Pumping Station Design
Second Edition

Pumping Station Design
Second Edition

Editor-in-Chief

ROBERT L. SANKS, Ph.D., PE
Consulting Engineer
and Professor Emeritus
Montana State University
Bozeman, Montana

Co-Editors

GEORGE TCHOBANOGLOUS, Ph.D., PE
Consulting Engineer and
Professor Emeritus
University of California at Davis
Davis, California

BAYARD E. BOSSERMAN II, PE
Principal Engineer
Boyle Engineering Corporation
Newport Beach, California

GARR M. JONES, PE
Senior Vice President, Design
Brown and Caldwell Consultants
Walnut Creek, California

BUTTERWORTH HEINEMANN

Boston Oxford Johannesburg Melbourne New Delhi Singapore

Disclaimer

All ideas, recommendations, methods, techniques, principles, training, or subjects mentioned are for information and educational purposes only. Because of the intangible and external influences that may bear on the use of these ideas and suggestions in any application and the absolute necessity for a totally integrated planning effort, the authors and publisher make no promises and accept no responsibility for the manner in which any of the information provided or contained herein may be applied, nor for the adverse effects that may result therefrom.

Recognizing the importance of preserving what has been written, Butterworth–Heinemann prints its books on acid-free paper whenever possible.

Butterworth–Heinemann supports the efforts of American Forests and the Global ReLeaf program in its campaign for the betterment of trees, forests, and our environment.

Library of Congress Cataloging-in-Publication Data

Pumping station design / editor-in-chief, Robert L. Sanks. — 2nd ed.
p. cm.
Includes bibliographical references and index.
ISBN 1-7506-9483-1 (alk. paper)
1. Pumping stations—Design and construction. 2. Water treatment plants. 3. Sewage disposal plants. I. Sanks, Robert L.
TD485.P86 1998
628'.29—dc21 97-15436
CIP

British Library Cataloguing-in-Publication Data
A catalogue record for this book is available from the British Library.

The publisher offers special discounts on bulk orders of this book.
For information, please contact:
Manager of Special Sales
Butterworth–Heinemann
225 Wildwood Avenue
Woburn, MA 01801–2041
Tel: 781-904-2500
Fax: 781-904-2620

For information on all Butterworth–Heinemann business books available, contact our World Wide Web home page at: http://www.bh.com

10 9 8 7 6 5 4

Printed in the United States of America

To the Memory of

Mary Clement Sanks
June 11, 1920–October 24, 1994

Musician of impressive accomplishments, organizer and director of the Annual Chamber Music Festival (1971–1994) for nearly 200 players from North America and Europe, inspiring teacher, enthusiastic chamber music player, and indefatigable worker. On the piano, she could switch keys or leave out a beat or a phrase to match others' mistakes so smoothly that the audience was unaware of a mishap. Generosity, kindliness, compassion, and service to others characterized her life. She once taught a university course for two students after hours for no compensation. A gentle, unassuming lady, she was loved by all who knew her.

Without her patient typing throughout many revisions for seven years, *Pumping Station Design* could not have been written at all. She was also our ultimate authority on grammar and clarity. If she, a non-engineer, could not understand something, it was recast until she could. She had much to do with the book's quality and readability. It is altogether fitting that our profession should know how much it owes to this one dynamic, dedicated musician.

Contents

Chapter 4
Piping 69

BAYARD E. BOSSERMAN II, JAMES C. DOWELL, ELIZABETH M. HUNING, AND ROBERT L. SANKS

Chapter 5
Valves 107

CARL N. ANDERSON, BAYARD E. BOSSERMAN II, AND CHARLES D. MORRIS

Chapter 6
Fundamentals of Hydraulic Transients 139

BAYARD E. BOSSERMAN II AND WILLIAM A. HUNT

Chapter 7
Control of Hydraulic Transients 153

BAYARD E. BOSSERMAN II

Chapter 8
Electrical Fundamentals and Power System Principles 179

STANLEY S. HONG, PHILIP A. HUFF, AND PAUL C. LEACH

Chapter 9
Electrical Design 219

STANLEY S. HONG, PHILIP A. HUFF, AND PAUL C. LEACH

Chapter 10 Performance of Centrifugal Pumps 241

GEORGE TCHOBANOGLOUS

Chapter 11 Types of Pumps 277

RICHARD O. GARBUS

Chapter 12 Pumps: Selection, Installation, and Intakes 317

GARR M. JONES AND ROBERT L. SANKS

Chapter 13 Electric Motors 401

PAUL C. LEACH

Chapter 14 Engines 423

GARR M. JONES

Chapter 15 Variable-Speed Pumping 443

MAYO GOTTLIEBSON AND ROBERT L. SANKS

Chapter 16 Pump-Driver Specifications 485

DAVID L. EISENHAUER AND GARR M. JONES

Chapter 17 System Design for Wastewater Pumping 495

GARR M. JONES

Chapter 18 System Design for Water Pumping 529

BAYARD E. BOSSERMAN II, RICHARD J. RINGWOOD, MARVIN DAN SCHMIDT, AND MICHAEL G. THALHAMER

Chapter 19 System Design for Sludge Pumping 573

CARL N. ANDERSON AND DAVID J. HANNA

Chapter 20 Instrumentation and Control Devices 601

ROBERT S. BENFELL

Chapter 26
Pumping Station Design Examples 803

GARR M. JONES, GARY S. DODSON, AND THEODORE B. WHITON

Chapter 27
Avoiding Blunders 827

ROBERT L. SANKS

Chapter 28
Contract Documents 841

JOHN E. CONNELL

Chapter 29
Costs 851

STEFAN M. ABELIN, MARC T. PRITCHARD, AND ROBERT L. SANKS

Appendix A
Physical Data 879

Appendix B
Data for Flow in Pipes, Fittings, and Valves 889

ROBERT L. SANKS

List of Examples

Locator for Equations, Figures, and Tables

Equations

Figures

Tables

Preface to the Second Edition

The reception of the first edition of this work by the engineering profession has indeed been gratifying. It seems to have become the standard reference for pumping station designers, and many have said it is the only reference they constantly use. In 1989, it received the "Excellence" award from the Professional and Scholarly Publishing Division of the Association of American Publishers. Each year a single engineering book is awarded this signal honor—a sort of Pulitzer Prize for engineering.

Matching that high standard with this second edition has been a challenge. Fortunately, most of the coeditors of the first edition again gave generously of their time, knowledge, and experience. Timothy Thor took the previous draftsman's place with equal artistry. Several experienced and competent authors and contributors joined the group to fill the omissions in the first edition. The absence of Mary Sanks to type and polish the manuscript left a gap that slowed the work and increased its difficulty.

This second edition is an improvement over the previous one in two major ways. First, every chapter has been examined and revised in some degree to reflect the best modern practice. Some changes are subtle—a word here and there, but many chapters were extensively rewritten. Second, a number of subjects, missing in the first edition, have been added. These include: (1) interviews with operators and supervisors of 15 utilities (that together manage 2700 pumping stations) to discover how to make operation better and maintenance easier and less expensive; (2) guidelines for troubleshooting existing vibration problems; (3) a straightforward explanation of how to avoid vibration problems in new stations; (4) objective, site-specific considerations in recommending whether large submersible pumps should be located in wet wells or dry pits; (5) directions for easily removing large submersible pumps from wet wells; (6) a comparison of life-cycle costs of constant-speed and variable-speed pumping stations; and (7) advice to utilities on how to choose a consulting engineering firm.

The eighth difference between the two editions is the addition of guidelines and worked examples for the design of modern pump intake basins for small to large pumping stations—especially self-cleaning basins for wastewater. In the first edition, wet wells for solids-bearing waters were limited to the few examples of Seattle Metro—now King County (Washington) Department of Metropolitan Services—pumping stations presented in Chapter 17. Other literature contained little of significance about this important subject, so a four-year period of development and research was immediately begun to improve the self-cleaning properties of the trench-type wet well and to develop guidelines for design. As a result, the self-cleaning properties were enhanced manyfold (as much as 50 or more), and the trench-type wet well, previously limited to variable-speed pumping, was adapted to constant-speed pumping—essentially made possible by the use of the sloping approach pipe described in Chapter 12. The inclusion of the results of this research and development is the most important improvement in the second edition.

Although the research was begun for the express purpose of improving this book, it was partly responsible for the appointment of the Committee on Pump Intake Design by the Hydraulic Institute. Following nearly three years of work by the committee, the standards for wet well design were extensively revised, and, at this writing, the draft is being circulated for public review as a step leading to approval by the American National Standards Institute. The trench-type wet well is included in the proposed new standards for both solids-bearing and clean waters. Other types are also allowed if provisions are made for cleaning those for solids-bearing waters. The new proposed standards are in consonance with the presentations in Chapters 12, 17, 26, and 29. The research has, furthermore, led to the construction of several successful trench-type wet wells, and more are being planned or constructed.

The co-editors join me in hoping that you find the second edition even more useful than the first. Suggestions for further improvements (other topics, elimination of errors, etc.) to make future printings or editions ever more valuable are welcome.

Acknowledgments

The co-editors, authors, and contributors have generously donated their time and effort to this second edition *pro bono*. All deserve the utmost appreciation from every reader. It seems unfair to single out specific contributions, but several have been extraordinary. They and their authors include: the concepts that make the trench-type pump sump adaptable to constant-speed pumping—Garr Jones; the hydraulics of sloping approach pipes—William Wheeler; the practical discourse added to vibrations—William Marscher; the survey of opinions held by operators of Seattle Metro pumping stations—Gary Isaac; the report on modern trends in the operation of 2700 pumping stations—B. E. Bosserman; and the comparative cost analyses of constant-speed and variable-speed pumping—Stefan Abelin and Marc Pritchard.

The presentation of trench-type sumps is based on surveys of many existing pumping stations and on research carried out by model studies at ENSR Laboratory (Redmond, WA) under the direction of Charles Sweeney and at Montana State University (Bozeman, MT) by Robert Sanks and by studies of a full-sized sump at Fairbanks Morse Corp. built by Virgil Beaty and tested by Beaty, Jones, and Sanks. Funds for the work were provided by a grant from the U.S. Environmental Protection Agency augmented by funds from Fairbanks Morse Pump Corp. (Kansas City, KS), The Gorman-Rupp Co. (Mansfield, OH), ITT Flygt AB (Solna, Sweden), and the Department of Civil Engineering and the Foundation of Montana State University (Bozeman, MT). Seattle Metro (now King County, Washington, Department of Metropolitan Services), the City of Steilacoom, the Town of Black Diamond (WA), G.A. Dodson & Associates (CA), and ITT Flygt AB (Sweden) arranged for investigative surveys of self-cleaning pumping stations.

Calculations in Chapter 12 were checked by Ned Paschke. All editors read, corrected, and finally approved every change made for the second edition. Hope Steele edited the manuscript. Brian Smith of Argosy was the manager for the conversion of manuscript into page proof. The staff of Butterworth–Heinemann was both encouraging and patient.

Thank you all.

Robert L. Sanks
Bozeman, Montana
January 1998

Preface to the First Edition

This book, *Pumping Station Design,* is unique in the following ways. It was written by consultants for consultants so as to be of the greatest practical use for designers. Each author is an expert whose writing is based mostly on personal experience. Little of it was obtained from, or based on, the existing literature. To make the book more usable and understandable, over 370 illustrations are included together with 59 design examples. Most design examples and all formulas are given in both SI and U.S. customary units. The book is complete enough for the novice and advanced enough to be useful to experienced designers and to those who direct or may be associated with design (such as utility managers, city engineers, or equipment suppliers). It is the only text available that deals comprehensively with the entire subject of how to design pumping stations. Finally, the book is unique in the number and expertise of its authors and contributors and in the meticulous care exercised during the seven years of its preparation (as described in the following paragraphs) to make it as easy to read as possible.

The first eleven chapters contain the fundamentals essential for effective design and include hydraulics, piping, water hammer, electricity, and theory and descriptions of pumps. The middle third is devoted to system design, including pump and driver selection and general piping layouts for water, wastewater, and sludge pumping. The last ten chapters contain supporting disciplines and subjects such as instrumentation and design, heating and ventilating, noise and vibration, comparisons of types of pumping stations and pumps, blunder avoidance, contract documents, detailing, and cost analyses. The appendices contain useful physical data, lists of codes and specifications, design checks, start-up checks, and addresses of all publishers given in the references. The tables of flow and headloss in pipes are compiled in a useful form not heretofore published. All of the work is extensively cross-referenced.

Perhaps never before has such a large, talented group of professionals been gathered to produce a book. The 132 expert contributors to this text provide broad and encompassing viewpoints gained from an aggregate of 20 centuries of practical experience. Each author was selected on the basis of specialized knowledge, past performance, experience, and commitment to the profession. Each produced one or more chapters (or parts thereof) based on detailed outlines suggested by the editorial board and improved by author and board as the rest of the book was developed. The other contributors, also selected on the basis of experience and competence, helped in the peer reviews and by supplying information.

Typically, I rewrote (or at least heavily edited) each chapter to conform to a uniform style and then sent it to from three to seven peer reviewers whose collected comments would be rephrased and given to the author with my own comments added. Following the author's reply, a second rough draft would be prepared and sent to author and reviewers. The returned comments would be recast into a third draft and again sent to the author. The fourth draft, usually called "final draft one," was sent to the co-editors. George Tchobanoglous checked every chapter for construction, clarity, and style. Garr M. Jones checked every chapter for practicality and good design practice. The other co-editors reviewed selected chapters for completeness and accuracy. Improvements, integration with other chapters, and nuances of wording often required as many as four subsequent "final drafts" until the chapter satisfied author, reviewers, and editors—a process that has taken seven years. As the book neared completion, new material was added and various subjects were sometimes shuffled between chapters for more logical presentation and cross-referencing. Alterations and improvements were continued through February 1989. Some idea of the effort taken can be appreciated by realizing that over 50,000 pages of review drafts have been distilled into this book. The result is considered to represent the state of the art (as of early 1989)—practical, authoritative, and essentially timeless. Consulting firms will find that this book can sharply reduce the time for an inexperienced engineer to become a competent pumping station designer. Project leaders will find the comprehensiveness, the checklists, and the list of blunders to be of great help. Utility managers will discover that selective reading of a few chapters will provide insights for directives that can produce better pumping stations for lower overall costs of construction, maintenance, and repair.

The work on this book was begun with a conference on pumping station design and a detailed proceedings outline, which served as a first approximation for the textbook to follow. Proceedings authors were selected on the basis of their experience records and were assigned chapters (or sections thereof) in strict adherence to the outline. The resulting *Proceedings,* published in 1981 in 4 volumes (1576 pages), are still available and valuable as an adjunct tools for design *[out of print in 1996—Ed.]*. Although the purpose of the conference was to make this new material immediately available to the profession, it also enabled us to find a group of experts and to gather resources for this book.

What prompted this project was the lack of a complete textbook about pumping station design in the United States (or in the English language insofar as we knew.) Of course, there were many books about pumps and pumping machinery and a few short manuals for designing pumping stations but there was no comprehensive, authoritative text or reference book dealing specifically with the design of all phases of water and wastewater pumping stations. Indeed, the literature about pumping station design has been fragmented, often superficial, sometimes wrong, and generally incomplete. One expert stated that 95 percent of all pumping stations he has seen contain serious design mistakes and that they occur in every category; if so there was a need for a book written by practicing engineers for consultants and other involved in decision making. Knowledge about the subject has been largely confined to consulting engineers, a few large public utilities, and to equipment manufacturers, so the overall purpose of this project was to gather, codify, and preserve the knowledge (much of which has never been printed) for the benefit of the public and the profession.

Carl W. Reh was the first co-editor appointed and, until his death in 1983, my chief proponent and supporter. The other co-editors, George Tchobanoglous, Donald Newton, B. E. Bosserman II, and Garr M. Jones (in order of appointment) have made this work possible. As technical advisor, Earle C. Smith provided much invaluable guidance and critiqued a large part of the work. All the authors and contributors have given a great deal of time to the project with no thought of reward beyond a desire to be of service to the profession.

Several consulting firms made extraordinary contributions of time, effort, and finances to the project, as follows: Greeley and Hansen Engineers, Chicago—six authors, including one editor, wrote four chapters, a part of another, and two appendices; Brown and Caldwell Consultants, Walnut Creek, California—three authors, including one editor, wrote six chapters and one appendix; Boyle Engineering Corporation, Newport Beach and Bakersfield, California—two authors, including one editor, produced five chapters and one appendix. Several firms, listed in Chapter 29, contributed cost data, an onerous task. Sincere appreciation is extended to all for this help, and, indeed, the engineering profession is indebted to all the contributing firms and personnel.

Mary C. Sanks patiently typed draft after draft and checked grammar, readability, punctuation, and spelling, and she assisted with galley and page proofs. Edimir Rocumback, student in architecture, drafted most of the figures. The entire project was made possible by the financial support of Montana State University. Officers directly involved included Theodore T. Williams, formerly Head, Department of Civil Engineering and Engineering Mechanics; Byron J. Bennett, formerly Dean, College of Engineering; and Lawrence T. Kain, formerly Administrator of Grants and Contracts.

Robert L. Sanks
Bozeman, Montana
March 1989

Authors and Contributors

*STEFAN M. ABELIN
Systems Engineering Manager
ITT Flygt Corporation
Trumbull, Connecticut

APPIAH AMIRTHARAJAH, PhD, PE
Professor of Civil Engineering
Georgia Institute of Technology
Atlanta, Georgia

*CARL N. ANDERSON, MS, PE
Consultant
Oakland, California

NICHOLAS J. ARHONTES
Collection Facilities O&M Manager
County Sanitation Districts of Orange County
Fountain Valley, California

RUSSELL H. BABCOCK, PE
Consulting Engineer
Westwood, Massachusetts

MICHAEL L. BAHM
Pump Maintenance Supervisor
Department of Public Works
Wastewater Treatment Division
City of Baton Rouge, Louisiana

VIRGIL J. BEATY (retired)
Formerly, Vice President, Engineering
Fairbanks Morse Pump Corp.
Kansas City, Kansas

*ROBERT S. BENFELL, PE
Chief Instrumentation Engineer
Brown and Caldwell Consultants
Seattle, Washington

HARRISON C. BICKNELL, PE
Product Application Engineer
General Electric Co.
Schenectady, New York

DONALD R. BJORK, MSEE, PE
Consultant
Bozeman, Montana

CHARLES T. BLANCHARD
CPC Corporation
Sturbridge, Massachusetts

KIRK BLANCHARD, PE
Manager of Manufacturing
Flygt Corporation
Trumbull, Connecticut

*BAYARD E. BOSSERMAN II, PE
Principal Engineer
Boyle Engineering Corporation
Newport Beach, California

PAT H. BOUTHILLIER, PEng
Professor of Civil Engineering
University of Alberta
Edmonton, Alberta, Canada

E. ROBERT BOUWKAMP, PE
Formerly, Electrical Design Engineer
Williams & Works, Inc.
Grand Rapids, Michigan

MEAD BRADNER
Consultant
Formerly, Applications Engineer
Foxboro Company
Foxboro, Massachusetts

ROBERT H. BROTHERTON, PE
Director of Public Works & Utilities
City of Dunedin, Florida
Formerly, Associate
Envirodyne Engineers
Knoxville, Tennessee

GEORGE R. BROWER, ScD, PE
Vice President
Barge, Sumner, and Cannon, Inc.
Knoxville, Tennessee
Formerly, Professor
University of Arkansas

*Asterisk denotes chapter author.

ROLAND S. BURLINGAME, PE (deceased)
Formerly, Senior Vice President
Camp Dresser & McKee, Inc.
Walnut Creek, California

FREDRIC C. BURTON, BCE, LS, PE (retired)
Formerly, Manager of International Operations
Flygt Corporation
Norwalk, Connecticut

CASI CADRECHA
Western Regional Marketing Manager
Golden-Anderson
Mars, Pennsylvania
Formerly, Manager Valve Division
Willamette Iron and Steel Co.

GEOFFREY A. CARTHEW
Vice President and Office Manager
Brown and Caldwell Consultants
Walnut Creek, California

A. L. CHARBONNEAU, PEng
Consulting Hydraulic Engineer
Kamloops, British Columbia, Canada

*JOHN E. CONNELL, PE (transferred)
Formerly, Senior Environmental Engineer
Christian, Spring, Sielbach and Associates
Billings, Montana

HARRY E. COVEY, PE
Chief Engineer
The Metropolitan District
Hartford, Connecticut

PATRICK J. CREEGAN, MCE, CE, SE, GE, PE
Vice President-Technical Manager
Water Storage and Conveyance Projects
Engineering-Science, Inc.
Berkeley, California

ROGER J. CRONIN, PE
Partner
Greeley and Hansen Engineers
Richmond, Virginia

ALFRED B. CUNNINGHAM, PhD
Professor of Civil Engineering
Montana State University
Bozeman, Montana

RODNEY L. CUSWORTH
Formerly, President
Mono Group, Inc.
Bensenville, Illinois

ROBERT A. DAFFER, JR., PE
Project Manager and Senior Engineer
Burns and McDonnell
Kansas City, Missouri

KUNDAN DESAI, MS
Zimpro/Passavant Inc.
Birmingham, Alabama

MICHAEL A. DEVINE
Engine Division
Caterpillar, Inc.
Peoria, Illinois

JOHANNES DEWAAL, PE
Formerly, Principal Associate
Clinton Bogert Associates
Fort Lee, New Jersey

JOHN L. DICMAS, PE (deceased)
Consulting Mechanical Engineer
Arcadia, California

*GARY S. DODSON, PE
President
G.S. Dodson & Associates
Walnut Creek, California

RICK A. DONALDSON
District Manager
Parco Engineering Corporation
Medfield, Massachusetts

*JAMES C. DOWELL, PE
Partner
Wilson & Company, Engineers & Architects
Phoenix, Arizona

RONALD W. DUNCAN
Senior Application Engineer
MagneTek Louis Allis Drives and Systems
Division of MagneTek
New Berlin, Wisconsin

*DAVID L. EISENHAUER, PE (retired)
Formerly, Manager
Engineering Standards Department
Brown and Caldwell Consultants
Walnut Creek, California

RAYMOND L. ELLIOTT, PE (retired)
Formerly, Director of Plan Review
Williams & Works, Inc.
Grand Rapids, Michigan

EDWARD J. ESFANDI
Senior Engineer
Los Angeles County Sanitation Districts
Compton, California

FRED A. FAIRBANKS, PE (transferred)
Formerly, President
Fairway Engineering
San Bernardino, California

*ERIK B. FISKE, MSEE, PE (retired)
Formerly, Director of Engineering
Byron Jackson Pump Division
BW/IP International, Inc.
Los Angeles, California

MAX FREY, PE
Consultant
Formerly, Vice President of Engineering
Cornell Pump Company
Portland, Oregon

*GEORGE FRYE, BS (deceased)
Formerly, Chief Engineer
Yeomans Brothers Company
Specialist
Greeley and Hansen Engineers
Chicago, Illinois

RAY A. GALL, MSME
Senior Development Engineer
Spicer Universal Joint Division
Dana Corporation
Toledo, Ohio

PAUL R. GALLO
Vice President, Engineering
Rodney Hunt Company
Orange, Massachusetts

*RICHARD O. GARBUS, MSME, MBA, PE (deceased)
Formerly, Senior Consulting Engineer
Fairbanks Morse Pump Corporation
Kansas City, Kansas

VICTOR G. GEREZ, PhD
Head
Department of Electrical Engineering
Montana State University
Bozeman, Montana

JAMES G. GIBBS, JR., PE
Partner
Wilson & Company, Engineers & Architects
Albuquerque, New Mexico

HAROLD D. GILMAN, BA, MA
Associate
Greeley and Hansen Engineers
Philadelphia, Pennsylvania

ROBERT C. GLOVER, PE
President
Fluid Kinetics Corporation
Ventura, California

HOWARD N. GODAT, PE
President
Howard Godat & Associates, Inc.
Olympia, Washington

*MAYO GOTTLIEBSON, BS, PE (deceased)
Formerly, President Dyna Systems Company, Inc.
Dublin, Ohio

WILLIAM F. H. GROSS
President
The Pitometer Associates, Inc.
Chicago, Illinois

L. V. GUTIERREZ, JR., DEng, PE
Public Works Engineer
City of Chandler, Arizona
Formerly, Vice President
Camp Dresser & McKee, Inc.
Boston, Massachusetts

*DAVID J. HANNA, PE
Associate Professor of Construction
Technology & Management
Ferris State University
Big Rapids, Michigan

GREG HARKER
Senior Applications Engineer, Flow Products
Endress + Hauser Instruments
Greenwood, New Jersey

ROBERT J. HART
Principal Engineer, Pumps & Pumping Systems
DuPont Engineering
Wilmington, Delaware

GEORGE E. HECKER
President
Alden Research Laboratory
Holden, Massachusetts

*EARL L. HECKMAN, PE (retired)
Formerly, Mechanical Group Head
Greeley and Hansen Engineers
Chicago, Illinois

*STANLEY S. HONG, BSEE, MSME, PE (deceased)
Formerly, Associate and Chief of Electrical Design
Greeley and Hansen Engineers
Chicago, Illinois

W. ERIC HOPKINS (retired)
Formerly, Project Manager
Gore and Storrie Limited,
Consulting Engineers
Toronto, Ontario, Canada

DAVID A. HOUSE
Engineering Product Manager
Dresser Pump Division
Taneytown, Maryland

*PHILIP A. HUFF, PE, R-M-S
Consulting Electrical Engineer
Acampo, California

*ELIZABETH M. HUNING, PE
Key Associate
Wilson & Company, Engineers & Architects
Phoenix, Arizona

*WILLIAM A. HUNT, PhD, PE
Consulting Engineer
HKM Associates
Bozeman, Montana
Formerly, Professor of Civil Engineering
Montana State University

GARY ISAAC (retired)
Formerly, Superintendent of Operations
Municipality of Metropolitan Seattle
(now King County Department of Metropolitan Services)
Renton, Washington

JOSEPH K. JACKSON
Vice President, Sales
Yeomans Chicago Corporation
Melrose Park, Illinois

CHARLES J. JECKELL, PE
Utility Engineering Administrator
Department of Public Utilities
City of Virginia Beach, Virginia

CASEY JONES
Manager
Application Engineering
Square D Company
Columbia, South Carolina

*GARR M. JONES, BSCE, BSIE, PE
Senior Vice President, Design
Brown and Caldwell Consultants
Walnut Creek, California

*GEORGE JORGENSEN, PE (retired)
Formerly, Chief Engineer
Salt Lake City Public Utilities
Salt Lake City, Utah

WILLARD O. KEIGHTLEY, PhD
Professor Emeritus
Montana State University
Bozeman, Montana

RONALD P. KETTLE
Superintendent of Desert Operations
Los Angeles County Sanitation Districts
Whittier, California

WILLIAM R. KIRKPATRICK, MSCE, PE
Project Manager
Engineering-Science
Berkeley, California

*FRANK KLEIN, BSAE, MSCE, Architect, PE (retired)
Formerly, President
Klein and Hoffman, Inc.
Chicago, Illinois

JOSEPH R. KROON, MSCE, PE
Director, Liquid Services
Stoner Associates, Inc.
Carlisle, Pennsylvania

MELVIN P. LANDIS, PE
Consulting Engineer
Carmichael, California

LONNIE LANGE
Maintenance Manager
Wastewater Operations
Division of Public Works and Engineering Department
City of Houston
Houston, Texas

R. RUSSELL LANGTEAU (retired)
Formerly, Pump Specialist
Black & Veatch, Consulting Engineers
Kansas City, Missouri

*PAUL C. LEACH, PE
Consulting Engineer
LaConner, Washington
Formerly, Chief Electrical Engineer
Brown and Caldwell Consultants

JOHN LEAK, BSc, ARCS
Instrumentation & Automation Engineer
Greeley and Hansen Engineers
Philadelphia, Pennsylvania

JOSEPH E. LESCOVICH
Chief Engineer
GA Industries
Cranberry Township, Pennsylvania

*JERRY G. LILLY, MS, PE
President
JGL Acoustics, Inc.
Bellevue, Washington

RICHARD R. MALESICH
Generator Sales Manager
A&I Distributors
Billings, Montana

RALPH E. MARQUISS, PE (retired)
Formerly, Partner
Rummel, Klepper & Kahl, Consulting Engineers
Baltimore, Maryland

*WILLIAM D. MARSCHER
President
Mechanical Solutions
Parsippany, New Jersey

COLIN MARTIN
President
Cham Engineering
Columbia, Connecticut
and Consultant to ABS Pumps, Inc.
Meriden, Connecticut

RHYS M. MCDONALD
Senior Scientist
Brown and Caldwell Consultants
Walnut Creek, California

M. STEVE MERRILL, PhD, PE
Project Manager
Brown and Caldwell Consultants
Seattle, Washington

WARREN H. MESLOH, PE
Director
Process Design & Equipment Selection
Wilson & Company, Engineers & Architects
Salina, Kansas

J. DAVIS MILLER, PE
President
White Rock Engineering, Inc.
Dallas, Texas

STEPHEN G. MILLER (transferred)
Formerly, Vice President
Komline-Sanderson
Peapack, New Jersey

ALOYSIUS M. MOCEK, JR., PE
Vice President of Engineering
Parrish Power Products, Inc.
Toledo, Ohio

JAMES L. MOHART, MSCE, PE, CVS
Director of Administrative Services
Black & Veatch, Consulting Engineers
Kansas City, Missouri

*CHARLES D. MORRIS, PhD, PE
Associate Professor of Civil Engineering
University of Missouri
Rolla, Missouri
Formerly, Principal Engineer
Camp Dresser & McKee, Inc.
Boston, Massachusetts

MICHAEL C. MULBARGER, MS, PE (retired)
Formerly, Vice President
Havens and Emerson, Inc.
Cleveland, Ohio

RICHARD L. NAILEN, PE
Project Engineer
Wisconsin Electric Power Company
Milwaukee, Wisconsin

TIMOTHY NANCE
Senior Engineer Specialist
Square D Company
Columbia, South Carolina

DENNIS R. NEUMAN
Research Chemist
Montana State University
Bozeman, Montana

*DONALD NEWTON, PE (retired)
Formerly, Partner
Greeley and Hansen Engineers
Chicago, Illinois

LORAN D. NOVACHEK
Applications Engineer
Waukesha Engine Division
Dresser Industries, Inc.
Waukesha, Wisconsin

ALAN W. O'BRIEN, PE
O'Brien & Associates
Albuquerque, New Mexico

MICHAEL R. OLSON
Senior Applications Engineer
Eaton Corporation, Electric Drives Division
Kenosha, Wisconsin

STEPHEN H. PALAC
Partner and Chief Electrical Engineer
Greeley and Hansen Engineers
Chicago, Illinois

CONSTANTINE PAPADAKIS, PhD, PE
President
Drexel University
Philadelphia, Pennsylvania

*RANDALL R. PARKS
Principal
Integra Engineering
Denver, Colorado

EVANS W. PASCHAL, PhD
Consultant
Whistler Radio Services
Sinclair Island
Anacortes, Washington

NED W. PASCHKE
Director of Engineering
Madison Metropolitan Sewerage District
Madison, Wisconsin

M. LEROY PATTERSON
Manager Electric Transmission Operations
Montana Power Co.
Butte, Montana

ALLAN W. PETERSON, MSc, PEng
Professor Emeritus
Department of Civil Engineering
University of Alberta
Edmonton, Alberta, Canada

ROBERT E. PHILLIPS, PhD (retired)
Formerly, Vice President of Applications and Sales
Turner Designs
Sunnyvale, California

WYETT C. PLAYFORD, PE (retired)
Formerly, Project Engineer
Williams & Works, Inc.
Grand Rapids, Wisconsin

JERRY P. POLLOCK
Engineering Manager
Johnson Power Ltd.
Toledo, Ohio

E. O. POTTHOFF, PE (retired)
Formerly, Application Engineer
General Electric Co.
Schenectady, New York

*MARC T. PRITCHARD
Manager, Planning and Control
Construction Management Division
Brown and Caldwell Consultants
Walnut Creek, California

RICHARD E. PUSTORINO, PE
President
R. E. Pustonno, PC
Commack, New York

EDGARDO QUIROZ
Engineer
Brown and Caldwell Consultants
Walnut Creek, California

SANJAY P. REDDY
Project Engineer
Carollo Engineers
Phoenix, Arizona

JOHN REDNER
Sewerage System Superintendent
Los Angeles County Sanitation Districts
Whittier, California

DAVID M. REESER, PE
Environmental Process Manager
RUST International Corp.
Portland, Oregon

CARL W. REH, PE (deceased)
Formerly, Partner
Greeley and Hansen Engineers
Chicago, Illinois

WILLIAM H. RICHARDSON, PE
Partner
Alvord, Burdick and Howson
Chicago, Illinois

*RICHARD J. RINGWOOD, PE
Consulting Engineer
Walnut Creek, California
Formerly, Manager of Environmental Engineering
Kaiser Engineers, Inc.
Oakland, California

RONALD ROSIE
Capital Projects Administrator
Seattle Metro
(now King County Department of Metropolitan Services)
Seattle, Washington

*ROBERT L. SANKS, PhD, PE
Consulting Engineer
and Professor Emeritus
Montana State University
Bozeman, Montana

*PERRY L. SCHAFER
Vice President
Brown and Caldwell Consultants
Oakland, California

JAMES W. SCHETTLER
Senior Project Manager, Mechanical
Brown and Caldwell Consultants
Walnut Creek, California

*MARVIN DAN SCHMIDT, PE
Principal Engineer
Boyle Engineering Corporation
Bakersfield, California

ARNOLD R. SDANO
Vice President, Engineering
Fairbanks Morse Pump Corporation
Kansas City, Kansas

DANIEL L. SHAFFER, PhD
Associate Professor of Chemical Engineering
Montana State University
Bozeman, Montana

RICHARD N. SKEEHAN, PE
Chief Electrical Engineer
Brown and Caldwell Consultants
Walnut Creek, California

W. STEPHEN SHENK, PE
Manager of Civil Engineering
Wiley and Wilson, Architects–Engineers–Planners
Lynchburg, Virginia

*LOWELL G. SLOAN
Chief Engineer
Prosser/ENPO Industries, Inc.
Piqua, Ohio

*EARLE C. SMITH (deceased)
Formerly, President
E.C. Smith and Associates
Upper Montclair, New Jersey

LARRY R. SMITH, PE
Civil Engineering Associate
Camp Dresser & McKee, Inc.
Dallas, Texas

ROBERT E. STARKE
Operations Engineer
Aurora/Layne & Bowler
North Aurora, Illinois

OTTO STEIN, PhD
Associate Professor of Civil Engineering
Montana State University
Bozeman, Montana

JOSEPH W. STEINER
Plants Superintendent
Public Utilities
City of Billings, Montana

DALE STILLER
Director of Maintenance
Madison Metro Sewerage District
Madison, Wisconsin

BRIAN G. STONE, PE
Consulting Engineer
Cottesloe, WA, Australia
Formerly, Vice President
James M. Montgomery Consulting Engineers
Pasadena, California

*SAM V. SUIGUSSAAR, PE
Senior Engineer
Greeley and Hansen Engineers
Chicago, Illinois

CHARLES E. SWEENEY, PE
Partner
Water Resources
ENSR Consulting and Engineering
Redmond, Washington

JAMES TAUBE
Electrical Designer
Florida Keys Aqueduct Authority
Florida City, Florida

HARVEY W. TAYLOR, PE (retired)
Formerly, Vice President
CRS Engineers
Portland, Oregon

*LEROY R. TAYLOR, PE (retired)
Formerly, Division Manager
CH2M-Hill
Boise, Idaho

WILLIAM R. TAYLOR, PhD, PE
Professor of Industrial and Management Engineering
Montana State University
Bozeman, Montana

*GEORGE TCHOBANOGLOUS, PhD, PE
Consulting Engineer and
Professor Emeritus of Civil Engineering
University of California, Davis
Davis, California

*MICHAEL G. THALHAMER, PE
Manager and Director of Water and Wastewater Engineering
Psomas and Associates
Sacramento, California

TIMOTHY A. THOR
Architect
KMD Architects and Planners
Portland, Oregon

PATRICIA A. TRAGER, WBE, PE
President
TRH Engineering
Chicago, Illinois

JERALD D. UNDERWOOD, PE
Public Utilities Director
City of Billings, Montana

R. DANIEL VANLUCHENE, PhD, PE
Associate Professor of Civil Engineering
Montana State University
Bozeman, Montana

ASHOK VARMA, MSME, PE
Vice President and Office Manager
Camp Dresser & McKee, Inc.
Dallas, Texas

LELAND J. WALKER, PE (retired)
Formerly, Chairman, Board of Directors
Northern Testing Laboratories, Inc.
(now Maxim Technologies, Inc.)
Great Falls, Montana

DAVID WALRATH, PE
Vice President
Hazen & Sawyer, PC
New York, New York

THOMAS M. WALSKI, PhD, PE
Professor
Wilkes University
Wilkes-Barre, Pennsylvania

HORTON WASSERMAN, PE
Senior Project Engineer
Malcolm Pirnie, Inc.
White Plains, New York

GARY Z. WATTERS, PhD, PE
Dean, College of Engineering, Computer Science and Technology
California State University
Chico, California

*THEODORE B. WHITON
Formerly, Project Manager
G.S. Dodson & Associates
Walnut Creek, California

THOMAS O. WILLIAMS
Technical Consultant on Gas Engines and Power Generation
Engine Products Division
Caterpillar, Inc.
Mossville, Illinois

ROY E. WILSON, PE
President
Wilson Management Associates, Inc.
Glen Head, New York

ERIC L. WINCHESTER, PEng
Senior Engineer
ADI Limited
Fredericton, New Brunswick, Canada

FREDERICK R. L. WISE, JR.
Vice President
Parco Engineering
Medfield, Massachusetts

JOHN E. WISKUS, MSSE, PE
Project Manager
CH2M-Hill
Boise, Idaho

FRANK A. WOODBURY, PE (retired)
Formerly, Senior Applications Engineer
Westinghouse Electric Corporation
Dallas, Texas

JAMES R. WRIGHT, PE
Senior Project Manager
Black & Veatch Engineers–Architects
Kansas City, Missouri

EUGENE K. YAREMKO, PEng
Principal
Northwest Hydraulic Consultants, Ltd.
Edmonton, Alberta, Canada

WILLIAM YOUNG
Operations Specialist
Brown and Caldwell Consultants
Walnut Creek, California

Pumping Station Design
Second Edition

Chapter 1

Introduction

ROBERT L. SANKS

CONTRIBUTORS
Roger J. Cronin
Marc T. Pritchard
Brian G. Stone
Roy L. Wilson

This book is written for a wide variety of readers: the expert and the beginner in a design office, the project leader of a design team, the city engineer or chief engineer of a water or sewerage authority (or their subordinates) who may review plans and specifications, and manufacturers' representatives who should know how their equipment is best applied to a pumping station. Recommendations for the utilization of the book by each group of readers are given in Section 1-7.

The aim of the volume is to show how to apply the fundamentals of the various disciplines and subjects into a well-integrated pumping station—reliable, easy to operate and maintain, and free from serious design mistakes.

To facilitate the selection of good design engineers, the publisher hereby gives permission to photocopy *Chapter 1 only* of this book for distribution to municipalities or utilities and their representatives.

1-1. Authors and Contributors

Each author or contributor is an expert with many years of experience in the subject discussed. Furthermore, all chapters were reviewed and critiqued by four to eight other equally qualified experts including the editors. (Editors' names are not listed unless an editor is a principal author.)

The reviewers are listed as contributors (except, of course, for the editors), but not all contributors are reviewers. A contributor is one who has helped in any way, from writing a short segment to giving advice.

Engineers do not always agree, and the viewpoints expressed do not necessarily reflect those of each individual contributor or even of the author. Where an unresolvable conflict occurs, both viewpoints are given. No effort has been spared to make this book the most authoritative possible. In spite of these efforts, there are more differences than can be encompassed and some errors may have occurred, so read thoughtfully and with care.

1-2. Responsibilities of Project Engineers

The design of a pumping station depends on several specialties, which are listed in Table 1-1 more or less in chronological order together with the approximate range of percentages of engineering or design costs. Not all specialties for all pumping stations are shown. For example, river engineering (not shown) might be a significant part of design costs for a raw water pumping station taking water from a river meandering in a sandy plain, and a raw sewage lift station at a treatment plant may be so closely tied to the treatment plant that it would be impossible to assign engineering costs to the lift station alone.

Table 1-1. Specialties and Disciplines in the Design of Pumping Stations

Specialty	Approximate percentage of design costs
Planning and water resources management	5–15
Surveying	0.5–1
Foundation and soils engineering	2–10
Hydraulic engineering	5–15
Analysis for transients	0–2
Structural engineering	20–35
Civil and mechanical engineering	15–40
Electrical engineering	10–30
Instrumentation engineering	2–5
Architecture	0.5–2
Specifications and contract administration	5–10
Construction engineering	15–50
Start-up, operations, and checkout	5–20

Although each discipline is important to the success of the design, several specialties do not require a detailed knowledge of hydraulics or pumps. To direct the design of a plant, it is not necessary to be a structural, soils, mechanical, or electrical engineer. But the project leader should be familiar with these disciplines and must be able to coordinate the work of the specialists.

Plant hydraulics requires great care because the hydraulic computations establish the ultimate capacity and, thus, the total capital costs. But the selection and specifying of reliable mechanical equipment is just as important, and an understanding of control methods, available monitoring equipment, and perhaps even display and data storage is fundamental for the design of a good and efficient plant.

It is the responsibility of the project leader to provide or coordinate all of the services in Table 1-1. It is bad practice to allow equipment suppliers to design custom plants in whole or in part. Many are interested in selling their equipment, and they cannot always be trusted to use the best. The project leader should ensure (1) that decisions are made by those who are completely objective and (2) that specifications are written so that inefficient and unreliable equipment with poor maintainability is eliminated.

1-3. Units

Metric (or SI) units are used worldwide and now take precedence in American technical literature. Although SI units will probably be used exclusively in the future, many American engineers still think in English units, plans for structures in the United States are still drawn in feet and inches, and pipes and machinery conform to inches (although many manufacturers now give dimensions in both units). Much of today's literature is written with SI units followed by English units in parentheses, a cumbersome practice but one that makes the work more universally appealing. That system is followed here, and formulas and many of the worked examples are also shown in both units. Abbreviations are defined in Chapter 2.

Engineering notation is typically used in tables and calculations. Exponents in multiples of three are useful for prefixing units with micro, milli, and kilo. Large and small numbers in calculations, for example, are written as 3.0×10^7 and 1.62×10^{-3}, whereas in tables the same numbers are designated in computer language as 3.0 E+7 and 1.62 E–3.

1-4. Standards and Codes

The use of standards in the specifications is extremely important. The quality of the product is largely dependent on a wide variety of standard codes and standard specifications—most of them the result of years of experience by both users and manufacturers. Because most standards and codes, updated from time to time, represent a consensus, they are an indispensable aid for quality control.

Readers, especially designers, should become familiar with the available standards (particularly those identified here) to reference exactly which standards (or portions of standards) apply to the project and to consider the advice contained in them intelligently. Never specify a standard unless you have read all of it and understand its implications. In today's litigious society, a specifier may be held legally responsible for a conflict or contradiction in a specification. Specifications must be accurate and clear (see "Specification Language" in Section 28-1).

Standards are referenced in contract documents and in this book by a simple code name such as "ANSI B16.1" or "ASTM B43." Such designations are entirely sufficient to identify the document, but nevertheless the documents are named, chapter by chapter, in Appendix E. Publishers' names and addresses are given in Appendix F.

1-5. Manufacturers' Recommendations

Be cautious when advised to consult manufacturers or to follow manufacturers' recommendations. Many manufacturers get business based on one consideration—low price—so they tend to stretch their criteria for use to the limit. Thus, their products perform as they say only if

- their instructions or recommendations are followed *exactly;*
- their product is used in a service *perfect* for its use.

These constraints are rarely met in field installations, so designers must consider the merit of the manufacturer's recommendations and often establish a more conservative design.

1-6. Safety

Many fatal accidents occur each year in the water and wastewater industries in the United States. Designers should be aware of the hazards and circumvent them insofar as possible both by good design and by adequate warnings and other instructions in the operation and maintenance (O&M) manual. Every designer should read *Life Safety Code* [1] and two of the NIOSH reports [2,3] as well as Sections 23-1 and 23-2 in this volume, in which some of the hazards in both water and wastewater pumping are explored.

1-7. How to Utilize This Book

Many of the facets of pumping station design are so interrelated that coherent discussions of different topics sometimes involve the same subject in several places and from different viewpoints. Hence, *none of the chapters should be studied in isolation. This is an integrated work and should be read as a whole.* For example, station head-capacity (H-Q) curves are customarily shown as single lines for simplicity. A reader who misses Figure 18-17 and the discussion of friction losses in Section 3-2 might not realize that H-Q curves should be considered as broad bands. Note that there is extensive cross-referencing; the locator in the prefatory section and the relatively complete index can be used to find the references to a given subject.

Recommended Minimum Reading

As a minimum, the user of this book should read Chapters 27 and 24 (particularly Sections 24-9 and 24-10). Those who are concerned with selecting consultants should read Section 1-8.

Beginners

Beginners should read the entire book and resist the temptation to study a single subject or a single chapter.

Project Leaders

Project leaders must have a working knowledge of all phases of the project and must be able to communicate with other members of the team to plan the project effectively. They must coordinate all phases, distinguish the good from the shoddy, and shoulder the responsibility for producing a well-designed facility. Hence, they too should be well acquainted with the entire book.

Experts

Experts are themselves the best judges of what to study. It would be wise to scan the entire volume to note the depth and thoroughness of coverage.

Public Utility Managers

Public utility managers and others who deal with or have control over the pumping station design would be wise to read the following:

- Chapter 25 from the beginning through Section 25-6 and, especially, Tables 25-3, 25-4, and 25-5 (for wastewater pumping) and Tables 25-6 and 25-7 (for water pumping)
- The first few pages of Chapters 17, 18, or 19 for wastewater, water, and sludge pumping, respectively
- Chapter 15, Sections 15-1 and 15-11 and, especially, Table 15-3 and the results of Example 29-1 if variable-speed operation is considered
- Before beginning to review plans, read Chapter 27, because much aggravation can be avoided and considerable funds may be saved by following the advice it contains.

Manufacturers

When advising consultants, it is not enough for manufacturers and their representatives to know their own products. They should also be able to help with the engineering involved in the incorporation of their products into the station. At a minimum, the chapters dealing with their products should be read together with Chapters 17, 18, and 19. Scanning the book may reveal other sections of importance.

1-8. How to Select Consulting Engineering Firms

The public at large tends to believe that registration of engineers ensures competency, but that is not true. Frankly, there just are not enough good engineers for every project. Many pumping stations (and treatment works) are flawed, and a distressing number are quite badly designed. So it is important to retain truly competent designers who will (1) produce a better facility than would mediocre or inexperienced engineers and (2) probably save the client significant life-cycle costs. Unfortunately, some public bodies are obsessed with the low-bid approach for choosing engineers in the mistaken thought that money is saved thereby—a penny-wise, pound-foolish notion that fosters hasty, ill-considered design, prevents adequate investigation of viable alternatives, and actually favors the inexperienced or incompetent who are willing to work for low wages. The nineteenth-century words of John Ruskin, *"There is hardly anything in the world that some man cannot make a little worse and sell a little cheaper and the people who consider price only are this man's lawful prey"* also apply to the services of any professional, including artists, engineers, lawyers, or surgeons. In fact, as engineering fees are typically only about one-eighth of the construction cost, a major saving in fees is insignificant compared with a minor saving in construction. The spread of construction bid prices is often greater than the entire engineering fee, so the place to save money is in construction. A thoughtful, resourceful designer might save more than the entire consulting fee in life-cycle cost.

Any organization that contemplates selecting an engineering firm on the basis of low bid should be aware of the inevitable results. As any private enterprise must make a profit to stay in business, the following disadvantages will occur when a firm is forced to compete with others for the lowest bid.

- The work will be bid exactly as written by the agency. If the scope fails to include all the tasks necessary for completing the project, those tasks will have to be completed by change orders that may negate the supposed low-bid savings.
- Options for long-term cost-saving and/or innovative, low-cost alternatives will not be considered. Instead, "cookbook" designs and copies of previous designs will be used. Although such designs may "work" (water may be pumped), the design is unlikely to be optimal or the most cost effective for the new project.
- The quality of plans and specifications will decline. Fewer hours will be allocated to coordination meetings, design reviews, and interdisciplinary coordination. Consequently, there will be more construction change orders (always expensive). The effort made in preparing detailed specifications will be the minimum possible. "Canned" specifications or specifications from a previous job will be used with a minimum of editing for current project needs. Construction inspection will often be under the direction of young and inexperienced engineers.

Low bid is *never* an adequate basis for selection *unless* the product can be defined and specified completely and accurately. Artistry and thoughtfulness in engineering or in any product of thought cannot be so specified.

Sometimes, there is an attempt to "save" money by dispensing with engineering services during construction, by in-house inspection, or by retaining the services of a separate consultant to provide engineering services during construction. A design is not really completed until after the project has entered service and the adjustments required during the commissioning period have been made. When any of the above arrangements are used, the inevitable result is that the designer is shielded from the day-to-day bidding and construction events that inevitably shape and refine the completed project. Loss of direct contact with the project by the designer as it progresses toward completion will inevitably mean the design will be compromised. Given this loss of contact, it is unreasonable for the owner to expect the completed project to be completely satisfactory, just as it was unreasonable for the owner to expect that the design, when it was completed for bidding purposes, is perfect in every way. The best construction can be obtained only by employing experienced inspectors under the control of (or at least answerable to) the design firm. Some requests for substitutions or change orders are inevitable, and it is better that the designer be in charge of these to protect the quality of the project and to avoid unnecessary added costs. Design services for a reasonably large project typically cost about 6 to 10% of project costs, and complete services (including design, inspection, and start-up) cost about 12 to 18%. Trying to save on this minor fraction of cost and major fraction of importance cannot be justified.

Good design begins with the selection of good designers by following the aphorism *"By their fruits, ye shall know them."* First, discard applications from firms that have not done related, recent work of roughly the same magnitude. This screening step reduces the applications by, perhaps, half. Second, conduct telephone interviews with the operation, maintenance, and supervisory staffs at representative facilities designed by the remaining applicants. Ask searching questions related to (1) the quality of design, (2) whether the specifications resulted in high-quality equipment, (3) ease of operation and maintenance, (4) whether the work space is pleasant and convenient, (5) whether the designers were both competent and cooperative during the project and after completion, and (6) the projects' success in meeting the purpose of the facility. These telephone interviews may well eliminate all but a very few applicants. Third, select an ad hoc committee of two or three members of the group that is responsible for choosing the consulting engineering firm and, accompanied by an engineering adviser who is both expert in the field and disinterested in the choice, visit a representative facility of each remaining firm. Allow a day for each visit. Tour the facility with the designer (or a representative) who will, of course, praise its virtues and suppress or minimize its faults. Dismiss the designer and tour it again with an operator and a maintenance worker to hear their objective views. Finally, tour it again with neither operator nor consultants so that the engineering adviser can frankly explain the good and bad features and, to a large degree, assess the competency of the designers. Such a visit is most revealing, and a wide disparity in the competence and thoughtfulness of designers for all but the simplest, smallest facilities will become apparent. Such visits may be expensive, but if the cost of the proposed facility is expected to exceed half a million dollars, the returns far outweigh the cost.

The above process may narrow the field to two or three applicants. Sometimes, none of the applicants appears qualified and it may be necessary to advertise for consultants again. Do not overlook the advantages of a consortium of a local firm and another firm with great experience. However, the experienced firm must have technical control over the project. Most engineers can do good work with only a minimum of guidance at the right times to the benefit of all concerned. In such situations, the above selection process, including visitations, should be redirected to the experienced firm.

By this time, the selection is usually apparent and the interview of firms on the "short list" is nearly a formality. Beware of "bait and switch" tactics wherein experienced engineers represent their firm at the interview and, thereafter, assign junior engineers to do the work. Make it plain that such practices will not be tolerated, and that those experienced engineers whose facilities were visited will themselves supervise the design. Follow up to be sure they do. Negotiate a contract for engineering services that will allow the designers enough leeway to do a proper job and to investigate promising alternatives. Cost plus a fixed fee with an upper limit protects both client and engineer. If no agreement can be reached, negotiate with the best of the other firms that made the short list or else advertise for consultants again.

The method outlined above is unsurpassed for getting the best possible facilities compatible with the lowest possible overall cost. Some states have recognized the need to discourage bidding for the selection of engineers by enacting statutes against the practice. The State of Nevada [4] requires selection based on qualifications. The state law of New Mexico encourages the qualification-based selection of engineers and also requires the services of a professional technical adviser to assist in defining the scope of work, evaluating proposals, and negotiating the contract. The price is determined by formal negotiations after selecting the consultant [5]. See Sanks [6], Shaw [7], and Whitley [8] for further information.

Once a selection has been made and a contract negotiated, be prepared to work closely with the consultant. A working team managed by an individual authorized to make budget and technical decisions for the owner should meet frequently with the consultant to monitor and advise as the project develops. With such a management, timely decisions affecting the project can be made and the finished product will more accurately reflect the owner's needs.

When the project is completed, put into service, and all adjustments made, the engineer should provide a set of "as-built" plans to the owner. These should be impressed with the responsible engineer's seal and signed. Some owners prefer the plans to be on diskettes, but such "electronic plans" should omit the seal and signature, because the engineer has no control over subsequent alterations to the diskettes.

1-9. Value Engineering

"Value engineering" is a formalized procedure for independently reviewing the proposed facilities before the design is completed. The Corps of Engineers and the U.S. Environmental Protection Agency are among those who either encourage or require these formalized studies of first costs together with operation, maintenance, and replacement costs, all of which are sometimes referred to as "life-cycle costs." If a value engineering study is required, the owner should pay for it as an extra.

A team of experienced engineering specialists in the fields appropriate to the study is assembled under the leadership of a "certified value engineer." The design engineer must assemble a presentation package for the reviewers. The assembled team normally spends about a week reviewing the designer's work. A team of experts who can give undivided attention to the broad aspects of a problem can sometimes save many times the cost of the study. Case histories abound where savings of large sums (up to 20% or even more) of the total project cost resulted from a peer review, but read these reports with caution. Many case histories give inflated values to the potential savings, and sometimes the recommendations are not consistent with overall project objectives.

To be most effective, such a study should be made as soon as the surveys are complete, the data are gathered, and a preliminary design concept is established, but before any detailed design has commenced—probably when the engineering is approximately 10 to 15% complete. The owner and the engineer review the report and may reject (with reasons) those recommendations that they consider inappropriate in their response to the funding agency.

Consulting engineers have often been forced to submit bids for their services and to reduce their costs to a minimum, with the result that adequate investigation of alternatives has been curtailed. If, however, the advice in Section 1-8 has been followed, and if the project has been subjected to careful in-house evaluation of alternatives, the redesign costs and delays brought about by the value engineering procedure will be minimized. If the work of the project engineer was well done, no significant changes will be needed.

The value engineering study is essentially a one-time review, whereas by contrast, the designer is concerned with costs throughout the project. The objective is that blend of economy in the cost of construction, operation, and maintenance together with intangibles (such as environmental effects and the ease of operation and maintenance) that can be termed the "most cost-effective" solution. A designer's attention is directed to increasingly smaller elements as the design proceeds. An important tool in the process is the "comparative cost estimates" of alternatives. Such estimates do not require detailed quantity takeoffs and can made before the design details are developed. Only those elements of the project that may vary because of the materials or equipment under evaluation are considered.

Every project of any size needs to be subjected to appropriate comparative cost estimates, regardless of whether a value engineering study is to be made.

1-10. Ensuring Quality and Economy

Pressure on engineers to achieve the lowest costs in the design of utility system installations has been increasing. Based on past experience, Jones [9] points out that high-quality design and construction features result in the lowest overall costs. The public has little patience for utility system failures—particularly for failures in water and wastewater systems. Consequently, features that incorporate a high degree of reliability are required. These features include:

- Heavy-duty construction of all equipment, not just the main pumps and drivers. See Sections 16-3, 28-5, and Appendix C for quality assurance for pumping equipment.
- Adequate ventilation to ensure a suitable environment for both equipment and workers. Today's electronic equipment requires far better environments than the mechanical relays and other devices of the past. Air conditioning to remove particulates and corrosive gases such as hydrogen sulfide and sometimes to cool the enclosure is often necessary.
- Support systems (such as on-site generators, sump pumps, service water, compressed air, and hoisting equipment) should provide the service necessary without fail.
- Features that make operation and maintenance simple and convenient include self-cleaning wet wells, simple control systems, and ample access for maintenance and operation.

Cost is the proper measure of success. The true cost of any facility should not only include capital, operation, energy, and maintenance costs but also the cost of failure to serve the public reliably. When all costs are properly assessed, the initial costs of design and construction are often a small part of the taxpayer's burden. Quality in design and construction is the best overall investment. Not only does it save money, but it makes life easier for management, operation, and maintenance personnel.

Joint Ventures

Joint ventures between small, local consulting firms and highly experienced firms result in several benefits including closer contact and cooperation with the client (at lower cost because of access and lower overhead), closer contact during construction, and the education of the local firm at no cost. The last benefit may make the local firm competent enough to design similar facilities with no future help. Note that the experienced firm must have control. A good start in the joint venture is a day-long retreat attended by the involved personnel from both of the firms and from the client to resolve major issues and to establish a close and cooperative relationship.

Another possibility is to invite an experienced construction manager to join the joint venture. The right person may save considerable expense by offering good advice on ease of construction and practicality.

Design-to-Cost

Design-to-cost is a proprietary, computerized management system used by one consulting firm for accurate cost estimating and scheduling at any stage of design completion. Because the potential for controlling costs is greatest in the early stages of design, the system is extremely useful. It enables (1) designers to focus immediately on the most cost-effective alternatives; (2) clients to see the consequences of incorporating special features; and (3) designers, construction contractors, and owners to share information that keeps costs under control. The system is made possible with a database of 20,000 items in which two subsystems (continuously upgraded costs and scheduling) are electronically linked. In early planning and design phases, estimators produce conceptual costs that are, on the average, accurate within an error range of +11 to –3%. As the design proceeds, new data are entered as soon as available, and the accuracy increases until the final estimate (based on quantities of materials, equipment, person-hours, location, and labor and market conditions) has an average error of only +7 to –1%. The key element is cooperation between all persons working on the project. The program can be used anywhere on a notebook computer [10, 11].

The concept can be used and a program developed by any firm that keeps good records of numerous and recent construction costs. Obviously, the capability of providing accurate costs from beginning to end helps to establish and maintain good relationships between the owner and designers and (during construction) between the contractor, owner, and design firm—particularly when change orders are processed.

1-11. Avoiding Litigation

In our litigious society, nothing can ensure complete protection from lawsuits, but the precepts that follow will reduce their incidence and severity.

Clear, Fair Contracts

Describe clearly both the rights and responsibilities of each appropriate party. Avoid ambiguity and one-sided requirements. Recognize potential problems, and provide for fair solutions openly.

More for Design, Less for Litigation

The intelligent application of money for design can promote better coordination between client and designer and between the various design disciplines. Proper funding also allows adequate investigation of viable alternatives and enough time for truly careful preparation and checking of plans and specifications with a probable decrease in the overall capital cost, with reduced chances for errors and expensive change orders and with improved client satisfaction.

Partnering

Partnering is a voluntary process that promotes nonadversarial relationships among the parties involved in a construction project. Partnering benefits all project participants (owner, contractor, and engineer) by reducing their exposure to disputes and litigation and by increasing their productivity. The teamwork during construction improves quality and the chance of timely completion and results in fewer lingering disputes. See "Partnering. A concept for success" [12].

Partnering between owner, engineer, and contractor has been tried, often with success but sometimes with mixed or poor results. Sincere efforts by each to be fair with the others are the keystone to success.

Resolve Potential Claims as Change Orders

Claims that lead to litigation are simply disputes that could not be resolved as change orders. The apparent ease of such a solution is deceptive if any of the following occurs:

- Ambiguous contract language
- Imprecise contract words or phrases such as "prompt," "reasonable," "workmanlike," and "to the satisfaction of the engineer," which will surely lead to different interpretations by different people
- Lack of communication and mutual respect.

Alternative Dispute Resolution

The construction industry has developed a variety of alternative dispute resolution (ADR) procedures to avoid the time, bitterness, and cost of resolving disputes by litigation. Those that should be considered include:

- Mediation, a formal method wherein both parties negotiate their differences before a neutral third party. Mediation may be mandated by the contract or adopted at the first irreconcilable dispute.
- Mini-trials wherein a third party acts as a judge and both sides present their positions and answer questions from the other side and the judge.
- Dispute review board named by the parties at the start of the project. The board thus becomes familiar with the project throughout its life.

Many other steps can be taken to prevent litigation, but the above are some of the most important. A more extensive discussion has been prepared by Wilson [13].

1-12. Library

Every designer should acquire an extensive library because no book, however complete, can contain adequate discussions of all topics. Read with caution and suspicion, however, because there are many mistakes in the literature, and some may lead to blunders or poor design. Just because a book is listed here or because it carries a prestigious label or a distinguished authorship, do not assume it to be universally authoritative or correct. Thoughtful reading of diverse opinions and consideration of the author's credentials coupled with your own experience and expertise can often distinguish between fact and fallacy, good and poor. The references in these chapters are publications often found to be useful, but by no means are they intended to include all of the useful literature. Furthermore, inclusion is not an endorsement of authority or assurance of freedom from error.

Readers must determine their own library needs. Pump sump design, pump testing, and vertical turbine and propeller pumps are discussed by Dicmas [14], and pumps, drivers, and pump systems are covered in *Pump Handbook* [15]. Storm water pumping (a subject omitted in this text) is discussed in the *Manual for Highway Storm Water Pumping Stations, Volumes I and II* [16] and *Design of Wastewater and Stormwater Pumping Stations* [17]. A compendium of standard design plans for storm water pumping is contained in the *Pump Station Design Manual* [18]. The number of times books or codes and standards are referenced, the title, and the context in which the references are given are clues to their importance. Consulting firms must have an extensive library of codes and specifications. Because these are voluminous and frequently revised, subscribing to a microfiche service is definitely worthwhile for a large office.

For your convenience, addresses for publishers are given in Appendix F.

1-13. Operator Training

Good operation is just as important as good design. A good crew can make a poor facility seem good, whereas a poor crew will make a good facility seem poor. Consequently, it is in the long-term interest of public utilities and agencies to spend the money necessary to send their operators and maintenance staff to seminars and other training sessions so that they can learn how to operate and maintain the pumping equipment properly. Training should be conducted (1) by qualified representatives of the equipment manufacturers, (2) at various seminars conducted by manufacturers, (3) by in-house training sessions, and (4) especially by the engineer responsible for the original design. This last individual is the only source of information on the operational concepts embodied in that design.

Purchasing expensive pumping installations with sensitive equipment and expecting untrained staff to maintain them is simply shortsighted. Indeed, over the long term, cutting back on training sessions will cost the utility more because: (1) more labor hours will be needed by poorly trained personnel to operate and maintain the station than would be required by properly trained staff; (2) equipment will have to be repaired more frequently than it would if it were properly cared for to begin with; (3) the station can be kept operating more efficiently and expensive, critical, and relatively fragile items such as mechanical seals and bearings for pumps are more likely to be properly installed and maintained by skilled maintenance workers, thereby reducing replacement parts costs; (4) skilled maintenance staff can repair equipment (for example, they can balance pump shafts and impellers) to a degree that minimizes problems such as vibration (which always reduces the longevity of rotating equipment); (5) properly trained staff will probably have a better attitude toward keeping the station neat and in good operating condition; and (6) good training will result in fewer accidents and missed work days—an important consideration because water and wastewater industries are among the most hazardous ones in the United States.

Large submersible pumps provide a good example of the cost-effectiveness of good training. These large pumps are out of favor with many public utilities, but a staff well trained in removing and repairing them can keep them operating with only ordinary effort and little down time *provided the equipment is of high quality and the facilities are tailored for easy pump removal and replacement* (see Sections 25-5 and 27-5). A utility with a goal of low capital expenditure would (in 1998) save an estimated construction cost of between 1 and 6 million dollars for a 1.3 m^3/s (30 Mgal/d) pumping station if three submersible pumps were used instead of three dry-pit pumps (see Figures 29-2 and 29-3). Obviously, the cost of training would be repaid many, many times.

Utilities should also realize that when the maintenance staff have become skilled in repairing complex equipment, their market value is much higher than that of "average" workers. If it does not offer a competitive salary appropriate to the level of skill, it will lose the benefits of its training program by becoming a training center for transferees to those private industries and other public agencies that do appreciate the worth of skillful mechanics.

1-14. References

1. NFPA 101, *Life Safety Code,* National Fire Protection Association, Quincy, MA.
2. *Papillon Creek Wastewater Treatment Plant, Omaha, Nebraska,* HETA 83-440-1537, NIOSH, U.S. Department of Health and Human Services, Atlanta, GA (1984).
3. *Criteria for a Recommended Standard. Working in Confined Spaces,* DHEW (NIOSH) Publication no. 80-106, U.S. Department of Health, Education, and Welfare, U.S. Government Printing Office, Washington DC (December 1979).
4. Nevada Revised Statutes, NRS 625, and Nevada Administrative Code Board Regulations, NAC 625 (1994).
5. Legislature of the State of New Mexico, "An act relating to public purchases, amending, repealing and enacting certain sections of the procurement code." Section 13-1-55 NMSA (1978).
6. Sanks, R. L., "How to choose a consulting engineer," *Public Works* **120** (10): 125–126 (September 1989).
7. Shaw, R. L., "Competing on price, again," *Civil Engineering* **59** (10): 6 (October 1989).
8. Whitley, F., "Consultant selection—controlling the process," *Public Works* **120** (4): 47–49 (April 1989).
9. Jones, G. M. "Municipal pumping stations, the case for quality and reliability," *Pumps and Systems Magazine,* pp. 32, 34, 36–37. (September 1993).
10. Pritchard, M. T., D. S. Parker, and A. Malik., "Design-to-cost with BACPAC." In-house publication, Brown and Caldwell, Pleasant Hill, CA.
11. Pritchard, M. T., D. S. Parker, and A. Malik., "Keeping costs from creeping," *Water Environment & Technology* **4** (12): 38–39 (December 1992).
12. AGC. "Partnering. A concept for success," Publ. No. 1205, Associated General Contractors of America, 1957 E Street N.W., Washington, DC 20006.
13. Wilson, R. L., President, Wilson Management Associates, Inc., Glen Head, NY. Personal communication (August 1995).
14. Dicmas, J. L., *Vertical Turbine, Mixed Flow & Propeller Pumps,* McGraw-Hill, New York (1987).
15. Karassik, I. J., W. C. Krutzsch, W. H. Fraser, and J. P. Messina, *Pump Handbook,* 2nd ed. McGraw-Hill, New York (1985).
16. Lever, W. F., *Manual for Highway Storm Water Pumping Stations, Volumes I and II,* Report no. FHWA-IP-82-17, Federal Highway Administration, National Technical Information Service, Springfield, VA (October 1982).
17. *Design of Wastewater and Stormwater Pumping Stations, Manual of Practice FD-4,* Water Environment Federation, Alexandria, VA (1993).
18. *Pump Station Design Manual,* Los Angeles County Flood Control District, Los Angeles, CA (n.d.).

Chapter 2

Nomenclature

MAYO GOTTLIEBSON
CARL W. REH
ROBERT L. SANKS

CONTRIBUTORS
Robert S. Benfell
Earl L. Heckman
Stanley S. Hong

2-1. Abbreviations

The abbreviations given here are the same for both singular and plural. The style follows common American usage. Abbreviations for some special disciplines are given near the beginning of the chapter about that subject.

A	Amperes (also amp).
A	Area, usually the cross-sectional area of a pipe.
a	Acceleration.
a	Acoustic wave speed.
@	At.
abs	Absolute.
ac	Alternating current.
ac/h	Air changes per hour.
acre-ft	Acre-foot (43,560 ft^3).
AF	Adjustable frequency.
AFD	Adjustable-frequency drive (for electric motors).
AM	Ante meridian (morning).
amp	Ampere (also A).
amp-h	Ampere-hour.
AN	As noted.
API	American Petroleum Institute.
ASHRAE	American Society of Heating Refrigerating and Air Conditioning Engineers.
atm	Atmosphere (= 14.7 lb/in.^{2})
avg	Average.
AWG	American wire gauge.
AWWA	American Water Works Association.
B	Bulk modulus of elasticity of water.

bec	Best efficiency capacity (discharge at bep).
bep	Best efficiency point.
BHN	Brinell hardness number (on the C-scale except as otherwise noted).
bhp	Brake horsepower.
bhp-h	Brake horsepower-hour.
BHRA	British Hydraulic Research Association.
Btu	British thermal unit (the heat required to raise the temperature of 1 lb of water by 1°F).
CADD	Computer-aided drafting and design.
c	One hundred.
©	Copyrighted.
C	Hazen–Williams roughness coefficient.
°C	Degree Celsius (centigrade) equals (°F – 32)$^5/_9$.
c–c	Center to center.
℄–℄	Centerline to centerline.
CFD	Computational fluid dynamics. A numerical means to predict flow patterns based on solving the governing equations in three dimensions at discreet nodes that represent the flow space. See Section 3-13.
cfm	Cubic feet per minute (also ft^3/min).
cfs	See ft^3/s.
ckt bkr	Circuit breaker.
cm	Centimeter.
cm^2	Square centimeter.
cm^3	Cubic centimeter.
coef	Coefficient.
conc	Concentration.
cond	Conductivity.
const	Constant.
COP	Coefficient of performance (a ratio of electrical energy input to heat energy output).
CPRV	Chlorine pressure reducing valve.
C/S	Constant speed.
CSO	Combined (wastewater and storm water) sewer overflow
CU	Coefficient of utilization used in lighting calculations.
cu ft	Cubic feet (also ft^3).
cu in.	Cubic inches (also in.^{3}).
cu yd	Cubic yards (also yd^3).
cwt	One hundred pounds.
d	Day.
D	Diameter.
dB	Decibel.
dBA	Decibel-A scale.
dc	Direct current.
dia	Diameter.
DIP	Ductile iron pipe.
do	Ditto.
E	Exponent (e.g., 3.6 E–6 = 3.6×10^{-6}).
E	Modulus of elasticity (see Table A-10, Appendix A).
E	Efficiency (sometimes followed by a subscript to indicate pump, motor, etc.).
ea	Each.
ed	Edition.
Ed.	Editor.
eff	Efficiency.
e.g.	For example.
EGL	Energy grade line.
El	Elevation. Also Elev.
emf	Electromotive force.
ENRCCI	Engineering News-Record Construction Cost Index.

E_p	Pump efficiency.
EPA	See U.S. EPA.
Eq	Equation.
est	Estimated.
et al.	And others.
ETM	Elapsed time meter.
F	Froude number.
°F	Degrees Fahrenheit.
ff	Following pages.
FFT	Fast Fourier transform (analyzer) for vibration analysis.
FLI	Full-load electrical current (see also I_{FL}).
FM	Force main.
fob	Freight on board.
fpm	See ft/min.
fps	See ft/s.
FRP	Fiberglass-reinforced plastic.
FSK	Frequency shift modulation.
ft	Foot or feet.
ft^2	Square feet.
ft^3	Cubic feet.
ft · cd	Foot-candles.
ft · lb	Foot-pound.
ft/min	Feet per minute.
ft/s	Feet per second.
ft^3/s	Cubic feet per second.
FVNR	Full voltage (nonreversing for magnetic motor starters).
g	Acceleration due to gravity (= 9.81 m/s^2 = 32.17 ft/s^2 = 386 $in./s^2$).
g	Gram.
g_c	32.17 $ft \cdot lb_m/lb_f \cdot s^2$
ga	Gauge.
gal	Gallon.
gal/d · cap	Gallons per day · capita.
gal/min	Gallons per minute.
GFCI	Ground fault circuit interrupter (a safety device).
GFI	See GFCI.
gpcpd	Gallons per capita per day (also gpcd; see gal/d · cap).
gpm or GPM	See gal/min.
Grd	Ground.
GTO	Gate turn-off thyristor.
h	Hour.
H or h	Head [usually in meters (feet) of water].
h_L	Headloss.
HDPE	High-density polyethylene (pipe).
HGL	Hydraulic grade line.
HOA	Hand-off-automatic (switch).
hp	Horsepower.
hp · h	Horsepower-hour.
H-Q	Head-capacity.
hr	Hour (see h).
HV	Heating and ventilating (also H&V).
HVAC	Heating, ventilating, and air conditioning (also HV&AC).
H–W	Hazen–Williams (Equations 3-8 and 3-9).
HWL	High water level.
Hz	Hertz (cycles per second)
I	Electrical current.
I_{FL}	Full-load current (amperes).
I_L	Line current (amperes).

I_Θ	Phase current (amperes).
I&C	Instrumentation and control.
ID	Inside diameter (mm [in.]).
i.e.	That is.
IEEE	Institute of Electrical and Electronics Engineers.
IGBT	Insulated-gate bipolar transistor (sometimes found in AF drives).
Imp gal	Imperial gallon.
in.	Inch.
in.^{2}	Square inches.
in. · lb	Inch-pound.
IR	Indicating and recording (instrument).
ISA	Instrument Society of America.
J	Joule.
J	Polar moment of inertia.
K	Bulk modulus of elasticity of water (*B* in Chapter 22).
K	Torsional rigidity (in Chapter 22).
°K	Degree Kelvin (= °C + 273.2).
kc	Kilocycles.
kcal	Kilocalorie.
kcmil	A unit of conductor area equal to the diameter of a wire in mils squared.
kg	Kilogram.
kip	Kilopound (1000 lb).
km	Kilometer.
kPa	Kilopascals (1000 N/m^2).
ksi	Kilopounds (1000 lb) per square inch.
kV	Kilovolt.
kVA	Kilovolt ampere.
kVAR	Reactive kilovolt ampere.
kW	Kilowatt (1000 watts).
kW · h	Kilowatt-hour.
L	Liter.
lb	Pound (means lb_f in this volume).
lb_f	Pound force = $lb_m\ g/g_c$; numerically (not dimensionally) equals lb_m. In this volume lb_f is written lb.
lb_m	Pound mass (= $lb_f \cdot g_c/g$).
lb/ft^2	Pounds per square foot.
lb/ft^3	Pounds per cubic foot.
L/d · cap	Liters per day · capita.
LEL	Lower explosion limit.
LF	Linear feet (also lin ft).
ln	Natural logarithm.
log	Logarithm to base 10.
LPG	Liquid petroleum gas (either propane or butane; specify which).
LWL	Low water level.
m	Meter(s).
m^2	Square meters.
m^3	Cubic meters.
mA	Milliampere.
max	Maximum.
MCC	Motor control center.
MCM	Now kcmil. A unit of conductor area equal to the diameter of a wire in mils squared.
MCP	Motor circuit protector.
Mfr	Manufacturer.
mg	Milligram.
Mgal	Million gallons.
Mgal/d	Million gallons per day.

mgd or MGD	See Mgal/d.
mg/L	Milligrams per liter.
mi	Mile.
mi/h	Miles per hour.
mil	1/1000 in. or 0.025 mm or 25 microns.
min	Minimum.
min	Minute.
ml	Milliliter.
mm	Millimeter.
mol	Mole.
mph	See mi/h.
mV	Millivolts.
N	Newton (the force required to accelerate 1 kg of mass 1 m/s^2).
n	Manning roughness coefficient (Equation 3-17).
NA	Not applicable.
NC	Normally closed (electrical switch).
n.d.	No date.
NEC	National Electrical Code.
NEMA	National Electrical Manufacturers Association.
NO	Normally open (electrical switch).
No.	Number.
NPS	Nominal pipe size.
NPSH	Net positive suction head (q.v. Section 2-2).
$NPSH_A$	Net positive suction head available (q.v. Section 2-2).
$NPSH_R$	Net positive suction head required (q.v. Section 2-2).
NRS	Nonrising stem (for valves).
n_s	Specific speed.
NTU	Nephelometric turbidity units.
OC	Overcurrent.
OD	Outside diameter.
O&M	Operation and maintenance.
OSHA	Occupational Safety and Health Administration.
OS&Y	Outside screw and yoke (indicates a rising stem in a valve).
oz	Ounce.
p	Pole (electrical).
p, pp	Page, pages.
Pa	Pascal (N/m^2).
PAM	Pulse amplitude modulated.
pcf	See lb/ft^3.
Pf	Power factor.
P&ID	Process and instrumentation diagram.
PIR	Peak influent rate.
PLC	Programmable logic controller.
PM	Post meridian (afternoon).
ppb	Parts per billion.
ppm	Parts per million.
PRV	Pressure reducing valve.
PS	Pumping station.
psi	Pounds per square inch (also $lb/in.^2$).
psia	Pounds per square inch absolute (pressure above vacuum, also $lb/in.^2$ abs).
psig	Pounds per square inch gauge (pressure above atmospheric, also $lb/in.^2$ ga).
PTC	Positive temperature coefficient thermistor.
PVC	Polyvinyl chloride.
PWM	Pulse width modulated.
Q	Flowrate, capacity, or discharge.
q.v.	Which see.

°R	Degrees Rankine (= °F + 491.7).
R	Electrical resistance (in ohms).
R	Hydraulic radius (wetted area/wetted perimeter).
R	Reynolds number.
®	Registered trademark (see also ™).
rad	Radian.
RCP	Reinforced concrete pipe.
RCR	Room cavity ratio used in lighting calculations.
rev/min	Revolutions per minute.
rev/s	Revolutions per second.
RMPDR	Recommended minimum pump discharge rate.
RMS	Root mean square (an average of 70.7% of peak value in a sinusoidal waveform).
rpm or RPM	See rev/min.
rps or RPS	See rev/s.
RS	Raw sewage (the older term, "sewage," is now replaced by "wastewater").
RTP	Reinforced thermosetting polyester (the modern acronym for fiberglass or FRP).
RVNR	Reduced voltage nonreversing (for magnetic motor starters).
RW	Raw wastewater.
s	Second.
SCADA	Supervisory control and data acquisition.
scfm	Standard cubic feet per minute.
Sch	Schedule (for pipe).
SCR	Silicon controlled rectifier.
SDR	Average outside diameter of a pipe divided by minimum wall thickness.
sec	Second (see s).
SF	Service factor.
shp	Shaft horsepower.
SI	System International (for metric units).
sp gr	Specific gravity.
sq	Square.
sq ft	Square foot or feet (also ft^2).
sq in.	Square inch or inches (also $in.^2$).
ss	Stainless steel.
Std	Standard.
t	Thickness.
t	Time.
TDH	Total dynamic head (= static head + friction head + other losses).
TDS	Total dissolved solids.
TEFC	Totally enclosed, fan cooled (motor).
temp	Temperature.
TIR	Totalizing, indicating, and recording (instrument).
™	As a superscript, trademark (see also ®).
Typ	Typical, typically.
UL	Underwriters Laboratory.
URV	Upper range value.
U.S.	United States or U.S. customary units.
U.S. EPA	United States Environmental Protection Agency.
V	Volts.
V	Volume.
V_L	Line-to-line voltage.
V_Θ	Phase voltage.
v	Velocity.
VA	Volt-ampere.
VAR	Reactive volt ampere.
VCP	Vitrified clay pipe.
VFD	Variable-frequency drive (also AFD [adjustable-frequency drive]).

viz.	Namely.
V/S	Variable speed.
vs.	Versus.
VTSH	Vertical turbine, solids-handling (pump).
W	Watt.
WC	Water column (pressure in meters [or feet] of water).
WEF	Water Environment Federation.
W · h	Watt-hour.
W · hM	Watt hour meter.
whp	Water horsepower.
WPCF	Water Pollution Control Federation (now Water Environment Federation).
WR^2	Inertia of moving parts in pump and driver.
wt	Weight.
XS	Extra strong (pipe).
XXS	Double extra strong (pipe).
yd	Yard.
yr	Year.
γ	Specific weight.
ϵ	Absolute pipe roughness (inches or millimeters).
μ	Dynamic viscosity.
μ	Poisson's ratio.
ν	Kinematic viscosity.
ρ	Density.
Σ	Sum, summation.
τ	Shearing stress.
ϕ	Phase (electrical current).
ω	Rotational speed (usually rev/min).
Ω	Ohms (resistance).
$\geq$	Equal to or greater than.
$\leq$	Equal to or less than.
$>$	Greater than.
$<$	Less than.
$\simeq$ or $\cong$	Approximately equals.
$\approx$ or ~	Approximately (prefix).
$\pm$	Approximately (suffix).
¢	Cent.
$	Dollar.
'	As a superscript, feet.
"	As a superscript, inches.
%	Percent.
·	Times (multiplication).

2-2. Definitions

The following list of terms is not intended to be complete but to contain only commonly used terms that might be unfamiliar to some readers. Some terms for special disciplines are contained at the beginning of the chapter about that subject.

Absolute pressure: Gauge pressure plus standard atmospheric pressure.

Affinity laws: A set of equations or laws governing the operation (discharge, head, power) of a single centrifugal pump operated at different speeds or, by close approximation, governing the relationship of operation of pumps or impellers of different sizes operated under dynamically similar conditions (see Section 10-3).

Air break: An air gap in a liquid system intended to ensure against back siphonage. The discharge into free atmosphere must be located well above the highest feasible water level. Air breaks are often required for positive protection of potable water.

Ampacity: Current-carrying capacity in amperes.

Archimedes screw: See *helical screw pump*.

Axial flow pump: A pump in which the impeller (somewhat similar to a motorboat propeller) moves the fluid parallel to the pump shaft.

Backflow preventer: A device installed in potable water piping to prevent the flow of nonpotable water into a potable system.

Background: The ambient condition before a change is made, such as the apparent concentration of a substance before the substance is added as a tracer.

Back head (also stuffing box head): The portion of the pump casing attached to the volute that houses the stuffing box and bearings on an end-suction pump.

Barrel (can) pump: A pump inside a pipe.

Best efficiency point: The discharge rate at which an impeller of a given diameter rotating at a given speed operates at maximum efficiency.

Bingham plastic: A material in which flow occurs in a straight-line relationship with shear stress, but only after a certain amount of shear stress is overcome.

Blank: A solution containing all the ingredients except the substance to be measured. Also the instrumental reading obtained from the foregoing.

Booster: A pump that takes suction from a pressurized piping system and discharges, at a higher pressure, to a second, isolated piping system.

Boss: Projection on a pipe fitting to provide extra thickness for a smaller threaded pipe.

Bowl: The casing that contains the impeller of a vertical turbine or vertical axial flow pump.

Brake horsepower: The power delivered to the pump shaft.

Bump: Momentary energization of a motor that causes it to rotate a few revolutions.

Bus: A bar (usually copper) that serves as the common connection for three or more electrical circuits.

Casing: The impeller housing that directs the fluid into the impeller and thence to the discharge flange (see also *scroll*).

Cavitation: Vapor bubbles formed on a solid surface (often an impeller) in contact with the liquid. The vapor bubbles occur when the pressure in the liquid falls below the vapor pressure.

Centrifugal pump: A rotodynamic pump in which the fluid is displaced radially by the impeller. *Colloquial:* Any rotodynamic pump in which the fluid is displaced radially, axially, or by a combination of both.

Characteristic curve: A plot of head versus discharge rate for a pump.

Clear well: A basin of filtered water in a water treatment plant.

Close coupled: A motor mounted close to the pump and in which the pump impeller is mounted on a specially lengthened motor shaft.

Closed impeller: An impeller with vanes enclosed by shrouds on both sides.

Closed system: A liquid piping system that does not include storage open to the atmosphere.

Coating: A liquid that cures after application to a surface. Example: paint (see *Lining*).

Column: The vertical piping through which a suspended pump discharges.

Comminutor: A machine designed to cut and shred stringy materials and large solids. It is sometimes installed before the wet well of a sewage pumping station to prevent clogging of the pumps.

Condition point: A point on the pump characteristic curve at which the pump delivers a given value of flow at a given head.

Conduit fill: The amount of wires allowed in a conduit. The allowable is 40% of the internal cross-sectional area.

Confined space: A space large enough and so configured that a person can bodily enter and perform assigned work; may have limited or restricted means for entry or exit (e.g., tanks, vessels, silos, storage bins, hoppers, vaults, pipes, trenches greater than 1.2 m [4 ft] deep, and not designed for continuous employee occupancy) (see *permit-required confined space*).

Control room: The superstructure of a pumping station, which usually includes electrical equipment and may also include motors for extended shaft or suspended pumps.

Corresponding point: One of two or more points related to one another by all of the affinity laws applied simultaneously.

Cutter pump: A centrifugal pump with a cutting blade (attached to the impeller) designed to cut stringy materials into short segments before entrance into the impeller eye.

Cutwater: The leading edge of the volute wall that separates the volute from the discharge nozzle. Minimum clearance between impeller and volute occurs at the cutwater.

Dashpot: A device in which a fluid escapement decreases the speed of closure (as of a valve disc).

Demand: A schedule of water requirements.

Developed head: The difference between the suction head and discharge head of an operating pump.

Diffusers: Stationary vanes (within the casing) to provide gradually expanding paths to change the direction of flow and convert liquid velocity into pressure.

Dirty water: Water that contains solids such as silt or sand.

Discharge flange: The flange at which liquid is expelled from the pump.

Discharge head: Static head plus friction headloss (in discharge piping) plus velocity headloss at the end of the discharge piping.

Double suction: A centrifugal water pump in which the liquid enters the impeller through eyes on both sides and thereby reduces the force on thrust bearings.

Downstream: In the direction of flow.

Drawdown: The vertical distance over which the surface of the pumped liquid is lowered during the pumping cycle.

Dresser® coupling: See *sleeve coupling*.

Dry pit: See *dry well*.

Dry well: The below-grade structure of a pumping station in which the pumped liquid is contained within piping, valves, and pumps.

Dry well pump: A pump designed to be mounted in a dry well and requiring a flooded suction pipe.

Ductile iron: A cast iron alloyed with magnesium and heat treated to concentrate the carbon into nodules to obtain ductility and high tensile strength. Uncontrolled high heating makes ductile iron revert to brittle gray cast iron.

Duty: Utilization and specific application of equipment, such as operation, standby, or emergency.

Duty cycle: Time-based utilization and specific application of equipment.

Effluent: Liquid discharged from a pumping station.

Extended shaft: Vertical pump and motor separately supported and connected by means of open shafting.

Fillet: Concrete in the bottom of the wet well shaped to smooth liquid flow into the pump suction openings and to prevent the accumulation of solids.

Finished water: Potable, treated water.

Firm pumping capacity: Capacity of the pumping station with the largest pump out of service or on standby.

Flexible coupling: A device to couple two shafts at a small angle.

Flexible pipe coupling: A coupling often used at walls, and sometimes elsewhere, to couple two slightly misaligned pipes.

Flooded suction: Condition in which the pump volute is below the low water level of the wet well.

Foot-candle: The amount of light falling on 1 sq ft of a spherical surface at a distance of 1 foot from a standard candle.

Force main: Piping, external to the station and filled with liquid under pressure, through which the station discharges.

Frazil ice: Crystals of ice formed in turbulent waters.

French drain: A rock-filled pit or trench designed to allow water to seep into surrounding soil.

Friction head: Equivalent head necessary to overcome friction losses through the piping, fittings, and valves.

Gauge pressure: Pressure above standard atmospheric pressure.

Gen-set: Engine generator set.

Gray iron: A brittle cast material principally of iron with 2 to 4% carbon and 1% silicon by weight and an ultimate tensile strength of 140,000 to 410,000 kPa (20 to 60 ksi), compressive strength 3 to 5 times greater than tensile strength, with negligible plastic deformation.

Grinder pump: A centrifugal pump with grinder assembly on an extension of the shaft, designed to grind solids larger than a fraction of an inch before they contact the recessed impeller.

Grooved-end pipe coupling: A pipe coupling used to make pipe fitting and disassembly easier. Often called a Victaulic® coupling (a trademark name).

Groundwater: Raw water obtained from wells.

Hazardous location: Defined in Article 500 of National Electric Code (see also Section 25-3).

Heat tracing: An electric heat tape wrapped around a pipe or tube (usually beneath insulation) to prevent freezing.

Hedstrom's number: A dimensionless number defined by Equation 19-5.

Helical screw pump (also Archimedes screw): A positive displacement, inclined, semi- or fully enclosed screw for lifting liquid and in which both suction and discharge are open to the atmosphere.

Hertz: Frequency in cycles per second.

High service pump: A water pump that discharges to a pressurized piping system, which may include nonisolated, elevated storage.

Home runs: Wires running directly back to the power source from field devices such as lights.

Horizontal pump: A pump mounted with its shaft horizontal.

Hydraulic radius: Cross-sectional area of water divided by the wetted perimeter of an open or closed conduit.

Impeller: A circular casting mounted on a rotating shaft with vanes to accelerate the fluid.

Impeller eye: An opening in the impeller of a centrifugal pump through which liquid enters the impeller.

Impeller hub: The outside portion of the impeller of a centrifugal pump that encloses the eye.

Impeller shroud: The outside part of the impeller of a centrifugal pump to which the vanes are attached.

Inrush (current): The temporary surge of electrical current to a motor that occurs with an across-the-line starter.

Intake: A structure from which the pumps take suction.

Intake basin: See *wet well.*

Invert: The inside bottom of a pipe.

Jockey pump: A small pump that maintains pressure in a distribution system during periods of low demand.

Knee: A reduction in the gradient of a pipeline or a change from a positive to a negative gradient.

Laminar flow: Flow with straight streamlines, i.e., without eddying or turbulence.

Laying length: The length of a piece of pipe or a fitting in a piping system.

Lining: A covering applied or attached to a surface. Examples: T-Lock® and Linabond®.

Lumen: The rate at which light from a one-candle source flows through a unit solid angle (1 m^2 at 1 m or 1 ft^2 at 1 ft).

Lux: A measure of luminous intensity equal to one lumen per square meter or 0.0929 foot-candles.

Mechanical bar screen: A screen of bars, through which the influent of a pumping system flows before entrance to the pumps, with a motorized cleaning mechanism that delivers the screenings to a collector.

Mechanical seal: An assembly consisting of one or two pairs of polished surfaces, one of each pair mounted on (and turning with) the shaft and the other connected to the casing to inhibit the leakage of liquid between the casing and the shaft.

Micron: One thousandth of a millimeter (1×10^{-6} m).

Mil: One thousandth of an inch.

Mixed-flow pump: A pump that produces a combination of centrifugal and axial flow (see Figure 10-8).

Multistage: Two or more impellers mounted on a common shaft housed within casings for each impeller and designed to deliver liquid from the discharge of one casing to the suction of the other.

Net positive suction head (also NPSH): Absolute (total) dynamic head of the pumped liquid at the suction eye of a pump (see also Section 10-4).

Net positive suction head available: The NPSH at which the pump in a given system operates at a given discharge rate.

Net positive suction head required: The minimum NPSH at which a pump can properly operate for a given discharge rate.

Newton: A unit of force in SI units that accelerates a 1-kg mass 1 m/s^2.

Newtonian fluid: A fluid (water, for example) in which shear flow is proportional to shear stress.

Nonclog pump: A centrifugal pump designed to pump liquids containing suspended solids and stringy material. (Ed. note: in spite of the name, they often do clog.)

Nonoverloading: The motor rating that is not exceeded by the brake horsepower at any point on the characteristic curve.

Open impeller: An impeller with vanes on only one side of a single shroud plate.

Open shaft: See *extended shaft.*

Open system: A liquid piping system that includes nonisolated storage open to the atmosphere.

Operating point (for pumps): The head and discharge at which a pump operates in a system; the intersection of the pump characteristic curve and the system curve.

Packing: Semiplastic material installed in a stuffing box to seal the shaft opening in the casing to restrict the leakage of liquid from the casing along the shaft.

Packing gland: See *stuffing box.*

Pascal: A unit of pressure or stress in SI units (= N/m^2).

Permit-required confined space: A confined space with one or more of the following characteristics: (1) contains or has the potential to contain a hazardous atmosphere such as lack of oxygen (<19.95%) or explosive or toxic atmosphere; (2) contains a material that has the potential for engulfing an entrant; (3) has an internal configuration such that an entrant could be trapped or asphyxiated by inwardly converging walls or by a floor that slopes downward and tapers to a smaller cross-section; or (4) contains any other recognized serious safety or health hazard.

Pig: A device for cleaning the inside of a pipe.

Point: Percentage efficiency (of a pump), e.g., 2 points equals 2%.

Poppet valve: A spring-loaded valve that operates automatically to relieve excessive pressure.

Potable water: Drinkable, finished water.

Pressure surges (also transients or water hammer): Rapid fluctuations in pressure in a closed piping system caused by a sudden change in the velocity of liquid in the pipe.

Primary element: An element that quantitatively converts measured variable energy into a form suitable for measurement.

Prime: Pump casing and suction piping completely filled with liquid.

Progressive cavity: A positive displacement pump with a helix rotor in a flexible, double helix casing for handling liquid with large amounts of solids in suspension.

Propeller pump: An axial flow pump suitable for pumping large volumes of liquids without stringy solids at low heads.

Pump: A machine that imparts kinetic and potential energy (from an external energy source) to a liquid to force a discharge from the machine.

Pump-down: An operation in which the water level in a sump is lowered as far as possible by the main pumps—generally to about $D/4$ above the last pump intake of diameter D.

Pumping station: A structure housing pumps, piping, valves, and auxiliary equipment.

Punch list: An itemized list of work the contractor must do to complete the job according to the plans, specifications, and change orders before the owner pays for substantial completion. Warranty repairs are not considered part of the punch list.

Radial vane: Impeller with vanes oriented radially to the shaft.

Raw sewage: Sewage that has not been settled or treated in any manner (except, possibly, for screening or grinding).

Raw water: Nonpotable water that must be treated to become finished water.

Reynolds number: A dimensionless number defined by Equation 3-11.

Root area: The cross-sectional area of a bolt calculated from the diameter at the root of the threads.

Rotodynamic pump: A mechanical device that pressurizes a fluid by means of a rotating impeller.

Runout: The operating condition for one pump, of a group of pumps, operating alone against reduced dynamic head. Hence, the pump operates beyond its bep.

Sail switch: A device used to indicate failure of the ventilation system. A light paddle (within an air duct) is kept pushed up by a normal flow of air, but when air ceases to flow, the paddle falls, thereby allowing a limit switch to open and trigger an alarm.

Salient pole: A rotor (in an electric machine) consisting of projecting elements surrounded by coils for energization.

Scroll: The casing enclosing the rotor of a water turbine.

Secondary element: An element that converts the output of a primary element (q.v.) into an electrical or pneumatic signal for monitoring or control, using switches for discrete signals and a transmitter for analog signals.

Self-priming pump: A centrifugal pump that normally retains prime or automatically reprimes itself when drawing suction from a liquid pool whose surface is below the casing of the pump.

Sequent depth: Depth following a hydraulic jump.

Service factor: A multiplier that, when applied to the rated horsepower, gives the load at which an electric motor can be operated.

Sewage: Domestic wastewater. May contain industrial wastewater.

Shaft sleeve: A replaceable cylinder mounted on (and rotating with) the pump shaft and extending through the stuffing box to protect the shaft from wear caused by the packing.

Shut-off head: The head developed by a centrifugal pump at zero discharge rate (against a closed discharge valve).

Sleeve coupling: A device to couple two plain-end pipes together (see Figure 4-3).

Sludge: Residue remaining after settling or treatment of sewage and raw water. Usually about 3% or more solids concentration in suspension.

Slug: The unit of mass that is accelerated 1 ft/s^2 by a force of 1 lb (~32.17 lb$_m$).

Slug: The instantaneous addition of a known (usually large) amount of a tracer added to a fluid.

Snubber: A device (such as a micropore filter) used to reduce pressure fluctuation to make a gauge indicate time-averaged pressure.

Soffit: The underside of a beam or the inside surface at the top of a pipe.

Soft foot: A machine foot (or support) with a bottom surface not coplanar with those of other feet and requiring compensation in shimming to avoid twisting the machine.

Soft start: Motor starting in which the inrush current is reduced.

Soldier beam: A vertical beam used to resist earth pressure.

Solids capability: The diameter of the largest sphere that can pass through a pump from the suction flange to the discharge flange.

Special: Any piece of pipe (except a full-length straight section), such as short pieces, manifold sections, elbows, reducers, adapter sections with special ends, sections with outlets.

Specific speed: A characteristic number for a geometrically similar series of pumps operating under similar conditions. Called the "type number" in Europe (see Table 10-1 and Figure 10-8).

Speed: Usually specific in this text for the velocity of propagation of a pressure wave.

Spike: A known addition to the concentration of a substance being analyzed; also a sudden change in a variable.

Split case: A two-piece casing bolted together for ease in providing access to the rotating assembly and all internal passages.

Spool: A section of flanged pipe shorter than a standard length.

Static discharge head: The vertical distance from the center of the pump casing to the surface of the liquid at the point of discharge.

Static head: See *total static head.*

Static suction head: The vertical distance from the centerline of the pump casing to the surface of the liquid being pumped; the casing must be below the liquid level (h_s in Figure 10-2).

Static suction lift: The vertical distance from the centerline of the pump casing down to the surface of the liquid being pumped; the casing must be above the liquid level (see h_s in Figure 10-3).

Steady bearing: A bearing between the pump and the motor for the lateral support of extended shafting.

Stuffing box: An assembly containing packing or a mechanical seal through which the pump shaft passes.

Submersible pump: A pump or pump and motor suitable for fully submerged operation.

Suction elbow: Any elbow that is connected to the suction flange of a pump or an elbow, bolted to the floor of a wet well, that engages a submersible pump.

Suction flare or bell: A flaring entrance fitting to the pump suction piping.

Suction head: See *static suction head* and *net positive suction head.*

Suction lift: See *static suction lift.*

Sump: See *wet well.*

Surface water: Raw water obtained from streams, lakes, or reservoirs.

Suspended: A pump in which the impeller is suspended below the liquid level on a vertical shaft supported from a motor above the liquid.

System head curve: Curve of total dynamic head (q.v.) versus flow for all flowrates within the capability of the pumping station.

Therm: 100,000 Btu.

Total dynamic head (TDH): The total head at which a pump operates at any given discharge rate (see Figure 10-2).

Total static head: The difference in elevation between the surface of the pool from which the pump draws water and the surface of the pool into which the outlet discharges (see Figure 10-2).

Trash basket: A basket with bars or screens through which influent of a pumping station flows before entering the liquid pool in the wet well to prevent large solids from entering the pumps.

Tremie concrete: Concrete placed under water through a tube that discharges under the surface of the concrete being deposited.

Tubercles: Rough deposits (formed inside of a pipe) that greatly increase resistance to flow.

Turbulent flow: Flow in which the streamlines are unsteady and the motion at a fixed point vanes in a random manner.

Type number: See *specific speed.*

Umbrella: An extension fastened to a pipe flare to form an entrance similar to a trumpet bell.

Upstream: Opposite to the direction of flow.

Varmeter: A wattmeter corrected for power factor.

Velocity: In pipe flow, the average velocity of the fluid.

Velocity head: The kinetic energy in a moving liquid; the average is $v^2/2g$.

Vertical pump: A pump with a vertical shaft.

Vertical turbine: A centrifugal pump usually with multistage impellers and bowls in a vertical column. It is suitable for pumping from wells and sumps.

Victaulic® coupling: See *grooved-end coupling.*

Volute: A centrifugal pump casing that provides a gradually expanding liquid path to change the direction of flow and convert velocity to pressure.

Vortex impeller: A recessed impeller that creates a vortex in the casing to induce flow. The liquid flows over the impeller, not through it.

Wastewater: Water-borne wastes (including sewage) from households, businesses, and industry.

Water hammer: Rapid, severe, and often destructive changes in pressure in a piping system caused by a sudden change of liquid velocity.

Water horsepower: The power delivered by the pump to the liquid being discharged (= pound of fluid per minute × feet of head/33,000).

Water lance: A long, light pipe used for directing a water jet at the desired point of delivery.

Water seal: Clear water supplied under pressure to the packing or mechanical seal of a pump to lubricate, cool, and flush the packing or seal. (Grease or oil is sometimes used in place of water.)

Wear rings: Removable, replaceable metallic rings on the casing and/or impeller at the impeller hub.

Wet pit: See *wet well.*

Wet well: The below-grade compartment of a pumping station into which the liquid flows and from which the pumps draw suction. Also "pump intake basin" and "sump."

Wet well pump: A pump designed to be directly immersed in the liquid.

Withstand rating: Capability of an electrical device to withstand a short circuit without damage long enough to trip the protective devices.

Wrought pipe fittings: Pipe fittings formed by forging, piercing, hammering, welding, rolling, bending, pressing, or extruding (or by a combination of these).

2-3. Symbols

Some of the symbols shown in Tables 2-1 to 2-6 are standard; others are typical, though not necessarily universal. Except for Table 2-4, which is reproduced from drawings by Brown and Caldwell Consultants, the tables are reproduced here by the courtesy of Greeley and Hansen Engineers.

2-4. Supplementary Reading

1. Lyons, J. L., and C. L. Askland, Jr., *Encyclopedia of Valves,* Van Nostrand Reinhold, New York (1975).

Table 2-1. Single-Line Piping Symbols

Pipe fittings			
Symbol	Description	Symbol	Description
	Cross	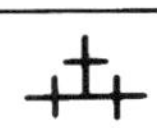	Tee
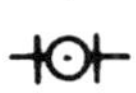	Cross (branch up)	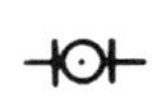	Tee (branch up)
	45° elbow		Tee (branch down)
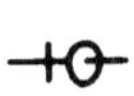	45° elbow (up)		Side outlet tee (up)
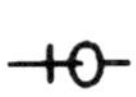	45° elbow (down)		Reducer
	Lateral		Eccentric reducer (Elevation)
	90° elbow	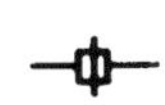	Union, screwed
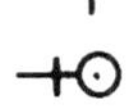	90° elbow (up)	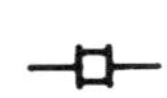	Sleeve coupling
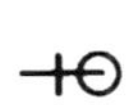	90° elbow (down)	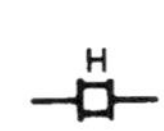	Sleeve coupling (harnessed)
	45° elbow (long radius)		Meter (identify type)
	90° elbow (long radius)		Venturi meter
	Base elbow		Expansion joint, metal bellows
	Side outlet elbow (up)		Expansion joint, rubber bellows
	Side outlet elbow (down)		Strainer

Pipe fittings

Symbol	Description	Symbol	Description
	Duplex strainer		Thermostat
	Flame trap		Pressure gauge
	Lube oil filter	MWL or LWL	Water level alarm
	Scale trape		
	Vent		Thermometer

Valves

Stem relative to plane: Perpendicular	Parallel	Description
		Gate
		Butterfly
		Eccentric plug
		Eccentric plug (alternate)
		Cone
		Ball
		Check, swing
		Globe
		Angle
		Three way
		Flap
		Diaphragm
		Auto air and vacuum release
		Auto air release
		Auto vacuum release
		Hose (bibb)
		Valve, manual operation
		Control valve with hydraulic, pneumatic or electric actuator

Pipe joints

Symbol	Description
	Flange
	Bell and spigot
	Mechanical joint
R	Mechanical joint, restrained
	Bell and ball

Wall fittings

Symbol	Description
	Wall sleeve, caulked
	Wall sleeve, annular seal
	Wall pipe, flange & flange, intermediate collar
	Wall pipe, bell & bell, intermediate collar
	Wall pipe, mechanical joint
	Wall pipe, mechanical joint, intermediate collar

Table 2-2. Mechanical Symbols[a]

Fixture symbols			
Symbol	Description	Symbol	Description
□ FD	Floor drain	○ WH	Water heater
□ AD	Area drain	□ L	Lavatory
○ RD	Roof drain	□ S	Sink
A	Compressed air outlet	☒ SH	Shower stall
G	Gas outlet	U	Urinal
VAC	Vacuum outlet	WC	Water closet
HB	Hose bibb	WC	Water closet, wall hung
□ DF	Drinking fountain	□ EWC	Electric water cooler

Piping symbols			
Symbol	Description	Symbol	Description
ST	Storm	F	Fire protection
S	Soil	SP	Fire sprinkler
		Plan El	
W	Waste	DT DT	Drum trap
D	Drain		P-trap
V	Vent	RT RT	Running trap
AW	Acid waste		Backflow preventer
AV	Acid vent	CO	Cleanout (exposed pipe)
SW	Service water	CO	Cleanout (floor or grade)
CW	Cold water	HWS	Heating water supply
EW	Effluent water	HWR	Heating water return
HW	Hot water (potable)	CWS	Cooling water supply
HWC	Hot water circulating (potable)	CWR	Cooling water return
A	Compressed air	CHS	Chilled water supply
VAC	Vacuum	CHR	Chilled water return

Symbol	Description	Symbol	Description
LPS	Low-pressure steam	* 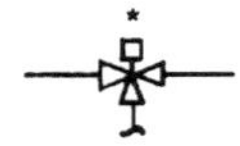	Three-way control valve * (P): pneumatic * (E): electric
LPC	Low-pressure condesate		
MPS	Medium-pressure steam	* 	Control valve * (P): pneumatic * (E): electric
MPC	Medium-pressure condensate		
HPS	High-pressure steam		Pipe guide
HPC	High-pressure condensate		Pipe anchor
FOS	Fuel oil supply	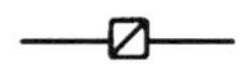	Float and thermostatic trap
FOR	Fuel oil return	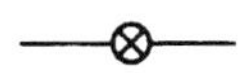	Thermostatic trap
BF	Boiler feed	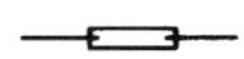	Expansion joint
NG	Natural gas		Expansion joint (harnessed)
LPG	Liquefied petroleum gas		Pipe flow direction
RL	Refrigerant liquid	DN	Pipe pitch down with respect to flow
RS	Refrigerant suction	UP	Pipe pitch up with respect to flow
CPD	Condensate pump discharge		Pipe capped

Miscellaneous symbols

Symbol	Description	Symbol	Description
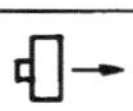	Unit heater	INV EL	Invert elevation
	Convection	℄ EL	Centerline elevation
	Cabinet heater	CFM	Cubic feet per minute
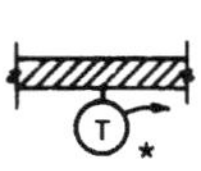	Thermostat * (P): pneumatic * (E): electric	MBH	Thousand Btu/hour
		VD	Volume damper
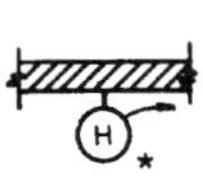	Humidostat * (P): pneumatic * (E): electric	TC	Temperature control
		NTS	Not to scale
BD/EL	Bottom of duct elevation	P	Pneumatic
TD/EL	Top of duct elevation	E	Electric

[a] After Greeley and Hansen Engineers.

Table 2-3. Heating, Ventilating, and Air Conditioning Symbols[a]

Symbol	Description	Symbol	Description
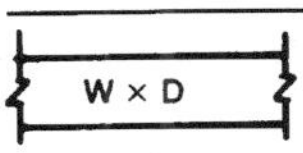	Duct width × depth (first dimension is dimension seen)	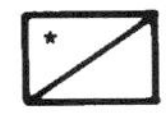	Negative pressure duct section; asterisk represents service designation
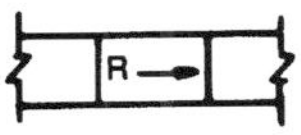	Inclined rise (with respect to air flow)	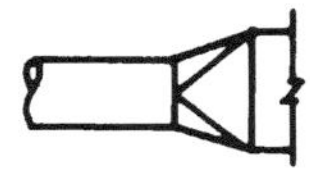	Round duct with transition to rectangular duct
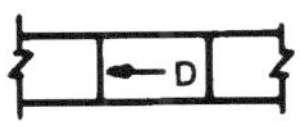	Inclined drop (with respect to air flow)	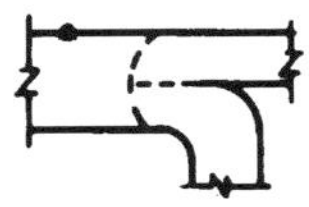	Splitter damper
	Volume damper (shaft parallel to paper)		Turning vanes
	Volume damper (shaft perpendicular to paper)	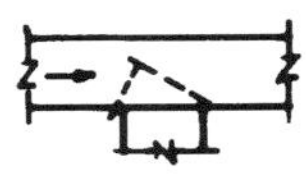	Deflecting damper with rod and lock
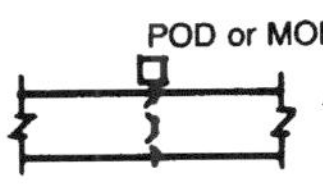	Automatic opposed blade control damper (POD) pneumatic; (MOD) electric	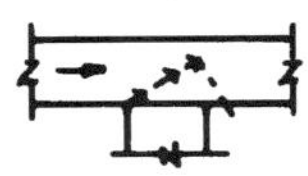	Extractor with rod and lock
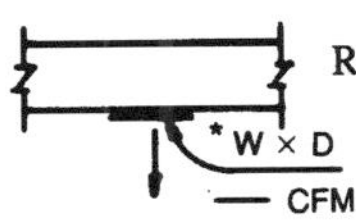	Register or grille (face perpendicular to paper); asterisk represents equipment designation	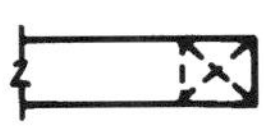	Supply duct down (positive pressure)
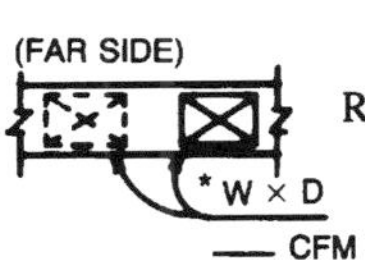	Register or grille (face parallel to paper); asterisk represents equipment designation		Exhaust or return duct down (negative pressure)
	Equipment designations		Flexible duct connection
R	Register		Access doors in duct
G	Grille	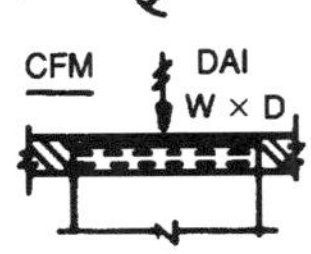	Outside air intake
CR	Ceiling register		
CG	Ceiling grille	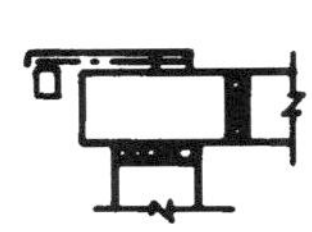	Single inlet fan (plan view)
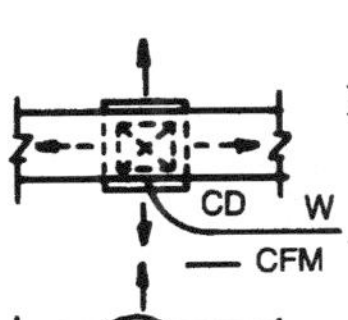	Rectangular ceiling diffuser (size given refers to neck size)	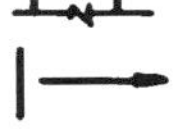	Outlet air direction
			Inlet air direction
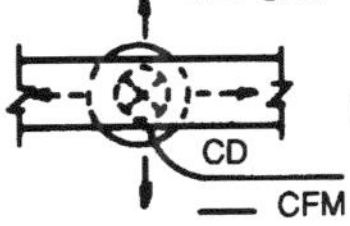	Round ceiling diffuser (size given refers to neck size)	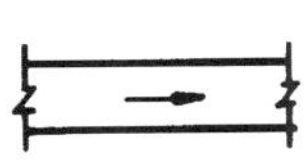	Direction of air flow in duct
			Service designations
		DA	Outside air
	Positive pressure duct section; asterisk represents service designation	S	Supply air
		E	Exhaust air
		R	Return air

[a] After Greeley and Hansen Engineers.

Table 2-4. Process and Instrumentation Diagram Symbols[a]

Miscellaneous mechanical equipment symbols					
Symbol	Description	Symbol	Description	Symbol	Description
	Centrifugal pump		Rotary lobe compressor		Spiral heat exchanger
	Submersible sump pump		Liquid ring blower or compressor		Shell-and-tube-type heat exchanger
	Vertical pump		Silencer	RAG	Right angle gear
	Gear pump		Inlet air filter silencer		Tank
	Rotary lobe pump		Mixer		Waste gas burner
	Progressive cavity pump	ASD	Adjustable-speed drive		Engine
	Diaphragm pump		Generator		Horizontal surface mixer
	Boiler		Grinder pump		Sight glass
	Blower or fan		Plate-type heat exchanger		
	Compressor				

Valve and actuator symbols					
Symbol	Description	Symbol	Description	Symbol	Description
N.O.	Normally open		Diaphragm valve	F.O.	Three-way valve (with typical fail position)
N.C.	Normally closed				
N.O. N.C.	Gate valve		Needle valve		
N.O. N.C.	Globe valve		Balancing cock	F.O. F.O.	Four-way valve (with typical fail position)
N.O. N.C.	Plug valve		Knife gate valve		Pressure-reducing valve (flow to right)
N.O. N.C.	Ball valve		Circuit-balancing valve		Back-pressure-reducing valve (flow to left)
	Butterfly valve		Telescoping valve		Valve with hand actuator
	Check valve		Relief valve	S	Solenoid-operated valve
	Ball check valve		Float valve	M	Electrical-motor-operated valve
	Pinch valve	F.O.	Fail open		Piston-operated valve
		F.C.	Fail closed		
	Angle valve				

Valve and actuator symbols

Symbol	Description	Symbol	Description	Symbol	Description
	Diaphragm-operated valve		Thermostatically controlled valve		Sluice gate (normally closed)
	Electrohydraulic-operated valve		Sluice gate (normally open)		

Instrumentation symbols for primary elements

Symbol	Description	Symbol	Description	Symbol	Description
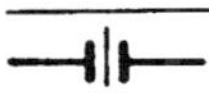	Orifice plate		Flume	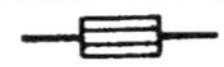	Flow-straightening vanes
	Venturi or flow tube	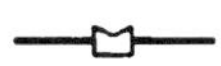	Weir	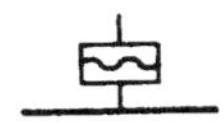	Chemical seal
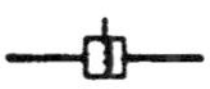	Averaging pitot tube	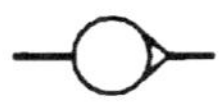	Variable-area flow meter (rotameter)		Concentric chemical seal
	Propeller or turbine meter		Rupture disc		Temperature well

Process and signal line symbols

Symbol	Description	Symbol	Description	Symbol	Description
	Main process flow (with typical direction of flow shown)		Electric signal (analog)		Electromagnetic or sonic signal (unguided)
			Pneumatic signal (discreet)		Software or data link
	Secondary process flow		Electric signal (discreet)		Mechanical link
	Instrument supply, process taps		Capillary tube		Hydraulic
	Pneumatic signal (analog)		Electromagnetic or sonic signal (guided)		

[a] Courtesy of Brown and Caldwell Consultants.

Table 2-5. Electrical Engineering Diagram Symbols[a]

SYMBOL	DESCRIPTION
	DC MOTOR WITH SERIES FIELD
	REACTOR (NONMAGNETIC CORE)
	REACTOR (MAGNETIC CORE)
	HALF WAVE RECTIFIER
	FULL WAVE RECTIFIER
	BATTERY
	CAPACITOR
	DIODE
R	RESISTOR
R	RESISTOR OR RHEOSTAT VARIABLE
R, M	RHEOSTAT-MOTOR OPERATED
	3-PHASE WYE (UNGROUNDED)
	3-PHASE WYE (GROUNDED)
	3-PHASE WYE (RESISTANCE GROUNDED)
	3-PHASE DELTA (UNGROUNDED)
	3-PHASE DELTA (GROUNDED)
	TWO MACHINES DIRECT CONNECTED
	MECHANICAL CONNECTION
	MECHANICAL INTERLOCK
	LIGHTNING OR SURGE ARRESTER

CONTROL AND INSTRUMENT SWITCHES

SYMBOL	DESCRIPTION
CS	CONTROL SWITCH
SL	SECONDARY CONTROL
GS	GOVERNOR SWITCH
SCS	SPEED CONTROL SWITCH
SYS	SYNCHRONIZING SWITCH
DT	DUTY TRANSFER SWITCH
VS	VOLTMETER SWITCH
AS	AMMETER SWITCH
SS	SELECTOR SWITCH
VR	VOLTAGE REGULATOR
ANN	ANNUNCIATOR
POT	POTENTIOMETER

SWITCHES

SYMBOL	DESCRIPTION

PROTECTIVE RELAYS

SYMBOL	DESCRIPTION
27	UNDERVOLTAGE
32	AC DIRECTIONAL OVERCURRENT
46	PHASE BALANCE
47	PHASE SEQUENCE VOLTAGE
48	INCOMPLETE SEQUENCE
49	THERMAL
50	INSTANTANEOUS OVERCURRENT
51	AC TIME OVERCURRENT
67	AC DIRECTIONAL OVERCURRENT
68	BLOCKING
81	FREQUENCY
86	LOCK OUT
87	DIFFERENTIAL
GSR	GROUND SENSING RELAY

CONTROL DEVICES AND SWITCHES

SYMBOL	DESCRIPTION
AR	AUXILIARY RELAY
CR	CONTROL RELAY
LS	LIMIT SWITCH
LL	LIQUID LEVEL SWITCH
DP	DIFFERENTIAL PRESSURE SWITCH
F	FLOAT SWITCH

CONTROL DEVICES

SYMBOL	DESCRIPTION
FW	FLOW SWITCH
P	PRESSURE SWITCH
V	VACUUM SWITCH
T	THERMOSTAT
TQ	TORQUE SWITCH
TR	TIMING RELAY
VIB	VIBRATION SWITCH

METERS AND INSTRUMENTS

SYMBOL	DESCRIPTION
A	AMMETER
V	VOLTMETER
W	WATTMETER
WH	WATTHOUR METER
DM	DEMAND METER
VAR	VARMETER
PF	POWER FACTOR METER
F	FREQUENCY METER
TG	TACHOMETER GENERATOR
TA	TACHOMETER
SY	SYNCHROSCOPE
TT	TOTAL TIME OR ELAPSED TIME METER
TMR	TIMER
T°	TEMPERATURE
GD	GROUND DETECTOR
SI	SPEED INDICATOR

* ⊗ – INDICATING LIGHT (* - INDICATES COLOR)

- G – GREEN INDICATES NON OPERATING CONDITION
- R – RED INDICATES OPERATING CONDITION
- B – BLUE INDICATES TROUBLE CONDITION
- Y – YELLOW INDICATES OPERABLE CONDITION
- A – AMBER
- C – CLEAR
- O – ORANGE
- W – WHITE

SYMBOL	DESCRIPTION
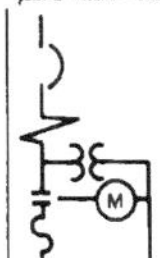	ACROSS-THE-LINE MAGNETIC CIRCUIT BREAKER OR MOTOR CIRCUIT PROTECTOR COMBINATION STARTER WITH CONTROL TRANSFORMER AND OVERLOAD RELAYS (M-OPERATING COIL)
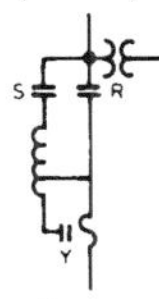	REDUCED VOLTAGE STARTER AUTOTRANSFORMER TYPE. CLOSED CIRCUIT TRANSITION WITH CONTROL TRANSFORMER AND OVERLOAD RELAYS (S-START CONTACTOR. R-RUN CONTACTOR. Y-WYE CONTACTOR)
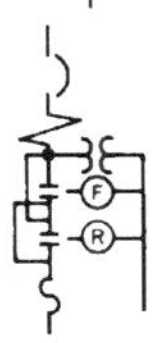	ACROSS-THE-LINE MAGNETIC CIRCUIT BREAKER OR MOTOR CIRCUIT PROTECTOR COMBINATION REVERSING STARTER WITH CONTROL TRANSFORMER AND OVERLOAD RELAYS (F-FORWARD. R-REVERSE)
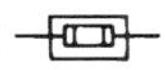	OIL FUSE CUTOUT (RATING)
	TRANSFORMER
	AUTOTRANSFORMER
	POTENTIAL TRANSFORMER
	CURRENT TRANSFORMER
	MANUAL CONTACTOR
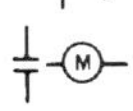	MAGNETIC CONTACTOR WITHOUT OVERLOAD PROTECTION (M-OPERATING COIL)
	THERMAL OVERLOAD ELEMENT (OL)
	SQUIRREL CAGE MOTOR
	WOUND ROTOR INDUCTION MOTOR
	SYNCHRONOUS MOTOR OR GENERATOR (G INDICATES GENERATOR)
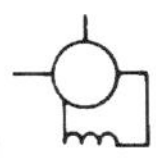	DC MOTOR WITH SHUNT FIELD OR DC GENERATOR (G INDICATES GENERATOR)

[a]Courtesy of Greeley and Hansen Engineers.

Table 2-6. Electrical Power and Lighting Symbols[a]

SYMBOL	DESCRIPTION
	SERVICE ENTRANCE (OVERHEAD)
	UNDERGROUND CABLE
	UNDERGROUND CONDUIT OR DUCT
	CONCEALED CONDUIT IN BUILDING
	EXPOSED CONDUIT OR DUCT IN BUILDING
	CONDUIT WITH ONE NEUTRAL AND ONE OR MORE HOT WIRES (LONG LINE IS FOR NEUTRAL, SHORT LINE IS FOR HOT)-LIGHTING
	HOME RUN TO PANELBOARD
	GROUNDING CONDUCTOR
	FLEXIBLE CONDUIT
	CONDUIT SEAL
	CONDUIT (UP)
	CONDUIT (DOWN)
	CONDUIT CAPPED
JB JB	JUNCTION BOX
PB PB	PULL BOX
TB	TERMINAL BOX
	GROUNDING ROD
	LIGHTING PANEL
	EQUIPMENT ENCLOSURE
	MANUALLY OPERATED SWITCH OR STARTER
M	MAGNETICALLY OPERATED STARTER FULL VOLTAGE
R	MAGNETICALLY OPERATED STARTER REDUCED VOLTAGE
C	MAGNETIC CONTACTOR
	THREE PHASE MOTOR (-HP)
	SINGLE PHASE MOTOR (-HP)
	PUSH BUTTON STATION
L	PUSH BUTTON WITH LOCKOUT ATTACHMENT
	CEILING OR PENDANT LIGHT
	BRACKET LIGHT
	COLUMN LIGHT
	EMERGENCY LIGHT
	EXIT LIGHT(BRACKET, PENDANT)
	FLOOD LIGHT
	FLUORESCENT LIGHT
	FLUORESCENT STRIP LIGHT
	FLUORESCENT LIGHT WITH EMERGENCY POWER SUPPLY UNIT
EM	EMERGENCY BATTERY UNIT
	EMERGENCY BATTERY LIGHTING HEAD
	UNIT HEATER
	RECEPTACLE (SINGLE AND DUPLEX)
XP	EXPLOSION-PROOF
WP	WEATHERPROOF

SYMBOL	DESCRIPTION
	RECEPTACLE - 3 PHASE
C	CLOCK
	SWITCH - SINGLE POLE
2P	SWITCH - 2 POLE
PL	SWITCH WITH PILOT LIGHT
T	SWITCH WITH THERMAL OVERLOAD PROTECTION
3W, 4W	SWITCH - 3 OR 4 WAY
	PLANT INTERCOM
	PUBLIC TELEPHONE
	PHOTOELECTRIC RELAY
SV	SOLENOID VALVE
K	SWITCH-KEY OPERATED
M	SWITCH-MOMENTARY CONTACT
	POWER POLE

SCHEMATIC AND WIRING DIAGRAM SYMBOLS

SYMBOL	DESCRIPTION
*	CIRCUIT BREAKER (-TRIP RATING)
	OPERATING COIL
	CONTACT - NORMALLY OPEN
	CONTACT - NORMALLY CLOSED
TC	NORMALLY OPEN CONTACT WITH TIME DELAY CLOSING
TO	NORMALLY CLOSED CONTACT WITH TIME DELAY OPENING
	3 - POSITION SELECTOR SWITCH
	NORMALLY OPEN PUSH BUTTON
	NORMALLY CLOSED PUSH BUTTON
	DOUBLE CIRCUIT PUSH BUTTON
	DOUBLE CIRCUIT MAINTAINED TYPE PUSH BUTTON
IC TO	NORMALLY OPEN CONTACT INSTANTANEOUS CLOSING AND TIME DELAY OPENING
	HORN OR SIREN
	BUZZER
	BELL
	CONDUCTORS CONNECTED
	GROUND CONNECTION

SYMBOL	DESCRIPTION
*	BUS (*-RATING)
	INCOMING FEEDER
	OUTGOING FEEDER
	POTHEAD
	DRAWOUT TYPE DEVICE
*	FUSE (*-RATING)
*	DRAWOUT TYPE FUSE (-RATING)
*	SINGLE-THROW DISCONNECT SWITCH (-RATING)
*S/*F	FUSED DISCONNECT SWITCH (*S-SWITCH RATING, *F-FUSE RATING)
	DOUBLE THROW SWITCH (-RATING)
	LOW VOLTAGE AIR CIRCUIT BREAKER WITHOUT TRIP DEVICE NON-AUTO (-FRAME RATING)
*F *T	LOW VOLTAGE DRAWOUT TYPE AIR CIRCUIT BREAKER (*F-FRAME RATING, *T-TRIP RATING)
F T	LOW VOLTAGE TYPE AIR CIRCUIT BREAKER (F-FRAME RAT'NG T-TRIP RATING)
*F S	LOW VOLTAGE AIR CIRCUIT BREAKER (*F-FRAME RATING, S-SOLID STATE SENSOR RATING)
	LOW VOLTAGE DRAWOUT TYPE AIR CIRCUIT BREAKER ELECTRICALLY OPERATED
	LOW VOLTAGE DRAWOUT TYPE AIR CIRCUIT BREAKER WITH ONE NORMALLY OPEN AND ONE NORMALLY CLOSED AUXILIARY CONTACT
52	MEDIUM AND HIGH VOLTAGE CIRCUIT BREAKER
52	MEDIUM AND HIGH VOLTAGE CIRCUIT BREAKER DRAWOUT TYPE
52	MEDIUM AND HIGH VOLTAGE CIRCUIT BREAKER WITH DISCONNECT SWITCH

[a]Courtesy of Greeley and Hansen Engineers.

Chapter 3

Flow in Conduits

ROBERT L. SANKS

CONTRIBUTORS
Appiah Amirtharajah
Alfred B. Cunningham
William F. H. Gros
George E. Hecker
James J. McCormick
Rhys M. McDonald
Charles D. Morris
Dennis R. Neuman
Constantine N. Papadakis
Robert E. Phillips
Sanjay P. Reddy
Earle C. Smith
Thomas M. Walski
Gary Z. Watters
William Wheeler

This chapter is a primer for those with little or no knowledge of hydraulics and a review for those with more experience. Even experts may find some of the material (such as that dealing with pipe flow tables and formulas, field measurements of flowrate and friction coefficients, and the use of tracers or models) to be new and useful.

Standards and specifications, such as ANSI B36.10, are commonly (and sufficiently) identified by the call number and, hence, are not referenced in Section 3-14. Double designations, such as ANSI/AWWA 104/A21.4, means that AWWA 104 is the same as ANSI A21.4. Titles and publishers are listed in Appendix E. Addresses of publishers are given in Appendix F.

3-1. Fundamentals of Hydraulics

The analysis of water flow in closed and open conduits depends on three fundamental principles—the conservation of (1) mass, (2) energy, and (3) momentum. Each principle is considered in terms of the equation(s) derived from its application.

Continuity Equation for Mass

Based on the conservation of mass, a material balance for steady, continuous flow in any conduit means that the flowrate of material weight in equals the flowrate of material weight out, or

$$\rho_1 Q_1 = \rho_2 Q_2 = \text{constant}$$

where ρ is density and Q is flowrate or discharge. However, Q equals average velocity times cross-sectional area, and for water $\rho_1 = \rho_2$, so

$$Q = A_1 v_1 = A_2 v_2 = \text{constant} \quad (3\text{-}1)$$

Energy Equation

Energy per unit mass in flowing water, expressed in kg · m/kg (or ft · lb/lb), is the summation of three quantities:

- Elevation head (potential energy) above a datum (an assumed level), usually termed z and expressed in meters (or feet);
- Pressure head (also potential energy). To put this into the same units as z (elevation), the pressure (p) is divided by the specific weight (γ) and is expressed in meters (or feet); pressure head is shown by the height to which the fluid would rise in a piezometer (an open tube), as shown in Figure 3-1;
- Velocity head (kinetic energy). To put this into the same units (meters or feet), it is expressed as $v^2/2g$, where v is velocity in meters per second (feet per second) and g is acceleration due to gravity, 9.81 m/s^2 (32.2 ft/s^2).

In equation form, for either open channels or closed and pressurized conduits, the total energy head is given by

$$E = z + \frac{p}{\gamma} + \frac{v^2}{2g} \tag{3-2}$$

In open channel flow, if the datum plane is taken at the invert (bottom of the conduit) and if the invert is level, z becomes zero and E is defined as the specific energy, E_s:

$$E_s = \frac{p}{\gamma} + \frac{v^2}{2g} = y + \frac{v^2}{2g} \tag{3-3}$$

where y is the depth of water.

Conservation of Energy

The energy of water flowing between two points in either an open channel or a closed conduit is derived from Equation 3-2 and becomes the Bernoulli equation

$$z_1 + \frac{p_1}{\gamma} + \frac{v_1^2}{2g} = z_2 + \frac{p_2}{\gamma} + \frac{v_2^2}{2g} + h_f + \Sigma h_m \tag{3-4}$$

where h_f is the energy per unit mass dissipated by friction between points 1 and 2 and Σh_m is the summation of headlosses due to turbulence in pipe fittings. The datum plane can be assumed at any level in Equation 3-4 because as z increases, p/γ decreases by the same amount. The application of Equation 3-4 to pipes is shown in Figure 3-1. Note how the pressure head, the height to which water would rise in a piezometer (an open vertical tube), decreases as the velocity increases (as it must) in the smaller pipe.

Energy and Hydraulic Grade Lines

The energy grade line is a plot of the energy head, E (as defined by Equation 3-2), vs. horizontal distance. In Figure 3-1, the energy grade line drops both gradually (due to friction head loss) and in steps (due to the turbulence in bends and discontinuities). The pressure at point 1 forces the water to rise in the piezometer. A pressure gauge at point 1 would read p_1, which, divided by γ, is the pressure head h_1. The locus of all pressure head values is the hydraulic grade line—always below the energy grade line by

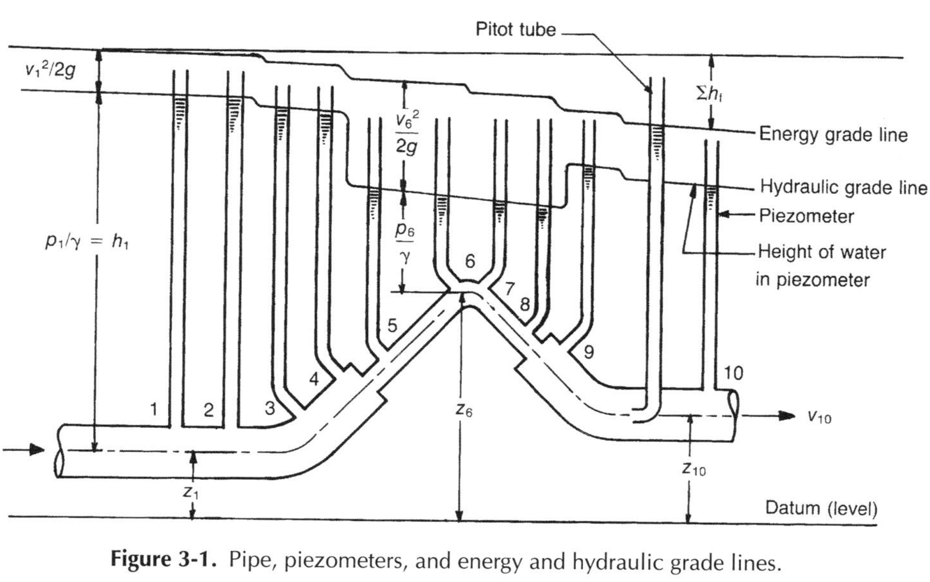

Figure 3-1. Pipe, piezometers, and energy and hydraulic grade lines.

the amount of velocity head, $v^2/2g$. At each bend or constriction there is a small energy loss due to turbulence. At expansions (point 8), the energy loss is substantially greater. Even if the pipe were straight and prismatic from point 1 to point 10, there would be an energy loss due to friction caused by (1) shear at the wall and in the fluid and (2) turbulence due to eddy formation. If the velocity in a medium-sized pipe (300 mm or 12 in.) is less than about 0.01 m/s (0.03 ft/s), there is no turbulence, friction depends only on shear, and the flow is called *laminar* or *tranquil*. The velocity distribution for laminar flow is pictured in Figure 3-2a. At the velocities usual in pipes, there are secondary, crosswise currents that cause mixing, the flow is turbulent, and the velocity distribution across a pipe appears as in Figure 3-2b. Between laminar and turbulent flow is a transitional zone where the flow might shift from laminar to turbulent and back again (or it might be a combination of both).

The pitot tube in Figure 3-1 penetrates the pipe wall and the probe is bent with its open end facing upstream. In this open end of the probe of the pitot tube, the velocity is zero because there is no flow in the tube. But a short distance upstream, the velocity in the center of the pipe is v_c. Between that upstream point and the end of the probe, the kinetic energy head, $v_c^2/2g$, is converted to potential energy head and the water in the tube rises to the energy grade line. Actually, the water level in the pitot tube rises slightly above the energy line because v_c is the maximum, not the average, velocity. An average of all possible pitot tube readings from wall to center would match the energy grade line.

Momentum Equation

The force acting on a mass accelerates it according to Newton's third law of motion:

$$\mathbf{F} = \frac{d(m\mathbf{v})}{dt} \qquad (3\text{-}5)$$

where **F** is force, m is mass, **v** is velocity, and t is time. **F** and **v** are vectors, and **F** can be considered the resultant (the vectorial sum) of any number of forces. If the mass remains constant, then

$$\mathbf{F} = m\frac{d\mathbf{v}}{dt} = ma \qquad (3\text{-}6)$$

where $d\mathbf{v}/dt$ is acceleration in the direction of **F**. In flowing water, it is convenient to rearrange Equation 3-6 to

$$\mathbf{F} = m\frac{d\mathbf{v}}{dt} = \rho Q\Delta\mathbf{v} = \rho Q(\mathbf{v}_2 \rightarrow \mathbf{v}_1) \qquad (3\text{-}7)$$

where **F** is the sum (resultant) of all forces acting, ρ is mass density, Q is flowrate or discharge, and the arrow sign indicates vectorial subtraction—change of velocity.

A force acting on a free jet to deflect it is applied by the curved vane in Figure 3-3. Vectorial subtraction is made by changing the sign of vector $\rho Q\mathbf{v}_1$ then adding it vectorially (as depicted in Figure 3-3b or 3-3c) to obtain **F**. The reaction of the jet against the vane is equal and opposite.

The momentum equation is needed to find the required strength of tie-downs, anchors, and thrust blocks to restrain piping at elbows, tees, etc. An example of the use of Equation 3-7 is given in Section 3-8.

3-2. Friction Losses in Pipe

The first well-known formula for flow in pipes was proposed by deChezy. The deChezy friction coefficient was given by a complicated equation developed

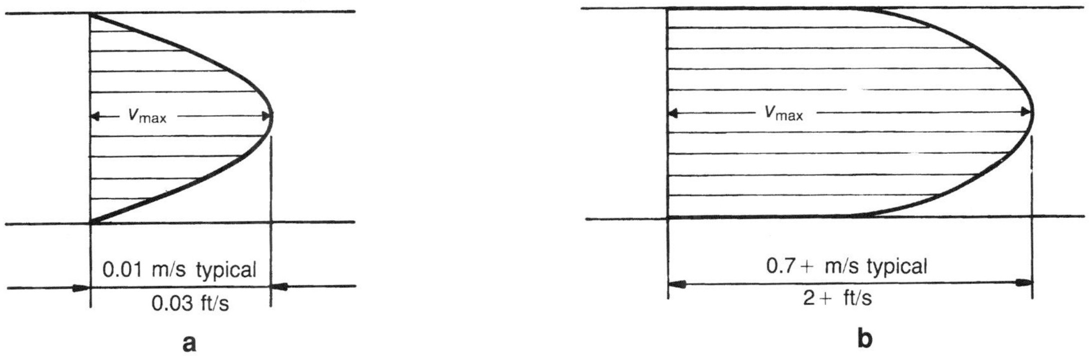

Figure 3-2. Velocity profile in a 300-mm (12-in.) pipe. (a) Laminar flow; (b) turbulent flow.

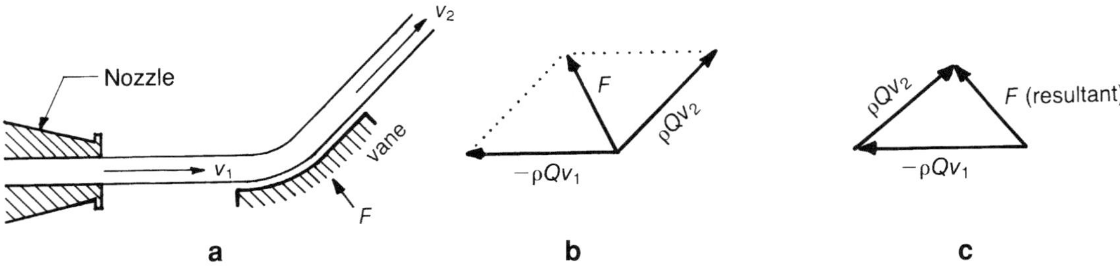

Figure 3-3. Impulse momentum. (a) Schematic diagram; (b) vector diagram; (c) equivalent vector diagram.

by Kutter. These formulas are no longer in common use. The Hazen–Williams (H–W) formula has been used in the United States for 90 yr. It is simple and easy to use, and it has been verified by many field observations for common sizes of pipes at conventional flowrates. The H–W formula is essentially required by the Ten-State Standards [1]. Unfortunately, the formula is irrational, it is valid only for water at or near room temperature and flowing at conventional velocities, the flow regime must be turbulent, and the C factor varies with pipe size. These disadvantages seem to be generally ignored, but errors are appreciable for pipes less than 200 mm (8 in.) and larger than, say, 1500 mm (60 in.), for very cold or hot water, and for unusually high or low velocities. Nevertheless, its continuous use has engendered a nearly blind faith in it.

The Manning equation (Section 3-5) is somewhat similar to the H–W formula and is subject to the same limitations. It is widely used in the United States for open channel flow, such as pipes that are partly full. Sometimes it is used for full pipes, but for this application it has no advantage over the H–W formula.

The Colebrook–White equation is more accurate than the H–W formula and is applicable to a wider range of flow, pipe size, and temperature. It is widely used in the United Kingdom and elsewhere in Europe.

The Darcy–Weisbach equation is the only rational formula, and it is applicable to turbulent, laminar, or transitional flow, all sizes of pipe, and any incompressible Newtonian fluid at any temperature. It has not been popular because, being an implicit equation, it must be solved by successive trials. It was therefore inconvenient to use, but now, modern computers (even programmable pocket calculators) can be used to solve the equation in seconds.

In this text, only the H–W, Manning, and Darcy–Weisbach equations are discussed. Benedict [2] has discussed formulas for pipe flow extensively.

Hazen–Williams Equation

The Hazen–Williams equation, developed from extensive reviews of data on pipes installed all over the world, was made public in 1905 [3]. The appeal of the equation is due partly to its simplicity and ease of use, partly to the source of the data (real pipes in the field—not just laboratory pipes), and, by now, to a tradition of nearly a century of use and faith in the results. The original form of the Hazen–Williams equation [3] was developed in U.S. units. In SI units, the equation is

$$v = 0.849CR^{0.63}S^{0.54} \tag{3-8a}$$

where v is velocity in meters per second, C is a coefficient ranging from about 80 for very rough pipes to 150 or more for smooth pipes, R is the hydraulic radius in meters, and S is the friction headloss per unit length or slope of the energy grade line in meters per meter. The hydraulic radius is defined as the water cross-sectional area divided by the wetted penmeter. For full pipes, R reduces to $D/4$ where D is the ID.

In U.S. customary units, the equation is

$$v = 1.318CR^{0.63}S^{0.54} \tag{3-8b}$$

where v is in feet per second, R is in feet, and S is in feet per foot.

Friction headloss is expressed more conveniently as the gradient, h_f, in meters per 1000 m (or feet per 1000 ft) instead of S. Velocity can also be expressed as discharge, Q, divided by water cross-sectional area, A. Substituting and rearranging Equation 3-8 yields another, somewhat more convenient form. In SI units,

$$h_f = \left(10{,}700\left(\frac{Q}{C}\right)^{1.85}\right)D^{-4.87} = \left(\frac{151Q}{CD^{2.63}}\right)^{1.85} \tag{3-9a}$$

where h_f is meters per 1000 m, Q is cubic meters per second, and D is pipe diameter in meters. Values of C are given in Table B-5 (Appendix B).

Equation 3-9a, expressed in U.S. customary units, is

$$h_f = \left(10{,}500\left(\frac{Q}{C}\right)^{1.85}\right)D^{-4.87} = \left(\frac{149Q}{CD^{2.63}}\right)^{1.85} \quad (3\text{-}9b)$$

where h_f is feet per 1000 ft, Q is gallons per minute, D is pipe diameter in inches, and C, again, is given in Table B-5 (Appendix B).

See Subsection "Friction Coefficients" for a discussion of the H–W C factor.

Darcy–Weisbach Equation

The equation for circular pipes is

$$h = f\frac{L}{D}\frac{v^2}{2g} \quad (3\text{-}10)$$

where h is the friction headloss in meters (feet), f is a coefficient of friction (dimensionless), L is the length of pipe in meters (feet), D is the inside pipe diameter in meters (feet), v is the velocity in meters per second (feet per second), and g is the acceleration of gravity, 9.81 m/s^2 (32.2 ft/s^2). The advantages of the Darcy–Weisbach equation are as follows:

- It is based on fundamentals.
- It is dimensionally consistent.
- It is useful for any fluid (oil, gas, brine, and sludges).
- It can be derived analytically in the laminar flow region.
- It is useful in the transition region between laminar flow and fully developed turbulent flow.
- The friction factor variation is well documented.

The disadvantage of the equation is that the coefficient f depends not only on roughness but also on Reynolds number, a variable that is expressed as

$$\boldsymbol{R} = \frac{vD}{\nu} \quad (3\text{-}11)$$

where $\boldsymbol{R}$ is Reynolds number (dimensionless), v is velocity in meters per second (feet per second), D is the pipe ID in meters (feet), and ν is kinematic viscosity in square meters per second (square feet per second) as given in Appendix A, Tables A-8 and A-9.

Determination of f

In the laminar flow region where $\boldsymbol{R}$ is less than 2000, f equals $64/\boldsymbol{R}$ and is independent of roughness. Between Reynolds numbers of 2000 and about 4000, flow is unstable and may fluctuate between laminar and turbulent flow, so f is somewhat indeterminate. When $\boldsymbol{R}$ is very large (greater than about 10^5), the flow is fully turbulent and f depends only on roughness. In the transition zone between turbulent and laminar flow, both roughness and $\boldsymbol{R}$ affect f, which can be calculated from a semianalytical expression developed by Colebrook [4]:

$$\frac{1}{\sqrt{f}} = -2\log_{10}\left(\frac{\varepsilon/D}{3.7} + \frac{2.51}{\boldsymbol{R}\sqrt{f}}\right) \quad (3\text{-}12)$$

where ε is the absolute roughness in millimeters (or inches or feet) and D is the inside diameter in millimeters (or inches or feet), so that ε/D is dimensionless. The Moody diagram, Figure B-1 (Appendix B), was developed from Equation 3-12 [5]. Note that the curves are asymptotic to the smooth-pipe curve (at the left). To the right, curves calculated from the Colebrook (also called Colebrook–White) equation are indistinguishable from the horizontal lines for fully developed turbulent flow given in Prandtl [6]. The probable variation of f for commercial pipe is about ±10%, but this variation is masked by the uncertainty of quantifying the surface roughness.

An explicit, empirical equation for f was developed by Swamee and Jain [7]:

$$f = \frac{0.25}{\left[\log_{10}\left(\frac{\varepsilon/D}{3.7} + \frac{5.74}{\boldsymbol{R}^{0.9}}\right)\right]^2} \quad (3\text{-}13)$$

The value of f calculated from Equation 3-13 differs from f calculated from the Colebrook equation by less than 1%.

Friction headloss can be determined from the Darcy–Weisbach equation in a number of ways:

- Use one of the Appendix tables (B-1 to B-4) to find the appropriate pipe size. Compute $\boldsymbol{R}$, find f from the Moody diagram, and compute an accurate value of h from the Darcy–Weisbach equation. Because f changes only a little for large changes of $\boldsymbol{R}$, no second trial is needed. Compare the h so obtained with the value in Tables B-1 to B-4 for an independent check.
- Program Colebrook's Equation 3-12 to find f as an iterative subroutine for solving Equation 3-10 with a computer. Once programmed (a simple task even for a hand-held, card-programmable calculator), any pipe problem can be solved in a few seconds.
- Use the Swamee–Jain expression for f in the Darcy–Weisbach equation. Equation 3-13 could even be used as a first approximation for iteration of Equation 3-12.
- Refer to the extensive tables of flow by Ackers [8].

- Program the Moody diagram on a computer by assuming it to consist of a family of short, straight lines [9].
- Guess the pipe size and thus estimate v, calculate $\boldsymbol{R}$, find f from the Moody diagram, and compute h from the Darcy–Weisbach formula. Revise v and D, if necessary, and recompute.

Other Pipe Formulas

There are many formulas for flow in pipes, but none is easier to use than the Hazen–Williams, and none is more accurate or universally applicable than the Darcy–Weisbach supplemented by the Moody diagram or the Colebrook equation. The limitation of the accuracy of all pipe formulas lies in the estimation of the proper coefficient of friction, a value that cannot be physically measured and, hence, is subject to large error (see Subsection "Friction Coefficients").

Comparison of *f* and *C*

The Darcy–Weisbach friction factor can be compared to the Hazen–Williams C factor by solving both equations for the slope of the hydraulic gradeline and equating the two slopes. Rearranging the terms gives, in SI units,

$$f = \left(\frac{1}{C^{1.85}}\right)\left(\frac{134}{v^{0.15}D^{0.167}}\right) \tag{3-14a}$$

where v is in meters per second and D is in meters. In U.S. customary units, the relationship is

$$f = \left(\frac{1}{C^{1.85}}\right)\left(\frac{194}{v^{0.15}D^{0.167}}\right) \tag{3-14b}$$

where v is in feet per second and D is in feet. For any given pipe and velocity, the relation between ε/D and C can be found by calculating f from Equation 3-14 and entering the Moody diagram with $\boldsymbol{R}$ and f to find ε/D.

Friction Coefficients

The major weakness of any headloss formula is the uncertainty in selecting the correct friction coefficient. The proper friction factor for new pipe is uncertain because of the variation in roughness of the pipe walls, quality of installation, effect of slight angular offsets in laying the pipe, and water quality. For example, the H–W C factor should be reduced by 5 units for pipe laid in hilly regions due to the angular deflection at joints.

Anticipating the friction factor after years of service is doubly difficult due to changes caused by corrosion, deposition of minerals or grease, or attachment of bacterial slimes. Estimation of the friction coefficient merits careful attention.

Half a century ago, unlined cast-iron water pipe (or cast iron with the then-common but short-lived bituminous linings) did indeed become coated with tubercles and, thus, became very rough in a few years. But the modern use of cement mortar or plastic linings for ductile iron and steel pipe has eliminated the devastating effect of rust and tuberculation on friction. Plastic pipe remains very smooth unless foreign matter collects on the walls. Water treatment and bacterial slime in water pipes and grease in wastewater pipes can greatly increase the interior pipe roughness independently of the pipe material. This subject is addressed in more detail below.

Hazen–Williams C

The basis of the Hazen–Williams C factor in Equation 3-8 has resulted in some confusion. The factor is a function not only of the smoothness of the pipe wall, but also of the difference between the actual ID of the pipe and the nominal pipe size. The calculation of C from field data is, by custom, based on the nominal diameter of the pipe. One could scarcely do otherwise, because finding the ID of a buried pipe is so difficult (see Section 3-9). Even if it is physically measured at one point, there is no guarantee of uniformity. The use of nominal diameters, however, leads to strange conclusions. For example, the C value of an uncoated, new Class 50 ductile iron pipe 300 mm (12 in.) in diameter is typically given as 130. A Class 56 pipe carries 94% as much water, so its C value should be 122, but such a listing is unlikely to be found. The ratio of actual to nominal diameter accounts for the difference between the published C values for ductile iron pipe (DIP) and steel pipe with its smaller bore when both are lined with cement mortar. The confusion over the proper C value to use is worsened because the nominal diameter of steel pipe 300 mm (12 in.) and smaller is the ID, whereas for larger pipe, the nominal diameter is the OD. To permit reasonably accurate estimation of friction losses, the C value ought to be selected for the type of lining thus allowing the true ID to be applied to the Hazen–Williams equation.

Matters are not improved by the apparent increase of C with diameter. According to AWWA Manual M11 [10], the average value of C for pipe with smooth interior linings can be approximated as $C = 140 + 0.17d$, where d is inside pipe diameter in inches. After a long term of lining deterioration, slime buildup, etc., $C = 130 + 0.16d$. However, above a diameter of about 900 or 1200 mm (36 or 48 in.), there is little increase in C values according to Gros [11], who has had many years of experience in measuring C values in the field. The values of C listed in the first part of Table B-5 reflect this experience.

In addition to the discussion above, there are other limitations on the value of C. Values of C less than 100 are only applicable for velocities reasonably close to 1 m/s (3 ft/s). At other velocities, the coefficients are somewhat in error. For water pipes, Lamont [12] advises the following:

- C values of 140 to 150 are suitable for smooth (or lined) pipes larger than 300 mm (12 in.).
- For smaller smooth pipes, C varies from 130 to 140 depending on diameter.
- C values from 100 to 150 are applicable in the transitional zone (between laminar and turbulent flow), but the scale effect for different diameters is not included in the formula.
- The formula is unsuitable and, hence, not recommended for old, rough, or tuberculated pipes with C values below 100.
- Force mains for wastewater can become coated with grease and C values may vary down to 120 for severe grease deposition.

In the past (before 1940 or 1950), it was common to line steel and cast-iron pipe with hot coal-tar dip, which provided poor protection and allowed C values to drop from 130 for new pipe to 100 or less for pipe in service for 20 yr or more [13]. The modern use of cement mortar or plastic linings makes pipe very smooth, prevents corrosion and tuberculation, and maintains its smoothness indefinitely. In field measurements [14] made all over the United States on new water pipe with diameters of 100 to 750 mm (4 to 30 in.) lined with cement mortar, the values of C varied from 134 to 151 (median = 149, average = 144). For 150 to 900 mm (6 to 36 in.) pipe in service for 12 to 39 yr, C varied from 125 to 151 (median = 139, average = 140)—a decrease of only about 5 units.

Water treatment often creates deposits that greatly increase friction in pipes. In one pipeline, lime incrustation reduced the measured value of C to only 80 downstream from the treatment plant. Pipe can, however, be cleaned and relined with cement mortar in situ and restored nearly to its original smoothness. Under some circumstances, deposits of bacterial slime in water pipes can change the smoothest pipe (whatever the material) into very rough pipe. Fortunately, chlorination destroys the slime and restores the former smoothness. In New York, for example, the C factor for a 1800-mm (72-in.) water main 7.7 km (4.8 mi) long drops from 140 to 120 about twice per year and is chlorinated to restore the C value to 140. Another example is a 1050-mm (42-in.) cement-lined steel cylinder prestressed concrete transmission main 48 km (30 mi) long. It develops a slime layer only about 3 mm (1/8 in.) thick every five years, but the thin slime is sufficient to decrease the C value from 140 to 100. A massive dose of chlorine restores its former smoothness. Instead of massive doses of chlorine at long intervals, however, the maintenance of a free chlorine residual of about 0.5 mg/L or a stronger (2 to 3 mg/L) dose for two hours twice per week in a southern California pipeline has been reported to maintain the original capacity. As bacteria do not develop immunity to chlorine, experimentation with doses, contact time, and time intervals between doses offers an opportunity to achieve overall economy [15]. Biofouling is far more prevalent than most realize, so chlorination facilities must be added for pipelines subject to slime buildup. The coefficient of friction should be determined when a new pipeline is first put into service to establish an irrefutable reference point for future cleaning needs and for evaluating cleaning procedures.

Sewers often become fouled with grease, and grease from industries (such as commercial laundries, slaughterhouses, or locomotive repair shops) can reduce the diameter of wastewater pipes by one-third or more. Their original size and smoothness can be restored, however, by cleaning the pipe in place. To prevent excessive buildup of grease, include pig launching and recovery stations (see Section 4-9).

The headloss in pumping station piping is usually small (about 2 m or 5 ft) and is largely related to valve and fitting losses (see Tables B-6 and B-7), so the selection of a C value for piping within the pumping station is of minor importance. If the static lift is the major part of the TDH and the transmission or force main is short, say 150 m (500 ft) or less, the C value is again of minor importance.

Long Force Mains

Friction coefficients for long force or transmission mains must be established with great care. Using the Hazen–Williams formula can lead to serious errors, particularly for (1) large pipes, (2) high velocities, or (3) water temperature that differs from 15°C (60°F) by more than about 11°C (20°F). For such situations, use the Darcy–Weisbach equation. If the energy loss is a vital design consideration, search the literature for tests on similar conduits instead of attempting to use the Moody diagram for the Darcy–Weisbach equation.

Pump and Impeller Selection

To base pump operating points on station curves drawn for an unrealistic roughness is a serious blunder. If some jurisdiction requires the use of some specific value of roughness deemed by the designer to be too great (for example, a C value of 120, as specified in the Ten-State Standards), use that value only to find the size of the pipe. Select the pump and its impeller for a rational envelope of curves that include the maximum and minimum limit of possible roughness, for example, $150 > C > 120$. By choosing a pump that can accommodate impellers of a substantial range of diameters, the pump can be modified to operate at its best efficiency point (bep) for any curve within the envelope. However, assuming an excessively rough pipe can be disastrous. One pumping station featured several sets of two pumps in series to develop the head calculated for a single C value of 100. To keep the pumps from vibrating, the operators partly closed a downstream valve. A better solution would have been to bypass the tandem pump and achieve a savings of $100,000 per year in electric power.

It is wise to use a calibrated pressure gauge in measuring the total dynamic head (TDH) during the start-up procedure so that the impeller trim can, if necessary, be refined with confidence.

3-3. Pipe Tables

So many materials, pipe diameters, wall thicknesses, and liner thicknesses can be used for pumping stations, yard piping, and transmission or force mains that complete tables of flow and headloss would have to be extensive indeed. Because interpolation between tabular values is onerous, calculating the flow and headloss with the Hazen–Williams formula is much quicker. Tables, however, can be used advantageously to find quickly the proper size of pipe, to approximate the friction headloss, and to provide an independent check on a solution by formula.

The purposes of the pipe tables in Appendix B (Tables B-l to B-4) are the following:

- For a quick, preliminary determination of pipe size, discharge, and headloss for a moderate friction coefficient (C = 120) and velocity (2 m/s in Tables B-l and B-3 and 5 ft/s in Tables B-2 and B-4);
- For finding the available sizes and weights of the thinnest (and most common) pipes used within pumping stations;
- For a quick, rough check of flow or headloss found by other means; and
- For providing useful data for both ductile iron pipe (DIP) and steel in both SI and U.S. customary units.

For final design, for different conditions, and for piping outside of the pumping station, calculate flow and headloss with the Darcy–Weisbach formula and consult the tables to check for blunders.

Pumping Station Piping

The joints in piping within a pumping station should almost universally be bolted flanges augmented with a few strategically placed grooved-end couplings (e.g., Victaulic®) or sleeve pipe couplings (e.g., Dresser®) to permit disassembly and allow for misalignment (see Chapter 4 for illustrations). Connections to equipment (such as pumps) should be flexible to prevent the transmission of undue forces, including sheer. Mason radial ply joints of resilient material (rubber, Viton, or Buna N) are excellent. Occasionally other joints are used, but only with many grooved-end couplings added for ease in disassembly. Ductile iron pipe (DIP) up to 750 mm (30 in.) in diameter seems to be preferred by Eastern designers both for water and sewage pumping, but steel pipe is often used in very large sizes and in all sizes on the West Coast where freight makes DIP very expensive. Some designers prefer steel pipe because the mitered and welded fittings allow greater flexibility in layout. A 48° bend, for example, is as easy to fabricate as a 45° bend. Steel pipe has the advantage of more convenient modification.

Flanges for steel pipe are always welded. Schedule 40, according to ANSI B36.10, is considered to be the "standard" wall thickness for pipe up to 600 mm (24 in.). But for 600-mm (24-in.) or larger steel pipe, most U.S. manufacturers produce selected OD cylinders with fractional inch plate. Deviations from a pipe maker's norms are expensive. To avoid blunders, the wise designer becomes familiar with the sizes of pipe readily available and with the many codes listed in Chapter 4 for pipe and connections. (For example, Schedule 40 pipe can be threaded if it is steel but not if it carries flammable fluids and never if it is polyvinyl chloride [PVC].) In the United States flanges for DIP are always screwed to the barrel by the pipe manufacturer, and because Class 53 is the thinnest DIP allowed by ANSI/AWWA C115/A21.15-83 for threaded flanges, it is therefore the most common thickness used within a pumping station. Grooved-end couplings also require Class 53 DIP for diameters of 400 mm (16 in.) or less, but even thicker pipe is required for larger sizes according to AWWA C151 and C606.

For yard piping or for transmission or force mains, mechanical joints or push-on joints would be preferred, and Class 50 DIP or any of the several other pipe materials would probably be used. At 2 m/s (6.5 ft/s) velocity, the discharge in Class 50 DIP is greater than in Class 53 by about 2 to 6% and the headloss is less by about 1 to 3%. Because the concern here is mainly with the pumping station proper, the tables in Appendix B are limited to standard weight for steel and Class 53 for DIP; see the manufacturers' literature for other piping.

Linings

Shop lining applied to DIP is usually one layer of cement mortar centrifugally cast with minimum thicknesses varying from 1.6 mm ($^1/_{16}$ in.) for small pipe to 3.2 mm ($^1/_8$ in.) for large pipe and given a thin asphaltic seal coat to control curing per ANSI/AWWA C104/A21.4. Optionally, a double thickness can be specified. The thicknesses of actual linings may vary considerably from those specified. These linings are so thin that many engineers specify double thickness to ensure adequate coverage, eliminate the danger of pinholes, and provide greater integrity. Shop linings applied to steel pipe can be coal-tar, enamel, thin plastic, or thick cement mortar varying from 6 to 13 mm ($^1/_4$ to $^1/_2$ in.) per AWWA C205 (see also Table 4-6).

Field-applied cement-mortar linings, according to Table 4-7, can vary from 3 to 13 mm ($^1/_8$ to $^1/_2$ in.). Considering all the possibilities for pipe thickness and for shop and field linings, inside diameters can vary substantially. Designers should determine the metal IDs from manufacturers' catalogs. Determine the probable net ID and the probable range of friction coefficients carefully. For final design, trust no table, but calculate the flow by formula. If the pipe is larger than about 600 mm (24 in.) or smaller than about 75 mm (3 in.), or if the temperature is less than about 10°C (50°F) or more than about 30°C (86°F), do not trust the Hazen–Williams formula. Use the Darcy–Weisbach formula instead.

Example 3-1
Designing Pipe with the Pipe Tables

Problem: Select the pipe for a water pumping station with a 15-km- (9.3-mi-) long transmission main. Maximum flow is 0.4 m^3/s (6360 gal/min or 14.1 ft^3/s).

Solution: One choice for the pumping station is DIP lined with cement mortar and sized for a velocity of about 2.5 m/s (8.2 ft/s), which is high enough to minimize the size and cost of valves and other fittings and low enough to prevent cavitation and excessive headloss.

Using Table B-l for SI units (or Table B-2 for U.S. customary units),

SI Units

Pipe size for v = 2 m/s: 500 mm

$$\frac{v_{\text{desired}}}{v_{\text{Table B-1}}} = \frac{2.5\,\text{m/s}}{2\,\text{m/s}} = 1.25$$

$$\text{Area required} = \frac{0.213\ \text{m}^2}{1.25} = 0.17\ \text{m}^2$$

Choose 450-mm pipe: $A = 0.172\ \text{m}^2$

$$v = Q/A = 0.4/0.172 = 2.33\ \text{m/s}$$

Friction headloss: use Equation 3-9a and $C = 120$

$$h_f = 10{,}700\left(\frac{0.4}{120}\right)^{1.85}(0.468)^{-4.87}$$

h_f = 11.3 m/1000 m

U.S. Customary Units

Pipe size for v = 5 ft/s: 24 in.

$$\frac{v_{\text{desired}}}{v_{\text{Table B.2}}} = \frac{8.2}{5} = 1.64$$

$$\text{Area required} = \frac{3.32\ \text{ft}^2}{1.64} = 2.02\ \text{ft}^2$$

Choose 18-in. pipe: $A = 1.85\ \text{ft}^2$

$$v = Q/A = 14.1/1.85 = 7.78\ \text{ft/s}$$

Friction headloss: use Equation 3-9b and $C = 120$

$$h_f = 10{,}500\left(\frac{6360}{120}\right)^{1.85}(18.4)^{-4.87}$$

h_f = 11.3 ft/1000 ft

Always check such calculations with the pipe table. Note that headloss is a function of Q or v to the 1.85 power. Hence,

$$h_{f_{\text{actual}}} = h_{f_{\text{table}}}\left(\frac{Q_{\text{actual}}}{Q_{\text{table}}}\right)^{1.85}$$
$$= 8.5\left(\frac{0.4}{0.344}\right)^{1.85}$$
$$= 11.2\text{m}/1000\text{ m}$$

$$h_{f_{\text{actual}}} = h_{f_{\text{table}}}\left(\frac{Q_{\text{actual}}}{Q_{\text{table}}}\right)^{1.85}$$
$$= 5.1\left(\frac{6360}{4160}\right)^{1.85}$$
$$= 11.2\text{ ft}/1000\text{ ft}$$

Entrance, fitting, and valve losses must be added (see Section 3-4).

At $C = 145$, the friction headloss is 8.4 m/1000 m, which, in the short length of pipe in a pumping station, would only be about 60 mm (0.2 ft). At 11.3 m/1000 m, the loss in head would be only about 40 mm (0.1 ft) more, which is insignificant.

For the transmission main, a velocity of about 2 m/s (6 ft/s) seems likely to be economical when the cost of pipe, valves and fittings, water hammer control methods and devices, installation, and energy are analyzed for, say, a 20-yr period. Using Table B-1 or B-2, a 500-mm (20-in.) pipe would fit the conditions. But flanged joints are not needed for the transmission main, where mechanical or push-on joints allow Class 50 DIP to be used. By referring to the DIPRA handbook [16], the wall thickness is 9.1 mm (0.36 in.). Let us assume the cement mortar is to be double thickness with negligible tolerance. The OD values in Tables B-1 and B-2 are correct for all classes of DIP, so

$$\text{ID} = 549 - 2(9.1 + 2 \times 2.38)$$
$$= 521\text{ mm}$$
$$h_f = 10{,}700\left(\frac{0.4}{120}\right)^{1.85}(0.521)^{-4.87}$$
$$= 6.69\text{ m}/1000\text{m}$$

$$\text{ID} = 21.60 - 2[0.36 + 2(3/32)]$$
$$= 20.5\text{ in}$$
$$h_f = 10{,}500\left(\frac{6360}{120}\right)^{1.85}(20.5)^{-4.87}$$
$$= 6.65\text{ft}/1000\text{ft}$$

Or, using Table B-1,

$$h_f = 7.5\left(\frac{0.4}{120}\right)^{1.85}$$
$$= 6.68\text{ m}/1000\text{ m. Good check.}$$

Or, using Table B-2,

$$h_f = 4.5\left(\frac{6360}{120}\right)^{1.85}$$
$$= 6.65\text{ ft}/1000\text{ ft. Good check.}$$

For 15 km (9.3 miles) of pipe, the total headloss at $C = 120$ is $6.7 \times 15 = 100$ m (330 ft) vs. 82.8 mm (232 ft) at $C = 145$. The difference in headloss and energy use is important and worthy of careful study. Check by using the Darcy–Weisbach formula; also consider the maximum and minimum water temperatures.

Air in Pipelines

Dissolved air (or other gas) is a serious problem in pipelines that have intermediate high points or are nearly flat. If air comes out of solution, it forms bubbles that, in turn, reduce the water cross-sectional area and increase resistance to flow—sometimes greatly—and the air-moisture environment is conducive to corrosion. Various ways to deal with air in pipelines include: (a) designing the pipeline profile to rise all the way to the exit, (b) installing air release valves at high points in the pipeline (or at frequent intervals for flat profiles), or (c) designing for velocities high enough to scour air bubbles to the exit. Obviously, the first is preferred if possible. Air release valves are risky because of uncertain maintenance. They should not be used at all on sewage force mains, because maintenance must be done so frequently (for example, monthly) and without fail. (See Section 7-1 for an exception.) If the valves are not maintained properly, they are worse than useless, because they then engender a false sense of security. Designing for scouring velocities may result in excessive head losses and energy needs. The required scouring velocities are given in Table B-9.

Some consultants customarily install manways at 450-m (1500-ft) intervals in water pipelines equal to or larger than 900 mm (36 in.) in diameter to permit worker entry and inspection of, and repairs to, the lining and to fix leaks. Air release valves are required in the manway covers to prevent the accumulation of air under them.

3-4. Headlosses in Pipe Fittings

Pumping stations contain so many pipe transitions (bends, contractions) and appurtenances (valves, meters) that headlosses due to form resistance (turbulence at discontinuities) are usually greater than the frictional resistance of the pipe. The simplest approach to design is to express the headlosses in terms of the velocity head, $v^2/2g$, usually immediately upstream of the transition or appurtenance. The equation for these losses is

$$h = K\frac{v^2}{2g} \tag{3-15}$$

in which K is a headloss coefficient (see Appendix B, Tables B-6 and B-7). The few exceptions to Equation 3-15 are noted in the tables.

The headloss coefficient, K, is only an approximation, and various publications are not always in agreement and may differ by 25% or more. The values in Tables B-6 and B-7 have been carefully selected from many sources and are deemed to be reliable.

In Equation 3-15, K varies with pipe size as noted in Table B-6. Furthermore, published values are for isolated fittings with a long run (for example, 20 pipe diameters) of straight pipe both upstream and downstream from the fitting. The headloss is measured between one point a short distance upstream from the fitting and another point at the downstream end of the piping system. This ensures symmetrical flow patterns. The difference in headloss with and without the fitting is used to compute K. Headlosses for a series of widely separated fittings are therefore directly additive.

Part of the headloss is due to the turbulence within the fitting, but probably about 30% (less for partially closed valves) is due to eddying and turbulence in the downstream pipe. So if one fitting closely follows another (as in a pumping station), the apparent K value for the first fitting is, probably, reduced to about 70%. For example, because K for a 90° bend is 0.25 (see Table B-6), K for two 90° bends would be 0.50 if the bends were separated by, say, a dozen pipe diameters. But if the bends were bolted together to make a 180° bend, K for the entire bend could be figured as $0.70 \times 0.25 + 0.25 = 0.43$, which is within 8% of the K value for a 180° bend in Table B-6. As another example, K for a 90° bend consisting of three 30° miters can be determined directly from Table B-6 as 0.30 or indirectly by adding reduced K values for each miter except the last: thus, $0.70\,(0.10 + 0.10) + 0.10 = 0.24$—an error of 20% (one publication lists the K for the mitered fitting as 0.20).

Pumps, especially when operating on either side of their best efficiency point, usually cause swirling (rotation) in the discharge pipe. Swirling sometimes also occurs in inlets and suction pipes. The effect of such swirling is to increase eddy formation and turbulence; consequently, the headloss in fittings can be doubled or even tripled. If swirling is likely to occur and if headloss within the pumping station is critical (which is often true in suction piping), the safe and conservative practice would be to design for headloss without swirling and again for headloss using, say, 200% of the fitting losses. Because there is no definitive body of literature about this complex subject, designers must either rely on experience or guess at headlosses.

Another method for computing headlosses is to use an "equivalent length" of straight pipe. This method is less accurate partly because, for example, 50 pipe diameters of a smooth pipe length would have less headloss than the same length of rough pipe. But, of course, corrections can be made by multiplying a tabular value of equivalent length by the ratio of $C_{actual}/C_{tabular}$. This method is used in pipe network analysis for simplification, but there is no reason for using it in pumping station calculations.

3-5. Friction Losses in Open Channel Flow

The most common equation used in the United States for open channel flow is the Manning equation. In SI units,

$$v = \frac{1}{n}R^{2/3}S^{1/2} \tag{3-16a}$$

where v is velocity in meters per second, n is Manning's friction coefficient (given in Appendix B, Table

B-5), R is the hydraulic radius in meters, and S is the slope in meters per meter. In U.S. units,

$$v = \frac{1.486}{n} R^{2/3} S^{1/2} \tag{3-16b}$$

where v is velocity in feet per second, R is hydraulic radius in feet, S is slope in feet per foot, and n is Manning's friction coefficient (in Table B-5); the constant, 1.486, converts SI to U.S. customary units. The Manning equation can be used for pipes flowing full, but has no advantage over the Hazen–Williams equation. For pipes flowing full and under pressure, the relationship between C and n is

$$n = 1.12 \frac{D^{0.037}}{CS^{0.04}} \tag{3-17a}$$

in SI units where D is the ID in meters. In U.S. customary units, the equation is

$$n = 1.07 \frac{D^{0.037}}{CS^{0.04}} \tag{3-17b}$$

where D is the ID in feet. (See Brater and King [17] for a more extensive discussion.)

Error in the Manning Equation

In spite of its common use for circular pipes flowing partly full, the Manning equation is valid only for full pipes. In extensive studies. Yarnell and Woodward [18] confirmed the exponents of $^2/_3$ and $^1/_2$ respectively for R and S in Equation 3-16 for full pipes and also proved that the apparent values of n vary considerably with depth of flow. At depths from 15 to 55% of the pipe diameter, the observed apparent n is about 25% greater than n for a full pipe. Camp [19] added the work of Wilcox [20] to that of Yarnell and Woodward to obtain the curves shown in Figure B-5. (Camp also included a graph of slopes required for keeping sewers clean at minimum discharge.)

Escritt [21] stated that the Manning equation was accurate within a few percent if half of the width of the free water surface were added to the wetted perimeter, P, of the pipe when computing a modified hydraulic radius, R_m. He also proposed an equation that Saatci [22] transformed into the expression

$$R_m = R_f \left[\frac{\theta - \sin\theta}{\theta + \sin\frac{\theta}{2}} \right] \tag{3-18}$$

where R_m is the modified hydraulic radius for a given depth, R_f is hydraulic radius ($D/4$) for a full pipe, and q is the central angle in radians. See Figure B-4 for relations between θ, area, perimeter, and depth of flow. Equation 3-18 fits the curves of observed data in Figure B-5, usually within an error of about 3%. Wheeler [23] reduced the error to less than 2% by introducing empirical coefficients into Equation 3-18 as follows:

$$R_m = R_f \left[\frac{\theta - \sin\theta}{\theta + 1.23 \sin\frac{\theta}{1.924}} \right] \tag{3-19}$$

Velocities, depths, and water cross-sectional areas corresponding to the Manning equation are compared in Table B-8, in which observed values are closely (within 2%) represented by Equation 3-19. The table makes it particularly easy to find accurate values of the hydraulic elements for a pipe flowing partly full. Wheeler also developed a computer program, PARTFULL®, that was used to find the conjugate and sequent (before and after a hydraulic jump) depths for the flowrates given in Tables 12-2 and 12-3.

Sewers constitute one kind of open channels in which stable, uniform flow rarely occurs because of discontinuities such as changes of gradient and because of constantly changing flow, all of which cause long backwater curves. For practical purposes, however, these effects are localized, and a steady-state equation is adequate if the equation itself can be verified by physical measurements.

Example 3-2
Design of a Sewer Pipe

Problem: Design a sewer pipe to carry a maximum design flow of 123 m^3/hr (541 gal/min). Select the pipe material and find the required diameter and slope.

Solution: Sewer pipe is usually clay, concrete, or plastic. Plastic is popular because of its light weight, ease of handling, tight joints, durability, and economy, but it is flexible and tends to flatten, so care must be taken to backfill it properly.

Size. Plastic sewer pipe is available in the following nominal diameters: 200, 250, 300, 375, 450, 525, 600, etc. mm (8, 10, 12, 15, 18, 21, 24, etc. in.). Below 375 mm (15 in.), some designers consider a sewer pipe flowing "full" when the water surface is at mid-depth. To pick up grit deposited at low flows and to scour the pipe clean, the velocity at maximum flow should equal or exceed 1.1 m/s (3.5 ft/s). In large (> 600-mm or 24-in.) pipe, keep the velocity greater than 1.5 m/s (5 ft/s) to inhibit septicity and odors (see Section 23-1).

SI Units	**U.S. Customary Units**
From Table A- 11 (Appendix A),	
$(123\ \text{m}^3/\text{hr})(2.78\times10^{-4}) = 0.0342\text{m}^3/\text{s}$	$(541\ \text{gal/min})(2.23\times10^{-3}) = 1.21\text{ft}^3/\text{s}$
The area required is, from Equation 3-1,	
$0.0342\text{m}^3/\text{s} = A\times1.1\text{m/s}$ $A = 0.0311\text{m}^2$	$1.21\text{ft}^3/\text{s} = A\times3.5\text{ft/s}$ $A = 0.346\text{ft}^2$
The area of a half-circle is $\pi D^2/8$, so	
$\pi D^2/8 = 0.0311\text{m}^2$ $D = 0.281\text{m} = 281\text{mm}$	$\pi D^2/8 = 0.346$ $D = 0.939\text{ft} = 11.3\text{in.}$
Choose a 300-mm pipe.	Choose a 12-in. pipe.

Slope. Plastic pipe is very smooth but it may become coated with grease, and Ten-State Standards [1] require n to be 0.013, which is rather rough (see Table B-5).

Because the pipe is larger than required, the water surface is below the center, so the perimeter is less than half the circumference. Tables for the area and perimeters of segments are available or the following formula can be used:

$$A = \frac{1}{2}r^2(\theta - \sin\theta) \tag{3-20}$$

where A is the area of the segment, r is the pipe radius, and θ is the central angle (in radians) of the sector enclosing the segment. Substituting,

SI Units	U.S. Customary Units
$0.0311 = \frac{1}{2}\left(\frac{0.300}{2}\right)^2(\theta-\sin\theta)$ $\theta = 2.95$ radians by trial	$0.346 = \frac{1}{2}\left(\frac{1}{2}\right)^2(\theta-\sin\theta)$ $\theta = 2.95$ radians by trial
The wetted perimeter is	
$P = r\theta = \frac{0.300}{2}\times2.95 = 0.443\ \text{m}$	$P = r\theta = \frac{1}{2}(2.95) = 1.48\ \text{ft}$
$R = A/P = 0.0311\ \text{m}^2/0.443\ \text{m} = 0.0702\text{m}$	$R = A/P = 0.346\ \text{ft}^2/1.48\ \text{ft} = 0.234\text{ft}$
Substituting in Equation 3-16a,	Substituting in Equation 3-16b,
$1.1\text{m/s} = \frac{1}{0.013}(0.0702)^{2/3}S^{1/2}$ $S = 0.0071\text{m/m}$	$3.5\text{ft/s} = \frac{1.486}{0.013}(0.234)^{2/3}S^{1/2}$ $S = 0.0065\text{ft/ft}$

The discrepancy is caused by inexact conversions of SI to U.S. units. The minimum slope from Ten-State Standards [1] is 0.0022 m/m.

3-6. Energy in Pressurized Pipe Flow

Many problems in hydraulics can be solved by equating the energy at two points along a pipe or a channel. Within a closed system (one inlet and one outlet) the total energy at point 1 equals the total energy at point 2 plus any losses due to friction, as shown by Equation 3-4. A typical example is the Venturi meter in Figure 3-4.

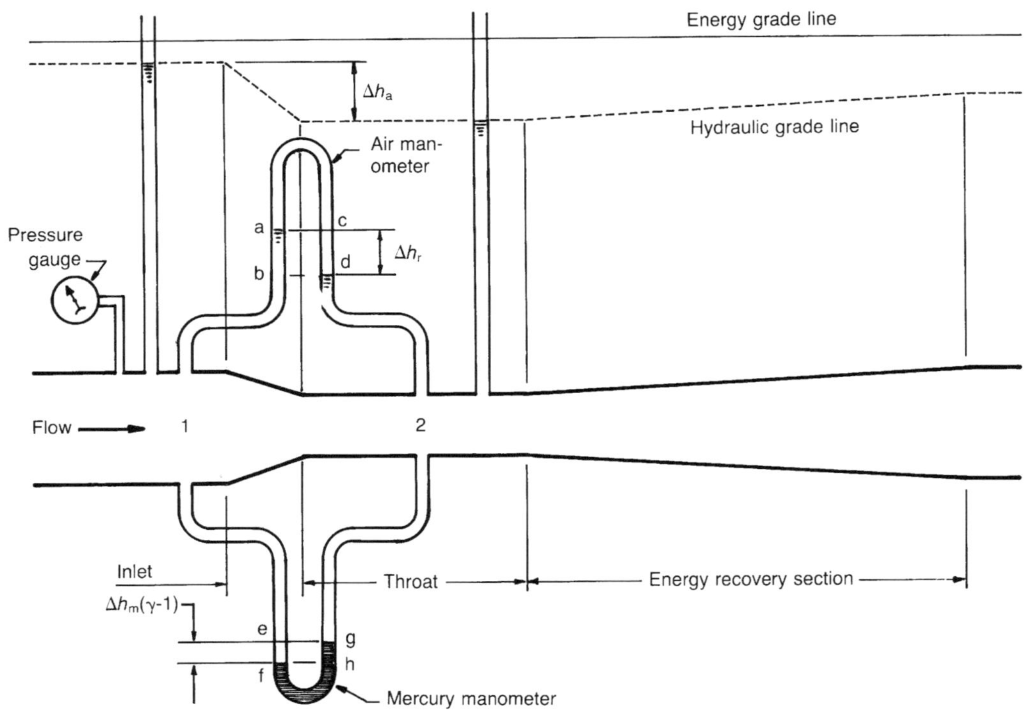

Figure 3-4. Venturi meter.

From the principles presented in Section 3-1, the pressure at point 1, as measured by a gauge or a piezometer, is greater than at point 2 by the amount $\Delta h_a = \Delta(p/\gamma)$ if the meter is horizontal. The differential pressure can be measured by a differential pressure gauge (not shown), by a mercury manometer, or by an air manometer. Differential pressure gauges or transducers are costly and delicate but necessary if a remote reading is needed. Mercury manometers are still made, but mercury is poisonous and an accidental spill, sure to occur eventually, is nearly impossible to clean up. The air manometer is sensitive, accurate (if purged properly), cheap, and excellent even when homemade. The column of air, even if highly compressed, is insignificant in mass, so the manometer reading, Δh_r, equals the true value, Δh_a, with negligible error.

The connecting tubing to any pressure gauge or manometer must be purged of air to ensure accuracy. In the air manometer of Figure 3-4, for example, there must be no air except between points a and d, and this requires purging petcocks or bleeders at strategic locations. A means of introducing the required bubble of air at the top of the manometer is also required.

Example 3-3
Venturi Meter in a Pipe

Problem: Assume the inlet diameter in Figure 3-4 to be 254 mm (10 in.) and the throat diameter to be 152 mm (6 in.) The air manometer reads 305 mm (12 in.). Find the flow, the mercury manometer reading, and the differential gauge pressure.

Solution: Assume the friction loss between points 1 and 2 to be zero (nearly true). If the datum plane is the pipe centerline, Equation 3-4 becomes

$$h_1 + \frac{v_1^2}{2g} = h_2 + \frac{v_2^2}{2g}$$

SI Units

From Equation 3-1, velocity v_2 must be

$$v_2 = \left(\frac{254}{152}\right)^2 v_1 = 2.79v_1$$

Substituting,

$$h_1 - h_2 = 0.305 = \frac{(2.79v_1)^2 - v_1^2}{2 \times 9.81}$$

$$v_1 = 0.939 \text{m/s}$$

$$Q = Av = \frac{(0.254)^2\pi}{4} \times 0.939$$

$$= 0.0476 \text{ m}^3/\text{s}$$

U.S. Customary Units

$$v_2 = \left(\frac{10}{6}\right)^2 = 2.78v_1$$

$$h_1 - h_2 = 1.0 = \frac{(2.78v_1)^2 - v_1^2}{2 \times 32.2}$$

$$v_1 = 3.09 \text{ft/s}$$

$$Q = Av = \frac{10^2\pi}{4} \times 3.09$$

$$= 1.69 \text{ ft}^3/\text{s}$$

In the mercury manometer of Figure 3-4, the pressure at f and h are equal (because the elevations are the same in a static continuum). The differential pressure between g and h equals the weight of mercury column $\overline{\text{gh}}$ minus the weight of water column $\overline{\text{ef}}$. From the terms of the problem, the differential pressure between points 1 and 2 (and, hence, between points e and g) is 305 mm (12 in.) WC. The specific gravity of mercury (Hg) is 13.6 and of water is 1.0, so

SI Units

$$305 = p_f - \overline{\text{ef}}(1.0) - [p_h - \overline{\text{gh}}(13.6)]$$

$$p_f = p_h \text{and } \overline{\text{ef}} = \overline{\text{gh}}$$

$$\overline{\text{gh}} = 305/(13.6 - 1) = 24.2 \text{ mm Hg}$$

U.S. Customary Units

$$12 = p_f - \overline{\text{ef}}(1.0) - [p_h - \overline{\text{gh}}(13.6)]$$

$$p_f = p_h \text{ and } \overline{\text{ef}} = \overline{\text{gh}}$$

$$\overline{\text{gh}} = 12/(13.6 - 1) = 0.952 \text{ in. Hg}$$

From Appendix Table A-11, the differential gauge pressure is

SI Units

$$p = 24.2 \times 1.33 \times 10^{-1} = 3.22 \text{ kPa}$$

U.S. Customary Units

$$p = 0.952 \times 4.91 \times 10^{-1} = 0.467 \text{ lb/in.}^2$$

3-7. Energy in Open Channel Flow

Water in open channels can flow in three regimes:

- subcritical or tranquil flow (waves can move upstream; example: sluggish stream);
- critical flow with standing waves (hydraulically unique; example: flow over a broad crested weir);
- supercritical or shooting flow (waves move only downstream; example: water flowing down a dam spillway).

Theory

To establish which of the three possible flow regimes occurs, consider Equation 3-3 and Figure 3-5 (review Section 3-1 if necessary).

$$E_s = y + \frac{v^2}{2g} \tag{3-3}$$

If the velocity is zero and the water is standing, the specific energy equals the depth, *y*, and plots as the straight line, oa, in Figure 3-5a. At a constant discharge, however, E_s is the curved line represented by $y + v^2/2g$. In tranquil flow, the velocity head is small, but as *y* decreases, the velocity (to produce the same discharge) increases—and very rapidly below the tranquil zone. Hence, there is a critical depth, y_c, at which specific energy is minimum. To find it, replace v^2 with $(Q/A)^2$, differentiate Equation 3-3 with respect to *y*, and set the resulting expression equal to zero; thus,

$$\frac{dE_s}{dy} = 1 + \frac{Q^2}{2g}\left(-2A^{-3}\frac{dA}{dy}\right) = 0$$

For a prismatic channel of any shape, as in Figure 3-5b, $dA = bdy$. Substituting bdy for dA, canceling dy/dy, and collecting terms yields

$$\frac{Q^2}{g} = \frac{A^3}{b} \tag{3-21}$$

at critical depth. Substituting $v_c = Q/A$ gives

$$\frac{v_c^2}{2g} = \frac{A}{2b} \tag{3-22}$$

at critical depth. Note that b is the width of the water surface—not the average width.

One illustration of the usefulness of Equations 3-21 and 3-22 is the application to open-channel Venturi flumes (Figure 3-6) for measuring discharge. Any obstruction in the channel that produces a short length of uniform flow with straight, level streamlines at critical depth can be a Venturi flume. One example is the Palmer–Bowlus meter (see Chapter 20). The Parshall is another popular measuring flume, but it does not have straight, level streamlines at the critical section and is not really a Venturi flume. In the Venturi flume of Figure 3-6, flow over the step meets the previous criteria. Strange though it may seem, the water surface drops between sections b–b and c–c, then rises again from sections c–c to e–e so that y_e nearly equals y_a. If either (1) the step, (2) a reduced throat width, or (3) a combination of (1) and (2) produces critical depth for a suitably wide range of flow, the Venturi flume can be used to meter the discharge with a basic uncertainty of only about ±3% [24]. Metering is easily accomplished because a wide range of dimensional changes between channel and throat—even a change in shape—can be used. For example, Palmer–Bowlus meters with trapezoidal throats are commonly installed in circular sewers.

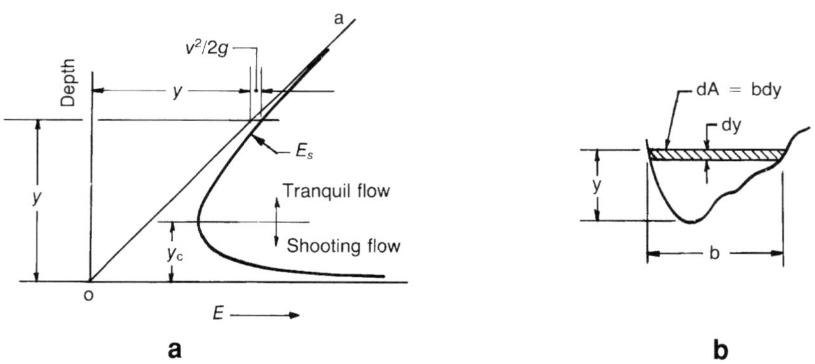

Figure 3-5. Energy in open channel flow. (a) Specific energy for constant discharge; (b) channel cross section.

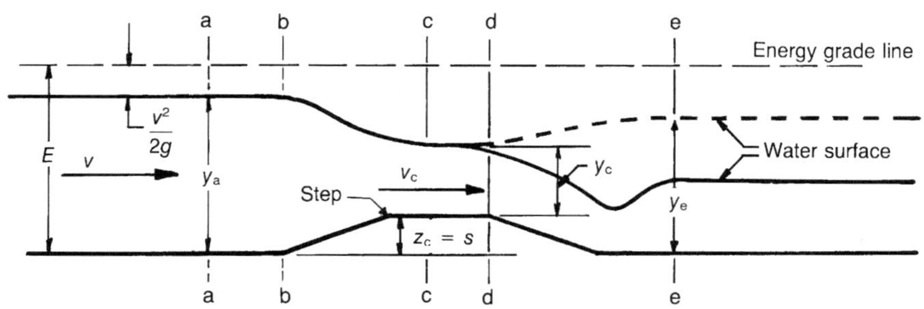

Figure 3-6. Longitudinal section through a Venturi flume in an open channel.

Discharge can be calculated by measuring y_c, using it to find A and b, and substituting into Equation 3-21. It is customary, however, to measure y_a (upstream) because it is greater than y_c and, hence, the accuracy is better. Methods for obtaining rating curves for y_a are given in Example 3-4.

Example 3-4
Venturi Flume in a Channel

Problem: A smooth, rectangular concrete channel is 610 mm (2 ft) wide and carries a flow of 0.170 m³/s (6 ft³/s) at a slope of 0.00124 m/m. A rectangular Venturi flume with a throat width of 457 mm (1.5 ft) is installed with its step 76 mm (3 in.) above the channel invert. Calculate (1) the depth of flow in the throat and (2) the depth in the channel in front of (upstream from) the flume, (3) compare that depth with the depth far upstream, and (4) find whether critical depth truly occurs in the measuring flume.

Solution: (1) Assume critical depth occurs in the throat and find the critical depth from Equation 3-21.

SI Units

$$\frac{0.170^2}{9.81} = \frac{A^3}{0.457}, A = 0.110\text{m}^2$$

$$y_c = \frac{A}{b} = \frac{0.110}{0.457} = 0.241\text{m}$$

U.S. Customary Units

$$\frac{6^2}{32.2} = \frac{A^3}{1.5}, A = 1.19\text{ft}^2$$

$$y_c = \frac{A}{b} = \frac{1.119}{1.5} = 0.793\text{ft}$$

From Equation 3-1, the throat velocity is

$v_c = Q/A = 0.170/0.110 = 1.55$ m/s

$v_c = Q/A = 6/1.19 = 5.04$ ft/s

(2) To find conditions in front of the measuring flume, use Bernoulli's equation (Equation 3-4) and assume h_f is zero (which is practically true).

$$z_a + y_a + \frac{v_c^2}{2g} = z_c + y_c + \frac{v_c^2}{2g} + 0 \tag{3-4}$$

Note that the velocity in the channel is

$Q/A = Q/by = (0.170/0.61)y_a = 0.279/y_a$

$Q/A = Q/by = (6/2)y_a = 3/y_a$

so Bernoulli's equation becomes

$$0 + y_a + \frac{(0.279/y_a)^2}{2 \times 9.81} = 0.076 + 0.241 + \frac{1.55^2}{2 \times 9.81}$$

$$0 + y_a + \frac{(3/y_a)^2}{2 \times 32.2} = \frac{3}{12} + 0.792 + \frac{5.04^2}{2 \times 32.2}$$

By trial,

$y_a = 0.416$ m

$y_a = 1.36$ ft

So the water surface falls

$y_a - (y_c + \text{step}) =$
$0.416 - (0.241 + 0.76) = 0.099$ m

$y_a - (y_c + \text{step}) =$
$1.36 - (0.793 + 0.25) = 0.317$ ft

(3) Use Equation 3-16 (Manning) to find the depth far upstream (or downstream). Choose $n = 0.011$ for smooth concrete.

SI Units	**U.S. Customary Units**
$v = \frac{1}{n}R^{2/3}S^{1/2}$	$v = \frac{1.486}{n}R^{2/3}S^{1/2}$
$v = Q/A = (0.170/0.610)y = 0.279/y$	$v = Q/A = 6/2y = 3y$
$R = A/P = 0.610y/(2y+0.610)$	$R = A/P = 2y/(2y+2)$
$\frac{0.279}{y} = \frac{1}{0.011}\left(\frac{0.610y}{2y+0.610}\right)^{2/3} \times (0.00124)^{1/2}$	$\frac{3}{y} = \frac{1.486}{0.011}\left(\frac{2y}{2y+2}\right)^{2/3} \times (0.00124)^{1/2}$
$y = 0.305$ m by trial	$y = 1.00$ by trial

So the measuring flume acts as a choke to raise the water surface from 0.305 m to 0.416 m (1.00 to 1.36 ft). The "backwater" created by the choke extends far upstream.

(4) If the total energy in the throat is greater (say, 5% greater to allow for friction losses) than the energy in the channel without the choke, critical depth must theoretically occur in the throat. Without the choke, the total energy (referenced to the channel bottom) would be, from Equation 3-2,

SI Units	U.S. Customary Units
$E = z + y + \frac{v^2}{2g}$	$E = z + y + \frac{v^2}{2g}$
$E = 0 + 0.305 + \frac{[0.170/(0.610 \times 0.305)]^2}{2 \times 9.81}$	$E = 0 + 1 + \frac{(6/2)^2}{2 \times 32.2}$
$E = 0.348$ m	$E = 1.140$ ft

With the choke, the total energy in the throat (again referenced to the invert of the channel) is

SI Units	U.S. Customary Units
$E = 0.076 + 0.241 + \frac{(1.55)^2}{2 \times 9.81}$	$E = 0.25 + 0.793 + \frac{(5.04)^2}{2 \times 32.2}$
$E = 0.438$ m	$E = 1.44$ ft

which is 26% greater than the energy without the choke, so critical depth does occur.

Field Determination of Critical Flow

The very uniqueness of critical flow makes it easy to determine if the metering flume is registering properly. A drowned meter usually has a nearly level surface from inlet to outlet. If a hydraulic jump occurs at the downstream end, the flow in the throat is certain to be critical. But a jump does not always occur, so another criterion is to compare the depths of flow upstream (at section a–a in Figure 3-6) and downstream (at section e–e). If the downstream depth is no more than 85% of the upstream depth, the flow in the throat is critical [25]. In Example 3-4, the upstream depth is 0.416 m and the downstream depth is 0.305 m or 73% of the upstream depth—another demonstration that critical flow does occur.

Arredi Diagram

Trial solutions are unsatisfactory for more than one or two points (unless programmed on a computer). To make a rating curve of head vs. discharge, an Arredi diagram [26] is preferable. Select several values of discharge (Q_1, Q_2, etc.) covering an appropriate range from small to large values. For each Q, plot a curve of throat area versus $v^2/2g$ as shown in the upper half of Figure 3-7. These curves are the same for all Venturi flumes regardless of size or shape. Now plot a curve of $v_c^2/2g$ versus throat area (curve OA) from Equation 3-21 in the upper half of Figure 3-7. This curve is specific for a particular throat. The points of intersections with the Q curves give velocity head energy for each value of Q.

In the lower half of the figure, plot (to the same scale) depth, d, versus area for the channel (curve OB) and plot depth versus throat area referred to the channel invert (curve CD). Note that OC is the height of the step, s, so throat depth is $d - s = y_c$.

Consider point e in Figure 3-7. The total energy in the throat is $\overline{eh}$, because $\overline{ef}$ is $v^2/2g$, $\overline{fg}$ is s, and $\overline{gh}$ is y_c. Set a pair of dividers to length $\overline{eh}$ and follow the curve of Q^2 to the right until the dividers intersect

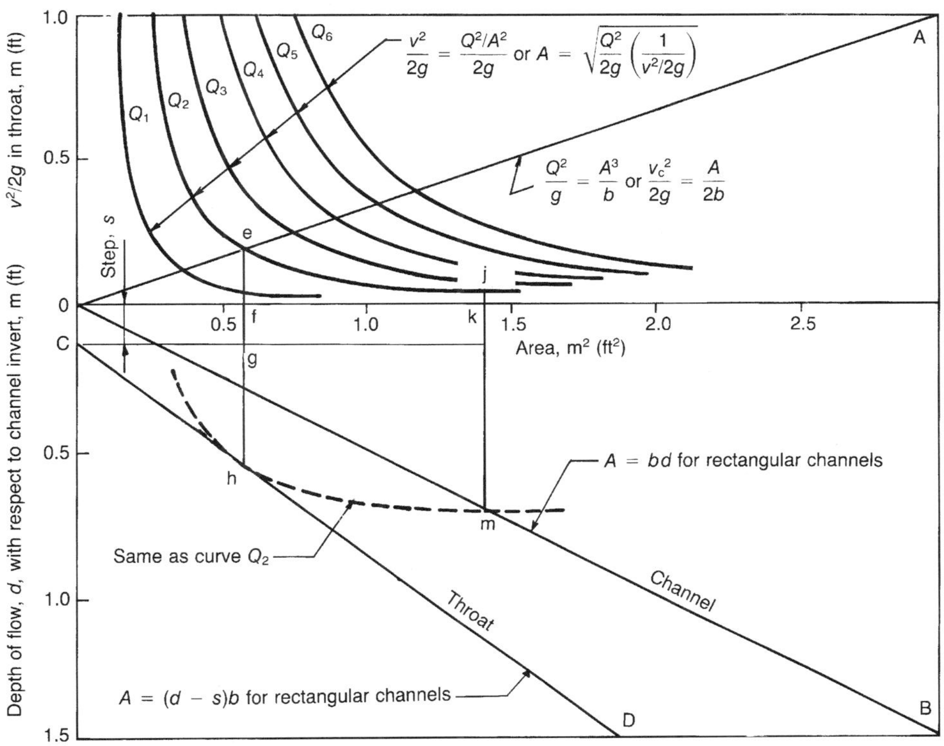

Figure 3-7. Arredi diagram.

curve OB so that $\overline{jm}$ equals $\overline{eh}$. The depth indicated by $\overline{km}$ is the upstream channel depth, y_a, for Q_1.

To develop a rating curve, repeat this procedure for several values of Q and construct a graph of y_a (as the ordinate) versus Q (as the abscissa). For shapes such as trapezoidal throats in semicircular flumes, lines OA, OB, and CD are curved.

The difference between these theoretical rating curves and discharges measured in the laboratory is less than about 4% of the maximum flow. In the field, the uncertainty of measurement would not exceed 4 or 5% if the flume is installed with reasonable care.

3-8. Unbalanced Hydraulic Forces

Adequate pipe support is important to protect flanges, bolts, and the pipe itself from destructive forces due to the weight of pipe and water or due to a change in fluid velocity. (Earthquake, vibration, and water hammer may also be destructive.) Valve bodies and pump casings are especially vulnerable, and undue strain can cause binding long before breakage occurs. Straight pipe needs support in the vertical plane (to resist gravity) and in the horizontal plane (to resist earthquake). Hangers or piers should be spaced to minimize vibration (see Chapter 22). Bends, tees, and wyes produce unbalanced thrusts in the plane of the fitting and, hence, require substantial anchors that must often resist large horizontal forces without appreciable deformation.

The purpose of the following example is to show how to compute the unbalanced forces. Review the fundamentals of momentum in Section 3-1, if necessary.

Example 3-5
Unbalanced Forces on a Wye

Problem: Compute the forces due to water velocity, static pressure, and water hammer on the horizontal steel wye shown in Figure 3-8 and determine whether an anchor is required.

Solution: Force due to velocity and pressure. In the force system of Figure 3-8, positive vectors act right or up. The net forces ($\mathbf{R}_x$ or $\mathbf{R}_y$) to produce the accelerations are assumed to be

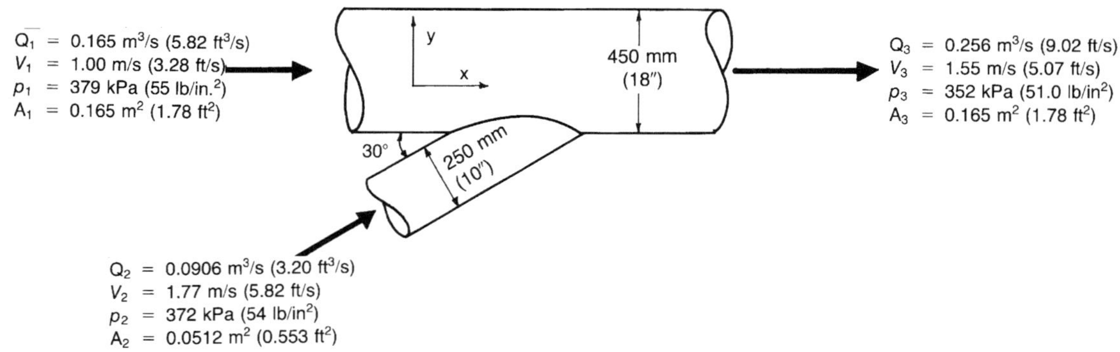

Figure 3-8. Flow in a wye.

positive (and, therefore, act up or right), so a negative answer indicates a force acting opposite (down or left). The x component of Equation 3-7 is

$$P_1A_1 + P_2A_2\cos 30° - P_3A_3 + \mathbf{R}_x = \rho Q_3 v_3 - \rho(Q_1 v_1 - Q_2 v_2 \cos 30°)$$

The y component is

$$0 + P_2A_2 \sin 30° + 0 + \mathbf{R}_y = 0 - \rho Q_2 v_2 \sin 30° + 0$$

Calculations for these terms are as follows.

SI Units	**U.S. Customary Units**
p_1A_1: $379 \times 0.165 = 62.5$ kN	$55.0 \times 1.78 \times 144 = 14{,}100$ lb
p_2A_2: $372 \times 0.0512 = 19.0$ kN	$53.9 \times 0.553 \times 144 = 4290$ lb
p_3A_3: $352 \times 0.164 = 58.1$ kN	$51.0 \times 1.78 \times 144 = 13{,}100$ lb
$\rho Q_1 v_1$: $1000 \times 0.165 \times 1.00 = 0.165$ kN	$1.94 \times 5.82 \times 3.28 = 37.0$ lb
$\rho Q_2 v_2$: $1000 \times 0.0906 \times 1.77 = 0.160$ kN	$1.94 \times 3.20 \times 5.81 = 36.1$ lb
$\rho Q_3 v_3$: $1000 \times 0.256 \times 1.55 = 0.397$ kN	$1.94 \times 9.02 \times 5.07 = 88.7$ lb

The x component is

SI Units:

$$62.5 + 19.0\cos 30° - 58.1 + \mathbf{R}_x = 0.397 - (0.165 + 0.160\cos 30°)$$

$$\mathbf{R}_x = -20.8 \text{ kN} = 20.8 \text{ kN (acting left)}$$

U.S. Customary Units:

$$14.100 + 4290\cos 30° - 13{,}000 + \mathbf{R}_x = 88.7 - (37.0 - 36.1\cos 30°)$$

$$\mathbf{R}_x = -4630 \text{ lb} = 4630 \text{ lb (acting left)}$$

The y component is

SI Units:

$$0 + 19.0\sin 30° + 0 + \mathbf{R}_y = 0 - 0.160 \sin 30°$$

$$\mathbf{R}_y = -9.58 \text{ kN} = 9.58 \text{ kN (acting down)}$$

U.S. Customary Units:

$$0 + 4290\sin 30° + 0 + \mathbf{R}_y = 0 - 36.1 \sin 30° + 0$$

$$\mathbf{R}_y = -2160 \text{ lb} = 2160 \text{ lb (acting down)}$$

Force due to water hammer. As shown in Chapter 6, water hammer pressure for the instantaneous (worst-case) closure of a downstream valve is

$$\Delta H = -\frac{a\Delta v}{g} \tag{6-2}$$

where ΔH is pressure rise in meters (feet), a is wave velocity in the pipeline in meters per second (feet per second), g is the acceleration due to gravity (9.81 m/s^2 [32.2 ft/s^2]), and v is the change in velocity in meters per second (feet per second), which is 1.55 m/s (5.07 ft/s) in this problem. Assuming some air is present in the steel pipe (see Equation 6-4 and Table 6-2), a is about 981 m/s (3220 ft/s), so

SI Units	**U.S. Customary Units**
$\Delta H = \dfrac{981 \times 1.55}{9.81} = 155 \text{ m}$	$\Delta H = \dfrac{3220 \times 5.07}{32.2} = 507 \text{ ft}$

and the pressure from Appendix Table A-11 is

$\Delta p = 155 \times 9.79 = 1520 \text{ kPa}$	$\Delta p = 507 \times 0.433 = 220 \text{ lb}$

This pressure increment can be assumed uniform throughout the wye, so the forces are

p_1A_1:	$(379 + 1520)0.165 = 313 \text{ kN}$	$(55.0 + 220)1.78 \times 144 = 70{,}500 \text{ lb}$
p_2A_2:	$(372 + 1520)0.0512 = 96.8 \text{ kN}$	$(53.9 + 220)0.553 \times 144 = 21{,}800 \text{ lb}$
p_3A_3:	$(352 + 1520)0.165 = 309 \text{ kN}$	$(51.0 + 220)1.78 \times 144 = 69{,}500 \text{ lb}$

Assume that the ρQv terms remain essentially unchanged because the water may still be moving when the shock wave arrives. Reinforcement by reflected waves could possibly occur in a complex situation, but the assumptions here are reasonable for most applications.

SI Units	U.S. Customary Units
$313 + 96.8\cos 30° - 309 + \mathbf{R_x}$	$70{,}500 + 21{,}800\cos 30° - 69{,}500 + \mathbf{R_x}$
$= 0.397 - (0.165 + 0.160\cos 30°)$	$= 88.7 - (37.0 - 36.1\cos 30°)$
$\mathbf{R_x} = -87.7 \text{ kN} = 87.7 \text{ kN (acting left)}$	$\mathbf{R_x} = -19{,}800 \text{ lb} = 19{,}800 \text{ lb (acting left)}$
$0 + 96.8\sin 30° + 0 + \mathbf{R_y}$	$0 + 21{,}800\sin 30° + 0 + \mathbf{R_y}$
$= 0 - 0.160\sin 30° + 0$	$= 0 - 36.1\sin 30° + 0$
$\mathbf{R_y} = -48.5 \text{ kN} = 48.5 \text{ kN (acting down)}$	$\mathbf{R_y} = -10{,}900 \text{ lb} = 10{,}900 \text{ lb (acting down)}$

and the forces acting on the anchor bolts are equal and opposite, i.e., same as force on water and opposite to forces on pipe.

The wye need not be anchored if the piping is able to resist the final resultant, which is

$\mathbf{R} = \sqrt{(87.7)^2 + (485)^2} = 100 \text{ kN}$	$\mathbf{R} = \sqrt{(19{,}800)^2 + (10{,}900)^2} = 22{,}600 \text{ lb}$

at a reasonable deflection (which should be calculated). Do not allow this resultant force to be transferred into valve bodies or pump casings, because they are not designed to resist such forces and are likely to bind or break. Anchorage for unbalanced forces is always required at flexible couplings.

3-9. Field Measurement of Friction Coefficient

Obstructions such as trash, valves partially closed through carelessness, reduction in pipe diameter due to mineral deposits, and the increase of roughness due to tuberculation or unevenness of mineral deposits are all manifested by an apparent decrease in the Hazen–Williams coefficient, C (or by an increase in the Darcy–Weisbach coefficient, f). The flow decreases markedly and the discharge pressure increases somewhat. Pumps run longer and consume more power unnecessarily—often at significant cost.

Field measurements to determine pipe roughness are needed whenever

- a new pump is to be used with an existing pipeline;
- the pumping station is to be remodeled;

- any expensive change is contemplated that involves existing pipe; and
- the capacity of the pump has declined without excessive impeller wear.

Other reasons for determining roughness include

- timing for cleaning and relining;
- calibration of a network model; and
- plans to increase the flowrate.

Using old plans and specifications as data for a new design without a confirming field survey is dangerous. According to some experienced consultants, the majority of record (or "as-built") drawings contain serious errors, such as wrong dimensions, alterations never recorded, significant mistakes in the static head, and an inability to achieve the design flow. Without a survey, trusting the plans is thus likely to lead to more blunders, embarrassment to both engineer and client, and an added, unnecessary cost for rectification.

The most dramatic result of a field survey is to find excessive roughness (low *C* value), which usually indicates a force main partially plugged with debris or heavily coated with grease, mineral deposits, or tuberculation. Mineral deposits over 50 mm (2 in.) in thickness were developed in one transmission main within two months [27]. Valves have been found partially closed, and improper repairs or additions long forgotten have collected sticks and rags. Camera inspection or forcing "pigs" through the line can disclose gross problems—once the need for such methods has been established by a field survey of pipe roughness. (A self-destructing pig that need not be recovered can be a loaf of hard bread, a block of ice, or a bag of ice cubes.)

Roughness is determined by measuring the friction headloss between two points separated by a known length of pipe of known internal diameter at a known rate of flow. Several runs (at least three) should be made, preferably at different flowrates, to obtain an average roughness coefficient. It is wise to demand two independent determinations of the roughness coefficient because either blunders or excessive errors can be so costly.

Pressure Gauging

Headloss can be measured with (usually) sufficient accuracy by installing a suitable pressure gauge downstream from the pump to determine the pressure drop to a free surface. The static head can be determined from a field survey or by halting the flow and reading the pressure gauge. To measure the friction head loss with enough accuracy, a "suitable" pressure gauge should have at least a 150-mm (6-in.) face, be accurate to 0.25%, and have a maximum pressure reading of not much more than 150% of the expected pressure; it should also have been calibrated recently.

Pressure gauges can be installed (without interrupting flow) by attaching them to corporation cocks, which in turn can be installed under pressure by the water utility for a nominal cost if the pipe is exposed. A short piece of flared copper tubing leading to pipe threads to match the gauge completes the installation. Some corporation cocks can be obtained with standard pipe threads or with standard pipe thread adapters (see Figure 20-6 for a good gauge assembly).

To obtain a static pressure reading, stop the flow in the force main and read the static gauge pressure, p_s. (It can also be calculated if the difference in elevation between the gauge and the free water surface is known.) After the flow is resumed, the gauge pressure is p_t and the pressure loss due only to friction head loss is

$$\Delta p = p_t - p_s \tag{3-23}$$

Convert differential pressure, Δp, to differential head, Δh, by using Table A-11, then calculate friction loss, h_f, in the usual units, meters per 1000 meters (or feet per 1000 feet), as

$$h_f = 1000\ \Delta h/L \tag{3-24}$$

where L is the length of pipe between gauge and free water surface in meters (feet).

If the transmission main branches or the diameter changes or if there is no free water surface, pressure must be obtained at a second point. If two pressure gauges are used, one at point A and another at point B, Equation 3-23 is modified to

$$\Delta p_{A-B} = (p_t - p_s)_A - (p_t - p_s)_B \tag{3-25}$$

Exposing buried pipe for the attachment of pressure gauges is very expensive, so connect the second gauge to a fire hydrant or a service line, if possible.

Unless the two pressure gauges are far apart, the error (even with gauges accurate to 0.25%) can become unacceptable. An alternative is to connect hoses or tubing at points A and B and lead them to a differential gauge or an air manometer such as that shown in Figure 3-4. The hoses must be purged, and a snubber may be needed to prevent rapid fluctuations.

Differential Pressure

Differential pressures must be measured when pitot tubes, Venturi meters, or orifice meters are used to determine flow. Usually, differential pressures range from 2 to 500 mm (0.1 to 20 in.) of water column. Pressure in the pipe may be 200 to 10,000 times greater, so it is futile to measure the differential pressure by trying to subtract gauge readings at the two pressure taps. Differential gauges are both expensive and delicate. Manometers that use mercury, carbon tetrachloride, or oil are, according to Murphy's Law, almost certain to blow out sooner or later and contaminate the pipeline—a serious disaster in a public water supply.

An inexpensive substitute is the air-filled, upside-down manometer in Figure 3-4. If the air blows out it can easily be replaced by a tire pump or a pressure tank, so a tire valve should be installed at the top. Petcocks to bleed air from the tubing leading to the manometer are also needed. Glass tubing is fragile, but plastic tubing is rugged. A satisfactory homemade device equipped with (1) 6-mm ($^1/_4$-in.) valves or petcocks, (2) a method for attaching it to the copper tubing from the meter (such as 6-mm [$^1/_4$-in.] heavy rubber tubing with hose clamps), and (3) a meter stick can be made for a few dollars from materials available at hardware stores or laboratories. Commercial models are available. Differential pressure measurement is discussed thoroughly by Walski [28, 29].

Pipe Diameter

The internal pipe diameter (ID) must be known with any of these methods, and the true ID must be known with reasonable accuracy to use a pitot tube, strap-on meter, or any device that measures only velocity in the pipe. Wall thicknesses of pipe are increased by decreasing the ID, so OD does not indicate ID with accuracy. Furthermore, coatings (with unreliable thicknesses) and mineral deposits reduce the ID. Hence, do not trust the plans or specifications. The most reliable means is to remove a section of pipe and measure the ID. If the pipe cannot be taken out of service, a "feeler" rod inserted through a packing gland fitted to a corporation cock gives the thickness, but only at one point and at a cost that is exorbitant for the purpose unless the cock is also needed for a pitot tube. (Note that mineral deposits may not be uniform.) Other nondestructive ways of measuring pipe thickness may exist or be developed, but thickness transducers (at the time of this writing) are inaccurate for pipe with a mortar lining or mineral deposits.

Calculation of Average Pipe Diameter

Equation 3-1 can be used to find the average pipe diameter if both average velocity and flowrate are known. A tracer can be used to find the average velocity as follows:

- Introduce the tracer as a slug at one point.
- Measure the average time for the tracer to pass a downstream point (actually, the time for half of the tracer to pass the point).
- Measure the length of pipe between the two points and calculate the velocity.

Tracers can also be used to find the discharge independently of the velocity as discussed at the end of this section. (Other methods discussed in the following subsection could also be used.) With very careful work, the average ID of a pipe can be found to about the nearest 2% of the pipe diameter.

Flowrate Measurement

Flowrate, in conjunction with friction headloss, is useful for constructing the existing system curve as well as the coefficient of friction. Flowrate can be measured in a variety of ways, generally dictated by the specific situations encountered.

- Obtain volumetric measurements using a stopwatch and the wet well, clear well, or the reservoir. Always make every effort to measure flow volumetrically by using any practical means available.
- Use the existing flowmeter in the station, but be suspicious if the calibration is not recent. Check the read-out system by manual readings, if possible, or by comparison with volumetric measurements. As a rule, errors in most flowmeter readings exceed 10%, and 50 to 200% errors are not uncommon [27].
- Temporarily install a Palmer–Bowlus flume in a manhole (preferably downstream) where there is open channel flow.
- Use the manufacturer's pump curves as a convenient check to confirm any of the other methods. When carefully done, the uncertainty of flowrate measurement can be kept to ±7%, if there is reasonable surety that (1) no wear has occurred, (2) the impeller has not been replaced with a significantly different one, and (3) the pump curves were developed from good data under conditions that can be reproduced in the field. Some consultants do not trust pump curves, but others have found them reliable.
- Install a temporary flowmeter (preferably in a long, straight section of pipeline), such as a multiport pitot tube or magnetic flow rod, or use a strap-on Doppler or ultrasonic meter. A traverse with a single port pitot tube is a useful substitute, but traversing is tedious, and within, say, 15 pipe diameters

downstream from a discontinuity (e.g., an elbow), accuracy suffers because of unsteady flow lines.

- Use tracers. The use of tracers provides the highest accuracy and reliability, but it requires either expensive equipment or a consulting specialist. Errors in repeatability can, with care, be kept within less than 2% of the particular flowrate being tested.

All of these techniques require resourceful, knowledgeable personnel who are either experienced or aggressive enough to learn how to make the selected tests with accuracy. If needed, pump consultants, some consulting engineers, or university staff members in environmental, civil, mechanical, or chemical engineering may be able to assist.

Volumetric Measurement

Volumetric measurement, the method of choice, is often the most convenient and reliable because many, perhaps most, pumping stations have associated wet wells, clear wells, or storage tanks. Volume computed from careful measurements can be used in conjunction with timed drawdown to determine flowrate with precision. Again, beware of trusting the construction drawings for calculating volumes; always make at least two confirming field measurements of the basin size. Under field conditions, it is difficult to time an event to the nearest second, so the running time should be at least 2 min; 5 min is better.

If the influent flow into a wet well or clear well cannot be halted for the duration of a run, a reasonable approximation of volume pumped can be made by first measuring inflow with the pumps off and then accounting for inflow while the pumps run. Because sewage inflow varies, several runs are necessary to get a reasonable, average result. Because the volume of a sewer pipe is not only very large but also indeterminate, be sure to keep the upper wet well level below the influent sewer pipe.

One method for timing wet well drawdown is to tie two short objects to a string at a convenient, measured distance apart. Attach a weight (a rock or brick will do) to the end of the string and suspend it in the wet well. As the water rises (or falls) the meniscus at the objects can usually be clearly seen by using a flashlight [30]. If visibility is poor, a pair of electrical contacts and a battery to operate low-voltage lights can be used.

Permanent Flowmeters

When present, a flowmeter may be the most convenient way to measure discharge, but care must be taken to ensure that the meter itself is not a source of error. Manufacturers' claims of accuracy are rarely realized in the field, and a surprising number of installations have shown excessive errors—usually more than 10%, often more than 20%, and occasionally as much as 200% [31]. (For example, the influent flowmeter at a water treatment plant in Montana was about 100% in error for nearly a year [32].) Reasons for poor accuracy include improper installation, clogged pressure taps, improper adjustment of the read-out system (mechanical or electrical), nonideal hydraulic conditions, incorrect flow coefficients, incorrect flowmeter gearing, and corrosion or moisture in electrical circuits.

The electronic read-out system of any differential pressure device, such as a Venturi or orifice, can sometimes be checked by temporarily installing a differential head device (differential pressure gauge or air manometer) that can be read manually. Unfortunately, only the differential pressure device is checked, whereas the meter itself may be at fault.

Temporary Meters in Open Channels

Sewage force mains and some water pumping stations take fluid from (or discharge into) open channels. Palmer–Bowlus flumes can be placed in sewer pipes at manholes. Weirs and Parshall or Venturi flumes can be installed in ditches. The water level can be read with a point gauge (equipped for convenience with a battery-operated electric contacter) in the upstream channel. Floats, battery-powered "dipper" needles, and sonic meters are also available.

Pump Curves

The use of pump curves is a convenient check on other methods. Pressure gauges are needed on both the suction and the discharge spools (but not in the pump casing). Worn impellers must not be used. (Of course, it is even better if the pump has been recently tested and certified test curves are available.) The accuracy of flowrate determination depends on knowing the conditions used by the manufacturer to derive the pump curves and avoiding operation where the pump curve is too flat.

Begin by operating the pump against a closed discharge valve and adjust (by calculation) the manufacturer's impeller head-discharge (H-Q) curve up or down to correspond to the differential pressure gauge reading across the pump at shutoff. If possible, operate the system at a point where the pump H-Q curve is reasonably steep by running only one of a bank of pumps. Calculate the flow from the adjusted pump curve at the proper differential pressure across the pump.

This technique and the probable errors associated with each component of the measurement are thoroughly discussed in DIN [33]. With care, the uncertainty of flowrate measurement can be kept within ±7% and, frequently, to within 5% [34].

Temporary Flowmeters in Pipelines

The recommendations for installing temporary flowmeters are the same as those for permanent ones. The manufacturer's advice must be followed, especially with regard to the required lengths of straight pipe before and after the meter. If there is insufficient length of straight pipe (which is often true in the cramped quarters of pumping stations), consider a meter installation in the yard piping.

Assuming a satisfactory location for a meter, several available types are compared in Table 3-1 (see also Chapter 20).

Tracers

The purpose of this subsection is to introduce a little-known, little-understood, but versatile and accurate method for measuring flowrate and for calibrating flowmeters in place. Errors of less than 5% of the measured flowrate are easily achievable, and with care, the errors can be reduced to less than 2%. There are few places where the use of tracers would be inappropriate (e.g., a closed or recirculating system). The method is useful for measuring flows in manufacturing plants, open channels, and pipelines of any size or of varying size. Engineers who have some background in wet chemistry can learn to use tracers in a couple of days.

Tracers can be used in three ways:

- Tracer velocity or time-of-travel,
- Slug addition, and
- Constant rate injection.

In the first method, a concentrated dose of tracer is dumped into an upstream point, and the time required

Table 3-1. Temporary Flowmeters

Type	Advantages	Disadvantages
Flow nozzle	Locate anywhere along pipe; only one pressure gauge required if at end of pipe.	Expensive to dismantle pipe.
Pitot tube, single port	Insert through corporation cock. Can install under pressure, inexpensive in exposed pipe, one pitot tube can serve all stations.	Unsuitable for raw sewage. Must take many readings to find average flow [35], flow must be steady while survey is in progress, expensive for buried pipe.
Pitot tube, multiport (integrating)	Single reading gives average flow; can follow changing flow, good as a permanent meter.	Suitable for only a single ID. Unsuitable for raw sewage.
Strap-on meters	Can apply to outside of pipe. Installation is inexpensive, can be used permanently.	Meter itself is expensive. Readings suspect without independent check, should calibrate meter on pipe.
Magnetic	For raw sewage or clean water.	
Time of flight	Accurate for clean water but only if installed far enough (2 blocks) downstream from the pump so that minute vapor bubbles are dissipated.	Not for raw sewage or dirty water.
Doppler	For raw sewage and dirty water.	Not for clean water.
Magnetic tube type	Same as multiport pitot tube.	Expensive. Requires corporation cock.
Orifice plate	Same as for flow nozzle; for grit-laden water, use segmental orifice plate to provide smooth invert.	Unsuitable for rags and stringy material.
Elbow meter	Inexpensive; can construct in the field using any elbow, fair accuracy but good precision (repeatability), stable operation, unobstructed flow.	Must calibrate in place; low differential pressure (use air manometer) (see Miller [36]).

for the peak concentration to pass a downstream point gives the average velocity. Salt is frequently used for the tracer and its concentration is measured with a portable conductivity meter. If the flow is steady and the channel is prismatic, an approximate flowrate can be determined by measuring the cross-sectional area. The accuracy is questionable. Errors may well exceed 15%.

The next two methods require precision. The general requirements for the use of tracers are:

- A precise method for introducing the tracer.
- Enough length and/or turbulence so that mixing is complete—a critical requirement.
- A tracer substance that is innocuous (nonhazardous and nontoxic), stable, and not significantly adsorbed on walls or particles.
- A tracer substance that occurs naturally only in very low or insignificant concentrations (low background).
- A means for obtaining small samples downstream.
- A method of chemical analysis that is sensitive, accurate, inexpensive, and free from significant interferences.
- Preliminary testing to (1) establish accuracy; (2) obtain calibration curves for the tracer diluted with the water or sewage; (3) quantify possible interferences, decay, or sorption on pipe walls or particles so that corrections can be made; and (4) determine the suitability of the means selected to preserve samples for shipment and analysis.

In the slug-addition method, a known mass of tracer is introduced at some upstream point in a manner to promote good mixing. At a point well downstream after thorough mixing, a series of grab samples is obtained and analyzed. Plotting the data yields a curve like that labeled "abd" in Figure 3-9. If the flowrate is constant over the short time needed for all the tracer to pass the downstream point, then tracer introduced equals tracer past the downstream point.

$$M = vC = Q\int_{t_1}^{t_2} (c - c_o)\,dt \tag{3-26}$$

where M is the mass of tracer, v is the small volume of tracer at high concentration, C, Q is the uniform flowrate, c is the measured concentration of tracer downstream, and c_o is "blank" or the background reading of water without tracer. Under ideal conditions, the error in determining Q can be less than 2%. In Figure 3-9, blank, c_o, is shown to be large for clarity. Actually, the peak concentration should be at least 100 times greater than blank.

Accurate results require a sufficient number of samples to identity clearly the time of first appearance and the time of complete passage and to define the shape of the curve. It is prudent to take many samples before the expectation of first appearance and many more samples after expected complete passage. The tracer concentrations should be determined with the same care as described below for the constant-rate injection method.

In the more versatile continuous constant-rate injection method, a small stream of concentrated tracer is added at the upstream point. If possible, the concentration of tracer past the downstream point should be monitored until uniform instrumental readings, as indicated by point "e" in Figure 3-9, are obtained. If monitoring is impractical, estimate the waiting period for the axial dispersion to stabilize and for a uniform concentration of tracer to invade all water in pockets (such as valve bonnets). Then take a series of grab samples for precise analysis. The remainder of this discussion applies to continuous constant-rate injection, although much of it would apply to slug addition as well.

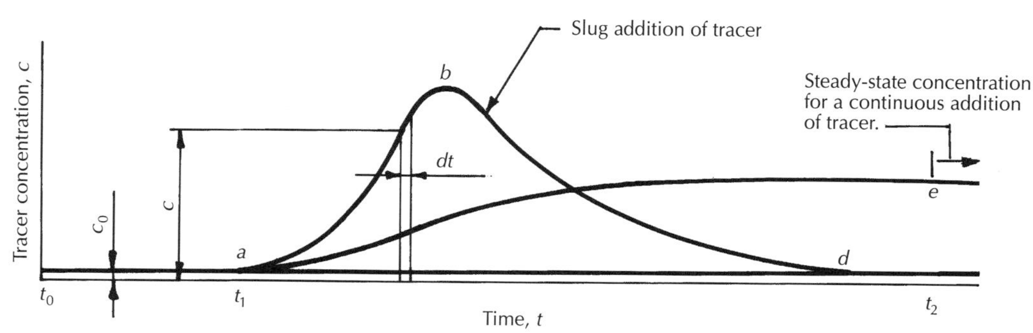

Figure 3-9. Downstream concentration distribution of tracer resulting from: (a) slug addition of tracer (curve abd) and (b) continuous constant-rate injection of tracer (curve ae) at an upstream point.

If the objective is to calibrate a flowmeter, obtain simultaneous readings of the flowmeter in engineering units and in raw sensor analog output. The latter can be measured by means of a digital meter with much greater accuracy than a visual reading of the flow indicator, and the multiplier and offset coefficients to convert sensor signal to flow can be calculated directly. Flowmeter signals typically fluctuate as much as 10% over short periods of time due to turbulence at the sensor, and, therefore, many data points are needed to establish a norm. If practical, obtain pump speed and pump motor current draw to compare pump performance with predicted performance and to check the reasonableness of the test results.

If the objective is to calibrate a flowmeter in an open channel, it is useful to obtain simultaneous depth measurements manually for comparison with the flowmeter measurements and to establish an independent flow rating curve. If a flow sensing element is replaced, the manual rating curve allows a quick check of the meter rating data. Record the measurement location on-site and document it in the report so the same point can be identified and used in the future.

The accuracy of the method depends directly upon (1) the accuracy and uniformity of tracer injection and (2) *complete* mixing at the second sampling point. Tracer can be metered with precision either from a positive displacement fluid metering pump or from a Marriot jar feeding an orifice, as shown in Figure 3-10. The Marriot jar is useful only for open channels, whereas the pump can inject tracer at high pressure. To ensure accuracy, however, the pressure must not change during injection. It is possible to inject tracer into the low-pressure side of a small pump that "chases" the dye to the point of application with a small stream of water. Constant speed metering pumps can be obtained from Fluid Metering Inc. [37] for 12 v battery operation, whereas variable speed pumps can be obtained for operation only on 110 v ac current. The flowrates can be adjusted by changing the piston angle to alter the piston stroke length. Even more adjustment in flowrate can be made by changing the pumping heads (on either kind of pump).The reservoir of tracer should be calibrated either volumetrically or gravimetrically so that the accuracy of dispensing can be measured with a stopwatch.

By the time the tracer reaches the sampling location, mixing must be complete so that the concentration of tracer is uniform. One hundred pipe diameters downstream of the injection point along a straight section of pipe is considered sufficient, but sometimes that distance can be reduced substantially. Dye can be injected at the inlet side of a centrifugal pump and sampled on the discharge side. Or it can be injected into a pipe through a ball valve or a corporation cock (wherein leakage between probe and valve is prevented by means of O-rings) and sampled downstream by the same means. It helps to have several ejection ports in the injection (probe) tubing. Always test for complete mixing by withdrawing samples from various points across the pipe and also by comparing consecutive samples. Extreme care must be taken to ensure that none of the tracer escapes by recirculation or by leaking through check valves. (Close all isolation valves for pumps that are not running during the test.) If the wet well is contaminated with tracer, it will take a long time to remove all of it, and it will compromise the data.

Of the recommended tracers shown in Table 3-2, Rhodamine WT is the best choice for municipal water and wastewater, because (1) flow-through fluorometers can be

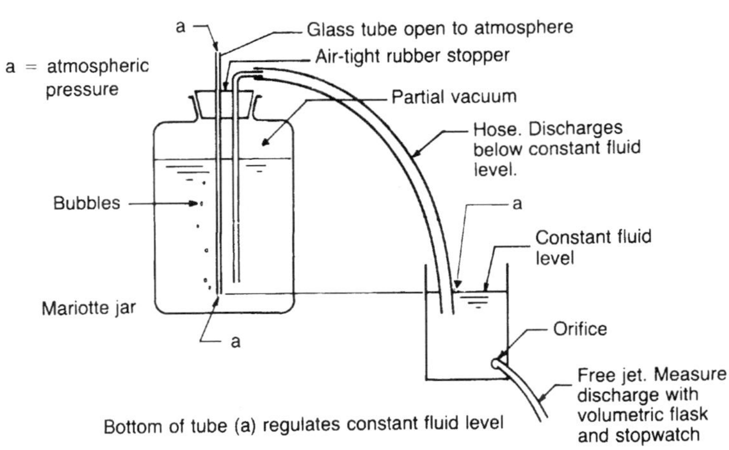

Figure 3-10. Constant head device for metering tracer.

Table 3-2. Recommended Tracers and Analyses

Substance	Use	Analytical method	Concentration (mg/L) Lower detection limit	Advisable range	Comments
Fluoride	Excellent for potable water.	Specific ion electrode.[a]	0.0500	0.5–2.5	If utility adds fluoride to water, no added spike needed, hence it may be the best field tracer. Calibrate fluoride feeder or substitute solution feed and metering jump or Mariott jar (Figure 3-9).
Glucose	Excellent for potable water.	Colorimetry by Park-Johnson method [38].	1	1–10	Occurs naturally in some fruits and is a food ingested by everyone. Sterilize any tracer solution to be injected into potable water. Little affected by chlorine. Protect samples from bacterial attack.
Rhodamine WT [39]	Excellent for sewage. Approved for drinking water by the EPA [40].	Fluorometry,[b] easiest of all tracer methods for use in laboratory or field.	0.00001	0.0002–0.4	Fluorometers uncommon in commercial laboratories. There is little background fluorescence. Adsorption on pipe walls and particles insignificant.
Lithium chloride	Good for manufacturing plants and water or sewage.	Electrothermal atomic adsorption spectroscopy, laboratory only.	0.001	0.005–0.025	Very low natural background, used medicinally (but in far larger quantities).
Sodium chloride	Sewage	Conductivity,[c] good for field use.	1	100–1000	High background (1000 μmhos) requires massive spikes (~100 mg/L NaCl) and large flow (q = 1/2% Q) of brine.
		Flame atomic adsorption.	0.002	0.03–1	Must dilute samples.
Chlorine	Water	DPD colorimetry [41], (can be used in the field).	—	0.2–4	Usually unacceptable due to rapid reaction and decay.

[a] SPADNS colorimetric method can be used, but it is less convenient, more expensive, and subject to more interferences.
[b] Temperature sensitive: variation of 0.4°C in samples gives 1% error; use a water bath or correct mathematically.
[c] Correct for temperature: use a temperature–conductivity calibration chart in the field.

used in the field to determine when stability has occurred and when the more accurate grab sampling procedure can start; (2) the tracer is innocuous, inexpensive, and provides linear instrumental response in concentrations from the limit of detection (about 0.01 ppb depending upon the fluorometer used) to 100 ppb; (3) sorption on walls or particles is usually very slight; (4) background fluorescence is low (usually less than 0.2 ppb); (5) determinations are quick and relatively convenient; and (6) determinations can be made in the field. The disadvantages are (1) fluorometers are expensive and rarely found in chemical laboratories (although they can be rented); (2) rhodamine is very temperature-sensitive and is attacked by chlorine; and (3) samples should not contain particles or air bubbles, so if the water is dirty, samples must either be centrifuged or tediously filtered through glass or Millipore filters.

If no tracer is lost (through leakage, reaction, absorption, etc.), a material balance yields:

$$\underbrace{Qc_o + qC}_{\text{tracer in}} = \underbrace{(Q + q)c}_{\text{tracer out}} \qquad (3\text{-}27)$$

where Q is flowrate in the pipe before the stream of tracer is added, c_o is blank or the apparent background concentration of tracerlike substance, q is the flowrate of tracer, C is the concentration of the tracer feed solution, and c is the final concentration of the tracer downstream. If Q is greater than $100q$ (as it should be), $Q + q$ approximates Q and Equation 3-27 can be modified to

$$Q = q\frac{C}{(c - c_o)} \tag{3-28}$$

The highest accuracy with Rhodamine WT cannot be achieved, however, without accounting for the effects of the flow stream on the dye (such as degradation or sorption) and on the actual fluorescent measurement itself (by masking the background fluorescence). Before and after every flowrate test, background samples should be collected upstream of the injection point and composited in sufficient volume (5 L) to produce a spiked background standard and a blank sample. The spike should be added in the field soon after collection at a concentration reasonably close to that expected in the downstream samples. Use only Class A volumetric glassware for dilution and pipetting.

Wastewater characteristics typically change under varying flow conditions. Solids increase significantly in raw domestic wastewater as the flow increases from early morning to midmorning, and they can significantly affect the measured dye concentration. Under certain conditions, background sampling is the only way to detect dye recirculation through, for example, leaking valves into the wet well. Furthermore, there may be some unknown material that degrades the dye or absorbs it over time in the flow stream.

These background samples should be analyzed in the same manner as the downstream samples and used to adjust the measured sample dye concentrations as shown by the following equation developed by McDonald [42]:

$$C_a = (C_m - b)\left[\frac{S}{(C_s - b)}\right] \tag{3-29}$$

where C_a is the adjusted concentration of the sample, C_m is the measured concentration of the sample, b is blank (or background concentration), C_s is the measured concentration of the spiked background, and S is the true concentration of the spiked background based on the volumetric dilution calculations. The true flowrate is calculated from

$$Q = q\left(\frac{C}{C_a}\right) \tag{3-30}$$

where q is the flowrate of the dye injected at concentration C.

As an absolute minimum, samples should be collected in groups of three for each flowrate tested. It is often difficult to predict how long it will take to reach steady-state conditions, so it is prudent to obtain extra grab samples to demonstrate that steady-state conditions were actually achieved. The concentration variability between samples in a group should be less than ±2% with complete mixing and steady-state conditions.

A flow-through fluorometer cell is useful for continuously monitoring downstream dye concentrations for determining completeness of mixing, equilibrium, and optimal dye concentration. But if precision is needed (for example, to calibrate flowmeters), use these data only for guidance. Use only grab samples, centrifuged and brought to uniform temperature, for precise results.

Although less than 20 ml of sample is required for measurement, the sample volume must be sufficient to allow for centrifuging or filtering the sample and for rinsing containers. Therefore, 250 to 500 ml of sample should be collected in groups of three and kept in borosilicate glass bottles. Decanting after settling by gravity instead of centrifuging is sometimes adequate, although it is time consuming. Borosilicate cuvettes (test tubes) should be acid-washed before each use and either dried or rinsed with sample prior to determinations. Mark cuvettes so each can always be inserted into the fluorometer with the same orientation, and test each for blank with distilled water. Use only those cuvettes with the same blank reading.

Fluorometers are now made with automatic temperature compensation for flow-through cells, and the flow-through cell can be used with large (1-L) grab samples by pumping the samples through it. But unless enough experience is gained to trust a particular instrument or procedure, it is wiser to put grab samples into small cuvettes and immerse them in a circulating water bath long enough (at least 20 min) for the samples to reach a predetermined stable temperature. After insertion into the fluorometer, obtain the instrument reading as soon as it stabilizes, because the temperature of the sample rises in the instrument due to the heat from the lamp, and readings change 2.6% per degree centigrade. A liquid-cooled cuvette holder (if one is available or can be custom made) is very desirable. Run all the samples without pause to obtain the best temperature control. Good advice on the details of fluorometric measurements is given by Turner Designs [43, 44]. This company also rents fluorometers and metering pumps.

Lithium chloride is also a good tracer. Lithium is rare in nature, so the background is low. It is not degraded by sunlight or chemicals, and sorption on walls and particles is less of a problem than it is with rhodamine. It can be measured by means of electrothermal atomic absorption spectroscopy or by inductively coupled plasma (ICP) emission spectroscopy. These instruments are common in chemistry laboratories, but (unlike fluorometers, which are easily operated) they require considerable skill for operation, and they are more expensive than fluorometers. The samples can, however, be measured by trained personnel in commercial or university laboratories. Usually, samples are acidified and digested, but it seems likely that for constant injection where the concentration is near optimum, the samples can be measured much more economically without the usual preparation. It would be wise to make preliminary tests to determine whether accuracy is thus compromised. The disadvantages include the necessity of taking samples to a laboratory for analysis, high cost per sample, and uncertainty or delay in the field for determining either contamination or stability.

Calculation of Pipe Friction Coefficient

With pipe diameter, headloss, and flow as known quantities, the Hazen–Williams roughness coefficient, C, can be calculated by Equation 3-9a. Rearranging it into a more convenient form gives

$$C = 151QD^{-2.63}h_f^{-0.54} \tag{3-31a}$$

where Q is in cubic meters per second, D is the pipe ID in meters, and h_f is meters per 100 meters. In U.S. customary units Equation 3-9b becomes

$$C = 149QD^{-2.63}h_f^{-0.54} \tag{3-31b}$$

where Q is in gallons per minute, D is in inches, and h_f is feet per 1000 feet.

It is desirable to make two or, better, three tests, preferably at different flowrates. The C values should be reasonably close. If C is less than about 130, investigate for excessive mineral deposits or obstructions. Note that buried pipe can be both cleaned and relined in place with cement mortar or plastic at a total of 20–40% of the cost for new pipe in a "turnkey" operation (which includes all facets of the job—even a temporary bypass). Cost data and practical field methods for cement-mortar lining have been discussed by Walski [45].

3-10. Flow of Sludges

The preceding equations for friction headloss in pipes (Equations 3-8 and 3-10, for example) are limited to Newtonian fluids—those in which shear stress is proportional to shear rate and thus to velocity (see Figure 3-11). But in thick sludge, the shear stress increases very rapidly at low shear rates and increases only moderately with an increase of shear rate at higher velocity. Consequently, at low velocity the headloss for sludge is many times the headloss for water, whereas at high velocity the headloss for sludge may be only marginally higher than that for water.

In general, the headloss for sludges with less than 2% solids is nearly the same as for water. Most secondary sludge and water treatment plant sludge would fall into this category. The headloss for thicker sludge (usually primary sludge) can be approximated as explained in Section 19-1 by calculating the headloss of water and then multiplying by a factor obtained from Figure 19-4 or 19-5. But sludges are unique and the headloss for a "worst case" could be double the factors shown in Figure 19-3. For long pipelines where pressure loss is important, it is wise to test the friction properties of the sludge. Sludge flow and pumping are considered in greater detail in Chapter 19 and by the EPA [46].

3-11. Unsteady Flow

In all of the preceding sections, it is assumed that the flowrate is steady and continuous and that Equation 3-1 (the continuity equation) is valid, so the flowrate

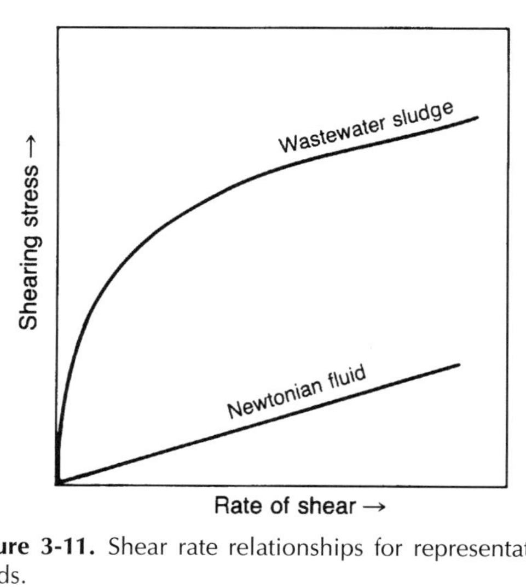

Figure 3-11. Shear rate relationships for representative fluids.

at one point equals that at another. In reality, this assumption is often only an approximation. Pipe material is not rigid and water is not incompressible. When valves are closed or opened quickly or when pumps are started or stopped, pressure waves are generated and travel back and forth in the pipe, which expands and contracts (as does the water) and causes the flow to become unsteady.

These violent pressure changes can have serious consequences on pumps, especially if the pressure changes continue for many cycles and if the head-discharge curve of the pump is relatively flat. The intersection of the pipe system curve and the pump curve is the operating point of the pump—but only during steady flow when the pressure at discharge is a function only of static and friction head.

If the pressure increases somewhat, for example, because of the partial closure of a valve, the flow decreases greatly. If the pressure is reduced slightly below normal, the flow increases substantially. Thus, a component of flow (in which the water sloshes back and forth) is superimposed upon the average discharge of the pump. If that component is in harmony with the natural frequency of the pressure waves in the pipe, resonance occurs, the pressure waves continue, and the pump can be destroyed in a few hours.

The problems associated with pressure waves can be even more severe and insidious with variable-speed pumps because the pumps operate over a considerable portion of the family of head-discharge curves.

The problem can be avoided or alleviated in a number of ways.

- Choose pumps that do not have a flat head-discharge curve at the operating point.
- Design the system so that the pump does not operate for more than a few seconds at any point where the phenomenon described can occur.
- Make a transient analysis (see Chapters 6 and 7) and design the entire system to limit the magnitude and repetition of pressure waves.

3-12. Model Studies

Some types of problems defy traditional mathematical analysis and must be solved by extraordinary means. One such problem type is the flow regime in forebays and pump intake facilities. For structures of reasonable size that are based upon proven, successful designs, it is usually sufficient to follow established rules or guidelines. For very large structures or those that differ significantly from designs known to be successful, a model study is usually the only means to ensure success.

Model studies of wet wells and/or forebays are indicated whenever:

- The design does not adhere in all respects (such as bay width, bell clearance, side slopes, piping, distance from obstructions, and change in direction of incoming flow) to the guidelines presented herein or to established designs proven to be successful.
- The flow is greater than about 9000 m^3/h (40,000 gal/min) per pump or 23,000 m^3/h (100,000 gal/min or 140 Mgal/d) for the station.
- Excavation cost for size or depth is at a premium, or where achieving maximum design flow is critical.
- The approach flow is asymmetrical or nonuniform or there is a change in direction of incoming flow due, for example, to some pumps not operating.
- An existing sump is malfunctioning, and retrofitting is required.

Conducted by experienced personnel, model studies can be used to evaluate measures for alleviating the undesirable hydraulic conditions described under "Pump Sump Problems" in Section 12-6.

Model Similitude

True similitude requires that (a) the Froude number ($\boldsymbol{F} = v/\sqrt{gL}$), a dimensionless function of gravitational and inertial forces; (b) the Reynolds number ($\boldsymbol{R} = vD/\nu$, Equation 3-11), a dimensionless function of gravitational and viscous forces; and (c) the Weber number ($\boldsymbol{W} = v^2D\rho/\sigma$), a dimensionless function of inertial and surface tensile forces be the same for both the prototype and a geometrically similar model. The velocity is v, g is the acceleration due to gravity, L is a length such as the flow depth, D is a length such as the intake diameter, ν is kinematic viscosity, σ is surface tension, and ρ is mass density. All dimensionless numbers cannot be satisfied simultaneously unless the model/prototype scale ratio is unity, so a compromise must be made. When a free surface exists, such as in a pump intake basin, flow patterns are influenced primarily by gravitational and inertial forces, and, therefore, similitude is based on the Froude number. Although the Reynolds number is much less important, it must be greater than 3×10^4 at the suction bell mouth to ensure minimum viscous effects. The Weber number should be greater than 120 for minimizing surface tension effects. For practicality in observing dye patterns and obtaining adequate measurements of velocity in the bell, the throat diameter of the suction should be at least 75 mm (3 in.). The linear scale ratio, L_m/L_p, of models is typically about $^1/_{10}$, but L_m/L_p for the models used for developing the guidelines in Section 12-7 was close to $^1/_4$, so that for equal model and prototype values of ***F***, other model/prototype scale ratios were close to:

- Velocity: $v = (L_m/L_p)^{0.5} = {}^1/_2$
- Time: $t = (L_m/L_p)^{0.5} = {}^1/_2$
- Volume: $V = (L_m/L_p)^3 = {}^1/_{64}$
- Flowrate: $Q = V/t = (L_m/L_p)^{2.5} = {}^1/_{32}$

Construction

The geometry of the model and prototype must be similar, and in particular, a sufficient amount of the approach structure or water body must be included to simulate accurately prototype flow patterns into the sump. At least one side of the model must be constructed of clear plastic or glass for observing flow patterns. Likewise, suction bells and pipes should also be formed from clear plastic. It is advantageous—in spite of extra cost—to construct the model so that selected dimensions, such as floor clearance for bells, depth of trench, or slope of floors, can be changed (possibly by adding or removing inserts) without rebuilding the model.

Impellers are not modeled, because the objective is to evaluate flow patterns approaching the impeller. Instead, the suction intake is equipped with a swirl meter consisting of four neutrally pitched vanes that rotate freely on a centered shaft. The outer diameter of the vanes should be 75% of the intake pipe diameter and the axial length of the vanes should be 0.6 pipe diameters. The swirl meter should preferably be located about four suction pipe diameters downstream from the mouth of the suction bell.

Measurements

Water level should be measured with a model accuracy of 3 mm ($^1/_8$ in.). Flowmeters should be calibrated to achieve an accuracy of at least ±2% of upper range value. Alternatively, a standard ASME orifice could be used without calibration. Very low approach velocities to pump intakes can sometimes be measured with dye by means of a stopwatch and a background grid. Dye is introduced through tubing small enough to be essentially nonintrusive. Higher velocities can be measured with a small propeller meter or by other suitable means at enough points to define the flow pattern and its stability.

For the final design, the distribution of axial velocity in the throat of the suction bell should be taken with velocity traverses along two perpendicular axes with a device capable of a repeatability of ±2% or better. Pitot tubes can be used, but a laser velocity meter with a computer readout, although an incredibly expensive device, provides much more extensive data and is completely nonintrusive.

The intensity of flow rotation, θ, is measured by the swirl angle computed from

$$\theta = \tan^{-1}\left(\frac{\pi d n}{u}\right) \tag{3-32}$$

where d is pipe diameter, n is revolutions per second of the swirl meter, and u is average axial velocity at the swirl meter. Swirl is unsteady. The meter may only rock slightly for short or long periods followed by slow, rapid, or even reversing rotation. Hence, average rotation is not a complete description, and data on the percentage of time versus swirl angle should be obtained. As the swirl meter may sometimes rotate too rapidly to count revolutions, a good system is to photograph the swirl meter for 10 min with a video camera that overlays time in seconds on the images. This permanent record can be analyzed (at slow speed when necessary) for any type of time-rotation rate distribution wanted.

Vortices are measured visually with the aid of dye and artificial debris by comparing them to Figure 3-12. Vortices are usually unsteady in both location and strength and intermittent in occurrence. Vortices should be observed at, for example, every 15 seconds during a 10-min interval to obtain vortex type versus frequency. Video recording is recommended.

Acceptance Criteria

The pump manufacturer, station designer, client, and research organization should agree on the acceptance criteria, because the criteria may vary with the size, application, and kind of pumps. In general, the following provisions are satisfactory.

- Surface vortices of Type 3 or more and subsurface vortices of Type 2 or more are unacceptable, except that in some stations they may be acceptable only if they occur less than 10% of the time or only for infrequent pump operating conditions. Be cautious, however, because a weak effect in a small prototype can become stronger in a larger prototype.
- Swirl angles should be less than 5°, except that 6° may be acceptable only if they occur less than 10% of the time or only during infrequent pump operating conditions.
- The time-average velocities at specific points in the bell throat should be within 10% of the area average velocity. Time-varying fluctuations should have a standard deviation of 10% or less.

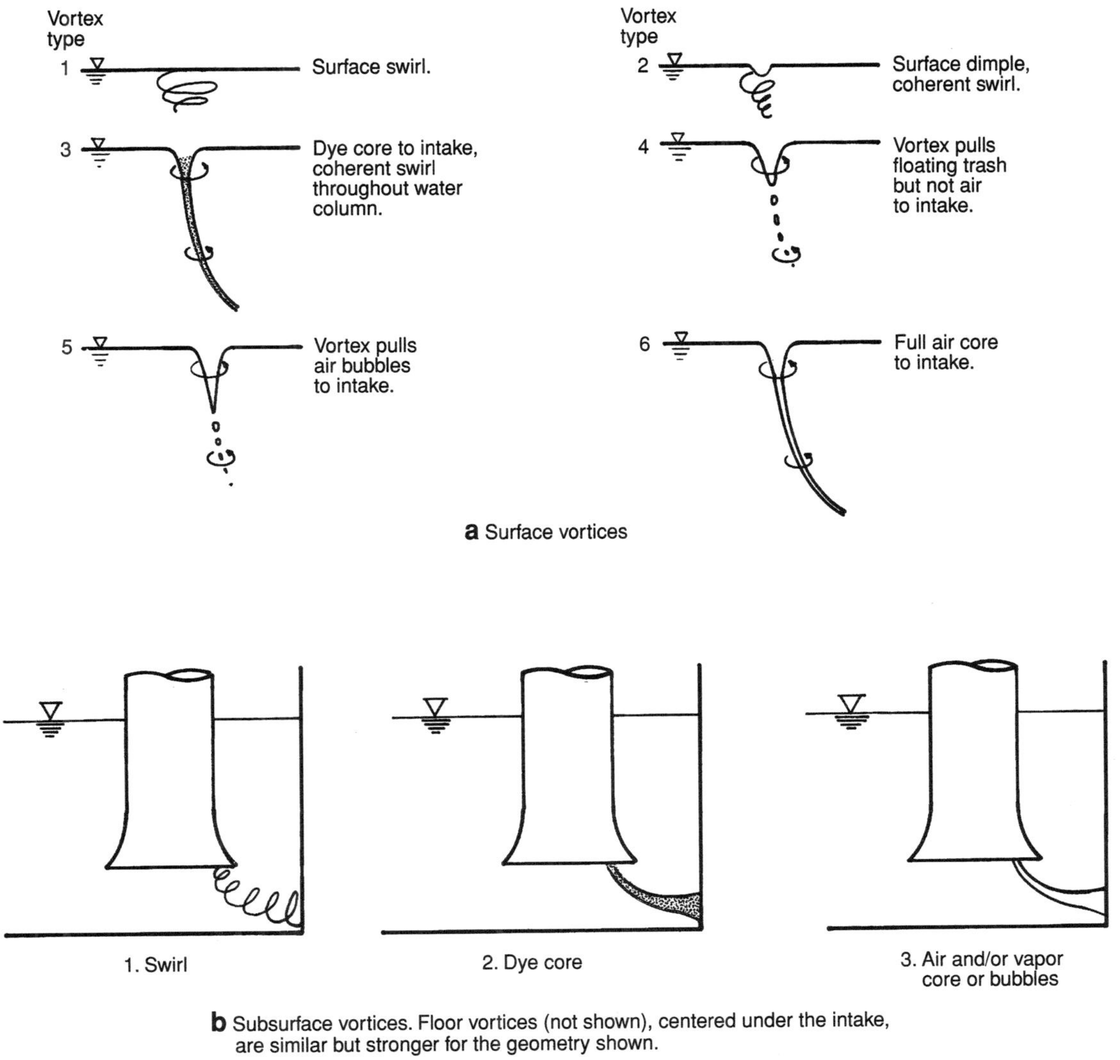

Figure 3-12. Classification of vortices. Courtesy of Alden Research Laboratory, Inc.

Tests

General observations should first be made on the initial design to identify any problems before making time-consuming detailed tests (unless detailed tests are wanted by the client for comparison with a final design). Modify the model until the general observations are consistent with satisfactory design. Then test operating conditions at minimum, intermediate, and maximum flows and water levels. Include all possible pump operating conditions and combinations of pumps. Selected tests for the formation of free-surface vortices should be made at 1.5 times the Froude number to compensate for possible scale effects. Sometimes strong vortices can organize unexpectedly—at the low velocities that occur with HWL and low flow, for example.

Vortices and swirl angles should be observed in all tests. Axial velocity determinations should be made for each pump—but only in the final design. Document tests of swirling and vortexing with a video camera and include a VHS cassette of the tests with the final report.

Comparison with HI Standards

Part of this section is adapted from (and all of it is compatible with) more detailed guidelines prepared by Hecker [47] for the next edition (circa 1998) of Hydraulic Institute Standards.

3-13. Computational Fluid Dynamics (CFD)

Computational fluid dynamics (CFD) is a powerful method for predicting fluid motion in a continuum by using numerical techniques. The Navier–Stokes equations governing fluid motion are assumed valid at all points, but, as it is not feasible to calculate velocities at every point, the equations are solved at a finite number of points called "nodes." Typically, nodes are closely spaced at discontinuities (such as walls and corners) where velocity gradients are large, whereas nodes are spaced farther apart where velocity gradients are low. Node spacing depends on the phenomena and geometry to be studied as well as on the numerical solver and the computer used. Computational time can vary from a few minutes to several days.

An increasing number of engineers are using CFD, because programs (called "codes") are being developed for high-speed, relatively low cost computers. Over 30 codes were available in 1996. Engineers with basic fluid mechanics backgrounds can, with minimal training, use these tools to analyze flow patterns. The most versatile codes include the ability to deal with several types of fluids (compressible, incompressible, Newtonian, non-Newtonian, etc.); change of phase, multiphase, steady and unsteady flow; shock waves; body forces; and even chemical reactions.

Fluid flow is described by partial differential equations of mass, momentum, and energy conservation. Equations that describe fluid properties such as viscosity and density are needed. Using the time-averaging technique, several terms that cannot be related to the mean flow quantities are generated. These quantities are called Reynolds stresses. It is impossible to solve the time-averaged equations unless certain assumptions are made for the Reynolds stresses. Various turbulence models are available depending upon those assumptions, and it is the responsibility of the user to select the appropriate turbulence model. After the set of differential equations is derived for a particular turbulence assumption, these equations are converted to a set of nonlinear equations by discretization. These simultaneous, nonlinear equations are then solved at all nodes to obtain the flow field by iteration to a predetermined level of accuracy.

A limitation of current (1996) codes is their inability to predict turbulence with accuracy. The energy in turbulence is contained in eddies whose sizes can range over ten orders of magnitude. To resolve the smallest eddies in three dimensions, about 10^{30} nodes would be required—well beyond the present limit of 10^6 nodes. Instead, special functions or turbulence models are used to estimate the production, transport, and dissipation of turbulent energy without dramatically increasing the number of nodes. The equations used to model turbulence contain constraints that are adjusted for the type of flow, for example, a free jet or flow over a backward-facing step. Turbulence levels and mean flow characteristics can be predicted for conditions near those for which the code was calibrated and verified. Research is currently being conducted to use direct numerical simulation to minimize assumptions for modeling turbulence.

CFD is useful for predicting overall flow distribution in pump sumps and forebays to within about one bell diameter of a pump intake. Approach flows to a pump intake (or to a pump bay from a forebay) can be influenced by withdrawals by other pumps, and such influences can be readily predicted by CFD. But flow details, such as free surface and subsurface vortices, cannot be predicted by CFD, nor can CFD predict objectionable phenomena such as swirling or asymmetrical flow within pump suction bells. Obtaining accurate data for such phenomena will require physical model studies until CFD codes are developed that can accurately model the smallest turbulent eddies that affect a flow field. The use of CFD in conjunction with physical model testing can sometimes reduce the total cost of modeling.

3-14. References

1. Ten-State Standards, *Recommended Standards for Sewage Works,* Great Lakes–Upper Mississippi River Board of Sanitary Engineers Health Education Service, Inc., Albany, NY (1978) (revised periodically).
2. Benedict, R. P., *Fundamentals of Pipe Flow,* John Wiley, New York (1980).
3. Williams, G. S., and A. Hazen, *Hydraulic Tables,* 1st ed., John Wiley, New York (1905); 3rd ed., 11th printing, Hazen and Sawyer P.C., New York (1965).
4. Colebrook, C. E., "Turbulent flow in pipes, with special reference to the transition region between the smooth and rough pipe laws," *Journal of the Institution of Civil Engineers,* London, **11,** 133 (1938–1939).
5. Moody, L. E., "Friction factors for pipe flow," *Transactions, American Society of Mechanical Engineers,* **66,** 671–684 (November 1944).
6. Prandtl, L., *Essentials of Fluid Dynamics,* Chapter 3, Hafner Publishing, New York (1952).
7. Swamee, P. K., and A. K. Jain, "Explicit equations for pipe flow problems," *Journal of the Hydraulic Division, Proc. of the American Society of Civil Engineers,* **102,** 657–664 (May 1976).
8. Ackers, P., *Tables for the Hydraulic Design of Storm-drains, Sewers, and Pipe-lines, Hydraulics Research Paper No. 4,* 2nd ed., Department of the Environment, Hydraulic Research Station, Wallingford, Berkshires, England (1969).

9. Jeppson, R., *Analysis of Flow in Pipe Networks,* Ann Arbor Science Publishers, Ann Arbor, MI (1976).
10. AWWA Manual M11. *Steel Pipe—A Guide for Design and Installation,* 3rd ed. American Water Works Association, Denver, CO (1989).
11. Gros, W.F.H., President, The Pitometer Associates, Inc., Chicago, IL. Private communications (February/March 1996).
12. Lamont, P. A., "Common pipe flow formulas compared with the theory of roughness," *Journal of the American Water Works Association,* **73,** 274–280 (May 1981).
13. Wolfe, T. F., "Advantages of cement linings for cast-iron pipe," *Journal of the American Water Works Association,* **38,** 11–15 (January 1946).
14. Miller, W. T., "Durability of cement-mortar linings in cast-iron pipe," *Journal of the American Water Works Association,* **57,** 773–782 (June 1965).
15. White, G.C. *The Handbook of Chlorination,* 2nd ed. Van Nostrand Reinhold Co., New York (1986), pp. 356–358.
16. DIPRA, *Handbook of Ductile Iron Pipe,* 6th ed., Ductile Iron Pipe Research Association, Birmingham, AL (1984) (revised periodically).
17. Brater, E. F., and H. W. King, *Handbook of Hydraulics,* 6th ed., McGraw-Hill, New York (1976).
18. Yarnell, D. L., and S. M. Woodward., "The flow of water in drain tile," Bulletin No. 854, Dept. of Agriculture Bulletin v 35, U.S. Government Printing Office (1922).
19. Camp, T. R., "Design of sewers to facilitate flow," *Sewage Works Journal* **18** (1): 3–16 (January 1946).
20. Wilcox, E. R., *A Comparative Test of the Flow of Water in 8-inch Concrete and Vitrified Clay Sewer Pipe,* Bulletin No. 27, Univ. of Washington Eng. Experiment Station (March 1924).
21. Escritt, L. B., *Sewerage and Sewage Treatment. International Practice*, edited and revised by W. D. Haworth, John Wiley, New York (1984).
22. Saatci, A., "Velocity and depth of flow calculations in partially filled pipes," *Journal of the ASCE,* Environmental Div., **116** (6): 1202-1208 (November/December 1990).
23. Wheeler, W., *Advancing Gravity Pipe Design,* unpublished. Available from William Wheeler, 683 Limekiln Road, Doylestown, PA 18901 (January 1994).
24. Kulin, G., *Recommended Practice for the Use of Parshall Flumes and Palmer–Bowlus Flumes in Wastewater Treatment Plants,* PB85-122745; EPA-600/2-84-186, U.S. Department of Commerce, National Technical Information Service, Springfield, VA (1984).
25. Grant, D. M., *Open Channel Flow Measurement Handbook,* 2nd ed., ISCO, Inc., Lincoln, NE (1981).
26. Wells, E. A., and H. B. Gotaas, "Design of Venturi flumes in circular conduits," *Transactions, American Society of Civil Engineers,* **133,** 749–771 (1958).
27. Jorgensen, G., private communication (November 1984).
28. Walski, T., "Measuring differential pressure accurately in water distribution systems," *WATER/Engineering & Management,* **131,** 28–30 (October 1984).
29. Walski, T. M., *Application of Procedures for Testing and Evaluating Water Distribution Systems,* U.S. Army Engineers Waterways Experiment Station, Technical Report, Vicksburg, MS (1984).
30. Smith, E. C., private communication (December 1984).
31. Shelley, P. E., and G. A. Kirkpatrick, "Sewer flow measurement—a state-of-art assessment." EPA-600/2-75-027, Contract No. 68-03-0426, Municipal Environmental Research Laboratory, Office of Research and Development, U.S. Environmental Protection Agency, 26 W. Martin Luther King Dr., Cincinnati, OH 45268 (November 1975).
32. Connell, J. E., private communication (January 1985).
33. DIN Number 1944, *Acceptance Test on Centrifugal Pumps (V.D. Rule for Centrifugal Pumps),* Deutsche Institut für Nommung (German Standards Institute), Berlin; available in English from Heyden & Son, Philadelphia, PA (October 1968).
34. Smith, E. C., private communication (February 1985).
35. Walski, T. M., *Analysis of Water Distribution Systems,* Van Nostrand Reinhold, New York (1984).
36. Miller, R. W., *Flow Measurement Engineering Handbook,* McGraw-Hill, New York (1983).
37. Fluid Metering, Inc., 29 Orchard St., P.O. Box 179, Oyster Bay, NY 11771.
38. Park, J. T., and M. J. Johnson, "A sub-micro determination of glucose," *Journal of Biological Chemistry,* **181,** 149–151 (November 1949).
39. Rhodamine WT, obtainable from Crompton & Knowles Corp., 7535 Lincoln Ave., Skokie, IL 60076.
40. "Flow measurements," *Fluorometric Facts,* FR-1084-IM, Tumer Designs, 845 W. Maude Ave., Sunnyvale, CA 94086 (n.d.).
41. *Standard Methods for the Examination of Water and Wastewater,* American Public Health Association, Albany, NY (revised periodically).
42. McDonald, R., Senior Scientist, Brown and Caldwell, Pleasant Hill, CA. Private communication (1995).
43. "A practical guide to flow measurement," *Fluorometric Facts*, Turner Designs, Inc., 845 W. Maude Ave., Sunnyvale, CA 94086 (Rev. November 1995).
44. "Flow measurements in sanitary sewers by dye dilution," *Fluorometric Facts,* SS-7-80, Turner Designs Inc., 845 W. Maude Ave., Sunnyvale, CA 94086 (n.d.).
45. Walski, T. M., *Cost of Water Distribution System Infrastructure Rehabilitation, Repair, and Replacement,* Technical Report EL-85-5, Department of the Army Waterways Experiment Station, Vicksburg, MS (1985).
46. *EPA Process Design Manual for Sludge Treatment and Disposal,* EPA 625/1-79-011, pp. 14-1–14-57, U.S. Environmental Protection Agency, Municipal Environmental Research Laboratory, Office of Research and Development, Center for Environmental Research Information, Technology Transfer, Cincinnati, OH (September 1979).
47. Hecker, G. E., President, Alden Research Laboratory, Inc., Holden, MA 01520 (May 1995).

Chapter 4

Piping

BAYARD E. BOSSERMAN II
JAMES C. DOWELL
ELIZABETH M. HUNING
ROBERT L. SANKS

CONTRIBUTORS
Earl L. Heckman
Charles D. Morris
W. Stephen Shenk

The types of pipe, fittings, and associated materials commonly used in pumping stations and force and transmission mains for the pressurized flow of water, wastewater, and sludge are discussed in this chapter. The emphasis is on pipe 100 mm (4 in.) in diameter or larger, but small pipe for fuel, seal water, and plumbing is included.

The chapter is conceptually presented in four parts:

- Selection of pipe material
  - Exposed piping (piping within the pumping station), Section 4-1
  - Buried piping (force mains and transmission pipelines), Section 4-2
- Descriptions of pipe, fittings, joints, gaskets, and linings
  - Ductile iron pipe, Section 4-3
  - Steel pipe, Section 4-4
  - Plastic pipe, Section 4-5
  - Asbestos cement pipe, Section 4-6
  - Reinforced concrete pressure pipe, Section 4-7
- Design of piping, Section 4-8
  - Tie rods for flexible couplings
  - Wall thickness in exposed pipe
  - Hangars, supports, and spacing
- Special piping and plumbing, Section 4-9
  - Water
  - Diesel fuel service
  - Sewers
  - Chlorine
  - Air
  - Design of plumbing systems

This organization makes it possible to select the type of pipe to be used with no need to read more than a fraction of the chapter. After the type of pipe is chosen, skip to the appropriate section (Sections 4-3 to 4-7) for a detailed description of the piping. Pipe design and the design examples in Section 4-8 emphasize exposed piping. The design of buried piping is limited to generalities because that subject is so well covered in the literature [1, 2].

References to a standard or to a specification are given here in abbreviated form—code letters and numbers only (such as ANSI B36.10). Double designations such as ANSI/AWWA C115/A21.15 indicate that AWWA C115 is the same as ANSI A21.15. Most standards are revised periodically, so obtain the latest edition. Titles of standards and specifications are given in Appendix E, and publishers' addresses are given in Appendix F.

There are many standards important in design that are not specifically mentioned in the text. They are listed under "Recommended Specifications and Supplementary Specifications" in Appendix E.

4-1. Selection of Exposed Pipe

Factors to be considered in the selection of pipe (whether exposed or buried) include the following:

- Properties of the fluid
  - Corrosive or scale-forming properties [3, 4, 5, 6, 7, 8]
  - Unusual characteristics, for example, viscosity of sludges
- Service conditions
  - Pressure (including surges and transients)
  - Corrosive atmosphere for exposed piping
  - Soil loads, bearing capacity and settlement, external loads, and corrosion potential for buried piping
- Availability
  - Sizes
  - Thicknesses
  - Fittings
- Properties of the pipe
  - Strength (static and fatigue, especially for water hammer)
  - Ductility
  - Corrosion resistance
  - Fluid friction resistance of pipe or lining
- Economics
  - Required life
  - Maintenance
  - Cost (fob plus freight to jobsite)
  - Repairs
  - Salvage value

Most of these factors are discussed here. If some that are not considered could influence selection, consult the literature and manufacturers' representatives.

Material

The great reserves of strength, stiffness, ductility, and resistance to water hammer; the wide range of sizes and thicknesses; and the wide variety of fittings available make either steel or ductile iron pipe (DIP) virtually the only logical choices for pump manifold piping. (See Section 2-2 for a definition of *ductile iron.*) The properties of these two pipe materials, compared in Table 4-1, are so similar that, in most situations, price is (or should be) the determining factor in the choice between DIP and steel. In the United States, ductile iron pipe is usually used on the East

Table 4-1. Comparison of Pipe for Exposed Service

Pipe	Advantages	Disadvantages/limitation
Ductile iron (DIP)	Yield strength: 290,000 kPa (42,000 lb/in.^{2}) Ultimate strength: 414,000 kPa (60,000 lb/in.^{2}) $E = 166 \times 10^6$ kPa (24×10^6 lb/in.^{2}) Ductile, elongation $\simeq$10% Good corrosion resistance Wide variety of available fittings and joints Available sizes: 100–1350 mm (4–54 in.) ID Wide range of available thicknesses Good resistance to water hammer	Maximum pressure = 2400 kPa (350 lb/in.^{2}) High cost, especially for long freight hauls No diameters above 1350 mm (54 in.) Difficult to weld Class 53 is the thinnest allowed for American flanged pipe (with screwed flanges in the U.S.)
Steel	Yield strengths: 207,000–414,000 kPa (30,000–60,000 lb/in.^{2}) Ultimate strengths: 338,000–518,000 kPa (49,000–75,000 lb/in.^{2}) $E = 207 \times 10^6$ kPa (30×10^6 lb/in.^{2}) Ductile elongation 17–35% Pressure rating to 17,000 kPa (2500 lb/in.^{2}) Diameters to 3.66 m (12 ft) Widest variety of available fittings and joints Custom fittings can be mitered and welded Excellent resistance to water hammer Low cost	Corrosion resistance low unless lined

Coast and in the Midwest, whereas steel pipe is usually chosen on the West Coast because of freight costs. Occasionally, steel may be preferred because it can be welded, which makes any configuration possible (see Figure 4-1). For example, mitered fittings can be fabricated to meet unique requirements, such as a 79° elbow. By comparison, DIP fittings are available only in standard configurations as described in AWWA C110.

Joints

The most practical connections are flanged joints (Figure 4-2) with enough strategically placed grooved-end couplings (such as Victaulic®), lugged (bridled or harnessed) gasketed sleeve-type compression couplings (such as Dresser®), and/or flanged adapters to make assembly and disassembly easy and convenient and to allow for small inaccuracies in the length of spools (pipes of less than standard length) (see Figures 4-3 and 4-4). Bolted flanges cannot be deflected and can resist a considerable amount of bending moment.

Inattention to the many and varied flange standards can result in misapplication with regard to pressure ratings and *costly,* embarrassing mismatches of flange bolt circles and bolt holes. The following is a summary of flange standards and designs encountered in the United States. Most of the following commentary concerns flanges with maximum designation of Class 300, although flange ratings as high as Class 2500 exist.

Standards

The flange standards most frequently encountered in water works piping designs are described in the following documents:

- ANSI B16.1, *Cast Iron Pipe Flanges and Flanged Fittings.* This standard covers Classes 25, 125, 250, and 800.
- ANSI B16.5, *Steel Pipe Flanges for Water Works Service—Sizes 4-Inch Through 144-Inch.* This standard covers Classes 150, 300, 400, 600, 900, 1500, and 2500.
- ANSI B16.42, *Ductile-Iron Pipe Flanges and Flanged Fittings, Classes 150 and 300.*

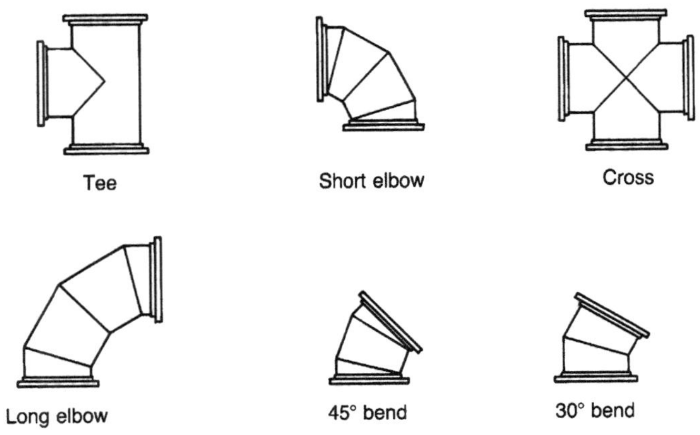

Figure 4-1. Typical mitered steel fittings.

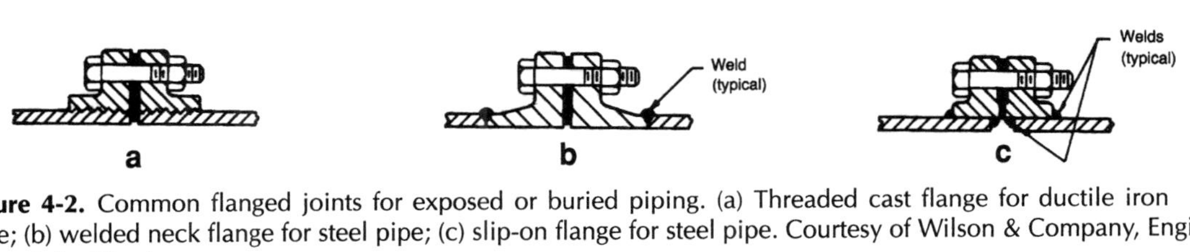

Figure 4-2. Common flanged joints for exposed or buried piping. (a) Threaded cast flange for ductile iron pipe; (b) welded neck flange for steel pipe; (c) slip-on flange for steel pipe. Courtesy of Wilson & Company, Engineers & Architects.

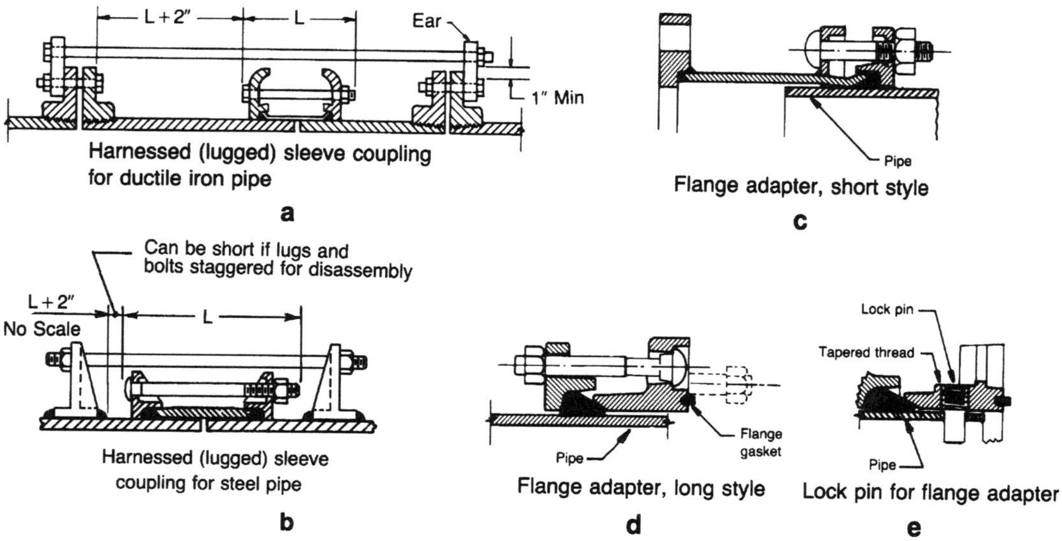

Figure 4-3. Sleeve couplings and flanged adapters. Unless restrained, the joints can separate. After Dresser Manufacturing Division.

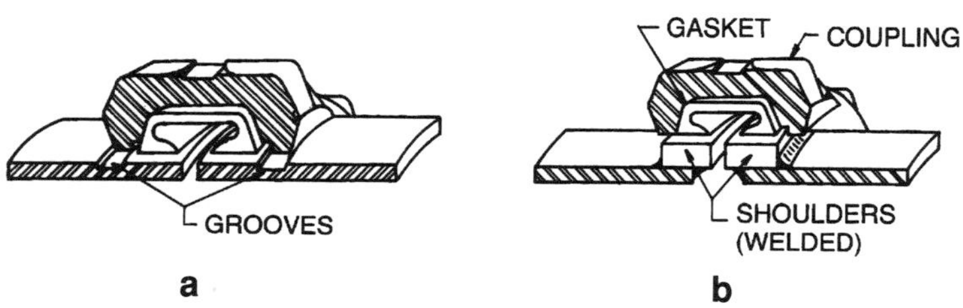

Figure 4-4. Grooved-end (Vitaulic®) couplings. (a) Grooved pipe; (b) welded shoulders.

- AWWA C207, *Steel Pipe Flanges for Water Works Service—Sizes 4-Inch through 144-Inch.* Four pressure rating designs, designated Classes B, D, E, and F, are described.
- AWWA C115, *Flanged Ductile-Iron Pipe with Threaded Flanges.*

Flanges constructed in accordance with ANSI standards do *not* match flanges constructed in accordance with AWWA C207 in terms of pressure rating. Some (but not all) combinations match in terms of bolt circle and bolt hole number and diameter.

To make matters more confusing, there is an API (American Petroleum Institute) flange standard—APO 605—that does not match ANSI or AWWA standards in any particular. Finally, there is a Manufacturer's Standardization Society (MSS) standard SP 44 that matches ANSI and AWWA in terms of flange OD, bolt circle, number of bolt holes, and bolt hole size but does not match in flange thickness. MSS-SP 44 does not match API 605 in anything. For the sake of brevity, the following discussion is limited to AWWA and ANSI flanges.

Facings

Important differences between AWWA C207 and ANSI B16.1 and B16.5 are:

- AWWA steel flanges are flat faced, while ANSI B16.5 steel flanges have a raised face. That is, the ANSI B16.5 flange has a projection inside the bolt circle that is about 1.5 mm ($^1/_{16}$ in.) higher than the portion of the flange outside the bolt circle.
- Above 1200-mm (48-in.) size, ANSI B16.1 and B16.5 and AWWA Class B, D, and E flanges do not

match in bolt hole size, but they do match in bolt circle and number of bolt holes.
- AWWA flanges have a serrated finish on the face whereas ANSI flanges are flat.

When mating steel flanges to cast-iron flanges, be sure the steel flange has the raised face removed. Steel to cast-iron flange connections should have flat-faced flanges. If the raised steel face is not removed, the stress induced by bolting the flanges together can crack the cast iron.

Dimensions (flange OD, bolt circle diameter, and number and size of bolt holes) of AWWA Class D flanges through 600-mm (24-in.) size are the same as those for ANSI B16.5, Class 150. AWWA Class D flange dimensions match ANSI B16.1, Class 125, cast-iron flanges in all sizes.

Cast-iron flanges ANSI B16.1, Class 125, have the same drilling and bolting dimensions as ANSI B16.5, Class 150, steel flanges in sizes 600 mm (24 in.) and smaller. Cast-iron flanges ANSI B16.1, Class 250, have the same drilling and bolting dimensions as ANSI B16.5, Class 300, steel flanges in sizes 600 mm (24 in.) and smaller. Class 125/150 flanges do *not* match Class 250/300 flanges in any size.

AWWA Class E flanges have a pressure rating of 1900 kPa (275 lb/in.^{2}), but they conform to the dimensions of ANSI B16.1, Class 125. Only AWWA Class F flanges conform to ANSI Class 250/300 dimensions (1200 mm or 48 in. and smaller).

The flange standards corresponding to various AWWA and ANSI pressure class ratings, materials, and pipe sizes are summarized in Table 4-2.

Pressure Ratings

Temperature is not mentioned in AWWA specifications for flanges, but there is a pressure–temperature relationship for ANSI flanges in which temperatures above 38°C (100°F) reduce the allowable pressure rating. In ANSI 16.5, a Class 150 steel flange has a pressure rating of 197 kPa (285 lb/in.^{2}) at 38°C (100°F). Hence, Class 150 flanges are adequate for any lower pressure at the temperatures usually encountered in traditional water works piping.

Some pressure ratings (static plus water hammer pressure) for various types of flanges usually encountered in water works piping are summarized in Table 4-3. Refer to AWWA C207, ANSI B16.1, and ANSI B16.5 for flange dimensions and to ASTM A307 and ANSI B31.1 and B31.3 for bolts and nuts.

Sleeve couplings (such as Dresser®) provide no longitudinal restraint (except for unreliable friction), so unless the pipe is otherwise anchored against significant longitudinal movement, the couplings must be harnessed as shown in Figure 4-3a and b. The number and size of tie rods are determined by the root area in the threads of the rods as demonstrated in Example 4-1. Harnessing adds to the laying length requirement and cost of the joint, so omit unneeded harnesses.

Flanged adapters (Figure 4-3c and d) accomplish the same purpose as sleeve couplings but are more economical for two reasons: (1) shorter laying lengths and (2) fewer flanges. Longitudinal movement can be restrained by the use of lock pins (Figure 4-3e), but if there are likely to be many repetitions of longitudinal force, the lock-pin holes in the pipe may wear and eventually fail. Flanged adapters, however, are useful and unquestionably excellent for externally restrained pipe.

Grooved-end couplings (such as Victaulic®) are the most economical of the flexible joints because (1) the laying length required is so short and (2) two flanges are omitted. Grooved ends can only be used in thick-walled pipe (see Tables 4-4 and 4-5), but thin steel pipe can be adapted for these couplings by

Table 4-2. Flange Standards

Pressure Class			Maximum pipe size		
kPa	lb/in.^{2}	Material	mm	in	Equivalent ANSI Standard[a]
860	125	C.I.[c]	2400	96	B16.1, CL. 125
1030	150	Steel[b]	600	24	B16.5, CL. 150
1730	250	C.I.[b]	1200	48	B16.1, CL. 250
2070	300	Steel[b]	600	24	B16.5, CL. 300
AWWA "D"		Steel[c]	3600	144	B16.1, CL. 125
AWWA "E"		Steel[c]		144	B16.1, CL. 125
AWWA "F"		Steel[c]		48	B16.1, CL. 250; and B16.5, CL. 300

[a] Equivalent in bolt circle and spacing and flange OD.
[b] Raised face.
[c] Flat face.

Table 4-3. Pressure Ratings of Flanges

Class	Standard	Material	Temperature range °C	Temperature range °F	Pressure rating kPa	Pressure rating lb/in.^{2}	Pipe diameter mm	Pipe diameter in.
125	ANSI B16.1	C.I.	–29 to +66	–20 to +150	1380 1030	200 150	300 350–1200	12 14–48
150	ANSI B16.5	Steel	–29 to +66	–20 to +100	1970	285		
250	ANSI B16.1	C.I.	–29 to +66	–20 to +150	3450 2070	500 300	≤500 350–1200	12 14–18
300	ANSI B16.5	Steel	–29 to +66	–20 to +100	5100	740		
"B"	AWWA C207	Steel	–29 to +66	—	590	86		
"D"	AWWA C207	Steel	–29 to +66	—	1210 1030	175 150	100–300 >300	4–12 12
"E"	AWWA C207	Steel	–29 to +66	—	1900	275		
"F"	AWWA C207	Steel	–29 to +66	—	2070	300		

Table 4-4. Minimum Wall Thickness for Grooved-End Couplings in Ductile Iron Pipe

Nominal size of pipe mm	Nominal size of pipe in.	Ductile iron pipe thickness class per AWWA C151 and C606
100–400	4–16	53
450	18	54
500	20	55
600	24	56

Table 4-5. Minimum Wall Thickness for Grooved-End Couplings in Steel Pipe per AWWA C606

Nominal size of pipe mm	Nominal size of pipe in.	Thickness[a,b] mm	Thickness[a,b] in.	ANSI B36.10 Schedule no.
75	3 ID	4.7	0.188	40[c]
100	4 ID	5.2	0.203	40[c]
150	6 ID	5.6	0.219	40[c]
200	8 ID	6.0	0.238	20[c]
250	10 ID	6.4	0.250	20
300	12 ID	7.0	0.279	30[c]
350	14 OD	7.1	0.281	20[c]
400	16 OD	8.0	0.312	20
450	18 OD	8.0	0.312	20
500	20 OD	8.0	0.312	20[c]
600	24 OD	9.5	0.375	30[c]

[a] Thin pipe can be used with shouldered couplings (see Figure 4-4b).
[b] Standard weight pipe up to 600 mm (24 in.) can be used.
[c] Nearest Schedule number.

welding shoulders to the pipe as shown in Figure 4-4. Grooved-end couplings allow expansion and contraction of ±3 mm ($^1/_8$ in.) or a total travel of 6 mm ($^1/_4$ in.) at each coupling. Consequently, special expansion joints can usually be eliminated. Standards for grooved-end fittings are contained in AWWA C606. Grooved-end couplings for steel and ductile iron pipe are manufactured of malleable iron, ASTM A 47 or A 536. Carbon steel bolts should conform to ASTM A 183. Stainless steel bolts should conform to ASTM A 193, Class 2. Gaskets should conform to ASTM D 2000.

Sleeve couplings, flanged adapters, and—to some degree—grooved-end couplings can be purposely deflected. However, use them only to provide for accidental misalignment, expansion, and ease of installation and disassembly. Use bends or other fittings for calculated deflection angles. To prevent strain from being transmitted to equipment (such as pump casings), use the types that do not transmit shear.

At times, transitions from one pipe material to another are required, for example, conversion from ductile iron for exposed service to polyvinyl chloride (PVC) for buried service. Adapters for this purpose are commercially available for almost any type of pipe material transition. Consult the pipe manufacturer for the particular application.

Gaskets

Gaskets required for ductile iron or gray cast-iron flanged joints are 3.2 mm ($^1/_8$ in.) thick, and those for steel flanges can be either 1.6 mm ($^1/_{16}$ in.) or 3.2 mm ($^1/_8$ in.) thick (see "Gaskets" in Sections 4-3 and 4-4 for details).

Thickness

Temperatures exceeding 871 to 927°C (1600 to 1700°F) erode the ductility and strength of ductile iron, so, if they are used at all, welds must be made with great care to avoid overheating. Hence, flanges for pipe made in the United States are screwed to the barrel by the pipe manufacturer. The thinnest allowable DIP for threaded barrels is Class 53. The thickness of Class 53 DIP is given in Tables B-1 and B-2 (Appendix B). Spools (short lengths of pipe) can sometimes be obtained with integrally cast flanges that may permit Class 50 pipe to be used. Flanges for ductile iron or gray iron valves and fittings are always cast integrally. Some European manufacturers do weld flanges to DIP barrels, and this practice permits the use of Class 50 instead of Class 53 pipe.

Steel pipe can be obtained in a very wide range of thicknesses and diameters [9]. Flanges are welded to the barrel of steel pipe. Thicknesses for carbon steel vary from Schedule 10 to Schedule 160 and from "standard" to "double extra heavy." Standard weight pipe is shown in Tables B-3 and B-4 (Appendix B). Standard weight and Schedule 40 are identical for diameters up to and including 250 mm (10 in.). All larger sizes of standard weight pipe have a wall thickness of 9.5 mm ($^3/_8$ in.)

Depending on pipe size and internal pressure, steel pipe may have to be reinforced at tees, wyes, and other openings. As discussed in Section 4-4, reinforcement can be one of three kinds: collar, wrapper, or crotch.

Linings and Coatings

Considering its low cost, long life, and sustained smoothness (see "Friction Coefficients" in Section 3-2), cement-mortar lining for both ductile iron and steel pipe is the most useful and the most common. Standard thicknesses for shop linings are given both in Table 4-6 and (for DIP) in Tables B-1 and B-2. Pipe can be lined in place with the thicknesses given in Table 4-7 after cleaning. Although cement mortar is normally very durable, it can be slowly attacked by very soft waters with low total dissolved solids content (less than 40 mg/L), by high sulfate waters, or by waters undersaturated in calcium carbonate. For such uses, carefully investigate the probable durability of cement mortar and consider the use of other linings. The marble test [10] is a simple, accurate, and inexpensive means for determining the calcium carbonate saturation. Sulfate and total dissolved solids are also easily tested in almost any water laboratory. Because the standard, shop-applied mortar linings are relatively thin, some designers prefer to specify shop linings in double thickness.

In specifying mortar lining, match the pipe ID with the required valve ID, particularly with short-body butterfly valves in which the valve vane protrudes into the pipe. If the ID is too small, the valve cannot be fully opened.

Table 4-6. Thickness of Shop-Applied Cement-Mortar Linings

Nominal pipe diameter		Lining thickness			
		Ductile iron pipe[a]		Steel pipe[b]	
mm	in.	mm	in.	mm	in.
100–250	4–10	1.6	$^1/_{16}$	6.4	$^1/_4$
300	12	1.6	$^1/_{16}$	7.9	$^5/_{16}$
350–550	14–22	2.4	$^3/_{32}$	7.9	$^5/_{16}$
600	24	2.4	$^3/_{32}$	9.5	$^3/_8$
750–900	30–36	3.2	$^1/_8$	9.5	$^3/_8$
1050–1350	42–54	3.2	$^1/_8$	12.7	$^1/_2$
>1350	>54	—	—	—	—

[a] Single thickness per AWWA C104. Linings of double thickness are also readily available.
[b] Per AWWA C205.

Table 4-7. Thickness of Cement-Mortar Linings of Pipe in Place per AWWA C602

Nominal pipe diameter		DIP or gray cast iron (new or old pipe)		Steel pipe			
				Old pipe		New pipe	
mm	in.	mm	in.	mm	in.	mm	in.
100–250	4–10	3.2	1/8	6.4	1/4	4.8	3/16
300	12	4.8	3/16	6.4	1/4	4.8	3/16
350–550	14–22	4.8	3/16	7.9	5/16	6.4	1/4
600–900	24–36	4.8	3/16	9.5	3/8	6.4	1/4
1050–1350	42–54	6.4	1/4	9.5	3/8	9.5	3/8
1500	60	—	—	9.5	3/8	9.5	3/8
1650–2250	66–90	—	—	12.7	1/2	11.1	7/16
>2250	>90	—	—	12.7	1/2	12.7	1/2

Table 4-8. Linings for Ductile Iron and Steel Pipe

Lining material	Reference standard	Recommended service
Cement mortar	AWWA C104, C205	Potable water, raw water and sewage, activated and secondary sludge
Glass	None	Primary sludge, very aggressive fluids
Epoxy	AWWA C210	Raw and potable water
Fusion-bonded	AWWA C213	Potable water, raw water and sewage epoxy
Coal-tar epoxy	AWWA C210	Not recommended for potable water
Coal-tar enamel	AWWA C203	Potable water
Polyurethane	None	Raw sewage, water
Polyethylene	ASTM D 1248	Raw sewage

Other linings and uses are given in Table 4-8. The plastics are usually about 0.3 mm (0.012 in.) thick. Glass lining is about 0.25 mm (0.010 in.) thick. In general, the cost of cement mortar is about 20% of that of other linings, so other linings are not justified except where cement mortar would provide unsatisfactory service. One exception is glass, which is supreme for raw primary sludge because it inhibits grease and solids buildup and is highly resistant to chemical attack. Applying the glass lining to the pipe requires a temperature of about 760°C (1400°F), which can safely be applied to DIP provided that the original thickness is at least Class 54 for 150- to 500-mm (6- to 20-in.) pipe or Class 56 for 100-mm (4-in.) pipe. The interior must be very smooth, so the pipe is either bored (from Class 56 to Class 53, for example) or blasted with abrasive grit. Manufacturers may differ in their thickness requirements, so consult them before writing a specification. Steel pipe should be seamless because weld seams cannot be covered. Welded pipe can, however, be used if the weld seams are machined smooth. Any end connection found on DIP can be used with glass lining, but only plain-end, flanged, or grooved-end connections can be used for glass-lined steel pipe. Steel bell and spigot joints deform under the heat required. Because glass lining becomes very expensive for pipes larger than 250 mm (10 in.), other linings (such as cement mortar) should ordinarily be considered.

Usually, ductile iron pipe is furnished with an exterior asphaltic coating. This asphaltic coating is only 1 mil thick and provides little or no corrosion protection. If the pipe is to be painted, either a bare pipe or a shop-applied primer coat compatible with a finish coat of enamel or epoxy should be specified.

Fittings

Some standard iron fittings are shown in Figure 4-5. A comprehensive list of standard and special fittings is given in Table 4-9. Greater cost and longer delivery times can often be expected for special fittings. Fittings are designated by the size of the openings, followed (where necessary) by the deflection angle. A 90° elbow for a 250-mm (10-in.) pipe would be called

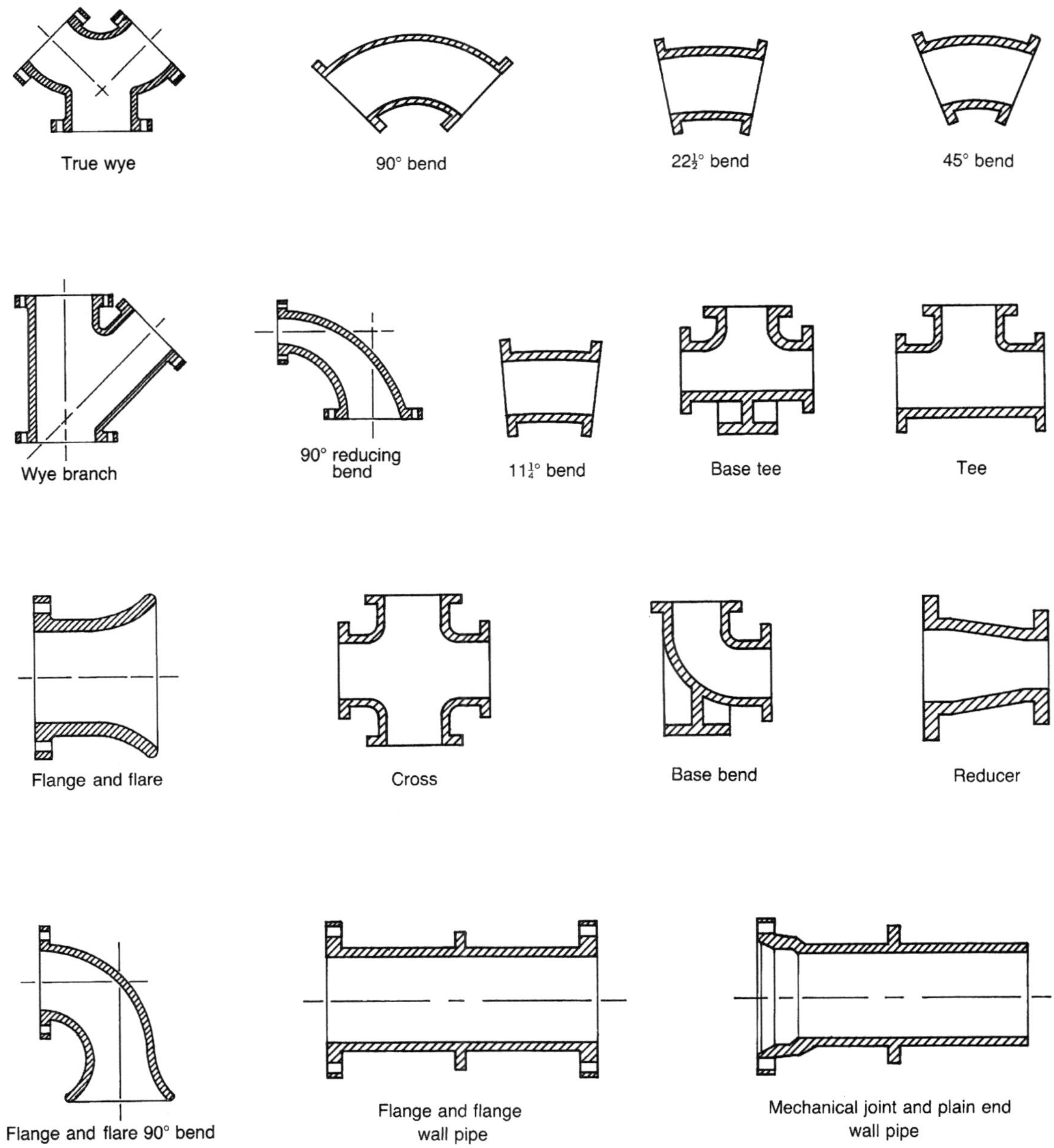

Figure 4-5. Ductile or gray iron flanged fittings.

a 250-mm (10-in.) 90° bend (or elbow). Reducers, reducing tees, or reducing crosses are identified by giving the pipe diameter of the largest opening first, followed by the sizes of other openings in sequence. Thus, the reducing outlet tee in Figure 4-6 might be designated as a 250-mm × 200-mm × 250-mm tee (10-in. × 8-in. × 10-in. tee).

Most steel fittings are mitered (Figure 4-1), but wrought fittings (Figure 4-6) are also used. A comprehensive list of customary mitered and wrought fittings is given in Table 4-10 (refer also to AWWA M11 [9]).

Expansion, Contraction, and Vibration

Pump vibration or expansion and contraction of the pipe may, at times, be significant. Compensation for these types of movement may be made by installing

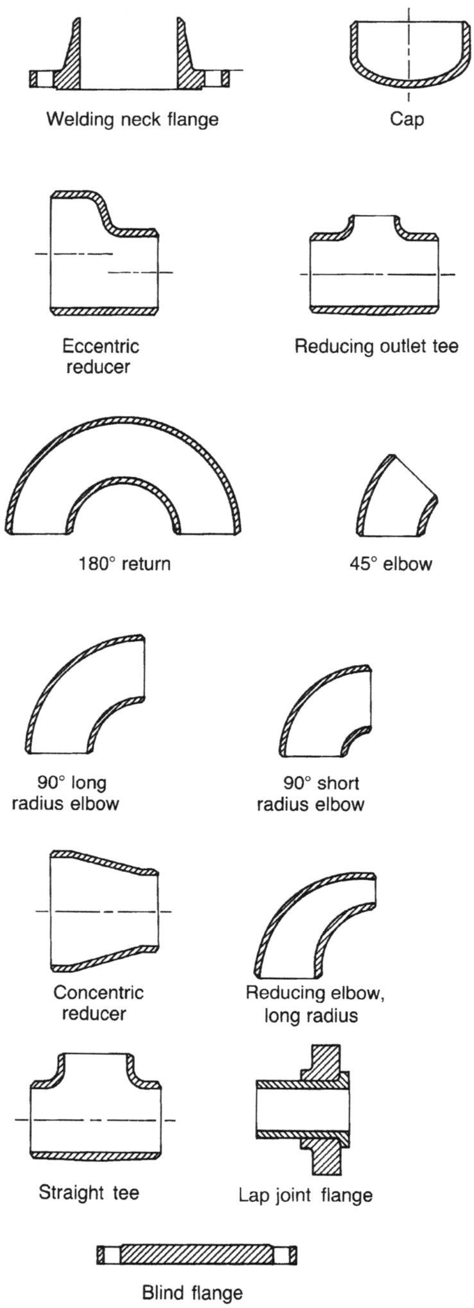

Figure 4-6. Wrought (forged) steel fittings for use with welded flanges.

flexible connectors. These connectors are usually constructed of rubber and may be bolted directly to the flanges of equipment and piping (refer to Chapter 22 and Figure 22-14). Many connectors for these applications are corrugated, but various shapes are available.

Table 4-9. Ductile Iron and Gray Cast-Iron Fittings, Flanged, Mechanical Joint, or Bell and Spigot[a]

Standard fittings	Special fittings
Bends (90°, 45°, 22½°, 11¼°)	Reducing bends (90°)
Base bends	Flared bends (90°, 45°)
Caps	Flange and flares
Crosses	Reducing tees
Blind flanges	Side outlet tees
Offsets	Wall pipes
Plugs	True wyes
Reducers	Wye branches
Eccentric reducers	
Tees	
Base tees	
Side outlet tees	
Wyes	

[a] Sizes from 100 to 1350 mm (4 to 54 in.).

Table 4-10. Steel Fittings

Mitered fittings	Wrought fittings
Crosses	Caps
2-Piece elbows, 0–30° bend	45° elbows
3-Piece elbows, 31–60° bend	90° elbows, long radius
4-Piece elbows, 61–90° bend	90° elbows, short radius
4-Piece, long radius elbows	90° reducing elbows, long radius
Laterals, equal diameters	Multiple-outlet fittings
Laterals, unequal diameters	Blind flanges
Reducers	Lap joint flanges
Eccentric reducers	Slip-on flanges
Tees	Socket-type welding flanges
Reducing tees	Reducing flanges
True wyes	Threaded flanges
	Welding neck flanges
	Reducers
	Eccentric reducers
	180° returns, long radius
	Saddles
	Reducing outlet tees
	Split tees
	Straight tees
	True wyes

4-2. Selection of Buried Piping

Buried piping must resist internal pressure, external loads, differential settlement, and the corrosive action of soils. The profile, the velocity of flow, and the size and stiffness of the pipe all affect water hammer; thus, they affect the design of the pumping station itself as well as that of the force main. General factors affecting piping selection are listed in Section 4-1.

Table 4-11. Comparison of Pipe for Buried Service

Pipe	Advantages	Disadvantages/limitations
Ductile iron pipe (DIP)	See Table 4-1; high strength for supporting earth loads, long life	See Table 4-1; may require wrapping or cathodic protection in corrosive soils
Steel pipe	See Table 4-1; high strength for supporting earth loads	See Table 4-1; poor corrosion resistance unless both lined and coated or wrapped, may require cathodic protection in corrosive soils
Polyvinyl chloride (PVC) pipe	Tensile strength (hydrostatic design basis) = 26,400 kPa (4000 lb/in.^{2}); E = 2,600,000 kPa (400,000 lb/in.^{2}); light in weight, very durable, very smooth, liners and wrapping not required, good variety of fittings available or can use ductile or cast-iron fittings with adapters, can be solvent-welded; diameters from 100 to 375 mm (4 to 36 in.)	Maximum pressure = 2400 kPa (350 lb/in.^{2}); little reserve of strength for water hammer if ASTM D 2241 is followed, AWWA C900 includes allowances for water hammer; limited resistance to cyclic loading as the fatigue limit is very low; unsuited for outdoor use above ground
High-density Polyethylene (HDPE) pipe	Tensile strength (hydrostatic design basis) = (8600 to 11,000 kPa (1250 to 1600 lb/in.^{2}); light weight, very durable, very smooth, liners and wrapping not required for corrosion protection; usually jointing method is thermal butt fusion, which develops the full strength of the pipe; flanges can be provided; diameters from 100 through 1575 mm (4 through 63 in.); low installed cost	Subject (as are many plastics, including PVC) to permeation by low molecular weight organic solvents and petroleum products, unsuited for manifold piping for pumping stations; scratches on the pipe wall can significantly reduce service life; requires careful bedding and compaction beneath the springline; cannot be solvent welded nor threaded
Asbestos cement (AC) pipe (ACP)	Yield strength: not applicable; design based on crushing strength, see ASTM C 296 and C 500; E = 23,500,000 kPa (3,400,000 lb/in.^{2}); rigid, light weight in long lengths, low cost; diameters from 100 to 1050 mm (4 to 42 in.), compatible with cast-iron fittings, pressure ratings from 1600 to 3100 kPa (225 to 450 lb/in.^{2}) for large pipe 450 mm (18 in.) or more	Attacked by soft water, acids, sulfates; requires thrust blocks at elbows, tees, and dead ends; maximum pressure = 1380 kPa (200 lb/in.^{2}) for pipe up to 400 mm (16 in.); brittle and requires care to install
Reinforced concrete pressure pipe (RCPP)	Several types available to suit different conditions (see Table 4-12); high strength for supporting earth loads, wide variety of sizes and pressure ratings, low cost, sizes from 600 to 3660 mm (24 to 144 in.)	Attacked by soft water, acids, sulfides, sulfates, and chlorides, often requires protective coatings; water hammer can crack outer shell, exposing reinforcement to corrosion and destroying its strength with time; maximum pressure = 1380 kPa (200 lb/in.^{2})

Material

As buried pipes are supported by the trench bottom, there are more possibilities for selecting piping materials and joints. As force mains are much longer than the pipe within a pumping station, substantial savings are possible by choosing the most economical pipe. Recommended materials are compared in Table 4-11. Plastics other than PVC are used for acids and chemicals. For sewage service, both acrylonitrile butadiene styrene (ABS) and PVC are commonly used, but PVC pipe is the only plastic commonly used for potable water service.

Reinforced concrete pressure pipe, generically described in Table 4-11, is divided into four types described in Table 4-12.

Practical Selection

Considering experience with, and characteristics of, the various pipe materials available for sewage force mains, the following materials are recommended. Use HDPE or PVC up to a diameter of 600 mm (24 in.) for wastewater force mains. Note, however, that if the pressure is 75% or more of the working pressure of

Table 4-12. Reinforced Concrete Pressure Pipe (RCPP)

Type	Specification	Description
Steel cylinder	AWWA C300	Steel cylinder within a reinforcing cage; walled inside and out with dense concrete, steel joint rings welded to cylinder, rubber gasket joint
Prestressed, steel cylinder	AWWA C301	Cement-mortar-lined steel cylinder, wire wrapped and coated with mortar or concrete; two types differ slightly
Noncylinder	AWWA C302	Circumferentially reinforced concrete pipe without a steel cylinder and not prestressed
Pretensioned, steel cylinder	AWWA C303	Cement-mortar-lined steel cylinder helically wrapped with continuous pretensioned steel bare and coated with mortar

the pipe or if there are repetitive pressure cycles, as with cycling constant speed pumps off and on, experience has shown that the material may fail. For larger pipe, use polyurethane or fusion-bonded, epoxy-coated ductile iron or steel pipe. Consider RCPP with a PVC lining above a diameter of 900 or 1050 mm (36 or 42 in.). If there is a possibility of air entering the force main, line the top of the RCPP with PVC to guard against the potential for acid corrosion. Careful evaluation with regard to internal pressures, external loads, corrosiveness of the wastewater, and corrosiveness of the soil is required in selecting the pipe material for a given project or application.

Joints

Because buried piping, such as a force main or water line, is fully laterally supported, the joints should be flexible to accommodate minor settlement; hence, less expensive joints, better suited for burial than bolted flanged joints, can be used (see Figures 4-7 and 4-8).

Many kinds of joints are available (see Figures 4-7 through 4-13), and several common ones are listed in Table 4-13. The check marks (√) show the type of pipe for which they are suited. For buried service, sliding couplings (such as Dresser®) need not be harnessed if thrust blocks are installed at bends or if there is a sufficient length of pipe on both sides for soil friction to resist the thrust (see DIPRA [11]). Grooved-end couplings (see Figure 4-4) allow for slight changes in alignment (about 5° per coupling), but do not deliberately make use of this flexibility; use bends or ball joints instead.

For ductile iron pipe, push-on and mechanical joints are the most commonly used for buried service. These joints allow for some pipe deflection (about 2 to 5° depending on pipe size) without sacrificing watertightness. Except in certain unstable soils, buried joints are not required to support the weight of the pipe.

Gaskets

Gaskets for various joints are described in Sections 4-3 to 4-5.

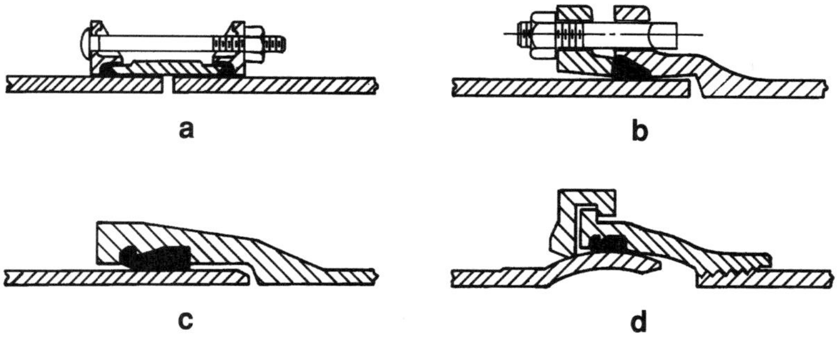

Figure 4-7. Couplings and joints for ductile iron pipe for buried service. (a) Sleeve (Dresser®) coupling (also for asbestos cement pipe); (b) mechanical joint; (c) push-on joint; (d) ball joint (also for exposed service). Courtesy of Wilson & Company, Engineers & Architects.

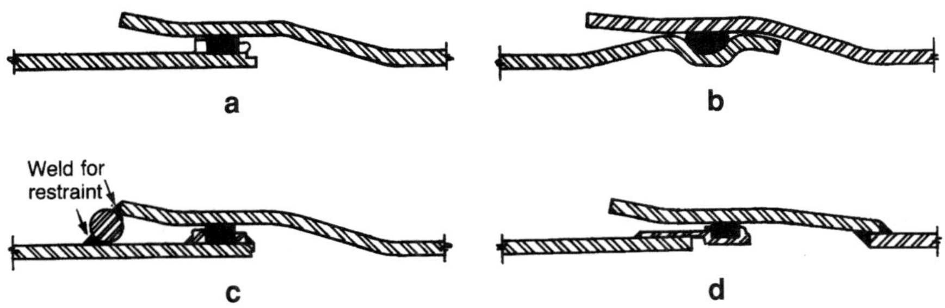

Figure 4-8. Rubber-gasketed push-on joints for steel pipe in buried service. (a) Fabricated joint; (b) rolled-groove joint; (c) tied joint; (d) Carnegie-shaped joint with weld-on bell ring. After AWWA M11 [9].

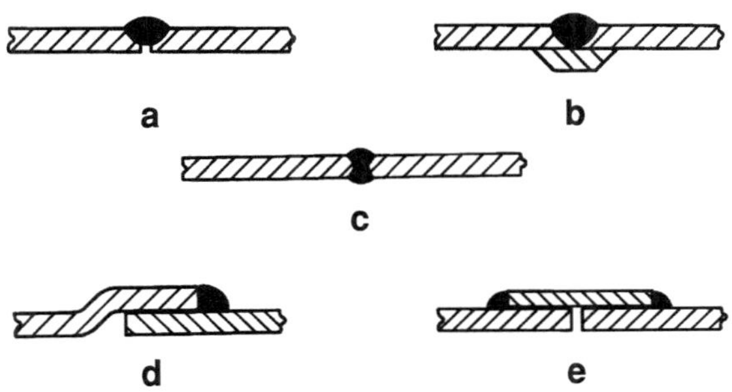

Figure 4-9. Welded joints for steel pipe in buried or exposed service. (a) Vee butt weld; (b) vee butt weld with weld ring; (c) double butt weld; (d) cup-type weld; (e) sleeve or butt strap weld. Courtesy of Wilson & Company, Engineers & Architects.

Figure 4-10. Couplings and joints for polyvinyl chloride (PVC) pipe in buried service. (a) Twin-gasketed coupling (also for asbestos cement); (b) bell and spigot joint. Courtesy of Wilson & Company, Engineers & Architects.

Thickness

The thickness of buried pipe must be sufficient to resist the external soil and overburden loads as well as internal pressure. If there is no traffic loading and the burial is shallow (say, 2 m or 6 ft), ductile iron pipe Class 50 can often be specified where internal pressure is less than the pipe pressure rating.

The stresses in (and deflection of) pipe walls due to soil and overburden loads are complex functions of (1) pipe material, diameter, and thickness; (2) bedding conditions; (3) soil properties; (4) trench width and depth; (5) frost; and (6) the overburden load and its distribution. Although not difficult, the calculations are too involved to be included here. Refer to

- AWWA M11 [9] for steel pipe
- DIPRA [11] for DIP
- ASTM D 2774 and Uni-Bell Handbook [12] for PVC pipe
- AWWA C401 and C403 for AC pipe
- AWWA M9 [13] for RCPP
- Spangler [14, 15] for the most extensive discussion

Linings, Coatings, and Cathodic Protection

The discussion of lining for exposed service applies to buried service as well. Buried glass-lined pipe should be DIP, but if steel is used, it should be plain-end pipe joined by mechanical couplings.

Steel and sometimes ductile iron pipe corrode on contact with some soils. Asbestos cement and concrete are attacked by acidic waters or by high sulfate soils. In aggressive environments the pipe can be coated (see Section 4-3 for details).

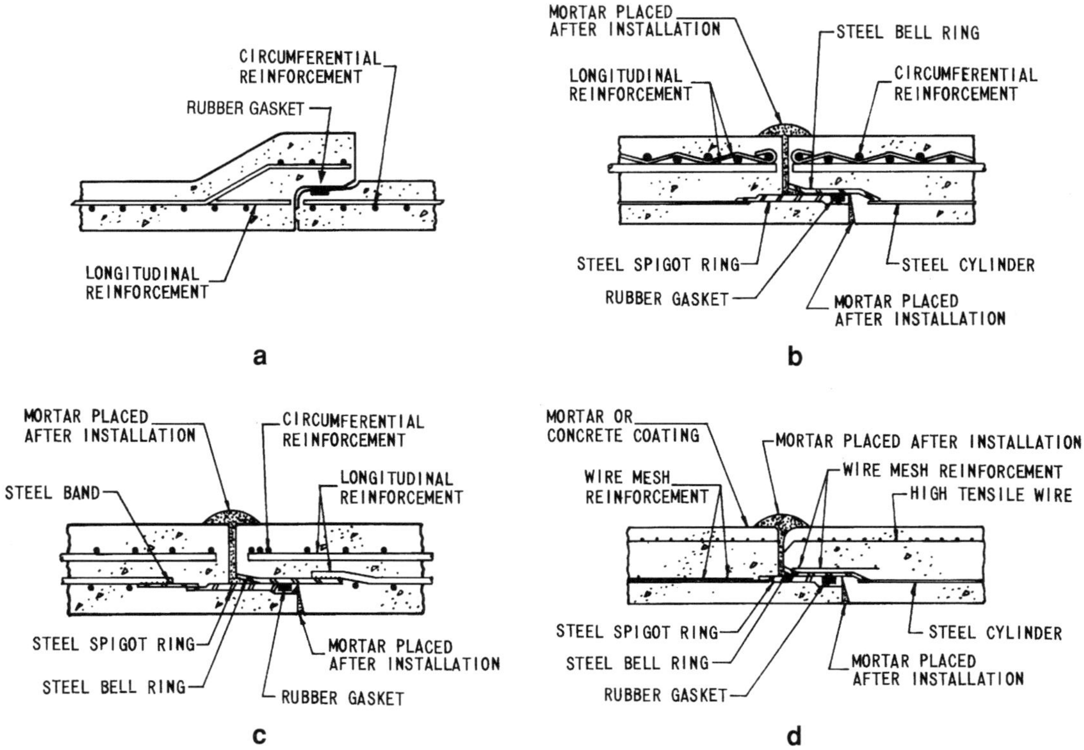

Figure 4-11. Push-on joints for concrete pipe in buried service. (a) Reinforced pressure pipe with rubber and conrete joint; (b) reinforced cylinder pipe with rubber and steel joint; (c) reinforced pressure pipe with rubber and steel joint; (d) prestressed embedded cylinder pipe with rubber and steel joint. Courtesy of Wilson & Company, Engineers & Architects.

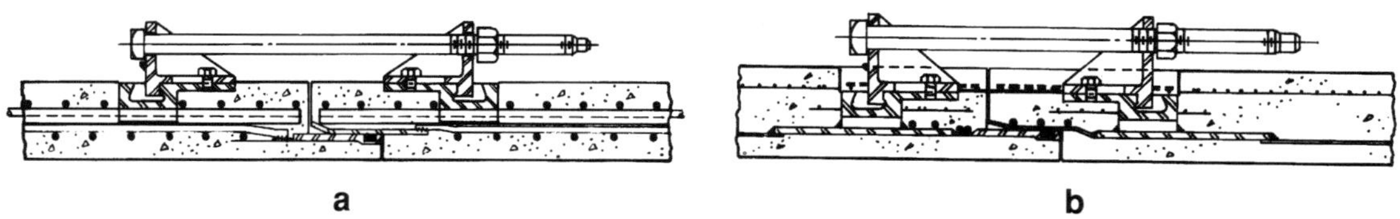

Figure 4-12. Restrained rigid joints for concrete pipe in exposed, buried, or subaqueous service. (a) Reinforced pressure pipe with rubber and steel joint; (b) reinforced cylinder pipe with rubber and steel joint. Courtesy of Wilson & Company, Engineers & Architects.

In addition to the protective coatings, cathodic protection may also be required. Corrosion of iron and steel pipe is an electrochemical reaction. Cathodic protection consists of introducing a controlled dc current to oppose the natural destructive current. An alternative method is to bury sacrificial anodes that also reverse the current. Cathodic protection of iron and steel pipes is commonly used in aggressive soils or where there are stray electrolytic currents. Stray electrolytic currents can occur near buried power transmission lines and near other pipelines equipped with cathodic protection systems. Most petroleum and gas transmission lines have impressed-current cathodic protection. Any ferrous metal in the vicinity becomes an anode, loses metal, and eventually fails unless it, too, is protected. If in doubt, consult a corrosion specialist. Minimum voltage requirements are determined by experience and field studies of ambient conditions such as moisture, chemical content, and resistivity of the soil. A more detailed discussion of cathodic protection is included in AWWA M11 [9] and NACE Standards RP-01-69 and RP 05-72.

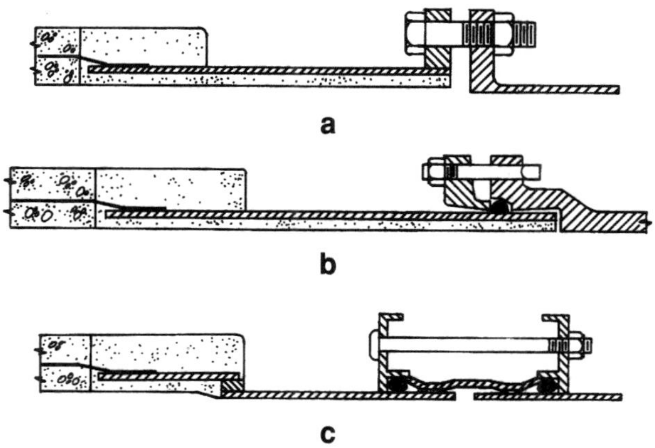

Figure 4-13. Adapters for connecting concrete pipe to ductile iron or steel pipe. (a) Flanged adapter; (b) mechanical joint adapter; (c) sleeve (Dresser®) coupling adapter. After Millcon Corp.

Table 4-13. Common Types of Joints for Buried Pipe

Joint	Figure	Type of joint[a]	Type of pipe: Ductile iron	Steel	PVC	AC	RCPP
Flanged joint	4-1, 2	R	√	√			
Lugged sleeve coupling (Dresser®), harnessed	4-3a, b	F	√	√			
Sleeve coupling (Dresser®), unharnessed	4-7a	S	√	√			
Mechanical joint	4-7b	F	√				
Push-on joint	4-7c	F	√				
Push-on joint	4-8	F		√			
Ball (also for exposed service)	4-7d	F	√				
Twin-gasketed coupling	4-10a	F			√	√	
Bell and spigot	4-10b	F, S			√		
Grooved-end coupling (Victaulic®)	4-4a	F	√	√			
Welded joint	4-9	R		√			
RCPP							
Push-on	4-11	S					√
Lugged, rigid	4-12	R					√
RCPP to iron or steel							
Flanged	4-13a	R					√
Mechanical or sliding	4-13 b, c	S					√

[a] R = rigid; S = sliding or slip; F = flexible.

Asbestos cement pipe is usually used without interior or exterior coatings. If corrosion can occur, consider other materials.

Reinforced concrete pressure pipe carrying corrosive water can be lined with coal-tar enamel or coal-tar epoxy. (Coal-tar enamels and epoxies are not used in potable water service.) These linings require maintenance, however, and should be used only where the pipeline can be removed from service periodically for inspection and where the pipeline is large enough in diameter to permit painting in place. Alternatively, the pipe can be lined by securing plastic sheets with T-shaped ribs to the pipe forms. The ribs lock the liner securely into the finished pipe. Field joints in the liner are made by heat-welding plastic strips after each joint of pipe has been installed in the trench. The pipe could also be encased in plastic to protect it from corrosive (high sulfate or acidic) soils.

Fittings

The ductile and gray cast-iron fittings shown in Table 4-9 are also available with mechanical joints. Fittings for other kinds of pipe are given in Table 4-13. Ductile iron fittings are used for asbestos cement pipe and PVC (AWWA C900). Fabricated steel fittings can also be used.

Economics

The cost of the pipe at the job site is only one factor of many. Others include

- Size
- Weight and length of sections, difficulty of handling, and machinery required
- Bedding. Flexible pipe, such as plastic sewer pipe, requires special bedding conditions (see Uni-Bell Handbook [12])
- Type of joint and amount of field labor
- Maintenance (frequency and difficulty of repair)
- Water hammer control (different for different materials)
- Friction coefficient
- Shoring requirement for laborers to enter trenches deeper (depending on soil stability) than 1.2 or 1.5 m (4 or 5 ft)—especially severe for trenches deeper than 3 m (10 ft). Consult OSHA regulations for shoring requirements.

Selection is further complicated by local conditions that significantly affect the cost of one material versus another, and these conditions may change radically with time. Carefully investigate the cost and availability of all types of suitable piping. For example, API steel pipe is made in great quantity and is frequently readily available and cheaper than other steel pipe. Coatings and linings can be applied to API pipe in the same manner as applied to any other steel pipe.

4-3. Ductile Iron Pipe (DIP)

Detailed descriptions of DIP, fittings, joints, installation, thrust restraint, and other factors relating to design as well as several important ANSI/AWWA specifications are contained in the DIPRA handbook [11] (see also Section 4-1 and Tables 4-1 and 4-11).

Materials

Cast-iron pipe is manufactured of an iron alloy centrifugally cast in sand or metal molds. Prior to the early 1970s, most cast-iron pipe and fittings were gray iron, a brittle material that is weak in tension. But now all cast-iron pipe except soil pipe (which is used for plumbing) is made of ductile iron in which the graphite is formed into spheroids by the addition of magnesium and heat treatment, which makes it about as strong as steel. Cast-iron fittings are still available in gray iron as well as ductile iron.

Tolerances, strength, coatings and linings, and resistance to burial loads are given in ANSI/AWWA C151/A21.51.

A special abrasion-resistant ductile iron pipe for conveying slurry and grit is available from several manufacturers. Regular ductile iron pipe (AWWA C151) has a Brinell hardness (BNH) of about 165. By comparison, abrasion-resistant ductile iron has a BNH of about 280. The sizes available are 150 through 600 mm (6 through 24 in.) per AWWA C151 in the standard AWWA wall thickness classes.

Available Sizes and Thicknesses

As shown in Tables B-1 and B-2, the available sizes range from 100 to 1350 mm (4 to 54 in.). The standard length is 5.5 m (18 ft) in pressure ratings from 1380 to 2400 kPa (200 to 350 lb/in.^{2}).

Thickness is specified by class, which varies from Class 50 to Class 56 (see the DIPRA handbook [11] or ANSI/AWWA C150/A21.50). Thicker pipe can be obtained by special order.

Joints

Buried joints should be of the mechanical joint or rubber gasket push-on type. Various types of restrained joints for buried service are also available.

Exposed joints should be flanged (AWWA C115 or ANSI B16.1) or grooved end (AWWA C606) (refer to "Joints" in Section 4-1).

Gaskets

Gaskets for ductile iron or cast-iron flanges should be rubber, 3.2 mm ($^1/_8$ in.) thick.

Gaskets for grooved-end joints are available in ethylene propylene diene monomer (EPDM), nitrile (Buna N), halogenated butyl rubber, Neoprene™, silicone, and fluorelastomers. EPDM is commonly used in water service and Buna N in sewage or sludge service.

Gaskets for ductile iron push-on and mechanical joints described in AWWA C111 are vulcanized natural or vulcanized synthetic rubber. Natural rubber is suitable for water pipelines but deteriorates when exposed to sewage or sludge.

Fittings

Dimensions of cast-iron and ductile iron flanged fittings are covered by ANSI B16.1 and AWWA C110, and fittings for abrasion-resistant pipe are generally furnished in one of the following categories:

- *Type 1:* Low alloy fittings with a minimum hardness of about 260 BNH, with flanged, grooved-end, or mechanical joints.
- *Type 2:* Special ductile iron fittings with a minimum hardness of about 400 BNH and flanged or mechanical joints.
- *Type 3:* Ni-Hard® fittings with a minimum hardness of about 550 BNH, with plain end or mechanical joints.

Linings and Coatings

For a discussion of linings, refer to Section 4-1.

Although ductile iron is relatively resistant to corrosion, some soils (and peat, slag, cinders, muck, mine waste, or stray electric current) may attack the pipe. In these applications, ductile iron manufacturers recommend that the pipe be encased in loose-fitting, flexible polyethylene tubes 0.2 mm (0.008 in.) thick (see ANSI/AWWA C105/A21.5). In some especially corrosive applications, a coating such as adhesive, hot-applied extruded polyethylene wrap may be required.

An asphaltic coating (on the outside) approximately 0.025 mm (0.001 in.) thick is a common coating for DIP in noncorrosive soils. In corrosive soils, consider the following coatings for protecting the pipe:

- Adhesive, extruded polyethylene wrap
- Plastic wrapping (AWWA C105)
- Hot-applied coal-tar enamel (AWWA C203)
- Hot-applied coal-tar tape (AWWA C203)
- Hot-applied extruded polyethylene [ASTM D 1248 (material only)]
- Coal-tar epoxy
- Cold-applied tape (AWWA C209)
- Fusion-bonded epoxy (AWWA C213).

Each of the coatings is discussed in detail in the AWWA specifications. Because each has certain limited uses, consider each specific installation and consult the NACE standards for the particular service.

4-4. Steel Pipe

The principal advantages of steel pipe include high strength, the ability to deflect without breaking, ease of installation, shock resistance, lighter weight than ductile iron pipe, ease of fabrication of large pipe, the availability of special configurations by welding, the variety of strengths available, and ease of field modification (see Section 4-1 and Tables 4-1 and 4-11).

Material

Conventional nomenclature refers to two types of steel pipe: (1) mill pipe and (2) fabricated pipe.

Mill pipe includes steel pipe of any size produced at a steel pipe mill to meet finished pipe specifications. Mill pipe can be seamless, furnace butt welded, electric resistance welded, or fusion welded using either a straight or spiral seam. Mill pipe of a given size is manufactured with a constant outside diameter and an internal diameter that depends on the required wall thickness.

Fabricated pipe is steel pipe made from plates or sheets. It can be either straight- or spiral-seam fusion-welded pipe, and it can be specified in either internal or external diameters. Note that spiral-seam, fusion-welded pipe may be either mill pipe or fabricated pipe.

Steel pipe may be manufactured from a number of steel alloys with varying yield and ultimate tensile strengths. Internal working pressure ratings vary from 690 to 17,000 kPa (100 to 2500 lb/in.^{2}) depending on alloy, diameter, and wall thickness. Specify steel piping in transmission mains to conform to AWWA C200, in which there are many ASTM standards for materials (see ANSI B31.1 for the manufacturing processes).

Available Sizes and Thicknesses

The available diameters range from 3 to 6000 mm ($^1/_8$ to 240 in.), although only sizes up to 750 mm (30 in.) are given in Tables B-3 and B-4. Sizes, thicknesses, and working pressures for pipe used in transmission mains range from 100 mm to 3600 mm (4 to 144 in.) and are given in Table 4-2 of AWWA M11 [9].

Manufacturers should be consulted for the availability of sizes and thicknesses of steel pipe (see also Table 4-2 in AWWA M11, which shows a great variety of sizes and thicknesses).

According to ANSI B36.10,

- Standard weight (STD) and Schedule 40 are identical for pipes up to 250 mm (10 in.). All larger sizes of standard weight pipe have walls 9.5 mm ($^3/_8$ in.)

thick [see Tables B-3 and B-4 for standard weight pipe from 3 to 750 mm ($^1/_8$ to 30 in.)]. For 300-mm (12-in.) pipe and smaller, the ID approximately equals the nominal diameter. For larger pipe, the OD equals the nominal diameter.

- Extra strong (XS) and Schedule 80 are identical for pipes up to 200 mm (8 in.). All larger sizes of extra-strong-weight pipe have walls 12.7 mm ($^1/_2$ in.) thick.
- Double extra strong (XXS) applies only to steel pipe 300 mm (12 in.) and smaller. There is no correlation between XXS and schedule numbers. For wall thickness of XXS, which (in most sizes) is twice that of XS, see ANSI B36.10.

For sizes 350 mm (14 in.) and larger, most pipe manufacturers use spiral welding machines and, in theory, can fabricate pipe to virtually any desired size. But in practice most steel pipe manufacturers have selected and built equipment to produce given OD sizes. For example, one major U.S. manufacturer uses a 578.6-mm ($22^{25}/_{32}$ in.) OD cylinder for a nominal 525-mm (21-in.) pipe. Any deviation from manufacturers' standards is expensive. To avoid confusion, either show a detail of pipe size on the plans or tabulate the diameters in the specifications. For cement-mortar-lined transmission mains, AWWA C200, C205, C207, and C208 apply.

Manifold piping in pumping stations should be considered a large special fitting or a series of fittings connected together. Dependence on AWWA C200 alone is inadequate for designing such headers or manifolds because it does not address the following:

- Reinforcement at openings (tees, laterals, branches) (see "Pipe Wall Thickness" in Section 4-8)
- Wall thickness for grooved-end couplings (see Table 4-5)
- Thrust harness lugs for flexible pipe couplings (see Example 4-1)
- Additional wall thickness required at elbows and other fittings because of stress intensification factor (see the following subsection on fittings).

Thus, the design of steel manifolds depends on a combination of a number of factors.

Steel pipe must sometimes either be reinforced at nozzles and openings (tees, wye branches) or a greater wall thickness must be specified. A detailed procedure for determining whether additional reinforcing is required is described in Chapter II and Appendix H of ANSI B31.3. If additional reinforcement is necessary, it can be accomplished by a collar or pad around the nozzle or branch, a wrapper plate, or crotch plates. These reinforcements are shown in Figure 4-14, and the calculations for design are given in AWWA M11 [9].

Steel manifold piping 500 mm (20 in.) and smaller must inevitably be made in accordance with ANSI B36.10. Material would usually conform to ASTM A 53, A 135, or API 5L. For pipe 600 mm (24 in.) and larger, a fabricator might elect to use pipe conforming to ASTM A 134 or A 139 as well. As shown in Tables B-3 and B-4, the size of the pipe (nominal diameter) approximates the ID for 300-mm (12-in.) pipe and smaller, but size equals the OD for 350-mm (14-in.) pipe and larger. For pipe larger than 500 mm (20 in.), steel pipe size can be specially fabricated to any size.

The ID of steel pipe should match the ID of iron valves, particularly butterfly valves. One way is to select pipe one size larger than the nominal pipe size and line it with cement mortar so that the ID of a mortar-lined pipe matches the nominal pipe size. Consider, for example, a requirement for a steel header with a nominal 400-mm (16-in.) ID. A pipe fabricator could use an ANSI B36.10 standard weight pipe with a true OD of 457 mm (18 in.). The wall thickness of the steel cylinder is 9.5 mm (0.375 in.), which gives an ID of 438 mm (17.25 in.). A 13-mm ($^1/_2$-in.) mortar lining provides a net ID of 413 mm (16.25 in.), which is close to the desired size, and if the mortar lining were 16-mm ($^5/_8$-in.) thick, the ID would be exactly 406 mm (16 in.). As shown in Table 4-6, the minimum thickness of the cement-mortar lining is 7.9 mm ($^5/_{16}$ in.).

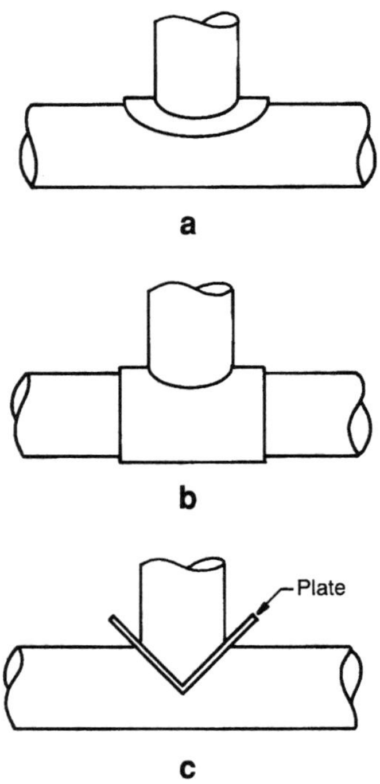

Figure 4-14. Reinforcement for steel pipe openings. (a) Collar plate; (b) wrapper plate; (c) crotch plates.

Obtaining internal diameters in even sizes for steel pipe smaller than 350 mm (14 in.) can be done by using plastic linings. For example, a 250-mm (10-in.) standard weight pipe has an OD of 273 mm (10.75 in.) and a wall thickness of 9.27 mm (0.365 in.); for a minimum cement-mortar thickness of 6.4 mm ($^1/_4$ in.), the ID is only 242 mm (9.5 in.). A practical alternative is to use a plastic lining, such as fusion bonded epoxy, that makes the ID equal to the nominal 250-mm (10-in.) diameter.

Some benefits of using standard weight ANSI B36.10 pipe are that (1) the pipe is readily available, and (2) the wall thickness is

- often sufficient, so that reinforcements at openings (wrappers, collars, or crotch plates) are unnecessary;
- often sufficient in pumping stations for use with thrust harnesses on flexible pipe couplings, so that additional wall thickness is unnecessary; and
- sufficient for use with AWWA C606 grooved-end couplings without additional reinforcement at the pipe ends.

Joints

Joints for steel pipe are listed in Table 4-13. For buried service, bell and spigot joints with rubber gaskets or mechanical couplings (with or without thrust harnesses) are preferred. Welded joints are common for pipe 600 mm (24 in.) and larger. Linings are locally destroyed by the heat of welding, so the ends of the pipe must be bare and the linings field applied at the joints. The reliability of field welds is questionable without careful inspection, but when properly made they are stronger than other joints.

Fittings

Specifications for steel fittings can generally be divided into three classes, depending on the joints used and the pipe size:

- Threaded (ANSI B16.3 or B16.11) but only for pipe 75 mm (3 in.) and smaller
- Flanged, welded (ANSI B16.9)
- Fabricated (AWWA C208).

Fittings larger than 75 mm (3 in.) should conform to ANSI B16.9 ("smooth" or wrought) or AWWA C208 (mitered); avoid threaded fittings larger than 75 mm (3 in.). The ANSI B16.9 fittings are readily available up to 300 to 400 mm (12 to 16 in.) in diameter. Mitered fittings are more readily available and cheaper for larger fittings.

The radius of a mitered elbow can range from 1 to 4 pipe diameters. The hoop tension concentration on the inside of elbows with a radius less than 2.5 pipe diameters may exceed the safe working stress. This tension concentration can be reduced to safe levels by increasing the wall thickness, as described in ANSI B31.1, AWWA C208, and *Piping Engineering* [16]. Design procedures for mitered bends are described in ANSI B31.1 and B31.3.

Gaskets

Gaskets for steel flanges are usually made of cloth inserted rubber either 1.6 mm ($^1/_{16}$ in.) or 3.2 mm ($^1/_8$ in.) thick and are of two types:

- ring (extending from the inside diameter of the flange to the inside edge of the bolt holes);
- full face (extending from the inside diameter of the flange to the outside diameter).

For mechanical and push-on joints, refer to "Gaskets" in Section 4-4.

Linings and Coatings

Cement mortar is an excellent lining for steel pipe (see Table 4-6 and AWWA C205 for the thickness required; refer to "Linings and Coatings" in Section 4-1 and to "Linings" in Section 3-3 for further discussion).

Steel pipe can also be protected with mortar, but soil conditions affect the necessary thickness of the mortar coating. Corrosive soils may require mortar coatings of 25 mm (1 in.) or more regardless of pipe size. Alternatively, a hot-applied extruded polyethylene coating with heat-shrink jackets for joints that complies with ASTM D 1248 is an excellent coating for both steel and ductile iron pipe.

As another alternative, epoxy-lined and -coated steel pipe could be used. Because this lining is only 0.3- to 0.6-mm (0.012- to 0.020-in.) thick, the ID of bare pipe is only slightly reduced by such linings (see Tables B-1 and B-2). Epoxy-lined steel pipe is covered by AWWA C203, C210, and C213 standards.

The supplier must be consulted to determine the limitations of sizes and lengths of pipe that can be lined with epoxy. Flange faces should not be coated with epoxy if flanges with serrated finish per AWWA C207 are specified.

4-5. Plastic Pipe

In the United States, where it is used in both water and sewage service, polyvinyl chloride (PVC) is the most commonly used plastic pipe. It is a polymer extruded under heat and pressure into a thermoplastic that is nearly inert when exposed to most acids, alkalies, fuels, and corrosives, but it is attacked by ketones (and other solvents) sometimes found in industrial wastewaters. It has a high strength-to-weight ratio and is durable and resilient, but it lacks the stiffness necessary for exposed service and is susceptible to flotation in groundwater conditions. Most types of PVC pipe should not be exposed to direct sunlight (see ANSI/AWWA C900). Some designers have had poor experiences with solvent-welded flanges and no longer use them on buried PVC pressure pipe in any size. PVC pipe conforming to AWWA C900 or C905 with ductile iron fittings can be used for buried service in sizes 1000 mm (24 in.) or smaller. For larger pipe, AWWA C905 may not be sufficiently conservative in some applications. A proper analysis of static and transient internal and external pressures is required. Note that the fatigue limit for PVC is very low, and the pipe could be vulnerable to rupture from excessive pressure cycles. HDPE pipe is better for cyclic loading.

High-density polyethylene (HDPE) pipe and polyvinyl chloride (PVC) are both suitable for use in potable water service and raw sewage service. HDPE, in general, is much less sensitive to surge pressures than PVC because of the long molecular chains in the plastic material. See AWWA C906 for the method of calculating the required pressure class. The heat fused butt joint system for HDPE pipe is a satisfactory joint that is easy to install in the field.

A key consideration in specifying either PVC or HDPE is their susceptibility to scratches in the pipe wall that can be caused by dragging the joined sections of pipe for long distances over the ground. Rocks usually cause scratches about 3-mm ($^1/_8$-in.) deep and sometimes much deeper. In HDPE, scratches 0.19-T (T is the pipe wall thickness) deep reduce the cyclic loading life span by 90%, whereas scratches 0.04 T deep have little influence on fatigue strength. Specifying that pipe must be dragged over smooth surfaces, such as railroad ties, is one way to avoid deep scratches.

In addition, careful attention must be paid in specifications and construction inspection to ensure that the pipe is fully bedded, with no voids or poorly compacted areas beneath the springline. Because of its permeability to organic solvents with low molecular weight (gasoline, for example) HDPE should not be used for potable water beside roads where gasoline could penetrate into the ground. Other than the above considerations, HDPE is ideal for systems with moderate pressures. Although the cost of the pipe itself is about the same as it is for steel, its installed cost is likely to be less than that of steel because of the ease and speed of handling.

Many designers refuse to specify PVC pipe for the small (75-mm or 3-in. and smaller) piping that, for example, conveys seal water inside pumping stations. The primary reason is due to the difficulty of getting the PVC pipe installers to do a proper job. Specific problems that occur include:

- Inadequate or excessive amounts of solvent cement applied to the joints. See ASTM D 2855.
- Inadequate curing time allowed before moving pipe with solvent-cemented joints. It is wise to specify eight hours of curing time before pipe can be moved (in spite of manufacturers' claims that two hours is adequate).
- Forcing and springing the pipe into position to make up for errors in initially installing the pipe. The overstress in the pipe can result in breakage after a time lapse of less than a year. Adequate inspection *during installation* could prevent the overstress, but inspection after installation is useless.

Because of these problems, copper, steel, or stainless steel for the small-size service piping is a better choice where applicable. These metallic piping systems are far more resistant to poor installation practices than is plastic piping. Furthermore, it is preferable to avoid exposed plastic pipe in any pumping station because of the hazard of melting (or even supporting combustion) in fires (see NFPA 820). Plastic pipe may, however, be the only practical selection for chlorine solution or other chemical services.

Other plastic piping materials include

- Acrylonitrile-butadiene-styrene (ABS)
- Chlorinated polyvinyl chloride (CPVC)
- Polypropylene (PP)
- Polyethylene (PE)
- Polyvinylidene fluoride (KYNAR™ or PVDF)
- Fiberglass-reinforced plastic (FRP).

Most of these materials are used for corrosive chemicals, but some may have special uses for water, sewage, or sludge. Consult manufacturers for properties, joints, and fittings. Investigate installations before specifying such materials. The impact strength of most plastics decreases when exposed to sunlight. Consequently, be wary of using plastic pipe in exposed outdoor service unless it is coated with an ultraviolet-resistant paint such as a polyurethane. However, FRP should be installed only where it is exposed and easily inspected;

if it is buried, it should be used with both very conservative safety factors and considerable caution.

Available Sizes and Thicknesses

The range of available sizes is given in Table 4-11, but consult the appropriate ASTM standards (D 1785 and D 2241) to find exact sizes, thicknesses, and pressure ratings.

Wall thickness design for PVC pipe is defined by two separate sets of nomenclature: (1) standard dimension ratios (SDR) and (2) schedules. The ratio of pipe outside diameter to wall thickness is called the "SDR." For PVC pipe, SDR is calculated by dividing the average outside diameter of the pipe by the minimum wall thickness. The available thicknesses are SDR 35 through SDR 14 (refer to ANSI/ASTM D 2241 for a complete discussion of SDRs and corresponding pressure ratings; refer to ANSI/ASTM D 1785 for a complete discussion of the wall thicknesses and pressure ratings for the various schedules of PVC pipe).

Joints

Joints of PVC can be solvent welded, flanged, push-on with rubber gaskets, or threaded. Threads should be used only for 100-mm (4-in.) pipe and smaller, and the thinnest threaded pipe should be Schedule 80 (see ANSI/ASTM D 2464). Solvent-welded Schedule 40 pipe is stronger than threaded Schedule 80 pipe, and solvent-welded Schedule 80 pipe is the strongest of all.

Gaskets

Gaskets for PVC flanges should be 3.2-mm- ($^1/_8$-in.)-thick ethylene propylene rubber (EPR), full faced, with a Durometer hardness of 50 to 70, Shore A. When connecting PVC flanges to raised-face metal flanges, remove the raised face on the connecting metal flange to protect the PVC flange from the bolting moment.

Valve Pressure Rating

The maximum allowable working pressure of PVC valves is even lower than that of Schedule 80 threaded piping. Most PVC valves are rated at 1040 kPa (150 lb/in.^{2}) at a temperature of 38°C (100°F). The maximum recommended pressure for any flanged plastic pipe system (PVC, CPVC, PP, PVDF) is the same.

Fittings

Threaded, flanged, or solvent-welded fittings are used in exposed and buried service for piping smaller than 100 mm (4 in.). Class 125 mechanical joint ductile or cast-iron fittings should be used in buried applications for pipes 100 mm (4 in.) and larger. The adapters must be installed in the manner prescribed by the manufacturer.

Fittings for PVC pipe include tees, crosses, wyes, reducers, and 22.5°, 45°, and 90° bends.

Criteria for Selection of PVC Pressure Pipe

Recommended criteria for using PVC pressure pipe is as follows:

- *Seventy-five mm (3 in.) and smaller, exposed and buried service:* Schedule 80 per ASTM D 1784 and D 1785. Other ASTM standards applicable to PVC pipe are D 2464, Schedule 80 Threaded Fittings; D 2467 Schedule 80 Socket Type Plastic Fittings; and D 2564 Solvent Cements for PVC Pipe. Based on the experience and observations of various engineers, this pipe should not be used in applications in which the operating pressure exceeds 550 kPa (80 lb/in.^{2}). Use copper or steel pipe for water service; use PVC pipe only for chemical service.
- *One hundred and 150 mm (4 and 6 in.), exposed service:* Schedule 80 per ASTM D 1748 and D 1785. Use solvent-welded—not threaded—joints. Based on the experience and observations of various engineers, this pipe should not be used in applications in which the operating pressure exceeds 345 kPa (50 lb/in.^{2}). Use PVC only for chemical service piping. Use steel or ductile iron for water piping and other services.
- *Two hundred mm (8 in.) and larger, exposed service:* PVC pipe should not be used at all in such sizes.
- *One hundred through 300 mm (4 through 12 in.), buried service:* AWWA C900. Fittings should be cast or ductile iron.
- *Three hundred fifty through 900 mm (14 through 36 in.), buried service:* AWWA C905. This standard may not be sufficiently conservative in its pressure ratings for some applications. For a pressure class of 1035 kPa (150 lb/in.^{2}), consider using an SDR of 18 for 350- to 600-mm (14- to 24-in.) pipe and an SDR of 21 for 750- to 900-mm (30- to 36-in.) pipe. These SDR values are based on a surge allowance of 345 kPa (50 lb/in.^{2}) over the pipe pressure class, with a safety factor of 2.5 for 350- to 600-mm pipe and a safety factor of 2.0 for 750- to 900-mm pipe.
- *Never use plastic pipe for air or compressed gas service.*

4-6. Asbestos Cement Pipe (ACP)

Asbestos cement pipe, available in the United States since 1930, is made by mixing portland cement and asbestos fiber under pressure and heating it to produce a hard, strong, yet machinable product. Over 480,000 km (300,000 mi) of ACP is now in service in the United States, and, according to a mid-1970s survey, more than a third of pipe then being installed was ACP [8].

In recent years, attention has been focused on the hazards of asbestos in the environment and, particularly, in drinking water. The debate continues with one set of experts advising of the potential dangers and a second set of experts claiming that pipes made with asbestos do not result in increases in asbestos concentrations in the water. Studies have shown no association between water delivered by ACP and any general disease, but fear may be as important as reality, so consult with owners and local health authorities before deciding whether to specify ACP.

In January 1986, the EPA published a proposed regulation banning further manufacture and installation of ACP, but it was made clear that the proposed action was based on the hazard of inhaling asbestos fibers during manufacture and installation of the pipe—not because it contaminated drinking water. For awhile after October 1987, the EPA had been reassessing the January proposal [8], but the proposed ban was overturned by a U.S. Federal Appeals court in 1991. Consequently, ACP is still being manufactured and used.

Available Sizes and Thicknesses

Available sizes are given in Table 4-11. Refer to ASTM C296 and AWWA 401, 402, and 403 for thicknesses and pressure ratings and AWWA C401 and C403 for detailed design procedures. AWWA C401-83, for 100- to 400-mm (4- to 16-in.) pipe, is similar to AWWA C403-84 for 450- to 1050-mm (18- to 42-in.) pipe. The properties of asbestos cement for distribution pipe (AWWA C400) and transmission pipe (AWWA C402) are identical. Nevertheless, under AWWA C403 the *suggested* minimum safety factor is 2.0 for operating pressure and 1.5 for external loads, whereas the safety factors under AWWA C402 are 4.0 and 2.5, respectively. So the larger pipe has the smaller safety factors.

Section 4 in AWWA C403 justifies this difference on the basis that surge pressures in large pipe tend to be less than those in small pipes. But surge pressures are not necessarily a function of pipe diameter (see Chapters 6 and 7). The operating conditions, including surge pressures, should be evaluated before the pipe class is selected. It is the engineer's prerogative to select which of the safety factors should apply. Per AWWA C400, safety factors should be no less than 4.0 and 2.5 if no surge analysis is made. The low safety factors given in AWWA C403 should be used only if all loads (external, internal, and transient) are carefully and accurately evaluated.

Joints and Fittings

The joints are usually push-on, twin-gasketed couplings (Figure 4-10), although mechanical and rubber gasket push-on joints can be used to connect ACP to iron fittings.

Ductile iron fittings conforming to ANSI/AWWA C110/A21.10 are used with ACP, and adapters are available to connect ACP to flanged or mechanical ductile iron fittings. Fabricated steel fittings with rubber gasket joints can also be used. ACP may be tapped with corporation stops, tapping sleeves, or service clamps.

4-7. Reinforced Concrete Pressure Pipe (RCPP)

Reinforced concrete pressure pipe can be made to meet special strength requirements by using a combination of a steel cylinder and steel cages, by using one or more steel cages, or by prestressing with spiral rods (as shown in Figures 4-11 and 4-12 and in Table 4-12).

A distressing number of failures of prestressed concrete cylinder pipe (AWWA C301) have occurred in the United States within the last decade. The outer shell of concrete cracks, which allows the reinforcement to corrode and subsequently fail. Therefore, do not depend on AWWA specifications or on manufacturers' assurances, but do make a careful analysis of internal pressure (including water hammer) and external loads. Make certain that the tensile strain in the outer concrete shell is low enough so that cracking either will not occur at all or will not penetrate to the steel under the worst combination of external and internal loading. Note again that strength is not the issue. The concern is for the tensile strain in the outermost concrete. When cracks do occur, they appear to penetrate to a depth of about 18 to 25 mm (0.75 to 1 in.), so a clear cover of at least 38 mm (1.5 in.) should be specified. Water hammer must be carefully analyzed and controlled. Wire-wrapped, prestressed concrete pipe has the worst history of failure. Reinforced concrete pipe is better, but if any significant water hammer can occur, ductile iron or steel is best. For salt or brackish water, PVC, HDPE, or ductile iron with cathodic protection should be used.

Sizes and Joints

Sizes range from 600 to 3600 mm (24 to 144 in.), as shown in Table 4-11.

Joints are shown in Figures 4-11 and 4-12. Joints for buried service include

- Rubber gasket and concrete (Figure 4-11a)
- Rubber gasket and steel (Figure 4-11b, c, d)
- Lugged rubber gasket and steel (Figure 4-12).

Note that rubber gasket and concrete joints (Figure 4-11) should be used only for pressures less than 380 kPa (55 lb/in.^{2}). The other joints can withstand pressures up to 2800 kPa (400 lb/in.^{2}), but consult manufacturers for joint pressure ratings.

Wall Thickness Design

The wall thickness design should be based on both external trench loads and on a detailed hydraulic analysis of the pumping system, including water hammer and surge pressure. Surge pressures should not be able to induce even hairline cracks in the external surface. A variety of wall thicknesses and reinforcing designs are available for each pipe diameter.

Some of these types of pipe have severe internal pressure limitations. The AWWA standards cited in Table 4-12 should be consulted.

Fittings

Standard fittings are tees, crosses, 45° wyes, eccentric reducers, concentric reducers, flange and mechanical joint adapters to connect concrete pipe to steel or ductile iron (Figure 4-13), and bends from 7.5° to 90° in 7.5° steps. RCPP can be tapped by drilling a hole into the pipe and then strapping a threaded or flanged tapping saddle to the pipe. Alternatively, a threaded steel outlet connection can be cast in the pipe wall during manufacture. Some pipe designers prefer to use fabricated steel specials for fittings and for any pipe segment containing an outlet or nozzle.

Linings

Pipe that is only partly full of sewage or in which air can enter by any means (temperature changes, vortices in the wet well, or leaks) requires protection against corrosion caused by bacterial action. Sometimes, pipe can fail in only a few years. Anaerobic, sulfate-reducing bacteria (such as *Desulfovibrio*) living in the slime below the water surface utilize the oxygen in sulfate and create hydrogen sulfide, which escapes from the water surface to the atmosphere above. Aerobic bacteria *(Thiobacillus)* on the sides and soffit of the pipe convert the hydrogen sulfide to sulfuric acid at a pH of 2 or even less. *Thiobacillus* cannot live in pipe that is always full, so it is important to keep force mains (whether concrete, steel, or ductile iron) full at all times. Therefore, where a force main terminates at a manhole, design the connection so that no part of the force main is exposed to air. Either connect the force main up through the bottom of the manhole, or if it enters from the side, either (1) set the invert of the downstream sewer above the crown of the force main or (2) use plastic pipe near the manhole.

Lining and coating systems for pipes include various brush- or spray-applied epoxies, resins, polyurethanes, and coal tars. Coatings such as coal-tar epoxies have a history of poor performance where hydrogen sulfide attack can occur. A more effective (and more costly) lining system consists of PVC liner sheets (such as Ameron Tee-Lock®) that are made with keys or ribs projecting from one side of the sheet. The smooth PVC face is attached directly to the forms prior to casting the concrete. The ribs project into the concrete and form a mechanical bond with it. PVC liners can be cast around the entire circumference of pipe and the longitudinal joint heat-fused to form a 360° liner, but it is cheaper to line only the portion above low-water level. These liners offer maximum protection for concrete pipe and manholes subject to corrosive environments, and they have a 40-year record of success. If the workmanship and inspection of the welding at the joints is good, the system will have a long, trouble-free life. The designer's problems are: (1) to write specifications that ensure workers are not hurried through this important task and (2) to ensure that inspection is of high caliber. PVC liners are usually applied only to pipes in sizes of 900 mm (36 in.) and larger and usually only to the upper 270° of the full circumference. Corrosion-resistant materials such as vitrified clay pipe, PVC, or HDPE should be used for smaller sewer pipes. See Section 25-11 for more discussion of concrete protection.

4-8. Design of Piping

The emphasis in this section is on the piping within the pumping station (i.e., exposed) and problems such as pipe thickness, flange bolts, and pipe supports. The design of external (i.e., buried) piping is limited to generalities because there is an extensive body of excellent literature on such problems [9, 11–15].

In selecting a pipe size, be aware that it is the outside—not the inside—diameter that is fixed. The inside diameter varies with the wall thickness, whereas the outside diameter does not. This is true for all sizes of ductile iron, copper, brass, and plastic pipe. It is also true for most steel pipe used in pumping stations.

Exposed Piping

The selection of pipe size governed by hydraulics is given in Example 3-1. Other practical considerations that depend on available pipe diameters and wall and lining thicknesses are discussed in Sections 4-3 and 4-4 under "Available Sizes and Thicknesses."

Tie Rods

Mechanical couplings, mechanical joints, and push-on joints (Figures 4-7, 4-8, 4-10, and 4-11) must be restrained from sliding apart either by soil friction (if the pipe is buried) or by tie rods (if the pipe is exposed) as shown in Figures 4-3 and 4-12. Always design rods and bolts for tensile stress at the net cross-sectional area at the root of the threads (see Table 4-14).

Table 4-14. Root Areas of Threaded Rods

Nominal rod diameter		Root area of coarse thread	
mm	in.	mm^2	in.^{2}
10	$^3/_8$	43.9	0.068
13	$^1/_2$	81.3	0.126
16	$^5/_8$	130.3	0.202
19	$^3/_4$	194.8	0.302
22	$^7/_8$	270.3	0.419
25	1	356.1	0.552
28	$1^1/_8$	447.1	0.693
31	$1^1/_4$	573.5	0.889
35	$1^3/_8$	679.4	1.053
38	$1^1/_2$	834.2	1.293
41	$1^5/_8$	977.4	1.515
44	$1^3/_4$	1125	1.744
47	$1^7/_8$	1321	1.048
50	2	1479	2.292
63	$2^1/_2$	2397	3.716
75	3	3626	5.672

Example 4-1
Design of Tie Rods for a Sleeve Pipe Coupling

Problem: Determine the number and size of tie rods required for a sleeve coupling in a 600-mm (24-in.) DIP under a maximum static and surge pressure of 1070 kPa (155 lb/in.^{2}).

Solution: The outside diameter of the pipe is 655 mm (25.8 in.); because the pressure acts on the gross face area of the end of the pipe, use the OD (not the ID) to compute thrust.

SI Units

$$\text{Pipe area} = \frac{\pi(0.655)^2}{4} = 0.337\ \text{m}^2$$

$$\text{Thrust} = 1070\ \text{kPa} \times 0.337\ \text{m}^2 = 361\ \text{kN}$$

U.S. Customary Units

$$\text{Pipe area} = \frac{\pi(25.8)^2}{4} = 523\ \text{in.}^2$$

$$\text{Thrust} = 155\ \text{lb/in.}^2 \times 523\ \text{in.}^2 = 81{,}100\ \text{lb}$$

Materials recommended in AWWA M11 [9] are (1) rods or bolts (ASTM A 193, Grade B7 or equivalent with a yield stress of 725,000 kPa [105,000 lb/in.^{2}]), (2) welded lugs (ASTM A 283, Grade B or ASTM A 285, Grade C or equivalent with a yield stress of 186,000 kPa [27,000 lb/in.^{2}] or ASTM A 36 steel with a yield stress of 248,000 kPa [36,000 lb/in.^{2}]) (see the ASTM standards for stresses).

Reduce the tensile yield stress at the root of the threads by a safety factor of 2 to obtain the allowable working stress. The total root area required is then

$$A = \frac{361\ \text{kN}}{362{,}000\ \text{kPa}} = 0.0010\ \text{m}^2$$

$$A = \frac{81{,}000\ \text{lb}}{52{,}000\ \text{lb/in.}^2} = 1.56\ \text{in.}^2$$

Many combinations of rod sizes and numbers can be used. Try several and choose a suitable one. (Note, however, that the resulting bending stresses in the pipe shell should be checked. Using two large-diameter rods, for example, causes greater bending stresses than using four smaller ones. Sometimes the pipe wall thickness must be increased to reduce the stress.) From Table 4-14 the root area of a 22-mm ($^7/_8$-in.) bolt is 270 mm^2 or 0.000270 m^2 (0.419 in.^{2}); thus, the number of rods required is

$$N = \frac{0.0010 \text{ m}^2 \times 10^6}{270 \text{ mm}^2} = 3.7 \qquad N = \frac{1.56 \text{ in.}^2}{0.419 \text{ in.}^2} = 3.7$$

Use four 22-mm rods. Use four $^7/_8$-in. rods.

Note that the rod material is high-strength steel (ASTM 193 Grade B7). If lower strength steel (carbon or stainless) were used, the number and/or size of rods would be quite different.

The rods can be supported by lugs welded to the pipe, which is excellent with steel pipe but can be done with ductile iron pipe only if great care is taken not to overheat the iron. An alternate detail for ductile iron pipe is to bolt an "ear" (as in Figure 4-3a) to the pipe flange so that the tie rod clears both the pipe flange and the pipe coupling. From the DIPRA handbook [11] and manufacturers' catalogs, the critical dimensions are (1) flange bolt circle [749 mm (29.5 in.)], (2) number and size of bolts [20 bolts 32 mm ($1^1/_4$ in.) in diameter], (3) flange OD [813 mm (32.0 in.)], (4) flange thickness [48 mm (1.88 in.)], and (5) coupling OD [782 mm (30.8 in.)]. If a clearance between the rod and flange of, say, 8 mm ($^5/_{16}$ in.) is chosen, the rod is centered 19 mm ($^3/_4$ in.) from the flange OD. The ear can be designed (with an adequate safety factor applied to the yield stress) to withstand the calculated force in the tie rod, but a more sensible design is to size the ear so that yield stress is reached simultaneously in the ear and tie rod. The bending moment in the ear is equal to the lever arm times the force in the tie rod.

$$M = 0.019 \text{ m}(2.7 \times 10^{-4} \text{ m}^2 \times 7.25 \times 10^8 \text{ N/m}^2) = 3.720 \text{ N} \cdot \text{m}$$

$$M = 0.75 \text{ in.}(0.419 \text{ in.}^2 \times 105{,}000 \text{ lb/in.}^2) = 33{,}000 \text{ lb} \cdot \text{in.}$$

The formula for bending stress is

$$s = \frac{Mc}{I} = \frac{M}{S} \tag{4-1}$$

where s is stress in newtons per square meter (pounds per square inch), M is moment in newton-meters (pound-inches), I is moment of inertia in meters to the fourth power (inches to the fourth power), and S is the section modulus in cubic meters (cubic inches). For rectangular cross-setctions, $S = bd^2/6$, where b is width in meters (inches) and d is depth in meters (inches). Rearranging Equation 4-1 and using the yield stress for A 36 steel gives

$$bd^2 = \frac{6 \times 3720 \text{ N} \cdot \text{m}}{248{,}000{,}000 \text{ N/m}^2} = 9.00 \times 10^{-5} \text{ m}^3 \qquad bd^2 = \frac{6 \times 33{,}000 \text{ lb} \cdot \text{in.}}{36{,}000 \text{ lb/in.}^2} = 5.5 \text{ in.}^3$$

Let b = 63 mm = 0.063 m. Let b = 2.5 in.

$$d = \sqrt{\frac{9.0 \times 10^{-5}}{0.063}} = 0.038 \text{ m} \qquad d = \sqrt{\frac{5.5}{2.5}} = 1.48 \text{ in.}$$

Use 63 mm × 38 mm plate. Use $2^1/_2$ in. × $1^1/_2$ in. plate.

Pipe Wall Thickness

The hoop (circumferential) tensile stress in metal pipe due to the working pressure should not exceed 50% of the yield strength. The working pressure plus surge pressure due to water hammer should not exceed 75% of the yield strength or the mill test pressure. Depending on the grade of steel used, the yield strength can lie between 248,000 and 414,000 kPa (36,000 and 60,000 lb/in.). The yield strength of DIP is more uniform—a minimum of 290,000 kPa (42,000 lb/in.^{2}).

The hoop tensile stress is given by the equation

$$s = \frac{pD}{2t} \tag{4-2}$$

where s is the allowable circumferential stress in kilopascals (pounds per square inch), p is the pressure in kilopascals (pounds per square inch), D is the outside diameter of the iron or steel cylinder in millimeters (inches), and t is the thickness of the iron or steel cylinder in millimeters (inches). (Theoretically, D should be the inside diameter, but the outside diameter is conservatively specified in most codes, partly because the ID is not known initially.) The longitudinal stress in a straight pipe is half of the circumferential stress.

Hangers and Supports

Supports or hangers must carry the weight of the pipe and fluid in exposed piping systems. The location of supports and hangers depends on the pipe size, joint systems, piping configurations, location of valves and fittings, weight of pipe and liquid, beam strength of the pipe, and the structure available to support the weight and all other static and dynamic forces, including expansion and contraction.

Supports or hangers must also carry the lateral forces due to earthquake. Either design vertical supports to resist the horizontal force of pipe and fluid or augment vertical supports and hangers with horizontal ones. Note that tension in wall anchors must resist the full horizontal force. Typical pipe supports are shown in Figure 4-15 and a few of the many hangers manufactured are shown in Figure 4-16.

Some general design precautions for pipe support system design include the following:

- Because pump casing connections to piping systems are rarely designed to support any load transmitted through the connection, supports should be provided on both the suction and discharge side of pumps to prevent pipe loads being transmitted to the pump casing.
- Flexible pipe couplings are recommended at pump inlets and outlets. They are useful because (1) they are "forgiving" and allow the contractor to level the pump and make minor horizontal adjustments; (2) properly selected, they isolate the pump from slight movements of the pipe; and (3) they reduce vibration and, to a limited extent, noise.
- If expansion and contraction of the piping could occur, the pipe supports should allow movement.
- Flexible joints, such as mechanical joints and couplings, must be supported on both sides because they are not designed to transmit loads. Flexible couplings must be constrained with lugs and lug bolts to prevent longitudinal movement caused by internal pipeline pressure.

ANSI B31.1 describes hanger and support spacings for steel pipe with a minimum wall thickness of standard weight. A maximum bending stress of 16,000 kPa (2300 lb/in.^{2}) and a maximum deflection (sag) of 2.5 mm (0.1 in.) for pipes filled with water is assumed in the ANSI spacings. The spacings given in Table 4-15 should not be exceeded.

Hanger rod sizes should be

- at least 9.5 mm ($^3/_8$ in.) for pipe 50 mm (2 in.) and smaller;
- at least 13 mm ($^1/_2$ in.) for pipe 63 mm ($2^1/_2$ in.) and larger.

Design hanger rods for a maximum working stress of 20 to 40% of the yield stress at the root area of the threaded ends (see Table 4-14). If the hangers are embedded in or supported by concrete, make sure the embedment is as strong as the threaded rod or limit the allowable load to the specifications in UBC Section 2624. Capsule anchors containing a two-part resin (such as Emhart Molly Parabond™) are superior to expansion anchors, which use friction to resist being pulled out. Some engineers do not trust glued anchors, however, because good and dependable workmanship is critical. Embedded eye bolts should be welded closed in addition to being welded to the reinforcing bars, as shown in Figure 4-16a and b.

Hanger rod sizing for plastic pipe should be the same as for steel pipe, but the spacing of hangers should be as recommended by the plastic pipe manufacturer. Spacings are typically half those for steel and ductile iron pipe.

In addition to installing supports or hangers for straight runs of pipe, provide additional hangers or supports at

- concentrated loads, such as valves;
- fittings;
- both sides of nonrigid joints; and
- pipe connected to suction and discharge casings of pumps.

A wide variety of standard hangers and supports is available [17]. Combining Equation 4-1 for stress due to bending ($s = Mc/I$) and bending moment for simple spans ($M = wL^2/8$) and rearranging, the equation for calculating pipe support spacings based on stress (for straight pipe runs) becomes

$$\text{Spacing} = L = \sqrt{\frac{8sI}{wc}} \tag{4-3}$$

where L is the spacing in meters (inches), s is the allowable stress in newtons per square meter (pounds

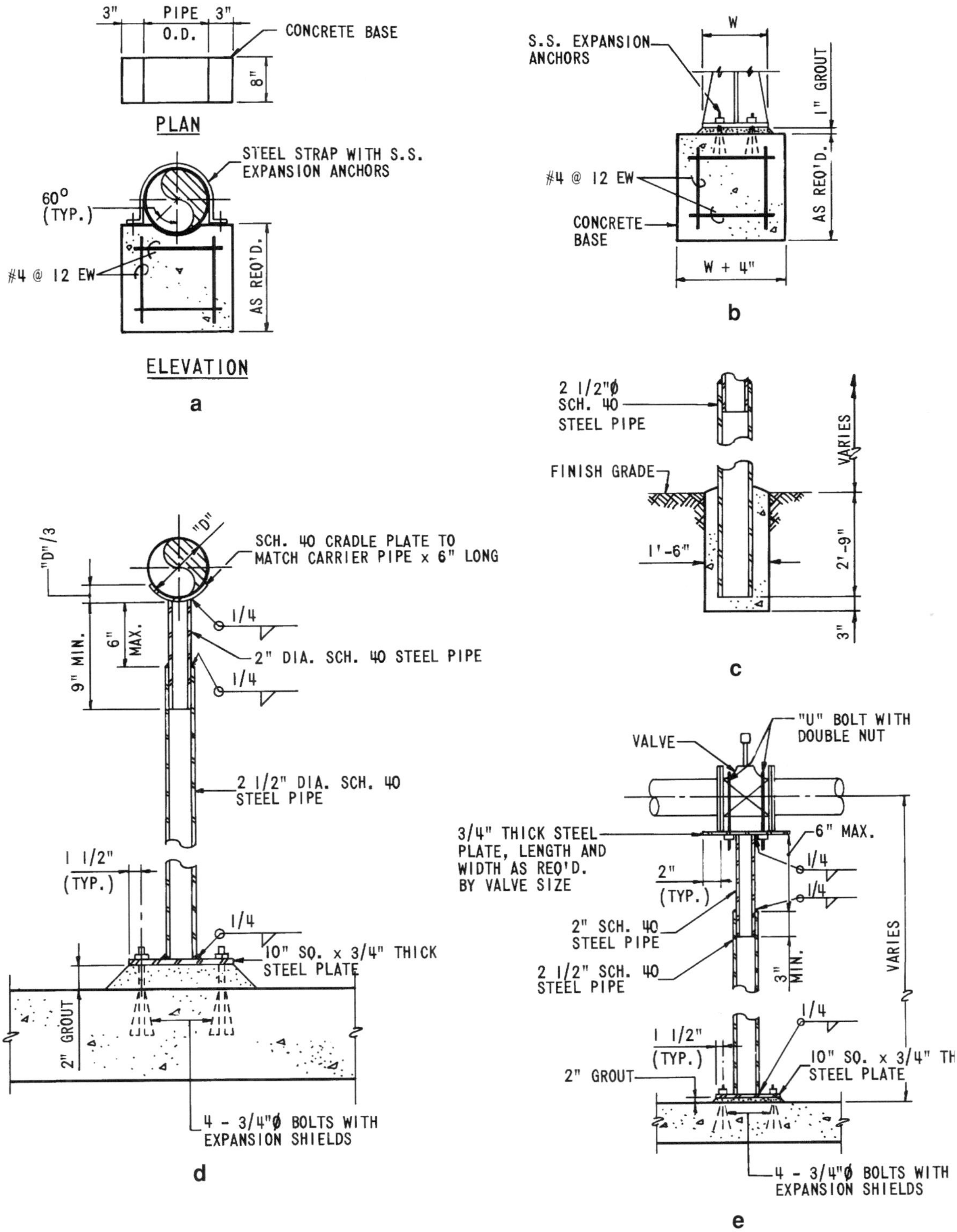

Figure 4-15. Pipe supports. (a) Concrete pipe support; (b) concrete base elbow support; (c) steel pipe support base in soil; (d) steel pipe support on floor; (e) steel valve support on floor. For floor supports, provide similar horizontal supports to resist seismic forces. Courtesy of Wilson & Company, Engineers & Architects.

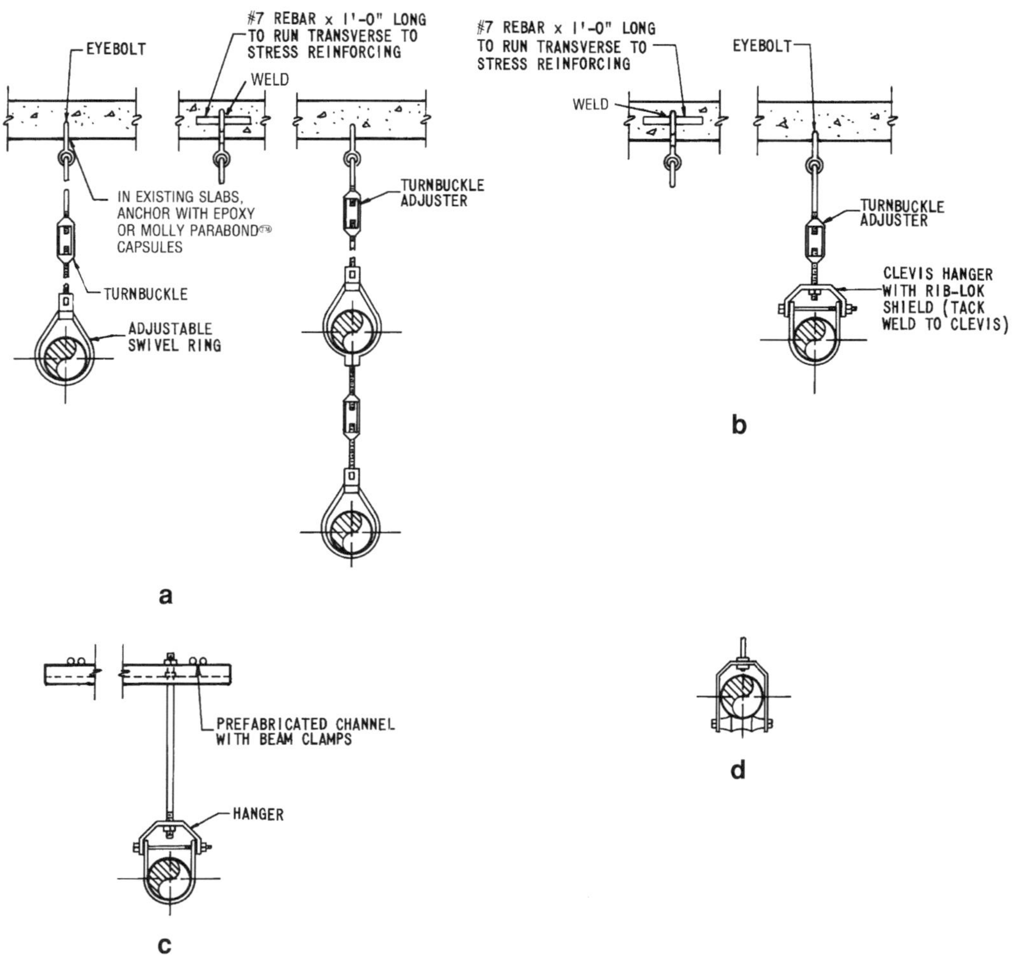

Figure 4-16. Pipe hanger details. (a) Swivel rings (for pipe that requires no protecting shield); (b) clevis hanger [(for flexible pipe, add a shield (e.g., a similar pipe split in half) between the pipe and hanger]; (c) clevis hanger (use for hanging from steel joists); (d) adjustable swivel roller (use for pipe that expands and contracts). Courtesy of Wilson & Company, Engineers & Architects.

Table 4-15. Support Spacing for Iron or Steel Pipe per ANSI B31.1

Nominal pipe size		Span			
		Water service		Steam, gas, or air service	
mm	in.	m	ft	m	ft
25	1	2.1	7	2.7	9
50	2	3.0	10	4.0	13
75	3	3.7	12	4.6	15
100	4	4.3	14	5.2	17
150	6	5.2	17	6.4	21
200	8	5.8	19	7.3	24
300	12	7.0	23	9.1	30
400	16	8.2	27	10.7	35
500	20	9.1	30	11.9	39
600	24	9.8	32	12.8	42

per square inch), I is the moment of inertia in meters to the fourth power (inches to the fourth power), w is the weight of pipe and water per unit length in newtons per meter (pounds per inch), and c is the distance from the neutral axis to the outer fiber (OD/2) in meters (inches). I for pipe is given by

$$I = \frac{\pi[(OD)^4 - (ID)^4]}{64} \tag{4-4}$$

The equation of sag for a uniformly loaded simple span is

$$\delta = \frac{5wL^4}{384EI} \tag{4-5}$$

and rearranging to find L gives

$$L = \left(\frac{384EI\delta}{5w}\right)^{0.25} \tag{4-6}$$

where E, the modulus of elasticity, is given in Table A-10. Be careful to use compatible units.

Example 4-2
Hanger Rod Sizing and Spacing

Problem: A 600-mm (24-in.) steel pipe of standard weight is filled with water. The allowable tensile stress (including the factor of safety) in the hanger rods is 62,000 kPa or 62×10^6 N/m^2 (9000 lb/in.^{2}). Assume that a factor of safety of 1.5 is to be applied to the ANSI maximum bending stress in the pipe, so that stress is limited to 1.07×10^7 N/m^2 (1500 lb/in.^{2}).

Solution: Calculate I from Equation 4-4 (see Table B-3 or B-4 for dimensions and weight of the pipe barrel).

SI Units

$$I = \frac{\pi}{64}[(0.610)^4 - (0.591)^4]$$
$$= 8.08 \times 10^{-4}\ \text{m}^4$$

U.S. Customary Units

$$I = \frac{\pi}{64}[(24)^4 - (23.25)^4]$$
$$= 2070\ \text{in.}^4$$

Calculate w, the total mass of pipe and water in kilograms per meter (pounds per foot) (see Tables B-3 and B-4 for the mass of the pipe barrel and for the inside cross-sectional area).

SI Units

$$w = 141\ \text{kg/m} + 0.273\ \text{m}^2 \times 1000\ \text{kg/m}^3$$
$$= 414\ \text{kg/m}$$

U.S. Customary Units

$$w = 94.6\ \text{lb/ft} + 2.94\ \text{ft}^2 \times 62.4\ \text{lb/ft}^3$$
$$= 278\ \text{lb/ft} = 23.2\ \text{lb/in.}$$

Transforming mass to force,

$$w = 414\ \text{kg/m}\ \frac{9.81\ \text{N}}{\text{kg}} = 4060\ \text{N/m}$$

From Equation 4-3 the maximum spacing is

SI Units

$$L = \sqrt{\frac{8 \times 1.07 \times 10^7\ \text{N/m}^2 \times 8.08 \times 10^{-4}\ \text{m}^4}{4060\ \text{N/m} \times 0.305\ \text{m}}}$$
$$= 7.47\ \text{m}$$

U.S. Customary Units

$$L = \sqrt{\frac{8 \times 1500\ \text{lb/in.}^2 \times 2070\ \text{in.}^4}{23.2\ \text{lb/in.} \times 12\ \text{in.}}}$$
$$= 299\ \text{in.} = 24.9\ \text{ft}$$

From Equation 4-6, the span for the allowable sag of 2.5 mm (0.10 in.) is

SI Units

$$L = \left(\frac{384 \times 2.07 \times 10^{11}\ \text{N/m}^2}{5} \times \frac{8.08 \times 10^{-4}\ \text{m}^4 \times 0.0025\ \text{m}}{4060\ \text{N/m}}\right)^{0.25}$$
$$= 9.35\ \text{m}$$

U.S. Customary Units

$$L = \left(\frac{384 \times 2.9 \times 10^7\ \text{lb/in.}^2}{5} \times \frac{2070\ \text{in}^4 \times 0.2\ \text{in.}}{23.3\ \text{lb/in.}}\right)^{0.25}$$
$$= 379\ \text{in.} = 31.5\ \text{ft}$$

So stress, not deflection, governs. Although the allowable span from Table 4-15 is 9.8 m (32 ft) based on an allowable stress of 15, 900 kPa (2300 lb/in.^{2}), use a lesser spacing, about 7.3 m (24 ft), for an allowable stress of 10, 400 kPa (1500 lb/in.^{2}) in this problem. The force per hanger is

$$F = 4060 \text{ N/m} \times 7.3 \text{ m} = 30{,}000 \text{ N} \qquad F = 23.2 \text{ lb/in.} \times 12 \text{ in./ft} \times 24 \text{ ft} = 6700 \text{ lb}$$

and the required area at the root of threads is

$$A = \frac{30{,}000 \text{ N}}{62 \times 10^6 \text{ N/m}^2} = 4.8 \times 10^{-4} \text{ m}^2 = 480 \text{ mm}^2 \qquad A = \frac{6700 \text{ lb}}{9000 \text{ lb/in.}^2} = 0.74 \text{ in.}^2$$

From Table 4-14, use rods 31 mm ($1^1/_4$ in.) in diameter. However, if the rods are to be anchored only by embedment in concrete, note that UBC Section 2624 allows a maximum load of only 14,000 N (3200 lb) per bolt. That requirement would permit a spacing of only

$$L = \frac{14{,}000 \text{ N}}{4060 \text{ N/m}} = 3.45 \text{ m} \qquad L = \frac{3200 \text{ lb}}{278 \text{ lb/ft}} = 11.5 \text{ ft}$$

Buried Piping

External Loads

Buried pipes must support external structural loads, including the weight of the soil above the pipe plus any superimposed wheel loads due to vehicles if the pipeline crosses a runway, railway, or roadway. The two broad categories for external structural design are rigid and flexible pipe. Rigid pipe supports external loads because of the strength of the pipe itself. Flexible pipe distributes the external loads to surrounding soil and/or pipeline bedding material. Consider DIP, steel, and PVC to be flexible, whereas AC and RCPP are rigid. (FRP is also rigid, but refer to Section 4-5 for a warning about buried service.)

Supporting strengths for flexible conduits are generally given as loads required to produce a deflection expressed as a percentage of the diameter. Ductile iron pipe may be designed for deflections up to 3% of the pipe diameter according to ANSI A21.50. Until recently, plastic pipe manufacturers generally agreed that deflections up to 5% of the diameter were acceptable. Some manufacturers now suggest that deflections up to 7% of the diameter are permissible. Many engineers, however, believe these values are much too liberal and use 2 to 3% for design.

Pipeline bedding conditions affect the safe supporting strength of both rigid and flexible conduits. Screenings, silt, or other fine materials are unsuitable for stable pipeline bedding and should be avoided particularly where the groundwater level may rise above the trench bottom. Better systems range from (1) merely shaping the trench bottom to (2) using select bedding material to (3) supporting the pipe on a monolithically poured concrete cradle. A stable, granular bedding material can be achieved with a well-graded crushed stone with a maximum particle size of 3/4 in. and containing not less than 95% by weight of material retained on the No. 8 sieve.

The design of pipe to resist external loads is involved because it depends on the stiffness of the pipe, the width and depth of trench, the kind of bedding, the kind of soil, and the size of pipe. Discussions are given in AWWA M11 [9], AWWA M9 [13], the DIPRA handbook [11], by Spangler [14, 15], and in many other publications.

Thrust Blocks

Where changes in flow direction occur, the exposed pipes must be restrained against the resultant thrust. Changes in flow direction occur at all bends, tees, plugs, caps, and crosses. Joint systems, such as flanges, welds, and grooved couplings, are designed to provide restraint up to the manufacturer's rating. But if the joint system is inadequate to contain the calculated thrust, external restraint is needed. Thrust calculations are illustrated in Example 3-5.

Buried pipes with mechanical or push-on joints require thrust blocking at deflections, bends, tees, plugs, or other changes in flow direction. Thrust blocks are constructed of cast-in-place concrete poured in the trench during pipeline installation. They are designed to act as horizontal spread footings that distribute the resultant force to the trench wall. The required area of the thrust block can be determined

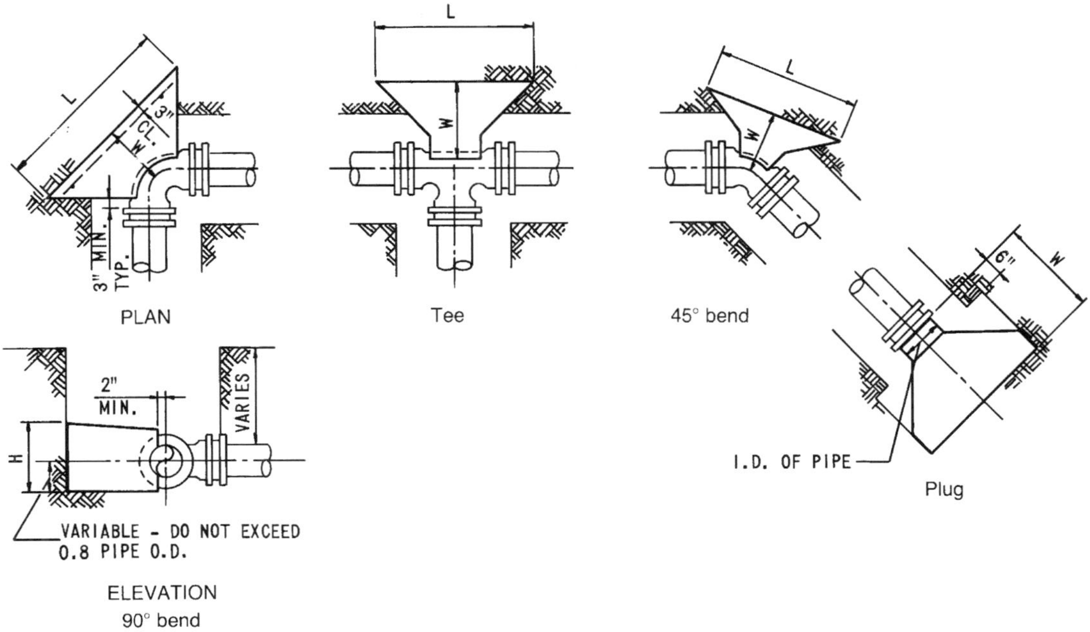

Figure 4-17. Typical underground thrust block details. Courtesy of Wilson & Company, Engineers & Architects.

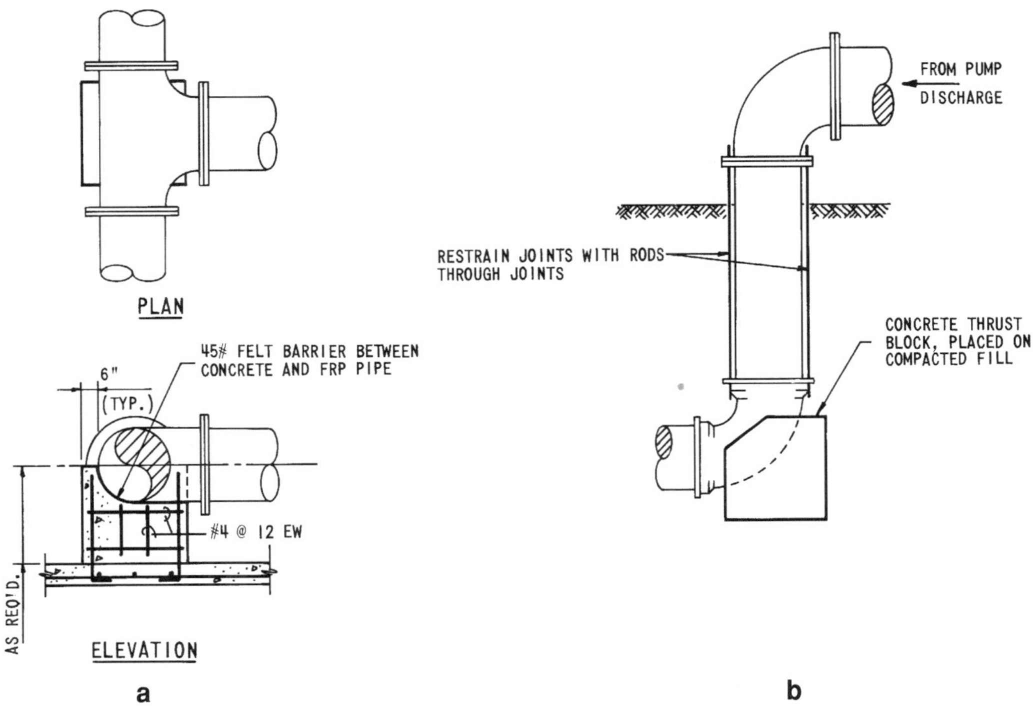

Figure 4-18. Typical thrust blocks for exposed pipes. (a) Combination thrust block pipe support; (b) thrust block/pipe support for emerging pipe. Courtesy of Wilson & Company, Engineers & Architects.

from the resultant force acting on the bend (see Example 3-5) and the allowable soil bearing pressure. For the resultant force, use whichever of the following is greater: (1) the total (working plus water hammer) pressure or (2) the pipeline test pressure. The allowable horizontal bearing pressure can best be found by calculating the Rankine passive pressure from the principles of soil mechanics, or it can be found in tables [9] (see also DIPRA [11] or UBC Table 29-B).

The concrete in thrust blocks should have a minimum compressive strength of 13,800 kPa (2000 lb/in.^{2}) at 28 days. The bearing area should be poured directly against undisturbed earth. Figures 4-17 and 4-18 show typical thrust block details.

The use of thrust blocks should be considered with great care. They are only as good as the stability of the soils used for reaction backing. In locations where the soils may be disturbed by future excavations (yard piping, treatment plant sites, busy streets, etc.), reliance on thrust blocks (particularly for large diameters and high-pressure pipeline systems) is not a very good idea. Instead, the use of restrained (lugged or harnessed) joints and trench friction is a better approach.

Cleanouts

Cleanouts should be installed in sludge and slurry lines that carry grease, grit, or other substances (such as lime) that may form deposits in the pipe (typical designs are shown in Figures 4-19 and 4-20). Lime-carrying waters are especially troublesome, so use troughs instead of pipe where possible.

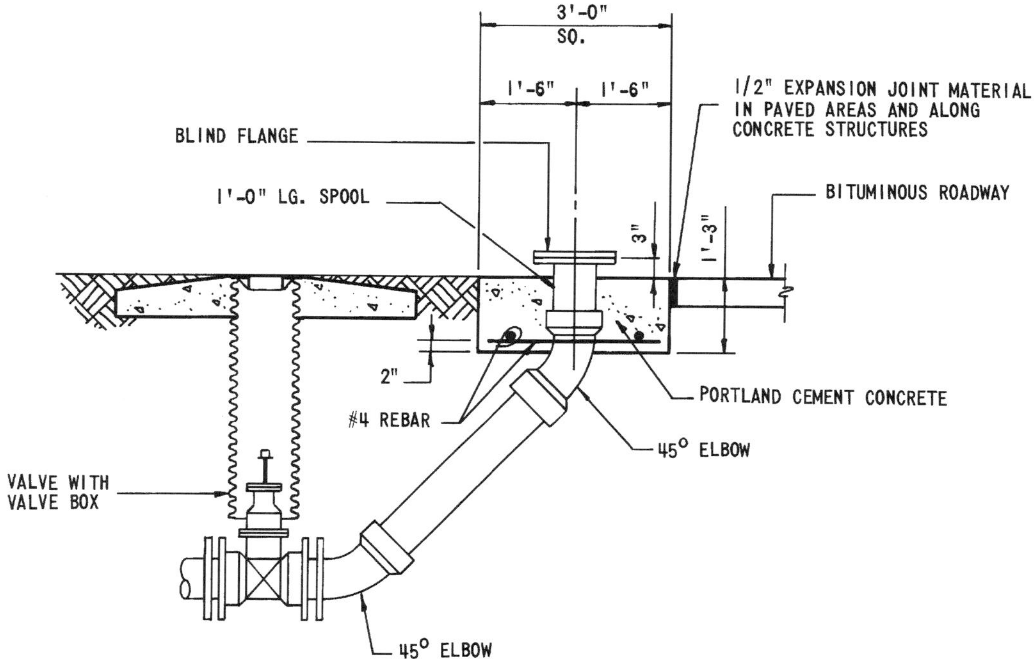

Figure 4-19. Underground pipe cleanout. Courtesy of Wilson & Company, Engineers & Architects.

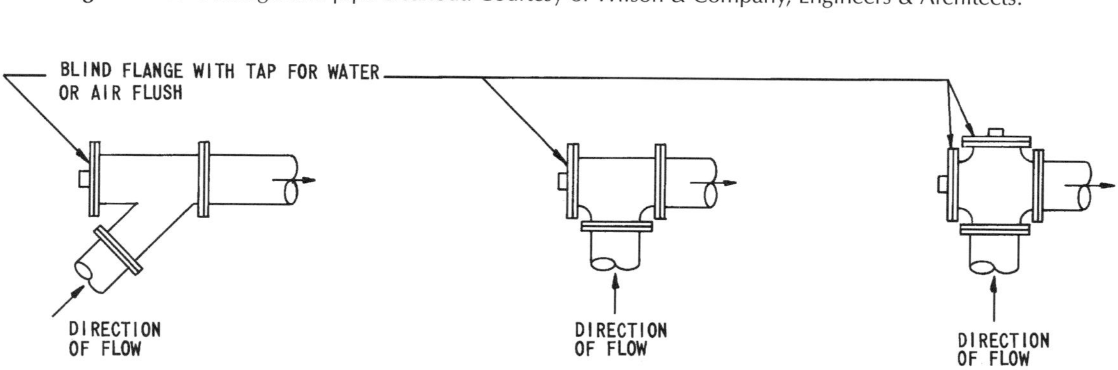

Figure 4-20. In-line cleanouts for exposed pipe. Courtesy of Wilson & Company, Engineers & Architects.

Special cleanouts are required for flowmeters, and a suitable system is shown in Figure 20-12.

Pig Launching and Recovery

Typical pig launching and recovery stations are shown in Figures 4-21 and 4-22. The pig stop, Figure 4-23, is placed in the pig retrieval pipe prior to the pigging operation. The $2^1/_2$ D barrel length is standard, but if space is tight, $1^1/_2$ D can be used (although not with deZurik pigs). Pressure gauges are unnecessary. The best way to determine whether the pig has gone is to isolate the launch barrel and insert a rod through the launching valve quick-connect. At the retrieval end, listen for the sound of the pig arriving.

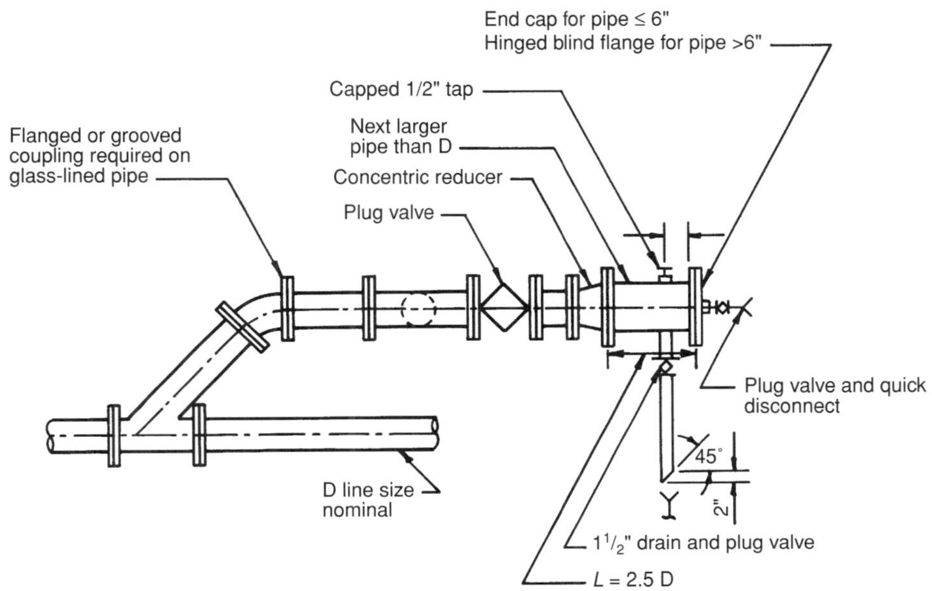

Figure 4-21. Typical pig launching station. Courtesy of Brown and Caldwell Consultants.

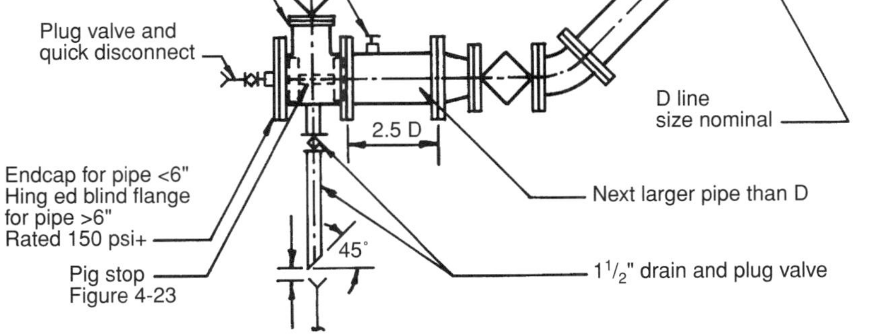

Figure 4-22. Typical pig retrieval station. Courtesy of Brown and Caldwell Consultants.

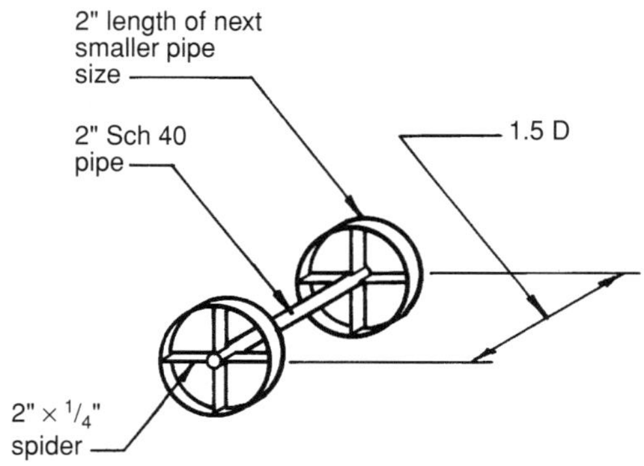

Figure 4-23. Typical pig stop. Courtesy of Brown and Caldwell Consultants.

4-9. Special Piping and Plumbing

The requirements for conveying small flows of wash-down water, fuel, cooling water, and sump drainage are entirely different from the foregoing. Pipes for fuel, wash water, seal water, and air are small and comparatively inexpensive even when constructed of materials such as copper or stainless steel. Special piping materials are listed in Table 4-16.

Water

If water is needed for lavatories or wash-down, galvanized steel with threaded connections is practical because the pipe would rarely exceed 38 mm ($1^1/_2$ in.) in diameter. Pipe fittings are usually made of malleable iron. Thread compound or Teflon™ tape should be applied to threads. There should be enough strategically placed unions to allow for dismantling and replacement. Threaded pipe must have a wall thickness no less than standard weight per ANSI B36.10.

Copper tubing, with joints soldered with tin-antimony (ASTM B 32, Grade SbS), is also practical. Because there is evidence that lead can leach into the water, avoid lead-based solder. Copper, however, is corroded by hydrogen sulfide. So copper or brass piping and tubing should be used with caution in areas of wastewater pumping stations exposed to a hydrogen sulfide atmosphere. Copper tubing also should not be allowed to come in contact with prefabricated wood trusses that have been treated with ammonium sulfate fire-retardant material, because the ammonium sulfate is very corrosive to copper. Require that the copper tubing be attached to wood trusses by pipe hangers, and not just attached directly to or laid directly on the wood.

PVC could be used in Schedule 40 for solvent-welded joints or Schedule 80 for threaded joints, but it requires closely spaced supports. Since galvanized steel pipe within corrosive atmospheres, such as wet wells, corrodes quickly, a more satisfactory material is stainless steel or PVC.

Stainless-steel pipe in most pumping station services is covered by ASTM A 312 for 3 through 750 mm ($^1/_8$ through 30 in.) and A 778 for 75 through 1200 mm (3 through 48 in.) for material; for dimensions, it is covered by ANSI B36.19 (not ANSI B36.10). Note that the minimum wall thickness for threaded pipe is Schedule 40S. Fittings 50 or 75 mm (2 or 3 in.) and smaller should be threaded and should conform with ASTM A 403 and ANSI B 16.3 (see the previous discussion on the applicability of threaded joints). Fittings larger than 75 mm (3 in.) should be butt welded, grooved end, or flanged and should conform to ASTM A 403 or A 774, and ANSI B16.9.

The wall thicknesses for the pipe schedules for stainless-steel pipe are as follows:

- Schedule 40S for stainless-steel pipe is the same as Schedule 40 for carbon steel pipe in sizes of 10 in. and smaller.
- Schedule 40S for stainless-steel pipe is the same as standard weight ($^3/_8$-in. wall) for carbon steel pipe in sizes of 12 in. and larger.
- Schedule 80S for stainless-steel pipe is the same as Schedule 80 for carbon steel pipe in sizes of 8 in. and smaller.

Note that the letter *S* always follows the Schedule number for stainless steel, for example, 40S.

Threaded Joints

Threaded malleable iron fittings and couplings (ANSI B16.3) can be used with steel pipe and are usually cheaper than forged steel fittings conforming to ANSI B16.11. Threading conforms to ANSI B1.20.1 for both malleable iron and forged steel fittings.

Forged steel fittings (90° and 45° elbows, tees, crosses, couplings, caps, plugs, and bushings) conforming to ANSI B16.11 are available in sizes of 3 through 100 mm ($^1/_8$ in. through 4 in.) Assembling and disassembling threaded fittings and joints larger than 50 to 75 mm (2 to 3 in.) is labor intensive, so only small pipes are connected by threaded fittings.

Unions at strategic locations are needed for disassembly. Steel unions conform to MSS SP-83; malleable iron unions conform to ANSI B16.39.

Table 4-16. Piping Materials and Standards

Abbreviation	Piping material	Standard
ABDI	Abrasion-resistant ductile iron	None
ABS	Acrylonitrile-butadiene-styrene	ASTM D 1527, D 2661, D 2751, D 2680, F 409, F 545
ACP	Asbestos cement pipe	AWWA C400, C402; ASTM C 296
ALUM	Aluminum pipe	ASTM B 241, B 361
ARCI	Acid-resistant cast iron	ASTM A 518
BR	Brass	ASTM B 43
CASST	Carpenter 20 stainless steel	ASTM B 464, B 474
CCP	Concrete cylinder pipe (pretensioned)	AWWA C303
CISP	Cast-iron soil pipe	CISPI 301; ASTM A 74, A 888
CML	Cement-mortar-lined steel pipe	AWWA C200, C205, C208; ASTM A 53, A 134, A 135, A 139, A 234, A 283, A 570, A 572
CMP	Corrugated metal pipe	AASHTO M36
CPE	Corrugated polyethylene	ASTM F 405, F 667, ASTO M252
CPVC	Chlorinated polyvinyl	ASTM D 1784, F 437, F 439, F 441, F 493
CTEL	Coal-tar enamel-lined steel	AWWA C200, C203; ASTM A 53, A 134, A 135, A 139, A 234
CTXL&C	Coal-tar epoxy-lined and -coated steel pipe	AWWA C200, C210; ASTM A 53, A 120, A 134, A 135, A 139, A 234
CU	Copper	ASTM B42 75, B 88
DIP	Ductile iron pipe	AWWA C150, C151, C1 10, C111, C115; ASTM A 395, A 436
DWV	Drain, waste, and vent copper tubing	ASTM B 306
FRP	Fiberglass-reinforced plastic pipe	ASTM D 2310, D 2992, D 2996, D 2997; PS-15-69; AWWA C950
FXL&C	Fusion-bonded epoxy-lined and -coated steel pipe	AWWA C213; ASTM A 53, A 120, A 134, A 135, A 139, A 324
GALVS	Galvanized steel pipe	ASTM A 53
HDPE	High-density polyethylene	AWWA C906
PCCP	Prestressed concrete-cylinder pipe	AWWA C301
PP	Polypropylene	ASTM D 2146, D 4101
PPLS	Polypropylene-lined steel pipe	ASTM F 492
PTFELS	Polytetrafluoroethylene- (Teflon™) lined steel pipe	ASTM F 423
PVC	Polyvinyl chloride (pressure) pipe	ASTM D 1784, D 1785, D 2464, D 2466, D 2467, D 2564; AWWA C900, C905
PVC (G)	Polyvinyl chloride (gravity) pipe	ASTM D 2688, D 3034
PVDF	Polyvinylidene fluoride (KYNAR™) pipe	None
PVDFLS	PVDF-lined steel pipe	ASTM F 491
RCP	Reinforced concrete pipe (gravity)	ASTM C 76, C 679
RCPP	Reinforced concrete pressure pipe	AWWA C300, C302; ASTM C 361
RPMP	Reinforced plastic mortar pipe	ASTM D 3517, D 3262; AWWA C950
SP	Steel pipe	ASTM A 53, A 134, A 135, A 139, A 324, A 283, A 570, A 572, A 795; AWWA C200
SST	Stainless-steel pipe	ASTM A 240, A 312, A 403, A 774, A 778
VCP	Vitrified clay pipe	ASTM C 301, C 700
WSP	Welded steel pipe	AWWA C200, C208
YOL	Yolloy™	ASTM A 714

Van Stone Flanges

Van Stone flanges are especially economical for stainless-steel piping. The flange can be carbon steel or ductile iron because it is not in contact with the liquid and, hence, does not have to be corrosion resistant unless the concern is for external corrosion. The joint is made by roll flaring the end of the pipe. A slip-on flange can then be placed over the pipe to make the connection. Such flanges are especially attractive with stainless-steel pipe, but where flanged joints would be exposed to corrosive conditions (as in a wastewater pumping station wet well), specify stainless-steel flanges.

Diesel Fuel Service

Both the Uniform Fire Code and the Standard Fire Prevention Code require that steel piping for flammable and combustible liquids have a wall thickness determined in accordance with ANSI B31.3 and B31.4 with a minimum wall thickness of standard weight. ANSI B31.1, B31.3, and B31.4 also contain design guidelines for flammable fluid piping, which are summarized as follows:

- *Prohibited pipe:* (1) furnace butt-welded steel, (2) cast iron, (3) copper, (4) brass, (5) aluminum, and (6) thermoplastic (aboveground).
- *Prohibited connections:* (1) cast, malleable, or wrought iron threaded couplings, (2) cast-iron flanges, unless integral with cast-iron valves, pressure vessels, and other equipment.
- *Required:* (1) double welding for slip-on flanges and (2) flat facing connecting steel flanges to cast-iron flanges.
- *Recommended materials:* (1) seamless steel such as ASTM A 53 Type S, (2) welded steel with straight seam conforming to ASTM A 53 Type E, ASTM A 134, ASTM A 155, or (3) API SL electric resistance or double-submerged arc welded with allowable stresses no higher than for ASTM A 53 Type S with an appropriate longitudinal joint efficiency factor included.
- *Recommended:* (1) welded joints between steel components where practicable, (2) where bolted flanged joints are necessary, gaskets suitable for the service, (3) where threaded joints are necessary, at least Schedule 80 (extra strong) pipe with extreme care in assembly to assure leak tightness, (4) steel or ductile iron for valves, fittings, and other piping components for systems within plants or buildings that contain equipment with open flame or parts that operate at temperatures over 260°C (500°F), and (5) black steel—never galvanized steel, because the zinc contaminates the fuel and the galvanizing sometimes flakes off and can clog fuel metering orifices.

The UFC Section 79. 305 requires that underground piping incorporate swivel joints where the piping leaves the dispensing location and just before connecting to the tank fittings. Swivel joints incorporate ball bearings and permit rotation and movement from one to many degrees of freedom.

Sewers

Cast-iron soil pipe (CISP) 50 through 375 mm (2 through 15 in.) and larger is suitable for use as drainpipe inside buildings if it is installed above the floor with no-hub ends and with neoprene sealing sleeves and Type 301 or 303 stainless-steel clamps (in accordance with CISPI 301). For buried service or service under slabs and buildings, use hub-and-spigot ends (see also ASTM A 74 standards). In lieu of CISP, ABS pipe is frequently used; for drainpipe smaller than 50 mm (2 in.) use steel, copper, or PVC.

Dry Chlorine Gas

As recommended in ANSI B31 and by the Chlorine Institute [18], make the piping arrangements as simple as possible. Keep the flanged or screwed joints to a minimum. Slope the pipe to allow drainage, avoid low spots, allow for expansion due to temperature changes, and be sure that the pipe is well supported.

Details of materials and construction can be obtained from Pamphlet No. 6 of the Chlorine Institute [18]. Other publications may also be applicable to pumping stations [19–25]. After assembly and pressure testing, chlorine gas piping must be thoroughly cleaned and *all* moisture must be removed.

Chlorine Solutions

Chlorine solutions can be carried in PVC piping. All fittings should be plastic, glass, or Hastelloy™. Most metals, including 316 stainless steel, corrode rapidly in concentrated chlorine solutions.

Air

Use stainless steel or copper. Do not use PVC because heat destroys plastic pipe. Do not use galvanized steel because moisture in the air corrodes it, and although dryers can be added at the compressor, maintenance cannot always be ensured.

Design of Plumbing Systems

Plumbing work in a pumping station usually includes roof drainage, toilet fixtures, floor drains, and sump pumps in addition to the necessary water, waste, and vent piping and a water heater.

Storm Drainage

The sizing of roof drains, horizontal conductors, and vertical leaders or downspouts (usually covered in the applicable plumbing code) is based on roof area and historical rainfall intensity. Storm drainage, including roof and area drains, catch basins, and foundation drains, should (1) connect to a storm drain, (2) discharge through a trap into a combined sewer, or (3) be drained to grade if neither (1) nor (2) is available. In addition to any sump receiving subgrade sanitary waste, a separate sump may be required by code for foundation drainage.

Roof drainage piping often conveys water at a temperature below the dew-point temperature of air in the building, so condensation tends to occur on the pipe. The horizontal conductors (at least) should be insulated to prevent dripping from such condensation, especially if the pipes pass over or near electrical equipment.

The materials usually used for storm drainage piping include galvanized steel pipe, galvanized cast-iron drainage fittings, cast-iron soil pipe and fittings, and polyvinyl chloride plastic pipe and fittings. Both storm and sanitary drain lines buried below floors should be encased in concrete when the floor is poured to protect them from corrosion and settlement.

Sanitary Drainage

Sanitary soil, waste, and vent piping sizes and arrangements must conform to the applicable plumbing code. Equipment drains receiving pump seal water can be enlarged hub drains raised above the floor or regular floor-mounted drains fitted with a funnel strainer. Cleanouts should be located on both sanitary and storm drainage lines at 15-m (50-ft) intervals to permit easy cleaning.

Cross Connections

Protection of the potable water supply against contamination is vital. Reduced-pressure principle backflow preventers, or an equivalent approved means of protection, must be installed in the branch pipe that supplies equipment connections, hose bibbs, or hose valves. Backflow preventers should be located above the flood level with a proper air gap in their drain connections. They must be tested at least annually to ensure safe operation, and the care of backflow preventers should be included in the O&M manual.

Some codes do not permit the installation of backflow preventers, but require potable water to be discharged through an air break into a tank, from which it may be pumped to potentially contaminating uses. Interior hose bibbs should be conveniently located for easy wash-down of the station, and exterior hose bibbs should be frostproof in cold climates.

Sumps

To prevent frequent pump starts, sumps for subgrade drainage should be sized to hold at least two to three times the flowrate capacity of the installed sump pump below the lowest inlet. Install duplex pumps and a sump high-level alarm if damage to other equipment could follow failure of a single pump. Each pump should discharge individually through a check valve and a gate valve before joining in a common discharge riser to the gravity flow waste line. If the piping system of a sump pump can be subjected to freezing temperatures, it should either (1) be designed to be self-draining or (2) be protected by heat tracing to prevent the formation of ice.

So that occasional solids can be passed, sump pumps should be of the semiopen impeller type without an inlet screen. Submersible pumps have the advantage of being inherently floodproof, provided their control panel is located above the flood level.

Mercury-type float switches are recommended in the sump. They are arranged (1) to alternate the operation of duplex pumps, (2) to start the second pump if the first does not handle the load for any reason, and (3) to energize local and remote alarms if the design HWL is exceeded. The pumps should be capable of continuous operation. Specify a manual "test" position on pump selector switches to permit periodic manual tests of pump operation, and state the recommended frequency of testing in the O&M manual. The pumps and their wiring, controls, and alarms must be explosionproof if open to a space classified as hazardous by Articles 500 to 502 of the NEC.

4-10. References

Anyone responsible for designing or selecting piping should have a minimum library of reference materials that includes design handbooks, standard codes and specifications, and one or more manufacturer's catalogs for each of the piping materials in the chapter. Manufacturers' catalogs often contain most of the needed design methods and useful excerpts from standard specifications.

Because some of these publications are revised occasionally, obtain the latest edition. Codes, standards, and specifications are listed in Appendix E. The addresses of publishers are given in Appendix F.

1. Crocker, S., M. L. Nayyar, and R. C. King, *Piping Handbook,* 6th ed., McGraw-Hill, New York (1992).
2. Deutsch, D. J., *Process Piping Systems,* Chemical Engineering, McGraw-Hill, New York (1980).
3. Langelier, W. F., "The analytical control of anticorrosion water treatment, " *Journal of the American Water Works Association,* **28,** 1500–1521 (October 1936).
4. Merrill, D. T., and R. L. Sanks, "Corrosion control by deposition of $CaCO_3$ films. A practical approach for plant operators," *Journal of the American Water Works Association,* Part I, **69,** 592–599 (November 1977); Part II, **69,** 634-640 (December 1977); Part III, **70,** 12–18 (January 1978).
5. Merrill, D. T., and R. L. Sanks, *Corrosion Control by Deposition of $CaCO_3$ Films. A Handbook of Practical Application and Instruction.* American Water Works Association, Denver, CO (1978).
6. Pisigan, R. A., Jr., and J. E. Singley, "Calculating the pH of calcium carbonate saturation," *Journal of the American Water Works Association,* **77,** 83–91 (October 1985).
7. Pisigan, R. A., Jr., and J. E. Singley, "Effects of water quality parameters on the corrosion of galvanized steel," *Journal of the American Water Works Association,* **77,** 76–82 (November 1985).
8. Pigg, B. J., "Asbestos cement pipe," *ASTM Standardization News,* pp. 56–57 (April 1988).
9. AWWA M11, *Steel Pipe—A Guide for Design and Installation,* 2nd ed., American Water Works Association, Denver, CO (1985).
10. Sanks, R. L., *Water Treatment Plant Design for the Practicing Engineer,* Ann Arbor Science Publishers, Ann Arbor, MI (1978).
11. DIPRA, *Handbook of Ductile Iron Pipe,* 6th ed., Ductile Iron Pipe Research Association, Birmingham, AL (1984).
12. *Handbook of PVC Pipe, Design and Construction,* Uni-Bell PVC Pipe Association, Dallas, TX (1991; updated frequently).
13. AWWA M9, *Concrete Pressure Pipe,* American Water Works Association, Denver, CO (1979).
14. Spangler, M. G., *Soil Engineering,* 2nd ed., International Textbook Company, Scranton, PA (1960).
15. Spangler, M. G., "Underground conduits—an appraisal of modern research" (with discussions), *Transactions of the American Society of Civil Engineers,* **113,** 316–374 (1948).
16. *Piping Engineering,* Tube Turns Division, Louisville, KY (1974).
17. *MSS, Pipe Hangers and Supports, Selection and Application,* Manufacturers Standardization Society, Inc., of the Valve and Fittings Industry, Arlington, VA (1976).
18. "Piping systems for dry chlorine," 13th ed., Pamphlet 6, The Chlorine Institute, Inc., Washington, DC (April 1993).
19. *The Chlorine Manual,* 5th ed., The Chlorine Institute, Inc., Washington, DC (1986).
20. "Non-refrigerated liquid chlorine storage," 5th ed., Pamphlet 5, The Chlorine Institute, Inc., Washington, DC (October 1992).
21. "Chlorine vaporizing systems," 4th ed., Pamphlet 9, The Chlorine Institute, Inc., Washington, DC (November 1994).
22. "Chlorine pipelines," 3rd ed., Pamphlet 60, The Chlorine Institute, Inc., Washington, DC (April 1990).
23. "Estimating the area affected by chlorine releases," 2nd ed., Pamphlet 74, The Chlorine Institute, Inc., Washington, DC (February 1991).
24. "Refrigerated liquid chlorine storage," 2nd ed., Pamphlet 78, The Chlorine Institute, Inc., Washington, DC (June 1994).
25. White, G. C., *Handbook of Chlorination,* 2nd ed., Van Nostrand Reinhold, New York (1986).

4-11. Supplementary Reading

1. Steele, A., *Engineering Plumbing Design,* Miramar, Los Angeles, CA (1977).
2. Lindsey, F. R., *Pipefitters Handbook,* 3rd ed., Industrial Press, New York (1967).
3. *Procedures for Pipewelding,* National Association of Plumbing, Heating, Cooling Contractors, Washington, DC (1983).
4. Singley, J. E., *et al., Corrosion Prevention and Control in Water Treatment and Supply Systems,* Noyes, Park Ridge, NJ (1985).

Chapter 5

Valves

CARL N. ANDERSON
BAYARD E. BOSSERMAN II
CHARLES D. MORRIS

CONTRIBUTORS
Casi Cadrecha
Joseph E. Lescovich
Harvey W. Taylor

Engineers typically lavish much attention on pumps but little on valves, which are just as important for the proper functioning of a pumping station. The discussion of valves and actuators in this chapter applies mainly to the control of the pumped fluid. Small valves for auxiliary purposes (e.g., seal water, fuel, and plumbing) are only briefly mentioned.

Most valves in a pumping station are for isolation service and, as such, are either open or closed. Actuators are usually manual for valves smaller than 600 mm (24 in.), and power-driven actuators are usually used for valves larger than 900 mm (36 in.). Check valves respond to flow direction and open and close automatically. Pump control valves serve a dual function as check valves, and the powered actuators are programmed to open and close slowly enough to control transient pipeline pressures within acceptable limits. If used at all, control valves are the most important valves in a pumping station. Flow-control valves (or valves that modulate to control flow or pressure) are used in small sizes for cooling-water or seal-water piping. Pressure-control valves are sometimes used in distribution systems to separate regions of two different pressures. In pumping stations, surge relief or surge anticipation valves arc occasionally used to relieve high-pressure surges.

The associated design considerations of cavitation, noise, actuator sizing, and vibration are specific for the brand and model of the valve used and, hence, are not discussed here. Designers should be aware, however, of the problems and explore them thoroughly with the manufacturer.

Additional information of value can be found in the literature and in the following references: Cook [1], O'Keefe [2], Deutsch *et al.* [3], AWWA M11 [4], the *ISA Handbook of Control Valves* [5], Lyons [6], and others [7–11]. Photographs and drawings that depict the various valves and show how they work are so readily obtainable from manufacturers that few are reproduced here.

References to a specification or standard are given in abbreviated form (such as ANSI B16.34) because such designations are sufficient for identification. The titles of references are given in Appendix E together with other standards that may not be referenced but, nevertheless, aid in the selection and specification of valves. Addresses of publishers are given in Appendix F. The dangers in referencing a standard without carefully reading the entire work are discussed in Section 1-4.

5-1. Designing for Quality

Choosing the right kind, style, and even make of valve in the right situation is vital to the proper functioning of the station. A valve proper for one installation may be improper for another. Style (and even model and maker) has a profound effect on satisfactory service. The problem of selection is complicated by the following considerations: (1) a valve satisfactory in one location may not be satisfactory in another location even if conditions are only slightly different; (2) makers of several styles of valves may make some good ones and some poor ones; (3) models are changed from time to time and a valve, once poor, may now be good; and, finally, (4) it is extremely difficult to write specifications to comply with the law, allow competitive bidding, and still obtain a satisfactory valve.

Familiarity with the various manufactured products, which is the key to good selection, can be achieved by

- Interviewing many manufacturers' representatives (but with critical skepticism), and
- Discussing valves with expert consultants and with users—the operators and the utility managers.

Note that many makes or models of valves look alike but differ significantly in quality. A valve is probably of high quality if a competitor agrees. The best valves are expensive, so a misplaced emphasis on low initial cost makes procurement of satisfactory valves difficult at best. Valves are the heart of the hydraulic system; if they fail, the system fails. In the long run, a cheap valve will have proved to be the most expensive. Skimping on valves is the wrong way to try to save money.

Good quality can be obtained by incorporating into the specifications such items or criteria as listed below.

- *Materials:* Abrasion-, corrosion-, and cavitation-resistant materials of construction—especially for seats (see Table 5-1).

Table 5-1. Typical Valve Seat Materials

Type specification	Life	Remarks
Resilient seats		
Buna N	Good	General-purpose elastomer for water and wastewater. Economical, suitable for most water and sewage uses.
Leather	Good	Usually impregnated with various waxes to improve qualities. Sometimes used for water, not sewage.
UHMW[a]	Very good	Very abrasion- and chemical-resistant; not expensive.
Teflon™	Poor	Impervious to chemical attack, creeps too much for normal use; expensive.
Viton™	Poor	Use only for aggressive liquids or high temperatures, creeps somewhat.
Natural rubber	Good	Suitable only for fresh water.
Rigid seats		
Bronze		
ASTM B 62, B 584 (34 alloys), B 16, B 371	Tremendous variation among alloys	Most common seat material, least resistant to erosion or corrosion.
Stellite		
SAE, J775, AMS 5373, 5375, 5378, 5380, 5385, 5387, 5788	Excellent	Expensive; best of all for resistance to both corrosion and erosion.

Stainless steel	Specification	Erosion[b]	Corrosion[b]	Remarks
440C	ASTM A276, alloy S44004	1	5	Most resistant to erosion, highest hardness.
420	ASTM A276 alloy S42000	2	5	
17-4 PH	ASTM A564, alloy S17400	3	3	
410	ASTM A276, alloy S41000	4	4	
405	ASTM A276, alloy S40500	5	4	
316	ASTM A276, alloy S31600	6	1	Highest resistance to corrosion.
304	ASTM A276, alloy S30400	7	2	Least resistant to erosion.

[a] Ultra-high-molecular-weight polyethylene.

[b] 1 signifies the best resistance.

- *Headloss:* Specify a price penalty based on life-cycle energy costs for more headloss than a stated value.
- *Proof-of-design tests:* Require certification of successful completion of proof-of-design testing conducted on a 6-in. (or larger) valve in accordance with AWWA C504, Section 5.5, altered as necessary to apply to the valve specified (e.g., "disc" in the standard means "plug" in a specification for a plug valve).
- *Massiveness of construction and conservatively designed bearings, shafts, and other moving parts:* Shafts, especially in cushioned-swing check valves, should be very large; compare the various makes and models to specify a high-quality product.
- *Service records:* Find a way to specify features that eliminate valves with poor records.
- *Complexity:* Specify valves and actuators that are simple, trouble free, and require minimum maintenance or the kind of maintenance within the capability of the workforce.
- *Resilient seat material that will not cold flow under differential pressures:* Look for well-designed mechanisms to retain seats in place (see Table 5-1); for wastewater and sludge, seat materials must be resistant to oils and solvents.
- *Responsibility:* Include a clause that involves the manufacturer in the responsibility for valve (and valve actuator system) performance. Warning: In some instances, manufacturers have inserted additional rubber shims or seals to the seats during some tests to meet the C504 testing requirement that the valve be drop-tight in both directions. Such practices should not be accepted.

Life-Cycle Cost

Quality might be said to be an inverse function of life-cycle cost, which is a combination of capital, maintenance, and energy costs. Headlosses can result in mind-boggling energy costs, as demonstrated in Example 5-1.

Example 5-1
Energy Penalties for Three Valves

Problem: Compare cone, butterfly, and globe valves for life-cycle costs of energy. Assume (1) electric power at \$0.05/kW · h, (2) a flow velocity of 3.05 m/s (10 ft/s), (3) headlosses based on the *K* factors of Table B-7, (4) a wire-to-water power efficiency of 75% for the pump, (5) interest at 8%, (6) a life of 20 yr, and (7) 300-mm (12-in.) valves wide open.

Solution: Calculate the annual cost of electric power, using headloss data from Table B-7, Equation 10-7 for the relation between head, power, pump efficiency, and flowrate, and Equation 29-4 for the present worth of an annual expenditure. From the results, it is evident that headloss is an important cost factor.

Valve	Headloss m	Headloss ft	Energy cost (in dollars) Annual	Energy cost (in dollars) Present worth
Cone	0.02	0.06	24	235
Butterfly	0.15	0.5	211	2,070
Globe	2.3	7.5	3,010	29,600

Location

Quality is a continuing concern and, hence, also a function of maintenance. Maintenance costs are elusive, but, whatever they are, they can be reduced by placing valves and actuators in locations that are easily accessible for servicing. Make it convenient to isolate and drain separate parts of the system, and write maintenance and exercising programs into the O&M manual. Note that, because of savings in parts and operation and maintenance labor, it may be more cost effective to install an expensive valve (such as a plug or a valve of corrosion-resistant construction) that works when needed than to install a cheap valve (such as a gate or a valve of lesser quality) that may not work until repaired.

High quality is achieved as much by good location and good piping layout as by specifying high-quality hardware. A good valve is cheapened by misapplication or poor location. Some considerations for proper location are

- In all but clean water service, install valves (such as gate or swing check) with bonnets upright so that (1) the bonnet cannot become clogged with debris that can render the valve inoperative and (2) the removal of the bonnet is easier and maintenance is less expensive (this means valves should not be located in a vertical pipe).
- Make sure that the connecting piping is large enough for projecting butterfly valve discs to rotate.
- Place a spool (at least two or, better, three pipe diameters long) between a butterfly valve and an elbow to prevent the diagonal streamlines from causing the vane to flutter, excessively wearing the bearings or even locking the vane; vane shafts should be horizontal so there is no bottom bearing to collect grit.
- Place all valves in locations such that operation, maintenance, and repairs can be comfortably accomplished.
- Locate sleeve, grooved-end, or flexible couplings nearby so that (1) valves can be easily removed for replacement or for factory repair and (2) movements and strain in the piping are not carried through the valve body (unless the valve body is designed to resist such loads).

Check Valves

In comparison with choosing other valves, the judicious selection of the right valve for check service is by far the most difficult and frustrating. Because there are many kinds and makes of check valves, choose the type and style only after careful study of Sections 5-4 and 7-1 and the literature of various manufacturers. Location is particularly important for check valves; for example, check valves on manifolds usually behave quite differently from isolated check valves. Check valves should, if possible, be placed no closer than four or five pipe diameters from a pump; otherwise turbulence from the pump tends to cause the valve disc to flutter and wear the bearings. (Flutter, however, is of less concern if the spring in a spring-loaded lever is stiff enough to prevent slam.) Furthermore, such a separation of pump and check valve allows the discharge pressure gauge to be placed farther from the pump where pressure fluctuation is much less violent, gauge wear is reduced (this is not a concern if the gauge piping contains a normally closed spring-loaded shutoff valve), and accuracy is greatly improved. The cost of the several fixed pressure gauges needed for several pumps can be greatly reduced by using a pair (one for suction, one for discharge) of portable gauges.

Safety

No automatic control system for valves is complete if it does not incorporate safety features to prevent damage from malfunctioning equipment. These safety features can be arranged to prevent the operation of an unprimed pump, to prevent the operation of a pump against a closed control valve that does not open on schedule, or to limit excessive pumping when discharging to a broken pipe. Most safety features include pressure-sensing devices and relays with timers that operate valves only when pressures are normal. When abnormal conditions occur, the valve should remain closed (or should be closed if open), and pumps should stop (after a suitable delay) to prevent damaging pumps and motors.

5-2. Isolation Valves

Isolation valves are either fully closed or fully opened. Valves that remain in one position for extended periods become difficult—even impossible—to operate unless they are "exercised" from time to time. Valves should be exercised at least once each year (more often if the water is corrosive or dirty), and the required exercise routine should be emphasized in the O&M manual.

Isolation Valves for Water Service

Isolation valves likely to be used for water service include

- Ball (these are expensive, so in sizes larger than, say, 100 mm [4 in.] they are used rarely for isolation only)
- Butterfly (these are popular in all sizes)
- Cone (these are also expensive, so in sizes larger than, say, 300 mm [12 in.] they are also used rarely for isolation only)
- Eccentric plug (these are excellent and useful in all sizes)
- Gate (these are popular in all sizes)

- Plug (these are either lubricated or nonlubricated, both of which are rarely used for only isolation in sizes larger than approximately 50 mm [2 in.]; the lubricated version is frequently used if the valve is to be closed for extended periods of time).

The gate valves most likely to be used are double disc or, for raw water with grit, resilient seat, but solid wedge, knife, or even sluice gates may be useful in some circumstances. The plug valve most likely to be used is the nonlubricated type with either a rectangular or a round port, but a lubricated plug valve would be used for higher pressures. Lubricants approved by the FDA are available for water service. Globe valves are not normally used (because of their high headloss) except in piping 50 mm (2 in.) and smaller in which an ability to regulate flow is desired. For clean water, the double disc gate and butterfly valves are the most frequently used. The more expensive ball or cone valves are used for flow control, pump control, or powered check service, usually in conjunction with another valve for isolation so that the control or check valve can be repaired.

Isolation Valves for Wastewater Service

Valves for wastewater service are more limited in type because stringy materials catch onto and build up on any obstructions to the flow and sticky grit tends to lodge in any pocket, such as a valve seat. Depending on the quality of treated wastewater, however, valves for water service might be used, albeit with some risk.

Isolation valves likely to be used for wastewater service include

- Ball (in unusual circumstances only)
- Cone (in unusual circumstances only)
- Gate (popular in all sizes)
- Eccentric plug (popular in all sizes).

The specific styles most often used are eccentric plug, knife gate, and resilient seat gate. Solid wedge gate valves with nonresilient seats might be used, but grit that can collect in the seat is often troublesome. Lubricated and nonlubricated plug valves and ball valves are used, especially if flow control is needed.

Description of Isolation Valves

The following descriptions are for both water and wastewater valves. The types of valves that are recommended or could be used for isolation service are given in Table 5-2.

Ball Valves

The rotor (round plug) in a ball valve rotates 90° from fully closed to fully open. A ball valve has very low

Table 5-2. Recommendations for Use of Isolation Valves

	Service usage[a]						
	Water		Wastewater				
Type of valve	Raw	Clean	Raw	Treated	Slurry	Gas	Fuel oil
Angle	G	G	X	G	X	G	G
Ball	E	E	E	E	E	E	E
Butterfly	G	G	X	G	X	—	—
Cone	E	E	E	E	—	—	—
Diaphragm	—	—	—	G	G	—	—
Gate							
Double disc	G	E	X–F	G	X	G	G
Knife	F–G	F–G	F–G	F–G	F	X	X
Sluice	G	G	G	G	—	—	—
Resilient seat	G	G	F–G	F–G	F	X	X
Solid wedge	G	G	F	G	X	G	G
Globe	X	G	X	F	X	G	G
Pinch	G	G	G	G	G	G	G
Plug							
Eccentric	E	E	E	E	F	—	—
Lubricated	G	G	G	G	G	G	G
Nonlubricated	G	G	G	G	—	F	F

[a] E, excellent; G, good; F, fair; X, do not use; —, use is unlikely.

headloss in the fully open position because the bore of the pipe is carried straight through the ball, which results in headloss nearly the same as in a straight piece of pipe of the same laying length as the valve. Ball valves are of two basic types: (1) seat supported, usually for valves smaller than 150 mm (6 in.) and (2) trunnion supported, usually for valves 150 mm (6 in.) in size or larger as described in AWWA C507.

Ball valves in water, wastewater, and sludge pumping stations are not ordinarily used as isolation valves. Their laying length, weight, and cost are much greater than those of gate and butterfly valves. Seat-supported ball valves are widely used in auxiliary piping in such services as seal water, fuel oil, and natural gas as well as in isolating pressure gauges and air and vacuum valves because such piping is smaller than 75 mm (3 in.). An advantage of a ball valve is that it offers a relatively leak-free seal.

Ball valves for severe duty service (wastewater, storm water, surge control, and pump control) are usually of the trunnion type and should be selected with a great deal of care, especially with respect to materials for seats and bearings, because ball valves vary widely in quality. Seats are subjected to wear and tear from grit and tramp iron (nuts, bolts, scraps) in wastewater and storm water service, particularly in check and surge control usage. Bearings and shafts must be designed to withstand unbalanced forces during rapid closure on pump failure and during surge control episodes. Ball valves are often reserved for severe service conditions, and some engineers have experienced difficulties with cold-flow of resilient seat materials under high differential pressures. In some seat designs or under some operating conditions, those materials tend to escape the mechanism intended to retain them in the plug or body castings. Stainless steel and stainless steel against monel metal are excellent alternatives for seats in such service—although these are not a universal panacea, and problems have been reported with metal-seated ball valves as well.

There are two kinds of mechanisms for operating ball valves. In most valves, a shaft allows the ball to rotate about a fixed axis so that the seat is wiped by the turning of the ball. In some metal-seated valves (e.g., Figure 5-1a), a loose-fitted trunnion allows pressure from the force main to push the ball against its seat. When the pump starts against the closed valve, the pressure moves the ball away from its seat and allows the valve to open freely. If the valve is closed before the pump is stopped, the action is reversed, and when the pump stops, the manifold pressure again seats the ball tightly. In effect, the action is like that of a cone valve but without its complex mechanism. The seat is thus never wiped except on power failure.

An alternative mechanism for accomplishing the same purpose (avoiding wiping the seat when the valve is opened or closed) is a movable retainer ring for holding resilient seat material. The inside surface of the ring is always exposed to the pressure in the force main. When the pump is started and its pressure exceeds force main pressure, the retainer ring retracts away from the seat and allows the valve to open freely. When the valve is closed and the pump subsequently stopped, pressure from the force main closes the seat and makes it driptight. The retainer ring mechanism can be adjusted for a wide range of differential pressures. Valves in service for 20 yr show no signs of seat wear.

Designers should carefully investigate the performance history of the ball valve under consideration and be satisfied that the valve will be satisfactory under the conditions that will prevail in service. In addition to considerations of cost and trouble-free operating life under imposed conditions, the cost, difficulty, down time, and probable frequency of repairs should be weighed. Some valves can be repaired in place, whereas others must be removed and dismantled for repairs.

Trunnion-supported ball valves should be installed with the shaft horizontal and should be located in horizontal—not vertical—pipes.

A ball valve for pump, check, and surge control in water or wastewater service is usually fitted with a worm gear and compound lever (or pantographic) operating mechanism, as shown in Figure 5-1. The operating mechanisms may be fitted with a variety of actuators. Such a valve is superior to all others in providing ideal opening and closing characteristics for minimizing surges caused by pump start-up and shutdown or loss of power. As shown in Figure 5-2, the last 10% of Cv (a measure of flow) requires about 50% of the stem travel, so the last portion of flow is choked off very slowly. The parameter Cv is more precisely defined by Equations 5-1 and 5-2.

In SI units

$$Cv = 2.919\frac{d^2}{\sqrt{K}} \qquad (5\text{-}1a)$$

where d is the diameter of the valve (approach pipe, actually) in meters and K is the dimensionless valve coefficient in Table B-7. At any differential pressure across the valve

$$Q = 0.3807 Cv\sqrt{\Delta P} \qquad (5\text{-}2a)$$

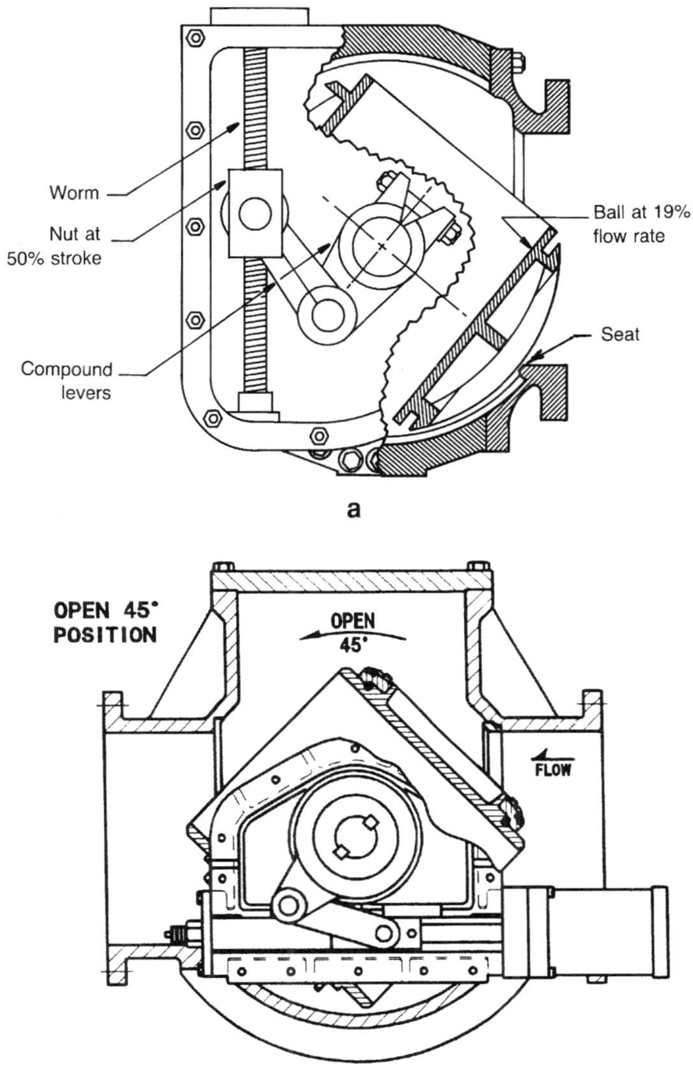

Figure 5-1. Ball valves with link and lever motion. (a) A metal-seated valve. Courtesy of APCO/Willamette Valve, Inc. (b) A resilient-seated valve. Note that removing the cover allows the seal ring to be replaced with the valve in situ. Courtesy of GA Industries, Inc.

where Q is in cubic meters per second and ΔP is differential pressure in kilopascals.

In U.S. customary units

$$Cv = 29.85\frac{d^2}{\sqrt{K}} \tag{5-1b}$$

where d is the diameter of the valve in inches and K is the valve coefficient in Table B-7. Flow through the valve is given by

$$Q = Cv\sqrt{\Delta P} \tag{5-2b}$$

where Q is in gallons per minute and ΔP is in pounds per square inch. One can think of Cv as the gallons per minute of water at 60°F that flow through the valve at a pressure difference of 1 lb/in.^{2} An inconsequential correction for density at other temperatures is omitted from Equation 5-2 because the correction applicable in pumping stations would never reach 2%. As the valve is closed, Cv is reduced, as indicated in Figure 5-2.

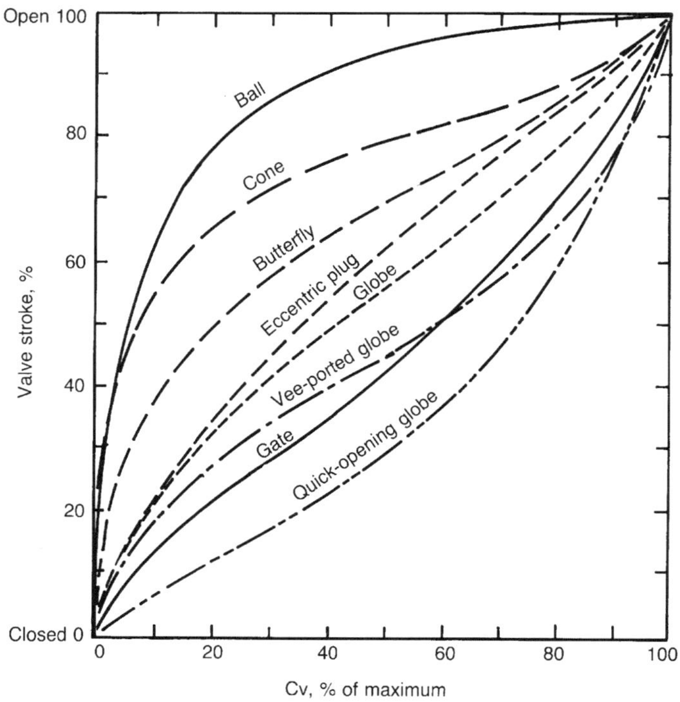

Figure 5-2. Stroke versus *Cv* for various valves. After GA Industries, Inc., DeZurik, A Unit of General Signal Corp., and Willamette Valve, Inc.

Butterfly Valves

A butterfly valve (Figure 5-3) is a quarter-turn valve in which a disc (or rectangular vane for square channels) is rotated on a shaft so that the disc seats on a ring in the valve body. The seat usually is an elastomer bonded or fastened either to the vane or to the body.

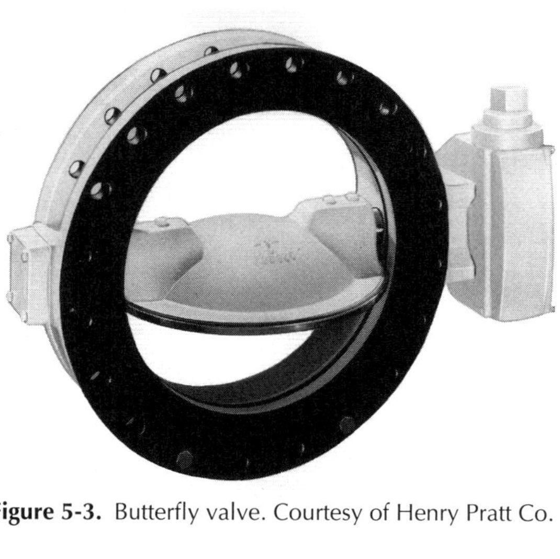

Figure 5-3. Butterfly valve. Courtesy of Henry Pratt Co.

Most, if not all, manufacturers have now standardized on the short body style (see AWWA C504). In the long body style, the vane is contained entirely within the body when the vane is in the fully open position. In the short body style, the vane protrudes into the adjacent piping when in the open position.

The wafer style, which is very thin, requires installation of the valve between pipeline flanges bolted across the valve body. When planning the use of short body or wafer-style butterfly valves, make sure the pipe on either side is large enough to accept the vane. Wafer valves are not recommended for isolating purposes where it may be necessary to remove the adjacent, connecting pipe spools. Some consultants refuse to use wafer valves under any circumstances.

Butterfly valves are used in both isolation and limited throttling service. Butterfly valves can be designed for leakproof shut-off, but leakage is significant without a resilient seat. Solids, wear, and scale buildup cause leakage even with resilient seats. In throttling service, the control range for the vane angle is about 15 or 20° to 60 or 70°. Because most butterfly valve designs are entirely unsuitable for throttling and some are prone to failure, select only valves recommended by the manufacturer for throttling. Then be suspicious. Investigate installations of valves proposed for throttling service. Pay particular attention to the seat design. The AWWA C504 standards alone do not ensure butterfly valve seats that are adequate for severe throttling (as in pump-control valves) where seats must be very rugged for longevity. Some valves have their rubber seats vulcanized into the body. Such valves are reliable, but when the seat needs to be replaced, the valve must be shipped to the manufacturer for repairs. Other valves have seats that are intended to be replaceable in the field, but after several years of service, corrosion may make it impossible to unscrew studs or nuts, and the valve must be sent to the shop for repairs anyway. So beware of sales claims concerning "easy replacement" in the field. Some (but by no means all) replaceable body seats are satisfactory.

Butterfly valves are used most frequently in water and air service. They are not suitable for sewage, sludge, or grit service because the disc collects stringy solids and the seating edge is easily abraded. When used in a "dirty" service, such as raw water, install the valves with the vane axles horizontal to prevent grit from settling in the shaft bearings and oriented so that the bottom part of the vane moves in the direction of flow to wash solids through the valve when it first opens.

Butterfly valves should be located at least three (and preferably five) pipe diameters from an upstream bend and at least two pipe diameters from a downstream bend to place the valves in regions of approximately symmetrical velocities and coaxial streamlines. Valves that are closer to bends may chatter, and if they are very close, excessive forces may be required to operate them. The excessive forces can be eliminated by mounting valve shafts in the plane of the bend, but it is best to orient shafts horizontally wherever possible.

Cone Valves

Cone valves, sometimes called "lift-action" plugs, are essentially plug valves that are positioned by first lifting the plug in the body before the operating mechanism moves the plug to its new position (see Figure 5-4). After reaching the new position, the plug is then reseated to provide a seal. The valve has excellent characteristics for surge-control and pump-control service. Cone valves have been used successfully in large water and wastewater applications. Compared with a ball valve, the axial motion in a cone valve reduces operating forces and seat wear and produces a tighter shutoff. However, because of the unusual stroking requirements, cone valve operating mechanisms require skilled maintenance personnel and can be complex and prone to failure.

A variant of the cone valve is the small taper-ground cock in which the plug is axially forced into tight contact by a spring. There is no mechanical lift, but a slight axial movement occurs anyway as the valve is turned. Cocks are useful for the isolation of taps, pressure gauges, and sampling ports.

Diaphragm Valves (Not Stem Guided)

A diaphragm valve contains an internal flexible elastomer diaphragm that presses down against the interior body wall (or, sometimes, against an internal weir

Figure 5-4. Cone valve. Courtesy of Allis-Chalmers Corp.

that is part of the body). The diaphragm seals the valve body from the stem, so the valve is leakproof and the stem requires no packing. This type of valve is useful in sludge and grit service because there is little obstruction through the body that could collect grit or solids. Cleaning tools such as pigs, however, cannot pass through them. These valves can be lined with various plastics such as PVC or polyethylene and are, therefore, often used in chemical piping (e.g., chlorine solution piping). They are not often encountered in raw sewage, water, and sludge pumping stations because other types of valves are cheaper. Sometimes a diaphragm valve or its close relative, the pinch valve, is used as a safety device on a bypass pipe around a positive displacement pump to prevent destruction of the pump or main pipeline if a downstream valve were to be mistakenly closed. The diaphragm or pinch valve is set to open automatically, either by a preloaded spring or by air pressure, at a suitable overpressure. The lack of a visible position indicator is a shortcoming.

Another type of diaphragm valve is a stem-guided globe valve with a diaphragm in the bonnet to actuate the valve (see Section 5-5). The two types are in no way similar and should not be confused.

Eccentric Plug Valves

In an eccentric plug valve (Figure 5-5), both the body and the plug seat are offset from the center of rotation so that when the valve is opened the plug-seat surface rotates away from the body-seat surface; this movement minimizes the scraping and deterioration of seats common in other valves. In the open position, at a quarter-turn from closed position, the plug rests against the side of the valve body. In some makes of valves, the port is either rectangular or, at least, not full ported. In others, the port is circular and the same size as the pipe so that a pig can pass through the valve. Some styles or makes are difficult to service in the field. Plug valves are especially attractive in wastewater and sludge applications because there is nothing to become jammed or clogged with solids. The plug is usually coated with an elastomer, such as neoprene, to obtain a resilient seating.

Eccentric plug valves used in sewage and sludge service, or for any fluid containing solids or grit, should be installed so that the plug rotates about a horizontal axis. The plug should be stored in the top when the valve is in the open position and should seat in the direction opposite the high-pressure side so that the pressure of the water forces the plug against the seat for a tighter seal. Because some designs hold pressure in only one direction, they must be installed to hold against the applicable pressure, which is often opposite to the direction of flow (as, for example, in the isolation valve on the discharge side of the pump). Debris problems are less severe if the body seat is upstream. The valve body may be marked to install the other way, so both the specifications and the O&M manual must clearly state both which way the valve is to be installed and the reason for the orientation.

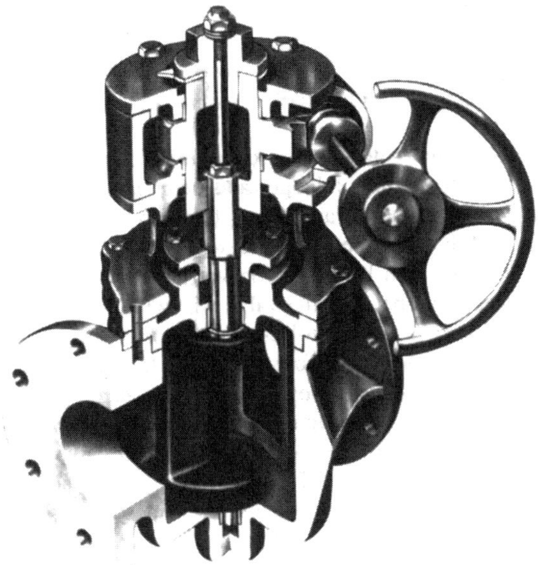

Figure 5-5. Eccentric plug valve. Courtesy of DeZurik, A Unit of General Signal Corp.

When these valves are installed, tighten the flange bolts only enough to stop leaks. Excessive tightening squeezes the gasket material against the plug and causes it to bind. This cautionary note should also be placed in both the specifications and in the O&M manual. Add the warning that ***all*** valves (not just eccentric plug valves) should be exercised on a regular schedule.

Gate Valves

A gate valve has a disc sliding in a bonnet at a right angle to the direction of flow. Gate valves are further divided into several subtypes:

- Double disc (Figure 5-6)
- Solid wedge resilient seated (Figure 5-7) or metal seated
- Knife (Figure 5-8)
- Rising stem (shown in Figure 5-7)
- Outside screw and yoke (OS&Y), which also is often a rising stem design (as shown in Figure 5-7)

- Nonrising stem (NRS) (as shown in Figure 5-6)
- Inside screw and outside stem thread
- Bolted, screwed, or union bonnet

A rising stem (as shown in Figure 5-8) allows the operator or observer to determine easily if the valve is open or closed. An NRS (as shown in Figure 5-6) allows the valve to fit into cramped areas where a rising stem would strike a ceiling or wall. The NRS style should be avoided wherever possible because when it is used there is no indication of whether the disc is fully seated. If a hard object is caught on the seat, workers may assume the gate is closed. Dismantling a pressurized pipe, thought to be isolated, can lead to flooding and even loss of life. Hence, some indication of valve position is desirable for all valves. Furthermore, in the O&M manual, caution workers always to back off nuts by two or three threads and crack the joint to determine whether there is pressure. If there is, the nuts can be retightened until the trouble is corrected.

Gate valves are suitable only for isolation (open/closed) service. They should not be used for throttling service because a relatively small valve opening allows a high flow capacity [1]. Furthermore, gate valves are subject to damaging vibration when partly open.

Double Disc Gate Valve

The double disc gate valve equipped with a rising stem is one of the most popular types for clean water.

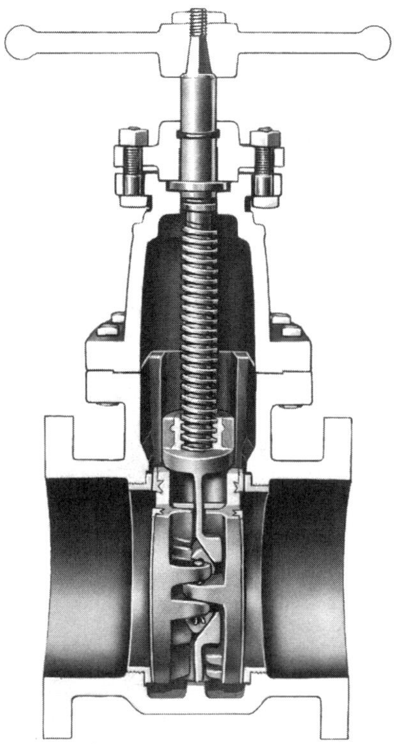

Figure 5-6. Double disc NRS (nonrising stem) gate valve. Courtesy of Mueller®.

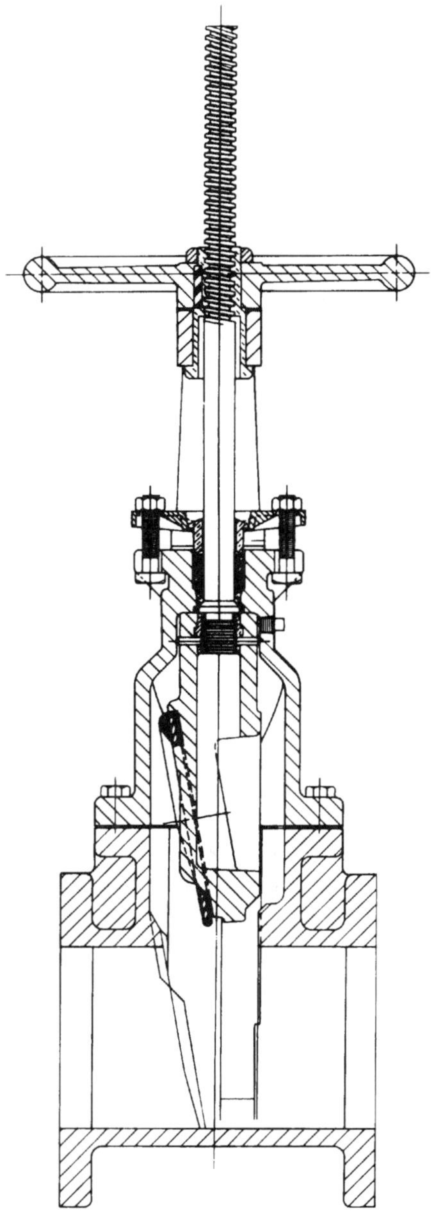

Figure 5-7. Solid wedge resilient seat OS&Y (outside screw & yoke) valve. Courtesy of Mueller®.

Figure 5-8. Knife gate valve. Courtesy of DeZurik, A Unit of General Signal Corp.

After the discs drop into their seats, further movement of the stem wedges the discs outward to produce a leakproof shut-off even at pressures exceeding 1700 kPA (250 lb/in.^{2}). Opening the valve reverses the procedure. Hence, the discs do not slide until the wedging is relaxed, and sliding and grinding between the disc rings and body rings are thus minimized.

This type of valve should not be used if the water carries a heavy load of grit or solids that would fill the seats (pockets) and prevent the discs from first dropping into place. For such service, use resilient seated gate valves instead.

Solid Wedge Gate Valve

Solid wedge gate valves are suitable for water containing grit, solids, sludge, and other matter. These valves offer the advantages of full port opening and limited throttling service, and, strange though it may seem, they can be placed on vertical pipes or even upside down because the valve will work even when the bonnet is full of solids. Confer with the manufacturer, however, before installing the valve on a vertical pipe or in any position other than near upright. Consider, instead, plug or ball valves for such applications. Solid wedge valves are less expensive than the double disc type. The wedge minimizes sliding and scraping as the valve opens, and solids in the pocket can usually be displaced by opening and closing the valve several times. If the grit is sticky and the valve is to remain open a long time, however, consider a resilient seated gate instead.

Resilient Seated Gate Valve

The seat of a gate valve is a pocket that can entrap solids and prevent the valve from closing fully. The resilient seat type greatly reduces this problem because it has no pocket in the body in which the gate seats (see AWWA C509). Instead, the rubber edge of the disc seats directly on the valve body, as shown in Figure 5-7.

Because there is no pocket for the disc at the bottom of the valve to collect grit and debris, the resilient seated gate valve is suitable for grit-laden waters and sewage as well as for clean water service. The disc is encapsulated with a resilient material (usually vulcanized rubber) that presses against the smooth, prismatic body of the valve. The valve is restricted to nearly horizontal pipelines and the bonnet must be oriented up or, at least, diagonally up. These valves can seal tight against working pressures up to 1380 kPa (200 lb/in.^{2}).

Knife Gate Valve

The knife gate valve (Figure 5-8) is lighter and more suitable for water carrying debris than other gate valve types, but it is more difficult to prevent leakage either through the closed valve or through the stem packing.

Unless some leakage is acceptable and the head is low (say, 6 m or 20 ft), another type of valve should be selected. The knife gate is adapted for pressures below approximately 170 to 350 kPa (25 to 50 lb/in.^{2}).

Pinch Valve

A pinch valve is essentially a rubber or elastomer tube closed pneumatically or by a screw, wedge, or lever. The valve is leakproof and requires no packing, but it is not suitable for high pressures and the tube weakens where it is compressed. When closed pneumatically, it is an inexpensive and suitable valve for a bypass pipe around a positive displacement pump for guarding against damage due to a blockage or to an inadvertently closed valve. But it is limited in size, the tube is relatively expensive and difficult to replace, and the laying length is greater than for other valve types.

Plug Valves (Lubricated and Nonlubricated)

A plug valve contains a cylindrical or tapered plug with an opening cast or cut into it. A 90° turn of the plug fully opens or fully closes the valve. Hence, plug valves are considered to be quarter-turn valves. The design of the plug seat is such that solids do not accumulate and cause the plug to jam or bind. Plug valves can be obtained with full-ported plugs or with reduced port areas.

Lubricated plug valves (Figure 5-9) contain a lubricating system in which a nearly solid lubricant is forced into the top of the plug and through a series of grooves into the bottom so that the faces of the plug and seat are wiped with the lubricant, which functions as a deformable sealant each time the valve is open or closed. This feature is attractive in applications where the valve may be fixed in the open or closed position for a long time. The valve is less likely to freeze in position, and, if necessary, a small amount of lubricant can be forced into it to unfreeze it. Many different lubricants, which allow the valve to be used in different fluid services, are available. Lubricant can also be used in slurry service because there are no spaces that can become packed with solids. On the other hand, some fluids can dissolve the lubricant off the plug face, which would cause the valve to seize and gall, but this usually is no problem with the fluids encountered in water, wastewater, and sludge pumping stations. A plug valve provides a very tight seal and is especially useful in service pressures greater than 1000 kPa (150 lb/in.^{2}). In the cylindrical style, excess lubricant is forced outside of the valve body where it can be examined for contamination, less torque is required to turn it, and the rotor has a 100% bore. Both styles are excellent for pump control service. Lubricated plug valves require occasional but simple maintenance.

Figure 5-9. Lubricated plug valve. Courtesy of Rockwell International.

Nonlubricated plug valves have become unpopular, and several former manufacturers no longer make them. They seem to have no advantages over the lubricated types.

5-3. Sluice Gates, Shear Gates, Flap Valves, and Stop Plates

The devices in this section are not actually valves, but they are used as a means of flow isolation in channels and on ends of pipes such as those entering wet wells.

Sluice Gate

Sluice gates (Figure 5-10) are used against a wall between two basins or between a pipeline and an open channel. Sluice gates are usually located where the influent channel or sewer enters the wet well and can be used if the pumping station

- is subject to flooding due to excessive influent flow or pumping failure;
- is critical, that is, if any major emergency repairs must be done as quickly as possible;
- has two wet wells; or
- has submersible pumps and the wet well must be dewatered when (or if) the pump lifting mechanism jams.

Figure 5-10. Sluice gate. Courtesy of Rodney Hunt Co.

Stems and actuators must be sized to overcome large friction forces for the following reasons:

- Breakaway forces are sometimes greater than the manufacturer's typical recommendation, particularly if the gate remains closed for a protracted period.
- Power actuators can develop very large torques and thrusting forces. Therefore, it is prudent to size operating stems in a way that will more than adequately sustain those forces, especially when the fluid contains debris that can be trapped underneath the sluice gate leaf when closing.

Once the gate has been broken loose, the operating forces are reduced greatly.

Shear Gates

Shear gates are mounted on the ends of open pipes. Unlike conventional valves, they cannot be mounted in a pipeline. One side of the shear gate contains a fixed bolt; on the other side is a lever, which is used to lift and rotate the gate about the fixed bolt on the other side.

Flap Valves

Flap valves (see Figure 5-11) for pump discharge are substitutes for check and isolation valves and are an economical, reliable method of preventing backflow through out-of-service pumps. With this type of device, no isolating valve is installed. Rather, the flap valve is installed on the individual pump discharge piping at the point of discharge to the receiving sewer, channel, or discharge structure. A prudent engineer makes some provision (slide gate or bulkhead slots) for isolating the flap valve for maintenance purposes, but no expensive, heavy-duty isolating valves are required. Be sure to provide a vent just upstream from the flap valve to drain the pump discharge and prevent slam. Use flap valves with a cushion design specifically intended for pump discharge service.

Stop Plates

A stop plate is a thin, vertical, rectangular plate used to form a temporary dam in open channel flow. It is

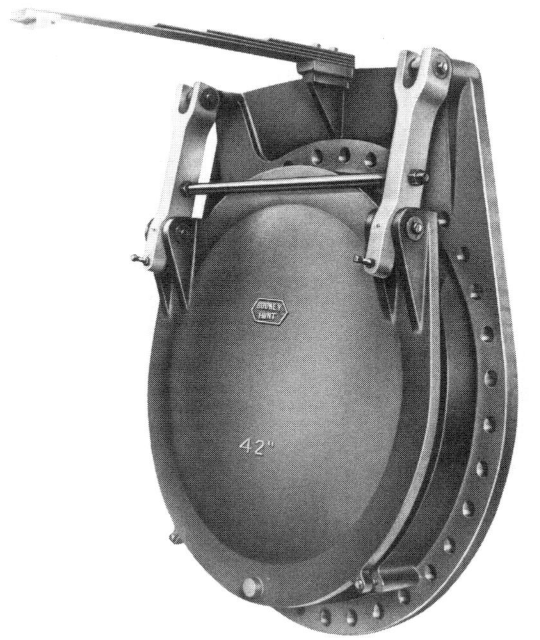

Figure 5-11. Flap valve designed for use with pump discharge. Courtesy of Rodney Hunt Co.

sometimes used in the wet wells of pumping stations to block flow to part of the wet well so that a pump and its suction piping can be dewatered for maintenance. The plate may have its own actuator or may be lifted by hand. Large plates can be lifted with a crane or hoist and stored on a rack or in a pit when not in service.

The plate is usually aluminum, but wood, fiberglass, stainless steel, and other materials are sometimes used. A local fabricating shop can make stop plates if supplied with detailed design information. Alternatively, a somewhat more sophisticated plate can be obtained from manufacturers, which means the engineer need not design such details as reinforcing.

Except for very small units, stop plates cannot be moved up or down when there is a substantial difference (more than about 0.2 m or 6 in.) in water level across the gate. If it is necessary to move the plate under such conditions, (1) a sluice gate may be used instead or (2) a valve (typically a 100- to 200-mm [4- to 8-in.] gate or butterfly) or small stop plate can be mounted in a larger plate to allow equalization of water levels before the larger plate is moved. Stop plates are inexpensive, are simple, are suitable for local fabrication, and take up little of the valuable space in a wet well, but moving them is awkward and the leakage is high.

5-4. Check Valves

A check valve is usually (but not always) required to (1) prevent reverse flow and prevent runaway reverse pump speeds when the pump is shut off, (2) keep the pipeline full of water to prevent the entrance of air, and (3) minimize water hammer and surges for pump start-up and shut-down.

Vertical pipelines are poor locations for check valves if the water contains grit or solids. For vertically placed valves in clean water service, special springs or counterweights may be needed. Manufacturers that state that a check valve can be placed in a vertical pipeline are referring only to the springs or counterweights and ignoring the danger of deposited grit and solids, which can (and will) jam the valve.

The designer's responsibility is the selection of a valve that will give good service in keeping with the pump selection, hydraulics, and size of the system. The first decision is whether a check valve will serve, or whether a more sophisticated pump-control valve is necessary to limit surges. Some insights for this decision are contained in Chapters 6, 7, and 26 as well as in Parmakian [12], but only a sophisticated mathematical model of the system solved by means of a computer can provide a rational analysis. Unfortunately, such modeling is too time consuming and expensive for common use.

Valve Slam

Check valves can be divided into two broad classifications: (1) those that are closed by the static pressure of water above the valve (mechanical checks) and (2) those held shut by an external actuator (pump control or controlled check valves). The latter do not slam, but swing checks do if, before the valve is fully closed, any substantial reverse velocity catches the valve disc and accelerates it until it strikes the body seat abruptly. The sudden stop of disc, lever, and counterweight (if there is one) plus the violent impact of the disc on the body seat (especially if the contact is metal to metal) causes an explosive noise and vibrations that shake the pipe and may shake the whole building. The real problem is the water hammer that results if the water column is flowing backward at a significant velocity when the valve closes. However, valve slam can occur without water hammer and vice versa.

At worst, valve slam can rupture water lines and pump casings. At best, it is annoying. In between, it pounds the system, can overstress pipes and joints, and may well result in eventual leaks and greatly increased maintenance. It is difficult to give advice on the best kinds of valves to specify because valve slam depends on many interrelated factors in addition to valve design. Other factors that are just as important include static head, friction head, the inertia and spin-down characteristics of the impeller and motor, size of pipe, and velocity of flow. Generally, valve slam is caused or aggravated in the following ways.

Low flywheel effect. The principal cause of valve slam is quick deceleration of the pump due to low angular momentum of the impeller, the driver, and the water within the casing. With enough inertia, valve slam can be prevented, but the necessary flywheels may be large and costly.

High proportion of static head. If the headloss is 70% static and 30% dynamic (due to friction), valves slam worse than if the headloss is 50% static and 50% dynamic. A simple vertical lift (e.g., into an adjacent elevated tank) is especially prone to valve slam.

Frequency of valve slam. Valve slam may stress material beyond yield strengths and cause permanent deformations. A few deformations of a given intensity may be acceptable, but numerous deformations eventually cause leakage or rupture. Even a single slam, if severe, is dangerous.

Large pipe diameter. As valve size increases, the resulting time to close it increases, the disc velocity increases, and the energy in the system increases.
Parallel pumps. If two pumps are connected to a header and pump 1 shuts off while pump 2 is operating, pump 2 may cause the short water column between the two pumps to reverse very quickly and, thus, cause the check valve of pump 1 to slam.
Column separation. Water column separation can cause valve slam in two different ways: (1) rapid reversal of flow through the check valve, even though flow in most of the pipeline reverses slowly, and (2) a fast-rising positive pressure surge due to the collapse of a vapor cavity if the surge arrives at the valve when it is not closed fully (see Chapter 6).
Air chambers and surge tanks. These units can prevent column separation, but at the same time they can cause rapid flow reversal at the valve and thus aggravate valve slam (see Chapter 7).
Insufficient closing force. If the closing force due to the disc weight and spring or counterweight is low, the valve operates too slowly. But the closing force should not be so high that the valve does not open fully under steady-state pumping conditions. If the valve is not fully open, the headloss increases and debris is more likely to hang up in the valve. Also, excessive closing force can cause the disc to bounce off the seat so that valve slam recurs, sometimes two or three times.
Constant-speed pumps. Constant-speed pumps have two features that aggravate valve slam: (1) they must be turned on and off at full capacity (unlike variable speed pumps), and (2) as they are turned off, their speed cannot be ramped down gradually. With variable-speed pumps, the speed can be ramped down during normal shutdown (although not when the power fails).
Friction in the hinge pin bearings. Friction is increased by dirt and corrosion. If the disc hesitates before moving, valve slam is almost certain to occur—mostly significant with tilting disc check valves.
Body shape. Details of check-valve design influence the closure operation. Because the disc must open wider in a valve with a straight body than in a valve with a bulbous body (Figure 5-12), the movement upon closing is correspondingly greater and the valve slam may be greater.
Inertia. Closing time increases with inertia of moving parts. A counterweight in a valve without a dashpot may therefore cause valve slam, and replacing such a counterweight with a spring sometimes minimizes slam if the pumps have no significant spin-down time.

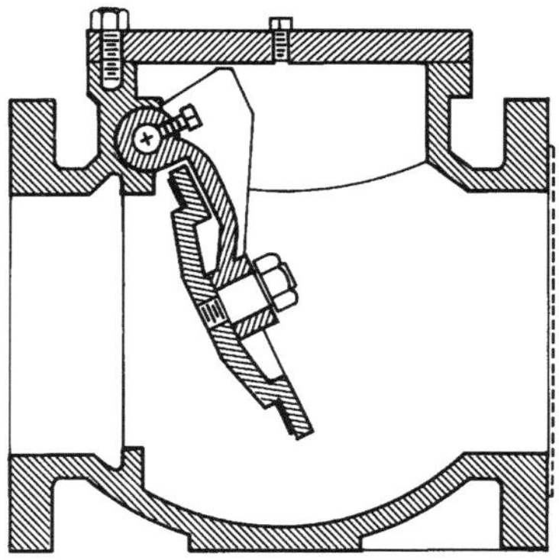

Figure 5-12. A swing check valve at full flow. After GA Industries, Inc.

Preventing Valve Slam

Valve slam can be prevented, or at least kept within bounds, by

- using a valve that closes quickly—before the flow can reverse by adding a heavy counterweight or a stiff spring to the external lever;
- adding a dashpot or buffer to make the disc seat gently but before the flow can reverse; or
- closing the valve with an external actuator so that the water column is gradually brought to rest without a significant increase in pressure.

The first two methods may prevent valve slam but do not necessarily prevent pressure surges.

Small Valves

For small valves, either confine swing checks to pipe less than, say, 250 mm (10 in.) in diameter or precede the valve with a reducer and follow it with an expander. Ordinary swing check valves are manufactured in large sizes, but there is a potential problem in using them in a low friction head system because they cannot close quickly enough.

Spring-Loaded Levers

Many engineers advocate the use of springs instead of counterweights to reduce the inertia of moving parts and thus speed the closure. However, as the valve closes, the spring tension relaxes and the torque on the disc shaft may decrease enough to be insufficient to prevent valve slam, so choose a design in which the combination of spring tension and the lever arm between the spring and shaft creates high torque at closure. A resilient seat aids in minimizing contact noise. These are the least expensive check valves.

Counterweight and Dashpot

A counterweight has the advantage of providing maximum torque on the disc shaft at closure, but it does not close the valve as quickly as a spring because of the inertia of moving parts. Some (especially manufacturers) think the valve should be equipped with either (1) a side-mounted or top-mounted oil-filled dashpot to cushion the movement of the lever at shutoff or (2) a bottom-mounted piston-type shock absorber that engages the disc before it closes. Air-filled dashpots are difficult—occasionally impossible—to adjust to prevent valve slam. The required massiveness of construction needed to resist the high force of water against the disc and the dashpot mechanism makes this valve more expensive, but it is the recommended style when the spring-loaded lever type is inadequate. But note, however, that a heavy counterweight or a stiff spring, properly adjusted, is cushioned by the water in the valve.

Pressure-Regulated Bypass Dump

A spring-loaded, pressure-actuated surge relief valve (Figure 7-8) with a pipeline returning the wasted water to the wet well can reduce the surge to an acceptable, preset level, but it does nothing to mitigate valve slam.

Actuator-Controlled Plug or Ball Valve

An actuator can be programmed to both open and close the valve slowly enough to prevent water hammer (see Figure 7-7). A stored energy system is needed to operate the valve when power failures occur.

Summary

Selecting a proper type of check valve and control mechanism is more art than science. Experience, not analytical theory, is a key consideration. Of course, a simple method to determine whether a conventional swing check valve can be used without excessive valve slam would be desirable, but unfortunately the complexity of the problem precludes a simple, accurate procedure. Complex computer programs can be used to predict with fair accuracy whether valve slam will occur. The cost of the analysis may be discouraging if the system is small, but large systems should always be so analyzed. If any general statement on check-valve selection can be made, it is probably this: use a swing check valve with an outside lever and spring. If that is inadequate, use a valve with a cushioned closure system such as a dashpot or bottom buffer. As a last resort, use a powered actuator. But note that even the experts disagree, and some prefer counterweights to springs.

Check Valves for Water Service

Check valves useful for water service include

- Swing check valves
- Center-post guided (or silent) check valves
- Double leaf (or double door, double disc, or split disc) check valves
- Foot valves
- Ball lift valves
- Tilting (or slanting) disc check valves.

The several styles of swing check valves can be divided into those with and those without an outside lever. Outside levers can be equipped with either springs or counterweights, and the levers can be cushioned or noncushioned. Bottom buffers can be used instead of dashpots affixed to the outside lever.

Check Valves for Wastewater Service

Valves for wastewater service must be capable of passing large solids and, as with isolation valves, must have no obstructions to catch stringy material. Valves likely to be used for wastewater are essentially limited to

- Swing check valves
- Flap valves, which might be used in special circumstances (for example, with combined sewers that contain stormwater and wastewater)
- Ball lift valves, which are useful in positive displacement sludge pumps as the ball can be lifted completely out of the flow path.

The rubber clapper swing check valve has no outside lever, is not fully ported, and, hence, should not be used for raw sewage or sludge.

Description of Check Valves

The following descriptions of check valves for water and wastewater offer some guidance and suggestions. A summary of recommendations for use is given in Table 5-3.

Table 5-3. Recommendations for Use of Check Valves[a]

Type of valve	Water		Wastewater		
	Raw	Clean	Raw	Treated	Sludge
Ball	E	E	E	E	E
Ball lift	F–G	F–G	F–G	F–G	G
Center-post guided	P	F	X	X	X
Double door	X	G	X	X	X
Flap	—	—	G	—	—
Foot	F	G	—	—	—
Swing					
No outside lever	P	P	P	P	P
Outside lever and counterweight	F–G	F–G	F–G	F–G	G
Outside lever and spring	F–G	F–G	F–G	F–G	F
Outside lever and air cushioned	F	F	F	F	P
Outside lever and oil cushioned	G	G	G	G	G
Slanting disc	G	G	X	F	X

[a] E, excellent; G, good; F, fair; P, poor; X, do not use; —, use is unlikely.

Ball Lift Check Valves

A ball lift check valve contains a ball in the flow path within the body. The body contains a short length or guide piece in which the ball moves away from the seat to allow the passage of fluid. Upon reverse fluid flow, the ball rests against an elastomeric seat.

These valves are often encountered in pumping stations in sizes of 50 to 150 mm (2 to 6 in.) or smaller for pump seal water or for wash water supply piping. They have also been successfully used by at least one major pump manufacturer for raw wastewater pump discharge piping up to 600 mm (24 in.). Except for small sizes, bodies are made of ductile iron. The ball is hollow with an external rubber coating resistant to grease and dilute concentrations of petroleum products, acids, and alkalies. The specific gravity of the balls can be adjusted to suit a wide range of operating conditions.

The valves are said to be self-cleaning, rugged, reliable, nonclogging and to be able to withstand repeated cycling, because each time the ball is reseated, a different part of the surface rests on the seat. In larger sizes (100 mm [4 in.] or more) for sludge pumping service, ball lift checks are part of the mechanism in plunger (piston) sludge pumps.

As check valves for wastewater pump discharge piping, the valves have these advantages: (1) the headloss is lower than it is for other types; (2) there are no external penetrations and no leakage to the outside (although good swing check valves properly set up do not leak either); and (3) stringy materials have nothing to wrap around and do not foul the valve. On the other hand, the standard valve (unlike swing check valves) gives no indication of whether water is flowing—a serious disadvantage. Ball lift check valves are, however, available with ball position indicator-proximity switches.

Decisions to use a ball lift check valve instead of, say, the faster-closing swing check valve with a spring-loaded lever should be based on a computer-aided dynamic hydraulic analysis of the system.

Center-Post Guided Check Valves

Center-post guided check valves are low cost and are called "silent check valves" by some manufacturers. They close more rapidly than any other check valve. As shown in Figure 5-13, the disc is held closed by a spring until the pump is started. The spring selection is very critical; it is the differential pressure across the valve (difference between static head and TDH) that must be specified and not the safety pressure rating of the system. An incorrect specification results in valve slam.

Three disadvantages of this valve type are that (1) the operating mechanism is enclosed so the valve must be removed for servicing, (2) there is no external indicator of the position of the disc, and (3) the headloss is high.

Double Leaf Check Valves

Double leaf (also double door, double disc, or split disc) check valves contain two hinged half-discs in a short body. The two half-discs are hinged in the middle and contain a spring that forces them closed. This type of valve has no connecting flanges of its own.

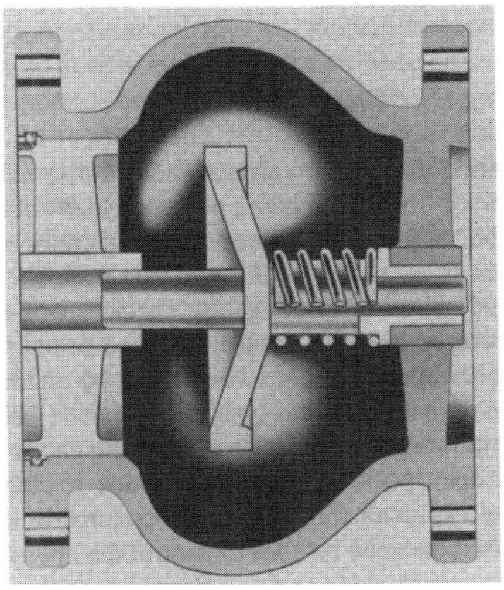

Figure 5-13. Center-post guided "silent" check valve. Courtesy of APCO Valve & Primer Corp.

Instead, it is inserted between two adjacent pipe flanges. It can be installed in either the horizontal or vertical position.

These valves should never be used in sewage or sludge service or in abrasive conditions because the hinge and discs can catch solids and the seat and discs would wear in abrasive service. Double leaf check valves are small, light, and inexpensive and have a short laying length. They close very quickly but they cannot be adjusted from the outside, nor can they be cushioned, so a sudden flow reversal can cause slam and water hammer. Shut-off is not leakproof. Other disadvantages are (1) the valve must be removed to service the mechanism, (2) there is no external indication of whether the valve is open or closed, and (3) they are subject to a fluttering motion caused by vortex shedding as the fluid moves past the valve plates. If the fluid velocity is less than 3.4 m/s (11 ft/s) and the valve is at least eight pipe diameters downstream from any source of flow disturbance such as a pump or a fitting, the problem is reduced [13].

Foot Valves

A foot valve is a special design of a lift check valve. It is used in the suction line of a sump pump to prevent loss of prime. It is designed for upflow and is attached to the bottom of a pump suction pipe.

Foot valves are prone to leakage, especially when used in fluids containing abrasives and solids, and they are difficult to service. Foot valves decrease the net positive suction head available ($NPSH_A$, see Section 10-4). In a raw sewage pumping station, a better choice would be to use a self-priming pump if a conventional wet well–dry well pumping station cannot be used or is not feasible.

Lift Check Valves

The body of a lift check valve is similar to that of a globe valve. A plug or stem moving within a guide lifts upward and allows fluid to pass through the valve. The plug seats when the flow reverses.

A lift check valve does not provide a tight shutoff. It cannot be used in fluids containing solids or abrasives, and gum-forming fluids can cause the stem to stick. Sudden flow reversal can cause water hammer.

This type of valve is normally encountered in pumping stations only in sizes 50 mm (2 in.) and smaller and in services such as utility water and compressed air.

Swing Check Valves

A swing check valve (Figure 5-12) contains a hinged clapper or disc that rests on a seat and prevents fluid from flowing backward through the body. A disadvantage of metal-to-metal design is the lack of a tight seal when the disc is seated, so a rubber seat is better. The disc is usually affixed to a hinge pin by means of an arm. The pin and arm allow the disc to move up and out of the flow path in the direction of fluid flow.

Swing check valves can be installed in both horizontal and vertical positions. In a horizontal position, the valve bonnet must be upright. In a vertical pipe, the valve must be installed so that fluid flow is in the upward direction, but never install swing check valves in vertical pipes in sewage, sludge, or slurry service because rags, debris, and grit would settle against the disc and eventually prevent functioning. Clearing the valve in this position is a messy, disagreeable task.

Slamming when a pump stops and the fluid reverses direction is a significant problem when using swing check valves of some designs. In general, swing check valves larger than 150 mm (6 in.) should have an outside lever and spring to close the disc quickly before the fluid can reverse direction. Note, however, that a commonly used check valve standard, AWWA C508, does not cover the outside lever and spring design. A disadvantage of this type of valve is that the outside lever and spring or counterweight can prevent the disc from opening fully, especially at low flow velocities—less than about 3 m/s (10 ft/s)—with a consequent increase of headloss. The swing check valve in Figure 5-12 is fully ported when open 20°, but most designs require a swing of 60° to open fully.

Headloss through the valve at low velocities is generally higher than the manufacturer's data, which are usually based on a fully open valve disc. If the valve is properly chosen for the specific application and the spring tension or the counterweight properly adjusted, however, the headloss should agree with the manufacturer's data. Headloss increase is often caused by increasing spring tension or the weight on the lever arm to reduce slam—a direct result of improperly selecting the valve. Swing check valves in pipes larger than 400 or 450 mm (16 or 18 in.) should be specified with caution, especially if the head exceeds about 15 m (50 ft), because the force on the disc is enormous.

Cushioned Swing Check Valves

Some check valve manufacturers offer pneumatic and/or hydraulic dashpots attached to the valve to regulate the speed of closure of the disc upon water column reversal. The cushioning system consists of

- a weighted lever arm attached to the disc pin or axle
- a piston mounted outside of the valve body and contained in a cylinder (dashpot) attached to the weighted arm.

As the velocity decreases, the weighted lever arm forces the disc to close, and the piston moves downward in the cylinder. The piston compresses the air (in a pneumatic system) or displaces oil through an orifice (in a hydraulic system). Adjusting the valves on the pneumatic or oil lines (or the orifices in the dashpot) controls the rate of closure.

The hydraulic system offers better control than the pneumatic system, which often does very little to reduce the slam. Sturdy valves can be closed quickly or slowly and can even be closed in two or three stages, such as quick closure to 50%, moderate speed of closure to 95%, and slow closure to shut-off.

Be very careful in selecting applications for these valves; close coordination with the manufacturer is necessary. In addition, field adjustment after installation is needed to set the closing controls properly.

Rubber Flapper Check Valves

The rubber flapper swing check valve is a swing check that is entirely enclosed. The seat is on a 45° angle and the steel-reinforced flapper need travel only about 35° to reach the fully open position. The short stroke and light weight of the flapper make it capable of very fast shut-off, which, combined with the resilient seat, reduces slam. The construction of the valve is simple, as is maintenance. There is no outside lever, no way of adjusting the closing force, and no way of determining whether the valve is open or closed. This type of valve should not be used in raw sewage service because debris can pack above the disc and prevent the disc from opening.

Slanting Disc Check Valves

A slanting disc check valve contains a disc balanced on a pivot. Instead of being perpendicular to the longitudinal axis as in conventional swing check valves, the seat is at an angle of 50 to 60° from the valve longitudinal axis. Slanting disc check valves should only be used in water service; rags and solids present in raw sewage and sludge would hang up on the disc.

The advantages of this type of valve are (1) headloss is low (although not as low as in a swing check valve) in the open position because the vane or flapper is designed as an air foil, (2) various pneumatic and oil-filled dashpots can be used to control the opening and closing speeds, and (3) the performance of these controls can be adjusted in the field. The disadvantages are (1) velocities less than 1.5 m/s (5 ft/s) do not fully open the vane; (2) the disc oscillates in the flow and the bearings wear on the bottom, so the valves begin to leak; and (3) the valve is not fully ported.

The two controls most frequently encountered are bottom buffers and top-mounted dashpots. These two systems are sometimes mounted together on one valve.

The bottom buffer consists of an oil-filled cylinder in which a piston is moved by the closing disc or vane. The disc moves freely for the first 90% of its closure, then strikes the buffer piston, which can be adjusted in the field to control the last 10% of disc travel.

The top-mounted oil dashpot system allows both the opening and closing speeds of the disc to be adjusted over the full range it travels. This adjustment can be especially valuable with pump start-up because the opening speed can be regulated to open the valve slowly, which greatly reduces hydraulic transient effects caused by pump start-up. A disadvantage is the high load exerted on the mechanical linkage when the pump reaches shut-off head. However, no electrical interconnections between the pump motor control center and check valve are needed.

5-5. Control Valves

Control valves are used to modulate flow or pressure by operating in a partly open position, thus creating a high headloss or pressure differential between upstream and downstream locations. Such operations may create cavitation and noise. If there is a large pressure differential and the limits of operation are approached or exceeded, the discs tend to flutter and bearings may wear quickly. Valve seats are especially vulnerable to wear because, if the pressure differential is high across the seat, small channels may be cut (called "wire drawing"), which prevents a tight seal, aggravates the wire drawing, and makes frequent replacement necessary.

To minimize wasting energy and to increase the life of the valve, it is desirable to minimize the time of operation at partly open positions. If the valve must throttle flow for extended periods, choose a style well adapted for the purpose and select hard materials for those parts that wear quickly.

Some control valves may be manually operated (for example, needle valves used to control the flow of a fluid in a valve actuator). Most control valves, however, are power-operated by programmed controllers. These valves are used for a variety of purposes: pump control, check valve control, control or anticipation of surges, or control of pressure or flow. The power source can be (1) hydraulic (usually oil), (2) pneumatic, (3) a combination of pneumatic and oil, (4) electric, or even (5) the pressure of the pumped water. All control methods feature some kind of adjustable-speed actuator, sometimes with three electric speeds that depend on the position of the valve mechanism. Whatever the power source, a backup is needed for power outages. The backup can be a pressure tank for pneumatic or hydraulic actuators or trickle-charged batteries for electric actuators (see Section 5-6).

Control valves are selected on the basis of the requirements of the hydraulic system and the characteristics of the pump. A major decision is whether to use a check valve that is controlled by the flow or a more sophisticated valve that itself controls the flow. The characteristics of the type—and even the brand—of pump-control or check valves are important. Every type of valve used as a check valve suffers some of the effects of cavitation, noise, and vibration while opening and closing, and some types are more vulnerable than others. Cavitation occurs at regions of large pressure drops.

Pump-Control Valves

Pump-control valves can be any type—angle, ball, butterfly, cone, globe, or plug—suitable for the liquid being pumped. Use angle and globe valves where high headloss can be tolerated or is desirable (as in bypass pipelines); and use ball, butterfly, cone, or plug valves where energy costs are important (see Example 5-1). Controls and electrical interlocks are provided so that the valve is closed when the pump starts. After the pump starts, the main valve opens slowly at an adjustable rate. When the pump is signaled to shut off, the valve slowly closes at an adjustable rate. When the valve is 95 to 98% closed, a limit switch assembly shuts off the pump.

Surges induced by start-up and shut-down of constant-speed water pumps can be effectively controlled by diaphragm- or piston-operated globe-type valves utilizing differential pressure to open and close the valve. Operation is usually initiated by activating solenoid valves that act on the trim piping controlling pressure on the diaphragm. The initiation of solenoid operation is usually linked electrically to the pump motor control circuit, and the speed of operation is controlled by adjusting needle valves in the trim piping. Variations of this basic type of valve include straight-through or angle bodies, surge relief valves, and head sustaining valves. To provide some assurance of reliability, the trim piping to the power side of the diaphragm must be fitted with a fine strainer to remove particulate material that might otherwise interfere with valve operation.

Piston-operated globe valves have an advantage over diaphragm-operated valves in that leakage from the valve occurs long before failure. Diaphragm-operated valves are completely sealed and do not leak, but, on the other hand, they give no warning of impending diaphragm rupture, which puts the valve out of service. Both valves are very effective in reducing surges due to pump start-up and normal pump shutdown, but they cannot prevent surges caused by power failure.

Power-actuated ball, butterfly, cone, and plug valves are more expensive to install but, when fully open, cause less headloss than other valves.

Control Valves for Water Service

The control valves likely to be used for water service include angle, ball, butterfly, cone, globe, needle (for fine flow regulation in control piping), and eccentric, lubricated, or nonlubricated plug valves. See Figure 5-14.

Control Valves for Wastewater

The only valves suitable for control of wastewater are ball, cone, long radius elbow, and eccentric, lubricated or nonlubricated plug valves.

Figure 5-14. Control valve for water service with external piping arranged for surge anticipation. Courtesy of Cla-Val Co.

Description of Control Valves

Except for globe and needle valves, all of the valves that can be used for control are described in Section 5-2. Recommendations for their use are given in Table 5-4.

Angle Valves

Angle valves and globe valves are similar in construction and operation except that in an angle valve, the outlet is at 90° to the inlet and the headloss is half as great as it is in the straight-through globe valve. An angle valve is useful if it can serve the dual purpose of a 90° elbow and a valve. Conversely, an angle valve should not be used in a straight piping run; instead, use a globe valve. As with globe valves, angle valves are best used in clear liquid service because fluids containing grit or abrasives cause severe seat erosion. Globe valves must never be used in sludge or raw sewage service because they are prone to becoming plugged with solids.

Globe Valves

As in the angle valve, a globe valve has a disc or plug that moves vertically in a bulbous body. Flow through a globe valve is directed through two 90° turns—upward and then outward—and is controlled or restricted by the disc or plug. The pressure drop or headloss is higher than in angle, gate, butterfly, or ball valves.

Because of this high pressure drop—even in the wide-open position—globe valves are not ordinarily used as isolation valves except in seal water, gas, and

Table 5-4. Recommendations for Use of Control Valves[a]

Type of valve	Water		Wastewater	
	Raw	Clean	Raw	Treated
Angle	F	G	X	X
Ball	E	E	E	E
Butterfly	F	G	X	F
Cone	E	E	E	E
Globe				
Diaphragm	G	G	X	F
Differential piston	G	G	X	F
Surge relief				
Diaphragm or piston	G	G	X	X
Angle valve (for water)	G	G	—	—
Long radius elbow valve (designed for sewage)	G	G	F	G
Surge anticipation				
Diaphragm or piston	G	G	X	X
Angle valve (for water)	G	G	—	—
Long radius elbow valve (designed for sewage)	G	G	F	G

[a] E, excellent; G, good; F, fair; P, poor, do not use; —, use is unlikely.

fuel oil pipelines. They are used in applications requiring throttling, such as pressure or flow control. Globe valves are suitable for clear liquid service, but not for fluids containing grit or abrasives, which cause severe seat erosion. Never use globe valves for sewage or sludge service because they are prone to becoming plugged with solids.

Globe or Piston Valves with Vee-Ports

Globe or piston valves containing vee-ports are made to eliminate seat wear by cavitation and to allow the flow to start in a controlled manner. Throttling is by the vee-ports as shown, for example, in Figure 5-15. The first 50% of the stroke allows only about 20% of full flow—a feature that minimizes the effects of surge caused by opening and closing pump-control valves.

Needle Valves

A needle valve is a special variation of a globe valve. The plug is a slender tapered needle. The flow annulus is easily fouled by particulates; otherwise, its characteristics and advantages and disadvantages are much the same as those of a globe valve.

A needle valve offers very fine pressure and flow control, even in a low flow range with the valve almost completely closed. It is often used in seal-water piping and, to control the speed of operation, in the control piping of other valve types.

Special Control Valve Functions

Some control valves regulate parameters such as pressure rather than flow, and these valves may operate by being either fully open or fully closed. Most special-purpose control valves are built on a single body design. Only the exterior piping to the hydraulic actuator (diaphragm or piston) in the bonnet is changed to effect the type of control wanted, whether it be constant flow, constant pressure, or proportional flow.

Make it easy to service these valves by incorporating enough isolating valves to close off the water supply to them.

Altitude Control Valves

Altitude control valves are used to add water to reservoirs and to one-way tanks used in surge control (see Chapter 7). The body design is usually of the globe type. Altitude control valves are made in many variations of two functional designs:

- One design in which the valve closes on high water level in the tank and does not open again until the water leaves through a separate line and the water level in the tank falls.
- A second design in which the valve closes on high water level in the tank and opens to allow water to flow out of the tank when pressure on the valve inlet falls below the reservoir pressure on the downstream side of the valve.

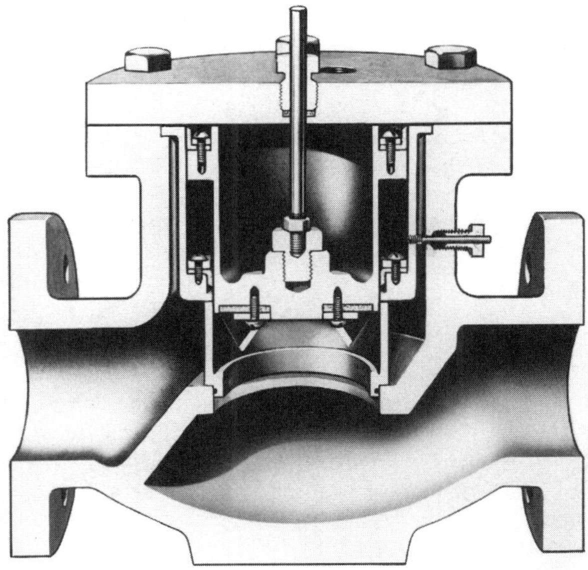

Figure 5-15. Vee-ported globe valve with differential piston. Flow is from right to left. Courtesy of GA Industries, Inc.

Pressure Relief Valves

Pressure relief valves are often of the globe type in terms of body design. A control system is added to establish how the valve operates. In the angle type (Figure 7-8), a direct-acting, adjustable spring is provided to open the valve and permit flow when the inlet pressure exceeds the spring setting. A common application is to place one of these valves on the branch of a tee on a pump discharge line. The valve then serves to release the fluid before a high pressure can develop and overstress the piping and valves. There is an inherent time lag in the opening of any valve, however, so the valve lags to some degree behind the actual pressure rise due to surges.

Surge Anticipation Valves

The entire surge anticipation system consists of a tee on the header, an isolation valve (to permit servicing the surge anticipation valve), the surge anticipation valve itself, a vent pipeline to waste, a pressure-sensing pipeline connected to the pump discharge, and a pilot system with an electronic timer. Following pump power failure, the pressure in the pump discharge drops, which opens the valve and vents water in anticipation of the subsequent high-pressure wave. The control system should be designed so that the valve does not open immediately after power failure but only after a timed delay (i.e., not until the pressure wave in the pipeline approaches the pumping station). The electronic timer keeps the valve open for a short period and then closes it slowly. If a second pressure wave follows, the valve reacts like a surge relief valve.

Although this type of valve can significantly reduce the return upsurge or high pressure after a pump power failure, it does nothing to control or reduce the effects of the initial downsurge or low-pressure wave (see Chapters 6 and 7).

5-6. Valve Actuators

Actuators (also called "operators") for valves can be manual or electrically, hydraulically, or pneumatically powered. Valve design (quarter-turn or lift type), valve size, operating pressures, and special requirements such as operation on loss of power or control of surge pressures determine the type and complexity of the valve actuating system. A comparison of valve actuators is given in Table 5-5.

Manual Actuators

Plug and butterfly valves 150 or 200 mm (6 to 8 in.) in diameter and smaller and ball valves 100 mm (4 in.) in diameter and smaller can be actuated with a simple lever attached to the plug shaft. Some lever-type actuators can be fitted with adjustable stops for balancing or throttling service. This feature, however, is rarely necessary for isolating valve service. Quarter-turn valves larger than the sizes indicated here should be fitted with geared manual actuators for two reasons:

- The torque required for actuation is too great for direct operation.
- Valves equipped with geared-type operators close slowly and thereby reduce the potential for damaging surge pressures.

Gate valves up to 300 mm (12 in.) in diameter can be actuated manually by handwheels acting directly through threaded nuts bearing on threads cut into the valve stem. Larger, manually operated gate valves should be fitted with gear reducers to reduce the force required to move the valve disc or plug to within reasonable limits. This accommodation is always provided at the expense of the time required to operate the valve. Thus, powered actuators are recommended for valves larger than 450 mm (18 in.) in diameter.

Manually actuated valves installed more than 2.1 m (6 ft 9 in.) above the operating floor should be equipped with chain actuators to permit operation without the need for a ladder, which is both inconvenient and dangerous. Small (100 mm [4 in.] in diameter and less) plug and butterfly valves can be equipped with extension arms and chains. Larger valves should be equipped with chain-wheel actuators. This type of actuator can be obtained with a hammer-blow feature to break loose valves that are hard to start.

The following are miscellaneous but important notes on specifying valves and designing installations:

- To avoid safety problems, valve stems and actuators should not protrude into walkways.
- If a valve or gate is below floor level and needs frequent operation, a floor stand is appropriate. A less expensive square nut, to be turned by a wrench, is sometimes a suitable substitute. Nuts should be standardized and, in the United States, the dimensions are given in AWWA standards.
- Ordinary chain wheels should not be used in a wet well because of the possibility of sparking. Nonsparking chain-wheel designs are available.
- Specifications and purchase orders for valves should state the direction of opening. The usual

Table 5-5. Comparison of Various Types of Valve Actuators

Type	Cost	Operating characteristics	Operation	Remarks
Manual, direct	Least	Relatively smooth.	Operator must be in attendance.	Suitable for smaller valves.
Manual, geared	Low	Smooth, slow.	As above.	Suitable for valves up to 450 mm (18 in.) in diameter.
Electric	Moderate to expensive depending on options	Smooth; can position only in increments unless provided with electronic positioner.	Avoid unless stored-energy (batteries) reserve system is provided.	Can be expensive depending on size, functions, and number of valves.
Pneumatic	Moderate	Tends to be jerky, difficult to position.	Easily provided with local receiver.	Corrosion, caused by water in compressed air, can be troublesome. These systems can freeze up at exhaust ports.
Hydraulic, water	Moderately expensive	Very smooth.	Easily provided with local hydropneumatic tank.	Corrosion and freezing can be problems.
Hydraulic, oil	Expensive	Very smooth.	Good, with precharged gas accumulators.	System can be complex. Reliability achieved only by using first-class components; recommend system pressures less than 14,000 kPa (2000 lb/ in.^{2}).

standard is for valve stems to turn counterclockwise to open. However, some valves open clockwise; if these must be used, paint handwheels, nuts, and levers red and/or mark the directions for operating them plainly.

- Valves are made with both rising and nonrising stems. The rising style is advantageous for indicating the valve opening. Geared actuators should incorporate position indicator dials.
- In some valves (especially butterfly valves), the flow through the valve tends to move the disc and may cause flutter. Such valves require locks. Worm gearing can be designed to be self-locking.
- Butterfly, plug, and ball valves 200 mm (8 in.) and larger in nominal diameter should be equipped with some type of actuator employing a mechanical advantage. Some engineers and pumping station workers prefer mechanical actuators for 150-mm (6-in.) valves as well. Worm gearing is the best because it is usually manufactured with greater precision than other types of actuators. Pantograph and traveling sleeve type actuators are often troublesome.

Powered Actuators

Powered actuators can operate directly on the valve shaft or stem or through gear reducers and special drive linkages. For rotary motion valves, such as ball or cone designs intended for surge control service and some butterfly valve actuating mechanisms, these special linkages serve two purposes: (1) conversion of linear motion to rotary motion and (2) special closing and opening characteristics to control pressure transients on pump start-up, shut-down, and power failure.

Electric Actuators

Electric actuators generally consist of electric motors driving through a gear train to power the valve stem or shaft. In general, the speed of operation and differential pressure at specified conditions determine motor power requirements. Motor operators have a hammer-blow feature to start hard-to-open valves (e.g., gate, nonlubricated plug). As with other actuators, a hammer-blow feature is important to start the valve in operation in either direction. Usually, this type of actuator is equipped with a handwheel for manual operation should the motor be disabled. It is important to specify a declutching mechanism that disengages the motor from the power train whenever the handwheel is being used—it can be motor preference or handwheel preference. Electric motor actuators can be specified, however, to accept remote commands, to telemeter position to remote locations, and to function with remote reversing starters. Electronic, modulating positioners are available, but these rarely are used in pumping station designs. To provide safeguards against potential damage, specify (1) torque-limiting switches for both open and closed positions and (2) four train limit switches to position the valve for seating. Specify integral, independent safety overrides.

Direct current power with battery support is recommended for all system control and monitoring functions where the actuating system must function during power failure. Batteries should be constantly trickle charged at low input with automatic switching to fast charging if the battery charge is low.

Hydraulic Actuators

Hydraulic actuators use fluid under pressure as a source of power, and both linear and rotary actuators are available. Hydraulic actuators (fluid power actuators) can be designed to use either oil under pressure from a self-contained system or water from the local potable supply, wherein the water is usually run to waste. However, because potable water supply systems must be designed to resist corrosion, specify stainless steel, bronze, or chrome-plated construction. Hydraulic actuators should be selected to provide sufficient power to break the valve loose. Once the valve is in motion, depending on the actuator linkage, a lower pressure differential may be required to move it from one position to another. One of the advantages of fluid power actuators is that fluid can be stored in pressure-charged accumulators or hydropneumatic tanks to provide a source of power under emergency conditions, such as commercial power failures. Another advantage is the ease of changing the speed of opening or closing the valve. Pressure should not exceed 14,000 kPa (2000 lb/in.^{2}) in all fluid power systems to limit leaks and joint failures, and a limit of 75% of that pressure is better. Specify premium components for fluid power systems. If operation of the equipment during emergencies is a prime concern, retain a specialist to design the system.

Pneumatic Actuators

Pneumatic actuators are available for both linear and rotary motions. The disadvantages of pneumatic operators include (1) noise; (2) poor operating characteristics because the powering fluid, a gas, expands on change of pressure; (3) a tendency to freeze because of expansion on release to atmospheric pressure; and (4) corrosion (with compressed air systems) because of water entrained in the gas.

A pneumatic actuator system generally has a lower initial installed cost than a motorized actuator system. However, the maintenance costs for the pneumatic actuators and associated equipment (compressors, receivers, traps, separators, filters, and piping) are usually much higher than they are for a motorized actuator system.

Pneumatic actuator systems are especially attractive for pumping stations because they can actuate valves when a power failure occurs. A receiver (tank) provides the compressed air to operate the actuator. A solenoid valve, energized to close (deenergized to open) is placed in the air line connecting the receiver to the pneumatically actuated valve. Upon power failure, the solenoid valve opens and the pneumatic actuator causes the valve to close. This system allows some control over the time of closure of the valve so that excessive surge pressures can be avoided. Size the receiver to hold twice as much air as needed to operate all of the valves through one cycle.

In most pumping stations requiring only a few powered valves (no more than three or four), an electric actuator system generally has the lowest installed cost. Hydraulic systems are usually the most expensive, with pneumatic systems in the middle. The cost of the hydraulic and pneumatic actuators themselves may be cheaper than the electric actuators, but the cost of the necessary auxiliary equipment—such as receivers, compressors, dryers, filters, and relief valves—rapidly increases the cost of small pneumatic systems. However, self-contained actuators that use the pumped water for power (so that auxiliary equipment is not required) are relatively inexpensive and low in maintenance labor.

Similarly, electric actuators require less maintenance than pneumatic and hydraulic actuators. Again, it is the maintenance associated with the auxiliary equipment that usually causes electric systems to be selected.

5-7. Air and Vacuum Valves

Air release and vacuum relief valves are often needed along transmission mains and may sometimes be unavoidable in sewage force mains. Air must be bled slowly from high points to prevent (1) "air binding" and (2) the reduction of the cross section of the pipe at high points. Vacuum conditions must be prevented when the pump head drops quickly (as in power failures) to prevent column separation. Vacuum relief valves can be as large as one-sixth of the diameter of the transmission main, whereas air release valves may be as small as one-fiftieth of the diameter of the pipe. Although most valves such as this are not within the pumping station, their presence in the transmission main has a profound effect on surge and, hence, on the whole system.

A pipeline designed for velocities high enough to scour air to the exit is an alternative approach that does not require the use of air release valves. Such velocities are within the normal design range for pipes 300 mm (12 in.) in diameter or smaller (see Table B-9). Air and vacuum valves are not objectionable and may be of some benefit (by eliminating air bubbles altogether), *but only if there is assurance that catastrophic failure is precluded by air-scouring velocities* in pipes on flat or downward slopes. Excessive headloss can be prevented by the use of larger pipe for upward slopes. Note, however, that air-scouring velocities must be reached frequently enough to prevent large air bubbles from forming. Also note that large pockets of air may greatly increase the head on the pumps. Design such combination systems on the assumption that the air release valves will sometimes fail to operate.

Air and Vacuum Valves in Water Service

Air release and vacuum relief valves (called "combination air valves") must often be used (sometimes at frequent intervals) along transmission pipelines. Wherever possible, select a pipeline profile that minimizes the number of these valves because they constitute an onerous maintenance problem. In water service, the short valve body (Figure 5-16) is appropriate for both air and vacuum valves. Lescovich [14] has discussed the use of air release valves in transmission mains.

Air and Vacuum Valves in Wastewater Service

Sewage grease, corrosive gases, solids, and scum in sewage combine to aggravate the maintenance problem and to reduce the reliability of valves in wastewater force mains (see Section 7-4). Use other control strategies, such as rerouting force mains, to avoid high points and the need for air or vacuum valves. If, for some reason, such valves are unavoidable, use the tall form or long body constructed of stainless steel or epoxy-lined iron in styles that permit the entire inside mechanism to be replaced quickly and easily in the field so that repair and cleaning can be done in the shop.

Because a valve that fails to operate might cause a pipeline rupture, reliability is mandatory. Some manufacturers recommend an overhaul every 6 months, but failures have occurred with such a schedule. To be safe, count on inspection and/or overhaul at frequent intervals (twice per week to be conservative or once per month for greater risk) and install both air valves and vacuum valves in pairs so that there is one for a backup.

One type of vacuum relief valve that is entirely appropriate for wastewater pipelines is discussed in Section 7-1 and shown in Figure 7-2.

Description of Air and Vacuum Valves

Air Release Valves

Air release valves slowly release the pockets of air that accumulate at high points in piping systems. In pumping stations, they are recommended on the discharge of vertical turbine pumps, especially when pumping from wells and sumps. This type of valve has a float that falls to vent air as the air accumulates in the body. Valves smaller than 19 mm ($^3/_4$ in.) usually have a float-activated compound lever with a linkage mechanism to provide a tight closure.

The valve body contains an orifice, usually 5 mm ($^3/_{16}$ in.) or smaller, through which the air escapes.

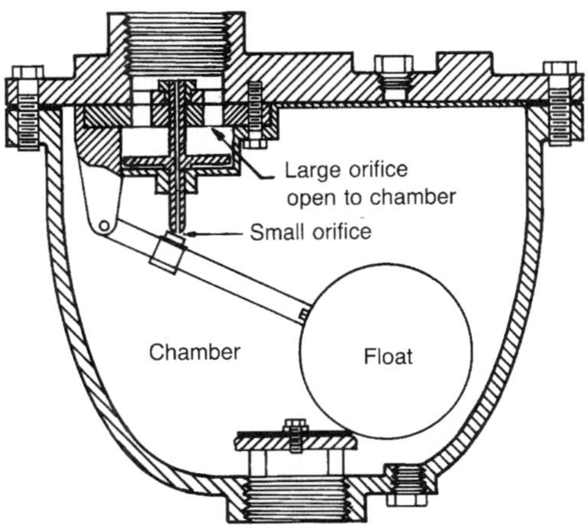

Figure 5-16. Combination air valve for water service. After APCO Valve & Primer Corp.

Combination Air Valves

A combination valve (Figure 5-16) consists of an air release and vacuum relief valve with an air release valve attached to it. It allows the use of one valve and one connection to the piping instead of two connections, and it fulfills two functions.

Slow-Closing Air and Vacuum Valves

This type of valve has a float assembly and large venting orifice to exhaust large quantities of air from pipelines when they are being filled and to admit large quantities of air when pipelines are being drained. The valve assembly includes a perforated water diffuser on the inlet to prevent the water column from rapidly entering the valve and slamming the float shut, which could possibly cause a severe water hammer problem.

Slow-Closing Combination Air Release Valves

A slow-closing air and vacuum valve with an attached air release valve allows the use of a single valve unit with one piping connection.

5-8. Materials of Construction

Bodies

Most valves in water, wastewater, and sludge pumping stations are not exposed to severely corrosive conditions. Bodies are usually cast iron (ASTM A48 or A126), cast steel (ASTM A216), or ductile iron (ASTM A395) for valves 100 mm (4 in.) or larger and bronze (ASTM B62 or B584, for which several alloys are available), for valves 75 mm (3 in.) and smaller. Fabricated steel (ASTM A36 or A516) is sometimes used in valves larger than 1800 mm (72 in.), especially in butterfly valves. In many locales, however, water and wastewater are indeed corrosive to iron and steel. For such liquids, the iron and steel bodies should be lined with epoxy (such as a product complying with AWWA C550) or other products. Also, some waters attack bronzes that contain high percentages of zinc and cause dezincification [15]. There is no universal agreement on an acceptable level of zinc in bronzes and other copper alloys. For example, AWWA C504 allows bronzes with zinc contents of up to 16%, but many engineers believe this is too high and allow no more than 5 to 7% zinc [16]. Some copper alloys (bronzes) frequently used in valves and having zinc contents of no more than 7% are alloys C83600, C87600, C90500, and C93700, defined in ASTM B 584.

A common phrase encountered in specifying valves is "iron body, bronze mounted (IBBM)." There is no universally accepted definition of this phrase. Some manufacturers provide a bronze disc and seat, while others provide only a bronze seat. Others furnish a bronze disc and seat up to a certain size, and then provide only a bronze seat in larger sizes. Be careful to define exactly what is meant when using a phrase such as "iron body, bronze mounted."

Seats

General

Seats in isolation and control valves are more subject to erosion and corrosion than bodies because the fluid velocity impinges most noticeably here. Bronze on bronze is the cheapest, and seats in metal valves 75 mm (3 in.) and smaller are frequently bronze (note the previous discussion of the zinc content in bronzes for valves used for some waters). Most seats or seat retention devices in valves 100 mm (4 in.) and larger are some grade of stainless steel. Stellite facing is much harder and more erosion resistant than stainless steel, but it is also much more expensive. Do not specify an exotic material unless there is a clear need for it. The most common materials for resilient seats are listed in Table 5-1, but there are numerous plastics available with special qualities. Buna N can be attacked by industrial chemicals, but unless there is an excess of illegal dumping of such solvents, it is a nearly ideal seat material. Teflon® is more resistant to attack, but it creeps.

Teflon®

Teflon® (polytetrafluoroethylene, or TFE) is suitable for both water and wastewater. It is hard, strong, and impervious to attack by nearly all chemicals, but it has shortcomings such as cold flow and creep. Other materials are usually more suitable.

Elastomers

Elastomers (natural or gum rubber and the synthetic rubbers, such as neoprene, Viton, and Buna N) are used for resilient seats, O-rings, and a few other parts. Natural rubber is suitable for clean water, but wastewater contains oil and grease products and organic solvents that attack it. The synthetic rubbers usually give good service in both water and wastewater systems. Elastomers are subject to wear in grit slurry service and where they rub against iron tubercles, hard scale deposits, or corroded surfaces. They are, however, suitable if the grit has been removed.

Engineers should be aware that rubber compounds—both natural and synthetic—are not uniformly and consistently resistant to some disinfecting agents commonly used in the water works industry, such as chloramines. There appear to be wide variations in resistance to attack on various elastomers and on differing formulations of the same elastomer. It should also be noted that various standards pertaining to valves (such as AWWA C500, C504, and C509) do not address the issue of resistance of the rubber seating material to either chloramines or free chlorine. The resistance to attack by disinfecting agents is addressed by Reiber [19]. As of 1995, the issue of developing a testing standard to specify the resistance of elastomers to disinfecting agents was being discussed and investigated by several AWWA standards committees.

Packing

Valve packing is as important in a valve as the packing gland is in a pump. It prevents leakage past the valve stem and damage to the valve housing. The seal is normally made by placing packing material around the valve stem and compressing it with a follower or gland, which is tightened by a packing nut. Although asbestos has historically been used as a packing material, more and more valve manufacturers are discontinuing its use and are switching to nonasbestos materials such as Teflon®, aramid fiber, acrylic fiber bound with nitrile, and Buna N. Note that some common standards specify packing material that is no longer even made. For example, AWWA C500 for gate valves specifies flax conforming to Federal Specification HH-P-106d, or asbestos conforming to Federal Specification HH-P34c. The flax specification HH-P-106d was discontinued in 1978 by the federal government because of insufficient usage. AWWA C504, pertaining to butterfly valves, still requires packing for stuffing boxes to be asbestos or flax. Most valve manufacturers do not comply (and, in fact, cannot comply) with these AWWA standards on packing materials.

Stems

Gate, globe, and needle valves have stems that rise in the body to seat or unseat. In both water and wastewater service, these stems are usually bronze, and the problem of dezincification is especially acute (see the discussion on zinc content in "Bodies" of this section).

5-9. Installation of Valves

Warping of the valve body due to pressure and thermal stresses in the connecting pipelines or lack of proper valve support can damage the valve enough to prevent it from functioning. The valve body should not be supported by the adjacent piping, nor should it support the piping. The following are some suggestions for valve support:

- For piping and valves supported on floors, provide separate bases or supports for valves 100 mm (4 in.) or larger.
- For piping and valves suspended from ceilings, provide separate hangers at valves—one hanger or support at each end of the valve body or on the connecting pipe within one pipe diameter of the valve end.
- Provide enough flexibility in connected piping so that thermal strains in the piping do not stress the valve.
- Install piping without springing, forcing, or stressing the pipe or any connecting valves.
- Some valves (such as AWWA C507 ball valves) are intended to be supported in a certain manner. Be sure to read the relevant standard before designing the support system.
- Always consult the valve manufacturer about proper support.

End Connections

End connections for valves can be

- screwed (ANSI B1.20.1);
- flanged (ANSI B16.1 for cast-iron valves, ANSI B16.5 for steel valves);
- grooved end (AWWA C606);
- butt welded (ANSI B16.34); or
- socket welded (most commonly found in plastic valves).

In general, valves 75 mm (3 in.) and smaller should have screwed ends, whereas larger valves have flanges or grooved ends because, in larger sizes, assembling pipes or valves with threaded ends becomes very laborious. Furthermore, bolted connections are much easier to disassemble than screwed connections, even if unions are installed at moderate spacing.

Butterfly, gate, and eccentric plug valves with grooved-end or flanged connections are probably the most commonly encountered valves in a pumping station. Butterfly valves must be of long body (not short body) style per AWWA C504 to accommodate grooved ends. The use of grooved-end connections provides greater ease than do flanges for removing valves from a piping manifold. Sleeve (Dresser®) couplings need not be provided adjacent to grooved-end valves, although such couplings may be needed for other reasons, such as thermal expansion, alignment, and differential settlement.

Locate valves and design the surrounding piping to prevent clogging with grit. Except for clean water service, bonnets must be within 45° of upright to keep out grit that can build up and prevent operation. If possible, avoid installing valves on risers, especially if there is a long section above the valve. If a valve must be placed on a riser, use a ball or plug valve that is wiped clean by operation and has no crevices to collect grit. Design risers to enter a header horizontally from an elbow to the tee so that grit and solids cannot block the riser. Even if valves are installed on horizontal pipes, locate them at least three (or, better, five) pipe diameters from the riser. Even in clean water service, butterfly valves should be at least two (three is better) pipe diameters from an elbow so that streamlines, entering from an angle, do not make the valve difficult to open or close and do not cause the vane to flutter. If the valve must be close to an elbow, orient the valve so the vane is not subjected to dynamic loading from the flow through the elbow.

5-10. Corrosion Protection

When used in water lines, sewage lines, sludge piping, or any other service in which the fluid is particularly corrosive, metal valves can be lined with epoxy. This epoxy lining can only be applied to valves 100 mm (4 in.) and larger.

In areas of corrosive soil, buried valves should be coated with coal tar, coal-tar epoxy, or a high-solid (usually 70% solids by volume or higher) epoxy.

Pay particular attention to the need for stainless-steel bolts in buried and submerged valves and valves exposed to H_2S atmospheres. In some soils and waters, galvanized steel bolts and nuts (ASTM A307) corrode readily. Therefore, type 304 or 316 stainless bolts (ASTM A193, Grade B8 or B8M) and nuts (ASTM A194, Grade 8 or 8M) may be required.

Low solids and low pH waters are very corrosive to brass [15], iron, and, to some degree, cement liners. Such waters can be rendered benign by chemical treatment [16, 17, 18].

Seats must neither corrode nor erode. The corrosion and erosion resistance of several seat materials are compared in Table 5-1 (see also Lyons [6]).

5-11. References

1. Cook, D. T., "Selecting hand-operated valves for process plants," *Chemical Engineering,* **89,** 126–140 (1982).
2. O'Keefe, W., "Pump controls and valves," in *Pump Handbook,* I. J. Karassik, W. C. Krutzsch, W. H. Fraser, and J. P. Messina, Eds., McGraw-Hill, New York (1976).
3. Deutsch, D. J., *et al., Process Piping Systems,* pp. 193–357, McGraw-Hill, New York (1980).
4. AWWA M11, *Steel Pipe—A Guide for Design and Installation,* 2nd ed., American Water Works Association, Denver, CO (1985).
5. Hutchison, I. W. (Ed.), *ISA Handbook of Control Valves,* 2nd ed., Instrument Society of America, Research Triangle Park, NC (1976).
6. Lyons, J. L., *The Valve Designer's Handbook,* Van Nostrand Reinhold, New York (1982).
7. Lyons, J. L., and C. L. Askland, Jr., *Lyon's Encyclopedia of Valves,* Van Nostrand Reinhold, New York (1975).
8. British Valve Manufacturer's Association, *Technical Reference Book on Valves for the Control of Fluids,* 2nd ed., Pergamon, London (1966).
9. Schweitzer, P. A., *Handbook of Valves,* Industrial Press, New York (1972).
10. Pearson, G. H., *Applications of Valves and Fittings,* Pitman, London (1968).
11. Zappe, R. W., *Valve Selection Handbook,* Gulf Publ., Houston, TX (1981).
12. Parmakian, J., "Water hammer," Section 9.4 in *Pump Handbook,* I. J. Karassik, W. C. Krutzsch, W. H. Fraser, and J. P. Messina, Eds., McGraw-Hill, New York (1976).
13. Collier, S. L., *et al.,* "Behavior and wear of check valves," *Journal of Energy Resources Management, Transactions of the American Society of Mechanical Engineers,* **105,** 58 (March 1983).
14. Lescovich, J. E., "Locating and sizing valves in transmission mains," *Journal of the American Water Works Association,* **64,** 457–461 (July 1972).
15. Jester, T. C., "Dezincification update," *Journal of the American Water Works Association,* **77,** 67–69 (October 1985).
16. "Valve dezincification prevented," *Heating/Piping/Air Conditioning,* **53**(11), 30 (November 1981).
17. Singley, J. E., *et al., Corrosion Prevention and Control in Water Treatment and Supply Systems,* Noyes Data, Park Ridge, NJ (1985).

18. Merrill, D. T., and R. L. Sanks, "Corrosion prevention by the deposition of $CaCO_3$ films," *Journal of the American Water Works Association,* Part 1, **69,** 592–599 (November 1977); Part 2, **69,** 634–643 (December 1977); Part 3, **70,** 12–18 (January 1978).
19. Reiber, S. H. *Chloramine Effects on Distribution System Materials*, American Water Works Association, Denver, CO (1993).

Chapter 6

Fundamentals of Hydraulic Transients

BAYARD E. BOSSERMAN II
WILLIAM A. HUNT

CONTRIBUTORS
Robert C. Glover
Joseph R. Kroon
M. Steve Merrill
Gary Z. Watters

The purpose of this chapter is to provide an overview of the problems caused by hydraulic transients and an insight into the circumstances that make a more thorough analysis necessary. The fundamental theory of hydraulic transient analysis is described simply, with no attempt to present rigorous mathematical or analytical methods. Simple numerical examples include surge pressure calculations, attenuation of surge pressure by programmed valve closure, and the design of pipe to resist upsurge and downsurge pressures.

For more complete discussions of hydraulic transients, see Parmakian [1], Rich [2], Wylie and Streeter [3], Watters [4], and Chaudhry [5].

6-1. Introduction

Hydraulic transients are the time-varying phenomena that follow when the equilibrium of steady flow in a system is disturbed by a change of flow that occurs over a relatively short time period. Transients are important in hydraulic systems because they can cause (1) rupture of pipe and pump casings; (2) pipe collapse; (3) vibration; (4) excessive pipe displacements, pipe-fitting, and support deformation and/or failure; and (5) vapor cavity formation (cavitation, column separation).

Some of the primary causes (and frequency of occurrence) of transients are (1) valve movements—closure or opening (often), (2) flow demand changes (rarely), (3) controlled pump shutdown (rarely), (4) pump failure (often), (5) pump start-up (rarely), (6) air venting from lines (often), (7) failure of flow or pressure regulators (rarely), and (8) pipe rupture (rarely).

The identification and calculation of pressures, velocities, and other abnormal behavior resulting from hydraulic transients make possible the effective use of various control strategies, such as the

- selection of pipes and fittings to withstand the anticipated pressures;
- selection and location of the proper control devices to alleviate the adverse effects of transients; and
- identification of proper start-up, operation, and shutdown procedures for the system.

The analysis of unsteady flow in pipe systems is generally divided into two major categories.

Rigid water column theory (surge theory). The fluid and pipe are inelastic, and pressure changes propagate instantaneously. These flow conditions are described by ordinary differential equations.

- *Solution:* closed-form integration or finite difference numerical integration.

- *Advantages:* the analysis can be applied by a person with little numerical analysis skill and with limited computational facilities.
- *Disadvantages:* the solutions, which are always approximations, are applicable only to simple pipelines. Considerable experience is required to know whether the results are applicable.

Elastic theory (water hammer). The elasticity of both fluid and pipe affect the pressure changes. Pressure changes propagate with wave speed, a, which varies from about 300 to 1400 m/s (1000 to 4700 ft/s). Flow conditions are described by nonlinear partial differential equations.

- *Solution:* arithmetic, graphical method or method of characteristics using finite difference techniques.
- *Advantages:* the theory accurately represents system behavior and is, therefore, applicable to a wide range of problems. Pipe friction, minor losses, and varying valve closure procedures can be incorporated.
- *Disadvantages:* applying the theory requires a substantial initial effort on the part of user to learn it, a digital computer programmed for the method of characteristics, and a knowledge of the operational characteristics of system components to set up the solution for the computer.

6-2. Nomenclature

In Chapters 6 and 7, "velocity" always means velocity of water and "speed" means the velocity of pressure waves. The symbols used in Chapters 6 and 7 are defined as follows.

- a Elastic wave speed in water contained in a pipe (in meters per second [feet per second])
- C Coefficient whose value depends on pipe restraint
- D Inside diameter of a pipe (in meters [inches or feet])
- e Wall thickness of a pipe (in meters [inches or feet])
- E_j Longitudinal joint efficiency in welded pipes (dimensionless)
- E Modulus of elasticity of pipe material (in newtons per square meter [pounds per square inch or pounds per square foot]; see Table 6-1).
- g Acceleration due to gravity (in meters per second squared [feet per second squared])
- h Head due only to surge (in meters [feet])
- Δh Change of head due to surge (in meters [feet])
- h_a Allowable head due to surge (in meters [feet])
- K Bulk modulus of elasticity of the liquid (in newtons per square meter [pounds per square inch or pounds per square foot]; see Table A-8 or A-9)
- L Length of pipeline (in meters [feet])
- ΔP Change of pressure due to surge (in newtons per square meter [pounds per square inch])
- ΔP_a Allowable pressure change due to surge (in newtons per square meter [pounds per square inch])
- P_{atm} Atmospheric pressure (in newtons per square meter (pounds per square inch); see Table A-6 or A-7]
- ΔP_c Difference between external and internal pressure on a pipe (in newtons per square meter [pounds per square inch])
- P_v Vapor pressure of water (in newtons per square meters [pounds per square inch]; see Table A-8 or A-9)
- SF Safety factor (dimensionless)
- s_y Yield stress (in newtons per square meter [pounds per square inch])
- t Time (in seconds)
- t_c Critical time ($2L/a$; in seconds)
- V Volume (in cubic meters [cubic feet])
- v Velocity (in meters per second [feet per second]; in this chapter, "velocity" is average velocity of fluid flow)
- Δv Change in velocity (in meters per second [feet per second])
- Δv_a Allowable change in velocity (in meters per second [feet per second])
- ρ Density (in kilograms per cubic meters [slugs per foot; see Table A-8 or A-9])
- μ Poisson's ratio (dimensionless; see Table 6-1)
- γ Specific weight of water (in newtons per cubic meter [pounds per cubic feet; see Table A-8 or A-9])

6-3. Methods of Analysis

Methods of analyzing pipelines for the effects of hydraulic transients, with or without various means of controlling them or reducing the severity, can be summarized as follows:

- Graphical [1]
- Arithmetic [2]
- Algebraic [3]
- Method of characteristics [3,4,5]
- Finite element
- Implicit differentiation

Table 6-1. Physical Properties of Pipe Materials

Material	Poisson's ratio	Modulus of elasticity: SI units (N/m²)	U.S. customary units: lb/in.²	U.S. customary units: lb/ft²
Aluminum	0.33	7.30 E + 10	1.05 E + 7	1.51 E + 9
Asbestos-cement	0.30	2.30 E + 10	3.40 E + 6	4.90 E + 8
Brass	0.34	1.03 E + 11	1.50 E + 7	2.16 E + 9
Copper	0.30	1.10 E + 11	1.60 E + 7	2.30 E + 9
Ductile iron	0.28	1.66 E + 11	2.40 E + 7	3.46 E + 9
Gray cast iron	0.28	1.03 E + 11	1.50 E + 7	2.16 E + 9
HDPE	0.45	1.0 E + 9[a]	1.5 E + 5[a]	2.2 E + 7[a]
PVC	0.45	2.70 E + 9	4.00 E + 5	5.76 E + 7
Steel	0.30	2.07 E + 11	3.00 E + 7	4.32 E + 0
Concrete	—	$4.73 \times 10^6 \sqrt{f'c}$ [b]	$57{,}000 \sqrt{f'c}$ [c]	

[a] At 16°C (60°F). Increases greatly with decreasing temperature and vice versa.
[b] $f'c$ is ultimate strength in newtons per square meter.
[c] $f'c$ is ultimate strength in pounds per square inch.

These are methods of *analysis,* not methods of *design*. In analysis, the system is described mathematically, and the behavior of the system is predicted by the analysis. In design, the desired physical results are described and alternative means for attaining these results are compared, which leads to a selection of one or more control measures. Analysis should be performed as a part of the design process.

All of the above methods involve the equations of motion and continuity used to describe the velocity and pressure variation in the pipeline. For computer modeling, the most widely used is the method of characteristics, in which the partial differential equations of motion and continuity are converted into four first-order equations represented in finite difference form and solved simultaneously with a computer. The method provides for

- the inclusion of many possible pipeline or system features, such as junctions, pumping stations, air chambers, air release valves, reservoirs, and line valves;
- the inclusion of fluid friction;
- the retention of small or secondary terms in the original equations so that accuracy is retained; and
- the computation of pressure and velocity as a function of time at various points throughout the entire pipeline system.

6-4. Surge Concepts in Frictionless Flow

Water hammer can occur in a pipeline flowing full when the flow is increased or decreased, such as when the setting on a valve in the line is changed. When a valve in a pipeline is closed rapidly, the pressure on the upstream side of the valve increases, and the pulse of increased pressure travels upstream at the elastic wave speed, a. This pulse (called an "upsurge") decreases the velocity of flow. Downstream from an in-line valve, the pressure is reduced and the wave of decreased pressure travels downstream, also at the elastic wave speed, a. This pulse (called a "downsurge") decreases the velocity of flow. If the velocity is reduced too rapidly and the steady-state pressure is low enough, the downstream pressure can be reduced to vapor pressure, which creates a vapor pocket. A large vapor pocket (called "column separation") can collapse with a dangerous explosive force produced by the impact between solid water columns and can cause the pipe to burst. This phenomenon can also occur upstream of the valve when the reflected positive wave returns to the valve.

The events following a sudden closure of a valve located a distance (L) downstream from a reservoir is described in Figure 6-1. Friction is neglected, and the energy gradeline (EGL) and hydraulic gradeline (HGL) are assumed to coincide because velocity heads are small compared with water hammer pressure heads. The steady-state EGL is called H, and the added-pressure head pulse is called h. The velocity of the fluid under steady-state conditions is v_0 just before the valve is closed (at $t = 0$).

The sequence of events between the valve and the reservoir occurs in a four-phase cycle; the duration of each phase is the time for the pressure wave to travel between the valve and the reservoir (the length of the

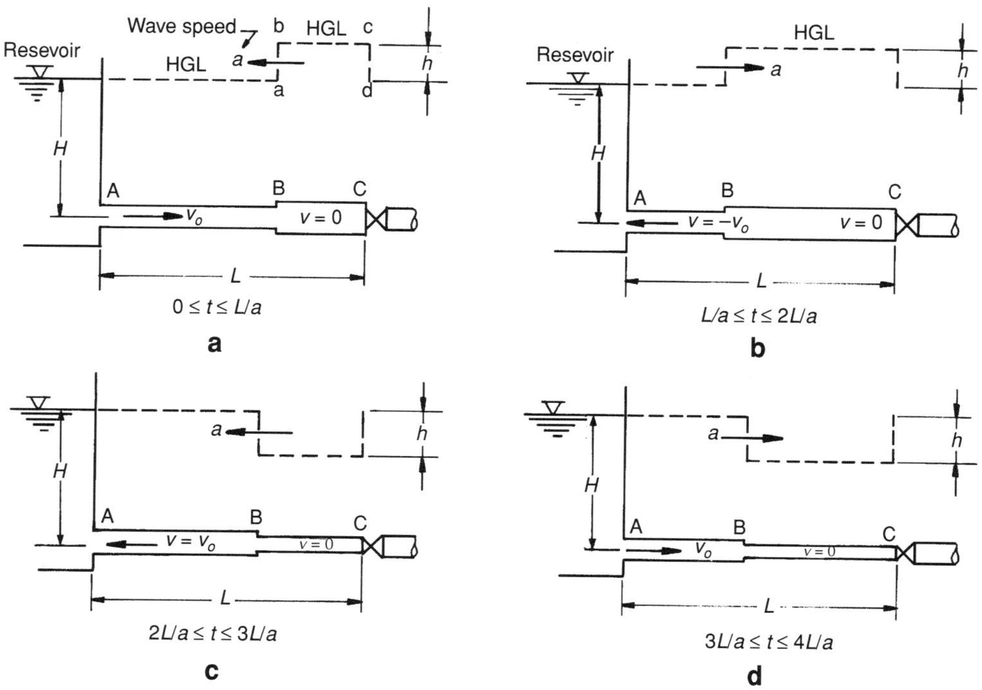

Figure 6-1. Sequence of events for one cycle after sudden valve closure.

pipeline divided by the elastic wave speed, L/a). The sequence occurs as follows:

1. $0 \leq t \leq L/a$. At $t = 0$, the fluid just upstream from the valve is compressed and brought to rest. Part of the pipe (section BC) is expanded and stretched, as shown in Figure 6-1a. This process is repeated for each successive increment of fluid as the pressure wave travels upstream. The fluid upstream from the wave front continues to flow downstream until it is stopped by the advancing pressure wave front. When the pressure wave reaches the reservoir at $t = L/a$, the fluid (at rest in the pipe) is under a total pressure head $H + h$, or h greater than the static head in the reservoir.

2. $L/a \leq t \leq 2L/a$. The pressure head difference at $t = L/a$ at the reservoir causes the fluid to flow from the pipe back into the reservoir with a velocity $-v_0$. The pressure along AB is reduced to the original steady-state level, H, and a negative wave producing normal pressure propagates back to the valve, as shown in Figure 6-1b. At $t = 2L/a$, the pressure is normal (equal to H) along the pipe but the velocity throughout the pipe is negative; that is, water is flowing away from the valve.

3. $2L/a \leq t \leq 3L/a$. At $t = 2L/a$, there is no fluid available to maintain the upstream flow at the valve, and the normal pressure head, H, is reduced by h to bring the fluid in section BC to rest (Figure 6-1c). This wave of reduced pressure propagates back toward the reservoir, all fluid comes to rest, and the reduced pressure allows the fluid to expand and the pipe walls to contract. At $t = 3L/a$, the reduced pressure, $H - h$, exists all along the pipe, the velocity is zero throughout the pipe, and the static pressure head in the pipe is less than the pressure head in the reservoir.

4. $3L/a \leq t \leq 4L/a$. At $t = 3L/a$, the unbalanced pressure head at the reservoir causes the fluid to flow back into the pipe at a velocity $+v_0$. A wave of pressure at the original static pressure head level propagates downstream toward the valve, but in section BC, the pressure is still reduced to $H - h$ (Figure 6-1d). At $t = 4L/a$, the pressure throughout the pipe is normal (equal to H), and the velocity is the same as the original $+v_0$ prior to the valve closure. But when the velocity, v_0, reaches the valve, the four-phase cycle repeats and continues to repeat periodically every $4L/a$ time period.

The variation of pressure with time during a $4L/a$ interval is shown at several points along the pipeline in Figure 6-2. Friction in a real pipe system eventually dampens the increased water hammer pressure head, h.

The principal concepts indicated by the events following a sudden valve closure are as follows:

- The time L/a is a significant parameter for water hammer analysis.
- The time $2L/a$ is critical because the pressure head at the valve reaches a maximum at $2L/a$. A valve closed in any shorter time produces the same maximum pressure head rise at the valve, whereas the

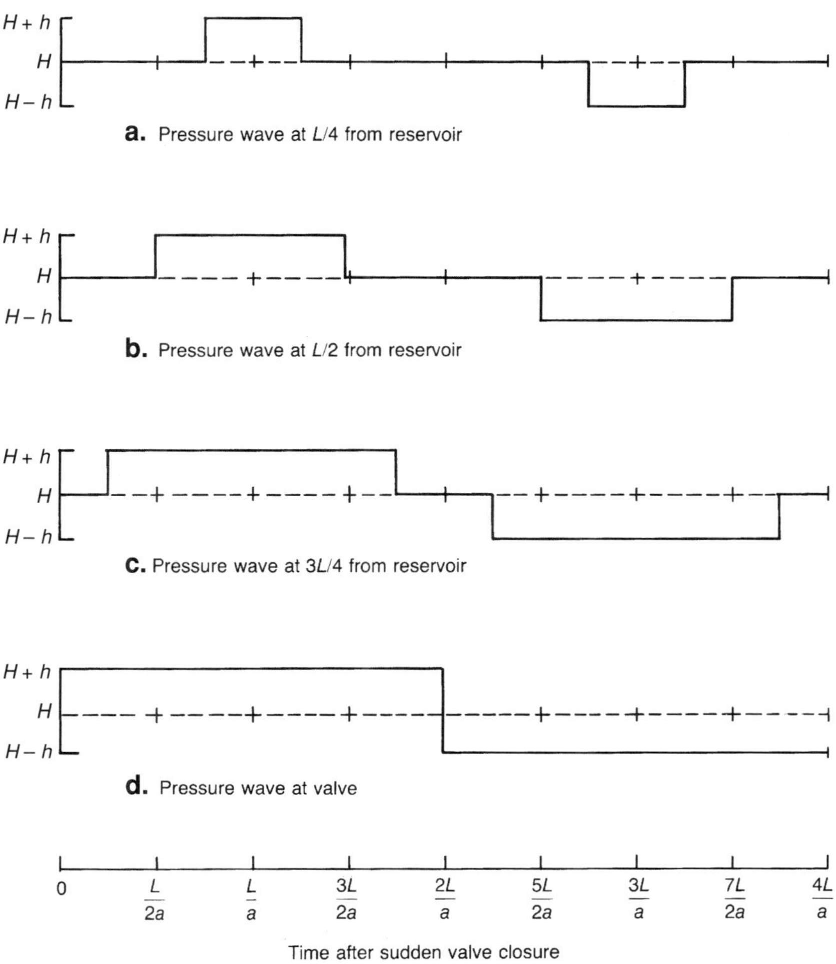

Figure 6-2. Head fluctuations at selected points in the pipeline of Figure 6-1 after valve closure.

pressure rise is reduced if the valve is closed in a longer time interval. Hence,

$$t_c = \frac{2L}{a} \tag{6-1}$$

where t_c is the critical time, L is the length of the pipe, and a is the elastic wave speed.

Pressure Head Change

A momentum analysis of the flow conditions for the valve closure shows the pressure head change, Δh, is a function of the change in flow, Δv.

$$\Delta h = \frac{-a\Delta v}{g} \tag{6-2}$$

where Δh is the change in pressure head in meters (feet), a is elastic wave speed in meters per second (feet per second), Δv is the change in velocity (of water) caused by the event in meters per second (feet per second), and g is the acceleration due to gravity in meters per second squared (feet per second squared). Use the negative sign for waves traveling upstream, use the positive sign for waves traveling downstream, and note that $\Delta v = v_2 - v_1$, where v_1 is the velocity prior to the change in flow rate and v_2 is the velocity following the change. If the flow is suddenly stopped, $\Delta v = v_1$ and $\Delta h = av_1/g$. Note, too, that Δh is positive if Δv is negative. If the valve on the downstream end of a pipe is closed incrementally, Equation 6-2 becomes

$$\Sigma\Delta h = -\frac{a}{g}\Sigma\Delta v \text{ (for } t < t_c) \tag{6-3}$$

Transient pressure heads due to valve closure can be reduced by slowly closing the valve over a time interval greater than t_c, as discussed in Section 6-5.

Elastic Wave Speed

The pressure head change due to a change in the flow rate requires the calculation of the elastic wave speed in the pipe. The wave speed depends on both the fluid properties and the pipe characteristics.

$$a = \sqrt{\frac{K/\rho}{1 + C\left(\frac{K}{E}\right)\left(\frac{D}{e}\right)}} \qquad (6\text{-}4)$$

where K is the bulk modulus of elasticity of the liquid in newtons per square meters (pounds per square foot), E is the modulus of elasticity of pipe material in newtons per square meters (pounds per square foot), D is inside pipe diameter in meters (feet), e is the pipe wall thickness in meters (feet), C is a correction factor for type of pipe restraint, and ρ is the density of the fluid in kilograms per cubic meter (slugs per cubic foot). The bulk modulus given in Table A-8 must be changed from kilopascals to newtons per square meter, and in Table A-9 it must be converted from pounds per square inch to pounds per square foot.

For thin-walled pipes (those with $D/e > 40$) the correction factor, C, varies according to pipe restraints. Values for three cases are

- $C_1 = 1.25 - \mu$ for pipes anchored at the upstream end only (Wylie and Streeter [3] use $1.0 - \mu/2$; the difference has little practical significance);
- $C_2 = 1.0 - \mu^2$ for pipes anchored against axial movement; and
- $C_3 = 1.0$ for pipes with expansion joints throughout.

The symbol μ is Poisson's ratio (see Table 6-1 for properties of pipe materials). Expressions for C_3 for thick-walled pipes ($D/e < 40$), which are more complex, are given by Wylie and Streeter [3]. Buried pipes are best represented by the factor C_2. Typical values for wave speed for water in pipes are given in Table 6-2. Because the elastic wave speed, a, for steel and ductile iron pipe is often close to 980 m/s (3200 ft/s), the change in pressure head from Equation 6-2 is, roughly, $\pm 100v$. For PVC, $h = \pm 34v$.

The speed of the pressure wave in a pipeline carrying liquids is greatly reduced if bubbles of free gas are entrained, as shown in Table 6-3. A detailed discussion of this phenomenon is given by Wylie and Streeter [3]. There seems to be no way of predicting air volume entrainment. Neglecting the effect of air entrainment provides a conservative analysis. However, as the wave speed decreases, L/a increases and the time for closure of valves must be increased.

Table 6-2. Typical Wave Velocities in Pipe for Water Containing Dissolved Air

Pipe material	Wave velocities m/s	Wave velocities ft/s
Asbestos cement	820–1200	2700–3900
Copper	1000–1300	3400–4400
Ductile iron	980–1400	3200–4500
HDPE[a]	180–370	600–1200
PVC	300–600	1000–2000
Steel	600–1200	2000–4000

[a] For a modulus of elasticity of 1.03 E + 9 N/m² (150,000 lb/in.²)

Table 6-3. Effect of Air Entrainment on Wave Speed[a]

V_{air}/V_{total}	Wave speed m/s	Wave speed ft/s
0	1200	4000
0.001	610	2000
0.002	460	1500
0.004	300	1000
0.008	210	700

[a] After Wylie and Streeter [3].

Example 6-1
Effect of Pipe on Wave Speed and Pressure

Problem: For 300-mm (12-in.) pipes of ductile iron, steel, and PVC, find the effects of pipe material, bedding, and joint conditions on wave speed and water hammer pressure. Assume the pipes are 3000 m (9840 ft) long and carry 0.15 m³/s (5.3 ft³/s) of water at a temperature of 15°C (59°F).

Solution: The wave speeds are based on Equation 6-4. The calculation of wave speed to the nearest 30 m/s (100 ft/s) is usually sufficient, because the accuracy of the data does not justify greater precision. The critical valve closure times, t_c, are based on Equation 6-1. Valve closure in any time interval less than t_c subjects the pipe to the maximum pressure, Δh, given by Equation 6-2. Note that $K = 2.15 \times 10^9$ N/m^2 (4.49×10^7 lb/ft^2) for water.

	SI Units			U.S. Customary Units		
	Ductile iron	Steel	PVC	Ductile iron	Steel	PVC
OD (m [ft])	0.335	0.324	0.324	1.10	1.063	1.063
e (m [ft])	0.0094	0.0048	0.0149	0.0308	0.0157	0.0259
E (N/m^2 [lb/ft^2])	1.66×10^{11}	2.07×10^{11}	2.70×10^9	3.46×10^9	4.32×10^9	5.76×10^7
μ	0.28	0.30	0.45	0.28	0.30	0.45
$C_1 = 1.25 - \mu$	0.97	0.95	0.80	0.97	0.95	0.80
$C_2 = 1 - \mu^2$	0.92	0.91	0.80	0.92	0.91	0.80
$C_3 = 1.0$	1.00	1.00	1.00	1.00	1.00	1.00
Wave speed, a (m/s [ft/s]), from Equation 6-4:						
$C_1 = 1.25 - \mu$, no air	1220	1150	280	4000	3770	930
$C_2 = 1 - \mu^2$, no air	1210	1130	280	3970	3710	930
$C_3 = 1.0$, no air	1190	1110	260	3920	3640	840
$C_3 = 1.0$, 0.2% air	490	490	230	1620	1600	760
For $C_3 = 1.0$ and no air:						
$t_c = 2L/a$ (in s)	5.0	5.4	23	5.0	5.4	23
Flow (m/s [ft/s])	1.8	1.8	1.9	5.9	6.0	6.2
h (in m [ft])	219	206	49.4	718	676	162
P (in kPa [lb/in.^{2}])	2150	2020	480	310	290	70

6-5. Slow Closure of Valves

To limit the pressure rise, the maximum deceleration of the water during the critical time period, t_c, must be limited. The maximum allowable deceleration can be calculated by using the ratio

$$\frac{\Delta P_a}{\Delta P} = \frac{\Delta h_a}{\Delta h} = \frac{\Delta v_a}{\Delta v} \tag{6-5}$$

where ΔP is pressure rise, Δh is the head rise, Δv is the change in velocity, and the subscript, a, means allowable. The terms in the denominator are for instantaneous valve closure.

Manufacturers can provide either (1) curves of the valve headloss coefficient, K, in Equation 3-15 wherein $h = Kv^2/2g$ (shown for a butterfly valve in Figure 6-3) or (2) values of flow coefficients, C_v, wherein C_v equals water flow in gallons/minute at 70°F and 1 lb/in.^{2} pressure differential shown for ball and eccentric plug valves in Figure 6-4 (see Appendix A for conversion to SI units.). Both K and C_v vary greatly with valve opening, so either is a convenient aid for solving valve problems.

Equation 6-5 can lead either to simplified equations or to computer solutions for determining the allowable rate of valve closure so that a given pressure is not exceeded. To be valid, a method of solution must include

- the change of valve coefficient, K, in the formula for headloss (Equation 3-15), or
- the shape of the valve closure curve, C_v/C_{vo}, and the nonlinear action of some valve stems, and
- the effect of the increasing pressure, which tends to sustain the original flow through the valve. The pressure, for example, depends on the shape of an associated pump curve and the location of the valve as well as on other characteristics of the system.

Because resistance to flow is relatively small in the initial stages of closure, a valve can be quickly closed to near shut-off if the final closure is slow. Of the three gate valve closure programs occurring in the same interval, $3t_c$, as shown in Figure 6-5, the least pressure is generated by a 95% closure in $0.5t_c$ with the remainder of the closure occurring in $2.5t_c$.

A valve poorly selected with respect to C_v closure characteristics can often be the difference between a problem situation and no problem at all. Ball valves (with their worm and compound lever actuators) have the ideal characteristic of closing rapidly at first followed by very slow closure without any complicated control gear. The capacity of the valve is reduced 80%

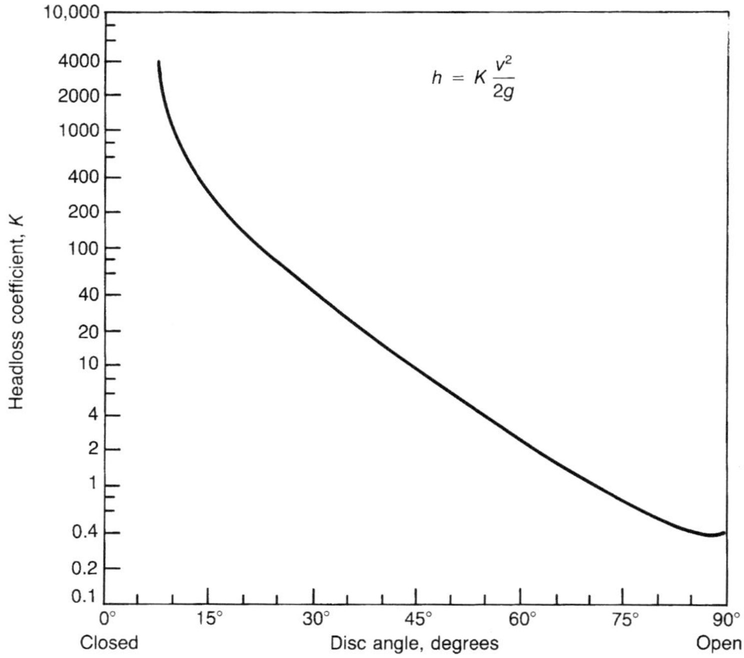

Figure 6-3. Headloss coefficient for a butterfly valve.

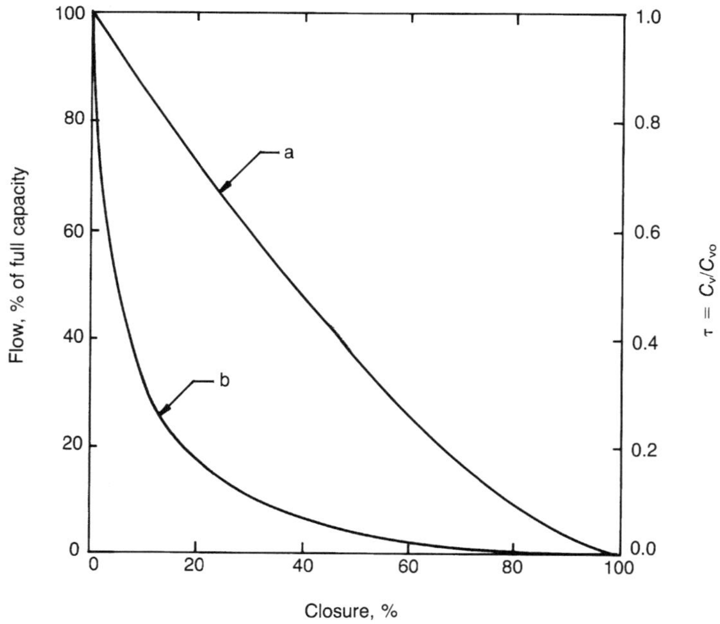

Figure 6-4. Valve closure functions. (a) Plug valve (DeZurik Series 100), the plug angle is linear with the time of closure; (b) ball valve (Williamette List 36), actuator travel is linear with the time of closure.

at about 20% closure (see Figure 5-2). In contrast, the eccentric plug with linear valve stem angle (or percentage of closure) has a near linear characteristic and would require either a much longer time for closure or a programmed closure to equal the ball valve's control of surge pressure.

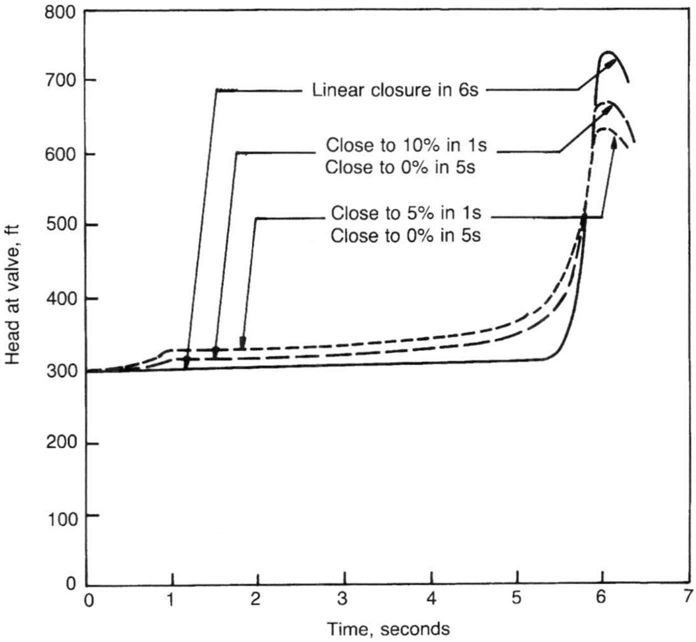

Figure 6-5. Water hammer caused by various gate valve closure programs. After Waters [4, p. 195].

As a rule, valves should not be closed in less than 2 to 10 times t_c. Insist on a computer solution wherever more certainty than this is needed, or when the problem is important because of the size of the pipe, the amount of flow, or the dynamic pressure involved.

Example 6-2
Seating a Valve

Problem: A 200-mm (actually, a 203-mm or 8-in.) butterfly valve discharges a flow of 12.6 L/s (200 gal/min) to atmosphere from a level pipe 4.83 km (3 mi) long with a gauge pressure of 690 kPa (100 lb/in.^{2}) just upstream of the valve. The valve design is such that it will close from its fully open position in 50 s.

Find the initial valve position (percentage of opening) and estimate the surge pressure when the valve closes.

Solution: First, find the percentage of valve opening from Equation 3-15 and a manufacturer's curve of K as shown in Figure 6-3. Then find the approach velocity in the pipe.

SI Units

$$v = \frac{Q}{A} = \frac{0.0126 \text{ m}^3/\text{s}}{(\pi/4)(0.200)^2} = 0.401 \text{ m/s}$$

U.S. Customary Units

$$v = \frac{Q}{A} = \left[\frac{2.00 \text{ gal/min}}{(\pi/4)(8/12)^2}\right]\left[\frac{0.00223 \text{ ft}^3/\text{s}}{\text{gal/min}}\right] = 1.28 \text{ ft/s}$$

The differential head across the valve is (see Tables A-8 and A-9):

$$h = \frac{P}{\gamma} = \frac{690{,}000 \text{ N/m}^2}{9789 \text{ N/m}^3} = 70.5 \text{ m}$$

$$h = \frac{P}{\gamma} = \frac{100 \text{ lb/in.}^2}{62.3 \text{ lb/ft}^3} \times \frac{144 \text{ in.}^2}{\text{ft}^2} = 231 \text{ ft}$$

SI Units

From Equation 3-16

$$K = \frac{2gh}{v^2} = \frac{2 \times 9.81 \times 70.5}{(0.401)^2} = 8600$$

U.S. Customary Units

$$K = \frac{2gh}{v^2} = \frac{2 \times 32.2 \times 231}{(1.28)^2} = 9080$$

The discrepancy between SI and U.S. units is due only to rounding off the difference between 200 mm and 8 in. From Figure 6-3, the valve is open about 7° (nearly closed).

The wave speed can be calculated from Equation 6-4 or simply estimated as about 100 g. For linear valve closure, the last 7° of valve movement would require (7/90)50 = 3.9 s.

$$a = 100 \times 9.81 \approx (980)$$

$$t_c = \frac{2L}{a} = \frac{2 \times 4830}{980} \approx 10 \text{ s}$$

$$a = 100 \times 32.2 \approx (3200)$$

$$t_c = \frac{2L}{a} = \frac{2 \times 3 \times 5280}{3200} \approx 10 \text{ s}$$

Because the valve will close in 3.9 s (less than the critical time, 10 s), it closes quickly enough to generate the full transient pressure given by Equation 6-2.

$$\Delta h = \frac{a \Delta v}{g} = \frac{980 \times 0.401}{9.81} \approx 40 \text{ m}$$

$$\Delta h = \frac{a \Delta v}{g} = \frac{3200 \times 1.28}{32.2} \approx 130 \text{ ft}$$

And the total pressure in the pipeline is

$$P = 690 \text{ kPa} + 40 \times 9.81 = 1080 \text{ kPa}$$

$$P = 100 \text{ lb/in.}^2 + 130 \times 0.433 \frac{\text{lb/in.}^2}{\text{ft}} = 156 \text{ lb/in.}^2$$

The assumption of a steady pressure upstream from the valve is often erroneous. The pressure usually changes with a change in flow as the system follows the characteristic pump H-Q curve. For pumps with type numbers greater than about 85 (specific speeds greater than about 4400), the head developed by a pump increases substantially as the flow decreases. If the type number is about 125 (specific speeds about 6600) or more, the increase in head is dramatic, as shown in Figure 10-16.

If the time of closure in Example 6-2 were greater than t_c, the resulting pressure surge would be less than the 40 m (130 ft) calculated, and a computer analysis would be used to reveal the pressure rise. Note that the shape of the valve closure curve would be very important in such an analysis.

6-6. Surge Concepts in Flow with Friction

The explanation of water hammer in Section 6-4 is simplified by omitting friction. Such simplification can be justified in some systems, such as those with short force mains in which friction head is a small part of the total design head. In most systems, however, friction contributes substantially to the total head, and it is important to understand its effect.

Figure 6-6 is similar to Figure 6-1a except for the hydraulic grade line, which slopes during steady-state flow. Shortly after sudden closure of the valve at point O, the wave front would arrive at point A. Consider the anomalies that would ensue by following the construction shown for frictionless flow in Figure 6-1a in which the pressure rise for the wave front at B forms rectangle abcd. Its counterpart is trapezoid oabc in Figure 6-6, but because the hydraulic grade line bc is sloping, water continues to flow past point A (corresponding to point B in Figure 6-1); therefore, v (and, hence, h) would be less in Figure 6-6 than in Figure 6-1a for the same original velocity. But flow continues only until the HGL becomes level. Meanwhile the wave front progresses to point B (Figure 6-6) and the same phenomenon occurs again, with more water flowing past point A, the pressure rising, the pipe expanding, and the water compressing. Wylie and Streeter [3] call this occurrence "packing." As the wave front approaches the reservoir, friction decreases v and h, a phenomenon called "attenuation." Eventually, the pressure builds to a maximum above the level of the reservoir and finally dies to the reservoir level.

Depending on the length of the pipeline, the pressure surges at the valve might appear, as shown in Figure 6-7. With each reversal of the pressure wave,

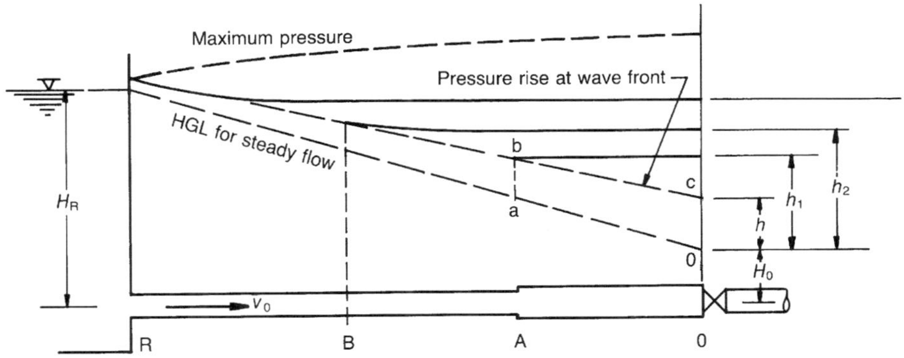

Figure 6-6. Packing and attenuation in a long pipe with friction.

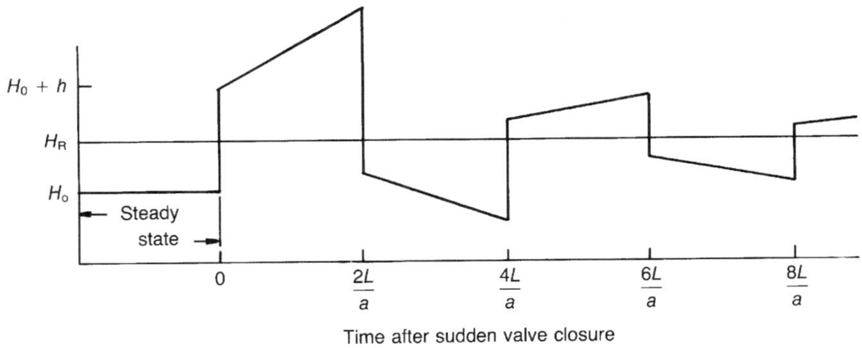

Figure 6-7. Surges at a valve in a pipe with friction.

friction decreases the pressure changes, so pressure eventually coincides with the reservoir level.

6-7. Column Separation

In the preceding sections, the pipelines are assumed to be level, but in real systems pipes may slope and downsurges result when power fails or when valves at the upstream end (the usual configuration) close quickly. Under some conditions, the downsurges can cause column separation—a condition to be avoided at any cost by a proper control strategy. As shown in Figure 6-8, a knee (a reduction in gradient or a change from a positive to a negative gradient) makes a pipeline especially vulnerable. If the power fails, the pumps stop quickly with an effect like closing a valve. The upstream flow (near the pumping station) stops whereas, due to inertia, flow continues at the downstream end (near the discharge). The static hydraulic gradeline begins to decay as shown by successive curves labeled $t = 1$ s, $t = 2$ s, $t = 3$ s, until, at $t = 4$ s, a slight negative pressure occurs between the pump and the knee. At $t = 4.5$ s, vapor pressure exists over a considerable length of the pipeline, and the water is boiling and forming large pockets of vapor. Column separation has occurred. On the upsurge of pressure that follows, the vapor pockets collapse and the two liquid columns can come together at literally express-train speed. Since the water is almost incompressible, the forces at impact can be enormous.

The knee makes the situation in Figure 6-8 worse, but note that column separation would occur with or without the knee and would occur even if the pipeline had a uniform gradient. However, if the pipeline profile were flat near the pump and the steep gradient occurred near the reservoir, column separation might be avoided—a useful control strategy that is maintenance free.

A computer analysis of a similar problem is given by Watters [4, pp. 235ff], and the results of still another somewhat similar problem with and without surge control devices are shown in Figures 7-12 and 7-13.

The hydraulic gradelines after power failure can sometimes be crudely estimated at several time intervals if the deceleration time of the pump is known or

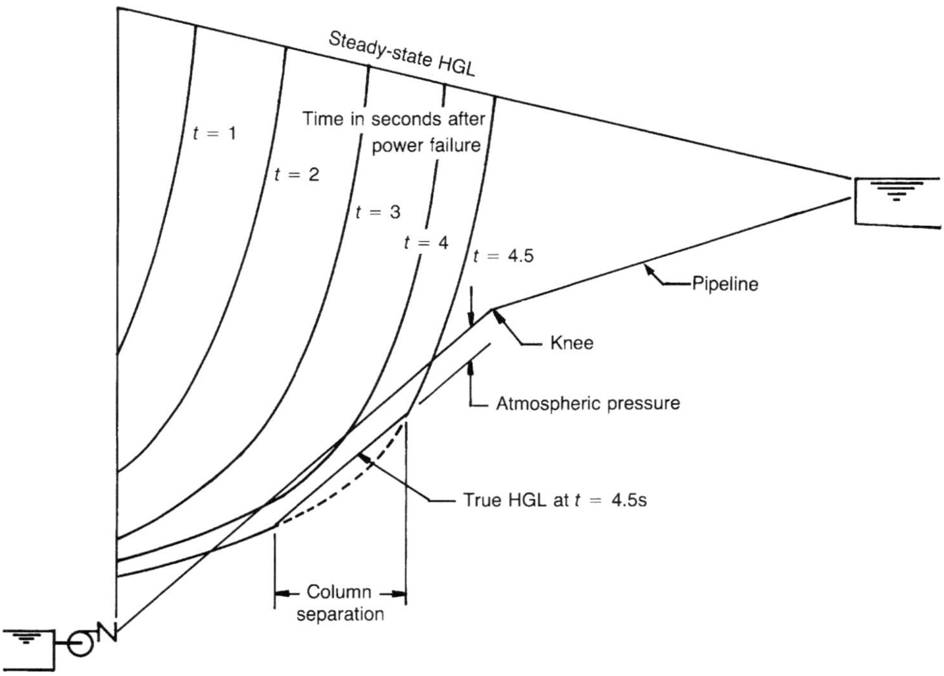

Figure 6-8. Successive hydraulic gradelines following power failure. Adapted from Watters [4, p. 271].

can be estimated. Flow through centrifugal pumps after power failure is a function of many variables, including

- inertia and speed of the pump, driver, and water within the casing;
- length and profile of the pipe;
- steady-state hydraulic gradeline;
- velocity of flow; and
- suction conditions in the wet well.

The true shape of the hydraulic gradelines requires solution by a computer.

6-8. Criteria for Conducting Transient Analysis

Every pump and pipeline system is subject to transient pressures, but in practice, it is impractical to spend the time and expense necessary to analyze all of them. The following empirical guidelines, which seem to be satisfactory in most (though certainly not all) situations, can be used to decide whether a complete transient analysis is required.

Do Not Analyze

- Pumping systems with flow less than 23 m^3/h (100 gal/min). Discharge piping is usually such that velocity is low and transient pressures are low. Even if transient pressures are high, small diameter (100-mm or 4-in.) piping has a high pressure rating and can usually withstand the pressures.
- Pipelines in which the velocity is less than 0.6 m/s (2 ft/s).
- Distribution systems or pipe networks (as in community potable water systems). The many junctions significantly dissipate the pressure waves.
- Reciprocating pumps, because virtually every reciprocating pump should have a pulsation dampener on the discharge (see Ekstrum [7] for methods of sizing such dampeners).
- Pumping systems with a static differential pressure between suction and discharge of less than about 9 m (30 ft).

Warning: it is possible that a very low static head coupled with a relatively high dynamic head could result in a column separation problem.

Do Analyze

- Pumping systems with a total dynamic head greater than 14 m (50 ft) if the flow is greater than about 115 m^3/h (500 gal/min).
- High-lift pumping systems with a check valve, because high surge pressures may result if the check valve slams shut upon flow reversal.

- Any system in which column separation can occur: (1) systems with "knees" (high points), (2) a force main that needs automatic air venting or air vacuum valves, or (3) a pipeline with a long (more than 100 m [300 ft]), steep gradient followed by a long, shallow gradient.
- Some consultants analyze any force main larger than 200 mm (8 in.) when longer than 300 m (1000 ft).

Checklist

Some additional insight can be gained from the following conditions, which tend to indicate the seriousness of surge in systems with motor-driven centrifugal pumps. A serious surge may well occur if any one of these conditions exists. If two or more conditions exist, a surge will probably occur with a severity proportional to the number of conditions met [8–11].

- There are high spots in pipe profile
- There is a steep gradient: length of force main is less than 20 TDH
- Flow velocity is in excess of 1.2 m/s (4 ft/s)
- Factor of safety (based on ultimate strength) of pipe (and valve and pump casings) is less than 3.5 for normal operating pressure
- There can be slowdown and reversal of flow in less than t_c
- There is check valve closure in less than t_c
- There is any valve closure (or opening) in less than 10 s
- There can be damage to pump and motor if allowed to run backward at full speed
- Pump can stop or speed can be reduced to the point where the shut-off head is less than static head before the discharge valve is fully closed
- Pump can be started with discharge valve open
- There are booster stations that depend on operation of main pumping station
- There are quick-closing automatic valves that become inoperative if power fails or pumping system pressure fails.

Criteria for determining whether to use simple hand calculations or a more detailed computer program are also given in *Pipeline Design for Water and Wastewater* [8, p. 65].

Shut-downs will occur, so plan for them. They can result in low pressures and column separation at knees in steep pipelines. Air venting valves that close too rapidly while an empty pipe is being filled also cause destructive hydraulic transients. Even on low-lift pumping stations, depending on the pipe profile, column separation can occur in the vicinity of the discharge header or farther downstream.

Computers

There is no simple, easy way to perform reliable transient analyses. Computer modeling is the most effective means available, but there are practical constraints on time and cost. Both computer time and the labor needed to analyze and review are expensive, so the extent of the analysis should be related to the size and cost of the project. For example, spending $1000 to $5000 to analyze transients for a $1,000,000 project is probably worthwhile even if no problem is found.

6-9. References

1. Parmakian, I., *Waterhammer Analysis,* Dover, New York (1963).
2. Rich, G. R., *Hydraulic Transients,* Dover, New York (1963).
3. Wylie, E. B., and V. L. Streeter, *Fluid Transients,* Feb Press, Ann Arbor, MI (1982, corrected copy 1983).
4. Watters, G. Z., *Analysis and Control of Unsteady Flow in Pipelines,* 2nd ed., Butterworths, Stoneham, MA (1984).
5. Chaudhry, M. H., *Applied Hydraulic Transients,* Van Nostrand Reinhold, New York (1979).
6. Wood, D. J., and S. E. Jones, "Water hammer charts for various types of valves," *Journal of the Hydraulics Division, Proceedings of the American Society of Civil Engineers,* 167–178 (January 1973).
7. Ekstrum, J. D., "Sizing pulsation dampeners for reciprocating pumps," *Chemical Engineering* (January 12, 1981).
8. *Pipeline Design for Water and Wastewater,* American Society of Civil Engineers, New York (1975).
9. AWWA MII, *Steel Pipe—A Guide for Design and Installation,* p. 54, American Water Works Association, Denver, CO (1985).
10. Kerr, S. L., "Minimizing service interruptions due to transmission line failures: Discussion," *Journal of the American Water Works Association,* **41,** 634 (July 1949).
11. Kerr, S. L., "Water hammer control," *Journal of the American Water Works Association,* **43,** 985–999 (December 1951).

Chapter 7

Control of Hydraulic Transients

BAYARD E. BOSSERMAN II

CONTRIBUTORS
Robert C. Glover
William A. Hunt
Joseph R. Kroon
M. Steve Merrill
Gary Z. Watters

The advantages, disadvantages, and typical uses of various kinds of control devices (such as tanks and valves) and control strategies (such as timing of valve operations, slow filling of empty pipelines, and adequate maintenance) used to limit transients are discussed in this chapter. Coping with transients is site specific, so the discussion is limited to examples that illustrate only a few of the many different problems designers may encounter. Chapter 6 is prerequisite to this chapter.

The approach to design is the same for all strategies. If surges are expected to be severe (see Section 6-8), the piping system is modeled for a computer solution and the magnitudes of surges for the critical conditions are determined. If they are excessive, a promising control scheme is selected and the system is then analyzed again by computer. The process of selecting an appropriate candidate solution and analyzing the results is continued until satisfactory control and reliability are achieved for minimum (or at least supportable) cost—a decision that usually requires considerable judgment.

Although the details of computer analyses are beyond the scope of this book, some of the factors involved in choosing a solution are included.

Standards and specifications given by a coded number (e.g., AWWA C401) are referenced in Appendix E. Addresses of the publishers are given in Appendix F.

7-1. Overview of Hydraulic Transient Control Strategies

Every piping system for water and wastewater should be evaluated for water hammer. From a review of the plan and profile of the pumping station and pipeline system as well as the operating scheme, it is frequently possible to determine where potential hydraulic transient problems may exist and what means might be taken to control them. For a quick check to determine the potential for column separation, draw a mirror image of the hydraulic profile, as in Figure 7-1. If the pipe lies well above the mirror image, column separation might occur or even be likely. If column separation can occur, or if the criteria for required analysis in Section 6-8 are met, use (or contract for) a trustworthy computer analysis of the system [1].

Make every effort to eliminate (or, at least, to mitigate) transients by avoiding knees, high spots, steep gradients near the pump, and air (or, much worse, vacuum) in pipes. When any of these conditions cannot be avoided, use a combination of pipe strength and

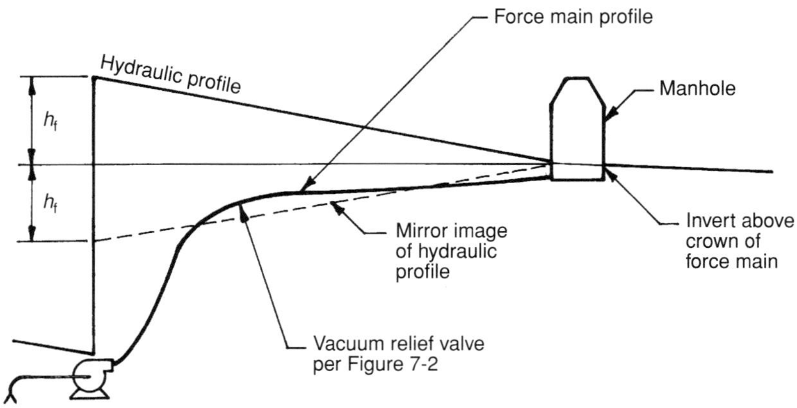

Figure 7-1. Construction for indicating the probability of water column separation in pipelines.

control strategies to provide adequate protection at reasonable cost.

Recommended Options for Wastewater Systems

Because sewage contains (1) microorganisms that form hydrogen sulfide and sulfuric acid; (2) grease (sometimes in large quantities) that floats, tends to stick to walls, and can eventually develop into a hard blanket; (3) stringy material (rags, paper) that tend to wrap around any protrusion; and (4) grit that sinks into any depression, only a few strategies can be used to control water hammer in wastewater pumping. See Section 7-7 for additioinal details. Briefly, surge control strategies are limited to the following, more or less in order of preference:

- Rerouting the force main to avoid knees or negative gradients so that air traps are eliminated and, potentially, the need for special surge protection can be avoided.
- Using an alternative source of power, such as an engine on some pumps, so that the entire system cannot fail at once. Properly designed, engine drives offer superior protection against power failure. Engines rarely fail or shut down suddenly even when starved for fuel, and—unlike motors in constant speed (C/S) operation—in normal operation, speed can be increased or decreased gradually.
- Using pump control valves that open and close gradually, or (on very long pipelines) using variable speed (V/S) drives for ramping speed up and down slowly. Either prevents the continual pounding of the system. The latter, however, offers no protection when power failure occurs.
- Increasing the rotating moment of inertia of the pump and driver by adding a flywheel.
- Using high-strength pipe. This option is limited to small systems, but if the force main is 200 mm (8 in.) in diameter or smaller, it may be the cheapest solution. But be aware that the continual pounding of surges may cause leaks and eventual failure.
- Providing air inlet (vacuum relief) valves at critical points to prevent the high surge that follows column separation. Use only swing check valves as in Figure 7-2, because other types of valves may clog with grease or particulate material. Install such valves only where the pipe slopes upward to its discharge (or at worst is level), so that the air admitted can be swept downstream to a free surface. Each valve should be sized to match the volumetric flow of water with the volumetric flow of air at a pressure drop of about 14 kPa (2 lb/in.^{2}) or less. Fleming [2] has described a successful installation of air inlet valves for a 7100-m^3/h (45-Mgal/d) pumping station.
- Providing air scouring velocities (per Table B-9) at knees (together with vacuum relief valves as per Figure 7-2) or in any pipe laid flat or at downward gradients. If required by the owner or a regulatory agency, adding air release valves could serve a useful purpose in preventing small accumulations of air and eliminating an environment conducive to corrosion. (Such valves should be of the tall body form and made of stainless steel to resist corrosion caused by hydrogen sulfide.) If, however, the air release valves are clogged, the high velocity will scour out the air (or gas) anyway and thus prevent a catastrophe. To minimize friction losses, only the pipe on a flat or downward gradient need be made small to obtain the high velocities required. (Note that the custom of never decreasing the diameter of

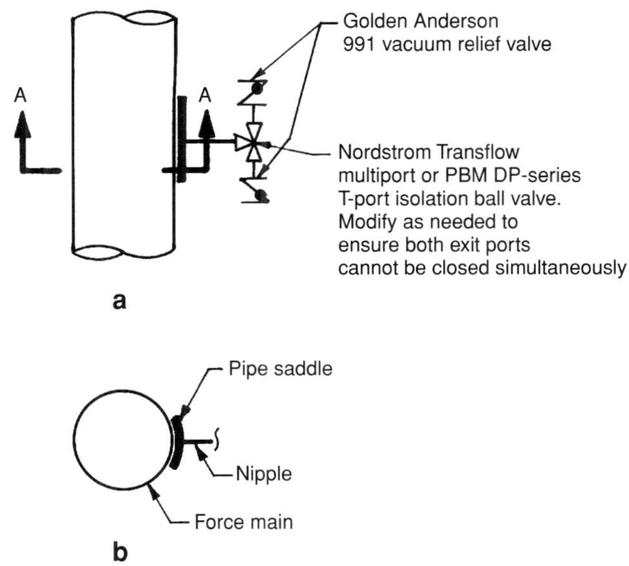

Figure 7-2. Vacuum relief (air inlet) valve scheme for sewage service (a) Plan; (b) Section A–A.

a sewer line does not apply to force mains.) Air-scouring velocities must be reached frequently—at least once per day.

In the device of Figure 7-2, the saddle is on the side of the pipe to inhibit the entrance of scum and grit. Install the check valves in pairs so there is always protection while one is being serviced. The clapper is held closed by a light counterweight or spring until a partial vacuum in the force main overcomes the counterweight or spring (usually at 7 to 14 kPa or 1 to 2 lb/in.^{2}) and opens the valve to admit large (equal in volumetric flow to the flow of water) quantities of air and thus prevent the formation of water vapor. The three-way valve must be of a type that can never close off both check valves at once.

The velocity of flow in wastewater force mains should reach 1.2 m/s (4 ft/s) once every day or so to ensure self-scouring. It is wise to include pigging facilities because pigging removes the bacterial slimes that produce hydrogen sulfide.

Options Undesirable for Wastewater Systems

Protection strategies that are undesirable or impractical in varying degrees for wastewater systems include:

- Air chambers. The use of air chambers is highly controversial, but in any situation, it is always a means of last resort. Opponents state that grease will enter the tank with each surge event, and eventually the grease will harden and can clog piping, thus eliminating the supposed protection. Some operators have stated that they immediately close valves to remove surge tanks from the system because of grief with them in the past. Other operators have not experienced the above difficulties and have not been upset over surge tanks in their pumping stations. A few have been enthusiastic, because the tanks effectively controlled surges. On the other hand, many have complained about the maintenance required for the compressors, level controls, pressure relief valves, and grease accumulations on sight gauges. Proponents also claim that running a stream of fresh water (with, of course, proper backflow prevention) into the tank keeps out the wastewater, although opponents reiterate that, as grease floats and surges introduce wastewater anyway, grease will accumulate and may eventually prevent proper functioning and expose the designer to lawsuits. In a water-short situation, a steady flow of fresh water into a surge tank at a rate needed to purge the tank effectively may be unacceptable.
- Air-vacuum valves. The protection these valves afford is most uncertain, and their life-cycle cost is exorbitant. But because their first cost is the lowest of all, they may sometimes be demanded by the owner. If they are to be used despite these objections, the owner should be warned in the preliminary design report and in the O&M manual that even the best of these valves will require a high level of maintenance—a minimum of servicing

monthly—wherein the best maintenance is replacement of the valve's interior mechanism and cleaning the replaced parts in the shop. Some engineers insist on placing such valves in pairs so that if one fails, the other might still protect the pipeline.

Recommended Options for Water Systems

All of the recommended strategies for wastewater also apply to water systems. Additional options for water systems include:

- The use of air chambers or standpipes (see Table 7-1).
- The use of one or more air release and vacuum relief valves—an inexpensive first-cost solution but one in which inspection and cleaning is required twice per year as a minimum One major pipeline failed because in 20 years of service, the air release and vacuum relief valves had never been maintained. Recognize that the failure of any means for protecting a pipe can lead to utter disaster.

Means for coping with transients are discussed in detail in Section 7-6.

Power Failure

Power failure will occur several to perhaps many times, so plan carefully for it. Power failure at a pumping station results in an initial rapid downsurge in the discharge header and piping close to the pumping station. The change in head that occurs in the pump discharge piping after power failure (or pump shut-off) is shown in Figure 6-8. The following methods can be used to reduce the magnitude of a downsurge.

- Increase the inertia of pump and motor by adding a flywheel. Engines have higher moments of inertia than do electric motors, and it is easy to increase the size of the flywheel. Flywheels control downsurge and can prevent column separation (which is especially important in sewage systems), whereas control valves (which control upsurge) cannot.
- Install an air chamber (see Figure 7-3) near the pumping station and, perhaps, at other points as well. Check valves near air chambers must be fast acting but slamproof because the short water column between the air chamber and the check valve can quickly accelerate.

Column Separation

The most serious consequence of downsurge is column separation, which must always be avoided. It is more likely to occur if there are knees in the pipeline (as in Figure 6-8) and is almost certain to occur at a high point. Column separation can also occur at other places along the pipe. The location and extent depends on the relative relationship between the dynamic head, the total head, and the profile of the pipeline (review Section 6-7). A quick way to determine whether column separation is likely is shown in Figure 7-1. If the mirror image of the hydraulic gradient intersects the pipeline profile, column separation will occur. Column separation may also occur as depicted in Figure 6-8. Various means for preventing column separation in raw sewage force mains include (but are not limited to) the following:

- Relocate the pumping station and the associated pipeline to avoid knees or high points.
- Reroute the pipeline or bury the pipe deeper to achieve a flat gradient near the pump so that successive hydraulic grade lines (as in Figure 6-8) do not intersect the pipe.
- Install one-way air valves or, preferably, a check valve to admit air into the pipeline on the downsurge and to trap the air on the upsurge.
- Add a flywheel sized to prevent the column separation.

In addition to these four methods, column separation in water transmission pipelines can be prevented by the following means:

- Install open surge tanks (standpipes) at knees or high points if the required height is not excessive. The stored water can control both high and low pressures (see Figure 7-4 and refer to Chapter 10 in Watters [3]).
- If the hydraulic gradeline (HGL) is too high for an open surge tank, a one-way (Figure 7-5) or a two-way (Figure 7-6) surge tank can be used to keep the HGL above the pipe. A check valve allows flow from the tank to the pipe but not from the pipe to the tank.
- Install air valves, more specifically called "vacuum relief and air release valves" or "combination air valves" (Figure 5-16), that admit air into the pipe and thereby prevent a vacuum. Upon the following upsurge, the air must be exhausted slowly enough to prevent overpressure when the two water columns meet.
- Install an air chamber (pneumatic surge tank). More than one might be required for very long pipelines.

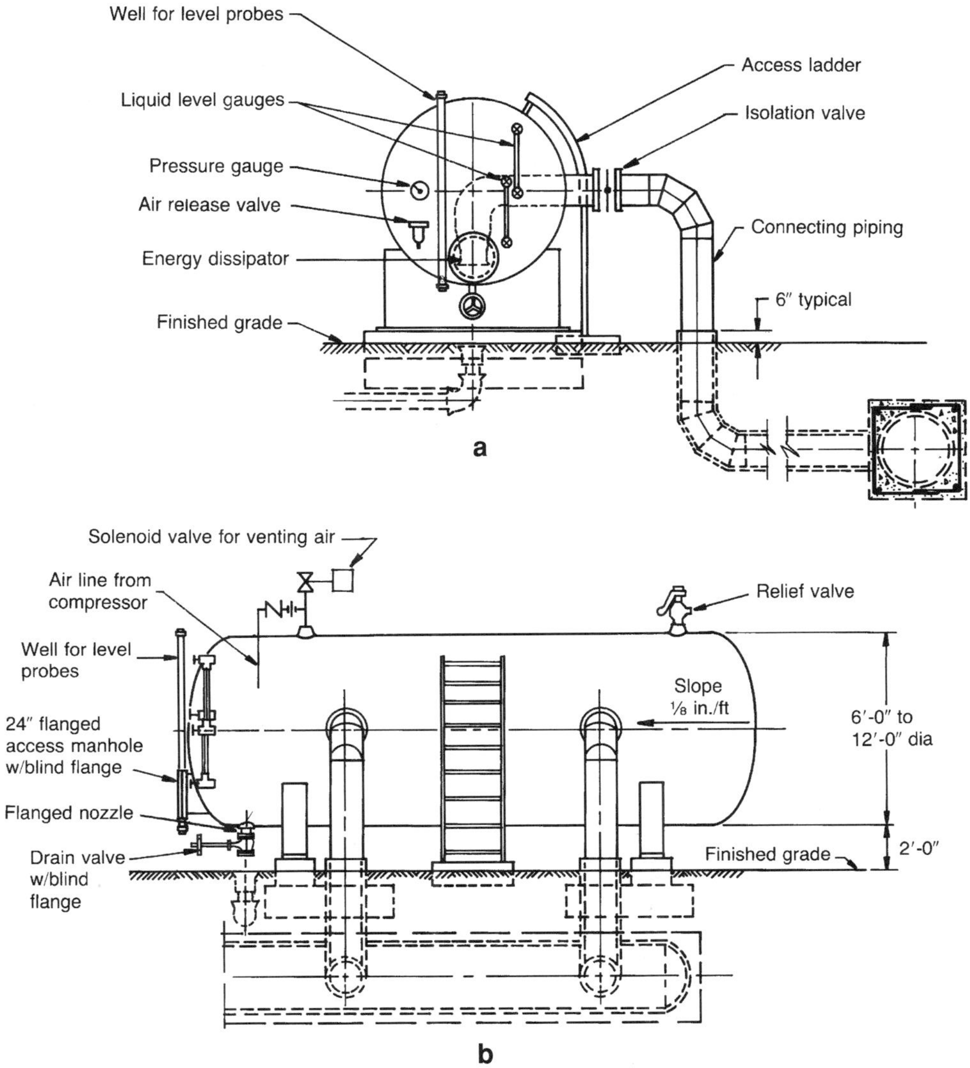

Figure 7-3. Horizontal air chamber for clean water service. (a) End elevation; (b) Side elevation.

Except for pipeline rerouting and the flywheel, all of these methods require routine maintenance, which cannot be guaranteed and is not always provided.

Start-Up

Pump start-up can cause a rapid increase in fluid velocity that may result in an undesirable surge, but usually it is not a problem unless the type number of the pump exceeds about 135 (specific speed exceeds approximately 7000). (Refer to Figure 10-16 for type number and specific speed.) Methods for controlling such surges, in order of increasing cost, are as follows:

- If there are several pumps, start them one at a time at intervals from four to ten times the critical period ($t_c = 2L/a$).
- Program a pump-control valve to open slowly (again from four to ten times t_c) after the motor starts. Pump motors typically reach full speed within 1 to 2 s after energization (see Figure 7-7).
- Use a variable-speed drive for each pump ramped up to full speed slowly enough (from four to ten times t_c) to avoid high surges.
- Add an air chamber (hydropneumatic tank), but only if quick-acting, slam-resistant check valves are used.

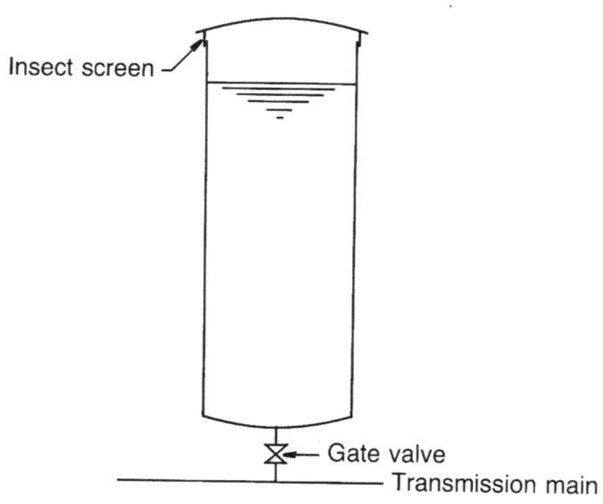

Figure 7-4. Open-end surge tank or standpipe.

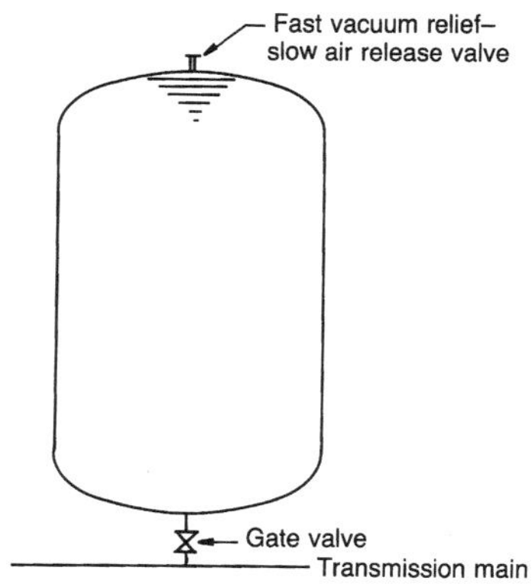

Figure 7-6. Two-way surge tank.

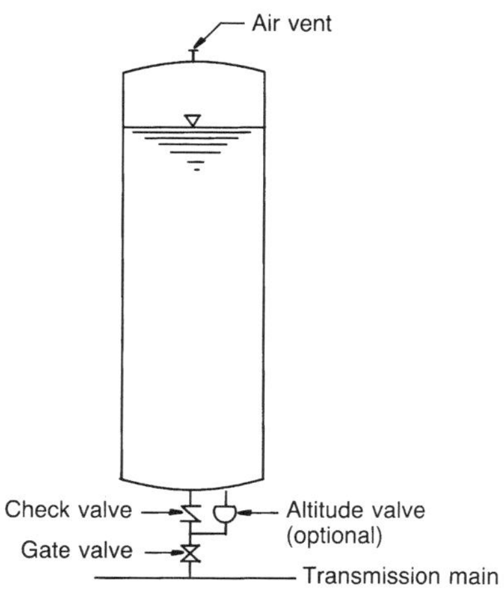

Figure 7-5. One-way surge tank.

Shut-Down

Normal pump shut-down may also cause surges. They can be controlled to remain within acceptable bounds by the following methods:

- Turn pumps off one at a time at intervals from four to ten times t_c.
- Program a pump-control valve to close slowly (four to ten times t_c) before the motor is stopped.
- If pumps are equipped with variable-speed drives, ramp down slowly.
- Increase the inertia of the motor and pump unit so that it coasts to a stop over a longer interval.
- Add an air chamber. The air chamber should not be needed for normal pump shut-down, but if one is installed for other reasons, it would be an effective control method.

Check Valve Slam

If the pumping station is small, or if there are several duty pumps, the sequencing of the pump shut-down is usually adequate to prevent valve slam. However, when the last pump shuts down, or when power fails, the liquid comes to rest and then reverses. An unrestrained check valve disc, seating after the fluid flow reverses, closes with a resounding and disconcerting slam that may shake the building. The slam may or may not be accompanied by a significant hydraulic surge. If slam does occur, there are several remedies that may, depending on circumstances, mitigate the problem.

- Ensure quick closure of the valve *before* the flow can reverse to cause a slam
- Ensure slow, gentle seating of the disc (especially during the last 5% or so of closure) by adding an oil-filled dashpot to an outside lever attached to the disc axle
- Substitute a pump-control valve that cannot slam. It must be equipped with a stored energy closure mechanism to ensure operation when power fails
- Equip the check valve with an oil-filled dashpot to allow the valve to close quickly but to seat gently
- Use rubber-cushioned flapper seats to prevent metal-to-metal contact and to cushion the closure
- Install a pressure-actuated relief valve (see Figure 7-8).

For a more extensive discussion, refer to "Slam" in Section 5-4.

Choosing Check Valves

Choosing the right check valve is vital. Often there are profound differences in the same kind and style of valve offered by different manufacturers. Headlosses

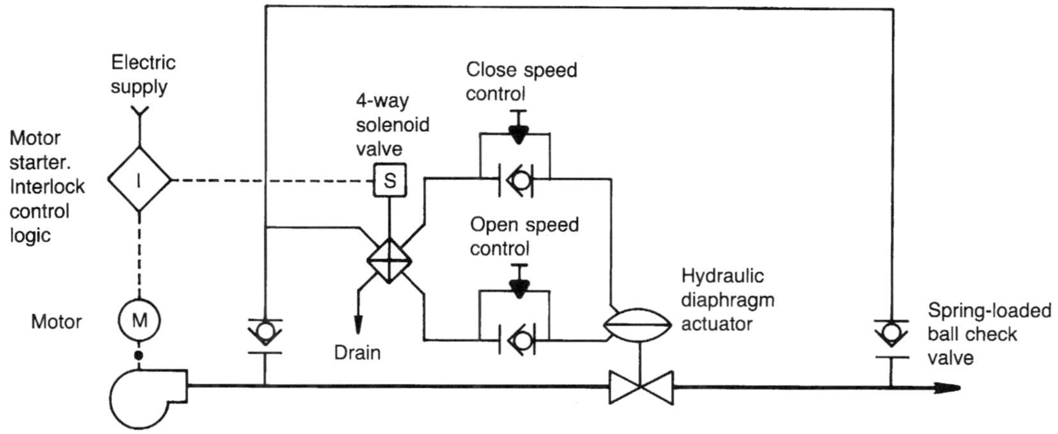

Figure 7-7. Pump-control-valve system. To ensure the highest available operating pressure, take the control water supply from both sides of the check valve. Provide separate speed-control needle valves on each side of the double-sided diaphragm to control the water flowrate out of the diaphragm actuator.

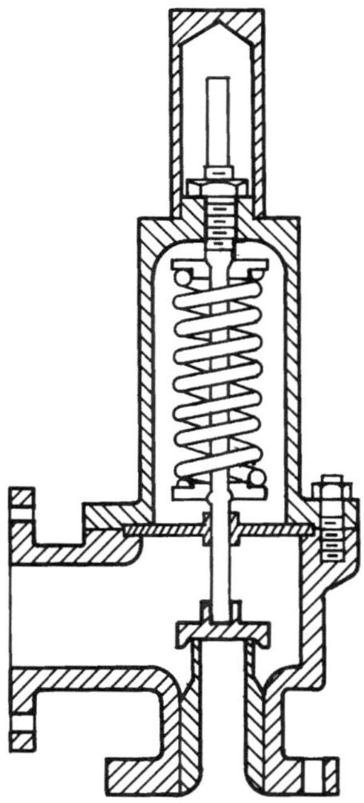

Figure 7-8. Pressure-actuated surge relief valve.

in some makes of swing check valves are twice as great as in others, and the massiveness of stressed parts varies widely.

Swing check valves should always be equipped with external levers, which are useful in several ways. The position of the lever indicates whether flow is occurring, and the lever can be equipped with an inexpensive switch that shuts off power to the motor (after a timed delay) if flow does not occur. The lever can be equipped with either springs or counterweights, which can be adjusted to mitigate slam or disc flutter. Dashpots can also be added to control slam. But whether counterweights, springs, dashpots, or pump-control valves are used, it is mandatory that the valves be properly adjusted in the start-up procedure and equally important that operators understand the valve operation and practice the necessary preventive maintenance.

Because many contrary opinions prevail, designers should be extremely careful in specifying check valves. Investigate them thoroughly by obtaining and analyzing advice from several sources.

Filling Empty Pipelines

Small air bubbles, which collect at summits of pipelines, can be bled off with air release valves without creating hydraulic transients. However, the initial filling of pipelines must be done cautiously with velocities kept below 0.3 m/s (1 ft/s) and with air release valves open to exhaust the air slowly. Avoid full-capacity start-ups until all of the large air bubbles are exhausted. Provide properly sized air and vacuum valves or slow-acting pump-control valves for pumps with long discharge columns. These measures complete air evacuation without sudden slamming of valves, which creates high transient pressures. Always include the start-up procedure for empty pipelines in the O&M manual.

Pipelines that are nearly flat may require air release valves spaced at intervals of approximately 400 to

800 m ($^1/_4$ to $^1/_2$ mi) to vent air properly during filling. Knees and high points require both air and vacuum valves. At best, air pockets in pipelines increase flow resistance by as much as 10% or even more. At worst, air pockets can generate pressures as high as ten times normal operating pressures, and air trapped at high points reduces the water cross-sectional area and, thus, acts like a restriction in the pipe.

Air release cocks or valves must be installed in pump casings (or at high points of pump manifolds) to prevent air binding. If the HGL can fall much below any part of the system on a downsurge, an air and vacuum valve is needed to prevent vapor cavities. Limit the vacuum to about half of an atmosphere, and exhaust the air slowly so that it acts as a cushion. Alternatively, especially for raw sewage, reroute the pipeline to produce a uniform or, better, an increasing gradient with no knees.

7-2. Control of Pumps

None of the methods for controlling water hammer is universally applicable. Some methods might control one cause of water hammer but leave the system unprotected from other causes. Some methods may be unacceptable for a variety of reasons, such as excessive maintenance or unreliability. Because a single device is often inadequate, several must be used to provide full protection. Several schemes for controlling pumps can be used to limit surges during start-up and shut-down. Some methods offer limited control of surges due to power failure.

Pump Sequencing

By controlling the sequence of pump start-up and shut-down so that the starting and stopping of several pumps are staggered, the magnitude of change in flow at any one time is reduced—often to acceptable levels for normal operation. Sequencing is automatic, reliable, and inexpensive.

Pump-Control Valves

By interlocking the pump with control valves in its discharge piping, transient problems caused by pump start-up and shut-down can be greatly reduced. The control valves are set to open and close slowly (4 to 10 times t_c). Upon start-up, the pump operates against a closed valve. As the valve opens, the flow into the pipeline gradually increases to the full pump capacity. Upon shut-down, the control valve slowly closes to decelerate the flow, after which power to the pump is shut off (but not until the valve is fully closed).

To circumvent power failure, the valves should be operated by a stored-energy auxiliary power source such as trickle-charged batteries for electric systems or a compressed air tank (with enough capacity to operate every valve through two cycles) and either a water or an oil valve actuator. Ball valves are excellent for either water or sewage service. For water service, butterfly valves can be used, but they have poor throttling characteristics when slow closing is required to control head rise following a power failure. Diaphragm-actuated globe valves operating off pipeline pressure with a check valve feature can also be used. For sewage service, eccentric plug valves can be used. Pump-control valves, however, cannot prevent downsurge on power failure.

Increasing the Rotational Inertia

The moment of inertia of the pumping system has an important effect upon hydraulic transients in the piping downstream from a pumping station. Upon pump power failure, the pump speed and head rapidly decrease and a negative pressure wave, or downsurge, is propagated down the pipeline. The greater the moment of inertia, the slower this decrease in head and pump speed and the lower the magnitude of the resulting downsurge becomes. Thus, adding a flywheel increases the moment of inertia of the system and slows the decrease in pump speed and head. A flywheel can be added to an electric motor by extending the housing or frame (usually toward the pump) to enclose the flywheel and to support bearings above and below the flywheel. When the conditions are suitable, high inertia is the most reliable method of all for control of surge in both normal operations and power failure, and, except for maintaining the bearings, it is maintenance free. An alternative is to use an engine. It is easy to enlarge the flywheel, and engines inherently take longer to come to a stop (even when starved for fuel) than do motors.

Variable-Speed Drives

A variable-speed drive that can be ramped up or down slowly has several advantages for controlling transients during normal operation: (1) enhancement of motor life due to infrequent starts and, for adjustable frequency drives, low inrush current; (2) flicker of lights (annoying to nearby residents) caused by inrush is avoided; and (3) the change of flow is gradual and does not upset primary sedimentation tanks or other processes in a wastewater treatment plant.

The disadvantages of a variable-speed drive are that (1) it does not provide protection from power failure, (2) such a drive for a motor is expensive (it costs more than the motor), (3) it may require special training for maintenance workers, (4) it adds complexity, and (5) it reduces reliability somewhat. If surge control is the only objective, variable-speed drives for electric motors are seldom the best answer.

Engines can be easily used for variable-speed drives unless there is prolonged operation at low loads. Note that engines are more reliable than electric power, although the maintenance required is high.

7-3. Control Tanks

Control tanks range from the simple, reliable standpipe to hydropneumatic tanks (air chambers). If all else fails, they are, *when properly maintained,* a positive and reliable means of control.

Avoid using surge tanks for sewage. Air chambers are sometimes used with sewage, but they should be flushed intermittently or continuously. Some installations are successful if well maintained, but some experienced engineers will not use air chambers for sewage under any circumstances. Air chambers are worse than useless if not maintained, because they become inoperable yet nevertheless give a false sense of security. The consequences of inadequate maintenance should be dramatized in the O&M manual.

Standpipes

Standpipes or open-end surge tanks (Figure 7-4) provide a free water surface at atmospheric pressure and act as small reservoirs to accumulate or supply water temporarily to control pressure variations. They are useful at high points where the hydraulic grade line is within 20 or 30 ft of the ground. They are simple to construct, easy to maintain, require no power or other utilities, and are completely automatic in operation.

Overflow drain, piping, and a water disposal area must be provided because upsurges often cause overflow. The height of the tanks may be aesthetically unacceptable. They cannot be used for pipelines with a large variation in HGL, and they must be protected from freezing in cold climates.

One-Way Surge Tanks

One-way surge tanks (Figure 7-5) are somewhat similar to standpipes, and, because they can be pressure vessels as well as gravity tanks, they are especially useful if the original HGL is too high to allow the use of a standpipe. If used as a pressure vessel, the tank must be closed with an airtight cover containing air release and vacuum relief valves. But unlike standpipes, they permit water to flow only from the tank into the pipeline. Therefore, a water-disposal system or storm drain (which is needed for a standpipe) is not required. One-way tanks are sometimes used at pumping station discharge headers in low head (15-m or 50-ft) applications. They are simpler in design and operation than hydropneumatic tanks, but are more complex than standpipes.

Electrical power is required to operate the solenoid valve on the fill line. Special attention to the selection and specification for the connecting check valve is required. The valves are usually in low-pressure service and require a soft seat material to prevent water from leaking past the disc.

Two-Way Surge Tanks

Two-way surge tanks are similar to one-way surge tanks except that they have no check valve and no external fill line (see Figure 7-6). Ordinarily, the tank is full of water. In this condition it is ineffective for an upsurge, but on a downsurge it releases water into the pipeline while a large vacuum relief valve allows air to flow into the plenum to avoid a vacuum. If there is a return upsurge, water flows into the tank against the increasing pressure of air in the plenum, which can escape (but only slowly) through a small air release valve. In this operational sequence, the two-way tank behaves much like an air chamber. Eventually, the surges dissipate and the tank gradually fills with water again.

Air Chambers

An air chamber (or hydropneumatic tank) is a pressure vessel about half filled with air and half filled with water and connected to the piping system on the discharge side of the pumps (see Figure 7-3). The air in the tank is compressed, which stores energy available to sustain flow after power failure. Pipe inlets to these tanks are usually fitted with differential orifices that allow water to exit the tank with low loss but cause high energy dissipation for flow into the tank.

The operation during a downsurge-upsurge sequence is as follows:

- Upon pump shut-down after power failure, the air pressure forces water out of the tank and into the pipeline. As water leaves the tank, the volume of air increases and the pressure decreases.
- The decreasing pressure causes the flow into the pipeline to decrease gradually. Eventually, the flow in the pipeline stops and then reverses, and the downsurge gives way to an upsurge.
- As water enters the tank, the volume of air decreases and the pressure increases.
- The increasing pressure causes the flow to decrease. Eventually, the flow in the pipeline stops and then reverses. The cycle then repeats until friction gradually halts it.

Because the tank must be large enough for a reserve of water to remain at the end of the downsurge and for a reserve of compressed air to remain in the plenum at the height of the upsurge, the computer modeling must be accurate. Air chambers are very effective and reliable for controlling both upsurge and downsurge and so versatile that they can be used in almost any water pumping system, although they are unusual in systems with force mains smaller than 250 mm (10 in.). They are commonly used on the pumping station headers and at high points. Air chambers require a fair amount of complex auxiliary equipment and controls and, hence, need frequent maintenance. For long, large transmission mains, the site area may need to be enlarged.

The relationship between air pressure and air volume within the air chamber can be described by the equation

$$P_1 V_1^{1.2} = P_2 V_2^{1.2} \tag{7-1}$$

The use of the empirical exponent 1.2 is standard practice and describes a gaseous expansion between isothermal (exponent = 1.0) and adiabatic (exponent = 1.43). Some analysts use an exponent of 1.3.

The following are the recommended design criteria and accessory equipment for air chamber systems:

- An air compressor to supply air automatically to the chamber. Oil-free air compressors are required for potable water systems by some health departments.
- A solenoid valve to vent excessive air from the chamber.
- Level probes, float switches, or capacitance probes in the chamber or in a separate probe well to start and stop the compressor and open and close the solenoid valve.
- Liquid-level gauges or sight glass to observe water level in the chamber. Equip the sight glass with stopcocks and cleanout plugs so that the glass can be cleaned while the tank is in service.
- A flanged access opening for inspection and maintenance.
- A flanged connection to the main pipeline.
- Design the air chamber in accordance with the ASME Boiler and Pressure Vessel Code, Section VIII [4]. The air chamber should bear the ASME Code stamp. Code vessels are, however, very expensive, especially in large sizes.
- Provide a poppet safety-relief valve on the air chamber.
- Specify slosh plates in horizontal air chambers. Use horizontal air chambers only for clean water—never for sewage.
- The ratio of air:water at normal operating (steady-state) pressure should be about 1:1.
- If an air chamber is used for sewage or dirty water service, use only a vertical tank with a hopper bottom (conical or elliptical) and a fresh water supply (for flushing the tank) pumped through a small (25- or 38-mm [1- or $1^1/_2$-in.]) pipe. The pump can run either constantly or intermittently. However, read the warning in Section 7-1 before deciding to use an air chamber.

7-4. Valves for Transient Control

Valves can be used effectively to control both upsurge and downsurge at pump start-up and shut-down, but they cannot prevent downstream column separation if that is the problem (as in Figure 6-8), nor can they prevent downsurge upon power failure.

Air and Vacuum Control

Vacuum Relief Valves

The only vacuum relief valve unequivocally recommended for wastewater service is shown in Figure 7-2. Although it is reasonably well protected from scum, grease, and stringy material, it must nevertheless be serviced on a regular schedule—at least once every four months.

Vacuum relief valves for water service are usually accompanied by air release valves, as described in the next subsection.

Air Release Valves

For bleeding air from pump casings or air pockets from other high points within a pumping station, a simple hand-operated cock is sufficient in small stations. For large stations and for long, flat pipelines, use air release valves that close automatically when air is expelled. Add a check valve so that air cannot be sucked back into the valve.

Air Release and Vacuum Relief Valves

Air release and vacuum relief valves (or "air/vac valves") are needed to remove all of the air during pump start-up and to introduce air to prevent a vacuum after pump shut-down. A critical consideration is sizing the air exhaust to prevent excessive shock when the water columns meet after column separation. Flexibility can be obtained by adding a throttling device for optimizing the release of air.

Air/vac valves may not be able to prevent vapor pockets at high points, because air may not be admitted quickly enough. The location of a vapor pocket in a pipeline is critical and cannot always be predicted accurately, so valves cannot ensure complete protection. Furthermore, the valves do nothing to relieve high upsurge pressures.

If air/vac valves are used on sewage force mains, specify stainless-steel trim and provide quick-connects for freshwater flushing. Instead of cleaning the float and linkage mechanism by flushing the valves in the field, it is preferable to replace the interior mechanism with a clean one and take the dirty one to the shop for a thorough cleaning and overhaul. According to Murphy's law, however, someone will eventually forget to open the isolation valve and render the valve useless unless the isolation valve functions like the one shown in Figure 7-2. It should be possible to design most force mains so that air/vac valves are not required. Some experienced engineers forbid the use of such valves for sewage under any circumstances (see Section 7-1) and have always managed to find a different control strategy.

Check Valves

Substantial control of surges is obtainable by selecting the correct valve. Some valves shut off very fast and might be preferred for a short (1-km or $^1/_2$-mi.) transmission line. Some can be adjusted to close very slowly or to close at different rates in three stages. Always choose a valve that closes automatically on stored energy when power fails.

Swing Check Valves

Swing check valves should be supplied with an outside lever with a weight or a spring adjusted to close the valve just before fluid reversal occurs following pump shut-off or power failure. Such check valves are especially useful in low head (15-m or 50-ft) pumping stations. Note that head loss depends less on flowrate than it does on the adjustment of the weight or spring (see Figures B-2 and B-3).

Cushioned Check Valves

An alternative to quick closure is cushioned closure. Cushioned check valves have either an air-filled or an oil-filled dashpot that can be adjusted for rate of closure. The oil-filled dashpot is much more positive in its action and more easily adjustable than the air-filled type.

Check valves for sewage must not obstruct the flow and must not have projections that could accumulate stringy materials. Consequently, such check valves are of the single-disc, top-pivot type. The angle-seated, rubber-flapper type closes the quickest, but its lack of an external indicator of valve position is a disadvantage. A variety of check valves, such as the double leaf and slanting disc types (which are suitable only for water), have varying degrees of resistance to slam.

If an air chamber is installed at a pumping station, the check valves must be compatible. The stored energy in the air chamber moves a short water column quickly, so a check valve must close very fast. Rubber seats or an oil-filled dashpot to cushion the seating may be helpful in preventing slam. Cushioned check valves are effective in limiting surges in both pump start-up and shut-down, but these valves cannot begin closing until after the pump stops.

Surge Relief Valves

Surge relief valves act to reduce upsurges. They do not control the initial downsurge that occurs on pump shut-down or power failure. Hence, they are most useful in short, steep pipe profiles where reversal of flow quickly follows power failure.

There is a wide variety of valves with guided discs, pistons, flappers, or membranes available in spring-actuated and diaphragm-actuated designs (controlled by springs, air pressure, or hydraulic pressure) that are kept closed until an upsurge arrives. The spring-actuated type is shown in Figure 7-8. Surge relief valves open quickly, remain open until the surge dissipates, and some then close slowly. Because upsurges travel at elastic wave speed, such a valve may not open quickly enough to prevent a very short surge of high pressure. Hence, insert the characteristics of a proposed valve into a computer model of the system to determine what pressure rise to expect. Very short surges are less important than longer ones and are difficult to model.

Surge Anticipation Valves

Surge anticipation valves (Figure 5-14) overcome the disadvantage of the surge relief valve by beginning to open before the upsurge arrives. The initial reduction of pressure following a power failure is sensed and a timer is then actuated to open the valve and release water before the anticipated high pressure arrives.

The sequence of operation is as follows:

- Pump power failure occurs. Water continues to flow into the pipe away from the pump, but the pressure downstream of the pump drops.
- The flow in the pipe halts, then reverses and flows back toward the pump and closes the check valve. Pressure begins to rise in the pipe next to the pump because the check valve acts as a closed-end pipe.
- The surge anticipation valve has already opened to release water and prevent high pressure. The valve then closes over a period of three to ten times t_c.

This type of valve reduces only the high transient pressures; it cannot control the initial low pressures that occur in the pump discharge piping upon pump power failure or shut-down. The conditions under which such valves can operate are restricted, so consult the valve manufacturer. Pressure relief valves without the "anticipation" feature may be preferable.

Surge anticipation valves can be used for water but not for sewage.

7-5. Containment of Transients

One method of coping with transients that is especially useful for pipes 200 mm (8 in.) or less in diameter is to specify thick-walled pipe and heavy valves, pump casings, and accessory equipment with higher pressure ratings than necessary for normal operating or test pressures. The disadvantage of such a method (beyond the obvious extra cost incurred) is that many pipe materials and valves may eventually fatigue and fail after several years of being subjected to intense cyclic pressures. Hence, portions of the force main may have to be replaced eventually, which would make this method costly indeed. On the other hand, it is a simple method to use, and the short-term maintenance cost is zero. This solution is common for small pumping systems.

For larger pipes, other devices are more effective and less costly. An overview of the various devices is given in Table 7-1.

7-6. Surge Control for Water Pumping Stations

The installations discussed in this section are more or less typical (or at least common), but they are by no means universal.

Deep Well Pump

The start-up of a deep well pump with a high static head poses a special problem. With the pump turned off, the pipe column between pump and check valve is empty. Unless air is admitted to the column, the fast-rising water may slam into the closed check valve (which is holding tons of water at rest) sometimes with a tremendous force that can displace motors and pipes. If such a problem occurs, it can be controlled by

- admitting air into the column (through an orifice) after shutdown so that, upon start-up, the air is exhausted slowly and acts as a cushion; plus
- adding a surge relief valve to bypass water until a pump-control valve (or a check valve) opens; plus
- constructing stout well system piping.

A deep well pumping system is shown in Figure 7-9 together with a typical arrangement of valves. It is assumed that a hydraulic transient analysis has been performed and the need for a surge relief valve established for the particular application. For clarity, the bypass piping containing valves E and D is shown schematically as vertical. Physically, this equipment would probably be installed horizontally. The function of the valves is discussed in the following subsections.

Air release and vacuum relief valve. Valve B removes air during start-up and introduces air after shut-down to avoid a vacuum. The valve must be sized so that optimum air cushioning prevents slam. A good way is to add a throttling device for flexibility in adjustment.

Check valve. Valve C prevents reverse flow if (1) the pump shaft breaks, (2) a failure in the electrical system shuts off the pump with the pump-control valve energized and open, or (3) the power fails. Applicable types of check valves include silent, double door, cushioned swing, and tilting disc. By placing the check valve at the location shown, the transmission main is protected from water hammer by the relief valve.

Table 7-1. Comparison of Devices for Controlling Hydraulic Transients

Method and application	Advantages	Disadvantages
Pump-control valve (electric, hydraulic, or pneumatic); for water and sewage	Effective in controlling surges due to pump start-up and shut-down. Allows automatic control system. Can control surges at any operating pressure for the pump and pipeline system.	Ineffective in controlling surges due to power failure. Must have auxiliary power source (e.g., batteries or compressed air) to operate valve when power fails; even then, it may not protect against downsurge and column separation.
Air chamber; primarily for water	Pressurized vessel can be used in almost any pipe and pumping system encountered in the waterworks industry; can be sized to control pressure changes to prescribed ranges; commonly used on pumping station discharge header, high points, or knees in pipeline; reliable. Can control both upsurge and downsurge.	Requires a fair amount of complex auxiliary equipment and controls. Requires frequent maintenance like any other mechanical-electrical system; large tanks can be expensive; additional land area required at site, especially for large tanks. Not recommended for sewage due to buildup of grease and settleable solids and production of gases; if used, specify frequent blow-down to minimize sewage in tank.
Standpipe; for water only	Nonpressurized tank or tower. Simple to construct and operate. Used at high points in pipelines to supply water to pipe to prevent column separation. Does not require electrical power at site.	Cannot be used if normal HGL at high point is more than about 9 m (30 ft) above ground level because tank would be uneconomical. Overflow drain, piping, and water disposal area must be provided if pressure cycles in pipe system can cause water to overflow standpipe. Requires tall tower, which may not be aesthetically acceptable at site; impractical in a pipeline that experiences large variations in the HGL.
One-way tank; for water only	Similar to standpipe in function; contains check valve on connecting pipe so tank can be used in system in which the HGL is much above the top of the tank; used at high points in pipelines to supply water to pipe to prevent column separation. Also sometimes used at pumping station discharge header in low head applications (usually discharge head < 15 m or < 50 ft); simpler in design and operation than a pressurized air chamber.	Not useful usually on pump discharge if discharge head is greater than about 15 m (50 ft). Requires tall tower, which may not be aesthetically acceptable at site. Requires special attention to selection and specification of the connecting check valve; requires separate fluid supply to maintain water in the tank.
Air release and vacuum relief valve; primarily for water, risky for sewage	Relatively simple devices to install; used at high point in pipeline to allow air to enter pipeline to prevent large negative (vacuum) pressures or reduce effects of column separation.	Requires regular maintenance so float and linkage mechanisms do not hang up or corrode. Should not be trusted to provide complete protection from column separation, because location in the pipeline is critical and cannot always be predicted accurately; does nothing to relieve high transient pressures (upsurges).

(Continued on next page)

Table 7-1. Continued

Method and application	Advantages	Disadvantages
Swing check valves; for water and sewage	Can dramatically reduce effect of flow reversal in pipeline after pump shut-off or power failure. Especially useful in low head pumping stations in which the discharge head is 15 m (50 ft) or less. No electrical power needed. Simple device to install and maintain.	Cannot prevent downstream column separation in the pipeline if that is the problem. Slam can be a severe problem (see Sections 5-4 and 7-1).
Cushioned swing check valve; for water and sewage	Same as swing check valves, except: slam can be reduced or eliminated; upsurge can be partially controlled.	Does not control downsurge. May cause pump to run backward.
Surge relief valves; primarily for water	Effective in reducing positive surges caused by reversal of water column in the pipeline after power failure.	Cannot prevent downstream column separation in pipeline if that is the problem. Requires drain piping and disposal area to dispose of the water that drains from the pipeline.
Higher pressure rated piping, valves, and equipment	Simple to use. Used in many small raw sewage pumping stations and force main systems; short-term maintenance cost is zero.	Pipe and other accessories may eventually fatigue after several years of being subjected to cyclic surge pressures. In the long term, portions of force main may fail and have to be replaced. Cost effectiveness may be poor.

Surge relief valve. Valve E limits excessive pressure during start-up or when the check valve closes with the pump-control valve open. Hence, lighter weight equipment and fittings can be used. The surge relief valve allows the first part of the well water (which may contain sand and entrapped air) to be exhausted. The valve supplies a free flow of water for cooling submersible motors if the pump-control valve fails to open, and also prevents free discharge and extremely high head, both of which are detrimental to the pump. The pressure setting should be about 10% higher than the normal pumping pressure, and the valve should be sized to handle the full pump flow. Either globe or angle body valves are suitable.

Pump-control valve. Valve F opens and closes slowly to limit surges in the transmission main. An electric timer (or an alternate system) should open it only after all of the air (or undesirable water) has been wasted. On shut-down, the pump is kept running until the valve is about 95% closed.

A globe valve provides good protection against surge and is easy to adjust and to service, but it has the greatest headloss (although it can be sized for "acceptable" headloss).

A butterfly valve has low headloss and is available in large sizes. If repairs are required, however, it must (unlike the globe valve) usually be removed and sent to the factory. Butterfly valves are not often used for pump-control service. Because the curve of flow versus stroke is poor, eccentric plug valves should not be used for controlled closure. Ball valves are excellent because of their good throttling characteristics and low headloss when wide open, but in large sizes they are expensive.

Shut-off valves. Always install enough shut-off valves (A, D, and G in Figure 7-9) so that other equipment can be repaired without draining the entire system. However, valves A and D in Figure 7-9 are unnecessary because, whenever the pump is off, all the piping upstream from the transmission main can be drained by stopping the pump and closing valve G. Valves A and D are actually hazardous (which should be noted in the O&M manual) because, if valve A is inadvertently left closed, the air and vacuum relief valve cannot function, and if valve D is closed, there is no protection against water hammer.

Safety features. Provide controls (1) to shut off the pump if the pump-control valve, F, does not open and

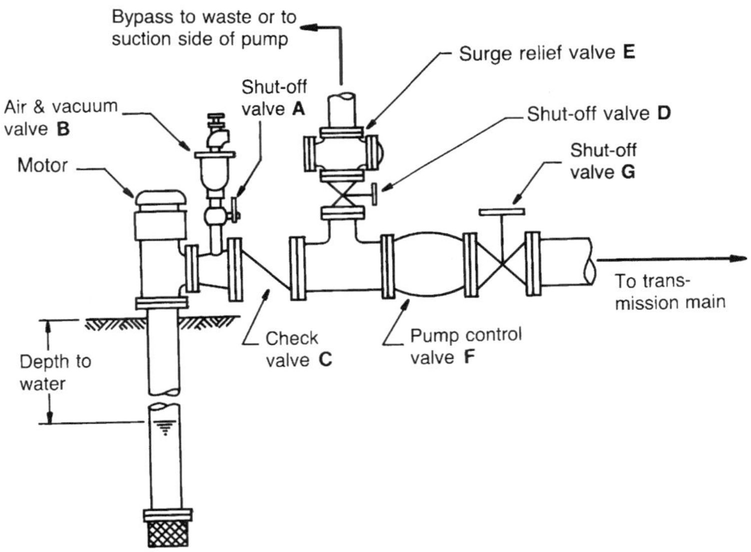

Figure 7-9. Valves for a deep well pumping station with an upstream bypass. After Hy/Con Valve Co. [5].

(2) to close the pump-control valve and shut off the pump if there is sustained low pressure, which would indicate a pipe break. Always include manual controls to override electrical malfunction of the automatic equipment.

Control-valve bypass. The valve arrangement of Figure 7-10 is superior because it provides better protection against upsurge than the valving of Figure 7-9. The surge protection on pump start-up is equally effective, and because the pump-control valve closes in anticipation of pump shut-down, the bypass arrangement limits the transmission main pressures to the relief valve setting. The configuration of piping and valves in Figure 7-10 is particularly preferred if the pump is to start and stop frequently (several times per day).

The pump-control valve, E, must have a fast-closing feature so that it acts as a check valve if the power fails or a pump shaft breaks. If a downsurge could occur as a result of power failure, another vacuum relief valve (with a large opening) may be needed between valves C and F. The orifice in the air and vacuum valve A must be carefully sized to prevent excessive upsurge during pump start-up.

The waste pipeline must have a positive air break before discharging into a sewer or drain line to prevent a cross-connection and possible contamination of the well.

The surge relief valve should be protected from excessive wear. So, if the well water contains sand, the system shown in Figure 7-11 is advantageous because the first portion of the discharge (which usually carries the most sand) is vented through valve B. As the surge relief valve, D, is never opened (unless a surge occurs as a result of power failure), the valve must be exercised periodically (say, weekly). The exercising should be part of the pump monitoring program and so specified in the O&M manual. The blow-off valve, B, can be a manual type, or a small pump-control valve can be substituted if pump starts are to be automatic. The pump-control valve, C, must be equipped with a fast-closing feature to prevent reverse flow following power failure.

A still simpler system is shown in Figure 18-13. Whether the protection provided by the valves of Figures 7-9, 7-10, and 7-11 is required depends on the static head, the length and size of the force main, the depth to groundwater, and the inclination of the designer toward conservatism.

Piping flexibility. Although not shown in Figures 7-9, 7-10, and 7-11, a sufficient number of flexible couplings specific to the installation are necessary for properly joining fittings and connections. Considerations include:

- Adjustment for potential misalignment and for preventing strain in rotating equipment
- Sufficient freedom to permit convenient removal of the pump and other high-maintenance equipment without the necessity for removing other valves or fittings
- Resolution of hydraulic thrusts generated during various operating modes.

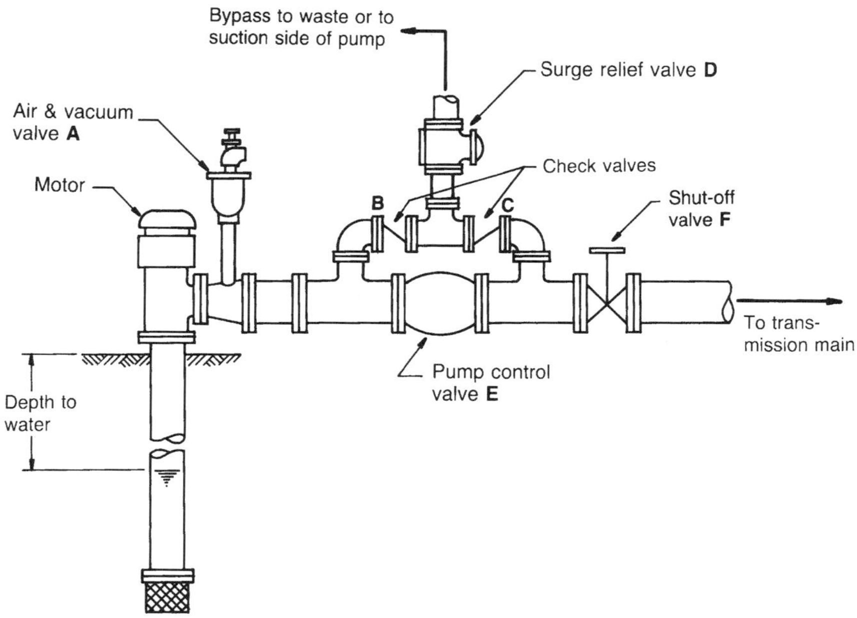

Figure 7-10. Upstream and downstream bypass of a pump-control valve. After Hy/Con Valve Co. [5].

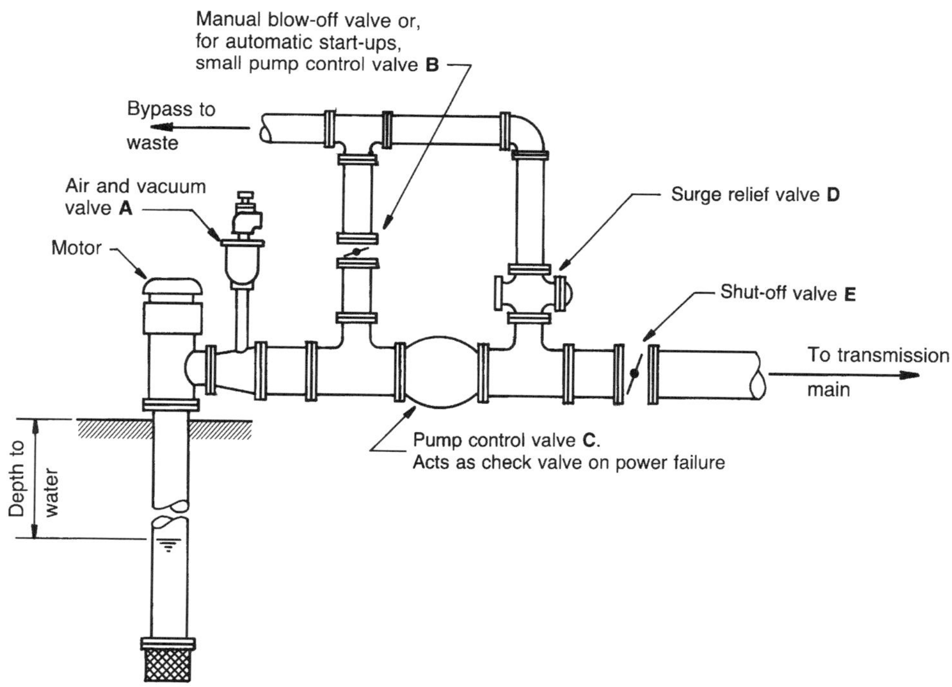

Figure 7-11. Bypass of pump-control valve for sandy water applications. Gate valves are preferred over butterfly valves in the smaller sizes (say, <250 mm or 10 in.). Butterfly valves should have gear actuators to prevent excessively rapid closure. Install enough flexible couplings for fitting up. After Hy/Con Valve Co. [5].

Flexible piping connections that can serve the above purposes include sleeve couplings (such as the Dresser® Style 38), the flanged coupling adapter (such as Dresser® Style 128), and the grooved-end coupling (such as Victaulic® Style 77). Designers should be aware that two couplings are usually required for effectiveness in coping with misalignment and eliminating piping strain.

Low Suction-Lift Turbine Pump

If the depth to water level is only a few feet (as it might be for a shallow well or for pumping from a clear well or forebay), the same arrangement of valves shown in Figure 7-9 can be used with the following modifications.

Air release and vacuum relief valve. Valve B can be smaller and the throttling device can be omitted because the volume of air to be exhausted is small. In fact, an air release valve might be used in lieu of the air/vac valve.

Pump-control valve. The opening of valve F need not be delayed because the volume of air to be exhausted is small and (if the supply is a clear well or forebay) the initial quality of water is good.

Safety feature. A low-level cutout switch should be located in a clear well or forebay.

Air Chamber with Turbine Pump

For wells without troublesome sand or silt in the discharge, an air chamber may sometimes be the appropriate surge-arresting facility. On pump start-up, air flows through an air vacuum valve and into the air chamber from which excess air is vented through an air release valve. When the pump is shut off, a vacuum breaker allows the water in the pump to drain into the well. If the depth to the water table is large enough, the air in the pump column can be used to maintain the air charge in the hydropneumatic tank, which eliminates the need for an air compressor and level controls.

Turbine Booster Pump

Because the major difference between a well pump and a booster pump is the continuously flooded suction of the booster pump, no air needs to be exhausted upon start-up.

Air release valve. Vacuum relief is not needed, so valve B in Figure 7-9 or 7-10 need only be a small air release valve.

Check valves. The purpose of the check valve is to prevent reverse flow through the pump in the event of a power failure while the pump-control valve remains open. However, pump-control valves that close quickly in the event of a power failure can also serve as check valves.

Relief valve. The relief valve is still needed for several of the reasons given in the subsection entitled "Deep Well Pump" in this section, but some of those reasons obviously do not apply to booster pumping.

Pump-control valve. It is necessary to open (and close) the pump-control valve slowly but, because the system is always full of water, unnecessary to delay its opening.

Safety features. In addition to those listed in the subsection entitled "Deep Well Pump," a pressure switch may be desirable to keep the pump-control valve shut until the pressure developed by the pump exceeds the pressure in the transmission main.

Centrifugal High Service or Booster Pump

The same valve arrangements can be used with centrifugal pumps with flooded suctions for both high service and booster pumping. However, the air release valve should be located at the top of the pump casing and should be equipped with a check valve to prevent air from returning to the casing. A manual petcock on the pump casing is usually adequate if the suction is always flooded.

7-7. Surge Control for Raw Sewage Pumping Stations

The strategies for controlling surges in raw sewage pumping are limited to those given in Section 7-1. Those transient control strategies are, however, adequate for nearly all situations. Of course, any proposed solution should be checked by a computer analysis.

Some additional comments on surge control are presented below.

Quick-Closing Check Valve

If the pump spin-down time as determined by computer analysis is sufficiently long (say, more than four or five t_c) and there are no knees, a quick-closing check valve is usually sufficient to prevent surges. Such conditions are common with force mains not much longer than 1 km (0.6 mi) and with uniform gradients of less than about 4%.

Pump-Control Valve

If a pump-control valve is provided so that the pump starts and stops against the closed valve, it can only be operated slowly on both closing and opening. A computer analysis should be used to program the closure time.

The pump-control valve can be an eccentric plug or a lubricated plug, but cone and ball valves have the best characteristics for this service and are preferred even at their higher cost.

If a power failure occurs, a stored-energy system must be used to activate the pump-control valve. But unless the valve closes quickly, the water runs backward through the pump, so the pump, impeller, and wet well systems must be able to handle the backward-flowing sewage.

Increasing Rotational Inertia

If column separation can occur, first investigate the increase of inertia (WR^2) of the moving parts. If the required additional WR^2 is not too great, this is a solution with the utmost reliability (see Section 7-2).

Other Control Strategies

If none of the foregoing can be used to control surge, consider the other strategies outlined in Section 7-1.

7-8. Pipeline Design

Selecting the wall thickness to withstand the pressures expected in the operation of a pipeline is crucial. Wall thickness depends on the pressures due to normal operation, upsurge, and downsurge as well as the pipe material and its safety factors. The safety factors are governed by user groups (such as the American Water Works Association), industries (such as the American National Standards Institute, Inc.), or by manuals of practice. In some standards, a specific overpressure allowance is required, whereas in others a safety factor to be applied to the yield strength of the pipe material is given.

Upsurge

The required wall thickness for positive line pressures for normal upsurge is

$$e = \frac{PD(SF)}{2s_y E_j} \qquad (7\text{-}2)$$

where e is pipe wall thickness in meters (inches), P is internal pressure in newtons per square meter (pounds per square inch), D is outside diameter in meters (inches), SF is the safety factor, s_y is the yield strength in newtons per square meters (pounds per square inch), and E_j is longitudinal joint efficiency associated with the effective strength of the weldment.

The allowable stress, s, may either be specified as a percentage of the yield strength or as s_y/SF. Typical values of SF for the water industry are given in Table 7-2. The longitudinal joint efficiency, E_j, is determined by the type of weld or it may be considered to be accounted for in the allowable stress specification (see ANSI B31.1). The joint efficiency factors in Table 7-3 are used primarily with welded steel pipelines.

Standards for materials of construction often include an allowance for pressure due to hydraulic transients. For example, "ordinary" surge pressures are incorporated in the safety factor of 4.0 for ACP, which conforms to AWWA C400 for internal pressure in combined loading, but the term "ordinary surge pressure" is not defined. A maximum value of 350 kPa (50 lb/in.^{2}) in addition to the design operating pressure is typical. For higher surge pressures, the pipe wall thickness should be increased. A design procedure for including such exceptional surge pressures in the design of ACP is described in AWWA C401.

Table 7-2. Typical Safety Factors (SF) for Pipe[a]

Pressure condition	Type of pipe: Ductile iron	Steel	PVC	AC
Maximum operating	2.0	2.0	2.0–2.5	2.0–4.0
Upsurge transient	*a*	1.5	*a*	*a*
Downsurge collapsing	4.0	4.0	None	None

[a] Include the upsurge transient in maximum operating pressure.

Table 7-3. Weld Joint Efficiencies for Steel Pipes (per ANSI B31.1)

Type of longitudinal joint	Weld joint efficiency factor (E_j)
Arc or gas weld	
Single butt weld	0.80
Double butt weld	0.90
Single or double butt weld with 100% radiography	1.00
Electric resistance weld	0.85
Furnace butt weld	0.60
Most steel water pipelines	0.85

According to AWWA C151, ductile iron pipe up to 450 mm (18 in.) is adequate for a rated working pressure of 2400 kPa (350 lb/in.^{2}) plus a surge allowance of 690 kPa (100 lb/in.^{2}). For larger pipes, the operating pressure varies with wall thickness class, although the surge allowance of 690 kPa (100 lb/in.^{2}) remains the same.

Downsurge

The minimum wall thickness required for protection against negative gauge pressures that tend to buckle the pipe often dictates the design in low-pressure systems where vapor cavities may form in high points of pipelines subject to downsurge conditions. The negative gauge pressure required to collapse a circular pipe of uniform wall thickness is

$$\Delta P_c = P_{atm} - P_v = \frac{2E}{(1-\mu^2)(SF)}\left(\frac{e}{D}\right)^3 \qquad (7\text{-}3)$$

where ΔP_c is the difference between the external and internal pressures on the pipe, P_{atm} is the atmospheric pressure, P_v is the vapor pressure of the liquid inside the pipe, E is the modulus of elasticity, μ is Poisson's ratio, SF is the safety factor, e is the wall thickness, and D is the outside diameter. Equation 7-3 is reasonably accurate for ductile iron and steel. However, because of end effects, wall thickness variation, lack of roundness, and other manufacturing tolerances in steel pipe, use

$$\Delta P_t = P_{atm} - P_v = \frac{C}{SF}\left(\frac{e}{D}\right)^3 \qquad (7\text{-}4)$$

where C is 3.45×10^8 kPa (5.0×10^7 lb/in.^{2}). Safety factors, SF, are given in Table 7-2.

Example 7-1
Determination of Minimum Pipe Wall Thickness

Problem: Find the required minimum wall thickness for a steel water transmission line 500 mm (20 in.) in diameter with the following characteristics:

- Pipe material specification: ASTM A 53, Type E, Grade A
- Yield strength = 210,000 kPa (30,000 lb/in.^{2})
- Maximum operating pressure = 1750 kPa (250 lb/in.^{2})
- Maximum pressure due to surge (static + dynamic + transient rise) = 2500 kPa (360 lb/in.^{2})
- P_{atm} = 94.5 kPa (13.7 lb/in.^{2})
- P_v = 1.7 kPa (0.26 lb/in.^{2})
- Longitudinal joint efficiency = 0.85

Solution: Find the required wall thickness based on three criteria: (1) maximum operating pressure, (2) maximum transient pressure, and (3) collapsing pressure.

Maximum operating pressure. From Equation 7-2 and for a safety factor of 2 from Table 7-2.

SI Units

$$e = \frac{1750 \times 500 \times 2}{2 \times 210{,}000 \times 0.85} = 4.90 \text{ mm}$$

U.S. Customary Units

$$e = \frac{250 \times 20 \times 2}{2 \times 30{,}000 \times 0.85} = 0.196 \text{ in.}$$

But the tolerance on wall thickness in ASTM A53 pipe is 12.5%, so increase e by 12.5%:

$e' = 1.125 \times 4.90 = 5.51$ mm

$e' = 1.125 \times 0.196 = 0.221$ in.

SI Units	**U.S. Customary Units**

Maximum transient pressure. The safety factor from Table 7-2 is 1.5 for the upsurge pressure.

$$e = \frac{2500 \times 500 \times 1.5}{2 \times 210{,}000 \times 0.85} = 5.25 \text{ mm} \qquad e = \frac{360 \times 20 \times 1.5}{2 \times 30{,}000 \times 0.85} = 0.212 \text{ in.}$$

Again, increase e by 12.5%:

$$e' = 1.125 \times 5.25 = 5.91 \text{ mm} \qquad e' = 1.125 \times 0.212 = 0.238 \text{ in.}$$

Collapsing pressure. Find the safety factor from Table 7-2 and rearrange Equation 7-4 to solve for e:

$$e = D\sqrt[3]{\frac{SF}{C}(P_{\text{atm}} - P_{\text{v}})}$$

$$e = 500\sqrt[3]{\frac{4}{3.45 \times 10^8}(94.5 - 1.7)} = 5.12 \text{ mm} \qquad e = 20\sqrt[3]{\frac{4}{5 \times 10^7}(13.7 - 0.26)} = 0.205 \text{ in.}$$

Increase e by 12.5%:

$$e' = 1.125 \times 5.12 = 5.76 \text{ mm} \qquad e' = 1.125 \times 0.205 = 0.231 \text{ in.}$$

The required wall thickness, 5.91 mm (0.238 in.), is governed by the maximum transient pressure of 2500 kPa (360 lb/in.^{2}). Pipe conforming to ASTM A53 is built according to the dimensions given in ANSI B36.10, and the thinnest applicable wall is Schedule 10—6.35 mm (0.250 in.).

Several other important pipeline design considerations that are omitted in Example 7-1 include

- Soil loading on buried pipes
- Additional reinforcement or thickness at branches and openings
- Internal or external corrosion allowance
- Temperature effects on the steel stress values
- Effect of column separation and rejoining.

7-9. Computer Analysis

The objectives of this section are (1) to show the value of computer analysis and (2) to provide insight into the effectiveness of various surge control devices. Refer to Watters [3] for extensive treatment of air chambers alone and in conjunction with one-way tanks, and see Watters [3], Chaudhry [6], and Wylie and Streeter [7] for the details of computer analysis, as these are beyond the scope of this book. Several commercial computer programs are available for hydraulic transient analysis.

An example of a water pumping system is shown in Figure 7-12a. The heavy line represents the profile of a ductile iron pipe 4285 m (14,060 ft) long with an ID of 420 mm (16.5 in.). Two pumps each discharge 386 m^3/hr (1702 gal/min) through 300-mm (12-in.) quick-closing check valves into the 420-mm (16.5-in.) manifold. The static lift is 57 m (186 ft) and the initial TDH is 93 m (305 ft). The system requires a computer analysis for several reasons (which are given in Section 6-8):

- The flow exceeds 6 L/s (100 gal/min).
- The head is greater than 9 m (30 ft).
- The system contains a knee and the initial gradient is excessively steep.
- The velocity exceeds 1.2 m/s (4 ft/s).

The system was modeled with a program utilizing the method of characteristics on a mainframe computer, which made a complete solution of pressure and velocity at 0.05-s intervals. Column separation would occur in the unprotected pipeline, so surge control measures are necessary.

During the filling of the pipe, an air release valve is needed at station 19+00 to expel trapped air and to release air bubbles that inevitably accumulate during operation. After the pipe is filled, it will stay filled (except for vapor cavities if column separation occurs) because the knee is below the static water level.

When power fails at time zero, the pumps stop quickly (within 6 s) because of the initially steep pipeline gradient. Successive hydraulic gradelines are shown at intervals labeled 0.2, 0.4, . . . , 3 s in Figure 7-12a. At 1.0 s, the HGL intersects the pipeline at sta-

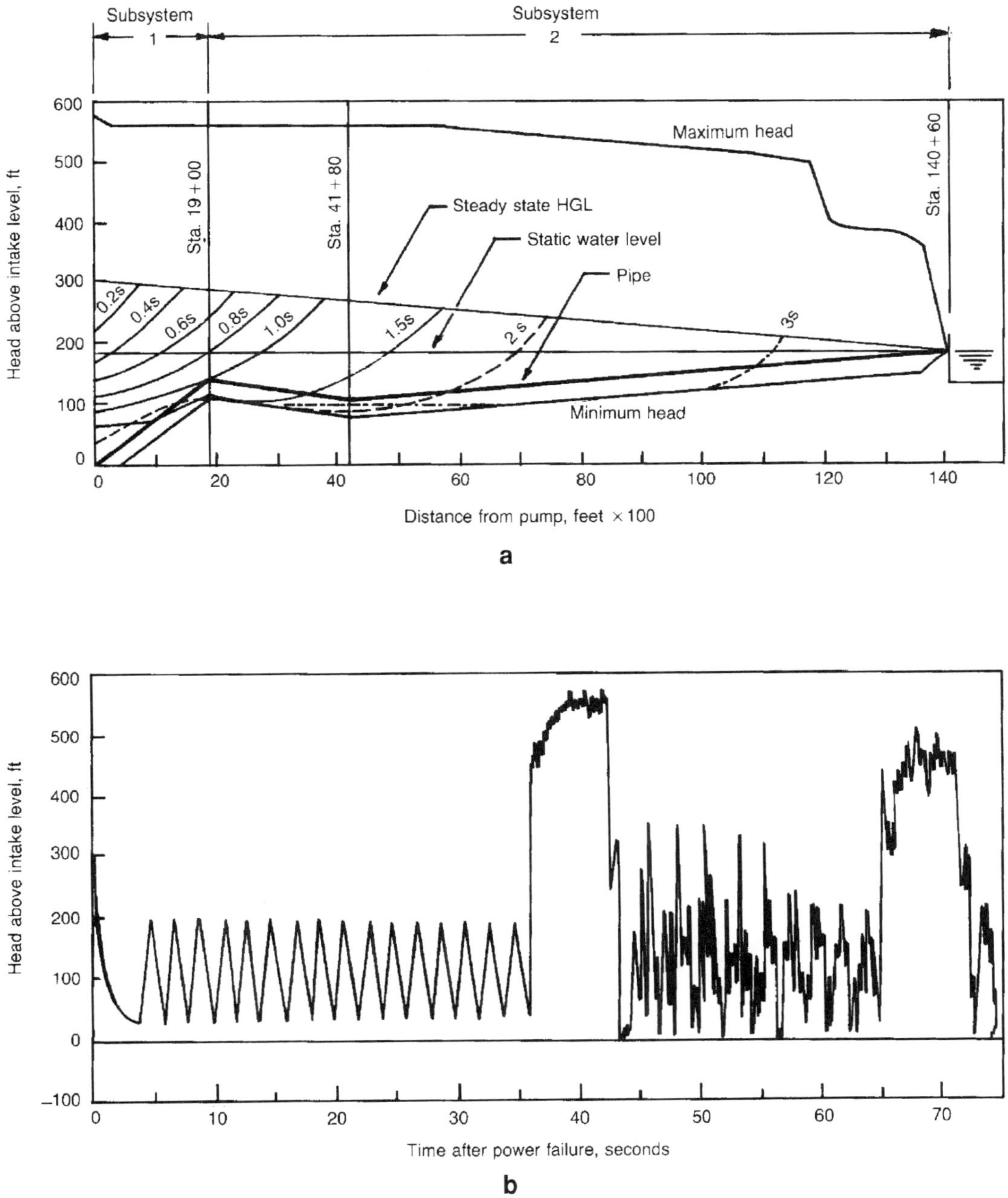

Figure 7-12. Effect of a power failure on an unprotected pipeline. (a) Pressure head along the pipeline; (b) Pressure head at pump (station 0+00). Courtesy of Stoner Associates, Inc.

tion 19+00, so the pressure drops to zero (note that a "station" is 100 ft). At 1.5 s, the HGL is below the pipeline for a considerable distance, and the gauge pressure falls to the minimum possible—vapor pressure at –291 kPa (–14.5 lb/in.^{2}) as referenced to zero at atmospheric pressure. (Depending on the elevation and water temperature, the negative pressure may be slightly less.) At 3 s, the water is literally boiling and forming vapor pockets over nearly all of the pipeline. The largest vapor pocket forms at station 19+00 because of the knee, and this splits the pipeline into two hydraulically separate subsystems: (1) stations 0+00 to 19+00 and (2) stations 19+00 to 140+00. For as long as a vapor cavity exists at station 19+00, it acts

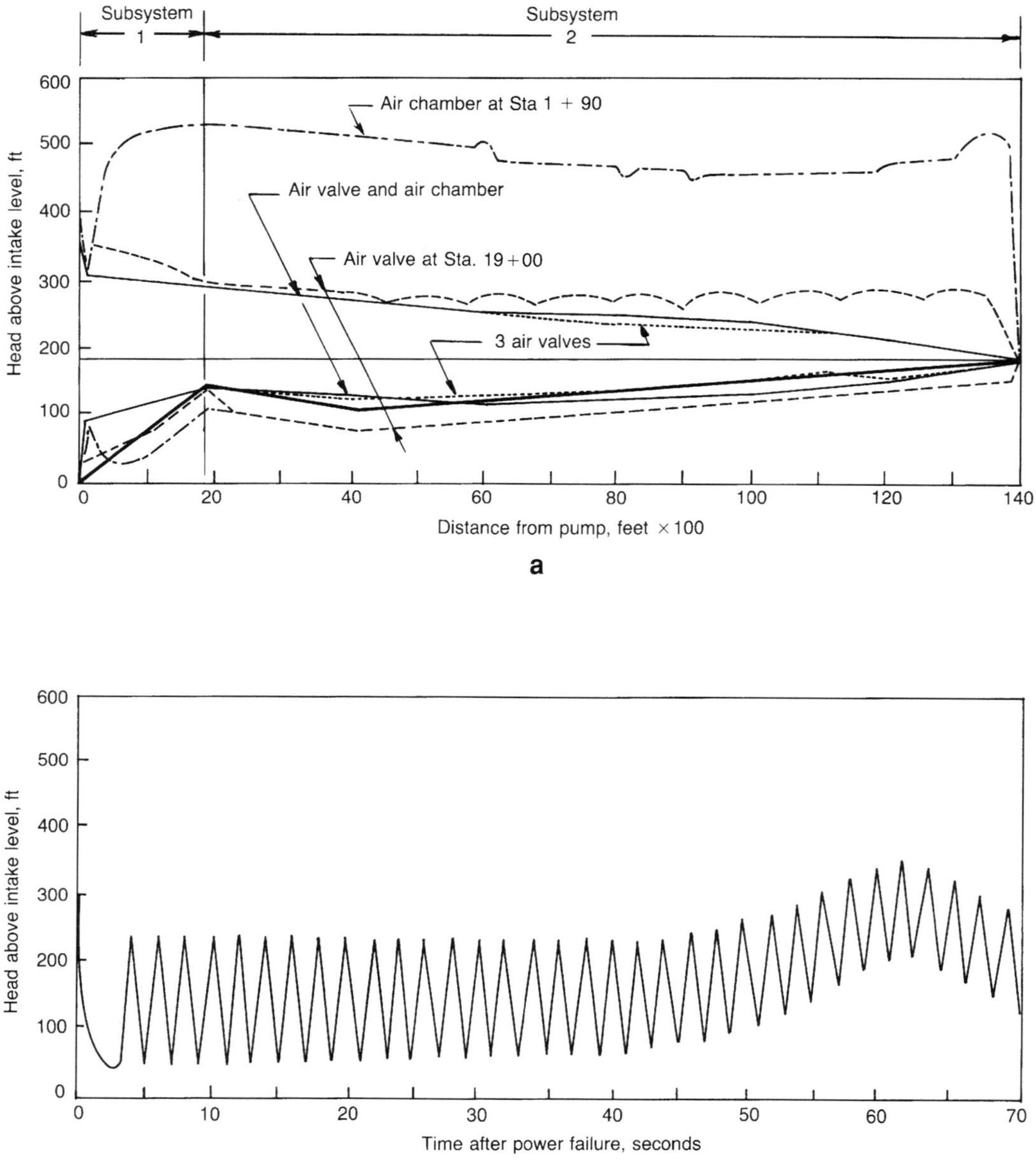

Figure 7-13. Effect of a power failure on a pipeline with surge control devices. (a) Maximum and minimum pressure heads along a pipeline with surge protection. (b) Pressure head at pump (station 0+00) for a pipeline with a $2 \times {}^1/_4$-in. air valve at station 19+00.

like a reservoir with a constant pressure of –101 kPa (–14.7 lb/in.^{2}). The steep gradient and high gravity force acting on subsystem 1 rapidly bring the column of water to rest and stop the pumps, and the spring-loaded check valves close before reverse flow can occur. Pressure oscillations travel back and forth between the closed check valves and the constant head source at station 19+00. The frequency of the oscillations should be $4L/a = 4 \times 1900/3800 = 2$ s, and, of course, that is evident in Figure 7-12b.

The slope of the pipeline in subsystem 2 is gentle, so there is less force to halt the fast-moving column. Flow continues for a relatively long period but, eventually, it does reverse, and when the large vapor

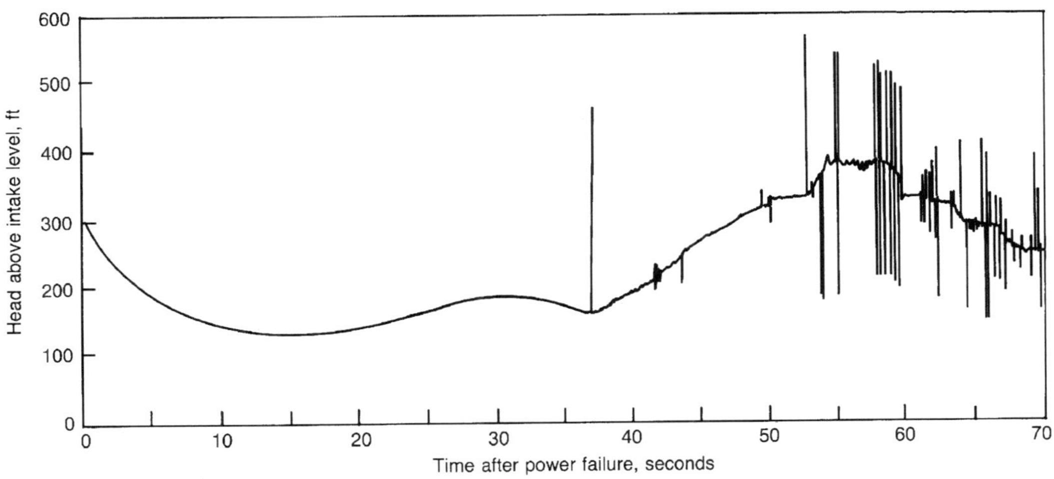

c

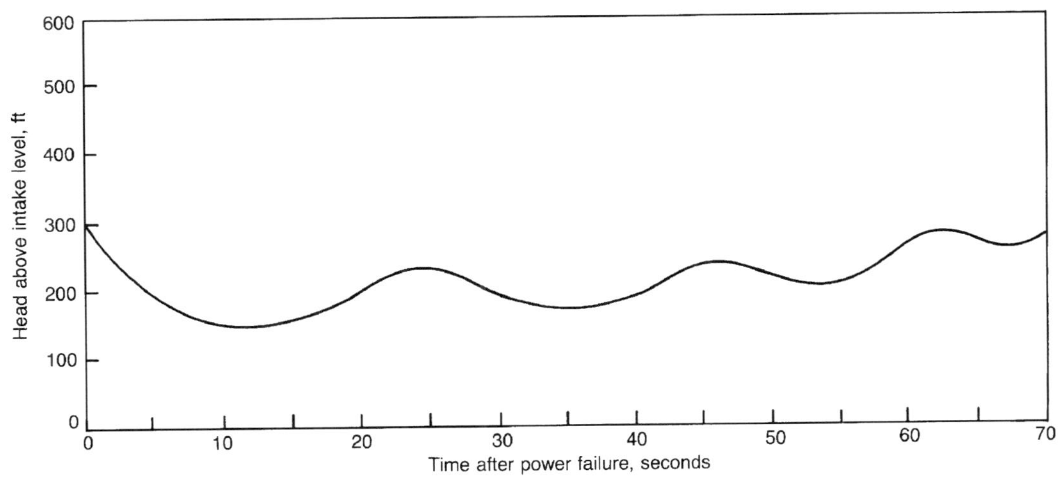

d

Figure 7-13. *(Continued).* (c) Pressure head at pump for a pipeline with an air chamber at station 1+90. (d) Pressure head at pump for a pipeline with an air chamber at station 1+90 and a 2 × $^{1}/_{4}$-in. air valve at station 19+00. Courtesy of Stoner Associates, Inc.

pocket at station 19+00 collapses, the reverse flow suddenly stops with a shock that increases the head to about 550 ft for 7.5 s—the time required for a wave to travel to the reservoir and be reflected back at the elastic speed of 1160 m/s (3800 ft/s) (see Figure 7-12b). Other vapor pockets also collapse and cause a jumble of peaks in the pressure trace. The energy generated by the vapor cavity collapse is large enough to create a rebound effect in subsystem 2 and open another vapor cavity at station 19+00. So, about 65 s after power failure, there is another sudden jump in pressure caused by the collapse of the new vapor cavity. Cavities may open and close several more times before friction dissipates the energy.

To minimize the high pressures, the violent collapse of the vapor cavity at station 19+00 must be prevented. The best defense is to prevent vapor formation in the first place, and the simplest means (assuming it is completely impractical to eliminate the knee) is to install a vacuum breaker valve and an air release valve at the knee. The sizes required are found by trial to be 50 mm (2 in.) for the vacuum breaker and 6 mm ($^1/_4$ in.) for the air release valve. The vacuum breaker valve admits air into the pipe when the gauge pressure becomes slightly negative and, thus, prevents the high negative pressure that produces vapor. Of course, a cavity still forms at the knee, but now it is composed of air and not vapor. Furthermore, in contrast to the unprotected pipeline—where the head to rejoin the two water columns is a constant differential of 14 m (46 ft) between reservoir and knee plus 10.2 m (33.5 ft) due to the vacuum at the knee—the head for flow reversal now only starts at this value and soon diminishes because as the cavity shrinks, the air is compressed and this air is emitted so slowly by the small air relief valve that contact is the soft, cushioned event shown in Figure 7-13b. Again, the periodic pressure oscillation with the 2-s frequency is caused by the reflection of the pressure wave between the closed check valves and the air pocket at station 19+00. The returning flow in subsystem 2 raises the oscillating pressure slightly with a peak at about 62 s.

The vacuum breaker valve at station 19+00 does not entirely eliminate downstream vapor cavities, although it does keep them small. To eliminate all downstream vapor cavities along flat pipelines requires vacuum breakers spaced at reasonable intervals. Note the further reduction of maximum pressures shown in Figure 7-13a with additional sets of vacuum breaker-air release valves at stations 81+70 and 110+20.

Although an air chamber at the pumping station is often effective, here it does not prevent formation of a vapor cavity and its subsequent collapse at station 19+00, as shown in Figure 7-13c. The air chamber does dampen oscillations between check valves and the knee, and it absorbs some of the shock so that the pressure spike is very narrow. But it does little for the pipe downstream from station 19+00, as can be seen by comparing Figures 7-12a and 7-13a.

In the mathematical model of this example, the air chamber can be placed no closer to the pumps than station 1+90 owing to the details and limitations of the model. In reality, the air chamber can be placed anywhere inside or outside of the pumping station with results little different from the model.

The use of both an air valve and an air chamber (Figure 7-13d) is beneficial in reducing the oscillations of Figure 7-13b. Together they offer very effective protection from water hammer.

An air chamber placed at station 19+00 might prevent column separation altogether, but because it would have to be very large, it might be uneconomical. Whether the air chamber would work at all, though, is uncertain and depends on the profile of the pipe. A computer analysis would have to be used to determine the size and the effectiveness. The profile in Figure 7-12a is such that the simpler and less expensive air release, vacuum relief valve offers a minimum of protection for a minimum cost.

7-10. Transients in Distribution Systems

As an example of how to identify and characterize possible transient conditions in a distribution system, consider the pumping station and pipeline system shown schematically in Figure 7-14. Each item, such as (1) a pump, (2) an interconnection of two or more pipes, (3) an end of a pipe, or (4) a reservoir is known as a boundary condition. Each boundary condition constitutes a node as shown by circles in the figure. The system shown consists of a source pump with forebay, two in-line booster pumping stations, and several turnouts with solenoid-operated valves through which water flows. The valves, which are of the energize-to-open, deenergize-to-close design, could close completely in 1 min. Each valve also has a manually operated isolation valve immediately upstream.

Possible causes of transient conditions are

- manual closure of one or more isolation valves,
- power failure at one pumping station or simultaneously at several pumping stations,
- shut-off of one solenoid valve at power failure or simultaneous shut-off of several (or all) solenoid valves, and
- combination of pump power failure plus a simultaneous closing of all solenoid valves.

Although any of these conditions—and more—may be possible, sometimes it is unlikely, or even impossible, for a particular condition to occur. For example, sudden, simultaneous closure of more than one isolation valve is extremely unlikely. Because neither time nor resources permit analyzing all the possibilities, the transient conditions *most likely* to cause the worst problems must be identified, as opposed to all *possible* conditions.

The profile of the transmission line can also give clues about possible transient conditions as discussed in Sections 6-7 and 7-9.

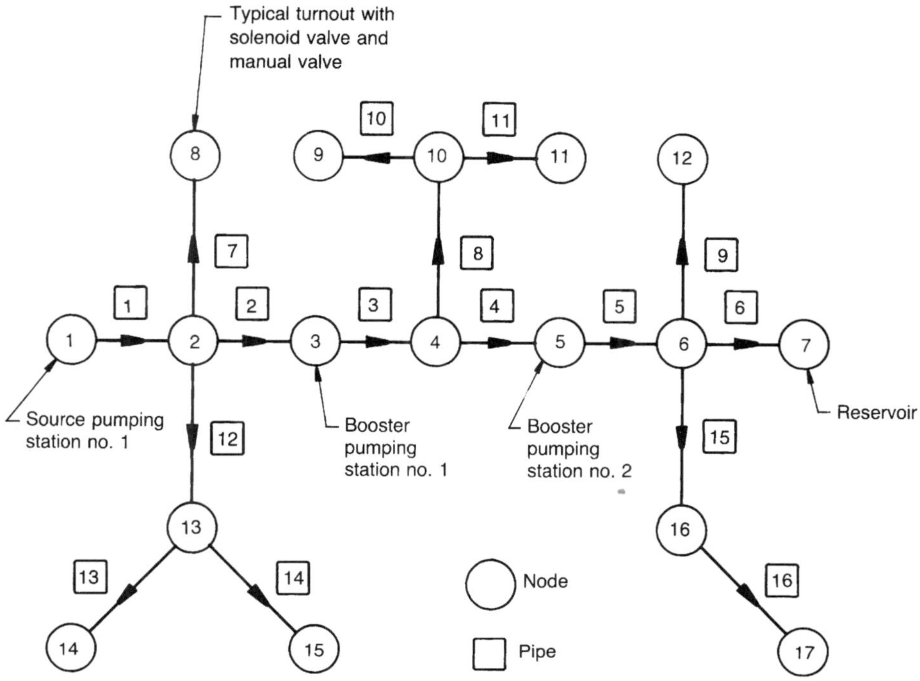

Figure 7-14. Schematic diagram of a pumping system.

7-11. References

1. Stoner Associates, P.O. Box 86, Carlisle, PA 17013-0086.
2. Fleming, A. J. "Cost-effective solution to a water hammer problem," *Public Works* **121**(8)**:** 42–44 (July 1990).
3. Watters, G. Z., *Analysis and Control of Unsteady Flow in Pipelines,* 2nd ed., Butterworth-Heinemann, Stoneham, MA (1984).
4. American Society of Mechanical Engineers, *ASME Boiler and Pressure Vessel Code, Section VIII,* United Engineering Center, New York (1983).
5. "Surge protection for water pump stations," Hy/Con Valve Co., Inc., Schenectady, NY (n.d.).
6. Chaudhry, M. H., *Applied Hydraulic Transients,* Van Nostrand Reinhold, New York (1979).
7. Wylie, E. B., and V. L. Streeter, *Fluid Transients,* Feb Press, Ann Arbor, MI (1982, corrected copy 1983).
8. AWWA M11, *Steel Pipe: A Guide for Design and Installation,* American Water Works Association, Denver, CO (1985).

Chapter 8

Electrical Fundamentals and Power System Principles

STANLEY S. HONG
PHILIP A. HUFF
PAUL C. LEACH

CONTRIBUTORS
Robert S. Benfell
Mayo Gottliebson
Stephen H. Palac
Evans W. Paschal
Richard N. Skeehan
Patricia A. Trager

This chapter is intended primarily for those project managers who have limited knowledge of electrical fundamentals but must nevertheless develop rapport and communication skills with electrical designers. The fundamentals of elementary electrical theory are reviewed in Section 8-2, power and control systems in Section 8-3, generators in Section 8-4, grounding and ground fault projection in Section 8-5, lighting, power outlets in Section 8-6, circuit diagrams in Section 8-7, and power and control system practice in Section 8-8. Although design principles are given herein, design per se is covered in Chapter 9.

References to books are given in standard form, but references to specifications, standards, and codes are given in abbreviated form, such as NEC (for National Electrical Code) or NFPA 78 (for National Fire Protection Association Lightning Protection Code). Titles of codes and standards are given in Appendix E and publishers' addresses are given in Appendix F.

Project engineers should realize that all electrical design must be done or supervised by a professional electrical engineer who is licensed in the state where the project is located. All work should be checked by another qualified electrical engineer. Furthermore, project engineers must be reasonably familiar with:

- Electrical codes
- Definitions (see Section 8-1, Chapter 2, and NEC)
- Electrical symbols (see Tables 2-5 and 2-6)
- Single line diagrams (see Section 8-7)
- Power and control system elements and practices (see Section 8-8)
- State and federal regulations pertaining to energy conservation—especially lighting
- Requirements of the local building codes
- EPA design guidelines.

8-1. Definitions and Code References

Refer to the latest revision of the NEC and ANSI/IEEE 100 for definitions of electrical terms. See also Section 2-2 and, for voltage terminology, Section 8-8.

The following are some commonly used terms in pumping station design. The sources of quoted definitions are given in brackets.

Ampacity: The current in amperes that a conductor can carry continuously under the conditions of use without exceeding its temperature rating [NEC].

Branch circuit: The circuit conductors between the final overcurrent device protecting the circuit and the outlet(s) and the utilization equipment such as motors, lights, etc. [NEC].

Circuit breaker: A device designed to open and close a circuit by nonautomatic means and to open the circuit automatically on a predetermined overcurrent without damage to itself.

Continuous load: A load in which the maximum current is expected to continue for three hours or more [NEC].

Controller: A device or group of devices that governs the electrical power delivered to the utilization equipment in a predetermined manner. A motor controller includes any device, such as a switch or contactor, normally used to start and stop a motor and may include equipment (such as a reactor, resistor, or solid-state system) that limits motor starting current [based on NEC].

Current withstand rating: The maximum allowable current, either instantaneous or for a specified period of time, that a device can withstand without damage [ANSI 100].

Fault: An unintended connection (a short circuit) between phases or between a phase and ground. A fault may result in an excessive current in the faulted equipment, in its supply conductors, and in the supply system itself.

Feeder: All circuit conductors between the service equipment or the source of a separately derived system and the final branch circuit overcurrent device [NEC].

Frequency: The number of periods (or cycles) per unit time.

Hertz: The unit of frequency in cycles per second, abbreviated ***Hz***.

Motor control center (MCC): An assembly of grouped control equipment used primarily for control of motors and associated power distribution applications.

Overcurrent: Any current in excess of the rated current of equipment or the ampacity of a conductor. Overcurrent may result from overload, short circuit, or ground fault [NEC].

Overload: Operation of equipment in excess of the normal full-load rating (or of a conductor in excess of rated ampacity) that would cause damage or dangerous overheating when it persists for a sufficient length of time. A fault, such as a short circuit or a ground fault, is not an overload [NEC].

Power factor: The cosine of the angle by which the current lags (due to inductance) or leads (due to capacitance) the voltage.

Service: The conductors and equipment for delivering electrical energy from the supply system (usually an electric utility) to the wiring system of the premises served [based on NEC].

Switching apparatus: A device for opening and closing or for changing the connections of a circuit. It includes switches, fuses, circuit breakers, and contactors [ANSI/IEEE Std 141-1986 Art 9.2].

8-2. Electrical Fundamentals

Electricity (the movement of electrons) flows easily in conductors such as copper and hardly at all in insulators such as rubber or glass. Electrons flow in the circuit conductors from the negative pole to the positive pole of a battery, but by custom, electrical current is said to flow from the positive to negative pole. In semiconducting media and in electrolytes both positive and negative charges flow to their respective attracting terminals.

Electricity flows when a voltage source providing electrical pressure (or force) to the electrons is in series with a complete circuit. Ordinarily, a complete circuit consists of the voltage source, conductors, and a load that transforms the electrical energy from the source into usable energy such as heat or mechanical work. A switch provides the means of making and breaking a circuit intentionally and, thus, controls the flow of electricity. If the circuit is broken at any point, electricity does not flow.

Electrical pressure (voltage) can be produced by: (1) chemical reactions (as in batteries), (2) movement of magnets near coils (as in generators), (3) contact between dissimilar metals (thermocouples, for example), (4) friction (lightning, for example), or (5) light energy impinging upon a thin semiconducting film (as in photovoltaic systems or solar cells).

Alternating current (ac) power is the most common, because it is easy to generate, transform (from one voltage level to another), and utilize. It is commonly available as single phase or three phase. Direct current (dc) is often used to power emergency lights and to provide control power in large systems. It is used to start engine generator sets, to power dc pump motors in adjustable-speed applications, and as field supply to synchronous motors and generators. But dc is not available from utilities. It must be produced on site.

The potential difference (the electrical equivalent of pressure) between any two points in an electric system is measured in volts (V). The unit of quantity of electrical charge is the coulomb (C). The ampere (A) is the measure of the rate at which charges move past a given point in coulombs per second. The energy or work required to move one coulomb through a potential difference of one volt is the joule (J) and the rate of expenditure of energy is measured in joules per second or watts (W).

In an alternating current system, an ampere of alternating current is defined as the measured value that causes the same heating effect as one ampere of direct current. The ampere is mathematically equal to 0.707 times the peak value of a sinusoidal alternating waveform, and it is sometimes referred to as the root-mean-square (RMS) value.

The measure of the opposition or resistance to the flow of electrical current is the ohm (Ω). In dc circuits and in unity power factor ac circuits, one volt of potential difference is required to cause a current of one ampere to flow through a resistance of one ohm—a relation known as Ohm's law.

$$I = V/R \tag{8-1}$$

where I is current in amperes, V is volts, and R is resistance in ohms. In dc circuits, the average values for current and voltage are used. In ac circuits with sinusoidal waveforms, RMS (root-mean-square) values are used for current and voltage.

Electrical power in dc and unity power factor ac circuits is the product of current and voltage.

$$P = VI \tag{8-2}$$

where P is power is watts. In ac circuits with a power factor that is not unity, the above relation becomes more complex (Equation 8-3). Nevertheless, power is equal to the product of resistance and the square of the RMS value of current. Power usage at the rate of one watt for a period of one hour is termed one watt-hour (or 3600 J) of electrical energy.

Generation of Electricity

An electrical source as described in the foregoing discussion generates a potential difference called "voltage." The voltage may be steady, pulsing (with constant polarity), or alternating in polarity. A battery generates a steady dc voltage, as do other static sources such as photovoltaic cells and thermocouples. An ac generator produces an alternating voltage with a sinusoidal waveform. A dc generator is an ac generator with either: (1) solid-state rectifiers or (2) a commutator that switches connections to coils on the rotating armature and thereby produces a pulsating dc voltage.

The movement of the bar magnet across the coil of wire in Figure 8-1 generates voltage because lines of magnetic force or flux move across the coil. Reversing the direction of movement reverses the voltage polarity, as from positive to negative. Alternatively, if the polarity of the magnet is reversed, the voltage polarity reverses. By rotating the magnet as shown in Figure 8-2 the voltage alternately reverses as the north and south poles of the magnet pass the coil. The voltage is produced as a sine wave, which is complete with one revolution of the magnet. If the magnet revolves at 60 revolutions per second, the voltage is produced at 60 Hz. Two magnets arranged with alternating polarity

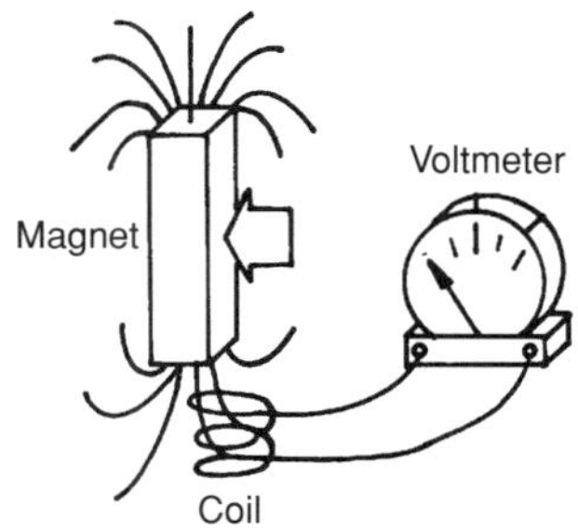

Figure 8-1. Producing potential (voltage) with a moving magnet.

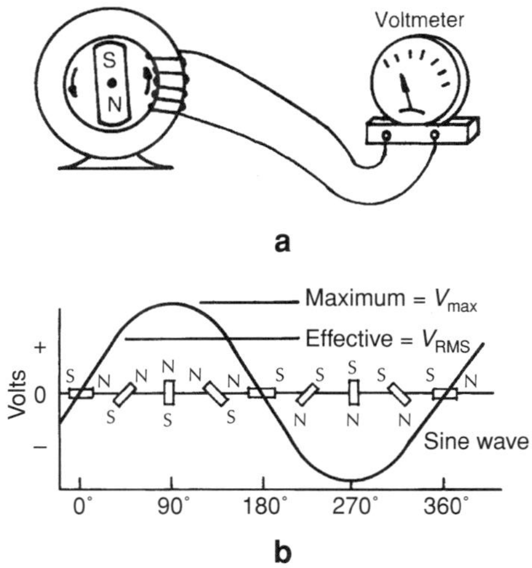

Figure 8-2. A simple ac generator. (a) Voltmeter and rotating magnet; (b) voltage produced.

likewise produce 60 Hz voltage when rotated at 30 revolutions per second. In a multi-pole generator, the required rotational speed in revolutions per minute to produce 60 Hz voltage is 7200/P, where P is the number of north and south poles (always an even number).

DC generators with commutators are always arranged with the magnets on the stationary member (the field) and the coils on the rotating member to facilitate commutator construction. Most ac generators are arranged with a rotating field, and, to eliminate the slip rings that would be required to conduct the generated current from the rotating member, the coils are on the stationary member. Slip rings and commutators are a significant source of wear and maintenance.

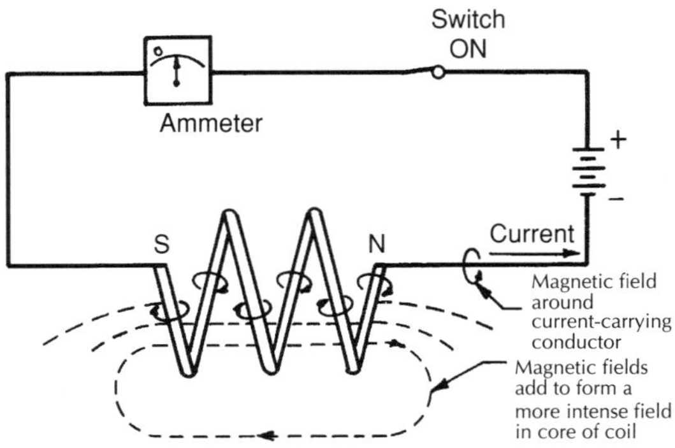

Figure 8-4. Magnetic field due to a coiled conductor.

Inductance

When current flows in a conductor, a magnetic field forms that envelops the conductor for its full length. If the current is steady, the field is static (see Figure 8-3). The presence of the field can be detected by a small magnet such as the needle of a compass. However, if the current starts to change, the magnetic field around the conductor also begins to change, and this changing field generates a voltage that tends to oppose the change in current. This opposition to a change in current flow in a conductor is termed an "inductive effect." The effect can be increased by coiling the conductor into a helix, which increases the magnetic field along the central axis of the coil as shown in Figure 8-4. If an iron bar or some other ferromagnetic core material is inserted inside the coil, a striking increase in the inductive effect occurs, because the iron creates an easy path for the magnetic field and enhances the magnetic flux created by the current.

The unit of inductance is the henry (H). It is the inductance required to generate one volt while a current change of one ampere per second occurs. In direct current circuits the inductive effect is noticed only during changes in the level of current. When the switch is closed in an inductive dc circuit, there is a slight time lag in the buildup of the current. Also, if a steady current is flowing in a circuit with a large coil, opening a switch and attempting to return the current to zero instantaneously causes a voltage to be induced in the coil. This voltage may be sufficient to break down the air gap insulation between the opening switch contacts and cause a momentary arc. Never be in contact with a dc circuit conductor that has a large inductive element in the circuit, because the voltage generated during switching may be several thousand volts and can cause a severe shock or burn.

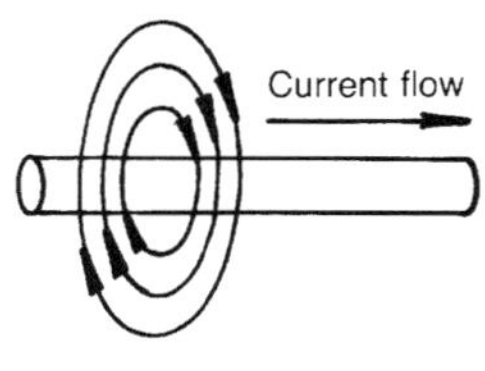

Figure 8-3. Magnetic field around a conductor.

In an alternating current system where significant inductance is present, the current is limited partly by the inductive effect, because the continuously changing of the value of the current generates a voltage in opposition to the supply voltage. From an examination of the sine wave shown in Figure 8-2, it can be seen that the voltage is changing at its most rapid rate as it goes through the zero axis and is not changing at all at the top and bottom of the cycle (90° and 270°). Thus the countervoltage generated by the inductive effect is maximum when the driving voltage is zero, and the current within the inductor is offset 90° from the voltage across the inductor.

Capacitance

Capacitance is a measure of the ability of a capacitor to store an electrical charge. One type of capacitor contains a series of metal plates separated by insulators (e.g., air gaps, mica, or other insulating materials called dielectrics). Alternate plates are connected to one or the other conductor. Another common type of capacitor is a sandwich of metal and insulator sheets rolled into a cylindrical form.

Capacitance is measured in units of farads (F). Common values are measured in picofarads (mmF or pF or 1.0×10^{-12} F); other applications call for microfarads (μF or 1.0×10^{-6} F). In power system work, the capacitance is usually designated in terms of the reactive power or reactive kilovolt amperes (kVARs) required from an alternating current source of a specific nominal voltage.

Assume that the switch in Figure 8-5a is closed. Because the two plates labelled "capacitor" are separated by a thin insulating medium, no direct current can flow through the circuit. However, there is a short-duration flow of current in the circuit as electric charges build up on the opposing plates. Thus, capacitors are charged by a short-duration current flow. The larger the plates, the greater is their ability to store charges. No current flows through the capacitance because of the air gap or insulator.

If a capacitor is placed in a simple alternating current circuit and the switch is closed, there is a flow of current in the circuit because, as the source voltage continually reverses, the current is made to reverse and the plates are each alternately charged in alternating polarity. Because current must flow before the charges build up and potential difference exists, the capacitive current is 90° out of phase with the applied voltage and leads it by 90°.

Never touch capacitor terminals after opening the switch. Because of the trapped charges on the plates, a voltage exists across the terminals that may equal line voltage. For safety, discharge the capacitor by shorting its terminals. Commercial units include a small resistor across the terminals to discharge a de-energized capacitor automatically, but the discharge takes several minutes.

Inductance, capacitance, and resistance form the principal electrical elements in power systems and are collectively termed "impedance." Resistance in a circuit results in a component of the total current that is in phase with the applied voltage. Inductive reactance in a circuit draws a current that lags the voltage and capacitive reactance draws a current that leads the voltage. Because the inductive current component lags by 90° and the capacitive leads by 90° they are opposite in effect, and if their respective currents are equal, they are cancelled. The result is a purely resistive circuit. Capacitors are used in practical power systems for cancelling inductive effects.

Power

Electrical power, the time rate of doing work, is defined in terms of the heating effect of electricity expressed in watts (W), and one W = 1.0 J/s (9.48×10^{-4} Btu). One horsepower (hp) equals 746 W. The total electrical energy used over a period of time is measured in watt-hours (W·h) or kilowatt-hours (kW·h). A kilowatt-hour equals 3413 Btu.

Direct Current Power

Direct current power is computed using Equation 8-2 from simultaneous readings of voltage and current. The average values of current and voltage are used.

Alternating Current Power

In ac circuits, power is the direct product of amperes and volts only when the load is a resistance (or a load that can be represented as equivalent to a resistance). Power can then be calculated from Equation 8-2 by using RMS values of voltage and current. In this situation, the current is in phase with the voltage and follows the same sinusoidal path with the same relative instantaneous values at all times, as in Figure 8-6.

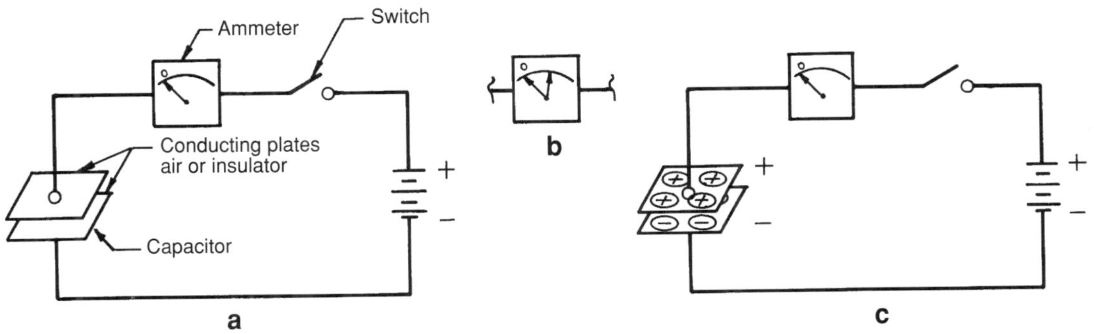

Figure 8-5. Basic capacitor. (a) Capacitor discharged; (b) switch closed and reopened, ammeter deflects during charge; (c) capacitor charged.

In a circuit containing pure inductance, the product of the voltage and current is termed reactive volt-amperes. The two values are 90° out of phase, so the real product (or power) averaged over a cycle is zero. In practical inductive circuits, there is always significant resistance, and the current lags the voltage by some angle less than 90° as illustrated in Figure 8-7. Total true (real) power is, then, the product of the voltage, the current and the cosine of the phase angle between the voltage and current (Equation 8-3).

$$P = VI\cos\theta = VI \times Pf \qquad (8\text{-}3)$$

where P is the true power, Pf is the power factor and equals cos θ, θ is the electrical angle by which the current lags or leads the voltage, and RMS values of current and voltage are used.

If the load on the circuit is resistance and capacitance, then the same equation is valid, but the current is leading the voltage, and the cosine of the angle between the capacitance current and the resistive current represents a leading power factor rather than a lagging power factor as in the inductive circuit (Figure 8-8).

In any ac circuit containing resistance, capacitance, and inductance, a phasor (scaled vector) can be laid

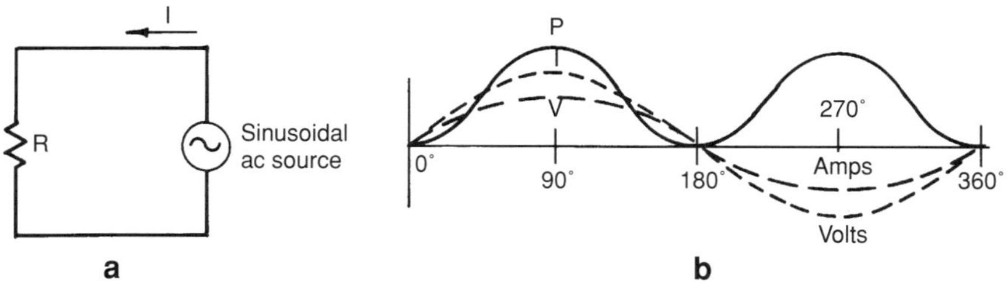

Figure 8-6. Basic resistance current. (a) The equivalent circuit; (b) waveform diagram; instantaneous values of *V, I,* and *P* are shown.

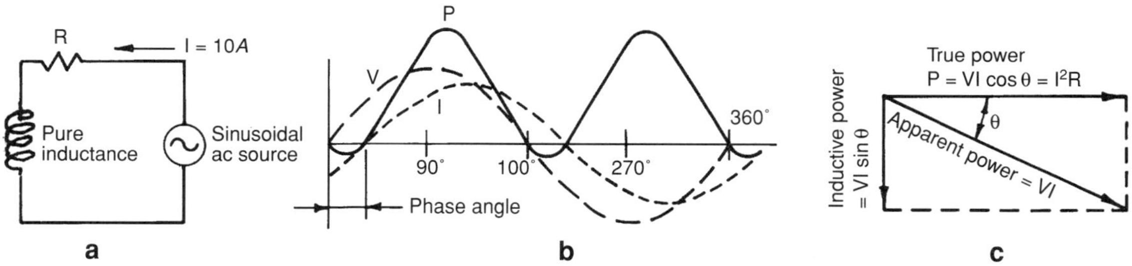

Figure 8-7. Basic inductance circuit. (a) The equivalent circuit resistance is in series with inductance; (b) waveform diagram. Instantaneous values of *V, I,* and *P* are shown; (c) phasor diagram.

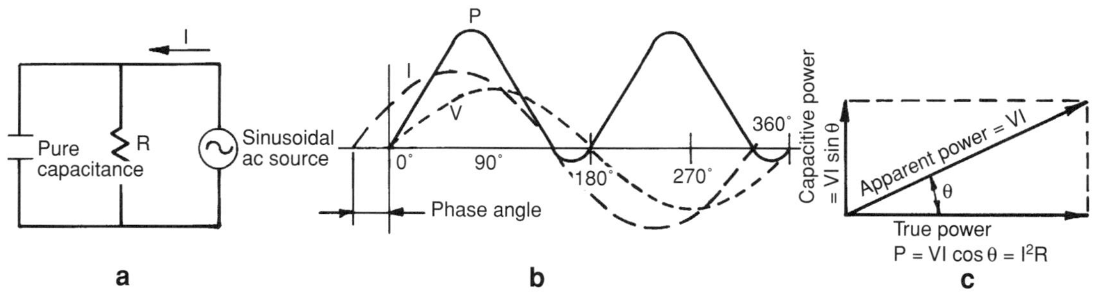

Figure 8-8. Basic capacitance circuit. (a) The equivalent circuit; (b) waveform diagram; instantaneous values of *V, I,* and *P* are shown; (c) phasor diagram.

out on the x-axis (representing true power), capacitive (volt amperes) can be laid out on the upward y-axis (representing the capacitive reactive power), and the inductive volt-amperes can be laid out on the downward y-axis (representing inductive reactive power) as in Figure 8-9a. The difference between the two oppositely directed phasors represent the total circuit reactive power, and this resultant reactive power and the real power are added vectorially to determine the power triangle values. These values are True Power, Reactive Power, and Apparent Power (the latter is the hypotenuse of the power triangle). True Power is illustrated in Figure 8-9c.

Three-Phase Power

Single-phase ac power pulsates. The use of three-phase alternating current smooths the power by adding overlapping power waveforms. Instead of using a single coil (as in Figure 8-2), three-phase power is generated by using three coils spaced 120° apart around the rotating magnet. Actually, the coils would be arranged in three sets of opposite pairs.

The three coils can be arranged in a "wye" connection, as shown in Figure 8-10a. The three sets of coils are labeled A, B, and C in Figure 8-10a. In a wye connection, the phase current (I_f) equals the line current (I_L) as shown in Figure 8-10b, but the phase voltages (V_f) must be added vectorially, as in Figure 8-10c, to obtain the line-to-line voltage (V_L). These statements applied to Figure 8-10 can be expressed as

$$I_\phi = I_L \tag{8-4}$$

$$V_L = V_\phi\sqrt{3} = 277\sqrt{3} = 480\ V \tag{8-5}$$

The power, computed from Equation 8-3, is

$$P_{3\phi} = 3V_\phi I_\phi Pf = \sqrt{3}V_L I_L Pf \tag{8-6}$$

Wye connected windings can be three- or four-wire. No common wire is used in the 3-wire connection.

Delta connected windings are usually three-wire, as illustrated in Figure 8-11. In delta connections, phase voltage equals line-to-line voltage, but line current must be obtained by the vectorial addition of the

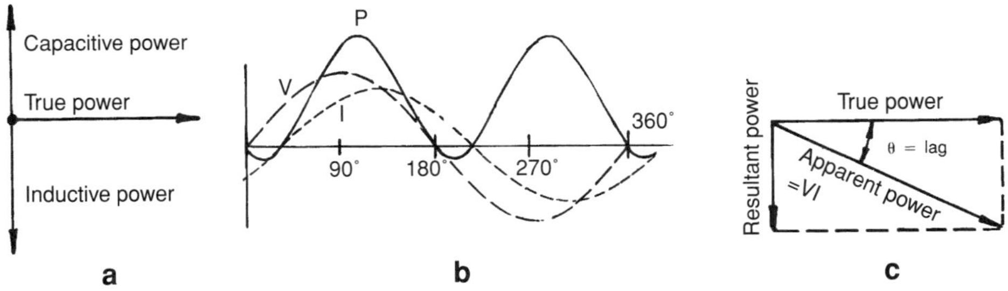

Figure 8-9. Basic circuit containing resistance, capacitance, and inductance. (a) Power in circuits with combined resistance, inductance, and capacitance; (b) actual power wave diagram; (c) phasor diagram.

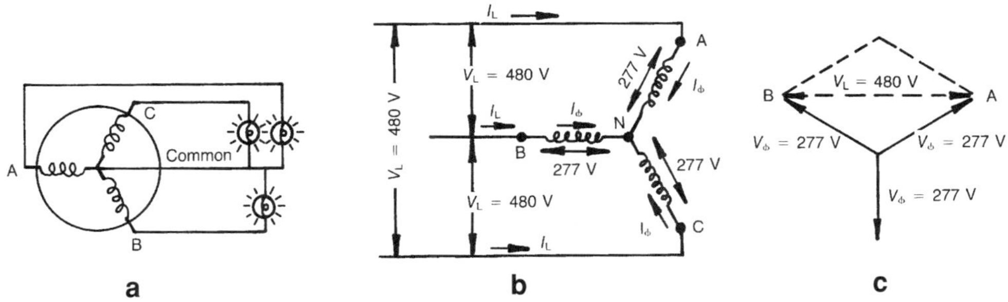

Figure 8-10. Example of generation of three-phase power. (a) The circuit; (b) voltage and current diagram for generator winding connected in wye; (c) phasor diagram.

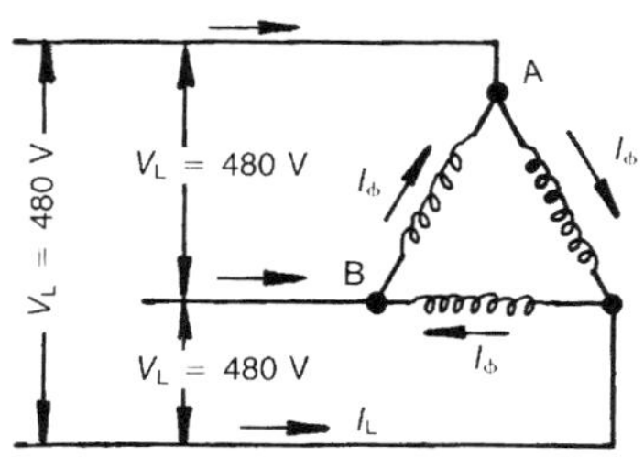

Figure 8-11. Voltage and current diagram for generator windings connected in delta.

phase currents. These statements for the delta configuration can be expressed as

$$V_\phi = V_L \quad (8\text{-}7)$$

$$I_L = I_\phi\sqrt{3} \quad (8\text{-}8)$$

As computed from Equation 8-3, power is

$$P_{3\phi} = 3V_\phi I_\phi Pf = \sqrt{3}V_L I_L Pf \quad (8\text{-}6)$$

Thus, the power is the same whether the windings are wye or delta.

Electrical Measurements

Electrical measurements of interest to the pumping station operator include volts, amperes, watts, power factor, and frequency. Voltmeters and ammeters usually have phase selector switches to minimize the quantity of meters needed. Metering in larger pumping stations, particularly at the highest voltage level, tends to be more complex and comprehensive than in smaller pumping stations on lower voltage levels. Service technicians have additional needs, such as resistance and recording instruments.

Electrical measurements primarily of interest to the electric utility include (kilo)watt-hours, watt-hour demand, and (kilo)var-hours.

Measurement errors may be classified into: (1) accidental or random errors and (2) such systematic errors as

- Instrumental errors
- Errors resulting from external conditions
- Errors caused by the observer.

Accidental errors are mistakes. Instrumental errors are the result of shortcomings of the instruments used or of attempting to make measurements for which the chosen instrument is unsuitable. External conditions that produce errors can be caused by magnetic and electrostatic fields, ambient temperature and pressure, humidity, and other factors. Observational errors are primarily the result of subjective interpolation of analog displays, but, as digital instruments have virtually replaced analog instruments, such errors are now of minor interest. Common full-scale errors are ±3 to 5% for analog instruments, ±0.5% for digital instruments, and ±1 to 2% for kilowatt-hour instruments.

The most serious measurement errors are caused by using unsuitable instruments and by stray magnetic fields. An example of the use of an unsuitable instrument is measuring distorted wave forms with an instrument that is incapable of giving useful readings under such conditions, and such errors can easily reach 50%. The most common error caused by stray fields is using a poorly shielded instrument such as a clamp-on ammeter in the vicinity of either an electric motor or a magnetic starter; such errors are typically about 20%.

Permanent-magnet moving-coil (PMMC) instruments are the most common type of analog unit in general use. Shaft torque is generated when current through the coil produces magnetism that reacts with the permanent magnet. Movement ceases when shaft torque is balanced with torque from the return spring(s) or taut band, at which time the instrument may be read. PMMC instruments are dc milliammeters. Rectifiers and thermo-elements are provided for ac operation, series resistors are provided for voltmeters, and parallel resistors are provided for ammeters. Panel-mounted PMMC ac meters are usually used with current and/or potential instrument transformers. These meters read average values, and ac scales are calibrated in RMS values and are, therefore, accurate only for sinusoidal waveforms.

Other analog instruments often used include the moving-iron radial vane, ac-dc units that respond to average values, use series and parallel resistors for range extension, and the dynamometer units, usually watt-meters, ac-dc units that respond to RMS values and use series and parallel resistors for range extension.

Digital instruments are becoming very popular in panel-mounted versions (and in portable versions for maintenance) due to such features as ruggedness, accuracy, automatic range changing, polarity indication, and reasonable cost. Averaging units may be calibrated for RMS ac values and are accurate only for sinusoidal waveforms. RMS units may indicate total RMS values, or RMS values of selected harmonics. Peak capture and hold is a feature useful in measuring motor-starting currents. Circuitry details are beyond the scope of this book.

Panel-mounted instruments used in ac circuits in new pumping stations should be digital units that respond to true RMS values.

Resistance is not ordinarily measured directly. Instead, the instruments have an integral voltage source (battery, hand-cranked generator, or other dc power supply), and current through the resistor is measured and displayed on the selected scale in terms of ohms.

The portable clamp-on ac ammeter has a movable half-jaw that enables the operator to slip it on over an insulated conductor for fast, easy, and safe current measurement. The jaws, when fully closed, form an iron core that encircles the conductor and intercepts the magnetic field generated by the current in the conductor. By transformer action, a voltage is induced in a secondary coil wound on the iron core. The voltage is measured and read in terms of amperes. Encircling more than one phase (or a phase and neutral) in the jaws cancels the magnetic field and causes a spurious reading or none at all. These instruments are usually multipurpose units that can also measure voltage and resistance.

Ac watt-hour meters are used to measure the flow or usage of electrical energy by means of the induction motor principle. Torque is produced on the aluminum rotor disk by means of magnetic fluxes from current and voltage coils. This torque, and thus disk rotational speed, is proportional to instantaneous power. The disk drives a mechanical or digital register that integrates disk speed with time, producing a watt-hour (or kilowatt-hour) reading. A three-phase meter uses two disks. Practical error limits are within about ±1 to 2%.

The maximum amount of electrical energy used in a given period is measured with a demand meter, often integral with pumping station kilowatt-hour meters. The maximum demand in any one year typically determines the power company's demand charge for the next year, because their distribution system must be able to provide the customer with that much energy, regardless of whether it is used at other times. Thus, for example, a storm water pumping station demand charge for an entire year may reflect the effect of one 3-day storm last year, even though the pumping station may be unused for 11 months.

Basic Electrical Calculations

The relation between resistance, volts, and current is given by Ohm's law, Equation 8-1.

$$R = V/I$$

where R is measured in ohms (Ω), V is volts, and I is current in amperes (A).

The total resistance of several resistances in series is the sum of the individual resistances.

$$R_T = R_1 + R_2 + \ldots + R_N \qquad (8\text{-}9)$$

The inverse of the total resistance of several resistances in parallel is the sum of the inverse of the individual resistances.

$$\frac{1}{R_T} = \frac{1}{R_1} + \frac{1}{R_2} + \ldots + \frac{1}{R_N} \qquad (8\text{-}10)$$

where R_T is the total resistance and R_1, R_2, and so forth are individual resistances, all in ohms.

Lest the reader think that all electrical calculations in the real world are as simple as presented above, one kind of calculation that an electrical engineer would make every day is to find the voltage drop through a simple electrical power feeder $E_D = E_S - E_L$ where the relation between E_S and E_L is given by Equation 8-11.

$$E_S = \left[(E_L\cos\theta + RI)^2 + (E_L\sin\theta + XI)^2\right]^{\frac{1}{2}} \qquad (8\text{-}11)$$

The voltage drop through the power feeder is E_D, E_S is the voltage at the source, E_L is the voltage across the load, θ is the angle whose cosine is the load power factor, R is the resistance of the circuit in ohms, I is the line current in amperes, and X is the reactance of the circuit (usually inductive) in ohms. The equation is best solved by means of a computer program.

8-3. Power and Control System Elements

Solenoids

A solenoid, illustrated in Figure 8-12, is an electromagnet arranged to produce linear mechanical motion. An electromagnet is formed by a number of turns of wire around an iron core, solid for dc but for ac, laminated to reduce eddy currents. An energized electromagnet draws an iron armature toward it by magnetic attraction. The armature may be connected by a suitable mechanical linkage to another object. The range of motion is small, perhaps 2 to 5 mm, except for long hollow core types where the core is drawn in, perhaps 25 mm (1 in.) or more. A common application in pumping stations is the solenoid valve. Energizing the solenoid causes the valve to operate from open to closed or from closed to open position. Other motions are possible, depending on the design of the linkage mechanism. Solenoids are also used to operate braking

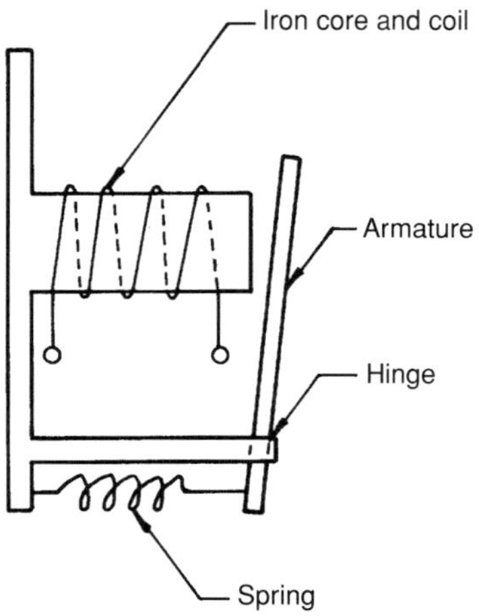

Figure 8-12. Basic solid-core electromagnet.

systems. Some electric motors are required to stop immediately when the power is turned off. A solenoid-operated brake is usually arranged to be held open by a solenoid that receives power when the motor is energized. Upon shutdown or loss of power, the brake is applied by the action of heavy-duty springs.

Solenoids are the basic operating elements of relays and contactors.

Relays

A relay is a solenoid-operated switch. A relay may have one or more switch contacts of various configurations, such as normally open (NO), or normally closed (NC), and various ratings, such as 10 A at 120 V ac. Operating coils may be ac or dc.

A popular relay type used in pumping stations is the industrial control relay. These relays are modular in construction, and switch contacts may be changed or added up to a limit. Other accessories such as timers and mechanical latches with unlatch solenoid coils may be added. NEMA has standardized contact ratings under the designation A600 (10 A continuous, 120 V, 60 A make, 6 A break). Other ratings are also given.

Another popular relay is the 2-pole or 3-pole double-throw plastic enclosed type. Many of these relays are fully interchangeable between manufacturers. Some are plug-in. Common contact ratings are 10 A or 15 A at 120 V or 240 V ac plus a small horsepower rating.

General-purpose power relays are often used to control nonessential pumping station loads such as air conditioning and heating. These relays are available in single-pole or double-pole versions. Single-phase contact ratings are 30 A, 300 V, 1.5 kW (2 hp), not necessarily concurrently. Such relays have very limited current-breaking capability. Plastic enclosed and general-purpose relays are often provided with equipment that is not in motor control centers, but they degrade the overall reliability of the pump and, therefore, should not be allowed. Only industrial control relays should be allowed. Thus, it is imperative that the proper relays, along with proper control wiring practices, be written into specifications for this equipment; that the shop drawings be checked (and most likely corrected!); and that a jobsite inspection be made to ensure that the specified relays were actually installed.

Timing Relays

Timing relays are delayed-action devices that are primarily relays. Output contact operation is delayed so that they open or close after a predetermined lapse of time after the relay is energized or de-energized.

There are several classifications, including: (1) ON Delay, where the relay contacts do not operate until after a timed period; and (2) OFF Delay, where the relay contacts operate instantaneously upon energization of the relay, but delay going back to their normal (shelf) position after de-energization of the relay.

Electromechanical or pneumatic relay timing is usually limited to a maximum period of about one minute, although they may be available to 60 minutes. Timing accuracy is ±10%, sufficient for most pumping station uses. Long operating times can be obtained from solid-state relays and from motor-operated timers. Both of the latter types can provide very sophisticated timing sequences and a wide range of timing. Timing error may be as low as ±0.1%.

Programmable Logic Controllers (PLCs)

The programmable logic controller (PLC) is a very useful form of a specialized digital computer that can provide a large number of timing arrangements and logical sequences to operate complex control functions. Isolated output contacts operate the various controlled devices. Because of their economical price and ease of programming, these units have replaced electromechanical relays on many pump sequence control applications. They are cost effective when compared with six to many hundreds of relays and timers. They should always be considered for pumping stations instead of large numbers of relays.

Transformers

Transformers operate on the principle of magnetic induction and are used to change voltage levels in ac systems and to isolate (electrically) one circuit from another. Two coils, illustrated in Figure 8-13, are wound around the legs of an iron core that forms a closed magnetic circuit. As the current begins to rise in the primary winding it causes a proportional change in the magnetic flux in and surrounding the core. This changing magnetic field, cutting through the secondary coil, induces a voltage that is proportional to the input or primary voltage multiplied by the turns ratio of the two coils.

If a transformer input voltage is 240 volts, the output voltage is 120 volts for a turns ratio of 2:1. Except for minor losses manifested in heat, the total power input equals the total power output and, therefore, the current in the secondary equals twice the current in the primary.

Although most transformer applications involve the changing of voltage levels in power circuits, some are used solely for isolation and instrumentation purposes. Current transformers are designed for instrumenting power systems and usually have a single-turn primary, which may in reality be a piece of buswork, carrying very large currents. The secondary is wound to feed instruments (including meters) that are rated for 5 amperes full scale.

Another special transformer that finds some application in pumping stations is the autotransformer, a single-winding unit used primarily for producing small voltage change ratios. The single transformer winding has a tap (or taps) to increase or decrease the output voltage, but it does not isolate the output from the input. Autotransformers are commonly used for reducing the voltage applied to a motor during starting. After the motor has accelerated to a desired speed, the connections are changed with contactors, and either another step of the autotransformer is provided or the motor is connected directly across the line for the balance of the acceleration time. This reduced-voltage starting limits the current required from the utility during the starting of the motor.

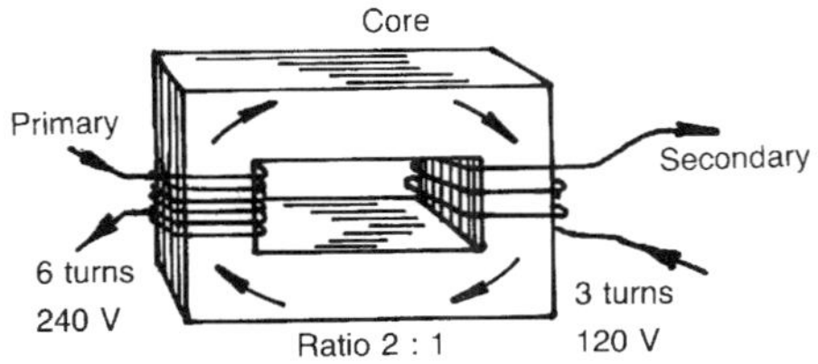

Figure 8-13. A single-phase transformer.

Power transformers are available in various types, sizes, and construction. They can be classified per ANSI/IEEE Std 141-1986 as follows:

- Distribution and Power—According to rating in kVA. The distribution type covers the range of 3 to 500 kVA, and the power type of all ratings above 500 kVA.
- Insulation—Classification as liquid or dry types. Liquid is further classified as mineral oil, nonflammable, and low flammable. There are dry-cast coil, totally enclosed, nonventilated, and sealed gas-filled types.
- Substation or Unit Substation—Substation usually denotes a power transformer with direct cable or overhead line terminations. A unit substation is designed for integral connection to primary and secondary switchgear.
- Primary and Secondary—A primary substation has a secondary voltage rating of 1000 V or higher. A secondary substation has a secondary voltage rating of less than 1000 V.

Transformers may be located indoors or outdoors. Dry-type units are best used indoors, where the NEC imposes few special requirements. Liquid-filled units are usually located outdoors, subject mainly to prescribed clearances from building walls and roof overhangs. In very quiet locations hum may be a problem.

Liquid-filled units may be located indoors because of aesthetic, accessibility, security, or noise requirements, but the NEC imposes some rather expensive room (vault) construction requirements with regard to fire resistance, fire suppression, ventilation, and liquid containment in the event of transformer enclosure rupture. These requirements vary with liquid type (oil, nonflammable, etc.) and size in kVA.

Utility-owned transformers are nearly always outdoors. Refer to the utility for their requirements regarding pads or foundations, conduit stub-ups, and guard posts. Coordinate the transformer location and screening with utility and with the appropriate legal authorities.

When selecting transformers, the following data and ratings should be specified:

- Rating in kilovolt-amperes or megavolt-amperes
- Single-phase or three-phase
- Frequency
- Impedance
- Voltage ratings
- Voltage taps
- Winding connections: Single-phase, single-coil secondary or double-coil secondary for dual voltage connection. Three-phase delta or wye

- Temperature rise
- Basic impulse level (BIL)
- K-rating, if applicable. K-rated transformers are specially designed to resist the heating effects of currents with harmonic frequencies (multiples of rated frequency). In pumping stations, adjustable-speed motor controllers are the major source of harmonic currents.

The desired construction details should include:

- Insulation medium—dry or liquid type
- Indoor or outdoor service
- Accessories—monitoring and safety devices, lightning arrestors
- Type and location of terminating facilities
- Sound-level limitations (if any)
- Manual or automatic load tap changing
- Grounding requirements
- Cooling or provisions for future cooling
- Energy conservation features.

In pumping stations of more than several kVA full-load rating, the transformers connected to the utility are ordinarily three-phase units. Some utilities serve small three-phase pumping stations at 240 V using three pole-mounted single-phase transformers. Two such transformers connected in open delta are sometimes used, but this practice should not be tolerated for fully loaded pumps, because the phase voltages are somewhat unbalanced and cause excessive phase currents and motor overheating. In larger installations, the transformer may be the property of the station owner and may be part of the switchgear lineup. Large transformer ownership can be negotiated with the utility.

Single-phase, dry-type transformers are commonly used to supply 120 V convenience receptacles, lighting, and small power loads, such as control and instrument and computer power. Three-phase transformers with a delta-wye winding connection shown in Figure 8-10 are also widely used. The wye connection can supply both three-phase and single-phase loads simultaneously. The neutral is grounded and results in a uniform potential relative to ground in all phases.

For long-term variations in supply voltage or for growth in plant load, power transformers should have two 2.5% primary taps above normal and two 2.5% taps below normal. These primary winding taps are usually specified for de-energized reconnection.

The impedance rating of a transformer is usually given in percent of full load ratings, and it is ordinarily specified at 5 to 8%. This impedance limits the current that would flow in the primary and secondary systems of the transformer if a short circuit occurs in the secondary system or at the terminals of the transformer. Therefore, the impedance (in conjunction with the supply system's available short circuit current) determines: (1) the required interrupting ratings of the load switching devices and fault current interrupting devices (fuses and circuit breakers) and (2) the bracing required in buswork. But the higher the impedance, the higher the voltage regulation, particularly during motor starting. Hence, carefully strike a balance between available short circuit current and voltage regulation.

Many pole-mounted and pad-mounted transformers furnished by the utilities have low impedances, such as 3%, and thus they allow relatively higher fault current during short circuit conditions than do higher impedance units. Low impedance transformers can result in more expensive switchgear, and thus the cheaper low-impedance transformer may be poor in overall economy. System voltage drop during motor starting is, however, less with low impedance transformers.

Transformers are rated in kVA at a specific temperature determined by winding resistance measurements in accordance with NEMA standards. The transformer temperature rating is a function of the insulation system used. The effects of excessive temperature rise on insulation in transformers and electric motors are identical. See Section 13-9 and Table 13-2 for more detail.

If proper allowance has been made in the specifications for a transformer, plant load increases may be accommodated by the addition of cooling fans and/or an oil circulation heat exchanger system to allow the transformer to withstand a larger load current safely.

Switches

The types of switches normally applied in power circuits include the following:

- Disconnecting switches
- Load interrupter (or load break) switches
- Safety switches
- Control switches.

A disconnecting switch is used for isolating a circuit or equipment from the source of power. It has no interrupting rating and is intended only for operation after the circuit has been de-energized by other means.

Load interrupter switches are intended for operating on circuits of 600 volts or above. They are of air or fluid-immersed construction. The load break switch is usually manually operated and has a quick-make, quick-break operating mechanism that operates at a speed independent of the speed of operation of the handle. When combined with a properly rated fuse, fast fault clearing as well as circuit isolation can be provided. This type of switch (in a metal clad or metal enclosed unit substation lineup) serves to disconnect and protect transformers and medium voltage pump motors. It is sometimes provided as a roll-out unit for safety and ease of maintenance. (See ANSI/IEEE 141.)

Safety switches are commonly used for isolating and disconnecting circuits of 600 V and below. These switches are enclosed and may be fused or unfused. A safety switch designed specifically for motor circuits is rated in horsepower and must be able to interrupt the maximum current drawn by the motor. The maximum current is the stalled or locked rotor current, which is 600% or more of the rated full-load current of the motor. (See ANSI/IEEE 141.)

Control switches are constructed in many types, sizes, and ratings. They are usually rated for a few amperes and for a few volts to about 250 volts in either ac or dc designs. They have large plated contacts in most circuits, and for milliampere control circuits, the contacts are usually gold plated.

Take precautions to ensure that control switches are operated within their voltage and current rating and that ac switches are not applied to dc circuits. The normal toggle switches for the control of lighting circuits are in a similar category and some are rated not only "AC only" but also for the type of lighting circuit.

Fuses

A fuse is an overcurrent protective device with a circuit-opening fusible link that is heated and melted by the passage of overcurrent through it (per ANSI/IEEE 100-1984).

Fuses are calibrated for a specific melting time for a specific current that persists for a defined time. Fuse time delay is approximately inversely proportional to the square of the current, but the exact nature of fuse operation must be determined by laboratory tests for each type and rating of fuse. The resulting time-current curves are plotted on graphs with special log-log scales and are used as fuse application guides.

Fuses are available in a wide range of voltage and current ratings and time characteristics. Fuses may be obtained with slow-blow characteristics for such applications as motor starting. These fuses must, without damage, pass up to 6 to 10 times the full-load motor current during the few seconds of acceleration time. Fuses are also available with fast-blow characteristics and with "current-limiting" characteristics. The current-limiting fuse is used in many applications where the downstream equipment is not fully rated for the short circuit currents available at the supply point. Such systems rated less than 600 volts ac can be interrupted in less than $^1/_4$ cycle and those above 600 volts in less than $^1/_2$ cycle.

Fuses are low-cost, fast-acting, and fail-safe operating devices. They are relatively small, simple and easy to install, and permit a close match to motor thermal characteristics. However, fuses are subject to corrosion and eventually to failure due to repeated heating and cooling, and must be inspected at regular intervals. In a three-phase system, if a ground fault occurs in one phase of a motor and the fuse in that phase opens, the motor will continue to operate as a single-phase motor with currents in the surviving phases increasing as much as 73%. Overload relays will trip off a fully loaded motor, whereas a partially loaded motor may operate to burn-up, because the overload relays would not sense overcurrent. In any event, a single-phased, three-phase motor will not restart.

Circuit Breakers

A circuit breaker is a mechanical switching device capable of making, carrying indefinitely, and breaking current under normal circuit conditions, and also making, carrying for a specified time, and automatically operating to break current (trip) under circuit overcurrent and downstream short circuit conditions. Most circuit breakers applied in pumping stations are enclosed in plastic cases, and are known as "molded-case circuit breakers." Also applied in large pumping stations are low-voltage power circuit breakers, both open and in plastic cases. Medium- and high-voltage circuit breakers, normally utility-owned, are not discussed herein.

The molded-case circuit breaker is enclosed in a ruggedly constructed molded plastic case. Smaller breakers are sealed. Larger breakers may be opened for maintenance. These breakers have external operating handles but are trip-free after automatic operation; that is, they cannot be reclosed into a fault. They may be obtained in single-pole, two-pole, and three-pole configurations. Handle ties are available for converting single-pole breakers into two- or three-pole configurations. But they are unreliable, because if one pole trips, the other(s) may not. Only multipole breakers with common internal trip bars should be used in pumping stations to avoid single-phasing three-phase motors and to ensure complete circuit de-energization for safety. Molded-case circuit breaker operating mechanisms wear if operated too often. They should not be used in lieu of motor contactors. Refer to ANSI/IEEE 141-1993.

Molded-case circuit breakers are available in several types. The thermal-magnetic type is the one most commonly used. This type has two automatic trip mechanisms, both of which operate to release the trip bar. The thermal or bimetal trip element operates in response to overload currents in an inverse time manner: the higher the current, the shorter is the response time. This element is designed to follow the heating characteristics of the conductor, and it allows for temporary overloads, such as for motor starting. Refer to Figure 8-14a. The magnetic trip element operates instantaneously (within a fraction of a cycle) upon occurrence of high-current short circuits, and the trip setting is usually eight to ten times the breaker-rated current. Many breakers have a magnetic trip adjustment of five to ten times the breaker rated current, and they should be adjusted to be as low as practicable for maximum short circuit protection (see Figure 8-14b).

Some circuit breakers have only magnetic trips. In pumping stations, most are used in combination motor starters where the overload function is provided in the overload relay. These circuit breakers are often known as motor circuit protectors (MCPs). MCPs have adjustable trips that should be set just above the motor inrush current. Refer to manufacturer's literature for available sizes and adjustment ranges.

Fused circuit breakers are hybrid circuit breakers with special fuses that open on very high fault currents that would destroy the breaker. Fuse operation releases the trip bar and opens the breaker to prevent single-phasing.

Molded-case switches are circuit breakers without trip elements and are used as manual disconnects. Some have nonadjustable magnetic trip units for self-protection. Molded-case switches must be applied with caution, because fault currents near their interrupting rating may force the contacts to part. Contacts in all circuit breakers are designed so that high currents aid in forcing the contacts apart. Therefore, they may have a withstand rating below their interrupting rating.

Electronic trip units are now available for molded-case circuit breakers. These are considerably more expensive than their thermal-magnetic counterparts, but they are available with a full range of features that increase pumping station reliability, because they can be adjusted to coordinate with other breakers in the circuit. Hence, a serious short circuit in a branch circuit would cause only the branch circuit breaker to trip, thereby sparing the main circuit breaker and per-

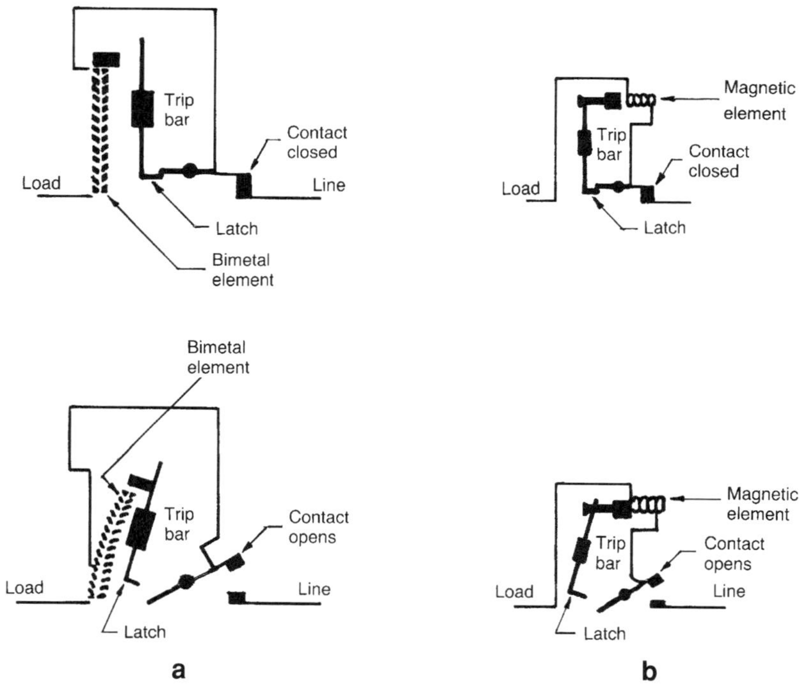

Figure 8-14. Operation of circuit breakers. (a) Thermal action; (b) magnetic action. For both types, the condition of the breaker before being tripped is at the top.

mitting the pumping station to continue in operation. Main and feeder thermal-magnetic circuit breakers do not coordinate with each other or with branch circuit breakers.

Molded-case circuit breakers have several ratings. Among them are:

- Rated current: most breakers carry 80% of rated current continuously (at least 8 hours). Refer to manufacturer's literature, and do not specify a breaker rated below 15 A. Use fuses.
- Voltage: 240 or 480: 480 V units may be applied on 240 V systems. Single-pole breakers are rated 120 or 277 V.
- Interrupting rating (RMS Symmetrical): a circuit breaker must be able to interrupt the maximum fault current available to it. Operators should always open and inspect a circuit breaker after such an operation, because it may need replacement.

Standard thermal-magnetic circuit breaker ratings are given in Table 8-1.

Because the utility may increase the available fault current without notice, designers should use discretion in circuit breaker selection. Do not apply those subject to utility fault current near their interrupting rating, and allow for motor contribution (4 times full-load current for common induction motors).

Most smaller molded-case circuit breakers are designed for use in circuit breaker panelboards. A circuit breaker panelboard is a factory assembly of circuit-breaker mounting arrangements and connecting buses, and may contain up to 42 single-pole circuit breakers. A main circuit-breaker may be included. The enclosure is of sheet metal for safety, and may contain a door (sometimes lockable) over the circuit-breaker operating handles. Circuit breakers may be either plugged into or bolted to the buses. Bolting is recommended for reliability, but bolting is somewhat more expensive.

Circuit breakers may be provided with a number of accessories, including lock-ons, padlock hasps, and alarm contacts. Refer to the manufacturer's literature. Ground fault relaying is available for use with main circuit breakers and with electronic trip units.

Low-voltage power circuit breakers are used in switchboards as the main and tie circuit breakers. They may have trip ratings to 4000 A or higher. Withstand and interrupting ratings are quite high, and they may be obtained with back-up fuses for use up to 200,000 A. Modern trip units are solid state with a full range of adjustments, and ground fault relaying is available. Breaker contacts are opened and closed by a stored energy system (springs) to provide quick-make, quick-break operation. Springs are compressed by electric motor devices. In the event of motor failure, springs may be compressed manually. Remote electrical opening and closing control is possible, such as for switching to alternate electrical sources or standby generators and tie-breaker operation. These breakers are easily inspected, maintained, and serviced.

Table 8-1. Standard Thermal-Magnetic Circuit Breaker Ratings

Continuous current rating	Interrupting rating	Continuous current rating	Interrupting rating
120/240 V		277/480 V	
15–100 A	10,000 A	15–100 A	14,000 A
15–150 A	22,000 A	15–125 A	18,000 A
100–225 A	42,000 A	15–100 A	25,000 A
15–30 A	65,000 A	70–250 A	25,000 A
		125–400 A	30,000 A
		300–1000 A	30,000 A
		15–125 A	35,000 A
		70–250 A	35,000 A
		125–400 A	35,000 A
		15–125 A	65,000 A
		100–250 A	65,000 A
		300–2000 A	65,000 A
		600–1200 A	100,000 A
		600–2500 A	100,000 A

Motor Branch Circuits and Controllers

The requirements for each component of a motor branch circuit are delineated in Article 430 of the NEC. A single-line diagram of a typical motor branch circuit is illustrated in Figure 8-15. The feeder or service disconnect and overcurrent protective device(s) and the feeder or service conductors are not part of the motor branch circuit, but their size calculations are influenced by the rules in Article 430.

The branch circuit disconnect and branch circuit overcurrent protection consist of either a circuit breaker or a fused switch. It must be convenient for a worker to padlock a switch in open position for safe motor or pump maintenance. Circuit breakers in motor control centers are often adjustable magnetic-only devices, often called "motor circuit protectors." They must be adjusted high enough to not trip on starting but to trip on a short circuit. Fuses are sometimes dual-element types selected for both branch circuit and motor overload protection. Fuses are low-cost, easily replaced devices that are fail-safe and easily coordinated with other protective devices. But they can allow single-phasing of the motor. Some fused switches are arranged to open the switch automatically if a fuse opens. Fused switches must be horsepower-rated below 200 hp, and equivalent measures must be taken above 200 hp. Large, important motors, particularly in sewage lift stations, justify the cost of a single-phase protective relay or of a ground fault protective relay. Single-phase monitors and protection devices cost about $700 per motor at 1997 prices, regardless of motor size. Installation would increase the cost by perhaps 50%. Where power is reliable, the frequency of a single-phasing event might be less than once per 20 years, whereas it might be once per decade in rural communities or where lightning is prevalent. Designers should assess the likelihood of burned-out motors and the loss of pumping against the cost of protection to balance risk with cost.

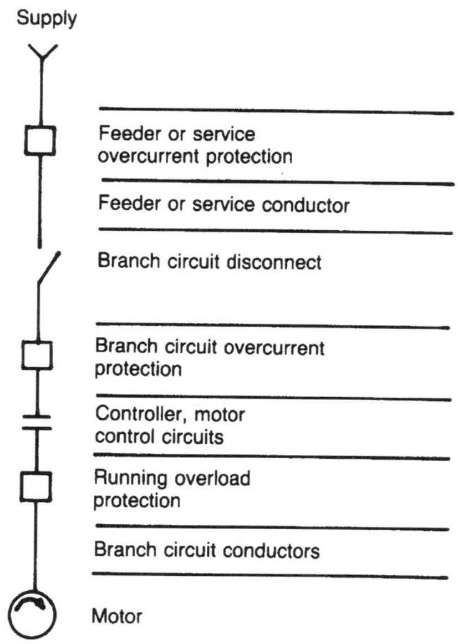

Figure 8-15. A typical motor branch circuit diagram per NEC.

Motor circuit conductors must be selected to carry at least 125% of motor-rated full-load current.

Controllers for AC Squirrel-Cage Motors

The controller must be capable of starting and stopping the motor and of interrupting the locked rotor current. For portable motors $^1/_3$ hp and smaller, a cord and plug assembly may serve as the controller, For stationary motors $^1/_8$ hp and smaller that are normally left running and cannot be damaged if stalled or jammed, the branch circuit protective device may serve as the controller. For attended motors, manually operated motor starters and some light switches of suitable ratings are available in single- and three-pole models, For unattended or automatically controlled motors, motor starters using magnetically operated contactors must be installed.

A contactor is a heavy-duty solenoid-operated switch used primarily as a controller for motor on-off control, but also for heaters and lights. When combined with an overload relay, the assembly is known as a "motor starter." A motor starter with a circuit breaker or fused switch is known as a "combination motor starter."

A motor contactor must be able to interrupt locked-rotor current. Locked-rotor current is drawn during the starting period or indefinitely should the pump or other driven device be jammed or stalled. Locked-rotor current is roughly six times full-load current or more for the high-efficiency line of motors. Contactors are not required to interrupt fault currents. The fuses or circuit breaker perform that function. The locked-rotor current rating determines the horsepower rating of the contactor.

Running overload protection is intended to protect motors, motor controllers, and motor branch circuit conductors from overheating due to motor mechanical overloads and from failure to start. It does not protect against short circuits and ground faults.

Most of the common single-phase motors larger than $^1/_{15}$ kW ($^1/_{20}$ hp) have an integral thermal switch (in series with the motor conductors) that will shut off the motor if it becomes overheated. Most three-phase motors are protected by thermal or solid-state overload relays (one in each phase) that are part of the motor starter. The following description applies to relays on size 5 motor starters and smaller.

Thermal overload relays of the melting alloy and bimetal types have been available for many years. Connected in series with motor conductors and on the load side of the contactor, they carry motor current and are heated thereby. Should excessive current flow for a length of time due to mechanical overload or to a locked-rotor condition, the relay (connected so as to de-energize the contactor coil when hot) is gradually heated. As the motor does not heat instantaneously, the relays are designed to have a time lag matching that of the motor, and this time lag allows the motor to start. Relays are available in quick trip (Class 10), standard trip (Class 20), and slow trip (Class 30) versions, and can be set for manual or automatic reset. Automatic reset should be selected for unattended pumping stations. Because ambient temperatures vary and significantly affect motor heating, noncompensated relays are provided as standard. Ambient-compensated relays are available for motors in a constant ambient condition, such as in a submerged application. Relays have heaters selected to match the motor and other conditions, and may be adjusted within limits as necessary for a closer match to motor-heating characteristics.

Solid-state overload relays are now available as direct replacements for the melting alloy type. They are somewhat more expensive than thermal relays. These solid-state relays have a large adjustment range and so do not need heaters. They may have other features, such as phase loss protection, phase unbalance protection, and electrical remote reset.

Across-the-Line Starters

The across-the-line (or full voltage) starter is the most commonly used starter in pumping stations because of its simplicity, ease of maintenance, and relatively low cost. It is nearly always chosen for motors of 55 kW (75 hp) and smaller. Motor starters are available in the following styles:

- Full Voltage Non-Reversing (FVNR) or "across-the-line"
- Full Voltage Reversing (FVR)
- Reduced Voltage Non-Reversing (RVNR).

Most of the above styles of motor starters are available in low-voltage starters (below 600 V) and in medium-voltage starters (2300 to 13,200 V). The ratings of motor starters are given in Table 8-2.

No motor starter smaller than Size 1 should be specified or permitted in pumping stations, and some designers say Size 2. The cost difference is not great, but the larger starters are considerably more robust and ensure greater reliability and longer service life.

Use caution in specifying motor starters. Thorough knowledge of the motor-operating conditions and

Table 8-2. Motor Starter Ratings

		Maximum hp			Maximum hp	
NEMA size	System voltage	Single phase	Three phase	NEMA size	System voltage	Three phase
00	120	$\frac{1}{2}$		3	208	25
	208		$1\frac{1}{2}$		240	30
	240	1	$1\frac{1}{2}$		480	50
	480		2			
0	120	1		4	208	40
	208		3		240	50
	240	2	3		480	100
	480		5			
1	120	2		5	208	75
	208		$7\frac{1}{2}$		240	100
	240	3	$7\frac{1}{2}$		480	200
	480		10			
2	120	3		6	208	150
	208		10		240	200
	240	$7\frac{1}{2}$	15		480	400
	480		25			
				7	240	300
					480	600

other constraints is required for selecting starter equipment for each application. Prices vary widely, so designers must not specify an expensive soft starter when a simple across-the-line unit would be adequate, nor should an overly expensive starter be specified because of a lack of design information. The decision should be made after sufficient information has been developed relating to the load conditions, accelerating time of the drive, and accelerating time allowed by the serving utility. Also, in pumping applications, avoid simple on-off controls for pumps when the pumping cycle time is likely to be very short (more than 3 starts/min). Otherwise, derate motor starters by using their jogging horsepower rating (see manufacturer's literature). Starting too frequently causes motors and motor starters to overheat due to the high inrush current. There have been situations where too-frequent starts have caused early failure of a motor. Failure is even more likely to occur if the motor is undersized for the peak load during the short-duration pumping periods, because both the load and the starting heating effects are additive.

Reduced-Voltage Starters

Use of a reduced-voltage starter should be considered: (1) if the power system cannot provide the high starting currents typical of larger motors, (2) if the power company limits starting current or kVA, (3) if a standby generator is used, (4) if the feeder and/or branch circuit is longer than 180 m (600 ft), or (5) if the voltage drop due to starting current would be 20% or more. Actually, the motor would start and run at a 30% drop, but the contactors might drop out.

When a motor is started at reduced voltage, the current to the motor is reduced in direct proportion to the voltage reduction, and the torque is reduced by the square of the voltage reduction. If a "typical" NEMA B motor is started at 70% of line voltage, the starting current would be 70% of the full voltage value (i.e., 0.70 × 600% = 420%). The torque would be then $(0.70)^2$ or 49% of the normal starting torque (i.e., 0.49 × 150% = 74% of full-load torque). In pumping stations, reduced-voltage starting effectively reduces inrush current, while the lower starting torque may pose no problem. Autotransformer starters are the most popular in pumping stations. Refer to manufacturers' literature for details of construction and operation of each type, and to Table 8-3 for their features, advantages, and disadvantages.

Motor Control Circuits and Devices

The simplest, most commonly used motor control circuit utilizes momentary start-stop push buttons and is illustrated in Figure 8-16. This control circuit is sometimes called "three-wire" control because three control conductors are run from the motor starter to the pushbutton station. Momentarily depressing the start button energizes the contactor M coil, which is sealed in through the normally open M contact (which closes when the contactor main contacts close). The start button may then be released without dropping out the

Table 8-3. Reduced-Voltage Starters for Motors

Type	Advantages	Disadvantages
"Soft start," solid state	Low starting line current. Frequent starting capability.	Most expensive. Sensitive to line transients, power quality, and high ambient temperatures.
Autotransformer	High torque. Low starting line current. Adjustable taps for reduced voltage.	Expensive.
Primary resistor	Multiple accelerating points. Frequent starting capability. High power factor.	Expensive. Resistors generate heat. Starting time less than 5 seconds.
Wye–delta	High torque and efficiency. Moderately priced. Low starting current.	Long acceleration starting time. Low starting torque. Motor must be suitable.
Part winding	Compact. Least expensive.	Low starting torque. Starting time less than 2 seconds. Not suitable for frequent starts. Nonadjustable. Motor must be suitable.

contactor, and thus the motor is energized through the closed main contacts of the starter. Depressing the stop button or tripping an overload relay de-energizes the M coil, opens the line contactor, and thus disconnects the motor from the line.

A similar control circuit using a maintained (a contact that remains closed after actuation) two-position (On–Off) or a three-position (On–Off–Auto) selector control switch can also be utilized for motor starting and running control. See Figure 8-17. The selector switch control circuit allows the motor to operate in either the manual or automatic mode. The manual mode consists of the On and Off positions. In the Auto mode (with the selector in the Auto position), the motor is started *or* stopped by the closing or opening of, for example, the wet well level control switch located remotely from the motor controller.

Control circuits for pumping applications may be designed to meet other requirements, such as:

- Pump sequencing, which reduces the frequency of motor starts, exercises rotating equipment and

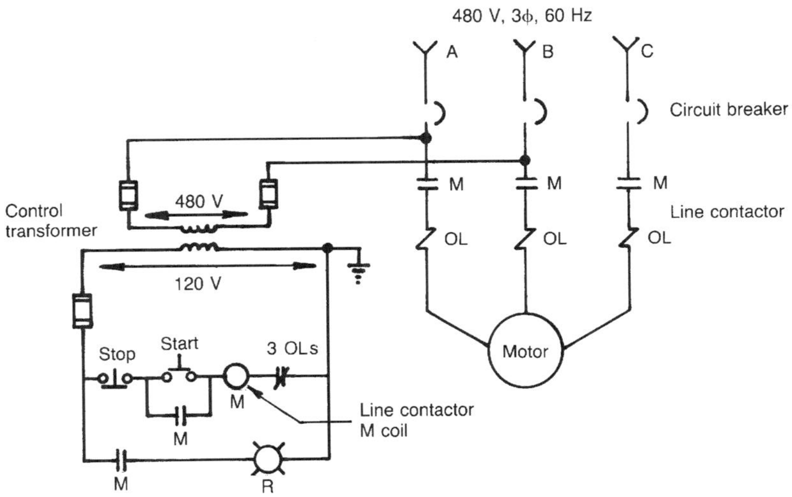

Figure 8-16. Pushbutton three-wire motor control circuit.

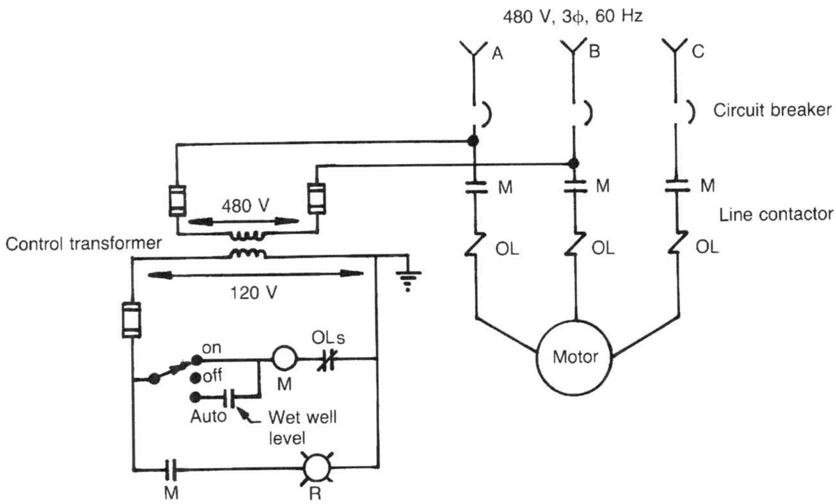

Figure 8-17. Three-position selector switch motor control circuit.

alternates the pumps to start on each successive operation. The pumps, however, tend to wear out at about the same time, which may or may not be acceptable to the owner.

- A time delay, which is used in a control circuit to delay the restarting of the pump after the restoration of power following power failure.
- A permissive condition that must be satisfied before control action can become effective, such as delay in shutting down an operating pumping unit until the replacement unit has reached a specific speed range and is thus capable of taking over the pumping duty.
- Electrical interlocks, which are usually a series of contacts on relays or other control devices used to enable or prevent the operation of equipment in a prearranged sequence.

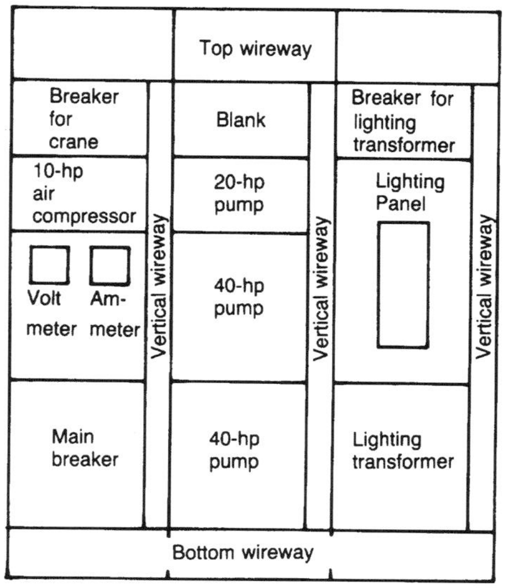

Figure 8-18. Elevation view of a typical low-voltage motor control center. Cubicles are labeled with name of load. Permanent name plates are recommended.

Motor Control Centers

A motor control center (MCC) is a unitized assembly of grouped motor controllers, interconnecting buses, wiring spaces, and other electrical controls and equipment associated with providing power and control for several motors. Motor control centers provide incoming line facilities, main circuit breaker space (if needed), horizontal bus, and a vertical bus in each section. MCCs may be obtained with plug-in cubicles for NEMA Size 1 through Size 4 motor controllers. Several classes of control wiring arrangements are identified by NEMA standards. Layouts of typical MCCs are shown in Figures 8-18 and 8-19. The MCC structure must have an enclosure and a seismic rating compatible with the location in which it is installed. The built-up motor control panel shown in Figure 8-19c is a low-cost alternative to an MCC—but only if labor costs are not considered.

It is important that the motor control center (bus bracing, combination starters, and individual circuit breakers) be rated for the short circuit duty predicted by the fault calculations. If a main circuit breaker is to be provided in the unit, then its setting should be coordinated with the supply overcurrent equipment as well as with the branch circuit breakers connected to the buswork. In the coordination with the branch circuit breakers feeding a number of motors, the breaker feeding the smallest motor has the highest interrupting duty, so be certain that protection for the smallest motor has adequate interrupting capacity. The reason for this high interrupting duty on the smallest circuit is that during the initial period of the short circuit, all other running motors contribute current to the fault, thus adding to the short-circuit current from the main power source. Large motors contribute larger short-circuit currents than do small motors.

NEMA Class II, Type B wiring is recommended in the MCC. In this wiring method, factory power and control wiring is terminated on terminal blocks in or adjacent to the cubicle. The terminal blocks form the interface between factory and field wiring. The factory furnishes complete layout, wiring, and control diagrams for each unit as well as for the entire assembly. All terminals are labeled and each wire is numbered and/or color coded. The manufacturer's diagrams should be required to include the entire motor branch circuit and controls.

It is advisable to require: (1) the circuit breaker (circuit disconnect) handle of each unit to be lockable in the open position and (2) provisions for at least two padlocks to be used for locking out the circuit. The door must be interlocked so that the circuit breaker is in the Off position before the door of the unit can be opened.

There must be enough space in each cubicle so that no device is mounted behind another device. All control equipment must be accessible for inspection when the front door of the cubicle is open. Empty spaces or spaces wired for "future" starters must have bus openings completely covered so that when the door is open, no live buses are exposed.

Motor control centers should be installed in easily accessible locations with adequate spacing from other equipment, in accordance with NEC and local and

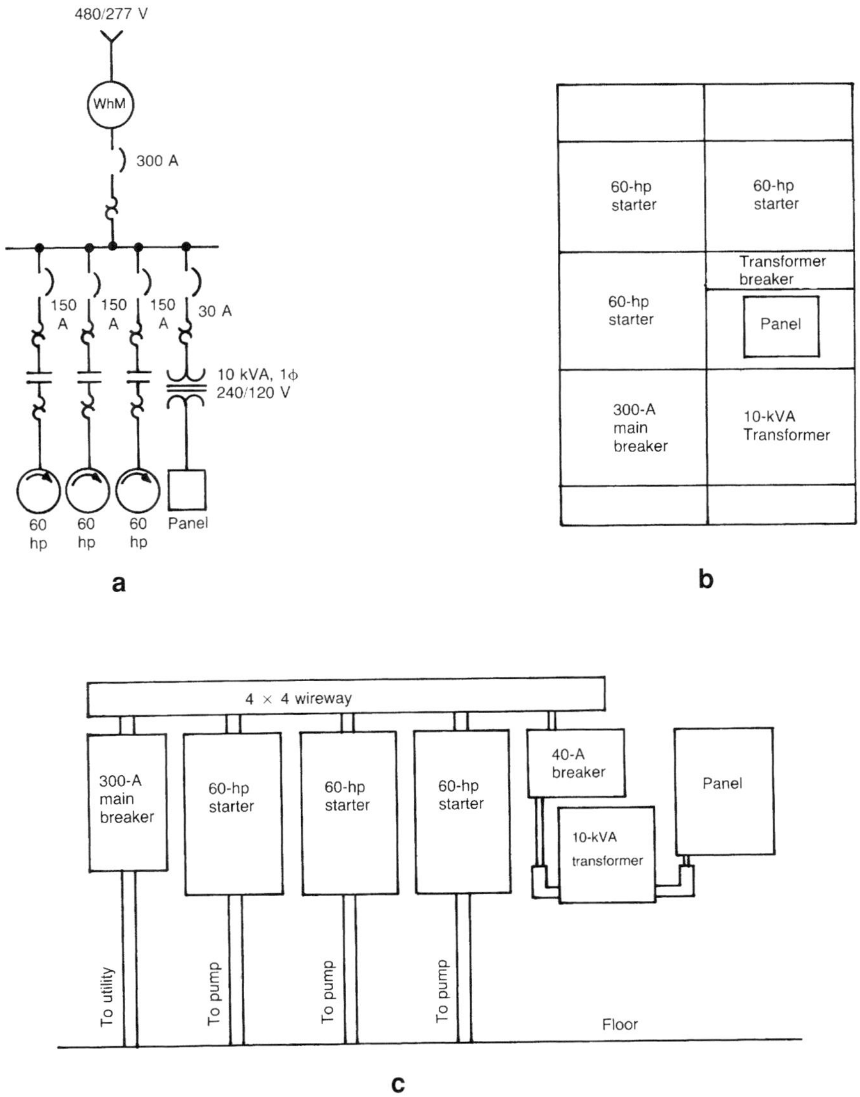

Figure 8-19. Examples of different types of motor control centers. (a) Single-line diagram; (b) motor control center; (c) built-up motor control panel.

state codes. They should be in a clean, dry location to avoid the need for specifying the more expensive gasketed enclosures. The space should be well lighted and ventilated (see Chapters 9 and 23).

Where control power transformers with fused primary and/or secondary are provided, it is advisable to install a blown-fuse indicator in the door of each unit for monitoring the control circuit continuity. It may also be desirable in some installations to include door-mounted pushbutton controls for the motor and indicating lamps to show: (1) power available, (2) motor running, (3) automatic or remote control indicators, and (4) running-time indicator. In large stations where computer control and monitoring is used, the running-time records may be more economically kept in the computer files.

Sometimes the plant control panel is installed in the motor control center when there is no other suitable location.

Insulated Cables and Conductors

Cables and conductors are not the same thing. A cable contains two or more conductors.

Cables

Insulated conductors used in electric circuits may be grouped together and surrounded by a jacket to form an insulated cable. Cable jackets must be suitable for the location and application, because they may be exposed to moisture, dirt, etc., without the protection of conduit.

Conductors and Terminations

Most conductors are of copper or aluminum. Both metals are considered to be excellent conductors of electricity. The conductivity of aluminum is only 61% of that of copper, with the result that aluminum conductors are approximately one size larger than the equivalent copper conductor. The specific gravity of aluminum is 30% of that of copper, so aluminum conductors are lighter and easier to handle.

Conductors may be solid or stranded. The NEC requires that conductors No. 8 AWG and larger be stranded. Smaller conductors may be either solid or stranded. A number of types of stranding are available, and the selection depends upon the application. Conductors No. 2 AWG and smaller are usually 7-strand rope-laid (center strand surrounded by 6 outer strands). In larger sizes, 12 strands surround the basic 7 to form a 19-strand conductor.

Oxide forms on the surface of both copper and aluminum—slowly on copper and rapidly on aluminum. Copper oxide is a relatively good conductor, but aluminum oxide has high resistance and must be removed in the termination process to prevent termination overheating and failure.

Conductor termination workmanship must always be of the highest quality to ensure a reliable, trouble-free electric system. But aluminum terminations are much more sensitive to workmanship than are copper ones. The aluminum must not be nicked when the insulation is cut off. The aluminum oxide must be removed by wire brushing, and the aluminum must be immediately coated with a conductive anti-oxidant compound. The connection must be torqued to manufacturer's specifications, retightened the next day, and retightened annually, because aluminum has a 31% greater thermal coefficient of expansion, which, together with its tendency to creep, loosens connections (see ANSI/AIEE 141). The only reliable alternative to torquing bolted terminals is the use of compression lugs, and these are recommended for high-quality work for both copper and aluminum conductors in Size No. 2 AWG and larger. Terminals for use with aluminum conductors must be so labeled. *Because of the termination problems, it is industry consensus that only copper conductors should ever be specified and used in pumping stations.*

Buses are electric conductors in the form of bars of various shapes and sizes. Buses are used in motor control centers, switchgear, and elsewhere for high amperage capacity. They may be insulated for safety. Aluminum buses are often tin-plated for corrosion resistance. Bolted connections are made with belleville washers, properly torqued.

Insulation

Most insulating material is composed of organic compounds. Two exceptions are magnesium oxide, used for mineral-insulated cables, and mica, used in some high-voltage cable construction. Insulation and cable jackets for pumping station applications should have physical and electrical characteristics such as:

- Resistance to moisture
- Resistance to heat
- Resistance to ozone
- High dielectric characteristics
- Resistance to abrasion
- Hardness at utilization temperature.

Examples of low voltage (600 V) conductor insulation are given in Table 8-4.

Medium-voltage (5 to 15 kV) cables may be constructed with semiconducting tapes on the inside and outside of the insulation with a conducting shield between the outer semiconducting tape and the jacket. Medium-voltage cable insulation materials commonly used are:

- Natural rubber
- Butyl rubber
- Cross-linked polyethylene (XLPE)
- Ethylene-propylene (EPR).

Shielded cables should always be used. The use of shielding in medium-voltage cables lowers the electrical stress on the cable insulation by providing a grounded surface equidistant radially in all directions from the conductor. When laid on a metallic surface, an unshielded cable allows a distorted field pattern within the cable structure that causes a high voltage gradient in a portion of the insulation. It shortens the life of the cable.

Current Rating

The ampacity (current rating) depends upon the conductor material, size, type of insulation, and the ambient conditions of its installation. Refer to NEC Tables

Table 8-4. Insulation Types for Low-Voltage Electrical Conductors

Type	Insulating material	Maximum operating temperature, °C [a]	Application
RHH	Heat-resistant rubber	90	Dry and damp
RHW	Moisture- and heat-resistant rubber	75	Dry and wet
THW	Moisture- and heat-resistant thermoplastic	75	Dry and wet
THHN	Heat-resistant thermoplastic	90	Dry and damp
THWN	Moisture- and heat-resistant thermoplastic, polyamid nylon jacketed	75	Dry and wet
XHHW	Moisture- and heat-resistant cross-linked polymer	75	Dry and wet
		90	Dry and damp

[a] Can be operated at these temperatures only if the circuit protective device is tested and rated by UL at these temperatures.

310-16 through 310-19 and 310-69 through 310-84 and the notes pertaining to them.

Raceways and Wireways

The NEC requires that all electrical conductors be enclosed and protected by some form of raceway. Jacketed cables, where permitted, are exempt. The most common forms of raceways in pumping stations are (1) conduit and (2) wireways.

Galvanized rigid steel is the most commonly used material for conduit systems. Lighter grades than schedule 40 steel do not generally exhibit the degree of corrosion resistance required for long life in the damp surroundings that are common in pumping stations. Threaded connections should always be used. The specifications should require full hot-dip galvanizing treatment after fabrication and threading. Field-cut threads should be painted with a zinc-bearing material.

On small projects, if high-grade hot-dip galvanized conduit is not available, consider schedule 40 or 80 rigid PVC conduit in wet places and where concrete encasement is required. If the pumping station is part of a large project, wherein office and shop space is allocated, intermediate-grade galvanized conduit may be specified for these areas. On all projects where PVC conduit is run through or embedded in concrete, special precautions must be taken to avoid breakage during the placement of the concrete.

Wireways are square or rectangular metal channels or troughs with removable covers designed for routing and connecting conductors. In pumping stations, use wireways in dry locations only. The main advantage of wireways is the ease of access for additions or removal of conductors and in adding circuits. A new conduit may be readily connected to a raceway and new power and control circuits added at minor cost compared to the cost of a new conduit run to the nearest power source.

If the mechanical specifications for the project require painting of all mechanical equipment and exposed metal in a particular area of the station, then the exposed conduit and wireways should be painted to conform with the other exposed metal. The zinc coating on galvanized metal must be properly pretreated and primed before painting. Any conduit entering a hazardous location must be provided with an explosion-proof seal fitting filled with a UL-listed sealing compound after all conductors are installed. Conduit runs must be sloped to drain, and breather/drain fittings (or, in nonhazardous areas, weep holes) should be provided at any low point in a run and prior to entering control cabinets, MCCs, and control panels.

All conduits below grade, whether buried, embedded in concrete, or exposed in underground rooms or spaces, are subject to infiltration of water and should be sealed inside the conduit around the conductors and/or cable(s) at each entrance into any enclosure below grade. O-Z/Gedney [1] has excellent sealing bushings and fittings that utilize sealing glands tightened by accessible set-screws that can be retightened during maintenance if necessary.

All conduits or cables entering a normally dry underground room or space in a pumping station and originating outside (buried) or in a room or space subject to possible flooding (such as a wet or clear well) should be installed in through-wall or floor sleeves equipped with sealing glands tightened by accessible set-screws that may be retightened. O-Z/Gedney [1] has suitable sealing sleeves.

Junction and Outlet Boxes

Junction and outlet boxes installed in pumping stations should be hot-dip galvanized cast iron. They should be surface-mounted, spaced a minimum of 6 mm ($^1/_4$ in.) from walls or mounting plates, and placed as high as possible in dry wells. Many pumping stations are subject to accidental flooding. Wet wells are even more subject to flooding, and, in addition, they are generally classified as hazardous areas because of the high probability of a large concentration of methane or other explosive gasses resulting from accidental or illegal dumping of gasoline and solvents into the sewer system.

Avoid placing junction boxes in wet wells (or in any other other moisture-laden locations) by making a concerted effort to feed electrical systems from junction boxes in dry locations through conduit sealed at the entry into the hazardous area. But if there is no alternative, specify nonwicking cable, watertight motor and device terminations, and junction boxes filled with a waterproof sealant. The box itself must be rated as waterproof when properly installed.

One example of the need for a junction box in a wet well occurs in pumping stations with submersible pumps fed by portable cables. A junction box is necessary to transfer from the normal branch circuit conductors to the heavy-duty flexible, multiconductor cable to the submerged motor. One solution is to mount the junction outside the wet well and feed the cable into the wet well through a sufficiently large conduit.

Switchgear and Switchboards

"Switchgear" as defined in ANSI 141 is a general term covering switching and interrupting devices alone or in combination with other associated control, metering, protective, and regulating equipment.

Metal-enclosed switchgear provides sheet metal covering on all sides and top with access via doors and/or removable panels. Other classifications are indoor, outdoor, and walk-in with an enclosed maintenance aisle, as defined in ANSI 141. Further classification provides for metal-clad switchgear that meets a number of specific safety requirements and is usually applied to systems operating at 1000 V or more.

Switchboards are not defined in ANSI 141 but, although of open construction, may be provided with equipment similar to that included in switchgear. Modern switchboards are of dead-front construction (no live parts exposed) for operator safety and to meet NEC requirements. Switchboards are usually an assembly of molded-case circuit breakers, whereas switchgear is ordinarily a grouping of power circuit breakers and may include large motor starters. Incoming and outgoing line-termination space must be provided, with particular attention to the required minimum bending radius of large-diameter power cables. Buswork must be braced for the calculated maximum fault current and the structure must be rated for the prevailing seismic conditions.

Service Entrances

Every building supplied with electricity directly from an electric utility must have a UL-listed service entrance. A service entrance usually consists of a utility-owned kilowatt-hour meter, any current or potential transformers (usually utility-owned), and a disconnect consisting of a fused switch or a circuit breaker. The neutral, if grounded, is grounded on the utility side of the disconnect. Service entrance equipment, including the meter sockets and instrument transformer provisions, must conform to utility requirements.

Meter and disconnect locations vary. For maximum security, both should be indoors, but utility and local regulations may dictate otherwise and must be followed. Sometimes other arrangements can be made.

A second building or structure on the same premises and under the same management must also have a UL-listed service disconnect. Neutral grounding can be at either building.

Equipment Installation

All electrical equipment must be equipped and installed per the NEC. Sufficient access and working space is required for ready and safe access for maintenance. Minimum working space in front of electrical enclosures not requiring rear access and facing metal or concrete walls is shown in Table 8-5. Illumination is required for all working spaces.

At least one adequate entrance must be provided for access to the working space in front of electrical enclosures. A continuous passageway is usually sufficient. For equipment over 1.8 m (6 ft) wide rated 1200 A or more and 600 V or less, and for all equipment over 600 V and 1.8 m (6 ft) wide, two entrances at least 0.61 m (2 ft) wide and 1.91 m (6.25 ft) high are required—one at each end. If, however, the depth shown in Table 8-5 is doubled, only one entrance is

Table 8-5. Minimum Working Space in Front of Electrical Enclosures

Nominal voltage to ground	Dimensions, m			Dimensions, ft		
	Depth[a]	Width	Height	Depth[a]	Width	Height
0–150	0.92	0.76	1.91	3.0	2.5	6.25
151–600	1.07	0.76	1.91	3.5	2.5	6.25
601–2500	1.22	0.92	1.98	4.0	3.0	6.5
2501–9000	1.53	0.92	1.98	5.0	3.0	6.5

[a] Or more if required for the access doors to swing 90°.

required, but the edge of the entrance nearest the enclosure must equal the depth shown in the table.

8-4. Standby Generators and Auxiliaries

Standby generators and their auxiliary equipment are needed in many pumping stations. Permanently installed units may be required by state or local laws or regulations, or they may be included in the station design because of the critical nature of the station, probability of loss of utility power, or the owner's or engineer's standards.

Portable generators provided with cord and plug connections are often used for backup power for small "package" pumping stations, and trailer-mounted engine-alternators of up to several hundred kilowatt ratings have been cord-connected with suitably rated cables and power plugs to serve medium-sized pumping stations during power outages. Bear in mind, however, that power outages occur in stormy weather that may make it difficult to reach a station, and if the power outage covers a wide area, many portable units would be required. Distance and time of response must also be considered. Finally, consider that legislation now allows competition between electric power suppliers—a policy that will inevitably lead to less reliable power because of the need to reduce costs to be competitive. Thus, the need for standby generators is likely to increase.

Codes and Legal Requirements

Generators in pumping stations are installed in accordance with NEC Article 701, Legally Required Standby Systems, or Article 702, Optional Standby Systems. If a local code requirement seems to require Article 700, Emergency Systems, be sure to negotiate a change with the authorities, and never call the generator an "Emergency Generator," because pumping station applications in no way involve life safety. Use Article 702, if possible, due to its minimal requirements and flexibility of operation. Use Article 701 only in case of a legal requirement to do so. See Article 701-2 (FPN), which refers to sewerage disposal and fire fighting applications, but take into account the wet well or clear well storage capacity, including any available storage in upstream piping.

Ratings

Continuous Rating

Two nameplate ratings are offered: prime and standby. The prime rating is that rating at which the engine-generator set is suitable for continuous operation at rated conditions. The standby (or continuous standby) rating is defined as that rating at which the generator set is suitable for continuous operation for only 30 days at the rated conditions. Continuous operation at the standby rating causes relatively high temperatures that age the generator insulation 4 to 8 times faster than at the prime rating. This aging is not considered serious for a standby generator because of its limited use.

Refer to Chapter 14 for engine ratings.

Motor-Starting Rating

Motor-starting ratings are limited by either the engine or the generator. The engine must supply the shaft power required by the actual maximum load applied to the generator. The generator size is normally chosen to match the engine output capability. Selecting an oversized generator where motor-starting kVA demands are high, however, may be advantageous. The running load and the motor-starting requirements should be specified. The largest motor should be started first, if possible. Consult the manufacturer for selection and sizing of the engine-generator set. A 30% voltage dip at motor start is usually tolerable, but 20% is a more common design parameter to prevent contactor opening.

When specifying a generator for use with AFDs, require the generator supplier to work with the drive supplier to ensure that the generator is suitable for the application. Loads with harmonic current content over about 10% of the generator rating cannot be supplied by standard generators.

Fault Rating

With a brushless exciter whose output is dependent on terminal voltage, the worst event is a three-phase bolted fault (three-phase line-to-line short) at the generator terminals, because (with greatly reduced excitation) the terminal voltage collapses to a low value and cannot supply sufficient fault current to trip the breaker on the faulted circuit. Field-forcing equipment should be specified to sustain the rated excitation and current up to three times the generator's rated current output—sufficient fault current to trip the generator breaker. Ensure that downstream and generator circuit breakers are coordinated so that the branch circuit breaker trips first.

An undervoltage relay is recommended for any generator unit, large or small. It is actually a voltage-restrained overcurrent relay that has a current setting low enough to trip breakers and shut down the engine if overcurrent at less than full voltage occurs for a predetermined length of time.

Derating

Refer to Section 14-6 for derating engines for elevation above sea level and for temperature. Derate the generator 1% for each 100 m (3%/1000 ft) above 900 m (3000 ft).

Engine-Generator Controls

The engine instrument panel should never be mounted on the engine or on the engine pad, because engine vibration causes instruments to deteriorate.The instruments for the engine should include:

- Oil pressure
- Oil temperature
- Water temperature
- Intake manifold pressure and vacuum gauges—especially for large units.

A battery-charging alternator and ammeter should be included. Require safety controls for engine shutdown to be manually reset. Such controls should include:

- High water temperature
- Low oil pressure
- Overspeed.

The following instruments and controls should be mounted on the generator control panel:

- An ammeter with current transformers, as needed
- A voltmeter with potential transformers, as needed
- An ammeter/voltmeter phase selector switch
- A frequency meter (45 to 65 Hz)
- A voltage-adjustment rheostat
- An elapsed-time meter
- An annunciator or monitoring panel with fault-indicating lights for low oil pressure, high water temperature, overspeed, overcrank, and generator undervoltage. Standard annunciator panels may be specified to perform all the foregoing functions, and they usually constitute a much more economical system than a hard-wired custom alarm system.
- A fault reset pushbutton
- A generator output circuit breaker
- A three-position (manual-off-auto) selector switch
- An engine-start switch
- An emergency stop pushbutton
- A voltage regulator
- An indicating-light test pushbutton.

Actuating the safety devices must shut down the generator set, indicate the cause of the shut-down by lighting the appropriate indicating light, and provide separate outputs for the remote alarm indication panel and the computer.

Automatic Transfer Controls

Code Requirements

Automatic transfer controls, per NEC 701-7 (1981), are needed for legally required standby systems. The interconnection of normal and standby power sources should not be possible in any mode of operation, except where parallel operation is intended and suitable automatic or manual synchronizing equipment is provided per NEC 230-83. In pumping stations, automatic transfer control equipment is commonly installed, except where portable generator sets are to be used.

Manual Transfer Switches

Manual transfer switches are sometimes used with portable generating equipment. Use kVA- (horsepower-) rated, quick-make/quick-break, heavy-duty switches or mechanically interlocked circuit breakers. If the transfer switch is part of the station service equipment, it must be UL-listed as service equipment or better, and it must have a separate UL-listed service disconnect (circuit breaker) ahead of it in a separate enclosure.

Automatic Transfer Switches (ATS)

Complete automatic transfer switches are available in suitable enclosures (e.g., NEMA 1) as well as in open style for installation in motor control centers. They must conform to all of the requirements of UL 1008 and be so listed and labeled. Bypass isolation switches (that allow the ATS to be removed for repairs) are desirable for avoiding shutdowns for repairs.

Automatic transfer switches include a switching element, relays, and controls and are available in these forms:

- Molded-case circuit breakers
- Contactors
- Double-throw switches.

Automatic transfer switches should have the following ratings:

- Continuous rating, which is the current rating on a 24-hr basis.
- Inrush rating, which is the ability to close a circuit with high inrush currents with minimum contact bounce and welding.
- Interrupting rating, which is the load- (not necessarily fault-) interrupting ability under the worst condi-

tions, e.g., locked rotor currents that occur when the motor rotor is locked in place due, say, to a clogged pump.
- Withstand rating, which is the ability to withstand the thermal and magnetic effects of downstream faults.

Automatic transfer switches should include a pause-in-neutral position (with an adjustable time delay that should be set for several seconds) that causes the motor to be disconnected from the power source during transfer and allows the motor voltage to collapse to a safe level prior to re-energization. Failure to allow the field to collapse can result in the motor field being out of phase with the new power source (which may well occur when transferring from a standby generator to the power utility). This out-of-phase condition can instantly destroy the motor by breaking the shaft or tearing out the windings. Because the pause-in-neutral feature may be unavailable (and adds complexity and reduces reliability), consider other methods such as motor shut-down before transfer and an in-phase monitor relay.

Batteries

Starting batteries for a standby generator are ordinarily lead-acid type with heavy-duty ratings. Cranking batteries for moderate-sized generators are usually size 4D or 8D diesel starting types. These may be series-connected for voltages greater than their 12-volt nominal rating. Cranking batteries should be located as close to the engine as possible—many times on a rack that is engine-mounted. Cranking currents up to 1000 A are not uncommon.

It is recommended that the batteries be specified as part of the engine-generator package, with the specifications written to give the manufacturer some freedom in sizing, while making certain that cranking capacity is adequate for ten attempts at starting and that the batteries are of the highest quality. Failure of an engine to start is usually a simple problem (e.g., fuel not reaching the cylinders), but one that cannot be resolved by remote control. Three failures to start should shut down the system and generate an alarm. After the operator has fixed the problem, enough cranking power must remain for a start with certainty.

Battery Chargers

Battery chargers vary widely, from simple half-wave rectifiers with a transformer and circuit overcurrent protection to precisely controlled units with voltage and current regulated and with timed alternate high-charging rate followed by a trickle-charging rate after initial recovery from a cranking cycle. Batteries are well charged by a few hours of generator operation.

A battery charger may be specified with the engine-generator equipment, but if the engineer wishes to include this item as part of the electrical contract, the specifications for the battery charger must be carefully coordinated with those for the engine-generator.

8-5. Grounding

Grounding is a conducting connection, whether intentional or accidental, made between an electrical system and earth or to some conducting body that serves in place of earth.

Two types of grounding related to pumping station design are (1) system neutral grounding and (2) equipment grounding. The system neutral ground is a connection to ground from the neutral point of a circuit, transformer, rotating machine, service, or system.

System Grounding

System grounding offers improved service reliability, improved fault protection, greater safety, and a reduction of transient overvoltages. The present trend is to use system neutral grounding at all voltage levels. Solid grounding of the system is used for systems operating at 600 V or lower, while at higher voltage levels a neutral resistance grounding unit is used between ground and the neutral of a power transformer or generator to limit the ground fault current to a low value. Grounding principles for ac systems are shown in Figure 8-20. The following methods can be used for system grounding:

- Solidly ground the system by connecting the neutral conductor to the grounding electrode conductor at the service.
- Solidly ground a separately derived system, such as a transformer, by connecting the neutral connector to the grounding electrode conductor. When the source is in the same building as the service, the grounding electrode conductor may be run back to the service.
- When the service conductors originate outside of the building and the neutral conductor is solidly grounded at the service, ground the neutral conductor at the transformer as well.

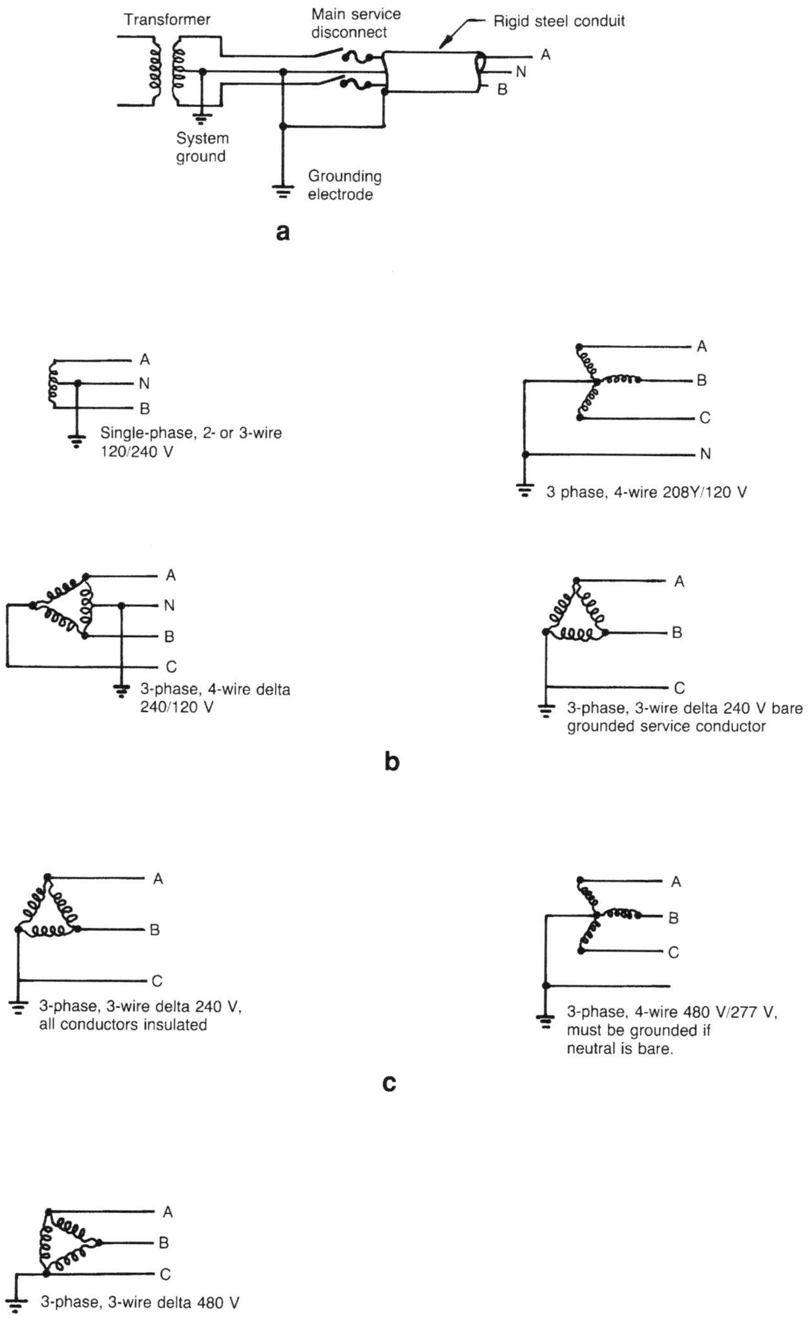

Figure 8-20. Grounding of low-voltage ac systems. (a) Grounding principles; (b) systems that *must* be grounded; 150 V or less to ground or bare service conductor used; (c) systems recommended to be grounded: not over 300 V to ground: (d) a system that may be grounded: over 300 V to ground.

- When resistance grounding a medium-voltage (600 to 15,000-V) wye system, ground the system at the service by connecting the neutral conductor to the grounding electrode conductor through the resistor. For low- (below 600-V) voltage systems, solidly ground to a grounding electrode.

Equipment Grounding

Equipment grounding is vital to protect people as well as the equipment by: (1) limiting the voltage between equipment and earth to a safe value and (2) providing a low-impedance return path for fault current to operate the protective devices quickly. The methods of equipment grounding include the following:

- Run the grounding conductor with the circuit conductors.
- Use the metal conduit (if it is electrically continuous) as a ground conductor. If it is not continuous, bond each section to the grounding conductor.

Ground Fault Protection

System Protection

The NEC requires ground fault protection of 277/480 V electrical systems with services 1000 A and higher. It is, however, good engineering practice to examine the possible consequences of ground faults in any polyphase system where an arcing ground fault may burn long enough to create major fire damage.

Ground Fault Circuit Interrupter

A ground fault circuit interrupter (GFCI)—or simply a ground fault interrupter (GFI)—is used on 120 V outlet circuits to interrupt the circuits when a fault current to ground exceeds a predetermined value but is less than that needed to operate the overcurrent protective device.

Class A GFCI units, designed to trip when line-to-ground current exceeds approximately 5 milliamperes (mA), protect personnel from electrical shocks. The average person can, without harm, withstand currents of 5 mA or more for the time needed to trip the unit. Hand-held power tools and their power cords can be hazardous if their insulation is defective, particularly if the worker is in contact with a conducting surface such as the ground, a concrete floor, or a metal pipe. In pumping stations, GFI units should always be required in outdoor and below-grade receptacles, in hazardous (per NEC 500) locations, in all areas with bare concrete floors and/or metal work benches, and where specifically called for in the NEC.

The ground fault interrupter device may be a part of the circuit breaker, or it may be inherent in a receptacle unit. Receptacles are cheaper and must be used if the circuit run is longer than about 30 m (100 ft). If circuit breakers are used, each circuit must have its own neutral because multiwire branch circuits do not work with a GFCI unit in the circuit breaker. The receptacle unit may be of the type that provides protection only for equipment plugged into it, or it may be a feed-through unit that protects all ordinary receptacles beyond it in the circuit. Two methods of ground fault protection are illustrated in Figure 8-21.

The NEC may require ground fault protection on heat trace circuits for water pipe freeze protection or for de-icing applications. Use Class B GFCI circuit breakers (which trip at 30 mA) for these applications, or ensure that the heat trace control package has this protection.

Surge and Lightning Protection

Lightning arresters and surge capacitors are used to limit the peak voltage of an impulse and to reshape the impulse to absorb the energy. They are a necessity in lightning-prone areas or if the pumping station is equipped with any electronic gear such as AFDs, and they are highly recommended in all pumping stations

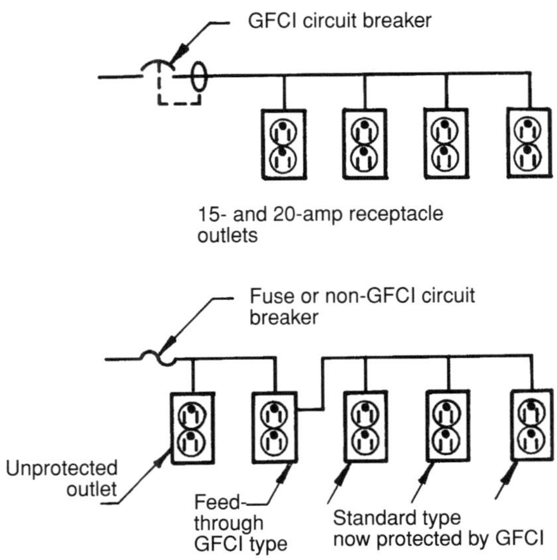

Figure 8-21. Ground fault circuit interrupter.

because of their very low cost (relatively speaking) and the excellent protection they afford motors. They are located at the service disconnect.

The following sources produce sustained overvoltages (surges):

- *Intermittent ground faults:* Intermittent arcing ground faults on 480-V ungrounded systems can alternately strike and clear, leaving dc charges on the line-to-ground capacitance. Surges or overvoltages of five to six times normal may be produced, but a properly sized grounding resistor can eliminate the overvoltages.
- *Series inductive capacitance faults:* The inadvertent connection of an inductance from line to ground on an ungrounded system can cause surges or voltage increases by resonance effect.
- *Circuit switching:* Circuit switching by an air contactor or circuit breaker does not cause overvoltages, because there is no trapped charge on the circuit when the arc is extinguished at a current zero. Vacuum contactors in circuit breakers or starters and current-limiting fuses cause overvoltages. Vacuum contactors force the current to zero before a natural zero occurs. Current-limiting fuses have built-in overvoltage surge suppressors.
- *Lightning:* Lightning is the discharge of a highly charged condenser. The clouds are one plate, the earth is the other.

Lightning Protection

The following discussion is related to basic design considerations for pumping station buildings of ordinary construction. Protection of high-voltage switchyards is usually the responsibility of the power company. The details of lightning protection are quite complex. An art rather than a science, they are best left to a specialist. The role of the pumping station designer is normally limited to specifying the quality of both the system and the materials.

Codes. In NFPA 78 (Lightning Protection Code) a distinction is made between the following building classes:

- *Class I:* An ordinary nonsteel building less than 23 m (75 ft) high.
- *Class II:* An ordinary building more than 23 m (75 ft) high or a building of any height with structural steel framing where the steel framing can be used for the down conductors. All steel is insulated from the ground by concrete.

Design Objectives. The design objectives are: (1) to provide a continuous low-impedance path for the discharge current to follow in preference to higher impedance paths such as wood, brick, or concrete and (2) to provide air terminals for building parts most likely to be struck, especially ventilators, gables, and roof edges.

Electrical Materials and Components. Electrical materials must be resistant to corrosion under the anticipated environmental conditions or must be suitably protected. The use of aluminum is restricted because of the problems it creates in all electrical work, and furthermore, it must not be used in contact with the earth or embedded in concrete or masonry. Always use copper or bronze. All materials used must be protected from mechanical injury. Typical components of a lightning protection system are air terminals, down conductors, secondary conductors, and ground electrodes (see Figure 8-22).

8-6. Lighting and Power Outlets

Interior Lighting Equipment Selection

Many factors (including room size, ceiling height, tasks performed, frequency and length of room occupancy, and special considerations such as damp, corrosive, or hazardous areas) determine the type of lighting fixtures and lamps to be used. Fixtures are varied and numerous, but the industrial type (ceiling or wall mounted) is the most suitable for pumping stations. Lamps are described by watts used and type (e.g., 100 W incandescent), but they are rated by lumen output. The number of lumens per watt is called the "efficiency of the light source."

There have been many changes in available lamps since 1985. High-efficiency lamps with novel shapes and bases and unusual wattages have been produced. Recent federal legislation has outlawed the production of a number of low-priced incandescent and fluorescent lamps that once were almost engineering standards. State energy conservation laws have proliferated. The net result is a greater choice of more expensive lamps producing light of higher quality than ever before. It is still possible, however, to make recommendations regarding pumping station lighting.

For new installations, 32 W T-8 fluorescent lamps should be used in industrial fixtures in such non-hazardous areas as control rooms, pump rooms, and dry wells. Hazardous-location fixtures are available but are expensive. Use high-power factor ballasts. Consider electronic ballasts where fixtures have high usage. These ballasts are now reliable, are cost effective, but still have a high first cost. Low-temperature magnetic ballasts are available for areas in which air

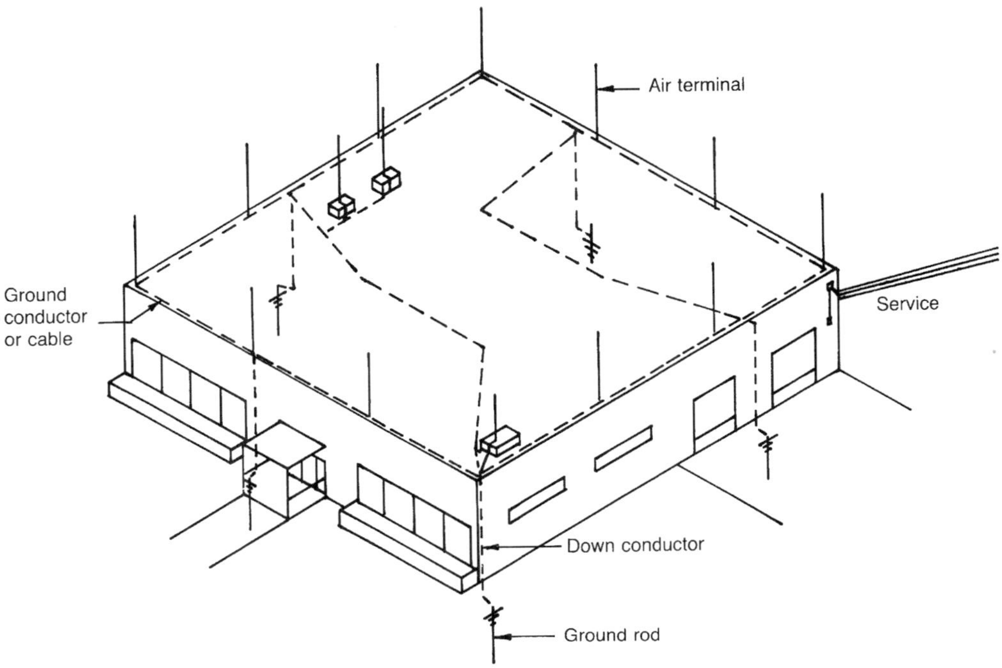

Figure 8-22. Lightning protection system for a building.

temperatures can fall below 10°C (50°F), but verify that lamps can be used with them. Fluorescent lamp life is 20,000 hours.

Use the new compact fluorescent lamps in sizes 7 to 26 W single or dual in small rooms with fairly high usage, corridors, and outside for security where the brighter, high-pressure sodium lamps are not wanted. Compact fluorescent lamps have excellent color and 10,000-h lives. Wattages higher than 26 are available.

Use incandescent or metal halide lamps in wastewater pumping station wet wells, in clear wells, and in small areas not often used. Incandescent lamps have low efficiency and short lamp life (750 to 2000 h), and several commonly used wattages have recently been outlawed. Use guards where needed. Metal halide lamps should have back-up quartz lamps where needed, and emergency lights require an off-delay to permit metal halide lamps to restrike. Use fixtures suitable for damp or wet locations where needed. In wastewater pumping station wet wells, use fixtures approved and labeled for use in Class 1, Division 1, Group D hazardous locations per NEC 500.

As alternatives to fluorescent lamps, use high-pressure sodium lamps outside as security lighting, in larger rooms with ceiling heights of 3 m (10 ft) or more where the glare and color is not objectionable, and in hazardous locations such as wastewater wet wells. Never use them where machinery is in motion, because there is a dangerous stroboscopic effect. These lamps are now available as small as 35 W. In larger sizes they are the most efficient commonly used lamps. Lamp life is 24,000 h, and color is constantly being improved. Do not use mercury vapor lamps in new construction. They are relatively inefficient and are now considered obsolete. Metal halide lamps have good color, are available in 32 W, 100 W, and in larger sizes, and have 10,000 hours of life, but they are less efficient than high-pressure sodium. Color is the only reason to choose them over high-pressure sodium. Low-pressure sodium lamps emit objectionable color.

Fixtures suitable for use in damp and/or corrosive areas are available with all types of lamps, and the selection depends on location and preference. The fixtures in damp areas must be vapor tight and those in corrosive areas should be made corrosion resistant with a coating of PVC (or the equivalent) on all metal parts.

Energy Conservation

Lighting energy conservation is mandatory in some states and may be imposed throughout the country by the federal government. Obtain requirements and forms from the appropriate regulatory bodies, and consult the local building inspector. Some states (California, for example) require two-level switching in heated or air-conditioned rooms larger than 9 m^2 (100 ft^2), containing two or more fixtures, and drawing 13 W/m^2 (1.2 W/ft^2). Such requirements might be waived for unattended pumping stations.

Fixture Location

Space fixtures relatively evenly with due regard for motor control centers and pipe risers. Pumps, motors, and auxiliaries such as chemical feeders and motor control centers need more light than piping areas. Do not obstruct pump and motor lift space, and do allow for bridge cranes and the like. In high-bay rooms such as motor and engine rooms, make certain that lamps in the fixtures can be replaced from the crane working platform.

Switching

Generally, locate a switch at each entry on the door latch side. Locate wet well or clear well switches inside the pump building (but not in the wet well or clear well) and include a labeled pilot light. Always provide unswitched lights at stairways and along corridors, especially in machine rooms and below-grade rooms where hazards may be present. Sketch a wiring diagram and carefully determine number of wires from the diagram. If the fixture is identified by the switch that controls it, these diagrams need not be included on plans.

Exterior Lighting

Entry or Security Lights

The following suggestions pertain to entry or security lights:

- Locate the lights at each principal entrance.
- Lights should be able to be switched or put on photocell control.
- Use vandal-resistant lenses for lights.
- Use 52- or 90-W incandescent lamps in smaller stations where fixtures are manually switched. Use high-pressure sodium fixtures in stations with photocell control.

Roadway and Parking Lots

There are many types of roadway and parking lot light fixtures utilizing both utilitarian and architectural designs. Use only high-pressure sodium lamps. Photocell or time-clock control is usual. For design information and selections, refer to manufacturers' catalogs and the IES handbook.

Emergency Lighting

Codes and Legal Requirements

Emergency lights are sometimes required by state and/or local law in structures similar to pumping stations where failure of power to the main lighting system would leave building exits in total darkness. Where required, they are subject to NEC Article 700 and NFPA Life Safety Code No. 101, Section 5-10.

Self-Contained Lighting Units

Approved lighting units are available complete with battery, charger, controls, and attached or separate lamp heads. Either unit-mounted or separate lamp heads are available, and the types of lamp heads include: (1) 8- to 25-W sealed beam models, (2) non-sealed beam models with automotive-type bulbs, (3) very high efficiency halogen models, (4) decorator models, and (5) explosion-proof (Class 1, Division 1, Group D from NEC Section 50) models.

Exit Signs

Incandescent and fluorescent exit signs are available with four lamps—two for line voltage and two for battery voltage. They are also available with an integral battery, charger, transfer controls, and down light. The atomic type of exit lights should be considered if the owners are made aware that there is a disposal problem for the luminous tubes at the end of their lives. They require no connections to a power source. LED (light emitting diode) exit signs are more expensive, and they have no down light, but with lamp life exceeding 20 years, and as they draw 2 W per face, they are very cost effective over their lifetimes.

Transfer Controls

Transfer controls consist of a relay with a voltage sensor that transfers to battery power and energizes lamp heads when line power fails. They are available with battery low-voltage protection that shuts off the lights to prevent deep battery drawdown.

The following are suggestions and recommendations for emergency lighting:

- If the pumping station is underground, use battery packs wired to the normal lighting circuit (but ahead of the light switch) to illuminate the stairs or ladder.

- Use at least two lamps at each lamp location so that no area is left in darkness if a lamp fails.
- Provide at least 10 lux (1 foot-candle) at floor level for each exit.
- Exit signs are not usually provided in pumping stations. If required, they may not have to be connected to the emergency lighting system.
- Use a sealed, maintenance-free battery rated for 10-yr life.
- Use a solid-state, automatic, two-rate trickle charger with trickle and high-charge indicators and a test switch.
- Use an automatic transfer relay with low-battery protection. Use delayed retransfer when the normal light source is metal halide or high-pressure sodium to allow for restrike time (i.e., the time required for lamp to relight).
- Use NEMA 4x enclosures in cold or damp areas.
- Except in hazardous areas, use sealed beam or sealed halogen lamp heads.
- Do not locate battery units in wet, corrosive, or hazardous areas. In these hazardous areas install only lamp heads connected to a battery that is located in dry, indoor, nonhazardous areas.

Convenience Outlets

Low Voltage

For 120 V outlets, use NEMA 5-20 receptacles, specification grade, on 20 A, single-pole circuits even though the NEC allows NEMA 5-15 if two or more are on a circuit. No more than five should be on a circuit. Locate them near pumps, electrical gear, and outside near pipe or wet well or clear well access. Generally place them so that any location that might require power can be reached with a 15-m (50-ft) extension cord. Do not locate them in a wet well or clear well. Install weatherproof covers if they are subject to weather or splashing, such as where equipment is hosed down.

Special Outlets

The following apply to special outlets:

- *Welding:* The voltage and ampere rating must be compatible with welding machine requirements.
- *Telephone:* Use NEMA 5-15 outlet on 15-A circuits where more than two telephone lines will be brought in.
- *Emergency lighting:* Where permitted by codes, use single NEMA 5-15 outlets on the same circuit as the room lighting. Permanent connection to the supply circuit is required for legally required emergency lighting.
- *Battery charger:* The charger should be sized (and specified) according to battery requirements.
- *Light fixtures:* Use twist-lock outlets near chain-suspended fixtures, and consider hard-wiring two fixtures together.

8-7. Electrical Circuit Diagrams

Single-Line Diagrams

A single-line diagram is used to show the electric power system from source to load in symbolic form. Conductors and their connections are omitted, and only circuit elements, their relationships, and the flow of energy through the system are shown. The diagram should be simple, utilize standard symbols, and present a complete picture that enables the observer to: (1) assess the blocks of power required, (2) identify the major feeders, and (3) ascertain the main protective devices responsible for plant outages.

An example of a single-line diagram of a radial circuit arrangement of a small pumping station is shown in Figure 8-23. The radial system is the simplest and most economical type of circuit arrangement. It offers good reliability and probably is the type most often used in small pumping stations. The drawback of the radial system is that a failure in the utility feeder, main breaker, or motor control center bus shuts down the whole station. The flexibility of being able to transfer manually or automatically to a standby power source is illustrated in Figure 8-24.

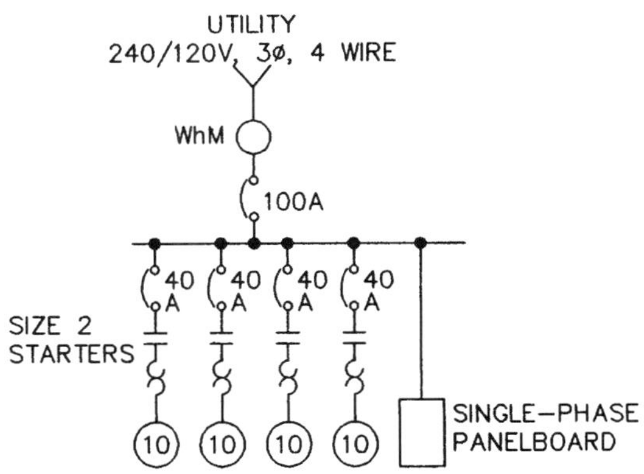

Figure 8-23. A typical single-line diagram for a small pumping station. Courtesy of Brown and Caldwell Consultants

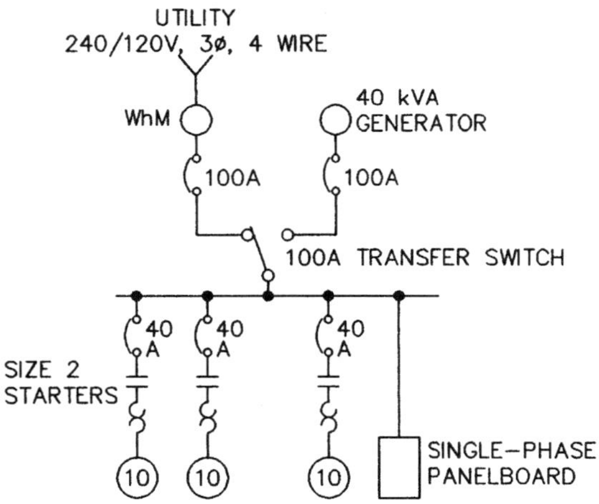

Figure 8-24. A typical single-line diagram for a small pumping station with standby power. Courtesy of Brown and Caldwell Consultants.

Similar flexibility is offered in Figure 8-25 with added reliability from using two utility power services. Each feeder is provided with a full-sized transformer and main incoming circuit breakers. The motor control center includes the two normally closed main breakers and a normally open tie breaker interlocked to prevent parallel operation of the two incoming lines. In the event of a power outage on a utility feeder, its associated main breaker can be opened (usually automatically) and the tie breaker closed, which allows the entire pumping station to be fed from the second source feeder. This pumping station includes two variable-speed pumps (one dc drive and one wound-rotor, slip-power recovery) and two constant-speed pumps. As all of the motors are of equal power (a blunder as explained in Chapter 15), the standby pump must be one of the variable-speed units, and it must be assumed that the other variable-speed pump is out of service. Hence, under a critical condition, there are one variable- and two constant-speed pumps opera-

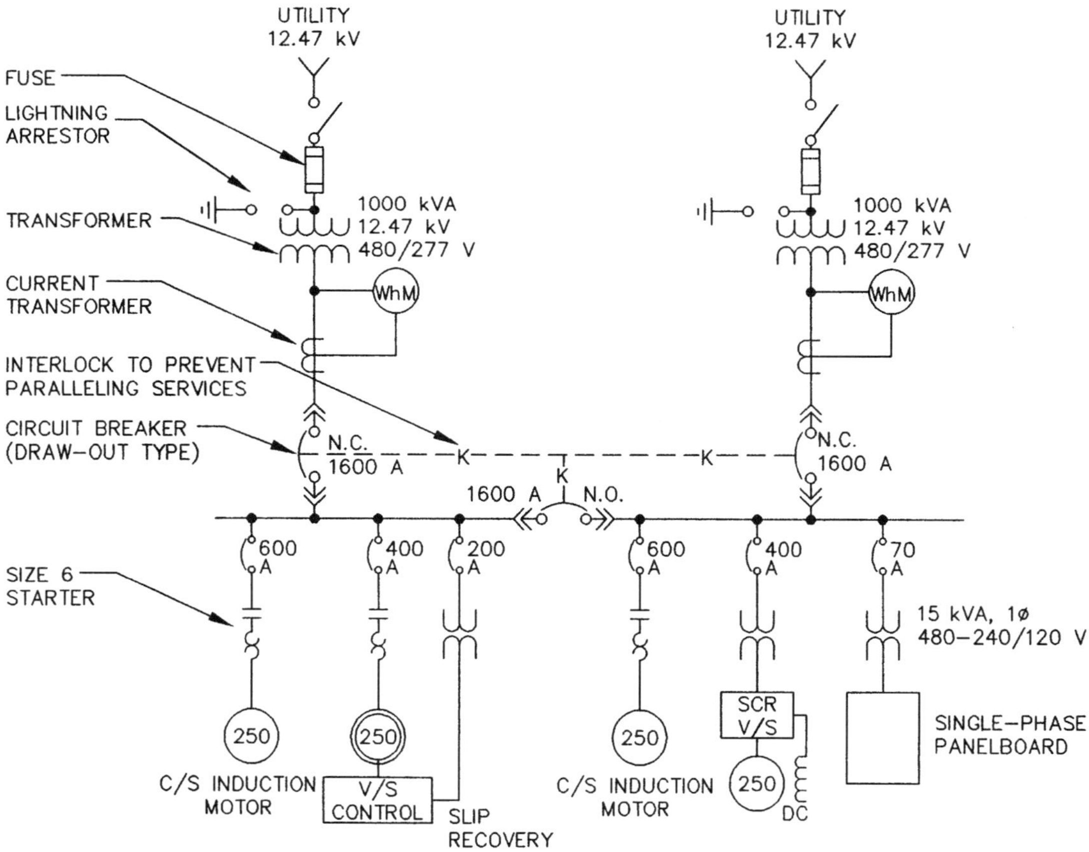

Figure 8-25. A typical single-line diagram for a large, low-voltage pumping station with dual incoming service. N.C., normally closed; N.O., normally open; SCR, silicon-controlled rectifier. Courtesy of Brown and Caldwell Consultants.

tional. A typical sequence of starting pumps as flow increases might be as follows:

- One operational variable-speed pump accommodates low flow.
- A constant-speed pump is added and the variable-speed pump operates at its lowest permissible speed (usually to discharge about 35% of its peak capacity) in an on–off mode until the flow increases to about 135% of a single pump's capacity. Now both pumps can operate continuously.
- The second constant-speed pump is added, and the variable-speed pump again operates at minimum capacity in an on–off mode until the flow requires it to operate continuously.
- When both variable-speed pumps are available, operational flexibility is improved so that on–off operation can be avoided.
- If the capacity of the variable-speed pumps were about 50% greater than that of the constant speed pumps, there would be no need for on–off operation (see Section 15-5).

The pumping station in Figure 8-25 is, however, an old-fashioned design. Wound-rotor and dc motors would nowadays be avoided in favor of induction motors driven by AF converters. Note, too, that the hydraulic efficiency of the system is less than it would be if all four drives were variable speed (see Figure 15-8 for proof). Furthermore, three variable-speed pumping units is a viable alternative with savings in: (1) size of wet well, (2) size of dry well, (3) number of pumps and amount of piping, (4) complexity, (5) upsets at the treatment plant, and, of course, and (6) cost of power.

Single-line diagrams are a very necessary and useful tool in short-circuit calculations and protective device coordination studies, because they are a skeleton representation of the system, and thus it is simple to follow through a number of levels of protection. Protective relay settings are derived from the diagram and the analysis of the time-current graphs of the various fuses and circuit breaker trip elements or relays.

Control Diagrams

Control diagrams represent the exact connection of all the elements of a control system. Interconnections of the elements with other systems that are diagrammed elsewhere are also shown. Simple motor control diagrams are shown in Figures 8-26 and 8-27. All the elements of the control system are usually shown horizontally on the diagram and the interconnections are drawn horizontally and vertically. The resulting diagram resembles a ladder and is thus termed a "ladder diagram" by many users.

All elements are identified and the function of each is either noted on the diagram or included in a supplementary explanation of the control element and its function. All lines of a ladder diagram should be numbered consecutively. Auxiliary relay contacts are usually shown to the right of the main diagram on or near the same line as the circuit element. The contacts are then individually identified with interconnection information. Relays are identified as CR1, CR2, and so forth, for control type units, and TR3, TR4, and so forth, for timing units. Timing relays are further identified at the coil symbol with the time setting and at the contact with information, such as TCE (time closing on energization), TOE (time opening on energization), and TCD and TOD for the respective operations on de-energization of a relay. The ladder diagram allows the engineer to follow the sequence of operation of a complex control system logically. Ladder diagrams are often used in training maintenance personnel, but their most important use is for analyzing systems that have failed to operate.

Many PLCs are programmed on a ladder diagram basis for the convenience of technicians who have no specialized training in computer programming.

Coordination

Coordination of overcurrent protective devices and of protective relays greatly improves the reliability of a pumping station, and it should always be required by the project engineer. In a coordinated system, the branch circuit protective device (usually a circuit breaker) trips off a faulted circuit. Should it fail to trip in a reasonable time, the upstream overcurrent device will trip, but the outage area will then be larger. Thus, all pertinent overcurrent devices have proper time delays and are said to be coordinated on a time basis. Fuses easily coordinate with each other and with a single downstream circuit breaker. Non-adjustable circuit breakers do not coordinate with each other. Short circuit current to a faulted branch circuit flows through the branch circuit breaker, the feeder circuit breaker, and the service circuit breaker, and they may all trip instantaneously, thereby shutting down the entire station. Fortunately, most faults are ground faults, and the ground-loop impedance keeps the short-circuit current to a sufficiently low value to prevent that scenario. Many commercial power systems are designed with 10,000 A branch circuit breakers and fully rated main circuit breakers

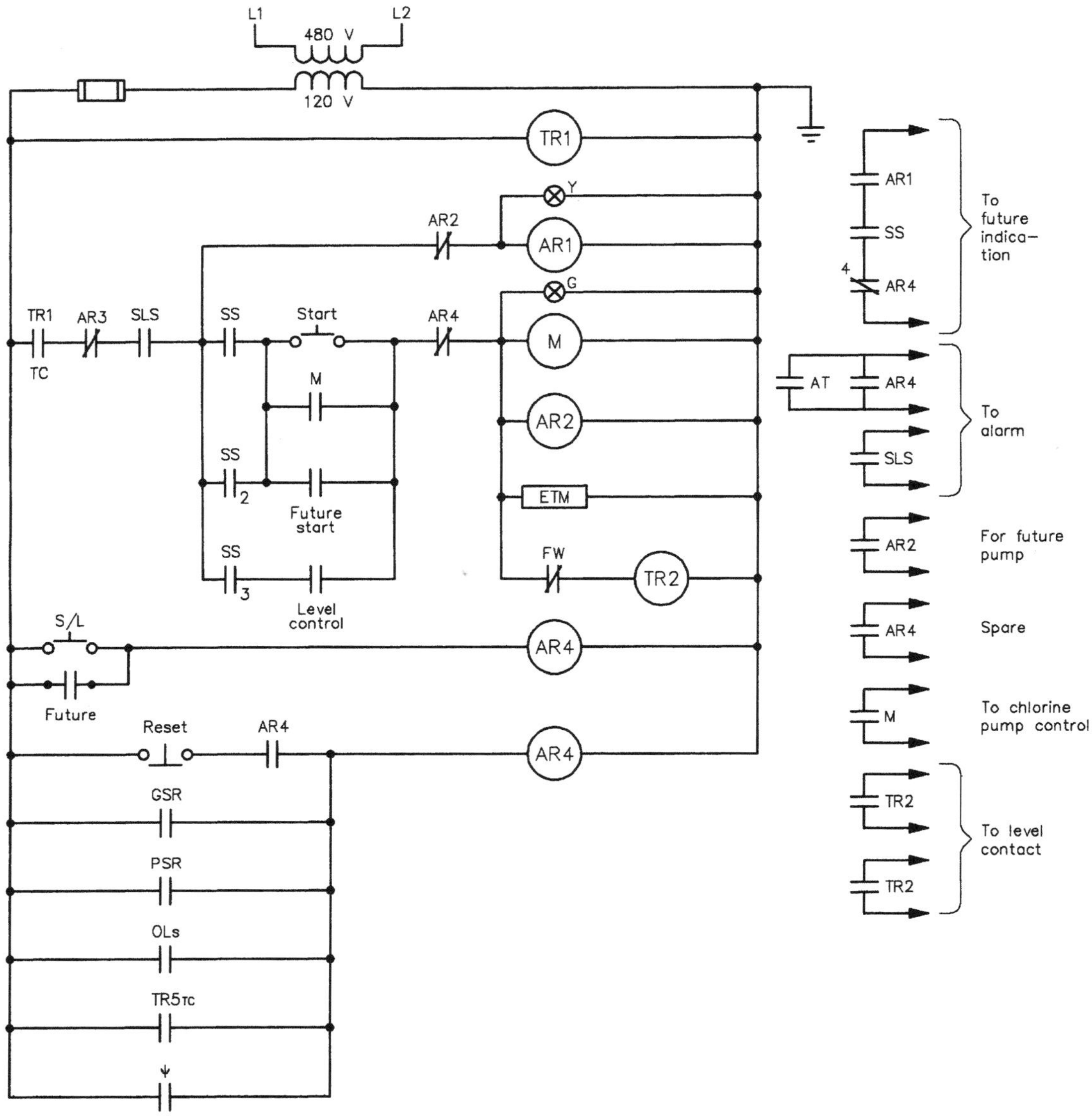

Figure 8-26. A typical schematic diagram for booster pump number 1. Booster pumps 2 and 3 are similar. L, line; OL, overload; PSR, power-sensing relay; SLS, suction valve limit switch; TC, time closing; (Ψ), motor temperature detector (see Figure 8-25 for other abbreviation). Courtesy of Brown and Caldwell Consultants.

for low first cost, and the main must trip to protect the branch circuit breaker for a branch circuit fault drawing more than 10,000 A. This cascade system is obviously unacceptable where a reliable power system is needed.

Coordination studies are performed to verify coordination using manufacturer-supplied time–current curves on log–log graph paper. The studies verify circuit trip and fuse ratings and predict settings for adjustable circuit breakers and protective relays.

8-8. Power and Control System Practices

Voltage Terminology

System voltage: The root-mean-square phase-to-phase voltage on an ac electric system. All subsequent defined voltages are in terms of root-mean-square phase-to-phase values unless noted otherwise.

Nominal system voltage: The voltage by which the system is designated and to which certain operating characteristics are related.

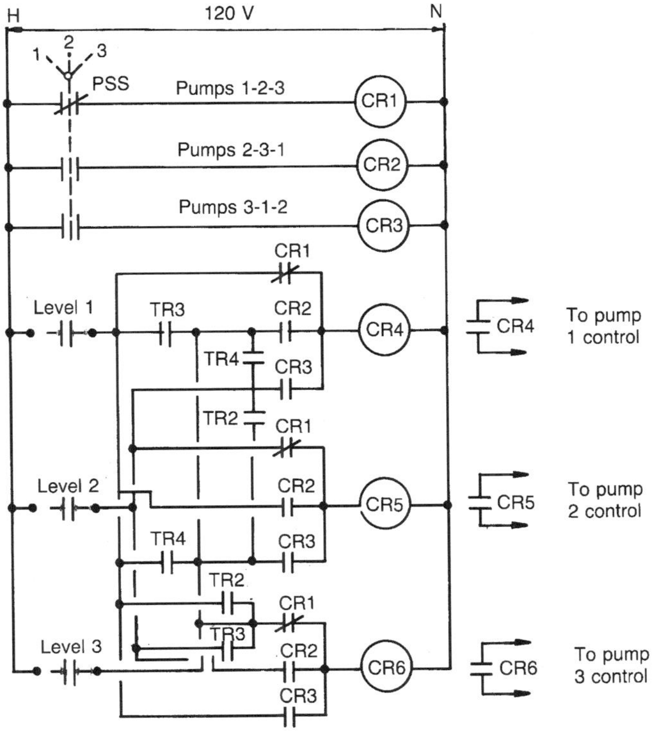

Figure 8-27. A typical schematic diagram for pump selection and level control. CR, control relay; H, hot; N, neutral; PSS, pump sequential selector switch (see Figures 8-25 and 8-26 for other abbreviations). Courtesy of Greeley and Hansen Engineers.

Maximum system voltage: The highest voltage that occurs on the system under normal operating conditions, and the highest voltage for which equipment and other system components are designed for satisfactory continuous operation without derating of any kind.

Service voltage: The voltage at the utility's point of service to the customer—ordinarily the metering point.

Utilization voltage: The voltage delivered to the line terminals of the utilization equipment.

Standard Nominal System Voltages

Standard nominal system voltages supplied by most electric utilities are given in Table 8-6. Utilities usually maintain them to within ±5%, but where energy conservation is of concern, the voltages are maintained to within +0, –5%.

For other system voltages and for exceptions to the values in Table 8-6, refer to ANSI/IEEE Std 141-1986. In some localities, the voltages offered may differ from the ANSI standard voltages.

Table 8-6. Standard Nominal System Voltages

Single-phase	Three-phase
120 V, 2-wire	208/120 V, 4-wire
240/120 V, 3-wire	240/120 V, 4-wire
208/120 V, 3-wire[a]	240 V, 3-wire delta[b]
	480/277 V, 4-wire
	480 V, 3-wire delta[c]
	2400 V
	4160 V
	12, 000 V

[a] Available in some inner-city areas with three-phase networks.

[b] Common in rural areas with pole-mounted transformers. Insist on three transformers to ensure balanced voltages.

[c] By special request. Be sure the utility does not ground the neutral at the transformer.

Station Voltage Selection

The preferred nominal system voltage for an electric service depends on the electrical load. For a pumping station where all electrical devices may operate simultaneously, the connected load is used. It is the sum of the ratings of all electrical equipment in the pumping station computed to the nearest hp, kW, or kVA. Motors are counted at their horsepower rating regardless of their actual loading, and all other devices are counted at their nameplate rating in kW or kVA. Conversions are usually made on a one-to-one basis (1 hp = 1 kVA). Welding machines are rated in hp at 1 kVA per hp.

Recommendations for pumping station service voltages are given in Table 8-7. Utility limitations may be different and must be checked.

Voltage Ratings for Utilization Equipment.

Nameplate ratings applicable to motors are listed in ANSI/IEEE Std 141-1986, Table 7. A few of these ratings are given in Table 8-8.

Motors used on 208-V systems should be rated 200 V. Motors rated 220 V or 230 V do not perform satisfactorily and should not be used.

Load Estimating

Load estimating evolves in several major stages. Preliminary load estimating is done during the pre-design stage of development of proposals for the pumping station design—usually with very preliminary estimates of the total plant pumping horsepower. It can be accomplished largely on the basis of discussions with the project managers and reference material from company design files on similar previous jobs.

The complexity of load estimating depends on the size of the station and the experience level of the project designers. For many years, successful professional design firms have developed estimating information from preliminary data sheets. These sheets allow for a variety of useful information relating to the various plant drives. The principal load is usually comprised of pumping units. Other important loads depend on whether the station is in an area of extreme summer heat or extreme winter cold, where extensive cooling or heating equipment may be required. The number of individual pumping units affects the size of the standby generator. The number of pumps also affects the complexity of the control and monitoring system of the station. All of these factors affect the electrical load estimate as well as the electrical system construction cost.

Data sheets should be generated in a computer system so that the data are accessible to all employees who are responsible for a portion of the total estimate.

Load estimation for small or "package" pumping stations is very simple, as there is frequently no special auxiliary equipment to complicate the station design.

Table 8-7. Pumping Station Service Voltage Recommendations

Load	Voltage
5 kVA (7 hp)	240/120 V, single-phase, 3-wire or 208/120 V, single-phase, 3-wire
40 kVA (50 hp)	240/120 V, three-phase, 4-wire or 208/120 V, three-phase, 4-wire or 240 V, three-phase, 4-wire
400 kVA (500 hp)	480/277 V, three-phase, 4-wire[a]
Above 400 kVA (500 hp)	4160 V, three-phase[b] or 2400 V, three-phase[b]

[a] Should be used for motors as small as 1/2 hp. Provide 240/120 V, single-phase, 3-wire or 208/120 V, three-phase, 4-wire for pumping station auxiliaries, derived from the 480/227 V system.

[b] Use for large motors. Provide 480/277 V three-phase, 4-wire for larger pumping station auxiliaries, derived from the high-voltage system. Provide also 240/120 V single-phase, 3-wire or 208/120 V, three-phase, 4-wire for smaller pumping station auxiliaries derived from the 480/277 V system.

Table 8-8. Nameplate Ratings for Motors

Nominal system voltage	Nameplate voltage	Power range kW	Power range hp
Single-phase motors:			
120	115	≤0.37	≤$^1/_2$
240	230	≤1.1	≤1.5
Three-phase motors:			
208	200	110	150
240	230	150	250
480	460	300	400
2400	2300	1500	2000
4160	4000	3000	4000

A few checklist items that apply to load estimation for many pumping stations follow:

- Number of pumps and required horsepower of each
- Number of pumps operating at any time and sequence of starting
- Special requirements, such as adjustable speed drives
- Auxiliary drives required
- Preliminary layout of structure and number of operating levels
- Special use areas (such as shops) and their electrical loads

- Heating, cooling, and ventilating requirements
- Standby power requirements (if any)
- Applicable electric utility rate schedules
- Available electric utility voltage levels and distance to their point of service
- Inside lighting requirements/outdoor lighting requirements
- Review local electrical codes and try to get information from the inspection authority on special interpretations that may affect the electrical construction costs. An example is the treatment of possible hazardous areas such as Class I, Division 1 versus Class I, Division 2.

8-9. Reference

1. O-Z/Gedney, Main Street, Terryville, CT 06786

8-10. Supplementary Reading

1. Beeman, D., *Industrial Power Systems Handbook,* McGraw-Hill, New York (1955).
2. *IES Lighting Handbook,* Illuminating Engineering Society, New York (latest edition).
3. *Lightning Protection, Institute Standard of Practice,* 3rd ed., Lightning Protection Institute, P.O. Box 458, Harvard, IL 60033-0458 (1987).
4. McPartland, J. F., *National Electric Code Handbook,* 22nd ed., McGraw-Hill, New York (1996).
5. Smeaton, R. W., *Switchgear and Control Handbook,* 3rd Ed., McGraw-Hill, New York (1996).
6. Catalogs of various manufacturers, such as General Electric, Square D, Westinghouse.

Chapter 9

Electrical Design

STANLEY S. HONG
PHILIP A. HUFF
PAUL C. LEACH

CONTRIBUTORS
Robert S. Benfell
Mayo Gottliebson
Stephen H. Palac
Patricia A. Trager

The primary purpose of this chapter, as with Chapter 8, is to provide project leaders with some understanding of electrical design, to aid communication with electrical specialists, and to help in coordinating electrical design with other disciplines. The chapter is appropriate for civil engineers, and it may also be appropriate for electrical engineers with limited experience in power engineering. A working knowledge of the principles of electricity, discussed in Chapter 8, is a prerequisite.

The topics discussed here include the coordination of electrical design both with other disciplines and with utility companies (Sections 9-1 to 9-4), the suitability of materials (Section 9-1), harmonics (Section 9-11), and construction services such as field testing (Section 9-12). Most of the essential electrical calculations for a pumping station of moderate size are given as worked examples. The examples include electrical load estimation, overcurrent protection, lighting, engine-generator sizing, and conductor, fuse, and breaker sizing for main and branch circuits.

Terminology is given in Sections 2-2 and 8-1, and abbreviations are defined in Section 2-1. The text and computations in the examples are in U.S. customary units for correlation with present codes and standards in the United States Conversions to SI units are given in Appendix A, and some are also given in this chapter.

References to a code, a standard, or a specification are given here in abbreviated form, such as NEC (for National Electrical Code) or NFPA Standard 493-1978 (for a National Fire Protection Association standard). The coded numbers are entirely sufficient for identifying the reference. Titles of the abbreviations are given in Appendix E, and addresses for publishers are given in Appendix F.

Much of the material in this chapter is referenced to the NEC, so read the referenced passages in sequence with this text. Keep in mind that the NEC is subject to local interpretation. Check with local authorities for applicable codes and regulations. The design must follow safety and protection standards established by OSHA and UL. Standards and recommendations published by organizations such as IES, IEEE, NEMA, ANSI, and NFPA should also be followed.

9-1. Final Construction Drawings

Long before final construction drawings are prepared—in fact, at the beginning of the project—the project leader should consult the electrical engineer to ensure that an adequate space is reserved for electrical equipment and to forestall interference with mechanical equipment and the building envelope. During the appropriate design stages, the project leader should also cooperate with the electrical engineers to ensure that the electrical drawings and specifications are complete and compatible with all of the other drawings and specifications.

List of Electrical Drawings

The following is a suggested list of electrical drawings required for a typical pumping station. The list would be modified according to the size and complexity of the job.

- Electrical legend
- Single-line diagram
- Control diagrams (schematics)
- Site plan (power and lighting)
- Building power plan (including control wiring)
- Building lighting plan
- Fire alarm, communications, and security plans as required
- Equipment elevations and schedules
- Conduit and wire schedules
- Details.

Symbols

In the electrical legend, always include a list of symbols used. See Chapter 2 for common electrical symbols.

Coordination with Other Pumping Station Drawings

The project engineer is responsible for coordinating the electrical design with the other disciplines (see Table 1-1)—a task vital for the continuity and completeness of the drawings. The project engineer should:

- Verify that electrical equipment, including light fixtures, does not conflict with structural elements or mechanical equipment, and that necessary space and mounting provisions exist.
- Check the plans and specifications of the other disciplines (particularly process and mechanical disciplines) for compatibility with electrical systems such as voltage, phase, alarms, and controls.
- Be sure that the electrical engineer checks every piece of equipment that requires electrical energy or control. Commonly missed items include instruments, limit switches, solenoid valves, and alarms.
- Check to see that conductor sizes, dimensions, etc. are shown only once. On other drawings, these items should be shown (if at all) as "reference-only" data to avoid conflicts.
- Perform a final complete check of electrical drawings against all other pumping station drawings and specifications just before the drawings are sent out.

Raceway System

For raceway systems,

- Show conduits only in schematic form except where noted specifically or detailed otherwise.
- Electricians often double-up home runs for lighting and receptacle circuits, control wire runs, and so on. If doubling-up is undesirable, so note or detail the home runs.
- A minimum conduit size of 19 mm ($^3/_4$ in.) is suggested.
- Detail and/or specify the mode of conduit entry into structures, including underground penetrations for the prevention of groundwater entry.
- Detail conduit duct banks carefully: include conduit layout, section, acceptable types of conduit, reinforcing bar sizes and extent, concrete encasement, and slope direction for drainage.
- Remember that PVC conduit deteriorates if exposed to sunlight or possible impact damage. Embedded PVC conduit may suffer severe damage during pouring of concrete in walls and particularly due to vibrators.
- The installation of spare (for example, not less than 10% of the total) conduits (capped at both ends) from panelboard and motor control center (MCC) or motor starter panel to exterior or to attic is suggested.
- In corrosive areas such as wet wells and chlorine storage rooms, use only PVC-coated, hot-dipped galvanized (inside and outside) rigid steel conduits for corrosion resistance. Require that all bare spots in the conduit system be recoated with PVC (or an equivalent).
- The trend is toward the use of aluminum conduit, but aluminum reacts quickly with the lime in damp concrete, so it must not be encased in concrete. For first-class construction, use only PVC-coated steel conduit in pumping stations.
- Allow contractors to size junction boxes and wireways in accordance with codes unless the designer requires oversize boxes for special use.
- Provide drainage for large handholes and manholes.
- Install pulling eyes on walls opposite every electrical conduit entry inside manholes and handholes to aid in pulling wires.

- Use cast, hot-dipped galvanized ferrous boxes for embedded systems—receptacle, switch, and junction boxes in all pumping station levels where embedment might be allowed or specified.
- NEC 370-3 restricts the use of nonmetallic boxes in metallic raceway systems. Such boxes cannot be used unless an internal ground jumper is provided between each metallic conduit. Ground jumpers are impractical with conduit bodies (condulets) and small boxes such as trade types FS and FD, so PVC-coated condulets and FS/FD boxes are recommended. For larger boxes, it is practical to field install an internal ground jumper, and non-metallic boxes can be used. Nonmetallic boxes should have nonmetallic or stainless-steel hinges, mountings, and clasps.

Wiring

For wiring,

- Show feeders, the pump motor branch circuit, and the principal ancillary equipment branch circuit conductor sizes either on plans, on the single-line diagrams, or on conductor schedules, but show them only once.
- Indicate number and minimum size of control, lighting, and outlet wiring on drawings or specifications. By note, direct installer to adjust wire size for the type of wire, conduit fill (defined in Section 2-2), voltage drop, and need for 60°C (140°F) ampacities at 15- to 100-A circuit breakers and fused switches. An installer can size conduit per the NEC, but it is better for engineers either to size conduit or to limit the percentage of fill if any conduit is to be sized by others.
- Coordinate process and instrumentation wiring carefully with respect to: (1) number of control wires serving each instrument and device, control requirements, and 120-V power; (2) shielded cable for 4- to 20-mA dc signals and other instrumentation signals (require the shield to be grounded at one end only) (if the cable is double shielded, follow manufacturer's directions); (3) for intrinsically safe wiring, refer to NFPA Standard 493-1978 for complete details; (4) combine circuits carefully to avoid exceeding the maximum voltage buildup; (5) use separate steel conduits for these circuits; and (6) in panels, separate wiring a minimum of 50 mm (2 in.) from all other panel wiring.
- Equipment ground-wire considerations: (1) A ground wire from the system ground to each 480-V motor and panel is recommended; (2) bond all ground wires to both ends of rigid metal conduit; (3) ground all metal enclosures, water pipes, and machines as required by NEC 250-42, 43, 44(d), and 80; and (4) provide a ground wire in every PVC conduit.
- Use duct seal in conduits where they emerge from ground and enter enclosures to prevent passage of water into the enclosures and building. Slope underground ducts to drain water to manholes.
- Number all equipment and devices. Require numbered wire tags at each end of all conductors.

Equipment Space Heaters

All space heaters should be powered from a 120-V panelboard so that equipment that is idle or down for repairs has condensation protection. At motors, control panels, and so on, provide labeled heater disconnects for safety during maintenance. Provide thermostats in equipment for operating space heaters and connect motor space heaters through a normally closed auxiliary motor starter contact so that the heater is on when the motor is off and vice versa. For large, important motors, consider a neon light across the contacts to indicate heater continuity when the heater is off. Consider low-voltage motor winding heating from a separate transformer in the motor starter cubicle (see Chapter 13), but note that there will be no heating when motor branch circuit is off.

Power and Telephone Utility Requirements

Instruct the contractor to contact the appropriate power and telephone companies for all service requirements and to conform to them. The engineer, however, must contact the utilities in the early stages of design to obtain their agreements to serve the facility adequately (see Section 9-3).

9-2. Specifications

Some specific hints for consideration in writing the electrical specifications are given in this section (see also "Specification Language" in Section 28-1).

In the "General Requirements" portion of the electrical specifications, provide an article titled "Electrical Devices Furnished with Mechanical Equipment" and have each specifier of equipment include a special callout of electrical contacts, alarm functions, and any auxiliary electrical power requirements. Reference these callouts to the "Electrical Devices" specification.

Information and sample specifications may be obtained from The Construction Specifications Institute [1]; their format is becoming popular and is increasingly being required by clients.

Coordination with Other Specification Sections

Coordination is very important to ensure the continuity and completeness of specifications. Wherever any equipment requires electric power, control, or instrumentation, check the following with the designer and be sure that this information is included in the equipment specifications:

- Voltage and phase
- Disconnects
- Enclosures
- Control philosophy, such as a local–remote selector switch with an On–Off pushbutton
- Alarm contacts
- Special features or circuits to ensure compatibility with other electrical equipment
- Any desired special component or wiring specifications.

In addition:

- Require that all conductors used for interwiring between panels and for input or output purposes be brought to numbered terminal strips and that all wiring (internal and external) be tagged at both ends with preprinted wire markers. Require that the wire marking code be submitted for acceptance prior to the manufacture of the equipment.
- Require that a copy of the final issue "As Delivered" control diagram be attached to the inside of the door of the equipment cabinet.
- Require the control diagrams to be drawn in the ladder-diagram format with all contacts and lines identified (numbered or coded) and all relays and control devices designated by appropriate names and numbers.
- If a PLC is to be provided, require that a ladder-diagram format control diagram be submitted for acceptance and require that a tape or disk copy of the final program accompany the equipment.

Seismic Requirements

All equipment must be attached or supported so that no hazard to personnel will result from any seismic force or motion. Contact the project structural engineer for help in fulfilling and enforcing this requirement. The seismic zone for every part of the country is designated in UBC Section 2332, Figure No. 1 (in 1996), and the contractor and suppliers must be instructed to follow all of the requirements. Zones vary from 1 through 4. The worst, Zone 4, corresponds to an earthquake with a magnitude of 8 on the Richter scale. Although equipment is usually not required to operate properly during a seismic event, specify that large equipment (such as a large circuit breaker) must not change position and that large engines and motors must not start or "bump" from momentary energization.

Equipment Labeling

Labeling, an indication of the testing and listing of the equipment by a recognized testing laboratory, is required by most local and state inspection authorities and must be required for all standard equipment. Custom equipment and control panels can be expensive if the authorities insist on a label. Whenever possible, design around equipment that is UL-listed as a complete unit (panel and all).

Shop Drawings

With the possible exception of standard, widely used materials, it is suggested that shop drawings be required for each item for which detailed specifications were written. Check the shop drawings thoroughly. Specifications should include (1) the equipment and in formation that must be submitted and (2) a submittal schedule.

9-3. Contacting Utilities

It is necessary to contact each utility early to obtain their requirements and the name of the person assigned as their representative. Back up all telephone conversations with confirmation letters, and visit the site with the utility representative.

Electric Power Utility Contact

An underground (instead of an overhead) service from the electric power utility is preferred because of reliability and appearance. Provide the service from the nearest pole or vault to the pumping station. In addition,

- Obtain all current rate schedules that might be applicable to the station. If an abbreviated rate schedule is given, ask for a complete rate book.
- Discuss the advantages and disadvantages of each schedule with the utility representative.
- Analyze demand charges, seasonal demand or energy charges, basic energy charges, power factor clauses, and the basis of the demand charge (15-min demand, 30-min demand, connected horsepower, etc.) and discuss them with the utility representative.
- Use an aerial photographic survey to mark the location of the pumping station and the preferred point of service. Review the proposed location in the field with the utility representative and send a marked copy to the utility.
- Obtain the utility's agreement on acceptable starting methods and the frequency of starts for the proposed pumping station drives. If they do not allow across-the-line starting for pump drives of the proposed size, discuss the alternatives, select the best one for reliability and economic factors, and inform the utility of the selection.
- Discuss transformer ownership and the utility's maintenance policy if the transformer is provided by the owner.
- Obtain the utility's transformer specifications and find the size they would likely provide, its impedance, and its primary fuse size and insulation (dry or oil).
- Check the utility's policy on overhead and underground service installation and determine whether they share costs or furnish the entire service. If they furnish it, determine whether underground services are encased in concrete. Cable for pumping stations should be in a concrete-encased conduit and not directly buried.

Telephone Company Contact

Obtain agreement with the telephone company on the following:

- Approximate required date of service
- Service point (e.g., pole)
- Routing of telephone company cables and the location of their equipment in the owner's facilities
- Clear definition of items (and their interface) furnished by telephone company and by owner (contractor)
- Telephone terminal facilities: type (plywood panel, enclosure, etc.) and size, 120-V power requirements, and grounding provisions
- Telephone extension conduit size and type
- Telephone extension outlet box size, type, location, and mounting height
- Pumping station equipment interface requirements, for example, for telemetry (space, proximity to 120-V sources, need for terminal block or surge suppressors).

9-4. Construction Information to Utilities

The following information should be sent to the electric power utility following the completion of the design.

- *Specifications:* special switchboard, substation, and special grounding specifications;
- *Drawings:* site plan, building power plan, single-line diagram, service gear elevation (switchboard, MCC, etc.), and substation plans and elevations; and
- *Other data:* the bid opening date and the approximate date service is needed.

The telephone company should be sent

- *Specifications:* terminal cabinet specification, including size;
- *Drawings:* site plan and building plan containing telephone provisions; and
- *Other data:* the bid opening date and the approximate date service is needed.

9-5. Load Estimation

To facilitate the design of the electrical power service entrance equipment, substation, motor control center, and so on, a load or motor list such as that shown in Example 9-1 should be utilized. The motor and equipment lists should be started at the first stage of electrical design. If no data sheets are available from the designers at the first meeting between the various design team members, a tentative list should be prepared by the electrical engineer and copies should be distributed for general discussion. The motor and equipment list must be kept up to date during the course of design as motors and loads change in rating or are added or deleted. One of the most subtly important points of the motor and equipment list is an early decision on a name for each equipment item exactly as the designer wants it inscribed on the nameplates. Even though the exact name of a particular item may change during the course of design, all parties involved will use the same name to describe a piece of equipment.

Following the motor and equipment list, the electrical calculations for the pumping station include the following:

- Sizing the branch circuit breaker or fused switch, starter, and conductors for each of the main pumps
- Sizing the circuit breaker, starter, and conductors for the sump pumps and other incidental motor loads
- Sizing the lighting and small power transformers, circuit breakers, and conductors
- Sizing the heating and ventilating equipment panel circuit breakers and conductors
- Sizing the main (service) overcurrent protection device, transformer, primary and secondary protection, and conductors
- Lighting calculations, layouts, and circuiting
- Sizing the power factor correction capacitor bank
- Sizing the standby engine-generator.

These calculations are illustrated in Examples 9-2 through 9-10.

Example 9-1
Electrical Load Estimation

Problem: Estimate the service electrical load for a pumping station with four constant-speed, 25 hp, 720 rev/min motors driving three duty pumps and a standby pump. The maximum planned capacity is 4500 gal/min at a maximum TDH of 40.7 ft for a Hazen–Williams C of 120 and 38.3 ft for a C of 145. The present required capacity of 3600 gal/min can be achieved with two pumps.

The floor elevations are: ground, 112.5 ft; basement, 102.5 ft; pump room, 90.0 ft; and the three floors are 21 by 42 ft in plan. The ground floor contains a trolley hoist, the motor control center, and an engine-generator. Air handling units are in the basement. The wet well floor is at elevation 89.25 ft, and the wet well plan size is 8 by 34 ft.

Solution:

Pump motors. According to NEC Table 430-150, the full-load current (I_{FL}) for a 25-hp, Code F. 720 rev/min, 460-V, three-phase, 60-Hz motor is 34 A. (Note: check the manufacturer's data for 720 rev/min motor to determine the actual full load current. Slow-speed motors [<1200 rev/min] typically have lower power factors and therefore draw more current than is shown in NEC tables, so compute with the manufacturer's data if they are available. If power factor correction at the motor terminals is included in the design, the full-load current may be reduced to the NEC value [or even less].)

Heating and ventilating. The power requirements for heating and ventilating (H&V) must come from the H&V engineer, who determines them on the basis of the size of rooms, insulation value of the building envelope, heat sources, and climate (as explained and calculated in Chapter 23). The H&V loads used in this example are higher than normal and would be appropriate for a pumping station in a severe climate. The H&V load is typicaly less than 10% of the total load.

Other electrical loads. Estimates for other loads can be made either by the project leader or by an experienced electrical engineer. As the design progresses and the loads become known more exactly, the estimates below may be modified.

Item		Power					Service			
				Full load						
Name	No.	hp	kW	A	V	Phase	Continuous	Intermittent	Design A	Essential load on generator, A
Pump	1	25		34	460	3	X		34	34
	2	25		34	460	3	X		34	34
	3	25		34	460	3	X	X	34	34
	4	25		34	460	3		X	Stdby	Stdby
Sump pump	5	0.5		1	460	3		X	1	1
	6	0.5		1	460	3		X	Stdby	
Instrument air compressor	7	0.5		1	460	3		X	1	1

Item		Power					Service			
				Full load						
Name	No.	hp	kW	A	V	Phase	Continuous	Intermittent	Design A	Essential load on generator, A
Pump room exhaust fan	8	0.5		1	460	3			1	1
Control room exhaust fan	9	0.5		1	460	3			1	1
Control room exhaust fan	10	0.5		1	460	3			1	1
Engine room exhaust fan	11	0.5		1	460	3			1	1
Air handling unit	12	0.5		1	460	3			1	1
Duct heater	13		16.5	21	460	3			21	
Pump room H&V control	14		0.5	1	460	1			1	1
Wet well										
intake fan	15	0.5		1	460	3			1	1
exhaust fan	16	0.75		1.5	460	3			1.5	1.5
duct heater	17		34.8	43.5	460	3			43.5	
Wet well H&V control	18		0.5	1	460	1			1	1
Lighting transformer	19	(9 kVA) 8.5		10.5	460	3			10.5	9
Totals		62.75		222.5					187.5	121.5

Maximum power required (for a corrected power factor of 95%) is

$$P_{3\phi} = \sqrt{3} \times 460 \times 187.5 \times 0.95/1000 = 14.9 \text{ kW}(149.4 \text{ kVA})$$

Demand factor: actual = 0.85. Demand factor used = 1.0. (The demand factor is an arbitrary percentage of connected, rated design loads that is used to determine actual operating loads. It allows for nonconcurrent loads and motors that are not running at full load.)

Minimum feeder to MCC = 187.5 A + 25% of largest motor = 187.5 + (0.25 × 34) = 196.0 A.

Minimum main circuit breaker = 187.5 A + 50% of largest motor = 187.5 + (0.5 × 34) = 204.5 A.

Minimum standard circuit breaker = 225 A.

Design load plus 20% "cushion" = 187.5 × 1.2 = 225 A.

Anticipated power needed (for a corrected power factor of 95%) is

$$P_{3\phi} = \sqrt{3} \times 460 \times 225 \times 0.95/1000 = 170.2 \text{ kW}(179 \text{ kVA})$$

Minimum feeder to MCC = 225 + (0.25 × 34) = 233.5 A. Minimum feeder conductor (NEC Table 310-16, 75°C column) = 250 kcmil (thousands of circular mils) rated for 255 A.

Minimum main circuit breaker = 225 + (0.5 × 34) = 242.0 A. Next smallest standard circuit breaker rated for 300 A. Minimum feeder conductor = 350 kcmil rated for 310 A.

Minimum standard transformer = 225 kVA.

Use: 225 kVA, 12.47 to 480/277-V transformer;
300-A three-phase main circuit breaker in MCC;
350-kcmil feeder conductors (rated for the main circuit breaker); and
#2 AWG (American Wire Gauge) ground conductor.

9-6. Overcurrent Protection and Conductor Sizing

The NEC has definite standards for selecting overcurrent protective device ratings and for sizing conductors. Conductor sizing is based on the load current of the equipment connected to the circuit and is influenced by circuit length (voltage drop) and ambient temperatures. It is usually wise to allow spare capacity. The NEC and local and state codes and rules must be followed as a *minimum* safety standard for design. Furthermore, consider designing for long-time service and load growth. Overcurrent protection is accomplished by applying current-sensing techniques at many points in the utility's distribution system as well as within the user's system. Sensing equipment may be self-destructive (e.g., fuses), resettable (e.g., circuit breakers), or remotely located (e.g., protective relay applications).

The electrical design engineer's role is to develop a system so coordinated that an overloaded motor will be removed from service before the motor insulation or bearings are damaged, and only the affected equipment (in this example, the overloaded motor) will be removed from the service. A properly coordinated overcurrent system will enable fault conditions to be isolated from the system without affecting the operation of other equipment. Coordination studies range from simple to very complex. A thorough discussion of the method applicable to system studies and coordination is contained in IEEE Standard 242.

A single-line diagram of the pumping station of Example 9-1 is shown in Figure 9-1. The utility fuses are pole mounted, and they protect (1) the utility from overloads (overcurrent, faults) within the pumping station electrical systems, (2) the transformer, and (3) the service cable connecting the transformer to the utility.

Fuses

Fuses are commonly used by the utility to protect their system from faults in the customer's electrical system.

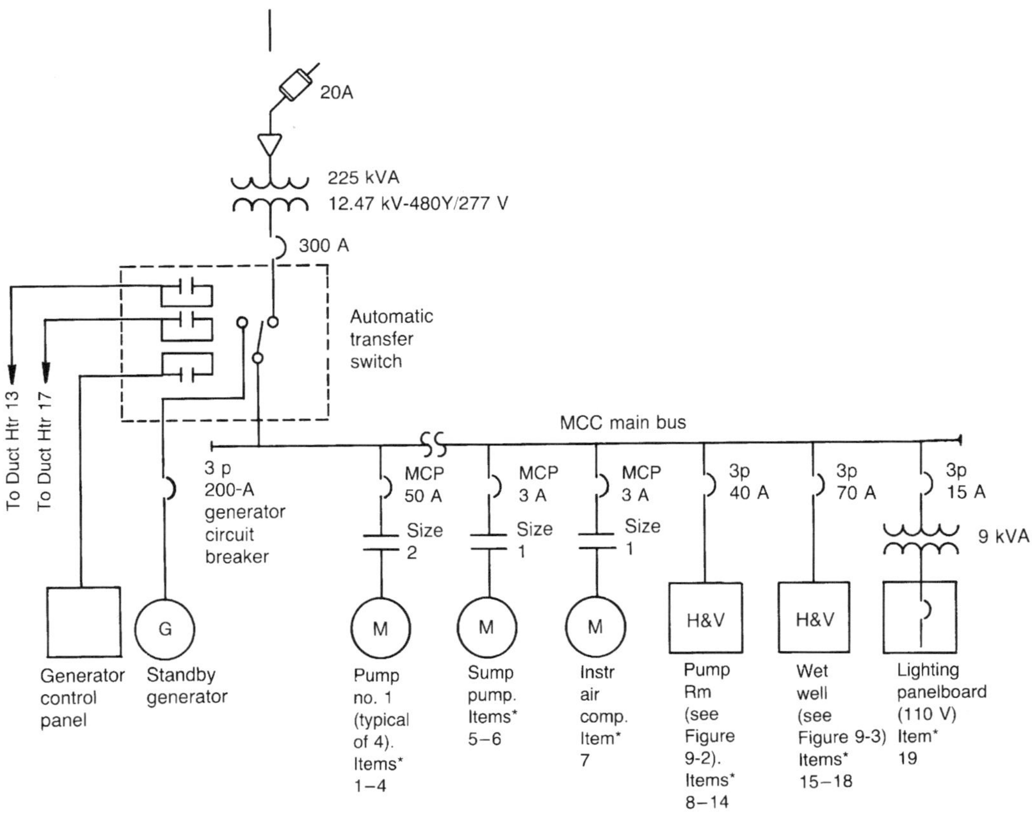

Figure 9-1. Pumping station electrical single-line diagram. Htr, heater; Inst, instruments; Comp, compressor; H&V, heating and ventilating; AHU, air handling unit; Cont Rm EF, control room exhaust fan; Int, intake. *See load estimates in Example 9-1.

Several classes of fuses are encountered in industrial design such as that for pumping stations. An example is pole-mounted fuses used by utilities. These may be of the expulsion type in which the gases generated by the melting fuse and the heated casing during an overcurrent condition cause a mechanical indicating device to function. The typical dropout fuseholder unlatches and the blown fuse remains with the hinged portion of the holder.

The fuses are generally selected by the utility, and so it is necessary to obtain the exact fuse data and available fault currents from the utility before starting a coordination check.

Other fuses commonly used in pumping stations include medium-voltage transformer primary fuses on large installations, medium-voltage current-limiting fuses on medium-voltage (2400 and 4160 V) motor starters, low-voltage current-limiting fuses in low-voltage switchgear (to reduce the fault current available to the loads), and low-voltage transformer primary fuses.

In the pumping station design examples in this chapter, fuses are used only to protect control transformers, and they are used in both the transformer primary and secondary sides. The electric power utility is to furnish the primary fuse at their service pole.

Circuit Breakers

Circuit breakers are electromechanical switching units that are designed to open a circuit when specified overload or fault conditions occur. Either thermal-magnetic or magnetic-only molded-case circuit breakers are normally used in pumping stations. The power circuit breaker type may be preferred for large or high-voltage motors. The type of circuit breaker chosen is based on current withstand, interrupting, and coordination ratings and on economic factors.

The thermal-magnetic circuit breaker is tripped by a magnetically operated element for severe faults and by a thermal element for long-duration overloads. Thermal-magnetic circuit breakers may be adjustable and nonadjustable and are used as main and branch circuit breakers and are used to protect almost all types of loads.

An adjustable-trip form of the magnetic-only circuit breaker is most often used in conjunction with a motor starter to protect motor branch circuits. This unit, called a "motor circuit protector (MCP)," is sized and adjusted to pass the starting current of a particular motor and will not trip when the sustained current is less than the setting (which can be as much as 13 times the full-load current of the motor). It will trip, however, under severe fault conditions. The overload relays required on the motor starter provide the required motor overload protection.

Example 9-2
Branch Circuit Calculations for Pump Motors

Problem: Size the branch circuit breaker, starter, and the branch circuit conductors for each of the main pumps of Example 9-1.

Solution: From Example 9-1, the motor full-load current, I_{FL}, is 34 A.

Size the branch circuit breaker. NEC 430-152 allows the circuit breaker to be sized to a maximum of 250% of full-load current, which is $34 \times 2.5 = 85$ A. Use a 70-A circuit breaker for a normal inertia system. If the inrush current (approximately $6 \times I_{FL} = 204$) does not persist more than about 10 s, the circuit breaker will not trip. Be careful to check the time–current curves of the circuit breaker.

If the drive inertia of the centrifugal pumps were higher than normal, the circuit breaker should be sized above the 250% value in accordance with NEC 430-52 (which limits the maximum setting to 400%) to accommodate the necessary longer acceleration time of the drive.

Size the motor circuit protector. For adjustable-trip, magnetic-only circuit breakers, NEC 430-152 allows the MCP to be set at 700% of full-load current, so $34 \times 7 = 238$ A. NEC 430-52 limits the maximum setting to 1300% of the full-load current, or $34 \times 13 = 442$ A. Select an MCP with a 50-A rating and an adjustable-trip range of 150 to 580 A. The initial setting should be approximately 200 A (see, for example, literature from the Square D Company; the ratings of other manufacturers are similar).

Size the disconnect switch at motor. If a disconnect switch is required at the motor, NEC 430-110 applies, and this requires the switch to be rated at least at 115% of the motor's full load current, or $34 \times 1.15 = 39$ A.

The next larger standard-size disconnect switch is rated at 60 A, so use a 60-A disconnect switch (in a housing suitable for the environment) with an auxiliary control contact that opens when the switch is in the open position. The contact should open before the main contacts open in order to drop out the motor starter.

Size the motor starter. The motor starter manufacturer's data shows that an NEMA size 2 starter is rated for 25-hp, 460-V, three-phase motor-starting duty, so use a NEMA size 2 starter. A size 3 starter would be specified if it is likely that a 30-hp motor will be substituted in the near future or the motor is of a type that draws more current than is shown in NEC tables. Low-speed motors often require larger starters; check with the starter manufacturer.

Size the branch circuit conductors. Branch circuit conductors must be rated at least at 125% of the motor's full-load current, or $34 \times 1.25 = 42.5$ A. Using a maximum conductor temperature of 60°C, NEC Table 310-16 lists No. 8 AWG copper at 40 A and No. 6 AWG copper at 55 A.

Use three No. 6 AWG TW or THW insulated copper conductors plus No. 6 or No. 8 AWG ground conductor (NEC Table 250-95 lists minimum-size grounding conductors). THW insulation is a thermal plastic material that softens when heated. Many designers prefer XHHW insulation, a cross-linked synthetic polymer, which is thermal-setting and hardens when heated.

Checking Branch Circuit Voltage Drop

NEC 210-19, footnote a, recommends that the branch circuit voltage drop be less than 3%, and the total drop in feeders and branch circuits must not exceed 5%. Good design limits the voltage drop to less than 2% each for feeders and branch circuits.

Another facet of voltage drop (which is not governed specifically by the NEC but is essential to good operation) is the motor-starting voltage drop in the branch circuits and feeders. An induction or synchronous motor draws about six times the full-load current at the moment of starting, and this decreases to the current necessary to drive the load during the remainder of the acceleration time and the normal running time. Ordinarily; a 13 to 15% voltage drop is acceptable and provides sufficient voltage for accelerating the drive. The starting voltage drop with No. 6 AWG copper conductors is approximately 1.4%, and the running drop is about 0.3%. The starting voltage drop is a problem only on long runs. The method for calculating the voltage drop is given in Example 13-1.

Miscellaneous Drives and Loads

Two important drives, the pump room sump pump and the instrument air compressor, are rated at $^1/_2$ hp each. These drives should be served by branch circuits of the motor control center at 460 V (three phase).

The exhaust fans and the duct heaters could also be served by individual motor circuits. Because of the interlocking requirements and to facilitate safe maintenance practices, two heating and ventilating panels should be provided. The arrangement of the panels is shown in single-line diagram form in Figures 9-2 and 9-3.

Each heating and ventilating control panel includes all of the power and control functions for the duct heater and its interlocked fans. The manufacturer of the panel should be given the option of providing a transformer large enough to serve the fans at 120 V or 208 V (single phase) from a 208Y/120-V panelboard in the H&V control panel.

Interlocks from the station's automatic transfer switch cut off the duct heater when power is being taken from the station standby generator. Fans and louvers are still required to operate under these conditions.

Example 9-3
Branch Circuit Calculations for Sump Pump

Problem: Size the circuit breaker, the starter, and the branch circuit conductors for the sump pump (and all other loads of identical horsepower rating) in Example 9-1.

Solution: One manufacturer's 1150-rev/min, 460-V, 1/2-hp motor is listed with a full-load current of 1 A and a locked-rotor current of 12.5 A.

Size the motor circuit protector. Size the MCP at 700% of full-load current per NEC 430-152, or $1.0 \times 7.0 = 7$ A.

Select an MCP rated at 3 A with an adjustable-trip range of 8 to 22 A—slightly above the 7 A required. Depending on the exact characteristics of the motor, the setting should be very close to (or slightly exceed) NEC limits.

Size the starter. Although the manufacturer's data show a size 0 starter, the starter should be an NEMA size 1, which is the minimum recommended by the authors.

Size the branch circuit conductors. These must be rated at 125% of the full-load current, or $1.0 \times 1.25 = 1.25$ A. The use of minimum-size wire, No. 12 AWG stranded copper conductor, is recommended; No. 14 AWG conductor is adequate.

The pumping station lighting and small ancillary loads operate at 120 V (single phase) or at 208 V (single and/or three phase). A step-down transformer and circuit breaker panelboard are used to feed this load. The calculations for feeding such equipment are illustrated in Example 9-4.

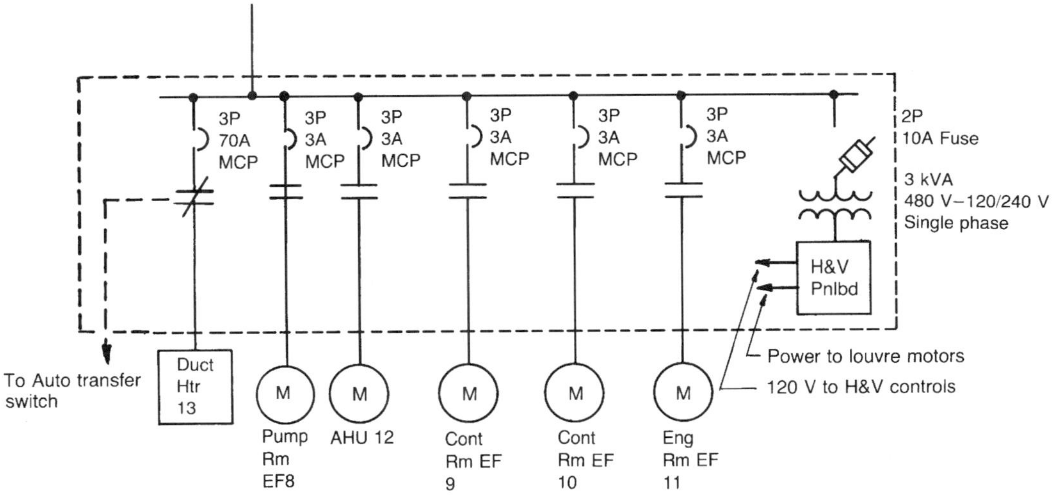

Figure 9-2. Pump room heating and ventilating panel.

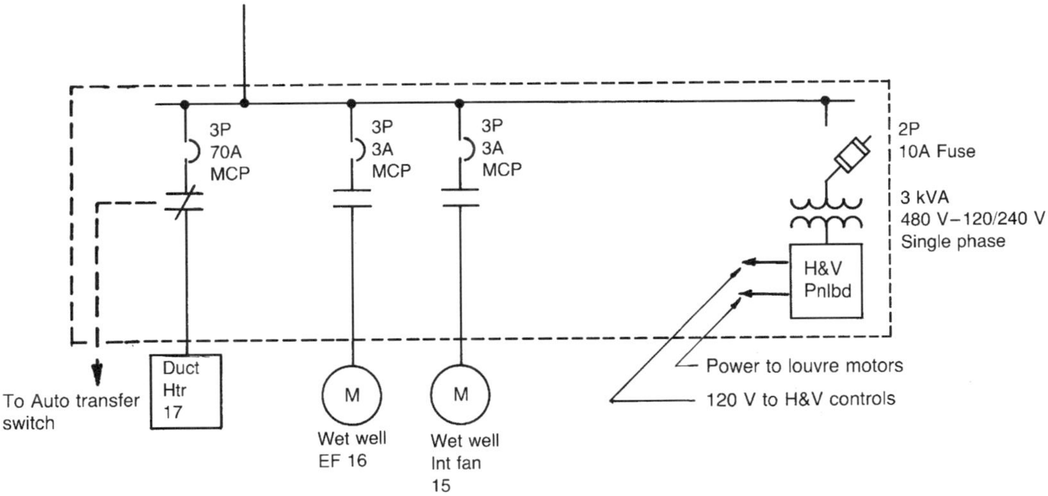

Figure 9-3. Wet well heating and ventilating panel.

Example 9-4
Lighting and Small Power Transformers

Problem: Calculate the size of the lighting and small power transformers in Example 9-1.

Solution: The principal loads on the lighting panelboard are the pumping station's lighting and receptacle circuits and the battery charger. Allow 1000 W for the battery charger. Allow 3000 VA for lighting in the pump building and 500 VA in the wet well. Allow 1500 W for receptacles. The calculated load is 600 VA or 6 kVA. The next larger size transformer is 9 kVA.

A 9-kVA, three-phase, 480-V delta primary, 208Y/120-V secondary transformer should be used.

Size the transformer primary circuit breaker. NEC 450-3b governs the sizing of transformer overcurrent protection. Where circuit breaker protection is provided in both the primary and secondary, the maximum primary overcurrent protection is 250% of full-load current. The maximum secondary overcurrent protection is 125% of the full-load current. From Equation 8-6 ($P_{3\phi} = \sqrt{3}V_L \times I_L \times Pf$),

$$I_{FL} = P_{3\phi}/(V_L \times \sqrt{3}Pf) = 9000/(480\sqrt{3} \times 0.95) = 11.4\text{A}$$

if a power factor of 0.95 for the lighting and small loads on this panelboard is assumed (usual for fluorescent lighting).

$$250\% \text{ of } I_{FL} = 11.4 \times 2.5 = 28\text{A}$$

This maximum must not be exceeded, so use the next lower size, which is a 20-A primary circuit breaker.

Size the secondary circuit breaker (panelboard main breaker). The full-load current is

$$I_{FL} = 9000/(208 \times \sqrt{3}) = 25\text{A}$$

$$125\% \text{ of } I_{FL} = 25 \times 1.25 = 31\text{A}$$

Use a 30-A secondary circuit breaker. Secondary protection should always be provided.

Size the primary conductors. The primary is protected by a 20-A circuit breaker (refer to NEC Table 310-16). Because the 60°C rating of a copper conductor at 20 A requires No. 12 AWG size, use three No. 12 AWG copper conductors plus one No. 12 AWG ground conductor.

Size the secondary conductors. The secondary is protected by a 30-A circuit breaker. Because the 60°C copper temperature rating at 30 A requires a No. 10 AWG conductor, use four No. 10 AWG copper conductors (three phase and one neutral conductor), plus one No. 10 AWG ground conductor.

Size the panelboard circuit breakers. The panelboard circuits should all be rated at 20 A, and all of the conductors should be No. 12 AWG copper.

The H&V panels require 460-V, three-phase power. The sizing of the feeder circuit breaker and feeder conductors is illustrated in Example 9-5.

Example 9-5
Heating and Ventilating Electrical Equipment

Problem: Determine the panel loads, the size of the circuit breakers, and the size of the conductors for (1) the pump building heating and ventilating and (2) the wet well heating and ventilating in Example 9-1.

Solution: (1) *Pump building H&V panel.* Per NEC code, for the largest motor add 25% of 1 A (= 0.25A) for minimum conductor size. The duct heater is rated at 16.5 kW, so the rated current is

$16{,}500/(460 \times \sqrt{3})$ =	21 A
Four $^1/_2$-hp motors at 1 A each =	4 A
Transformer, 1 kVA, single phase =	2 A
Largest motor × 25% =	0.25 A
Total H&V panel load =	27.25 A

Choose a circuit breaker for at least 125% of total calculated load:

$$27.25 \times 1.25 = 34 \text{ A}$$

Use a 40-A circuit breaker in the MCC.

Conductors must have an ampacity (current-carrying capacity) of no less than 40 A. In the 75°C column in NEC Table 310-16, No. 8 AWG conductors are rated at 40 A and No. 6 AWG conductors are rated at 55 A. Use three No. 8 AWG copper conductors and one No. 8 AWG ground conductor.

(2) *Wet well heating and ventilating panel.* The wet well H&V panel is calculated in a similar manner. A circuit rated at 60 A minimum is required. Use a 70-A circuit breaker in the MCC, and use three No. 4 AWG copper conductors and a No. 8 AWG ground conductor.

Service Load

Now that the branch circuits for motors and for heating and ventilating are sized, the total station (service) load can be determined, and the main transformer, its protection, and the conductors can be sized.

Example 9-6
Service Circuit Transformer, Protection, and Conductors

Problem: Find the sizes of the service circuit breakers, the main transformer, the primary and secondary protection for the transformer, and the service conductors for the pumping station in Examples 9-1 to 9-5.

Solution: Summarize all the pump station loads:

Pumps: 3 × 34 A =	102 A
Air compressor and sump pump:	2 A
Lighting load:	11 A
Pump building H&V:	27 A
Wet well H&V:	48 A
Total station load:	190 A

Treat all loads as "continuous duty." Per NEC 430-62(b), the rating of the main circuit breaker may be sized for possible future larger loads.

Size of main transformer. The transformer's kilovolt-ampere rating should exceed the calculated load of the station by 15 to 20% or more depending on several factors. The lack of a quickly obtainable replacement transformer may dictate operation at a somewhat lower total temperature than the full-load rating would produce. Predictions of substantial load growth within 10 yr make oversizing worth studying. A slight oversizing, such as 15%, allows for some adjustment in main pump sizes or in H&V or other auxiliary equipment without the necessity of rebuilding the entire station electrical system. For a station load of 190 A, $190 \times 480 \times \sqrt{3} = 158$ kVA and 15% oversizing adds 23.7 kVA for a total of 181.7 kVA, so use the next larger standard size unit, which has a rating of 225 kVA.

Size of transformer primary and secondary protection. Assume that the fuse will be provided by the utility. Although the utility is not required to design to NEC standards, the fuse size is likely be the same because it is based on load size, and utilities use conservative conductor current ratings for their equipment.

For transformers with less than 6% impedance and a primary voltage greater than 600, NEC allows a maximum fuse rating of 300% of the full-load current. The 480-V secondary circuit protection cannot be more than 250% of the full-load current. For a utility voltage of 12,500, the primary is $225{,}000/(12{,}500\sqrt{3}) = 10.39$ A, so the maximum fuse per the NEC is $10.4 \times 3 = 31$ A.

A 30-A fuse could be used, but a 20-A fuse designated 20T type provides transformer protection closer to the overloading point of the motor.

The secondary is $225{,}000/(480\sqrt{3}) = 270.6$ A. The maximum circuit breaker (main MCC circuit breaker) is $271 \times 2.5 = 677$ A; the minimum circuit breaker (the total connected load plus 50% of largest motor) is 207 A.

Use a 400-A frame circuit breaker with a 300-A trip.

Sizing the service conductors. The service conductors should be sized for any foreseeable load growth that can be accommodated without the necessity of replacement. The minimum size is that which will be protected by the main circuit breaker, namely, 300 A at 75°C (refer to NEC Table 310-16).

Use three 350-kcmil copper phase conductors plus one No. 2 neutral (grounded) conductor in a concrete-encased duct 4 in. in diameter. The neutral conductor must be sized according to NEC Table 250-94 as a minimum.

9-7. Lighting

Use the IES [2] lumen method for lighting calculations. Manufacturers' selection charts are too inaccurate for small buildings because they are based on somewhat different criteria. A lighting level of 20 foot-candles (ft · cd) is usually sufficient for pumping stations if there are adequate GFCI receptacles for work lights needed for repairs. The following lighting levels are recommended:

- Offices: 50–100 ft · cd
- Control areas: 30–75 ft · cd
- Equipment areas 15–50 ft · cd
- Outdoor areas: 1–5 ft · cd.

Many states now have a lighting limitation based on watts per square foot, and this limitation should be investigated prior to design.

Obtain fixture data from manufacturers' catalogs. These data are classified by several parameters based on Equations 9-1 and 9-2. The room cavity ratio, *RCR*, is defined as

$$RCR = 5H(L + W)/A \tag{9-1}$$

where H is the distance from the work plane (usually 2 ft 6 in. above the floor in offices and control areas, 3 ft above the floor in equipment areas, and the floor itself elsewhere) to the bottom of the light fixtures, in feet, L is room length in feet, W is room width in feet, and A is floor area in square feet.

From the IES handbook [2], the relationship between illumination and the number of fixtures is

$$\text{Number of fixtures} = \frac{\text{ft} \cdot \text{cd}}{(\text{number of lamps/fixture})(\text{lumens/lamp})} \times \frac{\text{area}}{(CU)(LLD)(LDD)(RSD)} \tag{9-2}$$

where area is in square feet, CU is a coefficient of utilization depreciation and varies from 0.01 to 0.99 (Table 9-1), LLD is lamp lumen depreciation, LDD is luminaire dirt depreciation (dimensionless), and RSD is room surface depreciation (dimensionless).

Table 9-1. Coefficient of Utilization (*CU*),[a] Manufacturer's Data for a Selected Two-Lamp, 40-W Industrial Fluorescent Fixture[b,c] (Spacing/Mounting height = 1.5)

Room cavity ratio	R_w (%) with R_{cc} = 80%			R_w (%) with R_{cc} = 50%			R_w (%) with R_{cc} = 10%		
	50%	30%	10%	50%	30%	10%	50%	30%	10%
1	0.68	0.65	0.62	0.62	0.60	0.58	0.55	0.54	0.52
2	0.59	0.54	0.50	0.54	0.50	0.47	0.48	0.46	0.43
3	0.52	0.46	0.41	0.48	0.44	0.39	0.43	0.39	0.37
4	0.46	0.40	0.35	0.43	0.37	0.33	0.38	0.34	0.31
5	0.40	0.34	0.30	0.37	0.32	0.28	0.34	0.30	0.27
6	0.36	0.30	0.25	0.33	0.28	0.25	0.30	0.26	0.23
7	0.33	0.27	0.22	0.30	0.25	0.22	0.27	0.23	0.20
8	0.30	0.23	0.19	0.27	0.22	0.19	0.25	0.21	0.18
9	0.27	0.20	0.17	0.25	0.20	0.17	0.22	0.18	0.15
10	0.24	0.18	0.15	0.23	0.18	0.15	0.20	0.16	0.13

[a] *CU* is for a 20% floor cavity reflectance (R_{fc}).

[b] R_{cc} = Effective ceiling cavity reflectance; R_w = wall reflectance.

[c] Coefficient of utilization varies depending on type of fixture, type of lamp, and manufacturer. Refer to manufacturer's data.

Example 9-7
Lighting a Small Pumping Station

Problem: Assume a pumping station control room is 21 × 42 ft with a 15-ft ceiling. Mount lamps 12 ft above the floor. Assume the work plane is 3 ft above the floor. Design lighting to provide 40 ft · cd of illumination on the work area.

Solution: Select 4-ft-long, two-lamp industrial fluorescent fixtures with 32 WT-8 lamps and an electronic ballast. The precalculation procedure outlined in the IES handbook [2] provides a detailed explanation of these parameters:

- *LLD* (lamp lumen depreciation): Use 0.90 for fluorescent lights.
- *LDD* (luminaire dirt depreciation): Use IES Category III and assume fixtures are cleaned annually: therefore use "medium" (0.87).
- *RCR* (room cavity ratio) from Equation 9-1:

$$RCR = H(L+W)/A = \frac{5\times(12-3)(21+42)}{21\times 42} = 3.2$$

- *RSD* (room surface depreciation): use 0.95.

Find the coefficient of utilization (*CU*) from the IES handbook [2], or use the manufacturer's data as shown in Table 9-1. For two-lamp industrial fluorescent fixtures, 2850 lumens per lamp, and typical room reflectance values of 50/50/20, *CU* = 0.46 by interpolation.

From Equation 9-2:

$$\text{Number of fixtures} = \frac{(40)(21\times 42)}{2\times 2850\times 0.46\times 0.90\times 0.87\times 0.95} = 18.0$$

Use 18 fixtures (2 rows of 9 fixtures). Fixture spacing must not exceed the fixture's designated spacing/mounting height ratio multiplied by the mounting height above the work plane in either direction. The illumination on the work area is found from Equation 9-2:

$$\text{Actual foot-candles} = \frac{18\times 2\times 2850\times 0.46\times 0.90\times 0.87\times 0.95}{21\times 42} = 39.8$$

Locating fixtures is best done by drawing the room to scale, making paper cut-outs of the fixtures to scale, and moving the fixtures around until the best effect is obtained. Assume the lamps line up with the long dimension of the room. Then:

$$\text{Transverse spacing} = (2^1/_2)/(12 - 3) = 1.2 < 1.5 \text{ OK}$$

$$\text{Longitudinal spacing} = [42 - (9 - 1)4]/(12 - 3) = 1.1 < 1.5 \text{ OK}$$

Similar calculations can be made for the other rooms in the pumping station. More than one light fixture and lamp combination may often appear appropriate for an area or space. The final selection may require: (1) calculations for each light fixture/lamp combination proposed, (2) trial layouts with consideration for room geometry and contents, and (3) consideration, perhaps, of state lighting limitations. For even lighting at the work plane, conform to the fixture manufacturer's designated spacing/mounting height limitations in locating fixtures. Ensure that fixtures are accessible for cleaning and lamp changing.

9-8. Power Factor

Capacitors are used to correct the power factor, as computed by Equation 9-3.

$$Pf = \frac{\text{actual power (kW)}}{\text{apparent power (kVA)}} = \frac{\text{active current}}{\text{total current}} = \cos\theta \qquad (9\text{-}3)$$

The relationship between power, current, and power factor in Example 9-8 is shown in Figure 9-4.

Example 9-8
Power Factor Relationship

Problem: Calculate the power, current, and power factor relationship for Pump No. 1 (see Example 9-1). The motor characteristics given by the manufacturer are 25 hp, 460 V, three phase, 34 A, and 0.75 *Pf*.

Solution: Use Equations 8-4 and 6 to calculate current and power values.

$$\text{Active (true) power} = P_{3\phi} = \sqrt{3}V_LI_LPf = \sqrt{3}\times 460\times 34\times 0.75 = 20{,}300 \text{ W} = 20.3 \text{ kW}$$

$$\text{Apparent power} = \sqrt{3}V_LI_L = \sqrt{3}\times 460\times 34 = 27{,}100 \text{ VA} = 27.1 \text{ kVA}$$

$$\text{Reactive power (see Figure 9-4a)} = \sqrt{27.1^2 - 20.3^2} = 17.9 \text{ kVAR}$$

Correction of Power Factor

Power company penalties for low power factor can be severe. It is often economical to raise the power factor to meet the power company's minimum limit. It is neither economical nor necessary to correct the power factor to 1.0, but raising the power factor to about 0.95 is likely to be economical.

Example 9-9
Capacitor Sizing

Problem: Calculate the capacitor sizes needed to raise the power factor of the pumping station from 0.75 to 0.95.

Solution: From Example 9-1, the design amperage is 187.5, so the normal operating station load is $\sqrt{3}(460)(187.5\text{ A})/1000 = 149.4\text{ kVA}$. At the present power factor of 0.75, the kilowatts (kW) used are $149.4 \times 0.75 = 112.0$ and the reactive kilovolt-amperes (kVAR) are $\sqrt{(149.4)^2 - (112.0)^2} = 98.9$. The vector diagram of power is shown in Figure 9-4b.

To increase the system power factor to 0.95, draw the vector diagram of Figure 9-4c. The kVA must remain constant. Therefore, $\text{kW} = (\cos\theta)(\text{kVA}) = (0.95)(149.4) = 141.9$ and kVAR =

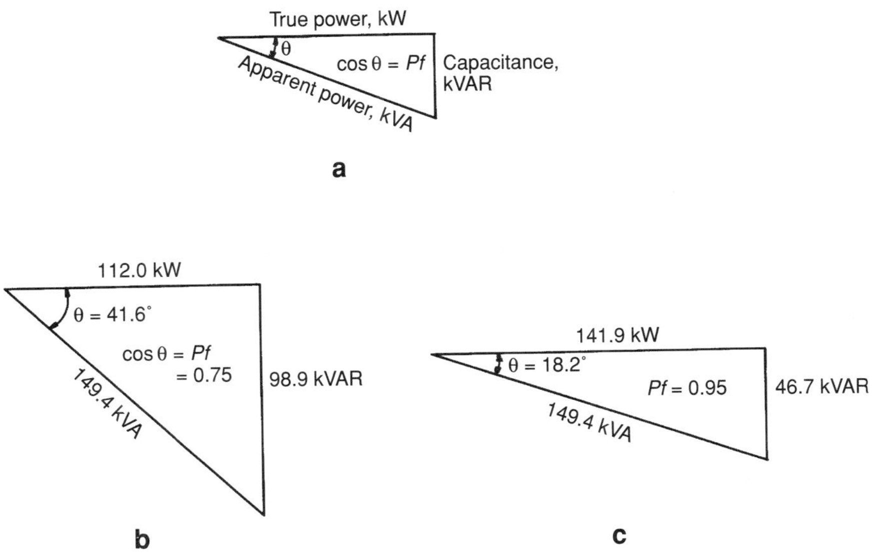

Figure 9-4. Vector diagrams for power factor correction. (a) vector diagram symbols; (b) power vector diagram without power factor correction; (c) power factor diagram with power factor correction.

$\sqrt{(149.4)^2 - (141.9)^2} = 46.7$. The rating of the capacitor must change the capacitance from 98.9 to 46.7 kVAR, so the capacitor rating must be 98.9 kVAR – 46.7 kVAR = 52.2 kVAR.

The 52.2 kVA is based on a station load of 187.5 A or 149.4 kVA. Because it is not economical to increase the power factor to a value greater than unity, less kVAR capacitor correction should be connected to the system at reduced station load. Therefore, it is desirable to distribute the capacitors to the system loads so that, as system load is turned on, additional capacitance is also added. The pump motors are typically the largest loads in a pumping station, so it is often advantageous to connect capacitors directly to each motor so as to switch them on with the motor. Calculations should be done so that the amount of capacitance added to each motor does not correct the motor *no-load* power factor to more than unity.

Install a total of 55 kVAR capacitance distributed as follows:

Each motor: 15 kVAR,

Bus capacitor: 10 kVAR (but sized with care to make up the deficiency after motor capacitors are finally chosen).

Capacitors

Obtaining satisfactory power factors by using capacitors is preferable to using more expensive synchronous motors unless synchronous motors are otherwise required. Switch capacitors with motors; that is, connect the capacitors to the motor terminals so that both units are switched on at the same time. Select capacitors to supply approximately (without exceeding) motor no-load magnetizing current or nearly no-load current. If the motor characteristics are not accurately known during design, check the shop drawings or confer with the manufacturer and resize the capacitor if necessary.

Conductors

The conductor ampacity should not be less than 135% of the rated capacitor current, nor should it be less than one-third of the ampacity of the motor branch circuit conductors (see NEC 460-8.a).

9-9. Engine-Generator Sizing

Preliminary calculations for determining the engine-generator size are shown in Example 9-10. The final size, however, should be determined by the manufacturer based on an even more careful consideration of station loading and starting requirements and the characteristics of the generator and engine.

Example 9-10
Size of the Engine-Generator Set

Problem: Determine the generator and the engine power rating requirements for the entire pumping station described in Example 9-1. Note that as more pumps come on line, the head on the pumps increases, but the incremental flowrate decreases, with the net result that (for this pumping station) one pump operates at 21.2 motor hp, two pumps operate at 20.0 hp each, and three pumps operate at 18.4 hp each.

Solution: At full load, the power factor of a 720-rev/min motor is about 0.75, but adding capacitors to the circuit, as shown in Example 9-9, increases the power factor to 0.95. The locked-rotor power factor of a standard 1800-rev/min, 25-hp motor is 0.44, and for a 720-rev/min motor, the power factor is estimated to be 0.40. The addition of a 15-kVAR capacitor (Example 9-9) would have little effect on the locked-rotor power factor. Starting a Code F motor requires 5.3 kVA for each horsepower.

In Example 9-1, the essential loads on the generator for heating and ventilating are 5 A in the pump room and 3.5 A in the wet well. During a power outage, lighting can be restricted to 11 A. The sump pump and instrument air compressor consume 1 A each, but these loads can be ignored during motor start-up because they are small, intermittent, and not likely to exist during the worst-case inrush, which persists only momentarily. Hence, the total load with no pumps operating is 5 + 3.5 + 11 = 19.5 A. Because personnel discomfort during power outages can be acceptable, but freezing (even of small pipelines such as for seal water supply) cannot, it may be necessary in very cold climates to include some of the heater loads in the sizing calculations.

Item	Starting		Running	
	kVA	kW	kVA	kW
Ventilation and lighting (V&L) loads				
$19.5 \times 460\sqrt{3}/1000$				15.5
15.5/0.95 *Pf*			16.3	
Start first pump				
25 hp × 5.3 kVA/hp	132.5			
132.5 kVA × 0.40 *Pf*		53.0		
Add V&L loads	16.3	15.5	16.3	15.5
Total	148.8	68.5	16.3	15.5
Run first pump at 21.2 motor hp				
21.2 hp × 0.746 kW/hp (1/0.88 efficiency[a])				18.0
18 kW at 0.95 *Pf*			18.9	
Add V&L loads			16.3	15.5
Total load			35.2	33.5
Start second pump				
25 hp × 5.3 kVA/hp	132.5	53.0		
Run 1st pump and V&L	35.2	33.5	35.2	33.5
Total	167.7	86.5	35.2	33.5
Run 2 pumps at 20 motor hp each				
20 hp × 2 × 0.746 kW/hp (1/0.88 efficiency[a])				33.9
33.9 kW at 0.95 *Pf*			35.7	
Add V&L loads			16.3	15.5
Total load			52.0	49.4
Start third pump				
25 hp × 5.3 kVA/hp	132.5	53.0		
Run 2 pumps and V&L	52.0	49.4	52.0	49.4
Total	184.5	102.4	52.0	49.4

Item	Starting kVA	Starting kW	Running kVA	Running kW
Run 3 pumps at 18.4 motor hp				
18.4 hp × 3 × 0.746 kW/hp (1/0.88 efficiency[a])			49.3	46.8
Add V&L loads			16.3	15.5
Total load			65.6	62.3

[a] Efficiency for high-efficiency motors is about 92 to 94%.

The critical generator loads are the kilovolt-amperes and kilowatts for starting the third pump. During the start, the first two running motors can act momentarily (for half a cycle) as generators and, hence, reduce the demand on the generator, but the effect lasts much less than a second and is difficult to assess, so the effect is ignored, especially as it may take up to 6 s to start the third pump.

Because the calculated size of the generator is 184.5 kVA, specify a 200 kVA standby duty rating. The future flow of 4500 gal/min requires about 15% more horsepower, but only about 5% more generator output because the inrush current for the 25-hp motor does not change. Therefore, a 200-kVA generator would be adequate.

The engine can be sized on the basis of running power, partly because the inertia is so great that the engine will maintain its speed during the few seconds of high inrush current and partly because the engine can be overloaded for a few seconds.

$$\text{Size of engine} = \frac{62.3 \text{ kW}}{0.85 \text{ generator efficiency}} = 73.3 \text{ kW} = 98.2 \text{ hp}$$

Use an engine of not less than 100 hp (continuous duty rating).

This approximate analysis may be refined by methods explained in manufacturer's literature [3], but always confirm engine-generator size calculations with the manufacturer. Manufacturers may calculate required engine and generator size somewhat differently, so supply all of the worst-case figures. Also supply generator site, altitude, and ambient temperature because these affect the generator rating.

The calculations for Example 9-10 are typical, but maximum flow might be a rare occurrence of short duration, and the diversity of loads (the probability that all demands may not occur at the same time) can often reduce a power requirement. On the other hand, some heating might be required. A more extensive analysis might result in a different engine-generator unit. A typical engine-generator set is shown in Figure 9-5.

9-10. Short Circuit Current Calculations

The purpose of short circuit current calculations is to find the maximum current available under short circuit conditions at each required location and to be certain that the circuit interrupting devices (circuit breakers and fuses) have the ability to interrupt this level of current. It is also important to be certain that other devices required to carry this high level of current have appropriate withstand ratings so that the device will not fail at the current level. Except in some medium-voltage circuits with local on-site generation, the maximum available short circuit current would occur if all phases or lines were connected together.

A preliminary calculation is needed during design for properly sizing, rating, and specifying equipment. The designer must use the best available data in making this calculation, although the equipment manufacturer and the exact data may be unknown.

If the expected fault currents approach equipment interrupting and withstand ratings (typically 22,000 A at 480 V), a final calculation is made during construction to verify the equipment ratings and set adjustable overcurrent protective devices. This work is often done by the equipment supplier using a computer for speed and accuracy. Make sure that the specifications detail this requirement properly. Spot check the calculations to ensure that the proper field data are used.

Fault (Short Circuit) Current Magnitude

The magnitude of a fault current depends primarily on the capacity of the supply system, but on-site

Figure 9-5. A typical engine-generator set. Courtesy of Waukesha Engine Div., Dresser Industries, Inc.

motors and generators can contribute to the fault current. In Figure 9-1, the characteristics of each contributor are:

- Utility. The fault current magnitude is assumed constant during the fault period.
- Local generators are not usually on line in pumping stations. If they are, the magnitude of the fault current decreases during the fault period. Practically, the fault current may be assumed to be constant in low-voltage systems.
- Synchronous motors act as synchronous generators driven by load and rotor inertias. The fault current magnitude decreases substantially during the fault period.
- Induction motors. Current is generated while the magnetic field is present. The induction motor's field decreases rapidly when the driving voltage drops to a small value (as in a fault or when the contactor or circuit breaker opens). The fault current magnitude is initially about that of the locked rotor current, but it decreases rapidly, usually over a few cycles. If the running power factor is corrected to near unity, the magnitude may not decrease as expected due to self-excitation.

The total or asymmetrical fault current is composed of an ac component plus a decaying dc component. The initial magnitude of the dc component depends on the circuit reactance/resistance (X/R) and on the instantaneous voltage of the faulted phase at the instant the fault occurs. To allow for the dc component, multiply the symmetrical value by 1.6 (for high-voltage fused starters, current-limiting fuses, and circuit breakers) or by 1.25 (for low-voltage fuses).

It is the asymmetrical current that must be interrupted by low-voltage breakers and fuses due to short (e.g., 1.5 cycles) interrupting times. Low-voltage breakers, however, are now rated in symmetrical current, so the asymmetrical current need be computed only for fuses. Medium-voltage breakers and fused switches have both withstand and interrupting ratings (which are lower due to their longer interrupting times), so the dc decay must be taken into account. For sample calculations, consult the literature [4, 5, 6].

Coordination

A coordinated system is one in which the overcurrent devices are rated and adjusted so that the device just ahead of a fault will safely clear the fault, dropping only the downstream part of the circuit. If that device should fail to clear a fault, the next upstream device will clear after a suitable time delay to give the proper device a chance to operate. In a coordinated system, the withstand ratings of equipment and cable are suitable for withstanding fault currents for the time needed to clear the fault without damage to the insulation (see the UBC for a detailed discussion).

9-11. Harmonics

In pumping stations with adjustable frequency drives, excessive harmonic contribution to the electrical power system must be prevented. Where the adjustable frequency drive load exceeds 25% of the station load or where adjustable frequency drives must be powered by an engine-generator, a harmonic calculation should be made and suitable filtering specified. IEEE 519 explains how to make harmonic calculations and specifies harmonic current limits for systems connected to electric power utilities. Where adjustable frequency drives are to be powered by an engine-generator, consult the manufacturer regarding limits for that specific equipment.

Power factor correction capacitors must not be connected to power systems with a harmonic voltage distortion in excess of 5% without suitable filters that exclude harmonic currents from the capacitors.

Past practice in the industry has been to make the drive vendors responsible for harmonic calculations. As with short circuit calculations discussed in Section 9-10, harmonic calculations are best performed using suitable computer software.

9-12. Construction Service

A thorough review of all submitted material is a time-consuming, unpopular, and unrewarding task, but *it is vital.*

Shop Drawing Review

The shop drawing review is the last chance to make a design change at a reasonable cost. Once construction has begun, changes are very expensive.

Allow no deviations without good reason and without a reduction in the contract price. Document all revisions, and keep the project manager and contractor fully aware of all conversations with suppliers that result in any revisions.

Obtain and review the electrical portions of submittals under other specification sections such as mechanical, process, or instrumentation.

Submittal Checklist

Check submittals for conformance to the drawings, specifications, applicable codes, addenda, and change orders. Checks of the following elements are a minimum:

- Voltage, phases, and other nameplate data
- Control (schematic) diagrams
- Wiring and interconnection wiring diagrams
- Control and alarm interfaces with other equipment
- Components
- Wiring details
- Inclusion and acceptability of required factory test reports.

Release the generator shop drawings to the general contractor only after including the following warning: "Accepted pending verification that *you* have coordinated the generator sizing with your pump and motor suppliers." If the warning is not included, the station's generator will be inadequate if the motor size is increased.

Inspection

Adequate inspection is required to assure the owner and authorities that the construction meets code, specification, and drawing requirements, so several inspectors, including city/county personnel in addition to the resident inspector or engineer, are likely to visit the job site. The designer should make spot checks, and there should be a final, detailed inspection. The following are suggestions for the resident inspector:

- Be as thorough as time allows. Have the contractor make the interiors accessible for inspection and observe the interior of as much equipment as possible. Note the condition of workmanship as well as conformance to drawings.
- Check all work against drawings, specifications, addenda, change orders, shop drawings, and codes.
- Make a written record of each deviation, whether it is to be corrected or left as is, and give the record to the project manager for transmittal to the contractor.

Tests

This section is limited to electrical system tests.

Factory Tests

Require the following factory tests:

- *Switchboard:* Test per NEMA Standard PB2.
- *Motor control center:* Test per NEMA Standard ICS.
- *Transformer, oil filled:* Test per ANSI C57.12.90.
- *Standby generator:* Run the unit at the rated load and power factor for at least 2 hr to demonstrate set performance; require the contractor to provide a load bank suitable for continuous duty, full-load operation of the engine-generator, and require complete mechanical and electrical records and fully describe witnessed testing.
- *Automatic transfer switch:* Test for proper operation in all modes.

Field Tests

The following tests are required:

- *480-V wiring:* With contactors closed, measure and record the insulation resistance to ground. Motors may be connected. Do not record or accept "infinity." Use a 500-V "megger" (megohm meter).
- *480-V motors and 230-V main pump motors:* Separately measure and record the insulation resistance to ground. Do not record infinity. Reject any measurements below 1 MΩ. Use a 500-V megger.
- *Circuit breaker solid-state trip units:* Have an independent, factory-certified testing organization field test and certify accuracy, operation, and adjustment.
- *Power sources:* Check the voltage at the service, at the lighting transformer, and at the generator terminals for amplitude and balance (loaded and unloaded). If the voltage is not within 5% of nominal or if the unbalance exceeds 1%, notify the power company, adjust generator voltage regulator, or reconnect taps as applicable.
- *Tighten all electrical connections:* Note that circuit breakers inherently protect against single phasing except in the case of a loose connection.
- *Motor phase currents:* Measure these currents for each pump motor at full load and check for overloading and imbalance. Note that a 1% voltage imbalance will cause an approximately 8% current imbalance. Excessive current imbalance will cause motor overheating. Consult with the equipment supplier when in doubt.
- *Motor and generator rotation:* Correct rotation, if necessary, by interchanging any two-line conductors on three-phase units.
- *Operational test:* Take nothing for granted. Check and test everything for proper operation: prevention of single phasing, lights, GFCI outlet circuit protection, each mode of the automatic transfer switch, standby generator alarms and shut-downs, motor space heaters, etc.
- *Motor protection list:* Record the following data and verify correctness to ensure proper motor protection for *each* motor on the job:

  (1) Unit name or designation
  (2) Motor nameplate data (horsepower, voltage, service factor, current, and temperature rating)
  (3) Circuit breaker or fuse rating
  (4) Circuit breaker setting, if any
  (5) Starter size
  (6) Overload relay manufacturer and heater catalog number
  (7) Capacitor size, if any.

Obtain a copy of the overload relay manufacturer's heater table with instructions for selecting heaters. Check each motor branch circuit for proper protection and adjustments.

9-13. References

1. *CSI Documents MP-2-1, Master Format, Master List of Section Titles and Numbers (1983 ed.)* and *MP-2-2 Section Format (1981 ed.),* The Construction Specifications Institute, Alexandria, VA.
2. Kaufman, J. E., and H. Haynes (Eds.), *IES Lighting Handbook,* Illuminating Engineering Society of North America, New York (1981).
3. *Generator Set Sizing Guide,* Bulletin LE BX 6220. Caterpillar Inc., Peoria, IL.
4. *Short-Circuit Current Calculation,* General Electric Co., Schenectady, NY.
5. IEEE Standard 242-1975, *Recommended Practice for Protection and Coordination of Industrial and Commercial Power Systems,* The Institute of Electrical and Electronic Engineers, Inc., Wiley-Interscience, Piscataway, NJ (1976).
6. IEEE Standard 141-1976, *Recommended Practice for Electric Power Distribution for Industrial Plants,* The Institute of Electrical and Electronics Engineers, Inc., Wiley-Interscience, Piscataway, NJ (1976).

Chapter 10

Performance of Centrifugal Pumps

GEORGE TCHOBANOGLOUS

CONTRIBUTORS
Richard O. Garbus
Robert J. Hart
Carl W. Reh
Lowell G. Sloan
Earle C. Smith

The purpose of this chapter is to introduce those fundamentals of centrifugal pump theory that are useful for background and sometimes necessary for selecting and specifying centrifugal pumps in water and wastewater pumping applications. A corresponding body of theory for positive-displacement pumps can be found in the literature [1, 2, 3].

This introduction includes (1) the general classifications of centrifugal pumps, (2) pump application terminology and usage, (3) pump operating characteristics, (4) cavitation and net positive suction head, (5) pump characteristic curves and operating ranges, and (6) an introduction to pumping system analysis.

10-1. Classification of Centrifugal Pumps

In colloquial usage in the United States, a "centrifugal pump" is any pump in which the fluid is energized by a rotating impeller, whether the flow is radial, axial, or a combination of both (mixed). Strictly defined (as in European practice), a centrifugal pump is a radial-flow pump only. But colloquial usage is followed here, and thus centrifugal pumps are divided into three groups:

- Radial-flow pumps
- Mixed-flow pumps
- Axial-flow or propeller pumps.

These classifications are derived from the manner in which the fluid moves through the pump (see Figure 10-1). Thus, the fluid is displaced radially in a radial-flow pump, axially in an axial-flow pump, and both radially and axially in a mixed-flow pump. The physical features of centrifugal pumps are described in detail in Chapter 11.

10-2. Pump Application Terminology, Equations, and Performance Curves

The basic terminology, equations, and curves for defining pump performance and solving pump problems are given in this section. The development of the pump performance curves, which are used to define the operating characteristics of a given pump, is reviewed briefly at the end of this section (this subject is covered in more detail in Chapter 12).

Capacity

The capacity (flowrate, discharge, or Q) of a pump is the volume of liquid pumped per unit of time, usually measured in SI units in cubic meters per second for large pumps or liters per second and cubic meters per

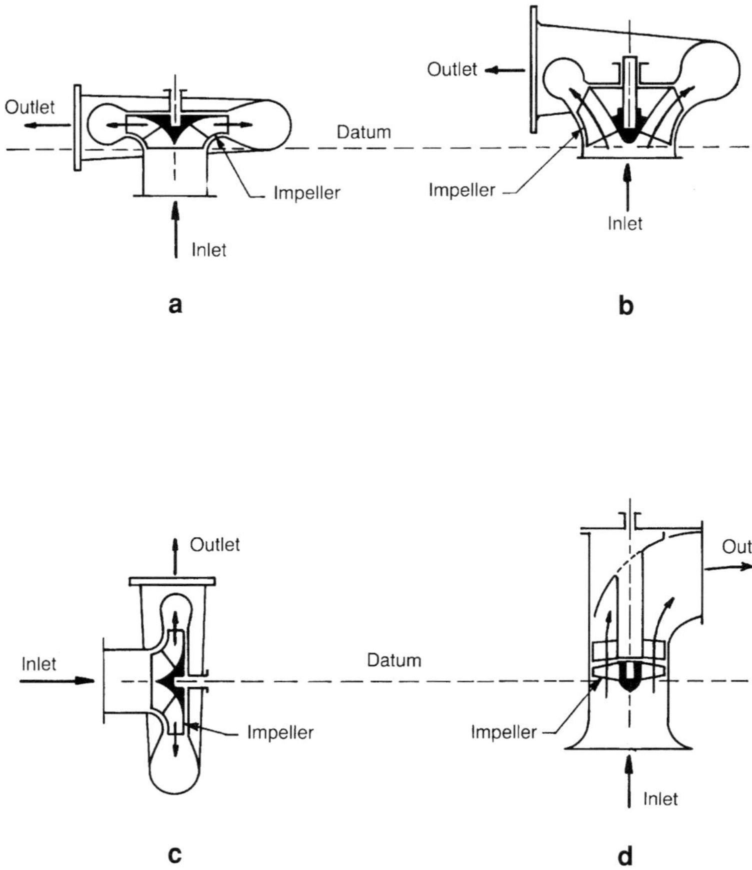

Figure 10-1. Typical flow paths in centrifugal pumps. (a) Radial flow, vertical; (b) mixed flow; (c) radial flow, horizontal; (d) axial flow.

hour for small pumps. In U.S. customary units, the capacity of a pump is expressed in gallons per minute, million gallons per day, or cubic feet per second. Equivalent units of measurement are given on the inside back cover and in Appendix A.

Head

The term head (h or H) is the elevation of a free surface of water above (or below) a reference datum (see Figures 10-2 and 10-3). For centrifugal pumps, the reference datum varies with the type of pump, as shown in Figure 10-1.

In accordance with the standards of the Hydraulic Institute [1], distances (heads) above the datum are considered positive and distances below the datum are considered negative. Each term, defined graphically in Figures 10-2 and 10-3, is expressed as the height of a water column in meters (feet) of water. H is used for total head, whereas h is used for head from the datum or for headloss. The subscripts s and d denote the pump suction and discharge, respectively. Other subscripts are defined as follows:

Total static head (H_{stat}): The total static head is the difference in elevation in meters (feet) between the water level in the wet well and the water level at discharge ($h_d - h_s$).

Static suction head (h_s): The static suction head is the difference in elevation between the wet well liquid level and the datum elevation of the pump impeller. If the wet well liquid level is below the pump datum, as in Figure 10-3, it is a static suction lift, so h_s is negative.

Static discharge head (h_d): The static discharge head is the difference in elevation between the discharge liquid level and the pump datum elevation.

Manometric suction head (h_{gs}): The suction gauge reading is expressed in meters (feet) measured at the

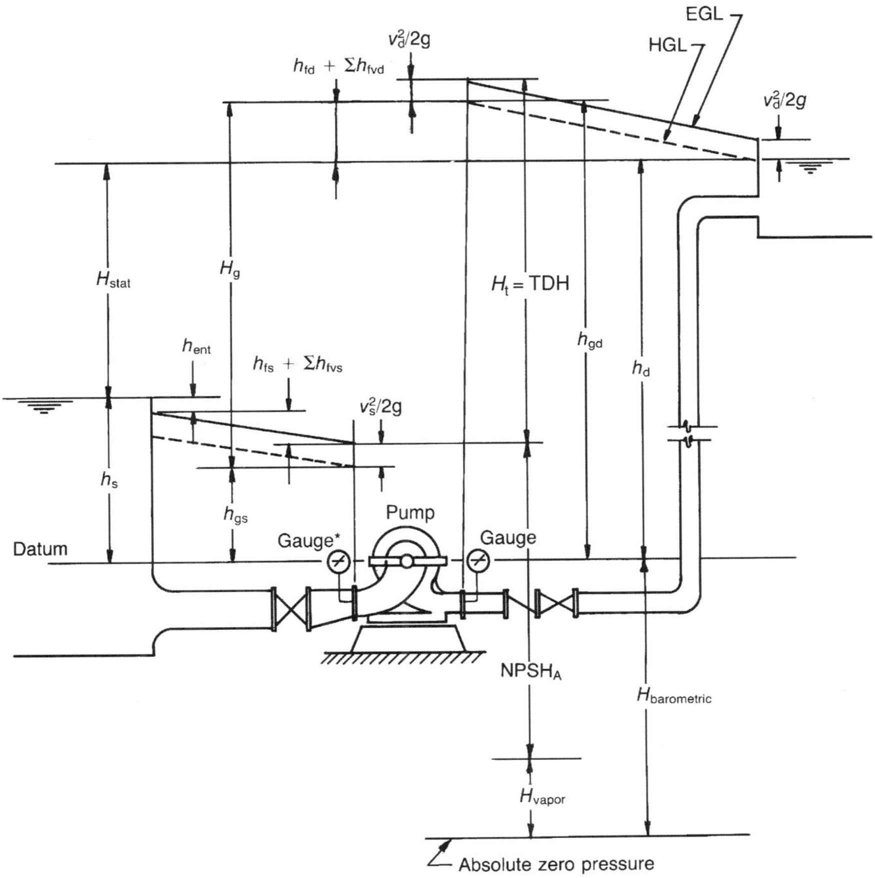

Figure 10-2. Terminology for a pump with a positive suction head. (*) The gauge is located to show theoretical pressures at the inlet and outlet flanges; see "Field Pump Tests" in Section 16-6 for practical gauge locations.

suction nozzle of the pump and referenced to the pump datum elevation and atmospheric pressure.

Manometric discharge head (h_{gd}): The discharge gauge reading is expressed in meters (feet) measured at the discharge nozzle of the pump and referenced to the centerline of the pump impeller. The gauge reading is the height that a water column would attain in a vertical pipe. It is also the distance to the hydraulic gradeline (shown dashed in Figures 10-2 and 10-3).

Manometric head (H_g): This is the increase of pressure head, expressed in meters (feet) generated by the pump ($h_{gd} - h_{gs}$).

Friction headloss (h_{fs}, h_{hd}): This is the head of water that must be supplied to overcome the frictional loss in the pipe. The frictional headloss in the suction (h_{fs}) and discharge (h_{fd}) piping systems can be computed with the Hazen–Williams or Darcy–Weisbach equations (Equations 3-9 and 3-10).

Velocity head ($v^2/2g$): The velocity head is the kinetic energy in the liquid being pumped at any point in the system. The energy gradeline (shown solid in Figures 10-2 and 10-3) is always above the hydraulic or piezometric or manometric gradeline (dashed line) by $v^2/2g$. The velocity head in the discharge pipe, $v_d^2/2g$ is lost if the pipe discharges freely in air or if it discharges abruptly below the surface of a reservoir. Some of the velocity head can be recovered if turbulence is inhibited by a gradually expanding section, but this is ordinarily impractical with the pipe velocities normally encountered (i.e., up to 2.5 m/s or 8 ft/s).

Fitting and valve losses (h_{fvs}, h_{fvd}): As a fluid flows through fittings and valves, energy is lost due to eddy formation and turbulence. Because the head lost in fittings and valves is small compared with the friction loss in long piping systems, the losses in fittings and valves are sometimes called "minor losses" and often

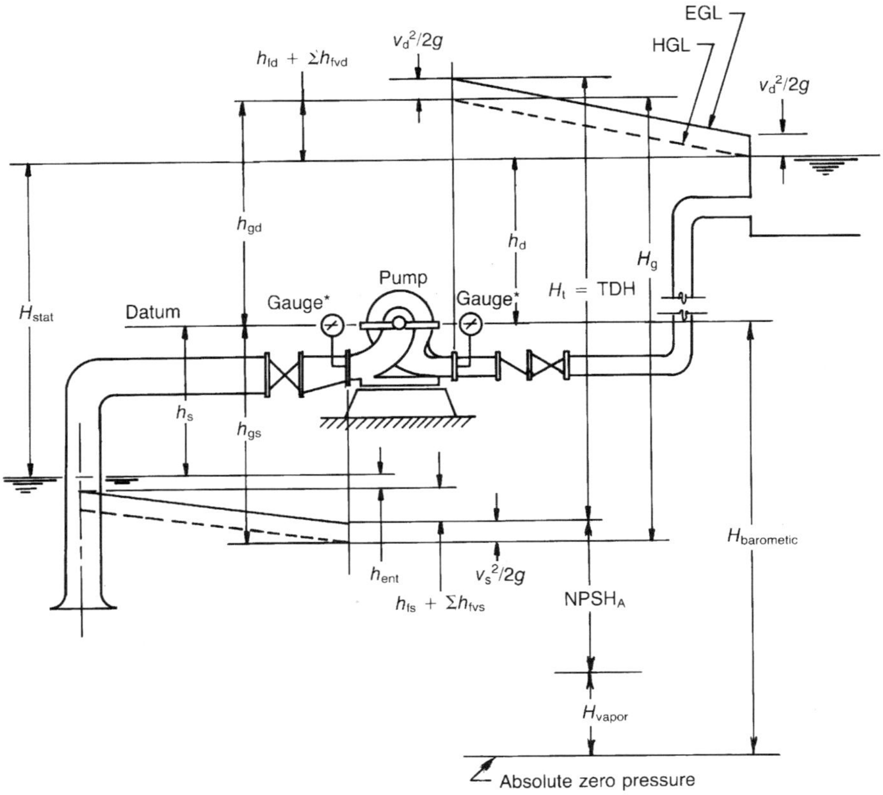

Figure 10-3. Terminology for a pump with a negative suction head. (*) The gauge location is to show theoretical pressures at the inlet and outlet flanges; see "Field Pump Tests" in Section 16-6 for practical gauge locations.

ignored. But the lengths of pipes within a pumping station are short, and the total headloss through the fittings and valves is likely to be greater than the pipe friction loss. Regardless of the shortness of pipe length, both the frictional headless and the "minor losses" should always be computed.

The loss of head through each individual fitting or valve is estimated by using Equation 3-15 ($h = Kv^2/2g$) with K values taken from Tables B-6 and B-7 (or from the literature). The total fitting and valve losses in the piping system are determined by summing the individual losses for each fitting and valve, but, as explained in Section 3-4, the total may in reality be somewhat less or much greater than this sum.

Total dynamic head (H_t or *TDH*): The total dynamic head is the head against which the pump must work. It is determined by adding the static suction and discharge head (with respect to signs), the frictional headlosses, the velocity heads, and the fitting and valve headlosses. The expression for determining the total dynamic head for the pumps shown in Figures 10-2 and 10-3 is given by Equations 10-1, 10-2, and 10-3.

$$H_t = h_{gd} - h_{gs} + \frac{v_d^2}{2g} - \frac{v_s^2}{2g} \tag{10-1}$$

where

$$h_{gd} = h_d + h_{fd} + \Sigma h_{fvd} \tag{10-2}$$

and

$$h_{gs} = h_s - h_{ent} - h_{fs} - \Sigma h_{frs} - \frac{v_s^2}{2g} \tag{10-3}$$

Substituting Equations 10-2 and 10-3 into Equation 10-1 and noting that $h_d - h_s = H_{stat}$,

$$H_t = H_{stat} + h_{ent} + h_{fs} + h_{fd} + \Sigma h_{frs} + \Sigma h_{fvd} + \frac{v_d^2}{2g} \tag{10-4}$$

Some designers consider all headlosses except pipe friction to be "minor" losses, h_m, and rewrite Equation 10-4 as

$$H_t = H_{stat} + h_{fs} + h_{fd} + \Sigma h_m \quad (10\text{-}5)$$

Power

Output Power

The power output of a pump is the energy delivered by the pump to the fluid. In SI units, the power output is defined as

$$P = \gamma QH = \frac{qH}{102} \quad (10\text{-}6a)$$

where P is the water power in kilowatts, γ is the specific weight of the fluid in kilonewtons per cubic meter (see Table A-8), Q is the flowrate in cubic meters per second, H is the total dynamic head in meters, q is the flowrate in liters per second, and 102 is a conversion factor for water at 15 to 20°C. In U.S. customary units, power output is defined as

$$P = \frac{\gamma QH}{550} = \frac{WH}{33{,}000} = \frac{qH}{3960} \quad (10\text{-}6b)$$

where P is the water power in horsepower, γ is the specific weight of the fluid in pounds per cubic foot (see Table A-9), Q is the flowrate in cubic feet per second, H is the total dynamic head in feet, 550 is the conversion factor from foot-pounds per second to horsepower, W is weight of water pumped in pounds per minute, 33,000 is the conversion factor from foot-pounds per minute to horsepower, q is flowrate in gallons per minute, and 3960 is used to convert gallon feet per minute to horsepower.

Input Power

Pump performance is measured in terms of the flowrate that a pump can discharge against a given head at a given efficiency. The pump capacity depends on the design, and design information is furnished by the pump manufacturer in a series of curves for a given pump. Pump efficiency, E_p, is the ratio of the useful power output (water kilowatts [wkW] or water horsepower [whp]) to the power input to the pump shaft. Hence, the brake power (bkW) that must be supplied by the drive is, in SI units,

$$\text{bkW} = \gamma QH/E_p = \text{wKW}/E_p \quad (10\text{-}7a)$$

The power input (bhp) in U.S. customary units is

$$\text{bhp} = \frac{\gamma QH}{550E_p} = \frac{qH}{3960E_p} = \frac{\text{whp}}{E_p} \quad (10\text{-}7b)$$

Pump efficiencies usually range from 20 to 85% and increase with the size of the pump (see Figure 10-4). Energy losses in a pump are volumetric, mechanical, and hydraulic. Volumetric losses are those of leakage through the small clearances between wearing rings in the pump casing and the rotating element. Mechanical losses are caused by mechanical friction in the stuffing boxes and bearings, by internal disc friction, and by fluid shear. Frictional and eddy losses within the flow passages account for the hydraulic losses.

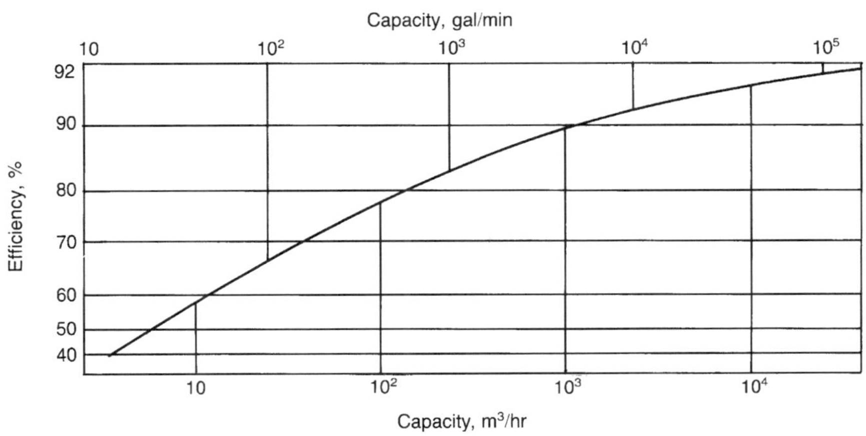

Figure 10-4. Maximum pump efficiency attainable at the best operating point.

Example 10-1
Evaluation of Pump Performance

Problem: Water is pumped by a radial-flow centrifugal pump at 20°C (68°F) through a new cement-mortar-lined ductile iron piping (DIP) system such as shown in Figure 10-3, except that there are two 90° elbows and a gate valve in the suction piping, and a gate valve and three 90° elbows in the discharge piping. Other data known, measured, or computed are

- Discharge, $Q = 0.1$ m^3/s (3.54 ft^3/s)
- Total static life, $H_{stat} = 14$ m (45.93 ft)
- Static suction lift, $h_s = 2$ m (6.56 ft)
- Power input to pump shaft, $P = 22.8$ kW (30.5 hp)
- Suction piping = 3.3 m of 300-mm (11 ft of 12-in.) pipe
- Discharge piping = 47.6 m of 250-mm (156 ft of 10-in.) pipe.

Evaluate: (1) the total dynamic head, (2) the power delivered to the water, and (3) the efficiency of the pumping unit.

Solution: Because pipe is exposed and joints are flanged within the pumping station, Class 53 is needed (see Tables B-1 and B-2 for dimensions). Outside, the pipe is buried, joints can be mechanical or push on, and the pipe can be Class 50 with a net ID of 264 mm (10.4 in.), which includes the lining (see the DIPRA handbook [4]). Because most of the discharge piping can be Class 50, ignore the slightly smaller diameter of the 250-mm (10-in.) Class 53 exposed piping within the pumping station.

Total dynamic head: Find velocity, velocity head, and friction losses in suction and discharge piping from Equation 10-4.

SI Units

For the suction piping:

$$\begin{aligned} ID &= 312 \text{ mm (Table B-1)} \\ A_s &= 0.0763 \text{ m}^2 \text{ (Table B-1)} \\ v_s &= 1.31 \text{ m/s} \\ v_s^2/2g &= 0.0875 \text{ m} \end{aligned}$$

U.S. Customary Units

$$\begin{aligned} ID &= 12.3 \text{ in. (Table B-2)} \\ A_s &= 0.822 \text{ ft (Table B-2)} \\ v_s &= 4.31 \text{ ft/s} \\ v_s^2/2g &= 0.288 \text{ ft} \end{aligned}$$

Find the entrance loss assuming a bell mouth entrance is used (see Table B-6).

$$h_{ent} = 0.05 \times 0.0875 = 0.004 \text{ m} \qquad h_{ent} = 0.05 \times 0.288 = 0.014 \text{ ft}$$

Find the friction headloss using Equations 3-9a and 3-9b.

$$\begin{aligned} h_{fs} &= 10{,}700(Q/C)^{1.85} D^{-4.87} \\ h_{fs} &= 10{,}700(0.1/145)^{1.85} \times (0.3121)^{-4.87} \\ h_{fs} &= 4.41 \text{ m/1000 m} \\ h_{fs} &= 4.41(3.3 \text{ m/1000 m}) = 0.015 \text{ m} \end{aligned}$$

$$\begin{aligned} h_{fs} &= 10{,}500(Q/C)^{1.85} D^{-4.87} \\ h_{fs} &= 10{,}500(1590/145)^{1.85} \times (12.3)^{-4.87} \\ h_{fs} &= 4.34 \text{ ft/1000 ft} \\ h_{fs} &= 4.34(11 \text{ ft/1000 ft}) = 0.048 \text{ ft} \end{aligned}$$

The "minor" losses are found using Equation 3-16 ($h = Kv^2/2g$). Note that there is an elbow as well as a gate valve in the suction piping. Obtain K values from Tables B-6 and B-7.

$$\begin{aligned} h(\text{bend}) &= 0.25 \times 0.0875 = 0.022 \text{ m} \\ h(\text{gate}) &= 0.2 \times 0.0875 = 0.018 \text{ m} \\ \Sigma h_{fvs} &= 0.040 \text{ m} \end{aligned}$$

$$\begin{aligned} h(\text{bend}) &= 0.25 \times 0.288 = 0.072 \text{ ft} \\ h(\text{gate}) &= 0.2 \times 0.288 = 0.058 \text{ ft} \\ \Sigma h_{fvs} &= 0.130 \text{ ft} \end{aligned}$$

For the discharge piping:

ID = 0.264 m	ID = 10.4 in.
A_d = 0.0548 m^2	A_d = 0.589 ft^2
v_d = 1.82 m/s	v_d = 6.01 ft/s
$v_d^2/2g$ = 0.169 m	$v_d^2/2g$ = 0.561 ft

Find the friction headloss from Equations 3-9a and 3-9b.

$$h_{fd} = 10{,}700(0.1/145)^{1.85}(0.264)^{-4.87} = 9.95 \text{ m}/1000 \text{ m}$$
$$h_{fd} = 9.95(47.6 \text{ m}/1000 \text{ m}) = 0.474 \text{ m}$$

$$h_{fd} = 10{,}500(1590/145)^{1.85}(10.4)^{-4.87} = 9.82 \text{ ft}/1000 \text{ ft}$$
$$h_{fd} = 9.82(156 \text{ ft}/1000 \text{ ft}) = 1.53 \text{ ft}$$

Find the "minor" losses using Equation 3-15.

$h(\text{check}) = 2.20 \times 0.169 = 0.372$ m	$h(\text{check}) = 2.20 \times 0.561 = 1.23$
$h(\text{gate}) = 0.20 \times 0.169 = 0.034$	$h(\text{gate}) = 0.20 \times 0.561 = 0.11$
$h(\text{bends}) = 2(0.25 \times 0.169) = 0.084$	$h(\text{bends}) = 2(0.25 \times 0.561) = 0.28$
$\Sigma h_{fvd} = 0.490$	$\Sigma h_{fvd} = 1.62$

The TDH computed using Equation 10-4 is

H_{stat} = 14.00 m	H_{stat} = 45.93 ft
h_{ent} = 0.004	h_{ent} = 0.014
h_{fs} = 0.015	h_{fs} = 0.048
h_{fd} = 0.474	h_{fd} = 1.53
Σh_{fvs} = 0.040	Σh_{fvs} = 0.130
Σh_{fvd} = 0.490	Σh_{fvd} = 1.62
$v_d^2/2g$ = 0.169	$v_d^2/2g$ = 0.561
TDH = 15.192 m	TDH = 49.833 m
Use 15.2 m	Use 49.9 ft

The slight discrepancies between SI and U.S. values are caused by rounding off coefficients and powers to three significant figures. All of the terms in Equation 10-4 are included in these additions. No allowance was made for the proximity of fittings or for swirling (see Section 3-4). The friction loss in the discharge piping is 0.474 m (1.53 ft), but only about 10% of it would be within the pumping station. The total pipe friction loss within the pumping station, therefore, would be about (0.1) (0.474) + 0.015 = 0.062 m (0.2 ft), whereas all the "minor" losses (due to fittings and valves) total about 0.53 m (1.75 ft)—almost ten times as much.

Power delivered to the water. Use Equations 10-6a and 10-6b.

$$P = \gamma QH = (9.77 \text{ kN/m}^3)(0.1 \text{ m}^3/\text{s})(15.2 \text{ m}) = 14.9 \text{ kW}$$

$$P = \gamma QH/550 = \frac{(62.25 \text{ lb/ft}^3)(3.54 \text{ ft}^3/\text{s})(49.9 \text{ ft})}{(550 \text{ ft}\cdot\text{lb/s})\text{hp}} = 20.0 \text{ hp}$$

Efficiency of the pumping unit. Use Equations 10-7a and 10-7b.

$$E_p = \frac{\text{wkW}}{\text{bkW}} \qquad E_p = \frac{14.9}{22.3} = 65\%$$

$$E_p = \frac{\text{whp}}{\text{bhp}} \qquad E_p = \frac{20.0}{30.5} = 65\%$$

Pump Performance Curves

The head that a pump can produce at various flowrates and rotational speeds is established in pump tests conducted by the pump manufacturer. During testing, the capacity of the pump is varied by throttling a valve in the discharge pipe, and the corresponding head is measured. The results of these tests and other tests with different impeller diameters are plotted to obtain a series of head-capacity (H-Q) curves for the pump at some given speed (see Figure 10-5). Simultaneously, the power input to the pump is measured. The efficiency at various operating points is computed, and these values are also plotted in the same diagram. Together, these curves are known as "pump characteristic curves."

10-3. Pump Operating Characteristics

The operating characteristics of pumps depend on their size, speed, and design. Pumps of similar size and design are produced by many manufacturers, but they vary somewhat because of slight design modifications. The basic relationships that can be used to characterize and analyze pump performance under varying conditions include

- Energy transfer in pumps
- Flow, head, and power coefficients
- Affinity laws
- Specific speed.

Energy Transfer in Radial Centrifugal Pumps

The transfer of energy in centrifugal pumps is based on the momentum principle, which, as applied to centrifugal pumps, can be stated as follows: "A net torque applied to a pump impeller causes a change in the angular momentum of the fluid." It is this change in angular momentum that develops head or pressure as the fluid passes through a pump impeller.

In Figure 10-6, the torque, T, applied to the pump impeller is the difference between the moment of momentum at the inlet and outlet of the impeller. For ideal conditions, the torque is given by

$$T = \frac{\gamma}{g}Q(r_2V_2 \cos \alpha_2 - r_1V_1 \cos \alpha_1) \qquad (10\text{-}8)$$

To obtain the power, the torque is multiplied by the angular velocity.

$$P = T\omega = \gamma QH \qquad (10\text{-}9)$$

where P is the power, T is the torque, and ω is the angular velocity (rotational speed) in radians per second.

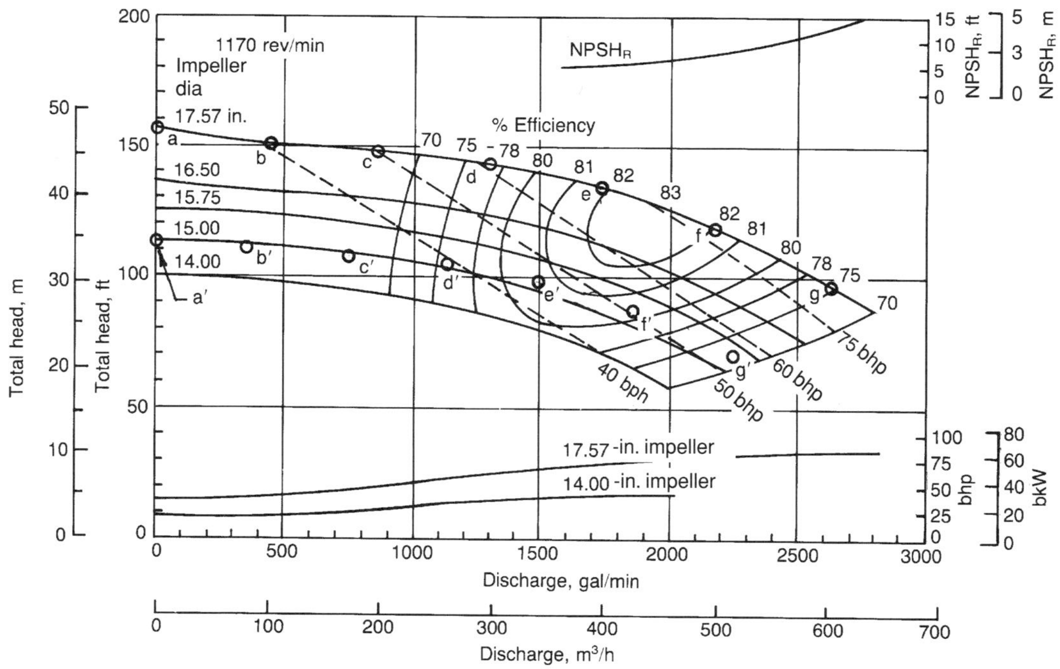

Figure 10-5. Typical pump characteristic curves. Double suction impeller.

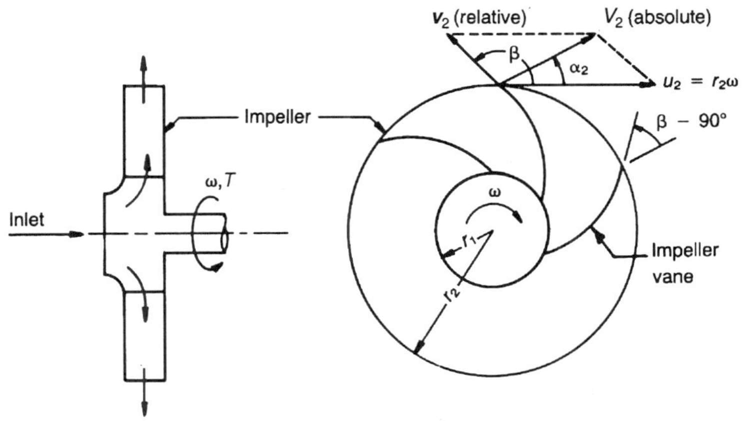

Figure 10-6. Velocity diagrams for a radial-flow pump impeller. *V*, absolute velocity; *v*, relative velocity; *u*, peripheral velocity; α_1, angle to V_1 at the hub; α_2, angle to V_2 at the periphery; β, angle to **v** at the periphery; ω, rotational velocity; *T*, torque.

The theoretical head, *H*, added to the fluid by the pump can be obtained by equating *T* in Equations 10-8 and 10-9

$$\frac{\gamma}{g}Q(r_2V_2 \cos \alpha_2 - r_1V_1 \cos \alpha_1) = \frac{\gamma QH}{\omega}$$

Substituting *u*/ω for *r*

$$H = \frac{(u_2V_2 \cos \alpha_2 - u_1V_1 \cos \alpha_1)}{g} \quad (10\text{-}10)$$

For radial flow, $\cos \alpha_1 = 0$ and

$$H = \frac{u_2V_2 \cos \alpha_2}{g} \quad (10\text{-}11)$$

A similar set of equations can be derived for mixed- and axial-flow pumps. From Equation 10-10, it can be seen that the theoretical head developed by a pump depends only on the increase of the moment of velocity within the impeller and is independent of specific weight and viscosity of the fluid (see the literature [5–11] for a more complete discussion of pump theory).

Flow, Head, and Power Coefficients

In centrifugal pumps, similar flow patterns occur in a series of geometrically similar pumps. By applying the principles of dimensional analysis, the following three independent dimensionless groups can be derived to describe the operation of centrifugal pumps and other rotodynamic machines. Note that the equations are the same in either SI or U.S. customary units. Only the value of the coefficient changes.

$$C_Q = \frac{Q}{nD^3} \quad (10\text{-}12)$$

$$C_H = \frac{H}{n^2D^2} \quad (10\text{-}13a)$$

or, for dimensional correctness,

$$C_H = \frac{gH}{n^2D^2} \quad (10\text{-}13b)$$

$$C_P = \frac{P}{\rho n^3D^5} \quad (10\text{-}14)$$

where *C* is a coefficient and the subscripts Q, H, and P correspond to capacity, head, and power. *Q* is capacity in cubic meters per second (cubic feet per second), *n* is speed in radians per second (revolutions per minute), *D* is impeller diameter in meters (feet), *H* is head in meters (feet), *g* is 9.81 m/s^2 (32.2 ft/s^2), *P* is the power input in kilowatts (horsepower), and ρ is the density in kilograms per cubic meter (slugs per cubic foot). Although Equation 10-13b is dimensionally correct, Equation 10-13a is commonly used in the United States for both SI and U.S. customary units.

Equations 10-12 through 10-14 apply only to "corresponding points"—the operating points at which similar flow patterns occur. Thus, every point on a head-capacity curve for a large pump corresponds to a point on the head-capacity curve of a geometrically similar but smaller pump that operates at the same speed. This correspondence can be corrected for different speeds provided that the speeds differ by no more than about 40%.

Affinity Laws

For a pump operating at two different speeds, the following relationships can be derived from Equations 10-12 through 10-14. (Note that neither diameter nor density changes.)

$$\frac{Q_1}{Q_2} = \frac{n_1}{n_2} \tag{10-15}$$

$$\frac{H_1}{H_2} = \left(\frac{n_1}{n_2}\right)^2 \tag{10-16}$$

$$\frac{P_1}{P_2} = \left(\frac{n_1}{n_2}\right)^3 \tag{10-17}$$

where Q is flowrate, H is head, P is power, n is rotational speed, and subscripts 1 and 2 are *only* for corresponding points. *Note that Equations 10-15 and 10-16 must be applied simultaneously to ensure that point 1 "corresponds" to point 2. Corresponding points fall upon parabolas through the origin. They do not fall upon system H-Q curves.* These relationships, known collectively as "the affinity laws," are used to determine the effect of changes in speed on the capacity, head, and power of a pump.

The affinity laws for discharge and head are accurate as found from actual tests for all types of centrifugal pumps including axial-flow pumps. The affinity law for power is not as accurate because efficiency increases with an increase in the size of the pump. But if the affinity law for power is used, the computed value of power can be corrected using the Moody equation [2, 8, 9], which accounts for efficiency as a function of pump size.

In applying these relationships, remember that they are based on the assumption that the efficiency remains the same when transferring from a given point on one pump curve to a homologous point on another curve. Because the hydraulic and pressure characteristics at the inlet, at the outlet, and through the pump vary with the flowrate, the errors produced by Equation 10-17 may be excessive, although errors produced by Equations 10-15 and 10-16 are very small. The application of these relationships is illustrated in Example 10-2 and in Section 15-3.

Example 10-2
Application of Affinity Laws

Problem: A pump with a normal operating speed of 705 rev/min and a 356-mm (14-in.) diameter impeller was tested at rotational speeds of 705, 625, 550, 450, and 350 rev/min. Using the head and capacity data collected at 705 rev/min, generate pump curves for rotational speeds of 625, 550, 450, and 350 rev/min and compare the curves generated with the measured data points.

	SI Units		**U.S. Customary Units**	
	Discharge (m^3/h)	Head (m)	Discharge (gal/min)	Head (ft)
705 rev/min				
	1596	5.91	7025	19.4
	1361	7.28	5994	23.9
	1140	8.66	5020	28.4
	1002	9.39	4410	30.8
	866	10.21	3550	33.5
	363	11.28	1600	37.0
	0	13.11	0	43.0
625 rev/min				
	1420	4.57	6250	15.0
	1136	6.25	5000	20.5
	908	7.32	4000	24.0
	681	7.92	3000	26.0
	454	8.63	2000	28.3
	227	9.45	1000	31.0
	0	10.36	0	34.0

	Discharge (m^3/h)	Head (m)	Discharge (gal/min)	Head (ft)
550 rev/min				
	1249	3.51	5500	11.5
	943	5.12	4150	16.8
	772	5.82	3400	19.1
	454	6.55	2000	21.5
	363	6.80	1600	22.3
	0	7.92	0	26.0
450 rev/min				
	1022	2.44	4500	8.0
	795	3.35	3500	11.0
	568	3.96	2500	13.0
	227	4.72	1000	15.5
	0	5.43	0	17.8
350 rev/min				
	568	3.13	2500	7.0
	472	2.41	2080	7.9
	363	2.59	1600	8.5
	0	3.05	0	10.0

Solution: Plot the pump head-capacity (H-Q) data obtained at 705 rev/min (shown as solid circles in Figure 10-7) and draw a smooth pump curve (shown as a solid line) through the plotted points.

Develop H-Q curves for the other rotational speeds from the smooth pump curve by applying the affinity laws for discharge and head simultaneously. For example, the discharge and head values at the point on the 625-rev/min curve that corresponds to point a on the 705-rev/min pump curve, with coordinate values of 1250 m^3/h and 8.10 m (5504 gal/min and 26.6 ft), are determined as follows.

Using Equation 10-15

$$Q_2 = Q_1/(n_1/n_2) = 1250/(705/625) = 1108 \text{ m}^3\text{/h}$$

Using Equation 10-16

$$H_2 = H_1/(n_1/n_2)^2 = 8.10/(705/625)^2 = 6.37 \text{ m}$$

So, point b with coordinates of 1108 m^3/h and 6.37 m (4875 gal/min and 20.9 ft) on the 625-rev/min curve *corresponds* to point a on the 705-rev/min curve. Enough other *corresponding points* are computed in the same manner to allow the entire curve to be drawn for 625 rev/min. The process is repeated for curves at 550, 450, and 350 rev/min, which pass through points c, d, and e, respectively. The curves are shown as dashed lines to indicate that they are derived mathematically from the curve at 705 rev/min. The measured values for head and capacity are plotted in Figure 10-7 as open circles. The correspondence between measured values and computed pump curves is excellent.

Points f, g, h, i, and j in Figure 10-7 are unique. Because the flow is zero at these points, Equation 10-15 can be ignored and Equation 10-16 can be used alone. However, the only time that Equations 10-15 and 10-16 need not be solved simultaneously is when the flow is zero. The errors in the four points derived from point f are 0.6, 0.8, 1.7, and 5.90%, respectively. In variable-speed pumping, pumps rarely operate at less than 60% of full speed, so in practice the errors in using the affinity laws for calculating head and flow would rarely exceed 2 or 3%.

Points a, b, c, d, and e lie on a parabola that passes through the origin. The equation for the affinity parabola can be found by solving Equations 10-15 and 10-16 simultaneously to eliminate n.

$$\left(\frac{Q_1}{Q_2}\right)^2 = \left(\frac{h_1}{h_2}\right)^2 = \frac{H_1}{H_2}$$

So $H_1 = H_2 (Q_1/Q_2)^2$, which is a parabola through point O. Hence, all *corresponding points* lie on parabolas that pass through point O, the origin. Station H-Q curves are also parabolas, but because such parabolas do not pass through the origin (unless the static head is zero), it follows that corresponding points *cannot* lie on a station H-Q curve. Expressed in the simplest terms, the foregoing statement means that *the flow from a pumping station (with a static lift) is not proportional to the speed of the pump*—a conclusion that might appear, superficially, to be at odds with Equation 10-15.

Approximate Relationships for Radial-Flow Pumps

To cover a wide range of flows with a minimum number of pump casings and impeller designs, manufacturers customarily offer a range of impeller diameters for each size of casing (see Figure 10-5). In general, these radial-flow impellers are identical as cast, but the size is reduced by machining the impeller to a smaller diameter. Equations 10-12 through 10-14, written for pumps of different sizes, can be modified into equations for impellers of reduced diameter in the same pump. Dividing Equation 10-13 for an impeller of diameter D_1 by the same equation for an impeller of diameter D_2 and canceling C_H and n^2 (because density, viscosity, and rotational speed are the same) gives

$$\frac{H_1}{H_2} = \left(\frac{D_1}{D_2}\right)^2 \tag{10-18}$$

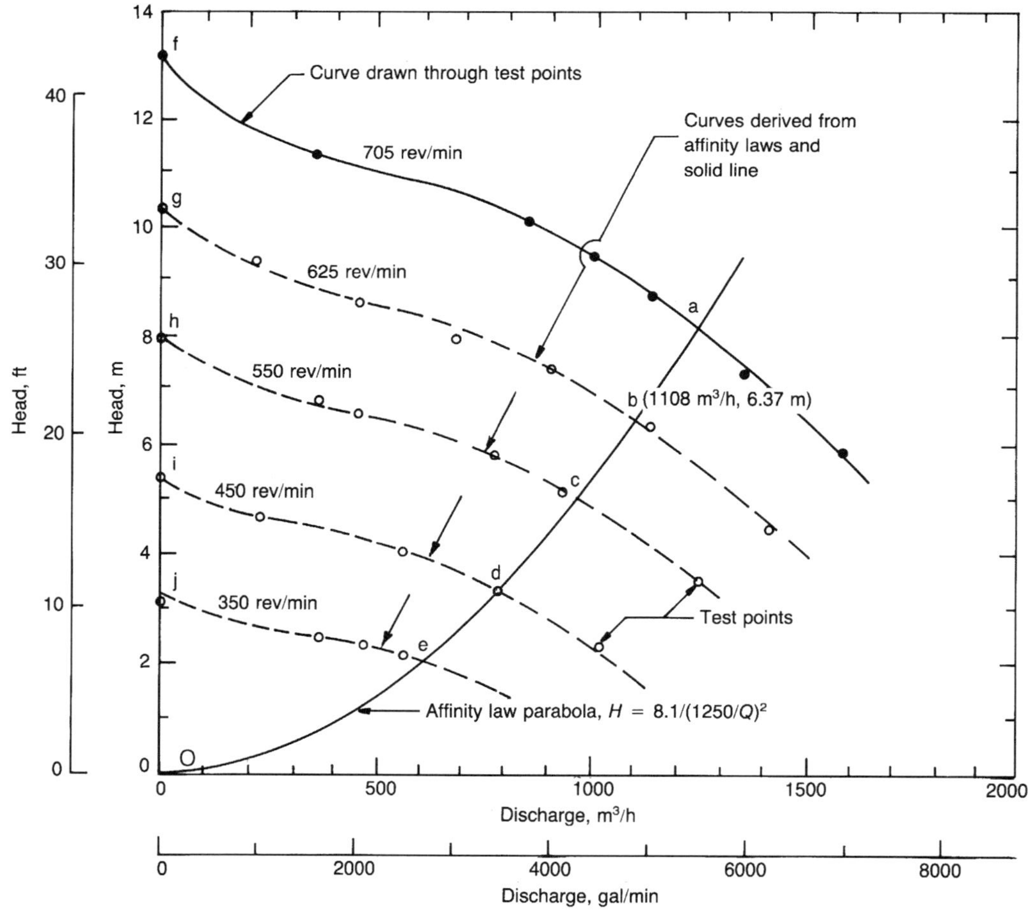

Figure 10-7. Pump curves developed from the affinity laws compared with test data.

The same analysis applied to Equation 10-12 gives

$$\frac{Q_1}{Q_2} = \left(\frac{D_1}{D_2}\right)^3 \quad (10\text{-}19)$$

Equation 10-19 applies only to two pumps of different size (i.e., the same scale-up in three dimensions). For a single pump, the discharge area for a radial-flow impeller is a function of impeller width and volute size, and both are nearly constant even if the impeller diameter is reduced. Because area is proportional to diameter squared, Equation 10-19 can be converted to an equation for a single pump by factoring out $(D_1/D_2)^2$, which leaves

$$\frac{Q_1}{Q_2} = \frac{D_1}{D_2} \quad (10\text{-}20)$$

Power is a function of head times discharge, so multiplying Equation 10-18 and Equation 10-20 gives

$$\frac{P_1}{P_2} = \left(\frac{D_1}{D_2}\right)^3 \quad (10\text{-}21)$$

These relationships, when used to predict the effect of changing the diameter of a radial-flow impeller, are somewhat less accurate than the affinity laws because the angle of the blade decreases slightly and the clearance between impeller and casing increases (which changes the geometry) as the diameter is reduced (see Figure 10-6). Two or more impeller designs are often available (each in a range of sizes) for the same casing, but if these impellers are not geometrically similar, the affinity laws and the previous relationships are invalid.

It should be noted that the modified affinity laws are less accurate for trimming impellers of mixed- and axial-flow pumps. In most designs, the pump is somewhat more effective when the pump impeller is trimmed than would be predicted using the modified affinity laws.

Example 10-3
Effect of Changes in Impeller Diameter and Speed

Problem: Data from Figure 10-5 for a pump with an impeller diameter of 0.4463 m (17.57 in.) operating at 1170 rev/min are tabulated as follows.

SI Units

Point	Discharge (m^3/h)	Head (m)
a	0	47.6
b	100	46.3
c	200	45.3
d	300	44.0
e	400	41.0
f	500	36.5
g	600	29.8

U.S. Customary Units

Point	Discharge (gal/min)	Head (ft)
a	0	156
b	440	152
c	880	149
d	1320	144
e	1760	135
f	2200	120
g	2640	98

(1) Develop a new head-capacity curve for the same pump fitted with a new impeller 0.3810 m (15 in.) in diameter. (2) Compare the results with the test curve for the new impeller shown in Figure 10-5. (3) Find the increase in rotational speed that would be required for the new impeller to match the performance of the original impeller at 1170 rev/min.

Solution:

H-Q for the smaller impeller. Compute this using Equations 10-20 [$Q_2 = Q_1(D_2/D_1)$] and 10-18 [$H_2 = H_1(D_2/D_1)^2$]. Sample calculations for point d are

$$Q_2 = 300(0.3810/0.4463) = 256 \text{ m}^3/\text{h} \qquad Q_2 = 1320(15/17.57) = 1130 \text{ gal/min}$$

$$H_2 = 44.0(0.3810/0.4463)^2 = 32.1 \text{ m} \qquad H_2 = 144(15/17.57)^2 = 105 \text{ ft}$$

Compare the results to Figure 10-5. The computed values of points a to g for the new 15.00-in. impeller diameter are summarized below.

SI Units

Point	Discharge (m^3/h)	Head (m) Computed	Head (m) Scaled
a'	0	34.4	34.4
b'	85	33.8	34.4
c'	171	32.9	33.2
d'	256	32.0	31.7
e'	341	29.9	29.0
f'	427	26.5	24.7
g'	512	21.3	—

U.S. Customary Units

Point	Discharge (gal/min)	Head (ft) Computed	Head (ft) Scaled
a'	0	113	113
b'	376	111	113
c'	751	108	109
d'	1130	105	104
e'	1500	98	95
f'	1880	87	81
g'	2250	70	—

Compare the computed with the scaled heads in the last two columns or compare calculated points a' to g' in Figure 10-5 with the 15.00 curve. Minor differences (up to 0.6 m or 2 ft) are due to errors in plotting and scaling. Larger differences are due to the fact that actual losses in the pump are not considered in Equations 10-18 and 10-20.

Rotational speed for the trimmed impeller. Use Equation 10-16 ($n_2 = n_1\sqrt{H_2/H_1}$) at zero flow to estimate the new rotational speed required to obtain the original flowrate with the smaller impeller.

$$n_2 = 1170\sqrt{47.6/34.4} = 1376 \text{ rev/min}$$

$$n_2 = 1170\sqrt{156/113} = 1375 \text{ rev/min}$$

Specific Speed

For a geometrically similar series of pumps operating under similar conditions, the diameter term in Equations 10-12 and 10-13a can be eliminated by dividing the square root of Equation 10-12 by the three-fourths power of Equation 10-13a:

$$n_s = \frac{C_Q^{1/2}}{C_H^{3/4}} = \frac{(Q/nD)^3}{(H_t/n^2D^2)^{3/4}} = \frac{nQ^{1/2}}{H_t^{3/4}} \qquad (10\text{-}22)$$

where n_s is the specific speed (also called the "type number" in Europe), n is in revolutions per minute (not radians per second), Q is discharge (customarily in cubic meters per second but sometimes in liters per second or cubic meters per hour), and H_t is the total dynamic head in meters. In spite of being dimensionally incorrect, it is this form of the equation that is used in the United States. (A dimensionally correct expression would be derived by using Equation 10-13b instead of 10-13a.) In U.S. customary units, n_s is the specific speed, n is the revolutions per minute, Q is discharge in gallons per minute, and H_t is the total dynamic head in feet. The relation between specific speeds for various units of discharge and head is given in Table 10-1, wherein the numbers in bold type are those customarily used.

Table 10-1. Equivalent Factors for Converting Values of Specific Speed Expressed in One Set of Units to the Corresponding Values in Another Set of Units

Quantity	Expressed in units of				
N	rev/min	rev/min	rev/min	rev/min	rev/min
Q	L/s	m^3/s	m^3/h	gal/min	ft^3/s
H	m	m	m	ft	ft
	1.0	**0.0316**	1.898	**1.633**	0.0771
	31.62	**1.0**	60.0	**51.64**[a]	2.437
	0.527	**0.0167**	1.0	**0.861**	0.0406
	0.612	**0.0194**	1.162	**1.0**	0.0472
	12.98	**0.410**	24.63	**21.19**	1.0

[a] For example, if the specific speed is expressed in metric units (e.g., N = rev/min; Q = m^3/s; and H = m), the corresponding value expressed in U.S. customary units (e.g., N = rev/min; Q = gal/min; and H = ft) is obtained by multiplying the metric value by 51.64.

For any pump operating at any given speed, Q and H must be taken at the point of maximum efficiency. When using Equation 10-22 for pumps with double suction impellers, one-half of the discharge is used unless otherwise noted. For multistage pumps, the head is the head per stage.

The variations in maximum efficiency to be expected with variations in size, capacity, and specific speed are shown in Figure 10-8 for single-stage, single-entry centrifugal pumps. The progressive changes in impeller shape as the specific speed increases are shown along the bottom of Figure 10-8. The efficiency increases with the size of the pump because of a reduction in the friction losses in the pump shrouds and a corresponding drop in the volumetric losses. The greatest efficiency is achieved in single-stage, single-entry pumps with a volute casing [7]. In general, for a given rotational speed,

- If the n_s value is low (specific speed <30, or in U.S. units, <1500), Q must be low and H must be high.
- If the n_s value is intermediate (specific speed of 30 to 80, or in U.S. units 1500 to 4000), Q and H must be of intermediate value.
- If the n_s value is high (specific speed >80, or in U.S. units, >4000), Q must be high and H must be low.

Example 10-4
Use of Specific Speed in Pump Selection

Problem: A flow of 0.01 m^3/s (158 gal/min) must be pumped against a head of 20 m (66 ft). The pump is to be driven with an electric motor at a speed of 1750 rev/min. What type of centrifugal pump should be selected, and what would be the corresponding efficiency?

Solution: From Equation 10-22 ($n_s = nQ^{1/2}/H^{3/4}$), the specific speed is

SI Units

$$n_s = \frac{1750(0.01)^{1/2}}{20^{3/4}} = 18.5$$

U.S. Customary Units

$$n_s = \frac{1750(158)^{1/2}}{66^{3/4}} = 950$$

From Figure 10-8, a radial centrifugal pump is needed and the expected efficiency is about 62%.

10-4. Cavitation

Cavitation is one of the most serious problems encountered in the operation of pumps because it may cause permanent damage and because the pump performance is reduced.

Cavitation and Its Effects

Cavitation is a potential danger, especially when pumps operate at high speeds or at a capacity much greater or much less than the best efficiency point (bep). Cavitation reduces pump capacity and efficiency and may damage the pump—sometimes very rapidly. It occurs in pumps when the absolute pressure at the inlet of the pump drops below the vapor pressure of the fluid being pumped. At first, air comes out of solution to form tiny bubbles, followed instantly by vapor as the water boils. As the vapor bubbles are transported through the impeller, they reach a zone of higher pressure where they collapse abruptly (see Figure 10-9a). If the collapse occurs on the surface of a solid, the liquid rushing in to fill the vacuous space left by the bubbles impacts tiny areas with tremendous, localized pressures and thereby pits and erodes the surface. Cavitation can also occur wherever local velocities are high, such as in vanes or nearly closed valves. The impeller shown in Figure 10-10a eroded because the absolute pressure at the pump inlet connection was too low. The damage to the suction bell in Figure 10-10b was caused by water recirculated by operating the pump at 35% of its bep discharge. It has been suggested that the pitting and erosion is accelerated by simultaneous chemical attack or that the high-impact pressure causes locally high temperatures that accelerate pitting.

In addition to the pitting and erosion, cavitation can also cause noise and vibration. Noise is produced by the collapse of the vapor bubbles as they enter the region of higher pressure. The vibration is due to the imbalance and surging caused by the uneven distribution of collapsing vapor bubbles.

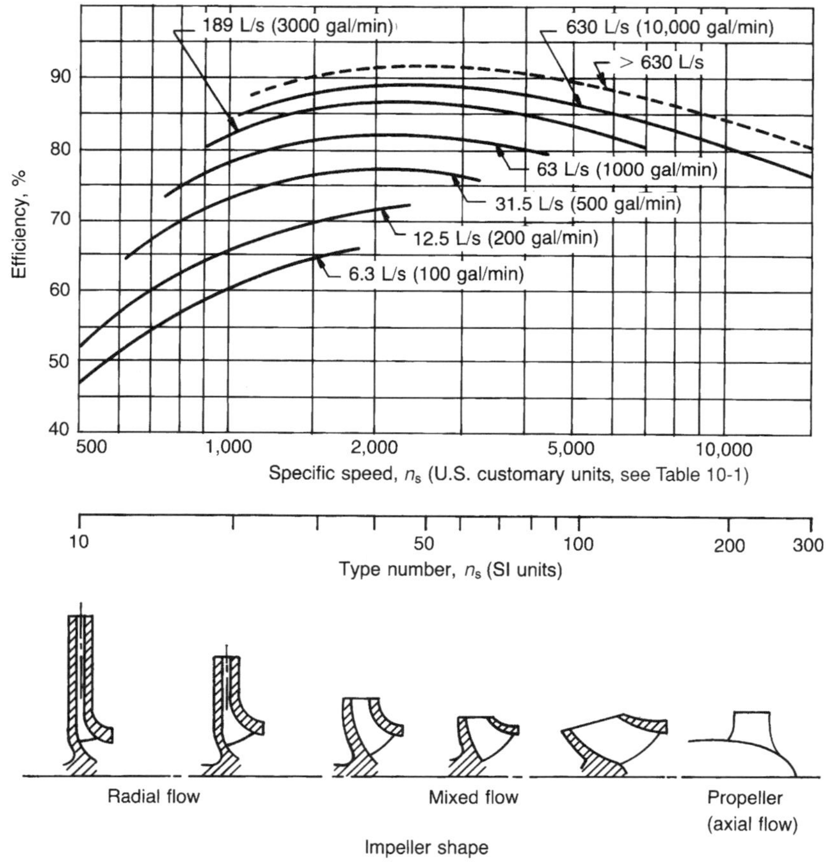

Figure 10-8. Pump efficiency as related to specific speed and discharge. After Dresser Pump Division.

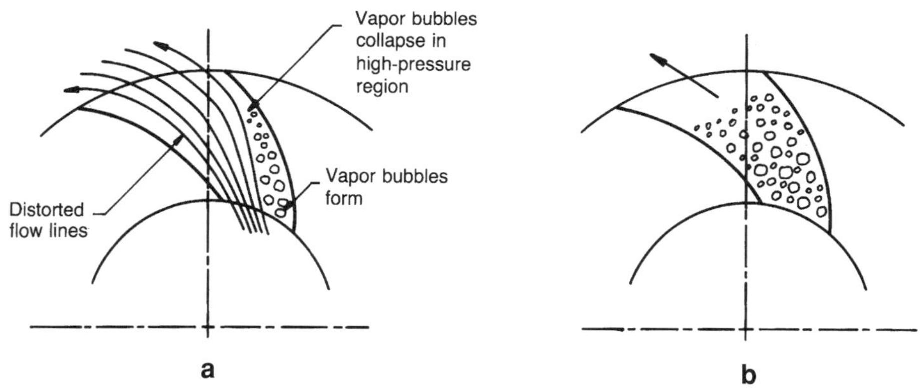

Figure 10-9. Formation of vapor bubbles in a pump impeller. (a) Partial cavitation; (b) full cavitation. After Addison [5].

Effects of Cavitation on Pump Performance

When cavitation occurs in centrifugal pumps with type numbers less than 30 (specific speeds ≤1500), there is a sharp drop or cutoff in H-Q and efficiency curves, as shown in Figure 10-11. Vapor bubbles begin to form at a lower discharge than the cutoff discharge where cavitation is fully formed. For pumps

a b

Figure 10-10. Cavitation damage in a vertical pump. (a) Impeller; (b) suction bell.

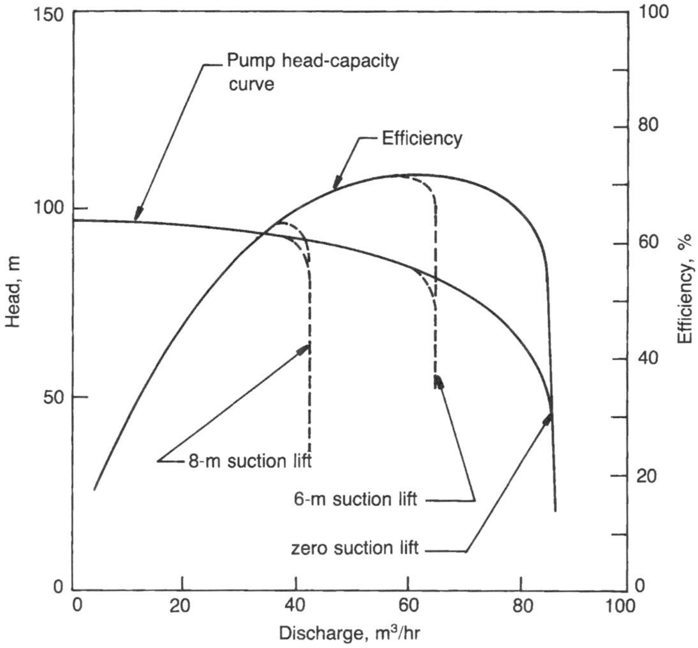

Figure 10-11. Pump performance curves at various suction lifts with cavitation. Note how rapidly cavitation begins.

with type numbers between 30 and 80 (specific speeds between 1500 and 4000 in U.S. units), the pump performance curves fall more gradually until the cutoff discharge is reached. For pumps with type numbers greater than 80 (specific speeds greater than 4000 in U.S. units), there is no distinct cutoff point.

Net Positive Suction Head (NPSH)

The absolute pressure plus the velocity head at the eye of the impeller converted to absolute total dynamic head is called the "net positive suction head" and abbreviated NPSH. Pump performance declines rapidly as the NPSH becomes less than the NPSH "required" ($NPSH_R$). The $NPSH_R$ is determined by tests of geometrically similar pumps operated at constant speed and discharge but with varying suction heads. The development of cavitation is assumed to be indicated by a 3% drop in the head developed as the suction inlet is throttled, as shown in Figure 10-12. It is known, however, that the onset of cavitation occurs well before the 3% drop in head [2, 10] and can, indeed, develop substantially before *any* drop in the head can be detected. Furthermore, erosion occurs more rapidly at a 1% change in head (where vapor bubbles are small but cause a higher unit surface implosion energy) than it does at a 3% change in head (where bubbles are large with a lower unit implosion energy). Injection of air reduces the hydraulic shock imposed on the mechanical equipment. Although air injection may improve mechanical stability by acting as a shock absorber, there are distinct disadvantages and seldom does it permit long-term operation, because: (1) oxygen content increases corrosion, (2) gas increases the $NPSH_R$, and (3) air causes noise and some vibration. In conclusion, serious erosion can occur as a result of blindly accepting data from catalogs because of the current standard (a 3% drop in head) used by most pump manufacturers. In critical installations where continuous duty is important or where the pump is to be operated at reduced speed, the manufacturer should be required to furnish the $NPSH_R$ test results. Typically, $NPSH_R$ is plotted as a continuous curve for a pump (see Figure 10-5). When impeller trim has a significant effect on the $NPSH_R$, several curves are plotted.

The NPSH "available" ($NPSH_A$) in the actual installation is calculated using Equation 10-23.

$$NPSH_A = H_{bar} + h_s - H_{vap} - h_{fs} - \Sigma h_m - h_{vol} - FS \qquad (10\text{-}23)$$

where:

H_{bar} is the barometric pressure in m or ft of water column corrected for elevation above mean sea level (see Table A-6 or A-7) Note that storms can reduce barometric pressure by 1.7%.

h_s is the static head of the intake water surface above the eye of the impeller. If the water surface is below the eye, h_s becomes minus.

H_{vap} is vapor pressure of the fluid at the maximum expected temperature. See Table A-8 or A-9 for the equivalent height of the water column.

h_{fs} is pipe friction in m or ft between the suction intake and the pump.

Σh_m is the sum of minor pipe friction losses such as entrance, bend, reducer, and valve losses. These "minor" losses are significant.

h_{vol} is the partial pressure of dissolved gases such as air in water (customarily ignored) or volatile organic matter in wastewater (customarily estimated to be about 0.6 m or 2 ft).

FS is a factor of safety used to account for uncertainty in hydraulic calculations and for the possibility of swirling or uneven velocity distribution in the intake. Also, the 3% drop in head allows some cavitation (that ought to be suppressed) to occur, and although a 3% loss in TDH is typically

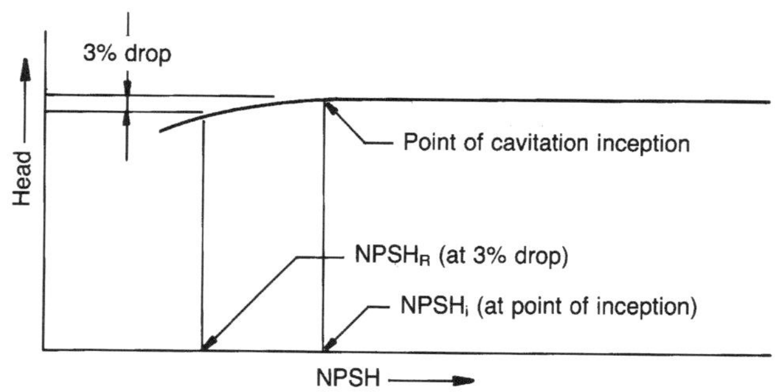

Figure 10-12. Net positive suction head criteria as determined from pump test results.

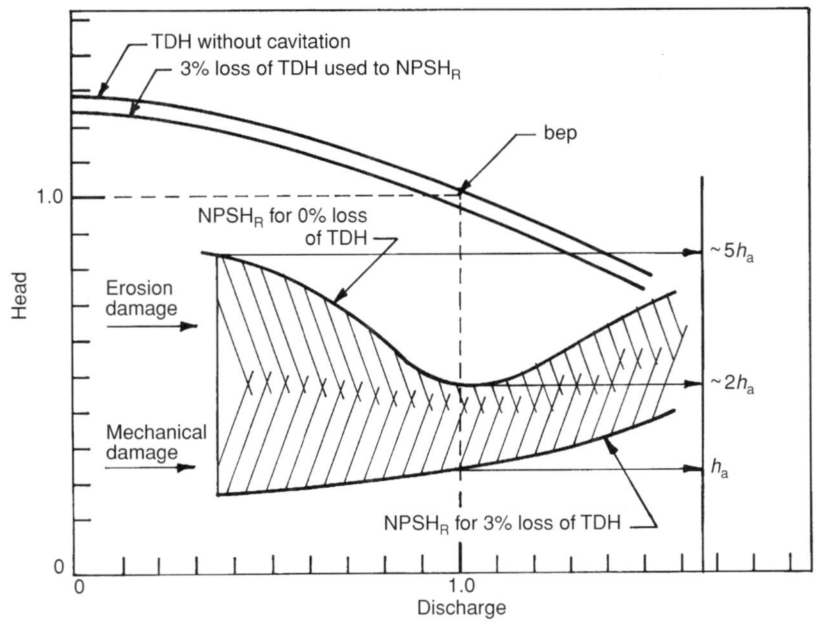

Figure 10-13. Effect of NPSH on TDH and pump damage. Head required at bep for 3% loss of TDH is termed h_a. Courtesy of DuPont Engineering [12].

insignificant, the shock on the equipment can reduce the life of bearings, seals, and impellers. Surprisingly large variations in $NPSH_R$ are sometimes observed in tests of supposedly identical pumps. Small pumps are more sensitive to casting imperfections than large ones. Now add the fact that the NPSH to avoid cavitation may increase substantially when the pump operates at any point but the bep, and the need for a generous factor of safety is apparent. In the past, it was customary to consider that 0.6 m or 2 ft (but no less than 20% of the $NPSH_R$) was adequate, and some very knowledgeable engineers continue to recommend such safety factors for water and wastewater pumps. *Others, however, warn that anything less than 1.5 m (5 ft) or less than 1.35 times $NPSH_R$ may be dangerous.* Consult the pump manufacturer to decide which rules to follow. But be aware that to eliminate cavitation and its effects on TDH entirely, the $NPSH_A$ must, depending on the operating flowrate relative to the bep, be 2 to 5 times the $NPSH_R$. The effect of NPSH on TDH is illustrated in Figure 10-13.

No term for velocity head is seen in Equation 10-23, because velocity head is part of the absolute dynamic head.

Example 10-5
Calculation of Net Positive Suction Head Available ($NPSH_A$)

Problem: Estimate the $NPSH_A$ for the pump system shown in Figure 10-3. Use the data given in Example 10-1. Assume the water temperature is 40°C (104°F) and the elevation is 500 m (1600 ft).

Solution: Use Equation 10-23 with data from Appendix Tables A-6 to A-9 and calculations in Example 10-1.

SI Units		**U.S. Customary Units**	
H_{bar} (Table A-6)	= +9.74	H_{bar} (interpolate Table A-7)	= +32.0
h_s (Example 10-1)	= −2.00	h_s (Example 10-1)	= 6.56

SI Units		**U.S. Customary Units**	
H_{vap} (Table A-8)	= −0.76	H_{vap} (interpolate Table A-8)	= −2.48
h_{ent}	= −0.011	h_{ent}	= −0.014
h_{fs}	= −0.015	h_{fs}	= −0.048
$\Sigma h_m = 0.044 + 0.018$	= −0.040	$\Sigma h_m = 0.144 + 0.058$	= −0.130
Subtotal	= +6.931	Subtotal	= +22.768
Safety factor[a]	= −1.5	Safety factor[a]	= − 5.0
$NPSH_A$	= 5.431	$NPSH_A$	= +17.768
Use	+5.4	Use	+18

[a] Note that, according to some engineers, the safety factor should be at least 1.5 m (5 ft) but not less than 35% of $NPSH_R$. According to others, 0.6 m (2 ft) but not less than 10% of $NPSH_R$ is adequate for water and wastewater.

Prevention and Control of Cavitation

The easiest, most direct, and best way to eliminate cavitation is to ensure that the internal pump pressure remains above the vapor pressure of the water. Where cavitation already exists, possible solutions might include

- Decreasing the suction lift (see Figure 10-3)
- Decreasing the suction losses
- Lowering the liquid temperature
- Reducing the impeller speed
- Changing the pump or the impeller
- Adding a booster pump or an inducer.

An inducer is an auxiliary axial-flow propeller attached to the pump impeller as in Figure 10-14 [10, 14]. It produces a small increase in the fluid pressure at the eye of the pump impeller, which reduces the risk of cavitation in the expensive impeller. Inducers are designed specifically for very low NPSHR and will not cavitate if operated at the design point. Unfortunately, the operating range that is free of cavitation is rather narrow. If the inducer erodes due to cavitation, it is easier and cheaper to replace than is the pump impeller.

Cavitation Constant

The ratio of the net positive suction head at the point of cavitation inception, $NPSH_i$ (see Figure 10-12), to the total dynamic head is known as Thoma's cavitation constant [9, 10, 11, 14].

$$\sigma = \frac{NPSH_i}{H_t} = \text{constant} \tag{10-24}$$

where H_t = total dynamic head in meters (feet). For multistage pumps, the total dynamic head is the total dynamic head per stage. In the literature, the $NPSH_R$ value for a pump is often used incorrectly in place of $NPSH_i$. The pump $NPSH_R$ cannot be used because cavitation is already occurring at that value. Furthermore, it should be noted that there is no known relationship between the $NPSH_R$ at the 3% level and the value of $NPSH_i$.

Because the specific speed is an indication of performance curve shape, it is possible approximately to

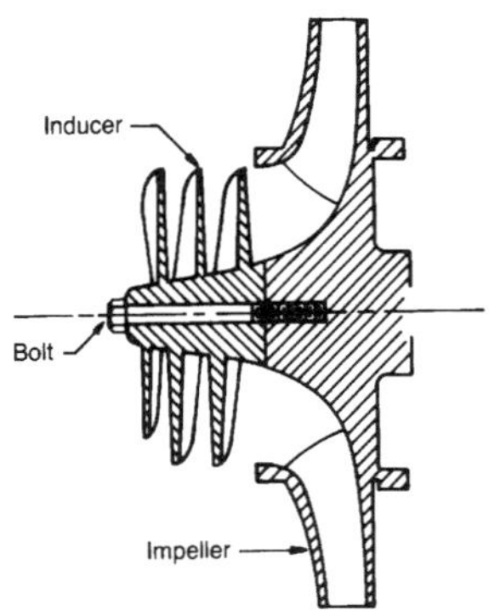

Figure 10-14. Typical inducer (axial-flow impeller) attached to a conventional pump impeller. After Turton [10].

correlate σ (and therefore $NPSH_R$) with the specific speed. The relationship between specific speed, Thoma's cavitation constant, and pump efficiency for single suction pumps as published by Rutschi [15] is illustrated in Figure 10-15. For single-suction pumps

$$\sigma = \frac{K \times n_s^{4/3}}{10^6} \tag{10-25}$$

where the values of K are given in Table 10-2. Equations 10-24 and 10-25 are presented for background only and should not be used for making design decisions. Recommendations for $NPSH_R$ should come from the pump manufacturer for the specific operating conditions and requirements.

Table 10-2. Values of *K* for Equation 10-25[a]

Pump efficiency (%)	*K*	
	SI units[b]	U.S. customary units[c]
70	1726	9.4
80	1210	6.3
90	796	4.3

[a] For double-suction pumps, use the same formula with Q equal to half of the actual value.

[b] Use with n_s values in cubic meters per second, meters, and revolutions per minute.

[c] Use with n_s values in gallons per minute, feet, and revolutions per minute.

Example 10-6
Determination of $NPSH_i$ and $NPSH_A$

Problem: At a total dynamic head of 36.5 m (120 ft) and a rotational speed of 1170 rev/ min., the pump described in Figure 10-5 delivers a flow of 500 m^3/h (2200 gal/min). Assume that the pump is to operate at sea level at a temperature of 15°C (59°F) and that the suction losses are 2.0 m (6.6 ft).

(1) Estimate the required $NPSH_i$, (2) compare the computed $NPSH_i$ to the $NPSH_R$ value given in the figure, and (3) estimate the allowable suction head.

Solution:

(1) *$NPSH_i$*. Equate Equations 10-24 and 10-25 to obtain

$$NPSHA_i = \frac{H_t K n_s^{4/3}}{10^6}$$

and calculate n_s from Equation 10-22 ($n_s = nQ^{1/2} H_t^{-3/4}$). Because the pump curves shown in Figure 10-5 are for a split-case pump with a double-suction impeller, one-half of the discharge is used in computing n_s. Use K values of 1140 and 6.0 at 82% efficiency (see Table 10-2).

SI Units

$$n_s = 1170\left(\frac{250}{3600}\right)^{1/2}(36.5)^{-3/4}$$

$$= 20.8$$

$$NPSH_i = \frac{36.5 \times 1140(20.8)^{4/3}}{10^6} = 2.38 \text{ m}$$

U.S. Customary Units

$$n_s = 1170(1100)^{1/2}(120)^{-3/4}$$

$$= 1070$$

$$NPSH_i = \frac{120 \times 6.0(1070)^{4/3}}{10^6} = 7.88 \text{ ft}$$

(2) *Compare $NPSH_i$ to $NPSH_R$*. The computed value is 4% greater than the value of 2.28 m (7.4 ft) given in Figure 10-5.

(3) *Estimate the allowable suction head.* Rewriting Equation 10-23 in terms of the suction head yields

$$h_s = NPSH_A - H_{bar} + H_{vap} + h_{vol} + h_{fs} + \Sigma h_m + FS$$

where, for this example, h_{vol} is assumed to be zero, and $h_{fs} + \Sigma h_m$ are given as 2 m (6.6 ft). Substituting the value of $NPSH_i$ for the $NPSH_A$, the value of h_s is

$$h_s = 2.38 - 10.33 + 0.17 + 2.0 + 1.5$$
$$= -4.28 \text{ m}$$

$$h_s = 7.88 - 33.9 + 0.58 + 6.6 + 5.0$$
$$= -13.8 \text{ ft}$$

The minus sign means that h_s can be a suction lift. The $NPSH_A$ should always be greater than the $NPSH_R$.

The accuracy implied by the results of Example 10-6 may be misleading. First, the data points in Figure 10-15 deviate by as much as 25% from the line of best fit. Second, the K coefficients in Table 10-2 vary more widely than do the efficiencies. (A change of 30% in efficiency reduces K by half.) Finally, the efficiency of pumps (upon which K is so dependent) can easily vary by 2%. Although the calculations cannot be trusted to yield errors of less than 30 or 40%, they nevertheless provide both insight and approximate results.

Cavitation at the Operating Point

If the pump operates at low head at a flowrate considerably greater than the capacity at the best efficiency point (bep), Equation 10-26 is approximately correct:

$$\frac{\text{NPSH}_R \text{ at operating point}}{\text{NPSH}_R \text{ at bep}} = \left(\frac{Q \text{ at operating point}}{Q \text{ at bep}}\right)^n \quad (10\text{-}26)$$

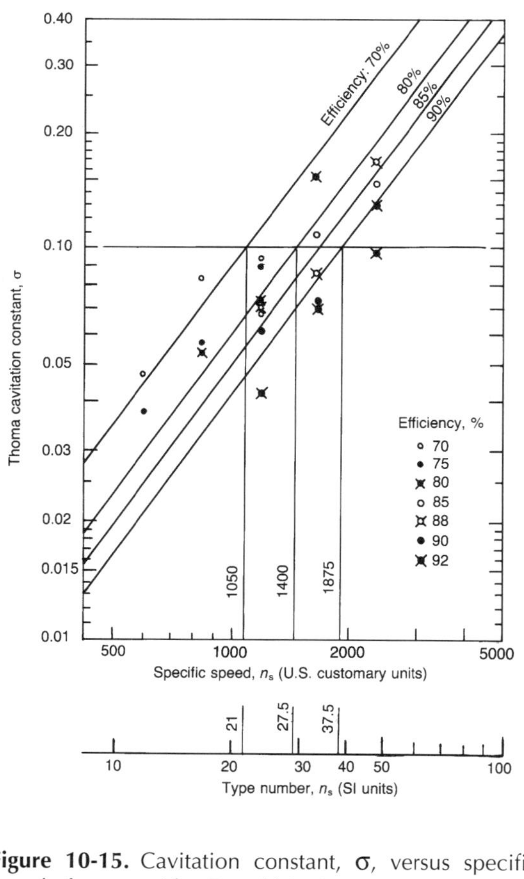

Figure 10-15. Cavitation constant, σ, versus specific speed of pumps. After Rutschi [15].

where the exponent n varies from 1.25 to 3.0, depending on the design of the impeller. In most water and wastewater pumps, n lies between 1.8 and 2.8. The NPSH_R at the bep increases with the specific speed of the pumps. For high-head pumps, it may be necessary either to limit the speed to obtain the adequate NPSH at the operating point or to lower the elevation of the pump with respect to the free water surface on the suction side.

10-5. Pump Characteristic Curves

In most pump curves, the total dynamic head H (in meters or feet), the efficiency E (a percentage), and the power input P (in kilowatts or horsepower), are plotted as ordinates against the capacity Q (in cubic meters per second or per hour or in gallons per minute or millions of gallons per day) as the abscissa.

Nondimensional Pump Characteristic Curves

Nondimensional pump curves are obtained by expressing head, capacity, power input, and efficiency as percentages of the corresponding values at the best efficiency point, and such curves are useful for (1) comparing the hydraulic properties of pumps belonging to the same type and (2) assessing the performance of pumps of different specific speeds for various applications [7]. The general shape of these curves varies with the specific speed, the form and number of blades on the impeller, and the form of the casing used. As shown in Figure 10-16, the slope of the head-capacity curve becomes steeper as the specific speed increases. Notice the shape of the power curves in Figure 10-16a. As head decreases below the bep, the power required increases. Thus, overloading the motor is most likely to occur in pumps with low specific speeds and with flat head-capacity (H-Q) curves. Unstable pump operating curves are typically encountered in such pumps. Furthermore, because of the shape of the characteristic curves in Figure 10-16, mixed-flow and axial-flow pumps are nonoverloading in their operating range. However, depending on the impeller blade design, a dip (and consequent instability) can occur as shown in H-Q curves for pumps of high specific speeds. Prolonged operation of high–specific speed pumps in this head-capacity range should be avoided.

As the specific speed increases, the effects on the power input are marked. With radial-flow pumps, where the specific speed is about 35 (about 1800 in U.S. units), the energy input decreases with a decrease in flow. For a mixed-flow pump with a specific speed of about 85 (about 4300 in U.S. units), the power input is nearly

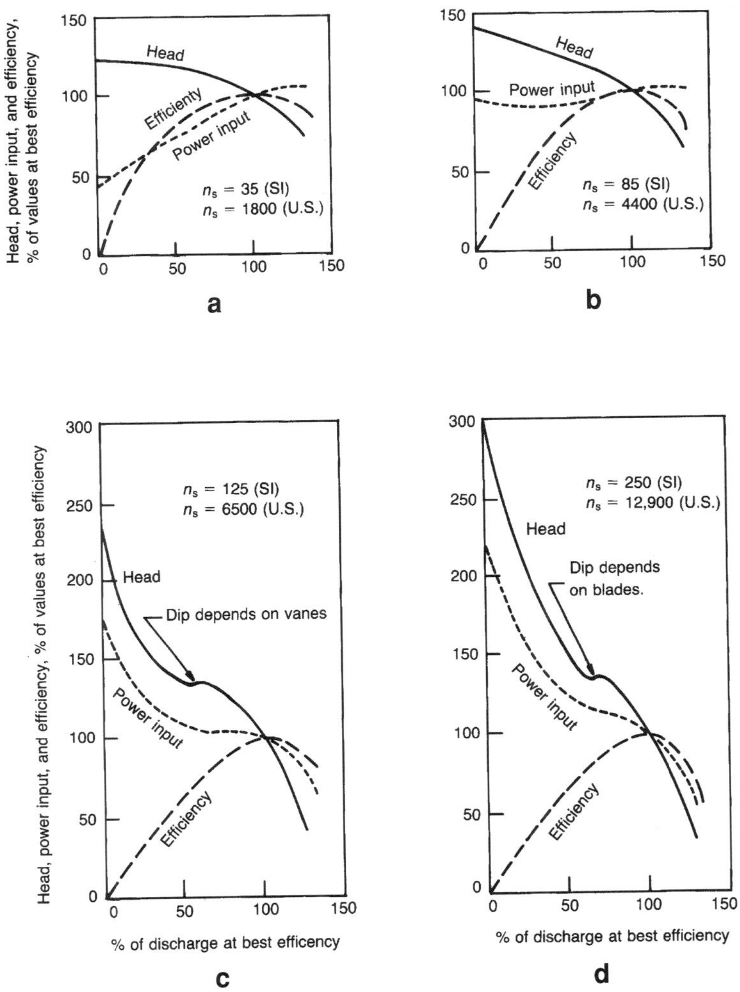

Figure 10-16. Typical dimensionless characteristic curves for centrifugal pumps. (a) Radial flow; (b) mixed flow; (c) mixed flow; (d) axial flow.

constant. In pumps with specific speeds greater than about 125 (greater than 6400 in U.S. units), the power adsorbed by the fluid rises steeply as the discharge is decreased to zero. Thus, with mixed- and axial-flow pumps, some means of unloading the pump (such as a bypass pipe with a pressure-activated valve) is needed as the head rises so as to avoid overloading the pump driver. Sometimes, it is more cost effective to oversize the motor than to install an unloading device. A comparative cost analysis should be made if there is doubt.

Example 10-7
Estimating the Size of the Pump Required

Problem: Water is to be pumped at a flowrate of 0.2 m^3/s (3166 gal/min) to a height of 30 m (98.4 ft) at a rotational speed of 1150 rev/min. Determine (1) the type of pump and (2) the impeller diameter.

Solution:

(1) *Type of pump.* Calculate the specific speed from Equation 10-23 [$n_s = n\,(Q^{1/2} H^{-3/4})$] and relate the type of pump to the specific speed.

SI Units	**U.S. Customary Units**
$n_s = 1150(0.2)^{1/2}(30)^{-3/4} = 40.1$	$n_s = 1150(3.166)^{1/2}(98.4)^{-3/4} = 2071$

From Figure 10-8, a pump with a mixed-flow (e.g., screw or helicoidal) type of impeller is needed.

(2) *Impeller diameter.* Referring to Figure 10-16, the normalized shut-off head for a pump with a specific speed of about 40 (2000 in U.S. units) is estimated to be about 1.4 times the value at the bep. At shut-off head, the head developed by a pump is proportional to $u^2/2g$, where u is the peripheral speed of the impeller (see Figure 10-6). Assuming that the head at the bep is 30 m (98.4 ft), the shut-off head is

SI Units	U.S. Customary Units
$H = u^2/2g = 1.4 \times 30 = 42$ m	$H = u^2/2g = 1.4 \times 98.4 = 137.8$ ft
$u = \sqrt{42 \times 2 \times 9.81}$	$u = \sqrt{137.8 \times 2 \times 32.2}$
$= 28.7$ m/s	$= 94.2$ ft/s
$u = \pi D\omega/60$	$u = \pi D\omega/60$
$D = 60u/\pi\omega$	$D = 60u/\pi\omega$
$= 60 \times 28.7/(3.14 \times 1150)$	$= 60 \times 94.2/(3.14 \times 1150)$
$= 0.477$ m (impeller diameter)	$= 1.56$ ft $= 18.8$ in. (impeller diameter)

Stable and Unstable Pump Curves

The head-capacity curves for radial-flow centrifugal pumps can be either stable (Figure 10-17a) or unstable (Figure 10-17b), whereas the head-capacity curves for mixed-flow and axial-flow pumps are always stable (except for a small area of operation). Unstable pump curves are usually limited to specific speeds of less than 20 (<1000 in U.S. units). With stable pump curves, there is only one flowrate for each value of head, whereas two discharge values are possible for a given value of head in unstable pump curves, as can be seen from Figure 10-17b.

The use of pumps with unstable pump curves can lead to pumping instabilities at heads greater than the shut-off head, especially where pumps are operated in parallel. It is best to avoid the use of pumps with unstable head-capacity curves entirely. In single-stage pumps, stable pump head-capacity curves can be obtained by reducing the number of impeller blades, changing the exit blade angle, and changing the configuration of the blades [7].

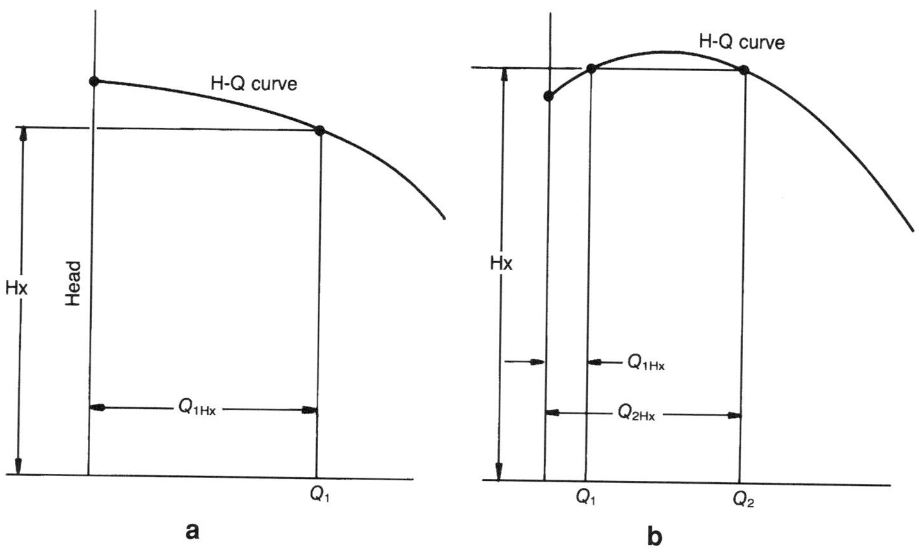

Figure 10-17. Stable (a) and unstable (b) pump H-Q curves.

10-6. Pump Operating Ranges

A pump operates best at its best efficiency point. Not only is the efficiency maximum, but radial loads on the impeller and the problems of cavitation are minimized. It is good practice to limit the operating range of pumps—especially radial-flow centrifugal pumps—to within approximately +20% and –40% of the discharge at the bep.

Radial Thrust

Pump volutes are so designed that at optimum discharge (at the bep) the average pressure throughout the volute approaches uniformity. At other discharges, the pressures around the volute vary considerably. The unbalanced radial forces are represented as the radial thrust, $P_2 - P_1$, in Figure 10-18a, where P is the pressure times the projected area of the impeller. Radial thrust always occurs in centrifugal pumps, but it is of greatest concern in end-suction overhung shaft designs as shown in Figure 10-18b.

The magnitude of radial thrust depends on such operating conditions as discharge, head, flowrate, and rotational speed. Radial thrust can be astonishingly large. For example, tests of a 350-mm (14-in.) pump rated at 0.5 m^3/s (7800 gal/min) at 50-m (163-ft) head showed that the radial thrust at shut-off was 36 kN (8200 lb) at 1160 rev/min and 24 kN (5400 lb) at a variable speed when operated along the station's system curve (see Figure 10-18c).

Be aware of the potential for damage by radial loads that may be too great for the load-carrying capacity of standard shafts and bearings. Pumps have occasionally been destroyed in a few days or even a

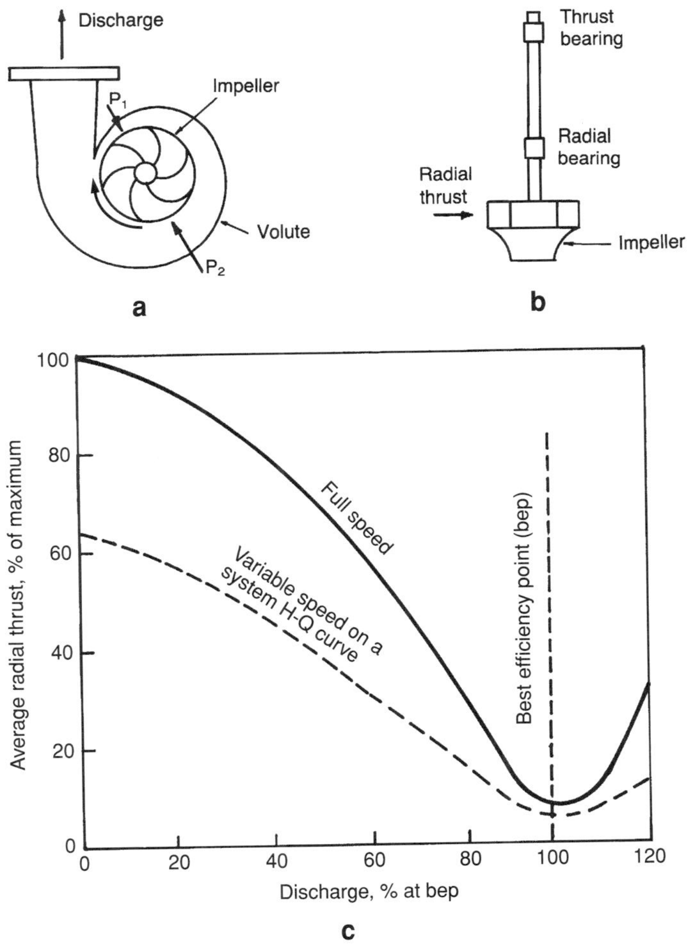

Figure 10-18. Development of radial thrust in a radial-flow centrifugal pump shaft. (a) Plan of pump; (b) impeller and shaft; (c) radial thrust versus flow.

few hours through excessive wear in bearings, seals, shaft sleeves, and wearing rings. Bearings, shafts, and casings must be designed to resist the anticipated radial loads (which can be approximated by the method given by the Hydraulic Institute [1]). Standard pumps can often be customized with larger shafts and bearings. One practical means of dealing with radial thrust is to define several critical operating points and require the manufacturer to furnish a pump with a specified life at such points (see Appendix C).

Cavitation

When the pump discharge rate increases beyond the best efficiency capacity (bec), the absolute pressure required to prevent cavitation increases so that cavitation is a potential problem. The $NPSH_R$ curves are the basis for selecting the maximum permissible discharge. When the pump discharge rate decreases toward zero at the shut-off head, the radial load increases and the recirculation of the pumped fluid within the impeller becomes a problem. This recirculation can cause vibration and may also result in cavitation that can ruin the impeller.

Pumps should not be operated at or near the shut-off head for extended periods of time because such operation also leads to heating of the liquid and can cause severe wearing-ring rub and other problems.

Fixed Efficiency Loss

For a given impeller diameter, the operating range of a pump can be established by (1) setting a limit on the minimum acceptable efficiency and (2) setting upper and lower limits on the allowable speed changes. Alternatively, if the diameter of the pump impeller is to be changed, the operating range of the pump can be established by (1) setting a limit on the minimum acceptable efficiency and (2) setting upper and lower limits on the allowable impeller diameters.

The operating range of a pump based on these criteria is illustrated in Figure 10-19.

Percentage of Capacity

An alternative approach used to establish the pump operating range is to set limits on the pump discharge as a percentage of the bec. An operating range of 60 to 120% of the bec is often recommended for pumps with specific speeds less than 40 (<2000 in U.S. units) [16]. Pump operating envelopes are defined using these values, as illustrated in Figure 10-20, for changes in speed as well as for changes in the impeller diameter.

10-7. Elementary Pump System Analysis

In any pump–force main system, the head developed by the pump must equal the total dynamic headloss in

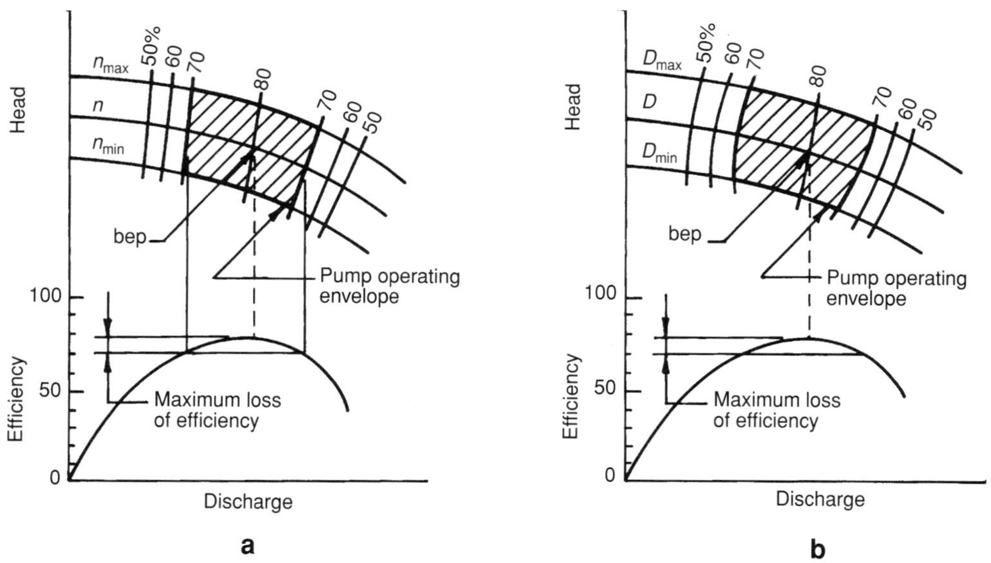

Figure 10-19. Pump operating envelopes based on a fixed efficiency loss. (a) Change of rotational speed; (b) change of impeller diameter.

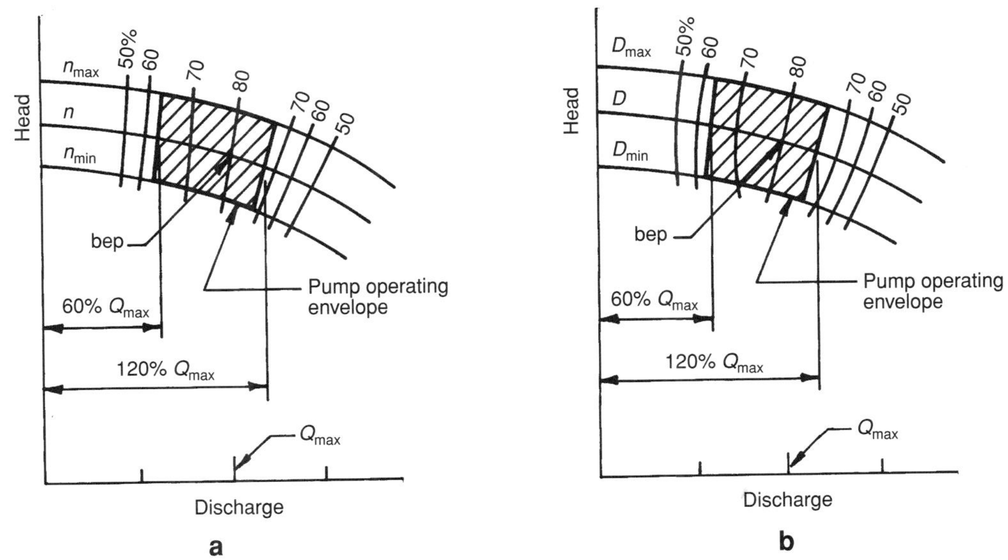

Figure 10-20. Pump operating envelopes based on the percentage of capacity at the best efficiency point. (a) Rotational speed; (b) impeller diameter.

the system, and, of course, the discharge in pump and pipe are equal. This relationship determines the "pump operating point," a single point on both the H-Q curve of the pump and the H-Q curve of the piping system. There are several ways of finding the pump operating point, including computer analysis and several graphical methods, but the simplest (for simple problems) and the most universally used method is given in this section.

System Head-Capacity (H-Q) Curves

The transmission pipeline H-Q curves in Figure 10-21 are plotted over a range of flows from zero (where the only head is due to static lift) to the maximum expected (which also includes all friction, fitting, and valve losses). Because transmission pipeline or force main head-capacity curves are approximate functions of $v^2/2g$, the curve is approximately a parabola with its apex on the zero Q line. Actually, headloss is a function of $Q^{1.85}$—not Q^2—in the Hazen–Williams equation. In the Darcy–Weisbach equation, the value of the friction factor, f, changes with the Reynolds number. However, the use of a parabola is close enough for practical purposes.

New pipe is expected to be very smooth for some indeterminate time. But bacterial slime, other deposits, or deterioration may reduce the H–W coefficient C to as little as 120 for pipes lined with cement-mortar (see Table B-5 and Section 3-2).

A simple envelope of transmission pipeline H-Q curves is shown in Figure 10-21.

Single-Pump, Single-Speed Operation

The point on any specific system head-capacity curve at which a single-speed pump must operate is determined by superimposing the pump H-Q curve on the transmission pipeline H-Q curve as shown in Figure 10-22. The point of intersection is the pump operating point. As the pipe ages and the pipeline H-Q curve rises, the operating point moves up and to the left along the pump H-Q curve.

As pumps start and stop (or as they change speed) because of the change in water level, the path of operating points is extended still further along the pump curve. Pumps and impellers must be chosen for a range of operating conditions. The efficiency curve is usually rather broad at the top. Hence, choose an impeller that operates from slightly left to slightly right of the best efficiency point (bep). If the range in operating points is too great to fit the efficiency curve nicely, choose an impeller that fits in the early years when the pipe is smooth. Then a new impeller of larger diameter can be installed when conditions warrant it.

Single-Pump, Variable-Speed Operation

The point on any head-capacity curve at which a variable-speed pump must operate is determined as described previously for the single-speed pump. The

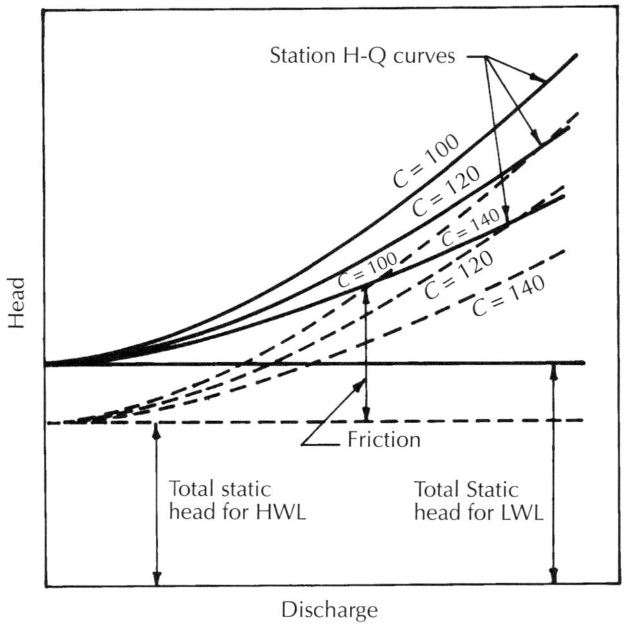

Figure 10-21. Typical envelope of transmission pipeline head-capacity (H-Q) curves.

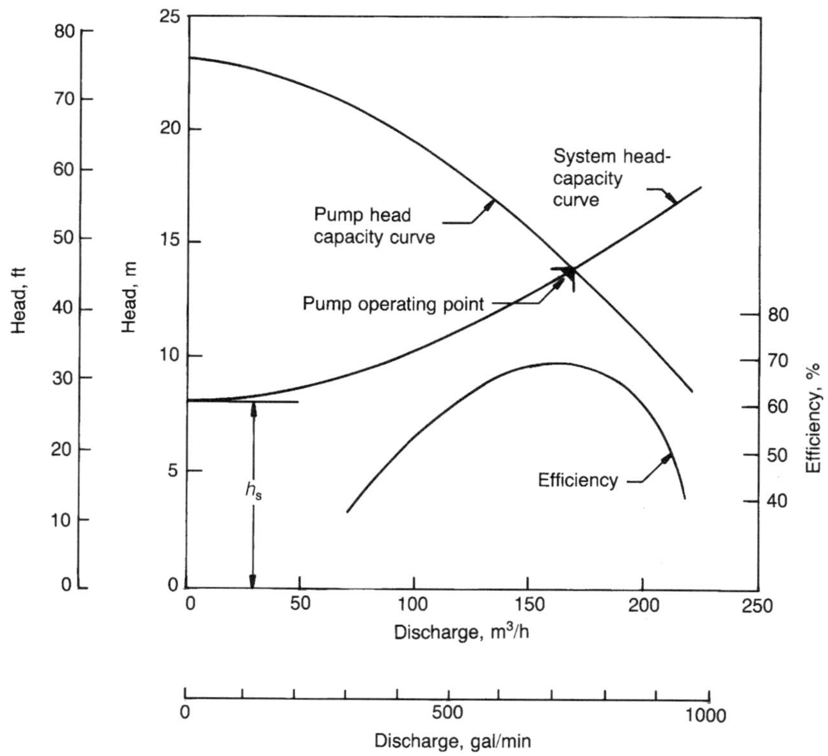

Figure 10-22. Determining the operating point for a single-speed pump with a fixed value of h_s.

affinity laws are used, as shown in Example 10-8, to determine the rotational speed at any other desired operating point on the system curve. Again, the operating point moves along the pump H-Q curve, although not so far, because the change between HWL and LWL is less than with single-speed operation.

Example 10-8
Application of a Variable-Speed Pump

Problem: The pump curve shown in Figure 10-23 is for a pump operated at 1150 rev/min that pumps 130 m^3/h (572 gal/min) at a head of 13.2 m (43.3 ft), which is shown as point a. If the pump must discharge 150 m^3/h (660 gal/min) at a head of 15.5 m (50.8 ft), which is shown as point b, what is (1) the required new operating speed and (2) the efficiency of the new operating point?

Solution: (1) *Operating speed.* Extend the system H-Q curve from point a to an intersection with the desired flowrate at point b. The new pump curve must pass through point b. Note that Equation 10-15 ($Q_1/Q_2 = n_1/n_2$) cannot be used to find the pump speed because points a and b are not corresponding points (see Example 10-2 and the text following Equation 10-17). One way to solve the problem is to create an "affinity law parabola" by solving Equations 10-15 and 10-16 simultaneously to eliminate n:

$$\left(\frac{Q_1}{Q_2}\right)^2 = \left(\frac{n_1}{n_2}\right)^2 = \frac{H_1}{H_2}, \qquad H_1 = H_2(Q_1/Q_2)^2$$

which defines a parabola passing through the origin and point b in Figure 10-23. Every point on the parabola is a corresponding point. Because point b corresponds to point c, Equation 10-15 can be used to find speed at point b after Q is scaled at point c.

SI Units	**U.S. Customary Units**
$Q_c = 136$ m^3/h (by scaling)	$Q_c = 598$ gal/min (by scaling)

From Equation 10-15 ($n_b = n_c\, Q_b/Q_c$),

SI Units	U.S. Customary Units
$n_b = 1150(150/136) = 1270$ rev/min	$n_b = 1150(660/598) = 1270$ rev/min

By scaling H at points b and c, the speed can also be found from Equation 10-16 ($n_b = n_c\sqrt{H_b/H_c}$).

SI Units	U.S. Customary Units
$n_b = 1150\sqrt{15.4/12.6} = 1270$ rev/min	$n_b = 1150\sqrt{50.5/41.3} = 270$ rev/min

(2) *Efficiency.* The efficiency of the new operating point is found by projecting a line from point c downward to point d on the efficiency curve, which gives an expected efficiency of 75%.

Multiple Pump Operation

In pumping stations where several pumps are installed, two or more pumps are usually operated in parallel. Occasionally, pumps are operated in series. The elements of the methods used to develop the combined H-Q curve for multiple pumps are outlined in the following subsections. A more extensive explanation for multiple pumps is given in Section 10-8.

Parallel Operation

When two or more pumps are discharging into the same header or manifold, the total flow is found by adding the individual flows at a given head. A correction (usually small) must be made, however, because the higher flow causes higher headlosses that slightly reduce the flow either pump would discharge if operated alone. One method for combining the flows is to subtract from the pump curve those station headlosses in the suction and discharge piping up to the intersection of the last pump discharge and the manifold (header). In the inset of Figure 10-24, the headlosses from a to b to c are subtracted from the H-Q curve for pump P1. Similarly, the losses from a to d to c are subtracted from the H-Q curve for pump P2. The modified pump curves are shown as dashed lines labeled MP1 and MP2 in the figure. Adding the abscissas (or discharges) of MP1 and MP2 produces the pump curve MP1 ‖ MP2, where the symbol "‖" means "in parallel with." Thus, he = hg + hf. The intersection of MP1 ‖

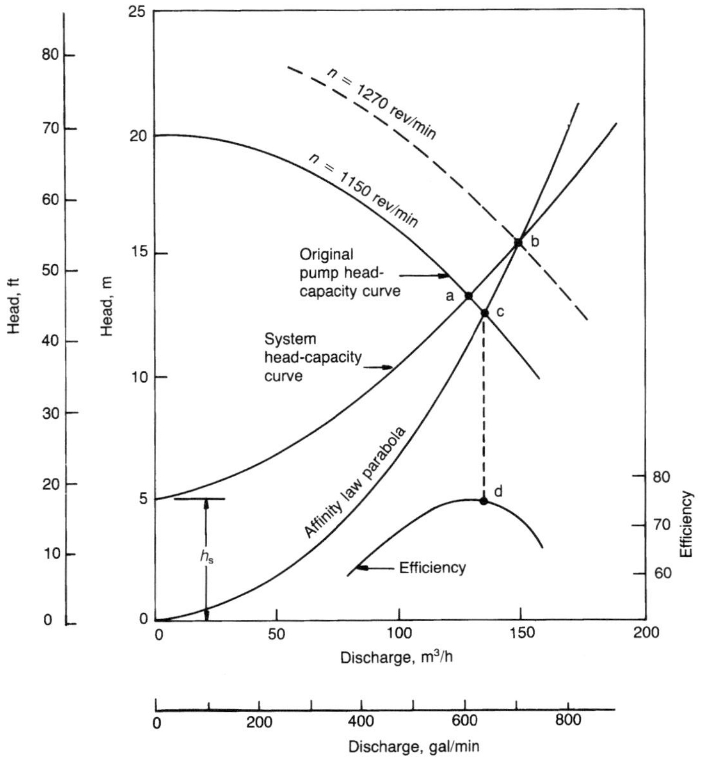

Figure 10-23. Determining the pump operating points for a single variable-speed pump and a system curve with a fixed value of static lift (h_s).

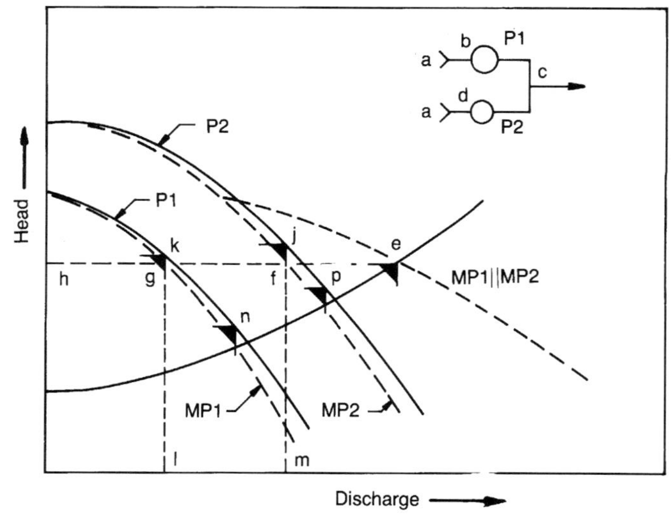

Figure 10-24. Operation of two pumps in parallel.

MP2 with the system curve for the manifold and force main downstream from point c is the operating point (point e) for the pumps when both are running.

To find the discharge contributed by each pump, draw the horizontal line, eh, which intersects the modified pump curves at f and g. The discharge from P1 is hg; from P2 it is hf. To find the total head at which each individual pump operates, project a vertical line from f to j and from g to k. The actual operating point of P1 is point k and the head on the pump is kl. The operating point of P2 is point j and the head is jm. With one pump off, the other operates at point n (for P1) or point p for P2. The pump specifications must be written for pump P1 operating either at point k or point n and for pump P2 at point j or point p.

Series Operation

Pumps are operated in series, as shown in Figure 10-25, when the head requirement is greater than can be obtained with a single pump. The combined head capacity curve, P1 + P2, is found by adding ordinates—that is, the heads developed at the same discharge. Thus, ad = bd + cd. The operating point is point c for pump P1 and point b for pump P2, and the operating point for the system is point a.

10-8. Practical Pumping System H-Q Curve Analysis

The fundamentals of pumping system analysis, explained in the previous section, should be reviewed before reading this section, in which additional details add practicality to this subject. The ultimate objective in analyzing a pumping system is to describe the required range of operating conditions in sufficient detail to be able to (1) understand the application yourself, and (2) with that understanding, describe the application to equipment suppliers in sufficient detail that they may intelligently apply their products.

With these objectives in mind, the full range of operating conditions must be examined to develop a complete understanding of the application. Constant speed (C/S) and variable speed (V/S) pumps are considered separately in the following discussion.

Constant-Speed (C/S) Pumps in Parallel

A typical schematic diagram of a pump and piping system for a pumping station with four assumed C/S pumps (one a standby) that discharge into a manifold and thence into the force main is shown in Figure 10-26. The system is divided into three basic components—force main, suction piping, and discharge piping. Head-capacity (H-Q) curves are drawn for each component. The H-Q curves for the pumps are then superimposed. Collectively, the curves are called "system curves." In Figure 10-27, dynamic head losses in suction and discharge piping are combined into Curve B + C.

Curves A, A', and A" (usually the first ones drawn) are combinations of static head plus the dynamic head losses from the manifold to the point of discharge. The curves must encompass the entire envelope of

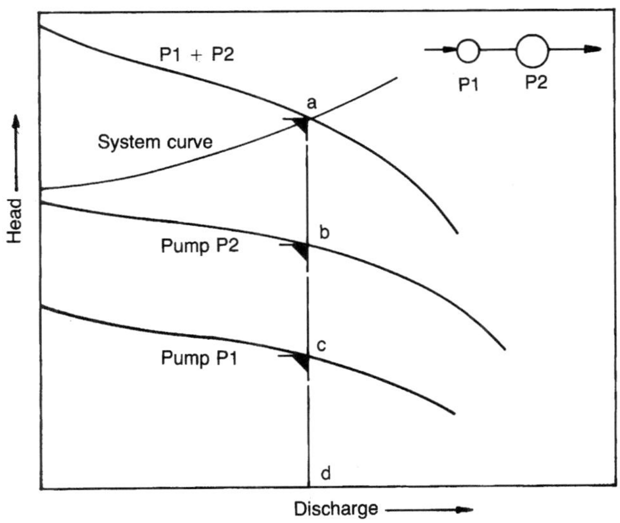

Figure 10-25. Operation of two pumps in series.

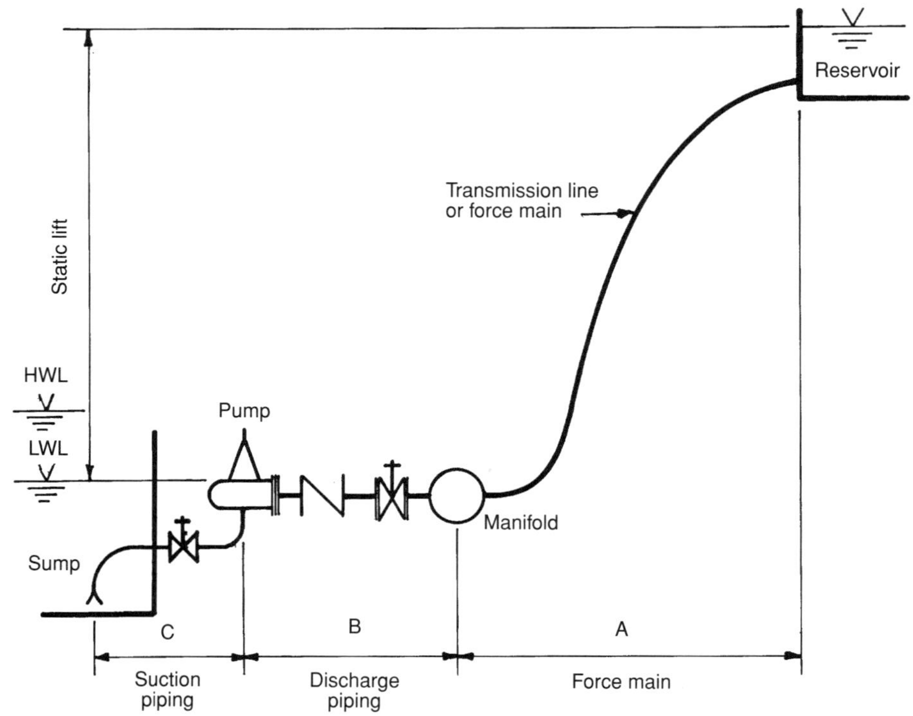

Figure 10-26. Schematic diagram of pump and piping system. One of four (three duty) pumps is shown.

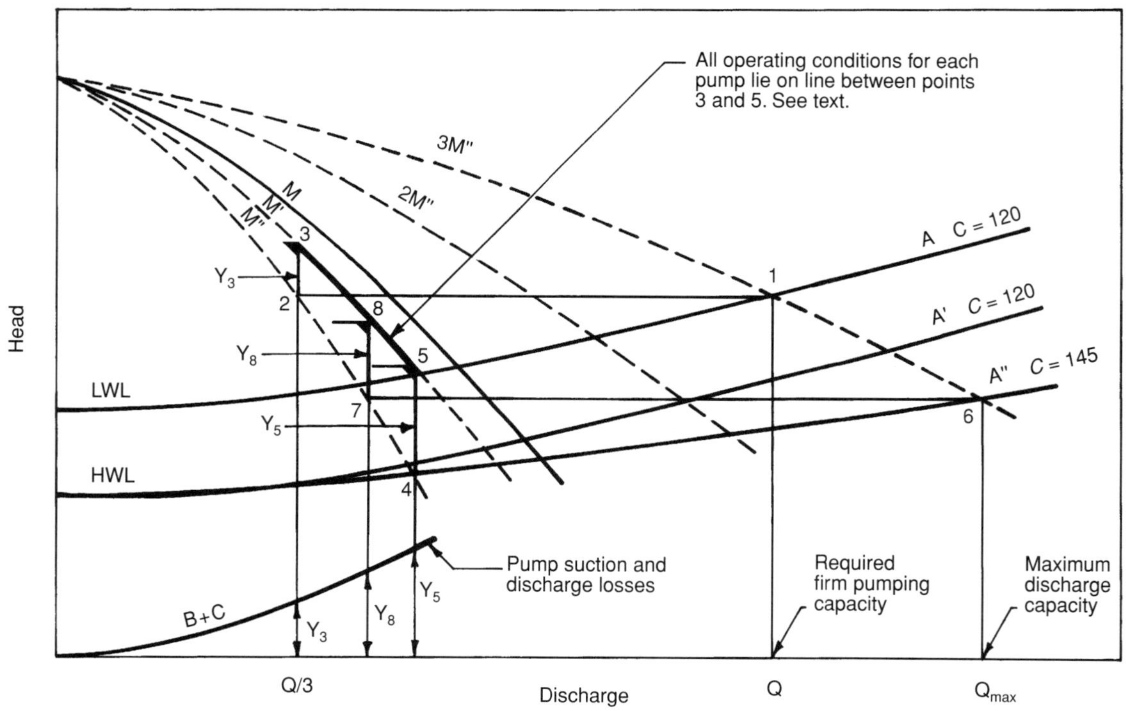

Figure 10-27. System H-Q curve analysis for three C/S duty pumps as per Figure 10-26.

operating conditions including both HWL and LWL in the wet well and the extremes of friction losses in the force main. (See Section 3-2, Tables B-5 to B-7, and Figures B-2 and B-3.) As shown, Curve A represents losses at the greatest static lift (LWL in the wet well) and the worst condition for pipeline resistance losses—in this example for Hazen–Williams $C = 120$. Curve A' represents the lowest static lift condition (HWL in the wet well) and similar pipeline losses, while curve A" represents the lowest static lift condition with the most favorable pipeline resistance losses—assumed here to be $C = 145$. Minor losses in the force main caused by valves and fittings can, perhaps, be ignored if the pipe is very long.

Curve B + C is the sum of friction and turbulence losses from the suction intake to the manifold. Losses on the inlet side of the pump are calculated and accumulated separately because they are needed to calculate $NPSH_A$. Here, turbulence losses in pipe fittings and valves are more significant than pipe friction losses. Typical loss coefficients are given in Tables B-6 and B-7. Draw Curve B + C for the maximum expected losses.

Begin the analysis by locating the operating point for the maximum head, so consider only Curve A and ignore (for the present) Curves A' and A". Determine the total required station discharge, Q, and draw a vertical line to intersect Curve A at Point 1. If the three duty pumps are of the same size, each must discharge Q/3. Draw a vertical line from Q/3 to intersect (at Point 2) the horizontal line through Point 1. Point 2 represents the discharge and head required of the pump system that includes intake and discharge piping. The pump itself must develop enough more head, y_3, to overcome the friction and turbulence losses in the intake and discharge piping, so plot y_3 from Point 2 to Point 3. Point 3 is the required operating point or rated condition for each of the three pumps. It is unique because it represents the highest head required. Note, however, that the pumps will rarely operate at this condition.

Now choose a pump with an operating curve that includes Point 3—in other words, above and to the right of Point 3 (refer to Sections 12-3 and 12-4 for selecting pumps). The manufacturer's pump curve, here called Curve M, can be made to fit Point 3 exactly by changing either the speed or the diameter of the impeller (or both) according to the affinity laws as defined by Equations 10-15 through 10-18, 10-20, and 10-21. *Note that these equations must be applied simultaneously. When an impeller is trimmed or the speed is changed, both head and discharge change.* The curve through Point 3 is called M', the adjusted pump curve. (If the manufacturer shows the curves for a synchronous speed, the curve must also be corrected to the speed of the induction motor to be used.) Subtracting the ordinates (y_3, y_5, etc.) of Curve B + C from the adjusted pump curve, M', gives M", the corrected pump curve. Think of M" as the curve if suction and discharge piping were considered to be part of the pump. This curve (M") must be extended to pass through Curve A".

Another operating point, unique because it represents the maximum discharge when one pump operates alone against reduced friction and turbulence losses, is found by intersecting Curves M" and A" at Point 4 and plotting y_5 above Point 4 to locate Point 5.

All operating points for the pumps (regardless of the wet well levels or the pipe roughness) are located along the Curve M' between Points 3 and 5. The pumps must be selected so that both points are within the manufacturer's recommendations.

It is interesting to plot Curves 2M" and 3M" to see what pump discharges and heads are developed under different conditions. For example, for HWL and minimum pipe friction, the flow and head for the pumping system is defined by the intersection of Curves 3M" and A" at Point 6. To obtain the head for the pump itself, add y_8 to Point 7 to obtain the total dynamic head at Point 8.

The conditions for both Points 3 and 5 *must* be published in the purchase order or project specifications, because this is the only legally defensible method for communicating performance requirements to the equipment manufacturer.

Always bear in mind that pump performance is worse, in terms of vibration, potential cavitation, and mechanical damage to the pump, at operating conditions different from the best efficiency point. As a general rule, off-the-shelf pumps perform best if operating conditions lie between 60 and 115% of best efficiency capacity, bec.

The next step *(which should not be omitted)* is to plot the hydraulic profile for the system curves against the profile for the pipeline under the intended modes of operation. Although extreme accuracy is not critical, it is important to assess the system as it is intended to function. Try to visualize how the system will behave under all operating conditions. Questions to be asked when performing this task are:

- Are all operating conditions properly defined by the system curves?
- What happens in the pipeline under each operating condition?
- Are there any peculiarities in the operation of the system that have not been considered in the construction of the system curves?
- Has *everything* that might influence pump operating conditions (and, hence, pump performance) been considered?

An example of where one can go astray is seen in Figure 10-28, in which a knee is shown in the pipeline profile. In this example, the operating point (Point 3 in Figure 10-27) for three pumps is properly defined. But when only one pump is running, the operating point (Point 5 in Figure 10-27) is wrong, because the hydraulic profile in Figure 10-28 is below the high point in the pipeline. The air-vacuum valve will open, and flow will stop until the pump increases the pressure enough to send water over the high point. So by plotting hydraulic profiles, it can be seen that the transmission pipeline H-Q curve must be raised, as shown in Figure 10-29, to accommodate a new static head. The true flowrate is less than the flowrate obtained from Figure 10-27, so the system will not perform as intended. Blunders of this sort are unlikely to inspire client confidence.

Variable Speed (V/S) Pumps in Parallel

The important difference between V/S and C/S pumping is that V/S pumps operate over a wide speed range, and care is required to make the pumps operate near their bep at all speeds (see Section 15-4). Other differences are: (1) storage is neither required nor desirable for V/S pumping; (2) fewer pumps are needed because of the V/S feature, so piping and space requirements are reduced; and (3) in properly designed systems, the difference between LWL and HWL in V/S pumping is small—slightly less than the diameter of the pipe discharging into the wet well, so the average static lift is less than that for C/S pumping.

For full-speed operation with all duty pumps (two in this example) operating, the analysis for V/S pumps, Figure 10-30, is nearly the same as that for C/S pumps, Figure 10-27. However, the pumps are programmed so that maximum discharge only occurs at HWL. Hence, Points 1 and 2 are constructed from Curve A' and not from Curve A. Note: two pumps including their appurtenant or station piping operate together at Point 1, whereas each pump including its station piping operates individually at Point 2 and discharges a flowrate of Q/2. By itself, the pump must operate at Point 3 to overcome B + C (station piping) losses. Curves M' (the manufacturer's curve adjusted for impeller trim), M", and 2M" can now be constructed and Point 4 located. Half the Point 4 discharge locates Point 5 and the pump operating point, 6, at y_5 above Point 5.

A single pump is programmed to turn at top speed when the water level is at the middepth of the inlet

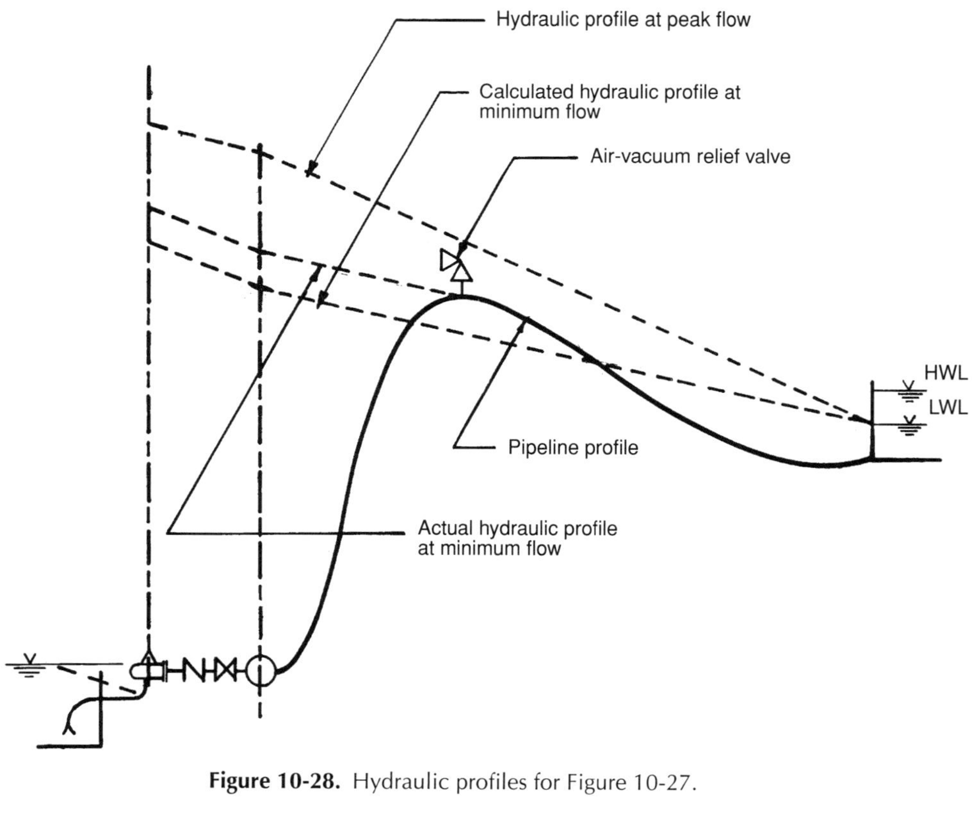

Figure 10-28. Hydraulic profiles for Figure 10-27.

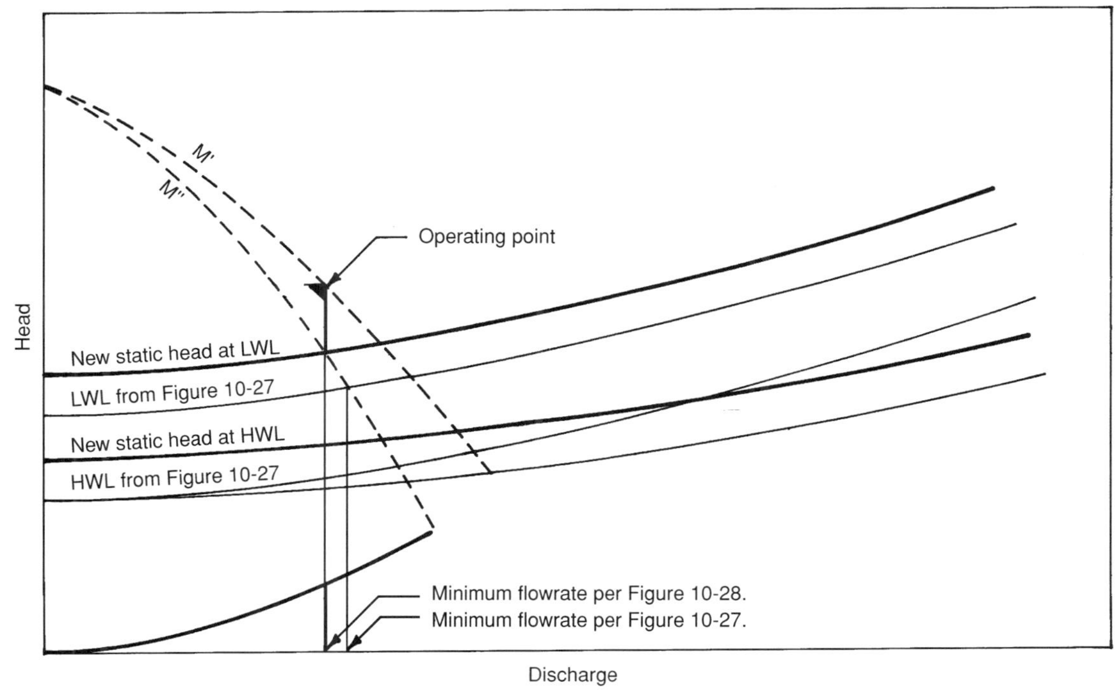

Figure 10-29. System curve analysis for one pump operating at minimum flow as per Figure 10-28. Compare with Figure 10-27.

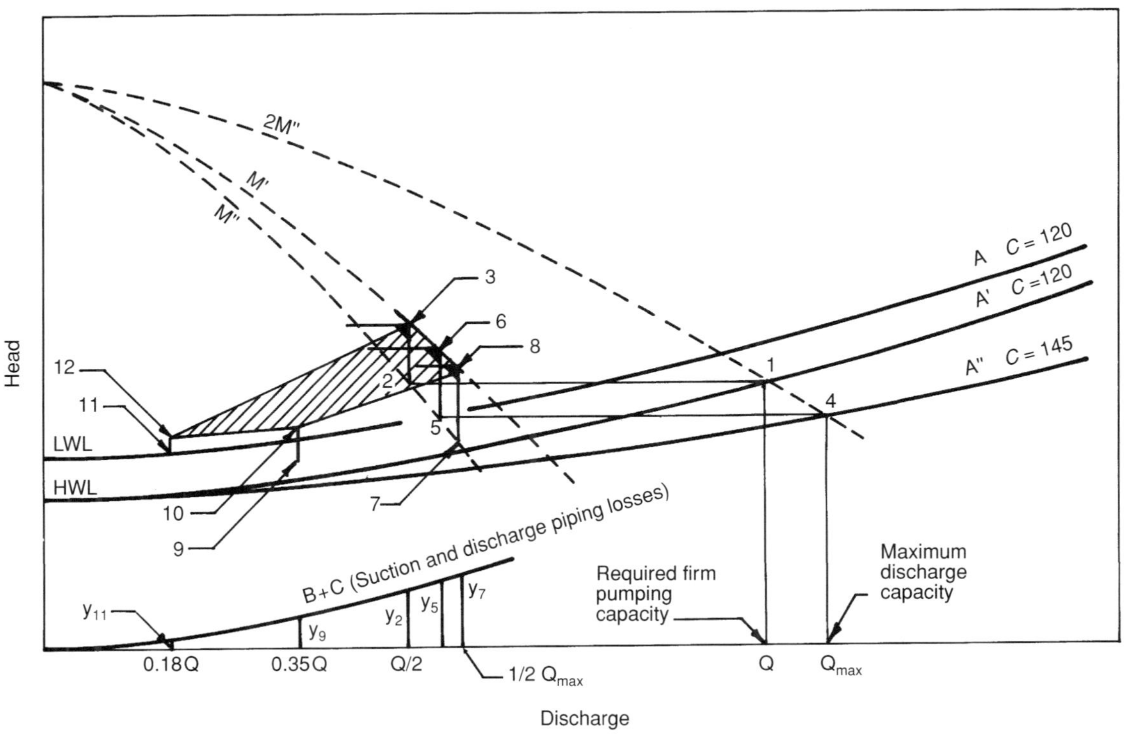

Figure 10-30. System H-Q curve analysis for two V/S duty pumps.

pipe—Point 7 in Figure 10-30, and the pump operating point is y_7 above—at Point 8. At minimum allowable speed, a single pump discharges (in this example) 35% of Q/2 or 0.18 Q and the pump operates at Point 12—y_{11} above Curve A, because minimum speed occurs when the water level is at (or near) the invert of the inlet pipe. Two pumps at minimum speed operate at Point 10—y_9 above the midpoint of Curves A and A'.

Trapezoid 3-8-10-12 encompasses all possible operating points for each pump. For all practical purposes, however, Point 10 can be ignored, and the operating points can be assumed to fall within Triangle 3-8-12.

Again, the extremes of the operating points (3-8-12) *must* be given in the specifications and to the vendors. Plotting the hydraulic profiles together with the profile of the piping system is even more important for V/S pumping because of the wider range of heads developed.

Computer Modeling

There are a number of digital computer programs that can be used to calculate pump performance curves and pump operating points. But none is known to be as flexible and to follow the principles stated herein as closely as PUMPGRAF2® [17].

10-9. Complex Pumping System H-Q Curves

The trend in engineering today is to depend on the newer and more advanced computer programs such as Cybernet® [18] or KY Pipe [19] to determine operating points in complex piping systems. Hence, the graphical method for solving complex problems described by Frey [20] in the first edition is omitted in this edition.

10-10. References

1. *Hydraulic Institute Standards for Centrifugal, Rotary, and Reciprocating Pumps,* 14th ed., Hydraulic Institute, Cleveland, OH (1983).
2. Karassik, I. J., W. C. Krutzsch, W. H. Fraser, and J. P. Messina (Eds.), *Pump Handbook,* 2nd ed., McGraw-Hill, New York (1985).
3. Warring, R. H., *Pumps: Selection, Systems, and Applications,* Trade and Technical Press, Morden, Surrey, United Kingdom (1979).
4. DIPRA, *Handbook of Ductile Iron Pipe,* 6th ed., Ductile Iron Pipe Association, Birmingham, AL (1984).
5. Addison, H., *Centrifugal and Other Rotodynamic Pumps,* 3rd ed., Chapman & Hall, London (1966).
6. Anderson, H. H., *Centrifugal Pumps,* 3rd ed., Trade and Technical Press, Morden, Surrey, United Kingdom (1980).
7. Lazarkiewicz, S., and A. T. Troskolanski, *Impeller Pumps,* Pergamon Press, Warsaw, Poland (1965).
8. Norrie, D. H., *An Introduction to Incompressible Flow Machines,* Edward Arnold Publishers, London (1963).
9. Stepanoff, A. J., *Centrifugal and Axial Flow Pumps,* 2nd ed., Wiley, New York (1957).
10. Turton, R. K., *Principles of Turbomachinery,* L. E. & F. N. Spon, London (1984).
11. Wislicenus, G., *Fluid Mechanics of Turbomachinery,* Vol. 1, 2nd ed., Dover, New York (1965).
12. DuPont Engineering, P.O. Box 80840, Wilmington, DE 19880-0840.
13. Cavi, D., "NPSHR Data and Tests Need Clarification," *Power Engineering,* **89,** 47–50 (February 1985).
14. Grohmann, M., "Extend Pump Application with Inducers," *Hydrocarbon Processing,* **59,** 121–124 (December 1979).
15. Rutschi, K., "Die fleiderer-Saugzahl als gutegrad der Saugfahigkeit von Kreiselpumpen," *Schweizerische Bauzeitung,* No. 12, Zurich (1960).
16. Metcalf and Eddy, Inc., *Wastewater Engineering: Collection and Pumping of Wastewater,* G. Tchobanoglous (Ed.), McGraw-Hill, New York (1981).
17. Wheeler, W., PUMPGRAF2®. For a free copy of this computer program with instructions, send a formatted 1.4 MB, $3^1/_2$-in. diskette and a stamped, self-addressed mailer to 683 Limekiln Rd., Doylestown, PA 18901-2335.
18. Cybernet®, Haestad Methods, Inc., 37 Brookside Rd., Waterbury, CT 06708.
19. KY Pipe, Dept. of Civil Engineering, University of Kentucky, Lexington, KY 40506.
20. Sanks, R. L., G. Tchobanoglous, B. E. Bosserman, D. Newton, and G. M. Jones, Frey's Method in Section 10-8 in *Pumping Station Design* (1st ed.), Butterworth-Heinemann, Newton, MA (1989).

Chapter 11

Types of Pumps

RICHARD O. GARBUS

CONTRIBUTORS
Kirk Blanchard
Rodney L. Cusworth
Kundan Desai
Mayo Gottliebson
David A. House
Joseph K. Jackson
R. Russell Langteau
Colin Martin
Stephen G. Miller
Lowell G. Sloan
Earle C. Smith
Robert E. Starke

The types of pumps commonly used in public utilities for the pumping of water, wastewater, and sludge are described in this chapter together with details such as impellers, casings, seals, and wear rings. A summary of pump sizes and applications is given in Section 11-11, and a general comparison of advantages and disadvantages of the pumps is presented in Section 25-6. A more complete listing of pump parts is given by the Hydraulic Institute [1, 2]. For pump dimensions and performance data, refer to the pump manufacturers' catalogs. Examples of pump selection are presented in Chapters 12, 17, 18, 19, 26, and 29.

References to standards are given in abbreviated form, such as ANSI B73.1. See Appendix E for a listing of titles of standards useful for selecting and specifying pumps and Appendix F for the addresses of the publishers of the standards.

11-1. General Classifications of Pumps

Pumps may be classified in many different ways. Classification by general mechanical configuration is used in this chapter. Figure 11-1 (an abridged version of the more complete listing given by the Hydraulic Institute [1]) is a chart of the pumps discussed here arranged according to their design. A slightly different version of Figure 11-1 is given in a later edition of Hydraulic Institute Standards [2], but it offers no advantage for this chapter. An overview of pumps is given in the following subsections. The pumps are described in more detail in Sections 11-3 to 11-10. There are two basic groups of pumps: (1) kinetic and (2) positive displacement.

Kinetic Pumps

Kinetic pumps impart velocity and pressure to the fluid as it moves past or through the pump impeller and, subsequently, convert some of that velocity into additional pressure. Kinetic pumps are subdivided into two major groups: (1) centrifugal (or volute) and (2) vertical (or turbine) pumps.

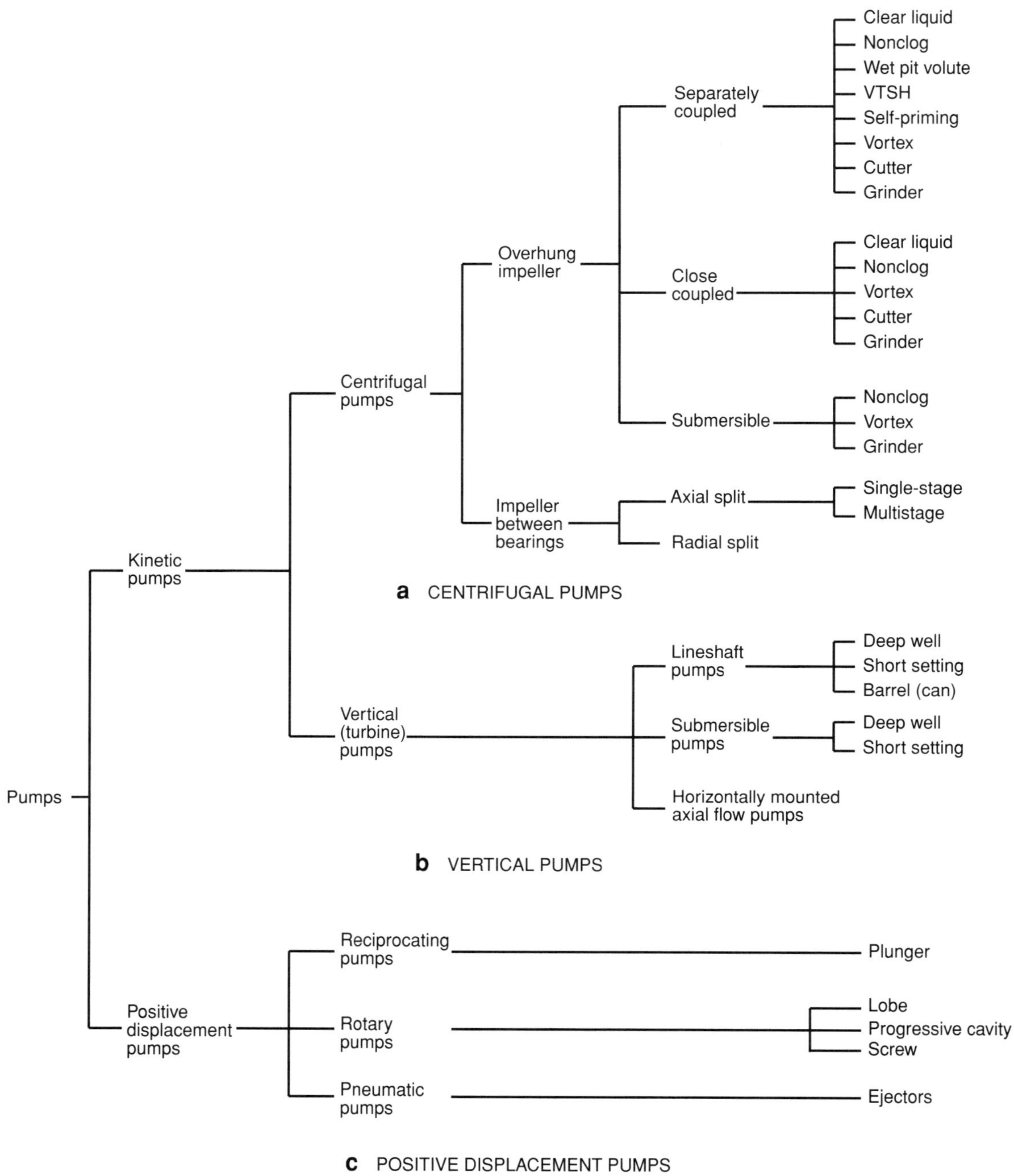

Figure 11-1. Classification of pumps. Courtesy of Hydraulic Institute Standards [1].

Figure 11-2. A centrifugal pump in a sewage pumping station. The outer shell is the volute or casing. Photograph by George Tchobanoglous.

Centrifugal Pumps

All centrifugal pumps have one common feature: they are equipped with a volute or casing (see Figures 11-2 and 11-11). The function of the volute is to collect the liquid discharged by the impeller and to convert some of the kinetic (velocity) energy into pressure energy.

Vertical Pumps

Vertical pumps are equipped with an axial diffuser (or discharge bowl) rather than a volute. The diffuser performs the same basic functions as the volute (see Figure 11-3).

Positive Displacement Pumps

In positive displacement pumps, the moving element (piston, plunger, rotor, lobe, or gear) displaces the liquid from the pump casing (or cylinder) and, at the same time, raises the pressure of the liquid. The three major groups of positive displacement pumps are (1) reciprocating pumps, (2) rotary pumps, and (3) pneumatic pumps.

Reciprocating Pumps

In a reciprocating pump, a piston or plunger moves up and down. During the suction stroke, the pump cylinder fills with fresh liquid, and the discharge stroke displaces it through a check valve into the discharge line. Such pumps can develop very high pressures. Plunger pumps are the only reciprocating type of interest here (see Figure 11-4).

Rotary Pumps

The pump rotor of rotary pumps displaces the liquid either by rotating or by a rotating and orbiting motion. Lobe pumps (Figure 11-5) and progressive cavity pumps (Figure 11-6) are the two rotary pump categories discussed in this chapter.

In an Archimedes' screw pump, a helical screw (Figure 11-7) rotates slowly and conveys water at atmospheric pressure trapped between the flights of the screw up a trough or a pipe.

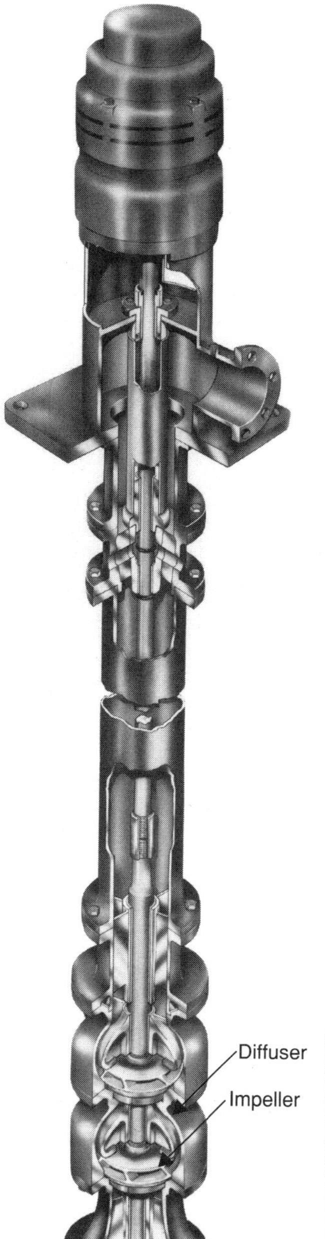

Figure 11-3. A vertical turbine pump. Courtesy of Johnston Pump Co.

Figure 11-4. A plunger pump. Courtesy of Komline-Sanderson Engineering Corp.

Figure 11-5. A rotary lobe pump. Courtesy of Mono Group, Inc.

Figure 11-6. A progressive cavity Moyno™ pump. Courtesy of Robbins & Meyers, Inc.

Figure 11-7. An open screw (Archimedes) Spiralift™ pump. Courtesy of Zimpro/Passavant, Inc.

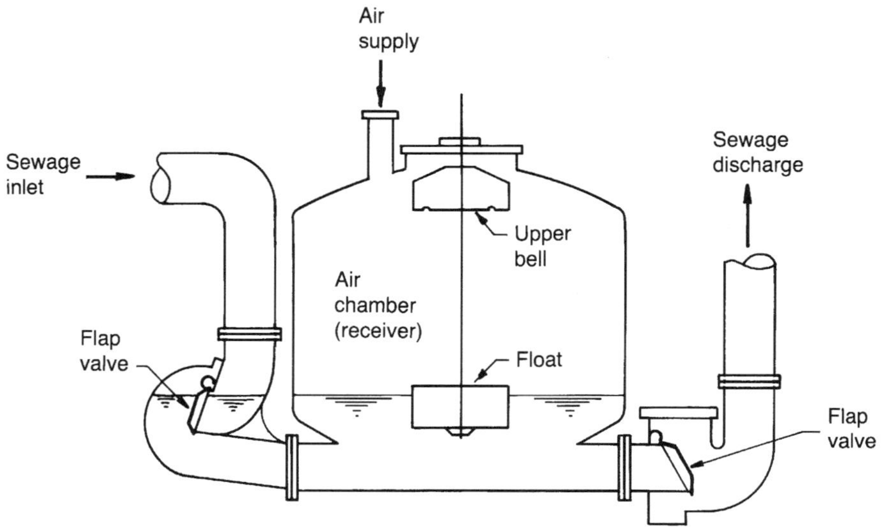

Figure 11-8. A pneumatic ejector. Adapted from Yeomans Chicago Corp.

Pneumatic Pumps

Compressed air is used to move the liquid in pneumatic pumps. In pneumatic ejectors, compressed air displaces the liquid from a gravity-fed pressure vessel through a check valve into the discharge line (see Figure 11-8) in a series of surges spaced by the time required for the tank or receiver to fill again.

The reduced density of a column of an air–liquid mixture is used to raise the liquid in an air lift pump (see Figure 11-9). But because the water is not pressurized, it is not really a pump, is not, therefore, included in Figure 11-1, and should be called a "water-lifting device."

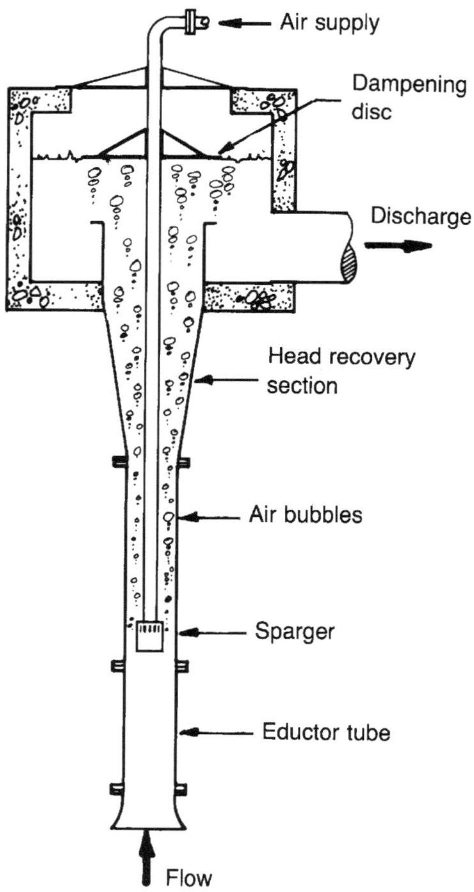

Figure 11-9. An air lift pump. Adapted from Walker Process Corp.

11-2. Classification of Centrifugal Pumps

Centrifugal pumps are subdivided into several categories, as shown in Figure 11-1a.

Overhung-Impeller Pumps

In an overhung-impeller pump, the impeller is mounted at the end of the pump shaft in a cantilever fashion. Both bearings are arranged on the same side of the impeller (see Figure 11-10).

Separately Coupled Pumps

Separately coupled pumps (also called "frame-mounted pumps") have their own shafts, which are coupled to and driven by the driver shaft (see Figure 11-10).

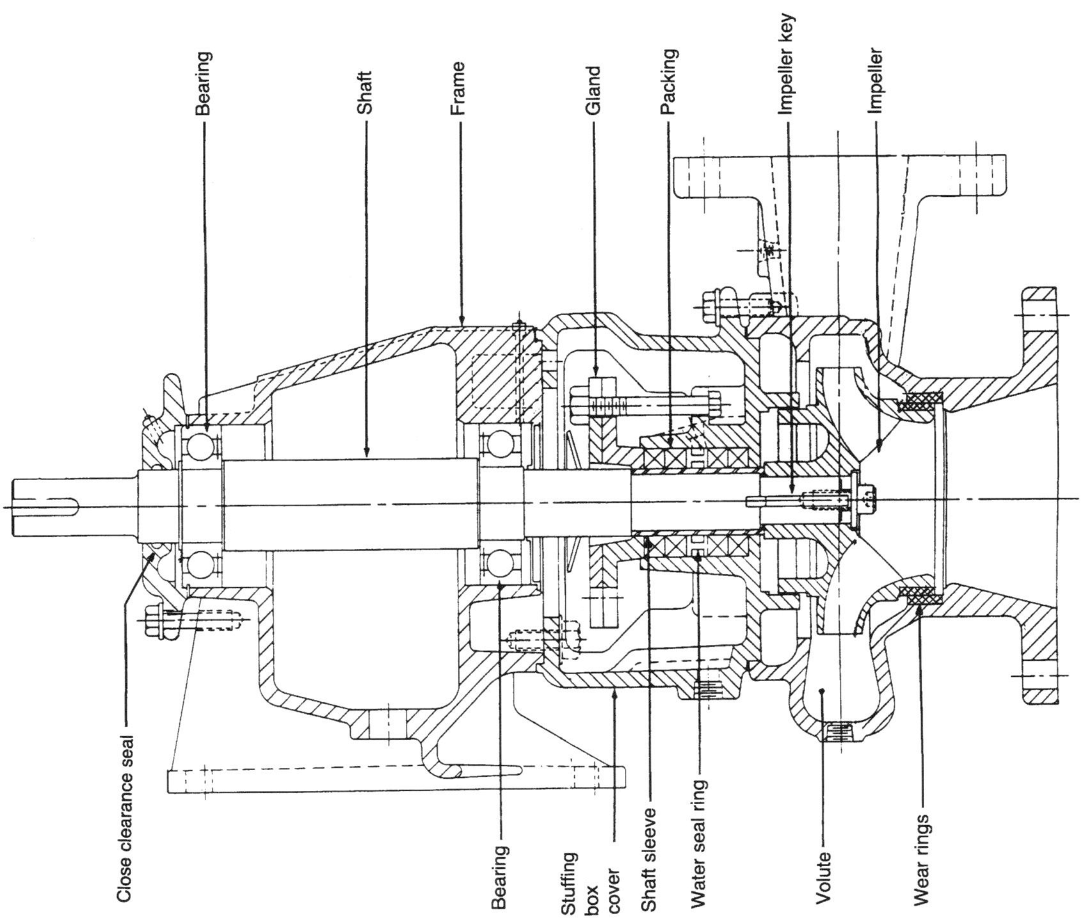

Figure 11-10. An overhung-impeller, separately coupled, end suction, clear-liquid pump. Courtesy of Fairbanks Morse Pump Corp.

Close-Coupled Pumps

The impeller of a close-coupled pump is mounted on the driver shaft. There is no separate pump shaft, and no coupling is required between the driver and the pump (see Figure 11-11).

Submersible Pumps

Submersible pumps are close-coupled pumps driven by a submersible motor and designed for submerged installation in a wet well (see Figure 11-12).

Impeller-between-Bearings Pumps

In impeller-between-bearings pumps, the bearings are mounted on each side of the impeller (see Figure 11-13). Only the axial split-case design is of interest here.

Axial Split-Case Pumps

Axial split-case pumps have a casing that is split along the (usually horizontal) centerline of the shaft. The impellers can be readily exposed for inspection and service by removing the upper half of the casing (see Figure 11-13).

11-3. Construction of Centrifugal Pumps

A representative centrifugal pump and its basic components are shown in Figure 11-10. The function of the components of the pump and the different options available with these components are discussed in the following subsections.

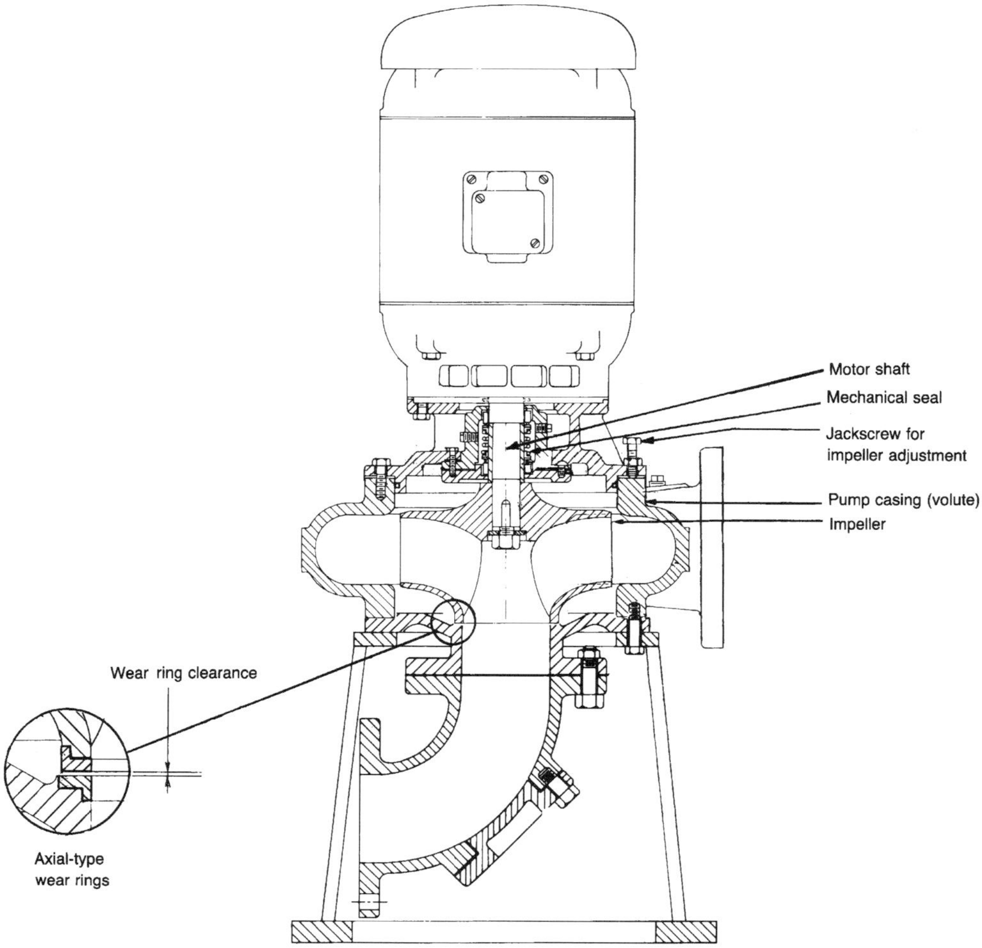

Figure 11-11. An overhung-impeller, close-coupled, end suction pump. After Fairbanks Morse Pump Corp.

Impeller

The impeller increases the velocity of the liquid and raises its pressure. The impellers of centrifugal pumps may be of the radial- or mixed-flow type. The pump characteristics are determined by the impeller type. The relationship between impeller type, specific speed, and pump characteristics is discussed in Chapter 10 and illustrated in Figure 10-8.

Centrifugal pump impellers may be enclosed, semiopen, or open (Figure 11-14). Enclosed impellers are by far the most common. Semiopen impellers are used less frequently; they have, in theory, slightly better efficiency and are easier to cast, but they require close tolerance of axial clearances between the blades and the front cover. The third type, the open impeller, requires close clearances on both sides simultaneously. Because such tolerances are difficult to maintain, open impellers are seldom used in centrifugal pumps.

The pressure of the liquid at the impeller discharge is higher than at the impeller inlet, so the liquid tends to circulate back to the impeller suction. Some provision must be made to contain this recirculation. This is the function of the impeller and casing wear rings in enclosed impellers and the wear plates in semiopen impellers.

The impellers are subjected to radial and axial thrust forces. Radial thrust is caused by unequal pressure distribution in the volute due to volute asymmetry. The theory of radial thrust is presented in Section 10-6.

The axial thrust is caused by the pressure differential between the front and back sides of the impeller. Axial thrust in multistage high-pressure pumps can reach high values. These pumps are sometimes equipped with hydrodynamically balanced impellers,

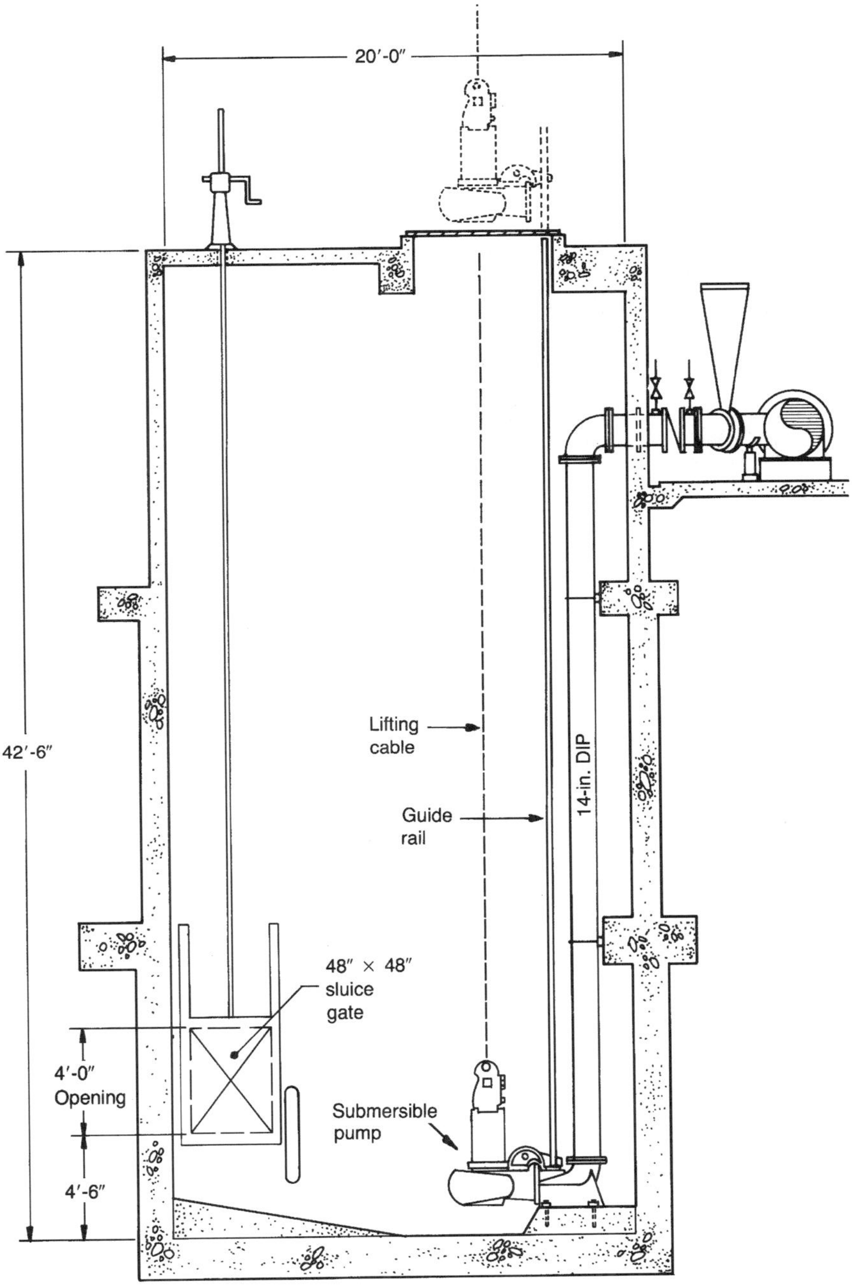

Figure 11-12. Submersible pump (pull-up design) in the Lancaster Reclamation Plant, Los Angeles County Sanitation District No. 14.

as shown in Figure 11-14e. This type of impeller has a balancing wear ring on its back shroud, which, combined with pressure relief (or pressure balancing) holes, equalizes the pressure on both sides of the back shroud and reduces the axial thrust. Although the back shroud wear rings reduce the axial thrust, they also have disadvantages: (1) they complicate the pump design, and (2) the pressure relief holes cause continuous recirculation of the pumped liquid with a resulting loss of efficiency. Consequently, hydro-dynamically balanced impellers are rarely used in the pumps discussed in this book. Instead, the thrust bearings carry

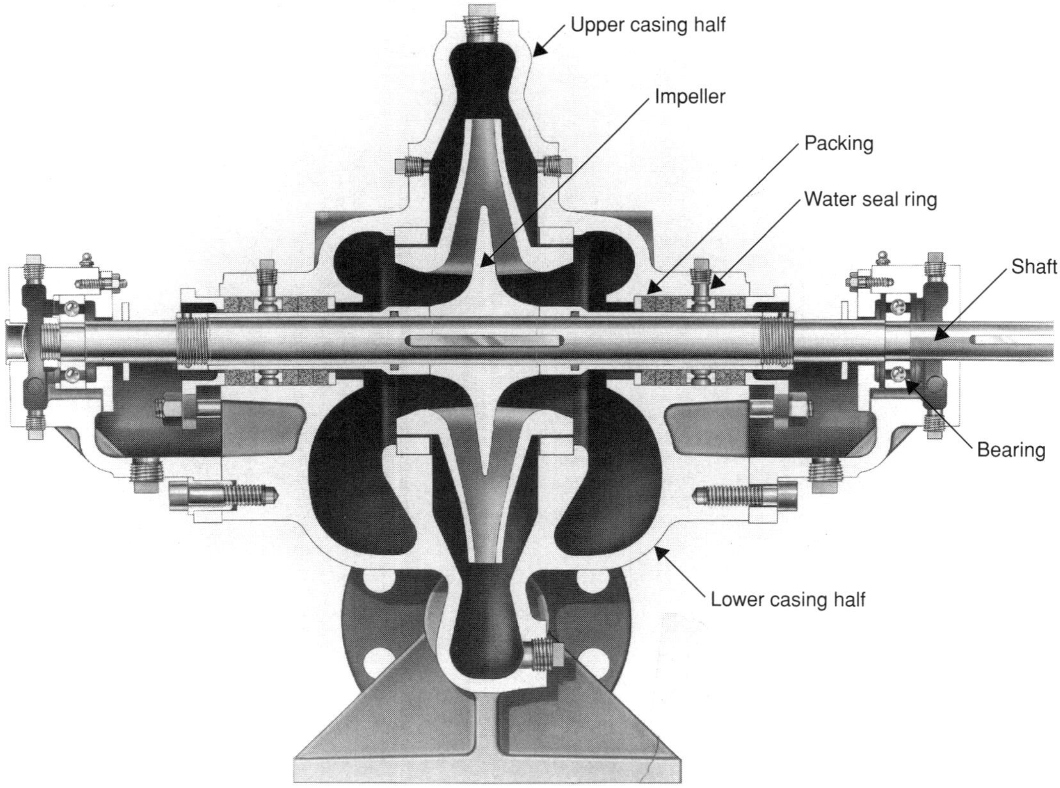

Figure 11-13. An impeller-between-bearings, axial split, single-stage pump. Courtesy of Dresser Pump Division.

the full axial load, which, from practical experience, has been shown to be the simplest technical solution.

Wear Rings

The wear rings control the liquid recirculation between the impeller discharge side and the inlet side. The wear of wear rings is caused by grit or other abrasive matter in the pumped liquid. Rubbing of wear rings also causes wear, but in well-designed pumps, the shaft deflection does not exceed wear ring clearance unless the pumps run outside of their design operating range (or close to shut-off). The wear rings may be separate or integral. Separate rings are machined independently from the impeller and the inlet cover and then pressed or fitted in place and locked by mechanical devices such as setscrews, pins, or locking compounds.

The phrase "integral wear rings" means that no separate wear rings are installed, but that a close clearance is arranged between the impeller and the inlet cover. Enough material is usually provided in the cover and on the impeller so that after the wear ring clearance exceeds the recommended limits, both can be remachined and separate replacement wear rings can be installed.

Wear rings can be of axial, radial, or some special design. Axial or "face-type" wear rings rely on the impeller setting relative to the suction head to maintain the wear ring clearance. They require an axially adjustable impeller or an axially adjustable suction cover (see Figures 11-11 and 11-15b). Axial wear rings have the advantage of being adjustable for wear in the field without the need for pump disassembly or wear ring replacement. They do, however, complicate the pump design. They are used primarily on pumps smaller than about 0.25 m^3/s (4000 gal/min).

Radial-type wear rings are not sensitive to the exact axial impeller location and, hence, do not require axial adjustment of the impeller (see Figures 11-10, 11-14a, 11-15a, and 11-15c). They cannot be readjusted for wear, however, so they must be replaced when leakage caused by wear becomes objectionable.

Some pumps designed for gritty liquids and very long wear ring life are equipped with flushed wear rings (see Figure 11-15c). Flushing liquid, free of abrasive particles and injected under pressure into the wear ring clearance, prevents the entry of the grit.

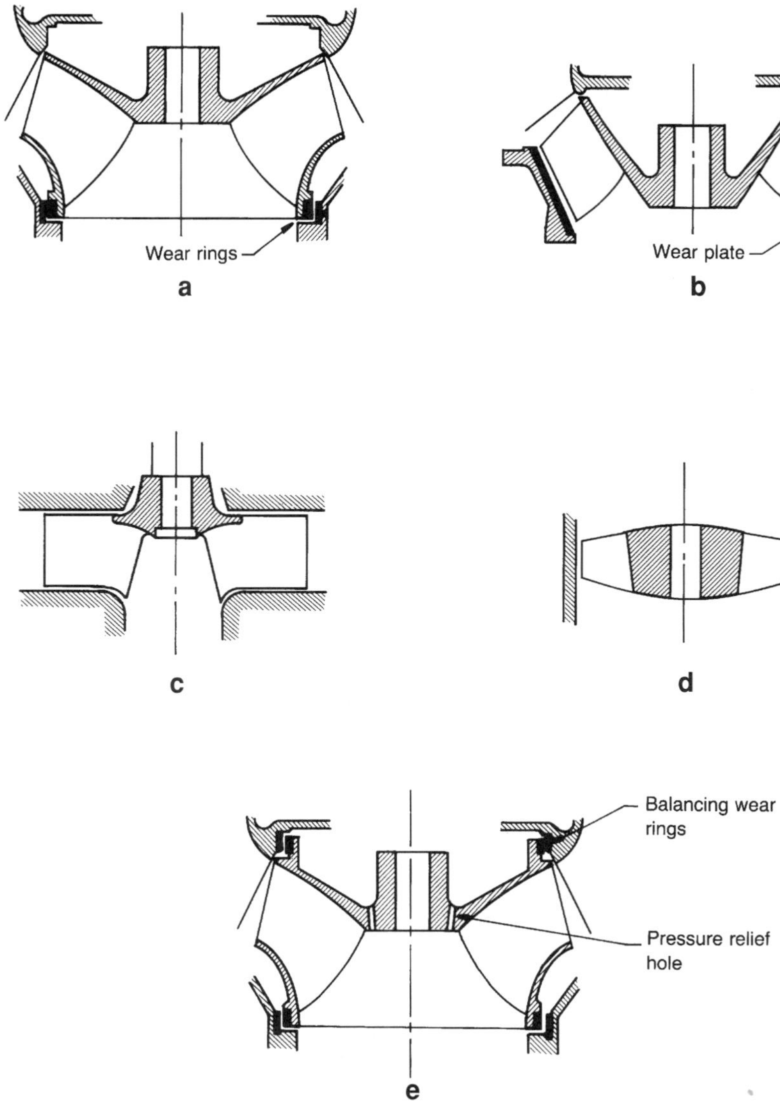

Figure 11-14. Impeller designs. (a) Enclosed impeller; (b) semiopen impeller; (c) open impeller, radial flow; (d) open impeller, axial flow; (e) hydrodynamically balanced impeller. After Fairbanks Morse Pump Corp.

L-shaped wear rings (Figure 11-15d) are another option designed to increase the length of the leakage path and to reduce the leakage rate. For gritty applications, they may be combined with the external flush option to conserve on flushing water requirements. They do require the same axial adjustment as the facetype rings.

Because the wear rings may contact each other under extreme operating conditions, the wear ring material must be both nongalling and compatible with the pumped liquid. A difference in ring hardness of Brinell hardness number 50 on the C scale (50 BHN) is generally recommended especially for stainless-steel rings. Front head rings are usually made of the harder material.

Wear Plates

Semiopen impellers have no wear rings but rely on the close clearance between the vane tip and the suction cover to control the leakage around the vane tips (see Figure 11-14b). When gritty liquids are pumped, wear of the blade tips and of the casing is to be expected. Hence, the suction cover is often equipped with a wear plate, which has a contour matching the blade tip contour and can be axially readjusted to compensate for wear. The wear plates are usually made of abrasion-resistant material such as Type 416, heat-treated stainless steel (e.g., ASTM A 743-CAIS, 350 BHN).

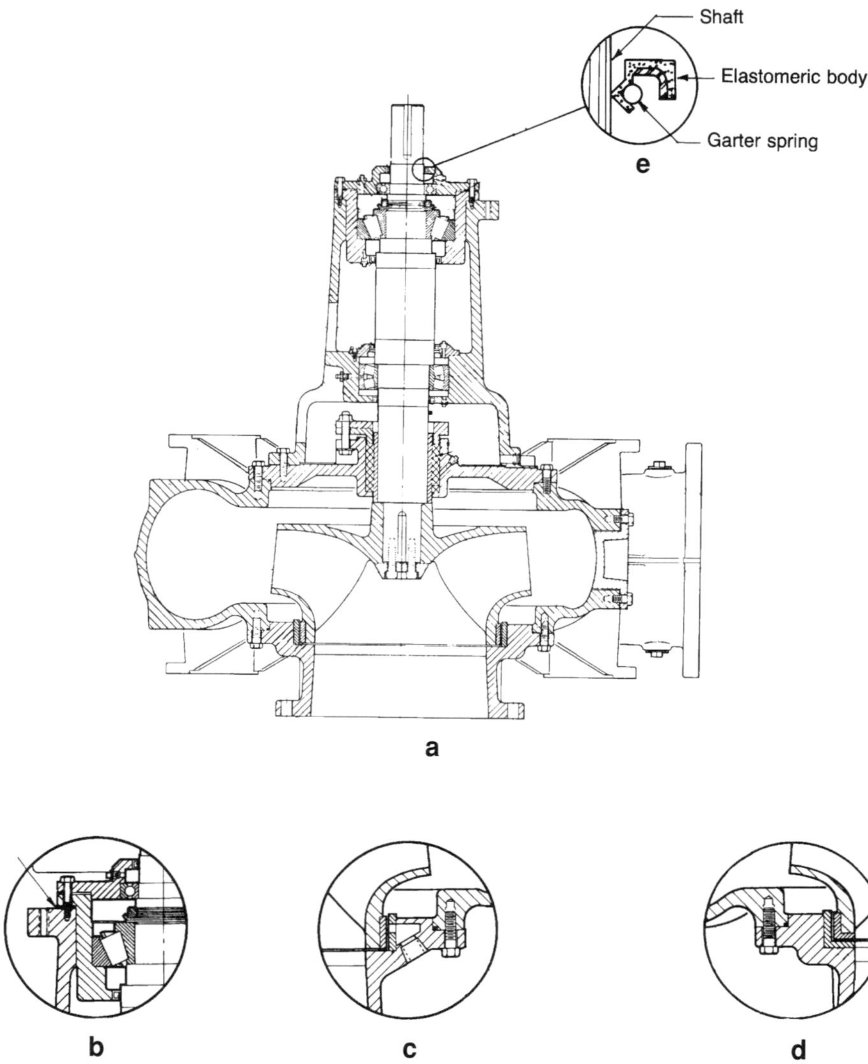

Figure 11-15. A solids-handling pump. (a) Pump; (b) impeller axial adjustment; (c) wear ring external flush; (d) L-shaped wear rings; (e) lip seal. Courtesy Fairbanks Morse Pump Corp.

Pump Shaft

In transmitting power from the driver to the impeller, the pump shaft must support all radial and axial impeller forces. It is subject to torsion, bending, and tension. Good shaft design is needed to ensure that the endurance limit of the shaft material is not exceeded under any anticipated operating conditions. Furthermore, the deflection of the shaft must be controlled to avoid wear ring rubbing and damage to the packing or the mechanical seals. A deflection of 0.05 mm (0.002 in. or 2 mils) at the shaft seal is often given as a recommended limit. Deflection at radial wear rings should be compatible with wear ring nominal clearance.

Shaft Sleeve

Most shafts are equipped with a replaceable shaft sleeve to protect the shaft from wear in the packing area (see Figure 11-10). Typical sleeve material is heat-treated type 416 stainless steel (ASTM A 754-CAIS, 300 to 350 BHN). The sleeve must be sealed to the shaft to stop leakage of the pumped liquid, and either O-rings or anaerobic sealing and locking compounds (which harden in the absence of oxygen) are used for this purpose.

Positive locking devices (such as pins, keys, or notches) are sometimes used to prevent sleeve rotation on the shaft, but the use of sealing compounds is also a successful locking method.

Pump Casing

The pump casing or pump volute encloses the impeller and contains the discharge and, sometimes, the suction nozzles (see Figure 11-10). The casing collects the high-velocity flow from the impeller and converts some of the velocity energy into pressure. The casing may be one piece or of axial- or radial-split design.

Suction Cover

The suction cover (or front head) encloses the suction opening of the casing and also contains the suction nozzle of the pump. On some designs, the suction cover is an integral part of the pump casing (see Figure 11-10).

Stuffing Box Cover

The stuffing box cover (also called the "back head" or "adapter") encloses the inboard opening of the casing and also contains the stuffing box (see Figure 11-10).

Frame and Bearing Housings

Separately coupled pumps with overhung impellers have a frame that contains the pump bearings (see Figure 11-10). The bearings may be mounted directly in the frame or in bearing housings mounted in the frame. Bearing housings facilitate the assembly of the bearings and are used in larger pumps. With horizontal pumps, the frame frequently supports the entire pump (see Figure 11-10).

Bearings

The pump bearings carry the shaft and absorb all of the radial and axial forces acting on the shaft. A great majority of the centrifugal pumps for pumping stations are equipped with antifriction bearings (rollers, tapered rollers, or balls in single or multiple rows). Journal bearings are used predominantly for high-speed and high-load applications (e.g., boiler feed pumps and multistage high-pressure chemical pumps) and are of little interest to pumping station designers.

The thrust bearing of the cantilever-type pump carries not only all of the thrust load but also its share of the radial load. The thrust bearing must be capable of supporting axial loads in both directions because thrust reversal can be expected with most pumps, especially during the starting process. In some designs, a separate (usually smaller) thrust bearing is installed for the thrust reversal. Typically, the thrust bearing is pressed on the shaft and locked in the bearing housing (or in the frame); thus, the thrust bearing determines the axial location of the shaft and the impeller. The radial bearing is also pressed on the shaft, but is free to slide in the frame or in the bearing housing. This protects the bearings from excessive axial loads generated by unequal thermal expansion of the shaft and frame.

The antifriction bearing design is of little consequence as long as the bearings have been selected by the manufacturer to deliver adequate bearing life and the design provides for proper bearing installation and lubrication. The bearing life can be expressed in statistical terms only. Commonly, the L-10 bearing life rating is used. It is defined by the AFBMA as the number of operating hours at a given load that 90% of a group of bearings will complete before the first evidence of fatigue develops. Average life is statistically three to five times the L-10 life (depending on whether the rating is for degassed or nondegassed steel). Average life is defined as the operating hours at which 50% of the group of bearings fail and the rest continue to operate.

The L-10 life must be specified and computed for a selected pump operating condition. The bearing life is a function of bearing load and speed. The impeller radial loads (and, to a lesser degree, the axial loads) vary considerably between the best operating point (bep) and shut-off of the pump (see Figure 10-18). As a result, *the computed L-10 life at shut-off for a typical, single volute pump can be as little as one-hundredth or less of the life at bep.*

A frequent specification is 40,000 h of L-10 life at the operating point for pumps in continuous, 24-h service. Specification of bearing life at shut-off or for the entire head-capacity range of the pump (which is the same as shut-off) should be avoided. Such a specification leads to oversized shafts and bearings and does not ensure the computed bearing life because, at shut-off, the bearings are exposed to shock loading, the effect of which cannot be readily predicted nor provided for. Operation at shut-off can be damaging to the shaft, impeller, and other structural components of the pump as well and, therefore, operation at shut-off for prolonged periods of time should be avoided.

L-10 life in excess of 100,000 h is of no practical value. It can be taken only as an indication that bearing fatigue failures are unlikely. Bearings fail anyway (as shown through experience) due to other causes such as corrosion caused by water contamination or deteriorated lubricants, rolling surface damage due to abrasive particles, and lack of lubricant.

Grease is the most common lubricant for pump bearings. It requires relatively little attention and service and usually gives long bearing life. Oil-bath lubrication is used much less frequently because retaining the oil in the bearing housing poses some technical difficulties, especially on vertical pumps. Force-fed oil lubrication is used very infrequently. Although it ensures the lowest bearing operating temperatures, provides the best protection from bearing contamination, and gives the longest bearing life, it requires expensive auxiliary equipment (which also needs service and attention). In the experience of pump operators, the benefits gained from force-fed lubrication do not justify the additional expense.

Bearing Seals

The bearings must be protected from contamination by abrasive particles or corrosive liquids (including water). Consequently, the bearing frame must be sealed. The following are typical seal arrangements.

Close-Clearance Seals

Close-clearance seals consist merely of a close clearance between the shaft and the frame or bearing housing (see Figures 11-10 and 11-15). They are very simple, have an unlimited life, and require no maintenance. However, they give only limited protection from water spray (e.g., hosing down the pumps) or from condensation during shut-down. Still, close-clearance seals are very successful on continuously operating and well-maintained pumps where the normal bearing operating temperature prevents any water condensation and grease acts as a barrier to outside contaminants.

Elastomer Seals

Elastomer (lip-type) seals are the most frequently used seals in current designs. They give good protection from contamination and moisture, but they also generate friction heat and cause higher bearing-operating temperatures with potentially reduced lubricant life (Figure 11-15e).

Labyrinth Seals

Labyrinth seals offer good bearing protection and generate no heat. They require more space and are relatively expensive. At present, they are offered as optional equipment only. They are suitable only for horizontal shafts.

Pump Shaft Seals

To contain the liquid in the pump, the pump shaft must be sealed against the pressure in the pump. The two basic types of seals are packings and mechanical seals, and both have advantages. The final selection of seal type depends on installation requirements and owner preferences, but packing is recommended for water and wastewater service in any installation where leakage from the stuffing box is not objectionable or on vertical turbine pumps that require dismantling for seal removal.

Packing

Packing is the simplest and most frequently used seal. The packing is usually made of braided synthetic fibers impregnated with special lubricating compounds. The packing rings require continuous lubrication and cooling, accomplished by using the pumped fluid or an external clean water source, so a properly adjusted packing must continuously leak and drip. All packings require periodic inspection and occasional adjustment of the gland (see Figure 11-10).

Packings are frequently equipped with a water seal ring (see Figure 11-10) so that a buffer liquid from an outside source (potable water or chlorinated final effluent for wastewater pumps) or from the pump discharge (for clean water pumps) can be injected into it. This ensures that the packing gets an adequate supply of lubricating and cooling liquid under all operating conditions, particularly when the pressure on the pump side of the packing drops below the atmospheric pressure. The injection of clean liquid is also used to prevent contaminants and gritty materials in the pumped liquid from entering the packing and causing accelerated shaft sleeve wear. If potable water is used as the buffer liquid, an air gap must be provided between the fresh water source and the packing to prevent contamination of the water supply with the pumped liquid.

Although most types of packing in wastewater service must be flushed with clear water to avoid long-term scoring or abrasion of shaft sleeves due to grit in the flow stream, some types of braided, graphite-lubricated and impregnated or Teflon-impregnated, non-asbestos packing have proven to work effectively in such service without any external clear water flushing.

Mechanical Seals

Mechanical seals have the advantages of no leakage and no need for periodic maintenance, but they are relatively expensive and require care in installation. Some are subject to catastrophic failure. For mechanical seal replacement, the pump must be partially disassembled, which is costly. Mechanical seals should therefore not be specified indiscriminately. Note that API 610 has an excellent discussion of mechanical seals.

A typical, simple, single-face mechanical seal is shown in Figure 11-16a. Mechanical seals rely on very

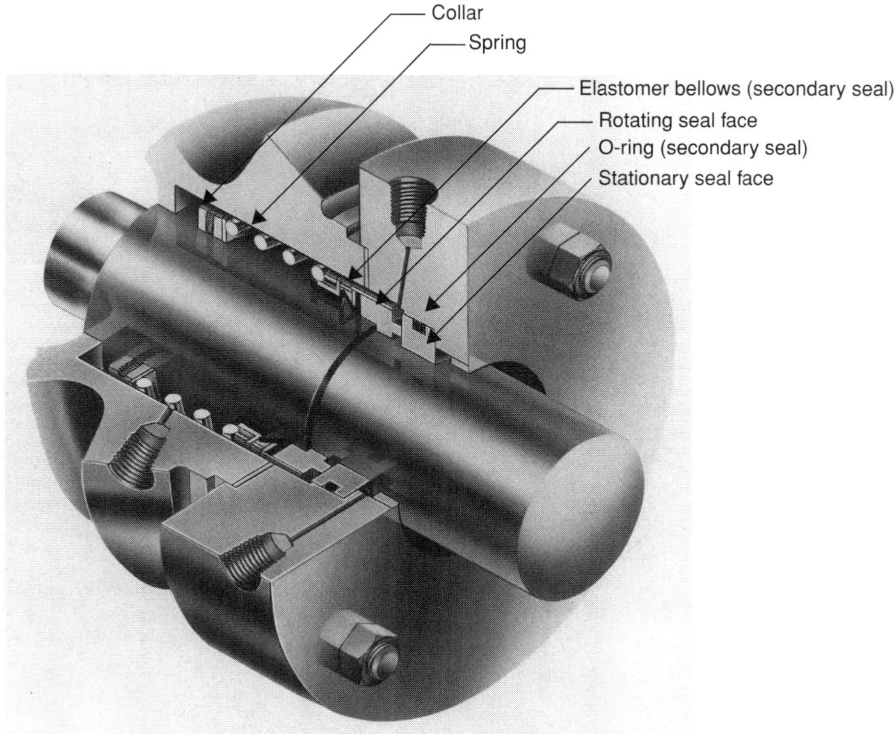

a

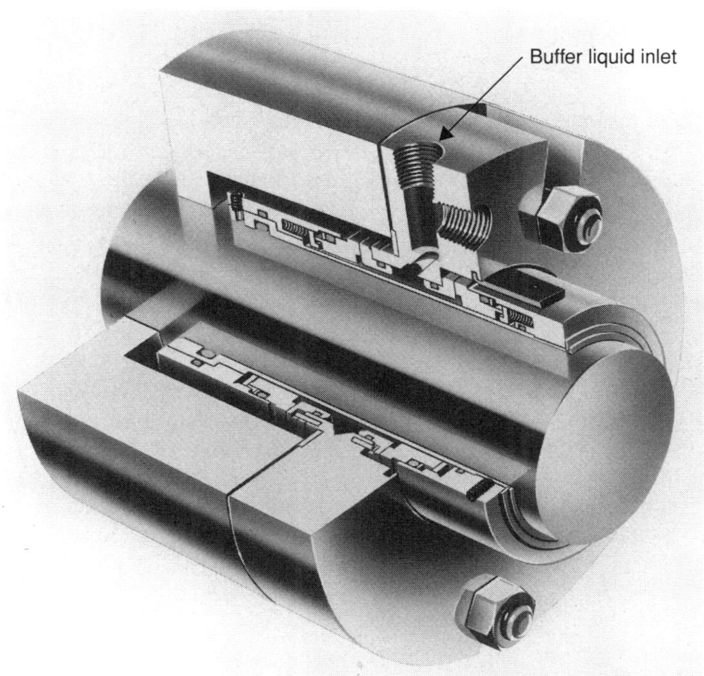

b

Figure 11-16. Mechanical seals. (a) Single-face mechanical seal; (b) double-face mechanical seal. Courtesy of John Crane, Inc.

close clearance between the stationary and rotating seal faces for sealing. Both faces are lapped flat to within approximately a micron. They are lubricated and separated from each other by an extremely thin film of the pumped liquid. The most frequently used sealing face materials are ceramic for the stationary seal and carbon (sintered graphite) for the rotating seal. The secondary seals that prevent leakage around the sealing faces are rubber or synthetic elastomer compound O-rings or bellows.

Double-face mechanical seals (Figure 11-16b) are installed when the pumpage contains abrasives (e.g., wastewater containing gritty material). Clean liquid is injected into the cavity between both seal faces at a recommended pressure of 70 kPa (10 lb/in.^{2}) above the sealed liquid pressure. It lubricates and cools both seals and prevents intrusion of the pumped liquid between the inner seal faces. A wide variety of seal designs, arrangements, and materials is available for corrosive or toxic liquids, high pressures, and high surface velocities, but these exotic designs are not required in water and wastewater service.

Separately Coupled Pump Design

A pump designed for a separately coupled driver is shown in Figure 11-10. The pump itself is totally self-sufficient and directly mounted on a base or a foundation. The driver is either separately mounted on the same base or foundation, or it can be mounted directly on the pump frame. A flexible coupling is usually used to connect the pump to the driver.

Close-Coupled Pump Design

Close-coupled pumps have the pump impeller installed on the motor shaft and the pump casing bolted directly to the motor (see Figure 11-11). Compactness and low initial cost are the prime objectives of this concept. The motor shaft of close-coupled pumps must resist all impeller radial and axial thrust forces. Hence, the motor bearings must carry the pump impeller loads in addition to normal motor electromagnetic and torsional loads. Standard flange-mounted motors are used for the close-coupled pumps when the shafts have adequate strength and rigidity and when the bearing life meets the specified requirements. Special motors with bigger bearings and shafts must be supplied when the loading exceeds standard motor capabilities.

Submersible, Close-Coupled Pump Design

Submersible, close-coupled pumps operate while immersed in the pumped liquid. The submersible motors are single-purpose machines designed specifically for this application (see Figures 11-12 and 11-17). The motor housing is hermetically sealed to prevent the intrusion of the pumped liquid. The motor is usually equipped with two mechanical seals, and the space between the seals is filled with oil. To compensate for the thermal expansion of the oil, the oil chamber must contain a pressure-limiting device, which is usually an air cushion or a sealed bladder. *The pressure-limiting device is important. Several serious accidents have occurred in which a pump exploded when lifted from the wet well soon after being stopped.* The cause of the explosion was the expansion of the oil due to stored heat that was no longer dissipated by water.

The motor housing itself may be filled with air or with oil. The oil-filled motor is less sensitive to the intrusion of pumped liquid, but the churning oil increases friction and lowers efficiency. Oil-filled motors are generally not supplied with larger pumps.

The motor junction box must also be hydraulically sealed at all joints, especially at the cable entry. Furthermore, the box is usually equipped with a hydraulically tight bulkhead for further protection of the motor windings from leakage.

Submersible motors should always be equipped with moisture-sensing devices. These devices are frequently installed in the seal oil chamber and can be used to initiate an alarm when the outer mechanical seal starts leaking. Some manufacturers install moisture probes in the motor itself to indicate moisture intrusion from any source.

The motors may be externally or internally cooled. Externally cooled motors rely on liquid in the wet well for cooling and must be at least partially immersed for continuous operation at full load. They can be operated in air only for short periods. Otherwise, they must be derated for continuous operation. Internally cooled motors circulate some of the pumped liquid through the motor-cooling passages or through a special motor-cooling jacket and can thus operate in the air without any restrictions.

11-4. Overhung-Impeller Pumps

Overhung-impeller pumps are the most frequently used type in pumping stations for pumping clear water, wastewater, and thin sludge. Different designs are

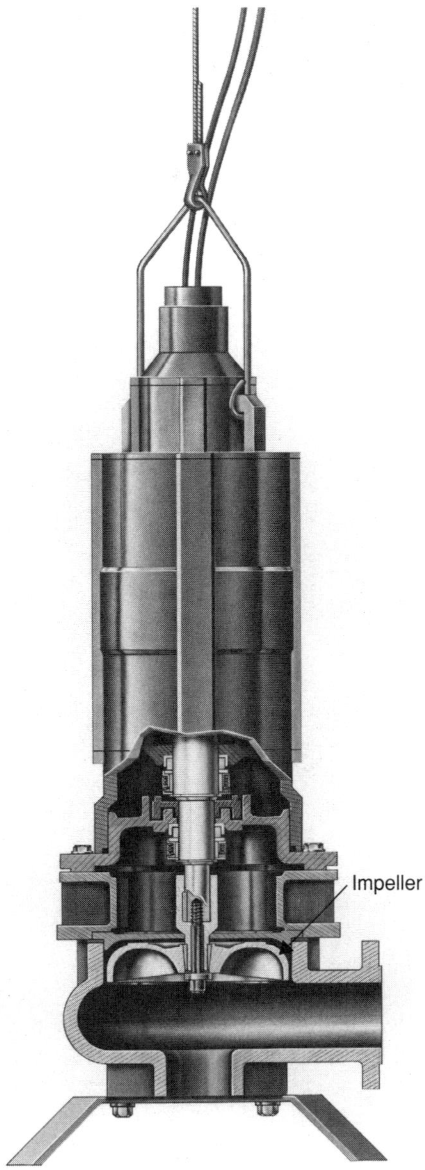

Figure 11-17. A submersible vortex pump. Courtesy of WEMCO.

required for different applications. The types of greatest interest to pumping station designers are pumps for clear liquids; nonclog pumps; wet well volute pumps; vertical turbine, solids-handling (VTSH®) pumps; self-priming pumps; vortex pumps; cutter pumps; and grinder pumps (see Figure 11-1a). Some of these pumps are available in separately coupled, close-coupled, and submersible designs.

Clear-Liquid Separately Coupled Pumps

The pump shown in Figure 11-10 is a clear-liquid, end-suction, separately coupled pump with an overhung impeller and is available in the horizontal model shown or in a vertical model. In vertical models, the driver may be mounted either on the frame (Figure 11-18a) or on a floor above the pump (Figure 11-18b) with a long line shaft. Line shafts usually require intermediate supports and bearings.

Impeller. The impeller of the pump typically has seven to nine vanes, the optimum for high hydraulic efficiency. Usually, the impeller is of the enclosed design.

Wear rings. Both types of wear rings—radial and axial—are common.

Shaft seal. Packing, as well as mechanical seals, are used for the shaft. Customer preference is the deciding factor.

ANSI B73.1 and B73.2 pumps. ANSI B73.1, "Horizontal End-Suction Centrifugal Pumps for Chemical Process," covers separately coupled pumps and basic design requirements. Most important, pump mounting dimensions are standardized, so that pumps made by different manufacturers can be interchanged without piping modifications.

ANSI B73.2, "Vertical In-Line Centrifugal Pumps for Chemical Process," covers vertical, separately coupled, and close-coupled pumps. It is similar to B73.1 in all other aspects. ANSI pumps can be an economical selection for clear liquids when head and capacity requirements fall within the ANSI standard range of 2.3 to 220 L/s (37 to 3500 gal/min) and 9.8 to 61 m (32 to 200 ft) head.

Clear-Liquid, Close-Coupled Pumps

Clear-liquid, close-coupled pumps are popular, especially for smaller sized pumps (Figure 11-19). These pumps have the advantages of relatively lower cost, compactness, and ease of installation. Close-coupled pumps are hydraulically similar or identical to separately coupled pumps and have the same options with respect to seals, wear rings, and materials. They are available in horizontal or vertical models. Both close-coupled and separately coupled pumps are described in ANSI B73.2.

A decided disadvantage of some close-coupled designs is that the motor must be removed to repack or replace mechanical seals, and the extra cost for maintenance, therefore, is likely to offset the slightly greater cost of a separately coupled unit.

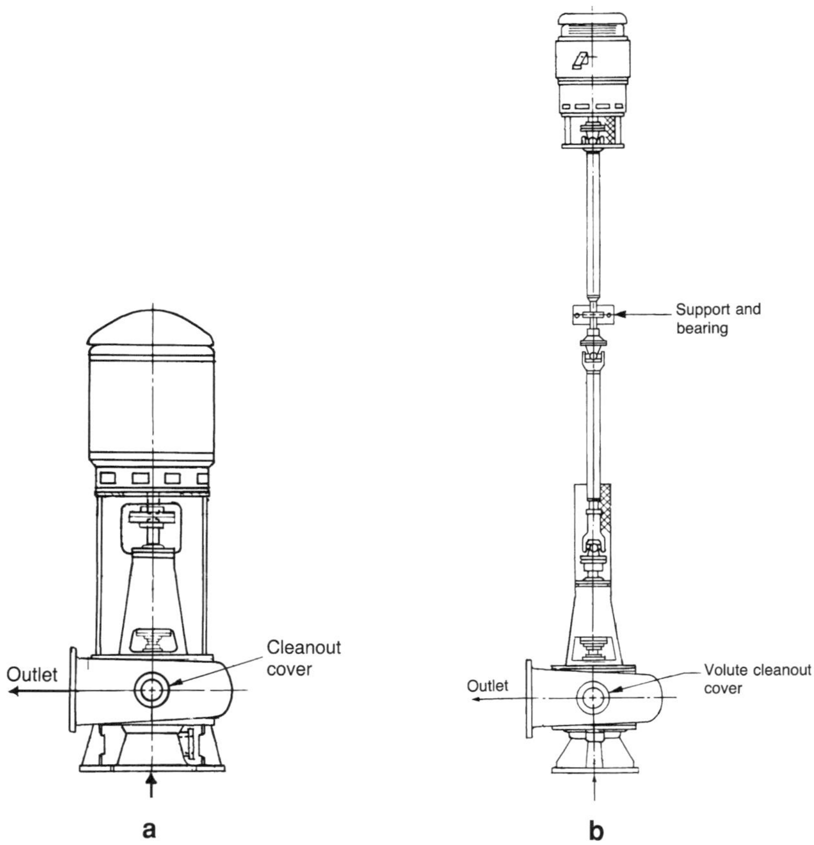

Figure 11-18. Vertically mounted, overhung-impeller pumps. (a) Frame-mounted motor; (b) pump with extension shaft. After Fairbanks Morse Pump Corp.

Separately Coupled, Nonclog Pumps

Nonclog (or solids-handling) pumps were developed to pump liquids containing soft solids and stringy material without plugging or needing frequent service and cleaning. Separately coupled, nonclog pumps are available for vertical mounting (see Figure 11-15) or for horizontal mounting (similar to the pump of Figure 11-10). Vertical pumps may be driven by a motor installed on the pump frame (Figure 11-18a) or by a line shaft to a motor installed on a floor above the pump (Figure 11-18b).

Externally, separately coupled, nonclog pumps are similar to clear-liquid pumps. Internally, there are some significant differences (compare Figures 11-15 and 11-19).

Impeller

A good, nonclog impeller has vanes with a hydraulic foil cross-section and, to prevent catching stringy material, blunt, well-rounded leading edges. Small pumps have only one or two vanes. Larger pumps are built with two or three vanes. A reduced number of vanes increases the size of flow passages and allows larger solids to pass.

The solids-handling capability of the pump is customarily defined as the solid-sphere size that will pass through the pump. It is accepted in the industry that passing a sphere 75 mm (3 in.) in diameter is required for smaller pumps and that a 100-mm (4-in.) sphere is acceptable for virtually all capacities. It should be noted that a restricting area may exist in the pump casing as well as in the impeller, and the pump specification should clearly indicate the minimum sphere diameter for the pump, not just for the impeller.

Wear Rings

Pumps can be equipped with axial, radial, or special design wear rings. Because grit is frequently present in wastewater, wear ring life is an important consider-

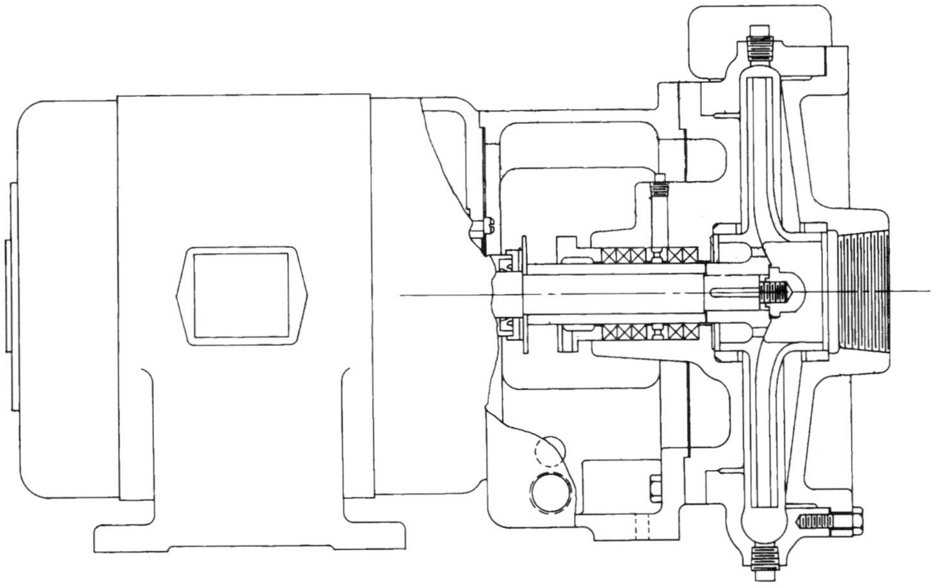

Figure 11-19. Close-coupled, clear-liquid pump. After the Hydraulic Institute [1].

ation. Optional design wear rings are sometimes specified for larger pumps to extend their life. These designs may comprise flushed rings, L-shaped rings, or rings made of special, abrasion-resistant ring materials, including special (sometimes ceramic) coatings. A typical mechanism for the axial ring adjustment is shown in Figure 11-15. The thrust bearing housing is equipped with jackscrews, which allow the bearing housing to move axially so that shims can be inserted between the frame and the housing flange. This moves the rotor assembly and sets the clearance of wear rings.

Cleanouts

The pump casing and suction nozzle are usually equipped with hand-hole covers for inspection of the impeller and easy removal of any trash caught in the pump. Cleanouts in the volute provide access to the volute cutwater, and cleanouts in the front head provide access to the impeller inlet—the two critical areas. The cleanout covers should not obstruct the flow and should be designed to be flush with the surface of the water passage (Figure 11-18).

Seals

The packings of solids-handling pumps should have a water seal ring with clean-liquid injection to prevent grit intrusion and shaft abrasion. Grease injection is sometimes used when clean water is not available.

Mechanical seals, of either single- or double-face design, are also frequently specified for nonclog pumps. Double-face mechanical seals usually have carbon and ceramic faces. They require clean-liquid injection into the seal cavity to prevent contamination of the seal faces with abrasive particles. When single-face mechanical seals are used, particular attention must be given to abrasion resistance of the seal faces, and high-grade and high-hardness materials should be chosen. Ceramics or tungsten carbide are frequently used. Seals must be arranged in the housing so that adequate flushing and cooling of the seal faces is ensured.

Close-Coupled, Nonclog Pumps

Close-coupled sewage pumps are designed for low initial cost and for compactness (Figure 11-11). The motor shaft supports the impeller just as in the close-coupled, clear-liquid pump shown in Figure 11-19. The impeller and volute, however, are designed as in Figure 11-15, so the hydraulic components of close-coupled pumps are identical or very similar to those of separately coupled pumps. In general, they have the same design and performance features and offer the same options as separately coupled pumps.

Customer resistance is sometimes encountered in proposals for the use of close-coupled pumps in sewage service because of the reputation of some of these pumps for being lightly built. Low initial cost may not be true economy if the life of various pump components is short and if downtime and maintenance costs are unusually high. Therefore, check the bearing life of the pumps for the actual operating conditions and compare the life with the anticipated service requirements (e.g., 24-h duty or 8-h duty). Also check the shaft stresses and shaft deflection both at the seals and at the wear rings to confirm that rubbing and rapid wear will not be a problem. Check the design of the stuffing box; it should be practical to remove the seal without disturbing either the motor, the volute, or the discharge piping. The maintenance cost and the downtime can thus be kept competitive with those of separately coupled pump designs.

Submersible Nonclog Pumps

Close-coupled, submersible nonclog pumps are equipped with motors designed to operate immersed in the wastewater. Because they are installed directly in the wet well (see Figure 11-12), pumping stations with submersible pumps require no dry well. They may have no aboveground construction whatever except for a concrete slab and a small housing for the control center. The construction cost is therefore less than that of the wet well-dry well stations as shown in Figure 29-9.

On the other hand, submersible pumps are not readily accessible for inspection and service, give little warning of incipient problems, and require shipment to qualified service centers for any kind of motor or seal repairs. This tends to lead to higher service and maintenance cost and longer turnaround times. Statements about low maintenance costs can be misleading when based on short-term (e.g., 2-yr) experience.

Pumps with externally cooled submersible motors must be immersed for continuous operation and can run only for short periods of time if exposed to air. Pumps with motors internally cooled by the pumped liquid can be run continuously at full load without immersion. The same is true for some (but not all) oil-filled submersible motors. Pumps that can operate dry continuously at full load are also available for dry pit installation. They are not damaged if the dry well is accidentally flooded, but this advantage comes at the expense of higher cost of the pump and motor, a slightly greater energy requirement (because of less efficiency), and a larger structure (because of the added dry pit). Submersible, nonclog pumps are frequently used for sump pumps.

Submersible pumps can be installed with fixed-discharge piping and can be supported by a tripod or a similar device mounted on the wet well floor. This type requires draining of the wet well for any kind of inspection, service, or maintenance. A more popular method for installing submersible pumps is the pull-up design (Figure 11-12) in which the discharge piping is connected to a special elbow that is permanently mounted on the wet well floor. The elbow and the pump discharge nozzle are equipped with a self-locking coupling. Because the pump-mounting bracket slides up and down on two rails, it can be raised from the wet well or lowered onto the elbow by means of a crane and a cable with no need for personnel to enter the wet well or to drain it. In some designs, the self-locking suction elbow of the pull-up design can allow considerable leakage, which reduces efficiency and capacity. Hence, *it is advisable to specify that pump tests include the suction elbow.* Such pump tests, however, do not guarantee efficient field performance because the joints can be subject to deterioration under actual operating conditions.

The pump ends of submersible pumps are very similar to those of the dry well solids-handling pumps described previously. Some manufacturers offer the same pump ends for dry well and wet well applications.

Wet Well Volute Pumps

The wet well volute pump is mounted on top of the wet well with the pump itself immersed in the liquid (Figure 11-20). The wet well pump concept eliminates the need for a dry well and provides significant savings in the pumping station cost and surface area requirements. Immersion of the pump in the wet well makes pump priming unnecessary and reduces NPSH requirements.

Impeller

The pump impeller design may be single or double suction (Figure 11-20), closed or semiopen. In all other respects, the impeller is similar to the clear-liquid or solids-handling pump impellers described previously.

Volute

Well-designed wet well volute pumps are equipped with double volutes. As discussed in Chapter 10, a standard volute with a single cutwater exerts radial thrust on the impeller, especially when the pump operates off the bep. The journal bearings of wet well pumps are lubricated by the pumped liquid and, therefore, have a very limited radial load capability. A dou-

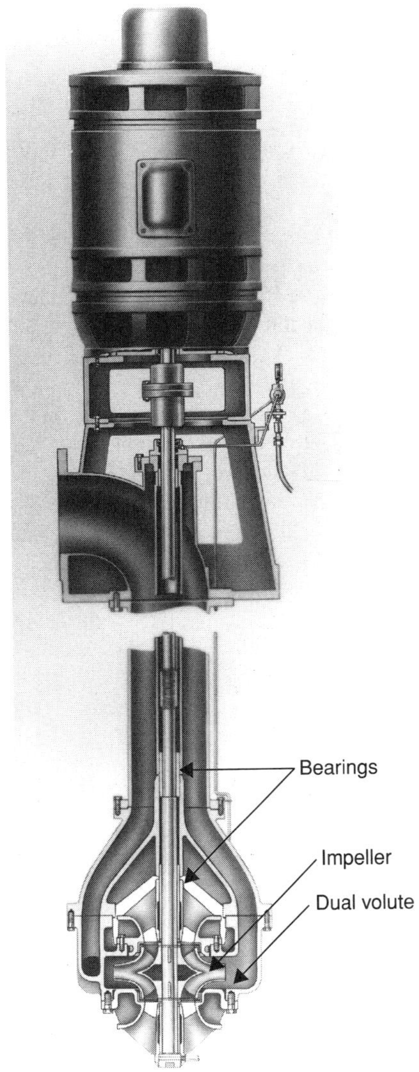

Figure 11-20. A wet well volute pump. Courtesy of Dresser Pump Division.

ble volute balances virtually all of the radial loads and resolves the bearing overload problem. The drawback of a double volute is that it significantly reduces the flow passage area, which except for large pumps normally prohibits using it for pumping wastewater.

Shafting

Wet well volute pumps are driven by lineshafting in an enclosing tube, which is similar to the shafting of vertical pumps (refer to Section 11-8 for a more detailed description).

Bearings

Journal-type, water- , oil- , or grease-lubricated bearings are used in a wet well volute pump. They are similar or identical to vertical pump bearings (for additional information, see Section 11-8).

Vertical Turbine, Solids-Handling (VTSH®) Pump

The VTSH® pump is a wet well, solids-handling pump that combines the advantages of the classical solids-handling pump with the well-proven vertical pump concept (Figure 11-21). The VTSH® pump is the latest arrival in the nonclog pump technology. It is a proprietary, patented design.

The pump is installed on top of the wet well and requires no dry well. The driver, either a motor or an angle gear, is mounted on top of the pump discharge head. Standard drivers with any desired design or control options are used. The driver is readily accessible for service or maintenance. The symmetrical bowl design eliminates radial thrust forces. Although the pump itself is expensive, the total cost of a pumping station may be reduced by using it because the pump concept eliminates the need for a dry well.

Impeller

The pump impeller is an enclosed, mixed-flow, solids-handling design with two vanes and a large solids-sphere diameter.

Bowl

The pump bowl replaces the volute of a standard nonclog pump. It contains a symmetrical, two-vane, nonclog-design axial diffuser that balances any radial loading on the impeller.

Bearings

The pump shaft is supported by two bearings located in the bowl. The bearings are rubber or bronze and are lubricated with water, oil, or grease. Because the radial thrust of the impeller is balanced and the axial load is carried by the motor bearings, there is no danger of bearing overload even with water lubrication.

Column, Lineshaft, Head, and Lineshaft Bearings

The remaining pump components are the typical vertical pump components discussed in Section 11-7.

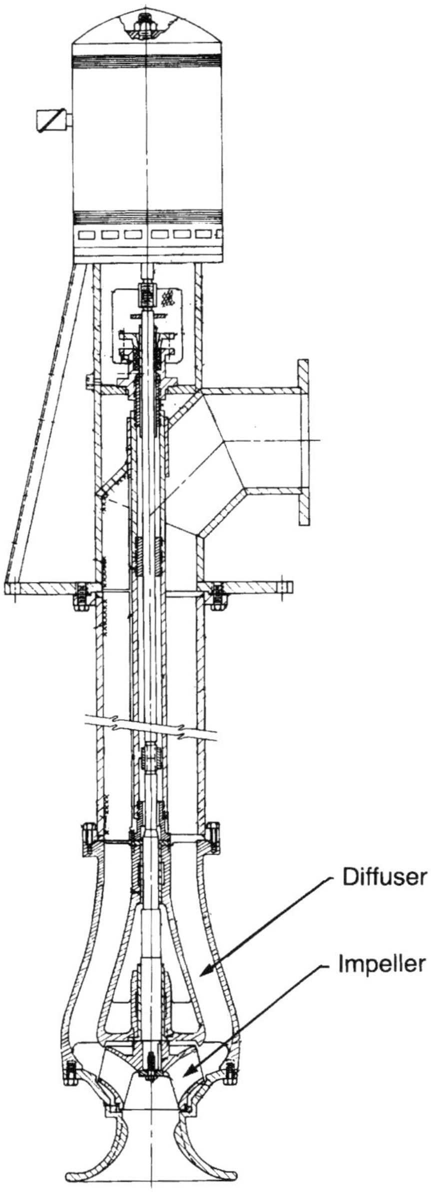

Figure 11-21. A water-lubricated, vertical turbine, solids-handling (VTSH®) pump. After Fairbanks Morse Pump Corp.

Self-Priming Pumps

Self-priming pumps are designed for automatic starting with suction lifts up to 7.6 m (25 ft). They require no external priming system. They are used for wastewater pumping in wet well stations where they can be installed above the wet well. No dry well is needed. Another popular application of self-priming pumps is the use of portable units for construction site dewatering.

Self-Priming Feature

A cross-section of a typical self-priming pump is shown in Figure 11-22. The pump has a large casing, which completely surrounds the volute. The lowest point of the volute contains an opening to the casing. The inlet and outlet openings of the casing are arranged above the impeller so that when the pump stops, the casing remains partially filled with the pumped liquid. A vent line to the wet well is provided in the discharge line (not shown).

At start-up, the liquid contained in the impeller inlet is displaced by the impeller into the volute and discharged into the casing. Some of the liquid returns through the volute opening into the impeller periphery and continues to circulate back to the casing. The recirculation of the liquid creates reduced pressure in the impeller eye, which draws the air into the suction line of the pump and mixes it with the recirculating liquid. The air separates from the liquid in the casing and is vented to the wet well or the atmosphere through the vent line. The priming action continues until all of the air is evacuated from the suction line and the liquid rises to the pump inlet level, at which time normal pumping operation begins.

The casing contains an inlet check valve, which helps to keep the suction and discharge lines filled with liquid upon shut-down and reduces the priming time for subsequent start-ups.

Impeller. The impeller of the pump is of the typical nonclog design. The pump characteristics with respect to performance and solid spheres are the same as for a standard nonclog pump of the same size. The pump efficiency is somewhat lower, due to the continuous loss of liquid through the vent line and due to the effect of the volute opening.

Wear rings. The impeller is usually equipped with axial design wear rings. The axial clearance of the wear rings is adjusted with shims at the thrust bearing.

Hand holes. Two hand holes with quick-access covers are provided for inspection and for removal of any debris from impeller intake and from the check valve.

Pump seals. Pump seals are usually of the double-face mechanical seal type. They are filled with filtered liquid from the pump discharge or with oil from an oil reservoir. This arrangement prevents dry seal operation and intrusion of air during the priming cycle.

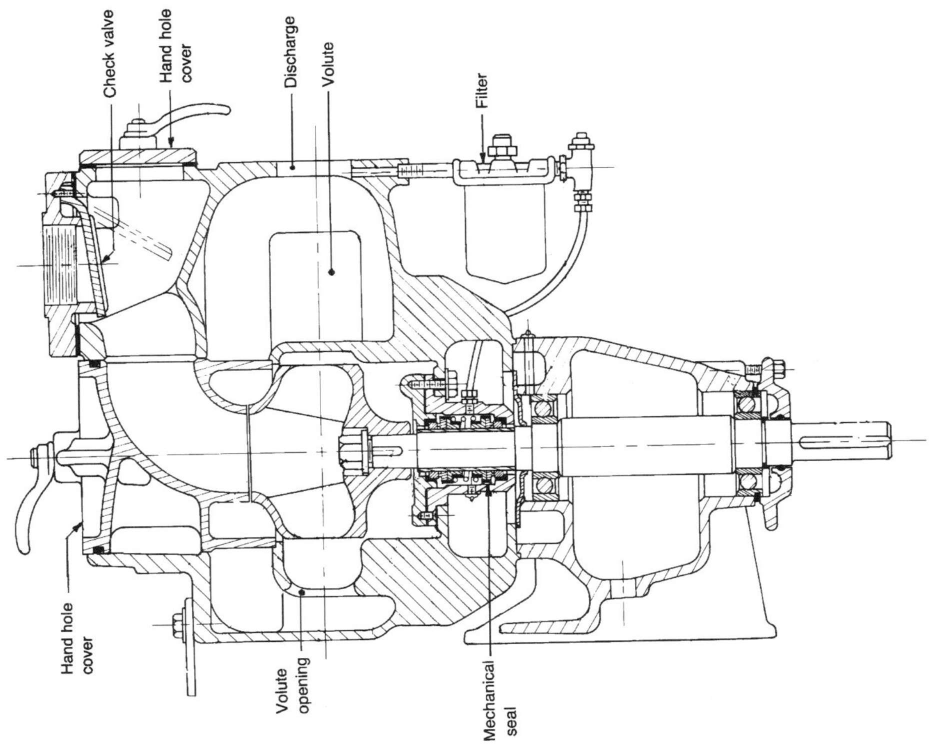

Figure 11-22. A self-priming pump. After Fairbanks Morse Pump Corp.

Vortex Pumps

Vortex pumps (Figure 11-17) are designed to pump large solids and gritty sludge or slurry. The impeller of the vortex pump is recessed in the stuffing box cover, and the impeller vanes do not extend into the pump casing. They induce a strong vortex in the space between the impeller and the suction cover. The pumped liquid does not have to pass through the impeller itself, but flows directly to the volute and discharge nozzle. Solids with a sphere size slightly smaller than the suction and discharge nozzle diameter will pass through the pump unimpeded. Vortex pumps also have good abrasion resistance due to the flow pattern through the pump and, therefore, are ideal for pumping gritty and abrasive sludge. The casing can be made of abrasion-resistant material such as Ni-Hard (ASTM A 532, Class I, Type A) or conventional cast iron (ASTM A 126).

Vortex pumps are frequently used for raw sewage in residential areas where the flows are low and where even the smallest single-vane, nonclog pump can no longer be selected to pass the required sphere size without being too big for the required flow and without being forced to operate in the low-flow, reduced-efficiency range near shut-off. Unlike other pumps, vortex pumps can be operated at shut-off without damage. The hydraulic efficiency of vortex pumps is low (about 35%), which restricts their use to small capacities. The H-Q performance curves of vortex pumps are very flat, and they can develop only low heads.

In other respects, the design of the vortex pump resembles the standard nonclog pump design. Vortex pumps are available for wet well installations (as shown in Figure 11-17) with submersible motors, but dry well installations are more prevalent.

Cutter Pumps

A cutter pump is a small overhung-shaft, close-coupled centrifugal pump in which an impeller is fitted with a special steel cutter cone whose two vanes are an extension of the impeller vanes (Figure 11-23). The cutter vanes and a cutter knife (both with tungsten carbide cutting edges) are adjusted to clear the vanes by 0.05 mm (0.002 in.) and, thus, create a shearing device that easily chops rags, disposable diapers, sanitary napkins, overalls, or even wire coat hangers into

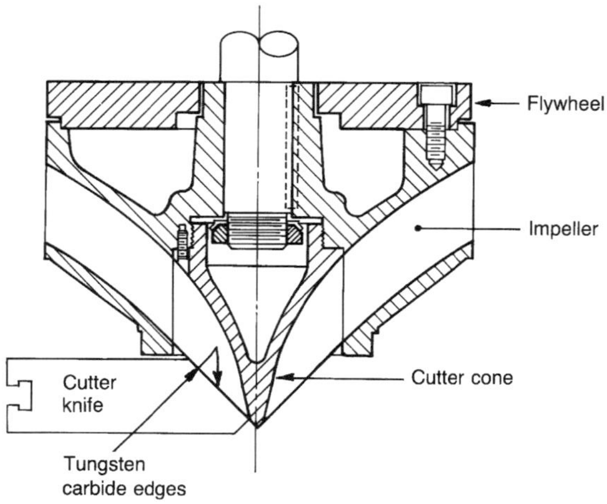

Figure 11-23. Impeller and cutter from a cutter pump. After Prosser-Enpo Industries, Inc.

25-mm (1-in.) pieces. Objects such as blankets, however, go about halfway through before the motor stalls, and pantyhose can clog the pump because the threads are too fine to be completely cut. Both submersible and dry well types with explosion-proof enclosures are available.

These units are for use in applications containing wastewater heavily laden with debris where some form of comminution is indispensable. Examples include raw household sewage; wastes from hospitals, prisons, and asylums; and industrial wastes from fabric mills, pulp and paper mills, and poultry-, fish-, and vegetable-processing plants. Cutter pumps do not develop high heads, but they can be used to precede a positive displacement pump (see also Section 19-2) to obtain a very high head.

Grinder Pumps

Grinder pumps are another solution to the clogging problem of small-capacity solids-handling pumps. The inlet of the grinder pump is fitted with a stationary, serrated, steel cutting ring. A close-fitting, lobed rotor is mounted in front of the impeller and shreds any solid material in raw sewage. The pump impeller and pump discharge are designed to match the flow and head requirements and are not affected by solids size.

The pump can be selected to match the application flow requirement and to operate at or near its bep. It is, therefore, of relatively small size and higher efficiency. It has the drawback of higher complexity and short life.

The pumps are usually submersible and motor driven, and their installations resemble those of small vortex or small standard nonclog pumps, except that they are seldom used as pull-up designs.

11-5. Impeller-between-Bearings Pumps

The impeller-between-bearings pumps have one bearing arranged on each side of the impeller, so the radial load of the impeller is shared equally by both bearings. The loads on the bearings are less, and the shaft has lower bending moments than has an overhung shaft. On the other hand, impeller-between-bearings pumps require shaft seals at two locations and two separate bearing housings.

Impeller-between-bearings pumps are available in axial-split and radial-split designs. Because radial-split pumps are not used in pumping stations, they are not considered here.

Axial-Split Pumps

A typical axial-split pump (also called "horizontal split" or "horizontal split-case pump") is shown in Figure 11-13. The pump casing is split along the centerline of the shaft. The lower half of the casing supports the entire pump and also contains the suction and discharge nozzles. The upper half of the casing can be removed for inspection, and the pump rotor can be removed for repairs without disturbing the suction or discharge piping. Axial-split pumps may be single stage (as in Figure 11-13) or multistage for higher pressures. The pumps are usually mounted with shafts in the horizontal position, but vertically mounted pumps for reduced floor space are also available.

Impeller. Single-stage pumps are equipped with double-suction impellers (Figure 11-13). Double-suction impellers are inherently axially balanced and, therefore, exert very low axial forces. They also have relatively large impeller intake eye areas and, therefore, low NPSH requirements. Multistage pumps usually have single-intake impellers that are arranged, when possible, in a back-to-back fashion to reduce the axial thrust. The impellers are equipped with radial wear rings.

Bearings. The pump bearings are mounted in two separate bearing housings. One of the two bearings is locked in the housing and carries the radial as well as the axial loads. The other bearing is for the radial load only. The bearings may be lubricated with grease or oil.

Seals. The two stuffing boxes contain either packing or single-face mechanical seals. For gritty liquids or for installations with low suction pressure, the packings are provided with lantern rings and buffer liquid injection.

11-6. Classification of Vertical Pumps

Vertical pumps were originally developed for well pumping. The bore size of the well limits the outside diameter of the pump and so controls the overall pump design. Vertical pumps are very versatile and are often used for installations not related to well pumping.

Vertical pumps can be subdivided into three major categories: (1) lineshaft pumps, (2) submersible pumps, and (3) horizontally mounted axial-flow pumps (Figure 11-1b).

Lineshaft Pumps

In lineshaft pumps (Figure 11-24), the driver is mounted on the discharge head. The lineshafting extends through the column to the bowl assembly and transmits torque to the pump rotor. Lineshaft pumps can be further subdivided into the following three categories:

- Deep well lineshaft pumps, which are used for the pumping of deep water wells
- Short-setting lineshaft pumps, which are used for the pumping of shallow wells or pump sumps
- Barrel pumps, which are short-setting pumps equipped with their own barrels or "cans" in place of pump sumps.

Submersible Pumps

Submersible pumps (Figure 11-25) are driven by a submersible motor. The motor is mounted below the bowl assembly and is directly coupled to the pump rotor shaft. The lineshaft is eliminated altogether and is replaced by a cable, which supplies power to the motor. Submersible pumps are available as

- Deep well submersible pumps, particularly useful for very deep (more than 180 m or 600 ft) settings or for crooked wells.
- Short-setting submersible pumps, used for pumping shallow wells and sumps or where the noise of a motor would be objectionable.

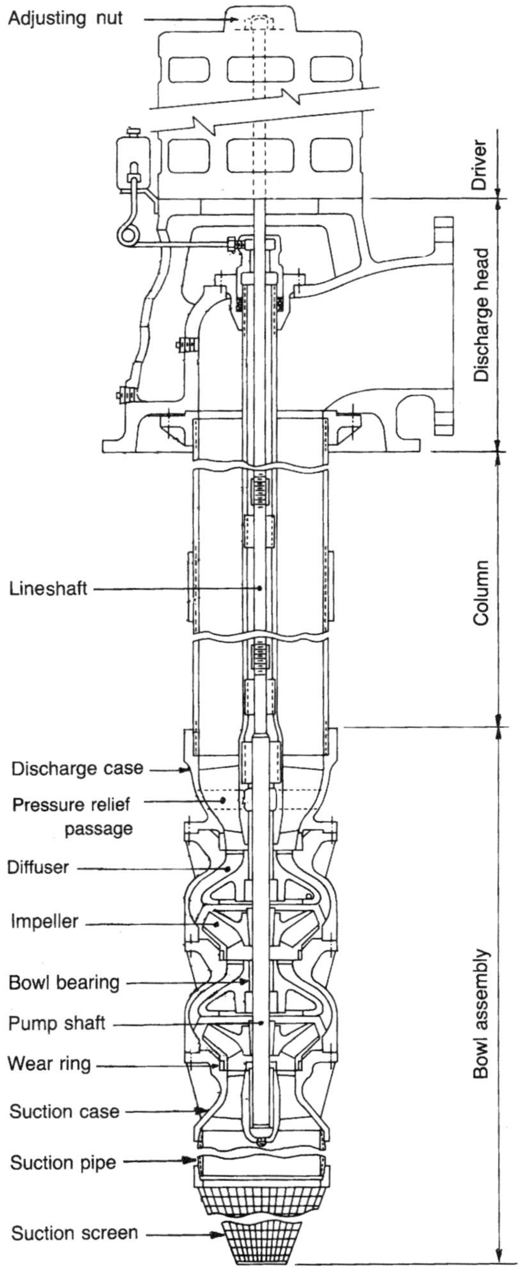

Figure 11-24. A lineshaft-driven vertical turbine pump. After the Hydraulic Institute [1].

Horizontally Mounted, Axial-Flow Pumps

Horizontally mounted, axial-flow pumps (Figure 11-26) are high-capacity pumps and are typically used for flood control and similar applications. They are often engineered for each particular installation and may have many different driver and bearing arrangements.

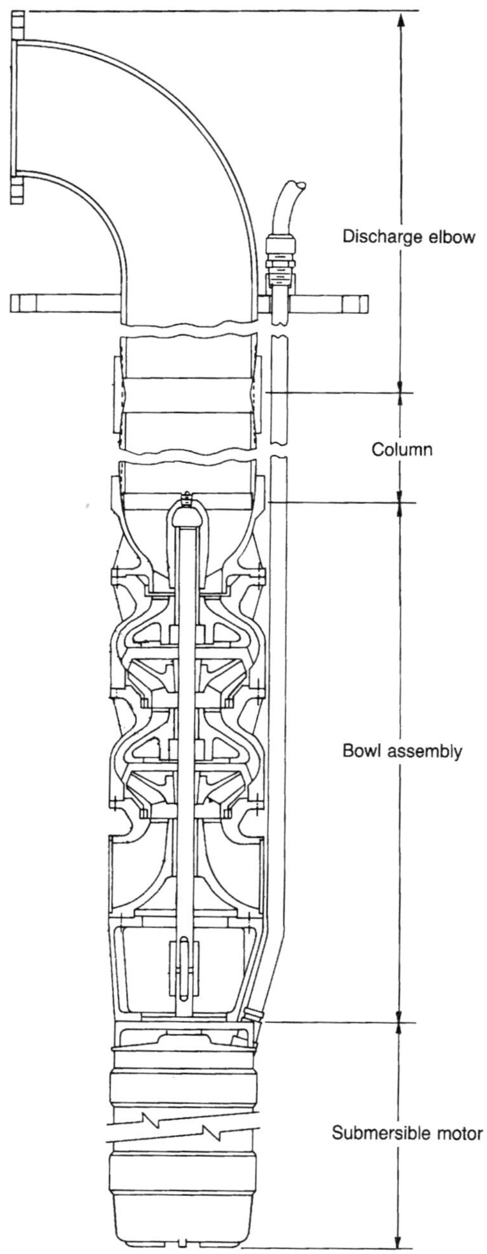

Figure 11-25. A vertical turbine pump driven by a submersible motor. After the Hydraulic Institute [1].

11-7. Construction of Vertical Pumps

A typical vertical pump consists of four major components: (1) the bowl assembly, (2) the column, (3) the discharge head, and (4) the driver. The function of these components and their design features are discussed in the following subsections.

Bowl Assembly

The bowl assembly of a vertical pump does the actual pumping. A typical multistage bowl assembly is shown in Figure 11-24. Each stage consists of one bowl with its impeller and bearing.

Shaft. All of the impellers of the bowl assembly are mounted on one common pump shaft.

Impellers. The impellers may be of the radial, mixed-flow (shown), or axial design. The impeller design determines the pump characteristics. The relationship between the impeller design, specific speed, and pump characteristics is discussed in Section 10-3 and is illustrated in Figure 10-8. The impeller design may be enclosed, semiopen, or open (Figure 11-14). The latter is typical of axial-flow or propeller pumps. The comments about impellers near the beginning of Section 11-3 apply to vertical pumps as well.

Pump bowl. The pump bowl contains the axial diffuser. Spiral diffuser vanes straighten the discharge swirl into axial flow and convert the kinetic velocity energy into pressure. Axial discharge from the bowl makes staging of the bowls possible without hydraulic losses. The bowls may be flanged and bolted together or machined for threading and screwed into each other.

Wear rings and wear plates. The bowl wear rings or wear plates, which are also called "liners" (Figures 11-14 and 11-24), are also mounted in the bowl (for the description of both and their function, see Section 11-3). The wear rings may be of integral or of separate design.

Bowl bearings. Each bowl also contains a bowl bearing. Because impellers with a diffuser are not subject to radial thrust, the bearings are not required to support any significant loads and act primarily as guide bushings for the shaft.

Suction case or suction bell. The suction case or the suction bell (Figure 11-24) is attached to the first bowl. The suction case is designed for the mounting of a suction pipe. The suction pipe is required in well pumping when the water level in the well is expected to drop below the bowl assembly inlet. The suction bell is commonly used in open pit installations and provides for an hydraulically smooth, unobstructed flow into the first stage. Both types of pump inlets contain straight axial inlet vanes, which help to guide the flow of liquid.

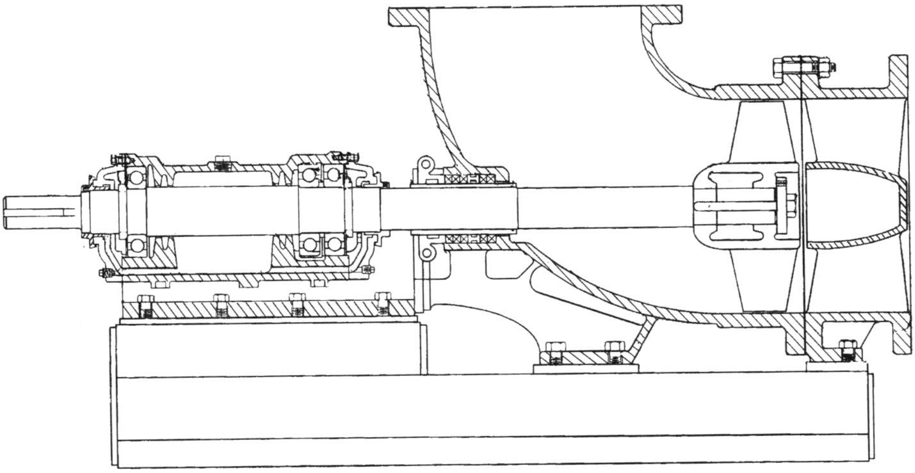

a

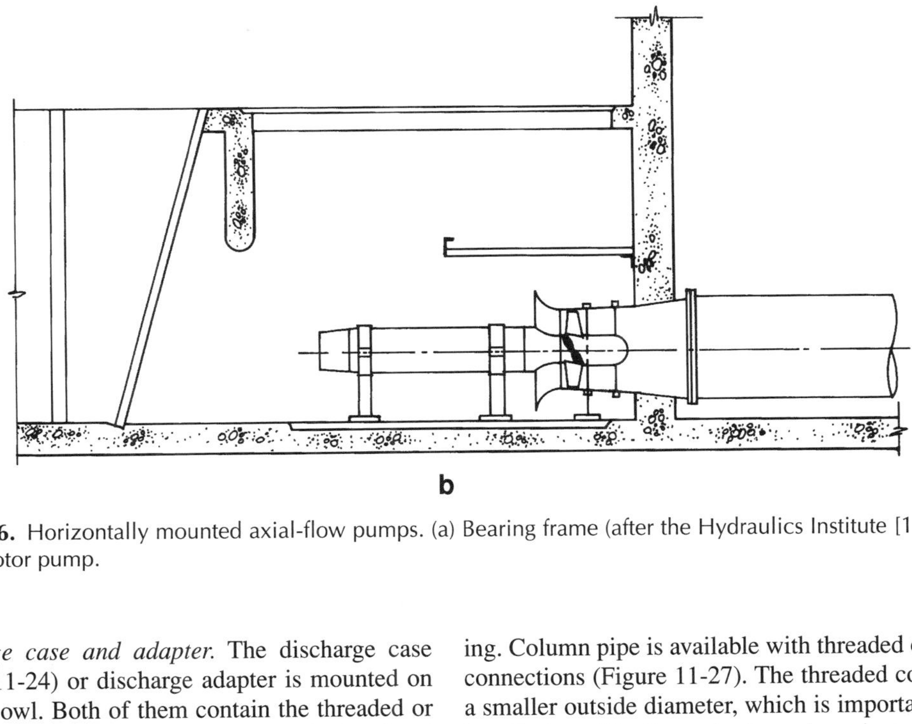

b

Figure 11-26. Horizontally mounted axial-flow pumps. (a) Bearing frame (after the Hydraulics Institute [1]); (b) submersible motor pump.

Discharge case and adapter. The discharge case (Figure 11-24) or discharge adapter is mounted on the top bowl. Both of them contain the threaded or bolted connection for attachment of the bowl assembly to the column. The discharge case contains a center hub to which the enclosing tube can be connected. The discharge adapter is fully open on the inside and does not accept an enclosing tube.

Column and Lineshafting

The column supports the bowl assembly, ducts the liquid to the discharge head, and encloses the lineshafting. Column pipe is available with threaded or flanged connections (Figure 11-27). The threaded column has a smaller outside diameter, which is important in well installations. The flanged column is easier to assemble and disassemble.

The lineshafting transmits the torque from the driver to the pump rotor. The lineshafting sections are connected to each other with threaded couplings or, for larger diameter shafts, with sleeve couplings.

Open lineshafting is exposed to the pumped liquid and is guided in lineshaft bearings, which are supported in bearing retainers. Lineshaft sleeves protect the shafting from wear in the bearings. With the enclosed lineshafting (Figure 11-27b), the enclosing

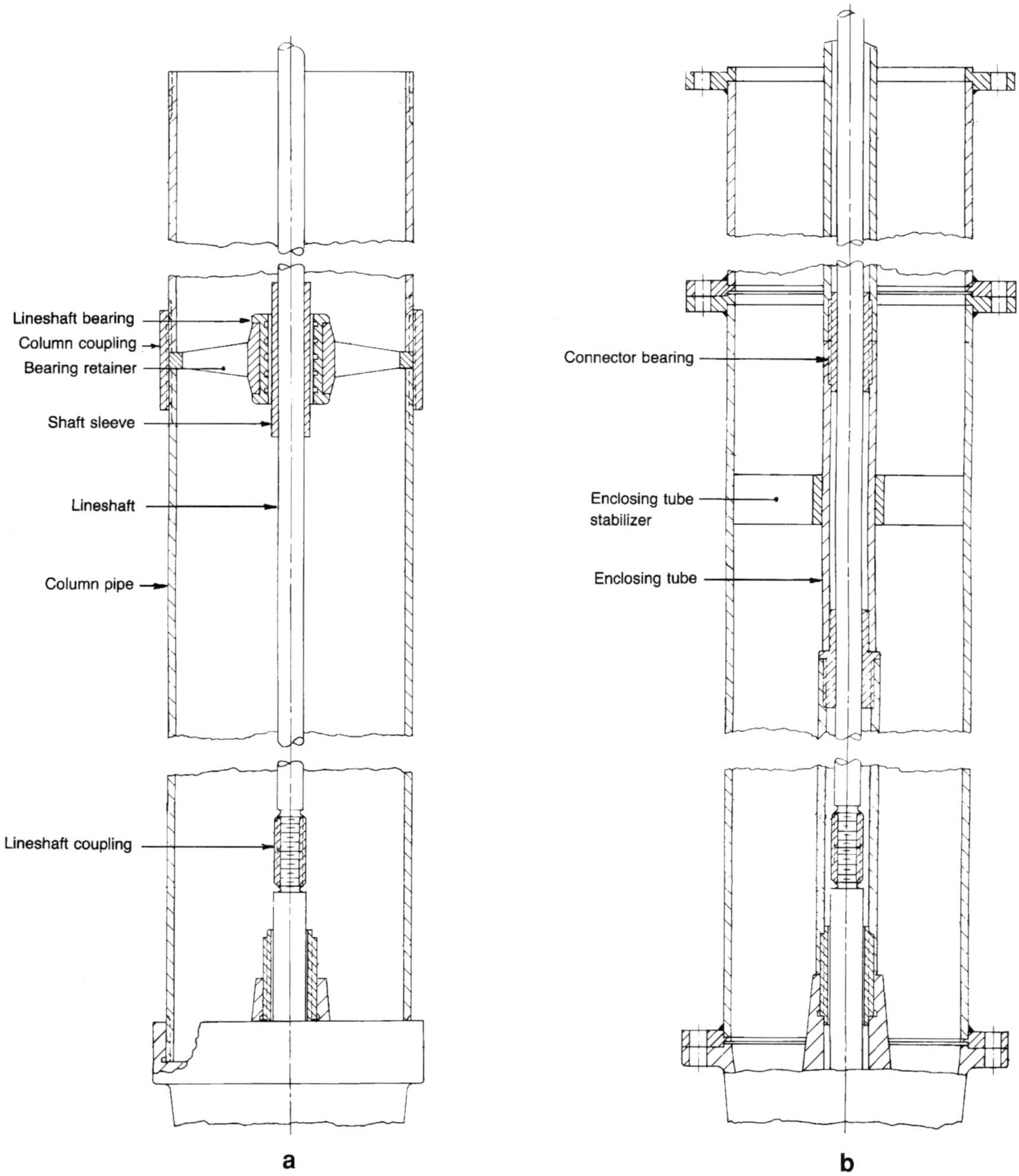

Figure 11-27. Column and lineshafting. (a) Open; (b) enclosed. After Fairbanks Morse Pump Corp.

tube isolates the shafting from the pumped liquid and provides a channel for the lubricating liquid. Externally threaded enclosing-tube connector bearings serve as tube couplings and shaft bearings at the same time. Enclosing-tube stabilizers support the tube in the column and protect it from lateral vibration.

Discharge Head

The discharge head directs the pump flow from the column to the piping system, provides for sealing of the shafting and the enclosing tube, and provides a base from which the pump is suspended and on which the driver is mounted (Figure 11-24). The discharge head may be a casting or a fabricated component.

The discharge head usually contains a separate packing box (Figure 11-28a) for the packing, which seals the lineshaft. Modified packing boxes for mechanical seals are also available (Figure 11-28b). The packing boxes for enclosed lineshafting are called tension nuts (Figure 11-28c). The top section of the enclosing tube is threaded into the tension nut, which is tightened to keep the tube under tension and to hold it straight.

Some pump installations require underground discharge piping. Underground discharge heads have an elbow mounted below the motor pedestal. The discharge head in submersible pumps is replaced by a simple discharge elbow (Figure 11-25).

Driver

The driver of lineshaft pumps is mounted on the discharge head. It can be an electric motor (Figure 11-24)

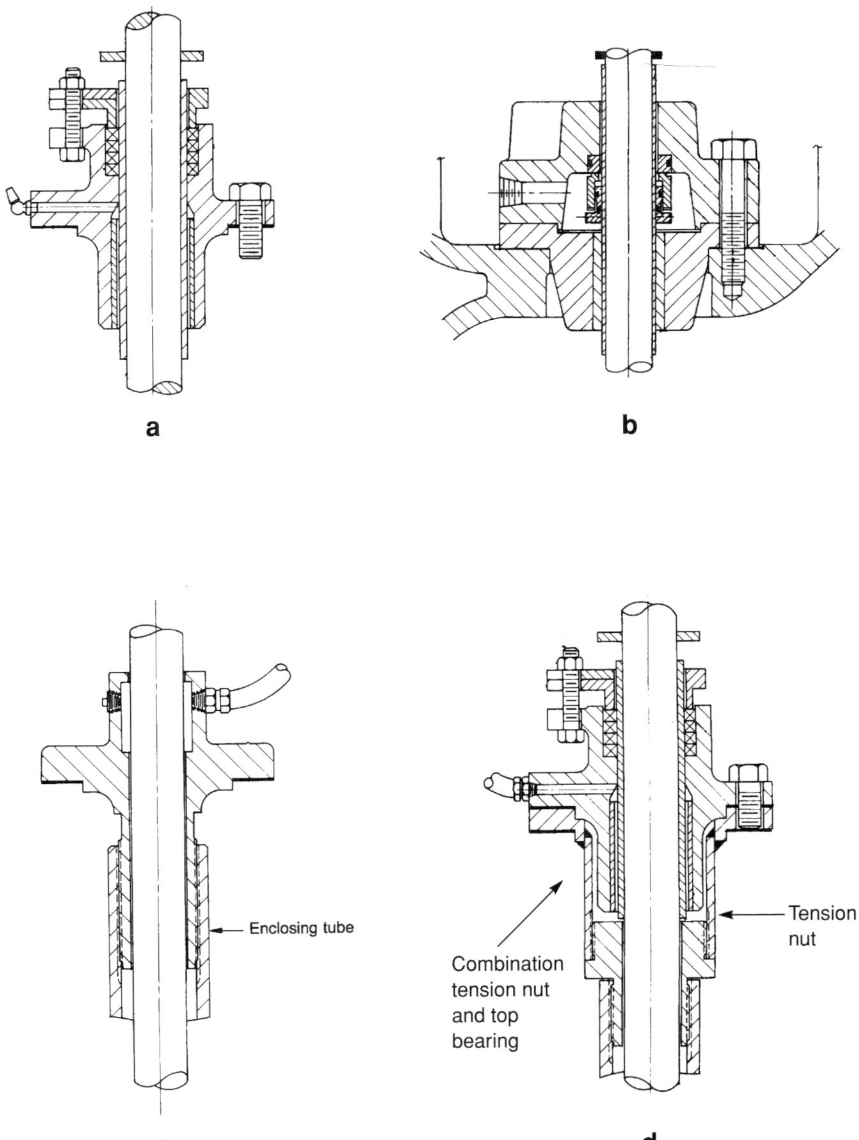

Figure 11-28. Head accessories. (a) Packing box; (b) packing box with mechanical seal; (c) tension nut; (d) packing box (water flush). After Fairbanks Morse Pump Corp.

or an angle gear with an engine. Because the pump bowl assembly or column contains no thrust bearings, the driver must be equipped with a thrust bearing capable of supporting the hydraulic axial thrust of the pump and the weight of the pump rotor with its lineshaft.

Solid shaft drivers require an axially adjustable coupling for the top section of the lineshaft (Figure 11-29). The adjusting nut of the coupling is threaded on the lineshaft and is used to set the pump rotor in the proper axial relation to the pump bowls. In hollow-shaft motors, the top section of the lineshaft extends through the hollow shaft to the top of the motor. An adjusting nut (Figure 11-24) with a coupling, which is installed at the top of the motor, provides for the proper setting of the rotor and for the transmittal of thrust and torsional loads.

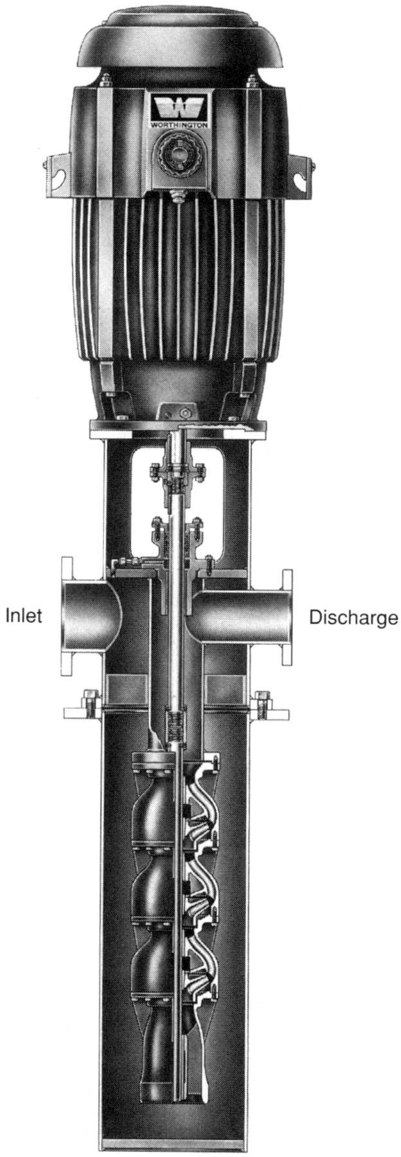

Figure 11-29. Barrel or "can" pump. Courtesy of Dresser Pump Division.

All submersible motors that drive submersible pumps are of a special hermetic design with a small outside diameter to fit the well. The motor may be filled with air, oil, or water. The first two types of motors are usually sealed with a double mechanical seal with an oil-filled chamber between them.

The thrust bearing of the motor must support the axial thrust of the pump as well as the weight of the motor rotor.

Lubrication of Vertical Pumps

In open lineshaft pumps, all lineshafting and bowl assembly bearings are flooded and lubricated by the pumped liquid. Enclosed lineshaft bearings require external lubricant supply and can be lubricated with water, oil, or grease. With water-lubricated and enclosed lineshaft pumps, clean water is injected under pressure into the tension nut (Figure 11-28d) below the stuffing box and forced through the lineshaft bearings. It then flows through the top bowl bearing and mixes with the pumped liquid. The injected water lubricates and cools the lineshaft bearings and protects them from any abrasives in the pumped liquid. It should be noticed that the flushing water does not reach the bowl bearings, which are, therefore, lubricated by the pumped liquid.

With oil-lubricated pumps, a drop-feed oiler is connected to the tap in the tension nut and feeds oil at a slow rate and at atmospheric pressure into the enclosing tube (Figure 11-28c). To allow the oil to flow through all of the lineshaft bearings, the discharge case of the bowl assembly is provided with two pressure-relief passages (Figure 11-24), which vent the cavity below the lowest connector bearing to the well and allow the oil to escape into the well. Here again, the bowl bearings remain lubricated by the pumped liquid.

With grease-lubricated pumps, external lines feed all of the pump bowl bearings. Individual grease-admission passages may be provided to each bowl, or the pump shaft may be gun drilled with cross bores to each bearing. The grease is injected under pressure by a manual or automatic lubricator. The grease also provides protection against the intrusion of abrasives.

11-8. Types of Vertical Pumps

The types of vertical pumps are shown in Figure 11-1b.

Deep Well, Lineshaft Pumps

A typical lineshaft-driven deep well pump is shown in Figure 11-24. The bowl assembly of the pump may contain as many as 60 stages, depending on the well setting and on the discharge pressure requirements. The bowls may be equipped with enclosed or semi-open impellers and optional wear rings or wear plates. Semi-open impellers offer the advantage of end-clearance adjustability for pumping liquids with a higher abrasive content, but they are difficult to adjust for deep settings.

Lineshafting. The lineshafting is usually open for settings up to 75 m (250 ft) and is usually enclosed for deeper settings. Enclosed shafting is drip fed and lubricated with oil. Open shafting is lubricated with pumpage but requires prelubrication with water before starting, because the lineshaft bearings must be provided with lubricant to avoid dry running until they become flooded with the pumped liquid. Grease lubrication is not very practical due to the requirement for long grease lines.

Column and head. The column and the head must be designed for the anticipated pressure and for high axial loads. For column diameters up to 300 mm (12 in.), columns are usually threaded. Larger columns are usually flanged for ease of assembly and disassembly.

Axial Thrust

The thrust due to the total dynamic head (including the pumped liquid contained in the column) combined with the weight of the bowl assembly, the lineshaft, and the column causes significant stretching of both the lineshaft and the column. This must be considered in component selection. The column and the lineshaft stretch by different amounts, and the final axial setting of the rotor requires particular attention with deep well pumps.

Deep Well Submersible Pumps

Submersible motor-driven pumps (Figure 11-25) are another popular configuration of deep well pumps. They have no lineshafting and, therefore, no related problems. The overall efficiency is improved somewhat because there are no lineshafting friction losses. These advantages become greater as the well depth increases. On the other hand, they must be equipped with a special submersible motor and a long power-supply cable. The motor is not accessible for inspection or maintenance without pulling the entire pump from the well (see Table 25-8). The reliability and life of a submersible pump depend greatly on the motor design and quality.

All bowl and column components of submersible pumps are very similar to those of lineshaft pumps.

Discharge Elbow

The discharge elbow (Figure 11-25) replaces the discharge head of lineshaft pumps. It is rather simple in design but must be capable of supporting the bowl assembly with the motor and the column.

Lubrication

The pump bearings are lubricated with pumpage because there is no practical way to supply lubricants of any kind from an outside source. The motor bearings are sealed in the motor and, depending on the motor design, are lubricated with water or oil. The hydraulic axial thrust and the weight of the pump and the motor rotor are carried by the motor thrust bearing.

Short-Setting Lineshaft Pumps

The basic components of a short-setting pump are the same as those of a deep well pump. The major difference is in the length of the column and in the lightweight design of the column and the head. The column of a short-setting pump is often limited to one section; occasionally, it is eliminated altogether with the bowl mounted to the head directly.

Bowl assembly. The bowl assemblies are one- or two-stage propeller bowls (Figure 11-30) or single- and multistage mixed-flow bowls.

Lineshafting and options. The lineshafting can be open or enclosed. The lineshafting and bowl assembly can be lubricated with pumpage, water flush, oil, or grease. The wear ring or wear plate, packing or mechanical seal, and different bearing material options are available and are of greater importance, because short-setting pumps are used to pump a great variety of liquids.

Installation. Short-setting pumps usually take their suction from a wet well. The performance of vertical pumps is rather sensitive to inlet flow conditions. The wet well should, therefore, be designed by a

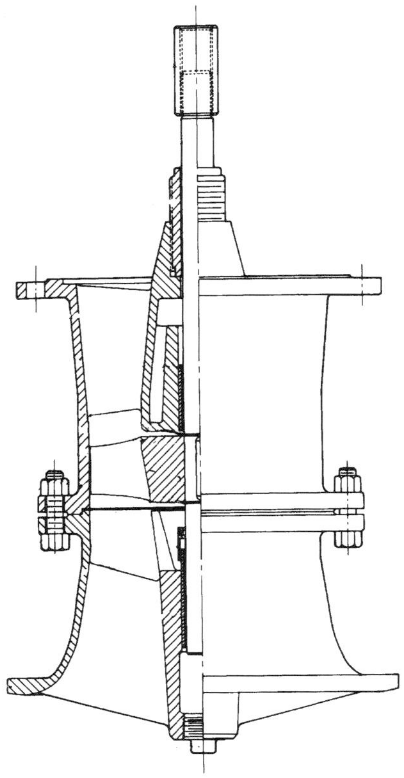

Figure 11-30. Propeller pump bowl assembly. Courtesy of Fairbanks Morse Pump Corp.

competent engineer. Particular attention must be given to (1) inlet flow velocities, (2) freedom from excessive turbulence and unsymmetrical flow, and (3) sufficient submergence to suppress vortexing and the intake of air (see Equation 12-1).

Short-Setting Submersible Pumps

Submersible motors are also used with short-setting pumps, but less frequently. They eliminate the need for a superstructure to protect the motor. Deeper sumps are required, however, to accommodate the motor length. Their main advantage (the elimination of the lineshafting) is of less significance. In all other respects, the short-setting submersible pumps resemble deep well submersible pumps.

Barrel Pumps

Barrel or "can" pumps are short-setting pumps arranged in a can or barrel (Figure 11-29). Used primarily for booster service, they are particularly attractive for installations where NPSH can be increased as needed by making the barrel deeper and lowering the bowl assembly. The expense of constructing a sump is thereby avoided. By specifying unit responsibility for pump, motor, and barrel, the manufacturer thus becomes responsible for barrel dimensions and hydraulic design.

The barrel inlet may be arranged in line with the pump discharge nozzle, as shown in Figure 11-29, or the inlet can intersect the barrel at any elevation. Underground or aboveground discharge heads are also available. Barrel pumps are otherwise very similar to short-setting pumps. Submersible motors are rarely used.

Sometimes rotating flow between the barrel and the pump causes poor performance. A plate between pump and barrel, extending from the bottom of the barrel to the water surface or to the inlet pipe, prevents rotation. A cross between pump bell and the bottom of the barrel is also helpful. A cone with fins might be better than a cross, but a cone is inappropriate if the bottom of the bowl assembly is like that shown in Figure 11-30.

Horizontally Mounted, Axial-Flow Pumps

Horizontally mounted, axial-flow pumps are typically low-head pumps, and larger sizes can have very high capacity—up to 30 m^3/s (500,000 gal/min) or more. Smaller pumps frequently have a bearing frame (Figure 11-26a) that contains radial and thrust bearings to support the pump shaft. The propeller is cantilever mounted without additional bearings. The frame bearings are lubricated with oil or grease.

Larger pumps are usually custom engineered, and their design varies from installation to installation. The thrust bearing is usually arranged outside of the pump and is lubricated with oil. Radial bearings can be mounted within the pump casing on one or both sides of the propeller. They are usually of underwater, grease-lubricated, journal-bearing design.

High-capacity pumps operating with suction lift and located above the high-water level may have a center hub that is equipped with water shaft seals and is accessed through a manhole. The hub is never flooded with water. The bearings are mounted inside the hub, are lubricated with oil or grease, and are always kept dry. They can be either of journal or antifriction design.

Submersible motor-driven, axial-propeller pumps are used in Europe for flood-control applications. The motor of these pumps may be filled with water and the bearings lubricated with water. The pump and motor are inherently flood resistant. Such pumping stations are compact, low in cost, and require no pump house. Only the controls and the power supply are installed above the surface (Figure 11-26b).

Submerged hydraulic motor-driven, axial-propeller, and mixed-flow pumps are also available. In these units the hydraulic fluid power unit is electric motor or engine driven and located away from the pump. The only connection to the pump is a pair of hydraulic hoses.

11-9. Positive-Displacement Pumps

When compared with kinetic pumps, positive-displacement pumps are inherently low-capacity, high-discharge-pressure pumps. Although they are used to pump a great variety of liquids, they are also capable of pumping slurries in small volumes or in consistencies that cannot be handled by centrifugal pumps—the basic reason for using positive-displacement pumps in water and wastewater pumping and treatment stations. Only designs suitable for sewage, slurries, sludge, and similar liquids are discussed in this chapter.

As shown in Figure 11-1c, positive-displacement pumps are subdivided into three categories: reciprocating pumps, rotary pumps, and pneumatic (ejector) pumps. Of all the various reciprocating pumps, only the plunger pump is of interest for pumping sludge and slurries. Two types of rotary pumps are of interest: lobe pumps and progressive cavity pumps.

The only type of pneumatic positive-displacement pump of interest is the pneumatic ejector. It is used for pumping raw sewage in small volumes and, sometimes, high heads.

Plunger Pumps

A representative plunger pump is shown in Figure 11-31. It can be used for all types of sewage, sludge, scum, slurries, and clarifier and thickener underflow. It can be applied for transfer and for metering service. Such pumps are available in single- and multicylinder models. The plunger contains the crosshead, driven by a camshaft arrangement. The capacity of the pump can

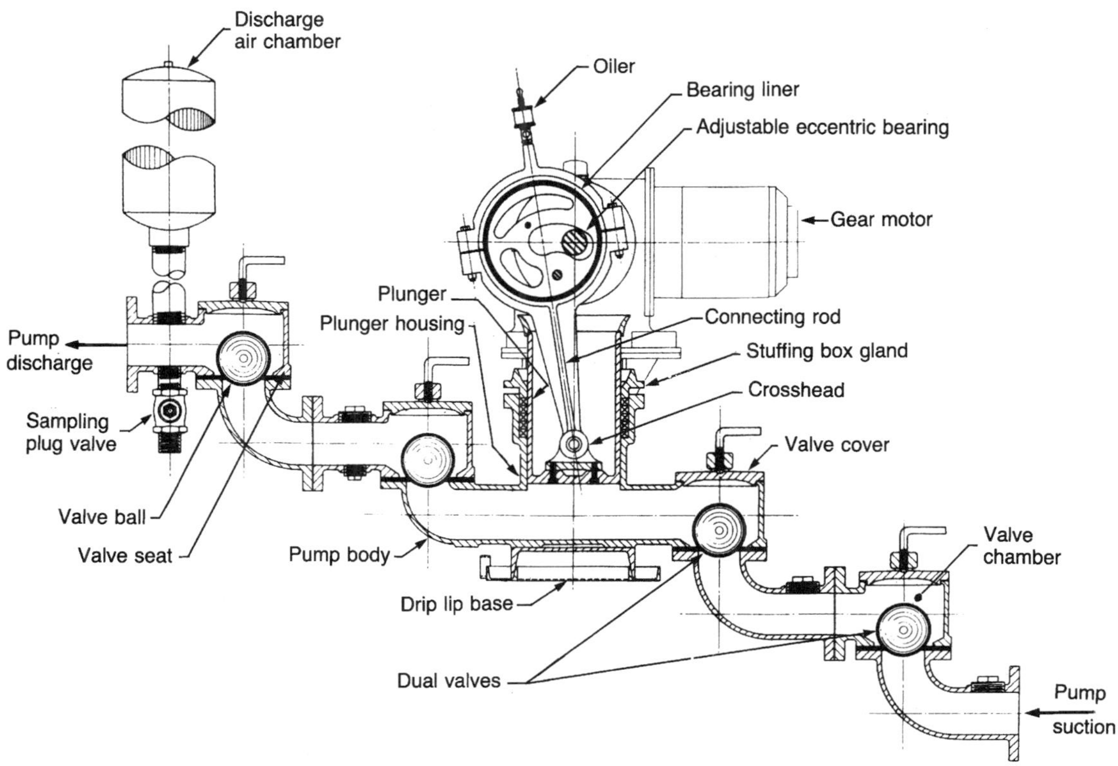

Figure 11-31. A plunger pump. Courtesy of Komline-Sanderson Engineering Corp.

be adjusted by changing the stroke, the rotating speed of the pump, or both. The stroke of the pump is changed by the eccentric pin setting.

A plunger pump is equipped with single or dual ball lift check valves. The dual design contains two ball check valves in series for each plunger on both the suction and the discharge side. Two valves rarely hang up on foreign matter at the same time, so if one valve is unseated the other continues to operate properly until the foreign matter is flushed through without affecting the pump operation. "Quick-opening" covers provide for easy access to the ball checks and seats for servicing or replacement.

The pump bodies and plunger housings are separate components, and the plunger can be removed for replacement without disturbing the shaft assembly, pump body, or piping. As with all positive-displacement pumps, plunger pump capacity is not altered by a changing discharge head. The positive pressure exerted by the plunger clears plugged lines. The pumps are, therefore, well suited for metering applications. The pumps are driven by constant- or variable-speed motors with gear reducers.

Rotary Lobe Pumps

A representative rotary lobe pump is shown in Figure 11-5. It contains two elastomer-coated rotors that are driven by an integral gear box and synchronized by timing gears. The rotors run without touching each other or the casing. The liquid is drawn through the inlet port into the pockets between the lobes and chamber walls. Because liquid cannot escape between the two rotors, it discharges in the direction of rotation of the outer lobes through the discharge nozzle.

The rotors may be cantilever mounted (as shown in Figure 11-5) or supported by bearings in each cover. Cantilever mounting allows a hinged cover assembly, which provides easy access to the pumping chamber for inspection and replacement of rotating components.

The pump discharges at a continuous and smooth flowrate and is relatively nonagitating and nonshearing. The pump is self-priming and can be run dry without damage from blocked or starved suction inlets. It is suited for handling a wide range of sludge viscosities and types. The smoothly rounded contours of the rotor and their full sweep within the pumping chamber provide a high tolerance to raggy sludges. No check valves are required (provided the gearing prevents backward rotation), and the pumping is not susceptible to rag buildup. Elastomeric coatings for the lobes have been developed to pass hard solids up to 120 mm (4 in.) in diameter, and the coating has good wearing life in mildly abrasive duty. Where there is a high content of debris, an automatic reverse mechanism can be provided to reduce operator attention.

At the head of the treatment plant (where sludges contain a low percentage of solids but a high grit content) urethane-coated rotors are recommended, and the pumps should be run at reduced speeds. Otherwise, Buna N–coated rotors are used. The pumps are used for pumping sludge with as much as 6% solids.

In summary, the initial cost is relatively high, but the advantages of (1) quick, easy, inexpensive replacement of moving parts, (2) compactness and space saving, (3) high tolerance for rags and large solids, (4) long life at low speeds, and (5) self-priming make the overall life cycle cost attractive (see also Section 19-2).

Progressive Cavity Pumps

A progressive cavity pump is designed specifically to transfer abrasive and viscous fluids with a high solid, fiber, and air content. A hard steel screw rotor rotates and orbits within an elastomer stator, as shown in Figure 11-6. The pitch of the stator is two times the pitch of the rotor. As the rotor turns, it contacts the stator along a continuous sealing line, creating a series of sealed cavities that progress to the discharge end. The cavity fills with liquid as it gradually opens and expands at the suction end of the rotor. The trapped liquid is transported to the discharge end and is then gradually discharged in an axial direction. Multistage pumps of up to four stages are available for reduced wear from abrasives. Some engineers require a minimum of two stages for pumping wastewater sludge.

The pump driveshaft is supported by bearings contained in the bearing frame. The shaft is sealed with a packing or a mechanical seal. The bearings are usually lubricated with grease. A Cardan shaft with universal joints of various designs turns the rotor and allows for orbiting motion at the same time. The universal joints may be lubricated either by the pumped liquid or by grease.

Progressive cavity pumps are used in wastewater treatment plants for transferring all types of slurries and sludge, and they can pass solids with a sphere diameter of up to 50 mm (2 in.). A "bridge breaker" can be added at the front end of the pump to reduce the size of large solids. These pumps are self-priming up to lifts of 8.5 m (28 ft), but they cannot be operated dry. The flow is even and the shear is low. The pump is relatively easy to service, but sufficient clear floor space must be provided for dismantling and access to the rotor and the Cardan shaft (see Figure 11-32).

Figure 11-32. Installation of a progressive cavity pump. Courtesy of Netzsch Inc.

Progressive cavity pumps are relatively low in cost, but stators and rotors may have to be replaced frequently, especially if grit is present in the fluid. To reduce wear, pump speed should be low (no more than 400 rev/min) when pumping sludge or raw sewage.

Pneumatic Ejectors

Pneumatic ejectors are used for pumping low flowrates of wastewater at high heads (flowrates to 140 m^3/h [600 gal/min] at heads to 90 m [300 ft]), especially if the flowrates are highly variable. The pump consists of a pressure vessel that is allowed to fill by gravity until a predetermined level is reached (Figure 11-8). Controls are then operated to admit compressed air to the vessel. The high pressure moves the liquid into the force main. When the chamber has been emptied, the controls close the air supply valve and vent the air in the tank to the atmosphere, which allows the next cycle to begin. The compressed air may be supplied from a plant air system or from compressors installed on location. Air receivers of adequate capacity for several cycles are sometimes installed to provide for limited continuing operation of the system during power outages.

Mechanical controls for the system have proven to be more durable than electronic controls. The check valves should be specifically designed for passing solids and stringy material. Because of the danger of freezing, the pneumatic ejector and the air supply system should not be exposed to low ambient temperatures.

11-10. Special Pumps

The pumps in this section are for the limited purpose of producing either high pressures at low flowrates or moderate to large flowrates at low lifts.

Regenerative Turbine Pumps

Regenerative turbine pumps derive their name from the many buckets machined into the periphery of the impeller. They have long been recognized for efficiently producing low flows at pressures much higher than those of centrifugal pumps of comparable size. They are excellent for seal water and washwater pumps.

Regenerative turbine pumps are unique in operation. The liquid circulates in and out of the impeller buckets many times on its way from the inlet to the outlet of the pump. Both centrifugal and shearing action combine to impart additional energy to the liquid each time it passes through the buckets. Pressure from the inlet in Figure 11-33a increases linearly to the discharge. The impeller runs at very close axial clearances with the pump channel rings to minimize the recirculation losses, so these pumps can only be used with clean liquids. The pump is damaged if operated against a closed valve, so a spring-loaded relief valve is always installed on the pump discharge.

A seal water pumping unit normally consists of a liquid holding tank with a float-operated makeup valve connected to the central water supply. To ensure reliability, two pumps, each alternating as duty and standby units, are connected to the tank. An air gap between the incoming water supply and the tank overflow prevents any possibility of backflow. Complete "package units" are available. Pumps are made with capacities ranging from 0.13 L/s to more than 4.4 L/s (2 gal/min to more than 70 gal/min) at heads ranging from 3 m to more than 120 m (10 ft to more than 400 ft).

Multistage Centrifugal Pumps

A multistage centrifugal pump is a viable alternative to a regenerative turbine for supplying seal water or wash water. The shaft and impellers in the pump shown in Figure 11-34 are fabricated from stainless steel. Sizes are available for pumping from 0.08 to over 22 L/s (1.2 to over 350 gal/min) at corresponding heads of 260 to 150 m (600 to 350 ft). An advantage of the multistage configuration is that pressures do not change very much with a change in flowrate.

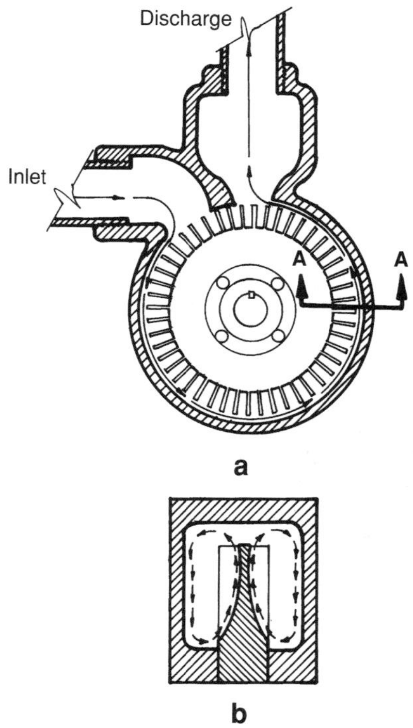

Figure 11-33. A regenerative turbine pump. (a) Impeller; (b) Section A-A. After Aurora Pump, a Unit of General Signal.

Screw Pumps

Screw pumps (or Archimedes' screws) are high-volume, nonclog, atmospheric-head devices that can pump a variety of solids and debris in raw wastewater without screening. There are two general types: (1) the open screw, which rotates in a trough (Figure 11-7), and (2) the enclosed screw, in which both the screw and an enclosing cylinder rotate (Figure 11-35). A major advantage of these pumps is variable pumping at constant speed, because the output (up to design capacity) is controlled by the sump level and equals the influent flowrate. The disadvantages are the inducement of turbulence, the release of odors and other volatile substances in wastewater, and the relatively high initial cost. But when comparing costs

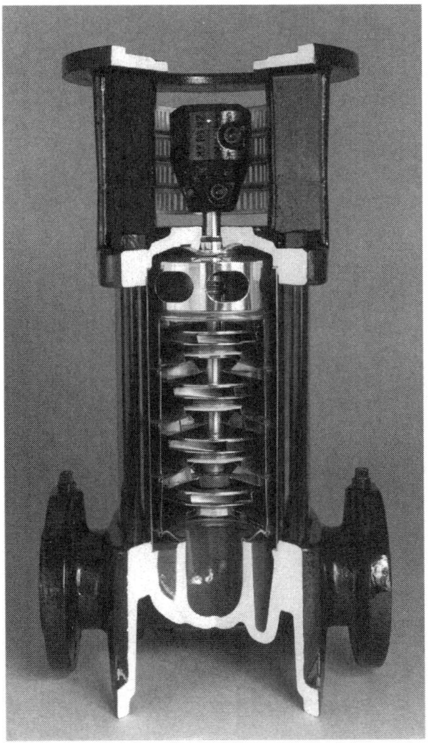

Figure 11-34. Multistage centrifugal pump for high-pressure discharge. Courtesy of Grundfos Pumps Corporation.

with those of other types of pumps, consider the cost of the total system, including all piping, wet or dry wells, screens, fittings, valves, variable-speed controls, and other accessories as well as operating and maintenance costs. Note that operators like screw pumps because the good ones, when properly installed, are so trouble free.

Open Screw Pump

The open screw pump consists of a torque tube to which spiral flights are attached, a lower submerged bearing, an upper radial and thrust bearing, a gear reducer (typically driven by an electric motor), and a trough in which the screw rotates at a constant speed. The limiting speed ranges from 20 rev/min in large pumps to 75 rev/min in small ones and cannot be exceeded without water spilling over the top of the screw. Pockets formed between the flights, torque tube, and trough trap the liquid and move it up the incline in a continuous flow. The steel screw can be protectively coated. Steel troughs can be used for smaller pumps, but concrete troughs are most common. Fabricated deflectors are usually installed in the pump trough along the uptake side of the spiral. Their concave surfaces extend the circular arc of the wall of the trough and improve pump efficiency.

The pump drive consists of a motor, vee belts [except for motors of more than 100 kW (150 hp)], and reducing gears. A backstop is usually furnished to prevent reverse rotation of the spiral when the unit is stopped.

If the headloss at the upper collector or chute point, which is about 0.3 m (1 ft) above the discharge pool, is ignored, the maximum overall system efficiency of the pump at design flow may reach 80%. At approximately 30% of design capacity, the efficiency drops to about 60% due to friction and "slippage" (backflow) of fluid between the flights and the trough.

Installation of the pump is relatively simple in concept, although there is some difficulty in placing concrete on such steep slopes. Once the pump is aligned, the radius of a concrete trough is grouted in place with a screed attached to the edge of the screw.

Enclosed Screw Pumps

Except for the following differences, the enclosed screw pump is similar to the open type:

- Because the flights are welded to the outer cylinder, there is no slippage (backflow). If the pump is stopped, the water is retained between the flights. (*In cold climates, provide for reverse operation after shut-down to prevent freezing.*)
- The lower bearing, a self-aligning set of rollers mounted above the high-water level, is easily accessible. The rollers carry most of the radial load.
- Because there is no slippage, the efficiency of the pump is very high and stays high even at low discharge.
- A massive concrete structure is not needed.

Air Lift Pumps

Air lift pumps are pneumatic devices that do not fit the categories of Figure 11-1. The pump consists of a simple tube immersed in the sump or wet well with a high volume of low-pressure compressed air admitted at the bottom of the tube. The reduced-density air–liquid mixture in the tube is raised by the static head in the wet well, and it overflows into the open discharge channel (Figure 11-9). The pump is extremely simple and can be constructed in the field. It is suitable for raw sewage, sludge, and sandy or dirty water.

The hydraulic efficiency is very low and rarely reaches 35%. This does not include the compressor

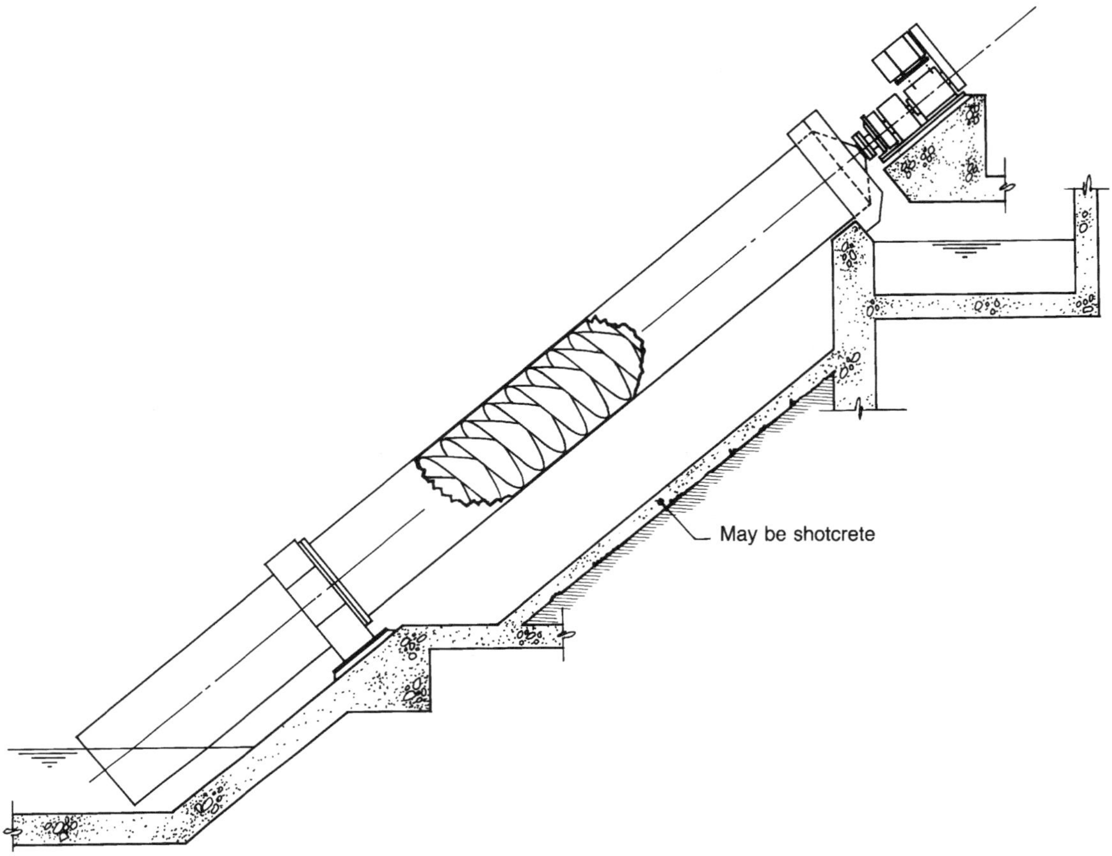

Figure 11-35. An enclosed screw pump. After CPC Engineering Corp.

efficiency, which reduces the overall efficiency even more. For all except very low heads, the pump requires large submergence.

11-11. Summary of Typical Pump Applications

The principal uses of the pumps shown in Figure 11-1 are given in Table 11-1. The listed discharge capabilities of pumps are intended to convey only a general

Table 11-1. Application of Pumps

Type of pump (see Figure 11-1)	Capacity		Service
	m^3/h	gal/min	
Overhung impeller			
Clear liquid	to 2300	to 100,000	Water
Nonclog	9–2300	40–100,000	Wastewater
Wet pit volute	50–9000	250–40,000	Water
VTSH®	450–5000	2000–22,000	Wastewater
Self-priming	23–910	100–4000	Wastewater, water
Vortex	10–230	50–1000	Sludge, wastewater
Cutter	11–230	50–1000	Wastewater
Grinder	<23	<100	Wastewater
Submersible nonclog	9–11,000	40–50,000	Wastewater
Submersible vortex	11–60	50–250	Sludge, wastewater
Submersible grinder	<23	<100	Wastewater

Table 11-1. Continued

Type of pump (see Figure 11-1)	Capacity m³/h	gal/min	Service
Impeller-between-bearings			
Axial split, single-stage	23–23,000	100–100,000	Water
Axial split, multi-stage	23–460	100–2000	Water
Vertical (turbine) lineshaft			
Deep well	9–2300	40–10,000	Water
Short setting	9–23,000	40–100,000	Water
Barrel pumps	9–9100	40–40,000	Water
Vertical (turbine) submersible			
Deep well	23–460	100–2000	Water
Short setting	23–1000	100–8000	Water
Horizontally mounted			
Axial flow	9–110,000	40–500,000	Storm water
Positive displacement			
Plunger	0–120	0–540	Sludge
Lobe	<450	<2000	Sludge
Progressive cavity	<100	<480	Sludge
Screw (open)	2300–14,000	10,000–60,000	Wastewater, storm water
Screw (enclosed)	110–8000	500–35,000	Wastewater, storm water
Ejector	0.5–140	2–600	Wastewater
Other (omitted in Figure 11-1)			
Air lift	11–570	50–2500	Water, activated sludge

idea of the available pump sizes. Some manufacturers may have standard designs smaller or larger than those listed. Custom-made pumps (which can be of any size) and European pumps (some of which are very large indeed) are excluded.

No distinction is made in Table 11-1 between clean water and dirty water (with gritty material such as sand in suspension), because a clean water pump can be made satisfactory for dirty water if it is constructed of abrasion-resistant materials and if it is equipped with, for example, the appropriate kinds of seals.

Wastewater is assumed to be unscreened and to contain both organic and inorganic solids (including stringy materials) that require a nonclog design. Wastewater from a treatment plant can be pumped with water pumps. Unless a sludge pump is specifically marked as suitable for pumping wastewater, it would not be used for such service even though such use is possible.

11-12. References

1. *Hydraulic Institute Standards for Centrifugal, Rotary, & Reciprocating Pumps,* 14th ed., Hydraulic Institute Parsippany, NJ (1983).
2. ANSI/HI 9.1-9.5-1994, *Pumps—General Guidelines.* Hydraulic Institute, Parsippany, NJ (1994).

11-13. Supplementary Reading

1. Pump manufacturers' catalogs, such as *Centrifugal Pumps* and *Vertical Pumps,* Fairbanks Morse Pump Corp., Kansas City, KS (n.d.).
2. Anderson, H. H., *Centrifugal Pumps,* 3rd ed. Trade & Technical Press, Morden, Surrey, United Kingdom (1980).
3. Pollak, F. *Pump Users' Handbook,* 2nd ed., Trade & Technical Press, Morden, Surrey, United Kingdom (1980).
4. Karrasik, 1. J., W. C. Krutzsch, W. H. Fraser, and J. P. Messina, *Pump Handbook,* 2nd ed., McGraw-Hill, New York (1985).
5. SWPA, *Submersible Sewage Pumping Systems Handbook,* Lewis Publishers, Chelsea, MI (1986).
6. Hicks, T. J., *Pump Selection and Application,* McGraw-Hill, New York (1957).
7. Stepanoff, R. J., *Centrifugal and Axial Flow Pumps,* Wiley, New York (1967).
8. Sarvanne, H., and H. Borg, *Submersible Pumps Handbook,* Oy E. Sarlin, AB, Helsinki, Finland (n.d.)
9. *PSI—Pump Selector for Industry,* Worthington Pump Division, Dresser Industries, Harrison, NJ (n.d.).
10. Matley, J., and the staff of Chemical Engineers (Eds.). *Fluid Movers: Pumps, Compressors, Fans, and Blowers,* McGraw-Hill, New York (1979).

Chapter 12

Pumps: Selection, Installation, and Intakes

GARR M. JONES
ROBERT L. SANKS

CONTRIBUTORS
Stefan M. Abelin
Virgil J. Beaty
Roger Cronin
John L. Dicmas
Rick A. Donaldson
Erik B. Fiske
Paul R. Gallo
Richard O. Garbus
James G. Gibbs, Jr.
Mayo Gottliebson
Alan W. O'Brien
Ned W. Paschke
Otto Stein
Charles E. Sweeney
William Wheeler

The purpose of this chapter is to describe: (1) how pumps are selected, (2) some important aspects of pump installation, and (3) superior pump sump designs.

The process of selection is difficult to explain. For seasoned engineers, it probably "just happens" because the knowledge gained from years of experience guides them through a winnowing process in which several possible selections are eliminated until only a few candidates remain. Up to this point the selection can be based on a set of guidelines (Section 12-4). But then it becomes a matter of relying on experience to select the best alternative from those that have survived this initial screening process.

Illustrative problems taken from practice in which the design was developed in U.S. Customary Units are presented in those units. Problems with no history are presented in SI units. But the use of both units simultaneously is avoided in this chapter.

12-1. Initial Screening

Before the process of selecting pumping equipment can begin, many factors relating to the application must be established. As these factors are defined, one inevitably begins to identify the type and size of equipment that will be most suitable. Initial selection factors, listed generally in order of precedence, include:

- Quality of the fluid to be pumped (clean or sandy water, wastewater)

- Required design capacity (maximum flowrate)
- Operating conditions (head, maximum and minimum flowrates, submergence, and/or NPSH).

Once these factors have been evaluated and the initial decisions have been made, the following factors must be considered:

- Mode of operation (such as in-line pumping, pumping from a well, or pumping from a clear well to a reservoir)
- Type of driver (motor or engine, constant or variable speed)
- Station location, configuration, and constraints (such as the number of pumps, horizontal or vertical pumps, units in parallel or series, wet well or dry well pumps, and submersible pumps).

The steps taken to complete the initial screening process are described in the following subsections. By necessity, this process is presented in a stepwise progression. In fact, it is an iterative procedure, with trade-offs from the ideal until the apparent optimum selection is found.

Fluid to Be Pumped

The characteristics of the fluid to be pumped have obvious effects on the pump selection process. The selection may not be so immediately obvious as one might think, however. Some raw water supplies may contain significant quantities of sand, grit, and floating or suspended material. For such waters, solids-passing capability and wear resistance (features commonly found in wastewater pumping equipment) may be appropriate. Pumps intended to function on unscreened wastewater should be of the type with appropriate eye inlet velocities, rounded and blunt impeller leading edges, and a configuration for nonclog service with rounded cutwater and adequate clearances to pass a desired minimum (specified) solids size. On the other hand, pumps intended for use with screened wastewater or treatment plant effluents of various qualities need not be selected to pass large solids. However, reconstituting of solids in the wet well, stringy materials, and abrasion due to sand and grit must be considered. Note that abrasive wear appears to vary with the third or fourth power of speed, and that unbalance (due, for example, to fouling by rags) is more serious at higher speeds. Consequently, the slower speeds are preferable for all applications but particularly for wastewater pumps.

Required Design Capacity

The required design capacity (both initially and at a future date), including the maximum, normal, and minimum flows to be pumped, must be considered when selecting the type and size of pumping equipment.

Unless the station is intended to accommodate a wide range of flows (caused, perhaps, by substantial further growth in the service area, by daily or seasonal changes, or by substantial storm inflow in a sewerage system), the following advice, derived from experience, is useful.

- Try to accommodate the peak demand with two or three duty pumps.
- Try to accommodate the normal demand with one duty pump.
- Try to limit the number of pump sizes. To reduce the inventory of spare parts, one size is best. Two sizes are acceptable.

These considerations, which keep the station size to a minimum, must be balanced against initial requirements. In some instances, the best solution may be to install small pumps initially that are to be replaced at some future date with larger equipment instead of more pumps. Note that the largest capital cost item is the structure itself (e.g., excavation, concrete, ventilation). In most instances, minimizing the number of pumps minimizes the capital cost of the station. If smaller pumps are used initially, the suction and discharge connections should be sized properly for the future units; use reducers as necessary for the smaller original units. Full-size drivers and starting equipment may be preferable for the initial pumping equipment. In any event, remember to allow space for future equipment and plan how it will be installed.

Once the preliminary pump sizes have been selected, the next step is to examine the complete range of operating conditions to be imposed on the equipment.

Operating Conditions

The full range of operating conditions should be examined to understand the application completely (see Section 10-8). Operating conditions (minimum flow, minimum/maximum discharge heads, NPSH and/or submergence limitations, and other requirements) may dictate pump selection. Some examples (drawn from actual experience) include the following:

- In a wastewater pumping station on a small site where over 70% of the tributary flow was delivered by off-site constant-speed (C/S) pumps at several pumping stations, large-capacity variable-speed (V/S) pumps were required. V/S equipment had to be capable of reacting quickly because little or no storage could be provided in the wet well.

- At a water booster station, C/S, vertical-turbine barrel pumps (described in Chapter 11) were selected because of the limited NPSH available at the pumping station site. The selection minimized the cost of the station by eliminating the need for a forebay and substructure.
- High static and total head conditions and the desire to use only one pumping stage at a wastewater pumping station limited the number of pump designs available for consideration.
- Widely fluctuating inlet head conditions at a water booster pumping station required pumps with a high shut-off head and a steep head-capacity curve.
- A nearly flat system curve at a V/S interstage pumping station in a treatment plant ruled out high specific speed pumps because a dip in the head-capacity curve (see Figure 10-16) would have caused unstable operation.
- A requirement for start-up and shut-down against a closed valve eliminated propeller pumps from consideration at a water pumping station because of the high power requirement while the valve is closed. Operation against a closed valve would require larger motors, switchgear, conductors, etc. (see Figure 10-16d). An alternative, however, would have been to bypass to the wet well during start-up. This alternative was not attractive because of the additional mechanical equipment, valve wear, and noise considerations.
- Plotting the system characteristics for *all* anticipated operating conditions is most helpful and is considered mandatory regardless of station size.

Station Location and Configuration

Various aspects of the site selected for a pumping station may force one pump option to be favored over all others. Site considerations may include the following:

- *Size:* A small site may require the use of vertical pumps (such as vertical turbines in barrels) to reduce the size of the station floor plan.
- *Hydraulic profile:* The required NPSH ($NPSH_R$) at the pump's (suction) inlet may result in a deep station, thereby discouraging the use of horizontal pumps in a dry well because of the cost associated with a large floor plan. Conversely, if a station can be shallow, horizontal pumps may be preferred because of the prospect for reduced: (1) vibration, (2) headroom required for a crane, and (3) height of superstructure.
- *Environment:* Weather conditions may rule out equipment that could otherwise be mounted outdoors. Concern over noise emissions may eliminate drivers or gear boxes mounted at or above grade, and submerged motors (which are quiet) might be favored. Pumping stations may, however, be soundproofed (see Chapter 22).

Mode of Operation

In most situations, pumps suitable for V/S applications are those with continuously sloping head-capacity (H-Q) curves, such as those shown in Figure 10-16b. The range of type numbers can be approximately 20 to 1250 in SI Units (specific speeds of 1000 to 7000 in U.S. Customary Units). Pumps with flat H-Q curves are usually unsuitable for V/S operation if two or more pumps operate in parallel or if a portion of the H-Q curve parallels the station system curve. Pumps with a dip in the H-Q curve are also unsuitable for V/S (Figure 10-16d) unless the pump H-Q curve intersects the system H-Q curve cleanly and well beyond the dip at all possible operating conditions. Pumps with a steep H-Q curve or a power curve that rises to the left (Figure 10-16c or d) are not an optimum choice if the pump must be started and stopped against a closed valve. Selection of pumps with such operating characteristics may require larger motors, starters, electrical conductors, and conduits to sustain the added load on start-up and shut-down.

Because of transient control measures or because of the details of the discharge piping, the pump may be required to run backward during the period shortly after shut-down. Operation of the pump at shut-off or at reduced capacity may increase vibration and significantly reduce bearing life.

Type of Driver

The type of driver influences pump selection. For example, horizontal pumps may be more appropriate than vertical units if engines are to be used, although note that right-angle gears can be used, as shown in Figure 12-1. Available motor speeds may also influence the pump selection process.

Miscellaneous Considerations

Other considerations that can influence the selection of pumping equipment include

- Client preference

Figure 12-1. Large engines driving pumps through right-angle gears. West Point Sewage Treatment Plant, Seattle Metro. Designed by Brown and Caldwell Consultants. Photo by George Tchobanoglous.

- Local custom (e.g., the use of vertical turbine pumps in the West vs. horizontal, double-suction, split-case pumps in the East)
- Budget and initial cost versus operation and maintenance cost
- Delivery time
- Experience (i.e., an engineer's previous experience with a particular manufacturer or pump type)
- Funding agency or statutory restriction (e.g., a prohibition against foreign manufacturers, proprietary specificiations, etc.)
- Local availability of parts and service.

12-2. Final Selection

Final selection involves a review in more critical detail of the considerations used in the initial screening process as well as detailed discussions with the selected manufacturers and the owner. Specific actions to precede a final selection include the following:

- Perform a detailed evaluation of system hydraulics and consider the performance of candidate pumps at all possible operating conditions.
- Furnish each candidate manufacturer with the details of the hydraulic requirements (including the system curve), the preliminary station layout, and preliminary specifications. Ask for written comments and a proposal specific to the project with budget equipment prices.
- Show the owner's representatives the station layout with a rough construction cost estimate and the preliminary equipment selections, preferably in a preliminary design report. Request their comments.

Once comments have been received from the owner and pump manufacturers, the final selection of equipment, layout of the pumping station, and writing of detailed pumping unit specifications can begin.

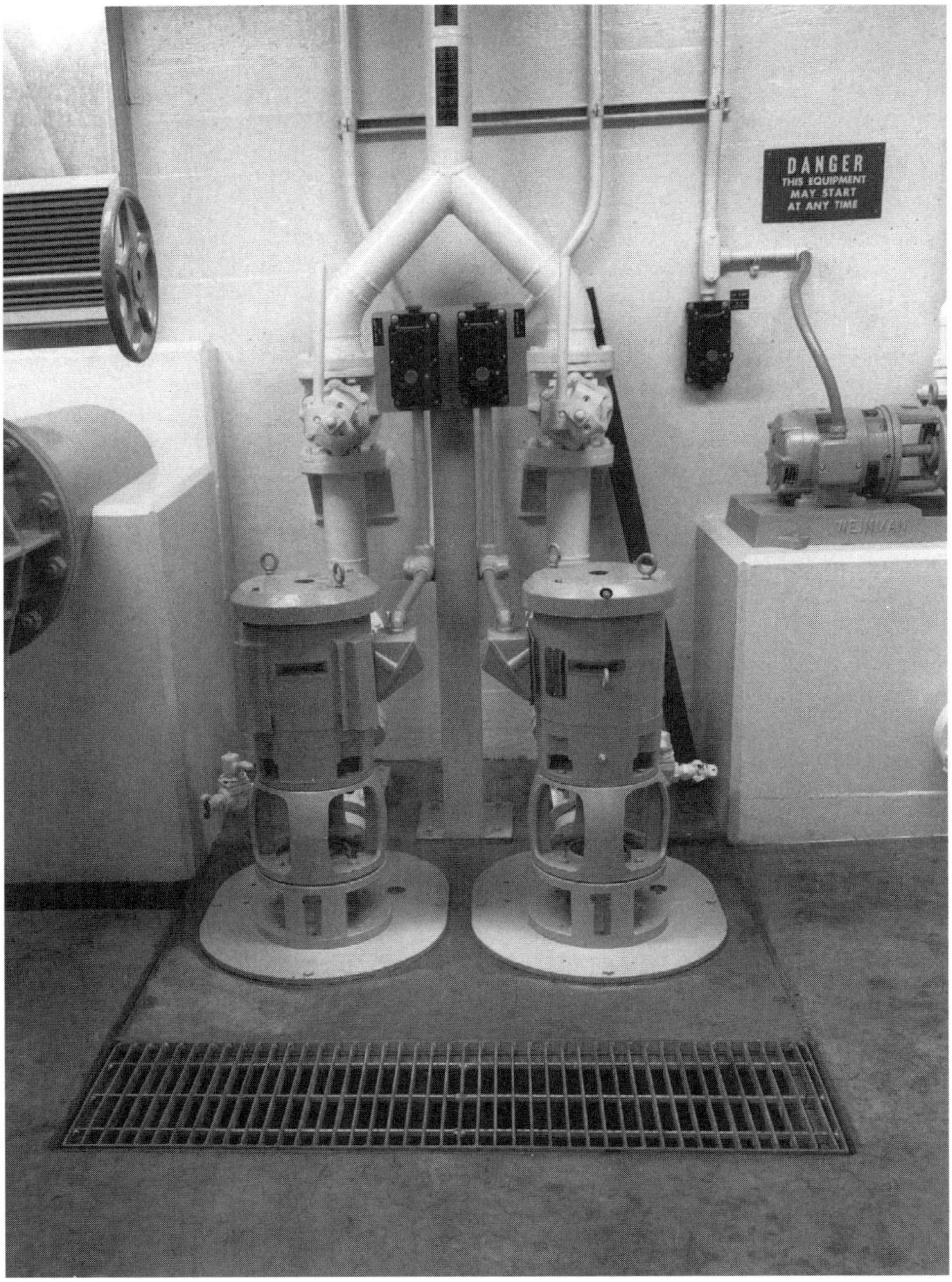

Figure 12-2. Sump pumps in Henderson Street Pumping Station, Seattle Metro. Designed by Brown and Caldwell Consultants. Photo by George Tchobanoglous.

12-3. Illustrative Examples

The examples in this section are presented to illustrate the pump selection process. Bear in mind that the process is one that cannot fully be described because a great deal of the selection effort is simplified through experience.

All too often, sump pumps (see Figure 12-2) are given too little attention by designers. In reality, these are pumping stations in their own right. The equipment should be selected and the installation should be designed with the same care as that afforded to the main pumping equipment.

Example 12-1
Sump Pumping System

A pumping station is designed to pump unscreened wastewater at a rated capacity of 20 Mgal/d at a total head of 67 ft. Each of the three main pumps is to be driven by a 175-hp (with nominal 200-hp motors) variable-speed system. The pumps are to be located in a room with a finished floor elevation of 95.33 ft. The motors are to be located at an elevation of 110.00 ft and the nominal finished grade outside the station is at an elevation of 128.75 ft. The maximum water surface elevation in the influent sewer system is to be limited by the elevation of overflows set to function at 121.00 ft. A typical station cross-section is shown in Figure 12-3. A hydraulically powered sluice gate isolates the pumping station upon power failure, abnormally high levels in the wet well, or other system malfunctions. The configuration shown has several advantages: (1) no intermediate shaft bearings (which increase cost and maintenance and often give trouble) are needed, (2) the station is quieter, and (3) the upper floor can be used for other purposes such as for the motor control center. As for the possibility of flooding, modern motors with encapsulated windings can withstand inundation (although not while running) without damage. Note that because the pump cannot be operated under water without damage, sump pumps or portable pumps must be used to remove water from a flooded station.

Problem: Select sump pumps and design a sump pump installation to protect the dry well, the motor room, and the pump room. The owner wants submersible pumps.

Solution: The purpose of the sump pump installation is (1) to accept liquids from incidental leakage from pumps, drainage from station maintenance and housekeeping operations, and

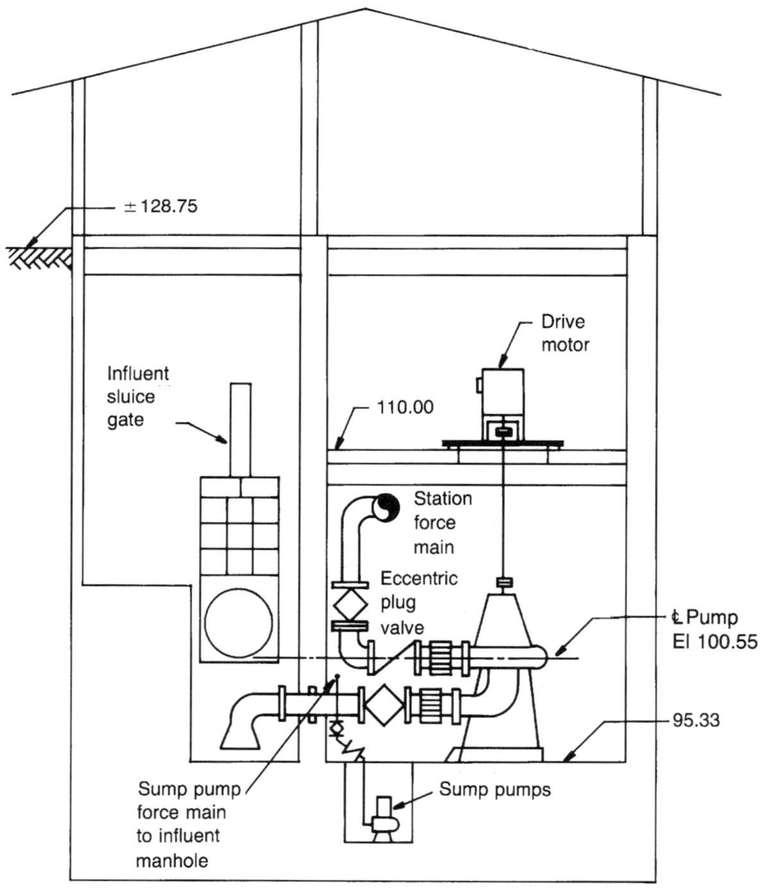

Figure 12-3. Cross-section of a typical pumping station. After Brown and Caldwell Consultants.

excessive inflow caused by broken or leaking couplings in the wastewater piping and (2) to discharge these liquids to a suitable location for disposal. Because the installation serves a wastewater pumping station and wastewater solids may be encountered from spillage during maintenance operations and at other times, the pumps must be selected to pass solids at least $1^1/_2$ in. in diameter. The pumps must discharge to the influent sewer upstream from the sluice gate (to provide a positive means for dewatering the station) at an elevation that protects the station against back siphonage.

The required discharge cannot be computed, but it is considered that a flowrate of 150 gal/min will protect the station. Under worst-case conditions (substantial inflow, perhaps caused by a partially separated pipe or pump casing joint), the maximum discharge required might be on the order of 300 gal/min. Such an unlikely event can be met by operating the standby pump as well.

Head-capacity curve. The influent manhole (upstream from the pumping station) is located approximately 25 ft away from the structure and the sump is to be at the opposite end of the structure, which is about 45 ft in length. The sump pump force main is to terminate in the influent manhole at a centerline elevation of 125.00 ft. The estimated length of piping between the sump and the point of discharge is 135 ft. The drain discharging to the sump is to have an invert elevation of 94.55 ft. The pump-control elevations are to be selected to maintain free drainage conditions in the drain system at all times during normal operation. The minimum static lift, therefore, is 125 – 94.55 = 30.45 ft when the sump water level is 94.55 ft (see Figure 12-4). The maximum static lift is 125 – 91.5 = 33.5 ft. and the average lift is (30.45 + 33.5)/2 = 32 ft.

The strategy that prompted the use of six ON–OFF switches in Figure 12-4 is the best possible for achieving the following. The follow pump will not start until after the high-water alarm is activated. Either excess flow into the sump or failure of the lead pump will initiate an alarm. The operator must go to the station to investigate the alarm condition and switch the follow

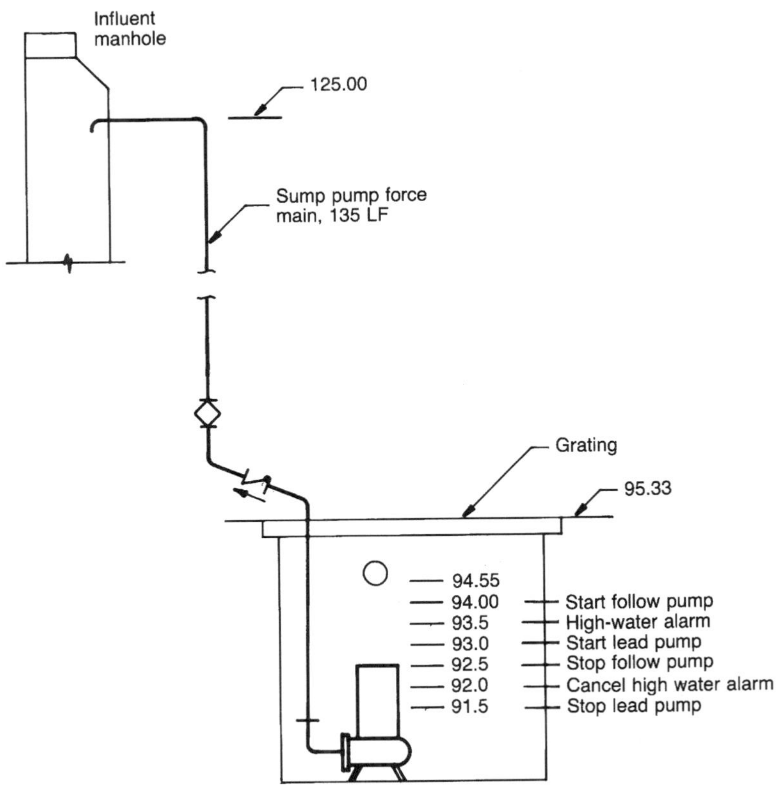

Figure 12-4. A sump pumping system. After Brown and Caldwell Consultants. See Figure 21-4 for a different strategy for ON–OFF switches.

pump to the lead position (with manual lead/follow selector) to clear the alarm. The problem with the lead pump can usually be corrected by the operator immediately.

For a discharge of 300 gal/min, the velocity (from Table B-4) in a 6-in.-diameter steel pipe, unlined or lined with plastic or coal tar, would be 5(300/447) = 3.4 ft/s, which is satisfactory. (If the lining were cement-mortar, the ID would be less, and the velocity would be 4.0 ft/s—which is also satisfactory.) In the system H-Q curves shown in Figure 12-5, the total dynamic head (TDH) for 150 gal/min is 32 to 35 ft; for 300 gal/min it is 36 to 39 ft.

Pump selection. Submersible pumps capable of passing solids and stringy material are essentially of the overhung-shaft, nonclog type (see Chapter 11). If the pumps are required to operate at near shut-off conditions, the pump vibrates and radial thrust forces (see Section 10-6 and Figure 10-18) acting on the impeller can be substantial. For this pump, the radial thrust forces would not only impose considerable operating loads on the shaft and bearings but would also cause appreciable deflection at the mechanical seal between the pump case and the motor casing. Failure of this seal would admit liquid to the motor windings and lead to motor failure. Hence, to achieve maximum reliability, the pumps should be selected to operate preferably

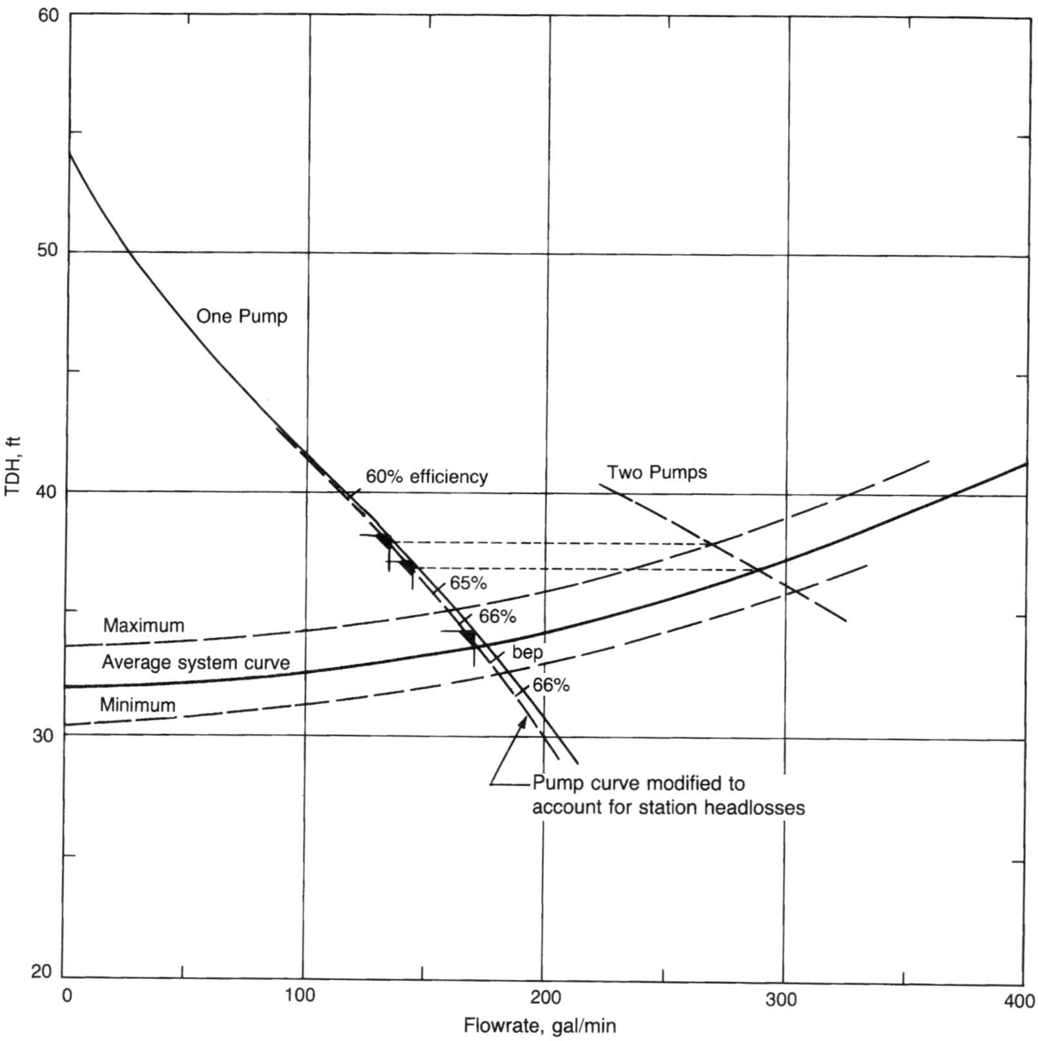

Figure 12-5. System and pump performance curves for sump pumps. Fairbanks Morse 5431 W submersible pumps (1750 rev/min, 3-in. suction, 7.00-in. impeller T3A1 X) were used.

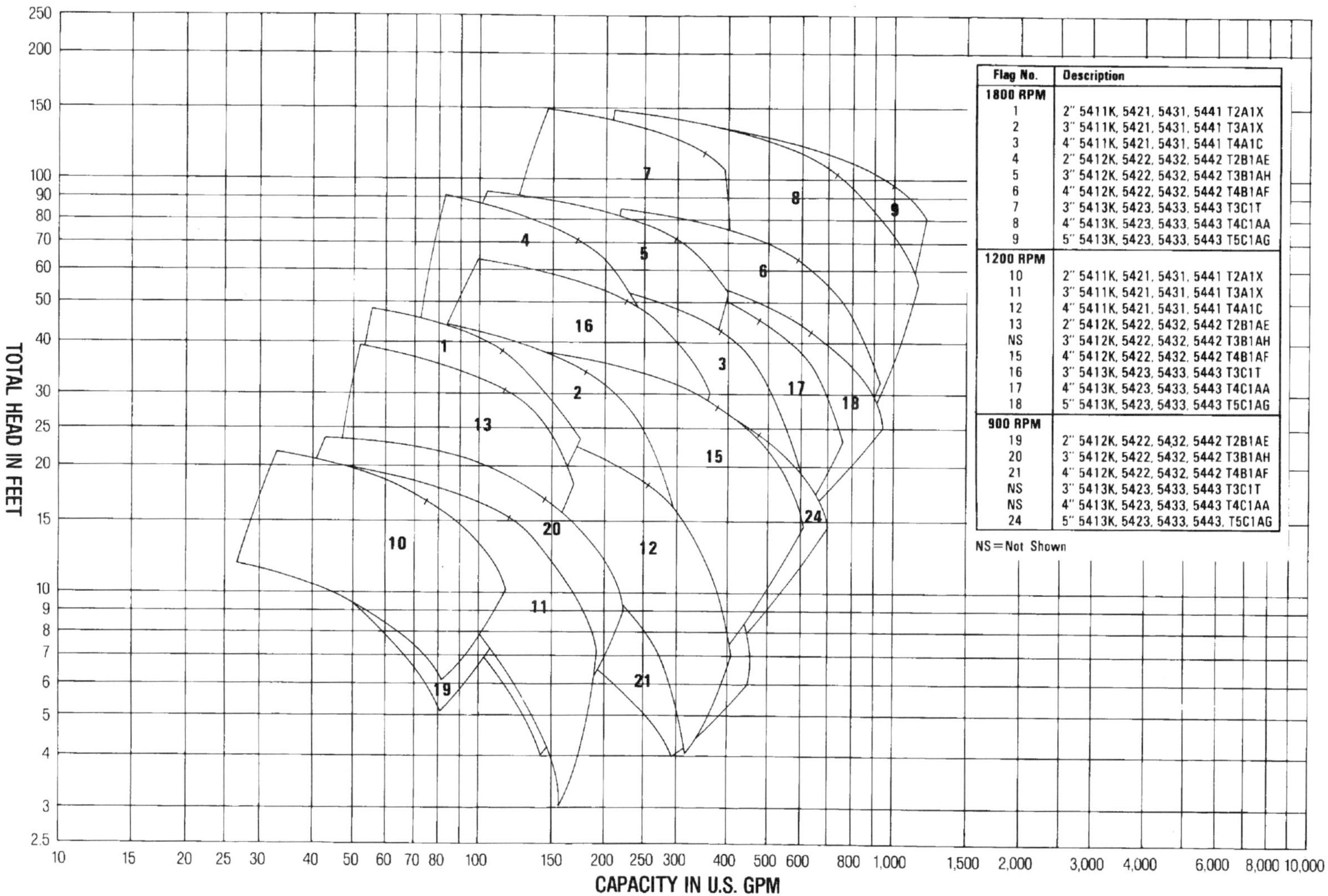

Flag No.	Description
1800 RPM	
1	2" 5411K, 5421, 5431, 5441 T2A1X
2	3" 5411K, 5421, 5431, 5441 T3A1X
3	4" 5411K, 5421, 5431, 5441 T4A1C
4	2" 5412K, 5422, 5432, 5442 T2B1AE
5	3" 5412K, 5422, 5432, 5442 T3B1AH
6	4" 5412K, 5422, 5432, 5442 T4B1AF
7	3" 5413K, 5423, 5433, 5443 T3C1T
8	4" 5413K, 5423, 5433, 5443 T4C1AA
9	5" 5413K, 5423, 5433, 5443 T5C1AG
1200 RPM	
10	2" 5411K, 5421, 5431, 5441 T2A1X
11	3" 5411K, 5421, 5431, 5441 T3A1X
12	4" 5411K, 5421, 5431, 5441 T4A1C
13	2" 5412K, 5422, 5432, 5442 T2B1AE
NS	3" 5412K, 5422, 5432, 5442 T3B1AH
15	4" 5412K, 5422, 5432, 5442 T4B1AF
16	3" 5413K, 5423, 5433, 5443 T3C1T
17	4" 5413K, 5423, 5433, 5443 T4C1AA
18	5" 5413K, 5423, 5433, 5443 T5C1AG
900 RPM	
19	2" 5412K, 5422, 5432, 5442 T2B1AE
20	3" 5412K, 5422, 5432, 5442 T3B1AH
21	4" 5412K, 5422, 5432, 5442 T4B1AF
NS	3" 5413K, 5423, 5433, 5443 T3C1T
NS	4" 5413K, 5423, 5433, 5443 T4C1AA
24	5" 5413K, 5423, 5433, 5443, T5C1AG

Figure 12-6. A manufacturer's range (coverage) chart. Courtesy of Fairbanks Morse Pump Corp.

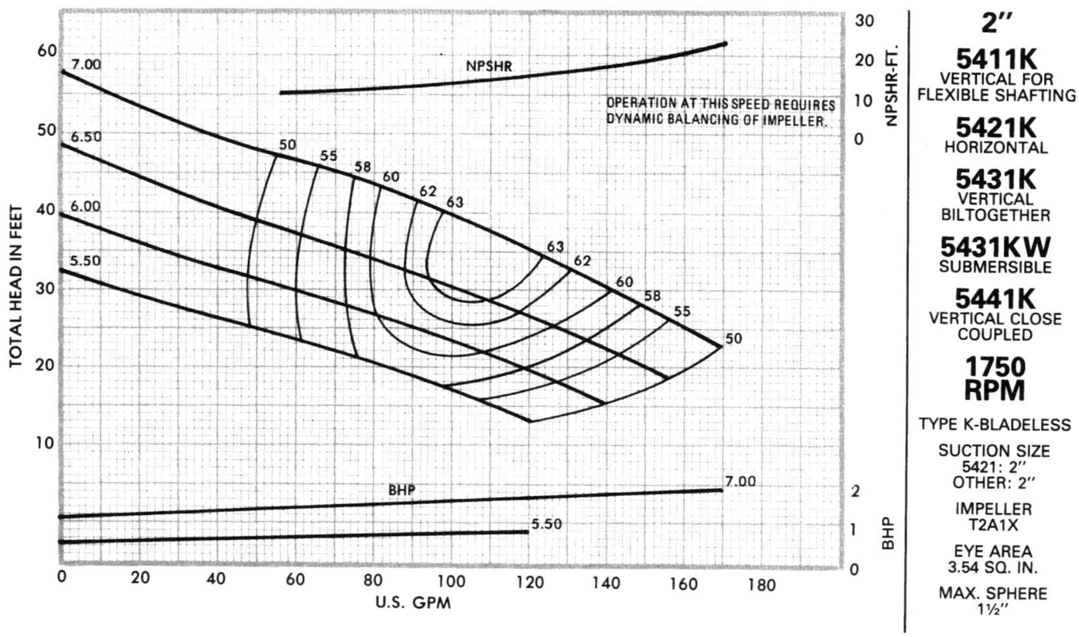

Figure 12-7. Flag 1. Performance curves for a 2-in. 5431 KW submersible pump. Courtesy of Fairbanks Morse Pump Corp.

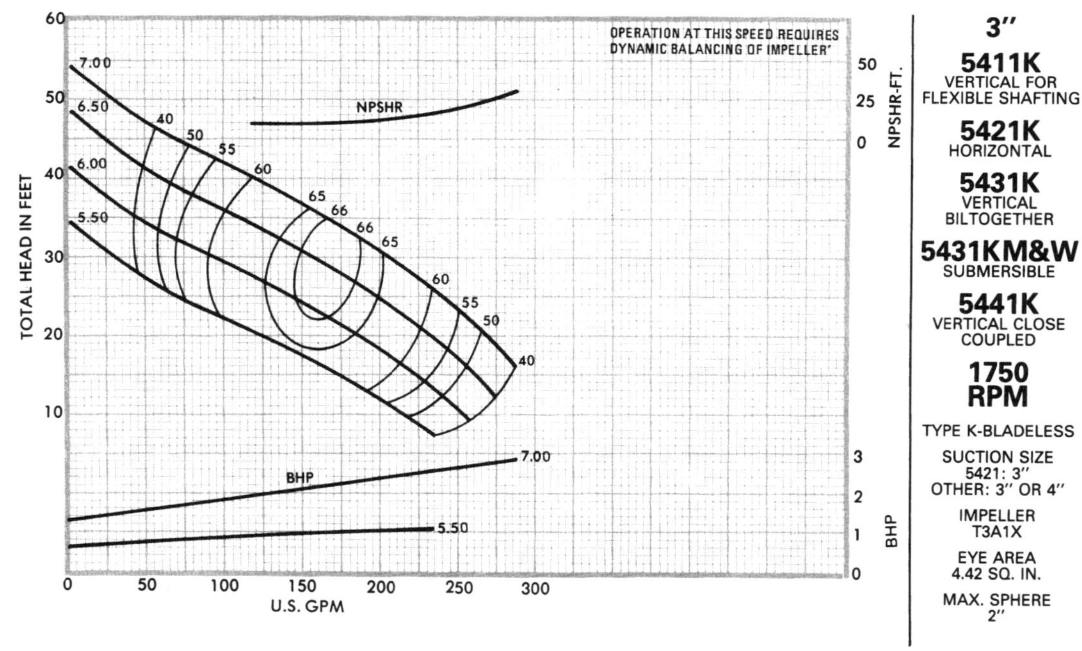

Figure 12-8. Flag 2. Performance curves for a 3-in. 5431 KM&W submersible pump. Courtesy of Fairbanks Morse Pump Corp.

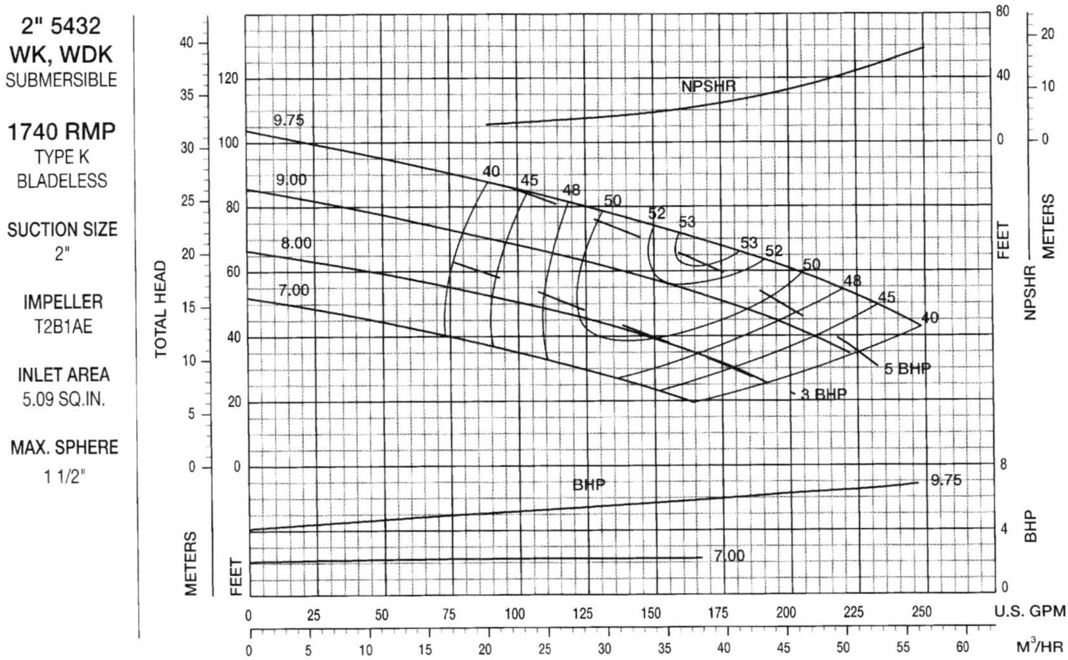

Figure 12-9. Flag 4. Performance curves for a 2-in. 5432 KW submersible pump. Courtesy of Fairbanks Morse Pump Corp.

within 60 to 115% of flow at the pump's best efficiency point (bep) where the magnitude of radial thrust forces is minimized.

Manufacturers' catalogs usually contain range or coverage charts such as the one shown in Figure 12-6. For a head of 37 ft and a flowrate of 150 gal/min for each of two pumps, envelopes (or flags), 1, 2, 4, 15, and 16 indicate pumps that might serve. The latter two are for 1200 rev/min pumps, which are unnecessarily massive for such intermittent duty. Pump characteristic curves for flags 1, 2, and 4 are shown in Figures 12-7, 12-8, and 12-9, respectively. The pump of flag 1 is limited to 110 gal/min at 37 ft of head and to 130 gal/min at 33 ft of head. This pump is adequate if the designer is willing to alter the original requirements.

The pump of Figure 12-8 would discharge 150 gal/min at 37 ft TDH and 170 gal/min at 33 ft TDH with a 7.00-in. impeller turning at 1750 rev/min. It is advisable to avoid the use of maximum-size impellers to allow for flexibility in meeting any unknown flow conditions. The flowrate required for the sump pump, however, is a very generous estimate, so the choice of impeller is justifiable. Furthermore, operating conditions are well within 60 to 120% of flow at bep. The $NPSH_A$ is not an issue with submersible pumps.

The pump of Figure 12-9 would meet the discharge requirements with near-minimum impellers, but it is larger and the efficiency is low (although for infrequent and short-term use, the efficiency is of little concern).

After reviewing the information, a good choice would be two (one duty, one standby) 3-in. 5431W pumps with 7.00-in. impellers as per flag 2, Figure 12-8. Similar performance can be obtained with pumps from competing manufacturers.

The power requirement (2.2 hp) can be read from Figure 12-8 or calculated using Equation 10-7b.

$$\text{hp} = \frac{170\ \text{gal/min} \times 33\,\text{ft}}{3960 \times 0.66} = 2.15$$

where 0.66 is the pump efficiency. A 3-hp motor is indicated.

Most of the basic procedures involved in selecting large pumps are illustrated in Example 12-2.

Example 12-2
Selecting Pumps for a Water Booster Pumping Station

The project consists of a booster pumping station to deliver water from an existing 84-in. filtered water aqueduct to a distribution reservoir. The total project capacity will be 25 Mgal/d, but the initial demand requires only 15 Mgal/d.

The station is to be located in a rapidly developing area and must be in service within 9 months. Environmental conditions include subfreezing temperatures during the winter and summertime temperatures exceeding 100°F. The temperature of the filtered water is expected to range between 50 and 75°F. The design data, shown in Figure 12-10, are as follows:

- The hydraulic gradeline (HGL) in the aqueduct varies from 6127 to 6132 ft.
- The suction pipe, 4000 ft long, is to be 48-in. reinforced concrete pressure pipe (RCPP).
- The elevation of the soffit (crown minus wall thickness) at the pumping station must be at least 2 ft (4 ft is better) below the minimum hydraulic gradeline to ensure that a full pipe is maintained at all times.
- The transmission pipe to the reservoir, 8500 ft long, is to be 36-in. steel lined with cement-mortar.
- The elevation of the water surface at the terminal reservoir varies between 6349 and 6357 ft.

Problem: Select the pumps. Before the pumps can be selected, however, the piping from the filtered water aqueduct to the reservoir must be designed to determine the dynamic headlosses. Optimizing the size of pipe to obtain a least-cost design is a duty of the project engineer but is not to be considered in this example.

Solution: A complete solution should follow the recommendations given in Chapter 17. A contour map would be used for a preliminary layout of the piping, pumping station site, and reservoir location as well as for the approximate elevations. The first calculations of headlosses, pump capacities, and motor sizes can be crude—accurate enough only for a guide in selecting the site, deciding on pipe sizes and materials, and choosing the type, number, and approximate sizes of pumps. The data given above summarize the results of such preliminary engineering.

For suction piping, the low elevation (6127 to 6132 ft) of the HGL in the filtered water aqueduct with respect to the ground elevation (6131 ft) presents some difficulty and must be considered in both the design of the supply pipeline to the pump and the selection of the pumps. To minimize losses to the pump, 48-in. ID pipe is selected. Because of frost penetration and traffic loads, the pipeline should have a minimum cover of 5 ft. The combination of low fluid pressure (only 6 lb/in.^{2}), high compressive strength of concrete to resist external loads, and low cost makes RCPP an ideal choice.

Suction piping losses. The Hazen–Williams (H–W) coefficient, C, for concrete pipe varies from about 135 for new pipe to 130 for old pipe (Table B-5), but, to be safe, a C factor of 120 is to be used for old pipe. Friction losses are calculated by using Equation 3-9b for both new pipe and old pipe at both 15 and 25 Mgal/d to obtain an "envelope" of high and low water elevations at the pump intake manifold. The use of the H–W equation for friction in pipe as large as 48 in. in diameter is justifiable in this example because the headloss is so low. Significant headlosses for large pipe should be verified by the Darcy–Weisbach equation (see Section 3-2). Headlosses in pipe fittings and valves are calculated using Equation 3-15 ($h = Kv^2/2g$). Hydraulic headloss coefficients, K, are listed in Table B-6 for fittings and in Table B-7 for valves. The velocities for the various expected sizes of pipe in both suction and discharge are as follows (see Tables 4-6 and B-4):

Nominal pipe diameter (in.)	Assumed ID (in.)	Area (ft^2)	Velocity (ft/s) 25 Mgal/d	15 Mgal/d	6.25 Mgal/d	5 Mgal/d
16	14.6	1.16	—	—	8.34	6.67
20	18.6	1.89	—	—	5.12	4.09
24	22.5	2.76	—	—	3.50	2.80
30	28.5	4.43	8.75	5.24	—	—
36	34.5	6.49	5.96	3.58	—	—
48	48	12.57	3.08	1.85	—	—

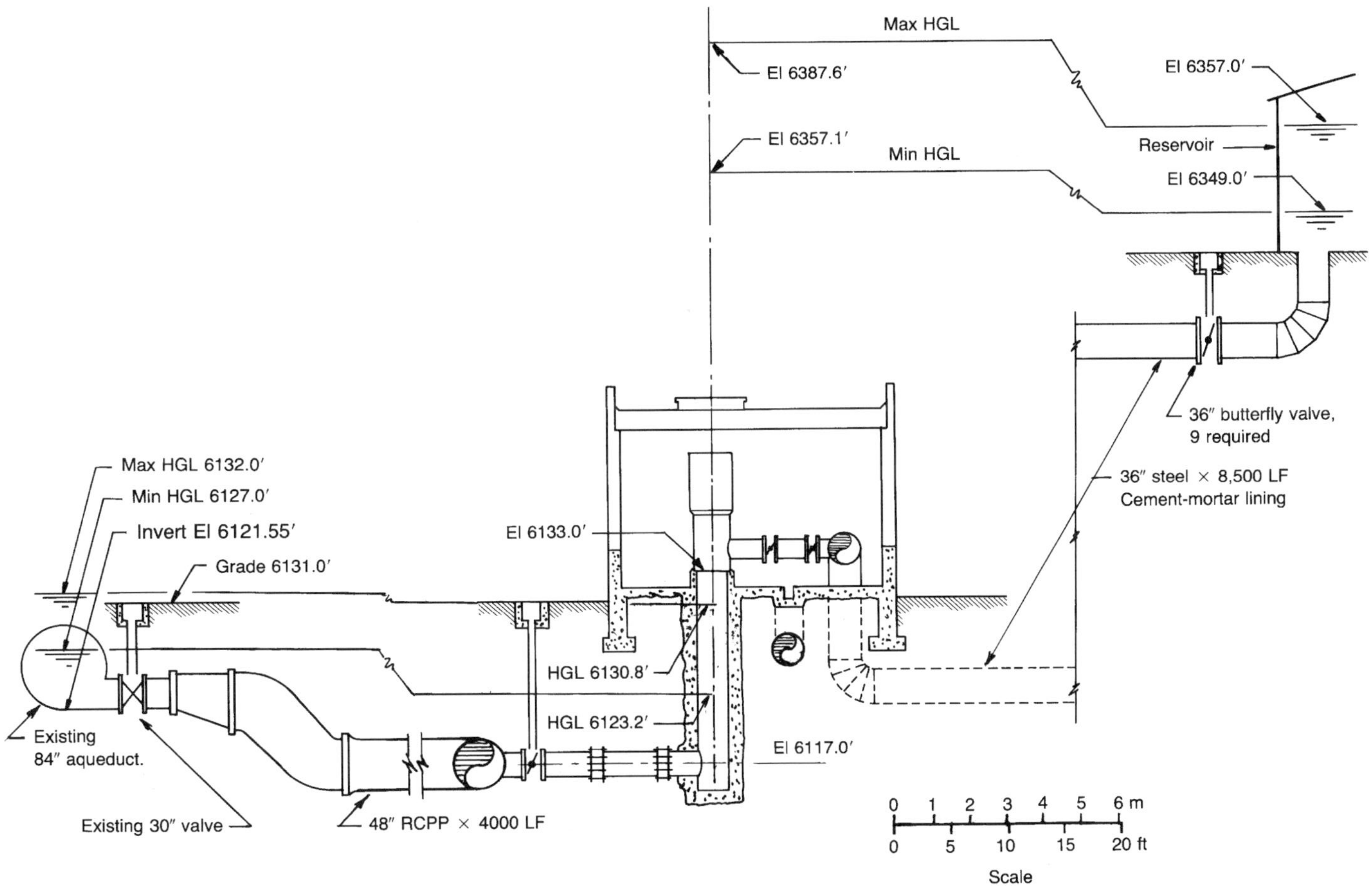

Figure 12-10. Profile of pipelines, Example 12-2. El, Elevation; LF, linear feet; RCPP, reinforced concrete pressure pipe.

The maximum suction pipe losses (in feet) are for old pipe, and the minimum losses are for new pipe. The calculations are as follows:

		Headloss			
		25 Mgal/d		15 Mgal/d	
Suction pipe losses	Calculations[a]	Max	Min	Max	Min
Entrance (30 in.)	$0.5v^2/2g$	0.59	0.59	0.21	0.21
Butterfly valve (30 in.)[b]	$0.16v^2/2g$	0.19	0.19	0.07	0.07
Short spool	Ignore	—	—	—	—
Increaser (30–48 in.)[c]	$K = 3.5\left(\tan\frac{\theta}{2}\right)1.22 = 0.45$				
	$h = K\left[1-\left(\frac{D_1}{D_2}\right)^2\right]^2 v_1^2/2g$	0.33	0.33	0.12	0.12
Two 45° bends (48 in.)	$h = 2\times 0.18v^2/2g$	0.05	0.05	0.02	0.02
Pipe friction	$4\times 10{,}500\left(\frac{\text{gal/min}}{C}\right)^{1.85}(48)^{-4.87}$	2.70	2.18	1.05	0.85
Total		3.86	3.34	1.47	1.27

[a] As shown in Appendix B, fitting and valve losses can vary ±30% or more from given values. For this problem such variations in losses are small and, therefore, are ignored to avoid complexity.

[b] For the low head in the suction pipe, a 25-lb Class valve is adequate (see Table B-7).

[c] The formula $h = 0.25(v_1^2 - v_2^2)/2g$ is simpler and sufficiently accurate.

Transmission piping losses. The (discharge) transmission pipe is to be 36-in. diameter steel with a cement-mortar lining. The H–W C factor for such large pipe when new would be 145 to 150 and might decrease to 130 in about 20 yr. To be safe, coefficients of 145 and 120 are used here for maximum and minimum headlosses, respectively.

		Headloss			
		25 Mgal/d		15 Mgal/d	
Transmission pipe losses	Calculations	Max	Min	Max	Min
Pipe friction	$8.5\times 10{,}500\left(\frac{\text{gal/min}}{C}\right)^{1.85}(34.5)^{-4.8}$	28.7	19.0	11.2	7.4
8 butterfly valves (36 in.)[a]	$8(0.35v^2/2g)$	1.55	1.55	0.56	0.56
90° bend (36 in.)[b]	$0.30v^2/2g$	0.17	0.17	0.06	0.06
Total		30.4	20.7	11.8	8.0

[a] Isolation valves at 1000- to 1200-ft intervals

[b] The two bends in the 36-in. pipe within the station are allocated to station losses.

It can be seen from these results that calculating some of the minor headlosses in long pipelines is often unnecessary because the losses are trivial. Some engineers just add 5% to the pipe friction instead (see Section 17-2).

Static head. The maximum static head (from Figure 12-10) is 6357.0 – 6127.0 = 230 ft and the minimum static head is 6349.0 – 6132.0 = 217 ft. The total head is

Static head + suction losses + station losses + discharge losses

If the station losses are assumed to be 5 ft (as recommended in Section 17-2), the TDHs are

$$\text{TDH}_{max} = 230 + 3.86 + 5 + 30.4 \simeq 269 \text{ at 25 Mgal/d}$$

$$\text{TDH}_{min} = 217 + 1.27 + 5(5/6.25)^2 + 8.0 \simeq 230 \text{ at 15 Mgal/d}$$

The ratio, 5/6.25, is proportional to velocities in the headers at 15 and 25 Mgal/d flowrates as explained below. The above data are used to construct the system H-Q curve in Figure 12-11.

Pump selection. If five pumps (four duty, one standby) are used to deliver 25 Mgal/d, each pump is rated at 6.25 Mgal/d (4340 gal/min). Pump No. 3, the middle pump, can be omitted at the initial flowrate of 15 Mgal/d, so the rating of the three duty pumps need only be 5 Mgal/d (3470 gal/min). If an end pump were omitted, water at that end of the manifold would stagnate.

To minimize construction costs and expedite the completion of the project, barrel-type or "can" vertical turbine pumps are to be used. Instead of coverage charts, Byron Jackson pumps are listed in a selection table as shown in Figure 12-12. Note that a Series 1700 pump (Figure 12-13) develops the required 3470 gal/min flowrate, but is already at bep, and higher flowrates can only be obtained at reduced efficiencies. A better choice is a Series 2000 pump. The pump performance curves are shown in Figure 12-14 for a single stage operating at 1770 rev/min. With a 13-in. impeller, the pump would deliver 4340 gal/min at a head of 134 ft. With two stages, the head is increased to 268 ft at an efficiency of 85%. The $NPSH_R$ is 28 ft, and the required horsepower is $2 \times 180 = 360$, so a 400-hp motor is needed.

As shown in Figure 12-10, the suction inlet must be at an elevation of 6117, which is 16 ft below the mounting plate at an elevation of 6133.0. The suction inlet in the standard pump is only 77 in. below the mounting plate, so the pump barrel must be extended. Priming must be ensured at all times, so the lower bowl should (as an estimate) be at an elevation of about 6115 ft. The pump manufacturer will be held responsible for pump barrel dimensions and for the depth of the lower bowl below the filtered water supply pipeline (submergence) as a part of a unit responsibility specification.

Manifold piping. The diameter of the suction inlet for the pump selected is 24 in., the barrel diameter is 36 in., and the discharge outlet is to be 16 in. in diameter because a butterfly-type pump-control valve will be used. A butterfly valve larger than 16 in. to control a flow rate from 0 to 5 or

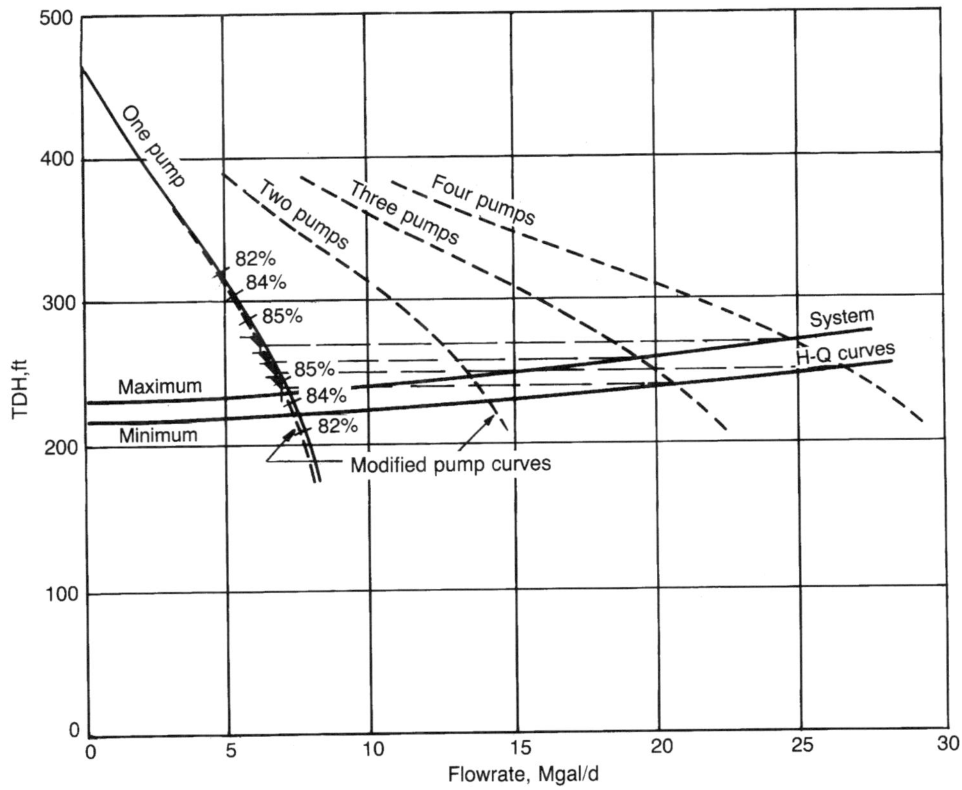

Figure 12-11. System head-capacity envelope and pump characteristic curves.

PUMP SERIES	SIZE			OPERATING LIMITS								SHAFT AREA IN.²	MAX. NET SLEEVE AREA IN.²
	DISCH. X SUCT. X "BD"	BARR. (IN.)	COL. (IN.)	MAX. CAP. GPM	MAX. SHAFT BHP 416 SS		MAX. SUCT. PRESS	MAX. DIFF. PRESS CAST IRON	MAX. DISCH. PRESS 300 LB FLANGES		MAX. DISCH. PRESS 600 LB FLANGES		
					RPM 3600	RPM 1800			SB	SH	SH		
100	2 x 3 x 12 3 x 4 x 12/16	8	4	100 160	60	30	285	650	740	N/A 740	N/A 1340	.52	.71
200	2 x 3 x 12 3 x 4 x 12/16 4 x 6 x 12/16	8 8 10	4	100 200 300	120	60	285	650	740	N/A 740 740	N/A 1340 1340	.79	1.29
600	3 x 4 x 12/16 4 x 6 x 12/16	8 10	4	200 350	120	60	285	660	740	740	1340	.79	1.29
700	4 x 6 x 12/16	10	4	600	120	60	285	610	740	740	1340	.79	1.29
800	6 x 8 x 20	12	6	1000	200	100	285	560	660	740	1130	1.11	.97
1000	6 x 8 x 20 8 x 10 x 20	16	6 8	850 1600	400	200	285	580	660	740	1130 1030	1.62	1.92
1100	6 x 8 x 20 8 x 10 x 20	16	6 8	850 1250	N.A.	300	285	580	660	740	1130 1030	2.24	1.31
1200	8 x 10 x 20 10 x 12 x 24	16 20	8 10	1500 1800	N.A.	300	285	570	660 440	740	1030 950	2.24	1.31
1300	8 x 10 x 20 10 x 12 x 24	16 20	8 10	1500 2100	N.A.	450	285	550	660 440	740	1030 950	2.95	1.96
					1200 RPM	1800 RPM							
1500	10 x 12 x 24	20	10	2400	450	700	285	430	440	740	950	3.76	3.91
1700	10 x 12 x 24 12 x 16 x 24/30 16 x 20 x 24/30	24 24 30	10 12 16	2300 3400 4600	800	1200	285	410	440 300 300	740 560 450	950 — —	5.67	2.00
2000	12 x 16 x 24/30 16 x 20 x 24/30 16 x 24 x 24/30 20 x 24 x 24/30	30 30 30 36	12 16 16 20	3400 5400 6500 7800	1100	1600	285 285 285 250	560	300 300 300 250	560 450 450 375	— — — —	6.78	2.17

Figure 12-12. Selection table for vertical turbine pumps. Courtesy of Byron Jackson Pump Division.

6.25 Mgal/d would be out of proportion; it would have to open and close too slowly, and the seats would wear quickly due to the cavitation that occurs when a butterfly valve is nearly closed.

The hydraulic losses through the suction and discharge headers and manifolds can now be computed (see Figures 12-15 and 12-16).

Suction header losses	Calculations	Headloss 6.25 Mgal/d	Headloss 5.0 Mgal/d
Tee, reducing to 24 in. and branch flow[a]	$0.5v^2/2g$	0.10	0.06
Valve, 24-in. butterfly[b]	$0.16v^2/2g$	0.03	0.02
17 LF of 24-in. pipe[c]	H–W equation	0.03	0.02
Exit velocity head (into pump barrel)	$v^2/2g$	0.19	0.12
Total		0.35	0.22

[a] Estimate loss for a sharp-edged intake.

[b] Use 25-lb Class valve because static head is so low (see Table B-7).

[c] LF, linear feet.

From these results it is seen that the minor headlosses (in valves and fittings) are considerably greater than pipe friction losses and must not be ignored in short pipelines.

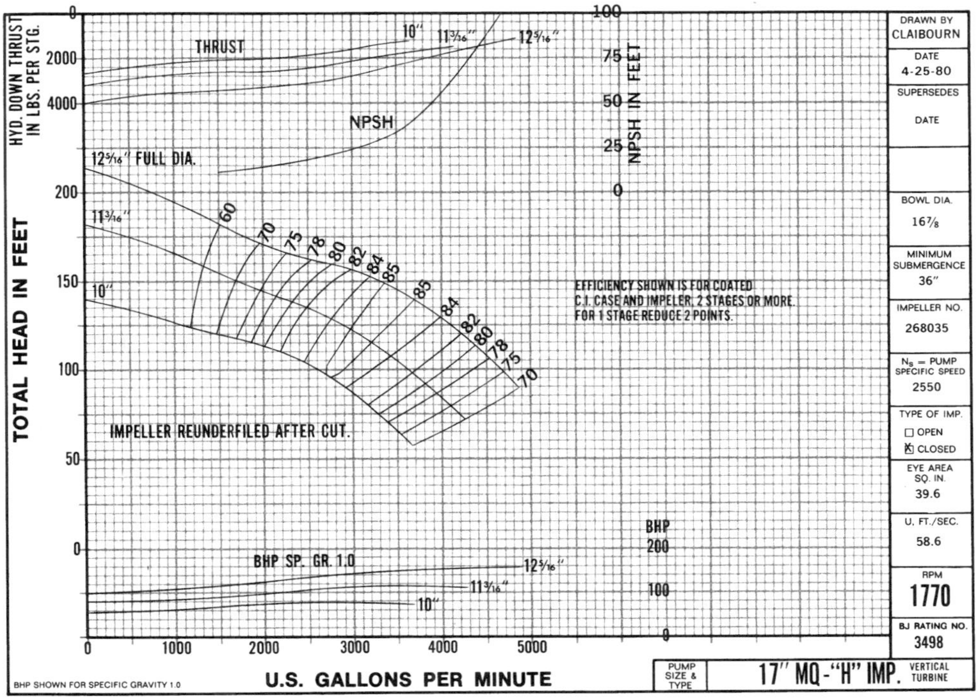

Figure 12-13. Series 1700 pump performance curves for a single stage. Courtesy of Byron Jackson Pump Division.

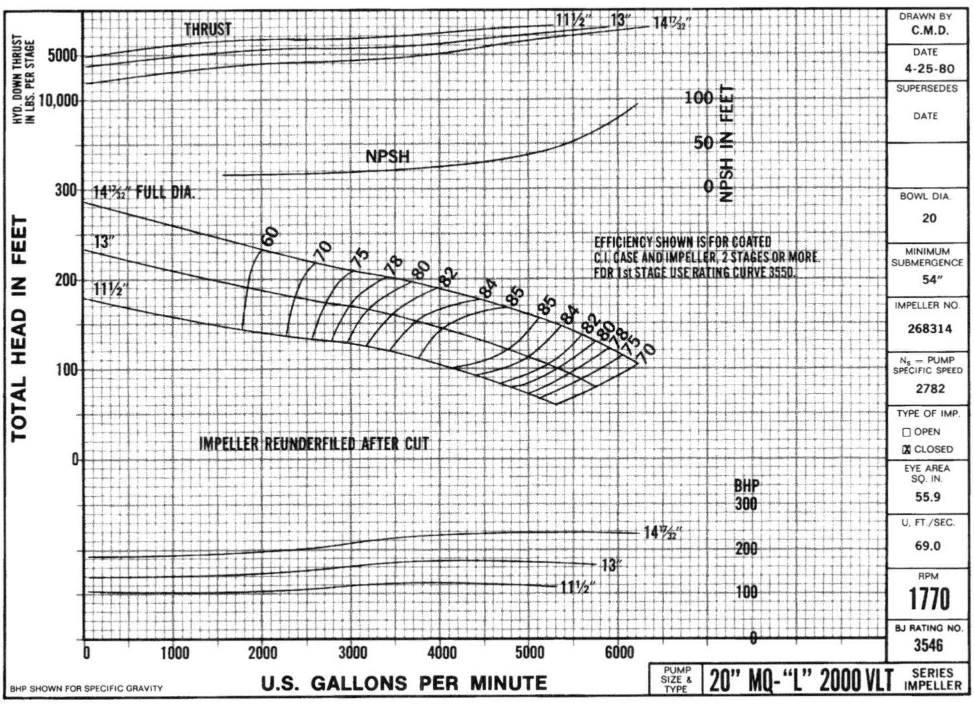

Figure 12-14. Series 2000 pump performance curves for a single stage. Courtesy of Byron Jackson Pump Division.

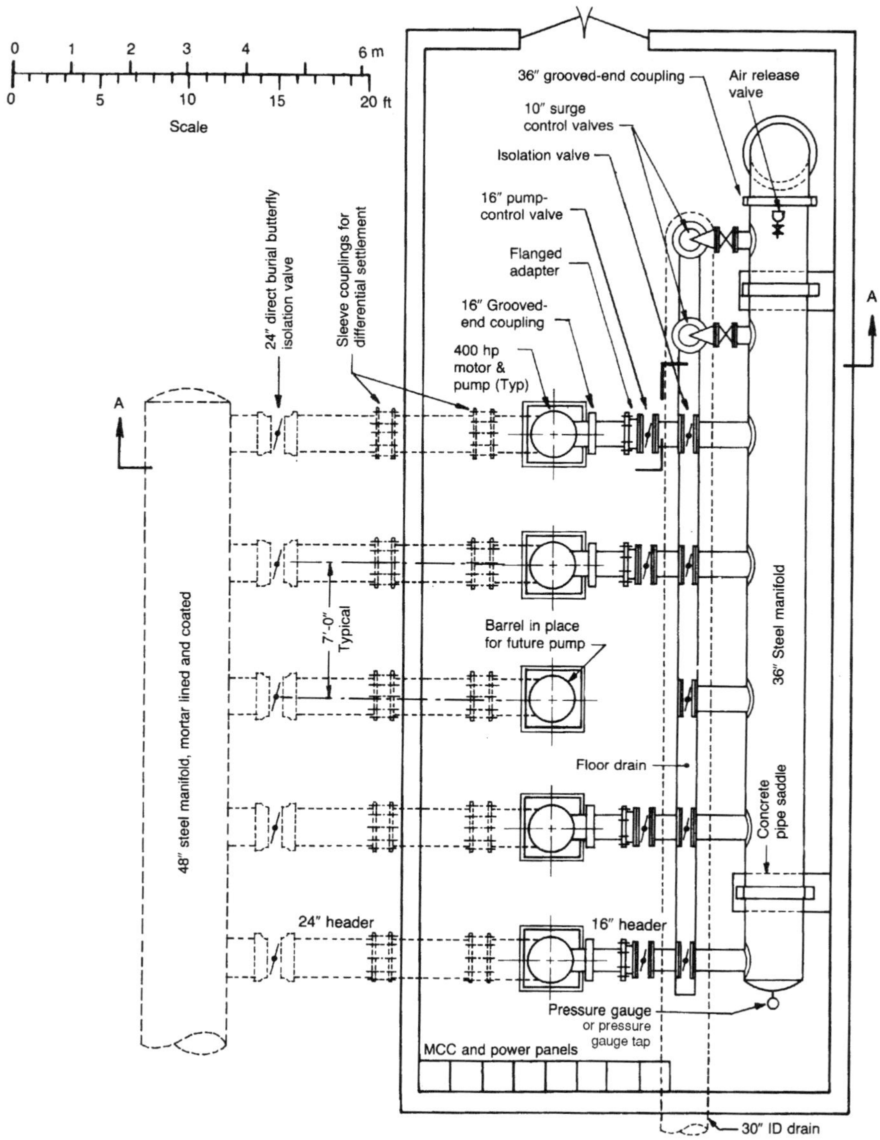

Figure 12-15. Plan of a booster pumping station.

Discharge header and manifold losses	Calculations	Headloss 6.25 Mgal/d	Headloss 5.0 Mgal/d
10 LF of 16-in. pipe	H–W equation	0.17	0.11
Pump-control valve, 16-in. butterfly[a]	$0.35v^2/2g$	0.38	0.24
Isolation valve, 16-in. butterfly[a]	$0.35v^2/2g$	0.38	0.24
Tee, branch flow, and sudden increaser (16 to 36 in.)[b]	$0.50v^2/2g$	0.54	0.35
	$(v_1^2 - v_2^2)/2g$	0.94	0.64
41 (say) LF of 36-in. pipe	H–W equation	0.14	0.04
Two 90° mitered bends	$2 \times 0.3v^2/2g$	0.33	0.12
Total		2.88	1.74

[a] Use 125-lb Class valve because of high (230 ft) static head (see Table B-7).

[b] Assume v_2 is half the normal velocity in the 36-in. pipe (for pump 2 or 4).

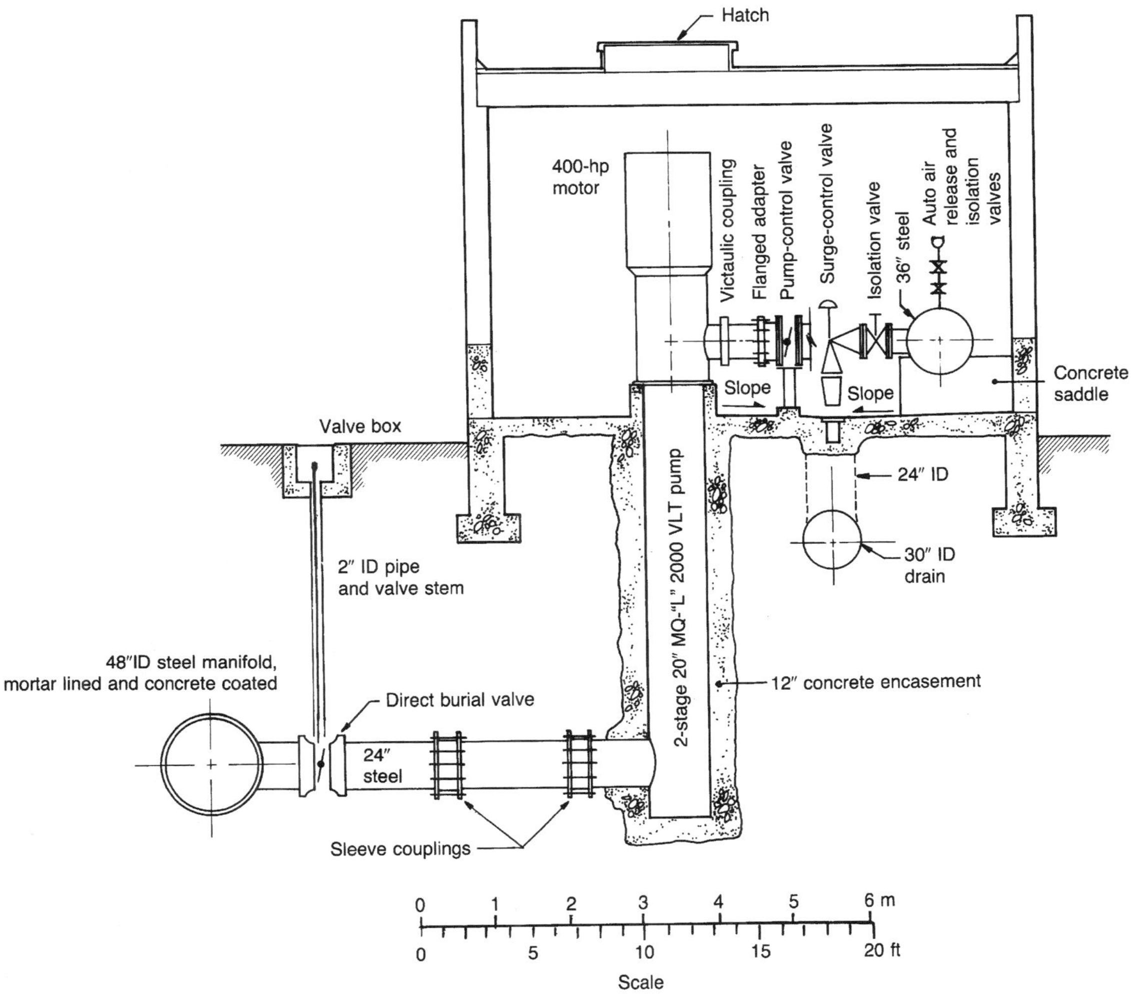

Figure 12-16. Section A–A from Figure 12-15.

Total dynamic head (TDH)

	Discharge headloss			
	25 Mgal/d		15 Mgal/d	
Loss	Max	Min	Max	Min
Suction pipe	3.86	3.34	1.47	1.27
Transmission pipe	30.4	20.7	11.8	8.0
Suction header	0.35	0.35	0.22	0.22
Discharge header and manifold	2.88	2.88	1.74	1.74
Static head	230	217	230	217
Total	267	244	245	228

Note that the system curve in Figure 12-11 needs no correction in this example. Many figures are carried out to 0.01 ft for clarity so that a reader can trace them through the calculations. Some values are so small that they can be safely ignored for an actual application. The pump selected (Figure 12-14) would be adequate with a 13-in. impeller, but note that larger impeller diameters or additional stages can be accommodated in the same barrel size.

$NPSH_A$. Assume that the first-stage pump bowl is at an elevation of 6115 (2 ft below the centerline of the suction piping). From Equation 10-23 and the explanation in Section 10-4, the $NPSH_A$ is found as follows:

Component	Heat (ft)
h_{bar} at 6000 ft altitude (Table A-7)	+27.2
h_{stat} (minimum) = 6127 – 6115 ft (Figure 12-10 and subsection "Pump Selection")	+12.0
h_{vap} (of water at 75°F) (Table A-9)	–1.0
$h_{ent} + h_{fs} + \Sigma h_m$ (3.78 + 0.35 = 4.1)	–4.1
Subtotal	+34.1
Safety factor (5 ft) According to recent data, the usual 2 ft is insufficient. See Section 10-4.	–5.0
Net $NPSH_A$	29.1

According to the pump manufacturer, 2 ft should be allowed for losses in the pump barrel. The $NPSH_R$ for the selected pump is 28 ft, so the pump setting is less than adequate; the first-stage impeller should be lowered at least 1 ft, to a maximum elevation of 6116. But specify the $NPSH_A$ (referred to datum in the 48-in. inlet piping) and let the pump manufacturer determine the pump setting elevations and the barrel and inlet sizes.

Surge protection. There is probably no need to protect the 48-in. suction pipe from surges because (1) the velocity is only 3 ft/s, and (2) the static plus dynamic head is only 15 ft. If necessary, however, the pipeline could be protected as described below for the transmission main.

The transmission main is protected against power failure by two pilot-regulated, diaphragm- or piston-operated, slow-closure, 10-in. surge relief valves (Figures 12-15 and 12-16). The valves are piloted shut during normal conditions, but upon power failure they are opened quickly by rising water pressure; then they close slowly. The valves depicted are 10-in. Parco angle needle surge anticipator valves, but 12-in. CLAVAL, Golden-Anderson, Bermad, or other valves would also work. By adjusting the closure mechanisms of the two valves to settings that differ by 5 to 10 lb/in.^{2}, the amount of water bypassed to waste can be reduced.

Surges during normal pump start-up and shut-off are prevented by the slow opening and closing characteristics of the pump-control valves. The valves should be designed to close quickly (within 5 or 6 s or, rarely, more than 25 or 30 s) by water pressure if the power fails. After (or while) the pump-control valves close, the surge relief valves, described above, open rapidly to dissipate any surges in the system. The water from the surge relief valves can be wasted to a local storm sewer. The water should not be wasted to the pump inlet conduit because of the potential for surge in that conduit.

Critique. Isolation valves on the suction and the discharge headers are always required because of the need to remove a pump occasionally. Butterfly valves are used for isolation because of their low cost. Underground valves should be direct burial valves (such as a Pratt Groundhog or a similar valve). The worm gear box (which is required for big valves) has a long stem (enclosed in a small pipe) ending in an operating nut in a small valve box at grade.

The two sleeve couplings on each suction header are needed to allow for differential settlement and for slight misalignment. The Victaulic® coupling and the flanged adapter allow for misalignment in the discharge header. Lock pins can be omitted from the flanged adapters because the discharge header is longitudinally restrained by the pump on one side and the manifold on the other. Because of the length and weight of the pipe and valves between the pump and the manifold, there must be intermediate support at the pump-control valve to prevent strain in the valve body.

Because pump-control valves limit surges to very low values for normal start-up and shutdown, they are much better in this station than check valves. But be cautious. Avoid designing a custom pump-control valve system unless the valves are so large that off-the-shelf, package systems are not available. Packages based on globe valves are proven, reliable designs, although the headloss can be high and the cost of energy therefore substantial (see Example 5-1). To save dynamic head, butterfly valve packages are used for pump control in this example. However, pump control is extremely severe service for a butterfly valve, so be sure the valve is specifically built for throttling. For throttling service, it is probably best to have seals that are cemented in place even though replacement is more difficult than for seals held in place by bolts, because bolts tend to loosen due to vibration.

After a short period, fixed-pressure gauges become unreliable because of vibration, corrosion, and other factors, so it is better to use portable gauges that can be calibrated before use. Usually, a pressure tap should be located on each discharge header well downstream from disturbances (about 6 pipe diameters) caused by the pump. But such locations in this design are so close to the manifold that the gauge readings would closely approximate the manifold pressure (within 0.33 m or 13 in. WC or less). Hence, a single-pressure gauge tap on the manifold is adequate.

The 36-in. discharge manifold is shown supported on concrete pipe saddles. Because of thrust generated in a surge, the saddle must be extended at least to the half-depth of the manifold on the wall side. Steel pipe supports could be used instead—probably at less cost—but horizontal supports between manifold and wall would also then be necessary. The bending stress with the manifold full of water is less than 2000 lb/in.^{2} (for a pipe wall $^3/_8$ in. thick), so neither stress nor buckling is a problem. If the manifold is too long for the fabricator, it can be built in two halves joined by a shouldered Victaulic® coupling and supported at the Victaulic® joint with either a pipe support or a concrete saddle. Note that the headers connect to the manifolds at 90° for ease in fabrication. Connections at 45° are sometimes used, but the headloss saved is negligible, the piping layout is not as neat, the floor space shape is less desirable, and the fabrication cost is greater. An air release valve at a high point in the piping is also always required, and in this station, the best location is at the first 90° bend where air would naturally accumulate.

The surge control valves discharge through a standard reducer (to prevent splashing) into a 24-in. downspout connected to a 30-in. sewer pipe designed to flow partly full to prevent "burping." The velocity from the surge control valves can reach 40 or 50 ft/s, so a large drain is necessary.

Note that there is 42 in. of clearance between pumps, a generous 6-ft clear walkway beside the pumps (which could be reduced to 5 ft or, if space were at a premium, even to a crowded 4 ft), and 42 in. of clearance in front of the 480-V electrical cabinets. As shown in Figure 12-15, there are five 20-in.-square cabinets for combination motor starters (at present, one is empty), another for the main breaker, and two for auxiliaries (such as heating and ventilating). If 4160-V motors were used, there would be seven cabinets 36 in. wide by 30 in. deep (five for combination motor starters, one for the main breaker, and one for a transformer) plus two 20-in.-square cabinets for auxiliaries as before; according to the NEC, the front clearance would have to be 4 ft. The clearances shown in the figure and the spacings between fittings on the discharge headers are reasonably minimum but adequate for ease of access and maintenance. The design is simple, compact, inexpensive, and practical. The station can be quickly fabricated and constructed and will be easy to maintain and operate.

12-4. Summary and General Considerations in Pump Selection

A step-by-step procedure for selecting pumps is as follows.

1. Identify the design point as well as all other operating points for which the pump must be selected.
2. Define any other operating conditions.
3. Compute and plot the station characteristics for the design and all extreme conditions.
4. Decide the number of pumps to be used and the type of pump (horizontal, vertical turbine, submersible, etc.).
5. From a manufacturer's coverage chart or selection table, find the pumps that operate in the desired range. Be sure to give all operating conditions and system H-Q curves to two or more manufacturers for independent evaluation and selection recommendations. However, always check the work, because your responsibility for selection cannot be shirked. If suitable pumps cannot be found,

contact the two manufacturers anyway, because pumps and impellers that are not shown in the catalogs are, nevertheless, often available.

The specified operating conditions must include:

- High and low static lifts and the envelope of maximum and minimum possible friction losses.
- For C/S pumps, the specified operating point should be within 5% of bep.
- If the pump must operate against a closed valve (or against water at rest in a long pipeline), furnish data on frequency and time length of such events. Check acceleration time for long pipelines to make certain the pump will not be subjected to unacceptably long periods of operation at high radial thrust loads.
- The $NPSH_A$, referenced to pump elevation.
- Unusual fluid characteristics. For example, excessive Cl^- (in salt water pumping) or grit. Stainless steel is attacked by Cl^-, so use Hastelloy or Monel metal. Erosion resistance of cast iron is greatly increased by adding 2 to 3% nickel.
- Service requirements (such as continuous or intermittent duty) and required service life.
- Special requirements such as heavy shafts, oversized bearings, and packing glands versus mechanical seals.

6. Examine the appropriate pump performance curve for the series of pumps selected in step 4 until a pump is found with a high efficiency at the operating point. Decide on the maximum permissible speed and the speed range for which the pump will be selected. Decide on the pump and pump component life to be selected. Whenever possible, choose larger, lower-speed pumps for resistance to abrasive wear (which varies with speed to an exponent of about 3 or 4), long life, and low maintenance. For sewage pumps, consider 1200 rev/min to be the maximum for any application, and above 0.2 m^3/s (3000 gal/min) capacity, think about limiting speeds to less than 1200 rev/min.
7. In most situations, the required capacity will increase somewhat over the years. There are several ways to accommodate this increase.
   - If practical, select a pump with a less-than-maximum-diameter impeller so that modest increases in flow can be obtained by changing to a larger impeller. In such instances, size the motor and all related equipment for the larger capacity.
   - Design the suction piping conservatively so that larger pumps can be substituted.
   - Leave room (and piping) for an additional pump, as in Example 12-2.
8. Plot the pump operating characteristics on the station H-Q curves. Make sure that the selection is compatible with the required operating conditions and pump life.
9. Compare the $NPSH_R$ with the $NPSH_A$.
10. Make sure that the pump is rugged enough for the intended service. For example, specify that the deflection of the shaft shall not exceed 0.05 mm (0.002 in.) at the packing gland and the impeller shall not be allowed to deflect more than 50% of the wear ring clearance under extreme conditions (e.g., at the minimum allowable discharge for variable-speed pumps). Require the pump suppliers to demonstrate compliance. Recheck the speed selection.
11. Repeat steps 5 through 10 using catalogs from at least two or three other reputable pump manufacturers until satisfied that a number of manufacturers can meet the operating conditions with standard pumps, or if not, by customizing standard pumps to accept larger shafts and bearings.
12. Be sure that all of the expected operating points of the pump were considered in the station and pump head capacity plot. Static heads may vary, pipe friction may differ from the assumptions, and variable-speed pumps operate through large variations of flowrate, so the selected pumps must be capable of meeting a variety of operating ranges. Conservative selections of pumps for future needs can cause problems under current conditions if not properly considered.

Designers must understand that published curves of pump efficiencies may or may not be attainable in reality. Some curves may be drawn for the maximum ideal efficiency of only a polished pump casing and its impeller with all other friction losses (in, for example, bearings, stuffing boxes, inlets, and outlets) deducted. The above warning applies especially to column type (axial flow and vertical turbine) pumps and to submersible sewage pumps where leakage may exceed 5% of the discharge—or more for old pumps. On the other hand, other curves (even in the same catalog) may indicate overall *attainable field* efficiencies. The difference in energy losses may be substantial and must be taken into account when comparing pumps from different manufacturers. In general, however, these differences have little meaning for designers in this initial selection process. But be aware that the size of the driver may be affected.

Once these initial selections have been made, make a closer examination of the application (such as variable-speed or off–on operation), the fluid characteristics (such as grit, solids, corrosivity), starting/stopping requirements, reliability, expected service life, and owner-dictated features.

13. Contact candidate manufacturers and request their *best* offering given the requirements established previously. If the application is unusual or demanding, contact the manufacturers' engineering departments directly. In some instances, it may be appropriate to visit the manufacturers' factories and confer with the pump engineers face to face. Note that equipment manufacturers' representatives may know shockingly little about their product lines other than how to price them.
14. With the manufacturers' best offerings in hand, determine which offerings meet the established criteria. Use a simple method for determining quality, such as size of shafts; bearing arrangement and size; provisions for maintaining shaft seals, sleeves, and wear rings; operating speed; and impeller eye diameter (to name a few). For example, one expert finds that massiveness usually accompanies quality, and he computes a relative shaft stiffness for overhung-shaft, nonclog pumps as L^3/d^4, where L is the pump shaft overhang and d is the shaft diameter at the radial bearing. The lower the computed result for a given pump, the greater the shaft stiffness. This value can be used in the final specifications to rule out pumps with more flexible shafts and, hence, greater deflection during operation. Specifications written for the most massive pumps eliminate light, flexible pumps that would otherwise be selected on the basis of low bid.
15. After each offering has been examined, weed out the weakest and least sufficient. Ask searching questions about user data and actual performance under similar conditions. Probe to obtain answers when the offered information is vague. Be firm; do not accept partial answers nor answers only partially understood. Finally, the selection depends on that hard-to-define-and-quantify process of professional judgment. It is in this step that the art and skill of the professional must finally be brought to bear to sort out quality and cost issues.
16. Obtain estimated fob costs from manufacturers and evaluate the cost and other desirable features (e.g., efficiencies, low maintenance, reliability, cost of repair, and availability of parts) to select the "best" unit for the owner—which may not necessarily be the cheapest pump. "Best" means the lowest life-cycle cost (capital cost, maintenance cost, and energy cost) for the entire station in conjunction with reliability and any other desirable attributes.
17. Write specifications to allow at least two manufacturers to bid on the pumps and to obtain the features necessary to provide reliable service. (For example, some experts require special alloys for parts in contact with water.) Be prepared to defend your specifications. Disappointed manufacturers will surely protest to your client if the pumping equipment package is of any size. If the evaluation process has been done properly, it will be possible to support professional judgments with facts that can be shown to lead to a well-considered conclusion.
18. Ship large pumps disassembled and with continuously recording accelerometers in X, Y, and Z planes.
19. Require the contractor to purchase pumps and driving units from one source (usually the pump manufacturer) to ensure the parts are compatible mechanically and electrically (see Sections 16-1 and 1.02 B in Appendix C).
20. To aid in ensuring an enforceable warranty, require the contractor to have the manufacturer certify that the installation is satisfactory.

With experience, these steps become automatic and the designer scarcely realizes they are being followed. The steps tend to blend together so that several may actually be under consideration at once. Although not indicated as one of the "steps," a designer should be thoroughly familiar with a manufacturer's equipment by reading the first part of the catalog where the features of pumps are described. If you have not used a particular catalog for a long time, read it again for the wealth of valuable information it contains.

The cost curves given in Chapter 29 may be helpful for evaluating the items in step 16. In general, the least expensive type of wastewater pumping station for small flows consists of self-priming pumps because (1) the only underground construction is the wet well and (2) all machinery is above grade where it is readily accessible for maintenance and requires only ventilation or, in cold weather, heating. For larger flows, stations with VTSH® pumps may be the lowest in cost despite the high cost of the pump, because underground construction is also limited to the wet well. Stations with submersible pumps are shown in Figure 29-9 to have the lowest initial cost, but maintenance involves removal and disassembly of the entire unit—a specialized, often expensive task. The prefabricated pumping station is next in cost, but access, ventilation, room for repairs, and safety considerations are often inadequate (see Table 25-3). The most expensive is the wet well–dry well custom station (many are poorly designed with inadequacies for maintenance and repair), although care and thoughtfulness in design can keep life-cycle costs comparatively low. On average, if the capital cost of a station with submersible pumps is taken as a base at 100%, a prefabricated station costs 120% and a wet well–dry well station costs 160% (Figure 29-9).

Life-cycle costs depend on maintenance costs in addition to capital and energy costs. Unfortunately, maintenance costs—even comparative ones—are quite uncertain because so many factors are involved (e.g., attitude of the utility; size, number, and kind of other pumping stations; and quality of equipment installed), none of which is quantifiable. But to ignore maintenance costs, however unpredictable they may be, is to skew selection toward low-initial-cost equipment—often to the disadvantage of the owner. Readers are cautioned to be skeptical regarding extravagant claims by salespersons or accounts in the literature about low-maintenance equipment that is new. Maintenance records of equipment used for less than, say, 15 yr are apt to be misleading.

Pumping stations should be visited daily (or at least every other day even with remote monitoring); depending on the location, the time allocated might vary from 365 to 730 h per year regardless of the type of station or equipment. Note, incidentally, that monitors for such parameters as bearing temperature and vibration are temperamental and expensive. In addition, the following worker-hours per year for maintenance might be allocated to sewage pumping stations of approximately 3 Mgal/d capacity.

Self-priming pumps:	400–1000 h
Submersible pumps:	300–600 h
Additional for pump repair estimated as once/30 months:	100–300 h
Wet well–dry well:	500–1500 h

Supervision, parts, tools, clerical, transportation, and overhead costs must be added to these hours and costs.

The above data are subjective and those for submersible pumps are apt to be misleading. Submersible pumps are operated to failure or to other signs of distress such as detected moisture. Then the pump must be lifted out, cleaned, dismantled, and usually rebuilt—all at a cost of about one-third that of a new pump. The preventive maintenance possible with a well-designed dry well pump installation—and consequently its longer life—may make the long-term cost of a wet well–dry well pumping station compare favorably with that of a submersible pumping station.

The maintenance of water pumping stations can be expected to be, perhaps, half to two-thirds of the above costs because the service is less severe.

12-5. Installation

For an installation to be successful, several aspects must be carefully considered, including (1) suitability of selected equipment for the application; (2) pump inlet conditions; (3) selection, arrangement, and support of connecting piping; (4) method of support for the equipment; (5) maintenance access; (6) method of controlling the pumping system; (7) environmental considerations; and (8) arrangement of service piping. The suitability of pumping equipment is discussed in Section 12-4.

Pump Inlet Conditions

Pump inlet conditions are among the most overlooked and misunderstood aspects of inlet design, yet they probably constitute the single reason most responsible for the success or failure of an installation. (The design of pumping station wet wells and forebays is a separate subject—see Sections 12-6 and 12-7.) Pump performance is entirely dependent on the quality of the effort to provide adequate conditions at the pump inlet. Regardless of the type of inlet (either pressurized, sump, or forebay), care must be exercised to avoid poor hydraulic conditions at the impeller. The pump intake design must satisfy the requirements for proper approach conditions by avoiding the following:

- Loss of efficiency and capacity
- Noise and vibration
- Unstable operation
- Damage to impellers, bowls, bearings, and shafts.

Undesirable features noted in many sump designs include:

- A free fall into open sumps from the inlet conduit into the pool below with the consequent entrainment of air in the liquid and (with wastewater) the release of odors. The air bubbles, easily captured by currents and carried into the pumps, cause loss of capacity and damage to the equipment.
- Piping with excessive velocities that cause excessive headloss and can lead to vibration problems caused by turbulence in fittings and valves.
- Abrupt changes in flow direction upstream from the pump inlet connection. In sumps, abrupt changes may cause vortices. In intake manifolds and pump inlet piping, abrupt changes in direction may cause flow to become asymmetrical and thus load pump shafts and bearings excessively. Abrupt changes in flow direction are acceptable when the pump manufacturer (1) supplies the fitting as a part of the

equipment or (2) does not take exception to the presence of the fitting. Operation is probably not adversely affected in such circumstances.

- Sump or inlet piping geometry that permits differential velocities and, thus, rotation of the water. With the slightest rotation, the spin increases as the water approaches the pump suction inlet. Swirling in the suction pipe may reduce the local $NPSH_A$ in the core to zero and thereby cause cavitation, noise, and rapid wear even though the average $NPSH_A$ is adequate. Preswirling at an impeller changes the angle of attack of the flow to the impeller blades and shifts the pump curve—often drastically.
- Horizontal velocities in sumps near the pump inlets that are too high. In general such velocities should be much less than 0.3 m/s (1 ft/s). Actual velocities usually differ greatly from calculated average velocities.
- Interference between adjacent intakes. Space intakes no closer than 2.5 *D* c-c (where *D* is intake bell diameter).
- Discontinuities such as corners without fillets and uneven distribution of currents caused by flow past pier noses that often result in the formation of air-entraining vortices. Although there is usually no surface indication, subsurface vortices may also occur, and they can be very damaging.
- Stagnant areas in wastewater pumping station wet wells where velocities are too low to prevent the deposition of putrescible solids from wastewater. Velocities of about 0.3 m/s (1 ft/s) are enough to keep organic solids moving whereas velocities in excess of 0.5 m/s (2 ft/s) are required to keep grit moving. After organic material and grit are deposited, velocities in excess of 1.6 m/s (5 ft/s) are required to move the deposits with appreciable effect.

Pump Suction and Intake Piping

Design considerations include:

- Velocities in pump intake manifolds should not exceed about 0.9 m/s (3 ft/s), and the manifolds should have a constant diameter as reducers increase cost and induce turbulence. Avoid a sudden acceleration into branch connections to pumps—if necessary, by installing an eccentric reducer in the branch at the manifold.
- The velocity of flow into the trench-type sumps should not exceed 1.2 m/s (4 ft/s). Use judgment in establishing velocity limits for other types of sumps.
- Limit velocities in constant-speed pump suction pipes to about 2.4 m/s (8 ft/s) and in discharge piping to about 3.7 m/s (12 ft/s) to keep headlosses within reason. When energy costs are considered, it may be more economical to reduce the above velocities by as much as 30%. For variable-speed pumps, however, these velocities can be increased by about 25%, because the discharge from a V/S pump averages about 75% of its maximum capacity.
- Carefully calculate the $NPSH_A$ under the worst possible combination of barometric pressure, fluid temperature, vapor pressure, and intake conditions. Allow a factor of safety of at least 1.5 m (5 ft) over the manufacturer's published $NPSH_R$ requirement for the pump. See Figure 10-13 and the accompanying discussion.
- Eliminate piping arrangements and any features (such as concentric reducers) that might trap air—especially in the inlet piping, but if that is not possible, provide a means to relieve trapped air to appropriate points of disposal.
- Use the room walls to support heavy valves and piping.
- Avoid placing check valves on vertical pipes (especially in wastewater pumping to prevent solids from jamming the valve). Provide convenient access to valve bonnets for maintenance.
- In wastewater stations, always connect pump discharge piping to the side of a manifold and never to the underside to prevent solids, moving along the manifold invert, from falling into the vertical pipe and clogging it.
- Arrange piping and supports to allow convenient access for maintenance on three sides of the equipment. Provide adequate clearance from all projections of at least 1.1 m (42 in.) on each of those three sides.
- Provide adequate room for the use of wrenches and to remove bolts. Pay particular attention to the removal of sleeves and bolts in sleeve couplings.
- There must be quick, unobstructed exit for people working around the pumps.
- Make certain all pipe and fittings are supported independently. Supports should be designed to carry the full weight of fittings and valves (and included water) to the structure. Plan piping and fittings to isolate strain from lateral misalignment at the pump connections. Strain isolation is best achieved by installing two flexible couplings (instead of one) at each pump connection.

Installation recommendations for the piping and piping supports for several kinds of pump installations are shown in Figures 12-17 to 12-25. Avoid piping layouts in which the entire width of a dry well is traversed by pipes just above the floor (as in Figure 12-17a) or at

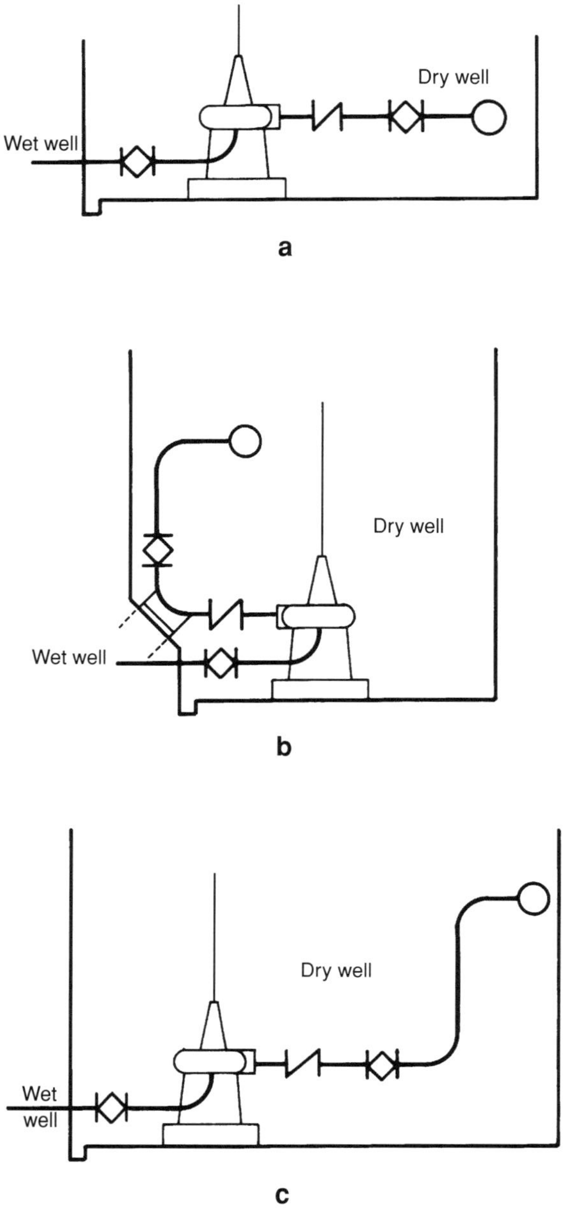

Figure 12-17. Schematic elevation views of pump room piping layouts. (a) Unacceptable piping layout; (b) Good piping layout; (c) Alternate piping layout (see text).

less than adequate headroom so that workers must either climb over them or stoop under them. Observe, in Figure 12-17b, how arranging the discharge piping on the same side of the pump as the inlet piping allows for maximum accessibility, minimum floor area requirements, and greatest convenience. Access to the entire pump room is provided at floor level. The large manifold pipes are installed overhead as indicated. Where discharge piping is larger than 250 mm (10 in.), the above arrangement may interfere with easy access to the fittings in the suction piping. One means to avoid the interference is to turn the pump so the discharge piping is at an angle (say, 45°) to the suction piping (and to the wall), but the spacing between the pumps and the overall length of the pump room may have to be increased.

Another alternative is shown in the piping diagram of Figure 12-17c. The elbow of the vertical pipe is best supported by a concrete pedestal. As no intermediate support is possible between the pedestal and the manifold (or the upper elbow), the unsupported length is substantial, and a careful analysis is required to ensure that resonating vibrations cannot occur. Of course, the width of the dry well must be substantially increased, but, unlike orienting the discharge pipe at an angle, the length of the wet well is not affected. The plan with the smallest footprint must, however, be determined by plotting both layouts. One of the largest utilities in the United States prefers this arrangement for discharge pipes larger than 250 mm (10 in.) in spite of the larger footprint and greater cost, because access for repairs or replacement is less obstructed, and the utility considers that safety is improved.

Avoid pipe fittings in grated trenches or pockets, because access for disassembly or repairs varies from inconvenient to difficult. Trenches collect dust and trash and become maintenance headaches. Advantages are, however, that access to the rest of the piping may be enhanced, and it gets some of the piping out of the way.

Use stainless steel for small pipes and bury them in the floor slab to prevent workers from tripping and falling. Run large pipes overhead as indicated in Figure 12-17b and c. Overhead clearance should be no less than 2.3 m (7.5 ft). Note how the piping diagrams to follow conform to these recommendations.

Piping diagrams for connecting two pumps in series are shown in Figures 12-24 and 12-25. Again, note how unobstructed walkways provide unhindered access to the entire pump room floor.

Both motors in Figure 12-24 are controlled by one AF converter so that both motors operate at the same speed. The drive shafts are offset at about a 3° angle to aid lubrication of the universal joints. However, universal joints operating at an offset angle produce a torsional excitation of twice (sometimes four times) the rotational speed. The excitation force increases with speed and with an increasing offset angle. Yokes out of phase plus incorrect and uncanceled universal joint angles have caused many vibration problems that can be solved by eliminating the angles entirely. Debates

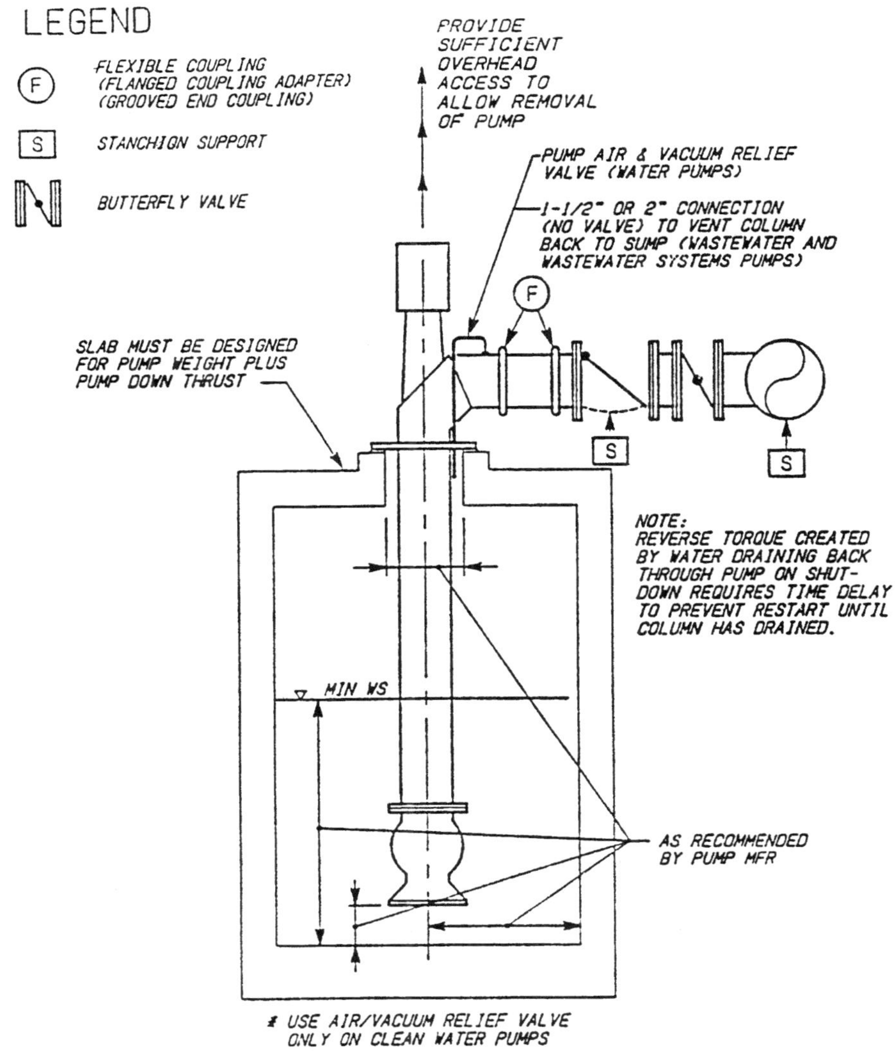

Figure 12-18. Installation recommendations for column-type (mixed-flow, propeller, and vertical nonclog) pumps.

about the need for offset angles led the Spicer Company to a test of two shafts—one at zero offset and one at a 3° offset angle. Both failed at about the same time. An offset angle of true zero cannot be achieved in practice, and the unavoidable tiny angle is sufficient to distribute grease to the needle bearings and to prevent "brinelling." Strong opinions pro and con still persist, but if the manufacturer prescribes offsetting the shaft, use the smallest angle acceptable.

Drive shafts operating between offset universal joints accelerate and decelerate twice per revolution. If the angles at each end are different, the shaft tries to accelerate at two different rates. If the yokes are not in the same plane, the shaft tries to start and stop accelerating at different times, thereby causing torsional and bending stresses in the shaft and vibration in the drive train. The worst situation is offsetting two drive shafts between a coaxial pump and motor. If upper and lower offset angles are 3°, the central angle is 6°, and the shaft bends in an "S" shape. The center universal joints will have a short life.

The pump and driver train in Figure 12-25 is unusually long because of the flywheel located between the motor and the pump on the right. The use of a single motor for both pumps ensures that both operate at the same speed.

Equipment Supports

Rotating equipment is best supported on massive concrete structures. Regardless, concrete and steel

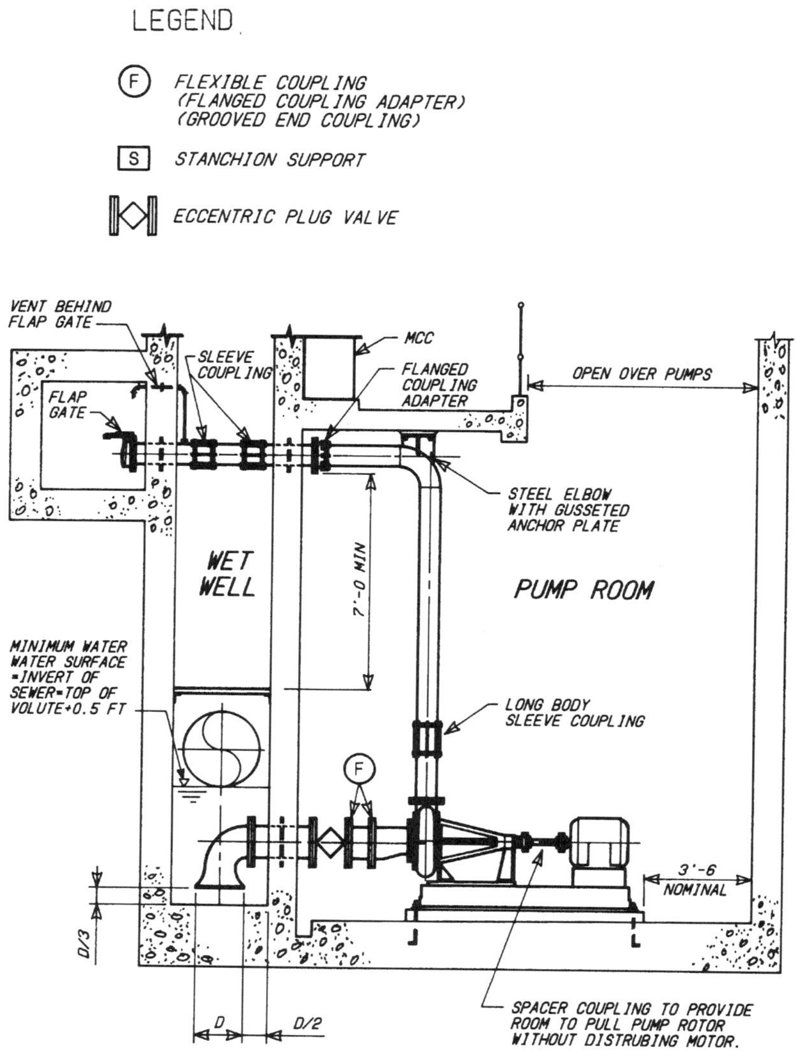

Figure 12-19. Installation recommendations for horizontal nonclog and mixed-flow pumps discharging to a canal or sewer.

supports should be designed for natural frequencies well above the highest operating speed. Always conduct a dynamic analysis of the structure as part of the design process. Don't forget that some equipment vendors have offerings that differ substantially in terms of operating speed, weight, number of impeller and diffuser vanes or cylinders and other important features. Make certain all variations are considered in the design.

All equipment should be placed on housekeeping pads. Housekeeping pads are concrete bases raised above the floor by a minimum of 100 mm (4 in.). The purpose of these pads is to prevent liquids on the floor from accumulating against ferrous metal and causing corrosion. All equipment should be grouted in place after shimming between the equipment and the pad to make certain it is supported and leveled properly. Large pieces of equipment should be installed on grouted-in sole plates furnished by the manufacturer. Only journeyman millwrights should be allowed to perform final alignment. Alignment methods should be those recommended and accepted by the manufacturer.

Maintenance Access

Access for maintenance is an important aspect of installation design. Ensure at least 1.1 m (3.5 ft) clear between adjacent pieces of equipment. Provide more clearance with large units. Vertical clearance over

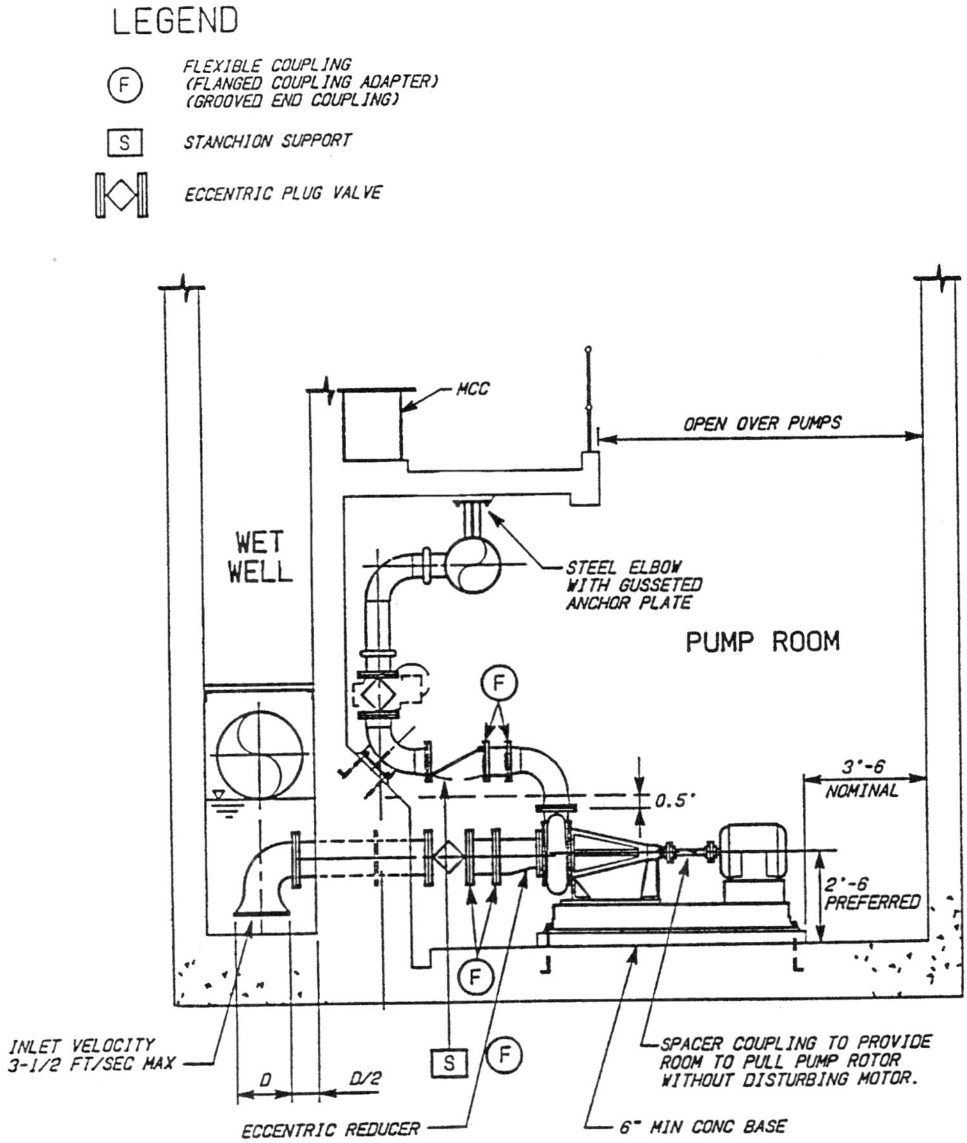

Figure 12-20. Installation recommendations for horizontal nonclog and small mixed-flow pumps discharging to a force main.

walkways and adjacent to major items of equipment should be not less than 2.3 m (7.5 ft). Make certain there is sufficient clear space around the equipment to remove subassemblies. Hoisting equipment should be provided in all stations where the largest individual equipment unit subassembly weighs more than 90 kg (200 lb). Electrified hoist, trolley, and bridge functions are recommended for all hoist capacities greater than about 900 kg (2000 lb).

Make certain the installation is designed to provide vertical hoist access to all large equipment components. Use lifting hooks over equipment only as a last resort. Cranes are much better than hoists and should be arranged to deliver the largest single equipment component directly to the bed of a service truck. Avoid arrangements where service personnel are required to place components on a dolly and manually move them to another position for later placement on a truck bed.

System Controls

Remember that the controls for a pumping installation are not for merely controlling a set of pumps, but

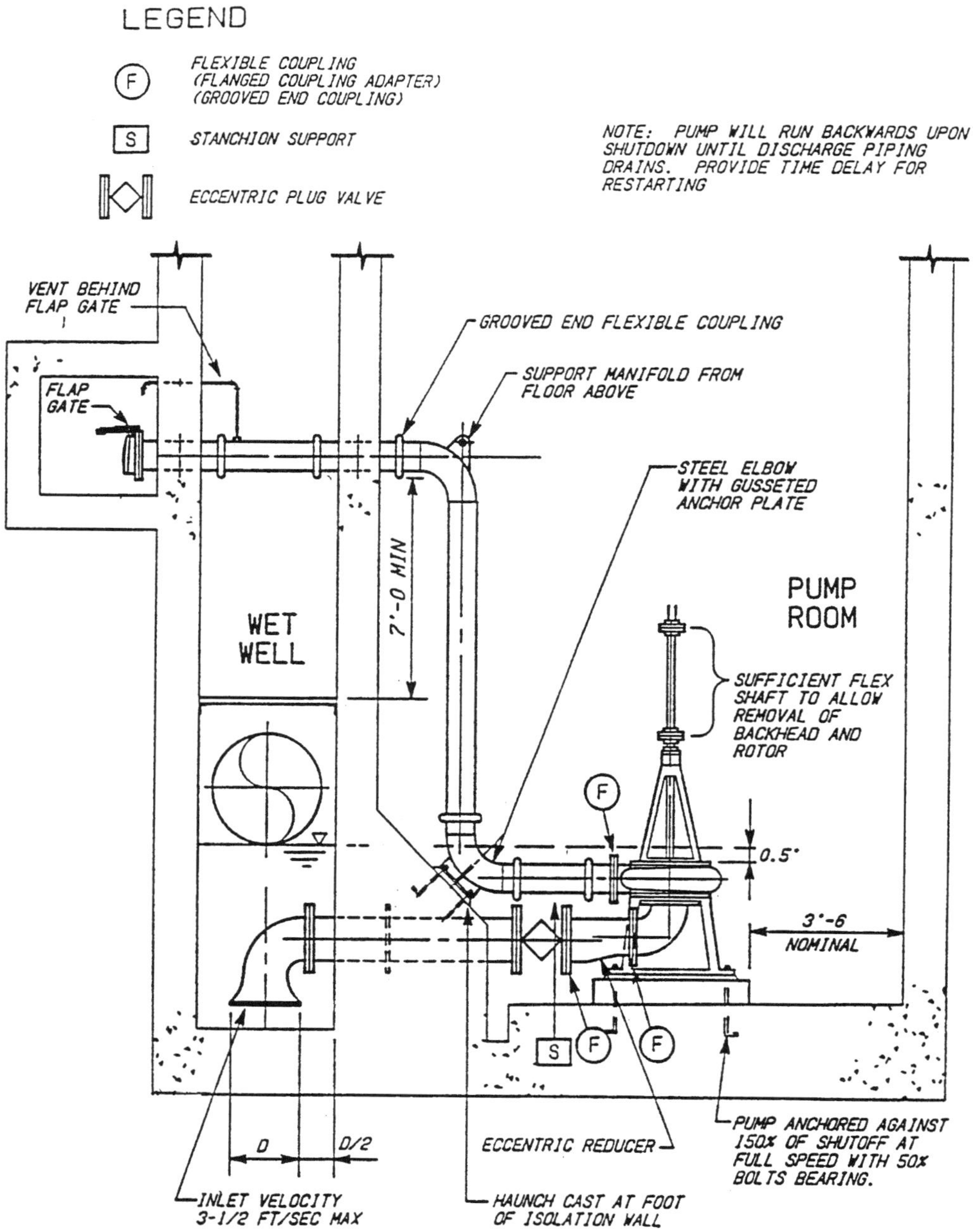

Figure 12-21. Installation recommendations for vertical nonclog and small mixed-flow pumps discharging to a canal or sewer.

rather a *whole system*. For that reason, it is important to consider the effects of operation of the equipment on other parts of the total installation, such as:

- Operation of too much pumping capacity (such as a standby pump along with all others) may overload a downstream system such as a receiving sewer or storm channel.
- Constant-speed operation may be inappropriate for the function of downstream treatment operations.
- If a pumping unit includes high-inertia components, quick restarting after loss of power while the equipment is still rotating may cause considerable damage to shafts, couplings, motor windings, and other components.

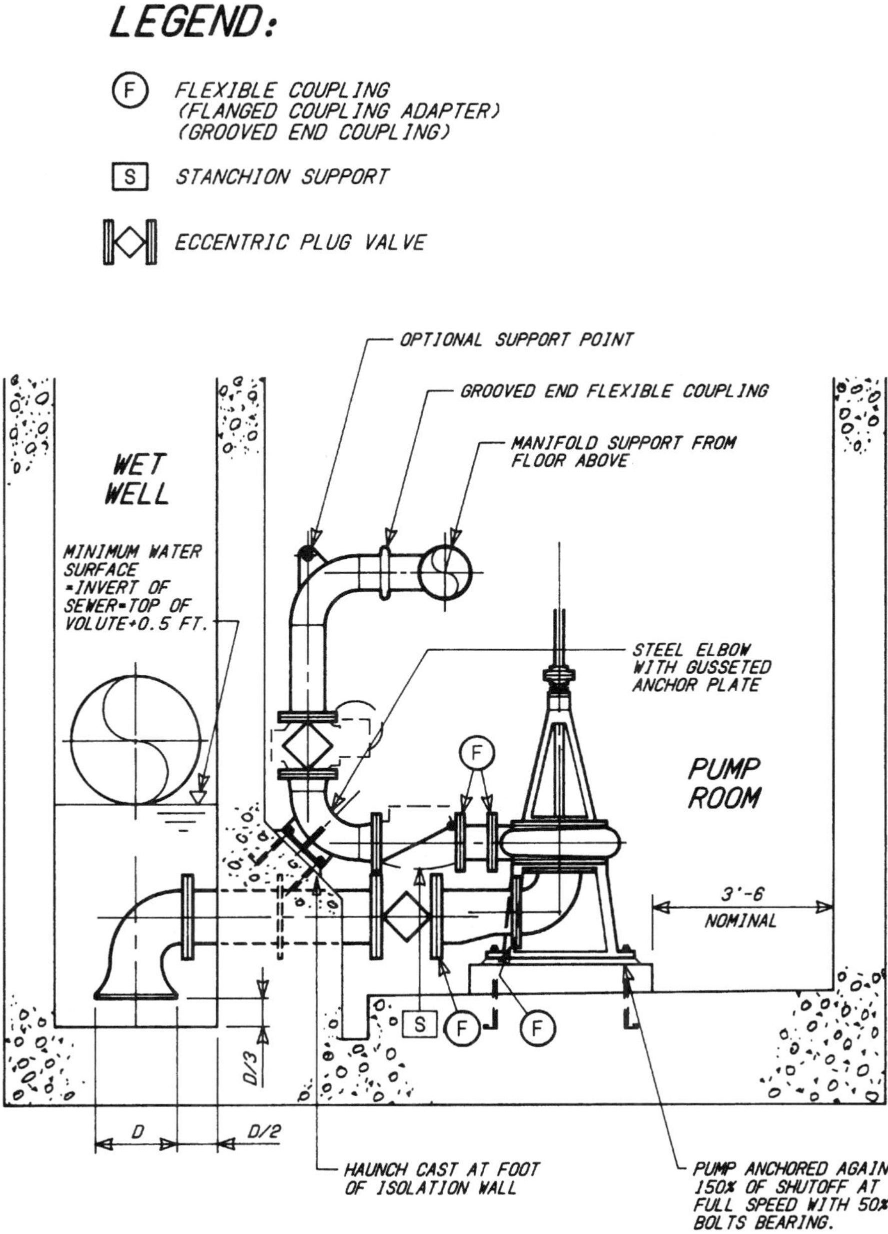

Figure 12-22. Installation recommendations for vertical nonclog and small mixed-flow pumps discharging to a force main.

- Some pumps, such as column pumps, will spin backward when stopped. Even if a nonreverse device has been installed, countertorque will act on the shaft until the column has been emptied. Starting the pump with a full column may cause considerable damage, so arrange the controls to lock out the starter until the column is empty. Confer with the manufacturer for finding means to circumvent these problems.

Environmental Considerations

For the most part, utility installations should not be seen, heard, or detected in any way that would degrade the local environment. The design of utility installations should include the following features:

- Architectural design that blends into the neighborhood and neither establishes a presence that

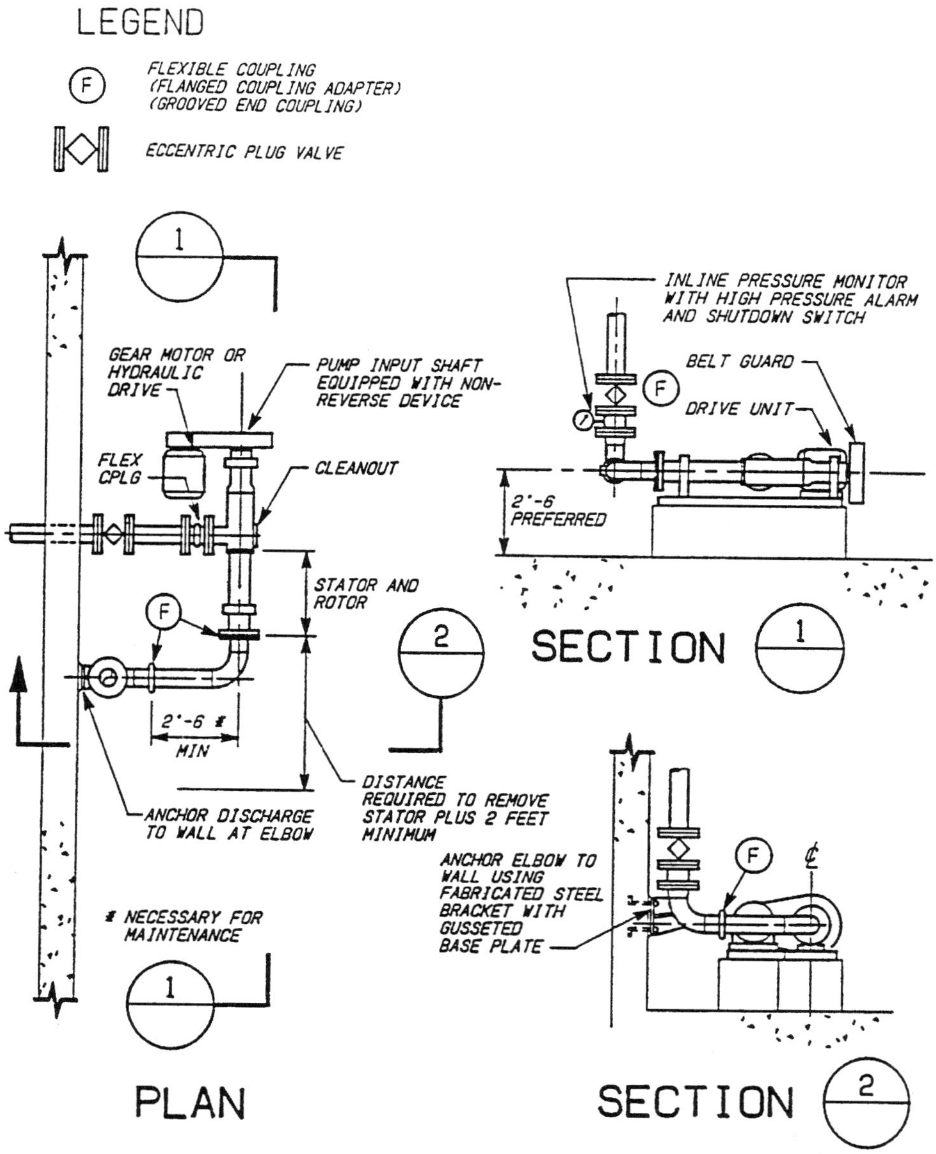

Figure 12-23. Installation recommendations for progressive cavity pumps.

declares the purpose of the installation nor degrades the value of its neighbors. Pumping station superstructures can, with minimal effort and expense, easily be crafted to appear to be homes, farmhouses, or restaurants.

- Noise and light emissions suppressed to avoid broadcasting the function of the installation. Special construction features can be included to eliminate virtually all of the high-frequency noise from motors. Hospital-grade silencers are available for engines at little additional cost. With diligence, the noise from ventilation equipment can also be suppressed. Lighting fixtures in a variety of architecturally acceptable designs are available to illuminate exterior areas without suggesting a commercial or industrial facility.
- Odor control and treatment for wastewater stations. Odor control is, perhaps, the most challenging environmental management task. The design should incorporate every technique for avoiding the conditions that generate odors. Ventilation system exhausts should be treated for odor removal. A variety of cost-effective and simple odor treatment techniques are available (see Chapter 23).

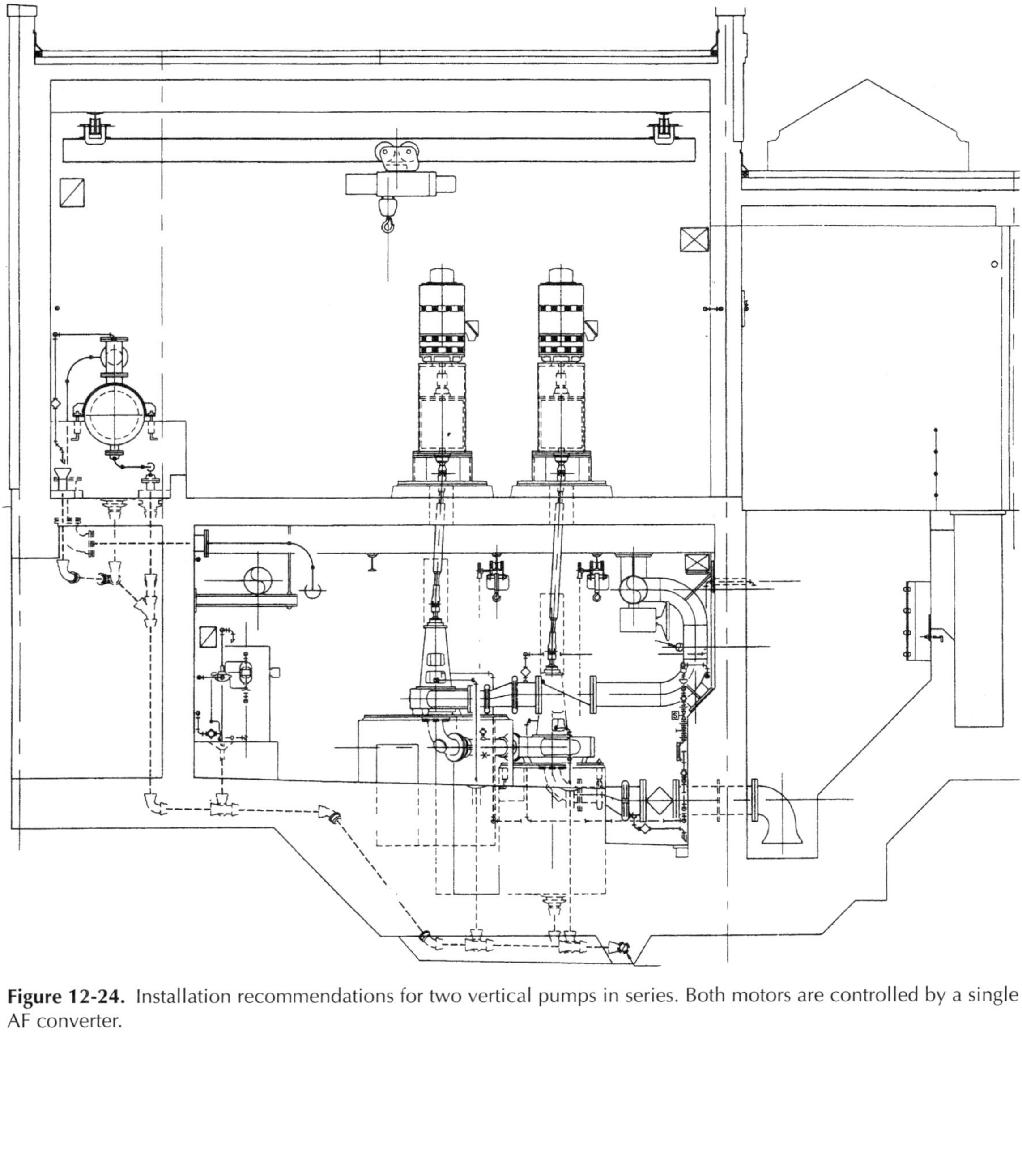

Figure 12-24. Installation recommendations for two vertical pumps in series. Both motors are controlled by a single AF converter.

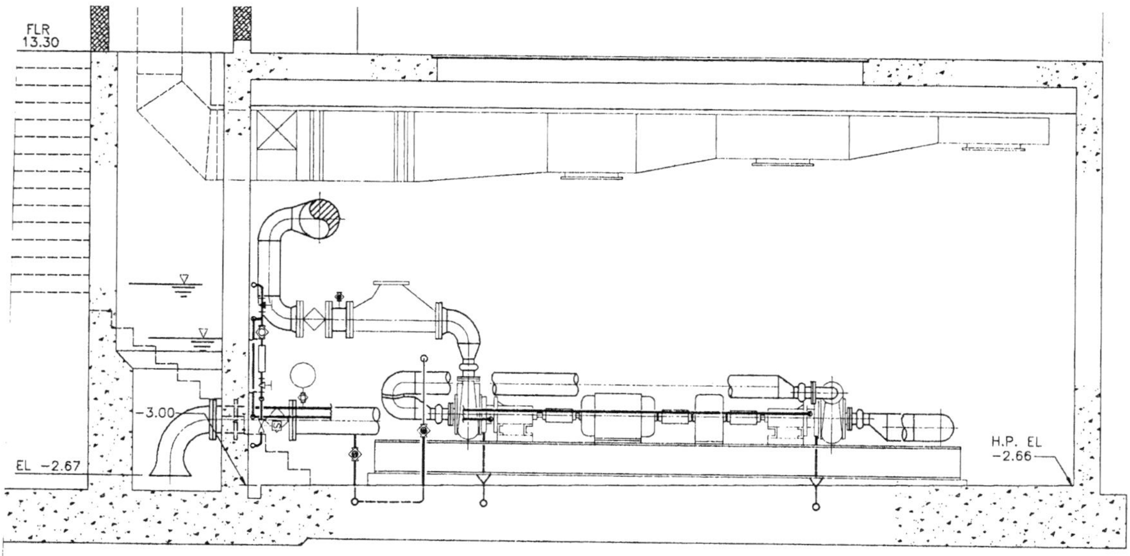

Figure 12-25. Installation recommendations for two horizontal pumps in series.

Service Piping and Appurtenances

Care should be taken to avoid cluttering equipment maintenance points with service piping, conduits, and other peripheral devices and fixtures. If gauges are necessary, they should be mounted on walls or gauge boards located away from the pump itself. Purge panels for packing boxes or mechanical seals should be located on walls or columns and the service piping (stainless steel) routed in slabs to the pump housekeeping pad. Every effort should be made to keep the equipment installation clean and free from items that could interfere with maintenance operations.

12-6. Pump Intake Basins: An Appraisal

The head-capacity curves for pumps developed by the pump manufacturer are based on tests of a single pump operating in a semi-infinite pool with no nearby walls or floors and no stray currents. Hence, flow into the pump suction is symmetrical with no vortices or swirling (prerotation). Pumping station designers rely on these curves to define the operating conditions for the pumps selected. But various constraints such as size, cost, limitations on storage time, and (for solids-bearing waters) deposition require walls, floors, and pump intakes to be in close proximity to each other. As the manufacturer's test conditions cannot be duplicated in the field, various problems arise and many designs, in spite of their common acceptance and continual usage, do not perform well.

The BHRA [1] recommendations and (in the United States) the Hydraulic Institute Standards [2, 3, 4] are considered to be the authoritative works for the design of pump intakes. However, up to and including the 1994 edition of the Hydraulic Institute Standards (as well as the BHRA publication), the focus has been largely confined to clear water, with discussion on solids-bearing water essentially limited to the advice that flat floor areas should be minimized, that designs proven to be successful should be followed, or that pump manufacturers should be consulted. In 1994, the Hydraulic Institute appointed a committee of users, engineering consultants, researchers, and pumping equipment producers to expand and improve the standards for clear water and, for the first time, to consider designs for solids-bearing waters in detail. The material in this chapter and the proposed standards are mutually compatible and complementary. Furthermore, the presentation in this chapter reflects: (1) experience gained through more than 35 years of designing pumping station wet wells; (2) information developed at the U.S. Corps of Engineers Waterways Experiment Station in Vicksburg; (3) investigations at Montana State University and ENSR Hydraulic Laboratory in Redmond, Washington, under financial assistance from the U.S. Environmental Protection Agency, The Gorman-Rupp Co., Fairbanks Morse Pump Corp., ITT Flygt AB; Montana State University, and R. L. Sanks; and (4) full-scale tests at the Fairbanks Morse Pump Corp. plant in Kansas City—all of which have led to several publications [5, 6, 7, 8].

Pump Sump Problems

The geometry of the pump sump and its associated piping and entrance conditions has a profound effect on pump performance. Hydraulic conditions detrimental to pump performance include:

- Asymmetrical velocities in the approach to the pump intake
- Nonuniform or asymmetrical velocity distribution in the pump intake
- Swirling or prerotation of flow at the impeller
- Surface and subsurface vortices.

Asymmetrical flow to a pump intake is likely to produce swirling and/or asymmetrical velocity distribution in the intake throat. Asymmetrical velocities in the intake throat create a higher load on one side of the impeller, bend the shaft, and put extra stress on bearings and couplings. They can cause rough operation, vibration, and loss of head and capacity. These effects worsen as pumps get larger and as the specific speed increases.

Swirling changes the angle of attack on the impeller vanes, reduces head and capacity, and decreases efficiency. An angle of 5° from axial is usually considered the maximum allowable. Swirling in the approach can degenerate into vortices. The pressure in the core of a vortex is reduced and air (or other gasses) coming out of solution can cause noisy operation and vibration. When a vortex is severe, it results in cavitation that quickly erodes metals.

Strong surface vortices entrain air bubbles that collapse as they pass from low to high pressure zones across the impeller vane, thus creating noise, uneven torque, undue stress on shafts and couplings, and erosion. A small amount of air can cause a large decrease of capacity, head, and efficiency. Milder surface vortices tend to become smaller but more intense as they enter the intake.

Strong subsurface vortices, formed as liquid separates from walls or floors, are often present and difficult or impossible to detect in the field. Normally, the bubbles in subsurface vortices consist of air that comes out of solution due to the decreased pressure in the core. Subsurface vortices are just as damaging as surface vortices—sometimes more so. If vapor bubbles can form, they cause rapid erosion. In fact, the rate of erosion can be reduced (although performance is also reduced) by introducing air deliberately to form large bubbles that do not completely collapse (see Section 10-4).

General Problems

Designers should try to visualize the flow patterns likely to develop in a sump. The mental picture may be somewhat inaccurate, but even so, it can serve to identify potential problems. For example, let us analyze several types of sumps by trying to visualize how water must behave in each. Bear in mind several phenomena that govern the flow of water including:

Phenomenon 1. Water emerging as a jet from a pipe into a relatively quiescent pool tends to continue to flow in a straight jet that spreads only about one unit per eight units of travel. The center of the jet tends to retain its original velocity for a considerable distance, and eddies (rotation) are generated along the edges of the jet.

Phenomenon 2. When a jet of water is forced to turn, swirls are created and some energy is lost. Swirls tend to persist and may develop into vortices at pump intakes.

Phenomenon 3. Vortices easily form in quiescent water both at the water surface and under water at walls or floors. Low velocities past pump intakes tend to prevent swirls from organizing into vortices.

Specific Examples

According to Hydraulic Institute Standards [2] or the ANSI/HI Standards [3, 4], the sump shown in Figure 12-26 is not recommended. The jet from the inlet pipe would continue with little loss of velocity until it caroms off the back wall. If Pump 1 is operating, most of the flow would turn toward that pump and create swirls on the inside of the turn. A general counterclockwise flow pattern would probably develop in the portion containing Pumps 1 and 2, and, aided by the

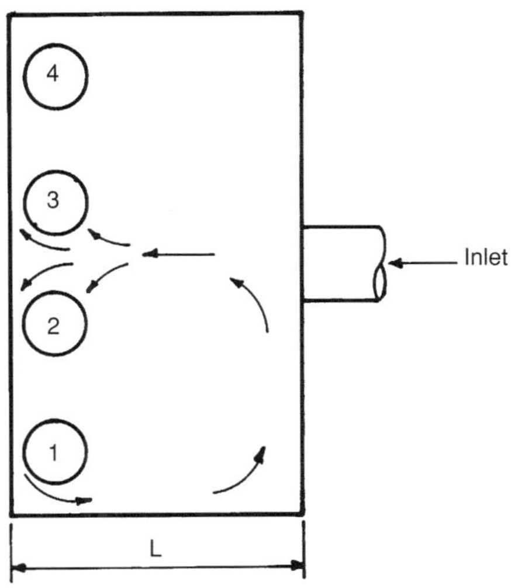

Figure 12-26. A sump labeled "Not recommended" in Hydraulics Institute Standards [2].

incoming jet, might become quite pronounced. On the other hand, if only Pump 4 were operating, the water would turn toward that pump and produce a clockwise current in the part of the sump containing Pumps 3 and 4. In either regime, however, flow to all pumps would be asymmetrical with respect to a line through the center of the pump and normal to the wall. Swirling would probably be excessive because of Phenomenon 2.

The sump shown in Figure 12-27 (but without the baffle) was used in a fire suppression system for two huge airplane hangars. The sump design was so poor that the pumps vibrated excessively and could not meet their rated capacity. The fire marshall refused to approve the system. *Against the express recommendation of the pump manufacturer,* a consultant (guided by 30 years of experience with trench-type wet wells) recommended a submerged baffle to place the pump intakes in a trench. As shown in the figure, the trench is fed by water flowing gently over the top of the baffle. Consequently, the circulation currents are destroyed, water now enters the pump intakes symmetrically, the pumps operate at rated capacity, and the installation was approved. *Nevertheless, the sump is deficient in several other respects (excessive floor clearance under pump intakes and poor inlet design, to name two) and is not recommended.*

Consider the sump shown in Figure 12-28. Some swirling along the edges of the jet may be expected and rotation due to the curvature of the jet as it strikes the end wall is not entirely suppressed. Flow, especially flow to the end pumps, is asymmetrical (although much less so than in Figure 12-26), and the asymmetry may induce swirling. In sumps of this type, baffle walls have sometimes been placed between the pumps to improve symmetry, but the noses of the baffle wall themselves may be a source of swirling. Furthermore, the submergence shown is inadequate for some installations. According to Sanks, Jones, and Sweeney [7], the submergence of the suction intake should be at least 2 D to prevent the formation of significant surface vortices at intake velocities up to 1.5 m/s (5 ft/s). Dicmas [9] has shown that the submergence should be 2.4 D in sumps 2 D wide and 1.9 D in sumps 3 D wide. According to Hecker [10], the intake of air from surface vortices can only be entirely eliminated at submergences of at least

$$S = (1 + 2.3\boldsymbol{F})D \qquad (12\text{-}1)$$

where S is submergence, D is the outside diameter of the suction bell, and $\boldsymbol{F}$ (the Froude number) is defined as $v(gD)^{-0.5}$ where v is the average velocity at the entrance to the bell. On the basis of Equation 12-1, the required submergence is usually 12 to 20% more than the criterion of 2 D. A significant disadvantage of the design in Figure 12-28 is its tremendous size. If the intake structure must be deep, such a sump would be costly indeed. It might well be improved by making the floor flat to within, say, D of the pump intake centers—a design (shown by the dashed lines) that puts the intakes in a trench. Then the flow to the intakes would probably become more symmetrical, and pump performance would likely be improved. The length could then be shortened for economy. Such a modified sump for pumping 37,600 m^3/h (238 Mgal/d) of settled wastewater at a treatment plant in Colorado performs very well indeed.

Another way to improve the performance of the sump in Figure 12-28 is to add long baffles between the pumps as described in the Hydraulic Institute Standards proposed for publication in 1998. The baffles extend from the end wall to a point at least 5 D from the pump. Thus each pump is located in an individual channel—a construction that requires a long sump. Such a sump is difficult to keep clean if the water contains solids.

Extensive research at the Hydraulics Laboratory of the U.S. Army Corps of Engineers Waterways Experiment Station at Vicksburg, Mississippi, resulted in a short but thorough and informative technical letter [11] for guidance in the design of wet wells of the configuration shown in Figure 12-28. The reason for troubles with pumps in such sumps is shown in one of the figures in the technical letter that illustrates a jet discharging into ponded water. Means for circumvention are shown in other figures in the technical letter.

Figure 12-29 represents the type of sump in which water is required to change direction by 90°, thereby running afoul of Phenomenon 2—a problem that can be alleviated by making the distance L sufficiently great. The baffles are intended to straighten the flow and make it symmetrical. Again, this design requires a large footprint and becomes costly.

Aping the geometry of traditional designs is no guarantee of success. Triplett, Fletcher, and Grace [12] emphasized the problems concerning some established pump sump design criteria with their statement in 1988 that *"approximately 50% of all sumps for flood-control pumping stations designed and constructed by the Corps [of Army Engineers] needed some sort of post-construction modification to improve the flow conditions to the pumps. These modifications were costly and in most cases did not correct the problem but improved the conditions only slightly."* Criticisms such as this should jar engineers out of any complacency toward sump design (see also Karassik *et al.* [13]). The problems did lead the U.S. Army Corps of Engineers to undertake extensive,

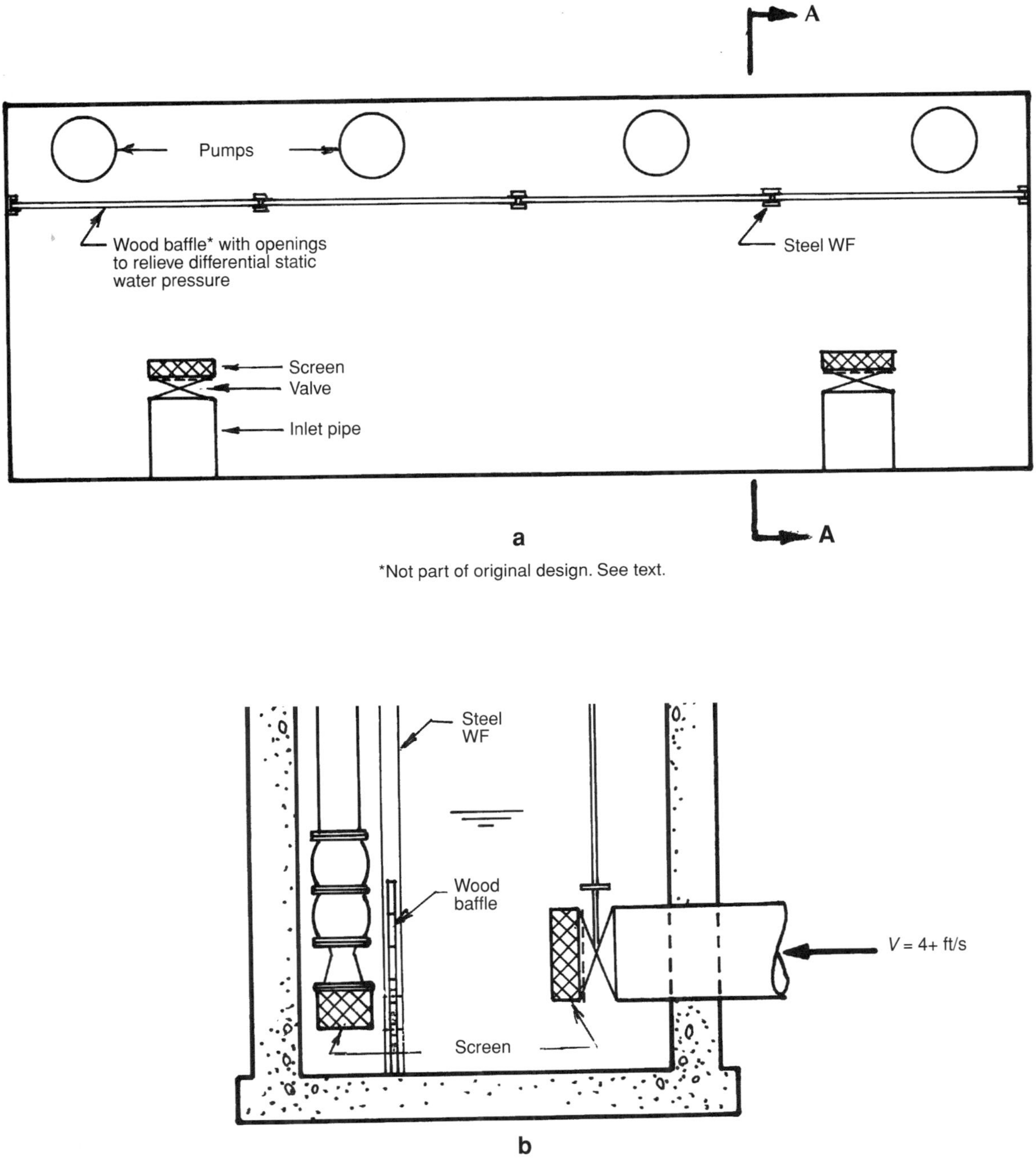

Figure 12-27. Sump for fire-suppression pumps at Portland, Oregon, airport. (a) Plan; (b) Section A–A. Not recommended (see text for partial remedy applied to original design).

large-scale model studies at the Waterways Experiment Station at Vicksburg, Mississippi, where an excellent formed suction intake (FSI) was developed for large pumps by Fletcher [14] and his co-workers. When the discharge for a pump is limited to $1.99\sqrt{gD^5}$, the hydraulic environment provided is so benign that the Corps has used this design extensively since 1990 without requiring model studies. Note that

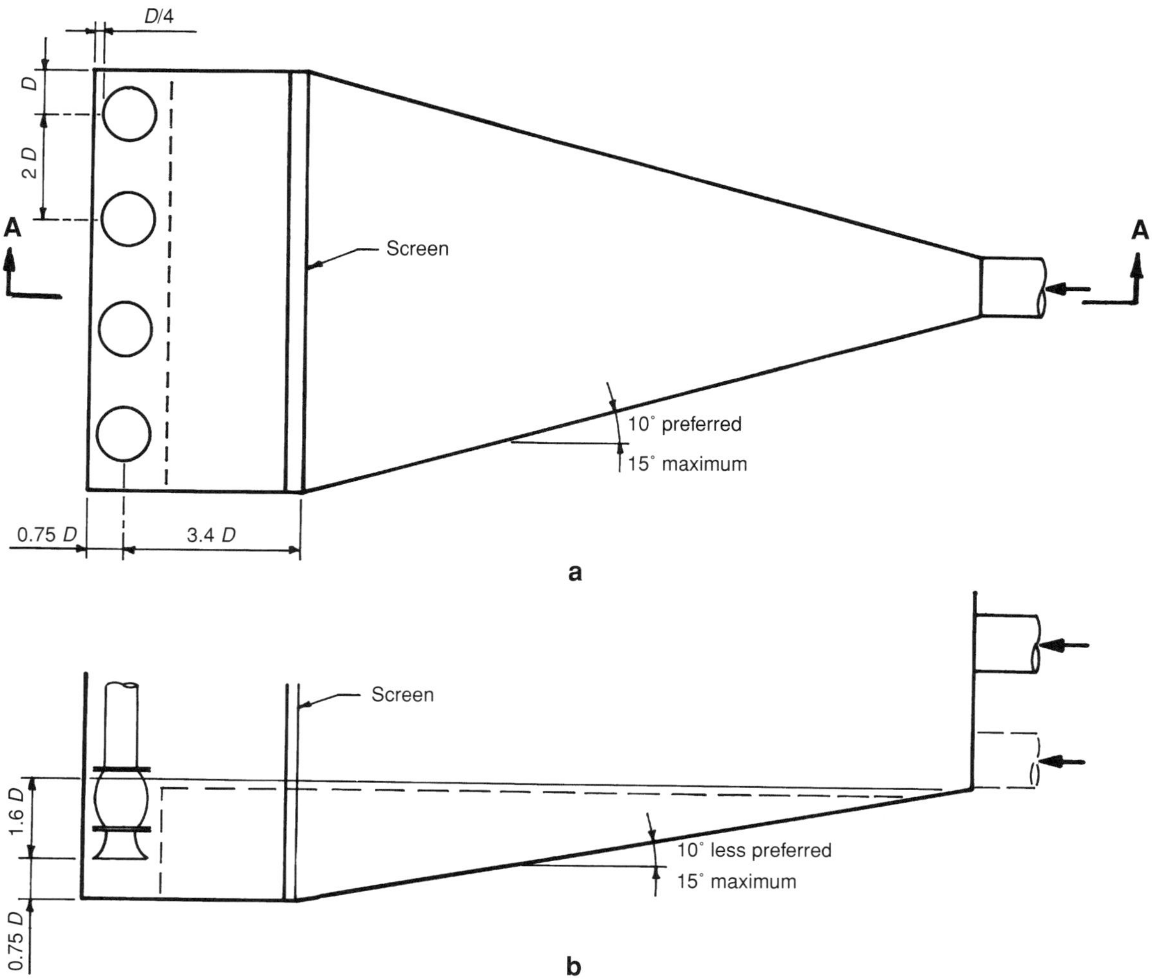

Figure 12-28. A sump recommended in the Hydraulics Institute Standards [2]. (a) Plan; (b) Section A–A. Improvement is shown by dashed lines.

D is the throat diameter of the intake (in either meters or feet), so for a throat of 1.0 m (3.28 ft), for example, the discharge is limited to 6.2 m^3/s (220 ft^3/s).

There are two designs of the FSI. The one with the lowest profile (and hence the unit of least cost) is shown in Figure 12-30 (page 356). Flow may approach the intake at any angle up to 90°. In studies of a model with a throat diameter of 200 mm (8 in.), forebay velocities of 1.2 m/s (4 ft/s), or even twice as much, caused no problems, according to Fletcher [15], and velocities in the throat met all requirements for uniformity and lack of swirling or vortices. Corresponding prototype velocities can be found using the model similitude relationships presented in Section 3-12. Caution with respect to high forebay velocities is, however, advised. Not every design with FSI intakes works equally well, however, and the configuration of the forebay does affect swirling and vortices, as demonstrated in model tests by Fletcher [16]. The dividing wall in the entrance is not a problem, because trash racks are assumed to be placed in the forebay to eliminate debris. Without trash racks, expect stringy material to be deposited on the nose of the dividing wall.

A schematic drawing of a wastewater pumping station wet well containing pumps originally designed to discharge a total flow of 6300 m^3/h (40 Mgal/d) is shown in Figure 12-31 (page 357). The pumps were very noisy and impeller noses were quickly eroded, so impellers had to be replaced frequently. When nearly double the capacity was required, a model study was made. The model proved that even a very short cascade (less than 0.3 m or 1 ft in the prototype) into the

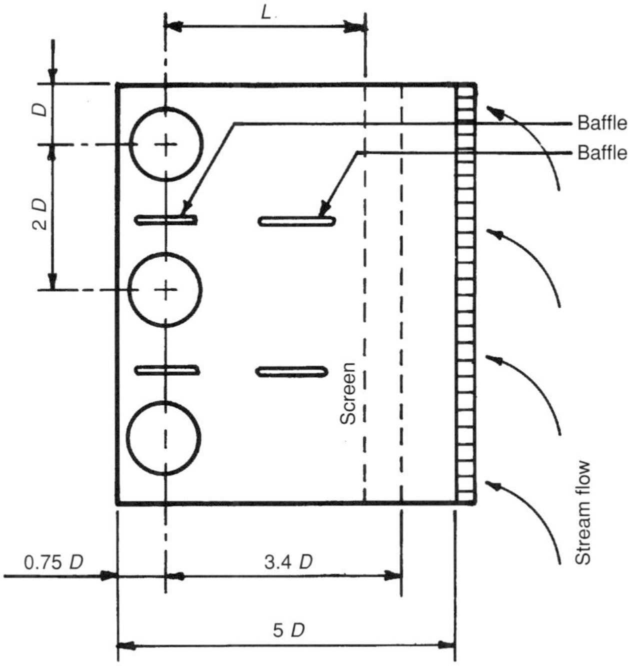

Figure 12-29. A sump with right-angle change of flow. Not recommended.

wet well caused great masses of air bubbles to be driven to the floor where the pump intake currents captured them, thereby creating noise and erosion. The poor characteristics were overcome by installing the baffle channel, shown by the dashed lines in the figure. Water from the cascade falls into the channel and must enter the sump horizontally. Hence, the bubbles are not driven to the floor of the sump and thus can float to the surface. No bubbles entered any pump intake after the baffle was installed even at the high discharge of 4700 m^3/h (30 Mgal/d) per pump, and the pumps ran smoothly even at a station discharge of 7500 m^3/h (118 Mgal/d). The wet well still traps sludge and scum, however, and remains unsatisfactory for that reason.

Air bubbles entrained by a free fall can persist for a surprisingly long time and are readily drawn into pump intakes with devastating effects on capacity, head, and efficiency. One % of air can, for example, reduce the efficiency and throughput of some pumps by as much as 15%. Pumps with a low specific speed (n_s) are more sensitive to air than those with a high specific speed. With a sufficient residence time in the sump and enough distance between the inlet to the sump and the pump intake, the air bubbles can be eliminated. Information given by Falvey [17] is useful for estimating this distance.

Solids-Bearing Waters

Many waters (raw water, stormwater, and wastewater) contain solids that settle rapidly in traditional wet well designs as the velocity at the exit of the inlet pipe falls to very low values in the sump. As a result, all of the above designs are generally unsuitable for such waters. The deposition of solids can change the hydraulic characteristics of the basin appreciably, and, if any organic material is present, its decomposition results in toxic, malodorous, and corrosive gases. Usually such solids can be removed from these sumps only with difficulty and at considerable expense. Designs for solids-bearing waters are specifically avoided in the Hydraulic Institute Standards [2] with the statement *"Figures apply to sumps for clear liquids. For fluid-solids mixtures refer to the pump manufacturer."* In a later edition [3], the wording is *"This standard applies to handling clear liquids,"* and these warnings preclude, for example, the use of such sumps as illustrated in Figures 12-26 to 12-29 for solids-bearing waters. When, despite the warning, traditional designs are used for water containing solids, sludge accumulates, often quickly. Sometimes sumps are divided (as in Figure 12-31) to allow one side to remain in service while the other is isolated and manually cleaned. But the manual cleaning of sumps is a

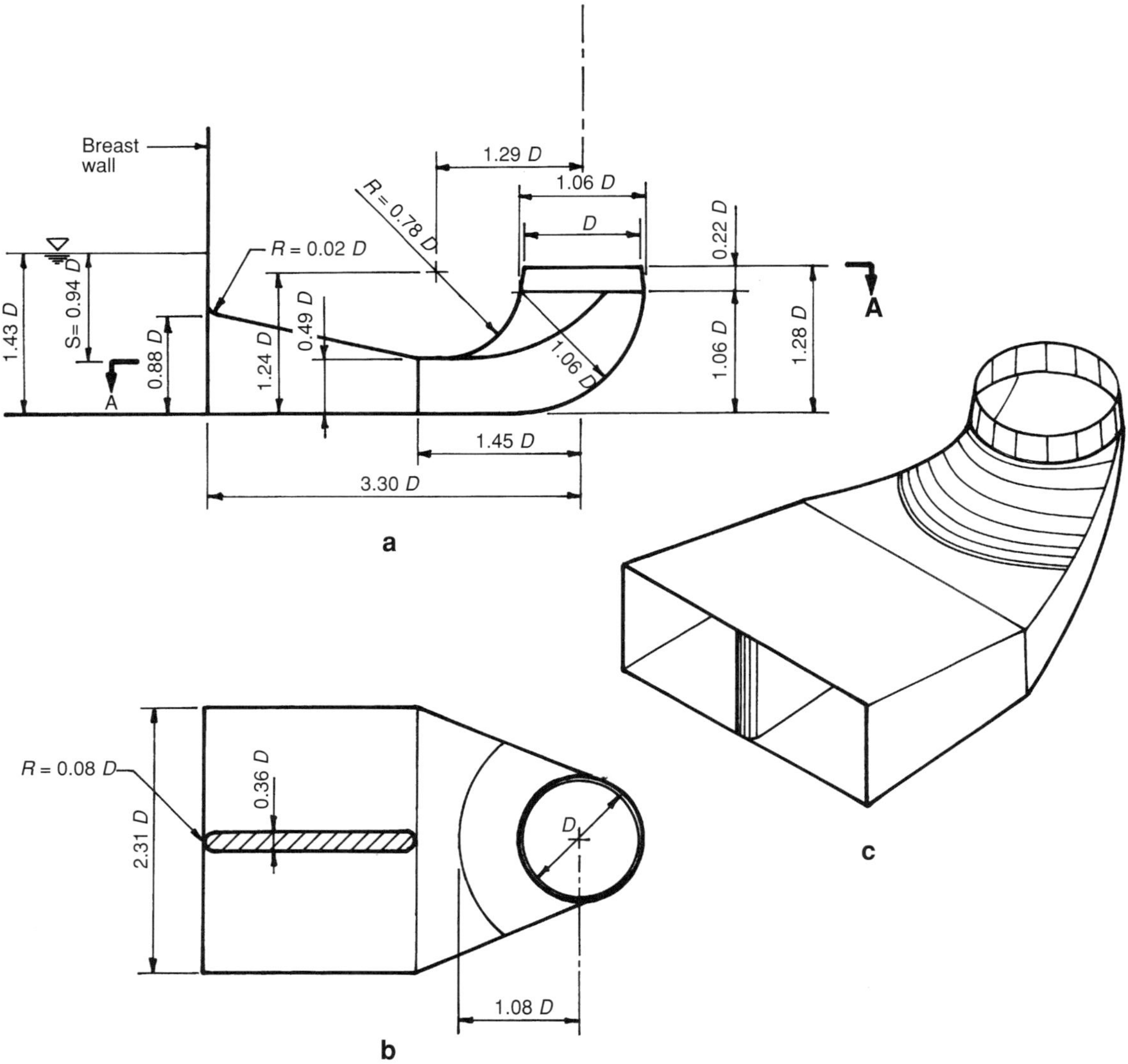

Figure 12-30. Formed suction intake (FSI). (a) Elevation; (b) plan; (c) isometric view. Courtesy of U.S. Army Corps of Engineers.

labor-intensive, disagreeable chore in a hazardous space. Cleaning by pumper truck is expensive. It is unlikely that such wet wells would be frequently (say, weekly) cleaned, so if the fluid is wastewater, odors are likely to be severe. Clearly, a design that permits quick, easy, inexpensive cleaning is needed.

Until 1997, the Hydraulic Institute Standards contained no advice about the design of pump intakes for solids-bearing waters beyond the statement that: (1) flat floor areas should be minimized, (2) approach currents should be kept high (above 1 m/s or 3 ft/s) to minimize deposition, and (3) previous successful designs should be followed. No means for accomplishing these goals were depicted. But from 1994 to 1997, the Intake Design Committee of the Hydraulic Institute addressed (1) improved designs for clear waters and (2) for the first time, specific and detailed design recommendations and geometries for pump intake basins for use with solids-bearing waters. The next edition of the Hydraulic Institute Standards will include the trench-type wet well and endorse the presentation in Section 12-7.

12-7. Pump Intake Basin Design

To be acceptable, a generic intake basin design must provide a good hydraulic environment for the pump

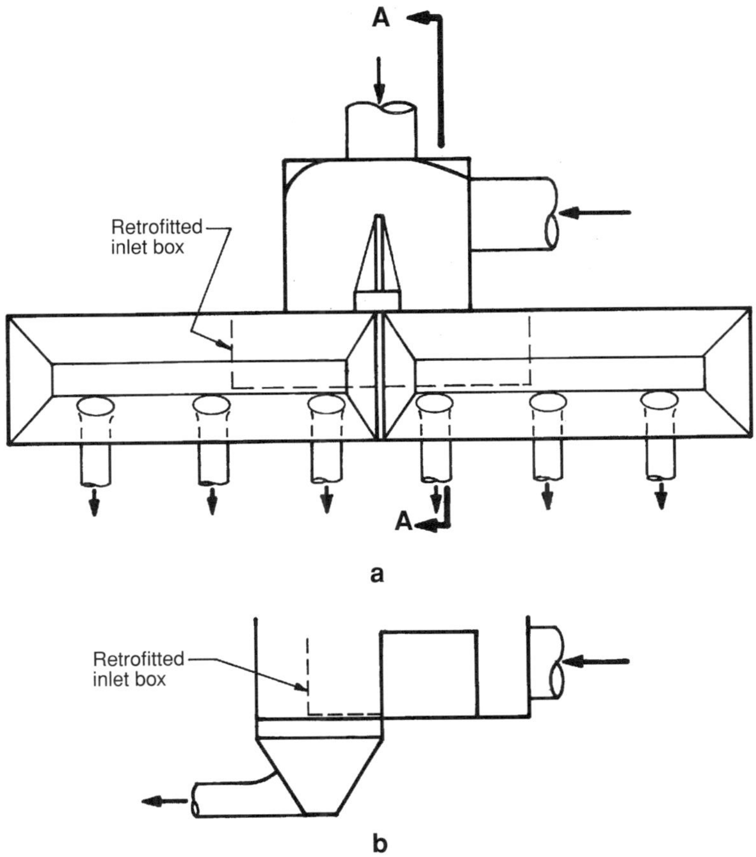

Figure 12-31. A divided sump. (a) Plan; (b) Section A–A. Not recommended (see text).

intakes. If the fluid contains solids, the design must allow ease in cleaning. Three kinds of sumps that can easily be kept clean are:

- The trench-type sump in which extreme drawdown creates an irresistible current that quickly sweeps all solids into the last pump suction. This design is suitable for medium to large nonclog column pumps, dry-pit pumps, and submersible pumps.
- Hopper-bottom sumps with steep sides and floors so small that settleable solids do not accumulate and in which occasional drawdown allows scum to be removed by the pumps. They are suitable for two or even three small submersible or self-priming pumps in a small, round wet well.
- Vigorous mixing of the contents during a short period of the normal pump operation cycle. Mixing can be accomplished by: (1) using a propeller (usually coupled to a submersible motor), (2) bypassing part of the pump discharge into the sump, or (3) bypassing some of the flow in the force main back into the sump just above the pump intakes.

Geometry is the favored method for facilitating cleaning, so the first two types are preferred for new construction. The third is useful for retrofitting existing sumps.

The concept of the trench-type pump sump was developed by Caldwell in the late 1950s. The design has been used at many locations. In particular, it was used in 27 pumping stations in Metropolitan Seattle by Metropolitan Engineers, a consortium of four consulting engineering firms managed by Brown and Caldwell Consultants. Six examples are described in Chapter 17. Note that trench-type sumps differ from traditional designs by having pump intakes in a deep, narrow trench and, in defiance of conventional wisdom, in tandem with the inlet so that water must theoretically pass one inlet to reach another. In spite of this violation of recommendations by the BHRA [1], Hydraulic Institute Standards [2], and ANSI/HI Standards [3, 4], these wet wells have been eminently successful as demonstrated in the field for four decades with pump sizes ranging from 63 L/s (230 m^3/h or 1000 gal/min) to 4.7 m^3/s (17,000 m^3/h or 75,000 gal/min). None of these sumps

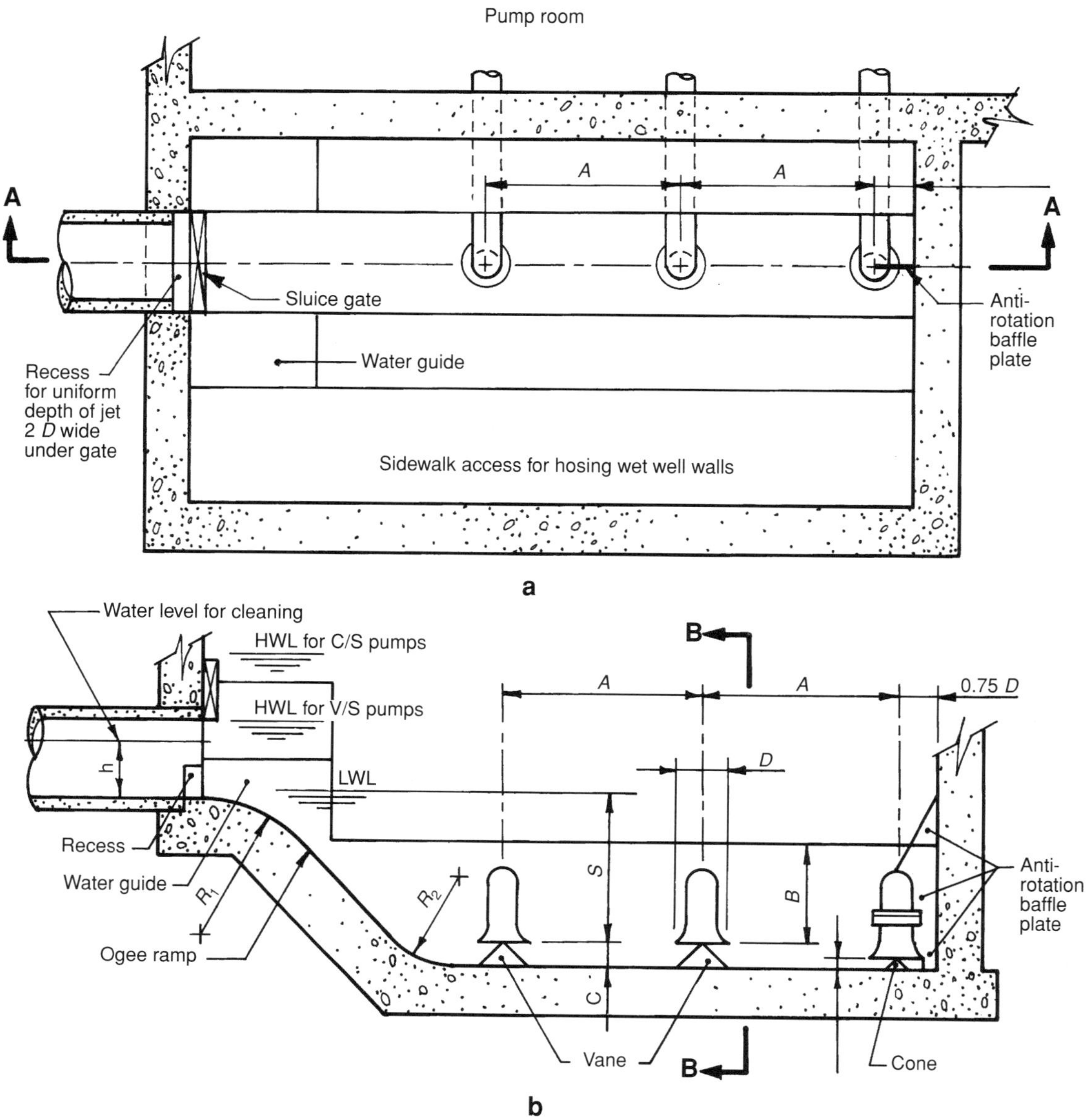

Figure 12-32. Trench-type sump for V/S or C/S pumps. (a) Plan view for solids-bearing water; (b) Section A–A for solids-bearing water, for clean water, omit ogee ramp, shorten sump, and use cones under all intakes; (c) Section B–B for solids-bearing water and dry pit pumps; (d) Section B–B for solids-bearing water and submersible pumps; (e) Section B–B for clean water and column pumps; (f) Floor flow splitter and fillets for suppressing floor and wall (subsurface) vortices at upstream pumps. (See Figure 12-33 for the end pump.)

has ever required retrofitting for the design flowrates, and no pump installed in one has failed to perform satisfactorily.

Trench-type pump sumps possess four distinct advantages over almost all other kinds of wet wells:

- Pump inlets are located near the bottom of a deep trench and far below the invert of the upstream conduit or channel. Consequently, currents near pump intakes are low—almost stagnant—and water tends to enter suction bells uniformly around their peripheries.
- Incoming flow is coaxial with the pump intakes, so Phenomenon 2 has no chance to develop.
- Cleaning is quick, easy, and essentially devoid of manual labor. Cleaning every week or more often is practical.
- The footprint (or plan area) required is minimal.

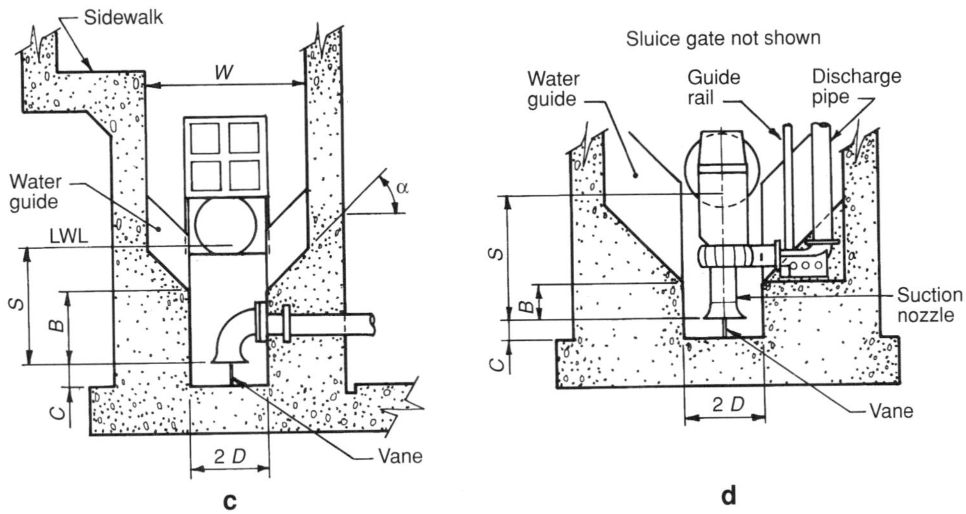

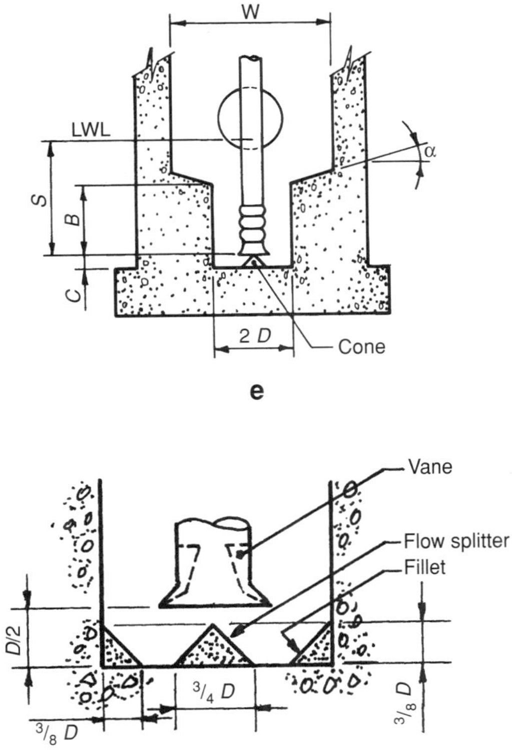

Recommended Dimensions of Trench-Type Sumps

Item	Dimension
	All Waters
A	$A \geq 2.5\ D$. Usually about 4.5 D unless suction pipes splayed to make room for pumps and motors. For submersible pumps, see Mfr.
B	$\geq 2\ D$
S	$(1 + 2.3\boldsymbol{F})D$ where $\boldsymbol{F} = v(gD)^{-0.5}$.
W	As small as possible but keep v_{avg} above trench ≤ 0.3 m/s (1 ft/s) for all flows and water levels.
	Wastewater
C	0.5 D. Last pump intake $\geq 0.25\ D$ below upstream intakes and $D/4$ above floor. Lower the floor if desired.
R_1	2.33 h where h is head on sluice gate.
R_2	0.67 R_1
α	$\geq 45^{\circ}$ for plastic lining; $\alpha \geq 60^{\circ}$ for concrete surfaces.
	Clean water
C	$0.25\ D \leq C \leq 0.5\ D$. Cone always preferred but is required for $C < 0.5\ D$.
α	$\geq 0^{\circ}$ but 45° preferred by some experts.

Figure 12-32. *(Continued).*

Research with models constructed to a linear scale of approximately 1:4 and tested between 1991 and 1995 at Montana State University and (under contract to the university) at ENSR Consulting and Engineering laboratories in Redmond, Washington, by Sanks, Jones, and Sweeney [7] plus assistance from Beaty [18] in testing a full-scale model at the Fairbanks Morse Pump Corp. plant has resulted in minor design improvements over the Metropolitan Seattle stations for normal operation and in major improvements for cleaning. An improved design, in which recommended dimensions are shown, is depicted for wet well–dry well pumps in Figure 12-32. Note that both plan and cross-section views are symmetrical below the water line, so asymmetrical flow to the intakes is avoided. During normal operation, the water jet from the inlet spreads slightly and the velocity is diminished somewhat before the jet strikes the rear wall and returns, thereby setting up a recirculation pattern above the trench from which water drifts slowly downward to the pump intakes.

For wastewater applications, the sidewalk shown in Figure 12-17a and c is strongly recommended for access when workers are hosing grease off the walls. Try to avoid any obstruction (beams or struts) across the wet well, because they interfere with water lances. Use the sidewalk as a beam to resist earth loads. Although the cross-section is massive, the actual volume of concrete is small when compared with the "typical" wet well. Furthermore, many pumping stations are relatively deep and well below the maximum groundwater level, so massive concrete volumes are required to prevent flotation.

The preferred position of a submersible pump, shown in Figure 12-32d, requires a suction nozzle appended to the pump. The disadvantage of the nozzle is its cost (because it may have to be custom fit, and the pump volute might have to be modified), the extra head space required to pull the pump out of the wet well, and the requirement of some sort of cradle or dunnage to support it on the floor or on a truck bed. (Without a nozzle, a submersible pump is stable on a floor.) When the pump volute is small enough, the pump can be lowered into the trench (by deepening the recess for the discharge elbow) so that no nozzle is required. After installing the discharge fitting, the recess can then be filled with lean concrete troweled into place with only enough room for the pump to be pulled out.

The trench-type wet well is readily adaptable for pumping clear water by omitting the ogee ramp, shortening the sump, and using a cross-section like that in Figure 12-32e. As there is no need for supercritical velocities for cleaning, the suction bells can be lowered to $D/4$ with a cone beneath the bell. The advantage of the trench-type sump is its small size, low cost, and the benign hydraulic environment for the pumps.

Caveat

Because of the success of the trench-type sump, it is to be preferred over all others wherever applicable. But unless the designer has had experience with installations of larger sizes, the trench-type sump should not be universally applied to pump sizes larger than about 1900 L/s (6800 m^3/h or 30,000 gal/min) without design-specific model testing.

Valves for suction or discharge pipes larger than 400 mm (16 in.) are large, heavy, and costly. Consider the use of draft tube inlets (as in Figures 17-9 and 17-12) with sluice gates for isolation for larger pump intakes. See subsection *"Sumps for Large Pumps"* near the end of Section 12-7.

Sumps for large pumps that run continuously (e.g., circulating water pumps in power plants) should be practically perfect as proven by model studies.

Trench-Type Sumps for Solids-Bearing Waters

The characteristics of a good sump for solids-bearing waters, as illustrated in Figure 12-32, are:

a. Water enters the sump horizontally with no free fall. LWL during normal operation is far enough above the pipe invert to limit the entrance velocity to about 1 m/s (3.5 ft/s). For V/S pumping, HWL is at the soffit of inlet pipe, but for C/S pumping, HWL can be anywhere above the inlet pipe.
b. A motorized sluice gate allows isolating the sump from the inlet pipe. The mechanism must allow setting the gate opening accurately (perhaps by means of a limit switch) even under static pressure from upstream storage.
c. The trench is 2 D wide by about 2.5 D deep. But actual depth is governed by: (1) required submergence of pump intakes (see Equation 12-1), (2) pump intake floor clearance (usually $D/2$), and (3) the required projected net cross-sectional area above the trench (see item e below). The top of the trench should, however, be no less than 1.9 D above any intake unless model studies indicate that less is allowable.
d. The invert of the influent pipe or channel is no less than 2.0 D above the highest bell mouth to keep incoming currents well above the pump intakes unless model studies indicate that less is allowable.
e. The projected net cross-sectional area (gross area minus area of column or submersible pumps) of the water above the trench is sufficient at any depth (and consequent flowrate) in the inlet pipe to maintain an average velocity (i.e., a nominal plug flow velocity) that never exceeds 0.3 m/s (1 ft/s).
f. The ogee ramp (used only during cleaning) extends from the invert of the inlet to the floor of the trench. The minimum radius of the upper ramp curve is given below by Equation 12-2. The radius of the reverse curve should be at least half as great. A short tangent between the two curves is desirable.
g. A water guide (shown in Figure 12-32) keeps water in a rectangular section from the inlet to the

trench. Without the guide, water climbs up the sloping sides, falls back into the trench, and interferes with the flow of water along the trench during cleaning.

h. Above the trench, the sides are sloped at least 45° if lined with, for example, PVC (or another plastic equally slick) or 60° if the surface is smooth concrete (steeper is better) to assist solids to slide into the trench. PVC makes it easier to hose grease and debris off the walls. For wastewater, the entire sump above LWL should be lined with a plastic such as Linabond® [19] to prevent corrosion.
i. Intakes may be spaced as close as 2.5 *D* center to center. Closer spacing may be used subject to model study. For intakes so close, splay the pump suction pipes apart so that pumps and valves have a clear space on three sides of no less than 1.1 m (42 in.).
j. Upstream intakes are no less than *D*/2 above the floor to avoid interference with the supercritical flow of water during cleaning. Interference patterns generated at the sides of the trench during cleaning may cause shock waves (or "rooster tails") that can reach a height of *D*/2. Their impingement on a suction bell can interfere with the flow of water underneath.
k. For flat floors as in Figures 12-32c to 12-32e, the entrance velocity at the mouth of the pump suction bell (based on the OD of the bell) should be 1.1 m/s (3.5 ft/s) or less for V/S pumping and 0.9 m/s (3.0 ft/s) or less for C/S pumping. To ensure effective disposal of the largest grit particles during cleaning, however, the entrance velocity at full discharge capacity should not be less than about 0.6 m/s (2.0 ft/s). Velocity in the suction pipe should not exceed 2.5 m/s (8 ft/s). If these restrictions are incompatible with standard bells, the bells can be machined to a smaller size.
l. The last pump intake clears the end wall by, preferably, no more than *D*/4 to inhibit surface vortices. The intake must be no more than *D*/4 above the upstream trench floor. The intake can be lowered by sloping the floor downward from the adjacent intake to a lower elevation and retaining a floor clearance of *D*/4. The lowered suction intake creates a more vigorous jump.
m. An anti-rotation baffle, as shown in Figure 12-33, must be installed at the last pump inlet, because otherwise the hydraulic jump, necessary for quick cleaning, may not reach the last pump. The cone is also necessary, and the attached vane is desirable.
n. Floor vanes as shown in Figure 12-34 are installed under all upstream suction bells to inhibit swirling caused by floor vortices.
o. Valve and connection is installed for a 38-mm ($1^1/_2$-in.) hose for washing walls with about 3 to 4.5 L/s (50 to 70 gal/min) of water at about 600 kPa (90 lb/in.^{2}) of pressure at the nozzle.

The trench-type sump of Figures 12-32a to 12-32e was developed from model studies made between 1992 and 1995 in which the maximum suction bell velocity was 1.07 m/s (3.5 ft/s). The adequacy of this design has been demonstrated by those model tests backed by the

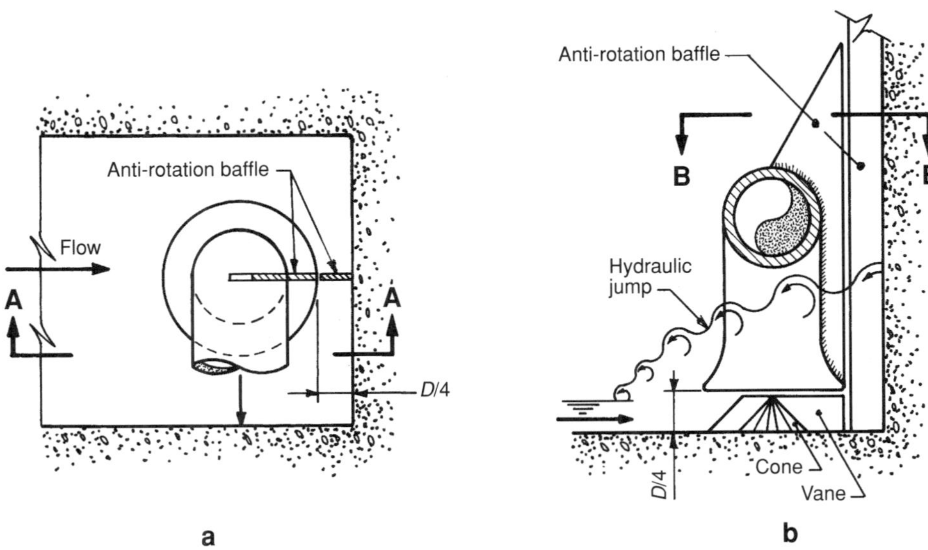

Figure 12-33. Details at last intake to ensure supercritical velocity along full length of trench. (a) Section B–B; (b) Section A–A.

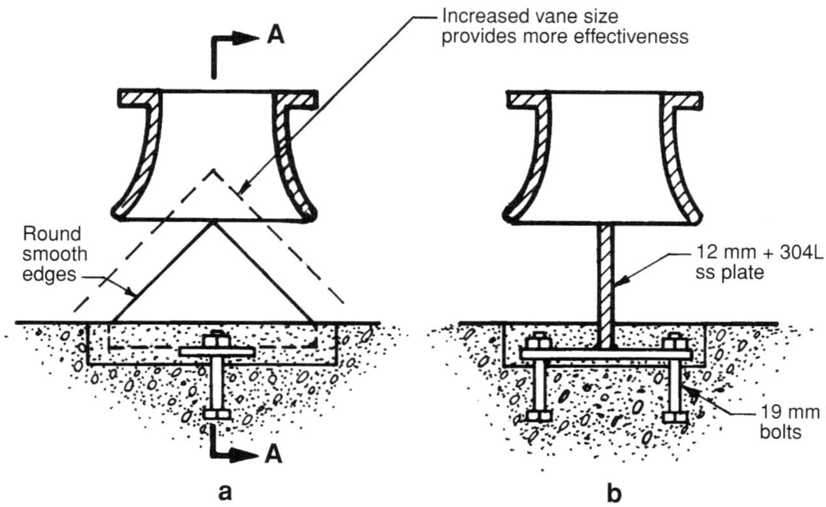

Figure 12-34. Floor vane for upstream pumps in sumps requiring periodic cleaning. (a) Longitudinal (axial) section; (b) Section A–A.

success of two-score wet wells of the type shown in Figures 17-13 through 17-17, which have operated at V/S without problems in the Seattle area for 10 to 30 yr. They have flat floors ranging from 1.88 to 2.6 D wide, suction-bell floor clearances of $D/2$, and no floor vanes or cones, but the intake velocity was nominally limited to 1.07 m/s. The addition of vanes and cones as in Figures 12-33 and 12-34 provides an added factor of safety against swirling and vortices. At higher intake velocities, however, the submerged floor and wall vortices are expected to worsen. Model tests reported by ENSR in June of 1997 [22] and discussed by Williams *et al.* [23] revealed that both subsurface vortices and excessive swirling at an intake velocity of 1.5 m/s (4.9 ft/s) could be controlled with a flow splitter and fillets as shown in Figure 12-32f. The flow splitter should extend from the ogee ramp to a point between the last two pumps. A cone per Figure 12-33 should be installed under the last pump. The fillets can extend to the end wall. The flow splitter eliminates floor vortices, and the fillets reduce wall vortices, but swirling is not controlled. Swirling can be reduced with two or more vanes inside the suction bell. The vanes should have smooth rounded edges at angles of no more than 15 or 20° to the streamlines to prevent the attachment of stringly material. The design shown in Figure 12-32f has not yet (in 1998) been tested in the field.

The acceptable velocity ranges for pump intakes proposed by the Intake Design Committee (consisting of users, engineering contractors, and producers convened and sponsored by the Hydraulic Institute) are given in Table 12-1. This proposal is expected to be published in the 1998 edition of ANSI/HI Standards. Note, however, that the velocities in Table 12-1 are likely to require special "fixes" (such as flow splitters and fillets) for proper pump operation, and even then the recommended velocities may be too high in trench-type wet wells.

The radius of the curve at the top of the ogee can be established by plotting the trajectory of the stream issuing under the partly open sluice gate. The coordinates of any point along the trajectory are $x = tv = t\sqrt{2gh}$, $y = gt^2/2$, where h is the depth of water on the upstream side of the sluice gate, t is time, and g is the acceleration due to gravity. The depth of water, of course, depends upon how much water must be stored

Table 12-1. Acceptable Velocities for Pump Intake Bells, Based on Bell OD

SI Units			U.S. Customary Units		
Flow, L/s	Range, m/s	Recommended[a] velocity, m/s	Flow, gal/min	Range, ft/s	Recommended[a] velocity, ft/s
<320	$0.6 \le v \le 2.7$	1.7	<5000	$2 \le v \le 9$	5.5
320–1300	$0.9 \le v \le 2.4$	1.7	5000–20,000	$3 \le v \le 8$	5.5
>1300	$1.2 \le v \le 2.1$	1.7	>20,000	$4 \le v \le 7$	5.5

[a] The editors recommend maximum velocities of 1.1 m/s (3.5 ft/s) for V/S pumping and 0.9 m/s (3.0 m/s) for C/S pumping. See text.

to complete the cleaning cycle (see Example 12-5). The trajectory is a parabola, and a circular curve fits it only approximately. For practical purposes, however, the radius, R, of a circular curve with its center on a point directly below the sluice gate is, with no factor of safety:

$$R = 2.33\ h \qquad (12\text{-}2)$$

where R is the radius and h is defined above. The upper curve should end in a short tangent no steeper than 45° before beginning the reverse curve. There seems to be no rational basis for determining a satisfactory radius for the reverse curve. In early tests at Fairbanks Morse Pump Corp. (described above), a short radius produced a high shock wave, which was substantially reduced by a radius approximately equal to that of the upper curve.

Note that items g and h above combine to establish the cross-section geometry, dimensions, and the water levels. The number of pumps and their spacing (see item i above) establish the length and, hence, the volume of the sump. For V/S pumping, the volume is irrelevant, because outflow always equals inflow. For C/S pumping, the volume of the sump plus the volume stored in upstream piping must be sufficient to limit the frequency of motor starts and the resting periods to safe values.

Controls for V/S Pumping

One advantage of V/S pumping (aside from the small size of the sump required) is that the influent sewer pipe can be used as part of the sump by keeping the water surface in the sump at the same elevation as the normal depth within the sewer pipe. For example, Figure 12-35 is drawn for three duty pumps operating at V/S. At 100% of full flow (Q_{max}), three pumps operate at V/S within Mode III. At 66% of Q_{max}, two pumps operate, and at 33% of Q_{max}, one pump operates. The modes (or ranges) are nearly equal in depth, so the pumping station control system can be programmed for a linear response. Hence, the normal depth of flow in the sewer is ensured at all times, which maintains carriage (scouring) velocities. Note that pump capacity does not vary linearly with speed. However, the difference in the relationship between sewer capacity and pump capacity is usually small, and refinements are not considered justified in terms of benefit versus additional cost for more sophisticated control. The P&ID diagram for Figure 12-35 is shown in Figure 21-6, and more explanation of the control strategy is given in Section 21-7.

Some engineers are concerned that the pump speed controllers may cause backup or surging in the sewer pipe and prefer to design more conservatively by setting the maximum height of Range 3 at the mid-depth of the sewer or even at the invert. If instability (or surging) is observed, however, the likely cause is rapid acceleration of the pumping equipment, so surging can be eliminated by decreasing the rate of acceleration. Backup is not a problem with properly set elevation controls. These adjustments should be part of the start-up procedure. If the HWL in the sump is lowered by, say, 0.3 m (1 ft) more than necessary in a 0.22 m^3/s (5 Mgal/d) pumping station where the average cost of electricity is 6¢/kW·h, the 20-year present-worth cost of such a decision is

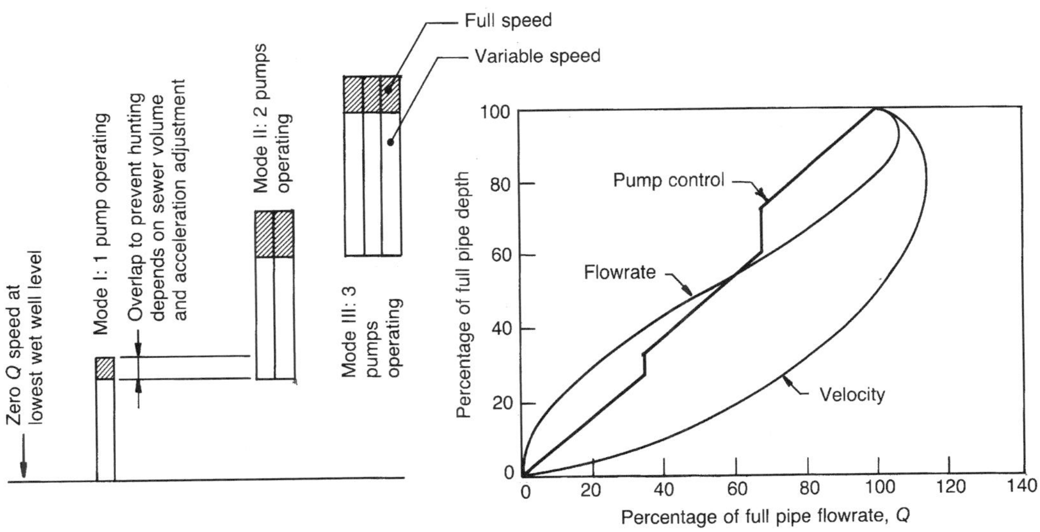

Figure 12-35. Design concepts for variable-speed pumping with the sewer used as part of the pump sump. After Brown and Caldwell Consultants.

nearly \$5000. The free fall also induces turbulence, odor release, and air entrainment in the sump, with the potential for damage to the pumping equipment.

Cleaning Trench-Type Sumps

The purpose of frequently cleaning pump sumps is to prevent odors, to protect the facilities (especially electronic and electrical gear) from corrosion caused by hydrogen sulfide in the air within the station, to eject scum before it can form hard rafts, and to prevent banks of sludge from interfering with flow patterns. To eject scum, the scum must be concentrated into the smallest possible area close (both horizontally and vertically) to a pump intake so that it can be quickly and easily sucked by a strong vortex into the pump intake during the cleaning process. Sludge and grit must be swept along the trench by a strong current to a pump intake in which the velocity is high enough to to lift sand and small gravel particles into the pump. An approximate relationship between the current and the movement of sand beds is shown in Figure 12-36. Note that moving a bed of sand 25 to 50 mm (1 to 2 in.) thick at 1.2 m/min (4 ft/min) requires a water velocity about 60 times greater than the sand velocity. Thus, at the above velocity, it would require about 4 min to eject such a bed of sand from a trench 5 m (16 ft) long—a long time to maintain stable flow conditions with a pump on the verge of losing prime. Light sludge with some grit moves more readily than sand, whereas consolidated, heavy, and sticky sludge is much more difficult to scour.

In conclusion, moderate water velocities per se are of limited effectiveness. For example, the trench in the Kirkland (Washington) Pumping Station sump is 5 m (16 ft) long by 750 mm (30 in.) wide, and the 400-mm (16-in.) suction bells are set 200 mm (8 in.) above the floor. The last pump has a maximum capacity of 12.6 L/s (4.68 ft^3/s). But that capacity is reduced to about 85%, due to air entrainment in a strong surface vortex at the minimum attainable suction bell submergence of about $D/4$. Hence, the current along a clean trench is limited to about 0.5 m/s (1.6 ft/s)—enough to remove light sludge (and essentially all odor) but not enough to remove all the grit. Obviously, better means are needed to remove all deposits.

An extraordinarily effective method of scouring a trench completely clean is to use an ogee ramp to preserve the potential energy of the influent by converting it to kinetic energy. A ramp only 1 m (3 ft) high results in a velocity at the toe of about 3.6 m/s (12 ft/s)—far above critical velocity. A hydraulic jump forms at the toe, and its turbulence immediately suspends all solids beneath it. The moderate current following the jump washes the suspension to the downstream pumps. The formation of the jump is sketched in Figure 12-37, and several stages in a pump-down are shown photographically in Figures 12-38 and 12-39.

It is easy to make the jump move upstream or downstream slowly or quickly by small adjustments of the sluice gate position. With either a dead-man switch and slow closing speeds or a properly adjusted limit switch,

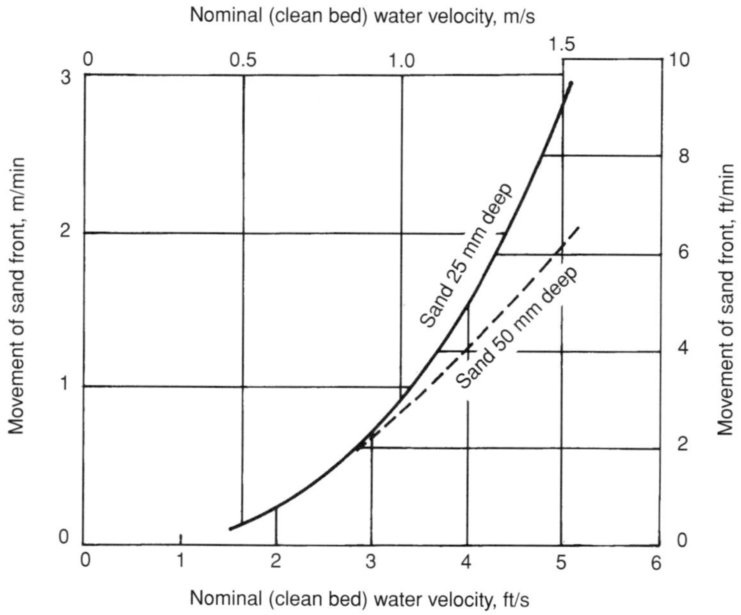

Figure 12-36. Average rate of sand movement as a function of fluid velocity. From Sanks, Jones, and Sweeney [7].

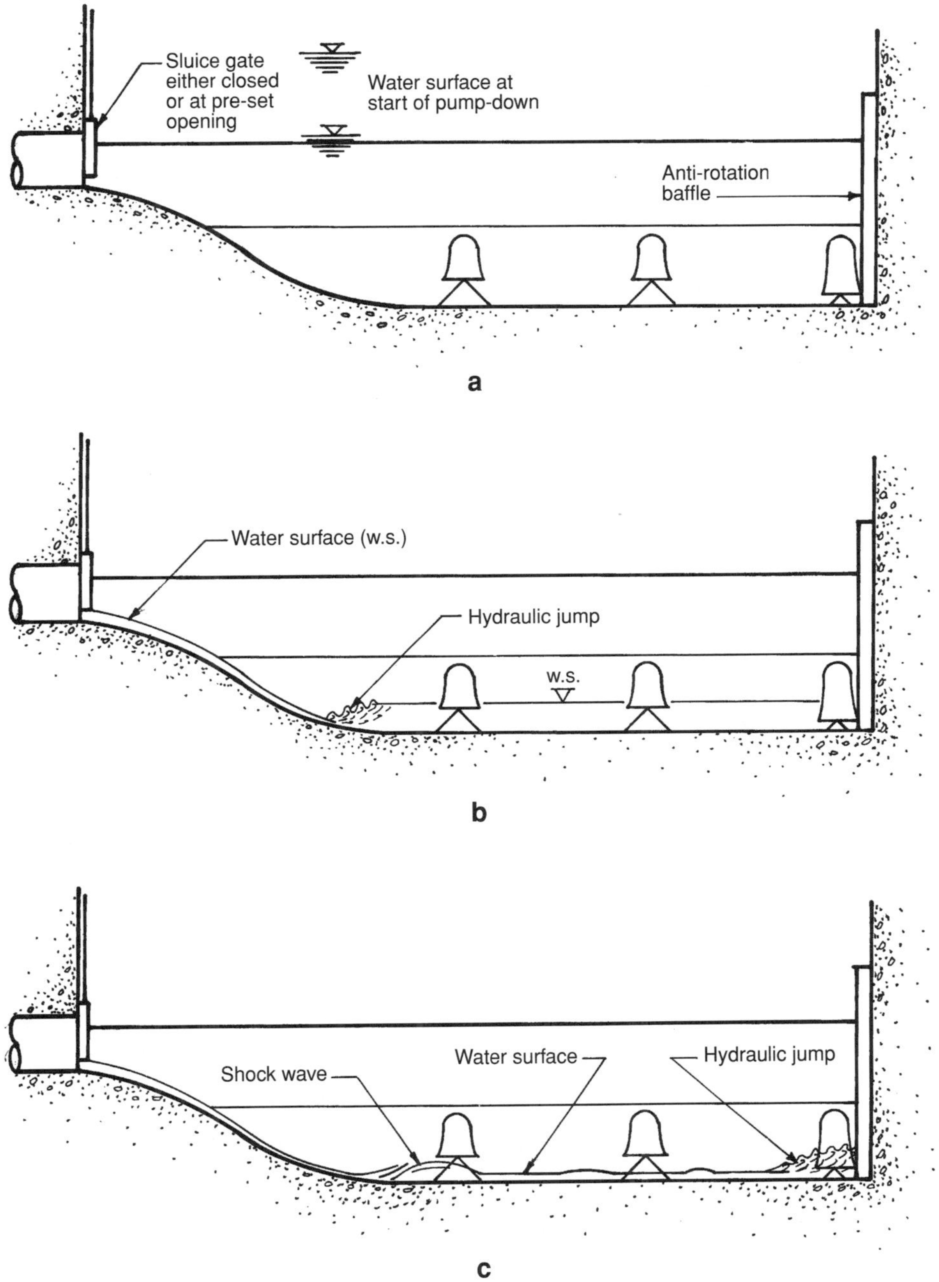

Figure 12-37. Stages in cleaning a trench-type sump. (a) Before pump-down; (b) intermediate pump-down; (c) end of pump-down and cleaning cycle.

the gate can be set very accurately (see the subsection "Sluice Gates"). After the jump is at the toe, a typical trench can be cleaned in less than half a minute provided: (1) the entering flowrate is about 50 to 60% of the maximum capacity of the last pump, and (2) the elevation of the last pump intake is no more than $D/4$ above the floor. (Greater clearance allows a higher water surface that may prevent the jump from reaching the pump.)

Because it is unlikely that the flow to the pumping station will be exactly what is required for cleaning (typically, it is insufficient), storing water upstream is often necessary to obtain enough water to complete

a

b

Figure 12-38. Demonstration model of trench-type, self-cleaning wet well. (a) Before pump-down with water level at top of influent conduit; (b) pump-down begins. Hydraulic jump on ogee ramp. Courtesy of Fairbanks Morse Pump Corporation.

a

b

c

Figure 12-39. Stages in the pump-down of the model in Figure 12-38. (a) Hydraulic jump at the toe of the ogee ramp. The model "sludge" (represented by the black particles) has been ejected, but in a prototype, most of the grit deposits would remain; (b) hydraulic jump under Pump Intake No. 1; (c) jump in front of Pump Intake No. 3 and end of pump-down. Except for grease on the walls above the trench, a prototype would now be clean. Courtesy of Fairbanks Morse Pump Corporation.

the cleaning cycle. There are two ways to store the water and to operate the pumps and sluice gate. The first way is to:

- Close the sluice gate to store water in the upstream piping.
- When enough water is stored upstream, dewater the wet well with the last pump.
- Open the sluice gate to deliver somewhat less flow than the last pump can discharge—easy to do with V/S pumps. With C/S pumps, this operation requires close timing and might be unsuccessful.
- Hose grease off the walls while the sump is dewatered.
- The trench is now clean. Open the sluice gate and restore the station to normal automatic operation. Reprime the pumps as necessary.

The second way is to:

- Lower the sluice gate until it reaches its proper, predetermined position.
- Simultaneously shut off all the pumps to allow the water level behind the sluice gate to rise to such a level that sufficient water is stored to complete the cleaning cycle.
- Turn on all the pumps (even the standby pump, if the facilities can withstand such a discharge without damage) at full speed to dewater the sump as rapidly as possible. Upstream pumps remove some of the sludge until they lose prime. Being set at a lower elevation, the last pump is the last to lose prime, and it does so only when the jump reaches it. Use the dead-man switch to jockey the gate so that the jump travels down the trench at the desired rate.
- Turn off the upstream pumps as they lose prime. A limit switch on the check valve lever or (better) a power monitor in the MCC is advantageous for doing so automatically. Continue to operate the last pump for a few seconds after the jump reaches it.
- Hose grease off the walls while the sump is dewatered.
- The trench is now clean. Open the sluice gate and restore the station to normal automatic operation. Reprime the pumps as necessary.

Sluice Gates

The sluice gate, its operating mechanism, and its installation are vital in ensuring ease and efficiency of cleaning. Select the best equipment, because the sluice gate is unlikely to cost more than about 0.6% of the station construction cost, whereas owner satisfaction will depend greatly upon the ease in operating and maintaining it.

Small (900-mm or 36-in.) sluice gates can be set manually. Larger gates require a mechanical operator, and limit switches on these larger gates are helpful. A sluice gate operator should be designed to move the gate at about 75 mm/min (3 in./min) to make it possible to set the gate accurately at the position wanted. A dead-man switch helps to jockey the gate to its exact position. If the time of closure for gates higher than 750 mm (30 in.) is considered to be excessive, plan for two-speed operation. Operators can be manual (usually operated with a portable power source), electric, and hydraulic. Electric operators are entirely satisfactory for gates up to about 750 mm (30 in.) in size. Hydraulic operators are more expensive and troublesome to maintain, but they make two-speed operation easy, the gate can be operated during a power outage (if equipped with an accumulator), and no electrical equipment is needed in the wet well. If desired, a limit switch can be installed to stop either operator at the proper gate setting, but, of course, it adds complexity.

Sluice gates, frames, and yokes are made of cast iron, mild steel, aluminum, and stainless steel. The latter is the most expensive but it is also the only material that is completely satisfactory. Be sure the yokes can resist about 50% more force than the gate operator can exert, and also specify deflections to be less than L/360 (or still less for large gates). Confer with a reputable manufacturer for features and specifications.

It is advantageous for water to flow under the sluice gate at a uniform depth from wall to wall of the ogee apron. That objective can be accomplished by forming a short (say, 100- to 150-mm or 4- to 6-in.) recess with a flat floor as wide as the trench on the upstream side of the sluice gate. The sluice gate frame would have to be recessed into the sides of the trench (see the water guides in Figure 12-32a). After installation of the frame, the wall recesses should be filled with grout for a hydraulically smooth entrance to the ogee apron. Although this construction is sure to produce much better results than discharge from a segment of a circle (as would occur without the recess), it has not been tested, and no quantitative measure of improvement can be given at present.

Wash Water

In large wastewater pumping stations, install a regenerative turbine pump or a multistage centrifugal pump with piping to a 38-mm ($1^1/_2$-in.) wash water hose to deliver 3 to 4.5 L/s (50 to 70 gal/min) of water at about 600 kPa (90 lb/in.^{2}) at the nozzle with a globe valve for controlling the output. A spring-loaded shutoff valve on the nozzle prevents the hose from becoming a flail if it is dropped. In small stations, the pressure could be lowered somewhat and the flowrate cut in half. If the supply is from a potable source, nothing less than a complete air break is satisfactory to guard against a cross connection.

Rectangular Sumps for Clean Waters

The trench-type sump is also a good design for clean waters because it furnishes such benign hydraulic conditions for the pump intakes. The only modification needed is the omission of the ogee ramp and alterations of some minor features as discussed below. A trench-type sump for pumping clean water with column pumps is shown in Figure 12-32e. The desirable characteristics are:

a. Water enters the sump horizontally with no free fall. The LWL during normal operation is slightly above the invert. For V/S pumping, HWL is at soffit of inlet pipe, but for C/S pumping, HWL can be anywhere above the inlet pipe.
b. A sluice gate allows isolating the sump from the inlet pipe.
c. The trench is 2 *D* wide and extends at least 2 *D* above the pump intakes.
d. Above the trench, the sides are sloped 45°. The slope could be less or even flat if model tests prove such geometry is acceptable.
e. The projected net area (gross area minus area of column or submersible pumps) of the water *above the trench* is sufficient at any depth in the inlet pipe to maintain an average velocity (i.e., a plug flow velocity) that never exceeds 0.3 m/s (1 ft/s).
f. The submergence of pump intakes is not less than given by Equation 12-1.
g. Performance is best when all intakes clear the floor by *D*/2. But *D*/4 is adequate if a cone or, better, if a cone and vane (as shown in Figure 12-40) is placed under each suction bell to eliminate floor vortices and reduce any tendency for swirling. If there is an objection to the use of a cone, the floor clearance should be increased to *D*/2, but then wall vortices, strong floor vortices, and swirling can occur. Wall vortices disappear when the floor clearance is less than *D*/3.
h. Intakes may be spaced as close as 2.50 *D* center to center, but the spacing must leave enough room around machinery for ease in maintenance. A clear space of at least 1 m (36 in.) on each of three sides is a minimum, and 1.1 m (42 in.) is better.
i. The last pump intake clears the end wall by, preferably, no more than *D*/4 to inhibit surface vortices. An anti-rotation baffle somewhat similar to that in Figure 12-33 is desirable.

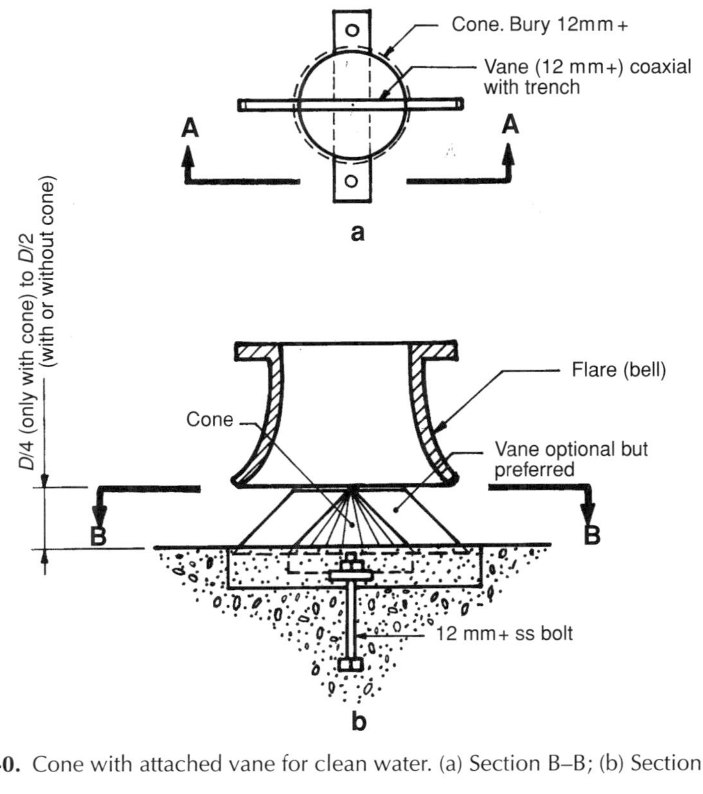

Figure 12-40. Cone with attached vane for clean water. (a) Section B–B; (b) Section A–A.

Active Storage Volume for C/S Pumping

The active storage (between HWL and LWL) in pumping stations for C/S pump motors must be sufficient to limit the number of starts and extend resting periods so as to avoid overheating and overstressing the motors and thereby reducing their life. The tendency of many engineers is to design oversized wet wells. Applying a safety factor that is too conservative to the wet well volume in wastewater pumping is, however, unjustified. At minimum flow, pump starts become infrequent, and the long storage times promote stagnant, anaerobic conditions that result in odors and corrosion. Some of the advice in the literature on allowable frequency of motor starts is based on the use of standard motors and is too cautious. Good judgment is needed to optimize life-cycle costs, so consult pump manufacturers rather than motor manufacturers, many of whom do not understand the overall picture and the need for objective compromise between motor and starter life, size and cost of sump, and number of pumping units necessary to achieve lowest life-cycle costs. Standard motors of moderate size with full-voltage starters can withstand about 4 starts/h with no effect on motor life. Special motors and/or starter systems may be needed to increase starting frequency to at least 6 starts/h for dry pit motors and 12 starts/h for submersible motors. See Section 13-11 for a more extensive discussion of motor-starting frequency.

Determination of Storage Volume

The required storage capacity can be calculated from Equation 12-3 derived as follows. If i is the inflow rate (variable), q is the pumping rate of a single pump (fixed), V is the active volume between LWL and HWL (fixed), and T is the allowable minimum cycle time between starts (time to fill plus time to empty), then:

$$T = \frac{V}{i} + \frac{V}{q-i} = \frac{Vq}{i(q-i)}$$

$$V = \left(i - \frac{i^2}{q}\right)T \text{ and } \frac{dv}{di} = \left(1 - \frac{2i}{q}\right)T = 0,$$

whence $i = q/2$

Substituting $q/2$ for i in the first expression yields:

$$V = \frac{Tq}{4} \tag{12-3}$$

This simple equation for a single pump can also be used for multiple pumps by using q as the increment of pumping capacity as a second (or third pump) is energized. In practice, however, a single pump (in a series of similar pumps) has more capacity than a succeeding pump, because friction head is less, and the single pump is said to pump at "runout." Hence the size of the wet well is governed by the first pump. Succeeding pumps are energized at, typically, 150-mm (6-in.) intervals of water level and de-energized the same way. Calculations are given in Part G of Example 12-3.

Given a known size of wet well and its volume, Equation 12-3 can also be solved for T, the time between pump starts. However, the solution is only for only one flowrate—the critical one that equals one-half of the pump capacity. To see the effect of other flowrates, examine the dimensionless plot of cycle time versus inflow rate in Figure 12-41. Note, for example, that for 50% of the time, the period between pump starts may be from 35% more than T in Equation 12-3 to infinity. In reality, inflow rates are constantly changing, so the period between starts is nearly always significantly greater than that given by Equation 12-3.

Graphical Solution for Pump Cycling Times

A graphical method for dealing with cycling frequency (given in Example 12-3, Figures 12-44 and 12-45) fosters insights not available from Equation 12-3. Goldschmidt [20] developed complex equations defining cycle times for stations with two or more duty pumps. But in a discussion by Wheeler [21], it was pointed out that Equation 12-3 is entirely sufficient for practical purposes.

Approach Pipe

In spite of the common configuration in lift stations whereby liquid is allowed to fall freely (cascade) from

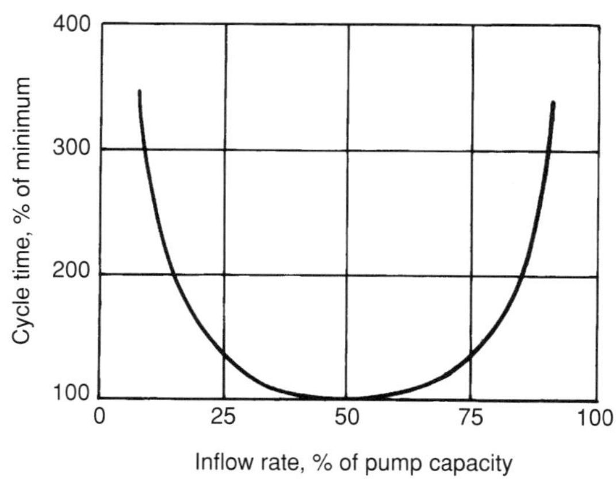

Figure 12-41. Pump cycle time versus inflow rate.

the influent conduit into the wet well pool, it is, nevertheless, poor practice. Even short free falls entrain air bubbles and drive them deep into the pool where they may be drawn into the pumps and reduce pump capacity, head, and efficiency. If the liquid is domestic wastewater, the turbulence sweeps malodorous and corrosive gasses into the atmosphere. The problem is (at least in 1997) universal in wet wells for constant-speed pumps where the active storage requirement makes it necessary to separate high- and low-water levels by, typically, about 1 m (3 or 4 ft) to avoid excessively frequent motor starts.

Objectives

Three desirable objectives in pumping station design are to: (1) eliminate any free fall whatever at all times, (2) supply some of the required active storage capacity in an appurtenant structure at low cost, and (3) discharge liquid horizontally into the wet well pool without turbulence at low velocities (less than 1.2 m/s or 4 ft/s).

One means for accomplishing all three objectives is to introduce the liquid into a sloping "approach pipe" at an upstream point and to discharge it into the wet well at low water level. Trench-type sumps in particular have limited capacity for storage (because of their sloping sides and need for a short trench), so active storage in the approach pipe is especially appealing. Low exit velocities can be obtained by setting the low-water level at an appropriate elevation above the invert of the approach pipe so that the turbulence from the hydraulic jump occurs in the pipe and not in the sump.

Precautions

The relatively steep slope of the approach pipe generates high (super-critical) velocities, so a hydraulic jump occurs when the swiftly moving liquid strikes the pool in the pipe at the level of liquid in the wet well. The sequent depth (depth after the hydraulic jump) must be less than the pipe diameter so that any air bubbles generated by the jump are not trapped. Consequently, it is necessary to use an approach pipe large enough to limit the depth and velocity to safe values, and it follows that smooth pipes and steep slopes require more stringent limitations than do rough pipes and flatter slopes. The steep slope and the sequent depth requirement create the apparent anomaly that a rough pipe can be allowed to carry more flow than a smooth one.

Engineers develop chest pains when told a rough approach pipe can carry more flow than a smooth one can. To illustrate, consider, for example, water flowing at its critical velocity of 2.17 m/s (7.11 ft/s) in a half-full 1219-mm (4-ft) pipe. (Note that a hydraulic jump cannot occur.) If the gradient is 2%, a Manning's n value of 0.030 is required to prevent acceleration. Although the above flowrate is not the maximum allowable, it is 50% more than the flowrate allowed for a Manning's n value of 0.010 (see Tables 12-1 and 12-2). Thus, a rough pipe can be allowed to carry more flow than a smooth one. QED.

Transition

The transition from the upstream conduit to the approach pipe should be designed to accelerate the liquid from the velocity in the upstream conduit to the terminal velocity in the approach pipe. One method for design is to plot the specific energy grade line, E_s, (see Equation 3-3) through the transition, allow for the drop in the E_s due to friction and turbulence, and position the approach pipe so as to have its E_s where it is wanted. However, if such a location results in a rising invert, arbitrarily slope the invert downward to make the exit from the manhole, say, 3 to 30 mm (0.01 to 0.10 ft) below the entrance.

As the sizes of the two conduits are different, it is necessary to hand form a suitable transition section in a manhole. A typical transition design is shown in Example 12-5.

Discharge

Horizontal flow into the wet well is desirable to keep the jet as far above all of the pump intakes as possible. Two ways to discharge the liquid horizontally into the wet well are: (1) beginning at a point 5 to 10 pipe diameters upstream from the wet well, bend the approach pipe from its normal slope to horizontal, or (2) install another manhole. (A manhole is useful as a discharge point for sump pumps.) The horizontal section makes it a little easier to confine the hydraulic jump in the pipe so that influent can enter the wet well smoothly at a velocity (for trench-type sumps) of 1.2 m/s (4 ft/s) or less.

If the slope of the approach pipe is shallow (less than 2%), it makes little practical difference whether the pipe enters the wet well at normal slope or horizontally.

Two Percent Slope

For a majority of installations, a slope of 2% is economical and close to the best compromise between length of approach pipe, active storage volumes between high- and low-water levels, and limitation of the super-critical velocities within the pipe. Tables 12-2 and 12-3 are designed for a slope of 2%, a Manning's n of 0.010, a sequent depth of 60% D_p (where D_p is the inside diameter of the approach pipe), and a useful active storage cross-sectional area of 72 to 81% of the total pipe cross-sectional area. The Froude numbers of the hydraulic jumps for the pipe sizes given are less than 2.5, so the jumps are mild and the turbulence and air entrainment are low. At the sequent depth of 60% of D_p, there is a free water surface 20 D_p long, to allow any entrapped air bubbles to rise to the surface and escape up the pipe. Allowable flowrates for pipe of a different roughness can be obtained using the multiplication factors given at the bottom of each table.

Deposition of solids in the approach pipe is not a problem because of the frequent high velocities and the scouring action of the hydraulic jump as it moves up and down the pipe.

Active Storage Volume

If the difference between LWL and HWL is 1.2 m (4 ft), an approach pipe at a 2% gradient is 61 m (200 ft) long plus the length of horizontal pipe. The active storage volume equals the volume above the water surface at maximum flow in the approach pipe at super-critical velocity and bounded between LWL and HWL. The approach pipe can usually hold about half of the required active storage volume.

The nonprismatic volume at the upper end of the approach pipe can be calculated using Equation 12-4, the prismoidal formula,

$$V = \frac{L(A_1 + 4A_m + A_2)}{6} \tag{12-4}$$

in which L is the length, A_1 and A_2 are the end areas, and A_m is the area at $L/2$ from either end. The formula is a reasonable approximation, but averaging only the end areas is grossly (as much as 50%) in error whenever the area at one end is zero or very small.

Corrosion

Owing to the conditions of wastewater service, corrosion in the approach pipe is likely to be severe. Reinforced concrete pipe 42 in. and larger can be protected with T-Lock® [23]. Consider plastic pipe (even smooth-wall polyethylene drainage pipe) for smaller sizes. A problem with plastic is that it is too smooth. Grease, however, would soon roughen it to some extent.

Table 12-2. Maximum Allowable Flowrates in Approach Pipes, in SI Units

Slope = 2%, Manning's n = 0.010,[a] sequent depth = 60% pipe diameter, D_p.

True Pipe		Flowrate		Before Jump				After jump	Jump energy
D_p, mm	Area, m²	m³/h	L/s	y/D_p,[b] %	Velocity, m/s	A_e/A,[c] %	Froude No.	y/D_p, %	loss, %
254	0.051	71	20	32	1.4	72	1.6	59	17
304	0.073	110	31	32	1.6	72	1.6	59	18
381	0.114	190	53	31	1.8	74	1.7	60	18
457	0.164	290	81	30	2.0	75	1.7	60	18
533	0.223	420	120	29	2.2	76	1.7	60	19
610	0.292	580	160	29	2.3	76	1.8	60	19
686	0.370	770	210	28	2.5	77	1.8	60	20
762	0.456	990	270	28	2.6	78	1.8	60	20
838	0.552	1200	340	27	2.8	78	1.9	60	20
914	0.657	1500	420	27	2.9	78	1.9	60	21
1067	0.894	2200	610	27	3.2	78	1.9	60	21
1219	1.17	3000	840	26	3.5	79	2.0	60	22
1372	1.48	4000	1100	26	3.7	79	2.0	60	22
1524	1.82	5100	1400	25	4.0	79	2.0	60	23
1676	2.21	6500	1800	25	4.2	81	2.1	60	23
1829	2.63	7900	2200	25	4.4	81	2.1	60	24

[a] For n = 0.009, multiply flowrates by 92%
For n = 0.011, multiply flowrates by 108%
For n = 0.012, multiply flowrates by 115%
For n = 0.013, multiply flowrates by 122%

[b] Depth divided by pipe diameter

[c] Empty area divided by total area

Table 12-3. Maximum Allowable Flowrates in Approach Pipes, in U.S. Customary Units.

Slope = 2%, Manning's $n = 0.010$,[a] sequent depth = 60% pipe diameter, D_p.

True Pipe		Flowrate		Before Jump				After jump	Jump energy
D_p, in.	Area, ft^2	M_{gal}/d	ft^3/s	y/D_p,[b] %	Velocity, ft/s	A_e/A,[c] %	Froude No.	y/D_p, %	loss, %
10	0.545	0.5	0.7	32	4.6	72	1.6	59	17
12	0.785	0.7	1.1	32	5.1	72	1.6	59	18
15	1.23	1.2	1.9	31	5.8	74	1.7	60	18
18	1.77	1.9	2.9	30	6.5	75	1.7	60	18
21	2.41	2.7	4.1	29	7.1	76	1.7	60	19
24	3.14	3.7	5.7	29	7.7	76	1.8	60	19
27	3.98	4.9	7.5	28	8.2	77	1.8	60	20
30	4.91	6.3	9.7	28	8.7	77	1.8	60	20
33	5.94	7.8	12.1	27	9.2	78	1.9	60	20
36	7.07	9.7	14.9	27	9.7	78	1.9	60	21
42	9.62	14.0	21.6	27	10.6	78	1.9	60	21
48	12.6	19.1	29.6	26	11.4	79	2.0	60	22
54	15.9	25.3	39.1	26	12.2	79	2.0	60	22
60	19.6	32.5	50.3	25	13.0	79	2.0	60	23
66	23.8	40.9	63.3	25	13.7	81	2.1	60	23
72	28.3	50.3	77.8	25	14.4	81	2.1	60	24

[a] For $n = 0.009$, multiply flowrates by 92%
For $n = 0.011$, multiply flowrates by 108%
For $n = 0.012$, multiply flowrates by 115%
For $n = 0.013$, multiply flowrates by 122%

[b] Depth divided by pipe diameter

[c] Empty area divided by total area

Warning

The approach pipe is needed (1) to limit the size and cost of the wet well for improved overall economy and (2) to prevent a free fall of wastewater into the pool below. For success, however, there are four criteria to be met.

- Deposition of solids in the sewer upstream from the manhole must be prevented by maintaining solids-carrying velocities. To prevent surcharging the upstream sewer at low flows (and thereby reducing the velocity), the wet well HWL at low flow should not be allowed to rise above the sewer invert by more than about $^1/_4$ of the pipe diameter.
- The invert of the manhole at the junction of the sewer and the approach pipe must be designed to accelerate the flow to supercritical velocity. Velocity always equals $\sqrt{2gh}$, where h is the vertical distance between the water surface and the specific energy grade line. For most situations, the invert should probably drop from 30 to 100 mm (0.1 to 0.3 ft) in the manhole. Make allowances for headloss due to turbulence in the manhole.
- To prevent excessive turbulence in the wet well, water should not exit from the approach pipe at the supercritical velocity should be reduced to subcritical velocity in the approach pipe immediately upstream from the wet well.
- Deposition of solids must be prevented in the approach pipe at any rate of flow. Therefore, ensure that each pump cycle produces a scouring velocity before the pump(s) is (are) stopped.

The last two criteria appear to be mutually exclusive, but they can be met by setting the pump start and stop cycles at elevations that (1) achieve the above objectives (as in Example 29-5), (2) provide sufficient active storage for proper cycle times, and (3) separate water levels for successive pump starts by a practical amount—usually 150 mm (6 in.). Thus, the LWL (or stop level) for any combination of operating pumps is based on an approach pipe exit velocity between 1 and 1.2 m/s (3 to 4 ft/s). The start levels can be adjusted to cope with the other considerations in preceding items 2 and 3.

The two criteria can also be met by using a smart controller (a PLC) to sense the approximate rate of flow (based on frequency of pump starts) and set both the LWL and HWL accordingly. PLCs are well understood, the programming is easy, and PLCs are inexpensive. They cost as little as, or less than, an ordinary hard-wired pump controller system.

The data in Tables 12-2 and 12-3 were calculated by Wheeler [25] using his PARTFULL® program and templates in Mathcad 4.0 to solve equations from *Design of Small Dams* [26] and Chow [27]. The required equations are as follows.

The distance from the center of a circle to the centroid (cg_c) of a circular segment is

$$cg_c = \frac{2\left(r \sin\frac{\theta}{2}\right)^3}{3A} \quad (12\text{-}5)$$

where θ is the central angle in radians, A is the area of the segment, and r is the radius of the inside of the pipe.

$$A = \frac{r^2(\theta - \sin\theta)}{2} \quad (12\text{-}6)$$

Substituting,

$$cg_c = \frac{4r\left(\sin\frac{\theta}{2}\right)^3}{3(\theta - \sin\theta)} \quad (12\text{-}7)$$

The distance, y, from the water surface (or from the chord) to the centroid is

$$y = cg_c - r + d \quad (12\text{-}8)$$

The central angle, θ, can be defined as

$$\theta = 2\cos^{-1}\left(1 - \frac{y}{r}\right) \quad (12\text{-}9)$$

where $\cos^{-1}$ is arc cosine. Solving for y gives

$$y = \left(1 - \cos\frac{\theta}{2}\right)r \quad (12\text{-}10)$$

The expression for the centroid to the water surface, y, obtained by combining the above equations is:

$$y = r\left[\frac{4\left(\sin\frac{\theta}{2}\right)^3}{3(\theta - \sin\theta)} + \left(1 - \cos\frac{\theta}{2}\right) - 1\right] \quad (12\text{-}11)$$

Equation 49, page 563 in *Design of Small Dams* [26], is

$$\frac{Q^2}{gA_1} + A_1 y_1 = \frac{Q^2}{gA_2} + A_2 y_2 \quad (12\text{-}12)$$

where subscript 1 is used before the jump and subscript 2 is used after the jump. Note that both y and θ are changed by the jump. Substituting the expressions for A (Equation 12-6) and y (Equation 12-11) with appropriate subscripts into Equation 12-12 results in the following expression:

$$\frac{2Q^2}{gr^2(\theta_1 - \sin\theta_1)} + \frac{r^3}{2}(\theta_1 - \sin\theta_1) \times \left[\frac{4\left(\sin\frac{\theta_1}{2}\right)^3}{3(\theta_1 - \sin\theta_1)} + \left(1 - \cos\frac{\theta_1}{2}\right) - 1\right] = \frac{2Q^2}{gr^2(\theta_2 - \sin\theta_2)} + \frac{r^3}{2}(\theta_2 - \sin\theta_2) \times \left[\frac{4\left(\sin\frac{\theta_2}{2}\right)^3}{3(\theta_1 - \sin\theta_2)} + \left(1 - \cos\frac{\theta_2}{2}\right) - 1\right] \quad (12\text{-}13)$$

The upper (left-hand) term is solved for a known Q and a θ at critical depth. Then the lower (right-hand) expression yields θ and d_2, the depth after the jump.

The values in Tables 12-1 and 12-2 can be approximated without a computer by assuming that the height of the energy grade line above the invert is 75% of the pipe diameter and that the energy loss in the hydraulic jump is 20%. The sequent depth (depth after the jump) is then 60% of the pipe diameter. A table of hydraulic properties or Figure B-4 is useful.

Examples of the Design of Pump Sumps

The following examples are exercises that demonstrate the differences between what the authors perceive as a poor approach and a more rational approach—the difference between wastewater or stormwater pump sumps that are designed for handling solids and those that are not. The design objectives are: (1) to use geometry that allows the pump sumps to be self-cleaning and (2) to minimize size to reduce costs.

The designs used in Example 12-3 is, in the authors' opinions, poor even though similar designs are commonly encountered. Sumps with dividing walls are intended to be cleaned by dewatering one side at a time and manually removing the solids—an expensive, costly, onerous task unlikely to be often repeated. Consequently, sludge will accumulate, become septic, and release odiferous and corrosive gases. Scum will also accumulate and may harden into rafts that can clog pumps.

The detailed designs of trench-type pump sumps in Examples 12-4 and 12-5 illustrate the use of geometry to make the removal of all accumulated solids easy and quick and to provide good hydraulic conditions for the pump suction intakes. The presentation is detailed and, consequently, tedious, but nevertheless, *first-time readers (even experienced engineers) are cautioned to follow the calculations step by step to avoid confusion.* An expert, who develops short-cuts and puts most of the calculations in tabular form, can design and detail an ordinary trench-type pump sump for variable-speed pumps in about half a day, even without a computer. Although the design of pump sumps for constant-speed pumps takes longer, the time required is certainly reasonable and not a deterrent. Do make sufficient, clearly understandable calculations and design notes to justify all important design features, because they are needed by reviewers, are filed in the company archives, and may be needed in court (see Section 17-1).

Example 12-3
Design of Typical Pumping Station Wet Well for C/S Wastewater Pumps

The pump sump for a C/S pumping station is to be rectangular with a modified V-shaped bottom and fed from a double channel with a comminutor in one side and a manually cleaned bar screen in the other side. Plans are shown in Parts B-4 and H of this example.

A. Design Conditions

1. Q_{max} = 220 L/s (5 Mgal/d). Q_{min} = 35 L/s (0.8 Mgal/d).
2. Typical wet pit–dry pit rectangular configuration.
3. Use three duty + one standby pump, each at 73.3 L/s (1165 gal/min), three × 73.3 = 220 L/s. If friction headloss is significant, a single operating pump would be subject to less head and would discharge more than 73.3 L/s.
4. Pumps: horizontal with horizontal motors.
5. Maximum water level change = 1.2 m (4 ft).
6. Frequency of motor starts = 6/h maximum = 600 seconds between starts.
7. Include comminutor and bar screen.

B. Size and Shape of Sump

1. Active volume required. See Equation 12-3: $V = Tq/4$

$$V = 600 \times 73.3/4 = 11\ \text{m}^3 (388\ \text{ft}^3)$$

2. Dimensions of sump. See Part H for final dimensions.
   a. Set pumps 1.8 m (6 ft) c-c for maintenance access.
   b. Center the outboard pump intakes 0.3 m (1 ft) from the end of sump.
   c. Length of sump = (3 × 1.8) + (2 × 0.3) = 6.0 m (19.7 ft) long.
   d. Make sump 2.5 m (8.2 ft) wide, because less width does not give enough room for sluice gate in dividing wall, as will be seen.
3. Water levels (see Figure 12-42).
   a. HWL to be at invert of inlet pipe, El 6.22 m (above mean sea level)
   b. Set pump start and stop levels 150 mm (6 in.) apart so that excessive sensitivity in switching is not required.
   c. Active depth (HWL – LWL per pump) = active volume/area sump. 11 m^3/(6.0 m × 2.5 m) = 0.73 m (2.4 ft); e.g., Pump 2 start to stop elevation difference = 0.73 m.
   d. Headloss in comminutor = 0.10 m (4 in.) from manufacturer's literature.
   e. Establish HWL at inlet pipe invert elevation—6.22 m. Arbitrary (many engineers worry about deposition in sewer if HWL is above invert).
   f. Then LWL in sump = 6.22 – (0.10 + 0.150 × 2 + 0.73) = 5.09 m.

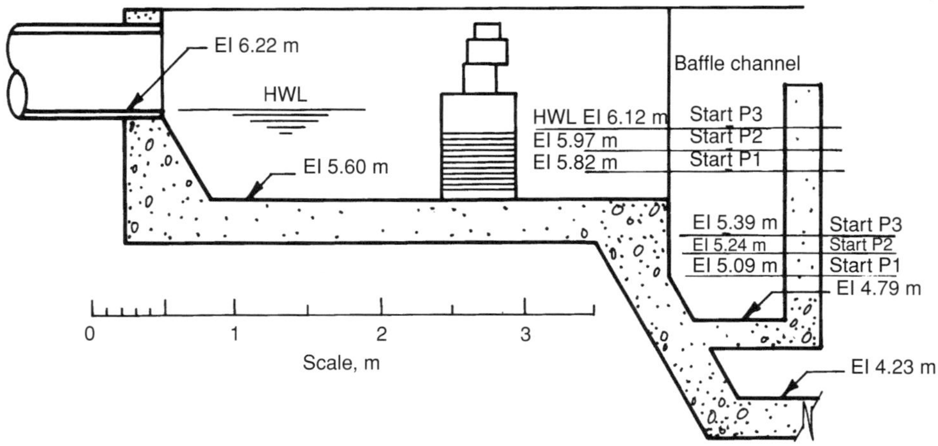

Figure 12-42. Inlet channel for "typical" sump. *Not recommended.*

4. Pump start-stop levels

	Start elevation, m	Stop elevation, m
Pump 3	6.22 – 0.10 = 6.12	6.12 – 0.73 = 5.39
Pump 2	6.12 – 0.15 = 5.97	5.97 – 0.73 = 5.24
Pump 1	5.97 – 0.15 = 5.82	5.82 – 0.73 = 5.09

C. Inlet Channel Design

1. Install comminutor in one side. Capacity = 220 L/s.
2. Install manual bar screen in other side. Add slides to direct water to either channel.
3. Water plunging into sump drives air bubbles into pumps. To prevent cascade, install baffle channel to discharge water horizontally into pool (see Figures 12-31 and 12-42 and Part H).
4. Velocity into pool from the baffle channel:

   Area = 0.7 m × 0.3 m (scaled from Figure 12-42).
   $v = Q/A = (0.220\ m^3/s)/(0.7 \times 0.3) = 1.04$ m/s (3.4 ft/s).
   OK because $v_{allowable} = 1.1$ m/s into any sump.

D. Circulating Currents in Sump

1. The above velocity (1.04 m/s) sets up an oblong current pattern as shown in plan of left cell in Part H.
2. Compute area available for the circulating current at LWL:

   a. Height = 1.0 m (scaled from Figure 12-46, shown later).
   b. Width = half-width of sump = (1.2 + 0.6)/2 = 0.9 m (also scaled from Figures 12-42 or 12-46).
   c. $v = Q/A = 0.22\ m^3/s/(1 \times 0.9) = 0.24$ m/s (0.8 ft/s). OK because current past pump intakes should be less than 0.3 m/s (1 ft/s).
   d. Above calculation is crude. Effect of pump inlets being below entering current would indicate lower velocity occurs, but using "average current" over half-width of sump indicates higher velocity along wall. It is best to be conservative in such situations.

E. Submergence for Pump Intakes

1. Intake $v_{max} = 1.1$ m/s (3.5 ft/s) at inlet entrance. (Hydraulic Institute Intake Design Committee allows 1.5 m/s, but less is better.)

2. Size of intake: $A = Q/v = (0.073\ m^3/s)/(1.1\ m/s) = 0.066\ m^2$.
3. Intake diameter: $D = \sqrt{4A/\pi} = \sqrt{4 \times 0.066/\pi} = 0.30$ m (12 in).
4. Suction pipe size: $v_{max} = 2.4$ m/s (8 ft/s) in suction pipe. (Approved by Hydraulic Institute Intake Design Committee.)
5. Suction pipe area: $A = Q/v = (0.073\ m^3/s)/(2.4\ m/s) = 0.030\ m^2$.
6. Suction pipe diameter = $\sqrt{4A/\pi} = \sqrt{4 \times 0.030/\pi} = 200$ m (8 in).
7. Submergence of center of intake. See Figure 12-43.

   a. Use Equation 12-1, $S = (1 + 2.3\boldsymbol{F})D$

      i. $\boldsymbol{F} = v/\sqrt{gD}$

      ii. $v = 0.073/[(0.30)^2\pi/4] = 1.03$

      iii. $\boldsymbol{F} = 1.03/\sqrt{(9.82\ m/s^2)0.3} = 0.60$

   b. $S = (1 + 2.3 \times 0.60)\ 0.30 = 0.71$ m (2.3 ft)

F. Sluice Gate in Dividing Wall

1. Need sluice gate to close off one cell for cleaning by half a dozen enthusiastic workers in moon suits, rubber boots, and perhaps breathing apparatus. Alternative is a pumper truck.
2. At LWL, area of gate = 0.95 m wide × 0.8 m high. Scaled from Figure 12-46.
3. $v = Q/A = (0.220\ m^3/s)/(0.95\ m \times 0.8\ m) = 0.29$ m/s (0.95 ft/s). OK, because less than 0.3 m/s, the maximum velocity past an inlet.

G. Pump Cycling Frequency

1. Use Equation 12-3 to compute the minimum required volume $V = Tq/4 = (600\ s)(0.073\ m^3/s)/4 = 10.95\ m^3$.
2. Alternatively, see derivation of Equation 12-3 and consider the following:

   a. Critical inflow (i_{crit}) equals half of pump capacity = $P/2$.
   b. $i_{crit} = 0.5\ P = 0.0365\ m^3/s$ for one pump cycling on and off. Filling rate = $0.0365\ m^3/s$. Emptying rate = $0.073 - 0.0365 = 0.0365\ m^3/s$ also.
   c. $i_{crit} = 1.5\ P\ (0.110\ m^3/s)$ for one pump always running and one cycling on and off. Filling rate = $0.110 - 0.073 = 0.037$. Emptying rate = $2 \times 0.073 - 0.11 = 0.036\ m^3/s$. (More exactly, both are 0.0365.)
   d. $i_{crit} = 2.5\ P\ (0.183\ m^3/s)$ for two pumps always running and one cycling on and off. Filling rate = $0.183 - 2 \times 0.073 = 0.037$. Emptying rate = $3 \times 0.073 - 0.183 = 0.036\ m^3/s$. Again, both are exactly 0.0365.

3. For calculating required storage, ignore storage in comminutor and bar screen channels. If included, volume in the wet well itself would be reduced.

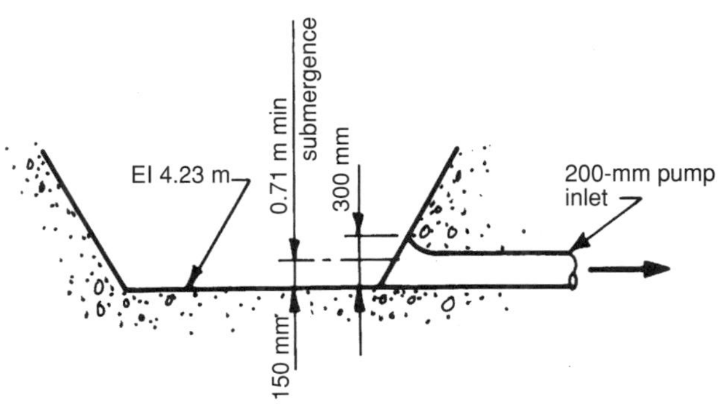

Figure 12-43. Pump inlet for Example 12-3. *Not recommended.*

4. No matter what critical flowrate is used (0.5 *P*, 1.5 *P*, or 2.5 *P*) filling rate is the same and emptying rate (in this example) is the same— 0.0365 m^3/s. Hence, for a filling time (or an emptying time) of 300 s, $V = qT = 0.0365 \times 300 = 10.95$ m^3 as before.
5. For $q = 0.0365$ m^3/s, see Figure 12-44.

   a. Without automatic sequencing of pumps:

      i. $a = b = c = d = V/q = 10.95/0.0365 = 300$ s = 5.0 min.
      ii. Pump starting frequency = 60 min/(2 × 5.0) = 6 cycles/h.
      iii. Resting time = 5.0 min.
      iv. If comminutor channel were included, cycling frequency < 6/h.

   b. With automatic sequencing:

      i. a = P1, b = P2, c = P3, d = P1
      ii. Cycling frequency = 6/3 = 2/h
      iii. Resting time = 5 × 5.0 = 25 min

6. For $Q = 0.183$ m^3/s, see Figure 12-45:

   a. Without automatic sequencing:

      i. $a = c = e = g = i$ = P1 + P2 on.
      ii. $b = d = f = h = j$ = P1 + P2 + P3 on.
      iii. P1 and P2 run continuously and P3 cycles at 6 starts/h. Compare to $Q = 0.037$ m^3/s.
      iv. Resting period for P3 = 5 min.

   b. With automatic sequencing:

      i. a = P1 + P2, b = P1 + P2 + P3 (all).
      ii. c = P2 + P3, d = all.

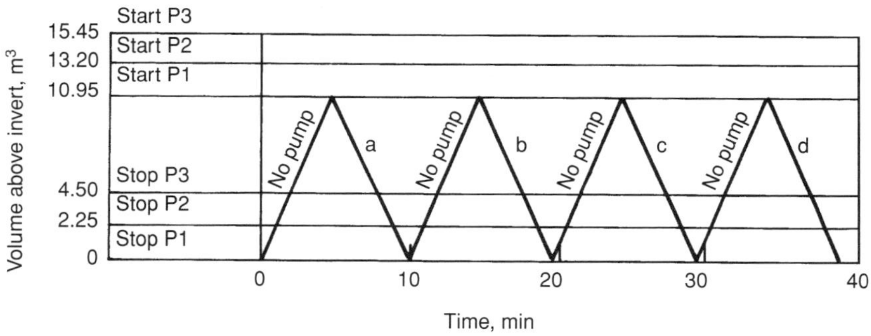

Figure 12-44. Cycling frequency at low flow.

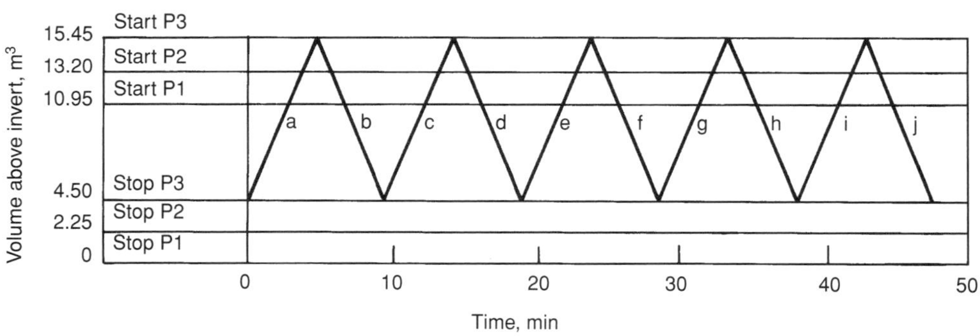

Figure 12-45. Cycling frequency at high flow. NB: If pipe friction is significant, the increment of flow produced by the last pump would be less, lines a to h would be flatter, and cycling frequency would also be less.

iii. e = P3 + P1, f = all.
iv. g = P1 + P2, h = all.
v. Starts/h = 6/3 = 2 for any pump.
vi. Resting time = 10.95/0.0365 = 300 s = 5 min for any pump.

7. Diagrams such as Figures 12-44 and 12-45 can easily be modified for pumps of different sizes and, of course, for any discharge or inflow rates.

H. Plans

Plans for the above example are shown in Figure 12-46.

I. Maximum Detention Time of Wastewater in Wet Well

1. To avoid septic conditions, wastewater should not be stored for long periods. Allowable storage periods are site-specific and depend on travel time to treatment plant, strength and freshness of wastewater, temperature, and other factors.
2. Maximum/minimum flows have been reported with considerable variation. At Sunset Beach station in Steilacoom, Washington, max/min = 197/13 L/s, so minimum flow (at 2 A.M. to 4:30 A.M. in dry weather) is only 7% of peak flow in wet weather.
3. Compute detention time as average liquid volume/min flow. (Not true detention time but rather a dilution factor, but, like air changes/h in a dry well, this computation gives some idea of detention time.)
4. Avg vol = 10.95/2 (see Figures 12-42 and 12-44) + vol below LWL
   = 5.48 + (5.09 – 4.23)(6.0)[(2.5 + 1.35)/2] = 15.41 m^3.
5. Flowrate = 7% max design flow = 0.07 × 0.220 m^3/s = 0.0154 m^3/s
6. For one dilution, T = 15.41/0.0154 = 1001 s = 17 min. Seems quite acceptable.

Critique of Example 12-3

The authors emphasize that this type of design is not recommended, although many stations similar to Example 12-3 are constructed each year. It is impossible to remove scum and sludge with the main pumps, because a pump intake can only suck up sludge within a distance of about $D/4$ from the intake. Dividing the sump in half so that half may be dewatered for cleaning implies entry by an enthusiastic crew in moon suits, rubber boots, and perhaps self-contained breathing apparatus working in a confined, hazardous space to muck out scum and sludge by hand. Entry and safety requirements combined with the revolting task would ensure infrequency of cleaning and, hence, an odiferous facility.

A far more acceptable alternate method is a pumper truck with a flexible suction hose for vacuuming the sludge under water, thereby making the dividing wall superfluous. Dividing walls with sluice gates are, nevertheless, common. Sometimes they are included to facilitate repairs.

Floor

The floor of the sump is much too wide, and deep banks of sludge will accumulate on it. The space under the baffle channel is inadequate to prevent it from being clogged with sludge. There are several remedies, such as a narrower sump, flatter slopes, or a deeper sump. For this facility, the best approach appears to be lowering the floor to Elevation 3.35 m as shown by the dashed lines in Figure 12-46. A downturned suction bell (actually "flange and flare 90° bend") with the mouth at elevation 3.52 m (also shown by dashed lines) makes a better intake, because it offers some protection against the entry of large heavy solids, and it allows the floor of the dry well to be 0.4 m higher than would a horizontal intake. Cleaning with a pumper truck would still be required, but the amount of sludge to be removed would be diminished.

Baffle Channel

In a similar, larger pumping station without a baffle channel, even a short cascade of water from the inlet channel into the pool at low-water levels generated large masses of air bubbles driven down to the floor and into the pump intakes, with the result that design pump capacity could not be reached, pumps were noisy, and repairs were frequent. The baffle channel was effective in discharging water horizontally, allowing bubbles to escape to the surface, and restoring full pumping capacity as shown by model studies.

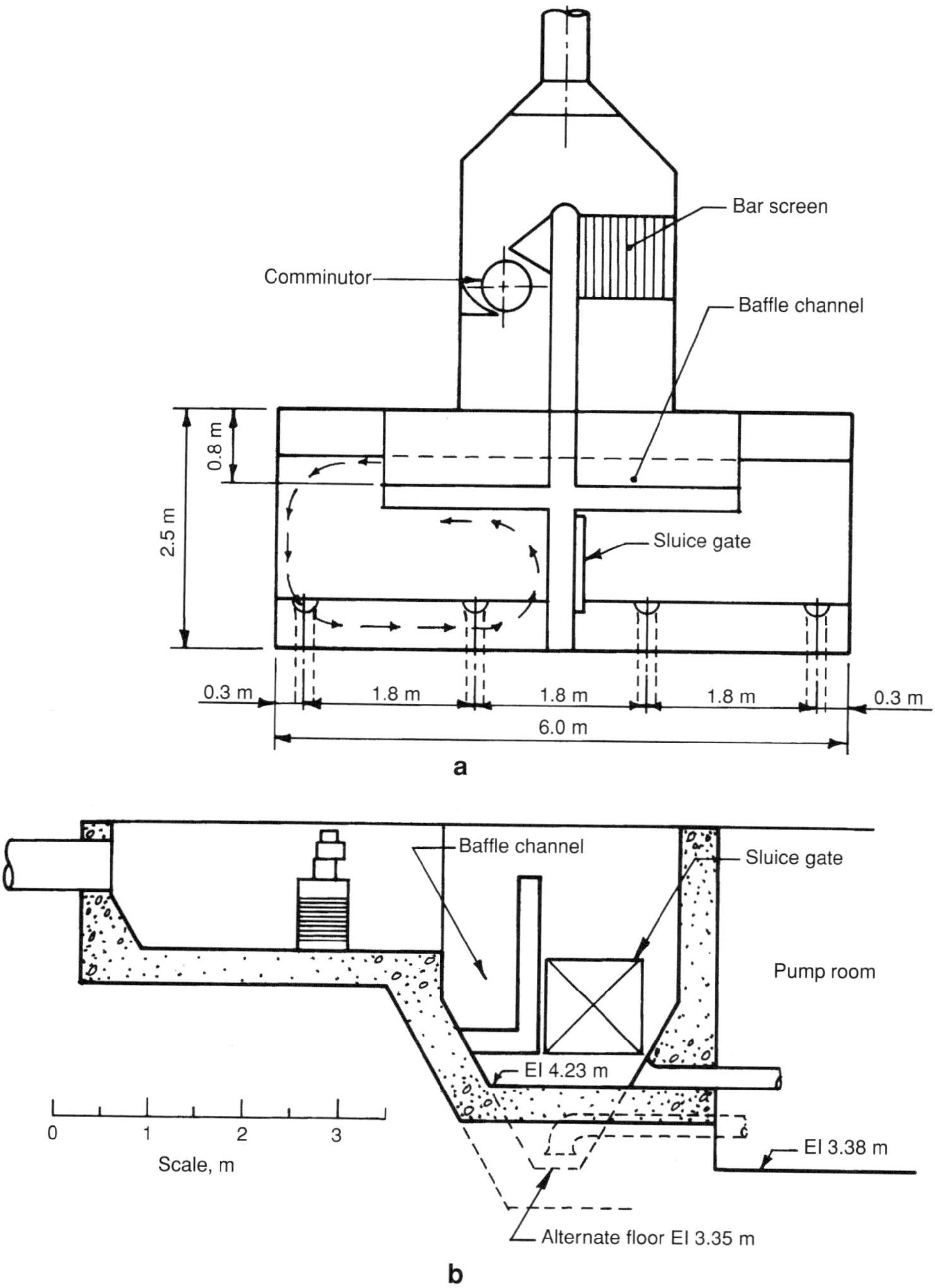

Figure 12-46. "Typical" wet well for C/S pumps. *Not recommended* (see text). (a) Plan; (b) longitudinal section.

However, turbulence in the baffle channel promotes the release of hydrogen sulfide from the entering liquid. Acid corrosion and odors are certain if the wastewater is stale.

Comminutor and Bar Screen

In some locations, there may be a need for a comminutor and a bar screen, but these facilities can be omitted in most pumping stations (see Section 24-11, Item 13). If the dividing wall is also omitted, only a single channel from inlet pipe to sump is needed, and that channel can be shortened but deepened to serve the same purpose as the baffle channel—thereby eliminating the need for the baffle channel. Care must be taken to keep the velocity less than about 1.1 m/s (3.5 ft/s) lest currents at pump intakes exceed 0.3 m/s (1 ft/s). On the other hand, the currents should be large enough to scour deposits from the trench frequently. See Figure 12-36 for the velocities required.

Pump Capacities

Unless the TDH is due almost entirely to elevation head and not to pipe friction, the capacity of one pump operating alone would be greater than one-third of the total capacity. Two pumps would have a greater capacity than two-thirds of the total capacity. (These capacities must be found from the pump and system H-Q curves.) More sophisticated expressions than Equation 12-3 can be used [20, 21], but Equation 12-3 can still be applied in at least two ways.

- Use the true capacity of P1 acting alone to find the volume required and to set stop and start levels. Then set stop and start levels for the follow pump at 150 mm (6 in.) higher. Do the same for the second follow pump, and so on.
- Find the sump volume for P1. Then find the additional volume for P2 by using the difference in pumping capacities between P1 and P1 + P2 for q in Equation 12-3. The difference in levels for starts or stops of subsequent pumps must still be 150 mm.

Graphical Analysis of Pump Cycling

In Figures 12-44 and 12-45, ordinates are wet well volumes with equal increments of volume shown by equal increments of ordinates. The abscissa is time. Beginning with an active sump volume determined by using Equation 12-3, start (and stop) elevations for lead and follow pumps are plotted at the proper volumes for elevation differences of—typically—150 mm (6 in.). For two duty pumps, critical inflow rates are $p/2$ and $(1 + {}^1/_2)p$. Both cycle time and resting time for an inflow of $p/2$ are shown in the figures. When one pump is always running, the cycle time can be doubled (for two duty pumps) by alternating lead and follow pumps in every cycle. The graph can be easily modified for a different active volume, for lag between starting and stopping a pump, for pumps of different sizes, and for changing discharge with changing TDH. Flexibility is greater than with mathematical analyses because, for example, different pump sequences, start-stop elevations, and smart controllers can be easily evaluated. Simplicity, ease of use, and more complete exposition are among the advantages of graphical analysis.

Example 12-4
Design of a Trench-Type Wet Well for V/S Wastewater Pumps

The design of a trench-type wet well with dry pit pumps operating at V/S is illustrated in this example. The design capacity is the same as in Example 12-3. See Section 26-2 for a somewhat different wet well design, for the use of computers to aid hydraulic calculations, and for the design of the entire station.

A. Design Conditions

1. Q_{max} = 220 L/s (5 Mgal/d). Q_{min} = 35 L/s (0.8 Mgal/d).
2. Trench-type wet well. Plans are shown later in Figures 12-47, 12-48, and 12-50.
3. Use three (two duty and one standby) pumps—always the best choice for V/S pumps if the flowrates can be met at all times.
4. Influent sewer pipe: 525-mm nominal (533-mm true = 21-in.) RCP.
5. LWL at pipe invert, HWL at pipe soffit.
6. Pump-speed controller programmed for 0 to 220 L/s with depth proportional to Q (see Figure 12-35).
7. No comminutor nor bar screen.
8. Influent or supply pipe is 533 mm true (21 in.) in diameter

B. Size and Shape of Sump

1. For simplicity, assume the system curve is flat and Q/pump = 110 L/s (2.5 Mgal/d). For a rising system curve, one pump alone would discharge more than 110 L/s (see Section 26-2).

2. Size the suction bell.

   a. Entrance velocity should be about 1.1 m/s (3.5 ft/s). According to Table 17-1, velocities can be 0.6 to 2.7 m/s (2 to 9 ft/s) but not for trench-type sumps.
   b. $A = Q/v = 0.110/1.1 = 0.1\ \text{m}^2$.
   c. $D = \sqrt{4A/\pi} = \sqrt{4 \times 0.1/\pi} = 0.357\ \text{m} = 357\ \text{mm}$.
   d. Suction bell (or flare) OD = flange diameter.
   e. From Table B-1,

Nominal Pipe dia	Flare OD	Entrance velocity	Pipe (see Table B-1) Area, m^2	Pipe (see Table B-1) Velocity, Table B-1
200 mm	338 mm	1.23 m/s (4.0 ft/s)	0.0341	3.23 m/s (10.6 ft/s)
250 mm	400 mm	0.87 m/s (2.85 ft/s)	0.0529	2.08 m/s (6.8 ft/s)

- To avoid excessive subsurface vortices in trench-type sumps, a maximum entrance velocity (into the pump suction bell) of 1.1 m/s (3.5 ft/s) is recommended for V/S pumping. That velocity should be reduced 15% for C/S pumping because C/S pumps always operate at nearly the same discharge, whereas V/S pumps rarely operate at full speed.
- The proposed (in 1998) revised Hydraulic Institute Standards recommend an entrance velocity of 1.67 m/s (5.5 ft/s), but such a velocity produces strong subsurface vortices unless extensive "fixes" (such as fillets and flow splitters) are used.
- Choose the 250 mm pipe. The entrance velocity is somewhat conservative for V/S pumping, but the pipe velocity of 2.08 m/s (6.8 ft/s) is nearly ideal. One method for controlling both the entrance velocity and the suction pipe velocity is to machine the bell to a smaller diameter.
- During cleaning, with only the last pump operating, the true intake velocity is a function of ID (not OD), a capacity of 85% of maximum (as discussed previously), and an increase of Q due to reduced friction losses in the force main. Without station and pump curves, an exact solution is impossible.

C. Design Trench (see Figure 12-32)

1. Trench = 2 D wide = 2 × 400 mm = 0.8 m (32 in.) wide.
2. Upstream bells to be $D/2$ = 200 mm (8 in.) above floor.
3. Last bell to be $D/4$ = 100 mm (4 in.) above floor.
4. Depth of trench to be 2.5 D = 1.0 m (39 in.).
5. To reduce the length of floor (to reduce cost and improve cleaning), set intakes at 2.75 D c-c = 1.1 m (3.6 ft). Must splay suction pipes at $11^1/_4°$ to obtain adequate clearance (1.85 m or 6 ft) around pumps. It might cost no more to make trench longer, set the bells farther apart, and give the contractor the luxury of rectilinear piping.

D. Submergence of Suction Bells

1. Use Equation 12-1. $S = (1 + 2.3\boldsymbol{F})D$, where $\boldsymbol{F} = v/\sqrt{gD}$
2. $\boldsymbol{F} = V/\sqrt{gD} = 0.87/\sqrt{9.82 \times 0.4} = 0.44$
3. $S = (1 + 2.3 \times 0.44)0.40 = 0.8$ m

E. Set Elevation of Top of Trench Relative to Inlet Pipe

1. Guideline: Avg (plug flow) velocity in prism of water above trench ≤ 0.30 m/s (1 ft/s) at all inflow rates.
2. First trial: Set top of trench at invert of inlet pipe.

   a. For $Q = 0.22\ \text{m}^3/\text{s}$, water level is at soffit of inlet.
   b. Scale widths of trapezoid in wet well between top of trench and top of inlet pipe from Figure 12-47.

      i. $A = H(W_1 + W_2)/2 = 0.533(0.8 + 1.90)/2 = 0.720\ \text{m}^2$.
      ii. $v = Q/A = 0.22/0.720 = 0.31$ m/s (1.0 ft/s). OK.

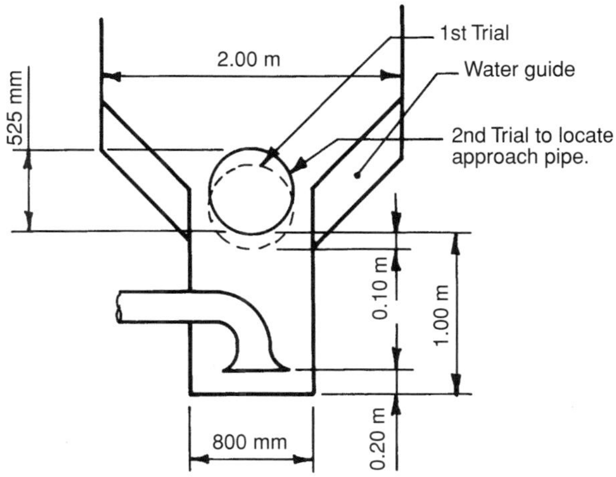

Figure 12-47. Cross-section through wet well of Example 12-4.

c. For $Q = 0.11\ m^3/s$, water level is at mid-depth of inlet pipe.

i. $A = 0.267(0.8 + 1.37)/2 = 0.29\ m^2$.
ii. $v = 0.11/0.29 = 0.38$ m/s (1.2 ft/s). Too great by 20%.

3. Second trial. Set top of trench at 100 mm below invert of inlet pipe.

a. For $Q = 0.11\ m^3/s$ and widths scaled from Figure 12-47,

i. $A = [(0.267 + 0.10)(0.8 + 1.50)]/2 = 0.422\ m^2$.
ii. $v = 0.11/0.422 = 0.26$ m/s (0.86 ft/s). OK.

F. Establish Radius of Ogee Ramp and Length of Wet Well

1. Read "Sluice Gates" in subsection "Cleaning Rectangular Sumps" in Section 12-7.
2. Assume P3 can discharge $0.11 m^3/s$ with adequate submergence and $85\% \times 0.11 = 0.094\ m^3/s$ with low submergence at end of pump-down.
3. Assume influent pipe full during cleaning. Depth = 0.533 m.
4. After jump forms at toe of ogee, assume flow under sluice gate equals, say, $(^2/_3)0.094\ m^3/s = 0.063\ m^3/s$.
5. Calculate sluice gate opening to discharge $0.063\ m^3/s$. This calculation not really necessary, because sluice gate opening can be determined by trial at start-up.

a. Install a recess with a flat floor at invert elevation for full width (0.8 m) of trench.
b. Equation for flow under sluice gate is

$$Q = \frac{WbC_cC_v\sqrt{2g(y_1 - y_2)}}{\sqrt{1-(C_cb/y_1)^2}}$$

where W is width of channel (0.8 m), b is gate opening, C_c (the contraction coefficient) is approximately 0.61 for large y_1/b values, C_v is approximately 1.0, g is $9.81 m/s^2$, y_1 is depth upstream from gate (0.533 m), and y_2 is depth downstream from gate ($= bC_c$). The denominator is very close to 1.0, so the approximate equation is

$$Q = 0.49b\sqrt{19.62(y_1 - y_2)}$$

c. Try $y_2 = 0.030$.

$$0.063 = 0.49b\sqrt{19.62(0.533 - 0.030)} = 1.54b$$
$$b = 0.041$$
$$y_2 = 0.041 \times 0.61 = 0.025. \text{ Close enough to } 0.030.$$

d. Velocity a short distance downstream from the gate is

$$v = 0.063/(0.025 \times 0.80) = 3.15 \text{m/s}$$

e. EGL = $0.025 + (3.15)^2/(2 \times 9.81) = 0.531$ m above apron, so very little head is lost as water flows under gate.

6. Radius of ogee from Equation 12-2:

$$R = 2.33\ h = 2.33(0.533 - 0.025/2) = 1.21\ \text{m}$$

7. See Figure 12-48 for dimensions.

G. Design Cleaning Cycle. Those experienced in this subject can avoid cleaning cycle calculations.

1. Read "Another Way" in subsection "Cleaning Rectangular Sumps" in Section 12-7.
2. Work backward from end to beginning of pump-down by using volumes of water expelled over time. Consider inflow (down the sewer to be 35 L/s) and outflow rates to get water levels and required volumes. See Figures 12-47 through 12-49.
3. Trial 1. Begin with hydraulic jump at P3 and upstream water depth at d_2 in Figure 12-49. Only P3 can expel water under these conditions.

a. d_2 (see Example 12-5) varies but averages about 0.030 m. Ignore it for volumetric and time calculations.

b. $d_1 \approx D/2 + (D/4 \text{ to } D/3) \approx 0.75 \times 400$ mm = 300 mm.

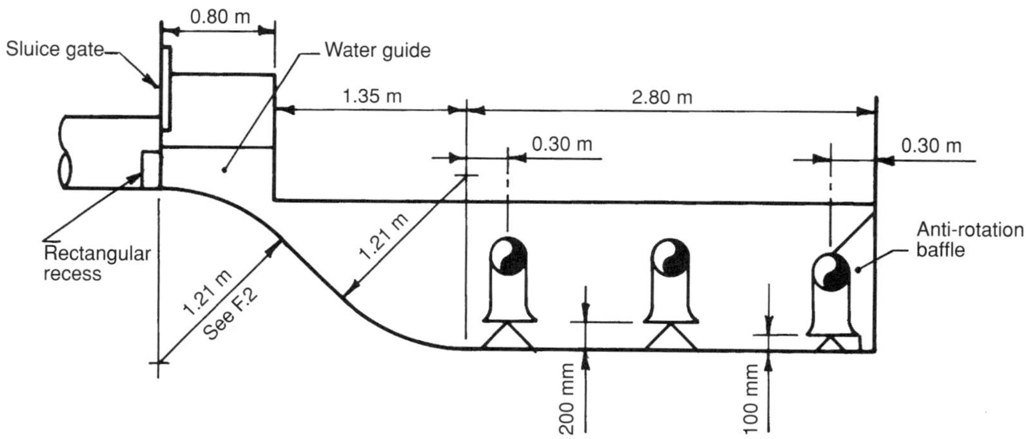

Figure 12-48. Longitudinal section through trench-type wet well for V/S pumps.

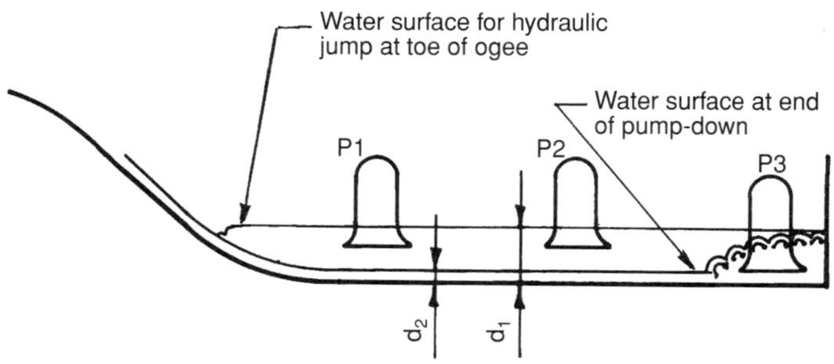

Figure 12-49. Water surface elevations during pump-down. P1 and P2 lose suction at submergences approaching *D*/3 or *D*/4.

c. Time to expel water volume between floor and d_1.

i. $V = 0.3 \times 0.8 \times 3.2 = 0.77$ m^3. Scale length, 3.2 m, from Figure 12-49.
ii. Last pump capacity = $Q_{P3} \approx 85\% \times 110$ L/s = 94 L/s.
iii. Inflow (under sluice gate) chosen (0.063 m^3/s) must be less then Q_{P3} to dewater trench in, say, 20 to 40 seconds, but enough for reasonably vigorous jump with good mixing, say, $F \geq 3$. But note that any velocity above 1 m/s moves grit well although subcritical velocities take longer than would a jump.
iv. $T = V/(Q_{P3} - Q_i) = 0.77$ m^3/(0.094 – 0.063) = 25 s. About right.
v. Ignoring friction along ogee ramp, the velocity of water at the toe of ogee for an EGL that is 1.3 m above water surface in trench (i.e., 1.3 m between the water surface upstream of the sluice gate and the water surface at the toe of ogee) is

$$v = \sqrt{2gh} = \sqrt{2 \times 9.82 \times 1.3} = 5.1 \text{ m/s (17 ft/s)}.$$

vi. Depth of flow in trench at toe of ogee is $d_2 = 0.063/(5.1 \times 0.8) = 0.015$ m = 15 mm.
vii. Friction reduces velocity and increases d_2 substantially. From approximate calculations, the 5.1 m/s velocity might be reduced to about 4 m/s at the toe of the ogee and to about 2 m/s at P3 (see Example 12-5, Figure 12-57).

d. Continue to go backward in time.

i. At depths above d_1, all three pumps operate at (or near) peak capacity. $Q = 110$ L/s $\times 3 = 0.33$ m^3/s. The force main should be investigated to determine whether it can accept so much flow.
ii. Volume from d_1 to top of trench.

$$V = (1.00 - 0.30)(0.8)(4.0) = 2.24 \text{ m}^3$$

Scale the 4.0 m from Figure 12-48.

iii. Volume from top of trench to HWL at soffit of influent pipe.

$$V = \left(\frac{0.8 + 2.0}{2}\right) \times 0.625 \times 4.95 = 4.33 \text{ m}^3$$

iv. Volume from d_1 to HWL = 2.24 + 4.33 = 6.57 m^3.
v. Set sluice gate to pass 0.063 m^3/s (see step G.3.c.3).

vi. Time for water level to drop from HWL to d_1 is, with three pumps running:

$T = V/Q = 6.57/(0.33 - 0.063) = 25$ s

vii. If force main can accept only two pumps pumps running,

$T = 6.57/(0.22 - 0.063) = 42$ s

e. Total time for cleaning is:

	For three pumps	For two pumps
From 3.c.iv	25 s	25 s
From 3.d.v	25 s	
From 3.d.vi		42 s
Total	50 s	67 s

f. In 1.1 minutes of cleaning, the volume used is 6.57 + 0.77 = 7.34 m^3. During this time, 0.035 m^3/s is assumed to be flowing toward the sump. The net volume change in the influent pipe is 7.34 – (0.035 × 67) = 5.0 m^3 (under the worst conditions). Obviously, the water level in the pipe would fall—but only by an insignificant amount. If cleaning required a long (say, 5 to 10 minutes or more) time, it would be necessary to determine whether sufficient water is available to complete the cleaning and to find (and cope with) the water level changes in the pipe.

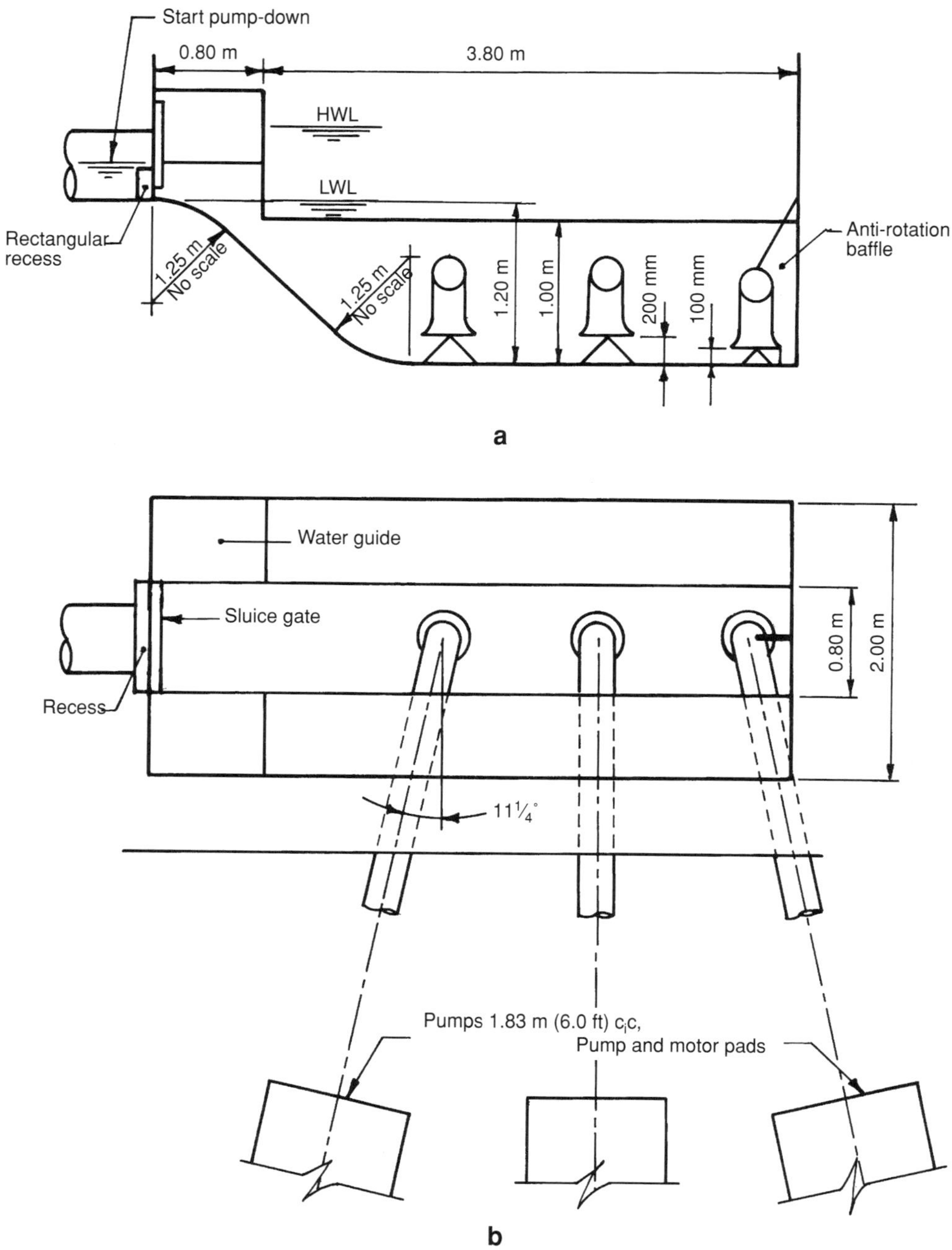

Figure 12-50. Final design of trench-type sump for horizontal dry pit pumps with V/S drives. Splayed pipes and short wet well versus rectilinear pipes and a longer wet well is designer's choice. (a) Longitudinal section; (b) plan.

H. Final Design

1. See Figures 12-47 and 12-50.
2. Note: the suction pipes are shown splayed at angle to illustrate how a long trench can be shortened to make it less expensive and quicker and easier to clean. Locating piping and machinery pads at these angles might be somewhat troublesome. Deciding between splayed and rectilinear piping (and a shorter or longer wet well) is a matter of judgment.

Critique of Example 12-4

Trench-type wet wells for V/S pumping are easy to design. The only concerns are:

- Size and submergence of the suction bell.
- Elevation of the influent pipe and the water level within it to keep the average (plug flow) velocity above the trench below 0.3 m/s (1 ft/s) at all flowrates.
- Enough water available for pump-down and cleaning.
- A reasonable radius for the ogee ramp.
- High velocity along the trench to the last pump to ensure rapid movement of grit. Froude numbers greater than about 3 suspend grit, with the result that the channel is swept clean many times faster than at lesser Froude numbers. The ideal is to produce a Froude number of about 3 at the last pump, but any velocity greater than 1.5 m/s (5 ft/s) moves grit reasonably well (see Figure 12-36).

A method for calculating Froude numbers by successive trial is given in Example 12-5 for general interest. It is not, however, necessary to calculate Froude numbers for wet wells of reasonably similar geometry and dimensions, because cleaning can be effective even if the Froude numbers vary substantially from those in the example. Furthermore, Froude numbers can be modified by changing the cleaning procedure—for example, by an increase in the sluice gate opening. The prudent engineer will, however, calculate Froude numbers for unusual circumstances, for large stations, for trenches that are relatively longer than the one in the example, and perhaps for the first trench-type station encountered.

If a piping arrangement similar to that in Figure 17-13 is used, the minimum satisfactory spacing for the pumps is about 1.8 m (6 ft). If suction piping were rectilinear (i.e., at 90° to the trench), the wet well would be 1.4 m (4.6 ft) longer and somewhat more expensive because of the additional PVC lining and form work. The shorter one would be easier and quicker to clean. On the other hand, the contractor might find nonrectilinear layout more difficult and charge somewhat higher unit prices. The choice is a matter of judgment and personal preference.

Example 12-5
Design of a Trench-Type Wet Well for C/S Wastewater Pumps

To prevent a cascade into the sump at LWL and to obtain more storage so that the sump can be smaller and cleaned more easily, an approach pipe larger than the sewer will be laid at a 2% gradient between an upstream manhole and the sump. Its free (above normal flow) volume is considered part of the active storage of the sump.

Again, the following presentation is detailed and, consequently, may seem tedious and time-consuming. Wheeler's PARTFULL® program [25] makes short work of many of the calculations. Note that operations can (and should) be adjusted during start-up to fit the design and the actual flowrates encountered.

A. Design Conditions

1. Q_{max} = 220 L/s (5 Mgal/d). Q_{min} = 35 L/s (0.8 Mgal/d).
2. Two duty pumps discharge 220 L/s together plus one standby.
3. Trench-type wet well.
4. Influent sewer pipe: RCP 525 mm nominal dia (535 mm true = 21 in.)
5. Maximum pump start frequency = 6/h, but less is better.

B. Select Size of Approach Pipe

1. Read subsection "Approach Pipe" in Section 12-7.
2. Convert % in Table 12-2 to numerical values (see also Table B-8).

True pipe			Before jump			After jump		
ID, mm	A, m²	Q, L/s	y, mm	v, m/s	A_e[a], m²	y, mm	A_w[b], m²	v, m/s
686	0.370	210	192	2.5	0.285	412	0.233	0.90
762	0.456	270	213	2.6	0.356	457	0.287	0.94

[a] A_e = area above water surface.

[b] A_w = area below water surface.

3. Q for nominal 675-mm (27-in.) pipe is 3% too low.
4. Q for nominal 750-mm (30-in.) pipe is 25% too high.
5. Choose nominal 675-mm pipe (actually 686 mm).

C. Size and Shape of Sump

1. Use three pumps—two duty (to pump 220 L/s together) and one standby.
2. Q/pump = 110 L/s (2.5 Mgal/d or 1740 gal/min) only when both pumps are operating. In a system with significant pipe friction, a single pump would, as noted previously, discharge more *because of reduced friction head*. However, for simplicity in this example, assume that a pump operating alone discharges 110 L/s.
3. Size of suction bell = 400 mm for 250-mm pipe. See Example 12-4.
4. Design trench to be the same as in Example 12-4, Parts C.1 through C.4.
5. Set elevation of trench relative to approach pipe entrance. See Figure 12-51.

   a. Water should not enter the sump at supercritical velocity because of the undesirable currents that would be created. Therefore, the hydraulic jump in the approach pipe should not be allowed to exit from the pipe.
   b. At Q = 220 L/s, LWL for follow pump (P2) should (see table above) be 412 mm (say 0.41 m or 16 in.) above invert to keep the jump in the approach pipe.
   c. Keep average plug flow velocity above the trench at or less than 0.3 m/s (1 ft/s). For Q = 220 L/s, the water surface must be 525 mm (say 0.53 m or 21 in.) above trench. See Example 12-4, Part E. Invert must be at least 0.53 – 0.41 m = 0.12 m (4.8 in.) above trench.
   d. At Q = 110 L/s, velocity in approach pipe should be about 1 m/s (3 ft/s) to scour solids from pipe.

      i. At v = 1 m/s, area is (0.11 m^3/s)/(1 m/s) = 0.11 m^2.
      ii. From Tables B-5 and 12-1, $A_w/A_t = 0.11/0.370 = 0.30$, so $y/d = 0.34$ and $y = 0.34 \times 0.686 = 0.23$ m (9 in.).
      iii. To keep plug flow velocity in sump to no more than 0.3 m/s (1 ft/s), area above trench must be at least (0.11 m^3/s)/0.3 m/s = 0.37 m^2.

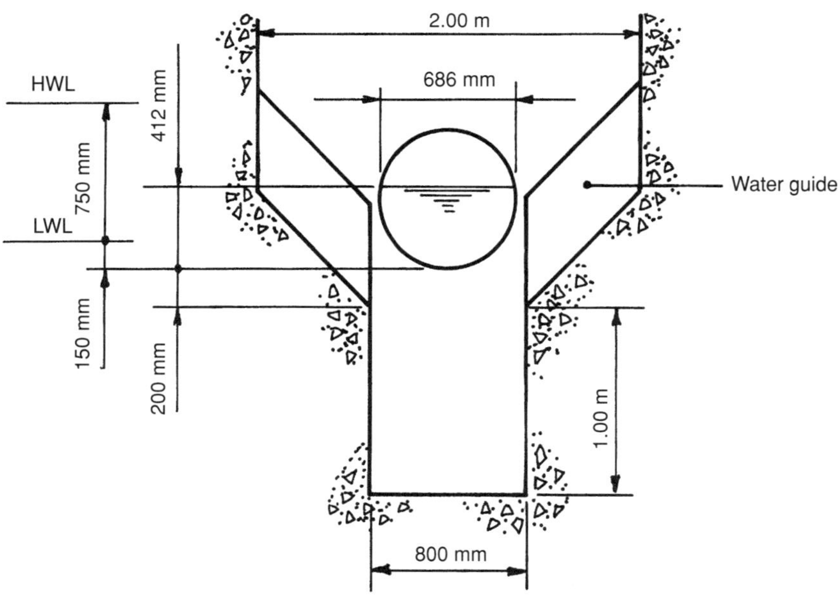

Figure 12-51. Cross-section through trench-type sump for C/S pumps.

iv. Place invert 0.20 m (instead of 0.12 m) above trench. Water depth above trench = 0.20 + 0.23 = 0.43 m, and wetted area above trench = (0.43)[0.8 + (0.8 + 2 × 0.43)]/2 = 0.53 m^2. $v = Q/A$ = 0.11/0.53 = 0.21 m/s. OK because < 0.3 m/s. (Actually, placing the top of the trench 0.12 m below the invert would have been satisfactory, but the added area and reduced currents provide more benign conditions for the pumps. Designing to the limit is not always best.)

6. Length of sump floor: set intakes 4.5 D = 1.8 m (6 ft) c-c. Floor is (1.8 + 0.3)2 = 4.2 m (13.8 ft) long (see Figure 12-52).
7. Length of ogee ramp.
   a. For flow under sluice gate, see Example 12-4, Part F.5.
   b. At beginning of pump-down, assume water level upstream from sluice gate to be at mid-depth of approach pipe.
   c. When water level in sump is at or below top of ogee, head on gate opening can be approximated as half the diameter of pipe less distance from invert to centroid of segment under sluice gate ≈ 685/2 – 57 = 285 mm. (The 57 mm value is an estimate.) A more rigorous analysis would involve a momentum and force balance. Unfortunately, little information is available on the downstream depth of flow from the circular segment.
   d. From Equation 12-1, ogee radius = 2.33 × 0.285 = 0.66 m.
   e. As in Example 12-4. Use at least 0.75 m.
   f. Determine dimensions graphically. Find total length (including ramp) to be 6.0 m (19.7 ft). Find final dimensions by trigonometry.
8. Check submergence of suction bells.
   a. See Example 12-4, Part D. Submergence = 0.8 m required.
   b. Actual submergence (from Figure 12-51) = 1.15 m (3.77 ft). OK.

D. Find Active Volumes Needed to Limit Frequency of Motor Starts

1. The storage volume required for either the lead (P1) or follow (P2) pump is given by Equation 12-3 as $V = Tq/4$ where q is the capacity of P1 alone or, for P2, the increase in capacity for P1 + P2. Pump starts are limited to 6/h or 600 s per cycle, and (in this example) q is 0.11 m^3/s, so $V \geq 600 \times 0.11/4 \geq 16.5$ m^3 between stop and start elevations for P1 or start and stop elevations for P2. Critical flowrates are 0.055 m^3/s for P1 and 0.165 m^3/s for P1 + P2.

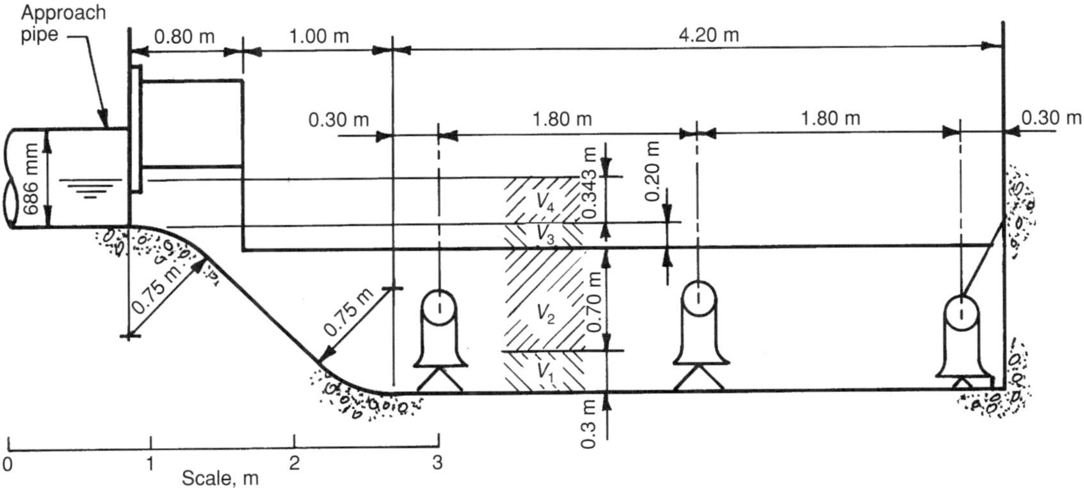

Figure 12-52. Longitudinal section through trench-type sump for C/S pumps.

2. Bear in mind that at any flow, the volume occupied by water flowing down the approach pipe and pooling before it enters the wet well is unavailable for active storage.
3. Also bear in mind that there should be about 150 mm (6 in.) difference in elevation levels for starting or stopping successive pumps (P1 and P2 in this example) to avoid spurious starts or stops caused by wave action in the sump.
4. From Part C.5.d.ii above, the level for stop P1 is 0.23 m (9 in.) above invert and from Part C.5.b, Stop P2 is 0.41 m (16 in.) above invert.
5. Compute active volumes in sump and approach pipe (by scaling distances in Figures 12-52 and 12-53) corresponding to these levels, add 16.5 m^3 to each and compute comparable start levels for these volumes (best done by computer, as calculations by hand are tedious). Adjust start P2 to make it 150 mm higher than start P1. Elevations above invert are:
   a. Stop P1 at 0.23 m above invert and stop P2 at 0.41 m. Difference = 0.18 m, which is > 0.15 m. OK.
   b. Start P1 at 1.00 m above invert and start P2 at 1.15 m. Difference = 0.15 m. OK.
6. See Figure 12-54 for graphical depiction of pump starting frequency.

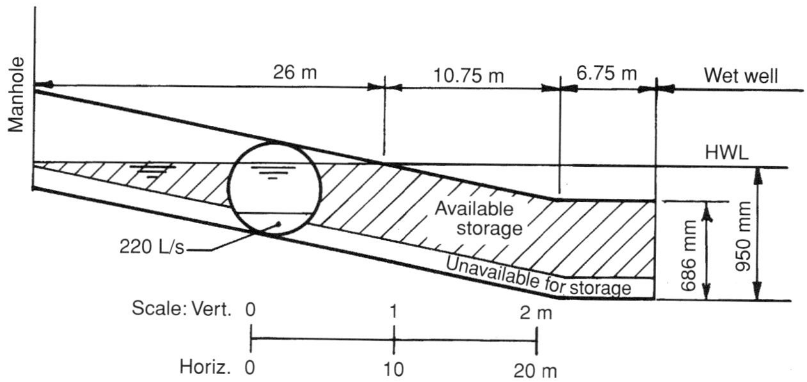

Figure 12-53. Sketch of approach pipe for finding volumes.

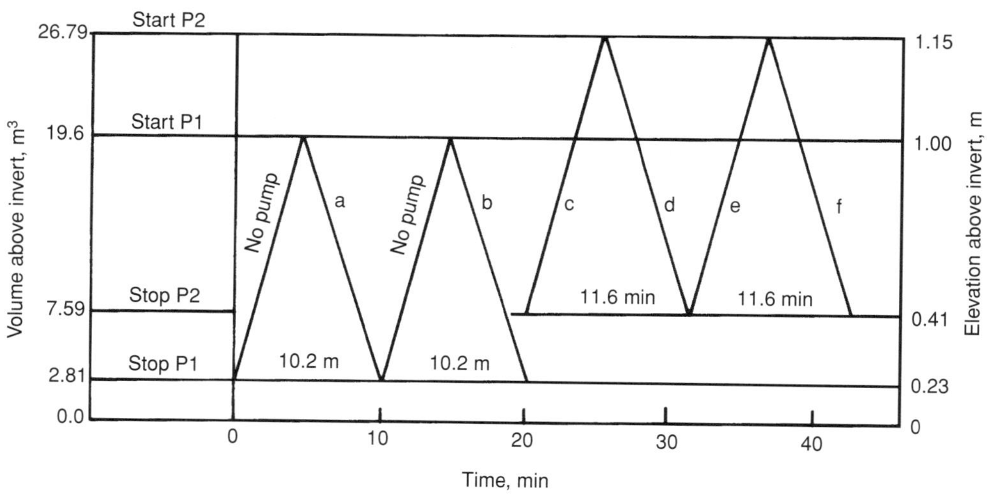

Figure 12-54. Pump cycling frequency. Low flowrate is shown at left and high flowrate at right.

7. With manual sequencing:

   a. $a = b =$ P1. Cycles = 5.9 times/h at $Q = 0.055$ m^3/s.
   b. $c = e =$ P1 and $d = f =$ P1 + P2. $V = 26.79 - 7.59 = 19.2$ m^3, and
      $T = 2[19.2/(0.165 - 0.110)] = 698$ s = 11.6 min,
      so frequency = 60/11.6 = 5.2 cycles/h at $Q = 0.165$ m^3/s.

8. With automatic sequencing, cycle frequencies are half of above.

E. Design the Transition between Sewer and Approach Pipe

1. Objectives:

   a. Prevent jump from reaching soffit of approach pipe and causing an air lock.
   b. Provide a reasonable length of free water surface to which air entrained by jump can rise and escape up the pipe.
   c. Objectives best reached by locating invert to be close to 75% d (approach pipe diameter) below the EGL so that initial velocity approaches the terminal velocity as per Table 12-2.

2. Flow in sewer.

   a. Assume sewer flows "full" at 220 L/s.
   b. From Figure B-5:

      i. $y = 91\%\ d = 0.91 \times 535 = 487$ mm.
      ii. $A = 95\%\ A_t = 0.95 \times 0.223 = 0.212$. See Table B-8.
      iii. $v = 0.220/0.212 = 1.04$ m/s.
      iv. Velocity head = $v^2/2g = (1.04)^2/(2 \times 9.82) = 55$ mm.
      v. EGL (energy grade line) = 487 + 55 = 542 mm above invert.

3. Flow in manhole.

   a. Water goes from low to high velocity somewhat as it does in a conical reducer (see Table B-6), but depth decreases and width increases, so K is much larger than 0.03
   b. Estimate $K = 0.25$, and $v = 2.5$ m/s (Table 12-2, 686-mm pipe).
   c. $h = Kv^2/2g = 0.25(2.5)^2/(2 \times 9.82) = 80$ mm (3.2 in.). See Figure 12-55.

4. Flow in approach pipe

   a. EGL to be 75% d above invert. See text and E.1.c above.
   b. Velocity head, $v^2/2g = (2.5\text{ m/s})^2/(2 \times 9.82) = 318$ mm.
   c. Water depth before jump (Table 12-2) = 28% $d = 0.28 \times 686 = 192$ mm.
   d. From Figure 12-55, invert of approach pipe is 487 + 55 – 80 – 318 –192 = 48 mm below sewer invert. Use 50 mm (2 in.) for ease in construction.

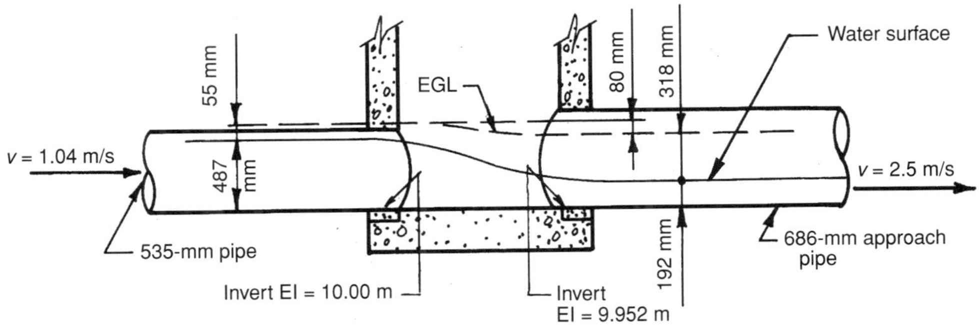

Figure 12-55. Manhole between sewer and approach pipes. Grade of approach pipe is 2%. *A larger drop in manhole invert recommended (see critique).*

F. Cleaning: Volumes and the Time Required

1. Volumes V_1 to V_4 are shown in Figure 12-52.
2. Volumes of water to be expelled:

   a. $V_1 = \mathrm{H} \times \mathrm{W} \times \mathrm{L} = 0.3 \times 0.8 \times 4.7 = 1.1\ \mathrm{m}^3$.
      Scale lengths from Figure 12-52.
   b. $V_2 = 0.7 \times 0.8 \times 5.2 = 2.9\ \mathrm{m}^3$.
   c. $V_3 = 0.2[(0.8 + 1.2)/2](5.6) = 1.1\ \mathrm{m}^3$.
   d. $V_4 = (0.686/2)\{[(1.2 + 1.9)/2][5.25 + (\text{between water guides})0.8 \times 0.75]\} = 3.1\ \mathrm{m}^3$.

3. Start at beginning of pump-down cycle with water level at mid-depth of approach pipe.

   a. Assume force main can accept discharge from all three pumps. $Q_{out} \approx 3 \times 0.11 = 0.33\ \mathrm{m}^3/\mathrm{s}$. (In a real problem, obtain from pump and system H-Q curves.)
   b. Q_{in} is flow under sluice gate. When sump is dewatered to invert of approach pipe, Q_{in} = 63 L/s, and gradually decreases as upstream water surface falls. Ignore this decrease.
   c. Time to expel a volume $= T = V/(Q_o - Q_i) = V/(0.33 - 0.063) = V/0.267$ seconds.
   d. $V_{4:}\ T = 3.1/0.267 = 12$ s.
   e. $V_{3:}\ T = 1.1/0.267 = 4$ s.
   f. $V_{2:}\ T = 2.9/0.267 = 11$ s.
   g. V_1: P1 and P2 lose suction. P3 discharges ~85% of $0.11 = 0.094\ \mathrm{m}^3/\mathrm{s}$,
      so $T = 1.1/(0.094 - 0.063) = 35$ s
   h. Total cleaning time = 62 s ≈ 1 min.

4. Storage volume required:

   a. During pump-down, 63 L/s flows under sluice gate while 35 L/s flows into approach pipe. Net approach pipe loss = 28 L/s.
   b. $V_{required} = (62\ \mathrm{s})(63 - 35) = 1740\ \mathrm{L} = 1.7\ \mathrm{m}^3$.
   c. The volume used causes the water level upstream to fall about 120 mm, so there is more than enough storage for cleaning.

G. Check Froude Number (***F***) at Last Pump.

1. Unusual conditions (such as long trenches) require the following calculations for velocities and Froude numbers. Those with little experience in designing self-cleaning, trench-type pumping stations should also complete them until experience can lead to short-cuts. *Note that failure to have neat, clear, and understandable calculations of every phase of a project leads to trouble for reviewers, for other engineers, and, of course, in court.* It is not necessary for engineers who have had considerable experience with trench-type wet wells to calculate Froude numbers for common heights of the ogee ramps and lengths of trench, because changing the mode of operation can accommodate a wide variety of dimensions. Note, too, that without the ogee ramp, most of the sludge in the original trench-type wet wells was ejected within a couple of minutes. The ramp simply makes cleaning much more effective and quicker. The following calculations are presented both for general interest and for guidance when they may be needed.
2. The dimensionless Froude number (***F***) should be about 3 or more for the jump to have enough energy to suspend immediately all particles within its influence. An ***F*** greater than 8, however, entrains a great deal of air, and so such large values should be avoided near the last pump. For the usual sump, a Froude number greater than about 5 at the last pump is unlikely.
3. One procedure for calculating velocities along the trench is the use of successive approximations made by dividing the ogee ramp and the trench into short sections. Begin the process by finding the sluice gate opening to pass $0.063\ \mathrm{m}^3/\mathrm{s}$ of flow, and find the average velocity of the flow under the gate. The flow from a circular segment splashing onto a flat apron is not uniform, and errors are likely to be large, so make calculations conservatively. *Note, however, that flow under the sluice gate provides much better hydraulic con-*

ditions along the trench when discharged through a slot (formed by the sluice gate and the apron) that is uniform in depth from wall to wall of the water guide. See Example 12-4. Flow through a circular segment is shown in Figure 12-56.

a. Formulas:
   Area pipe $= \pi r^2 = \pi\,(0.343)^2 = 0.370\ \text{m}^2$.
   $\theta = 2\cos^{-1}(b/0.343)$.
   A_s = area segment $= 0.370(\theta/360°) - 0.343\, b \sin(\theta/2)$.

$$cg_c = \frac{2\left(r\sin\frac{\theta}{2}\right)^3}{3A}$$

   (Equation 12-5 where cg_c is distance from center of circle to the centroid of the segment.)
b. C_c and C_v are ignored herein. Calculated velocity is slightly too large and the gate opening to pass the minimum Q (0.063 m³/s) is too small. Error is on the safe side.
c. For $Q = 0.063\ \text{m}^3/\text{s}$,

	1st trial	2nd trial
d, m	0.095	0.085
b, m	0.247	0.257
θ, degrees	87.52°	82.57°
0.370 (θ/360°)	0.08971	0.08463
0.342 b sin(θ/2)	0.05843	0.05799
A_s, m²	0.03128	0.02664
cg, m	0.2821	0.2875
$v = \sqrt{2g\ cg}$, m/s	2.35 m/s	2.38 m/s
$Q = A_s\, v$, m³/s	0.074 m³/s	0.063 m³/s
	Q too high	OK

4. EGL downstream from sluice gate must be somewhere between mid-depth of pipe and bottom of sluice gate. Assume it to be halfway. Erroneous but error is on safe side. (For a rectangular recess at sluice gate, assume the energy loss to be 10% or less.)

   a. EGL = Invert + (0.342 + 0.085)/2 = Invert + 0.214 m.
   b. From Equation 3-3,

$$E_s = y + v^2/(2g) = 0.214 = 0.085 + v^2/(2 \times 9.82), \text{ so } v = 1.59\ \text{m/s}.$$

   c. A velocity reduction of 2.38 to 1.59 m/s seems excessive. However, water at invert flows over a sill for the sluice gate, and edge water splashes down onto ramp with some turbulence and energy loss. Accept the 1.59 m/s until better data are available.

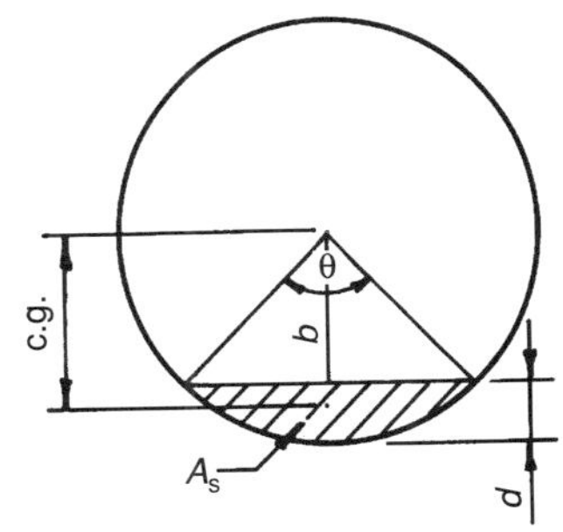

Figure 12-56. Sluice gate opening for a 675-mm pipe (686-mm true diameter). Note: a recess between pipe and sluice gate to produce a rectangular opening as wide as the trench is much better.

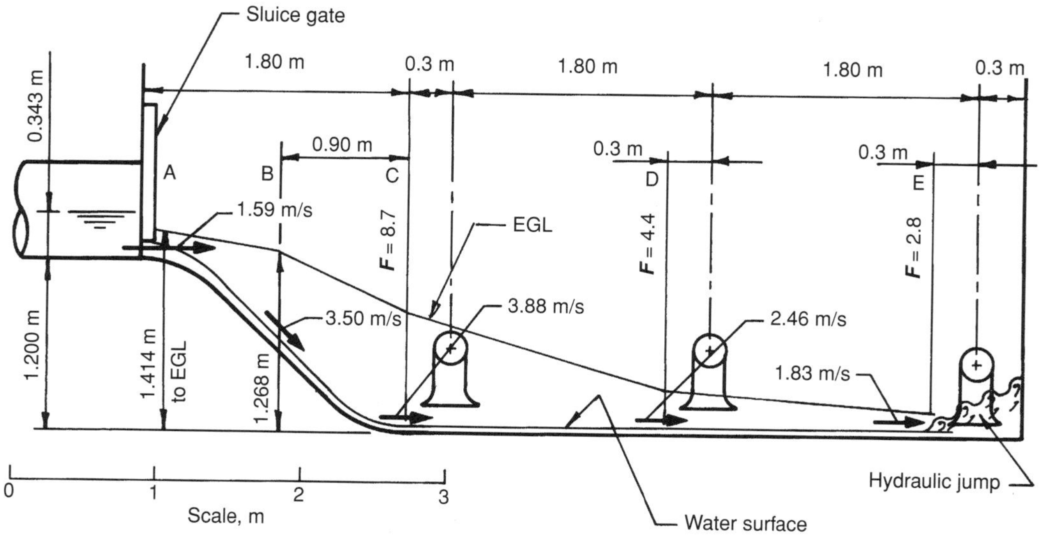

Figure 12-57. Energy grade line and velocities at pump-down.

d. EGL at top of ogee ramp = invert + E_s = 1.200 + 0.214 = 1.414 above trench floor (see Figure 12-57).

5. Divide the ogee ramp and the trench into four sections. Four are enough for preliminary results. For more accuracy, use six or eight and limit the allowable discrepancy between estimated and calculated values.

a. For any section, state the velocity, v_1, at the beginning of the section.
b. Compute depth, d_1, at the beginning of the section = $Q/wv_1 = 0.063/0.8\ v_1 = 0.07875/v_1$.
c. Compute EGL_1 at the beginning of the section = invert elevation$_1$ + d_1 + $v_1^2/2g$, where $2g = 2 \times 9.81 = 19.62$.
d. Estimate depth of flow, d_2, at the end of the section.
e. Compute velocity, v_2, at the end of the section = $Q/wd_2 = 0.07875/d_2$.
f. Compute EGL_2 = invert elevation$_2$ + d_2 + $v_2^2/2g$.
g. Compute $v = (v_1 + v_2)/2$.
h. Compute $d = d_{avg} = 0.063/0.8\ v = 0.07875/v$.
i. Compute R_m. The use of R_m in the Manning equation was proposed by Escritt [28], who stated that n is correct within a few percent if the perimeter includes half the width of the free water surface. Here, it is equivalent to increasing n from 0.010 to 0.013. See Section 3-5.

   i. $R_m = A/(P + w/2)$ where A is 0.8 d, P is $w + 2d = (0.8 + 2d)$, and $w/2$ is half the width of the free water surface, 0.40 m.
   ii. $R_m = 0.8\ d/(2d + 1.2)$

j. Compute $\Delta h = \text{L}(0.01\ v R_m^{-2/3})^2$
k. Compute $\text{EGL}_2 = \text{EGL}_1 - \Delta h$.
l. Compare EGL_2 in Step k with EGL_2 in Step f.
m. Repeat as necessary with a revised d_2.

6. Flow from A to B: L = 1.1 m. Invert_A = 1.20 m. Invert_B = 0.59 m.

Steps correspond to those of Part 5 above.

Step		Trial 1	Trial 2	Trial 3
a.	v_A, m/s	1.59		
b.	$d_A = 0.07875/v_A$, m	0.050		
c.	$\text{EGL}_A = \text{Invert}_A + d_A + v_A^2/19.6$	1.379		
d.	d_B, m (estimate)	0.022	0.023	0.0225
e.	$v_B = 0.07875/d_B$	3.580	3.42	3.50
f.	$\text{EGL}_B = \text{Invert}_B + d_B + v_B^2/19.6$	1.265	1.211	1.237
g.	$v = (v_A + v_B)/2$	2.585	2.510	2.545
h.	$d = 0.07875/v$	0.0305	0.0314	0.0309
i.	$R_m = 0.8\ d/(2d + 1.2)$	0.01935	0.01989	0.01959
j.	$\Delta h = \text{L}(0.01\ v\ R_m^{-2/3})^2$	0.141	0.1286	0.1349
k.	$\text{EGL}_B = \text{EGL}_A - \Delta h$	1.238	1.250	1.244
l.	EGL_B @ k – EGL_B @ f	–0.027	+0.039	+0.007 O.K.

7. Flow from B to C. L = 1.1 m.

Step		Trial 1	Trial 2	Trial 3
a.	v_B, m/s	3.50		
b.	$d_B = 0.07875/v_B$, m	0.0225		
c.	$\text{EGL}_B = \text{Invert}_B + d_B + v_B^2/19.6$	1.238		
d.	d_C, m (estimate)	0.020	0.021	0.0203
e.	$v_C = 0.07875/d_C$	3.938	3.75	3.879
f.	$\text{EGL}_C = d_C + v_C^2/19.6$	0.810	0.7377	0.788
g.	$v = (v_B + v_C)/2$	3.719	3.625	3.6895
h.	$d = 0.07875/v$	0.0212	0.0217	0.0213
i.	$R_m = 0.8\ d/(2d + 1.2)$	0.01364	0.01396	0.01374
j.	$\Delta h = \text{L}(0.01\ v\ R_m^{-2/3})^2$	0.4669	0.4300	0.455
k.	$\text{EGL}_C = \text{EGL}_B - \Delta h$	0.771	0.808	0.783
l.	EGL_C @ k – EGL_C @ f	–0.039	+0.070	–0.005 O.K.

8. Flow from C to D. L = 1.8 m.

Step		Trial 1	Trial 2
a.	v_C, m/s	3.879	
b.	$d_C = 0.07875/v_C$, m	0.0203	
c.	EGL_C	0.7889	
d.	d_D, m (estimate)	0.030	0.032
e.	$v_D = 0.07875/d_D$	2.625	2.461
f.	$\text{EGL}_D = d_D + v_D^2/19.6$	0.3812	0.3407
g.	$v = (v_C + v_D)/2$	3.252	3.170
h.	$d = 0.07875/v$	0.0242	0.0248
i.	$R_m = 0.8\ d/(2d + 1.2)$	0.01551	0.01588
j.	$\Delta h = \text{L}(0.01\ \text{v}\ R_m^{-2/3})^2$	0.4922	0.4532
k.	$\text{EGL}_D = \text{EGL}_C - \Delta h$	0.2966	0.3357
l.	EGL_D @ k – EGL_D @ f	– 0.085	–0.005 O.K.

9. Flow from D to E. L = 1.8 m.

Step		Trial 1	Trial 2
a.	v_D, m/s	2.461	
b.	$d_D = 0.07875/v_D$, m	0.032	
c.	$EGL_D = d_D + v_D^2/19.6$	0.341	
d.	d_E, m (estimate)	0.040	0.043
e.	$v_E = 0.07875/d_E$	1.969	1.832
f.	$EGL_E = Invert_E + d_E + v_E^2/19.6$	0.2376	0.2139
g.	$v = (v_D + v_E)/2$	2.215	2.146
h.	$d = 0.07875/v$	0.0356	0.0367
i.	$R_m = 0.8\ d/(2d + 1.2)$	0.0224	0.02306
j.	$\Delta h = L(0.01\ v\ R_m^{-2/3})^2$	0.1401	0.1263
k.	$EGL_E = EGL_D - \Delta h$	0.201	0.215
l.	$EGL_E\ @\ k - EGL_E\ @\ f$	–0.037	+0.001
			O.K.

10. Froude numbers

a. $v_E = 1.832$ m/s from Part 9.e.
b. $d = 0.063/(0.8 \times 1.832) = 0.043$ m.
c. $\boldsymbol{F} = 1.832/\sqrt{9.81 \times 0.043} = 2.82$
d. See Froude numbers on Figure 12-57

11. Comments

a. It was deemed desirable to carry some calculations to a greater number of significant places.
b. ***F*** can be increased in several ways, such as:

i. Increase the flowrate of water during pump-down. In this example, if the flow were increased 48% (about the maximum), the ***F*** would increase about 32%. However, as flow increases, it takes longer to dewater the trench. Of course, flow cannot exceed about 85% of the last pump's capacity.
ii. Improve the smoothness of the floor. For example, lining the floor with PVC reduces *n* to 0.009 and increases ***F*** by about 24%.
iii. Shorten the trench.
iv. Slope the trench floor to match the slope of the EGL beginning at the point where ***F*** becomes less than about 3.0. (It does no good to slope the floor upstream of such a point, because, as shown in Figure 12-57, the slope would have to be very steep.)
v. An incidental advantage of sloping the floor between the last two pumps is that the last pump intake can be lowered, thereby increasing its submergence during early stages of pump-down and increasing the discharge capacity of the last pump, because the surface vortex would not form so soon. The safety of the entire cleaning system is improved.
vi. It is not necessary for ***F*** to be as much as 3 at the last pump. As long as the velocity is above 1.4 m/s (4.5 ft/s), sludge and grit move reasonably rapidly (see Figure 12-36). An hydraulic jump at an ***F*** of 3 or more is, however, an extremely effective way to suspend solids and move them quickly.

Critique of Example 12-5

It is assumed that equilibrium prevails for the water surface shown in Figure 12-55. But it is doubtful whether equilibrium is reached when pumps are cycling rapidly. Furthermore, the S-shaped water surface curve would never be confined within the manhole. Instead, long backwater and tailwater curves would occur. It is likely that supercritical flow would occur first near the wet well and progress slowly up the approach pipe—with no positive assurance of reaching the manhole before the pump is stopped. Under such circumstances, dropping the invert only 48 mm may be risky. If the invert were dropped 92 mm (0.30 ft), the available velocity head would be about 318 + 92 = 0.41 m (1.34 ft). The resulting velocity would be 2.84 m/s (9.30 ft/s)—only 14% higher than the value in Table 12-2. As the sequent depth increases only one-third as much as the velocity increases, the depth would increase from 60% of the pipe diameter to 65% (still a conservative amount) with greater assurance of critical velocity in the manhole. If the crowns of the two pipes were at the same elevation, the velocity would approach about 3 m/s or 20% more than the nominal value in Table 12-2 and produce a sequent depth of about 67% of the pipe diameter. Such high velocities are quickly dissipated by friction. In conclusion, it seems wise to drop the invert by at least 100 mm (0.3 ft) to ensure supercritical flow at the exit from the manhole.

In calculations of the velocity increase in the manhole between sewer and approach pipe (Part E.3), the K-factor in the formula $h = Kv^2/2g$ is assumed to be 0.25, whereas in Example 29-1, K is assumed to be 0.10. There are no known data to support either assumption. Probably K lies between the two values and, perhaps, closer to 0.10. It should be noted that the choice of a K-factor is not of great practical significance, because the initial velocity in the approach pipe is insensitive to headloss (being a function of the square root of head) and the sequent depth is relatively insensitive to velocity (increasing only one-third as much as the velocity increases on a percentage basis.

The designer of a pumping station for C/S pumps must add the following to the concerns listed in Example 12-5:

- Storage volumes in both approach pipe and pump sump
- Setting elevations for starting and stopping pumps.

These tasks entail some tedious calculations, best done with computers. Note that the allowable active storage in the approach pipe depends on the influent flowrate, because the volume occupied by the influent flowing at supercritical velocity is not available for storage.

The trench might be made shallower, but consider that (1) the top of the trench must be positioned so that the cross-sectional waterway area above it is adequate to carry all of the inflow (at the water surface elevation, whatever it is) at an average velocity of less than 0.3 m/s (1 ft/s); (2) the requirement for adequate submergence of pump intakes (Equation 12-1) must not be compromised; and (3) there must be proper floor clearance for the pump intakes. These are the considerations that establish the elevation of the trench floor. Nevertheless, it is unwise to set the top of the trench at much less than 2 D above the intakes unless model studies indicate a shallower trench can be used.

The sluice gate in this example is poor. Flow through a circular segment results in large depths in the middle of the ramp and shallow depths near the edges. This condition is only partially overcome at the toe of the trench, and it results in excessive wave action ("rooster tails") that may interfere with the hydraulic jump. Discharge through a rectangular horizontal slot the full width of the ramp (as in Example 12-4) is preferred.

The radius of the ogee ramp curves are (at 0.75 m) rather small. The operators would be unable to start the cleaning routine with a higher water level than the mid-depth of the influent pipe. A radius of 1.25 m (as in Example 12-4) would improve the flexibility of operation for an increase in overall wet well length of only 0.4 m.

The calculations of velocity from the sluice gate to the last pump are only rough approximations. Greater accuracy can be obtained by dividing the sump into more segments. Without a computer program, such calculations are not justified unless Froude numbers obtained by the rough approximation are close to the limit.

For simplicity, the capacity of a single pump operating alone was taken to be 110 L/s. The capacity in a real problem must be the "runout" capacity found at the intersection of the pump and station H-Q curves, and the solution for Equation 12-3 must be based on the q at runout.

Round Sumps for Small Lift Stations

Round pump sumps for submersible pumps or self-priming pumps are popular for small flowrates and duplex pumps with motors of 3.7 to 15 kW (5 to 20 hp), although much larger pumps with motors up to 110 kW (150 hp) can be used. Most manufacturers' catalogs contain drawings of round sumps with duplex pumps in a flat-bottom sump—occasionally shown with corner fillets. The advantage of these sumps is their low cost. However, one overwhelming disadvantage is the odor from the deep, unstable sludge bank that develops and surrounds the pumps. A second disadvantage is the thick blanket of scum that develops, and a third is the free fall of water that cascades into the pool and may cause bubbles of air to enter the pumps. These disadvantages are eliminated in a hopper-bottom sump with a sloping-approach pipe discharging at LWL.

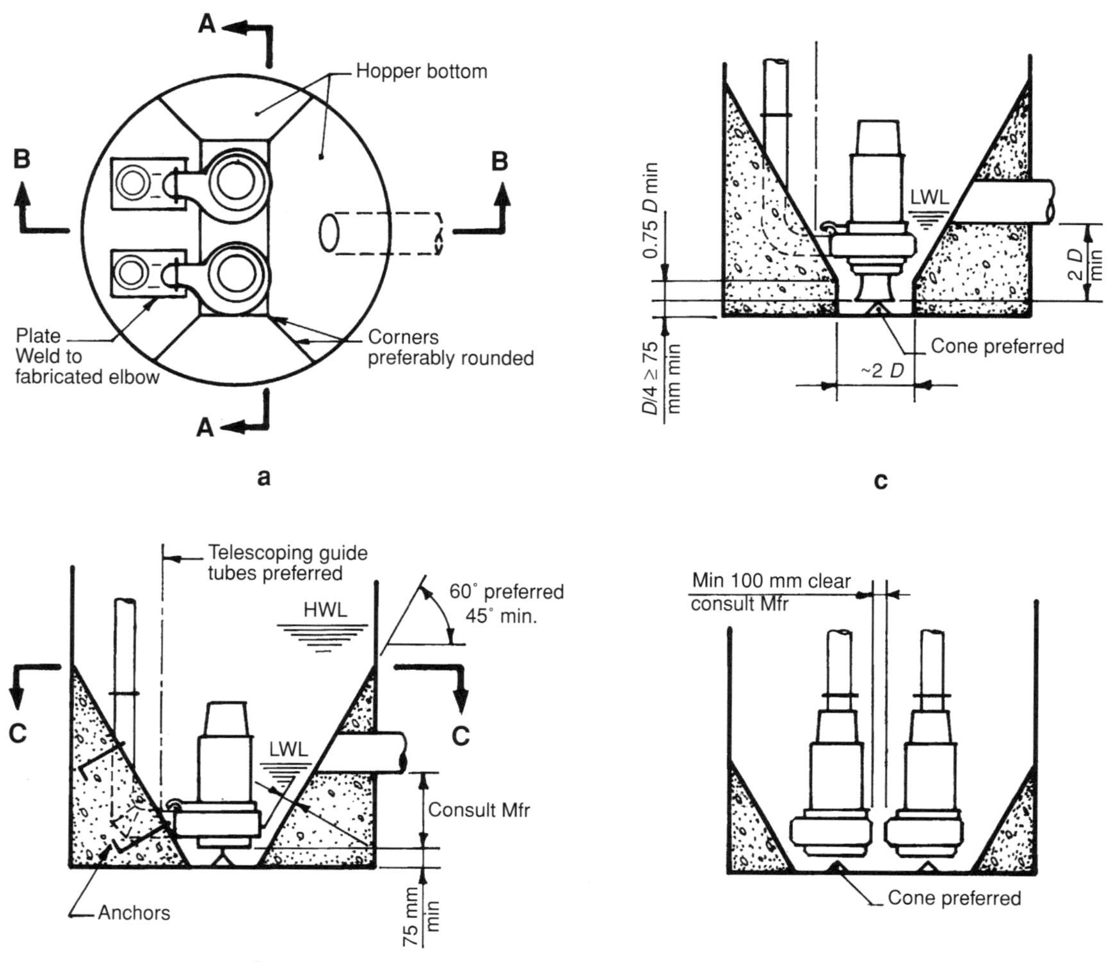

Figure 12-58. Duplex submersible pumps in round sump. (a) Section C–C; (b) Section B–B; (c) alternate Section B–B; (d) Section A–A.

Hopper-Bottom Pump Sumps

Hopper bottoms in round sumps (see Figures 12-58 and 12-59) prevent the accumulation of settleable solids, because the solids slide down the smooth, steep sides to the floor, and every time a pump is activated, the sludge is pumped out. Scum can be eliminated by a regular schedule of pumping until the pump loses prime (pump-down), thereby concentrating scum into a small area where a surface vortex sucks it into the pump. A PLC can be set for pump-down at regular intervals—say, once per day if the pump can reprime itself reliably.

The free water surface area at pump-down must be small if scum is to be completely discharged. As an example, the Black Diamond pumping station wet well (see Figure 17-11) does not hug the suction inlets closely enough to remove all the scum on the first pump-down. Note that pumps suck air and begin to lose prime when the submergence of the intake becomes much less than the intake diameter, *D*. Large hopper bottoms with pumps that lose suction when the water level falls to the top of the volute of submersible pumps are unlikely to be fully effective in removing scum. The smallest free water surface area at pump-down (and consequently, better and quicker cleaning) can be achieved by equipping submersible pumps with suction nozzles (a standard feature on some submersible pumps) set within a vertical-sided trench, as shown in Figure 12-58c.

For duplex or triplex pump installations, good cleaning can be obtained without pump-down (or

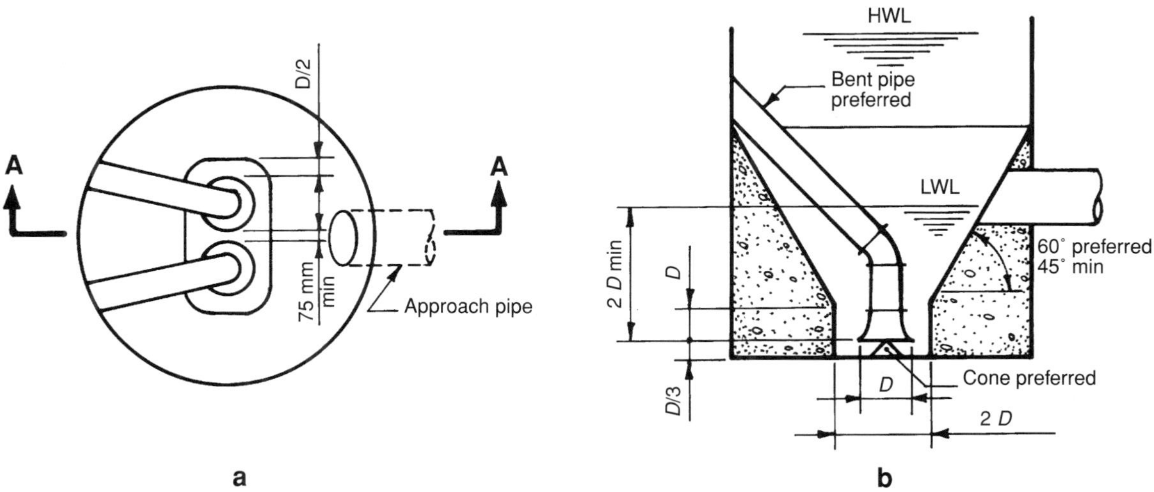

Figure 12-59. Sump for duplex, self-priming pumps. (a) Plan; (b) Section A–A.

with only partial pump-down) by operating the pump(s) while mixing the contents with either a mechanical mixer, a piping system such as shown in Figure 17-22, or a bypass in the pump that allows some of the pumped fluid to mix the contents. Mixers can be programmed to operate only a few minutes at the beginning of every pump start-stop cycle. If intakes are set as close together as shown in Figures 12-58 and 12-59, settleable solids cannot accumulate in significant amounts and removal of scum is the only reason for pump-down.

Sumps for Large Pumps

The most practical isolation valve for wastewater pump intakes is the eccentric plug type, but these valves become massive and expensive for pipes larger than 400 mm (16 in.). Solid-wedge gate valves have been used for this service, but they are less satisfactory and are not recommended. Furthermore, the larger pipes make the pump elevation above the dry well floor inconveniently high for the maintenance crew. The high pump pedestal exacerbates vibration and earthquake force resistance problems.

Draft Tube Designs

The foregoing considerations on size of pipe limits the size of pump for the configurations in Figures 12-19 through 12-22 to about 14 m^3/h (6000 gal/min). For larger pumps, consider draft tube designs such as those used for the Duwamish and Interbay Pumping Stations (Figures 17-4 to 17-9) for which an inexpensive sluice gate can be used for isolation. Such a configuration permits mounting the pump on a low housekeeping pad in the dry pit. The Duwamish Pumping Station has, incidentally, proven to be the best of the Seattle Metro stations for cleaning. Also consider using the FSI design shown in Figure 12-30.

Vortices are likely to form above a draft tube inlet unless a vertical vortex suppressor plate is installed in the mouth of the draft tube such as that shown in Figure 17-6. The forces acting on the plate are thought to be substantial, so use a thick (12- to 25-mm or $^1/_2$- to 1-in.) plate strongly anchored to the wall. Round the outboard edge so that rags will not be caught on it. For side wall entrances, the vortex suppressor can be mounted on the sluice gate.

12-8. References

1. Prosser, M. J. "The hydraulic design of pump sumps and intakes." British Hydromechanics Research Association, Cranfield, Bedford, England MK43 OAJ (July 1977).
2. *Hydraulic Institute Standards for Centrifugal, Rotary & Reciprocating Pumps,* 14th ed., Hydraulic Institute, Parsippany, NJ (1983).
3. ANSI/HI 1.1-1.5-1994, *Centrifugal Pumps for Nomenclature, Definitions, Application and Operation,* Hydraulic Institute, Parsippany, NJ (1994).
4. ANSI/HI 2.1-2.5-1994, *Vertical Pumps for Nomenclature, Definitions, Application and Operation,* Hydraulic Institute, Parsippany, NJ (1994).
5. Sanks, R. L., G. M. Jones, and C. E. Sweeney. "Designing self-cleaning wet wells for wastewater pumping," Conference Proceedings, American Society of Civil Engineers National Conference on Hydraulic

Engineering, Part 1, San Francisco, CA (1993), pp. 180–185.
6. Sanks, R. L., G. M. Jones, and C. E. Sweeney. "Self-cleaning wet wells: definitions and design for wastewater pumping," Proceedings of the International Conference on Pipeline Infrastructure II, San Antonio, TX. American Society of Civil Engineers, New York, NY (1993) pp. 102–114.
7. Sanks, R. L., G. M. Jones, and C. E. Sweeney. "Improvements in pump intake basin design." EPA 600/R-95/041, RREL-CI. Order No. PB95-188090, National Technical Information Service, 5285 Port Royal Road, Springfield, VA 22161.
8. Sanks, R. L., G. M. Jones, and C. E. Sweeney. "Better sumps for pumps," Engineering and Construction Conference, Proceedings, American Water Works Assoc., Denver, CO (March 1996), pp. 393–398.
9. Dicmas, J. L. *Vertical Turbine, Mixed Flow, & Propeller Pumps*. McGraw-Hill, New York, NY (1987).
10. Hecker, G. E., President, Alden Research Laboratory, Inc., Holden, MA (1996).
11. Kennon, H. H. "Design guidance for rectangular sumps of small pumping stations with vertical pumps and ponded approaches," Engineering Technical Letter No. 1110-2-313, Dept. of the Army, U.S. Army Corps of Engineers, Washington, DC (29 April 1999).
12. Triplett, G. R., B. P. Fletcher, and J. L. Grace. "Pumping station inflow—discharge hydraulics, generalized pump sump research study," Technical Report HL-88-2, Department of the Army, Waterways Experiment Station, Corps of Engineers, Vicksburg, MS (February 1988).
13. Karassik, I. J. , W. C. Krutzsch, W. H. Fraser, and J. P. Messina. *Pump Handbook*, 2nd ed., McGraw-Hill, New York (1986).
14. Fletcher, B. P. "Formed suction intake approach appurtenance geometry," Waterways Experiment Station, Corps of Engineers, 3909 Halls Ferry Road, Vicksburg, MS 39180-6199.
15. Fletcher, B. P. Private communication, July 21, 1995.
16. Fletcher, B. P. "Cypress Avenue pumping station: Hydraulic model investigations," Technical Report HL-94-9, U.S. Army Corps of Engineers Waterways Experiment Station, 3909 Halls Ferry Road, Vicksburg, MS 39180-6199.
17. Falvey, H. T. *Air–Water Flow in Hydraulic Structures,* U.S. Dept. of the Interior, Bureau of Reclamation, Engineering Monograph 41, Superintendent of Documents, Washington, DC (1980).
18. Beaty, V. J. Vice President and Chief Engineer (now retired), Fairbanks Morse Pump Corp., Kansas City, KS.
19. Linabond, Inc., 6922 Hollywood Boulevard, Ste. 303, Los Angeles, CA 90028.
20. Goldschmidt, G. "Mixing fixed-speed pumps to variable flows," *Journal Water Pollution Control Federation,* 50:1733–1741 (July 1978).
21. Wheeler, W. Discussion of "Mixing fixed-speed pumps to variable flows," *Journal Water Pollution Control Federation,* 51:2959–2960 (December 1979).
22. ENSR. "City of Portland, OR, Columbia Boulevard Wet Weather Treatment Facility Influent Pump Station Project Hydraulic Model Study." Prepared for Damon S. Williams Associates, L.L.C., Portland, OR. Document 2181-002-410, ENSR, Redmond, WA (1997).
23. Williams D. S., D. P. Sowles, S. P. Reddy, and T. C. Demlow. "Computer and physical modeling of a large hybrid CSO sanitary wastewater pump station," presented at WEFTEC '97, Chicago, IL (October 1997).
24. Linabond, Inc., 12960 Bradley Ave., Sylman, CA 91342. Phone (818) 362-7373; Fax (818) 362-5757; Web www.linkbond.com.
25. Wheeler, W. PARTFULL®. For a free copy of this computer program with instructions, send a formatted 1.4 MB, 3-1/2-in diskette and a stamped, self-addressed mailer to 683 Limekiln Road, Doylestown, PA 18906-2335.
26. *Design of Small Dams,* Revised Reprint, U.S. Department of the Interior, Bureau of Reclamation. US. Government Printing Office, Washington DC (1977).
27. Chow, V. T. *Open Channel Hydraulics*, Classic Textbook Reissue, McGraw-Hill, New York (1959).
28. Escritt, L. B. *Sewerage and Sewage Treatment: International Practice*, edited and revised by W. D. Haworth, John Wiley and Sons, New York (1984).

Chapter 13

Electric Motors

PAUL C. LEACH

CONTRIBUTORS
Robert S. Benfell
Harrison C. Bicknell
Donald R. Bjork
Victor G. Gerez
Mayo Gottliebson
Stanley S. Hong
Philip A. Huff
Richard L. Nailen
Stephen H. Palac
M. LeRoy Patterson
Richard N. Skeehan
Patricia A. Trager
Frank A. Woodbury

The focus of this chapter is the practical selection (from the myriad of special features and characteristics available) of the proper driver in a manner that will result in an economical, dependable drive unit. The discussion of motors includes an elementary explanation of the several types commonly used as pump drivers. Refer to standard texts on electrical fundamentals and alternating current machines for an in-depth theory of operation.

Before designing the electrical system for driving the pumps, it is important to be aware of choices, restrictions, and requirements, such as (1) the various motor characteristics available both as standard and as specialty or custom products, (2) the costs of the various types and configurations of suitable motors, and (3) the restrictions resulting from station design criteria or earlier decisions. For example, in rebuilding a station, there are usually severe restrictions on space—sometimes for the motor and pump combination and usually for motor controls.

References to a code or standard are given in abbreviated form (such as NEC for National Electrical Code). See Appendix E for the names of the abbreviations and Appendix F for the addresses of publishers.

13-1. General

Some of the definitions (given here for convenience) are shortened, simplified versions of the definitions in ANSI/IEEE Standard 100. For more definitions, see Chapter 2 (which also contains abbreviations), Chapter 8, and the latest edition of the NEC.

Definitions

Adjustable-speed drive: A drive designed so that speed can be varied throughout a considerable range (1) independently of the load and (2) as regulated by some external signal (see variable speed).

Armature: The member of an electric machine in which an alternating voltage is generated by virtue of relative motion with respect to a magnetic field. The armature is the stator in a squirrel-cage motor and the rotor in a dc motor.

Brush: A conductor (usually made of carbon) that maintains an electrical connection between the stationary and moving parts of a machine.

Collector ring (slip ring): A metal ring mounted on an electric machine that conducts current into or out of the rotating member through stationary brushes.

Commutator: A cylindrical ring or disc assembly of conductors, individually insulated in a supporting structure with an exposed surface for contact with current-collecting brushes. In effect, a commutator is a switching mechanism between the dc of the brushes and the ac in the armature of a dc machine.

Dripproof enclosure: An open machine with ventilating openings constructed so that drops of liquid or solid particles falling on the machine at any angle not greater than 15° from the vertical cannot enter the machine.

Explosionproof enclosure: An enclosure designed to withstand an explosion of a specified gas that may occur within it and to prevent the ignition of the same gas surrounding the enclosure. It must operate at a temperature that prevents the surrounding flammable atmosphere from being ignited.

Frame mount: A supporting structure that connects the motor to the pump with an intervening shaft coupling.

Framework (frame): The stationary supporting structure that contains the stator windings and the rotor.

Induction motor: An alternating current motor in which a primary winding (usually on the stator) is connected to the power source and a polyphase secondary winding or squirrel-cage secondary winding (usually on the rotor) carries induced current.

Inrush: The surge of electricity into a motor at rest when it is energized. The surge, which diminishes to the running load current at full speed in approximately 5 to 30 s, is typically about 6 times the full-load current, but it may be as much as 10 in some motors or it may be reduced in "soft starts" to 1.5 or even 1.0 times the full-load current.

Power factor: True power divided by apparent power. Lag of current behind voltage in inductive ac circuits reduces true power from VA to $V \times A \times \cos \theta$, where V is volts, A is amperes, and θ is the lag angle (see Equation 8-3).

Round (cylindrical) rotor: A cylindrical rotor in which the coil sides of the windings are contained in axial slots.

Salient pole: A field pole that projects from the yoke or hub toward the primary winding core. Salient pole and round (cylindrical) rotors ordinarily are used to describe the rotor design of synchronous motors; i.e., a salient-pole synchronous motor or a round- (cylindrical-) rotor synchronous motor.

Service factor (SF): A multiplier (applied to the rated power) that indicates the built-in capability of carrying a continuous load greater than the nameplate rating (but only at the rated voltage and frequency). Standard service factors are 1.0 and 1.15.

Slip: The difference between the rotational speed of the magnetic field in an induction machine and the rotor. An induction motor develops its ability to drive a load by virtue of its slip, whereas a synchronous motor operates at zero slip.

Soft start: A type of motor starting in which the inrush is greatly reduced. Soft start is generally applied to solid-state starting and acceleration control systems.

Splashproof enclosure: An open motor with the ventilating openings constructed so that drops or particles falling or coming toward it at any angle not greater than 100° downward from the vertical cannot enter. The splashproof enclosure is no longer generally available. Weather-protected Type I is an alternative enclosure type.

Stator: The portion of a machine that includes and supports the stationary portions of the magnetic circuit and the associated windings and leads. It may include a frame or shell, winding supports, ventilation circuits, coolers, and temperature detectors. A base is not usually considered to be part of the stator.

Submersible enclosure: An enclosure constructed so as to operate successfully when submerged in water under specified conditions of pressure and time.

Synchronous motor: A motor that has field poles excited by direct current so that the speed of the rotor equals that of the rotating magnetic flux.

Synchronous speed: The rotational speed of the magnetic field in an electric motor and, in synchronous motors, the rotational speed of the shaft and rotor.

Torque, accelerating: The difference between the torque produced within the motor and that required by the driven load at any particular shaft speed.

Torque, full-load: The torque required to produce the rated motor horsepower at the rated speed.

Torque, locked-rotor: The torque produced with the motor at a standstill but energized at the rated voltage. Locked-rotor torque is also called "starting torque." At the instant power is applied to an induction motor at rest, the motor is analogous to a transformer with its secondary short-circuited through a very low impedance.

Totally enclosed (TE) motor: An integral enclosure constructed so that, although not necessarily airtight, the enclosed air has no deliberate connection with the external air except as required for draining and breathing.

Totally enclosed, fan-cooled (TEFC) motor: A totally enclosed motor equipped for exterior cooling by means of fans integral with the motor but external to the enclosing parts.

Totally enclosed, nonventilated (TENV) motor: A totally enclosed motor that is not equipped for cooling by means external to the enclosing parts.

Totally enclosed, pipe-ventilated motor: A totally enclosed motor except for openings arranged so that inlet and outlet ducts or pipes may be connected to them for the admission and discharge of the ventilating air. This air may be circulated by means that are (1) integral with the machine or (2) external to and not part of the machine. The latter machines are called "separately ventilated" or "forced-ventilated" machines.

Variable-speed drive: A drive designed so that its speed varies through a considerable range as a function of load (see adjustable-speed drive).

Weather-protected (WP) machine: A guarded machine with the ventilating passages designed to minimize the entrance of rain, snow, and airborne particles to the electrical parts. There are two types, I and II. Type II is designed for more severe applications.

Wound-rotor motor: An induction motor with the rotor windings connected to collector rings. The brush terminals may be either short-circuited or closed through adjustable circuits.

Standards and Codes

The most authoritative information on standards relating to electric motors is NEMA MG-1. The Institute of Electrical and Electronics Engineers provides many sources of reference material in the *IEEE Standards* and in various technical articles in *IEEE Transactions on Industry Applications*. These references are recommended for in-depth information on motors, new developments in research, and motor manufacturing. Many manufacturers' catalogs and special brochures provide useful information on motor characteristics and pricing.

If motor characteristics are selected from a particular standard, there is no assurance that any manufacturer will build the motor as specified because each manufacturer has its own "standard" models and modifications (called "adders"). During early stages of design, therefore, check with several manufacturers whose representatives can supply current information.

13-2. Applications of Motors

Electric motors are the most frequently used drivers in pumping stations, primarily because of their versatility, compactness, and low maintenance. The most common machine is the polyphase squirrel-cage induction motor; these motors range in size from less than one to several thousand horsepower. In sizes larger than about 375 kW (500 hp), a wound-rotor motor (for variable-speed pumping) or a synchronous motor is sometimes more economical, although the controls are more complex. Induction motors up to about 600 kW (800 hp) are usually used for adjustable-speed drives, but larger drives tend to be more economical with a wound-rotor or a synchronous motor. Large squirrel-cage motors, however, have high efficiencies and power factors that make the operating costs approach those of synchronous motors without the high capital cost.

Almost all pumping stations also require electric motors to drive auxiliary equipment. These auxiliaries can range from fractional kilowatt or horsepower ratings up to 30 or 40 kW (40 or 50 hp) in the larger stations. The advantages of electric motors are summarized in Table 13-1.

Ambient Conditions

Electric motors are usually comparable in size to the pump itself—particularly in the small- to medium-sized stations. Therefore, the use of electric motors as drivers is not a major problem in pumping station space allotment. Enclosures for electric motors can be selected to suit a wide range of environments: (1) totally submerged, (2) humid, (3) explosive atmosphere, (4) clean and dry areas indoors, or (5) out-of-doors directly in intense heat, rain, and freezing temperatures. Although different environments require different enclosures, the basic operating characteristics are the same.

Mountings

Electric motors may be mounted in many ways. The motor can be close-coupled to the pump in either a horizontal or a vertical arrangement in which the pump impeller is fastened directly to the shaft of the motor. Wherever possible, however, avoid configurations that

Table 13-1. Electric Motors as Pump Drivers

Advantages	Disadvantages
Compactness and minimum space requirements Usually less expensive for installed cost than engines (engines may be more economical if electricity is costly and fuel is cheap or if inertia is needed for water hammer control) Reliable Usually easily removed for major maintenance or replacement Available in wide range of sizes Insulation systems available to suit variety of ambient temperature and environmental conditions A number of mounting configurations available A number of enclosures available to suit physical conditions	Subject to damage by flooding (most motors) High starting currents for most types Subject to utility outages Standby generation may be required for the full capacity of the system Control equipment and switchgear sometimes require considerable space Protection systems may be complex and sometimes expensive Most applications use purchased electrical power where sometimes engine fuels may be less costly Subject to possible overheating due to (1) nonsinusoidal supply as in adjustable-frequency drives using rectifier/inverter power sources (the harmonic component of current creates additional I^2R and core losses in the motor); (2) unbalanced phase voltages [1, 2] (see also NEMA MG-1); (3) overvoltage that causes overexcitation of magnetic materials and a consequent increase of core losses; (4) short cycling (too-frequent starting of motor by faulty ON–OFF controls or by careless design of pumping station or its control system); (5) excessive pump load conditions

lead to high maintenance or difficult replacement problems. Close-coupled pumps are more difficult to maintain, because most units must be disassembled to service the seals. Many units utilizing close-coupled designs have shafts and bearings that are inadequate for severe service in sewage pumping. If the motor is separated from the pump by a spacer or by a frame mount (which allows a coupling to be installed), seals can be changed without moving either the motor or the pump. Such arrangements are used in both horizontal and vertical configurations. Furthermore, for ease in maintenance the electrical system should not have to be disturbed to remove a pump, nor should the pump or piping have to be disturbed to remove a motor.

It is common to support the shaft and impellers for a deep well pump from a thrust bearing located atop a vertically mounted motor. Some of these motors have two thrust bearings. It is also common to mount a deep well turbine pump above a small-diameter motor and submerge the entire unit in the well.

In some sewage pumping stations, a vertical pump is located in a dry well. It is good practice to place the motor and other electrical equipment (such as starters and panelboards) on an operating floor that is

- above grade and safe if the pumping station is flooded;
- accessible only from the outside with no direct access to the wet well and completely sealed from the wet well to prevent contamination with sewer gases; and
- above the 100-yr flood level.

Any motors and electrical devices in locations classified as Class 1, Division 1, by NEC Subarticle 500-3a must be explosionproof. Motors and electrical devices in locations classified as Class 1, Division 2 must be rated for this classification and must be non-sparking, so there can be no open switches or open contacts; hence, wound-rotor motors with standard open enclosures are prohibited, although polyphase open or totally enclosed motors with no switches are permissible. Neither type of hazardous location should be allowed to occur in the design of new facilities. If an existing facility containing a hazardous area is to be remodeled, it is necessary to consult the proper inspection authority in the formative stage of design and obtain written agreement that the designer's interpretation of the classification is correct. Explosionproof motors and other electrical equipment are very expensive, but some savings are possible in the overall construction costs of equipment in Class 1, Division 2, locations compared to Class 1, Division 1, locations. There may be some savings in lighting costs, little or none in wiring methods, and a significant amount if there are a number of motors.

If the station is safe from flooding, an alternative arrangement for a vertical pump in a dry well is an open dripproof motor frame-mounted above the pump. This arrangement avoids a long shaft and intermediate supports. If flooding is a possibility, submersible motors can be installed, but the problem of ambient air cooling for normal operating conditions usually requires substantial derating of the motors. Note that the ordinary pump cannot be operated while submerged without damage to seals and bearings.

Still another alternative is the close-coupled submersible pump and motor mounted on slide rails in the wet well. These units are increasing in popularity because of the overall low initial cost due to savings in the superstructure. The motors, however, are subject to loss of seals, penetration of moisture, and overheating if the pumping level drops below the motor. Maintenance costs (1) depend on the size and operational procedures of the utility, (2) depend on the number of such pumping units, and (3) may be either less or more than those for the wet well–dry well configuration.

A careful study of the alternatives should be made before any of these configurations is selected.

13-3. Fundamentals

The discussion in this section is elementary, and only a hint of the complexity of electrical machinery is given. For a more extensive exposition, see Richardson [3, 4], or other good texts.

When a conductor is moved in a magnetic field, electricity (voltage) is induced in the conductor as shown in Figure 8-1. Because the voltage is able to force current through the conductor into a resistance (or other load), work is done, so energy is required to move the conductor in a generator. If a conductor is formed into a closed rectangular loop and mounted on end pivots so that it is free to rotate in a magnetic field, electrical energy can be converted back into mechanical energy, and the coil becomes a motor. In dc motors, a small part of the electrical input is used to create a strong magnetic field through which the conductor moves. When electrical connections are made to the conductor loop, there is a reacting force between the magnetic field of the conductor itself and the stationary magnetic field. Because the conductor is fastened to a shaft, it can move only by rotating and thus it becomes a motor, as shown in Figure 13-1. By means of a segmented commutator a number of coils are fed with current in a manner that produces a nearly constant rotational force.

The commutator type of motor with a series-connected field can operate on alternating current. The stator, as well as the rotating armature, must be laminated to reduce internal power losses. A constant positive torque is produced, because the current in the stator and armature coils reverse together with the alternations of the supply current. This kind of motor typically powers small hand tools.

Squirrel-Cage Induction Motors

The squirrel-cage induction motor operates on alternating current and has no physical electrical connections to the rotor. The rotor is constructed of steel laminations that have insulated slots for the conductors. It is common to use either copper or aluminum rotor bars cast in place in the slots with integrally cast end rings in motors less than 250 hp. Tangential bars, however, are used in large motors. There are many designs that provide the shaft-driven ventilation fan in the same integral casting, and one is shown in Figure 13-2.

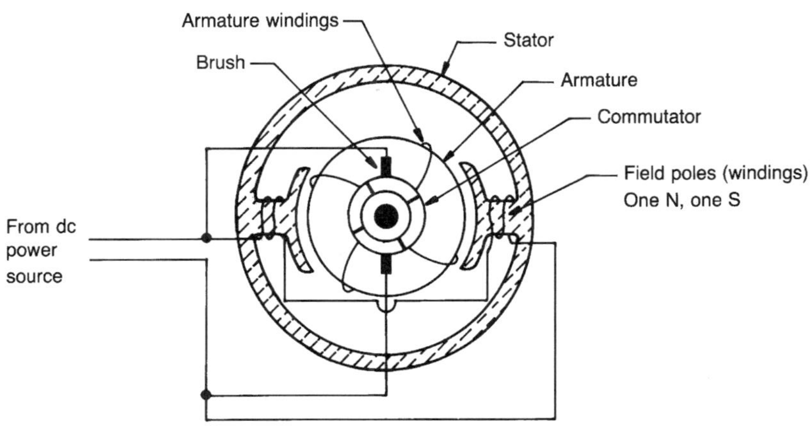

Figure 13-1. Schematic diagram of a direct-current motor.

Figure 13-2. Ventilation fan in a squirrel-cage motor. Courtesy of Louis Allis.

The stationary part of the motor (stator) is also made of laminations of special low-loss steel and is provided with slots to accommodate the stator windings. A small air gap for mechanical clearance separates the rotor and stator. The stator windings are arranged so that when three-phase voltage is applied, the magnetic field advances around the periphery at a rate proportional to the frequency of the ac voltage (see Figure 13-3). This advancing field induces voltages (and thus electric currents) in the rotor conductors. The magnetic field due to these currents reacts with the revolving magnetic field of the stator and produces torque on the rotor. At standstill, or start, the induction motor is identical in operation to a transformer with its secondary winding (see Figure 8-13) shorted together or connected through a very low impedance. But as rotation begins, the relative motion between the stator field and the rotor conductors diminishes, and the induced voltage and frequency decrease proportionately. If this machine could rotate at synchronous speed there would be no voltage induced in the rotor conductors, because there would be no relative motion between the field and conductors and, hence, no torque to maintain that speed. In practical motors, the difference in rotational speeds between the field and the rotor (slip) can be as low as 2% at the rated load, but 3 to 5% is common in motors smaller than 75 kW (100 hp). Higher slip motors are manufactured for special applications. The amount of slip depends on the load and on the rotor resistance.

When a squirrel-cage motor is started, the inrush current is 5 to 10 (typically, about 6) times the full-load operating current. Some approximate inrush values are given in NEC Table 430-151 for an assumed locked rotor current of about 6 times the NEC-listed motor full-load current values of the variously rated motors.

Wound-Rotor Motors

If the squirrel-cage rotor is replaced by a rotor with windings installed in the slots and if the winding ends are brought out through slip rings, then the resistance of the rotor circuit can be varied by means of switching arrangements external to the motor (see Figures 13-4 and 13-5). The rotor resistance is reflected to the primary (stator) by transformer action, which affects the motor current. Adding appreciable resistance in the rotor circuit at start is desirable to limit the inrush current to the motor while still producing a reasonably high torque, and it is practical to reduce inrush to low values such as 125% of full-load current.

Synchronous Motors

Synchronous motors represent the third type of ac motor widely used for large pumps (see Figures 13-6 and 13-7). A synchronous motor requires a rotating

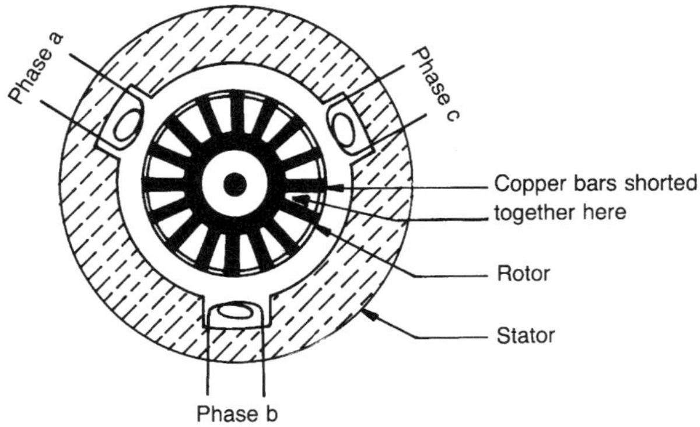

Figure 13-3. Schematic diagram of a squirrel-cage induction motor.

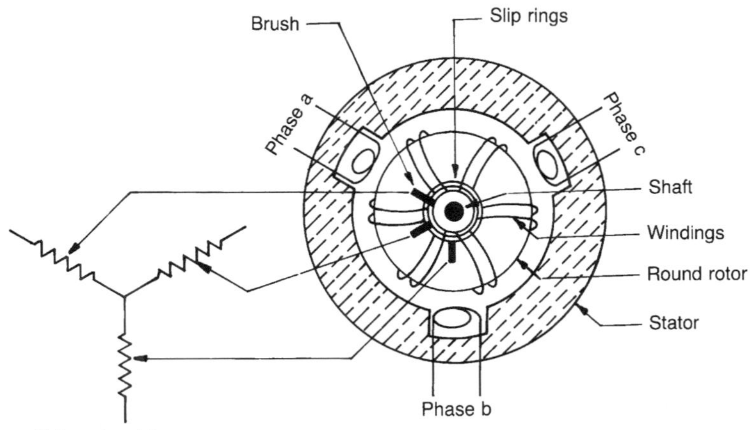

Figure 13-4. Schematic diagram of a wound-rotor induction motor.

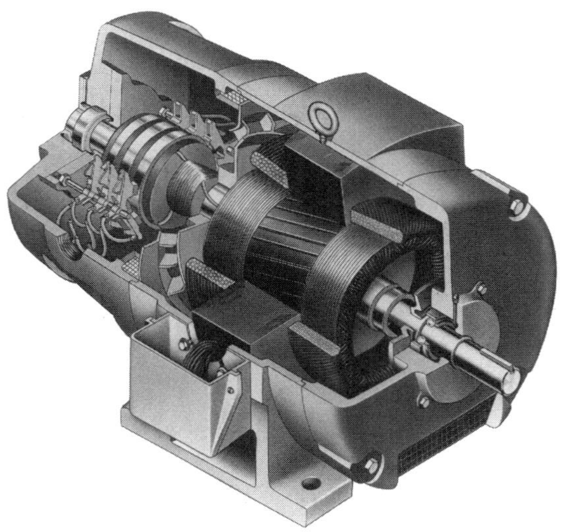

Figure 13-5. A wound-rotor induction motor. Courtesy of Marathon Electric Manufacturing Co.

field of fixed polarity, which is produced by a permanent magnet (in very small machines) or by a dc electromagnet (in larger machines). The direct current is fed to the rotor through slip rings and is usually derived from external rectifiers. In a more modern version of the synchronous motor (called a "brushless synchronous motor"), a small externally powered exciter field is used to control a shaft-mounted exciter–alternator. Integrally mounted diode rectifiers are installed to provide the direct current to the main motor field. The stator is then connected to the ac power source through contactors and switchgear.

In a typical salient-pole synchronous induction motor, squirrel-cage type windings are provided in slots in the pole faces (Figure 13-7). These windings serve two purposes: (1) to produce the starting torque required to accelerate the rotor from zero speed to near synchronous speed and (2) after synchronization, to produce a stabilizing torque that prevents "hunting" of the rotor, which is sometimes the result of changing or pulsating loads. The pole face winding is termed the "damper" (or "amortisseur") winding. A similar winding is also used in the synchronous alternator for stability purposes.

13-4. Types of Motors for Pump Drivers

Only the most common types of motors (squirrel-cage, wound-rotor, and synchronous) used in pumping stations are discussed.

Squirrel-Cage Induction Motors

Of the three types of motors, the squirrel-cage induction motor is by far the most commonly applied driver for pumping service because of its (1) simplicity, (2) ruggedness, (3) low maintenance, (4) relatively high efficiency (when well loaded), (5) low first cost, and (6) availability in many standard forms. Because the speed of rotation of the shaft of an induction motor varies only a few percentage points from no load to full load, it is considered a constant-speed driver.

The principal disadvantages of the standard induction motor are (1) the high starting current (usually about 600% of rated full-load current), (2) the low efficiency at light loads, and (3) the poor power factor, which worsens as the load is reduced. Because each start causes a voltage dip on the line and lights flicker, the high starting current (inrush) may be an annoying problem where a large motor is to be started frequently in a residential area.

Figure 13-6. Synchronous motor (600 hp) and eddy-current coupling for a raw sewage influent pump installed in 1962 at Renton, King County (Washington) Department of Metropolitan Services. Photograph by R. L. Niclas.

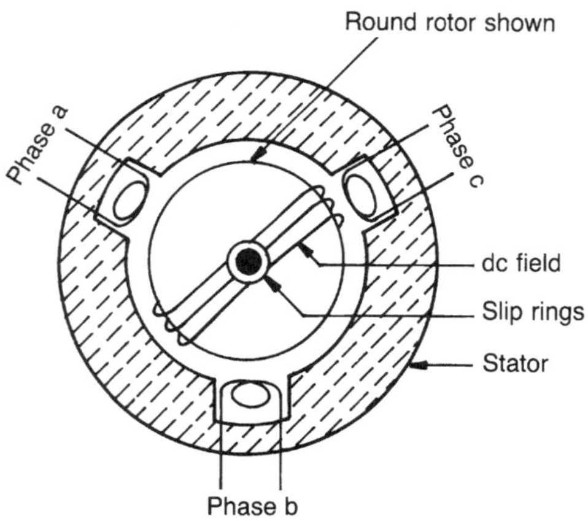

Figure 13-7. Schematic diagram of a synchronous motor.

Reduced-Voltage Starting

There are various forms of reduced-voltage starters available for motors. Reduced-voltage starters decrease the motor starting current during at least a portion of the acceleration time. Because the characteristics vary considerably, check with the utility for their approval of the proposed units before an application is made for a particular drive. Some types result in very high transition peak currents, which may cause lights to flicker momentarily in the immediate vicinity.

Part-winding starters have been widely applied in pumping stations in the past, but the motor must have two windings designed so that one winding is used for starting and both used for running. Of the older starting systems, the best unit has been the closed-transition, autotransformer, reduced-voltage starter. It is only slightly more expensive than a system without an autotransformer, but it takes up considerably more motor-control center space. It usually has three voltage taps on the autotransformer to allow for field adjustment of the voltage applied to the motor.

The advent of the solid-state, soft-start motor acceleration controller has made it possible to program different modes of acceleration control to match the requirements of the power system and load. Typically, a motor starting contactor is added to the solid-state unit to give complete electrical isolation when the motor is turned off. The contactor also provides motor-overload relays for running protection. The soft-start system can control the inrush current on the basis of a time limit, a current limit, or more complex programs. The costs of the various starting systems should be compared at the time of design. Roughly, in the 125-hp range, a NEMA 5 across-the-line motor starter costs about the same as a typical TEFC induction motor. The reduced-voltage starters (except, perhaps, for the wye–delta type) are about twice as expensive as the motor.

Efficiency

At the rated load, the efficiency of a standard squirrel-cage induction motor varies from about 85% at 3.7 kW (5 hp) to about 91% at 75 kW (100 hp). High-efficiency motors, which are now available (at higher cost) from most manufacturers, are made by using thinner laminations in the core, modifying the shape of the rotor bars, and using more iron. The efficiencies of these motors vary from about 93% at 3.7 kW (5 hp) to about 96% at 75 kW (100 hp), but may differ by several percentage points depending on the model and manufacturer. The life-cycle savings in power costs make high-efficiency motors cost effective in ratings from 1 to 150 kW (1 to 200 hp). Larger standard motors have efficiencies nearly as high as the high-efficiency type, so there is little difference. In fractional ratings, "high-efficiency" motors are not much of an improvement.

Wound-Rotor Induction Motors

Wound-rotor motors were common in the past when power systems were relatively weak, because the controllable starting and acceleration current were desirable. Today's systems can usually start squirrel-cage motors, but wound-rotor motors may still be advantageous when very large motors (more than 750 kW or 1000 hp) are required. The production of wound-rotor motors has been on the decline for years and will probably continue to decline. Some manufacturers no longer make them, and reduced demand has made them even more expensive.

The wound-rotor motor is now used principally where high starting torque requirements must be met. By providing a starter arrangement with a number of steps (each with progressively less resistance in the secondary windings), a wide variety of starting and accelerating currents and torques can be attained. The last step of the starter shorts out all external resistance, and this results in slip that is only slightly higher than that of an equivalent squirrel-cage motor.

Although the typical starting torque required for pumps is usually low, the wound-rotor motor can still be useful in two ways: (1) with the type of starter described, the starting current can be kept below a predetermined maximum value (such as 125% of full-load current), and (2) by leaving resistance in the external circuit, the starter can be used as a speed controller because the speed in any step varies considerably with load changes. An adjustable-speed drive is obtained merely by making the control system compensate for speed variations caused by the loading of the motor (see subsection "Liquid Rheostats" in Section 15-11).

Synchronous Motors

The availability of synchronous motors has been affected by a reduction in demand due to the price premiums required. Considering power factor, efficiency, and the more expensive synchronous motor starter, the life-cycle cost is usually less for squirrel-cage motors of up to about 750 kW (1000 hp). Consequently, synchronous motors are rarely applied in drive systems smaller than about 370 kVA (500 hp) because they are not cost effective in small ratings. Because the synchronous motor starts like an induction motor and the same problems of high inrush during the acceleration occur, the same solutions must be undertaken if high starting current is a problem.

The unique characteristic of a synchronous motor is that it runs at a constant speed that is determined by the supply frequency. Also, if the field strength (amperes) is externally controlled, the power factor of the line current of the motor with respect to the supply voltage can be varied over a wide range from lagging to unity to leading. Machines can be specified with unity, 0.8, or 0.6 leading power factors, for which the suitable machine current rating is provided by the manufacturer's design. In large pumping stations, the synchronous motor can sometimes be advantageous due to its ability to be operated with a leading power factor, which thus obviates the need for capacitor installation or at least reduces the size of the capacitor bank that may be required.

Multispeed Motors

Operating a pump at two different speeds allows the same unit to be used during dry weather flow as well as during normal- and high-flow periods. The 2:1 ratio that can be obtained by reconnecting a one-winding "variable torque" motor from a wye to a delta is a much higher speed ratio than can be utilized with centrifugal pumps. Reasonable speed ratios can normally be obtained with a two-speed, two-winding arrangement. For example, one winding may provide a four-pole speed of 1800 rev/min while the other provides a six-pole speed of 1200 rev/min with a speed ratio of 3:2. Closer ratios may be obtained by using a lower base speed, such as 1200 or 900 rev/min.

Constant- and Adjustable-Speed Systems

There are many ways to provide speed-control devices for constant-speed motors, including:

- Direct current drivers with field control
- Wound-rotor induction motors with rotor circuit resistance control (controlled slip)
- Variable-pitch sheaves
- Hydraulic couplings that produce controlled slip
- Eddy-current couplings that provide controlled slip within the coupling
- Hydraulic pump/motor combinations
- Adjustable-frequency drives.

All of the slip-producing devices, except the expensive wound-rotor slip-recovery system, result in considerable energy loss. In contrast, the energy losses in the adjustable-frequency and direct current drives are low. Reducing the speed of the pump by whatever means, however, often entails a loss of hydraulic efficiency unless the pump is selected according to the principles explained in Chapter 15, in which variable- and adjustable-speed systems are discussed at length.

13-5. Characteristics of Squirrel-Cage Induction Motors

A number of designs that have significantly differing starting current and torque characteristics are listed by NEMA. The two most useful in pumping stations are Design B and Design D motors. Design B is termed "normal starting torque, normal starting current, and normal slip." This is the most commonly applied motor design and is generally used for driving centrifugal pumps, fans, blowers, compressors with unloading start controls, and machine tools. The normal slip at full load is approximately 3 to 5%, and the starting current varies from 500 to 600%.

Design D is termed "high starting torque, normal starting current, and high slip." The rotor in Design D has a higher resistance than the rotor in Design B. The high resistance in the rotor produces high starting torque, but it also causes high slip (5 to 8% or even 8 to 13%) at full load; the inrush current is the same as for Design B. High slip in Design D causes greater internal loss and, therefore, less efficiency. Nevertheless, hard-starting loads (such as reciprocating pumps and progressive cavity pumps) may require a high-slip motor for satisfactory starting. A typical set of torque curves for various standard designs is shown in Figure 13-8. Manufacturers can frequently provide special designs to match a particular loading characteristic, but that may mean delayed delivery and troublesome replacement problems.

13-6. Motor Speed

The speed of the rotor depends on the speed of rotation of the magnetic field produced by the stator currents. The conductors can be wound in the stator slots to produce any desired number of resultant north-south pole pairs around the periphery of the stator. One pair of poles in a three-phase winding is illustrated in Section 13-3. One cycle of sinusoidal alternating current represents 360 electrical degrees and during this time the magnetic field rotates 360 mechanical degrees or completely around the stator. Because 60 Hz ac produces 60 revolutions of the field per second (or 3600 rev/min), a one-pole pair machine (commonly termed a two-pole machine) rotates 3600 rev/min at zero slip. An equation for synchronous speed in SI units is

$$\omega = \frac{2\pi(Hz)}{n} \tag{13-1a}$$

where ω is radians per second, Hz is current frequency in cycles per second, and n is the number of poles (which is twice the number of pole pairs). In U.S. customary units

$$\omega = \frac{120(Hz)}{n} \tag{13-1b}$$

where ω is revolutions per minute and the other terms are as defined above.

Preferred motor speeds for pumping station equipment depend on the characteristics of the driven equipment as well as the overall costs—in particular, the life-cycle costs of the various practical alternative speeds. As a general recommendation, the following guidelines apply to pumping station equipment:

- Blowers and fans: use 1800- or 3600-rev/min motors with speed increaser gears when higher speeds are required. Note, however, that high speed produces a high noise level.
- Water, sewage, or sludge pumping: use 900-, 1200-, or 1800-rev/min motors for medium to small pumps

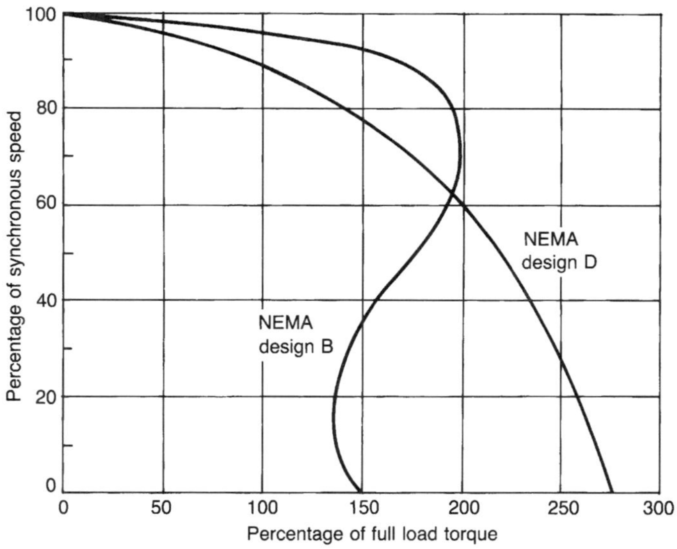

Figure 13-8. Speed-torque curves of NEMA design motors.

and speeds up to 600 rev/min or less for special, low-head, large pumps.

- Progressive cavity screw pumps: use 900-rev/min motors (or less) and consider connecting to the pump through a belt or gear speed reducer.

Note that European motors are designed for 50 Hz; operating these machines on 60 Hz will probably cause them to overheat. Likewise, do not operate a 60-Hz motor on a 50-Hz system unless the manufacturer specifically warrants the motor at the reduced frequency at some specific voltage and load.

13-7. Motor Voltage

Typical distribution system voltages are given in Table 8-6 as single phase (120 or 240 V) and three phase (208, 240, 480, 600, 2400, 4160, or 12,000 V). These voltages are the levels that the utility attempts to maintain at its point of service.

Voltage drops occur in the user's electrical system, and these are divided into two parts according to the NEC. The first drop is in the feeders and includes the service switchgear and the cables delivering electrical energy to local distribution centers, which are typically motor control centers (MCC) in pumping stations. The second voltage drop cited in the NEC is in the branch circuit, which includes all of the power wiring between the circuit protective equipment (typically MCC) and the motor. A 5% voltage drop is allowable in the total system and a maximum drop of 3% in either the feeder or the branch circuit is recommended by the NEC. For example, on a system rated at 480 V line-to-line, a 5% drop is 24 V and the net voltage delivered is 456 V line-to-line (at the full rated load). The designer may divide the drop in accordance with the NEC rules or choose to put in cables heavy enough to make the drop somewhat less. It is good design practice to keep the total calculated drop and the individual parts of that drop well below the specified levels discussed. At light loads, the terminal voltage at the motor is likely to approach the service voltage or to be approximately 480 V. The motor manufacturer usually rates the motor (full-load current and horsepower) at 460 V with a ±10% tolerance at full load. Thus, the motor guarantee would be valid at 506 V, but the life of the motor at 506 V may be shortened. A motor rated with a service factor may not be guaranteed at a voltage other than the nameplate voltage.

Typical rated voltages for motors are 115 or 230 V (single phase), and 200, 230/460, 460, 575, 2300, and 4000 V (three phase). The supply voltage for a very small pumping station may be 240 V, three phase, three wire from a utility delta-connected system, and a 120/240-V supply can be provided from a center tap between two phases on the utility's transformer. Sometimes, the utility may supply a 208Y/120-V, three-phase, four-wire system, particularly if the pumping station is located in a network area having this supply. It is also frequently available where an individual transformer is required for the pumping station. This multivoltage system is preferable to a 240-V three-phase system because it enables the load to be balanced on all three phases. Pumping stations of medium size—for example, 75 to 600 kW (100 to

800 hp)—need a full 480-V supply and usually derive their own 120-V circuits for lighting, small power outlets, and control devices from dry-type transformers located in the station.

Ordinarily, the service for a large station is most economical at the highest voltage available, and large pumping stations require 4160-V service. Because motors are usually rated at 4000 V maximum, however, there is no advantage in utility service above 4160 V unless the utility rate schedule for higher voltage service is particularly appealing to the owner. Higher service voltage in such situations would reduce cable size for power distribution, but transformers would be required to provide an operating voltage of 4160 or less, and they are costly. Note that some 230-V motors are guaranteed by the manufacturer for operation on 208 V, but some manufacturers then qualify this rating as not necessarily meeting NEMA standards. Therefore, use 200-V motors for 208-V systems.

A study of alternative voltage levels and sources should precede the selection of both service voltage and utilization voltage. For example, 480 V is nearly always the preferred and most economical level for both service and distribution in medium-sized pumping stations with short feeders and service runs. However, if the cable run from the service point to the pumping plant is, say, 150 m (500 ft) or longer, the excessive cable costs at 480 V can make 2400- or 4160-V service more economical. The cost of in-plant transformers versus utility-furnished transformers should be studied carefully on a life-cycle basis.

Service from a utility circuit is usually obtained through pole-mounted fuses, but in the largest stations, service may be obtained from a utility circuit breaker. In addition to the loading recommendations given here, make sure that the utility provides a balanced system voltage of rated value. In addition, require overload relays in each phase of the motor branch circuit to remove the motor from operation in the event of "single-phasing" (which occurs when a fuse on one leg blows). The prevention of single-phasing begins in the utility's system and carries throughout the pumping station electrical system.

Wherever single-pole switching equipment is used, install current-balance relays and circuit breakers with interpole tripping to cut station power in the event of an unbalance greater than the relay setting allows. Never install circuit breaker handle ties even in the smallest stations. Some unbalance is always present due to lighting or single-phase auxiliary equipment, and the current-balance relays must be set to accommodate this unbalance. If single-phase loads are large enough to cause voltage unbalance, distribute these loads between the phases as equally as possible.

Voltage unbalance causes excessive heating of motors. A 2% unbalance, according to IEEE Standard 141, adds about 8% to the heat in the motor windings and causes the temperature to rise from 80 to 86.4°C. A 3.5% unbalance causes an approximate 25% increase in heating, so the temperature rises (in a Class B insulation, T-frame motor) from 80 to 100°C.

13-8. Enclosures

The motor enclosure provides the environmental protection for the motor, a mounting space for the stator windings, the magnetic circuit for the stator field, a ventilation space, and a mounting for the bearings on most models. Motor enclosures, which are defined in Section 13-1, are

- Dripproof
- Splashproof (if not available, substitute weather protected Type I)
- Totally enclosed (TE)
- Totally enclosed, fan-cooled (TEFC)
- Totally enclosed, nonventilated (TENV)
- Totally enclosed, pipe-ventilated
- Weather-protected, Types I and II
- Submersible
- Explosionproof.

The open dripproof and totally enclosed fan-cooled enclosures are the ones most commonly used in pumping stations. Small- to medium-sized open motors may have epoxy-encapsulated windings, and they give excellent performance in quite severe conditions. They are the least expensive to put back into service in the event of flooding. Nonencapsulated dripproof motors are not usually applicable for pumping station environments. Horizontal machines are particularly vulnerable to hosing-down operations.

The totally enclosed fan-cooled motor usually has the fan mounted integrally with the motor under a shrouded-end bell arrangement. Air is drawn into the end of the motor and distributed over the outside of the motor enclosure by the fan. The cooling ability of a TEFC motor is less than that of a similar open dripproof motor. The TEFC motor may be used in both indoor and outdoor environments. Horizontal TEFC motors should be provided with drain holes or a breather drain in the bottom of the enclosure to permit the escape of condensation should it collect inside the enclosure. The shaft, of course, extends through and beyond the enclosure.

The totally enclosed, nonventilated motor is usually used only for small, noncritical loads.

Totally enclosed pipe-ventilated motors are likely to be used in adjustable-speed systems where, at low speeds, insufficient cooling would be available from the ordinary shaft-mounted fan. This type of ventilation may also be appropriate where the motor is subjected to higher than normal ambient temperatures but where cooler air is available.

Weather-protected Type I motors should always be considered for large drivers in ratings where TEFC is not a standard option. It is also applied indoors in some installations where pipe breaks may spray the equipment. Use weather-protected Type II motors in outdoor applications subject to high wind and rain.

Only explosionproof motors can be used in areas that are classified as Class 1, Division 1 hazardous locations in Article 500 of the NEC. These motors are similar to the TEFC in cooling, but are much more expensive than the TEFC motor.

Submersible motors are often used in water wells and in many sewage pumping stations. Most models must be totally submerged to cool properly, and because the motors are "ordinarily" submerged, a submersible rating is adequate. In the event of a malfunction and subsequent overheating combined with possible explosive gas accumulation in the manhole, however, an extreme hazard may be created. Some motors can be run intermittently or continuously in air. For additional discussion, see Section 13-2 and Chapter 25.

13-9. Insulation

Motor windings, by necessity, are closely packed into slots that are designed to accommodate them. The methods used to maintain conductor-to-metal, conductor-to-conductor, and phase-to-phase insulation vary greatly depending on the size and application of the motor. Pumping station designers and motor specifiers are more concerned with the class of insulation than with the details of packaging.

In addition to other considerations, the type of insulation that should be specified for a pump driver motor depends on its suitability in a possibly moist atmosphere, its ability to be dried readily in the event of a submersion, and the ambient temperature conditions that will prevail in its location. Insulation is ordinarily classified by letter designations that infer its ability to withstand continuous operation at a defined total temperature (or some rise above a standard ambient temperature). The letter designation is also defined in terms of the types of materials used in the insulation system.

The allowable temperature rises above a 40°C ambient temperature for the various standard letter designations and enclosure types (taken from NEMA MG-12.42) are summarized in Table 13-2. All of the temperatures in Table 13-2 are measured by resistance in accordance with NEMA standards. If the ambient temperature is higher than 40°C, a berating equation is given in NEMA MG-1 for the allowable temperature rise (see the latest revision of NEMA MG-1). The insulation materials for the several classes are:

- *Class A:* Combinations of materials such as cotton, silk, and paper impregnated or immersed in a dielectric liquid, no longer specified for pumping station applications and now encountered only in very old drivers
- *Class B:* Combinations of materials such as mica, glass fiber, asbestos, and so on with bonding substances
- *Class F:* Similar to Class B, but designed for a higher operating temperature at the same thermal life
- *Class H:* Similar to Class F, except for the use of component materials (such as silicone elastomers) rated for a higher operating temperature at the same thermal life. Note that Class H insulation may be unavailable, and its thickness causes design problems.

Epoxy insulation was used in the past to permit an inundated motor to be put into service almost immediately after the dry pit was dewatered. But epoxy is short lived because of the cracks that develop. Use Class F or H insulation and take the time to dry a previously submerged motor properly.

Although thermal life is not quantifiable by aging time alone, modern insulations (Class B, F, or H) can be expected to last 25 yrs at conservative temperatures unless they are subject to extreme vibration or extreme thermal cycling.

Table 13-2. Allowable Temperature Rise for Insulation in Motors[a]

Motor enclosure	NEMA insulation letter designation A	B	F	H
Totally enclosed, fan-cooled	60	80	105	125
Totally enclosed, nonventilated	65	85	110	135
Encapsulated with 1.0 SF	65	85	110	—
All other motors, with 1.15 SF	70	90	115	—
All other motors	60	80	105	125

[a] Degrees Celsius above 40°C ambient from NEMA MG-1 Table 12.42.1.4.

Always select water or sewage pump motors for long life. Furthermore, carefully control the temperature rise aspects of the insulation system by specifying the best of materials (within reason). Then take advantage of every means available to obtain an efficient motor with a low temperature rise at the specified loading conditions. A practical and economical way of buying a long-lasting motor is to specify Class B temperature rise but require Class F or H insulation.

Along with the insulation specification, the motor (if available in the enclosure specified) should have an applicable service factor (SF)—typically 1.15 for most sizes. Also specify that the maximum pumping horsepower load must not exceed 85% of the service factor rating (about 98% of the nameplate rating). If the motor type does not have a service factor rating available, then the maximum loading should not exceed 90% of the nameplate rating. This recommended maximum loading is not low enough to cause undue efficiency or power factor loss in a well-designed machine.

The overheating of insulation, whether due to overloading, insufficient ventilation, high ambient temperature, or too-frequent starting, can contribute rapidly to the shortening of insulation life. An old rule of thumb is that insulation life is halved for every 10°C increase above the rated value. The rule is not precise and may not apply to present-day insulation, but it is a good guideline when a decision must be made on whether to use a motor above its rating during an emergency or to replace it immediately with an adequately rated unit. Note that the rule applies to a continuous load. It does not mean that an occasional overload of short duration will halve the life of the insulation. Common sense dictates that conservative loading practices should be maintained in pumping station operation and design. See the IEEE standards for more information on insulation aging.

13-10. Service Factors for Squirrel-Cage Motors

Although the service factor of a motor implies that the entire design of the motor allows for the SF loading, the most sensitive factor is the stator winding temperature. Because the allowable temperature of the conductors is limited by the type of insulation used, the insulation system is the determining factor in the allowable temperature rise. A motor with a service factor rating of 1.15 (standard for most motors for which an SF is available) must therefore be designed for the nameplate load operation at an insulation temperature of somewhat less than the limiting value given in Table 13-2.

If the motor is selected according to the advice given in Section 13-9 and loaded only to 85% of the service factor rating, the operating temperature of the insulation will probably be low enough to allow a long life for the windings.

Insulation

Insulation systems and service factor are very closely related. The maximum loading of the motor can be easily controlled by the specifier and pump manufacturer. However, it is not ordinarily in the best interests of the owner to load motors too lightly under maximum operating conditions because of the inherent decrease in motor efficiency and power factor. The major considerations that provide long insulation life are those required to keep the temperature of the insulation below its maximum rated value.

Ambient Conditions

Do not place the motor in a closely confined space. It needs "breathing room." If the motor must be in an inherently dirty area, specify the proper motor enclosure for the conditions and describe the advantage and necessity of frequent inspection for clogged air passages and accumulations of dirt and debris around the motor in the O&M manual.

Responsibility

If a single manufacturer is responsible for the entire pumping unit (pump, frame, shafting, motor, and controller), the motor is selected on the basis of the complete operating conditions furnished by the project engineer (see Chapter 16 and Appendix C). But responsible project engineers must always make their own calculations, as shown in Example 13-1.

The maximum permissible length of the service feeder in Example 13-1 is about 300 ft if the voltage drop is limited to 2%. Because of the high cost of the cable and conduit (concrete-encased duct), it is wise to negotiate with the utility to place transformers closer to the load (say 30 m or 100 ft maximum) regardless of whether the utility or owner provides the service cable.

Starting a 25-hp motor is not likely to cause problems unless the utility's power lines are loaded to the limit. But if the motors were, say, 250 hp, the utility might object to frequent line starts, so consider reduced-voltage ("soft") starting (which might reduce the inrush to 300 or even 150% of the full-load current) or consider adjustable-speed drives.

Example 13-1
Motors for a Sewage Lift Station

Problem: Preliminary investigations for a sewage lift station have led to the following conditions:

- Flows: 800 gal/min minimum, 1600 gal/min design, 3600 gal/min maximum, and 4500 gal/min future maximum
- Static lift: 25 ft at low wet well level, 20 ft at high wet well level
- Friction head losses at 4500 gal/min: 15.7 ft at Hazen–Williams C = 120 and 13.4 ft at C = 145 (including "minor" losses within the pumping station).

Four (three duty, one standby) 705-rev/min pumps were selected to meet the various conditions of head discharge in the following tabulation. Each constant-speed pump is rated at 2100 gal/min for 30 ft TDH at 79% efficiency.

	Low wet well (C = 120)		High wet well (C = 145)		
Pumps operating	Head (ft)	Discharge (gal/min)	Head (h)	Discharge (gal/min)	Minimum cycle time (min.)
1	28.0	2300	24.0	2700	6.7
2	33.5	3500	31.0	4100	10
3	37.3	4100	35.3	4800	17

Select the drive motors, size the cables, and draw a single-line diagram of the installation.

Solution: For public works, design for a generic product. Standardization in manufacturing makes it easy to design and specify on the basis of NEC, NEMA, IEEE, and ANSI criteria.

Required power. From Equation 10-6b, the fluid horsepower is

$$hp = \frac{qH}{3960} = \frac{2300 \times 28}{3960} = 16.3$$

But the efficiency of the pump is 79%, so the output (shaft) motor horsepower must be

$$hp = \frac{16.3}{0.79} = 20.6$$

The motor power requirements for each of the conditions in this tabulation range from 16.6 to 21.2, so choose 25-hp motors, which can meet all conditions.

Required torque. Horsepower at full speed is only one criterion. The motor must develop enough torque at all speeds to exceed the resisting load by a comfortable margin; otherwise the motor will never reach full speed. The equipment suppliers know that and will furnish suitable motors provided they are supplied with the speed-torque characteristics of the pump.

To compare the torques, plot the speed-torque curves of the pump and the motor on the same graph, as in Figure 13-9. If motor torque everywhere exceeds pump torque by, say, 15%, the motor can develop full speed quickly.

Motor type. To determine whether a Code F motor is available as a standard unit (Code G may be standard), consult motor manufacturers. The motor is to be mounted high above the floor, so a dripproof enclosure is satisfactory. Alternatively, a TEFC motor might be preferred for some protection during washdown—certainly if the motor is mounted at floor level.

Starting. Obtain written approval from the electric utility for frequent line starting of one motor at a time (assuming there are controls for automatic sequencing). Consider the following alternatives:

- Line starting (unless the utility's branch circuit is loaded to the limit, line starting would be approved for such small motors)

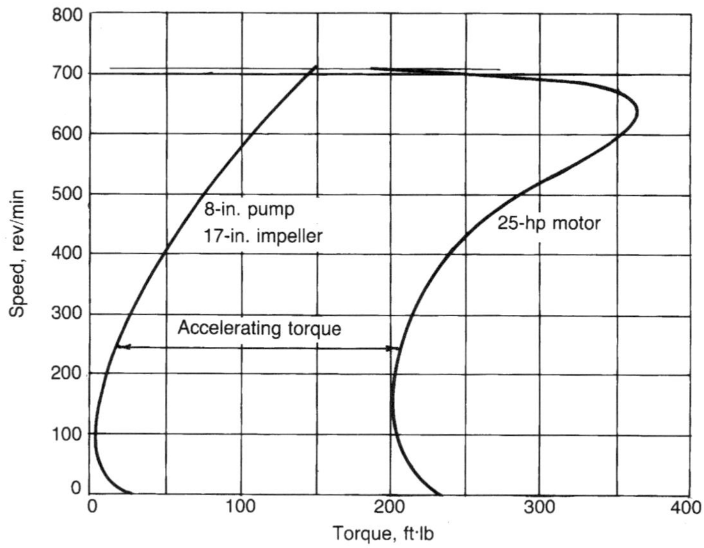

Figure 13-9. Torque curves of the motor and pump in Example 13-1.

- A limited number of line starts per day
- Line starts not permitted, but reduced-voltage starting allowed
- No line starts and only a limited number of reduced-voltage starts per day.

The last three alternatives may govern motors of 50 hp or more. If the final alternative applies, consider adjustable-speed drives.

Load calculation. Assume the current available is 480 V, three phase, and 60 Hz, and that the length of service run is 100 ft. A Code F motor has an inrush value in the range of 5.0 to 5.59 kVA/hp from NEC Table 430-7(b).

First, size the cable for current, then check the voltage drop. Assume (1) 5 kVA for miscellaneous (balanced three-phase) loads and (2) three 25-hp duty motors. For the 5-kVA load, the current is calculated from Equation 8-6. If the power factor is 1.0.

$$\begin{aligned} P_{3\phi} &= \sqrt{3} \times V_L \times I_L \times Pf \\ 5\text{kVA} \times 1000 &= \sqrt{3} \times 460 \times I_L, \text{ so } I_L = 6.3 \text{ A} \end{aligned} \tag{8-6}$$

From the same equation, the full-load current for a 25-hp motor with a power factor of 0.80 is, from Equation 8-6

$$\begin{aligned} 25 \times 0.746 \times 1000 &= \sqrt{3} \times 460 \times I_L \times 0.80 \\ I_L &= 29.3 \text{ A} \end{aligned}$$

But a 25-hp, 705-rev/min Code F motor has an efficiency of about 0.86, so the line current must be increased to

$$I_L = 29.3/0.86 = 34 \text{ A}$$

which can be verified by NEC Table 430-150. In accordance with NEC requirements, load calculations must be based on the referenced tabulated values as a minimum. If there is reliable information at the time of design that indicates that the motor to be purchased will have a higher current than listed in the NEC tables, the higher values must be used for all related calculations.

From the principles given in Chapter 9, the total branch circuit load is

• Miscellaneous:	6.3 A
• Motor No. 1:	34.0
• Motor No. 2:	34.0
• Motor No. 3:	34.0
• Motor No. 4 (standby):	0.0
• 25% of the largest motor:	8.5
Total:	116.8 A

The feeder would be designed for 125 A at full load if no expansion were contemplated, but a future expansion of 25% would increase the amperage to 116.8 A × 1.25 = 146 A, so select a cable for 150 A. Inrush persists for such a short interval that it is ignored in the load calculations. However, inrush does produce voltage drop and cannot always be ignored. Conservative engineers may limit the voltage drop at the motor terminals to 12 or 15% on starting, which provides a good, solid electrical system throughout. The drop is approximately 4% in this example.

Second, select a cable. From NEC Table 310-16, choose THW- or RHW-rated cable (75°C rating). The rating of I/O cable is 150 A.

Third, find the voltage drop from Equation 13-1 (a close approximation):

$$\Delta V_{\mathrm{L}} = \sqrt{3} \times I_{\mathrm{L}}(R\cos\theta + X\sin\theta)L \tag{13-1}$$

where ΔV_{L} is the line-to-line voltage drop, I_{L} is line current in amperes, R is resistance in ohms per 1000 ft, cos θ is the power factor, X is inductive reactance in ohms per 1000 ft, and L is length in 1000-ft units. If the power factor is 80%, cos θ is 0.80 and sin θ is 0.60. From Table 9 in Chapter 9 of NEC, R at 75°C is 0.12Ω/1000 ft and X is 0.05 Ω/1000 ft. Either 116.8 or 125 can be correctly substituted for I, but at the ultimate load of 150 A

$$\Delta V_{\mathrm{L}} = \sqrt{3} \times 150 \times (0.12 \times 0.8 + 0.05 \times 0.6) \times 0.10 = 3.24 \text{ V}$$

The voltage drop percentage = 3.24/480 = 0.67%.

Check using the IEEE Standard 141 to find 2.3 V × 10^{-4} × A × ft = 2.3 × 10^{-4} × 150 × 100 = 3.45 V (a fair check with 3.24 V).

The voltage drop of 0.67% is negligible compared with the NEC limit for feeder or branch circuits of 3%. Because a 2% voltage drop is the maximum for good design, the maximum length of the service run should not exceed 100 (2/0.67) ≃ 300 ft.

See Figure 9-1 for a single-line diagram of the electrical system.

If reduced-voltage starting is allowed by the utility with ON–OFF controls, compare costs and space requirements of autotransformer start with those of solid-state soft start controls. The newer controls have a much greater selection of starting currents than do other methods. The added cost for soft starting is only the installed cost of the added equipment, but adjustable-speed drives are likely to involve a complete redesign of the station with possibly fewer pumps (two duty and one standby, for example) and smaller wet and dry wells, with consequent savings that partly offset the cost of the adjustable-speed drives. Variable-speed operation might be more cost effective because it extends the life of the motors and less energy is used (see Example 29-1), but it is somewhat less reliable and there is increased maintenance of the adjustable-speed devices.

13-11. Motor Starting Frequency

Motor life is determined by the temperature of the stator winding insulation. From motor manufacturers' data, a motor operated at rated load with normal voltage and frequency in a 40°C (104°F) environment will last about 40,000 hours. A general rule is that motor insulation life is halved for every 10°C (18°F) rise in temperature. For example, a motor with Class F insulation is rated to have a temperature rise of 110°C (198°F) with a maximum ambient temperature of 40°C. If, however, the internal temperature rise is limited to 100°C, the insulation life is approximately doubled.

When an induction motor is started with an across-the-line starter, the stress on the windings is doubled, and the inrush current in the rotor is about six times normal current until operating speed is reached—usually 3 to 4 seconds if the motor is driving a pump. The inrush current heats the rotor and causes the stator winding temperature to increase. If the motor is started too frequently, the rotor heat causes the stator winding insulation to fail.

The frequency of motor starts is given by NEMA in both MG 1 and MG 10. The two publications differ radically. MG 1 is very conservative, whereas MG 10 allows more frequent starts—sometimes four times as many as MG-1. Because pumps are low-inertia machines, and because there must be a balance between first cost and service life, it seems reasonable to use about two-thirds of the starting frequency calculated from MG 10. The allowable starting frequency is, however, far more complicated than is indicated by such a calculation, because (1) an underloaded motor may be started more frequently than a fully loaded one; (2) the required frequency of starts in a pumping station with constant-speed motors is based on the assumption that inflow to the wet well is exactly half of the pump capacity, whereas half of the time, the required starting frequency is nearly 40% less (see Figure 12-41); (3) the specified severe ambient temperature conditions may persist for only short periods of time; and (4) motors can be custom designed and built to withstand many more frequent starts than standard motors can. The most important factor relating to the frequency of pump starts is whether (for multiple pumps in a station) automatic sequencing is used. Some engineers do not favor automatic sequencers and prefer manual selection of lead and follow pumps for better control of pump wear. Nowadays PLCs can be programmed to alternate the lead pump at every cycle reliably, and the reliability can be increased by automatic self-testing and switching to a backup PLC if a malfunction occurs.

Soft (reduced-voltage) starters typically induce an inrush current that is only about 150% of normal operating current. Solid-state soft starters distort the waveform and cause extra heating. The time required to accelerate the machine to normal operating speed is longer, so the heating effect is about the same as it is for across-the-line starters. Note that soft starters cost about half as much as adjustable-frequency drives, but adjustable-frequency drives not only provide soft starts but also make it possible to reduce the size of the wet well, usually reduce the power lost in overcoming pipe friction, and prevent the sudden changes of flow that tend to upset downstream treatment.

Motors 224 kW (300 hp) and larger are usually custom engineered to whatever requirements are specified. Hence, large motors can be specified to provide whatever frequency of starting is needed for the pumping station. However, if starting frequency is not specified, the motor will be designed to provide the same frequency of starts as small and medium motors.

Submersible motors, cooled by the pumped liquid, can usually withstand very frequent starts—sometimes much more frequently than 20 starts per hour. For such motors, it is the starter (not the motor) that limits starting frequency.

Because objective decisions concerning allowable frequency of motor starts must be based on such a myriad of site-specific factors, the pumping station designer's best course is to consult the pump manufacturer, who deals with such problems constantly and can temper theory and calculations with practical experience. The same cannot be said of motor manufacturers, because they rarely seem to be cognizant of the needs for pumping stations.

13-12. Miscellaneous Motor Features

Miscellaneous, but important, features to be considered in the selection of a motor include the shafts, bearings, space heaters or winding heating, temperature sensors, and (possibly) vibration monitors.

Shafts

Motor shafts are usually solid and extend beyond the enclosure in order to accept the coupling. The shaft may be ordered with standard NEMA dimensions in either short or long shaft extensions. For municipal work it is usually unnecessary to specify shaft length unless the installation is for a nonstandard arrangement.

Hollow shafts are usually supplied for the vertical motors for deep well pumps. The shafts extend through both ends of the motor with an adjustable thrust bearing above the motor frame to the deep well turbine pump below it. A nonreverse ratchet is required for deep well pumps where threaded-end sectional drive shafts are used. The ratchet prevents backspin caused by the discharge water column if the drive should ever stop against a high head and the check valve fail. Backspin could unscrew the connections between shaft sections. Nonreverse ratchets are also useful for other pumping arrangements, such as horizontal pump motors, because energizing a backspinning drive can break the shaft, tear windings loose, or overheat the motor. When a ratchet is set, the locked rotor causes large line currents to continue beyond the normal starting period, so the overload relays must operate to prevent damage to the motor.

Bearings

Antifriction ball bearings are provided in most motors. Recently, improvements in materials have added to the life expectancy of what was already an excellent product. In large horizontal motors, the tendency is to use sleeve bearings, but ball bearings may be preferable in some instances.

Bearings for vertical motors may be angular contact, grease lubricated, or oil lubricated for higher speed applications. Spherical roller bearings can withstand very high thrust. For the highest thrust application on very large drives, the Kingsbury thrust bearing [5] is available. It is expensive but has an extremely long life.

Space Heaters and Winding Heating

Space heaters are often applied to motors located in damp areas or outdoors where the ambient temperature and humidity vary over a wide range. The space heater is sized to maintain a reasonable temperature in the motor enclosure to ensure that the windings and insulation cannot collect moisture. A motor in standby service or one that operates only at infrequent intervals must have the heater on almost all of the time. Heater circuit continuity is seldom checked, and sometimes it seems that there are as many strip heaters that do not operate as ones that do.

An alternative to the space heater (but initially much more expensive) is a low-voltage heating circuit for the motor winding. The most modern motor winding heater is a solid-state device that does not use a transformer [6]. The safety of this type of circuit, however, may be questioned by inspection authorities, so consult these authorities prior to completing the design. If the low voltage is to be produced by a transformer, the transformer is placed within the motor starter enclosure, and a separate contactor and spare contacts on the motor starter are also required [7, 8]. A timing relay is needed to ensure that the motor voltage has had time to decay before connecting the low-voltage heating circuit to the stator of the motor. The temperature of the windings should be kept about 10°C above the ambient temperature. Low-voltage heating of the motor windings has proved to be very practical for motors less than about 150 kW (200 hp) where there is often little or no room for mounting a conventional space heater.

Temperature Sensors

Winding temperature sensors provide back-up protection for overtemperature conditions within the motor. The simplest kind is a direct-acting, bimetallic element that snaps from one position to another and either opens or closes a contact. Usually this contact is used as a control-circuit-stop contact (similar to the overload relay contact and stop pushbutton). This contact resets automatically on a decrease in temperature, so arrange the circuitry to prevent automatic restarting of the motor. The worst thing that can be done to a motor already in trouble from overtemperature is to cycle it on and off at short-duration intervals.

A more sensitive device that may be applied to any size of motor is the positive temperature coefficient (PTC) thermistor. It is a very small resistance element taped in the interstices of the end turns of the motor windings. It can be retrofitted in most motors—even those of small horsepower ratings [9, 10]. Several of the PTC units may be connected in series with only two leads brought out of the motor. The resistance of the PTC element is quite low up to some critical temperature (130°C, for example), at which point the resistance changes radically to a high value with rising temperature. A sensitive monitoring relay is provided in the control circuitry of the motor, and this relay is energized while the motor is in operation and the winding temperature is lower than the critical value of the sensor. The relay drops out, which opens the motor control circuit if the winding temperature exceeds the critical resistance temperature, and the control circuitry of the motor must be designed so that the motor cannot restart even after the windings are cool. A pump that stops due to the operation of any safety device (such as the overtemperature detectors or the overload relays of the motor starter) should *always* activate an alarm at the monitoring site—a fire or police station, a supervisor's home, or a central control station.

On large motors, it is more typical to use resistance temperature detectors (RTD). The RTDs have a linear resistance/temperature response and are monitored by an analog instrument. The temperature of the winding being monitored can be read directly, and various alarm points are usually set in the readout unit or in a separate monitor. Motor starters need thermal overload relays in each leg, so three are required.

Vibration Monitors

Vibration monitoring is considered applicable only on very large systems. To be effective, vibration monitoring systems are complex, and their cost must be weighed against the probable value of the information obtained. In high-speed machinery applications, the information obtained from sensors located radially on quadrature axes can be recorded and analyzed at intervals. Changes in these values, as well as changes in the axial movement of the shaft, may give an early warning of impending failure. In addition, the readout units provide an automatic alarm for out-of-limits radial or axial vibration. It is doubtful, however, that the relatively slow pump speeds in most pumping stations warrant the expense of the monitoring system.

In large pumping unit installations, particularly in unstaffed stations, there may be a good argument for a complete vibration monitoring and analysis system. The ultimate user of such a system must be committed to the regular collection and interpretation of the readout data to make the additional investment worthwhile.

In unattended stations that have smaller pumping units than inferred in the above discussion, simpler vibration monitors are recommended to shut down the unit and to signal an alarm. If the vibration is due to a bearing failure, the bearing is already lost, but early shut-down may prevent further damage to the unit.

The purpose of monitoring systems is to ensure the safety of the pumping station, but a complex system that is inoperable because of poor maintenance or lack of regular testing is a hazard in itself because it creates a false sense of security. Pumping stations should be visited regularly (daily, if possible) so that vibration can be easily heard or felt.

Moisture Sensors for Submersible Motors

The most common detector is the capacitance probe (or two-point electrode) in a moisture leak sensor assembly filled with transmission oil. Water forms an emulsion that conducts electricity and completes a circuit containing a warning light or other alarm. A resistor can be added in parallel with the electrode to make it possible to check the circuit.

Moisture sensors are always placed between the inner and outer seals of submersible motors, and sometimes they are also placed at the bottom of the motor housing. The probes are effective in detecting moisture leakage through the shaft seals, although they tend to give false indications of moisture intrusion. Moisture detectors do not indicate leakage through the power cable or its connection to the motor housing, and occasionally motors are burned out by such leakage—a hazard with any submerged motor.

13-13. Specifying Pumping Unit Drivers

To obtain a driver completely coordinated with the pumping equipment, it is advisable for a single responsible manufacturer to provide the motor and the driven equipment. Usually the pump manufacturer will take this responsibility and purchase a motor to meet both the owner's specifications and the load requirements and starting frequency of the pump. If an adjustable-speed system is specified, the speed-control equipment should be included in the drive package.

Motors for Water Pumping

Motors of about 10 to 60 kW (15 to 75 hp) are common in water pumping stations, but relatively large-size motors are not unusual. Even farm water pumping systems may be of several hundred horsepower if the irrigation system is extensive or the pumping head large. The electric utility may advise or require large motors to be provided with a reduced-voltage starting system. Either an autotransformer starter or a part-winding start motor can be satisfactory. Alternatively, the motor might be specified with lower-than-normal starting current or a standard motor with a soft start controller so that it can be line started within the utility's power supply capabilities. Motors for well pumps usually drive a multistage pump often several hundred feet below the motor. The hollow-shaft motor discussed in Section 13-12, as well as the nonreverse ratchet feature, is usually required. A high-thrust (or very high thrust) top bearing on the motor is required due to the high loading involved in supporting the long shaft, the impellers, and the TDH of the water column.

Other features to be considered for the specifications include

- Enclosure type
- Special cooling provisions
- Bearing type
- Bearing and winding temperature monitoring systems
- Special insulation system
- Special temperature rise limitations
- Stator winding temperature monitoring
- Voltage and frequency (for an adjustable-frequency drive, both the speed and frequency range and constant-volts-per-hertz ratio must be specified)
- Vibration monitoring, if applicable
- Brass nameplate
- Painting (manufacturer's standard, special, or prime coat only).

Motors for Sewage Pumping

Motors for sewage pumping range up to 600 kW (800 hp) or more and are likely to be indoors and even below grade where they may be subject to flooding.

Raw sewage is usually pumped at heads of less than 30 m (100 ft) by single-stage pumps with open impellers. Several (at least two) units are installed to provide back-up and to handle the widely fluctuating flow range. In motor sizes below 75 kW (100 hp), line starting is almost universal. Reduced-voltage or part-winding starting may be required in pumping stations located several miles or so from main electrical substations. Reduced-voltage starting may also be desirable to limit voltage dip. In adjustable-speed systems, raw sewage pumps usually require only a moderate speed range, which simplifies the speed-control equipment.

Features that should be considered when writing the specifications are the same as those listed for water pumping, but with special emphasis on the ability of the unit to be returned to service quickly after flooding.

13-14. Need for Engine-Generators

The need for standby power from engine-generators is not only site-specific but depends on the need for absolute reliability and what the owner is willing to pay. Diesel engines are usually preferred, because they are smaller than gas engines and hence less expensive. However, storing diesel fuel on site requires double-wall storage tanks and deterioration is a problem. On the other hand, if natural gas can be piped to the site, no storage is required. Gas may not be available at some sites, and in earthquake-prone areas, it is not completely reliable. Pollutant emissions from gas engines is minimal, whereas it is more difficult to control pollutant emissions from diesel engines. Furthermore, regulations tend to become more stringent with time (see Section 14-14). Altogether, the need for and choice of standby power generation requires the most careful consideration.

In the field of electric power supply, substations supplied from main transmission power lines are the most reliable. Substations rarely fail even when the power lines or even the substations themselves are struck by lightning. Generally, only one or two outages per year occur. The next most reliable component is the distribution substation, which operates at lower voltage. One to six outages per year might be considered average. The least reliable are long power lines to isolated communities where outages are more frequent and last for longer periods. These lower-voltage power lines suffer more frequent power outages because of accidents, lightning, construction, and so on.

Outages usually involve only a small area (for example, a quadrant of a small town), but occasionally an entire city or even several states are affected. On July 2, 1996, a flashover between a tree and a 345,000-volt transmission line triggered a chain of events that, due to record-setting demands for power, led to outages that lasted up to six hours in some areas throughout 15 states and affected approximately two million customers. The blackout of 1977 left New York City and Long Island without electric power for several hours. The blackout of November 9 and 10, 1965, left most of the northeastern United States and two Canadian provinces without electric power for most of two days. Natural disasters such as hurricanes, tornadoes, and earthquakes tend to create many hours or days of outages. Following the San Francisco (actually Loma Prieta) earthquake of October 17, 1989, power was not restored to all parts of the area for several days. The Oakland fire of 1991 burned the distribution power lines to the only water booster pumping station that supplied water to northeast Oakland, so there was no effective way to fight a grass fire that destroyed a large residential district.

Some pumping plants can be without power for as much as an hour (or even more) before flooding occurs, and some utilities have concluded that, for their situation, this grace period makes standby power unnecessary (see Section 24-11). On the other hand, the public has become more intolerant of utility failures. This attitude, coupled with the deregulation of power generation (a move that can only result in trimming costs by every producer to remain competitive) is likely to result in less reliable power in the future. Even though power in any given area has a history of good reliability, there is no guarantee that such reliability will continue in the future.

No blanket rules can be given. Investigate each project separately. To arrive at the most rational decision, examine the history of power failures, the consequences of down time, and the costs—both monetary and subjective—such as political concerns, customer exasperation, and so on.

13-15. Design Checklist

The following checklist contains aspects of the characteristics desirable in motors used for pumping service. The typical pumping station service is severe and warrants the expense of the more rugged motor designs. Such motors, manufactured in NEMA standard frames from size 286 through 445 (up to about 150 kW or 200 hp at 1800 rev/min), typically have cast-iron frames, whereas smaller motors with rolled steel or aluminum frames may be available from the same makers.

1. Motor frame (cast iron or [above 200 to 400 hp] fabricated steel)
2. Enclosure to suit application

3. Antifriction bearings (a Kingsbury bearing may be desirable on deep well pumps)
4. Copper stator conductors
5. Lifting eyes or welded-on hooks
6. All surfaces of the frame treated with corrosion-resistant epoxy paints
7. Nonsparking vent fan
8. Breather drain or (on small motors) two 1/4-in. drain holes in totally enclosed motors, explosion-proof breather drains on explosionproof units
9. Premium insulation (Class F or H materials)
10. Class B–rated temperature rise
11. All conductors brought out to the conduit box, lugged, and identified
12. Nonwicking insulation on motor leads
13. Large cast-iron or fabricated steel conduit box for motor leads
14. Grounding lug within conduit box
15. Separate cast-iron box for auxiliary circuits (temperature and vibration monitoring, etc.)
16. Special motor guarantees for severe applications:
    a. Submersibles
    b. Very high thrust vertical motors
    c. Nonsinusoidal voltage supply (rectifier/inverter supply)
    d. Unusual ambient temperatures or elevation
17. Submittal requirements for nameplate data
18. Submittal requirements for efficiency data at various loads and power factor at the same loads from "like-motor tests" or from factory testing of the first unit of the order
19. Space heater or a provision for winding heating of infrequently run motors and motors in damp locations
20. Overtemperature, vibration, and (for submersible motors) moisture monitors
21. Protection from dirt, rodents, and insects
22. Protection from moisture and flooding
23. Protection from vehicular traffic
24. Protection from sun and weather, if applicable
25. Balanced voltage supply at the rated value
26. Rated frequency supply (utility and local generation)
27. Protection from single-phasing for utility and in-plant systems
28. Rating of motor(s) and frequency of starts within the utility's system capabilities
29. Starting voltage drop reasonable
30. Adequate controls to limit the frequency of starts
31. Controls designed to prevent energizing a back-spinning motor
32. Prescheduled maintenance and frequency of inspections by adequately trained personnel written into the O&M manual.

13-16. References

1. Linders, I., "Effects of power supply variations on ac motor characteristics," *IEEE Transactions on Industry Applications,* **IA-8,** 383–400 (Jul./Aug. 1972).
2. Woll, R. F., "Effect of unbalanced voltage on the operation of polyphase induction motors," *IEEE Transactions on Industry Applications,* **IA-II,** 38-42 (Jan./Feb. 1975).
3. Richardson, D. V., *Handbook of Rotating Electric Machinery,* Reston Publishers, Reston, VA (1980).
4. Richardson, D. V., *Rotating Electric Machinery & Transformer Technology,* 2nd ed., Reston Publishers, Reston, VA (1982).
5. Karassik, I. J., *et al., Pump Handbook,* 2nd ed., McGraw-Hill, New York (1985).
6. Yuen, M. H., "Low voltage heating of motors in refineries and chemical plants," *IEEE Transactions on Industry and General Applications,* **IGA-5,** 300–309 (May/June 1969).
7. Dikinis, D. V., and M. H. Yuen, "Solid-state control—low voltage heating of motors," *IEEE Transactions on Industry Applications,* **IA-II,** 287–290 (May/June 1975).
8. Lukitsch, W. J., "Methods of protecting against damaging effects of moisture buildup in motors used in the petroleum/chemical industry," IEEE Conference Paper PCI-80-15, Petroleum and Chemical Industry Conference, September 1980, New York.
9. Sheffer, K. W., *et al.,* "Application of inherent thermal protection to industrial motor systems," *IEEE Transactions on Industry Applications,* **IA-II,** 14–23 (Jan./Feb. 1975).
10. Obenhaus, R. E., "Sensor retrofitting of motors for protection against overtemperature," *IEEE Transactions on Industry Applications,* **IA-II,** 24–32 (Jan./Feb. 1975).

Chapter 14

Engines

GARR M. JONES

CONTRIBUTORS
Robert A. Daffer, Jr.
Michael A. Devine
Stanley S. Hong
Philip A. Huff
Richard A. Malesich
Loran D. Novacek
James W. Schettler
Thomas O. Williams

Drives based on engines (either through generators or directly) should be considered for virtually every water, storm, and wastewater pumping station installation. Pumping stations are often provided for the protection of public safety, health, and property, and reliable operation under all conditions of weather and manmade or natural disasters is important, if not essential. This chapter is intended for those who may be unfamiliar with engine drives and their needs. It is an introduction—not a definitive work. Rely heavily on the advice of qualified manufacturers' representatives (especially in application engineering departments) and those with adequate experience in engine-based facility design, and consult them early in the design process.

References to a standard or code are given in abbreviated form, such as NFPA 37. Titles of standards and codes are listed chapter by chapter in Appendix E. Publishers' addresses are given in Appendix F.

14-1. Selecting an Engine Drive

Internal combustion engines are used only sparingly for pumping unit prime movers. To apply this type of equipment properly, a greater understanding of a wider variety of technical concerns is needed than for electric motor-driven installations. This section is intended only to outline the multitude of issues that must be addressed by the designer. To design the engine installation and specify the equipment properly, consult reputable engine manufacturers' engineering and sales representatives. Become completely familiar with the limitations, capabilities, and installation requirements for candidate equipment.

The following factors may lead to the selection of engine drives in lieu of electric motors.

- Greater reliability, if a reliable source of fuel is available and multiple engine-driven pumps are provided
- Economy of operation (especially at wastewater treatment plants where the availability of biogas and the need for process heat almost always makes cogeneration attractive)
- Improved protection against surge damage due to (1) higher rotating moment of inertia and (2) greater reliability
- Variable-speed operation for capacity control and slow, surgeless starts

- Remote location of installation and unreliable electrical supply
- Fuel (natural gas, diesel oil) available at prices competitive with electricity.

Factors that may cause rejection of engine drives include

- Concern over noise emissions
- Concern over air pollutant emissions
- Lack of adequate maintenance capability on the part of the owner
- Small (less than 60 hp) unit pump requirements.

To determine whether an engine drive is more economical than an electric motor, the best approach is to base the cost comparison on a unit of energy (such as kilowatt-hours or British thermal units). The costs must be modified to include the efficiency of the pump-driver combination. Do not forget to take into account obscure items such as transformation and motor efficiencies and slip, heat losses for motor drives, the cost of compressing fuel gas, gear losses, and power-absorbing accessories for engine drives. The costs for the various forms of energy depend on many factors, including location and, sometimes, time of day of consumption (particularly for electricity). These factors should be explored carefully and then base prices should be determined. Once this has been done, the following multipliers for U.S. customary units can be used to obtain the cost per British thermal unit:

- *Biogas*—free, except for the cost of compression and transportation and standby fuel, if any
- *Crude oil*—dollars per barrel: 19,000 Btu/lb, 7.5 lb/gal, 42 gal/bbl
- *Diesel fuel*—dollars per gallon: 19,000 Btu/lb, 7.5 lb/gal
- *Electricity*—dollars per kilowatt-hour: 3413 Btu/kW · h
- *LPG*—dollars per gallon: 2316 Btu/ft^3, 36.5 ft^3/gal
- *Natural gas*—dollars per therm: 1000 Btu/ft^3, 100 ft^3/therm.

Once the difference in energy costs has been calculated, it can be combined with adjustments in maintenance costs and a comparison can be made against capital cost difference to arrive at an initial assessment of economic viability. After an engine drive has been selected, a multitude of decisions is required. The primary categories for decisions include

- Duty cycle
- Fuel
- Aspiration
- Type of engine.

Secondary categories include

- Starting method
- Cooling method
- Controls
- Governors
- Accessories
- Combustion air
- Exhaust silencing
- Pollution control
- Vibration isolation.

Finally, the following peripheral systems must be designed:

- Lubrication oil storage and supply system
- Fuel oil or fuel gas storage and supply system
- Service piping
- Building envelope
- Ventilation systems.

Most of the decisions listed in the primary and secondary categories dictate the equipment to be supplied as a part of the basic equipment contract. Engine drives are complex because of the number of components that must be coordinated properly. A unit responsibility provision, with a single manufacturer responsible for providing the driven equipment, the engine, all engine-related accessories, any gear reducers or increasers, shafting, and equipment supports is a recommended approach for such systems. (See Section 16-1 and Appendix C, Section 1.02B for unit responsibility).

The third group of issues is designer oriented and relates to systems to be designed by the pumping station engineer and to be specified and supplied independently. In addition, be aware of code requirements governing stationary engine installations, particularly NFPA 37 and local fire codes.

14-2. Duty Cycle

The term "duty cycle" means both the specifics of the application (duty), such as generation, direct drive, emergency, or standby, and the time-based utilization of the equipment.

Direct Drive

Direct-drive systems require the smallest engine to drive a given load because the engine does not have to sustain the loads imposed by squirrel-cage induction motors' inrush currents. If an adjustment (gear reducer or increaser) is required between the engine and pump speeds, specify gears with a nonreversing mechanism to prevent reverse rotation on pump shut-down.

Generator Duty

Generator duty requires a careful analysis of the applied loads and generator response. Inrush currents require the generator to be sized to 120 to 150% or more of the nominal running load. The best approach is to apply loads in increments, thereby limiting the amount of the starting load to be sustained by the engine-generator at any one time. A careful analysis of load increments (type, size) is required to ensure that engine capability is not exceeded. A complete description of the conditions under which load will be applied to the standby generator must be furnished to the engine-generator set manufacturer. The information to be furnished should include

- Size and NEMA code letter for each pumping unit or other large motor to be started
- Magnitude of miscellaneous loads (electrical resistance loads, lights, small motors)
- Motor start sequence after the standby generator has started and reached stable operation.

Ask the manufacturer to consider this information and recommend specific equipment for the proposed application. The information thus received can be used for the *preliminary* allocation of space within the pumping station. Be sure to be generous with space requirements because manufacturers may change their minds in a bidding situation. If competitive bids are sought, include all of the information and the provision for a field test of the standby generator.

An analysis for the required size of generator and engine is given in Example 9-10.

Continuous Duty

Continuous-duty engines should be selected using conservative rating factors (Section 14-6). The principal objectives are reliability, low operating costs, and a long service life. Just as with electric-motor-driven units, continuous duty applications require multiple units to permit maintenance operations without a loss of capacity.

Standby Duty and Standby Generators

Standby duty engines used to drive pumps or generators should provide rapid starting and be able to assume the load soon after starting. Special starting aids, such as jacket water and lube oil heaters, should be provided. Consult the engine manufacturer for recommendations. Engines for this service should have a relatively high torque versus speed relationship.

Standby generators are considered in Chapter 9, which states that standby engine-generator sets need to meet and sustain the high starting loads imposed by squirrel-cage induction motors. One effective technique is to size the generator for the worst-case condition of running loads and start-up, to provide the generator with a field-forcing regulator, to size the engine for the maximum steady-state operating loads, and to equip the engine with a large flywheel to assist in sustaining the starting loads. It is important not to oversize diesel engines for standby generators. Underloaded diesel engines develop carbon deposits in the exhaust gas passages. These deposits reduce the reliability of the engine. Reduced reliability is particularly a problem with standby generators because of the following:

- Standby generators are typically selected for starting loads under worst-case conditions—not running loads.
- Standby generators are exercised frequently, usually under less than peak load conditions.

In pumping stations, solutions may include the following:

- Recommended operating procedures that force the station to operate under peak loads. These might include (1) shutting down a wastewater pumping station and storing the wastewater (if possible) until all pumps can be started and operated under power supplied by the standby generator for long enough to clear the carbon deposits from the engine or (2) operating a water pumping station at a maximum rate under standby power.
- Provision (in the station motor control center or at the standby generator breaker) for connecting a portable electrical load bank.

14-3. Fuel for Engines

Factors that govern the choice of fuel for an engine installation include

- Duty cycle.
- Application of the equipment (for example, diesel fuel may not be the best selection for a standby generator because of the duty cycle. Conversely, natural gas may not be available on a reliable basis).
- Economics.

Fuels available for fixed-engine installations are gaseous: (natural gas, liquid propane gas, biogas) and liquid (diesel, gasoline).

Combination fuel systems are available that allow some engines to function on more than one fuel, which improves economy or reliability. These include the following:

- Bifuel carburetion systems. These permit the utilization of more than one gaseous fuel. As a rule, these operate on an either/or basis. Blending is not possible. Each fuel has its own pressure regulator and carburetor. Switchover to the standby fuel occurs on low pressure of the preferred fuel.
- Blended fuel systems. These are external to the engine and require mixing equipment to blend two gaseous fuels together to meet engine fuel demand. Typically, air is used to reduce the fuel value of a standby, higher-value fuel to that of a lower-value primary fuel. The advantage is that the blending system can be used to maximize utilization of the primary fuel when engine demands are greater than its availability. Fuel blending systems can be costly. However, once set up they have proven to be reliable and economical to operate. Blended fuel systems can be used, with some limitations, on low exhaust emission applications.
- Dual fuel engines. This category is limited to relatively large [550 kW (750 hp) and larger], slow-speed (1200 rev/min or less), diesel engines. The engine is typically started as a diesel. Once in operation, up to 90% (heating value) of the fuel is provided in the form of gas (natural gas or biogas) with the remainder provided as diesel oil. Ignition is by compression, rather than by spark.

Natural Gas

Commercially available natural gas, with a net higher heating value (HHV) of 4.45×10^4 J/L or 12.4 kW · h/m^3 (1200 Btu/ft^3), is a nearly ideal engine fuel for the following reasons:

- Ignition takes place uniformly over time and avoids high momentary loads on engine parts.
- Gas has a nearly uniform fuel value of predictable quality.
- Gas is clean (no water vapor or contaminants).
- Gas is much more reliable than electricity (also, a limited fuel supply can be stored).
- Gas is usually available at a wide range of pressures.
- Engines are available from 20 kW (30 hp) to 3700 kW (5000 hp).
- Higher compression ratios for improved fuel consumption may be used.
- Clean-burning engine designs that meet air pollution code restrictions and that have substantially improved efficiencies are available.

Disadvantages include the following:

- Gas may be more expensive than alternative fuels, especially if a long transmission pipe is needed.
- Natural gas may not be available under emergency conditions if it is not stored on site. Thoroughly explore its availability with the gas utility.
- Gas may not be available with adequate pressure for new-technology clean-burning engines, and gas compressors may be required.
- Regulatory agencies may prohibit the use of gas for an engine fuel.
- Gas, in the right concentration in air, can create an explosive mixture.

Propane

Propane, available as a liquid under pressure (LPG), is useful as a reserve fuel for spark-ignited engines. The advantages of propane include the following:

- Propane has a relatively high fuel value of 9.32×10^4 J/L or 25.8 kW · h/m^3 (2500 Btu/ft^3).
- A large quantity of fuel can be stored in a relatively small volume 7.11 kW · h/L (91,900 Btu/gal).
- Propane can be stored for long periods without deterioration.
- Propane has a low liquid specific gravity (0.51).
- Propane vaporizes at –43°C (–42°F) and is, therefore, useful in cold climates.

The disadvantages of propane include the following:

- Unless the rate of fuel consumption is relatively low, liquid must be vaporized with an external heat source prior to transport to the engine carburetor.
- The high gaseous specific gravity (1.52 relative to air) is hazardous at low points in structures if leaks occur.
- The sharp peak in ignition characteristics may increase maintenance, and it limits the compression ratio to 8.0:1.
- Propane cannot be used in a compression-ignition engine.

Propane may be too expensive for continuous use, but given the advantages listed here, it can be an ideal reserve fuel or fuel for a standby generator.

Butane

Butane is somewhat similar to propane, but because its combustion characteristics are erratic, it is not acceptable for most modern engines. Butane does not vaporize below –0.5°C (31°F), which is a problem.

Biogas

Biogas, a product of anaerobic decomposition of organic materials, may be readily available for use as an engine fuel at wastewater treatment plants and at sanitary landfills. As a rule, the gas consists of 40 to 65% methane, 35 to 65% carbon dioxide, and small amounts of hydrogen sulfide, nitrogen, and halogenated hydrocarbons, and it leaves the digester or landfill saturated with moisture. It may also contain contaminants such as siloxane, arsenic, vanadium, chromium, and other substances. The moisture and, perhaps (depending upon the engine manufacturer's requirements), the contaminants must be removed before it is introduced into the fuel system. The fuel value varies from about 15,000 to 24,000 J/L or 4.1 to 6.7 kW/m^3(400 to 650 Btu/ft^3).

The advantages of using this gas for a fuel are as follows:

- The combustible constituent (methane) makes it a good fuel for internal combustion engines.
- The gas is a wasted byproduct of most wastewater treatment plants and sanitary landfills and, thus, is free except for the low capital investment required for the collection system.
- With the proper selection of engines and engine cooling systems and an appropriate gas system design, the only treatment required to make the gas suitable as a fuel is trapping and draining the condensate.

The following are the disadvantages of biogas:

- Only modest fuel pressures are available without resorting to gas pressurization systems.
- The low fuel value requires special carburetion systems.
- The saturated gas is warm as it leaves a landfill or a digester. As it cools, condensate forms and must be trapped and drained safely from all low points in the piping system.
- Contaminants such as particulates and hydrogen sulfide may cause corrosion and/or clogging of appurtenances.
- Most landfill gases contain a wide spectrum of corrosive constituents. Maintenance can be quite costly.
- The gas flow from wastewater treatment plants fluctuates greatly (3:1 peak to average or greater). This fluctuation nearly always mandates the provision of an alternate source of fuel (usually natural gas or diesel fuel, although propane is sometimes used).

Fuel supply systems must be designed and engines must be specified to accommodate the special characteristics of biogas. Some engineers provide scrubbing systems and fuel filters to protect biogas fuel engines against corrosion and contaminants. Some engine manufacturers have been able to apply naturally aspirated and some turbocharged engines to biogas fuels without conditioning the fuel to remove contaminants other than condensate. Turbocharged engines with draw-through (low fuel inlet pressure) systems have a good record. The only protection found to be necessary has been high-temperature cooling systems to guard against corrosion. Other engines, however, require extensive and costly contaminant removal systems. Engine manufacturers should be consulted and required to furnish guarantees regarding requirements for fuel quality before considering them for candidates for projects where biogas will be the fuel.

Diesel Fuel

Diesel oil is an excellent choice for standby systems. At current prices it is too expensive for continuous operation. Its advantages are:

- The high flash point and low volatility makes diesel oil safer to store than most fuels.
- Higher compression ratios compared to gas engines permit greater horsepower per cylinder and improved efficiency.
- It is common and easily obtainable.

The disadvantages of diesel fuel include the following:

- Diesel oil deteriorates in storage: at 6- to 12-month intervals, the tank must be drained and refilled. An alternative is to plan operating cycles and limit fuel tank size to ensure a full turnover of fuel inventory every 6 months.
- The fuel supply systems are more complex than most.
- Engines are harder to start than other types.
- Engine maintenance costs are high.
- Engines are costly (but no more so than gas engines).

Gasoline

Gasoline is often used for engines powering small- to medium-sized portable pumps, but the disadvantages for stationary gasoline engines are overwhelming.

- Gasoline is a fire and explosion hazard due to its low flash point and high vaporization rate.
- Gasoline engines generally are more temperamental than other types.

- Gasoline engines are relatively high-speed designs, and they are limited to approximately 375 kW (500 hp).
- The allowable storage period for gasoline is very short.

14-4. Aspiration

Engines may be either naturally aspirated or pressure-charged.

Pressure charging is typically accomplished by an exhaust gas-driven turbocharger or an engine accessory train-driven blower. For most engines, the combustion air is pressurized and the fuel is then added at the combustion chamber. This method of operation requires significant [140 to 420 kPa (20 to 60 lb/in.^{2})] pressure for the introduction of fuel into the combustion chamber. Diesel fuel is usually injected with an engine accessory train-driven pump. A variety of means, including fuel injection systems, are used for gas-fueled engines. A simpler system, available from some manufacturers, uses an arrangement wherein the fuel/air mixture is established at atmospheric pressure through a carburetion system, then compressed and delivered to the combustion chambers. This latter arrangement has proven particularly attractive for dirtier fuels and for control of exhaust emissions. Draw-through designs have also demonstrated much higher efficiency than other pressure-charged gas-engine designs. Turbocharged engines can more easily achieve exhaust emission requirements than naturally aspirated engines. The fuel gas passages and metering devices for naturally aspirated engines are large—an advantage when using biogas or other dirty gas fuels. Naturally aspirated engines may not be capable of meeting air pollution code restrictions in degraded air basins, even with currently available control technology.

Pressure charging is accomplished by an exhaust gas-driven turbocharger or by an engine accessory train-driven blower. With pressure charging, fuel gas pressures are high [from 140 to 550 kPa (20 to 80 lb/in.^{2})], engine efficiency is increased by 10% or more, and cooling requirements increase because more heat is wasted.

Pressure-charged designs that incorporate clean-burn technology are capable of achieving air pollution code restrictions without additional emission control technology.

14-5. Types of Engines

Engines useful in pumping stations can be categorized by ignition, cycle, and configuration.

Ignition

Based on ignition, engines can be subdivided into those using spark ignition and those using compression ignition.

With spark ignition, the fuel charge is ignited, as in most automobile engines, by a spark arcing between two direct current electrodes. Medium- or slow-speed spark-ignited engines should be specified with breakerless, electronic capacitor discharge-type ignition systems. If available, dual spark plugs should be specified for larger-bore engines.

Compression ignition engines employ the heat generated by the compression of the fuel-air mixture to ignite the fuel charge. Compression ignition engines include both straight diesel and diesel-gas (dual-fuel) engines.

Cycle

Stationary engines are available in both two- and four-stroke cycle designs. Two-stroke engines require a separate scavenging blower (usually powered from the engine accessory train) to remove exhaust gases from the cylinder and introduce combustion air into the cylinder. In general, two-stroke cycle engines are limited to compression ignition designs and are less efficient, more complex, lighter in weight, less expensive, and noisier than four-stroke cycle engines.

Four-stroke cycle engines are available as both spark ignition and compression ignition types.

Configuration

Stationary engines are available in either in-line or vee-block configurations. Compared with in-line engines on the same power basis, vee engines are generally more complex, more expensive to maintain, smoother operating, more compact, and require less floor space.

14-6. Application Criteria

Application criteria, the rules by which a designer establishes how engine manufacturers may apply their products to a given power requirement, include the following:

- Brake mean effective pressure
- Piston speed
- Rotational speed.

Recognize that not all engine manufacturers use the same criteria for furnishing the buyer with output power ratings for their products. Some manufacturers' ratings are very conservative, whereas others tend to overstress their products. Instead of accepting manufacturers' ratings, provide a set of criteria applicable to all manufacturers as a means of establishing a fair basis for bidding. The following recommended application criteria will produce a consistent approach to the selection of internal combustion engines.

Brake Mean Effective Pressure

Brake mean effective pressure (BMEP) is a widely used empirical measure of the load imposed on an engine. Although not universally accepted by all engine designers, it does provide a useful tool for controlling the size of the equipment to be used for a particular application. In SI units, BMEP is given by

$$\text{BMEP} = \frac{bkW}{LAn} = \frac{60\ bkW}{LAN} \tag{14-1a}$$

where the BMEP is in kilopascals, bkW is the brake kilowatts required at the output shaft, L is length of piston stroke in meters, A is net piston area in square meters, n is number of power strokes per second, and N is number of power strokes per minute. In U.S. customary units

$$\text{BMEP} = \frac{33{,}000\ bkW}{VN} \tag{14-1b}$$

where BMEP is in pounds per square inch, bhp is the brake horsepower required at the output shaft, L is piston stroke in feet, A is piston area in square inches, and N is number of power strokes per minute. Note that the coefficient 33,000 in Equation 14-1b reduces to 1.00 in Equation 14-1a.

When applied to two-cycle engines, Equation 14-1a reduces to

$$\text{BMEP} = \frac{60{,}000\ bkW}{VN} \tag{14-2a}$$

where V is the displacement in liters and N is number of power strokes (or engine revolutions) per minute. In U.S. customary units, Equation 14-1b reduces to

$$\text{BMEP} = \frac{396{,}000\ bhp}{VN} \tag{14-2b}$$

where V is the displacement in cubic inches. For four-cycle engines, the right side of Equation 14-2 must be doubled. A suggested set of BMEP limits for four-stroke cycle engines and various types of engine applications is given in Table 14-1.

Another approach advocated by some authorities is to limit BMEP to a percentage of the BMEP at the manufacturer's listed maximum power rating. The following are suggested limits to the engine's maximum power requirement.

- Continuous duty, relatively constant power demand—70%
- Continuous duty, variable power demand—75%
- Standby duty, incremental application of load—85%
- Standby duty, full-load application after starting—80%.

Derating for Altitude

Some engines, particularly turbocharged designs, may not be capable of achieving acceptable performance if the criteria in this section (14-6) are used. Note that the BMEP limitations should be adjusted downward for increasing altitude and increasing ambient temperature—1% for every 150 m above 460 m elevation (every 500 ft above 1500 ft elevation) and 1% for every 5°C above 32°C (every 10°F above 90°F).

The above rules are simplified. A better approach is to use all of the application criteria in this section.

Table 14-1. Recommended BMEP for Four-Cycle Engines[a]

	kPa			lb/in.^{2}		
		Turbocharged			Turbocharged	
Duty	Naturally aspirated	Gas	Diesel	Naturally aspirated	Gas	Diesel
Standby	660	1100	1600	95	160	170
Continuous						
Constant load	550	790	900	80	115	130
Variable load	590	860	930	85	125	135

[a] Sea level pressure and air temperature below 32°C (90°F)

Piston Speed

Piston speed, which is actually the *average* piston speed, is calculated as

$$v = 2LN = \frac{LN}{30} \qquad (14\text{-}3a)$$

where v is piston speed in meters per second and L, n, and N are as defined for Equation 14-la. In U.S. customary units

$$v = \frac{LN}{6} \qquad (14\text{-}3b)$$

where v is piston speed in feet per minute, L is stroke in inches, and N is number of power strokes per minute.

Piston speed is an indicator of (1) the rate of wear on cylinder walls and piston rings, (2) the magnitude of inertial forces at the top and bottom of stroke, and (3) the class of engine design—that is, "low," "medium," or "high" speed. Item 3 refers to rotating speed only indirectly. The important factor is the length of piston stroke at a given speed. The following limitations are suggested for average piston speed:

- Continuous duty—6.1 to 7.1 m/s (1200 to 1400 ft/min)
- Standby duty—6.6 to 8.1 m/s (1300 to 1600 ft/min).

Rotative Speed

Using rotative speed to compare competing engine designs can be somewhat misleading. Rotative speed does relate, however, to the frequency of piston reversals and, hence, to the rate of maximum stress occurrences.

The suggested limits for rotative speed are as follows:

- Continuous duty, 115 kW (150 hp) or less—20 Hz = 1200 rev/min
- Continuous duty, greater than 115 kW (150 hp)—15 Hz = 900 rev/min
- Standby duty, up to 115 kW (150 hp)—30 Hz = 1800 rev/min
- Standby duty, up to 375 kW (500 hp)—20 Hz = 1200 rev/min
- Standby duty, 375 kW (500 hp) or more—15 Hz = 900 rev/min.

Other Criteria

Other generalizations that may be useful for comparing competitive designs include the exhaust valve port area (a larger area requires less valve maintenance), the number of pistons (fewer pistons mean less complexity), and the experiences of other users, particularly in similar applications.

14-7. Starting Methods

Stationary engines may be started using one of the following stored-energy systems: compressed air, electricity (direct current from storage batteries), or hydraulic fluid.

When selecting a starting method, place substantial importance on storing sufficient reserve energy for a series of starting attempts without resorting to commercial power. A list of advantages and disadvantages of each starting system is given in Table 14-2. If either electric or hydraulic starting is chosen, the engine manufacturer usually supplies all of the starting equipment. The design of compressor, receiver, and piping systems for compressed air starting is usually the responsibility of the pumping station designer, however. Suggested design criteria for the various starting methods are given in Table 14-3.

14-8. Cooling Methods

Because the cooling requirements of each engine design are unique, the equipment should be furnished by the engine manufacturer. Cooling methods suitable for use with stationary engines include radiators, heat exchangers, and ebullient (boiling water) cooling. Air cooling, which is available on some smaller engines, is not a viable system for the engines used in most pumping stations.

The advantages and disadvantages of each method are given in Table 14-4 and the details of radiator, heat exchanger, and ebullient cooling are shown in Figures 14-1 through 14-3.

14-9. Controls

The engine manufacturer should be required to furnish all of the controls associated with the routine start-up and shut-down and emergency shut-down of the equipment. If remote start–stop initiation is needed, require interfacing relays for field connection to manufacturer-furnished logic circuits. Other important features of the control system include the following:

- Use direct current (24, 48, or 125 V) for all engine-related controls. Do not use alternating current because of power failures and loss of reliability.

Table 14-2. Comparison of Engine Starting Methods

Method	Advantages	Disadvantages
Electric		
DC starting motor with rectifier batteries	Lowest cost Simple system Can be mounted on engine base	Effectiveness and amount of stored energy falls with falling air temperature Batteries require considerable care Batteries subject to damage if rectifier fails or overcharges; use float-type charger Limited effectiveness on larger engines
Compressed air		
Air motor with compressors and receivers	Most reliable Can be used on all but largest engines Consistent power available until stored air is exhausted	Most costly Requires greatest building space More complex than electric Requires greatest air-storage volume Larger volume of air required per start
Direct injection with compressors and receivers	Very rapid start Requires less storage volume Smaller compressors	Higher pressures required Available on larger engines only Most applicable to diesel engines
Hydraulic		
Hydraulic motor with electric motor-driven pump, hand pump, and nitrogen-charged accumulators	Very rapid start High-speed cranking Requires least floor space May be appropriate for portable generators	System is complex Leaks in hydraulic lines can be a problem. System is costly to maintain

- Do not locate the control panel on or near the engine foundation because the vibration causes control system maintenance problems.
- Engine controls must be coordinated with controls for the driven machine. For example, if it is an engine-driven pump with power-operated valves, the valve controls must be coordinated with engine start-up/shut-down controls to avoid sudden surges or reverse engine rotation.
- Specify dust-tight construction for the control enclosure to protect the equipment.
- The recommended control and monitoring system features are listed in Table 14-5.

14-10. Governors for Engine Control

Two basic types of governors may be employed for engine control: (1) isochronous: (steady speed regardless of load) and (2) droop (speed varies slightly with engine load).

Isochronous governors are necessary where two or more units are operated in parallel and precise regulation and load sharing are required, such as when paralleling on an electrical bus. Droop mode is usually satisfactory for standby generators.

All governors use some type of detector (to monitor the engine speed) and a means of developing sufficient force to activate the throttle. Hydraulic governors are driven from the engine and use either a self-contained oil system or engine lubricating oil. Electric governors depend entirely on electronic/electric actuators. Pneumatic/hydraulic or electronic/hydraulic governors are the most suitable for variable-speed service. The recommended governor types for various applications are listed in Table 14-6.

14-11. Accessories for Engines

The recommended accessories for engines include the following:

- An electric or air motor-driven lubrication-oil priming pump mounted on the engine subbase. The pump should be timer controlled to operate hourly and prior to automatic engine start.
- Dual oil filters with quick shift-over valves for larger, continuous-duty engines.
- A lubrication-oil-level regulator (in all spark ignition and some diesel engines) to replace the oil consumed during operation automatically.

Table 14-3. Suggested Design Criteria for Engine Starting Methods

Starting method	Components	Design criteria
Electric	Rectifier	Floating-charge and quick-recovery features with automatic voltage regulation.
	Motor	12 or 24 V dc with Bendix release.
	Batteries	12 or 24 V nickel-cadmium type; sufficient storage capacity for at least three starts at 30°F (lower, if appropriate for project).
Compressed air	Compressor	Two- or three-stage air-cooled (liquid-cooled required for higher pressures). Provide two. Size to recharge receivers in 1 hr. Set start/stop controls to ensure receivers are fully charged. Allow no more than 10% loss of pressure before starting lead compressor alarm at 15% loss of pressure. Aftercooling is not necessary if discharge piping is insulated.
	Receivers	Provide at least two. Store air at 300 lb/in.^{2} (gauge) or greater to reduce size and provide better starting action. Provide sufficient storage for at least three successive starts—more if more than one engine. Size piping to engines to minimize pressure loss. Use copper piping to prevent corrosion.
	Starting motor (if required)	Provide pressure-reducing station at motor to adjust for motor limitations. Must be fitted with Bendix release. Provide lubricator and exhaust silencer. Solenoid valve located on receiver side of flexible hose connection to engine piping system.
Hydraulic	Pump (electric)	System pressure not more than 3000 lb/in.^{2} (gauge)
	Pump (engine)	Piston type with internal relief and check valves.
	Pump (hand)	Double-acting piston type with lever for hand operation.
	Accumulators	Nitrogen-precharged bladder type. Sufficient storage volume for at least four starting attempts. Not less than four accumulators. Independent means for draining each accumulator with pressure gauge on gas side. Filters required to protect seals and rings in motors, pumps, and accumulators.
	Starting motor	Multipiston type with Bendix release.

- Crankcase explosion vent valves (on large diesel and dual-fuel engines).
- A thermostatically controlled jacket water heater (on automatically started engines, especially diesels and in colder climates).
- Engine-driven circulating water pumps for external cooling water systems (heat exchanger and ebullient cooling).
- Blow-by emission absorbers for crankcase vents (specify cartridge type).

14-12. Combustion Air

Small- to medium-sized engines can be furnished with engine-mounted combustion air filters. If the

Table 14-4. Engine Cooling Systems

System	Equipment	Operation	Advantages	Disadvantages
Radiator, (Figure 14-1)	Engine jacket water circulating pump Thermostat Lube oil heat exchanger Radiator and fan (if located at engine, fan is usually engine driven. If radiator is remote from engine, fan is motor driven)	Water is circulated through engine, oil heat exchanger, and radiator by pump Thermostat maintains desired system water temperature Fan circulates air through radiator, thus cooling the water	Least costly system Simple to operate and maintain All engine heat, including oil, can be accommodated	Fan requires considerable power System is not readily adaptable for reclamation of waste engine heat Radiator is vulnerable to leakage (typical pressure rating 10–15 lb/in.^{2} (gauge) Fans make considerable noise Difficult to provide high temperature cooling needed when biogas is used as fuel Thermostat failure can cause engine overheating
Heat exchanger, (Figure 14-2)	Engine jacket water pump Thermostat Lube oil heat exchanger Waste heat exchanger External source of cooling water	Water is circulated through engine, oil heat exchanger, and waste heat exchanger by jacket water pump Water from an external source is circulated through waste heat exchanger to cool engine jacket water	Relatively simple system Modest cost Possible to reclaim waste heat for useful purposes Water or wastewater pumped by pumping station can be source of cooling water (through heat exchanger) All engine heat can be accommodated	Two pumped water systems Energy consumption can be significant System is not particularly suited for reclamation of exhaust heat Thermostat failure can cause engine overheating
Ebullient cooling, (Figure 14-3)	Engine jacket water pump may or may not be necessary Steam separator Exhaust boiler (if exhaust heat reclamation is desired) Source of softened make-up water Source of condensing water if reclamation of waste engine heat desired Source of cooling water (or radiator) required for engine oil	Water circulates through engine to steam separator (and exhaust heat boiler) thermally if no jacket water pump is required Jacket water flashes to steam. If no heat reclamation, make-up water is added to steam separator to replace lost water; if heat reclamation is desired, cooling water in steam condensing heat exchanger condenses steam; cooled water returns to engine; typical engine water inlet temperature is 225°F, outlet 228°F	Low temperature differential across engine reduces thermal stresses and wear Stable, simple, easy to maintain Ideal for maintaining high exhaust temperatures necessary for biogas fuels Particularly suited for reclamation of waste engine heat Quiet Power consumption nearly negligible; system is safe, impossible to overheat engine	Higher first cost Requires special engine construction (bearings, gaskets, and seals) Full advantage of system realized by utilizing thermal circulation Many engine designs are not suitable

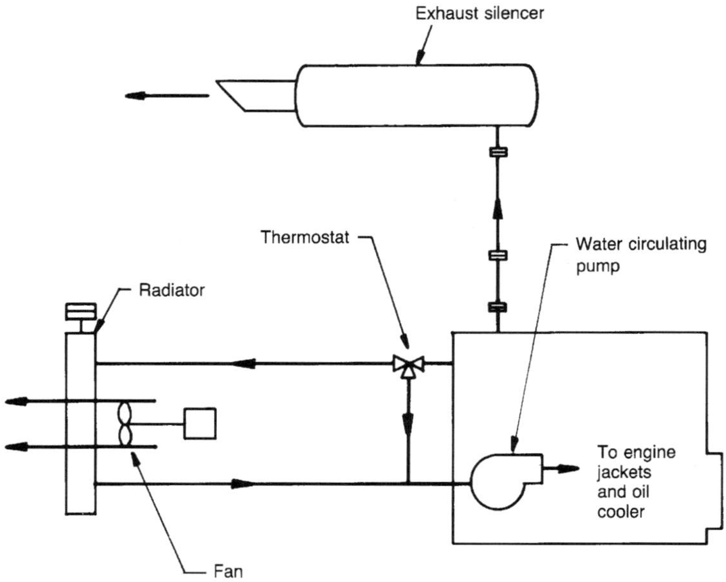

Figure 14-1. Radiator cooling. If the radiator is remote from the engine, the fan is motor driven; if the radiator is at the engine, the fan is engine driven. The thermostat controls the temperature of the water leaving the engine. The water circulating pump is engine driven. After Brown and Caldwell Consultants.

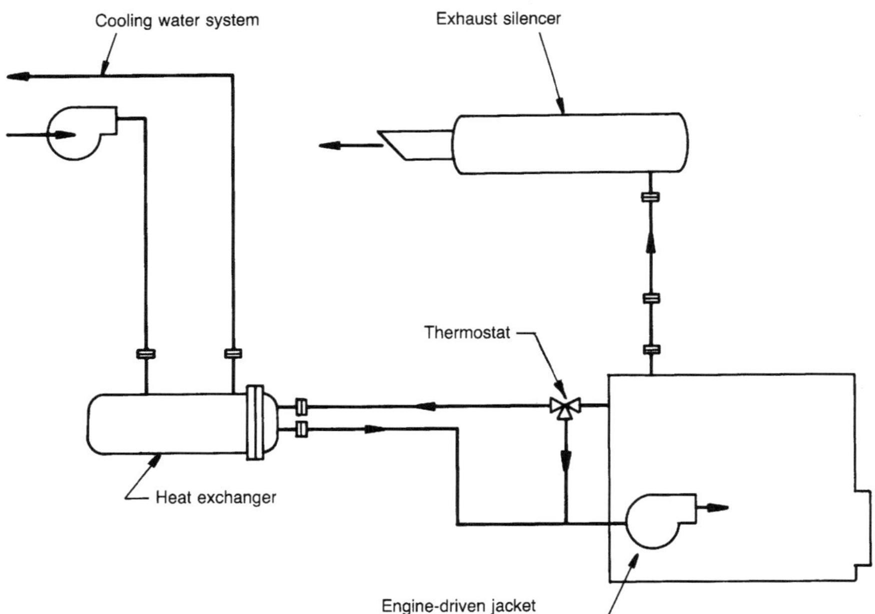

Figure 14-2. Heat-exchanger cooling. With the proper heat exchanger, the cooling water can be the flow from the pumping station and need not go to waste. The thermostat maintains the engine water outlet temperature. The exhaust silencer can be a heat exchanger or a boiler for heat reclamation. After Brown and Caldwell Consultants.

engine room ventilation air is filtered, specify dry-type filter elements. Larger engines require an air intake system ducted outside of the building to avoid upsetting the building ventilation system, and the pumping station designer should be responsible for such an air intake system. If turbocharged engines are installed, a separate silencer is needed to control intake noise.

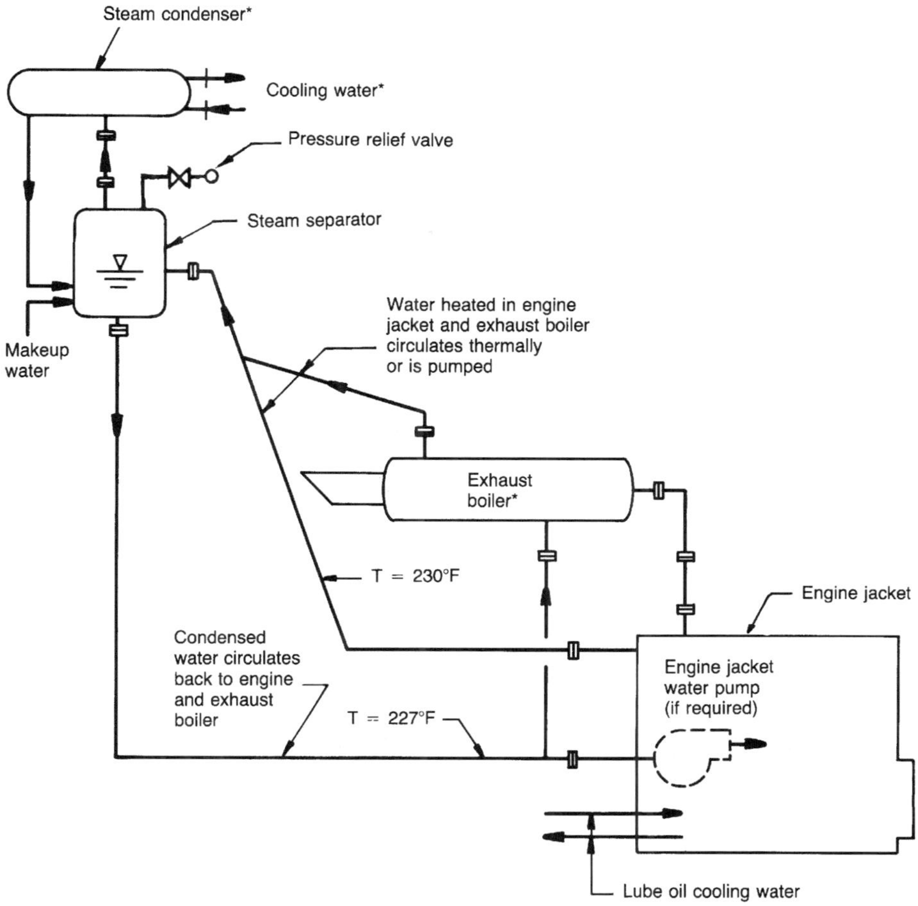

Figure 14-3. Ebullient cooling system. In the steam condenser, steam is condensed and returned to the separator (the conversion is 950 Btu/lb of steam). In the steam separator, heated water flashes to steam (the conversion releases 950 Btu/lb of water). The asterisks (*) indicate the components that are necessary to reclaim engine waste heat. If heat reclamation is not desired, the steam condenser, the steam separator, and the exhaust boiler can be eliminated, and the system will operate by boiling off water, but provide for blowdown of precipitated solids. After Brown and Caldwell Consultants.

14-13. Exhaust Silencing

Exhaust silencers must be selected specifically for the application because of engine back-pressure limitations and should, therefore, be furnished by the engine manufacturer. Chambered, ported types appear to be the best. There are four different grades of effectiveness:

- *Industrial*—minimal sound attenuation
- *Commercial*—suitable for nonresidential areas
- *Residential*—applications where a high degree of silencing is required
- *Critical*—maximum silencing for locations near hospitals and very quiet residential areas.

Do not rely solely on the silencer manufacturer's rating for the equipment. Instead, specify noise reduction by octave band. In critical applications, specify a result, that is, the noise level on the decibel-A scale (dbA) at a specific distance from the exhaust. An acoustical engineer should be consulted in such instances (see also Chapter 22).

Because of their design, exhaust heat boilers provide a high degree of silencing and approach residential quality. Specify inspection ports and drain plugs for all silencers and for the gas side of all exhaust heat boilers. (Davited inspection doors are preferred for boilers.) Stainless steel (type 347) may be advisable for silencers serving engines that operate only intermittently.

14-14. Pollution Control

Federal regulations and many local air pollution control codes require some form of exhaust emission control.

Table 14-5. Recommended Engine Controls, Monitors, and Alarms

Controls	Monitors	Alarms
Start-up logic	Oil temperature and pressure	Low oil pressure[a]
Cranking routine	Jacket water temperature	Jacket water temperature (radiator and heat exchanger cooling only)[a]
Actuation of associated remote devices	Engine speed	Low water level[a]
Status	Intake manifold vacuum (gas engines)	High oil temperature
Shut-down logic	Cylinder exhaust temperature (larger, continuous-duty engines)	Cylinder exhaust temperature (large, continuous-duty engines)
Actuation of engine-mounted and remote devices	Main bearing temperature (larger, continuous-duty engines)	Vibration (larger engines)[a]
Status	Vibration (larger engines)	Overspeed[a]
Emergency shut-down logic		
Actuation of engine-mounted and remote devices		
Status		
Manual start-up/shut-down		
Timer and logic for lube oil priming pump		
Timer and logic for remote devices such as cooling fans and pumps		

[a] Initiates emergency shut-down.

Pollutants of interest include particulates, and oxides of sulfur (SO_x) and nitrogen (NO_x). Many control technologies have been attempted—with mixed results.

Generally speaking, exhaust converters using noble metal catalysts have not worked well even when using a clean-burning fuel such as natural gas. Turbocharged engines, using a very lean mixture in the combustion chamber, have achieved better (but not reliable) performance. The reason is that many of the systems utilize sensors that are located in hostile environments such as combustion chambers and exhaust gas streams, and they depend on sensitive instrumentation systems. Maintenance costs are high because of the labor required to keep the equipment properly tuned for optimum emission performance.

Emerging technologies using parameters easily measured outside the combustion chamber and exhaust gas stream, such as fuel charge temperature and pressure, engine torque, and so on, promise improved performance and reduced maintenance costs. Some guidance regarding current (January 1996) emission control technology used by various manufacturers of turbocharged engines is given in Table 14-7.

Table 14-6. Recommended Governor Types for Engines

Application	Governor type
Generation	
Standby generator (single unit)	Droop, hydraulic
Standby generator (two, in parallel)	Isochronous, with synchronizing feature electronic/hydraulic
Continuous duty multiple units or cogeneration with utility	Isochronous, with synchronizing feature electronic/hydraulic
Direct drive	
Constant speed	Droop, hydraulic
Variable speed	Electrohydraulic or pneumatic/hydraulic

14-15. Vibration Isolation

Depending on the application, the subsoil conditions, and the nature of the building housing the engines, the following alternatives can be used to isolate engine vibrations from the building (in descending order of isolation efficiency). See also Chapter 22.

- Separate foundation, with vibration-absorbing filler between the building and the engine foundations
- Heavy inertia block, mounted on energy-absorbing material or spring-type isolators
- Spring-type energy-absorbing isolators mounted on a heavy engine subbase (the subbase may be filled with concrete to lower the center of gravity)
- Cork and elastomer type energy-absorbing sandwich material under the engine subbase.

Table 14-7. Fuel and Exhaust Emission Control Systems

Manufacturer	System/types	
	Fuel	Exhaust emission
Caterpillar	CI/OC	CS/ES
Cooper	CI	ES
Duete MWM	DT	FS
Jennbacher	DT	FS/T
Waukesha	CI/DT	ES

CI = Combustion chamber injection
DT = Draw-through
OC = Open chamber/carburetion
ES = Exhaust sensors
CS = Combustion chamber sensors
FS = Fuel charge sensors
T = Output torque

Before choosing which vibration isolation system to use, carefully evaluate the natural frequency of the structure and the effect transmitted vibration may have on sensitive electrical and electronic equipment. Because unbalanced forces in engines vary widely between types and manufacturers, the vibration isolation system should be made a part of the engine manufacturer's responsibility. Specify the maximum transmissibility and the maximum amplitude for the engine during operation and on start-up.

14-16. Lubrication Oil Storage and Supply

Oil should not be stored in large volumes inside buildings because of fire codes and insurance costs.

Automatic lubrication oil-makeup units are designed to sustain only modest pressures, so direct pumping is not possible. The solution is to use a small day tank similar to the 30-gal day tank for fuel, as illustrated in Figure 14-4. The system operates on a fill-and-draw basis. Waste oil is pumped to a storage tank by means of a portable, air-operated pump. For large engines, use a fixed pump and permanently connected oil drain piping. Do not use galvanized (zinc-coated) piping or appurtenances in any portion of the lubrication oil storage or supply system, because the zinc may contaminate the oil. The building design should incorporate provisions to contain any spills caused by day tank overflow or rupture.

14-17. Fuel Oil Storage and Supply

Diesel fuel injection systems are not designed for pumped supply but must, instead, draw fuel by gravity from a day tank. The system shown in Figure 14-4 ensures the constant supply pressure required for proper injector operation. A small quantity of excess fuel must be drained back to the fuel storage tank from the engine fuel oil system when the engine is in operation. Never use galvanized (zinc-coated) piping or appurtenances in any portion of the fuel oil storage or supply system.

14-18. Gaseous Fuel Storage and Supply

The design of storage and supply systems for gaseous fuels is governed by NFPA 37 and NFPA 58. In general, the local utility's storage and transmission system provides adequate service for natural gas and no separate storage system should be necessary. If a pressure-charged engine is allowed, however, a separate pressure-boosting system may be required.

Be particularly careful when designing gaseous fuel systems, because the effectiveness of carburetor systems depends on stable fuel supply pressures. If possible, design biogas systems that do not use compression and storage systems for the following reasons:

- The power required for compression and the initial cost of compressors and storage vessels reduces the economic advantages of biogas.
- Because biogas is contaminated with liquids, sulfur dioxide, and particulate matter, the condensate causes corrosion and other maintenance problems in compression equipment.

The gaseous fuel supply to naturally aspirated engines may be troublesome because of the relief valve required in paragraph 4-21 of NFPA 37. The relief valve is extremely sensitive and may leak, which results in the loss of fuel. Be aware of this particular provision and be prepared to deal with it. One obvious solution is to employ a fuel transport pressure of less than 3.5 kPa (0.5 lb/in.^{2}) gauge within the engine room.

14-19. Service Piping

Care must be exercised in the design of the service piping that connects the engine to various piping systems. Examples of service piping include exhaust, fuel, lubrication oil, starting air, and cooling water.

The engine is a source of vibration, so flexible piping sections (or flexible metal hose) should be provided at each engine connection. The piping on both sides of the flexible section should be firmly anchored to prevent the transmission of motion that could cause

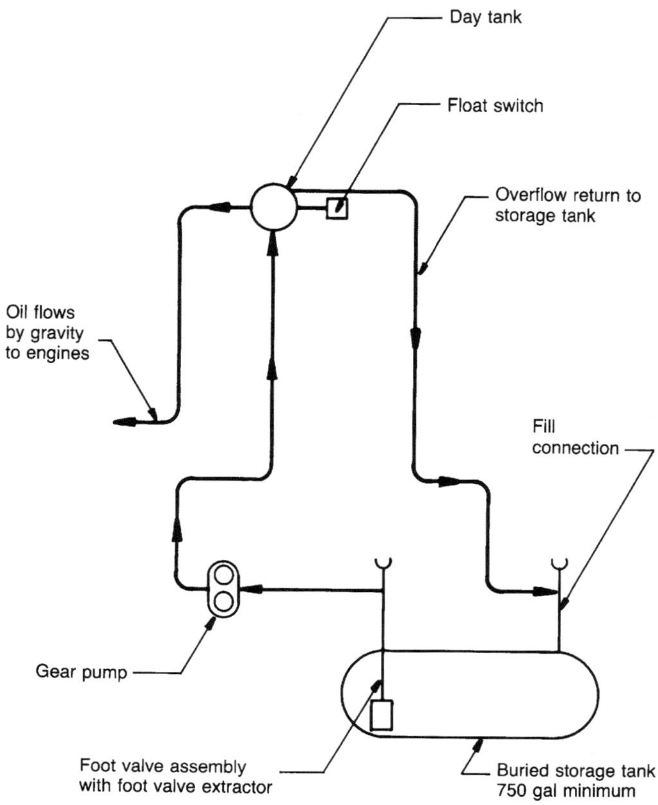

Figure 14-4. Schematic for a lubricating oil supply system. The day tank is less than 30 gal and is located high in the superstructure. The float switch controls the pump on a fill and draw basis. After Brown and Caldwell Consultants.

the joints or flexible hoses to fail. Also observe the following considerations:

- Install the flexible hose as close as possible to the exhaust outlet.
- Avoid fittings or lengths of pipe between the engine exhaust connection and the flexible hose; the weight could damage the exhaust manifold.
- Try to make flexible piping connections as close as possible to the engine crankshaft.
- Install flexible sections parallel to the engine crankshaft to increase their service life.
- Require flexible piping sections to be designed for a specific number of full displacement cycles (10,000,000 is suggested).
- Install isolating valves and automatically controlled valves, such as fuel and air valves, in the rigid gas piping upstream from any pressurized flexible section to avoid the escape of fuel to the engine room if the flexible section fails.
- Use braided stainless-steel hose rated for a working pressure at least twice as great as the system's maximum pressure.
- Provide flexible duct connectors between engine-mounted radiators and built-in exhaust louvers.

14-20. Building Envelope

When designing structures for engine installations, consider these caveats:

- Avoid the transmission of engine vibration by conducting a thorough analysis of each building element and eliminating harmonic natural frequencies.
- Provide a small room for engine maintenance tools adjacent to the engine installation.
- If permitted by the project budget, conduct a sound spectrum analysis and provide noise-control features. In critical applications where building noise emissions may be objectionable, employ an acoustical consultant. Coordinate the acoustical control efforts with the ventilation system and architectural design.
- Avoid crowding engines. A clear passage of 2 to 2.4 m (6 to 8 ft) between adjacent units is suggested.
- Provide some type of hoisting means (e.g., a bridge crane or a rolling gantry with chain hoist) to aid in engine maintenance.
- Be aware of special engine maintenance clearance requirements. One example is the vertical distance above the heads that is necessary to pull the pistons on some engine designs.

- Design for explosion venting (see NFPA 37).
- Except for gas compressors, explosionproof electrical devices for engine rooms are not required by any code. Engine rooms, however, should be designed for explosion venting when gas engines are used (see NFPA 37 and NFPA 68).
- Provide a means for draining the crankcase. A portable air-operated pump to remove the oil from the engine sump is preferred. If supplied from a fixed-service air system, the air can also be used for other purposes (e.g., torque wrenches and other air-driven tools).

14-21. Ventilation

Ventilation systems for engine rooms should be designed to provide the following:

- Balanced room pressure
- Powered intake and exhaust for continuous-duty applications
- A maximum room temperature of 43°C (110°F)
- Positive air flow along each side of each engine (which is necessary for personnel comfort)
- Air flow from driven equipment (pump, gear, generator) directed toward the engine
- 10% of the total air flow exhausted through the roof to clear out smoke and fumes; the rest should be exhausted at the wall at the rear of the engine
- Adequate engine combustion air if engine-mounted air intake systems are provided—a *variable* that depends on the engine load and the number of engines in operation
- Filtered air for continuous-duty applications, tempered air only to 13°C (55°F) in the winter; cooling is necessary only in hot climates.

14-22. Maintenance

The manufacturer's instructions for maintenance should, of course, be followed exactly, but a designer contemplating an engine drive should recognize the owner's involvement. The maintenance of engines depends on whether (1) the engine is used as the prime mover and is, thus, in regular operation or (2) the engine is on standby and (including exercising) operates, on average, only about 50 h per year.

Prime Movers

The life-cycle costs for a continuous duty diesel engine are compared with those for a gas engine in Tables 14-8 and 14-9. The total costs are based on an operating period of 20 years.

Standby Engines

Standby engines operate so seldom that maintenance depends on time as well as wear. Maintenance tasks for a 125-kW diesel engine-generator set are shown in Table 14-10. On a yearly contract, the total cost of maintenance labor and parts would be about $700/yr if an allowance of 2 h driving time is assumed per yearly visit.

There is a considerable difference between the maintenance of light, standby units and continuous-duty prime movers. For example, the service interval for valve overhaul in a heavy-duty engine is 20 times as long as the interval given for the light-duty engine in Table 14-10.

Exercising Engines

Instructions vary, but the following recommendations are good practice for modern engines.

Continuous-Duty Engines

At least once every 6 months, all engines should be operated at the rated load for at least 5 to 10 min after reaching operating temperatures to ensure that the engine is capable of developing the required rated power.

Standby Service

Every 6 months to a year, all engines should be full-load tested as follows: Operate at 1 or 2 min at no load, 5 min at 25% load, 5 min at 50% load, 5 min at 75% load, and 5 min at rated load.

Engines should be no-load exercised at least once each month. The engine should be run just long enough to ascertain that the starting and control systems are operating properly and that proper oil pressure is obtained. Excessive no-load operation causes the exhaust system to "slobber"—a problem that is mostly cosmetic (messy). Slobber, an oily discharge from the manifold stemming from fuel that is partly unburned, is especially noticeable in modern direct-injected diesel engines. The exercising can be done by an automatic timing device installed in the transfer switch. These devices are usually 1-week cycle timers. There are 2-week cycle timers, but they may not be available for most switches. Many engineers dislike automatic exercisers, because (1) an operator

Table 14-8. Data for Diesel and Natural Gas Engines[a,b]

Item	Diesel engine	Gas engine
Size of engine, hp	1150	1150
Engine output, kW	820	820
Engine speed, rev/min	1800	1200
Total engine price, $	158,000.00	268,000.00
Down payment, $	58,000.00	68,000.00
Loan interest rate, %	10.00	10.00
Inflation rate, %	4.00	4.00
Months payment due	120	120
Taxes, $	9,875.00	16,570.00
Insurance, $	0.00	0.00
Operation, h/yr	8000	8000
Fuel	No. 2 diesel	Dry natural gas
Fuel price	1.00 $/gal	0.35 ft^3
Fuel consumption at average load	59.06 gal/h	9500 ft^3/h
Oil	10W30	SAE 30
Oil consumption at average load	0.14 gal/h	0.05 gal/h
Oil change interval	500. h	1000. h
Sump size, gal	81.1	111.7
Overhaul interval		
Overhaul 1 (top end)	6530. h	16,006. h
Overhaul 2 (major)	13,061. h	48,017. h
Overhaul 3	———	80,002. h

[a] Courtesy of Caterpillar, Inc.
[b] Prices effective June 1997. ENRCCI = 5700.

Table 14-9. Financial Summary for Twenty Years of Operations[a,b]

Item	Diesel cost			Natural gas cost		
	$	$/h	¢/kW·h	$	$/h	¢/kW·h
Selling price	158,000	0.99	0.12	268,000	1.68	0.20
Interest	58,581	0.37	0.04	117,162	0.73	0.09
Taxes	9,875	0.06	0.01	16,570	0.10	0.01
Insurance (omitted)	—	—	—	—	—	—
Fuel	9,450,000	59.06	7.20	5,320,000	33.25	4.05
Oil	288,500	1.80	0.22	151,400	0.95	0.12
Preventive maintenance	241,500	1.51	0.18	153,900	0.96	0.12
Components	567,500	0.36	0.04	107,600	0.67	0.08
Overhauls	937,900	5.86	0.72	295,800	1.85	0.23
Total cost	11,202,000	70.01	8.54	6,430,000	40.19	4.90
With inflation	16,584,000	103.65	12.64	9,393,000	58.71	7.16

[a] Courtesy of Caterpillar, Inc.
[b] Prices effective June 1997. ENRCCI = 5700.

should be listening to an engine while it is exercised, (2) the maintenance crew may depend too much on automation and neglect the trouble-forestalling inspection of the running engine, and (3) very few automated exercising systems can adequately exercise the engine under load. The owner should be consulted about automatic versus manual control of exercising.

Manufacturers' Recommendations

Although the above exercising instructions comprise one manufacturer's recommendation for modern engines, there are many other makes and models and many older engines. Owners should be advised to follow the maker's recommendation for the particular engine model and the particular circumstances of service.

Table 14-10. Typical Maintenance Schedule for a 125-kW Standby Diesel Engine-Generator[a]

Maintenance task	Frequency Interval	
	Operating time	Calendar time
Inspect fuel, oil level, coolant	8 h	1 mo[b]
Inspect air cleaner, battery	50 h	1 yr
Clean governor linkage, breather, air cleaner	100 h	1 yr
Clean fuel filter, replace oil filter, change crankcase oil, check switchgear	200 h	1 yr
Clean commutator, collector rings, relays, cooling system; inspect brushes, valve clearances, starting and stopping systems, water pump	500 h	1 yr
Check injectors, grind valves (if required), remove carbon, clean oil passages, replace secondary fuel filter, grease bearings, clean generator,	1000 h	—

[a] Courtesy Onan Corporation.
[b] mo, Month.

Batteries

Starting batteries must be reliable and capable of five 10-s start attempts without a loss of cranking power. Maintenance-free batteries are recommended with either a two-stage trickle charger or, better, a float charger that is used to sense the charge on the battery and to limit the charging current accordingly.

Heaters

Moisture is the enemy of electrical equipment, including generators. Install space heaters inside the generators to keep them as dry as possible. Install heaters on the engine cooling water to keep the engine warm. Use a diesel fuel that is compatible with the ambient temperature at the storage location.

Chapter 15

Variable-Speed Pumping

MAYO GOTTLIEBSON
ROBERT L. SANKS

CONTRIBUTORS
Stefan M. Abelin
Robert S. Benfell
Harrison C. Bicknell
Ronald W. Duncan
Stanley S. Hong
Casey Jones
Paul C. Leach
Timothy Nance
Michael R. Olson
E. O. Potthoff
Arnold R. Sdano
Richard N. Skeehan
Earle C. Smith
Patricia A. Trager

In this chapter, the uses of variable-speed (V/S) pump operation in water and wastewater pumping stations are presented and discussed. The principal adjustable- and variable-speed drives are reviewed and compared.

The primary goal of V/S *wastewater pumping* is to keep the station discharge rate equal to the influent rate at all times. The station acts as though it were a sewer discharging by gravity. A pumping station that meets this goal requires very little (if any) storage in the wet well, does not rapidly cycle any of the pumps, and does not cause abrupt, large changes in the discharge rate.

The primary goal of V/S *booster pumping* is to maintain a nearly constant pressure at all times regardless of the demand. The discussion given here for V/S booster pumping also applies to high-service (high head) pumping for reducing water hammer in very long transmission mains. Otherwise, V/S is rarely used for water pumping.

In common usage, the term "variable speed" indicates that the speed of the pump is regulated to produce the desired discharge. Electrical engineers, however, consider "variable speed" to mean that the motor speed varies with the load, while the term "adjustable speed" means that the speed is nearly independent of load. An example of an adjustable-speed drive is an adjustable-frequency drive (AFD), sometimes mistakenly called a "variable-frequency" drive. An example of a variable-speed drive is a liquid rheostat in which, although speed varies with both load and control, the proper speed can be attained by the external control. Thus, pump speed can be varied at will by either an adjustable- or a variable-speed drive. In this chapter, both kinds of drives are called "variable speed" or "V/S" for convenience, but when communicating with electrical engineers, confusion can be avoided by strict usage of correct terminology.

Applications to both water and wastewater pumping are described in Sections 15-1, 15-3, and 15-11. Sections 15-2 and 15-4 through 15-6 are generally devoted to wastewater pumping, and Sections 15-7 through 15-10 are confined to water booster pumping.

15-1. Variable Speed versus Constant Speed

The difference between V/S and constant-speed (C/S) pumping is profound and has far-reaching consequences, so the choice should not be lightly made. Some reasons for the choice of V/S or C/S for a particular pumping station may be compelling while others, which may be significant for a different situation, may be trivial. Nevertheless, all of the reasons should be considered thoughtfully.

Advantages of Variable-Speed Pumping

The major advantages of V/S pumping, stemming from the objective that discharge always equals influent flowrate, are:

- No storage is required, so the wet well can be smaller, shallower, and less expensive than for C/S pumping, and size might compensate for the high cost of the V/S equipment.
- Fewer—although larger—pumps are needed in medium- to large-sized stations with savings in space for the pump room and less equipment to maintain.
- Downstream flowrates fluctuate slowly and do not upset treatment processes caused by surges nor do the sudden surges caused by C/S pumping pound downstream piping.
- The almost constant wet well surface and short residence time tend to reduce the production of odors and corrosive gasses.
- Energy savings are likely due to the higher average wet well surface elevation and the usually lower pumping rates and lower pipe friction losses.
- If adequate cooling is provided, motor life is prolonged because of the infrequent starts.
- The high inrush current that characterizes across-the-line starts of C/S drivers is greatly reduced in adjustable-frequency drives (AFDs) and liquid rheostats (although not in eddy-current couplings).
- The soft and reduced number of starts do not cause annoying power dips in nearby buildings. Power dips caused by across-the-line starts for C/S pumps can be avoided with soft start equipment, but the cost of such equipment makes adjustable-frequency drives more attractive economically.

Throttling the engine of a direct engine drive is a simple, inexpensive, and reliable means of achieving V/S, but read Sections 14-2 and 14-22 and consult the engine manufacturers regarding the problems of operating diesel or gas engines at less than full load.

Wastewater

The reasons for using V/S pumping for wastewater include the following:

- Continuous pumping and the very short liquid residence time in V/S operation eliminates or reduces (1) the *deposition* of organic solids—a problem in C/S pumping because of the difficulty of resuspension; (2) *putrefaction*—so the sewage is easier and less costly to treat; (3) the *production of odors* and corrosive, poisonous gases; and (4) *cyclic rise and fall of the water surface* with the resulting pumping of sewer gases into the atmosphere.
- V/S pumping eliminates the long, free fall of sewage and the entrainment of air bubbles that, if sucked into the pump, greatly reduce pump efficiency, although this advantage disappears in a comparison with C/S pumping if a sloping approach pipe is used (see Section 12-7).
- The average static head in C/S pumping can be reduced by 0.6 m (2 ft) or more in V/S pumping by keeping the wet well level the same as the normal level in the influent sewer (see Section 12-6 and Figure 12-35). Turbulence is thereby virtually eliminated, and the release of odorous gases is substantially reduced. The energy cost savings due to the consistently high wet well level may be significant.
- One important reason for using V/S pumping is to produce gradual changes in flow that do not upset (and thereby reduce the efficiency of) downstream processes such as sedimentation. Sudden flow changes caused by C/S pumps, even at significant distances from a treatment plant, can upset the entire treatment process. At one utility, it was found that such flow changes propagated through the entire primary/secondary treatment process and were upsetting the dechlorination system. Forcing the operators to use the V/S drives (that had been installed but, due to indifference, not used) solved the problem.

Water

Water is customarily pumped for long periods (of several hours) at uniform rates (often to fill reservoirs or tanks), so there are fewer reasons for choosing V/S for pumping water than for pumping wastewater, although they may be just as compelling.

- For booster pumping or pumping into a distribution system without a reservoir, V/S can provide constant water pressure during varying rates of usage (this is the most common application).
- Pumps can be ramped up (or down) slowly so that water hammer is reduced. In pipelines so long that $5t_o$ is more than 5 min (about 30 km or 20 mi), surges caused by the startup of C/S pumps can be serious. Solid-state starters or automatically operated control valves can also reduce water hammer, but only in shorter pipelines. The full-speed operation of C/S pumps for long periods (more than several minutes) at near shut-off head causes excessive (1) wear on pump bearings and (2) flexing of the shaft.

Disadvantages of Variable-Speed Pumping

The selection of V/S pumping units should only be made after due consideration of the disadvantages.

- Except for engine drives, V/S adds high cost and complexity, requires more equipment and more maintenance, and reduces reliability. Troubleshooting and repairs require personnel with on-the-job training. Each type of V/S drive requires a different kind of specialized training.
- If the operating personnel do not understand the equipment or find it difficult to maintain, they are likely to turn off the V/S drive, operate the pumps manually at C/S, and thus effectively lose the investment in expensive equipment plus the advantages it offers. Pump sumps designed for V/S pumping have insufficient storage for C/S pumping, and the frequent starting may burn out the motors and certainly reduces their life.
- V/S drives have less electrical efficiency than C/S motor drives (although V/S drives may use less energy because of lower pipeline friction losses).
- Avoiding vibration is more difficult with variable drive shaft speeds because the "windows" (the ranges of allowable supporting structure frequencies) for avoiding resonance are narrower. This avoidance should always be by design—never by luck (see Chapter 22).
- Instrumentation for regulating pump speed must usually be more refined, accurate, and costly than is required for C/S pumping. An exception is the pneumatically operated liquid rheostat, which requires no instrumentation.
- V/S drives are not well adapted to flat system H-Q curves because (1) good efficiencies cannot be obtained throughout the speed range; (2) small changes in speed produce large changes in discharge; and (3) energy losses (and costs) are high.
- Rapid changes in technology can make particular models obsolete.
- V/S drives may be much noisier than C/S drives.
- AFDs in particular are more vulnerable to lightning and electrical disturbances.

The disadvantages of V/S pumping are essentially advantages for C/S pumping. To complete the comparison, consider the following:

- The cost of the large wet wells required for C/S pumping may exceed the cost of V/S drives, but note that excavation costs are site dependent.
- Pump sizes and wet well design can be easily coordinated so that the number of pump starts per hour (for C/S pumps) is well within acceptable limits.
- A wide range in flows can also be accommodated by using several C/S pumps, and the programmers for alternating C/S pumps are more reliable and less expensive than control systems for V/S units. A penalty for using C/S pumps is a larger station, more pumping units, more piping and fittings, and discrete increments of flow.
- Surges due to starting or stopping one pump are not as severe as those caused by a power failure, which may occur when several pumps are running and for which the station and pipeline must be protected from water hammer anyway. Solid-state starters or pump-control valves can eliminate surges in usual circumstances. If the pumping station discharges a small proportion of the treatment plant flow, the sudden change of influent flowrate due to the starting and stopping of a single pump may be negligible. The effect cannot be quantified, so this is a matter of judgment.
- Relatively constant water pressure from a booster pumping station may be obtained with C/S pumps either with flat H-Q curves (if suction pressure is constant) or with elevated or pressurized tanks.

Summary

If objective evaluation of these advantages and disadvantages leads to a decision to use V/S, select the best drive for the particular application.

15-2. Design Considerations

Firm pumping capacity is the maximum station discharge with the largest pump out of service. Similarly, a station with V/S pumps must be able to accommodate any flow (from minimum to maximum) and operate in a normal manner when any one of the pumps or drives is out of service.

A V/S pumping station that includes C/S pumps can provide the same results as a station with all V/S pumps—but only if the relationship between the sizes of the two kinds of pumps is correct (see Sections 15-5 and 15-10) or if the sump is made large enough to permit V/S pumps at low speed to cycle on and off.

Pumps

The number of pumps to be installed in a station depends on (1) the range over which the influent flow rate varies, (2) the available pump selections, and (3) the initial and operating costs.

Wet Wells

For V/S pumping, the wet well need only be a sump for pumping at the instantaneous system flowrate. Because storage as such in the wet well is not needed at all, the wet well size can be reduced to the geometry necessary for adequate pump inlet conditions and access—nothing more. Furthermore, the change of water surface elevation can be reduced to the amount needed for regulating the speed of the pumps. Most V/S control systems start and stop the pumps and regulate speed over a range of about 0.6 m (2 ft) of change in the water elevations in the wet well. (An even smaller range can be used, but a more sophisticated control system may be required.) Hence, some static head can be saved as compared with a C/S pumping system in which the wet well level typically fluctuates over a range of 1.2 m (4 ft) or more.

It is easier to design good wet wells for V/S pumping than for C/S pumping because the free fall into the pool of a C/S pumping station wet well can be avoided together with the entrainment of air bubbles and the release of odors. The deposition of putrescible solids is virtually eliminated in V/S pumping, whereas deposition and long residence times of solids can be a problem in C/S pumping. The design of pump sumps is discussed in Section 12-7.

15-3. Theory of Variable-Speed Pumping

It may be helpful to review the fundamental theory for centrifugal pumps in Section 10-3, which was explained primarily by Equations 10-15 through 10-21 and by Examples 10-2, 10-3, and 10-8.

In pump manufacturers' published data, H-Q curves are usually shown for impellers of several diameters at two or more speeds. Such curves drawn for pump speeds of 880, 695, and 585 rev/min are shown in Figure 15-1. At variable speeds, the pump actually operates on an infinite number of speed curves between the maximum and minimum limits. The minimum speed may be less than the lowest speed for which a published curve is available.

Most nonclog, dry-well sewage pumps can be furnished with an impeller of any diameter desired between the minimum and maximum shown. For example, if the pump whose curves are shown in Figure 15-1 must deliver a maximum of 0.347 m^3/s at 15.2 m (5500 gal/min at 50 ft), the curve for that impeller would pass through those coordinates and would be represented by the dashed lined labeled 20.1 in. (impeller diameter).

Speed versus Flow Curves

When discharging into a hydraulic system, the pump must operate on the system head-capacity (H-Q) curve at all flows. In Figure 15-2, a typical system H-Q curve has been superimposed on the 20.1-in. diameter impeller curves. The pump delivers 0.133 m^3/s (2100 gal/min) at 10.8 m (35.5 ft) when it operates at 585 rev/min, 0.227 m^3/s (3600 gal/min) at 12.3 m (40.5 ft) at 695 rev/min, and 0.347 m^3/s (5500 gal/min) at 15.2 m (50 ft) at 880 rev/min. A curve of speed versus discharge can be plotted from these data, as in Figure 15-3. This curve is probably the most meaningful depiction of V/S pump operation for a single pump. Note that the pump efficiencies (obtained from Figure 15-1) vary from 70% at maximum discharge through 83% at midspeed to 77% at low speed—good efficiencies throughout the usual discharge rates.

Affinity Law for Speed

To determine a speed versus flow curve for any pump impeller, find the flowrates at which three or more of the pump H-Q curves (each at a different speed) intersect the system H-Q curve. If only one or two of the published pump curves intersects the system curve, points on another speed curve can be calculated from the affinity laws as shown in Example 10-2, Section 10-3. Note, incidentally, that points along the system H-Q curve are *not* "corresponding points" and, hence, discharge is *not* proportional to pump speed.

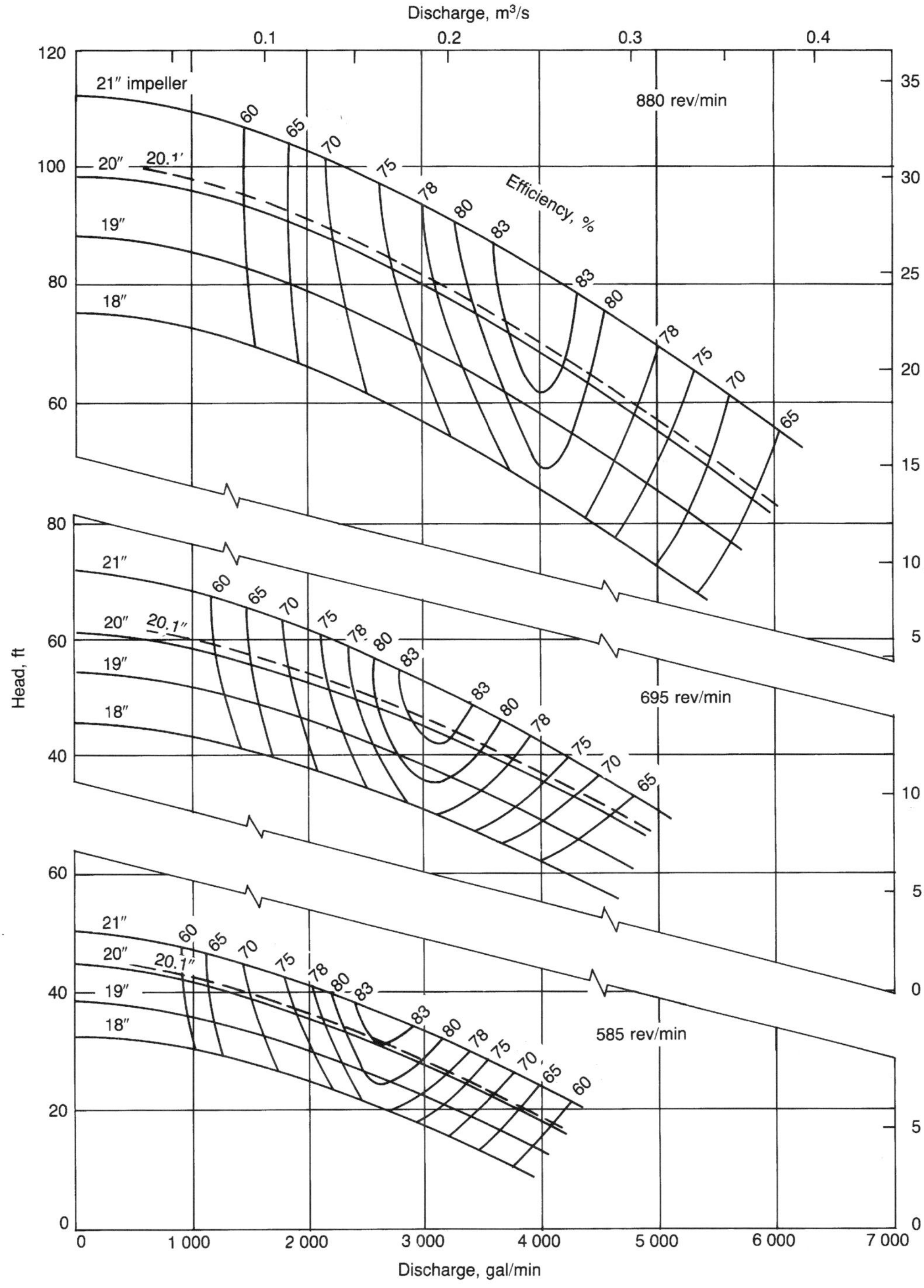

Figure 15-1. Pump characteristic curves at three speeds.

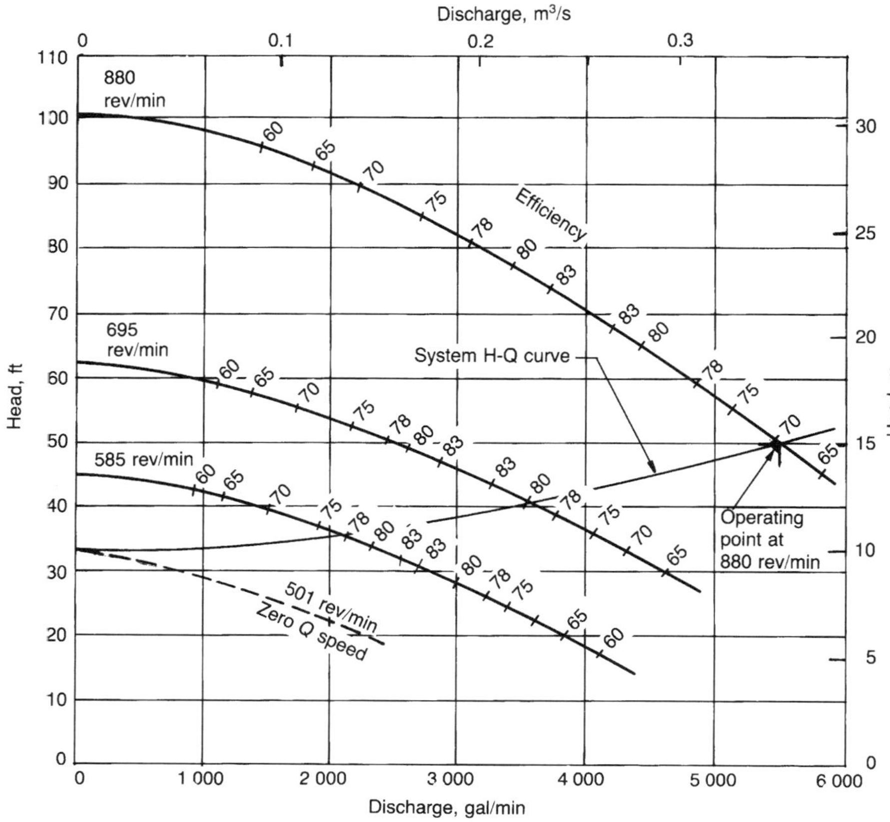

Figure 15-2. Pump characteristic curves for a 20.1-in. impeller.

Minimum Operating Speed

The minimum operating pump shaft speed (zero Q speed) is defined as the speed at which the pump is no longer able to maintain discharge against the static head. Hence it is unique for each pump and each pumping station, and it occurs at the speed for which the pump's shut-off head equals the static head.

At zero discharge (zero Q), and only at zero Q, the pump discharge head can change without a change in the flowrate. Hence, at this unique locus of points, Equation 10-16 is the only one needed for calculating the zero Q speed for a pump discharging into a hydraulic system. Solving that equation for the pump and system H-Q curve in Figure 15-2 gives

$$n_1 = n_2\sqrt{\frac{H_1}{H_2}} = 585\sqrt{\frac{33}{45}} = 501 \text{ rev/min} \qquad (15\text{-}1)$$

and a position of the curve for this speed is shown as a dashed line in Figure 15-2. However, most pumps, especially medium and large ones, should not be operated more than momentarily at zero flow. At zero flow and at low discharge rates, the pumped fluid recirculates within the volute, which causes an unbalanced radial thrust against the impeller that results in a flexing of the pump shaft, which can cause bearing failure or fatigue fracture of the shaft (see Section 10-6).

As a rule of thumb it is usually safe to operate a pump continuously at discharge rates of 30% or more of its best efficiency capacity (bec), which is its discharge capacity at the best efficiency point (bep) at the maximum recommended speed. The pump manufacturer's minimum discharge rate, however, should always be met; the minimum recommendation may be 50% or more of bec for some pumps. By requiring custom-engineered pumps (heavier shafts and bearings), the problem with the minimum recommended operating speed can be avoided with most pump makes. But a custom pump, especially in a nonclog design, might be expensive, so discuss the subject with one or more manufacturers. Perhaps a larger number of smaller pumps would be a good alternative.

Most V/S drives provide a method for preventing the pump from operating below a preselected minimum speed. When the influent rate is less than the dis-

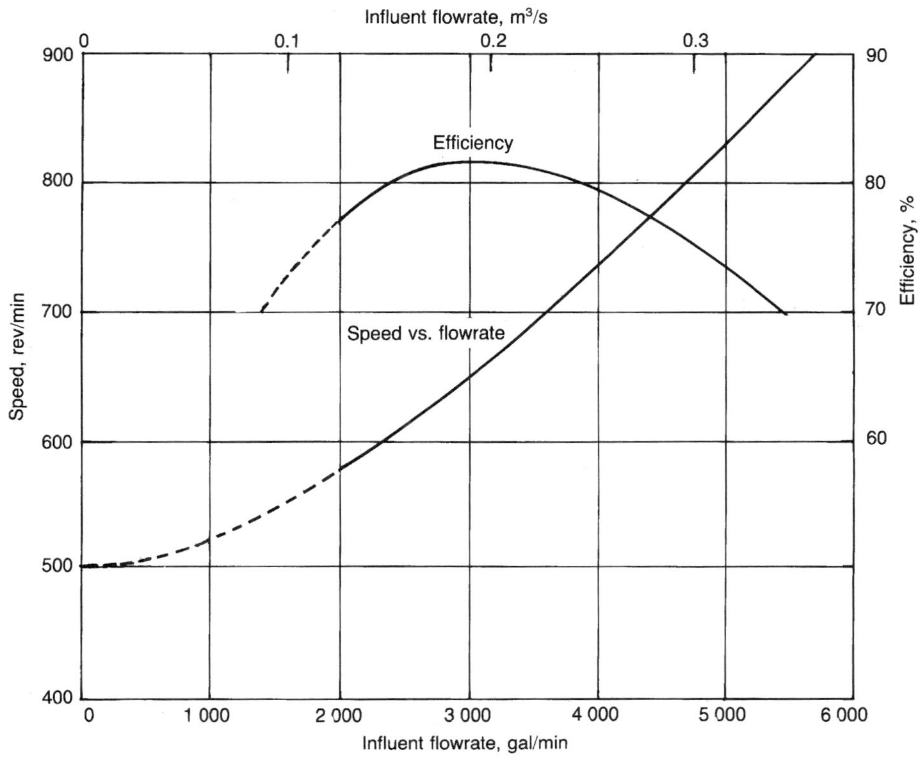

Figure 15-3. Speed versus flowrate and efficiency.

charge rate of the pump at its preselected minimum speed, the pump is cycled on and off. There is no speed variation during the on–off operation. If the minimum influent rate is less than the minimum recommended discharge rate of a single pump, the wet well (and the sewer pipe if its storage capacity is utilized) should include sufficient storage to prevent cycling the pumps too frequently at this rate. When the influent rate again reaches the discharge rate of the pump at its preselected minimum speed, the pump is returned to V/S operation by the control system.

The following two methods can be used to avoid cyclic action of the "lead" pump:

- Use a larger number of smaller V/S pumps. For example, use four 3000-gal/min pumps instead of two 6000-gal/min units. Select the pump capacities so that the recommended minimum discharge rate of each pump is less than the minimum influent rate.
- Add two auxiliary V/S pumps that operate only when the influent rate is less than the recommended minimum discharge rate of the main V/S pumps. Select the capacity of each auxiliary pump to deliver a maximum flow that slightly exceeds the recommended minimum discharge rate of one of the main pumps.

Pump Efficiencies

As speed changes, efficiencies follow parabolic curves with their apexes at the origin of the graph, as shown in Figure 15-4. Thus, except for a slight shift with Reynolds number, efficiency is independent of speed. Consequently, if pump H-Q curves are constructed for different speeds, the efficiencies can be plotted at the same time. By projecting efficiencies from a pump H-Q curve along a parabola to the system H-Q curve, the efficiency at any pump operating point can be found.

An alternative way to find pump efficiencies along a system curve is to superimpose characteristic curves for selected pump impeller curves at two or preferably three speeds (as in Figure 15-2) onto the system H-Q curve and connect points of equal efficiency. The pump efficiencies along the system H-Q curve can then be plotted as shown in Figure 15-3.

Power Required

When the pump discharges into a hydraulic system, the power delivered by the pump at any point on the system head curve is determined by the flow and the head at that point (see Equation 10-6).

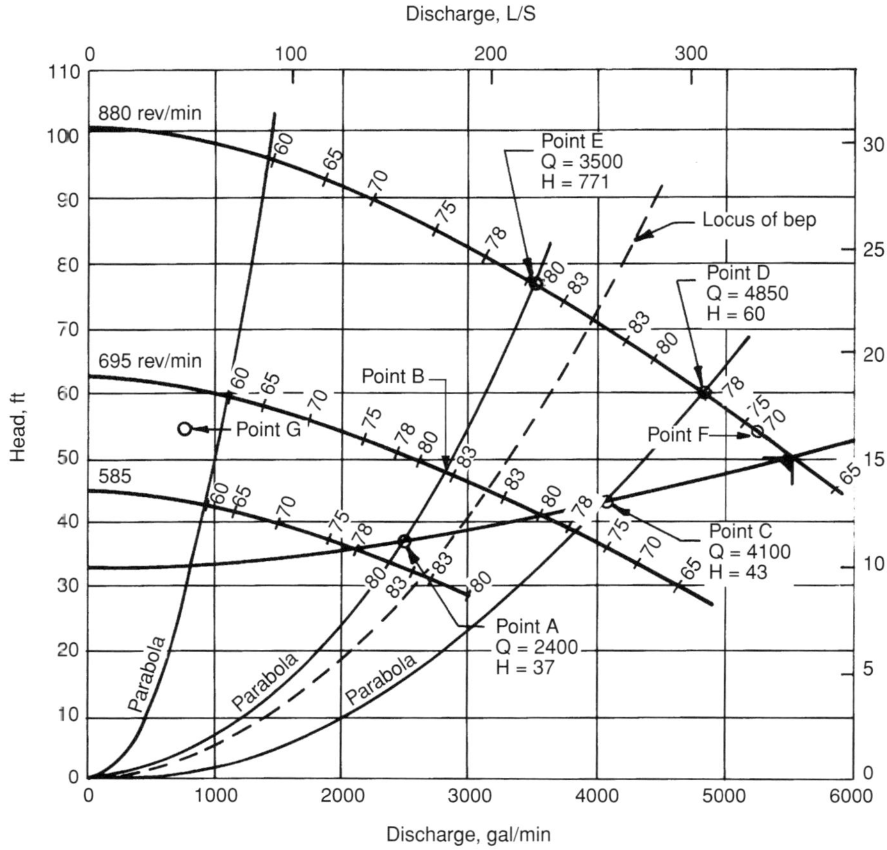

Figure 15-4. Characteristic curves for a single pump. Corresponding points A, B, and E lie on a parabola through the origin, as do corresponding points C and D. Points A and C (on the system curve) are not corresponding points.

According to the affinity laws, the ratio of power *delivered* by a pump (water power) at "corresponding points" (such as Points A and B in Figure 15-4) at different speeds equals the cube of the ratio of speeds. *But points along the system H-Q curve (such as Points A and C) are not "corresponding points," so the power delivered along the system curve does not vary as the cube of the ratio of speeds.* Using Equation 10-6 to calculate the ratio of power delivered at 695 and 585 rev/min along the system head curve in Figure 15-4 gives:

$$\frac{(3550)(40.5)}{(2100)(35.5)} = \left(\frac{695}{585}\right)^x$$

from which $x = 3.8$. But x varies along the system head curve. For the same pump, the exponent x can vary from less than 2 to more than 4 along the system curve.

Moreover, it is the *required* pump shaft or brake power that interests the designer. The power required by the pump at any point on the system curve is equal to the power delivered at that point divided by the pump efficiency at that point. Pump efficiency can vary considerably over the system curve, and this can cause the pump power requirements to further deviate from a cube relationship. Thus, drive manufacturers' efficiency data (which are based on a cube relationship of speed and required power) are not applicable.

Always compute power requirements from head, discharge, and efficiency (Equation 10-7)—not from the affinity laws.

15-4. Pump Selection

All of the arrangements described here include sufficient pumping capacity to discharge the peak influent rate when any single pump is out of service.

Two-Pump Facility

For an installation with two pumps, each capable of discharging the peak influent rate, a single duty pump with its peak discharge rate well to the right of the bep provides satisfactory efficiencies throughout the entire operating range and very good efficiencies in the center of the range. An example of the capability of a single pump is shown in Figure 15-4. Note that the efficiencies are greater than 70% over a range of flows equal to 3:1.

Three Variable-Speed Pumps

In the usual arrangement of a three V/S pump system, two pumps must be capable of discharging the design flowrate, while the third is idle as a standby unit. Unless the system curve is flat, a single pump must be capable of discharging somewhat more than half of the peak influent rate. The "lead" pump runs alone during periods of low flow. The "lag" pump is started when the influent rate slightly exceeds the maximum discharge rate of the lead pump, and both pumps operate until the influent rate again falls to slightly within the capability of the lead unit.

Assume, for example, that a pumping station must discharge flowrates from 0.13 m^3/s (2000 gal/min) to 0.50 m^3/s (8000 gal/min) at a static lift of 10 m (33 ft) and a TDH at peak discharge of 20 m (67 ft). The system H-Q curve is shown in Figures 15-4 and 15-5. The friction head is moderate and the curve is relatively flat, so a pump with comparatively steep H-Q curves should be selected. When two pumps are operating, the bep should be to the left of the operating point at peak flowrate so that the system H-Q curve sweeps across the curves of highest efficiency. Therefore, the bep of a single pump at top speed (880 rev/min in Figure 15-4) should be at the left of the midrange discharge (0.35 m^3/s [5500 gal/min]), and for a single pump operating, only the higher efficiency curves should straddle the system H-Q curve. The minimum discharge should be at least 30% of the maximum. The pump depicted meets all of these requirements and thus appears to be a good choice.

Load-Sharing Operation

When two of the above pumps operate in parallel at the same speed, the discharge (abscissa) is doubled with no change of head (ordinate), as is shown in Figure

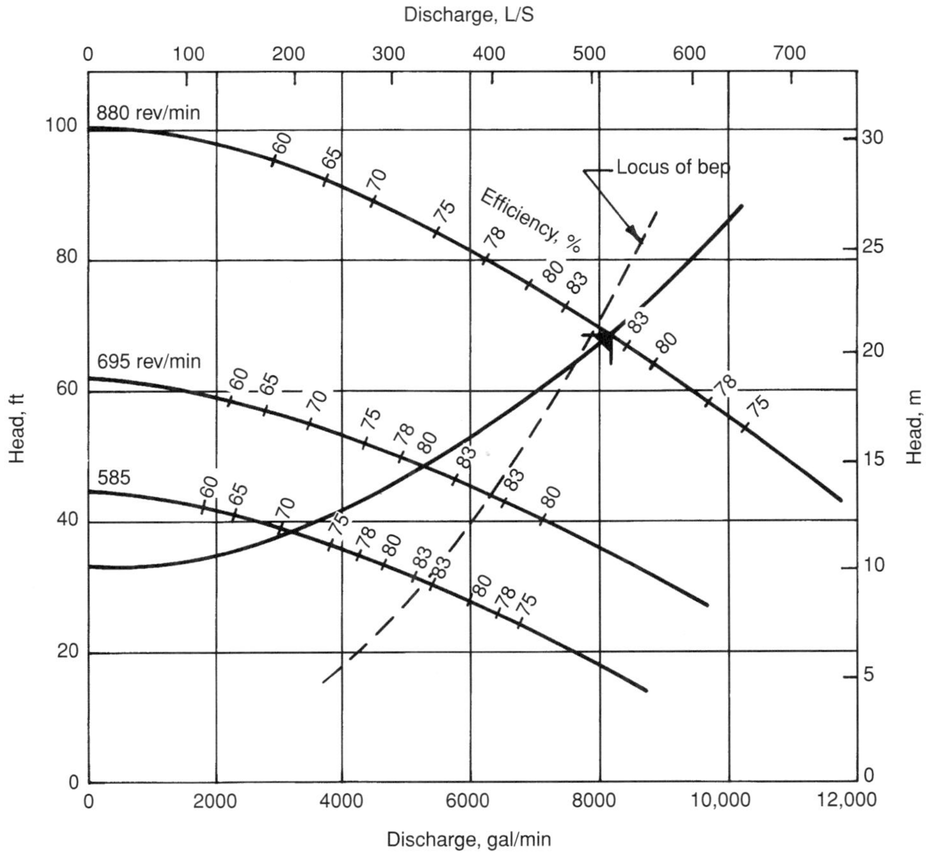

Figure 15-5. Characteristics for two pumps (per Figure 15-4) in parallel.

15-5. (Review Figure 10-24 and the accompanying text if an explanation is needed.) Note that the locus of bep for the two pumps at full speed is slightly to the left of the peak discharge. The efficiencies throughout the full range of flows from 0.13 m^3/s (2000 gal/min) to the peak discharge, shown in Figure 15-6, can be kept between 76 and 83%. Throughout the entire range of flowrates, the pumps operate at 36% or more of maximum capacity, and neither pump would cycle on and off frequently. Hence, the pump selection is indeed good.

Usually a lag pump is started when the influent rate slightly exceeds the capacity of the lead pump, and the two pumps then operate at the same speed. The lag pump runs continuously until the influent rate is again well within the capability of the lead pump. In the example of Figure 15-6, however, the lag pump could be stopped or started near the intersection of the efficiency curves for single and for double pump operation.

Staggered Operation

The staggered operation method is usually employed to permit the use of a common adjustable-speed controller for two or more V/S pumps. (Note that if there must be a standby pump, there should also be a standby V/S control system for the same reason—to ensure reliability.) The lead pump operates on its maximum-speed curve, while the lag pump operates on reduced-speed curves. Staggered operation is a poor way to operate pumping equipment and should not be used. The complexity of the controls is increased substantially, and the longevity of the pumping equipment is threatened. Unfortunately, staggered operation is sometimes advocated as a way to induce a designer to use variable-speed pumping at a "cost savings." Read Section 15-5 before deciding to use a common adjustable-speed controller for two or more pumps.

The discharge rates of the pumps at various influent rates can be calculated as follows:

- For any influent rate that exceeds the maximum capacity of a single pump, find the head, H', from the system H-Q curve.
- Determine the capacity of the lead pump at H' from its maximum speed curve.
- The discharge rate of the lag pump equals the influent rate minus the lead pump capacity.

A portion of Figure 15-4 is reproduced in Figure 15-7. As an example of staggered operation, consider a discharge of 0.38 m^3/s (6000 gal/min) shown as point E on the system H-Q curve. The head is 16.5 m (54 ft), and at this head the lead pump delivers about 0.33 m^3/s (5200 gal/min) at maximum speed (point F). The lag pump, therefore, must deliver 0.38 – 0.33 = 0.05 m^3/s (6000 – 5200 = 800 gal/min)—also at 16.5 m (54 ft) head, which is plotted as point G. By interpolating

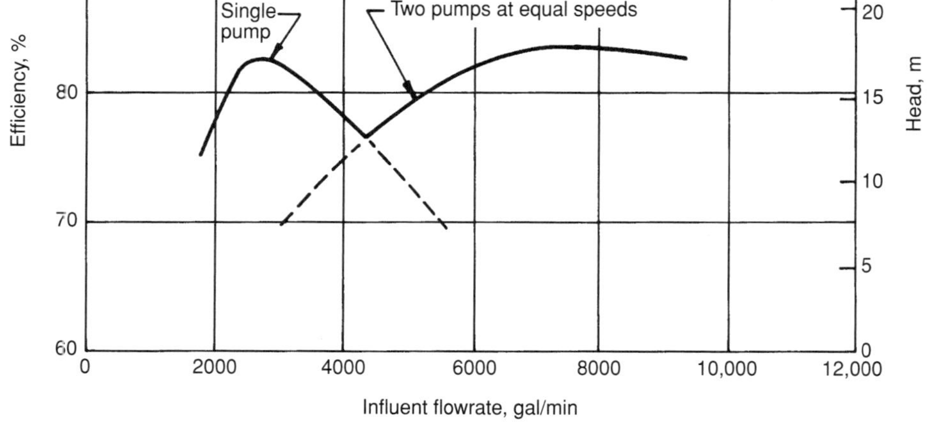

Figure 15-6. Efficiencies for one and two variable-speed duty pumps in load-sharing operation.

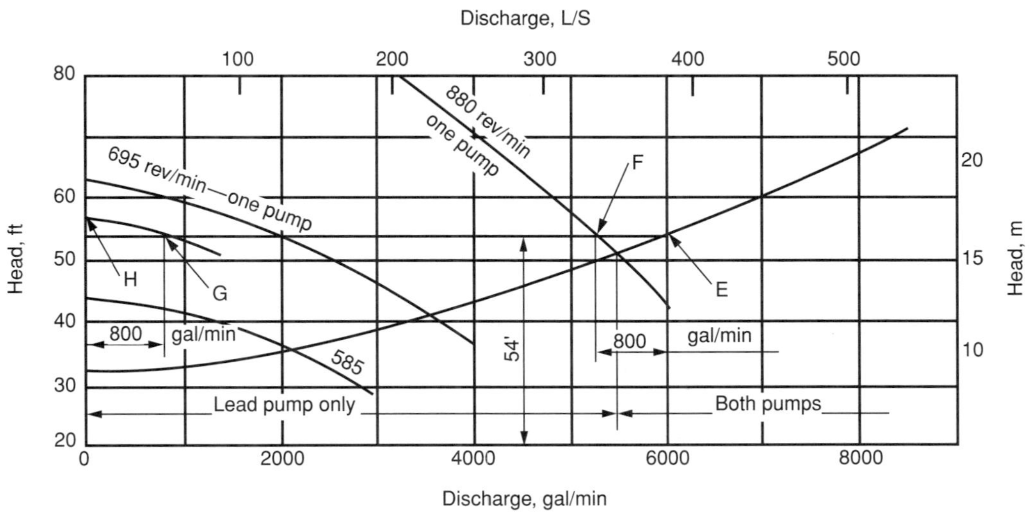

Figure 15-7. Two variable-speed pumps in staggered operation.

curve GH between the 585- and 695-rev/min curves, the shut-off head is found to be 17.4 m (57 ft). The shut-off head for 695 rev/min is 19.2 m (63 ft), so to produce a head of 17.4 m (57 ft), the speed of the pump would be, from Equation 10-16,

$$n_1 = 695\sqrt{17.4/19.2} = 662 \text{ rev/min}$$

or

$$n_1 = 695\sqrt{5.7/63} = 661 \text{ rev/min}$$

Interpolating between curves may be inaccurate where the curves are far apart (as in Figure 15-7), but accuracy can be improved by using the affinity laws (Equations 10-15 and 10-16 simultaneously).

The combined curve for the lead pump (at 880 rev/min) and the lag pump (at 661 rev/min) is found by adding the abscissas of the lag and lead pump curves (see Figure 10-24 for further explanation).

The efficiency of the lead pump operating alone is the same as that shown for the single pump in Figure 15-6. The efficiency of the lead pump when operating with the lag pump is found by projecting the head horizontally from point E on the system H-Q curve in Figure 15-7 to an intersection (point F) with the maximum speed curve (880 rev/min), transferring Point F to Figure 15-4 (note difference in horizontal scales), and interpolating between the efficiency lines to obtain an efficiency of approximately 73%. The lag pump efficiency for a total discharge of 0.38 m^3/s (6000 gal/min) is the efficiency at point G, which (transferred to Figure 15-4) is found to be below the pump manufacturer's data. It is probably about 55%.

Efficiencies at other discharges are found in the same manner. From these efficiencies and the heads and discharges, a plot of required pump shaft power versus discharge can be drawn as shown in Figure 15-8.

Comparison of Operating Modes

The power requirements for "load sharing" by the same pumps as above are also shown in Figure 15-8 for comparison. Staggered operation usually requires more power than load sharing whenever the flowrate varies between 101 and about 150% of the capability of a single pump. The influent rate could remain within this range for many hours at a time, with a consequent waste of considerable energy. Of even more concern is the operation of the lag pump at very low (and potentially damaging) discharge rates for long periods of time.

To avoid the problems of low discharge rates, the lag pump can be cycled on and off at a preselected minimum speed whenever the influent rate is between 101 and approximately 130% of the maximum output of the lead pump. (The minimum safe speed required for the lag pump to deliver its minimum discharge rate may be quite different from the minimum speed required when the same pump acts as the lead unit). But if the lag pump is cycled, the wet well and/or sewer must include sufficient storage between the proper elevations to avoid rapid cycling.

The only advantage of staggered operation is the cost saved by using one V/S controller for two or more pumps. The net cost reduction may be much less than the cost of one converter because contactors must be added to switch the converter to either motor and to

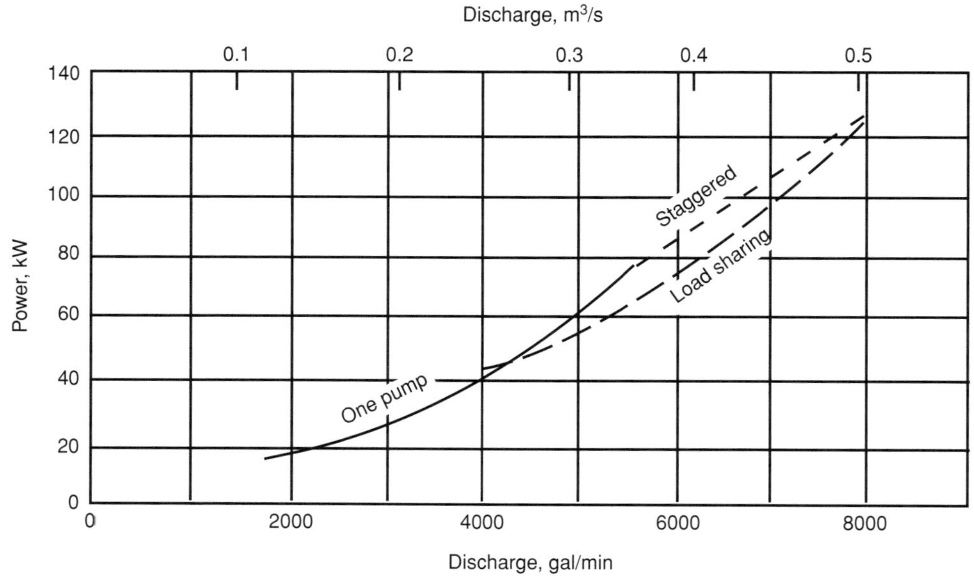

Figure 15-8. Pump shaft power consumption for two variable-speed pumps.

bypass the converter for C/S operation of each motor. When the lag pump is started, the lead pump is locked into full speed by an electrical contactor that bypasses the V/S controller, which is then switched to the lag pump. When stopping the lag pump, the lead motor should also be brought to a full stop—and quickly—before the V/S controller can be again connected to it. Meanwhile the water level in the wet well increases with neither pump running. The lead pump must be brought quickly to an intermediate or full speed to avoid too much increase in wet well level. Hence, switching from one pump to the other is complicated with one controller and staggered operation, whereas with load sharing there is a smooth transition without auxiliary switching equipment.

Selections for Steep System Curves

If the friction head is large in relation to the static head, the best pump selection would be to have the bep well to the left of the single pump maximum discharge rate. The pump shown in Figure 15-9 would be an excellent selection to deliver half of a peak influent rate of 0.58 m^3/s (9200 gal/min) with the system H-Q curve shown.

Four Variable-Speed Pumps

The same principles of selection apply to systems with four or more V/S pumps. A combination of V/S and C/S pumps, however, may provide all of the advantages of V/S drives at a much lower equipment cost.

15-5. Variable- and Constant-Speed Pumps in Simultaneous Operation

Throughout this section it is assumed that the system H-Q curve is flat (for clarity) and that the influent rate is always greater than the minimum recommended discharge rate of one pump (for brevity).

The goals of any V/S sewage pumping system are to ensure that

- The station effluent rate is always equal to the influent rate
- Proper operation still occurs with any single pumping unit out of service
- V/S pumps always discharge at or above their minimum recommended rates
- None of the pumps is cycled at influent rates above the minimum discharge rate of any V/S pump.

These goals can be attained with a combination of V/S and C/S pumps operating together effectively as one large V/S unit if:

- The "on-line" V/S capacity exceeds each step of C/S capacity to be added by 50% or more
- The standby pump has a complete V/S drive.

Many designers shudder at the cost of V/S drives for every motor and try to "save money" with a mix of

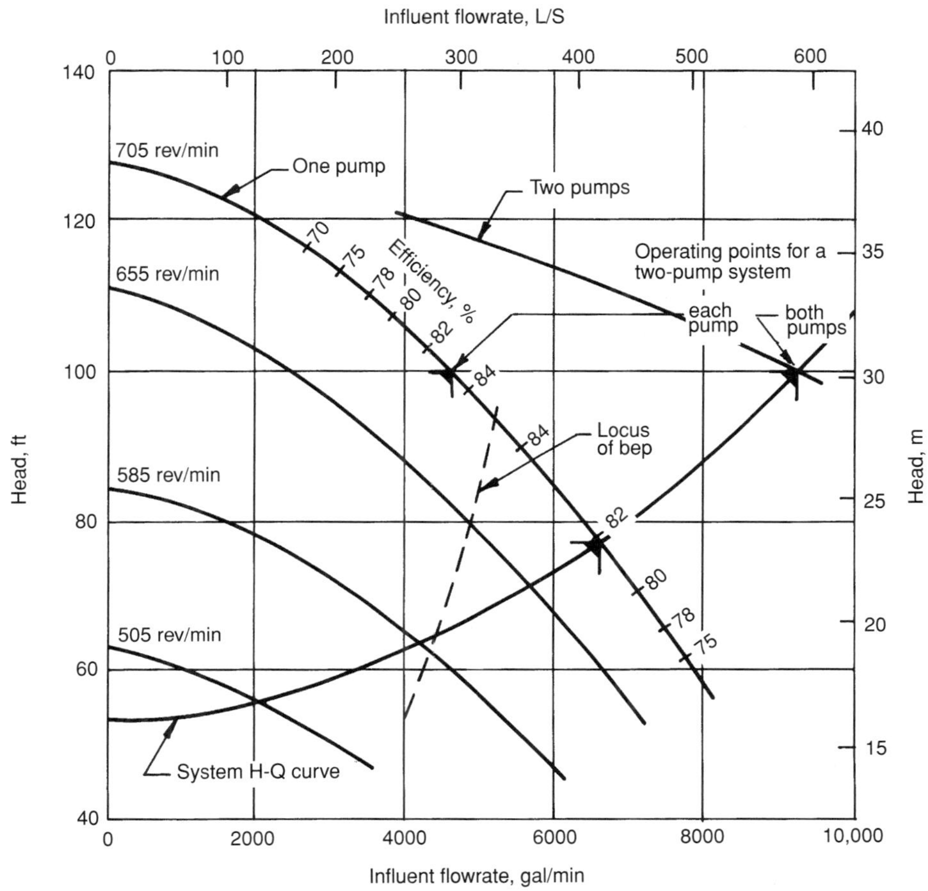

Figure 15-9. Variable-speed pumps for steep system H-Q curves.

C/S and V/S drives. Such a mix can be successful only if the V/S pumps are larger than the C/S pumps. The reasons are given in the following subsections. If, however, all of the pumps are to be of the same size, a mix will neither save money nor be successful, as explained by the following reasoning.

Many, perhaps most, pumping requirements can be satisfied with two duty pumps and a standby. If any of the pumps has a V/S drive, the standby must have a V/S drive also, so only one pump can be a C/S unit. Assume each pump has a capacity of 100 units. Below 100 units only a V/S pump can operate. Suppose the flow is slightly more than 100 (e.g., 102), and the C/S pump is started. The C/S unit can discharge nothing but 100 units, so the V/S pump must either slow to 2 units or cycle on and off at its minimum desirable speed. Centrifugal pumps cannot tolerate less than 30 or 35% of their maximum capacity without excessive vibration, cavitation, and damage to bearings and shafts. As the standby cannot be involved (consider it out of service), the only solution (for discharge to equal influent) is to add a fourth pump and only a V/S will do. Now we have three V/S drives and one C/S drive plus a larger dry well with more piping. No savings there! If the V/S pump is allowed to cycle on and off, the discharge surges and the wet well must be made large enough for adequate pump cycle periods.

In a second example, assume the peak influent rate is 300 units and that most of the time the influent rate ranges between one-ninth and one-fifth of the peak flow—a situation that could easily be met with four identical V/S pumps (one a standby). As the flow increases from 33 units, only one pump would be used until the flow reaches 100, and then a second pump would come on at 50 units and the first pump would drop to 50 units also. If the flow were to drop, the two pumps would continue running until the flow dropped to 95 (or perhaps to 90) to avoid switching pumps on and off too frequently. When the flow rises to 200, three pumps are activated at 67 units each and they remain on until the flow drops to 190 (or perhaps to 180). Except for very low flows, a pump is never required to operate below 45% of its capacity. The regime is shown in Table 15-1 in columns 1 through 3 and rows 2 and 3.

Now suppose one of the V/S drives (see columns 2 and 3) is replaced with a C/S drive (see columns 4 and 5) to make a mixed system. The mixed system would work, but it is not optimum, because when the C/S pump is running, the V/S pumps often operate at minimum discharge rates where their energy efficiency is lowest. Note, too, that it is simpler (and easier) to program for V/S-only drives. The C/S pump would rarely be used, and it might deteriorate through idleness. If the C/S drive were changed to V/S, the life of the pumps would be increased by 25%, and the operating staff would have much more flexibility in managing the station equipment.

Three Pumps

Because the standby must be a V/S pump, a three-pump system requires two V/S pumps and one C/S pump. The V/S pumps should each be capable of delivering about 60% of the peak influent rate (PIR), and the C/S should deliver no more than about 40% of the PIR. One of the V/S pumps operates as the lead unit. When the influent rate imperceptibly exceeds 60% of the PIR, the level in the wet well rises, which causes the C/S pump to start. The C/S pump delivers 40% of the PIR, while the speed of the V/S unit drops to pump 20% of the PIR. Thus, the lead V/S pump and the C/S pump operate together as one large V/S unit that discharges a combined total exactly equal to the influent rate.

Once started, the C/S unit operates *continuously* until the influent rate falls to less than 55% of the PIR, at which point the V/S unit is still operating at approximately 30% of its bec (the exact percent depends on the pump). After stopping, the C/S pump remains off until the influent rate again exceeds 60% of the PIR, which may be hours, days, or even weeks. This method provides the best possible automatic operation of the C/S pump. The V/S pump does not operate at discharge rates below the recommended minimum, and a V/S pump is always available as a standby to replace the lead V/S pump or the C/S pump if either fails.

Four Pumps

A four-pump combination can be configured either as (1) two V/S and two C/S pumps or (2) three V/S pumps and one C/S pump.

Table 15-1. Minimum, Minimum Optimum, and Maximum Relative Pumping Rates for a Mix of Drivers

Column	1	2	3	4	5
Pumps	1 V/S	2 V/S	3 V/S	1 V/S + 1 C/S	2 V/S + 1 C/S
Q_{min}	33	66	99	133	190
$Q_{optimum\ min}$	—	95	190	—	—
Q_{max}	100	200	300	200	300

Two V/S and Two C/S Pumps

Each V/S pump should be capable of discharging 45% of the PIR, and each C/S pump should discharge not more than 27.5% of the PIR.

Either of the V/S units operates as the lead pump. When the influent rate exceeds 45% of the PIR, a C/S pump is started and the V/S pump discharge drops to 17.5% of the PIR. These two pumps act together as one large V/S pump to discharge a total exactly equal to the influent rate.

The second C/S unit is started when the influent rate exceeds 72.5% of the PIR, at which point the V/S pump again discharges 17.5% of the PIR. The second C/S unit operates continuously until the influent rate falls to about 70% of the PIR, and the first C/S pump operates continuously until the influent rate falls to about 40% of the PIR. The minimum discharge rate of the V/S pump occurs just before either C/S pump is stopped, and this rate is at least 12.5% of the PIR.

The C/S pumps are usually automatically alternated at the end of each cycle of either or both. The V/S pumps are usually alternated every 24 h. The second V/S unit acts as a standby for the lead V/S pump as well as for either of the C/S pumps.

Three V/S and One C/S Pumps

In this combination, all pumps are of the same size, each with a capacity of 33.3% of the PIR.

One V/S pump is the lead unit. When the influent rate exceeds 33.3% of the PIR, a second V/S pump comes on line and the two operate in a load-sharing mode. The C/S unit is started when the influent rate exceeds 66.7% of the PIR. The V/S pumps discharge all of the influent above the capacity of the C/S unit. The C/S pump continues to operate until the influent rate falls to about 60% of the PIR, and the lag V/S pump is stopped when the influent rate falls to about 30% of the PIR. The third V/S pump is a standby to replace either the lead or lag V/S pump or the C/S unit.

Combination of Five or More Pumps

The capacity of each C/S pump should be limited by

$$Q_{C/S} \leq \frac{100}{N_{V/S} + 1.5} \tag{15-2}$$

where $Q_{C/S}$ is the capacity of a C/S pump as a percentage of the PIR and $N_{V/S}$ is the number of V/S pumps.

The capacity of each V/S pump should be

$$Q_{V/S} \geq \frac{1.5Q_{C/S}}{N_{V/S} - 1} \tag{15-3}$$

where $Q_{V/S}$ is the capacity of a V/S pump.

Consider a station with three C/S and three V/S pumps. From Equations 15-1 and 15-2, the capacity requirements are

$$Q_{C/S} \leq \frac{100}{3 + 1.5} \leq 22\% \text{ PIR}$$

$$Q_{C/S} \geq \frac{1.5(22)}{3 - 1} \geq 16.5\% \text{ PIR}$$

and the station would operate with one V/S pump as a standby. None of the C/S pumps is cycled at any influent rate, so wet well storage is not required.

In actual practice, the system curve is not at a constant head, and the pumps are selected to deliver their nominal ratings at the TDH of the maximum station discharge (PIR).

15-6. Special Design Considerations

Maximum Speed Limit

A pump must *never* be operated at a flowrate greater than that shown by the manufacturer's plotted curve. In other words, *never* extrapolate a pump H-Q curve. Hence, it may be necessary to limit the speed of a pump to ensure that the pump curve not only intersects the system curve but also overlaps it somewhat. Consider the range of possible system curves, and also note that pipe friction can (particularly when the pipe is new) be lower than anticipated, which may further limit the speed range of the pump.

A discharge pressure-control override can also be used when several stations discharge into a common force main where the head at any station can be very low at times.

Control of Pumping Units

A pump-control system should be designed so that the failure of any component or the trip of any single protective device (downstream of the main breaker) does not disable more than one pump. A complete and separate V/S drive for each V/S pump is strongly recommended to ensure that the facility can operate in a normal manner with any single V/S drive out of service.

Pump Failure Detection

Pump failure detection is a highly desirable feature that can usually be included in the pump-control system at little cost. If the pumps are equipped with outside-lever check valves, the valves can be fitted with limit switches to detect flow (or the lack of flow). The limit switch is actuated by the lever when the valve is fully closed; it is deactivated when the pump begins to deliver a small flowrate. Regardless of the cause, the check valve limit switch can detect the failure of the pump to discharge.

The electrical circuitry is designed to deenergize the motor of any pump whose check valve remains closed for a predetermined period of time (usually less than 3 min for V/S pumps) when the pump is required to operate. An alarm is signaled, and the standby pump replaces the unit that failed.

Check valve limit switches can also prevent energization of the motor of any pumping unit whose discharge check valve is not fully closed. Some drives can be severely damaged or destroyed and shafts can break if the motor is energized while the pump is backspinning.

If the check valves do not have outside levers, a pressure switch located between the pump discharge nozzle and the check valve is sometimes used as a flow indicator. Actually, the check valve only indicates pressure, not flow, and is misleading if flow is blocked by a downstream obstruction. A better flow indicator is a vane switch in clean water applications and, for sewage, a thermal-dispersion switch or a Doppler flow switch (see Section 20-5). When used for sewage or dirty water, any pressure device (such as a bourdon tube) should be filled with a clear liquid (glycerin, for example), which is separated from the dirty water by a flexible diaphragm.

Power-Operated Check Valves

If the pump discharge check valves are power operated (hydraulic or electric), it may be necessary to include a relatively sophisticated control system to protect the V/S pump drives against backspinning while the drives are energized. When a pumping unit is required to operate and the pump drive is activated, the check valve should not begin to open until the pressure on the upstream side of the valve exceeds the pressure on the downstream side. (The pressure can be sensed, as described previously.) When the pump is no longer needed, the check valve should close first, and the pump should not be stopped until the valve is at least 95% closed.

The pumping unit should be deenergized when

- The pump discharge pressure does not exceed the pressure in the force main within a short interval (say, 60 s) after the pump is started
- The check valve does not open fully within a few minutes after the pump discharge pressure exceeds the pressure downstream from the check valve
- The check valve does not close fully within a short interval (say, 3 min) after the pumping unit is signaled to shut down.

The control system should also prevent energization of the motor of any drive whose check valve is not fully closed.

15-7. Analysis of Variable-Speed Booster Pumping

The purpose of a V/S booster pump is to maintain a nearly constant discharge pressure while delivering the variable demand of a closed distribution network. Pump speed variation controls the pump discharge pressure, and the discharge rate is determined by the system demand.

To analyze the operation of a pump that discharges into a hydraulic network, both the V/S pump characteristic curves and the curve of required pump differential head versus demand must be considered. The pump must always operate at the intersection of the impeller's characteristic curve (at the speed of rotation) with the required differential head curve.

Pump Characteristic Curves

The curves for a typical horizontal split-case pump, with its impeller trimmed to deliver 681 m^3/h (3000 gal/min) at 33.5 m (110 ft) at maximum speed, are shown in Figure 15-10. Note the flatness of these curves.

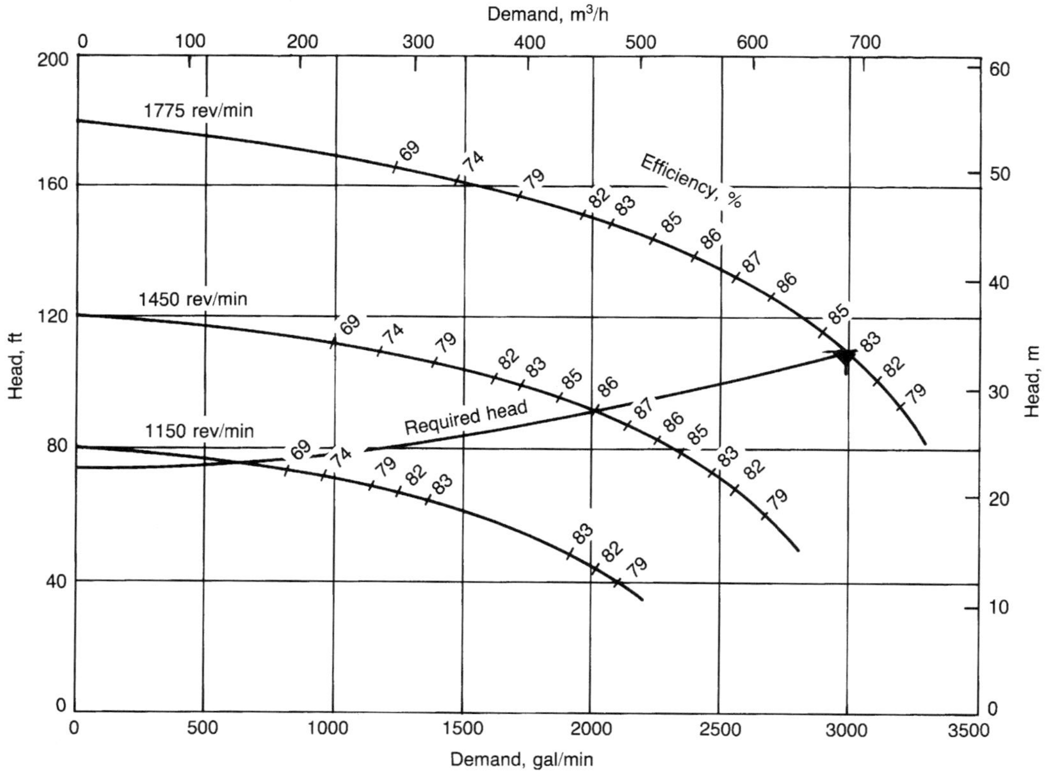

Figure 15-10. Characteristic curves for a booster pump.

Required Differential Head Curves

The significant contrast between wastewater and booster pumping is that the system H-Q curves for the two are derived differently and usually have different shapes. The system H-Q curve for a booster pumping facility is actually a "required differential head," which represents the differential head that must be added to the suction head to produce the desired constant discharge pressure at all demands.

There are four basic types of suction pressure variation.

- The suction pressure decreases with an increase of the pump discharge rate due to headlosses in the suction piping system.
- The suction pressure remains constant regardless of demand. An example is a pump that takes suction directly from a tank in which the water level is automatically maintained relatively constant by a float valve on the supply inlet.
- The suction pressure varies independently of demand. An example is a relatively small pump that takes suction from a large city main in which the pressure varies but is essentially unaffected by flow through the pump. Thus, there are an infinite number of suction head curves.
- The suction pressure decreases with an increase in the pump discharge rate and also varies independently of demand. Examples are (1) a booster pump in a city water main in which pressure varies due both to the pump discharge and to other demands and (2) an in-line transmission booster pump.

Suction Pressure Varies with Demand

The curve of required differential head that the pump must develop is shown by the solid lines in Figure 15-11 to be the desired discharge head minus the curve of suction head. But in actual practice, the booster pumps commonly used do not maintain a constant discharge pressure. Because of the requirement of the control system for a change in signal to the drive, the pump discharge pressure usually has a droop of about 10% from zero demand to maximum demand. The droop is not usually linear, but its actual curve is not important; a linear droop is used here for simplicity. The required differential head curve incorporating the droop is shown by the dashed line in Figure 15-11c, and it is also superimposed on the pump characteristic curves in Figure 15-10, where it is labeled "required head." Droop can be eliminated with a reset feature in the control system. Elimination of the droop, however, is complex, expensive, and not usually necessary because a 10% droop is not objectionable.

The curve of pump speed versus demand shown in Figure 15-12 can be obtained as explained for Figure 15-3. The pump can, of course, operate at any speed between the limits shown by the solid line and at speeds less than 1150 rev/min, as indicated by the dashed line. However, operation at zero flow, or even at low discharge rates, should be avoided as explained in Section 15-3. Again, the speed-demand curve is the most significant depiction of V/S booster pump operation.

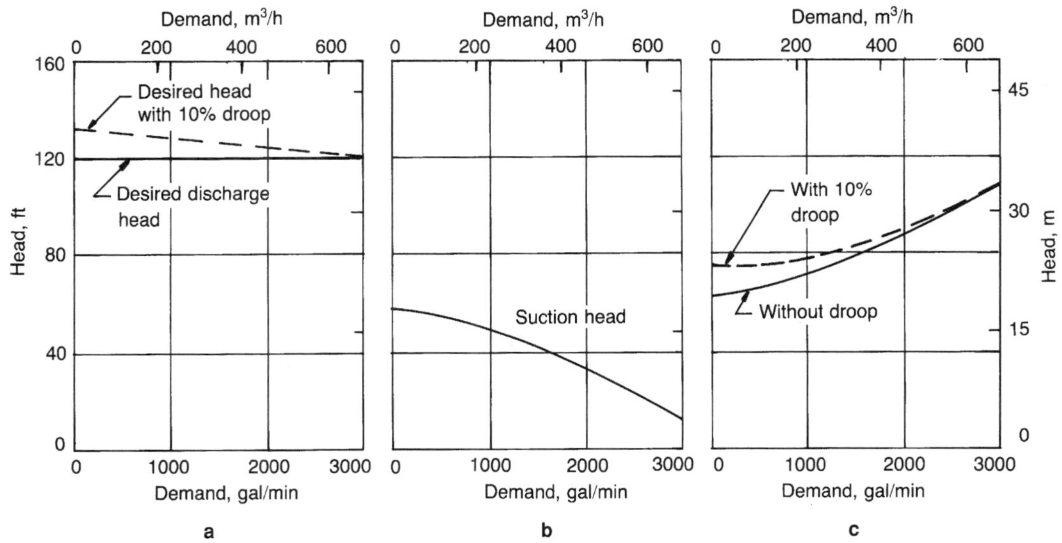

Figure 15-11. Differential head for a suction head varying with demand. (a) Discharge head; (b) suction head; (c) required differential head. Note: (a) – (b) = (c).

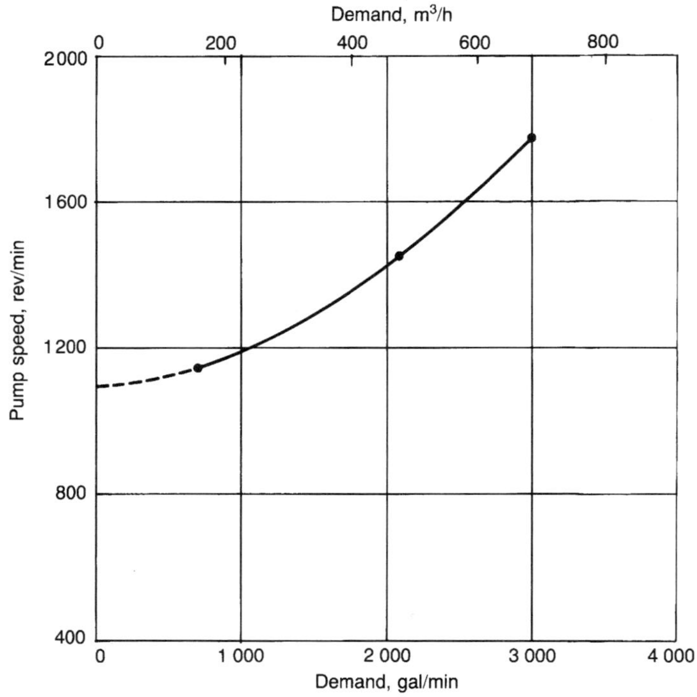

Figure 15-12. Speed versus demand for a booster pump.

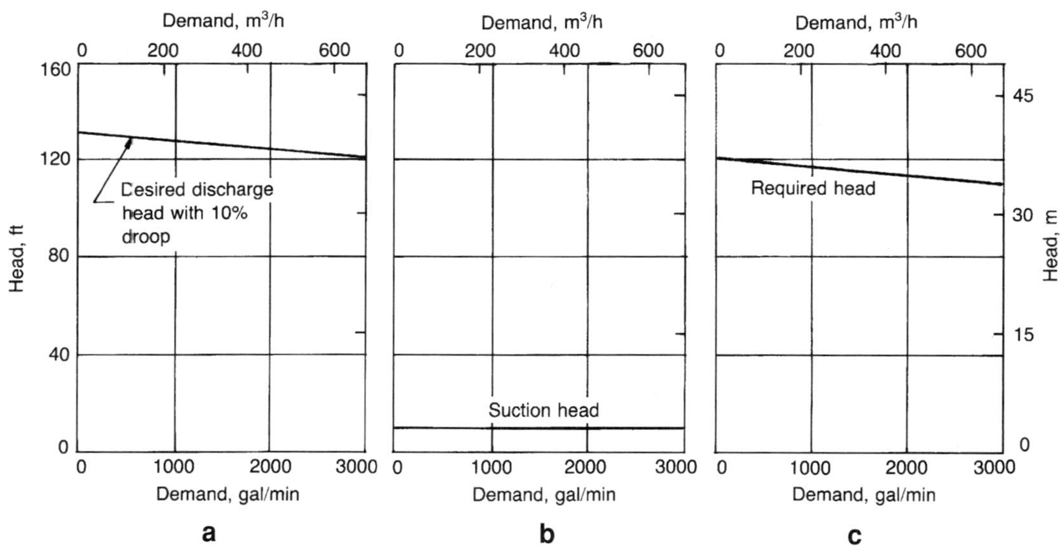

Figure 15-13. Differential head for a constant suction head. (a) Discharge head; (b) suction head; (c) required differential head. Note: (a) – (b) = (c).

Suction Pressure Constant

The development of the required differential head curve is shown in Figure 15-13. The superposition of this curve on the pump curves of Figure 15-10 is shown in Figure 15-14. The zero Q speed is found at the intersection of the 1450-rev/min pump curve with the required head curve, but if it were not, it could be found as explained in Section 15-3.

If a pump with a very flat curve is used as shown in Figure 15-15, the zero Q speed can be calculated to be 1760 rev/min and the pump would always operate

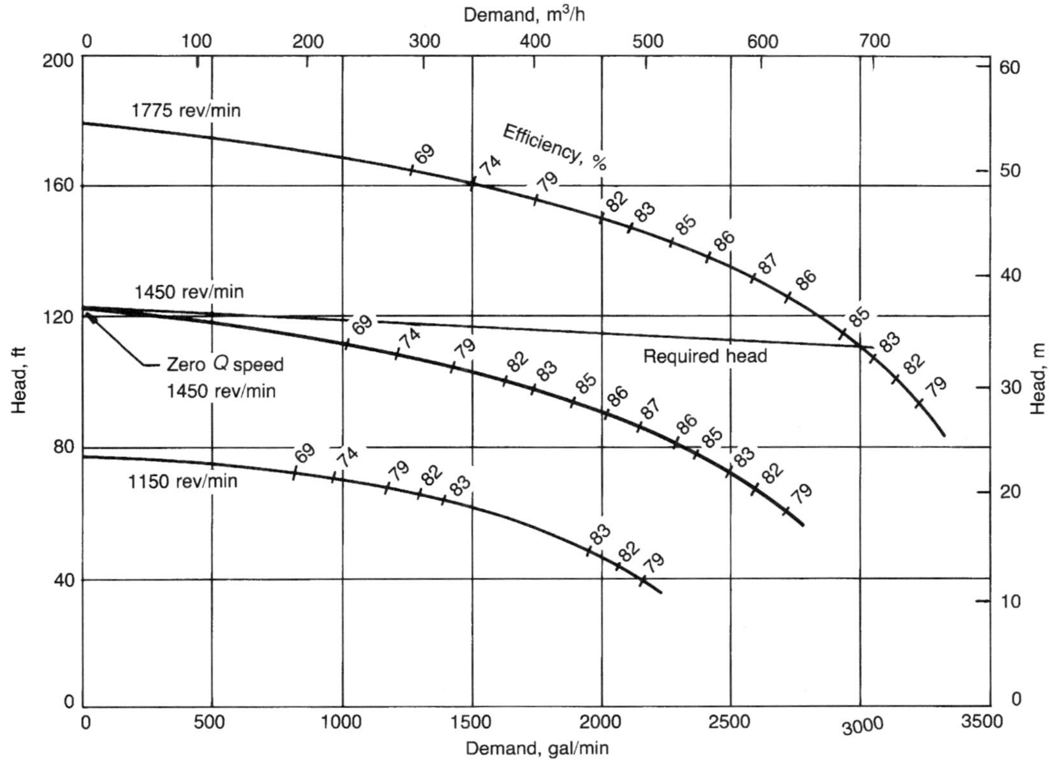

Figure 15-14. Booster pump and required differential head curve for a constant suction head.

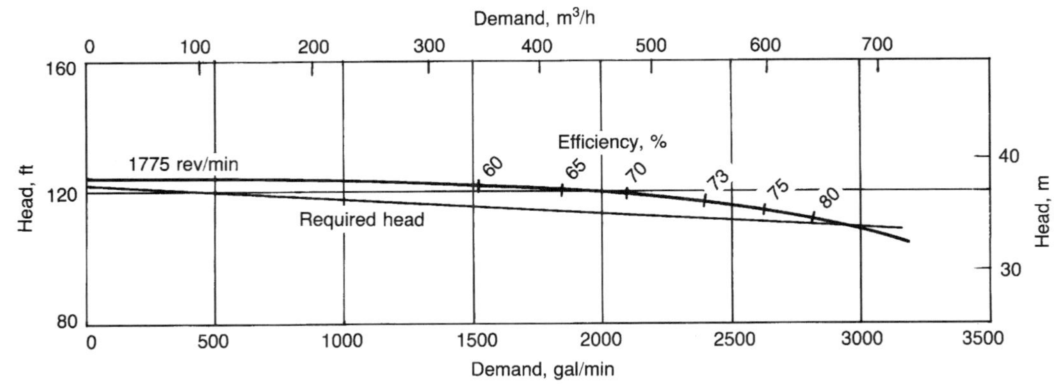

Figure 15-15. A flat booster pump curve.

within 1% of maximum speed regardless of demand. (All of the curves shown here are published curves for actual pumps offered by well-known manufacturers.) Hence *there is nothing to be gained by V/S operation of a flat-curve pump with a constant suction head.* The flat-curve pump provides good pressure regulation where the demand fluctuates rapidly. The use of steep-curve pumps is sometimes desirable because they permit some adjustment of the discharge pressure without changing the pump impellers. The discharge pressure of a steep-curve V/S pump fluctuates with abrupt changes in demand because the inertia of the rotating elements prevents instantaneous changes of pump speed.

Suction Pressure Varies Independently of Demand

The development of the required differential head curves is shown in Figure 15-16. The minimum suction head requires the maximum required differential head and vice versa. The pump and differential head curves are similar to those shown in Figure 15-14 except that there are an infinite number of required head curves as the suction head changes between the minimum and maximum of Figure 15-16b. The maximum speed pump curve and the required head curve for maximum suction head *must* intersect (which may require a steep-curve pump).

If the pump curve is very flat, the pump speed changes are almost entirely to compensate for suction pressure variations. But regardless of the manner in which the suction head varies, the drive automatically regulates the pump speed to deliver the desired discharge pressure at any demand within the capability of the pump.

Suction Pressure Varies both Independently and Dependently of Demand

The development of the required differential head curves is shown in Figure 15-17 for a minimum suction head of 12.2 m (40 ft) and a maximum suction head of 21.3 m (70 ft). The differential head is not a parabola with its apex at zero discharge because the droop causes it to be skewed.

The required differential head curves for the extremes of suction head in Figure 15-17c are superimposed on the pump curves of Figure 15-10 and shown in Figure 15-18. The pump operates on an infinite number of system curves between those for minimum and maximum suction head.

15-8. Minimum Discharge Rate

A minimum speed feature in a V/S drive *cannot,* by itself, prevent the operation of the pump at low discharge rates. Without a bypass, the pump cannot deliver more water than the amount that is being drawn from the distribution system. If the drive prevents operation at speeds lower than a preselected minimum, the pump must operate on its curve for that speed when the demand is less than the flowrate at which the minimum speed curve intersects the required head curve. If the pump with the characteristic curves of Figure 15-18 were prevented from operating at speeds less than 1300 rev/min, the pump would operate on the solid line ABCD shown in Figure 15-19. For example, at a demand (and discharge) of 114 m^3/h (500 gal/min), the pump would develop a head of 29 m (96 ft). If the minimum demand is less than the minimum recommended pump discharge rate, the pumps should be protected by one of the three methods discussed below.

Suction Pressure Varies with Demand

Flowmeter and Bypass

Include a flowmeter and a bypass with a modulating valve, as shown schematically in Figure 15-20a. The

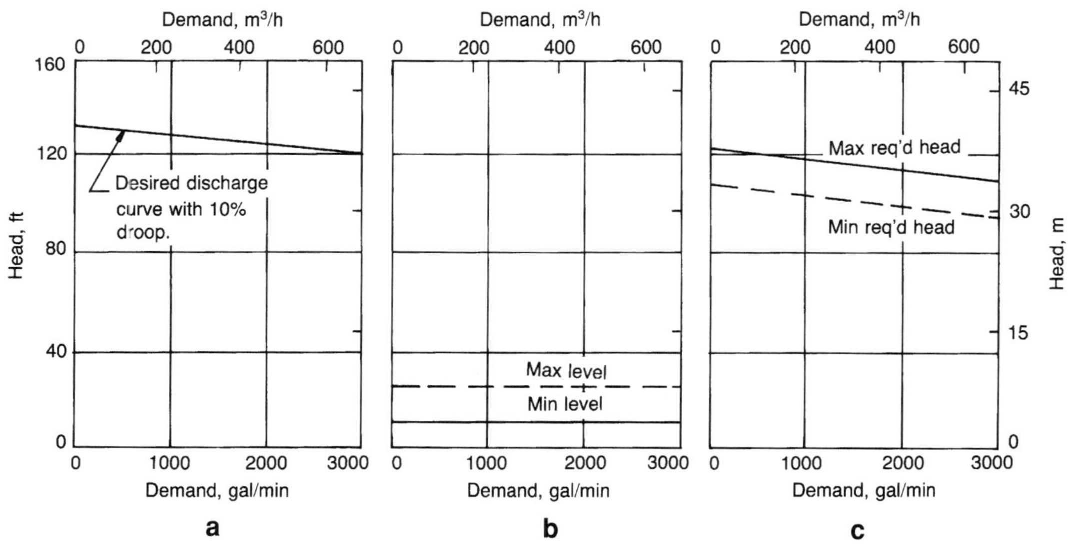

Figure 15-16. Differential head for a suction head that varies independently of demand. (a) Discharge head; (b) suction head; (c) required differential head. Note: (a) – (b) = (c).

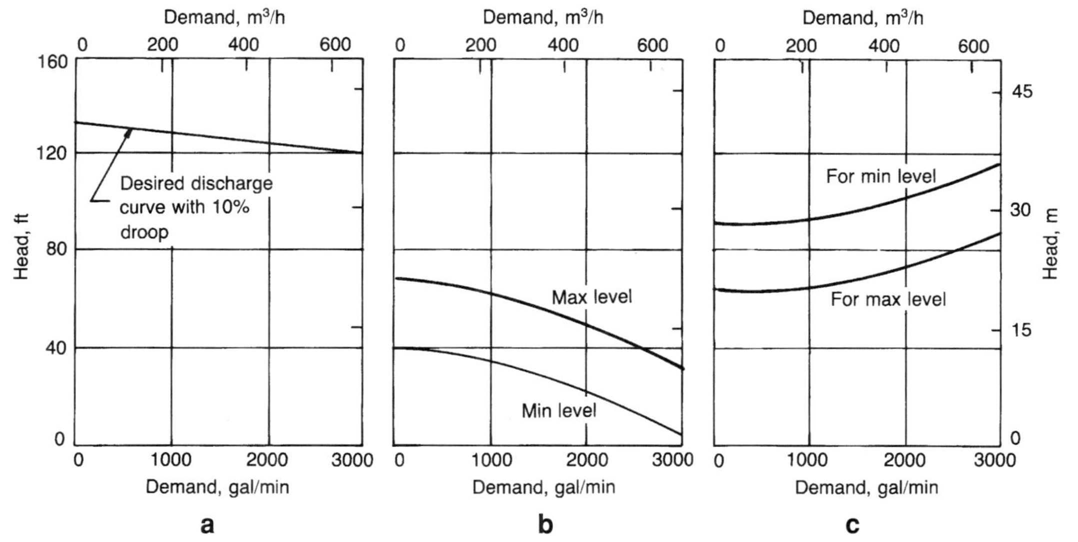

Figure 15-17. Differential head for suction pressure varying both dependently and independently of demand. (a) Discharge head; (b) suction head; (c) required differential head. Note: (a) – (b) = (c).

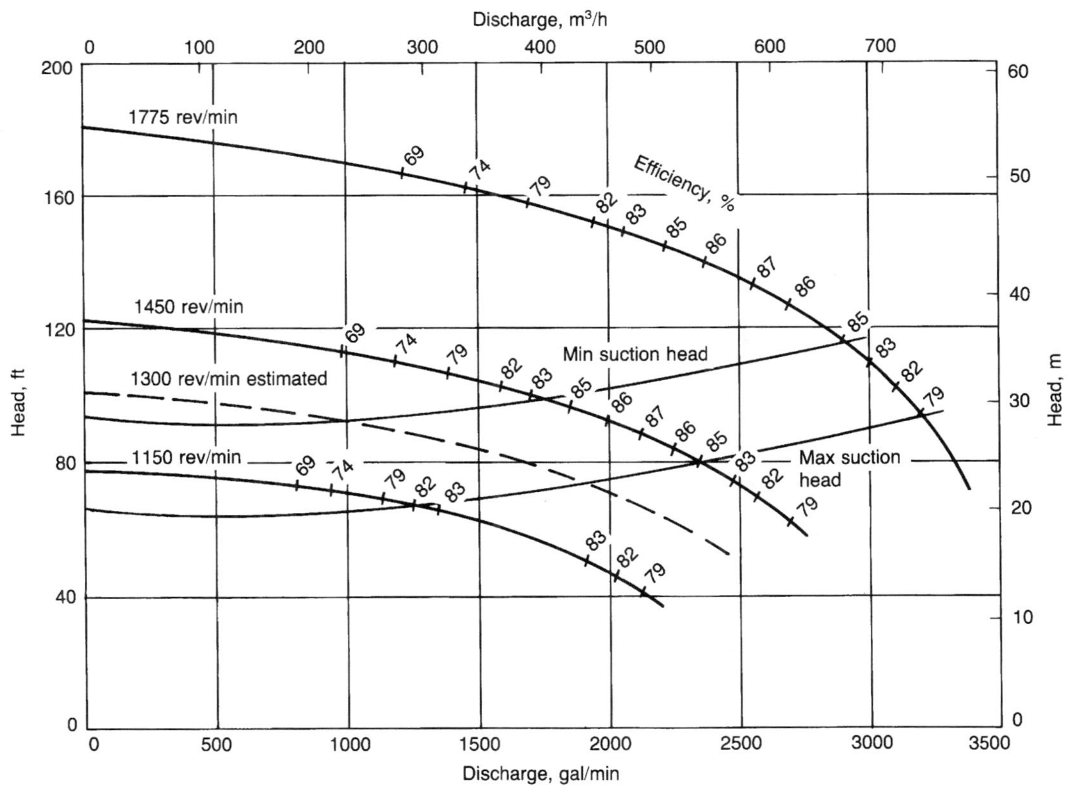

Figure 15-18. Suction varying both dependently and independently of demand combined with pump curves.

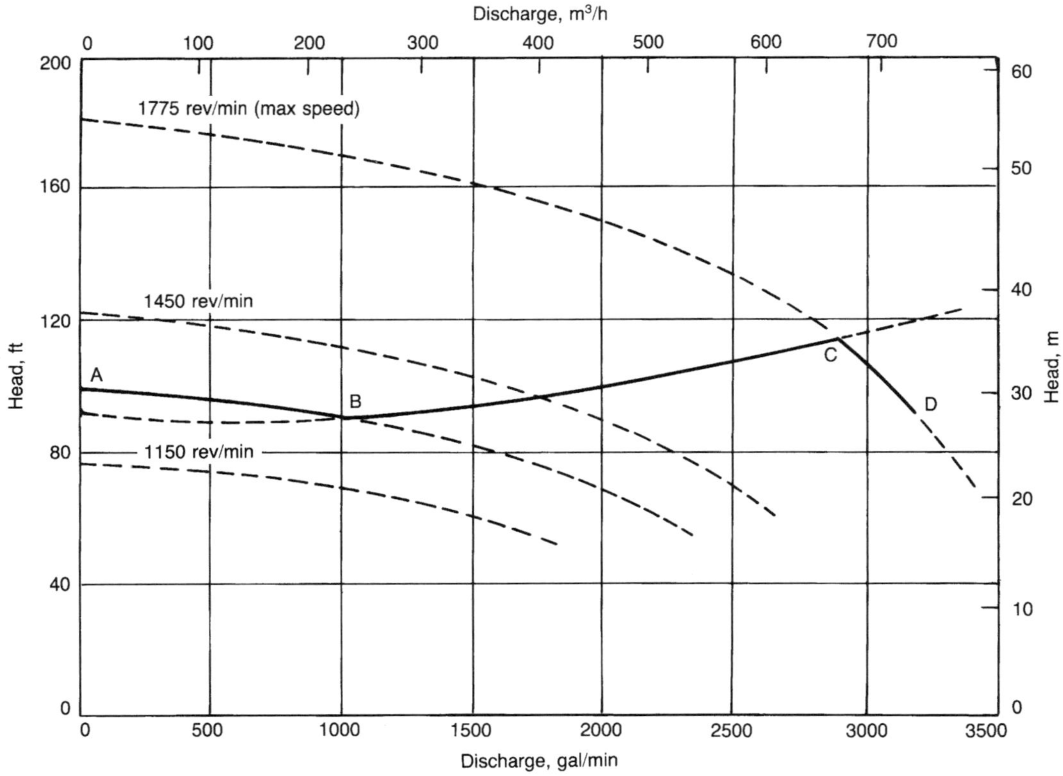

Figure 15-19. Booster pump operation at minimum suction head.

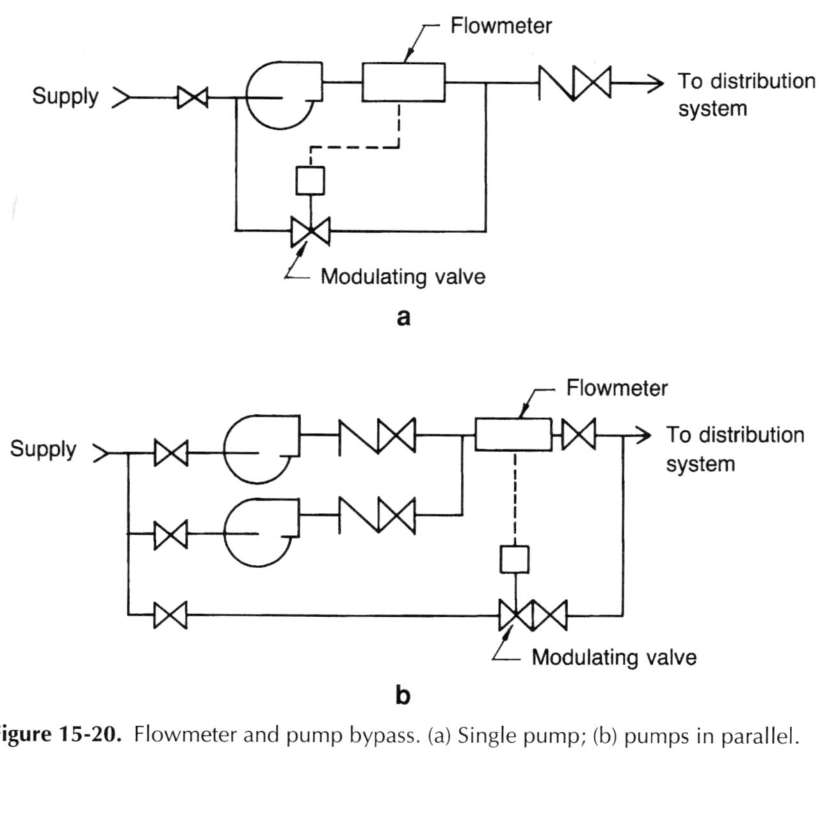

Figure 15-20. Flowmeter and pump bypass. (a) Single pump; (b) pumps in parallel.

flowmeter can be any type that has a 4- to 20-mA output signal. Accuracy is not usually important. Because the valve should modulate in response to the flowmeter output signal, a hydraulically operated valve is usually preferred.

The valve actuator is adjusted to maintain the recommended minimum pump discharge rate (RMPDR) through the valve when the demand is zero. The valve should be fully closed when the demand exceeds the RMPDR by 25 to 35%. At demands less than the RMPDR, the valve modulates to maintain the flow through the flowmeter equal to or slightly greater than the RMPDR.

If two pumps are to operate in parallel, as shown in Figure 15-20b, the signal from the flowmeter should be electronically converted to cause the valve to maintain a minimum combined discharge rate of twice the RMPDR when both pumps are operating. This method of operation, however, applies only to load-sharing parallel operation, as explained below.

A bypass around the flowmeter (not shown in Figure 15-20) is desirable so that the meter can be removed for repairs (see Figure 20-12).

The modulating or regulating valves shown in Figure 15-20 would nearly always operate as throttling valves subject to cavitation. The valve selected must be capable of withstanding such severe service for many years. It should be of the type in which metering occurs downstream from the seat, and the cavitation constant should be greater than 1.0. If metering occurs in the seat (as in an ordinary globe valve), the seat will be quickly cut. If the pressure drop is excessive for a single valve, two valves can be installed in series.

Jockey Pump

Include a relatively small, flat-curve, constant-speed jockey pump to operate during periods of low demand. The V/S pumps do not operate while the jockey pump is running. Because the head in the pump suction header is high when the flowrate is low, the jockey pump can be selected for a relatively low head.

The jockey pump is usually started and stopped by means of pressure switches with suitable time delays to prevent starting and stopping on pressure surges. The switches sense pressure in the station suction header. If the suction pressure variation at low flows is too little for reliable pressure switch operation, it may be necessary to use a flow-measuring device to start and stop the jockey pump.

Hydropneumatic Tank

A small hydropneumatic tank with an air compressor and an appropriate air/water ratio controller can be used with C/S pumps. It is rarely used with V/S pumps.

15-9. Operations in Booster Pumping

Booster pumps can be operated in parallel in much the same way as described in Section 15-4, except that the difference in power requirement is even greater. Typical, steep characteristic curves of a booster pump are shown in Figure 15-21 with the required head curve superimposed. When two of these pumps are operated in parallel, the total power required in both types of operation is as shown in Figure 15-22.

With a different combination of pump and/or required head curves, the difference in power requirements may be more or less pronounced than in this example, but a situation in which staggered operation would use only a little more power than load sharing throughout a significant portion of the discharge would be most unusual.

Sequencing

Starting Lag Pumps

A high-quality, mechanical pressure switch that senses pressure in the discharge header could be used to signal the second pump to start but only if the pump curve is steep. For starting either (1) flat-curve second pumps or (2) third (and subsequent) flat- or steep-curve pumps, the recommended method is to use signals from a flowmeter and a pressure switch. The lag pumps are normally started either (1) when the header flowrate *approaches* the maximum discharge rate of the on-line pumps *or* (2) when the pressure in the discharge header is low. Time delays should be included in pressure switch circuits to prevent action on momentary pressure fluctuations.

Pressure switches alone should not be used to provide starting signals for flat-curve pumps or pumps subsequent to the second pump, because the pumps on line may cavitate before the pressure in the discharge header falls enough (about 10% below the desired head) to actuate the pressure switch.

Stopping Lag Pumps

The preferred method of stopping lag pumps is always by means of signals from a flowmeter plus an electronic pressure switch. Each lag pump is stopped when (1) the demand falls to within the combined capacities of the remaining on-line pumps *and* (2) the pressure in the discharge header is at the desired value.

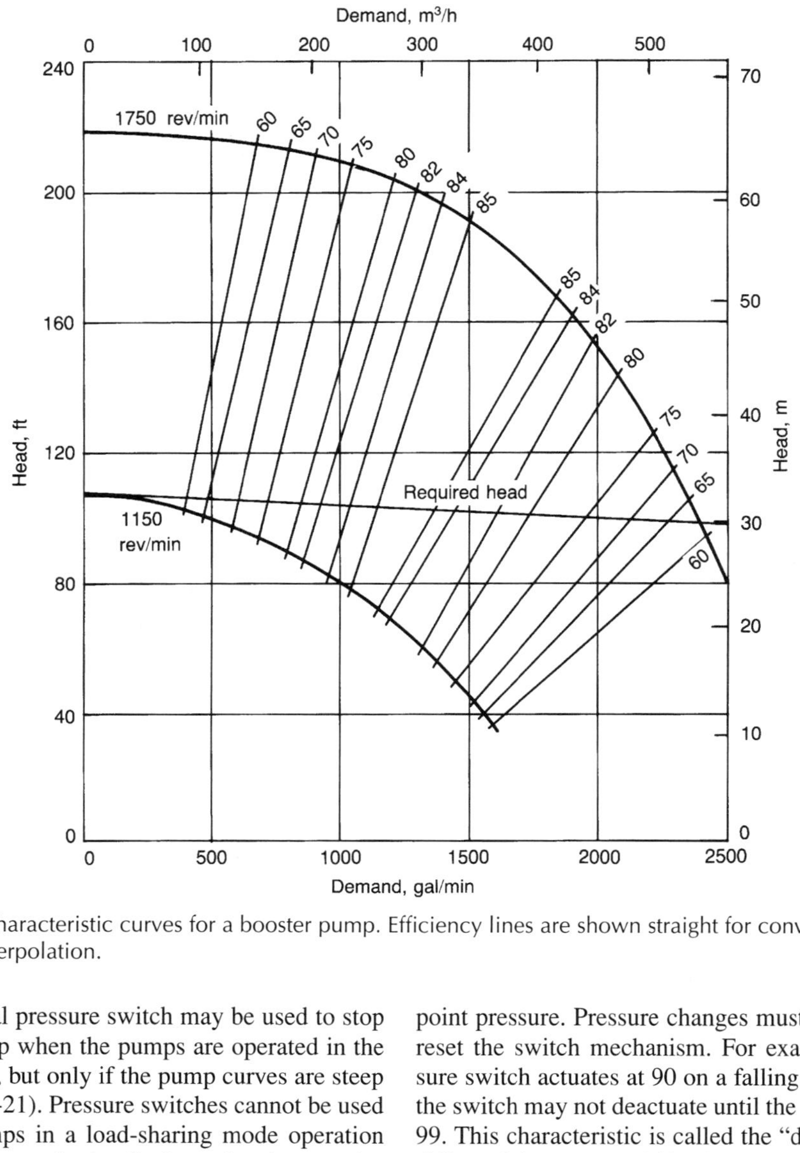

Figure 15-21. Characteristic curves for a booster pump. Efficiency lines are shown straight for convenient (although approximate) interpolation.

A mechanical pressure switch may be used to stop the second pump when the pumps are operated in the staggered mode, but only if the pump curves are steep (as in Figure 15-21). Pressure switches cannot be used to stop lag pumps in a load-sharing mode operation because the pressure in the discharge header remains relatively constant at all demands within the combined capacities of all on-line pumps.

For pumps operating in a load-sharing mode, the lag pump can be stopped by means of a running period timer. One timer is provided for each step of pumping capacity. A timer is actuated when its associated pump is started, and it runs for a preselected period of time. The associated pump is stopped at the end of the timer cycle, but if the demand still requires operation of the pump, the pump motor is restarted and operates through another timer cycle.

Mechanical pressure switches do not actuate or deactuate at the same pressure on every cycle. The band of actuation/deactuation may be ±5% of the set-point pressure. Pressure changes must be sufficient to reset the switch mechanism. For example, if a pressure switch actuates at 90 on a falling pressure signal, the switch may not deactuate until the pressure rises to 99. This characteristic is called the "differential." The differential span may drift with respect to the set point as the switch linkage wears.

15-10. Simultaneous Operation of V/S and C/S Booster Pumps

Combinations of V/S and C/S pumps are rarely used for water booster pumping. Such combinations could be used if *all* of the following conditions are met.

- The pump curves are steep.
- The capacity relationships are those discussed in Section 15-5 (i.e., the C/S pump has no more than two-thirds the capacity of any V/S pump).

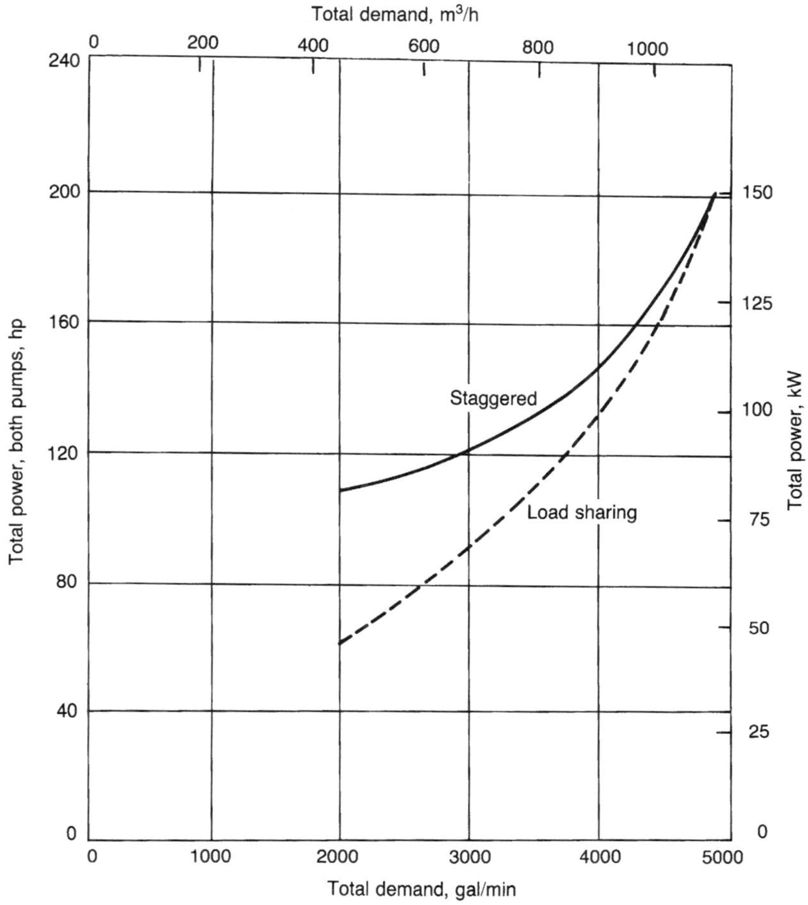

Figure 15-22. Power consumption in staggered and load-sharing operation.

- The demand does not change significantly over brief periods of time, so that rapid cycling of the C/S pump(s) is precluded.

Pumping Unit Failure Detection

A pressure switch connected directly between the pump discharge nozzle and the check valve can be used if the check valves do not have outside levers. Pressure switches are useful and can detect all internal pumping unit failures, but they cannot detect whether flow is occurring. Check valve limit switches are always preferred to pressure switches.

Other Considerations

The special design considerations discussed in Section 15-6 apply equally well to booster pumping.

15-11. Adjustable- and Variable-Speed Drives

The purpose of V/S drives is to regulate the flowrate, and various means for so doing are listed in Table 15-2. The last four types of V/S drives are those most commonly used for water and wastewater pumping. Because some other drives are common in other industries, a limited discussion of a few of them is also included. See Karassik *et al.* [1] for other and more extensive discussions of V/S drives.

A summary of the comparative advantages of the various types of V/S drives is given in Table 15-3. Descriptions and further details are given below.

Direct Engine Drives

Internal combustion engines (discussed in Chapter 14) are easily adapted for V/S pumping. All that is needed is a proportional linkage (mechanical, pneumatic, or

Table 15-2. Systems for Regulating Flowrate or Pressure

Motive unit speed	Pump speed	Regulating system	Notes and examples
Constant	Constant	Throttling valve	Pump may run near zero discharge; for water pumping only
Constant	Constant	Regulating valve	Excess pump discharge spilled to sump intake; for water pumping only
Constant	Constant	Variable-pitch propeller	Propeller pumps for pumping very large storm water flows
Constant	Constant	Intermittent operation and usually multiple pumps in parallel	Storage in wet well or clear well; variable number of duty pumps
Constant	Variable	Mechanical drive	Oil film cone rollers, automatic gears, vee belt
Constant	Variable	Hydraulic coupling	Hydrokinetic, hydroviscous
Constant	Variable	Hydrostatic drive	Hydraulic pump and hydraulic motor with squirrel-cage motor
Variable	Variable	Slip energy recovery	Wound-rotor motor
Variable	Variable	Engine drive	Throttle fuel
Adjustable	Variable	Electronic	Adjustable-frequency drive
Variable	Variable	Slip drive	Liquid rheostat with wound-rotor motor
Constant	Variable	Slip drive	Eddy-current coupling

electrical) between the sump level or a header pressure monitor and the governor. Note that prolonged operation at low loads causes carbon deposits in a diesel engine exhaust system (see Section 14-22).

Right-angle gears can be used to adapt the horizontally mounted engine to a vertical pump shaft. Nonreversing gears should be used. The connection between the driveshaft and a horizontal pump shaft should be flexible to limit vibration transfer.

Engine drives are ideally suited for the following:

- Pumping stations that, because of size or location, are a problem for electrical facilities. It may be more economical to use engines than to install long, large transmission power lines.
- In some water hammer situations where the required inertia of moving parts is greater than economically feasible for electric motor drives.
- Locations where gaseous fuels are available at low cost—near natural gas centers or large sewage treatment plants with surplus digester gas.

Engine-Generator Drives

If power must be locally generated by an engine, V/S pumping can be had for almost no added expense by controlling the generator speed with an asynchronous engine governor that responds to changes in the controlled variable (such as sump level or discharge pressure) to produce a variable-frequency power source. One or more pump motors can be powered by one generator in load-sharing operation. With a back-up engine-generator set (which is usually required anyway), this system is very reliable.

Combination Drives

If the utmost in reliability is required of the standby power system, consider using direct-drive standby engines for duty and standby pumps instead of an engine-generator set. For horizontal split-case pumps, the motor is mounted at one end of the pump shaft and the engine is mounted through an automatic clutch at the other end. If a vertical motor drives the pump, a combination right-angle gear and automatic clutch is available to provide a horizontal shaft for an engine and a mount for a vertical motor.

If power fails, the duty engine is started and the motor is allowed to spin freely. An interlock must disconnect the motor from the power source when the engine is started. When electric power is restored, the pump must first slow to well below synchronous speed and the engine must be disconnected before the motor is energized. Energizing a spinning motor

Table 15-3. Comparison of Variable Speed Drives

Drive	Advantages	Disadvantages
Direct engine drive	Very reliable when well maintained.	Engines expensive. Maintenance costs can be either high or moderate (depends on personnel).
Combination engine/motor drive	Utmost in reliability. Allows peak shaving to avoid electricity cost penalties (time of day, load factor, power demand). Emergency operation without costly switchgear, cabling, and generator. Engine can drive pump at higher speeds than motor if pump properly selected; greater capacity for storms, fire flow. Only a small engine-generator is needed for lighting, control, and—possibly—ventilation.	Practical, economical power range per engine is 75 to 200 hp. Speed range less than 1200 rpm. Multiple engines are expensive. Dedicated to one pump.
Adjustable frequency (AF)	Depending on motor and AF converter selected, peak wire to driveshaft efficiency ranges from 81 to 92% without a transformer and ~2% less with a transformer; efficiencies at lower speeds reasonably high. Efficiency is misleading; at reduced speeds, calculate power losses based on actual operating conditions; use life-cycle costs (see Example 29-1) for comparison with C/S drives. No motor starter needed unless bypass operation at full speed is required. Starting current is low. Has become very popular; most new V/S drives are AF drives (AFDs). Reliability good with modern equipment, if provided with adequate protection from electrical surges. Lowest life-cycle cost of any variable-speed drive system. Any practical maximum speed obtainable. Can produce frequencies slightly above 60 Hz. May permit better coordination between drive system and pump characteristics than is possible where maximum speed must conform to standard motor speeds such as 1170 or 1760 rev/min. Newest AF converters do not heat motors.	*Standard motors cannot be used with any AF drive*. To be satisfactory for AFD service, motors must be specially built with superior insulation and resistance to heating (see subsection "Motors" in Section 15-11). As a minimum, line side series reactors are recommended to reduce harmonic currents fed back into power supply system. Expensive harmonic filters may be required to control harmonics adequately and to meet utility requirements (see also Section 9-11). Filters may be needed to control harmonics in feedback and, if needed, may be very costly. Can be crippled by lightning or heavy power surges. Depending on lightning exposure, extensive lightning protection systems may be required. In all situations, add surge arrestors on incoming power. Adjustment and repair must usually be performed by factory representatives who may not be readily available in all geographical areas. For remote locations, specify one spare power module and one spare controller module. Owner can usually install either and send faulty unit to factory for repair. For owners with technicians trained in electronic maintenance, specify a comprehensive set of spare parts. Life expectancy: 8–15 years. Some may shut down frequently with minor power disturbances, so specify automatic flying restart. Motor efficiency reduced about 5% by the non-sine wave power input in older drives. The best new drives reduce efficiency by about 3%. The reduction causes increased heating of the motor.

(Table continues on next page.)

Table 15-3. Continued

Drive	Advantages	Disadvantages
Adjustable frequency (AF) *(continued)*		Standby generators must be greatly oversized and/or special designs are required. Canadian Standards Association and California Safety Order codes may require features not included in standard AFD to further increase cost.
Eddy-current coupling (a slip drive)	Depending on motor and coupling selected, peak wire to driveshaft efficiency ranges from 83 to 93%; as speed is reduced, efficiencies drop rapidly. Efficiency is misleading; at reduced speeds calculate power losses based on actual operating conditions; use life-cycle costs for comparison with other drives. Uses squirrel-cage induction or synchronous motor. Little maintenance: lubricate every 2–3 months, replace brushes, and dress slip rings. Proven life of 20 yr or more. Thousands in satisfactory use. Sizes available are 4–2200 kW(5–3000 hp). Longest mean time between failures and shortest mean time for repairs in most places. Most motor rewind shops can repair. Brushes and slip rings are eliminated in the stationary field design; maintenance is reduced to that of a squirrel-cage induction motor.	Maximum speed ≃ 98% of motor nameplate speed. Slip loss, dissipated as heat to station atmosphere (in air-cooled models), may require increased building ventilation. May require reduced-voltage starters for motors. Noisy with motors of 1200 or more rev/min. Replace brushes and dress slip rings every 2–4 yr (may require dismantling). Horizontally placed units increase the floor area needed; vertically mounted units may require extra headroom and/or special provisions to resist lateral seismic forces. Becoming obsolete. Fewer suppliers result in higher first cost than AFDs.
Liquid rheostat (a slip drive)	Efficiency is comparable to that of eddy-current coupling (q.v.). Pneumatically operated design has no moving parts in basic assembly. No electronics in pneumatic design Inrush current to motor less than full-load current, which may reduce size of standby engine-generator and increases life of motor. Nearly as reliable as eddy-current coupling; proven life of 20 yr or more. Relatively low maintenance (add distilled water monthly). Slip-loss heat can be discharged to pumped fluid at a saving in ventilation requirements. Sizes available are 3.7 to 7500 kW(5 to 10,000 hp).	Maximum speed ≃ 98% of motor nameplate speed. Competition limited; may increase cost. Requires external liquid–liquid or liquid–heat exchanger to remove slip-loss heat from electrolyte. Environment clean and cool: 29°C (85°F) or less, with cooling water or air at 27°C (80°F) or less. Movable plate designs with electrode positioning by a constant-speed motor inherently changes speed in steps and usually "hunts" continuously over small range. Wound-rotor induction motors expensive and have been discontinued by many former manufacturers. Replace motor brushes and dress slip rings every 2–4 yr. High cost.

Table 15-3. Continued

Drive	Advantages	Disadvantages
Fluid coupling (hydrokinetic) drive	Rugged machinery. Basic elements subject to little wear. Moderate cost.	For remote control, the scoop tube is usually driven by a small electric motor, which must have "dead zone" that results in continuous hunting over small range in water and sewage pumping applications and causes occasional failure of scoop tube motor and rapid wear of linkage. External heat exchangers are required. Reduced-voltage starters may be required. Low efficiency, about 5–7% less than eddy-current coupling at all speeds due to higher "windage" losses.
Slip recovery drive	Efficiency comparable to AFD. Can be closely sized to fit application Reliable May be more economical than AFD in ratings above 750 kW (1000 hp). Electronic conversion section requires 1/3 of the kVA rating of an AFD.	Solid-state components have caused maintenance problems. Unsuitable for less than 375 kW (500 hp). Requires wound-rotor induction motor. Poor power factor; requires large bank of capacitors for power factor correction. Repairs must be made by factory representative. Repair costs similar to those for AFD. Life expectancy: 10–15 yr. Reliability same as AFD.
Variable-ratio belt drive	High efficiency. Good reliability. Excellent for sludge pumping or other applications where speed adjustments are made manually and output speeds are in low ranges.	Ratings limited to 100 kW (125 hp). Most output speed ranges unsuitable for pumping sewage or water. Unsuitable for continuously fluctuating speed or for frequent starts and stops (causes rapid wear of belt), but chrome plating reduces wear. Belt replacement may require removal and disassembly; check with manufacturer. Cost: low from fractional to about 5.6 kW (7-$^1/_2$ hp); moderate from 7.5 to 19 kW (10 to 25 hp); high from 22 to 100 kW (30 to 125 hp).
Hydrostatic drive	Motor can be submerged indefinitely. Explosionproof; useful in hazardous atmospheres. Useful for constant-torque sludge pumps. Widely used in industry.	Limited power and output speed ranges. Pump *very* noisy. Overall efficiency of about 80% *at all speeds;* requires larger than normal electric motor. May require oil cooler. High cost.

might not harm it (unless it is spinning in reverse), but energizing a spinning motor with its residual field remaining from interrupted electrical power by connecting it to another, non-synchronized power source could displace the windings, break the shaft, and probably shear coupling bolts.

Peak Shaving

Significant savings in power costs can sometimes be obtained by using the motor to drive the pump for low and medium flows and by switching to engine drive for high flows. The advantages include

- A smaller electric motor
- Two or three small- to medium-sized engines and drive gears (plus a very small engine-generator set for controls and lighting) replace a large engine-generator set and expensive electrical transfer switch
- A lower electric power demand
- The engine need not be "exercised" if high flows occur with sufficient frequency
- The engine always operates at or near full load.

The design of a peak shaving system requires an extensive study of diurnal and monthly flowrates. A convenient and simple aid is the use of either arithmetic or log probability paper [2]. Fair *et al.* [3] have explained the use of probability paper.

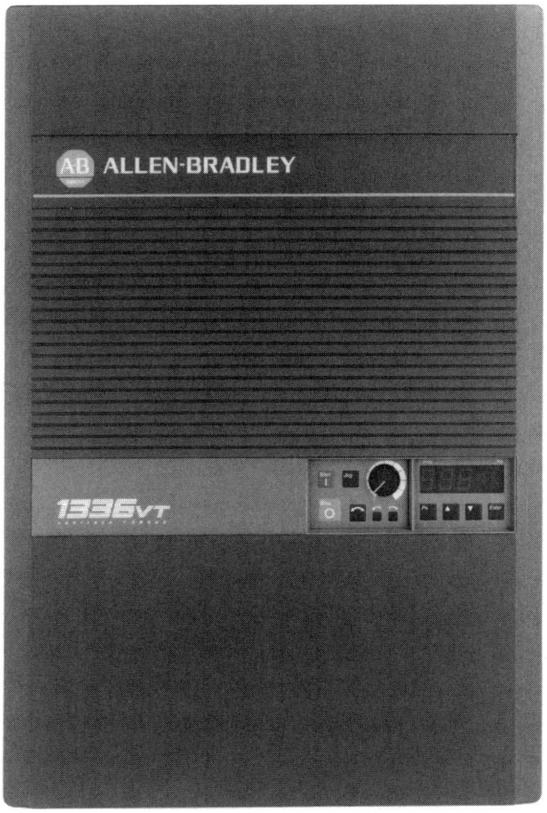

Figure 15-23. An adjustable-frequency converter for a 50-hp motor. Courtesy of Allen-Bradley Co., Inc.

Adjustable-Frequency Drives

The adjustable-frequency drive (AFD) consists of an adjustable frequency (AF) converter (Figure 15-23) and an ac squirrel-cage induction motor. The AF converter is a solid-state electronic assembly that contains

- A rectifier, which converts incoming ac power to dc power
- A filter to reduce the ripple in the dc power
- An inverter, which changes dc power to ac power at any selected frequency within the design range.

A control signal from external equipment is supplied to cause the inverter output frequency and voltage to vary as a function of some convenient variable, such as pressure for water pumping or wet well liquid level for wastewater pumping. The ac output of the converter powers an ac squirrel-cage induction motor, and the motor speed varies directly and linearly with the frequency of the applied power.

Almost all modern AF converters are pulse-width modulated (PWM) utilizing insulated gate bipolar transistors (IGBTs) in the output stage, which produces the output wave form shown in Figure 15-24c. Very large (over 300 kW or 400 hp at this time) AF converters may use older thyristor designs that produce a stair-step output or pulse amplitude modulated (PAM) (Figure 15-24a). The IGBT-type converter is less expensive than older designs and produces an output wave form that does not heat the motor as much as PAM output. The disadvantages of the IGBT-type are (1) a very high switching frequency that results in a high rate of change of voltage (dV/dt) and (2) high peak voltages applied to the motor windings. Older type converters were frequently designated as either (1) voltage source or (2) current source. Voltage-source converters (Figure 15-24a) have been replaced by PWM converters for drives to 300 kW (400 hp). The current source type (Figure 15-24b) still has adherents, is generally more expensive than voltage source for a given horsepower, and is usually used only for very large drives.

Current source converters must be closely matched to their load, but voltage-source converters can operate any size motor, even multiple motors, as long as

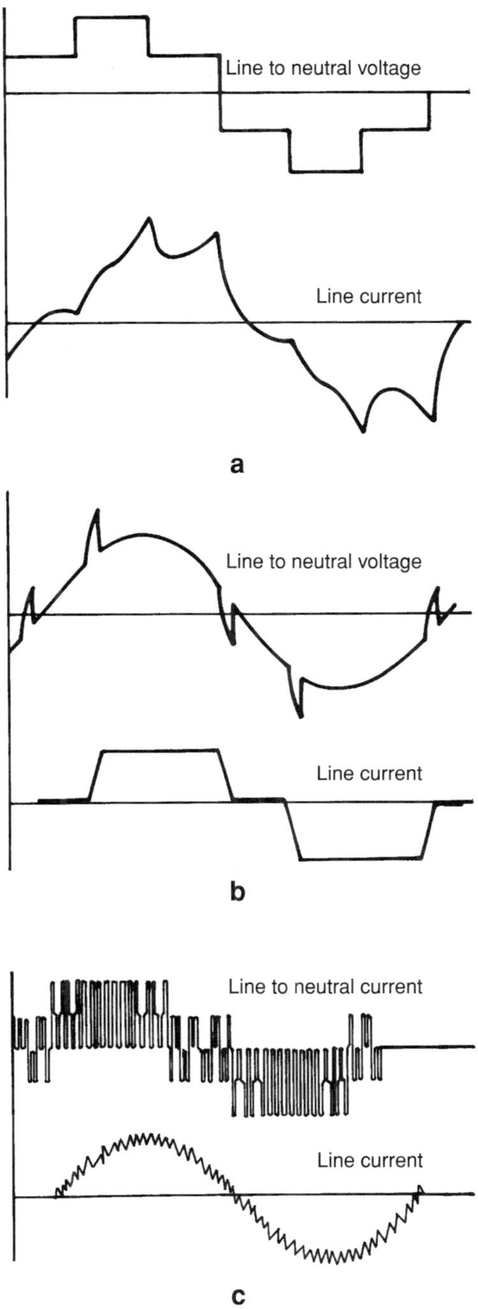

Figure 15-24. Output waveforms from adjustable-frequency converters. (a) Voltage source six-step; (b) current source; (c) pulse-width modulated waveforms. Courtesy of Cutler-Hammer Eaton.

the power requirement is within their rating. For drives over 600 kW (800 hp), consider the load-commutated converter, which either uses a synchronous motor or a reactive output network to accomplish commutation. Load-commutated converters are particularly suitable for motors over 1500 kW (2000 hp).

Harmonics

All AFDs draw harmonic currents (300, 420, 660, 780 Hz, etc.) from a 60-Hz power source. These harmonics distort the voltage supply and cause (1) excessive heating of voltage supply equipment such as transformers, (2) excessive heating or malfunction of other equipment connected to the voltage supply, and (3) interference with communication systems. Voltage distortion must be controlled to prevent difficulty with other plant equipment and the AF drives themselves. Harmonic currents must be controlled to a level acceptable to the power and telephone utilities. Generally, this level of control requires conformance to the requirements of IEEE 519, which limits both the harmonic current load and the amount of voltage distortion permitted (see Section 9-11).

Prospective harmonic distortion produced by AF converters can be readily predicted by power system analysts who use computer programs to simulate the behavior of the entire electrical system. Such an analysis should be performed on any power system where the converter load exceeds 25% of the total load. Power system analysts can prescribe various corrective measures to bring the installation within the requirements of IEEE 519. Typical corrective measures (in order of increasing cost) include:

- Series reactors
- Phase-shifting transformer and six-phase rectification
- Shunt filters
- Active filters.

These corrective measures increase the drive cost and may decrease the overall drive efficiency. Because of the space required for a typical filter, the need for filters *must* be determined early in the design of the building envelope.

Efficiency

The efficiency of a modern AF converter may approach 98% at full load and speed, but the effective efficiency is lower because harmonics increase losses in the motor and the power supply system, and harmonic mitigation schemes absorb energy.

A series reactor reduces drive efficiency by about 1%. A phase-shifting transformer reduces drive efficiency by 2 to 3%. Series reactors and phase shifting transformers usually remain connected to the power supply even when the drive is not running. Their efficiency loss, though small at no load, can have a significant effect on overall drive efficiency. Shunt filters absorb little power, but like series reactors and phase-shifting transformers, these units are frequently left connected to the power supply when the drive is not running, and they cause a significant energy loss. PWM drives have little effect on motor efficiency (less than 1%), but older drives with step-output wave forms reduce motor efficiency by 2 to 3%. Practically, AF drive system efficiency at full load cannot be quantified, but probably never exceeds 96% for PWM drives and 92% for step-output drives.

Efficiencies of AFDs vary as a function of both load and speed. The published efficiency data for AFDs driving centrifugal pumps are based on the assumptions that (1) the drive is fully loaded at maximum speed, (2) power varies as the cube of pump speed, (3) the motor efficiency at full speed is about 96%, and (4) the drive does not include transformers. Because (1) is usually incorrect, (2) is almost always incorrect, (3) is frequently incorrect, and (4) is often incorrect for wastewater pumping applications, the efficiencies of AFDs are usually significantly less than the published values of efficiency versus speed.

Efficiencies of electric motors fall rapidly at loads below about 25% full load. For this reason, the efficiencies of *any* variable speed drive that includes an electric motor is very poor at less than 20 to 25% of rated loads.

Note that efficiency is a misleading indication of operating costs. Low efficiency at low discharge is relatively unimportant because the power required is also low, and it is the power losses that are important. Typical calculated power losses in a complete driver system (AF converter and motor) for the 74-kW (99-hp) pump of Figure 15-4 are shown in Figure 15-25, where they can be compared with power losses of typical slip drives such as eddy-current couplings and liquid rheostats.

If dynamometer tests are specified, require that they be made *at the loads required by the pump* at several typical speeds and with instrumentation that maintains its accuracy in spite of any harmonics generated by the converter. If practical, insist that tests of the motor and the AF converter to be installed in the field be done together. It is even better, although expensive, to test the pump, motor, and AF converter as a unit at the pump manufacturer's test facility.

Operating Costs

The power required by the pump should always be calculated as a function of discharge, head, and pump efficiency and not by application (or, rather, misapplication) of the affinity laws. Affinity laws are too easy to misuse, and they are inaccurate when applied to pump efficiency. Overall (or wire-to-water) efficiency is thus obtained by calculating and plotting the following:

- Water power (kilowatts or horsepower) based on head and flow over the expected range of pump

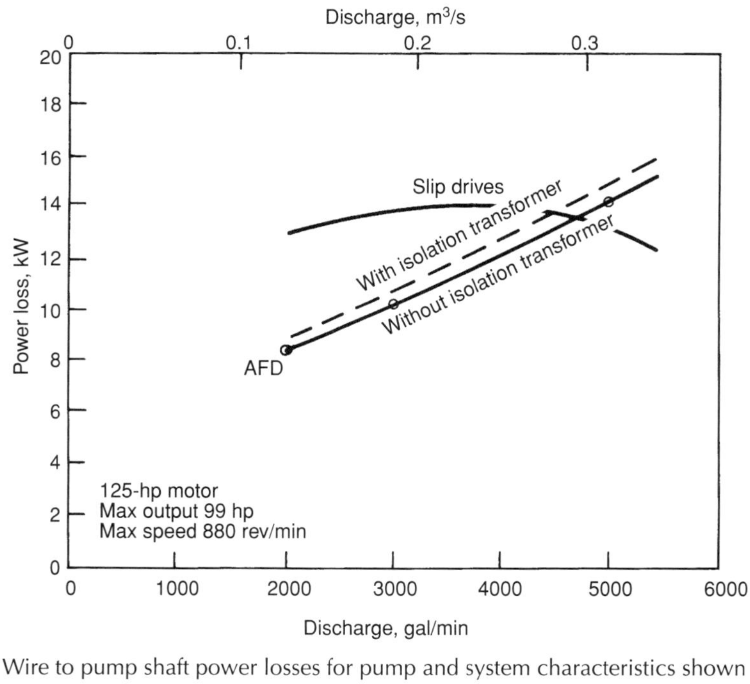

Figure 15-25. Wire to pump shaft power losses for pump and system characteristics shown in Figure 15-4.

speeds. Calculate the actual pump shaft input power (bhp) based on pump efficiencies

- Motor efficiencies reduced for (1) low speeds, (2) low torques, and (3) non-sine-wave input (~3%)
- Converter and transformer efficiencies for the actual loads
- Overall drive efficiency at each of several speeds obtained as the product of pump, transformer, converter, and predicted motor efficiencies
- Wire power (input power to the motor) obtained by dividing the bhp by the overall drive efficiency
- Power needs calculated from hours per year of operation at various discharge and pump speeds (see Example 29-1).

Compare various types of drives for each specific application on the basis of the present worth of cost, life and replacement, maintenance, and power (per Example 29-1). The results cannot be predicted reliably if any of these factors is ignored.

Motors

Standard motors cannot be used with any AF drive. All drive types impress a non-sinusoidal wave form on the motor. As motor speed decreases, motor cooling provided by shaft-driven fans decreases, but cooling is not the only consideration when applying adjustable-frequency drives. Rapid change in voltage versus time caused by some types of drives has been the source of rapid insulation failure in some motors. Furthermore, stray currents, apparently induced by some variable-frequency drives, have resulted in electrolytic corrosion of bearings and other components. To break the circuit of these stray currents, a thin, dielectric material can be specified for electrically isolating the shaft from the motor frame. Not all problems associated with this technology have been solved, but consensus standards for motors for application to AF drives are evolving.

The claim frequently made by drive manufacturers that standard motors can be used *is not true*. As it happens, AF converter application to existing motors is frequently successful, because motors are often oversized for the application, and older motors were frequently designed more conservatively than are modern motors.

Motors for use with AF drives must be specially designed for this service. For example, such motors are constructed with stator wires double-dipped in varnish (wires in standard motors are single-dipped); thin, paperlike insulation is added in the slots; and Class H insulation is used for Class A temperature rise. Also, as stated above, it is wise to insulate the bearings from the motor frame to interrupt stray currents and thus prevent electrolytic corrosion. *Be very conservative in specifying motors for use with AF drives.*

Step-output converters and PWM converters that do not use IGBT converters are generally satisfactory with existing 1.15 service factor motors driving centrifugal pumps if the motor is not loaded above the nameplate rating (service factor not used). However, both the foregoing AF converters are less efficient and more expensive than modern converters.

IGBT converters switch their output at a very high frequency (Figure 15-24b). They impress a high rate of voltage change (dV/dt), and the switching action also overshoots, thereby impressing high peak voltages on the motor. Both characteristics put greater stress on the motor insulation than does a sine wave. The motor insulation system must either be designed for this stress, or an output reactor must be inserted between the AF converter and the motor. Specially designed "power converter duty" motors are more desirable because a reactor reduces the efficiency of the drive system by about 1%. Also, always locate the converter within 45 m (150 ft) of the motor, because long motor leads magnify the peak voltages produced by the converter. If the drive cannot be located close to the motor, always use a reactor between the converter and the motor.

Power converter duty motors have insulation systems designed for the stresses imposed by AFDs. In addition to the usual motor parameters that must be specified when ordering a converter-duty motor, the speed range and load speed/torque characteristic must be specified. The motor manufacturer will then design the motor cooling for the application. Motor manufacturers catalog motors for speed ranges as great as 10:1 when driving constant torque loads—a much more severe service than driving a centrifugal pump over a typical 2:1 speed range.

Operational Problems

Any short interruption of power (sufficient to make a light flicker) can trip an older AFD, after which the motor must be allowed to come to rest before restarting—a requirement annoying in thunderstorms and windstorms, which could cause several interruptions per day. Some modern AFDs, however, are designed to operate through such momentary (less than 0.5 s) power losses, and AFDs can even be designed to synchronize with a spinning motor—but these features must be specified if they are wanted, and they increase the cost of the AFD.

Designing and/or specifying the interfacing between the motor and AFD is not a simple task. One way to promote compatibility is to specify "unit responsibility," in which the pump, motor, converter, and isolating transformer (if any) are all furnished by one supplier (the pump manufacturer) who has full responsibility for the proper operation of the entire pumping unit (see Chapter 16). The pump manufacturer is in no position to warrant the compatibility of the motor and the AF converter, so specify that the entire power package be supplied to the pump manufacturer by the AFD manufacturer, who must ensure the compatibility of the motor and the AF converter.

Because there are torsional pulsations in the output of a motor powered by an AF converter, it is strongly recommended that a torsional analysis be done for any drive of more than 300 kW (400 hp). Vertical motors supplied with variable-speed systems are subject to "reed frequency" vibration. The motor, the pump, and the control manufacturers must coordinate to eliminate this potential problem (see "Unit Responsibility" in Appendix C). Certain speeds may have to be blocked out of the variable-speed range. These speeds must, of course, be passed through during drive acceleration and deceleration, but the drive must not be allowed to operate there. A second problem that can occur if there is an anti-reversing ratchet in the drive shaft is that dwelling at certain speeds can cause an oscillation about the vertical axis, hammer the ratchet, and damage or destroy it.

Adjustable-Frequency Drive Specifications

A summary of specifications and requirements for a reliable, trouble-free AFD system includes:

- *A power system analysis.* A source voltage range of +10%, –5% from nominal (usually 460 volts) is specified in NEMA standards for AF drives. Most electrical equipment is designed for +10% from nominal. If the plant electrical system cannot meet the tighter tolerance required by AF drives, specify the voltage range that the drive must accept.
- *Voltage disturbances.* Modern drives have microprocessor controllers. These controllers are capable of maintaining drive operation through short power supply voltage sags and immediately and automatically restarting the drive if the sag exceeds the hold-up interval. Specify "flying restart."
- *Grounding.* Protect the microprocessors by specifying a ground grid in which each electrode, when isolated, has a maximum ground resistance of 2 ohms—a protection significantly better than that specified in the NEC.
- *Motor insulation and AFD compatibility.* Static power converters impose unusual stress on winding insulation. Early insulation failure will occur if the motor is not designed for this duty. For six-step inverters, specify conformance to NEMA MG 1, Part 30 (1000 volt peak with 12 μs rise time). For PWM inverters, specify conformance to NEMA MG 1, Part 31 (1600 volt peak with 0.1 μs rise time).
- *Drive tuning.* To ensure optimum performance, the AFD must be tuned to the motor and load characteristics. Some modern drives have a "self-tuning" function that permits the drive to adjust some parameters to motor characteristics, but no drive can analyze load requirements. Specify that the drive be set up by a factory-authorized technician. Do not rely on contractor electricians to set up AF controllers.
- *Voltage-vector control.* For best performance, specify drives with voltage-vector control to regulate both voltage and frequency to produce a sine wave output for optimum motor magnetization and performance.
- *Isolation.* A contactor to disconnect a drive physically from the motor is not necessary, but consider specifying contactors to isolate shunt filters when the drive is not running. If the drive is equipped with a phase-shifting transformer, specify a line contactor to eliminate the transformer losses when the drive is not running.
- *Harmonic currents.* Power factor correction is usually not required for AF drives. But if power factor correction is required because other loads on a power system have a lower power factor, traps must be provided to keep harmonic currents out of the capacitors. A power system analyst can design suitable trapped capacitor schemes for power systems with AF drives.
- *A temperature-controlled environment.* A maximum ambient temperature of 40°C (104°F) is specified in NEMA standards. Unless the space is air conditioned, the temperature requirement may not be met. AF drives generate substantial heat. Specify a realistic ambient temperature requirement. If the space is to be air conditioned (mandatory in many geographic areas), obtain realistic heat dissipation (including heat from transformers) values for the drive from the manufacturer, specify these as maximums, and provide the value to the air conditioning system designer.
- *Standby generators.* If the drive must be powered by a standby generator, harmonic currents drawn by the drive must be reduced to a minimum. Heavy harmonic filtration with shunt filters may result in a leading power factor. Use six-phase rectification, series reactors, and/or active filters in preference to shunt filters to reduce harmonics. Be sure the generator is capable of handling the harmonic current load and any leading power factor resulting from

any filters. Be aware that the generator must be oversized to meet these requirements.

- *Corrosive gases*. Hydrogen sulfide in the atmosphere must be limited to 3 ppb or less. Place the air intake on the upwind side of the building, and pass the air through a filter of, for example, activated carbon.
- *Noise*. An annoying hum (sometimes an ear-piercing high-frequency screech) is generated by AF controllers, and motors are also noisy, so consider sound-absorbing walls and ceilings to reduce the reverberation. IGBT controllers, however, are quiet and emit only a slightly noticeable hum.

In conclusion, power system design for AF drives presents problems that have been largely ignored in the past and have resulted in many unsatisfactory installations. Modern AF drives and proper power system design should result in a problem-free variable-speed pumping system with a life-cycle cost lower than that of any other V/S drive.

Emergency Generators

The harmonics generated by the converter are fed back to the standby generator and, unless the generator voltage regulators are specifically designed for compatibility with the AFDs, the generator may not operate the drive. The generator manufacturer must match the generator to the specific AFD to be used.

Maintenance

The problems of repairing AFDs are controversial. Some manufacturers claim that diagnostics and plug-in circuit boards make troubleshooting and most repairs relatively easy. Most circuit boards, however, are expensive, and untrained, nonelectrical personnel should not touch them. Moreover, replacing a thyristor requires a very skilled technician and the proper tools. Nevertheless, a competent maintenance electrician can service AFDs if

- Training in power electronics is provided at start-up and on a periodic basis.
- The proper tools are available.
- A spare of every type of printed circuit board and power electronic component is kept on hand. Many boards, even for different sizes of AFDs, are the same, so the inventory may not have to be large. (Circuit boards usually cost between $750 and $2500 each; the cost of repair is about half as much.) Power electronic components, though, are different for different sizes of drives. Another reason for keeping spare circuit boards is that some manufacturing firms continue to drop out of the solid-state business, and orphan electronics boards may be very costly to replace and slow in delivery.

Routine maintenance is low and consists only of keeping the equipment clean and all connections tight.

Other manufacturers state that all but simple problems require a factory-authorized technician, who is sometimes accompanied by an electrical engineer at a charge of about $700 per person-day portal to portal plus travel time and expenses. Repairs often take 2 or 3 days. A worst-case scenario might be a day for travel and diagnosis, a day waiting for parts, and a day for installation and start-up, which could total $4000 or $5000. Troubles might occur yearly, more often, or never. Typically, one might expect no trouble at all for 2 or 3 yr after start-up with only occasional problems later. Note, incidentally, that one lightning strike can zap all of the AFDs, so *make sure that the station is protected from lightning* (see Section 8-5).

Some expert consultants with considerable and lengthy experience have had nerve-racking problems with AFDs, and they approach decisions regarding their use with reservations and caution—especially for municipal projects. They point to the problem of competition with industry in hiring and keeping trained technicians and the difficulty in procuring spare parts for public utilities, and they urge that spares be included in the original bid package. Make sure that no critical items are omitted.

The industry is now more mature than it was a decade ago, the equipment continues to improve, reliability is better than it was in the past, and new components continue to reduce the price of AFDs. The improvements, however, tend to make electronic equipment obsolete and parts may become unavailable. It seems wise to plan on replacing electronic equipment every 10 to 15 yr, and this cost should be included in the cost analyses of alternatives. The cost of repairs is quite uncertain, but for economic analyses of alternatives, the assigned annual cost should probably lie between 5 and 10% of the fob price of the equipment. There are records of annual costs well above 10% and other records of almost zero.

If C/S drives are to be considered as an alternative to V/S drives, include a comparison of (1) the construction costs of the wet and dry wells, (2) energy usage, and (3) maintenance needs as shown in Example 29-1.

Summary

When evaluating V/S drives, avoid the entrapment of comparing only the efficiencies. Instead, compare power losses in kilowatts at several of the loads and speeds to be imposed. An alternative is to compute the annual energy requirement for various increments of discharge (as shown in Example 29-1), in which the proper power rate schedule is used. Then add maintenance, replacement, and capital expenditure and reduce all of the costs to present worth. Include isolation transformers in the capital cost.

Finally, consider the intangibles of simplicity and reliability together with the owner's situation, such as the number and size of pumps and pumping stations, the size of the public utility, the location (especially with respect to repair facilities), the cost of power, the abilities of the operators, the effects on downstream operations, and the qualifications of the supervisors. The energy savings in one or two small drives would not justify the training and spare parts inventory needed for AFDs. If, on the other hand, there are half a dozen large drives, justification for using AFDs (or any other V/S drive) is likely. The key is to find the overall economics for each situation.

Eddy-Current Couplings

The eddy-current coupling (Figure 15-26) is a unit of approximately the same size as the motor and is placed between the motor and the pump. The input shaft is coupled to the motor, which is usually a squirrel-cage induction motor. The output shaft, coaxial with the input shaft, is coupled to the pump. A small control panel is electrically connected to the eddy-current coupling.

An eddy-current coupling consists of two basic elements: (1) an iron ring (or drum) keyed to the input shaft and rotating with it at C/S and (2) a multipole electromagnet keyed to the output shaft (see Figure 15-27). The electromagnet pole faces are very close to the inner surface of the iron ring, and the magnets are energized by low-power dc current from the control panel through brushes mounted on the frame and slip rings mounted on the shaft.

When the driving motor is started and the iron ring revolves, the electromagnet is energized through the brushes and slip rings. Electrical currents (eddy currents) in the ring are induced by the field of the electromagnet, and they create magnetic fields. For brevity, it can be considered that the electromagnet is driven by the magnetic fields in the ring. The difference between the rotational speed of the ring and that of the electromagnet is called "slip." The amount of slip depends on the magnitude of electrical current through the electromagnet.

Efficiency

The efficiency of the drive is determined by three factors:

- Motor and coupling friction and windage losses
- Motor core, winding and stray losses
- Slip loss, which is by far the major loss
- Direct current supplied for coupling's field excitation.

The slip loss is the power lost in speed reduction due to slippage of the electromagnet with respect to the ring. Because of low friction and windage in the coupling, the output torque is only slightly less than the input torque. If friction and windage losses are neglected, the input and output torques are equal. Hence, the slip efficiency (power out/power in) is

$$e_s = \frac{2\pi n_o T}{2\pi n_i T} = \frac{n_o}{n_i} \tag{15-4}$$

where n_o is output shaft (pump) speed and n_i is input shaft (motor rated) speed. Thus, the slip efficiency depends only on the ratio of input and output speeds. It is linear and independent of load.

The loss of power due to slip is

$$P_s = P_p\left(\frac{1}{e_s} - 1\right) \tag{15-5}$$

where P_p is the power output to the pump. So the power lost due to slip depends on the ratio of speeds and the magnitude of the load at the lower speed. Furthermore, the slip loss is the same for any type of slip drive, including wound-rotor/liquid-rheostat slip drives.

The load for centrifugal pumps may vary between the third and fourth power of the speed over the useful speed ranges, so the power lost may be relatively small even though the efficiency may seem to indicate otherwise. For example, at 75% speed for a load that varies as the fourth power of the speed, the power lost would be

$$P_s = (0.75)^4\left(\frac{1}{0.75} - 1\right) = 0.105$$

or 10.5% of the power required at 75% maximum speed. Although these losses are not negligible, they should be considered objectively in a life-cycle engineering cost study that includes not only power costs but also the maintenance, reliability, and expected life of the equipment in comparison to other drives.

Typical power losses for an eddy-current coupling driving a pump are shown (as "slip drives") in Figure 15-25.

Stationary Field Eddy-Current Couplings

By altering the design of the unit, the rotating field of an ordinary eddy-current coupling can be converted to a stationary field, which eliminates the brushes and slip rings. Two U.S. companies offer this design. The efficiencies of the new and old designs are almost the

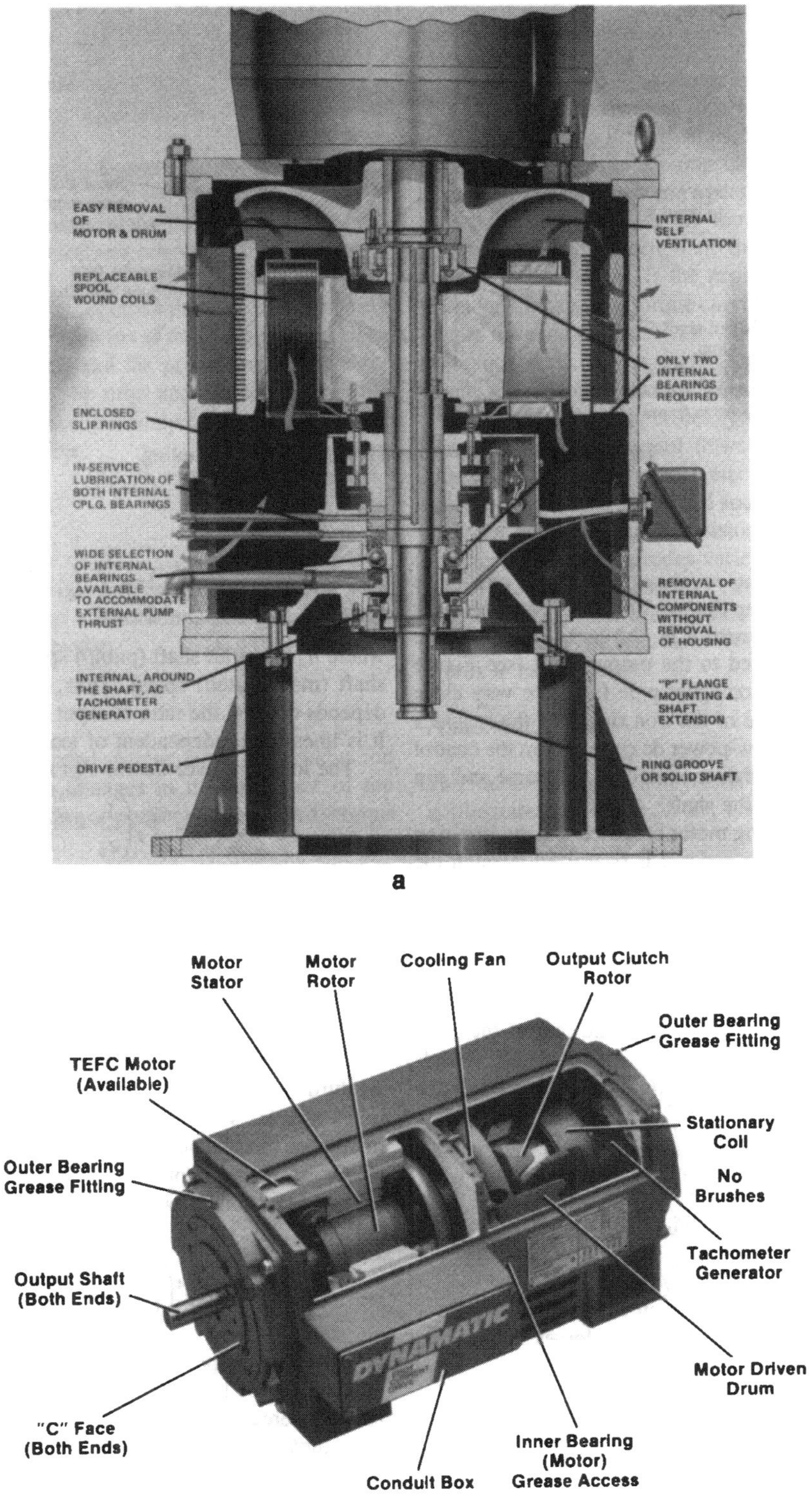

Figure 15-26. Air-cooled eddy-current pump drives. (a) Vertical mount, salient pole, rotating field (with brushes); (b) horizontal mount, stationary field (without brushes). Courtesy of Dynamatic Corp.

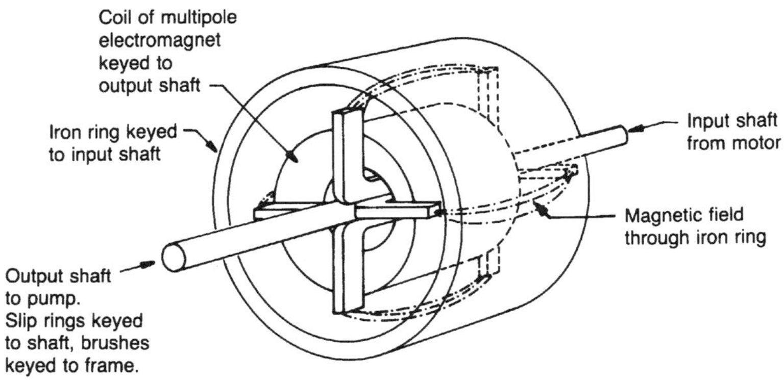

Figure 15-27. The principle of an eddy-current coupling.

same and the prices are comparable, but eliminating the slip rings and brushes reduces the maintenance to an amount comparable with a squirrel-cage induction motor.

Liquid Rheostats

A complete drive consists of a liquid rheostat with short electric power cables connected to the brushes of a wound-rotor motor.

Operation

The slope of the speed–torque curve of an induction motor varies with changes in the resistance of the rotor circuit. The lower portions of the speed–torque curves of any induction motor are shown in Figure 15-28 for many values of external resistance in the rotor circuit, and a portion of the speed–torque curve of a typical centrifugal pump in a typical system is superimposed on the motor speed–torque curves.

At any given speed, the motor and pump must operate at the intersection of the pump and motor speed–torque curves. If the resistance is decreased, the slope of the motor torque curve is increased, which causes the motor to accelerate on its curve (while the pump accelerates on its curve) until the motor output torque exactly equals the torque required by the pump.

A liquid rheostat is basically a large, three-phase resistor. Electrical current flows between copper or bronze electrodes through an electrolyte (a very weak solution of distilled water and sodium carbonate) that makes a good and stable resistor. The electrical resistance between electrodes varies inversely with the submerged area of the electrodes and also varies directly with the distance between the electrodes. The electrolyte is the resistor. The resistance of the electrodes themselves is insignificant.

Pneumatic Control

The model most commonly used for sewage pumping is illustrated in Figure 15-29. One rheostat electrode is connected to each rotor brush of its associated motor. The principle of operation is that of a manometer. The air pressure, P_a, in the sealed reservoir equals the air pressure, P_a, at the bottom of the bubbler. This air pressure raises the level of the electrolyte on the electrodes by an amount, D. The depth, D, in the rheostat equals D in the wet well and also equals P_a (expressed in height of water column). The air compressor is relatively small and (without storage) supplies air directly to the piping as shown. The three-way solenoid valve vents the sealed compartment to atmosphere when the drive is shut down.

The bubbler requires about 1.7 m^3/h (1 ft^3/min) of air at very low pressure, and an air compressor powered by a small (1/4-kW or 1/4-hp) motor is adequate for drivers up to 600 kW (800 hp). Each liquid rheostat includes an internal air compressor.

Mechanical Controls

Another type of liquid rheostat uses movable and stationary electrodes. Each of a set of upper, movable electrodes is connected to one of the motor rotor brushes and stationary, lower electrodes are connected together to form a wye-connected, three-phase resistor. The electrodes are fully submerged at all times, and resistance between the movable and stationary electrodes is varied by changing the level of the upper electrodes. Usually a small, single-speed motor and sprocket chains are used to raise and lower the movable plate assembly. The motor receives "up-hold-down" signals from external equipment.

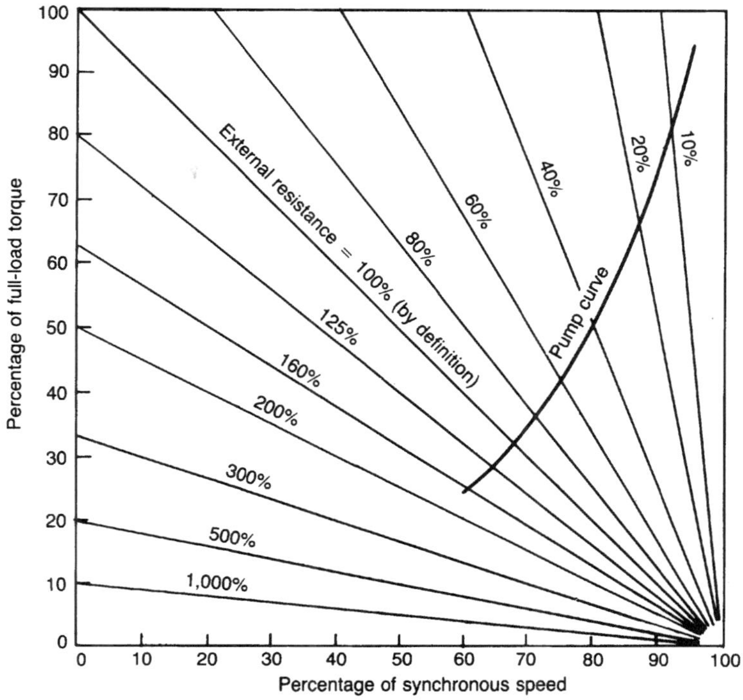

Figure 15-28. Speed–torque curves for a typical nonclog sewage centrifugal pump and a wound-rotor motor with external resistances.

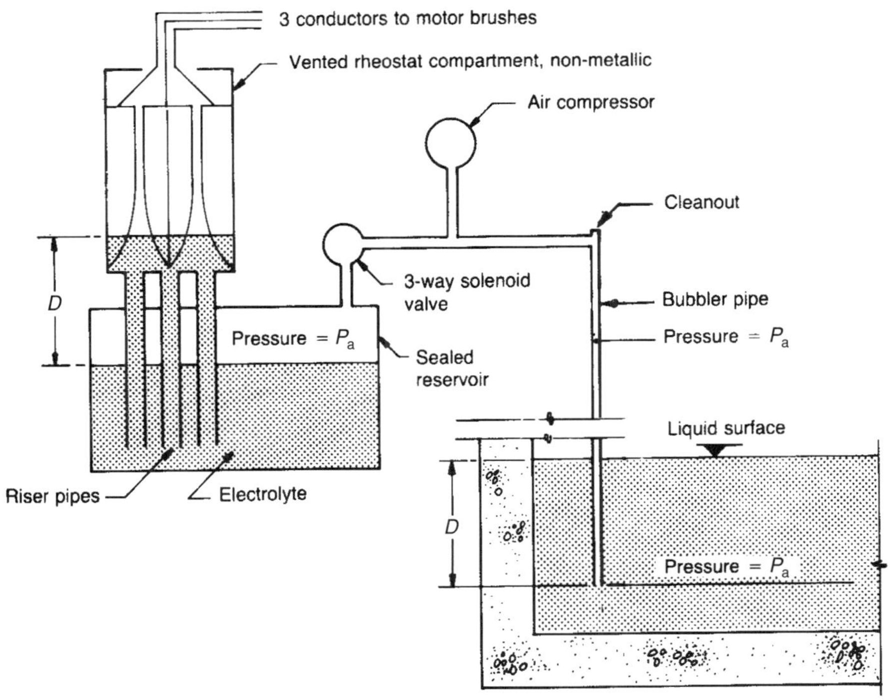

Figure 15-29. Pneumatic control system for a liquid rheostat.

Cooling

Because the electrolyte must be cooled, a heat exchanger is required. Up to ratings of about 19 kW (25 hp), a water-to-air heat exchanger can be incorporated into the same enclosure with the liquid rheostat if ambient temperatures permit. Larger units require an external water-to-water heat exchanger, and the most common is a baffled steel sleeve around the pump discharge piping. The annular clearance between sleeve and pipe is about 13 mm (1/2 in.) and the sleeve length is determined by the maximum heat losses that can occur based on the maximum slip loss. A small circulation pump (rarely exceeding 1/4 kW or 1/4 hp) is adequate.

Efficiency

The slip efficiency is

$$e_s = \frac{n_p}{n_m} = \frac{\text{motor shaft speed}}{\text{motor rated speed}} \quad (15\text{-}6)$$

in which n_p is the speed of the pump shaft and n_m is the full-speed rating of the motor. The slip efficiency is identical to that for an eddy-current coupling or any other type of clutch.

The total drive efficiency, determined by adding slip losses and motor losses, is about the same as that of an eddy-current coupling. The friction or windage losses (1 to 2%) in the coupling are absent, but in ratings below about 300 kW (400 hp), wound-rotor motor efficiencies at loads above a rating of approximately 25% are typically 1 to 2% less than those of squirrel-cage machines. The power required by air compressors and other accessories is negligible.

Fluid Couplings (Hydrokinetic)

In fluid couplings, the drive consists of a squirrel-cage induction or synchronous motor and coupling. The coupling, enclosed in an attractive housing, contains these major components:

- An impeller, which is mounted on and keyed to the input shaft
- A runner, which is mounted on and keyed to the output shaft
- A scoop tube
- A casing, which is mounted on and keyed to the input shaft.

A very small electrically driven pump is mounted on the outside of the housing. The pump delivers oil from a sump at the inside bottom of the housing to the casing. Oil is free to flow from the casing into the impeller and runner.

The impeller and runner have internal blades similar to those in a recessed impeller of a pump. Rotation of the impeller imparts energy to the oil in which it is partially submerged. The oil is "pumped" to the runner and imparts energy to the runner blades, thereby causing the runner and the output shaft to rotate.

The depth to which the impeller and runner blades are submerged in oil directly determines the torque that is transmitted to the runner and the output shaft. The depth of oil in the casing, and thus the submergence of the impeller and runner blades, is adjusted by a "scoop tube" that removes oil from the casing and returns it to the sump. The scoop tube can be positioned from the exterior of the housing through a linkage, and its position determines the oil level in the casing.

Slip-Recovery Drives

A slip-recovery drive consists of a wound-rotor induction motor and a solid-state controller.

The controller is electrically connected to the rotor brushes of the motor, and the effective resistance in the controller is electronically adjustable (see liquid rheostats). Rotor circuit current does not actually pass through resistors, but is rectified, inverted to 60 Hz, and returned to the motor stator terminals. The power losses that occur in resistor-type secondary controllers do not exist in the slip-recovery drive, which is slightly more efficient than the AFD. The controller should include all of the necessary capacitors for power factor correction and all of the circuit breakers required for line disconnect.

Variable-Ratio Belt Drives

The variable-ratio belt drive consists of a squirrel-cage induction motor and two variable pitch sheaves with a belt, shafting, bearings, a sheave-pitch-changing mechanism, and a large spring, all of which are enclosed in a housing.

One sheave (input) is either on a shaft directly coupled to the motor shaft or mounted on an extended motor shaft. The second sheave is mounted on the output shaft. Power is transmitted from the input to output sheaves by a heavy-duty vee-belt.

Output shaft speed changes are accomplished by changing the effective diameters of the sheave hubs by changing the spacing between the sheave discs. The discs are inside tapered so moving one disc (along the shaft) closer to the other disc of the sheave forces the belt to engage the disc at a greater distance from the hub. The position of one disc on the input sheave is changed by an external linkage. The position of the disc of the output sheave is changed by an internal spring to keep belt tension relatively constant regardless of changes in output speed. The discs should be chrome plated to reduce wear.

Hydrostatic Drives

A hydrostatic drive consists of the following:

- A squirrel-cage induction motor
- An adjustable displacement, multicylinder hydraulic pump
- A fixed displacement, multicylinder hydraulic motor
- An oil reservoir.

The hydraulic pump is flexibly coupled to the ac motor, and its shaft turns at C/S. The cylinder block (which is cylindrical in shape) is keyed to the shaft and contains the pistons. The entire assembly rotates with the shaft. The piston rods bear and slide on a "swashplate" that does not rotate. The swashplate can be tilted by an external device (through a linkage) to change the effective displacement of the pump.

The hydraulic motor is connected to the hydraulic pump by means of two hoses. The output of the pump is circulated through the motor and back to the pump intake. The shaft of the hydraulic motor is flexibly coupled to the load shaft. The motor construction is similar to the pump, except that its swashplate tilt is fixed. The motor output is connected to the shaft of the sewage pump. Changing the swashplate angle changes the discharge rate of the hydraulic pump thus controlling the speed at which the motor turns.

Refer to Table 15-3 for comparative advantages of hydrostatic drives.

Other Drives

Other drives are available but are not described here because they are not widely used for pumping water or wastewater. Obviously, this situation may change, so be alert to new products. Many types of rotating machinery that are quite satisfactory for some specific use, however, are unsuitable for municipal pumping. Be cautious in applying a new product by making a thorough investigation of installations and their maintenance records.

15-12. References

1. Karassik, I. J., W. C. Krutzsch, W. H. Fraser, and J. P. Messina, *Pump Handbook,* 2nd ed., McGraw-Hill, New York (1985).
2. Davis, C. V., and K. E. Sorensen, *Handbook of Applied Hydraulics,* pp. 40–16, McGraw-Hill, New York (1969).
3. Fair, G. M., J. C. Geyer, and D. A. Okun, *Water and Wastewater Engineering, Vol. 1, Water Supply and Wastewater Removal,* John Wiley, New York (1966).

Chapter 16

Pump-Driver Specifications

DAVID L. EISENHAUER
GARR M. JONES

CONTRIBUTORS
Richard O. Garbus
Philip A. Huff
Melvin P. Landis
Ralph E. Marquiss
Earle C. Smith
Horton Wasserman

Two approaches to writing specifications for pumps and drivers are (1) investing "unit responsibility" for the complete pumping unit (driver, shafting, pump) with a single manufacturer and (2) separating driver, shafting, and pump so that each can be furnished by a different supplier and assembled at the jobsite by the general contractor. These two approaches are discussed and compared in Section 16-1. The remainder of the chapter is confined to the first approach (unit responsibility) and an explanation of how to apply it to custom-engineered pumps and drivers for a high-lift sewage pumping station. "Custom engineered" means that most manufacturers must modify their "off-the-shelf" pumps with larger shafts, more rugged bearings, or some special materials to comply with the specifications. Other manufacturers may produce high-quality pumps that need no improvements.

This chapter is keyed to (and is an explanation of) Appendix C, which contains the specifications as written for two high-lift pumping stations. Each station has four pumps discharging into two force mains. Paragraph 1.02B of Appendix C shows how to vest the manufacturer with complete responsibility for properly coordinating sizes, system analyses, and procurement of the complete pump and driver system.

Specifications should be so clear that they cannot be misunderstood (see "Specification Language" in Section 28-1).

16-1. Comparison of Two Approaches to Writing Specifications

Many experts believe that the unit-responsibility approach is superior for public works projects. The benefits of this approach include the following:

- Fixing complete responsibility for the equipment with a single manufacturer
- Eliminating problems of misfit or mismatch in either size or performance (such as torsional resonance, reed frequency, and critical speed)
- Providing the owner with a single source for spare parts, troubleshooting, and future modifications
- Ensuring adequate design in subjects for which the consultant's personnel may be lacking in training or experience.

For the common, general-duty application of standard off-the-shelf pumps and motors of small pumping units—for example, 20 kW (25 hp) or less—the adoption of unit responsibility is less important but, nevertheless, has advantages over separate purchase specifications.

Some engineers prefer the second approach (the separation of specifications for each component of a pumping unit), but responsibility for properly mating all components then falls on the engineer. A list of the added responsibilities the engineer must assume includes the following:

- Coordinating the equipment to obtain correct overall shaft length
- Coordinating the compatibility of couplings
- Determining the critical speed for pump, motor, and shafting
- Analyzing the complete system for torsion (see Chapter 22)
- Coordinating the complete system to avoid dangerous resonance frequencies
- Selecting the motor starter
- Matching and coordinating the speed of the pump and driver
- Ensuring the compatibility over the entire speed range of variable-speed drives
- Coordinating the field testing of the entire system.

Consulting engineers normally do not have the resources necessary to perform these tasks because they do not manufacture the products. They are in no position to coordinate the multitude of mechanical, structural, electrical, and instrumentation details, and they rarely have the training or contractual relationship to do so.

Some contractors may prefer to purchase equipment components separately to obtain lower prices. In such cases, the owner loses value because of the lack of proper engineering in selecting the various pieces of rotating equipment and their appurtenances. The specification provision requiring unit responsibility is one of the best approaches available for controlling this type of activity. The owner always benefits from an open bidding process in which more than one manufacturer is allowed to meet the specifications. Specifications, however, should be clearly and properly drafted (consistent with applicable laws) to limit equipment to that most suited to the project. In many (if not most) pumping station projects, more than one manufacturer is capable of providing equipment within the limitations established by the owner's budget and project needs. It is the engineer's responsibility to provide the specification mechanism that will ensure good equipment and good mating of equipment components.

An example of the consequences of a contractor's evasion of the unit-responsibility clause (see Appendix C, Section 1.02B) was found in the construction of Sunset Beach Pumping Station in Steilacoom, Washington. Because the contractor purchased parts from different suppliers, the parts did not fit, completion was delayed by four months, and the contractor suffered significant losses due to penalties plus the cost of new equipment. Permitting such practice has no benefit to either contractor or owner and makes for poor coordination and higher cost. One means to forestall such noncompliance is to require an affidavit signed by both the contractor and an officer of the manufacturer, under the penalty of perjury, that the latter has accepted unit responsibility in accordance with the contract documents.

16-2. Methods for Specifying Quality of Equipment

Several methods for specifying the quality of products are compared in Table 16-1. More complete descriptions are given in Section 28-5 and in the CSI manual of practice [1].

16-3. Nonrestrictive Specifications

Appendix C is an example of a combination descriptive and performance approach to a nonrestrictive specification for pumps and drivers. The specification is a guide for (1) writing a pump specification and (2) the topics to be included. The format conforms to the CSI three-part section format [2].

The example in Appendix C illustrates how both performance and descriptive specification systems are combined in writing a pump specification. Brand names or manufacturers are seldom used in the example. The paragraphs on design requirements and operating conditions state the end results to be achieved and also include the necessary restrictions. Minimum salient characteristics are specified that limit the contractor's choice of equipment but that are sufficiently open to permit competition.

Research Needs

Except for the electric motor specification (which is basically oriented toward certain minimum performance criteria necessary for the project), most of the material requirements in Part 2 of the specification are listed in descriptive form. Preparing such descriptive portions requires researching pumps, pumping sys-

Table 16-1. Methods for Specifying Equipment

Method and description	Advantages	Disadvantages
Performance (functional results specified)	Common in building systems. Concept can be incorporated into any specification.	Burdens contractor with design. Acceptance not ensured. Cost is high.
Proprietary (name or brand and model specified)	Ensures quality desired. Short and easy to prepare.	Bidding noncompetitive. Requires assurance of fair price. Not allowed in public works without excessive justification.
Descriptive or generic (performance, materials, quality, etc. described in detail)	Excellent in principle	Considerable time to research and write. Often misused. Difficult to evaluate product submitted.
Nonrestrictive (also "or equal") (name two or more products and add "or equal")	Common method for complex products.	Evaluation is subjective, so documentation is required for rejects. Requires at least some performance or descriptive clauses.
Reference standards (reference to a standard specification)	Can be used in conjunction with the above methods. Familiar to contractors. Easy to write.	Limited to simple components or materials. Specifier must understand the entire reference.[a] Ease of use may lead to poor or incomplete specifications.

[a] Not necessarily disadvantageous. Specifiers have a duty (1) to understand fully each reference specification and (2) to include qualifying statements about which material does or does not apply to the equipment specification.

tems, and equipment to determine what is appropriate for the intended service and the critical features of materials and equipment required to perform under project conditions. This research consists of (1) developing and listing the performance requirements of the equipment; (2) identifying any special needs, such as special metallurgy or resistance to abrasion or corrosion; and (3) requesting preliminary proposals specific to the project requirements from candidate manufacturers. The final specification can then be drafted using the manufacturers' project-specific proposals as references. Descriptive requirements are also based on experience with specifications for similar equipment items as well as engineering judgment.

Effect of Size and Complexity

The content of any equipment specification varies greatly with the size and complexity of the project as well as with the equipment application and operating requirements. The example specification is written for wastewater centrifugal pumps of the end-suction, overhung-shaft design. Custom-engineered pumps are specified because, when operating at reduced speed against high discharge heads, there is a potential for damaging conditions caused by excessive vibration and unbalanced radial forces. Specifications for pumps operating at constant speed and low to moderate heads need far less detail and restrictive requirements, so they can be written around most manufacturers' standard pump designs. The example is not intended to exclude standard pumps if they comply with the specification. Instead, the special nature of the intended application is recognized in a specification that is designed to alert the manufacturer to these requirements.

In the following paragraphs, portions of the specifications in Appendix C are discussed and information is provided to assist in the preparation of a specification that will obtain the pumping unit required for a particular design or installation. Where proprietary specifications are permitted, the specification can be shortened considerably. But first read the warnings given in "Proprietary Specifications" in Section 28-5.

16-4. Operating Conditions

For pump installations—particularly variable-speed pumping installations—it is necessary that the pump manufacturer be aware of all operating conditions critical to pump performance and operation that are expected to occur over the life of the pump. A method of presentation of several operating conditions (described in the following paragraphs) is given in Paragraph 1.02D of Appendix C.

The purpose of "operating condition, point B" is to describe the range of full-speed conditions that will be imposed on the pump. Many pump manufacturers will guarantee only one point of operation, but it is nevertheless necessary to alert the manufacturer to other conditions that will occur over the life of the station. At each of two pumping stations, X and Y, the pumps discharge into two force mains. As more pumps are placed into service, the dynamic head increases, which alters the full-speed total dynamic head accordingly.

"Operating condition, point C" is selected judiciously to describe fairly the minimum sustained operation point for the equipment. In high head pumping applications, the selection of point C is critical because it represents the worst loading condition for determining the size of the radial bearings, the shafts, and, to a certain extent, the thrust bearings.

"Operating condition, point D" is provided to indicate the starting condition and to alert the manufacturer to a potential concern with motor starting power.

16-5. Mass Elastic Systems and Critical Speeds

For pumps 60 kW (75 hp) or larger and operating at variable speed, the analysis of critical speeds and the complete mass elastic system (specified in Paragraph 1.02E of Appendix C) is generally warranted and should be considered, particularly for custom-engineered units. The results guide the equipment manufacturer in avoiding combinations of rotating elements that may be subject to fatigue failure (see Section 22-11).

If critical speed falls within the operating range of a variable-speed unit, continuous operation at that speed must be avoided to prevent destructive vibration. Calculations can be made to predict the critical frequency of a unit, but variations in metal densities and minor casting irregularities make total accuracy impossible.

16-6. Pump Testing

The true test of pump performance is satisfactory operation in the intended installation, and it may be desirable or necessary to test a pump in the factory or in the field (or both) to provide assurance of the pump's capacity to meet project requirements. Consider, however, that the methodology used to determine flowrate is the result of projections based on unit demand factors or total consumption figures developed from historical data. These data, in turn, are based on flow records (which are usually inaccurate) and on population estimates, which in themselves are never more accurate than ±10%. Headloss figures are based on estimations of pipe wall roughness and data for turbulent losses through fittings, valves, and special conditions (such as dividing or uniting flows) in which each loss is within the "shadow" of another and losses may easily double by swirling. Some of these data were developed from model studies (1) without a reasonable allowance for the effects of scale; (2) that were performed many years ago, some under questionable assumptions; and (3) with metering and measuring devices less accurate than the apparent precision of the hydraulic computations in which they become an integral part. On the other hand, the range of static heads can be precisely determined, a range of pipe friction losses can be confined to an acceptable band with confidence, and the inaccurate losses in fittings are a small part of the total losses and, hence, of minor effect. In summary, pump performance requirements should be viewed objectively—not with the assumed precision many engineers assign to them.

Ensuring that the pump supplied will meet the required performance specifications requires a guarantee from the manufacturer, which, in turn, may require a factory test. Whether the test should be made at all (with the designer and client relying on previous tests of similarly sized impellers), whether a specific pump test should be witnessed, and whether factory tests must be followed with field tests depend on the cost–benefit ratio of testing and the need for positive assurance of performance. One way to approximate the cost–benefit ratio is to speculate about the effect of efficiency. Suppose, for example, the efficiency of a pump is 85% for a discharge of 0.38 m^3/s (6000 gal/min) at a head of 15 m (50 ft) and the efficiency of the motor is 95%. For a life of 20 yr at 8% interest with power at 6¢/kW · h, the present worth of a 3% change in efficiency is nearly $11,000 (see the interest formulas in Section 29-4). A factory test of a pump of this size might cost the owner $2500 (depending on the configuration of the unit). A witnessed test might cost nearly double plus the cost for the consulting engineer's on-site representative. Either a factory test or a witnessed factory test would establish not only the efficiency but also the head and discharge under test conditions and to some degree vibration and noise.

Test Requirements

At the outset, the designer should identify two or more manufacturers with the capability of conducting full-scale tests. Mid-sized equipment (up to 500 hp, 480 V, and 15,000 gal/min capacity) can be tested by most manufacturers. At least three manufacturers can test equipment up to 1500 hp and 4160 V. For larger equipment, it may be necessary to utilize an independent hydraulic laboratory. Despite these limitations and difficulties, some experienced engineers believe the cost of properly performed tests of each pump is money well spent. Consider, however, the following means of keeping costs at a minimum.

- Discuss the proposed tests with at least two manufacturers. Determine the probable cost of the tests and any viable modifications.
- Consider using a factory-calibrated motor or dynamometer. However, failure to use the actual production motor on large units leads to controversy if subsequent wire-to-water power field tests vary significantly from factory test results.
- Test at constant speed. But for variable-speed pumps, select four or five well-spaced operating points from minimum to maximum discharge.
- Consider testing only one of a group of identical units, but be aware that (1) the manufacturer may not guarantee any pump not tested, (2) the efficiency at the design point may vary as much as 3% and there may be significant variation in shut-off head and performance curve characteristics due to machining tolerance or errors, and (3) some experts insist on testing each pump in the group.
- Test only those pumps for which a cost–benefit ratio is reasonable. If the costs are too great, it may be more cost effective to oversize the pump and driver.
- Consider whether nonwitnessed pump tests are sufficient. Much depends on the manufacturer's reputation and standard methodology in testing.

Factory Pump Tests

If factory tests are made, the following are essential:

- Require the pump manufacturer (or the general contractor) to guarantee performance at the specified conditions of head, capacity, and efficiency. But note that unless the contract is a direct prepurchase one, the owner has no direct contractual relationship with the manufacturer, so obligate the general contractor, who in turn obligates the manufacturer.
- Require that the testing and the test setup conform to the Hydraulic Institute's standards [3] together with any special requirements [e.g., testing a submersible pump with its discharge elbow and a 1.5-m (5-ft) length of discharge pipe].
- Require that both test logs and curves be certified by an officer of the manufacturer's corporation. (Note that submittal of test logs is usually a special requirement.)

Performance requirements for the pumping station may not be satisfied if the field conditions cannot be reasonably met in the factory. For example, it may be impractical to test a long, deep well pump in a vertical position without reducing the length to a few bowls. Very large pumps may exceed the capabilities of the factory test site. Note also that the Hydraulic Institute's standard tolerances are +10% in capacity at the specified head and +5% in head at the specified capacity [3]. Be careful to ensure that these tolerances cannot overload the driver.

Both factory and field tests of pumps are discussed at length by Dicmas [4]. Groundwater pumping tests are extensively described by Walton [5], and a diskette programmed in BASIC is available from the publisher [5].

Witnessed Pump Tests

Owner-witnessed factory tests cost considerably more than factory tests because (1) the tests are made twice—once to make sure the pump will meet specifications and again when the owner's representative is present—and (2) the owner must pay for the engineer's attendance and travel. Still, witnessed factory tests are valuable because they tell the manufacturer that the owner is serious, which encourages honest performance in test setups. Anecdotes abound about the discovery of faults by a witness, but none, however, has been personally experienced by the authors of this chapter.

The benefit to the owner of witnessed tests depends on many factors, including the size of the pump, the projected hours of service, the cost of power, the reputation of the manufacturer, and the engineer's familiarity with the manufacturer's test methodology. In general, it is wise to confer with the owner to assess the potential benefits before specifying that tests be witnessed.

The following is a checklist for the engineer who is required to witness a pump test:

- Be thoroughly familiar with project specifications, any referenced standards, and the manufacturer's proposed test setup prior to arriving for the test. Be familiar with the test procedures and setups in the Hydraulic Institute's standards [3] and in Dicmas [4].
- Require submission and approval of the test setup prior to beginning the test.
- Develop a checklist based on project specifications and referenced test standards prior to arriving for the test.
- Be completely familiar with the test data compilation procedure prior to the test date.

- Be sure to arrive early enough the night before to be fully rested and alert.
- Be prepared to stay longer than scheduled if a test fails.
- Upon arrival, check the test setup against the agreed-upon testing procedure.
- Obtain photocopies of the calibration records for all test gauges, meters, motors, and dynamometers.
- Check the calibration curves against the serial number of each gauge, meter, motor, and dynamometer.
- Go over the test procedure with the manufacturer's test engineer to be certain all concerns have been addressed.
- Note the time testing begins for each pump.
- Make a note of the serial number of each pump. (On bowl-type pumps, the serial number is on the bowl.)
- Check all zero points or readings before starting the test.
- Once testing begins, observe the data recorded by the test technician, read gauges, manometers, and other instruments at the same time that the technician reads them, but do not record any data, only confirm. Be certain gauges and manometers are stable at the time readings are taken.
- *Subjectively* assess pump vibration by feeling each pump case. A pump usually vibrates more on the test stand than in the pumping station because the pumps are not mounted on a proper foundation and are not driven by the field drivers. On the other hand, resonance or equipment problems may occur in the field.
- Listen for any unusual noises (which may be difficult because of ambient noise at the test site).
- Note the time that testing is finished for each pump.
- Sign each test log and obtain photocopies of all test data as soon as they are developed.
- Spot check the calculated results by reducing a few data points yourself.
- Do *not* approve or sign developed performance curves until you have independently checked four or more data points.
- Be prepared for long periods of inactivity if several pumps are to be tested. It takes awhile to disassemble one test setup and assemble the next. Take your briefcase, work on the report, study this textbook, or tour the plant. This is an excellent opportunity for an experienced engineer to educate an intern who can then be the witness for other tests.

When the test is completed, the witness should prepare a report for the owner that describes the test, conditions, observations, and results. Appendices to the report should include copies of the test logs, generated curves, and any other information pertinent to the test procedure or results.

Variable-Speed Pumps

Many specifications require that the intended field drive equipment be used in the tests of variable-speed pumps, which doubles or triples the cost of the tests. The perceived benefits include (1) the evaluation of the pump's ability to perform at specified conditions, (2) the determination of wire-to-water power efficiency, (3) the determination of the compatibility of all components, and (4) the discovery of objectionable vibration. The setup in a manufacturer's test bed, however, is never the same as in the final installation. Although efficiency and compatibility can indeed be assessed, harmonics in AFDs are a function of the entire electrical system including the supply, which is different in the field. Observations of vibration under various operating conditions are qualitative at best because the vibrational characteristics of the supporting structure are missing in the field. Factory tests of a complete pumping assembly are particularly expensive if special test stands must be fabricated—for example, for motors mounted on a floor above the pumps. Shipping motors, controllers, and other parts of the driving units to the pump manufacturer as well as the assembly, disassembly, electrical work, coordination, and possible delays increase the costs for testing complete pumping units. In summary, engineers should have a clear idea of what is to be accomplished, exactly how the tests should be run, what added costs are incurred, and the benefits that accrue from such tests before specifying factory tests of complete pumping assemblies.

Example Specifications

Example specifications for hydrostatic tests and witnessed performances are given in Section 1.02F of Appendix C. Both of the testing procedures in Paragraphs 1.02F2 and 1.02F3 are desirable for custom-built pumping units.

Field Pump Tests

Field pump tests are of two general types: (1) *field operational pump tests* in which accuracy of flow and head are measured to within an error band (or uncertainty) of ±5 to 7% and (2) *pump acceptance tests* in which final acceptance may be based on satisfactory field performance with reward/penalty clauses written into the contract for failure of the pump to perform within specified limits. Such a specification requires expensive tests that may have limited practical value because manufacturers cannot be responsible for factors outside their control, such as improper sump design with intake problems, insufficient $NPSH_A$, system characteristics different from those submitted, and inadequate supporting structure. A good factory test is a great help in identifying and resolving such problems.

Field Operational Tests

Comprehensive field operational tests are important, and many engineers require such tests of every installed pump as a check in which the entire pumping unit (including instruments, controls, motor, pump, and valves) is observed for operational integrity and function. Testing flowrates and pressures at allowable errors (or uncertainties) of 5 to 7% is not difficult, time-consuming, or expensive. The methods given in Section 3-9 for measuring the roughness of pipe apply equally well to the testing of pump discharges. (Note, however, that a pump manufacturer's guarantee should apply only to the factory test.) Such tests are valuable for discovering trouble. For example, in a two-hour field trial, a single-stage vertical turbine pump delivered 30% less flow than expected. When the pump was disassembled, the impeller was found to have slipped down the shaft. Repairs restored the full discharge capacity. Without the test, the inefficiency would not have been discovered. A field test is, however, as much a test of the hydraulic design of the pumping station and of the technicians' skill as it is a test of the pump and driver. For example, vortices profoundly affect the pump performance, and these are, in turn, a function of the sump design and the suction intake velocity. Confer with the manufacturer to make sure the sump in the field test is acceptable. The field operational test is also valuable as the basis for the intelligent operation of the facility.

Before beginning field tests, observe the sump surface for vortex formation or signs of disturbance that would lead to rotation. Vortices may be intermittent on 15- to 30-s intervals, so allot sufficient time for the observation. Even with the use of clear water, subsurface vortices are not visible, but if they are suspected, dye (which is used in model studies) can also be used in field tests, although the water in the system must be changed when discolored. An alternative is to use colored plastic chips of polyethylene (which has a specific gravity of 0.92 to 0.96). Chips 25 × 25 × 1.6 mm (1 × 1 × 1/16 in.) are about the right size. They can be put into a small sack, which can be positioned at the suspected origin of a subsurface vortex and then broken to release the chips. (See various directories [6, 7] for suppliers of pigmented polyethylene chips.) The chips slowly float to the surface and can be netted. If vortices are detected, model studies are definitely required.

Flow can be measured with the pumping station flowmeter after calibration in situ by volumetric or tracer methods (see Section 3-9). If no permanent flowmeter is installed, a temporary flowmeter, such as a strap-on magmeter, orifice, pitot tube, or elbow meter, can be installed and also calibrated in situ. Alternatively, tracer and/or volumetric methods can be used for metering the flow throughout the test.

Because one purpose of an operational test is to provide the basis for facility operation, it is a *change* in the operational status—not absolute values—that is important. Consequently, the pressure gauges to be used by the operating personnel should be used for the test. The gauges should be calibrated and installed in accordance with the Hydraulic Institute's standards [3]. The discharge pressure gauge should be located at least three pipe diameters downstream from the pump discharge nozzle with no intervening flow disturbances such as those caused by valves or fittings. Electrical instruments in pumping stations are not adequate to obtain overall (wire-to-water power) efficiency. If the efficiency must be checked, special meters must be temporarily installed.

When running the tests, try to obtain two or three separated test points of head versus flow on both sides of the normal pump operating points.

Field Pump Acceptance Tests

Only with extraordinary care and special instrumentation can the errors in field testing be reduced almost to the same order of accuracy obtainable in factory tests (about 1%). The principal disadvantages of acceptance by precision field testing are the cost and the difficulty of rejecting a unit that fails the test. Claims (and, perhaps, lawsuits) will abound in regard to (1) the accuracy of the field test, (2) the validity of the design point, and (3) the reasonableness of the discharge requirements. Delays due to repairs, adjustments, or replacements defeat the project schedule and can force the engineer to a compromise adverse to the client. A comprehensive, witnessed factory test is not foolproof, but it can usually help to avoid such confrontations.

Consider, on the other hand, a factory test of a large, long, multistage deep well turbine. Some manufacturers might be able to test only a short column length consisting of, perhaps, only a single bowl (or four to five bowls at most). It might be necessary to operate at reduced speed because of power supply limitations. In the field, however, full-speed operation can be performed to evaluate (1) the suction well conditions, (2) the oscillation of the full assembly in its vertical configuration, (3) the actual hydraulic losses in the full column length, (4) the mechanical friction losses in additional shafting and intermediate bearings, (5) the true power requirements, and (6) the overall efficiency. The accuracy of such a field test might be superior to a factory test, provided that (7) the technicians are knowledgeable and skillful, (8) the pumping station was designed with field testing in mind (isolation valves and drain pipe from force main to sump for recirculating the flow), (9) pressure gauge taps are properly located, and (10) calibrated, sensitive instruments are used. But note that (11) the calibration of flowmeters to an accuracy comparable to a factory flowmeter (repeatability of, e.g., 0.5%) requires very great care, considerable skill, and meters that are installed in strict accordance with the manufacturer's recommendations; (12) pressure gauges and meters must be 5 or, better, 10 diameters downstream from disturbances; (13) the pressure taps should be located at quadrant points connected to a common collector; and (14) the pressure gauges should have a face at least 200 mm (8 in.) in diameter and must be readable and accurate to 0.25%. The other instrumentation must be of similar precision to approach the accuracy of the factory test (see Dicmas [4] and the Hydraulic Institute's standards [3]). Hence, specify factory tests where proper instrumentation is available followed by field operational tests unless circumstances are so unusual that such a procedure is inadequate.

16-7. Shipping Major Pumping Units

Damage to major equipment items during shipment is of concern on most construction work, particularly with custom-made units. To protect against potential damage from sudden impact, it is often desirable to specify shipping requirements and restrictions, as in Paragraph 1.02G of Appendix C. Recording accelerometers are available on a rental basis from most railroad and trucking firms at a slight additional cost. The presence of the device on a shipment usually provides some incentive on the part of the common carrier to exercise more than usual care. The 3.0 g value (of acceleration) is a commonly accepted criterion for indicating potential damage.

16-8. Submittals

If the product is specified by performance criteria or is to be used in a complex system, or if a variety of contractor options are permitted, submittals should be specified as in Paragraph 1.04 of Appendix C. Submittals are also required for equipment or materials that have long delivery times to reduce the chances of delays due to the rejection of the products after delivery.

Submittals listed in the specifications are designed to provide sufficient information to the engineer regarding the design to indicate that the contractor has properly interpreted the drawings and specifications. Because the submittal requirements are specific and limited, the contractor does not have to guess about the amount of information required and the design engineer does not waste time reviewing irrelevant or unnecessary documents. The submittal and the specification together should give a description sufficient for the engineer to determine whether the contractor's planned equipment or material will conform to design requirements. Specify only the information needed to make that determination. The submittal material requested in the pump and driver specification, Paragraph 1.04 of Appendix C, conforms to the above criteria.

16-9. Information to Be Provided

Equipment specifications also require that other information and product data be provided to meet a wide variety of owner needs. Much of the information goes into files as part of the project records, some is used by the engineer or construction manager in inspecting the contractor's work, and other data form the basis of the O&M manual. There must, however, be sufficient detail for the engineer or construction manager to determine the acceptability of the equipment. Examples of information to be provided with a pumping unit are listed in Paragraph 1.05 of Appendix C.

16-10. Seals

Pumps with a stuffing box, shaft packing, and clear-water seal are specified in Paragraphs 2.02B4 and 2.02B5 of Appendix C. Mechanical seals are often preferred over conventional packing because mechanical seals reduce maintenance. But mechanical seals are not only costly but are subject to failure on occasion, and the failure may be sudden. If packing fails, the pump can usually be kept running by temporary adjustments until it is convenient to shut it down. If a mechanical seal fails, the pump must be shut down at once. Some types of mechanical seals, however, fail gradually with ample warning.

There are a variety of proprietary seal designs and no industrywide standard exists as a convenient specification reference. Because seals are a continuing maintenance concern and interchangeability is desirable, it is recommended that, if possible, one manufacturer be required to supply all mechanical seals.

16-11. Pump Shafts

The extensive specification for the pump shaft (Paragraph 2.02B8 in Appendix C) is intended to require specific attention by the manufacturer to the need for heavy-duty shafts for applications where high discharge heads and operation at variable speed may result in substantial radial thrust loads. This provision can be modified to less stringent requirements if the pump is to operate at constant speed and moderate discharge heads.

The maximum shaft deflection of 6 mils (1.5 mm) at the worst radial-load condition applicable to the specified continuous operating condition is considered a reasonable objective. The value is typically 10 to 20% of the nominal wearing ring clearance for an end suction pump. The methodology specified as the basis for the calculations is presented in the Hydraulic Institute's standards [3].

16-12. Pump Shaft Bearings

For pump installations where radial thrust is a problem, bearings must be sized and rated to withstand successfully the worst loads applicable to the specified continuous operating conditions. To ensure compliance with this requirement, Paragraph 2.02B10 of Appendix C requires the submission of documentation to support the selection of bearing sizes.

As a guide, bearing design lives (based on the AFBMA L-10 rating method) for continuously operated pumps typically range between 40,000 and 60,000 h if good reliability is expected. If the operation of the equipment has to be extremely reliable or if the equipment is quite large, design lives of 100,000 to 200,000 h are not unreasonable. Check with equipment suppliers where extreme reliability is required to ascertain that the specified design life can be provided.

16-13. Vertical Drive Shafts

An intermediate bearing is required and specified for the vertical drive shafts in Paragraph 2.02C of Appendix C. If intermediate bearings are required, the design should provide both a rigid support for the bearing and convenient access for maintenance. The support can often consist of a reinforced concrete walkway that serves as lateral support for the wall (see Section 25-10).

16-14. Electric Motors

Standard electric motors less than 150 kW (200 hp) are normally covered by a separate specification. However, information such as voltage, power, speed, and enclosure type, plus all motor modifications including overtemperature devices and space heaters, should be specified in the specific equipment section.

Overtemperature or thermal protection for motors is often recommended for pumping units with variable-speed controllers to protect the motor against sustained overload conditions. Overtemperature sensing devices may be located either in the motor starter or in the motor winding. The sensing element specified in the motor winding (Paragraph 2.02D2, Appendix C) provides a more accurate and faster indication of winding temperature and is more reliable than a device in the motor starter.

Space heaters (to reduce condensation in the motor enclosure), thermal protection, and other motor modifications should be specified only when the value of the motor warrants the cost of modification. If the motor has epoxy-coated windings, space heaters are not usually required, but if the windings are unprotected, space heaters are normally specified whenever the motor is located (1) outside (especially if, as a standby unit, it sits idle for extended periods), (2) where humidity is high for long periods, or (3) below grade in a damp atmosphere.

16-15. Optimum Efficiency

An analysis of capital and operating costs (including the cost of energy) shows that the present worth of wire-to-water energy efficiency can have a profound effect on selecting the most advantageous bid. A purchase specification (for a pumping unit) that includes an efficiency evaluation is given in Paragraph 4.01 of Appendix C. The specification is simplistic, but more complete energy cost factors (such as maximum power demand, power factor, peak and off-peak rates, seasonal changes in rates, and other power-cost-modifying concerns) can be added as well as expected inflation and power escalation costs. The calculation of such costs can be imposed on the bidder if applicable formulas and data are given.

Although high efficiency is desirable, it is less important than reliability—a characteristic that is difficult to quantify. Service records of similar pumping units are sometimes used as a basis for judgment.

16-16. References

1. *CSI Document MP-FF/120, Methods of Specifying,* The Construction Specifications Institute, Alexandria, VA (1996).
2. *CSI Document MP-2-2, Section Format,* The Construction Specifications Institute, Alexandria, VA (1997).
3. ANSI/H1 1.6, *Centrifugal Pumps,* and 2.6, *Vertical Pumps,* Hydraulic Institute, Parsippany, NJ (1994).
4. Dicmas, J. L., *Vertical Turbine, Mixed Flow, and Propeller Pumps,* McGraw-Hill, New York (1987).
5. Walton, W. C., *Groundwater Pumping Tests, Design & Analyses,* Lewis Publ., Chelsea, MI (1987).
6. *Modern Plastics Encyclopedia,* McGraw-Hill, New York (updated annually).
7. *Plastic Directory,* Cahner's Publishing Company, division of Reed Publishing, Newton, MA (updated annually), with a subscription to *Plastic World.*

Chapter 17

System Design for Wastewater Pumping

GARR M. JONES

CONTRIBUTORS
Zbigniew Czarnota
Gary S. Dodson
James C. Dowel
William R. Kirkpatrick
William H. Richardson
Michael G. Thalhamer

The thought process involved in, and a formalized approach to, the design of wastewater pumping stations are described in this chapter. Although several aspects are unique to wastewater pumping, the process is similar for all types of engineering design. Regardless of the size of the project, engineers should school themselves to follow the procedure described here, which is equally applicable to the very smallest as well as the largest projects. The only difference is in the formalities used to ensure that nothing is omitted from the thought process. Novices are encouraged to follow the process formally without regard to project size. Experienced engineers may wish to take some shortcuts in the formalized process—especially for small projects.

17-1. Organization and Control of the Process

The development of a well-organized plan to complete the work is fundamental to the design process. Last-minute changes in project direction (often leading to an unsuccessful project) can be avoided by first thinking through and formalizing the individual steps required to complete the assignment. Because there is no substitute for an orderly planning process, set up a procedure for managing the documentation of project information. The documentation plan should provide for the filing and identification of the following:

- The project management plan
- Quality assurance plan
- Correspondence to and from:
  - The owner
  - Equipment manufacturers
  - Information sources
  - Consultants
  - Regulatory authorities
  - Others affected by or affecting the project
- Calculations
- Notes of meetings
- Reference documents such as utility maps and "as-built" drawings
- Reports from consultants
- Reports/recommendations to the owner
- Design memoranda.

All of these should be neatly written, dated, and entered into the record book of the project. Document control and prompt distribution of information to all members of the project team are basic requirements for a successful project. Recognize that, in future weeks or years, others will need to refer to the project record for information. Under the best of circumstances, the need will occur because the project has been eminently successful and others may wish to discover the secrets of success. Under the worst of circumstances, the records will be needed because contractor or client claims or litigation have made it necessary to determine how the project went awry. Either way, if the task was performed properly, there should be no fear of exposure.

Calculations should always be accompanied by an explanation of their significance, because otherwise, their meaning will make interpretation in future years almost impossible. Indeed, a good set of calculations usually involves more text than numbers. Engineers come and go, and the one who starts a job may not be the one to finish—another reason to make calculations perfectly understandable by any likely reader.

Another important element is the quality assurance procedures. Protocols should be established for

- The calculation format
- Checking the calculations
- Reviewing specifications and drawings
- Cross-checking between disciplines
- Peer review
- Final coordination reviews
- Final signing and sealing of contract documents.

A properly conceived and implemented quality assurance plan is the cornerstone of a successful project. If quality assurance is not considered and documented as an important facet, the result may be, at best, an inordinate number of costly change orders or, at worst, a disastrous failure.

Project events usually fall into three stages during the design process: (1) preliminary engineering, (2) detailed layout, and (3) detailed design. The general sequence of activities at each stage of design development is described in the following sections.

17-2. Preliminary Engineering

The purpose of preliminary engineering is to gather the information necessary to perform the design. If the preliminary engineering has been properly executed, all of the information required to complete the detailed design will be within arm's reach. Additional information and details on preliminary engineering are contained in Sections 25-1 to 25-4.

Need for Pumping Stations

Even small pumping stations represent a substantial investment over the life of the facility—particularly true with wastewater pumping stations where structure size, excavation depths, and environmental considerations generally result in comparatively greater costs than those of water pumping stations of an equivalent capacity. Therefore, deeper sewers, tunneling or jacking through hills, alternate pipeline routes, and other strategies should be considered on a present worth basis with generous allowances for the cost of operation and maintenance of the pumping station before deciding whether to pump.

Site Selection

Frequently, an engineer has little choice in site selection for wastewater pumping stations because of sewer system hydraulics, political considerations, and available land. Some aspects of site selection could have profound implications with respect to the cost of the project. These include subsurface conditions, aesthetic considerations, and the route of the force main.

The concerns and considerations associated with evaluating and selecting pumping station sites are discussed in Chapter 25 and are briefly summarized in the following subsections.

Subsurface Conditions

The characteristics of the underlying soils at a site often require some specific method of construction. For example, a caisson may be required because of soft soils and high groundwater; thus, the configuration of the pumping station is virtually dictated by the caisson and, in turn, by the subsurface conditions. This situation is illustrated in the Duwamish Pumping Station (Example 17-1), where a caisson was used. A caisson must be open at the bottom and top for excavating the material inside, so a cylinder (designed as a ring girder to resist wall loads) is more efficient than a box-type structure in which the necessary internal diaphragms or struts are obstructions. Once the structure has been sunk to the planned elevation, a tremie seal can be cast at the bottom. After curing, the structure can be dewatered and internal walls and slabs can be constructed. While this example is perhaps somewhat unusual in the United States, wastewater pumping stations are, by their nature, often deep structures. The location and thickness of floors and walls and the configuration of the structure are often determined by the external wall loads and buoyancy caused by both hydrostatic and passive soil pressures. Therefore, after a site has been selected, a comprehensive geotechnical investigation should have the highest priority.

Aesthetic Considerations

Most property owners agree that wastewater facilities are necessary and beneficial to the public good, but only when located "anywhere but in my backyard." As the sensitivity of sites to environmental conditions increases, provisions for architectural treatment, odor and noise control, landscaping, and similar concerns grow more costly. Pressure is often exerted to eliminate any visible presence of a pumping station, thereby requiring electrical and instrumentation equipment to be located in rooms below ground. Such an arrangement should be avoided (because of the potential for flooding) by selecting another site, if possible, or by appealing to reason. It usually requires many weeks to restore a station to operation after vital electrical and instrumentation equipment have been immersed in sewage.

Force Main Route

While not specifically a site issue, the selection of the force main route may have an impact on station cost and may require consideration of a different location for the pumping station. The identification and evaluation of hydraulic transients is considered in Chapters 6 and 7. Control of transients in wastewater systems is difficult because of the concern over the reliability of air and vacuum release valves and surge control valves when applied to a medium containing grease, grit, and rags. Thus, the relocation of the station or the force main or both to avoid knees, intermediate high points, and plateaus that could result in column separation is a high priority. Additionally, force main length should be short to reduce dynamic headlosses, the production of malodorous and corrosive gases, and cost.

Required Capacity

Because numerous texts [1, 2, 3] deal with the methodology for developing information on wastewater flows, the subject is not addressed here. It is sufficient to emphasize that an engineer must have reasonably well-developed projections of initial minimum, average, and peak flows before embarking on a design. Although predictions of future needs are often difficult, severe operational problems are likely to occur if initial loads are either under- or overestimated. The flow data must be competently developed and completely realistic. This information is essential when addressing issues such as

- Initial and final number and size of pumps and motors
- Initial size of the force main and the size of a future parallel force main, if any
- Schedule for a staged installation of pumps, motors, and force mains.

Alternatives include (1) installing minimum impeller sizes now and larger ones later, (2) installing two or three pumps now with provisions for more to be added later (see Example 12-2), (3) using small pumps initially with piping designed for larger pumps to be substituted later, and (4) installing a small force main initially with provisions for a larger one to be laid in parallel with the first and to be used alone in intermediate years (whereas both would be used at full development).

Wastewater Characteristics

In addition to quantity requirements, the characteristics of the wastewater must be understood. There is no such thing as "standard wastewater." If the wastewater system already exists, look for the following:

- Unusual quantities or size of gross solids. (Perhaps mechanical grinding or a screen will be required to protect the pumps, but try to pump the solids without screening if possible, because both screening and grinding equipment are notoriously maintenance intensive.)
- Unusual grit quantities or sizes. (Alternate metallurgy, such as Ni-Hard or, in a worst-case situation, an upstream grit removal system may be required.)
- Corrosive constituents. (Special metallurgy or chemical conditioning systems may be necessary.)
- The potential for toxic, explosive, or flammable materials. (Monitoring systems and scrubbers for ventilation exhaust may be needed.)
- Septic sewage and large quantities of hydrogen sulfide. (Special coatings in the wet well and variable-speed pumping to reduce residence time in the sump may be advisable.)

This information is needed in the equipment selection process, in the station layout, and in other aspects of design.

Mode of Operation

The mode of operation of the pumping station—constant speed (C/S) versus variable speed (V/S), the size of the pumping equipment, and the type of pump driver—should be settled as soon as required capacities have been selected. Issues to be considered in making such decisions include the following:

- *Impact on receiving facilities:* Constant-speed pumps may discharge slugs into a treatment plant and, thus, upset unit processes.
- *Cost:* Compared with V/S pumping equipment, C/S equipment is simpler to design and maintain, and the first cost of the entire station may often be slightly less, but the energy used, however, is usually greater. On a life-cycle cost basis, there is often no significant difference (see Example 29-1).
- *Production of odors:* Large wet wells and turbulence from slug discharges of anaerobic sewage may generate and release copious quantities of malodorous compounds.
- *Pump capacity:* If too large, V/S pumps may be forced to operate in regions of unfavorable radial thrust, which risks the possibility of damage and reduces efficiency (see Sections 10-6 and 15-3). In addition, start-up of large horsepower electric motors may induce unacceptable voltage fluctuations in the local electrical distribution system.

Preliminary Hydraulic Profile

A preliminary hydraulic profile should now be developed. It may not be necessary at this stage to refine the calculations, because they need be only accurate enough to ensure that equipment sizing and the layout for the preliminary system evaluations are valid for the final design. The following are some practical rules and shortcut hints for constructing a preliminary system head-capacity (H-Q) curve (hydraulic profile):

- In true pumping stations (where pumps discharge through isolation and check valves to a common header), use 1.5 m (5 ft) as the headloss from the pump inlet to the end of the header.
- Use street maps or even U.S. Geological Survey quadrangle maps both for force main length and location and for a preliminary estimate of static lift.
- Construct a system H-Q curve envelope using a Hazen-Williams C of 110 to 120 at a low wet well level to estimate the maximum total dynamic head (TDH) and a C of 130 to 140 at a high wet well level to estimate the minimum TDH. Add 5% for fitting losses if the force main is 250 mm (10 in.) in diameter or less or if the force main is less than 1000 ft long. Note that the above values are suggested unless a fitting-by-fitting takeoff is made, in which case calculate the turbulence losses in fittings and pipe friction separately (see Section 26-2 and Example 12-2).

The profile should be constructed for the pumps, the force main, the receiving sewer or structure, and the influent sewer(s) to provide a clear description of the operating conditions at initial and projected future conditions that include the extremes of maximum and minimum flows. The pumping station must work in harmony with its upstream and downstream hydraulic elements. The potential for damaging hydraulic transients should be included (see Chapters 6 and 7). Some shortcut methods for computing the effects of hydraulic transients are sufficiently conservative (especially for wastewater systems) to permit proceeding with reasonable confidence (see Parmakian [4]). Shortcut analyses are often sufficient to determine if conventional check valves can be used or if powered pump stop-and-check valves must be used to control head rise at the pumps upon power failure. Where doubt exists, a detailed computer analysis should be made in the final design stage.

Preliminary Equipment Selection

The selection of pumps is discussed in Chapter 12. However, consideration should also be given to the type of driver when making a preliminary selection of pumping equipment. Information on electric motors and internal combustion engines is contained in Chapters 13 and 14. In some circumstances, a combination driver (see Section 15-11) may be appropriate.

Unless the pumping station is to be constructed as a part of a treatment plant or the pumps must be quite small, try to select pumps that are true nonclog designs and pass solids at least 25 mm (2 1/2 in.) in diameter. It is better to avoid the need for influent screening or grit removal whenever possible because the best place for these labor- and maintenance-intensive operations is at the treatment plant. Grit or screenings stored at a pumping station almost always produce prodigious odors.

Once preliminary equipment selections have been made from catalog data, contact the candidate equipment manufacturers' representatives and describe (1) the preliminary operating requirements and criteria, (2) the preliminary equipment selection, and (3) any decisions regarding driver selection. Request written confirmation and any suggested alternatives.

Power Supply

After the preliminary selections of the main pumping equipment have been made, estimates of total station power requirements can be developed. The project electrical engineer should produce a preliminary estimate of station service requirements and discuss the availability of power for the project with the local electric utility. Service characteristics, reliability, and special concerns such as limitations on starting loads should be obtained at this time (refer to Chapter 9 for more information).

Owner Preferences

By now, a preliminary concept for the station is taking shape. It is time to discuss the project with the owner and, particularly, the owner's operating staff. The purpose of the discussion is to develop an understanding of the preferences and internal requirements for the following:

- Types of equipment
- Standards for reliability (such as more than one standby pump and back-up power supply)
- Station amenities (such as restrooms, offices, and shop)
- General arrangement
- Access to equipment (such as provisions for cranes, monorails, and vehicles)
- Controls (such as location of switches and type of variable-speed systems—see Chapters 9, 15, and 21)
- Monitoring systems (such as types of indicating lights, vibration monitors, and overload detectors—see Chapter 20)
- Telemetry (see Chapter 20)
- Aesthetics (see Chapter 25)
- Odor-control system (see Chapter 23).

Regulatory Agency Requirements

Regulatory agency requirements affect the equipment selection and the design of a variety of station features. Regulatory agencies to be contacted may include

- State and local water pollution control agencies
- U.S. EPA
- State health department, if different from the above state agency
- Building permit agency
- Planning commission
- U.S. Army Corps of Engineers
- Fire marshal
- Water utility
- Flood control district.

Public hearings, environmental impact statements, or use permits may be required. The subjects to be discussed (and resolved) with the regulatory agencies include

- Power supply reliability and the need for on-site standby power
- Potable water supply and backflow prevention
- Flood protection
- Requirements for station reliability
- Ventilation standards, including odor control
- Fire protection
- Architectural standards
- Building permit requirements
- Local design codes
- Access and egress. Local setback requirements, and on-site dedicated parking for O&M crews.

Station Utilities

The next step is to conceptualize the design of the station utility systems. Equipment need not be selected at this time, but the basic approach for all of the station needs should at least be resolved. Issues to be considered include the following:

- Source of water for pump seals and domestic and housekeeping purposes (local utility or on-site well)
- Ventilation system (heating, cooling, odor control)
- Compressed air (type, capacity, quality standards)
- Drainage (destination and connector limitations)
- Internal drainage (destination and method of control)
- Standby power.

17-3. Detailed Layout

If the preliminary engineering stage (which is now complete) is done properly, the project engineer has all of the information and tools to put the project together. What remains are the details that will make it work successfully. A generalized description of the sequence of considerations for a typical wastewater pumping station project is presented in the following subsections. Circumstances unique to an individual project may, however, require a different approach. Careful and thorough preliminary engineering work can provide the guidance for the specific sequence from this point forward.

Piping and Instrumentation Diagrams (P&IDs)

First and foremost is the production of piping and instrument diagrams (P&IDs), which are the most useful type of schematics. Examples of P&IDs for pumping stations are illustrated in Figures 17-1 to 17-3 as well as in Figures 18-14, 21-5, and 21-6. The P&IDs are used to

- Define the nature and control of each system;
- Show piping sizes;
- Locate piping devices, such as valves, pressure sensors, and meters;
- Show control methods and strategies;
- Specify the rated conditions and power requirements for all equipment.

The project leader, electrical engineer, and instrumentation engineer working together should develop the P&IDs. They are first sketched in pencil, then drafted when most details are worked out. Because many changes will occur as the project develops and because a P&ID drawn manually is initially so often large and unwieldy, it is advisable to use a CADD (computer-aided design and drafting) system for P&ID production and updating. Properly executed, P&IDs are useful for design in the following ways:

- As the source of project information for producing detailed drawings and specifications
- For control of project information and changes
- For cross-checking and final document coordination
- As a means of communication between project disciplines.

P&IDs are also useful (1) as control documents during construction, (2) as aids in personnel training, and (3) for the operation of the facility.

Structural Considerations

At this stage in the process, recommendations from the project geotechnical engineer should be available to the project structural engineer. It is now time to make important decisions regarding construction methods. A detailed station layout is not necessary at this time, but a rough idea of station dimensions and loads is necessary.

Later in the detailed layout stage, the structural engineer should again be consulted about the thickness and placement of walls and slabs, although structural design should not commence until the next stage.

Sump Design

Extreme caution should be exercised in the sump design, particularly as it relates to inlet conditions. A poor design will very likely result in an inability to develop the intended capacity of the pumps, contribute to or cause severe damage to the pumping equipment, and result in shortening the useful life of the station.

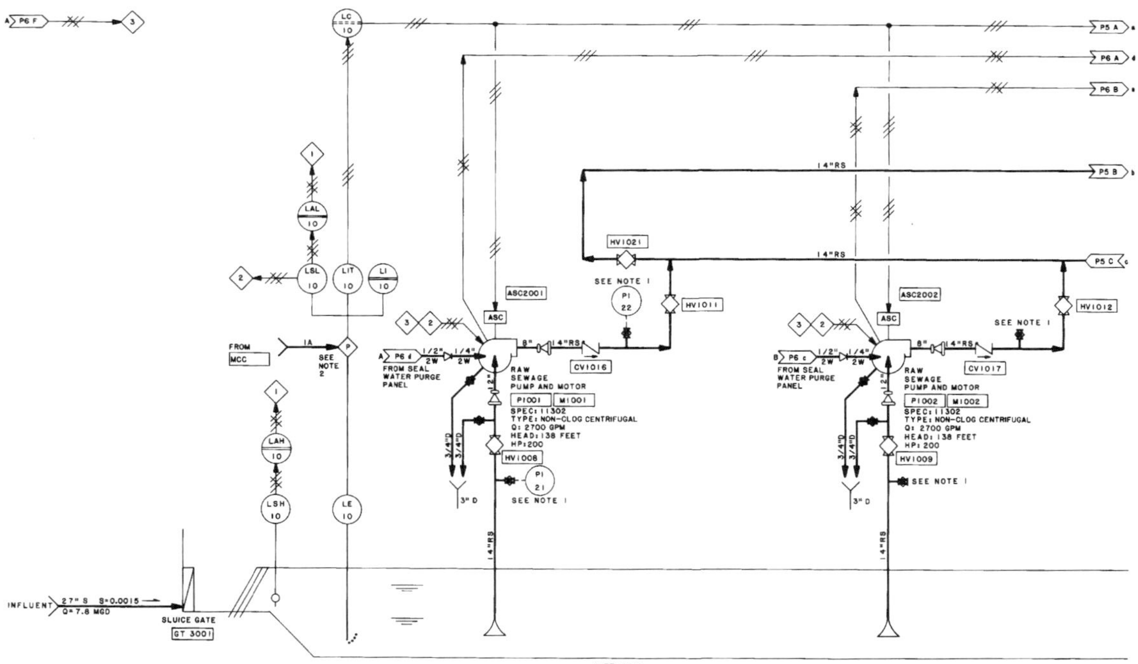

Figure 17-1. P&ID of main pumping units 1 and 2, Wastewater Pumping Station, Steilacoom, WA. Courtesy of Brown and Caldwell Consultants.

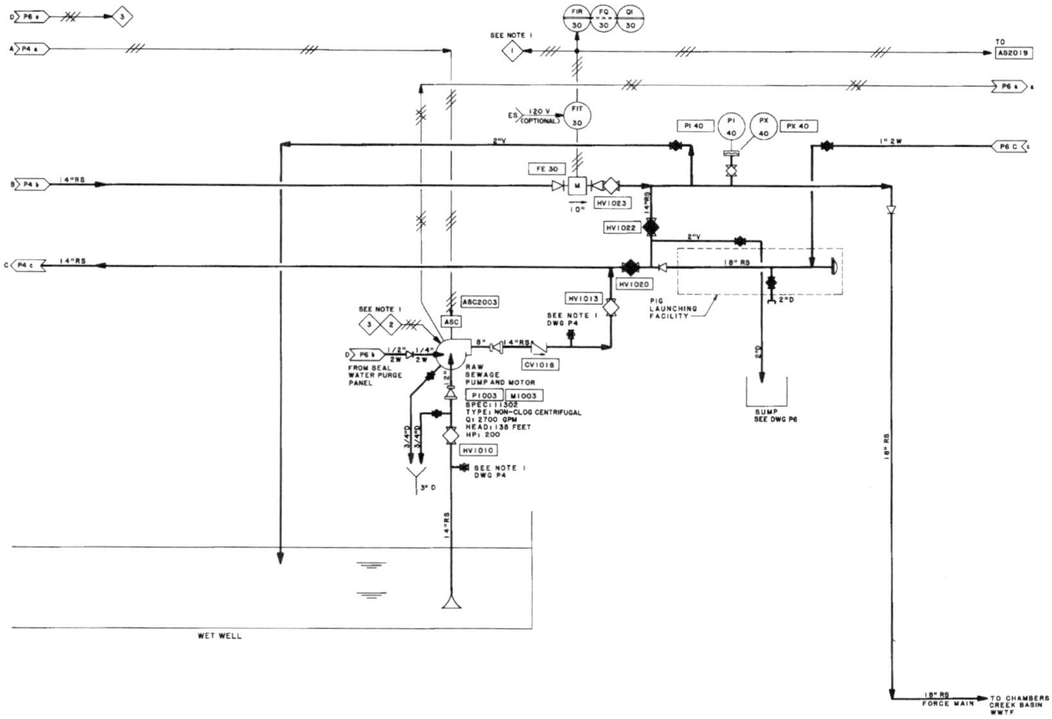

Figure 17-2. P&ID of pumping unit 3 and metering system, Wastewater Pumping Station, Steilacoom, WA. Courtesy of Brown and Caldwell Consultants.

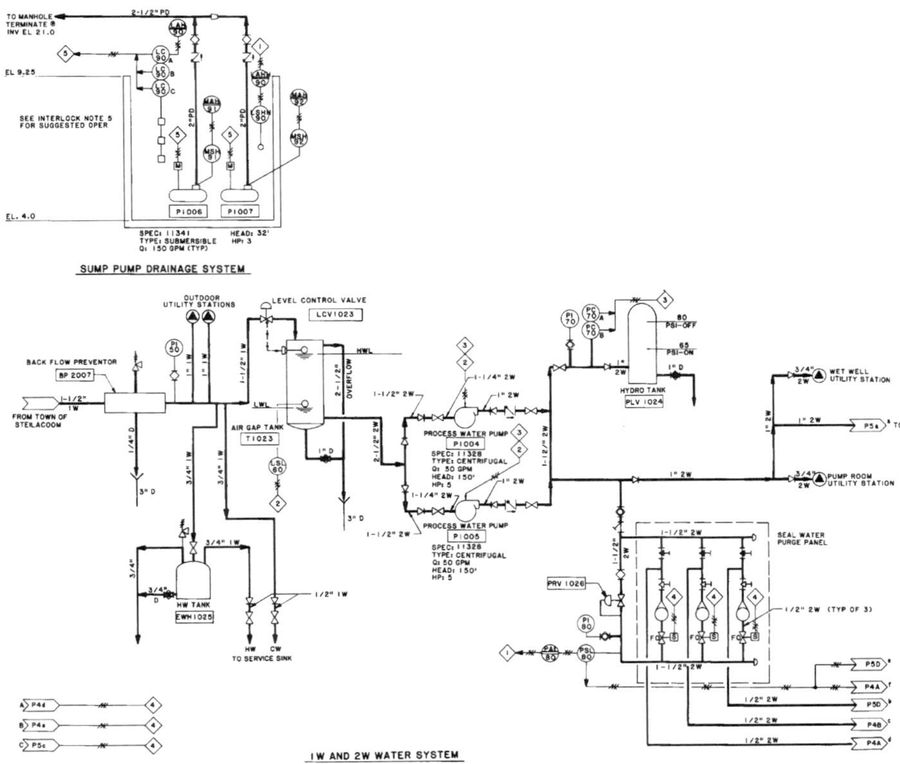

Figure 17-3. P&ID of station utilities, Wastewater Pumping Station, Steilacoom, WA. Courtesy of Brown and Caldwell Consultants.

Details of some successful sump designs for large, moderate-size, and small pumping stations are shown in Sections 17-5, 17-6, and 17-7, respectively. The method of construction may influence the sump design, for example, as in the Duwamish Pumping Station where a caisson was used to construct the station substructure. A good sump and inlet design must have the following:

- Minimal turbulence and no influence from the incoming sewer, such as a cascade that might entrain air in the liquid and, therefore, into the pumps and force main. If a long fall is unavoidable, there must be sufficient distance to the pump suction intake for entrained air to escape. The entrance velocity into the sump should not significantly exceed about 1.1 m/s (3.5 ft/s).
- Conservative values of $NPSH_A$ (see Figure 10-13 and text in Section 10-4) obtained by a suitable elevation of the pump relative to the LWL plus suction pipe velocities limited to about 2.4 to 3.4 m/s (8 to 11 ft/s) for keeping head losses low.
- Suppression of vortices by (a) good geometry, (b) sufficient pump intake submergence as determined by Equation 12-1 (see also Dicmas [5]), and (c) the addition of vanes under pump intakes to reduce floor vortices or cones to eliminate them (see Figures 12-32f, 12-33, 12-34, and 12-41 for design details).
- Adequate pump intake velocities: for removing solids and, at the same time, for obtaining a good hydraulic environment for the pumps. See Table 12-1, but be aware that the author and editors consider 1.5 m/s (5 ft/s) to be a near-maximum velocity and not a target. All of the pumping stations in Sections 17-5 and 17-6 were designed for a maximum of 1.1 m/s (3.5 ft/s) intake velocity, and they have performed very well.

Pipeline Orientation

The relative positions (vertical as well as horizontal) of the incoming sewer or sewers and discharge pipelines can influence the station layout. However, avoid undue influence in this regard. The principal cost savings in the project are realized by making the pump room and sump layout as efficient as possible, by saving excavation and structure costs, and by the smaller cost of ventilation and lighting. Thus, for the price of a manhole or bend or two, significant cost savings in the pumping station can often be realized. Efficient pump room layout is a key factor in owner satisfaction.

Pump Room Layout

The efficiency of the pump room layout, as suggested previously, has a great effect on the overall cost of the project. Used in this context, *efficient* or *efficiency* means conserving room or structure plan dimensions without sacrificing optimum machine performance or personnel access for operation and maintenance of the equipment. The following are some guiding considerations:

- Use the room walls to support heavy valves and piping.
- Avoid piping configurations that impede access to pumps, valving, and other equipment.
- Locate seal water valving and appurtenances, such as solenoids and pressure-regulating valves, rotameters, and isolating valves, on a wall adjacent to the pump and supply the seal water to the pump in copper or stainless-steel piping encased in the pump room floor slab. This arrangement avoids clutter around the pump and improves maintenance access.
- Never connect a pump discharge to a manifold or header from underneath. To do so invites the plugging of check valves and piping. Make the connection at the side of the manifold as shown in Figures 12-20 and 12-22.
- To avoid plugging check valves and the piping above them, locate check valves in horizontal pipes—never in a vertical pipe.
- For simple lift stations where the final discharge is not fully submerged, use a flap valve in a structure at the receiving sewer, which thereby avoids the space-consuming isolation and check valves (see Figure 12-21).
- Provide adequate clearance, such as 0.9 to 1.1 m (36 to 42 in.), from the outside edge of all piping flanges or other projections—not just the edge to edge of pump bases or pipe shells—on at least three sides of each pump and any discharge valving.
- Provide enough room to remove bolts from the thrust harnesses (if used) of sleeve couplings and to slide the coupling off the joint.
- Provide a quick, unobstructed exit for people working around the pumps.
- Don't forget ventilation. Be sure that air can be moved through the room efficiently to prevent the accumulation of odors and moisture condensation and to scavenge gases from the space adequately. Try to exhaust the room from a location near the drainage sump where the odorous gases tend to accumulate. Locate the fresh-air intake away from the exhaust and on the prevailing upwind side of the structure.

Some of these principles are illustrated in Figures 12-17 to 12-25 for different pump room layouts. Details of good piping layout are also shown in Figures 26-7 and 26-8.

Superstructure and Interior Spaces

The pump room layout should be developed with due consideration for other portions of the station. All spaces must work in harmony and must properly accommodate the function of all systems. Considerations in pumping station layout include the following:

- Access to and from all spaces for both personnel and equipment.
- Intermediate floors in deeper pumping stations can provide internal diaphragms to resist heavy wall loads and are excellent locations for utility systems such as water and compressed air equipment.
- Ventilation of all spaces. Now is the time to begin the preliminary sizing of ductwork and fans. If these features are left to the last, there will not be enough room and you may have to start all over.
- Location of utilities and support equipment. Some practical rules follow:
    - Water pumps, hydropneumatic tanks, and similar equipment can be located in the pump room and other spaces below grade.
    - Avoid locating fans in the pump room or in other areas below grade that are more prone to flooding.
    - Locate utility stations for housekeeping as close as possible to entrances to rooms below grade.
    - Do not provide washdown water services to engine rooms and electrical equipment (except motor) rooms.
- With the project electrical engineer's assistance, locate the motor control center and station switchgear in a roomy, dry area in the station superstructure. This equipment requires a source of clean, uncontaminated air and good air circulation. Cooling may be necessary. Be sure to be generous when estimating space requirements for electrical equipment. Add, say, 50 or even 100% to the space estimated by the electrical engineer. As the project progresses toward completion, new electrical loads will be added and initial load estimates will probably be too low. Also, plan for major conduit and cable tray runs to allow adequate provision in slabs or in passageways.

Architectural Treatment

Now that the station configuration and dimensions are roughly defined, consult the project architect about exterior treatment for the superstructure, site access, landscaping, and so on. These discussions should lead to final decisions on roof treatment, siding materials, and such other features as noise (see Chapter 22) and odor (see Chapter 23) suppression (see also Section 25-3 and Figure 25-1). The architect must bear in mind the clearances required for equipment operation, maintenance access, and removal.

Preliminary Sketches

With the foregoing completed, the engineer and architect can now produce preliminary sketches showing the location and arrangement for all station features. These should be produced to scale, although not in great detail. The purpose of this step is to provide a basis for discussion with the various discipline leaders. After review, the preliminary sketches are revised for use in the design report.

Design Report

Using the information and materials developed thus far, a draft design report is prepared. Included in the report are (1) the purpose of the project, (2) the design criteria, (3) illustrations of the proposed and alternative arrangements, (4) features of the pumping station, and (5) rough estimates of capital and operating costs. The intent is to produce a concise basis-of-design document for all project participants.

The draft design report is submitted to the owner as well as to the funding and appropriate regulatory agencies for comment. After receiving their comments, the engineer should meet with interested parties to discuss the comments and to seek agreement on revisions, if any. The design report is then revised and issued as the guidance document for final design.

17-4. Detailed Design

Once detailed design is authorized, the production of final construction documents can begin. By the time the work in this stage commences, the site has been selected, the property presumably has been acquired, the force main route is assured and rights of way (both temporary and permanent) are being secured, required permits are being obtained, and all legal impediments including public hearings are being resolved. But the project is not ready for drafting just yet. Before the final drawings can be prepared, the following steps must be completed.

Refine Hydraulics

Sufficient information is now available to refine the preliminary hydraulic calculations completed earlier.

Calculate $NPSH_A$ curves for all of the critical conditions in which altitude, sewage temperature, and low atmospheric pressure are considered. Draw accurate minimum and maximum system H-Q curves for new and old piping. Make or authorize a rigorous hydraulic transient analysis if the shortcut methods used earlier indicate one is necessary. Be sure to develop neat, legible calculations for all of the anticipated operating conditions.

Finalize Equipment Selection

Once final performance requirements are known, the draft specifications describing the operating conditions, the performance requirements, the required construction features (such as materials), and so on should be produced for the major pieces of equipment. Copies of these should be furnished to representatives of the candidate manufacturers for their comments and suggestions. Equipment manufacturers who should receive draft specifications include

- Pump manufacturers
- Motor manufacturers
- Engine manufacturers (for engine drives and standby generators)
- Makers of other important equipment packages.

In addition, final system calculations and equipment selection and control strategies should be developed for all peripheral and utility systems.

Revise P&IDs

Next, the P&IDs are revised to their final form, in which all monitoring and control elements, rated conditions for equipment, and power requirements are shown. The revised P&IDs can now be used as control documents for completing the construction documents.

Detailed Layout Sketches

Before preparing the final documents, the layout sketches are revised to show the final dimensions, including maintenance space requirements, code clearances, and final equipment outlines. By this stage, no more than 30 to 40% of the total design project effort should be expended.

Final Production

At this point, production of the final documents can begin. If the foregoing process has been followed with no significant revisions, the project should be ready for final production. All that remains is to follow the original plan and to coordinate the efforts of the various disciplines.

From now on, the project leader acts as a coordinator who makes certain that each member of the team performs according to plan at the proper moment and that all members are aware of project events. The quality assurance plan should be brought into full effect. Communication and cross-checking between disciplines is of utmost importance. Final specifications (see Chapters 16 and 28) should be developed at the same time that final drawing production begins. As the project reaches the 75% completion stage, interdisciplinary cross-reviews for both drawings and specifications should be well under way (see Sections 9-12, 13-15, Appendix G, and, by all means, Chapter 27).

Final Cross-Check

If the project has followed an orderly and carefully thought out plan, the final cross-check should not reveal any significant discrepancies. Regardless of project history, however, there is no substitute for an independent final detailed review by an experienced senior-level engineer.

The concept is simple. A cross-check consists of checking drawings against specifications and discipline against discipline to discover inconsistencies. The purpose is not to second-guess the design decisions; those are considered given. Rather, it is to uncover (if possible) those areas where one discipline is not properly coordinated within itself or where interdisciplinary communications have not been adequate. As a rule, at least an hour per drawing is required if coordination has been reasonably good. Well-coordinated documents require about half as much time, and poorly coordinated documents may require as much as five times as much effort plus commensurate efforts for document corrections.

17-5. Examples of Large Lift Stations

A great deal can be learned by studying the following examples of pumping stations selected to illustrate different problems and approaches to their solutions. All work well, meet their design objectives, and are neat, clean, and—even those in service for 15 to 25 yr—have the appearance of new stations.

All of the pumping stations in Sections 17-5 and 17-6 were designed to minimize deep excavation by using variable-speed pumping systems. The ventilation systems meet the recommendations in Section 23-2, in which air is powered in near ceiling and powered out from the lowest level to produce a slight vacuum in the wet well and ensure a safe, corrosion-free environment for personnel and equipment. Ease of access by operators and maintenance workers was obtained without unduly increasing the size of the structure by following the precepts embodied in Figures 12-20 and 12-22.

All of these stations have draft tube pump inlets somewhat similar to the FSI intake of Figure 12-30, because ordinary isolating valves (eccentric plug valves in smaller systems) become massive and costly for suction piping larger than 400 mm (16 in.) in diameter. Those big, expensive valves can be replaced by inexpensive sluice (or slide) gates. The configuration allows the pumps to be placed close to the floor of the pump room. In comparison with ordinary suction piping layout (as in Figures 12-20, 12-22, and 17-17), the pump support is far more rigid, and susceptibility to vibration and earthquake forces is greatly reduced. Access to machinery is improved by the comparatively low mounting.

Example 17-1
Duwamish Pumping Station

The substructure of Duwamish Pumping Station (see Figures 17-4, 17-5, and 17-6) was constructed as a caisson because of the poor quality of soils underlying the site and concern that conventional open cut construction, no matter how carefully accomplished, would result in the

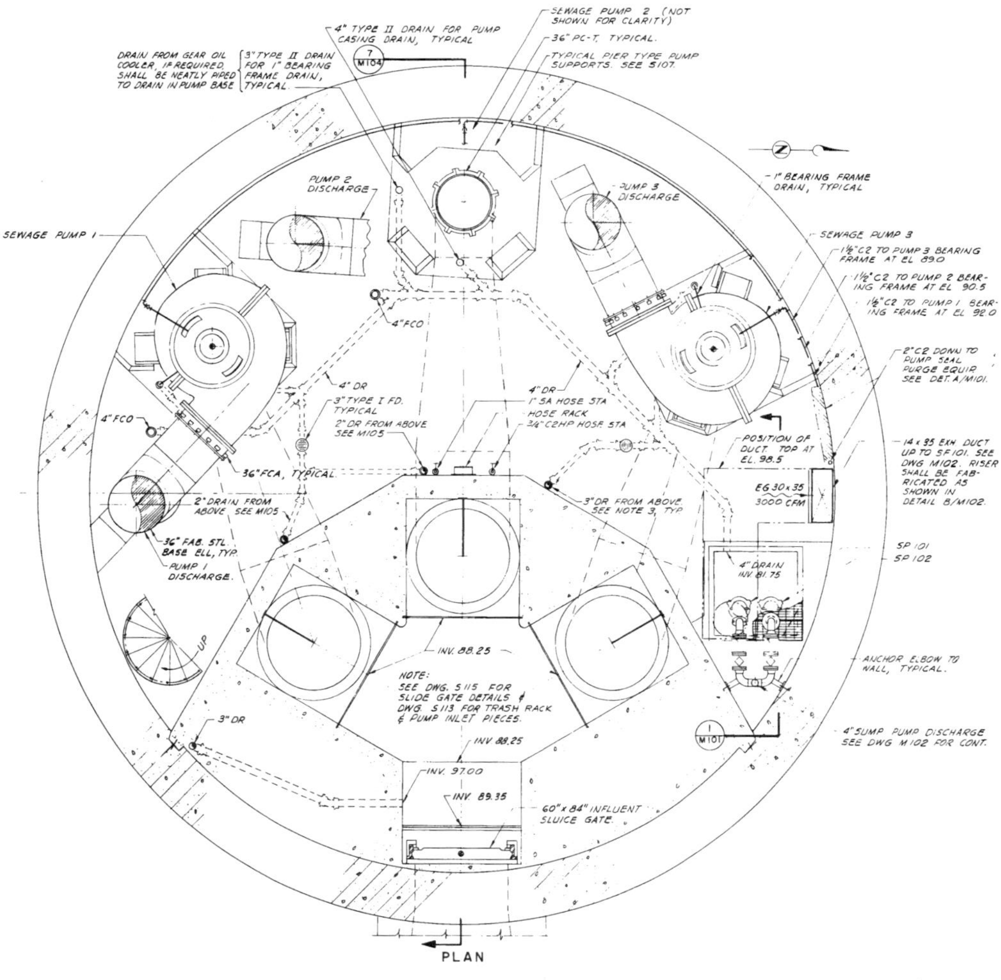

Figure 17-4. Pump room plan, Duwamish Pumping Station, King County (Washington) Department of Metropolitan Services. Courtesy of Brown and Caldwell Consultants.

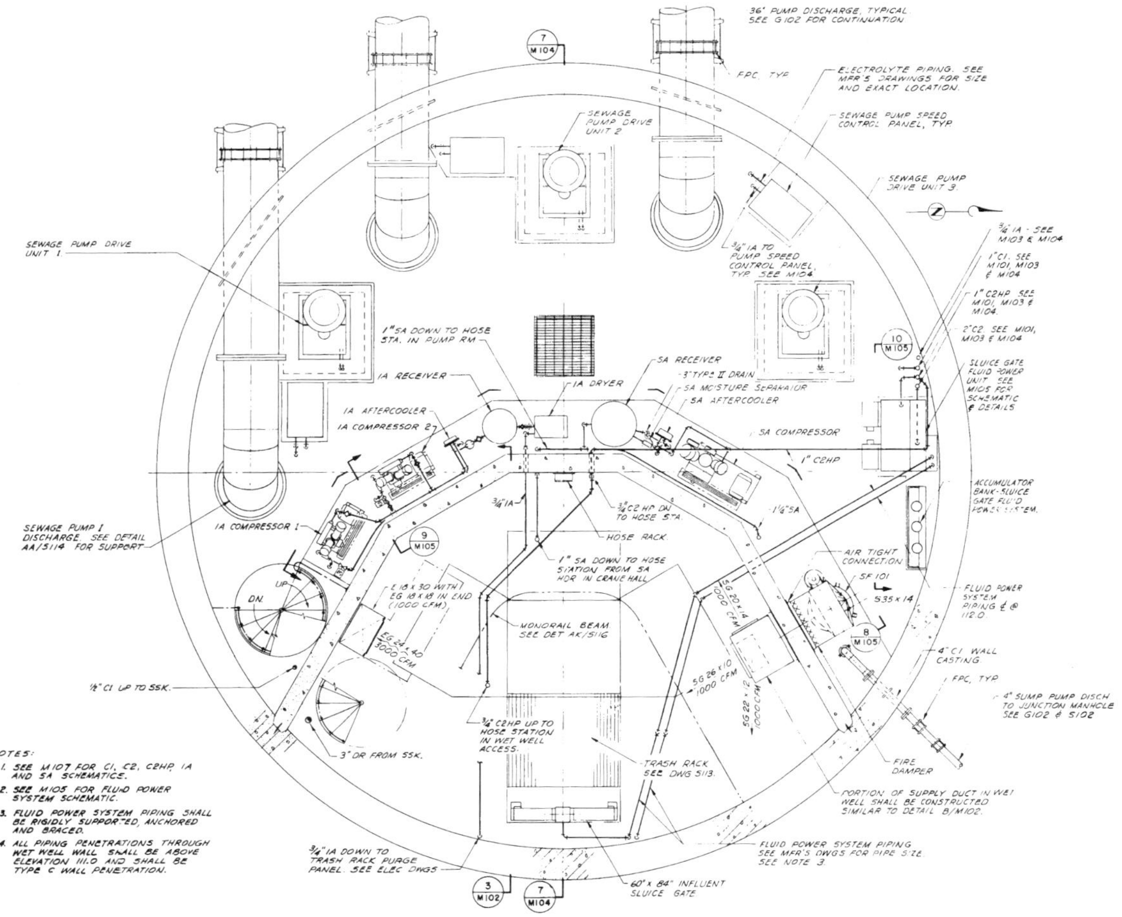

PLAN

Figure 17-5. Equipment room plan, Duwamish Pumping Station, King County (Washington) Department of Metropolitan Services. Courtesy of Brown and Caldwell Consultants.

settlement of nearby structures. Concrete for the substructure, 24.4 m (80 ft) in diameter and approximately 15.2 m (50 ft) deep, was placed in three lifts with the structure built at existing grade. After the concrete had cured, dunnage (cribbing) under the cutting shoe was removed, and the caisson was sunk into position within 2 days by excavating from within the structure. When the tremie seal was poured, the caisson was less than 25 mm (1 in.) below the planned elevation and only 0.25° out of plumb.

Each pump in this simple lift station (which was built in 1968) discharges through approximately 21 m (70 ft) of dedicated piping to individual flap valves in a structure at the receiving sewer. Expensive and space-consuming large-diameter isolating valves and check valves are thereby avoided. Each of the three 900-mm (36-in.) pumps is rated 2.2 m^3/s (50 Mgal/d) at a total head of 5.33 m (17.5 ft). The V/S pumps, which maintain normal levels in the upstream sewers, are driven by wound-rotor motors operating through a liquid rheostat for speed adjust-

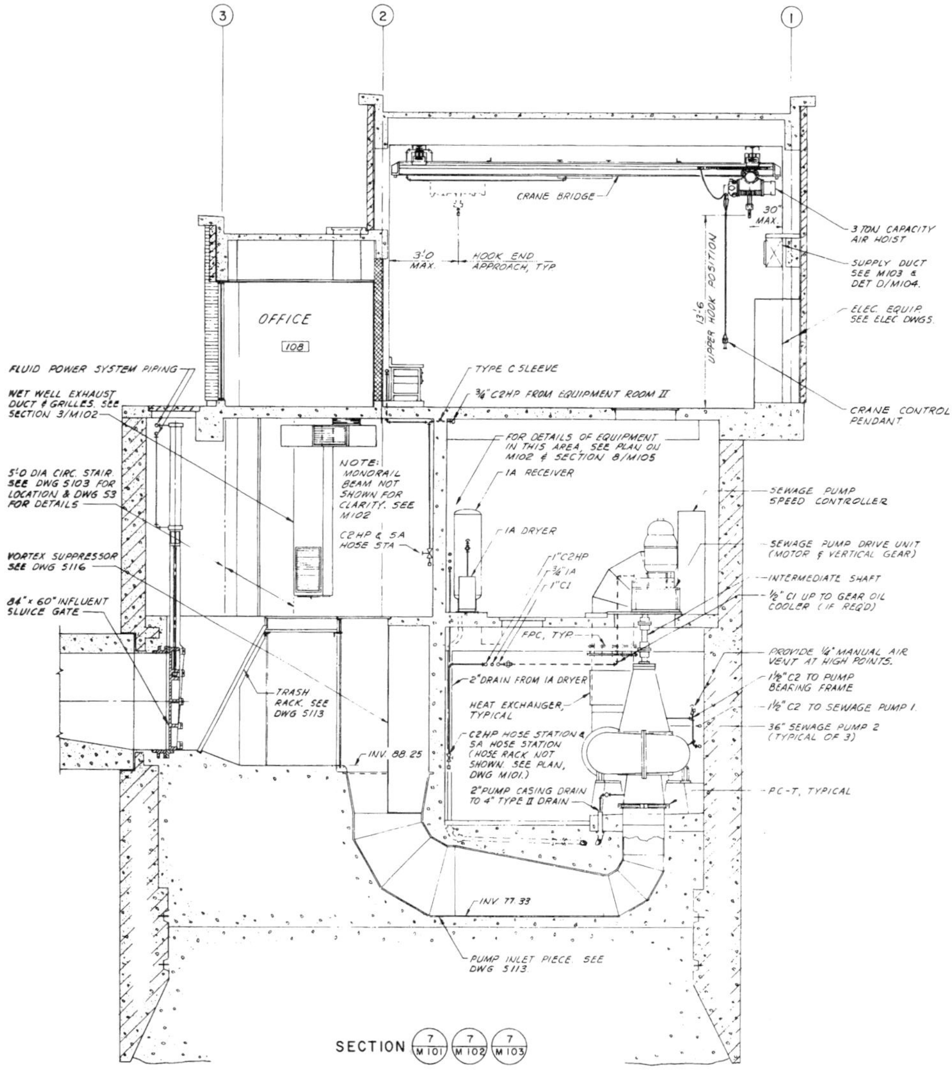

Figure 17-6. Typical section, Duwamish Pumping Station, King County (Washington) Department of Metropolitan Services. Courtesy of Brown and Caldwell Consultants.

ment. The maximum speed of the pumps at the rated performance requirement is approximately 300 rev/min. To reduce the cost of both structure and equipment, the pumps were specified with 1200 rev/min vertical motors to be mounted on specially designed parallel shaft gear reducers. An integral backstop prevents reverse rotation on pump shut-down. Each pump has a draft-tube-type inlet conduit fitted with a vortex suppressor. Each pump inlet can be isolated by means of a fabricated steel slide gate placed using a motorized hoist on a monorail. A hydraulically operated sluice gate, which is tripped by power failure or by float switches in the wet well and pump room, protects the station against flooding.

Example 17-2
Interbay Pumping Station

The Interbay Pumping Station (shown in Figures 17-7 through 17-9) was constructed in 1966 on a site in an ancient saltwater embayment. The underlying soils were quite soft, and the groundwater table reflected fluctuations in tidal level in an adjacent harbor. The site was very constrained, with important structures (some of them quite fragile) nearby. Because soils at the base of the structure lacked sufficient supporting strength, piles were favored over a caisson. The construction procedure was to (1) drive a sheet pile cofferdam, (2) excavate the site, (3) drive the pile foundation with the cofferdam flooded, (4) pour a tremie seal at an elevation below the planned foundation slab, (5) dewater the cofferdam after the tremie concrete had cured, (6) cut off the piles, and (7) construct the foundation slab. General dewatering at the site, which would have compressed the local soft soils and severely damaged many nearby structures and surface improvements, was thereby avoided.

Each of the three 900-mm (36-in.) pumps is rated to deliver 2.6 m^3/s (60 Mgal/d) against 12 m (39 ft) of total head at a maximum pump speed of 360 rev/min. The engines are rated for 370 kW (500 hp) at 820 rev/min. The twin 1200-mm (48-in.) force mains are each approximately

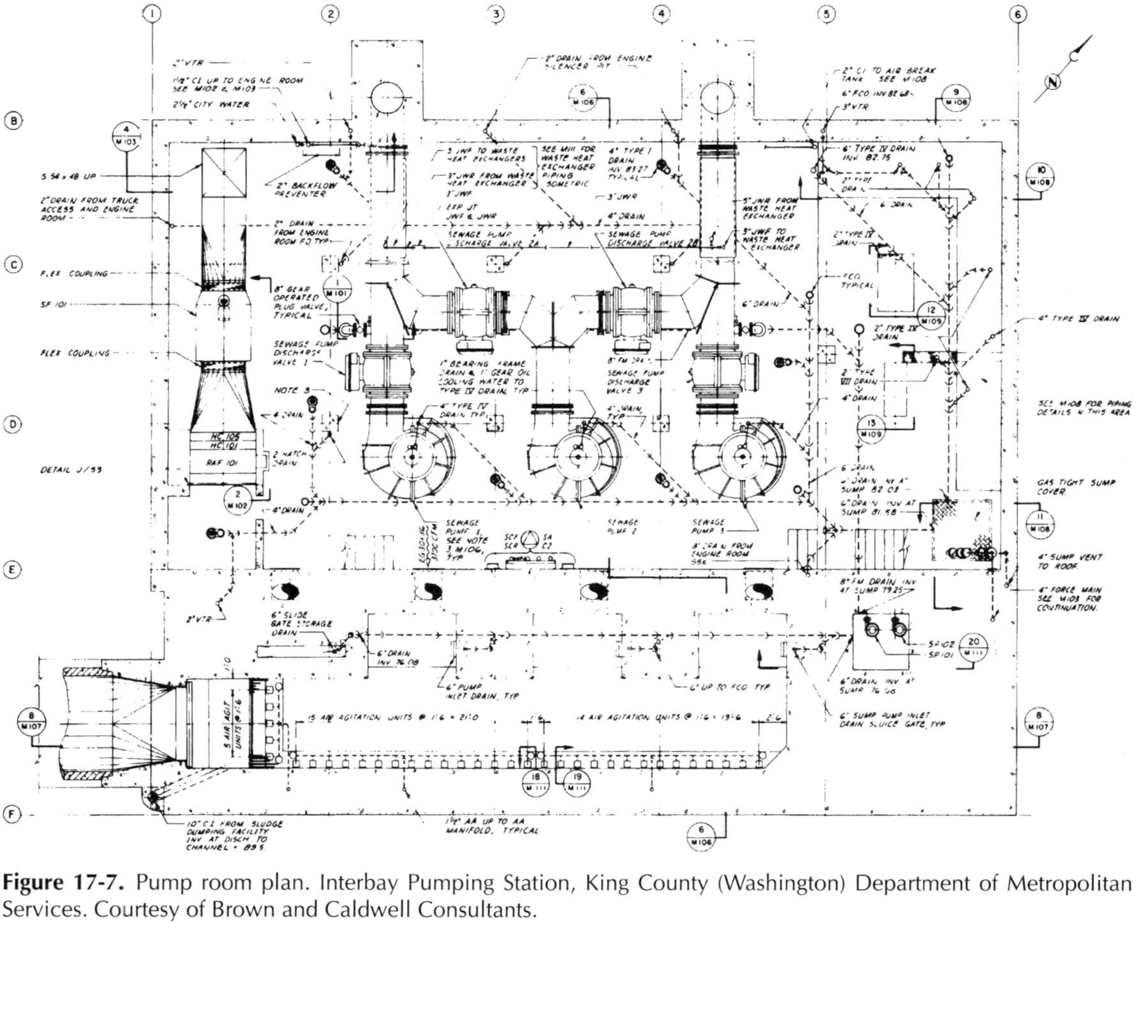

Figure 17-7. Pump room plan. Interbay Pumping Station, King County (Washington) Department of Metropolitan Services. Courtesy of Brown and Caldwell Consultants.

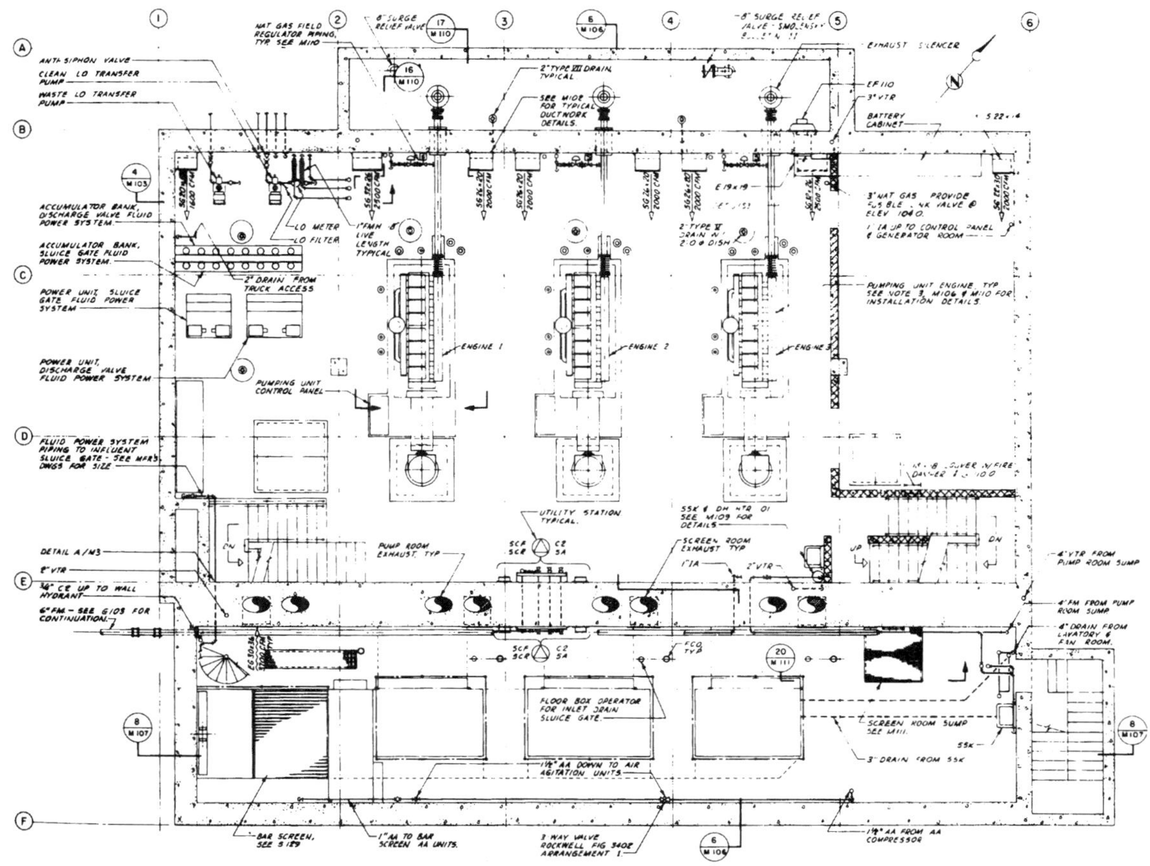

Figure 17-8. Engine room plan, Interbay Pumping Station, King County (Washington) Department of Metropolitan Services. Courtesy of Brown and Caldwell Consultants.

1100 m (3600 ft) long. Almost all of the static head in the system had to be gained at the pumping station itself because of construction difficulties associated with the elevation of nearby bridge pier foundations. On the basis of a transient analysis, column separation was likely to occur in the force mains at the pumping station on loss of power at virtually any flow—even when all reasonable measures were taken to avoid it. Mitigating means include (1) engine drives with dual fuel sources to provide a higher degree of reliability and (2) check valves installed on a branch of the force main at the probable point of column separation. These check valves act like vacuum-relief valves (see Figure 7-2) to admit a column of air that cushions the rejoining of the two columns of water (see Figure 17-13a). Hydraulically operated slow-closing pump stop-and-check valves on the pump discharges permit relief of pressure rise when the pipeline represurizes on the return wave.

To maintain normal depth in the influent sewer, the pumps operate at variable speed through control of the engine throttle setting. Engine fuel is natural gas with a liquid propane reserve supply. The propane is vaporized at the engine using the heat available in the engine's jacket water. A hydraulically operated sluice gate, triggered by float switches in both the wet well and pump room, protects the station against flooding.

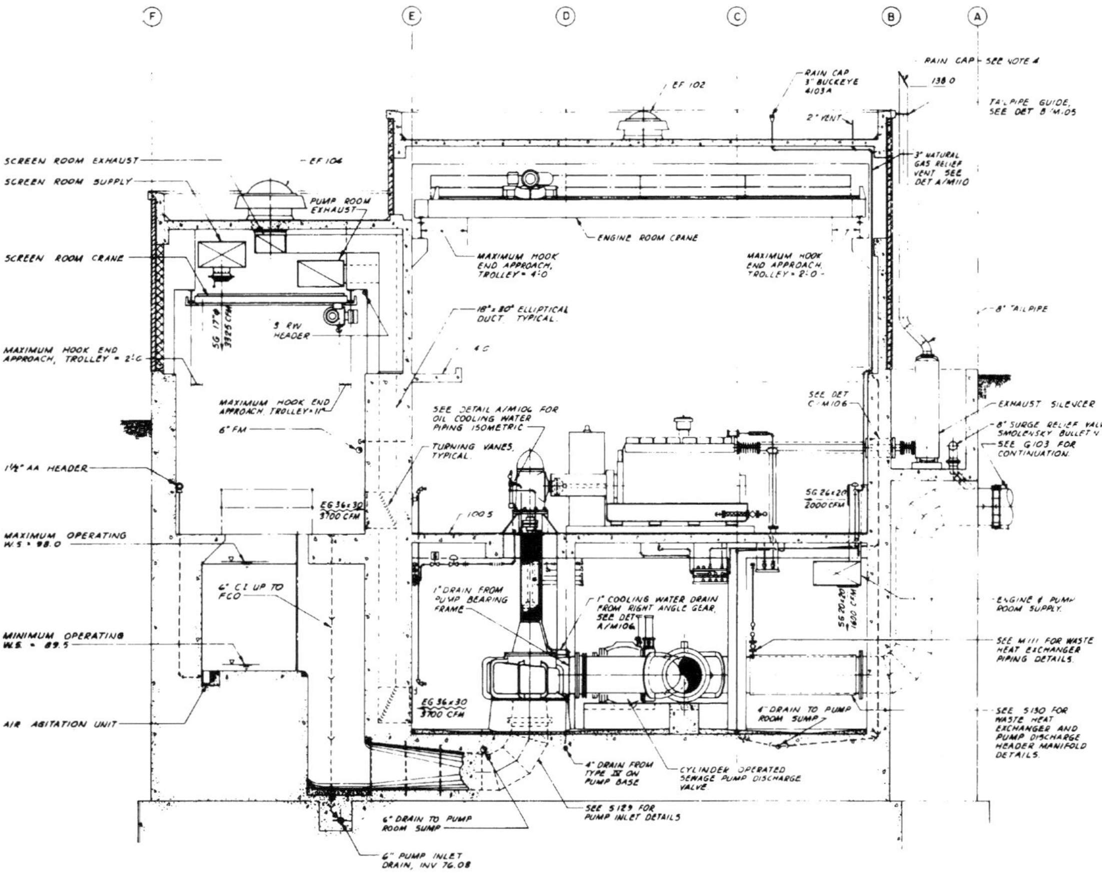

Figure 17-9. Typical section, Interbay Pumping Station, King County (Washington) Department of Metropolitan Services. Courtesy of Brown and Caldwell Consultants.

Example 17-3
West Point Influent Pumping Station

The West Point Influent Pumping Station (Figures 12-1 and 17-10 through 17-12), which was constructed in 1964, consists of four engine-driven 1350-mm (54-in.) pumps that lift wastewater to a large municipal treatment plant. The pumps, each rated to deliver 4.73 m^3/(108 Mgal/d) against a total head of approximately 5.3 m (17.5 ft), are operated at variable speed by adjusting the engine speed to maintain acceptable velocities through the bar screen channels and upstream sewers. The pumps discharge to a unique weir (see Figure 17-12), which is circular in plan, set 150 mm (6 in.) above the maximum liquid surface in the receiving channel. An expansion cone and deflector provide recovery of dynamic energy and maintain adequate carriage velocities—an arrangement that saves the cost of large-diameter isolating and check valves. The engines, rated at approximately 340 kW (450 hp) at 800 rev/min, are fueled with gas from the treatment plant's anaerobic digesters. Standby fuel is gaseous propane. Heat from the engines' jacket water and exhaust is reclaimed for process use and heating the building.

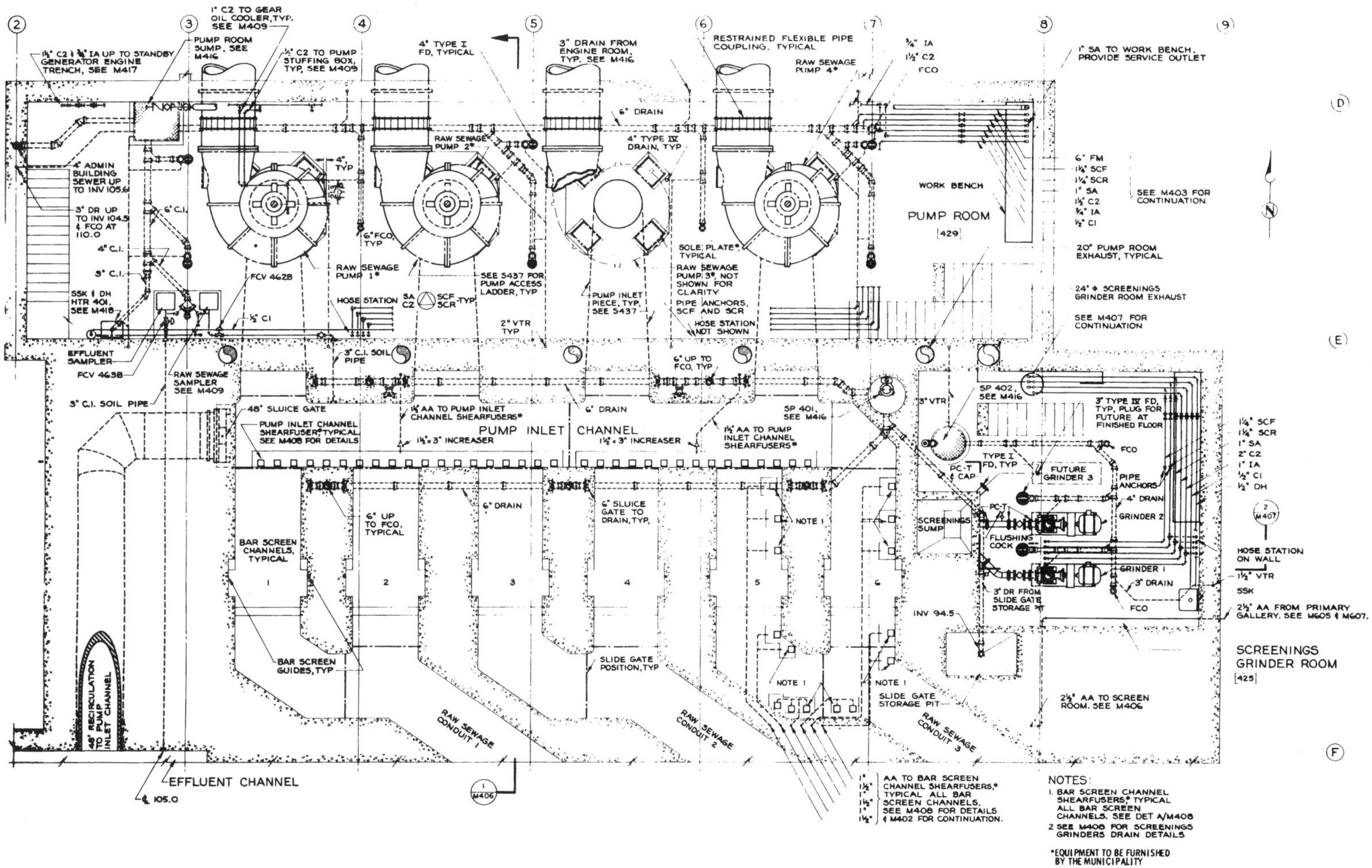

Figure 17-10. Pump room plan, West Point Influent Pumping Station, King County (Washington) Department of Metropolitan Services. Courtesy of Brown and Caldwell Consultants.

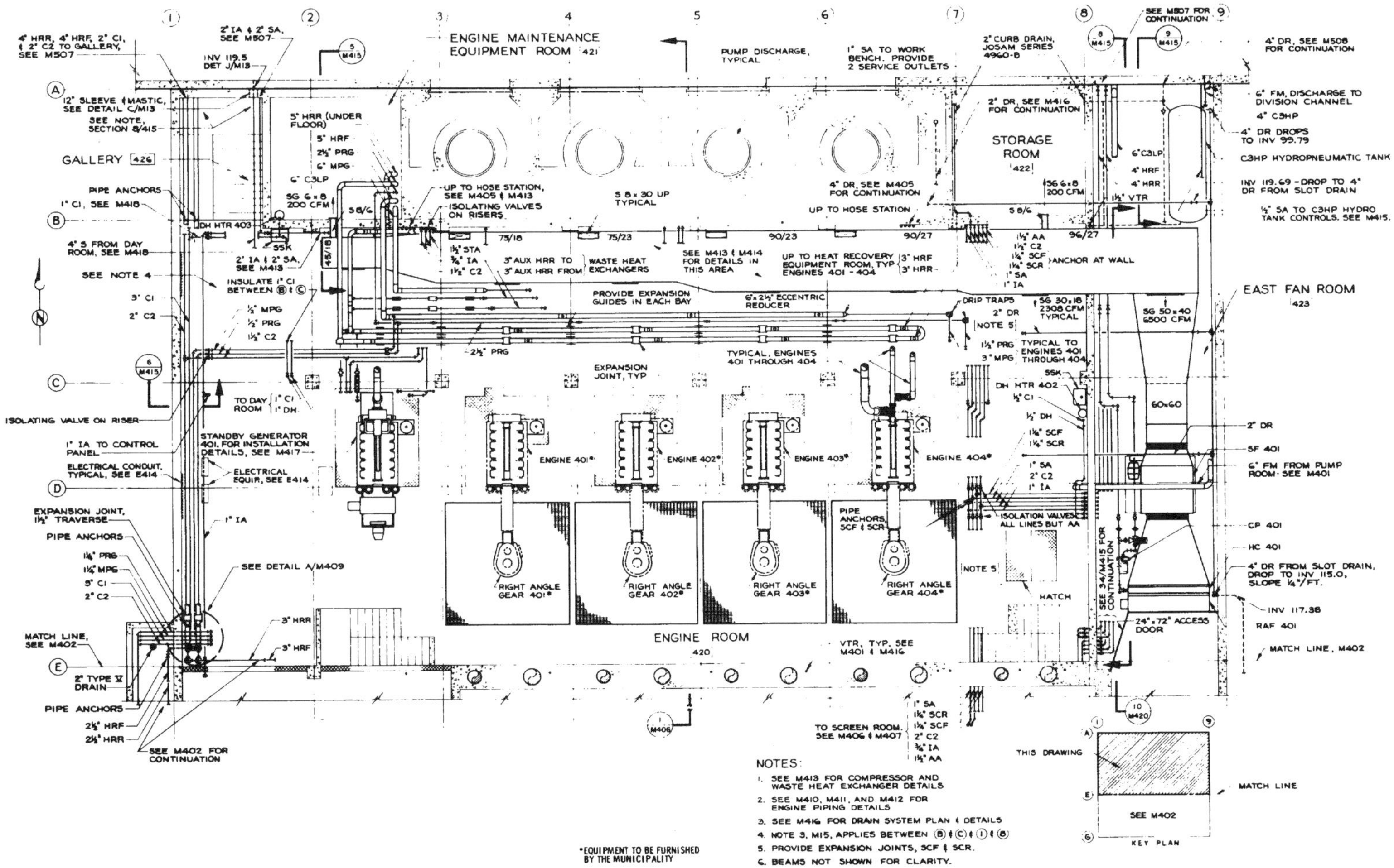

Figure 17-11. Engine room plan, West Point Influent Pumping Station, King County (Washington) Department of Metropolitan Services. Courtesy of Brown and Caldwell Consultants.

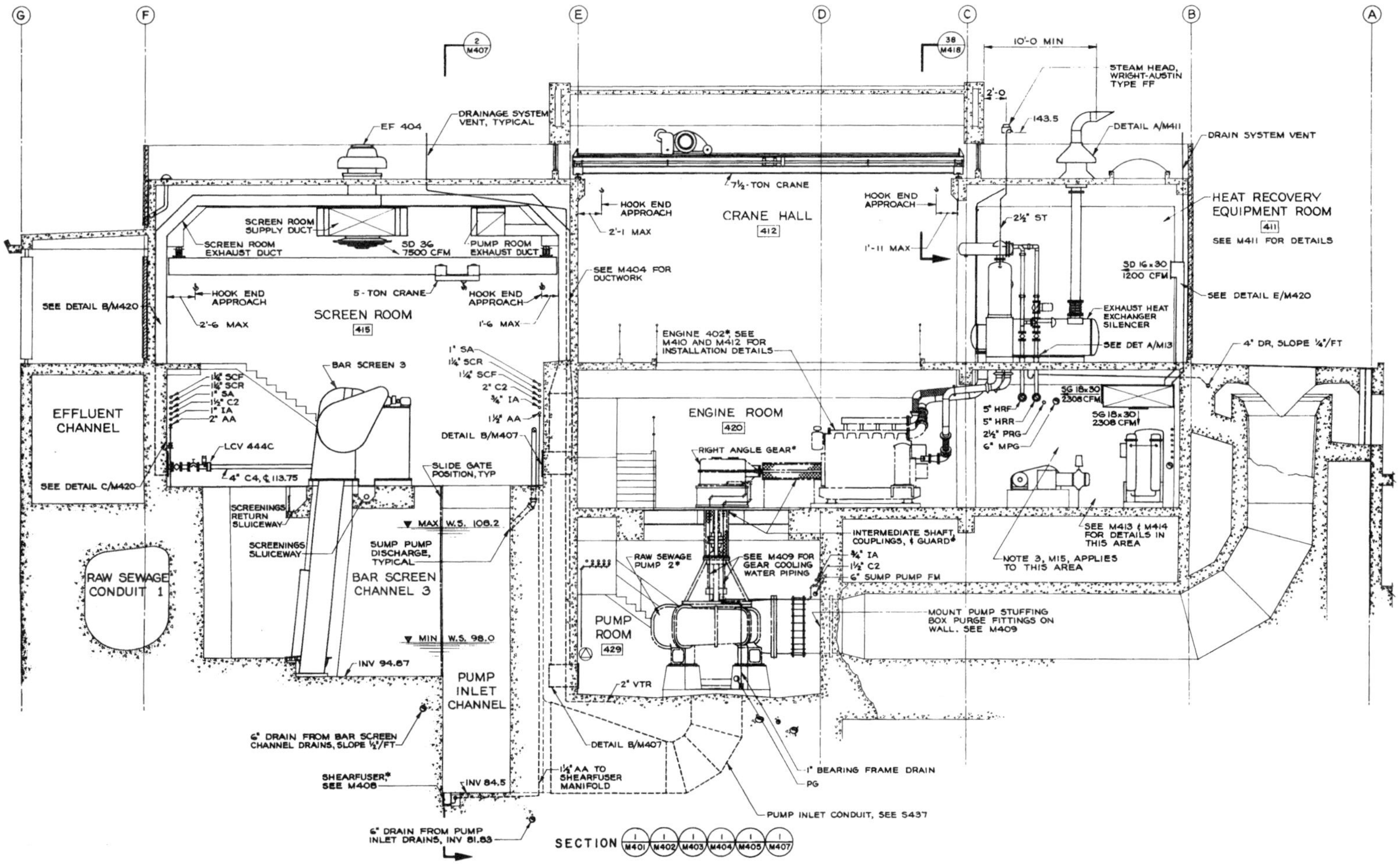

Figure 17-12. Typical section, West Point Influent Pumping Station, King County (Washington) Department of Metropolitan Services. Courtesy of Brown and Caldwell Consultants.

17-6. Examples of Medium-Size Lift Stations

As in Section 17-5, all of the pumping stations described in this section work well, meet the design objectives, and are neat, clean, and virtually odorless—even after 10 to 30 yr of service. All were designed to minimize the size and depth of the wet wells by incorporating V/S pumps. All have ventilation systems that meet the recommendations in Chapter 23. Air is blown in near the ceiling and removed by a powered exhaust system from the lowest level to produce a slight vacuum (about 6 mm of water column) in the wet well to ensure a safe and corrosion-free environment for both equipment and personnel. Note that ease of access, adequate room for maintenance and repairs, and a minimum size of footprint were obtained by following the equipment layout shown in Figures 12-20 and 12-22.

All of these station have trench-type wet wells designed before the advent of the ogee ramp, so they are cleaned by water coursing along the trench at only subcritical velocities. Consequently, not all sludge is usually removed, and there is an average residual depth of about 50 mm (2 in.) of hard sludge remaining after cleaning—a depth that could be reduced somewhat by modifying the operating procedure. (All scum, incidentally, is removed.) Nevertheless, the cleaning is effective in suppressing odors.

The Kirkland Pumping Station appeared to be one of the most typical or representative of the Seattle-area stations with the original trench-type wet well design. Thus it became the prototype for the model studies described in Chapter 12.

Example 17-4
Kirkland Pumping Station

The Kirkland Pumping Station (shown in Figures 17-13 and 17-14) was constructed in 1965 in the downtown area of a small city on the outskirts of a large metropolitan area. The station replaced a wastewater treatment plant as a part of a regionalization project. Because the incoming sewer would be only a few feet below the existing grade, horizontal pumps were selected over vertical units to keep the station profile as low as possible. By chance, the owner had two pumps (surplus from an abandoned station) equipped with two-speed motors that could satisfy project requirements. The existing impellers were replaced to improve solids-passing capability. The second (lower) motor speed, however, was too low and, therefore, was not used. Instead, the pumps were converted to V/S operation by using eddy-current couplings. A duplicate third pump was powered with a new single-speed motor to complete the main pumping units for the new station.

To avoid any outward indication of the station's purpose, the large transformer and high-voltage switchgear were enclosed within the station's superstructure. Each pump delivers 0.11 m^3/s (2.5 Mgal/d) against a rated head of 57.6 m (189 ft) at 1750 rev/min. The suction bells are 400 mm (16 in.) in outside diameter. At maximum flow for one pump, the intake entrance velocity based on the OD of the bell is 0.85 m/s (2.78 ft/s) and the suction pipe velocity is 2.1 m/s (6.8 ft/s).

Critique

The wet well is covered by a light grating that must be removed section by section for washing grease off the walls. Gratings were omitted in some of the other Seattle-area pumping stations in favor of a walkway beside the wet well for easy access for hosing walls—a feature much preferred by the operators.

The walls of the wet well were coated with coal-tar epoxy. After 30 yr of service, the coating is beginning to peel. Lining with PVC is much more expensive, but linings last far longer. Of course, it is also expensive to renew coatings in an operating wet well.

Observing the cleaning procedure at Kirkland and the water splashing onto the trench floor, losing energy that could be used to create supercritical velocities along the trench, prompted the idea that an ogee ramp would make cleaning more effective. From the model tests described in Chapter 12, the cleaning was at least 50 times more effective as measured by the time required to remove all solids. Regardless of whether time is the means used to rate cleaning effectiveness, the ogee ramp is a decided improvement.

The Kirkland Pumping Station serves as an outstanding example of the importance of good maintenance. Aside from some areas in the wet well where the coating has peeled, the station has the appearance of one built only a year or so ago.

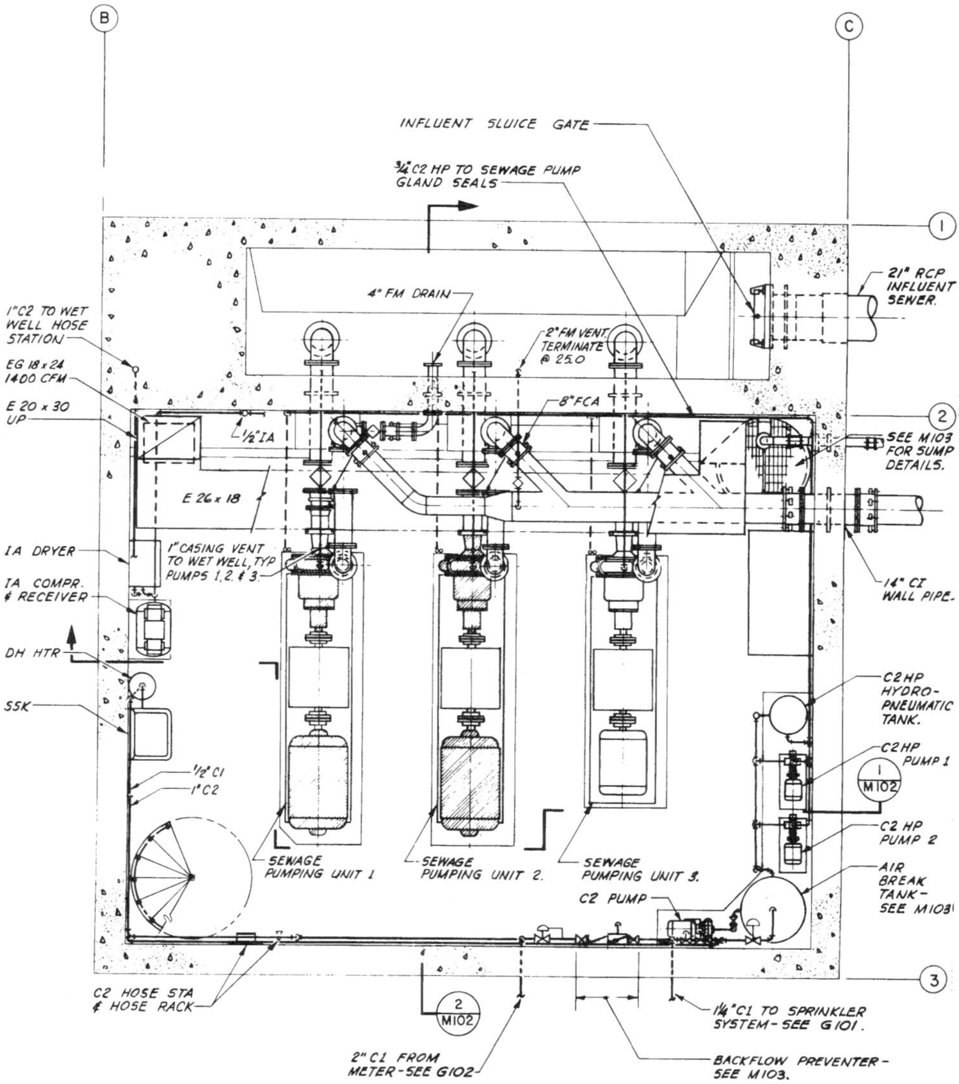

PUMP ROOM PLAN

Figure 17-13. Pump room plan, Kirkland Pumping Station, King County (Washington) Department of Metropolitan Services. Courtesy of Brown and Caldwell Consultants.

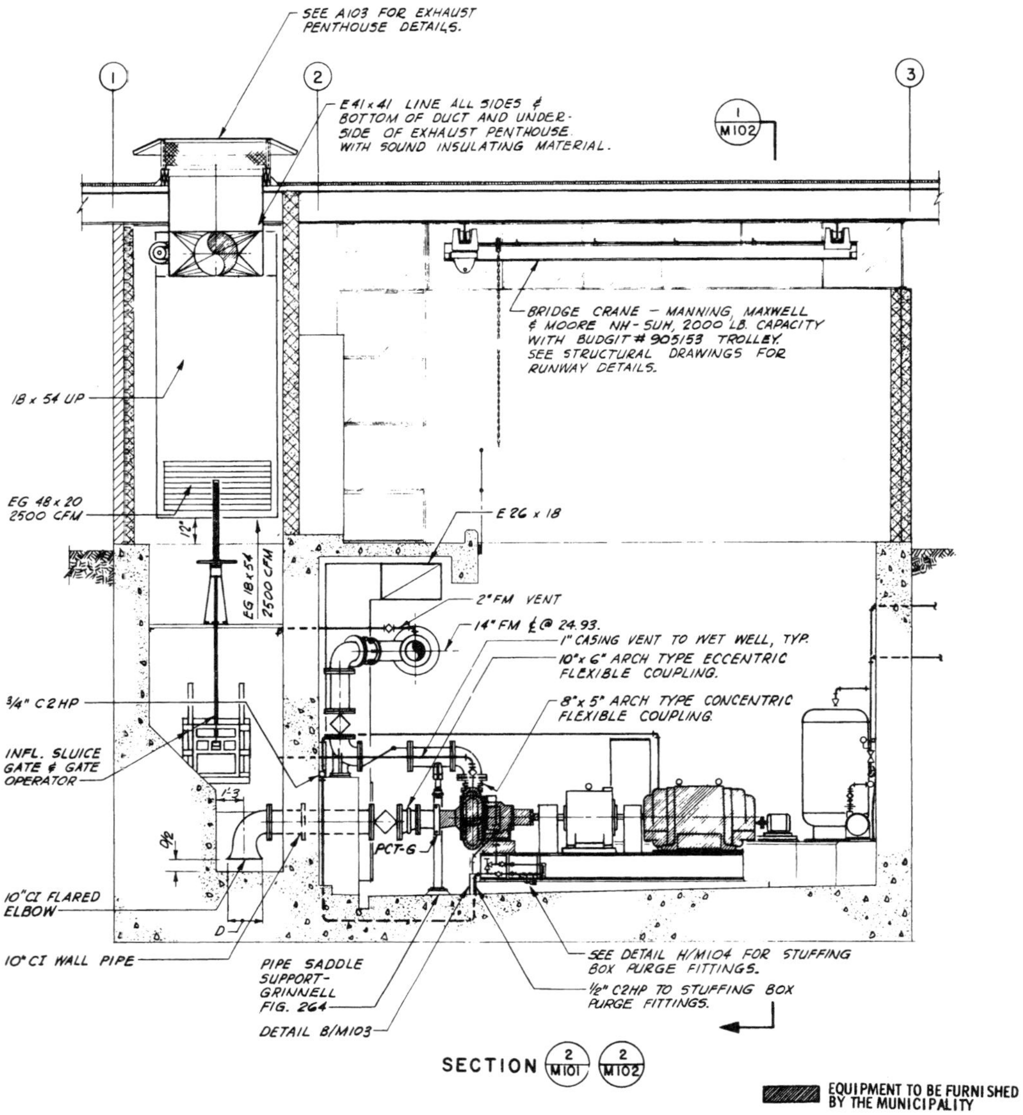

Figure 17-14. Section, Kirkland Pumping Station, King County (Washington) Department of Metropolitan Services. Courtesy of Brown and Caldwell Consultants.

Example 17-5
North Mercer Island Pumping Station

The North Mercer Island Pumping Station (Figures 17-15 through 17-17), which was built in 1969, is exceptionally deep (13.7 m or 45 ft). It is located in a quiet residential neighborhood where aesthetics (architectural treatment and control of odor, sound, and light emissions) all were of concern. The station had to have the appearance of a modern ranch house to match nearby structures. The typical cyclone fence, so common to most municipal wastewater pumping stations, had to be avoided at all costs. A small stream, which coursed through the site, had to be preserved. The sylvan character of the site, if anything, had to be enhanced by the proposed improvements.

Each of the pumps is rated 1.1 m^3/s (2.6 Mgal/d) against a total head of 43 m (140 ft). The force main route, the only economically feasible selection, featured a plateau at approximately the middle half of its 760-m (2500-ft) length. This plateau was at an elevation that corresponded roughly to two-thirds of the total static head. From a transient analysis, it was determined that column separation was likely to occur upon power failure at maximum pumping rates. The solution was to specify the pumps to be driven by wound-rotor, specially designed motors incorporating flywheels that have rotating moments of inertia sufficient to prevent column separation under the worst conditions. The motor speeds are controlled by a liquid rheostat and adjusted to maintain normal depth-of-flow conditions in the upstream sewer.

The noise suppression measures include the following:

- The pump drive motors are specially designed for quiet operation.
- The pump drive motors are located in a below-grade room with all access openings fitted with hatches or doors.

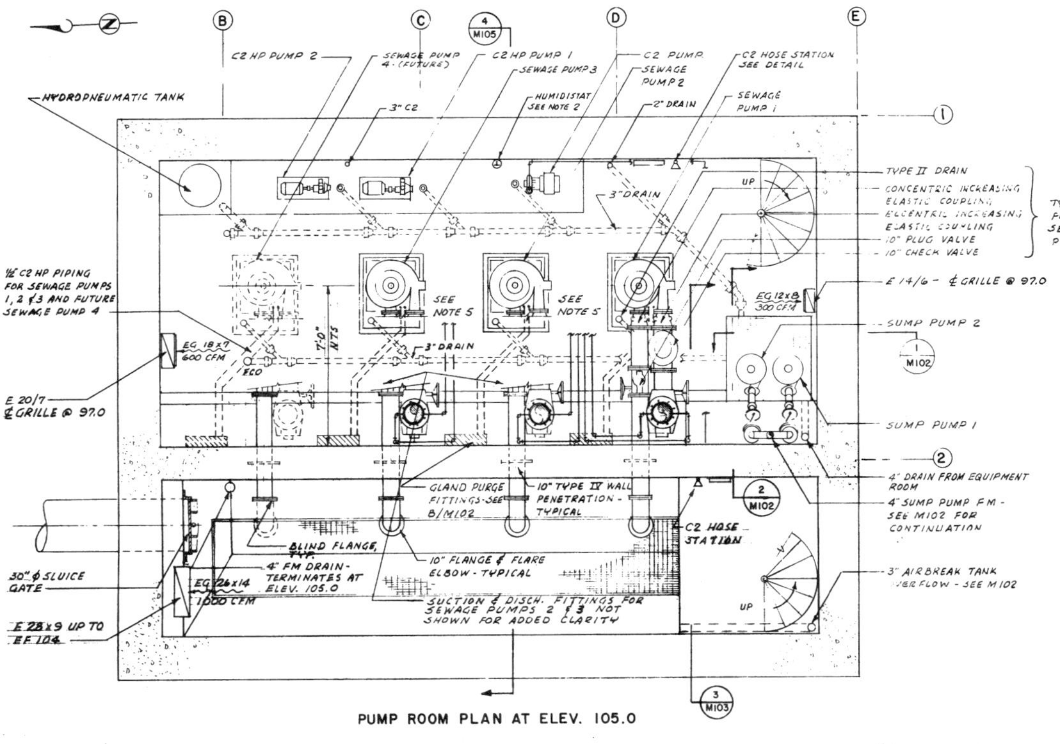

Figure 17-15. Pump room plan, North Mercer Island Pumping Station, King County (Washington) Department of Metropolitan Services. Courtesy of Brown and Caldwell Consultants.

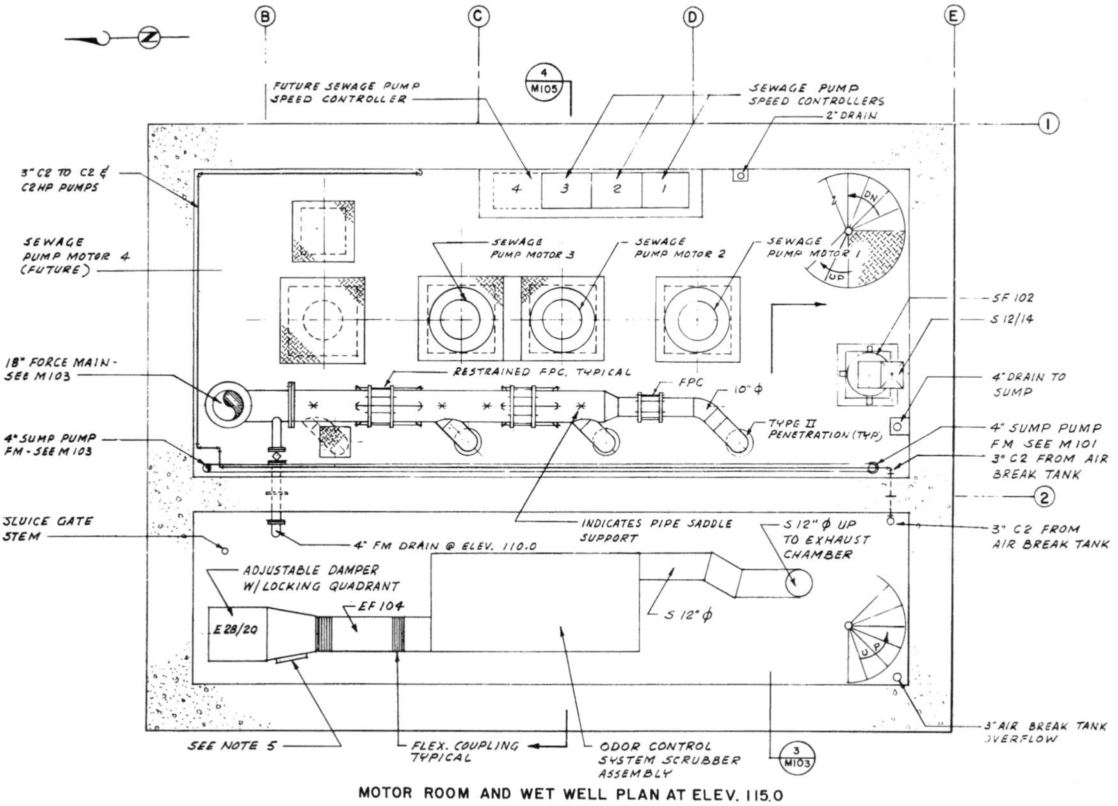

Figure 17-16. Motor room and wet well plan, North Mercer Island Pumping Station, King County (Washington) Department of Metropolitan Services. Courtesy of Brown and Caldwell Consultants.

- All walls above grades are of double-wall construction with a sound-absorbing filler between them.
- Double glazing (with nonparallel panes) is used for all windows. The panes of synthetic, bulletproof glass are resilient to absorb energy.
- The heavy transformer enclosure is located two-thirds of a wavelength from the transformer shell. The walls are high enough to focus a 60-Hz hum upward.
- A special sound-absorbing design is used for air inlet louvers.
- The design criteria included extra-heavy construction for all substructure and superstructure floors to prevent sound transmission to adjacent soils.
- All ventilation systems discharge upward with special enclosures to focus sound upward.

All above-grade lighting systems were designed to mimic residential lighting except when maintenance is required. Odor suppression measures included variable-speed pumping to maintain clean upstream sewers and a self-cleaning wet well. Originally, a dry scrubber with pellets of potassium permanganate and alumina was installed on the ventilation exhaust. However, material costs were very high, charging the reactors with pellets was labor intensive, and—worst of all—odor removal was poor. The system was replaced with a mist scrubber that removes 95% of the H_2S. *[Ed. note: a passerby would smell nothing, would hear nothing but the brook and the wind in the trees, and would assume the structure to be an attractive residence with an attached double garage and a wide driveway.]*

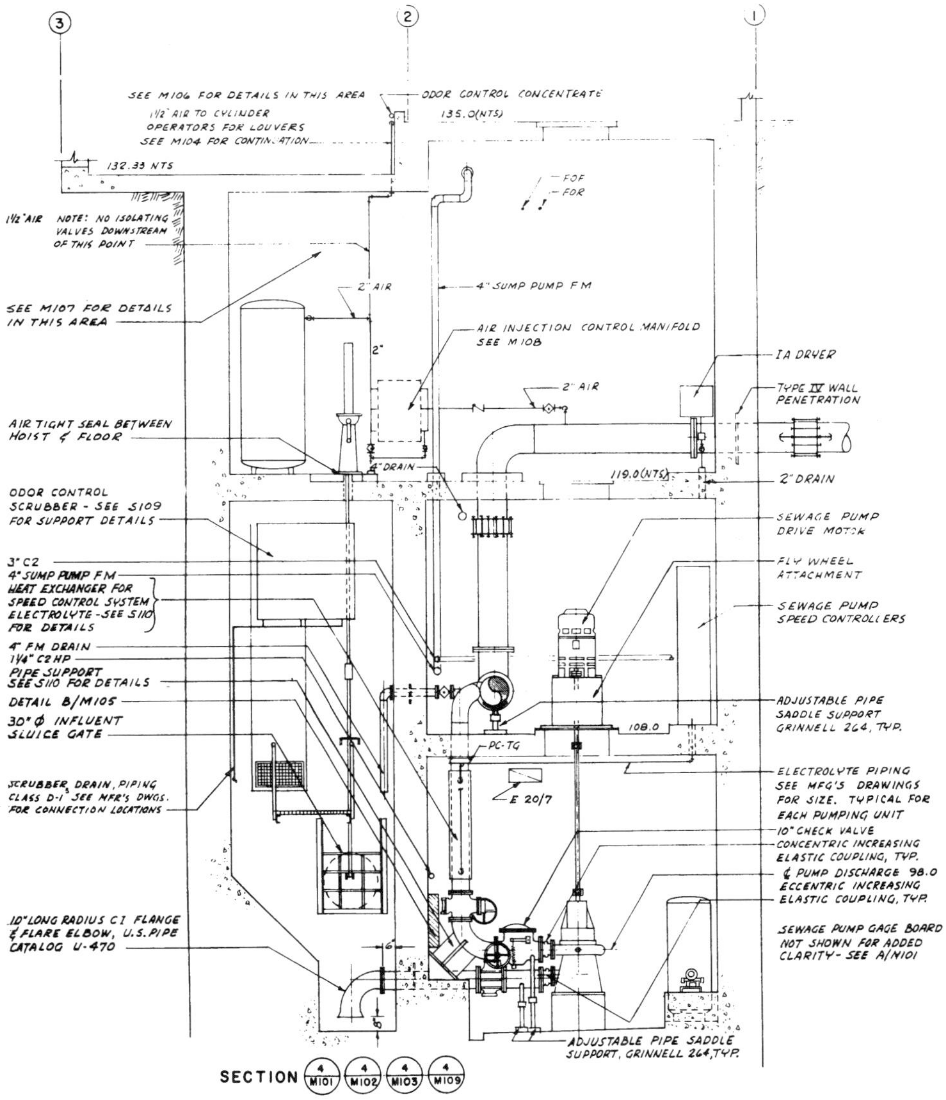

Figure 17-17. Typical section, North Mercer Island Pumping Station, King County (Washington) Department of Metropolitan Services. Courtesy of Brown and Caldwell Consultants.

Example 17-6
Sunset Pumping Station

The Sunset Pumping Station (see Figures 17-18 through 17-20), which was constructed in 1985, is one of two pumping stations that operate in series to lift wastewater over a high ridge. Each station is equipped with two pairs of pumps arranged to discharge through two parallel force mains. Each pump of the smaller pair is rated to deliver 0.17 m^3/s (3.8 Mgal/d) against a total head of 51.2 m (168 ft). Each of the large pumps is rated for a capacity of 0.48 m^3/s (11 Mgal/d) at 51.8 m (170 ft). Space is provided to permit the replacement of the small pumps with 11 Mgal/d pumps when required; thus, the future firm pumping capacity of the two-station system can be increased to 1.4 m^3/s (33 Mgal/d). All pumps are operated at V/S using variable-frequency drives to maintain the normal depth in the upstream sewers. The maximum motor power is 340 kW (450 hp).

As indicated from a transient analysis, there is a potential for a pressure rise of 2000 kPa (300 lb/in.^{2}) or more upon power failure, even at relatively moderate pumping rates. The

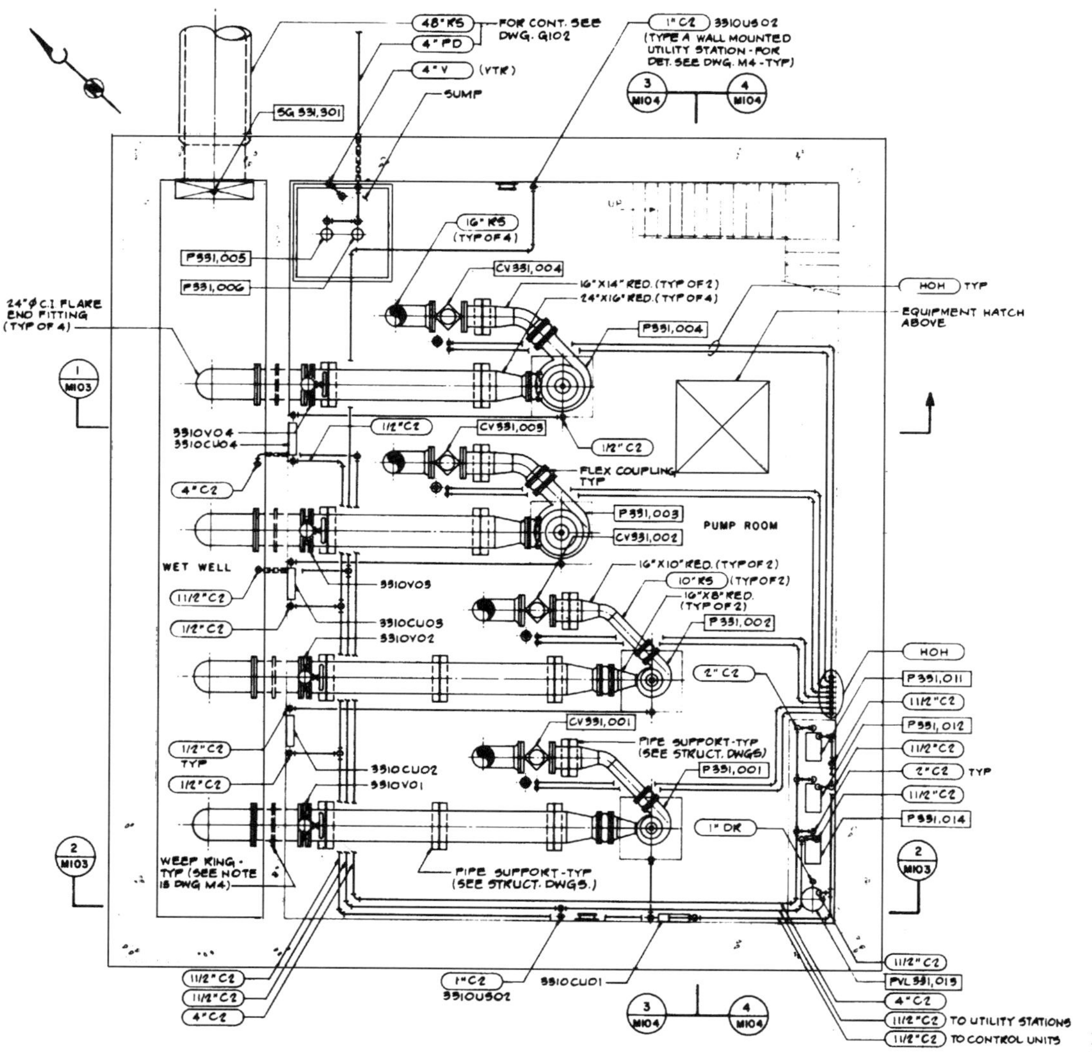

Figure 17-18. Pump room plan, Sunset Pumping Station, King County (Washington) Department of Metropolitan Services. Courtesy of Brown and Caldwell Consultants.

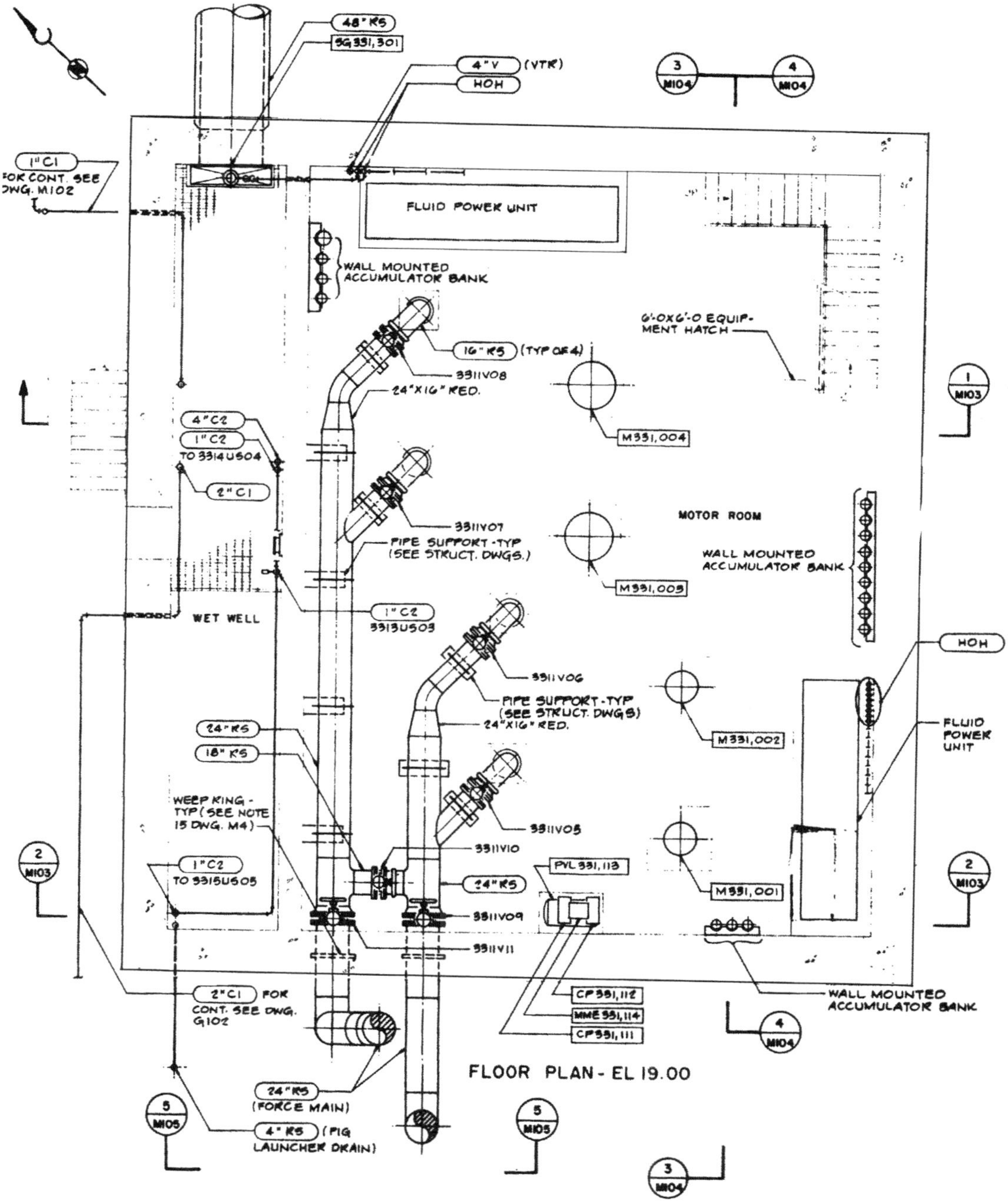

Figure 17-19. Motor room plan, Sunset Pumping Station, King County (Washington) Department of Metropolitan Services. Courtesy of Brown and Caldwell Consultants.

adopted transient mitigation strategy involves hydraulically activated, controlled-closure ball valves on the pump discharge to prevent damaging pressures. Normally, the pumps start and stop against a closed valve. Upon power failure, however, the valves operate in a timed cycle

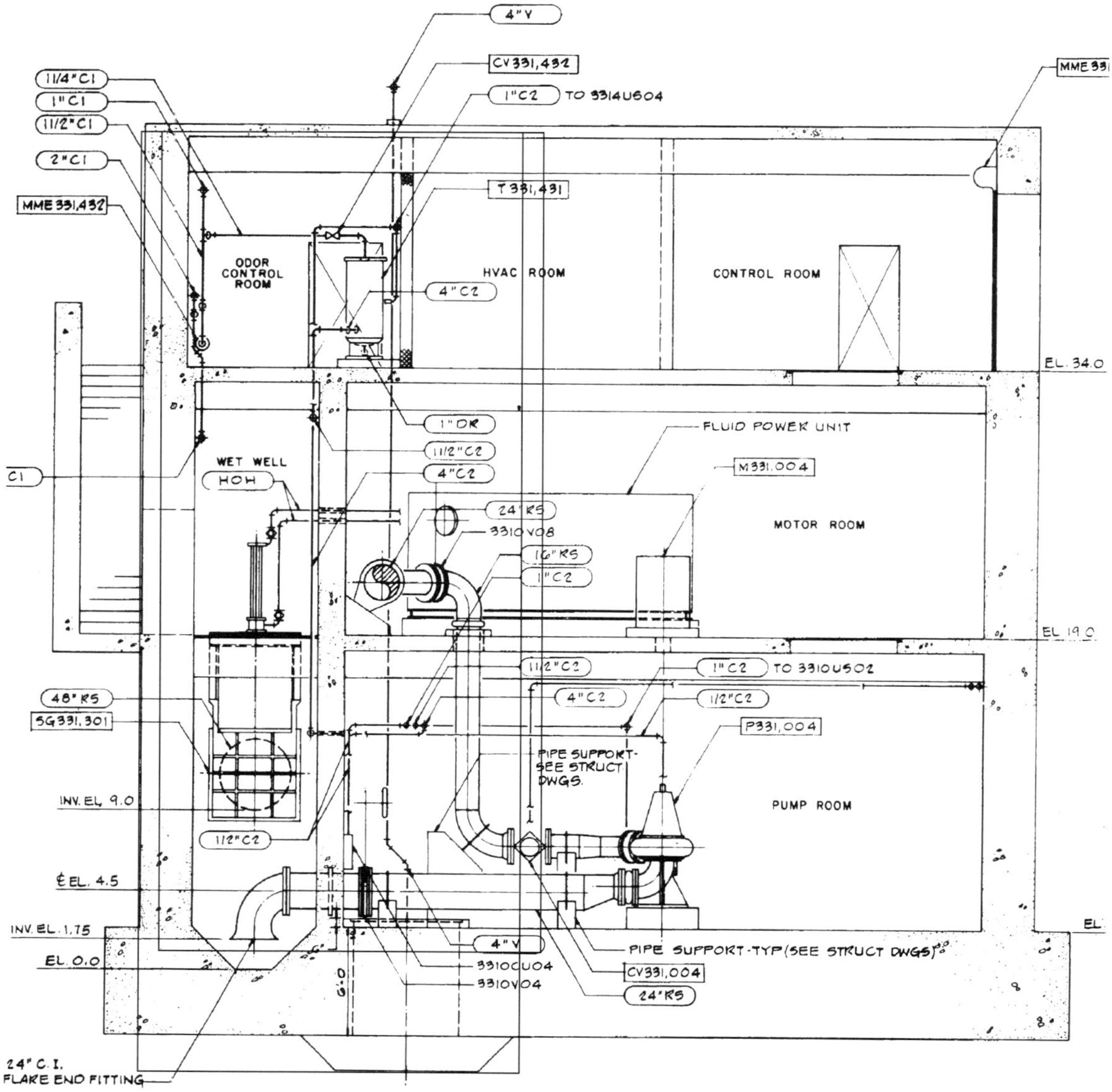

Figure 17-20. Typical section, Sunset Pumping Station, King County (Washington) Department of Metropolitan Services. Courtesy of Brown and Caldwell Consultants.

that results in the reverse rotation of the pumps for a considerable period (up to 5 min. depending on initial flow) until the valves finally reach the closed position.

Each station is served by a dual power supply. Space is allocated for switchgear and an engine-generator to provide an on-site standby power supply.

17-7. Examples of Small Lift Stations

Small pumping stations usually feature C/S duplex pumps in round wet wells. Some older designs have flat bottoms with either small fillets or none at all. Cleaning is given no consideration, and such wet wells are, typically, repositories of noxious sludge and scum. The facilities described herein were, on the other hand, designed to be either self-cleaning or cleaned with a minimum of effort. Consequently, all are successful, all meet their design objectives, and all are easy to clean and are kept clean. Passersby detect no odor whatsoever, and even when the wet well covers are opened, there is only the faint smell of fresh sewage. All of these pumping stations have been described by Sanks, Jones, and Sweeney [6].

Example 17-7
Vallby Pumping Station

Vallby Pumping Station is located in Sweden, not far from Stockholm. It has been so successful that a great many—perhaps most—of the small pumping stations built since are copies. It was designed by an experienced operator—not an engineer—who wanted a facility that could easily be kept clean and would require the absolute minimum of attention and time for maintenance and operation.

The hopper bottom shown in Figure 17-21 is made of 18-8 stainless-steel plate 12 mm ($^1/_2$ in.) thick and bent so that its top fits the round, vertical concrete pipe and its bottom conforms to a rectangle of minimum size with rounded corners. The sides are inclined at about 60°. Special discharge elbows were made and welded to a heavy plate for bolting to the hopper side. The space between the hopper bottom and the side of the round pipe is filled with concrete. The sump was perfectly clean at the time of inspection, and a demonstration of pump-down to the lowest achievable water level showed that cleaning would be very effective indeed.

The unique features include the following:

- By enclosing most of the pump discharge elbow, the flat floor can be made very small, so that virtually all sludge beneath a pump is ejected when the pump is turned on.
- A curb supports: (1) the guide rods, (2) the conduit through which the pump power cable and the level indicator cord reach the control box, and (3) the water lance and keeps it within easy reach.
- Guide rails are stainless-steel telescoping tubes that are raised out of the water except when a pump is to be reinstalled.
- Instead of floats for monitoring water level and controlling the pumps, a piezo-electric pressure cell is placed within an open 100-mm (4-in.) PVC pipe with its lower end beside the volute of the pump. These cells are reliable, are long lived, and are excellent for sophisticated systems because they can, unlike floats, provide input throughout the liquid level range to a PLC for activating the pumps for both normal operation and for pump-down. In the long run, they are probably less expensive than floats. (Some floats must be replaced yearly, although there are some that contain micro-switches instead of mercury switches and are supplied with more resistant electrical cables, which can last for many years.)
- A fresh water supply for washing is equipped with a quick-connect and a valve contained in a large pipe with a padlocked cover.
- Inside the wet well, there is a wash-down water lance that slides in a collar supported by a universal joint that permits freedom of direction. The water lance is equipped with a quick-connect at the top and a nozzle at the bottom. A short hose with mating quick-connects on each end is carried in the operator's truck. The system makes wash-down not only quick and easy, but the jet strikes the pump or hopper bottom so far away that splashing does not reach the operator.
- Under the hatch there are two hinged grates consisting of heavy rods spaced at about 150 mm each way to cover the opening, prevent large objects from falling in, and give workers a feeling of safety. Only one must be lifted to remove a pump. The curb is a support for the 150-mm (6-in.) pipe that carries the power cable from the control box to the sump. It takes only seconds to unplug the power cable and pull it out when removing a pump.
- Operating water levels can be easily changed at the control panel. A pushbutton is provided tor making a motor run backward to clear a clogged pump.

Critique

If the waterfall into the wet well were avoided by means of an approach pipe laid at a 2% grade and discharging at low water level, this wet well would conform to all the concepts for self-cleaning wet wells described in Chapter 12. At small flows, the jet from the inlet strikes the sloping stainless-steel plate, so the generation of bubbles is reduced.

The thickness of the stainless-steel hopper-bottom insert (12 mm) seems excessive and unnecessarily costly. The plate needs to be only stiff enough to allow concrete to be placed between it and the concrete pipe wall. Perhaps a thinner plate with interior stiffeners or a molded plastic shell would be less expensive and just as satisfactory.

The floor under the pumps is so small that significant amounts of sludge cannot accumulate. But where scum is a problem, the controls could be programmed for automatic pump-down at some suitable interval, say, twice per week. On the other hand, some engineers feel that any time a wet well is pumped down

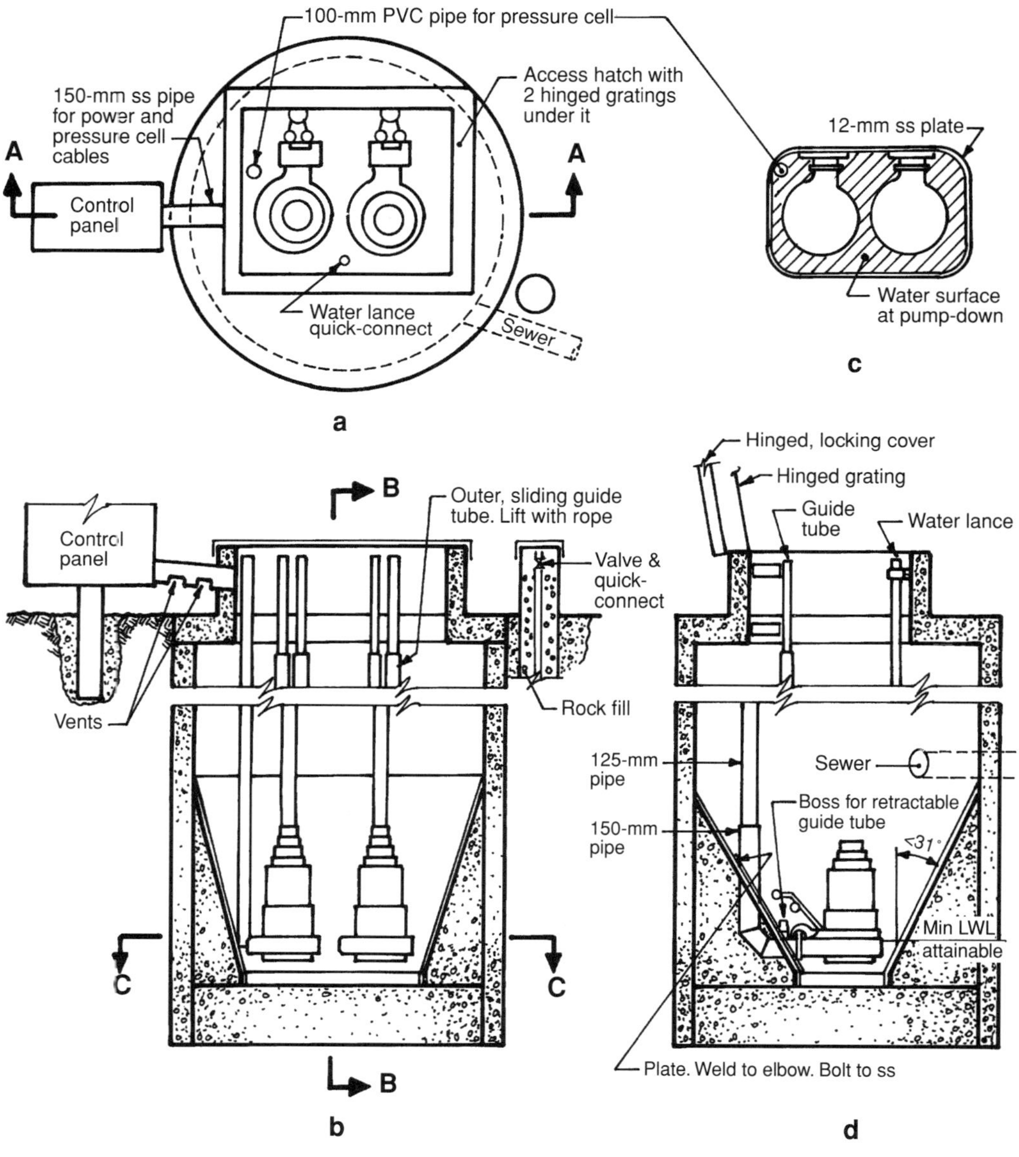

Figure 17-21. Schematic diagram of Vallby Pumping Station. (a) Plan; (b) Section A–A; (c) Section C–C; (d) Section B–B.

and the pump loses prime, an operator should be at hand for troubleshooting.

It is important to keep the area between the pump volutes and the side of the hopper bottom as small as possible. Otherwise, the vortex that forms at pump-down will not suck all the scum into the intake. Of course, it is not necessary to remove all the scum (more will accumulate anyway), but if cleaning is required to leave a spotless water surface, the smallest permissible water surface area at lowest water level is prerequisite (refer to Example 17-9).

Other Swedish Pumping Stations

A growing number of Swedish pumping stations are similar to Vallby Station not only in design but also operationally. One, intended to serve a future subdivision, was connected to only a few houses, so the detention time was very long, and the wet well produced overwhelming odors. The steep-walled sump is advantageous, because detention time and odors can be reduced by setting the HWL close to the LWL to increase the frequency of pump starts and keep the wastewater at least a little fresher. Adding a small flow of fresh water would also help, as would feeding iron chloride to sequester the sulfide ion. A small, obviously inexpensive hydro-pneumatic tank was installed in one station for controlling water hammer. Many such tanks have been in use for many years and are said to be quite satisfactory and devoid of excessive maintenance problems.

Example 17-8
Clyde Pumping Station

The pumping station at Clyde in Contra Costa County, California, is shown in Figure 17-22. The two 7.5 kW (10 hp) C/S pumps are each capable of discharging 26 L/s (410 gal/min) of wastewater at a TDH of 11.1 m (36.5 ft) at a minimum efficiency of 62%.

One of the main functions of the unique 3" piping system is to backflush the pumps to remove rags. Another function is to drain the short force main, and a third function is to promote the vigorous mixing of scum, sludge, and grit so the "homogenized" mix can be discharged to the force main and thus clean the sump. Smaller pipes might plug. The four eccentric plug valves allow complete flexibility of operation. Either pump can be used to unclog the other, and either can supply the mixing water while the other one pumps the mixture into the force main. Alternatively, the force main itself can be tapped for about 15% of its flow for recirculation as mixing water, while either (or both) pump(s) discharge the mixture.

The 3" pipe to the sump discharges just above LWL. At the time of inspection in 1993 (a few months after the station was put into service), the sump was remarkably clean with no material floating on the surface. There was no odor.

Although a little manual labor is required to manipulate the valves, the station can be cleaned with less effort than, say, Kirkland Pumping Station. So it is fair to describe it as a "self-cleaning" facility.

Critique

In comparison with the Vallby pumping station, the 30° bottom slopes allow more active storage volume, and the simpler bottom with its plane surfaces is much less expensive to construct. On the other hand, four extra plug valves are required in a valve vault that must be enlarged to accommodate them. The 3" piping system allows maximum versatility in mixing. If a less expensive system is wanted, the system could be reduced to one valve at the 4" force main. Flexibility would suffer, but there would be little loss of effectiveness. Discharging the 3" piping at or above the LWL drives both scum and bubbles into the pool, whereas discharge at a lower elevation would be just as effective for sludge and perhaps a little less effective for scum, but it would avoid the bubbles.

One advantage of the Clyde piping system is that one pump can be used to backflush the other to remove a blockage.

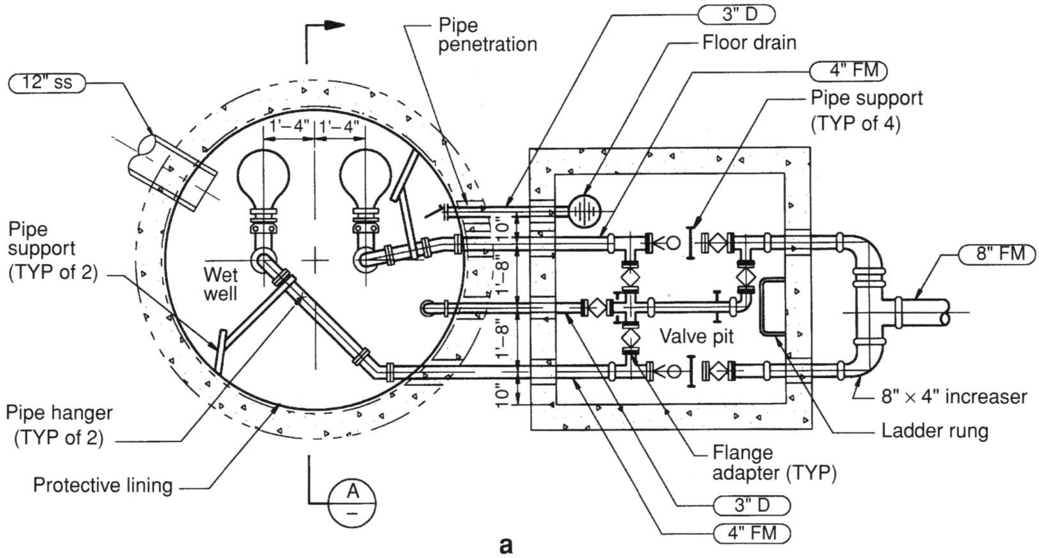

a

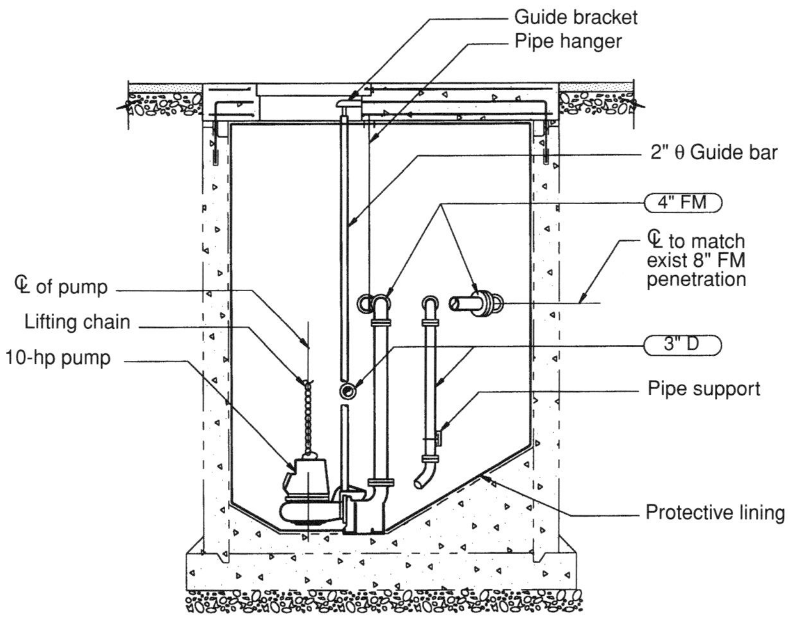

b

Figure 17-22. Clyde Pumping Station. (a) Plan; (b) Section A. Courtesy of G.S. Dodson & Associates, Consulting Engineers.

Unlike the Vallby pumping station, which ejects sludge whenever a pump is turned on and which can be programmed to eject scum automatically, an operator must visit the Clyde pumping station to clean it—no real disadvantage if the policy is always to have an operator present at pump-down.

At Clyde, the operator must climb down into the pump vault to operate the valves. An improvement would be to install a removable grating about 0.6 m (2 ft) below grade and extend the valve stems through the grating to just below the hatch. The operator could then manipulate the valves without entering the vault. This improvement is now (1996) being incorporated into the California Maritime Academy Pumping Station at Vallejo, California.

Example 17-9
Black Diamond Pumping Station

This station at Black Diamond, Washington, was designed to meet most of the concepts developed during this research project for C/S duplex pumps in a round, self-cleaning pump sump. The plans are shown in Figure 17-23. The 400-mm (16-in.) approach (influent) pipe slopes at a 2% gradient for 61 m (200 ft) but is horizontal for 10 pipe diameters before entering the wet well. There are two self-priming pumps of 63 L/s (230 m^3/h or 1000 gal/min) each housed in a nearby building with only the 250-mm (10-in.) pump suction pipes in the wet well. The diameter of each bell is 400 mm (16 in.), so the entrance velocity is only 0.49 m/s (1.6 ft/s).

The layout of the pump room (not shown) allows for the utmost flexibility of operation—a requirement of the owner. A diesel gen-set furnishes enough power to operate the station in the event of a power outage.

Critique

The behavior of the steeply sloping (2% gradient) approach pipeline is entirely satisfactory, but the LWL shown on the plans is too low to prevent the hydraulic jump in the approach pipe from reaching the sump. Of course, the LWL can be readily changed.

In the sump, the smooth sides sloping at 60° keep solids from sticking during pump-down. The pumps broke suction when the submergence of the bells was about 1 D. At this water depth, the area of the water surface was too large to confine the scum sufficiently so that all scum could be ejected on a single pump-down, and a second pump-down was needed. During an inspection, it appeared that operating the two pumps simultaneously would possibly remove the scum in a single pump-down, but that operation was not tested. The designer has stated that, in a redesign, the walls would be made vertical from the floor to 1 D above the pump intakes so as to restrict the area to which scum would be confined upon pump-down.

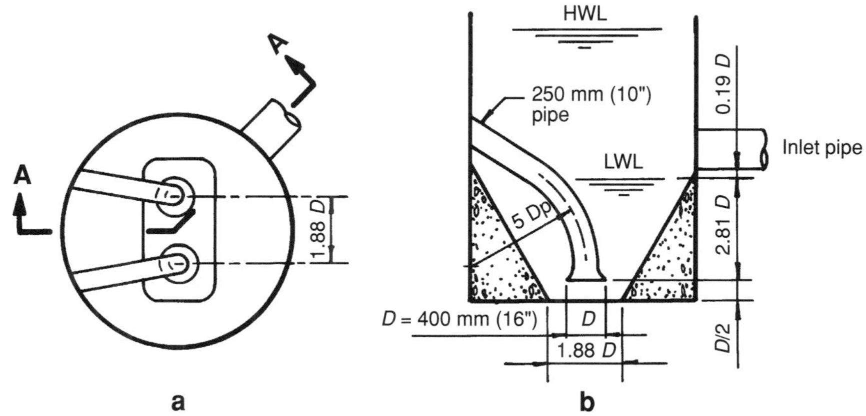

Figure 17-23. Black Diamond Pumping Station sump. (a) Plan; (b) Section A-A.

The pump entrance velocity is very low and might not suck out sand and gravel quickly. Nevertheless, cleaning is accomplished with reasonable dispatch. However, the designer stated he would follow the more restrictive dimensions of Figure 12-59 next time.

The success of the Black Diamond Pumping Station demonstrates the practicality and usefulness of the concepts explained in Chapter 12.

17-8. References

1. ASCE Manual 37, *Design and Construction of Sanitary and Storm Sewers,* American Society of Civil Engineers, New York (1970).
2. Metcalf & Eddy, Inc., *Wastewater Engineering: Collection and Pumping of Wastewater,* McGraw-Hill, New York (1981).
3. AISI, *Modern Sewer Design,* American Iron and Steel Institute, Washington, DC (1980).
4. Parmakian, J., *Waterhammer Analysis,* Dover Publ., New York (1981).
5. Dicmas, J. L., *Vertical Turbine, Mixed Flow, & Propeller Pumps,* Figures 6.6, 6.8, 6.9, and 6.10, McGraw-Hill, New York (1987).
6. Sanks, R. L., G. M. Jones, and C. E. Sweeney. "Improvements in pump intake basin design." EPA 600/R-95/041, RREL-CR. Order No. PB 95-188090. National Technical Information Service, 5285 Port Royal Road, Springfield, VA 22161 (1995).

Chapter 18

System Design for Water Pumping

BAYARD E. BOSSERMAN II
RICHARD J. RINGWOOD
MARVIN DAN SCHMIDT
MICHAEL G. THALHAMER

CONTRIBUTORS
Pat H. Bouthillier
Roland S. Burlingame
A. L. Charbonneau
Allen W. Peterson
David M. Reeser
Joseph W. Steiner
Jerald D. Underwood
Eric L. Winchester
James R. Wright
Eugene K. Yaremko

The purposes of this chapter are (1) to describe methods for determining the required capacity and pressure for a water pumping station, (2) to present recommendations on selecting and sizing the pumps, (3) to discuss control methods for operating pumps, and (4) to provide representative examples of pumping station design. Examples of pumping station design may be worked only in the units originally used, whereas some may be worked in both SI and U.S. customary units.

18-1. Types of Water Pumping Stations

For the purposes of discussion, water pumping stations are considered to fall into five general categories or systems:

- Source (such as a well) pump discharging into an elevated tank
- Raw water pumping from a river or lake
- In-line booster pumping into an elevated tank
- High service pumping of finished water at high pressure
- Distribution system booster without a storage tank in the piping system.

Functionally, there may be little difference between high service and distribution or in-line booster pumping or between low service (raw water pumping to a treatment plant) and booster pumping.

18-2. Pumping Station Flow and Pressure Requirements

Water pumping stations are fundamentally different from wastewater pumping stations because they do not have to be sized to pump at high peak flowrates.

A wastewater pumping station must be able to pump whatever sewage flow enters it, but a water pumping station can be designed to take advantage of water storage reservoirs in the system.

Flow Requirements

Water pumping stations are usually designed to supply water to an area in which the required demand is reasonably well defined or can be projected to a reasonable degree. In a water distribution system, the demand is a combination of customer needs and fire flow requirements. Average annual per capita water consumption, peak hour, and maximum daily demands vary widely depending on factors such as climate, income levels, population, and the proportions of residential, commercial, and industrial users. Typical water consumption values are 560 to 760 L/cap · d or 150 to 200 gal/cap · d. More detailed data on per capita water demand is readily available [1–5]. Most water supply utilities and cities also have extensive records on water consumption.

Fire flow requirements are usually dictated by the fire code adopted by the local jurisdiction. Formulas based on population or on the individual buildings [where area, height, type of construction, occupancy, installation of automatic sprinklers, and proximity (and thus exposure) to other structures] may be useful. For example, the fire flow requirement can be estimated using an early National Board of Fire Underwriters formula. In SI units, the formula is

$$Q = 232\sqrt{P}(1 - 0.01\sqrt{P}) \qquad (18\text{-}1a)$$

in which Q is the fire demand flow in cubic meters per hour and P is the population in thousands. In U.S. customary units, the formula is

$$Q = 1020\sqrt{P}(1 - 0.01\sqrt{P}) \qquad (18\text{-}1b)$$

where Q is fire demand flow in gallons per minute and P, again, is the population in thousands. Be wary of computing fire flows based on formulas that include only the population factor because they can give unrealistically low flow values in comparison to a fire flow demand that might actually occur. For example, in a town of 1000 people, P would be 1 and Q would be 230 m³/h (1000 gal/min), which may be inadequate. The presence of a lumber yard, for example, would require a high water flow allocated for fire. Furthermore, design flows for pumping stations and water mains depend on the reservoir storage capacity available. Generally, the design flow should be the larger of (1) peak hourly demand or (2) maximum daily demand plus fire flow.

For commercial areas, the flow demand is usually estimated as flowrate per area. Demands of 18.7 to 47 m³/ha · d (2000 to 5000 gal/acre · d) are not unusual in mixed commercial and residential areas [2,3,4,5]. Zoning classifications greatly affect these unit flows. In one large water agency on the East Coast of the United States, water consumption varies from 18.7 to 700 m³/ha · d (2000 to 75,000 gal/acre · d) depending on the zoning classification.

Pressure Requirements

Service connection pressures during normal system operations should lie between the following limits:

- Minimum pressure = 210–280 kPa (30–40 lb/in.²)
- Maximum pressure = 410–550 kPa (60–80 lb/in.²)

Some cities and agencies further define acceptable minimum pressures as, for example,

- 276 kPa (40 lb/in.²) for maximum daily flow
- 207 kPa (30 lb/in.²) for peak hourly flow
- 138 kPa (20 lb/in.²) for maximum daily flow plus fire flow.

A frequent practice in water districts is to have zones with a 45- to 60-m (150- to 200-ft) difference in elevation from top to bottom. This corresponds roughly to the maximum 410 to 550 kPa (60 to 80 lb/in.²) pressure range above. A major reason for the maximum pressure of 550 kPa (80 lb/in.²) is that household plumbing fixtures—especially water heaters—cannot withstand greater pressures.

In sparsely populated areas, it is possible to establish zones with a 90-m (300-ft) difference in elevation and to serve the lower areas through pressure-reducing valves. It is also possible to use booster pumping and storage in alternating zones (e.g., pump from zone 1 to 3 to 5) and to service intermediate zones (e.g., zones 2 and 4) through pressure-reducing valves until additional pumping and storage facilities can be justified.

Example 18-1
Flow Requirements in a Small Town

Problem: A town has a population of 10,000. Find the requirements of the water supply system if (1) an adequate reservoir is to be built, (2) an existing but somewhat inadequate reservoir exists, and (3) there is no reservoir.

Solution: Note that data and calculations are rounded off to the customary 2 (or sometimes 3) significant figures. The assumptions are as follows:

Item	SI Units	U.S. Customary Units
Population	10,000	10,000
Water consumption	758 L/d · cap	200 gal/d · cap
Fire flowrate (from Equation 18-1)	197 L/s	3100 gal/min
Fire flow time	8 h	8 h

So the demand flow rates are as follows:

Average daily	10,000 × 758/(1440 × 60) = 88 L/s	10,000 × 200/(1440 min/d) = 1400 gal/min
Maximum daily	2.5 × Avg daily = 220 L/s	2.5 × Avg daily = 3500 gal/min
Peak hour	4 × Avg daily = 352 L/s	4 × Avg daily = 5600 gal/min
Average daily + fire flow	280 L/s	4500 gal/min
Maximum daily + fire flow	420 L/s	6600 gal/min

(1) Storage reservoir. If a properly sized water storage reservoir is provided in the system, it would be reasonable to design the pumping station with a firm capacity equal to the maximum daily demand—in this example, 220 L/s (3500 gal/min). The reservoir would supply additional water to the system during peak hour and fire flow demands.

Although the detailed sizing and design of water storage reservoirs is beyond the scope of this book, some general criteria are as follows:

- Peaking storage equals 25 to 50% of the average daily demand
- Fire flow storage equals fire flow for 3 to 8 h duration
- Emergency storage required for loss of power supply with no standby power supply or alternative water supply available equals 2 to 3 days of maximum daily demand.

The total required storage, then, can range from the volume required by one of these criteria to the volume required by the sum of all of the criteria. The actual size of the storage reservoir depends on the pumping station capacity and design considerations, such as whether standby power is provided for the pumping station. In this example, if standby power is omitted, the required storage is

Storage	SI Units	U.S. Customary Units
For peaking	0.25(88 L/s)(86,400 s/d) × 10^{-3} m^3/L = 1900 m^3	0.25 (1400 gal/min)(1440 min/day) = 504,000 gal.
For fire flow	(197 L/s) (3600 s/h) (8 h) × 10^{-3} m^3/L = 5700 m^3	(3,100 gal/min) (609 min/hr) (8 h) = 1,500,000 gal.
For emergency (2 days of storage at maximum daily demand)	(2 d) (220 L/s) (86,400 s/d) × 10^{-3} m^3/L = 38,000 m^3	(2 d) (3500 gal/min) (1440 min/d) = 10,100,000 gal
Total	45,600 m^3	12,100,000 gal

Notice that by far the largest storage volume is due to the emergency storage requirement, which results from the assumption that the pumping station lacks standby power or that the water supply is temporarily lost.

Selection of pumps. When considering this reservoir, a reasonable design would be to provide four 320 m^3/h or 89 L/s (1400 gal/min) pumps identical for ease in stocking the same spare parts.

- One pump would normally operate for the average daily flow.
- The second and third pumps would come on as demands increase and the pressure consequently decreases up to the maximum daily flow. The nominal peak flow would be 3 × 320 = 960 m^3/h or 267 L/s (3 × 1400 = 4200 gal/min).
- The fourth pump would serve as standby when one of the pumps must be taken out of service for repairs.

In this example, the installed maximum operating capacity of 960 m3/h (4200 gal/min) is slightly greater than the required maximum daily flow of 220 L/s (3500 gal/min) because of the decision to use pumps of the same size. The actual total capacity would be less than the installed maximum because of the increased headloss at higher flows.

(2) Inadequate reservoir. Reservoir capacities are often marginal or even substandard. In a congested urban area, it may not be feasible to construct additional reservoir capacity. Even in relatively open areas, topography may be so flat that additional reservoirs are unwarranted. In such situations, constructing additional pumping facilities may be the only feasible alternative.

Consider the previous example, but assume it has been determined that sufficient reservoir capacity is available to satisfy peak hour and short-term maximum daily demands but not average daily plus fire flows. In such a situation, the pumping station must satisfy both (1) maximum daily flow and (2) average daily flow plus fire flow. A possible design might provide

Demand	SI Units	U.S. Customary Units
Average daily	One pump at 88 L/s	One pump at 1400 gal/min
Fire flow	One pump at 197 L/s	One pump at 3100 gal/min
Maximum daily	Both of the pumps operate	Both of the pumps operate
Standby	Another pump at 197 L/s	Another pump at 3100 gal/min

(3) No reservoir in the system. The difference between the first part of this example, 18-1(1), and this part, 18-1(3), is that in this part, the pumping station must supply all of the water demands—there is no reservoir to make up peak hour and fire flow demands. Thus, the pumping station should be designed to supply the larger of (1) a peak hour demand of 352 L/s (5600 gal/min) or (2) a maximum daily demand plus fire flow of 220 + 197 = 417 L/s (3500 + 3100 = 6600 gal/min).

A reasonable design might consist of the following number and sizes of pumps.

SI Units	U.S. Customary Units
One pump at 88 L/s	One pump at 1600 gal/min
Two pumps at 159 L/s	Two pumps at 2500 gal/min
One standby pump at 159 L/s	One standby pump at 2500 gal/min

In all of the examples, consideration must also be given to system operation at very low flows, such as might occur at night. In the first and second parts of this example, the pumps could be shut off and the reservoir could supply the small amount of water required. In this part, there is no reservoir, so a pump must continue to operate. It is usually not feasible to allow a pump to operate at flows less than, say, 33% of its optimum capacity, so there are three alternatives.

- Install a small pump that is sized to provide the low flow needed.
- Provide a variable-speed drive on the smaller pump to reduce its output. The variable-speed drive would be controlled to maintain a set minimum system pressure.
- Add a hydropneumatic tank (see Example 18-5).

Controls for the pumping station in all three parts of this example can be based either on the pressure sensed in the pumping station discharge or on the change in the level of a reservoir. Pressure at a pumping station could, depending on system geometry, be difficult for control. A control scheme might be as follows:

- Demand on the system causes a decline in water level in the reservoir or a decrease in the pressure in the system.
- As soon as the pressure at the pumping station falls to Pl, the first pump comes on.
- If the pressure continues to fall to P2, the next pump comes on.
- This procedure continues until (if necessary) all pumps are running.
- Eventually, the water level in the storage tank rises and fills the tank and the system pressure starts to rise.
- As the pressure rises, the pumps are stopped. The order in which the pumps are shut off can be as simple as shutting them off in the order in which they started. A more complicated scheme is to run combinations of pumps of various capacities to provide small changes in

flow. In this part of the example, the following combinations of pumps can be run to meet differing demands.

Average daily:	One pump at 100 L/s or one pump at 159 L/s	One pump at 1600 gal/min or one pump at 2500 gal/min
Maximum daily:	One pump at 100 L/s or one pump at 159 L/s Total of 259 L/s or two 159 L/s pumps for a total of 318 L/s	One pump at 1600 gal/min or one pump at 2500 gal/min Total of 4100 gal/min or two 2500 gal/min pumps for a total of 5000 gal/min
Peak hour or maximum daily plus fire flow:	Two 159 L/s pumps = 318 L/s plus one 100 L/s pump = 100 L Total of 418 L/s	Two 2500 gal/min pumps = 5000 gal/min plus one 1600 gal/min pump = 1600 gal/min Total of 6600 gal/min

Note that the lack of a reservoir results in operating inefficiencies. Large pumps must be provided for peak hour and fire flows and must consequently operate at lower flows and less efficiency at average and intermediate demands.

Utilizing either (1) another smaller pump (one-fourth to one-half the size of the smallest pump above) or (2) a variable-speed drive on the smaller unit can result in an even finer degree of control in response to system demand and pressure.

All these flows are nominal. Increases in friction head reduce the higher flows. Accurate flows can be calculated only by due consideration of the system H-Q curve.

18-3. Raw Water Pumping from Rivers and Lakes

Raw water may be pumped from a river, a natural lake, or an artificial reservoir. A potable (treated) domestic water supply may be pumped directly or indirectly through a distribution reservoir or into a water distribution system. A raw water supply may be pumped to a water treatment plant either from the source or before or after passing through desilting basins.

Depending on the source and ultimate use of the water, raw water pumping facilities are generally a combination of only three basic components: (1) the raw water intake structure, (2) the pumping facilities, and (3) the screening facilities, which may or may not be required. Design variations and arrangement of these basic components depend on the imagination, ingenuity, experience, and judgment of the design engineer.

Most raw water pumping facilities are shore installations. The intake may be placed below the lowest water level on record or the lowest level of reservoir drawdown determined by storage requirements. The intake may be placed (1) at a point in deep water such as a lake or impoundment, (2) near the shoreline in shallow water, or (3) directly on shore with an excavated channel to deeper water. Compared with wastewater pumping, there is greater flexibility in placing the pumping facilities for a water supply. The pumping station may be built in the intake structure at any of the points cited above, or it may be located remotely from the intake. The connection between the intake and the pumping station may be any of the following:

- A suction pipe in a shallow trench to a higher elevation and directly connected to the pumps. The pump would not be more than 20 ft (including pipeline losses) above the lowest water level at the intake. The pumps must be equipped with adequate priming devices.
- An excavated open channel leading to the pump suction well.
- A horizontal pipeline from the intake directly to the pumps.
- A tunnel (in larger installations) connecting the intake to a remote pumping station.
- Combinations of any of these.

It may be desirable to incorporate the intake works and/or the pumping and screening facilities into a water supply dam. In an earth dam, the intake may be

- A simple gate structure built into and parallel to the upstream slope, near the toe of the dam, with a conduit leading to a pumping station at or beyond the downstream toe.
- A tower inlet, located near the upstream toe, with an access bridge from the top of the dam. The tower can include pumps and have trash racks, screens, or multiple inlet ports. With this type of

inlet, a conduit may also be carried through the dam to a pumping station downstream.

- A concrete dam intake section may be in the dam's upstream face, with or without racks, screens, pumps, and multiple ports.

Because many options and alternatives are available, the choice is based on the siting considerations of the land and water environment. The most important factor influencing design is the depth of the raw water source. Depending on the design flows (pumping rates) and pumping heads, it may be difficult to deal with a large variation in water surface elevation due to reservoir drawdown or natural hydrological variations. Water quality variations at different levels in a deep lake or reservoir may require multiple port inlets to the intake structure to allow taking the best quality of water available at all times.

Deep reservoirs in temperate climates turn over (sometimes rapidly) in the fall or spring or both. Turnover is caused by cold and dense surface water overlaying warmer (less dense) layers. The unstable condition causes bottom layers to be suddenly brought to the surface, often with bottom mud. Normally, the water quality at or near the bottom of a deep reservoir is inferior to that of shallower depths. It may be devoid of or low in oxygen and high in manganese and/or iron due to decayed vegetation. Because the water quality at any given depth undergoes seasonal variations, multiple ports in the inlet works of deep reservoirs are desirable to draw the best water. Top and bottom waters can be continuously mixed to a uniform quality by extending a perforated air hose or pipe across the bottom of a reservoir and supplying air from a compressor on shore, an arrangement that may allow a single inlet in the intake structure. A sampling program should be carried out throughout the year to determine water temperatures and quality at various depths.

Surface conditions in lakes and impoundments influence the intake design. In warm, humid climates, aquatic plant growth, such as water hyacinth and duckweed, is a problem. A floating boom arrangement may prevent plant growths and surface trash from entering the intakes and pumps. In cold climates the intake works and pumping arrangements must also be designed to prevent damage from ice and freezing.

River Intake Design Considerations

This subsection is essentially an abridgement of pages 1173 to 1214 and 1247 to 1256 of *Proceedings* [6].

Low Water

The determination of design low water is important for an intake that must operate year-round. Design low water is an important factor in the design of a direct intake, in the layout of an infiltration gallery, and in the pump setting and wet well floor elevation. Design low flow can be obtained through a routine low-flow frequency analysis of water survey data. If there are no data for the stream, either a regional analysis or the analysis of transposed data may be required.

Cross-section data can then be used to calculate a design low stage. The calculations must begin at a downstream control section or at a control reach in the river. Consider the likelihood of channel changes that could significantly change low-flow levels over the expected life of the intake structure. The degradation of a rapids or a downstream control section or the migration of rapids from downstream to upstream of the site would result in lower water levels, as would the development of a downstream cutoff. The development of a large bar downstream, or a cutoff just upstream, could cause higher water levels. The likelihood of such an occurrence and the magnitude of the effect cannot be determined analytically. Any allowance made must be a matter of judgment based on a study of comparative aerial photographs and observations made in the field.

High Water

The elevation of design flood high water is also an important factor in determining the elevation of the working floor of the pumphouse and in determining the design of any river training works that might be associated with the intake. In many instances, design scour depths can also be determined from mean flood depths.

The discharge of the design flood can be estimated by analyzing the water survey data. The corresponding stage can be computed by using the cross-section data and the channel and flood plain roughness estimates. Direct stage measurements at one or two high flows of known magnitude are most useful for increasing the accuracy of the calculated design high water. High-water marks, information from local residents, and backwater computer programs are also very useful in simulating water levels once the channel characteristics have been determined.

The choice of design high- and low-water levels should be based on the aggradations of the river reach under consideration because estimated water levels and streambed levels can be dramatically reduced at an intake if the river reach is subjected to degradation. Alternatively, on an aggrading reach such as the backwater above a storage dam, water and streambed levels will rise, which results in more frequent flooding during periods of high flow and possible inundation of the intake with sediment.

Once the frequencies of the low- and high-water levels have been ascertained, the probability of levels above or below these values during the life of the structure can be determined. For a 50-yr design life of the structure and design high- or low-water levels with a return period of 100 yr, there is a 60% assurance that the high-water levels will be exceeded at least once in the life of the structure (see Figure 18-1). For the low-water-level condition (for a similar design life and return period), the assurance of lower water levels would also be 60%.

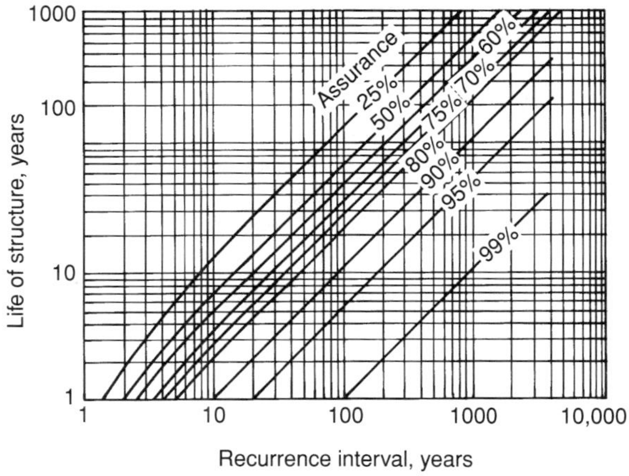

Figure 18-1. Probability of flood damage during the life of a structure.

Trash and Debris

Large amounts of trash and debris, floating or suspended, are likely to be carried by the stream at high discharges. River structures that project high above the bed can trap the debris and can sometimes cause the formation of an adjacent floating island that impedes the operation of the intake. Log booms can be constructed to deflect the floating debris away from the intake, but booms are effective only if the angle between the direction of the surface velocity and the boom is small. Suspended debris, such as waterlogged wood and coal, can be drawn into the system and can clog screens or trash racks. These materials can be removed by adequate backflushing, and frequent backflushing is usually required during periods of high discharge. Trash racks should be designed with horizontal bars exposed to the flow. The racks should be recessed into the face of the intake structure to minimize the hang-up of trash on the upstream side of the rack.

Ice Jams

Design high water in regions of below-zero weather may often result from ice jam flooding. For example, the water levels during spring break-up on the Athabasca River at Fort McMurray, Alberta, can be 20 ft above the summer high flood levels because of ice jams. Evidence for such high-water levels may be discovered during the field survey, from either ice scars on tree trunks or information from local residents. Ice-jam data can be found in the records of various government departments and public and private agencies, in the archives of local newspapers, in railway engineering and maintenance records, or even in old diaries of church missions.

Frazil Ice

Few water intakes in northern regions with freezing temperatures are immune from frazil ice problems. A length of river in which rapids or highly turbulent areas are combined with reaches of open water and subzero temperatures can generate large quantities of frazil ice that can result in very large ice thickness downstream with a high potential for jamming and possible ice damage to structures in this reach.

To combat active frazil ice, amorphous plastics, such as polyethylene and other special coatings on steel bars, or slightly heated surfaces can be used. Frazil ice does not form on wooden racks nearly as readily as on steel ones. A short frazil run and large rack openings will result in a small amount of ice deposit with no significant headloss.

Mechanical rack cleaning is successful if the frazil run is not heavy. A motor drives a device with teeth that mesh with the openings up and down the screen. But if the frazil run is heavy, the mechanical scrapers may seize.

Recent research has shown that an economical way to get rid of frazil ice problems is to use screens with large openings and to coat the steel bars with plastic resins, polyethylene coatings, or silicone grease.

Screens for Fish Protection

Some regulatory authorities have insisted that fish be protected by intake screens with small slots (2 mm or 0.08 in.) and approach velocities of no more than 0.15 m/s (0.5 ft/s). Such small slots are completely impractical in the presence of frazil ice. In severe climates, the only protective measure that works well appears to be electric shock.

During the preliminary stages of design, be sure to obtain approvals of conceptual plans from the U.S. Army Corps of Engineers, the regional U.S. EPA, the state department that deals with environmental protection, and, perhaps, the state fish and game authority.

Stable Channels and Mild Climates

If the river channel is stable and the water levels are controlled within reasonable limits either by an upstream dam or by a weir in the channel, if ice is not a serious problem, and if the intake capacity required is relatively small (say, no more than about 1 m^3/s or 20 Mgal/d), a relatively simple intake structure may suffice. Some of the types are

- Tower intake with multiple, screened ports on the downstream face
- Infiltration galleries along the bank
- Transverse perforated pipe under the streambed (protected with a rock overlay)
- Weir and side channel
- Forebay or lagoon constructed beside the river with a shore intake, as per Figure 18-2.

Lake Intake Design

Tower Intakes

Tower intakes are built in deep water, have a bridge to the shore or to the crest of a dam, and may or may not have screening or pumping facilities. These intakes are used when taking water from varying levels in a deep, newly constructed reservoir or in dewaterable reservoirs, so dry or shallow water conditions prevail throughout the construction period. Because most tower intakes are in deeper water, they are usually circular in plan. The configuration (circular, square, or hexagonal) is based on (1) structural considerations, (2) water depth, (3) the need for trash racks or screens, (4) pump space, if required, and (5) whether the tower may be dewatered under full or partly full reservoir conditions.

The cost of the access bridge may be a substantial part of the total intake works cost. The intake, therefore, should be placed as close to the shore as possible. If pumps are to be installed, the service bridge also provides a means of carrying the pipeline and power cables to shore, and it must be substantial enough to truck equipment (such as pumps, motors, and screening segments) from the shore to the intake and simultaneously carry the load of the discharge pipeline. High tower intake and bridge piers may impose relatively heavy, concentrated loads on the foundation—particularly if the reservoir is drawn down to the point where buoyancy is lost. The foundation conditions must be carefully investigated prior to design, and if the intake tower is to be dewatered under full reservoir conditions, the tower weight must be able to resist water buoyancy and prevent flotation.

A typical tower intake including vertical dry well pumps that take suction from a central well is shown in Figure 18-3. Water can be taken from four different elevations. Installing vertical wet well pumps would eliminate the dry well with a substantial cost savings

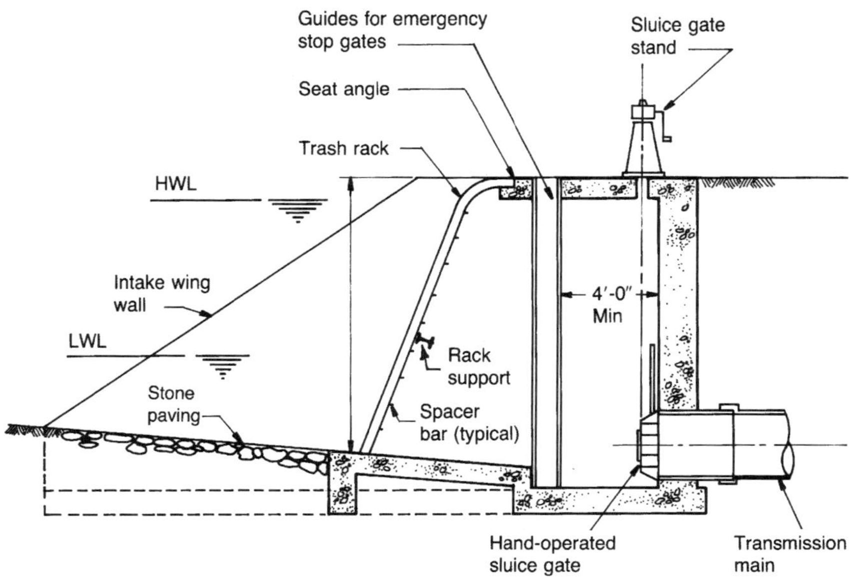

Figure 18-2. A typical shallow shore intake with a hand-cleaned trash rack. After Camp Dresser & McKee.

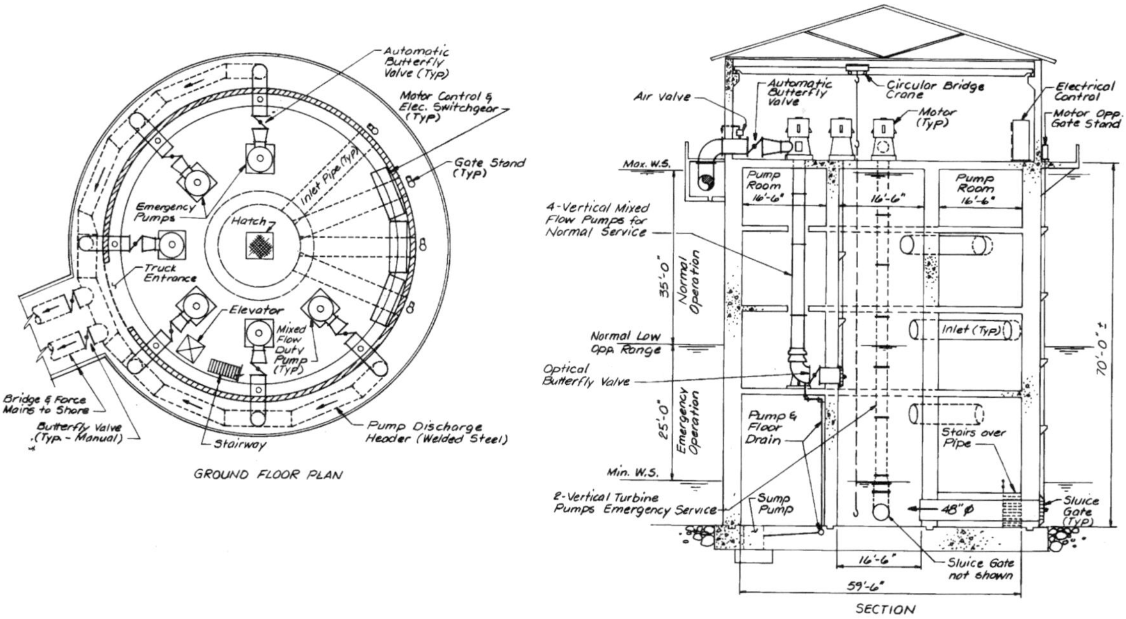

Figure 18-3. A tower intake. Courtesy of Camp Dresser & McKee.

even though the wet well would have to be enlarged to provide adequate spacing between the pumps. The advantage of installing wet well pumps is the reduced cost. The disadvantage is pump maintenance difficulties. If an impeller becomes clogged or damaged, the pump must be pulled to the motor room floor for servicing. In a shallow setting, it is relatively simple to pull the pump, but in a deep setting, such as the one shown in Figure 18-3, it is time consuming. If clogging or pump damage is periodic, the dry well configuration is advantageous because of the relative ease of operations and maintenance. For example, the pump impellers are readily accessible for inspection, maintenance, and repair because the pumps can be disassembled at the pump room floor level without disturbing either the motor or the pump column and without having to pull the entire pump.

A horizontal pump installation would require the exterior shell of the tower to be substantially larger. Furthermore, pump room floor-level motors could be subjected to flooding to the full depth of the intake—about 21 m (70 ft). The summary of the advantages and disadvantages of wet well versus dry well installations and of vertical versus horizontal pumping units is given in Table 25-6.

Alternative Tower Intake Designs

One alternative (if incorporated into a new dam) is to eliminate the pumps but install bar racks and extend a pipe from the bottom of the tower through the dam to a downstream pumping station. The pumps can be horizontal to make them more easily accessible. All piping would be protected from freezing. A valve or sluice gate in the outlet pipe within the intake structure can allow unwatering for periodic pipe inspection. If the dam is made of earth, a number of cutoff walls around the pipe as it passes through the dam should be arranged to prevent seepage, which, if not prevented, ultimately results in failure of the earth dam.

Another variation is to build an intake with two compartments into the upstream face of a concrete gravity dam. Trash racks are installed at the entrance of the first compartment, with openings at various elevations to admit water to the second compartment. A pipe at or near the bottom of the second compartment passes through the base of the dam to a pumping station either at the downstream face or at some convenient point below the dam. This arrangement (as with all intakes) may supply water by gravity to a treatment plant or distribution reservoir without using a pumping station. All intakes may be equipped with fine-mesh traveling water screens if required.

No raw water pumping station can be said to be typical, but many aspects of the concerns noted above are embodied in Example 18-2, in which there is both a side inlet and a tower intake in a river where ice creates problems.

Example 18-2
Raw Water River Intakes and Pumping Stations

Billings, Montana, with a population of 87,000 in 1988, lies in the Yellowstone River Valley at an elevation of 940 to 1100 m (3100 to 3700 ft). The climate is relatively severe with minimum and maximum mean temperatures of −11.2 and −1.2°C (11.8 and 29.9°F) in January and 14.5 and 30.4°C (58 and 86.6°F) in July. Recorded flows on the Yellowstone River have varied from 12 to 1970 m^3/s (430 to 69,500 ft^3/s). In 1987, the average flow was 138 m^3/s (4869 ft^3/s). The turbidity varies from 2 to 1500 nephelometric turbidity units (NTU), and the average in 1987 was 35 NTU. Spring floods occasionally carry large cottonwood trees, and summer flow carries leaves and algae. During the winter, heavy pack ice and frazil ice create serious problems. Pumpage in 1987 averaged 0.8 m^3/s (18.3 Mgal/d or 210 gal/cap · d) and a maximum of 1.79 m^3/s (40.9 Mgal/d—an average of 470 gal/cap · d). The 11 reservoirs have a total storage of 91,700 m^3 (24.2 Mgal).

The site of Billings' water supply is shown in Figure 18-4, an aerial photograph. Before 1954, however, the only supply was at Inlet No. 1, which then consisted of three fixed screens at the shore line.

Problem: The problems prior to 1954 were (1) to obtain a reliable supply of water at low river flow, (2) to be able to handle the high turbidities more effectively, and (3) to achieve an overall reliability greater than could be obtained with a single inlet.

Solution: The solution to these problems consisted of (1) constructing a tower intake (called Inlet No. 2) at the deepest portion of the river and (2) improving the original inlet.

Inlet No. 2. The construction of the intake tower is shown in Figures 18-5 and 18-6. The tower is connected to the shore by a walkway bridge with the pump suction pipes buried under the river bed, as shown in Figure 18-7a. The bedrock is shale soft enough for steel sheet piles to be driven into it. The low service pumping station is on the shore. The pump room floor is low enough to provide flooded suction for the horizontal split-case pumps. One of the smaller pumps is shown in Figure 18-7b.

Inlet No. 1. Some years later, a staged series of improvements were made at Inlet No. 1. The first was the construction of six 1200-mm (48-in.) reinforced concrete pipes (RCPs), which extended for some distance into the river. The second was the construction of the contact basin, the diffuser baffle, and the L-structure (Figure 18-8) designed to collect the flow and pass it through a rapid-mix basin consisting of two vertical mixers in tandem. Chemicals are piped to the L-structure from the treatment plant nearby. Potassium permanganate is added at the rapid-mix basin all year. Alum is also added from March through September, during which time the turbidity exceeds 20 NTU and complete treatment is desirable. Flocculation takes place in the contact basin, sometimes assisted by two horizontal propeller mixers driven by submersible motors carried on barges. Water is generally treated by direct filtration from October through February. The diffuser baffle in the contact basin inhibits short circuiting.

A third improvement at Inlet No. 1 was to drive sheet piling along the inlet and river both to protect the banks and to make it easier for a backhoe to remove sand and silt from the inlet. The next improvement was to cut off the six 1200-mm (48-in.) pipes and install canal gates at the cut ends. The final improvements were to construct the inlet baffle with three skimmer gates at the entrance of the inlet channel. The skimmer gates are raised and lowered in separate grooves by cables and hand winches so that a slot of any desired height can be positioned at any desired elevation to exclude sand and silt and to prevent the entrance of floating debris. Because there is almost no hydrostatic force on the skimmer gates, they are lightly built.

Operational experiences. Inlet No. 1 has been very satisfactory. Frazil ice is not a problem, and debris is kept out by positioning the three skimmer gates to draw water from below the water surface but above the river bottom to exclude heavy materials. Once a year, sand and silt are removed from the channel by means of a backhoe. The contact basin is cleaned with a floating dredge, a job that takes all summer to complete. The dredged material is stored in holding ponds.

Figure 18-4. Site of the City of Billings Intake No. 1 (upper left) and Intake No. 2 (right center). Aerial photograph courtesy of Christian, Spring, Sielbach & Associates.

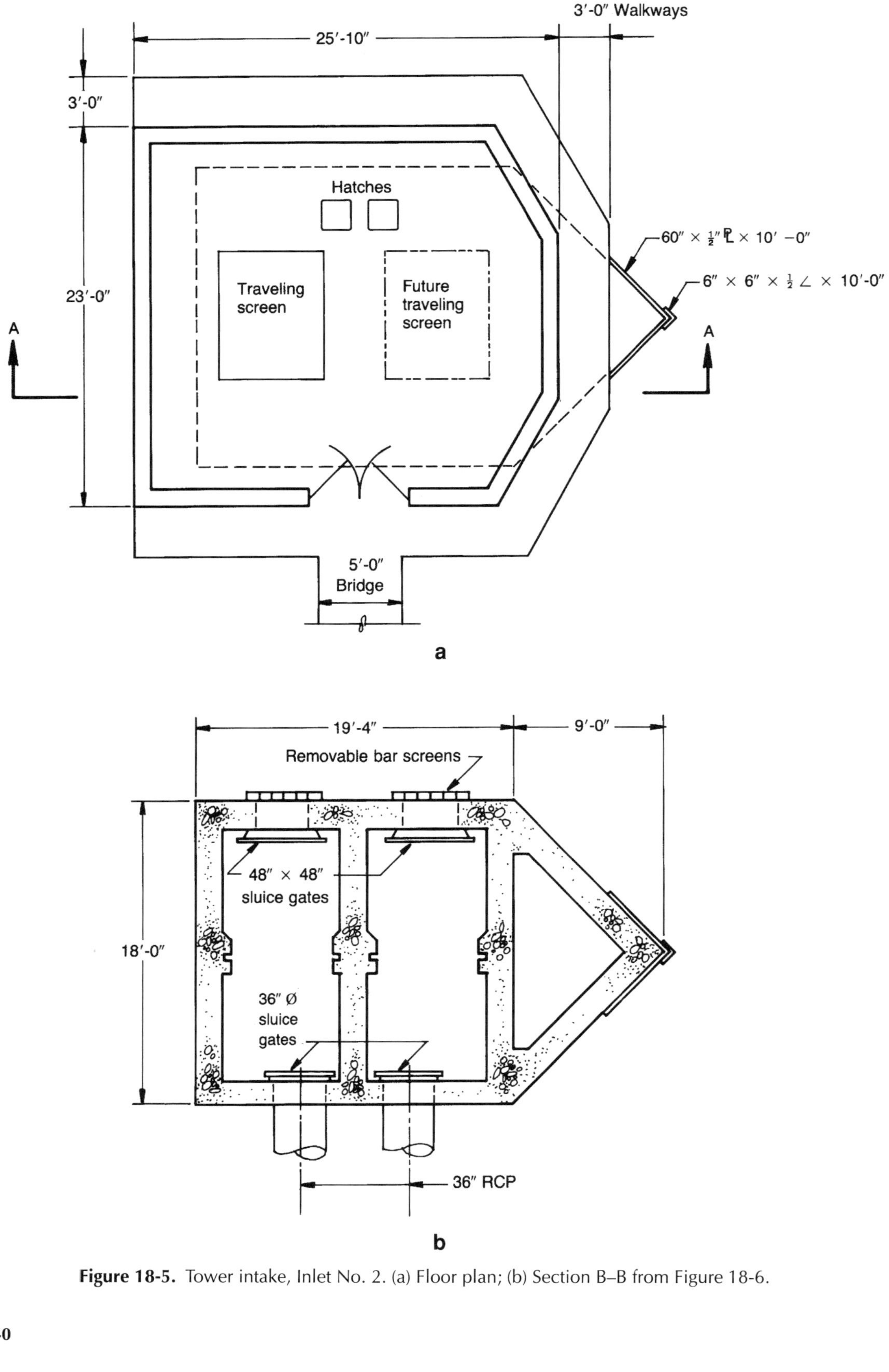

Figure 18-5. Tower intake, Inlet No. 2. (a) Floor plan; (b) Section B–B from Figure 18-6.

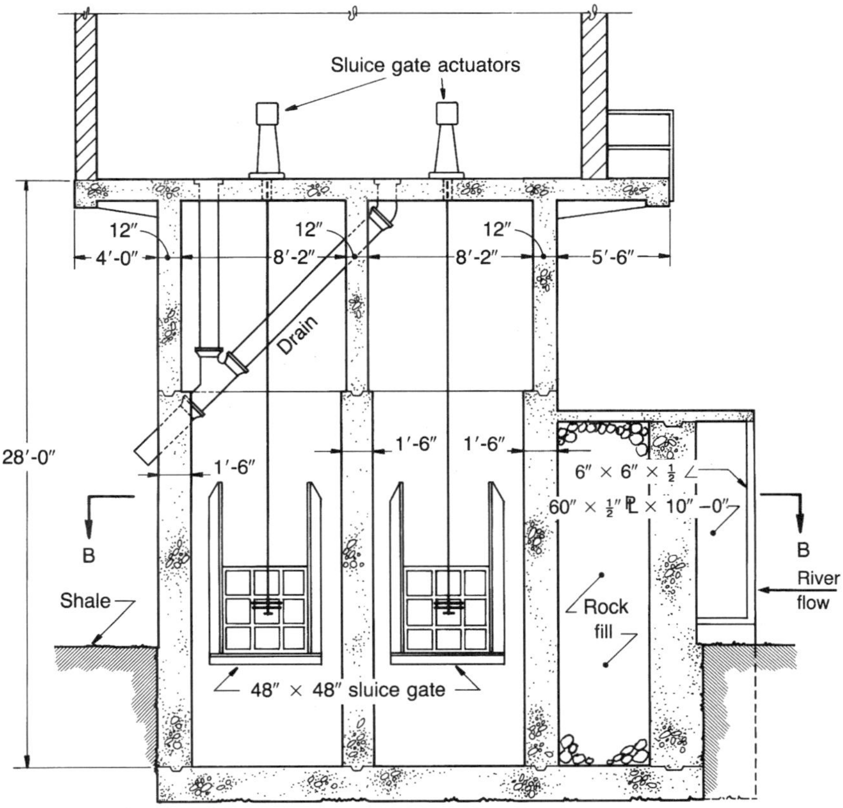

Figure 18-6. Section A–A from Figure 18-5.

Inlet No. 1 is used most of the time, partly because it is trouble free, but—more important—because it contains the L-structure, which is used as part of the treatment process.

Inlet No. 2. Inlet No. 2 is used for emergency service when Inlet No. 1 is closed for repairs; it is also used during periods of very low flow when the water can be treated by direct filtration. Frazil ice causes serious problems at times. Because it clogs the steel screens, which must be cleaned by hand (an onerous task), Inlet No. 1 is used (if possible) whenever frazil ice is forming.

Pack ice causes ice jams and, at times, a pileup of ice at both the intake tower and the walkway. Ice jams have formed downstream, and they have caused the water level to rise 11 ft in 25 min. Water has, at times, flowed over the bridge. Large cottonwood trees are sometimes caught either on the intermediate piers or on the walkway and can damage railings. To remove the trees, a cable must be attached and they must be winched upstream where the branches can be sawed off. Workers are reminded of the hazards of this duty by the shuddering of the bridge in such circumstances. A section of the bridge was once carried away. The walkway and one intermediate pier have since been repaired, and the nose of the intake tower has been sheathed with 12-mm (1/2-in.) steel plate.

Object lessons for designers. Designers of river works would do well to be guided by thoughtful contemplation of the three decades of operating experiences described above. Some conclusions are as follows:

- Side inlets are generally satisfactory if the problem of low water can be circumvented. Perhaps training dykes or a low weir would help to supply water during low flows.
- Any inlet may have to be closed for repairs. Either design so that an inlet can be partially closed or provide for an alternate emergency inlet.

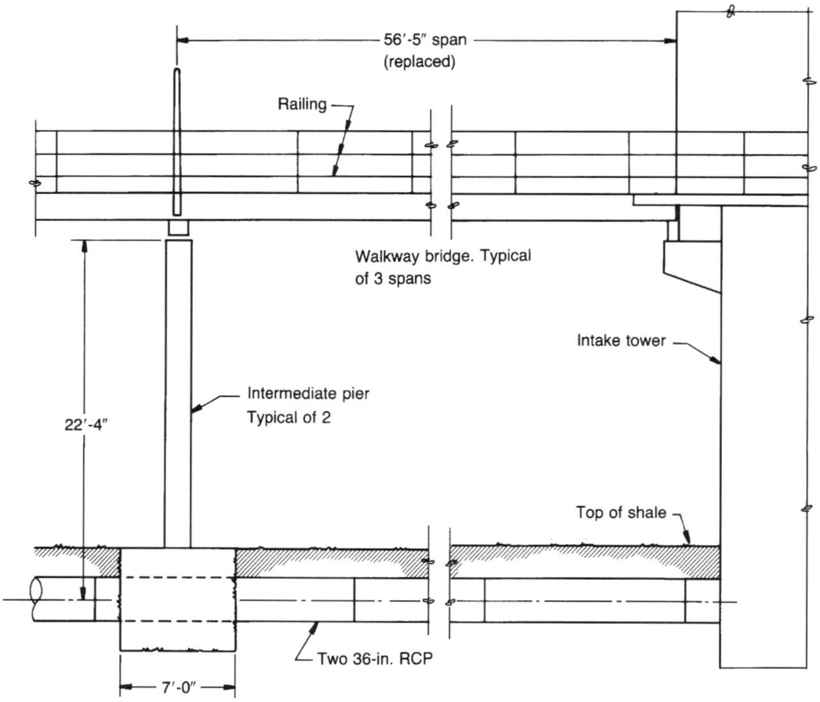

a

b

Figure 18-7. Bridge, suction pipes, and pump at Inlet No. 2, Billings Water Treatment Plant. (a) Bridge and pump suction pipes. (b) One of the four horizontal split-case pumps. Photograph by R. L. Sanks.

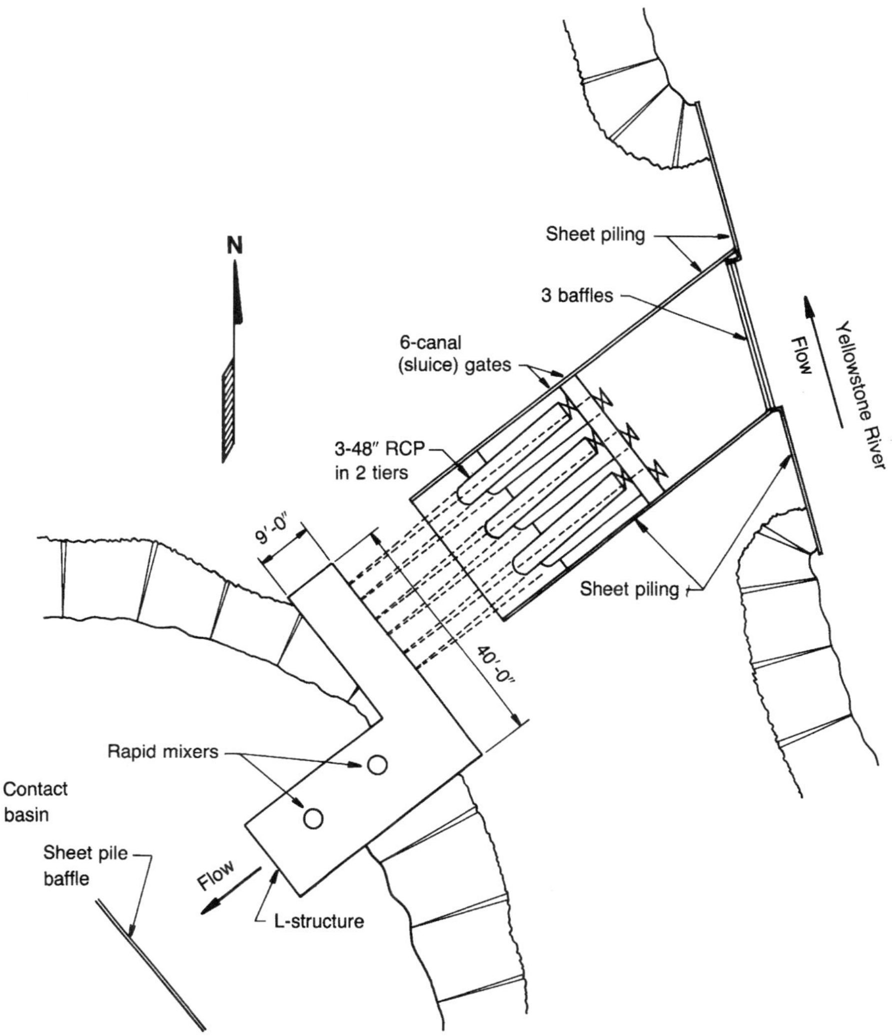

Figure 18-8. Inlet No. 1 at Billings Water Treatment Plant.

- Make a very thorough study of the river and its seasonal variations before beginning to design the river works. Studying records of low and high flow is not enough; frazil ice, pack ice, trees, debris, and fish can create serious problems.
- If pack ice or trees can be carried by the river, search for the highest ice jam that has occurred. Interview local residents and use ice scars on tree trunks for data. Set floor elevations of towers and walkways high enough to exceed the previous records, and then add another 3 m (10 ft) of elevation for additional safety.
- Consider designs that tend to keep pack ice away from the mouth of the intake. Ice tends to pile up on protruding objects in a river.
- Avoid intermediate piers in walkways. A single span for such light loads can easily bridge 60 m (200 ft).
- Sheath the nose of any towers or piers with steel up to the top of the highest ice jam that can occur.
- With reference to Figures 18-5 and 18-6, some engineers would prefer (1) thicker walls—particularly the nose walls, (2) anchoring the nose wall to the underlying shale with heavy reinforcing steel bars (or even railroad rails) grouted into deep [e.g., 6 m (20 ft)] holes to prevent overturning, and (3) using flanges or grooved-end couplings (for longitudinal restraint and to

allow for movement due to temperature) instead of bell and spigot joints in any pipe (such as the sloping floor drain in Figure 18-6) that is not buried.

- For gratings and screens, consider the use of solid plastics or plastic coatings on metal or other measures to combat frazil ice. At one time, excess steam from a nearby power plant was considered for keeping frazil ice off the screens, but the heating requirements were enormous. Locating the intake in a quiescent water zone and keeping the intake velocities very low is a good defense.
- Note that pack ice is unpredictable both in its ability to form ice jams and in the forces it imposes on structures. Because failures of bridge piers elsewhere in the United States have occurred with loss of life, river works that are subject to extraordinary forces of nature should be designed very conservatively indeed. Calculations are meaningless when the forces cannot be accurately quantified, so follow the old engineering adage to "build it Hell for stout."

18-4. Raw Water Pumping from Aqueducts

Functionally, low-service water pumping to a treatment plant from a raw water aqueduct is somewhat similar to booster pumping. However, the booster pump in an in-line pumping station must be coordinated with the supply pumps—not an easy problem. But with raw water pumping, only a portion of the flow is taken and the problems of coordination are less severe or even absent. Furthermore, the pump suction head in pumping from a canal or aqueduct does not usually fluctuate very much, whereas the suction head from a reservoir, for example, might vary by 30 m (100 ft) or more.

The design of a pumping station taking suction from an existing raw water aqueduct is shown in Example 18-3.

Example 18-3
Raw Water Pumping from an Aqueduct

The project, located in the north San Francisco Bay area in California, was needed to supply 5 Mgal/d of raw river water to an existing 5 Mgal/d municipal water treatment plant (operated 24 h/day) in which the process train consists of chemical mixing, flocculation, sedimentation, filtration, and chlorination. There are four filter beds, each with a sustained capacity of 875 gal/min, although the use of two beds (at 1750 gal/min) is average. The treated water enters a small clearwell with a capacity of 150,000 gal, and from there it is pumped to a series of distribution storage reservoirs on the basis of demand.

The raw water is high in turbidity, particularly during the spring and early summer period of maximum river flow. The treatment plant was previously supplied from the same raw water source via a 40-yr-old pressure pipeline 36-in. in diameter that passes by the treatment plant site (see Figure 18-9). In the new system, the source of water is a surge-controlling standpipe (see Figure 7-4 in Section 7-1) 150 ft in diameter and 22 ft high on a new 72-in. state aqueduct. A 14-in. isolation valve and a flow-metering station to supply the treatment plant is provided as a part of the state aqueduct system. The ground surface profile along the supply pipeline is shown in Figure 18-10.

The long-term future demand was not expected to exceed the previous plant capacity of 5 Mgal/d, but the treatment plant was then 35 yr old (in 1987) and due for rehabilitation, and the treatment process needed to be upgraded to accommodate revised U.S. EPA water quality standards.

Problem: Design a system to supply the municipal treatment plant with water from the standpipe on the 72-in. state aqueduct. The municipality insists on variable output at any flow from 875 to 3500 gal/min.

Solution:

Design decisions. The following are the major design decisions that were required:

- Pumping station location
- New pipeline requirements and size
- Type, number, and capacity of the pumps

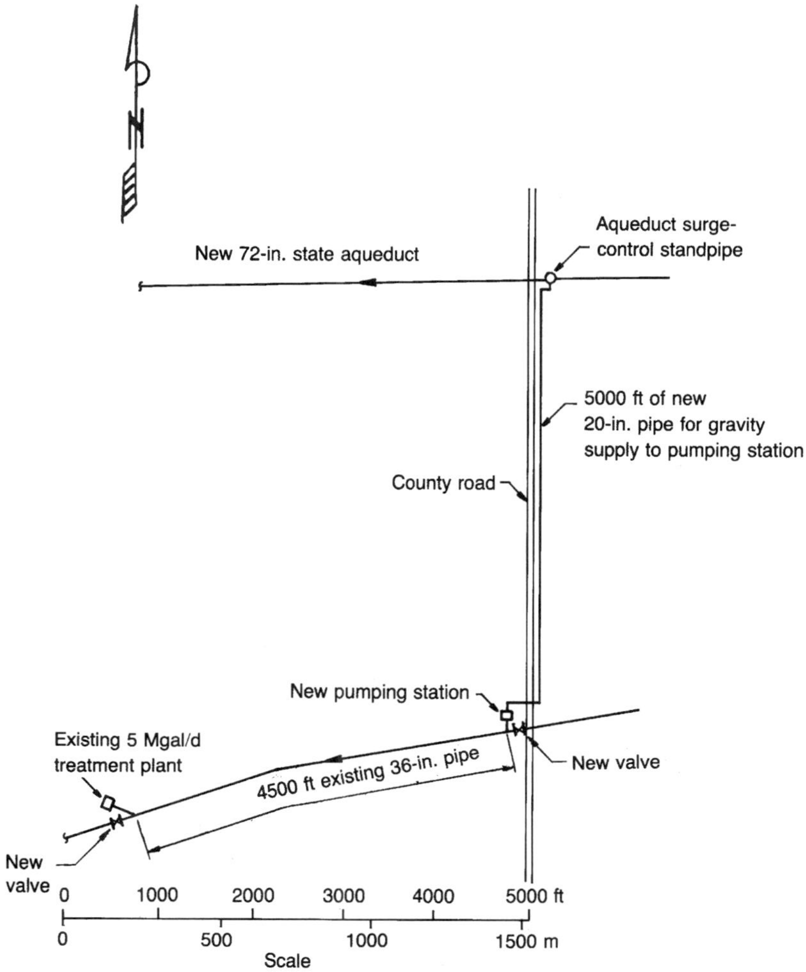

Figure 18-9. Site plan, Example 18-3.

- Type of driver
- Type of pumping station structure
- Pump-control system.

Pumping station location and pipeline route. The location considerations included:

- Gravity supply from the surge tank
- Site availability
- Pipeline right-of-way
- Access
- Power supply
- Environmental factors
- Security
- Minimum new pipeline construction.

These considerations led to the selection of the site shown in Figure 18-10. Any location between the standpipe and the selected site would have been satisfactory except for the availability of land. The alignment paralleling the county road fell within an existing pipeline easement and also permitted the use of the existing 36-in. pipeline to deliver the pumping station discharge to the treatment plant.

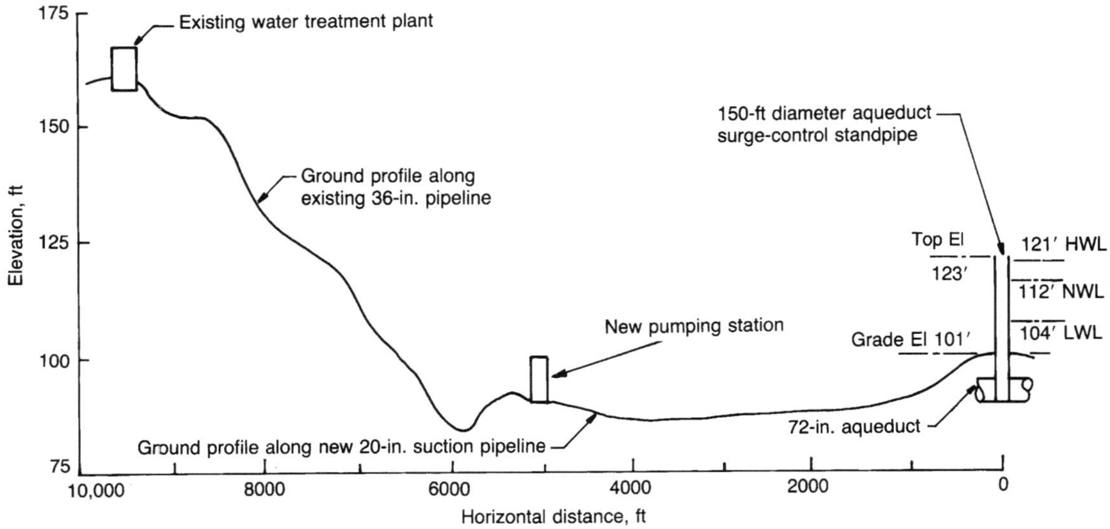

Figure 18-10. Profile of pipelines. For the existing water treatment plant, the influent channel HWL = 160 ft. For the new pumping station, floor elevation = 89 ft, grade elevation = 88 ft, and pump suction elevation = 76.5 ft.

A direct route from the standpipe to the treatment plant was discarded for several reasons:

- A greater length of new pipeline construction.
- Right-of-way acquisition through multiple property ownerships.
- To avoid environmental impact.

New pipeline. With the selected pumping station location, the new discharge piping is limited to the pump manifold and the connection to the existing 36-in. pipeline. Upstream and downstream valves are needed to isolate this portion of the 36-in. pipeline.

A present-worth analysis was performed on a range of pipe sizes for the (approximately) 5000 ft of suction pipeline from the 14-in. standpipe connection to the pumping station. Based on the installed cost of ductile iron pipe (the owner's preference), the power cost for average flow conditions at \$0.065/kW · h per 25-yr life expectancy, and interest at 8.0%, a pipeline 20 in. in diameter was determined to be the most economical size.

System curve. Pipeline friction losses were calculated by first determining the equivalent length of each size of pipe in the system to take into account system entrance and exit losses and losses due to fittings (elbows, tees, increasers, and reducers), flowmeters, and valves. With equivalent pipe lengths tabulated, system friction losses were calculated by a "pipe loss" computer program using the Darcy–Weisbach formula with the following relative roughnesses, ε/D:

Pipe	ε/D
Ductile iron (12, 14, and 20 in.)	0.003-in. (cement lining)
Steel (10 and 14 in.)	0.0005-in. (asphalt coating)
Old cement-lined steel (36 in.)	0.008-in. (bad shape)

The results of calculations for maximum static lift and total dynamic head (TDH) are tabulated below, and the system H-Q curve is shown in Figure 18-11.

Flow (gal/min)	Pumps operating	Friction loss (ft)[a]	Static lift (ft)	TDH (ft)
3500	2	48	56	104
1750	1	12	56	68
875	1	3	56	60

[a] For 5000 linear ft (LF) of new 20-in. pipe, 4500 LF of old 36-in. pipe, and all valves, meters, and fittings.

Pump suction conditions at maximum flow are as follows:

Minimum standpipe water level	+104 ft
Suction piping loss at maximum flow	−17 ft
Pump suction elevation	
(centerline of connection to pump barrel)	−76.5 ft
	Subtotal +11.5 ft
Barometric pressure head	+33.9 ft
Vapor pressure head at 70°F	−0.8 ft
	Subtotal +44.6 ft
Safety factor at 10%[a]	−4.5 ft[a]
$NPSH_A$ (worst condition)[a]	Total +40.1 ft

[a] However, readers are warned that a safety factor of less than 5 ft or less than 1.35 × $NPSH_R$ may be risky (see Section 10-4 and Figure 10-13).

Pump and driver selection. The most suitable type of pump for this application was considered to be a vertical turbine pump with the bowls encased in a vertical "can" or suction barrel with a flanged connection to the suction pipeline manifold. One alternative would be a horizontal

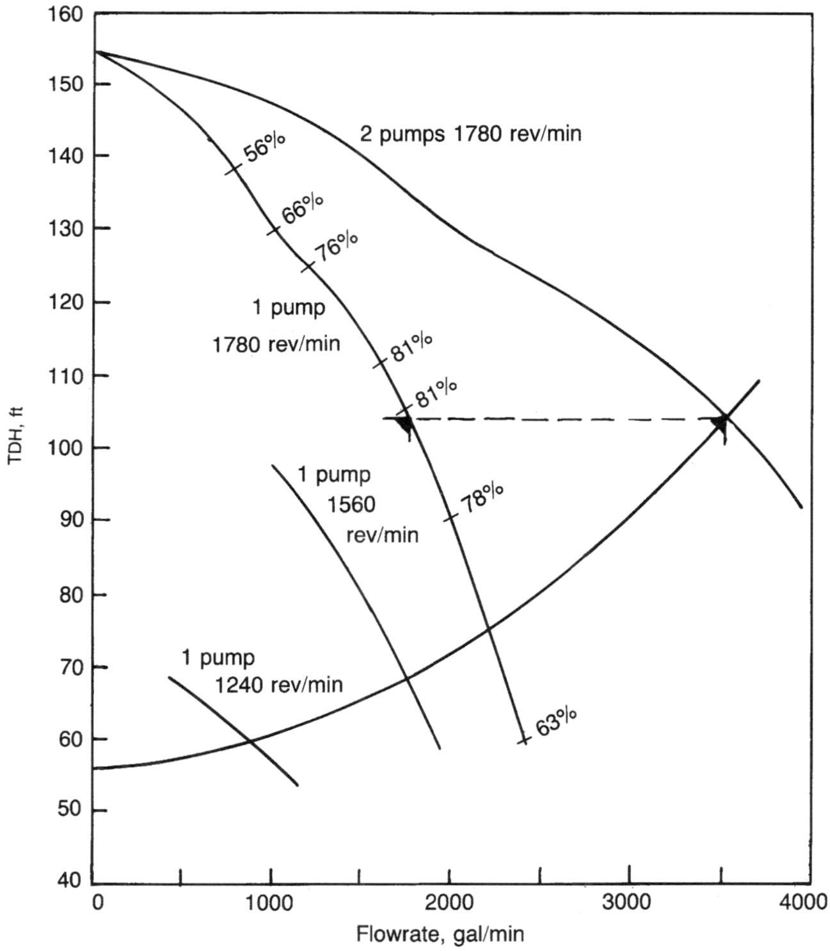

Figure 18-11. System head-capacity and pump characteristic curves.

split-case centrifugal pump, but the vertical turbine pump is lower in cost, occupies less floor space (thus reducing the cost of the building), and (with the concept of using multiple bowls) is more flexible.

The logical number of pumps to satisfy maximum and average flow requirements would be two units with a capacity of 1750 gal/min each to deliver the maximum flow of 3500 gal/min at a TDH of 104 ft. One pump operating alone would deliver more than 1750 gal/min due to the reduction in pipeline friction loss. If constant-speed pumps were used, the flowrate controller at the treatment plant could compensate for this difference, but to supply the minimum flow requirement of 875 gal/min would require either a third, smaller pump sized for this condition or excessive throttling (by the flowrate controller). In this application, however, the owner specified pump drives for operation at any flow between 875 and 3500 gal/min; hence, variable-speed (V/S) drives are required. The V/S drives permit the use of two identical duty pumps of 1750 gal/min capacity. A third unit is used for standby service.

Adjustable-frequency drives (AFDs) were selected because of their higher efficiency in comparison with eddy-current couplings or fluid-type V/S drives (but see Figure 15-24, Section 15-11, Table 15-3).

The pumps selected were two-stage, barrel-type vertical turbines to discharge 1750 gal/min at a head of 104 ft with 7.6-in. impellers at 1770 rev/min. The minimum permissible flow per pump is 600 gal/min. The pump performance curve is shown in Figure 18-11. Curves for reduced speeds were obtained by using the affinity laws (see Section 10-3), which are quite accurate for these pumps.

The power output of the pump is found from Equation 10-6b.

$$hp = \frac{qH}{3960} = \frac{1750 \text{ gal/min} \times 104 \text{ ft}}{3960} = 46$$

As noted in Table 15-3, the efficiency of AF converters is about 96%, the motor efficiency is also about 96%, but extra motor losses are about 3% and transformer losses are about 2%, so the net AFD efficiency is about 88%. Because the pump has an efficiency of 80.5%, the total motor output must be at least

$$hp = \frac{46}{0.88 \times 0.805} = 65$$

A 75-hp motor would be needed.

Pumping station structure. Although the pumping station is located in a temperate climate suitable for an outdoor installation, the owner elected an indoor installation to reduce and facilitate maintenance, to protect the equipment better, and to improve security. The foundation conditions at the site were favorable. The building is of concrete block construction with a pier-supported beam-and-girder concrete foundation and concrete floor and roof. Eight poured-in-place piers 2 ft in diameter by 20 ft deep were provided, one at each building corner and two along each side wall. Inside dimensions are 32 ft by 18 ft in plan and an interior height of 9 to 9.5 ft. Removable 6-ft-square roof hatches are provided over each pump to facilitate installation and removal. In addition to the pumping units, the station houses the electrical switchgear and control equipment, including the AFDs.

Ventilation is provided by air inlet screens and three 3500 ft^3/min wall-mounted exhaust fans. Building heating is not required. Pump suction piping connects to the pump barrels 12.5 ft below the floor level. The above-floor pump discharge passes through a slow-closing check valve and shut-off valve before turning down through the floor with a 90° bend for below-grade manifolding and connection to the existing 36-in. pipeline. The plan and a section are shown in Figures 18-12 and 18-13.

Pump-control system. The treatment plant is staffed 24 h/d, but the pumping station is visited only periodically for inspection and maintenance. The control system is therefore designed for remote control from the treatment plant. The treatment plant operator can select which pumps to operate and can set the desired flowrate. A "local/remote" switch is provided at the AFD cabinet at the pumping station. "Local" operation at the pump house is used only for maintenance and testing.

A multiplexing unit between the pumping station and the filter plant control room links the input–output signals over telephone lines for remote start–stop and analog display and control of the pumps. Analog signals for pressure (at each pump suction and at a common point on the pump discharge), level, flow, and speed are multiplexed for remote indication at the filter plant control room panel. Pressure gauges are also mounted on each pump discharge for local observation. At the control room panel, a three-position switch selects the pump group (1–2, 2–3, 1–3), and another switch selects the lag-lead pump. If pump No. 1 is selected as the lead pump and group 1–2 is selected to run, then pump No. 3 is automatically on standby. A start button will start pump No. 1 as the lead pump.

A flow input signal compares the flow set point against the actual flow and adjusts the pump speed until maximum flow is reached. If the flow requirement is greater than the flow output of pump No. 1, pump No. 2 will start automatically and share the load with pump No. 1 until flow requirements have been fulfilled. A reverse sequence occurs when flow requirements decrease. Overriding emergency controls to shut off the operating pumps automatically are

- A high water level in the treatment plant influent channel
- Low flows (which indicate possible pump failure)
- Low pump suction pressure and high pump discharge pressure.

A schematic process and instrument diagram (P&ID) of the system is shown in Figure 18-14.

Critique. In most circumstances, cement-mortar lining is superior to asphalt coatings and should always be considered (see Sections 3-2, 3-3, and 4-1). Some waters are aggressive toward cement mortar, so asphalt and other coatings (Table 4-8) are still used.

The use of variable-speed pumps in this example would normally be unjustifiable. Flows close to 100, 50, and 25% of a given flowrate can be achieved by driving the pumps selected (1750 gal/min at 104 ft TDH) at constant speed or by using constant-speed pumps of two sizes. The operating time per shift at the treatment plant could be adjusted to permit the plant to produce either 3500 gal/min with four filters in use or 2250 gal/min with three filters in use. Thus, the filters would always operate within 15% of rated capacity. If the owner requires V/S

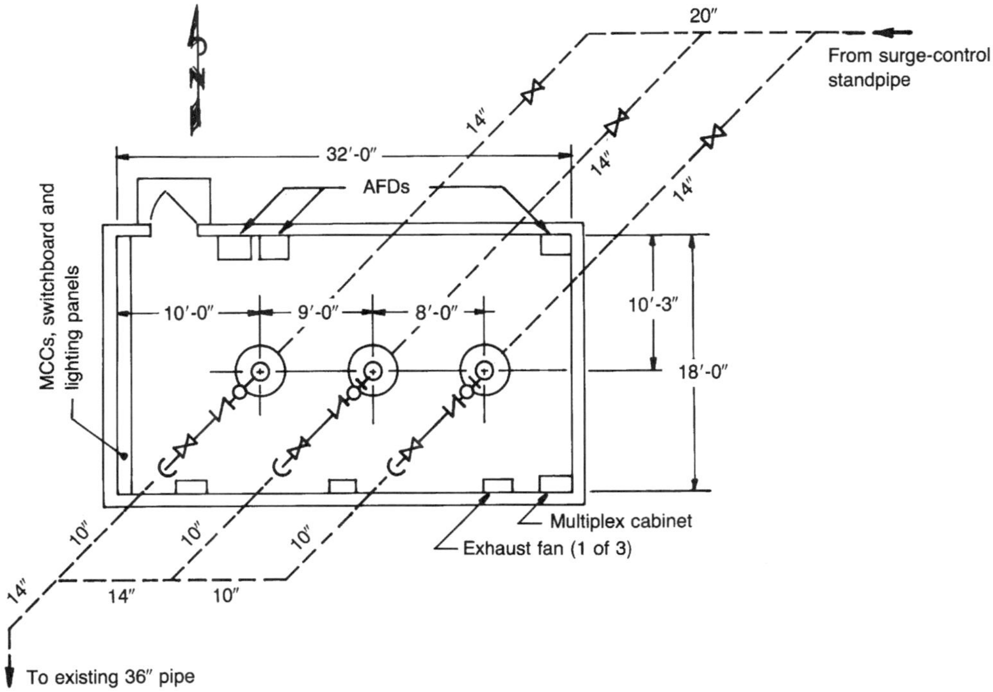

Figure 18-12. Plan of the pumping station.

drives despite their disadvantages (see Section 15-1), the engineer in a practical world has no choice.

The use of a slip drive (such as an eddy-current coupling) instead of an AFD may be worth an extensive investigation. In Example 29-1 of the first edition of this book, the eddy-current coupling was slightly more cost effective than an AFD despite the greater efficiency of the AFD. Comparisons depend on the effect of a frequency analysis of expected flowrates on the cost of energy; they also depend on capital cost (eddy-current couplings are becoming more expensive whereas AFDs are becoming comparatively cheaper), replacement cost of short-lived components, and considerations of reliability and maintenance.

Fittings in the discharge header are crowded. The pressure gauge location (so close to the discharge elbow) may induce some inaccuracy and fluctuation in the gauge readings. The combina-

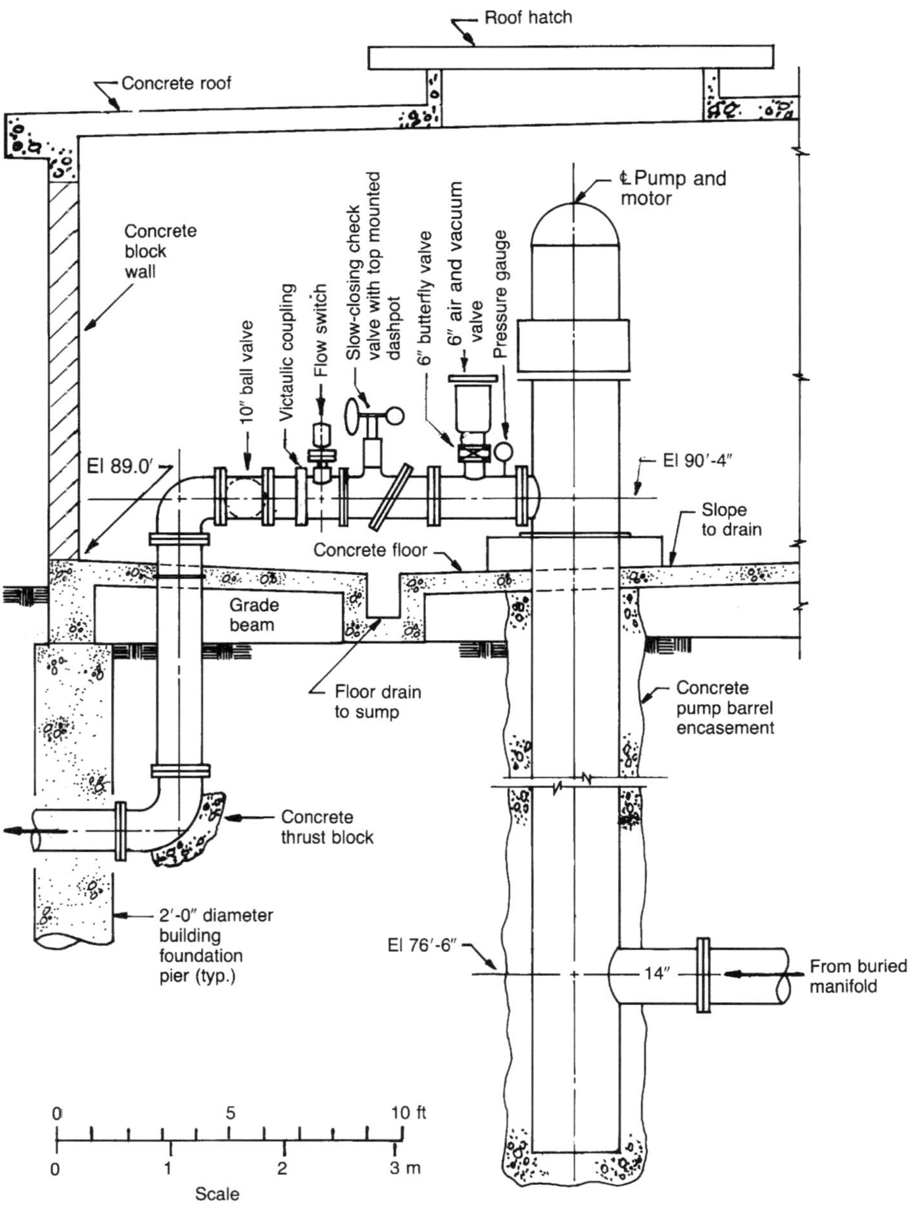

Figure 18-13. Section through the pump (typical of all three pumps).

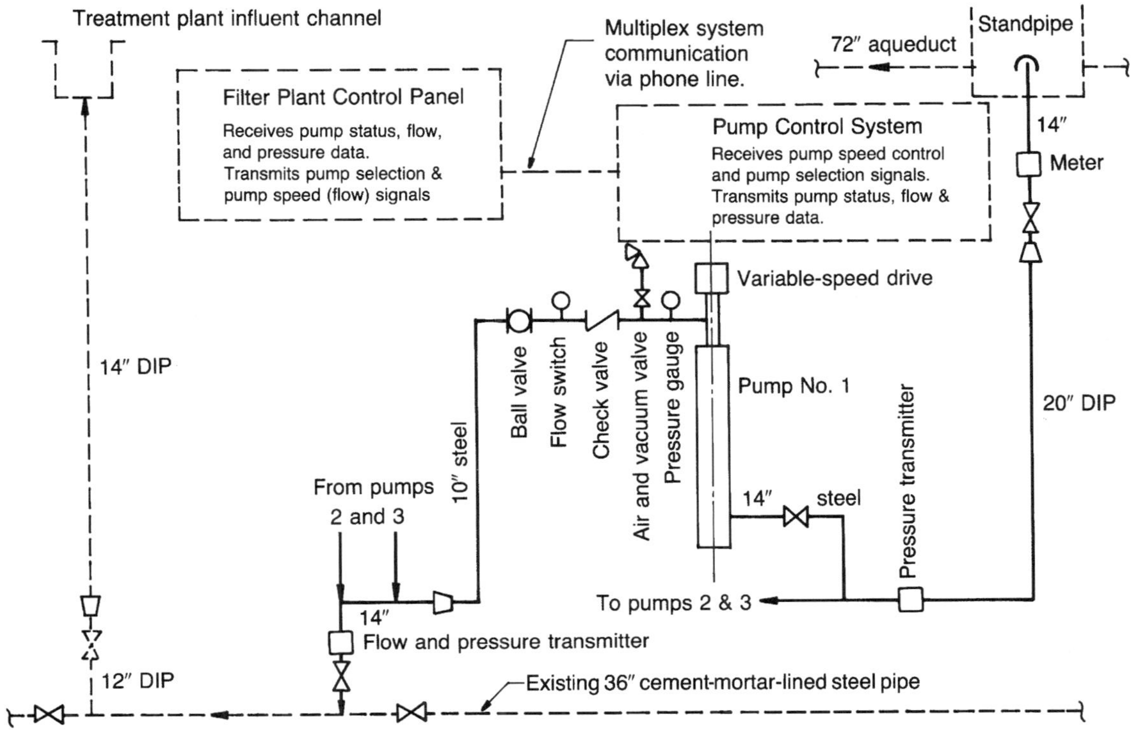

Figure 18-14. Piping and instrumentation diagram for Example 18-3.

tion of one grooved-end coupling and a slanting flange makes dismantling easy; another solution would be a grooved-end coupling on the discharge elbow and another in a horizontal plane just above the floor at the down-turned elbow. A pair of sleeve couplings (not shown on the drawings) separated by two or three pipe diameters is normally needed outside of (but adjacent to) the building to allow for differential settlement. As this building is set on piers, the sleeve couplings were not needed. Refer to Example 12-2 for a comparison of design approaches.

18-5. Well Pumps with Elevated Tanks

The most common well pumps are vertical turbines. High service pumps connected to a clearwell are usually of horizontal split-case, vertical split-case end-suction, or vertical turbine barrel design (see Chapter 11). Selection criteria and a discussion of the advantages and disadvantages of vertical turbine pumps are presented in Section 25-6.

As discussed in Chapter 4, cement-mortar-lined ductile iron or steel pipe is most frequently used. Isolation valves are usually gate or butterfly valves (see Chapter 5). Check valves should be either the swing type with outside lever and spring or outside lever and counterweight type with an oil-filled dashpot. The slanting disc type with a top-mounted dashpot and/or bottom buffer can also be used (see Chapter 5). If very quick closure is required, the center post "silent" check valve is a good choice. If valve slam or water hammer (or both) are likely to occur with a check valve (see Chapters 5, 6, and 7), a pump-control valve should be used instead. Pump-control valves are usually used if the piping is larger than 300 mm (12 in.). The purpose of these valves is to control surges in the discharge piping that occur when the pump starts and stops. The pump motor control center (MCC) is interconnected with the pump-control valve so the pump starts and stops against a closed valve. In discharge manifolds up to about 16 in. in size, these valves are frequently of the diaphragm-actuated or piston-actuated globe type. In larger installations, ball or cone valves equipped with motorized, pneumatic, or hydraulic actuators are frequently used.

Climate Considerations

Heat, cold, wind, dust, precipitation, and moisture are natural elements that must be considered in the design of any waterworks facility. Weatherproof enclosures for electrical equipment and devices and well-maintained paint, coatings, and linings for pumping units, piping, valves, and appurtenances may be adequate in moderate climatic conditions. More severe climatic conditions require additional protective measures. A good starting point is a field inspection of existing facilities in the proposed project area in which the type and adequacy of protection are being investigated.

High temperature is primarily a concern for electrical equipment such as switchgear, motor starters, and instrument control panels. Temperatures exceeding 30°C (85°F) tend to reduce equipment life and affect the accuracy of instruments. Increasing degrees of protection of equipment from excessive heat can be provided as follows:

- Provide sun shade
- Enclose in a vented building with an exhaust fan
- Enclose in an air-conditioned building.

Temperatures below freezing are of great concern. Short duration temperatures slightly below 0°C (32°F) require that small-diameter piping, valves, and water-holding appurtenances be protected against freezing. In colder climates, *all* of the piping, valves, and appurtenances must be protected against freezing and, in addition, structures and piping must be protected from damage due to frost heave. Piping, valves, and appurtenances that need not be accessible may be protected from both freezing and the effects of frost heave by being buried below the frost line. Piping, valves, and appurtenances that must be accessible may be given increasing degrees of protection from freezing as follows:

- Wrap with electrical "heat strips" and insulating materials (called "heat tracing"). Some engineers report poor experience with heat tracing and find that it usually lasts 3 yr or less.
- Enclose in a vault with or without insulating materials or space heaters.
- Enclose in a heated building.

Windblown sand and dust particles can damage painted or coated surfaces, bind moving parts, infiltrate electrical equipment cabinets and enclosures, and increase the maintenance. Electrical equipment must be maintained in a dust-free condition to provide quiet, reliable service. Enclosing station equipment and piping in an air-conditioned building offers the best protection from windblown sand and dust particles. Protection against precipitation and moisture is provided by

- Properly specified, applied, and maintained paints and coatings
- The use of weatherproof enclosures for exposed electrical equipment and devices
- The use of space heaters in vaults or motor enclosures
- Enclosure in a climate-controlled building.

Using buildings to protect pumping station equipment from the outside environment requires careful consideration of the resultant inside environment:

- Enclosing electrical equipment in a vented building provides a degree of protection from heat but not from dust.
- Enclosing electrical equipment in an air-conditioned building provides heat and dust control and a more pleasant environment for service or operating personnel.
- Enclosing pumping units in a heated building to protect them from winter cold requires proper ventilation and, perhaps, cooling to prevent temperature buildup in the summer due to motor or engine heat.
- Variable-speed motor controllers produce significant amounts of heat that must be properly vented if enclosed in a climate-controlled building; generally, most switchgear is rated for a 40°C ambient temperature, and it is difficult to obtain switchgear of higher ratings.

Wells

Location

The water supply of many (perhaps most) small communities comes from wells. Hydrogeologists, well drillers, and engineers share the responsibility of locating, testing, drilling, and casing wells. An engineer usually selects the pumps, designs the system, and assumes overall responsibility for the project. The location of wells depends on many factors, including

- The distance from other wells and the possibility of mutual interference of drawdown surfaces
- The distance from possible pollution sources, such as 30 m (100 ft) from a known septic tank or downstream from a chemical or solid waste dump site
- The land available for purchase or lease
- The location of a good groundwater aquifer; information can be obtained from local well drillers, previous well-drilling reports, health departments, or geological surveys.

Water Quality

A test well is normally drilled to determine the sustained yield, drawdown distances, and water quality. As the well is drilled, a soil log is kept to record soil type versus depth. This information, together with an electric well log, is used to determine the best water-bearing soil strata. The electric well log (a measure of the soil's resistivity) can be used by an experienced technician to determine at what soil strata levels a well screen should be positioned to provide the desired yield. In several states, such records must be submitted as part of the permit process for wells. A chemical analysis of the water in each aquifer should be made to select those aquifers with the highest water quality commensurate with suitable yield. Unwanted aquifers can be sealed.

Water Treatment

Well waters usually have a high content of dissolved solids and are usually "hard." Hard waters can be softened (and often demineralized to some extent) by lime-soda treatment, or they can be either softened or demineralized by such means as ion exchange or reverse osmosis [7]. All of these treatments are expensive. Well waters for potable use should be chlorinated even if the water from the aquifer contains no bacteria. Because both wells and distribution systems can become contaminated, residual chlorine is needed to protect the public health. Unless the well water is clean and free from sand and silt, a settling basin (or even a filter) is required for a municipal well supply. Many wells contain gases (such as carbon dioxide, methane, and/or hydrogen sulfide) that can be removed by aeration or, more completely, by vacuum. Iron and manganese can also be removed by aeration, lime-soda softening, or chlorination followed by filtration, or they can be removed by ion exchange.

Casing

Wells are always cased—usually with steel. The casing can be placed against the soil, or the bore hole can be enlarged to permit the placement of a layer of gravel, usually at least 150 mm (6 in.) thick, between the soil and casing. The "gravel pack" increases the yield of the well and reduces the velocity of inflow at the face of the soil, which thereby reduces the inflow of sand and silt. Samples of water and soil from the test hole can be used to design the gravel pack for maximum effectiveness in preventing the entrance of particulate In the aquifer, the casing is either perforated at closely spaced intervals by a special tool to allow water to enter or the casing is interrupted by well screens.

Development

The well is "developed" by pumping at a high rate (considerably higher than normal pumping rates) to wash fine particles out of the aquifer so that it can behave like a large, underloaded sand filter and, thus, produce clear water. Some wells produce sparkling, clear water; others never stop producing silty water. During a well's development, nearby wells or bore holes can be tested to establish the drawdown curve and to aid the hydrogeologist in predicting the safe yield. Walton [8] describes groundwater tests in detail, and his text is accompanied by a diskette containing test programs in BASIC.

Well Head

The design of the valving system at the well head is critical. When the pump starts after a period of rest, the water (at water table level) flows at high speed to the well head because the TDH is initially zero. If the speed is not reduced by the cushion of air in the well, the pipe or well casing may break and pumps may be displaced. The air must be allowed to escape slowly enough so that the moving column of water in the well strikes both the closed check (or pump-control) valve and the stationary column of water in the transmission main too gently to cause an undue pressure surge. If the first surge of water from the aquifer contains sand and silt, the water is usually wasted. Valving arrangements for well heads are shown in Figures 7-9, 7-10, and 7-11 in Section 7-6, as well as in Example 18-4.

Example 18-4
Design of a Deep Well Pumping Station

Problem: Maxwell (population 1800) needs a second, reliable water source at 500 gal/min minimum. The well location selected is 2000 ft from an existing 100,000 gal storage tank with a low water level 107 ft and a high water level 127 ft above the ground, which is level between the well and the tank. Based on the results of well tests, well screens were placed between depths of 595 to 605 ft and 720 to 724 ft. The drawdown curve is shown in Figure 18-15.

The problem is (1) to plot the system H-Q curve, (2) to select the pump, (3) to select the driver, (4) to design the pump layout at the well head, and (5) to plan the chlorination facilities. The transmission main has been selected by others.

Solution: As outlined in Chapter 17, a well-documented record of all verbal communications, letters, calculations, and equipment selections should be maintained.

(1) Transmission main. Ordinarily, comparative analyses of initial cost plus energy costs over a 20-yr (or longer) period would be made for two or three pipe diameters to select the optimum. The well, however, is the only water source on one side of a railroad, and because that area is growing the owner wants the largest economically feasible transmission main, which is an 8-in. diameter pipe. The profile of the pipe is shown in Figure 18-16. Data for TDH calculations at three points are as follows:

Item	Flow (gal/min)		
	300	500	700
Flow (ft^3/s)	0.67	1.11	1.56
Velocity [ft/s (8-in. ID)]	1.92	3.18	4.47
Velocity head ($v^2/2g$)	0.057	0.157	0.310
Dynamic losses, maximum			
287-ft lined steel pipe, C =140[a]	0.51	1.27	2.37
2000-ft PVC pipe, C = 150	3.03	7.79	14.51
TDH			
Maximum pumping head:			
Tank water level (above grade)	127	127	127
Well water level (below grade)	78	108	137
287-ft lined steel pipe, C = 140	0.5	1.3	2.4
2000-ft PVC pipe, C = 150	3.0	7.8	14.5
Minor losses: inlet, gate valve, check valve, 10 elbows, outlet, K = 7.3[b]	0.4	1.1	2.3
TDH	209	245	283
Minimum pumping head:			
Tank water level (above grade)	107	107	107
Well water level (below grade)	60	83	109
287-ft lined steel pipe, C = 140	0.5	1.3	2.4
2000-ft PVC pipe, C = 150	3.0	7.8	14.5
Minor losses: inlet, gate valve, check valve, 10 elbows, outlet, K = 5.2[b]	0.3	0.8	1.6
TDH	171	200	235

[a] Equation 3-9b.

[b] From Tables B-6 and B-7: K_{ent} (through screen) = 1; K_{gate} = 0.1 to 0.2; K_{check} = 0.6 to 2.6; K_{elbow} = 0.25; K_{exit} = 1. Equation 3-16: $h = Kv^2/2g$.

Plot the system curve. The system H-Q curve is a band as shown in Figure 18-17. Hence, the pump can operate anywhere along the pump curve between points A and B.

(2) Select pump. Consult a manufacturer's catalog for a vertical turbine pump that will discharge at least 560 gal/min at a TDH of 255 ft and 630 gal/min at a TDH of 220 ft (from the system curve in Figure 18-17). Make a tentative selection from a pump catalog "coverage chart." At six bowls, the head per stage for a candidate pump would be about 43 ft, which is reasonable. The head for a single stage of the turbine pump is multiplied by 6 to obtain the total pump H-Q curve, as shown in Figure 18-17. Refinements in the calculation of total head or energy due to water density, derating for the number of stages, whether bowls are enameled, impeller material, and so on can be found in the manufacturer's catalog. The pump operates between points A (560 gal/min) and B (630 gal/min).

The $NPSH_R$ at 630 gal/min is 16 ft from the manufacturer's catalog. The $NPSH_A$ is calculated as follows:

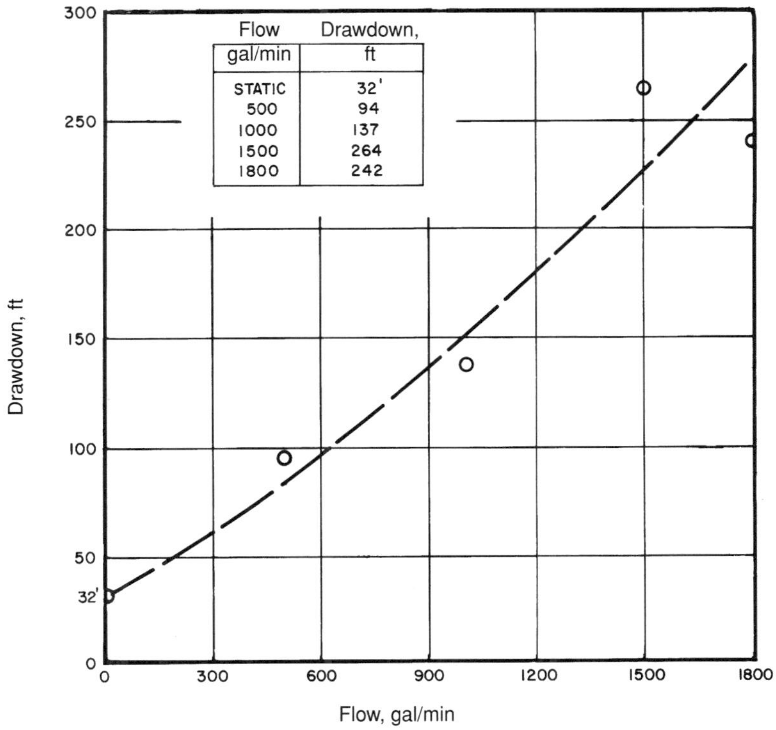

Figure 18-15. Pump drawdown curve for the well. Courtesy of Psomas and Associates and Raymond Vail and Associates.

- Lowest bowl elevation = 150 ft below grade
- Drawdown at 630 gal/min = 126 ft below grade (worst case)
- Net submergence = +24 ft
- Atmospheric pressure head = +33.9
- Vapor pressure head (60°F) = − 0.6
- Entrance loss = − 0.25
- Safety factor (for wells) = − 5
- $NPSH_A$ = 52 (obviously adequate)

The safety factor usually suggested for $NPSH_A$ is 2 ft, but due to the uncertainty of drawdown for a well, larger safety factors should be used. A safety factor of 20 ft might be too little in some circumstances.

Other considerations in pump selection are allowable internal pressure in the bowls, shaft stretch, and hydraulic losses between casing and pump. Ordinarily, the manufacturer should be consulted for these concerns, but for this problem, such concerns are trivial because the pump is not deep and the head is not large.

The total pumping head is 255 ft, and about 5 ft per 100 ft of pump column should be added to allow for friction and turbulence losses.

The bowl head pressure = 255 + (5 × 160/100) = 263 ft = 114 lb/in.^{2} (also satisfactory).

Shaft stretch, thrust-bearing load, and bearing life should be calculated by the manufacturer; the calculations should be certified and given to the designer for checking. Pumps from several manufacturers should be investigated and the selection made on the basis of efficiency as well as on the quality of the pump. Specifications can be written to include a penalty/reward clause for efficiency. Quality can be attained by writing specifications as explained in Section 12-4.

(3) Select the driver. As shown in Figure 18-17, the operating point depends on the drawdown, the piping losses, and the elevated tank water level, so operation could occur at any point

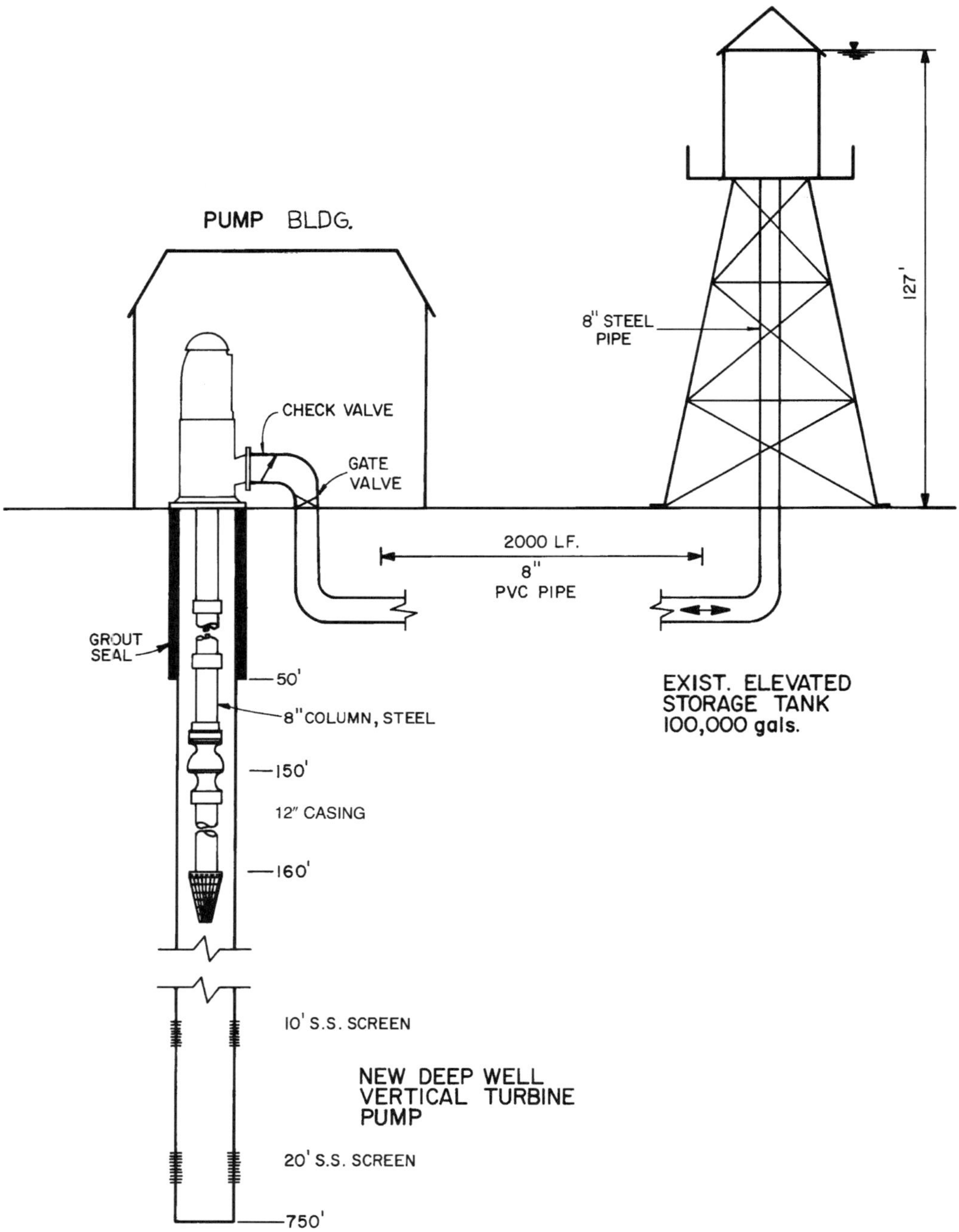

Figure 18-16. Well and pipe profile for Example 18-4. Courtesy of Psomas and Associates and Raymond Vail and Associates.

along the pump curve between points A (560 gal/min at 255 ft) and B (630 gal/min at 220 ft). From Equation 10-6b, the water power is

$$hp = \frac{560 \times 255}{3960} = 36$$

$$hp = \frac{630 \times 220}{3960} = 35$$

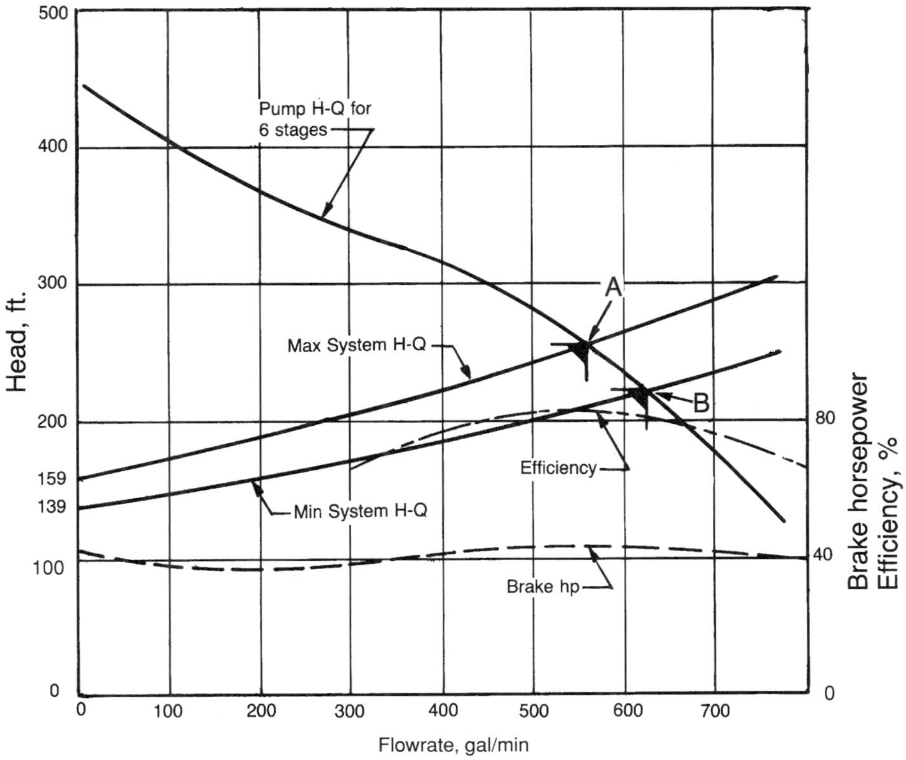

Figure 18-17. System head-capacity and pump performance curves for Example 18-4. Courtesy of Psomas and Associates and Raymond Vail and Associates.

The pump efficiency is 82%, so the brake horsepower at the pump shaft is

$$hp = \frac{36}{0.82} = 44$$

which is confirmed by the plot of horsepower in Figure 18-17.

The motor selected should be the next higher motor size manufactured, or 50 hp. A check of the pump curve shows that a 50-hp motor can power the pump along the full range of the pump curve. Note that there are minor losses caused by shaft bearing and thrust bearing friction. Such losses should be computed by the manufacturer, certified, and transmitted to the designer for checking.

(4) Pump layout. The pump and piping layout is shown in Figure 18-18, and well details are shown in Figure 18-19. Items of special note are as follows:

- An air release valve to vent air on pump start-up. The venting is critical. It must cushion the shock of opening the check valve and starting the movement of water in the transmission main. Data for calculating the size of the orifice must be obtained from the valve manufacturer.
- A check valve to minimize water hammer (see Chapter 7). A "silent" (center-post-guided globe-style) check valve or a swing check could be used.
- An isolation gate or butterfly valve.
- An advantage of the low velocity in the transmission main is that the simple valving system shown in Figure 18-18 was found by analysis to be adequate for water hammer control.
- A pump bearing lubrication system. Use water lubrication in most potable water installations.

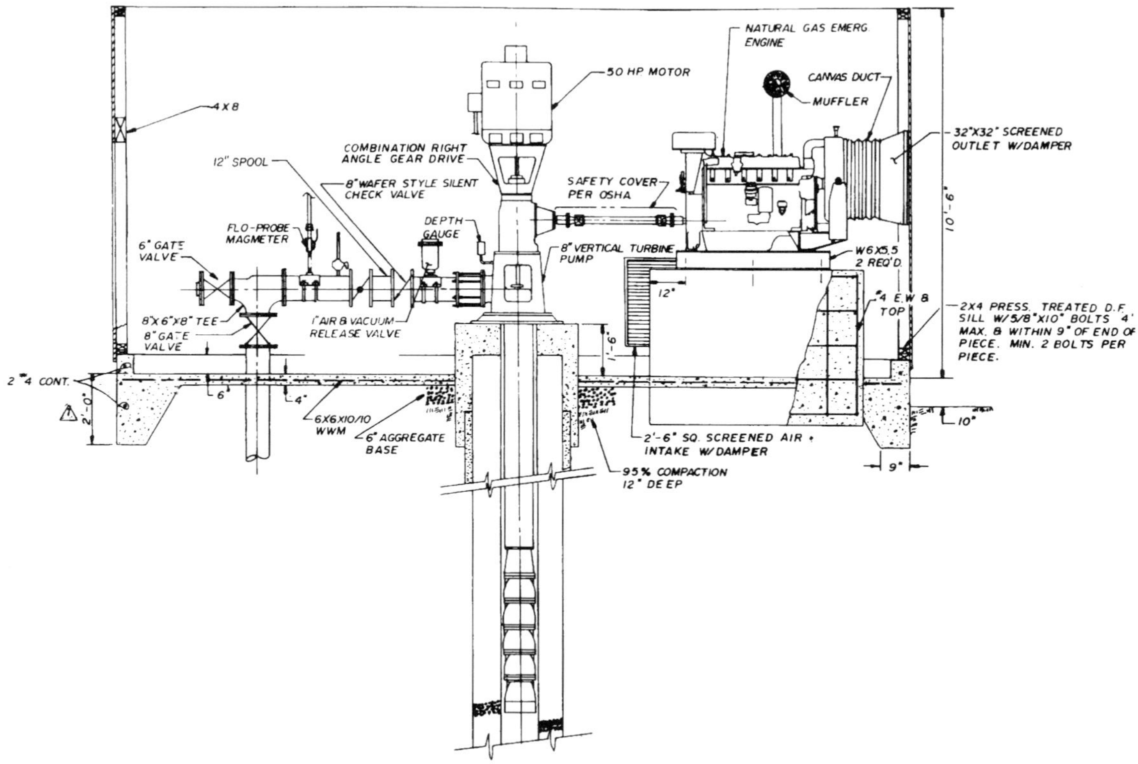

Figure 18-18. Pump and piping detail for Example 18-4. Courtesy of Psomas and Associates and Raymond Vail and Associates.

- A pump packing seal or gland or stuffing box.
- A $^{1}/_{2}$- to $^{1}/_{4}$-in. tap for a pressure gauge (see Figure 20-6).
- A 1-in. tap for chlorine feed addition.
- A direct-coupled, hollow-shaft motor with a top adjusting nut to adjust for shaft stretch or abrasive wear.
- Other pump features are available and important for particular installations and should be discussed in detail with the pump manufacturer. Some features include thermal protection and winding heaters.
- The pump motor is energized by pressure switches set to respond to the level of water in the elevated reservoir.
- If an extended power outage is likely, auxiliary power is required. Because the system has storage sufficient for 3 h of supply at peak day demand, a manually started engine directly connected to the pump through a right angle gear is satisfactory. If an engine is installed, use it to pump for a 1-h period at least twice per month (once per week is better).

(5) Chlorination facilities. The State Department of Health required the well water to be chlorinated. If the well were operated continuously at 560 gal/min, the chlorine gas use at 2 ppm (a relatively low dosage) would be

$$560 \text{ gal/min} \times \frac{1 \times 10^6 \text{ gal/d}}{694 \text{ gal/min}} \times \frac{8.34 \text{ lb}}{\text{gal}} \times \frac{2 \text{ lb}}{10^6 \text{ lb}} = 13 \text{ lb/d}$$

This well, however, is one of three available and is used only part time, so the amount of chlorine used would be small. Hypochlorite is more expensive than compressed chlorine, but it is safer and the equipment is less expensive, so a batch hypochlorination system was installed. The system, shown schematically in Figure 18-20, consists of a 60-gal PVC batch tank and a metering pump connected to the well pump discharge pipeline. The pump controller also starts the metering pump.

Critique. The pump setting, 150 ft is relatively shallow, so a lineshaft pump is proper. For deep wells with pump settings greater than 600 ft for crooked wells, or for wells where noise must be suppressed, consider the use of submersible pumps.

The simple valving system at the well head is adequate for this particular well. Neither valve slam nor excessive surge occurs because the water velocity is very low. It is important to make a water hammer analysis. If valve slam or excessive surges can occur, use a more sophisticated valve system such as one of those discussed in Section 7-6.

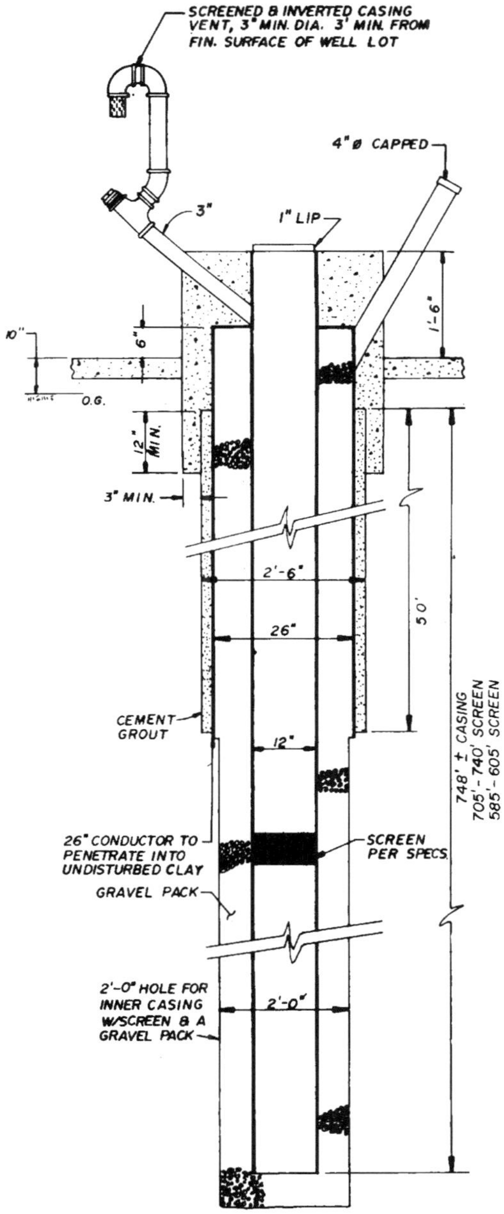

Figure 18-19. Well detail for Example 18-4. Courtesy of Psomas and Associates and Raymond Vail and Associates.

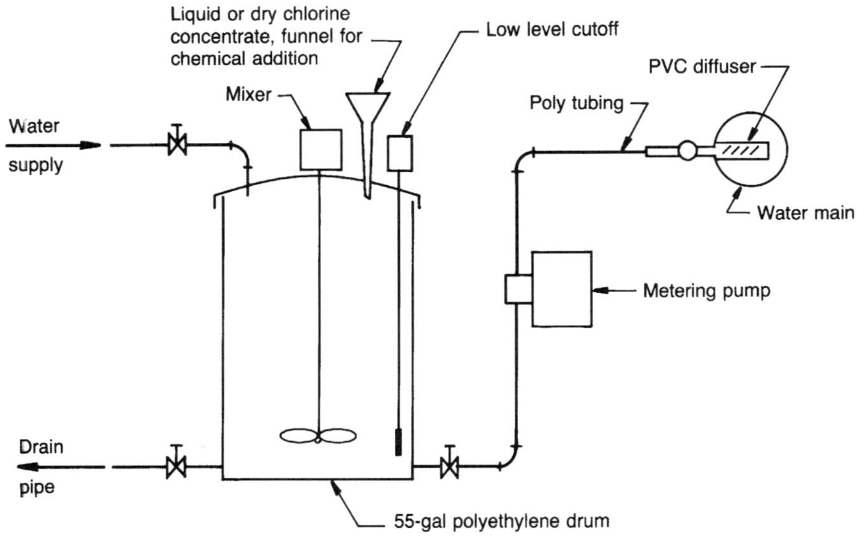

Figure 18-20. Batch hypochlorinator schematic.

Some engineers would specify a harness for the sleeve coupling at the pump discharge elbow. Others would note that if the 8-in. gate valve body is not overstressed by the shear load imposed, nothing can move and no harness is needed. The sleeve coupling is crowded in its location, and two alternatives are (1) a flange adapter and (2) a grooved-end coupling. Some engineers would also provide an intermediate pipe support for the discharge header, but the header is only 6 ft long and can bridge that distance.

The device to the right of the Flo-Probe Magmeter in Figure 18-18 is a sampling tap. Taps for a pressure gauge and chlorine addition are included but not shown on the figure. When the pump starts, a water-air mixture explodes out of the air release valve, so drain piping (also not shown) leads from the valve past the pump and engine.

The space between the 6-in. gate valve and the wall is only about 2 ft. More clearance (at least 30 or 36 in.) should be provided.

18-6. Booster Pumping Stations

Booster pumps can be divided into two types: in-line and distribution. In-line boosters take suction from an incoming pipeline, pressurize all the water, and discharge it into another pipeline. Distribution boosters take suction typically from storage (although sometimes from a portion of the flow in a pipeline) and maintain a given pressure (within limits) for supply in a distribution system at wide ranges of demand.

In-line Booster Pumping Stations

The potential advantages of an in-line booster pumping station include (1) the pipeline on the suction side of the booster station can be designed for a lower pressure rating, which thereby reduces the pipeline construction costs; (2) all of the water in the system need not be pumped at maximum system pressure, so energy costs are reduced; and (3) the primary pumping station (e.g., the source pumping station at a clear well or reservoir) need not be designed for the high-pressure conditions that are necessary for only a part of the entire water system.

The potential disadvantages include (1) additional pumping station construction cost, unless the cost of the primary pumping station can be reduced; (2) additional pumping station O&M costs; (3) increased operational complexity; (4) additional electric power substation required; (5) complicated analysis and control of system hydraulic transients; and (6) the possible need for other facilities, such as access roads and power lines.

General comments concerning the typical transmission main in-line booster pumping station, illustrated schematically in Figure 18-21, are as follows:

- Operationally, the booster pumping station could be located anywhere between point A and point C. The relationship of the minimum suction side hydraulic grade line (HGL) to the ground is the limiting factor. The critical HGL (based on the lowest reservoir water level and the peak flowrate) should exceed ground level by at least 3 to 6 m (10 to 20 ft).
- The operating pressure in pipeline segment AC is significantly reduced compared with the alternative of placing the pumping station at point A instead of at point C.
- Energy can be saved if the portion of flow delivered through the turnout at point B does not require boosting.
- The booster pumps must be capable of producing the required flowrate at the maximum TDH, and they must operate without cavitation at the minimum TDH.
- Both suction and discharge pipelines should be analyzed for transients, and facilities should be designed as necessary to prevent water hammer.

The addition of a primary pumping station (Figure 18-22) creates a system that is hydraulically and operationally more complex than the one in Figure 18-21. General comments about Figure 18-22 are as follows:

- Capital and annual O&M costs of the booster pumping station must be offset by the reduction in the costs of the primary pumping station and the pipeline segment AB due to a lower HGL.
- The HGL between points A and B depends on the pump H-Q curves selected for each station.
- The rate of flow produced at both stations must be equal at all times, so careful matching of the pumping units at the two locations is required.
- The pipeline transients must be analyzed and a surge control system must be designed.
- Provision must be made to shut down each pumping station on high or low pressure if a power outage occurs at the other station.
- Depending on pump selection, it may be necessary to allow gravity backflow from point B to point A to produce the proper back pressure for the pumping units at point A to start against.

Distribution Booster Pumping Stations

Examples of booster pumping station installations that provide pressure to a municipal water supply upper service zone are schematically shown in Figures 18-23 and 18-24.

The following comments relate to the system shown in Figure 18-23:

- The system hydraulics are controlled by the water level in the elevated reservoir.
- The pumps are started by pressure switches located at the booster station as pressure drops due to customer demand.

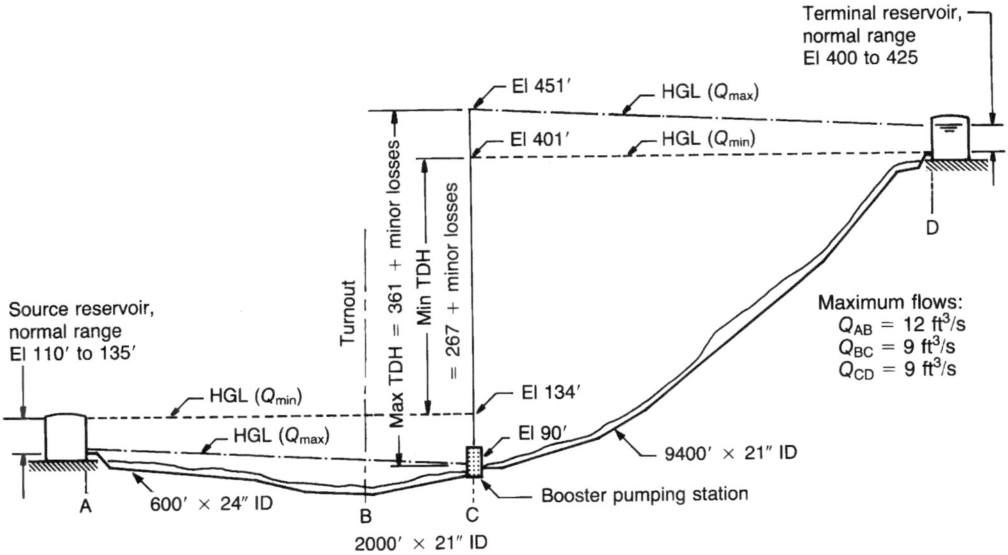

Figure 18-21. Transmission main with an in-line booster pumping station. Gravity flow out of the source reservoir. After Boyle Engineering Corp.

- Pumping in excess of demand refills the elevated reservoir. When the reservoir is full, the altitude valve closes and causes a pressure rise, and a pressure switch sequentially shuts off the booster pumps at set time intervals.
- The discharge of each booster pump is equipped with a pump-control valve or some other device to limit the transient pressures of start-up and shutdown.
- Rapid pressure fluctuations in this type of system often cause customer dissatisfaction. Service can be improved by adding a hydropneumatic tank at the booster.

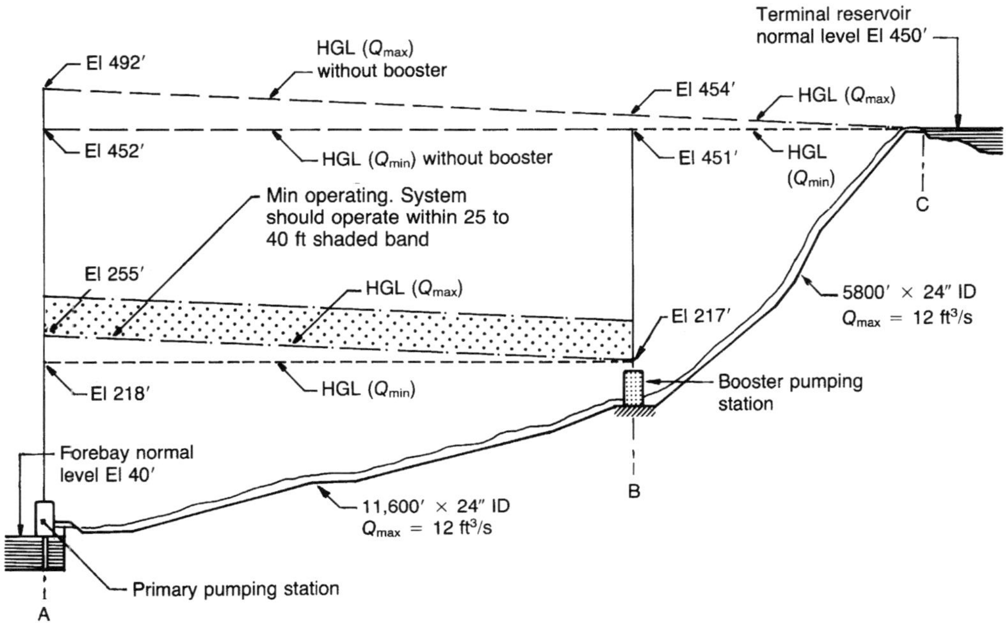

Figure 18-22. Transmission main with a primary pumping station and a booster pumping station. After Boyle Engineering Corp.

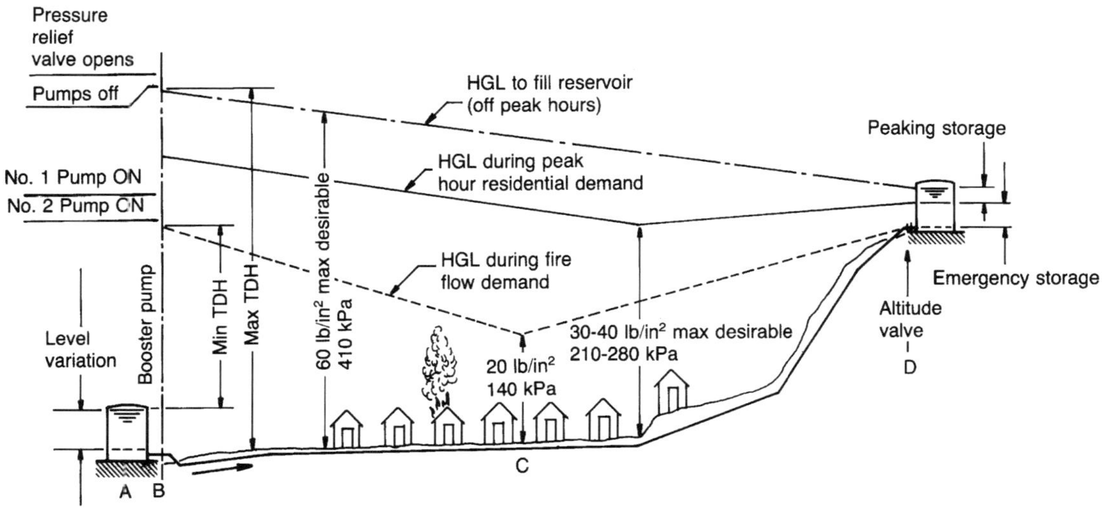

Figure 18-23. Distribution system booster with reservoir control of the hydraulic gradeline. After Boyle Engineering Corp.

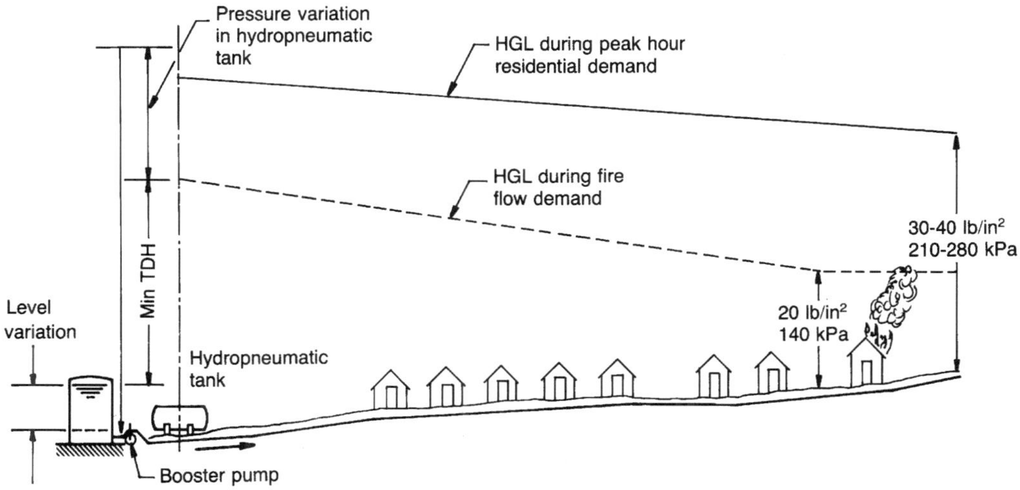

Figure 18-24. Distribution system booster without reservoir control of the hydraulic gradeline. After Boyle Engineering Corp.

Comments on the system in Figure 18-24 are as follows:

- It is assumed that an elevated reservoir is not economically justified, due to a small service area and flat terrain.
- A hydropneumatic tank maintains system pressure within prescribed limits. The pumps are started and stopped based on water level and pressure in the hydropneumatic tank.

Example 18-5
Small Distribution System Booster without Reservoir

Problem: Design a booster pumping station for a small residential distribution system pressure zone serving 200 persons. The booster is to be located adjacent to and pump from an existing ground-level welded steel reservoir. It has been determined that an elevated storage reservoir cannot be justified economically. A hydropneumatic tank is to be included in the design to maintain system pressure within prescribed limits. Because the climate is mild with temperatures rarely below freezing, an outdoor installation is acceptable.

Solution: Note that differences in results between SI units and U.S. customary units are due to the number of significant digits used.

Data and assumptions.

Item	SI Units	U.S. Customary Units
Population	200	200
Water consumption	946 L/d · cap	250 gal/d · cap
Fire flowrate		
(minimum rate per local code)	31.5 L/s	500 gal/min
Demand flowrates		
Average daily	200 × 946 L/d = 2.2 L/s	200 × 250 gal/d = 35 gal/min
Maximum daily	2.5 × avg daily = 5.5 L/s	2.5 × avg daily = 87 gal/min
Peak hour	4.0 × avg daily = 8.8 L/s	4.0 × avg daily = 139 gal/min
Suction reservoir operating elevations		
Maximum water surface	103.9 m	341 ft
Minimum water surface	100.0 m	328 ft

Item	SI Units	U.S. Customary Units
Ground-level elevations		
At booster station	98.1 m	322 ft
At system high point	104.9 m	344 ft
At system low point	89.9 m	295 ft
System pressure requirements		
Maximum pressure	517 kPa	75 lb/in.^{2}
Minimum pressure		
Peak hourly flow	207 kPa	30 lb/in.^{2}
Max daily flow plus fire flow	138 kPa	20 lb/in.^{2}

Define the booster peak design flow rate. Because the distribution system is without storage, a reasonable peak design flowrate for the booster station is the sum of the maximum daily average flowrate and the required fire flowrate.

SI Units	U.S. Customary Units
(5.5 + 31.5) L/s = 37 L/s	(87 + 500) gal/min = 587 gal/min

Define the allowable pressure variation in the hydropneumatic tank. The maximum operating pressure in the hydropneumatic tank should be the pressure that, during static conditions, produces the maximum allowable system pressure at the lowest customer service in the pressure zone. Estimate the water surface in the hydropneumatic tank to be 1.5 m (5 ft) above ground level.

The maximum operating pressure at the hydropneumatic tank is

SI Units	U.S. Customary Units
(517 kPa + 89.9 m/0.102 m/kPa) – [(98.1 m + 1.5 m)/0.102 m/kPa] = 1398 kPa – 976 kPa = 422 kPa	[75 lb/in.^{2} + 295 ft/2.31 ft/(lb/in.^{2})] – [(322 ft + 5 ft)/2.31 ft/(lb/in.^{2})] = 203 – 142 = 61 lb/in.^{2}

The minimum operating pressure in the tank must be at least as great as the larger of the following:

- The pressure required to deliver the peak hourly flowrate with at least 207 kPa (30 lb/in.^{2}) throughout the system, or
- The pressure required to deliver the maximum day average flow plus the required fire flow at a pressure of 138 kPa (20 lb/in.^{2}) or more to the most critical hydrant location.

These determinations require a system network analysis, which is typically accomplished using computer programming. For a small system, however, it can readily be done by hand using the Hardy Cross method. For this problem, the following assumptions have been made:

- The latter criterion governs, i.e., the delivery of maximum day average flow plus fire flow.
- The critical hydrant is located at the system high point.
- The drop in system pressure from the hydropneumatic tank to the critical hydrant due to pipeline friction and other system losses is 96 kPa (14 lb/in.^{2}).

The minimum operating pressure at the hydropneumatic tank is

SI Units	U.S. Customary Units
(138 kPa + 104.9 m/0.102 m/kPa) + 96 kPa – [(98.1 m + 1.5 m)/0.102 m/kPa] = 286 kPa	[(20 lb/in^2 + 344 ft/2.31 ft/(lb/in.^{2})] + 14 lb/in^2 – [(322 ft + 5 ft)/2.31 ft/(lb/in.^{2})] = 41 lb/in.^{2}

The allowable pressure variation in the hydropneumatic tank can thus be between 286 kPa (41 lb/in.^{2}) and 422 kPa (61 lb/in.^{2}), or a 136-kPa (20-lb/in.^{2}) differential. Differential pressures from 136 kPa (20 lb/in.^{2}) to 207 kPa (30 lb/in.^{2}) are considered normal for hydropneumatic systems.

Pumping unit selection. The following subjects must be addressed relative to pumping unit selection:

- Horizontal split-case pump versus vertical turbine pump.
- Electric motor versus natural gas (or diesel) engine driver.
- The number and capacity of pumping units.

Considerations that influence the choice between horizontal and vertical pumps include the suitability of available pump curves to meet the full range of system head-capacity requirements, pump speed, pump efficiency, construction costs, space requirements, available suction pressure, and owner preference. Pump curves from several manufacturers of both vertical turbine and horizontal split-case pumps were compared with the system requirements. The primary problem in pump selection was finding a pump that would operate at both the upper and lower head conditions with reasonable efficiency. The pump curve that best met the conditions was that of a five-stage vertical turbine pump operating at 1760 rev/min. A vertical turbine barrel pump has another advantage compared with a split-case pump: it can operate safely with a lower suction hydraulic gradeline.

Factors influencing the choice of an electric motor or an engine drive include equipment costs, maintenance costs, space requirements, the availability and dependability of electric power, the availability of natural gas, the costs of these alternative energy sources, and, again, owner preference (see Chapters 13, 14, and 25). These factors vary from area to area and from time to time. For this problem, it is assumed that an economic evaluation of initial installation costs plus present worth of long-term energy costs favors electric motors. A power outage in excess of a few minutes, however, would mean *no water* because the system has essentially no storage. (Consideration will be given to minimal emergency storage in the hydropneumatic tank later in this example problem.) It is thus vital to contact the utility supplying electric power to obtain an outage history. If outages have been significant, then one of the following options should be selected:

- Natural gas (or diesel) engines as main drivers
- Electric motors with direct-connected natural gas or diesel engines as backups
- Electric motors with a natural-gas-powered or a diesel-engine-powered standby generator.

For this problem, the outage history is assumed to be very good with no outages longer than 5 min experienced within the last 2 yr. Thus, electric motors are to be used, but only after careful review with and approval by the owner of the proposed booster and distribution system.

The number and capacity of pumping units are determined on the following basis: (1) the pumps normally have to produce flows only up to the peak hour demand flowrate of 8.8 L/s (139 gal/min) and (2) the pumps must also be capable of occasionally producing the peak design flowrate of 37 L/s (587 gal/min). Because these flowrates are likely to occur for durations of an hour or more, they must be supplied by the pumps because hydropneumatic tank storage is inadequate for such periods. The pumps should also be selected so that, when new, they include a "wear allowance" of 5 to 10%. In effect, the pumps should initially produce excess flow at the design head such that after 5 yr or so of operation and the resultant loss of capacity due to wear, the pumps will still produce the design flowrate.

Standby (backup) units for each size of pump are desirable for domestic water supply systems and probably mandatory where the system does not include storage. For a system of this size, the following tentative combinations should be considered:

- One primary pumping unit plus an identical standby unit (a total of two units)
- Two identical primary pumping units plus one identical standby unit (a total of three units).

The selection of the pumping unit combination to be utilized in the booster station design should be based on the following considerations:

- Costs of pumping units, electrical switchgear, piping, and valving—which favor fewer, larger units
- Operational flexibility and system reliability—which favor more, smaller units
- System surge control—which favors more, smaller units, but the fundamental problem is always power failure
- Energy cost—which favors more, smaller units if a "demand charge" is a significant element of the power rate schedule
- Hydropneumatic tank size and cost—which favor more, smaller units
- Space requirement—which favors fewer, larger units.

For this example, the preferred pump combination is found to be three identical pumping units—two duty pumps and one standby unit.

Develop system head-capacity and pump characteristic curves. It is necessary to establish the full range of pumping conditions to select pumping units properly. As shown below, the total dynamic head on the pumps varies. Assume minor losses between the reservoir and the hydropneumatic tank at 1.8 m (6 ft). (The minor loss estimate should be verified during the design of the booster station piping.)

The maximum hydropneumatic tank pressure and minimum suction reservoir level is $TDH = HGL_{tank} - HGL_{res}$ + minor losses.

$$TDH = (98.1\text{ m} + 1.5\text{ m} + 422\text{ kPa} \times 0.102\text{ m/kPa}) - 100.0\text{ m} + 1.8\text{ m} = 44.4\text{ m}$$

$$TDH = [322\text{ ft} + 5\text{ ft} + 61\text{ lb/in.}^2 \times 2.31\text{ ft/(lb/in.}^2)] - 328\text{ ft} + 6\text{ ft} = 146\text{ ft}$$

This maximum TDH is plotted as a horizontal line in Figure 18-25.

The minimum hydropneumatic tank pressure and maximum suction reservoir level is $TDH = HGL_{tank} - HGL_{res}$ + minor losses.

$$TDH = (98.1\text{ m} + 1.5\text{ m} + 286\text{ kPa} \times 0.102\text{ m/kPa}) - 103.9\text{ m} + 1.8\text{ m} = 26.7\text{ m}$$

$$TDH = [322\text{ h} + 5\text{ ft} + 41\text{ lb/in.}^2 \times 2.31\text{ ft/(lb/in.}^2)] - 341\text{ ft} + 6\text{ ft} = 87\text{ ft}$$

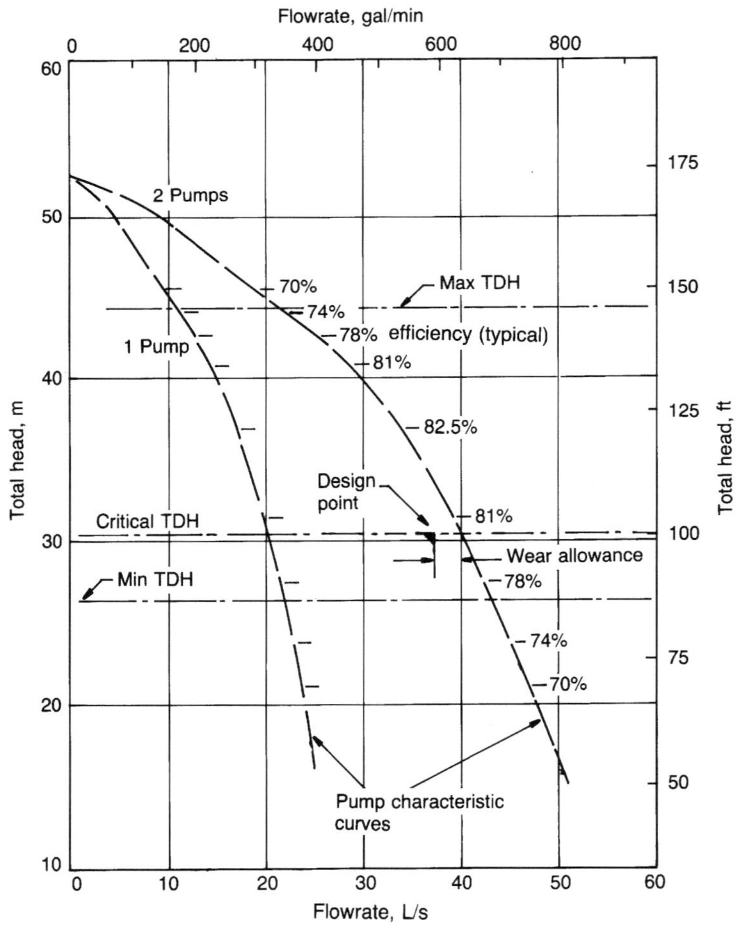

Figure 18-25. Booster pump head-capacity curves. The pump characteristic curves are based on identical pumps selected from the manufacturer's data: vertical turbine, five-stage, 1760 rev/min, 73/4-in. bowls. After Boyle Engineering Corp.

This minimum TDH is shown as a horizontal line in Figure 18-25.

The minimum hydropneumatic tank pressure and minimum suction reservoir level is critical because the suction pressure is minimum. TDH = $HGL_{tank} - HGL_{res}$ + minor losses.

TDH = (98.1 m + 1.5 m + 286 kPa × 0.102 m/kPa) – 100.0 m + 1.8 m = 30.6 m	TDH = [322 ft + 5 ft + 41 lb/in.^{2} × 2.31 ft/(lb/in.^{2})] – 328 ft + 6 ft = 100 ft

This critical TDH (30.6 m or 100 ft) is also shown as a horizontal line in Figure 18-25. The pumps must be able to discharge the peak design flowrate at this critical TDH, so the "design point" for the system is 37 L/s at 30.6 m (587 gal/min at 100 ft).

The next step is to find a pump curve that satisfies the following requirements when two such pumps are operating:

- Operate with acceptable efficiency (70% or better) at both maximum and minimum system heads.
- Operate without cavitation at minimum system head. Thus, $NPSH_A$ must equal or exceed the $NPSH_R$ shown on the manufacturer's pump curves.
- Operate at the "design point" increased by a flowrate of 5 to 10% for wear allowance.

Available pump characteristic curves are presented graphically in pump manufacturers' catalogs, and the manufacturers' sales representatives should be consulted to assist designers in the selection of the best-suited pumps. A sample pump characteristic curve that satisfies the design requirements is plotted in Figure 18-25 for one pump and for two pumps operating in parallel.

The electric motor must be able to drive the pump at any combination of head and discharge from the maximum to the minimum TDH. The maximum required power can be found by trial, but pump manufacturers' catalogs usually show the point where the power demand peaks. For the pumping unit selected, the peak occurs at a flowrate of 18.3 L/s (290 gal/min) and a TDH of 34.1 m (112 ft) at a pump efficiency of 82%. From a combination of Equations 10-6 and 10-7, the motor output power is

$$P = \frac{qH}{102\ E_P} = \frac{18.3\ \text{L/s} \times 34.1\ \text{m}}{102 \times 0.82} = 7.46\ \text{kW} \qquad P = \frac{qH}{3960\ E_P} = \frac{290\ \text{gal/min} \times 112\ \text{ft}}{3960 \times 0.82} = 10.0\ \text{hp}$$

Determine the required hydropneumatic tank capacity. The operation of a hydropneumatic tank is based on the following relationship (if constant temperature is assumed):

$$P_1 V_1 = P_2 V_2 \tag{18-2}$$

where P is the absolute pressure, V is the volume of air, and the subscripts 1 and 2 indicate initial and final, respectively.

A hydropneumatic tank is typically designed to be at least 10% filled by water at the minimum station pressure. The tank volume is normally selected to limit pump cycling to 4 to 6 cycles per hour. For this problem, assume a minimum volume of 10% and a maximum of 5 cycles per hour.

Select the following pump operational criteria:

- First pump on when tank pressure falls to 314 kPa (45 lb/in.^{2}) or 32.0 m (104 ft) WC
- Second pump on when tank pressure falls to 286 kPa (41 lb/in.^{2}) or 29.2 m (95 ft) WC
- Second pump off when tank pressure rises to 394 kPa (57 lb/in.^{2}) or 40.2 m (132 ft) WC
- First pump off when tank pressure rises to 422 kPa (61 lb/in.^{2}) or 43.0 m (141 ft) WC.

Both one- and two-pump operations need to be analyzed to determine which governs the size of the hydropneumatic tank. The minimum hydropneumatic tank volume for the one-pump operation is found as follows:

Determine the percentage of tank water volume used in the one-pump operation, that is, the pressure change from 314 kPa (45 lb/in.^{2}) gauge to 422 kPa (61 lb/in.^{2}) gauge. The tank water volume at minimum pressure 286 kPa (41 lb/in.^{2}) gauge was previously assumed at 10%. Atmospheric pressure (see Table A-6 or A-7) is 101 kPa (14.7 lb/in.^{2}) at the site. From Equation 18-2,

SI	US
(286 + 101) kPa × (100 – 10) = (314 + 101) kPa × (100 – % water)	(41 + 14.7) lb/in.^{2} × (100 – 10) = (45 + 14.7)1b/in.^{2} × (100 – % water)
At 314 kPa, water = 16.1%	At 45 lb/in.^{2}, water = 16.0%
(286 + 101)kPa × (100 – 10) = (422 + 101)kPa × (100 – % water)	(41 + 14.7)1b/in.^{2} × (100 – 10) = (61 + 14.7)1b/in.^{2} × (100 – % water)
At 422 kPa, water = 33.4%	At 61 lb/in.^{2}, water = 33.8%
Volume used = 17.3%	Volume used = 17.8%

Determine the average capacity of the single-pump operation varying between its on and off points. Assume the suction reservoir is at its maximum operating level [water surface elevation = 103.9 m (341 ft)], because this results in increased pump discharge (reduced TDH) and is therefore a critical case for hydropneumatic tank sizing. The hydropneumatic tank pressure for the one-pump operation varies from 314 kPa (45 lb/in.^{2}) to 422 kPa (61 lb/in.^{2}). The average tank pressure is 368 kPa (53 lb/in.^{2}). The average one-pump TDH is calculated as follows (see Figure 18-26):

SI	US
TDH = 98.1 m + 1.5 m + (368 kPa × 0.102 m/kPa) – 103.9 m + 1.8 m = 35.0 m	TDH = 322 ft + 5 ft + [53 lb/in.2 × 2.31 ft/(lb/in.2)] – 341 ft + 6 ft = 115 ft

From the one-pump curve (Figure 18-25), the pump capacity (new) for a TDH of 35.0 m (115 ft) is 18.0 L/s (286 gal/min).

Determine the hydropneumatic tank minimum volume using the following relationship: minimum volume = $^1/_2$ pump capacity × $^1/_2$ cycle time/portion of tank used (expressed as decimal):

SI	US
Minimum volume = $^1/_2$ × 18 L/s × $^1/_2$ × 720 s/0.173 = 18,700 L	Minimum volume = $^1/_2$ × 286 gal/min × $^1/_2$ × 12 min/0.178 = 4820 gal

Determine the percentage of tank water volume used in the two-pump operation:

SI	US
(286 + 101) kPa × (100 – 10) = (394 + 101) kPa × (100 – % water)	(41 + 14.7) lb/in.^{2} × (100 – 10) = (57 + 14.7) lb/in.^{2} × (100 – % water)
Percentage water at 394 kPa = 29.6%	Percentage water at 57 lb/in.^{2} = 30.1%
Volume used = 19.6%	Volume used = 20.1%

Determine the average capacity of the second pump operating between its on and off points with the first pump operating continuously. Again assume the suction reservoir is at its maximum operating level [water surface elevation = 103.9 m (341 ft)]. The hydropneumatic tank pressure for the second pump operation varies from 286 kPa (41 lb/in.^{2}) to 394 kPa (49 lb/in.^{2}). The average second pump TDH is calculated as follows (see Figure 18-26):

SI	US
TDH = 98.1 m + 1.5 m + (340 kPa × 0.102 m/kPa) – 103.9 m + 1.8 m = 32.2 m	TDH = 322 ft + 5 ft + [49 lb/in.^{2} × 2.31 ft/(lb/in.^{2})] – 341 ft + 6 ft = 105 ft

From the one-pump curve (Figure 18-25), the capacity of the second pump for a TDH of 32.2 m (105 ft) is 19.2 L/s (304 gal/min).

Determine the minimum hydropneumatic tank volume for the operation of the second pump:

SI	US
Minimum volume = $^1/_2$ × 19.2 L/s × $^1/_2$ × 720 s/0.196 = 17,600 L	Minimum volume = $^1/_2$ × 304 gal/min × $^1/_2$ × 12 min/0.201 = 4540 gal

Thus, the minimum hydropneumatic tank volume is 18,700 L (4820 gal) as determined by single-pump operation.

Because the system is vulnerable to power outages, reliability relative to short-duration power outages and for average day system demands could be improved by increasing the tank

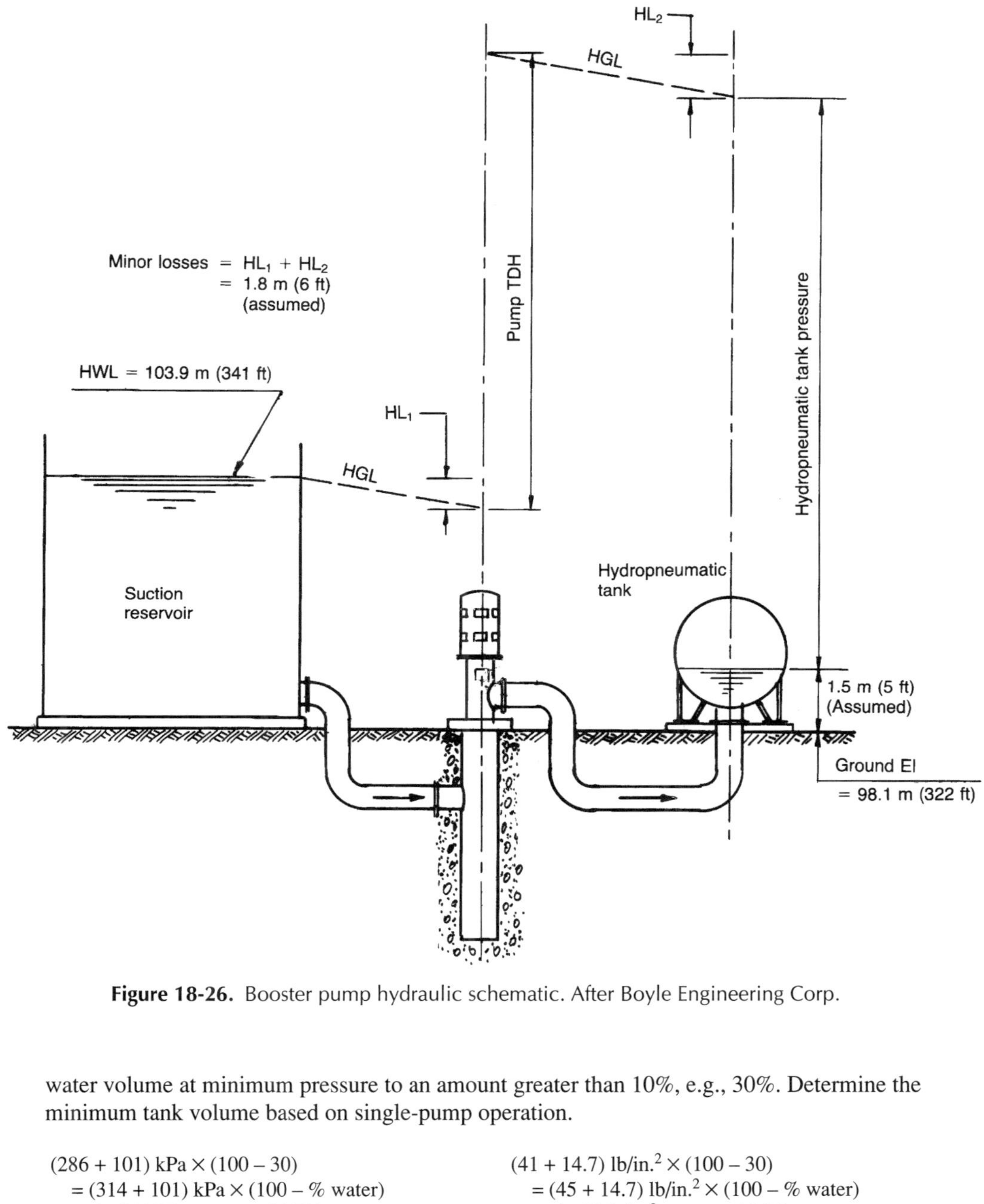

Figure 18-26. Booster pump hydraulic schematic. After Boyle Engineering Corp.

water volume at minimum pressure to an amount greater than 10%, e.g., 30%. Determine the minimum tank volume based on single-pump operation.

(286 + 101) kPa × (100 – 30)	(41 + 14.7) lb/in.^{2} × (100 – 30)
= (314 + 101) kPa × (100 – % water)	= (45 + 14.7) lb/in.^{2} × (100 – % water)
At 314 kPa, water = 34.7%	At 45 lb/in.^{2}, water = 34.7%
(286 + 101) kPa × (100 – 30)	(41 + 14.7) lb/in.^{2} × (100 – 30)
= (422 + 101) kPa × (100 – % water)	= (61 + 14.7) lb/in.^{2} × (100 – % water)
At 422 kPa, water = 48.2%	At 61 lb/in.^{2}, water = 48.5%
Volume used = 13.5%	Volume used = 13.8%
Minimum volume = $^1/_2$ × 18 L/s × $^1/_2$ × 720 s/0.135 = 24,000 L	Minimum volume = $^1/_2$ × 286 gal/min × $^1/_2$ × 12 min/0.138 = 6220 gal

Thus, the hydropneumatic tank would be about 29% larger than the minimum required. It would, however, provide additional reserve capacity at slightly reduced pressure in the event of power outage. This reserve capacity would amount to the volume between 30 and 10% water,

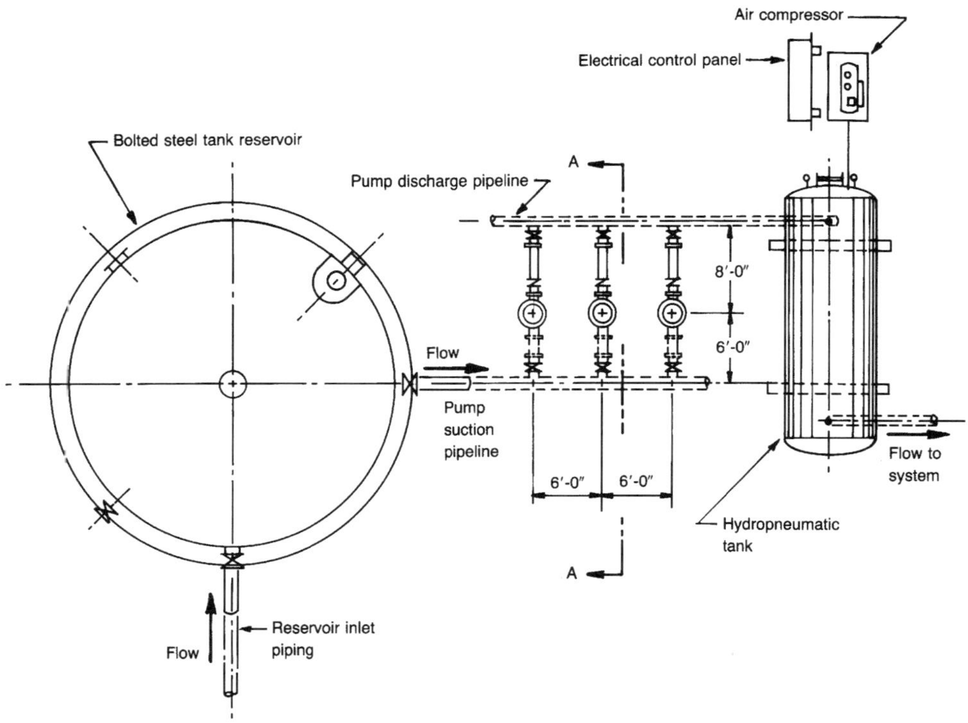

Figure 18-27. Booster pumping station plan. After Boyle Engineering Corp.

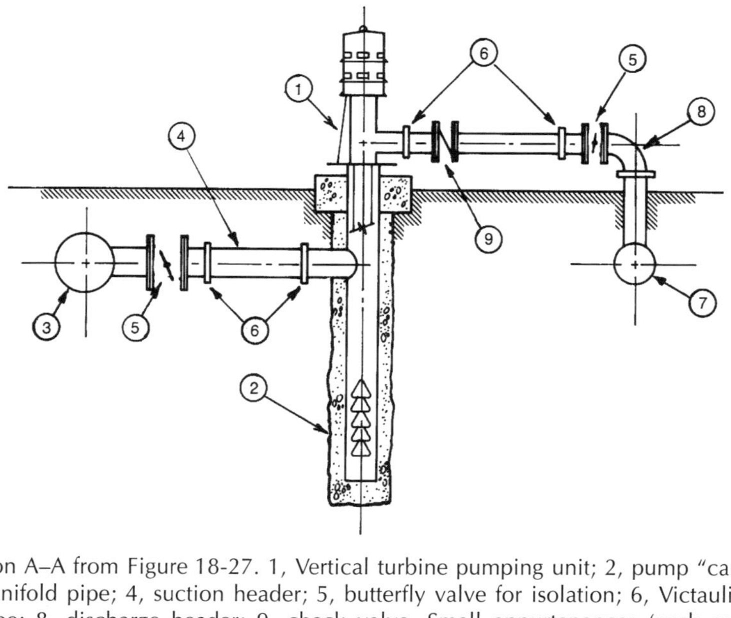

Figure 18-28. Section A–A from Figure 18-27. 1, Vertical turbine pumping unit; 2, pump "can" enclosed in concrete; 3, suction manifold pipe; 4, suction header; 5, butterfly valve for isolation; 6, Victaulic couplings; 7, discharge manifold pipe; 8, discharge header; 9, check valve. Small appurtenances (such as air release valves, pressure switches, valve supports, motor connection boxes, and pressure gauges) are not shown. After Boyle Engineering Corp.

or 4800 L (1244 gal)—equivalent to 35 min reserve capacity at average daily flowrate. This additional investment in tank size would probably avoid an occasional depressurization of the system and time-consuming repressurization.

System layout. A sample layout of the required facilities is illustrated in Figures 18-27 and 18-28.

Packaged Booster Pumping Stations

At least one manufacturer (see Section 11-10, Subsection "Multistage Centrifugal Pumps") markets packaged booster pumps with built-in variable speed controls designed to satisfy demands for very small to rather large flows for houses, high-rise buildings, or subdivisions at little change of pressure. Some (but not all) of the foregoing calculations can be avoided because (1) integrated diaphragm tanks are sized correctly for the pumps and the application and (2) the speed controls ramp the pump speed up or down automatically as more pumps are switched on and off.

18-7. References

1. Metcalf and Eddy, Inc. *Wastewater Engineering: Collection, Treatment, Disposal, and Reuse,* 2nd ed. McGraw-Hill, New York (1978).
2. "Water supplies—fire flow tests." Special Bulletin No. 42, National Board of Fire Underwriters, New York (July 1958).
3. Fair, G. M., I. C. Geyer, and D. A. Okun. *Water and Wastewater Engineering, Vol. 1, Water Supply and Wastewater Removal,* pp. 5-1–5-21. John Wiley, New York (1966).
4. Babbitt, H. E., J. J. Doland, and J. L. Cleasby. Water *Supply Engineering,* 6th ed. McGraw-Hill, New York (1967).
5. Peavy, H. S., D. R. Rowe, and G. Tchobanoglous. *Environmental Engineering.* McGraw-Hill, New York (1985).
6. Sanks, R. L., C. W. Reh, and A. Amirtharajah (Eds.), *Conference Proceedings, Pumping Station Design for the Practicing Engineer,* Vol. III, pp. 1173–1214 and 1247–1256. Department of Civil Engineering and Engineering Mechanics, Montana State University, Bozeman, MT (1981). Available (1) as photocopies or (2) as Vol. III on interlibrary loan from the university library.
7. Sanks, R. L. *Water Treatment Plant Design for the Practicing Engineer.* Ann Arbor Science Publishers, Ann Arbor, MI (1978).
8. Walton, W. C. *Groundwater Pumping Tests, Design & Analysis.* Lewis Publishers, Chelsea, MI (1987).

18-8. Suggested Reading

1. Walski, T. M. *Analysis of Water Distribution Systems.* Van Nostrand Reinhold, New York, NY (1984).

Chapter 19

System Design for Sludge Pumping

CARL N. ANDERSON
DAVID J. HANNA

CONTRIBUTORS
Robert H. Brotherton
George R. Brower
Geoffrey A. Carthew
Michael C. Mulbarger
Wyett C. Playford

The purpose of this chapter is to introduce the key issues involved in the design of sludge pumping systems. A number of significant problems, not usually met in the design of other pumping systems, are encountered in pumping sludge. These problems include:

- Variability in the nature of the pumped material
- Resulting variability in the frictional headloss characteristics
- Sludge behavior (as a non-Newtonian fluid) that makes traditional design (such as that for water pumping) inappropriate
- The need to understand complex fluid mechanics and not rely on "rule-of-thumb" approaches.

Due to the unusual and complex fluid mechanics associated with the wide varieties of sludges, designers are cautioned (1) to treat each sludge pumping application as a unique design problem and (2) to develop site-specific design criteria based on detailed evaluation of the specific sludge characteristics. The discussion in this chapter is limited to such problems of sludge systems as the selection of pumps, pipe, valves, and cleaning and flushing stations. Problems associated with scum, screenings, grit, ash, and chemical slurry pumping are not addressed, but note that these materials behave, in general, entirely differently from wastewater sludges.

The key to pumping sludge is to use (1) a pump properly sized to develop sufficient head and (2) a smooth pipe sized to produce the proper velocity (neither too high nor too low) without constrictions or projections and with as few bends as possible. The key to maintaining such a system is to have large, easily opened cleanouts on the pump, at any elbows on the suction side of the pump, and (where possible) at all elbows on the discharge side of the pump. Quick-disconnect air and/or water hose connections on both the suction and discharge sides of the pump are desirable where possible. Sludge lines should rarely be smaller than 150 mm (6 in.) and should preferably be larger. Glass-lined pipe is superior where high concentrations and large amounts of grease are present. Glass lining is expensive, however, whereas cement-mortar lining—which is also satisfactory—is not. The elimination of elbows and constrictions is of more practical importance than extreme smoothness for reducing friction.

The procedures presented in this chapter are applicable only to short pipelines. Long pipelines [1.6 km (1 mi) or longer] present a host of complex problems, briefly stated in Section 19-5. In general, extensive field testing and a thorough understanding of the literature and past experience with sludge pumping to develop engineering judgment is required for the successful design of systems for transporting sludge over long distances.

19-1. Hydraulic Design

Some of the traditional, well-known difficulties in evaluating the hydraulics of sludge flow [1] are as follows:

- Sludge is nonhomogeneous and has variable, peculiar properties.
- Parameters useful for water (such as Reynolds number) do not directly apply to sludge, which is a non-Newtonian fluid that may or may not behave as a Bingham plastic. Consequently, viscosity cannot be treated as a constant in pressure-drop calculations, and special methods must be used to calculate the friction loss.
- Friction losses decrease with decreasing solids concentration, a lower proportion of volatile solids, and increasing temperature. The suspended solids concentration is generally accepted as the most dominant variable.
- Due to the nature of biological solids, fresh undigested sludges and sludges from combined sewage behave more erratically than digested sludges.
- Sludge flow can be either laminar or turbulent. Both flow regimes, shown in Figure 19-1, are widely used.

Flow in Pipelines

Sludges are most efficiently moved within a treatment system or between locations by pumping through pipelines. The unique flow characteristics of sludge present unusual difficulties in estimating the frictional flow characteristics, and the literature contains some information on hydraulic design parameters for several types of sludges, the bulk of which is on digested and raw sludges [3–8]. Before 1970, designers typically relied on empirical rules and unreliable methods for predicting frictional headloss, although the fundamental theory was available [4]. Better understanding was provided by the literature of the 1970s [9–12], and still more valuable knowledge has been gained from later studies [2, 13–18]. A brief, noncomprehensive summary of fluid mechanics applicable to sludge systems is presented below.

Headloss in Laminar Flow

Sludges generally behave as non-Newtonian fluids. Because frictional headloss depends on the fluid rhe-

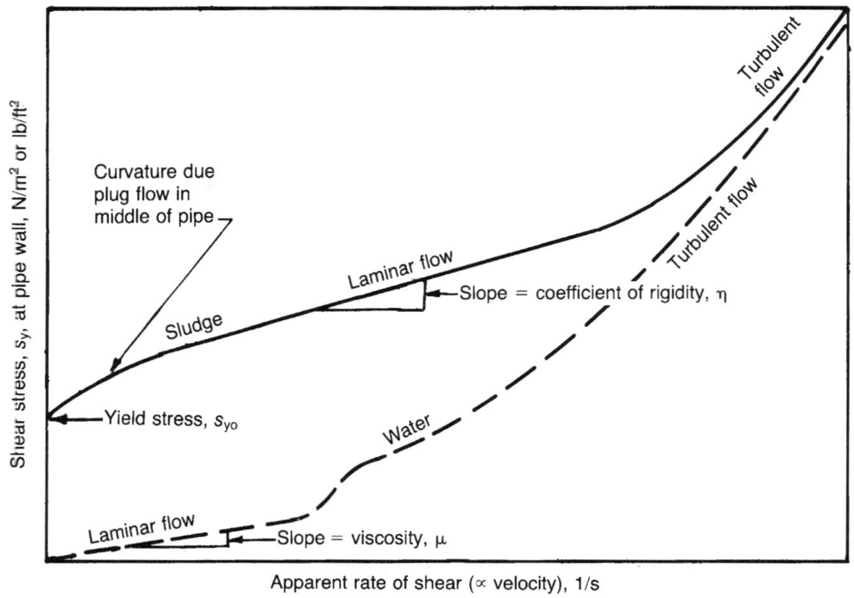

Figure 19-1. Comparison of wastewater sludge and water flowing in pipes. After EPA [2].

ology (viscosity, elasticity, plasticity) as well as on pipe diameter and flow velocity, it can be many times the headloss for water. Thixotropic behavior (resistance to flow until the shearing force is significant), grease accumulation on pipe walls, the increase of friction with an increase of solids, and (to a lesser extent) a high proportion of volatile solids contribute to the uncertainty of predicting headloss.

For water (a Newtonian fluid), pressure drop due to flow is directly proportional to the fluid velocity and viscosity under laminar flow conditions. When critical velocity is reached, the flow becomes turbulent. Critical velocity is a function only of Reynolds number (see Chapter 3 and Figure B-1). Unlike water, sludges often move in the laminar flow region where, for non-Newtonian fluids such as sludge, the pressure drop is not proportional to flow. The precise Reynolds number at which turbulent flow characteristics are encountered is uncertain for sludges. Convention has evolved to an accepted definition of two critical velocities for sludge: a lower critical velocity (below which flow is laminar) corresponding to a Reynolds number (***R***) of 2000 and an upper critical velocity (above which flow is turbulent) corresponding to an ***R*** of 3000. As discussed below, there is some debate about the best way to evaluate the laminar-turbulent transition. At any rate sludge behaves much like a Bingham plastic, a substance with a straight-line relationship between shear stress and flow only after flow begins. A Bingham plastic is described by two constants: (1) the yield stress (Figure 19-2) and the coefficient of rigidity (Figure 19-3).

An efficient procedure for using both constants is presented in depth by Mulbarger [14] (with comments by Anderson [14]), whose curves are included here as Figures 19-2 through 19-5. Designers are urged to provide additional data from known projects and to perform the suggested testing to develop site-specific data, particularly when pumping sludge long distances. Once the two parameters, the yield stress and the coefficient of rigidity, are obtained, laminar flow headlosses and the two critical velocities can be obtained from Equations 19-1 through 19-4. Babbitt and Caldwell performed early pioneering work in this study of hydraulics using a method by Bingham [3, 4, 6]. The Bingham equation is applicable only under laminar flow, and its derivation includes an approximation that makes it conservative at very low velocity. The Buckingham equation [15] may be used to avoid this approximation, but it is more difficult to solve. Further difficulties arise in measuring the two parameters, s_y and η, and because sludge is not exactly a Bingham plastic. Even so, the Bingham equation is useful for estimating laminar flow headlosses. In SI Units, the equation is

$$\frac{H}{L} = \frac{16 s_y}{3 D \rho g} + \frac{32 \eta v}{\rho g D^2} \tag{19-1a}$$

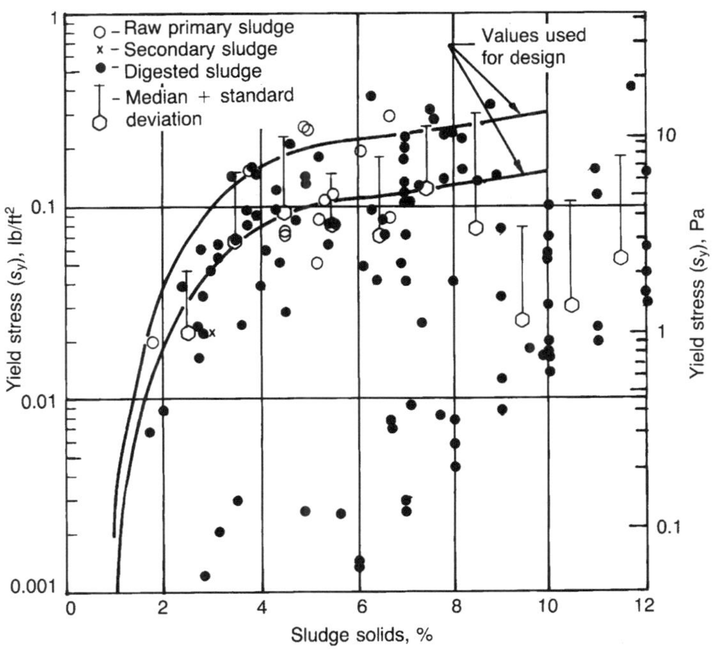

Figure 19-2. Yield stress versus sludge solids.

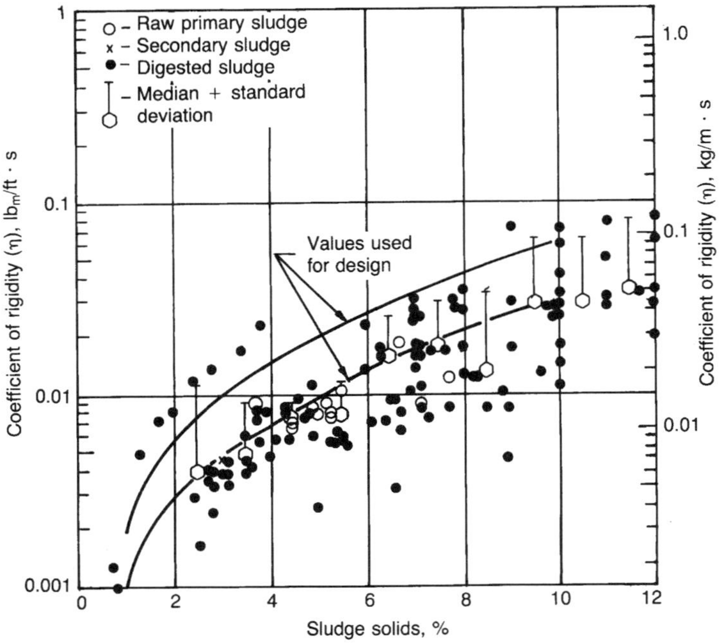

Figure 19-3. Coefficient of rigidity versus sludge solids.

where H is the headloss in meters, L is the length of pipe in meters, s_y is the yield stress in Pascals, ρ is the density in kilograms per cubic meter, D is pipe diameter in meters, η is the coefficient of rigidity in kilograms per meter-second, v is mean velocity in meters per second, and g is 9.81 m/s^2.

In U.S. Customary Units,

$$\frac{H}{L} = \frac{16s_y}{3\gamma D} + \frac{32\eta v}{g\gamma D^2} \tag{19-1b}$$

where H is the headloss in feet, L is the length of pipe in feet, s_y is the yield stress in pounds per square foot, γ is specific weight in pounds mass per cubic foot, D is the inside pipe diameter in feet, η is the coefficient of rigidity in pounds mass per foot-second (Figure 19-3), v is mean velocity in feet per second, and g is 32.2 ft/s^2. Although Equation 19-1 correlates well with field data, an approximation is included that could sometimes cause errors exceeding 10%. The approximation can be eliminated by solving the Buckingham equation [15] by successive approximations—a tedious task without a computer or a programmable pocket calculator. The development of site-specific data and a comparison with published information are recommended, and if the pipeline is longer than approximately 1 km (0.6 mi), site-specific data are vital. It is acknowledged that specific data are an ideal that is not likely to blanket all conditions to be actually experienced. However, it is far better to collect specific data for tempering both the designer's personal experience and the published information.

It is possible to define an apparent viscosity, μ. However, the viscosity of Bingham plastics is not a characteristic of the fluid alone, as it is with Newtonian fluids. Instead, viscosity depends on the relationship shown in Equation 19-2. The dynamic viscosity in SI Units is

$$\mu = \frac{Ds_y}{6v} + \eta \tag{19-2a}$$

where μ is the coefficient of viscosity in Pascal-seconds and other terms are as previously defined. In U.S. Customary Units,

$$\mu = \frac{g_c Ds_y}{6v} + \eta \tag{19-2b}$$

where μ is the coefficient of dynamic viscosity in pounds mass per foot-second and g_c is 32.2 lb$_m$ · ft/lb · s^2. The other terms are as previously defined.

The dynamic viscosity of Newtonian fluids can be measured with a viscometer. For sludge, however, the dynamic viscosity depends on the pipe diameter and velocity of flow, so it cannot be measured directly with a viscometer.

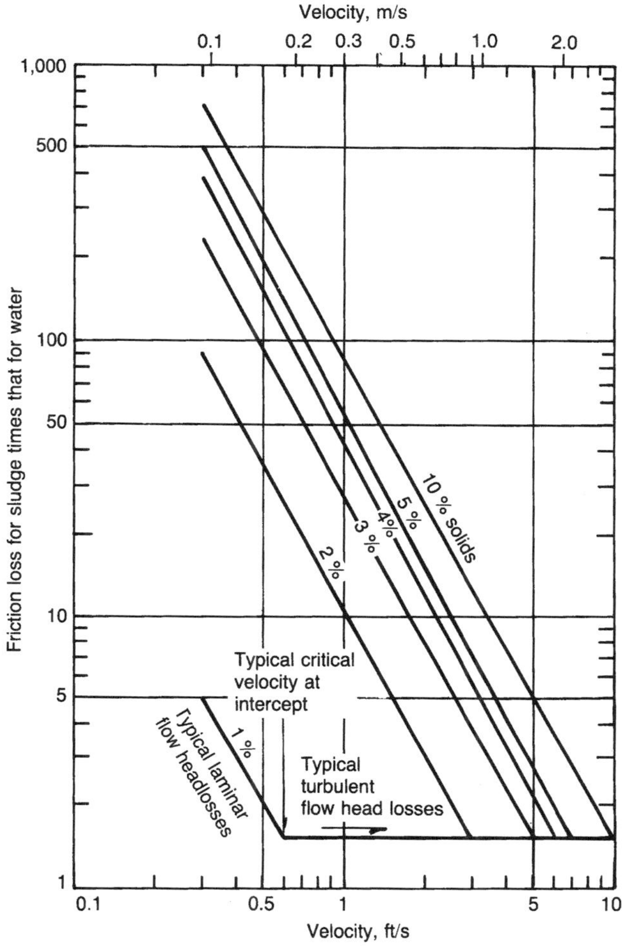

Figure 19-4. Recommended design curves. Headloss prediction for worst-case design.

Detailed examples of calculations using this development are presented by the U.S. EPA [2].

Data for Different Sludge Types

There is now a growing data base for different types of sludge for estimating sludge coefficients [16–18]. These data indicate that an increase in the content of unstabilized (undigested) biological solids in a sludge causes an increase in the coefficient of rigidity and hence an increase in the dynamic headloss in sludge pipelines. The headlosses may be considerably higher than those indicated for the worst case in Figure 19-4. Considering the variability of the data in Figures 19-2 and 19-3, it is wise to be conservative when designing for unstabilized sludges or sludges containing metal salts, and the values indicated in Figures 19-2 and 19-4 should probably be adjusted upward (tempered with the designer's experience) by an additional 50% or more.

Laminar-Turbulent Transition

Two general approaches have been developed for the transition from laminar to turbulent flow. The first approach was developed by Babbitt and Caldwell [3, 4, 6] and many authors since then, including Mulbarger *et al.* [13, 14] and Carthew *et al.* [18]. Equations 19-3 and 19-4 present upper and lower critical velocities based on a Reynolds number between 2000 and 3000. The lower critical velocity in SI units is

$$v_{\mathrm{lc}} = \frac{1000\eta + 1000\sqrt{\eta^2 + s_y \rho D^2/3000}}{D\rho} \qquad (19\text{-}3a)$$

and the upper critical velocity is

$$v_{uc} = \frac{1500\eta + 1500\sqrt{\eta^2 + s_y \rho D^2/4500}}{D\rho} \qquad (19\text{-}4a)$$

where both critical velocities are in meters per second.

In U.S. Customary Units the lower critical velocity is

$$v_{lc} = \frac{1000\eta + 1000\sqrt{\eta^2 + s_y \gamma g D^2/3000}}{D\gamma} \qquad (19\text{-}3b)$$

and the upper critical velocity is

$$v_{uc} = \frac{1500\eta + 1500\sqrt{\eta^2 + s_y \gamma g D^2/4500}}{D\gamma} \qquad (19\text{-}4b)$$

where both critical velocities are in feet per second. Equations 19-2, 19-3, and 19-4 have been used by several authors and work fairly well for most sludges.

Hanks and Dadia [19] criticized using only the Reynolds number and suggested the addition of the Hedstrom number for a more rigorous determination of the transition turbulent flow of a Bingham plastic. Although somewhat complex, the approach requires only the yield stress, coefficient of rigidity, density, and pipe diameter. The Hedstrom number is defined in SI units as

$$\boldsymbol{H} = \frac{D^2 s_y \rho}{\eta^2} \qquad (19\text{-}5a)$$

where $\boldsymbol{H}$ is the dimensionless Hedstrom number and other terms are as defined above. In U.S. Customary Units,

$$\boldsymbol{H} = \frac{D^2 s_y g_c \gamma}{\eta^2} \qquad (19\text{-}5b)$$

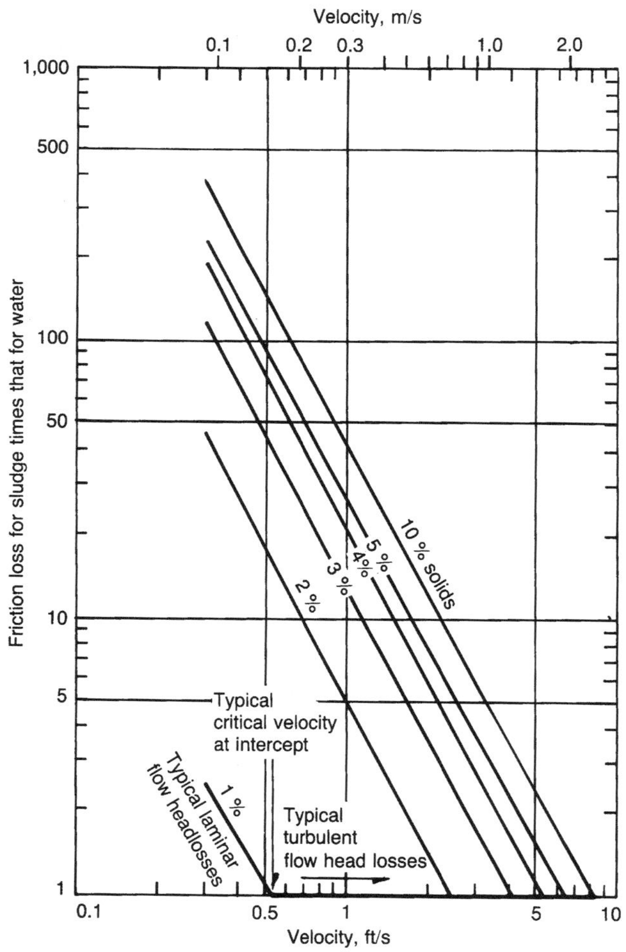

Figure 19-5. Recommended design curves. Headloss prediction for routine operation.

where $\boldsymbol{H}$ is the dimensionless Hedstrom number and other terms are as defined above. A special Reynolds number is also calculated (see Figure 19-6) and the plotted position of $\boldsymbol{H}$ and $\boldsymbol{R}$ on Figure 19-6 indicates laminar versus turbulent flow.

For sludge pipelines, the Hedstrom number is usually about 10^5 to 10^6. In this range, Figure 19-6 implies critical velocities not too different from those of Equations 19-3 and 19-4. If it is important to know whether flow will be laminar or turbulent (as in a sludge heat exchanger, because laminar flow would virtually prevent heat transfer), check both methods. Allow a generous safety factor because a turbulent zone near the pipe wall can coexist with a plug flow zone near the center of the pipe [20].

Headloss in Turbulent Flow

It is often practical to provide sufficient velocity to produce turbulent flow, especially for sludge with a low solids content. Under turbulent conditions, headloss vanes with velocity raised to an exponent of 1.7 to 2.0 [18]. Pipe roughness can be highly significant, and if so, procedures such as Hazen–Williams (Equation 3-9) may be used for sludge. In a smooth pipe, turbulent flow of sludge produces a headloss that is similar to or somewhat higher than the headloss with water. Major references include papers by Caldwell and Babbitt [4, 6] and Hanks and Dadia [19] and encyclopedia articles by Hanks [20] and Darby [21]. The test pipe data of Carthew *et al.* [18] cover turbulent as well as laminar flow, and the EPA manual [2] provides example calculations.

Some engineers select velocities to avoid the laminar-turbulent transition. The laminar-turbulent transition of Newtonian fluids causes unstable, unpredictable headlosses. Sludge, however, is non-Newtonian, and data [18] indicate only a change in trend at the critical velocity—not the abrupt jump that occurs in the headloss of Newtonian fluids.

There are a number of procedures and equations available for the design of either water or sludge pipelines. In general, it is just as important for a designer to understand the practical significance of pipe roughness and interior pipe conditions as it is to select the most appropriate design equation or procedure [22]. The approach of Hanks and Dadia [19] is probably the most accurate method to predict headloss under turbulent conditions. To use this method calculate the Hedstrom number from Equation 19-5 and the Reynolds Number as presented in Figure 19-6 and read the friction factor from Figure 19-6. Losses in SI Units are computed from

$$\Delta p = \frac{2f\rho L v^2}{D} \tag{19-6a}$$

where Δp is differential pressure in Pascals and f is found from Figure 19-6. In U.S. Customary Units,

$$\Delta p = \frac{2f\gamma L v^2}{g_c D} \tag{19-6b}$$

where Δp is differential pressure in pounds per square foot and f, again, is found from Figure 19-6. All other terms are defined above.

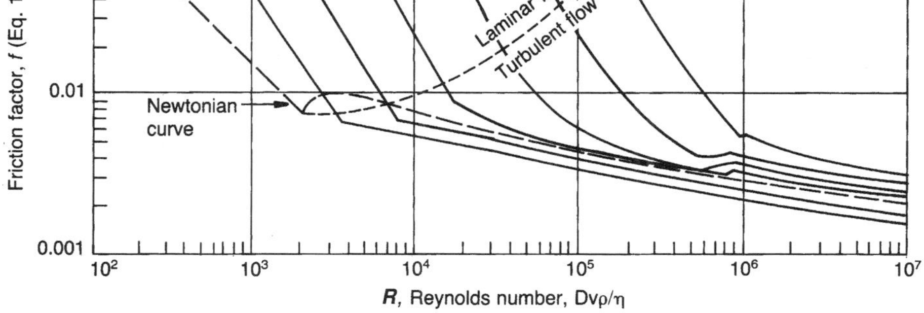

Figure 19-6. Theoretical friction factor for sludge analyzed as a Bingham plastic. After EPA [2].

As with water under turbulent conditions, the friction loss in sludge flow is approximately proportional to velocity squared but modified by a multiplier found by using Figure 19-4 or 19-5.

Mulbarger [13, 14] suggested a 50% allowance for worst-case design for headloss under turbulent flow conditions. The 50% allowance is equivalent to using a Hazen–Williams C of about 110 instead of 140. The difference between sludge and water increases with an increase of solids under both laminar and turbulent flow as shown by recent experiments [18]. If the two Bingham parameters (s_y and η) can be determined with accuracy, the procedure of Hanks and Dadia [19] may be a more exact method. It is certainly wise to allow for an increase in pipe roughness due to moderate deposits of grit and grease. The above safety factor, however, may not suffice under some conditions, especially at higher solids concentrations. Wherever more accuracy is needed (as in all but short pipelines), the graph of Figure 19-6 should be used.

Thixotropy

Sludges sometimes exhibit characteristics of thixotropy; at other times they display characteristics similar to those of a Bingham plastic. In thixotropic behavior, flow resistance depends on the time at rest, so high pressure is needed to start the fluid moving after it has been at rest. Because some studies have shown that sludge is markedly thixotropic, it is good practice to assume that thixotropy may occur and, hence, raise the friction loss in suction piping. Suction piping should therefore be as short as practical. After passing through a pump, thixotropic effects are unlikely to be important, except when restarting a pipeline that has been shut down while full of concentrated sludge.

Simplified Headloss Calculations

Most of the above information has been available for many years but has not been widely used. A recent advance has been the compilation of Figures 19-4 and 19-5. The use of these figures is demonstrated in Example 19-1, in which sludge is considered to be a Bingham plastic with values of s_y and η, as shown in Figures 19-2 and 19-3. The factors obtained from Figures 19-4 and 19-5 are compared to losses for water using the Hazen–Williams equation with a C value of 140. Choose velocity and percent solids, compute the friction headloss for water, and multiply the headloss by the factors from Figure 19-4 (worst case) or Figure 19-5 (routine design). These figures cover laminar and turbulent flow conditions, both of which are widely used. At lower velocities, sludge friction decreases only slowly as velocity drops. With water, however, friction drops sharply as velocity drops. Hence, the sludge/water headloss ratio increases as velocity decreases. Above a critical velocity, sludge flows more like water and friction varies as velocity to an exponent of about 1.85. Figure 19-4 includes a factor of 1.5 in this region. Headloss also varies with solids content, especially at the lower velocities.

Consider the following when using Figures 19-4 and 19-5:

- Figures 19-4 and 19-5, supplemented with data published since 1980, are far more accurate than popular references published before 1970. These figures were proposed by Mulbarger [14] to represent the general case(s) based on further examination of data derived in his earlier work [13].
- *Figures 19-4 and 19-5 must be used correctly.* The water headloss is computed as a basis for the sludge headloss. Most engineers are comfortable with the approach of (1) calculating headlosses for water for a selected pipe roughness and then (2) multiplying the headloss by a selected sludge factor. The factors in Figures 19-4 and 19-5 are based on an H–W C factor of 140 for water. Do not use a different C factor even if you would do so for water. Minor losses should never be computed as equivalent pipe lengths but should be computed separately. Minor loss coefficients may be taken as roughly the same for sludge as for water.
- Some sludges may exceed the "design worst-case" curves of Figure 19-4. For undigested activated sludge, the design worst-case should be raised by a factor of 1.5 until more data are available. A similar factor may apply to primary and secondary sludges that contain aluminum or iron salts.
- The charts should be used with great caution for pump suction piping in the calculation of available net positive suction head ($NPSH_A$). Even if $NPSH_A$ appears to be adequate, long suction pipelines represent poor practice and should not be allowed because of possible thixotropy.
- Lower velocities and larger pipe may or may not reduce total friction headlosses. Calculate headlosses for at least two pipe sizes and evaluate capital, operating, and maintenance (pipe cleaning) costs before deciding the pipe size (see Example 19-1).
- Figures 19-4 and 19-5 are inadequate for highly accurate results. Accuracy is especially important when the pumped distance is large or when significant variations in sludge are expected. A strongly

suggested guideline is to conduct a specific testing program whenever (1) the pipeline is more than 1.6 km (1 mi) long, (2) the sludge is expected to be at or over 7% solids, (3) a fluctuation in fluid characteristics of the sludge is likely to occur, or (4) calculations based on Figure 19-4 indicate a friction head greater than 15 m (50 ft).

- Do not assume that resistances obtained by testing will never be exceeded. Design the facility to allow for diluting the sludge to ensure an adequate installation.

19-2. Types of Pumps

Depending on the application, sludge pumps can be centrifugal, vortex, combined screw-centrifugal, air lift, or positive displacement pumps such as plunger (or piston), progressive cavity, diaphragm, rotary lobe, and high-pressure piston. These pumps are described in Chapter 11 and compared in Table 25-10. In general, centrifugal pumps are most suitable for pumping large volumes of sludge at low (2 or 3%) concentrations, whereas higher concentrations and intermittent pumping warrant the use of a positive-displacement pump.

Centrifugal Pumps

The preferred choice of centrifugal pumps for sludge applications is the nonclog pump because of the need for larger passages through vanes to prevent obstructions. In general, pumps for moving sludge fall in the low end of the mixed-flow range of specific speeds. Sludge pumps must be sturdier with larger, more reliable bearings, shafts, seals, and other internal components. Reliability is preferred over pumping efficiency.

One of the major problems in selecting a centrifugal pump for sludge is the difficulty in obtaining the proper size for both flow and head conditions. (An example of the range of design considerations is presented in Example 19-1.) Due to the wide range of expected flow conditions, variable-speed (V/S) control is usually used. The discharge valve should not be throttled in an attempt to control flow because throttling promotes increased valve wear, plugging of the pipeline, and wasted energy. Velocity criteria in the sludge piping system are often compromised when the pump capacity is reduced and the head is raised by throttling a discharge valve.

The impeller can be specified to include one or more vanes and to be enclosed, semi-enclosed, or open. Efficiencies tend to be higher for enclosed impellers and lower for open impellers. Closed impellers are best suited for large return activated sludge pumps or raw waste pumps with suction and discharge pipes of 200 mm (8 in.) or larger. Open or recessed impellers are better for smaller-volume pumping applications involving stringy material and for suction and discharge piping sizes of 100 to 150 mm (4 to 6 in.). Centrifugal pumps are not recommended for very small volumes of sludge with suction and discharge piping less than 100 mm (4 in.). Note that many designers object to pipe smaller than 150 mm (6 in.) and refuse to use pipe smaller than 100 mm (4 in.) for sludge in any circumstances.

Vortex Pumps

The impeller of a vortex pump (also called a "recessed impeller pump") develops a vortex in the incoming fluid so that most of the solids never touch the impeller. The openings through the pump are large because the impeller is almost entirely out of the liquid flow path, so the plugging often encountered with nonclog pumps is eliminated. (Despite the name, nonclog pumps often clog.) Hence, the size of solids is limited mostly by the size of the suction and discharge nozzles. Vortex pumps have the advantages of less maintenance, greater reliability, and greater ease of parts replacement. But the sizing of vortex pumps is critical because they have very flat system head curves—a major complication when pumping thicker sludges (because of the highly variable pressure drop). Internal fluid recirculation is a concern, the efficiency is very low (only 45% or less), and in some types the impellers cannot be trimmed. In summary, the vortex pump is a desirable selection when the sludge pumping system requirements are such that reliability is paramount in the pumping of sludge with a low (say, 4% or less) concentration of solids. Vortex pumps can pump up to about 6% solids, but a variable-speed drive and a reliable flowmeter are required to allow process control at higher solids contents. Rugged mechanical construction also becomes critical as varying process conditions may require the pump to operate away from its best efficiency point (bep). A positive means of moving sludge through the suction piping into the pump must be available if pumping is to be reliable.

Horizontal mounting is recommended, because vertical installations can result in noise and/or cavitation.

Restrictions are often built into the discharge passages of the pump to steepen the flat system head curves, so do not fail to specify the size of sphere that must pass entirely through the suction, discharge, and internal passages of the pump.

Combined Screw-Centrifugal Pumps

Recognizing the inherent problems in pumping sludge, pump manufacturers have periodically offered innovative designs. One type (such as the Chicago Pump Scru-Peller®), used for half a century, combines a two-flight screw conveyor with an open-type, two-blade centrifugal nonclog impeller. It was designed and applied particularly for primary sludge pumping. The sharp stellite edges of the screw conveyor and impeller and the stellite shear bars in the screw housing cut or shred stringy materials.

The advantages of the Scru-Peller® pump are its ability to (1) cope with stringy materials, (2) produce a steady discharge, (3) develop heads up to 50 m (160 ft) at 160 m^3/h (700 gal/min), and (4) minimize clogging effects and associated downtime and maintenance. The disadvantages are primarily higher first cost and operating costs as compared with costs for a centrifugal nonclog pump but less as compared with costs for a vortex pump. The pump is best applied for intermittent pumping instead of continuous duty.

A pump that has become almost standard for sludge pumping and has often been used to replace an inefficient vortex pump is the combined screw-centrifugal impeller pump (such as the WEMCO Hidrostal®). This type of pump has been applied to sludges for about 16 years in the United States. The design was developed to pump live fish without damage for canneries. The impeller has two sections (screw and centrifugal) that combine the clog-free features of a vortex pump with the gentle action of a screw pump at the higher efficiencies of a centrifugal pump. The large open channels in the fluid path provide pumping without the abrupt changes in direction that capture stringy materials. It is claimed that the corkscrew action of the screw section helps to start thicker sludges moving. The advantages of this pump include (1) higher (up to 80%) efficiencies, (2) nonclog operation, (3) gentle fluid action without high turbulence, (4) a steep head-capacity curve, (5) low NPSH requirements, and (6) a nonoverloading power curve. The disadvantage is that for a given flow rate, it is larger and more expensive than a centrifugal pump.

Both of these pump types may be useful when the combination of steady discharge with clog-free operation is desired. The pumping of primary sludge and of sludges with stringy and fibrous materials are said to be good applications for both of these pumps.

Air Lift Pumps

Air lift pumps are sometimes used for the thin sludges produced at oxidation ditches or other small extended aeration activated sludge plants. Because the control of pumping rate is limited for a single pump, they are often installed in multiples. Air lift pumps are used for activated sludge return where the lift is low—typically less than 2 m (6 ft)—and the pipeline is short [23]. The pump itself is almost indestructible and, except for the splashing and the daily need for cleaning, requires virtually no maintenance. The air blowers, however, do require maintenance. The pump has several disadvantages: (1) it is difficult to regulate flow (so process control is also difficult); (2) the pumpage changes erratically with small variations in the air delivered; and (3) the efficiency is low (typically less than 30%). The best way to regulate discharge is to maintain the maximum air pressure entering the pump and close (or open) the air exhaust ports at the bottom of the pump column. Flushing connections are recommended.

Plunger Pumps

The plunger (or "piston") pump has long been a workhorse and is one of the most commonly used types for pumping sludge. The piston is driven by an exposed drive crank whose eccentricity is adjustable and, thus, provides a stroke length that can be varied to suit the desired output. The discharge from a plunger pump is therefore a series of interrupted sinusoidal pulses, so the "pulsed" flow (at the peak of the sine curve) is greater than the average flow. The pulsing effect is reduced with multiple piston (duplex or triplex) pumps and/or with air chambers. Design calculations must be based on the peak of the pulsing flow. The approximate flow variations as a percentage of the average flow are shown in Table 19-1. Because the tabular values are based on the assumption of a rigid machine and an incompressible fluid, the observed effects may be somewhat less. Nevertheless, discuss this aspect of pump performance with the pump manufacturer. Air chambers should be specified on the suction side when the suction head is more than 4.6 m (15 ft), and they should always be furnished on the discharge side. Air chambers, or pulsation dampeners, induce a more uniform flow. The volume of an air chamber or dampener for simplex pumps should be six to eight times the displacement of one plunger per stroke (V_s); for duplex and triplex pumps, the volume should be three to four times V_s [25]. The kinematics of reciprocating pumps produce periodic variations in

Table 19-1. Approximate Flow Variations as a Percentage of the Average Flow in Plunger Pumps[a]

Pump configuration	Number of plungers	Flow variations as a plus or minus percentage of average flow
Simplex	1	320
Duplex	2	160
Triplex	3	125
Two duplexes	4	133

[a] After Henshaw [24].

flow, velocity, and head conditions, so design the system to prevent the coincidence of harmonics with the fundamental harmonics of the suction or discharge piping [26].

The advantages of plunger pumps are (1) superior reliability, (2) ease and simplicity of maintenance and repair, and (3) low cost of replacement parts. But the most important advantage is their constant capacity when the head varies or when the head cannot be reliably estimated. The pump motor must, of course, be large enough for the maximum pressure conditions. A procedure for calculating the head for a plunger pump is given in Example 19-2.

The disadvantages of plunger pumps include: (1) pulsating flow, which tends to induce "rat-holing" in the sludge sump, (2) noise and vibration, (3) filth and smell, (4) frequent replacement of the balls in the ball check chambers, and (5) the reluctance of some operators to maintain or repair them. The piston frequently tends to tilt during the downstroke, thus causing rapid wear. Piston guides are recommended to reduce this tendency.

Plunger pumps should be designed and operated at the maximum discharge stroke setting for the best operation. Two in-line ball check valves on both sides of the plunger pump are also recommended (1) for better backflow protection if back siphoning from the discharge to the suction tank is possible and (2) to make the operation of the pump smoother if the suction head is high. The use of two ball check valves along with other design features should be discussed in detail with the pump manufacturer. Ball check valve seats should be stainless steel or bronze and should be replaceable without disturbing the valve chamber piping.

If a pulsating discharge rate is acceptable, positive-displacement reciprocating pumps are recommended as the primary means of sludge pumping due to their reliability in pumping the thicker and heavier sludges.

Progressive Cavity Pumps

Progressive cavity pumps are advantageous if a smooth, steady, predictable discharge is required. The pump produces a relatively constant output of flow because as one cavity reduces in size, the opposing cavity is increasing at the same rate. Flow capability is a function of rotor speed, rotor diameter, rotor eccentricity, and helix pitch. Pressure capability is a function of the number of rotor/stator stages [2]. These pumps are available with a wide variety of flow and head ratings and are very useful when pumping thicker sludges if they do not contain grit, debris, or abrasives.

Because a progressive cavity pump acts as a check valve when inactive due to the minimal clearances between the stator and rotor, check valves may be eliminated, but isolation valves must always be provided. If static back pressure is high, however, the pump might rotate backward unless either a check valve or antireverse gearing is used. The size of solids that can pass through a progressive cavity pump is limited to less than 45 mm ($1^3/_4$ in.) in diameter, so the sludge must not contain large solids. Unless sludges are relatively free from grit and other abrasives, wear is accelerated, which causes costly and frequent maintenance. Operating speeds are limited to about 300 rev/min or less to minimize wear. However, the pump clogs at very low speeds. A general recommendation is to provide an adequate velocity through the system piping while (at the same time) pumping at 200 rev/min or slightly more. Pump speed should be discussed in depth with the pump manufacturer.

Progressive cavity pumps have a reputation for a high rate of wear and the requirement of more maintenance labor than any other widely used type (see also rotary lobe pumps in this section). Wear of the metal rotor on the elastomeric stator increases clearances and reduces flow and pressure, so rotors and stators must be replaced frequently—typically once a year and sometimes more often. Replacement parts are expensive and difficult to install. Ample room is needed to slide internal pump parts out of the casing, so the pump requires significantly more floor space than other pumps.

Diaphragm Pumps

Diaphragm pumps, especially the spring-assisted, air-operated type (such as the Dorr-Oliver type ODS), have been successfully applied to low head sludge pumping service for at least 20 years. The spring acts on the diaphragm to draw sludge into the pumping chamber, so the pump is capable of a modest suction lift although high suction lift cannot be sustained reliably [27]. Compressed air forces the sludge out of the pumping chamber. Check valves on the inlet and discharge connections prevent backflow. Stroke adjustment and timer control of the air supply solenoid permits a wide range of pumping rates. Nonassist and pneumatic-assist types are available. The pump can be furnished with ball-type check valves in either singles or pairs with quick opening covers (recommended). Capacities range up to 34 m^3/h (150 gal/min) as water (and 60% or more of such values as sludge) at 350 kPa (50 lb/in.^{2}) gauge. Compressed air requirements are significant, however, and the air exhaust can be quite noisy. Diaphragm pumps have a good reputation, problems are few, and they are suitable for small treatment plants.

The disadvantages include: (1) pulsating flow, (2) vibration, (3) requirement for compressed air and consequent inefficiency, and (4) high cost per unit of flow. The type with a single valve for suction and discharge is not suitable for stringy materials, but valves in tandem pairs can be obtained or the pump can be preceded by a grinder.

Membranes typically last about two years if the air supply and the solenoid air valve are properly adjusted. Cracks in the membrane can be readily detected, so the membrane can be replaced before it ruptures entirely. A ruptured membrane allows the air lines to fill with sludge to the muffler, but the muffler liner quickly clogs so that little sludge can spill. Cleaning the air lines is a chore; membranes should be replaced well before cracks develop. The O&M manual should contain a warning about the danger of flooding the station with sludge if a muffler and liner are not in place. Better protection is afforded by mounting the muffler high enough to prevent any possible spill.

Another diaphragm pump options consists of a low-pressure pump using dual free or unbalanced diaphragms driven by an eccentric shaft. The features of the pump include a mechanical drive, self-priming characteristics, ability to operate at speeds reduced to a "snore" condition, the elimination of check valves and moving seals, and an unusually low number of moving parts. The pump is valveless and glandless and is able to operate dry for a short time without damage. It has been available since the mid-1980s and has been successfully applied in over 2000 sludge pumping applications. The pump can reportedly handle up to 8.5% solids [28]. This type of diaphragm pump is limited to a TDH of 100 feet (or less for the largest model). Maximum solids-handling capacity is 1 in. The pump tends to retain its performance over its operating life because disc and seal wear is low. Be cautious, however, in specifying this pump, because, when overpressurized, it can break internally without warning. Applications should be reviewed with the manufacturer in detail.

Rotary Lobe Pumps

The rotary lobe pump has been applied to wastewater sludges only since 1970 but with promising results. Its advantages are (1) positive displacement with little agitation or shear, (2) high efficiency, (3) self-priming, (4) compactness, (5) good accessibility for maintenance, and (6) sizes up to 0.12 m^3/s (2000 gal/min). The disadvantages are (1) the lack of an extensive operating history and (2) close clearances that are subject to wear and abrasion (which makes sludges containing much grit, abrasives, debris, and rags unsuitable).

Difficulties have been experienced with a number of rotary lobe pumps (installed between 1980 and 1988) in which rapid wear of lining materials and sometimes catastrophic loss of liners have occurred, according to a survey conducted in late 1988 [29]. Some operators reported rapid wear of end plates and a resulting loss of volumetric capacity. The causes were sorption of water by the liners and failure of the adhesive. The problems have since been corrected [30]. Rotary lobe pumps are so compact that it might be difficult to replace an unsatisfactory unit with another type because of space limitations.

High-Pressure Piston Pumps

Belt conveyors and screw conveyors are usually used to move dewatered sludge to truck loading points or incinerators. An alternative recently developed is the use of specially designed piston pumps to force dewatered sludge through pipes.

The basic pump design is derived from concrete pumps. A high-pressure piston pump typically has two pumping cylinders of small diameter and long stroke. Each cylinder is driven by the pressure of hydraulic oil in an adjacent drive cylinder. Hydraulic oil is also used to power-operate the valves, because ordinary check valves are not satisfactory in plastic fluids with high yield stress. At the pump intake, a short screw conveyor may be used to force sludge into the pump. The installation is mechanically complex and requires considerable floor space, but other conveying systems have the same drawbacks.

Very high pressure drops occur although data are sparse. In one unpublished test of a sludge containing

36% solids, pressure drops of about 3000 kPa (400 lb/in.^{2}) occurred after thixotropic breakdown in a pipe only 61 m (201 ft) long [31]. The pressure needed to achieve thixotropic breakdown after a day of rest was 4000 kPa (600 lb/in.^{2}). The pipe diameter was 130 mm (5 in.). In spite of the drawbacks, a pumping system may sometimes be more cost effective than a conveyor belt.

Pump Selection

For sludge service, maintenance-free operation with pumps that rarely clog is preferable to the operation of more efficient units in which these advantages are compromised. Should the process design require a centrifugal pump, however, it is good design practice to add a standby positive-displacement pump to ensure that thicker sludge can be moved.

19-3. Pumping System Design

The most important criterion for design is to have sludge pumps capable of transferring the design quantity of solids throughout the range of expected solids concentrations within the required time interval. In most treatment plant sludge pumping, the consistency of the sludge changes during pumping. At first, the most concentrated sludge at the bottom of the basin is pumped. After most of that sludge is removed, a more dilute sludge with characteristics ranging from semisolids to those essentially of water is pumped. This variability in the fluid characteristics causes the pump to operate at different points along its design curve. Designers must take this variability into account and size the pump and motor for all expected service conditions. If the pump motor is not sized correctly, the pump may become overstressed or the motor damaged.

It is desirable to pump as uniformly as practical, but with small systems it is better to pick larger pumps and pipelines than are needed and pump intermittently with timers. The pump should operate long enough (approximately 10 to 15 min) to heat the motor to the operating temperature. The pump cycle time (subject to review with the manufacturers of the pump, motor, and motor starter) should be at least 15 min. The velocity of pumping is important. The friction in starting the sludge movement is much greater than in keeping the sludge moving once it is started. In pumping thick sludges, it may be desirable to flush the line with a more dilute concentration of sludge before the pump is turned off.

Calculation of System Curves

Consider (1) the worst possible combination of variability of solids concentrations and (2) discharge to the several possible outlets. If several pumps are to discharge to a common pipe, each pumping route must be considered for sizing the piping and determining the dynamic headlosses. Two typical sludge pumping system schematics for an activated sludge plant are shown in Figures 19-7 and 19-8. Note the various pumping routes available in both figures. Furthermore, the static head may change substantially during pumping. Consequently, four system curves are essential for analysis.

- *Maximum system curve:* The pressures that occur with maximum solids concentration. The system operates along this curve when the sludge is most concentrated and when it is being pumped through the piping system route that results in the maximum static and dynamic loss. This route usually has the greatest length and the smallest pipe diameter, but differences in the static lifts and heads between alternative routes must also be considered.
- *Average system curve:* The pressures that occur at average operating solids concentration.
- *Minimum system curve:* The pressures that occur at minimum solids concentration.
- *Water system curve:* The lowest pressures that can occur when water without solids is pumped through the critical route of the piping system. The critical route is the one with minimum static and dynamic loss, which is usually the pipe with the shortest length or the largest diameter, but differences in static lifts between alternative routes must also be considered.

In the design of centrifugal pumps, the motor power required is maximum for the water system curve because the flowrate is maximum. However, in the design of positive displacement pumps, the maximum system curve is the most critical because the pump must deliver its unvarying flowrate with the most concentrated sludge.

Design Procedure for Centrifugal Pumps

A design procedure for selecting a centrifugal (or vortex) sludge pump and motor [14] is outlined below for the plant shown in Figure 19-7. For illustrative purposes, it is assumed that only one pump discharges at a time and that each outside duty pump is

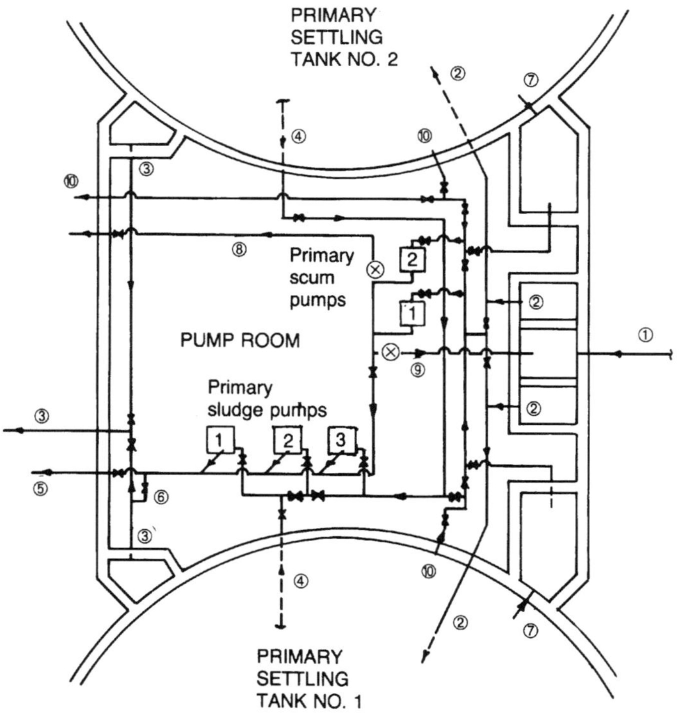

Figure 19-7. Primary sludge and scum piping. 1, raw sewage from pumping station; 2, raw sewage to settling tanks; 3, settled effluent to aeration tanks; 4, sludge drawoff; 5, sludge to sludge thickener (normal route) or to digesters (alternate route); 6, sludge bypass to aeration tanks; 7, scum removal; 8, scum pumped to truck for disposal; 9, scum semiliquid to influent well; 10, tank drains to plant sewer; ⊗, air-operated valves. Courtesy of Stearns & Wheler Engineers and Scientists and J. Kenneth Fraser & Associates.

dedicated to a separate settling tank. The middle pump is the common standby and is set up to take over the duty of either pump 1 or pump 3. Were several pumps to operate simultaneously, the total flow pumped through the system must be considered in determining total head. Note also that the scum pump is a positive-displacement pump that can also serve as a backup to the centrifugal pumps.

1. Plot the maximum, minimum, average, and water system curves (see Figure 19-9 in Example 19-1).
2. Examine the curves in step 1, choose a candidate pump, superimpose a plot of the pump H-Q curves, and locate the operating points on each of the system curves.
3. If a constant-speed (C/S) pump is to be used, select the smallest impeller size that intersects the three sludge system curves beyond the required flow. For vortex pumps with a choice of impeller sizes, use the largest impeller diameter and be sure the pump has a positive suction head (including allowances for headlosses at design flowrates).
4. If a variable-speed (V/S) pumping unit is to be used, select the maximum pump speed required for the three sludge system curves.
5. If a C/S pump is to be used, the intersection of the pump H-Q curve for the impeller found in step 3 with the water system curve determines the power required. Either calculate the power or find the smallest horsepower-designation curve beyond this intersection. A C/S pump installation makes flow difficult to control. A V/S pump should ordinarily be used.
6. If a V/S drive is to be used, consult the pump manufacturer to determine the maximum speed of the drive. The water power required is determined by the intersection of the maximum speed curve and the water system curve. Add corrections for the efficiency of the pump and the net efficiency of the driver to convert water power to the required power of the motor.
7. The manufacturer's pump performance curves must not be extrapolated because cavitation or vibration problems could occur at flows higher than those covered by the manufacturer's curves. If

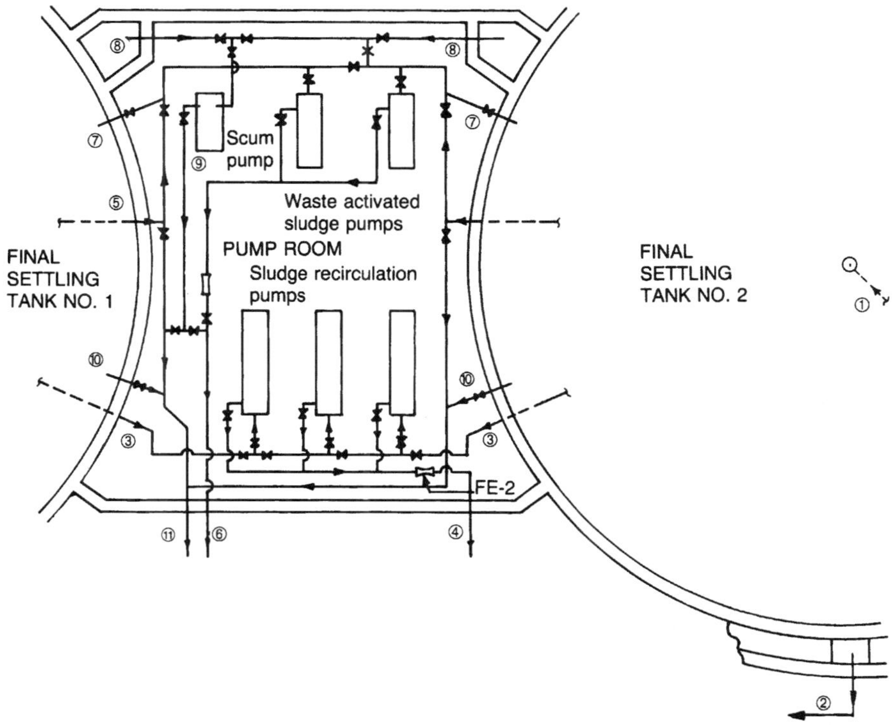

Figure 19-8. Return and waste activated sludge piping. 1, influent from aeration tanks; 2, settled effluent to chlorine contact tanks; 3, recirculation sludge drawoff; 4, recirculation sludge to aeration tanks; 5, waste activated sludge (WAS) drawoff; 6, WAS to thickener via control building; 7, sidewall drawoff; 8, scum drawoff; 9, scum to pumping station via plant sewer; 10, tank drain; 11, drain to plant sewer. Courtesy of Stearns & Wheler Engineers and Scientists and J. Kenneth Fraser & Associates.

the minimum system curve extends into the region beyond that covered by the pump curves, select a larger pump. If no suitable centrifugal pump can be found, consult the manufacturer. Always verify the selection of sludge pumps with several manufacturers.

Example 19-1
Design of a Vortex Pump System for Sludge

Problem: In the process design of a municipal activated sludge plant, a maximum of 13,600 kg/d (30,000 lb/d) on a dry basis of primary sludge solids (with no industrial component) must be transferred. The layout consists of two duty pumps and a standby pump, as shown in Figure 19-7. Note that there are several piping discharge routes, but for the purpose of this example, consider a system of 107 m (350 ft) of 150-mm (6-in.) piping and a static head of 3.2 m (10.5 ft) at average design conditions.

For this example, (1) plot the system curves, (2) select pumps for the V/S drives, (3) select the driver, and (4) check the computed headloss by a different method.

Solution: For problems of this kind, consider

- Several discharge piping routing options with varying pipe sizes, lengths, and static heads
- Varying static heads on any or all piping routes due to operating conditions
- The likely design feature of different-sized suction and discharge piping
- The varying mass and concentration of solids that must be transferred
- Variations in flowrates over the design life from initial operating conditions to final design conditions. (Consider ranges in solids concentration at both initial and final design conditions.)

Determine pump capacity. The required pumping capacities to transfer 13,600 kg (30,000 lb) of dry sludge solids per day are found from the following expressions:

SI Units

V_s = volume of sludge (m^3/d)
M_s = mass of dry solids (kg/d)
ρ = density of water (1000 kg/m^3)
S_s = specific gravity of sludge
P_s = percentage of solids (as a decimal)

U.S. Customary Units

V_s = volume of sludge (gal/d)
M_s = mass of dry solids (lb/d)
γ = specific weight of water (62.4 lb/ft^3)
S_s = specific gravity of sludge
P_s = percentage of solids/100

Solve the expression for cubic meters per day (or gallons per day) for a reasonable range of solids (10, 5, 3, and 1%). For the purpose of this example, assume the specific gravity of the sludge is 1.00.

10% solids:

$$V_s = \frac{13{,}600}{1000(1.0)(0.1)} = 136 \text{ m}^3/\text{d} \qquad V_s = \frac{30{,}000 \times 7.48}{62.4(1.0)(0.1)} = 36{,}000 \text{ gal/d}$$

5% solids:

$$V_s = \frac{13{,}600}{1000(1.0)(0.05)} = 272 \text{ m}^3/\text{d} \qquad V_s = \frac{30{,}000 \times 7.48}{62.4(1.0)(0.05)} = 72{,}000 \text{ gal/d}$$

3% solids:

$$V_s = \frac{13{,}600}{1000(1.0)(0.03)} = 453 \text{ m}^3/\text{d} \qquad V_s = \frac{30{,}000 \times 7.48}{62.4(1.0)(0.03)} = 120{,}000 \text{ gal/d}$$

1% solids:

$$V_s = \frac{13{,}600}{1000(1.0)(0.01)} = 1360 \text{ m}^3/\text{d} \qquad V_s = \frac{30{,}000 \times 7.48}{62.4(1.0)(0.01)} = 360{,}000 \text{ gal/d}$$

Estimate two duty pumps, each dedicated to a separate settling tank (including the third pump as a common standby) and each operating 6 h/d and at different times. The discharge rate required of each pump is

10% solids:

$$\left(\frac{136 \text{ m}^3/\text{d}}{2 \text{ pumps}}\right)\left(\frac{1 \text{ d}}{6 \text{ h}}\right) = 11.3 \text{ m}^3/\text{h} \qquad \left(\frac{36{,}000 \text{ gal/d}}{2 \text{ pumps}}\right)\left(\frac{1 \text{ d} \times 1 \text{ h}}{6 \text{ h} \times 60 \text{ min}}\right) = 50 \text{ gal/min}$$

5% solids:

$$\left(\frac{272 \text{ m}^3/\text{d}}{2 \text{ pumps}}\right)\left(\frac{1 \text{ d}}{6 \text{ h}}\right) = 22.7 \text{ m}^3/\text{h} \qquad \left(\frac{72{,}000 \text{ gal/d}}{2 \text{ pumps}}\right)\left(\frac{1 \text{ d} \times 1 \text{ h}}{6 \text{ h} \times 60 \text{ min}}\right) = 100 \text{ gal/min}$$

3% solids:

$$\left(\frac{453 \text{ m}^3/\text{d}}{2 \text{ pumps}}\right)\left(\frac{1 \text{ d}}{6 \text{ h}}\right) = 37.8 \text{ m}^3/\text{h} \qquad \left(\frac{120{,}000 \text{ gal/d}}{2 \text{ pumps}}\right)\left(\frac{1 \text{ d} \times 1 \text{ h}}{6 \text{ h} \times 60 \text{ min}}\right) = 167 \text{ gal/min}$$

1% solids (essentially water):

$$\left(\frac{1360 \text{ m}^3/\text{d}}{2 \text{ pumps}}\right)\left(\frac{1 \text{ d}}{6 \text{ h}}\right) = 113 \text{ m}^3/\text{h} \qquad \left(\frac{360{,}000 \text{ gal/d}}{2 \text{ pumps}}\right)\left(\frac{1 \text{ d} \times 1 \text{ h}}{6 \text{ h} \times 60 \text{ min}}\right) = 500 \text{ gal/min}$$

Fluid velocities at pump capacities. Find the velocity in 150-mm (6-in.) pipe at the pump discharge rates above for maximum (10%), average (5%), and minimum (3%) solids concentrations.

Sludge concentration	Discharge (m^3/h)	Velocity, m/s 150-mm pipe	Velocity, m/s 100-mm pipe	Discharge (gal/min)	Velocity, ft/s 6-in. pipe	Velocity, ft/s 4-in. pipe
10%	11.3	0.18	0.40	50	0.6	1.3
5%	22.7	0.35	0.80	100	1.1	2.6
3%	37.8	0.59	1.34	167	1.9	4.3

The velocities for 150-mm (6-in.) pipe are somewhat low for sludge service, so tabulations for 100-mm (4-in.) pipe are added.

(1) Plot system curves. Plot system curves for 10, 5, 3, and 1% solids flowing in 150-mm (6-in.) and 100-mm (4-in.) pipe by determining headlosses on the basis of Hazen–Williams C = 140. Then multiply the headlosses by the friction factors from Figure 19-4 (for worst case) or from Figure 19-5 (for routine operation). Headloss predictions for worst-case design are recommended for the maximum system curve. Headloss predictions for routine operations are recommended for all other system curves. The calculation for one point on the 5% solids, routine curve for 150-mm (6-in.) pipe is given below to illustrate the procedure.

Item	SI Units	U.S. Customary Units
Velocity	0.35 m /s	1.14 ft/s
Friction head, C = 140	0.91 m/1000 m	0.91 ft/1000 ft
Friction factor (Figure 19-5)	21	21
Friction loss	0.91 × 21 = 19.1 m/1000 m	0.91 × 21 = 19.1 ft/1000 ft
Pipe length	107 m	350 ft
Friction head	(107/1000)19.1 = 2.0 m	(350/1000)19.1 = 6.7 ft
Static head	3.2 m	10.5 ft
Velocity head	0.01 m	0.02 ft
TDH	5.21 m	17.22 ft

The curves of TDH for 150-mm (6-in.) pipe are shown in Figure 19-9 and those for the smaller pipe, in Figure 19-10. In comparing the curves of the two figures, the following are evident:

- In both sets of system curves the headlosses at high flowrates approach those of water. Note that some conservative designers would always provide some multiple of headloss above that of water. However, in this example such high flowrates are beyond those normally encountered.
- The 10% solids *worst-case* curve has a 50% safety factor over the *routine operations* curve. The safety factor approximates a deterioration of pipe from C = 140 to C = 110. If the designer believes that the pipe will deteriorate more than this or prefers an additional allowance for headlosses, then the system curves should be modified accordingly. Note that flexibility in pump drive speed may allow for future adjustments to higher head conditions.

Based on a review of the system curves, 150-mm (6-in.) piping is selected because of (1) lower TDH and horsepower requirements, (2) better size selection for maintenance and cleaning, (3) lower operating costs, and (4) future operational flexibility.

(2) Pump selection. A vortex pump is a good choice for this application. A 100-mm (4-in.) Model C WEMCO vortex pump is selected. Performance curves for the pump combined with the developed system head curves for 150-mm (6-in.) piping are shown in Figure 19-11. A review of Figure 19-11 indicates:

- Speeds of about 400 to 800 rev/min are adequate for a wide range of operating conditions.
- A speed of 710 rev/min is required to pump 10% solids at worst-case conditions with low velocity. A speed of 800 rev/min gives better velocity.
- A speed of about 640 is required to pump 1% solids.

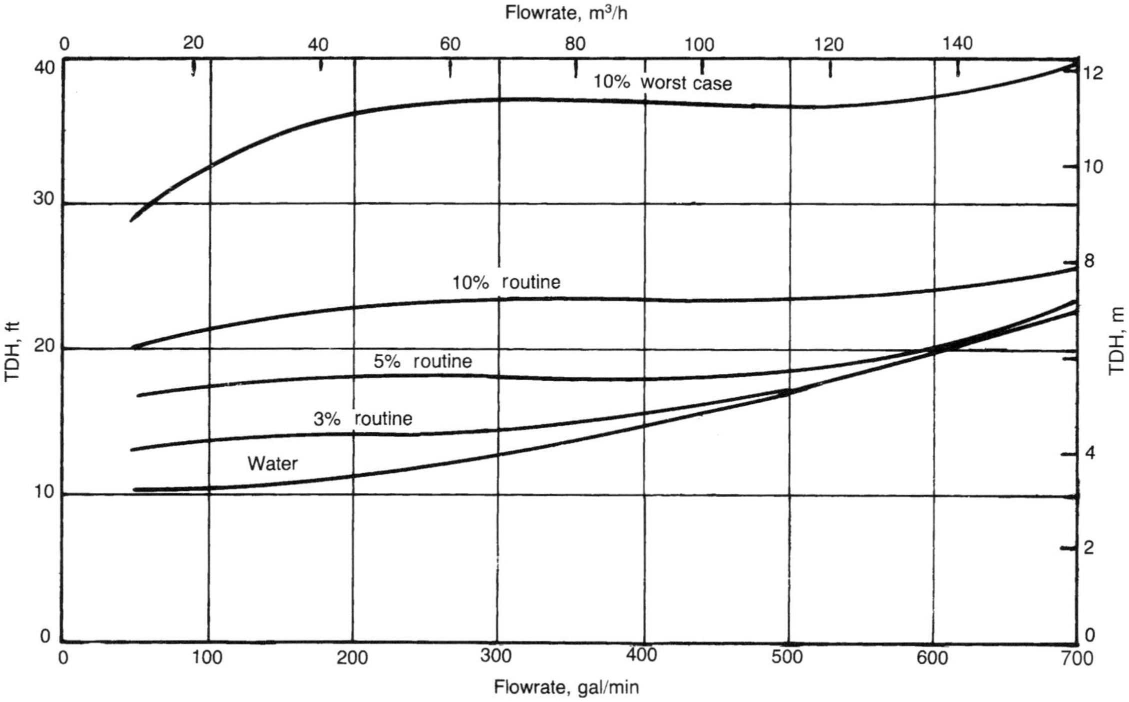

Figure 19-9. System H-Q curves for sludge in a 150-mm (6-in.) pipe.

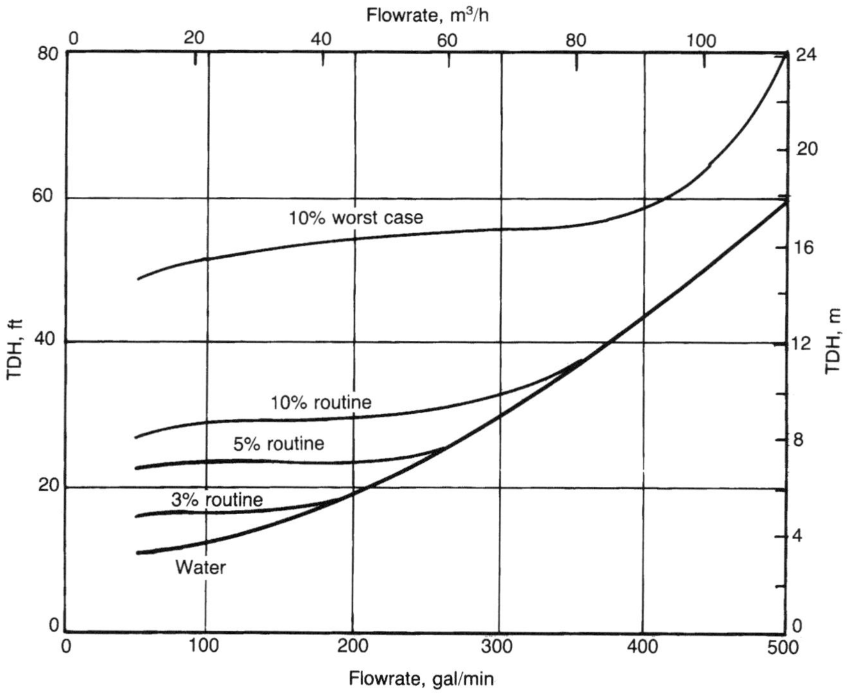

Figure 19-10. System H-Q curves for sludge in a 100-mm (4-in.) pipe.

Solids (%)	Speed (rev/min)	Flow (m^3/h)	TDH (m)	Flow (gal/min)	TDH (ft)
10[a]	710	11.3	8.81	50	28.9
10[b]	600	11.3	6.13	50	20.1
5[b]	570	22.7	5.67	100	18.6
3[b]	525	37.9	4.27	167	14.0
1[b]	640	113.3	5.27	500	17.3

[a] Worst-case.
[b] Routine operation.

(3) Select the driver. Following the pump selection, a decision can now be made to limit the drive speed to desirable values. The motor horsepower is determined by the intersection of the water system curve with the maximum drive output speed. In this example, if the maximum drive output speed is chosen to be 900 rev/min, the motor horsepower must be 20. Select the best practical motor drive and pump speed combination based on costs, sizes, and equipment

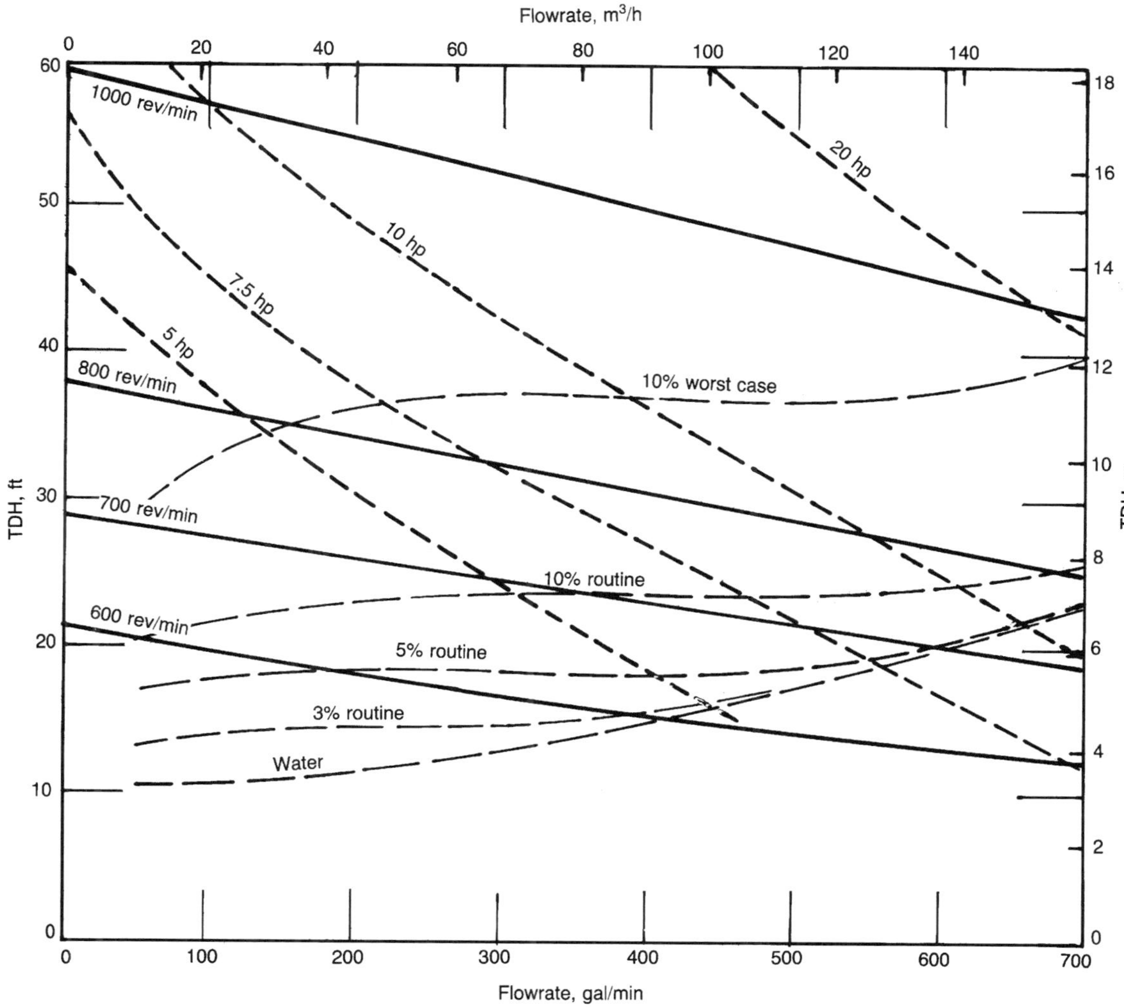

Figure 19-11. Performance curves for a 100-mm (4-in.) vortex pump and system curves for a 150-mm (6-in.) pipe. After WEMCO.

compatibility, and adjust the pump flowrates and times of operation to meet process transfer requirements.

Possibilities for drives include:

- Constant speed at 900 rev/min, with the time period of pumping adjusted accordingly.
- Variable speed at 900 to 450 rev/min to allow either a slip drive or an AFD (see Section 15-11) with appropriate running time adjustments. (One method is the use of time clocks.)
- Variable speed at 900 to 450 rev/min, a turndown ratio of at least 2:1 to allow a variable-ratio belt drive to be used.

Without more knowledge of the owner's circumstances, there is no answer better than the one given here.

(4) Compare headloss by a different method. Always cross-check both the design method and the actual calculations when designing sludge systems, because an error can have disastrous consequences. For example, use Equations 19-3 and 19-4 to confirm that the flow will be laminar, then use Equation 19-1 to check the headloss calculated in Part (1). Consider a sludge with a specific gravity of 1.02, containing 5% solids, flowing at a rate of 22.7 m^3/h (100 gal/ min) with a velocity of 0.35 m/s (1.14 ft/s) in a 150-mm (6-in.) pipe.

SI Units

Equation 19-3a

$$v_{lc} = \frac{1000\eta + 1000\sqrt{\eta^2 + s_y \rho D^2/3000}}{D\rho}$$

η = 0.015 kg/s · m

s_y = 4.8 Pa = 4.8 N/m^2

ρ = 1020 kg/m^3 (specific gravity = 1.02)

D = 0.15 m

g = 9.81 m/s^2

$$v_{lc} = 1000 \times 0.015/(0.152 \times 1020) + 1000 \times \frac{\sqrt{(0.015)^2 + (4.8)(1020)(0.152)^2/3000}}{(0.152)(1020)}$$

v_{lc} = 1.35 m/s

Equation 19-4a

$$v_{uc} = 1500 \times 0.015/(0.152 \times 1020) + 1500 \times \frac{\sqrt{(0.015)^2 + (4.8)(1020)(0.152)^2/4500}}{(0.152)(1020)}$$

v_{uc} = 1.69 m/s

U.S. Customary Units

Equation 19-3b

$$v_{lc} = \frac{1000\eta + 1000\sqrt{\eta^2 + s_y \gamma g D^2/3000}}{D\gamma}$$

η = 0.010 lb/s · ft

s_y = 0.10 lb/ft^2

γ = 63.65 lb$_m$/ft^3 (specific gravity = 1.02)

D = 0.5 ft

g = 32.2 ft/s^2

$$v_{lc} = 1000 \times 0.010/(0.5 \times 63.65) + 1000 \times \frac{\sqrt{(0.010)^2 + (0.10)(63.65)(32.2)(0.5)^2/3000}}{(0.5)(63.65)}$$

v_{lc} = 4.43 ft/s

Equation 19-4b

$$v_{uc} = 1500 \times 0.010/(0.5 \times 63.65) + 1500 \times \frac{\sqrt{(0.010)^2 + (0.10)(63.65)(32.2)(0.5)^2/4500}}{(0.5)(63.65)}$$

v_{uc} = 5.52 ft/s

Both lower and upper critical velocities are well above the actual velocity, so the Bingham equation for laminar flow is applicable.

Equation 19-1a

$$\frac{H}{L} = \frac{16 s_y}{3D\rho g} + \frac{32\eta v}{\rho g D^2}$$

$$\frac{H}{L} = \frac{16(4.8)}{3(0.152)(1020)(9.81)} + \frac{32(0.015)(0.35)}{(1020)(9.81)(0.152)^2}$$

Equation 19-1b

$$\frac{H}{L} = \frac{16 s_y}{3D\gamma} + \frac{32\eta v}{\gamma g D^2}$$

$$\frac{H}{L} = \frac{16(0.10)}{3(0.5)(63.65)} + \frac{32(0.010)(1.14)}{(63.65)(32.2)(0.5)^2}$$

SI Units	**U.S. Customary Units**
$\frac{H}{L} = 0.0176 = 17.6 \text{ m}/1000 \text{ m}$	$\frac{H}{L} = 0.0175 = 17.5 \text{ ft}/1000 \text{ ft}$
From Part (1) the friction slope is	
$\frac{H}{L} = 19.1 \text{(good check)}$	$\frac{H}{L} = 19.1 \text{(good check)}$

Critique

In the pump layout for the primary sludge pumps shown in Figure 19-7, there are two duty pumps with a common, identical standby pump (the one in the middle). This design allows a duty pump to be dedicated to each settling tank, allows separate process control for each tank, and gives the operator much more flexibility. The sludge flow characteristics are better with one pump for each tank than with one pump to serve two tanks, because there are no imbalances in friction head from each tank's suction piping to cause different sludge levels in the tanks.

The use of 150-mm (6-in.) piping requires confirmation by the manufacturer for applying the pump below the speeds shown on the catalog curve. This pump (4-in. Model C WEMCO) has been applied successfully for this particular example on the basis of eight years of operating experience at a speed of 300 to 350 rev/min [32]. Some engineers might favor the 100-mm (4-in.) piping because of better scouring due to greater fluid velocities, but objects such as toothbrushes or popsicle sticks can clog such small pipe unless grinders are used, so other engineers invariably avoid 100-mm (4-in.) pipe for sludge service. Note that operational velocities of 0.3 to 0.6 m/s (1 to 2 ft/s) occur in the 150-mm (6-in.) pipe at the expected solids concentrations of 3 to 5%. Such velocities are satisfactory in sludge pipelines if the grit content is not excessive. However, a variable-speed drive gives the operators the capability of pumping at greater velocities to clear the pipe. Adequate flushing water combined with the use of the plunger pump (see Figure 19-7 and Example 19-2) provide additional assurance for clearing obstructions in the 150-mm (6-in.) pipe. Hot water is helpful in flushing grease accumulations and is a desirable maintenance feature.

It is unlikely that primary sludge concentrations would reach 10%, although 7 or 8% may be attainable. However, the data for parameters s_y and η for undigested sludge are not extensive, so the use of parameters for 10% sludge is justifiably conservative.

If a more concentrated sludge at lower volumetric throughput is desirable, a different type of pump would probably be required. Options include: (1) a plunger pump, (2) a progressive cavity pump (if preceded by screening and grit removal), and (3) the combined screw-centrifugal impeller pump (such as a WEMCO Hidrostal®). An advantage of a positive displacement type pump is that process control measurements need not depend on flowmeters. Flowmeters may be unreliable and inaccurate in measuring sludge at low velocities.

The selection of a pump operating speed range of 450 to 900 rev/min allows operators to cope with varying sludge properties. Low pump speeds result in significantly less wear and reduced maintenance costs but require sturdier construction.

Design Procedure for Positive-Displacement Pumps

The design procedure for positive-displacement pumps is as follows:

1. Determine the maximum design pumping capacity required at the minimum sludge solids concentration.
2. Determine which combinations of simplex, duplex, and triplex pumps are feasible using the capacities found for each type of pump in a manufacturer's catalog. (Capacities vary from manufacturer to manufacturer).
3. Plot the sludge system head curve for the maximum percentage of dry solids sludge concentration.
4. Calculate the actual pulsing flow for each of the pumps under consideration by using the percentages shown in Table 19-1.
5. From the system head curve, read the total head corresponding to the maximum pulsed flow for each type of pumping combination. Each type of

pump specified must meet this total head requirement, which represents the most critical condition against which the plunger pump must operate.

6. Study manufacturers' publications to determine the motor horsepower required. Consult with the manufacturer to confirm the selection.
7. Select the type of pump most applicable with consideration for the following factors:

- Because the single cylinder, simplex pumps are the least balanced of the pumps, this type exerts the maximum loads on pump bearings and gears, and causes the highest friction in the pipe. These considerations are of the greatest concern at high heads.
- When design conditions approach 690 kPa (100 lb/in.^{2}), it is preferable to select the smallest possible bore and multiple cylinders.

Example 19-2
Design of a Plunger Pump Installation

Problem: Design a plunger pump backup for the vortex pumps of Example 19-1. Assume 100-mm (4-in.) piping.

Solution: As a backup, the plunger pump would be used only when a pipeline is clogged or if all three vortex pumps were being repaired, so concern is limited to concentrations of solids higher than about 3% and, hence, to flows of approximately 11.3 to 37.7 m^3/h (approximately 50 to 167 gal/min).

Pump. Find the pump scheme that approximates the design flow. The capacities per plunger per number of cylinders for one brand of pump are as follows.

Pump scheme	Average rated capacity for various cylinder diameters			
	225 mm	275 mm	9 in.	11 in.
1 simplex	19.3 m^3/h	28.4 m^3/h	85 gal/min	125 gal/min
2 simplexes	38.6 m^3/h	56.8 m^3/h	170 gal/min	250 gal/min
1 duplex	34.1 m^3/h	49.9 m^3/h	150 gal/min	220 gal/min
1 triplex	52.2 m^3/h	68.1 m^3/h	230 gal/min	300 gal/min

One duplex pump with the larger cylinder size is adequate.

Variable-speed drive. Variable speed gives better control over pumping conditions than does constant speed. At high sludge concentrations, the high flowrates delivered by a constant-speed pump may result in "keyholing" in which the sludge cannot enter the suction pipe at high rates. Power is based on the head at pulse flowrates. The ratio of pulse to average flowrate for a duplex pump is 1.60 (see Table 19-1). Find the head corresponding to the pulse flow rate from the curves in Figure 19-10.

Curve in Figure 19-10	SI Units			U.S. Customary Units		
	Average flow (m^3/h)	Pulse flow (m^3/h)	Pulse TDH (m)	Average flow (gal/min)	Pulse flow (gal/min)	Pulse TDH (ft)
10% worst case	11.3	18.2	14.3	50	80	47
5%	22.6	36.3	7.3	100	160	24
3%	37.7	60.6	7.6	167	267	25

In Figure 19-12, find the intersection of pulse TDH and average flow. For 3% sludge, the shaft output is about 5 hp (by interpolation). The speeds should vary about 167:50 or 3:1, and the maximum speed is 50 rev/min. A variable-ratio belt drive (see Table 15-2 in Section

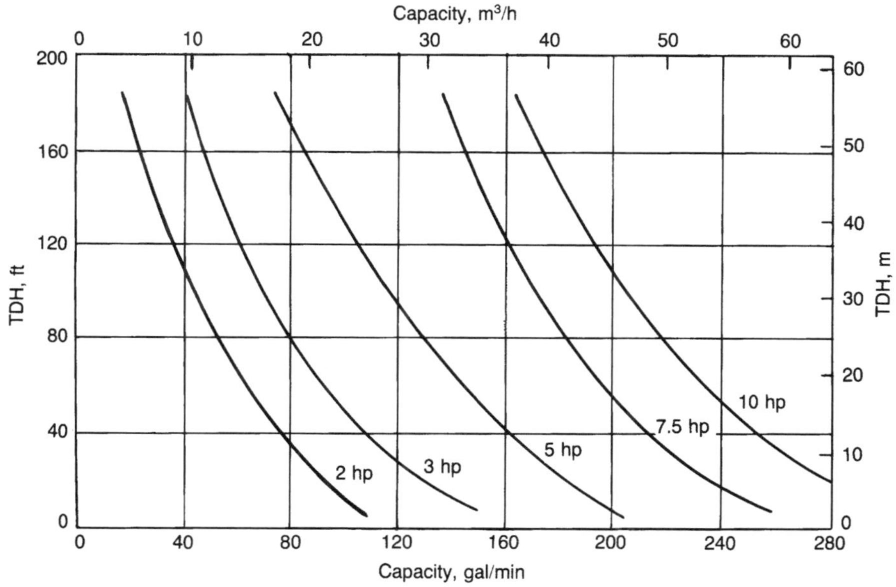

Figure 19-12. Performance curves for a duplex, heavy-duty, dual-valve, 275-mm (11-in.) plunger pump with a maximum speed of 50 rev/min and a 3-m (10-ft) suction lift (for pulse TDH and average flow). After Komline-Sanderson Engineering Corp.

15-11) is the recommended choice. Consult with the manufacturer to determine the efficiency and, thus, the size of motor required. Note that some manufacturers' curves are based on a "nonpulsating" flow and corresponding head due to probable inclusion of air chambers. Be cautious; use experience and judgment to interpret the basis of manufacturers' curves and design accordingly.

Critique

There are many other choices that might be made, because the operating conditions for a backup positive-displacement pump are a matter of judgment. For example. smaller pumps can be operated for longer times. Typically, the primary objective (pumping scum) determines the sizing of the plunger pump, and then the size is double checked against the pump's standby function.

19-4. Piping System Design

Due to the nature of sludge, the requirements of sludge piping systems differ in many ways from those of water piping systems. The following discussion relates specifically to sludge as an addendum to the general design practices for water and wastewater presented elsewhere in this text.

Pipe Materials

The most common pipe materials for sludge piping systems are ductile iron and steel; regional preferences are toward ductile iron in the eastern United States and steel in the western United States. The choice depends primarily on costs and availability because the performance of either, when properly protected from corrosion, is satisfactory.

Designers differ on the minimum allowable size of pipe. Some limit sludge piping to a minimum of 150 mm (6 in.), and some require no less than 200 mm (8 in.). Some do not object to 100 mm (4 in.) with glass or cement-mortar lining. Pipe used in sludge piping systems should be lined to resist corrosion and/or abrasion or to resist grease buildup. Grease is a significant concern with scum and primary sludges, and a smooth surface in piping for digested sludge inhibits the formation of struvite crystals [2]. The most common lining for ductile iron and steel pipe is cement mortar, but an even smoother lining, such as glass, is desirable. The economics and long-term reliability of glass-lined pipe should be evaluated carefully. It may be possible to achieve the desired goal of reducing maintenance on sludge pipelines to the same level that is projected for glass-lined pipe by judiciously selecting the allowable velocities, eliminating bends and pipe constrictions or "bottlenecks," and providing adequate flushing and cleaning facilities in pipes lined with cement mortar. The use of polyethylene-lined ductile iron pipe is a viable, economic alternative to glass-lined pipe. If the grit content of the sludge is low, some plastic pipes exhibit good resistance to wear and maintain clean internal surfaces.

Valves

The discussion in Chapter 5 should be modified by the following general comments and tempered with the designer's experience. Note that any valve should be both accessible and operable from the pump room floor. If pigs or other cleaning devices must pass through an open valve, select a ball, cone, or eccentric plug valve, preferably with a full, round port to favor clog-free operation.

Eccentric Plug Valves

One of the most common types of valves, the nonlubricated, eccentric plug valve, is usually specified to have a port area of at least 80% of the full pipe area to provide clog-free operation. These valves can be obtained with a full, round port opening, which is advantageous if the line is to be rodded or pigged. However, some pigs can pass through square ports.

Typically, valves smaller than 200 mm (8 in.) are equipped with 50-mm (2-in.) nuts for wrench operation, whereas larger valves have worm gears and hand wheels. Synthetic rubber covering is usually specified for eccentric plug valves to avoid wear of and damage to the seating surface. Eccentric plug valves that provide tight shut-off with pressure in either direction (a distinct advantage in sludge pipelines with multiple flow routings) are offered by at least one major manufacturer.

Ball Valves

Consider full-bore ball valves instead of plug valves on applications where a stoppage could require a major process tank (such as a digestion tank) to be dewatered to service the valve. In general, three-way valves are not used routinely because they usually have smaller openings and increase the likelihood of clogging. Also, if the valve fails both piping routes are out of service until the problem is corrected. Ball valves are sometimes preferred on corrosive service because the relatively small body size and weight mean less of an expensive alloy. Some ball valves are commonly made of thermoplastic with tension seats [33].

Check Valves

Check valves add to the design problem on sludge systems if they are installed on a vertical pipe because solids pack on the discharge side. Install check valves (if used at all) on horizontal pipes.

Ball Check Valves

Ball check valves are used in (and recommended for) plunger and diaphragm pumps (see also Section 5-4). The balls, made of lead-impregnated synthetic rubber, are contained in a chamber with a quick access door to facilitate the frequent replacement required.

Stringy materials tend to prevent proper seating of the balls, so plunger pumps are commonly supplied with a pair of check valves in tandem on both suction and discharge sides of the pump (see Figure 11-31). Diaphragm pumps can also be obtained with pairs of ball check valves. The reliability of a plunger or diaphragm pump can be improved by installing a grinder or macerator ahead of the pump. If grinders are placed too far upstream, they are likely to be less effective because separated and ground stringy material tends to become reconstituted.

Pinch Valves

Pinch valves are also widely used in sludge service, especially as relief valves where there is a need to protect a positive-displacement pump (or other downstream equipment) from an inadvertent valve closure in the discharge line. The system consists of an air-loaded pinch valve at a pipe tee that discharges fluid back to the pump suction source. A pinch valve system is relatively maintenance free, but it does require an air source and a pressure regulator.

Cone Valves

Cone valves are designed for throttling control and have been used successfully on water and raw sewage lines to regulate valve closures for elimination of water hammer effects. Cone valves also replace both the pump discharge check and isolation valves. None of these aspects, however, routinely applies to sludge service. Apparently, cone valves are not used in sludge systems and one manufacturer's representative stated that the valve is not intended for such service [34].

Gate Valves

Gate valves are seldom used today in sludge service because of their tendency to collect debris and solids as compared to the preferable plug valve [27].

Pump Seal Systems

Most sludge pumps are specified with seal water systems rather than mechanical seals due to the nature of pump service. The seal water flow requirement should be reviewed with the pump manufacturer. Seal water is taken from a water system but never from the pumped fluid discharge. Potable water must be separated completely from seal water by an air gap in the seal water tank or an acceptable backflow preventor system.

Seal water lines should have a pressure-reducing valve on the seal water header supplying seal water to a series of pumps. The valve prevents the seal water from entering the pump casing at an excessive pressure that could dislodge the packing and accelerate packing and shaft sleeve wear. Seal water leaving the pump should be conveyed by piping to the floor drain system and never be allowed to overflow the pump base or pump room floor. Pump bases should be curbed with drain outlets. A typical water seal system schematic is shown in Figure 19-13 and a suggested pump drain system is shown in Figure 19-14.

Flowmeters

The nature of sludge makes it necessary to use flowmeters that do not obstruct the flow. Sludge metering systems are expensive, often inaccurate, and always require preventive and corrective maintenance. Unless positive-displacement pumps are used, meters are nevertheless necessary to measure the flow of sludges. It is also desirable to measure solids content continuously so that the operators can determine the mass of (dry) solids transferred to track process control. If some means of sampling solids concentrations is not included, the flow data will only track volumes of fluids and not mass of (dry) solids.

Venturi, magnetic, ultrasonic, or in-line density flowmeters can be used. The first three are typically preferable to the fourth due to pipeline obstruction considerations. One advantage of a positive-displacement pump is that the pump itself can be used as a meter if the number of revolutions or strokes over a specific time period is recorded.

A concern about meters in sludge pipelines is pipeline cleaning. Flowmeters should be installed in the piping system with a bypass for routine maintenance,

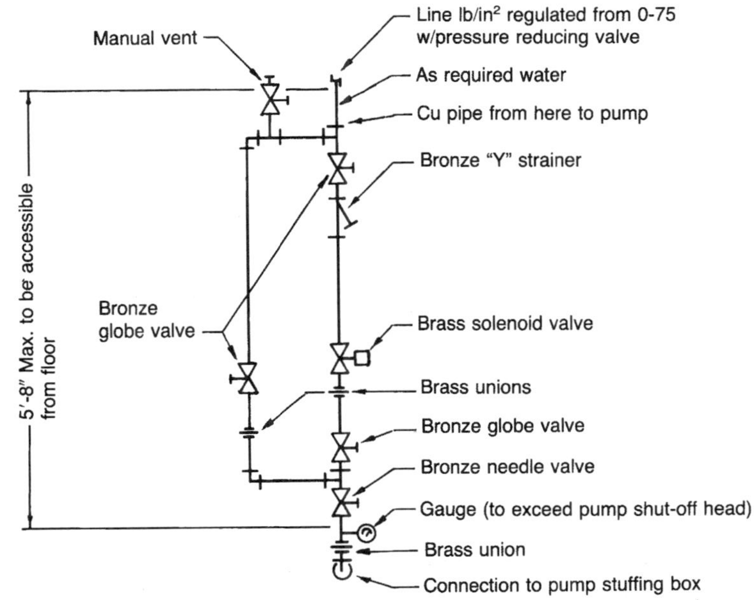

Figure 19-13. A typical water seal connection. Courtesy of Stearns & Wheler Engineers and Scientists and J. Kenneth Fraser & Associates.

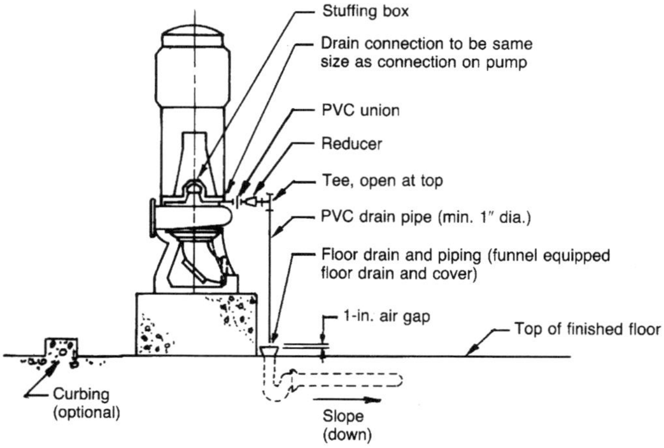

Figure 19-14. Pump drain system. This arrangement is also applicable to horizontal pumps. Courtesy of Stearns & Wheler Engineers and Scientists and J. Kenneth Fraser & Associates.

calibration, and (if needed) a water purging system (see Figure 20-12). A pipe spool piece should be available to replace the meter when it is removed for servicing or when the line must be pigged.

The relative accuracy of each type of sludge meter is debatable. Many engineers believe the Venturi system is the best overall choice when accuracy, maintenance, and long-term reliability are considered. Others prefer magnetic or ultrasonic meters.

Grinders

Grinders or macerators are sometimes provided in sludge piping systems for shredding solids, rags, and other debris to reduce wear on downstream pumps and sludge handling or processing equipment.

When an upstream sludge grinder is used to protect the pump, add an interlock between the sludge grinder and the pump to keep the grinder operating whenever the pump is running. However, a manual bypass should be installed around the grinder so it can be repaired.

Pipe Flushing and Draining

Flushing sludge piping systems is a common maintenance chore in water and wastewater treatment plants. Flushing requires an adequate velocity and pressure of water to dislodge solids, move obstructions, and generally clean the pipe. Flushing connections should be liberally provided to allow convenient operations. Flushing water should be available to the piping system on both the suction and discharge sides of sludge pumps.

Pipe Cleaning Stations

A well-designed system for pipe cleaning allows the piping system to be maintained in mint condition and reduces the pumping power requirement. Two systems used for cleaning pipe are (1) rodding out each straight section of piping and (2) cleaning the entire pipe with pipe-cleaning pigs.

Rodding

Rodding requires specifying cleanouts on all elbows, tees, and crosses within the pumping station. Require pipe-sized cleanouts for diameters up to 100 mm (4 in.) and cleanouts one-half the pipe size on larger pipe at bosses on all bends or other angled fittings. Additionally (or alternatively), bends can be replaced by crosses with removable blind flanges for cleanout or rodding. All cleanout plugs should be fitted with 25-mm (1-in.) or larger plug valves with hose adapters. The sludge piping can then first be drained or cleared by high-pressure water in preparation for rodding. Cleanouts for rodding must be located no farther than 15 m (50 ft) apart unless special rodding tools are available.

Pigging

The main advantage of cleaning pipe with pigs is that the pipe length is practically unlimited. Cleaning pigs range from soft swabs to flexible, abrasive pigs to rigid steel scraper pigs. Block ice or bagged ice cubes can also be used; with this method, retrieval is unnecessary and permanent jamming cannot occur. Flexible pipe-cleaning pigs can clean several pipe sizes, but are limited in turning around pipe crosses. Transmitter-cleaning pigs enable the maintenance crew to locate underground pipe failures accurately.

Cleaning the piping with pigs requires provisions for the insertion and retrieval of the pigs. The pigs are inserted in an isolated length of pipe one size larger than the piping to be cleaned and located as close to the primary pumps as possible. The pig launcher is fitted with a pressure gauge, a drain, a high-pressure water fitting, and an easily removed blind flange or mechanical coupling. The pump discharge and the pig launcher pressure gauges should be easily read from one point. Ideally, this control point is protected from a pipe rupture and includes other controls as necessary. The pressure gauges last longer if they are filled with glycerin and protected by an isolation diaphragm and a snubber (glycerine-filled capillary tubes) to keep sludge out of the gauge and attenuate sharp pressure spikes. The details of cleaning pipe with pigs are discussed by Playford in the *Proceedings* [14, pp. 878–879].

The ease of cleaning pipes with pigs and the increasing cost of power warrants the installation of pipe-cleaning, launching, and receiving stations on most pipe systems and networks.

Safety Factors for Sizing

Sludge piping systems should be as short as possible with a minimum number of bends. Bends (unless replaced by tees or crosses) should be the long radius type to minimize the headloss. In addition, the headloss contributed by check valves, flowmeters, and other piping accessories should be considered critically in the analysis of the system. If at all possible, omit check valves and fittings that would cause plugging of solids or add unnecessary headloss.

In each sludge pumping system, at least two pumps sized at full capacity should be provided so that at least one full-capacity standby pump is available. It may well be appropriate to use a positive-displacement standby pump because of its reliability and ability to move heavy sludge.

19-5. Long-Distance Pumping

Pumping wastewater sludges through pipelines 1.6 km (1 mi) long (or longer) requires greater attention to issues that are not as critical in pipelines of shorter length. Small variations in the unit dynamic headloss are magnified by the length of piping in a long pipeline. Dynamic headlosses are significantly affected by

- Concentration of the sludge
- Type of sludge (raw, secondary, or digested)
- Variability of blends if a mixture of sludge is pumped
- Additives used in the wastewater or sludge treatment process (polymers or metal salts).

Other design and operating considerations for long sludge pipelines include

- Periodic venting of gases released from solution or generated by the digestion process
- Correct sizing of the pipeline to allow for large flowrate variation over the system lifetime
- Providing sludge storage or a backup pipeline to enhance reliability and allow for pipeline maintenance
- Selection of pumps that can perform satisfactorily over a wide range of discharge pressures as concentrations and sludge blends change.

The designer of a long sludge pipeline must give serious consideration to measuring the characteristics of the range of sludge blends, solids concentrations, chemical additives, and sludge conditions (including temperature) likely to be encountered. The need for site-specific sludge testing is highlighted by the variability of the published data [13, 18] on sludge characteristics—particularly when major investments are to be made in pumping transport and storage systems. Extensive testing is even more necessary when undigested secondary sludges are to be pumped. These complex problems are covered in more detail by Carthew *et al.* [18].

The use of specially designed diaphragm pumps capable of large flowrates at pressures up to 2400 kPa (350 lb/in.^{2}) has been successful. The application of such pumps is unusual, but several advantages include (1) low maintenance, (2) high reliability, (3) excellent protection against overpressuring the pipeline, (4) clean, quiet operation, and (5) separation of sludge from most of the pump. Diaphragm pumps not specifically manufactured for sludge or slurry service do not have these advantages.

19-6. References

1. Metcalf, L., and H. P. Eddy, *American Sewage Practices. Volume III: Disposal of Sewage,* 3rd ed., McGraw-Hill, New York (1935).
2. *EPA Process Design Manual for Sludge Treatment and Disposal, EPA 625/1-79-011,* pp. 14-4–14-17, 14-29–14-37, U.S. Environmental Protection Agency, Municipal Environmental Research Laboratory, Office of Research and Development, Cincinnati, OH (September 1979).

3. Babbitt, H. E., and D. H. Caldwell, "Laminar flow of sludges in pipes with special reference to sewage sludge." *University of Illinois Engineering Experiment Station, Bulletin Series, No. 319* (1939).
4. Caldwell, D. H., and H. E. Babbitt, "The flow of muds, sludges, and suspensions in circular pipe," *Trans. of the American Institute of Chemical Engineers,* **37,** 237 (April 1941).
5. Wolfs, J. R., "Factors affecting sludge force mains," *Sewage and Industrial Wastes,* **22,** 1 (January 1950).
6. Babbitt, H. E., and D. H. Caldwell, "Turbulent flow of sludges in pipes," *University of Illinois Engineering Experiment Station, Bulletin Series, No. 323* (1940).
7. Brisbin, S. G., "Flow of concentrated raw sewage sludges in pipes," *Journal of the Sanitary Engineering Division American Society of Civil Engineers,* **83(SA3),** 1274 (June 1957).
8. Chou, T. L., "Flow of concentrated raw sewage sludges in pipes," *Journal of the Sanitary Engineering Division American Society of Civil Engineers,* **84(SA1),** 1557 (February 1958).
9. Sparr, A. E., "Pumping sludge long distances," *Journal of the Water Pollution Control Federation,* **43,** 1702 (August 1971).
10. Dick, R. L., and B. B. Ewing, "The rheology of activated sludge," *Journal of the Water Pollution Control Federation,* **39,** 543 (1967).
11. Bourke, J. D., "Sludge handling characteristics in piped systems," *Proceedings of the Northern Regional Conference of the California Water Pollution Control Association,* Monterey, California (October 19–20, 1973).
12. Vesilind, P. A., *Treatment and Disposal of Wastewater Sludges,* 2nd ed., Chapter 4, Ann Arbor Science Publ., Ann Arbor, MI (1979).
13. Mulbarger, M. C., *et al.,* "Pipeline friction losses for wastewater sludges," *Journal of the Water Pollution Control Federation,* **53,** 8, 1303–1313 (August 1981).
14. Sanks, R. L., C. W. Reh, and A. Amirtharajah (eds.), *Conference Proceedings, Pumping Station Design for the Practicing Engineer,* pp. 854–899. Montana State University, Bozeman, MT (1981). Available through interlibrary loan.
15. Govier, G. W., and K. Aziz, *The Flow of Complex Mixtures in Pipes,* Van Nostrand Reinhold, New York (1972).
16. Cain, C. B., *et al.*, "Design of 90-mgd wastewater reclamation plant," *Journal of the Environmental Engineering Division American Society of Civil Engineers,* **107(EEI),** 29 (1981).
17. Rimkus, R. R., and R. W. Heil, "The rheology of plastic sewage sludge," in L. K. Cecil, ed., *Second National Conference on Complete Water Reuse, American Institute of Chemical Engineers,* Chicago, IL (May 4–8, 1975).
18. Carthew, G. A., C. A. Goehring, and J. E. van Teylingen, "Development of dynamic headloss criteria for raw sludge pumping," *Journal of the Water Pollution Control Federation,* **55,** 472–483 (May 1983).
19. Hanks, R. W., and B. H. Dadia, "Theoretical analysis of the turbulent flow of non-Newtonian slurries in pipes," *Journal of the American Institute of Chemical Engineers,* **17,** 554 (May 1971).
20. Hanks, R. W., "Principles of slurry pipeline hydraulics," in N. P. Cheremisinoff (ed.), *Encyclopedia of Fluid Mechanics, Vol. 5, Slurry Flow Technology,* Gulf Publishing, Houston (1986).
21. Darby, R., "Laminar and turbulent pipe flows of non-Newtonian fluids," in N. P. Cheremisinoff (Ed.), *Encyclopedia of Fluid Mechanics, Vol. 7, Rheology and Non-Newtonian Flows,* Gulf Publishing, Houston (1988).
22. *Pipeline Design for Water and Wastewater,* American Society of Civil Engineers, Report of the Task Committee on Engineering Practice in the Design of Pipelines, New York (1975).
23. Karassik, I. J., W. C. Krutzsch, W. H. Fraser, and J. P. Messina, *Pump Handbook 2nd ed.,* Chapters 2 and 9, McGraw-Hill, New York (1986).
24. Henshaw, T. L., *Reciprocating Pumps,* Chapter 6, Van Nostrand Reinhold, New York (1987).
25. Baumeister, T. (ed.), *Standard Handbook for Mechanical Engineers,* 8th ed., Chapter 14, McGraw-Hill (1978).
26. Wylie, E. B., and V. L. Streeter, *Fluid Transients,* Chapter 14, McGraw-Hill, New York (1978).
27. Water Pollution Control Federation and the American Society of Civil Engineers, *Wastewater Treatment Plant Design,* WPCF Manual of Practice No. 8. Lancaster Press, Lancaster, PA (1977).
28. Burrage, K., President, Penn Valley Pump Co. Private communication (September 1997).
29. Jones, G. M., Vice President, Brown and Caldwell Consultants. Private communication (February 1989).
30. Svaras, J. C., Application Engineer, Alfa Laval Pumps, Inc. Private communication (September 1997).
31. Caballero, R. C., County Sanitation District of Los Angeles. Private communication (undated; tested March 1983).
32. Cunningham, J., Chief Operator, Clinton Wastewater Treatment Plant. Private communication (January 1989).
33. O'Keefe, W., "Valves for Corrosives and Slurries," *Power,* Special Report, pp. 51–516 (May 1986).
34. Allis-Chalmers Valve Division. Private communication (January 1989).

Chapter 20

Instrumentation and Control Devices

ROBERT S. BENFELL

CONTRIBUTORS
Russell H. Babcock
Mead Bradner
Harold D. Gilman
Mayo Gottliebson
Paul C. Leach
John Leak
Earle C. Smith
James Taube
Horton Wasserman

The purpose of this chapter is to introduce the basic instrumentation and control devices that are needed and used in pumping stations. The information presented should be of value to (1) designers, who must plan for and specify equipment, (2) project engineers, who must decide how a pumping station will operate and what kind of equipment will be needed, and (3) owners, who must maintain the equipment selected. The types of instruments and their principles of operation are discussed in detail by Lipták [1]. Specific control devices and their advantages can be found in manufacturers' literature. However, what is needed (and presented in this chapter) is a careful analysis of the *relative* merits of instruments specifically for use in pumping stations.

20-1. Reliability

Instruments for pumping stations should be selected to provide long life, low maintenance, and high reliability in damp, corrosive environments. These requirements are not necessarily unique to pumping stations, because many industrial applications have similar requirements. But industrial users are frequently less concerned with extended life, and most such users maintain highly skilled, full-time maintenance staffs to service sophisticated equipment. Pumping station operators often have limited sophisticated maintenance skills and usually rely on outside maintenance services. Furthermore, most pumping stations are designed to operate for 20 yr or more without major reconstruction. Accordingly, the simplest instrument that will perform the required function should be selected for pumping stations, and premium materials, such as stainless steel, should be used wherever possible to provide maximum resistance to environmental conditions.

Because many pumping stations are operated unattended and the potential property damage is significant if a serious malfunction occurs, instruments should be selected for their inherent reliability. Backup systems should be provided, and the consequences of component failure must be carefully considered.

Instrument design is a highly specialized field that requires knowledge of both instrumentation equipment and fluid mechanics. Any engineer who designs instrumentation and control systems must carefully consider the application limitations of the systems to be specified and should seek the services of qualified specialists if unusual and/or complex systems are required.

20-2. Instrument Selection

There are many instruments and instrument systems on the market that were designed specifically for water and wastewater facilities. But only some of this equipment was designed with serious consideration given to the requirements of such facilities. Unfortunately, much of it was designed with low cost as the primary consideration to compete in a market in which much equipment (1) is bought from the "low bidder" and (2) is probably inferior to equipment that was designed primarily for the industrial marketplace. Every instrument or functional instrument system must be evaluated solely on the basis of its suitability for the application. Instruments and instrument systems designed for the water and wastewater marketplace are not inherently better or worse than instruments and instrument systems designed for industrial applications for service in pumping stations.

Environmental Conditions

The environmental conditions to which the various instruments in a pumping station are subjected vary considerably from station to station and from location to location within a pumping station. Most pumping stations have areas of high humidity, many have areas of extreme temperature, and most wastewater pumping stations have areas that are corrosive or classified as hazardous by the NEC [2]. Instruments that are resistant to these conditions are generally available. However, instrument life, reliability, and/or maintainability are adversely affected by service under such conditions. Effects of corrosive gasses and high humidity can be reduced or eliminated by adequate ventilation or by removing as much of the instrumentation equipment as possible to more hospitable areas of the station. Electrical instruments in hazardous atmospheres can be put into explosionproof cases, but pneumatic instruments and intrinsically safe electronic instruments are easier to maintain.

Reliability and Maintainability

In many instruments, the "motion-balance" principle is used for operation. These instruments have springs, racks and pinions, and bearings that are all subject to wear. A simplified motion-balance Bourdon-tube mechanism used for pressure measurement is shown in Figure 20-1. An increase in pressure within the Bourdon tube straightens it slightly, which lifts the linear voltage differential transformer (LVDT) core. The primary coil is excited with a high-frequency alternating current, and the secondary coils are connected in series opposition so that the two voltages produced are of opposite phase. As the core moves up, the magnetic coupling between the lower secondary coil and the primary coil decreases and the magnetic

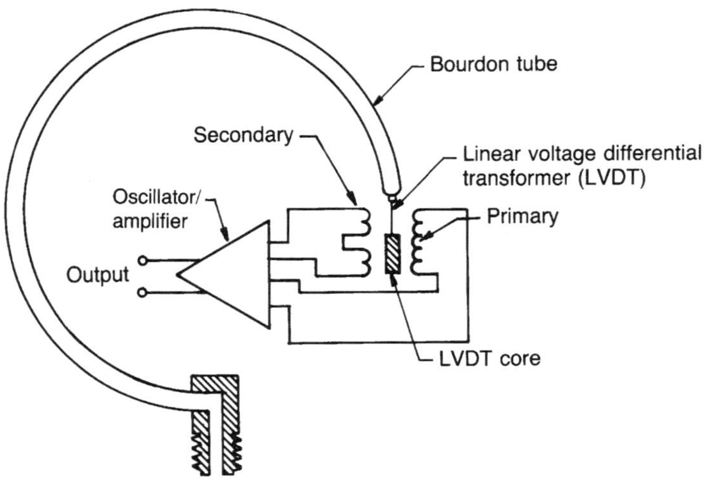

Figure 20-1. Motion-balance mechanism.

coupling between the upper secondary coil and the primary coil increases. Suitable phase-sensitive electronics are used to produce a direct current output proportional to core position.

"Force-balance" instruments reduce the amount of movement and, therefore, reduce wear. A simplified force-balance mechanism used for pressure measurement is shown in Figure 20-2. An increase in pressure within the housing displaces the diaphragm to the left, which rotates the force beam clockwise about the fulcrum and unbalances the null detector. The amplifier detects this unbalance and increases its output, which tends to rotate the force beam counterclockwise until balance is again established. The current passed through the solenoid force motor is also the output current, which is typically 4 to 20 mA. By designing the solenoid to provide a linear response and using a high gain null detector and amplifier, the motion required to provide a signal is reduced to a very small value. The only parts actually strained are the diaphragm and the seal at the fulcrum. The stiffness of these parts is made insignificant, which results in a very stable pressure-measuring device.

Many modern instruments utilize solid-state circuitry with no moving parts. Ordinarily, force-balance and solid-state instruments are more reliable and require less maintenance than motion-balance instruments. However, motion-balance instrument mechanisms are more easily understood by maintenance personnel familiar with mechanical equipment. Consider who is going to maintain the instruments before selecting exotic modern equipment that may require highly skilled specialists for routine servicing.

Utilities

Many instruments require either electrical power or instrument air for their operation, and such facilities constitute a major cost of instrumentation systems. Electric power must be (1) conditioned to remove surges, which may interfere with proper instrument operation or even damage sensitive electronic circuitry, (2) converted to the required voltage levels, and (3) regulated to ensure stable operation. In many pumping stations, provision must be made to ensure that electric power for instruments is maintained during service power outages.

Pneumatic instruments require clean, dry air except for bubbler systems connected to pressure switches. If the instrument air supply is not of adequate quality, instrument reliability suffers and maintenance requirements increase very significantly. Because the cost of purchasing and maintaining adequate instrument air equipment is significant, electronic instrument systems are most often selected for modern systems. However, pneumatic instruments have significant advantages in high temperature, corrosive, and hazardous locations. Also, level measurements are more

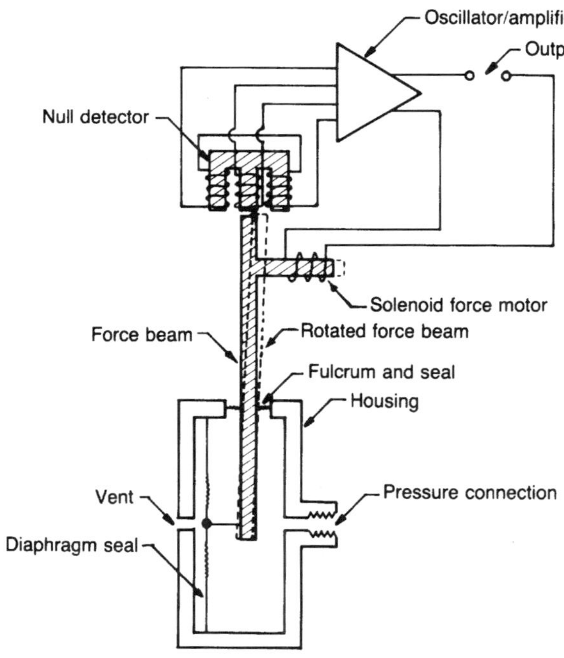

Figure 20-2. Force-balance mechanism.

reliably made with pneumatic bubbler systems than with electronic systems. In many pumping stations, particularly large wastewater stations, pneumatic instruments can be used advantageously if the maintenance workers are familiar with this type of equipment.

It is true that explosionproof cases are available for field instruments, but there is no way to service or calibrate them without opening the case, which makes them "nonexplosion" proof. Intrinsically safe circuitry sounds easy but is difficult to accomplish. In warm climates, pumping stations are often outdoors and sunlight can heat the cases to temperatures that electronic devices cannot withstand. Chlorine and hydrogen sulfide can quickly destroy electronic circuitry.

The choice of electrical or pneumatic equipment may depend on the skills of the service personnel. A skilled technician can disassemble, clean, and reassemble pneumatic equipment quickly (often in much less than an hour) in the field. Electronic equipment usually has to be removed and returned to the factory.

Accuracy

Accuracy requirements for instrument systems are generally not severe. The exception to this rule is "custody transfer" flow measurements (situations in which money passes from one entity to another based on flow quantities) that occur in water or wastewater applications. Instrument systems usually consist of a number of interdependent elements, each of which contributes to the total measurement error—errors that are always greater than the ones published in manufacturers' literature. A major source of error is the piping configuration at the point of measurement. It seldom resembles the piping in the hydraulics laboratory that was originally used to establish instrument performance.

The error of many instruments is stated as a percentage of span. Most instruments ordinarily operate at a point considerably below the maximum calibrated range. If high accuracy (such as in custody transfer) is required, extreme care must be taken to be sure that the instruments are properly applied and installed and that unnecessary error is not introduced by less than optimum configurations of multiple instruments in a system. Where instrument systems are used for custody transfer, it must be recognized that a certain amount of error is inevitable, and contracts must be written to recognize and accept the error.

Package Systems

Package systems are available to perform many instrumentation functions commonly used in pumping stations. Common examples include pump sequence control and chlorination control. The obvious advantage of a package is that it can be applied by a designer who is not skilled in instrumentation system design.

The following are disadvantages of package systems:

- It may be impossible to find one that does exactly what is desired, which may lead to compromises in control strategy.
- Packages may be difficult or even impossible to modify if the control strategy needs to be changed.
- Nonstandard signals between elements are frequently used, so all replacement parts must be obtained from the original vendor, who may discontinue the model or even go out of business.
- Packages tend to be designed so that only the original vendor can service them.
- Many are designed without adequate consideration of the environment in a pumping station or the need for reliability and long life.

This is not to say that packages should never be used. They frequently provide cost-effective solutions, particularly in relatively simple applications. But take the same care in selecting packages as in selecting components for custom-designed instrument systems.

Measurement Types

There are two basic types of measurement: analog and discrete. Analog measurements produce a signal or readout proportional to the value of a certain parameter, such as flow or level. Discrete measurements produce a true/false indication, such as valve "open/closed" or pump "running/stopped."

Signal Transmission

Many instruments provide only a local indication of a measurement. A pressure gauge is a common example. But, frequently, signal transmission to another location is required. Wherever possible, use standard signal transmission systems, such as 4 to 20 mA for electrical systems and 20 to 100 kPa (3 to 15 lb/in.^{2}) for pneumatic systems. These systems allow unrestricted interconnection of instruments of different manufacturers. Many nonstandard systems have been used over the years in water and wastewater facilities. Some are well established and conversion devices are readily available to incorporate such signals into standard systems, but such conversion devices (1) add more equipment to maintain, (2) reduce reliability, and (3) increase system error.

Variables Measured

Variables or parameters frequently measured or sensed are listed in Table 20-1. Not all of them would be used in any but very large stations. The purpose of a pumping station is to pump—not to collect data. If there is not a well-defined use and need for data, do not collect them. If a device is not essential, omit it.

20-3. Level Measurements

Level is one of the most common measurements made in pumping stations. Instrument types frequently used for this measurement are summarized in Table 20-2 more or less in order of popularity.

Floats

A float is a buoyant body that rides on the surface of a liquid and includes a method of monitoring its float elevation. A float may be suspended from a rod or a cable that mechanically links it to an elevation monitoring device, or the float may contain a magnet and simply slide up and down on a rod containing reed switches that are magnetically actuated. Another type of float, sometimes called a "tethered float," consists of a spheroidal, hollow vessel with a mercury switch inside. When lifted (or buoyed), the float changes position and actuates the switch. Tethered floats (Figure 20-3) are frequently used in wastewater applications with some degree of success. Floats have these disadvantages:

- They are difficult to adjust.
- Access to the wet well is required for servicing.
- The electric cable flexes and wears out, so the float must be replaced periodically (2 to 3 yr).

Float devices, which are available in a wide variety of materials, are simple and inexpensive. They are not well suited to analog transmission applications because of the complications involved in converting the motion produced to a transmission signal, but they are good for discrete applications and local indication applications. Because the presence of floating debris in wastewater may foul the floats, use only large floats and call for periodic washdown in the O&M manual in such service.

Although popular, tethered floats have sometimes given trouble and they do not carry a UL label for use in Class I, Division 1 hazardous areas such as sewage wet wells. The hazard can be circumvented by using very low energy, intrinsically safe control circuitry, but, because this adds another level of design complication, some engineers avoid them.

Displacers

Displacers differ from floats in that the buoyant body has a density greater than that of the liquid being monitored. The displacer, shown in Figure 20-4, is a cylindrical body whose height is at least equal to the maximum level span to be monitored. As the displacer is immersed, it loses apparent weight by the weight of the liquid displaced (in accordance with Archimedes' principle), and this apparent weight loss stretches the spring slightly and actuates the mercury switch used to indicate liquid level and start or stop pumps. Built-in limit stops prevent overstretching of the spring. Displacers do not produce any significant mechanical motion, but rather work on the force-balance principle. They are very suitable for both analog and discrete measurements, but do not produce a direct readout of level. The useful measurement range is limited by the practical cylinder length—less than 10 ft. Displacers are better than floats in wastewater applications because the liquid rises and falls on a smooth cylinder producing a washing action, although fouling is still possible.

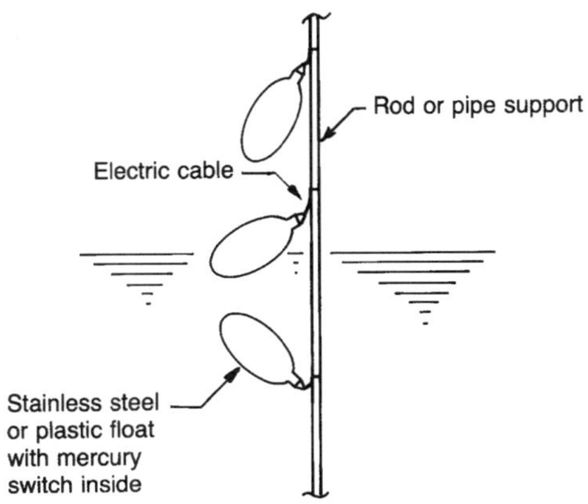

Figure 20-3. Tethered floats.

Table 20-1. Parameters Frequently Monitored

Parameter	Station type: Clean water	Station type: Wastewater	Reasons
Liquid level			
Wet well		X	Pump control Maintenance dispatch
Suction storage	X		Pump control Maintenance dispatch
Distribution storage	X		Pump control Maintenance dispatch
Dry well		X	Maintenance dispatch
Pressures			
Distribution system	X		Pump control
Pump suction	X	X	Troubleshooting
Pump discharge	X	X	Troubleshooting
Flow			
Influent or effluent	X	X	Pump control Troubleshooting Billing Chemical feed control
Pump status			
Running	X	X	System operation
Speed		X	System operation
Available	X	X	System operation
Maintenance dispatch			
Primed	X	X	Pump control
Dry suction		X	Equipment protection
Bearing temperature	X	X	Equipment protection
Motor temperature	X	X	Equipment protection
Operation time	X	X	Maintenance dispatch
Time start/stop	X	X	Troubleshooting Litigation defense
Start/stop sequence fail	X	X	Maintenance dispatch
Vibration	X	X	Equipment protection Maintenance dispatch
Water in motor	X	X	Maintenance dispatch
Valve and gate positions	X	X	System operation
Chemical feed systems			
Storage weight or level	X	X	Inventory Maintenance dispatch
Feed rate	X	X	System operation
Chlorine residual	X	X	Feed rate control
Feed pumps running	X	X	Maintenance dispatch
Utilities and environment			
Voltage	X	X	Maintenance dispatch Alternate source control
Amperage	X	X	Troubleshooting Expansion planning
Wattage	X	X	Power cost control
Power factor	X	X	Power cost control
Temperature	X	X	Equipment protection Maintenance dispatch
Explosive atmosphere		X	Personnel protection Equipment protection
Chlorine gas leak	X	X	Personnel protection Maintenance dispatch Public protection

Table 20-1. Continued

Parameter	Station type: Clean water	Station type: Wastewater	Reasons
Sulfur dioxide gas leak	X		Personnel protection Maintenance dispatch Public protection
Hydrogen sulfide gas		X	Personnel protection
Ventilation equipment status	X	X	Equipment protection
Fire	X	X	Equipment protection Maintenance dispatch Fire brigade dispatch
Unauthorized intrusion	X	X	Equipment protection Public protection
Air pressure	X	X	Maintenance dispatch
Fuel level	X	X	Inventory
Control power voltage	X	X	Maintenance dispatch
Battery chargers operating	X	X	Maintenance dispatch
Sample pumps running		X	Maintenance dispatch
Engine systems			
Engine running	X	X	System operation
Engine available	X	X	System operation
Engine malfunction	X	X	Maintenance dispatch

Admittance Probes

The admittance system is a modern improved version of the older capacitance probe. It consists of a slender, usually Teflon®-coated rod that extends into the monitored liquid. The rod is excited with a low-power, high-frequency voltage. An electrical current flows between the rod and a ground reference, which may be either the fluid itself (if it is a conductor) or a ground plane on the wall of the tank (if the fluid is essentially a nonconductor). The current flow varies as the fluid level changes and provides a measurement of the tank level. The system is totally solid state, which produces neither motion nor mechanical force. Solid-state systems are most useful where a transmitted signal is required. They do not produce a direct mechanical readout, but they can be fitted with a local electronic indicator to display the level.

Ultrasonic Measurement

The ultrasonic level measurement system consists of a transducer (suspended above the fluid) and associated

Table 20-2. Level Measuring Elements[a]

Element	Clean water	Wastewater	Discrete	Analog	Comments
Air bubbler	E	E	E	E	High cost, general application
Float	E	P	E	F	Low cost
Ultrasonic	E	E	E	E	High cost, general application when instrument air is not available
Immersible head transmitter	E	F	NA	E	Replaces the diaphragm box
Differential pressure	E	G	NA	E	High cost, limited application
Conductivity	E	F	E	NA	Low cost
Admittance probe	G	G	E	G	Moderate cost
Displacer	E	F	E	G	Limited range
Diaphragm box	G	F	G	G	High cost, obsolete
Microwave	E	E	E	E	High cost, use instead of ultrasonic for difficult conditions (see text)

[a] E, Excellent; G. good; F, fair; P. poor, NA, not applicable.

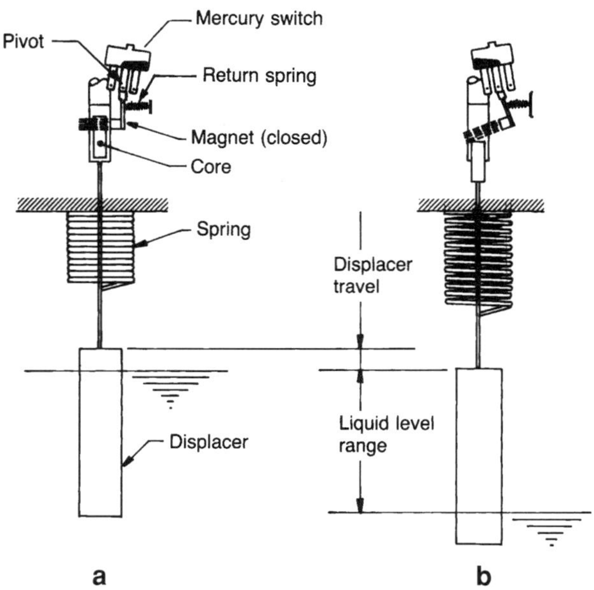

Figure 20-4. Displacement-type liquid level control. (a) Mercury switch closed; (b) mercury switch open.

electronics. The transducer periodically emits a pulse of sonic energy toward the liquid, and the pulse is reflected by the liquid surface back to the transducer. The transit time of this pulse is approximately proportional to the distance between the transducer and the liquid surface. The speed of the sonic pulse is affected by the density of the gas above the liquid and, hence, by both barometric pressure and temperature. Good-quality ultrasonic level systems are fully compensated and provide very accurate level measurement. But many systems are either uncompensated or only temperature compensated and cannot measure with accuracy. Readings may also be affected by foam, which causes either a false reflection or a loss of reflection. Discontinuities in the vessel walls may also cause false reflections.

The better quality ultrasonic level systems have microprocessors that reject spurious reflections. Nevertheless, when applying ultrasonic level measuring devices, check the angle of the emitted sonic pulse to ensure that the space within the transmission cone is clear of obstructions. Manufacturers rate ultrasonic level transducers for operation from – 40 to +90°C (–40 to +200°F), but field installations have not been satisfactory below freezing or in windy locations. The ultrasonic system is totally solid state and does not provide a direct mechanical indication of level.

Modern microprocessor-based ultrasonic level systems are accurate and reliable. They have supplanted bubbler systems in many facilities where maintenance of instrument air supply equipment has become uneconomical or onerous.

Microwave

The microwave level measurement system is similar to the ultrasonic, except that a microwave frequency pulse is emitted toward the liquid surface. Microwave pulses penetrate foam, and the speed of microwaves is affected very little by the density of the gas above the liquid or by very low or high temperatures. Thus, although considerably more expensive than the ultrasonic system, the microwave system operates reliably under much more difficult conditions.

Immersible Head Transmitter

Immersible electronic head transmitters are designed to be completely immersed in the liquid being monitored. A cable containing both the signal leads and a pressure reference tube connects the transmitter to a junction box above the vessel or channel. The pressure element is typically the strain gauge type described under pressure transmitter. Immersible head transmitters were originally designed for oil field borehole use. They are economical, rugged, and reliable. They are cylindrical in shape, and some users drop them into the liquid through a pipe such that the sensing diaphragm, which forms the bottom end to the transmitter, is flush with the bottom of the pipe. This convenient installation protects the transmitter from gross solids present in wastewater applications. These units have been used as a complete replacement for air bubblers in clean water applications. However, the typical size (25 mm or 1 in.) is not suitable for narrow spans (less than 700 mm or 2.3 ft). With 75-mm (3-in.) diaphragms, they can be used for controlling V/S pumps where narrow spans are required. With reasonable frequency of cleaning, they are suitable for wastewater applications if the enclosing pipe is extended below LWL to protect the transmitter from scum.

Conductance Probes

Conductance probes utilize the ability of virtually all liquids to conduct electricity. When the probe is immersed, a small electric current flows between probes or between a probe and the tank ground, and this current is detected by a sensitive relay. Conductance probes are suitable only for discrete measurements. They are used primarily in clean water applications. In wastewater applications, grease deposits and the necessity of using low electrical energy levels to prevent ignition of hazardous atmospheres generally rule out these devices.

Air Bubblers

The bubbler is probably the most common liquid level measurement device used in wastewater applications. It consists of a dip tube into which a small amount of air is metered. The back pressure produced is a measure of the liquid level above the tube outlet. Any pressure-measuring device can then be used to produce either an analog or a discrete signal, and a pressure gauge can be used for a direct readout. The bubbler works well with any liquid including sludge and scum. Its advantages also include safety in explosive atmospheres and longevity in corrosive environments. The perceived disadvantages are the complexities, continual maintenance, problems with leaks, and high cost. But if the advice in this section is followed, air bubbler systems are easily maintained, leaks are easily fixed, and the system is the most reliable of all. Some bubbler systems have been replaced by tethered floats (sometimes not the best of choices), by ultrasonic measurement systems, or by immersible head transmitters. Consider the latter two if the bubbler is the only device requiring instrument air.

A complete air bubbler assembly for level measurement is shown in Figure 20-5. Instrument air is the most vulnerable of all systems, so do not use a cheap installation. Oil-free air (which cannot be obtained with oil-pumped and filtered air) must be supplied by oil-less compressors. Air is most commonly supplied by dual-redundant, piston-type compressors derated from 550 to 690 kPa (80 to 100 lb/in.^{2}) to 400 kPa (60 lb/in.^{2}) discharging to an oversized receiver so that the compressor does not cycle frequently. Use refrigerated air dryers and specify a hot gas bypass to prevent freezing. Provide for air purging of the dip tube at no less than 200 kPa (30 lb/in.^{2}). Neither compressor should have to operate more than 5 min/h to produce 28 L/h(1 ft^3/h) of air.

The pressure regulating valve, shown in Figure 20-5, reduces supply pressure to a level low enough to prevent possible damage to the pressure element if the dip tube plugs but high enough to ensure purge flow at maximum liquid level. If high accuracy is required, a constant-flow regulator is needed to prevent variations in purge flow from introducing dynamic headloss variations. The regulator is set to maintain a constant differential pressure, usually 20 kPa (3 lb/in.^{2}), across the needle valve so that the valve becomes a constant-flow regulator. The flowrate

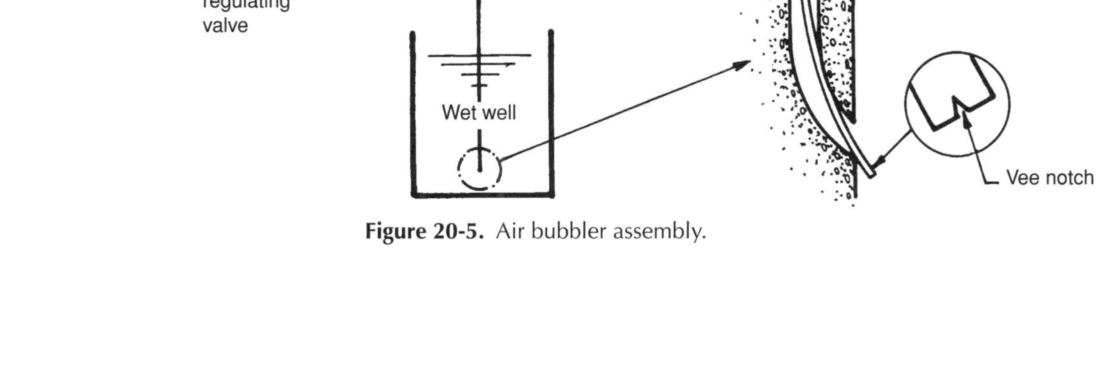

Figure 20-5. Air bubbler assembly.

is measured by the rotameter. Usually, the differential pressure regulator, needle valve, and flow indicator are furnished as a factory-assembled unit. Air flow is typically adjusted to about 0.025 m^3/h (1.0 ft^3/h), but the flowrates are sometimes a third as much.

Dip tubes should always be installed so that they can be easily rodded out if they become plugged. The bottom of the dip tube is notched to keep bubble size as small as possible to maintain the highest accuracy. Some engineers (and owners) prefer straight dip tubes (for easy rodding) made of 13-mm ($^1/_2$-in.) Schedule 40 PVC or stainless-steel (ss) pipe attached to the wet-well wall with pillow blocks. Others prefer to install soft ss tubing in 50-mm (2-in.) conduit of harder ss or galvanized steel encased in the concrete. The conduit must be bent so that the dip tube can enter the pool, but the bends must have a long radius to make it easy to insert the dip tube. Rodding must be done with a flexible rod (preferably furnished by the contractor). In either installation, make all joints, especially the rodding port, easily accessible for finding and fixing leaks. Joints should *never* be encased in concrete.

Pressure instruments used with bubbler systems should be installed adjacent to the bubble pipe (dip tube) to keep tubing lengths as short as possible because long tubing runs (e.g., 30 m or 100 ft) can add significant time delays to the measurement, increase the possibility of leaks that disable a bubbler system, and, in closed-loop control systems, cause instability. One way to eliminate these problems is to mount the purge panel and pressure instruments on a wall or hand rail adjacent to the bubble pipe to keep tubing runs to less than about 3 m (10 ft).

High accuracy is not usually required for pump control—especially if the range of wet well levels is 1 m (3 ft) or more—so the high cost can be somewhat reduced by simplifying the system. The differential pressure regulator is often omitted even though this causes the purge flow to vary with liquid level, and small errors result from changes in system dynamic pressure losses and changes in bubble size. If an analog signal is required, a bubbler with an electronic change-of-pressure transmitter is satisfactory. Providing a different (or even an additional) system for backup is good practice, so use a float switch for the high-level alarm.

Diaphragm Boxes

A diaphragm box consists of a closed box with an elastomer diaphragm forming one wall. The system is obsolete and has been supplanted by immersible electronic pressure elements, which may be installed in the same fashion and are much more accurate.

Differential Pressure Elements

Differential pressure units may be flange mounted to a tank to expose the pressure element directly to the fluid.This approach is practical only when (1) freezing does not occur or can be prevented and (2) access to the side of the tank is readily available. If the tank is not at atmospheric pressure, a second connection must be made above the maximum fluid level. The pressure exerted on the pressure element is then a direct indication of the fluid head above the connection point. This system, which is simple and very accurate, is the favored level measuring system in industrial applications, but it does not work if the specific gravity of the fluid in the tank is unpredictable. An isolating valve can be placed between the pressure unit and the tank to facilitate instrument maintenance without emptying the tank, but this modification entails a length of piping between the pressure element and the tank that is subject to clogging in dirty water applications.

Install a water connection for convenient flushing of any pipe used between the vessel and the diaphragm. Flushing can done be at regular intervals or a continuous small flow can be used to isolate the pressure cell from contaminated fluid.

20-4. Pressure Measurements

Pressure is a basic measurement that is made not only for itself but also for converting many other measurements into readable parameters. For example, many level and flow measurement systems produce a pressure that can infer a value for another parameter. Pressure is, in fact, almost always measured as a differential pressure. Atmospheric pressure is normally the reference, but virtually all pressure-measuring elements can be arranged for connection of an independent reference pressure. The various types of pressure elements are summarized in Table 20-3.

Pressure gauges should be installed on both the suction and discharge spools (short pipe section) of every pump (see Figure 20-6). Instead of fixed gauges, however, pressure taps with short, 12-mm ($^1/_2$-in.) pipes terminating in shut-off cocks and quick-connects allow more accurate portable gauges to be used. Portable gauges are advantageous because only two are required and they can be kept in calibrated condition. Suction and discharge pressures are valuable checks on the condition of the station because they can be used to diagnose such troubles as blockages, deposits in the pipe, and worn impellers. In conjunction with the manufacturer's pump curve, pressure

Table 20-3. Pressure-Measuring Elements[a]

Element	Clean water	Wastewater	Discrete	Analog	Comments
Bourdon	E	G	E	G	Low cost, general application
Diaphragm	E	G	E	G	Low cost, low pressure
Bellows	G	P	E	G	Special applications
Force balance	E	E	NA	E	Becoming obsolete
Strain gauge	E	E	NA	E	Moderate cost, general application
Capacitance	E	E	NA	E	Moderate cost, general application
Manometers	E	NA	G	E	High cost, special application

[a] E, Excellent; G, good; F, fair; P, Poor; NA, not applicable.

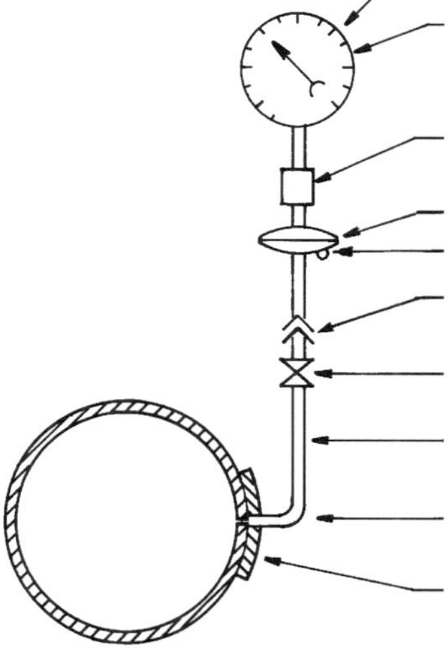

Figure 20-6. Proper Bourdon pressure gauge assembly.

measurements can be used to measure flow with an error of no more than about 7% (refer to Section 3-9).

An alternative to the installation of a pressure gauge on the discharge of every pump is a single pressure gauge installed on the manifold; this is advantageous because only one gauge is needed and the flow and pressure are likely to be more stable than at a gauge near the pump. The disadvantages are the need to shut off all of the pumps except the one under observation and the loss of head between pump and gauge.

Bourdon Tubes

The Bourdon tube is the most common of pressure-measuring elements. It is used in virtually every pressure gauge and in many pressure switches. It consists of a semicircular length of flattened tubing. When pressure is applied to the inside of the tube it tends to straighten, and the mechanical motion can be coupled to indicators, switch mechanisms, and transmitter mechanisms. It is a motion-balance device. The tube can be fabricated from copper-based and stainless-steel alloys, which offer moderate corrosion resistance. Sewage and dirty water must be excluded from the Bourdon tube by a diaphragm seal, as shown in Figure 20-6. All of the tubing (including the Bourdon tube) above the diaphragm seal is filled with glycerin. The case is also filled with glycerin, thereby immersing the rack and pinion gears both to prevent corrosion and to dampen rapid pressure fluctuations. The snubber (a micropore filter) can be omitted, particularly if the shut-off cock is normally closed. Bourdon tube pressure elements are suitable for moderate to high pressures—generally more than 100 kPa (15 lb/in.^{2}) although low-pressure gauges are available.

For hard-to-gauge liquids such as sludge, one form of Bourdon gauge installation consists of an annular, flexible, impermeable tube or liner that is recessed so that its inner wall is flush with the pipe. The tube is connected to the Bourdon gauge, and both are filled with a sensing fluid. Various fluids and tube materials are available. The advantages are a protected gauge, responsive pressure readings, and protection from clogging. The disadvantages are cost (perhaps five times the cost of a Bourdon gauge), a small amount of hysteresis, and some loss of accuracy.

Diaphragms

Diaphragm pressure elements are used for lower pressure spans than those of Bourdon tubes. A variety of materials can be used to provide corrosion resistance, and mechanical configurations are available that permit mounting the diaphragm flush with a vessel wall to permit use with solids-bearing fluids. Housings can be designed to protect the diaphragm from high overpressures. Large sizes can be made to provide sufficient sensitivity for measuring very low pressure spans. The limited amount of mechanical motion is sufficient for direct actuation of precision switches and transmission mechanisms. Diaphragms are also coupled to Bourdon tubes by liquid-filled capillary tubes to produce a common type of chemical seal.

Bellows

A bellows is a stack of diaphragms. Bellows elements can measure very low pressure spans and, at the same time, produce large amounts of mechanical motion for operating indicators and switch and transmission elements. Bellows elements are not available in corrosion-resistant materials, their configuration is unsuitable for solids-bearing fluids, and they cannot be protected against overpressure.

Force-Balance Transmitters

The force-balance mechanism is frequently coupled to a diaphragm to produce a very rugged pressure transmitter (see Figure 20-2). Mechanical power available from the transmitter pneumatics or electronics is used in a null-balance circuit to oppose movement of the diaphragm. As a result, mechanical motion and, hence, wear is limited to a very small amount. Force-balance transmitters can be built in very low pressure spans and can withstand high surges and overpressures without damage or alteration of the calibration. The pneumatic force-balance mechanism is the standard industrial pressure transmitter, but the electronic versions have become obsolete.

Strain Gauge and Capacitance Transmitters

These transmitters have replaced the electronic force-balance transmitter. Modern electronic circuitry permits the mechanical deformation of a diaphragm element to be limited to a very small amount and still provide sufficient movement to produce a usable output. These transmitters are extremely rugged and, in addition, provide error values of less than 0.25% of span.

Manometers

A manometer is simply a U-tube (see Figure 3-4), which contains an indicator fluid (such as mercury, air, and carbon tetrachloride) of known density. If a pressure difference exists between the two ends of the U-tube, the indicator fluid elevation in the two legs is different, and this difference is a measure of the pressure. If any but the lowest pressures are to be measured, either the U-tube must be very tall or the density of the indicator fluid must be great. Mercury has been used frequently, but it is a very poisonous substance that must *never* be used for potable water and should be avoided for any application.

Manometers are useful for direct readout of very low pressures. Mechanisms for the conversion of indicator fluid position to useful motion for switch and transmitter operation are complex and expensive. Manometers are very accurate, however, and are sometimes used in conjunction with air bubblers to monitor reservoir levels where very small changes in level represent large quantities of water.

20-5. Flow Measurements in Pipes

Flow is an easy measurement to make in clean water applications, but very difficult to make in wastewater applications. Most flow elements convert flow to some other variable that is more readily measured, such as pressure, level, or an electrical signal. The accuracy of flow measurement is frequently of considerable concern, and sustained accuracy is difficult to achieve without giving great care to the installation and the continued maintenance of the instruments (see ASME MFC-3M [3]). Instrument types frequently used for flow measurement are summarized in Table 20-4 and discussed by Miller [4]. The relative costs for some of the more common types of flowmeters, including the

Table 20-4. Flow-Measuring Elements[a]

Primary element[a]	Clean wastewater[b]	Wastewater[b]	Secondary element[a]	Comments
Orifice plate	G	X	Pressure	High headloss, range 8:1, error ±1–2%[c]
Venturi meter[d]	G	F	Pressure	High cost, range 8:1, error ±1–2%
Flow tube	G	F	Pressure	Medium cost, range 5:1, error ±1–2%
Elbow meter	F	P	Pressure	Cheap, range 4:1, error ±2%[e]
Magnetic	E	E	Electronic	High cost, accurate, range 10:1, error ±0.5–1%
Doppler	P	G	Electronic	Low cost, no intrusion into pipe range 10:1, error ±2–20%
Transit-time ultrasonic	E	F	Electronic	Special applications, error ±1%.
Propeller	E	X	Mechanical	Low cost, accurate, range 10:1, error ±2%
Turbine	E	X	Mechanical	Accurate, range 100:1, error ±1%
Weir	E	X	Level	High headloss, standard open channel device, fair accuracy, range 20:1, error ±2%
Flume	G	G	Level	High cost, fair accuracy, range 20:1, error ±4%
Vortex	G	X	Pressure change frequency	Range 20:1, insensitive to fluid properties, error ±1%

[a] See Chapter 2 for definition.

[b]E, Excellent; G, good; F, fair; P, poor; X, *not to be used.*

[c]Typical error to expect under "best" field conditions with premium quality instruments and frequent, careful recalibration. Expect double the stated error if recalibration is infrequent.

[d]Venturi meters are considered to be the standard flow measuring device for custody transfer.

[e]Only if accurately calibrated in situ.

4- to 20-mA output but excluding installation, are shown in Figure 20-7.

The accuracy of flowmeters is (for most types) significantly affected by piping configurations in the vicinity of the flow element. Because designing the approach conditions to match factory or laboratory flow test facilities is ordinarily impossible, the manufacturers' claims of accuracy are rarely realized in practice. The effects of nonstandard approach conditions on orifice and Venturi meters have been thoroughly explored and illustrated graphically by Starrett *et al.* [6]. Investigations of field installations show that errors greater than 10% are the rule in most installations and errors from 50 to 200% are not uncommon [7]. *In situ verification of calibration is essential for confidence and accuracy.* Calibrations with the "meter provers" (often used in the process industries) is impractical for the flows typically encountered in pumping stations. Adequately sized tankage associated with a pumping station can often be isolated and used for an approximate volumetric calibration, but much more accurate and reliable results can be obtained with tracer techniques (see Section 3-9). Pipeline traverses with either an inserted magnetic probe or a pitot tube should be considered only as a last resort. Accurate traverses are difficult to accomplish in the presence of swirling and/or pulsating flow conditions caused by flow profile disturbances.

Orifice Plates

An orifice plate is a simple, flat restriction (in a pipeline) that produces a differential pressure proportional to the square of the flow velocity in accordance with Bernoulli's principle (see Figure 20-8). Small reductions in flow cause a great reduction in the differential pressure, so orifice plates should not be used for flows that vary over a range greater than 8:1. Because the abrupt restriction does not pass solids, this instrument is unsuitable for sewage, and head recovery downstream is poor so there is a high permanent headloss. Orifice plates require a fully developed flow profile to produce accurate measurements, and, depending on pipe and orifice size, a straight pipe 6 to 45 diameters long is required upstream. The straight pipe requirement downstream is 5 diameters. The orifice plate system becomes unpredictable for Reynolds numbers below 10,000.

Venturi Tube

The Venturi tube (shown in Figure 20-9) also produces a differential pressure proportional to the square of the velocity. With its smooth approach and recovery cones, the Venturi tube works well in clean water applications and reasonably well in wastewater

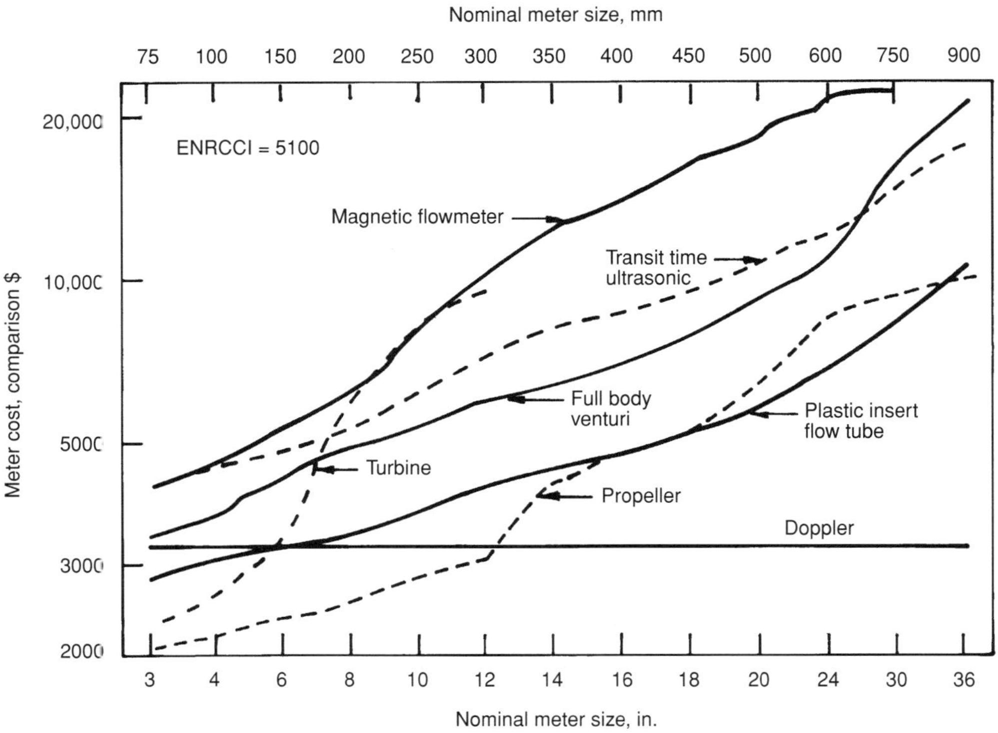

Figure 20-7. Relative 1993 costs (ENRCCI = 5050) for flowmeters with 4- to 20-mA outputs but no processor or installation costs. Courtesy of *WaterWorld Review* [5].

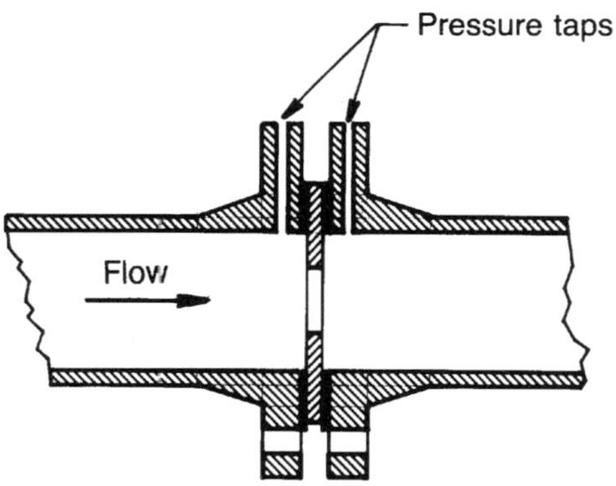

Figure 20-8. Orifice meter.

applications. Head recovery is excellent. The approach cone tends to correct aberrations in flow profile caused by poor approach conditions. Some error is introduced by irregular flow profiles, but the error is much less than with orifice plates. Standard Venturi tubes are acceptable as custody transfer flow-metering devices and are frequently used in clean water applications as revenue meters.

The principal disadvantage of a standard Venturi meter is the high cost for large sizes and long laying length (see Starrett *et al.* [6]). A number of proprietary designs (see, for example, Figure 20-10) have been developed to shorten the length and reduce the cost. The performance of these proprietary designs is generally excellent, but the volume of laboratory data available to substantiate their performance may not be as great, and their use as custody transfer meters may not be fully acceptable.

When Venturi tubes are used in dirty water or wastewater applications, solids must be excluded from the differential pressure measurement connections. The most common method is to install a water purge system, but even this is sometimes troublesome with raw sewage. Diaphragms, flush-mounted in the throat wall, may also be used, but some difficulty is encountered in maintaining measurement accuracy with such arrangements.

Elbow Meter

A 90° bend can be converted to an elbow meter by installing pressure taps on the inside and outside of the bend at either 22.5° or 45° [4] and measuring the differential pressure. Pressures at taps installed at

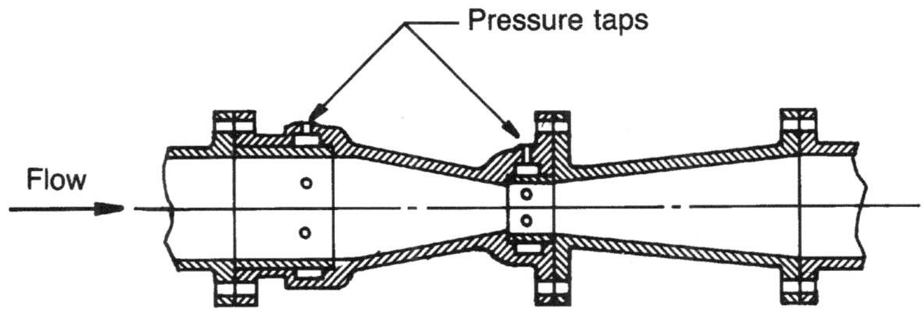

Figure 20-9. Venturi meter.

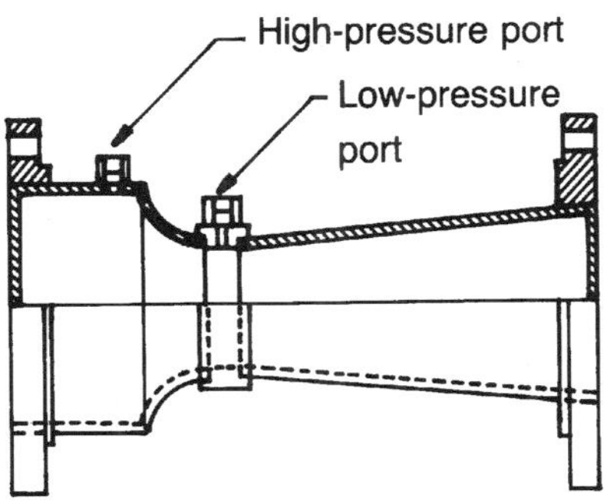

Figure 20-10. A typical proprietary flow tube.

angles greater than 45° are erratic. Because the differential pressure is very low, an air manometer (Figure 3-4) is, probably, the best type of measuring device. The low cost and ease of installation (or the use of an existing elbow) make it a useful flowmeter for field testing in stations not otherwise equipped with a flowmeter.

Magnetic Flowmeters

The magnetic flow tube (shown in Figure 20-11) is an open cylinder that produces practically no headloss and readily passes solids. It functions with any conductive liquid, such as water or sewage. It consists of a flow tube that develops a magnetic field across the flow profile. When a conductor passes through a magnetic field, an electrical potential proportional to the conductor's velocity (in accordance with Faraday's law) is produced. With proper design and modern electronics, very accurate flow measurements (which are virtually immune to distorted flow profiles) can be obtained.

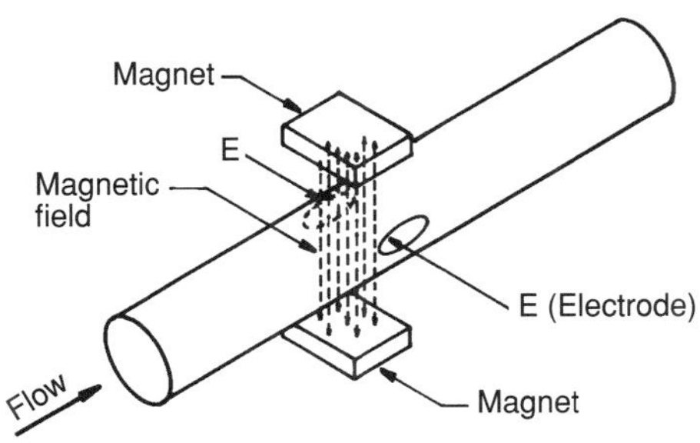

Figure 20-11. Magnetic flowmeter.

The principal difficulty is with wastewater containing grease, which can coat the electrodes and interfere with the potential measurement. This problem has been overcome somewhat by high-quality electronic circuitry that can measure the potential across very poor conductors. Nevertheless, magnetic flowmeters for wastewater applications should be installed to facilitate cleaning. A piping configuration that allows the electrodes to be cleaned without removing the flow tube from the pipe is shown in Figure 20-12. Alternatively, the meter can be installed in the straight pipe with a Dresser coupling on one side and a Victaulic® coupling on the other side (or with Victaulic® couplings on both sides) to allow the meter to be removed while the flow goes around the installation in a bypass pipe. Another cleaning method sometimes used consists of pig launching and removal stations. Use only soft pigs for magnetic flow tubes. Always arrange the magnetic flow tube so that (1) no stress is placed on the flow tube and (2) mating flanges can be separated to permit easy removal of the tube.

The magnetic flow tube liner must be both nonconductive and nonmagnetic. Polyurethane, Teflon®, and ceramic are the most commonly used liners. Polyurethane provides good abrasion resistance and is the standard liner material. Teflon® is used for meters in sewage sludge service for its superior resistance to grease, but it has poor abrasion resistance and must never be used if the sludge contains much grit. Ceramic liners are available for meters up to about 150 mm (6 in.) and are resistant to both grease buildup and abrasion. For meters larger than 150 mm (6 in.), used with liquids that contain significant amounts of grit,

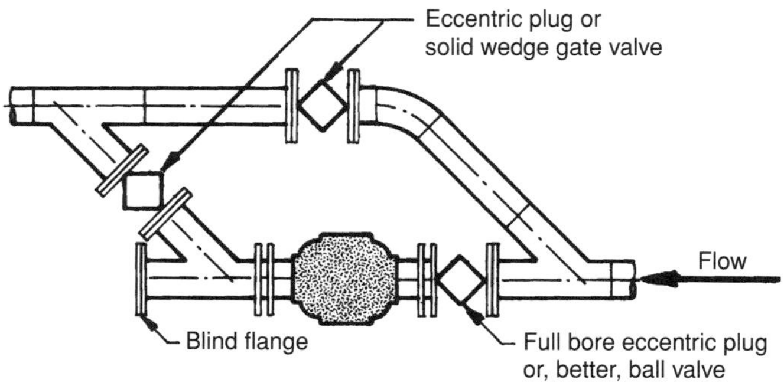

Figure 20-12. Piping plan to allow cleaning a magnetic meter in situ. With permission of The Foxboro Company.

use polyurethane and limit the velocity to 3 m/s (10 ft/s). The grit content of raw sewage is usually low enough to allow a peak flow velocity of 4.5 m/s (15 ft/s) and the grease content is usually low enough to allow a minimum velocity of 0.3 m/s (1 ft/s), so the usable range is 15:1. With sludge, the flow velocity range is about 1.0 to 3.0 m/s (3 to 10 ft/s) unless grit is effectively removed before entering the meter. Always provide a means for easily cleaning the electrodes. Failure to observe these limitations has frequently resulted in unsatisfactory service in wastewater applications. Nonetheless, the magnetic flowmeter remains the most satisfactory wastewater flowmeter. In clean water flow applications, the magnetic flowmeter can function from 0.03 to 6.0 m/s (0.1 to 20 ft/s), but the electronics generally limit the usable range to about 20:1.

Probable errors in magnetic flowmeters are typically stated as 0.5% of flow for liquid velocities greater than 1.5 m/s (5 ft/s) and 0.0075 m/s (0.025 ft/s) for velocities less than 1.5 m/s. Most vendors can furnish meters with half of these errors, but typically there is extra cost for custom calibration. The magnetic flowmeter is one of the most accurate meter types available. The initial cost may seem high, but the simplicity, reliability, and low maintenance makes its life-cycle cost competitive with other meter types up to about 400 mm (16 in.) in diameter. For applications other than sludge and larger than 400 mm, consider the transit-time ultrasonic meter.

Sonic Meters

There are two types of sonic flowmeters: (1) Doppler meters and (2) transit-time meters. Other than their common use of ultrasonic pressure waves, the operating principles of the two are completely different, and their applications are different. The Doppler meter is useful for measuring flows in fluids containing at least 100 ppm of suspended solids or air bubbles for which the transit-time meter is useless. Unlike the Doppler meter, the transit-time meter is very accurate.

Doppler Meter

The Doppler meter (see Figure 20-13) consists of a pair of piezoelectric transducers, which are usually contained in a common case and mounted on the outside of the pipeline. One transducer injects an ultrasonic pressure wave at frequency f_1 into the pipeline, and this pressure wave is reflected by solids, gas bubbles, or other discontinuities in the liquid back to the second transducer. The reflected frequency, f_2, varies by an amount proportional to the reflecting surface velocity. According to the principle of Christian Doppler,

$$f_a = f_1\left(\frac{v}{v - s}\right) \tag{20-1}$$

where s is the velocity component axial to the normal pressure wave propagation and v is the velocity of sound in the fluid. The practical execution of this principle is fraught with difficulties. The sound pressure waves must be precisely controlled, and the reflecting particles must be moving coaxially with the pipe at a speed representative of the flow (which varies from wall to center).

There are a large number of error sources in Doppler flowmeters: (1) the reflecting surfaces may not be moving at the same speed as the average velocity of the liquid; (2) they may not be moving parallel to the main liquid flow; (3) the frequency shift is affected by the speed of sound in the liquid, which is a function of temperature, pressure, and density and cannot be fully compensated; and (4) the angle of injection of the pressure wave is difficult to control accurately. In summary, the Doppler meter is inaccurate. Errors as great as 20% occur regularly in practical installations. Doppler meters are used only in dirty water applications because clean water does not contain reliable reflective surfaces.

Doppler flowmeters do have the outstanding advantage that there is nothing to foul or corrode. Users consider them to be maintenance-free devices. They are frequently used on sludge pipelines where accuracy is not particularly important and the flow is very difficult to measure at all with other devices. The maximum velocity is limited only by what can be practically pumped through a pipeline, and the minimum velocity is limited only by the resolution of the frequency measurement circuitry—about 0.3 m/s (1 ft/s) for most units. In sludge service, however, the minimum velocity must be enough to prevent a buildup on the pipeline walls, which would change the effective pipe diameter.

The Doppler flowmeter is also available as a moderately priced, nonintrusive flow switch. Accuracy is not critical in this application. The Doppler flow switch is particularly useful as a pump discharge flow detector, where limit switches cannot be used on a discharge check valve. If the Doppler flow switch is located immediately downstream of a pump, reliable signals are produced even with clean water, because the bubbles from the pump impeller reflect the sonic pulse.

Transit-Time Ultrasonic Flowmeters

The transit-time ultrasonic flowmeter (see Figure 20-14) consists of at least one pair of piezoelectric transducers mounted on opposite sides of the pipeline and offset from each other by 30 to 45°. Additional transducer pairs are often used to provide the ability to average the flow profile. An ultrasonic pressure pulse is injected by one transducer and the period of time for it to reach the second transducer is measured. The process is then reversed. The difference in time required for the pulse to propagate upstream and downstream through the fluid is proportional to the average fluid velocity between the transducers. This metering system is capable of very high accuracy, and when multiple transducer pairs are used, it can average out irregularities in the flow profile. The process of reversing the signal direction cancels errors that would be caused by variations in the speed of sound through the fluid. This meter does not perform well in dirty water or wastewater applications. The same solids, gas bubbles, and other fluid discontinuities that make the Doppler meter work diffract the pressure pulse so that it arrives at the receiving transducer as a burst of noise, not as a single pulse. High-quality

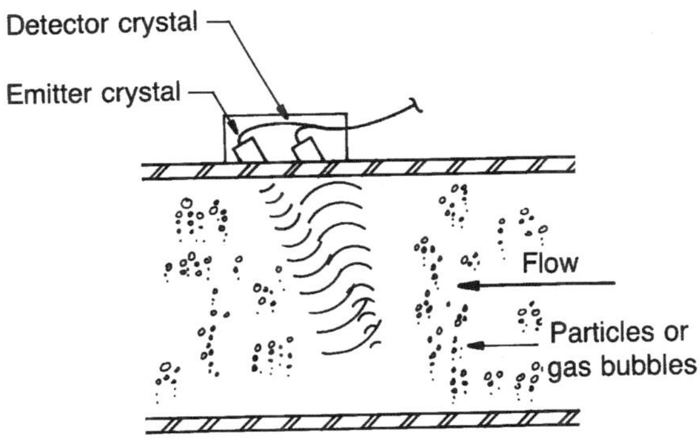

Figure 20-13. Doppler flowmeter.

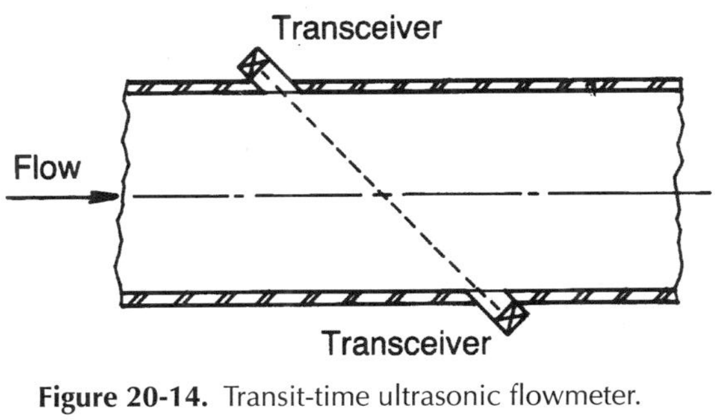

Figure 20-14. Transit-time ultrasonic flowmeter.

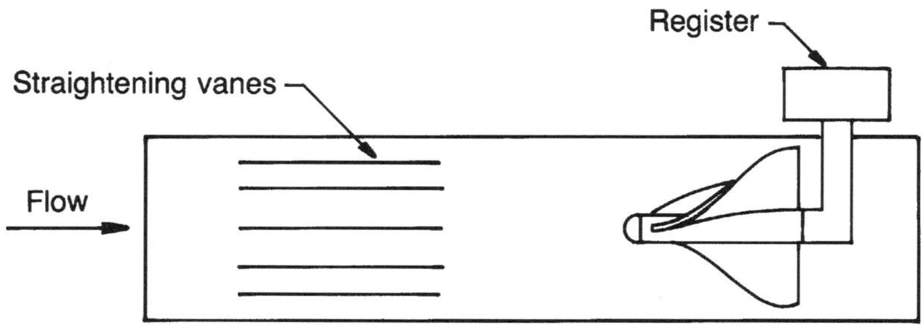

Figure 20-15. Propeller meter.

electronics can deal with the noise to a certain extent, but they increase the cost of the meter substantially.

The transit-time ultrasonic flowmeter is especially useful on very large pipes where cost becomes less of a factor. The maximum velocity is limited only by what can be pumped through a given size pipe, and the minimum velocity is limited by the resolution of the timing circuits—0.3 m/s (1 ft/s).

Propeller Meters

As its name implies, a propeller meter contains a propeller installed coaxially in the center of the pipe, as shown in Figure 20-15. Flow turns the propeller at a speed proportional to the liquid velocity. It is a low-cost device used primarily in clean water applications where it gives excellent service. Depending on the manufacturer and the meter size, the mechanical nature of the equipment limits the usable flow range to about 0.15 to 3.0 m/s (0.5 to 10 ft/s). Its only disadvantage (for clean water service) is the need to replace bearings from time to time.

Turbine Meters

Turbine flowmeters have many blades on a rotor whose speed is a linear function of fluid flow velocity within an error of about ±0.5% over a very wide flow range of 10 or even 20:1. The blades are vulnerable to particulate matter, so the meter is used only for clean water.

Vortex Flowmeters

In vortex flowmeters, a transverse, flat bar (called a "bluff body") across the pipe splits the flow into two paths and causes vortices to shed alternately from the sides of the bar at a frequency that is (1) linearly proportional to velocity and (2) independent of fluid density over flow ranges up to 30:1. The meters can be used for liquids, gases, and cryogenic fluids. There are no moving parts, and the construction allows sand in dirty water to pass without obstruction. The shedding rate for vortices is linear above a Reynolds number of 10,000. The meters are available in sizes from 38 to 400 mm (1.5 to 16 in.), and their use in chemical and industrial fields is increasing because of their accuracy (which is slightly greater than the accuracy of a Venturi meter), range, and insensitivity to fluid properties. These excellent characteristics are no less applicable to the measurement of water.

Summary

For clean water applications, the propeller meter is a good choice for all sizes of pipes. For sewage, the Venturi meter is widely used even though keeping the pressure ports clean is troublesome. Venturi meters for pipes larger than 600 mm (24 in.) become comparatively costly and require straight-approach pipes that may be inconveniently long. Install a freshwater purge system and a pipe configuration that allows the Venturi meter to be removed for cleaning. The magnetic meter is much preferred by many experts because of its superior accuracy and reliability, although maintenance is still a problem. Both Venturi and magnetic meters are heavy in sizes larger than 300 mm (12 in.), so hoisting beams should be installed overhead. Consider the Doppler meter for measuring the flow of sludge. The accuracy of Doppler meters is not good (but usually adequate), maintenance is negligible, and operators like them. Except for raw wastewater service, consider the vortex meter.

20-6. Open Channel Flow Measurement

Weirs and flumes are used for open channel flow measurement. Weirs are applicable only to clean water and treated wastewater because they cannot pass heavy solids. Flumes may be used in both clean and wastewater applications.

Weirs

A weir is simply an obstruction across a channel, and weirs are built in a number of configurations, including rectangular, vee-notch, and trapezoidal. Vee-notch weirs typically have an included angle of 30°, 60°, or 90°. Weirs are fabricated in the field, usually of concrete or steel, and the care used in their construction is a major limitation on their accuracy. Because the flow downstream from a weir must fall freely away from the crest, weirs introduce substantial headloss into the channel.

Depending on the weir configuration, the flow varies as a power (1.5–2.5) of the upstream level, so the change in level is not great with respect to the flow, and level instruments must be extraordinarily accurate. The usable range is typically 20:1, but the accuracy is limited because errors often exceed 5%.

Flumes

Flumes are built in a number of configurations including Parshall (Figure 20-16), Palmer–Bowlus (Figure 20-17), parabolic, and cutthroat. Except for the Palmer–Bowlus flume (see Section 3-7), they are purely empirical devices and their accuracy is limited by the degree to which the field installation resembles the laboratory model. Approach conditions that produce irregular flow profiles and/or flows greater or less than that specified for the size give poor and unpredictable results. Properly sized and installed, flumes provide reliable measurement of clear water and sewage—even with gross solids. The head change developed by a flume with a rectangular throat is proportional to the flow to approximately the 0.65 power. Flumes have a wide usable range—typically 20:1.

20-7. Chlorine Residual Measurement

Water is often chlorinated in water pumping stations. Standard analyzers in "package" units are available to

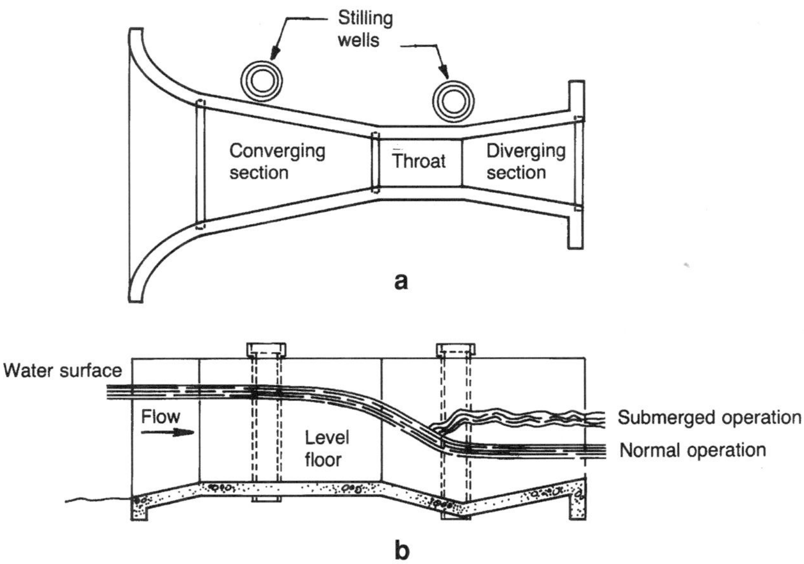

Figure 20-16. Parshall flume. (a) Plan; (b) longitudinal section.

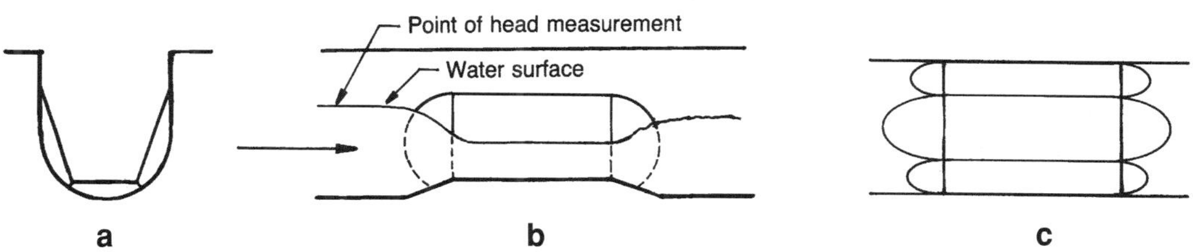

Figure 20-17. Palmer–Bowlus flume. (a) Cross-section; (b) longitudinal section; (c) plan.

control the dose or to measure the residual chlorine. Refer to the treatise by White [8] for a complete discussion of chlorination.

20-8. Utility and Environmental Measurements

Utility and environmental measurements include a large variety of miscellaneous measurements that provide information about and protection of the pumping station itself. These measurements are not unique to pumping stations. They might be made in any facility containing valuable equipment.

Electrical Parameters

Electrical parameters such as volts, amperes, watts, and power factor are easily measured either with direct reading instruments or with transducers that convert these readings to standard transmission signals for activating alarms or telemetering to a distant location. Most electric power–monitoring instruments operate on rather simple electromagnetic principles and are very accurate, reliable, and virtually maintenance free.

Temperature

There are a great many ways to measure temperature. Mercury thermometers should be confined to the laboratory because of the extreme hazard of a mercury spill. In field applications where only a discrete signal is required, the most common device is a closed chamber in which a fixed amount of gas or liquid produces a variation in pressure with changes in temperature.

A bimetal element consists of two metal strips with different thermal expansion coefficients formed into a spiral. As the temperature changes, the spiral tends to wind tighter or unwind. Bimetal elements are commonly used in residential applications both to indicate the temperature and to control on-off devices such as furnaces and air conditioners. These devices are low in cost but do not provide the long-term accuracy of liquid- or gas-filled systems. Where electronic transmission signals are required, coils of platinum, copper, or nickel alloys (which change resistance as the temperature changes) are usually used for temperature transmitters. Nickel alloy is low in cost and commonly used for heating, ventilating, and air conditioning, but it has a nonlinear temperature-resistance coefficient. Platinum has a linear coefficient, is extremely stable, and is commonly used where accurate, long-term measurement is required.

Environmental Safety Instruments

Methane, hydrogen sulfide, and oxygen depletion monitoring are required in wastewater pumping stations. Instruments for measuring these parameters are costly and require substantial maintenance—particularly when installed in wet wells. Only fixed hydrocarbon monitors are required in wet wells by the NFPA. Hydrogen sulfide detectors are too unstable for permanent installation and these monitors should be carried by the personnel. Oxygen depletion is not generally considered to be a problem in wet wells.

The calibration of any of these instruments should be checked at least monthly. For small facilities, portable instruments carried by maintenance personnel may be a better solution if acceptable to the fire marshal. Portable instruments should be carried by maintenance personnel anyway, because they frequently must enter collections system manholes where permanently installed equipment is never present.

Explosive Atmosphere

Methane (an explosive gas) occurs in underground water sources and forms naturally in wastewater facilities, and fuel vapors sometimes occur due to accidental leakage or intentional dumping into sewers. The most common explosive-gas-monitoring device is the catalytic detector, which consists of two heated elements. One is exposed to the ambient gas and the other is isolated. If the ambient gas is combustible, it burns on the exposed element, which raises its temperature above that of the isolated element. The difference in temperature measures the combustible gas present. Because catalytic elements are easily poisoned by the hydrogen sulfide frequently present in wastewater facilities, detector life is limited. Some manufacturers produce catalytic detectors with improved resistance to hydrogen sulfide poisoning and typically guarantee the detector for one year. In extreme conditions infrared detectors, which are totally immune to poisoning, are available, but in 1997 these cost about $2500 per point—four times as much as the improved catalytic type.

Chlorine Gas Leak Detectors

Chlorine gas leak monitoring is required in either water or wastewater pumping stations if chlorination equipment is present. Permanently installed monitoring equipment is necessary at any chlorination location, because chlorine leaks can present a hazard to the public outside of the station. Where permanently installed monitoring systems are used, audible and visual warning devices must be installed at all entrances to the monitored space.

Hydrogen Sulfide Gas Detectors

Hydrogen sulfide is a ubiquitous, dangerous gas as deadly as hydrogen cyanide. Because it is so common around sewage works (who has not smelled a faint odor of rotten eggs?), it is often ignored and many people have been killed or have suffered permanent brain damage from this insidious gas. Below ground, sample for hydrogen sulfide continuously. The effects of various concentration are given in Table 20-5. Detectors are seldom installed permanently in pumping stations because they should be carried at all times by personnel in wastewater facilities when working in any confined space where sewage is present.

Table 20-5. Effects of Hydrogen Sulfide

Concentration in air	Effect
3 ppb	Max concentration for electronic systems per ISA
10 ppb	Max concentration for electrical equipment per NEMA
100 ppb	Threshold concentration for smell
1 ppm	Offensive odor of rotten eggs
5 ppm	Deadens olfactory senses
10 ppm	Max 24-h exposure limit per OSHA
15 ppm	Max 8-h exposure limit per OSHA
50 ppm	Max 30-min/d exposure per OSHA
300 ppm	Instant death for sensitive people
500 ppm	Instant death for everyone
46,000 ppm	Lower explosive limit (LEL)

Oxygen Depletion

Oxygen depletion monitors are available from the same companies that manufacture explosive gas and hydrogen sulfide detectors. A common type of unit has a galvanic cell containing two dissimilar electrodes in a basic electrolyte. Oxygen diffusing through the sensor cell face initiates a redox reaction that generates a minute current proportional to the oxygen partial pressure. The typical range is 0 to 25% oxygen concentration.

Fire

Severe fire hazards do not normally exist in water and wastewater facilities because they are usually constructed of noncombustible materials and very little of the contents can burn. In very large facilities, the control equipment may present a significant fire hazard, and prompt detection of a fire may reduce the damage and facilitate repair. Because these electrical fires produce little visible smoke or heat, "products-of-combustion" detectors are the most suitable type. In some facilities, the local fire marshal may require a fire alarm system conforming to applicable building codes to protect personnel. The design of such systems is usually outside of the competence of instrumentation engineers, and specialists should be retained.

Unauthorized Intrusion

Intrusion alarm systems are frequently needed in urban areas not only to protect the pumping station, but also to protect the intruders from injury. Most pumping stations have a limited number of possible entry points, and simple door switches are adequate. If there are windows, passive infrared detectors may also be needed. To be effective, intrusion alarms must be transmitted to a continuously staffed facility, such as a police or fire station.

Air Pressure

Where instrument air and/or starting air systems form an essential part of a pumping station, pressure alarm systems are required. Simple pressure switches are adequate for this purpose.

Fuel Level

Where fuel is stored on-site for engine-driven pumping units or emergency generators, the fuel supply must be monitored. A simple staff gauge is frequently used. If remote monitoring is required, some type of level measurement system is used. Automatic monitoring of diesel tanks can be accomplished with bubblers, floats, radio-frequency probes, or ultrasonic units. Buried liquid fuel tanks must be furnished with leak detection systems. Package tank monitoring units, which include the leak detection and level monitoring, are available and recommended.

Propane tanks are more difficult. Pressure measurement, which is often used for monitoring propane tanks, does a better job of measuring tank temperature than liquid level until the tank is almost empty. A differential pressure element connected to the liquid withdrawal nozzle on the bottom of the tank with a compensating connection to the gas withdrawal nozzle on the top of the tank provides a simple, accurate measure of tank level. Such transmitters are readily available with the static pressure rating of 4140 kPa (600 lb/in.^{2}) required for this application. Alternatively, the tank can be mounted on platform scales to give a direct measure of propane quantity.

Control Power Availability

Where possible, control circuitry should be designed so that it is not dependent on a single control power supply. If a single control power supply is used, it should be monitored by an undervoltage relay to provide an alarm if failure occurs.

Battery Charger Operation

Chargers designed for constant-voltage-float charging of stationary batteries are available with built-in alarm circuitry that provides a warning when the charger fails.

Sample Pumps Running

Sample pumps are used to transfer sample flows to chlorine residual analyzers and/or containers for subsequent laboratory analysis. In noncritical applications, a motor-starter auxiliary contact or control relay connected across the motor gives a reasonable indication of pump operation. A running pump, however, is not a sure indication of flow. Because a valve might be closed or the pipe might be blocked, a flow switch is needed in critical applications. Vane switches installed in a pipeline can be used in clean water applications. In wastewater applications, a thermal-dispersion flow switch (which is a heated temperature-sensitive element encased in a smooth cylinder that is inserted into the pipeline) is better. When flow is present, the element temperature stays near the fluid temperature, but when flow stops, the element temperature rises because the heat is not dissipated effectively. This temperature change can be detected by suitable electronic circuitry. The thermal-dispersion flow switch is much more expensive than vane switches, but the system is fairly resistant to fouling.

20-9. Pumping Unit Monitors

Most pumps are driven by constant-speed electric motors of moderate (less than 110 kW or 150 hp) size. Such units require only the standard protective devices provided with any electric motor, such as the replica-type thermal overload protection normally furnished in standard motor controllers. "Replica" means that the thermal overload device, which is located in the starter where it is heated by motor current, duplicates motor heating. It is most common in motors up to 150 kW (200 hp). In larger motors, it is better practice to add temperature monitors in the windings. Temperature monitors should also be used on most adjustable-frequency drives. Always equip engine drives with cooling water temperature monitors and lubrication oil pressure monitors.

Bearing Temperature

Bearing housings of larger (greater than 150 kW or 200 hp) units may be provided with temperature monitors. Frequently, these are filled-system temperature switches and may be specified to be furnished with the pumping unit. Alternatively, resistance temperature detectors (RTDs) or thermocouples (TCs) may be installed in bearing housings and connected to external electronic monitoring equipment. Both the RTDs and the TCs are more accurate and easier to adapt to remote monitoring.

Motor Winding Temperature

Motor winding temperature monitors provide much better overtemperature protection than replica-type thermal overloads. Three types are commonly used: (1) end turn snap switches (Klixons™), (2) positive temperature coefficient (PTC) thermistors installed in the winding slots, and (3) resistance temperature detectors (RTDs) installed in the winding slots. Klixons™ are not adjustable, may produce a winding hot spot where attached, and are not recommended. The

PTC thermistors are recommended for motors up to about 750 kW (1000 hp), and RTDs are recommended for larger motors.

Pump Speed

In many applications, a motor-speed meter is an adequate measure of pump speed and can be obtained as an accessory with adjustable-speed drive equipment. Where an actual measure of pump shaft speed is required (either for positive interlocking purposes or because the adjustable-speed drive equipment cannot provide a speed signal), either a belt-driven tachometer or a reluctance pickup device may be used. Belt-driven tachometers are mechanical devices with bearings and brushes that require significant maintenance. The belt cannot be replaced without opening the coupling between the driver and the pump. A reluctance pickup consists of a split gear clamped to the pump shaft and a magnetic pickup coil assembly mounted close to (but not in contact with) the gear. As the gear teeth pass the pickup coil, a small alternating current voltage is generated that can be detected by suitable electronics. These devices are essentially maintenance free and are very accurate.

Vibration

Vibration monitoring is a specialized field, and most engineers defer to the recommendations of the machinery builder. There are three parameters that may be monitored: (1) acceleration, (2) velocity, and (3) shaft displacement. The choice depends on machinery characteristic dynamics such as speed, mass, and bearing type. In general, low-cost accelerometers utilizing a spring-suspended mass attached to a machine housing give a little protection. Elaborate electronic systems costing thousands of dollars may be able to provide a degree of warning of incipient failure, but nothing substitutes for periodic inspection by a competent operator.

Engine Driver Systems

Engine drivers of all sizes require extensive instrumentation (see Chapter 14). The instrumentation is usually furnished as part of the engine-control equipment. Electric or electronic sensors should be used on the engine itself with signals wired to a separate, vibration-free panel. Only sensors designed specifically for engine service should be mounted on the engine.

20-10. Control Equipment

Control equipment used in pumping stations has advanced from crude, motor-driven cam stacks to the compact, electronic, stored-program microprocessor controllers used in the new designs. The older systems were readily understood by personnel accustomed to maintaining mechanical devices, but many operators now can deal with modern control equipment. When there is a failure, either the device must be returned to the vendor or the vendor must come to the site to make repairs. This situation is unacceptable for equipment that must be kept operating. The overall reliability of modern solid-state controllers, however, is much better than that of the older mechanical devices, so modern devices are now frequently applied to pumping stations, even in locations far from skilled mechanics. If solid-state controllers are to be used, it is essential that emergency manual control systems be provided, that spare control devices be stocked by the operator, and that operators have or gain adequate skill levels to make repairs by replacing defective modules.

Motor Controllers

A motor controller is required for every motor in a pumping station. Article 430 of the National Electrical Code [2] establishes the minimum requirements for (1) motor controllers, (2) motor and branch-circuit overload protection, (3) motor branch-circuit, short-circuit, and ground-fault protection, (4) control circuit characteristics, and (5) disconnecting methods. Beyond these code requirements, a great number of variations in motor controllers are established by the characteristics of the power source and the requirements of the driven equipment.

Motor Branch-Circuit, Short-Circuit, and Ground-Fault Protection

Either a fused switch or an automatic circuit breaker is used for branch-circuit, short-circuit, and ground-fault protection. Fuses provide higher short-circuit interrupting capacity than circuit breakers, but they must be replaced after a circuit interruption. A single fuse may blow, which causes damaging single-phase power to be fed to a three-phase motor. Circuit breakers always open all conductors feeding a motor circuit and may be reset following a circuit interruption. Where the available short-circuit current exceeds the interrupting capacity of circuit breakers, fuses may be used in conjunction with the circuit breaker to provide the advantages of both. But this combination system is objectionable in terms of cost (30% added to the cost of the motor and starter), space requirements, and convenience. Although standard fuses are readily available, the special fuses used in conjunction with a circuit breaker may be difficult to obtain in an emergency. The type of short-circuit and ground-fault protection to be used is not a clear-cut decision, and the selection should be made in conjunction with the preferences of the facility operator.

Motor Starters

Motor starters typically provide motor branch-circuit overload protection as well as a means of starting the motor. Both magnetic and solid-state motor starters are available, but, at this time, solid-state units are still very expensive (they add about 50% to the cost of the motor and starter) and are seldom used except in special applications.

Full-Voltage, Nonreversing (FVNR) Magnetic Starters

The FVNR starter is the most commonly used type of magnetic starter. It connects and disconnects an electric motor with the power source in response to a relatively low energy control signal.

Reduced-Voltage, Nonreversing (RVNR) Starters

The RVNR starter must be used where the capacity of the electrical power source is limited in comparison with the size of motor. This is a common limitation in pumping stations located far from other plant facilities and in residential districts. Electric power utilities frequently stipulate that reduced-voltage starters be used under these conditions. Reduced-voltage starters are also helpful in getting a motor started on an emergency generator. The various types of RVNR starters, along with their advantages and disadvantages, are listed in Table 20-6 (see also Section 8-3).

Multispeed Motor Starters

Multispeed motors are sometimes used to provide two or more pumping capacities from a single pumping unit. Multispeed motors may be of the two-speed/one-winding, two-speed/two-winding, or four-speed/two-winding types, depending on the required speed ratios. The type of multispeed starter used depends on the type of motor to be started. The design of the control circuitry used with multispeed starters must ensure that the motor is deenergized for a sufficient time to allow the motor magnetic flux to decay during speed changes. If the time for decay is too short, extremely high transient torques may be generated when the motor is reenergized. Magnetic flux decay time varies from 0.5 s for small motors to over 1 s for large motors.

For adjustable-speed motor drives, refer to Chapter 15.

20-11. Control Logic

Control logic is required when equipment operates automatically. Although control logic is trivial in many small pumping stations, it can become quite complex in larger pumping stations with multiple pumping units and/or numerous auxiliary devices. For example, some pumping stations have several pumps of different sizes and require control logic to sequence them in a way that matches the pumping demand as closely as possible. Pumping stations discharging into long force mains frequently require power-operated check valves that are opened only after an oncoming pumping unit has been started (and are closed before the pump is stopped) to reduce surges that can occur when a pump suddenly starts or stops.

Table 20-6. Reduced-Voltage, Nonreversing (RVNR) Motor Starters

Starter type	Characteristics[a]
Autotransformer	Moderate cost, good starting current reduction, uses standard motors
Primary resistor	Moderate cost, poor starting current reduction, uses standard motors, not applicable to high-voltage (over 600 V) motors
Primary reactor	Same characteristics as primary-resistor starter, but may be used with high-voltage motors
Part winding	Low cost, poor starting current reduction, requires a special motor
Wye–delta	High cost, moderate starting current reduction, requires a special motor
Solid state	High cost, excellent starting current reduction, presently not applicable to high-voltage motors

[a] Moderate cost is 100%, low cost is 95%, and high cost is 140%.

Control logic can be provided either by electromechanical devices, such as relays and drum programmers, or by solid-state devices, such as programmable logic controllers and microcomputers. Mechanical devices are usually more expensive for all but the simplest control systems, and the reliability of large numbers of mechanical devices in complex control systems is poor because dirt on a single relay contact can cause a malfunction. Solid-state systems have become very reliable if properly designed and installed, but they are more difficult to repair when a malfunction occurs.

Relays

Two types of relays are available: machine tool and miniature plug in. Machine-tool relays are comparatively large and are capable of operating large motor starters. Because a large armature and contact mass must be moved quickly when a machine-tool relay operates, these relays are subject to mechanical failure in high duty-cycle applications. Miniature plug-in relays have a limited contact rating and are prone to early failure if applied to switching heavy loads. Miniature relays, however, may be quickly and easily replaced. Therefore, miniature plug-in relays should be used for logic level circuits and machine-tool relays should be used for switching large motor starters.

Timers

Most machine-tool timing relays provide time delays by forcing air through an adjustable orifice. Time delays of up to approximately 1 h can be reliably obtained in this fashion, but the accuracy is not particularly good. Where longer time delays or precise timing is required, synchronous-motor-driven timers are used. Solid-state timers are also available in either large-frame or miniature plug-in configurations. Solid-state timers can provide high-accuracy timing with delays of many hours.

Drum Programmers

Complex sequential logic requires large numbers of relays. Drum programmers provide a much simpler method of accomplishing such logic. Drum programmers can be actuated by either solenoids or motors. Either type is suitable for low duty-cycle applications, but the motor-actuated type is more reliable in high duty-cycle applications. The cam stacks that are used in many existing pumping stations are a form of motor-driven drum programmer.

Programmable-Logic Controllers

Programmable-logic controllers (PLCs) were originally developed for the automotive and machine tool industries. These units are industrial-grade computers that have been designed to emulate conventional relay logic. Reliability and maintainability have been developed to a very high level. Programming can be accomplished by any competent control electrician, and repairs are facilitated by extensive diagnostic systems and extensive use of plug-in components. PLCs are less expensive and more reliable than discrete components for all but the simplest of logic systems, and they are greatly preferred for pumping station control systems if control logic is needed.

Microprocessor Controllers

Microprocessor controllers are similar to programmable-logic controllers, but (1) they require computer programming skills to alter their logic and (2) they ordinarily lack diagnostic facilities and plug-in components. These controllers frequently appear in pre-programmed vendor packages.

This type of equipment is useful in applications such as constant-speed, multipump wastewater lift stations and multipump water system booster stations, but it should not be used for overall control of a pumping system unless the facility operator has, or is willing to develop, the requisite maintenance skills. When compared with programmable-logic controllers, the simpler design of a microprocessor controller does provide high performance at low cost.

20-12. Altitude Valves

Altitude valves (see Section 5-5) are sometimes used with water system storage to control pumps indirectly. The pumps are typically started by a time clock. When the storage tank is full the altitude valve closes, and a discharge pressure switch located at the pump detects an increase in system pressure and shuts the pump off. Because no electrical connections are required between the reservoir and the pumping station, this pump control is simple and economical.

The operation of pressure-controlled valves is explained in manufacturers' literature.

20-13. Monitoring and Data Acquisition

Providing information about the operation of pumping station equipment is particularly useful for large, unattended pumping stations. The simplest systems consist of only an event recorder (to record pump operation) or an annunciator with memory (to give visiting operators an indication of malfunctions). Telemetry systems may be added to transmit station status to an attended location for immediate indication of failure. Full supervisory control and data acquisition (SCADA) systems are capable of monitoring pumping station operation and controlling station operation from a distant location.

Annunciators

An annunciator monitors one or more discrete contacts and activates a warning light or alarm on malfunction. The addition of memory causes the annunciator to retain transitory alarms to inform a visiting operator of something gone wrong in the worker's absence. Recording annunciators provide a printed record of each alarm occurrence as well as the time, data, acknowledgment, and return to normal condition.

Recorders

Recorders produce a permanent record of analog or discrete variables on charts, which are frequently retained for many years. Flows and/or water levels are recorded in virtually all but the smallest pumping stations. The major disadvantages of recorders are as follows: (1) using information from recorder charts is tedious; (2) it takes considerable effort to keep pens working and charts replaced; and (3) the maintenance of recorders is particularly poor in many pumping stations. Recorders, therefore, should be sparingly used.

Telemetry

Transmitting data to an attended receiving station makes them more (and immediately) useful as compared with data stored in a remote pumping station. Telemetered alarms provide immediate warning of malfunctions. Centralized recorders are easier to maintain, and trends can be monitored continuously.

Communication channels are a major part of any telemetry system. They may be leased from common carriers (such as local telephone companies), or the pumping station utility may construct its own. Historically, the performance of leased facilities has been less than satisfactory. Data channels require high reliability—particularly during storms when leased facilities are least reliable. Leased channels are primarily designed for audio (voice) transmissions, which are not seriously impaired by intermittent noise. But noise is a severe problem for data transmissions. Consequently, dedicated communication facilities are frequently constructed even though they are almost always more expensive than leasing.

The various types of communication channels used for telemetry and SCADA systems are listed in Table 20-7. The type of telemetry equipment required depends on the type and amount of data to be transmitted and whether remote control capability is desired. In the simplest system, the data are simply displayed on annunciators, indicators, and recorders. In larger systems, video display terminals and data-logging typers are frequently used.

Tone Equipment

Tone equipment is a system for modulating discrete contact signals into the voice-frequency range for transmission over voice-grade communication channels. It permits multiple signals to be transmitted over a single communication channel by using different carrier or center frequencies for different signals. By utilizing multipoint communication channels, this system is suitable for multiple signals from a single station and for multiple signals from different stations.

For best performance, the number of different carrier frequencies used on one channel should be limited to about ten. Two types of tone equipment are available: amplitude modulation (AM) and frequency-shift-keyed (FSK) modulation. Amplitude modulation has very poor resistance to communication channel interference and should never be used. Frequency-shift modulation has some resistance to communication channel interference, but is still generally unsatisfactory. Tone systems are ordinarily capable only of transmitting data in one direction—from the remote station(s) to the central station.

Digitally Encoded Tone Equipment

Tone equipment manufacturers have generally recognized the interference problem, and most now offer digital encoding systems that permit the tone equipment to reject errors. Digital encoding also permits several pieces of information to be transmitted using a single carrier frequency, which substantially increases the data capacity. With multipoint communication channels, one carrier frequency can be assigned to each station connected to the channel and, thus, permits about ten stations to be connected to the channel.

Supervisory Control and Data Acquisition Systems

Supervisory control, a logical progression from simple telemetry, infers that data can be transmitted not only from the remote station to the central station but also from the central station to the remote station. Originally, SCADA systems were implemented with miniature mechanical relays for transmitting data over direct-current telephone lines to control and monitor remote gas and electric utility systems. Modern SCADA systems utilize microcomputer technology and voice-grade communication channels. The development of the "personal computer," which is used as the master station, has reduced the cost of very sophisticated SCADA systems to a relatively low figure. Remote terminals may be implemented either with programmable controllers or with microprocessor controllers. In either, it is practical to provide both (1) local automatic control at the remote station with the SCADA equipment and (2) a method of overriding the local controls from the master station.

20-14. Design Considerations

The performance of an instrumentation and control system can be affected more by the way the equipment is installed than by any other factor. Manufacturer installation and maintenance manuals provide extensive information about how an instrument should be installed. The API RP550 [9] is an excellent guide for the installation of many instruments used in pumping stations.

Location and Access

Instruments and control devices require special maintenance. The location of process sensors is generally dictated by process considerations, but considerable flexibility is available in the location of other instrument and control devices. Wherever possible, process sensors should be installed so that they can be removed without draining pipelines. Bubble pipes should be installed so that they can be easily rodded. Magnetic flow tubes should be installed so that the electrodes can be wiped off without removing the flow tube from the pipeline. Sufficient flexibility must be designed into pipelines so that in-line devices can be removed without having to disassemble whole piping systems. If possible, valving and connections should be provided to permit in-place instrument calibration.

Table 20-7. Telemetry Communication Channels

Channel type	Characteristics
Voice-grade telephone	Lowest cost and very good, but reliability may be a problem in storms. Similar to conventional telephone service except that a dedicated circuit is leased; signals must be modulated into voice-frequency range of 300 to 3000 Hz; moderate amounts of data can be transmitted over unlimited distance. Channels can be "conditioned" to improve the performance, and with a high degree of conditioning, large amounts of data are transmittable. Multipoint configurations permit several pumping stations to be connected to the same channel, but telemetry equipment must be able to separate the station signals.
VHF and UHF radio	Best system if a license can be obtained. Voice-grade communication channels functionally similar to voice-grade telephone; VHF radio frequencies are virtually all taken, but UHF frequencies (especially 928 to 952 megahertz) are still available for water systems (although not for wastewater systems) in many areas.
Microwave radio	Signals bounce off temperature inversion layer, but by using both high and low antennas (the principle of space diversity) signals go through temperature inversions 99% of the time but at twice the cost. Microwave-radio frequencies are available, but cost and transmission capability are beyond most budgets and needs. Birds may cause spurious signals (but not if the system is properly designed for error rejection).
Underground cables	Use pipeline easements for underground cables; such cables may be either buried directly or in conduits. Communication is reliable and of high quality, but initial costs are very high.
Overhead cables	Practical either on leased pole space or on privately owned poles, but, because they are subject to damage by storms and automobiles, reliability is inadequate.
Fiber optics	Impractical; very expensive to repair damage; suited to transmitting huge quantities of data in infinitesimal time periods—completely unnecessary for pumping stations.

Protective Enclosures

Some instruments are designed for installation under field conditions and adverse environments. Many instruments and control devices, however, are designed for installation only in a clean, dry, air-conditioned environment. If such an environment does not exist in the pumping station, install a suitable protective enclosure for these instruments and, if necessary, use force ventilation or even air conditioning in instrument enclosures. Because many electronic control instruments designed for the process industries require rear access for electrical connections, give careful consideration not only to panel design but also to providing the ready access needed.

Equipment Handling

Instruments and control devices are fragile, and some (e.g., magnetic flow tubes) are also heavy. Provide a method of removing and installing instrumentation and control equipment that permits the weight to be handled easily. These methods frequently entail special access hatches or doors and permanently installed hoisting beams or brackets.

Safety

Most modern instrumentation and control systems are electrically operated and must operate continuously, so they cannot be shut down for maintenance. Use low voltages (24 V dc) in instrumentation and control systems as much as possible. Make it easy to disconnect higher voltages quickly and safely from individual instruments for servicing. Use plug-in connections wherever possible so that the instrument can be easily replaced by a spare and taken to a shop for servicing.

20-15. References

1. Lipták, B. G., *Instrument Engineers' Handbook, Process Measurement,* Chilton Book Company, Radnor, PA (1982).
2. NEC, *National Electric Code,* National Fire Protection Association, Boston (updated periodically).
3. ASME MFC-3M, "Measurement of fluid flow in pipes using orifice, nozzle, and Venturi," American Society of Mechanical Engineers, New York (1985).
4. Miller, R. W., *Flow Measurement Engineering Handbook,* 2nd ed., McGraw-Hill, New York (1989).
5. Moorman, P. "Flow meters for water- and wastewater treatment plants," *WaterWorld Review,* pp. 25–27, Jan/Feb 1994.
6. Starrett, P. S., H. B. Nottage, and P. F. Halfpenny, "Survey of information concerning the effects of nonstandard approach conditions upon orifice and venturi meters," 65-WA/FM-5, presented at the winter annual meeting of the American Society of Mechanical Engineers, Chicago (November 7–11, 1965).
7. Shelley, P. E., and G. A. Kirkpatrick, "Sewer flow measurements—a state-of-the-art assessment," EPA-600/2-75-027, November 1975.
8. White, G. C., *Handbook of Chlorination,* 2nd ed., Van Nostrand Reinhold, New York (1986).
9. API RP550, *Manual on Installation of Refinery Instruments and Control Systems,* Part 1, Sections 1–13, and Part 11, Sections 7–10, American Petroleum Institute, Washington, DC (revised periodically).

20-16. Supplementary Reading

1. Bean, H. S., *Fluid Meters,* American Society of Mechanical Engineers, New York (1971).
2. Considine, D. M., *Process Instruments and Controls Handbook,* McGraw-Hill, New York (1974).
3. Hayward, A. T. J., *Flowmeters,* Macmillan, New York (1979).
4. Shinskey, F. G., *Process Control Systems,* McGraw-Hill, New York (1979).
5. U.S. Bureau of Reclamation, *Water Measurement Manual,* Government Printing Office, Washington, DC (1971).

Chapter 21

Instrumentation and Control Applications

ROBERT S. BENFELL

CONTRIBUTORS
Mead Bradner
Horton Wasserman

Pumping stations of different types and sizes have very different instrumentation requirements. Large utilities with many facilities to operate and maintain generally need more extensive instrumentation than smaller utilities, partly because smaller utilities usually lack the staff specialists necessary to maintain instruments. Several typical pumping station configurations and the instrumentation and control devices that might be required for that type of station are considered in this chapter.

21-1. Process and Instrumentation Diagrams

Control systems are frequently depicted on process and instrumentation diagrams (P&IDs). The level of detail shown on P&IDs varies considerably with their purpose. The figures included in this chapter are very simplified and show only the pumping station main process flow and the location of major monitoring instruments. No attempt is made to show control equipment. The following balloon symbols (circles) appear on the figures, as exemplified in Figures 21-1 and 21-2:

A, B	Force main A, force main B
AE	Analytical element
ARC	Analytical recorder/controller
FC	Flow controller
FE	Flow element (such as a magnetic flow tube or Venturi meter)
FIQ	Flowmeter with instantaneous and totalized indication
FQ	Flow integrator
FQR	Flow recorder with totalizer
FR	Flow recorder
FT	Flow transmitter
HS	Hand switch
KS	Timer or time clock
LAH	Level alarm high
LAHH	Level alarm high–high (back-up alarm)
LAL	Level alarm low
LC	Level controller
LCV	Level control valve (altitude valve)
LG	Level gauge
LR	Level recorder
LS	Level switch
LSH	Level switch high
LSHH	Level switch high–high (back-up alarm)
LSL	Level switch low
LSLL	Level switch low–low (interlock to shut down pumps)
LT	Level transmitter
M	Motor
PC	Pressure controller
PI	Pressure indicator or gauge
PSH	High-pressure switch

PSL	Low-pressure switch
PT	Pressure transmitter
SS	Speed switch
ZS	Position or limit switch
○	Field mounted
⊖	Mounted on front of panel
⊖	Mounted inside or behind panel

Note that a transmitter (such as FT and PT) may be designated as equipped with a local indicator by inserting the letter "I" in the symbol: FIT and PIT. The order in which output functions (such as Q or I) are listed is not important.

Other acronyms (not used in balloons) include

AS	Air supply
ASD	Adjustable-speed drive
CNTL	Control
MAG	Magnetic
P	Purge
P&ID	Process and instrumentation diagram
PLC	Programmable logic controller
S	Solenoid-operated
VSD	Variable-speed drive

The balloon symbol as used on P&IDs represents the function of an instrument, not the actual hardware used to perform the function. Thus, a physical pressure switch might be used to measure a level and (because the primary purpose takes precedence) would have LS as the functional symbol. The distinction between a switch and a controller is also important. A switch is a device the normal function of which is to detect an abnormal condition. A controller may be the exact same physical device, but its normal function is actual control. Thus, a pressure switch used to actuate a low-pressure alarm has the function symbol PSL; a pressure switch used to start and stop a pump in order to maintain pressure in the system has the functional symbol PC.

21-2. Well Pump with Hydropneumatic Tank

A well pump with a hydropneumatic tank (shown in Figure 21-1) is common in small water systems—frequently with a single well.

Pumping Unit Control

While there are differing theories about how to operate a hydropneumatic tank, the one proposed here ensures trouble-free operation. The maximum water level in the tank is typically set at one-third to one-half full to provide an adequate air-pad volume for reasonably constant-pressure operation, and air is added or vented as required only when the tank is at or close to the maximum water level. The operating sequence is as follows:

- As water is drawn from the tank, pressure (as measured by PSL) will fall, which causes the well pump to start and refill the tank.
- When the water level reaches the maximum operating level (LC), the pump is stopped.

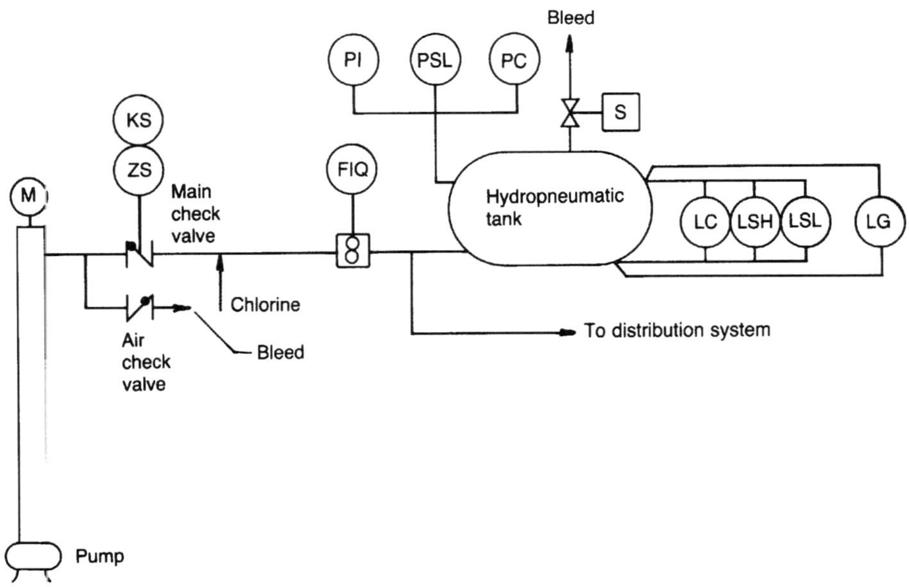

Figure 21-1. Well site with a hydropneumatic tank.

- Air is added or vented as necessary to adjust tank pressure (PC) to the maximum desired value.

In those installations where adequate air volume can be trapped in the well pump discharge piping, provision is required only for air venting. Because excess air can be vented easily and very quickly, the controls are quite simple. In installations where air must be added, an extra level switch is required to prevent the addition of air (which would air bind the tank) as the tank water level falls. Level switches on hydropneumatic tanks should always be installed in an external "cage" with isolating valves between the cage and the hydropneumatic tank so that the switches can be serviced without venting the tank.

In deep well applications, provide positive interlocking to eliminate any possibility of restarting a pump during backspin. A limit switch (ZS) on the check valve and time-delay relay (KS) or a zero speed switch can provide this protection. Also, flow detectors are sometimes installed on the intermediate shaft lubrication system to prevent the pump from being started before lubrication is established.

Chlorination Control

In most well pumping stations, chlorination equipment is operated at a constant rate. The chlorine injector pump is simply turned on at the same time that the well pump is turned on. At well sites that have multiple well pumps, step-rate control can provide different chlorine feed rates for different pump combinations.

Alarms

The following malfunction-monitoring devices and alarms may be furnished:

- Tank pressure low (PSL)
- Tank water level high (LSH)
- Tank water level low (LSL)
- Shaft lubrication failure
- Well water level low
- Pump discharge flow failure (ZS)
- Intrusion
- Chlorine leak (if applicable)
- Power failure.

Other Monitors

A totalizing water flowmeter (FIQ) is desirable on all wells and is frequently mandated by regulatory agencies. A pressure gauge (PI) and a level gauge (LG) should be furnished on the hydropneumatic tank for monitoring control system operation. Note that the level gauge is connected to the hydropneumatic tank separately from the switch cage so that it can remain in service if the switch cage is isolated.

21-3. Booster Stations

Booster stations are commonly used to lift water from one distribution zone to a higher distribution zone. The configuration shown in Figure 21-2 is controlled by the level in the zone 2 reservoir, but requires no

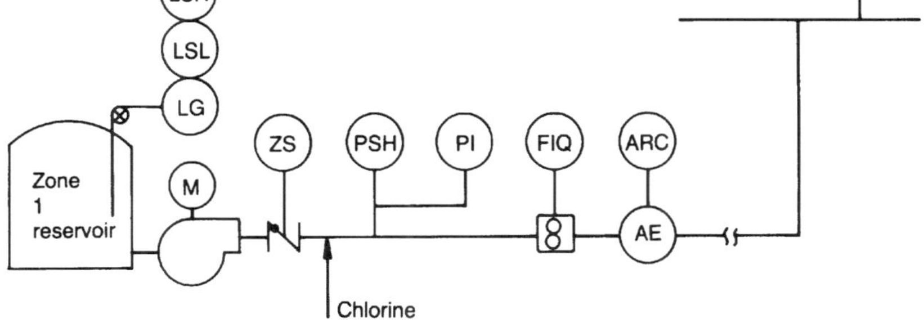

Figure 21-2. Booster pumping station.

telemetry between the pumping station at the zone 1 reservoir and the zone 2 reservoir.

Pumping Unit Control

The booster pump(s) is started by the time clock (KC)—if possible, in the evening to take advantage of off-peak power rates. Water enters the zone 2 reservoir until the reservoir is full and the altitude valve (LCV) closes. This causes the discharge pressure on the pump as measured by a high-pressure switch (PSH) to increase and shut down the pump.

For such a system to operate properly, the booster pump(s) must be of significantly greater capacity than system demand and the zone 2 storage reservoir must have ample capacity to supply the distribution system until the time clock initiates the next pumping cycle. With this type of system, it is frequently necessary to start the booster pump(s) manually or reset the time clock during high demand periods. It may also be necessary to start the pumps manually in case of unusual demand, such as could be caused by a fire.

A level gauge with low level switch (LG/LSL) is furnished on the zone 1 reservoir to shut down the pump(s) in case there is insufficient water in zone 1.

Chlorination Control

Where the zone 1 reservoir is large, water may remain in the reservoir for a long enough period to lose some or all of its chlorine residual. In this situation, chlorination equipment must be installed, and residual control is recommended because the residual chlorine in the water drawn from the reservoir is unpredictable.

A water sample from the discharge of the booster pump is passed through the residual analyzer/controller (AE/ARC), which compares the actual residual with the desired residual and adjusts the chlorine feeder until the difference between the actual and desired residuals is essentially zero.

Alarms

The following malfunction-monitoring devices and alarms may be furnished:

- Zone 1 reservoir level high (LSH)
- Zone 1 reservoir level low (LSL)
- Pump discharge flow failure (ZS)
- Chlorine leak (if applicable)
- Intrusion
- Power failure.

Other Monitors

A totalizing flowmeter (FIQ) is frequently furnished to record the amount of water transferred between zones, and a pressure gauge (PI) used in conjunction with the flowmeter can indicate any changes in pump performance. A chlorine residual recorder is shown as part of the chlorine residual controller.

21-4. High-Service Pumping Station

Ideally, storage reservoirs would be built for all service zones. In actual practice, however, many service zones do not have storage reservoirs, and the pumps supplying such zones must operate as necessary to furnish instantaneous demand. Multiple pumps are frequently used for this service, and they may either be selected to have a relatively flat head versus flow curve or be operated with a variable-speed drive to maintain constant pressure. A common arrangement is to have one small, constant-speed pumping unit to maintain pressure during low flow periods and two or more larger, variable-speed units (as shown in Figure 21-3) for normal or high demand.

Pumping Unit Control

The following control strategy might be used. The small pumping unit runs constantly until the pressure falls below a minimum set point as measured by an electronic trip (PC1) connected to a pressure transmitter (PT). A large pumping unit is then started and the small unit is shut down. A second electronic trip (PC2), which is set at a slightly lower pressure, is used to start the second large pumping unit if the first unit fails to maintain the required pressure. The speed of the large pumping units is controlled by the PIC, which compares the desired pressure with the actual pressure and adjusts the variable-speed drives as necessary. Because the large units simply slow down when water demand decreases, pressure cannot be used to shut the large units down when less pumping capacity is required. Therefore, electronic trips (FC1 and FC2) connected to the flow transmitter (FT) must be used to shut down the large units when demand can be safely handled by fewer pumps.

Chlorination Control

Chlorine demand varies over a wide range in this application, and compound loop control is recom-

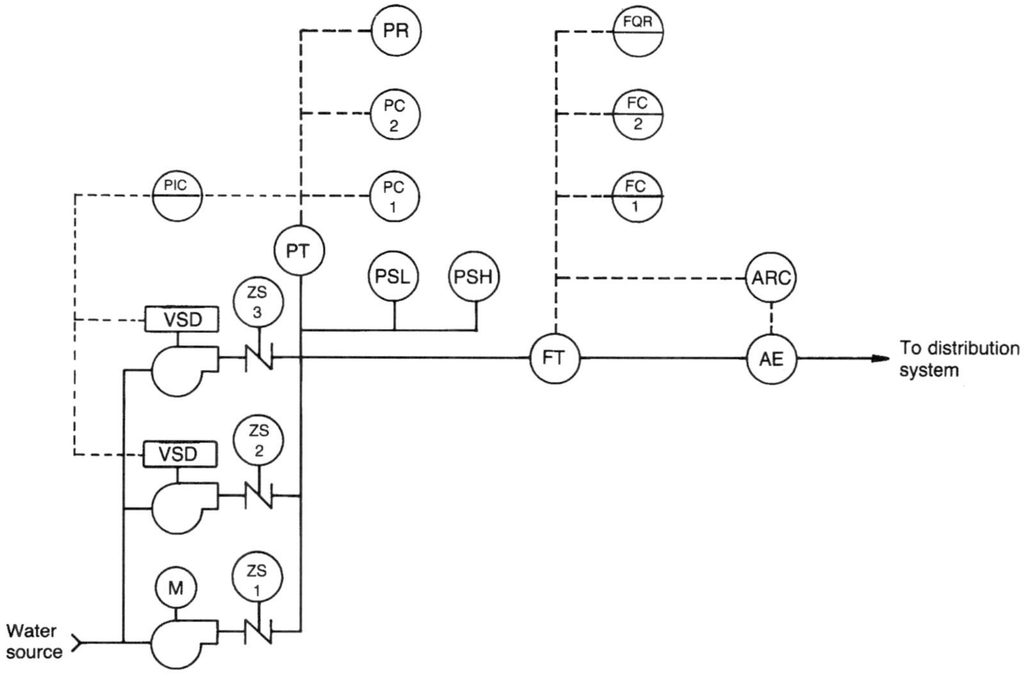

Figure 21-3. A distribution service pumping station.

mended. The output of the chlorine residual analyzer/controller (ARC) is multiplied by the flow to determine the chlorine feed rate. The action of the chlorine residual controller is the same as described previously for booster stations. By multiplying the controller output by flow, the feed-rate corrections are automatically made small at low flowrates and larger at higher feed rates.

Alarms

The following malfunction monitoring devices and alarms may be furnished. Note that the electronically transmitted pressure signals used to control the pumps are not used for alarms. Separate pressure switches are used to protect against the failure of the pressure transmitter or its associated electrical circuitry.

- Discharge pressure low (PSL)
- Discharge pressure high (PSH)
- Pump discharge flow failure (ZS1, ZS2, or ZS3)
- Loss of water source (ZS1, ZS2, and ZS3)
- Intrusion
- Chlorine leak
- Power failure
- Variable-speed drive malfunction (produced by the drive controller).

Other Monitors

A pressure recorder (PR) and flow recorder with an integral totalizer (FQR) are shown in Figure 21-3. A chlorine residual recorder is shown as part of the residual controller.

21-5. Small Wastewater Lift Station

Small lift stations and drainage sumps within larger facilities typically consist of a pair of submersible pumps—one duty and one standby. Such a station can be monitored and controlled by four float switches, as shown in Figure 21-4.

With an increasing water level, the duty pump is set to start at about 80% of the available fill/draw range (LC-B). The follow pump is started at 90% of the available fill/draw range (LC-C). A high-level alarm (LAH) is generated by the controller any time the follow pump is called. The high-level alarm should be electrically locked such that it can be reset only at the pumping station (HS-A). Both pumps are stopped at 0% of the available fill/draw range (LC-A).

Provide a high–high level alarm at 100% of the available fill/draw range (LSHH). The float should be set at such an elevation that it will not be fouled by floating debris in normal operation. A duty pump

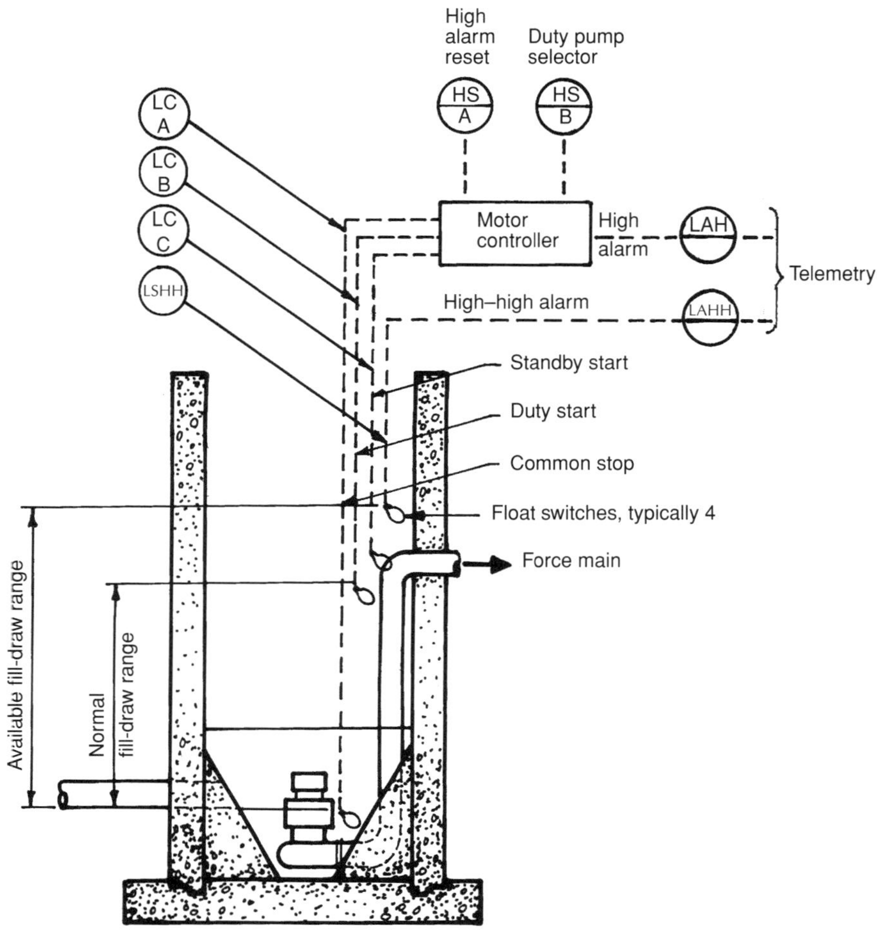

Figure 21-4. Controls for a small wastewater lift station with two C/S pumps.

selector switch is required (HS-B). Manual alternation is recommended to avoid the added complication of automatic alternators. If automatic alternation is used, however, a three-position selector switch (A-B-AUTO) should be provided.

One advantage of manual alternation, in addition to operator control of equipment wear, is the ability to provide an alarm-before-standby-start feature. With this arrangement, the high-water alarm device has a lockup feature that does not allow reset until the water level is lowered below the level at which the follow, or standby, pump shuts down. The alarm is set to initiate prior to starting the standby pump. This feature possesses the following advantages:

- It gives warning that something unusual has occurred and operation of the standby pump is required. The difficulty could be excessive inflow into the wet well, failure of the duty pump, or partial clogging of the duty pump.
- It gives operation and maintenance staff an opportunity to determine the cause before a genuine problem (flooding, backed up sewer, etc.) causes damage.
- The alarm condition will remain as long as the duty pump cannot successfully control liquid level, and thereby it produces a signal that does not appear to the casual observer as a spurious alarm.

The above arrangement requires only four level devices, as shown in Figure 21-4, and provides almost the same functionality as the seven floats shown in Figure 12-4. Furthermore, almost the entire fill/draw range can be utilized for each pump cycle. If the lead pump fails, the operator must visit the site to cancel and reset the high-level alarm—an important first step to ensure reliability. The second step is the immediate repair or remedy for the problem. Note that a careless operator could circumvent the reliability of the station and prevent alarm recurrence by simply exchanging duty and standby pumps. Utilities should be aware of, nurture, and continually assess the sense of duty and responsibility in its workers and, for critical tasks, assign only those who demonstrate these virtues.

Float switches, although widely used in small lift stations, have a number of serious disadvantages: (1) when untethered, they tangle badly even in stilling wells; (2) if tethered to prevent tangling, the power cords are severely flexed by the turbulence in the wet well, cable life is shortened, and adjustment is inconvenient; and (3) they become quickly coated with grease that alters their buoyancy. The life of float switches is only about 2 yr. A float is, however, unsurpassed as an independent switch for the high-water alarm. To facilitate testing a high-water alarm float, attach a light nylon cord to the free end so the float can be tilted by lifting the cord.

Two superior systems for small stations include (1) packaged bubbler systems and (2) pressure cells. A pressure cell can be suspended in a PVC pipe for protection from grease, sludge, and currents. The bottom of the pipe is set at about the mid-height of the submersible pump volute—above possible sludge banks and (to protect the cell from floating grease) below the lowest water level. Both the pipe and the pressure cell can be cleaned if necessary, although maintenance personnel usually dislike doing so (see Section 24-11).

21-6. Moderately Sized Lift Station

A lift station of moderate size, as shown in Figure 21-5, may have three constant-speed pumping units that are typically each less than 60 hp in size. This type of pumping station differs from the small lift station described in Section 21-5 in that it has a separate dry well for the pumping units and may have a small surface structure to house the control equipment.

Pumping Unit Controls

Pressure switches LC-A, LC-B, and LC-C are connected to a bubbler system to control the pumps through very simple relay logic. Ordinarily, the pumps are sized so that two pumps can handle the peak flow and one pump is a standby unit. The amount of control logic required in this type of station can typically be handled by less than half a dozen relays. If remote monitoring of the station is required, however, a solid-state programmable controller or packaged microprocessor controller can provide both the control and telemetry functions at a lower cost than mechanical relays and a separate telemetry unit.

A major consideration for the control strategy of this type of station is limiting the number of pump

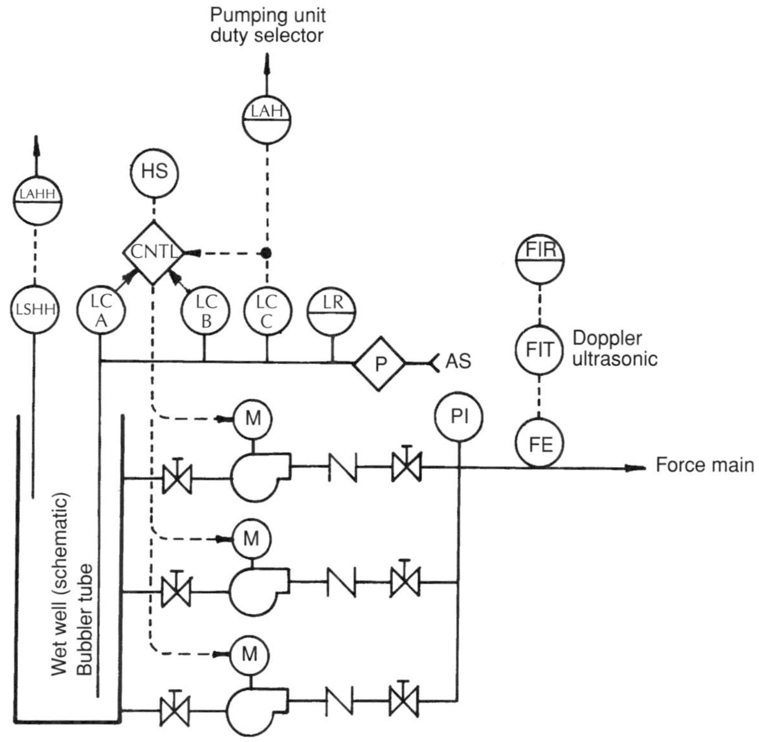

Figure 21-5. Moderate-sized wastewater lift station.

start/stop cycles. The lead pump might be set to start at 80% of the available fill/draw range and stop at 5% of this range (LC-A). The follow pump would be set to start at 85% of the available fill/draw range and stop at 10% of this range (LC-B). If the standby pump is configured for automatic start, it would be set to start at 90% and stop at 15% (LC-C). Regardless of whether the third pump is configured to start automatically, LC-C should generate a high-level alarm that requires manual resetting at the station (LAH).

A manual duty selector switch is shown (HS). Motor duty-cycle considerations frequently require automatic alternation of the lead and first follow pumps driven at constant speed. For this situation, the duty selector switch requires three positions:

Pumps 1 and 2 are duty pumps: switch marked 1/2
Pumps 2 and 3 are duty pumps: switch marked 2/3
Pumps 1 and 3 are duty pumps: switch marked 1/3.

Auxiliary Systems

The auxiliary systems in this type of station include an air compressor for the bubbler and a dry well sump pump. Package equipment with built-in controls is suitable for both units.

Alarms

The following malfunctions should be monitored and alarmed to a central continuously staffed facility:

- Wet well high level
- Third pumping unit started (if automatic)
- Dry well high level
- Instrument air pressure low
- Power failure.

21-7. Large Wastewater Pumping Station

The instrumentation and controls shown in Figure 21-6 are installed in an actual large pumping station operated by a sophisticated wastewater utility with a large number of pumping stations, many of which are larger than the one shown. This station was selected because it represents a complex instrumentation and control system that might well be used in other, simi-

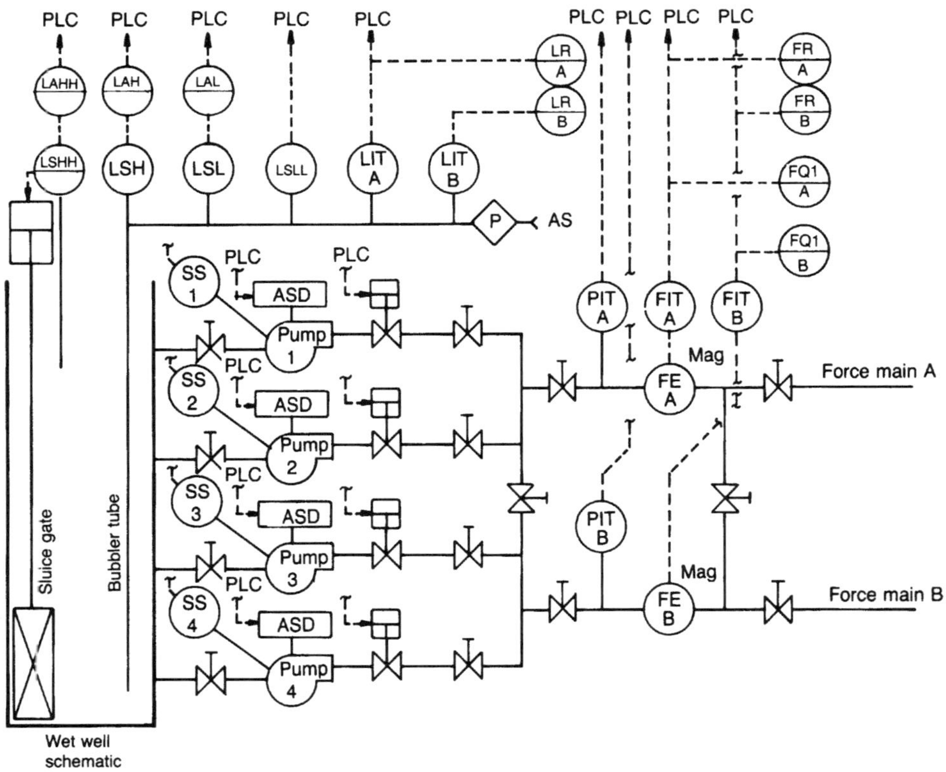

Figure 21-6. Simplified P&ID of the Sunset Pumping Station (see Figures 17-18 to 17-20). After Brown and Caldwell Consultants.

lar stations. The station has four 335-kW (450-hp) adjustable-speed pumping units and a maximum discharge capacity of 1.4 m^3/s (32 Mgal/d) at a head of 52.4 m (172 ft). The instruments installed are listed in Table 21-1.

A solid-state programmable controller is used for the execution of all control logic, including pumping unit sequencing, speed control, and control of all station auxiliaries. The programmable controller also serves as the remote terminal unit for remote monitoring and control from the treatment plant. Accordingly, all signals generated by the instrumentation system are interfaced to the programmable controller whether needed for control or only for monitoring. About 130 relays and considerable analog instrumentation (which are normally required for a station of this type) are replaced by the programmable controller.

Pumping Unit Controls

The wet well is operated with a varying level to take advantage of the influent sewer head/flow characteristic and to maintain scouring of the upstream sewer at low flows (see Figure 12-35). All pump sequencing is handled as a function of the wet well level, and all pumps running in a given level zone operate at the same speed and are linearly controlled by the water level within the zone.

The electronic level transmitter (LIT-A) produces a 4- to 20-mA signal proportional to wet well level. It is used to drive a recorder (LR-A) and the programmable logic controller (PLC) to provide pump speed control and sequence functions. A second electronic level transmitter (LIT-B) permits monitoring a much larger level range than is used for control. LIT-B is connected to a second recorder pen (LR-B). Pressure switches, independent from the electronic level transmitters, are connected to the bubble tube to provide high-level alarm (LSH/LAH), low-level alarm (LSL/LAL), and low–low level alarm (LSLL/LALL). A high–high level float switch acts as a backup to the bubbler system.

As illustrated in Figure 12-35, the level controller must be adjusted to drive the lead pump from minimum pumping speed to full speed as the wet well level increases from minimum to approximately 33% of influent pipe level so as to match the Mode I control curve shown. The follow pump is then required.

The PLC now activates the follow pump and must also change the speed-control adjustments to cause the pumping units to follow the Mode II curve. Both the control gain and set point must be changed to match the new curve. The results of changing the control adjustments are to slow the lead pump to a speed that discharges about 50% of its capacity and to accelerate the follow pump to the same speed as the lead pump. The result of these adjustments is that two pumps now pump the same flow discharged previously by one pump. The switchover is quite smooth. As flow continues to increase, full-speed operation will again be attained at approximately 66% of influent pipe level.

The switchover to three-pump operation is similar. The PLC must activate the second follow pump and readjust the speed control gain and set point to follow the Mode III curve.

Operation on decreasing level is reversed. The level switching points from Mode III to Mode II are delayed to approximately 60% of influent sewer level and from Mode II to Mode I, the delay is to approximately 30% of the influent sewer level so as to eliminate hunting at the switch points.

The need to readjust both the speed-control set point and gain may not be immediately obvious. The controller set point represents the vertical position,

Table 21-1. Instruments in a Large Wastewater Pumping Station

Instrument	Parameter	Instrument Type	Purpose
LSHH	Wet well level	Float switch	Alarm; close influent gate
LSL	Wet well level	Pressure switch on bubbler	Alarm
LSH	Wet well level	Pressure switch on bubbler	Alarm
LSLL	Wet well level	Pressure switch on bubbler	Stop pumps
LIT-A	Wet well level	Pressure transmitter on bubbler	Pump sequence and speed control; monitoring
LIT-B	Wet well level	Pressure transmitter on bubbler	Wide-range monitoring
SS	Pump shaft speed	Reluctance pickup	Pumping unit and discharge valve control
PIT	Force main pressure	Pressure transmitter with diaphragm seal	Monitoring
FIT	Force main flow	Magnetic	Monitoring

and the gain represents the slope of the pump control line segments in Figure 12-35. Control loop gain includes not only the control system gain but also the gain of the pump and associated drive system. If one pump is running, the gain of the pump and drive system may be considered to be G. If two pumps are running, the gain of the pump and drive system becomes $2G$. Therefore, the gain of the speed controller must be halved to maintain the same total gain or slope for the pump control line. Similarly, with three pumps running, the gain of the pump and drive system becomes $3G$, and the gain of the speed controller must be $^1/_3$ of the gain with one pump running to maintain the same slope. Executing such a sophisticated control strategy is difficult in conventional control systems, but it is easily executed in PLCs.

Due to the high discharge head of this pumping station, pumps must be started and stopped against a closed discharge valve in much the same way as required for high service water pumping stations. After a pumping unit is started and a speed switch indicates that the pump shaft is rotating at a preset minimum speed, the discharge valve is opened. When a pumping unit is to be stopped, the discharge valve is first closed. When the valve actuates the closed-position limit switch, the pump motor is stopped. The pump shaft speed switches are also used to ensure that the motor is never energized while the shaft is rotating in reverse, as might happen if the discharge valve failed to close completely.

Auxiliary System Controls

As might be expected, a pumping station of this size has a large number of auxiliary systems that are essential to station operation. All auxiliary system control logic is executed by the programmable controller system.

Dry Well Drainage Sump Pumps

Two pumps are furnished and periodically alternated by the operators. Because there is always a possibility of raw sewage getting into drainage sumps in wastewater pumping stations and an instrument air system is available, a bubbler is used in the drainage sump with pressure switches to control the pumps and generate a low-level alarm. An independent radio-frequency probe is used for the high-level alarm.

Instrument Air Compressors

Two compressors are furnished and periodically alternated by the operators. Receiver pressure switches provide control. A system alarm is generated whenever the alternate compressor is needed.

Air-Gap Tank

An air-gap tank provides positive isolation between domestic water systems and station service water. Simple float switches are used to control make-up water and to generate high- and low-level alarms.

Seal Water Pumps

As with the instrument air system, two pumps are furnished and controlled by simple pressure switches. An alarm is generated whenever the second pump is required. A bladder-type accumulator (a sealed pressure tank with a rubber bladder to separate a precharge gas from the water and, thus, provide a variable-volume storage for the water) is used on this system to eliminate the need for hydropneumatic tank controls, as described in Section 21-2. The limited volumetric capacity of the expansion tank requires automatic alternation of the seal water pumps to limit the motor starting frequency. An interlock between the air-gap tank low-level alarm and the seal water pumps is needed to protect against loss of water supply.

Fluid Power Pumps

The station is furnished with two complete fluid power systems: one for the influent sluice gate and one for the pumping unit discharge valves. Each fluid power system has two pumps that are used for the normal operation of valves or gates and for charging accumulators for emergency operation of the valves or gates. Pressure switches are used to control the pumps. Reservoir level float switches and temperature switches are also furnished to protect the systems from malfunction.

Utility Power Supply Monitoring

The pumping station is furnished with two electric utility power services and an emergency generator. Phase-sensitive relays are installed in the electrical switchgear to detect undervoltage, phase unbalance, and phase reversal on each incoming power supply. If the power supply is not within specifications, these relays (1) prevent actuation of the large, solid-state, adjustable-frequency drives; (2) signal the emergency generator to start; and (3) generate appropriate alarms.

Alarms

The following malfunction monitoring devices and alarms are provided:

- Dry well sump level high
- Seal water pressure low
- Pump discharge valve fluid power pressure low
- Wet well level high
- Wet well level low
- Wet well level high–high
- Influent sluice gate closed
- Air-gap tank level high
- Air-gap tank level low
- Odor control system failure
- Ventilation system failure
- Instrument air system trouble
- Sluice gate fluid power pressure low
- Control power system trouble
- Programmable controller trouble
- Raw sewage pumping unit trouble (four alarms)
- Utility power failure (two alarms)
- Emergency generator failure.

All of the alarms are locally annunciated and interfaced to the programmable controller system for relaying to the supervisory control system.

Chapter 22

Vibration and Noise

JERRY G. LILLY
WILLIAM D. MARSCHER

CONTRIBUTORS
Roger J. Cronin
Ray A. Gall
Richard O. Garbus
Willard O. Keightley
J. Davis Miller
Aloysius M. Mocek, Jr.
William E. (Ed) Nelson
Randall R. Parks
Jerry P. Pollack
James W. Schettler
Theodore B. Whiton

Vibration is one of the most vexing problems with pumping machinery, and it is the cause of considerable altercation and litigation. Noise can become a significant, annoying problem. Excessive vibration from primary equipment can be transmitted directly to the building structure, which causes uncomfortable (and sometimes dangerous) structural vibration levels. Excessive vibration of equipment and piping can destroy portions of the equipment (such as drive shafts and seals), loosen or break pipe anchors, and even cause pipes to burst under certain conditions.

To make this chaper easier to use, the presentation is divided into four parts:

- Avoiding vibration problems. Sections 22-1 and 22-2.
- Troubleshooting excessive vibration. Section 22-3.
- Vibration analysis. Sections 22-4 to 22-12.
- Noise analysis. Sections 22-13 and 22-14.

The first two parts require no background knowledge of the subject and are easy to read. The third part is written for those who wish to delve more deeply into the subject of vibration. The fourth part is, of necessity, a simplified mathematical presentation of noise but with enough worked examples to make it easy to follow.

22-1. Problems of Vibration and Noise

For the reader's convenience, the problems of vibration and noise in pumps and drivers are treated separately.

Pumps

Sources or causes of vibration include:

- The reaction of the impeller vane as it passes the casing cutwater.
- Pumps operating off the best efficiency point (bep) and thereby creating eddies within pump.

- Eddies created by water flowing through bends, valves, and other obstructions.
- In every rotating (e.g., motor) or reciprocating (e.g., engine) machine, some imbalance and shaft misalignment always exists.
- Resonance. If the natural frequency of the machine or system is nearly equal to the frequency of excitation, however small, the resulting vibration can become destructive.

Much of this chapter is devoted to the problems of lateral or translational vibrations, where the vibrating element actually moves back and forth. Rotating shafts may vibrate in both translational and/or torsional modes. In the torsional mode, the centerline of the shaft may not physically move in space, but the shaft does rotate at slightly different instantaneous rotational speeds with different phases at various points along the length of the shaft, which causes an oscillating torque. This torsional (twisting) vibration of the shaft can develop stresses high enough to cause the shaft to fail without any audible or visible warning signs prior to the failure. The subtle character of torsional vibration makes it all the worse.

The impact of a new pumping station on environmental noise levels in the community should be evaluated prior to final site selection, because noise adversely affects residential property values, causes sleep interference and general annoyance, and may lead to legal action against the owner of the plant. Building codes often include noise level limits at property lines. Noise problems are most likely to occur when facilities are located within 300 m (1000 ft) of a residential neighborhood. To protect the owner from legal challenges should a subsequent complaint develop into a court case, documentation of environmental noise levels by a qualified acoustical consultant prior to the start of construction in these instances is vital.

All too often, nothing whatever is done to reduce the noise level within the pumping station. Excessively high noise levels here can impair speech intelligibility and cause temporary or even permanent hearing loss to operating personnel exposed over extended periods of time. It is usually impractical to enclose a noisy machine in a sound-absorbing housing, partly because of the need to service the machine and partly because such measures are relatively ineffective. Common practice, therefore, is to enclose all the noisy machinery within thick concrete walls and ceilings. Requiring workers to wear ear protection and designing for reduced maintenance time and worker occupancy is helpful. The noise level in the reverberation field can be reduced by acoustical treatment (see "Indoor Sound Propagation" in Section 22-13). Although the noise level cannot be reduced by much more than 6 to 8 decibels, this reduction is often enough to permit conversation and substantially increase the level of comfort. At 1996 prices, the cost of acoustical treatment varies from \$40–130/m^2 (\$4–12/ft^2).

Drivers

Sometimes electric motors are the cause of excessive noise or vibration. Vibration tends to have unusually high spikes at one and/or two times the line frequency (60 and/or 120 Hz). If there is a broken rotor bar or bad stator winding, a high-pitched noise (and accompanying vibration acceleration) is often evident at the "slot pass" frequency (about 100 times running speed). In motors driven by AF controllers (particularly the older models), strong vibration pulses are evident even for healthy motors at 6×, 12×, 18×, and sometimes 24× the frequency with which the controller feeds the motor at any given speed. These frequencies seldom cause dynamic or structural resonances of any significance. However, these harmonics can sometimes excite linked-rotor system torsional critical speeds that are difficult to track down with standard vibration equipment that cannot directly sense torsional oscillations, even if shaft stresses are becoming excessive. If shaft failures are occurring at the shaft shoulder near the pump impeller or coupling hub, this situation should be investigated.

Engines

If the driver is an internal combustion engine, the primary driving frequencies (besides 1 times and 2 times running speed and the pump vane passing frequency) are the cylinder stroke frequency (i.e., the number of cylinders times the engine speed) and its first ten harmonics. For most engines, the second harmonic (i.e., 2 times engine speed times number of cylinders) is the strongest, and no natural frequency of the engine-support floor structure or the engine-gear-pump train should be allowed to exist anywhere in the typical operating speed range. Resonances of the other first ten harmonics cannot, however, be ruled out, and they should be investigated if there is excessive noise or vibration in such installations. Offset or right angle gear boxes can also be a torsional excitation source at the gear mesh frequency (number of gear teeth times running speed). Gear mesh excitations can also cause gear tooth damage and/or excessive noise due to axial resonances of the drive shafts, particularly if the drive shafts are hollow. Filling a hollow drive shaft with grease may eliminate the problem by the energy-absorbing damping so provided.

For every engine-driven pump—whatever its size—designers should specify that the pump manufacturer shall make (or have a qualified analyst make) at least a torsional vibration analysis of the pump-driver system. A translational vibration analysis should also be made for the engine-pump-support structure. Many low-cost pump bidders are not qualified for such work, so prequalify bidders to protect both the client and your reputation as a designer.

Active Noise Control

The use of electronically generated sound waves to cancel or reduce unwanted sound (noise) is called "active noise control." This technique has been known for decades, but it has not been widely implemented because of practical limitations in its application. Recent advances in signal processing hardware and software have improved the situation considerably, but active noise control is still generally restricted to low-frequency applications (typically below 200 Hz) and to noise contained inside a duct or a pipe. Once the noise has escaped into the listening environment, this technique is no longer a viable means of achieving a significant noise reduction over an extended region of space.

22-2. Avoiding Vibration Problems

By following a few simple rules, a nonspecialist can—fortunately—avoid about 90% of all vibration problems. The presentation herein is focused on minimizing vibration in pumping station machinery and its related systems. Necessary concepts are explained in simple terms, and their practical use is described. The information is relevant to small- or medium-sized centrifugal pumps of all types, reciprocating pumps, ventilation fans, and electric motor and diesel drivers. For stations with pumps larger than 75 kW (100 hp), or with engine or variable-frequency drives, review the more detailed procedures presented in later sections of this chapter or have a formal vibration analysis made.

Pump Selection

Select a pump that will operate close to its best efficient point (bep). Plot the maximum and minimum headloss envelope, and choose a pump suitable for the entire range. Then, as conditions change (flowrate increases, pipe becomes rougher), it is necessary only to change the impeller to meet the new conditions. Beware of applying large factors of safety to the flowrate, because this blunder could result in an oversized and therefore unreliable pump.

Balancing

Adequately balance all rotating components of diameter equal to at least $^1/_3$ of the pump impeller diameter. Balance all components (in two planes) that have a length of at least $^2/_3$ their diameter. The entire rotor system should be check-balanced in the assembled condition. A balance standard shown to be conservative for pumping station equipment is $e = kW/N$, where k is a constant equal to 0.423×10^{-3} for SI units, e is the imbalance in k · m, W is the mass of the balanced component in kg, and N is the peak operational speed in Hz. In U.S. customary units, the equation is $e = 16W/N$, where e is in oz · in., W is weight in lb, and N is speed in rev/min. Reports to demonstrate compliance with the balancing standards should be submitted as product data. Do not accept any loose rotating component fits where the mass of the component multiplied by the radial clearance approaches the above criterion. Check the shafts to ensure that, if shaft bow is detectable, the amount of bow times the supported component weight also does not exceed the above criterion.

Accurate alignment is necessary to make the centerline of the driving and driven equipment coincide as closely as possible both in concentricity and parallelism. To specify adequate coupled shaft alignment for a rigid coupling or one with one axial location of flexibility ("single engagement"), a good rule is to maintain an error of no more than 37 to 50 microns (1.5 to 2 mils) in the concentricity between coupling hub centerlines. This concentricity criterion may be loosened by an additional 1 to 2 mm per meter (1 to 2 mils per inch) of length between coupling hubs for double engagement or "spacer" flexible couplings, depending on the coupling type. The maximum permissible angle between shafts at the point of coupling engagement is 5 to 10 min, which is automatically maintained if the above offset specifications are maintained. Coupling manufacturers often quote larger amounts of offset and parallel misalignment than the above values, but these quotes account only for coupling survivability, not equipment bearing survivability, and they also do not account for the increased equipment vibration that misalignment might cause.

Pumps with stuffing boxes (packing glands) are less sensitive to misalignment than those with mechanical seals, and the maximum criteria above are satisfactory where stuffing boxes are used. Use the minimum criteria for pumps with mechanical seals.

Alignment should be performed only by qualified technicians or millwrights. Alignment is best determined by the reverse dial indicator method or its modern optical equivalent (although the latter is best performed only by an expert) as described in detail on page 2.284 in *Pump Handbook* [1]. Keep in mind that the above alignment criteria can be applied with the equipment cold, but steps must be taken to ensure that they also apply to the "hot" alignment—i.e., with the pump or fan running, after warm-up. Also, be sure to account for significant machining inaccuracies in whichever coupling faces or rims are used for measurement (reverse dial indicator largely accounts for that). Furthermore, compensate for gauge support arm sag, because the sag can be larger than the alignment specification level. Once alignment is achieved, the pump or fan feet should be doweled to the baseplate to avoid loss of alignment from slip at bolted joints. Grout baseplates to the foundation pad with nonshrinking epoxy grout (per manufacturer's recommendations), and take care to leave no air pockets of significance between the grout and baseplate by the liberal use of vent holes—some located by tapping the baseplate top and listening for hollow sounds to find voids.

Pump Support

The suction and discharge piping end flanges and the opposing pump nozzles must be properly supported. Avoid the use of unrestrained pressure-bearing "expansion" or "flexible" joints at pump nozzles for pipes equal to or larger than 150 mm (6 in.) in diameter unless the contained pressure times the nozzle cross-sectional area is within the manufacturer's nozzle load limits, and the pump and piping natural frequencies are well removed from the range of vane pass frequencies. Although such joints relieve any piping thermal expansion or Bourdon-tube effects from "loading" the equipment nozzles, they do not allow the piping to absorb the cross-sectional nozzle hydraulic load (i.e., the pressure times the open area). This nozzle hydraulic load can produce a large thrust perpendicular to the nozzle opening and severely load the pump casing. Failure to account for such loads has caused serious operating alignment problems, casing rubs, and system damage in many installations.

Natural Frequencies of Pump and Piping Supports

Design piping and pump supporting structures (including floors) to have natural frequencies that are at least 25% less or 25% more than the operating range of the key excitation frequencies in the pump. Typically, these excitation frequencies equal (1) the running speed and (2) the number of vanes times running speed. Loaded-floor natural frequencies can be significantly different from those of the bare floor. In addition, the floor stiffness can influence the natural frequency of the pump, because the manufacturer calculates the natural frequency on the basis of an infinitely stiff floor. For horizontal and vertical non-clog pumps, include the pump mass (plus the water contained) and pedestal stiffness when calculating loaded-floor natural frequencies. For vertical turbine pumps, include the motor and pump bowl assembly masses, and the discharge head and column piping stiffness when calculating loaded-floor natural frequencies. The component masses and stiffnesses are available from the manufacturers upon request.

These calculations can be made by the finite element method, but it is possible that the same models used for the structural design can be used to calculate the building floor and pipe structural natural frequencies after bulk elements are included for the pump and motor, as described above. Natural frequencies for the pump itself with added mass and stiffness of the floor and piping added are best done with the finite element method.

Pipe Support Spacing

Use a liberal number of unequally spaced pipe supports. This rule includes discharge piping for submersible pumps. To avoid vibration problems, piping must be well supported in three perpendicular directions (axes x, y, and z)—not just in the vertical direction. To avoid or suppress vibration, both the weight of the piping and adequate stiffness must be considered in designing proper supports (see Section 22-10).

Drive Shaft Offsets

Do not offset drive shafts at an angle. In the past, drive shaft manufacturers have often recommended at least a 3° offset to help lubricate needle bearings, but it has been found that an offset is not needed for bearing lubrication. Furthermore, an offset excites vibration at a frequency of 2 and 4 times the running speed.

Baseplates and Foundation Pads

Foundation pads should extend beyond the machinery baseplate by at least 75 mm (3 in.) or half of the floor-plus-pad thickness, whichever is larger. The effective thickness from subfloor beams and gussets need not be taken into account in this calculation. The mass of machinery foundations should be 5 to 10 times that of the machine itself. The foundation should have an adequate aspect ratio. That is, the "footing" lines from the centerline of the pump inclined 30° to the vertical should pass through the bottom of the foundation mass—not through its sides. The centerline of the pump is considered to be at the center of the impeller hub except for vertical turbines.

Pump Suction Piping

If possible, use straight suction piping up to the pump suction nozzle, and keep any required bends in a single plane. As a minimum, avoid piping reducers or bends within 5 to 10 suction pipe diameters of the pump inlet, unless the pump has been specially designed to accommodate such fittings closer to the pump (as in some vertical nonclog pumps). If in doubt, seek the pump manufacturer's advice.

Single and Double Volutes

The vane passing force is generally 2 to 3 times higher for single volute pumps than for "twin" or "double" volute pumps. Where practical, choose twin volute pumps (preferably with an odd number of impeller vanes), because they minimize the vane passing force and maximize both bep and, particularly, off-design efficiency. Single volutes are often necessary because of cost or solids-passing capability. The hydraulic imbalance present in single volutes is minimized, at some decrease of efficiency, by increasing the clearance between impeller and volute tongue.

Pump Suction Conditions

Ensure adequate pump suction conditions with sufficient $NPSH_A$ including all significant effects (see Figure 10-13) and a sump design in which surface and subsurface vortices are suppressed.

Recommended Detailed Design Practices

The potential for vibration problems is often given insufficient attention at the beginning of the pumping station design process. By following appropriate design rules, and with a bit of luck, it is likely that the vibrational behavior of the installed equipment will be acceptable. However, this is a risk not worth taking for pump drivers of more than 56 kW (75 hp), and the cost to all parties in those situations where vibration problems do occur quickly overshadows the "savings" of not performing analyses and test procedures such as those recommended below.

Controlling Excitation Forces

The first step in controlling excitation forces is to limit mechanical operational forces that can act to drive machinery vibration. The primary forces that must be controlled are caused either by imbalance of large diameter rotating components such as coupling hubs or pump impellers, or by misalignment of coupled shafts. These issues are discussed above.

Although mechanical forces may be the most common cause of excessive vibration, another common problem is the use of a pipe roughness coefficient that is too conservative and a resulting TDH that is too high. A pump selected on such a basis will operate well to the right of its bep, will discharge a higher flowrate than it should, and will increase the hydraulic excitation forces. The motor will draw more current and may be overtaxed. Eliminate this problem by ensuring the proper value of C is used for selecting both pump and impeller.

The primary frequency of hydraulic excitations is the vane-passing frequency. The forces and the resulting vibration depend upon several factors, controllable at the design and specification stage. The most important method of minimizing vane-passing vibration is to design proper suction piping, to use double volute or centered volute pump designs wherever possible, and—most important—to operate the pump at or near the bep.

For all types of volutes, it is possible to lower vane passing forces at some expense of efficiency by opening up the clearance between the impeller vanes and the volute tongue or diffuser vanes. This opening is called the "B-Gap." Designers should, in general, be wary of excessive vane-passing vibrations from any pump with a diametral B-Gap of less than 4% of the impeller diameter, and a B-Gap of 6 to 10% or more is preferable, although excessive B-Gap can encourage discharge recirculation, as discussed below, when operating below the bep. A sound approach is to follow manufacturer guidance on the setting of the B-Gap, while enforcing appropriate vibration specifications (discussed below) and performance guarantees throughout the operating range.

Reduced Flow

For all types of centrifugal pumps, the vibrations, shaft deflections, and bearing loads usually increase as the pump is run at lower flows. They do not decrease as is commonly assumed. In fact, if centrifugal pumps are operated well below bep (say, at 50% of the design flow at any given speed), suction and/or discharge recirculation will probably occur, accompanied by possibly large excitation forces at vane pass and at frequencies below running speed. The problem in both suction and discharge recirculation is that stalling is induced on the vanes or tongue similar to the stalling that occurs around an airplane wing at a bad angle of attack. This condition causes strong eddies, which resemble and act like miniature tornadoes, to be spawned and "kicked upstream" of the stalled passage. These "stall cells" often subsequently rotate near the impeller at a speed somewhat less than running speed (usually about 10 to 40% in the diffuser or volute or about 60 to 90% at the inlet of the impeller). The high local velocity and low local pressure associated with these cells (keep in mind the analogy of a tornado) interact with the vanes of the impeller and cause new and usually unexpected excitation force frequencies. In summary, *running a centrifugal machine at flows (loads) below its rated capacity (i.e., well below its bep) typically causes more, not less, stress and vibration, and decreases machine life and reliability.* (Of course, vibration due to unbalance or misalignment decreases at lower speeds.) These consequences should be considered in the lifetime cost and factored into design and purchasing decisions. Pumps can be selected for both present and future (larger) capacities by choosing a model that can meet both requirements merely by changing the impeller. The motor must be capable of operation at both discharge rates as well.

Selection of Machines

For reliable operation and long life, select centrifugal machines that will operate as close to their bep as possible throughout their operating range, and keep the flowrate within the manufacturer's maximum and (especially) minimum limits. Because (1) flow resistance can increase with time due to increasing pipe roughness (particularly in long force mains), (2) the pump capacity can decrease significantly as clearance around the wear rings increases, and (3) erosion further decreases discharge capacity, designers should pay attention to system changes with time. The effect of system changes on suction line losses, and therefore on $NPSH_A$, must also be analyzed and compared to the manufacturer's requirements. In reciprocating pumps, the effect of "acceleration head" due to the unsteady pulsing flow in long suction lines must be considered (see pages 25 and 26 in the ANSI/HI standards [2]).

Acoustic "organ pipe" resonances must be avoided (as discussed later in this chapter), or they may become exciting forces. Concern with acoustic resonance is particularly apt in reciprocating machines. Their strong flow and pressure pulsations at piston or plunger frequency are able to excite certain lengths of manifold or piping, and they are often run across a broad speed range. Accumulators are often needed near the inlet or discharge manifolds to detune and/or dampen potential acoustic resonances.

Net Positive Suction Head

To minimize hydraulic forces for all types of pumps, it is important to operate the pump with sufficient $NPSH_A$ (see Section 10-4). In addition to causing cavitation, inadequate $NPSH_A$ can excite unexpectedly high vibrations at various natural frequencies. From Bernoulli's equation, any increase in local velocity decreases static pressure and thus increases the cavitation potential, so it is important to keep the pump inlet flow velocities low (preferably below about 2 m/s or 6 ft/s) and evenly distributed. Hence, a large suction pipe (at least equal to the pump suction flange ID) should be used. Pipe reducers should preferably be at least five pipe diameters upstream of the pump flange. Maintaining this distance from more aggressive suction disturbances such as valves and elbows is even more important. Besides avoiding cavitation, this practice also ensures well-distributed flow velocity and pressure at the inlet and throughout the pump, and thus minimizes the hydraulic excitations in general. High velocity in the discharge is not usually important because the pump increases the static pressure well above the cavitation point.

Wet Wells

Minimize the possibility of vortex formation and discourage whirling flows around column piping by using care in designing the wet well. Follow the advice in Chapter 12 (preferably) or in pages 90–100 in ANSI/HI 1.1–1.5 [3] and pages 50–67 in ANSI/HI 2.1–2.5 [4].

Avoidance of Resonance

Resonance may cause excessive vibration even when mechanical and hydraulic forces are properly controlled. To avoid this problem, distribute the installed system (pump, piping, pedestal, and floor) mass and stiffness in such a manner that no natural frequencies are close (within at least 25%) to excitation frequencies, such as 1 or 2 times the running speed or 1 times the vane-passing frequency. This analysis is done in the design stage by successive trials, using appropriate computer models, such as finite element analysis. In such a model, pump and motor component drawings of sufficient detail must be obtained from the manufacturers—at least overall external dimensions, weight, and center of gravity (relative to the mounting flange)—and pedestal or base (including gussets), floor (including subfloor beam and column dimensions and locations), piping (if rigidly connected, including joint directional rigidity), and water mass must be included. Simple manual calculations are not likely to be sufficiently accurate to avoid resonances because of the complexity of these interlinked components. So much effort may be overkill for small stations, but it should be included in the design of large stations.

A common misconception is that if only the pump is designed stiffly enough, then natural frequency resonances will be avoided. Because the lower natural frequencies of assembled pumping and compression systems usually depend strongly on the base and floor design as well, as discussed by Reichert [5], any reliable analysis must include the pump, base, and floor together. Similarly, any test for the lower natural frequencies of installed machines and possible resonances is unlikely to be valid in the pump manufacturer's shop (although hydraulic/aerodynamic performance tests and rotordynamic testing should be valid). Only when all equipment is installed on its final foundation, filled with water, and with all bolts tight, will representative natural frequencies be present in the lower frequency ranges and detectable by a shaker or impact test.

Avoiding all matching between significant forcing frequencies and all natural frequencies may not be feasible with some pumps, particularly those operating at variable speed with a turn-down ratio of 30% or more. For example, if the first natural frequency occurs at 70% of the running speed in a vertical pump with a variable frequency 2:1 turn-down range, it is unlikely that mechanical design changes can be made that would move such a natural frequency up enough or down enough to be out of the speed range added to the resonance avoidance margin of 25%. Although vibration isolation pads can be used to decrease the reed frequency as much as necessary, the resulting flexibility of the revised system is likely to cause alignment problems, and also is likely to decrease the second natural frequency into the running speed range, thereby trading the original resonance for a new one. If resonance occurs in such instances, there are several alternative solutions:

- Tighten the balance specifications, and tighten the maintenance schedule and procedures, for example, by field trim balancing in which mass is added to the coupling hubs without disassembly.
- Reduce the natural frequency as much as possible by, for example, installing vibration isolation pads under foundation bolts until alignment approaches a potential problem due to static or low frequency hydraulic forces. Unfortunately, reducing the natural frequency is accomplished by successive trials until the second natural frequency is within 25% of the running speed range.
- Use "lockout speed range" settings on the speed controller to avoid operation within ±25% of the problem natural frequency. The use of lockout settings may cause a larger "hole" in the operating speed range than is tolerable from an operational standpoint. If the amplification factor at the edges of the compromised range results in acceptable vibrations at maximum imbalance and misalignment and at minimum flow, the lockout range can be relaxed to ±15% and even perhaps to ±10%. The natural frequency sometimes shifts with ambient temperature or with water level in the sump.

Rotordynamic Analysis

Rotordynamic analysis of the pump, compressor, or fan rotating system should be performed by the manufacturer or a third-party consultant. This analysis requires specialists, and details are not presented here. Some of the items that should be requested, however, are discussed below.

Usually the rotordynamic analysis performed by the manufacturer is for lateral vibrations only, not for torsional or axial vibrations. All three analyses should be performed, particularly for variable-speed systems. It should be made clear in the request for quotes (or in the equipment specifications) that a complete analysis is required so that all manufacturers are bidding on an equal basis and will include the cost of the analysis in their bid.

Regardless of whether a structural vibration and rotordynamic analysis is performed on a pumping station machine and drive train, acceptance should be based upon the concurrence with the specifications of the results of a vibration test performed on-site in the final, fully assembled and running condition. If the vibration is found to be excessive, the manufacturer should be required either to fix the problem or prove that the problem is not related to the equipment. If the latter is true, a knowledgeable consultant should be retained to ensure that the manufacturer is correct and to help find a fix to the problem.

Lateral Rotordynamic Analysis

The minimum rotordynamic analysis should consist of the determination of rotor critical speeds (i.e., noncritically damped rotor natural frequencies) and mode shapes from a frequency of zero up to the maximum excitation frequency. If the bearing stiffness has a large tolerance or might shift significantly with time, such as from the Lomakin effect (see subsection "Description of Basic Vibration Terms, Concepts, and Equipment") in centrifugal pumps (as the clearances increase due to wear in normal service), a critical speed map should be constructed in which the value of these frequencies versus the range of possible bearing stiffness values is plotted. The predicted critical speeds need to be compared with the excitation frequencies to determine whether a resonance is possible. If so, the offending natural frequency must be moved or sufficiently damped by, for example, using a different style of bearing.

If any of the rotor natural frequencies might (with reasonable allowance for system degradation) match the frequency of an excitation force within the intended speed range, a "forced response" analysis should be performed. Sometimes this analysis is done only for 1 times running speed excitation. It is then called an "unbalance response" analysis. In such an analysis, it is assumed that the rotor is driven by a worst-case excitation and the amount of vibration that can occur at the various critical locations along the rotor (e.g., at bearings, wear rings, and mechanical seals) is calculated. These results can be compared with available clearances to determine whether rubbing is likely and whether the bearings might become overloaded.

Vertical turbine pumps exhibit nonlinear shaft dynamics because of the large shaft excursions that occur in the lightly loaded long length/diameter ratio bearings. Given the flexibility of the lineshaft and the weak support provided by the pump casing column piping, and given the relatively large assembly tolerances and misalignments in the multiple lineshaft bearings of these machines, the contribution of each bearing to the net rotordynamic stiffness is nearly random and is constantly changing, as explained by Marscher [6]. As a result of the changing stiffness there is no single value for each of the various theoretically predicted natural frequencies. Instead, the natural frequencies of the lineshafting and shaft in the bowl assembly must be considered on a time-averaged and location-averaged basis and they constantly vary between two limiting states. A useful simplified method of predicting lineshaft reliability with a worst-case model is known as the "jump rope" model, and is documented by Marscher [6]. Another important vibration problem in certain vertical pumps is the vibration of the shaft enclosure tubes that provide a jacket of contaminant-free lubricating fluid to the lineshaft bearings. The natural frequencies of these tubes (supported by "spider" assemblies) can be analyzed for various end supports using finite element analysis or by using the multiply supported beam models of Blevins [7], pp. 134–148. Vertical pump vibration analysis of the stationary structure, the lineshafting, the shaft enclosure tubes, and the pump and motor rotors should be done simultaneously by means of finite element analysis.

The possibility of rotordynamic instability due to lateral motion needs to be analyzed in centrifugal and axial flow machinery, particularly in large fans and certain motors wherein cross-coupling forces (those induced by rotor motion perpendicular to the motion, as discussed in the subsection "Basic Vibration Terms, Concepts, and Equipment") are often large enough to overwhelm the damping. Rotordynamic instability refers to phenomena whereby the rotor and its system of reactive support forces are able to get "out-of-synch" with each other and become self-excited. Potentially catastrophic vibration levels ensue even if the original excitation forces are quite low. The characteristic of rotordynamic instability, sometimes called "shaft whip," are that it is whirling at about half running speed and begins when running speed exceeds twice the first bending natural frequency of the shaft. Very few pumps have natural frequencies in the running speed range with damping low enough that this process can occur, but motors and fans sometimes do.

If an unstable machine is encountered, a typical design modification to reduce the tendency for rotordynamic instability involves bearing changes. The type of bearing most likely to participate in instability problems is the plain journal bearing, because it has very high cross-coupling, although it also has high beneficial damping. Bearings that discourage whirling lubricant flow tend to decrease cross-coupling—dramatically so in terms of the axially grooved and tilt pad style bearings described in the *Pump Handbook* [1].

Torsional Vibration Analysis

Unlike lateral rotordynamic analysis, which usually can be performed independently by various manufacturers, torsional analysis is valid only when performed for the entire linked-up drive train. Torsional analysis of only the driver or the pump or compressor alone is without value. Also, the flexibility, backlash, and gear ratio of couplings and gear assemblies, if any, must be estimated and included in any accurate model. Methods of manually calculating the first several torsional natural frequencies for simple rotor systems are given in Blevins [7], pp. 187–195, but a finite element analysis is generally needed for complicated pumping station systems.

Forced response torsional analysis is necessary if any of the torsional natural frequencies coincide with a strong excitation frequency. To determine the frequencies at which large values of vibratory excitation torque are expected and the value of the torque occurring at each of these frequencies, the pump torque at any given speed and capacity can be multiplied by a "per unit factor." The per unit factor at important frequencies can be obtained from pump, compressor, motor, and control manufacturers for a specific system, and it is typically in the range of 0.01 to 0.1 zero-to-peak for important excitations. The most important torsional excitation frequency from an electric motor is the motor rotating speed times the number of motor poles. Unsteady hydraulic torque from the pump is also present at a frequency equal to the running speed times the number of impeller vanes, and the maximum intensity equals the delivery torque divided by the number of impeller vanes, although typically the per unit factor is 0.03 to 0.1. The gear mesh frequency (number of teeth times the rotational speed for a given gear) is usually a strong torsional excitation frequency with a typical per unit factor of about 0.02.

The worst-case torsional vibrations in pumping station rotors often occur due to temporary excitations during start-up, trip, and motor control transients. Therefore, it is wise to make a time-integration analysis to determine the transient peak stresses caused by these transients.

Particular care should be taken with systems involving adjustable frequency drives (AFD). Besides sweeping the excitation frequencies through a large excursion and therefore increasing the chances of a resonant encounter (Marscher [8]), AFD controllers provide new control pulse excitations at multiples of the motor running speed, commonly at 6 ×, 12 ×, and 18 ×, and often at whole-fraction submultiples as well. The controls manufacturer can predict these frequencies and their associated per unit factors.

Judgment of the acceptability of the assembly's torsional vibration characteristics should be based on whether the forced response shaft stresses are below the fatigue limit by a sufficient factor of safety at all operating conditions. The recommended factor of safety is 3 if all stress concentrations (such as keyways) are taken into account. If specific stress concentration factors are not known, assume that they have a value of 3.0 and use a revised factor of safety of 9.0 at any shaft location that has a step, attached part, thread, or keyway.

Vibration Specifications

Specifications concerning acceptable vibration levels, the method and frequency range of measurement, and the method of interpretation of the results should be clear and reasonable. Vibration-minimization responsibilities, such as who will do the pre-design analysis and who will do the post-installation testing, must be delineated and the responsible parties must be identified. There should be a list of items required, such as results from certain types of validated computer programs or testing procedures. Responsibility for each item should be assigned in the bid request so that bids are on an equal basis. The inclination or ability of low-bid suppliers to provide an appropriate level of technology should be specifically required—not assumed—in the bid and contract.

Injury to machines due to excessive rotor vibration consists of wear or fatigue damage to the pump internal components, such as bearings, annular seals, mechanical seals, and shafts. Most of this damage depends on the total displacement associated with the vibration. However, as machines are made faster, they become smaller, and hence the amount of vibration displacement they can tolerate decreases proportionally to the machine speed. Therefore, the allowable running speed vibration velocity (displacement times running speed) is roughly constant regardless of the running speed of the machine. Historically, machine survival versus failure data support the use of constant vibration velocity versus speed as an acceptance criterion in assessing vibration severity (see Rathbone [9], Blake [10], Baxter [11], and Hancock [12]). However, the raw information in these references was based on measurement equipment that could not distinguish between various frequency components. Therefore, vibration severity could be plotted only as unfiltered (total vibration including all frequencies) displacement readings versus machine running speed and not filtered (individual values at specific frequencies) velocity values versus frequency. Unfortunately, in many specifications (with the notable exception of the ANSI/HI standards [2]), it is assumed that the original data can be interpreted as velocity versus frequency, and the specifications are written on that basis. Be very cautious in using such specifications, because instability and hydraulic problems that cause rubbing at low frequencies tend to be overlooked, and the specifications may require unnecessarily small vibration displacements at high frequencies. Such specifications cause the rejection of good equipment at the expense of all involved, as discussed by Marscher [13].

Typical specifications for pumping stations to establish vibration test types, measurement locations, and acceptance criteria are contained in the ANSI/HI standards [2], API 610 [14], Mil-STD-167-1 (Ships) [15], and various ISO machinery vibration specifications, such as ISO 2372 [16]. These specifications require the monitoring of bearing housing vibration by using an accelerometer or velocity probe in three perpendicular directions at the location of each of the bearings in the drive train. The acceleration readings are typically integrated to obtain vibration displacement and/or velocity values—the typical terms for the criteria. Occasionally, the specifications contain procedures for the installation and evaluation of shaft proximity probes mounted in some bearing housings to measure shaft versus housing displacement.

Acceptability criteria varies between the various specifications. A conservative approach, consistent with all of the quoted specifications, is to require that vibration does not exceed any of these three criteria at any frequency: 0.05 mm (2.0 mils) displacement peak-to-peak, 6 mm/s (0.25 in./s) zero-to-peak (or just peak) velocity, and 1.0 g peak acceleration.

In essence, vibration velocity can be thought of as displacement times the frequency at which it occurs times a scaling constant. Likewise, acceleration can be thought of as displacement times frequency squared, times a scaling constant. The value of scaling constants is not a key to the discussion here, although knowledge that the above relationships exist brings out the important point that vibration displacement, velocity, and acceleration are all measurements of the same quantity, but with different emphasis given to vibration that occurs in different frequency ranges. At low frequencies, for example, displacement is the most stringent criteria, while at high frequencies acceleration is the most stringent. Depending on the type of equipment, displacement may be more likely to reject unreliable machines and pass reliable machines than acceleration would be. On the other hand, for machines prone to problems that show up at high frequencies (usually not true of pumping station equipment) the opposite would be true. Vibration velocity is used in many specifications as a "middle-of-the-road" single acceptance criterion, but this practice is not recommended, because it tends to be blind to potential rubs at low frequencies and is too stringent as a general criterion for most machinery at high frequencies. In practice, the 1.0 g peak acceleration specification given above is less stringent than the displacement and velocity specifications except for high frequencies. Generally speaking, high acceleration measurement is useful only as a flag that there is something unusual (possibly harmless) in the system, and by itself does not necessarily indicate that there is a problem with the machine.

For vertical pumps, the maximum allowable displacement may be significantly higher than 0.05 mm (2 mils) due to the geometrical lever principle acting to magnify the motion associated with a tall motor or pump bearing tower (see pages 120 and 121 in the ANSI/HI Standards [2]).

Piping vibration is in a special class. Pipes can be allowed to vibrate up to roughly three times that allowed for the machine to which it is attached (see Wachel [17]).

22-3. Troubleshooting Excessive Vibration

Vibration problems may sometimes occur in existing stations, in new stations because of corners cut during the design process, or in new stations in spite of following recommended practice. These problems can often be solved by a combination of inspection and common sense. If the problems persist, then vibration experts with appropriate vibration measurement equipment can be called in. Make sure that these experts focus on solving the problem at hand instead of just gathering data. Prior to calling in consultants, the following procedures can serve as a guide, but be prepared to modify them to fit the particulars of your situation.

Basic Troubleshooting Procedures

Obtain a history of the problem: is it recent or has it always occurred? Has the system or its operation recently changed? If the problem is recent in an existing installation, it is likely to be caused by looseness, clogging, erosion, or wear. Pump disassembly may be required for answers, but even so, delay for the moment. Before dismantling the pump for inspection, follow the procedures given below as far as conditions permit, and answer the following questions: (1) Does the problem relate to flow, the sump level, or some other circumstance? (2) Is the vibration constant or intermittent? (3) Have components been failing? (4) Which locations shake beyond the specifications, and which do not? (5) Is there a pattern to the locations that shake and/or make noise? A list of checking procedures follows, more or less in the order to be performed.

- A common fault is a pump operating off the bep. Install a calibrated pressure gauge on the discharge and another on the suction pipe. Read static pressures and total dynamic pressures when the pump(s) is (are) running. It is helpful to obtain the flowrate (see Section 3-9). Compare with the manufacturer's curve for the impeller. Check for a worn impeller.

- Check for loose or broken connections, and faulty or improperly set valves. Check all bearing housing, foundation, and piping bolts, gasketed joints, and isolation pads for proper installation and tightness.
- Check for pump baseplate "soft foot" (bottoms of all feet not coplanar) by following the procedures on page 2.288 in the *Pump Handbook* [1].
- Inspect all baseplates and foundation blocks for significant cracking that might decrease machine support stiffness.
- Have a millwright check driver/pump shaft alignment.
- Check the shafts for straightness, for example by rotating the shaft by hand while a dial indicator is mechanically held against it.
- Listen near the pump inlet for the crackling sound associated with cavitation. Crackling noises heard near the pump suction suggest inlet cavitation due to insufficient $NPSH_A$, severe impeller vane stalling, internal recirculation due to swirling or skewed flow at the inlet, operation too far from bep, or submerged vortices in the sump. If any noise is detected, place a calibrated pressure gauge of the proper range close to the pump suction. Ensure that there is sufficient $NPSH_A$ by noting static pressure at a tap at the pump inlet flange, and compare it with the manufacturer's $NPSH_R$ versus capacity curve. Is $NPSH_A$ adequate? If not, perhaps the suction line is clogged or there is a sump problem. If, in spite of inlet crackling noise, $NPSH_A$ is adequate, recirculation or sump problems are suggested. Other clues are: (1) inlet cavitation noise is relatively constant, whereas (2) recirculation noise tends to be throbbing, and (3) sump noise tends to come and go without following any pattern.
- Place a calibrated pressure gauge of the proper range at the discharge, preferably about 3 to 5 diameters downstream. Use a certified pump system head/capacity curve or a flow meter in the discharge line to determine flowrate. For variable-speed systems, use a tachometer or strobe to determine pump speed. At a given speed and flow, compare the discharge head minus the suction head with the value obtained from the last test performed, and to the manufacturer's head/capacity curve. If the total discharge head (TDH) is larger than expected and/or flow is lower than expected, some form of discharge blockage is possible. If the TDH is lower than expected, wear of the wear rings or erosion of the impeller vanes or volute tongues is possible. If the TDH indicates that the operating point lies outside the range of 75 to 110% of bep, especially after years of service, the impeller or the volute tongue may be excessively worn. If the station is new or if the vibration has always occurred, the designer may have used an unrealistic roughness coefficient for the piping.
- Ensure that expansion joints and flexible joints such as Dresser® couplings are constrained properly, especially if the pressure is high.
- What are the bearing shell or lubricant temperatures, at least approximately? Bearing lubricant temperatures are typically 60 to 82°C (140 to 180°F), and the shell and housing temperatures are usually 11°C (20°F) lower. The loss of oil viscosity at elevated temperatures can reduce bearing support film thickness and allow scuffing and excessive bearing wear to occur, and thermal growth mismatch between rotating and stationary parts may allow radial or axial rubbing. Excessive housing temperatures indicate severe pump/driver misalignment, need for lubricant addition or change, or (contrary to intuition) excessive grease in a packed bearing cavity.
- Are any wear particles or pumpage contamination visible in samples taken from the lubricant? The most common cause of bearing failure in pumps is lubricant contamination—usually by water passing a leaky seal or packing. Condensation from humid air onto the cool walls of an oil sump or inside a grease-packed bearing with badly worn lip seals that allow the free ingress of air is often overlooked. Sometimes the contamination is varnish, carbon deposits, or metal flakes from bearing surface fatigue. If the lubricant change interval has passed or if the bearing was overloaded or overheated for even a few minutes, the effect snowballs to more degradation.
- Following disassembly, if rubbing was involved, was it on just one side of the rotating or stationary components, or full circumference on each? Does corrosion appear to be a factor? If fracture occurred, are fatigue striations evident? If so, what does a high-power microscope indicate concerning how many cycles it took from crack initiation to failure? Such information is useful in determining whether the problem was some sort of brief transient event or is inherent in the pump or system. If fatigue occurred in relatively few cycles, there was a severe overload condition such as recirculation caused by running the pump below its recommended minimum continuous flowrate. Millions of cycles indicate gradual degradation, such as gradually worsening

rotor imbalance or pump/driver misalignment. Microscopy performed by a qualified metallurgist can often determine the root cause of failure so that it can be avoided in the future.

- Check the balance and fit of the disassembled pump impeller and coupling hub. Poor fit can be remedied by reaming the bore and installing a sleeve of proper fit or, sometimes, by chrome plating or knurling the shaft followed by grinding to the proper dimensions. The impeller/shaft assembly should then be rebalanced.

More Detailed Test Procedures

If the above approach does not lead to a solution of the problem, detailed vibration testing may be required. In the most common type, "signature analysis," an accelerometer output is sent to a fast Fourier transform (FFT) analyzer to document the amount of vibration at each frequency within a tested range. Typically, this range is from several Hz to beyond the pump vane pass frequency. The frequencies at which most of the vibration is occurring and the locations where the vibration is the greatest are used as clues to determine the cause of the vibration.

Vibration testing often ends here. However, it is recommended that vibration testing include experimental modal analysis (EMA). EMA involves artificially exciting a machine or structure (e.g., with an impact hammer), preferably while the machinery is running so that all bearing stiffnesses are representative (see Marscher [18]). The purpose of EMA is to determine the natural frequencies of a pump (or other machines), its rotor system, and the attached system components. These frequencies can be compared with excitation frequencies to ensure that resonance will not occur. They can also be used to confirm that all equipment and supporting structures have adequate separation margins between the excitation frequencies and natural frequencies.

Recognizing Vibration Problems

The great majority of pump vibration problems can be solved by: (1) re-balancing the rotor assembly, (2) alignment of the couplings when the system is at its rated conditions—especially if it is hot (see Dodd [19]), and/or (3) running the pump within the bounds of its specified head versus capacity curve. Remaining vibration problems are usually caused by a resonance of a pump internal natural frequency or a systemwide natural frequency. During resonance, the rotor vibrations can exceed internal clearances, and excessive bearing loads can occur, even if loads such as imbalance are within normally acceptable limits.

In performing vibration troubleshooting, generalized charts (such as those in the *Pump Handbook* [1]) for matching symptoms to possible causes can be useful for many typical or simple problems. However, do not rely too heavily on such lists, especially if their initial application does not lead to an immediate resolution of the problem. Persistent pump vibration problems are usually due to an unexpected combination of factors, some of which are specific to the particular pumping system, such as mechanical or acoustical piping resonances, or hot running misalignment of the pump/driver due to thermal distortions of the piping or baseplate. With this warning, a list of some of the more common pump vibration symptoms and their causes is given in Table 22-1. The following predictive maintenance and troubleshooting list can be used for interpreting the reason for many common vibrations and therefore could be the kernel of any future predictive maintenance software. The list is *not* meant to be all-inclusive and is in the order of the observed frequency—not in the order of likelihood or importance for reliability.

Typical Fixes for Vibration Problems

Although many pump vibration problems are caused by operation off the bep, most machinery vibration problems are traceable to excessive imbalance or misalignment. Determination of the reason for the imbalance or misalignment (and its subsequent elimination) is the cure in these cases. Most vibration problems that persist after proper balance and alignment, however, are related to some form of resonance.

When resonance occurs, simple trial-and-error field fixes, such as addition of gussets at appropriate locations for stiffening, can often be effective. Such "fixes" may, however, only shift the problem to a slightly different frequency. If "cut-and-try" methods cannot be made effective within a reasonable time span, it is usually cost effective to bring in a vibration expert from either the manufacturer or from a specialty consulting firm. The main priority in selection should be competence and experience. Fee differential is unlikely to compensate for your own personnel time, the cost of hardware changes, and lost operating time wasted by an inexperienced consultant.

In solving resonance problems, vibration experts may employ one of the following:

- Change the stiffness of the resonant component and/or its support (possibly including the floor). An increase in stiffness raises the natural frequency—

Table 22-1. Troubleshooting Tips for Identifying the Source of Vibration Problems

Frequency of vibration (in terms of multiples of running speed)	Other symptoms	Probable causes
0.05× to 0.3× (may be broadband)	Vibration response is unsteady and fluctuates somewhat in frequency.	Diffuser or volute wall stationary passage stall.
Exactly $^1/_2$×, $^1/_3$×, or $^1/_4$×	Vibration may increase dramatically shortly after the above (0.05× to 0.3×) appears, except in vertical pumps with four or more lineshaft bearings, where it is common, and generally not harmful.	Light rub or unexpected looseness combined with shaft of bearing support natural frequency.
0.42× to 0.48×	Near shaft natural frequency, and orbit "pulses," forming an inside loop. Vibration onset is sudden at a speed roughly twice the excited natural frequency, and locks onto the natural frequency in spite of speed increase.	Rotodynamic instability due to fluid whirl in close clearances, e.g., "shaft whip."
0.6× to 0.93×	Smaller peaks at f× and at $\pm(1-f)$× "sidebands" of the first several multiples of running speed where f is the frequency of the amplitude modulation caused by pressure pulsations. Amplitude modulation is manifested as a beat frequency f. Often accompanied by rumbling noise and beating. Occurs at capacities below bep, but improves at very low capacity. Independently, depends on speed and flow.	Internal flow recirculation in the impeller, probably at its suction, but possibly at the discharge.
Less than 1×	Increased broadband vibration and noise level below running speed as $NPSH_A$ decreases, especially at high flows. Often accompanied by decreased vibration and noise above 1×.	Cavitation without recirculation.
1×	(1) Stronger on shaft than on housing. Hydraulic performance and/or suction pressure normal. Axial vibrations within normal limits and vibrations above runout increase roughly with speed squared.	Imbalance in rotating assembly.
	(a) Vibration highest on drive and pump IB (inboard) housing.	Coupling imbalance.
	(b) Vibration highest on driver IB housing.	Driver rotor imbalance.
	(c) Vibration high on machine inboard (IB) or outboard (OB) housing, low on driver.	Machine imbalance.
	(d) Natural frequency, 1×.	Resonance.
	(2) Axial vibration is over $^1/_2$ of radial vibration, or/and vibration increases much slower than the square of the speed. Also, bearing temperature is high.	Machine/driver misalignment at the coupling.

(Table continues on next page)

Table 22-1. Continued

Frequency of vibration (in terms of multiples of running speed)	Other symptoms	Probable causes
1× *(continued)*	(3) Discharge pressure pulsations are strong at 1× but not at impeller vane pass in a single volute machine.	Clogged or damaged impeller passage.
	(4) Same as (3), but vane pass also strong, especially at flows far above or below the design point.	Volute tongue designed too close to impeller OD, or clogged or damaged volute, or excess impeller eccentricity.
2×	(1) Axial vibrations are low.	
	(a) Both shaft and housing vibrations are strong, and discharge pressure pulses are strong, but impeller vane pass vibrations are low in a twin volute pump.	Clogged or damaged impeller passage.
	(b) Same as (a), but with unusually high vane pass vibration and discharge pulsations.	Volute vanes designed too close to impeller OD, clogged or damaged volute, or excessive impeller/volute eccentricity.
	(c) Shaft vibrations stronger than housing vibrations, maybe exceeding clearance. Decrease in shaft first natural frequency. Multiples of running speed may be strong.	Looseness in bearing retainer, or cracked shaft.
	(d) Vibration of one bearing housing high, shaft may be quiet.	Loose or cracked bearing housing.
	(e) Shaft vibrations stronger than housing.	Torsional excitation.
	(f) Motor driver frequency equals electrical line frequency and highest vibration on motor.	Electrical problem with motor.
	(2) Axial vibrations are over $^1/_2$ of horizontal, and 1× vibrations are also high, and bearing oil temperature is high.	Pump/driver misalignment at the coupling.
Number of impeller vanes × running speed	(1) Discharge pressure pulsations reasonably high and both shaft and housing vibrations high.	Volute too close to impeller OD due to design or excessive rotor eccentricity.
	(2) Same as (1), but housing vibrations much higher than shaft vibrations.	Piping mechanical resonance at vane pass or loose bearing housing.
	(3) Discharge pressure pulsations high at vane pass frequency but suction pressure pulsations reasonably low.	Acoustic resonance in discharge pipe.
	(4) Suction pressure pulsations high.	Acoustice resonance in suction pipe.
	(5) Rotor or casing natural frequency close to vane pass.	Resonance.

Table 22-1. Continued

Frequency of vibration (in terms of multiples of running speed)	Other symptoms	Probable causes
Several multiples of running speed, including 1×, 2×, 3×, 4×, and possibly higher	(1) Orbit shows sharp angles or shows evidence of "ringing," and/or spectrum shows exactly $^1/_2$ or $^1/_3$×. Grinding noises may occur.	Internal rub or poorly lubricated gear coupling.
	(2) Orbit is fuzzy but does not pulse or ring. Seal coolant flow is unexpectedly high or low, and may exhibit high temperature. Spectrum may also exhibit $^1/_2$ or $^1/_3$×.	Jammed, clogged, or damaged seal.
	(3) Orbit pulses, usually in one direction much more than the other and shaft vibrates more than the housing. May also observe at exactly 0.5×, 1.5×, 2.5×, etc.	Shaft support looseness, especially in bearing insert or cap retention.
	(4) Shaft orbit fairly steady, and housing vibrates more than shaft. Often combined vibration over a broad range of low frequencies.	Looseness in casing, pedestal, or foundation or bearing housing.
Non-integer multiples of running speed	(1) Vibrations stronger at problem frequency at certain points in the piping or on the foundation than on the pump.	System structural resonance.
	(2) Vibrations high in piping but not foundation. Suction or discharge pressure pulses as strong or stronger relative to background of spectrum than the vibration is in the piping or the machine.	Hydraulic piping dynamics.
	(3) Vibrations low in piping and foundation at a fluid machine natural frequency.	Machine resonance, excited by turbulence.

possibly enough to drive it out of the resonance range. However, it is possible that the increase leaves the natural frequency still within the operating speed range but now excited by larger forces at the top of the speed range. It is also possible that a lower natural frequency, not previously a problem, can be moved into the operating speed range. For these reasons, it might be better to decrease rather than increase stiffness to drop the natural frequency below the operating speed range. Drawbacks to this latter approach are: (1) a higher natural frequency that was not previously resonant might drop into the operating speed range and (2) decreasing stiffness may make static forces (such as weight) and alignment more difficult to manage.

- Add weight to locations of the largest vibratory motion, thus decreasing the problem natural frequency. Measurements need to be taken to confirm whether the new frequency is out of the operating speed range. Potential drawbacks are greater sensitivity of the revised structure to seismic loads and the possibility of reducing a previously higher natural frequency into the operating speed range.
- Install vibration isolators between the vibration excitation source and the problem vibration location by adding flexibility between the two. Be careful to ensure that the vibration is sufficiently diminished in all potential problem locations and that the isolation does not hide an excessive excitation force that could cause other problems if untreated. The use of isolators is not often practical.
- Design and attach a dynamic absorber (made of a dead weight and a suspending rod) to a point of

maximum motion for the problem natural frequency's mode shape. The mass and adjusted rod stiffness is chosen so that the mass/rod reed frequency equals the problem natural frequency. If the system is tuned just right, the motions of both natural frequency/mode shapes cancel each other at the running speed and dramatically reduce the vibration level at the absorber attachment point. The disadvantages of this approach are: (1) it can be unsightly, (2) if the absorber mass is not large enough it can be difficult to maintain proper adjustment, and (3) the concept works only for constant-speed applications. The use of dynamic absorbers is tricky and best left to vibration experts.

Vibration problems that are not related to imbalance, misalignment, or resonance are usually caused by excessive hydraulic forces. If the cause is cavitation from insufficient $NPSH_A$, the $NPSH_A$ should be increased to the manufacturer's requirement or even more. If vibration is caused by excessive vane pass interaction, then the impeller/stator vane gap may be too small, and the gap should be opened to about 10% of the impeller diameter (if acceptable to the manufacturer and if the required discharge and head can be met). If a centrifugal pump is being run at relatively low flow, suction or discharge recirculation may be at fault—possibly with accompanying rotating stall. If the flow cannot be increased, then a bypass line can be placed between the pump discharge and a location in the suction plenum or suction line at least 5 pipe diameters upstream of the pump suction flange. A valve can be placed on the bypass line, and bypass flow can be increased gradually until the vibration problem ceases. Alternatively, a new impeller less sensitive to low flow can be installed. Although efficiency may decrease several percent, the operating cost is usually less than it is with a bypass line.

Description of Basic Vibration Terms, Concepts, and Equipment

Vibration analysis of machinery and its piping and foundations may be divided into two parts:

- *Structural dynamics:* the vibration of stationary components such as housings, piping, and supporting structures. An important factor in considering structural vibrations is that the amount of vibration at the point of interest may be strongly related to other system components far removed.
- *Rotordynamics:* the vibration of the rotor assembly. The rotor vibration is influenced by all factors relevant to structural vibration but, in addition, may be strongly affected by unexpected reaction forces and gyroscopic effects.

Concepts of key importance for both structures and rotordynamics in all types of rotating machinery are defined and explained below:

- *Excitation forces* supply the energy in a vibrating system. They include imbalance, at a frequency of 1 times running speed and misalignment, usually at both 1 and 2 times running speed. In machinery with vanes or lobes, vane-passing pulsations occur at the number of impeller vanes times running speed. In centrifugal pumps, the level of vane-passing pulsations (and the vibrations they cause) are dependent on the ratio of the capacity at the bep to the actual pump discharge. As the departure from bep increases, the flow path directions inside the pump become less well matched with the impeller vane and volute tongue angles. Running a centrifugal machine at flows below its rated capacity at a given speed causes increased stress and vibration, and decreased machine life and reliability as dramatically stated by Agostinelli [20]. In contrast, the pulsations of screw and reciprocating machines usually increase with load. Strong pulsation frequencies in screw pumps occur at 1 and 2 times the number of screw lobes, whereas in reciprocating machines, strong pulsations occur up to 4 or more times the plunger or cylinder stroke frequency (a value equal to the number of cylinders times running speed).
- *Natural frequencies* are those frequencies at which mechanical systems vibrate freely after energy is imparted, such as after striking a tuning fork. Machinery systems have not only one, but an infinite number of natural frequencies. Only natural frequencies close in frequency to strong excitation forces are of practical significance and merit the designer's or user's attention.
- *Mode shapes* are the vibrating movements associated with a given natural frequency. Each natural frequency has a unique mode shape. The mode shape becomes more complicated as the natural frequency becomes higher. At lower frequencies, simple mode shapes are most likely to possess motion consistent with the action of exciting forces, and these are the most troublesome to the plant designers and maintenance engineers. The lowest natural frequency of vertical pumps and motors resembles a reed blowing in the wind and is, therefore, called the "reed frequency" [2]. The second natural frequency has an initially similar motion that starts at the base in one direction, but then it turns and finishes with reed motion at the top in the opposite direction. The lowest natural frequency in horizon-

tal pumps is usually a swaying back-and-forth motion of the pump casing allowed through flexure of its support base, and a second natural frequency is a see-saw motion of the pump casing (as viewed from above). Both sets of these vertical and horizontal pump mode shapes and natural frequencies are often strongly influenced by (1) the stiffness and/or mass of the foundation and building floor and (2) the added mass of rigidly attached piping. Hence, the floor and piping should be considered together with the pump as a total, connected system. Remember these statements during the design process and during troubleshooting.

- *Damping* is the absorption of vibration energy. The strongest damping typically occurs in components where there are tight fluid-filled clearances such as in journal bearings and impeller wear rings. Handbook charts and readily available computer programs exist to estimate and optimize the damping of bearings and seals.
- *Resonance* occurs when the frequency of an excitation force equals the natural frequency. Resonance stores vibration energy from cycle to cycle. Very large vibrations build up during resonance even if the excitation forces are reasonably low—especially if the force is applied near a point of maximum vibration in the natural frequency's mode shape and if the damping is small. The affected machine is then observed to be "balance sensitive" or "alignment sensitive," because the resulting high vibration can be temporarily eliminated by very fine balancing or alignment. However, as these vibrations degrade slightly after brief operation, the vibration problem returns.
- *Amplification factor* refers to how much more (or less) a pump or system component moves during vibration than it would move if the force causing the vibration were static instead of oscillating. If the excitation frequency is well below the system's first natural frequency (<75%), the amplification factor is about 1—that is, the structure's maximum motion in response to the force is essentially the same as if the slowly oscillating force were static. Systems responding as though to a static force are known as "stiffness dominated" or "rigid structures," and they are unlikely to have balance-related resonances (by far the most common type of resonance problem). At the other extreme are systems with important excitation frequencies well above (by at least 25%) those natural frequencies that have simple, easily excited mode shapes. These are known as "mass dominated" or "flexible structures," and the amplification factor for them is less than 1. Hence, there is even less vibration motion than would be expected if the excitation force were applied statically. Although amplification factors less than unity sounds like an ideal situation, the high degree of flexibility implied in such systems makes them very prone to distortion from static forces and may lead to serious misalignment between the driver and the pump. The distortion may be accompanied by internal rubs and possible large dynamic misalignment in spite of the low amplification factor. Between the rigid and flexible structures are "resonant" structures, in which the excitation force frequency and natural frequency are close to each other—a situation to be avoided because of the high amplification factor and consequent high vibration levels.
- *Parametric resonances* refer to large nonlinear vibrations that can occur in response to looseness or rubbing. Such resonances are typically caused by bearing support looseness or a rub at a bearing, seal (or other close running clearance), or slip in a gear or spline coupling. The symptoms are a pulsating orbit, with a large amount of vibration at exact whole fractions of running speed, such as $^1/_2$, $^1/_3$, $^1/_4$, etc. In addition to the above, rotor systems may be strongly affected by more subtle phenomena. The most important of these phenomena are listed below.
- *Gyroscopics* describes the reluctance, due to conservation of angular momentum, of a rotating object of large radius (such as an impeller) from changing the direction of its axis.
- *Cross-coupling* is a force that develops perpendicularly to rotor motion when the rotor moves to an eccentric position within its bore clearances. It is caused by the higher pressure of the flow dammed up upstream from the minimum clearance pinch point, and it occurs in combination with the restoring bearing force that acts to support the rotor opposite in direction to the motion. The non-intuitive cross-coupling force acts in the opposite direction of the damping vector and can become an unexpected factor to cause decreased energy dissipation due to the vibration—in some circumstances to the point that the vibration becomes self-excited and unstable. After a certain threshold speed is passed, the self-excitement produces "rotor dynamic instability," in which the rotor orbit takes up the entire clearance, whirls at about half rotor speed, and rapidly wears out the machine. Rotordynamic instability is extremely destructive, but, fortunately, it is not common in the type of equipment used in pumping stations.
- *Added mass* refers to the effective inertia of the fluid surrounding the rotor. It is developed from three sources: (1) the fluid trapped in the impeller

passages adds mass directly; (2) the fluid displaced by the presence of the impellers and shaft adds its mass because, as the rotor vibrates within the fluid, it must displace this mass; and (3) the fluid in close clearances must accelerate off to the sides much faster than the rotor vibration acceleration to make room for the rotor motion. Item (3) can, potentially, add mass many times its displaced mass.

- *Lomakin effect* is an unexpected support force that occurs in pumps at annular seals such as wear rings due to the action of Bernoulli's effect during the normal leakage process. This effect, discussed in detail by Black [21] and Marscher [22], can change the rotor support stiffness dramatically and hence the rotor natural frequencies. The effect thereby either avoids or induces possible resonance between strong forcing frequencies at 1 and 2 times the running speed and at one of the lower natural frequencies. Although the effect is usually beneficial, it strongly depends on the annular seal diametral clearance, and loss of this clearance due to erosion and wear in service can lead to loss of Lomakin effect and the appearance of a resonance problem where there was none before.

Vibration Measuring Equipment

Modern vibration test equipment can be divided into three levels of detail and sophistication:

- *Hand-held sensor/meter packages.* The advantages of this equipment are that it is easily portable, easy to use, inexpensive, and the sensor is relatively lightweight and rugged. The disadvantages are (1) the frequency information it provides is not very distinct; (2) it is difficult to distinguish hydraulic forces and mechanical resonances from excitation harmonics of multiples of running speed caused by imbalance and misalignment; (3) only one probe at a time can be used, so the powerful modal testing discussed by Marscher [18] cannot be performed; and (4) the velocity probes used in these units rapidly lose accuracy outside of the range 10 to 17 Hz.
- *Single-channel FFT or real time analyzers.* These devices are rather complicated and best used by experts. Velocity probes can be used with them, but accelerometers are better because they (1) have a broader accurate range of frequency (2 to 17,000 Hz) and (2) are robust. FFTs may also take input from eddy current shaft proximity probes that sense the displacement of a rotating shaft relative to some stationary point on the pump housing. The advantages of the single-channel FFT are (1) it is moderate in cost; (2) it gives accurate plots of running vibrations versus frequency (thereby allowing the broader frequency range taken up by resonances and hydraulic forces to be detected relative to the narrow frequency range vibration response of mechanical forcing frequencies); and (3) it provides enough frequency resolution to distinguish subtle but important frequency differences between resonances, hydraulic forces, and mechanical forces. An important disadvantage of single-channel FFTs is that experimental model analysis (EMA) testing and recording of shaft position versus time "orbits" cannot be obtained. Another disadvantage is that the timing relationship or frequency correlation between pressure pulsations or vibrations in a possible "problem source" area and vibrations in the problem "symptom area" cannot be obtained. These procedures often play an important role in vibration problem solving.
- *Multi-channel FFT or real time analyzers.* As with single channel FFTs, these are also based on the fast Fourier transform. They are heavier (7 to 36 kg or 15 to 80 lb) and cost more than single-channel units, but they allow measurements in two or more channels at once, so that the EMA, orbit, and correlation measurements mentioned above can be performed. As with single-channel FFTs, any type of voltage-output transducer can be used, including velocity probes. The more accurate accelerometers and displacement proximity probes are, however, preferred.

Data Types and Formats

Vibration test data are usually plotted in three different forms:

- A Cartesian plot of vibration amplitude versus frequency ("signature plot" or "spectrum"). Sometimes this plot may be combined with a plot of the phase angle (the "lag angle" from application of the exciting force to the responding vibratory motion) versus frequency. It is then called a "Bode plot."
- A Cartesian plot of vibration amplitude versus time, similar to a typical oscilloscope trace ("time" plot).
- A polar plot of vibration versus time in a plane perpendicular to the shaft axis (an "orbit").

The amplitude scales may be linear or dB (base 10 logarithm). A dB scale is often used to improve the resolution of natural frequency peaks. The amplitude scale generally represents rms (root mean square) values unless specifically scaled otherwise. To convert rms values at a specific frequency to peak to peak, multiply by 2.83, and to convert from peak to peak to zero to peak, divide by 2.

22-4. Introduction to Vibration and Noise Calculations

The goals of the following 11 sections are to provide for a deeper understanding and appreciation of vibration, to present means for evaluating a particular situation, and to offer suggestions for reducing vibration or noise to an acceptable level.

The calculation procedures presented here are generally simplified expressions that are meant only to give an idea of the severity of a given situation. If accurate predictions of noise and vibration levels are required, a qualified specialist should be consulted. It may not be easy to find qualified specialists, but a good place to start may be either the National Council of Acoustical Consultants, the Institute of Noise Control Engineers, or the Vibration Institute (see Appendix F for the addresses of these organizations).

Nomenclature

Many of the following symbols are unique to this chapter and, hence, are omitted from Chapter 2.

Symbols

a	Acceleration amplitude of a vibrating body [m/s^2 (in./s^2)].
B	Bulk modulus of fluid [Pa (lb/in.^{2})].
c	Speed of sound in air or speed of wave propagation in fluids or solids [m/s (ft/s)].
cm	Centimeters
d	Static deflection of vibration isolator [mm (in.)].
dB	Decibel.
dBA	A-weighted decibel.
D	Diameter of pipe or resonator [mm (in.)] (see *ID* and *OD*).
D_n	Noise dose (dimensionless).
e	Wall thickness of pipe [mm (in.)].
e_o	Radius of motion of eccentric mass [mm (in.)].
E	Modulus of elasticity [Pa (lb/in.^{2})].
E_t	Total vibrational energy [N · m (ft · lb)].
f	Frequency (Hz).
f_d	Pump driving frequency (Hz).
f_n	Resonance frequency of a multiple degree-of-freedom system (Hz).
f_o	Natural frequency of a single degree-of-freedom system (Hz).
F	Dynamic force [N (lb)].
g	Acceleration of gravity [9.81 m/s^2 (386 in./s^2 or 32.17 ft/s^2)].
G	Modulus of elasticity in shear [Pa (lb/in.^{2})].
Hz	Hertz (frequency in cycles per second).
I	Moment of inertia [m^4 (in.^{4})].
ID	Inside diameter (of a pipe).
J	Mass moment of rotational inertia [kg · m^2 (lb$_m$ · in.^{2})].
k	Spring stiffness constant [N/m (lb/in.)].
K	Torsional rigidity [N · m/rad (lb · in./rad)].
L	Length of pipe or shaft between support points [m (in.)].

lb — Pounds force.

lb_m — Pounds mass = lb/g = lb · s^2/32.2 ft.

L_p — Sound pressure level (dB or dBA).

$L_{p(rev)}$ — Reverberant field sound pressure level (dB or dBA).

L_{dn} — Day–night sound level (dB).

m — A small mass in kg (lb_m).

m — Mass per unit length [kg/m (lb_m/in.)].

M — Mass [kg (lb_m)].

NR — Noise reduction (dB).

NRC — Noise reduction coefficient (dimensionless).

NRC_a — NRC of added acoustical material (dimensionless).

OD — Outside diameter (of a pipe).

p — Acoustic pressure (μPa).

P — Static pressure of fluid in pipe [Pa (lb/in.^{2})].

R — Radius of gyration [m (in.)].

rad — Radian (1 rad = 57.3°).

rms — Root mean square.

S — Surface area [m^2 (ft^2)].

S_o — Cross-sectional area [m^2 (in.^{2})].

S_a — Surface area of added acoustical material [m^2 (ft^2)].

STC — Sound transmission class (dB).

STC_c — Composite (or effective) STC (dB).

T — Repetition time (period) of a vibrating object.

T — Temperature.

T — Transmissibility—the ratio of transmitted vibrational force with isolators to that without isolators.

v — Velocity amplitude of a vibrating body [m/s (in./s)].

w — Weight of pipe or shaft per unit length [N/m (lb/in.)].

W — Weight of a vibrating body [N (lb)].

x — Displacement amplitude of vibrating body [m (in.)].

γ_n — Coefficient (dimensionless) for bending resonances in piping (the subscript n is the resonance order).

δ — Damping coefficient [kg/s (lb_m/s)].

δ_c — Critical damping coefficient—the minimum damping required to allow a displaced system to return to the rest position without undergoing oscillations [kg/s (lb_m/s)].

δ_r — Damping ratio—the ratio of the damping coefficient to the critical damping coefficient (dimensionless).

ε — Efficiency of vibration isolator (percentage).

λ — Wavelength, m (ft).

μPa — Basic unit of acoustic pressure (1 μPa = 1.45 10^{-10} lb/in.^{2}).

ρ — Fluid density [kg/m^3 (lb_m/ft^3)].

ω — Angular frequency [rad/s ($\omega = 2\pi f$)].

Definitions

A-weighted: A commonly used frequency weighting that closely approximates the frequency response of the human ear.

Absorption: Absorption refers to the conversion of acoustical energy into other forms of energy (usually heat). Sound-absorbing materials are usually rated by the noise reduction coefficient (NRC), a numerical value (usually between 0.1 and 1.0) that approximates the ratio of acoustical energy absorbed to the energy incident upon the material.

Acoustic pressure: The small pressure fluctuations that occur about the static atmospheric pressure (101.3 kPa or 14.7 lb/in.^{2} at sea level) that is sound.

Amplitude: A quantitative descriptor of the level or strength of a noise or vibration signal or waveform.

Attenuation: The reduction in amplitude of a noise or vibration signal.

Blade passage frequency: The frequency associated with the motion of individual blades of a fan or pump. It is the same as the pump frequency for a centrifugal pump.

Critical damping: The minimum damping required to prevent oscillation in a displaced mechanical system.

Critical speed: Any rotational speed of a shaft that excites a resonance.

Damping: The mechanism by which energy is removed from a vibrating system.

Day–night sound level: The energy average of the A-weighted sound pressure level measured over a continuous 24-h period with the sound levels biased upward 10 dB between the hours of 10:00 P.M. and 7:00 A.M.

Decibel: A dimensionless unit used to quantify a relative amplitude of an acoustic or vibration signal. Decibel is also used as the basic unit for the sound pressure level that is relative to a standard reference pressure of 20 μPa.

Direct field: The region near the source of a sound that is not influenced by the acoustical characteristics of other surroundings (e.g., room boundaries). In this region, the sound pressure level decreases approximately 6 dB with every doubling of distance (valid only for sources that are small in size compared with the distance).

Directivity: Directional radiation properties of a source that may make more noise in one direction than in another.

Displacement: The physical change in the position or angle of a body or particle as measured from its normal rest position.

Driving frequency: The frequency of forced vibration.

Dyne: The unit of force that causes 1 g to be accelerated at 1 cm/s^2 (= gram/9.81 m/s^2).

Frequency: The inverse of the time required for a body or particle in a vibrating system to go through a full cycle of motion.

Fundamental: The mode of vibration with the lowest natural frequency.

Harmonic: A signal that is usually generated along with the fundamental. The frequency of a harmonic is always an integer multiple of the fundamental frequency.

Inertia base: A heavy mass (usually concrete) added to the frame of a piece of vibrationally isolated equipment.

Mode: A natural, repetitive pattern of harmonic motion in which every particle moves at the same frequency.

Natural frequency: The frequency of the natural or normal modes of vibration of a vibrating system. A system at rest will vibrate only at its natural frequencies after being displaced by an impulsive force.

Node: Any point in a vibrating system where the particle motion is zero for a given mode of vibration.

Period: The time required for vibrating particles to complete a full cycle of motion for a given mode.

Phase: The relative time or displacement between two points in a vibrating system. Any two points are said to be "in phase" (i.e., the phase shift is 0°) if their relative motion is always in the same direction at the same time.

Pump frequency: The primary frequency generated by the pump in hertz (cycles per second). It equals $n \cdot$ rev/s where n is the number of pressure pulses generated in the fluid with each full rotation of the shaft.

Resonance: A condition in which a vibrating system responds with maximum amplitude to a driving force. This condition occurs only at the natural frequencies of the vibrating system.

Reverberant field: The region in a room (usually far from the source) where the sound pressure level is reasonably constant with position and distance from the source.

Rotor frequency: The frequency (in hertz) of the shaft of a pump or other rotating machine.

Short circuit: Any mechanism that allows vibrational energy to bypass a vibration-isolating device, thus rendering it ineffective (e.g., a spring compressed until the coils touch).

Sound pressure: A synonym for acoustic pressure. The sound pressure level is defined as 10 times the common logarithm of the square of the ratio of the weighted or unweighted acoustic pressure to the standard reference pressure (20 μPa).

Sound transmission class: A single-number rating for a material that describes the ability of the material to resist sound transmission.

Spectrum: The frequency distribution of a sound or vibration signal.

Standing wave: A periodic wave that does not appear to propagate but remains stationary. The wave is actually formed by two or more progressive (moving) waves of equal frequency traveling in opposite directions.

Stiffness: The ratio of the change in force to the change in displacement or deflection of an elastic element.

Static deflection: The deflection of an elastic element caused by the static (dead) load of the mass it supports.

Transmission loss: A value (usually expressed in decibels and a function of frequency for a given material) proportional to the ratio of incident to transmitted energy in a sound- or vibration-isolation system. A wall or barrier provides a transmission loss of sound. A spring or rubber mount is usually involved in the transmission loss of vibration.

Wavelength: The distance between "in phase" positions on an acoustic or vibration wave at any given instant. The wavelength is related to frequency, f, via the wave propagation speed, c, as $\lambda = c/f$.

Wave resonances: Resonances in vibration isolators that can permit the transmission of noise through the vibration isolator with little or no transmission loss. Wave resonances usually occur only at frequencies much greater than the fundamental resonance frequency of the vibration-isolation system.

22-5. Vibration and Noise Characteristics

To develop a good foundation for discussing vibration and noise issues in a pumping station, it is important to understand the characteristics and sources of vibration and noise as well as the techniques for measuring these quantities.

In reciprocating machinery such as engines, vibration is primarily due to the dynamic forces exerted on the machine by the reciprocating and rotating internal parts (pistons, crankshaft, etc.). The vibration levels of this equipment can be relatively high. The spectrum is typically dominated by the fundamental frequency (e.g., the cylinder firing rate in a diesel engine) and usually contains several (3 to 10 or more) prominent harmonics.

In centrifugal equipment, vibration is caused by unbalanced rotating parts, misaligned shafts, and fluctuations in the load. The vibration of centrifugal equipment is typically lower in level than comparable reciprocating machinery, and the spectrum usually contains only the fundamental shaft frequency, the blade passage frequency, and one or two harmonics.

Piping vibration is usually caused by the transmission of vibrations from the primary equipment to which it is connected. Piping can be excited by the pipe wall contact with the pump inlet or discharge (even through a flexible connection) or by the pressure fluctuations within the fluid in the pipe. In addition to the frequencies associated with the pump, pipes also vibrate at their own natural or resonance frequencies.

Building vibration results from dynamic forces within the primary vibration elements (i.e., equipment and piping). In addition to vibrating at the frequencies associated with the equipment and piping, the structural elements of the building also have their own natural frequencies. Excessive vibration of the building's structural elements (floors, walls, etc.) is usually a potential problem only when a structural resonance frequency is very nearly equal to one of the driving frequencies associated with the primary equipment. It is also possible to transmit vibration into the earth and, eventually, into adjacent structures. Aspects of this problem are beyond the scope of this chapter.

Translational Vibration

Translational vibration is usually measured with a contact sensor (an accelerometer or velocimeter), which is a small device rigidly attached to the vibrating body. The accelerometer produces an electrical output (voltage) proportional to the instantaneous acceleration of the body, whereas a velocimeter produces a voltage proportional to the velocity. Translational vibration problems in pumping stations almost always occur at frequencies below 100 Hz, and usually the problem is associated with a resonance frequency.

The waveform describing the translational motion of a vibrating body at resonance is usually sinusoidal, as shown in Figure 22-1. The three curves describe exactly the same motion (displacement, velocity, and acceleration) of a single point on the object that is vibrating—all as a function of time. All curves have the same shape and repetition period of T seconds. The only difference between these curves is the amplitude (vertical) scale and the relative phase (shift in time).

The acceleration, velocity, and displacement amplitudes of a sinusoidal signal are related to each other by

$$x = \frac{v}{2\pi f} = \frac{a}{4\pi^2 f^2} \tag{22-1}$$

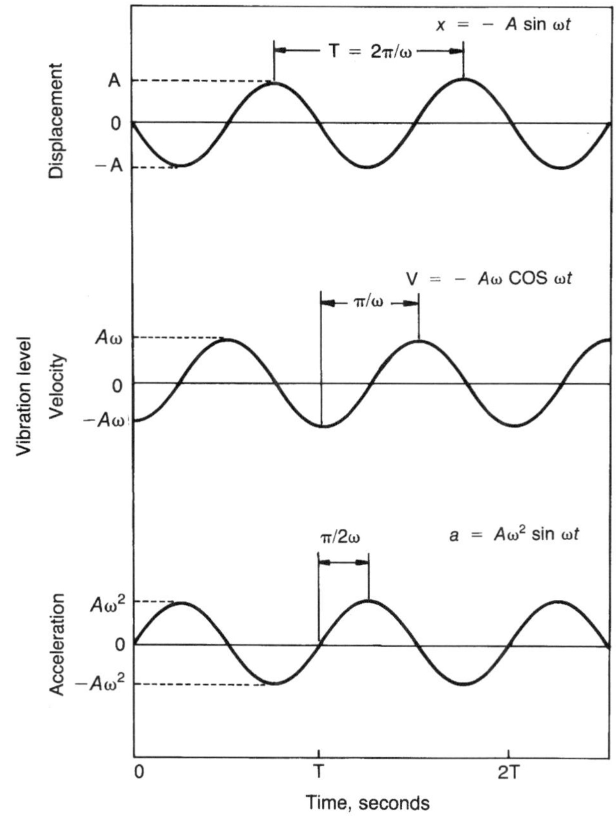

Figure 22-1. Displacement, velocity, and acceleration waveforms of a particle in simple sinusoidal motion.

where x is the displacement amplitude in meters (inches), v is the velocity amplitude in meters per second (inches per second), f is the frequency in hertz, and a is the acceleration amplitude in meters per second squared (inches per second squared). Vibration levels are sometimes reported as peak levels (the maximum deviation from the zero or "at rest" value), as peak-to-peak levels (the maximum total swing from positive to negative values), or the root mean square (RMS) value. The RMS value is 0.707 times the amplitude (or peak value) for sinusoidal waveforms (as shown in Figure 22-1).

For example, an object vibrating sinusoidally with a peak displacement of 1 mm at a frequency of 10 Hz (600 cycles/min) could be described accurately by any of the following (from Equation 22-1):

- 62.8 mm/s peak velocity at 10 Hz
- 3.95 m/s^2 peak acceleration at 10 Hz
- 0.0444 m/s root mean square velocity at 10 Hz
- 7.90 m/s^2 (0.805 g) peak-to-peak acceleration at 10 Hz.

An important concept to keep in mind when analyzing vibrating systems is the buildup and dissipation of energy in the system. As a power source begins operation from rest, the system vibration rapidly builds up to a steady-state level. After the system reaches the steady state, energy is dissipated at a rate equal to the rate at which energy is put into the vibrating system. This rate is usually only a very small fraction of the total power consumption of the source. The energy dissipation is in the form of conversion to heat via various damping mechanisms as well as energy transmission to remote systems (such as building structure and the earth).

The total vibrational energy present in the system can be determined by computing the instantaneous potential energy at peak displacement (when the instantaneous velocity is zero) or by computing the instantaneous kinetic energy at peak velocity (when the instantaneous displacement is zero). Either way, for a point mass the result yields

$$E_t = 2\pi^2 M x^2 f^2 \tag{22-2}$$

where E_t is the total vibrational energy in the system in newton-meters (inch-pound force), M is mass of the system in kilograms (pounds mass), x is the displacement amplitude in meters (inches), and f is frequency in hertz (see Section 22-4).

Solving Equation 22-2 for the dynamic displacement yields Equation 22-3, which demonstrates that, for a fixed amount of vibrational energy, the displacement is inversely proportional to the frequency and the square root of the mass. Thus, if the mass and energy remain constant, doubling the frequency results in a 50% reduction in the dynamic displacement.

$$x = \frac{1}{2\pi f}\sqrt{\frac{2E_t}{M}} \qquad (22\text{-}3)$$

By combining Equations 22-1 and 22-2, it can be seen that the vibration velocity, v, is independent of frequency at constant energy and the acceleration, a, is directly proportional to frequency at constant energy.

Torsional Vibration

Torsional vibration is difficult to measure because the shaft must be rotating during the measurement. The technique usually employed is to mount a lightweight, multitoothed gear near each end of the shaft at an accessible location and to install eddy-current sensors on a rigid foundation in close proximity to the gears. As the shaft rotates, the eddy-current sensor produces a pulsed output signal whose frequency is equal to the instantaneous shaft rotational speed in cycles per second (Hz) times the number of evenly spaced teeth on the gear. Torsional vibration of the shaft is detected as a periodic modulation of the output frequency, which indicates that the shaft speed is fluctuating at the sensor location. A frequency-to-voltage converter then transforms this pulsed signal into a low-frequency (usually below 100 Hz) waveform proportional to the instantaneous angular displacement in degrees or radians (or angular velocity in degrees per second or radians per second). This waveform can then be processed by conventional spectrum analysis techniques to obtain the torsional vibration levels at the frequencies of interest. Of course, for sinusoidal torsional waveforms, Equation 22-1 can also be used to convert angular displacement to angular velocity and angular acceleration if a consistent system of units is used.

Noise

Contrary to the characteristics of vibration, noise is usually only of concern at frequencies greater than 100 Hz. Noise problems are also almost never single-frequency problems, and, as a result, some form of spectrum or frequency analysis is usually required to solve acoustical problems.

Noise is usually measured with a sound level meter, which is a hand-held instrument incorporating a microphone, an amplifying circuit, a frequency-weighting network, a display, and, perhaps, a set of filters (octave or one-third octave) for spectrum analysis. The microphone generates an electrical output proportional to the instantaneous acoustical pressure oscillations, and the sound level meter displays a frequency-weighted (or unweighted) average sound pressure level.

Noise levels are reported as a sound pressure level, L_p (in decibels), which is computed from the acoustic pressure, p, by

$$L_p = 20 \log_{10} (p/p_{ref}) \qquad (22\text{-}4)$$

where p_{ref} is the standard reference pressure of 20 μPa (0.0002 dynes/cm^2). The standard reference pressure is the approximate threshold of hearing for young adults in the ear's most sensitive frequency region (500–2000 Hz).

Subjectively, a noise level increase or decrease of 1 dBA is difficult to detect with the ear even though it does represent a 12% change in the acoustic pressure amplitude. A 10-dBA increase is generally perceived as a doubling of the loudness; conversely, a 10-dBA decrease is perceived as a 50% reduction in loudness. This nonlinear dependence is primarily due to the logarithmic response characteristic of the human ear.

Some typical A-weighted noise levels are given in Table 22-2. Noise levels can vary substantially with time, with location, and, in some instances, with changes in weather conditions. In reporting noise levels from an actual noise source or from an actual environment, it is extremely important to document the relevant conditions of the measurement, including the exact location of the microphone, the time and duration of the measurement, and the atmospheric conditions (primarily temperature, atmospheric pressure, wind speed,

Table 22-2. Typical Noise Levels

Noise source or environment	Noise level (dBA)
Quiet suburb	40–55
Busy city	60–75
Gasoline-powered lawnmower[a]	80–90
Diesel truck (at 15 meters)	75–95
Chain saw[a]	95–120

[a] At operator position.

and direction). Without this back-up information, the validity of a noise measurement can be challenged.

Noise in and around pumping stations is usually broadband (meaning that it contains sounds at all frequencies simultaneously). A measure of an existing noise level is not complete unless the frequency weighting is also given. The most common frequency weighting is A-weighting, which closely approximates the frequency sensitivity of the human ear. Noise levels measured with this frequency weighting should be dimensioned as dBA for clarity.

To perform design calculations to predict noise levels with a reasonable degree of accuracy (±5 dBA), it is necessary to measure or have information concerning the frequency content (spectrum) of the noise sources under consideration. An explanation of this level of analysis is beyond the scope of this chapter; if this level of accuracy is required, consult a professional acoustical engineer.

22-6. Applicable Codes

Vibration

There are no established legal limits for vibration levels of equipment and piping associated with pumping stations, but there are guidelines for normally acceptable vibration levels of major pieces of equipment. Below 600 rev/min, the vibration criteria are based on peak-to-peak displacement; above 600 rev/min, the criteria are based on peak velocity (see Figure 22-1). The relationship between displacement and velocity is given by Equation 22-1. The Hydraulic Institute [2] has published acceptable field vibration limits for clear liquid and nonclog horizontal and vertical centrifugal pumps. Vibration limits for reciprocating and rotary pumps are not available. Table 22-3 contains recommended vibration criteria (compiled from various sources) for major equipment used in pumping stations.

The vibration criteria listed in Table 22-3 refer to vibration levels at the shaft bearing housings. Vibration levels at other, less rigid portions of the machine may be higher. If the vibration level of a particular machine exceeds these values by more than 25% (particularly if the vibration level increases steadily with time), the unit and the system should be analyzed carefully to solve the problem. If the excessive vibration is present while the equipment on a rigid foundation is operating independently from the rest of the system, it should be returned to the manufacturer for balancing and/or alignment to correct the problem. If the vibration levels are only exceeded under load, the problem may be due to an interaction with another portion of the system.

Vibration criteria for piping systems are discussed in detail by Wachel and Bates [17] and are shown in Figure 22-2. The main danger of excessive piping vibration is failure due to repeated stress from distortion of the pipe. Stress in a piping system is primarily a function of the maximum displacement of the pipe as well as its vibratory mode shape. The allowable pipe displacement decreases with increasing frequency, as illustrated in the curves of Figure 22-2, because increasing frequency

Table 22-3. Equipment Vibration Levels[a]

Source	Peak-to-peak displacement below 600 rev/min		Peak velocity above 600 rev/min[b]	
	mm	in.	mm/s	in./s
Centrifugal pumps[c]				
Clear liquid	0.125	0.005	7.8	0.31
Nonclog	0.20	0.008	12.5	0.50
Electric motors[d]	0.08	0.003	5.0	0.20
Fans				
Centrifugal	0.15	0.006	9.5	0.38
Axial	0.10	0.004	6.3	0.25
Generator sets				
Diesel	—	—	20	0.80
Gasoline	—	—	15	0.60
Air compressors				
Reciprocating	0.4	0.016	25	1.00

[a] Levels refer to filtered vibration amplitude at the vibration frequency and at the running speed (forcing frequency).

[b] See Equation 22-1 for the relationship to displacement.

[c] See the Hydraulic Institute's standards [2] for recommendations on vertical pumps.

[d] Limits apply to base-mounted motors. Use pumps for pump-mounted motors.

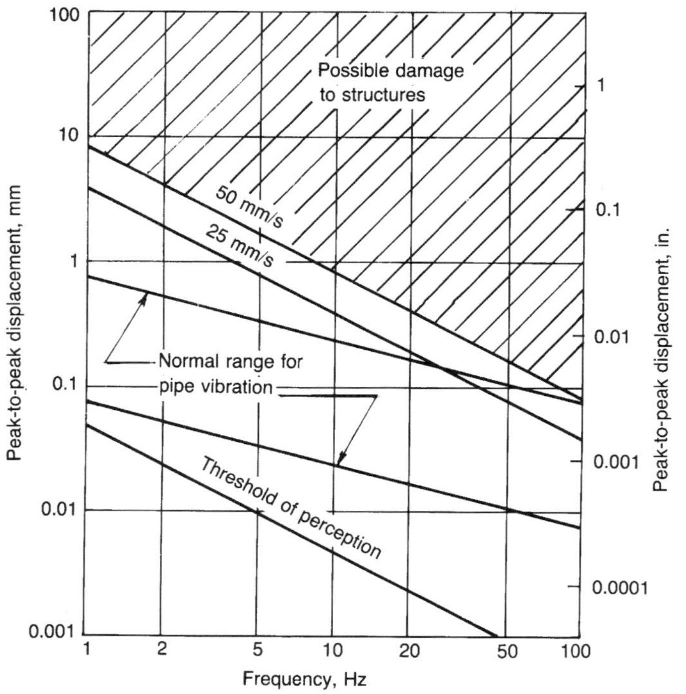

Figure 22-2. Vibration displacement criteria.

implies more vibrational energy input into the system and a greater number of stress cycles over the lifetime of the piping system. An alternative peak-to-peak pipe velocity criterion of 25.4 mm/s (1.0 in./s) has also been suggested by Wachel and Bates. This velocity criterion, which includes allowances with a nominal safety factor for typical conditions found in real piping systems, is also shown in Figure 22-2 along with the approximate threshold of human perception.

Vibration criteria for structures are discussed briefly by Harris and Crede [23]. They state that a peak-to-peak level of 25.4 mm/s (1.0 in./s) is generally regarded as an upper limit for safe, continuous vibration levels of structural floors. This is not to say that structural failure will result at higher levels. During earthquakes, peak-to-peak vibration levels sometimes exceed 50 mm/s (2 in./s) without damage. But the probability of eventual damage increases as the vibration level increases. Continuous vibration levels exceeding 50 mm/s (2 in./s) present a serious threat to the safety of most structures.

Noise

Legal limits on noise generally fall into two categories: (1) occupational exposure and (2) community exposure. Occupational noise exposure limits have been established by the federal government and are administered by OSHA [24]. The federal noise exposure regulations apply to employers who engage in interstate commerce, but most states have passed similar legislation extending these limits (some states are even more restrictive) to virtually all occupational environments. Compliance requires substantial design and investigative effort. The intent of the law is to prevent hearing loss that can be induced by exposure to high noise levels over extended periods of time. Reducing noise levels in the work environment is also advantageous for improved speech intelligibility, which in turn improves worker safety. The relationship between background noise level and the vocal effort required for adequate voice communication at various distances is shown in Figure 22-3.

In essence, the law restricts the length of time an employee can spend in a given noise environment during any 24-h period. Maximum allowable exposure times for various noise levels are listed in Table 22-4.

In most work environments, employees are exposed to noise that varies continuously throughout the day. The noise exposure under these conditions can be estimated by computing the noise dose, D_n,

$$D_n = \sum_{i=1}^{n} \left(\frac{C_i}{T_i} \right) \tag{22-5}$$

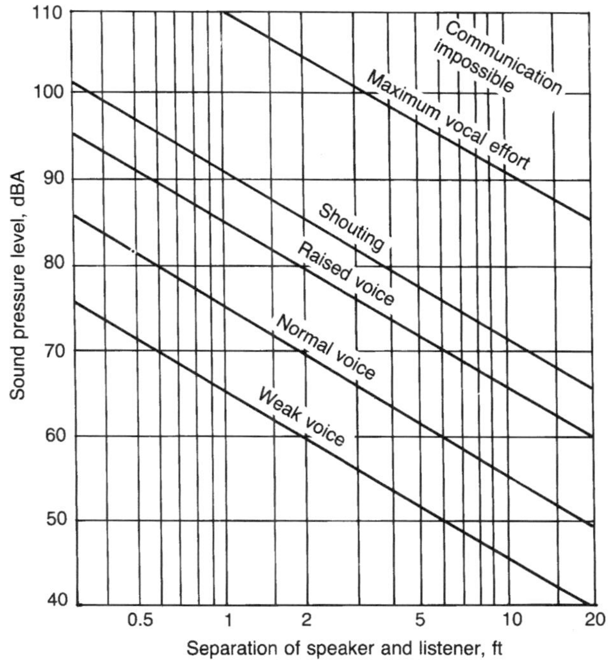

Figure 22-3. Relationship between voice and sound levels.

Table 22-4. Maximum Allowable Employee Noise Exposure

Exposure time (h/d)[a]	Noise level (dBA)
8.0	90
7.0	91
6.2	92
5.3	93
4.6	94
4.0	95
3.5	96
3.0	97
2.6	98
2.3	99
2.0	100
1.7	101
1.5	102
1.4	103
1.3	104
1.0	105
0.87	106
0.76	107
0.66	108
0.57	109
0.50	110

[a] Exposure times for all noise levels in the 80- to 130-dB range can be computed from the equation $T = 8/2^N$, where the exponent $N = (1/5)(L_p - 90)$.

where C_i and T_i are the cumulative exposure time and maximum allowable exposure times for the *i*th noise interval. If D_n exceeds 1.0, the employee is overexposed. Sometimes, it may be necessary to install portable dosimeters on the employees to monitor the actual noise level and compute the noise dose on a continuous basis.

If working conditions in any facility exceed the local or federal regulations, each employee should wear approved hearing protectors (earplugs or earmuffs) whenever the noise exceeds 90 dBA to avoid permanent hearing loss. Good-quality hearing protectors provide an insertion loss (effective noise reduction to the inner ear) between 10 and 20 dB if they are installed properly. It is important to emphasize that extended exposure to high levels of noise can *permanently* damage hearing.

If the noise levels in a pumping station are expected to exceed federal or local regulations, steps should be taken during the design of the facility to mitigate the problem. These steps include the following:

- Design the facility in such a way that workers do not have to spend much time in noisy areas.
- Specify and/or select less noisy equipment.
- Install acoustical lining in noisy areas.
- Isolate noisy equipment with acoustical enclosures.
- Provide sound-absorbing barriers around noisy machinery.

Although there is no federal legislation affecting noise radiated into the environment, many local and state governments have such legislation to protect the welfare of the population living near permanent noise sources. Typically, these limits are much lower than the occupational noise exposure levels because the intent is not to protect against hearing loss but to preserve a quality community environment.

A typical example of a community noise ordinance is illustrated in Table 22-5. The maximum allowable noise level is often a function of how the source and receiver property is zoned (not necessarily how the property is being used). Most ordinances also have allowances for somewhat higher noise levels for short periods of time as well as a requirement for lower noise levels (as much as 10 dB lower) on residential property during sleeping hours. The ordinance should also address the situation where existing ambient noise levels already exceed the ordinance requirements.

In critical situations, it may be necessary to monitor existing noise levels in the surrounding community prior to completing the design of the pumping station. This information on background noise can be used to assess the acoustical impact of the facility and make the appropriate design decisions to minimize the impact. The preferred descriptor of environmental noise levels is the day–night sound level, L_{dn}. The L_{dn} is a time-weighted energy average (over a 24-h period) of the continuously varying A-weighted sound pressure level at a point. The L_{dn} is usually determined by direct measurement with a special instrument called a "community noise monitor," which is designed for computing noise statistics at permanent or semipermanent outdoor installations. In special circumstances where the noise level does not vary with time over the 24-h day, the L_{dn} is approximately equal to the continuous A-weighted sound pressure level plus 6 dBA.

The way in which noise is perceived by average residential communities is shown in Figure 22-4. For reference, typical L_{dn} values for various areas are given in Table 22-6.

In addition to meeting the legal requirements of any state or local noise ordinances, it may be necessary to provide additional noise control measures to prevent any adverse community noise impact. Such measures would particularly apply to communities with nonexistent or poorly written ordinances, or where the ordinance requirements were inadequate to meet the needs of the community. To avoid an adverse community reaction, design pumping stations to meet the following four criteria:

- Do not increase the existing L_{dn} more than 3 dBA.
- Do not increase the existing L_{dn} more than 1 dBA if the existing L_{dn} is more than 60 dBA.
- Do not increase average hourly nighttime maximum noise levels more than 5 dBA in residential areas.
- Do not increase average hourly nighttime maximum noise levels more than 2 dBA in residential areas if the existing hourly nighttime maximum levels exceed 60 dBA.

The reason for restricting the hourly maximum noise level on residential property during the nighttime hours (usually defined as 10:00 P.M. to 7:00 A.M.) is to minimize sleep disturbance. Noise-induced sleep interference is more dependent on maximum noise

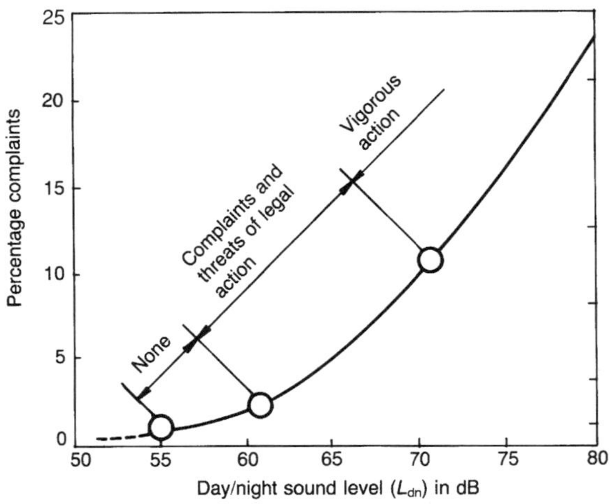

Figure 22-4. Average community reaction to noise.

Table 22-6. Typical Day–Night Sound Levels at Various Locations

Location	L_{dn} (dBA)
Near major international airport	75–85
Major city in downtown area	70–80
Noisy urban residential	60–70
Suburban residential	50–60
Quiet rural	45–55

Table 22-5. Commonly Used Maximum Permissible Community Noise Levels (dBA)

	Receiving property		
Source property	Residential	Commercial	Industrial
Residential	55	57	60
Commercial	57	60	65
Industrial	60	65	70

levels than on average noise levels [25]. Maximum nighttime noise levels are quite volatile, that is, they usually vary by as much as 10 dB or more from hour to hour as well as on a daily basis. Consequently, it is best to base a design on an average noise level taken over several nights. These four basic criteria can usually ensure little or no acoustical impact on community noise levels.

22-7. Equipment Vibration

The vibration limits recommended in Table 22-2 for major items of equipment in pumping stations apply to equipment mounted directly to a rigid foundation. Vibration levels are usually higher if the equipment is mounted on a flexible structure because both the support and the machine contribute to the total vibration level (ANSI/HI standards [2]). All structures are flexible to some degree, but a rigid structure as defined in the ANSI/HI standards has a fundamental natural frequency higher than 125% of the highest rotating speed of the machine.

When vibrating equipment is rigidly mounted to a massive structure (such as a building floor slab) the majority of the vibrational energy generated by the machine is transmitted into the structure. Much of this energy is absorbed by the building (which causes vibration of the building's structural elements), and some is dissipated into the earth via the foundation. If the building is structurally capable of handling this vibrational energy, and if the resulting noise radiated by the vibrating structure is within acceptable limits, the rigid mounting of equipment is preferable for most pumps because, unless the pump or motor excites a building resonance, it puts less stress on the machine.

If the rotational speed of the equipment happens to match one of the resonance frequencies of the supporting structure, the structure may vibrate at excessive levels. Obviously, if the structure supporting the equipment vibrates at an excessive level, this vibration also contributes to the vibration observed at the machine. Often, the equipment is thought to be out of balance when this occurs; in fact, it may be within acceptable tolerances if removed and tested on a rigid foundation. There are three potential solutions to the problem of excessive structure vibration caused by rigidly mounted equipment:

- Avoid machine speeds that excite structural resonances.
- Modify the structure to increase the resonance frequencies so that they are above the highest machine frequency.
- Isolate the equipment from the structure with vibration isolators.

Avoiding the resonance condition altogether is the preferred technique, but when this is not feasible consider vibration isolation of the equipment.

Vibration isolators reduce the amount of vibrational energy transmitted to the structure supporting the equipment and, hence, can reduce the vibration levels of the structure. Vibration isolators do not reduce the amount of vibrational energy generated by the machine, and, in fact, the vibration level of a machine mounted on them is usually greater than that of a rigidly mounted machine, particularly at low frequencies (i.e., in the region of the machine's shaft speed). The amount of additional vibration is primarily a function of the total mass of the machine. Equipment that is very massive (such as a large diesel generator set) usually has vibration levels that are almost independent of mounting, while lighter equipment (such as an air compressor) is subject to a significant increase in vibration level when moved from a rigid to an isolated mounting.

Because vibration isolation of equipment is an added cost, it should be considered only when there is a need for reduced noise or vibration. In general, vibration isolation of equipment does not reduce the airborne noise radiated by the equipment. Although vibration isolation can significantly reduce structure-borne noise, its effects are not noticeable unless the airborne noise is also controlled. In most centrifugal pumps, the vibrational energy is primarily low frequency, and structure-borne noise is not usually a problem. But structure-borne noise may be a problem for reciprocating pumps and rotary pumps if they have fundamental pump frequencies greater than 100 Hz. Therefore, most pumps do not require vibration isolation in well-designed pumping stations.

Equipment that should be vibration-isolated in most pumping stations includes generator sets, fans, and air compressors. This equipment usually contains sufficient low- and high-frequency energy to cause excessive structure-borne noise in other areas of the building. Structural vibrations from this equipment may also be a problem for computer equipment and other sensitive electronic instruments.

22-8. Vibration Isolation Theory

Equipment vibration in its simplest form can be modeled as a small mass, m, rotating about the center of gravity of a rigid body with mass M at a distance, e_0, from the center of gravity of mass M at a fixed rotational speed. The rotating mass produces an oscillating force, F_o, whose magnitude is

$$F_o = me_o\omega^2 \tag{22-6}$$

If the equipment is rigidly attached to the supporting structure, all of this force is transmitted into the structure, which causes it to vibrate at the frequency of rotation of the eccentric mass. The transmitted force (and resulting vibration) can be reduced by reducing (1) the eccentric mass, (2) the eccentricity, e_o, or (3) the rotational speed. Unfortunately, there are practical limits to equipment balancing (to reduce e_o), and equipment speed cannot always be reduced. The alternative method of reducing vibration transmission into the structure is by isolating the equipment with a resilient element such as a spring or a compliant pad. This isolating element, if properly designed and installed, can reduce the force transmitted into the structure and the resulting vibration dramatically.

The natural frequency of a simple, single degree-of-freedom (one mass) isolation system mounted to a rigid structure (in the absence of damping) can be expressed as

$$f_o = \frac{1}{2\pi}\sqrt{k/M} = \frac{1}{2\pi}\sqrt{kg/W} = \frac{1}{2\pi}\sqrt{g/d} \tag{22-7}$$

where f_o is the natural frequency of vibration in hertz, k is the spring constant of the isolator in newtons per meter (pounds force per foot), M is the total mass supported by the isolator in kilograms (pounds mass), W is the total weight supported by the isolator in newtons (pounds force), g is the acceleration of gravity [9.81 m/s^2 (32.2 ft/s^2)], and d is the static deflection of the isolator due to the weight of the equipment in meters (feet). At frequencies much less than f_o the isolators provide no isolation and the system acts as though the isolators were not present. Within ±50% of the resonance frequency, the isolation system reinforces and amplifies the vibration of the equipment as well as the vibration of the supporting structure. Only at frequencies much greater than f_o does the isolation system reduce the vibrational energy injected into the structure.

The efficiency of a simple, single degree-of-freedom vibration isolation system is defined as

$$\varepsilon = 100[1 - (F_2/F_1)] = 100[1 - (x_2/x_1)] \tag{22-8}$$

where ε is the efficiency (a percentage), F is the vibrational force, x is the displacement, the subscript 2 means transmitted to the structure with isolators, and the subscript 1 means transmitted to the structure with rigid connections.

In general, the efficiency is a function of frequency, but usually the efficiency at the driving frequency or the resonance frequency of the supporting structure is the main concern. The ratio F_2/F_1 is often called the transmissibility, T, of the system. At the lowest driving frequency, which is the rotational speed of the machine, the transmissibility is expressed as

$$T = \frac{F_2}{F_1} = \sqrt{\frac{1 + (2\delta_r f_d/f_o)^2}{\left(1 - f_d^2/f_o^2\right)^2 + \left(2\delta_r f_d/f_o\right)^2}} \tag{22-9}$$

where f_d is the driving frequency, f_o is the natural frequency of the isolation system, and δ_r is the damping ratio, δ/δ_c. Steel spring vibration isolators typically have damping ratios between 0.01 and 0.03 (1 to 3%). The transmissibility can be evaluated at any frequency, f, by substituting f for f_d in Equation 22-9. The frequency and damping ratio dependence of the transmissibility of the single degree-of-freedom system is illustrated in Figure 22-5.

In addition to reducing the vibration levels of structures that support equipment, it is also important to reduce the vibration levels of the equipment itself. The general expression for the displacement ampli-

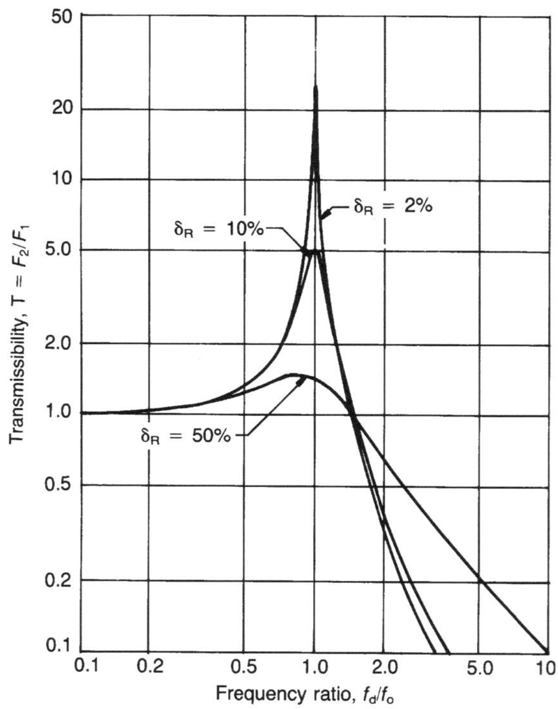

Figure 22-5. Transmissibility of a simple vibration isolation system.

tude of a rigid mass mounted on vibration isolators with a spring stiffness constant, k, is (for $f = f_d$)

$$x = \frac{F_o/k}{\sqrt{\left[1-(f_d/f_o)^2\right]^2 + \left[2\delta_r(f_d/f_o)\right]^2}} \qquad (22\text{-}10)$$

which is presented graphically in Figure 22-6 as a function of frequency and damping ratio. At low frequencies ($f << f_o$) the displacement amplitude asymptotically approaches the value F_o/k. At high frequencies ($f >> f_o$) the displacement amplitude approaches the value $F_o/(M\omega^2)$. In the resonance region the displacement amplitude is controlled entirely by the damping factor.

Damping is the process by which vibrational energy is converted into heat. The damping ratio, δ_r, is the ratio of the damping factor, δ, and the critical damping factor, δ_c. The damping factor, δ, is a physical property of the isolator that is relatively constant (there is a slight temperature dependence) with frequency. Materials with high damping factors do not readily vibrate because the energy is rapidly converted into heat and it is not allowed to build up to high levels. Critical damping is the amount of damping required to prevent a displaced mechanical system from oscillating during its return to a rest position and is a function of the system parameters. For viscous damping, δ_c equals two times the mass times the natural frequency in radians per second. Consequently, there are no fixed values of δ_r for a given material. Most spring-mounted vibration isolator systems, however, have damping ratios less than 5%. Systems with cork or rubber isolators may have δ_r values in the 10 to 20% range.

Although Figures 22-5 and 22-6 are similar in appearance, a completely separate but related phenomenon is described in each. Note that both curves show comparable response at low-frequency ratios. At high-frequency ratios (e.g., 2 or more), however, an increase in the damping ratio slightly reduces the vibration amplitude (Figure 22-6) of the equipment but increases the transmissibility (Figure 22-5) and, hence, increases the displacement amplitude of the building. Thus, depending on whether the equipment or the building is of greater concern, the addition of damping may be detrimental except at f_d/f_o ratios between 0.4 and 1.5.

For high-efficiency vibration isolation, the lowest driving frequency should be 5 to 10 times the natural frequency of the isolation system, and the damping factor, δ_r, should be low. The relationships found in Equations 22-7 through 22-9 are summarized in Figure 22-7. The required isolator static deflection needed to achieve a particular isolation efficiency at a given rotational speed is shown for a damping ratio of

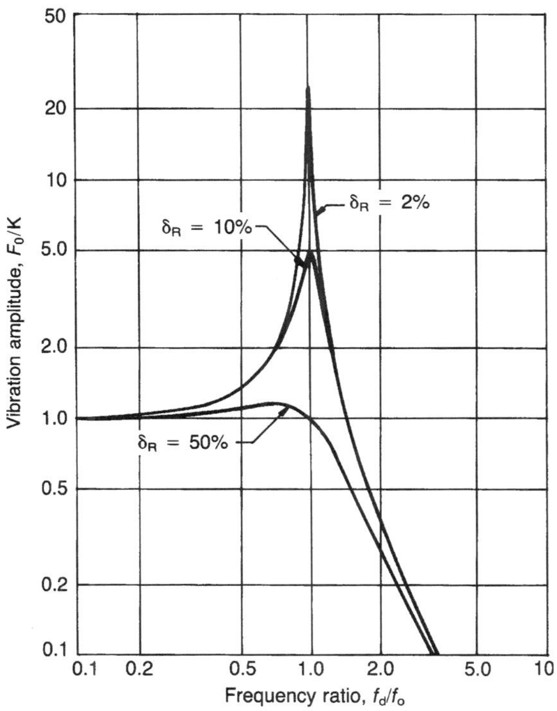

Figure 22-6. Amplitude response of a simple vibration isolation system.

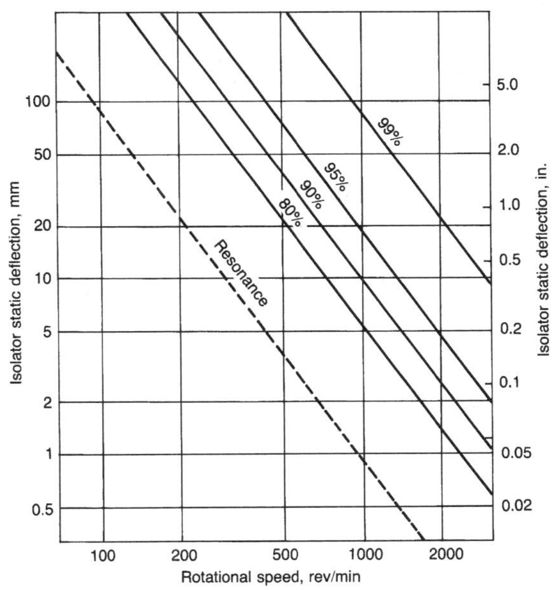

Figure 22-7. Vibration isolation efficiency.

0.02. Be cautious because the assumption in the preceding analysis is a linear single degree-of-freedom system (i.e., a rigid body mounted to a resilient element obeying Hooke's law with viscous damping) supported from an infinitely rigid structure. In reality, none of these assumptions is strictly true, but they are approximately valid for frequencies less than $10f_o$ when the natural frequency of the supporting structure is much greater than $10f_o$. At higher frequencies, other factors (such as wave effects in which energy is transmitted directly into the structure via the isolation element) take over to make the isolation efficiency much less than would be predicted with the above equations. In practice, the vibration isolation efficiency very rarely exceeds 95% at any frequency.

Warning: do *not* be confused with the isolation efficiency figures, which are a function of the linear ratio of the applied to the transmitted dynamic force (see Equation 22-8). Human perception of noise and vibration levels follows a logarithmic response, and the perceived reduction is always less than that derived directly from the isolation system efficiency.

22-9. Vibration Isolators

There are three major types of vibration isolators. The least expensive is the pad type, which is usually made of rubber, cork, fiberglass, or a combination thereof. These isolators generally provide static deflections of less than 5 mm (0.2 in.) and, hence, are effective (efficiencies greater than 90%) only for equipment with driving frequencies greater than 25 Hz (1500 rev/min).

The most common vibration isolator is the coil steel spring. Springs are manufactured in a variety of designs that provide static deflections from 10 to 100 mm (0.4 to 4.0 in.). Springs are suitable for isolating all types of equipment with driving frequencies greater than 300 rev/min. A typical open spring isolator (so called because the spring is not enclosed in a metal housing) is illustrated in Figure 22-8. Metal housings are often used to provide lateral restraint to resist forces due to earthquakes. The threaded rod in the top plate has height adjustment nuts for leveling the equipment after mounting. The neoprene rubber pad under the base plate reduces the transmission of high-frequency energy (via wave resonances in the steel coil) into the structure. Anchor bolts in the base plate prevent equipment creep, but they must be isolated from the base plate with a neoprene washer and grommet to prevent high-frequency energy transmission into the structure via the anchor bolt. For stability reasons, the coil diameter of open springs should be at least 0.8 times the operating height of the spring.

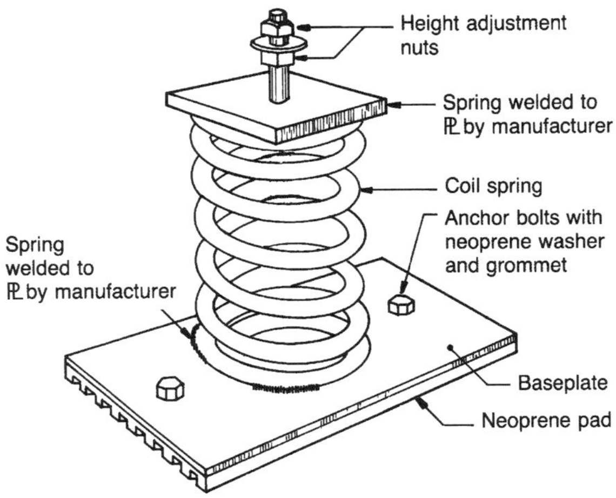

Figure 22-8. Open spring vibration isolator.

For extremely sensitive installations (which are not anticipated in pumping stations) or for very low-frequency equipment (less than 300 rev/min), air springs may be necessary. Air springs have an enclosed volume of air to support the equipment at each support point. Air springs do not exhibit a linear load deflection curve (Hooke's law), and as a result, the simple theory presented above does not apply. Air springs do, in general, provide superior vibration isolation, partly because of the resulting low natural frequency and also because of the absence of high-frequency wave resonance effects. Air springs should be used only with the guidance of a qualified vibration consultant.

The recommended method of isolating a centrifugal pump from its supporting structure is illustrated in Figure 22-9. The machine is rigidly mounted to an inertia base, a concrete-filled frame designed to add mass to the equipment. The frame is then supported 50 mm (2 in.) above the housekeeping pad with steel springs. Seismic restraints should be provided at each support point with cushioned bumpers to limit motion to no more than 8 mm (0.3 in.) in any direction. Discharge piping and electrical power should be connected with flexible fittings.

The primary purpose of the inertia base in Figure 22-9 is to reduce the amplitude of the vibration of the pump, not to reduce the vibration level of the supporting structure. Adding mass to equipment that is already supported by vibration isolators may improve the theoretical efficiency slightly because it lowers f_o. However, if enough mass is added to make a significant improvement in the isolation efficiency (by increasing the ratio f_d/f_o), the springs would probably overload, because springs typically can provide a safety factor of only 20 to 50% of load capacity before

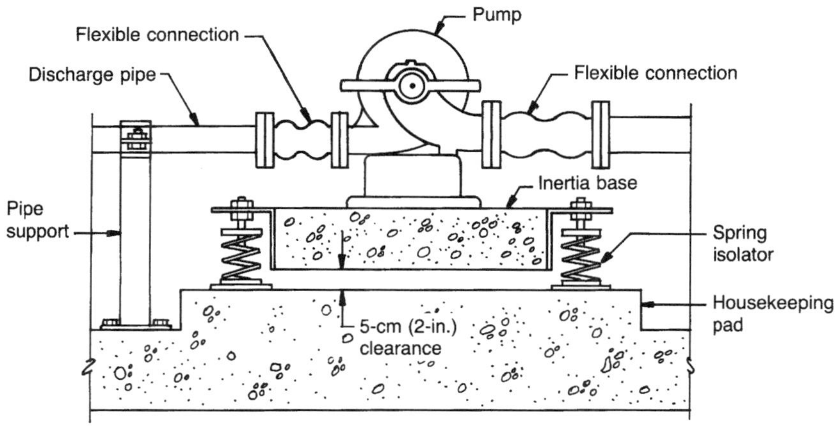

Figure 22-9. Vibration isolation of a centrifugal pump.

"short circuiting." The inertia base does not substantially decrease the vibratory force transmitted into the structure because it does not change the transmissibility. Furthermore, the inertia base does not reduce the driving force, because from Newton's first law ($F = ma$), the decrease in acceleration, velocity, or displacement is only due to a corresponding increase in mass. Thus, a 20% increase in mass results in a 20% decrease in equipment displacement (or velocity or acceleration) amplitude with no change in transmissibility or transmitted force into the structure. The efficiency of the isolation is controlled entirely by the selection of the springs (the static deflection), as shown in Figure 22-7.

In pumping stations, reciprocating air compressors and large, high-pressure centrifugal fans and pumps are usually the only machines that require an inertia base. The amount of additional mass required varies from 1.5 to 5 times the equipment weight and depends on the level of the vibration forces in the machine and the total weight of the machine. Heavy equipment (such as diesel generators) has sufficient mass and rigidity to make inertia bases unnecessary. The only penalty for overdesigning the mass of the inertia base is the increased cost of the base and isolators.

If it is necessary to isolate a pump from the structure, it is absolutely mandatory that the pump, drive shaft, and motor (or engine) all be mounted on a common, rigid foundation (preferably a concrete inertia base) to avoid vibration-induced misalignment between these components. This isolation requirement automatically eliminates the possibility of vibration isolation for some pumps (such as vertical turbines) with motors that must be supported by a separate structural element. So as not to degrade the overall isolation provided by the isolation mounts, piping and electrical connections must also be flexible. It is difficult to specify exactly how much relative motion must be allowed, but about 5 to 10 times the displacement amplitude of the pump is recommended. This displacement tolerance can usually be achieved by using flexible rubber hose (with nylon cord reinforcement) connectors.

Example 22-1
Vibration Isolation of a Centrifugal Pump

Problem: A 150-kW (200-hp) close-coupled centrifugal water pump is driven by an induction motor at 900 rev/min. The pump/motor assembly is rigidly mounted to a concrete pad 20 cm (8 in.) thick. The supporting floor vibrates at a peak-to-peak displacement of 1.0 mm (0.04 in.) at 15 Hz while the pump is running. During the test, the unit was operated without fluid in the connected piping, and it was found that the presence of the fluid in the system had little effect on the structural vibrations of the floor.

Design an isolation system for the pump-motor assembly that reduces the structural floor vibrations to meet the safe limit for the structure with a safety factor of 2 and provides reasonable equipment vibration levels. The total mass of the pump–motor assembly is 800 kg

(1760 lb_m), and the manufacturer states that the out-of-balance force should not exceed that of an eccentric mass of 1 kg (2.204 lb_m) with an eccentricity of 25 mm (0.984 in.).

Solution: The recommended safe limit for structures is 25 mm/s (1 in./s) peak to peak. Compute the required efficiency of the isolation system. Begin with Equations 22-1 and 22-8.

SI Units

$$d = \frac{25}{2\pi 15} = 0.265 \text{ mm (peak to peak)}$$

$$d_{\text{desired}} = \frac{0.265}{2} = 0.132 \text{ mm}$$

$$\varepsilon = [1 - (0.32/1.0)] \times 100 = 87\%$$

U.S. Customary Units

$$d = \frac{1}{2\pi 15} = 0.0106 \text{ in. (peak to peak)}$$

$$d_{\text{desired}} = \frac{0.0106}{2} = 0.0053 \text{ in.}$$

$$\varepsilon = [1 - (0.0053/0.04)] \times 100 = 87\%$$

From Figure 22-7, the required isolator static deflection is about 12 mm (0.47 in.). This deflection cannot be achieved with most pad-type isolators but is easily obtained with conventional steel spring isolators. By definition, the spring constant equals force divided by deflection so the spring constant ($k = F/d$) for each of four springs must be about

$$k = \left(\frac{800}{4}\right)kg \times \frac{9.81 \text{ m/s}^2}{0.012 \text{ m}} = 163{,}500 \text{ N/m} \qquad k = \frac{(1760/4)}{0.47} = 936 \text{ lb/in.}$$

From Equation 22-6 ($F_o = me_o\omega^2$), the maximum out-of-balance force for a mass of 1 kg at 25 mm (2.204 lb_m at 0.984 in.) is

$$F_o = (1)(0.025)(2\pi 15)^2 = 222.1 N \qquad F_o = \frac{2.204}{32.17}\left(\frac{0.984}{12}\right)(2\pi 15)^2 = 49.9 \text{ lb}$$

From Equation 22-7, the natural frequency of the isolation system is $f_o = (1/2\pi)\sqrt{k/M}$, where M is the pump mass supported by each spring and k is the stiffness of each spring:

$$f_o = \frac{1}{2\pi}\sqrt{\frac{163{,}500}{(800/4)}} = 4.55 \text{ Hz}$$

$$M = W/g = \frac{(1760/4)}{32.17 \times 12} = 1.140 \text{ lb} \cdot \text{s}^2/\text{in.}$$

$$f_o = \frac{1}{2\pi}\sqrt{\frac{936}{1.140}} = 4.56 \text{ Hz}$$

The ratio of driving frequency to natural frequency is $f_d/f_o = 15/4.56 = 3.29$. From Equation 22-10, $x = F_o/k[(1 - f_d^2/f_o^2)^2 + (2\delta_r f_d^2/f_o^2)^2]^{-1/2}$ and assuming a 2% damping ratio, the maximum displacement amplitude of the pump is

$$x = \frac{222.1}{163{,}500}[(1 - 3.29^2)^2 + (2 \times 0.02 \times 3.29)^2]^{-1/2}$$
$$= 1.38 \times 10^{-4} \text{ m}$$
$$= 0.138 \text{ mm peak to zero}$$

$$x = \frac{49.9}{936}\left[\left(1 - 3.29^2\right)^2 + (2 \times 0.02 \times 3.29^2)\right]^{-1/2}$$
$$= (0.053)/92.69 + 0.017$$
$$= 0.0054 \text{ in. peak to zero}$$

From Equation 22-1 ($x = v/2\pi f$),

$$v = 2\pi f x = 2\pi(15)(0.138) = 13.0 \text{ mm/s (peak)} \qquad v = 2\pi(15)(0.0054) = 0.51 \text{ in./s (peak)}$$

The computed value is almost double that allowable for clear liquid pumps given in Table 22-3. Hence, the vibration level of the machine must be reduced, so an inertia base must be added. If the inertia base has an added mass of 1000 kg (2204 lb_m) and if the spring constant is nearly doubled to 300,000 N/m (1713 lb/in.), the new static deflection of each of the four isolators will be

$$d = \frac{(1000 + 800)}{4} \times \frac{9.81}{300{,}000}$$
$$= 0.0147 \text{ m} = 14.7 \text{ mm}$$

$$d = \frac{(2204 + 1760)}{4 \times 1713} = 0.579 \text{ in.}$$

which still meets the 12-mm (0.47-in.) minimum. From Equation 22-7, $f_o = (1/2\pi)\sqrt{k/M}$, the new frequency would be

$$f_o = \frac{1}{2\pi}\sqrt{\frac{300{,}000}{1800/4}} = 4.1 \text{ Hz}$$

$$f_o = \frac{1}{2\pi}\sqrt{\frac{1713 \times 386}{(2204 + 1760)/4}} = 4.1 \text{ Hz}$$

The new dynamic displacement is calculated from Equation 22-10.
Note that $f_d/f_o = (900/60)/4.1 = 3.659$.

$$x = \frac{222.1}{300{,}000}\left[\left(1 - 3.659^2\right)^2 + (2 \times 0.02 \times 3.659)^2\right]^{-1/2}$$
$$= 5.98 \times 10^{-5} \text{ m}$$
$$= 0.060 \text{ mm peak to zero}$$

$$x = \frac{49.9}{1713}\left[(1 - 13.385)^2 + 0.04(3.659)^2\right]^{-1/2}$$
$$x = 0.00233 \text{ in. peak to zero}$$

From Equation 22-1 ($x = v/2\pi f$), the peak velocity is

$$v = 2\pi 15(0.060) = 5.63 \text{mm/s (peak)}$$

$$v = 2\pi(15)(0.00233) = 0.220 \text{ in./s (peak)}$$

which meets the criteria listed in Table 22-2.

22-10. Piping Vibration

The piping system attached to any pump vibrates as long as the pump is moving fluid inside the piping. The amount of vibration present in the pipe depends on many factors, including (1) the size, type, and rotational speed of the pump; (2) the fluid characteristics; and (3) the size, construction, support, and layout of the piping system. The purpose of this section is to present and discuss the basic design information needed to avoid vibration problems in piping systems.

To understand the nature of pipe vibrations, it is important to realize that vibration can occur in many forms; the most common are shown schematically in Figure 22-10. In general, all three modes—compression, pure bending, and radial (as well as other, less important modes)—coexist in varying degrees in all piping systems.

In the compressional mode, the vibration displacement of the pipe wall is along the length of the pipe, and the instantaneous density of the pipe wall material oscillates slightly with time and distance along the pipe. This form of vibration does not tend to stress the pipe unduly, and, consequently, it is of less importance.

In the pure bending mode, the entire pipe displaces in a direction perpendicular to the length of the pipe, much like that of a violin string when plucked. This action can put significant stress on the pipe wall if the vibration amplitude is large enough.

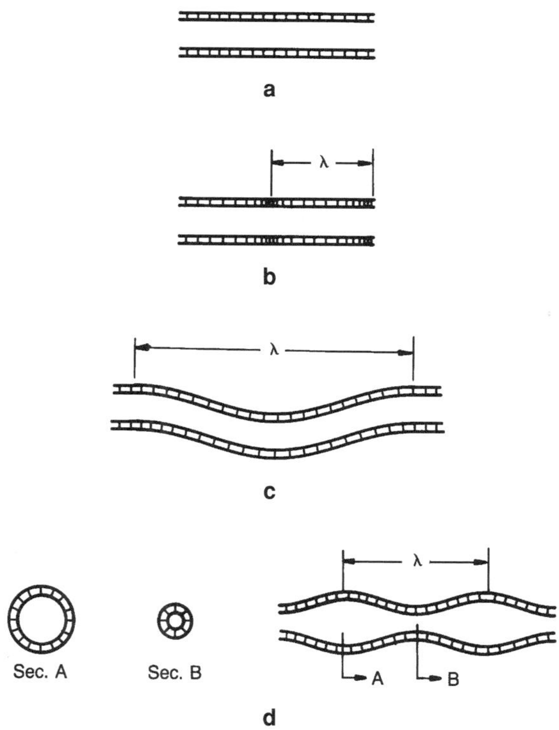

Figure 22-10. Modes of pipe vibration. (a) Pipe at rest; (b) compression; (c) bending; (d) radial.

In the radial mode of vibration, the pipe wall motion is perpendicular to the length of the pipe, but the pipe wall at a given location moves radially inward and outward. This mode of vibration is naturally induced by pumps that discharge the fluid into the piping system as a series of pulses that cause a sinusoidal variation of fluid pressure in the piping.

These three forms of vibration, called the "natural modes of free vibration," occur only at particular discrete frequencies, called "resonance frequencies." Resonance frequencies are determined by the physical characteristics of the piping system.

Although each section of pipe has an infinite number of resonances in each of several forms of vibration, only a few of these are of concern to the designer because of the following:

- Vibrational energy from the pump is usually concentrated at the low-order pump frequencies, which are usually less than 100 Hz.
- Resonances occurring at low frequencies are the only ones that might cause large structural displacements and corresponding high stress.

In most situations, it is adequate to consider only the first two resonance orders for piping systems. The first-order resonance is called the "fundamental frequency." The second-order resonance is defined in Figure 22-11.

Bending Modes in Piping Systems

At each bending mode resonance frequency of the piping system there is a characteristic shape that the pipe assumes. The pipe oscillates back and forth resuming the same shape at periodic time intervals equal to the inverse of the resonance frequency. The exact shape of the pipe for a given mode is slightly different from the theoretical shape because the pipe also simultaneously vibrates (to a lesser degree) at other modes.

The mode shape for a given resonance is determined by the boundary conditions (i.e., how the pipe section is supported) and the order of the mode. The lowest resonance frequency for a given system is called the "fundamental" or "first-order resonance." This resonance is usually the most important because it has the greatest potential for large pipe displacements. The second-order resonance occurs typically at a frequency two to four times the fundamental and, therefore, has the potential for only about one-half to one-fourth of the displacement of the fundamental mode of vibration if equal energy input into each mode is assumed.

The boundary conditions describe how the pipe is supported (restrained from motion) at the ends of a pipe section. There are numerous ways of supporting pipe sections, but four of the more common are given in Table 22-7.

Mode shapes and corresponding natural frequencies for various combinations of the first three boundary conditions have been determined analytically [7, 26], and some are shown in Figure 22-11. The analysis of the vibration-isolated support does not seem to have been published to date, but the mode shapes and natural frequencies are expected to fall somewhere between the free and simply supported conditions.

The general equation for computing bending mode resonance frequencies for a piping system of uniform diameter and physical properties can be expressed as

$$f_n = \frac{\gamma_n^2}{2\pi L^2}\sqrt{\frac{EIg}{w}} \tag{22-11}$$

where f_n is the resonance frequency in hertz, γ_n is the coefficient of bending resonance, L is the length of

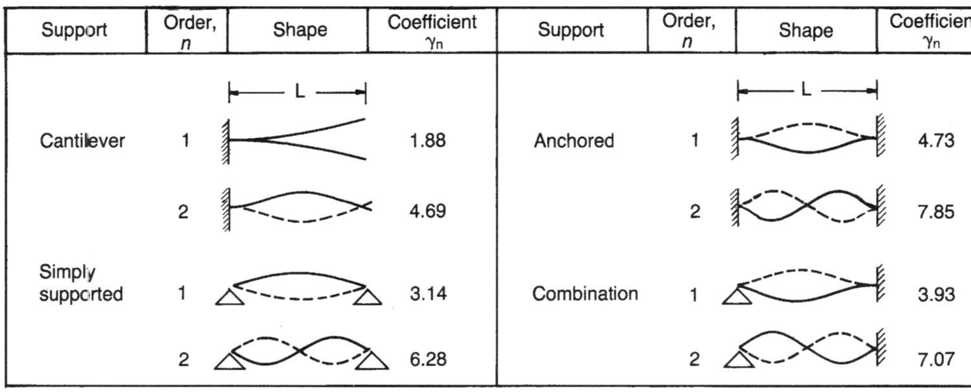

Figure 22-11. Bending mode resonances.

Table 22-7. Boundary Conditions for Pipe Bending Modes

Support	Boundary conditions
Free	Pipe is unrestrained to any motion (moment and shear loads are zero)
Simply supported	Pipe is supported to prevent displacement but allow rotation
Clamped	Pipe is clamped to prevent both displacement and rotation
Vibration isolated	Pipe is resting on a compliant support (e.g., a spring) that allows both displacement and rotation

pipe between supports in meters (inches), E is the modulus of elasticity in Pascals (pounds force per square inch), I is the moment of inertia in meters to the fourth power (inches to the fourth power), g is the acceleration due to gravity [9.81 m/s^2 (386 in./s^2)], and w is the weight (including fluid) per unit length in newtons per meter (pounds force per inch). In Equation 22-11, it is assumed that the pipe is straight and has a constant diameter. Values for γ_n are given in Figure 22-11 for the fundamental and second-order resonance for various boundary conditions. The moment of inertia for pipe is

$$I = \frac{\pi}{64}\left[(OD)^4 - (ID)^4\right] \tag{22-12}$$

where OD and ID are the outside and inside diameters of the pipe. Weights of pipe are readily available from the pipe manufacturers (see also Tables B-1 to B-4).

Another condition of interest to designers is a straight section of uniform pipe with an added center mass, which may be a valve or some other device supported by the pipe. Blevins [7] has published an expression for the approximate natural frequency for simply supported boundaries:

$$f_1 = \frac{2}{\pi}\sqrt{\frac{3EI}{L^3(M_o + 0.49M_p)}} \tag{22-13}$$

where M_o is the central concentrated mass in kilograms (pounds mass) and M_p is the total mass of the pipe and fluid between the support points. The other terms are as defined previously.

An approximate solution for clamped boundaries and noncentered masses is also given by Blevins, but it is too lengthy to reproduce here. Techniques for computing the natural frequencies of piping systems containing elbows and other complexities are addressed by the Kellogg Company [26].

Forced Vibrations

The source of vibrational energy in a typical piping system is usually the pump (sometimes augmented by an engine drive). The energy can be transmitted to the pipe either through the coupling between the two or through the fluid in the pipe. The coupling between the pump and the pipe wall transmits vibrations from the pump housing, which is usually dominated by the rotational frequency of the pump shaft and the pump frequency (the shaft frequency times the number of blades, teeth, or pistons). The vibration of a pipe due to a continuous energy source (e.g., a pump) is called "forced vibration." The term "forced vibration" is used because the system vibrates at the driving frequency of the source in addition to its own natural frequencies.

The amplitude of the forced vibration of a pipe depends on the amount of energy and frequencies introduced by the pump as well as the physical characteristics of the piping system. If the forced vibration occurs at one of the natural frequencies of the pipe, system vibration can be amplified to extremely high levels. This condition is called "resonance." The amplification of vibration near resonance follows the same basic laws discussed in Section 22-7.

Most metallic pipes possess values of δ_r that are much less than 0.05 and even as low as 0.02. Consequently, during resonance ($f_d = f_n$) the vibration amplitudes, depending on the damping factor, can be as much as 10 to 25 times that of the nonresonant condition. Most vibration-induced piping failures are due to resonance. Resonance can best be avoided by designing the piping support system so that the driving frequency is either less than $f_1/2$ or greater than $3f_1/2$.

The spacings between pipe supports should be approximately equal, but not necessarily exactly equal. Unequal support spacings simply add additional natural frequencies to the system, which reduces the frequency spacing between the driving frequency and the natural frequencies of the piping system. However, there is some merit to varying the pipe support spacing slightly (e.g., 10% or less) in an irregular manner, particularly if the system must operate near resonance. Adjacent sections of piping are more likely to reinforce vibration at resonance if the support spacings are identical. The damping factor is so low for most pipes that spacing variations as small as 2% should be adequate to minimize this problem.

If it is necessary to minimize bending mode vibrations in piping systems, observe the following design guidelines:

- Anchor the pipe to the structure at interval spacings that avoid natural frequencies in close proximity to forcing frequencies.

- Eliminate sharp elbows in the piping system—especially in the vicinity of pumps—since the fluid acceleration due to the change in direction at elbows helps transfer the energy of fluid pressure pulsations into bending mode vibrational energy in the pipe.
- Install flexible connections between the pump and piping at the inlet and discharge.

Example 22-2
Pipe Vibration with a Variable-Speed Centrifugal Pump

Problem: An end-suction centrifugal pump discharges clean water into a Class 53 ductile iron pipe 600 mm (24 in.) in diameter. The pump has a twin vane impeller that rotates at speeds between 560 and 800 rev/min. Design a piping support system free of bending mode resonance problems.

Solution: The rotating frequency of the pump shaft and impeller varies from $f_{min} = 560/60 = 9.3$ Hz to $f_{max} = 800/60 = 13.3$ Hz, and the fundamental blade passage frequency varies from $f_{min} = 2(9.3) = 18.6$ Hz to $f_{max} = 2(13.3) = 26.6$ Hz. The second and third harmonics (two and three times the fundamental, respectively) of the blade passage frequency vary as follows: $f_2 = 37 - 53$ Hz and $f_3 = 56 - 80$ Hz.

Support spacing requirements. Select the supporting spacing to avoid the resonance condition (within 50% as recommended previously). Because a variable-speed pump is used, virtually all frequencies above 18.6 Hz are possible driving frequencies or harmonics thereof. There is an available window between 13.3 and 18.6 Hz, but the window is not wide enough to meet the 50% clearance requirement. Because there is no practical method of completely avoiding the pipe resonances, select a support spacing that puts the fundamental pipe resonance above 100 Hz. The resonance problem would thus be moved outside of the frequency region where vibration problems are most common, thus reducing the amplitude of vibration because displacement is inversely proportional to the frequency.

Resonance frequencies. Compute these by using Equation 22-11, which requires the determination of I (from Equation 22-12 and Tables B-1 and B-2), E (from Tables 4-1 or A-10), w (from Tables B-l and B-2), γ_n (from Figure 22-11), and L.

SI Units

$$I = \frac{\pi}{64}\left[(0.655)^4 - (0.629)^4\right] = 0.00120 \text{ m}^4$$

$$E = 166 \times 10^6 \text{ kPa} = 1.655 \times 10^{11} \text{ N/m}^2$$

$$w = w_{water} + w_{pipe}$$

From Tables B-1 and A-8, for one meter of length:

w_{water}	$= (0.308\text{m}^2)(998.2 \text{ kg/m}^3)$ $(9.81 \text{ m/s}^2) =$	3020 *N*
w_{pipe}	$= (170 \text{ kg/m})(9.81 \text{ m/s}^2) =$	1670 *N*
w_{liner}	$= (2.38 \times 10^{-3} \text{ m})\pi(0.627 \text{ m}) \times$ $(2400 \text{ kg/m}^3)(9.81 \text{ m/s}^2) =$	110 *N*
w	per meter =	4800 *N*
γ_n	= 3.14	
Try L	= 3 m	

$$f_1 = \frac{(3.14)^2}{2\pi(3)^2}\sqrt{\frac{(1.655 \times 10^{11})(0.00120)(9.81)}{4800}} = 111 \text{ Hz}$$

$$f_2 = \left(\frac{\gamma_2}{\gamma_1}\right)^2 f_1 = \left(\frac{6.28}{3.14}\right)^2 \times 111 = 444 \text{ Hz}$$

U.S. Customary Units

$$I = \frac{\pi}{64}\left[(25.8)^4 - (24.9)^4\right] = 2880 \text{ in.}^4$$

$$E = 24 \times 10^6 \text{ lb/in}^2$$

$$w = w_{water} + w_{pipe}$$

From Tables B-2 and A-9, for one inch of length:

w_{water}	$= (3.32 \text{ ft}^2)(62.3 \text{ lb/ft}^3)$ $(1 \text{ ft})/12 =$	17.2 lb
w_{pipe}	$= (114 \text{ lb/ft})(1 \text{ ft})/12 =$	9.5 lb
w_{liner}	$= (3/32 \text{ in.})\pi(24.7 \text{ in.})(150 \text{ lb/ft}^3)$ $(1/1728) =$	0.7 lb
w	per in. = 27.4 lb	
γ_n	= 3.14	
Try L	= 118.1 in.	

$$f_1 = \frac{(3.14)^2}{2\pi(118)^2}\sqrt{\frac{(24 \times 10^6)(2880)(386)}{27.4}} = 111 \text{ Hz}$$

$$f_2 = \left(\frac{\gamma_2}{\gamma_1}\right)^2 f_1 = \left(\frac{6.28}{3.14}\right)^2 \times 111 = 444 \text{ Hz}$$

If these guidelines are followed, the worst situation that could result is a pipe system resonance excited by one of the higher harmonics of the pump. The displacement amplification may reach 20 or even 50, but the displacement of the pipe would also be reduced by the increase in vibration frequency and the decrease in energy available from the pump at higher harmonics. By increasing the natural frequency of the piping system from 15 Hz (the center of the available window between 13.3 and 18.6 Hz) to 112 Hz, the displacement amplitude is reduced by a factor of 15/112 or 0.134 if equal energy in all harmonics is assumed. If harmonics above f_3 are assumed to have less than 1/10 the energy of the fundamental (a safe assumption for centrifugal pumps), then, even with a displacement amplification of 50 at 112 Hz (assuming the worst condition), the amplitude of the pipe vibration would be less than 0.134 times 50 divided by 10. This displacement is less than 70% of the value that would be expected at 15 Hz if a spacing of 10 m (30 ft) had been selected to fit into the 13.3- to 18.6-Hz window.

Another solution to Example 22-2 would be to switch to a constant-speed pump. If the specified pump were changed to a triple vane impeller centrifugal pump driven at a constant speed of 1200 rev/min, the driving frequencies would be 20 Hz (the shaft speed), 40 Hz (2 times the shaft speed), 60 Hz (the pump frequency), 120 Hz, 180 Hz, and so on. A pipe spacing of 3 m (118 in.) would not be satisfactory because its resonance frequency (112 Hz) is close to the second harmonic of the blade passage frequency (120 Hz). With this pump, pipe resonance should ideally be about 90 Hz. Although the 90-Hz pipe resonance would provide only 30% clearance to the nearest driving frequencies, it would be preferred because the nearest driving frequencies are higher harmonics (which typically have less energy than the fundamental) and because the higher frequency would produce lower displacement amplitudes. Equation 22-11 can be rearranged to solve for L, which yields 3.33 m (10.9 ft) for 90 Hz.

The problems introduced by variable-speed pumps are demonstrated in Example 22-2. The variable-speed pump can also excite structural resonances of the building, a problem discussed in detail in Section 22-12.

Acoustic Resonances

Periodic pressure fluctuations caused by the pump can also excite acoustic resonances within the piping system. Just as there are an infinite number of bending mode resonances in a piping system, there are also an infinite number of acoustic resonances. An acoustic resonance is created when a discontinuity in the piping system (e.g., a termination, elbow, or branch) reflects a pressure wave. When a pressure wave propagating down the pipe reaches a discontinuity, a portion of the energy in the wave is reflected back toward the source while the rest of the energy in the wave continues down the pipe. If the length of the pipe and the speed of wave propagation in the fluid are such that the initial and reflected waves are in phase, the system operates at an acoustic resonance. In actuality, the pressure wave reflects back and forth at each discontinuity several times, creating what is called a "standing wave" in the pipe. The number of times a wave is reflected before it dissipates depends on the type of discontinuity (i.e., the strength of the reflection at each end) and the viscous losses in the system. Terminations or abrupt changes in area provide the strongest reflections. Reflections from small branches and 45° elbows are generally weak enough to be disregarded.

What occurs in the fluid when a system operates at an acoustic resonance is graphically illustrated in Figure 22-12. The instantaneous fluid pressure (with the movement in the pipe wall greatly exaggerated) is shown in a fixed section of pipe at four different times. At time $t = 0$, the pressure is unevenly distributed along the pipe in a sinusoidal manner with the peak

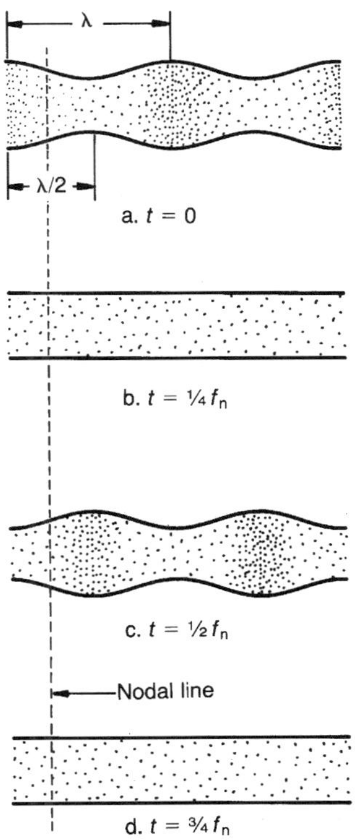

Figure 22-12. Internal pressure distribution at acoustic resonance.

pressure locations separated spatially by a distance, γ, the wavelength of the acoustic resonance in the fluid. At $t = f_n/4$, the pressure distribution and pipe shape are uniform. At $t = f_n/2$, the uneven distribution returns, except that the location that was at high pressure (at $t = 0$) is now a low-pressure zone, and vice versa. It is important to note that the average pressure in the pipe (averaged over a time interval of $1/f_n$ or its multiple) is the same at all points.

There are fixed locations along the pipe where the internal pressure does not oscillate about the average pressure. These points are called "nodes," and they also are separated by a distance, $\lambda/2$, along the pipe. Note that the nodes and antinodes (which are regions of high-pressure fluctuations) of acoustic resonances are at fixed locations that are determined by the pipe system geometry. They do not move with the fluid in the pipe.

The natural acoustic frequencies of a straight, uniform length of pipe are determined by the pipe length, the boundary conditions at each end, and the compressional wave propagation speed in the fluid (speed of sound). Depending on the impedance presented by the discontinuity, the acoustic wave reflected by a discontinuity may be reflected with or without a 180° phase reversal. In general, a pipe opening into a large volume with an abrupt termination creates a large reflection with a 180° phase shift. This termination is called an "open-end" condition. At the resonance frequency of the open-end condition, the system node is located at the termination. On the other hand, a pipe section that terminates with a blocked end cap, a tee connection, or a 90° elbow is said to have a "closed-end" condition, which provides no phase reversal for the reflected wave. Consequently, the dynamic pressure is greatest at the termination and the nodes occur at other positions along the pipe.

For straight runs of uniform pipe, the acoustic natural frequencies for pipes open at both ends or closed at both ends can be computed as follows:

$$f_n = \frac{nc}{2L} \tag{22-14}$$

where n is the resonance order ($n = 1, 2, 3, \ldots$), c is the wave propagation speed of fluid, and L is the length of the column of fluid between discontinuities where a reflection is anticipated.

For pipes open at one end and closed at the other end,

$$f_n = (2n-1)\frac{c}{4L} \tag{22-15}$$

Note that L has nothing to do with the spacing of pipe supports or the spacing of couplings.

In addition to individual sections of water columns, it is also possible to encounter acoustic resonances that incorporate one or more individual water columns, nonuniform sections of pipe, and other complexities. The equations for computing these frequencies are very complex—well beyond the scope of this text (see *Design of Piping Systems* [26]).

Speed of Sound

The speed of sound in liquids is

$$c_o = \sqrt{\frac{\gamma B}{\rho}} \tag{22-16}$$

where c_o is speed of sound in a free liquid in meters per second (feet per second), γ is the ratio of specific heats (1.0 for fresh water and 1.01 for sea water), B is the isothermal bulk modulus of the fluid in pascals (pounds force per square foot), and ρ is the density of the fluid in kilograms per cubic meter (slugs per cubic foot). In general, these quantities vary somewhat with temperature and pressure in a complex manner as shown in Tables A-8 and A-9. γ is further defined as c_p/c_v, where c_p is specific heat capacity of the fluid at constant pressure and c_v is the specific heat capacity at constant volume. In SI units, an empirical expression for the speed of sound in fresh water is

$$c_o = 1403 + 5.02T - 0.055T^2 + 0.0003T^3 + 0.00173P \tag{22-17a}$$

where c_o is the speed of sound in meters per second, T is the temperature in degrees Celsius, and P is the gauge pressure in kilopascals.

In U.S. customary units,

$$c_o = 4248 + 13.23T - 0.072T^2 + 0.00016T^3 + 0.0392P \tag{22-17b}$$

where c_o is speed of sound in feet per second, T is the temperature in degrees Fahrenheit, and P is the gauge pressure in pounds force per square inch.

The speed of sound in salt water is slightly higher (by approximately 3%) than it is in fresh water due to the effects of salinity. The speed of sound can be lowered dramatically by air bubbles in the fluid. As long as the concentration of air is less than 0.01% by volume, the effect is insignificant. However, when the air concentration reaches 0.1%, the speed of sound is reduced by nearly 50% (see Streeter and Wylie [27] for more information on the effects of air bubbles).

In addition to its dependence on temperature and pressure, the speed of sound is also influenced by the pipe. The finite mass and compliance of the pipe wall decreases the speed of sound as the pipe wall thickness decreases and the pipe diameter increases. The expression for the speed of sound within a pipe is

$$c = \frac{c_0}{1 + B(ID/Ee)} \tag{22-18}$$

where c_0 is given by Equation 22-17, B is the bulk modulus of the fluid, ID is internal pipe diameter, E is the modulus of elasticity of the pipe, and e is the thickness of the pipe wall. The speed of sound in fresh water within several Class 53 ductile iron pipes is shown in Figure 22-13 as a function of water temperature.

Acoustic resonances do not necessarily cause visible or perceptible vibration in the wall of the pipe, but they can cause very high dynamic pressures inside the pipe at the antinodes. It should be emphasized that the pressure rating for piping is based on internal static (steady) pressures. Safe limits for dynamic pressures are typically less than one-half of the static limits because of pipe fatigue. For most piping systems, this limits safe dynamic pressures to less than 690 kPa (100 lb/in.^{2}) peak to peak.

Acoustic resonance may also increase the noise level radiated by the pipe if the frequency is high enough. Depending on the ambient noise level in the plant, and if the frequency is lower than 250 Hz (as it would be for most pumping station systems), radiated noise from an acoustic resonance should not be perceptible.

Suppression Techniques

Once a potential or existing piping vibration or acoustic resonance problem is discovered, one is usually

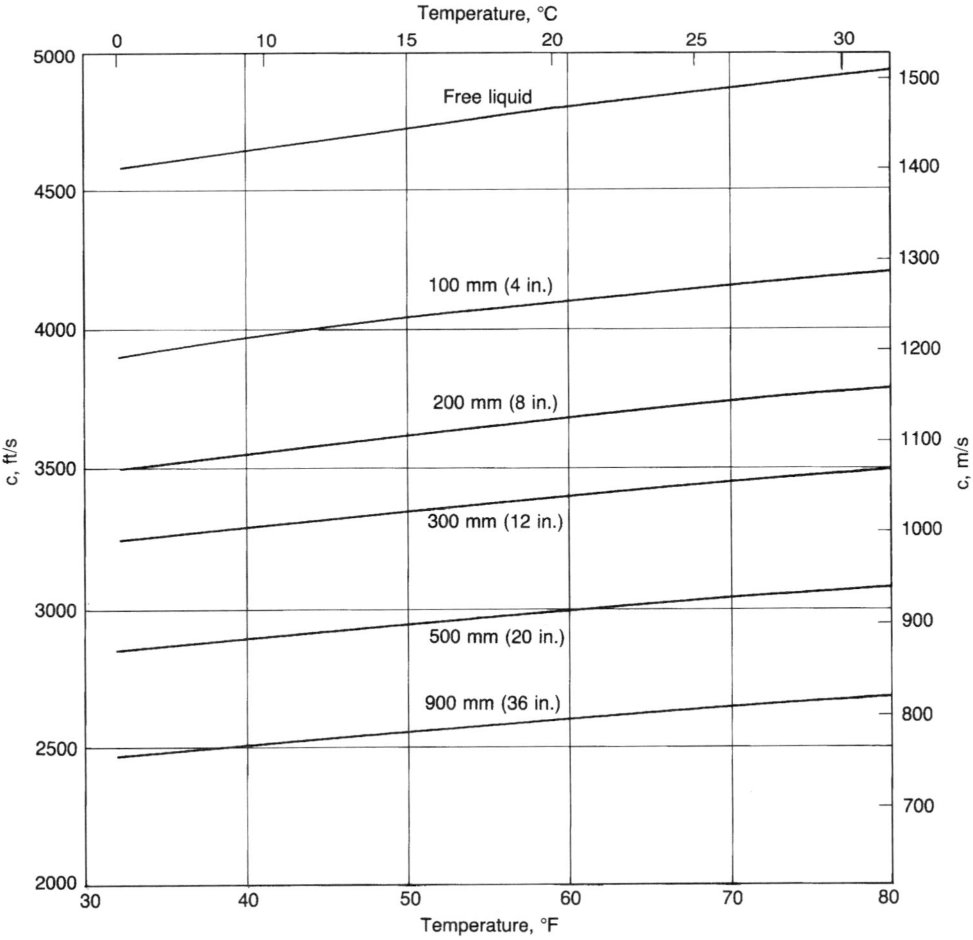

Figure 22-13. Speed of sound in fresh water at atmospheric pressure in Class 53 ductile iron pipe.

interested in finding a remedy—or at least a method of improving the situation with as little effort and expense as possible. The solution, of course, requires knowledge of the nature of the problem as well as some specifics concerning the frequencies and levels of vibration. For existing problems, this information is best obtained by direct measurement with an accelerometer and a spectrum analyzer. These measurements usually require the expertise of a qualified vibration consultant. Once the specifics of the problem are identified, a proper suppression technique can usually be applied.

Detuning

Because most vibration problems are associated with the mechanical resonance of one or more elements in the system, detuning is the simplest and most effective method of solving a piping vibration problem. Detuning merely implies changing either the driving frequency of the pump or the natural frequency of the piping system. Prior to detuning any problem system, it is important to determine the exact driving frequency and natural frequency of the installed piping system, particularly if the anticipated solution is to make a modest change in pump speed. Without this information, it is possible to make the situation even worse if the new pump speed turns out to be closer to the piping's natural frequency than the original pump speed.

Flexible Connections

A flexible connection between the pump and the piping can be an effective means of reducing pipe vibration if the pump is vibration-isolated from the structure and if the flexible connection is truly flexible. When the pump and piping are not isolated from the structure, it is possible for this energy to get back into the piping via the common structure. As shown in Figure 22-14, the most effective flexible connectors are made from molded neoprene rubber. A series of "accordion"

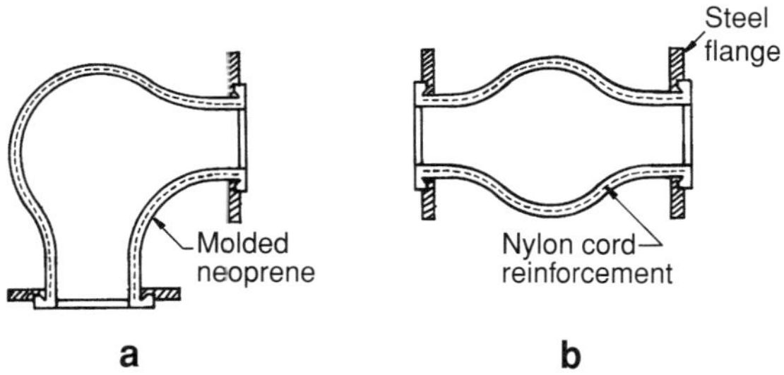

Figure 22-14. Vibration isolators for pipe. (a) Elbow coupling; (b) straight coupling.

corrugations is even more effective. In addition to isolating the piping from the structural vibrations of the pump, this particular design also effectively reduces pulsations in the fluid by virtue of the expansion capabilities of the rubber walls. The use of neoprene rubber is limited to temperatures below 100°C (212°F) and pressures below 1000 kPa (150 lb/in.^{2}).

The conduit for the motor should likewise be flexible to prevent transmitting vibration into the structure via the electrical conduit.

Resilient Supports

Resilient supports for piping systems are used in some industries to provide for expansion and contraction of the piping system as well as to reduce structure-borne noise from the pipe. In these devices a resilient element (e.g., neoprene rubber or fiberglass) is placed between the pipe wall and the pipe clamp, which allows the pipe to flex slightly and independently of the structure. Theoretically, these clamps also add some additional damping to the bending modes of pipe vibration, but they cannot reduce pressure pulsations in the fluid. Because structure-borne noise is not usually a big problem in pumping stations, the use of these devices is probably not warranted.

Dynamic Absorbers

A dynamic absorber is a passive mass attached to a vibrating element or system with a resilient component (e.g., a spring). The auxiliary mass and spring are attached to the vibrating element and are designed (or "tuned") to vibrate out of phase with the primary system. This results in significant vibration, which is significantly reduced at the resonance frequency. The tuning of the auxiliary mass and spring is critical because these auxiliary elements create a two-degree-of-freedom system that may cause an increase of vibration level at other nearby frequencies. The use of a dynamic absorber is applicable only for a system operating at or near resonance and should be attempted only with the assistance of a qualified vibration consultant.

Expansion Chambers

Expansion chambers are simply enlarged sections of pipe with abrupt discontinuities at each end, as shown in Figure 22-15. The expansion chamber provides a loss of dynamic pressure (i.e., reduced pressure pulsations) primarily through the interaction of reflections within the chamber. The dynamic pressure losses occur at all frequencies except at 0 Hz and at the reso-

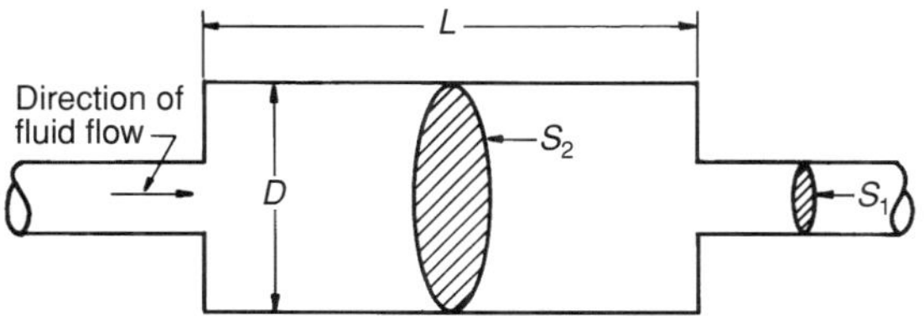

Figure 22-15. Expansion chamber.

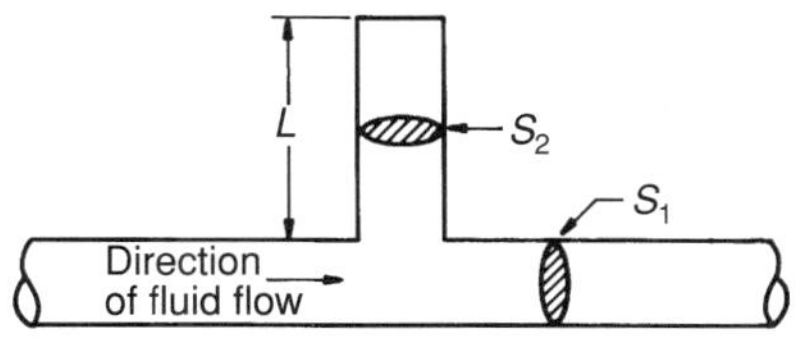

Figure 22-16. Side branch resonator.

nance frequencies of the expansion chamber (when the length, L, equals an integer multiple of one-half of the wavelength) where no losses occur. The expansion chamber cannot amplify the fluid pulsations in the downstream pipe.

The effectiveness of an expansion chamber can be described by the ratio of the outlet to inlet dynamic pressures:

$$P_{out}/P_{in} = \frac{1}{\{1 + [0.25(S_2/S_1 - S_1/S_2)^2]\sin^2(2\pi fL/c)\}} \qquad (22\text{-}19)$$

where S_1 and S_2 represent the cross-sectional area of the inlet pipe and the expansion chamber, respectively. Note that from this expression there is no pulsation reduction whenever the argument of the $\sin^2$ function is an integer multiple of π. The maximum pulsation reduction occurs at the frequency where the argument of the $\sin^2$ function is equal to $\pi/2$, or

$$f = (2n - 1)c/4L \qquad (22\text{-}20)$$

Thus, maximum reduction occurs when the length of the expansion chamber equals one-fourth of the wavelength, three-fourths of the wavelength, five-fourths of the wavelength, and so on.

In most pumping applications, the concern is with low-frequency performance where the expansion chamber length is only a fraction of the wavelength of the pump frequency. Under these circumstances, the expansion chamber is designed to ensure that the ratio S_2/S_1 and the pump frequency are high enough to produce the necessary pulsation reduction.

Tuned Resonators

Tuned resonators are sections of pipe that are added to the system in order to remove dynamic pressures in a very narrow frequency band. The simple example of a side branch resonator (a closed-end tube attached to the main pipe) is shown in Figure 22-16. If the effective length of the side branch resonator is exactly equal to one-fourth of the wavelength of the pump frequency, it can reduce the dynamic pressure in the downstream piping dramatically—by as much as 99%. It is important to realize that this device works only at the following particular frequencies:

$$f_n = (2_n - 1)c/2L \qquad (22\text{-}21)$$

where $n = 1, 2, 3, \ldots$. The length of the resonator must equal one-half of the wavelength, three-halves of the wavelength, and so on. These lengths differ from those of the expansion chamber discussed previously, because the side branch resonator is closed at one end, whereas the expansion chamber is open at both ends. The sound wave in the expansion chamber can propagate the distance L in a single pass. But the sound wave in the side branch resonator travels the distance L in one direction, is reflected and travels back again, and gives rise to the different factors ($2L$ vs. $4L$) in Equations 22-20 and 22-21.

At other frequencies the resonator is transparent to the system; that is, it acts as though it were absent or "invisible." Therefore, consider this device only for single-speed pumps moving fluids with temperatures that remain constant. Side branch resonators are not economically feasible for low-speed pumps because the resonator length becomes very long.

Pulsation Dampers

A pulsation damper reduces the dynamic pressure fluctuations within a piping system, and it can also reduce the bending mode vibration levels of the piping system if the pipe vibration is induced by the pulsating fluid. Two types are commonly used. The most common pulsation damper is a fluid-filled spherical vessel mounted in line with the piping at the discharge of the pump. A sphere is commonly utilized to achieve maximum volume and maximum pressure rating with minimum weight and cost. The general design concept is the same as that of the expansion chamber discussed above. Typically, pulsation dampers reduce dynamic pressures by 75% and more.

The other type of pulsation damper is the air chamber (see the hydropneumatic tank in Chapter 7), which also functions as an accumulator. This device is essentially a pressurized expansion tank (partially filled with gas) connected to the piping near the pump discharge. The accumulator adds and accepts fluid from the piping through an orifice as the dynamic pressure fluctuates. The accumulator is not an in-line device; therefore, its effectiveness at reducing high-frequency (50 Hz and above) dynamic pressures in the downstream piping is limited. Neither the accumulator nor the pulsation damper has a sufficient volume of gas to control water hammer in most instances. Pulsation control devices should be considered for both the suction and discharge flow lines because these units protect the pump as well as the piping; usually, they also improve pump performance.

Example 22-3
Dynamic Pressure from a Plunger Pump

Problem: A reciprocating (single-acting plunger duplex) pump delivers fresh water at 3.15 L/s at 2413 kPa (50 gal/min at 350 lb/in.^{2}) working pressure from a 5°C (50°F) reservoir at 400 rev/min into a Class 53 ductile iron pipe 300 mm (12 in.) in diameter that terminates in a large holding tank. Assume that the peak dynamic pressure generated by the pump is 25% of the static head (50% peak to peak). Compute the acoustic resonance frequencies and determine if the system is in danger of fatigue fracture if the pipe is relatively straight, has a total length of 64 m (210 ft), and has a yield strength of 290,000 kPa (42,000 lb/in.^{2}).

Solution: The speed of sound in the pipe can be computed to be 1007 m/s (3303 ft/s) from Equations 22-17 and 22-18. If the effect of pressure is ignored (because it is so small), the speed of sound in the pipe can also be estimated from Figure 22-13. The driving frequency of the pump is $f_d = (400)\,(2)/60 = 13.3$ Hz, and the harmonics of the driving frequency are 26.6 Hz, 40 Hz, etc. The pipe is open at one end, and the pump itself closes the other end. From Equation 22-15 [$f_n = (2n - 1)\,c/4L$], the acoustic resonances are as follows:

SI Units	**U.S. Customary Units**
$f_1 = (2 \times 1 - 1)1007/(4 \times 64) = 3.93$ Hz	$f_1 = (2 \times 1 - 1)3303/(4 \times 210) = 3.93$ Hz

The second- and third-order acoustic resonances are

$f_2 = (2 \times 2 - 1)1007/(4 \times 64) = 11.8$ Hz	$f_2 = (2 \times 2 - 1)3303/(4 \times 210) = 11.8$ Hz
$f_3 = (2 \times 3 - 1)1007/(4 \times 64) = 19.7$ Hz	$f_3 = (2 \times 3 - 1)3303/(4 \times 210) = 19.7$ Hz

The second-order acoustic resonance (11.8 Hz) is close (within about 10%) of the fundamental driving frequency of the pump ($f_o = 0.89 f_d$), which could excite the first-order acoustic resonance within the pipe. A pulsation damper is required in this installation to reduce the dynamic pressure, but the pulsation damper will not change the resonance match between the driving frequency of the pump and the water column in the pipe.

Effect of damping. If the pulsation damper reduces the dynamic pressures in the discharge pipe by, for example, 75% and the pressure amplification for the acoustic resonance is estimated to be about 5 (from Figure 22-6, for 2 to 10% of critical damping), the maximum dynamic pressure can be computed as working pressure times 25% times (1 – pulsation damping reduction) times the acoustic resonance amplification factor, or

$(2413)(0.25)\,(1 - 0.75)(5) = 754$ kPa	$(350)(0.25)(1 - 0.75)(5) = 109$ lb/in.^{2}

These pressures meet the 50% of rated value criterion, but they are higher than the 690 kPa (100 lb/in.^{2}) guideline. If the damping ratio were as low as 0.02, amplification factors could reach 20, which could bring the dynamic pressures inside the pipe to as high as 3020 kPa (440 lb/in.^{2}), and the total (working plus dynamic) pressure would be 5440 kPa (790 lb/in.^{2}). From Equation 4-2 ($s = PD/2e$), the hoop tensile stress in the pipe at peak pressure would be

$s = \dfrac{5440 \times 0.335}{2 \times 0.0102} = 89{,}300$ kPa	$s = \dfrac{790 \times 13.2}{2 \times 0.40} = 13{,}000$ lb/in.^{2}

This seems safe enough, but the maximum yield strength of 290,000 kPa (42,000 lb/in.^{2}) is applicable only to static stress levels. Dynamic stress levels must be substantially lower to avoid fatigue failure.

Changing operating speed. In addition to the pulsation damper, the operating speed of the pump should be changed to avoid operation near the acoustic resonance frequencies and, thus, reduce the internal dynamic pressure. A suitable choice might be a smaller duplex pump running at 480 rev/min, which would yield driving frequencies of 16 Hz, 32 Hz, etc., and would result in an estimated pressure amplification of less than about 2.0 for all values of the damping ratio (see Figure 22-6) for $f_d/f_o = 16/11.8 = 1.356$. The maximum dynamic pressure at the driving frequency of the pump would then be (2420)(0.25)(1 – 0.75)(2) = 302 kPa (44 lb/in.^{2}), which is safe by both criteria.

22-11. Vibration of Drive Shafts

Drive shaft vibration is a complicated phenomenon composed of two independent forms of vibration—translational and torsional. The approximation methods presented here are intended to provide only (1) a reasonable idea of safety margins and (2) sufficient background for pumping station designers to judge properly the results presented by the shaft, engine, or (usually) pump manufacturer, who should be responsible for the design of the entire rotating system (see Appendix C).

Translational Vibration

Translational vibration (which is sometimes called "lateral vibration") involves the oscillatory motion of the drive shaft about its ideal "at rest" centerline. The shaft has natural frequencies and assumes related mode shapes just as piping systems do, and all of the equations and analyses on piping vibration in Section 22-10 also apply to drive shafts.

Drive shafts are much more sensitive to vibration problems than are piping systems, because the drive shaft rotates and any unbalance in the drive shaft or misalignment in the coupling causes the shaft to vibrate or deflect laterally, which results in "shaft whip." This vibration is very difficult to measure directly, but it can be measured indirectly via sensors at the shaft support bearings. Avoiding rotational speeds that match or excite the natural or resonance frequencies of the translational vibration of the drive shaft is imperative. These speeds are called "critical speeds," and they are determined entirely by the physical properties of the shaft.

If the drive shaft is assumed to be a uniform, straight, steel tube with a circular cross-section (or a solid cylinder if $ID = 0$), Equation 22-11 can be rearranged into a somewhat simpler form for a quick evaluation of the lowest natural frequency.

$$f_1 = \frac{\beta}{L^2}\sqrt{(OD)^2 + (ID)^2}$$
$$\beta = \frac{\gamma_1^2}{8\pi}\sqrt{\frac{E}{\rho}} \qquad (22\text{-}22)$$

where f_1 is the first-order shaft resonance in hertz, OD is the outer diameter of the shaft in millimeters (inches), ID is the inner diameter of the shaft in millimeters (inches), L is the effective length of the shaft in millimeters (inches), and β is 2.03×10^6 mm/s in SI units for universal (pinned) joint couplings or 4.6×10^6 mm/s for direct-coupled (fixed) shafts. In U.S. customary units, β is 7.9×10^4 in./s for pinned joint couplings and 1.79×10^6 in./s for fixed shafts. A drive shaft also has other modes of lateral vibration at higher frequencies, but these are usually not problem frequencies because the rotational speed of shafts is usually below the first resonance.

When using Equation 22-22 to compute the resonance frequencies of the drive shaft, the length, L, of the shaft should be measured from the center of the universal joints on both the driving and the driven ends (see Figure 22-17). Steady bearings may be used as intermediate supports to reduce L and increase the resonance frequency out of the region of concern. To avoid operating at a critical speed, the drive shaft should be designed to ensure that the highest rotational shaft speed is less than 75% of the lowest natural frequency. Furthermore, rotational speeds that are nearly equal to one-half of the lowest bending mode frequency of the shaft should be avoided because secondary moments from universal joints produce an excitation frequency at twice the shaft speed. Combining these two criteria gives two acceptable ranges for the lowest shaft frequency: (1) 1.4 to 1.6 times the

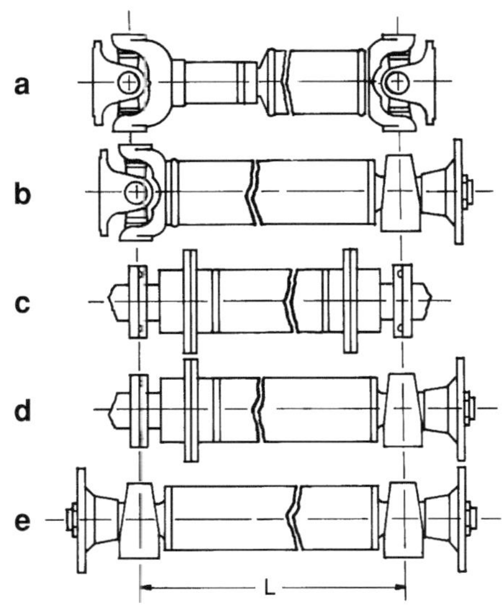

Figure 22-17. Typical drive shaft couplings. (a) Universal joint drive shaft. Vibration is accentuated if the two shafts are not exactly coaxial; (b) universal joint coupling shaft with right end flanged to connect with another shaft; (c) solid coupled shaft; (d) solid coupled coupling shaft; (e) middle shaft (with two support bearings) in a series of three shafts.

shaft rotational speed or (2) at least 2.5 times the shaft rotational speed.

Although the diameter of a drive shaft can be determined by the length of the shaft and the maximum operating speed, it is good design practice to limit the length to no more than 30 times the shaft diameter. Extremely long shafting must be divided into sections supported by steady bearings mounted on cross beams at the joints. If multiple sections are required, each section should, if possible, be of approximately equal length. The resonance frequencies of, say, four sections of nearly equal length are almost the same, but, if the lengths are substantially different, the additional resonance frequencies complicate the problem.

Support of Steady Bearings

When a drive shaft is divided into more than one section, the intermediate bearings must be supported adequately to resist not only the static load but also the dynamic load induced by the shaft rotation. Auxiliary steel members attached directly to the building structure usually support the intermediate bearings. A common rule of thumb is to make sure that the lowest natural frequency of the support structure is at least four times the shaft rotational speed. If the lateral support of the steady bearings does not meet this criterion, the boundary conditions will not approximate the "pinned" or "fixed" conditions and the first resonance frequency of the shafts will be lower than the calculated value.

The natural frequency of a single span support beam is given by Equation 22-11. The values of γ_n for clamped–clamped and pinned–pinned beams are given in Table 22-8 and, for modes 1 and 2, in Figure 22-11.

The length of the beam is often such that its lowest natural frequency is less than the recommended minimum of four times the shaft rotational speed. If so, it may be necessary to provide intermediate supports for the beam. If the main beam is supported at the center of the span, the same equation can be used (with modified values of γ_n) to compute the new resonance frequencies. Blevins [7] gives modified values of γ_n for various boundary conditions and positions of the intermediate support. In the special case of a center support with pinned boundary conditions (no displacement, but unrestrained rotation), the modified values of γ_n are given in Table 22-9. Most real-world boundary conditions for "rigidly" attached beams are such that some displacement and rotation occur at the ends of the beam no matter how they are fastened to the structure. For this reason, a more suitable choice for the frequency coefficient, γ_n, is a number somewhere between the theoretical values for the clamped–clamped and pinned–pinned boundaries.

Table 22-8. Frequency Coefficients for Bending Modes of Straight Beams

Mode (n)	Clamped–clamped (γ_n)	Pinned–pinned (γ_n)
1	4.73	3.14
2	7.85	6.28
3	11.00	9.42
4	14.14	12.57

Table 22-9. Frequency Coefficients for Bending Modes of Straight Beams with Center Supports[a]

Mode (n)	Clamped–clamped (γ_n)	Pinned–pinned (γ_n)
1	7.8	6.4
2	9.5	7.8
3	14.2	12.6
4	15.7	14.2

[a] The distance L is the full length of the beam—not the distance from the end to the center support.

Torsional Vibration

Torsional vibration problems occur less frequently than translational vibration problems, but they can be just as troublesome and sometimes catastrophic. Because they are not detected by our senses until failure occurs, it is important to select equipment components that preclude this possibility.

Torsional vibration can be magnified significantly if a torsional exciting frequency is close to the torsional natural frequency. The disturbing force in Figure 22-18 is the dynamic torque that oscillates about the average torque. At all frequencies less than 1.3 times the natural frequency, some magnification of the disturbance occurs, whereas at greater frequencies the disturbing force is damped. The maximum magnification occurs at resonance, and the value 3.3 shown is an arbitrary choice that is nevertheless typical. The maximum magnification can be calculated, but it depends on the damping factor, which is difficult to ascertain. The simplest and best procedure for designing shafts is to position the torsional natural frequency well outside any exciting frequencies. Typical exciting frequencies include (1) the shaft rotational speed; (2) the pump vane passage frequency and its harmonics; and (3) the universal joints, which generate a frequency twice that of the shaft speed.

To avoid torsional resonances, keep the fundamental exciting frequencies less than 50% of torsional natural frequency. At times, it is necessary to run the system at speeds where harmonic exciting frequencies are above the natural frequency. This mode of operation is not a problem (1) if the exciting frequencies are more than 1.5 times the natural frequency and (2) if the pass through the natural frequency range is accomplished quickly (which is normally true in electric motor pumping installations). If the exciting frequency is at least 1.5 times the torsional natural frequency, the effective disturbing system forces are less than the driver input disturbing forces. If the exciting frequency is less than 0.5 times the torsional natural frequency, the magnification factor will not exceed 1.3. Steel drive shafts typically have a minimum torsional elastic limit of 235 kPa (34,000 lb/in.^{2}) or an endurance limit (infinite life) of 117 kPa (17,000 lb/in.^{2}). The 1.3 stress magnification factor combined with the endurance limit means that a maximum allowable design torsional stress of $117/1.3 = 90$ kPa (13,000 lb/in.^{2}) ensures safety if torsional exciting frequencies are less than 0.5 or more than 1.5 times the torsional natural (resonance) frequencies.

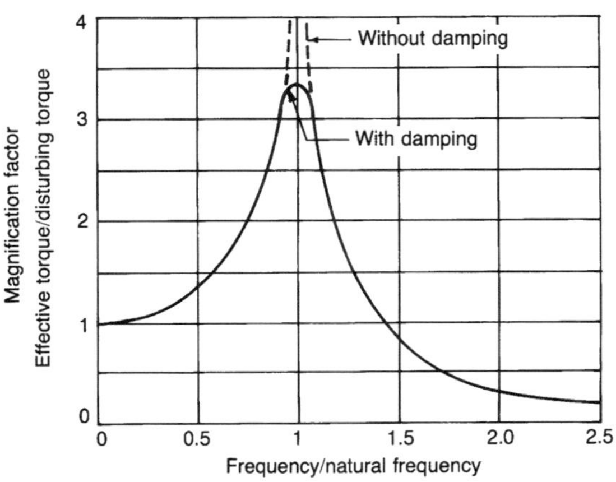

Figure 22-18. Magnification of a disturbing force in torsional vibration.

In the simplest of systems (a pump driven by an electric motor with intermediate shafting), the torsional natural frequency of the system can be computed in SI units as follows:

$$f_o = \frac{1}{2\pi}\sqrt{\frac{K(J_m + J_p)}{J_m J_p}} \qquad (22\text{-}23)$$

where f_o is the torsional resonance frequency in hertz, K is the torsional rigidity of the shaft in newton-meters per radian, J_m is the mass moment of inertia of the motor in meters squared times kilograms, and J_p is the mass moment of inertia of the pump in meters squared times kilograms.

In U.S. customary units, K has units of pounds force-inches per radian and J_m and J_p have units of inches squared times pounds mass. The mass moment of inertia of a motor rotor is approximated by that of a cylinder of equal weight and diameter in which

$$J = \frac{D^2 M}{8} = \frac{D^2 W}{8g} \qquad (22\text{-}24)$$

where D is the diameter in meters (inches), M is the mass in kilograms (pounds mass), and W is the weight in newtons (pounds force). Approximating the mass moment of inertia of the impeller and the water within the casing is neither easily done nor likely to be accurate, so obtain it from the equipment manufacturer. Accurate drive shaft torsional rigidity values can be obtained from computer programs (usually proprietary) or calculated directly by analyzing the shaft in sections and combining the rigidity constant, K, for each individual section to obtain the torsional rigidity of the entire shaft. The torsional rigidity of the stubs (the connecting link between two shafts), however, depends on the design—which is different for each manufacturer—so the gross torsional rigidity of shaft and stub must also be obtained from the maker.

The torsional rigidity of a uniform, hollow, circular shaft of length L is expressed as

$$K = \frac{\pi G}{32L}\left[(OD)^4 - (ID)^4\right] \qquad (22\text{-}25)$$

where G is the modulus of elasticity of steel in shear, which is 82.7×10^9 Pa (12×10^6 lb/in.^{2}), and linear dimensions are in meters (inches). For solid shafting, use $ID = 0$. If a shaft is made up of two or more sections of shafting with different physical properties, the torsional rigidity of the entire shaft can be computed from the torsional rigidity of the individual sections as

$$K = \frac{1}{\Sigma_{i=1}^{n}(1/K_i)} \qquad (22\text{-}26)$$

Example 22-4
Torsional Vibration in a Shaft

Problem: A two-vane, end-suction centrifugal pump is driven by a 75-kW (100-hp), variable-speed, direct-drive electric motor at speeds varying from 840 to 1200 rev/min (14 to 20 Hz). The power shafting is a universal joint, 3.353-m (132-in.) drive shaft (see Figure 22-17a) with 2.67 m (105 in.) of 101.6-mm (4-in.) OD tubing and a wall thickness of 2.108 mm (0.083 in.). The manufacturer gives the mass moment of inertia as $J_p = 1.729$ kg · m^2 (15.3 lbm · in.^{2}) and $J_m = 1.345$ kg · m^2 (11.9 lbm · in.^{2}). The torsional rigidity of the drive shaft stubs (excluding the tubing between the stubs) is given as 31,400 N · m/rad (277,802 lb · in./rad).

Evaluate the translational and torsional resonances of the shafting and determine if the system is within acceptable guidelines. If intermediate support bearings are required, size the structural members within recommended guidelines.

Solution:

Lateral resonance. The rotating frequencies of the shaft are 14–20 Hz (fundamental) and 28–40 Hz (second harmonic and pump vane passage frequency). From Equation 22-22, the first-order lateral resonance of the shaft is

SI Units

$$f_1 = \frac{2.03 \times 10^6}{(3353)^2}\sqrt{(101.6)^2 + (97.384)^2}$$
$$= 25.4 \text{ Hz}$$

U.S. Customary Units

$$f_1 = \frac{7.99 \times 10^4}{(132)^2}\sqrt{(4)^2 + (3.834)^2}$$
$$= 25.4 \text{ Hz}$$

The maximum shaft speed is 20 Hz, which is 78% of the fundamental resonance of the shaft. Ideally, the shaft speed should not exceed 75% of the first resonance, so this shaft speed is a marginal condition.

Torsional resonance. To compute the torsional resonance frequency, first determine the shaft rigidity, K_s, from Equation 22-25.

$$K_{\text{tubing}} = \frac{\pi(82.7 \times 10^9)\left[(0.1016)^4 - (0.097384)^4\right]}{32(2.67)}$$
$$= 50{,}526 \text{ N} \cdot \text{m/rad}$$

$$K_{\text{tubing}} = \frac{\pi(12.0 \times 10^6)\left[(4)^4 - (3.834)^4\right]}{32(105)}$$
$$= 447{,}935 \text{ lb} \cdot \text{in./rad}$$

From Equation 22-26:

$$K_{\text{shafting}} = \left[(31{,}400)^{-1} + (50{,}526)^{-1}\right]^{-1}$$
$$= 19{,}368 \text{ N} \cdot \text{m/rad}$$

$$K_{\text{shafting}} = \left[(277{,}802)^{-1} + (447{,}935)^{-1}\right]^{-1}$$
$$= 171{,}463 \text{ lb} \cdot \text{in./rad}$$

Determine the torsional resonance frequency from Equation 22-23:

SI Units

$$f_o = \frac{1}{2\pi}\sqrt{\frac{(19{,}368)(1.729 + 1.345)}{(1.729)(1.345)}}$$
$$= 25.4 \text{ Hz}$$

U.S. Customary Units

$$f_o = \frac{1}{2\pi}\sqrt{\frac{(171{,}463)(15.3 + 11.9)}{(15.3)(11.9)}}$$
$$= 25.4 \text{ Hz}$$

This is below the recommended minimum of 1.5 times 20 Hz (the highest shaft speed) and is, thus, unacceptable.

Increasing the shaft OD to 127 mm (5 in.) and the ID to 120 mm (4.72 in.) would increase the first-order lateral resonance to 31.6 Hz and the torsional resonance to 29.7 Hz. This modification would nearly meet the torsional criteria, but the lateral resonance may be excited by the second harmonic of the shaft speed at 948 rev/min (or 15.8 Hz). An alternative solution would be to retain the original shaft diameters and split the shaft into two equal sections with an intermediate support bearing. The fundamental lateral shaft resonance then becomes 101.8 Hz, which is acceptable, because $f_1/f_d = 5.09$. The new torsional spring constant of each section is 101,052 N · m/rad, which yields a total spring constant of 23,956 N · m/rad and a torsional resonance of 58.2 Hz, which is also acceptable.

If the intermediate shaft bearing is supported by a section of steel angle 20.3 cm × 20.3 cm × 1.9 cm (8 in. × 8 in. × $^3/_4$ in.) thick that weighs 568 N/m (38.9 lb/ft), has a moment of inertia about both axes of 2.9×10^{-5} m^4 (69.7 in.^{4}), and spans 6.10 m (20 ft) with fully restrained end points, the fundamental resonance frequency of this beam is, from Equation 22-11 and Figure 22-11,

$$f_1 = \frac{(4.73)^2}{2\pi(610)^2}\sqrt{\frac{(2 \times 10^{11})(2.9 \times 10^{-5})(9.81)}{568}}$$
$$= 30.3 \text{ Hz}$$

$$f_1 = \frac{(4.73)^2}{2\pi(240)^2}\sqrt{\frac{(29 \times 10^{6})(69.7)(386)}{(38.9/12)}}$$
$$= 30.3 \text{ Hz}$$

The computed resonance frequency of the beam is insufficient to meet the conservative criterion of four times the maximum operating speed, but if a center support (with pinned boundary conditions) were added to this beam, the lowest beam frequency would be increased by a factor of $(7.85/4.73)^2 = 2.75$. The fundamental beam frequency would be increased to 2.75 × 30.4 = 83.6 Hz, which is 4.18 times the maximum operating speed of 20 Hz and does meet the recommended criterion. If the bearing were to be located at the center of the main beam (where the center support beam is attached), this problem would not arise because the vibrating force would be applied at a node of this first beam resonance, and it would not be excited.

22-12. Vibration of Structures

Pumping station equipment can cause structures to vibrate at levels exceeding normally acceptable criteria [23], particularly if the equipment is operating at a speed that matches one of the low-order resonance frequencies of the structure. In the design phase, it is too difficult to predict accurately the natural frequencies of supporting structures (building floor slabs, pedestals, etc.), but there are some simple calculations that can be made to determine whether a potential problem exists.

The lowest natural frequency of a supporting structure can be estimated from basic information concerning the static load-carrying capability of the structure. This information is often available directly from the structural engineer. Assuming that the structural member undergoes a static deflection directly proportional to the dead load of the equipment, the dynamic behavior of the structural member is similar to that of a steel spring (for small deflections). If the maximum static deflection of the structure (due to the weight of the equipment), d, is expressed in millimeters, the natural frequency of the structure in hertz is approximately

$$f_o = 15.8\sqrt{1/d} \tag{22-27a}$$

If d is expressed in inches,

$$f_o = 3.13\sqrt{1/d} \tag{22-27b}$$

This expression is approximately valid (±25%) only when $d < L/500$, where L is the smallest cross dimension (width or height) of the structure, but this condition should be satisfied by most pumping station buildings. Although the correlation between the vibration of a point mass on a spring and the distributed mass of a concrete floor may seem unreasonable, the two systems vibrate at nearly the same frequency, because the use of deflection in Equation 22-27 tends to compensate for the differences between the two systems.

Much more accurate determinations of structural resonance frequencies can be obtained by modal analysis and other more elaborate techniques, but these methods are usually beyond the scope of most pumping station projects. Determining the maximum structural deflection caused by the static weight of the equipment is sometimes complicated, but approximations that provide an upper limit for d are usually adequate. Practical solutions lie in the close cooperation between the project leader and the structural engineer, which illustrates why a project leader must be much more than an administrator who assigns tasks and responsibilities to others.

Ideally, vibrating equipment should not operate at a rotational speed near the fundamental frequency of the supporting structure. Although problems are unlikely if the rotating speed is 5% greater than or 5% less than the resonance frequency, a 25% safety margin on each side is recommended.

Natural frequencies of a variety of structural members are given by Blevins [7]. In general, the natural frequency of any structure is proportional to the square root of the composite moment of inertia, I, divided by the total mass of the system.

$$f_o \propto \sqrt{I/M} \tag{22-28}$$

Any efforts to change the natural frequency of an existing system, therefore, should be devoted to these two parameters. Notice that doubling the mass of the system (equipment plus structure) reduces the natural frequency by 29% $(1 - \sqrt{0.5})$ if it is assumed that mass can be doubled with no change in the moment of inertia. However, structural modifications usually affect both M and I, so the modifications should be evaluated carefully prior to making the change.

Example 22-5
Vibration of a Floor

Problem: A 300-kVA (400-hp) variable-speed induction motor (900 to 1200 rev/min) is to be mounted on a 100-mm- (4-in.-)thick housekeeping pad over a 200-mm- (8-in.-) thick concrete slab that is continuously supported by the perimeter wall as well as by a concrete beam that divides the slab in the center. The motor is mounted with the shaft vertical and is rigidly bolted to the housekeeping pad. The drive shaft penetrates the floor beside the beam and in the center of the pad. The structural engineer estimates the static deflection of the floor slab (due to the weight of the motor) to be 1 mm (0.04 in.). Evaluate the potential for a floor vibration problem and provide a solution if necessary.

Solution: The driving frequency range of the motor is 15 to 20 Hz for the fundamental and 30 to 40 Hz for the second harmonic. Note that because vibrating machinery produces excitation forces in all directions, the orientation of the motor is of little practical significance.

From Equation 22-27 the estimated natural frequency of the floor is as follows:

SI Units

$$f_o = 15.8\sqrt{\frac{1}{1}} = 15.8 \text{ Hz}$$

U.S. Customary Units

$$f_o = 3.13\sqrt{\frac{1}{0.04}} = 15.7 \text{ Hz}$$

which is in the operating range of the pump's fundamental frequency.

The natural frequency of the floor should be at least 25 Hz (1.25 times the maximum fundamental driving frequency of 20 Hz). Solving for d in Equation 22-27, the maximum allowable deflection is

SI Units

$$25 = 15.8\sqrt{\frac{1}{d}}$$

$$d = 0.4 \text{ mm}$$

U.S. Customary Units

$$25 = 3.13\sqrt{\frac{1}{d}}$$

$$d = 0.016 \text{ in.}$$

This would require a reduction of more than 50% from the original conditions, and it might be very difficult for the structural engineer to design for so small a deflection. As an alternative, consider reducing the natural frequency of the floor so that the lowest driving frequency is about 25% above the natural frequency of the floor. This modification would require a floor frequency of 15 Hz/1.25 = 12 Hz, which (again from Equation 22-27) corresponds to a static deflection of

$d = 1(15.8/12)^2 = 1.73$ mm $\qquad$ $d = 0.04(15.8/12)^2 = 0.07$ in.

By directing the structural engineer to alter the supporting structure (decreasing the slab thickness, modifying the beam supports, etc.) to meet the desired static deflection, the floor resonance problem can be avoided. It should be noted that the floor would have other natural frequencies higher than the fundamental frequency approximated by Equation 22-27, and these natural frequencies may become resonant with the higher harmonics of the pump. Evaluating such a problem is so complex that it is beyond the scope of this chapter.

22-13. Noise

The intent of this section is to provide project leaders with some elementary tools to evaluate the severity of sound propagation through the air, including transmission through barriers. A discussion of structure-borne sound (i.e., structural vibrations resulting in subsequent noise radiation from solid surfaces) is of little importance in most pumping stations and, hence, is not included. Heed OSHA regulations and local ordinances, as explained in Section 22-6.

Outdoor Sound Propagation

Noise radiates from a source in all directions, and sound propagates in air at a nominal velocity of 344 m/s (1130 ft/s) at 20°C (70°F). The actual velocity varies somewhat with temperature and barometric pressure, but for practical purposes this dependence can usually be ignored. The intensity of sound decreases with the square of the distance from a source (much like the intensity of light from a light bulb) as the acoustical energy spreads out over a larger and larger surface area. In addition to this "spreading loss" there is also a slight loss due to atmospheric absorption (conversion of acoustical energy into heat), but this can be neglected in most applications.

The sound pressure level, L_p, at a point can be expressed as a function of the distance to the source, x, and the sound pressure level from the source at a reference distance, x_o, in the same direction. This spreading loss can be expressed as

$$L_p(x) = L_p(x_o) + 20 \log_{10} (x_o/x) \qquad (22\text{-}29)$$

The effects of partial or full barriers, the reflections from surrounding surfaces, and the directivity of the source are not considered in Equation 22-29, and in an enclosed room the equation may be valid only within a few feet of the source. For an outdoor source, the expression may be reasonably accurate for hundreds of meters provided that (1) there are no barriers or major reflecting surfaces (such as buildings) and (2) the direction of the reference measurement is the same as the direction of interest. In the special case of $x = 2x_o$ (where the location of interest is twice as far from the source as the reference position), the difference in the noise levels at the two positions is 6 dB. Thus, whenever the distance to the source is doubled, the sound intensity drops by 6 dB. Likewise, whenever the distance to the source is reduced by 50%, the sound intensity increases by 6 dB. These general rules are only strictly valid in a free field (i.e., an environment without reflective or absorptive surfaces), but they are approximately valid in many typical situations.

Indoor Sound Propagation

In an enclosed or partially enclosed room, noise radiating from equipment is reflected from the room boundaries and builds up to a higher noise level than if the equipment were located outside. This buildup of noise caused by acoustical energy being contained within the room creates a "reverberant area or field" in the room.

The reverberant field concept is illustrated in Figure 22-19, where noise level is plotted as a function of distance from a noise source. In the region near the source (the direct field), the noise level falls off at the 6 dB per double distance rate because the acoustical energy radiating directly from the source overpowers the reverberant field. As the listener moves farther from the source, the noise level reaches a sound pressure level that is relatively constant. It is this region that is called the reverberant field. In SI units, the noise level in the reverberant field can be approximated by the following expression:

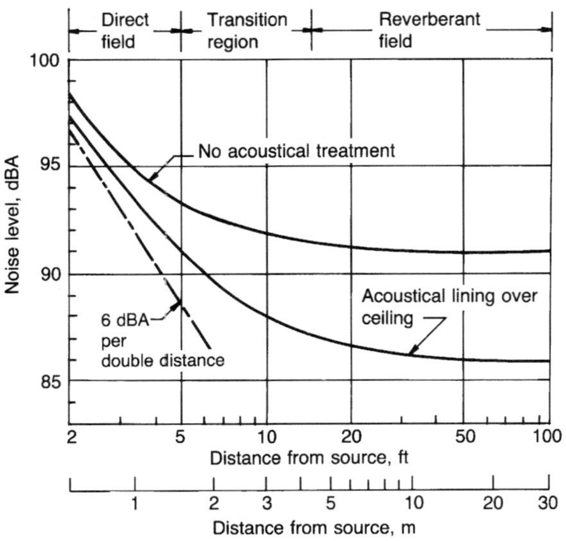

Figure 22-19. Noise level in a room.

$$L_{p(rev)} = L_p(1\text{ m}) - 10\log_{10}(NRC \times S) + 17 \quad (22\text{-}30a)$$

where $L_p(1\text{ m})$ is the sound pressure level 1 m from the source, S is the total surface area of the interior surfaces of the room in square meters, and *NRC* is the average noise reduction coefficient of the room surfaces exposed to the noise.

In U.S. customary units, the expression is

$$L_{p(rev)} = L_p(3\text{ ft}) - 10\log_{10}(NRC \times S) + 27 \quad (22\text{-}30b)$$

where S is in square feet.

The noise reduction coefficient is a single number (between 0 and 1.0) that approximates the ratio of the absorbed to the incident acoustical energy. For example, a material with an *NRC* rating of 0.60 would absorb 60% of the incident sound and reflect the other 40% of the sound back into the room. A list of common building materials and approximate *NRC* ratings is shown in Table 22-10. Note that most hard, smooth surfaces have low *NRC* ratings, while most soft surfaces have high *NRC* ratings.

The noise level in the reverberant field can be reduced by adding materials with high *NRC* values. The noise reduction achievable depends primarily on the product of the surface area and the *NRC* rating of the added acoustical material, but it also depends on the average *NRC* rating and the total surface area of the existing room. For pumping stations with little or no existing acoustical materials, the noise reduction by adding acoustical materials can be approximated by

$$NR = 10\log_{10}[1 + 10(S_a/S)NRC_a] \quad (22\text{-}31)$$

Table 22-10. Values of *NRC* for Common Building Materials Based on Total Surface Area (All Sides of the Absorber)

Material	Approximate *NRC*
Concrete (poured or sealed)	0.05
Steel (and other hard metals)	0.05
Concrete block (porous)	0.05–0.10
Wood (all types)	0.05–0.10
Glass	0.05–0.10
Gypsum board	0.05–0.10
Floor tile (ceramic, vinyl, etc.)	0.05–0.10
Acoustical masonry blocks	0.35–0.75
Acoustical ceiling tile (mineral fiber)	0.45–0.75
Carpet	0.45–0.65
Acoustical roof decking	0.45–0.85
Fiberglass insulation board	
25 mm (1 in.) thick	0.75–0.85
Fiberglass insulation board	
50 mm (2 in.) thick	0.85–1.00
Acoustical baffles (hanging)	0.65–0.95
Acoustical panels (wall or ceiling mounted)	
12 mm (1/2 in.) thick	0.35–0.60
25 mm (1 in.) thick	0.65–0.80
50 mm (2 in.) thick	0.85–1.00
Spray-on acoustical system	
12 mm (1/2 in.) thick	0.35–0.60
25 mm (1 in.) thick	0.65–0.80
37 mm (1-1/2 in.) thick	0.85–0.95

where S_a and NRC_a are the surface area and *NRC* rating of the added acoustical material. When a material is placed over an existing material (shielding it from the noise field), the NRC_a value used in Equation 22-31 should be the net increase in *NRC*. For example, if a 5-cm- (2-in.-) thick fiberglass insulation board is placed under a steel deck roof, the NRC_a should be 0.95 – 0.05 = 0.90 (see Table 22-10). Hanging baffles (or unit absorbers) are sometimes rated in terms of metric or English sabins per unit rather than by *NRC*. These numbers can easily be converted to *NRC* values by dividing the sabins per unit by the total surface area of the absorber to yield *NRC*. Equation 22-31 is represented in graphical form by Figure 22-20.

Noise Sources

There are numerous noise-generating devices typically found in pumping stations, but the most common noise problems stem from engine generators, air compressors, fans, electric motors, and pumps.

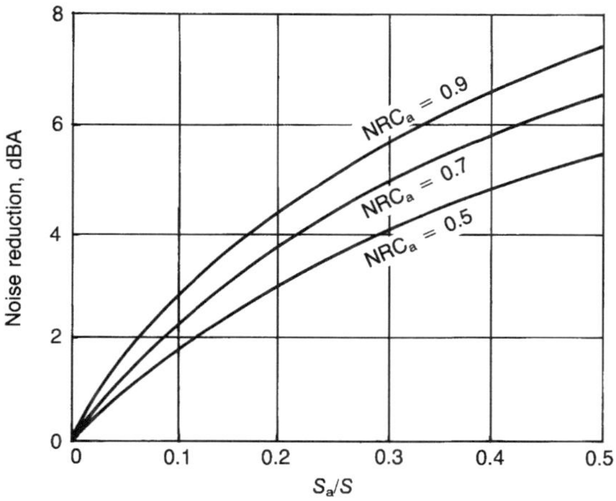

Figure 22-20. Reverberant field noise reduction with acoustical treatment on walls and ceiling.

Equipment manufacturers should be able to provide reasonably accurate noise data for a particular piece of equipment. The noise (sound pressure) levels in Tables 22-11 and 22-12 apply to the normal operation of equipment in good repair, but they are approximate and should be used only when manufacturer's data are unavailable or, for some reason, unreliable. Equipment with worn bearings or damaged internal parts may generate higher noise levels. Avoid comparing noise data in different formats. For example, sound power levels are not equivalent to sound pressure levels, even though they may appear to have the same units. (Actually, they do not have the same units.) Also be aware that sound pressure levels may be A-weighted, B-weighted, C-weighted, D-weighted, or even unweighted. They may also be octave band, one-third-octave band, or even narrow band. To be applicable to the expressions presented in this chapter, all noise levels must be A-weighted sound pressure levels at some identified direction and distance from the source.

Table 22-11. Typical Noise Levels from Diesel Engines[a]

		A-weighted noise level, [dBA at 1 m (3.28 ft)]	
Horsepower	(rev/min)[b]	Casing radiated	Exhaust radiated
under 50	1200	94	107
50–100	1200	96	109
100–200	1200	99	112
200–400	1200	102	115
400–800	1200	105	118
800–1600	1200	108	121

[a] Use the same data for spark-ignition engines. They are quieter, so these data are more conservative.

[b] For other engine speeds add 30 $\log_{10}$ [(rev/min)/1200].

Table 22-12. A-weighted Sound Pressure Levels of Electric Motors [dBA at 1 m (3.28 ft)]

		Speed (rev/min)	
Power (kVA)	Horsepower	450–900	900–1800
7.5–37	10–50	88	92
37–75	50–100	92	96
75–150	100–200	95	99
150–300	200–400	98	102
300–600	400–800	101	105

Engine-Powered Generators

Most of the noise generated by engine-powered generators is caused by the engine firings within the individual cylinders. This acoustical energy escapes via direct transmission through the engine block and the exhaust stack into the surrounding environment. Although some noise exits the combustion air intake, it is usually much lower in intensity because it is blocked off during the ignition while the gases in the cylinder are exhausting. Mechanical noise from the internal rotating parts in the engine and the generator are usually insignificant. The radiator fan is the only other significant noise source on the machine (unless it is remotely cooled). Noise levels from these fans can be found in the subsection entitled "Fans."

Approximate A-weighted sound pressure levels for diesel engines are given in Table 22-11. Casing-radiated and exhaust noise levels are shown separately because the exhaust stack outlet is usually remote from the engine. The conditions for the exhaust noise data are predicated on (1) no muffler in the exhaust pipe, (2) an exhaust directed upward, and (3) a measurement point 1 m to the side (90° from the exhaust direction) at the elevation of the opening. Under these assumptions there is very little directionality to the source (i.e., the noise level is independent of azimuth).

The noise levels given in Table 22-11 are for steady operation at fixed speed. Start-up noise may be higher, particularly if compressed-air starters are used. Compressed-air starters should be avoided in residential areas because noise levels inside the plant can exceed 120 dBA.

Mufflers are nearly always required to bring the exhaust stack noise levels down to meet environmental noise ordinances. Depending on the muffler design, the engine, and the operating speed, mufflers usually reduce noise by 10 to 40 dBA. Most manufacturers provide exhaust noise data for their engines, with a variety of mufflers, matched to the engine operating at fixed-governed speeds. In general, it is best to locate the muffler close to the engine exhaust with a flexible coupling between the two, as shown in Figure 22-21. This location minimizes the length of exhaust pipe with high internal noise, some of which radiates outward through the pipe wall. The preferred technique of silencing a diesel generator is illustrated in the figure.

Electric Motors and Pumps

Electric motors can be a significant source of noise in pumping stations. In most applications, the noise radiated by a motor-driven pump is generated by the electric motor, not the pump. There are some instances where the pump can be noisier than the motor, but these are usually limited to high-speed (pump frequencies greater than 200 Hz), high-pressure, low-volume pumps such as gear pumps, which are not common in pumping stations. Approximate A-weighted noise levels from induction motors are shown in Table 22-12.

Air Compressors

Noise levels produced by air compressors vary widely with design and application. Very large machines that are designed to deliver volumes of compressed air greater than 50 standard L/s (100 scfm) can generate noise levels in excess of 120 dBA near the source. Air compressors common in pumping stations require only a fraction of this capacity and, as a result, are much quieter (but still rather noisy). Typical noise levels from these machines at 1 m (3.28 ft) are in the 90- to 100-dBA range.

Fans

The noise levels produced by ventilating fans are a function of the fan blade and housing design, the number of blades, the tip speed of the blades, and the volume flow and static pressure created by the fan. The most popular types for use in pumping stations are propeller fans and centrifugal fans. Propeller fans are commonly used in an open condition (no housing or ductwork connected to the fan) to ventilate spaces through an opening in the wall or roof of the space. Under these conditions the static pressure requirements of the fan are minimal—usually less than 25 mm (1 in.) of water. Centrifugal fans, on the other hand, are commonly used in ducted applications where static pressure requirements are more substantial. Noise levels radiated from centrifugal fans are computed as shown in the ASHRAE handbook [28].

The approximate noise levels from propeller fans of various diameters and speeds are given in Figure 22-22. The blade design (airfoil, flat, etc.) also has some effect on the resulting noise level, but the noise produced by most propeller fans should fall within 5 dBA of these curves.

Multiple Sources

When two or more sources of noise operate simultaneously, the total noise level at a point from all sources can be computed from the following expression:

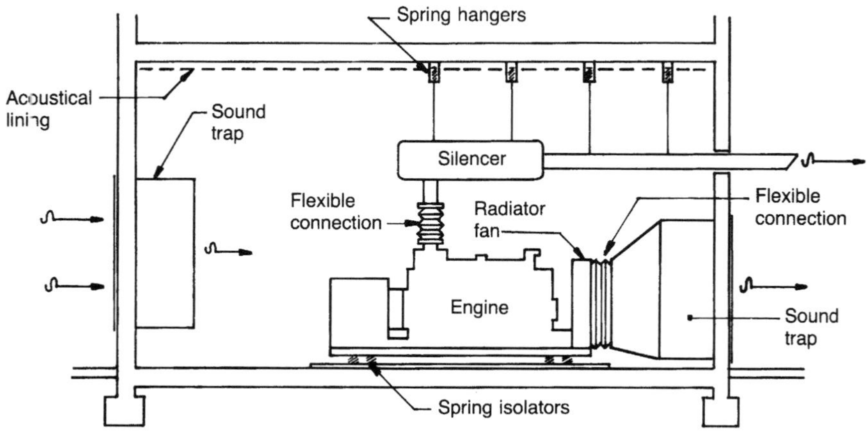

Figure 22-21. Noise control for a diesel generator.

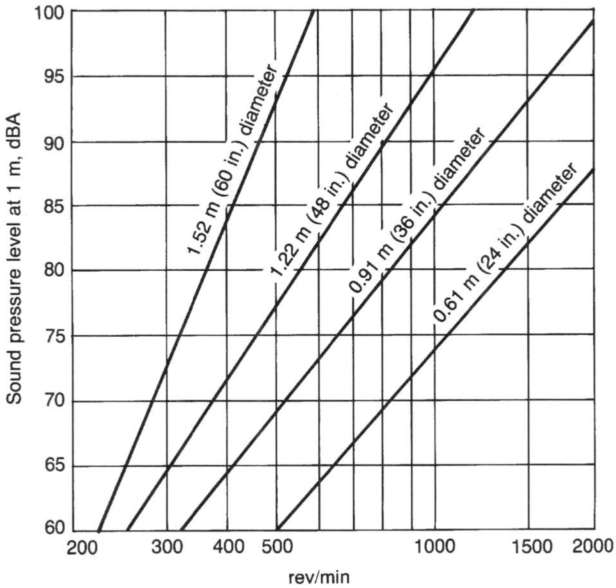

Figure 22-22. Noise level of propeller fans.

$$L_p = 10 \log_{10} \sum_{i=1}^{m} 10^{N_i} \tag{22-32}$$

where L_p is the total noise level in DBA, N is $L_i/10$, L_i is the noise level of the ith source, and m is the total number of sources. This expression is valid for multiple sources in a free field (i.e., outdoors) or within a room as long as the noise level from the direct field and the reverberant field are each evaluated independently as separate sources. If all sources have the same noise level at a point, Equation 22-32 reduces to

$$L_p = L_i + 10 \log_{10} m \tag{22-33}$$

Noise Reduction by Enclosures

Noise levels from mechanical equipment can be reduced by (1) modifying the source, (2) adding acoustical materials to the room, or (3) isolating the source within a complete or partial enclosure. Techniques for modifying the source (e.g., reducing the speed or changing the equipment design) can often be very effective, but they may be costly and are usually undesirable for one reason or another. Adding acoustical materials to the room is somewhat expensive and only partially effective because it does not reduce the noise near the source. Usually a 5- to 8-dB noise reduction is the most that can be achieved (see Figure 22-20). Consequently, the most common method of reducing noise from machinery is to enclose it within partitions and/or barriers.

The noise reduction, NR, provided by a partition or wall separating two spaces is defined as the difference between the noise level on the source side, L_{p1}, and the receiver side, L_{p2}, of the partition. In general, the noise reduction is primarily a function of the transmission loss of the partition, the total area of the common partition through which the noise radiates, the physical size and acoustical properties of the rooms in question, and the frequency distribution of the noise on the source side. A complete description of the equations required to compute the noise reduction accurately is well beyond the scope of this text. However, the following simple equation is usually accurate to within ±8 dBA:

$$NR = L_{p1} - L_{p2} = \text{STC} - 5 \tag{22-34}$$

The STC term in Equation 22-34 refers to the sound transmission class, STC, of the partition. The STC rating of a partition is a single number descriptor determined from laboratory measurements of the sound transmission loss as a function of frequency. The STC ratings of some common building materials are listed in Table 22-13.

For materials not listed in the table, the STC ratings can be obtained from data in Harris [29] or

Table 22-13. Approximate STC Ratings for Common Building Materials

Material	STC[a]
Poured concrete, 150–300 mm (6–12 in.) thick	50
Hollow concrete block, 200 mm (8 in.) thick, unpainted	40
Steel acoustical panel, 100 mm (4 in.) thick	40
Drywall partition,[b] 150 mm (6 in.) thick	35
Drywall partition,[c] 150 mm (6 in.) thick	40
Safety glass, 6.35 mm ($^1/_4$ in.) thick	30
Hollow metal door, without seals	20
Metal louver, more than 50% open area	2

[a] Note that the STC values are *not* additive for composites: two partitions each with an STC of 35 *do not* yield an STC of 70. Each composite must be separately tested.

[b] Metal studs with 1 layer of gypsum board on each side.

[c] Batt insulation between metal studs and gypsum board on each side.

approximated from the surface weight of solid materials using the following formula:

$$\text{STC} = 8 + 17 \log_{10} w \qquad (22\text{-}35a)$$

where w is the surface mass of the material in kilograms per square meter.

In U.S. customary units, the expression is

$$\text{STC} = 20 + 17 \log_{10} w \qquad (22\text{-}35b)$$

where w is the surface weight in pounds mass per square foot.

In most applications, an enclosure is made from more than one material. For example, the door and roof are usually constructed differently from the walls, and there may be openings for ventilation. To evaluate this complex situation, calculate the composite sound transmission class, STC_c. The composite STC can be determined from substituting the exposed areas and STC of each component of the barrier system into the expression

$$\text{STC}_c = -10 \log_{10}\left[1/s \sum_{i=1}^{n} A_i 10^{-N_i}\right] \qquad (22\text{-}36)$$

where N is $\text{STC}_i/10$, A_i and STC_i are the exposed surface area and STC rating of the ith material, and s is the total exposed surface area of all components in the direction of the listener. In most situations, the isolation provided by a composite system is only as good as its weakest component, unless that component has a surface area less than one-tenth of the other components. A unique but common example is the enclosure with an opening because openings have an STC of 0. The dependence of the composite STC on opening size is illustrated in Figure 22-23. Note that even small leaks can cause serious degradation of performance, particularly for partitions with high STC ratings.

Sound Traps

In most practical applications, enclosure openings are required for ventilation. Louvered openings are typically more than 50% open area, so these openings present a major acoustical problem—a problem usually solved with sound traps. A sound trap or duct silencer is usually a rectangular steel box (but sometimes round for circular ducts) open at both ends with a packing of acoustical materials that create parallel baffles, as in Figure 22-24. The air passes straight through most units without changing direction. These devices are prefabricated in a variety of sizes and are furnished with laboratory-tested acoustic and aerodynamic performance data. The STC of an opening can, therefore, be significantly increased by inserting one or more sound traps into the opening in the barrier and sealing the perimeter airtight with acoustical (nonhardening or resilient) caulking.

The acoustical performance of a sound trap depends on its size and internal design. The length

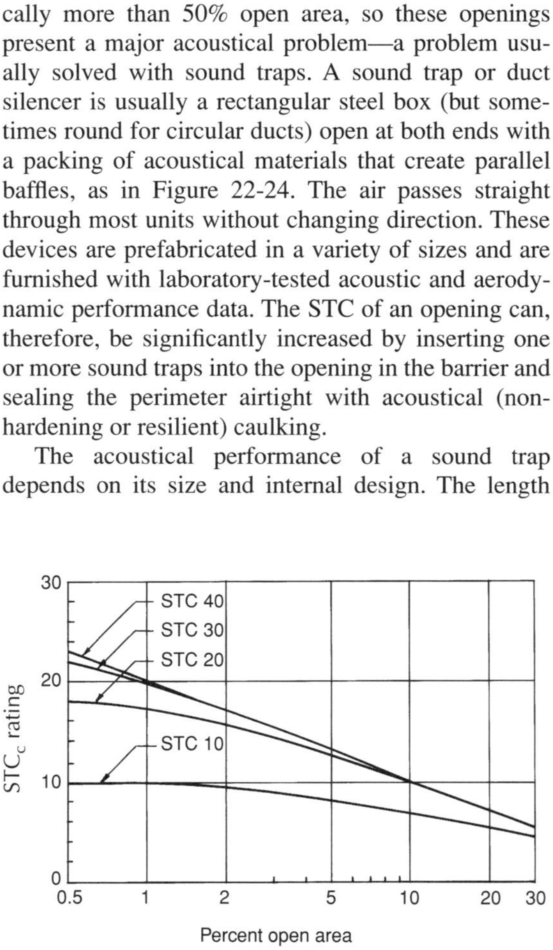

Figure 22-23. Dependence of composite STC rating on openings in partitions or enclosures.

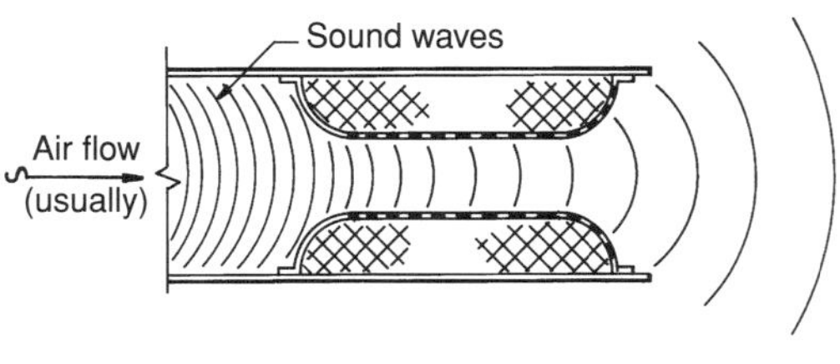

Figure 22-24. A typical duct silencer.

(the dimension in the direction of air flow) of the sound trap is the most important factor affecting performance. Sound traps are available in lengths ranging from 0.3 to 3 m (1 to 10 ft). Within each length, most manufacturers produce several designs that essentially vary in the size of the air passage (i.e., the percentage of open area). Units with larger air passages have lower acoustical ratings and lower aerodynamic pressure drops. Sound traps are rated by the dynamic insertion loss (in decibels) at a variety of frequencies (and not by STC values). The dynamic insertion loss is the difference in noise level with and without the sound trap in the system (with air flowing through the silencer). Aerodynamic and acoustic performance data for a few typical sound traps are given in Table 22-14. The STC of a sound trap can be approximated roughly by the manufacturer's dynamic insertion loss rating in the frequency band nearest the primary frequency of the noise source. The primary frequency would be the blade passage frequency for a fan or the firing rate frequency for a diesel engine. If the primary frequency of the noise is unknown, the insertion loss at 250 Hz may be used to approximate the STC of the sound trap.

Table 22-14. Performance Characteristics of Sound Traps, Face Velocity of 5.08 m/s (1000 ft/min)

Length		Pressure drop		Approximate
m	ft	mm Hg	in. H_2O	STC rating
0.3	1	1.40	0.75	6
1.0	3	0.09	0.05	10
1.0	3	0.19	0.10	15
1.0	3	0.37	0.20	25
1.5	5	0.15	0.08	18
1.5	5	0.28	0.15	25
1.5	5	0.37	0.20	30
2.1	7	0.19	0.10	25
2.1	7	0.28	0.15	30
2.1	7	0.48	0.25	35

Noise Reduction of Barriers

It is often neither practical nor economical to enclose a noise source completely. An often-used compromise is to erect a barrier between the noise source and the receiver to block the direct transmission of noise to the listener. For the barrier to be effective, it must (1) be solid (nonporous) and (2) completely block the straight-line path between the source and the listener. Chain-link fences, shrubs, or rows of trees are virtually useless as noise barriers, but earth berms and concrete walls have been used successfully. Barriers do not provide as much noise reduction as enclosures because the noise tends to go around the barrier.

The noise reduction provided by a barrier is primarily dependent on (1) the material of construction, (2) the height of the barrier, and (3) the presence of other adjacent reflecting surfaces. For an infinitely long solid wall blocking the line-of-sight transmission between two points (as shown in Figure 22-25) with no nearby reflecting surfaces, the noise reduction is approximated by the following dimensionless equation from Baranek [30]:

$$\mathrm{NR} = 20\log_{10}\left(\frac{\sqrt{2\pi N}}{\tanh\sqrt{2\pi N}}\right) + 5 \qquad (22\text{-}37)$$

where N is the dimensionless Fresnel number, which is defined by

$$N = \frac{2f}{c}(A + B - r) \qquad (22\text{-}38)$$

in which f is the primary frequency of the noise source in Hz and c is the speed of sound in air in meters per second (feet per second). See Figure 22-25 for the definitions of A, B, and r, which are in meters (feet). Although very high barriers might reduce noise by more than 20 dBA, they usually do not because of sound transmission either through the barrier material

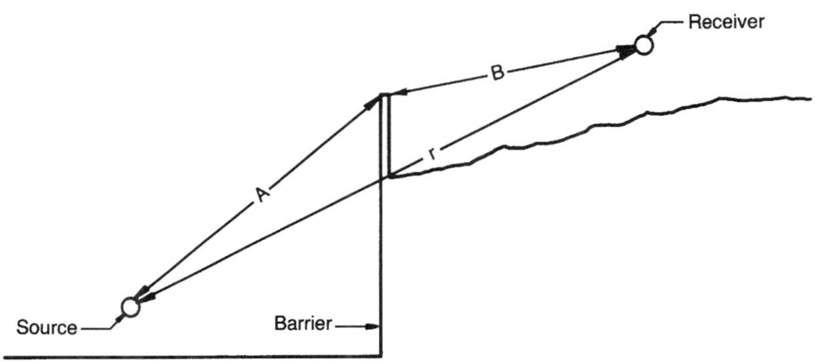

Figure 22-25. Noise reduction of barriers.

or by another flanking path. In general, the barrier material should have an STC rating about 15 dB greater than the noise reduction predicted by Equation 22-37. Note that the noise reduction given by Equation 22-37 represents the change in sound level at the receiving position with and without the barrier.

Example 22-6
Reducing the Sound Level at a Pumping Station

Problem: A pumping station with concrete exterior walls 200 mm (8 in.) thick houses two 500-kVA (670-hp), 1800-rev/min standby diesel generators that supply power to six 200-kVA (270-hp), 600-rev/min direct-drive centrifugal pumps, two of which are backups. The layout of the pumping station is such that the nearest residential property is 60 m (197 ft) from the edge of the building, as shown in Figure 22-26. Compute the expected noise levels inside and outside of the building with no special acoustical considerations, then determine the appropriate measures needed to reduce the acoustical impact to an acceptable level. Assume that the NRC for the untreated interior spaces is 0.10. An NRC of 0.05 is not realistic for real-life conditions because of absorption from miscellaneous fixtures, etc.

Solution:

Interior noise levels in pump room and office. The main noise sources in the pump room are the six electric motors that power the pumps. From Table 22-12, the expected noise level from each pump is 98 dBA at 1 m (98.8 dBA at 3 ft). The noise level in the reverberant field from each pump is evaluated from Equation 22-30. Assume the room dimensions to be 13 × 26 × 8 m (43 × 85 × 26 ft). The total surface area is S = 1300 m^2 (14,000 ft^2). Use Equation 22-30 and Table 22-10.

SI Units

$$L_{p(rev)} = 98 - 10 \log_{10} [(0.10)(1300)] + 17$$
$$= 93.8 \text{ dBA} \simeq 94 \text{ dBA}$$

U.S. Customary Units

$$L_{p(rev)} = 98.8 - 10 \log_{10} [(0.1)(14{,}000)] + 26.5$$
$$= 93.8 \text{ dBA} \simeq 94 \text{ dBA}$$

The total noise level in the pump room is evaluated by using four noise sources for m in Equation 22-33. (Two of the six pumps are always off.)

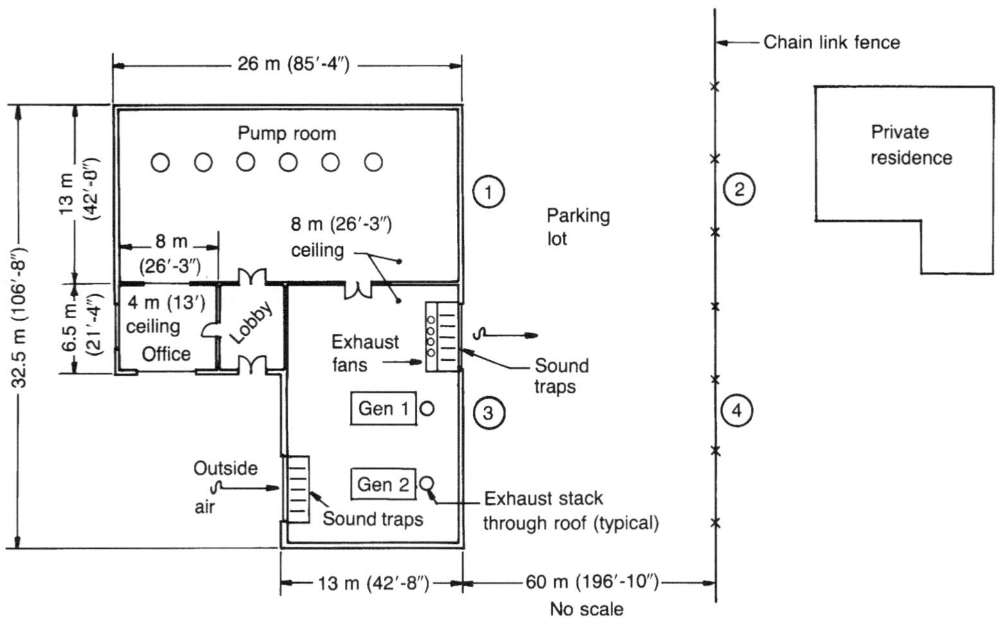

Figure 22-26. Pumping station for Example 22-6.

SI Units	**U.S. Customary Units**
$L_p = 94 + 10 \log_{10} 4 = 100$ dBA	$L_p = 94 + 10 \log_{10} 4 = 100$ dBA

The noise level in the office can be computed by estimating the sound transmission through the 32 m^2 (345 ft^2) gypsum board partition with a 3 m^2 (32.4 ft^2) safety glass window separating the office from the pump room. Assuming the partition has an STC of 40 and the window has an STC of 30 (from Table 22-13), the composite STC is determined from Equation 22-36

$$STC_c = -10 \log_{10}\left\{\frac{1}{32}\left[(29)(10)^{-4} + (3)(10)^{-3}\right]\right\} = 37.3$$

$$STC_c = -10 \log_{10}\left\{\frac{1}{325}\left[(313)(10)^{-4} + (32)(910)^{-3}\right]\right\} = 37.3$$

From Equation 22-34, the noise reduction is

$NR = 37.3 - 5 = 32.3 \simeq 32$ $\qquad$ $NR = 37.3 - 5 = 32.3 \simeq 32$

and the noise level in the office is approximately

$L_p = 100 - 32 = 68$ dBA $\qquad$ $L_p = 100 - 32 = 68$ dBA

Interior noise in generator room. Noise data from the manufacturer indicate that the casing-radiated noise level from each unit (including the radiator fan) is 98 dBA at a distance of 3 m (9.84 ft), which can be converted to 1 m (3 ft) by Equation 22-29.

SI Units	U.S. Customary Units
$98 = L_p(1\text{ m}) + 20 \log (1/3)$	$98 = L_p(3\text{ ft}) + 20 \log (3/9.84)$
$L_p(1\text{ m}) = 98 + 9.5 = 107.5$ dBA	$L_p(3\text{ ft}) = 98 + 10.3 = 108.3$ dBA

The reverberant field level is computed using $S = 13 \times 19.5 \times 2 + (13 + 19.5)2 \times 8 = 1027$ m^2 (11,055 ft^2), which yields (from Equation 22-30)

$$L_{p(rev)} = 107.5 - 10 \log_{10} (0.10 \times 1027) + 17 = 104.4 \text{ dBA for each unit.}$$

$$L_{p(rev)} = 108.3 - 10 \log_{10} [(0.1)(11{,}055)] + 26.5 = 108.3 - 30.4 + 26.5 = 104.4 \text{ dBA for each unit.}$$

When both units are operating, the noise level (from Equation 22-33) increases to

$L_p = 104.4 + 10 \log_{10} (2) = 107.4$ dBA $\qquad$ $L_p = 104.9 + 10 \log_{10} (2) = 107.4$ dBA

The other major noise sources in the generator room are the six propeller fans 1 m (36 in.) in diameter located in the louvered opening. These fans rotate at 600 rev/min, and each fan generates a noise level of 75 dBA at 1 m, as shown by Figure 22-22. (For U.S. customary units, first use Equation 22-29 to obtain the noise level at 3 ft). From Equation 22-30 the reverberant field noise level from each fan is

SI Units:

$$L_{p(rev)} = 75 - 10 \log_{10} (0.10 \times 1027) + 17$$

$$L_{p(rev)} = 75 - 20.1 + 17 = 71.9 \text{ dBA} \simeq 72 \text{ dBA}$$

U.S. Customary Units:

$$L_p\,(3\text{ ft}) = 75 + 20 \log_{10} (3.28/3) = 75.8$$

$$L_{p(rev)} = 75.8 - 10 \log_{10} [(0.1)(11{,}055)] + 27$$

$$L_{p(rev)} = 75.8 - 30.4 + 26.5 = 71.9 \simeq 72 \text{ dBA}$$

and from all six fans, Equation 22-33 gives

$L_p = 72 + 10 \log_{10} 6 = 79.8$ dBA $\qquad$ $L_p = 72 + 10 \log_{10} 6 = 79.8$ dBA

From Equation 22-32, the total noise from the six fans and the two generators is

$$L_p = 10 \log_{10} [(10)^{7.98} + (10)^{10.74}] = 107.4 \text{ dBA}$$

$$L_p = 10 \log_{10} [(10)^{7.98} + (10)^{10.74}] = 107.4 \text{ dBA}$$

Exterior noise levels. Independently compute the noise levels from each source type and add them together by using Equation 22-32. First, consider the pump room. The interior noise level is 100 dBA (the reverberant field level) because the wall is not within the direct field of any of the pumps (see Figure 22-19). The STC rating of the partition is 50 (see Table 22-13), so if there are no penetrations or openings in the wall, the exterior noise level at position 1 (Figure 22-26) is, from Equation 22-34,

SI Units

$$L_p = 100 - NR = 100 - (50 - 5)$$
$$= 55 \text{ dBA}$$

U.S. Customary Units

$$L_p = 100 - NR = 100 - (50 - 5)$$
$$= 55 \text{ dBA}$$

Estimating the average distance from position 1 to the four motors as 12 m (40 ft) and from position 3 to the four motors as 72 m (235 ft), the pump noise level at position 3 is, from Equation 22-29,

$$L_p(1) = 55 + 20 \log_{10} (12/72) = 39 \text{ dBA}$$

$$L_p(2) = 55 + 20 \log_{10} (40/235) = 39 \text{ dBA}$$

The generator room noise is evaluated in a similar manner. The main difference here is the fact that the generator room has a louvered opening containing six propeller fans in the wall facing the receiving property. The total area of the opening is 14 m^2 (151 ft^2), or approximately 9% of the 156 m^2 (1680 ft^2) exterior wall area. From Equation 22-36, the STC-50 partition is reduced to a composite (effective) STC$_c$ rating of

$$\text{STC}_c = -10 \log_{10}\left\{\frac{1}{156}\left[14(10)^{-0.2} + 142(10)^{-5.0}\right]\right\}$$
$$= 12.5 \text{ dBA}$$

$$\text{STC}_c = -10 \log_{10}\left\{\frac{1}{1680}\left[151(10)^{-0.2} + 1529(10)^{-5.0}\right]\right\}$$
$$= 12.5 \text{ dBA}$$

The STC$_c$ rating from Figure 22-23 is 11, a satisfactory check. From Equation 22-34 again, the generator casing noise level at position 3 with both generators running is

$$L_p(3) = 107.4 - NR = 107.4 - (12.5 - 5)$$
$$= 99.9$$
$$\simeq 100 \text{ dBA}$$

$$L_p(3) = 107.4 - NR = 107.4 - (12.5 - 5)$$
$$= 99.9$$
$$\simeq 100 \text{ dBA}$$

Assuming the distance from position 3 to the generators to be 5 m (16.4 ft), the generator casing noise level at position 4 is, from Equation 22-29,

$$L_p(4) = 99.9 + 20 \log_{10} (5/65) = 77.6 \text{ dBA}$$

$$L_p(4) = 99.9 + 20 \log_{10} (16.4/213)$$
$$= 77.6 \text{ dBA}$$

The propeller fan noise is evaluated from Figure 22-22. As shown, each of the fans is expected to generate a noise level of 75 dBA at a distance of 1 m (3 ft). Thus, at position 4, each fan yields a noise level of

$$L_p(4) = 75 + 20 \log_{10} (1/60) = 39.4 \text{ dBA}$$

$$L_p(4) = 75 + 20 \log_{10} (3.28/197) = 39.4 \text{ dBA}$$

The noise level from all six fans (from Equation 22-33) would be

$$L_p = 39.4 + 10 \log_{10} (6) = 47 \text{ dBA}$$

$$L_p = 39.4 + 10 \log_{10} (6) = 47 \text{ dBA}$$

The manufacturer lists the exhaust stack noise from each unit to be 62 dBA at a distance of 15 m (49 ft) when equipped with the standard residential-grade silencer. Using Equation 22-29, the noise level from one of the silenced exhausts at position 4, which is 63 m (207 ft) distant, would be

$$L_p(4) = 62 + 20 \log_{10} (15/63) = 49.5 \text{ dBA}$$

$$L_p(4) = 62 + 20 \log_{10} (49/207) = 49.5 \text{ dBA}$$

From Equation 22-33, the total noise level from both stacks would be

$$L_p(4) = 49.5 + 10 \log_{10} (2) = 52.5 \text{ dBA}$$

$$L_p(4) = 49.5 + 10 \log_{10} (2) = 52.5 \text{ dBA}$$

The total exterior noise level at the residential property from all sources originating from the pump station property is greatest at position 4; from Equation 22-32, it is

$$L_p = 10 \log_{10} [10^{7.76} + 10^{4.7} + 10^{5.25}]$$
$$= 77.6 \text{ dBA}$$

$$L_p = 10 \log_{10} [10^{7.76} + 10^{4.7} + 10^{5.25}]$$
$$= 77.6 \text{ dBA}$$

which is well above the maximum allowable noise standards for residential communities (see Table 22-5).

Potential interior noise reduction measures. Noise levels in both the pump and generator rooms may exceed OSHA standards [24] for 8 h/d employee noise exposure. Assume that on an average day an employee is expected to spend time as follows:

Time (h)	Place	Noise level (dBA)	Allowable time per Table 22-4	D_n from Eq. 22-5
1.0	Pump room	100	2.0	0.5
0.25	Generator room	108	0.66	0.38
6.75	Office	68	No limit	0.00
Total				0.88

The computed D_n value is below the allowable limit of 1.0, and there should be no problem complying with OSHA noise laws provided that all employees spend a considerable amount of time in the quiet areas. But the anticipated noise levels would make vocal communication difficult between co-workers in the pump room and the generator room.

The generator room has a total surface (floor, ceiling, and walls) of 1027 m^2 (11,055 ft^2), and if the ceiling, which has an area of 253.5 m^2 (2728 ft^2), were entirely covered with 5-cm- (2-in.-) thick fiberglass insulation board, the reverberant field noise level would be reduced by (from Equation 22-31 and Table 22-10)

$$NR = 10 \log_{10} [1 + 10(253/1027)(0.95)] = 5.2 \text{ dBA}$$

$$NR = 10 \log_{10} [1 + 10(2728/11{,}055)(0.95)] = 5.2 \text{ dBA}$$

This 5-dBA noise reduction would (from Figure 22-3) nearly double the distance for which voice communication is possible. If the pump room ceiling were also covered with fiberglass insulation board, the D_n noise dose would be reduced as follows:

$$NR = 10 \log_{10} [1 + 10(13 \times 26/1300)(0.95)] = 5.4 \text{ dBA}$$

$$NR = 10 \log_{10} [1 + 10(42.7 \times 85.3/14{,}000)(0.95)] = 5.4 \text{ dBA}$$

Time (h)	Place	Noise level (dBA)	Allowable time per Table 22-4	D_n from Eq. 22-5
1	Pump room	95	4.0	0.25
0.25	Generator room	102	1.5	0.17
Total				0.42

22-14. Reducing Exterior Noise

As shown in Example 22-6, the exterior noise radiating from a pumping station may stem from several sources, such as motors, pumps, engine generators, and cooling fans.

Noise is attenuated by (1) walls, (2) partitions or other enclosures (Table 22-13), (3) distance, and (4) silencers and sound traps. Reducing reverberant noise by acoustical treatment [acoustical tile, masonry blocks (if specially manufactured to do so), or thick fiberglass] also reduces radiating noise unless the source is near an opening in a wall. Placing motors below grade or using submersible pumps and motors is very effective. (But engines should not be below grade.) Fan noise can be reduced by using (1) larger diameter blades at lower rotational speeds, (2) more effective sound traps, and/or (3) exhausting through the roof. Reducing fan noise by using larger blades and lower air velocity carries the penalty of a larger louver, which decreases the STC of the wall (or roof), but the penalty can be overcome by permitting a higher pressure drop in the sound trap, thus making it more effective for suppressing both fan noise and generator-casing noise. In some situations, it may be desirable to provide cooling by refrigeration—an expensive solution.

When designing for noise suppression, contributing sources with noise levels that are 10 dBA or more below the total noise level allowed can usually be ignored because (from Equation 22-32) the effect on total noise level is less than 1 dB.

Example 22-7
Reducing Exterior Noise from a Pumping Station

Problem: The pumping station of Example 22-6 must meet a community noise ordinance that limits the maximum noise level at adjacent residential property to 55 dBA. Find and evaluate the modifications necessary to comply with the local ordinance.

Solution: All three major noise sources (generator casing, engine exhaust, and fan noise) must be reduced. From Example 22-6, the noise level at position 3 (Figure 22-26) emanating from the pump room is only 39 dBA, so further reduction of the noise from this source is not necessary.

The following noise levels at position 4 are obtained from Example 22-6, for which there are no special noise control features:

Generator casing:	77.6 dBA
Propeller fans:	47.0 dBA
Generator exhaust:	52.5 dBA
Total noise level:	77.6 dBA

To meet the maximum noise level of 55 dBA requires a reduction of 23 dB. Try STC-30 sound traps in the louver. The pressure drop for these 1.5-m- (5-ft-) long traps is about 0.37 mm Hg (0.20 in. WC) at 5.08 m/s (1000 ft/min). To reduce this pressure drop to the recommended 0.09 mm Hg (0.05 in. WC), the velocity must be reduced. Because the pressure drop is proportional to the square of the velocity,

$$\frac{\text{new pressure drop}}{\text{original pressure drop}} = \left(\frac{\text{new velocity}}{\text{original velocity}}\right)^2$$

SI Units

$$\frac{0.09 \text{ mm Hg}}{0.37 \text{ mm Hg}} = \left(\frac{v}{5.08 \text{ m/s}}\right)^2$$

from which $v = 2.51$ m/s

U.S. Customary Units

$$\frac{0.05 \text{ in.}}{0.20 \text{ in.}} = \left(\frac{v}{1000 \text{ ft/min}}\right)^2$$

from which $v = 500$ ft/min

The original 14-m^2 (151-ft^2) louver sized for a face velocity of 3.81 m/s (750 ft/min) must be enlarged to

$A = 14 \text{ m}^2(3.81/2.51) = 21 \text{ m}^2$ $\qquad$ $A = 151(750/500) = 226 \text{ ft}^2$

The net wall area is

$A = (19.5 \text{ m})(8 \text{ m}) - 21 \text{ m}^2 = 135 \text{ m}^2$ $\qquad$ $A = (64 \text{ ft})(26.25 \text{ ft}) - 226 \text{ ft}^2 = 1454 \text{ ft}^2$

From Equation 22-36,

$$STC_c = -10 \log_{10} [(1/156)(21 \times 10^{-30/10} + 135 \times 10^{-50/10})] = 38.4 \text{ dB}$$

$$STC_c = -10 \log_{10} [(1/1680)(226^{-30/10} + 1454^{-50/10})] = 38.4 \text{ dB}$$

Recompute the noise level at position 4. Noise from the generator casing is reduced by an STC_c 38.4-dB barrier instead of the original STC_c 12.5-dB wall (without sound traps). The net noise reduction is 38.4 – 12.5 = 25.9 dB.

The propeller fan noise is reduced by the STC rating of the sound traps (30 dB) minus the STC rating of the louver (2 dB) for a net noise reduction of 28 dB. The components at position 4 are, therefore,

Generator casing (77.6 – 25.9)	= 51.7 dBA
Propeller fans (47 – 28)	= 19 dBA
Generator exhaust	= 52.5 dBA

and the total noise from Equation 22-32 is

$$L_p = 10 \log_{10} [10^{51.7/10} + 10^{19/10} + 10^{52.5/10}] = 55.1 \text{ dBA}$$

$$L_p = 10 \log_{10} [10^{51.7/10} + 10^{19/10} + 10^{52.5/10}] = 55.1 \text{ dBA}$$

This sound level is just slightly above the 55-dBA limit. To make further reductions, choose critical-grade exhaust silencers for an additional 5-dB noise reduction on the exhaust noise. Note that both engines must be silenced with critical-grade silencers to achieve this result.

With the critical-grade exhaust silencers and the STC 30 sound traps, the noise level at position 4 is

Generator casing:	51.7 dBA
Propeller fans:	19.0 dBA
Generator exhaust:	47.5 dBA
Total noise level:	53.1 dBA (from Equation 22-32)

This meets the noise ordinance with a small margin of safety. For an increased margin of safety, STC-35 sound traps could be used for the louver opening. This application would require sound traps that are 2.1 m (7 ft) long, which would require considerable floor space—one of the several reasons for avoiding retrofit in favor of planning for noise reduction during the first stages of layout.

An additional margin of safety is possible, but it may become quite expensive. Generally speaking, noise control measures tend to follow the law of diminishing returns; that is, the cost of the first 10 dBA of noise reduction is generally much less expensive than the second 10 dBA, and so on. In situations where noise is important, it is usually cost effective to hire a consultant who can take the responsibility of performing a detailed frequency analysis and specifying exactly what is required to meet code requirements without excessive safety margins.

22-15. References

1. Karassik, I. J., W. C. Krutzsch, W. H. Fraser, and J. P. Messina. *Pump Handbook,* 2nd ed. McGraw-Hill, New York (1986).
2. American National Standard for Reciprocating Pumps for Nomenclature, Definitions, Application and Operation, ANSI/HI 6.1–6.5-1994. Hydraulic Institute, Parsippany, NJ (1994).
3. American National Standard for Centrifugal Pumps for Nomenclature, Definitions, Application and Operation, ANSI/HI 1.1–1.5-1994. Hydraulic Institute, Parsippany, NJ (1994).
4. American National Standard for Vertical Pumps for Nomenclature, Definitions, Application and Operation, ANSI/HI 2.1–2.5-1994. Hydraulic Institute, Parsippany, NJ (1994).
5. Reichaert, F. E., J. R. Hall, and R. D. Woods. *Vibrations of Soils and Foundations.* Prentice-Hall, Englewood Cliffs, NJ (1970).
6. Marscher, W. D. "Determination of pump rotor critical speeds during operation through use of modal analysis." Proceedings ASME 1986 WAM Symposium on Troubleshooting Methods and Technology, Anaheim, CA (December 1986).
7. Blevins, R. D. *Formulas for Natural Frequency and Mode Shape.* Robert Krieger Publishing Co., Malabar, FL (1984).
8. Marscher, W. D. "The effect of variable frequency drives on vibration problems in vertical pumps." Proceedings of the Water & Wastewater 1990 Conference (Publ. Walpole Productions, Houston), Barcelona, Spain (April 24, 1990).
9. Rathbone, T. "Vibration tolerance." *Power Plant Engineering,* **43**:721–724 (November 1939).
10. Blake, M. "New vibration standards for maintenance." *Hydrocarbon Processing and Petroleum Refining,* **43**:111–114 (January 1964).
11. Baxter. R. L. and D. L. Bernard. "Vibration tolerances for industry." American Society of Mechanical Engineers, 67-PET-**14**:1–8 (1967).
12. Hancock, W. P. "How to control pump vibration." *Hydrocarbon Processing,* **53**:107–113 (March 1974).
13. Marscher, W. D. "The relationship between pump rotor system tribology and appropriate vibration specifications

for centrifugal pumps." Proceedings Institute of Mechanical Engineers (London) IMechE 3rd European Congress on Fluid Machinery for the Oil and Petrochemical Industries, The Hague, Netherlands (May 1987).

14. API 610, 8th ed. American Petroleum Institute, Washington, DC (1995).
15. MIL-STD-167-1(SHIPS), U.S. Dept. of Defense. (1 May 1974).
16. ISO 2372 Mechanical Vibration of Machines (1974).
17. Wachel, J. C., and C. L. Bates. "Techniques for controlling piping vibration and failures." 76-Pet-18, American Society of Mechanical Engineers, New York (1976).
18. Marscher, W. D. "How to use impact testing to solve pump vibration problems." Proceedings Electrical Power Research Institute Power Plant Pumps Symposium, Tampa, FL (June 1991).
19. Dodd, V. R. *Total Alignment.* The Petroleum Publishing Co., Tulsa, OK (1974).
20. Agostinelli, A., D. Nobles, and C. R. Mockridge. "An experimental investigation of radial thrust in centrifugal pumps." *Journal Engineering for Power Transmission.* American Society of Mechanical Engineers **82**:120–126 (1960).
21. Black, H. F. "Effects of fluid-filled clearance spaces on centrifugal pump vibrations." 8th Turbomachinery Symposium, Texas A&M Univ. (1979).
22. Marscher, W. D. "Analysis and test of multistage pump at critical speeds." Society of Tribologists and Lubrication Engineers/American Society of Mechanical Engineers Joint Tribology Conference, Ft. Lauderdale, FL (October 1989).
23. Harris, C. M., and C. E. Crede. *Shock and Vibration Handbook,* 2nd ed. McGraw-Hill, New York (1976).
24. OSHA. "Safety and health standards for federal supply contracts." (Walsh Healey Public Contracts Act, U.S. Department of Labor). Federal Register, 34, 7948, Washington, DC (1969).
25. Miller, J. D. "Effects of Noise on People." *Journal of the Acoustical Society of America,* **56**: 746 (1974).
26. M. W. Kellogg Company. *Design of Piping Systems,* John Wiley, New York (1956).
27. Streeter, V. L., and E. B. Wylie. *Fluid Transients.* FEB Press, Ann Arbor, MI (1982).
28. *ASHRAE Handbook, Systems Volume.* American Society of Heating, Refrigerating and Air Conditioning Engineers, Inc., Atlanta, GA (1984).
29. Harris, C. M. (Ed.). *Handbook of Noise Control.* McGraw-Hill, New York (1979).
30. Beranek, L. L. (Ed.). *Noise and Vibration Control.* McGraw-Hill, New York (1971).

Chapter 23

Heating, Ventilating, and Cooling

EARL L. HECKMAN
PERRY L. SCHAFER

CONTRIBUTORS
Richard E. Pustorino
Ashok Varma

The needs, criteria, and design procedures for ventilating, heating, and cooling pumping stations are discussed in this chapter. Worked examples are used to illustrate the principles presented. Because these examples are drawn from practice, only U.S. customary units are used. The design of air conditioning systems is complex enough to require a specialist, so the discussion of air conditioning is limited to elementary considerations.

References to codes and standards are given in abbreviated form—code letters and numbers only (such as ASHRAE Standard 90-80). The titles are given in Appendix E, and publishers' addresses are given in Appendix F.

23-1. Need for Heating, Ventilating, and Air Conditioning

Heating and ventilating (H&V) with either air conditioning or cooling (HVAC) are included in pumping stations for the following reasons:

- To facilitate the safe and efficient performance of operating and maintenance personnel
- To minimize the deterioration of the equipment, controls, and structure
- To promote community acceptance of the station by helping to control noise and odor emissions.

Spaces within a pumping station are divided into two broad categories:

- The "wet well" or "sump" is a pump intake basin containing the fluid. In wastewater pumping stations, the term often includes wastewater channels, grit and screen rooms, and the pump suction chamber. If the fluid is potable water, it is also called a "clear well," "forebay," or "suction well."
- The "dry well" or pump room is a term also loosely used to include ancillary spaces for pumps, motors, and auxiliary equipment.

Hazardous Environments

Hazardous environments can be defined differently in different codes. In general, it is any environment that can be dangerous because of the possible presence of toxic, flammable, or explosive gases or liquids. As defined by NFPA 820, it is an environment that could contain explosive gases. Toxic gas is not covered. Regarding pumping stations, NFPA 70 classifies hazardous environments for electrical equipment as

- Class 1, Group D, Division 1 where there is a *high likelihood* of an explosive hazard. Group D environment is one where hydrocarbon vapors may be

present. There are about a dozen hazardous environment groups. All equipment in Group D must be explosionproof—an expensive requirement for polyphase motors.
- Class 1, Group D, Division 2 where there is the *possibility* of an explosive hazard. Only single-phase equipment (switches, lights, motors) must be explosionproof.

A Division 1 classification is assigned to a wet well either not ventilated or only intermittently ventilated in accordance with NFPA 820 criteria. If the pump intake basin is well ventilated at all times, the likelihood of an explosive atmosphere is greatly reduced and the space could be reclassified as Division 2.

Personnel Safety

There have been too many incidents of death and injury to ignore good practice for workers entering any room containing wastewater or the possibility that sewer gases could enter a room. Some jurisdictions require permanently installed meters for explosive gas (hydrocarbons), hydrogen sulfide, and low oxygen levels. However, the chance that such equipment will not be calibrated with adequate frequency or that detectors will become fouled with hydrogen sulfide poses a risk. It is better that workers be equipped with portable monitors regularly calibrated and tested. The monitors are expensive, but there is no good substitute for the protection they offer.

Confined Spaces

OSHA defines confined spaces as follows:

- A "confined space" is an area that is large enough and so configured that an employee can bodily enter and perform assigned work, has limited or restricted means for entry or exit [for example: tanks, vessels, silos, storage bins, hoppers, vaults, trenches greater than 1.2 m (4 ft) in depth, and pits] and is not designed for continuous occupancy.
- A "permit required confined space" is defined as a confined space with one or more of the following characteristics: Contains or has the potential to contain a hazardous atmosphere such as lack of oxygen (less than 19.5%), or explosive or toxic gasses; contains a material that has the potential for engulfing an entrant; has an internal configuration such that an entrant could be trapped or asphyxiated by inwardly converging walls or a floor that slopes downward and tapers to a smaller section; or contains any other recognized serious safety or health hazard.

Merely providing ventilation per codes, such as NFPA 820 or the Ten-States Standards, does not by itself change the classification of a confined space to a nonconfined space. The criteria regarding entry, exit, and continuous occupancy must also be addressed. In practical terms, it would seem that the following features must be present in a wastewater pumping station wet well to avoid classification as a confined space:

- Stairway access. Ladder access probably would not be considered to meet the criteria of eliminating "limited or restricted means for entry or exit."
- Permanently installed, continuous ventilation sufficient to control the accumulation of any hazardous (toxic or explosive) gasses and prevent oxygen depletion below concentrations necessary to support life.
- Permanently installed lighting.
- Permanently installed or (if always used) portable detectors for combustible gas, hydrogen sulfide, and oxygen concentrations.

Even with the above features, there are some sanitation agencies that do not allow individuals to enter wet wells alone. They must always be accompanied by an observer who does not go with them into the wet well.

Hazards in Wet Wells

Proper ventilation of pumping stations is an often misunderstood and neglected subject. Enclosures below grade, such as wet wells and vaults, are considered by OSHA to be confined spaces, and there are stringent requirements for access to them and elaborate protective measures for persons entering them. Many deaths in both water and wastewater pumping stations could have been avoided with good ventilation. Although ventilation by itself does not change the classification of a "confined space" to an "unconfined space," good ventilation goes a long way in improving the safety of wastewater pumping stations. In an accident in England, methane gas leaked from the forebay into a water pumping station and collected over a weekend. The station was not continuously ventilated. When a party of visitors entered the station, a spark from the lighting system caused an explosion that killed 14 people. In another incident, an operator on his normal rounds breathed a fatal concentration of hydrogen sulfide gas in a wet well inadequately ventilated by an inappropriate design. In subsequent air sampling, hydrogen sulfide concentrations up to 20 times greater than the recommended limit were found. These examples, along with many others, underscore the need for proper ventilation design. In a matter of life and death, there should be no compromise to save costs. Decision makers and designers should be well aware of the hazards (see especially the two NIOSH reports [1, 2]) and the methods to overcome them.

Good ventilation is also important in maintaining a benign, dry environment for mechanical and electrical equipment. The cost of the ventilating equipment may be partially or completely offset by the reduced deterioration of equipment and improved efficiency of personnel due to better working conditions. Without ventilation, explosive gases (from illegally dumped flammable liquids) may accumulate in wastewater wet wells and result in devastating explosions. The formation of explosive concentrations of methane through bacterial action is also possible.

Bacterial growth under anaerobic conditions also can produce hydrogen sulfide, which is more poisonous than cyanide. At trace concentrations, it has the smell of rotten eggs, but at high and dangerous concentrations the gas overwhelms the olfactory senses. Because it then cannot be smelled, it is the more insidious and deadly. Victims overcome but not killed outright often suffer irreparable brain damage. The depletion of oxygen in an enclosed space is also a possible danger.

Hazardous conditions in the clear wells of water pumping stations are less frequent than in the wet wells of wastewater stations, but they nevertheless exist (as cited at the beginning of this section). Ventilation is necessary, and heating may be required to protect equipment from corrosion and freezing.

Summary of Good Ventilation Practice

- Intermittent ventilation systems are inherently hazardous and may be a death warrant, because workers will not wait for the lengthy scavenging period even of high-speed fans. For example, at 60 air changes/hour (AC/h), it takes 2 hours to reduce 2000 ppm H_2S to 50 ppm at 100% efficiency, because an AC only dilutes the air and does not actually change it.
- NFPA 820 recommends 12 AC/h continuously for wastewater wet wells and 6 to 10 AC/h continuously for pump rooms.
- Some states require considerably greater ventilation rates—up to 30 AC/h.
- Prudent engineers design for 20 to 45 AC/h in wet wells and 10 AC/h in dry wells.
- Continuously force air in at the ceiling through industrial diffusers, and force air out at the floor with fans at both inlets and outlets.
- Supply enough air to cool the motors.
- Use either stainless-steel or fiberglass fans and shafts with permanently lubricated bearings.
- Use fiberglass or aluminum ducts.
- Use run-around circuits and heat exchangers to recover heat.

An Alternative to Ventilation

Neither OSHA nor NFPA 820 require ventilation to be provided in wet wells. If ventilation is not provided, however, the wet well must be classified as a "permit required confined space" per OSHA and as Class I, Groups D, Division 1 locations per NFPA 820. Entry into such spaces requires:

- A permit
- Atmospheric testing, purging, and monitoring for oxygen, flammability, and toxicity
- Personnel training
- Established procedures for working, standby, and rescue
- Safety equipment and clothing to protect the head, body, hands, and feet plus safety harnesses and life lines
- Self-contained breathing apparatus
- Record keeping
- Competent persons at the site as observers and supervisors.

Augment these requirements by consulting the literature [2] and local authorities. Include the required safety procedures and measures in the operating and maintenance (O&M) manual.

Temperature, Humidity, and Condensation

The life expectancy of equipment and controls decreases at excessive temperatures and humidities. Most electrical gear and electronics are rated for use in surrounding air temperatures up to 40°C (104°F). High humidity increases the likelihood of moisture condensation, which occurs when air is cooled below its dew point; rust, mildew, and deterioration of electrical and thermal insulation follow, particularly if corrosive gases are present. Blowing warm air over cooler pipe and wall surfaces can raise their surface temperature sufficiently to prevent condensation, thus preserving them.

Dry Wells

Heating and ventilating requirements for pump rooms and dry wells are similar for both water pumping stations and wastewater pumping stations. Maintaining a reasonable minimum temperature in cold weather and removing excess heat in warm weather are the principal reasons for ventilating a dry well or pump room. The following are sources of unintentional heat gain:

- Motors of pumps and other equipment
- Variable-speed drives and inverters
- Motor controls
- Solar gain and lights
- Personnel (negligible).

The sum of unintentional heat gains often entirely offsets building heat losses due to transmission through the structure and infiltration of colder outside air. Any remaining net heat loss must be made up by additional heat supplied to the space. In cold climates, building insulation and double glazing can be used to minimize structure heat losses and reduce condensation.

Heat gain in pumping stations usually exceeds their heat loss in moderate weather. The excess heat must be removed by ventilating with cooler outdoor air; if the ambient air temperature is too high, it must be removed by (1) direct evaporative cooling, (2) indirect evaporative cooling, or (3) refrigeration. Direct evaporative cooling is generally ineffective if the relative humidity of outdoor air exceeds about 75%. Even then, the nearly saturated air supply may promote rusting. Spaces with either high heat gains or a noise level that must be contained may justify the higher costs of refrigerated cooling or the use of some of the pumped water as the heat transfer medium in an air cooling system.

Aesthetics

Pumping stations, especially those in residential neighborhoods, must be acceptable to the community. Acceptability may require (1) noise control, (2) a building in harmony with other structures in the area, and (3) for stations pumping wastewater, control of odors. The superstructure of some stations has been designed to resemble a residence. Zoning ordinances may affect station height, setback, and appearance. Local noise ordinances may require sound attenuation at openings in the building envelope, particularly if an emergency generator is installed (see Chapter 22 for a discussion of noise control).

HVAC Complexity Related to Station Type and Location

The required extent of HVAC, its controls, and its cost increase when:

- Pumping capacity and station size increase
- Auxiliary spaces are added
- Both heating and cooling are required
- Distance to a residential area decreases
- Variable-speed pumps are used
- Stations are automated
- Continuous attendance is required.

23-2. HVAC Design Criteria

Codes and guidelines for good engineering practice are presented in this section with emphasis on safety for workers.

Codes, Ordinances, and Standards

Confer with the jurisdictional authorities to learn which codes, ordinances, or adopted standards apply at the site. Examples of documents that may govern pumping station design include

- City ordinances
- State or local plumbing codes
- State or local fire codes
- State or local building construction codes
- State or local mechanical codes
- State or local energy codes
- Federal or state EPA regulations
- Occupational Safety and Health Act (OSHA) [3]
- National Electric Code (NEC)
- National Fire Protection Association (NFPA) codes and standards
- Ten-State Standards [4]
- ASHRAE Standard 90-80 energy code, which includes recommended heat-transfer coefficients for buildings and rules for energy expenditure and recovery.

Codes and standards, whether mandatory or not, generally represent minimum requirements. Good engineering practice and safety considerations often require adherence to more stringent criteria.

Wet Well Design Guidelines

There is a profound distinction between wastewater wet wells designed for frequent entry (accessible) and those designed for nonentry (sealed).

Accessible Wastewater Wet Wells

Wastewater wet wells containing either bar screens or mechanical equipment must be accessible to workers for maintenance and servicing, so they must be ventilated mechanically. Continuous ventilation with air forced in and forced out is the best safeguard against the buildup of hazardous gases. Section 32.7 of the Ten-State Standards [4], however, only requires forcing the air into the wet well; multiple inlets and outlets for wet wells over 5 m (15 ft) deep are also recommended. Of course, interconnections between wet well and dry well ventilating systems are not allowed.

Fan wheels and shaft seals should be rated nonsparking in accordance with standards of the Air Movement and Control Association. Electric motors, wiring, controls, and monitors must meet NEC requirements for Class 1, Group D, Division 1 hazardous areas. Both automatic heating and dehumidification should be considered for worker comfort and safety as well as for protection from corrosion.

According to the Ten-State Standards, ventilation can be either

- continuous with 12 complete air changes per hour (based on wet well volume above high water level), or
- intermittent with 30 air changes per hour with the high-speed fan ventilation switch interlocked to the wet pit lighting system. However, intermittent ventilation is prohibited by NFPA 820 if the space is to be a Division 2 location.

Ventilation at a low rate can be automatically increased by gas sensors that (1) detect the presence of combustible gases or hydrogen sulfide and (2) either increase fan speed or start an auxiliary fan. (The incremental airflow is normally unheated.) Methane detectors are usually set for 20% or less of the lower explosive limit (LEL), which is the lowest concentration of a combustible gas in air at which an explosion can occur (approximately 5% by volume for methane). Hydrogen sulfide gas detectors can be set for 10 ppm, which is considered a safe level for 8-h exposure [5]. If such detectors are to be dependable, they must be recalibrated and tested at least monthly, and this maintenance should be specified in the O&M manual. Consider the possibility of intermittent ventilation design giving a false sense of security because maintenance may become unreliable and sensors may fail. Personnel should always carry portable detectors when entering potentially hazardous enclosures.

Good ventilation is not easily achieved because the purging of gases (that may be heavier than air) is by dilution and not by a "clean sweep," as is suggested by the phrase "air changes." Thus, even with the best supply air distribution, the purging of dangerous gases from odd-shaped areas is far from perfect. Opening the door to a wet well ventilated only with forced air inflow with gravity relief defeats the ventilation system because the airflow escapes through the door. The best practice for hazardous spaces, such as wastewater wet wells, is to blow air into the chamber at or near the ceiling and to exhaust air at or near the lowest level at a rate of approximately 5% more than the air intake rate, thus producing a partial vacuum of 30 to 60 Pa ($^1/_8$ to $^1/_4$ in.) of water column (WC). Even though permitted by some codes, such practices as providing air supply or exhaust alone at a continuous, but minimum, airflow rate that is to be switched to a high rate only at the time of entry should be avoided. It compromises safety for a relatively small reduction of cost. Minimum or intermittent air changes may not prevent explosions and do not adequately protect impatient workers who may not wait for the required period of high-speed scavenging.

According to NFPA 820, the *minimum* recommendations for ventilation are as follows:

- All ventilated spaces are to be served by both supply and exhaust fans and powered from two independent sources to ensure operation during power failure of a single source.
- Continuous ventilation at the rate of 12 air changes per hour with combustible gas detectors is required. A two-speed ventilation system is recommended with high-speed operation initiated at the warning level of gas concentration.
- Equipment rooms and other spaces below grade containing gas piping are to be (1) ventilated at the rate of 12 air changes per hour and (2) equipped with combustible gas detectors and, preferably, two-speed fans.
- Galleries and tunnels are to be treated the same as below-grade spaces with a 0.38 m/s (74 ft/min) air velocity.
- Below-grade spaces without gas piping are to be ventilated at the rate of 10 air changes per hour; galleries are to be ventilated at that rate or by an air velocity of 0.25 m/s (50 ft/min), whichever is greater.

The airflow pattern can be controlled only by using a combination of supply and exhaust ducts. Without ducts, air follows the path of least resistance and leaves stagnant areas. Suggested duct design velocity and friction ranges are given in Table 23-1. Ducts must be routed to clear equipment access and removal space, hoist rails, and hoistways. The air quantity supplied should be based either on the recommended air change rate needed or on the required heat removal, whichever is larger. Outside ventilation air should be filtered for cleanliness and insect control. Arrange insect screens for easy removal for cleaning. In the

Table 23-1. Air Velocities and Friction Losses in Duct Design

Item	Air velocity		Friction loss	
	m/s	ft/min	mm WC/100 m	in. WC/100 ft
Supply duct	5.1–9.1	1000–1800	6.7–10	0.08–0.125
Exhaust duct	4.1–7.6	800–1500	5.0–8.3	0.06–0.10
Registers, grilles	3.0–5.1	600–1000	4.2–6.7	0.05–0.08 net
Intake louvers	1.3–2.0	250–400	2.5–5.0	0.03–0.06 net

O&M manual, alert maintenance workers to the need for filter replacement. Consider installing a filter gauge to show the pressure drop across the filter or, preferably, use a differential pressure switch to energize a "clogged filter" signal at the control panel. In cold climates, select a storm louver type and net air velocity to prevent the entry of snow; provide heat at louvers to keep them from being clogged with ice.

Sealed Wet Wells

"Sealed" (not easily entered) wet wells require an adequate vent (perhaps only a manhole cover) to accommodate air displacement due to changes in the liquid level. Except in very cold climates, heating or heat tracing is rarely required because the temperature of the moving water is sufficient to prevent freezing. However, if the water temperature is near freezing (especially if it contains frazil ice) and the pumps run only intermittently, some form of heating is needed. Explosionproof electrical equipment is required for sealed wastewater wet wells just as it is for accessible wet wells.

Heat

If no equipment is located in a wet well, heating the airstream is frequently omitted, which allows the wet well temperature to reach an equilibrium between the water temperature and outdoor air temperature. Pipes that are filled with motionless water for long periods and exposed to subfreezing air can be "traced" (wrapped) with electric heat tape beneath their insulation to prevent freezing. Panel heating of stairs and walkways can be used to prevent ice formation in cold climates. Wet wells with screens or other equipment that require inspection and maintenance should be heated to approximately 10°C (50°F) for worker comfort, safety, and efficiency as well as for corrosion protection. Any heat source open to wastewater wet well air must have neither an open flame nor a temperature above 260°C (500°F) to prevent the ignition of combustible gases.

Explosionproofing

All electrical equipment (including motors) in or open to the wet well must be explosionproof and should be located above the flood level. Even submersible pump motors should be explosionproof because they may not always be submerged. Their control panels should not be in the wet well but in a nonhazardous location. Some engineers recommend explosionproof equipment regardless of location.

23-3. Odor Control

Odors at wastewater pumping stations constitute one of the worst difficulties that agencies encounter. If not resolved quickly, they can take inordinate amounts of staff time in dealing with irate neighbors. A more advisable and cost-effective approach is to conduct appropriate evaluation of any potential odor problems during planning and design of the pumping station, and to install all necessary odor control measures along with the construction of the pumping station itself. There are sufficient valid engineering and scientific tools available today to allow fully workable solutions to be determined during the project design.

Odor Control Evaluations

Odor control evaluations need to be conducted within the broadest context possible. Too many engineers believe odor control is synonymous with "foul air treatment." Actually, foul air treatment is often the most costly type of odor control and should be avoided unless absolutely necessary. Other types or categories of odor control should normally be evaluated first to determine if foul air treatment can be avoided.

Considerable information is needed to conduct a proper odor control evaluation, and information about the wastewater entering the pumping station is crucial. The details of the upstream collection system (including operation of other pumping stations); sources, kinds, and amounts of wastewater; industrial contributions; and other information is vital. Specific sampling and testing of existing wastewater flows in the same area as the future pumping station are likely to be helpful. Several chapters in WEF Manual 22 [6] are helpful in conducting such an evaluation. See also ASCE Manual 69 [7].

Four odor control strategies are defined and discussed herein. In order of effectiveness, they are:

- Minimizing or preventing production of odorous compounds
- Treating odorous compounds within the liquid phase
- Containing and treating foul air
- Enhancing atmospheric dispersion of foul air.

A well-organized evaluation of these categories of odor control almost always results in a successful project. Operation and maintenance costs as well as first cost must be evaluated for each strategy. Odorous substances include a large variety of compounds. The reduced sulfur family of compounds is the major problem in most wastewater systems, and hydrogen sulfide is the most common offender. Various effects and standards relating to hydrogen sulfide are described in Table 23-2. But other sulfides, disulfides, and mercaptans are also frequent problem compounds, because their odor thresholds are almost all in the part per billion range or less. Reduced sulfur compounds, amines, aldehydes, ketones, and various organic acids can cause problems. Ammonia is only rarely a problem, because its concentration is typically low compared with its odor threshold. The concentration of odorants can be measured in the liquid phase as well as the gas phase for almost all potential odorous compounds. Liquid phase analysis is often easier to complete and more accurate, whereas gas phase testing can be costly, especially when scanning for large numbers of potentially odorous organic compounds.

Minimizing Odorous Compounds

The first line of defense against odor problems is to design the entire system to produce the absolute minimum quantity of odorous compounds allowed to enter the pumping station via the influent water. Upstream controls may be beyond the purview of the pumping station designer, but they may need to be explored, because it could be less costly to solve the problem upstream. Control measures could include the following:

- Further pretreatment of specific industrial discharges to the system.
- Minimizing slug loads of wastewater from industries or other point sources.
- If hydrogen sulfide is the significant problem, keeping wastewater pH well above 7 to minimize hydrogen sulfide off-gassing. A pH of 8 would usually be adequate, but pH 9 might be required sometimes.
- Designing upstream sewers to maintain aerobic conditions in the wastewater—usually by keeping velocities above 1.5 m/s (5 ft/s).
- Adding chemicals (such as ferrous sulfate or ferric chloride) to precipitate hydrogen sulfide.

At the pumping station itself, there should be minimum turbulence of wastewater, because turbulence

Table 23-2. Effects and Standards of Hydrogen Sulfide Gas in the Atmosphere[a]

Concentration, ppm by volume	Effect
0.0005	Olfactory detection threshold
0.003	Max concentration for electronic equipment per ISA
0.002–0.008	Practical odor threshold range
0.010	Max concentration for electrical equipment per NEMA
0.030	Ambient Air Quality (odor-based) Standard in California
1	Offensive odor; rotten egg smell
5	Deadens olfactory senses
10	Max 24-h exposure per OSHA
15	Max 8-h exposure per OSHA
10–50	Headache, nausea, and eye, nose, and throat irritation
50	Max 30-min exposure per OSHA
50–300	Eye and respiratory injury
300–500	Life threatening; pulmonary edema
700	Immediate death for everyone
46,000	Lower explosive limit

[a] Adapted from ASCE Manual 69 [7] and industry standards.

promotes off-gassing of odorous compounds. Drop inlets into the wet well can and should be avoided. In stations with constant-speed pumps, use a sloping approach pipe with its invert at or slightly below the low water level even though its crown may be submerged at the high water level (see Section 12-7). Variable-speed pumping is most desirable (especially in larger stations where odor problems are more likely and the investment more easily justified), because matching water elevations in the sewer and wet well allows smooth, nonturbulent entry into the wet well. Refer to Section 12-7 for design details.

Wet wells should be kept small enough to minimize stagnation and the settling of solids. These deposits are anaerobic and produce odorous compounds that diffuse into the liquid above and thence into the air. Slime layers form on submerged walls of wet wells and also produce odors. Such wall areas should also be minimized. Wet wells should be frequently (say, weekly) cleaned. Refer to Section 12-7.

The velocities needed to keep domestic wastewater aerobic, promote scour, and eliminate odor-producing deposits in pipes are given in ASCE Manual 69 [7]. In general, force main velocities of 1.1 to 1.2 m/s (3.5 to 4.0 ft/s) occurring at least once per day are advisable (depending on pipe diameter) to minimize problems.

Treating Odorous Compounds in the Liquid Phase

There are a host of chemicals that can be added to wastewater to inhibit or treat odorous compounds, thus minimizing off-gassing and subsequent odor problems. Only the broad categories of such chemicals are defined here:

- Oxidants, such as chlorine, sodium hypochlorite, and hydrogen peroxide. These chemicals oxidize sulfide (and perhaps other compounds) that already exist in solution, and they minimize additional generation for a limited time downstream.
- Precipitants (such as ferrous or ferric chloride) that precipitate sulfide as insoluble, dark-colored compounds. Reactions usually take 15 to 20 minutes.
- Inhibitors, such as high pH slugs (pH greater than 12 for 20 minutes) or anthroquinone slugs, greatly inhibit sulfate-reducing bacteria densities for a 1- to 2-week period. Continuous addition of nitrate compounds is sometimes advantageous to promote nitrate reduction and minimize sulfate reduction.
- Aerobic conditioning by the addition of air or high-purity oxygen. If more than about 0.5 to 1.0 mg/l of dissolved oxygen is kept in the wastewater, little anaerobic activity can take place to form the reduced, offensive compounds. In gravity sewers, it is difficult to add dissolved oxygen except by natural re-aeration of wastewater flowing at velocities in excess of 1 m/s (3.3 ft/s). In some force mains (with certain rising profiles, sufficient pressure, and compatible pipe materials), it is feasible and cost effective to add high-purity oxygen.
- Bases such as lime or caustic added continuously to keep pH above 8.5 minimizes hydrogen sulfide production and off-gassing.

Chemicals, their reaction times, and transit times through the facilities must be carefully evaluated [6, 7, 8]. Chemicals can be added far upstream, at the wet well, or at the discharge of the force main.

Containing and Treating Foul Air

The first step in designing a foul air treatment system is to develop a reliable containment and ventilation system that brings all foul air to the treatment device. Containment of foul air is not always easy with sewers bringing gases into the wet well, and doors or hatches being opened. But regulating fans on both inlet and exhaust to maintain the slight vacuum described in Section 23-2 helps, as does proper dispersal of incoming air at the ceiling and collection of foul air at the floor.

Various ventilation, corrosion, safety, and related issues are discussed separately in this chapter. A large variation in pollutant concentration over the course of the day sometimes results in poor treatment, unless the system is specifically designed for it. Therefore, details about upstream system characteristics and operation are critical.

Biological Foul Air Treatment

Of the various types of biological foul air treatment systems, the one most applicable to pumping stations is bulk media biofiltration. That system involves (1) adsorbing odorants onto the surface of organic media and (2) biologically oxidizing the odorous compounds over time via the aerobic action of micro-organisms that live on the media surfaces. The media is usually a mixture of wood or bark chips, compost, soil, sand, peat moss, and/or other materials. The moisture content of the media is critical for the biomass, and the retention time of the foul air in the media must be sufficient—usually 30 to 120 seconds of empty bed contact time.

Biofilters can be built within closed containers or built in the ground with open-air vertical discharge at the bed surface. The latter usually consists of perforated pipe buried about 0.6 m (2 ft) deep in a bed of gravel, covered with a thin layer of sand overlaid with loamy soil, seeded with grass or other plants. The trench is designed for 0.3 m^3/min of gas per m^2 (1 ft^3/min of gas per ft^2) of ground surface area, so the area requirements can be significant. Good performance is reported for a wide variety of odorants with proper design and operation. Hydrogen sulfide in the foul air produces sulfuric acid and causes the media to become acidic, so underdrains and leachate piping must be corrosion resistant. PVC is a good choice. Maintenance costs are relatively low, although the media may have to be changed from time to time. Fortunately, the exhausted media is not classified as a hazardous material. For further information see the literature [6, 7, and 8].

Adsorption

If odorant concentrations are low, foul air can be effectively treated with activated carbon, potassium permanganate, or other adsorption materials. When the adsorptive material becomes exhausted, it must be either regenerated or removed and replaced. In-place regeneration of activated carbon has proven difficult—practically unworkable at most sites. Disposal of spent material has been a problem at some locations. The high humidity of foul airstreams sometimes causes poor treatment, and large amounts of carbon are needed and quickly exhausted when odors are very bad.

Granular activated carbon treated with caustic for extra hydrogen sulfide absorption is highly reactive, and instances of overheating and fires have occurred when moisture conditions become critical (see Appendix I).

Activated alumina impregnated with potassium permanganate has also been used for light-duty foul air treatment. Although cost-effective, removal of spent material is an extremely onerous task.

Wet Chemical Scrubbing

Wet chemical scrubbing is in widespread use at wastewater treatment facilities for odor control and is space-conserving for large foul air flowrates. The scrubbing solution is maintained at a high pH with caustic, to which an oxidant such as sodium hypochlorite or hydrogen peroxide is added. The solution is sprayed over the packing media, and blowdown to the sewer is required. Demisting of exhaust removes most mist particles from the airstream, but some can remain in the treated gas discharge.

Chemical mist scrubbing is another wet chemical technique. The same chemicals are used and a fine mist of the chemical solution (particles of 10 to 20 microns) is sprayed into a reaction chamber with the foul air. Mist particles are contained in the exhaust, which is often visible. Good treatment performance is possible with both systems. See WEF Manual 22 [6] and ASCE Manual 69 [7] for more details.

Other Treatment Systems

High-temperature oxidation of foul airstreams is rarely used at pumping stations because of its high cost. Achieving consistent performance with ozone systems has proven difficult. Neutralizing chemicals and counteractants are also sometimes used. Two-stage systems are occasionally used when a very high degree of foul air treatment is required.

Enhancing Atmospheric Dispersion of Foul Air

Occasionally, elevating the foul air discharge point to about 30 or 40 feet above ground level can solve a local odor problem. A discharge stack may be aesthetically objectionable, but with small foul airflows, a hollow light pole 200 to 250 mm (8 to 10 in.) in diameter can be used as a disguised stack.

Discharging gas at the roof line of the pumping station building should be avoided because downwash from wind blows the discharge to ground level. Carefully evaluate how local wind conditions will disperse odors. In critical situations, atmospheric odor modeling may be necessary to determine downwind effects [6, 7].

Sealing wet wells, with the intention of avoiding any gas discharge, is not normally recommended. Maintenance personnel need access to the wet well on occasion and resealing is difficult. Also, so-called gas tight structures often have leaks. Safety and corrosion issues normally demand some degree of ventilation. Small wet wells need only a ventilation pipe to allow gases to be discharged either close to the ground ar at an elevation of at least 3 m (10 ft) as the situation requires (see OSHA requirements [3] and NFPA 820).

Controlled Atmospheres for Sensitive Equipment

Motor control centers should be placed in a separate room with an atmosphere controlled for temperature and contaminant level.

Sensitive electrical and electronic equipment (such as PLCs or AF controllers) require highly controlled atmospheres to avoid corrosion and to keep temperatures at least below 30°C (85°F). ISA-S71.04 (1986) defines four "Severity Levels": G1 is mild, G2 is moderate, G3 is harsh, and GX is severe. Environments in wastewater pumping stations are typically G3-harsh, an environment in which there is a high probability that corrosive attack will occur. HVAC systems for electronic equipment such as AF controllers should be designed and specified to produce a G1 environment, as partially described in Table 23-3.

In addition, temperature should be maintained as low as possible (22°C, 72°F or lower if possible). Rooms or cabinets are often pressurized to 2.5 mm (0.1 in.) of WC to minimize leaks of outside contaminated air. Rooms or cabinets need to be sealed to achieve this limitation. Clean, filtered air is required to eliminate dust and other airborne particles.

To achieve these standards, air cooling is almost always required, along with filtration for particulate control. At wastewater stations, hydrogen sulfide is the primary contaminant of concern, although ozone concentrations in urban areas are often excessive, and various hydrocarbons can also be present in the ambient air. Use three-stage filters (prefilter at 10 microns, followed by a carbon filter, followed by a 1-micron high-efficiency filter) on air supplied from the upwind side of the pumping station. Relative humidity control is also required for many portions of the country. Chemical filtration media are often used to adsorb pollutants, and sometimes to oxidize them. Activated carbon media, along with potassium permanganate, are often used [9].

Equipment for integration into HVAC systems, or separate, stand-alone equipment is available. Equipment is also available that performs these functions on a small scale for individual cabinets or for banks of cabinets. These facilities pressurize the cabinet to allow it to operate within a room containing harsh environmental conditions, and they include filtration, pollutant adsorption filters, air conditioning, and, sometimes, relative humidity control. Airflow capacities in the 2.8 to 5.7 m^3/min (100 to 300 ft^3/min) range are typical for cabinet-type systems.

Table 23-3. ISA G1 Environmental Conditions

Atmospheric contaminant	Maximum concentration
Relative humidity	50%
H_2S	3 ppb
SO_2, SO_3	10 ppb
Cl_2	1 ppb
NO_X	50 ppb
HF	1 ppb
NH_3	500 ppb
O_3	2 ppb

23-4. Dry Well Design Guidelines

The following guidelines are recommended for designing the heating, ventilating, and cooling systems of dry wells. They supplement those briefly outlined in the Ten-State Standards [4]. Other suggestions can be obtained from *Design of Wastewater Treatment Plants* [10].

Heating

There are three basic heating techniques:

- *Space heaters (gas, oil, or electric) with a thermostat and a summer fan-only switch.* Heater locations and airflow patterns should be chosen to produce thorough circulation through the space. In milder climates this may be the simplest solution. A separate fresh-air heater may also be provided. This approach has the *least* temperature control, but usually the stations do not require precise temperature control.
- *Infrared radiant heating.* With either gas or electricity, this is a simple and effective way of heating large, spacious working areas. Comfort conditions are achieved at a lower air temperature with radiant heat than with any other heating method.
- *Ducted systems with temperature controls.* Mixing dampers can be used to control the proportion of outdoor air to recirculated air in response to room temperature. A supply-air low limit control is recommended.

Building heat loss in cold climates should be minimized by using wall and roof insulation, double glazing, and air dampers properly installed with blade and jamb seals to limit unwanted air infiltration. Follow a locally adopted energy conservation code or ASHRAE Standard 90-80. Maintain adequate space temperatures to facilitate essential activities and prevent freezing (see Table 23-4). Install a manual-reset, capillary-type low-temperature thermostat downstream from the preheating or heating coils and set it at 3°C (37°F) to stop the fan and energize a remote warning for 100% outside air units in unattended sta-

Table 23-4. Recommended Temperatures

	Minimum		Maximum	
Space type	°C	°F	°C	°F
Continually occupied areas	20	68	26	78
Occasional work areas	13	55	35	95
Unoccupied areas	>4	>40	43	110

tions. For systems with a recirculating air damper, an automatic reset low-temperature thermostat may be used. In addition to energizing the warning signal, it repositions the dampers for recirculation only. It may either stop the fan or leave the fan running at the designer's option. The fan and dampers automatically resume their normal operation when the freeze potential is no longer sensed. Provide extra heating capacity controlled by a thermostat (or use portable heaters) that can raise the temperature in the work area to a reasonable, comfortable level when maintenance and repairs must be done in the coldest expected weather. Locate air outlets and inlets to take advantage of the natural upward convection currents from hot equipment as well as to offset the heat losses at their source. Introduce heat near the bottom of exposed walls. If heated air is supplied from overhead (either from ducts or from unit heaters), its downward velocity must be sufficient to force it down to the occupied level. Consider central heating systems if heated ventilation air is to be supplied to multiple spaces totaling more than about 300 m^2 (3000 ft^2) of floor area. Specify industrial-quality equipment for long service with low maintenance.

Ventilating

The primary function of dry well ventilation is the removal of excess heat for the protection of equipment (specifically, motors and controls) and for the comfort of personnel. Secondary functions are moisture removal, corrosion prevention, and odor reduction. The outdoor air ventilation rate can be varied in cold, dry weather by using thermostatically controlled mixing dampers to blend sufficient outdoor air with return air to match the need for heat removal. In SI units the required ventilation is

$$Q = \frac{G - L}{(1208)\Delta T} \tag{23-1a}$$

where Q is the airflow rate in cubic meters per second, G is the heat gain in watts, L is the heat loss in watts, $G - L$ is the net heat gain, and ΔT is the temperature difference in degrees Celsius.

In U.S. customary units,

$$Q = \frac{G - L}{(1.08)\Delta T} \tag{23-1b}$$

where Q is the airflow rate in cubic feet per minute, G is the heat gain in British thermal units per hour, L is the heat loss in British thermal units per hour $G - L$ is the net heat gain, ΔT is the temperature difference in degrees Fahrenheit, and 1.08 = (0.075 lb/ft^3)(0.2395 Btu/lb · deg)(60 min/h).

Cold Weather

The recommended minimum air temperature in an occasionally occupied space is 13°C (55°F) (see Table 23-4). If the supply air is thoroughly mixed with warmer room air near the ceiling, the supply air temperature can be reduced to as low as 7°C (45°F) when necessary for heat removal.

Heat recovered from the dry well exhaust air can be used to help heat the wet well by using any of the heat exchangers discussed in Section 23-5. There must, however, be complete separation of the two airstreams to preclude any leakage of toxic or explosive gases from the wet well to the pump room. The heat pipe coil is the safest exchanger in this regard. The heat recovery device should be protected against corrosion.

Hot Weather

The required summer air-handling capacity of ventilating equipment for heat removal is determined for a given heat load by the difference between the outside air temperature and the allowable maximum indoor temperature (ΔT in Equation 23-1).

The heat gain to the space includes:

- Heat generated by pump motors, drivers, and controls
- Heat from other miscellaneous motors
- Heat from lights and occupants
- Inward heat transmission through walls, windows, and roof
- Direct solar heat gain through windows and skylights, and indirect solar gain through walls and roof previously heated by the sun
- Heat from emergency generators (obtain the data on heat lost to the room and the recommended airflow rate for this equipment from the engine manufacturer).

In hot climates the rate of outdoor ventilation needed to remove excess heat may be exorbitant due to the small difference between the outdoor temperature and the allowable indoor temperature. Try first to reduce the required ventilation rate by reducing the heat gain to the space. For example, cooling air that leaves a pump motor is well above room temperature. If that warm air can be immediately removed by a hood or by direct-ducted exhaust before mixing with room air, the total room heat gain and required ventilation are less. Of course, hoods and ducts must not unduly interfere with pump motor access and removal.

Evaporative Cooling

In dry climates, cooling the air supply by evaporation is effective. Direct evaporative cooling occurs when unsaturated air passes through a water spray or a wet filter. The heat required to evaporate part of the water is extracted from the air and, thus, lowers its dry bulb (sensible) temperature. The added moisture increases the total heat content of the air as measured by its increased wet bulb temperature. Less of the more humid but cooler air is required to remove the sensible heat from the space at a given rate.

Indirect evaporative cooling occurs when evaporatively cooled air is used in turn to cool warmer air by passing both airstreams through an indirect heat exchanger.

"Indirect–direct" evaporative cooling units are even more effective. The supply air is first cooled by passing it through an indirect heat exchanger in counterflow to a second (waste) airstream previously cooled by direct evaporation. The supply airstream is then further cooled by direct evaporation. In this way, both the dry bulb and wet bulb temperatures of the original airstream are reduced.

Due to the small ΔT usually available between supply and room air temperatures, the evaporatively cooled supply air rate is usually in the range of 15 to 30 air changes per hour. Such rates require larger fans and ducts and better air diffusion than a refrigerated air conditioning system with a larger ΔT. Recirculation of evaporatively cooled air is impractical because of the resulting buildup of humidity.

Be warned that evaporative cooling systems increase the humidity, may cause condensation, make rusting more likely, and increase maintenance. Unless winterized by complete draining in severe climates, the evaporative equipment can be badly damaged by freezing.

Refrigerated Cooling

Refrigerated cooling may be necessary where outdoor temperature and humidity are high and where internal heat gains are relatively large, as in a control room containing inverters for variable-speed motors. Other cooling methods should be investigated first because of the comparatively high operating cost of refrigeration compressors.

Chlorination Rooms

Chlorination and chlorine storage rooms require special ventilation. Both the *Chlorine Manual* [11] and the Ten-State Standards [4] recommend that these rooms be heated to 16°C (60°F) and have the exhaust fan capacity required for 60 air changes per hour. The exhaust fan should be energized by an automatic door switch and a manual switch (located outside the door) that simultaneously open an air intake damper. Ventilation air should be introduced at the ceiling and exhausted at floor level because chlorine is heavier than air.

23-5. Energy Use and Conservation

The high energy prices of the recent past years reinforce the importance of energy conservation. The adoption of energy-saving methods is at least partially responsible for the present trend toward increased energy availability. Awareness of the irreplaceability of fossil fuel resources warrants continued efforts by designers to avoid its waste.

Energy Sources for Heating

Frequently available heat energy sources include electricity, natural gas, propane, and fuel oil. Compare the probable life-cycle costs and reliability of each before making a choice. For stations in warm to moderate climates, the air-to-air heat pump may be an option for both heating and cooling as long as these loads are reasonably in balance. In an electric motor-driven air-to-air heat pump, a refrigerated coil is used to extract heat from outdoor air. That heat, plus the heat of compression, is then delivered to the space by a condensing coil in the supply airstream. The "coefficient of performance" (COP) of such a system is the ratio of electric energy input to heat energy output. With an outdoor temperature of 7°C (45°F), for example, a COP of about 2.5, which is not unreasonable, means that the heat energy delivered is 2.5 times greater than would be obtained from an electric resistance heater with the same electrical input. The greatest economy results when the summer cooling load also approximates the cooling capacity of the refrigeration unit selected, thus allowing year-round operation of the unit.

Solar heating is generally unsuitable because a full back-up heating system (or very large solar heat storage) is necessary to meet demand at the times when solar heat is unavailable. Furthermore, there is no compatible use for the greater heat collection in summer that may require separate expensive disposal.

Major Energy Uses and Conservation

The major HVAC energy use is for heating the station and its ventilation air in cold weather and for cooling the station and its ventilation air in hot weather. Continuously operating fans handling large quantities of air against substantial static resistances are the next largest energy users. Energy economy results from (1) a well-insulated structure, (2) prudent selection of air quantities and indoor design temperatures, and (3) minimized air-path resistance. Both the maximum and minimum indoor design temperatures should be balanced between operator efficiency and operating cost. Select air conditioning equipment to be fully loaded at peak summer weather conditions so that it can perform more acceptably at partial load conditions. Multiple or two-speed fans allow operation at reduced capacity and cost at light load conditions. Multistage thermostats or step controllers can actuate additional fan capacity as the space temperature approaches the design maximum.

Introducing ventilating air at lower levels (in dry wells only) and relieving or exhausting it at the ceiling augments natural convection flow. Early opening of properly sized inlet and outlet dampered louvers allows excess heat at partial loads to be removed by convection, which delays the need for mechanical ventilation. If the design of the station allows, consider passive cooling and heating by (1) bringing air through ducts buried below grade to take advantage of the earth's temperature, which usually ranges from 10 to 16°C (50 to 60°F) or (2) using skylights and other glass openings arranged to admit direct sunlight only in the winter months.

Automatic backdraft dampers are necessary on multiple supply and exhaust fans to prevent short-circuiting the air. Intake louvers should preferably be located on the windward side of the building so that prevailing wind pressure assists station ventilation. Wall-type propeller exhaust fans should exhaust to the leeward side of the building. Propeller fans move more air per unit of power than other fan types if the air resistance is 125 Pa (0.5 in. WC) or less, and they offer low resistance to natural airflow when not running. Propeller fans are usually less efficient than centrifugal fans against the higher resistances of long duct runs or heat-recovery devices.

Where wet wells must be heated for extended periods, heat-recovery devices should be considered. Their efficiency is the recovered fraction of the heat content difference between the two airstreams. Much less "new" heat needs to be added to the makeup airstream when such devices are used. The following are commercially available types of air-to-air heat-recovery devices. Recovery efficiency, ranging from 70 to 80% for the first three types, varies with the temperature difference between the warm and cool airstreams.

- *Plate-type heat exchangers.* Warm and cool airstreams pass on opposite sides of metal plates and exchange heat by conduction and convection. Cleaning is difficult, condensate drainage is a problem, and duct arrangements are sometimes awkward.
- *Heat wheels.* These rotate through the counterflowing airstreams. The wheel absorbs heat from the warmer airstream and releases it to the cooler airstream. The heat-recovery rate varies with the speed of rotation.
- *Heat pipe coils.* These consist of a number of individual, sealed, finned tubes containing a metered amount of refrigerant. One half of each tube is placed in the warmer airstream where it absorbs heat and vaporizes the captive refrigerant. The refrigerant then condenses in the opposite half of the tube and releases heat to the cooler airstream. The advantages include no moving parts and the ease of corrosion protection by applied coatings. Face and bypass dampers and other methods of heat-recovery control are used.
- *Coil energy-recovery loops.* Coils both in the exhaust and intake ducts are interconnected with piping and a pump to transfer heat from the exhaust air to the outside makeup air. A glycol solution is used as the heat-transfer medium if the coils are subject to freezing. A three-way bypass valve on the intake coil is controlled to keep the air temperature leaving the exhaust coil above 0°C (32°F) to prevent the freezing of condensate on the exhaust coil in cold climates. The exhaust coil, at least, should be corrosion protected. Overall efficiency is about 50%. This heat-recovery method is useful where supply and exhaust fans are widely separated.

These devices can recover a significant portion of exhaust heat that would otherwise be wasted, but their air resistance, which is in the range of 125 to 250 Pa (0.5 to 1.0 in. WC), penalizes fan energy. Filters are desirable in both airstreams to prevent dirt from clogging the closely spaced fins of the heat transfer coils.

In determining the cost effectiveness of such devices, consider (1) the installed first cost, (2) the purchased energy cost, (3) the actual energy savings, (4) the additional space required, (5) the added complexity of the system, (6) the increased maintenance costs, (7) the reliability, and (8) the higher resistance against which both the supply and exhaust fans must operate. When used in air conditioning systems, such energy-recovery devices can also precool the hot makeup air by thermal exchange with cooler exhaust air. This dual use makes them more economical in applications requiring heating in cold weather and refrigerated air conditioning in summer.

23-6. Corrosion Protection

Metallic equipment, especially ductwork, is susceptible to deterioration in varying degrees due to moisture, acid (from hydrogen sulfide and moisture), and salt often contained in the air in wet wells. Raw edges of cut metal sheets and surface scratches provide a starting point for oxidation to spread under coatings of metals such as the zinc on galvanized steel. Type 316 stainless steel is generally corrosion resistant, but its discoloration may be unacceptable aesthetically. Copper and copper-bearing materials are highly susceptible to attack by hydrogen sulfide and should be avoided or protected in wastewater stations. Aluminum should not be used in damp chlorine rooms. Where metals are the logical choice for items such as louvers, dampers, and grilles, they should be type 316 stainless steel, aluminum, or galvanized steel, or be protected with an epoxy or baked phenolic corrosion-resistant coating applied by the manufacturer.

Polyvinyl chloride (PVC) or fiberglass-reinforced plastic (FRP) are more corrosion resistant (and more expensive) than type 316 stainless steel, but their detracting characteristics—such as fire and smoke ratings, structural strength, greater weight and support requirements, and the greater difficulty of on-site modification—must be given due consideration. PVC and FRP ductwork are usually factory-fabricated from sheets of material 4.8 mm (3/16 in.) or more in thickness. Joints in either type of duct can be made by solvent welding. FRP joints also can be formed with multiple layers of glass cloth laid up by hand in a resin binder. Duct accessories used, such as dampers, turning vanes, hangers, and fasteners, should be equally corrosion resistant.

Fans, tanks, and piping may also be fabricated from PVC and FRP. Heat exchangers (which must be metal for good heat transfer) should be protected by corrosion-resistant coatings, such as thermosetting phenolics. The metal is dipped and then oven baked for several hours. Two or more coats may be used for severe applications. Epoxy compounds and some phenolics can be applied cold by painting or spraying. The resistance of the coatings to the various causes of corrosion can be obtained from their manufacturers. The heat-transfer capability of corrosion-protected heat exchangers may be decreased by 15% or more, so compare cost, energy, and permanence in making a selection.

The types and severity of corrosion to be expected should be ascertained from previous installations at the same or at a similar site, if possible. Also consult experienced coating vendors. The difficulty and cost of replacing unprotected components should be weighed against the cost for protection. Note that corrosion protection increases first cost from 20 to 400% but replacement costs can be even higher.

Electrical switches and control components should also be protected from corrosion. Such protection can be provided in a number of different ways.

- Control hydrogen sulfide at the source to the greatest extent possible. A complete odor-control system mitigates the problem except for electrical equipment within wet wells or under tank covers.
- Isolate control equipment in a separate room and design the ventilation to exclude sulfides or, if that is impossible, treat the air supply for the entire room. If electronic equipment is installed, pressurize the room slightly and eliminate chlorine and other corrosive gases as well.
- Hermetically seal relays and switches in plastic boxes attached to plastic conduit, seal the boxes as completely as possible, and put potassium permanganate pellets in the bottom of the boxes.
- Pressurize the boxes or panels to maintain 25 kPa (0.1 in. WC) with either uncontaminated air or with air filtered through potassium permanganate and activated carbon. Such pressurizing in accordance with NEC 500 also reduces enclosure explosion hazard ratings from NEC Class 1, Division 1 to Class 1, Division 2. Adequate differential pressure can be monitored by an attached pressure switch and alarm. To guard against excessive compressed air consumption if the bolts are loose, the gaskets are old, or the panel is warped, a limited amount of air can be supplied by a small, inexpensive, diaphragm-type fish tank air pump.

23-7. Sequence of Design Steps

Pumping station design is a cooperative effort among the various design disciplines that must be coordinated as the design progresses. Each design step affecting other disciplines should have confirmed agreement before proceeding to avoid later redesign.

Step 1: Write and distribute a design memorandum defining the applicable codes, the indoor and outdoor design temperatures, the building materials and orientation, the sources and amounts of heat gain (especially from motors and engines), and the design air change rate for each space. State the energy source to be used and describe the proposed ventilating, heating, and cooling systems.

Step 2: Calculate the heating and cooling loads for each space. Measure and record wall, glass, and roof areas from the architectural plans. List exposure directions and calculate space volumes. Obtain or calculate heat transmission coefficients (U values) for wall, roof, and window construction. Use the methods in the *ASHRAE Handbook of Fundamentals* [12] to calculate the heat loss and the heat gain for each space. Sum the heating or cooling loads in watts (British thermal units per hour) for each air handling unit and obtain the grand totals.

Step 3: From the appropriate designer, obtain the locations of the main pumps and piping, cranes or hoists, columns, stairs, beams, and other obstacles to air distribution. Also obtain the heat emissions from simultaneously operating equipment. Calculate the initial ventilation rates starting with 12 air changes per hour for wet wells (more in some states or for severe conditions) and 6 (or more) air changes per hour for dry wells. Make preliminary equipment selections. Estimate the outside air infiltration to each space by either the crack or the air change method, and calculate the expected outdoor air heating and cooling loads. See the *ASHRAE Handbook of Fundamentals* [12] for details of the crack and air change methods.

Step 4: From space volumes, heat gains, and design air change rates, calculate the normal and required purge air supply quantities for each space. If the summer airflow for heat removal exceeds 30 air changes per hour, consider evaporative or refrigerated cooling. Choose supply and return/exhaust register locations, assign the proper air quantity to each, and, from catalog data, select register sizes to give desired air throw (for supply) or velocity (return/exhaust). Note the corresponding pressure drops.

Step 5: Make an initial estimate of HVAC motor number and sizes for the electrical engineer. Fan motor horsepower can be approximated in SI units by

$$W = \frac{QR}{E} \tag{23-2a}$$

where W is brake watts, Q is the airflow in cubic meters per second, R is the static resistance in pascals, and E is the fan static efficiency as a decimal. In U.S. customary units,

$$bhp = \frac{QR}{6356E} \tag{23-2b}$$

where bhp is brake horsepower, Q is the airflow in cubic feet per minute, R is the static resistance in inches of water column, and E is the fan static efficiency as a decimal.

Lacking more precise information, use the fan resistance and efficiency assumptions given in Table 23-5 for the initial motor size calculations. Then, for safety, select the next larger standard motor size. Size all fans for at least 60 Pa (0.25 in. WC) static resistance. For propeller fans more than 0.6 m (2 ft) in diameter, use a motorized shutter to reduce resistance and to give more positive closure.

Step 6: Calculate the final heating and cooling loads and modify the initial air distribution rates as required to meet the allowable difference between the temperatures of the supply air and the room. Maximum temperature differences of 60°F for heating and 25°F for cooling are recommended.

Step 7: Determine feasible types and locations for air-handling equipment and lay out tentative duct runs. Avoid obstructions and consider how the ducts should be supported.

Step 8: Beginning with the outlets farthest from the fan, assign cumulative air quantities to each duct section. Using air friction charts or calculators, size the ducts (1) to fit the available space, (2) to give a streamlined airflow, (3) to look presentable, and (4) to simplify duct fabrication.

Step 9: Obtain gross face areas using velocities of (1) 1.3 m/s (250 ft/min) for wall intake and relief louvers, (2) 2.5 m/s (500 ft/min) for wall exhaust louvers, and (3) 5 m/s (1000 ft/min) for openings to roof exhaust fans. Confirm opening sizes and locations with the architect and structural engineer.

Table 23-5. Fan Static Pressures and Efficiency

Static Resistance[a]			
Without heat recovery		With heat recovery	
Pa	in. WC	Pa	in. WC
Directed supply			
250	1.0	620	2.5
Directed exhaust			
125	0.5	500	2.0

[a] Unless more accurate information is available, use in Equation 23-2(a) or (b) with an assumed fan static efficiency of 65% as initial assumptions for estimating the size of motor required.

Step 10: Draw the duct systems on the plan. From the resistances for registers, straight lengths of ducts, and fittings, cumulatively sum the actual friction drop of the longest duct run from the farthest point back to the fan. Duct friction can be calculated by the Darcy–Weisbach equation, which for air takes the form

$$p = f_D(cf \times L/D)p_v \quad (23\text{-}3)$$

where p is the total pressure friction loss in pascals (inches of water column), f_D is the friction factor (dimensionless), cf is the conversion factor, L is the duct length in meters (inches), D is the duct diameter in meters (inches), and p_v is the velocity pressure in pascals (inches of water column). The friction factor is a function of Reynolds number, however, and can be calculated only by iteration. Air resistance for straight ducts is more easily calculated with a slide rule made especially for this purpose, with a programmable calculator, or with a computer. Duct friction tables and charts and a listing of fitting friction losses are available in the *ASHRAE Handbook of Fundamentals* [12].

Step 11: Make final best selections of the heat transfer and air-handling equipment based on the design flowrate and the calculated total air resistance, including the friction drops for louvers, dampers, filters, coils, ducts, and registers. The friction drop for clean, pleated glass fiber filters is about 25 Pa (0.1 in. WC), but the fan motor must be sized to supply the required air at a dirty filter resistance of 120 to 160 Pa (0.5 to 0.65 in. WC). Filter and coil resistances are usually considered to be part of the external resistance—separate from the internal resistance of the air-handling unit given by the unit manufacturer.

Step 12: Show the selected air supply and exhaust equipment on the plans. Provide (1) raised housekeeping pads for floor-mounted units, (2) vibration-isolated supports for suspended and wall-mounted units, (3) flexible connectors for duct connections, and (4) ample access for filter, motor, belt, and bearing maintenance. Allow space for replacing heating and cooling coils and filters.

Provide direct access to the equipment. Floor-mounted units are, therefore, preferable to ceiling-hung equipment. Rooftop equipment is more likely to be maintained if it is accessible by stairs.

Step 13: Select the additional air-handling equipment for purging and summer heat removal. The equipment required may be only strategically placed high wall or roof exhaust fans and low dampered weatherproof wall intakes. Nonducted intakes and exhaust fans should be selected to result in no more than 60 Pa (0.25 in. WC) negative pressure within the space. A force of about 120 N (27 lb) is required to open a standard outwardly swinging door against such a vacuum.

Step 14: Recheck the actual motor kilowatts (or horsepower) needed for the designed systems and correct the motor list as required. Make sure motors are selected for the electrical characteristics available. Coordinate motor, starter, and control locations with the electrical engineer so that all of the necessary wiring is provided. Motors and controls should be listed by Underwriters Laboratories (UL). High-efficiency motors are cost effective for continuously operating fans and pumps. A specification requirement that air-handling equipment shall be rated in accordance with the AMCA No. 99 Standards gives increased assurance that it will meet the rated capacity.

Step 15: Specify each component of the system, stating each important criterion of construction and performance necessary to determine the acceptability of items subsequently submitted by the successful bidder. Equipment schedules are a simple way of summarizing required equipment designations, capacities, and special features. Write a performance description of (1) the intended sequence of operation and (2) the operating temperatures. Require a submittal of control components and schematic and wiring diagrams. Require acceptance tests to ensure that the installed system functions as intended and meets specification requirements.

23-8. Ventilating System Design

Pump rooms located where the outdoor temperature range is between freezing and 35°C (95°F) and in areas where noise need not be suppressed can often be ventilated without ductwork. Those rooms located below grade, however, require ductwork. For pump rooms at or above grade, louvered air intakes (with dampers and filters) and roof exhaust fans (with gravity shutters and sequencing thermostats) are sufficient. For lower outdoor temperatures, heating is usually required. For higher temperatures, some form of cooling is normally necessary.

Heating and cooling calculations are always necessary to determine the required extent of heat removal or heat addition. The air quantities to be handled, as derived from the heat load calculations, must be compared to those determined by required air change rates or other criteria, and the larger quantity must be used as the design basis.

Heating Load Calculations

Structure heat load is composed of (1) transmission loss through walls, windows, and roof and (2) infiltration of cold outside air, which must be heated to the design space temperature.

The transmission loss through any portion of the structure envelope is

$$H = U \times A \times \Delta T \qquad (23\text{-}4)$$

where, in SI units, H is the heat loss in watts per hour, U is the overall heat transmission coefficient in W/($m^2 \cdot h \cdot °C$), A is the surface area in square meters, and ΔT is the inside-to-outside temperature difference in degrees Celsius. In U.S. customary units, H is in British thermal units per hour, U is in Btu/($ft^2 \cdot h \cdot °F$), A is in square feet, and ΔT is in degrees Fahrenheit. Transmission coefficients for typical walls, windows, and roofs as well as methods of calculating the coefficient for the combinations of materials are given in Chapter 25 of the *ASHRAE Handbook of Fundamentals* [12].

Cooling Calculations

Heat gain throughout the structure, which makes up the external cooling load, is more difficult to calculate than heat loss because of the effects of radiated heat from the sun. This additional heat source must be considered in addition to the heat gain by conduction due to the difference between outdoor and indoor temperatures. Sunlight does not heat space air directly, but it raises the temperature of sunlit surfaces as a result of absorbed radiation.

Many factors influence the amount and timing of solar heat reaching the space. The intensity of incident solar radiation depends on latitude, time of year, time of day, cloud cover, and atmospheric pollution. The time lag between solar heat input and interior air temperature rise may vary from a few minutes to several hours. The lag is affected by the surface color of the exterior, the heat storage capacity and the insulating value of the construction, and the daily outdoor temperature range, among other factors.

The combined heat from both the higher outdoor temperature and solar radiation moves progressively through walls and roof, finally raising the interior surface temperatures. Loss from those surfaces occurs by radiation to cooler surfaces and by convection to the adjacent air. Air warmed by convection expands, becomes lighter, rises, and is replaced by cooler air, which continues the convective process.

Except for minor reflected losses, sunlight passes directly through clear glass and is absorbed by the interior surfaces it strikes. Their increased temperature transfers heat to other cooler surfaces by reradiation and to the air by convection.

Simply stated, solar heat gain is taken into account by using a higher outdoor temperature than actually would exist at the time considered for each heat gain calculation. The *ASHRAE Handbook of Fundamentals* [12] contains tables of cooling load temperature differential (CLTD) data (for use in calculating conduction heat gain through sunlit walls and roofs) and cooling load factors (CLF) (for calculating the solar radiation through glass). Both sets of data include the effect of time delay due to thermal storage. The total resistance to heat transfer for each type of construction is found by adding the resistances of its components, as illustrated in Example 23-1.

Peak Heat Gain

The peak heat gain to a space is the largest sum of external and internal heat gains that occur simultaneously. Heat removal capacity, in the form of ventilation or cooling, equal to the peak heat gain must be available to maintain the design space temperature. External heat gain results from an outdoor temperature that is higher than the indoor temperature as well as from solar radiation. Internal heat gain comes primarily from operating motors and engines.

Air Intake and Exhaust Openings

Intakes for ventilation air should be through screened louvers that exclude rain, snow, birds, and insects. Ducted intakes that are connected to air-handling units with filters that exclude insects need only bird screens. Nonducted intakes should include insect screens or filters. All screens and filters must be readily accessible for cleaning or replacement. Bird screens do not need mesh openings smaller than 25 mm ($^1/_2$ in.). The air resistance of filters used at unducted wall intakes should not exceed 24 Pa (0.1 in. WC) when clean to limit negative pressure in the space (created by exhaust fans) to 63 Pa (0.25 in. WC); low negative pressure makes doors easier to open and close. Both the louver and screen should be constructed of a corrosion-resistant material or should have a corrosion-protective coating. Provide an intake damper at each louver in cold climates, preferably with positive closure on failure of its pneumatic or electric actuator. Hurricane-prone locations may warrant an additional manually operated damper (on the outdoor air intake) that can be locked closed when a storm approaches.

Air exhaust openings should be protected from the weather by a wall louver, hood, penthouse, or weather cap. The discharge openings should be located so as to minimize unwanted recirculation into air intakes.

A high-velocity vertical discharge of untreated wet well exhaust air—about 20 m/s (4000 ft/min), which requires 250 Pa (1 in. WC) static pressure—may reduce odor near the station (but with attendant noise). A thorough model study of the likely results (which are not readily predictable) should precede any use of this approach for odor control.

Ducted Ventilation

A ducted air supply is usually necessary to achieve the desired air distribution pattern. Once delivered to the right location, the air can be directed upward or downward by registers (grilles with directional vanes and attached balancing dampers), or it can be more thoroughly mixed with room air by diffusers that give high entrainment of room air. The distance an airstream travels after leaving the outlet (the "throw") depends on its velocity, temperature, and any obstructions to straight-line flow. Catalog data on grilles, registers, and diffusers give the throw and pressure drop for a given flowrate of standard air. The throw is longer for an outlet discharging horizontally within about 0.6 m (2 ft) of a smooth ceiling due to the low induction of room air.

Install a manual volume damper in each main duct branch and at each inlet or outlet grille to allocate or "balance" the total air quantity among the various points of supply and exhaust. Such dampers are furnished as a standard part of registers, but must be separately specified for grilles and diffusers. Complete testing and balancing of ducted systems should be specified to deliver the design air quantity within 10% at each register and diffuser.

Wherever an air duct penetrates a rated fire separation required by code, install an accessible fire damper of equal fire rating. Codes generally require rated fire separations between floors, around vertical shafts and stairs, at the envelope enclosing life safety exitways, and, in some instances, around mechanical equipment rooms and storage rooms. The fire damper is held open by a fusible link that melts and allows the fire damper to close automatically if the air temperature rises above its fusion point 52 or 74°C (125 or 165°F). The fusible link is selected on the basis of location and normal air temperature. Such dampers are required by NFPA Standard 90A (which is included by reference in many other codes) to prevent fire and smoke from spreading through the air ducts. High-temperature thermostats or smoke detectors (as required by applicable codes) should be provided and interlocked to stop both the supply and the exhaust fans if fire is detected. Some codes require tight-closing, remotely operable smoke dampers (which are actuated by smoke detectors) at certain locations.

Detailed duct design methods are given in Chapter 33 of the *ASHRAE Handbook of Fundamentals* [12].

Ventilation Controls

Excessive and intricate controls for ventilation systems (or any HVAC system) are likely to result in more operational problems than they were intended to overcome. The rule is to keep controls as simple as feasible while still maintaining the temperature within an appropriate range. A temperature variation from above freezing to 43°C (110°F) would not be serious in an unmanned pumping station, but in an occupied office or control room the temperature variation should be held to a range of 6°C (10°F) or less.

The following are suggested ventilation control functions:

- Energize controls when the supply fan starts.
- Control outside-return air mixing dampers to use at least 10% outside air initially and modulate them to increase the proportion of outdoor air as the space temperature rises, subject to a discharge low-limit control set at 7°C (45°F) minimum.
- Interlock the exhaust fan to run with the supply fan, and the exhaust damper to operate in conjunction with the outside air damper.
- Interlock the supply fan to stop if the supply air high-limit thermostat or smoke detector functions, or if the supply air low-limit thermostat senses a temperature below 3°C (37°F).
- For a wet well, start a supplementary exhaust fan and open an air intake damper to purge the space if a gas detector senses 20% of the LEL of a combustible gas.
- For a dry well, start an additional exhaust fan and open an air intake damper when a space high-temperature thermostat senses a temperature above 30°C (85°F).

Example 23-1
Design of a Ventilating System

Problem: Plans for a water pumping station are shown in Figures 23-1 through 23-4. There are currently four 200-hp pumps (three duty, one standby), but six 350-hp pumps (five duty, one standby) will be used in the future. Although the office is currently occupied, the station will

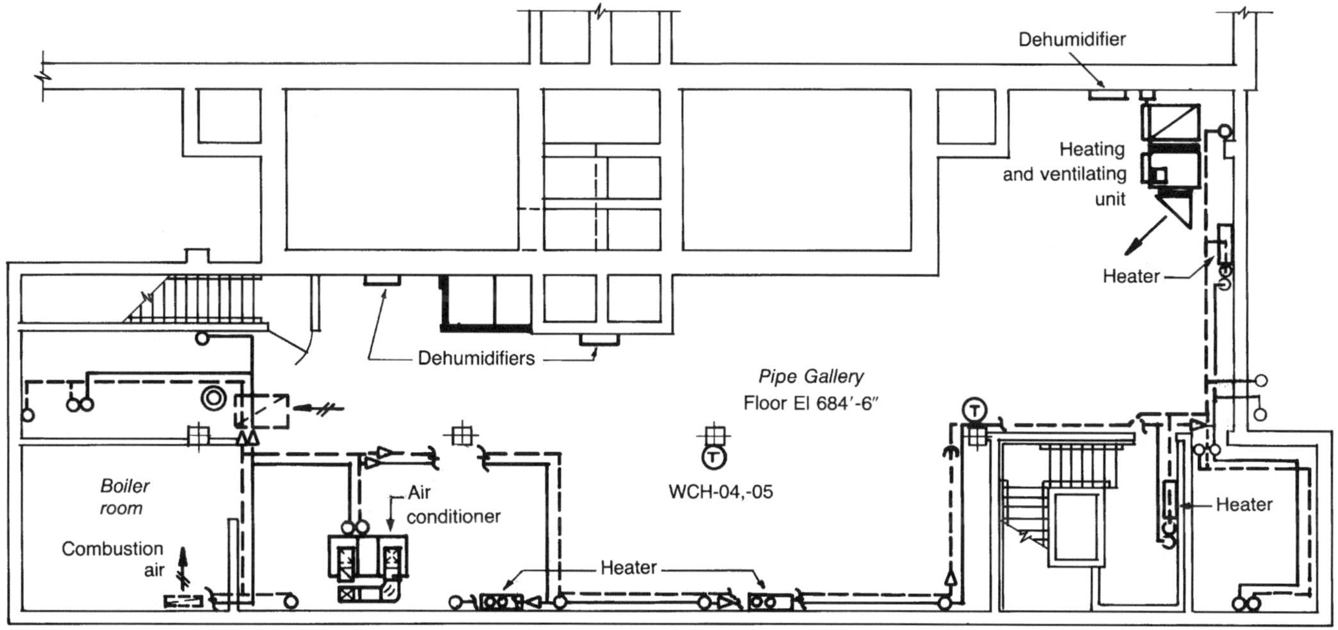

Figure 23-1. Basement plan.

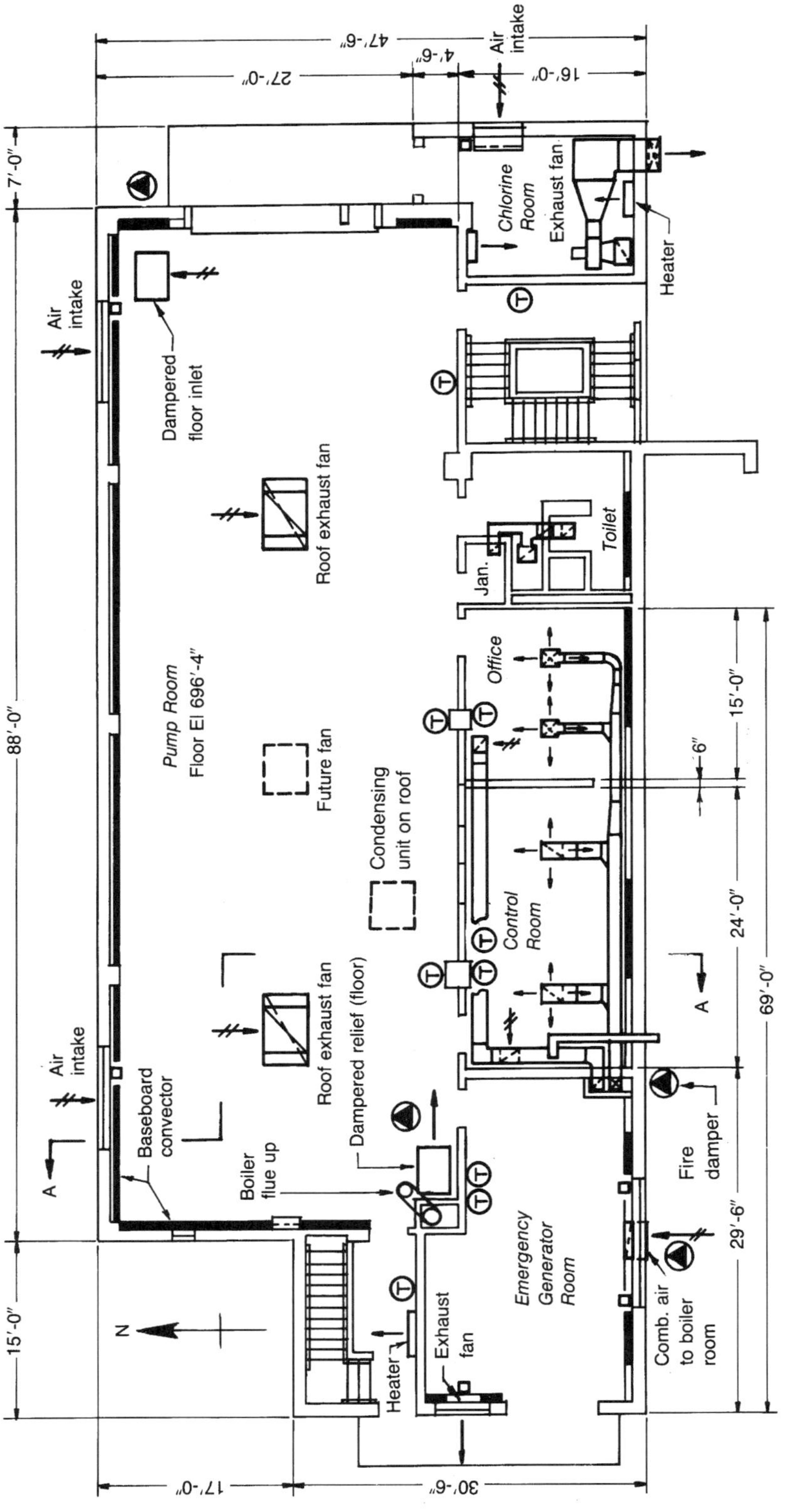

Figure 23-2. Ground-level floorplan.

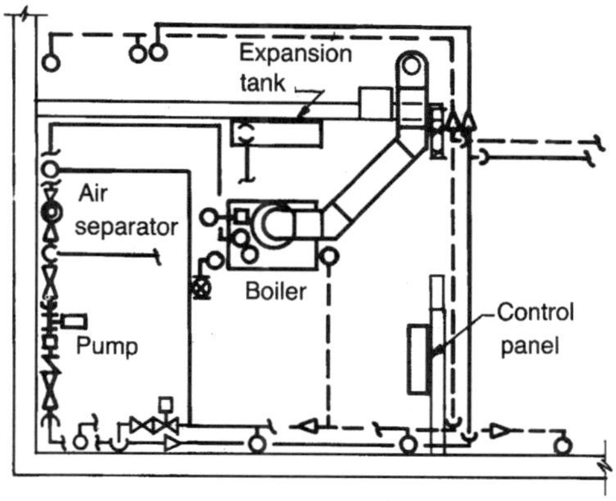

Figure 23-3. Boiler room plan.

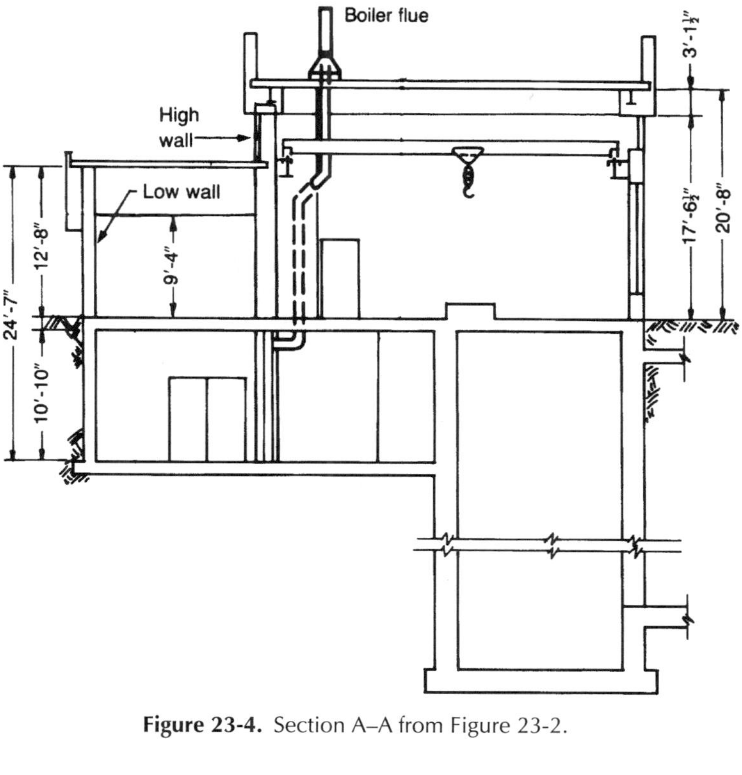

Figure 23-4. Section A–A from Figure 23-2.

also be unattended in the future. The latitude is 41.5° north, the elevation is 660 ft, and the daily temperature range is 20°F. Other temperature data are as follows:

Design conditions	Winter	Summer
Time equaled or exceeded	99%	1%
Outside temperature, dry bulb	–20°F	94°F
Outside temperature, wet bulb	—	74°F
Inside temperature (occupied)	70°F	78°F
Inside temperature (unoccupied)	60°F	105°F

Calculate (1) the heat transmission coefficients for superstructure walls, roof, and glass; (2) the heat loss for the pump room; (3) the present heat gain for the pump room; (4) the future heat gain for the pump room; and (5) the requirements for heat and ventilation.

Solution:

(1) Heat transmission coefficients. Refer to the *ASHRAE Handbook of Fundamentals* [12], pp. 23.12–23.20 and Tables 1, 3A, and 3B. Sum the component thermal resistances (R) for each surface of the enclosure and invert ($1/R$) to obtain the heat transmission coefficients (U) values (i.e., $U = 1/R$). An example follows.

Lower walls		Roof		Glass
				Double solar bronze, nonreflective
Outside surface:	0.17	Outside surface:	0.17	
4-in. face brick:	0.44	Built-up roof:	0.33	
1-in. styrofcam:	4.0	3-in. lightweight fill	4.69	
1-in. air space:	0.94	1-in. polyurethane	6.25	
6-in. concrete block:	0.91	6-in. flexicore	0.66	
2-in. glazed tile:	0.60			
Inside surface:	1.35	Inside surface:	0.61	Winter $R = 2.05$
	$R = 8.41$		$R = 12.71$	Summer $R = 1.79$
	$U = 1/8.41 = 0.12$		$U = 1/12.71 = 0.079$	$U = 0.49$[a]
				$U = 0.56$[a]

[a] Factors for heat transfer through glass differ for winter and summer conditions (see ASHRAE [12], p. 27.10, Table 13).

(2) Heat loss for pump room. Calculate the heat losses using Equation 23-4 ($Q = A \times U \times \Delta T$). (Also see the *ASHRAE Handbook of Fundamentals* [12], p. 25.1ff.) For algebraic consistency, heat gains are shown as positive and heat losses as negative. See Figure 23-4 for high and low walls. In tabular format, heat losses are as follows:

Low walls above grade: $(88 + 17 + 27)12.7 \times 0.12(60 + 20) =$	–16,094
High walls: $2(31.5 + 88) \times 3.3 \times 0.28 \times 80 =$	–17,667
Glass: $2(85 + 30)3 \times 0.49 \times 80 =$	–27,048
Doors: $[(12 \times 10) + (3 \times 7)](0.59 - 0.28) \times 80 =$	–3,497
Roof/ceiling: $30 \times 88 \times 0.079 \times 80 =$	–16,685
Infiltration (based on 0.75 AC/h at 80°F ΔT):	
$(20.7 \times 88 \times 30 \times 0.75 \times 1.08 \times 80)/60 =$	–59,020
Winter pump room heat loss (Btu/h):	–140,011

(3) Pump room present summer heat gain. Refer to ASHRAE [12], pp. 26.8, 26.19, 26.29, 26.32, and Tables 5A, 11A, and 24. Note that heat gain due to motors is

$$\text{Heat gain (Btu/h)} = bph\left(\frac{1-E}{E}\right)2545 \quad (23\text{-}5)$$

where bhp is the brake horsepower, E is the overall motor efficiency as a decimal, and 2545 is the British thermal units per hour · brake horsepower.

For the cooling load temperature difference (CLTD) adjustment for roof and walls in the following example, see ASHRAE [12], pp. 26.8 and 26.10, Tables 5 and 7, Note 2. For glass transmission factors, see Tables 10, 11, and 13, pp. 26.14–26.17, and Table 29, p. 27.29 together with their accompanying explanations and example. For our example, maximum heat transfer occurs at 4 P.M. on July 21. The pump room wall is Group No. C. Directions are identified using capital letters (N = north, etc.).

Structure solar and transmission heat gain

The heat gain is determined as follows:

Roof with a suspended ceiling (see ASHRAE [12], p. 26.8, Table 5, Roof 12):

$0.079 \times 30 \times 88[(30 + 1) + (78 - 104) + (84 - 85)] =$	+834
N wall:	
$0.12 \times 15.3 \times 88[(12 - 0)0.83 + (78 - 104) + (84 - 85)] =$	–2,753
E wall:	
$0.12 \times 15.3 \times 28[(29 - 0)0.83 + (78 - 104) + (84 - 85)] =$	–151
W wall:	
$0.12 \times 15.3 \times 18.6[(16 - 0)0.83 + (78 - 104) + (84 - 85)] =$	–469
S wall (upper):	
$0.12 \times 85 \times 1.5[(20 - 1)0.83 + (78 - 104) + (84 - 85)] =$	–172
N glass, solar:	
$3 \times 86 \times 38 \times 0.70 \times 0.55 =$	+3,775
E glass, solar:	
$3 \times 30 \times 216 \times 0.24 \times 0.55 =$	+2,566
W glass, solar:	
$3 \times 30 \times 216 \times 0.49 \times 0.55 =$	+5,239
S glass, solar:	
$3 \times 86 \times 109 \times 0.43 \times 0.55 =$	+6,651
Glass conduction:	
$3(86 + 30)2 \times 0.56(-12) =$	–4,677
Total summer structure external heat gain (Btu/h) =	+10,843

Internal heat gain

Pump motor heat gain in Btu/h (from Equation 23-5):	
$3 \times 200[(1 - 0.9)/0.9]\ 2545 =$	+169,667
Estimated auxiliary motor heat:	
$20[(1 - 0.67)/0.67]\ 2545 =$	+25,070
Heat from lights:	
$30 \times 88 \times 2 \times 3.413 =$	+18,021
Heat added by people:	
$2 \times 350 =$	+700
Total internal peak heat gain in Btu/h:	+213,458

Present total internal heat gain: Adding the structure heat gain to the internal peak heat gain yields the pump room's present maximum heat gain:

+213,458 + 10,843 =	+224,301

(4) Pump room future internal heat gain.

Pump motor heat gain:	
$5 \times 350\ (10/90)\ 2545 =$	+494,861

Auxiliary motor heat gain:

$300 \times 0.67\ (10/90)\ 2545 =$	+56,838

Lights:

$30 \times 88 \times 2 \times 3.413 =$	+18,021

People:

$2 \times 350 =$	+700
Total internal heat gain =	+570,420
Total summer structure external heat gain [from part (3) above]:	+10,843
Pump room future maximum heat gain:	+581,263

(5) Ventilation requirements. The present maximum summer ventilation requirement for a 10°F space temperature rise above the outdoor temperature is as follows: The present exhaust in cubic feet per minute is $224{,}301/(1.08 \times 10) = 20{,}769$ ft^3/min. The total future exhaust in cubic feet per minute is $581{,}263/(1.08 \times 10) = 53{,}821$ ft^3/min.

Currently use two exhaust fans at 17,940 ft^3/min each and include a capped curb for a third identical fan in future. Thermostats are provided to turn on the three fans and open corresponding outdoor intake dampers in sequence at 85, 90, and 95°F, respectively.

23-9. Design of Heating Systems

Any structure for which the heat loss exceeds the simultaneous heat gain for a period of time will experience a drop in indoor temperature. To maintain the design indoor temperature, a heating system of some type must add an amount of heat equal to that net heat loss at any given time.

Heating System Types

Heating systems can be categorized in the following ways:

- By heat source—steam, hot water, warm air
- By energy source—gas, oil, electricity
- By distribution method—central system, unitary

Any combination of the three categories is possible. In stations with few individual spaces and those in warm climates, only unitary equipment (such as cabinet heaters or unit heaters) is needed. For small stations, an oil- or gas-fired warm-air furnace is frequently applied and supplemented by electric unit heaters or baseboard heaters in spaces such as chlorine rooms and toilets from which code prohibits return air. Small stations require minimal or no ducting, although a return duct taking air from the lowest level prevents the pooling of heavy cool air in the bottom space.

Duct distribution systems are used with furnaces or other air-handling units to distribute warm air (including ventilation air) to multiple rooms and separate floors. In the simplest system, the heat source is cycled on and off by a single thermostat located in the most representative space. The temperature is controlled in that space only, and some variation in the temperature of other spaces may be expected.

Heating Load

The extent of heat loss depends primarily on the insulating value of the building envelope and the inside-to-outside temperature difference, and secondarily on wind velocity and outdoor air infiltration. In addition to the structure heat loss, heat also is required to raise the outdoor ventilation air to room temperature. The indoor design temperature is set by the designer. The setting should be high enough to keep the temperature of interior surfaces above the interior air dew point to prevent condensation. An indoor dry-bulb temperature of 13°C (55°F) usually meets that criterion for a well-insulated building and allows for normal maintenance activities without excessive discomfort. A lower indoor temperature (but well above freezing) may be acceptable in an unstaffed station if (1) the heating system has the capacity to raise the space temperature as needed for extended maintenance work or (2) portable auxiliary heating equipment is available.

The design outdoor temperatures for many locations are listed in the *ASHRAE Handbook of Fundamentals* [12]. If continuous operation of the pumping station is critical, use the 99% winter design outdoor dry-bulb temperature. Lower temperatures normally occur no more than 1% of the time. Additional weather data can be obtained from the National Climatic Center of the National Oceanic and Atmospheric Administration in the Department of Commerce and from Engineering Weather Data [13].

To calculate the heat loss from a structure, determine the U value, area, and temperature difference for each surface exposed to the outdoors and substitute those related values in Equation 23-4 to obtain the heat loss. The sum of the losses through all such surfaces in each room is the room transmission heat loss, and the total for all rooms is the building transmission loss. To this must be added the heating loads due to the infiltration of outside air, outdoor air introduced to ventilate the space, and the evaporation of moisture occurring within the space. The infiltration heat requirement can be decreased or disregarded if a positive pressure is continuously maintained in each space.

Basic heat load calculations for a small hot-water heating system are given in Example 23-2, and methods for designing warm-air, steam, and hot-water central systems are detailed in Chapters 12, 13, and 15, respectively, of the *ASHRAE Handbook: Systems* [14].

Example 23-2
Design of a Heating System

Problem: Calculate the pumping station heating load and select a hot-water boiler for the water pumping station in Example 23-1.

Solution: The method of calculating the heat loss for the pump room is shown in Example 23-1. Use the same method to calculate the heat loss for other rooms and sum all of the losses to arrive at the total heat loss for the structure.

Because perimeter heat is desirable to offset wall heat losses and to control the temperature in several spaces independently, a system of hot-water heat using a gas-fired boiler is chosen. Natural gas is preferred over fuel oil or electric heat at this site primarily because (1) it is available at a firm rate; (2) it meets the fuel needs for the building heating system; (3) it can power the required standby generator; and hence, (4) it offers low initial cost, simple controls, and low maintenance. A gravity-fired, cast-iron boiler is selected for its simplicity and expected long life (refer to AMCA No. 99 Standards, p. 24.3).

To minimize the danger of freezing in the event of a power or control outage, a 20% solution of ethylene glycol in water is used for the heating medium (see p. 15.19, AMCA No. 99 Standards). Combustion air is supplied by gravity flow at a net face velocity of 200 ft/min to the boiler room through a wall louver, which is sized to pass 12 times more air than the natural gas to be burned. This quantity of air is 20% greater than the stoichiometric volume for natural gas with a heat content of 1000 Btu/ft^3 (high heat value).

The current winter heat gain and loss in British thermal units per hour (for the pump room only) are summarized as follows:

Three 200-hp pumps running (maximum):	
Total winter pump room heat gain (from Example 23-1, part 3)	+224,301 Btu/h
Less structure heat loss (from Example 23-1, part 2)	–140,011
Excess heat to be removed	+84,290
One 200-hp pump running:	
Heat gain from one pump only: 169,667/3	+56,556
Auxiliary motor heat: 25,070 × 0.5	+12,535
Lights and people	+18,721
Total internal heat gain	+87,812
Structure heat loss (Example 23-1, part 2)	–140,011
Minimum ventilation load: 1.08 (–1000) 70 (see below)	–75,600
Total internal heat gain	–87,812
Operating net heat deficiency	–127,799

With all pumps off, the station not operating, no ventilation, no lights, and no people, the building heat loss is 140,011 Btu/h.

The minimum outside air rate for ventilating this city water pumping station was selected as 1000 ft^3/min, or roughly double the calculated infiltration. The ventilation air heating load in British thermal units per hour is

$$H = W \times c \times \Delta T \tag{23-6}$$

where W is the weight of air in pounds per hour (60 γ ft^3/min = 1.08 lb/h), c is the specific heat in British thermal units per pound-degree (0.2395 from Table A-3), and ΔT is the difference between outside and inside temperatures in degrees Fahrenheit. Note that W can be expressed as $Q \times 60 \times \gamma$, where Q is the flowrate in cubic feet per minute and γ is the specific weight of air in pounds per cubic feet (0.0752 at sea level from Table A-7). Collecting and multiplying the constants gives

$$H = 1.08Q \times \Delta T \tag{23-7}$$

The outside air (ventilation) heat load in British thermal units per hour is

$$H_{OA} = 1.08(-1000)[60 - (-10)] = -75{,}600$$

Heat equal to the greatest operating net heat deficiency (127,799 Btu/h from the previous calculations when only one pump is running) must be added to the pump room to maintain the design temperature. Heating equipment capacity should be adequate to maintain room temperature at 45°F or higher under any condition.

$$\text{Boiler input} = \frac{(HD + H_{OA})1.2}{0.8} \tag{23-8}$$

where HD is the net heat deficiency in British thermal units per hour, 1.2 is a "piping and pickup" allowance for generated heat not conveyed to the heated space and for extra capacity for quicker warmup from a reduced setback temperature, and 0.8 is the boiler thermal efficiency. Note that flue gas heat loss (including heating of the combustion air) is included in the thermal efficiency factor. Calculations (in Btu/h) for the boiler size are:

• Operating pump room net heat deficiency (1 pump)	–127,799
• Calculated remainder of building heat loss (not shown)	–77,201
• Minimum ventilation heat load	–75,600
• Total building heating load	–280,600
• Boiler input 280,600(1.2)/0.8	420,900

Use the next larger standard boiler with an input of 450,000 Btu/h.

Required combustion air (12 times the gas to be burned) is

$$\frac{(450{,}000 \text{ Btu/h})(12)}{(1000 \text{ Btu/ft}^3)(60 \text{ min/h})} = 90 \text{ ft}^3\text{/min}$$

The louver face area is

$$(90 \text{ ft}^3\text{/min})/(200 \times 0.25 \text{ free area}) = 1.8 \text{ ft}^2$$

Specify an automatic vent damper in the boiler flue to minimize convection losses from the hot boiler when the burner is not firing (see p. 16.5 of the *ASHRAE Handbook: Equipment* [15]).

Extend a Type B double-wall or insulated full-size steel flue through the roof to a height that is at least 2 ft above a parapet (or other obstruction within 10 ft horizontally). Provide a ventilated sleeve and weatherproof flashing at the roof and an approved weatherproof flue cap at the top.

Additional Required Design Steps

Other design steps are necessary to complete Example 23-2.

Size the Auxiliary Heat-Transfer Equipment

Select finned radiation, a unit heater, or a cabinet heater with the capacity to offset the calculated heat loss for each room. Base the calculation on an average water temperature of 93°C (200°F) and a temperature drop of 11.1°C (20°F) (see pp. 9.3 and 27.4 of the *ASHRAE Handbook: Equipment* [15] and p. 30.5 of the *ASHRAE Handbook: Systems* [14]).

Water Flow

Calculate the required heating water flow for each heating unit from Equation 23-9. In SI units,

$$Q = \frac{\text{Heat load}}{4184(\Delta T)} \tag{23-9a}$$

where Q is the liters per second, the heat load is in kilowatts, ΔT is the temperature difference in degrees Celsius, and 4184 is the specific heat times mass in watt-seconds per kilogram. In U.S. customary units,

$$Q = \frac{\text{Heat load}}{500(\Delta T)} \tag{23-9b}$$

where Q is in gallons per minute, the heat load is in British thermal units per hour, 500 is 8.34 lb_m/gal × 60 min/h, and ΔT is the temperature difference in degrees Fahrenheit.

Plans

Sketch the proposed heating circuits including the boiler, the pump, the compression tank, and the supply and return piping for each heating unit. A reversed return piping arrangement (which gives equal circuit lengths through all heating elements) is desirable to simplify the balancing of system water flow (see p. 16.3 of the *ASHRAE Handbook: Systems* [14]). Connecting the compression tank to the system main pipe just upstream of the heating pumps ensures that the pumping head is positive throughout the system. This prevents air from leaking into the system, which causes air binding (flow impedance due to the accumulation of air at high points). Any leaks can be located by water drips.

Sizing the Heating Pipe

Size the heating water piping based on the cumulative flowrate for each section of piping within a velocity range of 1.2–2.1 m/s (4–7 ft/s) and an approximate flow resistance of 0.4 kPa/m (4 ft WC/100 ft) of pipe. Begin by sizing the piping at the most remote unit supplied and at the first connection to a reverse return main and, accumulating the flowrate progressively, work back to the pump through each main.

Pressure Drop

Total the calculated pressure drops through the longest circuit of piping, including the fittings, valves, coils, and other system components (see p. 34.1 of the *ASHRAE Handbook of Fundamentals* [12]).

Heating Pump

Select the heating pump needed to deliver the total cumulative flowrate against the total calculated flow resistance by using a pump manufacturer's catalog data (see p. 30.5 of AMCA No. 99 Standards). The pump should be located downstream of the boiler with the compression tank connection at the pump inlet to avoid adding the pump head to the static pressure on the boiler.

Air Elimination

An air separator with an automatic float-type air relief valve should be connected on the discharge side of the pump. Manual or automatic air vents should also be located at the high points of the piping system and at each piping drop in the direction of flow. It is important that all air be removed from the heating water to prevent air binding and to minimize internal corrosion.

Water Volume

Calculate the total water volume in the system by adding the individual volumes of piping sections, coils, convectors, boiler, air separator, and compression tank (see p. 15.40 of the *ASHRAE Handbook: Applications* [16]).

Compression Tank

A compression tank is needed to accommodate the expanding volume of system water as it is heated. A diaphragm type of tank, in which a flexible elastomeric diaphragm separates a closed air compression space from the heating water, is recommended. The flexing diaphragm compresses the contained air as the water expands and vice versa. Size the compression tank to accept expansion of the system water as it heats from, for example, 10°C (50°F) to the design supply temperature without exceeding the allowable pressure setting of the boiler pressure relief valve, which is 186 kPa (27 lb/in.^{2} ga) for a low-pressure boiler rated for 207 kPa (30 lb/in.^{2} ga) operating pressure.

Maintenance

On a circuit diagram, locate the isolating, balancing, drain and control valves as well as thermometers and other accessories needed for maintenance and service of the system.

Gas Piping

In accordance with the chapter on pipe sizing in the *ASHRAE Handbook of Fundamentals* [12], size the gas piping for a gas flow equal to the boiler input rating at the pressure required by the boiler burner. This pressure is normally about 1.0 to 1.3 kPa (4 to 6 in. WC) for a gravity-fired natural gas burner. The gas piping installation should be specified to meet NFPA Standard 54.

Safety Provisions

Select and specify required operating and safety controls for the hot water heating system (see p. 24.3 of AMCA No. 99 Standards). The following controls are typically included:

- Boiler water temperature operating control to operate the burner
- Boiler water high-temperature limit control with manual reset to shut off the burner upon excessive temperature
- Boiler water low-level cutoff control to shut off the burner if the system water falls to an unsafe level
- Boiler overpressure relief valve rated for maximum fuel input to the burner in accordance with ASME Code VIII
- Boiler water make-up provision. If make-up water is from a potable source, it must be connected (1) through a backflow preventer or a break tank in accordance with the applicable code and (2) through a pressure-reducing station that includes a gate valve, strainer, and pressure-reducing valve. Set the pressure-reducing valve to maintain the minimum boiler pressure needed to give at least 1.5 m (5 ft WC) pressure at the highest point in the system piping with the heating pumps off. This minimum pressure ensures positive venting of air from the system.

Chemical Water Treatment

A means of introducing water treatment chemicals into the system is needed to prevent internal corrosion or scale formation. The simplest device is a chemical feed pot with isolating and drain valves. This permits adding antifreeze and corrosion inhibitors to be added when needed as shown by testing heating water samples. The pH of the heating system water should be maintained at about 9.0.

For steam boilers having direct steam usage and resulting condensate loss and makeup, an automatic chemical feed system should be used to supply the required chemicals at the rate recommended from tests conducted by a competent chemical company to prevent boiler corrosion and scaling.

Temperature Controls

Select and specify automatic temperature controls to operate the heating system (see p. 15.16 of the *ASHRAE Handbook: Applications* [16]).

23-10. Design of Building Cooling Systems

The removal of excess heat can be accomplished by moving outdoor air through the building as long as the temperature of the outdoor air is sufficiently below the design indoor temperature. If not, evaporative cooling can reduce the dry-bulb temperature if the initial wet-bulb temperature of the air is well below the desired indoor temperature. Refrigeration, either as an additional stage of the evaporative cooling process or as the sole cooling method, may be necessary if still more cooling is required.

Objectives other than cooling the air may include reducing its moisture content, minimizing its corrosiveness and odor, removing dust and aerosols, and achieving an acceptable air movement and sound level in the conditioned space. The term "air conditioning" encompasses all aspects of air enhancement. The extent of the air treatment applied varies with the requirements of the application. For example, air conditioning for an industrial process may be quite different from air conditioning for human comfort. Where both types of air conditioning are involved, some compromises may be necessary.

Evaporative Cooling

Evaporative air cooling (especially of equipment spaces) is often practical where the design summer wet-bulb temperature is lower than about 24°C (75°F), as is true in most of the Western states. Wet-bulb temperatures for many locations are listed in the *ASHRAE Handbook of Fundamentals* [12]. The heat content of air at any combination of dry-bulb and wet-bulb temperatures can be read from a psychometric chart for the site altitude. Humid air (with its higher wet-bulb temperature) has a higher heat content than dry air at the same dry-bulb temperature and is less susceptible to evaporative cooling. An indirect evaporative cooling first stage, followed by direct evaporation (called "indirect–direct"), is a more effective method. A high-limit humidistat or some other means of stopping or reducing evaporation (thereby also limiting the cooling effect) may be necessary to prevent excessive humidity.

Due to the smaller temperature differential between supply and room air temperatures, the evaporatively cooled airflow rate may often need to be 20 to 30 air changes per hour as compared with 8 to 12 changes for refrigerated cooling. Such higher rates require larger fans and ducts and better air distribution than the lower rates. The larger airflows also tend to create more noise and drafts. Recirculation of evaporatively cooled air is not practical due to the resulting buildup of humidity.

Example 23-3
Design of an Evaporative Cooling System

Problem: Design an evaporative cooling system for the unoccupied pump room of Example 23-1. Assume that a bridge crane in the pump room leaves too little headroom for ductwork below the roof, but that roof-mounted equipment is aesthetically acceptable.

Solution: As shown in part (4) of Example 23-1, the sensible heat load to be removed after the future increase in station capacity is 581,263 Btu/h. An indirect–direct evaporative cooling system is used to obtain maximum cooling with the lowest air supply quantity.

Supply air. The wet-bulb depression (the difference between outdoor dry-bulb and wet-bulb temperatures) is 94°F – 74°F = 20°F. Manufacturer's data for an indirect evaporative cooler shows that a unit rated at 10,000 ft^3/min can directly cool the waste airstream to within 5 degrees (called the "approach") of its wet-bulb temperature, or to 79°F dry bulb. The evaporatively cooled waste airstream goes through one path of the indirect cooler and flows across the supply airstream in the other path. According to the manufacturer's literature, the supply air is thereby cooled to 85°F dry bulb, and its wet-bulb temperature is reduced to 71.3°F. In passing through the second (direct) stage cooler, the supply air dry-bulb temperature is reduced to 76.3°F, or within 5 degrees of its new wet-bulb temperature. Simultaneously, however, the addition of moisture increases its total heat content by an equivalent amount, which leaves its wet-bulb temperature (the measure of total heat content) unchanged at 71.3°F. Direct evaporative cooling is an adiabatic process (or one in which total heat content remains unchanged).

Supply air equipment. The dry-bulb temperature difference between the evaporatively cooled supply air and pump room design temperature is 105°F – 76.3°F = 28.7°F. The airflow rate required to remove the calculated future pump room heat gain is, from Equation 23-lb,

$$Q = \frac{581{,}263}{28.7 \times 1.08} = 18{,}753 \text{ ft}^3\text{/min}$$

Two indirect–direct cooling units will be needed, one initially and another in the future, each supplying 9500 ft^3/min to the four diffusers located symmetrically in the room length. Each supply fan is selected for a capacity of 9500 ft^3/min at 1.25 in. WC. Using Equation 23-2b, the size of the motor is

$$\text{bhp} = \frac{9500 \times 1.25}{6356 \times 0.65} = 2.9$$

Use a 5-hp, 3-phase, 60-Hz motor. From a manufacturer's literature, select four round supply diffusers to deliver 4750 ft^3/min each that are 42 in. in diameter and have adjustable cones to give a downward projection. Specify installation of the first rooftop unit on a prefabricated curb with weatherproof counterflashing for the roofing. Specify a similar prefabricated curb with an insulated weatherproof cap for an equal fan unit to be installed in the future, when the station pumps are increased.

Water supply. Pipe city water through a backflow preventer to preclude contamination of the city water supply. Connect a $^3/_4$-in. city waterline through a ball valve, strainer, and three-way solenoid drain valve within the building to the float valve on each cooler. Provide a solenoid drain valve between the cooler pan drain and the common drain line. The solenoid drain valves are activated by an outdoor thermostat and they shut off the water supply and drain the supply pipe and cooler pan when the outdoor temperature drops to 60°F. The makeup water float valve in each unit is set to a level slightly above the overflow standpipe to prevent the buildup of minerals from water evaporation by allowing a small amount of water to be continuously bled off during operation.

Air controls. Install control dampers at the inlet of each supply air evaporative cooler to permit air recirculation from the pump room ceiling in cold weather when at least one of the supply fans runs without evaporation. Enough outdoor air is admitted (under the control of a room thermostat) to remove any net heat gain. Insulate and weatherproof the supply unit casings and ductwork above the roof to minimize heat loss in cold weather.

Cooling controls. The supply fans are manually started. One could be shut off automatically by an outdoor thermostat set at 50°F, below which a single unit is sufficient. A space thermostat modulates the inlet mixing dampers of the operating unit(s). Motorized dampers at the two relief louvers modulate in conjunction with the outdoor dampers on the supply units. The relief louvers are sized on the basis of a 250 ft/min face velocity or 38 ft^2 each. When the outside air damper is fully open and the space temperature rises above 85°F, the evaporative system water solenoid valves are activated and the waste airstream fans are started. A discharge low-limit control in the supply air from each unit takes control of the mixing dampers if the supply air temperature drops below 45°F. A smoke detector in each supply airstream, as required by NFPA Standard 90A, shuts off the fans and energizes a remote alarm if combustion products are sensed. For an unattended station, provide pump room high- and low-temperature warning signals to an attended station.

23-11. Design of Refrigerated Cooling Systems

Both the temperature and the humidity of air affect the comfort of occupants because they both affect the rate at which the body loses heat. High humidity causes discomfort even at normal air temperatures, so both humidity and temperature are controlled in an air conditioning system. Because of its power demand and relatively high operating cost, refrigerated air conditioning is mainly used where (1) human comfort or reduced humidity are prime considerations or (2) the maximum allowable ambient temperature for installed equipment cannot otherwise be maintained.

Thus, room comfort is affected by (1) "sensible" heat gain (from solar radiation, thermal transmission, internal motors, and lights or other dry heat sources) and (2) "latent" heat gain (from water evaporation and moist ventilation air). Both should be evaluated and controlled.

Different types of refrigerated air-conditioned equipment can be contrasted as follows:

- Packaged units versus custom-designed systems
- Direct expansion coils versus chilled-water coils
- Air-cooled refrigerant condensers versus evaporative or water-cooled types.

Heat Gain Calculations

Determine the maximum sum of simultaneous heat gains to the space or "peak heat gain" for the year. This maximum load is used to establish the required capacity of the refrigerating equipment. It is usually necessary to calculate the heat gains for air-conditioned spaces at more than a single time. Heat gain varies with the time of year, time of day, orientation and construction of the building, and the internal heat gains present (such as from occupants, motors, engines, and lights). The procedures for these calculations are outlined on p. 26.3 of the *ASHRAE Handbook of Fundamentals* [12].

Latent Heat Gain

Latent heat gain in pump rooms usually is of concern only if the pumped water temperature drops below the indoor air dew point temperature, which results in moisture condensation on piping, valves, and walls. In wastewater pumping stations, such occurrences usually are brief and infrequent. In city water stations, condensation can be a severe problem during the spring, which may make a separate dehumidification system advisable. Coating piping and valves with up to 1/8 in. of a sorbent material minimizes the effects of condensation.

Humid outside air introduced for ventilation is a major source of latent heat. Standard air conditioning equipment readily handles that heat if the proportion of outside air is limited to about 10% of the total under peak load conditions. The extent of both sensible and latent ventilation loads can be determined by means of a psychometric chart or psychometric tables. For the detailed methods and the extensive data needed, see Chapter 26 of the *ASHRAE Handbook of Fundamentals* [12].

Supply Air

Supply air quantities for refrigerated air conditioning systems are based on the calculated sensible cooling load and an acceptable temperature differential—limited to a maximum of 17°C (30°F)—between supply air and room air. Air changes ranging from 8 to 12/h are typically required. The total air-handling capacity of a constant-volume air conditioner is the sum of the required design air quantities for all of the spaces supplied. For variable air-volume systems, however, peak loads for all of the various spaces supplied are unlikely to be concurrent due to differences in solar exposure or occupancy. Thus the required total air quantity is determined by the largest sum of simultaneous sensible heat gains in the spaces supplied.

Refrigeration Equipment

The design cooling load for the refrigeration equipment is the highest sum of simultaneous sensible and latent cooling loads, including ventilation air, as calculated for all of the air-conditioned spaces served. The capacity of the refrigeration compressor should be somewhat less than the maximum cooling load because any rise in space temperature under brief peak loads will be minimal. Compressor performance at partial load (which is most of the running time) is better if the compressor is not oversized.

Select the refrigeration compressors in conjunction with the cooling coils to obtain the proportion of sensible to latent cooling capacity best matching the load. Equipment suppliers can assist with the choice.

Air Conditioning Units

Air conditioning equipment is rated by its cooling capacity at industry standard conditions, and 1 ton of refrigeration is equal to 3.52 kW (12,000 Btu/h) cooling capacity, which is equivalent to melting 1 ton of ice.

The simplest and least expensive of the available types is usually a prepackaged air-cooled conditioner. Either it must be installed outdoors or it must circulate an outdoor airstream to which the heat removed by the refrigerant plus its heat of compression is rejected. Package air conditioners of this type are designated as (1) rooftop, (2) at grade pad-mounted, (3) through-the-wall, and (4) above-ceiling units. If lack of space precludes these types of air conditioners, a "split" system (separate air handling and remote condensing units) is necessary.

Cooling Coil Drain

Most refrigerated air conditioning coils sometimes condense moisture from the circulated air. The condensate must be conducted to a satisfactory disposal point. The condensate drain line should be at least 25 mm (1 in.) in diameter to minimize clogging. A trap should be installed in the drain line near the coil drain pan and sized to prevent air leakage at the design static pressure. The direct connection of the drain to a plumbing waste pipe is prohibited by code, so terminate the drainpipe to leave an air gap of at least 65 mm ($2^1/_2$ in.) above a floor drain.

Noise Considerations

Air conditioning equipment in or near occupied rooms or outside of a pumping station in a residential area must be acceptably quiet. Many cities have noise ordinances defining sound levels permitted at property lines. Most compressors and condenser fans are direct driven, so choose a speed low enough to meet the allowable sound rating. However, supply and exhaust fan speeds must be adequate to develop the required static pressure. The fan with the highest efficiency is usually the quietest. Because sound level diminishes with distance, place fans as far from the occupied space (or from neighbors) as practical. Where a noise ordinance limit cannot be met by equipment selection or location, a masonry sound barrier or other sound attenuating means may be needed (see Chapter 22).

The air velocity in ducts and outlets also contributes to the noise level. Duct liner, consisting of specially treated insulation attached to the inside duct walls, helps to meet room sound criteria. The sound level in an office, for example, should not exceed about 45 dBA (decibels on the A-scale of a sound meter).

Example 23-4
Design of a Refrigerated Cooling System

Problem: Design an air conditioning system for the control room and the present office for the pumping station shown in Figures 23-1 through 23-4. See Example 23-1 for the design conditions.

Solution: Calculate the heat gain for each room at the peak time, which is at 8 P.M. on July 21, because it takes solar heat an additional 4 h to penetrate the south wall and roof.

Load	Control room (Btu/h)	Office (Btu/h)
Roof	692	496
South wall	674	421
Interior wall	735	830
Interior glass	3,079	1,703
Infiltration	2,401	1,087
People (2)	1,300	650
Lights	2,457	1,536
Electrical equipment	8,400	—
Total heat gain	29,838	6,723

System selection. A "split system" (the refrigeration condensing unit remote from the air-handling unit) is selected because it fits the available space and meets aesthetic requirements. It is comprised of an air-handling unit in the basement, an air-cooled refrigeration condensing unit on the roof, return and supply air ductwork, and controls. The air-handling unit includes mixing dampers for outside and return air, filters, a direct-expansion cooling coil, a hot-water heating coil, and the supply fan, all enclosed in a metal casing. The supply fan capacity is based on meeting the total sensible cooling load rounded to 36,500 Btu/h at a temperature difference (between the supply and room air) of approximately 20°F. From Equation 23-lb, the required airflow is

$$Q = \frac{36{,}500}{20 \times 1.08} = 1690 \text{ ft}^3/\text{min}$$

which is divided into 1380 and 310 ft^3/min for the two rooms, respectively. The thermostat, which controls operation of the compressor and the heating coil valve in sequence, is located in the control room. The cooling load there is continuous, although it varies with the number of pumps operating. A self-contained thermostatically controlled supply diffuser in the office varies the air quantity supplied to match the actual load. In winter, a heating thermostat in the office modulates heating water flow to the baseboard convector to maintain a room temperature of 70°F. The damper in the outside air duct is adjusted so that the minimum make-up air quantity matches the toilet room exhaust rate.

Basement dehumidification. Assume the pumped water temperature in the basement mains and adjacent reservoir varies from about 40°F in winter to about 75°F in summer. In the spring, water temperature is below the dew point temperature of the outdoor ventilation air and would tend to condense moisture from the ambient air on the piping. To prevent rusting and wet floors caused by such condensation, install three small refrigerated dehumidifiers in the basement to reduce the air dew point to below the cold surface temperatures. In cold weather, warm dry air from the upper level automatically is circulated through the basement by a transfer fan. For additional information on dehumidification, see p. 41.8 of AMCA No. 99 Standards.

23-12. References

1. *Criteria for a Recommended Standard, Working in Confined Spaces,* DHEW (NIOSH) Publication No. 80-106, U.S. Department of Health, Education and Welfare, U.S. Government Printing Office, Washington, DC (December 1979).
2. *Papillon Creek Wastewater Treatment Plant, Omaha, Nebraska,* HETA 83-440-1537, NIOSH, U.S. Department of Health and Human Services, Atlanta, GA (1984).
3. *Occupational Safety and Health Standards, General Industry Standards and Interpretations.* Vol. 1, OSHA 2077, U.S. Dept. of Labor, Superintendent of Documents, Washington, DC (1977).
4. Ten-State Standards, *Recommended Standards for Sewage Works.* Great Lakes–Upper Mississippi Board of Sanitary Engineers, Health Education Service, Inc., Albany, NY (updated periodically).
5. WEF. *Operation of Wastewater Treatment Plants,* Vol. I—MOP-11, Water Environment Federation, Washington, DC (1996).
6. WEF and ASCE. *Odor Control in Wastewater Treatment Plants*—MOP-22. Water Environment Federation, Washington, DC (1995).
7. ASCE. *Sulfide in Wastewater Collection and Treatment Systems,* Manuals and Reports on Engineering Practice No. 69 American Society of Civil Engineers, New York (1989).
8. Wolstenholme, P. *et al.* "Comprehensive odor and VOC performance tests on biofilters." *Proceedings*, Odor and VOC Emission Control Specialty Conference in Jacksonville, FL, Water Environment Federation. (April 1994).
9. "Design Standards for Controlled Environments and Selection of Gas-Phased Filtration Systems," Technical Brochure 300, Purafil, Inc., Doraville, GA (undated).
10. *Design of Municipal Wastewater Treatment Plants, MOP 8.* Joint Committee of the Water Environment Federation and the American Society of Civil Engineers, New York (1992).
11. *Chlorine Manual,* 4th ed. The Chlorine Institute, Washington, DC (1976).
12. *ASHRAE Handbook of Fundamentals.* American Society of Heating, Refrigerating and Air Conditioning Engineers, Inc., Atlanta, GA (1985, updated quandrennially).
13. *Engineering Weather Data.* Departments of the Air Force, the Army, and the Navy, U.S. Government Printing Office, Washington, DC (1978).
14. *ASHRAE Handbook: Systems.* American Society of Heating, Refrigerating and Air Conditioning Engineers, Inc., Atlanta, GA (1982, updated quandrennially).
15. *ASHRAE Handbook: Equipment.* American Society of Heating, Refrigerating and Air Conditioning Engineers, Inc., Atlanta, GA (1984, updated quandrennially).
16. *ASHRAE Handbook: Applications.* American Society of Heating, Refrigerating and Air Conditioning Engineers, Inc., Atlanta, GA (1983, updated quandrennially).

23-13. Supplementary Reading

1. *Dodge Construction System Costs.* McGraw-Hill, Princeton, NJ (updated annually).
2. *Means Mechanical Cost Data.* Robert Snow Means Co., Inc., Kingston, MA (updated annually).
3. Strock, C., and R. L. Koral. *Handbook of Air Conditioning, Heating and Ventilating.* Industrial Press, New York (1965).
4. *Pocket Manual on Heating.* Dunham Bush, Inc., West Hartford, CT (1978).

Chapter 24

Designing for Easy Operation and Maintenance

BAYARD E. BOSSERMAN II
GEORGE JORGENSEN
GARY ISAAC
ROBERT SANKS

CONTRIBUTORS
Nicholas J. Arhontes
Roger J. Cronin
Ronald P. Kettle
Earle C. Smith
Dale Stiller
John A. Redner

This chapter is confined to those design considerations that facilitate operations and maintenance by improving labor efficiency and thereby reducing down time and the cost of replacing parts. Operations and maintenance procedures themselves are not considered. See also Chapters 12, 25, and 26, which contain much advice relative to designing for ease in operations.

Sections 24-9 and 24-10 are especially valuable. Section 24-9 contains the opinions of many operators with an aggregate of two centuries of experience with wastewater pumping stations. Most of their comments also apply to water pumping stations. Section 24-10 is a distillation of knowledge gained from telephone interviews with 15 public sanitation agencies that operate more than 2700 pumping stations and site visits to more than 50 of these stations.

24-1. Site Selection

When selecting a site for a pumping station, give careful consideration to access, not only for operating personnel but also for equipment. Equipment can vary from a pickup truck for small stations to a crane for large stations. Weather conditions may influence access requirements. For example, road gradients may be limited in areas of heavy snowfall.

Route the force main to minimize the possibility of column separation, even if it means relocating the pumping station.

24-2. Landscaping

Landscaping the pumping station to fit or to blend into the surroundings is desirable, but design the landscaping for low maintenance by using ornamental woody perennials and ground cover that require little attention. Install automatic sprinkler systems activated by moisture probes and a timer.

24-3. Hydraulics

The hydraulics of pumping systems are covered in several other chapters, but it is worth repeating that if the analyses are improper, inadequate, or do not include *all* of the operating conditions (including start-up, shut-down, and power failure), excessive maintenance may result.

Head-Capacity Curves

A careful study of the full range of conditions under which pumping will occur is essential so that the proper pump, impeller, and motor can be chosen. Such a study is particularly needed where a battery of pumps discharges into a manifold. As additional pumps come on line, the total flow and head increase, but as the head increases, the flow from each individual pump decreases. In a smaller pump (or one with a lower shut-off head than the others), the discharge may drop to zero, which can cause excessive radial loads. If the analysis is incomplete, the pumping conditions may be outside the limits of the pump curve for a particular pump (refer to Section 10-6).

Pressure Gauges

Taps (in bronze pipe saddles secured to the pipe with double bronze or stainless-steel straps) for screw connections of pressure gauges are useful on both the intake and discharge pipelines. Permanently mounted pressure gauges can be used, but portable gauges, which can be used when needed, are usually better because they can be easily recalibrated and can serve for many locations (see Chapter 20). Fixed gauges are subject to equipment vibration and, unless periodically recalibrated, will not remain reliable. The maintenance schedule should require pressure readings at regular intervals (e.g., monthly) along with simultaneous flowmeter readings. These records can be used to detect abnormalities such as clogging of the force main or impeller wear.

Flow Measurement

To check the efficiency of the pumping system, a flowmeter in conjunction with the pressure gauges facilitates checking the power consumed against the water output. It is also a good idea to make it easy to calibrate the flowmeter in place; for water systems, a means of inserting a pitot tube is a good way to do so. Time-volume measurements using the wet well, a storage tank, or a clear well are reliable and can often be used. The use of tracers (dye dilution) is accurate, almost universally applicable, and easy to use if taps (one to introduce the tracer and another at a point far downstream for withdrawing samples) are installed (see Section 3-9).

The lengths of straight inlet and outlet pipe recommended by the manufacturer must be observed for all types of meters; otherwise the meter readings will probably be erroneous. The errors increase exponentially as the inlet and outlet pipes are shortened.

Surge Control

Water hammer need not instantly rupture a pipe to be damaging. Surges that continue for many years gradually destroy a pipe's integrity through material fatigue, so surge control is required in most pumping stations to prevent damage to the piping system. Several methods of surge control can be employed:

- Pump-control valves (which are popular) for start-up
- Motor starters that ramp up and down when pumps are started or stopped
- Flywheels to control the rates of acceleration or decay of liquid system velocities.

These and other methods are discussed at length in Chapter 7. Neither of the first two methods, however, provides any surge control for power failure. As explained in Chapter 7, if the flowrate exceeds about 30 L/s (500 gal/min) and the total dynamic head exceeds 12 or 15 m (40 or 50 ft), a surge control analysis is needed and other surge control devices or systems may be required.

If surge can occur in the intake pipe, consideration must be given to surge control both in the discharge and intake piping.

Air–Vacuum Release Valves

Try to avoid the need for air–vacuum release valves in water service (and make every effort to avoid them in wastewater service) because they become unreliable if not properly maintained. Because these valves seldom receive the maintenance required to ensure 100% reliability and because no designer can guarantee proper maintenance in the future, relying on air–vacuum release valves to positively prevent water hammer is unwarranted. If a water hammer control device can fail, it will do so sometime and is, therefore, inadequate protection. Use other defensive strategies, such as alternate pumping locations and/or a suitable force main profile obtained by relocation, deep trenching, or even tunneling. If necessary, increase the rotational inertia of the pump and motor by adding a flywheel.

If alternatives are completely impractical, conduct a careful analysis of the system for the proper size of air–vacuum release valve openings. Always install such valves in pairs, particularly in sewage systems, to provide a backup if one fails. Use the tall type and stainless-steel trim for sewage service. If iron valve bodies are used, require an epoxy coating to prevent corrosion. Provide flushing facilities for all air and vacuum release valves and specify weekly maintenance in the O&M manual. Note that only an air or a vacuum release valve might be needed instead of the combination.

Intake Sumps

Follow the recommendations given in Sections 12-6 and 12-7 in designing clear water and (especially) wastewater wet wells. These designs provide nearly ideal hydraulic environments for pump intakes, and with only the added cost of an ogee spillway in storm water stations plus (for wastewater stations) a small (1.5 to 4.5 L/s or 25 to 70 gal/min) washwater hose, they can be easily and quickly cleaned.

For clean waters, some engineers may prefer to follow the designs in Hydraulic Institute Standards of 1983 [1] or in the later ANSI/HI Standards [2,3], but beware! Those designs are unsuitable for water containing either settleable or floating debris, because, without modification, they cannot easily be cleaned. (Mixing the contents during pumping is one modification that is sometimes effective, but designers should contact manufacturers that make equipment specifically for that purpose.)

If there are unusual conditions or if the flow is very large, model studies should be used to optimize the geometry and the need for baffles and suppression of swirling and vortices.

If practical, provide enough storage in the wet well and influent sewer so that short-term (e.g., 1 h) power outages can, at least for dry-weather flows, occur without damage.

Heat generated by inrush current during start-ups can damage motors. The sump should be of adequate size to limit the starts per hour to the manufacturer's recommendation.

24-4. Mechanical Considerations

One of the most costly elements over the life of the facility is maintenance labor. Hence, developing a concept or method for the disassembly of piping, pumps, check valves, gate valves, traps, flowmeters, and so on for maintenance or removal and replacement is important, even if three-dimensional drawings or spatial models are necessary. Include such plans in the O&M manual. The following guidelines make the maintenance or replacement of pipe, fittings, and pumps easy:

- Locate these elements at a convenient height.
- Orient them so that removing the covers does not drench the workers.
- Include a drain and a vacuum cock for emptying the pipe.
- Install strong, safe, lifting eyes in the ceiling for a chain hoist in small stations, monorails and electric hoists for small to medium stations, and traveling bridge cranes for large stations. Even in small stations, consider a trolley rail for moving the heaviest load to a hatch accessible to the upper-level hoist. Hoists should travel over all heavy equipment and terminate at a doorway. Alternatively, especially for vertical turbine pumps, install roof hatches for access by a mobile crane parked outside. The hatches and floor space must be adequate for removing the largest equipment. It is bad practice to build the upper part of the station around equipment already in place.
- Provide plenty of clearance [0.8 m (30 in.) minimum, 1 m (36 in.) recommended] between the piping and pumps or other equipment for workers and their wrenches and other tools. Make sure that nuts and bolts are accessible. Increase the clearance for larger pumps. There must be adequate space under the crane or hoist to lay removed parts on the floor.
- Design the piping with enough sleeve and/or grooved-end couplings to permit the easy removal of pumps, valves, etc. Blocking, tie downs, or bridles are required for sleeve couplings.
- Provide valves to isolate pumps so that any pump can be removed without draining the header and force main.
- Specify spacer shafts between pumps and motors to allow for the quick and easy removal of backheads and rotors of end suction pumps to make replacement of seals, shaft sleeves, and wearing rings easy without moving the motor or pump.
- Provide a spool of the same length as the flowmeter and shut-off valves on the header to permit the removal of the flowmeter. An alternative arrangement is depicted in Figure 20-12.

- Provide adequate lighting for safe working conditions.
- Install GFCI protection on all service receptacles.

Locate the following items in places that are easily accessible for all routine maintenance:

- Traps that must be periodically cleaned
- Pump vent piping
- Grease fittings (especially intermediate bearings on long shafts)
- Intermediate bearings on long shafts (safely accessible)
- Oil reservoirs
- Dry-break, quick-connect couplings on hydraulic hoses
- Dehydrating canisters
- Mechanical or electrical devices that need periodic adjustment.

If necessary, provide permanent steps and/or platforms for access.

Miscellaneous considerations include the following:

- Keep walking areas clear of overhead obstructions to a height of at least 2.2 m (7 ft). If obstructions cannot be eliminated, cover them with padding and hang warning devices (heavy strings or light chains, for example) on both sides.
- Install convenient hose bibbs. A proper backflow preventer must be installed on the potable water connection to the station.
- Install a lavatory or at least a wash basin.
- Provide access that complies with OSHA requirements to ladders and manhole steps.
- Provide adequate storage space for spare lubricants, parts, and special tools.

Nine power plants were surveyed by the Electric Power Institute to identify problems that affect productivity and the safety of maintenance personnel [4]. The suggestions made are also useful for pumping station designers. They include the following:

- Provide easy access to adjustment points, test points, and filling and draining locations on all equipment.
- Arrange equipment so that access to the malfunctioning unit does not require the disassembly of adjacent units.
- Provide sufficient clearance to allow for effective personnel interaction when a team effort is required.
- Ensure that equipment clearances in hazardous areas allow access by personnel encumbered by protective garments and associated gear.
- Provide a clearance of at least 1.1 m (42 in.) at the front or back of electrical panels.
- Code or key interchangeable units that are functionally dissimilar to prevent their insertion in the wrong unit.
- Provide places to put and support components being removed or installed.
- Use hinged doors rather than cover plates (which are time consuming to remove) where frequent access is needed.
- Place controls and associated instrumentation within easy visual and manual reach of normal work positions.

A universal complaint was the difficulty of accessing equipment that required attention. About 30% of the maintenance time could have been saved had ideal or unrestricted access been possible.

Remember that some types of equipment (e.g., submersible well pumps and submersible wet pit sewage pumps) are maintenance intensive and must be removed and reconditioned at regular intervals, and this may cost about one third as much as a new pump. Parts for some kinds of equipment (e.g., variable-frequency drives and some foreign-made pumps) are very expensive or, worse yet, may be available only after long delays. Investigate this aspect before using such equipment and conduct lifetime present-worth analyses of the maintenance labor and parts as well as of the costs for energy. Avoid considering only the first cost.

24-5. Electrical Considerations

According to the NEC, electrical control panels must have a minimum clearance in front of at least

- 0.91 m (3 ft) for 0 to 150 V
- 1.07 m (3 ft, 6 in.) for 151 to 600 V
- 1.22 m (4 ft) for 601 to 2500 V
- 1.52 m (5 ft) for 2501 to 9000 V.

The same clearance is required at the back if the panel is a free-standing unit. Free-standing panels are recommended over wall-mounted panels.

A lock-out stop should be installed at each motor (or at any rotating equipment) that cannot be seen from the control panel to prevent accidental energization while the equipment is being serviced. A time-running meter for each pump can furnish useful information for scheduled maintenance. Include heaters or desiccant containers (to keep electrical equipment dry) and put control panels in a location free from dust, fumes, and H_2S.

For large pumping stations, consideration should be given to computerization for monitoring and controlling the system.

Make certain that the electrical engineer provides a liberal number of convenience outlets for tools and maintenance equipment, such as lights and fans (120 V, single-phase), and for welders (480 V, three-phase). All outlets should be GFI protected.

Use hermetically sealed switches and relays in wastewater pumping stations. Use NEMA 7 equipment for hose-down spaces.

Post a single-line diagram of the power system—especially if there is auxiliary power.

24-6. Architectural Considerations

The entire layout, including housing, drainage, storage for equipment, and safety, is very important to the maintenance operation. Give proper consideration to the following items.

For the housing:

- Access for the installation and removal of equipment (for example, use double doors or roll-up doors to remove large pieces of equipment such as pumps, motors, or engines)
- Heating and air conditioning, which may be necessary depending on weather conditions
- Convenience for the cleaning and painting of the facility.

For drainage:

- The drainage system and sump for wash-down water
- Drainage from the sump of water pumping stations piped to a storm drain or sewer (if this is not possible and a French drain is used, consider the groundwater table for all seasons of the year; if a submersible sump pump is used in a wastewater pumping station and the discharge is piped to an adjacent downstream manhole, it may be possible to dewater a small flooded station)
- A drip line for water-lubricated packing and a drain system to the sump
- Sloping the floor to floor-drains to eliminate puddles.

For storage:

- Cabinets to store records, spare parts, and tools
- Record drawings (plans) should be stored at each station.

For safety and security:

- Sewage pumping stations must be equipped with an adequate forced-air ventilating system
- Adequate lighting to supply at least 200 to 300 luxes (20 to 30 ft · cd) at the floor or, alternatively, about 100 luxes (10 ft · cd) with auxiliary portable lights and GFI receptacles
- Carefully consider security and the risks of vandalism
- Intrusion protection with a remote alarm.

Miscellaneous. Include the following:

- There should be a means of checking the water level in water wells because it helps in determining the proper pump head, drawdown, or changes in the water level in the aquifer.
- It may be desirable to have a color-code system for the piping, particularly if a water station pumps into more than one pressure zone.

24-7. Standby Facilities

Consider using standby pumping facilities for important pumping stations. Most states require a second source of power for sewage pumping. Possible sources include

- Another utility source (feeding from two separate substations is almost always adequate)
- A nearby major industrial plant with a large electrical generating capability
- A fixed, dedicated engine-generator set
- A dual engine/motor drive with an appropriate, automatic clutch that disengages the engine when the motor is energized and engages the pump shaft, thus allowing the motor to spin freely when the engine is running
- A portable engine–generator set that is readily accessible and well maintained.

Standby diesel or natural gas engines should be exercised for at least 20 to 30 min after reaching operating temperature once each month at nearly full load (see also Section 14-22). Low-load operation results in carbon deposits in the exhaust of diesel engines, and inadequately exercised engines require periodic dismantling for cleaning. An engine sized for pumping peak flows or operating under the worst conditions means that the engine may operate at one-third or less of its capacity when exercised. One solution is to design the electrical system so that a portable load bank can be used to exercise engines under full-load conditions. Reliability can be improved by electrically heating the engine, by trickle charging the batteries, and by vigilant maintenance—all of which should be included in the O&M manual.

24-8. Specifications

Include provisions in the specifications requiring the supplier of pumps and other equipment to furnish pump curves, maintenance data, parts lists, and the manufacturer's recommendations for service and maintenance. The names and locations of parts suppliers should be a part of the records.

Designers should recognize that many old, poorly designed pumps are still being sold (some by well-known manufacturers). Specifications obtained from the manufacturers (or any organization representing them) coupled with low bids may lead to the use of these shoddy pumps. For normal service, specify a limit for shaft deflection of no more than 0.15 or 0.18 mm (0.006 or 0.007 in.) at the wearing ring and 0.05 mm (0.002 in.) at a mechanical seal. The allowable deflection at stuffing boxes is greater, but, because operators may want to replace a stuffing box with a mechanical seal, the more restrictive deflection is preferable. Interviews with manufacturers at pump exhibits and with pump maintenance crews or supervisors of several utilities provide valuable instruction in maintenance economy, and a present-worth analysis of the expected future costs of maintenance, labor, parts, and power is revealing.

24-9. Operators' Preferences

The opinions of operators are invaluable. This section contains a potpourri of selected opinions on a variety of subjects. Although these comments come from operators of wastewater pumping systems, most of them apply to fresh water systems as well. Comments by non-operators are given in italics, and the commentators are named.

Los Angeles County Sanitation Districts

The following comments are paraphrased from conversations with Kettle [5].

Wet Wells

The Lancaster Reclamation Plant wet well in Figure 11-12 is very poor, because it is so large, and because the floor is so flat. Large quantities of grit and scum accumulate. Serious consideration was given to rebuilding it with floor slopes of 45° and pumps or pump inlets set into a small pocket so that scum could be readily sucked out, but the cost of rebuilding was prohibitive.

Valves

Trouble was experienced with high-quality eccentric plug valves in digester lines. They were replaced with solid wedge gate valves exercised weekly, and there have been no problems with them. In fairness, however, the plug valves had not been exercised. *Sanks: Others have used eccentric plug valves with complete success. Lescovich [6] warns that excessive tightening of the flange bolts on eccentric plug valves squeezes the rubber gasket out of the flanges and binds the plug. Bolts should be only tight enough to prevent leakage.*

Bubblers

Air tubing was plagued by plugging with grease and with leaks that could not be fixed because the tubing was buried in concrete. These ills, plus the maintenance of compressors and associated mechanical equipment, makes the air bubbler system irritating. Pressure cells were substituted at all pumping stations by attaching a 150-mm (6-in.) flange to the dry pit wall below LWL, installing a 150-mm gate valve and drilling through the wall into the wet well. A short spool is attached to contain the pressure cell and a fresh water line is fitted to flush the cavity continuously with 0.06 L/s (1 gal/min) of fresh water. An electronic control box completes the installation. *Sanks: Leaks can be fixed when air tubing is installed in the open or in conduit and all joints are accessible. Suitably derated air compressors are trouble-free.*

Packing Glands versus Mechanical Seals

According to Redner [7], new pumps are always required to have mechanical seals, and the seals that come with the pump are immediately replaced with the owner's standard tungsten carbide-faced seals. To fit mechanical seals to their older pumps that were equipped with packing glands required machining the pumps for installing larger shafts and bearings. The maximum allowable deflection at the end of the shaft is (and was) 0.025 mm (3 mils). More deflection would quickly destroy a mechanical seal.

Packing glands leak, of course, so housekeeping is a problem, they must be adjusted weekly (a continual source of maintenance), and shaft sleeves eventually become scored and must be replaced at considerable expense. Mechanical seals are preferred, because they require no maintenance, last for many years (10 or more), and eventually pay for themselves, although their first cost is many times greater than the modest cost for packing glands. *Sanks: The costly mechanical seal can be reconditioned to be like new for about a third of the cost for a new one. A mechanical seal can be reconditioned several times.*

County Sanitation Districts of Orange County, CA

Fresh water for seal water systems is avoided in most pumping stations. By using graphite-teflon packing, the wastewater itself, without filtration, becomes the seal water. This scheme has worked well for a decade, and, according to Arhontes [8], the wastewater does not cause excessive wear. This system is economical and more reliable because there is no dependence on a city water supply that might be interrupted by an earthquake. When pumps are dismantled to replace worn impellers, new shaft sleeves are also installed.

In contrast to the antipathy to air bubbler systems in the Los Angeles County Sanitation Districts (see above), air bubbler systems are preferred, and the Orange County engineers have designed a standard package for the entire system. The air tubing is installed in stainless-steel conduit to allow the bubbler system to be easily removed for fixing leaks.

Seattle Metro

The operational friendliness of 29 Seattle Metro—now King County (Washington) Department of Metropolitan Services—pumping stations with V/S drives and two older "traditional" designs with C/S pumps were carefully studied after two decades of operation by 10 senior operators with an aggregate total of 200 years of operating experience. Most of the 29 pumping stations have trench-type sumps of the type illustrated in Figures 17-14 to 17-20. In addition, there are two "traditional" designs with large flat floors. The following conclusions were reached by the operators in a day-long seminar held in 1989 and attended by Garr Jones (one of the original principal designers of these stations) and Gary Isaac (Superintendent of Operations).

Access

1. We recommend catwalks of fiberglass or aluminum gratings with rails (for safety) to all parts of the wet well and for all out-of-reach locations where routine maintenance is required.

Alarms

2. Warning systems (whether internal or transmitted elsewhere for monitoring) should be limited to those critical for station operation. Too many alarms can cause operator complacency and failure to respond when response is really needed. We have eliminated station-occupied and ventilation-system failure alarms in an effort to reduce troublesome calls.
3. Having early warning of high water followed by flood warning usually gives operators enough time to respond.
4. It should be impossible to clear alarms at the treatment plant when the acknowledge button is pressed. Alarms should be cleared only at the remote station, but when two stations are interconnected, it should be possible to clear alarms at either one.
5. There should be alarm lights for all significant water levels in the wet well.
6. Explosive gas detectors are desirable if they are maintained properly. *Jones: NFPA 820 mandates the use of explosive-gas detectors.*

Auxiliary Power

7. All stations should have back-up power, either engine-generators or dual power feeds.
8. Dual power feeds are preferred over engine-generator sets and seem to be more reliable. High winds can sometimes disrupt both feeds, so at least one should be underground. *Sanks: The reliability of dual feeds is suspect because they offer no protection in the event of an area-wide power failure.*
9. We prefer natural gas for engine-generator sets. *Jones: The use of gas is site-specific. In the South, where it is common practice to shut down gas producing and distributing installations for days at a time when a hurricane threatens, diesel is the fuel of choice. In the Pacific Northwest, gas may be the best choice because buried utilities are secure against windstorm and most floods. However, leaks are very hazardous. As the engine is presumably in an unattended station, a hydrocarbon gas detector is a better choice for shutting off fuel than a fuel pressure detector. The latter gives rise to spurious shut-downs. Cronin: Gas may be unavailable if power failure is due to earthquake. Also, gas engines are larger than diesel engines.*
10. Portable engine-generator sets are of no value in snow or for remote stations.
11. Stationary engine-generator sets may not start automatically. *Jones: For reliability, exercise them with regularity and long enough at or near full load operating temperatures to cook off moisture in the crankcase and elsewhere. See also the advice in Section 14-22.*
12. We think engine-generators should shut down automatically when commercial power is restored. *Isaac: I disagree. Operators should be*

dispatched to check all systems. Jones: There is great potential for damage if control circuits are not reset and full power is not restored to divided busses as a consequence of any power failure, no matter how brief.

13. Adequate noise control is needed for engines. *Sanks: See Chapter 22 for measures to reduce noise.*
14. Storage tanks for diesel fuel should be large enough for at least one day of operation. *Jones: Typically, power failures rarely last longer than one hour, although storms, earthquakes, and other disasters have interrupted power in every section of the country for several days, thus prompting engineers to design for excessive storage. Diesel fuel deteriorates, so the exercise program should be able to recycle the fuel inventory every 6 to 12 months.*
15. There should be load banks or some other means to exercise engines under no less than about 90 percent of full load.
16. The transfer switch and the station's main breaker should be sized to take all of the station's load.
17. PLCs (programmable logic controllers) should be programmed to start pumps one at a time.

Auxiliary Water Systems

18. Solenoid valves to main pumps should all be on essential power.
19. In addition to a backflow preventer, we [in Seattle] must also have a water storage tank with an air gap, and we use a regenerative turbine pump to pressurize the system. *Sanks: Multistage centrifugal pumps are also excellent for either seal water or wash water (see Figure 11-34).*
20. Make sure that water pressure is available in the station while the tank is filling.

Controls

21. Keep controls as simple as possible.
22. We prefer the control room to be isolated from the pumps. *Sanks: All controls should be above the flood plain. Jones: Modern electronic controls should be located in air-conditioned rooms with air treated to remove hydrogen sulfide and particulates.*
23. PLCs work well but need a manual backup for those times when the system goes down. *Cronin: All stations should have duplex PLCs. Sanks: Some operators would like a pump-down selector switch with a timer to return the switch to automatic operation after a preset time. Jones: Timers are not recommended. Cleaning operations should be supervised and the station returned to normal operation with operating personnel present.*
24. Duty switches need to be set so operators can select any combination of lead/follow pumps. *Sanks: Light switches and other switches in dry wells should be waterproof so that the entire dry well can be hosed down.*
25. It should be possible to control the pump either at the control panel or at the pump unit. *Jones: This seems to me to be a bad idea, because it causes an unnecessary complication in controls. Worse, it can lead to damage by making it possible for an operator to lock out equipment or leave it in the manual mode when leaving the station. Furthermore, with hand-held, low-power radios so readily available, one operator can be at the machine and another can be manipulating the controls. Cronin: If the client insists on a control at the pump when in manual mode, install an operation light to remind operator to reset the controls to "Auto." Sanks: Do so regardless of the location of the manual control.*
26. It is nice to have an indicator of the water level in the pump room for stations with deep dry wells. *Jones: The operator should also have the wet well level instruments available when operating the pump in manual mode.*
27. Bubbler systems are still our standard, but sonar has improved in reliability recently. Back-up floats for high-water alarms are also needed. Otherwise we do not recommend float switches except for constant speed pumps. *Sanks: See Air Bubblers, Section 20-3, for proper design. Jones: Bubblers are immune to power failure whereas electronic instruments are not. Float switches are mechanical devices and subject to wear. Bubblers and pressure cells (with virtually no moving parts) are inherenelty more reliable.*

Cranes, Hoists, Lifting Eyes, and Access Hatches

28. All equipment weighing more than 45 kg (100 lb) should be accessible by crane.
29. Bridge cranes, properly located to allow direct vertical access to major station equipment, are absolutely necessary.
30. The crane hoist should be electrically operated. Removing a heavy object from far below grade with a manually operated hoist is a waste of a journey-level worker's valuable time.
31. Include the "inching" mode of operation—a feature (available from most manufacturers) that allows slow and precise positioning of large items such as motors.
32. Extend crane rail girders outside the building if possible. *Jones: A better design allows trucks to move*

into the building and under the crane for loading heavy equipment directly onto the truck bed.

33. Lifting eyes should be an integral part of station design and need to be strategically located over critical equipment.
34. Hatches or floor covers with plenty of clearance for pump removal should be located over each pump. They should be light in weight (preferably aluminum) and either spring-loaded or easily removable.
35. Openings around hatches need proper barricades built into the hatch design to ensure a safe working area.
36. When lifting eyes are needed, locate them properly and design them with an adequate reserve of strength. [See Figure 25-6.]

Custom-Engineered Pumps

37. We are firmly committed to high-quality equipment and custom-engineered pumps with larger shafts and bearings.
38. The best metallurgy for impellers and pump casings (particularly for raw wastewater) is cost-effective in prolonging useful life. For example, our pumps have generally lasted 20 yr or more. Pumps in four stations finally wore out, and our engineers, unfamiliar with the thought that went into the original designs, replaced them with standard pumps. Troubles developed at the outset, and the standard pumps lasted only 4 yr or less.
39. We now replace custom-engineered pumps with units of the same quality. Where grit is a problem, pump casings should be hard cast iron.

Design Implications

40. Use common sense in laying out piping and valves so stations can be operated safely and with easy access to all equipment.
41. Good hydraulics is important.
42. Design for the lowest head in pumping. Do not allow a free fall of wastewater into a pool below.
43. Involve the operators in the design of the station.
44. Work with the neighbors so that they can claim ownership in the final station with regard to colors, landscaping, and noise and odor control.
45. Success requires good design, a good contractor, and good inspection.

Doorways

46. Doors should allow heavy equipment to be loaded easily onto maintenance vehicles for transport to the shop.
47. Measure the utility's trucks to be certain doors are wide enough. At two of our stations, the entrance door had to be modified to allow adequate clearance for maintenance vehicles.
48. Roll-up doors are preferred over double doors.

Drains

49. All areas should have adequate drains and all floors should slope about 1% to the drain to avoid standing water.
50. In the immediate vicinity of the drain, the slope should be increased to 2%.
51. Drains should be large and must not have check valves.
52. Pipe drains are preferred to an open rectangular drain along one wall.

Floor Finishes and Coatings

53. Use non-skid coatings where water is prevalent.
54. Use finishes that are easy to keep clean. Urethane finishes are satisfactory and control concrete dust.
55. Do not paint floors; paint peels in a matter of months.
56. The newer sealers have not been holding up well.

Force Main Draining and Cleaning

57. Keep force mains small to maintain adequate velocity. They must be large enough to accommodate only the sewage volumes to be removed.
58. Sump pumps and drains should be plenty large enough to take care of accidental spills. [See Section 12-3.]
59. Low turbulence discharge structures reduce hydrogen sulfide damage.
60. Pig launching facilities should be provided for all force mains. Pigs should not be larger than 600 mm (24 in.). Use dual force mains if necessary to avoid larger pipe. Use dual headers for dual force mains.
61. Design the force mains for ease in accessibility in the station, along the alignment, and at the discharge structure.

Gear Boxes

62. We dislike gear boxes because of the heat, noise, and oil leakage. *Sanks: They also tend to induce vibrations.*

Gratings

63. Removable gratings over the wet well (e.g., Kirkland Pumping Station, Figure 17-14) are a nuisance.

To facilitate hosing the side walls, design a catwalk along the full length of the wet well.

64. If gratings cannot be avoided, use non-skid fiberglass easily lifted. Even aluminum is too heavy.

Horizontal versus Vertical Pumping Units

65. Pumps and motors in a horizontal configuration are excellent, because alignment is better than in vertical units, but they require a larger footprint area and are more susceptible to flooding.
66. Avoid the use of intermediate shafts if possible. Vertical units require shafts, and vibration problems tend to develop.
67. *Isaac: We have some units with flywheels. They have not been a problem, but nevertheless they should be avoided unless necessary to solve a hydraulic problem. Jones: To the best of my knowledge, they have worked well.*

Instrument Air Systems

68. Air tanks (receivers) should have hand bleed-offs. Automatic bleeds are not necessary.
69. Dual compressors should be valved so that one can be removed while keeping the second in service.
70. Both compressors should not be on the same switch gear.
71. Purge valves should be spring-loaded to return to normally closed position.
72. Air dryers are recommended for all instrument air.
73. Design for high flowrates at low pressure: 240 to 350 kPa (35 to 50 lb/in.^{2}) is adequate.
74. Lead/follow compressor selection is better than automatic alternating compressor operation. *Jones: We have not advocated automatically alternating lead sequencers. We think it is just one more thing to go wrong. Instead, we furnish the operating staff with hour meters on the MCC cubicle doors and a lead/follow selector switch. This approach leaves it up to the operator to decide when to select the other unit to function as the lead unit in a lead/follow automatic starting sequence.*
75. Some new level control systems eliminate the need for instrument air: sonar and pressure cells are two examples.

Landscaping, Irrigation, and Shrubs

76. Many of our stations are located in areas of heavy public use and high visibility, so good landscaping is necessary. Landscaping should blend with surrounding area and be kept simple with ease of maintenance in mind. Low, slow-growing, disease-resistant plants should be specified. Avoid ivy and thorny bushes.
77. Lawns should be kept to a minimum. Consider grasses that need a minimum of weeding and mowing.
78. Automatic sprinkler systems with timers should be installed. *Sanks: Moisture probes are also useful for irrigating when necessary.*

Lighting

79. Good lighting is very important. Interior lights should come on instantly.
80. Where gases can exist, lighting should be explosionproof.
81. Pumping stations should be well lit outside for protection from vandalism.
82. Include emergency back-up lights by battery for each pumping station floor.
83. Some lights take too long to light (high-pressure sodium) and take a very long time to turn on right after they have been turned off, so choose other types. *Sanks: Sodium vapor lights are all right for outside lights that turn on automatically.*
84. Locate fixtures so that access by ladder for changing bulbs is easy.

Noise Control

85. Noise control is an important consideration for employee health and public relations.
86. Engine-generator sets may require noise control.
87. Variable frequency drives are also noisy and may require special treatment. *Sanks: AF converters are sensitive to dust, hydrogen sulfide, and temperatures higher than 32 °C (90 °F), so an air-conditioned enclosure should be considered.*
88. Consider the use of noise absorption walls and ceiling.
89. Odor control systems will probably require noise abatement.
90. Rubber-mounted equipment helps to absorb vibration and reduce noise.

Odor Control Facilities

91. Facilities should be built to suit the environment.
92. The discharge of air from the wet well should be diffused. Consider mixing wet-well air with fresh air from a booster fan.
93. Consider noise problems from odor control or ventilation fans.
94. Packed wet chemical scrubbers work well.

95. Carbon towers work well, but they still allow the escape of a slight odor that is not always acceptable to the neighbors. On heavy odor days, however, carbon is reasonably satisfactory. The carbon needs to be changed about twice a year, and it is dirty to handle.
96. Hypochlorite generating towers require lots of servicing time and the hypochlorite generators are too inconsistent. They need to be recharged often, and on heavy odor days the odor control is erratic. Because of the salt, these units corrode rapidly.
97. Potassium permanganate/alumina pellets require too much maintenance. They are expensive and do not last. *Jones: North Mercer Island pumping station was originally designed with "filters" for these pellets, but the system did not work well (an experience we and others have had) and was replaced with a hypochlorite wet scrubbing system.*
98. Masking agents are ineffective.
99. Proper maintenance equipment must be included.
100. Provide a proper storage area for replacement chemicals.

Overflow Structures

101. Structures should be kept as inconspicuous as possible but easily accessible.
102. Use flap valves rather than hydraulic or electrically actuated valves.
103. A measuring weir with recorder would be useful together with an alarm to signal when overflow is occurring.
104. A catch basin to collect solids would be desirable to reduce clean-up problems.

Potable Water Supply

105. No significant problems have been identified with potable water systems.
106. All outlets should be above flood level with no possibility of cross-connections.
107. Non-potable outlets should be clearly signed.

Raw Wastewater Pumps

108. Provide easy access to pumps and a clear working space around them.
109. Pumps are much more trouble-free when speed and head are low.
110. Impellers should be nickel iron or stainless steel for better wear. *Jones: Some types of stainless steel have excellent wearing properties but come at an exorbitant cost premium. We prefer cast iron with 2 to 3% nickel.*
111. Provide guards for all drive shafts.
112. We do not like mechanical seals and are replacing the few we do have with packing glands. *Jones: Mechanical seals generally do not function well where the shaft may deflect because of high radial thrust conditions.*
113. Plumb seal water pressure lines and drain lines to keep them out of the way of the operators. Provide seal water drains at the bottom of pump casings.
114. Valves should be located to make operation easy. Make sure manual actuators have enough mechanical advantage to make valves easy to open and close.
115. Bleed-offs are required. They should be large and plumbed into a drain.
116. Provide covers for volutes during repairs.
117. Provide spare parts.

Security

118. Provide vandal-proof doors, vandal-proof lights, and unbreakable windows.
119. Fencing may be required in some areas.
120. Provide moderate outside lighting switched on and off by a timer.
121. For stations in remote locations, it is wise to put special security locks on station doors.
122. Avoid long driveways to facilities.
123. *Isaac: We have "locked covers" for manholes that make it difficult for amateurs to stuff debris into the sewer. Sanks: For discouraging vandalism and midnight dumpers, some manholes in Europe have a special, heavy (140-kg or 300-lb) cast iron hatch (below the manhole cover) that can be lifted only with a hoist .*

Stairwells

124. Stairwells located on the outer perimeter of the station leading to the various doors are preferred.
125. Fiberglass treads are preferred.
126. Aluminum handrails are preferred over galvanized iron.
127. Provide adequate headroom over stairs.
128. Circular stairs are better than ladders or ship-type companionways, but straight stairs with intermediate landings are best. *Sanks: Design stairs so that it is easy to carry an injured worker or a box of tools.*
129. We prefer non-skid spiral stairwells for smaller stations but like wide straight steps if there is enough room.

Sump Pumps

130. Sump pumps should be industrial-grade units of relatively large size capable both of passing wastewater solids and of dewatering an entire flooded dry well.
131. There should be at least two pumps.
132. Check valves should have an external arm with a spring or counterweight to prevent slam and to indicate whether the pump is pumping.
133. Ease of access is critical. The pumps should be conveniently located in the dry well for ease of cleaning and pump removal.
134. Gratings over sumps should be light in weight and readily removable. Pumps should be easy to flush out. Grease collection on the surface of sumps is a problem that takes either manual removal or a lot of hosing down.
135. Built-in floor drains are preferable to open channels along one wall. The floor drains should be of relatively large size.
136. Lead/follow switches should be provided.
137. Add an alarm so that operators can tell when the follow pump is activated.
138. Either air bubblers or float controls are satisfactory.
139. We have numerous submersible sump pumps as well as submerged extended-shaft pumps with encapsulated motors located above the pumps. Although both types are acceptable, we prefer the latter. Our experience with submersibles has been less satisfactory, possibly because the pumps have not been rated for the intended service. *Isaac: Many submersible pumps are not designed for passing wastewater solids—many are too light for this duty, and seals tend to fail and cause motor shorts. Jones: Once the shaft seal fails, the motor windings are sure to go next and that is the first indication to the operating staff that there is a problem. After much trouble with submersibles, we switched to wet pit pumps with the motors above grade where they are less vulnerable and the operator can service them. According to NFPA 820, submersible pumps in sumps receiving wastewater-contaminated drainage must have explosionproof enclosures. Sanks: Submersible pumps can be satisfactory if they are carefully selected for the service conditions, are of rigid design, and have moisture probes to warn of seal failures.*

Support Systems

140. We are firmly convinced that support features of a pumping facility are of paramount importance. Some support systems that can make a major difference in ease of maintenance (but are sometimes given too little attention) include: (1) sump pumps; (2) cranes, hoists, lifting eyes, and access hatches; (3) alarms; (4) doorways; and (5) landscaping. *Sanks: These topics are discussed herein under separate subheadings.*

Switches

141. MANUAL/OFF/AUTO switches are too easily left on MANUAL with the possibility of damaging the equipment.
142. Either use PUSH-TO-TEST/ OFF/AUTO or add a timer to the MANUAL position that automatically turns the switch to AUTO after a preset period.

Telephones

143. For safety and convenience, every station should have a telephone—particularly in deep stations. *Jones: Portable phones (often carried in the maintenance vehicles anyway) are good substitutes.*

Toilets

144. Every station should have a toilet and washing facilities for the convenience and safety of operators and maintenance personnel.

Valves

145. Locate for easy isolation of pumps and force mains.
146. It should be possible to close all valves manually if necessary, and overhead valves should be activated with a chain. Large valves should have a setup for operation with an air wrench.
147. Do not use knife valves on discharge lines, because they tend to leak. *Isaac: Some manufacturers make quality knife valves that are adequate for this service, and specifications can be written to ensure this quality.*
148. Use gate valves for pig launchers.
149. Force mains discharging into channels should have flap gates instead of valves. [See Figures 12-19 and 12-21.]
150. Valves over walkways or located in walking areas should have a clearance of 2.1 m (7 ft).
151. Avoid cheap gate valves on discharge lines. *Isaac: Avoid cheap valves everywhere.*
152. We have no problems with hydraulic systems to operate valves. *Jones: Hydraulic systems unquestionably require more maintenance and are messier than electric actuators, but hydrau-*

lic actuators are better for protection against hydraulic transient conditions and flooding.

Variable-Speed (V/S) versus Constant Speed (C/S) Pumps

153. Variable-speed pumps are preferred and seem to do a better job of keeping the wet well clean.
154. Our types of V/S drives include liquid rheostats, eddy-current couplings, and adjustable-frequency drives (AFDs). All have been satisfactory, but each has shortcomings.
155. Liquid rheostats are relatively slow in response to wet well level changes. They have more maintenance problems. They can become fouled by electrolyte leaks. Changing electrolyte should be done with care and only by trained technicians.
156. Eddy-current couplings are the most maintenance-free system in use. They run well with very few problems and have proven to be very reliable. But we have had difficulty recently in obtaining parts from certain manufacturers who no longer build them in the smaller sizes, 56 to 112 kW (75 to 150 hp).
157. AFDs are quick in response, but they are subject to nuisance shutdown from small changes in incoming line voltage. All of ours are sensitive to voltage dips. Overheating is a problem. In general, AFDs give more trouble than other types of variable-speed drives. *Sanks: AFDs have recently become more versatile and robust. Automatic restart with power line synchronization is obtainable and should always be specified. Some drives now produce a nearly perfect sine wave without the spikes that cause heating. The overheating problem at low speed can be solved by adding a fan driven by a small motor directly connected to 60 Hz, ac power.*
158. The control system must be tailored for the application and flowrate.

Ventilation

159. Good ventilation is very important for both equipment and operators.
160. No less than 10 air changes per hour (ac/h) for dry wells and 20 ac/h for wet wells is required to prevent corrosion and odors. *Jones: We recommend 10 ac/h for pump rooms and a minimum of 20 ac/h for wet wells and screen rooms (40 ac/h in southern states where the wastewater is warmer and the need for hydrogen sulfide production is greater).*
161. The general design principle of using both inlet and exhaust fans to introduce air at the ceiling and discharge air into ducts at the floor has been effective in providing a pleasant work environment and protection of the equipment. For example, there has never been condensation on any equipment—even on tanks. *Sanks: Proper ducting is just as important as the number of air changes per hour.*
162. If the dry well is vented to the wet well, set up the duct work so that a high wet-well level cannot flood the dry well. A few problems in early designs were corrected in subsequent designs. The dry well in the Bellevue Pumping Station was flooded shortly after being placed in service. Power failure caused a high wet well level and allowed wastewater to flow through a ventilation duct between the wet and dry wells. *Jones: Any ducting whatsoever between wet and dry wells is now prohibited by NFPA 820.*
163. Sail switch alarms in the ventilation ductwork systems were provided in many of the early stations and were subsequently connected to the main treatment plant control centers over leased telemetry lines. These caused numerous nuisance calls to check stations that could have been serviced during the next routine inspection.
164. The high (10 to 20 ac/h) air exchange rates specified in early stations has created some difficulties when trying to retrofit facilities with odor control equipment. Replacing fans or adjusting sheave sizes has largely alleviated this problem.
165. If exhaust gases must be scrubbed for odor control, high-volume ventilation becomes very expensive. *Jones: This consideration puts the engineer at a decided disadvantage. Ventilation at lesser rates [than above] creates, in effect, confined spaces that require three-person entry teams and safety equipment. It also exposes the engineer to the potential for liability claims if an accident occurs. Given the value placed on human life, the added expense of an odor control system for the greater ventilation rates seems a justifiable, needed expense.*

Wet Wells

166. The long, narrow self-cleaning [trench-type] pump intake basin works well in practice. All of our future pumping stations will be of this type.
167. Seattle Metro's wet wells are pumped and hosed down at least weekly and even more frequently in troublesome locations.
168. The older "traditional" wet wells for constant-speed (C/S) pumps were dangerously odiferous and huge piles of grit accumulated rapidly.

169. We prefer suction bells that terminate near the floor over other types of intakes.
170. Provide high-pressure "hose-down" facilities. *Sanks: Supply at least 1.6 L/s (25 gal/min) at a pressure of no less than 550 kPa (80 lb/in.^{2}) at the nozzle.*
171. Smaller wet wells, where storage can be provided in the sewer lines (and not in the wet well) have fewer problems in general.
172. Minimize control instruments in wet wells.
173. Eliminate heavy gratings that must be lifted to gain access to the wet well for hose-down and/or maintenance. [See *Gratings*]
174. In other cities, wet wells are often dark, dismal, and poorly ventilated—a scourge for operations. Yet operators must enter them frequently to control odors and remove grease and accumulated solids. Some of the features that make frequent wet well entry less objectionable are:

    (a) Easy-opening, counterbalanced access hatches.
    (b) Ventilation systems that provide enough air exchanges (15–20 ac/h) to prevent condensation and remove potentially harmful gases.
    (c) Wash-down stations with local explosion-proof START/STOP switches to provide large volumes of water at high pressure.
    (d) Flood- and explosionproof lighting that provides good visibility for any operation chores.
    (e) Wet wells with no bar screens. Our experience has been that with V/S pumping, flow and debris come into the station constantly and uniformly, so they move through the system with minimal clogging. Our experience with inlet bar screens has not been very satisfactory. When operators rake the screens, a considerable amount of debris reaches the pumps in bunches and tends to clog them. Most stations are not equipped with facilities to handle removed screenings except to hand-carry them in buckets—a sure way to make operators unhappy. *Bosserman: Provide screenings removal equipment at stations where bar screens are deemed necessary. Jones: If screenings removal equipment is provided, then screening treatment equipment (washing, dewatering, and pressing) equipment and facilities for storage and removal are a must to prevent odors and minimize labor-intensive operation.*

24-10. Survey of Two Thousand Wastewater Pumping Stations

During 1996, one of the editors [9] investigated the performance of a great many raw wastewater pumping stations in several large sanitation agencies in the United States [10,11]. The initial part of the project (February and March) consisted of telephone interviews with operation and maintenance supervisors and chief engineers at 15 large public sanitation agencies. These 15 agencies operated a total of more than 2700 raw wastewater pumping stations. Based on the interviews, four agencies (City of Houston, Texas; City of Baton Rouge, Louisiana; City of Nashville, Tennessee; and King County Department of Metropolitan Services, Seattle, Washington) were visited in April. These agencies operate a total of more than 1000 pumping stations. Interviews were conducted with O&M personnel to determine what problems they frequently encountered and to establish their preferences for improving O&M characteristics of the pumping stations. Finally, more than 50 pumping stations were visited. Operations were observed, and the actual station O&M personnel were interviewed. The following is a short summary of some of the key issues that were identified.

1. How Not to Design a Wet Well

The design engineers for several pumping stations built within the past five years in one agency insisted that baffles per the Hydraulics Institute Standards [1,2,3] needed to be designed into the wet wells to provide quiescent conditions for the pumps and to provide greatest pump efficiency. The resulting baffle system did just that. It provided *very* quiescent conditions. As a result, the grit and solids, settling to the bottom of the wet wells, caused severe odor problems and sometimes clogged the intake pipes to the pumps. It takes an eight-person crew to remove the grit from the wet wells, because the wet wells are considered to be confined spaces. Willing to accept a little reduction in pump efficiency in exchange for eliminating the cost of removing the grit that accumulates in wet wells with baffles, the agency no longer installs baffles and has begun to demolish the existing baffles. *Sanks: The designers did not heed the warning in the Hydraulic Institute Standards: all H.I. Standard wet wells, (up to and including the 1994 edition) are for clear water only. If they are nevertheless to be used for solids-bearing waters, provisions for cleaning accumulated grit and scum must be made. Deposition occurs wherever currents decrease much below about 0.6 m/s (2 ft/s). Demolishing the baffle would make vigorous mixing (as a means of solids suspension and removal) more effective. See Sections 12-6 and 12-7 for viable alternative designs.*

Bosserman: If a designer uses conventional wet well design, consideration must be accorded to the removal of scum from the water surface and grit and other settleable solids from the bottom of the sump. Unfortunately, only recently (with some exceptions) have designers and published design standards or guidelines giving this factor the serious consideration that it deserves. Access must be convenient for bringing in cleaning equipment such as hoses to wash grease off the sump walls and vacuum hoses or other means to remove the accumulated sediments. The wet well walls and ceiling should be lined with smooth PVC to prevent the attack on concrete by sulfuric acid (derived from hydrogen sulfide) above the water line and to make it easy to wash grease off the walls. Gratings, sidewalks, or landings that do not interfere with hoses should be provided for the convenience of the cleaning crews.

2. More Comments on Wet Well Design

The vast majority of the dry well pumping stations in these agencies have sealed wet wells without ventilation. They contain a "breather" vent pipe, typically 150 mm (6 in.) in diameter, coming out of the wet well roof. When O&M personnel have to enter the wet wells (which is not very often), they use portable blowers and ducting to supply the necessary ventilation. In the latest designs, the wet wells are lined with PVC. One agency (King County Department of Metropolitan Services) that does desire routine and convenient personnel access to the wet well used the following design in their most recent pumping station (Interurban), started up in March, 1996:

(a) A catwalk of FRP grating over the entire length of the wet well about 0.6 m (2 ft) above the high-water level), so that personnel can walk over the entire wet well. The clearance between the catwalk and the walls is about 0.3 m (1 ft) to permit unobstructed wash-down of the walls.
(b) Variable-speed pumping, so that the wet well water level does not fluctuate very much.
(c) Explosionproof (NEMA 7) lighting fixtures, accessible from the grating for replacement.
(d) Walls and ceiling are lined with PVC.
(e) No screens or grinders.
(f) The ventilation system is so effective (the editor entered the wet well) that there was hardly any "sewage" odor, in spite of the fact that raw wastewater was just 2 ft below the grating. There are several ventilation ducts with diffusers on each, to distribute the air into the wet well.

3. Example of Unwarranted Optimism in the Design Engineer

A consulting engineering firm decided to use close-coupled vertical wastewater pumps, because the wet well was of a design that could not possibly be flooded. There was a solid, unpierced concrete wall between the wet well and the dry well from the bottom slab of the structure all the way up to the roof slab of the dry well containing the pumps. The following sequence of events occurred one day:

(a) There was a storm, during which a power failure occurred.
(b) In accordance with Murphy's Law ("If anything can go wrong, it will"), the standby engine generator failed to start.
(c) There was a rest room with a toilet on the roof slab over the dry well. The toilet was connected via a pipe to the wet well. There was no check valve in the pipe. The wastewater backed up in the wet well, flowed up the pipe, over the toilet, across the floor, and down the stairway into the dry well that couldn't be flooded.

4. Maintaining Dry Pit Pumps versus Submersible Pumps

All the O&M personnel at all the agencies visited felt that it is much easier to maintain a dry pit pump, unless the submersible pump was manufactured by one specific manufacturer. This one manufacturer (a large corporation) was almost unanimously acclaimed as having the most reliable pump on the market, with O&M requirements that were comparable to those for conventional dry pit pumps. Three other manufacturers were frequently cited as providing good pumps, although not quite as good as Manufacturer No. 1.

(a) Their main reason for stating that it is much easier to maintain a dry pit pump is that with a dry pit pump, the O&M staff can inspect it and tell when something is amiss and repair it before the problem becomes too serious. Such inspection is impossible with submersible pumps, because they cannot be observed routinely and must be hauled out of the wet well for repairs. Consequently, there is a greater tendency to let the submersible pump run until it simply fails. Then the cost of repair is usually very high.
(b) Submersible pumps are quite variable in frequency of maintenance required. Some

agencies have had to pull submersible pumps as frequently as every 6 months. Others have gone as long as 6 or 7 yr before having to pull the pumps for maintenance. The typical interval seems to be about 2 yr. Seal and bearing failures are the most common problems with submersible pumps. Damage to power cables and motor windings occur less frequently.

(c) It can take a crew of 6 to 10 workers to remove a submersible pump from a wet well. Depending on the size of the pump, the crew typically consists of the following workers:

   (i) 1 or 2 crane operators
   (ii) 2 or 3 electricians
   (iii) 3 to 5 other persons actually to remove the pump.

(d) Most agencies have discovered that O&M costs for submersible pumps larger than 75 to 150 kW (100 to 200 hp) become extremely high. Generally speaking, the agencies have discovered that:

   (i) Submersible pumps and motors larger than 75 kW (100 hp) break down more often than smaller ones.
   (ii) Removal of large submersible pumps takes a lot more effort than they had originally thought. In general, it requires more personnel to remove submersible pumps than conventional dry pit pumps. With large pumps (greater than 75 kW or 100 hp) workers in at least one large public utility usually must enter the wet well to disconnect the electrical power cable from the pump, unseat or disconnect the pump from the connecting piping, and help guide the pump out of the wet well. When the pump is reinstalled after being repaired, workers must again enter the wet well to reconnect the power cable and reconnect the pump to its discharge piping. As the wet well is classified as a confined space, this effort must be made by workers wearing rubber suits and air masks and accompanied by observers and supervisors above. *Sanks: Contrast that scenario with the experience of a large utility that manages 425 pumping stations of which 170 have submersible pumps including many large ones. No worker ever goes down into a submersible pump wet well except to work on guide rails or brackets. The pumps can be pulled up or replaced in half an hour by a crew of three. Minor service (such as changing oil or checking shafts for deflection) is completed on the pump while it rests horizontally in a cradle beside the wet well hatch. The cable is disconnected at the motor starter and accompanies the pump and motor to the shop for major repairs such as the replacement of seals and bearings. The power cable is not disconnected from the motor. The difference between the two kinds of operations lies in the training of the workers, attitude (do the best you can with what you have), and designs that make these operations easy (see Section 25-6).*
   (iii) The frequency of repair of submersible pumps is greater than that for conventional dry pit pumps. One agency thinks (although exact records are not available) that submersible pumps must be repaired 20 to 40% more often than dry pit pumps.

(e) One agency reported that it takes a six-person crew up to half a day to remove a submersible pump. If maintenance then has to be performed, additional time and labor is needed to transport the submersible pump to the maintenance shop, repair it, transport it back to the pumping station, and reinstall the pump. A conventional dry well pump can be repaired with a fraction of this time and effort. However, this agency also felt that good-quality submersible pumps (of which there are not very many manufacturers, in their opinion) are more reliable than the conventional dry well pumps in the sense that they do not fail as often as dry well pumps.

(f) A majority of the agencies send all their submersible pumps to private shops for repair. The most common repairs are replacing bearings, replacing seals, and rewinding motors. Typically, it cost $4000 to $8000 to repair a submersible pump and sometimes much more. For example, it could cost as much as $27,000 to replace the rotor and stator on a 160 kW (215 hp) submersible pump. Use an ENRCCI of 5600 for these costs (see Chapter 29).

(g) It can cost $600 to purchase a new upper bearing assembly for a submersible pump. A lower seal assembly for a submersible pump can cost $2400. By comparison, the seals (conventional packing) and bearings for a dry pit pump can be purchased for only $200.

(h) In the opinion of the O&M staffs at most of the agencies, the degree of effort required to maintain various types of wastewater pumping stations (from easiest to most difficult) is as follows:

  (i) Submersible pumps in a dry well.
  (ii) Conventional dry well pumps in a dry well
  (iii) Submersible pumps in a wet well.

(i) The sales representatives for submersible pump manufacturers tend to be very optimistic regarding the effort that it will take to maintain a submersible pump. A careful and realistic life-cycle cost analysis should be performed before deciding to use submersible pumps, especially if the contemplated pumping station is fairly large—greater than 440 or 660 L/s (10 or 15 Mgal/d) capacity.

5. Depth of Wet Wells for Submersible Pumps

It was observed and reported that in deep wet wells, the heavy electrical power cables to the pump motors can sometimes stretch and crack. The turbulence in deep wet wells can also cause the long cables to whip and bend, thereby damaging the cover, the power wires, or especially the connection at the motor cap. When a crack develops in the cable cover, hydrogen sulfide can enter the cable and corrode the power wires. It is expensive to purchase a replacement power cable and it sometimes takes four to six weeks to obtain a new cable from the manufacturers, most of which are overseas. *Sanks: Stretching can be controlled by fastening the power cable to a stainless-steel wire rope designed to carry the tension. Some manufacturers regard cable replacement as emergency repair and keep cable on hand so that replacement may require as few as three work days.*

Another factor in wet well depth, especially important for removing large pumps, is that the long, thick power cables cannot be bent to a short radius without damaging the cable. Consequently, it requires much care and effort to lift both the pump and the cable together without overstressing the latter. It is for this reason that some agencies disconnect the cable from the pump before lifting the pump out of the wet well. Agencies with experience in removing large submersible pumps prefer to limit the maximum allowable wet well depth to about 12 m (40 ft). *Sanks: Excessive flexure near the motor connection can be controlled by a cable saddle or by the use of semi-flexible stainless-steel cable sheathing that controls the radius of bending. It may be preferable for the cable to exit downward from the motor caps of large pumps to prevent excessive bending when the pump (with cable attached) is lifted. Cable sheathing or a saddle is still required because the cable must be tight during normal pumping operation to keep it from swinging with the currents and thus wearing out the cable connection at the motor cap.*

6. Discharge Elbows for Submersible Pumps

A major problem that has sometimes been encountered (fortunately, not very often) with submersible pumps is the discharge elbow in the wet well tearing loose from its anchor bolts. When that happens, it requires a major effort to repair it. The wet wells are classified as confined spaces per OSHA, so it takes an eight-person crew to do the work in compliance with all the confined-space observation and supervision requirements. It is believed that the cause of this problem is as follows:

(a) A pump ingests a large rock or some solid.
(b) The pump then vibrates as it tries to expel the solid.
(c) This vibration creates forces and movements in the pump and its discharge elbow that are much greater than those the elbow anchor bolts and associated concrete anchoring system were designed to withstand.
(d) Each time such an event occurs, it weakens the concrete and anchoring system a little more, and eventually, the bolts tear out of the concrete or the concrete breaks, and the agency has a major repair problem. *Sanks: Beware of light anchor bolts. A bolt smaller than 22 mm ($^7/_8$ in.) is a toy. Heavier bolts cost very little more than smaller ones. Bolts should be (1) buried deep enough to develop their ultimate tensile strength and (2) surrounded by a strong cage of reinforcement to prevent cracking of the concrete.*

7. Variable-Speed Pumping

Many agencies use variable-speed pumping extensively. Those with significant experience feel it is a very effective and efficient means of operating raw wastewater pumping stations. Most new variable-speed units are adjustable-frequency drives (AFD). One AFD manufacturer in particular was cited as providing especially reliable equipment and excellent service and training support. The drives feature insulated gate bipolar transistor (IGBT) technology, which results in an extremely quiet design compared with that of other older units. There is only a slight 60-cycle background hum, compared with the ear-piercing noise sometimes produced by older designs. In addition, the IGBT design runs cooler and does not emit nearly as much heat, thereby reducing the air-conditioning requirement for the control room.

8. Standby Engine-Generators

Although standby power is usually thought to be essential for raw wastewater pumping stations, there are sometimes exceptions to this rule. One agency (City of Houston) has none in any of their 330 pumping stations. Another agency (City of Baton Rouge) has none in more than 400 pumping stations. Both have concluded that their power supply is very reliable, and there is no need for standby generators. Some of their older stations have connections for use with portable generators, but the newer stations do not. These agencies have determined that the hydraulics of their systems allows them to withstand power outages of 3 to 5 hours duration before any wastewater spills occur, and the power failures have never lasted that long. Both have portable generators (up to 400 kW) that they can use when needed, but they are typically only used 10 to 12 times per year. Not having to provide permanently installed standby generators or dual power feeds to these several hundred pumping stations has saved considerable construction cost. *Sanks: Each of the above agencies may have both a fortunate and a unique situation. Each facility should be evaluated for standby power need on the basis of size and the consequences and probability of power failure. See Section 13-14 for developing a more balanced viewpoint and for examples of power outages that have caused disasters.*

9. Odor Control

Most O&M staff stated that they prefer packed carbon towers, because they are easy to maintain. Both the activated carbon type and the impregnated carbon type have been used. *Jones: The impregnated carbon type has a well-established poor safety record (fires, hazardous chemicals) and is not recommended.* The carbon systems last a long time. At some pumping stations, the carbon has not had to be changed in over 3 yr. Wet scrubbers (such as caustic soda and sodium hypochlorite) are high-maintenance systems. Sodium hypochlorite corrodes everything it touches when it spills and is a hard system to maintain.

10. Flow Metering

Four agencies (City of Houston, Texas; Ann Arundel County, Maryland; Orlando, Florida; and Frederick County, Department of Public Works, Maryland, with a combined total of more than 900 raw wastewater pumping stations) have no flowmeters at any of their pumping stations. They just read the elapsed time meters (ETMs) for the pumps, and knowing the approximate capacity of each pump, they can then approximate the flowrate. They see no need for greater accuracy. Flowmeters are installed at the treatment plants where the flow data are needed. Typically, flow measurement devices of some sort are needed only at pumping stations to indicate flow build-up over time to allow the O&M staff and engineering staff to determine when they need to add additional pumps or to expand a pumping station. In the opinion of these agencies, time meters furnish adequate data.

11. Pump Speeds

Most agencies prefer no more than 1200 rev/min speed for any pump (dry pit or submersible) or perhaps even no more than 900 rev/min maximum speed. Some of the pumps run at speeds up to 1800 rev/min. These higher-speed units wear out much more quickly than pumps operated at 600 to 1200 rev/min. Packing has to be replaced more often, and the bearings and shaft sleeves wear out faster and must be replaced more often. In general, pumps with operating speeds above 1200 rev/min have significantly higher O&M costs than slower speed units.

12. Cranes and Monorails

Almost all agencies have permanently installed bridge cranes or monorails in their conventional dry well/wet well pumping stations. In the opinion of the O&M staffs at several of the agencies, design engineers need to do a better job of considering the crane layout and how the cranes will be used. The O&M personnel also stated that they need more access space around equipment to be able to use the crane systems effectively. In some of the pumping stations, the crane layout was so poor that rework was necessary to make them function properly. For example:

(a) Some cranes were designed so that that the crane blocks had to be operated at an angle to remove pieces of equipment—a violation of OSHA crane safety regulations.
(b) In another pumping station, a bridge crane was supposedly designed so the trolley could be centered over the motors and pumps for removing them. Unfortunately, the designers placed a 90° elbow and discharge riser directly on the pump discharge nozzle. When the trolley was moved toward the motor, it struck the riser pipe and could not be centered over the motor or pump.
(c) In several pumping stations at four agencies, the overhead bridge cranes could not be easily used because of interferences from structural beams. Better coordination is required between the mechanical engineers who lay out the pumps, piping, and crane systems, and the structural engineers who design the structural beams and supporting columns. *Sanks: Project leaders, take note!*

13. Accessibility of Equipment

A nearly universal complaint of the O&M staffs was that more space or clearance needs to be provided around equipment for maintenance access. In some of the pumping stations visited, the access was so cramped that maintenance personnel could not even remove the bolts to the pump flanges without a major team effort. In others, access to the stuffing boxes was restricted so that changing the packing was difficult. Designers need to give much better thought to important issues such as:

(a) How will packing or mechanical seals be changed?
(b) How will pump shafts be removed so that the shaft sleeves can be replaced?
(c) How will impellers be removed and replaced?
(d) How will flanges be disassembled and reassembled?
(e) How will the top covers on check valves be removed?

Designers need to keep in mind that all these maintenance activities require a team effort of two to four people to be carried out, and there must be sufficient space around the equipment so they can work together effectively.

14. Screens and Grinders

Generally speaking, most agencies have no screens or grinders in any of their pumping stations. Typically, O&M staffs see no need for them and consider them to be high-maintenance items. At some older stations, the existing screens have been taken out of service, because they served no real purpose and required maintenance. Of the 15 agencies investigated, only one (City of Nashville, Tennessee) required screens or grinders at all of their stations. That agency reported they consistently get large quantities of construction debris (pieces of brick and masonry, lumber, drill bits, bolts, nails, etc.) in their sewage and they need to have screens or grinders ahead of the pumps to avoid excessive wear of the pumps. This problem does not, however, appear to be a typical or common occurrence in municipal sewage systems.

15. Check Valves

Most agencies normally use hydraulically buffered swing check valves or at least require springs or weights on the lever arms of the check valves. In pumping stations built within the past 5 to 10 yr, the trend toward hydraulically buffered controls on the swing check valves was especially evident. *Sanks: See also Section 5-4. With a heavy spring or counterweight, the liquid in the valve is a hydraulic buffer as effective as a dash pot (although the spring increases the headloss).*

16. Mechanical Seals

Another definite trend was the use of mechanical seals in the pumps for new pumping stations. Agencies that have investigated mechanical seals most extensively prefer the split-seal type with tungsten carbide (and not ceramic) faces. Split seals are easy to install and disassemble. Leakage is much less with a mechanical seal than with packing, and the mechanical seal lasts much longer than packing. One agency reported that in their life-cycle cost analysis, they concluded that a mechanical seal pays for itself in 3 yr because of the reduced maintenance requirements. The mechanical seal lasts 10 yr—an excellent bargain, in their opinion.

17. Packing

Another trend observed, especially in those agencies that have not switched to mechanical seals (Orange

County Sanitation District and City of Baton Rouge), was the use of packing materials that do not require an external clear water flush. At least two agencies have acquired extensive experience with such materials. One agency (Orange County Sanitation District) has been using one of them [12] for more than 10 yr, and they report excellent success. The packing lasts two to three times as long as conventional packing. The sealant material is so fine that it screens out wastewater solids that leak up the shaft, so that the solids cannot come in contact with and abrade the shaft sleeve. These agencies no longer even have the flushing water systems installed in their new wastewater pumping stations.

18. Level Sensors in the Wet Wells

Bubbler systems are by far the most popular wet well level sensing devices at all of the agencies. The bubbler systems have the least problems and are the easiest to maintain. Various agencies have also used floats, pressure transducers, and ultrasonic level sensors.

(a) Floats were generally disliked. They become caked with grease and scum, and were ruined by the turbulence in the wet wells. Even placing the floats in stilling wells or pipes in the wet wells does not solve the turbulence problem. Floats were regarded as high-maintenance items and were considered to be much less reliable than bubbler systems. Most agencies used floats only for emergency high-level alarm service but no longer for controlling pumps.

(b) Only one agency (L. A. County Sanitation Districts) reported that they preferred pressure transducers. This agency has developed a design that consists of a pipe spool mounted against the dry side of the wet well wall near the dry well floor. A gate valve is attached to the spool and the transducer is attached to the gate valve. A 12-mm ($^1/_2$-in.) water flushing line is connected to the spool, and the water stream keeps the pipe spool clear of grease and solids. The design is reported to be extremely effective and almost maintenance free. The other agencies have not reported success with their designs, but none of them used the wall spool mount with the water flushing system as did the one successful agency. See Section 24-9.

(c) One agency (King County Department of Metropolitan Services, Washington) has tried several manufacturers of ultrasonic level devices to monitor the water level in their sewage pumping station wet wells. They were able to find only one manufacturer whose product worked reliably in such service.

19. Coatings for Impellers, Pump Volutes, and Shaft Sleeves

One agency (City of Baton Rouge, Louisiana) reported that they coat the pump impellers and volutes with IPI Fluid Ceramic or Cerami-Tech C.R. [13] in two equal layers to a total thickness of 0.50 mm (20 mils). These coatings are very hard, offer excellent abrasion resistance, and increase the life of the impellers and volutes by making them more resistant to abrasion by grit.

(a) They also coat the shaft sleeves by first machining about 1.25 mm (50 mils) off the sleeve and then applying Cerami-Tech E.G. [13] to build the sleeve slightly oversize. The coating is so hard that, after it cures, it can be machined smooth to its original size only with a diamond-tipped cutting tool.

(b) They apply these coatings in their maintenance shop. Their pump specifications do not require the coatings to be applied as part of the original equipment manufacturer's scope.

20. Electrical Conduit Installation

All the O&M personnel reported that they prefer conduit to be exposed and not embedded in the walls or floors. Over the life of a pumping station, changes are inevitably made to the electrical and control systems, and it is easier to make these wiring changes if the conduit is exposed.

21. Air Release Valves on Sewage Force Mains

One agency has over 600 air release valves installed in their force mains. These valves are extremely high-maintenance items. They have two teams (of two persons each, plus a fifth person serving as a supervisor) whose only duty is to remove air release valves from the force mains and bring them back to the maintenance shop for cleaning and repair. Additional staff then do the cleaning and repair work. The most common problem is that the valve outlets or connections become clogged with grease and scum, thus making them useless.

(a) Each valve is sandblasted to remove build-up of grease and crud. The internal parts are then cleaned or replaced.
(b) The staff can remove, clean, and reinstall 10 valves per day. Hence, each valve is cleaned every 60 working days (about once every three months). Even this level of effort does not keep all the valves in operating condition 100% of the time.

Bosserman: Any agency contemplating the use of air release valves in wastewater force mains should consider the high O&M effort that will be required to keep them operating. If this effort is not provided, the valves quickly become useless and were better not installed at all, because they give a false sense of security.

22. Materials of Construction

All the agencies are much more aware of corrosion than they used to be, and now insist on corrosion-resistant materials. The following materials are typically used by all the agencies for various pieces of equipment or components in pumping stations constructed within the past 5 to 10 yr:

(a) Stairway treads, landings, grating, access platform decks, etc.: fiberglass reinforced plastic (FRP) or aluminum.
(b) Outdoor electrical cabinets: stainless steel.
(c) Handrails, access hatches, HVAC ductwork, and doors: aluminum.
(d) Odor control system ductwork: FRP.
(e) Chain-link fencing: PVC-coated for corrosion protection. Galvanized fencing near wastewater pumping stations eventually corrodes because of the minute concentrations of hydrogen sulfide. The PVC is typically green, although red is available (and definitely eye catching!)

23. Lighting

The O&M staffs typically now prefer halide lighting. Fluorescent lighting is too dim for dry wells. Some of the O&M staffs also felt that the designers underestimate the level of lighting required for them to see easily what they are doing during maintenance. Newer pumping stations typically have the walls and ceilings in the dry wells painted off-white or beige, although one agency has been doing this for almost 30 years. The light color gives a noticeable improvement in the overall brightness of the room and helps to provide better lighting for maintenance.

24. HVAC Design for Control Rooms Containing AFD Equipment

Control rooms containing adjustable-frequency drive (AFD) control panels in four pumping stations in two cities had heating, ventilating, and air-conditioning (HVAC) systems so poorly designed that extensive rework was required after construction. Apparently, the HVAC designers did not consider the heat loads emitted by the AFD equipment. In one station, the rework consisted of adding two outdoor air-conditioning units and additional ductwork to connect them to the control room. In another station, the cooling system had to be more than doubled in capacity. Temperatures as high as 61°C (142°F) occurred prior to the rework. In all four pumping stations, the AFD equipment was not of the IGBT design discussed in Subsection 7.

25. The State of the Art in Sewage Pumping Station Design

A majority of the agencies investigated prefer submersible pumps installed in a dry well. Their most recent pumping stations use this concept for the following reasons:

(a) Submersible pumps eliminate the lineshafts associated with conventional dry well pumps. These lineshafts can require a lot of maintenance, especially with pillow block bearings. When the pillow block bearings fail and are replaced, extraordinary effort is required to reinstall and realign the lineshafts.
(b) Submersible pumps eliminate the O&M costs associated with shaft and equipment vibration, which is never zero regardless of what anyone says.
(c) Submersible pumps eliminate the initial construction cost of catwalks for access to the intermediate bearings, lineshaft supports, etc.
(d) Submersible pumps allow much better access from the overhead bridge cranes because there are no interfering lineshafts, catwalks, or bearing support structures in the dry well.

(e) Submersible pumps are noticeably quieter than conventional dry well pumps.

(f) *Sanks: In addition to being vibration-free, submersible pumps are more compact and cleaner. They eliminate the need for a seal water supply. On the other hand, frame-mounted WP-1 motors and dry pit pumps are less expensive in both first cost and maintenance. Even if a dry well is flooded (a rare occurrence), the motors can be dried out in a shop, the bearings replaced, and the motors put back into service in (usually) a week. Balance the risk of flooding and its consequences with cost.*

24-11. References

1. *Hydraulic Institute Standards for Centrifugal Rotary & Reciprocating Pumps,* 14th ed. Hydraulic Institute, Cleveland, OH (1983).
2. American National Standard for Centrifugal Pumps, ANSI/HI 1.1-1.5-1994. Hydraulic Institute, Parsippany, NJ (1994).
3. American National Standard for Vertical Pumps, ANSI/HI 2.1-2.5-1994. Hydraulic Institute, Parsippany, NJ (1994).
4. Parris, H. L., and J. L. Seminara. "Factoring human needs into power plant maintainability." *Power,* **127,** 82–85 (January, 1983).
5. Kettle, R. Superintendent of Desert Operations, Los Angeles County Sanitation District. Private communication, August 10, 1989.
6. Lescovich, J. Chief Engineer, Golden Anderson Industries. Private communication, 1989.
7. Redner, J., Sewerage System Superintendent, Los Angeles County Sanitation Districts. Private communication, October 3, 1995.
8. Arhontes, N. Staff Engineer, Maintenance, County Sanitation Districts of Orange County. Private communication, October 3, 1995.
9. Bosserman II, B. E., Principal Engineer, Boyle Engineering Corp., Newport Beach, California (May 1996).
10. Bosserman, B. and S. Collins. "Pumped-up pumping stations." *Civil Engineering* **67** (7) 65–66 (July 1997).
11. Bosserman, B., A. Will, and S. Collins. "Current concepts in wastewater pumping station design," presented at the 28th Joint Annual Conference of the Chesapeake Water Environment Association and the Water and Waste Operators Association of Maryland, Delaware, and District of Columbia, Ocean City, Maryland (July 9–11, 1997).
12. TP 5400 made by Tom Pac Inc., 7575 Trans-Canada Highway, St. Laurent, Quebec H4T 1Z6.
13. Cerami-Tech C.R. and Cerami-Tech E.G. Thortex Division of E. Wood Ltd., Standard Way, Northallerton, N. Yorks, UK DL6 2XA.

Chapter 25

Summary of Design Considerations

GARR M. JONES
RANDALL R. PARKS
ROBERT L. SANKS

CONTRIBUTORS

Stefan M. Abelin
Michael L. Bahm
Charles T. Blanchard
Pat H. Bouthillier
Roland S. Burlingame
Fredric C. Burton
A. L. Charbonneau
Harry E. Covey
Roger J. Cronin
Patrick J. Creegan
Johannes de Waal
James C. Dowell
Edward J. Esfandi
Richard O. Garbus
Harold D. Gilman
Mayo Gottliebson
L. V. Gutierrez, Jr
Stanley S. Hong
W. Eric Hopkins
George Jorgensen
Frank Klein
Lonnie Lange
R. Russell Langteau
John Leak
Ralph E. Marquiss
Warren H. Mesloh
Stephen G. Miller
Allen W. Peterson
Edgardo Quiroz
John Redner
David M. Reeser
Marvin Dan Schmidt
Earle C. Smith
Brian G. Stone
LeRoy R. Taylor
R. Daniel VanLuchene
Leland J. Walker
David Walrath
Horton Wasserman
Frederick R. L. Wise
Frank A. Woodbury
Eugene K. Yaremko

The many considerations involved in the design of pumping stations make this summary chapter desirable for project engineers, for designers, for city engineers, for utility managers, and for both beginners and (as a partial checklist with helpful hints) for experienced designers. Topics discussed in other chapters are treated briefly, whereas topics such as structural engineering not mentioned elsewhere are covered in more detail. As in other chapters, published codes and specifications are referenced by number only. See Appendix E for full titles.

The purpose of the preliminary engineering phase of a pumping station design project is to make the major decisions that establish such things as

- Whether a pumping station can be avoided in favor of an alternative solution
- The best site (with all factors considered)
- The type of pumping station and pumps
- The kind of drivers, constant or variable speed, electric or engine
- The need for (and type of) standby power
- The type of construction: excavation with sloping sides, sheeting, cofferdam, or caisson
- Extent of automation, control, and data recording and/or transmission
- Force main route and profile
- Auxiliary systems such as cranes and chlorination.

Good "front-end" engineering is vital. No astuteness in subsequent work can overcome poor front-end decisions, so this stage of design should be supervised by the wisest and most experienced engineers in the firm. One pump expert has stated that 95% of all pumping stations contain significant design blunders and that these mistakes occur in every conceivable category including hydraulics, the mechanical system, the electrical system, and common sense. Blunders are expensive in construction, operation, or maintenance. At best, they occasion minor annoyance and embarrassment. Make yours one of the elite 5% with sound planning, forthright conferences with clients and their operators, periodic design reviews and design checklists, and thoughtfulness and the application of common sense to reduce errors and omissions to an acceptable level.

Some decisions can be reached only after cost comparison studies are made (see Example 29-1), but nearly all decisions should be made with the coordinated help of supporting professionals, whether in-house or carefully selected outside consultants, in the following disciplines and subdisciplines:

- Civil engineering
    - Surveying
    - Soils (or geotechnical) engineering
    - Hydraulics (including transient analysis)
    - Structural engineering
- Mechanical engineering
    - Heating, ventilating, and/or air conditioning
    - Noise
    - Vibration
    - Odor control
    - Pumps and piping
    - Engines
- Electrical engineering
- Instrumentation and control
- Architecture.

The project leader must be able to communicate with the professionals in these disciplines, and to do so effectively requires familiarity with the language, symbols, and (to some degree) the problems involved with each discipline.

After reaching the major decisions that affect the basic design criteria, the requirements of each system must be clearly conveyed to each designer. The duty of the project leader is to coordinate information transfer between support disciplines to ensure an efficient design and an economical pumping station devoid of blunders caused by the interference of one discipline with another. The importance of keeping a complete and legible set of records in, for example, a three-ring binder cannot be overemphasized. The records should include memoranda of important conferences and telephone calls, design memoranda to individuals in support disciplines, design calculations, and sketches. All records should be indexed, and they should be self-explanatory and easily understood because (1) accidents and opportunity frequently change the complement of an office, (2) design commenced by one engineer may have to be completed by another, and (3) a lawsuit may hinge on adequate records analyzed and interpreted by someone other than either author.

Augment this chapter with Sections 6-8, 15-1, 16-1, the first third of Chapter 17, Sections 22-1 and 22-2, and (as all of them apply to preliminary design) Chapters 24, 26, and 27. Codes and standards are identified only by their abbreviations, which are defined in Appendix E.

25-1. Need for Pumping Stations

Pumping stations are expensive to construct, maintain, and operate. They should be avoided whenever practical. Consider the following factors when deciding whether to install a pumping station:

- Topography, excavation, elevations, and capacity of existing water distribution and treatment systems or sewage collection and treatment systems
- Capital, operation, and maintenance costs along with the possibility that additional skilled personnel may be needed
- Problems such as odor or noise and other adverse aesthetic effects.

In some circumstances, odor control may represent the major cost of operation. In 1981 regulatory agencies forced the owner of a 1.1 m^3/s (25 Mgal/d) sewage pumping station in California to install an odor-control scrubbing system that cost \$200 to \$300/d for operation. Because there is no scientific way to evaluate odors, the owner and the engineer may be at the mercy of hostile residents.

Possible alternatives to installing a pumping station include the following:

- Increasing the head at an existing water pumping station by changing impellers, pumps, or drivers
- A reservoir at a critical elevation that could be filled when water demands (and transmission losses) are low
- A deep, gravity-flow interceptor sewer, perhaps installed in a tunnel
- Individual, on-site, or community underground sewage disposal systems at a lower elevation, especially when the area served is small.

In some situations it may be desirable to construct a bare-bones design for a short service life pending the eventual construction of a long-term (and more costly) conveyance facility.

An economic analysis of alternatives should include

- Capital cost
- Annual operation and maintenance costs
- Spare parts inventory
- Service life and replacement cost recovery
- Power costs, including heating and ventilating
- Annual lubrication and parts such as seals, bearings, packing, and gaskets
- Capital and annual costs for odor-control facilities and the additional treatment and odor-control costs related to the septicity of wastewater detained several hours in force mains.

Equation 29-6 is useful for evaluating the economics of alternatives. An engineering comparison of alternative designs is presented in Example 29-1.

25-2. Site Selection

Designers may have little choice in selecting the sites for wastewater pumping stations, which must often be located in low-lying areas characterized by poor soils and high groundwater. The deep excavations for wet wells create difficult problems requiring expertise and care. The designer of a water pumping station has a better choice of sites. The location of sites for raw water intake pumping stations usually depends on the requirements of the intake structure rather than the pumping station. Sites for lake or reservoir intakes are governed by such factors as shoreline topography, unstable soils or rock, depth of water and need for multiple port inlets, aquatic growth, fish protection, surface trash, and ice. For rivers, the factors also include the scour or deposition of bed or bank material, the distance from unstable channel reaches, the adverse effect of nearby bridge piers or other hydraulic structures, the stability of river banks, the location of river bends, the character of the thalweg, and the hazards from floods. Attempts to improve poor river intakes by adding groins, training walls, or other works is not only expensive but often ineffective. Even the U.S. Army Corps of Engineers has experienced unsuccessful attempts to make a poor site satisfactory.

Factors to Consider for Site Selection

The following factors should be considered when selecting a site.

- *Land availability and cost.* Allow for construction activities, future expansion, and parking for maintenance vehicles and even mobile cranes. Except in wet, unstable soils or very deep pits, it is less expensive to excavate side slopes than to use sheet piling.
- *Topography.* Land should be flat enough to minimize construction problems but have enough slope for surface drainage.
- *Soils.* Complete the subsurface survey prior to land acquisition.
- *Protection from flooding.* Some states require designs based on the "hundred year" flood. Flood elevations may be obtained from (1) the U.S. Geological Survey, (2) the Soil Conservation Service, (3) the U.S. Corps of Engineers, (4) the Department of Housing and Urban Development Flood Insurance Studies, (5) state agencies, or (6) by observing the high-water marks pointed out by the owner or local residents. Note that changes in upstream or downstream developments may alter future flood levels.
- *Availability of utilities*
    - *Water and fire protection.* Field check the available pressure and obtain the local fire department requirements for minimum flow and pressure.
    - *Power.* Check the available voltage and capacity. If electricity from a second, separate substation is used for standby power, the owner may have to pay for installation. Find the costs of extending water and other utilities to the site.
    - *Gas.* Heating and standby power may be operated with natural gas or with biogas from digesters.
- *Access.* Roads adequate for required construction and future maintenance are necessary.
- *Security.* Consider the potential for theft and vandalism. Avoid remote sites that are visible and readily accessible from a road.
- *Aesthetics.* Highly visible sites or sites near other structures may require special architectural treatment and/or provisions to minimize odors and/or noise.
- *Multiple use.* Obtain owner input on site use and combining other facilities with the pumping station.
- *Local land use and zoning ordinances.* Are there any special restrictions imposed by planning agencies?
- *Transmission pipelines.* Minimize profile elevation changes and the length of pipelines. Consider the problems and costs of rights of way, eminent domain, and construction in busy neighborhoods.

Subsurface Investigations

It is vital that a qualified geotechnical engineer make subsurface (soil) surveys and prepare an engineering report. The geotechnical engineer must inform the structural engineer about expected soil conditions and describe their probable effect on the various options available for the structural design engineer. It is the structural engineer's responsibility to evaluate the geotechnical engineer's report and select the most suitable option. It is often advantageous to involve the soils engineer during the construction phase to ensure that construction procedures conform to design assumptions regarding the magnitude of earth pressures and the sequence of construction. A sufficient number of borings should be made to determine the stability of the excavation. The following information is required:

- The stability of slopes
- The possibility of heave in the bottom of the excavation
- The high and low groundwater levels
- The best dewatering procedure: (1) pumping from open sumps, (2) well points, (3) deep wells, (4) ejectors, (5) freezing, or (6) sheeting driven to an impervious stratum and caulked to prevent leakage
- The need for permanent underfloor drainage
- The best method of excavation: (1) open pit (include the allowable slope), (2) sheeting (steel sheet piling or soldier beams and lagging), (3) the need for bracing (either cross-lot bracing or prestressed tie-backs), (4) freezing the embankment (to eliminate the need for sheeting and bracing), (5) coffer dams, or (6) caissons
- The best method for resisting uplift due to groundwater: (1) weight, (2) tension (uplift) piles, (3) prestressed drilled anchors, or (4) a coffer dam driven to impervious substratum with a permanent sump pump as backup for any slight leakage
- The lateral earth pressure (1) on rigid walls with passive earth pressure or (2) on flexible walls for both moist soil and saturated soil
- Soil properties at all changes in strata: (1) natural moisture content, (2) soil unit weight, (3) Atterberg limits for cohesive soils, (4) unconfined compression tests in cohesive soils, (5) standard penetration tests in sand, (6) cores in rock, (7) coefficient of friction between the bottom of the slab and soil, and (8) shear strength
- The probable effect of excavation, dewatering, and/or pile driving on nearby structures
- Any unusual conditions such as variable rock formations and elevations, springs, quick conditions, the sinkhole potential in deposits over limestone, the potential for "boil" in the bottom of the excavation, and perched water table
- Feasible foundations: (1) base slab on grade; (2) base slab on piles with treated or untreated wood piles, steel H piles, concrete-filled steel pipe piles, concrete-filled and mandrel-driven tube piles, precast concrete piles, or mandrel-drilled, concrete-filled, step-tapered piles; or (3) base slab on drilled piers
- If piles are required, the types, required depth of embedment, capacity, whether batter piles are needed, and whether pile load tests are required
- If drilled piers are feasible, the elevation of the bottom, whether the drill shaft needs to be lined, and whether the piers should be belled
- If a base slab on grade is the solution, (1) allowable bearing pressure, (2) probable settlement, and (3) coefficient of friction between the base slab and soil
- The possibility of unequal lateral forces and the resultant tendency for sliding if adjacent areas are excavated in the future
- The water table: install a permanent well point in one boring to determine the static water levels because levels obtained when borings are made are unreliable.

The foregoing information is the responsibility of the geotechnical engineer. The investigation is made with the help of preliminary design information furnished by the structural engineer.

Hydrogen Sulfide Investigation

Hydrogen sulfide (H_2S), a product of stale sewage, is a deadly, odiferous (it has a rotten-egg smell) gas slightly heavier than air. Certain bacteria convert H_2S to sulfuric acid, which is very corrosive to electrical equipment and to concrete, iron, and steel. Large wet wells exacerbate the H_2S problem because the quiescent conditions allow solids to settle, and, because they are not readily resuspended, the long retention time increases H_2S production. Always investigate the concentration of H_2S at the beginning of the project because a substantial (more than, say, 1 mg/L) concentration of H_2S or a combination of high (>22°C or >72°F) sewage temperature and long (>5 h) residence time in the sewers, may:

- Be a deciding factor in choosing variable-speed drives
- Require a sealed wet well or pump sump
- Alter the location of the pumping station site
- Require odor-control units on the ventilation exhaust system to meet state requirements (in California, the concentration of H_2S in air at the property line is limited to 30 ppb)
- Increase the capital and operation costs of the ventilating system
- Influence construction details, such as the kind of cement used or the type and lining of the wet well and sewers
- Require a specific regimen of wash-down operations to keep acid-producing bacteria in check.

25-3. Architectural and Environmental Considerations

Before the pumping station design is begun, make sure all the environmental considerations are met and public hearings have been held. The site should be checked for archaeological findings, and local and state regulatory agencies that may have jurisdiction should be consulted. Become acquainted with the pertinent codes (see the subsection on Chapter 25 in Appendix E).

Local planning agencies are likely to require review and approval of plans. Discussions with the owner, the local residents, and the planning agency are useful in determining the style of building to be used. Consider using a consulting architect if one is not available in house. In general, the cost of architectural treatment is minor; one expert estimated the cost to be less than 2% of the construction cost. Others maintain that it is often difficult to determine whether architectural treatment costs anything.

Aesthetics

For the most part, utility installations should not be seen, heard, or detected in any way that would degrade the local environment. To that end, the design should include the following features:

- Archetectural design that blends into the neighborhood and neither establishes a presence that declares the purpose of the installation nor degrades the property value of its neighbors. Pumping station superstructures, with minimal effort and expense, can be easily be crafted to appear to be homes, farmhouses, or restaurants (see Figure 25-1).
- As alternatives, pumping stations can either be concealed by burial (with the exception of electrical and instrumentation equipment, which must be protected from flooding), partly concealed by landscaping, or made into attractive architectural features. Landscaping plans that minimize maintenance should be developed by a landscape architect.
- Noise and light emissions should be suppressed to avoid broadcasting the function of the installation. Special construction features can be included to virtually eliminate the high-frequency noise from motors. Hospital-grade silencers are available for engines at little additional cost. With diligence, the noise from ventilation equipment can be suppressed. Lighting fixtures in a variety of architecturally acceptable designs are available to illuminate exterior areas without suggesting a commercial or industrial facility.
- For wastewater stations, odor control and treatment are perhaps the most challenging environmental management tasks. The design should incorporate every technique for avoiding the conditions that generate odors. A variety of cost-effective and simple treatment techniques are available.

Hazardous Areas

An unventilated or sealed wastewater pumping station wet well is rated as a Class A Confined Space. Because of the extreme danger of explosion, flammability, toxic gases, and oxygen deficiency, it can be entered only under stringent regulations (see Section 23-1). If the wet well contains electrical equipment, it is rated by NEC standards as a Class 1, Division 1 hazardous space when flammable or explosive gases are normally present, as in an unventilated space. However, if flammable gases are absent except under abnormal conditions (e.g., failure of ventilation system), the wet well is rated as a Class 1, Division 2 space. All electrical equipment and devices in Class 1, Division 1 spaces must be explosionproof, whereas in Class 1, Division 2 spaces, only arc-producing equipment (e.g., switches with external contacts) must be explosionproof. (A squirrel-cage motor in a Class 1, Division 1 space must be in an explosionproof enclosure, whereas it can be in an open, dripproof enclosure in a Class 1 Division 2 space as long as it is not equipped with switches.) Division 1 spaces in wet wells can be avoided (except below the grating) by installing a ventilation system that always operates at 12 air changes per hour or more (see Section 23-2). The ratings of the various areas must be decided in conjunction with the authority having jurisdiction—the local or state electrical inspector.

A dry well with access to a wet well also becomes a Class 1, Division 1 or Division 2 hazardous space, and such access should never be tolerated in new construction nor should it be allowed in an existing pumping station being remodeled. Access to a wet well should be possible only through an outside door.

Ventilation

Sewer gas is deadly. The safety of personnel who must enter pumping stations should not be compromised, especially if they must enter wet wells. Codes require a certain number of air changes per hour, and minimums are shown in Table 25-1. However, some states require a higher rate of ventilation and, if the wastewater is stale, so should prudent engineers.

Figure 25-1. Seattle Metro sewage pumping stations designed by Brown and Caldwell Consultants. (a) Trees and shrubbery conceal the station from the road above; (b) architectural enhancement of a pumping station; (c) a pumping station in a residential district. Photography by George Tchobanoglous.

Some codes allow a lower rate of ventilation while the station is unattended if two-speed fans are provided for ventilating a hazardous area at high rate before a worker enters. This practice should be discouraged because it does not guard against explosions, corrosion, or the impatience of workers. Air to wet wells should be *both supplied and exhausted* through ducts by powered blowers so that a slight negative pressure is maintained (refer to Sections 23-1 and 23-2).

c

Figure 25-1. *(Continued)* (c) a pumping station in a residential district.

Table 25-1. Suggested Minimum Ventilation Criteria for Wastewater Pumping Stations

Space	Ventilation criteria
Wet wells with access for maintenance	For continuous operation, 12 to 30 air changes per hour. Supply air at the ceiling. Withdraw air at the floor near sewage channels. Use powered blowers on both intake and exhaust.
Motor rooms	Sufficient air to control the buildup of heat.
Engine rooms	Sufficient air to control the buildup of heat of the air supplied. Exhaust at least 10% at the roof to control buildup of smoke and fumes.
Pump rooms	Use 10 air changes per hour. Supply air at the ceiling; withdraw air at floor.
Motor control centers	Provide 6 air changes per hour of filtered air, and air condition if heat would otherwise be a problem.
Dry wells	For continuous operation, 6 air changes per hour.

In stations housing large motors (whether for pumping sewage or water), the rate of ventilation required for cooling the motors may be a controlling factor, especially in hot climates where the ventilation rates can be surprisingly high. Heat dissipation in stations equipped with engine drives can be of even greater magnitude. The ventilation problem is one to be solved by an experienced heating and ventilating engineer.

Requirements for the design of ventilation systems for wastewater installations are provided in NFPA 820. Designers should be particularly aware that control and electrical equipment require an environment that has proper temperature control and protection against atmospheric conditions such as hydrogen sulfide gas. Because overall station reliability greatly depends on the proper operation of these devices, environmental control is of vital importance to the success of the installation. See Chapter 23 for information on proper design for purging and scavenging toxic, dangerous, and corrosive gases from an enclosure.

Odors

Unpleasant odors are emitted from some sewage pumping stations—a serious problem for stations in or near a residential area. Bad odors are often caused by grease and scum buildup on the wet well walls, so good housekeeping practices may eliminate the nuisance. Steps can often be taken upstream from the station to minimize hydrogen sulfide and reduce odor problems. By sealing the wet well, odors can be largely confined, particularly if the drivers are variable-speed (V/S) units that eliminate the "pumping" action produced by the rise and fall of the wet well levels that occurs with constant-speed (C/S) pumping units. If it is impossible or impractical to eliminate odors, then air-scrubbing units may be required in the air exhaust system—an expensive solution. If freezing can occur, dry scrubbing units may require thermostatically controlled heaters to avoid plugging by frozen condensate.

Access

Provide safe, adequate access for personnel and adequate space and facilities for replacing any or all machinery. The considerations include the following:

- Pump rooms (dry wells) and wet wells must be completely separated. Therefore, provide separate ground-level access to each. *Never* interconnect wet and dry sides even above the flood level.
- Entrance depths of 3.6 m (12 ft) or more should have stair landings or ladder platforms.
- Ladders should have nonslip, serrated, or splined-edged rungs of aluminum and should be surrounded with a cage. Aluminum can corrode at the concrete interface, so coat the aluminum at the point of contact or consider plastic-coated steel or stainless steel.
- Conventional stairways and landings are by far the best. It is difficult to carry tools and nearly impossible to carry injured workers up ladders or circular stairways.
- Hatches, doors, and other openings must be large enough to remove the largest piece of equipment.
- Large access covers are difficult to lift, so specify standard spring-loaded covers made skidproof by using checkerplate or an equivalent.
- Individual hatches over each submersible pump or a duplex split hatch over two pumps are preferable to a single large lid over two or three pumps. One edge of each hatch opening over a submersible pump should be rounded and smooth so that the electrical power cable cannot be cut by a sharp edge as the pump is pulled out and swung inboard.

Storage

In larger pumping stations, provide space for the storage of spare parts inventory, tools, and maintenance equipment.

Lighting

Uniform lighting of about 200 to 300 lux (20 to 30 ft · cd) at the floor is adequate except for repairs, which require about twice as much light. Install additional ceiling lights that can be switched on when needed (required in some states for energy conservation). An alternative is GFCI outlet receptacles for portable lights. Outlet receptacles are also required for small power tools. Supply 120-V power for lights, outlets, and small loads from a small, dry-type transformer (25 kVA or less) supplying a single panelboard.

Personnel

Personnel requirements for pumping stations vary over the broad range given in Table 25-2. Some owners allow major pumping stations to be unstaffed, while others may require 24-h two-person operation for a similar facility. Factors to be considered for operational personnel requirements include (1) the capital investment, (2) the degree of reliability needed, (3) security and the risk of vandalism, (4) safety, (5) union requirements, and (6) the owner's policy. The extent of automation and remote telemetry required as well as many of the amenities of the station depend on whether it is to be attended or unattended.

Scheduled daily inspection, including maintaining a daily log, is normally provided for unattended pumping plants. The facilities should be adequate for regular scheduled maintenance as well as for correc-

Table 25-2. Operational Requirements of Pumping Stations

Type of Facility	Operational personnel	Comments
Water pumping stations of less than 750 kW (1000 hp); wastewater lift stations of less than 0.2 m^3/s (5 Mgal/d)	Unattended	Daily inspection, automatic controls, alarm and fail-safe features.
Major water and wastewater stations with engine drives, screens, chlorination facilities, or odor-control facilities.	Staffed, 8 h/d, 5–7 d/week	Automatic controls, local and remote alarms, fail-safe features.
Major regional water and wastewater stations for communities of more than 250,000.	Staffed, 8–24 h/d, 7 d/week	Automatic controls, local and remote monitoring.

tive and emergency unscheduled maintenance, which may require working crews of several persons for several days. To avoid problems later, discuss the station layout with the owner's maintenance personnel before beginning the design.

Convenience

Items providing for convenient operation and maintenance include the following:

1. The building:

- Materials on the walls, floors, and ceilings easy to maintain with no rough finishes
- Coves at the base of all walls
- Good floor drainage (slope the floor to the drains so that water cannot form puddles)
- Easy access to all windows, louvers, and equipment to be serviced
- Sufficient space to service and maintain equipment
- Loading platform and ramps
- Restrooms, a slop sink, and a janitor's closet for all but the smallest stations
- Room for storing miscellaneous small tools and maintenance items such as impellers, wearing rings, packing, and other materials as specified by the owner or manufacturer
- Workbench, tools, and log books (for all but the smallest stations).

2. Access:

- Convenient roads
- Convenient hatches for equipment removal
- Pipe basements instead of crawl spaces

3. Equipment:

- Cranes or eyes for handling the debris from screening equipment
- Drainage to sump pumps for dewatering and washdown
- Eyes, monorails, or cranes for equipment handling.

4. Miscellaneous:

- Hose valves (bibbs) with a backflow preventer for cleaning and for yard irrigation (freezeproof them if necessary); check state regulations for the separation of potable water from wash water
- A desk, chairs, and file cabinets for storing records, maintenance manuals, and plans
- An O&M manual for reference and troubleshooting
- Special tools
- Site maintenance equipment and storage rooms
- A drinking water fountain or bottled water and drinking cups.

Safety

Safety considerations include the following:

- Nonslip walkways of checker or diamond plate or covered with nonskid "sanded" paints
- Handrails and kick plates around openings and stairways
- Safety cages on ladders and intermediate platforms for ladders longer than 3.6 m (12 ft); straight stairways are far more convenient and safe than are ladders
- Air-purging systems
- Chlorine detectors and gas masks
- A self-contained air-breathing tank and mask apparatus
- Explosive air detectors
- Hydrogen sulfide detectors
- Eye washers and showers if hazardous chemicals are used
- Personnel equipment such as hard hats, safety harnesses, first-aid equipment, ropes, and life vests or rings
- Fire extinguishers
- Telephone or two-way radio communication—unnecessary if service trucks are always so equipped
- Grounding of all electrical equipment including GFCI receptacles for hand tools
- Backflow (cross-connection) preventers or an air gap between the potable water supply and the pumping station water supply systems
- No open-style personnel lifts; all lifts and elevators to have solid walls, ventilation, roof escape hatch, telephone, and no gates.

Security

Consider the following as security items:

- Fencing and exterior lighting
- Unauthorized entry alarms with signal transmission to some authority (e.g., the police).

25-4. Future Expansion

Designers must have the vision and presence of mind to plan for expansion beyond the design life. If a future demand for increased flow is likely (which is nearly always true), consider provisions for future expansion and weigh the cost of such provisions against the future cost of remodeling or abandonment. The size and importance of the pumping station, the owner's requirements, and the uncertainties of need, time lapse, and expected flow increase must be taken as intangible but essential and sometimes overriding considerations.

The method for achieving the future increase in capacity should be included in the specifications and the plans. Future expansion can be facilitated in a number of ways that include:

- Pumps with medium-sized impellers that can be changed for larger ones
- Generous space for each pump so that a larger pump can be installed; the extra space will be appreciated by the maintenance staff
- Suction and discharge piping large enough to permit the increased flow
- Space for additional pumps; terminate suction piping and pipe connections to the header with valves and blind flanges
- Provision for enlarging the building by adding steel waterstops and dowels for connecting future walls and floors; bend the dowels to the face of the concrete and protect exposed steel with lean mortar (note that rubber or plastic waterstops can be too easily damaged by future construction)
- An electrical system designed so that the entire system will not have to be rebuilt; use larger conduit, for example, and leave extra space where needed.

Plan the operations necessary to allow the changes to be made while the station remains in operation. Consider the following:

- The pump manufacturer should know whether larger impellers are to be used so that bearing frames, bearings, motor, and other associated parts are sized properly.
- The removal of blind flanges on the discharge side of a future pump requires either (1) a means for quickly draining pressurized pipelines or (2) a valve at the flange. Note that valves are likely to become inoperative unless regularly exercised. Gate valves are the worst, lubricated plug valves are good (and the most expensive), and stop plates or slide gates are the best, the most reliable, and the least expensive when they can be used for isolating pump inlets.
- The removal of blind flanges on suction pipes requires dewatering the wet well (or a portion thereof) in a way that minimizes (or eliminates) downtime or prevents overflows in wastewater pumping stations. The wet well may be compartmentalized by inserting stop plates into grooves provided for that purpose.
- Considerable storage is available in sewers upstream from most wastewater pumping stations—storage that is sufficient to make the needed connections if planned properly and executed at the lowest flow.
- Because blind flanges are heavy, provide lifting handles to speed removal.
- Ensure convenience for the lifting equipment needed to install new or replacement facilities, and provide access for the installation of new equipment.

The capacity of an existing pumping station can often be increased. To avoid blunders, first make field measurements of flow, static head, friction head, true pipe diameter, and roughness coefficient. Trust no plans—not even record drawings. Measure the station, especially the piping connections, so that the new facilities will fit exactly. Various methods of increasing the capacity are

- Cleaning and relining transmission pipelines in place
- Laying new transmission lines or force mains
- Substituting larger impellers in existing pumps
- Substituting larger pumps and motors if the suction and discharge piping permits
- Adding submersible pumps in the existing wet well
- Converting the dry well to a wet well and adding submersible or VTSH pumps
- Adding a new wet well with submersible pumps
- Constructing a new pumping station
- Adding a booster pumping station.

25-5. Hydraulic Constraints

The hydraulic constraints include the flow capacity (and the range of capacities), the discharge characteristics required of the pumps (the range of both the flow and the head), the suction head, and the type of fluid pumped.

Wastewater Pumping

Pumping stations should be designed to handle flows from the entire service area. Obtain owner input regarding the size of the service area. Install a pumping capacity that is adequate for the fully developed service area, or install a pumping and force main capacity that can handle both the current needs and those of the immediate future with an adequate provision for economically adding to the capacity in the future. Consider both average and peak flows, which may be based on (1) existing and future land use, (2) per capita flow estimates, or (3) actual field flowrate measurements. Consult the literature before making flow projections [1-5].

- Choose a realistic design period, but make it possible to meet flow demands beyond the design period.
- Design the station layout to allow for planned increases in the capacity at specific times within the design period.
- Always use the larger pump size when choosing between two sizes.
- Optimize the size of force mains. The velocity should be at least 0.76 m/s (2.5 ft/s) to keep grit moving. At such a low velocity, a daily flush at 1.2 m/s (4.0 ft/s) for an hour or more is desirable. The maximum velocity should not exceed about 2.4 m/s (8 ft/s) because of high head losses and the possibility of water hammer.

Water Pumping

For the probable population to be served and the estimated flow needs, choose a realistic design period and allow for future expansion beyond the design period.

- The station layout should permit plans for periodic increases in flow capacity.
- Carefully consider the maximum daily and hourly flowrates, the requirements for fire flow, and a combination of the two.
- If head and flow are expected to change significantly over the years, choose pumps with medium-sized impellers so that impellers only—not pumps—need to be changed to match the flow.
- Choose the correct pipe roughness (or range of roughnesses) and not a single "conservative" value (see Section 3-2).
- Optimize the size of transmission mains. Where velocities exceed about 2.4 m/s (8 ft/s), water hammer may become a concern and may require additional, expensive control equipment and structures.

Sludge Pumping

Sludge is usually confined to treatment plants and transported for only short distances—a few hundred meters or yards at most. Centrifugal pumps can be used for thin sludges in large quantities. But positive-displacement pumps are usually required for thick sludges, because these sludges are thixotropic and, after immobilization, high pressures are required to start their movement. For wastewater sludges, consider:

- Sludge properties, which are entirely different from those of water (see Chapter 19)
- Sludge pipelines longer than 1.6 km (1 mi) must be designed on the basis of *measured* friction. (The associated problems are difficult and beyond the scope of this book. See Section 19-5.)
- Comparisons of plunger, progressive cavity, and lobe pumps based on both first cost and maintenance
- Pipelines of no less than 150 mm (6 in.) in diameter, although glass-lined piping down to 100 mm (4 in.) may be suitable
- Glass lining for sizes below 250 mm (10 in.) and cement-mortar lining for any size (but particularly for larger pipe).

Water plant sludges are of two general types: (1) those resulting from sedimentation and flocculation and (2) those resulting from lime softening. The former are gelatinous, bulky, and difficult to dewater. It may be desirable (although difficult) to decant them, but they are not difficult to pump. The characteristics of softening sludges vary greatly, and it is difficult to predict a conservative design that operates well over the range of necessary conditions. It is wise to

- Choose pumps that can be easily and relatively quickly disassembled and cleaned and locate them for ready accessibility.
- Use straight pipelines that can be rodded out, or, better, use open troughs that are easily accessible for cleaning.
- For a more extensive discussion of sludge pumping, refer to Chapter 19.

25-6. Types of Pumping Stations

The following discussion about the types of pumping stations is necessarily abridged. Greater detail is given in Chapters 11–15, 17–19, 26, and 29 of this volume, so this section is limited to a few selected essentials relating to choice. Develop alternatives only to the point where realistic, comparative cost estimates can be made (e.g., as shown in Example 29-1).

Wastewater Pumping Stations

Decisions should be based on the required capacity, cost, reliability, the owner's preference, and aesthetic considerations. Reliability is more important than efficiency. Sewage pumps must pass stringy materials and other solids, and it is frequently specified that they must be able to pass a 75-mm (3-in.) sphere. But because many relatively large pumps contain restrictions, always require certification on the size of solids to be passed *entirely through* the pump.

Before choosing the type of pumps and pumping stations, consider the following:

- Minimum 100-mm (4-in.) piping for sewage may dictate the pump size required to maintain minimum velocities. Many authorities and some codes allow 75-mm (3-in.) piping, but if this small pipe is used, make certain the pumps (or screens) cannot pass solids 75-mm (3-in.) in diameter.
- Provide enough headroom, floor space, and access to replace the pumps, motors, valves, and piping.
- Single volute pumps and single vane impellers are asymmetrical and cannot be hydraulically balanced. Many manufacturers cannot trim single-vane impellers. If a single-vane impeller is trimmed, it should be dynamically balanced (often by using attached counterweights) after trimming.
- A V/S drive allows substitution of a small pump sump for a large wet well, but V/S pumping may increase capital cost, does increase complexity and maintenance, and may decrease efficiency. Many V/S installations seem to have been misapplied, so read Sections 12-5 and 15-1 and Example 29-1 before deciding to use V/S.
- For flat system head curves, multiple C/S pumps alternated "first on–first off" may have most of the advantages of both V/S and C/S pumping.
- Provide one spare standby unit of the largest capacity, and consider the need for other spares depending on the critical effects of station outage.
- Supply standby power. (See Section 8-4 for engine-generator requirements and Example 9-10 for engine-generator sizing. Engine requirements are discussed in Chapter 14. Note that a separate electrical utility service can also be used for standby.)

Submersible Pumps

Large (75 to 750 kW or 100 to 1000 hp) submersible pumps are gaining market acceptance, although a number of utilities seem to agree that (due to unfortunate experiences and possibly inadequate facilities) they should be installed only in dry pits. In one large utility, no submersible pumps larger than 75 kW (100 hp) are allowed in wet wells. In another large utility, it required a dozen people working nearly a whole morning to remove a large submersible pump from a storm water pumping station because it was thought necessary to disconnect the power cable from the motor before pulling the pump. Hence, two workers were required to enter a wet well—a task that is not recommended by any pump manufacturer. As the wet well is a confined space, the large crew was necessary for safety.

Contrast the above scenario with the experience of still another large utility that likes large submersible pumps. The pump maintenance supervisor at that utility stated that 190 kW (250 hp) submersible pumps can be lifted and removed in half an hour by a crew of three (boom operator, electrician, and mechanic) using a truck-mounted boom. Larger pumps require a mobile crane, but the job is otherwise just as easy. (Workers never enter the wet well except to work on guide rails or brackets.) The pumps are cradled on their sides for access to impellers. The same utility has mechanics who are certified for repairing submersible pumps and who can even balance impellers to tolerances more precise than factory specifications. Minor servicing (such as changing the oil and checking the shaft and bearings for play) can be performed while a pump is in the cradle. When a pump is to be sent to the shop for major repairs, such as replacement of seals and bearings, the power cable is disconnected at the motor starter and not at the motor. The cable is pulled out of the conduit and accompanies the pump. This procedure allows the connection to the motor to be tested for shorts. After repairs and before approval for returning the pump to the wet well, the motor is run for a final check. Thus, large submersible pumps in wet wells *can* provide good service when the facilities are specifically tailored for their use and maintenance crews are well trained and motivated. Handling these large pumps is, however, entirely different from handling small ones because of their weight and the weight, size, and stiffness of their electrical power cables. Both cable and pump must be maneuvered with the assistance of cranes, whereas the cables for smaller pumps can be handled manually.

Special care must be taken in both the design and handling of large submersible pumps to prevent cables from being cut or rubbed on projections. At least one edge of the hatch should be round and smooth so the pump can be swung inboard without damage to the cable. Cables should not be allowed to swing freely in currents, because the movement may eventually crack the insulation or wires. Pulling the slack out of the cable is usually sufficient, but discuss the problem with the pump manufacturer,

One technique for removing large submersible pumps is to lift only the pump and allow the cable to hang freely in a "U"-shaped loop like an elevator cable. After the pump is swung inboard and placed on the floor (or in a cradle), the cable can be retrieved with the aid of the crane or boom. If the cable is to be disconnected at the motor cap (cable entry), the disconnected end can be held at floor level with, for example, a Kellems® Support Grip. This method eliminates the need for separate handling of the cable and for special arrangements for storing the cable while the pump is being repaired. On the other hand, disconnecting the cable at the motor starter does not disturb an attachment that must remain absolutely waterproof.

When a pump is pulled up in the above manner, the cable bends in an arc above the motor cap. This bending does not harm short cables for small pumps, but it may overstress long, heavy cables for large pumps. The cable can be protected (1) by means of semiflexible stainless-steel cable sheathing designed to control the radius of the curvature, (2) by the use of cable saddles, or (3) by emerging vertically downward from the motor cap. The need for special protection should be discussed with pump manufacturer. An alternative method is to lift both pump and cable together, thereby eliminating bending near the motor cap. A motor-driven cable reel mounted close to the wet well is of great help for handling and storing cable. The cable is unwound from the reel when the pump is installed, and a predetermined tension keeps slack out of the cable. When the pump is removed, both it and the cable should be washed with a high-pressure hose for sanitation.

The tension in long, heavy electric power cables for large pumps may require relief. One relieving method is to clamp the cable at intervals to a stainless-steel wire rope that carries the weight of the cable. Another method is to use a Kellems® Support Grip to attach the midpoint of the cable to a stainless-steel wire rope held by a hook device at the hatch of the wet well.

The unprotected end of the power cable is not submersible and should not even be exposed to water. Submersible pump control and power cables should never be spliced. The pump should be ordered with sufficient cable to avoid the need to splice. Water that does enter the ends of a control cable can cause false signals in the control panel and possibly enter the motor itself. Connecting a flexible cable to another in a junction box, however, is acceptable if the junction box is properly located. Water has entered power cables through junction boxes located improperly (for example, recessed into floors or mounted in regions of condensing humidity).

Electric power cables for submersible pumps in trench-type wet wells are particularly vulnerable to movement, because the pumps (being on the axis of the trench) are located directly in the path of the strong currents from the inlet. The cables will also collect stringy material. Exposure to currents and rags can be a serious problem that deserves careful evaluation by both the designer and the pump manufacturer. If the distance from the top of the motor to HWL is short, a rigid conduit rigidly fastened to the motor and extending to HWL would shield the cable from all currents. The cable emerging from the conduit must, however, be protected from excessive bending.

The above method does not apply to deeply submerged pumps. A large (150-mm or 6-in.) plastic conduit buried in the concrete wall or a stainless-steel conduit attached to the wall and ending near the discharge coupling is a possible solution if the cable can slide easily up and down the conduit. Again, excessive cable flexure at the ends of the conduit must be prevented. Note that the cable would have to be very long and be payed out from a cable reel as the pump is lifted. Note also that this method is not cheap. Another method is to attach the power cable at intervals to sliders or rollers held in a track (like fastening sails to a mast) that is bolted to the wet well wall out of the way of strong currents. These methods have not been tried, however, and there is no guarantee of success, so a full-scale test should be made.

Designers who have any doubts should describe the cable environment to manufacturers of cables and seek their advice. There may be no problem at all, or perhaps simple measures will suffice. Otherwise a summary of options is (1) to employ one of the methods above; (2) to invent some adequate alternative way to protect the cable; (3) to do nothing special and replace cables from time to time; (4) to locate the pumps in a dry pit—expensive in first cost but less costly to maintain, so the life cycle cost may be reasonable; or (5) to substitute a different type of self-cleaning wet well wherein cables are not subjected to strong currents.

Refer to Tables 25-3 and 25-4 and Sections 24-10 and 27-6 for further discussion.

Comparison of Different Types of Wastewater Pumping Stations

Different types of pumping stations (Table 25-3) are interrelated with different types of pumps (Tables 25-4 and 25-5), so all three tables must be considered together. Be alert, because a comment in Table 25-3 is not repeated in Table 25-4.

Water Pumping Stations

All pumping stations (except wells) require redundant pumping equipment so that while the largest pump is removed for maintenance, the remaining pumps meet all the demands on the station.

Raw Water Pumping

Some influencing considerations that are not under the designer's control include variation in the water level, potential flooding, trash and debris, fish protection, aggressive water, silt and turbidity, microorganisms and slimes, and ice—frazil, surface, and anchor. Head and discharge requirements influence the type of pump and, perhaps, the type of station.

Table 25-3. Comparison of Wastewater Pumping Station Types

Advantages	Disadvantages
Dry well–wet well	
Easy access for maintenance.	Greater cost due to excavation and building below grade—expensive if groundwater is high, if soils are very poor, or if blasting is required.
Wider range of head and capacity.	Greater risk of outage due to flooding; dry well must be kept dry.
Wider choice of driver arrangements.	Flood-protected motors (in the dry well) are expensive.
Possible to use flood-protected motors.	Long leads to the motor (in the dry well) from the control panel if the motors are frame-mounted to the pumps.
Flooded suction improves reliability.	
With long shaft and flood-protected motors at grade, electrical leads are short.	
Wet well pumps with above-grade drivers	
Less excavation (eliminates the dry well).	Difficult to remove for servicing, especially the pumps with separate discharge pipe.
Small superstructure.	Difficult to service in place, wet well must be drained or pump removed for major disassembly to reach many parts.
Above-grade drivers protected from flooding.	High superstructure needed to permit the removal of the pump.
	Submerged bearings are subject to overloads and frequent failures.
	Difficult to keep lubricated and protected from grit.
	Should not be used in raw sewage (clogging) applications.
Wet well with self-priming pumps above grade	
Least construction cost, easiest maintenance.	Reduced reliability due to the need for priming.
Convenient access to pumps above ground.	Distance from the low wet well level to the pumps is limited by NPSH and the priming system to about 8 m (25 ft).
Eliminates dry well.	If operated infrequently, putrefaction with gas production can occur to break prime and/or to make sewage boil at 60°C (140°F), so an external priming system must then be used.
No flood hazard to motors.	
Self-priming pumps do not have trimmable impellers, so they are usually belt driven to reach the design point.	
Belt drives are easily changed to meet a new condition point, but V-belts require proper adjustment and periodic replacement.	
Wet well submersible pumps and motors	
No dry well; excavation and concrete reduced.	Valves and headers must be accessible (1) in an adjacent vault, (2) in a small above-grade superstructure, or (3) by exposing the header above grade.
No superstructure required except for engine–generator or cabinet for motor controls.	Pump must be removed and disassembled for inspection and maintenance. Requires a hoist or crane and specially trained mechanics
No seal water system, no long shafts with steady bearings required.	Hazard of pumps jamming on guide rails or not seating properly.
Reduces the land area needed.	Often more difficult to remove pumps than manufacturers admit.
All of the above reduces construction costs (see Figure 29-9).	Pumps larger than 75 kW (100 hp) especially difficult to handle (see text above).
Quick removal and replacement in emergencies.	Special motors, seals, and moisture monitoring required (but moisture probes are useless for leaks via power cable).
Well adapted for increasing the capacity of a pumping station using existing wet and dry wells.	Makes differ greatly with respect to satisfactory performance, and quality of pump/driver unit has a very high impact on maintenance and life-cycle cost.
No daily nor weekly maintenance (but overhaul needed every 1 to 5 yr).	Vibration has occurred with some makes of pumps larger than 22 kW (30 hp).
Units removable for shop servicing, minimizes field work.	Guarantees often are valid only if repairs are made by authorized service centers.
Quiet operation.	At least one uninstalled spare unit (in addition to standby units) in each size is needed to permit shop servicing, which adds to the cost trade-off for submersibles.
Safety from flooding.	Pumps and motors are generally not suited mechanically for V/S operation (although AFDs can be used).
On a life-cycle basis, balance the lower first cost of submersible pumping station with its lack of regular, frequent maintenance chores against the cost of complete overhauls by specially trained mechanics (or at manufacturer's service center).	Inaccessibility of the pump and motor for routine preventive maintenance and the need for occasional expensive overhauls lead some engineers to refuse to countenance their use.
Elimination of (1) the dry well, (2) frequent preventive maintenance, and (3) the seal water systems that, coupled with simpler design and low first cost, lead some engineers and many operators to prefer submersible pumps.	

Table 25-3. Continued

Advantages	Disadvantages
Wet well submersible pumps and motors (continued)	
	Large units tend to break down more often than do small ones and tend to require high maintenance costs.
	Choice of pumps *for wet well service* should depend almost entirely in the utility itself—attitude, experience, and commitment to train and to support maintenance workers (see Sections 1-13, 24-11, and 27-6).
Horizontal vs. vertical configuration	
Horizontal pumps	
Some double-suction pumps available in large sizes.	Requires a dry well.
Easy access for maintenance.	Requires a large floor area.
Motors less costly.	Long leads from the control panel to the motor.
More convenient for belt drive.	
High headroom not required.	
Permits a high head design with a double-ended shaft motor and two pumps coupled with series piping.	
Direct-connected vertical pumps	
Less floor space and a smaller superstructure (see also Table 25-4).	Same as for horizontal pumps (above) plus.
	Double suction not available.
	High headroom required for lifting motor.
	Supporting the motor by the pump casing is poor practice in seismic areas.
Extended shaft vertical pump	
The same as direct-connected vertical pumps plus Avoids flood hazard to motor. Permits work on pump without disturbing motor. Short electrical leads.	Long shaft requires a structure to support the intermediate bearings plus added lubrication and maintenance. Consider a stiff, hollow shaft without intermediate bearings (which may be impractical because limitations are severe).
	High headroom needed for replacing motors, shafts, and pumps.
	Flexible couplings required at pump and motor.
	Must ensure that manufacturer analyzes shaft for torsional vibration—especially for variable-speed drivers.
Field construction vs. package plants	
Field construction	
No limitations of head or capacity.	Higher construction cost.
Complete flexibility of layout.	More engineering time and skills.
Special features can be included, such as bar screens, comminutors, better access and working room, overhead hoists, and spare parts storage.	More difficult to avoid blunders.
In smaller sizes, may be faster to build than package stations.	
Greater ease of maintenance due to less crowding.	
Package types—general	
Low construction cost (see Figure 29-9).	Alarming number of fatalities and injuries in can-type prefabs due to asphyxiation, H_2S, flooding, and the difficulty of escape.
Consulting engineers' cost is lower for "off-the-shelf" units.	Access is often poor and rescue almost impossible.
Equipment can be factory assembled and tested prior to shipment.	Ventilation is usually woefully inadequate.
Standardization reduces chances for blunders.	Layout design is inflexible and special features are limited.
Especially suitable for small stations.	Cramped working space makes maintenance and repairs difficult and more costly because of added labor.
Available in following types: underground wet well–dry well, wet well with suction pumps, submersible pumps in wet well, and pneumatic ejector.	Capacity is usually limited to about 110 m^3/h (500 gal/min), although larger ones have been built.
	Corrosion and buckling are potential problems with steel shells, which often corrode quickly (in 2 yr) due to stray electrolytic currents, and hence cathodic protection is required.
	FRP is difficult to use where soil loadings are significant; it is also vulnerable to corrosion if the protective resin coating cracks.
	Discuss hazards and disadvantages with owner.

(Continues on next page)

Table 25-3. Continued

Advantages	Disadvantages
Package type: wet well–dry well	
The same as the dry well–wet well type above plus it can be less conspicuous with only hatches aboveground.	Inconvenient access.
	Layout design and special features are limited.
	Crowded space adds maintenance difficulty.
	Forced draft ventilation is often inadequate, so the atmosphere is somewhat corrosive.
Package type: wet well with self-priming pumps	
Same as the wet well with suction pump above plus it is less expensive than the package wet well–dry well.	Maximum capacity is about 90 m^3/h (400 gal/min).
Package type: pneumatic ejector	
Suitable for basements and for small capacities and high heads.	Relatively high construction cost.
Can provide scouring of force main at low flows.	Low efficiency.
	Consider cutter pump (for low head) followed by a centrifugal pump (for medium head) or by a positive displacement lobe or progressive cavity pump (for high head).
	Maintenance of auxiliary equipment is expensive.
Package type: submersible pumps in wet well	
Low cost.	Pump characteristics somewhat limited.
	See wet well submersible pumps and motors above.

Table 25-4. Comparison of Non-Clog Wastewater Pumps

Advantages	Disadvantages
Self-priming centrifugal pumps	
Sizes to 250-mm (10-in.) suction/discharge.	Typical efficiencies are 10–20% below other non-clog pumps.
Capacities to 0.15 m^3/s at 24 m (2400 gal/min at 80 ft) TDH.	Priming depends on the volume of suction line, which must be short.
Heads to 38 m (125 ft) at less discharge.	High dynamic suction loss.
Low required NPSH.	
Suction lift to 7.6 m (25 ft).	
Dry well is unnecessary if suction lift is not excessive.	
Quick, easy access for impeller cleanout.	
Close-coupled centrifugal pumps (common pump and motor shaft)	
Lowest first cost.	Motor bearings must carry the radial and thrust loads of pumping. Some pumps require special motors with a 3-month delay after ordering.
Most compact.	Maintenance problems arise because the motor must be removed to replace packing or mechanical seals (not true with some models).
Conditions favoring use	
Limited finances.	
Temporary stations.	
Operating time less than 2000 hours/yr.	
Horizontal frame-mounted centrifugal pumps	
Pump bearings can be selected for long life and suitability for continuous operation with high radial and thrust loads.	Higher cost than close-coupled pumps.
Does not need a special motor.	
Coupling between the motor and pump reduces lateral vibrations, permits work on the pump without disturbing the motor.	
Wide range of pump sizes and characteristics is available.	
Vertical frame-mounted centrifugal pumps in dry well	
Same as for horizontal frame-mounted pumps (see also Table 25-3).	Same as for horizontal frame-mounted pumps plus motor must be lifted to work on the pump or to replace mechanical seals (but not to replace seal packing).

Table 25-4. Continued

Advantages	Disadvantages
Vertical frame-mounted centrifugal pumps in dry well with long shafts for motors on floor above	
Same as vertical frame-mounted pumps (see also Table 25-3).	Highest cost; shafting adds 5–7% to pump cost. Must analyze the shaft for torsional vibration. Impractical for large (1.3 m^3/s or 20,000 gal/min) pumps.
Wet well submersible pump and motors	
Wide range of sizes and characteristics—up to 5 m^3/s (80,000 gal/min) for a vertical propeller pump. Conditions favoring use Need for low first cost. Many other units in same system. See also Table 25-3.	V/S control limited to AFD. Pump and motor are more expensive than dry well types. Maintenance requires removal (a messy operation) from wet well and disassembly by trained workers, so factory service may be required at substantial expense.
Dry well submersible pumps	
Excellent reliability. Maximum protection against dry well flooding. No seal water. No intermediate shafting to disassemble for work on pumps and no steady bearings nor structure to support or gain access to them. Eliminates maintenance of such bearings. No daily maintenance. No oiling, greasing, or leaking packing glands. Reduced housekeeping. Air fin or water jacket cooling permits much greater pump-cycling frequency. Above advantages may make submersibles the most cost effective. Minimal or no vibration. Quieter than ordinary dry pit pumps. No long shafts to interfere with cranes. Becoming popular. Conditions favoring use High risk of dry well flooding. Deep stations. Reduced maintenance requirements.	More costly than pump and separate motor, although elimination of seal water systems and reduction of wet well size because of more frequent allowable starts may make price competitive. Water jackets require frequent flushing, occasional disassembly, and they are often clogged with solids in pumped fluid at low speeds. Fresh water cooling expensive. Repairs in place may void manufacturer's guarantee. If delays are likely for shop repairs, must have uninstalled spare unit. Air cooling is inadequate for many models (requires cooling water jacket, fins, or large fan on motor). Variable-speed control limited to AFD, which may increase cooling problems especially at lower speeds (see Section 15-11). Access to the pump (the part of the unit requiring the most maintenance) is costlier than with a conventional pump.
Propeller pumps	
High efficiency. Large discharge, up to 57 m^3/s (2000 ft^3/s). Low cost, low maintenance. Available with variable pitch blades, which reduces motor horsepower and starting torque. Conditions favoring use Large discharge at low head (9 m or 30 ft) per stage for axial flow and up to 24 m (80 ft) per stage for mixed flow. Screened sewage or storm water.	Unsuited for raw sewage flows smaller than 1 m^3/s (15,000 gal/min) and unsuitable for high heads. Sewage should be screened for small pumps. Variable pitch blades may not be reliable. Check with users.
Vertical column, wet well pumps	
Axial- or mixed-flow pumps permit high capacity and, with multiple stages, reach high head. Avoid centrifugal volute-type pumps for wet well applications and use submersible or vertical turbine, solids-handling pumps instead. Conditions favoring use Large flows following primary treatment or screening. Propeller pumps for intermittent service on screened storm water.	Non-clog volute pumps limited to small capacities. Radial loads on submerged bearings limit bearing life. Diffusion vaned pumps cannot pass stringy solids, not useful for raw sewage in smaller sizes. Limited stable capacity range. Entire pump with column must be removed for service.

(Continues on next page)

Table 25-4. Continued

Advantages	Disadvantages
Vertical turbines for handling solids (VTSH®)	
Eliminates need for a dry well. True non-clog design; reliably pumps long stringy solids and passes 75-mm (3-in.) spheres (see Section 11-4). Flowrates of 0.13–1.3 m^3/s (2000–20,000 gal/min.) at 6–21 (20–70 ft) TDH. Motor above ground floor is protected from flooding. Capacity of existing pumping stations can be increased by converting dry wells to wet wells. At least two manufacturers. High efficiency over a broad operating range. Ideal for increasing the capacity of existing pump stations and for V/S. Can operate at very low speed.	Pump is expensive. Sizes smaller than 750 mm (10 in.) not available. Repairs to pump require lifting entire unit out of wet well. Hence, the superstructure must be very high or a hatch must be located so the unit can be lifted by a mobile crane. When a pump is removed for service, the wet well is interconnected with the pump room which, depending on the ventilation system (see NEC Art. 500), becomes a Class 1, Group D, Div. 1 or 2 hazardous space. TEFC motors are acceptable if the room is ventilated at 15 air changes per hour. At lower ventilation rates, explosionproof motors are strongly recommended (and may be required in the future). All electrical equipment should be protected against corrosion and labeled for hazardous environments. All electrical panels must be in a separate room. As with any column pump, the column must be vented; a conventional air release valve is unreliable. Use open vent with return to the sump or a solenoid-controlled, pilot-operated sleeve or diaphragm valve.

Table 25-5. Comparison of Special Design Wastewater Pumps

Advantages	Disadvantages
Open Archimedes screws	
Available in diameters of 0.6 to 3.7 m (2 to 12 ft) and flows to 3.8 m^3/s (60,000 gal/min). Maximum typical efficiency is 75% over wide delivery range. Constant-speed drive produces equivalent of V/S pumping; self-adjusts to match inflow. Conditions favoring use Low lift, as at treatment plants. Little variation in flow level at discharge. Can handle large solids so screening and grit removal unnecessary. Need for V/S pumping with C/S drive. Adequate area outdoors.	Does not pressurize fluid. Pumps only to a fixed discharge level. Requires critical clearance between trough and screw. Efficiency (1) depends on slip due to clearance (so it cannot be pretested), (2) is reduced by *cascade* at exit receiver, and (3) decreases at lower flows due to slip. Maximum lift 9 m (30 ft). Lower bearing is submerged in the sump and inaccessible unless the sump is drained. Concrete trough is difficult to construct. Large units and high lift are subject to binding and excessive backflow. Requires more space than other lift stations. Ice is troublesome at the inlet. Pump may require an insulating cover in cold climates. Releases odor if sewage contains H_2S. High cost.
Enclosed Archimedes screws	
Same as open screws except Guaranteed efficiency is 85% and only decreases to 80% at low flow. No concrete trough. No leakage. All bearings are in the open and readily accessible. Flowrates from 115 to 8000 m^3/h (500 to 35,000 gal/min).	In freezing weather the pump must be run nearly dry (in reverse) if it goes on standby. Largest size is 3 m (10 ft) in diameter. High cost.

(Continues on next page)

Table 25-5. Continued

Advantages	Disadvantages
Pneumatic ejectors	
Capacity range is 0.5 to 140 m^3/h (2 to 600 gal/min). High head. Automatically variable volume. Batch discharge velocity scours force main. Operation is usually trouble free in small sizes even without screening or comminution. Rugged, reliable, long life. Conditions favoring use Where centrifugal pumps unsuitable. An air supply is already available. Low (~11 m^3/h or 50 gal/min), inconsistent flow and high head. Single receiver is acceptable and a standby is not needed (but avoid unless fully justified).	Maximum efficiency is 60%—typically less due to compressor efficiency and blow-down losses. Some designs are unsuitable for raw sewage due to the use of conventional swing check valves. No suction lift; fills by gravity alone. Maintenance (for some makes) includes unclogging check valves (which are susceptible to plugging) and cleaning electrodes for level controls. Noisy and creates high ambient temperatures in an enclosed space. High capital cost, especially in large sizes where a standby receiver is included. Settling in the discharge pipeline may occur unless the unit is sized properly. Electrode-controlled units corrode. Two compressors and two pressure vessels are needed to provide a semblance of continuous flow. Flow cannot be increased in the future except by adding ejectors. Adequate basement required for installation.
Grinder pumps (centrifugal or positive-displacement types)	
Many pumps can be connected to a single, small force main. Positive-displacement types can produce flow at heads to 36 (120 ft) or more with capacities up to 1.8 m^3/h (8 gal/min) and motors up to 7.5 kW (10 hp). Conditions favoring use Residences below the sewer main with a long force main. High head, low flow.	Costly. Unsuitable for municipal application. Usually installed one per residence, which mandates prompt repair service at any hour in any weather. Reliability is improved and high pump cost is slightly reduced by a single sump containing two pumps serving two or more residences. Centrifugal types may not pump when the force main is pressurized.
Cutter pumps	
Sizes available are 75–150 mm (3–6 in.). Capacities are 11–340 m^3/h (50–1000 gal/min. Power range (dry pit) is 2.3–30 kW (3–40 hp). Power range (submersible) is 2.3–15 kW (3–20 hp). Comminutors and bar screens may be eliminated. Very effective on all (e.g., overalls and disposable diapers) but the most unusual solids (e.g., panty hose, which are too thin to cut). Long life (many years) in severe service. Effective with relatively low horsepower motors.	Only three-phase motors are available at 1150 or 1750 rev/min. Cost is double that of a comparable centrifugal pump. Efficiency is low (30 to 60%) because of the cutting required. Maintenance of submersible styles requires disconnection and removal, or work must be done in a hazardous area.

Prior to selecting the pumps, a basic decision must be made on the use of the following:

- Vertical wet well pumps
- Horizontal dry well pumps in a dry well–wet well design
- Vertical dry well pumps in a dry well–wet well design
- Horizontal pumps that are floor-mounted above the suction well, with a priming system.

In some installations, the most economical design is obvious, but in others, alternative designs must be prepared on the basis of first cost and operation and maintenance costs.

- If a deep pump setting is required to reach the water surface and if construction is difficult due to poor soils or groundwater problems, a wet well station design with vertical pumps is always the most cost effective.

- If the setting is shallow and if the design of the intake works eliminates the need for a separate suction well, a horizontal pump installation with the pumps (below low water) in a dry well is usually the most economical.
- If the distance between the pump room floor and the low water surface is such that a horizontal pump can handle the required suction lift, an economical design is horizontal pumps installed at the ground-floor level. Complete reliability depends on the proper functioning of the pump priming system—particularly in an unattended station.

A small pumping station may well use a single operating pump and, to reduce the spare parts inventory, an identical standby unit. The electrical design should preclude the two pumps operating together and should also ensure that each operates alternately. In multiple pump stations, pumps should be started one at a time. The hydraulic transient or surge analysis should be based on the simultaneous shutdown of all operating units as a result of power failure.

Most solutions can be met with the variety of pumping stations compared in Table 25-6 and pumps compared in Table 25-7.

Well Water Pumping

Pumps used for well water are almost always vertical mixed flow or radial flow (compared in Table 25-8). If flooding can occur and station reliability is essential, all electrical machinery must be above grade, which might require vertical pumps. If space is limited, vertical turbines can be mounted directly above a clear well or reservoir, although protection against contamination is a significant factor for pumps in a clear well.

The types of wells include

- Shallow or caisson wells for depths up to 60 m (200 ft)
- Horizontal collectors and infiltration galleries
- Deep, cased (drilled) wells for depths of 6 to 600 m (20 to 2000 ft) and diameters of 100 to 750 mm (4 to 30 in.).

Table 25-6. Comparison of Water Pumping Station Types

Advantages	Disadvantages
Vertical wet well pumps	
Wide selection is available: axial flow (single stage), mixed flow (single and multistage), Francis turbine (single and multistage), and radial flow (turbine in single or multistage).	High superstructure to pull pumps or, at greater expense, pumps can be pulled through a hatch in the roof of a low superstructure by a truck crane parked outside—a method that may be cheaper than inside crane.
Small floor area and small superstructure.	Requires disconnecting the motor and pulling the pump for inspection or repair.
Ground floor pump motors above flooding.	Requires more shaft bearings.
Pump suction always flooded; no priming.	Priming may be required if air in the pump column causes problems.
NPSH easily met.	If idle pumps collect air in pump column, either (1) an automatic air vent valve at the discharge elbow or (2) priming must be installed.
Ideal for deep installation and large variations in water level.	
Vertical turbines are especially flexible in meeting head requirements by adding stages.	
Wet well–dry well, horizontal pumps below water level (flooded suction)	
Pump types available include split-case centrifugal, end-suction. overhung-shaft, and horizontally mounted axial-flow (propeller and mixed-flow) types.	Large pump room floor area.
Eliminates priming and air problems.	Greater excavation.
Low headroom requirement.	Motors are subject to flooding.
Easy maintenance and accessibility.	Longer electric conduits.
Service can be accomplished without disconnecting motor.	Pumping stations are more costly due to additional floor area, access, lighting, ventilation, etc.
	Greater forced ventilation needed to cool motors.
Horizontal pumps on floor above suction well	
Easy maintenance and accessibility.	Requires priming equipment.
Short electric conduits.	Dependability (due to priming equipment) is reduced.
Motors above flood level.	Limits the choice of pumps to those for negative suction.
Ventilation minimized.	
Reduced excavation and below-ground construction. Similar to wet pit.	

Table 25-7. Comparison of Water Pumps

Advantages	Disadvantages
Axially split-case pumps vertically or horizontally mounted	
Easy to maintain and to inspect. Axially balanced, so there is no axial thrust.	Requires a housing to prevent freezing and protect bearings and packing from dust.
Vertical axial flow, mixed flow, and radial flow (turbine) pumps	
Can be housed in low, flat-roofed buildings with roof hatches for pulling the pump with a crane parked outside. Quiet, especially with submersible motors. Submersible motor style needs no housing except for the control panel and valve vault. Virtually no maintenance except for pulling the entire unit every 2 to 5 yr.	If outdoors, standby units may freeze, noise is a problem in residential areas, and the station is easier to vandalize.

Table 25-8. Comparison of Well Pumps

Advantages	Disadvantages
Self-priming centrifugals	
Capacity to 0.25 m^3/s (4000 gal/min).	Limited to caisson, gallery, or wells with a water table less than 4.6 m (15 ft) below the pump
Vertical turbines (V/T)	
Most common type of well pump. Motor or engine driven. Diameter: 50 to >1200 mm (2 to >48 in.). Capacities exceeding 0.6 m^3/s (10,000 gal/min). Use for finished water and booster pumping is increasing. Can be tailored for specific head by adding bowls.	Efficiency as low as 50% depending on the size and application.
Vertical turbines, lineshaft	
Minimum maintenance. Driver accessible. Excellent for wells less than 90 m (300 ft) deep.	Unsuited to crooked wells. Somewhat noisy. Suitability is marginal at depths over 300 m (1000 ft). Even at depths less than 180 m (600 ft), shaft stretch wears impellers and bowls unless the thrust bearing is kept carefully adjusted.
Vertical turbines, submersible	
Quiet, so they are suitable near hospitals, schools, and residences. Only solution for crooked wells. Practical at depths over 210 m (700 ft). Water cooling is very effective. No long-shaft problems.	Maintenance requires pulling the entire unit. Regular maintenance is required. Seal problems may be severe. Long electric cables.

Booster Pumping

Selecting the type of booster pump is somewhat complicated. The following discussion is oversimplified and should be supplemented by reviewing Chapter 18.

Booster pumping stations fall into two general classes: (1) distribution boosters, which increase pressure to serve a water distribution system at a higher elevation, and (2) in-line boosters on a transmission main.

Boosters should be constructed only when justified by sound economic and operational reasons. The advantages and disadvantages of booster pumping stations make it necessary to conceive alternative system layouts, some of which might exclude a booster. Make a choice on the basis of a present-worth analysis of capital and O&M costs as well as adaptability for meeting the requirements of the situation.

The comparative advantages of the different types of pumps for booster stations are the same as for raw and finished water stations. In general, most booster pumps are best served by either a horizontal split-case double suction pump or a vertical turbine can pump. The personal preference of the engineer and owner is an important factor. During the past three decades, the trend has been toward the increasing use of turbine pumps, especially in the western United States. However, the selection of the pumping units requires considerable analysis because of the many factors that must be weighed against each other. Some of these are as follows:

- A smaller number of larger pumps is less costly.
- More favorable flowrates with less storage are obtained with more small pumps.
- Pumps of different sizes (proportional to flow ratios of 1:2:4) give the greatest number of flow combinations, but they (1) violate the desirability of standardization, (2) cause one pump to do most of the work, and (3) increase the cost of the standby pump.
- The use of V/S pumping improves flow matching, but it (1) adds complexity, (2) increases the difficulty of the engineering, and (3) reduces reliability.
- Before resorting to V/S, explore all other possibilities first, such as (1) balancing storage at either or both ends of system, (2) using complementary C/S pumps of two or more sizes, (3) bypassing a portion of the discharge, or (4) adding a throttling valve.
- Because of maintenance, capital cost, and inefficiencies, V/S is justified only (1) when feeding large systems with inadequate storage and large demand fluctuations or (2) in an intermediate zone where both the upper and lower zones contain sources of supply and variable speed is needed to balance pressures.
- Two philosophies in selecting the size of booster pumps are (1) the use of pumps of the same size (which is often better for small stations), and (2) the use of two groups of pumps: small ones for average demands and large ones for maximum demand (which is often better for large stations with heavy fire demands).

A summary of advantages and disadvantages of booster pumping stations is given in Table 25-9.

Sludge Pumping

Sludge pumping is usually a part of a water or wastewater treatment plant. Hence, sludge pumping stations per se are uncommon and pumps are usually housed in the dry well. Pumps are compared in Table 25-10. Sludge pumping evaluation and design is complex and not straightforward due to variations in the characteristic behavior of sludge as a non-Newtonian fluid in addition to other factors that render traditional design approaches inappropriate. For an extensive discussion, refer to Chapter 19.

25-7. Power, Drivers, and Standby

Early decisions about drivers and standby units affect the type, configuration, and physical size of the pumping station. The preliminary design considerations should include plans for the electrical system, such as load estimating (power and lighting), load data collection, load characteristics, selection of the power source, plans for load growth and change, selection of the best voltage, and selection of the best distribution system for reliability, flexibility, safety, and maintainability (see Chapters 8 and 9).

Motors

A generalized guide for selecting C/S drivers is presented in Table 25-11, and a guide for selecting V/S drivers is given in Table 15-2. About 90% of all drivers are electric motors, and the squirrel-cage induction motor, by far the most common, is available in a wide variety of casings, windings, insulation, allowable

Table 25-9. Comparative Advantages of Booster Pumping Stations

Advantages	Disadvantages
Allows suction side pipeline to be designed at a lower pressure rating and may reduce pipeline construction cost.	Additional construction costs, O&M, and replacement costs.
May avoid designing primary pumping station for abnormally high discharge pressure with resultant cost savings.	Increases operational complexity.
May reduce maximum system pressures over large service areas and reduce energy costs and leakage.	Offsite facilities (access roads and power lines) may be required.
Useful in increasing flow in existing pipeline systems.	Complicates analysis and control of water hammer.
Useful in eliminating substandard pressures in zone extremities.	Requires matching the flows of primary and in-line booster stations. Complicates design because head-capacity curves cannot be independently established for either station.
	Large fire flow requirements and small minimum domestic flow needs require a wide range in pumping capability.

temperature rise, shafts, and bearings. Synchronous motors are occasionally used for very low-speed applications or for drivers larger than 375 kW (500 hp). Wound-rotor motors are sometimes used for V/S applications, but they are expensive and maintenance is greater than with squirrel-cage motors. Direct current machines are rarely used in pumping stations.

NEMA Designations

NEMA *design letters* A, B, C, and D are used for designating the starting torque and starting current of general-purpose motors up to 150 kW (200 hp).

- Design A (never used in pumping stations) is normal starting torque and normal starting current.

Table 25-10. Comparison of Sludge Pumps

Advantages	Disadvantages
Air lifts	
Suitable for return activated sludge, raw sewage, and sandy waters. Also used for pumping shallow wells. Maximum flow is about 550 m^3/h (2500 gal/min). Simple and cheap. No moving parts, very little maintenance. Conditions favoring use Return activated sludge. Decanting digester supernatant. Sometimes for shallow wells. Close control of flow is not critical. Unusual cost for alternative system. Air supply is already available. Digester gas mixing systems.	Limited to low lifts—about 2 m (7 ft). Does not pressurize fluid, so they are limited to free discharge. Maximum efficiency is 35% or less. Difficult to regulate discharge, and pumping is unreliable if air is throttled for low flowrates. Requires a large submergence. Sensitive to small changes in head and viscosity. Suitable only for thin sludges without large solids. Use for flow measurement is unwieldy and inaccurate. Requires air compressor or blower.
Plunger pumps (positive displacement)	
The most reliable of the positive-displacement types. Discharges to 120 m^3/h (540 gal/min) at pressures up to 13,800 kPa (2000 lb/in.^{2}) or more. Can pump very thick sludge, large solids, stringy materials, and grit. Slow speed and long life. Suitable for suction lift. Major repairs can be made in a local machine shop. Capacity is constant despite large changes in head. Conditions favoring use Sludge pumping for digesters. High and/or variable head. High suction lift. Viscous fluids.	Maximum efficiency is typically 40 to 50% depending on the type. Needs extra attention for lubrication and routine maintenance. Requires sizable floor space. Solids size is limited by check valve clearance. Check valves must be dismantled to locate trapped solids; dual checks are helpful. Check valve spheres are short lived. Low capacity; primarily useful only for sludge. Air chamber is usually required to prevent water hammer. Requires bypass pressure relief or shear pin protection. Head depends on peak of sinusoidal flow characteristics. Air chambers are needed to smooth out flow pulsations.
Rotary lobe pumps (positive displacement)	
Efficiency near maximum (~75%) at all speeds and pressures. Quick, easy replacement of moving parts (in the cantilever type). No check valves are required. Pressures to 690 kPa (100 lb/in.^{2}). Capacity to about 450 m^3/h (2000 gal/min). High tolerance for rags and stringy materials and can pass solids to 120 mm (4 in.) in diameter; easy access for cleanout. Long life at low speed. Meters the flow accurately despite changes in TDH. Good for viscous sludge; creates the least shear rate. Can run dry or in reverse without damage. Smooth, relatively nonpulsating flow. Conditions favoring use Where trouble-free operation and low maintenance are more important than capital cost. Where floor space or building cost is at a premium.	High first cost (about 130% of the cost of a progressive cavity pump). Efficiency is low for thin liquids. Comminution, cutting, or grinding solids and grit removal (for moderate to high grit content) is an essential pretreatment. In usual service, replace lobes yearly, seals every 2 yr at an approximate annual cost, depending on the model, of \$1200 to \$3000 (in 1997). Suction lift due to slip is limited to 3 m (9 ft). Side clearances are abraded by grit, so avoid where grit loads are heavy. Bypass relief pipe is required to prevent damage if discharge is plugged or valve is closed. Limited to a maximum solids content of about 5%.

(Continues on next page)

Table 25-10. Continued

Advantages	Disadvantages
Diaphragm pumps (positive displacement)	
Simple and easy field repairability. High pressure. Suitable for suction lift. Moderate shear rate. Low cost. Can meter flow. Handles viscous liquids, sandy and muddy waters. Conditions favoring use Temporary use, especially with a gasoline engine outdoors. Low flows. Dewatering excavations. Slurry pumping. High grit loads.	Instant shut-down on component failure. Check valves are required. Solids handling ability is limited by ball check. Poor for rags, sticks, and string.
Progressive cavity pumps (positive displacement)	
Typical maximum efficiency is ~75% . Good solids handling for abrasive and viscous liquids. Generates pressures to 7000 kPa (1000 $lb/in.^2$). Maximum capacity is 110 m^3/h (480 gal/min). Rotor acts as a check valve. Smooth, relatively nonpulsating flow. Low capital cost. Conditions favoring use High head, variable head. High suction lift. Grit-free sludge with a high solids content. Handles gas inclusion.	High maintenance cost. Typically, stators replaced yearly, rotors rebuilt every 2 yr at about $400/yr for 20 gal/min pump to $3300/yr for 450 gal/min pump in 1997. Large floor space and clearance are required—especially to pull rotor. Unsuited to heavy grit loads. Cannot pump solids larger than 45 mm (1.8 in.). May require in-line grinding as pretreatment. Cumbersome piping. Instant failure when run dry or if the discharge plugs. Safety devices for (1) high pressure and (2) no flow are recommended. Elastomer (material in stator) limits the type of liquids pumped. Slow speed drive required.
Non-clog centrifugal	
Wide selection of flow and head applications. Vertical and horizontal configurations are available. If applied accurately, pump efficiency is high and pumping is economical. Conditions favoring use Thin sludges without debris (waste and return-activated sludge). High volume of sludge to be pumped. Need for economic pumping outweighs maintenance costs.	NPSH considerations may be critical. May be difficult to size large enough to pass solids without clogging, yet small enough to avoid dilution by drawing in overlying liquid (rat holing). Requires capacity adjustment (variable-speed drives). Needs separate flowmetering for process control and speed control. Pump clearances must be adequate to pass the material typically contained in sludge.
Vortex pumps (recessed impeller, centrifugal)	
Readily passes solids (debris, grit, rags, stringy materials). Good longevity. Materials for gritty slurries available. Excellent reliability. Conditions favoring use Primary sludges Undegritted sludges. Need for compactness.	Efficiency only 35 to 50%. Very flat H-Q curve. Usually requires variable-speed drive. Must be accurately sized to avoid excessive recirculation. Cannot trim impellers of U.S. makes. Increase of head greatly reduces capacity

- Design B is normal starting torque and low starting current (the most common type in pumping stations).
- Design C is high starting torque and low starting current.
- Design D is high starting torque, low starting current, and high slip.

NEMA *code letters* are different from NEMA design letters. The code letter identifies the starting conditions in "kilovolt-amperes per horsepower" when the motor is started on full voltage. These data are primarily used to determine if the capability of the electric supply or on-site generator is sufficient for starting and picking up the load.

Starting in-rush current requirements are needed for sizing the emergency generators, conductors, and controllers and are of interest to the power company. The motor controls should permit starting only one unit at a time, and if drivers of different sizes are used, the emergency generator size can be minimized by starting the largest motor first.

Hazardous Locations

A Class 1, Division 1 wastewater wet well requires full containment of electrical conductors and equipment. A non-explosionproof submersible motor can be made compliant by using a redundant low-level float switch to disconnect the motor and sound an alarm when the liquid level falls to the top of the motor. Class I, Division 2 allows the use of non-explosionproof wiring and equipment provided it does not produce arcs (see NEC 501-3).

Close-Coupled Motors

Close-coupled motors have a long shaft that carries the pump impeller. Radial loads for multivaned impellers pumping clear water are relatively small, but radial loads for impellers pumping raw sewage are large, they fluctuate (twice per revolution for a two-vane impeller), and with large solids present, the fluctuations are irregular. Shafts or bearings and seals in a close-coupled unit pumping sewage may be short lived, particularly at high discharge heads, unless the unit is specially designed for severe service.

Motor Specifications

Motors located indoors and above grade can have open dripproof enclosures. These are the least expensive, and they allow motors to run cooler than other enclosures. Some designers suggest splashproof enclosures for some environments so that the entire area can be washed with a hose. To provide protection from flooding, specify installations below grade to be either (1) open, dripproof enclosures with encapsulated motor windings or (2) totally enclosed, fan-cooled motors. Outdoor installations should be specified as Weather-Protected II enclosures. For Class 1, Division 1 hazardous locations, specify that "motors shall be rated Class 1, Division 1," which is a totally enclosed, explosionproof motor. For Class 1, Division 2 locations, the open, dripproof enclosure is acceptable for squirrel-cage motors if there are no switches or other arc-producing components in the motor.

Other specifications for motors should include:

- Class B temperature rise
- Class F insulation
- 480-V, three-phase power for motors from 2 to about 60 kW (2 to about 75 hp)
- 2300-V three-phase power for motors of 75 kW (100 hp) or more
- A service factor of 1.15
- High energy efficiency and a high power factor (because these terms have no accurate definitions, state the actual values in the specifications)
- Blower, fan, and centrifugal pump motor torque characteristics that fit NEMA Standards Design B
- Antifriction ball bearings with a life of 100,000 for motors up to 300 kW (400 hp); ball or roller bearings can be used
- NEMA code letters F or G for motors of 150 kW (200 hp) or less (see Chapter 13).

Variable Speeds

Use V/S drives only if necessary and do not use complex control systems (such as AFD) for remotely located pumping stations except with the client's full understanding of the problems stated in the subsection entitled "Adjustable-Frequency Drives" in Section 15-11. Specify unit responsibility (see Chapter 16) for the drive and pump, but, if AFD is to be used, either (1) specify also that guaranteed service and critical parts will be available within a reasonable distance (say, 160 km or 100 mi) or (2) make sure that the client will have trained electronic technicians on the staff or that there is a reliable, qualified service organization nearby. If V/S is necessary and AFD cannot be justified and if slightly higher power losses can be tolerated, consider the simpler slip drives such as the eddy-current coupling (see Sections 15-1 and 15-11). One way to "improve" the efficiency of slip drives is to design so that a pump rarely operates at less than about 85% of full speed. Also consider the almost 100%-reliable combined squirrel-cage motor/diesel engine drive where the engine with a larger rating is used both for peak flow and for standby power. If V/S is used for one pump, there must be a V/S standby. A combination of a V/S pump and a C/S pump is satisfactory *only if the discharge capacity of the V/S pump exceeds the capacity of the C/S pump by at least 50%.* If the two pumps have equal capacities, the V/S pump will "starve" on too little flow and wear out quickly, for the reasons given in Sections 10-6 and 15-5.

Flat H-Q curves for both the pump and the system must be avoided because, with an acute angle of intersection, small changes in head (due to air in the force

Table 25-11. Guide for Selecting Drivers

	Application criteria		
Type	Power limits	Reliability	Capital cost
Electric motor (direct drive)	No practical limit	Good, depending on utility	Low
Electric motor (belt drive)	600 hp	Good to poor depending upon frequency of belt and pulley replacement	Low
Electric motor (gear drive)	No practical limit	Good (depending upon utility), but has additional maintenance cost	Low
Engine (direct drive)	No practical limit	Excellent, especially with on-site fuel storage or production	Moderate to high
Engine (gear drive)	No practical limit	Same as direct-drive engines	High
Engine/motor (with clutch or declutching)	Generally less than 200 hp	Superior, especially with on-site fuel storage or production	High
Engine-generator standby duty	No practical limit	Superior	Very high

main, changes in the wet well level, or a change in speed due to fluctuating voltage) causes highly unstable operation. The best operational approach is to start a second V/S pump when the lead V/S pump capacity is slightly exceeded and to run the two pumps at the same speed in a "load-sharing" operation, which is more efficient than operating the two pumps at different speeds in "staggered operation" (see Figure 15-8).

Miscellaneous but Important Considerations

The following miscellaneous considerations are also important:

- For flood protection in dry wells, consider (1) submersible motors, (2) vacuum and pressure epoxy encapsulation of the motor windings with a flood level float to disconnect the motor automatically before inundation, or (3) vertical pumps with extended shafts.
- For high-service duty, select the slowest speed that can perform satisfactorily. Slow speed motors and pumps cost more, but wear is a function of speed to approximately the third power.
- Consider slip or high slip (8–13%) motors with positive-displacement pumps for high static heads or for long force mains.
- Keep the motor control panels in a clean, dry environment above grade and above the flood level, but locate them as close to the motor as possible to keep the leads short.
- Calculate the number of motor starts per year and obtain motors that will last for many years by optimizing life-cycle costs. Specify the minimum cycle times and the minimum life required for motors and obtain the manufacturer's guarantee in writing (never verbally). The best economy is often found in a special, more expensive motor.
- Limit the permissible deflection of the impeller shaft at the wearing rings under the worst specified continuous operating condition (see Sections 10-6, 11-3, and Appendix C).
- Specify the bearing grade. Some designers require certified calculations of pump bearing life at the design operating condition.

Engines

Internal combustion engines are used sparingly as prime movers. Nevertheless, they are more reliable than electric motors, often more economical (especially if operating on biogas), offer improved protection against surge dam-

Table 25-11. Continued

Application criteria		
Operating cost	Speed limits	Complexity
Moderate to high	Essentially limited to 1800, 1200, 900, or 720 rev/min for 60 Hz current; variable speed possible	Varies depending upon size and system configuration; synchronous drives can be complex
Moderate to high below 1720 rev/min	No practical limit to mechanically adjustable speed	Moderate, but if speed changes constantly, belts wear quickly. If speed is seldom adjusted, grooves worn into cone pulleys make adjustment impossible, so specify chromium plated pulleys
Moderate to high	No practical limit either up or down Variable speed (through motor speed adjustment)	Moderate
Moderate to low, depending upon fuel cost. Can be very low if engine waste heat can be recovered for heating or cooling	Limited to available engine speeds; variable speed easily achieved	Moderate to very complex
Same as direct drive engine	No practical limit either up or down Variable speed easily achieved	Moderate to very complex
Moderate to low, depending upon fuel cost	Practical limit is 1200 rev/min; variable speed is easily achieved with engine	Moderate to very complex
Moderately low in standby duty. Must exercise frequently	Same as electric motor	Very complex

age (because of the increased inertia of moving parts), and are easily controlled in V/S operation. But they are noisy, complex, and require skilled mechanics for maintenance and repairs. If engines are seriously considered, consult the application engineering departments of equipment manufacturers—certainly before proceeding with the design. *Never* install engines below grade.

Fuel

Consider diesel, natural gas, biogas, and LPG (propane) fuels. For engines over 560 kW (750 hp), consider dual fuel. Gasoline is unsatisfactory because its storage properties are very poor, and it has a low flash point, which increases the danger of explosion and fire. Diesel is much less hazardous, and it can be stored for approximately a year, but double storage tanks are required to guard against leakage. An advantage of LPG is its unlimited storage life, but it is heavier than air, very explosive, and the building envelope must positively prevent leakage from accumulating in low areas.

Duty

The term "duty" means the time-based utilization of the driver and the specifics of the application such as direct drive or electrical generation and emergency or standby duty.

- Direct drive is the most efficient. Gear reducers permit any pump speed at the best engine revolutions per minute. If gears are required, they should have a nonreverse mechanism to prevent reverse rotation of the engine and the interruption of lubrication.
- Continuous duty engines should be designed with conservative rating factors to meet the objectives of reliability, low operating costs, and long service life.
- Standby duty engines should be selected for rapid starting and an ability to develop full load quickly. Specify starting aids, such as two-stage trickle chargers for batteries, jacket water and lube oil heaters, and high torque versus speed.
- Electrical generation requires a careful analysis of the loads (see Example 9-1) and thoughtful application to the generator (see Example 9-10). The manufacturer should confirm the analysis and make the final equipment selection.

Standby

Standby may be supplied in four ways:

- Dual electrical power systems. Dual power does not guarantee that a general failure of interlocked systems (which has occurred over very wide areas) or that a failure in the main switchyard (as by a lightning strike) will not occur.

- Standby engines with either clutches and direct drives or an engine–generator.
- Use of only engines as prime movers with one engine and pump standby. Diesel fuel provides the utmost in reliability. Natural gas is much more reliable than electrical power. For example, the only interruption in natural gas to an area in Southern California occurred in the 1932 Long Beach earthquake.
- Storage tanks or reservoirs for the time anticipated to restore power. The time cannot be known, so consider a suitable safety margin.

If standby electrical generation is required for a pumping station with electric motors and V/S drives, using engines and direct drives (with microprocessors to trace the system head curve) is more economical in both capital and operating costs (assuming a 56-kW (75-hp) station and energy at 6¢/kW · h for electricity and 36¢/therm for gas). Maintenance for engines is much greater than for motors, but the savings in capital and operations is more than adequate for the maintenance and ultimate replacement of engines (see Section 8-4 for engine–generator requirements, Example 9-10 for engine–generator sizing, and Chapter 14 for engine requirements).

Number of Pumping Units versus Capacity of the Station

Optimize the size of the wet well and the number of pumping units so that the permissible number of starts per hour is never exceeded. Consider these factors:

- Present average and peak flows
- Ultimate design average and peak flows
- Time period until the ultimate design flows are obtained
- Desirability of adding pumping units in the future
- Size and number of pumping units required
- Size of standby power unit(s) needed to operate the station.

In general, the minimum number of pumping units is as follows:

- One pump and one standby for small stations, e.g., <160 m^3/h (<700 gal/min)
- Two or three pumps plus a standby for medium stations, e.g., 160–450 m^3/h (700–2000 gal/min)
- Three to five (or more) pumps plus standby for large stations. Consider V/S or a combination of V/S and C/S pumps (see Chapter 15).

25-8. Station Auxiliaries

Just as much care is required in the engineering and equipment selection for station auxiliary systems as for the main pumping equipment. In retrospect, station reliability, as determined by the ability of the main pumping equipment to perform their intended function, is greatly dependent on these systems. A host of benefits flow from properly engineered station auxiliary systems:

- A wholesome environment, free from condensation and dangerous gasses, and affording protection for personnel, equipment, and structure alike (ventilation system, Section 25-3).
- Reliable protection against flooding (sump pumping system).
- Convenient removal and replacement of heavy equipment components (hoisting equipment).
- Reliable supply of air (if needed) for instruments (compressed air systems).
- Reliable supply of housekeeping and shaft seal water for cleanliness and protection of equipment (utility water supply).
- Warning of probable system trouble before it becomes a crisis (system controls and alarms).

Some insights into the considerations that enter into design and equipment selection for these systems to provide the best investment are considered in the following.

Ventilation

Guidance for the design of ventilation systems for wastewater pumping stations is provided by Chapter 24 and by NFPA 820. Hence, this chapter contains no further discussion of the subject.

Sump Pumping System

Often given short shrift in the design of many installations, sump pumping stations provide a vital service: *they keep the floor dry.* To perform this function satisfactorily, they must be able to accomplish the following:

- The sump pumping station force main must discharge with an air break above the highest possible water surface at the point of disposal for station drainage. The best place is at some location where isolation from the station is possible—for example, upstream from the influent manhole, where a sluice gate has been provided, or into the main pumping station discharge manhole.

- Sump pumping stations should contain two pumps, and each should be capable of accommodating all normal operating conditions.
- The pumps must be capable of passing solids of the size commonly encountered. For an example, in a wastewater pumping station, the sump should be capable of passing a 50-mm (2-in.) solid.
- The operating conditions should be close to the bep to avoid damaging vibration and radial thrust.
- Large rates of flow should be able to enter the sump through a grated opening. Avoid sumps sealed with a solid cover plate.

If all of the above rules are followed, the sump pumps will have a capacity of 34 m^2/h (150 gal/min) or more, because the requirements to pass large solids and to discharge to a higher elevation dominate all other considerations. However, as the pumps will not be required to operate very often (leakage and other contributions under most conditions are quite small), the sump itself need only be of sufficient size to accommodate the installation and removal of the pumping equipment. If a grated sump cover is used, ample capacity is available if a pipe coupling comes loose or breaks and large quantities of liquid enters the room.

Hoisting Equipment

Electrified hoisting equipment is recommended for all but the very smallest stations. Large items of equipment or subassemblies from large units are awkward to move and place. As the persons charged with this work are typically unskilled in using hoisting equipment, specify crane trolleys, bridges, and hoists with no more than the following speeds:

Motion	Recommended maximum speed, ft/min
Hoist	10
Trolley	20
Bridge	30

Note that the above speeds are considerably slower than those provided with commercially available cranes intended for operation by skilled personnel. The added costs for slower-speed motors and larger gears is justified for safety. For a rated capacity of more than 3600 kg (4 tons), specify an inching hoist feature to lift or lower the maximum load at no more than 1 ft/min for the precision alignment needed for couplings.

Cranes in wastewater pumping stations are subjected to varying concentrations of hydrogen sulfide. Such cranes should be constructed of corrosion-resistant material, and all the equipment should be explosionproof if located in the wet well area within an enclosure. Power to all electrical devices must be supplied by festooned cables.

Cranes should have full, vertical access to heavy equipment in lower rooms through removable cover plates or supports for equipment located overhead (such as motors). Do not expect that items of equipment can be jockeyed under an opening from an equipment position using tie-back tackle or "coffin hoists"—a dangerous practice. Design the doors to be wide enough and high enough for trucks to be driven under the crane for removal of the bulkiest equipment. *Do not* assume equipment can be placed on wheeled carts moved by hand outside the structure for mobile cranes or jib cranes to hoist onto trucks. Cranes require rigid frames or diagonal bracing to withstand lateral and longitudinal forces.

Specify all cranes and hoists to conform to the standards of the Crane Manufacturers Association of America (CMAA), and specify that the installation must be tested and licensed by the local state entity providing this type of service. In addition, the specifications should include crane operation and safety training to be conducted by the crane vendor.

For small stations, lifting eyes (or hooks) over the equipment are feasible (although awkward and somewhat dangerous) for all items capable of being lifted by manually operated portable hoists. Design and specify lifting eyes to be strong and substantially anchored into the overhead structure, as shown in Figure 25-2. Be sure to design adequate headroom for the hoist and the machinery so that dismantling is unnecessary. The design maximum capacity should not exceed 900 kg (2 tons) and should preferably be no

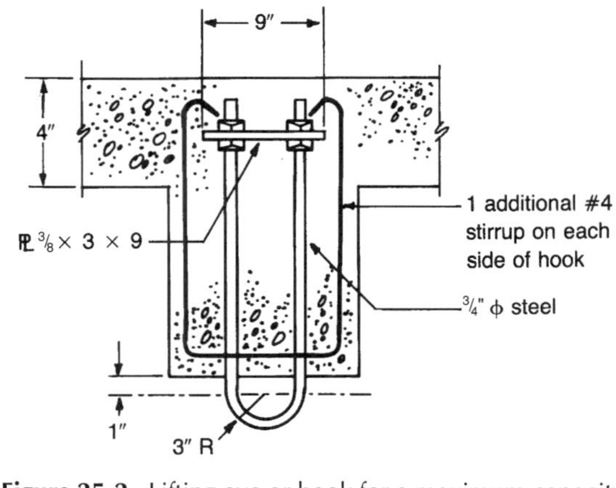

Figure 25-2. Lifting eye or hook for a maximum capacity of 2 tons and a desirable capacity of half a ton.

more than 25% of that capacity. Lifting eyes are completely inadequate for medium and large stations.

Monorails, which may be adequate in medium-sized stations, are often supported by bolts in concrete roof slabs. Design the bolts and the slab to resist impact as well as static loads. Overloading the anchorages is frequently a problem that violates OSHA requirements (see ANSI B30.2.0, B30.11, B30.16, and B30.17). Avoid the use of threaded inserts (into which bolts are screwed) whether cast in place or drilled. Instead, use bolts with adequate mechanical anchorage. An alternative is a low superstructure with a large hatch in the roof for access by a mobile crane parked outside (see CHMA standards).

The lifting effort of cranes or truck-mounted booms actually used in lifting submersible pumps should preferably not exceed about 150% of the weight of the pump and must not exceed the allowable strength of the discharge elbow and its eccentrically loaded connection to the concrete base. In at least one known accident, the discharge elbow and pipe were pulled up with the pump, thus leaving the owner with a very expensive repair job. Anchors for discharge elbows should be buried deep enough to develop the ultimate tensile strength of the anchors, and the anchors should be surrounded with a strong cage of reinforcing bars to prevent the gradual cracking and deterioration of concrete caused by vibration of the pump.

Compressed Air Systems

Air compressors, particularly the positive displacement type, can be high-maintenance items. Be conservative about estimating air demand, then double it for leakage at joints, etc. Select compressors on the basis of at least 10 times that figure. Be generous when calculating distribution system losses, and avoid selecting a compressor at the extreme limit of its range. For example, do not select an 80 ft^3/min, 100 lb/in.^{2} compressor for an application where the calculated figure is 75 ft^3/min at 90 lb/in.^{2}

Furnish at least two compressors and plenty of receiver capacity. All distribution piping should contain sediment and water traps at low points. If it is a large system, consider using terminal receivers in addition to the control receivers at the compressor location.

Instrument air systems require nonlubricated compressors and some means of removing water vapor from the compressed gas stream. Refrigerated dryers have worked well for this purpose as long as the piping system will not be exposed to freezing temperatures. If freezing temperatures may be encountered, use heat reactivated desiccant type downstream from the refrigerated dryer to provide a pipeline dewpoint of –73°C (–100°F) (see also Section 20-3).

Water Systems

Usually two water systems are required in a pumping station: one to provide a pressurized supply for shaft seal purge and lubrication, and the other for housekeeping purposes. Usually the seal water supply requirements (small demand, high pressure) can be satisfied by using a pair of regenerative turbine pumps (two are required for redundancy) operated on the pressure in a hydropneumatic tank. To avoid air binding, use a diaphragm-type tank for control and storage. The multistage centrifugal pump is an alternative to the regenerative turbine (see Section 11-10).

Housekeeping water supply requirements of 2 to 4 L/s (30 to 60 gal/min) at 600 kPa (90 lb/in.^{2}) gauge pressure or more) can be satisfied by providing start/stop controls at each utility station. An overriding timer can be provided to shut the pump down after a timed interval (say, 15 min), thereby avoiding the possibility the pump may be operated against a closed valve.

State regulations for protection of potable water systems against backflow require at least the installation of a reduced pressure principle backflow prevention device on the water supply to the station premises. Some states require an air break tank for absolute protection. Locate these in an area with good free drainage to the dry well. When these devices fail, they frequently release considerable quantities of water.

Alarms

Alarm functions should be from devices independent from controls systems, because the problem could very well be caused by a malfunctioning control system. Alarms should be configured to provide warning of abnormal system behavior before the problem becomes a crisis. For example, in a two-unit system designed to provide ample performance with one unit in service, furnish the following features:

- Use a broad-band alarm switch with two pairs of contacts.
- Set the alarm to initiate at a value less than that selected to start the follow unit.
- Set the second pair of contacts to open at a value below that which will shut down the follow unit.
- Wire the switch contacts to maintain the alarm condition until the second set of contacts have opened.
- Provide the system with a manual lead/follow selector switch.

With the above arrangement, if the lead unit should fail, or an unusual demand is placed upon the system, an alarm will be initiated. The second unit will operate, but it will not be possible to cancel the alarm condition until the controlled variable is within the range of the lead unit. The alarm condition serves as a reminder to the operating staff that something is wrong and encourages investigation. An example of this arrangement, portrayed as the controls for a sump pumping station, is shown in Figure 12-4.

25-9. Instruments and Control

A good basic rule is to avoid too much instrumentation because it increases costs and adds complexities that can compromise the entire system.

Typical Minimum Instrumentation and Control

Every pump should have a pressure gauge on the discharge piping. It is the "stethoscope" that is used to check the condition of the pump and force main or transmission pipeline—conditions such as worn impellers, blockage in the force main, deposits in the transmission pipeline, and partly clogged pumps. The gauges can be permanently mounted (which is convenient but costly if there are many pumps) or portable and attachable with quick-connects (which is less expensive, more easily calibrated, and less subject to wear, but less convenient and less apt to be frequently used). A good gauge installation is shown in Figure 20-6.

Other instruments and controls include the following:

- *Incoming power feeders:* voltmeter, ammeter, and watt-hour meter; the watt-hour meter readings may differ somewhat from those of the utility's meter.
- *Motors:* elapsed time meter; on large motors [>400 kW (>300 hp)], individual ammeters and thermal protectors.
- *Engines:* instruments and controls as recommended by the manufacturer for the particular engine, and its auxiliary systems such as fuel, cooling system, and storage tanks.
- *Wet well:* level sensors (bubbler or float switches) with low- and high-water pump controls and low- and high-water alarms.
- *Wet well storage tank:* level sensor (pressure transducer, capacitance, or ultrasonic) with high and low levels and alarms.
- *Dry well:* water level sensor for flood alarm.
- *Boosters:* pressure switches.
- *Water wells:* pressure switch, automatic timer, or on-off switch actuated from a central station.
- *Chlorination equipment:* if necessary.
- *Flowmeter (optional):* choose reliability and simplicity and allow enough room for installation according to maker's specifications. Most flowmeters require a straight entrance of 15 pipe diameters, and even magnetic flowmeters require 2 to 5 pipe diameters.
- *Multiple pumps:* controllers to start one at a time on a "first off–first on" basis; some experts object to automatic duty rotation unless the number of starts per hour would otherwise be excessive. Automatic alternators add unnecessary complexity. Wear can be equalized (if desired) on a periodic basis by operating personnel.
- *Variable speed:* controls and instruments as required by manufacturer.

Automatic Control

Entirely manual control is not economically efficient. Properly designed automation

- Reduces operator workload
- Optimizes operational efficiency
- Avoids potentially dangerous occurrences.

Automatic controls can be anything from a simple float switch that turns motors off and on to a microprocessor or even a computer. Standardized (or "off-the-shelf") microprocessors are efficient, effective, reliable, and relatively inexpensive. Like computers, they are getting better and less expensive. Microprocessors, in particular, have several advantages over simple switches even for simple process controls such as the level in a wet well or storage tank (see Section 20-11 and Chapter 21 for additional discussions of automatic controls).

Recording

Recording instruments vary from simple, circular chart recorders to strip chart recorders that document many operational variables at once, either at the station or at a centralized location. Consult the owner to determine whether recorded data are needed. Unless the data will be used, recorders should be omitted. However, consider using digital data storage (tape cassettes) that can be recycled, or consider digital recorders that automatically produce a properly annotated chart when abnormal conditions occur. Some information could be invaluable if a malfunction occurs in an unattended station.

Complex Instrumentation

Sophisticated instrumentation can take many forms, such as: (1) local monitoring and control, (2) local control with central monitoring, (3) local analog control with digital monitoring, and (4) distributed digital control. The total initial costs vary from about

$300,000 to about $1,000,000. The total cost per month (including amortization, energy for instruments and control, and salaries) ranges from about $20,000 to $40,000. Thus, sophisticated control instrumentation is very expensive (see Chapter 21).

25-10. Structural Design

Many parameters differentiate the design of a pumping station from other commercial or industrial-type structures. The two most important are:

- The structure must be watertight
- The structure will be exposed to a potentially corrosive environment.

Pumping stations, as with most sanitary structures, are usually required to be continuously operated for their entire service life of 30 to 50 years. Hence, durability and serviceability are overriding considerations. Conventional concrete structures are designed to crack—the very assumption upon which strength design is based. Although it is virtually impossible to design conventionally reinforced concrete structures without cracks, crack depth, width, and number can be reduced by modifying the design procedures of ACI Building Code 318. When cracks are reduced, the permeability of the concrete is also reduced and the durability of the structure is increased. To achieve these ends, ACI 350, (Concrete Sanitary Engineering Structures), produced in 1983, should be used instead of ACI 318 for pumping station concrete design. As compared with ACI 318, ACI 350:

- Lowers the allowable tensile stress of reinforcing by requiring either the use of working stress design or by requiring higher load factors for strength design
- Emphasizes serviceability requirements and lowering of the "Z" factor
- Suggests guidelines on mix design components and proportions
- Increases minimum reinforcement requirements
- Provides guidelines on limitations for construction and expansion joint spacings
- Provides guidance on concrete coating use and selection
- Provides discussion and guidance on design for impact and dynamic loading.

Abbreviations and Definitions

Consult ACI Building Code 318 for a more complete list of abbreviations and definitions. Those given here are primarily for the figures in this chapter.

A_s	Area of tension reinforcement.
A_{vf}	Area of shear friction reinforcement (in.^{2}).
b	Breadth.
C_3A	Tri-calcium aluminate.
f_c	Specified compressive strength of concrete (lb/in.^{2}).
f_y	Specified yield strength of reinforcement (lb/in.^{2}).
h	Height or total thickness (in.).
ℓ_d	Development length; equals (in. = $\ell_{db} \times$ modification factors).
ℓ_{db}	Basic development length (in.).
U	A factor (multiplier) used on the live load for ultimate strength design.
V_n	Nominal shear strength (lb/in.^{2}).
V_u	Factored shear force at a section (lb.).
Z	A distribution factor for flexural reinforcement. See text.
ϕ	Strength reduction factor.
μ	Coefficient of friction.

1 k/in.^{2} = 1000 lb/in.^{2}= 6.9 MPa = 6,900 kPa
1 k/in. = 1000 lb/in. = 0.175 kN/mm
1 in.^{2} = 645 mm^2.

The factor Z is defined by ACI 318 as

$$Z = f_s\,(d_c\,A)^{1/3} \tag{25-1}$$

where f_s is the computed tensile stress in a steel reinforcing bar, d_c is the depth of cover over the bar, and A is a cross-sectional area of concrete equal to 2 $d_c \times$ bar spacing.

Geotechnical Considerations

The purpose of most wastewater pumping stations (raising the hydraulic grade line to minimize pipe depth) means that these structures are usually founded quite deep—frequently below groundwater. The importance of obtaining good geotechnical information prior to designing the structure cannot be overemphasized. A geotechnical engineer should be consulted, and the following information and recommendations should be solicited.

- Safe construction slopes for excavation
- Design high groundwater level including seasonal and tidal effects
- Recommended dewatering methods
- Feasible foundation types
- Characterization of site soils
- Active, at-rest, and passive lateral soil pressures
- Seismic loadings including dynamic soil loading (if UBC indicates site is in Zones 3 or 4)

- Corrosivity of soil or groundwater (water-soluble sulfates and chlorides)
- Soil unit weight and buoyant soil unit weight

The determination of type of excavation should be based upon both cost and feasibility. Open cutting with a 2:1 slope (2 horizontal to 1 vertical) is the cheapest method where it can be used, but it is usually impractical in areas with high groundwater, or where there are adjacent structures or property line restrictions. These and other factors may require sheeted excavation with bracing or tie-backs.

Buoyancy can be a major factor in the determination of both foundation types and structural section. The groundwater level may be subject to seasonal or tidal variation, and the level at the time that the geotechnical investigation was performed may be inappropriate. The design high groundwater level should be used in designing the structure for buoyancy. The factors that can be used for resisting buoyancy are

- The weight of the structure
- The uplift resistant of piers or piles
- The buoyant weight of soil acting on the footing.

The latter two factors require evaluation by the geotechnical consultant. The proportion of soil mobilized by the footing to resist uplift can be determined by taking the buoyant weight of the prism of soil over the toe of the footing and adding the effect of the soil shearing stresses acting on this prism. A simpler and more common practice is to use the buoyant weight of the soil contained in an envelope defined by the toe of the footing and a 30° projection (from the vertical) to the ground surface. When conventional spread footings are used, it is usually more economical to generate uplift resistance by extending the toe of the footing than to thicken the walls or slab. Depending upon the confidence in the design high groundwater level, a factor of safety of 1.25 to 1.5 should be used for buoyancy.

A mat foundation is usually the preferred foundation type unless conditions exist that would preclude its use or make the mat foundation cost-prohibitive. Unstable soils, expected high settlements, expansive soils, or high groundwater with a limited footprint are all conditions that may lead to consideration of alternative schemes such as:

- Piers
- Piles
- Caisson.

A caisson is a heavy structure consisting of the outer walls of the pumping station built above ground and sunk under its own weight, often with the aid of water jets. By excavating soil from the inside of the caisson, a controlled soil failure occurs under the bottom "cutting edge" accompanied by a subsidence of the caisson. After sinking the structure to the required depth, a concrete floor is poured to form the wet well and pump room floor. A caisson can be an economical pumping station design in certain soil conditions or where adjacent structures or property lines are a problem. However, this type of construction is somewhat inflexible for layout, and it must allow for some larger-than-normal tolerances on plumbness.

Design Criteria and Analysis

Loads

The following loads should be considered in the design of pumping stations:

1. *Lateral earth.* Either at-rest or active lateral earth loads should be applied to the below-grade walls of the structure. Values and criteria for determining which to use should be obtained from the geotechnical consultant.
2. *Hydrostatic.* Lateral loads from water contained in the wet well and/or from groundwater.
3. *Seismic.* In addition to seismic forces applied to the above-ground structure (dictated by UBC), hydrodynamic and dynamic soil lateral loading should be evaluated for structures in Seismic Zones 3 or 4. Dynamic soil loading should be obtained from the geotechnical consultant. Hydrodynamic loading should be calculated as described in Chapter 6 and Appendix F of *Nuclear Reactors and Earthquakes* [6].
4. *Buoyancy.* The buoyant forces created from the volume displaced by the portion of the structure below high groundwater (see subsection "Geotechnical Considerations").
5. *Surcharge.* If traffic or construction vehicles are anticipated adjacent to the below-ground structure, the associated lateral loads should be applied to the walls.
6. *Floor live load.* Such loading is usually higher than that specified by building codes. Depending on the intended activities, a live load between 4.8 kN/m^2 (100 lb/ft^2) and 4 times as much should be used. Operation and maintenance activities involving equipment break-down, replacement, or use of portable gantry cranes can easily generate the higher loads.
7. *Equipment.* Most large pumps and motors generate dynamic or vibration loading in addition to their own dead weight. These dynamic loadings must be carefully evaluated to ensure that no unforeseen amplification occurs due to resonant frequencies. In general, structural natural frequencies must be at least 25% above equipment excitation frequencies.

These loads must be combined so that the result is a conservative design in which both the construction sequencing and the potential future conditions are recognized. The following design load combinations are suggested:

- Load combination I: full lateral earth + surcharge + hydrostatic high groundwater + dynamic soil + floor dead load. This loading condition should produce maximum negative vertical moment at the base of the walls and maximum positive vertical moment near the midpoint of the walls.
- Load combination II: same as above + floor live load. This loading should produce maximum negative vertical moment at the top of the wall as well as maximum negative moment in the floor.
- Load combination III: one-half lateral earth + low lateral groundwater + floor dead load + floor live load. This loading should produce the maximum positive floor moment. Using half of the lateral earth load is rational, because the development of full design earth pressures may take a long time or may never occur at all.
- Load combination IV: no lateral earth pressure but full hydrostatic loading from the wet well. Such a loading should produce maximum positive and negative wall moments in the opposite faces to that produced by soil and groundwater loading. This load combination is meant to account for the hydrotest of the wet well prior to backfilling the excavation. Designers should determine whether this test can be performed with or without the top slab of the wet well and note any restriction accordingly on the drawings.

Wall Analysis

Pumping stations are often quite deep and the loading of the below-grade walls is much greater than in conventional buildings. With these high loadings, one way to limit wall thickness is to recognize the potential for two-way action (wall spanning both horizontally and vertically). If the aspect ratio of the below-grade wall is 3 to 1 or less, there is probably an economic advantage in designing it as a two-way wall.

Even if walls are designed as two-way structures, shear instead of flexure can often become the controlling parameter for determining wall thickness, because walls are not traditionally reinforced for shear. With deep walls, especially in high groundwater, this mindset can result in extraordinarily thick walls. If, without reinforcing, shear criteria results in a wall thickness that is over 25% thicker than is required for flexure, shear ties in the walls should be considered.

There are several ways that walls can be analyzed as two-way structures. One way is to use a structural analysis computer program that permits element loading perpendicular to the element face. If this program is not available, use *Nuclear Reactors and Earthquakes* [6], "Rectangular concrete tanks" [7], Moody [8], and Housner, [9, 10].

Engineers should pay special attention to the support conditions at the wall perimeter. As discussed above, the floor slab may not be in place at the time the structure is backfilled or hydrotested, so there may be a free condition on the top of the wall for some loading conditions. Additionally, caution should be used in assuming fixed conditions where rotation at the support could occur and result in underestimation of the mid-height, positive moment.

Dynamic Analyses of Buildings

The presence of rotating machinery requires an analysis of the effects of dynamic loading or vibration on buildings—especially of equipment supported by elevated floors. The principal focus of this analysis is to determine the natural frequencies of the structural support system. The lowest fundamental natural frequency must to compared with the excitation frequency of the machinery to guard against the potential for resonance. As a rule, resonance is unlikely if the fundamental natural frequency falls outside a range of 0.5 to 1.5 times the operating frequency of the equipment.

Most finite element structural analysis programs now have the capability of calculating the natural frequencies and corresponding mode shapes of the structural support system. These programs can be used to find the potential resonant conditions or to aid in detuning the structure to avoid resonance. Avoiding resonance problems by addressing these issues during design is considerably easier than trying to solve them once they have arisen in a constructed facility. An extensive discussion of vibration theory is presented in Chapter 22.

ACI 350 Code

The use of either working stress design or a modified strength design is allowed by ACI 350. An additional "service factor" by which the normal load factors in ACI 318 are multiplied is used in the modified strength design. The result is a larger, more heavily reinforced section than if designed per ACI 318. However, the modified strength design usually results in a somewhat more economical section (thinner with less reinforcement) than one designed using working stress.

ACI 350 also prescribes an additional 12-mm ($^1/_2$-in.) concrete cover over the reinforcing steel, because it is recognized that these structures function in a potentially aggressive environment.

Crack control and serviceability (receiving greater attention in ACI 350) is accomplished by reducing the value of the reinforcing distribution factor, Z, from a maximum of 25 kN/mm (145 k/in.) for exteriors to a maximum of 20 kN/mm (115 kips/in.) for a normal sanitary exposure—even down to 17 kN/mm (95 k/in.) for severe sanitary exposure. The practical effect is that, for a given required area of steel, smaller reinforcing bars at closer spacing are required. In SI units, for example, if the required area of steel in a flexural section is 37.5 cm^2/m, the use of number 35M bars at 265 mm spacing would satisfy the strength requirement, but d_c is 71 mm, so $Z = 22.2$, a number that exceeds the maximum of 20. However, for number 30M bars at 180 mm spacing, d_c is 68 mm , so $Z = 19.0$ and, therefore, the second choice for reinforcement satisfies both provisions. In U.S. customary units, the required area of steel is 1.78 in.^{2}/ft., and although number 10 bars at 8.5-in. spacing satisfies the strength requirement, d_c is 2.72 in., and Z = 118.4—more than 115. Number 9 bars at 6.5-in. spacing with d_c at 2.62 in. yield a Z of 103, which satifies both requirements.

One other modification in ACI 350 aimed at crack control is a significant increase in minimum temperature and shrinkage reinforcing. In ACI 350 the minimum reinforcing is correlated to the spacing of movement joints by allowing less reinforcing where joints are spaced more closely. The primary effect of this provision would occur in sections where flexure is not a controlling parameter. As shown in Figure 25-3, even with movement joints at a 9-m (30-ft) spacing, 50% more reinforcing steel is required than in nonhydraulic structures.

Concrete Mix

The concrete mix design for any hydraulic structure subject to the same environment as in pumping stations should be modified over that specified for building construction. Design 28-day compressive strengths should be 28,000 kPa (4000 lb/in.^{2}) minimum for concrete exposed to freezing and thawing cycles and 24,000 kPa (3500 lb/in.^{2}) minimum for concrete not exposed to these cycles. Again, the goal of these modifications is to produce a more durable and watertight structure.

Cement

In structures exposed to wastewater or wastewater effluent, the cement should have a C_3A content of less than 8%. Where possible, therefore, ASTM C 150, Types II or V Portland cements should be used. In some parts of the country, requiring Type II or V cement can significantly increase the concrete cost. Substituting a pozzolan, ASTM C618 (fly ash), for a portion of the cement so that the total C_3A content does not exceed 5% can offer similar sulfate resistance with Type I cement. The pozzolan should not exceed 20% by weight of the cement plus pozzolan.

Engineers should be aware that the use of Type V cement or the addition of fly ash frequently suppresses

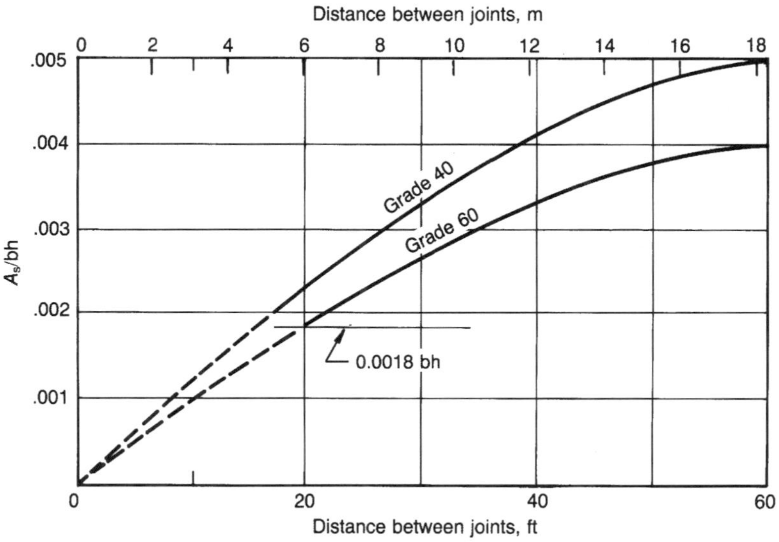

Figure 25-3. Wall shrinkage reinforcement.

the early strength gain of concrete and is therefore not always popular with contractors. The long-term concrete strength is not affected. Additionally, a fly ash mix can sometimes be more difficult to trowel finish than a normal cement-only mix. These measures should be used only where they are warranted for sulfate resistance, and other mix designs should be allowed elsewhere.

Admixtures

An air-entraining admixture conforming to ASTM C260 should always be specified. Additionally, it is frequently desirable to use a water-reducing admixture and/or a superplasticizer to maintain a workable mix with the reduced water–cement ratio required in hydraulic structures. These admixtures should conform to ASTM C494.

Because of the concern for watertightness, a water reducer-retarder is sometimes specified to reduce the potential for a cold joint during a large concrete pour. However, the retarding effect can become a problem for contractors during cold weather—air temperatures of less than 4°C (40°F). A concrete mix using a retarder with set times of a couple of hours in the summer can increase to 8 to 10 hours on cold days. When temperature becomes a problem, allow for the use of an alternative mix with the water reducer but without the retarder.

Mix Proportioning

The two most important parameters for producing watertight, durable concrete are the cement content and the water–cement ratio.

A minimum cement content should always be specified for concrete used in hydraulic structures. Guidelines for minimum cement content are given in ACI 350. These cement contents frequently produce concrete strengths significantly exceeding the specified 24.000 to 28,000 KPa (3500 or 4000 lb/in.^{2}). Bear in mind, however, that the high cement content is not only for strength but also helps produce a denser, more chemically resistant concrete.

Controlling the water–cement ratio is paramount in producing watertight structures. Generally, the less water used, the less shrinkage cracking will occur in the cured concrete structure. The maximum water–cement ratio should not exceed 0.45. Water–cement ratios of 0.43 to 0.38 are not uncommon when water-reducing admixtures or superplasticizers are used. When pozzolans are used, the water–cement ratio should be calculated as the water weight divided by the cement plus pozzolan weight.

It is desirable to use the largest aggregate size practical—up to 38 mm (1.5 in.) for concrete in hydraulic structures. Large size aggregate also helps to reduce shrinkage. As aggregate size decreases, the cement content should increase and the water–cement ratio should decrease. The maximum size of aggregate allowed should be determined by the following:

- One-fifth the width of the wall, or
- Three-fourths the clear spacing between reinforcing bars, or
- One-third the depth of the slab.

Watertightness

In addition to the modified concrete design and mix design, other elements are required to produce a watertight structure. These are discussed in this section.

Waterstops

Waterstops (see Figure 25-4) are often a troublesome source of problems because an amazing number of contractors and workers do not realize that, to be effective at all, waterstops must be fully continuous with all splices carefully welded and not simply lapped. Corner splices and intersections should be shop made. Waterstops between floors and walls should not be placed in a groove but in an upturned key as shown in Figure 25-4a. Waterstops should never be permitted to be bent out of shape by reinforcement. Waterstops can be made of rubber dumbbells, ribbed plastic, or steel plate.

In addition to the traditional rubber, plastic, or steel strip–type waterstops, there are numerous "second generation" waterstops available. Rather than being a strip that is cast into the first and then the adjacent pour, they are usually placed on the face of the joint between pours by adhesive or mechanically fastening. They rely on either adhesion to the concrete or expansive compression to produce a watertight joint. These types of joints are slowly gaining acceptance. Their advantages are that:

- They are often the only practical waterstop for tying into existing structures with a watertight joint.
- Some have a lower installed cost than traditional strip waterstops.

Their disadvantages are that:

- Those that rely solely upon adhesion may leak if any subsequent movement occurs in the joint.
- Those that rely on their expansive properties must be confined adequately by the concrete. At least 3 inches of cover must be provided or the expansion of the waterstop may break the concrete.
- The adhesives used to attach the waterstop to the first pour face tend to perform poorly in cold or wet weather.
- Some have a *higher* installed cost than traditional strip waterstops.

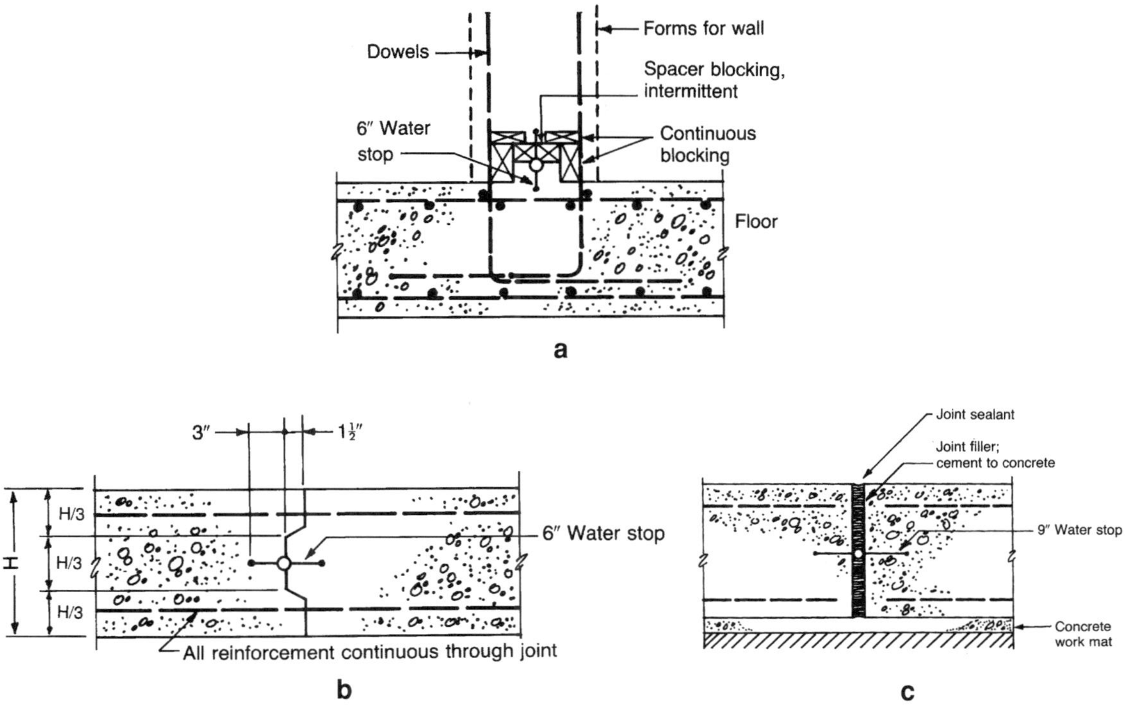

Figure 25-4. Details of construction joints with waterstops. (a) Upturned key; (b) wall joint; (c) expansion joint in a base slab.

Joints

As discussed above, location and spacing of construction and expansion joints affects the amount of cracking that occurs. The locations of these joints should either be shown on the plans or parameters for their placement should be given in the specifications. Typical details are shown in Figures 25-4 and 25-5.

Walls exceeding 15 m (50 ft) in length should have construction joints spaced at no more than 12 m (40 ft). It is desirable to wait at least 5 days between pouring adjacent panels. This waiting period does not delay progress unduly if alternating panels are poured. Walls or slabs longer than about 37 m (120 ft) should have expansion joints spaced no more than 31 m (100 ft) apart.

Slabs should be cast in panels not to exceed 12 m (40 ft) in length or 1000 square feet in area. Furthermore, the panels should be cast in a checkerboard pattern with at least 3 days between adjacent panel pours. An exception to these rules are slabs that can be cured by water-ponding. Good success has been obtained with very large monolithic pours cured by ponding water at least 50 to 75 mm (2 to 3 in.) deep for at least 6 days. Shrinkage appears to be minimized by the unlimited supply of free water during hydration.

If concrete is placed from a height, a tremie should be used to limit the free drop to 1.5 m (5 ft) maximum to reduce the potential (1) for segregation of the concrete and (2) for dislodging waterstops or other inserts. Pumping, however, is preferable to placing concrete by tremie, because the result is likely to be more uniform, with fewer rock pockets.

Exterior Coatings for Groundwater Exclusion

Exterior coatings are used in pumping stations for two reasons: (1) corrosion protection and (2) waterproofing for keeping interior walls dry. Many of the corrosion coatings can serve both purposes. See Section 25-11 for the application of coating and linings inside the wet well.

The three main types of waterproof coatings are:

- Sprayed, brushed, or rolled-on coatings
- Membrane systems
- Cementitious crystalline waterproofing.

With new regulations on volatile organic compounds, the sprayed or rolled-on coating field is changing rapidly. Spray-on coal-tar coating has traditionally been the low-cost damp-proofing used where groundwater was either not present or only intermittently

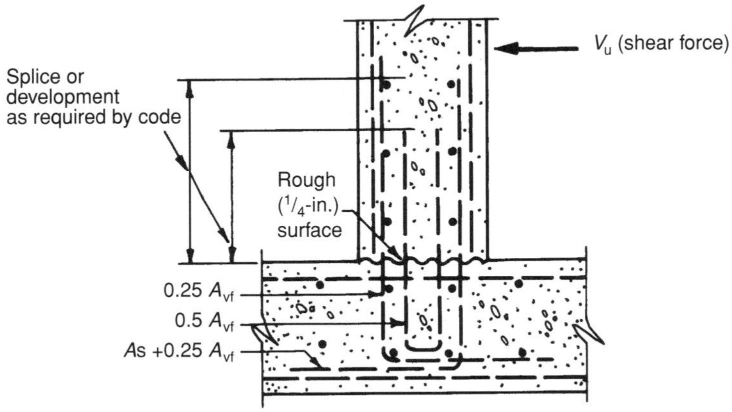

a

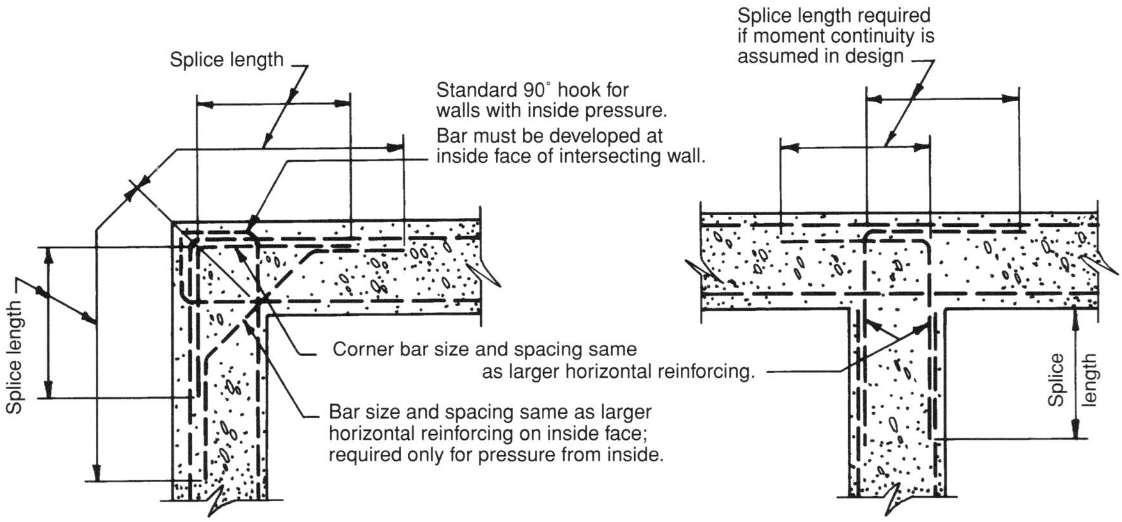

b

Figure 25-5. Concrete joint details. (a) Shear friction joint; (b) horizontal corner bars for walls; $V_u \le \phi V_R \le 0.2 f'_c A_c \le 800\ A_c$; $\phi = 0.85$ for shear and tension; $V_n = A_{vf} f_y \mu$; $\mu = 1.0$ for rough surfaces.

present. The new higher-solids epoxies that have taken their place are of higher quality at higher cost. Unless there is a potential for corrosion, other methods for waterproofing are desirable.

Cementitious crystalline waterproofing is sprayed or brushed onto new concrete prior to complete curing. It reacts with free water and forms a crystalline matrix that seals the pores and capillaries of the concrete. These systems perform quite well in most conditions at relatively low cost.

Membrane systems consist of sheets or panels that are applied to the concrete surface to act as an impermeable barrier for groundwater. Some systems actually employ clay panels that swell with contact to water providing an impermeable barrier. In general, membrane systems are more expensive but more effective than crystalline waterproofing.

Details

This section contains miscellaneous aspects of details common to pumping stations. As with any facet of structural design, there are many variations of these basic details, and designers should use only those that are appropriate for the specific application.

Detailing Reinforcement

One of the more important considerations in designing reinforcing at corners and other intersections is maintaining adequate clearance between bars to permit concrete placement that produces a watertight structure. Honeycombing or rock pockets promote leakage. Vertical construction joints should usually be located at least 1 to 1.5 m (3 to 5 ft) from corners to allow good bar placement without interference between waterstops and corner reinforcing. Splice and development lengths from ACI 318 are applicable.

The reinforcing for walls designed as two-way elements must reflect the support conditions assumed at the corners. Support conditions are especially critical for walls resisting inside pressure, because the interior horizontal reinforcing must be developed at the face of the intersection wall. See "Rectangular Concrete Tanks" [7] and Figure 25-5 for typical corner reinforcing at intersecting walls.

Anchors

Anchoring mechanical equipment to concrete is one of the most troublesome tasks in construction, and it requires careful detailing and superior inspection during construction to avoid problems. Nowhere else are there greater rewards for care or costlier penalties for carelessness. The environment for anchorages is often hostile due to submergence, high humidity, and the corrosiveness of sewage and some waters. Generally, it is more economical to use stainless steel or appropriate coatings on mild steel than to oversize the material to allow for rust and corrosion. The labor of replacement far exceeds the cost of prevention. Anchors are compared in Table 25-12. Refer to ACI's *Guide to the Design of Anchor Bolts and Other Steel Embedments* [11] for design procedures.

Table 25-12. Anchorages

Type	Advantages	Disadvantages
Physical embedment of equipment	Justified only for pipe sleeves in walls	Cannot be realigned nor replaced Use only if unavoidable.
Welding the equipment to embedded plates	Rigidity and strength	Difficult to realign or replace Field welds in cramped places and hostile environments are unreliable
Anchor bolts welded to embedded bars or shapes	Location easily changed Can be fitted to replacement equipment Rusted bolts replaceable	Suitable only for mild steel bolts Field welds unreliable
Embedded bolts	Slight realignment possible when set in sleeves Can use stainless steel or other corrosion-resistant materials	Must be precisely located in plan and elevation. Use templates
Bolts in holes through thin slabs	Good for thin slabs such as roofs Can develop full strength of bolt	Usable only if there is access to both sides of the slab Protruding bolt head may be unacceptable
Bolts with expansion shields	Inexpensive. Suitable for concrete already cast	Suitable only for light loads and no vibration Questionable reliability for tension loads
Bolts set with two-component systems (glue) in drilled holes	Suitable for concrete already cast Easily located and easily changed Reliability of two-component systems is excellent Excellent for tension and shear Inexpensive due to ease of installation Stainless steel (or other material) is suitable	Drilled hole may strike reinforcement Hole must be (a) thoroughly cleaned of all dust and (b) dry; otherwise, failure is likely. Some experienced engineers do not trust these fastenings at all because of the dependence upon responsible workmanship. If used for critical service, bury bolts deeper than required by manufacturer and specify proof tensile tests for, say, 10% of the bolts

Adjacent Structures

There are often ancillary structures on shallow foundations adjacent to the pumping station. They might include small generator rooms, hydropneumatic tanks, or valve vaults. Their foundations should be designed to accommodate the mechanical or piping connections between the two structures. The pumping station is usually deep, whereas the adjacent structures usually rest upon backfill, thus creating a perfect scenario for differential settlement. Consider either (1) designing flexible mechanical or piping connections between the structures or (2) supporting the ancillary structure with the pumping station itself by using, for example, a cantilevered counterfort.

Provision for Future Expansion

When designing pumping stations, always consider the likelihood of future expansion and discuss the potential for increasing the size of pumping equipment with the owner. Provisions can be made in the original design that greatly simplify expansion.

Motor or pump floor systems can be readily designed for greater future loads, whereas retrofitting an existing marginal support system for higher loading can be costly and can disrupt the operation of the pumping station.

If walls or slabs will be extended in a planned expansion, waterstops and dowel inserts to facilitate the new connection can be cast into the original structure and protected with lean concrete. Sliding can be a significant problem if the expansion involves excavation of soil at the wall face while the remaining structure remains backfilled. A shear key (like a retaining wall) cast below the floor prevents sliding. An alternative is to excavate both ends of the below-grade structure to equalize the soil loadings.

25-11. Concrete Protection: Coatings and Linings

Concrete structures exposed to the atmosphere of wastewater, such as the area above low water in sewers and wet wells, are attacked and corroded by sulfuric acid generated by bacteria. The mechanism is described in Section 4-7. The extent of attack is indicated by the pH of the surface. Surface scrapings of new concrete in distilled water yield a pH range of 11 to 13. As concrete weathers, calcium hydroxide is converted to calcium carbonate, which then slowly dissolves to calcium bicarbonate. Carbon dioxide in the atmosphere can lower the pH to less than 7. Sulfate-reducing bacteria in stale sewage reduce the sulfate ion in the water to hydrogen sulfide, and sulfide-oxidizing bacteria on the walls above the water convert the hydrogen sulfide to sulfuric acid, which dissolves components in the portland cement. The surface pH can be as low as 1, but any pH of 4 or less indicates active corrosion. A protective covering of inert and impenetrable material is required for longevity. It is better and far cheaper to apply such protection to new structures than to repair existing ones.

The County Sanitation Districts of Los Angeles [12, 13] have tested a wide variety of protective systems for many years. Systems are screened by (1) allowing a maker or supplier to coat or line a short, vertical concrete pipe with a product; (2) partly filling the pipe with 10% sulfuric acid; and (3) observing the effects after a long period of time. Products that pass this test are then tested under actual field conditions.

Concrete Pipe

See Section 4-7 for a discussion of protection for concrete pipe.

Wet Wells

Coatings are painted on wet well walls, but most coatings, such as coal-tar epoxy, have not been successful. Because wet wells have many corners that often meet at odd angles, it is difficult to apply linings. Lining new concrete with PVC is, however, about the only system that has provided any degree of success.

New Construction

Unbonded, ribbed liners (such as T-Lock®) mechanically locked into the concrete during casting are effective for pipe, but in wet wells, the many joints in straight walls and in corners as well as the holes created by the ties that hold formwork in place require a great deal of welding and patching—the weak points of the system. The high quality of workmanship required cannot be assured today. For wet wells, a bonded liner is better.

The most important aspect of bonding liners to concrete is the surface preparation. New concrete is sand blasted in a pattern sweep to remove laitance, open the surface, and remove all debris and form oil. In the Linabond® system, for example, a two-component primer is applied followed by trowelling a 1.5-mm (60-mil) layer of polyurethane mastic. An activator is applied to a sheet of PVC and allowed to cure for a few minutes.

The sheet is then applied to the mastic, and air bubbles are rolled out. Joints are overlapped 100 mm (4 in.).

Rehabilitation

It is necessary to remove all damaged concrete to hard gray concrete with a surface pH of at least 7 by using a water blast at preferably 55,000 to 69,000 kPa (8000 to 10,000 lb/in.^{2}) pressure or by sand blasting. If there is a delay after water blasting, a sweep sand blast is needed. The surface can then be rebuilt to its original contour with, for example, a polymer grout that is chemically impervious to acid. With Linabond®, the polyurethane mastic is applied in a layer twice as thick as for new construction to allow for a wavy surface.

Other Products

There are a myriad of products for protecting concrete, and engineers should be alert for good, new ones. Be wary and very skeptical of sales claims, however, because systems that are excellent in the laboratory may be disappointing in the field. Furthermore, the traditional low-bid approach for choosing a contractor is not conducive to good field work. However contractors are chosen, competent inspection is important in producing quality work.

It is best to investigate field installations that have been in place several years before specifying a product.

Inspections of field applications by the County Sanitation Districts of Los Angeles indicate there are no fail-safe coatings, even though some (polyester mortar, polyurea, and a sulfur concrete) have withstood the 10% sulfuric acid test well. The only system that has a long-term (more than four decades) history of protection in a corrosive environment is formed-in-place PVC liners, but proper welding of the seams is critical for success. Combinations of both coating (polyurethane mastic) and PVC lining have a shorter history of success, but mastic with polyethylene linings failed in bond. Evaluations are continuing, and it would be premature to draw irrefutable conclusions yet.

25-12. Force Main Design

The force main is not part of a pumping station, but because the station curve depends on the friction head as well as the static head, the discharge piping system cannot be ignored. Furthermore, the design of the force main affects water hammer or hydraulic transients, which, in turn, have an influence on the design of the pumping station.

Hydraulic Transients

Transient control receives far too little attention. Hence, there have been many dramatic disasters from the cracking of valves or pump casings to the rupture of pipes resulting in flooding and, at times, loss of life. Pump start-up with a force main full of water causes surges that may become major for pumps with a high shutoff head. Broken mains, damaged check valves, distorted pump shafts, or a multitude of small leaks caused by loose joints and cracks resulting from repeated water hammer impacts are the result.

Transient analyses are usually required when the following conditions exist:

- The TDH is greater than 12 to 15 m (40 to 50 ft) and the flow exceeds about 110 m^3/h (500 gal/min).
- Pipelines have high points or "knees" near the midpoint. Power failures can cause a partial vacuum at the knee or can even result in column separation (with only water vapor in the pipe at the knee)—a problem to avoid at all costs.
- The static pressure differential for an in-line booster pump is more than 12 m (40 ft).

Transient analyses are not usually required if (1) the total TDH is less than 12 m (40 ft), (2) the flow is less than 23 mm^3/h (100 gal/min), (3) the pipe velocity is 0.6 m/s (2 ft/s) or less, or (4) the pumps discharge into a domestic water network. Such networks have many branches and loops to dissipate the transients. See Chapters 6 and 7 for more explanation.

Simplified transient analyses may be incorrect or misleading. Computer modeling is the most effective method, and the cost for an experienced consultant can range from about $1000 for a simple system to $10,000 or more for a complex problem. Some manufacturers and suppliers offer computer modeling as a free service, but as such modeling may be incomplete or biased, the user had better know how to review the assumptions and the results.

There are many methods for controlling transients, but none can be used indiscriminately. For example, the pipe profile dictates both the kind and placement of hardware. Only an expert should make the analysis and design the transient control system.

Size

Optimize the size of the pipe based on the life-cycle cost, including power and capital costs. A practical maximum velocity is about 2.4 m/s (8 ft/s). Higher flow results in greater headlosses and may result in excessive water hammer. The lowest design velocity that should be used for raw sewage is 0.6 m/s

(2 ft/s) to keep grit moving, and a peak daily velocity of 1.1 m/s (3.5 ft/s) is desirable to resuspend settled solids.

Friction Coefficient

Use the correct friction coefficients for pipes. Excessively rough friction factors, although conservative with respect to the carrying capacity of the pipe, are dangerous for the selection of motors and pumps (see Section 3-2). When determining the carrying capacity, a conservative practice is to use $C = 120$ for lined or plastic pipe (or $C = 100$ for unlined pipe), which conforms to Ten-State Standards; then redraw the system curve for $C = 145$. Make sure that the system can operate at both conditions or at any intermediate condition.

25-13. References

1. Crites, R., and G. Tchobanoglous. *Small and Decentralized Wastewater Management Systems,* McGraw-Hill, Inc. New York (1998).
2. Tchobanoglous, G., and F. L. Burton. *Wastewater Engineering: Treatment, Disposal, Reuse,* 3rd ed., McGraw-Hill, New York (1991).
3. Metcalf and Eddy, Inc. *Wastewater Engineering: Collection and Pumping of Wastewater,* edited by G. Tchobanoglous, McGraw-Hill, New York (1981).
4. WEF. *Design of Municipal Wastewater Treatment Plants, Manual of Practice No. 8, Vol. 1,* Water Environment Federation, Alexandria, VA (1991).
5. ASCE. *Gravity Sanitary Sewer Design and Construction, ASCE Manuals and Reports on Engineering Practice No. 60,* American Society of Civil Engineers, New York (1982).
6. *Nuclear Reactors and Earthquakes.* Chapter 6 and Appendix F. Technical Information Document 7024, Lockheed Aircraft Corp. (1963).
7. "Rectangular Concrete Tanks," *Information Sheet* ISOO3.03D, Portland Cement Association, Skokie, IL (1969 Rev. 1981).
8. Moody, W. T. "Moments and reactions for rectangular plates," *Engineering Monograph No. 27.* U.S. Bureau of Reclamation, Denver, CO (1960 Rev. 1963).
9. Housner, G. W. "The dynamic behavior of water tanks," *Bulletin,* Seismological Society of America, **53** (2): 381–387 (1963).
10. Housner, G. W. *Earthquake Pressures on Fluid Containers,* California Institute of Technology, Pasadena, CA (1954).
11. ACI. *Guide to the Design of Anchor Bolts and Other Steel Embedments,* AB-81, American Concrete Institute, Farmington Hills, MI (1981).
12. Redner, J., R. Hsi, and E. Esfandi. "Evaluating coatings for concrete wastewater facilities: An update,"*Journal of Protective Coatings and Linings,* **11** (12): 50–161 (December 1994).
13. Redner, J., R. Hsi, and E. Esfandi. "Evaluating coatings for concrete exposed to sulfide generation in wastewater treatment facilities," *Journal of Protective Coatings and Linings,* **8** (11): 48–56 (November 1991).

Chapter 26

Pumping Station Design Examples

GARR M. JONES
GARY S. DODSON
THEODORE B. WHITON

CONTRIBUTORS
Roger Cronin
Philip A. Huff
Paul C. Leach
Ralph E. Marquiss
M. Steve Merrill
Earle C. Smith
Patricia A. Trager
William Wheeler

Chapter 12 contains details for selecting pumping equipment and installing it in properly designed wet wells with piping arrangements that allow for easy access in a minimum of space. The principles for the layout of the basic elements of wastewater and water pumping stations respectively are given by example in Chapters 17 and 18. These principles should be followed with an overall objective to keep the installation as simple as possible. The guiding rule should be to provide protection for the equipment and installation and, as much as practicable, protection against the prospect of catastrophic failure. Care should be exercised to follow the requirements of appropriate code documents and industry standards.

This chapter contains examples of both actual and hypothetical designs that reflect improvements in technology that have occurred since publication of the first edition of *Pumping Station Design* in 1989. The examples of pumping station design include:

- Section 26-1. Duplex submersible pumps for domestic wastewater in a hopper-bottom sump. Capacity: 25.9 L/s (410 gal/min). Improvement of original design.
- Section 26-2. Three dry-pit, V/S pumps for domestic wastewater in a trench-type sump. Capacity: 219 L/s (5 Mgal/d). Improvement of original design.
- Section 26-3. Three four-stage, two-speed vertical turbines for raw water in a trench-type sump. Capacity: 920 L/s (21 Mgal/d). Designed in 1993, and began service in 1995.
- Section 26-4. Three submersible pumps with long suction nozzles in a trench-type sump for combined sewage overflow (CSO) service. Capacity: 227 L/s (5.2 Mgal/d). Construction virtually complete in January 1998.

Dimensions and values in the plans and text for each section are, for simplicity, written in either SI units or in U.S. customary units but not usually in both.

Readers should have no difficulty in transforming one set of units into the other.

26-1. Redesigned Clyde Wastewater Pumping Station

The Clyde Wastewater Pumping Station in Contra Costa County, California, was rebuilt in 1991 to feature a self-cleaning sump. It is cleaned by pumping the water level down while vigorously mixing the contents with water from the force main. In the as-built plans, shown in Figure 17-22, eccentric plug valves in the valve vault can be regulated to take water from either the force main or from either of the two pumps. The water is discharged under considerable pressure at the surface of the lowered water level in the sump while a pump discharges the mixed liquid to the force main. The system works well for removing both scum and sludge and leaves the wet well remarkably clean.

In this section, the station is described as it might be designed in 1997 with the technology developed since the original plans were drawn. These changes consist of (1) steeper slopes to allow sludge to slide down to the pump intakes so that sludge is removed with every motor start and (2) a sloping approach pipe for introducing the inflow without a cascade and for supplying added storage to reduce the size of the wet well. In other respects, the design approach closely follows that of the existing Clyde pumping station except that fewer valves are used in the valve vault.

The actual design was carried out in U.S. customary units, so those are the units used in this example. The original sewer design studies, surveys, and discussions with operating and maintenance staff established the following general requirements for this sewage lift station:

- *General:* submersible pumps were preferred because of overall low cost, low maintenance, simplicity in operation, and minimizing visual impact on the neighborhood.
- *Flowrates:* Present average dry-weather flow: 30 gal/min.
  Present peak wet-weather flow: 236 gal/min.
  Future peak wet-weather flow: 410 gal/min (equals the capacity of one pump).
- *Ground elevation:* 13.2 ft. Pumping station site is relatively flat.
- *Force main:* An existing 8-in. cement-lined ductile iron pipe 2750 ft long was available.
  Invert elevation: 6.6 ft at the pumping station and 20.6 ft at the discharge.
  Slope: constant.
- *Reliability:* ability to pump future peak wet-weather flow with either of the two pumps out of service.
  Hook-up for portable engine-generator due to lack of space for permanent engine-generator.
  High wet well power-failure and intrusion alarm hooked up to an auto-dialer.
- *Location:* on shoulder of narrow residential street.
  Considerations include space, visibility, odors, noise, and security.

Station Siting

Station siting is established by the low point in the tributary area as well as access, availability of property, proximity to residents (i.e., farther is better), and the cost of piping to and from the site. The low point in the tributary area usually dictates the general location. Access is important because operation and maintenance staff must be able to visit the facility at any hour of the day and under adverse conditions. Access by public roads (paved, if possible) without the need to traverse private property or move parked automobiles is required. It is also preferable to provide room for maintaining the station without obstructing traffic or endangering workers.

Property and easement acquisition begins immediately after selecting the preferred site and before design on the pump station begins. Many projects have been delayed and/or designs changed because the site acquisition process did not begin soon enough. Such delays and changes will result in significant costs to the owner of the facility.

Hydraulic Design

Hydraulic design includes sizing the force main and developing the system curves, which are then used to select the number and size of the pumps. The rest of the facility is designed around the pumps. The force main invert elevation at the pumping station should, if possible, be set to allow for a constantly rising slope. High spots (knees) in a wastewater force main are to be avoided if at all possible, because knees require air release valves, which clog with grease and require constant maintenance. (Some utilities regularly replace all working parts with shop-cleaned parts every month.) Force main installation costs increase with depth, so it is best to keep the pipe as shallow as possible. The discharge end of the force main is susceptible to hydrogen sulfide corrosion and should be protected by using corrosion resistant piping (PVC, HDPE, VCP) where exposed to air or else be submerged to prevent corrosion. Corrosion resistant piping. as shown in Figure 26-1, should begin 10 ft before the point where the static water level contacts the soffit of the force main.

Force mains should be sized to provide a minimum velocity of 2.5 ft/s at present flows and a maximum velocity of 6 to 8 ft/s at future peak wet weather flows. The minimum velocity ensures that most solids will be moved through the force main. The maximum velocity is set to minimize headless and reduce surge pressures in long force mains. A 6-in. force main would meet this criteria for the stated flow conditions. However, an 8-in. mortar-lined ductile iron pipe was already in place and was, therefore, used. The velocity in the force main for this pumping station with its one duty pump is 2.5 ft/s (see Table B-2 for the cross-sectional area).

System curves are developed to define the design operating point and extreme operating conditions for the pumps. Computations for these conditions at the future peak wet weather flowrate are given in Table 26-1. Note that K-values are not absolutes. Different engineers may elect to use different values. Those in Table 26-1 differ somewhat from those in Table B-6. Minimum losses are given for a pipe roughness corresponding to C = 145, and the maximum is for $C = 120$. The results are shown graphically in Figure 26-2. Point A is the normal flow and head condition, and Point B is the extreme flow and head condition at which the pump may operate.

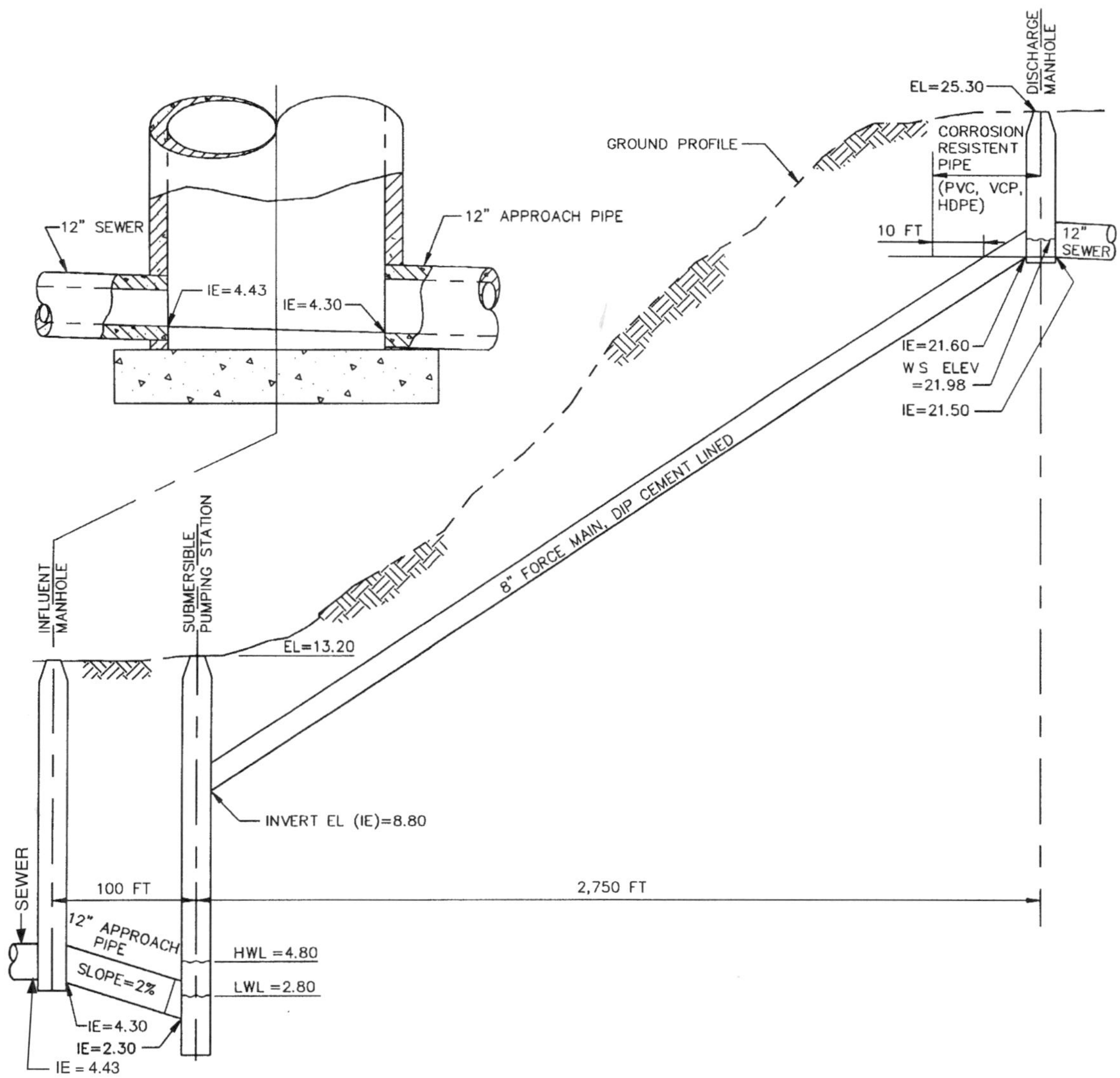

Figure 26-1. Piping profile.

Table 26-1. Total Dynamic Head Losses for Q = 410 gal/min (0.91 ft³/s)

Descript on		Head losses, ft	
		C = 145 Minimum	C = 120 Maximum
Station losses (4-in. DIP, v = 10.3 ft/s)	K-value		
Entrance	0.50	0.82	0.82
90° bends, 2 at 0.25	0.50	0.82	0.82
45° bend	0.20	0.33	0.33
Ball check valve	1.35	2.22	2.22
Eccentric plug valve	0.50	0.82	0.82
Tee, line flow	0.30	—	0.49
90° bend 4 to 8-in. expanding	0.50	0.82	0.82
Tee, branch flow, v = 2.49 ft/s	0.75	0.07	0.07
Force main losses			
2750 ft 8–9 in. DIP, lined (v = 2.49)		6.93	9.83
Minor losses, valves, discharge. ΣK = 2.5		0.24	0.24
Static lift		8.78[a]	19.18
Total design head		21.93 (≈22)	35.75 (≈36)

[a] Wet well is assumed to be filled to ground level

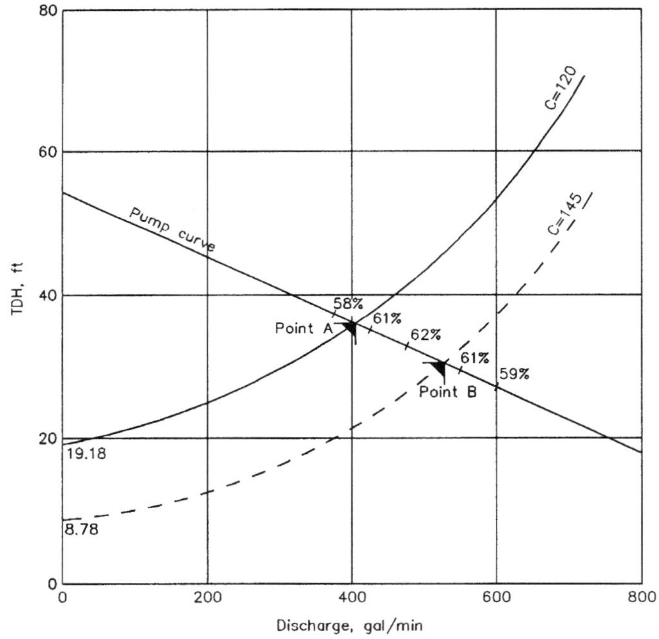

Figure 26-2. Pump and system head-capacity curves.

Not all submersible pump manufacturers include the entrance and discharge elbow losses in their pump curves. The specified design point should be clear on what has been included for losses. The minor losses are not very important for pumping stations with long force mains, but they could be significant for a lift station with low head requirements.

Pump Selection

Operating and maintenance personnel prefer equipment with which they are familiar. They do not like "experimental" applications. Pump selection starts by soliciting input from the people who operate them. Identical pumps are used for multiple pump applications whenever possible to simplify maintenance and provide interchangeability of parts. It is advantageous to have service and parts available from a nearby source.

The selected pump curve and impeller diameter should meet the design point near the best efficiency point. The selected pump should also operate at the low head and high head extremes without cavitating or vibrating. Pump curves with steep slopes are better than those with flat slopes, because there is less variation in capacity with varying head conditions. (Flat spots or dips in the pump curve are therefore undesirable.) Wherever possible choose impellers of intermediate size so that a larger impeller can be substituted for larger future flows. The motor is sized for the worst possible operating point (which is often the low head extreme with one pump operating). For multiple pumps, the best pump efficiency should be at normal operating conditions and not at the ultimate peak flows. However, the emphasis should be on finding pumps that can operate without vibration or cavitation at *all* anticipated service conditions.

Wet Well

The wet well is designed to (1) provide adequate space for the pumps; (2) facilitate cleaning; (3) contain sufficient storage volume; (4) limit pump starts; and (5) minimize installation costs. For a small duplex submersible pump station the most economical wet well is often a reinforced concrete pipe 1.8 or 2.4 m (6 or 8 ft) in diameter standing on a cast concrete bottom.

To improve solids removal, the pumps are confined by close-fitting, nearly conical, smooth walls sloping at 45° to 60° (preferably the latter). Sludge and grit slide down the walls to the pump suction. If the floor is a minimum size (about 0.6 × 1.2 m or 2 × 4 ft), the sludge is so confined that most of it is pumped out on each pump cycle.

The sloping walls may be constructed by placing a custom form in the wet well and injecting concrete behind the form. Fiberglass reinforced plastic or stainless steel can be used as the form and can then remain in place as a liner to provide a smooth, corrosion-resistant surface to facilitate cleaning. Holes are required in the form for the pump discharge elbows, which are cast into the walls. If a disposable form is used, the concrete should be covered with a protective liner such as PVC. The vertical walls above the "cone" should also be protected with a PVC liner.

Active Volume

The active or working volume of wet well and approach pipe must be adequate to limit the frequency of pump starts to a safe value. Conical bottoms reduce the volume available in the wet well, but the addition of an approach pipe laid on a 2% gradient, as described in Section 12-7, supplies additional volume.

The normal LWL is set above the invert of the inlet so that there will be no free fall into the wet well. Consult the manufacturer about the minimum submergence for the motor and the maximum frequency of pump starts. Typically, the motor should be about half submerged at LWL, although under some circumstances the allowable submergence might be less. Starting frequency for small to medium-size submersible pumps may be expected to vary between 6 and 15 starts/h. Using an alternator in stations with multiple pumps to switch lead pumps after each pump cycle reduces the starting frequency. However, one pump may be out of service, so duplex stations must be operable without alternation. Nevertheless, when both pumps are serviceable, alternation could be used to extend the life of the starters and motors.

Calculate the required active volume from Equation 12-3. If the allowable starting frequency for one pump is 10 cycles/h, each cycle takes 6 min and the required volume is

$$V = \frac{TQ}{4} = \frac{(6\text{ min})(410\text{ gal/min})}{4} = 615\text{ gal} = 61.8\text{ ft}^3$$

The wet well and approach pipe are laid out so that the working volume is 615 gal. The existing wet well has limited volume, which requires an approach pipe 100 ft long by 12 in. in diameter. It is laid on a 2% slope with its invert elevation at the wet well at 2.30 ft. The LWL is set at 2.80 ft to force the hydraulic jump to occur in the pipe and not in the wet well. The HWL is set at 4.80 ft. The wet well furnishes about 490 gal of active storage and the approach pipe supplies another 380 gal of storage. The total is 41%% over the requirement.

At the upstream manhole, the invert elevation of the approach pipe is 4.30 ft. On a rising grade of 2%, the invert on the upstream side of a 48-in. manhole would be 0.08 ft higher, but there is some form and friction loss in the transition from half-filled sewer to quarter-filled approach pipe, so an increase of 0.05 ft brings the sewer invert to an elevation of 4.43, as shown in Figure 26-1.

Standby Power

The site cannot accommodate the installation of a permanent engine-generator for standby power. The operator has chosen to store a portable engine-generator at a site about a mile from this facility. A power failure and high water alarm provide indication of potential overflow at the station. During dry weather flow, more than one hour is available to transport and connect the portable engine-generator. However, during wet weather periods, there may be less than 10 min available for that task. Wastewater pumping stations should normally be provided with permanent, on-site standby power to reduce the exposure to wastewater overflows.

Station Piping

Station piping within the wet well is limited to the two pump discharge lines and a force main drain line. Piping within the wet well is as simple as possible with few fittings and no valves. All flange bolts within the wet well are 316 stainless steel, as are the pipe supports and hardware. The piping design within the wet well minimizes items that corrode, that require regular maintenance, or that may catch floating debris. Valves and fittings are contained in a separate valve pit next to the wet well.

The valve pit contains the pump isolation and check valves on the two pump discharge lines and their connection to the force main. A valved force main drain line that discharges back into the wet well is also included. It can be used to agitate the contents of the wet well vigorously so that scum, mixed into the contents, is ejected with the wastewater. Such mixing eliminates the need to pump the water level down to the pump volute to develop vortices for engulfing the scum—an operation that subjects the pump to vibration, stress and wear of the mechanical seals, and possible air binding. Valve stems and nuts are extended nearly to grade to permit operation without the need to enter the vault. The plans are shown in Figure 26-3.

An alternative is the use of a pump that ejects part of its discharge into the wet well during the first minute or two after the pump is switched on. The water ejected through a flush valve mixes the solids so that most are discharged to the force main during each pump cycle. The advantage is that the wet well is kept continuously clean automatically. The disadvantages are a small loss of efficiency and the added mechanical device located in the wet well.

Controls and Alarms

The pumps are set up in a lead/lag arrangement with automatic alternation after each pump cycle to balance run times and minimize starts per hour. HAND/OFF/AUTO operator selector switches are provided for each pump at the control panel located with a view of the wet well. Pumps are operated by the wet well water level indicated by a pressure transducer/transmitter hanging in a PVC pipe in the wet well. Key pad controllers are generally preferred because they are easy to operate, do not have pins (easy to lose), and can be programmed with a security code. The high water level alarm consists of a float switch connected to an auto-dialer that pages the on-call operator. A stainless-steel chain supported by the hatch frame is connected to the float switch so the operator can periodically test the switch. A nylon cord would serve the same purpose at less cost.

Operation and Maintenance

Double-leaf, spring-loaded aluminum hatches rated for an H-20 loading are installed over the wet well for access to the pumps. Similar hatches are installed over the valve vault. Safety chains are provided to create a barrier when the hatch is open. Hatches are equipped with padlocks for protection against vandalism.

Telescoping stainless-steel tubes are used for guide rails for installing the pumps. Except during times needed, they are lifted out of the water and hence do not collect debris. Stainless steel cables are attached to the pumps for removal and installation. The owner uses a truck-mounted boom and winch for this service, but, alternatively, a fixed crane or hoist could be provided on site. Winches are equipped with ratchets for use in both directions.

The site is equipped with overhead lights with electric power supplied from a nearby power service pole. Lights are also provided within the outer door of the control panel. A weather guard extending 18 in. in front of the panel is installed at the top front of the control panel. A hook-up with a manual transfer switch makes it easy to use the owner's trailer-mounted engine-generator.

Wash-down water is obtained on-site from the potable water supply. The equipment consists of an air-gap tank, water pump, and hose bibb contained in a locked steel cabinet. The wash-water pump, equipped with a preset timer for automatic shutoff, also has an automatic recirculation line with a pressure regulating valve. Steel traffic bollards are placed around the water and electrical control cabinets. No fencing is provided around the site.

Final Check

After completing the pump selection and piping layouts, the system hydraulics are checked again to see that the pump selected is the best fit, and that the motor and electrical gear are sized adequately. The drawings and specifications are reviewed by the owner, and the operators are walked through the design system by system. Final revisions are made before bidding the project.

Critique

Compare Figures 26-3, 17-21, and 17-22. The choice of pumping station configuration for each site should be based on sound judgment in which first cost is balanced against cleanliness and

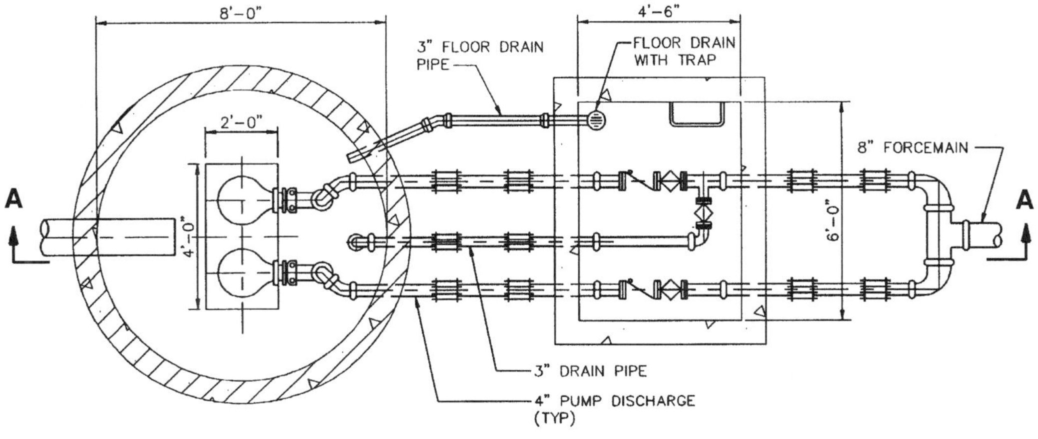

a

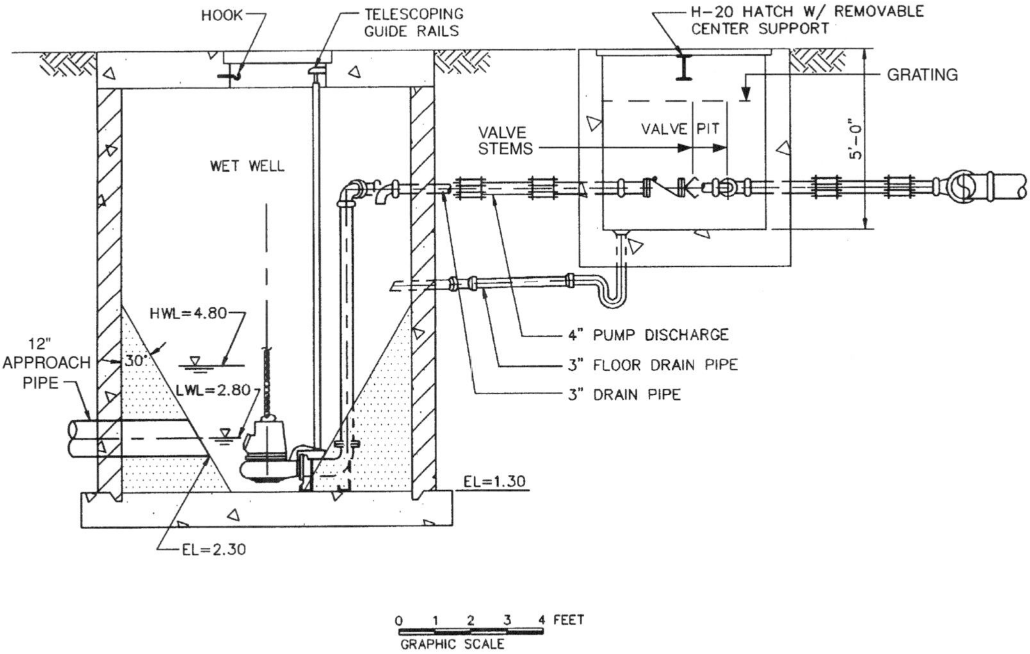

b

Figure 26-3. Layout of redesigned Clyde pumping station. (a) Plan view; (b) Section A–A.

desired control of odors and the ease and cost of maintenance.

Some operators prefer that the pump discharge lines be cross-connected with each other upstream from the check valves (with a valve on each branch) and connected to the force main drain line, as in Figure 17-22. Although such piping increases the number of valves, the size of the valve vault and the project cost, it allows the operator to agitate the wet well contents with one pump before pumping to the force main with the other. It also allows the operator to backflow one pump with the other to remove clogs without removing the pump.

The pumps could be equipped with nozzles as in Figure 12-58c, but these pumps are so small that such refinements are of debatable value.

The plans in Figure 26-3 indicate a pyramid-shaped hopper bottom. To keep the water surface area as small as possible so that scum will be readily drawn into the pump intakes at pump-down, the sides of the sump should hug the pump volutes with no more than 4 in. of clearance. Instead of being a 2-ft-by-4-ft rectangle, the hopper bottom would be improved if rounded ends were used and if the clearance between pump volutes were reduced to 4 in.

26-2. Redesigned Kirkland Wastewater Pumping Station

Designed originally in 1965, the Kirkland Pumping Station is shown in U.S. customary units in Figures 17-13 and 17-14, so the same units are used in this section. The station has been in continuous operation since 1967. The following example is a revised version of the existing station, modified to reflect: (1) wet well design developments that provide the self-cleaning features described in Chapter 12; (2) the best in current technology; and (3) more recent and stringent reliability standards. The station consists of three 2.5 Mgal/d pumps operating against a total dynamic head of 189 feet.

The top of the influent sewer is no more than 6 ft below finished grade. For such a shallow site, horizontal pumps were selected for both the 1965 and the current designs. Because they are less prone to vibration, horizontal pumps are preferred when they can be justified by little increase in structure cost. Instead of the eddy-current couplings used in the original station, the revised design has 125 hp adjustable-frequency drives. A 300 kW standby generator is provided for protection against power outages.

The force main terminates at an interceptor sewer 3150 ft from the pumping station site at an elevation 123 ft above the soffit of the influent sewer. Calculations were performed by using PUMPGRAF© [1], a computer spreadsheet program configured specifically for pumping station design work.

Individual Hydraulic Losses

Calculations for individual hydraulic losses from the pump inlet to the connection with the discharge manifold were performed first and are shown in Table 26-2. The pump inlet bell diameter in the wet well was selected on the basis of a conservative limiting velocity of 4.5 ft/s even though the HI Standards [2] proposed in 1997 allows 5.5 ft/s. Maximum velocities in pump connecting piping were considered acceptable for a variable speed station, where higher losses are only realized when the equipment operates at full speed. Upon completion of these calculations, they were automatically loaded into the program for the calculation of station system losses.

Station System Losses

Station system head losses in the force main (including static lift) were calculated from the manifold to the point of discharge and are shown in Table 26-3. Instead of the asbestos cement pipe used in the original project, HDPE was the material selected for the force main. Losses were initially calculated for a Hazen–Williams *C* of 140 and then recalculated for a *C* of 120. Individual losses for the pump inlet and discharge piping in Table 26-2 were not included in these calculations.

Pump Selection

Station system losses were then transferred through the program to the pump selection program (see Table 26-4), and a pump was selected from a previously entered library of pump manufacturers' catalog information. A plot of the selected performance curves against station system curves is shown in Figures 26-4 and 26-5. The pump performance curves plotted on the figure have been adjusted for individual pump inlet and discharge piping losses of 9.2 ft at 2.5 Mgal/d. These values must be added to the information on the plot to arrive at the correct rating for the pump. In this example, the pumps are to be rated at 2.5 Mgal/d at a total head of 180 ft. The pump selection is considered acceptable, because the intersection between the pump performance curve and the expected range of operating conditions lies well within the manufacturer's published data. Note that the intersection between the manufacturer's curve and the station system curves

Table 26-2. Pump Intake and Discharge Head Losses

2.5	= Starting Flow	– mgd	Title:	Revised Kirkland Pumping Station Brown and Caldwell Consultants
119.98	= Elev @ Start	– Feet		
140	= Default C-Value	>25	New C:	140 HiGrf LoGrf PmpCorr
14	= Default Diam	– Inch	New D:	14 53.34 = Target Head Loss
304.8	= Default Pipe Length	– Feet		9.185 = Total Head Loss – Feet
Date:	06-July 96	ParaEq =		3169.6 = EqivLnth (Std=def) – Feet

Flow, mgd	Item Description of Friction Loss	Diameter, in.	K or Cval	Fixed Loss	Length, ft	Item No.	Loss, ft	Hyd Grad, ft Pmp -> Dch	Vel, fps	V-Hd, ft
2.5	Bell Mouth Entrance	14	0.05			1	0.01	119.97	3.6	0.2
2.5	90° Elbow	10	0.3			2	0.23	119.74	7.1	0.78
2.5	Plug Valve	10	0.6			3	0.47	119.27	7.1	0.78
2.5	Reducer	10	0.01			4	0.01	119.26	7.1	0.78
2.5	PUMP	14	0.01			5	0.00	119.26	3.6	0.2
2.5	Increaser	6	1			6	6.03	113.23	19.	6.03
2.5	90° Elbow	10	0.6			7	0.47	112.76	7.1	0.78
2.5	Check Valve	14	3.5			8	0.71	112.05	3.6	0.2
2.5	Plug Valve	10	0.6			9	0.47	111.58	7.1	0.78
2.5	Branch Flow Tee	10	1			10	0.78	110.80	7.1	0.78
2.5						11				

Table 26-3. Force Main Head Losses for $C = 140$

5	= Starting Flow –	mgd		Revised Kirkland Pumping Station
119.98	= Elev @ Start –	Feet	Title:	Brown and Caldwell Consultants
120	= Default C-Value >	25	New C:	120 HiGrf PmpCorr
14	= Default Diam –	Inch	New D:	14 = Target Head Loss
1000	= Default Pipe Length – Feet			169.49 = Total Head Loss – Feet
Date:	06-July-96	ParaEq=		3341.6 = EqivLnth(Std=def) – Feet

Flow, mgd	Item Description of Friction Loss	Diam, in.	K or Cval	Fixed Loss	Length, ft	Item No.	Loss, ft	HydGrad, ft Pmp -> Dch	Vel fps	V-Hd, ft
5	Fixed Static head			122		1	123.00	289.48		
5	90° Elbow	14	0.6			2	0.49	288.99	7.2	0.8
5	Straight Pipe	14	120		25	3	0.35	288.64	7.2	0.8
5	90° Elbow	14	0.3			4	0.24	288.40	7.2	0.8
5	Straight Pipe	14	120		140	5	1.95	286.45	7.2	0.8
5	45° Elbow	14	0.25			6	0.20	286.25	7.2	0.8
5	Straight Pipe	14	120		1040	7	14.47	271.77	7.2	0.8
5	90° Elbow	14	0.3			8	0.24	271.53	7.2	0.8
5	22° Elbow	14	0.1			9	0.08	271.45	7.2	0.8
5	Straight Pipe	14	120		975	10	13.57	257.88	7.2	0.8
5	90° Elbow	14	0.3			11	0.24	257.64	7.2	0.8
5	Straight Pipe	14	120		995	12	13.85	243.79	7.2	0.8
5	Outlet Loss	14	1			13	0.81	242.98	7.2	0.8
5						14				

Table 26-4. Pump Selection for Two Pumps Operating

ENTER PLOT DATA RIGHT: Title --> Revised Kirkland Pumping Station
ENTER PUMP DATA BELOW: Brown and Caldwell Consultants

Name——— Fairbanks Morse
Nonclog, 4″ 5414/5424
Crv/Imp—T4D1B, 3″ solids

Impeller Rng 12.0 to 15.5
Speed Rng 1150 to 1785
15.5 = Impeller for Curve Points
1785 = Speed for Curve Points

Point/ Eff/NPSH	Head, ft	Flow, gpm	Head/Stg
	290.00	0.00	
	267.00	400.00	
62.00	248.00	800.00	
70.00	232.00	1200.00	
71.50	222.00	1400.00	
72.80	212.00	1600.00	
72.30	200.00	1800.00	
70.50	188.00	2000.00	
67.00	170.00	2200.00	
64.00	158.00	2300.00	

List Up To 12 Points In Any Order.
Put 1st Pump Curve Labels First
Column (Optional).
Delete or Add -->
Curve Labels -->
To Suit -->

Imported Data->Hi-Grf Lo-Grf
Pump Curve Calculations
2 = Number Pumps
For Fixed Speed Pumps,
Omit 2nd, 3rd & 4th Speeds
Data For Plot
14.70 = Impeller for Plot
1785.00 = Speed for Plot
= Optional 2nd Speed
= Optional 3rd Speed
= Optional 4th Speed

Pump Curve Correction, 0 if not used
9.18 feet Pmp Pipg Loss
2.50 mgd Flow Used

Overriding System Head Curve Data
0.00 mgd =Design Flow
100.00 = C-Value for Friction Loss
11.01 feet =Frtn Loss @ DgnFlo
50.00 feet =High Static Head
50.00 feet =Low Static Head
120.00 =Low C-Value
140.00 =High C-Value
1.40 =Peak Factor Sys Head
C = 120.00 Hi-SysHd Curve Label
C = 140.00 Low-SysHd Curve Label
1785.00 1st Pump Curve Label
2nd Pump Curve Label
3rd Pump Curve Label
4th Pump Curve Label

lies to the right of the pump's best efficiency point. As speed is reduced in variable speed operation, the point of intersection of the curves passes through the pump's zone of best efficiency.

After an acceptable pump selection was found, the program was used to evaluate pump performance at variable speed, and these data are shown in Table 26-5 and in Figure 26-5. As indicated, the minimum operating speed is approximately 1300 rev/min, which corresponds to a rate of discharge of 0.6 Mgal/d.

Hydraulic Profile

Next, a plot of the force main and hydraulic profiles was constructed, as shown in Figure 26-6. The purpose of this exercise was to provide a visual portrayal of hydraulic conditions in the system at both minimum and maximum system capacity. From an inspection of the plot, it was found that the hydraulic profile remains well above the force main profile from the pumping station to the point of discharge under all operating conditions. In addition, there appeared to be no high points or knees that would be the location of column separation in a hydraulic transient condition. Finally, the pipe line profile does not touch the mirror image line, so column separation does not appear to be a problem (see Section 7-1). For a 21-in. force main constructed of HDPE, the wave propagation velocity is about a third of that for ductile iron or cement asbestos (refer to Chapter 7). Based on inspection of the force main/hydraulic profile plots and the low wave propagation velocity, no formal transient analysis was judged to be necessary.

Geometry

The station capacity is 5 Mgal/d or 7.75 ft^3/s. From the principles presented in Chapter 12, the cross-sectional area of the wet well above the trench should be sufficient to limit the average forward (plug flow) velocity above the trench to 1 ft/s for any inflow. At the maximum inflow of 5.0 Mgal/d a cross-sectional area of 7.75 ft^2 is required, and at half the flow with the inlet pipe half full, the required cross-sectional area above the trench must be no less than 3.88 ft^2. In this design, the second criterion is the more critical and results in locating the top of the trench 4.0 in. below the invert of the influent pipe. If the sides are allowed to slope at 45°, the wet well above the inlet pipe can be made 5 ft wide. Note that installing the sluice gate requires a flat surface at least 12 in. wider than the diameter of the inlet pipe.

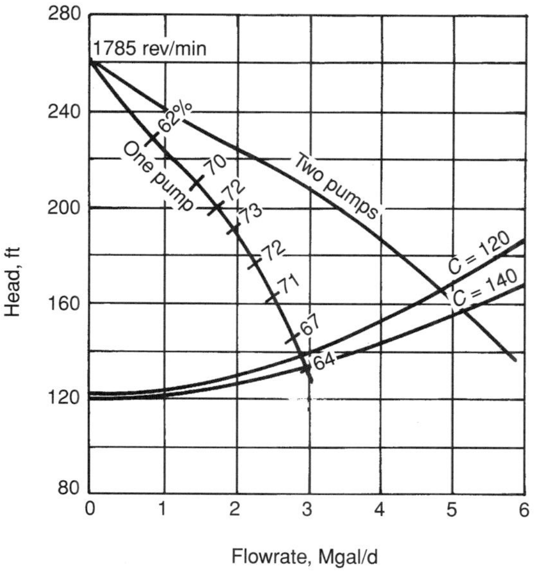

Figure 26-4. Pump and system curves for two pumps in parallel. Revised Kirkland Wastewater Pumping Station.

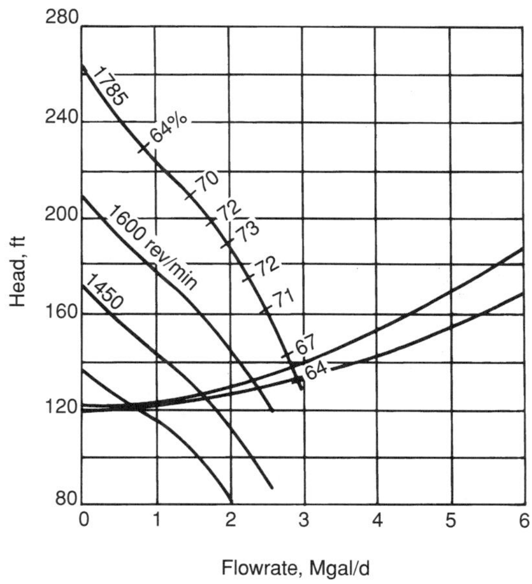

Figure 26-5. Pump and system curves for a single variable-speed pump. Revised Kirkland Wastewater Pumping Station.

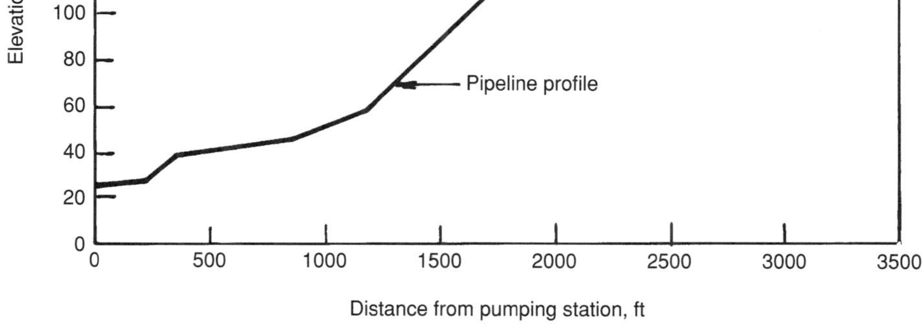

Figure 26-6. Plot of force main and hydraulic profiles. Revised Kirkland Wastewater Pumping Station.

Table 26-5. Operation of a Single Pump at Variable Speed

ENTER PLOT DATA RIGHT: Title --> Revised Kirkland Pumping Station
ENTER PUMP DATA BELOW: Brown and Caldwell Consultants

Name——— Fairbanks Morse

Nonclog, 4" 5414/5424

Crv/Imp—T4D1B, 3" solids

Impeller Rng 12.0 to 15.5

Speed Rng 1150 to 1785

15.5 = Impeller for Curve Points

1785 = Speed for Curve Points

NmbrStgs = 1.00

Point/ Eff/NPSH	Head, ft	Flow, gpm	Head/Stg
	290.00	0.00	
	267.00	400.00	
62.00	248.00	800.00	
70.00	232.00	1200.00	
71.50	222.00	1400.00	
72.80	212.00	1600.00	
72.30	200.00	1800.00	
70.50	188.00	2000.00	
67.00	170.00	2200.00	
64.00	158.00	2300.00	

List Up To 12 Points In Any Order.
Put 1st Pump Curve Labels First
Column (Optional).

Delete or Add -->
Curve Labels -->
To Suit -->

Imported Data->Hi-Grf Lo-Grf
Pump Curve Calculations
1 = Number Pumps
For Fixed Speed Pumps,
Omit 2nd, 3rd & 4th Speeds
Data For Plot
14.70 = Impeller for Plot
1785.00 = Speed for Plot
1600.00 = Optional 2nd Speed
1450.00 = Optional 3rd Speed

1300.00 = Optional 4th Speed

Pump Curve Correction, 0 if not used
9.18 feet Pmp Pipg Loss
2.50 mgd Flow Used

Overriding System Head Curve Data
0.00 mgd =Design Flow
100.00 = C-Value for Friction Loss
11.01 feet =Frtn Loss @ DgnFlo
50.00 feet =High Static Head
50.00 feet =Low Static Head
120.00 =Low C-Value
140.00 =High C-Value
1.40 =Peak Factor Sys Head
4.50 =Curvature factor
C = 120.00 Hi-SysHd Curve Label
C = 140.00 Low-SysHd Curve Label
1785.00 1st Pump Curve Label
1600.00 2nd Pump Curve Label
1450.00 3rd Pump Curve Label
1300.00 4th Pump Curve Label

Pump Intakes

The dimensions of a ductile iron bend and flare were judged to be unfavorable (see Example 12-4, Part B), so a special fitting consisting of a short radius steel 90° bend welded between a 14 × 10 reducer (with an OD of 14 in. or 1.17 ft) and a steel flange was selected. The trench was made 2.33 ft wide—equivalent to 2 bell diameters.

Intake Submergence

Submergence, governed by Equation 12-1, was calculated to be 3.0 ft. (Details of a somewhat similar calculation are given in Example 12-4, Part D.) As LWL is to be set at the invert of the inlet pipe, the pump intakes should be at least 3.0 ft below the invert or 2.72 ft below the top of the trench. The two pump intakes nearest the influent sewer should be placed D/2 (7 in.) above the floor. The depth of the trench should be 2.5 D or 2.92 ft.

The pump farthest from the influent sewer should have its inlet located D/4 or 3.5 in. above the floor of the trench. The minimum center-to-center spacing of pump intakes should be at least 2.5 D or 2.92 ft.

Plans

The plans, shown in Figures 26-7 through 26-9, illustrate the conceptual design for the revised station. Because the station must be located on a very small site, the pump inlets were located at the minimum permissible separation so as not to encroach on the space needed for the engine-generator. The pump inlet piping for Pumps 2 and 3 were skewed to permit the

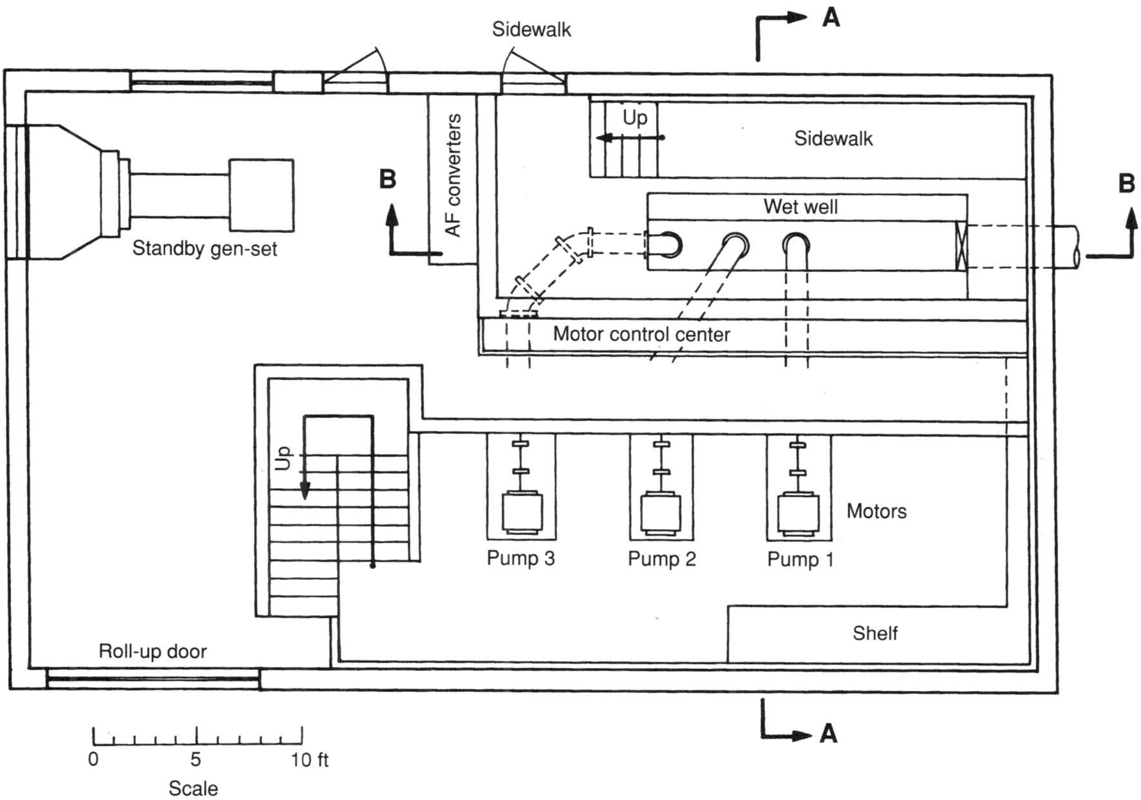

Figure 26-7. Ground floor plan. Revised Kirkland Pumping Station.

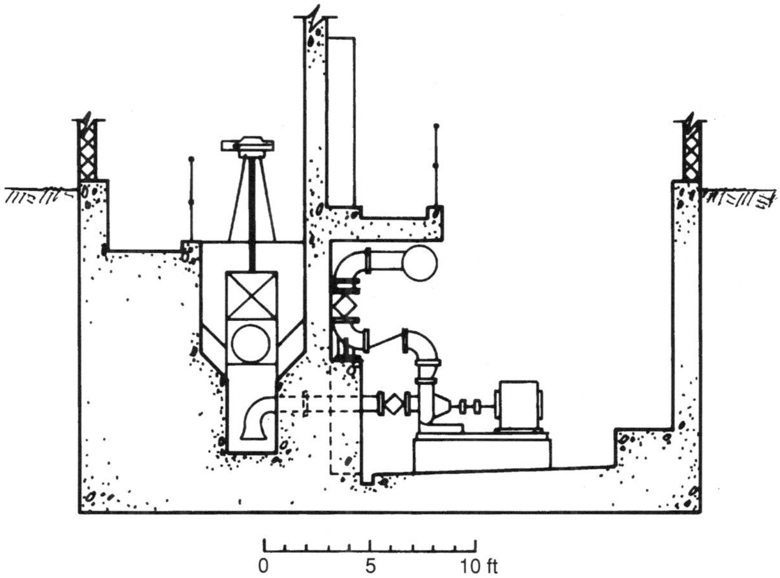

Figure 26-8. Section A–A. Revised Kirkland Pumping Station.

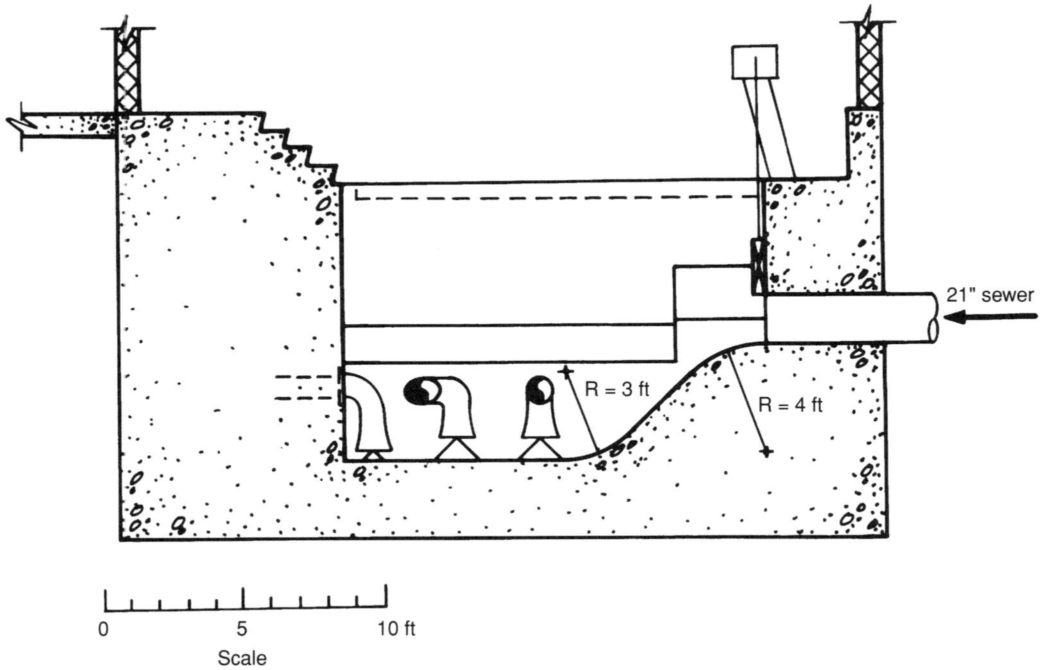

Figure 26-9. Section B–B. Revised Kirkland Pumping Station.

pumps to be located with a clear separation of 3.5 ft between bases.

The wet well has been revised to include an ogee ramp and a motor-operated sluice gate to improve the effectiveness of the cleaning operation. Instead of locating a grated walkway over the pump inlet channel as in the original design, a walkway alongside the channel makes access far more convenient for housekeeping chores (such as hosing grease off the walls).

Instead of a spiral stair (permissible under then-prevailing codes), a conventional stairway for access from grade level to the pump room is used in the revised design. All station auxiliaries (water pumps, hydropneumatic tank, air compressors) are located on an L-shaped bench along the south and east walls. To conserve space at grade level, all cabinets for electrical equipment (instead of being located in a separate room with a controlled climate) are to be furnished with an internal cooling system.

Critique

The suction bells, spaced at 2.5 D on centers, could, perhaps, be spaced at 2.0 D on centers (as shown for unconfined wet wells in an older edition of the HI Standards [3]) and thereby save 14 in. of wet well length. But the greater spacing is more conservative, and still keeps the wet well very short. Whether to keep this spacing and splay or bend the suction pipes or to use rectilinear piping (as in the original shown in Figure 17-13) and a much longer wet well is a matter of engineering judgment. One disadvantage of the plans shown is that if the suction piping for Pump No. 3 ever clogs, it would be more difficult to rod out debris than it would be for a straight pipe. The need for rodding, however, is rare because of the protection afforded by setting the pump intakes at relatively small floor clearances. The pump intake piping will be fitted with flush-out connections to permit clearing the pipe with high-pressure water. Mitered bends of a radius long enough to pass rodding equipment could be used as an alternative. Another disadvantage is that, because of the elbow connection, the distance from the back wall to the center of the pump intake is 1.0 D instead of the preferred 0.75 D. The latter disadvantage can be overcome by adding vertical fillets at the corners of the back wall to inhibit vortex formation in the stagnant water downstream of the intake.

26-3. Jameson Canyon Raw Water Pumping Station

The Jameson Canyon Pumping Station was constructed to replace an existing installation at the Cache Slough Reservoir, which receives water from the

Sacramento River system and delivers it to the City of Vallejo, California, Fleming Hill Water Treatment Plant. In addition to the new pumping station, the project included construction of a parallel pipeline from the reservoir to the treatment plant on an alignment through Gordon Valley. The original station, which has been retained and modified to function as a reserve facility, was found to be troublesome because of capacity-limiting inlet conditions and other defects. In addition to other problems, the station was sometimes troubled by accumulations of silt in the reservoir.

Vertical turbine pumps were selected for the new station, because it could be constructed without the cost of a pump room below ground level. Three pumps were installed initially, although there is a position for a fourth. Each pump has a nominal capacity of 416 L/s (9.5 Mgal/d).

The new pumping station can receive water from three sources, all with differing energy gradients. To limit costs and provide acceptable sump levels for operation with any source, a modulating butterfly valve (installed upstream from the station but downstream from the connection to the three sources) operates to maintain a fixed level in the pump sump. The siltation problem noted in the operation of the existing station dictated a sump configuration that could avoid accumulations of the silt that hampered the operation of the existing station. The plan and cross-sectional views of the finished design are shown in Figures 26-10 and 26-11.

An evaluation of operational considerations showed that two-speed pumps afforded considerable savings in operation and maintenance costs over variable-speed drives, which would have no significant advantages for this facility. The sump was designed with sufficient volume to accept a two- to three-minute fluctuation in water delivery rate without upsetting the pump control system. Power-operated 400-mm ball valves on the pump discharges were specified to control surges in the station sump on pump start-up and shut-down.

An overflow consisting of a weir, trough, and three 600-mm (24-in.) pipes (Figures 26-10 and 26-11) from the sump to the reservoir is for relieving surges and backflow to the sump on power failure.

After a careful economic analysis of the needs and options for service during a power failure, an on-site emergency generator was eliminated in favor of reconditioning a 597 kW (800 hp) pump with a combination motor and engine drive installed in the existing station. Piping from the new station to the existing station a short distance away included power-actuated valves that can be operated from a remote control center.

Using the Stoner Associates LIQT-PC [4] program, a transient analysis of pumping operations revealed that a high point in the transmission main approximately

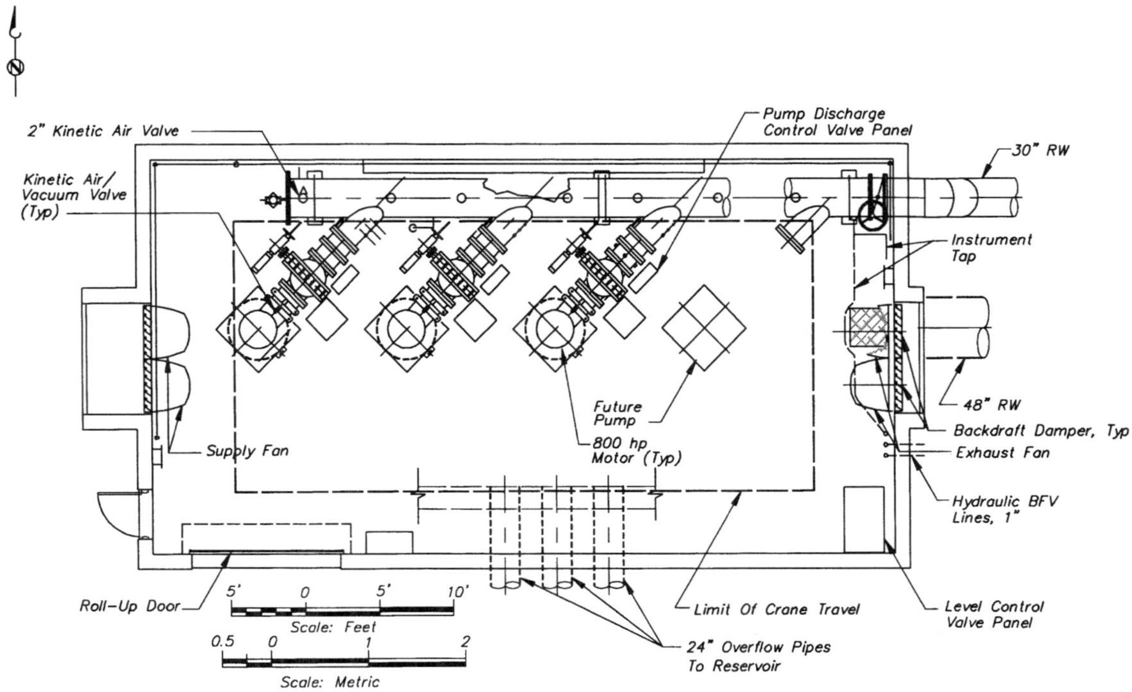

Figure 26-10. Plan of Jameson Canyon Pumping Station. Courtesy of Brown and Caldwell Consultants.

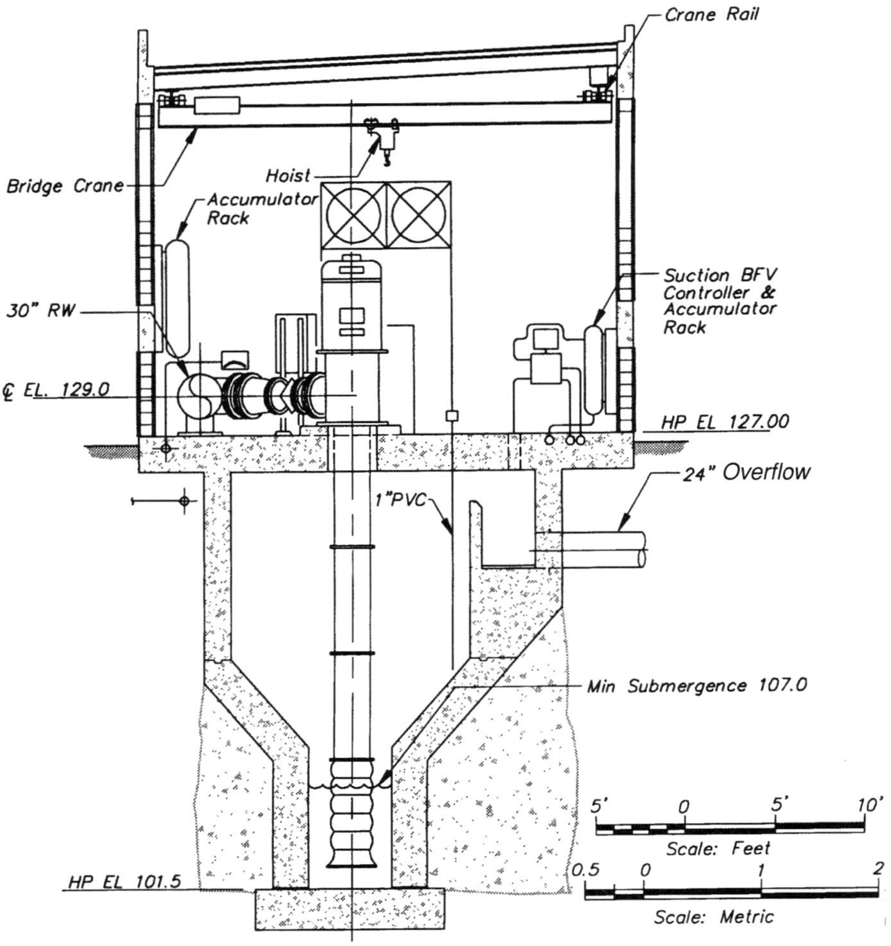

Figure 26-11. Cross-section of Jameson Canyon Pumping Station. Courtesy of Brown and Caldwell Consultants.

5500 m from the pumping station effectively limited station discharge head at lower flows. Depending upon the operational mode at the water treatment plant (throttling or free discharge), the hydraulic profile was dominated by the high point for flows less than 657 L/s (15 Mgal/d) for throttling or 920 L/s (21 Mgal/d) for free discharge (see Figure 26-12).

The high point in the transmission main also would be the site of a vapor cavity of damaging proportions on power failure at maximum flowrates. The transient analysis showed that the stop-and-check ball valves at each pump must be controlled to close in 8 min on power failure to keep the pressures in the transmission main under control (see Figure 26-12). During power failure, water will bleed back through the pumps as the ball valves close. The pumps had to be specified to accept this reverse flow and spin backward without damage. Time delays are provided to prevent restart until the pump stops turning. A large number of air-vacuum valves were installed in the transmission main.

Critique

The decision to omit the ogee spillway for removing sediment is a matter of balancing the extra cost of a longer wet well against the possible deterioration of water quality caused by possible organics deposited with the solids. The likelihood of problem sediments was judged to be remote for Jameson Canyon, but for other raw water pumping stations, the inclusion of an ogee spillway should be considered. If the water is clean, some cost could be saved by substituting a flat bench for the sloping side walls.

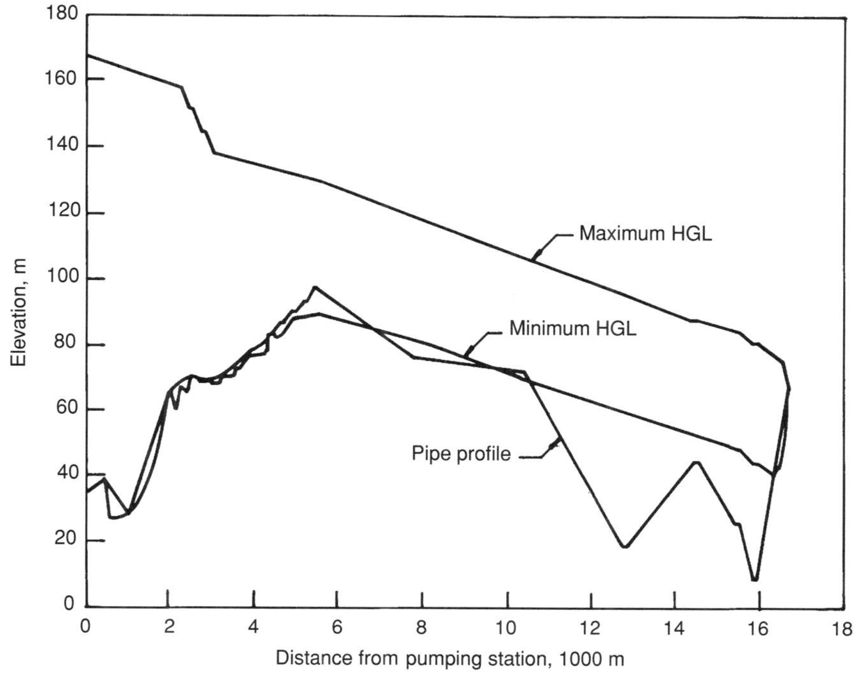

Figure 26-12. Minimum and maximum hydraulic grade lines and pipeline profile. From a Stoner Associates LIQT-PC transient analysis.

26-4. Albany Combined Sewer Overflow Pumping Station (CSO PS 88)

The combined sewer overflow (CSO) project in Albany, Georgia, is intended to ensure that most storm water in the existing system is intercepted and conveyed to a new regional treatment plant. Pumping Station 88, one of several under construction or being upgraded as a part of the project, was nearly ready for service in October 1997.

The station is designed for a maximum capacity of 227 L/s (5.2 Mgal/d). Constant-speed submersible pumps are used. Owing to an upstream interception project, the only flow in an existing local 750-mm (30-in.) sewer at the site is sanitary wastewater. The pumping station will intercept wastewater currently contributing to an overloaded interceptor and convey it through a 2100-m (6900-ft) long force main to a new interceptor.

Static lift on the station is only 3 m (10 ft). Because of the long force main, however, the total head at peak pumping capacity is 18 m (60 ft). Refer to Tables 26-6 through 26-8 and Figure 26-13 for the hydraulic calculations and pump selection.

The existing sewer was already at a slope consistent with a self-cleaning wet well designed for constant-speed pumps (see Chapter 12). The wet well and approach sewer are sized for the volume required to limit the pump starting frequency to 12 cycles/h in accordance with the manufacturer's recommendations. From the plans, the wet well contains approximately 10.6 m^3 of storage for every meter of depth (114 ft^3/ft). The critical storage requirement occurs when a single pump operates alone and discharges 175 L/s (4 Mgal/d). The volume required is given by Equation 12-3

$$V = \frac{Tq}{4} = (300\ \text{s})\frac{0.175\ \text{m}^3/\text{s}}{4} = 13.1\ \text{m}^3(464\ \text{ft}^3)$$

The station will have a 7.6-m (25-ft) length of 750-mm (30-in.) diameter sewer constructed at a slope of 0.1% upstream from the wet well, and then a new approach pipe at 2% grade will be constructed to intercept the existing sewer (also at a 2% grade). As more than adequate approach piping is available to develop the storage volume, the station pump controls will be set to provide reasonable regulation without incurring excess storage time, which could lead to septic wastewater and corrosion and odor problems.

As the above calculation need not be precise, the use of the average end area formula for obtaining the needed change in liquid level is sufficient. With the invert of the sewer at the station at elevation 52.3 m

Table 26-6. Suction Piping Losses for Albany CSO PS 88

114	= Starting Flow – Lps		Title:	Albany PS 88, pump correction						
52.288	= Elev @ Start – Mtrs			Brown and Caldwell Consultants						
120	= Default C-Value > 25		New C:	140	HiGrf		LoGrf	PmpCorr		
300	= Default Diam – mm		New D:	300		=Target Head Loss				
304.8	= Dflt Pipe Lnth – Mtrs				0.855	=Total Head Loss		– Mtrs		
Date:	06-Jul-96 ParaEq=				89.96	=EqivLnth (Std=def)		– Mtrs		
Flow Lps	Item Description of Friction Loss	Diam, mm	K or Cval	Fixed Loss	Length, m	Item Number	Loss, m	HydGrad, m Pump -> Dch	Vel, mps	V-Hd, m
114	Bell Mouth Entrance	350	0.05			1	0.00	52.28	1.2	0.07
114	PUMP	406.4	0.001			2	0.00	52.28	0.9	0.04
114	90° Elbow	300	0.6			3	0.08	52.20	1.6	0.13
114	Straight Pipe	300	120		3.048	4	0.03	52.18	1.6	0.13
114	Swing Check Valve	300	4			5	0.53	51.65	1.6	0.13
114	Gate Valve	300	0.2			6	0.03	51.62	1.6	0.13
114	Branch Flow Tee	300	1.4			7	0.19	51.43	1.6	0.13
114						8				
						9				

(171.6 ft), the low level shutdown for the lead pump will be set at elevation 52.7 m (172.8 ft), corresponding to the normal depth for a 750-mm (30-in.) sewer at a slope of 0.1% flowing at 176 L/s (4 Mgal/d), the capacity of one pump. Using the elements of a conduit flowing partly full, it can be calculated that 0.07 m^3 of storage per m of length (2.5 ft^3/ft) is available above the normal depth when the sewer is flowing part full at a slope of 0.1%. The sewer, however, will rise 8 mm (0.025 ft) in this 7.6-m (25-ft) distance.

The difference in elevation will have a negligible effect on storage. Thus, if the water surface elevation is allowed to rise from the low level shutoff to the soffit (elevation 53.06 m (174.1 ft) of the influent sewer before starting the lead pump, the total storage will be 5.7 m^3 (201 ft^3).

The plans for the pumping station are shown in Figures 26-14 through 26-17.

Using the elements of a circular conduit flowing part full and the previously calculated storage volume in the wet well itself and in the relatively flat (0.1% grade) section of the approach sewer, 7.4 m^3 (261 ft^3) of storage must be provided in the approach sewer at 2% slope to meet the manufacturer's requirements. Approximating this storage volume by using average end area, the program PARTFULL© [5] shows the depth in the sewer at 88 L/s (2 Mgal/d) to be approximately 112 mm (5.4 in.) with a depth after the hydraulic jump of 325 mm (12.8 in.). The 325-mm depth after the jump yields a cross-sectional area of 0.137 m^2 (1.47 ft^2). The average area available for

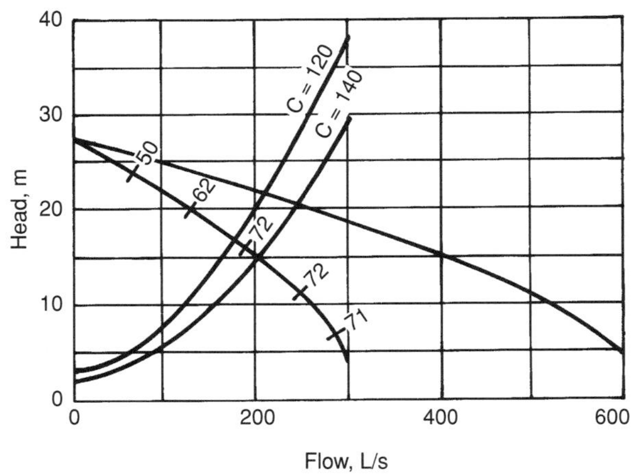

Figure 26-13. Pump and system curves for Albany CSO PS 88.

Table 26-7. Force Main Friction Losses for Albany CSO PS 88

228	= Starting Flow	– Lps	Title:	Albany CSP Project/Pumping Station 88			
52.288	= Elev @ Start	– Mtrs		Brown and Caldwell Consultants			
120	= Default C-Value	> 25	New C:	140	HiGrf	LoGrf	PmpCorr
400	= Default Diam	– mm	New D:	400		=Target Head Loss	
304.8	= Dflt Pipe Lnth	– Mtrs			23.456	=Total Head Loss	– Mtrs
Date:	06-Jul-96 ParaEq=				2415.6	=EqivLnth (Std=def)	– Mtrs

Flow Lps	Item Description of Friction Loss	Diam, mm	K or Cval	Fixd Loss	Length, m	Item Nmb	Loss, m	Hyd Grad, m Pmp -> Dch	Vel Mps	V-Hd, m
228	Fixed Static Head	400		3.04		1	3.05	75.74		
228	Passing Tee	400	1.4			2	0.23	75.51	1.8	0.17
228	Straight Pipe	400	120		1.524	3	0.01	75.50	1.8	0.17
228	Passing Tee	400	1.4			4	0.23	75.26	1.8	0.17
228	90° Elbow	400	0.3			5	0.05	75.21	1.8	0.17
228	Straight Pipe	400	120		99.06	6	0.84	74.37	1.8	0.17
228	45° Elbow	400	0.25			7	0.04	74.33	1.8	0.17
228	Straight Pipe	400	120		36.576	8	0.31	74.02	1.8	0.17
228	90° Elbow	400	0.3			9	0.05	73.97	1.8	0.17
228	Straight Pipe	400	120		457.2	10	3.86	70.11	1.8	0.17
228	45° Elbow	400	0.25			11	0.04	70.07	1.8	0.17
228	Straight Pipe	400	120		172.21	12	1.46	68.61	1.8	0.17
228	45° Elbow	400	0.25			13	0.04	68.57	1.8	0.17
228	Straight Pipe	400	120		4.572	14	0.04	68.53	1.8	0.17
228	45° Elbow	400	0.25			15	0.04	68.49	1.8	0.17
228	Straight Pipe	400	120		7.62	16	0.06	68.43	1.8	0.17
228	45° Elbow	400	0.25			17	0.04	68.38	1.8	0.17
228	Straight Pipe	400	120		10.058	18	0.08	68.30	1.8	0.17
228	90° Elbow	400	0.3			19	0.05	68.25	1.8	0.17
228	Straight Pipe	400	120		117.34	20	0.99	67.26	1.8	0.17
228	45° Elbow	400	0.25			21	0.04	67.21	1.8	0.17
228	Straight Pipe	400	120		13.716	22	0.12	67.10	1.8	0.17
228	45° Elbow	400	0.25			23	0.04	67.06	1.8	0.17
228	Straight Pipe	400	120		117.95	24	1.00	66.06	1.8	0.17
228	90° Elbow	400	0.3			25	0.05	66.01	1.8	0.17
228	Straight Pipe	400	120		266.7	26	2.25	63.76	1.8	0.17
228	90° Elbow	400	0.3			27	0.05	63.71	1.8	0.17
228	Straight Pipe	400	120		491.33	28	4.15	59.55	1.8	0.17
228	90° Elbow	400	0.3			29	0.05	59.50	1.8	0.17
228	Straight Pipe	400	120		35.052	30	0.30	59.21	1.8	0.17
228	90° Elbow	400	0.3			31	0.05	59.16	1.8	0.17
228	Straight Pipe	400	120		71.628	32	0.61	58.55	1.8	0.17
228	90° Elbow	400	0.3			33	0.05	58.50	1.8	0.17
228	Straight Pipe	400	120		22.86	34	0.19	58.31	1.8	0.17
228	90° Elbow	400	0.3			35	0.05	58.26	1.8	0.17
228	Straight Pipe	400	120		16.764	36	0.14	58.12	1.8	0.17
228	45° Elbow	400	0.25			37	0.04	58.07	1.8	0.17
228	Straight Pipe	400	120		74.676	38	0.63	57.44	1.8	0.17
228	90° Elbow	400	0.3			39	0.05	57.39	1.8	0.17
228	Straight Pipe	400	120		68.58	40	0.58	56.81	1.8	0.17
228	90° Elbow	400	0.3			41	0.05	56.76	1.8	0.17
228	Straight Pipe	400	120		38.1	42	0.32	56.44	1.8	0.17
228	90° Elbow	400	0.3			43	0.05	56.39	1.8	0.17
228	Straight Pipe	400	120		60.96	44	0.52	55.88	1.8	0.17
228	90° Elbow	400	0.3			45	0.05	55.83	1.8	0.17
228	Straight Pipe	400	120		38.1	46	0.32	55.50	1.8	0.17
228	Outlet Loss	400	1			47	0.17	55.34	1.8	0.17
228						48				

Table 26-8. Pump Characteristics for Flygt 3300 Pump

ENTER PLOT DATA RIGHT:	Title ---->	Albany CSO, PS 88
ENTER PUMP DATA BELOW:		Brown and Caldwell Consultants

Name----- Flygt CP3300

Curve 63-6

Crv/Imp--Imp 370 mm

Implr Rng 370.0 to 370.0

Speed Rng 1150 to 1150

370.0 = Implr for Curve Points

1150 = Speed for Curve Points

Point/Eff/ NPSH	Head Head-mtr	Nmbr Stgs = Flow Flow-Lps	Head/Stg
0.00	27.43	0.00	0.00
50.00	23.77	63.09	0.00
62.00	20.12	126.18	0.00
72.00	15.85	189.27	0.00
72.00	11.58	252.36	0.00
71.00	6.10	283.91	0.00

List up to 12 points in any order.
Put 1st pump curve labels first column (optional).

Delete or add -->
Curve Labels -->
To suit-->

ImprtdData->Hi-Grf Lo-Grf

Pump Curve Calculations

2 = Number Pumps

For Fixed Speed Pumps,
Omit 2nd, 3rd & 4th Speeds

Data For Plot

3.70 = Implr for Plot
1150.00 = Speed for Plot
= Optional 2nd Speed
= Optional 3rd Speed
= Optional 4th Speed

Pump Curve Corrctn, 0 if not used

0.80 Mtrs Pmp Pipg Loss
113.91 Lps Flow Used

Overriding System Head Curve Data

0.00 Lps = Design FLow
100.00 = C-Value for Friction Loss
3.36 Mtrs = Friction Loss @DgnFlo
15.24 Mtrs = High Static Head
15.24 Mtrs = Low Static Head
120.00 = Low C-Value
140.00 = High C-Value
1.40 = Peak Factor Sys Head
4.50 = Curvature Factor
C = 120.00 Hi-SysHd Curve Label
C = 140.00 Lo-SysHd Curve Label
1200.00 1st Pump Curve Label
2nd Pump Curve Label
3rd Pump Curve Label
4th Pump Curve Label

storage between the downstream end of the sewer (which is full) and the area above the wastewater at the jump, is, therefore, $(0.46 - 0.19)/2 = 0.14\ \text{m}^2$ ($1.45\ \text{ft}^2$). An equation can be written to solve for the required rise in liquid level above the soffit of the sewer to obtain the required volume

$$7.4 = 10.6y + 0.14x$$

where y = the rise in wet well elevation above elevation 53.1 (soffit of sewer at the station), and x = the length of sewer required to provide the required storage. Because $y = 0.02x$, the equation can be rewritten:

$$7.4 = (10.6)(0.02x) + 0.14x = \text{volume in wet well} + \text{volume in sewer}$$

Whence

$$x = 21\ \text{m (70 ft) and } y = 0.4\ \text{m (1.3 ft)}$$

The pump control program is, therefore, as follows:

Project elevation, m			
Rising level		Falling level	
Second follow on:	54.6	Second follow off:	53.3
High water alarm:	54.4	Cancel high alarm:	53.2
First follow on:	53.8	First follow off:	53.0
Lead pump on:	53.5	Lead pump off:	52.7

The pumps must be able to accelerate the liquid column downstream from the station each time the

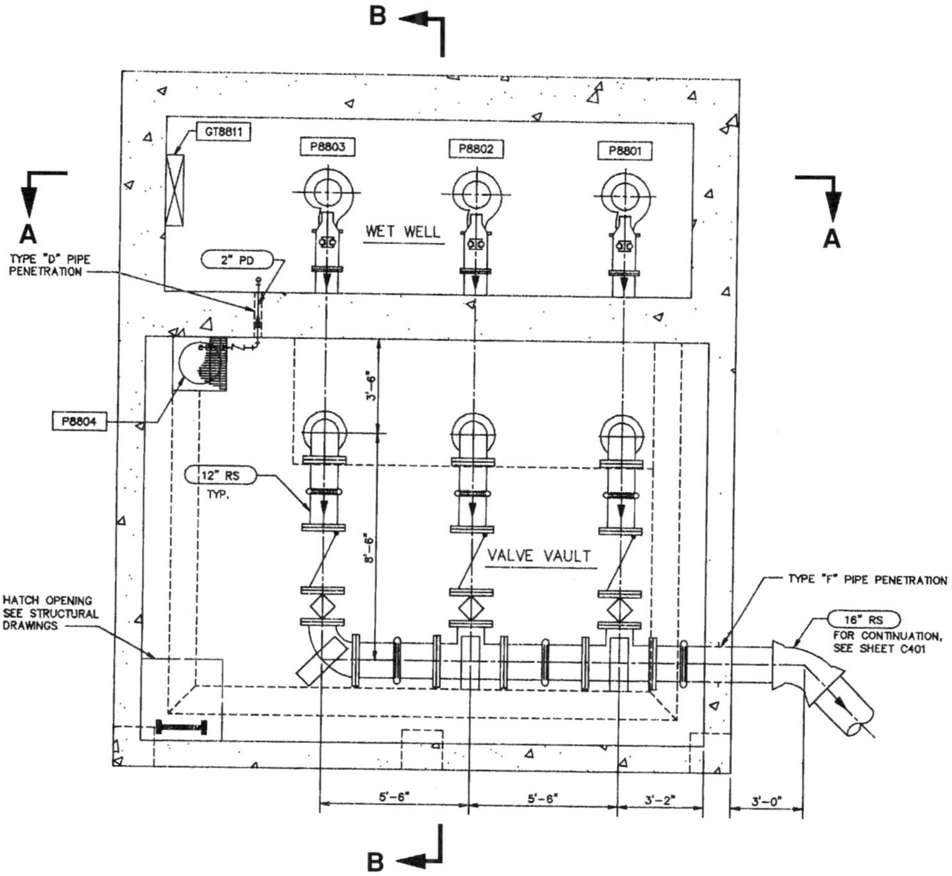

Figure 26-14. Plan at elevation 57 m (187 ft). Albany CSO PS 88. Courtesy of Brown and Caldwell Consultants.

pumps start without damage to the pump shaft seal. According to Krebs [6], the formula for calculating acceleration is:

$$\Delta T = 0.031\left[\frac{L\ v}{H_{\text{so}} - H_{\text{s}}}\right]$$

where ΔT is acceleration time in seconds, L is pipeline length in meters or feet, v is velocity in the pipeline at full pump discharge in m/s or ft^3/s, H_{so} is the pump shutoff head in m or ft, and H_{s} is the static lift on the system in m or ft.

From the information presented previously, the calculated acceleration time for the system is approximately 13 s—sufficient to cause some concern. Candidate pump manufacturers were contacted to discuss this problem and determine whether they were willing to warranty their products for this duty. If not, the furnished equipment may suffer from excessive maintenance costs associated with premature failure of the shaft seal.

26-5. References

1. Wheeler, W. PUMPGRAF© For a free copy of this computer program with instructions, send a formatted 1.4 MB, $3^1/_2$-in. diskette and a stamped self-addressed mailer to 683 Limekiln Road, Doylestown, PA 18906-2335.
2. Hydraulic Institute. "HI 9.8 Pump Intake Design." Hydraulic Institute, Parsippany, NJ (1997).
3. *Hydraulic Institute Standards for Centrifugal, Rotary & Reciprocating Pumps,* 14th ed. Hydraulic Institute, Parsippany, NJ (1983).
4. Stoner Associates. PO Box 86, Carlisle, PA 17013.
5. Wheeler, W. PARTFULL©. For a free copy of this computer program with instructions, send a formatted 1.4 MB, $3^1/_2$-in. diskette and a stamped self-addressed mailer to 683 Limekiln Road, Doylestown, PA 18906-2335.
6. Krebs, R. "Have you broken any shafts lately?" *Pumps and Systems,* **2**:18–19 (June 1994).

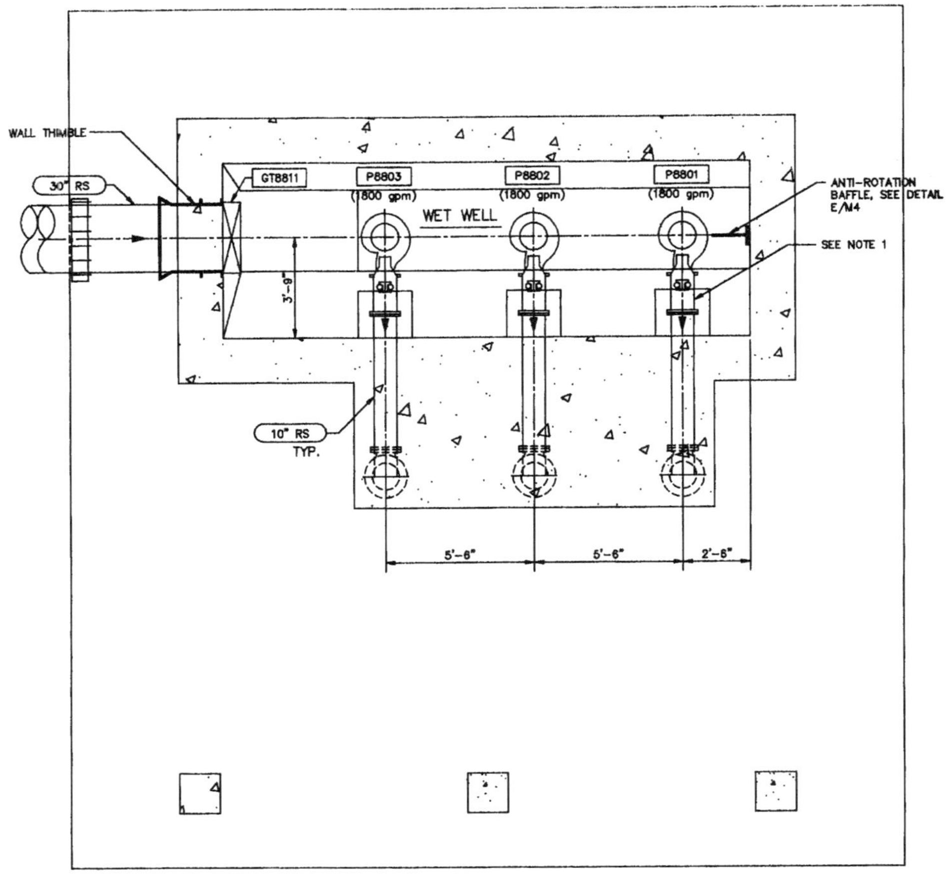

Figure 26-15. Plan at elevation 54 m (177 ft). Albany CSO PS 88. Courtesy of Brown and Caldwell Consultants.

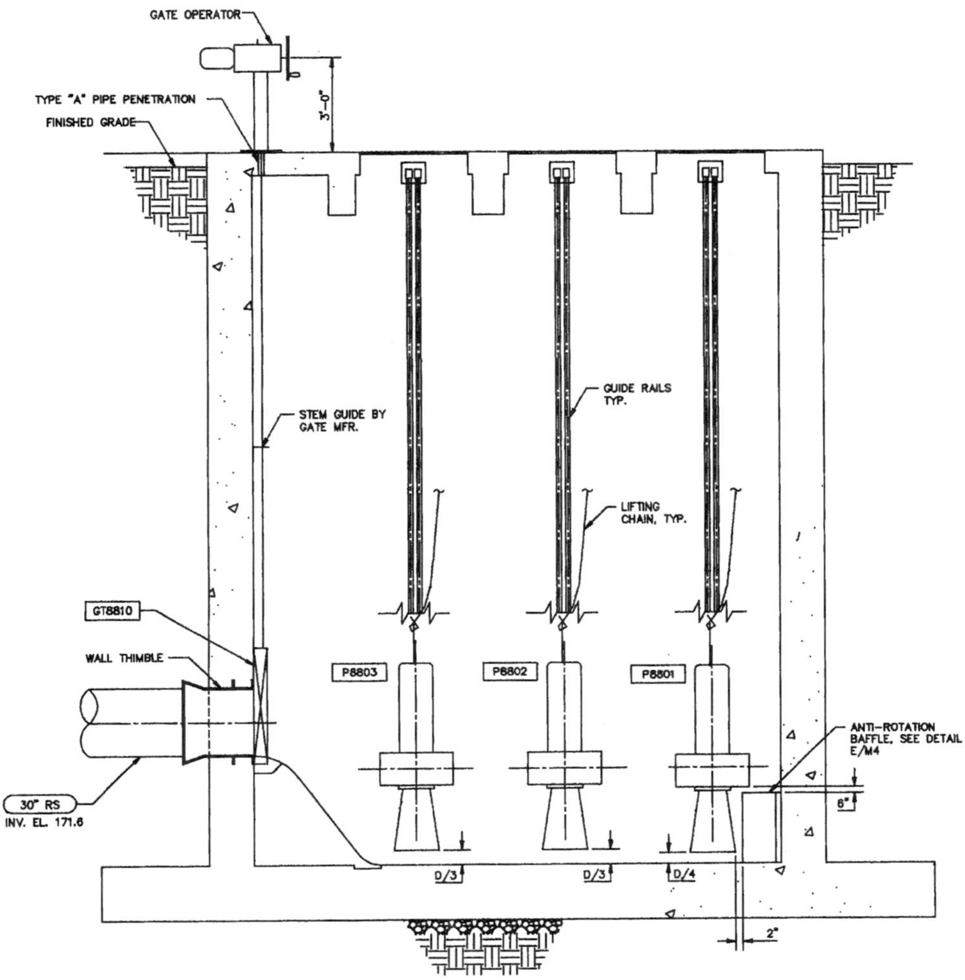

Figure 26-16. Section A-A. Albany CSO PS 88. Courtesy of Brown and Caldwell Consultants.

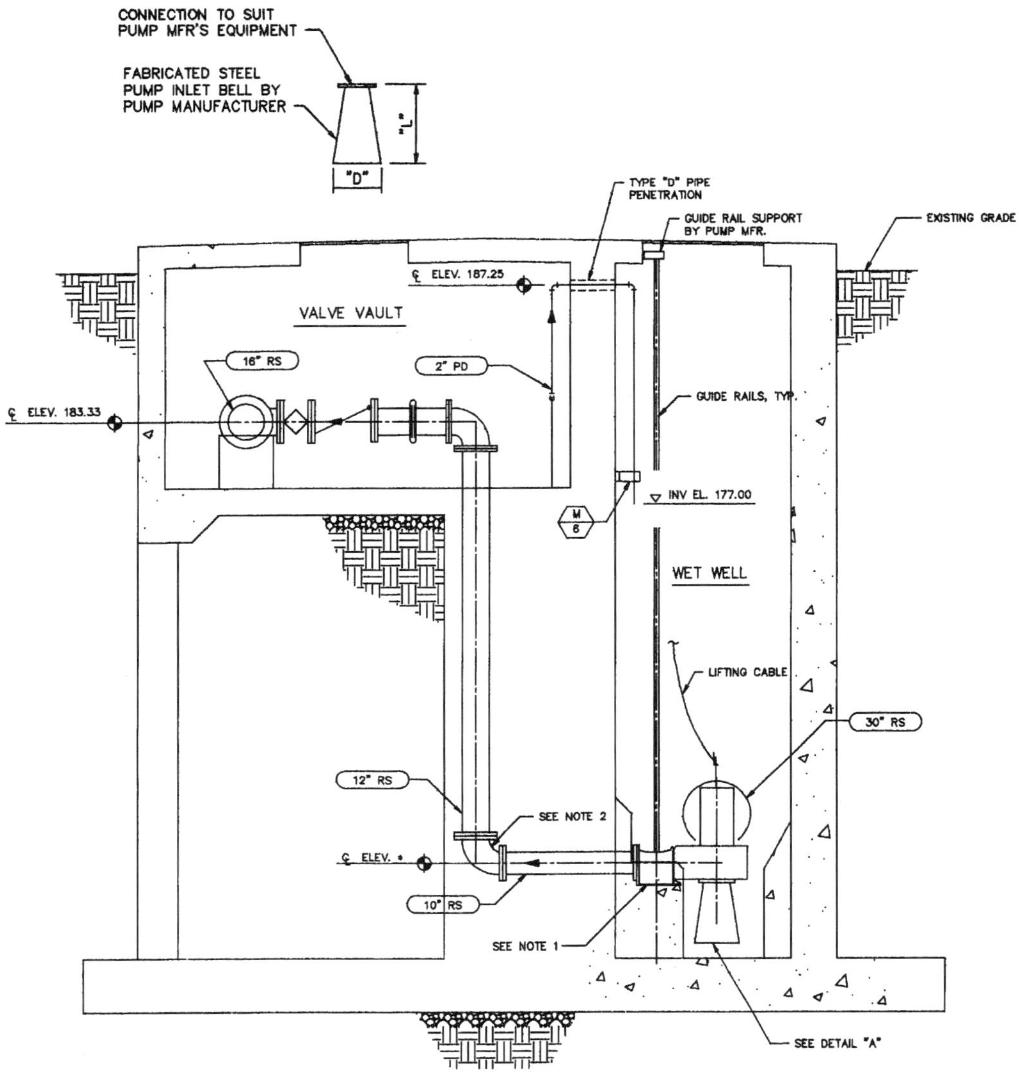

Figure 26-17. Section B-B. Albany CSO PS 88. Courtesy of Brown and Caldwell Consultants.

Chapter 27

Avoiding Blunders

ROBERT L. SANKS

CONTRIBUTORS
Stefan M. Abelin
Robert H. Brotherton
Fredric C. Burton
Patrick J. Creegan
Johannes deWaal
James C. Dowell
Fred A. Fairbanks
Philip A. Huff
William R. Kirkpatrick
Frank Klein
R. Russell Langteau
Jerry G. Lilly
Ralph E. Marquiss
Charles D. Morris
Constantine N. Papadakis
Carl W. Reh
William H. Richardson
Marvin Dan Schmidt
Earle C. Smith
Charles E. Sweeney
LeRoy R. Taylor
Michael G. Thalhamer
David Walrath
Gary Z. Watters
Frederick R. L. Wise, Jr.

Mistakes in pumping station design are sometimes disastrous and always costly and troublesome. They can be minimized by (1) knowing and frequently reviewing the most common blunders so as to avoid them; (2) making periodic design reviews (see Appendix G) at various stages of design—in-house if experienced personnel are available but, if not, by engaging outside, experienced consultants; (3) acquiring extensive knowledge about the manufactured products and pumping station design (by attending equipment shows, examining many stations, and interviewing managers and operators); (4) using a good deal of common sense and care; (5) properly coordinating the work of the different disciplines and professionals; (6) making knowledgeable inspections during construction; and (7) giving warnings against hazardous operation in the O&M manual.

27-1. General

Some of the most common blunders and some ways to avoid them are given in this chapter. The list of blunders is abridged, however, so read Chapter 24 (especially) as well as Appendices G and H for more completeness.

One of the commonest blunders—one that fits no particular category—is failing to maintain adequate written records of all significant conversations, letters, decisions, and calculations that pertain to the project design. The records must be neat, indexed, complete, understandable, and dated so that any engineer can interpret them; careful record keeping is desirable for others who may wish to discover the secrets of a successful design and a necessity when personnel changes are made or when there is a subsequent arbitration or a lawsuit. The records should be kept neatly bound in, for example, one or more three-ring binders.

Another of the most common blunders—one that seems to appear in most or at least many pumping stations—is inadequate room for heads, hands, feet, wrenches, and rumps. Much of the maintenance and repair work requires space for a team of workers—not just one individual. An absolute minimum of clear space free from all protuberances is 0.92 m (36 in.), preferably on three sides of the machine, and 1.1 m (42 in.) is better.

Many blunders seem so obvious. It does not even take an engineer to avoid them. It just takes thoughtfulness and a sense of responsibility. It helps to cultivate an attitude of asking, "How can this piece of equipment be maintained, disassembled, fixed, and reassembled, or replaced?"

Standard specifications and codes are identified only by their common abbreviations (see Appendix E for a complete listing).

27-2. Site

Existing utilities and obstructions not shown on the plans. Obtain utility company information for the specific site. Prepare a good site plan and field check the design.

Location in floodplain without adequate protection. Interview old timers about high-water marks, obtain government flood-hazard maps, and, if critical, compute flood elevations.

Inadequate street access and parking. Plan for access and enough parking space at the site (closed to public parking) for maintenance vehicles, including, if appropriate, a crane.

Nonconformity with setback and other planning and zoning ordinances. Review the regulations before selecting the site.

Site plan scale inadequate to show all details of site work. Normally, choose a scale of 1 in. = 50 ft or larger.

Wells located too close to sanitary hazards or property lines. The site plan should show the existing facilities. Follow the state sanitary codes.

Elevation of finished work incorrect. Insist on ties to at least two benchmarks. Field check the contractor's work prior to placing concrete for critical items.

Ground floor too low or floors without slope. Flooding or drainage problems are created and may violate sanitary codes. Have a good site survey and site plan, establish benchmarks, slope the floors to drains, and follow the state sanitary codes.

27-3. Environmental

Inadequate silencers on engines. Specify an acceptable noise level, say, 55 dB, at the lot line. Use a residential-quality silencer.

Fans too noisy. Use larger fans at lower speed.

Noisy machinery located within, say, $^1/_2$ km ($^1/_3$ mi) of inhabited areas. The alternatives are (1) relocating the pumping station; (2) using storage (except for sewage) instead of standby diesel power; (3) obtaining independent, supplementary electrical power as a standby power source; (4) using submersible pumps; or (5) planning to soundproof the station at the outset of the design. Take ambient noise-level readings before construction for protection in possible lawsuits.

Sewer inlet with a free-fall to the water surface. A free-fall of sewage into a wet well might (1) enhance odors and (2) promote foaming and air entrainment. Consider Examples 12-4 and 12-5 for preventing ingestion of air into pipes, and, for both controlling air ingestion and reducing odors and release of toxic, corrosive gases, consider an approach pipe as shown in Figure 12-53.

Odor production within about $^1/_2$ km ($^1/_3$ mi) of residential areas. The alternatives are (1) relocating the pumping station, (2) pretreating the sewage, (3) planning for frequent housekeeping by adding hose valves (bibbs) for washing out odor-producing deposits, (4) sealing the wet well with manhole covers (see Section 23-1 for details and safety precautions for entry), or (5) adding odor-control facilities on the wet well exhaust.

Odor control. Near residential areas, all vented air may require odor control (see Section 23-2), which may well be the most expensive part of O&M. *Always* investigate present and future needs for odor-control facilities if the station is (or will be) near inhabited areas, and design it so that odor control can be added. When using odor-control systems such as carbon sorption, consider thermostatically controlled heating to prevent the freezing of moisture in the exhaust.

27-4. Safety

It is poor economy to skimp on items that have a strong potential for life-threatening or hazardous situations.

Ventilation

Inadequate ventilation system for wet well. For wastewater wet wells that must be entered frequently, 12 air changes per hour (ac/h) is a minimum to meet most codes. But 12 ac/h can be hazardous, and an exchange rate of 20 ac/h is a more reasonable minimum. The California State Department of Health recommends 25 and the Arizona Department of Health requires 30 ac/h. Intermittent ventilation, which is allowed by some codes, including Ten-State Standards [1], is considered here to be entirely inadequate where human life is at stake. The scavenging of wet wells is imperfect at best, and the energy cost saving is not worth the risks. Consider omitting powered ventilation for wet wells by ensuring that there is no need for personnel to enter routinely (but note that a vent is required).

Wet well fans. Power air into the wet well at the top with high-energy diffusers, and power the exhaust air out at a low elevation using slightly more exhaust volume than supply volume to create a small negative pressure (see Section 23-2). (Opening or closing doors defeats the purpose of ventilation *by a single fan* whether the air is powered in or out.) Design for Class 1, Division 2 (NEC Code), and use corrosion-resistant materials (such as fiberglass) and explosion-proof construction in the exhaust system. Wet well exhaust is very damp and corrosive. Protect screens from freezing.

Inadequate ventilation for dry well. Ventilation in dry wells of either water or sewage pumping stations is required to limit the humidity and accumulation of toxic or explosive gases. (Explosions have occurred in water pumping stations.) Six air changes per hour for continuous ventilation or 30 complete air changes/h for intermittent ventilation is a practical minimum. Intermittent ventilation may be hazardous because it might not prevent the accumulation of explosive or toxic gases.

Other Considerations

Inadequate headroom. Design for 2.1 m (7 ft) clearance under all pipe flanges, hand wheels, and other protuberances. But if such a clearance cannot be met, pad the protuberance and hang a row of light chains around it as a warning. However, locate oil or water traps low enough so they can be serviced easily.

Obstructions in walkways. Lay out dry well and other floors to provide easy access to all maintenance points (such as packing glands, oil and grease points, and pressure gauges) so that workers need not climb over piping or other obstructions. A good rule is to require a 1.07 m (42 in.) clearance (horizontally) between pumps, motors, valves, etc., and any obstruction.

Ladders and circular stairs. Avoid ladders if possible, because (1) emergency egress is impeded, (2) personnel can carry tools only with difficulty, and (3) disabled workers cannot be carried at all. Observe OSHA rules for landings (required at maximum intervals of 3 m or 10 ft), ladder cages, nonskid treads, and toe plates. Make ladders easy and safe to enter.

Circular stairs are only marginally better. Straight stairways with landings conforming to local building codes are justifiable, considering safety, convenience, and daily use for 20 years or more. If a compromise must be made, consider ship stairs for minimum run length (1 horizontal to 2 vertical). Check local building codes for double-access requirements and for stairs to augment elevators.

Guards. Install guards around dangerous rotating machinery, belt drives, and other moving machinery.

Rescue. Provide a means for moving a helpless worker to the building exit. Consider the need for gas masks and self-contained breathing apparatus (always in pairs) at sewage pumping stations.

Codes. Become familiar with the safety regulations of such codes as OSHA, NFPA, and NEC.

27-5. Hydraulics

Sumps and wet wells inadequate or mismatched with pumps. Review Section 12-6. Unless the design is similar in geometry, size, and flowrates to an existing successful design, consider model tests. Sumps or wet wells for systems larger than about 3 m^3/s (50,000 gal/min) or those with unusual approach conditions should always be designed from model tests because the rules published in the various standards and design guides can (and often do) lead to poor hydraulic conditions. For small- to medium-sized inlets, refer to Prosser [2] and to the text by Dicmas [3].

Pipe roughness overestimated. If the actual friction headloss is less than the estimate, the actual discharge will be greater than predicted. The greater discharge increases the horsepower requirement and, thus, overloads the pump and motor. Design for the code-specified roughness—e.g., in the Ten-State Standards [1], $C = 120$ for lined and plastic pipe or $C = 100$ for unlined pipe—or design for the roughness expected after, say, 20 yr of service. Then redraw the station curves for new pipe (e.g., $C = 145$), find the new operating point, and specify a pumping unit that can handle all of the pumping conditions between the two extremes. For final calculations, use the actual, true inside pipe diameter for headloss and flow calculations.

Prerotation in wet wells. The vortexing that results from prerotation entrains air, reduces flow, and leads to vibration and noise in pumps. See Sections 12-6 and 12-7 for advice about wet wells.

Suction inlets too small for rated flows. Cavitation and vibration will occur and the pump will not discharge the designed capacity. The "average" velocity at the bell mouth of the inlet should not exceed 1.7 m/s (5.5 ft/s), and the velocity in the inlet pipe should not exceed 2.4 m/s (8 ft/s), but see also Table 12-1 (and nearby text) and Section 17-3.

Pump volutes inadequately submerged. Always place the low water cutoff above the pump volute or, more conservatively, above the stuffing box.

Suction inlets inadequately submerged. The required submergence is based on empiricism, and opinions vary. Probably the best rule is given by Equation 12-1. Gratings and other devices above the inlet can reduce the required submergence, but they collect debris in wastewater wet wells. Be careful and conservative. For authoritative and extensive discussions, refer to Prosser [2] and Dicmas [3]. See also the Hydraulic Institute's standards [4].

Vertical pipes connected to invert of manifolds. The solids in raw water and wastewater can collect in the riser and, given time, plug it. Connect the riser to the side of the manifold with a short length of horizontal pipe.

Float switches untethered in turbulent wet wells. Untethered float switches tangle, so attach them to a weighted chain (which permits easy removal) or to a vertical pipe. Also consider other sensors described in Section 20-3.

Pipe profile incompatible with hydraulic profile. Plot both profiles on the same drawing for comparison. Avoid high spots in pipe profile—especially in wastewater service where the height of the pipe should never be above the mirror image of the hydraulic profile (see Figure 7-1). In water service, the pipe profile should never be above the hydraulic profile by more than 7 m (23 ft.), and then control equipment (e.g., air-vacuum valves) must be installed at high points. Use the services of experienced personnel or consultants for air release needs and detailed surge analysis. Prevent column separation in the force main at all costs. A complete transient analysis should always be made for all systems exceeding, say, TDH = 12 m (40 ft), Q = 30 L/s (500 gal/min), and v = 1.5 m/s (5 ft/s). Beware of water hammer with an expulsion of air at the top of wells. Make certain that thrust blocks and anchors at bends and elbows can resist both static head and surges.

Sewage systems are especially vulnerable to water hammer because of buildup of grease and foam in control devices such as air valves. Pilot-operated surge-control valves for sewage are unreliable due to the potential clogging of pilot lines, so avoid them in wastewater pumping stations. Avoid water hammer in sewage force mains by controlling the pipeline profile and/or by adding flywheels to drivers.

Air chambers close to swing check valves. Never put swing check valves close to air chambers, because the short water column can accelerate very quickly and cause valve slam with exceedingly high pressures that can explode pipe or break valve casings.

Backflow Prevention

Cross connection installed. Use reduced-pressure-principle backflow preventers or, better, completely separate potable water from the station (seal water, washdown) water supply with a positive air gap.

Applicable plumbing code not checked. Under some codes, a reduced-pressure backflow preventer is not an acceptable substitute for an air gap.

Backflow preventer not equipped with drain. Under some conditions, the backflow preventer can release a large stream of water, which requires a high-capacity drain that can inundate the dry well, so exhaust the vent pipe outside the building above flood level.

Backflow preventer installed at too low an elevation. Always install backflow preventers above any possible flood level.

Potable water. Water downstream from a backflow preventer is subject to contamination and is not potable. Provide only potable water at sinks.

Potable water faucets with hose threads. Hoses are possible sources of cross connection and contamination. Do not install hose valves (bibbs) on potable water pipelines.

Hose valves (bibbs) without warning signs. Place warning signs indicating that the water must not be drunk.

27-6. Pumps

Improper selection of the type of pumps and pumping station. A design concept that may be entirely suitable for one agency or application may be entirely unsuitable for another, even if the stations happen to serve the same hydraulic conditions. For example, the use of a submersible sewage pump design may be appropriate for an agency that already has a significant number of such facilities and has the attitude, experience, personnel, and equipment to maintain them. Such a pumping station may be quite unsuitable for an agency that has no experience with such a facility and lacks the equipment and personnel to service and maintain it. A dry well may be the best option for the second agency in spite of the additional first-time construction cost. In summary: *all aspects of the facility*—initial construction cost, power costs, ease of operation and maintenance, availability of equipment and skilled maintenance personnel, and so on—must be considered in evaluating the various types of stations and selecting the most appropriate.

Large submersible pumps. Large [more than 75 kW (100 hp)] pumps may have very high operation and maintenance (O&M) costs if the pumping station is not tailored for ease in handling big pumps. Ease in handling large submersible pumps is particularly important for wet well installations due to the long, heavy power cables that (because they cannot be handled manually) make large submersible pumps entirely different from smaller ones in their requirements for trouble-free removal. There are many horror stories of power cables cut on sharp edges or flexed beyond their limit, pumps jammed on guide rails that are thereby bent, the need for massive mobile cranes to pull pumps located an excessive distance from a parking area, long discharge pipes that vibrate excessively, pumps dropped by inexperienced crane crews, and even of a discharge elbow pulled out with the pump. In addition to these scenarios of woes, large pumps (especially those of moderate quality) tend to break down more often than do smaller pumps. Many utilities seem to agree that large submersibles should be confined to dry pits where they can be serviced in place. In at least one large utility, however, wet well installations of submersible pumps are favored regardless of size (see the subsection "Submersible Pumps" in Section 25-6).

Cable attachment details omitted. Lack of foresight in planning for the removal of submersible wet pit pumps (especially those over 75 kW or 100 hp) can make removal difficult and expensive and may even require entry in confined hazardous spaces (with all the associated personnel and safety requirements). Discuss the removal of the pump with the manufacturer and the utility, decide what method is to be used, and tailor the facilities to make the job as easy as possible. Detail the method in the O&M manual, and describe the cable attachments in the specifications and show them on the plans. See subsection "Submersible Pumps" in Section 25-6. It is attention to details such as these that makes the difference between easy and difficult maintenance and between good and bad engineering.

Unreasonably conservative estimates of flowrates and pipe roughness. Fear of insufficient capacity for future flows often leads to the most frequent and worst blunder of all—choosing pumps and pipes too large for the actual flows to be encountered. The results are (1) pumps that operate too far to the right of the bep, exceed their capacity, vibrate, and cavitate; and (2) motors that are overloaded and therfore overheat. Pumps and motors have short lives, and parts must frequently be replaced.

Avoid this blunder by choosing the pumps and impellers for stable operation at low (as well as high) flows, for smooth pipe as well as aged pipe, and for flooded wet wells (with lowered static lift) as well as for low water levels. In lift stations with multiple pumps, for example, characteristically only one pump operates most of the time—and it does so at substantially reduced total head. So the pump should operate near its bep for that condition. Also select pumps and motors to operate safely over the full range of conditions to be encountered. Low flows are, if anything, more critical than high ones. See Figures 10-27 and 10-30. If future flows will be considerably larger than present ones, try the following in the order listed: (1) trim impellers now and replace them with larger ones when needed, (2) leave space and hook-ups for another, future pump, or (3) plan on replacing present pumps and motors with larger ones when needed. Piping must be satisfactory for both the present (with velocities high enough to transport solids) and the future (with friction headloss within bounds). Hazen–Williams C values for lined steel, ductile iron, and plastic pipe should be in the range of 130 to 150 for normal operation (see Sections 3-2 and 3-3). Future flows might require a second, parallel force main.

During start-up, always test both suction and discharge pressure at each pump for the widest range of conditions encountered. If necessary, change or trim impellers to suit the true system requirements and to reduce noise and vibration (an embarrassment to the designer). Test motors for current draw to be sure they are not overloaded at any condition. Make sure the ventilation system can adequately cool the motors.

Never allow variable speed units to operate at speeds lower than recommended by the pump manufacturer. If flows are too low, add a smaller unit or program the controller to operate on a start-and-stop, fill-and-draw basis while making sure the electrical equipment can withstand the frequency of pump starts.

Net positive suction head (NPSH). Always make sure $NPSH_A$ is at least 1.35 times $NPSH_R$, and also never less than 1.5 m (5 ft) more than $NPSH_R$ (see Section 10-4 and Figure 10-13).

Pump efficiency poor. Reliability and ease of maintenance are of prime importance, but search diligently for the best pump because an increase of only 2 or 3% in pump efficiency returns a startlingly high present-worth cost saving.

Pump and impeller selection poor. Avoid the largest impeller in a family of impellers because of vibration and reduced bearing life. Try to use a diameter that is no more than 90% of the largest diameter. Select a pump to provide good efficiency above the expected head range so that the heads "bracket" the highest efficiency of the pump.

For V/S units, specify (or select) pumps so that the operating point lies to the right of the best efficiency point (bep). Thus, the pump selected will be the smallest applicable and will operate through the zone of best efficiency at the various normal (reduced) speeds. But specify (or select) pumps with the operating point at or near the bep for C/S pumps. If a greater capacity is anticipated, specify the future operating conditions.

Relationship between impeller, speed, flow, head, and power incorrect. When calculating flows (which vary as revolutions per minute) for reduced speeds, also calculate the head (which varies as revolutions per minute squared) to find a new point on the pump H-Q curve. Draw the new pump H-Q curve through several such points to an intersection with the station H-Q curve to find the true flow and head at the reduced speed. To avoid mistakes in applying the affinity laws, calculate the power from head, flow, and pump efficiency—not from the relationship between power and revolutions per minute cubed. Note that manufacturers of V/S pump drives generally assume—erroneously—that power varies as revolutions per minute cubed, and, hence, their curves of driver efficiency are inaccurate (see Section 10-3, Example 10-3, and, especially, Figure 15-4 and Section 15-3).

Excessive radial loads. Avoid applications where pumps operate for extended periods near the shut-off head. Radial forces developed in the casing and acting on the impeller at low speeds (see Figure 10-18) cause rapid bearing wear and damage the wearing rings, shaft sleeves, and packing. If such operation is necessary, specify custom-made pumps with heavy-duty shafts and bearings.

Machinery vibration. Test pumps for vibration during factory pump testing, but be aware that foundation or supports in the field are different and affect vibration. Support vibrating machinery (such as engines) on spring isolators. (It is a fallacy that massive foundations can control vibration.) Isolate piping from vibrating machines with vibration isolators (see Chapter 22 for a discussion).

Inadequate or unclear pump performance specifications. Manufacturers differ in their presentation of published pump performance data. Hence, it is important to define the conditions of performance clearly so that competing products can be compared on the same basis. One basis for comparison is to specify energy, head-capacity, and efficiency criteria for the complete pump, for example (1) flange to flange for dry well pumps, (2) pool to discharge flange for wet well pumps, or (3) pool to discharge elbow outlet flange for submersible pumps.

Single-vane impellers. A single-vane impeller is asymmetrical, and no amount of machining can balance it dynamically in water. Some manufacturers can and do trim these impellers; others do not. As alternatives to trimming, (1) consider V/S drives unless an unaltered impeller can be found to match the flow conditions or (2) use a symmetrical impeller, which can be trimmed and balanced more easily.

High pump speeds. Avoid high speeds (especially for sewage pumps) because wear is a function of the second or third power of speed. Use the slowest practical speed (such as 705 to 1150 rev/min) to reduce maintenance costs.

Impellers improperly chosen for deep well turbine pumps. Because shaft stretch becomes important for deep pump settings greater than about 30 m (100 ft), specify closed impellers and consult the pump manufacturer or a specialist.

Oil-lubricated deep well pumps. In general, oil lubrication should not be used in pumps for potable water.

Prelubrication system omitted in lineshaft deep well turbine pumps. A system for supplying water to the bearings during starts is required. Consult the pump manufacturer for assistance in the design.

Light-duty pumps. Many pumps with attractive prices are too light for constant or severe service. Require rigid construction. For example, shaft deflection at the packing gland should not exceed 0.05 mm (0.002 in.), nor should deflection at wearing rings (outboard side of impeller) exceed about 0.20 mm (0.008 in.). A custom-designed pump may be needed to meet these requirements. Even sump pumps should not be light-duty machines (see Example 12-1).

Mechanical seals on pump shafts with a long overhang from bearing to impeller. If pumps are variable speed or operate frequently at low flow (or near the shut-off head), high radial loads, which bend the shafts and cause excessive wear on mechanical seals, are produced. Most seals cannot withstand such shaft distortions. Pumps must be engineered for such applications. Specifying a maximum shaft deflection of 0.20 mm (0.008 in.) at the impeller and a B-10 bearing life of 50,000 to 100,000 h at the worst combination of operating conditions promotes longer life, low maintenance, and life-cycle economy.

Bearings and seals or packing difficult to replace. Packing and bearings are the most frequently replaced items on pumps. Make seal replacement easy by ensuring easy access to the seal (such as avoiding close-coupled pumps and motors) and, for bearing replacement, provide a convenient means for lifting and storing motors and pump parts.

Bearing life not specified. Specify bearing life and insist that the manufacturer furnish certified bearing life calculations for the design point, pump speed, and impeller selected for the worst combination of specified continuous operation conditions. The worst combination of operating conditions is particularly critical for V/S pumps. Be especially careful with high radial load conditions when operating near shut-off, especially if the head exceeds about 15 m (50 ft).

Positive-displacement pumps unprotected. A clogged line can break a positive-displacement pump unless it is protected by (1) a bypass line with a pressure-relief valve such as a simple air-loaded pinch valve or (2) a pressure-actuated switch to stop the motor.

Poor or nonexistent pressure gauge installations. Every station should either have gauges or outlets for gauges on the pressure side (and also on the suction side, if practical) of a pump or pipe (see Figure 20-6). By using quick-connects with double shutoffs, one gauge can serve several locations. Ease of frequent calibration and inconsequential wear are two advantages of portable gauges.

Sewage pumps inadequate for passing solids. Specify that the pump must be certified by the manufacturer to pass a particular size of sphere. Furthermore, test the pump at the site before installation. The Ten-State Standards [1] requires at least 75 mm (3 in.). Many 100-mm (4-in.) or even 150-mm (6-in.) pumps cannot pass a 75-mm (3-in.) sphere.

Cleanouts not specified. Some pumps are built without suction elbows or suction piping containing hand holes for removing rags or other debris from non-clog pumps, and such pumps cannot be cleaned without dismantling. Always specify hand holes for unclogging sewage pumps and consider how to drain the pump and its piping conveniently (such as a drainpipe to the sump).

Use of main pumps for scum control. The buildup of scum in wet wells can be controlled with main pumps by manually bypassing the low-level pump control and by operating one or more pumps at full speed until the pumps break suction prime. The short-term vortexing sweeps the scum into the pumps. Equip the manual bypass feature with a spring return-to-automatic switch to restore the water level to the standard LWL and to avoid the hazard of operator error (see also Section 27-16).

Variable-speed drives in low friction head systems. Variable-speed drives should not be used if the system can operate at constant speed. Furthermore, V/S should not be used for centrifugal pumps in systems where (1) the slope of the head-capacity curve of the pump is nearly flat or (2) simple, four-position wet well level controls are used.

Variable-speed is appropriate for (1) positive-displacement pumps; (2) centrifugal pumps in systems where an instantaneously variable demand will occur, such as a tankless water system; (3) situations where relatively precise water level regulation is required; or (4) situations where sudden changes of discharge due to start-up cannot be tolerated. But be careful, because V/S design in low friction head systems requires a skillful design of pumps and controls because a very small change in speed may produce a very large change in discharge.

Use of pressure gauges at start-up. Impellers may not be compatible with the system hydaulics because of errors or even blunders in determining TDH. To avoid this one source of vibration and noise, always use calibrated pressure gauges on both suction pipe and discharge pipe for testing the pump. Compare with manufacturer's curve and trim or change impellers if necessary to keep the pump operating point within –15% to +10% of the bep.

27-7. Valves

Check Valves Improperly Selected

Be careful to match the valve in type and size to the system and to the flow.

Swing checks. Oversized valves operate improperly, but reducers and smaller valves can be used for low flows to obtain sufficient velocity to open the flapper fully. An external arm or a position indicator is desirable to show the position of the flapper.

Slam. Check valves may slam and cause disastrous water hammer. Avoid slam by using valves that close either very quickly (so that no reverse flow can occur) or very slowly (so that the water column comes gradually to rest). Quick-closing valves are preferable. Slow-closing valves tend to promote water hammer unless the closing mechanism (such as an external dashpot) is adjusted carefully and matched to the system. A water hammer analysis and an expert to adjust the closing mechanism is required (review the subsection entitled "Slam" in Section 5-4).

Quick closure. The quickest closure is provided by center-guided lift (or "silent") check valves followed by double-disc (or "double-leaf" or "double-door") check valves, but these types are for use only in clear water service. Closing characteristics of swing check valves with counterweighted versus spring-loaded lever arms is controversial (see "Slam" in Section 5-4). Carefully select check valves for the specific system.

Slow closure. Reverse flow will occur and the pump will run backward unless restrained. Impellers must be keyed, not screwed to the shaft. The wet well must be able to accept the reverse flow. Rags in sewage flowing backward may jam swing checks, so use ball, lubricated plug, or cone valves for slow closure in wastewater service. One requirement is a stored-energy closure system that operates the valve quickly through the first stage of closure, then slowly to shut-off so as to minimize water hammer. Swing check valves are satisfactory for water service if equipped with dashpots that can be regulated to meet system requirements.

Swing check valves without cover plates. In wastewater service, the lack of a cover plate means that the entire valve must be removed to extract debris, so require removable cover plates.

Other Aspects of Valves

Improper location. Location is often critical for any type of valve. Never allow swing checks or gate valves in a vertical pipe carrying raw sewage or grit-laden raw water. When the flow stops, grit settles and clogs the valve (or the valve bonnet of a gate valve), and the valve will eventually fail. If valves must be placed on a vertical pipe carrying dirty water, use only eccentric plug, ball, cone, or pinch valves. Never install butterfly valves near (within 2 pipe diameters of) elbows or pumps because the curved streamlines may cause so much torque on the vane that the valve actuator will not function. Never install valves in a wet well.

Air- and vacuum-release valves in sewage service. Go to any lengths to avoid using air-release and vacuum-relief valves for sewage service. Relocate the force main or the pumping station. Dig deep trenches or tunnel. If the use of such valves is unavoidable, realize that they are high-maintenance items, that they frequently clog, and that they are unreliable in wastewater service. Specify them in pairs to provide a backup, and specify sewage types with stainless-steel trim and flushing connections. Require not less than monthly cleaning in the O&M manual. Removal of all internal parts and shop cleaning is more effective than field flushing.

Throttling. Except for slow starts and stops to control water hammer, seek methods other than throttling to control flow (such as variable speed, spillage, a bypass, or a smaller impeller) because throttling wastes energy. Gate valves should not be throttled for extended periods because they are quickly ruined by vibration. Low velocity through a throttling valve results in poor control, so if throttling is unavoidable, consider using a reducer, a smaller valve, and an expander.

Nearly all butterfly valves are unsuitable for severe throttling service (as in pump-control valves). AWWA C504 specifications alone are insufficient to ensure the adequacy of valve seats. Some (but not all) replaceable body seats are satisfactory. Some manufacturers of pump-control systems who either use top-of-the-line makes or buy customized valves of special design may be able to offer advice.

Air-release valves in transmission and force mains. Air-release valves can cause water hammer (see Section 7-4) unless appropriately sized.

Vacuum-relief valve danger. Be extremely wary of using vacuum-relief valves in a pipeline if the pump valve is of the slow-closing type. If the water column reverses its direction and the velocity increases while the check valve is still mostly open (and unable to prevent high velocities), an explosive force can be produced that is sufficient to throw slabs of concrete into the air. Hydraulic analyses amd consultation with the vacuum-relief valve manufacturers are necessary to determine whether such a problem can occur.

Air in piping. Preferably, arrange piping so that air is automatically expelled as the wet well or sump fills. Air entrapped between the pump inlet and a check valve can air bind the pump, so install manual or automatic (or both in sewage service) air-release valves at all possible points of air entrapment, such as the volutes of horizontal pumps, pressure gauges, suction piping, and both upstream and downstream from the check valve. A manual air release should have, say, a 12- or 19-mm (1/2- or 3/4-in.) gate valve (for water) or a ball valve (for water or sewage) followed by an air break and drainpipe to the floor sump.

Natural rubber seats. Because natural rubber in wastewater valves is attacked by petroleum products, specify a resistant synthetic elastomer.

Thin elastomer seats. Elastomers may be cut by grit in slurry service, so thick layers should be specified. Urethane is preferable to either natural or synthetic rubber.

Zinc and aluminum bronze. Avoid zinc and untempered aluminum bronze in valves for waters that cause dezincification (see C508-93, 2.2.2.4 and C504, Foreword 111.7 and main text 2.2.11.2).

Valve handles. Chain wheels in sewage wet wells are an explosion hazard. Hand wheels should be placed within easy reach and away from walls or other knuckle-cracking obstructions. Use either hand wheels or chain wheels with worm gears instead of levers on plug valves 150 mm (6 in.) or larger.

Valve stems protruding into walkways. Protruding valve stems constitute a safety hazard in walkways, so relocate them.

Power-actuated valves without manual actuators. The valve cannot be moved if the power actuator does not work. Always specify an auxiliary manual closure.

Power actuators not designed for continuous throttling service. Valves designed for throttling should have actuators that are specified accordingly; otherwise, maintenance is excessive.

Hand wheel with nonstandard direction. Valves should either open when the hand wheel is turned counterclockwise or the hand wheel should be permanently marked with the direction for opening.

Valve permits cavitation. Cavitation causes excessive noise, wear, or premature failure of valves. Select and install valves so that cavitation does not occur. Valves must operate within the limits recommended by the manufacturer.

Valve fails in the wrong manner. Design the valve and appurtenant piping so that the valve acts in a safe manner when power fails (as it will).

Valve cannot be exercised without raw wastewater discharge. Redesign the system so that valves can be exercised with no resulting problems.

Improper support. Improperly supported valves or pipe can be distorted, which causes binding. Use an adequate number of supports.

Water valves in wastewater service. Use only valves designed for wastewater in wastewater service. Rags can wrap around shafts or catch on protuberances.

Freezing. Valves must be protected from freezing. If the fluid is warm enough, insulating the valve housing may be sufficient unless there is a possibility of zero flow.

Plug valve installed backward. Eccentric plug valves seat in only one direction. Analyze the requirements carefully and indicate the seating direction. In sewage service, install eccentric plug valves with the shaft horizontal so that the plug is in the upper portion of the valve in its open position.

Cleaning tools will not pass. Select valves that allow passage of a cleaning "pig." Otherwise, the piping may have to be disassembled for cleaning—a serious blunder.

Isolating valves added to pressure relief or other control valves. It must be possible to repair a valve by isolating it. Always design the system so that safety is not compromised by the forgetfulness of a careless worker (in reopening a valve, for example).

Control piping of incorrect diameter. It is important to use the correct diameter with respect to the size and speed requirements of valve actuators.

27-8. Mechanical

Cranes and hoists. Cranes and hoists must have access to the equipment with vertical falls to meet OSHA requirements. Piping and other obstructions above pumps or heavy valves must be located so as not to interfere. Design the facilities to permit lifting all heavy equipment straight up and depositing it directly on a truck bed.

Lifting eyes. Lifting eyes are an abomination, but if (for some reason) they must be used, at least design them so that stresses (in steel) do not exceed 69,000 kPa (10,000 lb/in.^{2}) at the smallest cross section. Never use cast iron or malleable iron for any fittings to hold lifting eyes (see Figure 25-2).

Obstruction in walkways. Avoid them. For example, seal water can be carried in ss piping buried in floors. Overhead piping and other objects are not obstructions if the headroom is at least 2.1 m (7 ft.)

Unharnessed sleeve couplings. Sleeve (Dresser®) couplings, flanged adapters, bell and spigot joints, and mechanical joints can blow apart at any time unless they are longitudinally restrained. The restraint can be provided by burial in soil, confinement between rigidly restrained piping or equipment, or harnessing with enough bolts to resist the longitudinal forces due to water pressure (plus water hammer) in the joint.

Long shafts. Either avoid the use of long shafts or else plan for them carefully. Provide sufficient headroom for disassembly. *Always* balance long shafts dynamically, and *always* investigate torsional vibration, which is not apparent to the senses and is therefore more dangerous. Provide access for lubrication and consider framing the concrete structure for both the support of and access to any intermediate steady bearings. Also confer with the shafting manufacturer when considering the use of tubing large enough to allow for the omission of intermediate bearings.

Pipe flanges of different pressure class than pump, valve, or fitting flanges. Coordinate the plans and specifications to ensure flanges will fit.

Piping system inflexibility. Flexibility must be sufficient for the replacement of valves and fittings. Consider flange adapters or grooved-end (Victaulic®)or sleeve (Dresser®) couplings at strategic locations.

Sludge lines clogging. Sludge or concentrated raw sewage tends to clog when at rest or when pumped at velocities that are too low. Design for velocities above the critical velocity, which is about 1 m/s (3 ft/s) for 2% sludge, 1.5 m/s (5 ft/s) for 4% sludge, and 2.4 m/s (8 ft/s) for 10% sludge. Refer to Chapter 19 for the calculation of headloss. Design sludge systems for easy cleaning—for example, by using crosses instead of elbows to make rodding easier.

Water meter inaccurate or readout and capacity improper. Water meters are inaccurate unless the entrance pipe is straight for 15 to 20 pipe diameters for "head" meters and 2 to 5 pipe diameters for magnetic meters. Downstream flow disturbances also affect meter accuracy. Do not "fudge" on the manufacturer's recommendations for the required straight inlet and outlet distances. Choose capacity and readout units for present (not future) flow variations. Calibrate *in situ*.

Inadequate well head design. Many problems can occur, such as improper well casing ventilation, depth gauge, tap for raw water samples, clearance of top casing above the floor, and pressure gauge for pump testing. Review sanitary codes and engage a well pump specialist.

Flywheels used for water hammer control. Flywheels impose very severe service for bearings. Use oversize bearings (100,000 hour, L-10 life) and flexible couplings to the flywheel shaft.

27-9. Electrical

Control system too complex for contractor or operators. Assess and do not exceed the level of sophistication of local contractors and the client's personnel. Keep it simple.

Corrosion in control panels. Locate control panels in a clean, dry environment. If they must be in a damp environment, use NEMA 4 or 12 panels and purge them with dry air. NEMA 4 panels require thermostatic heaters in most of the United States because of nighttime moisture condensation in the cabinet. Use moisture-absorbing cartridges for corrosion control in all panels.

Excessive starts per hour. Starting motors and motor starters too frequently reduces their life. Contact the manufacturers—not the sales representative—of both the motor and the starter regarding the number of allowable starts per hour and obtain guarantees in writing. Consider (1) "soft-starting" with reduced-voltage, solid-state starters for unlimited starts per hour or (2) V/S pumping.

Big motors. Big motors may not start on the first try. Then the operators must wait for hours for a second try. To avoid the delay, specify enough thermal capacity for at least two starts. An operator, who should be standing by, should stop the motor if it fails to reach full speed in 10 seconds. Correct the cause of the malfunction before attempting a restart.

Junction boxes in wet wells. Avoid placing junction boxes in wet wells or at any location where water could enter the box. Boxes in moisture-laden atmospheres accumulate water that can travel along conductors and cause serious faults. In a wet well, a worker must enter a hazardous, confined space to disconnect, for example, a submersible pump. One means to keep junction boxes out of wet wells is shown in Figure 17-21. If there is no alternative, specify watertight motor and device terminations. Keep moisture from entering by filling the junction box with potting compound and, for submersible motors, also specify non-wicking cable.

Inadequate lighting of work areas. Review lighting needs with the operating personnel and the electrical engineer. Operating panels should be lighted with a minimum of 500 lux (50 ft · cd), but 200 lux (20 ft · cd) is adequate for the pump room. Portable work lights are cheaper than fixed lights and are actually better. Check the hazard rating. Install guards on lamps and ensure easy access for replacement. Add a few incandescent lamps to provide instantaneous light to augment lighting that requires a long warm-up time before reaching full brilliance.

Inadequate or dangerous outlet receptacles. Always specify GFCI receptacles and always test them. Plan enough receptacles for power tools, electric hoists, and lights.

Phase not protected. Use breakers, not fuses, on motor-control and other three-phase circuits. Specify a three-phase monitoring instrument on the incoming power line to protect against undervoltage or a loss of phase.

Improperly sized circuit breakers. Specify properly sized circuit breakers to protect against (1) short circuits, (2) overloading over a short time interval, and (3) nuisance tripping. All electrical equipment must have adequate withstand ratings.

Improper grounding. All electrical (and mechanical) equipment must be properly grounded (see "Grounding" in Section 8-5).

Power failures. Power failures will occur, so design the system to prevent column separation and other disasters. Add backup in the form of stored energy to operate slow-closing valves.

Power supply inadequate. Check with the power company before beginning the design.

Positive-displacement pumps with high static heads or long force mains. These sometimes require high-slip (8 to 13%) "D" motors to accelerate properly the tons of liquid in a force main. Check the pump and operating characteristics with the motor manufacturer.

Transfer switches without time delay in neutral. The field voltages on motors driving loads with a large WR^2 do not collapse rapidly. If the motor is transferred quickly from one power source to another (say, to an emergency generator or vice versa) and if the two sources are out of sync, mechanical forces that are large enough to twist shafts, break couplings, and tear windings out of motors are produced. Specify a transfer switch with an adjustable time delay in neutral. Usually 20 to 30 cycles is a sufficient delay.

Wet well electrical equipment not explosionproof. Always classify sewage wet wells as Class I, Group D, Division 1 or 2 in accordance with the NEC. Electrical equipment should be explosionproof. Use nonsparking tools and safe maintenance practices when working in or around wet wells.

27-10. Structural–Architectural

Anchor bolts incorrectly located or with inadequate projection. Anchor bolts out of place is an all-too-frequent problem that can be solved by (1) using sleeves to permit minor adjustments of bolts and (2) fastening bolts to a plywood template before casting the concrete to ensure exact placement. Admonish the field inspector to check and recheck to ensure correct placement. An alternative is to place anchors in accurately located drilled holes with epoxy (or a similar quick-setting, two-component glue) in breakable capsules, but some experienced engineers will not use them, because the strength of the anchor depends so much on the quality of the field work. In corrosive environments, use extra-heavy bolts, or use stainless steel or nonferrous material.

Inadequate detailing. Show details of such elements as water stops, roof flashing and drainage, louvers, weather stripping, clearance around equipment, and provisions to guard against freezing. Detail roof and wall penetrations for adequate clearance between engine exhaust pipes and combustible materials. Ensure that the work of structural, mechanical, electrical, instrumentation, and architectural disciplines is closely coordinated.

Differential settlement between structure and exterior piping. Be sure that the exterior pipeline has enough flexibility in its joints adjacent to the wall or support both the structure and the pipe on piles. Always use wall sleeves. Because of the high probability of differential settlement or frost heave, interrupt plastic electrical conduit with metal conduit through exterior walls.

Foundation problems. Obtain the services of a qualified geotechnical engineer and adequately investigate the site. Include field borings or test pits (see "Subsurface Investigations" in Section 25-2 and "Geotechnical Considerations" in Section 25-10).

Inadequate hatches. Make hatches large enough (at least 700 mm × 700 mm or 28 in. × 28 in.) for a worker with a belt of tools to enter. In sewage wet wells, allow at least 700 mm × 900 mm (28 in. × 36 in.) for a self-contained breathing apparatus in a backpack. Consider aluminum checkerplate with stainless-steel hardware, positive latch openers, and inside grab rails for easy access to the ladder or stairs. Check OSHA standards for hatches and their locations.

Level floors and, consequently, puddles. Slope floors at least 1% to a floor drain (2% is better). Concrete floors can be cast level if a wearing slab of variable thickness is applied later, but there is considerable difficulty in obtaining a good bond.

Inadequate monorails or cranes. Preferably, arrange the station to provide monorails (for small stations) or cranes (for medium-sized to large stations) above all heavy equipment, and make it easy to set heavy items on a truck in a doorway. Consider motorizing all motions for cranes and hoists with capacities greater than 680 kg ($^3/_4$ tons). Use slow speeds, because pumping station operators are rarely skilled in the operation of hoisting equipment. Make access hatches light (for easy lifting) and large enough to remove equipment without dismantling it. At the very least, small stations should have lifting eyes for portable hoists (see Figure 25-2).

Inadequate access and clearances. Allow enough room for the removal (and replacement) of every piece of equipment without disassembly or with minimum disassembly. Avoid cramped working areas by allowing at least 1.1 m (42 in.) around all equipment (including appurtenant piping) for heads, hands, feet, rumps, wrenches, and lights for maintenance repairs and operations. Carefully draw plans for complicated equipment such as engines or generators together with all attached auxiliary or appurtenant equipment to ensure that there is about 3 m (10 ft) of clearance for ease in dismantling for repair.

Inadequate strength, stiffness, and watertightness. Avoid using the minimum design requirements in the ACI 318 Code. Heavier construction is required so use ACI 350 code.

Inadequate ventilation for motor cooling. Provide for close coordination between mechanical and electrical disciplines. Consider air conditioning, especially if noise must be suppressed.

Vibration. Avoid resonance frequencies by supporting pipes at frequent intervals and using deep beams and thick floor slabs. Isolate diesel engines by setting them on spring supports (see Chapter 22).

Wet pit atmosphere seeping into pump room. Completely sequester wet pits from the pump room so that neither gas nor sewage (even if the wet well overflows) can enter the pump room. Seal wires within interconnecting conduits placed at the highest practical elevation. Sump pumps should always discharge above the highest possible water level to a sewer *outside* the pumping station.

27-11. Specifications

Equipment specifications not to code. Equipment specified by a single named standard or by sole-source procurement must meet federal and state requirements.

Pump testing. Require witnessed factory tests of all large and important pumps, but use judgment before requiring witnessed tests for small pumps. Require the field operational tests described in Section 16-6.

Incomplete or conflicting specifications. The project engineer must ensure that specifications for all materials, workmanship, and submittals are complete and must coordinate them with the plans (see Chapter 17 and Appendix G).

Specifications not incorporating unit responsibility. Specifications should require that a factory-certified representative of the manufacturer be responsible for the complete coordination of multiple-component equipment packages (such as pumps, V/S driving equipment, drivers, and associated controls) when furnished by a single manufacturer. Otherwise, you must be prepared to coordinate all the parts and assume responsibility for such diverse aspects as motor size, motor heating characteristics, control system characteristics, pump torque requirements versus motor torque output at variable speed, coupling sizes, coupling bore sizes, shaft length, torsional analysis, shaft guard design, and a host of other concerns over which—not being the manufacturer—you have little or no control.

27-12. Economics

Failure to select an appropriate pump. Thoroughly investigate a pump's efficiency, ease of maintenance, energy costs, and compatibility with other pumps in the system.

Reliance on first cost for bid selection. A penalty/reward specification can be added in a bid clause to include the present worth of anticipated energy use (see Chapter 29). Pump specifications can be written for low maintenance (see Chapters 11, 12, and 16).

Obsession with economy. How well the station works and how easily and conveniently it can be maintained is vital. Two or three points of efficiency or a 4 to 5% savings in construction cost is trivial in comparison with a design that ensures a maximum of convenience and a minimum of maintenance and repairs.

27-13. The Future and Remodeling

Future expansion not considered. Design the station to allow for an increased discharge by (1) specifying pumps that can accommodate larger impellers, (2) adding another intake with a blind flange, or (3) choosing oversized piping (especially on the suction side) so that pumps and motors can be replaced with larger ones. Fasten a lifting eye on large blind flanges because they are otherwise difficult to handle.

Develop a strategy for adding or changing pumps and piping so that the station need not be shut down for the change. A valve with a blind flange is one way, but valves not exercised will freeze and, thus, be worse than useless. The strategy must be safe and sure and should be written into both the specifications and the O&M manual.

Reliance on record drawings and specifications. Never trust record drawings. Always make an accurate field survey (or engage an expert to do it) to determine the *actual* static head, TDH, flowrate, pipe roughness coefficient, possibility of obstructions in the force main, wire horsepower, exact locations of pipe and flanges, and any other measurements that may affect the new plans, before beginning the design.

27-14. Illustrative Examples

Schematic diagrams of two pumping stations, each containing at least 10 blunders, are shown in Figures 27-1 and 27-2. Many of these blunders are discussed in this chapter, but others are found elsewhere. Sharpen your wits by finding them before turning to the answers in Appendix D.

27-15. Design Reviews

A design review strategy should be imposed prior to beginning the design. Ideally, review teams should be made up of the personnel listed in Table 27-1.

For a typical large project, reviews might well take place on the following schedule:

- Predesign review
- Intermediate reviews (at 10, 50, and 90% completion plus a field check)
- Regulatory agency review
- Postconstruction review.

For smaller projects, the reviews might be made less often (or at different times), but project leaders should realize that reviews by those not directly associated with the detailed design are vital if blunders, errors, and problems of inadequate design are to be minimized. The cost of such reviews is small compared with the cost of field correction of mistakes, the value of the firm's reputation, and malpractice insurance premiums.

Appendix G contains checklists for the following disciplines:

- Civil design
- Structural/architectural design

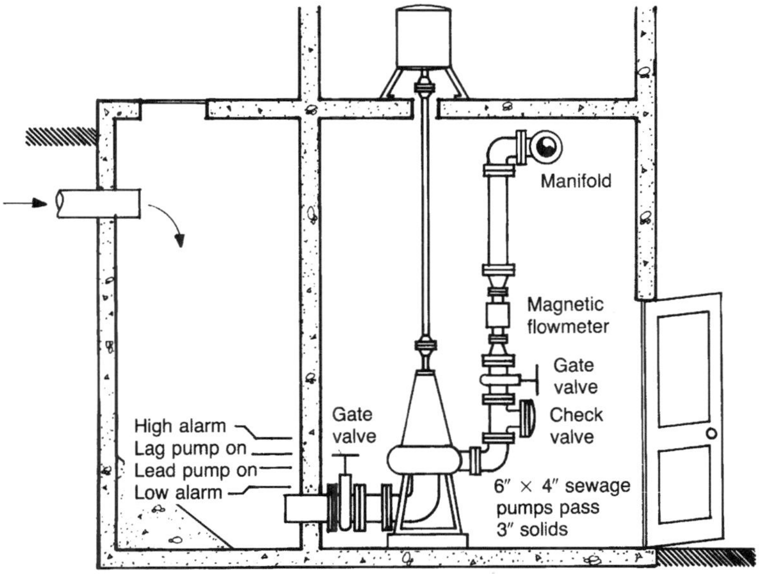

Figure 27-1. A composite of several sewage pumping stations. Find at least 10 blunders before looking at the answers in Appendix D.

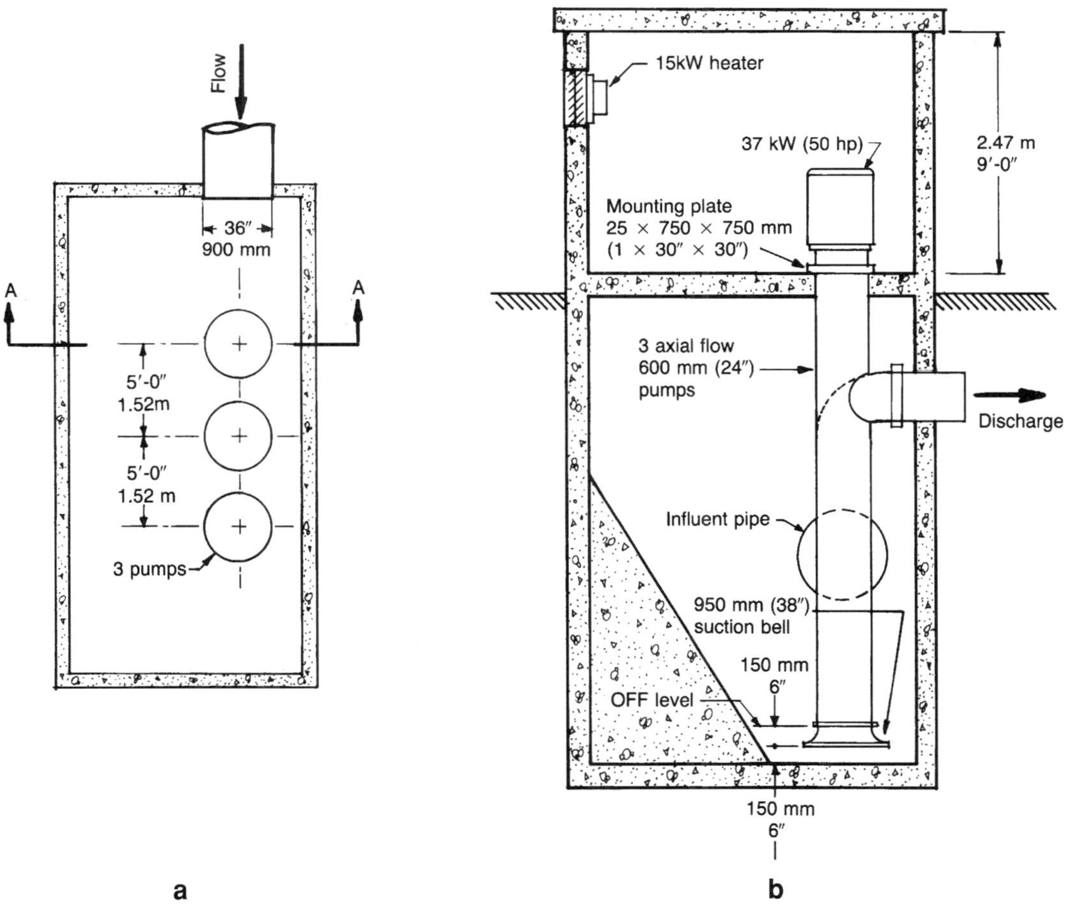

Figure 27-2. An actual low head sewage pumping station. The station has one 60-kW standby generator. (a) Plan; (b) section A–A. Find 10 blunders before looking at the answers in Appendix D.

- Electrical design
- Instrumentation and control
- Cross-connection control, especially for wastewater plants
- Mechanical design.

27-16. Operations

O&M manual not supplied. Always supply an O&M manual. If the owner objects to paying for it, do your best to overcome the objection by explaining it is a savings—not an expense. No pumping station should be without an O&M manual in which inherent pitfalls and dangers are identified clearly, and how to get the best from the equipment is stated in simple language. State the design assumptions and considerations, especially with respect to meeting future flows (by changing impellers or adding pumps). Put schematics on one page for the operator's use. Avoid engineering jargon in the O&M manual and use simple, clear English with (if necessary) sketches for clarity. Furnish several copies of the O&M manual and at least one copy of record plans and specifications. Furnish a weekly/monthly inspection form developed in cooperation with the maintenance supervisor. During the design, write portions of the O&M manual to clarify your thinking and to produce a better, more easily maintained pumping station.

Operating pumps in flooded station. Operating pumps when submerged is hard on seals and bearings and may ruin the motor. Warn operators in the O&M manual to dewater first.

Entry into confined spaces. In the O&M manual, state the safety precautions for entry into wastewater wet wells and other hazardous confined spaces. The precautions include pretesting for explosive gases before removing manhole covers, using portable forced-air ventilation, testing for toxic gases and oxygen depletion before entry, using a safety harness hooked to a winch, and using trained support personnel.

Pumping scum with main pumps. Give proper directions in the O&M manual for the removal of scum with main pumps; operate all pumps to draw water surface down to suction inlets quickly where the low submergence causes vortexing and draws scum into pumps. As soon as air is drawn into the pumps (which is signaled by an increase in vibration), stop the pumps and make sure that the switch is returned to automatic and that the pumps are reprimed. Operation of submersible motors with water level below the motor for more than a few moments may overheat and damage the motor. Verify the allowable time for out-of-water operation with the manufacturer. If it is necessary to remove scum more frequently than once a week, investigate the source of the scum and grease and have it controlled or use other methods to pump scum from the wet well.

Warnings buried in the O&M manual. Prominently display a summary of hazards and warnings at the beginning of the manual. Give safe operating procedures for emergencies such as chlorine leaks, fire, flooding, power outages, and injury to workers.

Disassembly not given in the O&M manual. Furnish a strategy for the disassembly of piping and the removal and replacement of machinery.

Table 27-1. Personnel on the Review Team

Small pumping stations	Large pumping stations
Project leader	Project leader
Engineer from each discipline[a]	Lead engineer for each discipline[a]
Owner's representative	Project and/or technical consultants with extensive experience in each major discipline involved
At 50% completion of detailed design, an engineer with broad and extensive experience (in- house if available, or an independent consultant if not)	Owner's representative
	Owner's operating personnel
	Regulatory agency personnel

[a] See the listing of disciplines at the beginning of Chapter 25.

27-17. References

1. Ten-State Standards. *Recommended Standards for Sewage Works,* Great Lakes–Upper Mississippi River Board of Sanitary Engineers, Health Education Service, Inc., Albany, NY (updated periodically).
2. Prosser, M. J. "The hydraulic design of pump sumps and intakes." British Hydromechanics Research Association (BHRA)/Construction Industry Research and Information Association, Cranfield, Bedford, United Kingdom (July 1977).
3. Dicmas, J. L. *Vertical Turbine, Mixed Flow, & Propeller Pumps.* McGraw-Hill, New York (1987).
4. ANSI/HI 1.1-1.5-1994, "Centrifugal Pumps for Nomenclature, Definitions, Application and Operation." Hydraulic Institute, Parsippany, NJ (1994).

Chapter 28

Contract Documents

JOHN E. CONNELL

CONTRIBUTORS
David L. Eisenhauer
Earle C. Smith

This chapter contains a brief description of what is included in the project documents, how they should be prepared, and a recommended format to be used. The information given here in conjunction with the references should be sufficient for guidance and assistance in the development of clear, concise contract documents.

28-1. General

Project specifications constitute the formal written agreement between the owner, who is paying to construct a facility, and the contractor, who has agreed to construct the facility. The legal relationship between the primary parties to the contract and the relationship of third parties such as the engineer and the construction observer are defined in the written agreement. For the engineer, the term "specifications" is apt to indicate the detailed description of materials, products, performance, quality, and testing of those components to be incorporated in the project. But specifications also describe the entire bidding procedure and, thus, serve a function prior to the existence of a formal agreement.

Purpose

The purpose of the project specifications can be divided into three parts as follows:

- Bidding documents
- Contract documents
- Technical specifications.

Thus, project specifications take on a much more general role than the technical description that accompanies the drawings.

The specifications are the official or legal communication between the owner and the contractor. They must, however, address others in addition to the contractor. The specifications first address bidders. These bidders are the prime contractors, the subcontractors, and the suppliers of products and materials. The contract agreement addresses the successful bidder—the prime contractor. The technical specifications address the prime contractor (the party legally responsible for their execution) and, in practice, the suppliers and manufacturers. Addressing so many parties presents one of the difficulties in communicating the technical part of the specifications. A specification for a pump, for example, clearly addresses all bidders, the prime contractor, and the successful subcontractor who will supply the product. Furthermore, the specification must ensure both the performance of suppliers and the performance of the product through the prime contractor, who is the legally responsible party.

The specifications may be outlined as follows:

I. *Bidding*. Addresses prospective contractors, subcontractors, and suppliers.

   a. Advertisement for bids
   b. Information for bidders

c. Bid form
d. Bid bond

II. *Contract.* Addresses the prime contractor.

a. Agreement form
b. Payment bond
c. Performance bond
d. Notice of award
e. Notice to proceed
f. Change order form
g. General conditions
h. Special conditions
i. Addenda

III. *Technical specifications.* Addresses the contractor directly and addresses the suppliers and manufacturers through the contractor.

Specification Language

The specifications are a complex and extensive set of instructions that direct the contractor. The writing style must reflect this objective by using clear, positive statements. The indicative mood is most effective in giving this direction to the contractor. Language that only suggests or implies an action is likely to cause misunderstandings or misinterpretations of the true intent of the statement. Engineers should be precise in specifying what is required in the manner of materials of construction and workmanship. Phrases such as the following should not be used in plans and specifications:

- *Suitable*
- *Stainless steel* (without stating the AISI grade)
- *Steel angle* (without stating the ASTM A 6 designation)
- *As required*
- *As necessary*
- *Appropriate*
- *In a workmanlike manner*
- *Of the best quality*
- *Heavy duty*
- *Industrial grade*
- *Commercial grade.*

For accurately describing what is to be provided in the contract documents, the importance of careful reading before referencing current standards cannot be overemphasized.

Format

Choose a format for the specifications that is clear, logical, and consistent from section to section. The format for some jobs is established by the owner. Federal and state agencies often require their format be used in writing specifications. Many consulting companies have their own specification format, which uses a standard approach and is set up on a wordprocessing system. There are also professional societies, such as the Construction Specification Institute (CSI) [1], that support a specific style and format (exemplified in Appendix C) or system for writing specifications. These societies have large libraries of standard specifications for bidding, contract agreements, and technical specifications.

28-2. Contractual or Legal Documents

Although the contract documents include the entire set of drawings and specifications, it is common to think of documents such as the bid, agreement, and general and special conditions as the actual contract documents. These documents are used to set out the basis for the owner–contractor relationship in a binding contract agreement. The documents listed in Section 28-1 under "Bidding" and "Contract" are generally considered the contractual or legal documents. They must be coordinated so that they do not conflict, and they must be properly referenced.

Special articles and paragraphs must be added to the bidding and contract documents for projects funded by the U.S. Environmental Protection Agency (EPA) to satisfy agency requirements. Become familiar with agency requirements and include the additional documentation when required.

Standard bid documents, agreements, general conditions, and other documents are published by various professional engineering societies. Perhaps the most useful to engineers are the documents prepared by the Engineers' Joint Contract Documents Committee: (EJCDC) [2] and issued and published jointly by the National Society of Professional Engineers (NSPE), the American Consulting Engineers Council (ACEC), the American Society of Civil Engineers (ASCE), and the Construction Specification Institute (CSI). The use of such documents is beneficial because contractors, engineers, and owners become familiar with their contents. Owners such as state agencies, federal agencies, cities, or industries may require their own set of documents. Standard forms and documents recommended by the American Institute of Architects (AIA) [3] are used for most private building construction. Engineering firms often develop their own standard documents. There should always be a review by an attorney to ensure that the content is appropriate for the contract.

Advertisement for Bids

The advertisement for bid (also known as the "invitation to bid") should be concise and must contain

- The owner's name and address
- A brief statement of work
- The time, place, and method of placing bids
- The locations where bid documents may be obtained and information on plan deposit can be found
- The amount and type of bid surety
- Special qualifications that may be required of bidders

This document is usually published in the construction or trade journals in the region where the project is to be constructed. Government projects usually require a legal advertisement in a local paper, and the advertisement for such bids must be written to comply with these requirements.

Information for Bidders

Information for bidders (also known as "instructions to bidders") is a summary of the bidding procedure and the requirements of the bid. It is more detailed than the advertisement and contains information on preparing and submitting the bid, on bid bonds, on performance and payment bonds, and on the methods of evaluating responsive bids as well as a schedule and procedure for awarding the contract and proceeding with the work. Available information on existing conditions (such as buildings and soil information) and special requirements (such as license requirements or other regulations affecting the contractor) are often mentioned in the information for bidders. Care must be taken in preparing this and all other documents so that they do not conflict with other parts of the specifications. To reduce the chance of a conflict, do not repeat items that are addressed in detail elsewhere in the documents. Examples of information for bidders are given in EJCDC and AIA documents [2, 3].

Bid Form

The bid form contains elements that are the essence of the agreement. These elements are the agreed compensation for the work described, the time for completion of the work, and the agreed damages for failure to complete the work on time. All prices, whether for a lump sum or a unit-price contract, are given in words and figures. If there is a discrepancy, the amount in words governs.

The time of completion and damages are set by the owner. The completion time must be reasonable for the work, and the damages must also have a reasonable basis. Unreasonable damages will likely not be awarded if challenged in court by the contractor. Reasonable damages may include additional engineering fees, administrative costs, loss of revenue to be generated by the project, and fines or penalties assessed by another authority (such as fines for a violation of the National Pollution Discharge Elimination Permit).

A bid bond is normally attached to the bid form. Other attachments often include bidders' prequalification data and a list of subcontractors. Examples of such attachments are included in the EJCDC and AIA documents [2, 3].

Bid Bond

The bid bond may be issued by a bonding company; the alternative of presenting a certified check in the amount of the bid bond is sometimes allowed. The purpose of this bond is to protect the owner if an apparent low bidder defaults and refuses to enter into a contract agreement. The bid bond is usually 10% of the bid price. The bonds for the three lowest bidders are held until a contract is awarded.

Contract Agreement

This brief document is the legal instrument used as the formal contract. The contracting parties, the time frame of the work, and the amount of compensation are identified or established in the agreement, which also incorporates all of the contract documents by reference (see the EJCDC and AIA documents [2, 3]).

Performance and Payment Bond

These bonds are provided by the contractor from a surety company in the amount of 100% of the contract price for each bond. These bonds protect the owner if the contractor fails to complete the work properly or if the contractor fails to pay those who worked on the project. Examples are listed in the EJCDC documents [2].

Notice of Award

The selected bidder (usually the low bidder) is notified in writing that his bid was accepted, and the contractor is given a specified time to present the necessary bonds, insurance, and executed contract agreement to the owner. An example of the notice of award is given in the EJCDC documents [2].

Notice to Proceed

Upon review and acceptance of the contractor's bonds, insurance, and agreement, the owner executes the agreement and provides written notice for the contractor to begin work within a specified time and to complete the work within the agreed completion time (see the example in the EJCDC documents [2]).

Change Order Form

Changes in the work, completion time, and contract price all require a "change order" to the contract. A specific form for this purpose is sometimes used as part of the formal contract documents. It is typical to add attachments in which any changes are described in detail. An example of a change of order form is contained in the EJCDC documents [2].

General Conditions

The general conditions are the focal point of the contractual documents. Documents and terms are defined. The authority of the owner and the relationship of the engineer and construction observer are outlined as well as the duties and obligations of the contractor. ("Construction observer" or "resident project representative" are the preferred terms because the words "inspector" or "supervisor" may communicate an unintended meaning—and liability—to the courts.) Method of payment, insurance requirements, project completion, guarantee of the work, and methods of resolving differences are examples of the content of the general conditions. The standard general conditions as published by the professional societies are commonly used in specifications (see examples in the EJCDC and AIA documents [2, 3]). All other documents must be consistent with the content of these general conditions.

Special Conditions

Special conditions are used to add, expand, or alter the general conditions. Special conditions are normally written for a project and incorporate requirements that are specific to the project. When the owner requires his or her own special conditions be used in the documents, it may be necessary to have an attorney review both the general conditions and the special conditions to avoid conflicts. When altering a set of standard general conditions, make it clear that it is a change and reference the section to be altered. Avoid intermixing technical specifications with the special conditions. Also avoid the tendency to repeat special conditions from previous project specifications—a common source of conflicting requirements or unnecessary specifications.

Addenda

Addenda are either written or graphic instruments issued prior to the bid opening to clarify, revise, add to, or delete from the original bidding documents or previous addenda. Like change orders, addenda are often prepared in a special format that later becomes part of the formal contract documents.

28-3. Technical Specifications

Technical specifications provide a detailed description of the scope of work, type and quality of materials, performance of equipment and systems, and the level of workmanship expected of the contractor. The drawings, which are also part of the contract documents, must be coordinated with the technical specifications. Drawings illustrate construction and provide a graphic means of showing the work to be done. Specifications should supplement, but not repeat, the information shown on the drawings. Unless otherwise stated, specifications normally take precedence over the drawings, and if a conflict arises the specifications govern.

The greatest difficulty in writing technical specifications is to decide when a given item is adequately described. Obviously, each and every move required by workers in carrying out the work cannot be described. The underlying presumption is that contractors execute work at a level that is consistent with their particular trade. The specifications are not to instruct carpenters, pipefitters, plumbers, or electricians in their trades. Industry standards of practice, however, are frequently referenced as part of the technical specifications. Building codes, electrical codes, plumbing codes, and so on are typical of such standards of performance.

Pitfalls: Who Should Write and Coordinate the Specifications?

Obviously, one person cannot write all the technical specification required in a pumping station project. Such projects are complex—sometimes *very* complex—and several disciplines are involved, among them civil, structural, architectural, electrical, instrumentation, and so forth. Various specialties, such as acoustics, vibration, model study, and so on, are sometimes needed. All these disciplines or specialties (and more) may be involved in preparing technical specifications for a project, and hence, various parts of the specification must be written by those experienced in such specialties.

A key issue is: who should coordinate and review the technical specialists' work? Uncoordinated specifications can result in problems ranging from duplicate—and different—specifications for a topic to a subject not covered at all (because the specialists assumed someone else was writing the specification).

Usually, the project manager should be responsible for the review and coordination of the technical specifications. However, one person physically cannot manage both the nontechnical and technical aspects of very large, complex projects. Considerable time is required to examine and coordinate the design drawings and specifications page by page. The task requires someone with experience in all aspects of design or at least with enough experience to know where the interdisciplinary problems usually occur and to know how to resolve the problems. Consequently, a technical assistant project manager or chief project engineer may be required to review and coordinate plans and specifications continuously during the project and to perform final review at the end of the design period. Some large organizations have a department of specification writers who become very skilled at this activity.

Problems such as the following are frequently found and must be resolved:

- Geotechnical specialists providing standard or "canned" specifications that do not match the recommendations of the geotechnical report.
- Mechanical and civil (sometimes called "process") specifications for valves, pipe hanger spacings, pipe pressure ratings, pressure test requirements, etc. that are inconsistent.
- Electrical wiring that does not match the requirements of the instruments.
- Materials for corrosion control that are inconsistent, e.g., stainless steel specified by one discipline while galvanized steel is specified in the same area by another discipline.
- Specifications for packaged equipment (e.g., pumps, compressors) that are not coordinated with instrumentation and control specifications and P&IDs.

The inexperienced pumping station designer may think such coordination is simple and easily resolved by a day-long conference of each of the discipline leaders—a naive supposition.

Addressees

Technical specifications must address a wide range of parties. Although the contractor is the primary audience, the technical specifications must also address bidders, subcontractors, suppliers, manufacturers, workers, and the construction observer.

Each technical section is used for:

- Bidding
- Product and material submittals, standard of quality, and acceptance
- Product description, materials of construction, and performance
- Product installation, start-up, and functional demonstration.

However, the design engineer writes the specifications to the contractor, who will sign the agreement with the owner. Subcontractors are not legal parties to the agreement. Consequently, the design engineer should not write specifications stating things such as "the concrete contractor shall . . ." or "the mechanical contractor shall. . . ." A phrase to be used carefully is "by others." The intent of the engineer, for example, may be to tell the concrete subcontractor that he does not provide the monorail support beams. But, because the contract documents are addressed to the general contractor—the entity with whom the owner has the legal relationship—the actual effect is to tell the general contractor that he does not have to provide monorail support beams at all. The phrase "by others" should only be used to denote equipment or work that is being done under another set of contract documents.

Most projects involve a wide variety of trades, products, and materials that are needed to complete a pumping station. Such complexity requires a systematic approach to writing technical specifications. The systems used may be developed by an engineering firm, by the owners (such as federal specifications), or by technical or professional societies.

The system or approach should not be confused with specification content. The system is the method that is used to break the complex project into meaningful sections or parts. These parts may then be described in detail, and their relationship to other parts may also be described. An excellent system of logical divisions, which is widely accepted in architectural and engineering projects, has been developed by the AIA [4]. A similar approach is contained in the CSI literature [5], which includes a more detailed breakdown into three parts: (1) general, (2) products, and (3) execution. The content of each of these parts is further described in the CSI format for writing specifications.

28-4. Source Material

Numerous sources of information are available for specifying an element of work or a product to be incorporated in the work. In general these sources are as follows:

- Manufacturers or suppliers of products and materials
- Regulatory requirements (such as building and plumbing codes) or owner-required specifications (such as federal or military specifications)
- Professional or trade organizations that have set standards for material composition, performance, and product standards.

As a designer or specification writer, you will encounter numerous representatives of equipment and materials. Most representatives will provide typical specifications for their products. Although such information is very useful for keeping informed on the current competitive market, be cautious when using a representative's standard specifications. The product may meet the regulatory requirements and the trade standards, but it may also contain elements placing it in an unnecessarily favorable bidding position that excludes other acceptable products. Review several such specifications and then edit or rewrite the specifications to ensure two or more sources that would be acceptable for the application. Avoid specifying elements that are not of standard manufacture unless there is a special reason for such a choice. Remember, special items may be difficult to maintain or replace.

Regulatory requirements affecting the work should be referenced in appropriate specification sections. Although the various trades working on a project may be required by law to perform work under a given code, it is best to state this fact in the general portion of the specification. Do not attempt to repeat or paraphrase such codes because that could lead to conflicts and misinterpretations.

There are numerous professional or trade organizations that have developed product and material standards and quality tests. The given trade or industry uses these standards as a means of self-regulation. Many have become national standards that are often recognized by regulatory agencies. The use of such standards is commonplace in technical specifications. Materials may often be specified adequately by simply referencing the appropriate trade standards. A more complex item, such as a valve, may be specified by reference to an appropriate AWWA specification. However, the reference alone may not be adequate because the referenced standard usually has selection options that must also be identified. Thus, such a reference must also identify the options allowed by the referenced standard.

The use of standards is extremely important. Unfortunately, the engineering industry in general and design engineers in particular are all too often increasingly unfamiliar with standards such as ASTM and ANSI, and the inexperienced engineer has, at best, a superficial knowledge of standards. Unfortunately, it is also common for the design or specifying engineer or architect to be unfamiliar with these referenced standards and, frequently, not to have read them at all. Many of these referenced standards, if blindly incorporated into a project contract document by reference, will drastically alter the relationship between the design engineer, the owner, and the general contractor. Other referenced standards may require the design engineer to take actions that may be unwanted. Therefore, read every part of a referenced standard carefully and thoughtfully and refer exactly to those portions that apply to the project (see Section 1-4).

Limitations of Published Standards

Published standards provide the specifier with convenient means for incorporating recognized benchmarks of quality into the detailed requirements for contractor or manufacturer performance. Most published standards are the product of countless hours of effort provided by volunteers interested in improving their industry. Today, after legal decisions have caused a re-examination of the process, most (but not all) standards-setting organizations use balanced committees (memberships representing manufacturers, users, and consultants or specifiers) to develop consensus documents. Once a document has been developed, it is then published for public comment. Public comments are then considered by the committee and the document is adjusted if necessary before final publication. Standards organizations usually have an oversight process to ensure the documents contain no biased requirements

These consensus documents *usually* provide the specifier with sound advice on the basic requirements for a product and options for enhanced quality or for alternative features appropriate for special applications. Many standards contain options for quality assurance reporting and for user-nominated requirements for construction options. No standard, however, is perfect. It is unlikely that any standard will be entirely applicable to a given application without some modification. Consequently, it is incumbent upon the specifier to read and understand every standard completely with a watchful eye for any deficiencies such as omissions, poor or weak practice, or inconsistency with other standards. The following shortcomings are a few examples taken from commonly used standards.

- *Omission.* Omission of requirements for surface preparation for coatings, film thickness testing, or frequency of testing.
- *Poor practice.* Allowing threaded joints in Schedule 30 pipe.

- *Inconsistency.* Referencing a specification at odds with the main body of the standard.
- *Weak practice.* Allowing excessive pipe hanger spacing that results in excessive pipe sag.

Many other examples can be cited, but the message is clear. *Referenced standards must be read completely and with care (1) to discover and correct omission, contradictions, weaknesses and (2) especially to ensure conformance with the objectives of each project.*

A list of organizations publishing standard references that are commonly used in specifications is given in Section F-l. Every specification writer should have access to the *ASTM Standards* [6] (in 66 volumes, revised annually), the *AWWA Standards* [7], and the AIA *MASTERSPEC* [8] (as well as the other references listed in Section 28-7).

28-5. Specifying Quality

Five basic methods are used to specify the quality of products or materials:

- Performance
- Nonrestrictive (also "or equal")
- Proprietary
- Generic
- Reference standards.

Occasionally, modifications are made by combining two or more methods (see also Table 16-1).

Performance Specification

The performance specifications describe the functional result that a product must achieve without identifying any specific material or product. This approach usually places the burden of some design work on the contractor. Acceptance of the product may not be ensured until it has been demonstrated that it can perform the intended function. The cost for such a specification may be high because the contractor must include design costs as well as costs to cover unforeseen problems should the products fail to perform as specified. Performance specifications are more typically used in conjunction with other methods of specifying a product.

Nonrestrictive (or Equal)

The competitive bidding requirements of most governmental agencies have resulted in an approach to specifications in which the writer identifies by name two or more products that meet the requirements of the design. These names are followed by the words "or equal." If a contractor submits a product from another manufacturer, the item is evaluated against those named in the specification. The engineer must then determine whether the item is "equal." Such a determination is somewhat subjective because the items are not expected to be *identical*. Thus, the "or equal" specification is usually combined with a certain amount of performance and product description information (such as a listing of the salient features of each unit with particular regard for maintenance, life, and efficiency) that aids both bidders and engineers in making the "or equal" decision and settling disputes. The evaluation should ensure that all specified performance and quality objectives are being satisfied. This approach has become one of the most common methods of specifying the more complex products.

Federal requirements for naming two products are eased somewhat by Public Law 97-117, which requires naming only one product or equal. However, some EPA regions and some states and local governments still require naming two or more products, so investigate the local requirements.

Proprietary

In proprietary specifications, the product is defined by name. However, only one product is allowed. Obviously, there is no competitive bid in this approach. When a design requires the use of a single source, care must be taken to establish a reasonable price for the product prior to bid. The bid form is set up in a manner that discloses the value of the proprietary item, thereby allowing the owner to check prices and ensure that the supplier did not take unfair advantage of the noncompetitive situation.

Proprietary specifications are banned under laws and regulations governing public works projects for most federal, state, and local government agencies. However, under Public Law 97-117, Section 204(a)(6) of the Federal Water Pollution Control Act was amended to permit the grantee to use single "brand name or equal" specifications. But be aware of the following restrictions:

- PL 97-117 applies exclusively to projects financed under the Federal Water Pollution Control Act (PL 92-500).
- The permission to use a single brand name may be superseded by the provisions of state and local laws.
- To qualify, the grantee (owner) must agree that a single-source specification is acceptable "when [in the words of the law] in the judgment of the grantee, it is impractical or uneconomical to make a

clear and accurate description of the technical requirements. . . ."

In most instances, an engineer would be hard pressed to justify an exclusive specification for the equipment commonly found in municipal water and wastewater pumping stations.

Generic (or Descriptive)

The generic specification requires the product be described in sufficient detail using, for example, performance, materials, quality control, and tolerances without naming a product. To have a basis for the specification, the specification writer must have identified at least one or more products that will meet the specification. This type of specification is usually lengthy and difficult to write for complex products. It also presents problems in evaluating the product submitted for approval by the contractor.

Reference Standards

The use of reference or industry standards is one of the most common methods of specifying materials or products. As mentioned in Section 28-4, referring to a standard specification may be sufficient for single items. Reference standards for more complex items, such as valves, require more description than a simple reference because there are often options that must be defined in order to complete the referenced specification. Reference specifications tend to be limited to materials of construction and simple components. Such references, however, are used extensively to describe the quality of more complex items. Thus, references are used in conjunction with the previously described methods.

Be sure to read subsection "Limitations of Published Standards" in Section 28-4 before using reference standards.

28-6. Submittal Requirements

The specifications must maintain a certain level of quality and performance in addition to allowing competition so that the owner can get the best price. The determination of acceptable products and materials is often difficult and controversial. One method of handling this situation is to require the prequalification of manufacturers. Prequalification adds an extra step to the bidding process but provides a very good method of controlling the quality of the products. This approach also eliminates controversy during and after bidding.

The more common acceptance method is through the shop-drawing procedure. Prequalifications do not eliminate the shop-drawing step. Generally, the shop-drawing review is used to ensure that the details of the specifications are being followed. This review may also be used as a basis for rejecting the entire product as submitted by the contractor. Complete rejection of a product at the shop-drawing-review stage can create scheduling and financial problems for a contractor, particularly if the products are a major portion of the contract. The specification should identify products adequately so that contract bids are not based on unacceptable products that must be rejected.

In addition to shop-drawing reviews, other quality-control tests and certifications may be required. These controls represent another method of checking product quality. Use care in warranty requirements, because unreasonable requirements unnecessarily increase prices.

28-7. References

1. *CSI Document MP-I,* The Construction Specifications Institute, Alexandria, VA:
   Document MP-1-1, "Construction Documents and the Project Manual" (1996).
   Document MP-1-2, "Bidding Requirements" (1996).
   Document MP-1-3, "Types of Bidding and Contracts" (1996).
   Document MP-1-4, "The Agreement" (1996).
   Document MP-1-5, "Conditions of the Contract" (1996).
   Document MP-1-6, "Division 1, General Requirements" (1996).
   Document MP-1-7, "Relating Drawings and Specifications" (1996).
   Document MP-1-8, "Changes to Bidding and Contract Documents" (1996).
   Document MP-1-9, "Specification Writing and Production" (1996).
   Document MP-1-10, "Specification Language" (1996).
   Document MP-1-11, "Methods of Specifying" (1996).
   Document MP-1-12, "Performance Specifications" (1996).
   Document MP-1-13, "Procurement Specifying" (1996).
   Document MP-1-14, "Civil Engineering Applications" (1996).
   Document MP-1-15, "Mechanical and Electrical Engineering Applications" (1996).
   Document MP-1-16, "Preparation and Use of an Office Master Specification" (1996).
2. EJCDC, Engineers' Joint Contract Documents Committee, The Construction Specifications Institute (CSI), Alexandria, VA:
   Document No. 1910-8, "Standard General Conditions of the Construction Contract" (1996).
   Document No. 1910-8-A-1, "Standard Form of Agreement between Owner and Contractor on the Basis of a Stipulated Price" (1996).

Document No. 1910-8-A-2, "Standard Form of Agreement between Owner and Contractor on the Basis of Cost-Plus" (1996).
Document No. 1910-8-B, "Change Order Form" (1996).
Document No. 1910-12, "Standard Form of Instructions to Bidders" (1996).
Document No. 1910-18, "Bid Form" (1996).
Document No. 1910-22, "Notice of Award" (1996).
Document No. 1910-23, "Notice to Proceed" (1996).
Document No. 1910-28A, "Construction Performance Bond" (1996).
Document No. 1910-28B, "Construction Payment Bond" (1996).

3. AIA, The American Institute of Architects, Washington, DC:
Document A201, "Owner–Contractor Agreement Form, Stipulated Price" (1987).
Document A111, "Owner–Contractor Agreement Form, Cost Plus Fee" (1987).
Document A201, "General Conditions of the Contract for Construction" (1987).
Document A701, "Instructions to Bidders" (1987).
4. AIA. *Uniform Construction Index.* The American Institute of Architects, Washington, DC (latest edition).
5. CSI Documents MP-2-1. *Master Format, Master List of Section Titles and Numbers* (1995); MP-2-2 *Section Format* (1997). The Construction Specifications Institute, Alexandria, VA.
6. *ASTM Standards,* 66 vols. American Society for Testing and Materials, Philadelphia, PA (revised annually).
7. *AWWA Standards.* American Water Works Association, Denver, CO (latest edition).
8. AIA. *MASTERSPEC* (Vols. I–V). The American Institute of Architects, Washington, DC (updated periodically).

Chapter 29

Costs

STEFAN M. ABELIN
MARC T. PRITCHARD
ROBERT L. SANKS

CONTRIBUTORS
Kirk Blanchard
Robert H. Brotherton
James C. Dowell
Ronald W. Duncan
Howard N. Godat
Charles J. Jeckell
James L. Mohart
Michael C. Mulbarger
Donald Newton
Michael R. Olson
Carl W. Reh
William H. Richardson
Ronald Rosie
Larry R. Smith
LeRoy R. Taylor
William R. Taylor
Michael G. Thalhamer
Patricia A. Trager
Eric L. Winchester
John E. Wiskus
Alvord, Burdick & Howson
Black & Veatch, Engineers-Architects
Brown and Caldwell Consultants
Camp Dresser & McKee, Inc.
CH2M-Hill, Inc.
City of Virginia Beach
Flygt Corporation
Havens and Emerson, Inc.
King County (Washington) Dept. of Municipal Services
Raymond Vail and Associates
Wilson & Company, Engineers & Architects

The cost of a project must be estimated at several stages: (1) at the time the project is conceived and before any funds have been expended, (2) during the planning stage while the facility plan report is being prepared and the best two or three alternatives must be selected, (3) during the design period when the most cost-effective plan must be chosen, and (4) after completion of the plans for use as a basis for both informing the owner of probable cost and judging bids.

Construction cost curves based on data for several types of pumping stations obtained from the contributors to this chapter are presented for rough cost estimating. The construction costs are keyed to a cost index to avoid obsolescence and to facilitate estimating future costs. Formulas for dealing with interest, escalation, and inflation are given, and an example of the economic comparison of alternatives is worked in detail. The level of detail in this chapter is necessarily limited, so a review of the literature before undertaking complex cost analysis is recommended.

29-1. Cost Indexes

Unfortunately, much of the cost data reported in the literature is worthless because of incomplete descriptions or a failure to reference the cost data to a cost index. Some form of cost index must be used to account for inflation. Those of merit include the Engineering News-Record Construction Cost Index (ENRCCI) [1], the Handy–Whitman Index [2], several EPA cost indexes [3], and the Richardson Index [4]. The ENRCCI, which is the oldest, is regularly updated, easy to find, and in common use. If the extreme variations in the costs of pumping stations similar in size and type are considered, it seems futile to attempt to improve on the accuracy of the ENRCCI.

Engineering News-Record Cost Indexes

The ENRCCI, which begins with an index of 100 for the year 1913, is based on constant quantities of structural steel (weighted 15%), portland cement (2%), lumber (10%), and common labor (73%) in 20 cities. The average of these is considered to be the national average, and plots of yearly national averages are shown in Figure 29-1.

The Engineering News-Record Building Cost Index (ENRBCI) [1] was introduced in 1928 to include the impact of skilled labor, which is weighted 55%, with the remainder assigned as 25% for structural steel, 17% for lumber, and 3% for portland cement.

Both cost indexes are updated weekly in the "Market Trends" section of the *Engineering News-Record* (now called "ENR"). Historical curves of ENRCCI and ENRBCI and the materials component are shown in each mid-March "Quarterly Cost Report" along with forecasts to the end of the year. Indexes 12 months ahead are predicted in the mid-December "Quarterly Cost Report." Reprints of "Quarterly Cost Reports" are readily available [1].

Richardson Construction Cost Trend Reporter

As part of the Richardson Rapid Estimating System [4], the *Reporter*, which is published quarterly, contains the Davis–Bacon and union wage rates and escalation indexes for 15 categories of laborers in 120 U.S. and 7 Canadian cities. The *Reporter* contains many cost indexes, including the following:

- *Engineering News-Record:* ENRCCI and ENRBCI
- EPA: sewers and wastewater treatment plants
- Boeckh: residences, apartments, hotels, offices, commercial buildings, and factories
- Handy–Whitman: buildings and electric light and power
- Department of Commerce Composite Cost Index
- Bureau of Reclamation.

Prices of materials used in construction and interest rates for various types of loans or bonds are also included.

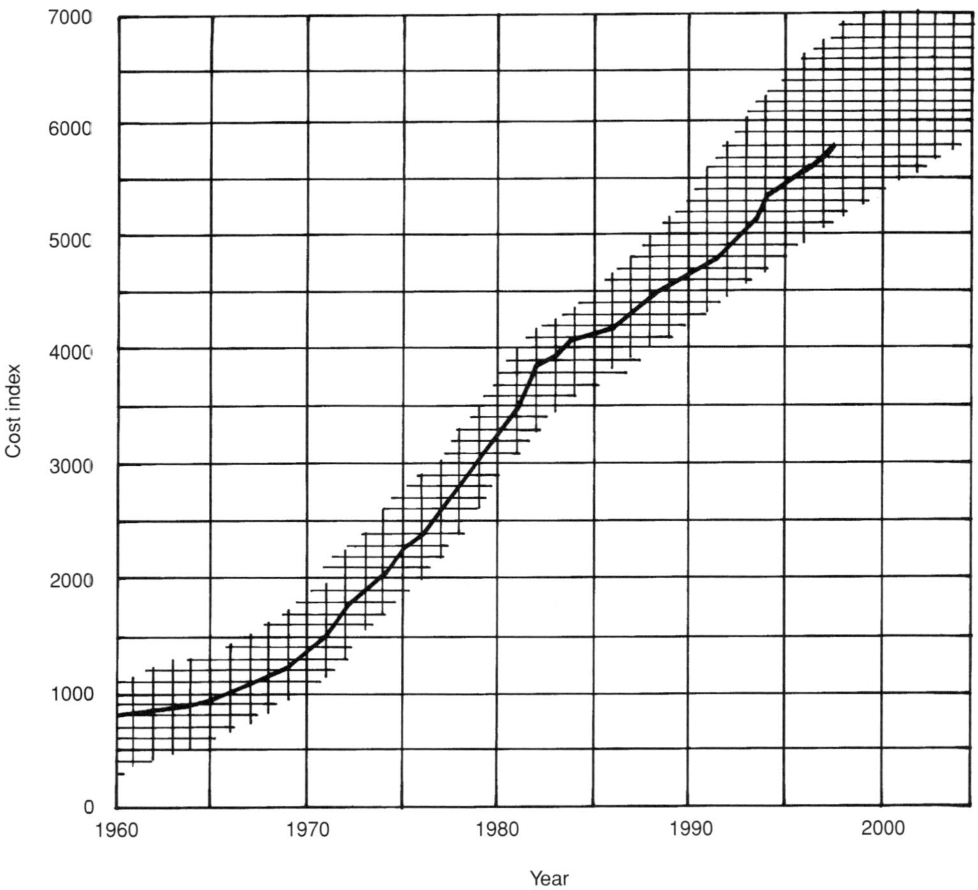

Figure 29-1. Inflation as measured by the ENRCCI.

Using the ENRCCI

To use the ENRCCI (from Figure 29-1), follow these steps:

- Estimate the date of construction and extrapolate the curve to that date.
- If desired, correct for the region of the site by choosing the ENRCCI ratio (F' in Equation 29-1) for the nearest city as compared to the national average.
- If the construction site is more than 30 mi from an urban area, labor cost increases about 10% for each additional half hour of driving time. Investigate local union rules for more accuracy. As labor is roughly half of the total cost of construction, add about 5% for each additional half hour of driving time beyond the 30-mi limit (F'' in Equation 29-1). If the site is in a congested area (such as in the center of a city), the cost index increases greatly due to traffic interference, limited work area, and underground and overhead interferences. The advice of experienced local contractors is helpful in estimating the impact on costs.
- Calculate the final cost index as follows:

$$C_f = C_p \frac{\text{ENRCCI}_f}{\text{ENRCCI}_p} \times F' \times F'' \qquad (29\text{-}1)$$

where the subscript f represents a future date, the subscript p represents a past date for which the construction cost is known, and F' and F'' are correction factors for the region and locale. The terms F' and F'' indicate a supposed accuracy that is entirely overshadowed by the construction conditions, the designer's concept of appropriate design, the amount of instrumentation, the addition of standby power, and, especially, the bidding climate factors that combine to make prices scatter far more than the effect of terms F' and F''.

29-2. Cost Curves

The data for the cost curves presented in this section were obtained from a survey of the contributors to this chapter. The cost data represent construction between 1966 and 1987, but more than 90% of the data are more recent than 1974. All costs were corrected to an ENRCCI of 4500 by means of Equation 29-1; the trivial terms, F' and F'', were omitted in consideration of the scatter that often exceeded 300%. Only data for "typical" pumping stations are included. Data for "atypical" pumping stations (such as those excessively deep or those built to unusual specifications) were either discarded or are explained in the text. All costs are contract prices for construction plus costs for extra work, and all are limited to the construction allocated only to the pumping station (and excluding costs of other works such as force mains and treatment plants). To obtain the total cost of a pumping station, charges for engineering and legal fees, land, administration, and interest during construction must be added.

No distinct pattern was found to explain the scatter—neither the number of pumps, variable- versus constant-speed drivers, the presence or absence of standby power, high versus low head, nor difficult foundation conditions. The factors discussed at the end of Section 29-1 account for the variations, and these factors are not quantifiable. Deep foundations, high head, standby power, and variable speed, however, tend to increase costs toward the upper limit lines in the figures.

The data are limited, and the cost envelopes in the following figures should be used with caution. In particular, it should not be assumed that the cost "curves" continue in a straight line beyond the limits of the data points. It seems logical to suppose that the cost curves become flatter as the size of the pumping station decreases below about 23 m^3/h (100 gal/min), although no data are presented here to support such a conclusion. Costs in Figures 29-2 to 29-9 are given in k$, thousands of dollars.

Wastewater Pumping

Pumping stations equipped with standby power are shown by open circles; those without standby power are indicated by smaller solid dots. Of stations for which complete data on standby power were given, 35% had diesel engine–generators, 30% had dual electricity sources, 15% had gas engine–generators, 10% had propane and diesel engine–generators, and 10% depended on portable engine–generators. The lines depicting upper and lower limits are merely estimates that indicate a likely range of probable costs.

Custom-Built Wet Well–Dry Well Stations

The numbers adjacent to plotted points are TDH in feet. Although there is no clear correlation between head and cost, data for pumping stations with a TDH above 21.3 m (70 ft) tend to lie along the upper limit of costs in Figure 29-2. Of the three points well above the upper limit, one is for a station (TDH = 47) that required automatic screens, odor control, and, during construction, blasting. Another is for a station (TDH = 55) that included comminutors, chlorination equipment, and provision for future telemetering. The third station (TDH = 38) had odor-control equipment and room to increase the number of pumps from four to six. The cluster of 25 points at the lower left is for one location, Virginia Beach, where nearly all of the pumping stations were built by developers to city requirements.

Submersible Pumps

The TDH for half of the stations is less than 15.2 m (50 ft). Data points for all but two stations with a TDH of more than 15.2 m (50 ft) lie close to or above the upper limit line shown in Figure 29-3.

Self-Priming Pumps

No data are shown for self-priming pumps at grade, but construction cost is slightly less than for submersible pumps. The maintenance of self-priming pumps is lower than that for submersible pumps because the equipment is at grade, the access is easy, the wet well can be sealed so that there are no fumes to corrode electrical equipment, and there is no need for ventilation other than louvers for cooling the motors [5].

Prefabricated Pumping Stations

Prefabricated pumping stations can be obtained in a number of types: (1) wet well–dry well with flooded suction—that is, the dry well is as deep as the wet well; (2) wet well–shallow dry well with suction lift; (3) wet well with a submerged pump connected to the motor with a long shaft; (4) wet well with suction lift

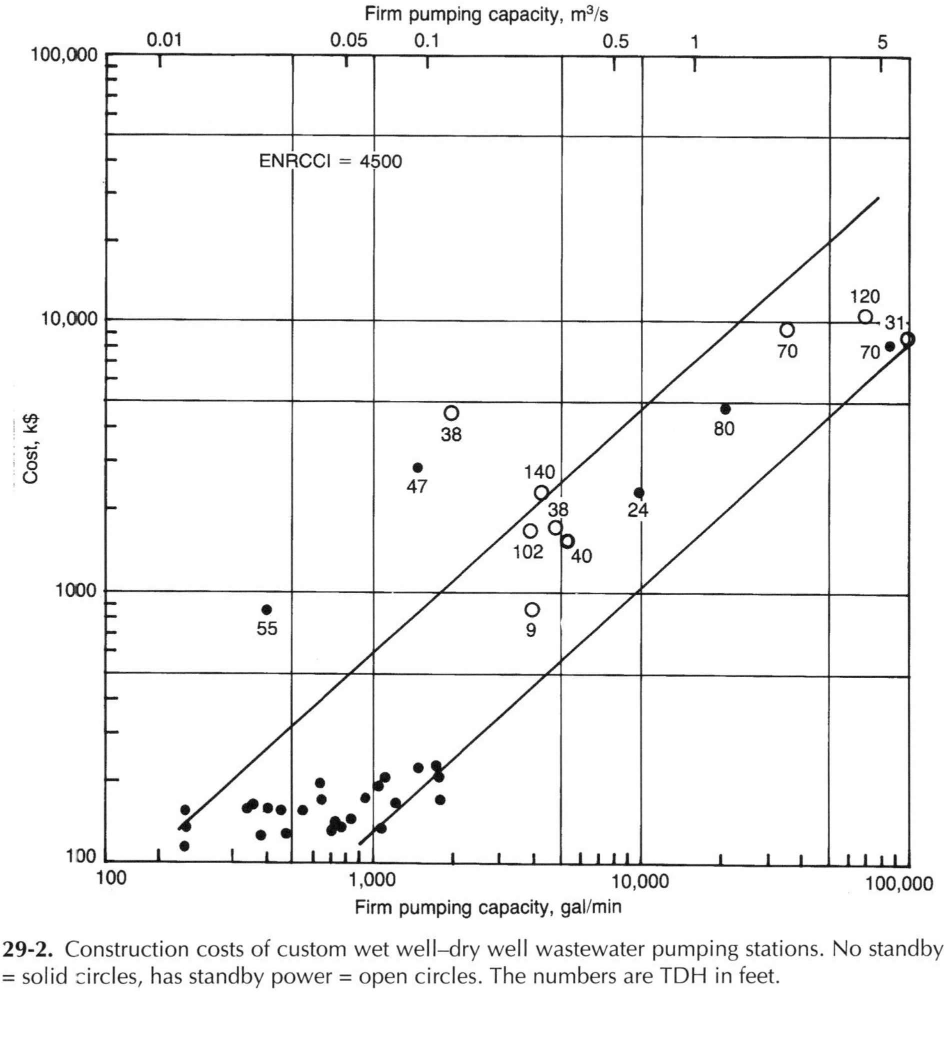

Figure 29-2. Construction costs of custom wet well–dry well wastewater pumping stations. No standby power = solid circles, has standby power = open circles. The numbers are TDH in feet.

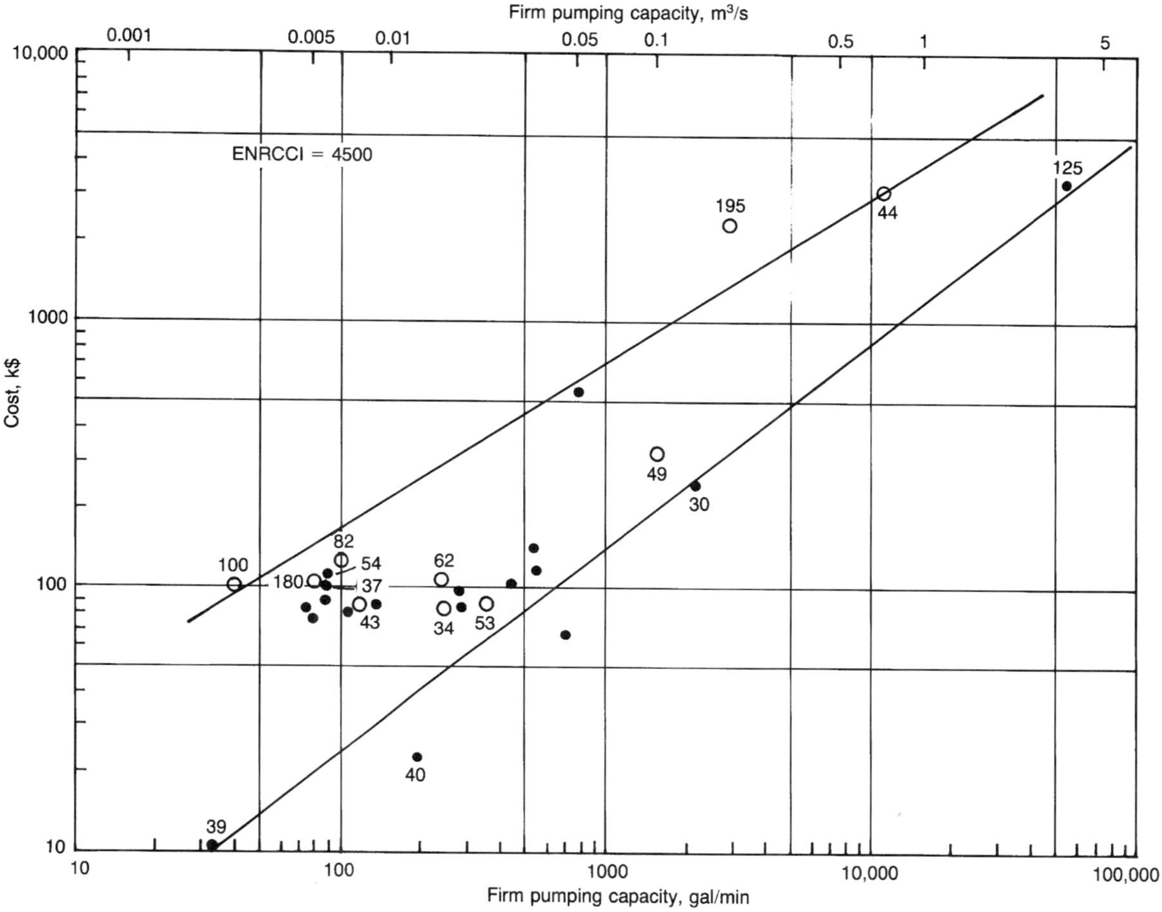

Figure 29-3. Construction costs of submersible-pump wastewater pumping stations. No standby power = solid circles, has standby power = open circles. The numbers are TDH in feet.

to self-priming pumps at grade; and (5) submersible pump and close-coupled (submersible) motor. The data for prefabricated pumping stations in Figure 29-4 are for the wet well–dry well type except for two, which utilize self-priming pumps, and one, which is an enclosed (Archimedes) screw pump. With one exception, the variation in heads does not affect the costs. The high cost of one pumping station is due to the inclusion of headworks (comminutor, Parshall flume, manual screen bypass). The Archimedes screw pump is included with prefabricated stations because it is an enclosed screw and, thus, "prefabricated" to a degree.

Water Pumping

In general, high speed (1800 rev/min) is common up to 150 kW (200 hp), and 1200 rev/min is usual for larger pumps.

Raw Water Pumping

All of the pumping stations shown in Figure 29-5 were constructed in lakes or on the banks of rivers where deep structures, sheet piling, or coffer dams were required. The high cost of pumping station no. 1 is attributed (1) partly to facilities designed for two pumps with space for four more and a dual electrical feeder for standby power and (2) partly to minimal bid competition and the construction of a lake intake. Pumping stations nos. 3 and 4 have traveling screens, no. 3 has a massive (3.3-m- or 11-ft-thick) foundation of tremie concrete, and no. 4 has prestressed rock anchors.

The TDH varied from 15.8 to 73 m (52 to 240 ft). The stations with the lowest relative cost (nos. 2 and 6) had nearly the highest and lowest heads, whereas the stations with highest relative costs (nos. 1, 3, 4, and 5) also had the lowest (no. 1) to the highest (no. 5) heads. Head is evidently a minor consideration in raw water pumping.

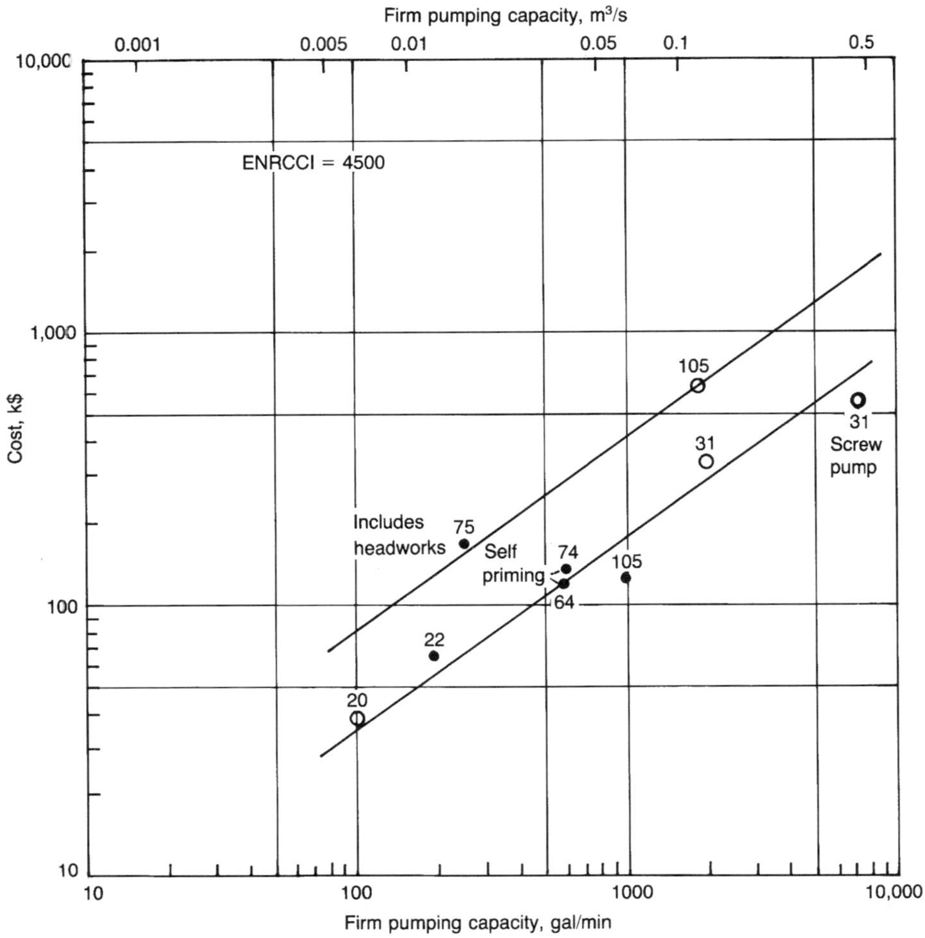

Figure 29-4. Construction costs of prefabricated wastewater pumping stations. No standby power = solid circles, has standby power = open circles. The numbers are TDH in feet.

Service Pumping

The TDHs for the finished water pumping stations shown in Figure 29-6, which vary from 43 to 114 m (140 to 375 ft), seem to have no correlation with cost. The station with the highest head was moderate in relative cost, and the station with the lowest head was the most expensive. The remaining data fit a similarly random pattern, and costs do not appear to correlate with other features such as standby power, ventilation, or foundation problems.

Booster Pumping

The TDHs for the booster pumping station costs in Figure 29-7 vary from 9.1 to 95 m (30 to 310 ft). Again, cost appears to be unrelated to either head or the inclusion of standby power. Seven pumping stations are in-line boosters, and five are distribution boosters. The type of booster also seems to have no effect on cost.

Well Pumping

The total costs of pumping stations, including the pumps and drilling and casing the well, are shown in Figure 29-8. Adequate data for the costs excluding the well were not available. Standby power in pumping station no. 5 consists of a natural gas engine–generator. Standby power for no. 3 is a direct-drive natural gas engine with a manual switchover. In general, any correlation between depth and cost is poor.

Summary

Median lines for each type of pumping station shown in Figures 29-2 through 29-8 are shown for comparison in Figure 29-9. If the upper and lower limits in Figures 29-2 through 29-8 were plotted, the costs for all of the pumping station types would overlap; a custom station, therefore, could be less expensive than a prefabricated or submersible station.

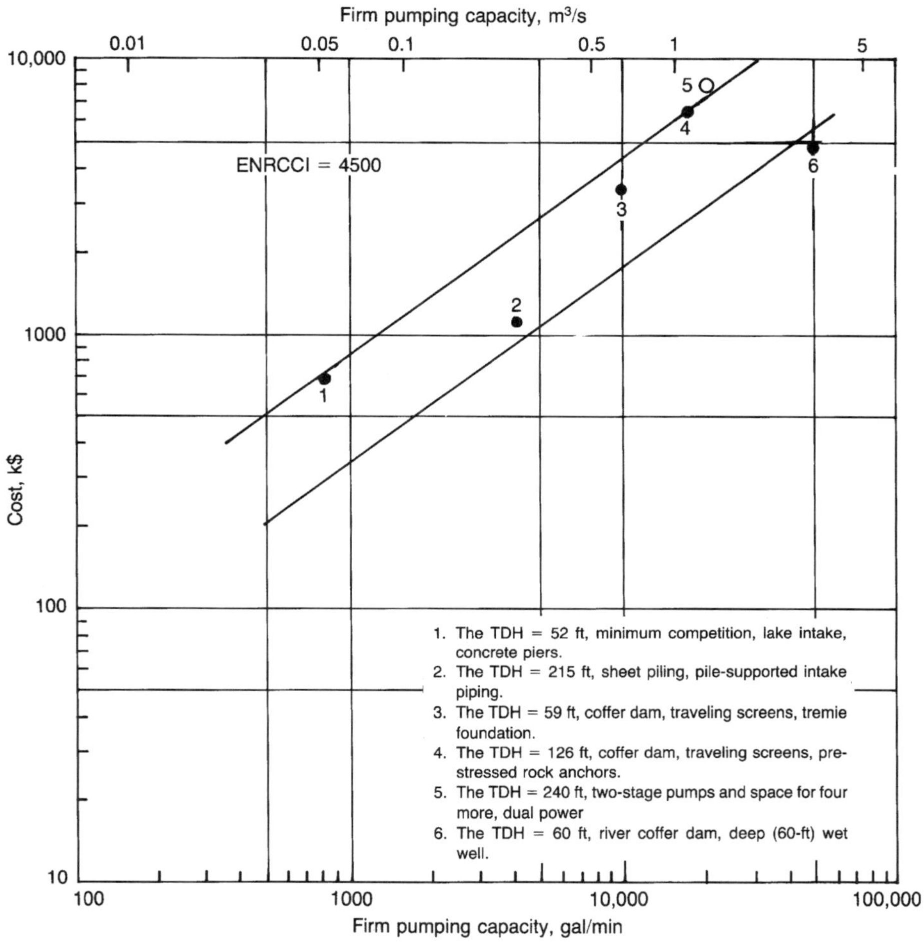

Figure 29-5. Construction costs of prefabricated wastewater pumping stations. No standby power = solid circles, has standby power = open circles. The numbers are TDH in feet.

If the cost for one size (capacity) of a pumping station is known, the cost for another size (of the same type) can be found by means of the "six-tenths factor." Simply multiply the known cost for one facility by the ratio of sizes raised to the exponent 0.6. For most of the plots of Figure 29-9, however, a more realistic exponent is 0.75. For water wells, the exponent is about 0.42.

29-3. Maintenance and Energy

The cost for maintenance is elusive partly because (1) it depends upon the owners' policies, which vary greatly from utility to utility (and also with time and personnel) and (2) so few owners have reliable, accurate records of maintenance labor and supplies for each pumping station. Because pumping stations differ considerably in maintenance requirements, records for several stations lumped together are of little use. The true cost of labor should include fringe benefits, supervision, costs for support (tools, vehicles, insurance, etc.), and the expenses associated with maintaining support facilities such as shops, clerks, records, etc. Costs of operating and maintenance labor, material, and supplies for wastewater pumping are shown as curves by Patterson and Banker [6], but these data should be used with caution because (1) labor time can easily be double the values shown and (2) there is no correlation with owners' policies or types of stations.

Maintenance of equipment is also equipment specific. For example, the maintenance for a wound-rotor motor with its brushes and slip rings is much greater than for a squirrel-cage motor. A self-priming pump can be repaired by the maintenance crew, but, to prevent voiding the contractor's guarantee, a submersible pump might have to be shipped to a service center with shipping costs added to the high (perhaps 30% of the original purchase price) cost of repair. Repairs made to

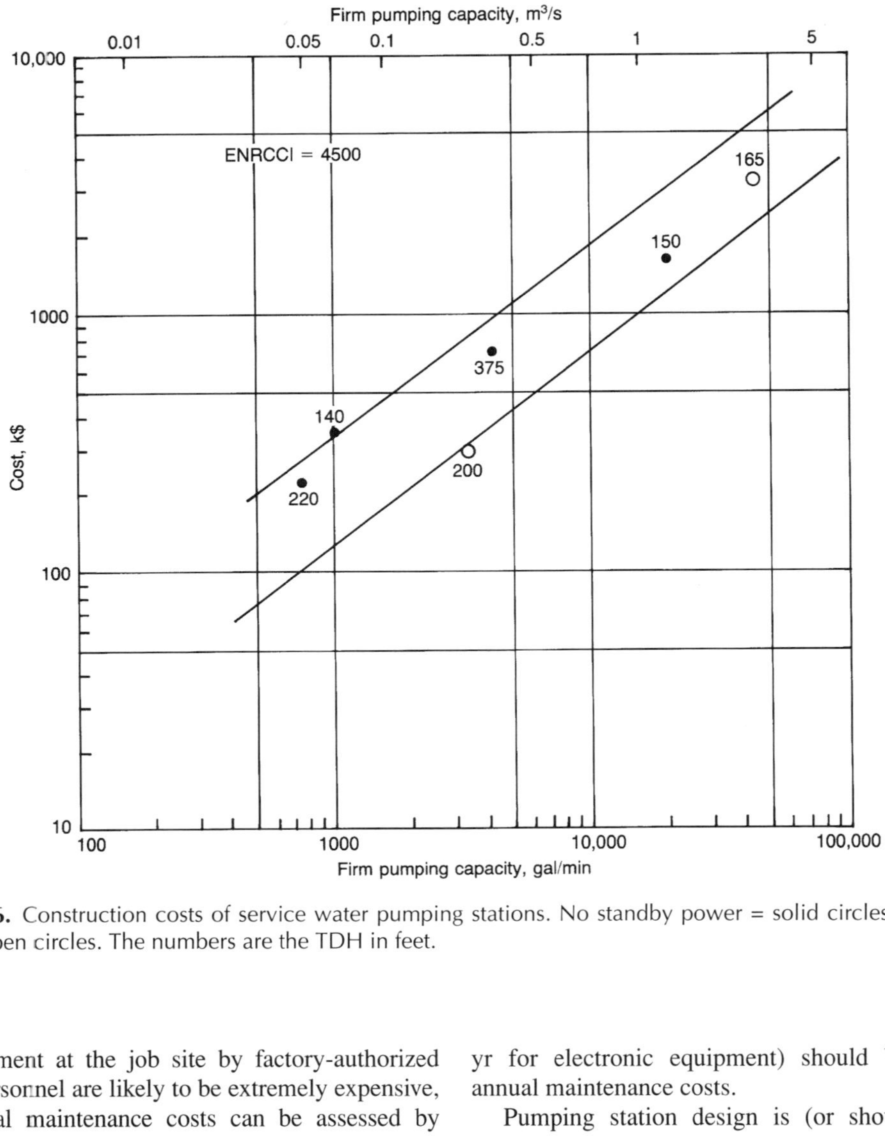

Figure 29-6. Construction costs of service water pumping stations. No standby power = solid circles, has standby power = open circles. The numbers are the TDH in feet.

any equipment at the job site by factory-authorized service personnel are likely to be extremely expensive, but rational maintenance costs can be assessed by investigating service contracts.

Excessive dependence on the effect of either size or purchase price in estimating maintenance costs is unwise. For example, the maintenance costs for a large pump might actually be less than for a small, less expensive pump because of better accessibility (and, hence, reduced labor time, e.g., for replacing packing) and because of the lower speed of the larger pump. Finally, two identical machines in similar circumstances may have, inexplicably, quite different maintenance needs and costs. However, to ignore maintenance costs because they are difficult to evaluate is to skew cost comparisons and decisions. To put it simply, investigate as necessary and do the best you can. Some equipment (such as solid-state electronic controls) become obsolete, so a sinking fund to replace such items in a reasonable period (say, 8 to 15 yr for electronic equipment) should be added to annual maintenance costs.

Pumping station design is (or should be) profoundly influenced by the cost of power. Dealing with energy costs can be very complex. Electric power rates vary with the classification of the user, the amount of power used, and, often, with the time of day, the season of the year, the power factor, and the power demand. The rates differ by as much as 500% or more depending on location. Furthermore, the escalation of cost may exceed, keep pace with, or even lag behind inflation. The costs of natural gas and diesel fuel are less complex and subject to fewer influences, but they still depend on location and politics. Electric power and gas utilities are the most reliable source of information on past and present costs of electricity and gas. Some utilities make careful predictions of future power costs. For critical problems, a good procedure is to present the appropriate utility with a complete spectrum of power needs and let them calculate

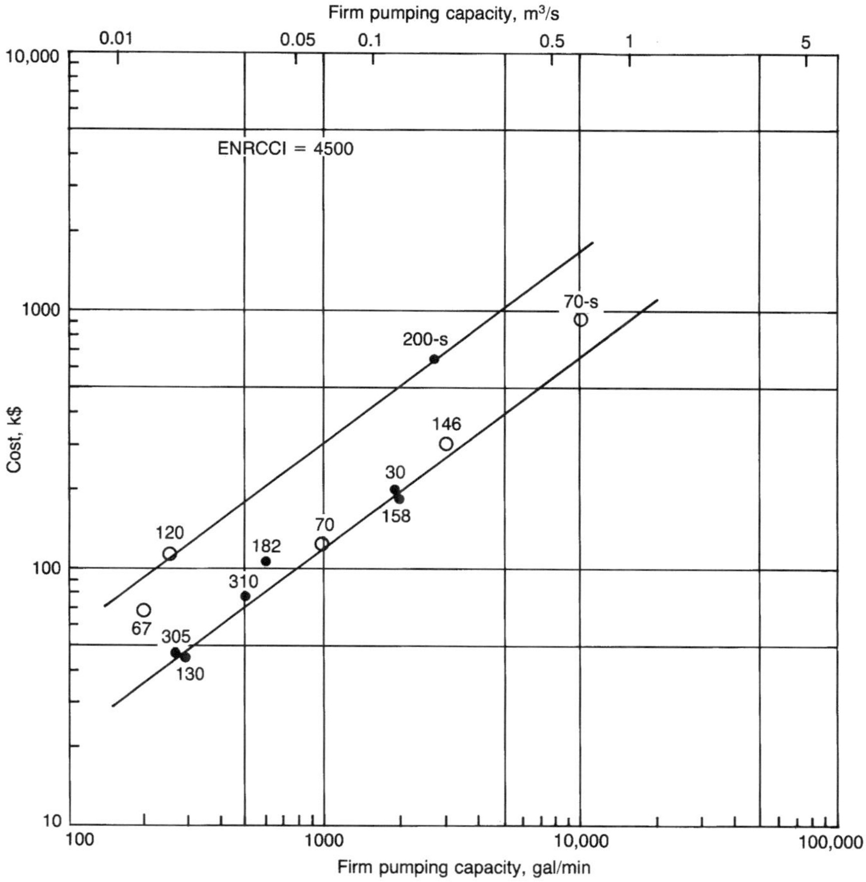

Figure 29-7. Construction costs of booster pumping stations. In-line booster = solid circles, residential booster = open circles, standby power = -s. The numbers are TDH in feet.

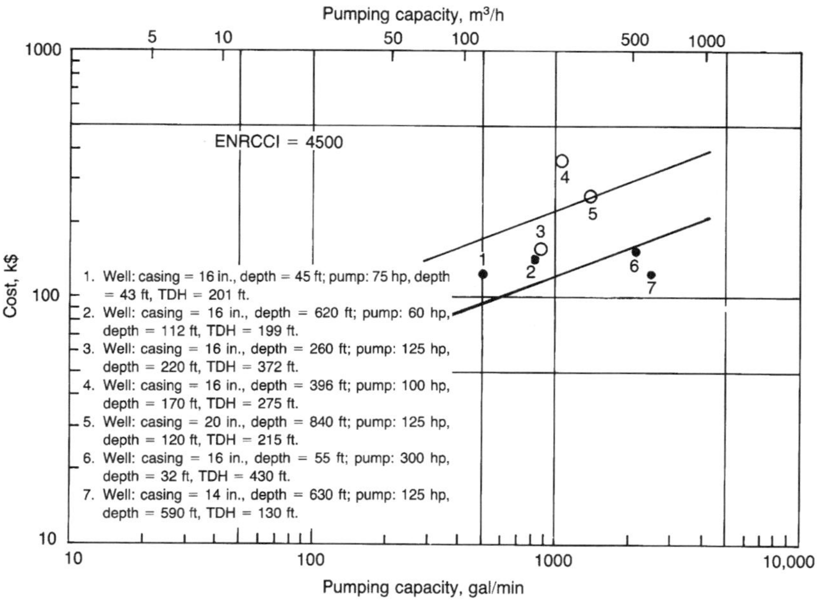

Figure 29-8. Construction costs of water wells and pumping stations. No standby power = solid circles, has standby power = open circles.

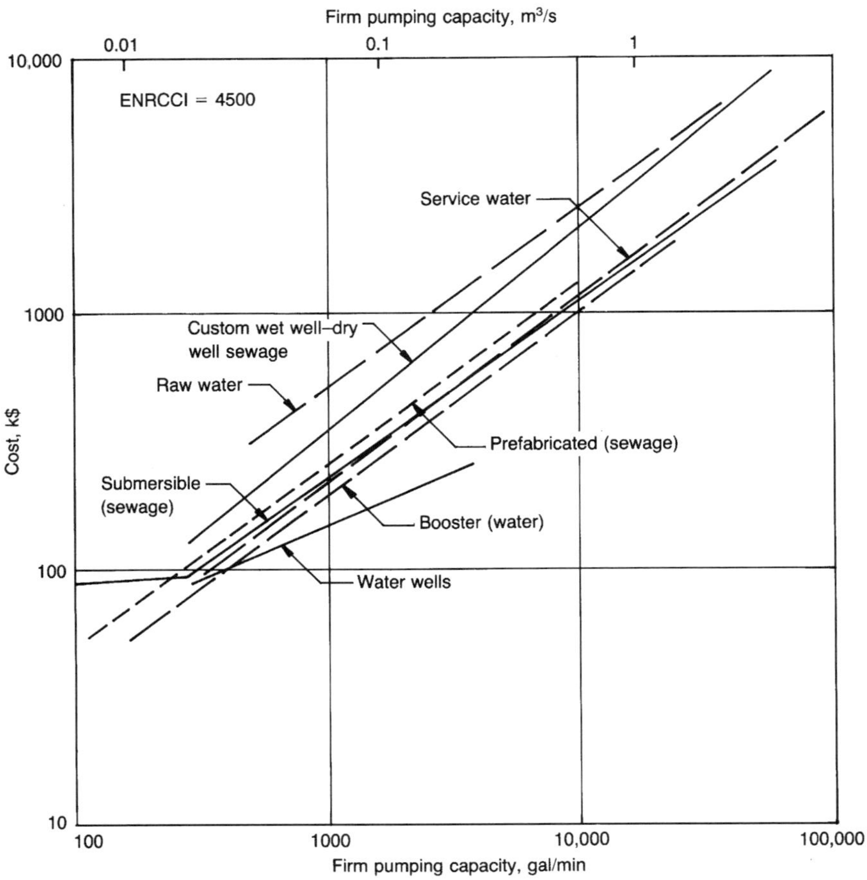

Figure 29-9. Comparison of average construction costs for pumping stations.

the annual cost. By following their method of calculation, the designer can then accurately evaluate power costs for alternatives.

29-4. Interest Formulas

The comparative worth of two or more alternative plans can be evaluated on the basis of either annual cost or present worth.

Equivalence

Costs can be compared only on some equivalent basis such as present worth or a uniform series of annual payments. The two basic formulas for the relationship between present worth, future sum, and periodic payments are

$$F = P[(1+i)^n] \tag{29-2}$$

$$F = A\left[\frac{(1+i)^n - 1}{i}\right] \tag{29-3}$$

where i is the interest rate (in decimals or percentage/100) per payment period, n is the number of payment periods, P is a present (lump) sum of money, F is the future (lump) sum of money at the end of n periods equivalent to P at interest rate i, and A is the uniform end-of-period (usually annual) payment equivalent to a present (lump) sum. If the payment period is, for example, monthly, the annual interest rate is divided by 12 and n is multiplied by 12.

Other formulas can be derived algebraically from Equations 29-2 and 29-3 as follows:

$$\begin{aligned} P &= \frac{F}{(1+i)^n} \\ &= A\left[\frac{(1+i)^n - 1}{i(1+i)^n}\right] \end{aligned} \tag{29-4}$$

$$A = P\left[\frac{i(1+i)^n}{(1+i)^n - 1}\right]$$

$$= P\left[i + \frac{i}{(1+i)^n - 1}\right] \tag{29-5}$$

The bracketed terms in these equations are named as follows:

- *Equation 29-2:* single-payment compound-amount factor (spcaf)
- Inverse of spcaf: single-payment present-worth factor (sppwf)
- *Equation 29-3:* uniform-series compound-amount factor (uscaf)
- Inverse of uscaf: sinking-fund deposit factor (sfdf)
- *Equation 29-4:* uniform-series present-worth factor (uspwf)
- Inverse of uspwf: capital-recovery factor (crf).

The annual cost, A, in Equation 29-5 is limited to the debt service (retiring the bonded indebtedness), but it can be expanded to include annual operation (O) and maintenance (M) costs as follows:

$$A = P\left[i + \frac{i}{(1+i)^n - 1}\right] + O + M \tag{29-6}$$

and this is a most useful formulation for comparing alternatives.

Arithmetic Series Gradients

Costs (or quantities such as power usage) sometimes increase uniformly with time. Let the increase in cost (or quantity) be G and the constant cost be C. Then the total payments for each period are

- C at the end of year 1
- $C + G$ at the end of year 2
- $C + 2G$ at the end of year 3
- $C + (n - 1)G$ at the end of year n.

The total equivalent uniform annual series of payments or quantities is

$$A = G\left\{\frac{1}{i} - \left[\frac{n}{(1+i)^n - 1}\right]\right\} + C \tag{29-7}$$

The term in braces is called the arithmetic-series factor (asf). The quantity A can be converted to present worth by means of Equation 29-4.

Geometric-Series Factor

If the costs or quantities increase geometrically with time, the total payments for each period are

- C at the end of year 1
- aC at the end of year 2
- a^2C at the end of year 3
- $a^{n-1}C$ at the end of year n.

The geometric increase ($aC + a^2C + \ldots + a^nC$) can be converted to present worth by using the equation

$$P = C\left[\frac{a^n - (1+i)^n}{a - 1 - i}\right] \tag{29-8}$$

The term in brackets is called the geometric-series, compound-amount factor (gscaf). The present worth, P, can be converted to a uniform annual amount, A, by means of Equation 29-5.

The derivations of these equations, along with a complete discussion of their use, are given by Taylor [7].

Inflation and Escalation

Inflation is the decrease in the value of money. Escalation is the increase in price of a particular commodity of concern, such as energy. Although both are common experiences, there is some uncertainty about how they should be treated (if at all) in an economic analysis. Equations 29-2 through 29-6 can be modified for inflation or inflation can be calculated from Equation 29-7 or 29-8, but predictions of the inflation rate are apt to be faulty, and, in many kinds of problems, the results are not altered by inflation.

If the escalation of energy costs equals the inflation rate, it, too, can be ignored. But if it is expected to exceed inflation, there is an undue penalty on a more costly but more efficient pump. White *et al.* [8] proposed the concept of "constant-worth" dollars expressed by the equation

$$C_t = \frac{T(1+e)^t}{(1+j)^t} \tag{29-9}$$

in which C_t represents purchasing power at time t relative to year 0, T is the fixed, constant-worth dollar amount, e is the escalation rate in percentage/100, and j is the inflation rate in percentage/100. Note that if escalation equals inflation, C_t equals T.

Dell'Isola and Kirk [9] proposed modifying Equation 29-5 to include escalation as follows:

$$A = P\left\{\frac{\frac{1+m}{1+i} - 1}{\left(\frac{1+m}{1+i}\right)\left[\left(\frac{1+m}{1+i}\right)^n - 1\right]}\right\} \tag{29-10}$$

where $m = e - j$ and e and j are defined above. If the escalation rate equals the inflation rate, Equation 29-10 reduces to Equation 29-5. Consult the literature for more complete discussions.

For finding the present worth of energy (or, more exactly, a "discounted break-even value") in an escalating market, Wood [10] proposed

$$P = E\left[\frac{1 - x^n}{(1 - x)y^n}\right] \tag{29-11}$$

where $x = (1 + e)/(1 + j)$, $y = (1 + i)/(1 + j)$, and E is the present annual cost of energy. If the escalation rate equals the inflation rate, the equation becomes indeterminate. The formula is not strictly comparable to Equation 29-10 because the derivation is based on payment at the beginning of each pay period, whereas all previous formulas are based on payment at the end of each pay period. Instead of Equation 29-10 or 29-11, however, Equation 29-7 or 29-8 can be used to account for inflation and escalation as well as for an increase in energy needs provided that the increases occur in either a straight-line or a geometric progression. For irregular increases in the quantity of energy to be used and to account for inflation, escalation, or both, set up a table of computations in which all quantities are reduced to an equivalent present worth.

For projects funded or controlled by the U.S. EPA, comparisons of alternatives may be required to follow a specified format. For example, inflation (except for land and energy) should be ignored [11]. Historical bases for the escalation of energy costs can be obtained from public utility commissions and from public relations personnel of power utilities.

The cost of energy over the life of a pumping station is surprisingly high and, hence, has a profound influence in any objective analysis on the selection of pumps and drivers. In today's labor market, maintenance is also likely to be a major cost. To ignore either is to skew the results of a comparative cost analysis. The cost of a pump is roughly proportional to the square of the impeller diameter, but, because energy is approximately proportional to the fifth power of the impeller diameter, the ratio of energy to capital cost increases rapidly as the pump becomes larger. Evidently, first cost is more important for small pumps, whereas for large pumps the cost of power may be many (up to 50) times the cost of the pump itself.

29-5. Cost Estimates

As defined by the American Association of Cost Engineers, there are three types of cost estimates:

- Order-of-Magnitude estimates
- Budget estimates
- Definitive estimates

Order-of-Magnitude Cost Estimates

There are various methods available to create an Order-of-Magnitude estimate for a pumping system. The use of Figures 29-2 through 29-9 is one.

Budget Cost Estimates

At the selection phase of a proposed pumping system project, a more detailed budget estimate is created to help identify the most cost-effective system. The types of estimating methods used in the past to establish these budgetary costs have had a history of minimal accuracy. With today's modern computers, the means to achieve a higher level of accuracy in establishing the budgets for various types of pumping systems is possible. Accuracy is important, because owners use the costs estimated at the early stages of the project to establish the budgets for the project. By using the latest computer technology, the level of accuracy that can be obtained in the budgetary phase can be surprising. For example, a computer-based program called BAC-PAC [12, 13] has been used to obtain budget estimates that (since 1991) have had an accuracy between +5% and –15% of the final, in-place construction costs.

Definitive Cost Estimates

After the budgetary costs have been established and the design becomes more definitive, the cost estimates also become more definitive and more accurate. Using the latest computer technology, it is relatively easy and quick to upgrade the budget cost estimates and to use these improved editions to help keep the design within the budget or to determine the effect on costs that various alternatives can have. For example, based on various soils conditions, BACPAC computes the thickness and cost of concrete walls together with the amount and cost of reinforcement automatically—the reason for omitting the thickness of walls in Figures 29-12 and 29-15. BACPAC cost estimates for final designs have had an historical accuracy (since 1991) between +2% and –5% of final, in-place costs.

During the design process, it is important to make cost estimates from time to time to inform both owner and designer of the consequences of various design decisions on the cost of the project and so avoid overruns. One system for coping with this objective is the "Design-to-Cost" system described in Section 1-10.

Regardless of the system used, the final cost estimate is based on a complete listing of the quantities of all materials described in the plans and specifications multiplied by the prices for materials and labor plus an allowance for the contractor's profit and overhead. Even with a computer program, good estimating requires great skill and an intimate knowledge of construction practices and market conditions. An apprentice, however, if careful to miss nothing, can make reasonably acceptable estimates with the help of current publications that are updated annually. Useful and authoritative publications include Richardson standards [4], Means [14, 15], and the Dodge Cost Systems [16]. Means is particularly useful for buildings, and the Dodge Cost Systems includes heavy construction and remodeling. Richardson (1) has the most complete and extensive cost estimating publication [4]; (2) has developed a rapid estimating system; and (3) provides 3- and 5-day seminars in cost estimating. Walski [17] discusses costs for planning. European engineers may find the publications of the Water Research Centre [18] useful. Articles such as that by Hall et al. [19] can remain useful for many years if the costs are updated by the ENRCCI [1].

Obtain prices from manufacturers wherever possible. List prices are usually reduced by large percentages that depend on the customer. The "customer" for a pumping station is the general contractor or a subcontractor. Shipping and installation costs (including profit and overhead) must be added to the manufacturer's cost. Installation involves (1) anchoring the unit in place; (2) "hooking up" to pipes, conduits, and electricity; and (3) start-up and testing. Installation costs are often surprisingly high even for a skid-mounted unit. Depending on the size of a machine or device and the complexity of the hook-up, typical installation costs might normally range from 10 to 40% of the purchase price, but the cost can be either less or more (see Richardson [4], Means [14, 15], or Dodge [16]). Costs are affected by many factors, including

- Competition and bidding climate
- Accessibility to a labor market [remote locations cause an increase in labor prices—about 10% for every 48 km (30 mi) of travel]
- Foundation conditions (dewatering, sheet piling, unstable soils, rock excavation)
- Earthquake resistance
- Surrounding conditions (such as nearby buildings, buried conduits, etc.)
- Need for standby power
- Whether the owner is a private or public agency.

Local contractors can be of great help in assigning values to these considerations—if their cooperation can be secured, but there is no substitute for developing some appreciation for construction practices and the problems confronting a contractor who must do a specified job for a fixed price. Because of planning, land acquisition, legal fees, owner's involvement, interest during construction, administrative costs for the project, start-up, operational training, and the costs for spare parts, the total cost to the owner is about twice the contract bid price.

Comparative Cost Estimates

A rational choice between alternative designs can be made only when life-cycle cost estimates are compared. *Basing decisions on capital (construction) costs only is misleading*. To obtain a direct measure of the relative costs between different design alternatives, it is not necessary, however, to include features common to all alternatives. Finally, to reach a decision on the best alternative—not necessarily the cheapest—the intangibles (which cannot be quantified in dollars) must also be considered. Intangibles include (but are not limited to) historical reliability, construction time, operational ease, simplicity, esthetics, compatibility with other facilities, familiarity, and the preferences of the owner, operators, and designer.

Life-Cycle Cost

Life-cycle cost is the combined cost of engineering, design, construction, and equipment procurement to construct and activate a pumping station during year 1 plus the accumulated operational costs of energy, maintenance, and repair for the established life of the pumping station. Depending upon the project circumstances, the life of the pumping station can be based on either economic or technical considerations. Important factors include cost of money (interest), the need for equipment repair, inflation, and future energy costs.

The initial cost together with each of the annual costs (corrected to present worth value by means of the equations in Section 29-4) yields the total life-cycle cost. Although the life-cycle cost may not be exact, it is the only way in which the relationship between different alternatives can be objectively assessed.

Unfortunately, owners are seldom willing to pay for a life-cycle cost analysis that is truly complete, and it is customary to consider capital costs only and, even so, to ignore the effects of the project on the cost of the whole system. (For example, savings using C/S instead of V/S pumping may stress—and hence increase the cost—of the treatment process downstream.) The designer, as a consequence, is often required to guess at answers, and such guesses can be misleading. In Example 29-1 of the original (1989) edition of this book, stations with constant speed (C/S), variable speed (V/S) with AFDs, and V/S with eddy-current couplings for pump drivers were compared for an 880 L/s (20 Mgal/d) wastewater pumping station. In view of the inefficiency of eddy-current couplings, the results were surprising. The comparative costs were 126, 107, and 100 respectively. Thus, the lowest cost was for eddy-current couplings. The results might have been different if energy costs had been higher than those assumed—essentially a peak power rate of 6.37 ¢/kW · h and an off-peak rate of 3.15 ¢/kW · h. Nowadays, the prices of eddy-current couplings have increased whereas the prices of AFDs have decreased, so in today's market, the AFDs would probably be the most cost effective.

Example 29-1
Life-Cycle Cost Comparison of Three Alternative Wastewater Pumping Stations

Two common problems today involve: (1) the choice between using a few large pumps and more smaller ones (with the consequent increase in piping and space requirement) and (2) the choice between using C/S and V/S drives. Both problems as applied to submersible pumps are illustrated in Example 29-1. The example is hypothetical, but the life-cycle cost analyses are realistic and supportable. Three options are compared:

- Three V/S pumps (2 duty, 1 standby) of equal size.
- Three C/S pumps of the above size plus a smaller jockey pump.
- Three C/S pumps of equal size.

To avoid tedious repetition in tables, this example is worked only in U.S. customary units.

Problem: A new pumping station with a trench-type sump is to replace an old, outdated station in a fully built-up older residential neighborhood. Any potential increase in inflow is expected to be offset by water conservation measures. Therefore, it is assumed that the flow duration data given in Figure 29-10 will essentially hold steady for the life of the station.

Flowrates, Heads, and Controlling Elevations

- Minimum: 1.9 Mgal/d (1300 gal/min). See also Figure 29-10.
- Average: 5.4 Mgal/d (3750 gal/min)
- Maximum: 13.3 Mgal/d (9230 gal/min or 20.6 ft^3/s) for 2 duty pumps
- Sewer: 36 in. at 0.1% slope. Invert elevation 24.05 ft at 60 ft from wet well
- Station piping: typical piping to manifold. Parabolic curve for friction head loss = 9.0 ft at 7000 gal/min
- Force main: 20-in. mortar-lined, ductile iron 4000 ft long. Discharge elevation: 60.10 ft.
- Friction headloss based on $C = 135$ is 50 ft at 10,000 gal/min.

Other Data

- Design life: 25 yr
- Interest: 7%
- Inflation: 4.5%
- Energy: $0.06/kW · h with annual inflation increase of 5%.

Design Option 1 (Three V/S Pumps)

Three 12-in. Flygt CP3312, 175 hp close-coupled submersible pumps are selected. The system curve for the force main is shown in Figure 29-11 together with pump characteristic curves.

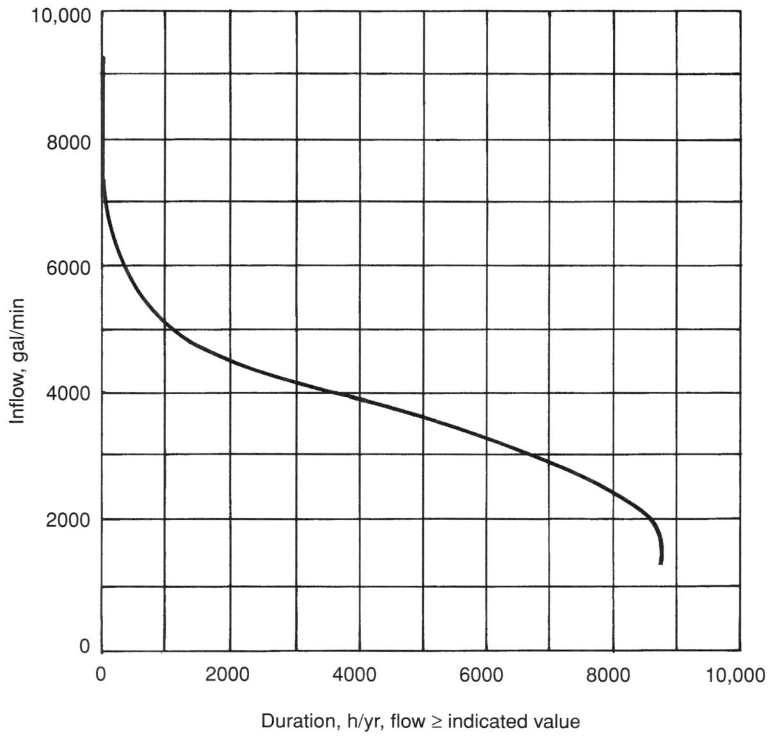

Figure 29-10. Inflow duration curve.

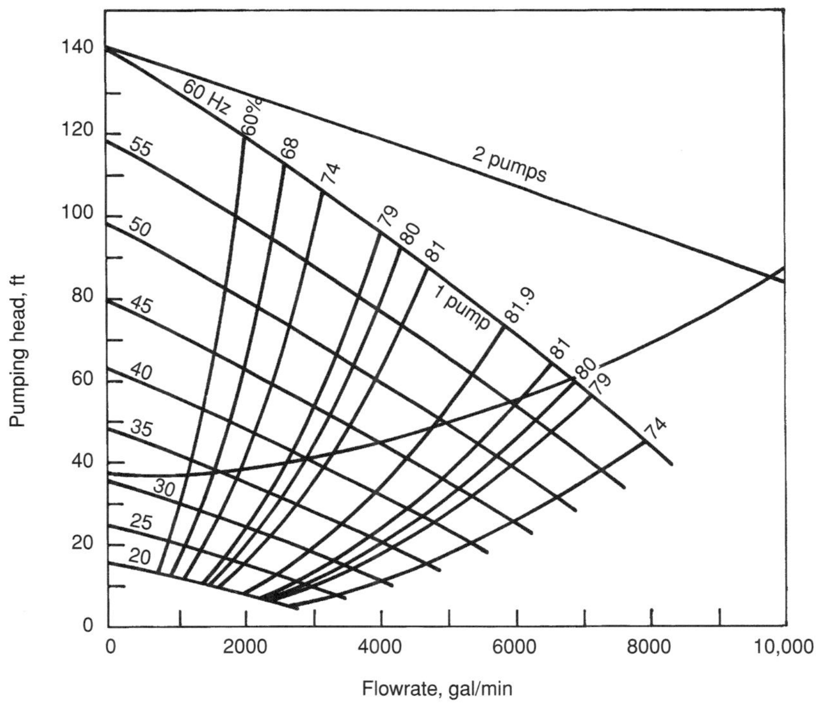

Figure 29-11. Pump and system curves for V/S pumping.

Station piping headlosses for a pump are assumed to follow a parabola to a loss of 7 ft at a discharge of 6000 gal/min.

To make sure that the station can meet its requirements, the H–W *C* is often considered to vary between 145 or 150 and 120. For a life-cycle cost analysis, however, the use of excessive safety factors is misleading, and the best—not the safest—values are needed, so the chosen value of *C* is 135. Two pumps together can discharge 9860 gal/min at 89.5 ft of head whereas a single pump can discharge 6790 gal/min at 69.6 ft of head.

Each pump is equipped with a top-of-the-line pulse width modulated (PWM) type of AFD to produce the highest drive efficiencies available to date. Because of their sensitivity to heat, fumes, and dust, the AFDs are housed in a small air-conditioned building where incoming air is filtered to remove all dust and fumes, particularly H_2S. Motors controlled by AF converters are subjected to severe service and are consequently derated. Despite the derating, they require more maintenance than do motors for C/S pumps.

The principal advantage of V/S pumping is that discharge equals inflow at all times, and thus there are no sudden changes of the inflow to the treatment plant—an advantage that may be overwhelming or negligible depending on the relative sizes of pumping station and treatment plant and the sensitivity of the treatment plant facilities to shock loads. The other advantages are the increased operational flexibility, the reduced energy consumption, the smaller size of the wet well, and a shallower excavation if the wet well must be self-cleaning. The disadvantages are the substantial costs of the AFDs, the increased control system complexity, the potential problems with harmonics and power quality, plus the increased maintenance both for the AF converters themselves and for the air-conditioning unit required to protect them from dust, fumes, and heat. Furthermore, AFDs require highly skilled personnel for servicing and repairs. Overall station reliability is reduced somewhat, but AFDs have become more dependable in recent years. High-quality equipment is reliable when well maintained and housed in an air-conditioned space.

The design follows the guidelines in Chapter 12. At an entrance velocity of 5.5 ft/s, the diameter of the suction bell for 6800 gal/min (Figure 29-11) would be 1.88 ft, but the next standard flange and flare is 1.96 ft in diameter. The maximum entrance velocity is 5.03 ft/s. The trench width is 2 D = 3.92 ft. and the depth is 2 D + floor clearance = 4.90 ft. Submergence required (Equation 12-1) is 4.8 ft. The top of the trench was set 0.95 ft below the invert of the sewer so that at neither peak inflow nor at half the peak inflow (at a lower elevation, of course) would the average velocity in the wet well above the trench and around the motors exceed 1.0 ft/s. The actual submergence is about 5.2 ft at LWL and 7.5 ft at HWL.

The pumps can be centered in the trench by recessing the discharge elbow (see Figure 29-12b). The pumps chosen are furnished with short suction nozzles. The floor is lowered below the last pump to ensure strong currents along the trench and minimum air ingestion during cleaning, so its nozzle is longer (see Figure 29-12c). The pumps are set on 4.66-ft centers as required by the manufacturer. Closer spacing might be tolerated pending model studies. The plans, excluding the valve vault and a small building for the AF converters, are shown in Figure 29-12.

Design Option 2 (Four C/S Pumps)

The parallel pump array consists of three C/S 175 hp Flygt CP3312 pumps and a C/S 75 hp Flygt CP3300 jockey pump selected for flows below 3600 gal/min. Compared with V/S pumping, the advantages of C/S pumping include lower equipment cost, reduced maintenance, simplicity, and somewhat greater reliability, whereas the disadvantages include a deeper wet well with an approach pipe to augment the storage volume so as to allow pumps to cycle on and off at intervals long enough to avoid overloading motors and motor starters. The surges accompanying essentially instantaneous pump starts may or may not be a significant disadvantage. The characteristic pump curves and an average system curve are shown in Figure 29-13.

In C/S pumping stations, the advantage of a four-pump facility that includes one relatively small pump is greater flexibility and the improvement of efficiency at lower flowrates because of the reduced friction losses. The disadvantages are the somewhat longer wet well and valve vault needed for the extra pump and its discharge piping, and an increase in mechanical equipment, maintenance, and repair costs.

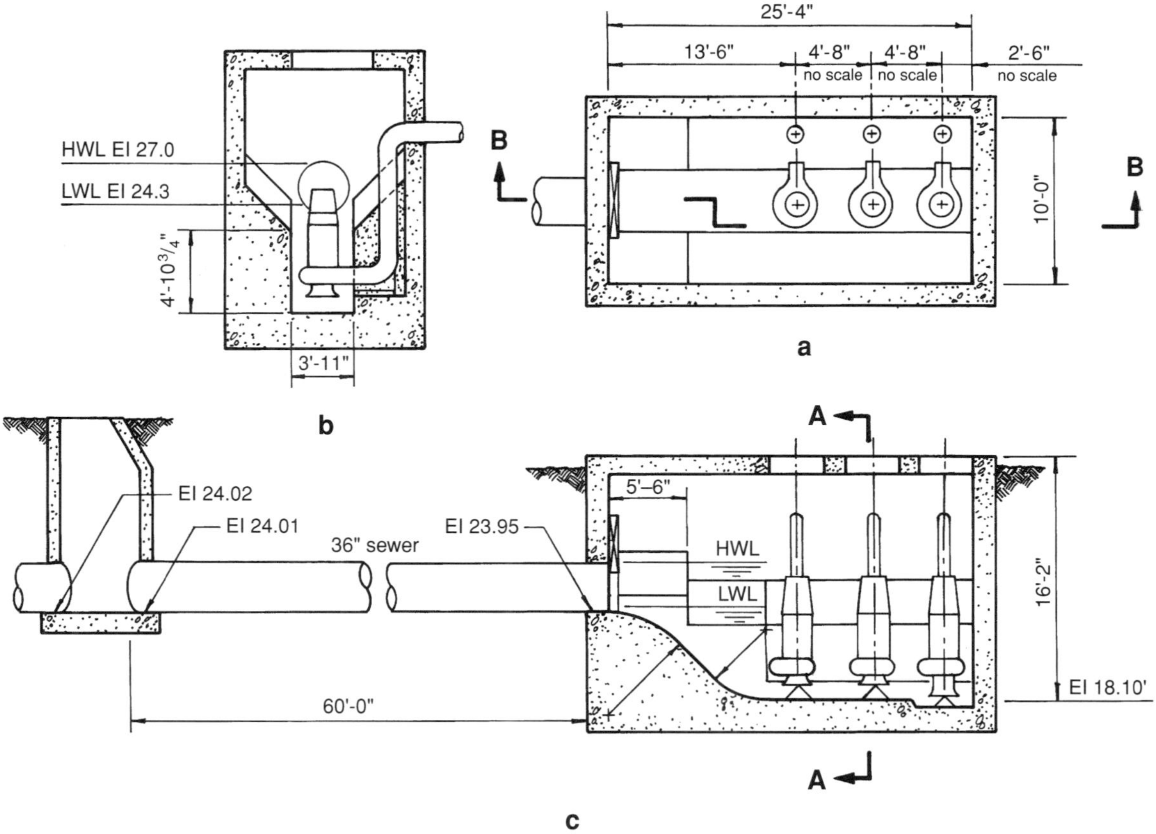

Figure 29-12. Plans for a pumping station with three V/S pumps. (a) Plan at Elev. 28.0 ft; (b) Section A–A; (c) Section B–B.

To keep excavation cost to the absolute minimum, the wet well elevation was set as high as possible by making the approach pipe as short as possible and allowing the HWL for any pump or pump combination to reach the nominal water level in the upstream sewer. At the transition manhole (Figure 29-14), the crown of the approach pipe was set at the same elevation as that of the sewer to ensure supercritical velocities at both intermediate and maximum flowrates. Little is known about headloss coefficients for flows in transition manholes where the shape of the water cross-section changes from nearly a full circle to a shallow segment of a larger circle. A K-factor of 0.1 applied to the exit velocity seems to be a reasonable assumption from which the drop in the energy grade line (EGL) is $0.1(v_2)^2/2g = 0.25$ ft, as shown in Figure 29-14. Fortunately, an error in calculating headloss is of little consequence, because velocity varies only as the square root of the head (water surface to EGL) and the increase in sequent depth (after the hydraulic jump) is only about $^1/_3$ as much as the percentage change in velocity. Hence, at the maximum flowrate, the exit velocity could be 21% too high with a sequent depth of, therefore, 67%—both acceptable values.

The design closely follows that of the foregoing V/S station except for the need for fluctuating water levels as pumps cycle on and off. An approach pipe is needed to prevent a free fall of water and to contribute to the active storage volume. The cross-section (Figure 29-15b) is the same as shown in Figure 29-12b, but the wet well was lengthened to accommodate an extra jockey pump, chosen to discharge 3500 gal/min. The complete plans are shown in Figure 29-15.

The volume of the wet well was found in preliminary calculations to be almost sufficient, so the approach pipe was made only 60 ft long with its last 10 ft level. To obtain a graph of pump-cycling frequency (similar to Figure 12-54), a plot of storage volume versus elevations above the approach pipe invert was made by dividing the volume into separate sections: (1) the prismatic part of the wet well; (2) the part occupied by water guides; (3) the level part of the approach pipe;

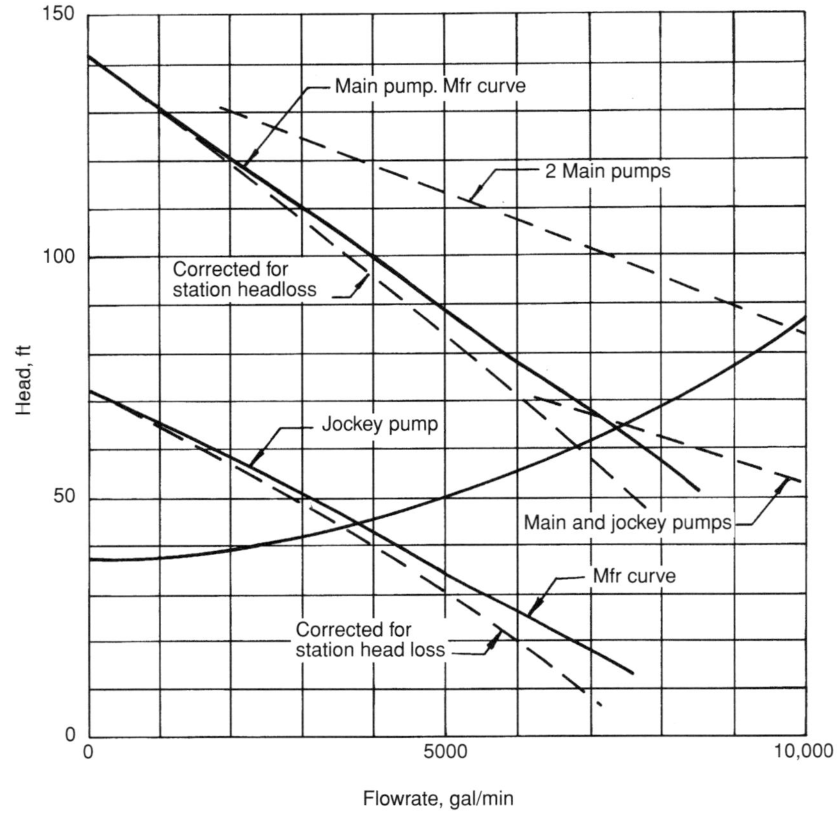

Figure 29-13. Pump and system curves for a station with constant speed pumps.

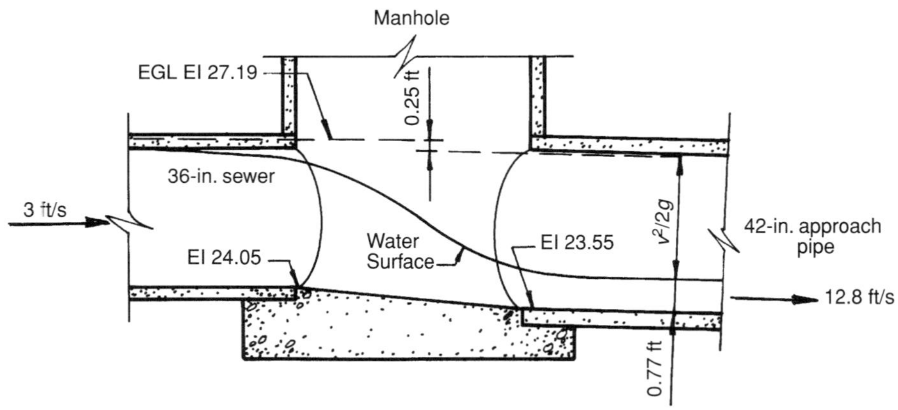

Figure 29-14. Transition manhole at the junction of sewer and approach pipes.

and (4) the sloping part of the approach pipe. Each section was divided into horizontal segments by selected horizontal planes, and the volumes of the segments were added to obtain abscissas as a function of elevation. (The volume occupied by the inflow moving at supercritical velocity in the approach pipe was not included.) Graphically adding the curves for each section produced a curve of total water volume versus height, and these values were transferred to Figure 29-16. A computer program would shorten the above tedious process to a few moments.

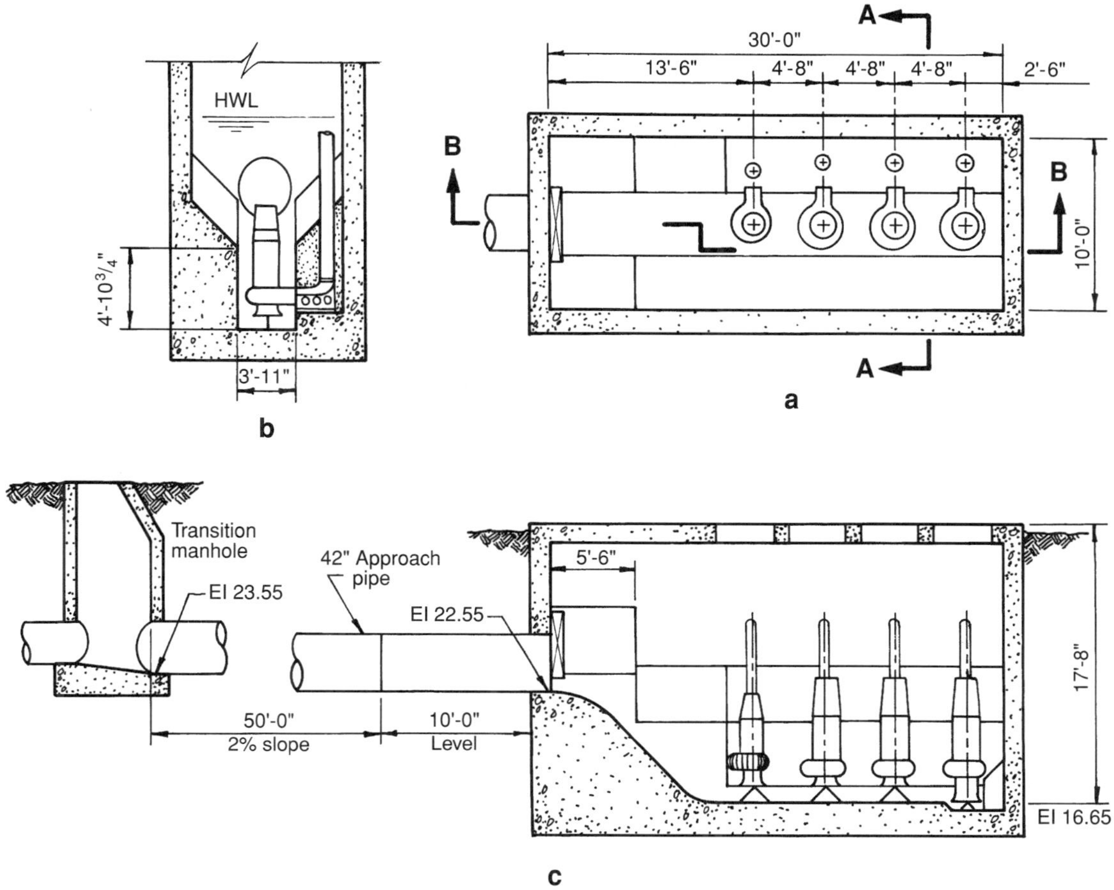

Figure 29-15. Plans for a pumping station with four C/S pumps. (a) Plan at Elev. 28.0 ft; (b) Section A–A; (c) Section B–B.

The entrance velocity into the wet well at the end of every pump cycle should, ideally, be more than 3 ft/s (to scour the approach pipe) but less than 4 ft/s (to avoid strong currents). With fixed stop-level controls for each pump (or combination thereof), some compromise may be necessary. For the jockey pump, the stop level was set at 0.45 ft above the invert for end-of-cycle entrance velocities varying from 2 to 4.6 ft/s. For 97% of the time, however, the minimum velocity is 3 ft/s. The stop level for P2 was set at 1.47 ft to produce velocities of 2.1 to 4.0 ft/s, and the stop level for P2 + P3 was set at 1.86 ft to produce velocities of 3 to 4 ft/s (see Figure 29-16).

Start levels were set to yield a minimum cycle time of 6 min and for minimum successive start levels differing by at least 0.5 ft. The corresponding volumes can be found by Equation 12-3 ($V = Tq/4$) or by plotting the first leg of the cycle time at a slope of $Q_{critical}$ to an intersection with half the cycle time (note the first and third legs in Figure 20-16). $Q_{critical}$ is half of the pump capacity or the average capacity of P1 + P2 or P2 + P3, etc.

The start and stop levels were not allowed to interfere with the velocities in the upstream sewer.

Design Option 3 (Three C/S Pumps)

Three pumps constitute the simplest type of installation—typical of many, especially those of small and medium sizes. The principal attraction is the perceived low first cost, but the drawbacks include higher energy consumption and less operational flexibility than either a four-pump C/S station or a V/S station.

The pump and station curves are similar to those of Figure 29-15 except that there is no jockey pump. Omitting the jockey pump would allow the wet well to be shortened. But, as the

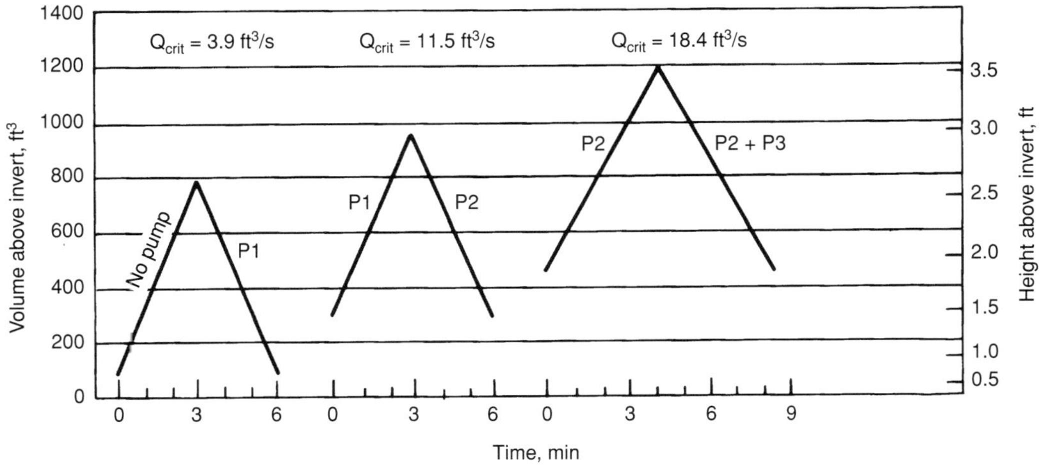

Figure 29-16. Wet well plus approach pipe volumes and pump-cycle times. Jockey pump (7.80 ft³/s) is P1, main pumps (15.1 ft³/s each) are P2, P3, or P4.

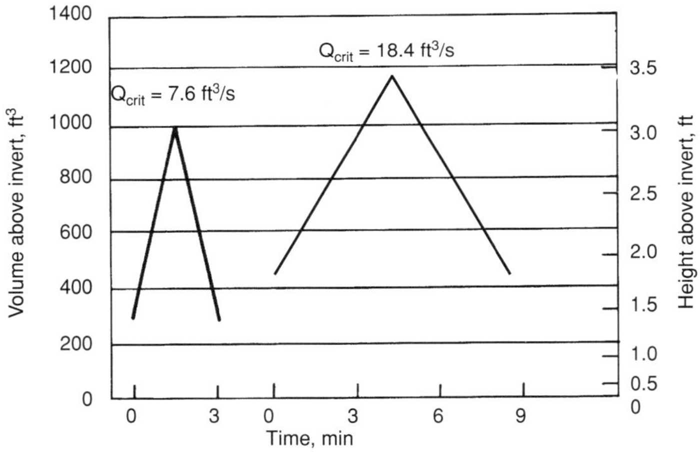

Figure 29-17. Wet well plus approach pipe volumes and pump cycle times for two main duty pumps (15.1 ft³/s each) alternated.

active storage volume required was no less, the needed volume could be obtained with either (1) a longer approach pipe and a shorter, deeper wet well or (2) by retaining the same dimensions. The latter option was chosen, and the plans are the same as those of Figure 29-15 except that there are three pumps instead of four.

A diagram of the pump-cycling frequency is shown in Figure 29-17. The start and stop elevations had to be changed to accommodate the omission of the jockey pump, but the greater volume required for the larger starting pump could be reduced by alternating the pumps in successive cycles—a common expedient that cuts the required cycle time in half.

Table 29-1. Estimate Summary Sheet for a Pumping Station in Poor Soil. Three V/S Pumps. Courtesy of Brown and Caldwell Consultants, Construction Management Division.

ESTIMATE SUMMARY SHEET

Project:	Pumping station cost, poor soil 3 pump V/S system	Estimate #:	1111C1
		Job #:	Figure 29-12
		Est date:	6-17-97
Made by:	Pritchard/Abelin	PRT date:	10-11-97
Checked by:	________________	PRT time:	8:49 pm

Summary totals

Labor:	$105,933	Sales Tax:	$ 14,449
Material:	$240,832	Markup:	$ 66,730
Subs:	$386,253	Subtotal:	$836,082
Equipment:	$ 21,885	Bond:	$ 8,361
		Revised Subtotal:	$844,443
		Contingency:	$126,666
		Grand Total:	$971,109

Markup data

Labor M/U:	15.00%	Sales Tax (Mat'l) :	6.00%
Material M/U:	12.00%	Sales Tax (Equip):	.00%
Subs M/U:	5.00%	Bond Rate:	1.00%
Equipment M/U:	12.00%	Contingency:	15.00%

GRP #	Process Area	Labor M/C Hrs[a]	Labor Amount ($)	Material Amount ($)	Sub Amount ($)	Equipment Amount ($)	Sales Tax ($)	Markup ($)	Total ($)	Job %
010	Contractor Indirects	1,384	65,394	608	2,450	14,490	36	11,744	94,722	11.3%
015	Yard Development	192	3,955		6,570	480		981	11,986	1.4%
018	Sewer Pipe	84	2,494	7,040	5,645	1,230	422	1,648	18,479	2.2%
020	Pump Station	1,042	25,125	215,997	111,588	4,616	12,962	35,822	406,110	48.6%
022	Valve Vault	354	8,965	17,187		1,069	1,029	3,535	31,785	3.8%
176	Elect/instr				260,000			13,000	273,000	32.7%
	Estimate Subtotal:	3,052	$ 105,933	$ 240,832	$ 386,253	$ 21,885	$ 14,449	$ 66,730	$ 836,082	100.0%
	Plus Bond, if Required:								$ 8,361	1.0%
	Revised Subtotal:								$ 844,443	101.0%
	Contingency:								$ 126,666	15.0%
	GRAND TOTAL:								$ 971,109	

[a] Man-h/Crew-h

One unresolved difficulty is that the use of fixed stop levels produces end-of-cycle entrance velocities that, for P1, would vary from 4.0 to less than 1 ft/s. To avoid the buildup of deposits not expelled at velocities much less than about 2.5 ft/s, a PLC could be programmed to lower the stop level for low flows. Inflow can be approximately quantified by programming the computer to measure the time period for the up leg of the pump cycle.

Construction Costs

Construction costs estimated for the three design options above are based on the prevailing wage rates for the Atlanta, Georgia, area, with an ENR 20-cities average of 5798. For a graphic presentation of the effects that specific ground and soils conditions can have on the costs of a particular pumping station, two estimates were made for each design option: (1) one for good soil conditions and (2) one for poor soils with high ground water. The difference in costs is due to the need for water-tight sheet piling and dewatering in poor soils.

A single cost estimate, even at this budget stage, requires seven pages of fine print and is thus too lengthy to include herein, but a typical summary is presented in Table 29-1. Grand totals for all six cost estimates are given in Table 29-2.

Table 29-2. Grand Total Capital Costs

Option	Grand total capital cost, $	
	Good soil	Poor soil
1 (3 V/S pumps)	859,910	971,109
2 (4 C/S pumps)	920,160	1,054,875
3 (3 C/S pumps)	782,184	914,209

It is of considerable interest to see how each option affects the costs attributable only to the pumps and unaffected by all of the rest of the pumping station costs. These costs, shown in Table 29-3, are included in the total construction costs of Table 29-2.

The costs for installation, piping, vaults, conduit, wiring, contractor indirects, overhead, profit, bonding, and insurance must be added to the costs in Table 29-2. These additional costs are significant. For example, the fob cost of the jockey pump, $31,000, increases by a factor of 4.5 to $140,000 when the foregoing components of costs are included.

Energy Usage, Design Option 1 (Three V/S Pumps)

Adjustable frequency drives cause a loss of efficiency as the frequency and speed decrease. A plot of the representative combined efficiency of the AF converter and the induction motor are shown in Figure 29-18. The graph represents a modern pulse-width modulated (PWM) type of adjustable frequency drive. These drives feature a linear volts-to-hertz ratio and typically achieve high efficiencies at full speed. The efficiency curve in the figure differs from that commonly published as AFD efficiency, because it accounts for (1) the lower motor efficiency resulting from a non-sinusoidal wave form emitted from the AF converter and (2) the usual motor loadings at lower speeds. Published data, on the other hand, are based on the assumption that power is proportional to speed cubed—a good assumption for fans and blowers but quite erroneous for pumping stations where static head is a substantial portion of the total design head. The curve of Figure 29-18 is based on actual testing of pump motors driven by PWM-type AFDs. The losses are manifested in the form of additional heat in the motor and in the AFD.

Specific energy (E_s) is a useful concept for comparing energy usage for various selections of pumping equipment. The term is a measure of the energy used per unit volume pumped, and in this example is expressed as kilowatt hours per million gallons pumped (kW · h/Mgal). The calculations used to obtain E_s are shown in Table 29-4.

The total energy use per year is 462,000 kW · h. As the inflow increases from a minimum, a single V/S pump is operated until its capacity is exceeded. Then two scenarios are possible:

- The first pump continues to run at top speed while the second discharges the excess inflow, called "staggered operation" and shown as a dashed line in Figure 29-19. See "Conclusions and Critique."
- The first pump is ramped down as the second pump is ramped up until both are running at equal speeds. "Load-sharing operation" is shown as a heavy solid line in Figure 29-19 (see also Figure 15-8).

Table 29-3. Capital Costs (fob) of Pumps, Motors, Guide Rails, and Controls

Option	Pump and accessory cost, $	Control cost, $	Total, $
1 (3 V/S pumps)	165,000	160,999	325,000
2 (4 C/S pumps)	191,000	55,000	246,000
3 (3 C/S pumps)	165,000	50,000	215,000

Table 29-4. Computation of Specific Energy for One V/S Pump and (in Staggered Operation) Two V/S Pumps

CP3312 pumps operating at:		Total head, ft		Discharge, gal/min			Total pump	Specific energy
V/S	C/S	Pump 1	Pump 2	Pump 1	Pump 2	Station	kW	kW · h/Mgal
1	0	37.8	0	1000	0	1000	20.1	335
1	0	38.6	0	1500	0	1500	23.5	261
1	0	39.8	0	2000	0	2000	27.2	227
1	0	41.4	0	2500	0	2500	32.3	215
1	0	43.2	0	3000	0	3000	38.2	212
1	0	45.5	0	3500	0	3500	45.6	217
1	0	48.5	0	4000	0	4000	53.4	223
1	0	51.0	0	4500	0	4500	62.9	233
1	0	54.3	0	5000	0	5000	73.5	245
1	0	57.9	0	5500	0	5500	85.1	258
1	0	61.8	0	6000	0	6000	98.6	274
1	0	66.2	0	6500	0	6500	113.1	290
1	1	69.7	61.6	6594	406	7000	144.9	345
1	1	72.7	65.6	6274	1256	7500	153.5	341
1	1	75.8	70.0	5942	2058	8000	163.7	341
1	1	79.1	74.9	5576	2924	8500	177.5	348
1	1	82.5	80.3	5162	3837	8999	193.3	358
1	1	86.6	86.6	4750	4750	9500	212.8	373

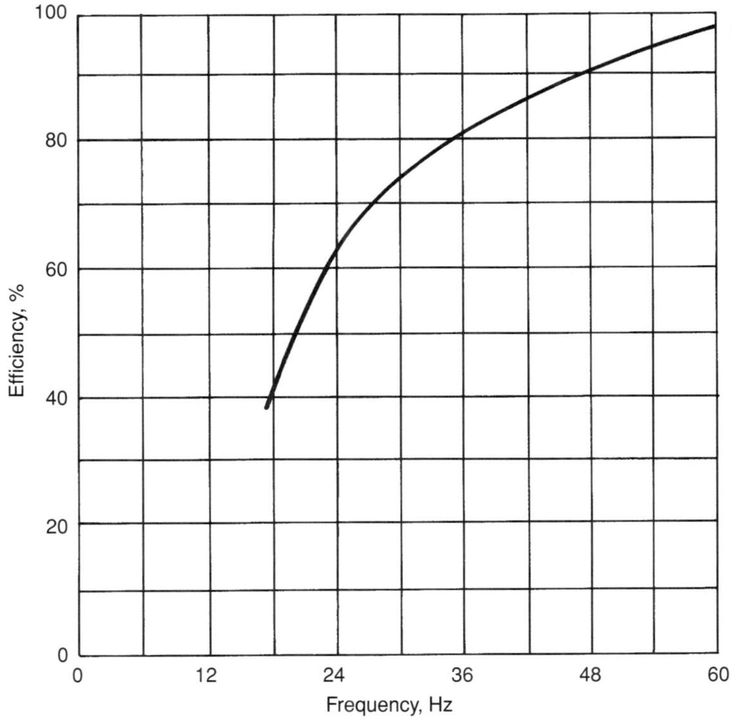

Figure 29-18. Representative efficiency of AFD/motor systems.

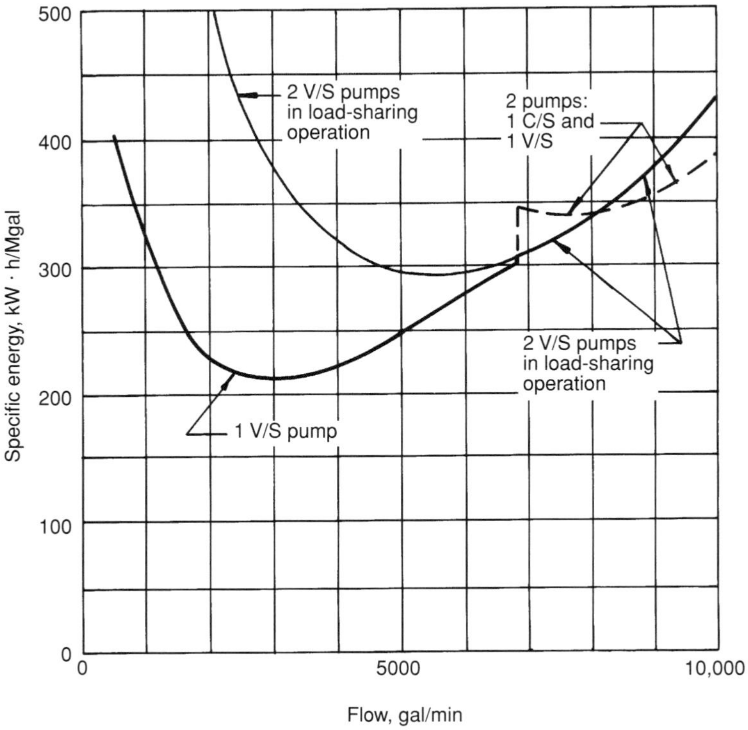

Figure 29-19. Specific energy for two V/S pumps versus one V/S and one C/S pump.

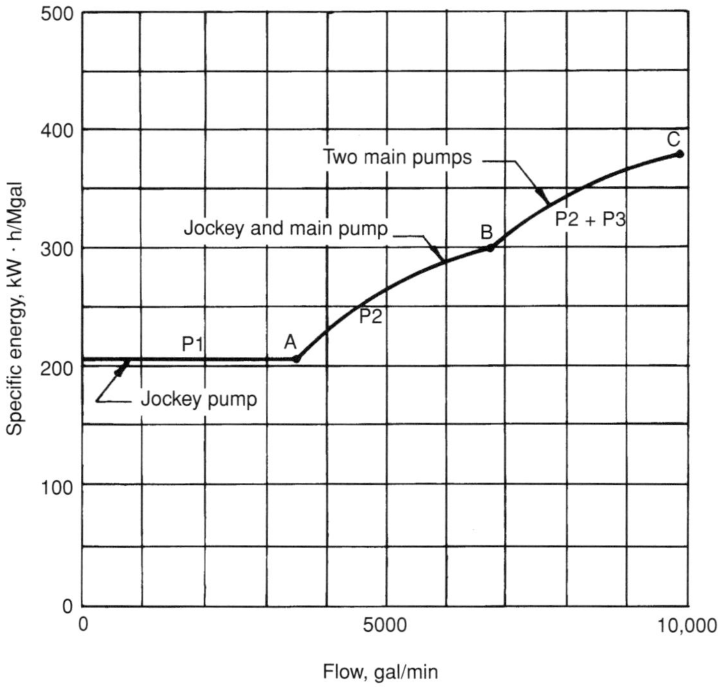

Figure 29-20. Specific energy for a small (jockey) pump and two main constant-speed pumps.

Energy Usage, Design Option 2 (Three C/S Pumps + One C/S Jockey Pump)

Only the jockey pump operates until the flow exceeds 3500 gal/min at Point A in Figure 29-20. The E_s is constant in this range, because the jockey pump, operating on and off, always pumps against nearly the same head. From 3500 to 6800 gal/min (Points A to B), a large pump alternates with the jockey pump. Because the E_s of the jockey pump alone is only 210, the average E_s in this operating range (between Points A and B) becomes a slightly curved line.

At Point B, the jockey pump is no longer useful and only the large pump with its E_s of 300 is required. Above a flow of 6800 gal/min, one large pump operates continuously while the other cycles on and off. Again a curved line is developed between Points B and C. Instead of cycling one pump on and off, the two can be alternated to double the time for a cycle. The data for calculating E_s for Figure 29-20 are given in Table 29-5.

Energy Usage, Design Option 3 (Three C/S Pumps)

The E_s curve between Points B and C in Figure 29-21 is exactly the same as in Figure 29-20. However, at flowrates smaller than 6800 gal/min (Point B), only a 175 hp pump is available for

Table 29-5. Computation of Specific Energy for a C/S Jockey Pump and Two C/S Main Pumps

Pumps operating	Head, ft			Flowrate, gal/min		Total kW	Specific energy, kW · h/Mgal
	Piping	Main	Total	per pump	Station		
1-CP3300	2.6	44.4	47.0	3600	3600	48.6	225
1-CP3312	8.6	61.4	70.0	6786	6786	122.0	300
2-CP3312	4.5	85.4	89.9	4930	9860	221.7	376

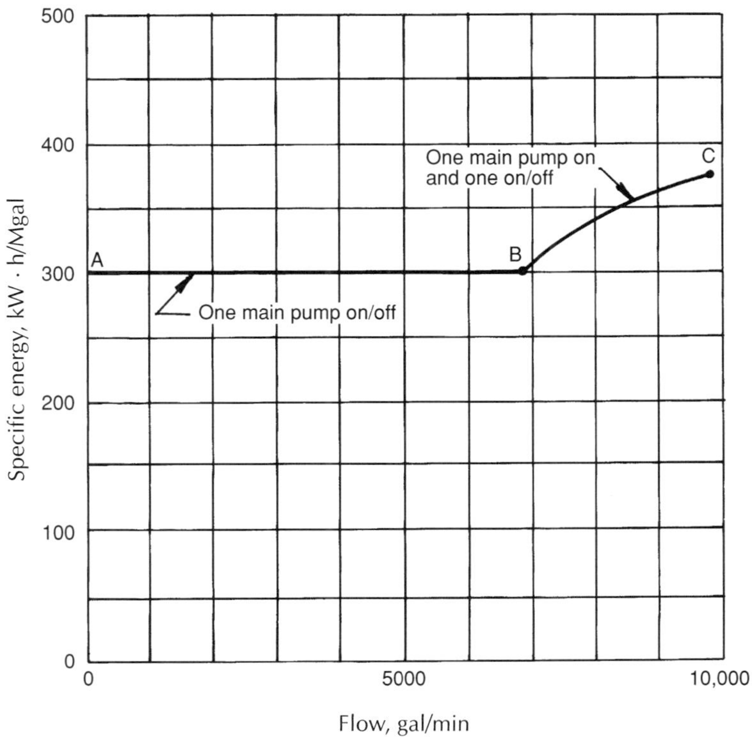

Figure 29-21. Specific energy for two main C/S pumps.

Table 29-6. Annual Energy Consumption and Cost of Operations

Option	Energy: Use, kW · h	Energy: Cost, $	Maintenance and service, $	Annual operation cost, $
1 (3 V/S pumps)	498,041	28,882	16,250	46,132
2 (4 C/S pumps)	519,986	31,199	12,300	43,499
3 (3 C/S pumps)	635,086	38,105	10,750	48,855

Table 29-7. Total Life-Cycle Costs Based on Present Worth (PW) for 25 Years

Option	Operation, $ (Present worth)	Capital, $	Total life-cycle, $[a] (Present worth)	Comparison, %
		Good soil		
1 (3 V/S pumps)	871,210	859,910	1,730,000	102
2 (4 C/S pumps)	821,481	920,160	1,740,000	102
3 (3 C/S pumps)	922,629	782,184	1,700,000	100
		Poor soil		
1 (3 V/S pumps)	871,210	971,109	1,840,000	100
2 (4 C/S pumps)	821,481	1,054,875	1,880,000	102
3 (3 C/S pumps)	922,629	914,209	1,840,000	100

[a] Rounded to three significant figures.

operation and the E_s is constant at 300. Consequently, the overall specific energy required is higher, and pumping is, on the average, less efficient.

Annual Cost Analysis

The total annual costs for energy and maintenance (including repairs and service) are given in Table 29-6. Maintenance is taken as 5% of the capital cost of pumps, accessories, and control given in Table 29-5. The 5% is an estimated value that may not be completely realistic. Objective figures, however, are not only difficult but probably impossible to obtain. There are many circumstances where identical pumps in identical circumstances have a history of decidedly different repair costs for no discernible reason.

Conclusions and Critique

The cost differences for the three options are less than the errors in cost estimating. There is, in other words, no practical difference at all. Many engineers have assumed that a V/S pumping station is more costly than a C/S station because of the added cost of the AF converters. Such short-sighted reasoning ignores the comparative sizes and depths of wet wells and cost of electricity. Furthermore, none of the costs herein include the added cost of treatment operations because of the added burden of meeting effluent standards due to the effect of hydraulic shock loads generated by C/S pumps going on and off. If the entire life-cycle system costs were evaluated carefully, it seems likely that V/S pumping (at least for similar pumping stations) is actually less expensive than C/S pumping in the long run.

Treatment systems simply work better when changes in hydraulic loading vary slowly instead of abruptly. For example, in 1986, the chlorination system in the Lompoc (California) Wastewater Treatment Plant could not meet the required discharge standards. An examination of the plant revealed that the operators did not like (or, perhaps, did not understand) the V/S drives and had switched off the AF converters to make the pumps operate as C/S units. Returning the station to V/S pumping solved the chlorination problem.

Owners should be advised to avoid consideration of first cost only. If capital cost were the only criterion, Options 1, 2, and 3 would, in the same order, compare as 110, 118, and 100 in good soil and 106, 115, and 100 in poor soil—a poor presentation for the owner.

Example 29-1 is representative of a rather typical pumping station, but the results should be viewed with objective skepticism when applied to markedly different circumstances. In general, however, small differences in dimensions or elevations would have a negligible effect on cost comparisons.

Location

Great care should be taken to ensure that comparative cost estimates are made with the very best available information. The ENRCCI is a reliable method for comparative cost estimating using historical data. Note that the region of the country has a profound effect on the costs of a pumping station. The costs in Example 29-1 would be 75% of those given if the station were to be built in New Orleans and 113% if built in San Francisco.

Matching Pump and System Curves

It is important in V/S pumping to select the highest flow duty point well to the right of the bep so that, as the flow decreases, the operating point does not stray too far from the high-efficiency zones (see also Section 15-3). By following the above guideline, the hydraulic efficiencies can be kept at or above 80% for flows from 3000 to 7000 gal/min. Note as well that the high-efficiency zone coincides with the zone of the most stable pump operation. If the guideline is followed, the pump is less likely to suffer cavitation, vibration, and radial thrust damage. It is also important to analyze the movement of duty points when pumps are turned on or off, or when their speed is varied. Two significant considerations are (1) ensure operation within the recommended operational range of the pump, i.e., avoid approaching shut-off head or run-out; and (2) pumping economy. With respect to pumping economy, the flow duration curve should be carefully considered, and the duty points with the best efficiency should be chosen to coincide with flows of the longest duration. When duty points for long operating hours have been selected close to the bep, the pump will run smoothly and give the best possible service life.

Staggered Versus Load-sharing Pump Operation

Staggered operation seems to be widely used—usually to save the cost of one or more AF converters by providing a mix of V/S and C/S drives—common practice in Europe. To prevent operation at excessively low discharges, the station is designed to operate in C/S mode until inflow increases to the minimum allowable pumping capacity of one V/S unit. Consider the cost data, the statements above, and Section 15-1. You should form your own conclusions about the advisability of a mix of V/S and C/S pumps after considering the cost data and statements above and Section 15-5.

Note that staggered operation is somewhat less efficient than parallel operation at flows slightly greater than the capacity of a single pump, as shown in Figure 29-19. Above 8300 gal/min in this example, however, staggered operation is more efficient, and the pump controller could be programmed to switch from load sharing to staggered operation. But flows above 8000 gal/min occur for such a negligible period of time (see Figure 29-10), that switching to staggered operation is excessively fussy. Note, however, that a small energy savings could be obtained if flows between 7000 and 8000 gal/min were pumped with 2 V/S units in load sharing operation instead of staggered operation with 1 V/S and 1 C/S unit.

29-6. References

1. *ENR*, published weekly by McGraw-Hill, New York. (Available from Construction Economics, Engineering News-Record.)
2. Whitman, Requardt & Associates, Engineers-Consultants, Baltimore, MD.
3. U.S. Environmental Protection Agency, Office of Municipal Pollution Control, Municipal Utilities Division, Washington, DC.
4. *Process Plant Construction Estimating Standards,* Vols. I, II, III, and IV. Richardson Engineering Services, Inc., San Marcos, CA (updated annually). Also *Reporter* (published quarterly).
5. Sloan, L., Chief Engineer, ENPO Pump Co. Private communication (April 1987).
6. Patterson, W. L., and R. F. Banker. *Estimating Costs and Manpower Requirements for Conventional Wastewater Treatment Facilities.* U.S. EPA, 17080 DAN, Washington, DC (1971).
7. Taylor, G. A. *Management and Engineering Economy, Economic Decision-Making,* 3rd ed. Brooks/Cole Engineering Division, Monterey, CA (1980).
8. White, J. A., M. H. Agee, and K. E. Case. *Principles of Economic Analysis,* 2nd ed. John Wiley, New York (1984).

9. Dell'Isola, A. J., and S. J. Kirk. *Life Cycle Costing for Design Professionals.* McGraw-Hill, New York (1981).
10. Wood, K. Private communication (January 3, 1986).
11. *Construction Grants 1985 (CG-85).* U.S. EPA 430/9-84-004, Office of Water Program Operations (WH-546), Washington, DC (July 1984).
12. Pritchard, M. T., D. S. Parker, and A. Malik. "Design-to-Cost with BACPAC." In-house publication, Brown and Caldwell Consultants, Pleasant Hill, CA.
13. Pritchard, M. T., D. S. Parker, and A. Malik. "Keeping costs from creeping." Water Environment & Technology **4**(12): 38–39 (December 1992).
14. Means, R. S. *Means Facilities Cost Data.* R. S. Means Co., Inc., Kingston, MA (updated annually).
15. Means, R. S. *Means Mechanical Cost Data.* R. S. Means Co., Inc., Kingston, MA (updated annually).
16. Dodge Cost Systems, Vol. 1, *Dodge Assemblies Cost Data; Vol. 2, Dodge Unit Cost Data; Vol. 3, Dodge Square Foot Cost Data; Vol. 4, Dodge Heavy Construction Cost Data. Vol. 5, Dodge Remodeling & Retrofit Cost Data.* McGraw-Hill, Princeton, NJ (updated annually).
17. Walski, T. M. (Ed.). *Proceedings of the Symposium on Cost Estimating for Water Supply Planning Studies.* Miscellaneous Paper EL-83-6, U.S. Army Corps of Engineers, Waterways Experiment Station, Vicksburg, MS (September 1983).
18. Water Research Centre. *Cost Information for Water Supply and Sewage Disposal.* TR 61, Water Research Centre, Medmenham, Marlow, Buckinghamshire, SL7 2HD, United Kingdom (1977).
19. Hall, R. S., J. Morley, and K. J. Naughton. "Current costs of process equipment." *Chemical Engineering,* **89:** 80–116 (April 1982).

Appendix A

Physical Data

CONTRIBUTORS
Carl N. Anderson
W. Eric Hopkins
Constantine N. Papadakis
Daniel L. Shaffer
Earle C. Smith

In this volume, SI means System International (a metric system) and U.S. means U.S. customary units (an English system). The notation 6.22 E−2 means 6.22×10^{-2}. Most tabular entries are shown to only three significant digits because the errors in engineering data are rarely less than 1%. If greater accuracy is needed, use the conversion factors in Table A-3.

By convention, lb_f (pound force) is written as lb, but lb_m (pound mass) is always written lb_m.

Table A-1. Base SI Units

Quantity	Name	Symbol
Length	Meter	m
Mass	Kilogram	kg
Time	Second	s
Electric current	Ampere	A
Thermodynamic temperature	Kelvin	°K
Amount of substance	Mole	mol
Luminous intensity	Candela	cd
Plane angle[a]	Radian	rad
Solid angle[a]	Steradian	sr

[a] Supplementary units.

Table A-2. Derived SI Units and Physical Quantities

Quantity	Symbol	Name	Formula
Acceleration, angular	—	Radian/second^2	rad/s^2
Acceleration, linear	—	Meter/second^2	m/s^2
Angle	rad	Radian	
Area	—	Meter^2	m^2
Capacitance	F	Farad	C/V
Charge	C	Coulomb	A · S
Concentration, molar	—	Molecular mass/meter^3	mol/m^3
Concentration, gravimetric	—	Milligram/liter or	mg/L
	—	Gram/meter^3	g/m^3
Conductance	S	Siemen	A/V
Density, mass	—	Kilogram/meter^3	kg/m^3
Electric field intensity	—	Volt/meter	V/m
Electric flux	C	Coulomb	A · s
Electric flux density	—	Coulomb/square meter	C/m^2
Electric potential	V	Volt	
Electric resistance	Ω	Ohm	V/A
Energy	J	Joule	N · m
Force	N	Newton	kg · m/s^2
Frequency	Hz	Hertz	cycle/s
Impedance	Ω	Ohm	V/A
Inductance	H	Henry	Wb/A
Luminous flux	lm	Lumen	cd · sr
Magnetic field intensity	—	Ampere/meter	A/m
Magnetic flux	Wb	Weber	V · s
Magnetic flux density	T	Tesla	Wb/m^2
Moment of force	—	Newton-meter	N · m
Power	W	Watt	J/s
Power, apparent	—	Volt-ampere	V · A
Power, reactive	—	Volt-ampere reactive	var
Pressure	Pa	Pascal	N/m^2
Reactance	Ω	Ohm	V/A
Resistance	Ω	Ohm	V/A
Resistivity	—	Ohm-meter	W · m
Stress	—	Newton/meter^2	N/m^2
Velocity, angular	—	Radian/second	rad/s
Velocity, linear	—	Meter/second	m/s
Viscosity, dynamic	—	Pascal · seccond	Pa · s
Viscosity, kinematic	—	Meter^2/second	m^2/s
Voltage	V	Volt	W/A
Volume	—	Meter^3	m^3
Volume, specific	—	Meter^3/kg	m^3/kg
Work	J	Joule	N · m

Table A-3. Physical Constants

Name	SI units	U.S. customary units
Acceleration due to gravity[a]	9.80665 m/s^2	32.174 ft/s^2
Atmosphere (standard)	101.325 kPa	14.696 lb/in.^{2}
Atmosphere (standard)	10.333 m of WC	33.899 ft of WC
Bar[b]	100.0 kPa	14.504 lb/in.^{2}
Mass density of water[b]	1000 kg/m^3	1.94 slugs/ft^3
Specific heat of air[c]	1003 N · m/(kg · °C)	0.2395 Ftu/lb$_m$· °F = 6000 ft · lb/(slug · °F)

[a] Standard value, changes with latitude and elevation.
[b] At 5°C (41°F).
[c] At 20°C (68°F) and 1 atm.

Table A-4. SI Prefixes

Multiplication factor		Prefix[a]	Symbol
1,000,000,000,000	E+12	tera	T
1,000,000,000	E+9	giga	G
1,000,000	E+6	mega	M
1,000	E+3	kilo	k
100	E+2	hecto[b]	h
10	E+1	deka[b]	da
0.1	E−1	deci[b]	d
0.01	E−2	centi[b]	c
0.001	E−3	milli	m
0.000 001	E−6	micro	μ
0.000 000 001	E−9	nano	n
0.000 000 000 001	E−12	pico	p

[a] The first syllable of every prefix is accented so that the prefix retains its identity. Therefore, the preferred pronunciation of kilometer places the accent on the first syllable, *ki*-lometer.
[b] The use of these prefixes should be avoided, except for the measurement of areas and volumes and for the nontechnical use of centimeter.

Table A-5. Conversion Factors, U.S. Customary Units to Commonly Used SI Units

Multiply the U.S. customary unit		by	To obtain the SI unit	
Name	Symbol		Symbol	Name
Acceleration feet per second squared[a]	ft/s^2	3.0480 E−1[a]	m/s^2	Meters per second squared
Angle	degrees (°)	1.745 E−2	rad	Radian
Area				
Acre	acre	4.047 E−1	ha	Hectare
Square foot	ft^2	9.2903 E−2	m^2	Square meter
Square inch	in.^{2}	6.4516 E−4[a]	m^2	Square meter
Concentration (for gases), parts per million	ppm	[b]	mg/m^3	Milligrams per cubic meter
Energy				
British thermal unit	Btu	1.0551 E+0	kJ	Kilojoule
Foot-pound (force)	ft · lb	1.3558 E+0	J	Joule
Horsepower-hour	hp · h	2.6845 E+0	MJ	Megajoule
Kilowatt-hour	kW · h	3.600 E+3[a]	kJ	Kilojoule

[a] Exact value or exact conversion.

(Table continues on next page.)

[b] $\text{mg/m}^3 = \frac{(\text{ppm})(\text{grams/mol})}{22.4\ \text{L/mol}} \times \frac{273}{273 + °\text{C}}$ Note: Regulations use °C = 25, but source measurements are referenced to °C = 21.1.

Table A-5. Continued

Multiply the U.S. customary unit		by	To obtain the SI unit	
Name	Symbol		Symbol	Name
Force				
Pound force	lb	4.4482 E+0	N	Newton
Flowrate				
Cubic feet per second	ft^3/s	2.8317 E−2	m^3/s	Cubic meters per second
Gallons per minute	gal/min	6.3090 E−2	L/s	Liters per second
Gallons per minute	gal/min	6.3090 E−2	L/s	Liters per second
Gallons per minute	gal/min	2.2712 E−1	m^3/h	Cubic meters per hour
Million gallons per day	Mgal/d	4.3813 E+1	L/s	Liters per second
Million gallons per day	Mgal/d	3.7854 E+0	ML/d	Million liters per day
Length				
Foot	ft	3.0480 E−1[a]	m	Meter
Inch	in.	2.5400 E+1[a]	mm	Millimeter
Mile	mi	1.6093 E+0	km	Kilometer
Luminous intensity				
Foot-candle	ft · cd	1.076 E+1	lx	Lux
Mass				
Pound	lb_m	4.5359 E−1	kg	Kilogram
Slug (32.174 lb_m)	slug	1.4594 E+1	kg	Kilogram
Moment				
Pound-inch	lb · in.	1.1298 E−1	N · m	Newton-meter
Power				
Horsepower	hp	7.457 E−1	kW	Kilowatt
Pressure (force/area)				
Atmosphere (standard)	atm	1.0133 E+2	kPa	Kilopascal (kN/m^2)
Pounds (force) per square inch	$lb/in.^2$	6.8948 E+0	kPa	Kilopascal
Specific heat	$Btu/(lb_m \cdot °R)$	4.188 E+0	J/(g · K)	Joules per gram-degrees kelvin
	ft · lb/(slug · °R)	1.672 E−1	N · m(kg · K)	Newton-meters per kilogram-degrees kelvin
Stress				
Pounds (force) per square inch	$lb/in.^2$	6.8948 E+0	kPa	Kilopascal
Kips (force) per square inch	$k/in.^2$	6.8948 E+0	MPa	Megapascal
Temperature				
Degrees Fahrenheit	°F	0.5556 (°F − 32)	°C	Degrees celsius
Degrees Fahrenheit	°F	0.5556 (°F + 459.67)	°K	Degrees kelvin
Degrees Rankine	°R	0.5556 (°R)	°K	Degrees kelvin
Velocity				
Feet per second	ft/s	3.048 E−1[a]	m/s	Meters per second
Miles per hour	mi/h	1.6093 E+0	km/h	Kilometers per hour
Volume				
Cubic foot	ft^3	2.8317 E+1	L	Liter
Cubic foot	ft^3	2.8317 E−2	m^3	Cubic meter
Gallon	gal	3.7854 E−3	m^3	Cubic meter
Gallon	gal	3.7854 E+0	L	Liter
Weight, specific				
Pounds per cubic foot	lb/ft^3	1.5708 E−1	kN/m^3	Kilonewtons per cubic meter

[a] Exact value or exact conversion.

[b] $mg/m^3 = \frac{(ppm)(grams/mol)}{22.4\ L/mol} \times \frac{273}{273 + °C}$. Note: Regulations use °C = 25, but source measurements are referenced to °C = 21.1.

Table A-6. Atmospheric Pressure (SI Units)

Elevation above sea level (m)	Atmospheric pressure[a] kPa	Water column (m)	Mercury column (mm)	Specific weight (g) of air at 20°C (kN/m³)[b]
0	101.3	10.33	760	1.18 E−2
500	95.6	9.74	717	1.11 E−2
1000	90.1	9.19	676	1.05 E−2
1500	84.8	8.64	636	9.87 E−3
2000	79.8	8.13	598	9.29 E−3
2500	73.3	7.47	550	8.53 E−3
3000	70.3	7.17	527	8.19 E−3
3500	66.1	6.74	496	7.70 E−3

[a] Storms commonly reduce atmospheric pressure by about 1.7%.

[b] At other temperatures and pressures, use $p_1v_1/K_1 = p_2v_2/K_2$ where p is pressure, v is volume, and K is degrees Kelvin (°C + 273).

Table A-7. Atmospheric Pressure (U.S. Customary Units)

Elevation above sea level (m)	Atmospheric pressure[a] lb/in.²	Water column ft	Water column in.	Mercury column (mm)	Specific weight (g) of air at 68°F (lb/ft³)[b]
0	14.7	33.9	29.92	760	7.52 E−2
1,000	14.2	32.7	28.90	734	7.26 E−2
2,000	13.7	31.6	27.87	708	7.00 E−2
3,000	13.2	30.4	26.81	681	6.73 E−2
4,000	12.7	29.4	25.90	658	6.51 E−2
5,000	12.1	28.2	24.92	633	6.26 E−2
6,000	11.8	27.2	24.01	610	6.04 E−2
7,000	11.4	26.3	23.19	589	5.83 E−2
8,000	10.7	24.8	21.89	556	5.50 E−2
9,000	10.5	24.2	21.38	543	5.37 E−2
10,000	10.1	23.4	20.63	524	5.18 E−2

[a] Storms commonly reduce atmospheric pressure by about 1.7%.

[b] At other temperatures and pressures use $p_1v_1/T_1 = p_2v_2/T_2$, where T = °F + 460, or use the general formula for atmospheric pressure

$$\frac{p_b}{p_a} = \exp\left[-\frac{gM(z_b - z_a)}{g_c RT}\right]$$

where $g_c = 32.2\ \text{ft} \cdot \text{lb}_m/\text{lb} \cdot \text{s}^2$, $g = 32.2\ \text{ft/s}^2$, $M = 29\ \text{lb}_m/\text{lb}_{mol}$, $R = \dfrac{1545\ \text{ft} \cdot \text{lb}}{\text{lb}_{mol} \cdot T}$, and $T = 460 + $ °F.

Table A-8. Physical Properties of Water (SI Units)[a]

Temperature (°C)	Specific weight[b] γ (kN/m^3)	Density ρ (kg/m^3)	Bulk modulus of elasticity[c] K (kPa)	Dynamic viscosity μ (Pa · s)	Kinematic viscosity ν (m^2/s)	Surface tension[d] σ (N/m)	Vapor pressure p_v (kPa)	WC (m)
0	9.805	9.998 E+2	1.98 E+6	1.78 E−3	1.79 E−6	0.0765	0.61	0.06
5	9.807	1.000 E+3	2.05 E+6	1.52 E−3	1.52 E−6	0.0749	0.87	0.09
10	9.804	9.997 E+2	2.10 E+6	1.31 E−3	1.31 E−6	0.0742	1.23	0.13
15	9.798	9.991 E+2	2.15 E+6	1.14 E−3	1.14 E−6	0.0735	1.70	0.17
20	9.789	9.982 E+2	2.17 E+6	1.00 E−3	1.00 E−6	0.0726	2.34	0.24
25	9.777	9.970 E+2	2.22 E+6	8.90 E−4	8.93 E−7	0.0720	3.17	0.32
30	9.764	9.957 E+2	2.25 E+6	7.98 E−4	8.00 E−7	0.0712	4.24	0.43
40	9.730	9.922 E+2	2.28 E+6	6.53 E−4	6.58 E−7	0.0696	7.38	0.76
50	9.689	9.880 E+2	2.29 E+6	5.47 E−4	5.53 E−7	0.0679	12.33	1.27
60	9.642	9.832 E+2	2.28 E+6	4.66 E−4	4.74 E−7	0.0662	19.92	2.07
70	9.589	9.778 E+2	2.25 E+6	4.04 E−4	4.13 E−7	0.0644	31.19	3.25
80	9.530	9.718 E+2	2.20 E+6	3.54 E−4	3.64 E−7	0.0626	47.34	4.97
90	9.466	9.653 E+2	2.14 E+6	3.15 E−4	3.26 E−7	0.0608	70.10	7.41
100	9.399	9.584 E+2	2.07 E+6	2.82 E−4	2.94 E−7	0.0589	101.33	10.78

[a] Adapted from Vennard and Street [1].

[b] Specific weight, γ, is force per unit volume. The relationship between γ, ρ, and the acceleration due to gravity, g, is $\gamma = \rho g$. See Table A-3 for the value of g.

[c] At atmospheric pressure.

[d] In contact with air. Note 1000 N/m = dynes/cm.

Table A-9. Physical Properties of Water (U.S. Customary Units)[a]

Temperature (°F)	Specific weight[a] γ (lb$_m$/ft^3)	Density[b] ρ (slug/ft^3)	Bulk modulus of elasticity[c] K (lb/in.^{2})	Dynamic viscosity μ (lb × s/ft^2)	Kinematic viscosity ν (ft^2/s)	Surface tension[d] σ (lb/ft)	Vapor Pressure p_v (lb/in.^{2})	WC (ft)
32	62.42	1.940	2.87 E+5	3.746 E−5	1.93 E−5	0.00518	0.09	0.2
40	62.43	1.940	2.96 E+5	3.229 E−5	1.66 E−5	0.00614	0.12	0.3
50	62.41	1.940	3.05 E+5	2.735 E−5	1.41 E−5	0.00509	0.18	0.4
60	62.37	1.938	3.13 E+5	2.359 E−5	1.22 E−5	0.00504	0.26	0.6
70	62.30	1.936	3.19 E+5	2.112 E−5	1.06 E−5	0.00498	0.36	0.8
80	62.22	1.934	3.24 E+5	1.799 E−5	9.30 E−6	0.00492	0.51	1.2
90	62.11	1.931	3.28 E+5	1.595 E−5	8.26 E−6	0.00486	0.70	1.6
100	62.00	1.927	3.31 E+5	1.424 E−5	7.39 E−6	0.00480	0.95	2.2
110	61.86	1.923	3.32 E+5	1.284 E−5	6.67 E−6	0.00473	1.27	2.9
120	61.71	1.918	3.32 E+5	1.168 E−5	6.09 E−6	0.00467	1.69	3.9
130	61.55	1.913	3.31 E+5	1.069 E−5	5.58 E−6	0.00460	2.22	5.2
140	61.38	1.908	3.30 E+5	0.981 E−5	5.14 E−6	0.00454	2.89	6.8
150	61.20	1.902	3.28 E+5	0.905 E−5	4.76 E−6	0.00447	3.72	8.8
160	61.00	1.896	3.26 E+5	0.838 E−5	4.42 E−6	0.00441	4.74	11.2
170	60.80	1.890	3.22 E+5	0.780 E−5	4.13 E−6	0.00434	5.99	14.2
180	60.58	1.883	3.18 E+5	0.726 E−5	3.85 E−6	0.00427	7.51	17.9
190	60.36	1.876	3.13 E+5	0.678 E−5	3.62 E−6	0.00420	9.34	22.3
200	60.12	1.868	3.08 E+5	0.637 E−5	3.41 E−6	0.00413	11.52	27.6
212	59.83	1.860	3.00 E+5	0.593 E−5	3.19 E−6	0.00404	14.70	35.4

[a] Specific weight (γ) is the weight per unit volume. The relationship between γ, ρ, and the acceleration due to gravity, g, is $\gamma = \rho g$. See Table A-3 for the value of g.

[b] At atmospheric pressure, 1 slug = 32.2 lbm × s^2.

[c] In many formulas, K must be in pounds per square foot. Multiply by 144 in.^{2}/ft^2.

[d] In contact with air.

Table A-10. Physical Properties of Pipe Materials

Material	Poisson's ratio, μ (dimensionless)	Modulus of elasticity (*E*) SI units (N/m^2)	U.S. customary units ($lb/in.^2$)	U.S. customary units (lb/ft^2)
Aluminum	0.33	7.30 E+10	1.05 E+7	1.51 E+9
Asbestos-cement	0.30	2.30 E+10	3.40 E+6	4.90 E+8
Brass	0.34	1.03 E+11	1.50 E+7	2.16 E+9
Copper	0.30	1.10 E+11	1.60 E+7	2.30 E+9
Ductile iron	0.28	1.66 E+11	2.40 E+7	3.46 E+9
Gray cast iron	0.28	1.03 E+11	1.50 E+7	2.16 E+9
HDPE[a]	0.45	1.03 E + 9	1.50 E+5	2.16 E+7
PVC	0.45	2.70 E+9	4.00 E+5	5.76 E+7
Steel	0.30	2.07 E+11	3.00 E+7	4.32 E+9

[a] Varies

Table A-11. Equivalent Weights and Measures

Length

U.S. customary units				SI units		
Miles	Yards	Feet	Inches	Kilometers	Meters	Millimeters
1	1.76 E+3	5.28 E+3	6.36 E+4	1.61 E+0	1.61 E+3	1.61 E+6
5.68 E−4	1	3.00 E+0	3.60 E+1	9.14 E−4	9.14 E−1	9.14 E+2
1.89 E−4	3.33 E−1	1	1.20 E+1	3.05 E−4	3.05 E−1	3.05 E+2
1.58 E−5	2.78 E−2	8.33 E−2	1	2.54 E−5	2.54 E−2	2.54 E+1
6.21 E−1	1.09 E+3	3.28 E+3	3.94 E+4	1	1.00 E+3	1.00 E+6
6.21 E−4	1.09 E+0	3.28 E+0	3.94 E+1	1.00 E−3	1	1.00 E+3
6.21 E−7	1.09 E−3	3.28 E−3	3.94 E−2	1.00 E−6	1.00 E−3	1

Area

U.S. customary units				SI units		
Square miles	Acres	Square feet	Square inches	Square kilometers	Hectares	Square meters
1	6.40 E+2	2.79 E+7	4.01 E+9	2.59 E+0	2.59 E+2	2.59 E+6
1.56 E−3	1	4.36 E+4	6.27 E+6	4.05 E−3	4.05 E−1	4.05 E+3
3.59 E−8	2.30 E−5	1	1.44 E+2	9.29 E−8	9.29 E−6	9.29 E−2
2.49 E−10	1.59 E−7	6.94 E−3	1	6.45 E−10	6.45 E−8	6.45 E−4
3.86 E−1	2.47 E+2	1.08 E+7	1.55 E+9	1	1.00 E+2	1.00 E+6
3.86 E−3	2.47 E+0	1.08 E+5	1.55 E+7	1.00 E−2	1	1.00 E+4
3.86 E−7	2.47 E−4	1.08 E+1	1.55 E+3	1.00 E−6	1.00 E−4	1

Volume

U.S. customary units					SI units		
Cubic yards	Cubic feet	Imperial gallons	U.S. gallons	Cubic inches	Cubic meters	Liters	Milliliters
1	2.7 E+1	1.68 E+2	2.02 E+2	4.67 E+4	7.65 E−1	7.65 E+2	7.65 E+5
3.70 E−2	1	6.23 E+0	7.48 E+0	1.73 E+3	2.83 E−2	2.83 E+1	2.83 E+4
5.95 E−3	1.61 E−1	1	1.20 E+0	2.77 E+2	4.55 E−3	4.55 E+0	4.55 E+3
4.95 E−3	1.34 E−1	8.33 E−1	1	2.31 E+2	3.79 E−3	3.79 E+0	3.79 E+3
2.14 E−5	5.79 E−4	3.60 E−3	4.33 E−3	1	1.64 E−5	1.64 E−2	1.64 E+1
1.31 E+0	3.53 E+1	2.20 E+2	2.64 E+2	6.10 E+4	1	1.00 E+3	1.00 E+6
1.31 E−3	3.53 E−2	2.20 E−1	2.64 E−1	6.10 E+1	1.00 E−3	1	1.00 E+3
1.31 E−6	3.53 E−5	2.20 E−4	2.64 E−4	6.10 E−2	1.00 E−6	1.00 E−3	1

(Table continues on next page.)

Table A-11. Continued

Discharge (flowrate)						
U.S. customary units			SI units			
Million gallons per day	Cubic feet per second	Gallons per minute	Cubic meters per second	Cubic meters per hour	Liters per second	Millions of liters per day
1	1.55 E+0	6.94 E+2	4.38 E−2	1.58 E+2	4.38 E+1	3.79 E+0
6.46 E−1	1	4.49 E+2	2.83 E−2	1.02 E+2	2.83 E+1	2.45 E+0
1.44 E−3	2.23 E−3	1	6.31 E−5	2.27 E−1	6.31 E−2	5.45 E−3
2.28 E+1	3.53 E+1	1.58 E+4	1	3.60 E+3	1.00 E+3	8.64 E+1
6.34 E−3	9.81 E−3	4.40 E+0	2.78 E−4	1	2.78 E−1	2.40 E−2
2.28 E−2	3.53 E−2	1.59 E+1	1.00 E−3	3.60 E+0	1	8.64 E−2
2.64 E−1	4.09 E−1	1.83 E+2	1.16 E−2	4.17 E+1	1.16 E+1	1

Force					
U.S. customary units				SI units	
Tons	Kilopounds	Pounds	Grains	Kilonewtons	Newtons
1	2.00 E+0	2.00 E+3	1.40 E+7	8.90 E+0	8.90 E+3
5.00 E−1	1	1.00 E+3	7.00 E+6	4.45 E+0	4.45 E+3
5.00 E−4	1.00 E−3	1	7.00 E+3	4.45 E−3	4.45 E+0
7.14 E−8	1.43 E−7	1.43 E−4	1	6.35 E−7	6.35 E−4
1.12 E−1	2.25 E−1	2.25 E+2	1.57 E+6	1	1.00 E+3
1.12 E−4	2.25 E−4	2.25 E−1	1.57 E+3	1.00 E−3	1

Mass				
U.S. customary units		SI units		
Slugs	Pounds mass	Kilograms	Grams	Milligrams
1	3.22 E+1	1.46 E+1	1.46 E+4	1.46 E+7
3.11 E−2	1	4.54 E−1	4.54 E+2	4.54 E+5
6.85 E−2	2.20 E+0	1	1.00 E+3	1.00 E+6
6.85 E−5	2.20 E−3	1.00 E−3	1	1.00 E+3
6.85 E−8	2.20 E−6	1.00 E−6	1.00 E−3	1

Pressure or stress (force/area)						
	U.S. customary units			SI units		
Standard atmospheres	Pounds per square inch	Hg (in.)	Water (ft)	Water (m)	Kilopascals	Hg (mm)
1	1.47 E+1	2.99 E+1	3.40 E+1	1.04 E+1	1.01 E+2	7.60 E+2
6.80 E−2	1	2.04 E+0	2.31 E+0	7.04 E−1	6.90 E+0	5.17 E+1
3.34 E−2	4.91 E−1	1	1.14 E+0	3.46 E−1	3.39 E+0	2.54 E+1
2.94 E−2	4.33 E−1	8.81 E−1	1	3.05 E−1	2.98 E+0	2.24 E+1
9.66 E−2	1.42 E+0	2.89 E+0	3.28 E+0	1	9.79 E+0	7.34 E+1
9.87 E−3	1.45 E−1	2.95 E−1	3.35 E−1	1.02 E−1	1	7.50 E+0
1.32 E−2	1.93 E−2	3.94 E−2	4.47 E−2	1.36 E−2	1.33 E−1	1

Power				
U.S. customary units				SI units
British thermal units per second	Horsepower	Foot-pounds per second	Foot-pounds per minute	Kilowatts
1	1.41 E+0	7.78 E+2	4.67 E+4	1.06 E+0
7.07 E−1	1	5.50 E+2	3.30 E+4	7.46 E−1
1.29 E−3	1.82 E−3	1	6.00 E+1	1.36 E−3
2.14 E−5	3.03 E−5	1.67 E−2	1	2.26 E−5
9.48 E−1	1.34 E+0	7.38 E+2	4.43 E+4	1

Energy				
U.S. customary units			SI units	
Horsepower-hours	British thermal units	Foot-pounds	Kilowatt-hours	Joules
1	2.54 E+3	1.98 E+6	7.46 E−1	2.68 E+6
3.93 E−4	1	7.78 E+2	2.93 E−4	1.05 E+3
5.05 E−7	1.29 E−3	1	3.77 E−7	1.36 E+0
1.34 E+0	3.41 E+3	2.66 E+6	1	3.60 E+6
3.73 E−7	9.48 E−4	7.38 E−1	2.78 E−7	1

A-1. References

1. Vennard, J. K., and R. L. Street. *Elementary Fluid Mechanics,* 5th ed. John Wiley, New York (1975).
2. Metcalf & Eddy, Inc. *Wastewater Engineering, Treatment, Disposal, and Reuse,* 3rd ed. Revised by G. Tchobanoglous. McGraw-Hill, New York (1991).

Appendix B

Data for Flow in Pipes, Fittings, and Valves

ROBERT L. SANKS

CONTRIBUTORS
William F. H. Gros
Charles D. Morris
William Wheeler

The use of Tables B-1 through B-4 is explained in Section 3-3. Headlosses for pipe diameters 100 mm (4 in.) and larger are given for Hazen–Williams (H–W) C of 120 given by Ten-State Standards [1] for cement-mortar lined pipe. In Tables B-3 and B-4, headlosses for smaller pipe were calculated by the Darcy–Weisbach equation in which a value of ε (0.23 mm or 0.008 in.) was found that resulted in the same headloss for 100-mm (4-in.) pipe as was found using an H–W C of 120. For smaller pipe, the more accurate Darcy–Weisbach equation gives higher headlosses than the H–W equation. Therefore, using the H–W equations for pipe smaller than 100 mm (4 in.) leads to large errors on the unsafe side. The pipe roughness coefficients in Table B-5 do not include safety factors.

The energy loss coefficients, K, in Tables B-6 and B-7 were taken from a variety of sources. Published values from any source should be used conservatively unless the test procedure used to the values is known. In general, the true value may well be 20% less or 30% more than the tabulated value, but some published data differ by more than 100%. Note that K values depend partly on size. For a crude rule, increase K by 5% for each 25-mm (1-in.) decrease in size below 300 mm (12 in.). When fittings are grouped close together, as in a pumping station, the extra turbulence in (and possible rotation of) the fluid may further increase the headlosses shown by 30% or more. If the velocity in check valves is low (see the manufacturer's data), the coefficient K may be several times the values given in Table B-7.

The Moody diagram for determining f in the Darcy–Weisbach formula is shown in Figure B-1. Headlosses in valves are often assumed to equal a constant coefficient times the velocity head, but this is not true for spring-loaded and counterweighted swing check valves. The headlosses in such valves based on tests of Mueller valves are shown in Figures B-2 and B-3.

Table B-1. Mortar-lined Class 53 Ductile Iron Pipe [2], Hazen–Williams $C = 120$, SI Units. For checking or preliminary use. Use Darcy–Weisbach formulas and exact dimensions for critical problems.

Nominal size (mm)	OD (mm)	Thickness (mm)		Inside	
		Iron wall	Mortar lining[a]	Diameter (mm)	Area (m^2)
75[e]	101	7.87	1.59	82	5.24 E–3
100	122	8.13	1.59	102	8.25 E–3
150	175	8.64	1.59	155	1.88 E–2
200	230	9.14	1.59	208	3.41 E–2
250	282	9.65	1.59	259	5.29 E–2
300	335	10.2	1.59	312	7.63 E–2
350	389	10.7	2.38	363	1.03 E–1
400	442	10.9	2.38	415	1.35 E–1
450	495	11.2	2.38	468	1.72 E–1
500	549	11.4	2.38	521	2.13 E–1
600	655	11.9	2.38	627	3.08 E–1
750	813	13.0	3.18	781	4.79 E–1
900	973	14.7	3.18	937	6.90 E–1
1050	1130	16.5	3.18	1090	9.35 E–1
1200	1290	18.3	3.18	1250	1.22 E+0
1350	1450	20.6	3.18	1550	1.55 E+0

[a]Single thickness. For the same discharge, double thickness increases velocity by about 2% and increases headloss by about 6%.
[b]Calculated from Equation 3–9a. For any other velocity or C factor, multiply the headloss by $[(v/2)(120/C)]^{1.85}$
[c]Of metal only.
[d]Flanges screwed to barrel by the pipe manufacturer.
[e]May not be available.

Table B-2. Mortar-lined Class 53 Ductile Iron Pipe [2], Hazen–Williams $C = 120$, U.S. Customary Units. For checking or preliminary use. Use Darcy–Weisbach formulas and exact dimensions for critical problems.

Nominal size (in.)	OD (in.)	Thickness (in.)		Inside	
		Iron wall	Mortar lining[a]	Diameter (in.)	Area (ft^2)
3[e]	3.96	0.31	1/16	3.22	5.64 E–2
4	4.80	0.32	1/16	4.04	8.88 E–2
6	6.90	0.34	1/16	6.10	2.03 E–1
8	9.05	0.36	1/16	8.21	3.67 E–1
10	11.10	0.38	1/16	10.2	5.69 E–1
12	13.20	0.40	1/16	12.3	8.22 E–1
14	15.30	0.42	3/32	14.3	1.11 E+0
16	17.40	0.43	3/32	16.3	1.46 E+0
18	19.50	0.44	3/32	18.4	1.85 E+0
20	21.60	0.45	3/32	20.5	2.29 E+0
24	25.80	0.47	3/32	24.7	3.32 E+0
30	32.00	0.51	1/8	30.7	5.15 E+0
36	38.30	0.58	1/8	36.9	7.42 E+0
42	44.50	0.65	1/8	43.0	1.01 E+1
48	50.80	0.72	1/8	49.1	1.32 E+1
54	57.10	0.81	1/8	55.2	1.66 E+1

[a]Single thickness. For the same discharge, double thickness increases velocity by about 2% and increases headloss by about 6%.
[b]Calculated from Equation 3–9b. For any other velocity or C factor, multiply the headloss by $[(v/5)(120/C)]^{1.85}$
[c]Of metal only.
[d]Flanges screwed to barrel by the pipe manufacturer.
[e]May not be available.

For v = 2 m/s				Weight[c]		
Discharge, Q			Headloss,[b] h_f			
m³/s	m³/h	ML/d	(m/1000 m)	Pipe[c] (kg/m)	Flange,[d] each (kg)	Flange or flare OD, mm (true)
1.05 E–2	3.77 E+1	9.05 E–1	66	16.2	3	191
1.65 E–2	5.94 E+1	1.43 E+0	50	20.6	6	229
3.76 E–2	1.36 E+2	3.25 E+0	31	31.9	8	279
6.82 E–2	2.46 E+2	5.89 E+0	22	44.8	12	343
1.06 E–1	3.81 E+2	9.14 E+0	17	58.3	17	406
1.53 E–1	5.50 E+2	1.32 E+1	14	73.3	26	483
2.06 E–1	7.43 E+2	1.78 E+1	12	89.5	33	533
2.71 E–1	9.76 E+2	2.34 E+1	9.8	104	41	597
3.44 E–1	1.24 E+3	2.97 E+1	8.5	120	41	635
4.26 E–1	1.54 E+3	3.68 E+1	7.5	136	52	699
6.17 E–1	2.22 E+3	5.33 E+1	6.1	170	73	813
9.57 E–1	3.45 E+3	8.27 E+1	4.7	229	109	984
1.38 E+0	4.96 E+3	1.19 E+2	3.8	313	159	1168
1.87 E+0	6.73 E+3	1.62 E+2	3.2	408	227	1346
2.44 E+0	8.80 E+3	2.11 E+2	2.7	517	284	1511
3.09 E+0	1.11 E+4	2.67 E+2	2.4	652	345	1683

For v = 5 ft/s				Weight[c]		
Discharge, Q			Headloss,[b] h_f			
gal/min	ft³/s	Mgal/d	(ft/1000 ft)	Pipe (lb/ft)	Flange,[d] each (lb)[e]	Flange or flare *OD*, in.
1.27 E+2	2.82 E–1	1.82 E–1	39	10.9	10	7.50
1.99 E+2	4.44 E–1	2.87 E–1	30	13.8	13	9.00
4.55 E+2	1.01 E+0	6.55 E–1	19	21.4	17	11.00
8.24 E+2	1.84 E+0	1.19 E+0	13	30.1	27	13.50
1.28 E+3	2.85 E+0	1.84 E+0	10	39.2	38	16.00
1.84 E+3	4.11 E+0	2.66 E+0	8.2	49.2	58	19.00
2.49 E+3	5.56 E+0	3.59 E+0	6.9	60.1	72	21.00
3.27 E+3	7.29 E+0	4.71 E+0	5.8	70.1	90	23.50
4.16 E+3	9.27 E+0	5.99 E+0	5.1	80.6	90	25.00
5.15 E+3	1.15 E+1	7.42 E+0	4.5	91.5	115	27.50
7.45 E+3	1.66 E+1	1.07 E+1	3.6	114	160	32.00
1.16 E+4	2.58 E+1	1.66 E+1	2.8	154	240	38.75
1.67 E+4	3.71 E+1	2.40 E+1	2.3	210	350	46.00
2.26 E+4	5.03 E+1	3.25 E+1	1.9	274	500	53.00
2.95 E+4	6.58 E+1	4.35 E+1	1.6	347	625	59.50
3.73 E+4	8.32 E+1	5.38 E+1	1.4	438	760	66.25

Table B-3. Standard Weight (ANSI B36.10) Steel Pipe [3], Hazen–Williams $C = 120$, SI Units. For checking or preliminary use. Use the Darcy–Weisbach formula and exact pipe dimensions for critical problems.

Nominal size (mm)	OD (mm)	Thickness (mm) Steel wall	Thickness (mm) Plastic lining[a]	Inside Diameter (mm)	Inside Area (m^2)
3	10.3	1.73	None	6.8	3.67 E–5
6	13.7	2.24	None	9.2	6.71 E–4
9	17.1	2.31	None	12.5	1.23 E–4
12	21.3	2.77	None	15.8	1.96 E–4
18	26.7	2.87	None	20.9	3.44 E–4
25	33.4	3.38	0.305	26.0	5.32 E–4
32	42.2	3.56	0.305	34.4	9.32 E–4
38	48.3	3.68	0.305	40.3	1.28 E–3
50	60.3	3.91	0.305	51.9	2.12 E–3
63	73.0	5.16	0.305	62.1	3.03 E–3
75	88.9	5.49	0.305	77.3	4.70 E–3
88	102	5.74	0.305	89.5	6.29 E–3
100	114	6.02	0.305	102	8.12 E–3
125	141	6.55	0.305	128	1.28 E–2
150	168	7.11	0.305	153	1.85 E–2
200	219	8.17	0.305	202	3.21 E–2
250	273	9.27	0.305	254	5.06 E–2
300	324	9.53	0.305	304	7.27 E–2
350	356	9.53	0.305	336	8.86 E–2
400	406	9.53	0.305	387	1.17 E–1
450	457	9.53	0.305	438	1.50 E–1
500	508	9.53	0.305	488	1.87 E–1
600	610	9.53	0.305	590	2.73 E–1
750	762	9.53	0.305	742	4.33 E–1

Larger pipe is available. Consult the manufacturers.

[a]Mortar linings can be substituted in 100-mm and larger pipe. Thicknesses for shop applications are 6.4 mm for 100- to 250-mm pipe, 7.9 mm for 300- to 550-mm pipe, and 9.5 mm for 600- to 900-mm pipe.

[b]For any other velocity and/or C factor, multiply the headloss by $[(v/2)(120/C)]^{1.85}$.

[c]Darcy–Weisbach equation and Moody diagram for $\varepsilon = 0.23$ mm, $T = 15°C$ (see Table B-5).

[d]Welded slip-on type.

For $v = 2$ m/s				Weight		
Discharge, Q						
(m^3/s)	(m^3/h)	(ML/d)	Headloss[b] h_f (m/1000m)	Pipe (kg/m)	Flange,[d] each (kg)	Flange or flare OD, mm (true)
7.33 E–5	2.64 E–1	6.34 E–3	1900[c]	0.36	NA	NA
1.34 E–4	4.83 E–1	1.16 E–2	1200[c]	0.63	NA	NA
2.46 E–4	8.87 E–1	2.13 E–2	800[c]	0.85	NA	NA
3.92 E–4	1.41 E+0	3.39 E–2	570[c]	1.19	0.5	89
6.88 E–4	2.48 E+0	5.95 E–2	400[c]	1.68	0.7	99
1.07 E–3	3.83 E+0	9.20 E–2	300[c]	2.50	0.9	108
1.86 E–3	6.71 E+0	1.61 E–1	200[c]	3.38	1.1	117
2.55 E–3	9.18 E+0	2.20 E–1	170[c]	4.05	1.4	127
4.23 E–3	1.52 E+1	3.66 E–1	120[c]	5.44	2.3	152
6.06 E–3	2.18 E+1	5.23 E–1	95[c]	8.62	3.6	178
9.39 E–3	3.38 E+1	8.11 E–1	71[c]	11.3	4.1	191
1.26 E–2	4.53 E+1	1.09 E+0	59	13.6	5	216
1.62 E–2	5.84 E+1	1.40 E+0	51	16.1	6	229
2.56 E–2	9.20 E+1	2.21 E+0	40	21.7	7	254
3.70 E–2	1.33 E+2	3.20 E+0	31	28.3	8	279
6.42 E–2	2.31 E+2	5.54 E+0	23	42.6	13	343
1.01 E–1	3.65 E+2	8.75 E+0	17	60.3	18	406
1.45 E–1	5.23 E+2	1.26 E+1	14	73.8	28	483
1.77 E–1	6.38 E+2	1.53 E+1	13	81.3	38	533
2.35 E–1	8.46 E+2	2.03 E+1	11	97.6	48	597
3.01 E–1	1.08 E+3	2.60 E+1	9.2	105	49	635
3.75 E–1	1.35 E+3	3.24 E+1	8.1	117	67	699
5.47 E–1	1.97 E+3	4.72 E+1	6.5	141	93	813
8.66 E–1	3.12 E+3	7.48 E+1	5.0	177	136	984

Table B-4. Standard Weight (ANSI B36.10) Steel Pipe [3] Hazen–Williams $C = 120$, U.S. Customary Units. For checking or preliminary use. Use the Darcy–Weisbach formula and exact pipe dimensions for critical problems.

		Thickness (in.)		Inside	
Nominal size (in.)	OD (in.)	Steel wall	Plastic lining[a]	Diameter (in.)	Area (ft^2)
1/8	0.405	0.068	None	0.269	3.95 E–4
1/4	0.540	0.088	None	0.364	7.22 E–4
3/8	0.675	0.091	None	0.493	1.32 E–3
1/2	0.840	0.109	None	0.622	2.11 E–3
3/4	1.050	0.113	None	0.824	3.70 E–3
1	1.315	0.133	0.012	1.03	5.79 E–3
1 1/4	1.660	0.140	0.012	1.36	1.01 E–2
1 1/2	1.900	0.145	0.012	1.57	1.34 E–2
2	2.375	0.154	0.012	2.04	2.27 E–2
2 1/2	2.875	0.203	0.012	2.45	3.27 E–2
3	3.500	0.216	0.012	3.04	5.04 E–2
3 1/2	4.000	0.226	0.012	3.52	6.77 E–2
4	4.500	0.237	0.012	4.00	8.74 E–2
5	5.563	0.258	0.012	5.02	1.37 E–1
6	6.625	0.280	0.012	6.04	1.99 E–1
8	8.625	0.322	0.012	7.96	3.45 E–1
10	10.75	0.365	0.012	10.0	5.45 B–1
12	12.75	0.375	0.012	12.0	7.82 E–1
14	14.00	0.375	0.012	13.2	9.54 E–1
16	16.00	0.375	0.012	15.2	1.26 E+0
18	18.00	0.375	0.012	17.2	1.62 E+0
20	20.00	0.375	0.012	19.2	2.02 E+0
24	24.00	0.375	0.012	23.2	2.94 E+0
30	30.00	0.375	0.012	29.2	4.66 E+0

Larger pipe is available. Consult the manufacturers.

[a] Mortar linings can be substituted on 4-in. and larger pipe. Thicknesses for shop applications are 1/4 in. for 4- to 10-in. pipe, 5/16 in. for 12- to 20-in. pipe, 3/8 in. for 24 to 30 in. pipe.

[b] For any other velocity or C factor, multiply the headloss by $[(v/2)(120/C)]^{1.85}$

[c] Darcy–Weisbach equation and Moody diagram for $\varepsilon = 0.0035$, $T = 60°F$ (see Table B–5).

[d] Of metal only.

[e] Welded slip-on type.

For v = 5 ft/s				Weight[d]		
Discharge, Q			Headloss,[b]	Pipe	Flange,[e]	Flange or
(gal/min)	(ft^3/s)	(Mgal/d)	(ft/1000 ft)	(lb/ft)	each (lb)	flare OD, in.
8.88 E–1	1.98 E–3	1.28 E–3	1120[c]	0.24	NA	NA
1.62 E+0	3.61 E–3	2.33 E–3	750[c]	0.42	NA	NA
2.96 E+0	6.60 E–3	4.27 E–3	450[c]	0.57	NA	NA
4.76 E+0	1.06 E–2	6.85 E–3	340[c]	0.80	1	3.50
8.31 E+0	1.85 E–2	1.20 E–2	240[c]	1.13	1.5	3.88
1.29 E+l	2.87 E–2	1.85 E–2	180[c]	1.68	2	4.25
2.27 E+l	5.05 E–2	3.26 E–2	120[c]	2.27	2.5	4.62
3.01 E+l	6.70 E–2	4.33 E–2	100[c]	2.72	3	5.00
5.10 E+l	1.14 E–1	7.37 E–2	73[c]	3.65	5	6.00
7.36 E+l	1.64 E–1	1.06 E–1	57[c]	5.79	8	7.00
1.13 E+2	2.52 E–1	1.63 E–1	43[c]	7.58	9	7.50
1.52 E+2	3.39 E–1	2.19 E–1	36	9.11	11	8.50
1.96 E+2	4.37 E–1	2.82 E–1	30	10.8	13	9.00
3.09 E+2	6.88 E–1	4.45 E–1	23	14.6	15	10.00
4.47 E+2	9.95 E–1	6.43 E–1	19	19.0	17	11.00
7.75 E+2	1.73 E+0	1.12 E+0	14	28.6	28	13.50
1.22 E+3	2.72 E+0	1.76 E+0	10	40.5	40	16.00
1.76 E+3	3.91 E+0	2.53 E+0	8.4	49.6	61	18.00
2.14 E+3	4.77 E+0	3.88 E+0	7.5	54.6	83	21.00
2.84 E+3	6.32 E+0	4.09 E+0	6.4	65.6	106	23.50
3.63 E+3	8.09 E+0	5.23 E+0	5.6	70.6	109	25.00
4.52 E+3	1.01 E+l	6.51 E+0	4.8	78.6	148	27.50
6.60 E+3	1.47 E+1	9.51 E+0	3.9	94.6	204	32.00
1.04 E+4	2.33 E+1	1.51 E+1	3.0	119.0	300	38.75

Table B-5. Probable Coefficients for Pipe Friction for the Design of Pipes Flowing Full for Nonaggressive Waters and Good Pipeline Maintenance. For critical problems consult the literature [4, 5, 6].

Note: Modify all tabular entries to obtain maximum and minimum values with a wider range to ensure an envelope of piping headloss curves that includes all possible values of roughness. See Section 3-2.

Material	Hazen–Williams C	Manning n[c]	Moody Diagram ε mm	Moody Diagram ε in.
Probable range of coefficients				
Plastic, FRP, and Epoxy[a]				
≤ 400 mm (16 in.)	140–130	0.010–0.010	Smooth–0.20	Smooth–0.008
= 600 mm (24 in.)	145–135	0.009–0.010	Smooth–0.20	Smooth–0.008
≥ 900 mm (36 in.)	150–140	0.009–0.010	Smooth–0.20	Smooth–0.008
Cement mortar lining[a]				
Centrifugally spun				
≤ 400 mm (16 in.)	135–125	0.010–0.011	0.13–0.33	0.005–0.013
= 600 mm (24 in.)	140–130	0.010–0.010	0.13–0.33	0.005–0.013
≥ 900 mm (36 in.)	145–135	0.009–0.010	0.13–0.33	0.005–0.013
Trowled in place[a]				
≤ 400 mm (16 in.)	130–120	0.010–0.011	0.20–0.50	0.008–0.020
= 600 mm (24 in.)	135–125	0.010–0.011	0.20–0.50	0.008–0.020
≥ 900 mm (36 in.)	140–130	0.010–0.010	0.20–0.50	0.008–0.020
Values below taken from the general literature[b]				
Ten-State Standards [1]				
Cement mortar or plastic	120	0.011	0.41	0.016
Unlined steel or ductile iron	100	0.013	1.5	0.060
Old pipe or lining in service for 20 yr or more and nonaggressive water[b]				
Smooth glass or plastic	135	0.010	0.13	0.005
Cement mortar,				
centrifugally spun	130	0.010	0.19	0.0075
troweled	125	0.011	0.28	0.011
Asbestos cement	125	0.011	0.28	0.011
Centrifugally cast CPP	130	0.010	0.19	0.0075
Wood stave	110	0.012	0.89	0.035
Riveted steel	80	0.016	5.6	0.22
Concrete, formed	80	0.016	5.6	0.22
Clay, not pressurized	100	0.013	1.5	0.060
Wrought iron	100	0.013	1.5	0.060
Galvanized iron	90	0.014	3.0	0.12

[a]For use with true ID of pipe.

[b]For Hazen–Williams C, use with nominal diameter of pipe. Headlosses predicted with the Hazen–Williams equation are too low for small pipe and/or low velocities.

[c]The Manning equation is incorrect for partly full circular pipes. See Figure B-5.

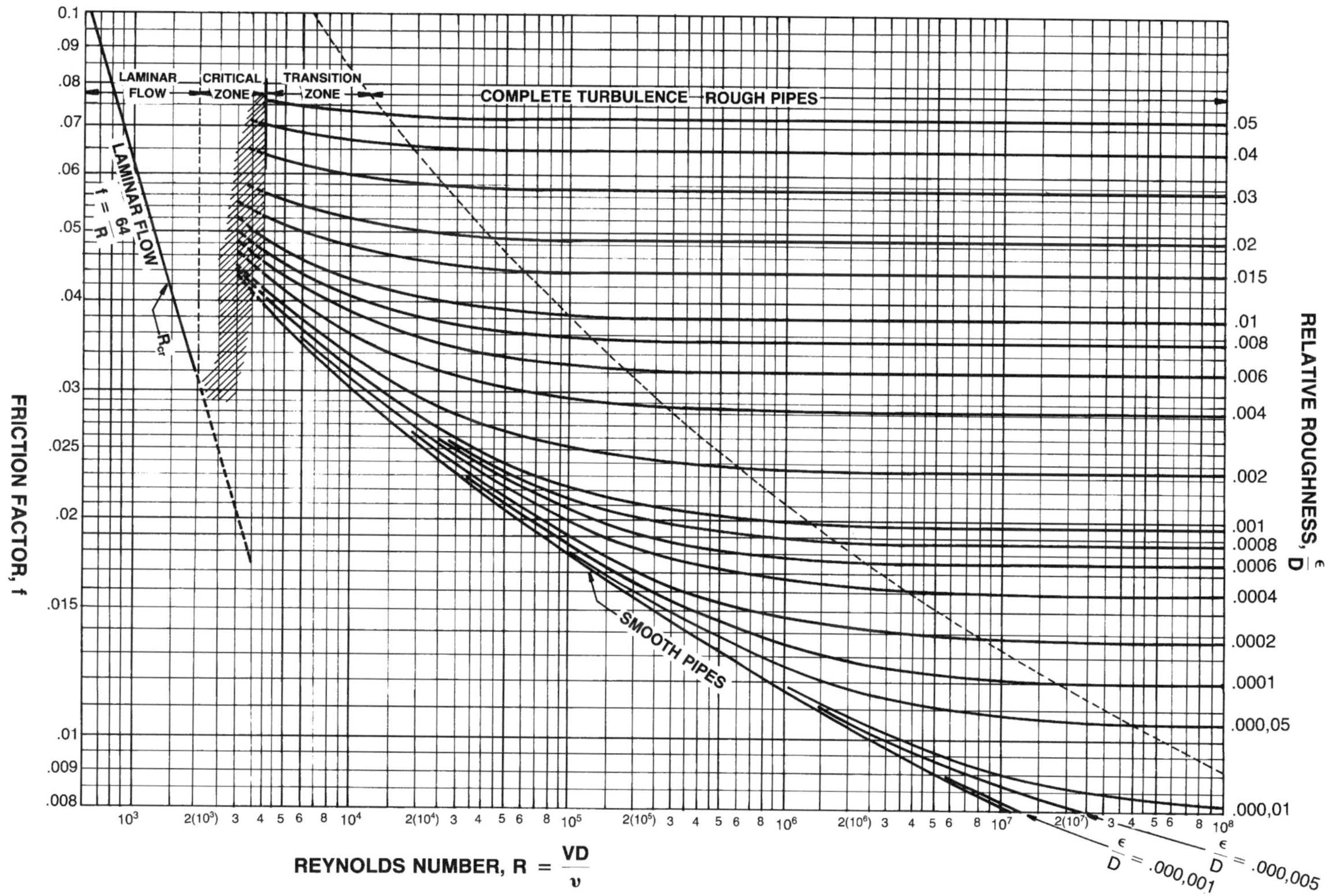

Figure B-1. Moody diagram for determining *f* in the Darcy–Weisbach formula: a Graphical solution of the Colebrook equation 3-12. After Moody [7].

Table B-6. Recommended Energy Loss Coefficients, *K*, for Flanged Pipe Fittings[a]

Fitting		*K*
Entrance	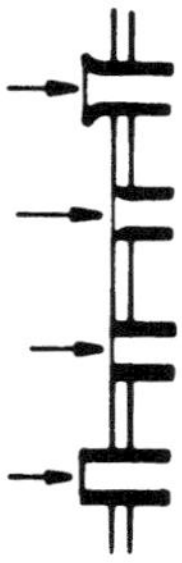	
Bellmouth		0.05
Rounded		0.25
Sharp-edged		0.5
Projecting		0.8
Exits		
All of the above		1.0
Bends, mitered		
$\theta = 15^\circ$		0.05
$\theta = 22.5^\circ$		0.075
$\theta = 30^\circ$		0.10
$\theta = 45^\circ$		0.20
$\theta = 60^\circ$		0.35
$\theta = 90^\circ$		0.80
90° bend $3 \times 30^\circ = 90^\circ$ $4 \times 22.5^\circ = 90^\circ$		0.30
Forged or cast fittings		
90° elbow, standard		0.25
90° elbow long radius		0.18
45° elbow		0.18
Forged or cast fittings		
Return bend, $r = 1.4$ D		0.40
Tee, line flow		0.30
Tee, branch flow	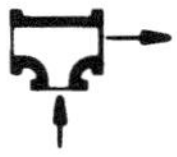	0.75
Cross, line flow		0.50
Cross, branch flow	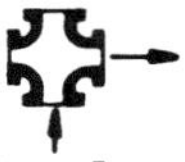	0.75
Wye, 45°	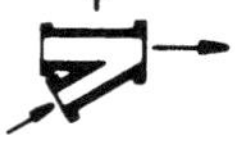	0.50
Increasers		
Conical	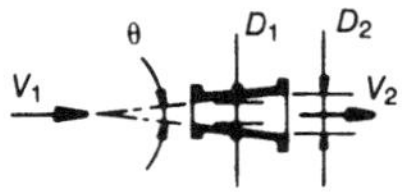	$h = K\left[1-\left(\frac{D_1}{D_2}\right)^2\right]v_2^2/2g$ $K = 3.5(\tan\theta)^{1.22}$
Conical (approximate)		$h = 0.25\left(v_1^2 - v_2^2\right)/2g$
Sudden	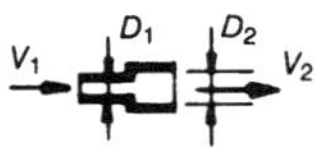	$h = \frac{v_1^2 - v_2^2}{2g} = \left[\left(\frac{A_2}{A_1}\right)^2 - 1\right]\frac{v_2^2}{2g}$
Reducers		
Conical		$h = Kv_2^2/2g$ $K = 0.03 \pm 0.01$
Sudden	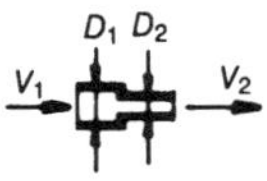	$h = \frac{1}{2}\left[1-\left(\frac{D_2}{D_1}\right)^2\right]v_2^2/2g$

[a] $h = Kv^2/2g$, where v is the maximum velocity in nonprismatic fittings. Increase *K* by 5% for each 25-mm (1-in.) decrement in pipe smaller than 300 mm (12 in.). Expect *K* values to vary from –20 to +30% or more.

Table B-7. Recommended Energy Loss Coefficients, *K*, for Valves Fully Open[a,b,c]

Valve type	*K*
Angle	1.8–2.9
Ball	0.04
Butterfly	
25-lb Class	0.16
75-lb Class	0.27
150-lb Class	0.35
Check valves	
Ball	0.9–1.7 but see Mfr's data for specific size and flowrate.
Center-guided globe style	2.6
Double door	
8 in. or smaller	2.5
10 to 16 in.	1.2
Foot	
Hinged disc	1–1.4
Poppet	5–14
Rubber flapper	
$v < 6$ft/s	2.0
$v > 6$ ft/s	1.1
Slanting disc[d]	0.25–2.0
Swing[d]	0.6–2.2, but see Figures B-2 and B-3.
Cone	0.04
Diaphragm or pinch	0.2–0.75
Gate	
Double disc	0.1–0.2
Resilient seat	0.3
Globe	4.0–6.0
Knife gate	
Metal seat	0.2
Resilient seat	0.3
Plug	
Lubricated	0.5–1.0
Eccentric	
Rectangular (80%) opening	1.0
Full bore opening	0.5

[a] $h = Kv^2/2g$, where v is the velocity in the approach piping.

[b] For 300-mm (12-in.) valves and velocities of about 2 m/s (6 ft/s). Note that *K* may increase significantly for smaller valves. Consult the manufacturer.

[c] Expect *K* to vary from –20 to +50% or more.

[d] Depending on adjustment of closure mechanism, velocity may have to exceed 4 m/s (12 ft/s) to open the valve fully. Adjustment is crucial to prevent valve slam.

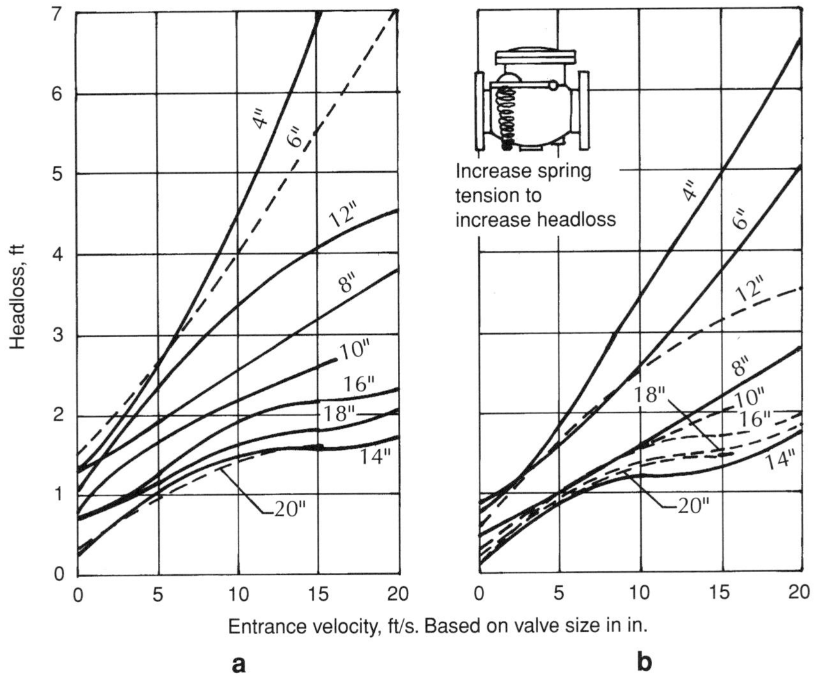

Figure B-2. Headlosses for Mueller swing check valves with spring-loaded levers. (a) Maximum headloss; (b) minimum headloss.

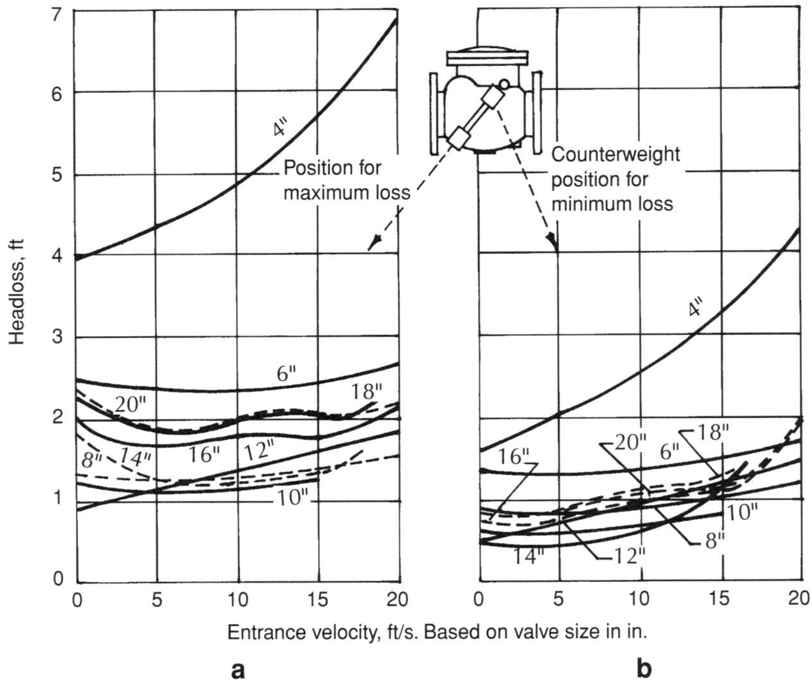

Figure B-3. Headlosses for Mueller swing check valves with counterweighted levers. (a) Maximum headloss; (b) minimum headloss.

Geometric properties of circular segments are shown in Figure B-4. As explained in Section 3-5, the Manning equation, although accurate for pipes flowing full, is in error by as much as 25% for pipes flowing partly full. The comparison between Manning's data and extensive field data as depicted by Camp [8] is shown in Figure B-5. Wheeler [9] developed an empirical equation that fits Camp's curve of field data within an error of less than 2%, and comparative values of Manning's data and observed (actually Wheeler's) data are given in Table B-8.

The velocities required to clear or scour air pockets from a pipe that slopes downward, given in Table B-9, were computed by Wheeler [10] from an equation (developed by Wisner, Mohsen, and Kouwen [11]) that can be rearranged to

$$v = \sqrt{gD}(0.825 + 0.25\sqrt{\sin\theta}) \tag{B-1}$$

where v is the velocity required, D is the ID of the pipe, and θ is angle of the downward gradient of the pipe profile. The use of velocity as a mechanism for air bubble removal was not common in 1996. Until it has become common practice, users are advised to study the literature—at least Edmunds [12], Walski et al. [13], and Falvey [14] to gain confidence in the method.

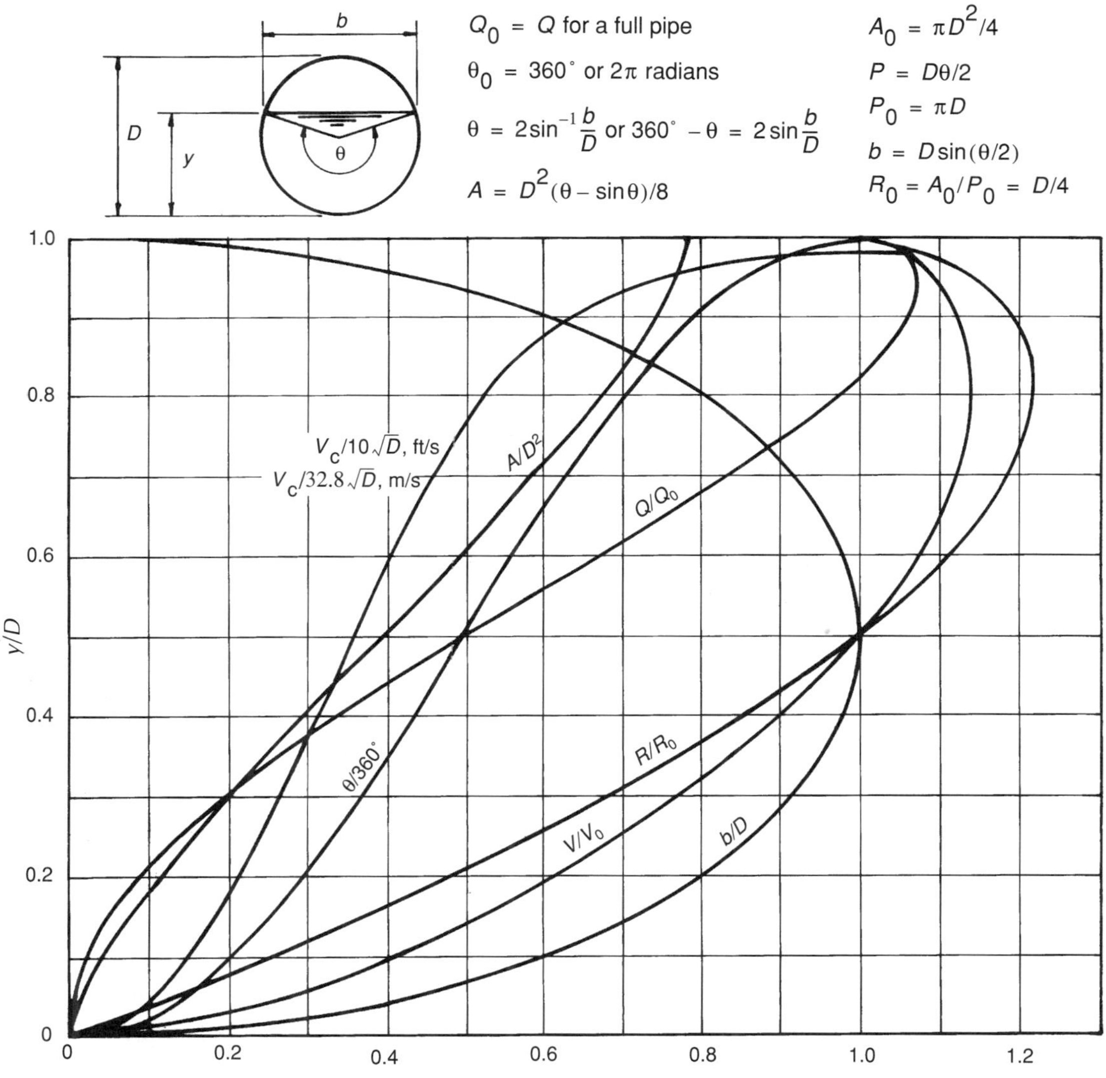

Figure B-4. Geometric properties of partly full pipes and velocity and discharge based on the uncorrected Manning equation.

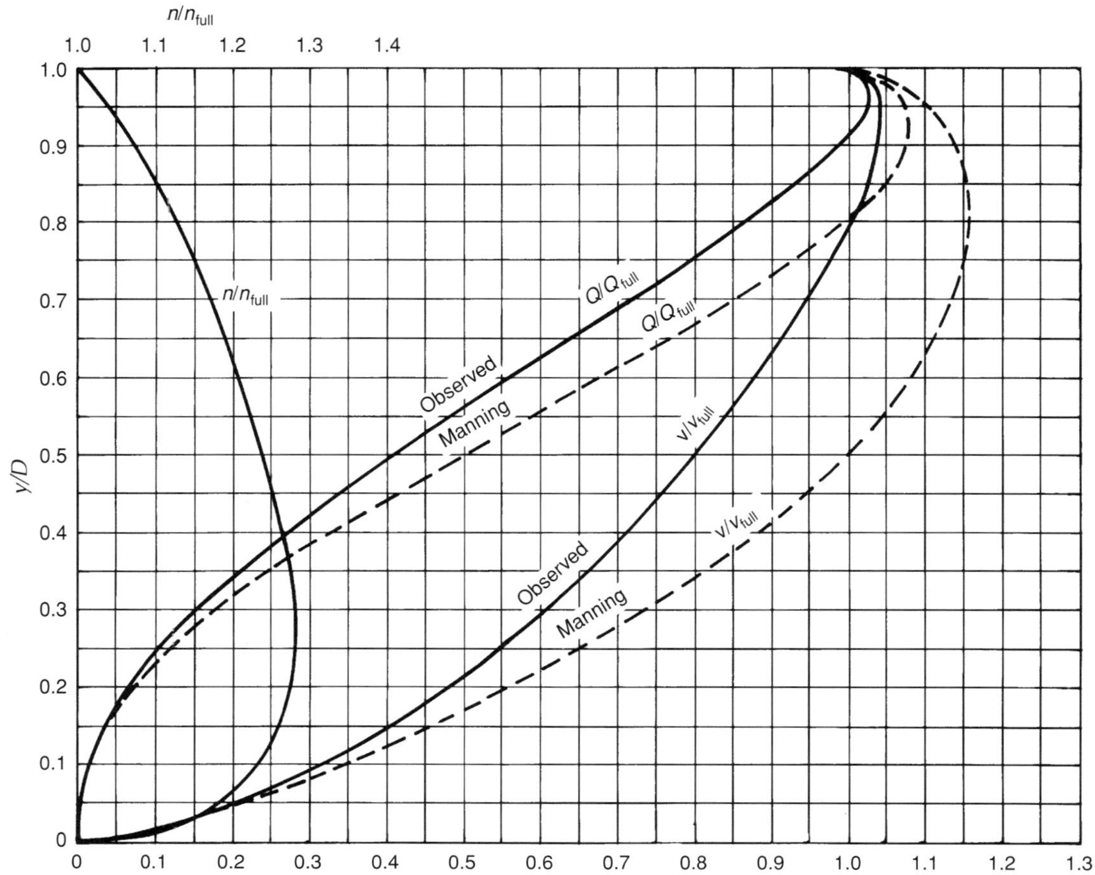

Figure B-5. Hydraulic elements of circular pipes as open channels. Observed versus Manning values. After Camp [2].

Table B-8. Comparison of Computed Values of Velocity, Depth, Area, and Flow in Open Channels. Manning versus Observed Values. After Wheeler [10]. All values in percent of those for a full pipe. Q = flow, y = depth, V = velocity, A = wetted area.

	Manning			Observed				Manning			Observed		
Q	y	V	A	y	V	A	Q	y	V	A	y	V	A
0	0	0	0	0	0	0	50	50	100	50	57	86	58
1	7	32	3	8	26	4	51	51	101	51	57	86	59
2	10	40	5	11	32	6	52	51	101	52	58	87	60
3	12	45	7	14	36	8	53	52	102	52	59	87	61
4	14	49	8	16	40	10	54	52	102	53	59	88	62
5	15	52	10	18	42	12	55	53	102	54	60	88	62
6	17	55	11	19	45	13	56	54	103	55	60	89	63
7	18	58	12	21	47	15	57	54	103	55	61	89	64
8	19	60	13	22	49	16	58	55	104	56	62	90	65
9	20	62	15	23	51	18	59	55	104	57	62	90	66
10	21	64	16	25	52	19	60	56	105	57	63	91	66
11	22	66	17	26	54	20	61	56	105	58	64	91	67
12	23	67	18	27	55	22	62	57	105	59	64	92	68
13	24	69	19	28	57	23	63	58	106	60	65	92	69
14	25	71	20	29	58	24	64	58	106	60	65	92	69
15	26	72	21	30	59	25	65	59	107	61	66	93	70
16	27	73	22	31	61	26	66	59	107	62	67	93	71
17	28	75	23	32	62	28	67	60	107	63	67	94	72
18	29	76	24	33	63	29	68	61	108	63	68	94	72
19	30	77	25	34	64	30	69	61	108	64	69	94	73
20	30	78	26	35	65	31	70	62	108	65	69	95	74
21	31	79	27	36	66	32	71	62	109	65	70	95	75
22	32	80	27	36	67	33	72	63	109	66	71	96	75
23	33	81	28	37	68	34	73	63	109	67	71	96	76
24	33	82	29	38	69	35	74	64	110	68	72	96	77
25	34	83	30	39	69	36	75	65	110	68	72	97	78
26	35	84	31	40	70	37	76	65	110	69	73	97	78
27	36	85	32	41	71	38	77	66	110	70	74	97	79
28	36	86	33	41	72	39	78	66	111	71	74	98	80
29	37	87	34	42	73	40	79	67	111	71	75	98	81
30	38	87	34	43	73	41	80	68	111	72	76	99	81
31	38	88	35	44	74	42	81	68	111	73	76	99	82
32	39	89	36	44	75	43	82	69	112	74	77	99	83
33	40	90	37	45	76	44	83	70	112	74	78	100	83
34	40	90	38	46	76	45	84	70	112	75	78	100	84
35	41	91	38	46	77	46	85	71	112	76	79	100	85
36	42	92	39	47	78	46	86	72	113	77	80	101	86
37	42	93	40	48	78	47	87	72	113	77	81	101	86
38	43	93	41	49	79	48	88	73	113	78	81	101	87
39	43	94	42	49	80	49	89	74	113	79	82	101	88
40	44	94	42	50	80	50	90	74	113	80	83	102	89
41	45	95	43	51	81	51	91	75	113	80	84	102	89
42	45	96	44	51	81	52	92	76	114	81	84	102	90
43	46	96	45	52	82	53	93	76	114	82	85	103	91
44	46	97	46	53	83	53	94	77	114	83	86	103	92
45	47	97	46	53	83	54	95	78	114	84	87	103	92
46	48	98	47	54	84	55	96	79	114	84	88	103	93
47	48	99	48	55	84	56	97	79	114	85	89	104	94
48	49	99	49	55	85	57	98	80	114	86	90	104	95
49	49	100	49	56	85	58	99	81	114	87	91	104	95
50	50	100	50	57	86	58	100	82	114	88	92	104	96

Table B-9. Velocities Required[a] to Scour Air Pockets from Pipelines. Values computed by Wheeler [4] using Equation B-1 developed by Wisner, Mohsen, and Kouwen [5].

Pipe diameter, mm	Velocities, m/s Slope 0%	5%	25%	45°	90°	Velocities, ft/s Slope 0%	5%	25%	45°	90°	Pipe diameter, in.
25	0.4	0.4	0.5	0.5	0.5	1.4	1.4	1.6	1.7	1.8	1
50	0.6	0.6	0.7	0.7	0.8	1.9	2.0	2.2	2.4	2.5	2
75	0.7	0.8	0.8	0.9	0.9	2.3	2.5	2.7	2.9	3.1	3
100	0.8	0.9	0.9	1.0	1.1	2.7	2.9	3.1	3.4	3.5	4
150	1.0	1.1	1.2	1.3	1.3	3.3	3.5	3.8	4.2	4.3	6
200	1.2	1.2	1.3	1.5	1.5	3.8	4.1	4.4	4.8	5.0	8
250	1.3	1.4	1.5	1.6	1.7	4.3	4.6	4.9	5.4	5.6	10
300	1.4	1.5	1.6	1.8	1.9	4.7	5.0	5.4	5.9	6.1	12
350	1.6	1.6	1.8	1.9	2.0	5.1	5.4	5.8	6.3	6.6	14
375	1.6	1.7	1.8	2.0	2.1	5.2	5.6	6.0	6.6	6.8	15
400	1.6	1.8	1.9	2.1	2.1	5.4	5.8	6.2	6.8	7.0	16
450	1.7	1.9	2.0	2.2	2.3	5.7	6.1	6.6	7.2	7.5	18
500	1.8	2.0	2.1	2.3	2.4	6.0	6.5	6.9	7.6	7.9	20
525	1.9	2.0	2.2	2.4	2.5	6.2	6.6	7.1	7.8	8.1	21
600	2.0	2.2	2.3	2.5	2.6	6.6	7.1	7.6	8.3	8.6	24
675	2.1	2.3	2.5	2.7	2.8	7.0	7.5	8.1	8.8	9.2	27
750	2.3	2.4	2.6	2.8	2.9	7.4	7.9	8.5	9.3	9.6	30
825	2.4	2.5	2.7	3.0	3.1	7.8	8.3	8.9	9.7	10.1	33
900	2.5	2.7	2.8	3.1	3.2	8.1	8.7	9.3	10.2	10.6	36
1050	2.7	2.9	3.1	3.4	3.5	8.8	9.4	10.1	11.0	11.4	42
1200	2.9	3.0	3.3	3.6	3.7	9.4	10.0	10.8	11.8	12.2	48
1500	3.2	3.4	3.7	4.0	4.1	10.5	11.2	12.0	13.1	13.6	60
1800	3.5	3.7	4.0	4.4	4.5	11.5	12.2	13.2	14.4	14.9	72

[a]See first page of Appendix B.

Practical considerations:

- Air problems do not occur where the pipe gradient is positive in the direction of flow [5].
- Avoid excessive headloss by using smaller diameter pipe (to obtain higher velocities) only where gradient is flat or slopes downward.
- For air scouring to be effective, the tabular velocities must occur frequently (e.g., daily or more often).
- Air release valves in small pipes may be of little or no value.
- Blowback from clearing air in large pipes may cause surges that cannot be estimated. See Wisner, Mohsen, and Kouwen [5].

Before designing piping systems for air scouring, it is advisable to read "Air binding in pipes" by Edmunds [6], the chapter on closed conduit flow in Falvey [14], and, for wastewater, "Hydraulics of corrosive gas pockets in force mains" by Walski *et al.* [13] .

B-1. References

1. Ten-State Standards. *Recommended Standards for Sewage Works*. Great Lakes-Upper Mississippi Board of Sanitary Engineers, Health Education Service, Inc., Albany, NY (1978).
2. DIPRA. *Handbook of Ductile Iron Pipe*, 6th ed. Ductile Iron Pipe Research Association, Birmingham, AL (1984).
3. ANSI B36.10. "Welded and Seamless Wrought Steel Pipe." American National Standards Institute, New York (updated periodically).
4. Lamont, P. A. "Common pipe flow formulas compared with the theory of roughness." *Journal of the American Water Works Association*, **73,** 274–280 (May 1981).
5. Peggs, L. A. *Underground Piping Handbook*. Robert E. Krieger Publishing Company, Malabar, FL (1985).
6. Miller, W. T. "Durability of cement-mortar linings in cast-iron pipe." *Journal of the American Water Works Association*, **57**, 773–782 (June 1965).
7. Moody, L. F. "Friction factors for pipe flow." *Transactions of the American Society of Mechanical Engineers*, **66**, 671–684 (November 1944).
8. Camp, T. R. Discussion of "Determinations of Kutter's *n* for sewers partly filled," *Transactions of the American Society of Civil Engineers*, **71**(3) Part 2:240–245 (1944).
9. Wheeler, W. PARTFULL®. For a free copy of this computer program with instructions, send a formatted 1.4 MB, $3^1/_2$-in. diskette and a stamped self-addressed mailer to 683 Limekiln Road, Doylestown, PA 18906-2335.
10. Wheeler, W., Consulting Engineer, Doylestown, PA. Private communication (November 1995).
11. Wisner, P. E. F. N. Mohsen, and N. Kouwen. "Removal of air from waterlines by hydraulic means." *Proceedings of the American Society of Civil Engineers, Journal Hydraulic Div.* HY2 **101,** 243–257 (February 1975).
12. Edmunds, R. C. "Air binding in pipes." *Journal of the American Water Works Association*. **71**(5), 272–277 (May 1979).
13. Walski, T. M., T. S. Barnhart, J. M. Driscoll, and R. M. Yencha. "Hydraulics of corrosive gas pockets in force mains." *Water Environment Research* **60** (6):772–778 (September/October 1994).
14. Falvey, H. T. *Air-Water Flow in Hydraulic Structures* (Chapter on Closed Conduit Flow). U.S. Dept. of Interior, Bureau of Reclamation Engineering Monograph 41. Superintendent of Documents, Washington, D.C. (1980).

B-2. Supplementary Reading

1. Benedict, R. P. *Fundamentals of Pipe Flow*. John Wiley, New York (1980).
2. Culp/Wesner/Culp. *Handbook of Public Water Systems*, 834–870. Van Nostrand Reinhold, New York (1986).
3. *Flow of Fluids Through Valves, Fittings, and Pipe*, Technical Paper No. 410. Crane Co., New York (1981).
4. *Hydraulics and Useful Information*. FMC Bulletin 9900, Clow Corp., Melrose Park, IL (1973).
5. *Hydraulic Institute Engineering Data Book*, 2nd ed. Hydraulic Institute, Parsippany, NJ (1990).
6. Idel'Chik, I. E. *Handbook of Hydraulic Resistance*. Gosudarstvennoe Energeticheskoe Izdatel'stro, Moskva-Leningrad (1960); English translation, Israel Program for Scientific Translations, Ltd., U. S. Department of Commerce, Springfield, VA (1966).
7. Miller, D. S. "Internal flow systems," *BHRA Fluid Engineering Series*, Vol. 5, British Hydromechanics Research Association, Cranfield, Bedford, United Kingdom (1978).
8. Chow, V. T. *Open-Channel Hydraulics*. McGraw-Hill, New York (1959, reissued 1988).

Appendix C

Typical Specifications for Pumps and Drivers

DAVID L. EISENHAUER
GARR M. JONES

CONTRIBUTORS

Richard O. Garbus
Philip A. Huff
Melvin P. Landis
Earle C. Smith
Horton Wasserman

[The format, headings, and numbering system in this appendix generally follow the three-part section format in the recommendations of the Construction Specification Institute Manual of Practice Document MP-2-2. *SI units were not incorporated in this document. The following is an actual specification quoted from a project for which bids were received in the spring of 1985. Four qualifying proposals were received for the specified equipment. None of the candidate suppliers took serious exception to the specification requirements.*

Codes and standards are given here in abbreviated form. Titles are listed in Appendix E, and publishers' addresses are given in Appendix F.]

SECTION 11302
CUSTOM ENGINEERED HEAVY-DUTY VERTICAL NONCLOG
CENTRIFUGAL PUMPING UNITS

PART 1—GENERAL

1.01 DESCRIPTION

A. SCOPE:

This section specifies custom engineered heavy-duty wastewater pumping units. Each pumping unit consists of a vertical nonclog centrifugal pump, vertical drive shaft and couplings, squirrel-cage induction motor suitable for variable-frequency speed control, and all necessary appurtenances to provide complete pumping systems. The contractor shall require that the pumping units specified herein are supplied by a single manufacturer.

1.01B

B. TYPE:

Each pump shall be of the vertical, dry pit, bottom suction volute casing, enclosed impeller, nonclog centrifugal type, designed so that the impeller shaft and bearings can be removed without disturbing the connecting piping or casing. The pump casing shall be pedestal mounted on a common cast-iron or fabricated steel base with sufficiently large openings to permit access to the pump suction line.

C. EQUIPMENT LIST:

	Pumps	Motors
Pumping Station X		
Sewage pump 1	P331,001	M331,001
Sewage pump 2	P331,002	M331,002
Sewage pump 3	P331,003	M331,003
Sewage pump 4	P331,004	M331,004
Pumping Station Y		
Sewage pump 1	P330,001	M330,001
Sewage pump 2	P330,002	M330,002
Sewage pump 3	P330,003	M330,003
Sewage pump 4	P330,004	M330,004

D. REFERENCE SPECIFICATIONS:

This section contains reference to the following specifications for more detailed requirements which must be complied with in providing the equipment covered under this section. For reference, the Contractor's attention is directed to the following sections included as part of this project manual:

Section	Title
10030	Noise requirements and Controls
1100	General requirements for Equipment
11020	Vibration and Dynamic Requirements
16158	Adjustable Frequency Drives

1.02 QUALITY ASSURANCE

A. REFERENCES:

This section references the following documents. They are a part of this section as specified and modified. In case of conflict between the requirements of this section and those of the listed documents, the requirements of this section shall prevail.

Reference	Title
	Standards of the Hydraulic Institute, latest edition
AFBMA Std 11-78	Load and Fatigue Life for Roller Bearings
ASTM A48-83	Gray Iron Castings
ASTM A276-83	Stainless and Heat-Resisting Steel Bars and Shapes
ASTM A36-84a	Structural Steel
ASTM A322-82	Steel Bars, Alloy, Standard Grades
NEMA MGI-78	Motors and Generators
UL 1004-84	Motors, Electric
ANSI/IEEE 112-84	Test Procedures for Polyphase Induction Motors and Generators

1.02B

B. UNIT RESPONSIBILITY AND COORDINATION:

The contractor shall cause all equipment specified under this section to be furnished by the pump manufacturer who shall be responsible for the adequacy and compatibility of all pumping unit components. Any component of each complete pumping unit not provided by the pump manufacturer shall be designed, fabricated, tested, and installed by factory-authorized representatives experienced in the design and manufacture of variable-speed pumping equipment. This requirement, however, shall not be construed as relieving the Contractor of the overall responsibility for this portion of the work.

Additionally, the Contractor shall cause the pump manufacturer and its suppliers to coordinate design of the pumping units with the manufacturer of the adjustable-speed drive equipment and controls such that all equipment is compatible and capable of achieving the performance requirements specified below.

[Author's note: Some contractors will try to circumvent the Unit Responsibility requirement in an effort to save costs. To forestall such a move, the Unit Responsibility specification should be accompanied by a requirement for the contractor to confirm by affidavit that the specified arrangement has been provided. The affidavit should be signed by the authorized representatives of the contractor and the manufacturers of the principal equipment items under the penalty of perjury.]

C. DESIGN REQUIREMENTS:

The arrangement shown on the drawings is based upon the best information available to the Engineer at the time of design and is not intended to show exact dimensions peculiar to any specific equipment unless otherwise shown or specified. Therefore, it may be anticipated that the structural supports, foundations, connected piping and valves shown, in part or in whole, may have to be changed in order to accommodate the pumping equipment furnished. No additional payment will be made for such changes. All necessary calculations and drawings for any related redesign shall be submitted to the Engineer for his approval prior to beginning the work.

Pumping units shall be designed to operate at variable speed without cavitation or damaging vibration over the entire specified range of flow and head conditions. Pumping units shall not produce undue noise or vibrations during reductions in flow from the specified operating capacity range to zero flow. *[Authors' note: Noise cannot be quantified as it has more to do with quality than magnitude. An expert, used to listening to similar pumps, can identify "undue" noise. Vibration is also complex and is, in addition, a function of the support. Again, an expert can determine whether vibration is "excessive." Specifying magnitude of either noise or vibration is futile.]*

All components shall be designed to safely withstand forces resulting from flow reversals, up to 125 percent of maximum speed, within the pump during shutdowns caused by power failure. The Contractor shall be responsible for this coordination and shall ensure that the component parts do not interact on each other to produce unacceptable vibrations, stresses, or undesirable conditions.

The pump base shall be designed for anchor bolting to a concrete foundation, assuming that the pump, without restraints at the suction and discharge connections, is subjected to a displacing force equal to that developed by an internal pressure equal to three times shut-off head at maximum operating speed. The motor shall be supported independently on the motor room floor above and shall be connected to the pump with two intermediate shaft lengths as shown on the drawings. No vertical hydraulic thrust shall be transmitted to the floor above from the pumps below.

The complete pumping unit shall be designed to operate without overload on any component at any point along the pump's entire full-speed operating curve.

D. OPERATING CONDITIONS:

[Authors' note: although this specification is silent on allowable maximum speed, the authors prefer to limit speed to 1150 rev/min whenever possible.

1.02D

Each pumping unit shall be capable of operating at infinitely variable speeds to meet the following operating conditions outlined below:

Pumps P331,001, P331,002, P330,001 and P330,002

Operating Condit on	Pumping Station X	Pumping Station Y
Maximum Speed Operation		
A[1] Capacity, mgd	3.8	3.7
Total head, feet	166	169
B[2] Capacity, mgd	2.7	2.9
Total head, feet	202	192

Operating Condition	Pumping Station X	Pumping Station Y
Reduced Speed Operation		
C[3] Capacity, mgd	0.5	0.5
Total head, feet	150	154
Pump speed	reduced	reduced
D[4] Capacity, mgd	0	0
Total head, feet	149	153
Pump speed	minimum	minimum
Minimum NPSH available at any specified conditions	31	30
Characteristics		
Minimum pump efficiency at best efficiency points, percent	71	71
Minimum spherical solids-passing capacity diameter, inches	3-1/2	3-1/2
Maximum full-load horsepower required	200	200

Pumps P331,003, P331,004, P330,003 and P330,004

Operating Condition	Pumping Station X	Pumping Station Y
Maximum Speed Operation		
Al Capacity, mgd	10.8	11.0
Total head, feet	172	169
B[2] Capacity, mgd	9.0	9.9
Total head, feet	194	183
Reduced Speed Operation		
C[3] Capacity, mgd	4.4	4.3
Total head, feet	153	154
Pump speed	reduced	reduced
D[4] Capacity, mgd	0	0
Total head, feet	149	153
Pump speed	minimum	minimum
Minimum NPSH available at any specified condition	32	31

1.02D

Operating Condition	Pumping Station X	Pumping Station Y
Characteristic		
Minimum pump efficiency at best efficiency point, percent	82	82
Minimum pump solids-passing capacity diameter, inches	4	4
Maximum full load motor horsepower required[5]	450	450

Notes (see superscripts):

1. Condition A shall be taken as the rated, continuous-duty operating condition.
2. Condition B is presented to indicate the expected head capacity condition with pumps operating in parallel.
3. Condition C represents the expected minimum sustained flow, continuous-duty operating condition for one pump operating at reduced speed.
4. Condition D represents the expected momentary flow conditions for one pump operating at minimum speed. The pumps will be started and stopped against a closed discharge valve. This condition will occur for a short time only—typically less than 45 seconds.
5. The variable-speed control system will employ variable-frequency drives to vary the rpm of the squirrel-cage induction motor. Motors shall be selected such that along the pump's full-speed curve the motor is not loaded at more than 90 percent of the motor's nameplate rating as a result of the pump's brake horsepower requirement and other power-absorbing accessories included as a part of the complete pumping unit.

E. MASS ELASTIC SYSTEMS AND CRITICAL SPEEDS:

Each complete system, including pump, intermediate shafting, motor, and all appurtenances, shall have no dangerous critical or resonance frequencies or multiples of resonance frequencies within 20 percent above and 35 percent below the speed range required by the pump to meet the specified operating conditions.

For the purposes of design, a dangerous vibratory critical speed shall be defined as one which produces a torsional stress exceeding 3500 psi. The Contractor shall be responsible for the analysis of critical speeds and the complete mass elastic system, which shall be analyzed and certified by a registered professional engineer regularly engaged in this type of work. Calculations shall be submitted to the Engineer for review prior to construction of the equipment. Nothing in this provision shall be construed as relieving the Contractor of his responsibility for this portion of the work.

F. FACTORY TESTING:

1. MATERIALS: Melt and strength tests of the nickel cast iron used in the manufacture of the pumps' major components shall be performed in accordance with all applicable ASTM standards. The Contractor shall furnish the Engineer with five certified copies of the results of all tests. The composition and relative physical properties of all other component materials shall be submitted.

2. HYDROSTATIC TESTS: Each pump shall be hydrostatically tested. The test pressure shall be not less than twice the shut-off head as shown on the approved head-capacity curve passing through or above the rated point (point A). The test procedure shall be as follows:

Condition	Time, minutes
A. Test pressure	180
B. Atmospheric pressure	5
C. Test pressure	15
D. Atmospheric pressure	5
E. Test pressure	30

At no time during this test shall the casing show undue deflection or signs of weakness at any point, nor shall the external surfaces of the casing show sweating through porous metal or leaking through gaskets or cracks or other defects.

The Contractor shall furnish the Engineer with five certified copies of the results of the tests.

3. PERFORMANCE TESTS: Each pump shall be subject to witnessed performance tests to determine the head, capacity, speed, and brake horsepower at each condition point specified in paragraph 1.02D. Tests shall be conducted in accordance with the test code of the Standards of the Hydraulic Institute except that predicted per-

formance from model tests or from the performance at a single test speed will not be permitted. Should the manufacturer's testing facilities be limited to constant-speed pump drivers, a test curve may be developed utilizing the constant speed closest to the rated condition point on the test curve. In any event, test data shall be sufficiently comprehensive to produce guaranteed performance curves showing head versus capacity, efficiency, and brake horsepower for the rated speed plus or minus 10 percent and at one other lower speed, preferably at condition point C. The test data shall be corrected to the rated speed and the speed at condition point C by the methods outlined in the Hydraulic Institute Standards. Five certified copies of all original test data, calculations, and final performance curves shall be furnished to the Engineer for acceptance before shipment of the equipment.

The Contractor shall submit a sketch of the proposed test setup, along with a description of the proposed testing procedure to the Engineer for his review at least 10 weeks in advance of the proposed testing date. No tests shall be performed until the test procedure meets with the Engineer's approval. In addition, the Contractor shall furnish the Engineer with at least 4 weeks advance written notice of the date and location of the witnessed performance tests. The results of the factory tests shall be considered official and conclusive for the purpose of determining whether or not the equipment complies with the performance specifications. Final acceptance of the equipment will depend on satisfactory operation after installation.

4. MOTOR TEST: Motors shall be tested in accordance with NEMA and IEEE procedures. The test shall include the following:

a. Routine test:
   1) Running—no-load amperes
   2) Locked rotor amperes
   3) Winding resistance, DC
   4) High-potential test at twice rated voltage plus 1000 volts, with a minimum of 2200 volts for 1 minute, winding to ground
b. Complete test:
   1) Rated load temperature rise
   2) Slip in percent
   3) Locked rotor amperes (3 phase, full voltage)
   4) Locked rotor torque
   5) Breakdown torque
   6) High-potential test (see 4 above)
   7) Efficiencies tabulated at 100, 75, and 50 percent of full load
   8) Power factor tabulated at 100, 75, and 50 percent of full load

The Contractor shall furnish the Engineer with five certified copies of the results of all tests.

G. SHIPMENT OF PUMPING UNITS:

Pumping units to be furnished under this contract shall be shipped to the site in enclosed, weather-tight, sealed containers in a manner designed to protect the pumping units against damaging stress caused by sudden acceleration or deceleration. A recording accelerometer, designed to record the magnitude of any sudden impact in all three directions (X, Y, Z) on continuous strip charts, shall be shipped with and fixed to each assembly or its packing crate. Upon arrival of each shipment, the accelerometer shall be removed in the presence of representatives of the Engineer, the Contractor, and the equipment manufacturer. If the magnitude of the maximum acceleration exceeds 3.0 times the acceleration due to the force of gravity (3.0 g), the assembly or subassembly shall be dismantled and completely inspected. All damage shall be corrected before the assembly is incorporated into the work. The Contractor shall bear all costs arising out of dismantling, inspection, repair, and reassembly.

1.03 ENVIRONMENTAL CONDITIONS

The equipment to be provided under this section shall be suitable for installation and operation at elevations of about 10 and 150 feet above sea level inside weather-protected structures. Outside ambient temperatures range between 10 and 95°F, and reported wastewater temperatures vary between 47 and 65°F. Relative humidity is expected to range between 5 and 100 percent.

1.04G

1.04 SUBMITTALS

The information listed below shall be submitted to the Engineer for review in accordance with Section 01300. The submittal shall, as a minimum, include the following data drawings and other descriptive material:

A. Shop drawings to describe and show pump construction and materials.
B. Pump performance curves showing the head, capacity, speed, efficiency, and brake horsepower required when operating at specified conditions. Pump curves shall include a minimum of four pump curves approximately equally spaced between minimum speed and full operating speed with a minimum of four points each including shut-off head and maximum capacity.
C. Reed frequency for motor and pump rotors.
D. Mass elastic system calculations.
E. Complete calculations for bearing and shafting selection necessary to determine size, material, and styles.
F. All motor information as outlined by this specification.

1.05 INFORMATION TO BE PROVIDED

1. Certified copies of all material test reports.
2. Certified copies of all performance tests on both pump and motor.
3. Manufacturer's certification that the pumping units and variable-speed control equipment are fully in conformance with the requirements described and depicted in this project manual.
4. O&M manual information in accordance with Section 01730.

PART 2—PRODUCTS

2.01 MATERIALS

Component	Material
Suction nozzle, volute casing, back head, impeller	Cast iron, ASTM A48, Class 30C, close grained with 2 to 3 percent nickel
Bearing frame and housing	Cast iron, ASTM A48, Class 30C
Shaft	Steel, ASTM A322, Grade 4140, or equal
Wear rings	Stainless steel, ASTM type 405; minimum Brinnell hardness 450, casing; 350, impeller
Shaft sleeve	Stainless steel, ASTM A276, type 405; minimum Brinnell hardness 500
Bolts, studs, nuts	Stainless steel, ASTM A276, type 316

2.02 EQUIPMENT

A. GENERAL:

Each pumping unit shall consist of a custom engineered heavy-duty vertical nonclog pump, vertical alternating current induction motor, and all necessary shafting, couplings, and support bases. The pumping unit shall be capable of operating at variable speed without cavitation or damaging vibration over the specified range of flow and head conditions. The pumping units shall also be suitable for adjustable-speed operation by means of a variable-frequency drive as specified in Section 16158.

B. PUMPS:

1. SUCTION HEAD AND NOZZLE: Each suction head shall be cast separately from the casing and shall be of one piece close-grained cast iron containing 2 to 3 percent nickel and conforming to ASTM A48, Class 30. Each suction elbow shall be fitted with a 4-inch hand hole for access to the impeller, shall be tapped and fitted with a 1-inch drain cock and shall be drilled and tapped for a 1-inch gauge connection. *[Author's note: The manufacturer's standard construction provides a gauge connection adjacent to the suction flange.]* The inner contour of the hand hole shall exactly match the contour of the suction elbow. The drain cock shall be located in the suction nozzle below the hand hole. Each suction nozzle shall be equipped with 125-pound ASA standard flange as follows:

Pumps	Size, inches
P331,001, P331,002, P330,001, and P330,002	8
P331,003, P331,004, P330,003, and P330,004	16

2. BASE: A rigid steel fabricated base shall be firmly bolted to the casing. Large openings on all sides shall permit free access to the suction and hand hole, drain and gauge fittings. Anchor bolts shall be embedded in the concrete floor in accordance with the pumping unit manufacturer's recommendation. Bases for pumps P331,001, P331,002, P330,001, and 330,002 shall be designed such that the anchor bolt locations match those recommended for the bases of pumps P331,003, P331,004, P330,003, and P330,004.

2.02B

3. CASING: Each casing shall be made of close-grained cast iron containing 2 to 3 percent nickel and conforming to ASTM A48, Class 30. A 4-inch hand hole shall be provided on the discharge nozzle side of each casing to permit inspection of the impeller. The inner contours of each hand hole cover shall exactly match the contours of the casing in which it fits. Each casing shall be provided with lifting eyes or other means for conveniently attaching hoisting bridles at not less than three points. The surfaces of all water passages shall be smooth and free from all pits and projections which might cause undesirable turbulence. Each casing shall be equipped with a 1-inch drain and plug. Each discharge nozzle shall be drilled and tapped for a 1-inch gauge connection, and a valved 1-inch air vent shall be fitted to the top of each discharge nozzle. Each discharge nozzle shall be equipped with a 125-pound ASA standard flanged discharge fitting of at least the following size:

Pumps	Size, inches
P331,001, P331,002, P330,001, and P330,002	6
P331,003, P331,004, P330,003, and P330,004	12

4. BACK HEAD: The back head shall be so constructed as to allow the complete removal of the impeller, shaft, and bearings without disturbing the pump suction or discharge piping. The back head shall be accurately machined to provide a self-centering fit with the volute casing and frame and shall be drilled and tapped for the installation of a 1-inch seal water drain designed to minimize standing seal water within the back head.

The back head shall contain a stuffing box of sufficient depth to ensure trouble-free operation. Large *[Authors' note: This depends on size of pump and manufacturer.]* openings shall be provided adjacent to the stuffing box to permit packing replacement and adjustment. The gland shall be of the split removable type, constructed of bronze. The stuffing box shall contain a split bronze lantern ring and shall be drilled and tapped for a clear water seal. Each lantern ring section shall be drilled and tapped for a 1/4-20 thread to permit removal by means of a special threaded tool to be furnished with the spare parts.

5. FRAME: The frame shall be of cast iron and shall be designed to carry both radial and thrust bearings. Large openings shall be provided adjacent to the stuffing box to permit packing replacement and adjustment. The frame shall be provided with accurately machined self-centering and self-indexing fits with their back heads to ensure proper alignment of bearings and stuffing box.

6. IMPELLER: Each impeller shall be of the one-piece, single suction, two-vane or three-vane, enclosed, nonclog centrifugal or nonclog, mixed-flow type. All exterior surfaces shall be accurately machined, and all water passages shall be smooth and free from hollows, cracks, pinholes, and projections which might incite or encourage cavitation. Each impeller shall be keyed to the shaft and locked in place with a threaded fastener designed to prevent accumulation of stringy material.

7. WEARING RINGS: Removable wearing rings shall be provided for both the impeller and the suction head of each pump. Wearing rings shall be nongalling, heat-treated stainless steel with a chrome content of not less than 11 percent. Impeller rings shall have a Brinell hardness of not less than 300. Casing ring hardness shall exceed impeller ring hardness by not less than 100 points on the Brinell scale. The impeller and suction head rings shall be so designed as to be easily and readily replaceable. Wearing rings shall preferably be attached by screwed fasteners.

8. SHAFT: Each pump shaft shall be of heat-treated, high-strength steel, turned, ground, and polished. Shafts shall be of ample strength and stiffness to operate without distortion or vibration throughout the range of service specified. The section of shaft fitting between radial and thrust bearings shall be of sufficient section to limit deflection at the impeller centerline to not more than 6 mils when the pump is operating at the minimum sustained flow, continuous-duty point, as defined under paragraph 1.02D. Shaft deflection shall be calculated using the following relationship:

2.02B

$$Y_{max} = \frac{W_r}{3E}\left(\frac{b^3}{I_b} + \frac{a^3 - b^3}{I_a} + \frac{a^2 c}{I_c}\right)$$

where

Y_{max} = deflection, inches

E = modulus of elasticity, psi
30×10^6 psi for carbon steel
28×10^6 psi for 316 stainless steel

a = shaft length, inches, from impeller centerline to the centerline of the radial bearing

b = shaft length, inches, impeller centerline to shaft sleeve shroud at radial bearing

c = shaft length between bearings, inches

I_a = moment of inertia of the shaft just outboard of the radial bearing, in.^{4}

I_b = moment of inertia of the shaft under the shaft sleeve, in.^{4}

I_c = moment of inertia of the shaft between the radial and thrust bearings, in.^{4}

W_r = radial force, pounds = KHD_2B_2

where

H = total head, pounds per inch

D_2 = impeller diameter, inches

B_2 = impeller width, including shrouds, inches

$$K = 0.36\left[1 - \left(\frac{Q}{Q_n}\right)^2\right]$$

where

Q = pumped flow at condition C

Q_n = capacity at the pump's best efficiency point

9. REPLACEABLE SHAFT SLEEVE: The section of each shaft or impeller hub which extends through or into the stuffing box shall be fitted with a replaceable, hardened chromium stainless-steel sleeve with a Brinell hardness of not less than 500. The sleeve shall be held to the shaft to prevent rotation and shall be gasketed to prevent leakage between the shaft and the sleeve. Minimum shaft sleeve thickness shall be $^1/_2$ inch.

10. BEARINGS: Each pump shall be fitted with two sets of shaft bearings conservatively designed to withstand all stresses of the service specified herein. Bearings shall be of the antifriction grease-lubricated type. The bearing nearest the impeller shall be capable of carrying all radial thrust loads, while the outboard bearing shall be capable of carrying some radial load as well as the unbalanced hydraulic thrust. All bearings shall be rated in accordance with AFBMA L-10 for a continuous (24 hours/day) duty life of not less than 50,000 hours at the most severe loads imposed by the specified continuous-duty conditions (condition points A and C, paragraph 1.02D). Calculations substantiating the bearing selections shall be submitted to the Engineer with the pump submittals. Pressure grease fittings shall be provided for bearing lubrication. Only standard commercial bearing sizes shall be used. Grease fittings shall be Alemite hydraulic type.

11. BALANCE: Each rotating assembly, including coupling, shaft impeller with wearing rings, and impeller nut, shall be dynamically balanced prior to final assembly. *[Editor's note: This stringent specification requires the manufacturer to assemble and disassemble the rotor several times, requires the use of balancing machines not usual in a production line, and results in only a little added precision. Individual balancing of components (not assemblies) and dynamic balancing only for the impeller and coupling may be adequate. Mass unbalance forces are insignificant compared with hydraulic unbalance forces.]*

When properly balanced, each rotating element shall meet the following minimum criteria at speeds varying from 50 percent of the maximum up to maximum operating speed. When properly balanced, each rotating element shall meet the following at every point throughout this range:

2.02B

Maximum speed, rpm	Maximum amplitude peak to peak, inches
600	0.0035
720	0.0033
900	0.0025
1200	0.002
1800	0.0013

The Contractor shall furnish five certified copies of all logs to demonstrate that all rotating elements have been balanced in accordance with these specifications. All balancing logs shall be submitted to and accepted by the Engineer as a condition precedent to shipment of the equipment from the factory.

C. VERTICAL DRIVE SHAFTS:

Each pump shall be provided with two sections of tubular shafting for connection to the motor driver. Shafting shall be complete, with flexible couplings sized to carry the loads of the intended service and a shaft guard which shall extend 3 feet above the pump frame or 7 feet above the floor, whichever is greater. The bearing shall be of the heavy-duty, self-aligning spherical-roller, expansion type 3 capable of a free vertical movement of $^1/_2$ inch and shall be designed for AFBMA L-10 continuous-duty life of not less than 50,000 hours with $1^1/_2$ degrees misalignment. All grease fittings shall be mounted in an approved readily accessible location. The shaft-to-hub connection of each end shall be by means of a keyless mechanical device which exerts an external clamping force on the connections and which does not require the application of heat for removal of the hub from the shaft.

Shafts shall employ universal-type flexible couplings and shall have grease-lubricated needle bearing universal joints and splined slip joints. Universal-joint-type shafts shall be offset to the manufacturer's recommendations and shall be as manufactured by the H.S. Watson Company, Twin Disc, Inc., Parrish or equal. *[Authors' note: Balancing is specified in 2.02B.II and 2.02D.7.]*

D. ELECTRIC MOTOR DRIVERS:

1. GENERAL: Each motor shall be designed for variable-speed service in conjunction with the variable-frequency drive controllers as specified under section 16158. Motors shall be designed with torque characteristics equal to those required for the pumping units, and total load applied to the motor by the driver equipment and variable-frequency drive system shall not exceed 90 percent of the motor's nameplate rating. Each motor shall be designed to withstand a reverse overspeed of 150 percent of rated synchronous speed.

Motors shall be three-phase, squirrel-cage induction types designed for 460 volts, 60 Hz, and maximum kVA inrush current shall be NEMA Code F. Each motor shall be of vertical, solid-shaft, "P" base construction and shall, as a minimum, meet the requirements of NEMA M61-1978. Temperature rise shall not exceed 80 degrees C above a 40 degrees C ambient when operated at full load. Insulation shall be NEMA Class F, and the motor shall have a 1.15 service factor. Enclosures shall be NEMA WPI. Each motor shall be provided with a minimum of two lifting eyes.

2. THERMAL PROTECTION: Thermal protector sensing elements shall be of the same manufacture and shall be coordinated with the thermal protecting relay. The sensing elements shall be embedded and sealed in the end winding of each stator phase. The sensing elements of all three phases shall be connected in series and the end leads brought out to a conduit fitting as shown.

The thermal protector relay contacts shall be of ample capacity to operate the motor starter control circuits as shown. The protector relay shall drop out on overtemperature.

Information shall be supplied to the Engineer regarding the characteristics of the thermal device and its required control circuit protective device. The thermal devices shall be designed to operate at 10 degrees C above the total temperature specified in paragraph 2.02D.1.

3. SPACE HEATERS: Each motor shall be provided with a space heater. Heaters shall be carriage type or flexible wraparound type installed adjacent to the core iron. Heaters shall be rated 120 volts, single phase. The

2.02D

space heater terminals shall be separately wired to a terminal block or pigtails in the power conduit box. Space heater rating in watts and volts shall be noted on the motor nameplate.

4. BEARINGS: Motor bearings shall be of high-precision manufacture, antifriction type designed for an AFBMA L-10 continuous-duty life of 70,000 hours. Bearings shall be provided with pressure-type lubrication fittings with removable plastic cap protective covers and grease drain with threaded, removable drain plugs.

5. CONDUIT BOX: Motors shall be fitted with a cast-iron conduit box provided with threaded hubs. Conduit boxes shall be provided with gaskets at the base and between the halves of the box and shall be capable of 360 degree rotation for mounting and for bonding of a ground conductor employing a stud or grounding pad inside the box. A separate cast-iron signal conduit box shall also be provided for temperature detection leads.

6. MOTOR SUPPORT PEDESTAL: Each motor shall be mounted on a support pedestal designed to carry the motor weight and the weight of the vertical shafting plus all dynamic loads associated with the operation of the unit. The pedestal shall be designed with adequately sized openings to permit access to the upper universal joint of the drive shaft. The openings shall be protected by easily removable stainless steel guards. The guards shall be of comparable quality to the air grille on the motor.

7. BALANCE: Each impeller and each rotating assembly, including coupling half, shaft and rotor shall be dynamically balanced prior to final assembly.

Dynamic balancing shall be accomplished by supporting the rotating element in its bearings and turning the element at speeds varying from 50 percent of maximum up to maximum operating speed. When properly balanced, each rotating element shall meet the following criteria at every point in this range:

Maximum operating speed, rpm	Minimum amplitude peak to peak, inches
600	0.0010
720	0.0010
900	0.0010
1200	0.00075
1800	0.0005

The Contractor shall furnish five certified copies of all logs to demonstrate that all rotating elements have been balanced in accordance with these specifications. All balancing logs shall be submitted to the Engineer for his review and acceptance as a condition relevant to final shipment of the equipment.

E. SPARE PARTS:

Furnish to the Owner the following spare parts for each size pumping unit:

1 each, each type of coupling
2 sets, all gaskets
1 each, impeller hub or shaft sleeves
1 each, impeller
2 sets, complete replacement all packing
1 each, lantern rings as specified
1 set, each type of bearing
1 each, stuffing box gland with bronze bolts and nuts
1 set, special tools as required for pumping unit disassembly

PART 3—EXECUTION

3.01 INSTALLATION

The pump supplier shall provide the complete pumping system and factory-trained personnel to supervise installation and initial operation of all components. The pumps shall be aligned, connected, and installed at the

3.01E

locations shown and in accordance with the manufacturer's recommendations. Pump supplier shall certify that the equipment is installed in a manner to ensure proper operation.

3.02 CERTIFICATION

Manufacturer shall supply certified pump performance curves demonstrating compliance with the performance specified herein.

3.03 TESTING

After the completion of installation, each pumping unit shall be field tested to ensure compliance with the performance requirements as specified.

END OF SECTION

Note: *The following is not part of the original specification. It is offered as an example of how competitive equipment could be evaluated. Strictly speaking, the following provision should be located either in Part I of the section or in the general provisions of the contract.*

4.01 EVALUATION OF BIDS

Evaluation and Selection. Each proposal will be evaluated for selection by the Owner based on the information given in the proposal and data submitted in the brochure. The proposed pumping unit will be evaluated on the basis of first cost and on present worth of estimated operating cost. Operating costs will be determined by computing the cost of power as detailed below:

A. Life of pumping unit: 25 years
B. Interest rate: 8 percent
C. Power cost: $0.06/kWh
D. Annual operation
   1. Design point: 2200 hours
   2. 2nd point: 1300 hours
   3. 3rd point: 900 hours

The guaranteed wire-to-water efficiency at the design point and the wire-to-water efficiencies at the 2nd and 3rd design points stated in the brochure will be used to estimate the operating costs. *[Authors' note: The stated energy rate should be the real cost per unit for power and should, therefore, include the transformer losses, the power factor and time of usage penalties, and the demand charges averaged over the estimated monthly (or annual) energy consumption.]*

Additional factors affecting selection will be delivery time, operating curve characteristics, NPSH requirements, and compatibility with existing plant equipment to reduce need for additional spare parts, availability of service, and familiarity of plant operators with the equipment.

The Purchaser reserves the right to select any proposal or alternative he deems in his best interest.

The Bidder shall allow the OWNER or his representative to witness the acceptance tests. *[Authors' note: The specifier must key this statement to the acceptance test procedures.]*

Appendix D
Common Blunders

EARLE C. SMITH

D-1. Figure 27-1

Figure 27-1 is a composite of several sewage pumping stations constructed between 1970 and 1980. The drawing represents a classic case of "file-drawer" engineering. A standard design was used in which the motors are raised to prevent flood damage. But this pumping station is on a hillside and has a door and windows, so motor flooding is not a problem. The blunders are real and actually were made. They include:

1. The design of the suction piping in the wet well is the worst possible. The entrance headloss is high, and the configuration tends to cause vortexing. An elbow and a bell entrance turned down should be added.
2. The "pumps off" level is too low, which causes vortexing and cavitation.
3. The lead pump "on" level is below the shaft seal, which tends to draw air through the stuffing box and causes the pump to "air bind."
4. The inlet allows the sewage to drop from a great height, which increases the static head, wastes energy, and causes foaming and odors (by sweeping out H_2S). Bubbles caused by the cascade may air bind the pump and, perhaps, the force main.
5. The check valve has no external lever. The operator cannot tell whether it is open or closed. A lever can indicate clogging and can sometimes be manually operated to dislodge trash.
6. The check valve is on a vertical pipe. Grit and rags settling on it when the pump is off may prevent it from opening fully. When the cover is removed for cleaning, the operator will probably get a bath of filthy water. This type of installation is prohibited by the Ten-State Standards [1].
7. The gate valve is on a vertical pipe. The bonnet will fill with grit, and the valve will eventually fail.
8. The magnetic flowmeter is in the wrong location. A straight section of pipe, five diameters in length, should be placed between the flowmeter and the reducer. The flowmeter should be on the manifold to measure the flow from all the pumps.
9. The flowmeter will probably clog because the pump passes a 3-in. sphere.
10. There is no shut-off valve after (above) the flowmeter to allow its removal for repair.
11. The long shaft has no steady bearing and is too slender to be without an intermediate support.
12. There is no lifting eye, hoist, or other lifting apparatus to aid in dismantling the pump.

D-2. Figure 27-2

An actual pumping station in the United States (similar to several constructed in Saudi Arabia), shown in Figure 27-2, is so badly designed that the equipment had to be installed before all the concrete could be cast! The blunders include:

1. The height in the pump room is inadequate for lifting the hollow-shaft motor off the pump shaft.
2. There is no removable hatch in the roof for a crane to lift the pump.

3. The mounting-plate hole is not large enough to allow the removal of the pump.
4. The pumps are mounted in line with the inflow—a poor arrangement. (Exceptions are given in Sections 12-6 and 12-7 and Chapter 17.)
5. The suction bells are too close together [2].
6. The 150-mm (6-in.) clearance between the suction bell and the floor is too small [3] as indicated by the then-prevailing Hydraulic Institute Standards [2, 3].
7. The "off" level is too low, which causes vortexing.
8. There is no ladder or steps for access to the sump.
9. There is no hatch for access to the sump.
10. The generator is too small to start even one pump.

D-3. References

1. Ten-State Standards. *Recommended Standards for Sewage Works,* Great Lakes–Upper Mississippi River Board of State Sanitary Engineers. Health Education Service, Inc., Albany, NY (updated periodically).
2. *Hydraulic Institute Standards for Centrifugal, Rotary & Reciprocating Pumps,* 13th ed. Hydraulic Institute, Parsippany, NJ (1974).
3. *Hydraulic Institute Standards for Centrifugal, Rotary & Reciprocating Pumps,* 14th ed. Hydraulic Institute, Parsippany, NJ (1983).

Appendix E

Codes, Specifications, and Standards

Reference to and use of standard codes and specifications are indispensable for designing, preparing specifications, providing work of high quality, and reducing labor and uncertainty while improving construction practices. In this appendix, the codes and specifications referenced throughout the book are listed chapter by chapter together with the specifications or codes considered to be important but not specifically referenced. The names, abbreviations, and addresses of the organizations that publish the codes and standards are given in Appendix F.

Chapter 3. Flow in Conduits

ANSI B36. 10	Welded and Seamless Wrought Steel Pipe.
ANSI/AWWA C104/A21.4	Cement-Mortar Lining for Ductile-Iron and Gray-Iron Pipe and Fittings for Water.
ANSI/AWWA C1 15/A21. 15	American National Standard for Flanged Cast-Iron and Ductile-Iron Pipe with Threaded Flanges.
ANSI/AWWA C205	Cement-Mortar Protective Linings and Coatings for Steel Water Pipe—4 In. and Larger—Shop Applied.
AWWA C151	Ductile Iron Pipe, Centrifugally Cast in Metal. Molds or Sand-Lined Molds for Water or Other Liquids.
AWWA C601	AWWA Standard for Disinfecting Water Mains.
AWWA C602	AWWA Standard for Cement-Mortar Lining of Water Pipelines—4 In. (100 mm) and Larger—in Place.
AWWA C606	Grooved and Shouldered Type Joints.

Chapter 4. Pipes and Fittings

Recommended Standards and Specifications

The minimum of specifications needed is given here. These publications are frequently revised, so obtain the latest revision.

Ductile Iron Pipe

ANSI B16.1	Cast-Iron Pipe Flanges and Flanged Fittings.
ANSI/AWWA C104/A21.4	Cement-Mortar Lining for Ductile Iron and Gray-Iron Pipe and Fittings for Water.
ANSI/AWWA C105/A21.5	Polyethylene Encasement for Ductile Iron Piping for Water and Other Liquids.
ANSI/AWWA C110/A21.10	Ductile-Iron and Gray-Iron Fittings, 3 In. through 48 In. for Water and Other Liquids.
ANSI/AWWA C111/A21.11	Rubber Gasket Joints for Ductile Iron and Gray-Iron Pressure Pipe and Fittings.
ANSI/AWWA C115/A21.15	Flanged Ductile Iron and Gray-Iron Pipe with Threaded Flanges.

ANSI/AWWA C150/A21.50	Thickness Design of Ductile Iron Pipe.
ANSI/AWWA C151/A21.51	Ductile Iron Pipe, Centrifugally Cast in Metal Molds or Sand-Lined Molds for Water or Other Liquids.
ANSI/AWWA C600	Installation of Ductile Iron Water Mains and Their Appurtenances.
AWWA C602	Cement-Mortar Lining of Water Pipelines—4 In. and Larger—in Place.
AWWA C606	Grooved and Shouldered Type Joints.

Steel Pipe

ANSI B36.10	Welded and Seamless Wrought Steel Pipe.
AWWA C200	Steel Water Pipe 6 In. and Larger.
AWWA C203	Coal Tar Protective Coatings and Linings for Steel Water Pipe—Enamel and Tape—Hot Applied.
AWWA C204	Chlorinated Rubber-Alkyd Paint System for the Exterior of Aboveground Steel Water Piping.
AWWA C205	Cement-Mortar Protective Lining and Coating for Steel Water Pipe—4 In. and Larger—Shop Applied.
AWWA C206	Field Welding of Steel Water Pipe.
AWWA C207	Steel Pipe Flanges for Waterworks Service—4 In. through 144 In.
AWWA C208	Dimensions for Steel Water Pipe Fittings.
AWWA C209	Cold-Applied Tape Coatings for the Exterior of Special Sections, Connections, and Fittings for Steel Water Pipelines.
AWWA C210	Liquid Epoxy Coating System for the Interior and Exterior of Steel Water Pipelines.
AWWA C213	Fusion-Bonded Epoxy Coatings for the Interior and Exterior of Steel Water Pipelines.

Polyvinyl Chloride Pipe

ANSI B31.8	Gas Transmission and Distribution Piping Systems.
ANSI/ASTM D 1785	Poly(Vinyl Chloride) (PVC) Plastic Pipe, Schedules 40, 80, 120.
ANSI/ASTM D 2241	Poly(Vinyl Chloride) (PVC) Plastic Pipe (SDR-PR).
ANSI/ASTM D 2464	Threaded Poly(Vinyl Chloride) (PVC) Plastic Pipe.
ANSI/ASTM D 2466	Socket-Type Poly(Vinyl Chloride) (PVC) Plastic Pipe Fittings, Schedule 40.
ANSI/ASTM D 2467	Socket-Type Poly(Vinyl Chloride) (PVC) Plastic Pipe Fittings, Schedule 80.
ANSI/ASTM D 2672	Bell-End Poly(Vinyl Chloride) (PVC) Pipe.
ANSI/ASTM D 2740	Poly(Vinyl Chloride) (PVC) Plastic Tubing.
ANSI/ASTM D 2749	Standard Symbols for Dimensions of Plastic Pipe Fittings.
ANSI/ASTM F 412	Definition of Terms Relating to Plastic Piping Systems.
ANSI/AWWA C900	Polyvinyl Chloride (PVC) Pressure Pipe 4 In.
ASTM D 2774	Underground Installation of Thermoplastic Pressure Piping.
UNI B-1	Thermoplastic Pipe Joints, Pressure and Non-pressure Application.
UNI-B-3	Installation of Polyvinyl Chloride (PVC) Pressure Pipe.

Asbestos-Cement Pipe

AWWA C400	Asbestos-Cement Distribution Pipe, 4 In. through 16 In., for Water and Other Liquids.
AWWA C401	The Selection of Asbestos-Cement Distribution Pipe, 4 In. through 16 In., for Water and Other Liquids.
AWWA C402	Asbestos-Cement Transmission Pipe, 18 In. through 42 In., for Water and Other Liquids.
AWWA C403	The Selection of Asbestos-Cement Transmission and Feeder Main Pipe, Sizes 18 In. through 42 In.
AWWA C603	Installation of Asbestos-Cement Pressure Pipe.

Reinforced Concrete Pressure Pipe (RCPP)

ASTM C361	Reinforced Concrete Low-Head Pressure Pipe.
AWWA C300	Reinforced Concrete Pressure Pipe, Steel Cylinder Type, for Water and Other Liquids.
AWWA C301	Prestressed Concrete Pressure Pipe, Steel Cylinder Type, Pretensioned, for Water and Other Liquids.
AWWA C302	Reinforced Concrete Pressure Pipe, Noncylinder Type, for Water and Other Liquids.
AWWA C303	Reinforced Concrete Pressure Pipe—Steel Cylinder Type, Pretensioned, for Water and Other Liquids.

Supplementary Specifications

The following specifications contain a wealth of practical information valuable for those designing piping systems and selecting and specifying piping materials. ANSI B31.4 and B31.8 are referenced in municipal codes, such as the Uniform Fire Code and the Uniform Mechanical Code, and, thus, can be considered as having at least some legal force in engineering design. Others, though titled for power or petroleum piping, contain guidelines useful for pumping station piping.

Design Practice

ANSI A13.1	Scheme for the Identification of Piping Systems.
ANSI B31	Corrosion Control for ANSI B31.1 Power Piping Systems.
ANSI B31.1	Power Piping.
ANSI B31.2	Fuel Gas Piping.
ANSI B31.3	Process Piping.
ANSI B31.4	Liquid Transportation Systems for Hydrocarbons, Liquid Petroleum Gas, Anhydrous Ammonia, and Alcohol.
ANSI B31.8	Gas Transmission and Distribution Piping Systems.
ANSI B31.9	Building Service Piping.
ANSI B31.11	Slurry Transportation Piping Systems.
ANSI B31.G	Manual for Determining the Remaining Strength of Corroded Pipelines.
ANSI Z53.1	Safety Color Code for Marking Physical Hazards.
API 686	Recommended Practices for Machinery Installation and Installation Design (see Chapter 6).
ASME MFC-1M (1995)	Measurement of Fluid Flow in Pipes Using Orifice, Nozzle, and Venturi.
MFC-5M (1994)	Measurement of Liquid Flow in Closed Conduits Using Transit-time Ultrasonic Flowmeters.
MFC-16M (1995)	Measurement of Fluid Flow in Closed Conduits by Means of Electromagnetic Flowmeters.
NACE Std RP-01-69	Control of External Corrosion on Underground or Submerged Metallic Piping Systems.
NACE Std RP-01-75	Control of Internal Corrosion in Steel Pipelines and Piping Systems.

Dimensions and Design

ANSI B1.20.1	Pipe Threads.
ANSI B16.1	Cast Iron Pipe Flanges and Flanged Fittings.
ANSI B16.3	Malleable Iron Threaded Fittings.
ANSI B16.4	Cast Iron Threaded Fittings.
ANSI B16.5	Steel Pipe Flanges and Flanged Fittings. [*Covers Classes 150, 300, 400, 600, 900, 1500, and 2500—Ed.*]
ANSI B16.9	Factory-Made Wrought Steel Butt Welding Fittings. [*Covers "smooth" elbows, tees, caps, etc., as opposed to the mitered fittings—Ed.*]
ANSI B16.10	Face-to-Face and End-to-End Dimensions of Ferrous Valves. [*Cast-iron and steel gate, plug, globe, angle, and check valves—Ed.*]
ANSI B16.11	Forged Steel Fittings Socket Welded and Threaded.
ANSI B16.12	Cast Iron Threaded Drainage Fittings.
ANSI B16.14	Ferrous Pipe Plugs, Bushings, and Locknuts with Pipe Threads.
ANSI B16.15	Cast Bronze Threaded Fittings.
ANSI B16.18	Cast Copper Alloy Solder-Joint Pressure Fittings.
ANSI B16.20	Ring-Joint Gaskets and Grooves for Steel Pipe Flanges.
ANSI B16.21	Nonmetallic Flat Gaskets for Pipe Flanges.
ANSI B16.23	Cast Copper Alloy Solder Joint Drainage Fittings DWV.
ANSI B16.24	Bronze Flanges and Flanged Fittings 150 and 300 lb.
ANSI B16.28	Wrought Steel Buttwelding Short Radius Elbows and Returns.
ANSI B16.29	Wrought Copper and Wrought Copper Alloy Solder-Joint Drainage Fittings.
ANSI B16.31	Non-Ferrous Pipe Flanges.
ANSI B16.34	Steel Valves.
ANSI B16.39	Malleable Iron Threaded Pipe Unions.
ANSI B36.10	Welded and Seamless Wrought Steel Pipe.
ANSI B36.19	Stainless Steel Pipe.
ANSI B40.1	Gages, Pressure and Vacuum.

Steel Piping

ASTM A 53	Pipe Steel Black and Hot-Dipped Galvanized Zinc-Coated Welded and Seamless.
ASTM A 106	Seamless Carbon Steel Pipe for High-Temperature Service.
ASTM A 134	Electric-Fusion-Welded Steel Plate Pipe (Sizes 16 In. and Over).
ASTM A 135	Electric-Resistance-Welded Steel Pipe.
ASTM A 139	Electric-Fusion-Welded Steel Pipe (Sizes 4 In. and Over).
ASTM A 587	Electric-Welded Low-Carbon Steel Pipe for the Chemical Industry.
AWWA C208	Dimensions for Steel Water Pipe Fittings.
AWWA C214	Tape Coating Systems for the Exterior of Steel Water Pipelines.

Steel Fittings

ASTM A 105	Forgings Carbon Steel for Piping Components.
ASTM A 181	Forgings Carbon Steel for General Purpose Piping.
ASTM A 182	Forged or Rolled Alloy-Steel Pipe Flanges, Forged Fittings and Valves and Parts for High-Temperature Service.
ASTM A 216	Carbon Steel Castings Suitable for Fusion Welding for High-Temperature Service.
ASTM A 234	Piping Fittings of Wrought Carbon Steel and Alloy Steel for Moderate and Elevated Temperatures.

Other Process Piping

AWWA C902	Polybutylene (PB) Pressure Pipe Tubing and Fittings 1/2 In. through 3 In. [*Note: Has generated many lawsuits because the fittings fail; the pipe is permeable to gasoline—Ed.*]
AWWA C950	Glass-Fiber-Reinforced Thermosetting-Resin Pressure Pipe.

The American Petroleum Institute (API) has several publications relevant to piping applications:

API Bull 5C3	Formulas and Calculations for Casing, Tubing, Drilling Pipe, and Line Pipe Properties.
API Spec 5L	Specification for Line Pipe.
API Spec 5LE	Specification for Polyethylene Line Pipe.
API Spec 5LP	Specification for Thermoplastic Line Pipe (PVC and CPVC).
API Spec 5LR	Specification for Spiral-Weld Line Pipe.
API Spec 5LX	Specification for High-Test Line Pipe.
API Spec 5L4	Recommended Practice for Care and Use of Reinforced Thermosetting Resin Line Pipe (RTRP).

Chapter 5. Valves

Each number below identifies a specific alloy in terms of chemical composition and physical strength properties. All are aerospace materials—steels, cast, corrosion and heat resistant.

AMS 5373	
AMS 5375	
AMS 5378	
AMS 5380	
AMS 5385	
AMS 5387	
AMS 5788	Aerospace materials—steels, wrought, corrosion and heat resistant.
ANSI B16.10	Face-to-Face and End-to-End Dimensions of Ferrous Valves. (Includes the laying lengths of most valves except ball and butterfly.)
ANSI B16.34	Steel Valves, Flanged and Buttwelding Ends.
API 593	Ductile Iron Plug Valves.
API 594	Wafer Type Check Valves.
API 595	Cast Iron Gate Valves, Flanged Ends.

API 598	Valve Inspection and Test.
API 599	Steel Plug Valves, Flanged or Buttwelding Ends.
ASTM A 36	Specification for Structural Steel.
ASTM A 48	Specification for Gray-Iron Castings.
ASTM A 126	Specification for Valves, Flanges, and Pipe Fittings.
ASTM A 193	Specification for Alloy Steel and Stainless Steel Bolting Materials for High-Temperature Service.
ASTM A 194	Specification for Carbon and Alloy Steel Nuts and Bolts for High-Pressure and High-Temperature Service.
ASTM A 216	Specification for Steel Castings, Carbon, Suitable for Fusion Welding for High-Temperature Service.
ASTM A 276	Specification for Stainless and Heat-Resisting Steel Bars and Shapes.
ASTM A 307	Specification for Carbon Steel Bolts and Studs.
ASTM A 395	Specification for Ferritic Ductile Iron Pressure Retaining Castings for Use at Elevated Temperatures.
ASTM A 516	Specification for Pressure Vessel Plates, Carbon Steel, for Moderate and Lower Temperature Service.
ASTM A 564	Specification for Hot-Rolled and Cold-Finished Age-Hardening Stainless and Heat-Resisting Steel Bars, Wire, and Shapes.
ASTM B 16	Specification for Free-Cutting Brass Rod, Bar, and Shapes for Use in Screw Machines.
ASTM B 62	Specification for Composition Bronze or Ounce Metal Castings.
ASTM B 371	Specification for Copper-Zinc-Silicon-Alloy Rod.
ASTM B 584	Specification for Copper Alloy Sand Castings for General Applications.
AWWA C500	Gate Valves.
AWWA C504	Rubber-Seated Butterfly Valves.
AWWA C506	Backflow Prevention Devices—Reduced Pressure Principle and Double Check Valve Types.
AWWA C507	Ball Valves—6 In. through 48 In.
AWWA C508	Single Swing Check Valves for Ordinary Waterworks Service.
AWWA C509	Resilient-Seated Gate Valves for Water and Sewage Systems.
AWWA C550	AWWA Standard for Protective Interior Coatings for Valves and Hydrants.
AWWA C606	AWWA Standard for Grooved and Shouldered Type Joints.
MSSSP-70	Cast Iron Gate Valves, Flanged and Threaded Ends.
MSSSP-71	Cast Iron Swing Check Valves, Flanged and Threaded Ends.
MSSSP-78	Cast Iron Plug Valves, Flanged and Threaded Ends.
MSSSP-80	Bronze Gate, Globe, Angle and Check Valves.
MSSSP-88	Diaphragm Type Valves.
SAE J775	Engine Poppet Valve Materials.

Chapter 8. Electrical Theory

ANSI/IEEE Std 100	Standard Dictionary of Electrical and Electronics Terms.
CSA	Canadian Standards Association.
CSI	Construction Specifications Institute.
IEEE Standards	The Institute of Electrical and Electronics Engineers.
IEEE Std 141	Recommended Practice for Electric Power Distribution for Industrial Plants.
IEEE Std 242	Recommended Practice for Protection and Coordination of Industrial and Commercial Power Systems.
IEEE Std 446	Emergency and Standby Power.
IES	Lighting Handbook.
NEC	ANSI/NFPA 70 National Electrical Code (NEC).
NEMA Std ICS	Industrial Control and Systems.
NEMA Std MG I	Motors and Generators.
NFPA Standards	National Fire Protection Association.
NFPA 78	Lightning Protection Code.
NFPA 101	Life Safety Code.
UBC	Uniform Building Code.
UL Standards	Underwriters Laboratory.

Chapter 9. Electrical Design

ANSI C57.12.90.	Transformers
IEEE Std 242	Protection and Coordination of Industrial and Commercial Power Systems.

IES	Illuminating Engineering Society.
NEC	National Electric Code.
NEMA Std ICS	Industrial Control and Systems.
NEMA Std PB2	Deadfront Distribution Switchboards.
NFPA Std 493	Standard for Intrinsically Safe Apparatus and Associated Apparatus for Use in Class I, II, and III, Division 1 Hazardous Locations.
NFPA 820	Standard for Fire Protection in Wastewater Treatment and Collection Facilities.
UBC	Uniform Building Code.

Chapter 11. Types of Pumps

The following list of standards for pumps frequently encountered in water and wastewater was developed primarily for the petrochemical industry. However, these standards contain a wealth of experience and information relevant to the water and wastewater industry.

ANSI B73.1	Specifications for Horizontal, End-Suction Centrifugal Pumps for Chemical Process.
ANSI B73.2	Vertical In-Line Pumps for Chemical Process.
ANSI/AWWA E101	Deep Well Vertical Turbine Pumps—Line Shaft and Submersible Types.
API 610	Centrifugal Pumps for General Refinery Service.
API 674	Positive Displacement Pumps—Reciprocating.
API 675	Positive Displacement Pumps—Controlled Volume.
API 676	Positive Displacement Pumps—Rotary.
ASME Power Test Code 7.1	Displacement Pumps.
ASME Performance Test Code PTC 8.2	Centrifugal Pumps.
ASTM A 126	Specification for Gray Iron Castings for Valves, Flanges, and Pipe Fittings.
ASTM A 532	Specification for Abrasion-Resistant Cast Iron.
ASTM A 743	Specification for Castings, Iron-Chromium, Iron-Chromium-Nickel, and Nickel-Base, Corrosion-Resistant, for General Application.

Chapter 13. Motors

ANSI/IEEE Std 100	Standard Dictionary of Electrical and Electronics Terms.
IEEE Standards	The Institute of Electrical and Electronic Engineers
IEEE Std 141 (Red Book)	Recommended Practice for Electric Power Distribution for Industrial Plants.
NEC	National Electrical Code.
NEMA MG 1	Motors and Generators.
NEMA MG 10	Energy Management Guide for Selection and Use of Polyphase Motors

Chapter 14. Engines

NFPA 37	Standard for the Installation and Use of Stationary Combustion Engines and Gas Turbines.
NFPA 58	Standard for the Storage and Handling of Liquefied Petroleum Gases.
NFPA 68	Guide for Explosion Venting.

Chapter 20. Instrumentation and Control Devices

ASME MCF-3M	Measurement of Fluid Flow in Pipes Using Orifice, Nozzle, and Venturi.

Chapter 23. Heating, Ventilating, and Cooling

AMCA No. 99	Standards Handbook
AMCA 210	Laboratory Methods of Testing Fans for Rating Purposes.
AMCA 261	Directory of Products Licensed to Bear the AMCA Certified Ratings Seal.
AMCA 300	Test Code for Sound Rating. Air Moving Devices.
ASHRAE Std 90	Energy Conservation in New Building Design.
ASME CODE VIII	ASME Boiler and Pressure Vessel Code, Section VIII: Pressure Vessels.
NEC	ANSI/NFPA 70 National Electric Code (NEC).
NFPA	Codes and Standards. [*Some of the 243 included standards are updated each year—Ed.*]
NFPA Std 90A	Standard for the Installation of Air Conditioning and Ventilating Systems.
NFPA Std 493	Standard for Intrinsically Safe Apparatus and Associated Apparatus for Use in Class I, II, and III, Division 1 Hazardous Locations.
NFPA Std 820	Recommended Practice for Wastewater Treatment and Transmission Facilities.
Ten-State Standards	Recommended Standards for Sewage Works.

Chapter 25. Summary of Design Considerations

ACI 318	Building Code Requirements for Reinforced Concrete.
ACI 350	Recommended Practice for the Design of Concrete Sanitary Structures.
ANSI MH 27.1	Specifications for Underhung Cranes and Monorail Systems.
ANSI B30.2.0	Overhead and Gantry Cranes (Top Running Bridge, Multiple Girder).
ANSI B30.11	Monorails and Underhung Cranes.
ANSI B30.16	Overhead Hoists (Underhung).
ANSI B30.17	Overhead and Gantry Cranes (Top Running Bridge, Single Girder, Underhung Hoist).
ASTM C494	Specification for Chemical Admixtures for Concrete
CMAA Spec No. 70	Specifications for Electric Overhead Traveling Cranes.
CMAA Spec No. 74	Specification for Top Running and Underrunning Single Girder Electric Overhead Traveling Cranes.
HMI 100	Standard Specifications for Electric Wire Rope Hoists.
UBC	Uniform Building Code.

Chapter 26. Station Layout

NEC	ANSI/NFPA 70 National Electric Code (NEC).

Chapter 27. Avoiding Blunders

AWWA C504	AWWA Standard for Rubber-Seated Butterfly Valves.
AWWA C508	AWWA Standard for Swing Check Valves for Ordinary Waterworks Service.

Chapter 28. Contract Documents

See Section 28-7.

Appendix C. Typical Specifications for Pumps and Drivers

CSI MP-2-2	Section Format.
AFBMA Std 11	Load and Fatigue Life for Roller Bearings.
ANSI/IEEE 112	Test Procedures for Polyphase Induction Motors and Generators.

ASTM A 36	Structural Steel. Carbon Steel.
ASTM A 48	Gray Iron Castings.
ASTM A 276	Standard Specifications for Stainless and Heat-Resisting Steel Bars and Shapes.
ASTM A 322	Steel Bars, Alloy, Standard Grades.
CSI Doc MP-2-2	Section Format.
NEMA MG 1	Motors and Generators
UL 1004	Motors, Electric.

Appendix H. Start-Up

IEEE 81	Guide for Measuring Earth Resistivity, Ground Impedance, and Earth Surface Potentials of a Ground System.

Appendix F

Publishers

F-1. Abbreviations

AASHTO	The American Association of State Highway and Transportation Officials.
ACEC	American Consulting Engineers Council.
ACI	American Concrete Institute.
ACPA	American Concrete Pipe Association.
AFBMA	Anti-Friction Bearing Manufacturers Association.
AGC	Associated General Contractors of America.
AGMA	American Gear Manufacturers Association.
AIA	The American Institute of Architects.
AISC	American Institute of Steel Construction.
AISI	American Iron and Steel Institute.
AMCA	Air Movement and Control Association.
ANSI	American National Standards Institute, Inc.
API	American Petroleum Institute.
ASA	See ANSI.
ASCE	American Society of Civil Engineering.
ASHRAE	American Society of Heating, Refrigeration and Air Conditioning Engineers.
ASME	American Society of Mechanical Engineers.
ASTM	American Society for Testing and Materials.
AWS	American Welding Society.
AWWA	American Water Works Association.
BHRA	British Hydromechanics Research Association.
CISPI	Cast Iron Soil Pipe Institute.
CMAA	Crane Manufacturers Association of America, Inc.
CSA	Canadian Standards Association.
CSI	The Construction Specifications Institute.
DIPRA	Ductile Iron Pipe Research Association.
EJCDC	Engineers' Joint Contract Documents Committee.
FEDSPEC	Federal Specifications.
FMC	FMC Corporation.
HI	Hydraulic Institute.
HMI	Hoist Manufacturers Institute.
IAPMO	International Association of Plumbing and Mechanical Officials.
ICBO	International Conference of Building Officials.
IEEE	The Institute of Electrical and Electronics Engineers, Inc.
IES	Illuminating Engineering Society.
INCE	The Institute of Noise Control Engineers.
ISA	Instrument Society of America.
MILSPEC	Military Specifications.
MSS	Manufacturers Standardization Society, Inc.
NACE	National Association of Corrosion Engineers.

NBS	National Bureau of Standards.
NEC	National Electrical Code (Publisher: National Fire Protection Association).
NEMA	National Electrical Manufacturers Association.
NESC	National Electrical Safety Code (Publisher: American National Standards Institute).
NFBA	National Board of Fire Underwriters.
NFMA	National Fan Manufacturers Association (superseded by AMCA).
NFPA	National Fire Protection Association.
NIOSH	National Institute for Occupational Safety and Health.
NTIS	National Technical Information Service.
OSHA	Occupational Safety and Health Administration.
SAE	Society of Automotive Engineers.
SAMA	Scientific Apparatus Manufacturers Association.
SBCCI	Southern Building Code Congress International, Inc.
SSPC	Steel Structures Painting Council.
UBC	Uniform Building Code (Publisher: ICBO).
UL	Underwriters Laboratories, Inc.
UPC	Uniform Plumbing Code (Publisher: IAPMO).
WEF	Water Environment Federation.
WPCF	Water Pollution Control Federation (now WEF).

F-2. Addresses of Publishers

These addresses of publishers are the places to which requests for literature should be sent. They may change with time.

Acoustical Society of America
335 East 45th Street
New York, NY 10017

Air Movement and Control Association
30 West University Drive
Arlington Heights, IL 60004

The American Association of State Highway and Transportation Officials
444 North Capitol Street N.W., Suite 249
Washington, DC 20001

American Concrete Institute
P.O. Box 9094
Farmington Hills, MI 48333

American Concrete Pipe Association
8615 Westwood Drive, Suite 105
Vienna, VA 22180

American Gear Manufacturers Association
1500 King Street, Suite 201
Alexandria, VA 22314

The American Institute of Architects
1735 New York Avenue N.W.
Washington, DC 20006

American Institute of Steel Construction
1 East Wacker Drive, Suite 3100
Chicago, IL 60601

American Iron and Steel Institute
1133 15th St. NW
Washington, DC 20005-2701

American National Standards Institute, Inc.
11 West 42nd Street, 13th Floor
New York, NY 10036

American Society for Testing and Materials
1916 Race Street
Philadelphia, PA 19103

American Society of Civil Engineers
1801 Alexander Bell Drive
Reston, VA 20191

American Society of Heating, Refrigeration and Air Conditioning Engineers
United Engineering Center
345 East 47th Street
New York, NY 10017

American Society of Mechanical Engineers
345 East 47th Street
New York, NY 10017

American Water Works Association
666 West Quincy Avenue
Denver, CO 80235

American Welding Society
550 LeJuene Road, N.W.
Miami, FL 33126

Anti-Friction Bearing Manufacturers Association
1200 19th Street N.W., Suite 300
Washington, DC 20036

British Hydromechanics Research Association
Cranfield
Bedford MK 43 OAJ, England

Brooks/Cole Engineering Division
555 Abrego Street
Monterey, CA 93940

Building Officials and Code Administrators International, Inc.
4051 W. Flossmoor Road
Country Club Hills, IL 60477

Bureau of National Affairs, Inc.
1231 25th Street N.W.
Washington, DC 20037

Canadian Standards Association
235 Montreal Road
Rexdale, Ontario KIL 6C7
Canada

Cast Iron Soil Pipe Institute
5959 Shallowford Road, Suite 419
Chattanooga, TN 37421

Caterpillar, Inc.
100 NE Adams Street
Peoria, IL 61629

Chemical Manufacturers Association
2501 M Street N.W.
Washington, DC 20037

Chilton Book Company
Chilton Way
Radnor, PA 19089

The Chlorine Institute
201 L Street N.W.
Washington, DC 20036

Clow Corporation
Water Reclamation Group
Pump Division
1999 N. Ruby Street
Melrose Park, IL 60160

or

Water Systems Group
1211 West 22nd Street
Oak Brook, IL 60521

Columbia Books, Inc., Publishers
1212 NY Avenue N.W., Suite 330
Washington, DC 20005

The Construction Specifications Institute
601 Madison Street
Alexandria, VA 22314-1791

County of Los Angeles
Department of Public Works
P.O. Box 4089
Los Angeles, CA 90051

Crane Co.
300 Park Avenue
New York, NY 10022

Crane Manufacturers Association of America, Inc.
8720 Red Oak Boulevard, Suite 201
Charlotte, NC 28217

Department of the Army Waterworks Experiment Station
Corps of Engineers
P.O. Box 631
Vicksburg, MS 39180-0631

Department of Civil Engineering
(Formerly Department of Civil Engineering and Engineering Mechanics)
Montana State University
Bozeman, MT 59717

Department of the Environment
Hydraulic Research Station at Wallingford
Her Majesty's Stationery Office, Government Bookshops
49 High Holborn
London WCIV 6HB, England

Ductile Iron Pipe Research Association
245 Riverchase Parkway East
Birmingham, AL 35244

Dunham-Bush, Inc.
101 Burgess Road
Harrisburg, VA 22801

Engineering News-Record, now ENR
P.O. Box 950
New York, NY 10020

FEB Press
P.O. Box 2431
Ann Arbor, MI 48106

Federal Specifications
General Services Administration
Specification and
Consumer Information
Distribution Branch
Washington Navy Yard, Bldg. 197
Washington, DC 20407

FMC Corporation
Environmental Equipment Division
2240 W. Diversey Avenue
Chicago, IL 60647

or

Water Treatment Dept. 2240
2050 N. Broad Street
Lansdale, PA 19446

General Electric Co.
1 River Road
Schenectady, NY 12345

Health Education Service, Inc.
P.O. Box 7126
Albany, NY 12224

Her Majesty's Stationery Office,
Publications
51 Nine Elms Lane
London SW8 SDR, England

Hydraulic Institute
9 Sylvan Way
Parsippany, NJ 07054-3802

Hydraulics Research Station
Wallingford, England
(See Her Majesty's Stationery Office
or Department of the Environment.)

Illuminating Engineering Society
Publications Office
120 Wall Street, 17th Floor
New York, NY 10005

Industrial Press
93 Worth Street
New York, NY 10013

or

200 Madison Avenue
New York, NY 10016

Ingersoll-Rand Company
c/o Thomas Associates, Inc.
1300 Sumner Avenue
Cleveland, OH 44115

The Institute of Electrical and
Electronics Engineers, Inc.
345 East 47th Street
New York, NY 10017

or

Wiley-Interscience of
John Wiley & Sons
445 Hoes Lane
(IEEE Service Center)
Piscataway, NJ 08854

The Institute of Noise Control
Engineers
P.O. Box 3206
Poughkeepsie, NY 12603

Instrument Society of America
P.O. Box 12277
Research Triangle Park, NC 27709

or

400 Stanwix Street
Pittsburgh, PA 15222

International Association of
Plumbing and Mechanical Officials
5032 Alhambra Avenue
Los Angeles, CA 90032

International Conference of
Building Officials
5360 South Workman Mill Road
Whittier, CA 90601

Robert E. Krieger
Publishing Company
P.O. Box 9542
Melbourne, FL 32901

Lewis Publishers, Inc.
121 South Main Street
Chelsea, MI 48118

Los Angeles County
Flood Control District
Department of Public Works
P.O. Box 4089
Los Angeles, CA 90054

Manufacturing Chemists Association
(See Chemical
Manufacturers Association.)

Manufacturers Standardization Society, Inc. of the Valve and
Fittings Industry
127 Park Street NE
Vienna, VA 22180

McGraw-Hill, Inc.
1221 Avenue of the Americas
New York, NY 10020

McGraw-Hill Information
Systems Co.
P.O. Box 28
Princeton, NJ 08540

R.S. Means Co., Inc.
100 Construction Plaza
Kingston, MA 02364

Military Specifications
Naval Publications and Forms Center
5801 Tabor Avenue
Philadelphia, PA 19120

Miramar Publishing Company
2048 Cotner Avenue
Los Angeles, CA 90025

Montana State University
(See Department of
Civil Engineering.)

National Association of
Corrosion Engineers
P.O. Box 218340
Houston, TX 77218

National Association of Plumbing-
Heating-Cooling Contractors
180 South Washington Street
P.O. Box 6808
Falls Church, VA 22046

National Board of Fire Underwriters
85 John Street
New York, NY 10038

National Council of
Acoustical Consultants
66 Morris Avenue
Springfield, NJ 07081

National Electrical
Manufacturers Association
1300 North 17th Street, Suite 1847
Rosslyn, VA 22209

National Fan Manufacturers
Association (Superseded by Air
Movement and Control
Association.)

National Fire Protection Association
1 Batterymarch Park
P.O. Box 9101
Quincy, MA 02269

National Institute for Occupational
Safety and Health
U.S. Department of Health and
Human Services
Public Health Centers for Disease
Control
1600 Clifton Road N.E.
Atlanta, GA 30333

National Technical Information
Service
U.S. Department of Commerce
5285 Port Royal Road
Springfield, VA 22161

Noyes Publications
Noyes Data Corporation
Mill Road at Grand Avenue
Park Ridge, NJ 07656

Occupational Safety and Health
Administration
The Bureau of National Affairs, Inc.
1231 25th Street N.W.
Washington, DC 20037

Plastic Directory
Cahner's Publishing Company,
Division of Reed Publishing
275 Washington Street
Newton, MA 02158

Reston Publishing Co., Inc.
P.O. Box 500
Englewood Cliffs, NJ 07632

Richardson Engineering Services, Inc.
909 Rancheros Drive
P.O. Box 1055
San Marcos, CA 92069

Oy E. Sarlin, AB
P.O. Box 750
SF-001 01 Helsinki 10
Finland

Scientific Apparatus
Manufacturers Association
225 Reinekers, Suite 625
Alexandria, VA 22314

Southern Building Code
Congress International, Inc.
900 Montclair Road
Birmingham, AL 35213-1206

Trade & Technical Press, Ltd.
Crown House
Morden, Surrey SM4 SEW, England

Tube Turns Division
Allegheny International, Inc.
2900 W. Broadway
Louisville, KY 40211

Turner Designs
920 West Maude
Sunnyvale, CA 94086

Underwriters Laboratories, Inc.
333 Pfingston Road
Northbrook, IL 60062

Uni-Bell PVC Pipe Association
2655 Villa Creek Drive, Suite 155
Dallas, TX 75234

Uniform Building Code
International Conference of
Building Officials
5360 South Workman Mill Road
Whittier, CA 90601

U.S. Environmental Protection
Agency
Office of Municipal Pollution Control
Municipal Utilities Div. (WH 546)
401 M Street S.W.
Washington, DC 20460

U.S. Government Printing Office
732 North Capitol Street N.W.
Washington, DC 20401

Van Nostrand Reinhold Company, Inc.
7625 Empire Drive
Florence, KY 41042

The Vibration Institute
6262 S. Kingery Highway, Suite 212
Willowbrook, IL 60514

Water Environment Federation
601 Wythe Street
Alexandria, VA 22314

Water Pollution Control Federation
(See Water Environment
Federation.)

Water Treatment Department
2050 N. Broad Street
Lansdale, PA 19446

Westinghouse Electric Corporation
Printing Division
Forbes Road
Trafford, PA 15085

Whitman, Requard & Associates
1304 St. Paul Street
Baltimore, MD 21202

John Wiley & Sons, Inc.
605 Third Avenue
New York, NY 10158

Wiley-Interscience
A division of John Wiley & Sons, Inc.
IEEE Service Center
445 Hoes Lane
Piscataway, NJ 08854

Worthington Pump Division
Dresser Industries, Inc.
401 Worthington Avenue
Harrison, NJ 07029

F-3. Reference

1. Colgate, C., Jr., J. J. Russell, and R. Germain. *National Trade and Professional Associations of the United States,* 32nd ed. Columbia Books Inc., Publishers, Washington, DC (1997).

Appendix G

Checklist for Project Reviews

LEROY R. TAYLOR

CONTRIBUTORS
Raymond L. Elliott
Horton Wasserman

The purpose of this appendix is to present a series of checklists as a guide to ensure the consideration of important aspects of design. It does not, however, include all aspects for every kind of pumping station, so add to it as necessary.

G-1. Civil Design Checklist

This list should be reviewed by a civil engineer.

1. Construction facilities.
   a. Access, staging area, borrow area, special erosion-control facilities.
   b. Waste area.
2. Fence location, details.
3. Grading and drainage.
   a. Existing and finish grades and two benchmarks shown.
   b. Corrective measures for undrainable areas.
   c. Flood elevation shown and accommodated.
   d. Off-site irrigation and drainage maintained.
   e. Culvert location, size, and material.
   f. Storm sewers at large sites.
4. Irrigation plan.
5. Landscaping plan: planting and seeding.
6. Location, size, and material of interfering utilities, power poles, and structures.
7. Paving.
   a. Thickness design.
   b. Area, elevations.
   c. Drainage.
   d. Cross-sections.

8. Property lines: bearings, distances, ties, future additions.
9. Specifications: cover all materials and workmanship.
10. Test hole information shown: location, materials, depth to water, date measured.
11. Walkway locations, details.
12. Handicap parking and access.
13. Plan review by civil and geotechnical engineer.
14. Cost estimates.

G-2. Structural/Geotechnical/Architectural Design Checklist

This list should be reviewed by a structural engineer, a geotechnical engineer, and an architect.

15. Buoyancy checked.
16. Soils.
    a. Bearing, subsidence, compaction.
    b. Lateral pressure.
    c. Pile bearing.
17. Clearance around and over equipment adequate for maintenance and/or replacement.
18. Door details: louvers, weatherstripping, sills, sections.
19. Door and window schedules—no windows at sites where vandals have easy access.
20. Finish details and color schedule.
21. Footing depth below frost line.
22. Gutter, downspout, and roof drainage details.
23. Hardware schedule and keying.
24. Leakage tests of water-holding basins required.
25. Orientation to sun.
26. Provisions made for dewatering, foundation drains, etc.
27. Roof and flashing details; slope roof to gutters or drains.
28. Disinfection of potable water-holding basins required.
29. Handrailing.
    a. Around hazardous areas (open water basins, walkways, stairs, floors, etc.).
    b. Cages and landings for ladders.
30. Specifications cover all materials and workmanship.
31. Structural continuity and integrity.
32. Traffic patterns analyzed.
33. Waterstops called out and detailed.
34. Expansion joints in masonry walls (include corner provisions).
35. Coping and flashing on parapet walls.
36. Flashing and drainage for flat roofs.
37. No interior wall penetrations into hazardous areas.
38. Plan review by structural engineer and architect.
39. Cost estimates.

G-3. Electrical Design Checklist

This list should be reviewed by an electrical engineer. Consult Section 9-12 in addition to the following.

40. Conduit.
    a. Material, size, concealed versus exposed.
    b. Control and power circuit separation.
    c. No junction boxes in wet well or other places of condensing humidity.
41. Drivers.
    a. Characteristics suitable for proposed service.
    b. Nonreverse ratchet if required by pump.

c. Lubrication, bearing type, and life.
d. Explosion and weather hazards considered in choosing motor type.

42. Explosion hazards and grounding.
43. Illumination.
    a. Fixture type.
    b. Photocell control for outside lighting.
    c. Switching locations.
    d. Work areas covered.
44. Specifications: all materials are covered.
45. Switchgear and motor starters.
    a. Sizing, fused on all three legs, clearance from waterlines, adequate spare circuits.
    b. Corrosion potential.
    c. Protection from lightning.
46. Wiring: material, size, number of conductors.
47. Review by electrical engineer.
    a. Concept.
    b. Constructability.
    c. Plans. Adequate space in front of and behind panels and around other electrical gear.
    d. Specifications.
    e. Schedule.
    f. Cost estimates.

G-4. Instrumentation and Control Checklist

This list should be reviewed by an expert in instrumentation and control.

48. Control cable.
    a. Provisions for protection from lightning and voltage spikes.
    b. Provisions for damping inductance.
    c. Telephone cable of proper grade.
49. Instruments and controllers.
    a. Proper range, units, visibility.
    b. Environment controlled-temperature range, dust, humidity, etc.
50. Power supply.
    a. Adequacy, reliability, back-up power.
    b. Failsafe provisions.
51. Process and instrumentation diagrams.
    a. All functions covered, logic correct.
    b. Alarms and alarm silencer for critical functions.
    c. All interconnects covered.
    d. Control stability, time delay requirements.
    e. Interlocks not too complex.
    f. Provisions for expansion.
52. Specifications cover all materials.
53. Plan review by instrumentation and control engineer.
54. Transmitters and receivers.
    a. Freeze-up, heat, and humidity problems considered.
    b. Lightning protection.
55. Cost estimates.

G-5. Cross-Connection Control

This list may be reviewed by either a civil or a mechanical engineer.

Water Plants

56. Cross-connections in water plants can result in contamination of potable water supplies by untreated or partially treated water, chemicals, sanitary wastes, drainage water, and similar hazardous sources. During design reviews, a checklist should be used to spot hazards. Include the following:
    a. Inadvertent pipe connections between systems eliminated.
    b. Avoid potential cross-connections due to hoses.
    c. Back siphonage hazards eliminated.
    d. Backflow prevention for underground sprinkler systems, pump water seal systems, cooling water systems, etc.; provide air gap on a receiving tank, if possible.
    e. Water and sewer line crossings meet state standards.
    f. Positive intake pressure is maintained on booster pumps.
    g. Pump suction lines are not buried (or at least negative pressure is prevented); burial is satisfactory for raw water before treatment.
    h. Backflow preventers in floodproof location (above flood elevation).
    i. Operation and maintenance instructions include cross-connection-control program.

Wastewater Plants

57. Cross-connection hazards in wastewater plants are similar to those in a water plant, but sanitary hazards are much more dangerous. A backflow preventer should be installed on the main water supply line to the plant and an air gap should be provided on hose bibbs where the hose may inadvertently be left in a tank or other sanitary or chemical hazard. Extra vigilance must be used to spot actual pipe cross-connections during both design and construction. Because there are often a number of grades of water (such as potable and plant effluent), pipe-marking systems are especially important. Keep backflow preventers above flood level and vent to the outside of the building. Angele [1] and AWWA M-14 [2] give more complete discussions.

G-6. Mechanical Design Checklist

This list should be reviewed in general by a civil or mechanical engineer. Parts of it require specialists in heating and ventilating, engines, pumps, control and instrumentation, water hammer analysis, and hydraulics.

58. Auxiliary power.
    a. Air cooling systems adequate.
    b. Alarm systems for critical functions.
    c. Automatic start system and timer for critical systems, marked to indicate hazard of automatic starting.
    d. Coupling on shaft shielded for safety.
    e. Engine horsepower adequate for start-up and continuous duty after derating for elevation.
    f. Exhaust system meets requirements for neighborhood noise levels, is insulated from flammable materials, and is installed above snow level.
    g. Fuel systems and capacity.
    h. Water cooling systems protected against freeze-up and separated by air gap from potable water systems.
    i. Heat recovery feasibility.
    j. Chlorine gas: design feed rates.
59. Drivers—see Electrical Design Checklist.
60. Flowmeter.
    a. Clearance from obstructions and flow disturbances (adequate length of straight pipe on both ends).
    b. Sizing and proper units for readout.
    c. Volumetric accuracy. Calibrate *in situ* for guaranteed accuracy.
    d. Bypass for removal or cleaning.
61. Heating, ventilating, and air conditioning.
    a. Air inlet, exhaust, and fan adequate for motor cooling.
    b. Floor level exhaust for chlorine room with outside switch and high level air inlet.

- c. Positive diffusion of emergency vented chlorine gas.
- d. Heating and insulation adequate for freeze-up protection.
- e. Wet well ventilation and odor control.
- f. Heat recovery facilities.
- g. Fuel or electricity for heating and heating method (unit heaters versus heated ventilating air).

62. Piping.
    - a. Corrosion review—see Sections 4-9 and 5-10.
    - b. Cross-connections: backpressure, back siphon.
    - c. Floor drainage, traps, vents, wall penetrations.
    - d. Drains for suction and discharge piping for valve or pump repair; large drain to pump sump for field testing pumps.
    - e. Freeze-up hazards considered.
    - f. Pipe cover adequate for frost protection.
    - g. Pressure gauges and appropriate water-sampling taps.
    - h. Size, material, pressure class, joint type, flanges, gaskets, taps, fittings and couplings, wall sleeves, flexible joints at walls, vibration sources.
    - i. Thrust restraint.
    - j. Spacing of pipe supports adequate; supports cushioned.
    - k. Water hammer analyzed for normal operation and power loss conditions.
    - l. Air and gas traps: positive means to remove trapped air (or other gases) at high points—manual for start-up and automatic where gas may accumulate; cross-connection control required with automatic air release valves on potable water systems; avoid high points entirely (especially in sewage systems) if practical.
    - m. Adequate Victaulic® or Dresser® couplings for dismantling and flexibility.
63. Pumps.
    - a. Anchorage and grouting.
    - b. Bearing type, life, prelubrication, lubrication.
    - c. Clearance from walls, need for vortex breaker.
    - d. Materials choices.
    - e. NPSH adequate for pumps.
    - f. Provisions for field and factory tests.
    - g. Pump curves, system curves, capacity, impeller size and ability to pass solids, capacity for start-up, normal and emergency (flooded suction) operating conditions.
    - h. Facilities for removal for maintenance.
    - i. Guards for and accessibility of bearings and supports for extended shafting.
    - j. Isolation from or designed to withstand water hammer and surge pressures.
64. Valves.
    - a. Air and vacuum valve sizing, type, location, closing characteristics, method of isolating and draining.
    - b. Check valve size, type, location, and closing characteristics.
    - c. Isolation valves adequate in number, location, type, and size.
    - d. Globe, plug, ball, cone or butterfly valve location, size, type, and suitability for throttling use.
    - e. Pressure relief valve size, type, drainage.
    - f. Pump-control valve sizing, sequencing, failsafe on power failure.
    - g. All installations allow exercising valves.
65. Wells.
    - a. Well vent and drawdown gauge provided.
    - b. Proper sealing.
66. Safety considerations.
    - a. Showers and eyewashes if chemicals are used.
    - b. Gas masks or self-contained breathing apparatus if required.
    - c. Clearances around machinery, piping, and cabinets.
    - d. Labeling and color coding for piping.
    - e. Adequate signs.
    - f. Ventilation.

67. General review by project engineer (civil or mechanical engineer, probably).
    a. Concept.
    b. Mechanical calculations.
    c. Constructibility and bid clarity.
    d. Plans.
    e. Specifications complete and cover all materials and products.
    f. Schedule.
    g. Cost estimates.

G-7. References

1. Angele, G. J., Sr. *Cross-Connection and Backflow Prevention.* American Waterworks Association, Denver, CO (1974).
2. AWWA. *Recommended Practice for Backflow Prevention and Cross-Connection Control,* M 14. American Water Works Association, Denver, CO (1990).

Appendix H
Start-Up

ERIK B. FISKE
GEORGE FRYE
LOWELL G. SLOAN
SAM V. SUIGUSSAAR

CONTRIBUTORS
Robert S. Benfell
E. Robert Bouwkamp
LeRoy R. Taylor
William Young

At the conclusion of the pumping station design and while details are still fresh in the mind, the project leader (aided by the various designers who represent all the systems involved, including the start-up engineer) should prepare a start-up checklist for each system. The purpose of the checklist is to make sure that all equipment and subsystems are functional and operate as the designer intended. Do not assume that the installation will be flawless. There have been many instances of incomplete or loose electrical connections, valves installed backward, motors that run in reverse, and even the substitution of check valves for gate valves. So take nothing for granted. Make the start-up check thorough and complete to avoid malfunction, damage to equipment, or future operating problems.

Plant start-up should always be the responsibility of the contractor. Some equipment is almost always damaged at start-up. Often, if not usually, the damage is the result of installation errors. Even the most thorough checking by the most knowledgeable and conscientious resident engineer cannot possibly forestall every error. A start-up team should be organized with the following suggested personnel:

- *Start-up coordinator*—an operations specialist provided by the designer.
- *Electrical engineer*—provided by the designer.
- *Instrumentation engineer*—provided by the designer.
- *Resident engineer*—provided by the designer, the client, or the construction manager.
- *Mechanical contractor*—provided by the general contractor.
- *Electrical contractor*—provided by the general contractor.
- *Instrumentation contractor*—provided by general contractor.
- *Operators*—provided by the owner.

The duty of the start-up coordinator is to implement the testing plan devised by the consultant, the contractor, and the owner. During all phases of testing, the start-up team members provide both expertise in their respective fields and verification checks. After start-up, the team should remain at the job site to facilitate fine tuning and debugging.

The following list of items is based on hypothetical pumping stations of various types. The list is intended as a reminder and does not encompass every possible pumping station or subsystem. Several subsystems are, in fact, omitted. For example, ventilation, gas detection (Cl_2, H_2S, explosive gases), emergency generator, odor control, and fluid power systems are so specialized or specific as to type and manufacturer that a generalized list might be inappropriate. Some parts of the following list are generic, and some are detailed to emphasize the complexity and the completeness needed for start-up. The point to be emphasized is that start-up checklists should be made with meticulous care and must be specific to each individual pumping station and each individual system.

The distinction among design, start-up, and troubleshooting checks is often ill-defined. The troubleshooting tips given on pages 145–148, 197–207, and 288–292 of *Hydraulic Institute Standards* [1], in Table 9-1 of the classic *Pumping Manual* [2], or on pages 2.294–2.301 and 12.16–12.8 of Karrasik *et al.* [3] may well be adaptable to start-up.

H-1. Pre-Visit Check

Before visiting the pumping station, review the specifications, plans, O&M manual (control section), and manufacturers' submittal data. Review the following application data.

Item	Notes
1. Curves. System H-Q and pump performance	Speeds, maximum heads and capacities, and design conditions.
2. Liquid pumped	Chemistry of liquid, temperature, specific gravity, pH, solids.
3. Capacity	Station influent flow rates, discharge rates, and pressure heads.
4. Pump mounting	Horizontal, vertical, or suspended.
5. Number of pumps and operation strategy (from O&M manual)	Sequencing, types of drivers (C/S or V/S or combinations such as 2 V/S and 1 C/S).
6. Pump suction	Wet well head or lift referred to pump datum point. $NPSH_A$ versus $NPSH_R$.
7. Pump discharge	Total head in feed and pounds per square inch or the SI equivalents. Obtain the required time to close discharge valve to avoid water hammer.
8. Power transmission	Direct shaft-mounted coupling, intermediate shaft, vee-belt drive, etc.
9. Other power considerations	Speed, motor overloading, motor efficiency and power factor, station voltage, station electrical systems (one or two incoming lines or engine-generator power), switchgear, transformers, induction or synchronous motors. Compare drive starting torque to pump shut-off torque, and effective pumping volume versus motor starts/hour (responsibility of designer).
10. Piping	Flexible connections, base elbow, pipe supports, discharge air-release valve with drain and air gap.
11. Discharge check	Air, oil, air-oil, or water with speed control, swing with valves spring or weight and lever, etc.
12. Shop tests	Certified or witnessed. Shop test performance data that indicates speeds, safe (limit lines) operating conditions, and $NSPH_R$.
13. Wet well	Low wet well level must be higher than the pump casing and there must be no vortexing or influent waterfall effect. Otherwise, furnish an automatic primer system. [*Ed. note—including such a feature is the designer's responsibility.*]

H-2. Pre-Start-Up Check

Before start-up of any equipment, inspect the pumping station with the contractor, and the mechanical, electrical, and instrumentation personnel. The start-up of complex equipment (such as engine-generator sets and variable-speed controls) requires a factory representative, so coordinate his visit, and let him take the responsibility for his equipment (to keep manufacturers' warranties in force). Make sure the following items have been accomplished.

General

14. Review and verify the resident engineer's inspection checksheets.
15. All electrical equipment has been installed and adjusted by the supplier as specified.
16. All mechanical equipment specified has been installed properly and anchored. All bolts and nuts are tight.
17. Ventilation system is operational.
18. Heating system is operational.
19. Sump pumps are operational.
20. Instrumentation system is operational.

Wet Well

21. Wet well is clean. All debris has been removed.
22. Wet well water level sensing system (e.g., floats, bubbler) has been installed properly and controls have been adjusted for operating levels as specified.

Piping and Valves

23. Suction pipe, unless submerged, is vacuum tight. Fasteners are tight at all pipe connections.
24. Pipe supports are installed properly and strain in the equipment (particularly in valve bodies) is eliminated.
25. Piping is level or continuously rising to prevent formation of air pockets, and air-relief valves and air bleeds are installed where specified.
26. Harness bolts are installed properly in sleeve connections (unless pipe is externally restrained so that joint cannot separate).
27. The piping has been pressure tested and all pipelines to all equipment have been flushed.
28. Valves operate properly. Check to be sure valves (especially check valves) are installed in the right flow direction. Test limit switches and pressure switch operation. Bolts on eccentric plug valves should be only tight enough to prevent leakage.
29. Valve springs or lever weights are properly adjusted. Dashpots are properly timed. (Do during operational test.)
30. Pump control or check valves are operational and properly timed or adjusted. (Fine tune during operational test.)
31. Both suction and discharge valves are closed.
32. Automatic valves operate and air receivers (if any) are fully charged.
33. It is always good procedure to close all valves prior to beginning operational tests.

Pumps

34. Motor and pump shafts are disconnected and tested for alignment. Leave disconnected (see "Motors," Items 53–63).
35. Pumps are level and there is no strain in the pump casing.
36. Packing glands are adjusted with nuts finger tight for a small, steady stream of water for a 24-h break-in period.
37. Shaft seals are lubricated as required.
38. The rotation arrow on the pump is in place and correct, and relate arrow to the direction of the pump discharge nozzle.

Seal-Water System with Open Tank Reservoir

39. Incoming float control valve is mounted properly.
40. Float can move freely.
41. Incoming liquid line to the tank is furnished with a shut-off valve.
42. Shut-off tank drain valve is closed.

43. Seal-water discharge piping is connected to the main pump.
44. Seal-water system piping is furnished with a discharge relief to reservoir for excessive pressure.
45. Piping from the control valve in the tank is higher than the projected tank liquid level (as determined by the float level).
46. Piping from the drain is directed to the sump pump.
47. A needle valve and a flow indicating device are furnished in the seal-water piping at the main pump.
48. Push the STOP button and LOCKOUT button to the STOP position at the seal-water equipment.
49. Rotate the seal-water pump shaft by hand.
50. Reassemble the coupling and replace the coupling guard.
51. Set the pressure switches in the discharge line as specified.
52. Clean and fill the reservoir and adjust the float switch so water will not overflow. Confirm there is an air gap between the incoming water pipe and the top of the tank (for cross-connection prevention).

Motors

53. Motor nameplate data (horsepower, speed, and electrical characteristics) agree with the specifications.
54. The equipment (including coupling and gearboxes) has been lubricated as specified by the supplier.
55. Confirm that the coupling between the motor and pump is disconnected (Item 34). "Bump" the motor to make sure the direction of rotation is correct. Reconnect the coupling.
56. Overload protectors are sized properly.
57. Wiring connections are tight.
58. Temperature overload device is operational.
59. Guards are installed around rotating machinery.
60. Notched belts are installed with proper tightness and alignment in belt drives.
61. Alignment between pump and motor shafts is checked.
62. Variable-speed units: manufacturer's representative **MUST** be on site for start-up.
63. Engines and engine-generator sets: the factory representative must start up, operate, and troubleshoot.

Submersible Sewage Pumps

In addition to the applicable checks above, make sure

64. Return elbows are securely bolted in place and fasteners are *tight*.
65. Slide rails are installed properly. A midrail guide is in place if the depth exceeds 6 m (20 ft).
66. Pumps seat accurately.
67. Lifting device (if included to remove pumps) is operational. Check insertion and removal of pumps. Ensure there are no sharp edges to cut power cables.
68. There are **NO** electrical connections, splices, or junction boxes in the wet well or in locations of condensing humidity
69. The leak sensor is operating.
70. Oil is at the proper level in the rotating assembly.

H-3. Electrical Systems

Pre-Energizing Inspection

Electrical systems for all pumping stations with either low, medium, or high voltages are very dangerous. It is therefore extremely important that all electrical checks be made by qualified and competent electrical inspectors and/or engineers. The statement in the introduction that even the most conscientious resident engineer cannot possibly prevent every error is particularly true of electrical work, where hundreds (or even thousands) of wires must be properly terminated. Nevertheless, the engineer can reduce start-up problems by being sure that the following electrical tests are made.

Before energizing anything, the following should be done:

71. The grounding system should be tested. The best way to accomplish this test is specified in IEEE 81. The 5-ohm value suggested meets code but is not very good for wet-process equipment. Generally, values of less than 1 ohm can be obtained without difficulty. The actual value required depends on the capacity and voltage levels of the power system and should be determined by the electrical design engineer. In situations where a low enough ground resistance cannot be obtained, such as in desert areas, the design engineer should have known about it and designed a ground mat for the facility that would, at least, hold potential gradients within values safe to human beings. Check switchgear, yard transformer, pump and motor control centers, motor feeder cables, pumps and equipment at all electrically connected panels, and switchgear ground bus for grounding.

72. Electrical insulation should be tested. For low-voltage (<600 V), this is easily accomplished with meg-ohm meters that typically apply 1000 V to the circuitry under test. For systems rated over 600 V, "high-potting" is required. Only personnel trained in these procedures should attempt high-voltage insulation testing.

73. Protective devices, such as fuses and circuit breakers, should be checked at least visually for conformance to design documents. Circuit breakers should be operated manually to ensure that the mechanisms are not binding. Adjustable circuit breakers and protective relays should be checked for correct settings. Checking low-voltage motor overload heaters and motor starter fuses is specifically omitted at this point. Overload heater and motor starter fuses should be removed from all starters before energizing the system. High-voltage starters and circuit breakers should be racked out to the test positions.

In power systems rated more than 1200 A at 480 V and all power systems operating at over 600 V, it is strongly recommended that all protective devices be tested by passing current through their sensing elements with a high-current low-voltage testing machine—a task that should normally be done by contractors specializing in electrical testing.

74. Electrical gear must be vacuumed out thoroughly to remove construction dust and debris.

75. Every circuit and breaker and disconnect switch should be opened before the utility turns on the power.

Some of the detailed checks to be made before the utility turns on the power are as follows:

76. Cable and circuit are identified per contract specification requirements.

77. Switchgear key interlock scheme prevents paralleling feeders (if there is more than one feeder).

78. Switchgear door interlock functions are

a. Door cannot be opened with switchgear closed.

b. Switch cannot be closed with door open.

c. Access to all power fuses is possible only when switches are open.

79. All switchgear secondary and low-voltage compartments are isolated by barriers from high-voltage compartments.

80. Switchgear current-limiting fuses are rated in accordance with the contract documents.

81. Switchgear safety-glass window makes disconnect *blade* visible. Stored-energy disconnect switch mechanism operates.

82. Switchgear weatherproofing (if outdoor type), ventilation, thermostat, and heaters are operational.

83. Switchgear protective relay settings are in accordance with manufacturers' recommendations and coordination/short-circuit protection limits. Each incoming line has time-delay overcurrent and one time-delay ground fault protective relay.

84. Potheads or other cones on all power cables are installed properly.

85. Nameplates, warning, and "danger—high voltage" signs are installed.

86. Perform high potential (voltage) tests at switchgear for 1 min on primary and secondary circuits. Factory test reports on high potential tests are acceptable.

87. Switchgear schematic control diagrams are correct.

88. See that transformer nameplate data, high-voltage warning signs, and high- and low-voltage compartment isolation are in place.

89. Transformer door interlocks: assure

a. Low-voltage compartment door can be opened first.

b. High-voltage compartment door cannot be opened first.

90. Transformer high-voltage compartment switch fuse is operational.

91. Transformer low-voltage compartment has
   a. No-load tap changer.
   b. Liquid-level gauge on liquid-fill transformer.
   c. Dial thermometer.
   d. Level and temperature alarm contact.
   e. Ground pad and proper grounding.

Motor Control Centers (MCC)

92. Control centers are NEMA class as specified.
93. Bus rating conforms to specifications.
94. The circuit breaker is sized properly, of the proper type, and is adjusted as specified.
95. Motor starters are sized properly and provided with proper thermal overload relays with heaters on each leg as specified.
96. Wiring and conduit from the control to the motor is sized per NEMA standards as a minimum.
97. Automatic and manual switches are furnished and operational as specified.
98. If combination motor-circuit protectors and magnetic starters are furnished, determine that the motor-circuit protectors are sized in accordance with manufacturer's recommendations. Adjust the current limiters properly.
99. Feeder circuit breakers are installed, sized, and adjusted as specified.
100. Main and tie circuit interrupters are rated as specified and capable of carrying full-rated current to feed the breakers as specified.
101. Current and potential transformers are sized and electrically connected as specified.
102. All control switches are furnished and operating as required.
103. Relay panels are furnished with subpanels, control relays, timing relays, repeat cycle timers, protective relays, phase-failure relays, timers, and latching relays and are adjusted as required. Terminal block wiring and all necessary appurtenances are complete.
104. Fuse blocks are furnished and sized properly.
105. Elapsed-time meters are furnished and can be adjusted as specified.
106. Remote control (stop–start) devices are installed at the equipment and are properly connected to the MCC.
107. Manual motor starters are furnished for the low-voltage, low-horsepower, single-phase equipment and have proper current protection.
108. Line disconnect devices or circuit breakers are furnished for manual starters.
109. The spare parts for switchgear and MCC are supplied as specified.

Switchgear Operations for Two Independent Power Lines (If Furnished)

110. Coordinate energizing of switchgear with client personnel, power company, and contractor. Arrange for power outage testing as required.
111. Coordinate switchgear automatic transfer control voltage and time-delay relay settings with power company. Review power company recommended settings with client personnel and make setting adjustments as directed by the client.

Energizing the Station

112. Now, stand as far back as possible and have the utility turn on the power. In any event, make sure that nobody is standing in front of any electrical equipment when it is energized for the first time or, for that matter, reenergized at any time.
113. Close the main circuit breaker or fusable switch. This will be the first time that the switchboard bus is energized since the high-pot test (on-site or at the factory). *Never* stand in front of a circuit breaker or switch when operating it.

114. Test for correct voltage range. Do not use a multimeter for this purpose. Do use a solenoid tester with fused leads (such as manufactured by Ideal or Square D). Once the presence of the correct voltage range has been established, exact voltage can be measured with a multimeter. Utilities typically are required to set voltages within ±10% of nominal. At the no-load condition, the voltage for a 480-V system should be between 480 and 506 V.
115. Test for correct phase sequence. Most engineers specify phase sequence as left to right when viewed from the front and front to back on switchboards.

At this point, testing may diverge into two separate paths. The contractor is likely to want to start "bumping" motors because it is easy. However, it is preferable to start testing the control system. Because overload heaters and fuses have been removed from the starters, a complete functional test can be run on the control system without endangering mechanical equipment.

116. Check calibration of all instrumentation equipment. Instrumentation may well have been calibrated before it left the factory, but since then it has suffered vibration and shock from transportation and dampness from sitting in a nonenergized state for several months.
117. Test manual control devices. Operate start, stop, jog, etc. devices to determine whether the contactors and circuit breakers respond correctly.
118. Test safety devices. Close power devices manually and manually actuate the safety device to determine if the power device trips.
119. Test automatic control systems by placing all equipment in the automatic mode and simulating operation of the various control devices to determine whether the power devices function correctly. Simulation of control device operation should be accomplished as close to the process as possible. For example, it is better to apply a pressure signal to transmitter process input with a pneumatic test set than to simulate the pressure transmitter output with an electronic test set. As another example, it is better to move a valve manually to test limit switches than simply to move the limit switch arm by hand.

At this point, much of the power system will have been energized in a no-load state for several days. Insulation failures that are likely to occur will probably have occurred with a minimum of inconvenience to the start-up process. It is time to start energizing power equipment.

120. Check actual motor nameplates against motor starter application data and install correct motor overload heaters. Follow the same procedure for power fuses on high-voltage motor starters.
121. Rack in high-voltage starters and circuit breakers to the operating position.
122. Bump motors. Before bumping any motor, verify that reverse rotation will not damage mechanical equipment. If mechanical equipment can be damaged by reverse rotation, open coupling or remove drive belts as applicable.

Tests of the electrical and control systems have now been taken as far as practical without actual operation. In wastewater pumping stations particularly, it is certainly prudent to test the entire station with clean water before filling the station with sewage, but if the electrical designer has done his job and the start-up engineer has performed the above steps, the station electrical system should be in reasonably good condition. Once the station is actually started, it will be necessary to adjust the control system for stable operation under dynamic conditions. The time delay cannot be set until the station is operating normally.

Energizing a Station with Two Independent Power Sources

123. Energize both power company lines and transformers. Establish normal switchgear operation with two nonparalleling sources.
124. See that switchgear metering operates properly and that the meter registers accurately with reference to known existing loads.
125. Test automatic transfer operation with controller in live test mode. Simulate loss of power at one power line by pushing and holding loss-of-voltage test pushbutton. Note indicator lights that indicate opening of line switch followed by closing of tie switch. Note timing of both switch indicator lights. Note that switches do not actually operate in this test. Note annunciator alarms at MCC or control panel (if any are furnished).

126. Simulate return of power by release of test button. Note that tie switch light indicates opening condition followed by the line switch light indicating closed position. Note that switches actually do not operate. Acknowledge and reset annunciator.
127. Repeat the above test on the other power company line and note the results.
128. With switchgear returned to normal, place automatic transfer control switch on automatic position. Simulate loss of voltage on one phase of power company line by other than the test pushbutton. Note time to actual opening of the line switch followed by closing of tie switch. Note annunciator alarms.
129. Simulate loss of second power company line. Note that no switch position changes take place.
130. Simulate return of power to power company first line.
131. Note time to open power company second-line switch, followed by time to close first-line switch. Note that the tie switch remains closed.
132. Simulate return of power to power company second line. Note particularly the time between opening of the tie switch and closing of the second power company line switch. Reset annunciator alarm.
133. Repeat the above test by simulating simultaneous loss of both power company lines. Note that no switch transfer takes place. Return the system to normal.
134. Simulate phase and ground overcurrent on each power company line by protective relay action. Note opening of affected line switch. Note automatic transfer control lockout, tie switch remaining open, and annunciator alarm.
135. Reset the lockout at the switchgear, place the selector switch on manual position, and close the tie switch.
136. Place the automatic transformer control switch in manual position and repeat simulated loss and return of power. Note that the switches respond only to manual controls.
137. Return the switchgear control and plant equipment to normal.

Automatic Transfer Switch (If Furnished)

138. Note that the automatic transfer scheme of the motor control center is similar to the switchgear automatic transfer scheme. Differences include the use of circuit breakers instead of high-voltage disconnect for voltage transfer, and utilization of control time setting longer than the switchgear setting. Transfer and return time delay should be set approximately 2 seconds longer than the switchgear setting to avoid out-of-phase transfer and damage to equipment.
139. MCC automatic transfer test setting (automatic mode) can therefore be coordinated with the switchgear testing and with testing of standby generator.
140. Be sure MCC incoming voltage and control voltage are within acceptable limits for both primary and alternate power sources.
141. Indicating lights are operational for motor off or running, valves open or closed, power on, ready, and trip.
142. Control relays and timers are properly sized and adjusted as specified.
143. Alternators are operational.
144. All indicating voltage, current, and power devices are registering properly.

H-4. Simplified Operational Checks for Small Stations

First read Items 71 through 144 and 187 through 336. Follow any procedures that apply. Make the following tests for each main pump.

145. Start the main pump with the discharge isolation valve closed and the discharge pressure gauge valve closed. Read ampere draw.
146. Slowly open the discharge valve and slowly fill the transmission pipe or force main. Bleed any entrapped air.
147. When pipe is full, open discharge isolation valve wide open. Open valve to pressure gauge and bleed air.
148. Compare discharge pressure head and flowrate with design operating point. Read ampere draw and compare with calculated wattage. If necessary, trim impellers to make pump operate within, for example, ±10% of bep.

149. Repeat the above steps for other pumps.

150. Run through cycles and observe level controls, operation of pump sequencer, pressure, and flowrate for all combinations of pumps operating. (Can be accomplished by introducing false signals into the control system.)

151. Be alert to any undue noise, vibration, or other operational problems.

H-5. Well Pumps

Pumping from a well is usually done with a multistage, vertical turbine pump, either of the lineshaft or the submersible type. The pump setting (the distance from the pump suction to the well head) can, depending on the water table, vary from a few feet to several hundred feet. Although lineshaft and submersible units each have distinct features and advantages, lineshaft pumps are typically used for shorter settings, and submersible pumps are often preferred for settings greater than 500 to 700 ft. Lineshaft pumps with substantially deeper settings are used where electric power is not available.

Before installing any well pump, the following should be checked:

152. The well casing must be reasonably straight, preferably within 25 mm per 30 m (1 in. per 100 ft) depth and without double bends to avoid binding the equipment. Straightness may be checked with a dummy assembly, slightly longer and larger in diameter than the actual pump, lowered on a cable down the well to the actual pump setting.

153. New wells should be checked for the release of excessive amounts of sand for the first few days after start-up. Because sand is harmful to close-running clearances within the pump, wells suspected of being sandy should first be "developed" by a pump other than the production unit.

154. Make sure the perforated section of the well casing is well below the operating water level to avoid cascading and air entrainment (that causes reduced pump output). For the same reason, return lines should not be connected at the top of the well casing.

155. With the pump installed and operating, check the "drawdown" with an air line down the well to make sure the pump still has adequate $NPSH_A$. ("Drawdown" is the amount the water level drops in the well from static condition to full operating flow.)

Submersible Well Pumps

Submersible pumps are directly connected and driven by a submersible electric motor. Two types of motors are used: wet winding and dry winding. In wet winding motors, well water is permitted to enter the motor and surround the heavily insulated stator winding, or the motor is filled with a clean, aqueous solution. The dry winding motor is either sealed and completely filled with oil, or only the stator winding is sealed off from the well water.

Because the pump and motor are close-coupled on submersible units, shaft critical speed is typically not a concern. The unit can therefore be operated at the optimum rotational speed.

In addition to the general precautions and checks above, make sure of the following:

156. The motor shaft sealing device varies significantly from manufacturer to manufacturer, so make sure the proper prestart instructions are followed.

157. Check the liquid level if the motor is fluid filled, and refill as required.

158. If pump and motor are shipped separately, make sure the shafts are aligned carefully when the two assemblies are connected.

159. When the power cable is shipped separately, make sure a watertight connection is made at the motor terminal box.

160. Just before lowering the unit into the well, take a megger (ohm meter) reading on the motor-cable assembly to ensure electrical integrity.

161. When the motor is submerged to the normal operating level, repeat the megger test of the motor-cable. If the megger reading is below the minimum specified by the manufacturer, the cause must be found and corrective measures taken before start-up.

162. Check the starting panel to be sure that all electrical controls and instruments are functioning properly.
163. Check the unit for rotation by comparing the closed valve-head reading with the manufacturer's rating curve. Centrifugal pumps develop substantially lower head at the closed valve when run in reverse. Switch two of the leads in the starting panel if necessary.
164. Repeat the megger reading after the motor has reached operating temperature and has been running for a few hours.

Lineshaft Pumps

Lineshaft pumps are most commonly driven by a vertical, hollow-shaft motor, which is mounted on top of the pump discharge head and connected to the pump bowl assembly by a lineshaft housed within the discharge column. A drive clutch at the top end of the motor provides for substantial vertical shaft adjustment. The adjustment is important, because the hydraulic thrust developed in the pump must normally be carried by the thrust bearing in the motor. Elastic elongation of the lineshaft occurs while the pump is operating, and this elongation can be substantial for deep setting pumps.

Lineshaft pumps can also be driven by a vertical solid-shaft motor, connected to the pump lineshaft with a conventional, rigid coupling, or the pump can be driven by a horizontal gasoline, diesel, or gas engine connected to a hollow shaft through a right-angle gear.

If reverse rotation would cause equipment damage (by overspeed or to engine drives) the drivers may be fitted with a nonreverse ratchet. However, it should be noted that when a pump spins in reverse due to backflow, the direction of torque in the shafting is the same as for normal operation, so pump damage is usually not a concern.

The lineshaft runs in bearings housed within the discharge column. When clean water is pumped, the bearings are lubricated by the pumped liquid. The bearings are usually of rubber or bronze material and housed in spiders at 3-m (10-ft) centers. When water containing abrasive material is pumped, the lineshaft is located within an enclosing tube, and externally threaded bronze bearings are used to connect the enclosing tube sections. The bearings are lubricated with oil or clean water injected into the enclosing tube at the discharge head. Piping and valve arrangements at the well head may vary greatly depending on the application. Some typical arrangements of piping are shown in Figures 7-9 to 7-11, and 18-18.

In addition to the general precautions recommended above, note the following points:

165. Before connecting the driver to the lineshaft, check the driver for correct rotation as indicated by the arrow on the pump discharge head.
166. See that the discharge head is grouted properly at the well head and that a well seal is provided if required by local code.
167. Deep setting, open lineshaft pumps normally require pre-lubrication of the lineshaft bearings. Make sure the proper connections are made and the pre-lube system is functioning.
168. For pumps with fresh water or oil injection to the inner column, see that the lubrication reservoirs are filled, the proper piping connections are made, and electrically operated valves are functioning.
169. See that all pump discharge connections are made and pressure tested properly, flexible pipe couplings have been correctly harnessed with tie-bars, pipe anchors are in place, and no undue pipe stress is being imposed on the pump discharge head.
170. Check that the air and vacuum release valve on the pump discharge is connected and functioning correctly.
171. Make a final wiring and instrument check at the starting panel and include connections for electrically operated control valves.
172. Check that the pump rotating element has been given the proper lift as specified by the manufacturer. (The vertical adjustment is made at the driver coupling clutch.)
173. Follow the procedure outlined in the main pump section for filling and venting the discharge pipeline.
174. Start the unit and slowly open the discharge control valve. Adjust the shaft seal at the discharge head as required.
175. With the pump operating at design flow, check general mechanical functioning (e.g., noise, vibration).
176. Double check controls and alarms for functioning. Make sure the manufacturer's minimum flow requirements will be met.
177. Take pump head readings with the pump operating at design flow and also against the closed valve, and compare with the manufacturer's performance curve. With good field instruments, the readings should be within 5% of the factory test.

H-6. Chlorination

178. See that the gaskets on the doors of the chlorination room fit tightly.
179. See that air packs are furnished and are operational.
180. Verify the proper operation of the gas alarm system.

The manufacturer's representative should perform the necessary checks and tests for the chlorinator equipment and installation. These systems are critical, complex, and require more detail than the abridgement listed in Items 181 through 186.

181. Check the chlorinator equipment and installation with respect to
- O-rings
- Inlet gas heater
- Venting
- Gas inlet filter
- All lines and valves
- Flowmeter
- Leaks in chlorine lines, valves, and fittings from chlorine tanks to chlorinator
- Clean chlorine pressure reducing valve (PRV) cartridge
- Clean and operational vacuum relief valve
- Clean and operational trimmer drain relief valve.

182. Check the chlorine evaporator with respect to
- Proper operation of temperature and pressure gauges
- Leaks
- Clean chlorine PRV filter
- Clean chlorine chamber. A dirty or oily one is very dangerous.
- Clean liquid line from tank car. A dirty or oily one is very dangerous.

183. Proper operation of ejectors.
184. Check the chlorine solution pump with respect to
- Items 34 through 38
- Lubricated pump and motor bearings.

185. Proper size, type, material, and installation of all valves.
186. Check the chlorine tank connections with respect to
- Proper size, type, material, and installation of hose lines
- Leaks in hose lines
- Leaks in pipe nipples and valve installed on the chlorine tank.

H-7. Complex Drives

Engines, engine-generator sets, variable-speed drives, and controls **MUST** be checked and tested by a factory representative.

H-8. Control Panel and Electrical Systems

These tasks should be performed by an electrical inspector or engineer and the electrical contractor foreman. Supplement Items 187 through 196 with "Field Tests" in Section 9-12.

187. There is proper voltage on each leg into the panel.
188. The voltage on the control side of the panel is proper.
189. The pump sequencer is operational.
190. Overload protectors are properly sized.
191. All alarms are operational.
192. All limit switches and pressure switches are installed and operating.
193. The telephone dialer is operational.

194. All nuts, screws, and instrument and junction box inspection covers are tight.
195. Make certain that all GFCI outlets have been tested.
196. Set all STOP/LOCKOUT buttons on the equipment to the STOP position.

Automatic Controls, Devices, and Monitors

197. If practical, measure the wet well (do not trust the plans) to determine volume used for *in situ* testing of the flowmeter, if furnished (see Item 204). Make sure the flowmeter is installed properly and the recording mechanism is operating. Following the main pump start-up (Items 321 through 336), check the accuracy of the flowmeter volumetrically. If volumetric calibration is impractical or if greater accuracy is needed, use tracers (see Section 3-9).
198. Any other devices not mentioned above should also be tested for proper operation.

H-9. Bubbler Systems

199. Construction debris and all foreign materials are removed from liquid channels and wet wells.
200. Check wet well and channel bubbler pipe for extension. The bubbler stilling well is constructed in accordance with the contract plans.
201. All indicating devices on the bubbler panel are furnished and adjusted properly.
202. All connections from the bubbler panel are properly made to the MCC and/or control panel.
203. Operate the compressed air system. Adjust the airflow to the wet well.
204. If practical, admit known amounts of liquid to the main pumping station well to calibrate it volumetrically. Otherwise, measure the wet well accurately and calculate the volume at HWL and LWL (for calibration check of the flowmeter).
205. Make bubbler system calibration adjustments as necessary.
206. Allow channel and wet well levels to rise. Check level indication at the liquid level control and verify accuracy by comparing to liquid level as measured with the calibrated rod.
207. Allow the liquid to rise to the maximum level. If the bubbler system calibration requires readjustment, reverify the entire range of bubbler system calibration. Refer to the specifications for proper level setting.

H-10. Vacuum Priming Systems

There are three types of vacuum priming systems. The first is a simple eductor type used with water or air to remove the air from the main pump casing and to draw liquid from the wet well.

The second type is a dry vacuum pump with a protective tank installed between the vacuum pump and the main pump. The tank is equipped with an electrode. Before starting this system, water must be emptied from the protective tank. When the vacuum pump is energized, a vacuum is created in the tank, so air and then liquid is drawn from the main pump casing. When the protective tank fills with liquid from the main pump casing to the low level of the electrode, the vacuum pump is automatically deenergized and the main pump is primed.

The third type is a package system that basically consists of one or two water-displacement, wet-type vacuum pumps, vacuum tank, vacuum valve, vacuum relief valve, water level switches, and pressure switches and gauges. The "wet-type pump" is dependent on a continuous supply of cool, clean "service water," which enters the vacuum pump on the suction side, is discharged with the compressed gas, and thus draws air and water from the vacuum tank, main pump casing, and wet well. The device then shuts down automatically.

Eductor Type Systems

208. Main pump casing nozzle should have shut-off valve and a quick-disconnect fitting installed.
209. The quick-disconnect fitting on the water or air priming hose must match casing nozzle disconnect.

210. Priming hose must be pressure rated for the water or air used and must be in good condition.
211. Priming hose must have a manual shut-off valve and sufficient pressure to evacuate the pump casing air and draw liquid from the wet well.

Dry Vacuum Pump Systems

212. Separate the pump coupling.
213. Energize the control and local START button to bump the motor. Compare the motor rotation with the arrow on the vacuum pump.
214. Deenergize the system.
215. Check vacuum pump lubrication.
216. Manually turn the vacuum pump in the direction shown by the rotational arrow.
217. Check motor–vacuum pump alignment.
218. Reconnect the vacuum pump coupling.
219. Replace the coupling guard.
220. Inspect piping between vacuum pump and protective tank. Piping should be installed near or at the top of the protective tank.
221. Inspect piping between protective tank and main pump.
222. Inspect the electrical connection to the transformer required for the electrode installed in the protective tank. The electrode system operates on low voltage.
223. The protective tank must have a drain with a shut-off valve. If the system is specified as an automatic drain type, a normally closed solenoid must be furnished on the drain piping and must be electrically connected. The solenoid opens when the motor is deenergized, which drains the vacuum tank.
224. Inspect the protective tank liquid sight glass and determine whether its elevation corresponds with the bottom of the electrode.
225. Open all vacuum pump suction and discharge valves.
226. Fill the protective tank with water so that the bottom of the electrode is submerged.
227. Energize the vacuum pump. The unit should *not* operate. If it does, immediately deenergize the vacuum pump and correct the wiring.
228. After all corrections and trial runs are completed, open the main pump suction shut-off valve.
229. Do not open *any* pump discharge valves.
230. Energize the vacuum pump system, and check the vacuum gauge reading. If the gauge reading is above 635 mm (25 in.) of mercury and the pump is a rotary vane type, immediately shut down the system and see whether all required valves are open. *Note:* Above 686 mm (27 in.) this type of unit will destroy itself, and high vacuum is very dangerous to the piping and the protective tank.

Wet-Type Vacuum Systems

231. Separate the pump coupling.
232. Energize the control and local START button to bump the motor. Compare motor rotation with the arrow on vacuum pump.
233. Fill the pump with water up to the shaft level. Never start the motor with liquid level above this shaft line because of the high power consumption.
234. Rotate the pump shaft by hand and check for binding. If the pump does seize, gently tap the shaft end to free the impeller. If the trouble persists, fill the pump with a solvent approved by the manufacturer to loosen scale binding and rotate the pump shaft by hand.
235. Check alignment of the shaft.
236. The equipment must be level because of the water-filled pump.
237. Packing-equipped pump glands must be finger tight only. Adjust after the unit is operating as specified.
238. Reassemble the coupling and guard.
239. Check motor and vacuum pump lubrication.
240. All gauges and switches should be made operational. Adjust switches and valves as specified.

241. Determine that the piping is connected properly.
242. Disconnect the piping to the main pump, energize the system, and operate for several hours. Make adjustments to find the optimum operating conditions.
- Packing gland drip and temperature.
- Motor amperes are below motor nameplate rating.
- Pump temperature is acceptable.
- Amperes used versus motor nameplate amperes rating.

243. Deenergize the system.
244. Reconnect the piping to the main pumps.
245. Reenergize the system. Again inspect the packing, amperage, motor temperature, vacuum gauge reading, and pump temperature. The pump must be able to reach the design vacuum in the time specified without a motor overload.
246. If the pump cannot reach design vacuum, it may be necessary to increase the water flow into the pump slightly. Check motor amperage during the time this is done.

H-11. Compressed Air Systems

Compressed air may be used for air-operated check valves, air-to-oil cylinder check valves, and bubbler systems. The air lines must be furnished with shut-off valves, pressure relief valves, and pressure-reducing valves as specified and required. Then follow the procedure below:

247. Separate the driven equipment coupling.
248. Check alignment, grouting, and doweling of components.
249. See that all piping is properly supported and under no strain.
250. Rotate the compressor shaft several times by hand to check for rubbing or binding. If an automatic belt-operated lubrication system is furnished, push each lubricator site feed plunger a number of times and rotate the shaft one-quarter turn. Repeat this procedure until the shaft has been turned a full 360°.
251. Energize the motor and check the direction of rotation against the rotational arrow.
252. Reconnect the coupling.
253. Fill the oil bath open-type intake air filter with oil to the level indicated with a lubricant recommended by the manufacturer. If a dry-type air filter is used, clean the medium.
254. Check the compressor and motor for lubrication.
255. Set pressure switches on the air receiver as specified for the lead compressor, the lag compressor, and the alarm.
256. If the compressor is a water-cooled unit, open the bypass line around the automatic water solenoid valve and water temperature flow regulator. Turn on the cooling water supply and fill the jacket and cooler completely. Bleed the system of air and close the bypass line.
257. Open all drain valves in the air and gas piping to remove any liquid. Close the drain valves.
258. Check the operation of all manual valves. Adjust packing glands (if furnished) as necessary. Close the bypass valve at the air dryer.
259. Start the compressor in automatic mode.
260. Check and adjust pressure-regulating valves as specified for regulators to the wet-bulb bubbler system, regulators to HVAC equipment, air-operated cylinder check valves, and air-to-oil cylinder check valves as specified.
261. Check the pressure on the compressor loader.
262. Check air receiver pressure gauges, condensate site glasses, drain valve, and automatic moisture eliminator.
263. Check operation of the air dryer, cooling dry filters, automatic condensate discharge valve, and the compound gauge indicating pressure and temperature of refrigerant.
264. Check all air lines for moisture and drain them.
265. Turn the compressor control to OFF, then decrease the system pressure to less than design conditions and note annunciator alarm.
266. Return the compressor to automatic operation. Observe the automatic starts and shutoffs of the compressor at design pressures.

267. Check and adjust the temperature switches. The switches should be set at approximately 6 to 8°C (10 to 15°F) above expected normal operating temperatures.
268. Observe the discharge air temperature. It should be approximately equal to the expected discharge temperature for the operating pressure as shown on the manufacturer's data sheet.
269. Observe the water outlet temperature and wait for it to stop rising before making corrections.
270. Adjust water flow through the aftercooler (if installed) to maintain approximately 38°C (100°F) outlet air temperature.
271. Reduce the pressure and note the lead compressor's automatic start.
272. Reduce the pressure and note the lag compressor's start.
273. Check operation of the duty transfer switch.
274. After a unit has been operating for approximately 2 h, slowly lower the setting of each temperature switch until the switch opens to shut down the driver. Readjust the switch to 6 to 8°C (10 to 15°F) above this shutdown point and restart the unit. Repeat this procedure with each temperature switch.
275. After all the checks have been made and the equipment is operating smoothly, the unit is ready for service in the system.

H-12. Hydropneumatic Tank Systems

The optimum water–air ratio for any size tank and operation condition can be found by applying Boyle's law, which is expressed mathematically as

$$P_1V_1 = P_2V_2 \tag{H-1}$$

where P is pressure and V is volume. The design engineer has selected the hydropneumatic tank for a specific pressure-level relationship. Obtain the information and be sure it is understood before beginning a check of the system. Follow the procedure below:

276. Check the water receiving tank and clean it. Flush the incoming water lines. Check the backflow preventer.
277. Fill the water receiving tank. Check the drain, overflow, and float-operated fill valve. Adjust as necessary.
278. Check the flow-through chemical feeder (if installed) for seal-water treatment.
279. Check the duplex strainer, seal-water pumps, hydropneumatic tank, and main pump seal-water piping and valves.
280. Check the hydropneumatic tank. Flush the tank and piping with water.
281. Check seal-water pump and motor bearing lubrication. Check pump packing.
282. Disengage the pump coupling.
283. Energize the motor and compare the direction of rotation with rotational arrow on the pump.
284. Turn the pump by hand in the direction of the arrow and check for binding.
285. Reconnect the pump coupling.
286. Fill the hydropneumatic tank to specified maximum water level.
287. Check the hydropneumatic tank relief valve against specification (usually factory-set to tank pressure rating).
288. Inspect check valve and hydropneumatic tank air lines for proper functioning. Open air valve and raise tank air pressure to maximum operational pressure. Close air valve.
289. Set pressure switches as specified (by draining tank) to:
- Set automatic-start seal-water pumps
- Stop seal-water pumps under manual control
- Set seal-water pressure alarm.

290. Operate seal-water pumps manually to check operation. Refill the tank.
291. Override the seal pump and manual control pressure interlock by holding the pump start pushbutton. Note water elevation, operating pressure, and annunciator high water alarm.
292. Drain the tank and note the high-water-alarm clearing level.
293. Place the seal-water pump in automatic mode. Continue to drain the tank until the lead pump starts. Allow the lead pump to refill the tank. Note pressure and level at shut-off.
294. Repeat the automatic test with the second pump in the lead position.

295. Repeat the automatic test and simulate lead pump failure-to-start. Note standby substitution for lead pump and annunciator alarm. Return to normal.
296. Repeat failure-to-start test with the second pump in the lead position. Note annunciator alarm. Return to normal.
297. Check the tank air-add system for panel function.
298. Adjust the seal-water regulator to the main pump to regulate the pressure as required.
299. See that all manual valves to main pumps and auxiliary equipment are open.
300. Check main pump packing, seal-water solenoid, sight flow indicator, control circuit timer, and auxiliary equipment as part of the main pump test sequence. *Main pump packing gland bolts should be only finger tight.*

H-13. Main Pumps, Final Pre-Start-Up Checks

The procedure depends to some degree on whether the pumps are constant speed (C/S) or variable speed (V/S). Variable-speed equipment should be checked only by the manufacturer's representative. Pumps furnished with wearing rings should never be operated dry. Read the manufacturer's instructions for the type of lubricants and for cooling (if required) and make sure the bearings are lubricated properly. Follow the procedure below:

301. Energize the seal-water system and be sure that water is flowing to the stuffing box or seal box.
302. Stuffing box gland for packing is only finger tight.
303. Air, oil, air-oil, or water is flowing to the check valve cylinder.
304. Manual check valve weight or spring is adjusted, and fine tune after operation.
305. Remove the coupling guard, separate the pump coupling half, and check the alignment between the pump and the driver.
306. Check motor and pump lubrication.
307. Energize the master control panel.
308. Energize the local control button and equipment and bump the motor to determine pump rotation relative to the pump arrow and pump nozzle.
309. If pump rotation is incorrect, lock out the pump at the local control, deenergize the pump control, and correct.
310. Reenergize the pump control and push START button to bump the motor and recheck rotation.
311. Rotate the pump shaft manually.
312. Reassemble the pump coupling and replace the coupling guard.
313. Shut off the suction and discharge pressure gauges.
314. Repeat Items 302 through 313 for each main pump.

H-14. Wet Well and Testing of Main Pumps

315. Refer to bubbler system start-up.
316. Check controls for START/STOP levels. Readjust as required.
317. Electrically jumper out (bypass) the actuator for the discharge control valve to determine whether the valve is open and note opening and closing times. Readjust speed controls as required.
318. Close all pressure gauges to the liquid being pumped.
319. Energize the seal-water system and determine that the water is flowing to the stuffing or seal box.
320. Energize the pump control.
321. Energize the START button of local control. (Note that the pumps are noisy at this shut-off condition.)
322. Open the manually operated discharge valve slowly, and fill the discharge header and the force main or transmission pipeline slowly. Check for undue motor and pump noise and vibration.
323. Open shut-off valves for discharge and suction pressure gauges; bleed all air from the pressure gauge piping or tubing; and compare head at shut-off and flow, head, and power at operating conditions with the pump performance curve.
324. If the pumping station has no flowmeter, flow can be measured by the methods described in Section 3-9.
325. If the stuffing box is leaking excessively, tighten the gland slowly. Be sure that a small, steady stream is leaking from the packing during a 24-h break-in period.
326. Close the discharge shut-off valve slowly enough to avoid water hammer.
327. Deenergize the pump.

328. Remove the jumper from cylinder solenoid check valve. The valve should close.
329. Determine that check valve limit switches are operating properly.
330. Set the check valve differential pressure switch as required.
331. Open the discharge shut-off valve.
332. Energize the pump and note the time of check valve opening. Readjust if required.
333. Deenergize the pump and close suction and discharge shut-off valves. Close all valves to pressure switches.
334. Repeat Items 317 through 333 for each pump.
335. After each pump has been tested individually, open all the suction and discharge valves and operate the station as specified.
336. In a V/S pumping station with wet well levels set at the normal level of flow in the incoming sewer pipe (see Figure 12-35), adjust the acceleration of the pump speed control (usually by reducing it) so that surging of flow in the influent sewer does not occur.

H-15. Cleaning Wet Wells

The operational and timed sequence used for ejecting scum and settled solids in wet wells for solids-bearing waters must be established. Either clear or dirty water can be used.

Hopper-Bottom Sumps

337. Set the pump control to MANUAL (or lock out the low level STOP switch) and pump the water level down as far as it can go with either one or both pumps operating. Look for a number 4 or 5 surface vortex that would (or does) draw floating material into the pump(s). The vortex should persist for 5 or 6 seconds—long enough to engulf most of the scum. In a properly designed sump, the hopper bottom concentrates sludge so near to the pump intake that sludge should be ejected in every pump cycle.
338. Test the wash water hose to see that flowrate and pressure are sufficient to scour grease off the walls of the sump.

Trench-Type Sumps

Depending on the circumstances, it may take several trials to reach a satisfactory operating sequence of storage, sluice gate setting, and pump operation. Three scenarios are possible: (1) inflow rate, Q, is greater than the capacity, Q_p, of the last pump; (2) Q equals or is somewhat less than Q_p; and (3) Q is much less than Q_p. There is no need to test the first two scenarios, and the sluice gate is not needed. The operator simply runs the last two pumps at full speed for a short time (a minute or two) after the hydraulic jump reaches the toe of the ogee ramp. The third scenario requires: (1) water storage sufficient to complete the cleaning cycle and (2) the release of the stored water under the sluice gate at a rate of about 50% to 80% of the capacity of the last pump.

339. Obtain from the designer the sluice gate setting required to release an adequate flow and operate the sluice gate to obtain that setting before testing with water. Adjust the automatic stop (if there is one) at that setting. If no stop is installed, close the sluice gate, then open it for as many seconds as are required to reach the proper setting.
340. Obtain from the designer the time required to store a sufficient volume of water for various influent flowrates. Water can be stored by closing the sluice gate or by shutting off the pumps. In either way, the pumps must be stopped for awhile, because a trench-type wet well can usually be dewatered in about half a minute, whereas storage (depending on influent flowrate) takes much longer.
341. For a range of influent flowrates, the designer must calculate the sequential times to activate the sluice gate, shut off the pumps, and start the pumps again (see Examples 12-4 and 12-5).
342. Test items 339 to 341 with either clear or dirty water. Adjust sequential times and sluice gate settings so that the hydraulic jump travels from the toe of the ogee ramp to the last pump in a reasonable length of time (15 to 40 seconds). Correct calculated values by observations as necessary and supply operators with instructions and timing sequences for the various influent flowrates to be expected.

343. If there is one, adjust the power monitor in the MCC (or the limit switch on the check valve lever) to stop each pump as it loses suction near the end of the cleaning cycle.

344. Test the wash water hose to see that flowrate and pressure are sufficient to scour grease off the walls of the sump.

345. If the station does not have a flowmeter, supply operators with charts or tables relating flowrate to speed (for V/S pumps) or to time between pump start and stop (for C/S pumps).

H-16. References

1. *Hydraulic Institute Standards for Centrifugal, Rotary & Reciprocating Pumps,* 14th ed. Hydraulic Institute, Cleveland, OH (1983). (See also later editions.)
2. *Pumping Manual,* 6th ed. Trade & Technical Press, Ltd, Morden, Surrey, England (1979).
3. Karrasik, E. J., W. C. Krutzsch, W. H. Fraser, and J. P. Messina. *Pump Handbook,* 2nd ed. McGraw-Hill, New York (1986).

Appendix I
Addenda

GARR M. JONES

CONTRIBUTORS
Robert S. Benfell
Arnold R. Sdano

I-1. Fire Safety in Pumping Stations

Several codes (NFPA 820, NFPA 70, NFPA 101, to name a few) provide guidance for fire protection for installations such as pumping stations. In addition to these documents, the designer should consider the following:

Choice of Materials

- Be careful in the choice of materials that could add to the fuel load (the amount of burnable material present) in a structure. Plastic, although useful for corrosion protection and resistance, will burn if ignition occurs. The products of combustion from plastics are often toxic and corrosive and become more so when they contain fire retardants.
- Be suspicious of materials that could auto combust, such as some forms of activated carbon. If these materials are used, make certain that they are isolated from other combustible materials.

Fire Suppression Systems

- Storage areas, such as those for records, lubricants, and fuels, should be separated from other areas of the structure by two-hour rated fire walls and should be served by independent ventilation systems. Appropriate fire suppression systems should be considered for these areas.
- Cable trays and other locations where electrical conductors are not protected by conduit should be monitored by rate of temperature rise, particle ion, and smoke detection systems.
- In large installations, consider dividing the electrical systems into identical halves and placing the switchgear in separate fire-rated rooms to avoid the prospect that an electrical fire will disable the entire facility.
- In critical facilities, install sprinklers over cable trays.

I-2. Application-Engineered Equipment

Consider specifying application-engineered equipment for those facilities considered to be critical to the overall function of the installation. Included in this category are the main pumps and motors, engine-generators, and engine drives for pumps. The hallmark of application-engineered equipment is the design of machines for the specific installation and application. Generally, a greater level of conservatism is employed to provide equipment more robust (and therefore more reliable with longer life) than that obtainable from standard machines. The following are suggestions for improving the performance of typical machines.

Pumps

- A maximum permissible pump shaft deflection of 0.05 mm (0.002 in.) at the pump seal is suggested for any operating condition specified. Pumps can usually be modified to accept larger bearings and shafts.

- Specify bearings with L-l0 ratings of 100,000 hours for any loads imposed by any specified operating condition. Some engineers insist that life is actually increased by specifying an L-10 rating of 50,000 hours because of the reduced tendency for skidding in the more heavily loaded bearing. Others maintain that skidding is not a problem (or at least unlikely) in bearings turning at 1725 rev/min or less.
- Specify either austenitic cast iron (ASTM A439) or that nickel be included in iron castings (2–3% is suggested) for improved resistance to corrosion and erosion. Alternatively, consider applying ceramic coatings such as Thortex Cerami-Tech [1] to internal pump surfaces to reduce wear and erosion (see page 758).
- Specify cast stainless steel or aluminum bronze for impellers where $NPSH_A$ margins are borderline to reduce erosion due to cavitation. The more flexible ceramic coatings (such as Thortex Cerami-Flex [1]) on ordinary impellers might also serve at less cost.
- Require the pump rotor unbalance to be no greater than grade G2.5 per ISO Standard 1940/1.
- Require the pump and driver supplier to perform a torsional analysis for systems with engine drivers of 250 kW (335 hp) or larger or for synchronous motors of 500 kW (670 hp) or larger.

Motors

- Specify NEMA Class H insulation, but require the motor be designed for a Class B temperature rise. Note, however, that the thicker insulation requires wider slots and thereby reduces the amount of iron and the efficiency.
- Require a 1.15 service factor over nameplate rating, but prohibit the use of the service factor for driving the machine at any specified operating condition.
- Specify inverter duty motors for all variable speed applications.
- Specify IEEE efficiency compliant motors for cooler operation.
- Consult the vendor for derating motors at altitudes above 1000 m (3000 ft).

Engines

- Derate engines to allow no more than 85% of the manufacturers continuous duty rating.
- Pay attention to elevation and temperature conditions to make certain engines are selected properly.
- Insist upon a thorough mass elastic system analysis.
- Do not allow controls to be mounted on the engine.

Generators

- Specify the same construction as detailed above for motors.
- Require solid state field-forcing regulators.
- Specify generator loading criteria specific to the application—that is, if the generator is to start 75 kW (100 hp) motors in sequence, so state. Require the manufacturer to produce a voltage transient study to demonstrate that the generator and engine have been selected for the actual starting duty.

I-3. Air-Vacuum Valves for Wastewater Service

Historically, air release and vacuum relief valves for wastewater service have been expensive to maintain and are prone to jamming either open or closed due to grease and debris. In side-by-side tests in a system known for considerable difficulty with traditional air-vacuum valves, a new design, Vent-O-Mat® series RGX [2], was found to be much more reliable and easier to maintain. After testing, a large number of these valves were purchased for installation in the system tested. Note that the period between cleanings of conventional air-vacuum valves is typically one to three months whereas the permissible period between cleanings of the RGX valve is said to be at least six times longer. A somewhat different style, Series RBX, is intended for water service. Product information comes with software to help the engineer select the proper size valve for a specific application.

I-4. Machine Foundations and Installation

Experience has shown that commonly accepted practices for preparation of machine foundations and for machine alignment have not been adequate in many instances. Lack of adequate care in preparation of the foundation has resulted in poor support of the equipment, progressive misalignment of the machines, and high maintenance costs. Excessive equipment vibration can often be traced directly to an inadequate foundation. Often, equipment manufacturers are not fully aware of the need for the care required to construct a proper machine foundation and thus their own recommendations for equipment supports are found wanting. API standard 610 provides an excellent specification for baseplate construction details. Although a rigid baseplate or soleplate is important, it must be installed level and with sufficient care to ensure that it is fully supported by a long-lasting grout base. Key elements required to achieve this objective include:

- Anchor bolt sleeves are filled with wax or similar compounds to prevent grout from bonding to the anchor bolts. The wax filling assures that the full clamping force required to anchor the machine is transmitted to the structural foundation and not to the grout.
- The concrete foundation pad is roughened by chipping the concrete to a depth of 50 to 100 mm (2 to 4 in.) to reach solid aggregate. The baseplate or soleplate is roughened by abrasive blasting. Both concrete and baseplate are primed with an epoxy bonding compound prior to grouting.
- Soleplates or baseplates are leveled on shims and steel blocks or jackscrews (not leveling nuts) against the anchor bolt nuts to a tolerance of 0.17 mm/m (0.002 inches/ft) prior to grouting.
- Epoxy grout, poured under a head of not less than 150 mm (6 in.) is used rather than cementitious grout. Care is taken to ensure complete contact with baseplate or soleplate.
- Once the grout has fully cured and reached its design strength, the leveling devices are removed and the anchor bolts are torqued sequentially by increments to develop the required clamping force and ensure complete contact with the grouted foundation.
- Pockets left for the removal of shimming devices are filled with grout and pointed.

Although the cost of the procedures outlined above is somewhat greater than that with commonly accepted practices, the results can be startling in terms of reduced vibration and lengthened equipment service life. One company was able to reduce the magnitude of equipment vibration by more than half by using these procedures [3]. Another company credits the major portion of an increase in mean time between failures of 950 percent (from 6 months to 57 months) in pumping equipment in their plant to improved equipment support requirements [4].

In addition to improved equipment support design, owners have found that advanced alignment techniques, using laser- and computer-based equipment, have resulted in improved power train alignment. This equipment is also capable of detecting support imperfections on both driving and driven machines. A study reported by the University of Tennessee demonstrated that even small amounts of misalignment result in significant increases in the loads applied to bearings, shafts and seals [5]. Machine alignment should be the responsibility of specialists using a variety of tools to align all elements of the power train accurately.

I-5. References

1. Maintenance Protection Systems, 43317 Moody Dixon Road, Prairieville, LA 70769, or Thortex America, 420 Babylon Road, Horsham, PA 19044, or Thortex America West, 600 S.E. Maritime Ave., Ste. 250, Vancouver, WA 98661, or Mountain West Thortex, 6969 N. 16th., Coeur d'Alene, ID 83814.
2. International Valve Marketing, Inc., P.O. Box 56, Plainfield, IL 60544. Phone (815) 439-0588; Web www.ventomat.com.
3. Monroe, P.C. "Pump baseplate installation and grouting." Proceedings of the Fifth International Pump Users Symposium, Turbomachinery Laboratory, Texas A & M University, College Station, TX, pp 117-128 (1988).
4. Hrivnak, S.J. "In-plant perspective: Eastman Kodak Company." Pumps and Systems (January 2000).
5. Piotrowski, J. "Understanding and using shaft-to-shaft alignment measurement systems." Pumps and Systems (April 1999).

Index

Q

R

S

T

W

Equivalent Weights and Measures

Length

U.S. customary units				SI units		
Miles	Yards	Feet	Inches	Kilometers	Meters	Millimeters
1	1.76 E+3	5.28 E+3	6.36 E+4	1.61 E+0	1.61 E+3	1.61 E+6
5.68 E−4	1	3.00 E+0	3.60 E+1	9.14 E−4	9.14 E−1	9.14 E+2
1.89 E−4	3.33 E−1	1	1.20 E+1	3.05 E−4	3.05 E−1	3.05 E+2
1.58 E−5	2.78 E−2	8.33 E−2	1	2.54 E−5	2.54 E−2	2.54 E+1
6.21 E−1	1.09 E+3	3.28 E+3	3.94 E+4	1	1.00 E+3	1.00 E+6
6.21 E−4	1.09 E+0	3.28 E+0	3.94 E+1	1.00 E−3	1	1.00 E+3
6.21 E−7	1.09 E−3	3.28 E−3	3.94 E−2	1.00 E−6	1.00 E−3	1

Area

U.S. customary units				SI units		
Square miles	Acres	Square feet	Square inches	Square kilometers	Hectares	Square meters
1	6.40 E+2	2.79 E+7	4.01 E+9	2.59 E+0	2.59 E+2	2.59 E+6
1.56 E−3	1	4.36 E+4	6.27 E+6	4.05 E−3	4.05 E−1	4.05 E+3
3.59 E−8	2.30 E−5	1	1.44 E+2	9.29 E−8	9.29 E−6	9.29 E−2
2.49 E−10	1.59 E−7	6.94 E−3	1	6.45 E−10	6.45 E−8	6.45 E−4
3.86 E−1	2.47 E+2	1.08 E+7	1.55 E+9	1	1.00 E+2	1.00 E+6
3.86 E−3	2.47 E+0	1.08 E+5	1.55 E+7	1.00 E−2	1	1.00 E+4
3.86 E−7	2.47 E−4	1.08 E+1	1.55 E+3	1.00 E−6	1.00 E−4	1

Volume

U.S. customary units					SI units		
Cubic yards	Cubic feet	Imperial gallons	U.S. gallons	Cubic inches	Cubic meters	Liters	Milliliters
1	2.7 E+1	1.68 E+2	2.02 E+2	4.67 E+4	7.65 E−1	7.65 E+2	7.65 E+5
3.70 E−2	1	6.23 E+0	7.48 E+0	1.73 E+3	2.83 E−2	2.83 E+1	2.83 E+4
5.95 E−3	1.61 E−1	1	1.20 E+0	2.77 E+2	4.55 E−3	4.55 E+0	4.55 E+3
4.95 E−3	1.34 E−1	8.33 E−1	1	2.31 E+2	3.79 E−3	3.79 E+0	3.79 E+3
2.14 E−5	5.79 E−4	3.60 E−3	4.33 E−3	1	1.64 E−5	1.64 E−2	1.64 E+1
1.31 E+0	3.53 E+1	2.20 E+2	2.64 E+2	6.10 E+4	1	1.00 E+3	1.00 E+6
1.31 E−3	3.53 E−2	2.20 E−1	2.64 E−1	6.10 E+1	1.00 E−3	1	1.00 E+3
1.31 E−6	3.53 E−5	2.20 E−4	2.64 E−4	6.10 E−2	1.00 E−6	1.00 E−3	1

Discharge (flowrate)

U.S. customary units			SI units			
Million gallons/day	Cubic feet/second	Gallons/minute	Cubic meters/second	Cubic meters per hour	Liters per second	Millions of liters per day
1	1.55 E+0	6.94 E+2	4.38 E−2	1.58 E+2	4.38 E+1	3.79 E+0
6.46 E−1	1	4.49 E+2	2.83 E−2	1.02 E+2	2.83 E+1	2.45 E+0
1.44 E−3	2.23 E−3	1	6.31 E−5	2.27 E−1	6.31 E−2	5.45 E−3
2.28 E+1	3.53 E+1	1.58 E+4	1	3.60 E+3	1.00 E+3	8.64 E+1
6.34 E−3	9.81 E−3	4.40 E+0	2.78 E−4	1	2.78 E−1	2.40 E−2
2.28 E−2	3.53 E−2	1.59 E+1	1.00 E−3	3.60 E+0	1	8.64 E−2
2.64 E−1	4.09 E−1	1.83 E+2	1.16 E−2	4.17 E+1	1.16 E+1	1